ADVANCED BIOTECHNOLOGY

For B.Sc. and M.Sc. Students of Biotechnology and Other Biological Sciences

Dr. R.C. Dubey
Ph.D.
Professor
Department of Botany and Microbiology
Gurukul Kangri University, Haridwar, Uttarakhand

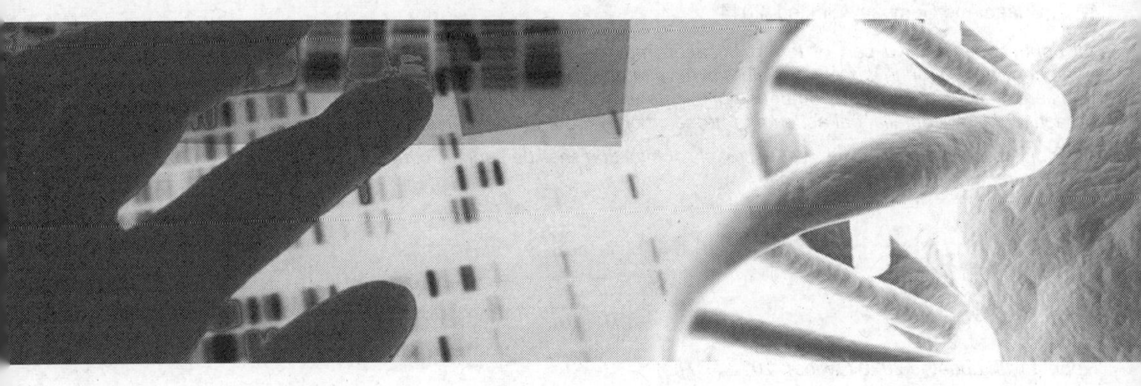

S. Chand And Company Limited
(ISO 9001 Certified Company)

S. Chand And Company Limited

(ISO 9001 Certified Company)

Head Office: D-92, Sector–2, Noida – 201301, U.P. (India), Ph. 91-120-4682700

Registered Office: A-27, 2nd Floor, Mohan Co-operative Industrial Estate, New Delhi – 110 044, Phone: 011-49731800

www.**schandpublishing.com;** e-mail: **info@schandpublishing.com**

Marketing Offices:

Chennai	:	Ph: 23632120; chennai@schandpublishing.com
Guwahati	:	Ph: 2738811, 2735640; guwahati@schandpublishing.com
Hyderabad	:	Ph: 40186018; hyderabad@schandpublishing.com
Jalandhar	:	Ph: 4645630; jalandhar@schandpublishing.com
Kolkata	:	Ph: 23357458, 23353914; kolkata@schandpublishing.com
Lucknow	:	Ph: 4003633; lucknow@schandpublishing.com
Mumbai	:	Ph: 25000297; mumbai@schandpublishing.com
Patna	:	Ph: 2260011; patna@schandpublishing.com

First Edition 2014
Reprints 2018, 2019
Reprint with updates 2020 (Twice), 2022 (Twice)

Reprint 2023

ISBN: 978-81-219-4290-4 **Product Code:** H6BTC68BTCH10ENAA14O

PRINTED IN INDIA

By Vikas Publishing House Private Limited, Plot 20/4, Site-IV, Industrial Area Sahibabad, Ghaziabad – 201 010 and Published by S. Chand And Company Limited, A-27, 2nd Floor, Mohan Co-operative Industrial Estate, New Delhi – 110 044.

To

The Sweetest Memory of

My Mother

PREFACE

Biotechnology has emerged as a fascinating interdisciplinary subject during the last couple of decades. It has been possible only due to rapid research work done in this area throughout the world. Moreover, there are many separate books in specialised areas of biotechnology. Today some of biotech products are available in the national and international markets. Many more industries have also been established and producing genetically, engineered products including enzymes, detergents, proteins, antibiotics, hormones, vaccines, and animals as well. DNA fingerprinting is also being applied in forensics for identifying the suspects. In the recent years, biotechnology has been included in the syllabi of almost all Indian Universities, in various subjects of biology or as an independent subject.

Being an interdisciplinary area, this subject is very vast. Though my earlier published book entitled **A Textbook of Biotechnology** has been benefitting the students of biotechnology at the undergraduate and postgraduate levels since 1993, yet much was needed to incorporate the recent developments in the former book which was not possible at the author's end. Hence, keeping the pace with the demand of more materials this book '**Advanced Biotechnology**' is written so that the demand of the readers can be fulfilled. Though some of the chapters are overlapping, yet those have been described in rather detail.

The book embodies 22 chapters covering various important disciplines of biotechnology, such as cell biology, molecular biology, molecular genetics, biophysical methods, genomics and proteomics, metagenomics, enzyme technology, immune-technology, transgenic plants and animals, industrial microbiology and environmental biotechnology. The book is illustrative. It is written in a simple language so that the students can easily grasp the matter. Possible questions have been written at the end of each chapter so that the matter can be thoroughly understood. Glossary of some important technical terms has been incorporated. Subject index has been given at the end of the book so that the students may search the related topics with complete case.

I acknowledge the publishers and authors of books and journals from where information in the form of matter and figures have been retrieved and used in this book. The author expresses his sincere thanks to the management and the editorial team of S. Chand & Company. Pvt Ltd. for help and support in bringing out this book.

Last but not least, I extend my thanks to Poojya Ma Bharti (Mataji) for her constant inspiration and blessings, my wife (Mrs. Malti Dubey) and son (Sujaan Dubey) for their endurance during the course of writing of the text. Be it appreciation or criticism from the readers (via e-mail), I shall welcome their suggestions equally. This shall help me in improving the subject matter in future editions.

Dr. R.C. Dubey
Email: profrcdubey@gmail.com

CONTENTS

Chapter 1: Introduction 1–21

History of biotechnology- from 'zymotechnology' to 'biotechnology'; timeline of biotechnology, *traditional vs. modern biotechnology*; *the world of biotechnology* - red biotechnology (= pharmaceutical biotechnology) (gene splicing, genetic engineering, pharmacogenomics and medicines, cloning, genetic testing, gene therapy, stem cell therapy), green biotechnology (plant tissue culture, plant genetic engineering, plant molecular marker-assisted breeding, biofertilizers, biopesticides, global area of biotech crops), white biotechnology (= industrial biotechnology), blue (marine) biotechnology*; bioinformatics; synthetic biology; public perception of biotechnology; biotechnology in India*- bio-business (biotechnology market) in India, (biotech turn-over, booming biotech market, success story of biotech market, consumption of biotech products, policy initiatives), biotechnology institution in India (top biotech schools (public sector), top biotech schools (private sector)*; ethical issues*- stem cells, cloning, food and agriculture, use of animals in research; *questions*.

Chapter 2: Cell Biology 22–70

Cell structure and function - journey from molecules to a living cell- polymerisation of nucleotides, natural selection of self-replicating molecules, catalytic RNA molecules (ribozymes), evolution of self-replicating system of RNA molecules, flow of genetic information, evolution of first cell, the heredity material, *Prokaryotic and eukaryotic cells*- structure of bacterial cells (capsule, flagella, pili and fimbrae, cell wall, plasma membrane, cytoplasm), evolution of eukaryotes, compartmentalization of organelles, Eukaryotes (cell structure and components)- cell wall (structure of cell wall, cell wall chemistry, function of cell wall), cell membrane (lipids, proteins, function of cell membrane), protoplasm, cell organelles (nucleus, endoplasmic reticulum, ribosomes, Golgi complex, lysosomes, peroxysomes, mitochondria, plastids; *transport across cell membrane*- membrane transport proteins, active and passive transport systems, Carrier protein and active membrane transport- Na^+/K^+–ATPase, Ca^{2+}–ATPase pumps; *transport systems in prokaryotes and eukaryotes*- ABC transport super family; *ion channels*- measurement of ion flow; *cell signalling and signal transduction*- recognition of extracellular signal molecules, forms of signalling (paracrine signalling, synaptical signalling, endocrine signalling, autocrine signalling); *types of receptors*- cell surface receptors, classes of cell surface receptor proteins involved in signaling [G-protein linked receptors, enzyme-linked receptors (e.g. receptor tyrosine kinase), integrins, toll-like receptors, ligand-gated ion channel receptors], intracellular receptors; *signal transduction pathway*- role of second messenger molecules in signal transduction pathway, disease and cell surface receptors; *signal amplification*- signal amplification at the transmembrane hormone receptor; *secondary messengers*- cyclic adenosine monophosphate (cAMP), cyclic guanosine monophosphate (cGMP), inositol triphosphate (IP3) and diacylglycerol (DAC) (inositol-1,4,5 triphosphate (IP3), diacylglycerol (DAC), calcium ions; *questions*.

Chapter 3: Molecular Biology 71–127

Nature of nucleic acids (DNA and RNA)- deoxyribonucleic acid- chemical composition (five carbon ring, nitrogenous base, phosphate group, phosphodiester bond), Chargaff's rule of equivalence, X-ray crystallographic studies of DNA, consideration of Watson and Crick model of DNA, Watson and Crick's Model for DNA; *Conformation of nucleic acids*: *torsion angles*-different torsion angles, sugar rings (endocyclic torsion angles, sugar puckering, furanose pseudorotational angles, *anti-syn* orientation); *conformational variation in double helical structures*, alternative form of right handed DNA- different forms of DNA (A, B, C, D, and Z forms of DNA, single stranded DNA; circular and super-coiled DNA), palindromic DNA (palindromes); genome organisation (genome organisation in prokaryotes, genome organisation in eukaryotes), *structure of ribonucleic acids-* ribosomal RNA (rRNA), messenger RNA (mRNA) (5′ cap, non-coding regions, initiation codon, coding region, termination codon, poly (A) sequence), transfer RNA (tRNA) or soluble RNA (sRNA) (structure of tRNA, clover leaf model of tRNA); *DNA replication: Watson and Crick model for DNA replication-* Meselson and Stahl's experiment for DNA replication by semi-conservative method, enzymes involved in DNA replication (nucleases, DNA polymerases, DNA ligases, DNA helicase, topoisomerase, DNA gyrase), mechanism of DNA replication (unwinding of double helix, DNA replication, removal of RNA primer and completion of DNA strand, ioining of fragments), models for DNA replication (Cairns model for DNA replication, rolling circle model, replication in eukaryotic chromosome); *gene expression,* central dogma; *transcription-* RNA polymerases (types of RNA polymerase, site of transcription, progress of transcription), chain initiation (promoter recognition, binding of RNA polymerase to promoter, unwinding of DNA double helix, synthesis of first base of RNA chain), chain elongation (release of sigma (σ) factor, direction of transcription, proof reading of mRNA, protein-protein interaction with RNA polymerase), chain termination (Rho-dependent termination, Rho-independent termination); *translation (protein synthesis)-* charging of tRNA, initiation of polypeptide synthesis, chain elongation, translocation, termination of polypeptide chain, post translational processing, signal hypothesis (protein export), *question*.

Chapter 4: Molecular Genetics 128–181

Chromosomal basis of Mendelian genetics- Mendel's laws of inheritance (law of segregation, law of independent assortment), chromosomal theory of inheritance (verification of chromosome theory, sex chromosome and chromosomal theory), crossing over, (synaptonemal complex), linkage, genetic recombination, gene mapping, (recombination frequency, LOD score method for estimation of recombination frequency); *cytoplasmic inheritance-* maternal effects, extranuclear inheritance (chloroplast inheritance, mitochondrial inheritance); *microbial genetics: transformation of genetic materials in prokaryotes-* transformation (identification of transforming substance, factors affecting competence, mechanism of transformation, Hershey and Chase experiment), conjugation (role of surface protein, F factor, genetic map of F plasmid, conjugal transfer process, barriers to conjugation, high frequency recombination strains, formation of F′ prime), transduction (generalised transduction, specialised transduction, use of transduction in genetic mapping in viruses); *transposable elements-* nomenclature of transposable elements, classes of transposable elements (insertion sequence element, transposons) *mutation and mutagenesis: spontaneous mutation*- random nature of mutation, evidence for spontaneous mutation, origin of spontaneous mutation; substitution, spontaneous deamination of 5-methylcytosine, frame shift mutation; *induced mutation*- base analogue, alkylating agents, intercalating agents; *radiation as mutagens*- UV radiation, X rays; *expression of mutation*; *detection and isolation of mutants*- replica plating technique, resistance selection method, substrate utilisation method, carcinogenicity test; *DNA repair systems*- mechanisms of DNA repair (photo reactivation, excision repair, recombination repair, mismatch repair, SOS repair); *questions*.

Chapter 5: Biophysical Methods (= Biophysics) 182–244

Chromatography: *gel permeation* (determination of molecular weight of macromolecules, application of gel permeation); *ion exchange chromatography* (anion exchangers, cationic exchangers, matrix quality, separation of solutes, application of ion exchange chromatography); *electrophoresis* (factors affecting migration of substances), *types of electrophoresis*-paper electrophoresis, agarose gel electrophoresis, polyacrylamide gel electrophoresis, SDS-PAGE, pulse-field gel electrophoresis, 2D gel electrophoresis (isoelectric focussing), *electroblotting*; *gene sequencing*: *DNA sequencing*- Maxam and Gilbert's method (cleavage of purine, cleavage of pyrimidine), Sangar and Coulson's method (dideoxynucleotide chain termination method), automated DNA sequencer; *RNA sequencing, restriction mapping, site-directed mutagenesis*- method of SDM; *spectroscopy*- colorimetry, UV-VIS spectrophotometry, fluorescence spectroscopy; *polymerase chain reaction*-working of PCR (denaturation, primer annealing, primer extension), application of PCR (diagnosis of pathogen, diagnosis of specific mutation, prenatal diagnosis, DNA fingerprinting, in research, in molecular archaeology, diagnosis of plant pathogens); *type of PCR*- reverse transcription PCR, real time PCR (RT-PCR) (fluorescent reporter probe method, application of RT-PCR), arbitrary primed PCR, quantitative PCR, miniprimer PCR; *PCR-based molecular markers*- RAPD marker (step of RAPD PCR and application of plant DNA), DNA amplification fingerprinting (DAF); *radioactivity*- ionizing radiation (types of ionizing radiation, effects of radiation on health), radioactivity (isotopes, beta decay, half-life); *tracer techniques (= radiotracer techniques)*- tracer isotopes, detection of radioactivity (Geiger-Müller counter or Geiger counter, scintillation counter, neutron rem meter (with proportional counter, proportional counter), applications (archaeological dating (= ^{14}c dating), metabolic research, medical applications, in *autoradiography*- method of autoradiography, replication of circular bacterial chromosomes; *radioimmuno assay* (RIA)- method of radioimmunoassay; *immunohistochemistry* (IHC); *laws of thermodynamics*- zeroth law of thermodynamics (thermodynamic equilibrium, conditions for equilibrium), first law of thermodynamics, second law of thermodynamics, third law of thermodynamics; *questions*.

Chapter 6: DNA Modifying Enzymes and Restriction Endonucleases 245–280

Exonucleases- lambda exonuclease, exonuclease I, exonuclease III, Bal 31 exonuclease; *endonuclease*- S1 nuclease, mungbean nuclease, deoxyribonuclease I (DNase I), ribonucleases; *restriction endonucleases: restriction-modification system, types of restriction endonucleases* - type I, type II (recognition sequences, properties of type II restriction enzymes, frequency of restriction enzyme sites, action of *Eco*RI, nomenclature of restriction enzymes), type III endonucleases, other restriction enzymes, factors influencing restriction enzyme activity (buffer composition, incubation temperature, effect of DNA methylation, star activity, digestion with multiple enzymes); *DNA polymerase- E. coli* DNA polymerase, Klenow (large) fragment of *E. coli* DNA polymerase I, polymerisation activity, application of Klenow fragment, T4 DNA polymerase, T7 DNA polymerase, thermostable DNA polymerases; *terminal transferase-* application of terminal transferase (labelling of 3' ends of DNA, addition of complementary homopolymer tails to DNA); *reverse transcriptase-* activities of reverse transcriptase (DNA polymerase activity, RNase H activity), uses of reverse transcriptase (copying RNA into DNA, generating DNA copies of RNAs prior to amplifying that DNA by PCR, creation of cDNA library, preparation of antiviral drugs); *bacteriophage RNA polymerase; other DNA modifying enzymes-* DNA ligases (types of DNA ligases- *E. coli* DNA ligase, T3 DNA ligase, T4 DNA ligase, T7 DNA ligase; ligation mechanism), alkaline phosphatase (sources of alkaline phosphatase, uses of alkaline phosphatase), polynucleotide kinase (forward reaction, exchange reaction); *questions*.

Chapter 7: Cloning Vectors **281–332**

Bacterial plasmid vectors- some considerations of plasmid vectors (plasmid numbers, amplification of plasmids, examples of plasmids), pSC101, pBR322, pUC, pUC57, pACYC184, pUN121, *Agrobacterium*–based plasmid vectors (Ti-plasmid, Ri-plasmid), properties of Ti-plasmids, pTi-based vectors (co-integrate vectors, binary vectors); *bacteriophage vectors-* phage λ vectors (why phage lambda as cloning vector, elimination of restriction sites), cosmid vectors, phage M13 vectors (advantages of using M13 vectors, M13mp18/ M13mp19 vectors), phage P1 vector; *phagemid (phasmid) and fosmid vectors-* phagemid pBluescript II KS (+/–), phagemids pTZ19R and pTZ19U, fosmid vectors; *bacterial artificial chromosome (BAC)-* cloning of BAC vector, selection of transformed bacteria; *yeast plasmid vectors-* yeast vectors, yeast integrative plasmid (YIp), yeast episomal plasmid (YEp), yeast replicating plasmid (YRp), yeast centromeric plasmid (YCp) or yeast centromere (CEN), yeast linear plasmid (YLp); *yeast artificial chromosomes (YAC); plant and animal viruses as cloning vectors-* plant viral vectors (caulimoviruses, geminiviruses, tobamoviruses), animal viral vectors; *transposons as vector-* Ac-Ds elements, P element; *shuttle vectors and expression vectors-* shuttle vectors, expression vectors (features of *E. coli* expression vector); *questions*.

Chapter 8: Methods of Gene Manipulation **333–382**

Host cells- prokaryotic cells, eukaryotic host cells; *gene cloning in prokaryotes: strategies of recombinant DNA technology-* basic technique of gene cloning, pros and cons; *DNA library or gene bank-* types of DNA libraries (how many clones are required for a library, size of gene library), genomic library, cDNA library (isolation of mRNA, reverse transcription, oligo-dC tailing, alkali hydrolysis, addition of oligo-dG primer, synthesis of second strand of cDNA, cleavage of hairpin loop, generation of cohesive ends on double-stranded cDNA), screening of cDNA clone; *insertion of a foreign gene (DNA insert) into a cloning vector-* blunt-end ligation by linkers and adaptors, homopolymer tailing; *transformation of bacterial cells by recombinant DNA-* transformation, transfection; *selection (screening) of recombinants-* direct selection of recombinants, insertional inactivation method, blue-white screening method, screening by colony hybridisation (nucleic acid hybridisation) technique

(DNA probes, colony (or plaque) hybridisation technique, plaque lifting method), screening by *in vitro* translation method, direct immunological screening technique; *blotting techniques*- Southern blotting techniques, northern blotting technique (applications of the northern blot, advantages of northern blotting, disadvantages of northern blotting, reverse northern blot), western blotting techniques (protein blotting or electroblotting technique), dot blot technique (dot blot analysis); *recovery of cells containing recombinant plasmid : expression of cloned DNA*- Shine-Dalgarno sequence , expression vectors; *detection of nucleic acids*- radioactive labelling (nick translation, random primed radiolabelling of probes, probes developed by PCR), non-radioactive labelling (horseradish peroxidase (HRP) method, digoxigenin (DIG) labelling system, biotin-streptavidin labelling system, fluorescent labelling with quantum dots); *gene cloning in eukaryotes- gene cloning in yeasts and molds*- gene cloning in yeasts, transformation in filamentous fungi (application of transformation of fungal protoplasts); *gene cloning in plant cells*- gene transfer in dicots by using *Agrobacterium* Ti-DNA as vector, gene transfer in monocots, plant cell transformation (plant cell transformation by co-cultivation method, co-integration method, and ultrasonication, liposome-mediated gene transfer, electroporation, particle bombardment gun, fibre- mediated DNA delivery); *gene cloning in animal cells*- animal viruses, microinjection; *questions*.

Chapter 9: Gene Therapy 383–420

History of gene therapy, genetic diseases- Alzheimer disease, sickle-cell anaemia, thalassaemia, Duchenne muscular dystrophy, severe combined immunodeficiency; *background of gene therapy, types of gene therapy*- germ-line gene therapy, somatic gene therapy; *methods of gene therapy; Viral vector-mediated gene therapy* (gene therapy of SCID) – retroviruses, adenoviruses, adeno-associated viruses; *non-viral methods*- direct injection of naked DNA, lipoplexes and polyplexes, receptor-meditated gene delivery, embryo therapy through IVF technology, magic bullets—cells as drug delivery system [genetically modified mesenchymal stromal cells (MSC) for therapy, bone marrow-derived mesenchymal stromal cells], virosomes in drug targeting, chemically modified viral vectors and synthetic viruses (optimisation of synthetic viruses, bioresponsive elements); *recent development in gene therapy*- gene silencing (RNA interference), sense RNA and antisense RNA, Dicer enzyme, antisense therapy (example antisense therapies); *current status of gene therapy research; problems and consideration*- problems related to gene therapy, ethical considerations, social considerations, metaphysical considerations, *governmental oversight and public participation*; *questions*.

Chapter 10: DNA Profiling (DNA Fingerprinting) 421–450

DNA fingerprinting (= DNA profiling)- biological evidence, variable number of tandem repeats (VNTR), microsatellites, minisatellite; *method of DNA fingerprinting*- RFLP analysis, PCR analysis, short tandem repeat (STRs) analysis, amplified fragment length polymorphism (AFLP), Y-chromosome analysis, mitochondrial DNA (mtDNA) analysis; *application of DNA profiling (DNA fingerprinting) technology*- setting up of a national DNA databases, reuniting lost children, identifying bodies or body parts, establishing paternity, solving of baby swapping case, application in forensic diagnostics (establishing identity of criminals and rapists- sexual escapades, tandoor case), solving immigrant dispute, DNA ancestry testing, DNA testing in IVF technology; application of DNA profiling in live stock industry; *hurdles to DNA profiling technology*; *disease diagnosis: diagnosis of parasitic diseases*- diagnosis of parasitic diseases through DNA probe (identification of African trypanosomes); *genetic counselling*- indirect gene tracking, antenatal diagnosis of haemoglobinopathies, foetal cell/DNA prenatal paternity test; *questions*.

Chapter 11: Proteins: Structure, Engineering and Design **451–523**

Chemical bonds- ionic bond, covalent bond, hydrogen bond, van der Waals force, peptide bond; *protein structure : amino acids*- structure of amino acids (isomerism, ionic forms of amino acids, zwitterions, polypeptide chains), classification of amino acids (non-polar amino acids, polar amino acids with no charge, charged polar amino acids), polymerisation of amino acids, nomenclature (IUPAC system), chemical properties of amino acids (calculation of pI values, reaction with ninhydrin); *conformation analysis of proteins*- special features of proteins, primary structure of proteins (post-translational modifications), secondary structure of proteins (alpha helix- Ramchandran plot, β-pleated sheet), tertiary structure of proteins, quaternary structure of proteins, protein sequencing (Sanger's method, Edman degradation method, enzymatic method for fragmentation of polypeptide chains, chemical method); *X-ray crystallography*- X-ray diffraction analysis of crystals, methods of X-ray diffraction (crystallisation, data collection, data analysis); *nuclear magnetic resonance (NMR)*- theory of NMR (nuclear spin and magnet, spinning in a magnetic field, magnetic resonance by nuclei, nuclear shielding), working principle of NMR, types of NMR spectroscopy (NMR spectroscopy, continuous wave spectroscopy, fourier transform spectroscopy, multi-dimensional NMR spectroscopy), applications of nuclear magnetic resonance (NMR in structural biology, medicine, chemistry, non-destructive testing); *mass spectrometry (MS)*- principle of working of mass spectrometry, components of mass spectrometry, uses of mass spectrometry; *matrix-assisted layer desorption-ionisation (MALDI)*- ionisation method, application (in biochemistry, organic chemistry, polymer chemistry); *surface-enhanced laser desorption-ionisation (SELDI)*- SELDI-TOF-MS, sample preparation and data processing; *electrospray ionisation (ESI), desorption electrospray ionisation (DESI), protein-DNA interac*tion- types of DNA-binding domains (helix-turn-helix domain, helix-loop-helix domain, zinc finger domain, leucine zipper domain, HMG-box domain, coiled coil motif), unusual DNA-binding domains (immunoglobulin fold, B3 domain, TAL effector DNA-binding domain); *protein engineering and protein design*- differences between protein engineering and protein design, strategies for protein engineering (rational design of proteins, directed evolution, examples of engineered proteins), steps of protein engineering; *methods in protein engineering and design*- physical and computational techniques (X-ray crystallography, nuclear magnetic resonance, spectroscopic technique, prediction of protein structure, computer modelling), biochemical techniques and molecular biology (site-directed mutagenesis, recombinant DNA techniques); *structure-function relationships*- models of protein structure and function used in protein design; *some achievements in protein-engineering and design*- how does serine protease (subtilisin) work ?, protein engineering in subtilisin, recombinant vaccines, protein engineering in enzymes (triosephosphate isomerase, beta-lactamase), protein design, improvement in nutritional quality of seeds (essential amino acid profile); *questions*.

Chapter 12: Bioinformatics **524–569**

Historical background, what is bioinformatics ?, databases- types of databases (primary databases, secondary databases, composite database, other databases-flat-file database, relational database, hierarchial database), database structure, database management systems (DBMS) (components of DBMS, current trends), database models (network model, dimensional model, object-oriented database models), model-view-controller (MVC) (model, view, controller, variation of MVC); *sequences and nomenclature-* IUPAC symbols, nomenclature of DNA sequences, nomenclature of protein sequences, directionality of sequences, types of sequences used in bioinformatics (genomic DNA, cDNA, organellar DNA, expressed sequence tag (EST), gene sequence tag (GST), other biomolecules; *information sources-* National Centre for Biotechnology Information (NCBI) (function of NCBI, GenBank, mouse genome database (MGD); *data retrieval tools-* ENTREZ [the ENTREZ databases, online Mendelian inheritance in man (OMIM)- MIM code], PubMed, Taxonomy browsers, LocusLink, sequence retrieval system; *database similarity searching-* basic local alignment search tool (BLAST) (uses of BLAST), FASTA (FASTA format, method of searching); *resources for gene level sequencing-* UniGene database, HomoloGene database, RefSeq database; *sequence alignment-* global and local alignments, pair-wise alignment, dot matrix methods, dynamic programming, word methods, multiple sequence alignment [dynamic programming, progressive methods, iterative methods, motif finding (profile analysis)], structural alignment (DALI, SSAP, combinatorial extension); *use of bioinformatics tools in analysis; questions.*

Chapter 13: Genomics and Proteomics **570–621**

The era of genomics and proteomics; genome sequencing projects, microbial genome sequencing projects- viral genomics (standards for viral genome and annotation data, biotechnological applications); *bacterial genome sequencing projects- Escherichia coli* genome project, *Bacillus* genome project*; eukaryotic genome sequencing projects*; *human genome project (HGP-* donors of genome, benefits of human genome project, cost of human genome project, implications of genomics on medical science, ethical, legal and social issues (ELSI)*; methods of gene sequencing-* Sanger sequencing method, pyrosequencing (= next generation sequencing) (limitation of the method, commercialisation, applications of pyrosequencing), DNA nanoball sequencing, hierarchical sequencing (clone-by-clone sequencing) method (construction of physical map, shotgun cloning, automated DNA sequencing), whole-genome shotgun sequencing (library construction, random sequencing, fragment-alignment and gap closure, proof reading), expressed sequence tag (EST) method*; genome assembly and finishing-* trimming of vector sequences, sequence quality determination (Phred quality scores, sequence assembly, assembly view), problems related to draft sequence and genome finishing (gap closure, validation and refinement), assembly verification*; gene prediction and gene counting-* methods of gene prediction (content-based method, pattern-based method, comparative methods), gene prediction algorithms (homology-based gene prediction, *Ab initio* gene prediction, systenic gene prediction), accuracy and validity of gene prediction algorithms, gene prediction software*; genome similarity and single nucleotide polymorphisms (SNP)-* genome similarity, single nucleotide polymorphisms (SNPs)*; types of genomics-* structural genomics, functional genomics (functional genomics toolbox, determination of function of unknown genes, patterns of gene expression), comparative genomics (example of comparative genomics, databases for comparative genomics), DNA chip (DNA microarray) technology (types of DNA chips, comparative microarray hybridisation, application of DNA chips); *proteomics*- relation between gene and protein, approaches for study of proteomics, types of proteomics (expression proteomics, structural proteomics, functional proteomics), protein structure database; *questions*.

Chapter 14: Metagenomics **622–654**

Historical background, about metagenomics- biological insights from metagenomics; *from genomics to metagenomics; different metagenomics projects*- environmental microbial metagenomics project, microbial metagenomics project, hydrocarbon metagenome project, red sea marine metagenomics; *metagenomics at work, technologies for metagenome study*- metagenomic nucleic acid extraction and enrichment strategies (GC enrichment by ultracentrifugation, bromodeoxyuridine (BrdU) enrichment, stable-isotope probing, suppressive subtractive hybridisation, differential expression analysis, phage display, affinity capture, microarray, enrichments for metagenomics clones forming metabolic consortia, metagenomic cDNA (transcriptomic) libraries); *metagenome gene discovery tools*- gene-specific PCR, RNA as biomarker, integrons, homologous recombination cloning, immobilised oligonucleotides, microarrays, fluorescent *in situ* hybridization (FISH), substrate-induced gene expression screening (SIGEX), pre-amplification inverse-PCR (PAI-PCR), metagenomic DNA shuffling, function-driven analysis, sequence-driven analysis; *metagenomics via next generation sequencing*- 454 sequencing technology, genome Sequencer™ FLX system, true single-molecule sequencing technology (tSMS), single molecule real time technology, Oxford nanopore technologies; *metagenomics via bioinformatics*- CAMERA, CARMA, CLC bio workbench, DOTUR, ∫-LIBSHUFF, galaxy, IMG/M, metagenomics RAST server (MG-RAST), MetaSim, MEGAN, MetaStats, STAMP, SONS, SmashCommunity, TreeClimber, Treephyler; *application of metagenomics*- agricultural application, human health, environment, industrial application; *questions.*

Chapter 15: Enzyme Technology **655–714**

Enzyme structure- cofactors and coenzymes (cofactors, coenzymes); *properties of enzymes*- presence of species specificity, variation in activity and ability (effects of pH, effect of temperature, effect of ions, effect of other factors), substrate specificity, activation and inhibition (competitive inhibition, non-competitive inhibition, feedback inhibition (allosteric inhibition), regulation (control) of enzyme activity (allosteric control, proteolytic activation, isoenzymes, reversible covalent modification, control by limiting amount of enzyme); *novel enzymes*- abzymes (naturally occurring abzymes, application of abyzme technology), synzymes, extremozymes; *classification and nomenclature of enzymes*- nomenclature committee of the International Union of Biochemistry and Molecular Biology (NC-IUBMB), list of recommended names for enzymes, glossary, reaction pathways and links to other databases (enzyme subclasses); *mechanism of enzyme catalysis*- induced fit theory, mechanisms of transition state stabilisation (catalysis by bond strain, catalysis by proximity and orientation, acid/base catalysis- catalysis involving proton donors or acceptors, covalent catalysis, electrostatic catalysis, quantum tunneling); *enzyme kinetics*-

principles of enzyme kinetics, enzyme assays, single-substrate reactions, Michaelis–Menten kinetics (linear plots of the Michaelis–Menten equation, significance of kinetic constants), multi-substrate reactions (ternary-complex mechanisms, ping–pong mechanisms), non-Michaelis–Menten kinetics*; immobilization of enzymes and whole cells-* carrier matrices and its properties (properties of carrier matrix, carrier binding), methods of enzyme immobilization (adsorption, covalent bonding, cross-linking or co-polymerisation, entrapment, encapsulation), advantages and disadvantages of immobilised enzymes (advantages, disadvantages), immobilisation of whole cells (immobilisation of animal cells culture)*; application of enzymes, immobilised enzymes and whole cells- application of enzymes and immobilised enzymes-* applications in pharmaceutical industry (uses in semi-synthetic penicillin production, immobilisation of amino acid acylase, production of high fructose syrup, enzymes used in production of glucose syrup), therapeutic uses (= medical and clinical uses) (enzymes used in cancer therapy), uses of foods and beverages, uses in textile industry, use in removal of stickies, use in detergents (uses of lipases in detergent industry), application in dairy industry, uses in food industries, uses in biosensors; *application of immobilized whole cells-* enzyme produced by immobilised cells, alcohol production by immobilised cells, use of cell immobilisation in meat industry; *biosensor : characteristics and types of biosensors-* components of a biosensor, characteristics of biosensors, type of biosensors (glucose-electrode biosensor, electro-chemical biosensor, amperometric biosensor, thermistor-containing biosensor, split-flow enzyme thermistor (cholesterol biosensor), bioaffinity biosensors, calorimetric biosensors, opto-electronic biosensor, piezo-electric biosensors, whole cell biosensors (microbial biosensors)*; applications of biosensor-* uses in medicine and health, uses in industry, biosensor in military, uses of biosensors in environmental monitoring (gas biosensors, immunosensors, BOD biosensor, luminescent bacterial biosensors, electrochemical biosensors, enzyme electrodes, whole cell biosensors); *enzyme engineering; questions*.

Chapter 16: Immunology **715–766**

History of immunology; branches of immunology; antigen- types of antigens (T-independent, T-dependent), origin of antigens [exogenous antigens, endogenous antigens, autoantigens, tumour antigens, native antigen (= nativity)]*; innate immunity, cells of innate immune response-* erythrocytes, leukocytes, mast cells, phagocytes, basophils and eosinophils, natural killer cells, gamma-delta (γδ) T cells*; responses of innate immune system-* inflammation, complement system, cytoplasmic viral receptors, antimicrobial peptides, apoptosis (programmed cell death) (innate immune response, adaptive immune response); *adaptive or acquired immunity; humoral immunity, antibody (structure and diversity)-* antigen binding, structure of antibodies, immunoglobulin fragments- structure:function relationships (Fab, Fc); *immunoglobulin isotypes-* immunoglobulin gamma (IgG), immunoglobulin delta (IgD), immunoglobulin alfa (IgA), immunoglobulin mu (IgM), immunoglobulin epsilon (IgE)*; antigen determinants of immunoglobulins-* isotypes, allotypes, idiotypes; *antibody kinetics-* regulation of humoral response (antigen blocking and receptor cross-linking, idiotypic networking); *cell-mediated immunity; antibody diversity-* genetic basis of antibody diversity- light chain (kappa or lamba) formation, heavy (H) chain formation), isotype switching, germline diversity : multigene families [diversity due to DNA rearrangement (combination diversity), diversity as a result of somatic hypermutations]*; complements-* activation of complement cascade (classical pathway, alternative pathway, lectin pathway *i.e.* mannose-binding lectin

(MBL) pathway); *histocompatibility- Major Histocampatibility Complex-* MHC Class I broteins, MHC Class II proteins, MHC Class III proteins; *hypersensitivity-* type I, type II, type III, and type IV hypersensitivity; *autoimmunity-* causes of autoimmunity, tolerance (origin of immunological tolerance, genetic factors), treatment of autoimmune diseases; *questions.*

Chapter 17: Immunotechnology **767–827**

Hybridoma technology- Monoclonal antibodies (production of monoclonal antibodies) (purification of monoclonal antibodies, production of recombinant monoclonal antibodies), applications of monoclonal antibodies (diagnostic tests, monoclonal antibodies for cancer treatment), monoclonal antibody therapy (origins of monoclonal antibody therapy, function of human and therapeutic antibodies), mechanism of action, antibody-directed enzyme prodrug therapy (ADEPT), drug and gene therapy (immuno-liposomes), approved therapeutic antibodies, problems with successful therapy; *interferons-* types of interferons (type I and type II), cell activation by interferon-alpha and interferon-beta, role of STAT protein (STAT family (activation, function)), therapeutic uses of interferons, pegylation of interferon, production of recombinant interferon (alfa-2a); *vaccines, types of vaccines-* live attenuated vaccines- inactivated (killed) vaccines, subunit vaccines, toxoid vaccines, conjugate vaccines (component vaccines), DNA vaccines (construction of DNA vaccines, mode of administration, mechanisms of action, advantages of DNA vaccines over classic vaccine), recombinant vector vaccines (recombinant vaccine for hepatitis B virus, indigenous hepatitis-B vaccine); *polio vaccine-* oral polio vaccine (OPV) (advantages of OPV, disadvantages of OPV), inactivated polio vaccine (IPV) (advantages of IPV, disadvantages of IPV), vaccination schedule; *edible vaccines (plant-derived vccines)-* edible vaccines, edible antibodies (plantibodies), edible interferon; *transplantation (transplant immunology)- organ transplants-* types of transplants (autograft, allograft, isograft, split transplants, domino transplants, xenograft or xenotransplant), types of donor [lving or deceased, deceased (formerly cadaveric)], reasons for donation (living related donors, compensated donation, forced donation), organ transplant laws, ethical concerns; *haematopoietic stem cell transplantation-* types of HSCT graft (autologous graft, allogeneic graft), sources of HSCT (bone marrow, peripheral blood stem cells, amniotic fluid, umbilical cord blood), complications and side effects [veno-occlusive disease, mucositis, infection, graft-versus-host disease (GVHD)]; *transplant rejection-* types of rejection (hyperacute rejection, acute rejection, chronic rejection), rejection mechanisms, treatment of rejection; *tumour immunology- tumour antigens-* tumour-specific antigens, tumour-associated antigens [tumour-associated transplantation antigens (TATA), tumour-associated developmental antigens or onco-fetal antigens], tumour-associated transplantation antigens on viral tumours, importance of tumour antigens; *cancer immunotherapy-* active immunotherapy, passive immunotherapy (specific, and non-specific passive immunotherapy) Monoclonal antitumour antibodies *questions.*

Chapter 18: Plant Biotechnology **828–884**

History of plant cell, tissue and organ culture- research work on plant tissue culture in India; *laboratory techniques: a tissue culture laboratory-* basic facilities (washing and storage, media preparation room, transfer area); *nutrient media-* complex mixture of salts (*macroelements, microelements,. iron source*), organic supplements (carbon source, vitamins, amino acids), growth hormones (cytokinins, auxins, gibberellins, *abscisic acid, ethylene*), antibiotics, solidifying agents, pH; *maintenance of aseptic environment during culture*- sterilisation of glassware, instruments, culture room and transfer area, and nutrient media; *methods of plant cell, tissue and organ culture*- basic steps (preparation of suitable nutrient medium, selection of explants, sterlisation of explants, inoculation of explant, incubation, regeneration, hardening, plantlet transfer to greenhouse and field); *meristem culture- production of disease-resistant plants*- production of virus-free plants from meristem culture; *callus culture*- factors affecting callus culture; *plant regeneration*- organogenesis (cytokinin–auxin control of shoot meristem development, auxin–cytokinin regulation of *in vitro* organogenesis); *root culture*- hairy root culture- application of hairy root culture; *cell (suspension) culture*- cell culture in mass (batch culture, continuous culture, benefits from cell culture), culture of isolated single cells, application of cell suspension culture (production of secondary metabolites, production of stress-tolerant plants); *protoplast culture*- isolation of protoplasts (surface sterilisation of leaf samples, rinsing in suitable osmoticum, plasmolysis of cells, peeling of lower epidermis, purification of protoplasts, protoplast culture and regeneration); *protoplast fusion and somatic hybridisation*- hybridisation and cybridisation, methods of somatic hybridisation, selection of somatic hybrids and cybrids (biochemical methods, auxotrophic mutant selection method, morphological or visual selection method, cytometric method), applications of somatic hybridisation; *somatic embryogenesis and encapsulated seeds*- somatic embryogenesis (types of encapsulated seeds, steps of somatic embryogenesis, factors influencing somatic embryogenesis), encapsulated artificial seeds (methods of production, advantages of artificial seeds); *micropropagation; somaclonal variation*- causes of somaclonal variation (physiological causes, genetic causes, biochemical causes), application in crop improvement, examples of somaclonal variants; *anther and pollen culture*: *haploid production*- method of pollen culture (*in vitro* androgenesis, acclimatisation, factors affecting androgenesis, limitations in haploid production, use of haploids); *production of monoploid plants*- uses of monoploids; *triploid production*- seedless watermelon, triploid papaya; *mentor pollen technology*- advantages of mentor pollen technology; *ovary culture and embryo culture: ovary culture, embryo culture*- culturing method; *embryo rescue*- method of embryo rescue, application of embryo rescue; *cryopreservation*- method of cryopreservation (selection of materials, addition of cryoprotectors, freezing, storage in liquid nitrogen, thawing, washing and reculturing, regeneration of plantlets), risks-associated with cryopreservation, achievement through cryopreservation; *questions*.

Chapter 19: Transgenic Plants **885–940**

Gene construct: organisation of a typical plant gene- promoters (TATA box, CAAT box), transcription start site, enhancers/silencers, introns and exons, reporter genes (marker genes), scorable (screenable) markers, selectable markers); *methods for developing GM plants* steps for developing GM plants, agroinfection and trasformation, satellite RNA and its use in transformation; *herbicide-resistant transgenic plants*- EPSP synthase, EPSP synthase (*aro A*) gene, microbial EPSP synthase (*aro A*) genes, EPSPS-resistant GM plant (Roundup Ready); *virus- resistant transgenic plants: viruses as cloning vector*- caulimoviruses (35S promoter of CaMV, production of new promoters), geminiviruses, cucumovirus (CMV); *expression of viral gene*- expression of coat protein (CP), expression of other genes, anti-viral proteins of other plants; *transgenic plants*- GM papaya, GM tomato; *fungus- resistant plants*- pathogenesis-related proteins (PRs), Barnase–barstar component system (mechanism of catalysis, electrostatic optimisation of binding complex, transgenic plants)*; insect-resistant plants- Bacillus thuringiensis* (Bt) (mechanism of action of Cry protein crystals, genetics of *Bacillus thuringiensis,* promoters used with *cry* gene, transformation of *A. tumifaciens* cells and creation of transgenic plants), Bt cotton [Bt cotton in India, use of *cry* gene(s) in cotton, biosafety tests and assessment of toxicity to non-target organisms, economic impact of Bt cotton in developing countries], Bt brinjal (methods of transformation, mode of action of Cry1A(c) protein, field trials, Bt brinjal controversy)*; modification in nutrient quality*- Golden rice (*Oryza sativa*) (protest against release of Golden rice)*; molecular pharming: molecular of plantbodies (antibodies in plants)*- different plant-based expression systems, downstream processing, regulatory landscape*; edible vaccines*- methods of production of edible vaccines (*Agrobacterium* –mediated gene transfer method, virus- mediated gene transfer method), advantages of edible vaccines, advancements made toward plant-based vaccines (hepatitis B, influenza, rotavirus, tuberculosis), commercial use of transgenic plants; *improvement of seed protein quality; questions.*

Chapter 20: Transgenic Animals (= Animal Biotechnology) 941–991

Gene construct- promoter/enhancers, reporter (= marker) gene [chloramphenicol acetyltransferase (CAT), thymidine kinase (TK), dihydrofolate reductase (DHFR) (resistant to methotrexate), bacterial xanthine-guanine phosphoribosyl transferase (XGPRT), neomycin phosphotransferase (NPT) (resistant to neomycin, G-418), transcriptional and translational reporters]; *Cloning vectors*- baculovirus expression vectors (Bac-to-Bac method)*; mammalian virus vectors*- SV40-based vectors (SV40 transducing vectors, SV40 plasmid vectors), bovine papilloma virus (BPV) vectors, mammalian artificial chromosome (MAC) vectors; *transfection methods*- types of transfection (stable transfection, transient transfection), methods of transfection (calcium-phosphate precipitation, DEAE-dextran mediated transfection, viruses-mediated transfection, liposome-mediated transfection (lipofection), electroporation, impalefection, microinjection), production of transgenic cattle*; embryonic stem cell-mediated gene transfer*- embryonic stem cell transfer technology (generation of embryonic stem cells), transfection of gene construct, production of transgenic animals by using transfected ES cell clones; *gene targeting (targeted gene transfer)*- targetted gene transfer method (gene disruption, gene replacement, tag and exchange), identification of homologous recombination (negative selection method, PCR and Southern blot methods, recombineering method, reference-query pyrosequencing (RQPS) method, bioinformatics method), site of chromosomal integration (integration sites, expression of transgene, effect of transgene on function of other genes, determination of transgene integration sites)*; rescue and recovery of transgenes from animal cells: detection of transgenic animals*- Southern blot analysis of transfected gene construct, northern blot analysis of cytoplasmic RNA, translation analysis*; breeding with transgenic livestock*- advantages of transgenic breeding over selective breeding, applications of transgenic animals (agricultural application, medical application, industrial applications); *transgenic animals produced*- mouse (Alzheimer's mouse, OncoMouse™), fish (GloFish®), cows (low cholesterol cow milk, disease-resistant cows, human breast milk in cows), goat (spider goat, therapeutic milk proteins), sheep (Tracy, Dolly and Bonny, Polly and Molly, Geep (sheep-goat chimera), 'Peng Peng' sheep, organ donor sheep), pigs (green pig (= FluoroPig), EnviroPig™), dogs [Snuppy, Ruppy (glowing dogs)], chickens*; molecular pharming; bioethics; questions.*

Chapter 21: Industrial Microbiology **992–1043**

Microbial growth- growth parameters, viable plate count, cell counting by direct microscopic count method, viable counting (plate counting) technique, measurement of cell mass, absorbance (optical density turbidity) measurement method, growth in batch culture (lag phase, log phase, deceleration phase, stationary phase, death phase); microbial growth kinetics and specific growth rate, synchronous growth diauxic growth, growth in fed-batch culture, growth in continuous culture (chemostat for growth measurement)*; technique of microbial culture-* sources of nutrition, culture media (types of media, preparation of media)*; fermentor: designing and construction-* basic functions of fermentors, construction of fermentors (control of temperature, aeration and agitation, design and operation), types of fermentors (fluidized bed, air-lift, membrane, pulsed column, bubble column, photo- and packed tower bioreactors), use of computer in fermentor, maintenance of aseptic conditions (sterilisation, containment)*; fermentation processes-* batch, fed-batch, and continuous fermentation (single stage, multiple-stage fermentation, turbidostat)*; stoichiometry, process flow, material balance (mass balance)* (some calculations); *scale-up of fermentation-* sterilisation of gases and nutrient solutions, stock cultures; *production of microbial products* : *strain improvement-* methods of strain improvement (plasmid improvement, protoplast fusion, mutation), isolation of mutants and their use in microbial processes (localised mutagenesis and computation, sexual and parasexual processes)*; fermentation process-* upstream process, downstream process, product recovery (cell harvesting, broth conditioning technique, sedimentation, centrifugation, filtration, disruption of microbial cells)*; alcohol production-* preparation of medium, fermentation, recovery*; malt beverages-* beer production (malting process, mashing process, finishing), wine production (microbial process, wine from grapes, fermentation, recovery), microbial deterioration and spoilage of wine*; organic acid production-* vinegar production (substrate, methods of production), lactic acid production (fermentation, medium and manufacturing process, recovery, uses), citric acid (Koji process or solid state fermentation, liquid surface culture process, submerged culture process, recovery, uses)*; antibiotic production-* penicillins (fermentation, biosynthesis of penicillin G (benzylpenicillin), recovery, units of antibiotics), streptomycin (structure of streptomycin, production, recovery, unit of streptomycin)*; amino acid production-* L-lysine production (fermentation, recovery, strain improvement for lysine production), L-glutamic acid (fermentation, recovery, strain improvement for glutamic acid production, uses); *questions*.

Chapter 22: Environmental Biotechnology **1044–1101**

Biosensors in environmental monitoring: bioremediation- Types of bioremediation- in situ bioremediation (intrinsic bioremediation, engineered *in situ* bioremediation), *ex situ* bioremediation (slurry phase bioremediation, solid phase bioremediation), microorganisms involved in bioremediation (aerobic microbes, anaerobic microbes, lignolytic fungi), bioremediation process (bioremediation of heavy metals, polycyclic aromatic hydrocarbons), factors affecting bioremediation (bacterial chemotaxis, biofilm, biosurfactants, abiotic factors), mass balance, superbug, monitoring of bioremediation, biosafety from GMOs, phytoremediation by superfund superplant;

composting- methods of composting, factors affecting composting (C : N ratios, urine, microorganisms), types of composting (biodung composting, compost tea, vermicompost, bokashi compost, industrial composting system), *vermicomposting technology*- working mechanism of vermiculture, importance of vermicompost, physical quality of vermicast, vermicomposting by using earthworms (use of local varieties, use of exotic variety, collection of native earthworms, preparation of vermipit, preparation of vermibed, harvesting of compost), benefits of vermicompost, vermicompost manufacturers and suppliers, *biogas* (*gobar* gas)- biogas technology in India (benefits from biogas plant, feedstock material, construction of a biogas plant), biogas production (solubilisation, acidogenesis, acetagenesis, methanogens), biogas production from different feedstock (Salvinia, water hyacinth, municipal sewage), microbial diversity in biogas digesters, factors affecting methane formation, uses of biogas and sludge, *biomining (bioleaching)*- identification of bacteria (immunofluorescence, dot immunoassay, dot-blot hybridisation) methods of biomining (*in situ* leaching, damp leaching, heap leaching, vat leaching), bioremediation of metals by bacteria, iron bioleaching, copper bioleaching (mechanism of copper leaching, microbiology of copper bioleaching, biomining of copper, resistance mechanism in biomining bacteria), gold biomining (biooxidation of gold, future application), uranium recovery, metal bioleaching with fungi; *questions*.

Glossary **1102 – 1139**

Index **1140 – 1161**

CHAPTER

Introduction

A. History of biotechnology
B. Traditional vs. modern biotechnology
C. The world of biotechnology
D. Bioinformatics
E. Synthetic biology
F. Public perception of biotechnology
G. Biotechnology in India
H. Ethical Issues

The term biotechnology is believed to have been coined in 1917 by a Hungarian engineer Karl Ereky (1878- 1952). In 1919, Ereky first used the word 'biotechnology' in his book *Biotechnologie der Fleisch-, Fett- und Milcherzeugung im landwirtschaftlichen Grossbetriebe* (Biotechnology of Meat, Fat and Milk Production in an Agricultural Large-Scale Farm). In this book he described a technology on the basis of conversion of raw materials into a more useful product. Ereky built a slaughterhouse for a thousand pigs and a farm for 50,000 pigs, and produced over 100,000 pigs in a year. This enterprise was one of the largest and profitable meat and fat operations in the world. Further he developed a theme that biotechnology can provide solutions to shortage of food and energy. For Ereky, 'biotechnologie' was the process to upgrade the raw materials biologically into socially useful products. Therefore, Ereky is regarded as the 'father of biotechnology'. Thus the term 'biotechnology' has its roots into *bio* (the use of biological processes) and *technology* (to make useful products applying techniques).

Thus biotechnology is the use of biological processes, organisms or systems to manufacture products for improvement in quality of human life. In recent years scope and applicability of biotechnology has expanded Therefore, application of technical advances in life science for development of commercial products is called biotechnology.

DEFINITION OF BIOTECHNOLOGY

Development of biotechnology has been possible due to revolutionary discoveries made in biology and technology. Biotechnology is not a pure science, but an integrated effort of these two different disciplines, the root of which lies in biological science. Biotechnology is defined by different organizations in different ways. It has been broadly defined as, 'the development and utilization of biological processes, forms and systems for obtaining maximum benefits to man and other forms of life'. Biotechnology is 'the science of applied biological process' (*Biotechnology* : A Dutch Preservative, 1981). Following are few definitions given by other organisations:

1. *Any technological application that uses biological systems, living organisms, or derivatives thereof, to make or modify products or processes for specific use* (United Nations Convention on Biological Diversity).

2. *The application of biochemistry, biology, microbiology and chemical engineering to industrial process and products and on environment* [International Union of Pure and Applied Chemistry (IUPAC), 1981].
3. *The integrated use of biochemistry, microbiology and engineering sciences in order to achieve technological application of the capabilities of microorganisms, cultured tissue, cells, and parts their of* [(European Federation of Biotechnology (EFB), 1981].
4. *Application of scientific and engineering principles to the processing of material by biological agents to provide goods and service"* [The Organisation for Economic Cooperation and Development (OECD), 1981].
5. Biotechnology is the "*integrated use of biochemistry, microbiology and engineering sciences in order to achieve technological applications of capabilities of microorganisms, cultured tissues, cells and part there of*" [The European Federation of Biotechnology (EFB), 1981].
6. *The formation of new combination of heritable material by the insertion of nucleic acid molecules produced by whatever means outside the cell, into any virus, bacterial plasmid or other vector system so as to allow their incorporation into a host organism in which they do not naturally occur by in which they are capable of continued propagation* (Smith, 1996).

A. HISTORY OF BIOTECHNOLOGY

About 10,000 years ago humans started growing crops and raising animals for stable supply of food and clothing. *Vedas* describe the daily practice of '*agnihotra*' (oblation) using cow milk-based fat prepared after churning the curd. Several useful products, such as bread and cheese could be made by using the biological processes of microorganisms about 6,000 years ago. We cannot forget '*makhan chori*' of Lord Krishna about 1,500-2,000 B.C. Science made progress gradually; moreover, during 1960s the result of biological research reached a level when we started to use the biological molecules (*i.e.* the smallest parts of organisms called 'biomolecules'), rather than using whole organisms.

1. From 'Zymotechnology' to 'Biotechnology'

The term 'zymo' refers to enzyme 'zymase', which is used in fermentation yeast or other fungi. Zymotechnology is a subset of biotechnology. Really, biotechnology arose from the field of zymotechnology, which began as a search for a better understanding of industrial fermentation, particularly beer. Beer was an important industrial commodity. In late 19th century in Germany, brewing processes contributed as a national product and a significant amount was collected as revenue by the government. In 1875, Emil Christian Hansen from Carlsberg Institute pioneered the pure yeast process for the production of beer. The private consultancies advised the brewing industry. A German-born chemist John Ewald Siebel established the Zymotechnic Institute at Chicago.

C. Weizmann (1874 – 1952)

Biological industries were encouraged at the end of World War I to create opportunities for a market of new fermentation products, especially non-alcoholic drinks. Emil Siebel established his own institute called the 'Bureau of Biotechnology', which offered expertise in fermented non-alcoholic drinks.

The zymotechnology expanded during World War I to support the war and to fulfill the industrial need. During the war yeasts were grown on a large amount to meet out about 60% needs of animal feed of Germany. A Russian chemist Chaim Weizmann produced acetone by fermenting maize; acetone was used as a raw explosives material that solved the problem of acetone shortage in Britain. Thus the industrial potential of fermentation technology formed a base to the development of zymotechnology that gave a way to biotechnology.

Alexander Fleming

In the 1927, Alexander Fleming discovered penicillin in England. In 1940, penicillin was produced industrially in the U.S.A. by using a deep fermentation process that was originally developed in Peoria, Illinois. A radical shift in the standing of the pharmaceutical industry was resulted in due to enormous profits and the public expectations of this miracle drug, penicillin. The public penicillin represented the perfect health. During 1950s, steroids were synthesised using fermentation technology. Like penicillin, cortisone also had revolutionary ability.

During 1970s, the biotechnology industry originated based on recombinant DNA technique details of which were published in 1973 by Stanley Cohen (Stanford University) and Herbert Boyer (University of California). Several proteins were made using recombinant DNA techniques, for example human insulin, and cultured cells under controlled manufacturing conditions. Boyer became the co-founder of Genentech, which is the largest biotechnology company today. Biotechnology has contributed significantly in modernisation of the European industries. Enhancements in competitiveness, increased economic growth, and improvement in welfare of European citizens have been made due to broad range of high-tech applications of biotechnology. Moreover, application of modern biotechnology can be classified into three broad groups : healthcare and pharmaceutical, industrial processes and manufacturing, and agriculture (livestock, veterinary products and aquaculture).

2. Timeline of Biotechnology

The important events of biotechnology are summarised in brief in timeline of biotechnology (Table 1.1)

Table 1.1. Timeline of biotechnology.

Year	*Major discoveries / events*
Around 8000 BC	Yogurt and cheese made with lactic acid-producing bacteria by various cultures
7000	Brewing beer, fermenting wine, baking bread with help of yeast
4000	Egyptians used yeast to make bread rise
2000	Biotechnology used to leaven bread and ferment beer using yeast (Egypt)
500	First antibiotic: moldy soybean curds (tofu) was used to treat boils (China)
1590 AD	The microscope was invented by Sacharias Jansen
1675	Microorganisms were discovered by Antonie van Leeuwenhoek
1797	First vaccination was done; Edward Jenner took pus from a cowpox lesion, inserted it into an incision on a boy's arm
1830	Proteins were discovered
1856	Gregor John Mendel discovered the laws of inheritance
1857	Louis Pasteur proposed that microbes cause fermentation. He later conducted experiments that supported the germ theory of disease
1915	Bacteriophages were discovered
1919	Karl Ereky, a Hungarian agricultural engineer, first used the word biotechnology
1928	Alexander Fleming noticed that a certain mould could stop the duplication of bacteria, leading to the first antibiotic, penicillin
1944	DNA was proven by Oswald Avery, Colin MacLeod and Maclyn McCarty to carry genetic information

1953	James D. Watson and Francis Crick described the structure of DNA
1955	The amino acid sequence of insulin was discovered by Frederick Sanger
1972	DNA composition of chimpanzees and gorillas was discovered to be 99% similar to that of humans.
1973	Stanley N. Cohen and Herbert Boyer performed the first successful recombinant DNA experiment, using bacterial genes
1975	Method for producing monoclonal antibodies developed by Kohler and Cesar Milstein
1978	North Carolina scientists Clyde Hutchinson and Marshal Edgell showed it is possible to introduce specific mutations at specific sites in a DNA molecule
1980	– The U.S. patent for gene cloning was awarded to Cohen and Boyer – In the landmark case of *Diamond v. Chakrabarty,* the U.S. Supreme Court approved the principle of patenting organisms, which allows the Exxon oil company to patent an oil-eating microorganism.
1981	The first transgenic animals were produced by transferring genes from other animals into mice
1982	Humulin, Genetech's human insulin drug produced by genetically engineered bacteria for the treatment of diabetes, is the first biotech drug to be approved by the Food and Drug Administration
1983	Kary Mullis discovered the polymerase chain reaction (PCR) technique which makes unlimited copies of genes
1986	First recombinant vaccine for hepatitis B was approved for human
1987	First approval for field tests of a genetically modified food plant, *i.e.* virus-resistant tomatoes
1990	The Human Genome Project was launched
1994	The United States Food and Drug Administration approved the first GM food: the 'Flavr Savr' tomato
1997	British scientists led by Ian Wilmut from the Rosalin Institute reported cloning Dolly sheep using DNA from two adult sheep cells
1998	Human embryonic stem cell lines were established
2000	– Completion of a 'rough draft' of the human genome in the Human Genome Project. – The first entire genome of a plant, *Arabidopsis thaliana* was sequenced. – Golden Rice that makes vitamin A was produced to cure blindness.
2001	Bush allowed for a full federal funding for research on adult and umbilical stem cells, but no federal funding for research on human Embryonic Stem cells from Donor Embryos created specifically for developing stem cells or for research in therapeutic cloning
2002	The draft sequence, of the most important pathogen of rice, a fungus that destroys enough rice to feed 60 million people annually, was completed.
2003	– The SARS (severe acute respiratory syndrome) virus was sequenced three weeks after its discovery. – Farmers in 18 countries plant GM crops on 67.7 million hectares
2004	The first cloned pet, a kitten (CopyCat or Cc) was delivered to its owner

2006	A recombinant vaccine against human papillomavirus (HPV) received FDA approval
2007	President Bush issued his second veto of a measure lifting his restrictions on human embryonic stem cell experiments
2009	Master heart stem cell was identified by the Harvard Stem Cell Institute
2009	Elizabeth H. Blackburn, Carol W. Greider and Jack W. Szostak won the Nobel Prize in Physiology or Medicine 2009 *'for the discovery of how chromosomes are protected by telomeres and the enzyme telomerase.*
2010	– Researchers at the J. Craig Venter Institute created the first synthetic cell. – The Nobel Prize in Physiology or Medicine 2010 was awarded to Robert G. Edwards *'for the development of in vitro fertilization'*.
2011	The Nobel Prize in Physiology or Medicine was divided, one half jointly to Bruce A. Beutler and Jules A. Hoffmann *'for their discoveries concerning the activation of innate immunity'* and the other half to Ralph M. Steinman *'for his discovery of the dendritic cell and its role in adaptive immunity'*.

B. TRADITIONAL VS. MODERN BIOTECHNOLOGY

Traditional or classical biotechnology has its root in ancient time, while transition of humans was taking place from hunter-gatherers to farmers. Humans used to collect the wild plants as farmers, cultivated them and selected the best yielding strains for mass production of agricultural products. Therefore, yielding the new products or modifying the existing ones by using living organisms by applying number of ancient ways is called 'traditional biotechnology'. Archaeological evidence suggests that the early Romans recovered copper leached by bacteria from natural copper sulphide deposits. In the early 1700s, the first record of large-scale bio-mining process was started in Rio Tinto, Spain.

However, biotechnology was applied empirically but it was not well understood. During ancient times, animal and vegetable species were selected and crossed to make them more productive; strains of yeast were used to ferment bread, beer and wine, or as compressed yeasts. These products were used to feed the growing population, or to prepare and preserve certain foods. There is no reflection of biotechnology in the traditional biotechnology as we know it today. In traditional biotechnology the processes were experimental that were run in uncontrolled ways.

The diversity of microorganisms and development of genetics largely expanded the potential of traditional biotechnology, and led to the development of modern or new biotechnology. The modern biotechnology refers to a number of techniques that involve the purposeful manipulation of genes, cells and living tissue in a controlled manner to alter the genetic make-up of an organism, for example recombinant DNA (rDNA) technology or genetic engineering), plant cell and tissue culture. Modern biotechnology was started with the discovery of the structure of DNA in 1953, and unraveled the mystery of passing of genetic information over generations. James Watson and Francis Crick published the double helical structure of DNA, for which they won a Nobel Prize in 1962 along with Maurice H. F. Wilkins. Two strands of DNA are composed of pairs of chemicals, adenine (A), guanine (G), thymine (T) and cytosine (C). DNA is present within the cells of all organisms is called the genetic blueprint that determines the cells to store, duplicate, and pass information from generation to generations. A segment of DNA that encodes protein is called a gene. The order of DNA base pairs that determines specific genes encoding specific proteins, controls the ch[illegible]teristics of an individual.

[illegible] the start of molecular genetics *Escherichia coli* was commonly used as a 'host' for gene [illegible]lso *E. coli* was able to incorporate foreign DNA 'vehicles' in the form of phages and plasmids. One of the main tools of modern biotechnology is the ability to manipulate living organisms at the genetic level.

In 1970, Hamilton Smith isolated microbial enzymes called restriction endonucleases that, along with DNA ligase, caused gene manipulation and inter-species gene transfer. During 1973, methods for isolating individual genes could be developed; by the 1980s, single genes could be transferred from one organism to another.

In 1972, Paul Berg created the first recombinant DNA molecules by combining DNA from the monkey virus SV40 with that of the lambda virus. In 1973, Herbert Boyer and Stanley Cohen created the first transgenic organism by inserting antibiotic resistance genes into the plasmid of an *E. coli* bacterium. A year later Rudolf Jaenisch created a transgenic mouse by introducing foreign DNA into its embryo, making it the world's first transgenic animal. In the beginning the new biotechnology focused on the development of methods to clone, over express, purify, and administer biopharmaceuticals, such as insulin, growth hormone, blood factor VIII (required for haemophilic patients), erythropoietin, etc. These discoveries revolutionised the potential of biotechnology in the health-care, agrifood, industries, environment, and other sectors. Biotechnology has also been a new source of innovations in bioinformatics and nanotechnology.

In 1990, the Human Genome Project was launched to map and sequence all human genes; this project was supposed to take 15 years costing about $3 billion. In the beginning of the project, less than 5% of all human genes could be identified; it was hoped that new drugs for the diagnosis, treatment, and prevention of human disease could be done with the discovery of new genes. These drugs included therapeutic proteins, diagnostics, gene therapy reagents, and small molecules. Besides, a significant proportion of the human genome has been sequenced and many new human disease genes have been characterised. These advances enabled biotechnologists to measure disease potential and expand the genomic diagnostics, and devised new therapeutic methods. Different branches of modern biotechnology are given in Fig. 1.1.

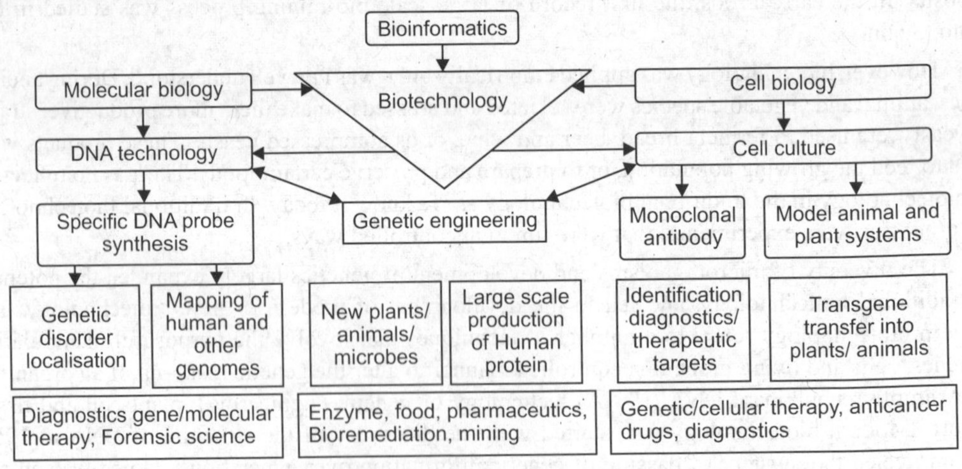

Fig. 1.1. Different branches of modern biotechnology.

The umbrella of biotechnology encompasses a broad array of technologies, including both 'traditional' and 'modern'. The term biotechnology has become synonymous with genetic engineering to the general public. Genetic engineering is also referred to as recombinant DNA technology, genetic modification, and gene technology and/or gene manipulation. Using genetic engineering DNA can be cut into genes, and recombined the cleaved genes into other DNA molecules. Th ability of an organism can be enhanced to produce a specific product, or to inhibit the producti a product. Thus, a genetically engineered (GE) or genetically modified (GM) organism can a new purposeful product that had not produced before.

C. THE WORLD OF BIOTECHNOLOGY

Different disciplines of biology and technology that enriched the wealth of biotechnology and caused its development are : biochemistry, cell biology, genetics, molecular biology, embryology, robotics, information technology, computer science, bioinformatics, chemical engineering, animal husbandry, research assistance, and production management, etc. In recent years, biotechnology has touched almost every aspect of human life. Biotechnology has helped in the area of medical sciences, industries, environment, agriculture, forensics sciences, etc. directly touching the lives of human beings and is making a significant contribution. The area of biotechnology that deals with medical and health care is called *red biotechnology*. When biotech area makes concerned with agriculture, it is known as *green biotechnology;* the area related to industrial processes is called *white biotechnology*. The area of biotechnology concerned with the application of molecular biological methods to marine and freshwater organisms is known as *blue biotechnology*.

1. Red Biotechnology (= Pharmaceutical Biotechnology)

Biotechnology applied to the pharmaceutical, medical and healthcare field is termed as 'red biotechnology', which deals with pharmacogenomics, designing organisms to produce antibiotics and vaccines, clinical research and trials, gene therapy and diagnostics. Red biotechnology is also useful in veterinary science and poultry farming. Moreover, the production of active pharmaceutical ingredients using biotechnological processes, including genetic engineering, is a young. Intensive research in this field has assured hopes for various life threatening diseases and enhanced the quality of life. Various areas of red biotechnology are briefly described below:

(*a*) Gene Splicing : Gene splicing is a tool of recombinant DNA technology for transfer of gene(s) from one organism to the other. Changes in genetic make-up of organisms resulted in the development of recombinant DNA; development of such recombinant DNA molecules and its transfer into a desired host (*E. coli*) has proved a milestone for production of insulin for type-2 diabetes. Red biotechnology has helped in the development of vaccines; genetically altered cowpox is used against influenza, herpes and hepatitis.

(*b*) Genetic Engineering : Genetic engineering has been used to produce chemicals and hormones required by our body. Such application of biotechnology is like a boon for healthcare. It cures by altering the genetic material of an individual. Gene therapy involves manipulation of DNA by replacing the defective genes or supplementing the normal genes in *ex vivo* or *in vivo* conditions. For detail see Chapter 9 : *Gene therapy*.

(*c*) Pharmacogenomics and Medicines : Pharmacogenomics is the study of pharmaceuticals and genetics that makes possible to design and produce drugs as per needs of the patients. Possibly the dosage of drugs according to the genetic information and body response of a patient can also be determined. Moreover, the better quality of medicines according may also be developed by a pharmaceutical industry. Biotech drugs or biotherapeuticals derived from genetic engineering of proteins in organisms are used specifically.

(*d*) Cloning : Cloning technique became popular in 1997 with the popularisation of the cloned sheep 'Dolly' at Roslin institute. Cloning method has made a platform for successful development of human clones also. However, human cloning has been banned in most of the countries; only therapeutic cloning is permitted for saving human lives. In this method, a nucleus is taken out from one cell, inserted into an unfertilized egg and grown for further development identical to the nucleus donor. For detail see Chapter 20, *Transgenic Animal (= Animal Biotechnology)*.

(*e*) Genetic Testing : Thousands of human genetic diseases can be identified by detecting mutations in a single gene by using genetic tests.

(*f*) Gene Therapy : Identification of genes responsible for genetic diseases and its treatment through insertion of functional genes have been described in detail in Chapter 9, *Gene Therapy*.

(g) Stem Cell Therapy : Really stem cell therapy has much potential in future. It can be very helpful in treating deadly diseases such as human cancers. The cells of initial stage of growth, unspecialised to develop into some specific cells, are called stem cells. The stem cells can get renewed for long time through cell division; they can differentiate and grow into specific cells under certain biochemical conditions. Such specific cells can be inserted in place of abnormal cell for treatment of the damaged tissues. Consequently, the self- renewal property of cells results in generation of tissues, which in turn replace the tissues of the damages areas. In recent years, stem cells from endangered species have been successfully produced that may hopefully be used to save the endangered animals.

2. Green Biotechnology

The use of biotechnology for agricultural processes and products is called 'green biotechnology' or plant biotechnology, for example an organism designed to grow under specific environmental conditions or in the presence (or absence) of certain agricultural chemicals. In recent years, you can hear about genetically modified (GM) fruits, vegetables available in any season that have specific nutritional values. Transgenic plants modified for increased resistance to pests and diseases, improved flavour and enhanced growth in adverse weather conditions have been produced. Green biotechnology is a rapidly increasing field within modern biotechnology. Green biotechnology has been possible due to the following research areas :

Green biotechnology

(a) Plant Tissue Culture: Methods used to grow or maintain plant cells, tissues or organs under sterile conditions on artificial nutrient medium is called 'plant tissue culture'. Whole plants can be produced from a piece of plant parts, such as roots, leaves, stems, a single plant cell or cells without cell walls (protoplasts) on culture media consisting of nutrients and plant hormones. Using plant tissue culture methods disease-free planting material and genetic resources can be preserved rapidly. For detail, see Chapter 18, *Plant biotechnology*.

(b) Plant Genetic Engineering: Some genetically engineered crops (such as cotton, sweet potato, etc.) for enhanced features have been produced through deliberate transfer of genes. Environmental stress-resistant crops have also been developed, such as drought-resistant crops, salt-resistant crops, etc.

Several plants having property of herbicide tolerance or resistance to insects and disease are available in market. Farmers are benefited from such plants in the form of reduced use of pesticides and labour. GM plants with improved nutrients such as improved amino acid composition or increased content of a certain vitamin (e.g. Golden Rice, with an increased provitamin A content) have been produced that have improved food quality and prevent dangerous diseases like childhood blindness caused by vitamin A deficiency. Improvement in flavour of fruits and vegetables (FlavrSavr tomatoes), and reduction in allergen content in foods have also been done in some GM plants. Such modification has directly benefited to consumers. Transgenic plants with an altered pathway for fatty acid synthesis are being applied for production of industrial oils; they can also produce biopolymers to replace petroleum-derived compounds. For detail see Chapter 19, *Transgenic plants*.

(c) Plant Molecular Marker-assisted Breeding : Biotechnology has shortened the time from 20-25 years (in conventional breeding) to 7-10 years for new crop varieties to enter into the market. Marker-assisted selection (MAS) is one of the tools which make it easier and faster to select plant traits. Now MAS plays a key role in the field of plant breeding. The main uses of molecular markers in crop genetic studies are : (*i*) assessment of genetic variability and characterisation of germplasm, (*ii*) identification and fingerprinting of genotypes, (*iii*) estimation of genetic distances between population, inbreeds, and breeding materials, (*iv*) detection of monogenic and quantitative trait loci (QTL), (*v*) marker-assisted selection, and (*vi*) identification of sequences of useful candidate genes.

In 2000, R.C. Grube and collaborators have described the comparative genetics of disease resistance within the family solanaceae. In recent years, molecular markers are now being widely used for breeding the tomatoes. More than 40 genes conferring resistance to major pathogens have been mapped, cloned, and sequenced. Currently, tomato breeding through MAS has resulted in varieties with resistance or tolerance to one or more specific pathogens.

***(d)* Biofertilizers :** Insects and pests are a headache for farmers; however, their crops can be made immune employing biotechnological methods, for example using the biofertilizers and pesticides. A biofertilizer is a formulation consisting of living microorganisms which promotes plant growth by increasing the supply of primary nutrients to the host, when applied to seed, plant surfaces or soil through root and/or rhizosphere colonisation. Microorganisms carryout nitrogen fixation, phosphorus solubilisation, and plant growth stimulation through the synthesis of growth-promoting substances, resulting in a reduction in chemical fertilizers and pesticides. Microorganisms used as biofertilizers are : *Rhizobium, Azotobacter, Azospirillum* and blue green algae (BGA) or cyanobacteria such as *Nostoc, Anabaena, Anabaena, Tulypothrux,* and *Aulosira*.

The phosphate-solubilizing bacteria, such as *Pantoe agglomerans* strain P5 or *Pseudomonas putida* strain P13, solubilise the insoluble phosphate from organic and inorganic sources of phosphate. The rate of available phosphate (Pi) in soil is well below the requirement of plants due to immobilization of phosphate by mineral ions such as Fe, Al and Ca or organic acids. Soon the inorganic P(i) are also immobilized in soil, therefore, plants absorbe less than 20% of added fertilizer. On the basis of many roles played by these microorganisms, the beneficial bacteria have been termed as 'plant-growth promoting rhizobacteria' (PGPR) by Kloepper and coworkers in 1979.

Microorganisms are symbiotically or asymbiotically associated with plant roots, convert complex organic material in simple compounds (to which plants easily take), increases crop yield by 20-30%, replaces chemical nitrogen and phosphorus by 25%, stimulates plant growth, improve soil fertility. Interestingly, they cast residual effects and provide protection against drought and some soil-borne pathogens. Biofertilizers are cost-effective as compared to chemical fertilizers having low manufacturing costs, and are eco-friendly. For detailed description see *A Textbook of Biotechnology* by R.C. Dubey.

***(e)* Biopesticides :** According to the US EPA, biopesticides are 'the naturally occurring substances that control pests (biochemical pesticides), microorganisms that control pests (microbial pesticides), and pesticidal substances produced by plants containing added genetic material (plant-incorporated protectants) or PIPs'. Therefore, biopesticides are of three different types: microbial pesticides, biochemical pesticides, and plant-incorporated protectants (e.g. GM plants). For detailed description see *A Textbook of Biotechnology* by R.C. Dubey.

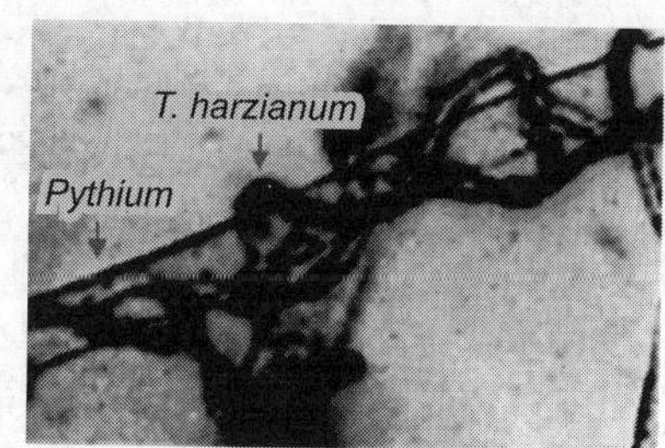

Mycelium of *T. harzianum* parasitising of mycelium of *Pythium*.

***(f)* Global Area of Biotech Crops :** It is hoped that in poor countries the genetically engineered crops can help the farmers by giving them access to new varieties of crops that will resist disease better and adapt to climate change. According to the annual biotech crop report of the ISAAA (International Service for the Acquisition of Agri-biotech Applications), a record 16.7 million farmers planted biotech crops in 160 million hectares (395 million acres) in 29 countries in 2011 (Fig. 1.2). This is a sustained increase of 8% or 12 million hectares (30 million acres) over 2010. Most of that growth of 8.2 million hectares came from the developing world, led by Brazil and India. The growth rate for biotech crops in developing countries was 11%, which is twice as fast and twice as large as industrial countries at 5% or 3.8 million hectares (James, 2011).

The top six mega countries that have cultivated biotech crops are : the U.S., Brazil, Argentina, India, Canada and China; each country planted more than 10 million hectares of the GM crops. Of the

16.7 million farmers who grew biotech crops, about 14 million were small, resource-poor farmers in China and India; most of them planted pest-resistant Bt cotton. In Africa, three countries viz., South Africa, Burkina Faso and Egypt have commercialised biotech crops; the others, such as drought-tolerant maize, are being tested. However, if biotech crops do not make farmers more productive or save time or money, they will not spread around the world as consistently as they have been planting it.

In 2011, herbicide-tolerant soybean had been the dominant biotech crop that occupied 75.4 million hectares (= 47%) of global biotech area. Biotech maize was the second most dominant crop with stacked traits, which occupied 37.3 million hectares (or 23%) of the global biotech area. The third most dominant crop was Bt cotton, which was planted in more than 17.9 million hectares in India, China, Pakistan, Myanmar, Burkina Faso, Brazil, USA, Argentina, Australia, Colombia, and Costa Rica. This is equivalent to 11% of the global biotech area (James, 2011).

The Indian farmers need such crops that should be resistant to pest, pathogens and weeds, and that can withstand flood, drought and salinity. They require more nutritious food grains. Agricultural biotechnology has the potential to offer viable solutions to these problems.

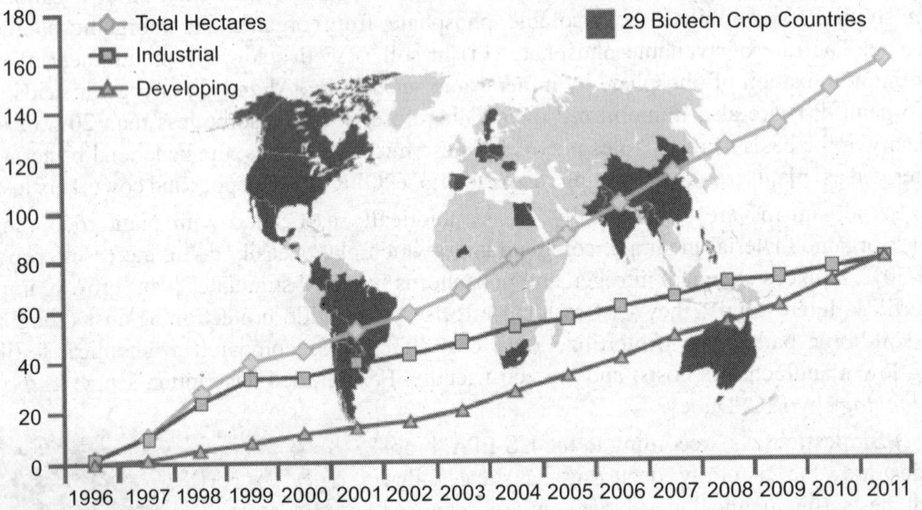

Fig. 1.2. Global area of biotech crops (million hectares) (1996-2011) (*Source*: Clive James, 2011).

3. White Biotechnology (= Industrial Biotechnology)

Application of biotechnology for industrial purposes, including manufacturing, alternative energy or bio-energy, and biomaterials is known as 'white biotechnology' mainly in Europe. Industrially useful products are produced by using cells or cell components such as enzymes. It involves fermentation and enzymatic processes that are the best alternatives to physicochemical and mechanical applications due to their economic environment and easier to use.

White biotechnology is also applied in water purification by involving bacteria. Several products such as biofuels, biodiesel, bioplastics, etc. have a promising future and opened research opportunities. Use of enzymes is more competitive and eco-friendly. For example, in textile industry the cost and the consumption of energy and water have been reduced up to 50% by using enzymes for the purification of tissues during cleaning processes. White biotechnology employs living cells of yeasts, moulds, bacteria and plants, and their enzymes to synthesise products. The bacterial enzymes have been used in manufacturing of food and washing powders (as an active ingredient that reduces the amount of artificial surfactants). Human insulin (Humulin®) has been produced in large-scale fermentation tanks by using transgenic *Escherichia coli*. During early of 1988, the first designed enzyme was introduced in detergents to break down fat.

White biotechnology is already bringing many innovations to the chemical, textile, food, packaging and health care industries by using alternative substrates and energy sources. Obviously researchers, industry and policy makers are interested in white technology with a hope that it may contribute for sustainable development. In place of traditional fossil fuel-based methods, white biotechnology is affecting chemical, textile, paper, food, mining and cosmetics industries by introducing eco-friendly biological processes. White biotechnology has also concerns with the alternative energy resources. Ethanol production from starch and carbohydrates began the era of biofuels which is a substitute of gasoline. This has started changing the 'hydrocarbon economy' to 'carbohydrate economy'.

4. Blue (Marine) Biotechnology

According to the OECD definition, blue biotechnology involves the use of valuable substances produced by living marine organisms, as well as parts, products and models thereof, to alter living or non-living materials for the production of knowledge, goods and services. This term also refers to the application of molecular biological methods to marine and freshwater life using marine organisms, and their derivatives, to increase seafood supply and safety, control proliferation of noxious water-borne organisms, and develop new drugs. Marine organisms used in blue biotechnology are bacteria, fungi, and microalgae. Their ingredients may be biological active compounds, pigments, antioxidents, vitamins, polysaturated fatty acids, enzymes, polymers and biomaterials. However, the potential for developing marine active ingredients remains mainly unveiled due to the diversity of marine microorganisms.

High value marine products have a wide range of applications. They offer some shorts of answers to global challenges related to health, food demands and environmental concerns. Human health and genetic engineering are the fields proposed for patents associated with genes from marine organisms. However, the marine resources are largely untapped so far. Hopefully, they can provide a new resource for the identification of high value ingredients. At present, there are about 20,000 marine-derived natural products as compared with >155,000 natural terrestrial products.

C-nucleosides were isolated from the Caribbean sponge (*Cryptotheca crypta*) about 40 years ago, which provided the basis for the synthesis of cytarabine, the first marine-derived anticancer agent to be developed for clinical use. Trabectedin (a metabolite of the sea squirt *Ecteinascidia turbinata*) is an anti-tumour drug which is marketed under the trade name Yondelis. It is approved for the treatment of advanced soft tissue sarcoma in Europe, Russia and South Korea. Its clinical trials are undergoing for the treatment of breast, prostate, and pediatric sarcomas.

In 2004, the Food and drug Administration (FDA) approved ziconotide when delivered as an infusion into the celebrospinal fluid. Ziconotide is an active ingredient of the commercial pain-killer Prialt. Ziconotide is a synthetic derivative of ω-conotoxin produced by a cone snail species, *Conus magus*. It alleviates severe chronic pain. Its prialt does not cause respiratory depression.

Extreme marine environments represent a source for new enzymes called extremozymes isolated from extremophiles. Laccases (the oxydizing enzymes) have much potential in various fields in industries including effluent detoxification, kraft pulp and dye bleaching, polymer synthesis, cosmetics, bakery, wine and beverage stabilization, manufacture of anticancer drugs, and as part of biosensor for immunoassays.

Flavours and fragrances from marine microorganisms offer a new field of marine biotechnology. Now companies are able to produce a variety of compounds that impart flavours that cannot be artificially produced in laboratory. The future of flavour production lies in pathways of new metabolic or fermentation methods. Several enzymes from marine organisms are certainly responsible for producing tastes and fragrance. This source has not been much explored so far.

D. BIOINFORMATICS

In 1979, Paulien Hogeweg invented the term 'bioinformatics' to study the processes of information technology into biological systems. Basically, it is a branch of science which deals with the applications of information technology and computer science into the field of molecular biology. Initially the biological data are in raw form; therefore, certain storage houses are required in which the data can be stored, organized and manipulated. Algorithms and biological software of computer are developed by bioinformatics to analyze and record the data related to biology, for example genes, proteins, drug ingredients and metabolic pathways. Hence, the data can be procured from these databases and used according to the requirement.

The terms bioinformatics and computational biology are sometimes interchangeable. Bioinformatics deals with the creation and advancement of algorithms, computational and statistical techniques, and theories to solve problems and analyse the biological data. On the other hand, computational biology refers to investigation of a specific biological problem using computers based on the hypothesis by using the experimental data. Bioinformatics refers to the creation and maintenance of databases of biological information. In recent years, advancement in molecular biology and modern equipments has eased the rapid sequencing of genes. Bioinformatics plays a key role in finding the DNA sequence of organisms, predicting the structure of discovered proteins, clustering protein sequences and developing protein models. Major research areas of bioinformatics are structural genomics, genetic engineering, DNA finger printing, functional genomics and evolution modelling. Bioinformatics has many applications and research areas where it can be applied.

The genes which encode regulatory sequences or peptides are determined by sequence analysis using the information of sequencing. There are many powerful tools that analyse sequence, and computers analyse the genome of various organisms. The DNA mutations in an organism are identified by these computers and tools, and related sequences are also detected. Sequence analysis of several DNA fragments is carried out by shotgun sequence techniques. However, overlapping fragments and their assembly are observed by special software.

In many organisms the expressed sequence tag (EST) analysis is widely applied for the discovery of gene transcripts. In the dbEST NCBI (National Center of Biotechnology Information) 73,360,923 ESTs entries were recorded up to July 1, 2012. Most of these are generated from laboratory models and economically important organisms. The 10 most frequently sequenced organisms are given in Table 1.2.

Table 1.2. The first 10 organisms ranked by amount of EST entries in dbEST.

Rank	*Organisms*	*Amount of ESTs*
1	*Homo sapiens* (human)	8,692,773
2	*Mus musculus* + *M. domesticus* (mouse)	4,853,570
3	*Zea mays* (maize)	2,019,137
4	*Sus scrofa* (pig)	1,669,337
5	*Bos taurus* (cattle)	1,559,494
6	*Arabidopsis thaliana* (thale cress)	1,529,700
7	*Danio rerio* (zebrafish)	1,488,275
8	*Glycine max* (soybean)	1,461,624
9	*Triticum aestivum* (wheat)	1,286,173
10	*Xenopus (Silurana) tropicalis* (western clawed frog)	1,271,480

Source: http://www.ncbi.nlm.nih.gov/dbEST/dbEST_summary.html; dbEST release 120701, July 1, 2012.

The regulatory sequences and protein coding are determined by marking the genomes in *genome annotation*. It determines the regulatory sequences; therefore, it is a very important part of the human genome project.

The branch of bioinformatics that determines the genomic structure and function, and relation between different biological species is known as *comparative genomics*. Therefore, inter-genomic maps are constructed which help to trace the evolution processes occurring in genomes of different species. The information about the point mutations and duplication of large chromosomal segments can be found out in such maps.

The primary structure of proteins in the form of amino acids present on the DNA molecule can easily be determined. However, determination of secondary, tertiary or quaternary structures of proteins is difficult. Therefore, the complex structures of protein can be predicted by using the method of crystallography or bioinformatics tools.

The bioinformatics tools also help in *drug discovery*, and disease diagnosis and management. Complete sequencing of human genes has made possible to prepare medicines and drugs targeting of above 500 genes. Drugs can easily be delivered employing different computational tools and drug targets because only diseased or mutated cells can be targeted.

E. SYNTHETIC BIOLOGY

In 1974, a Polish geneticist Wacław Szybalski (professor of oncology at the University of Wisconsin) used the term 'synthetic biology', which is also known as *Synbio*. After the development of molecular biology and the discovery of restriction enzymes, he thought about possibility of building of some new or parts of genomes. Synthetic biology is a new area of biological research and technology that combines science and engineering together. Synthetic biology possesses different methodologies and disciplines with a common goal to design and construct the new and unnatural biological functions and systems. Most definitions of synthetic biology have two parts, the construction of completely novel biological entities, and the re-design of already existing ones.

Waclaw Szybalski

In 1978, when the Nobel Prize in Physiology or Medicine was awarded to Arber, Nathans and Smith for the discovery of restriction enzymes, W. Szybalski wrote in an editorial comment in the journal *Gene*: *'the work on restriction nucleases not only permits us easily to construct recombinant DNA molecules and to analyze individual genes, but also has led us into the new era of synthetic biology where not only existing genes are described and analyzed but also new gene arrangements can be constructed and evaluated'*.

The creation of new biological system is focused on exploring the use of synthetic biology for benefit of society. The biological approach exploits the insertion of synthetic DNA into a living cell, and chemistry on gene synthesis.

In 1962, Szybalski inserted DNA into human cells for the first time, and created functional DNA in a Petri dish. He invented a technique for separation of genetically modified cells on HAT (hypoxanthine-aminopterin-thymidine) selection medium. After a decade, HAT selection medium was used for expression of monoclonal antibody, which became a Nobel-winning invention. Monoclonal antibodies can recognize specific proteins and cell types; it can be used in several areas of medicine, such as cancer, rheumatoid arthritis and even identification of pathogens.

There are two approaches of the *Synbio,* top-down and bottom-up. The top-down approach starts from a whole organism and then its genome is reduced until it provides a specific function. The second bottom-up approach is the opposite of the first one that creates an organism just with biobricks (the trademarked words), a system developed by the MIT (Massachusetts Institute of

Technology). Just add the genes to the genome of an organism that may carry out the desired function. In this frame three methods have been developed : minimal genome, artificial cells and synthetic biomolecules. Moreover, development of synthetic biology is closely related to the development of information technology and nanotechnology (Fig. 1.3).

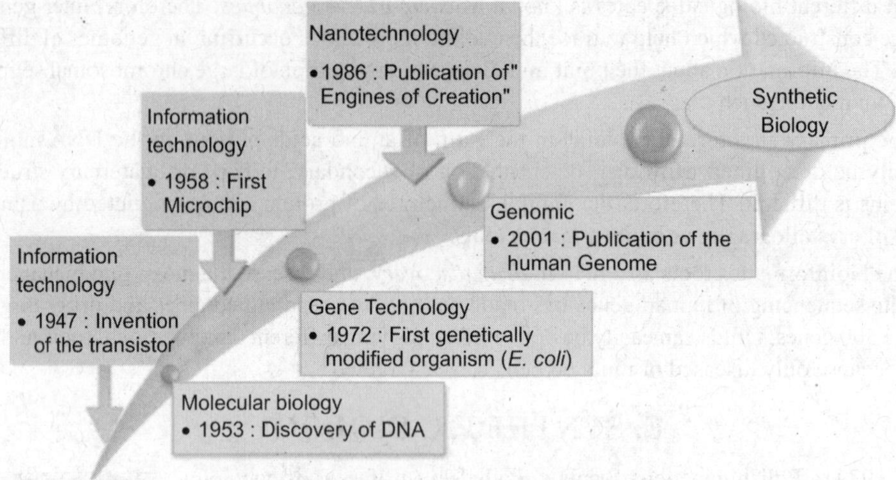

Fig. 1.3. Development of synthetic biology.

In 2007, several companies were offering the synthesis of genetic sequences up to 2000 bp long within two weeks, expending about $1 per base pair. However, some improvement was done in the method; hence the price had dropped to less than $0.50 per base pair by September 2009. In 2012, George Church encoded one of his books about synthetic biology in DNA. The data of the book are 5.3 Mb, which is >1000 times greater than the previous largest amount of information stored in synthesized DNA.

Synthetic life is artificial life created *in vitro* by using biochemicals. Experiments done on synthetic life searches out the origins of life, study the properties of life, or recreate life from abiotic materials substances. In 2010, Craig Venter and his collaborators announced for assembling a complete genome of millions of base pairs, insert it into a cell, and start of replicating. The DNA code was transcribed as a computer file, edited with new code, sequenced by the Blue Heron Bio Company, put together by yeast and other cells, and finally transplanted into a cell after removing the all genetic material. Consequently, the cell was totally controlled by the new genome, which also divided further. Craig Venter called this cell as the 'first synthetic cell'; over $40 million dollars were spent to create such cell. Craig Venter Institute uses the term 'synthetic bacterial cell' because it has been created out of already existing life using synthetic DNA.

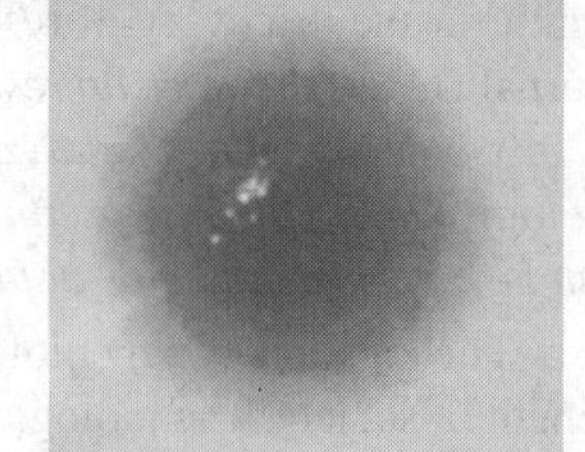

Fig. 1.4. The world's first synthetic life form.

Synbio is the multidisciplinary workforce, which needs biologists, engineers, chemists, information technology specialists, and social-, ethical- and economic experts. The essential tools of synthetic biology are : DNA sequencing, gene fabrication, modelling for behavior of synthetic genes, and measuring of gene behaviour.

Similar to the other types of engineering, the applications of synthetic biology is potentially very broad, including microbial biofuel production, bioremediation, biosensors, new drug development pathways and synthetic vaccines. However, there are social, political and ethical issues of synthetic biology.

In 2011, J Y. Zhang and coworkers in his book have described about 39 reports on synthetic biology and the social and ethical issues that have been raised since 2004. The issues associated with synthetic biology are: (*i*) biosafety (prevention of release of synthetic organisms into the environment), (*ii*) biosecurity (misuse of synthetic microorganisms in bioterrorism), (*iii*) intellectual property (use of synthetic biological parts by private companies or freely available for all), and (*iv*) creation of new life (artificial creation of life is unnatural, *i.e.* 'playing God'). Moreover, there are other issues also such as democratisation, standardisation, and inter-disciplinarity research on synthetic biology.

F. PUBLIC PERCEPTION OF BIOTECHNOLOGY

In 1985, the U.N. assembly constituted Brundtland Commission, which is also known as the World Commission on Environment and Development (WCED). The WCED submitted its report based solely on 'sustainable development'. The WCED defined it as the *development that meets the needs of the present without compromising the ability of the future generations to meet their own needs*. It has been approached by two ideologies: the *holistic ideology* and the *reductionist ideology*. The demand of holistic ideology is a return to pre-industrial, traditional, social and economic organisation. It may result in unequal production of food on the earth, and consequently unequal distribution. On the other hand the reductionists discuss the population problem and its effect on food deficits in many countries of the world. Will the earth be capable of feeding 10 billion populations by 2050 with the limited resources ? There shall be less availability of land and water per capita. The mankind would look up on the tools of biotechnology to meet their demand.

Implication of the biotechnological processes such as potential for healthcare and production, processing and quality of food go well beyond the technical benefits offered, and in turn will depend upon their acceptance by consumers. It is possible by genetic engineering of crops, fertilizers, pesticides, vaccines, etc. Public perception of biotech products will have a major influence on rate and direction of developments.

Public support is a crucial component for the introduction of new technologies and products into the society. Strong public support has a major impact on investment and marketing of companies. In this respect, consumer's attitude acts as a factor that directly affects the competitiveness. Eurobarometer survey of 2005 shows a growing public support regarding biotechnology and life sciences applications with the exception of food linked with genetically modified organisms. After a period of decline between 1991 and 1999, hope for contribution of biotechnology has increased between 1999 and 2005. This decline may have been due to the difficulties in communication, *i.e.* benefits and potential risks of genetically modified organisms.

More than 75% people are of the opinion that biotechnology will improve their quality of life in the next 20 years. Since 1999, such trust in biotechnology has been shown in universities, scientist and industry. There has been a growing public knowledge about biotechnology and genetics since 2002.

Medical biotechnology has received much strong support from gene therapy and pharmacogenetics. Biotechnology firms active in applications of health are significant as compared to the firms involved in food and beverage processing, agriculture or the other fields.

Europeans also favour embryonic stem cell research with the condition of strongly regulation and timely information about the risks and benefits of stem cell research. They also accept the use of genetic data for personal medical diagnosis and gene banks for research associated with diseases. But, they strongly oppose granting government or non-government agencies to access genetic information about individuals. Moreover, Europeans extensively support the industrial biotechnology. More than 70% of them favour the development of bio-fuels and bio-plastics. They broadly accepted the use of biological processes to produce pharmaceuticals.

However, a majority of people are ready to buy GM food if it is healthier, contained less pesticide residues, or was more environment-friendly. Respondents in Europe and Canada have similar opinions on average regarding GM food, whereas US respondents think that GM food is useful for society, morally acceptable and does not involve risks.

Due to the eco-friendly features white biotech has received a positive social acceptance throughout the whole world excluding some countries. New materials and fuels not obtained from petrochemical processes are provided by biotechnology; moreover, it also improves and enhances the bioremediation of water, soils and uses less fossil-fuel energy. However, people are aware of harmful impacts of white biotech on the biological diversity in many of the defined processes and release of genetically modified organisms into the environment. Strict bio-safety regulations are applied in factories, bioreactors and greenhouses.

Bt cotton was the first transgenic crop to be released in India in the year 2002. There has been a lot of controversy surrounding Bt cotton after its introduction. Its performance, impact on the environment, biodiversity and health of cattle has been widely debated. In September 2007, the GEAC announced the approval for large scale field trials for Bt brinjal in India. Some analysts opined that Bt brinjal would have a significant economic impact on farmers, while some raised doubts on its safety and environmental implications.

The Union Minister of Environment and Forest faced a huge outcry from activists, scientists, farmers, and general public. Finally, in March 2010, he put a moratorium on the release of this genetically modified vegetable. It shall be released after more scientific studies and field trials. It shows the crucial failure of the Indian biotechnology sector due to its inability to communicate to the public the pros and cons of the GM technology in an unbiased manner.

G. BIOTECHNOLOGY IN INDIA

In 1986, the Department of Biotechnology (DBT) was established by the Ministry of Science and Technology. This gave new impetus to the development of biotechnology in India. Several centers of excellence have been established by the DBT in the country. These centers are responsible for the generation of skilled workers, the development of research initiatives and opportunities, and sustenance of its R & D in the private sector. This facilitated the interaction between academia and industry that led to the real estate entrepreneur and a number of initiatives to take root and grow biotechnology in India.

The Indian government has developed bio-security and helped to establish the patent rules. He also participated in technology transfer and international cooperation. For promotion of small and medium biotech companies, the center has introduced the venture capital funds in line with its Technology Development Fund (TDF).

A decent regulatory framework for approving GM crops and products of recombinant DNA for human health has been set up by the Government of India. A proactive government policy allows purposeful stem cell research in the country with its code of ethics. In 2005, the patent system came into force and gave a message to the world and the Indian industry that India supports the framework of rules and rewards of research and new initiatives. The second amendment was done in the Indian patent law which makes provision of a patent period for 20 years after the filing of patents. This bill is similar to WTO and makes Indian laws compatible with the framework of multilateral negotiations.

In India several states like Andhra Pradesh, Karnataka, Gujarat, Maharashtra, Kerala, Tamil Nadu, Uttarakhand and Himachal Pradesh have taken their own initiatives for developing Biotech Parks, for encouraging research and establishing links between research institutions and industry.

The Ministry of Science and Technology has established several centers of excellence in various areas of biotechnology. These centres have a world-class infrastructure and fully developed research centres; some of them are : Plant Genomics Center (New Delhi), Center for Human Genetics (Bangalore), National Institute of Biologicals (New Delhi), Centre for Cellular and

Molecular Biology (CCMB) (Hyderabad), National Facility for Macromolecular Crystallography (BARC, Mumbai), Tata Institute of Fundamental Research (TIFR) (Mumbai), Central Institute of Drug Research (CDRI) (Lucknow), National Brain Research Centre (New Delhi), Central Institute of Medicinal and Aromatic Plants (CIMAP) (Lucknow).

In 1995 there were about 400 companies in India related to biotechnology activities. This number got increased to about 850. Indian biotechnology industry is small as compared to its pharmaceutical industry (about 16,000 drug makers big and small). Biotech in India started in the mid-1980s and has focused in large national or multinational drug. Many small biotechnology companies have been created since 1986 only after the birth of the Department of Biotechnology (DBT), Ministry of Science and Technology, Govt. of India.

The industry consists mainly of pure biotechnology startups and established pharmaceutical and agri-business companies. The major players in this industry are : Biocon India, Shantha Biotechnics, Bharat Biotech, Wockhardt, Dr. Reddy's Laboratories, Serum Institute of India Ltd, Zydus Cadila, Aventis Pharma and Reliance Life Sciences. Moreover, the key markets in Indian biotech industry are : industrial biotech (e.g. enzymes), vaccines (e.g. recombinant Hepatitis B), diagnostics (e.g. immunology kits), veterinary products (e.g. animal health products), agri-biotech products (e.g. genetically modified seeds), etc.

1. Bio-business (Biotechnology Market) in India

The sequencing of genes and dissemination of genomic information for drug companies is the size of the next boom industry. India has the potential to become one of the strengths of choice in the development and manufacture of genomic medicine with a good atmosphere and well-directed research initiatives. Currently, there are about 190 biotech companies in India. India is a huge market for products based on biotechnology. However, most of these products are not developed here but are imported by domestic and foreign companies. According to the analysis of capitalmarket.com, the Indian biotech industry is expected to reach Rs 4,40,000 crore in 2020.

Biotechnology has yielded fruit in terms of profit-making business enterprises. The success of the biotechnology industry has led to sustained economic growth in the developed nations like US, European countries and Canada. India is also fast emerging as a key player in biotechnology related activities and investments. Although the biotechnology market in India contributes only a 2% share in the global market, it is on the threshold of a colossal growth in the coming decade.

The importance of the biotech market in India and its prospective growth can be grasped from the steps taken by the Indian government to further development of research and entrepreneurship initiatives in the country. The Department of Biotechnology (DBT) and the various Centres of Excellence provide the backbone of furthering research activities and laying the foundation for generating knowledge and trained manpower to assist entrepreneurship programmes and the biotech private industry in India. Several industry houses from the 'old industry' have already jumped in the Indian biotechnology market to take advantage of the opportunities to this new and exciting field.

(*a*) Biotech Turn-over : Biotechnology industry in India has a strategic role to play in addressing various developmental challenges of the nation, such as healthcare both in terms of affordable medicines and therapies and their delivery models, as well as meeting the challenges of the food and energy security of the country.

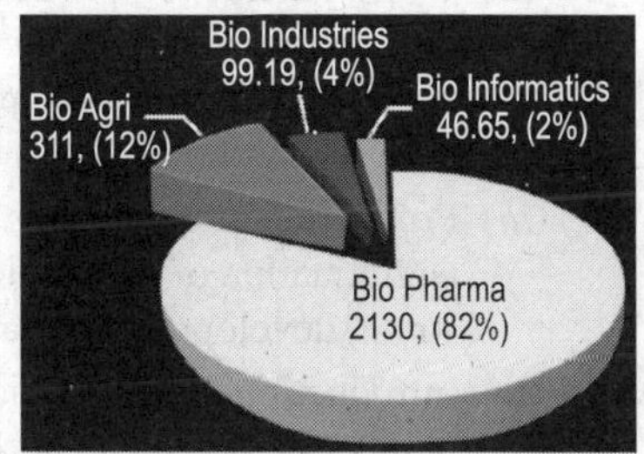

Fig. 1.5. Indian biotech sector (US $ million).

In year 2010-2011 the Indian biotechnology industry revenues touched ₹ 15,000 crores, despite the global downturn in 2009 and slow recovery in 2010. Over the last decade the industry has grown consistently in double digits with average revenue growth figures greater than 20%. It is estimated to reach a turnover of around Rs 45,000 crores by 2015 which would require growing at about 30% year on year. To achieve this target a synergistic support, both in the short as well as long term, is needed from the central and state governments. Contribution from different sectors of Indian biotech is shown in Fig. 1.5.

(*b*) **Booming Biotech Market:** Biotechnology has provided a national and international market, good career to researchers, new jobs to youth, new life to patients and new possibilities in many areas of life sciences and more. Marketing of biotech products is called bio-business. In the beginning, Indian biotechnology industries were slow but now are gaining good momentum. Although, the current share in the global market is just 2 % percent; but India is today among the top five countries in the Asia Pacific region. India shows immense potential not only as a destination for new generation pharmaceuticals, biotech products and diagnostics but is also becoming an important hub for outsourcing of clinical trials and contract research.

Biotechnology in India can be divided into three broad areas, human and animal health care, agricultural, and industrial sectors. The key success factors for India in the biotech industry are: (*i*) a large and strong pool of qualified scientists and engineers, (*ii*) existing and upcoming research facilities and institutional networks, and (*iii*) rich biodiversity that are advantageous for biotech companies to carry on their effective research and drug discovery. Secondly both the state and central governments provide strong support to the Indian biotech industry in the form of different incentives. In India, the top five biotech companies (of Indian origin) have performed well in the international arena.

In India, biotechnology market is one of the world's most potential markets. The Indian biotechnology market, valued at US$ 2.86 billion in 2007-08, is projected to hit a 'compound annual growth rate' (CAGR) of 30% in near future. Present exports dominate the Indian biotechnology market with bioservices and bioinformatics in particular being export driven industries. Biopharma remains the India's biggest sector accounting for nearly 70% of the total revenues in 2007-08. By 2012-13, this sector is expected to be worth US$ 5.7 billion.

The bio-agriculture sector, driven by the sales of Bt cotton, is the industry's fastest growing sector. Commercial approval of more transgenic plants including Bt brinjal is likely to push further growth in this sector. Strong R & D capability and a large patient pool for clinical trials are expected to make the bioservice sector amongst the most potential sectors in the Indian biotechnology market.

Bharat Biotech is one of the fastest growing biotechnology companies of India with a rich base of technical and commercial expertise. Apart from the Indian market, Bharat Biotech has also entered in various International markets. Currently Indian biotech companies are exporting its products to Kenya, Zambia, Peru, Bolivia, Philippines, Bangladesh, etc.

(*c*) **Success Story of Biotech Market:** The success of biotech marketing in India is given below:

(*i*) Development and production of indigenous recombinant Hepatitis B vaccine by a number of companies, including Shantha Biotech Private Limited, Hyderabad.

(*ii*) World Patents for a solid state fermentation called plafactor, US patents for F-virosome, drug delivery system for gene therapy, US and European patents for refamicine shuttle vectors in *E. coli*, US patent in the production of hydrophilic nanoparticle for drug delivery system, etc.

(*iii*) Production of indigenous HIV I and HIV II detection rapid test kit using naked eye visible agglutination of the whole blood and a confirmatory Western blot test kit utilizing Indian strain, development of various other indigenous diagnostic kits and transfer to Industry.

(*iv*) Introduction of transgenic Bt cotton and successful trial under contained conditions of a number of vegetable, horticulture and crop transgenics, such as Bt brinjal.

(*v*) Development of a number of indigenous rDNA products to be launched shortly like human insulin, growth hormone, G-CSF, interferons (α-2a and α-2b), erythropoetin, etc.

(*vi*) Initiatives for the establishment of various biotechnology parks in different states of the country, besides pilot plant facilities for the small entrepreneurs.

India has many advantages in terms of knowledge, skills, R & D facilities and costs. In India, the institutional infrastructure provides the basic foundation to translate into business opportunities. Biotechnology in India can be divided into three broad areas : human and animal health care, agricultural and industrial.

The core competence of Indian biotech industry exists in : (*i*) handling of fermentation-based products, (*ii*) handling of labile fermentation-based compounds efficiently in downstream processing methods, (*iii*) use of plant and animal parts for extracting value added products of high purity, (*iv*) use of cell/microbial culture techniques, and (*v*) plant cell/tissue culture, etc.

(*d*) Consumption of Biotech Products: The estimated average consumption of biotech products in India during 1997 was US$ 1,798 million. The areas of main competence include : (*i*) capacity in bioprocess engineering, (*ii*) skills in gene manipulation of microbes and animal cells, (*iii*) capacity in downstream processing and isolation methods, (*iv*) skills in extraction and isolation of plants and animal products, (*v*) competence in recombinant DNA technology of plants and animals, (*vi*) excellence in traditional and molecular marker assisted breeding of plants and animals, and (*vii*) infrastructure in fabricating bio-reactors and processing equipment.

(*e*) Policy Initiatives: The Union Government as well as the State Governments have taken various initiatives to boost biotechnology in India. Several state governments such as Karnataka, Tamil Nadu, Andhra Pradesh, Maharashtra, Uttarakhand, Uttar Pradesh, Himachal Pradesh and Delhi have taken initiatives to encourage entrepreneurs to set up biotech industries in their states. India's biotech sector is expected to grow at a robust 20 per cent per annum in the near future, given the growing demand for biopharmaceuticals, biosimilars and vaccines, says a veteran in the field.

According to *Padma Bhushan* Kiran Mazumdar Shaw, the Chairman & Managing Director of Biocon Limited (a biotechnology company based in Bangalore), and a member of the board of governors of the prestigious Indian School of Business, the size of the country's biotech sector had reached US$ 3.5 billion in 2010 and is poised for robust growth in all segments of biotechnology. India is already a world leader in vaccine production, Bt cotton and bio-pharmaceuticals, especially bio-similars. India is also a large producer of industrial enzymes for green technologies. Biofuels would also offer a huge growth opportunity in biotech. Kiran Mazumdar Shaw said *"we are entering the era of bioeconomy, where biotechnology can provide powerful solutions to some of the grave challenges that we face today such as food scarcity, energy deficit, environmental damage, unmet medical needs and industrial pollution."*

Kiran Mazumdar Shaw

- The 10th annual BioSpectrum-ABLE Biotech Industry Survey brings all round cheers within the industry. The biotech industry has performed creditably to cross the ₹ 20,000 crore revenue for the first time ever. The industry, buoyed by healthy vaccine exports and Bt cotton sales, has grown by 18.5% in 2011-12 over the previous year to record revenues of ₹ 20,440 crore.
- Many Indian companies have introduced products of original research through technology transfer from R&D institutions in the field of vaccines, diagnostics and reagents. Some others have teamed up with foreign companies for sourcing technologies and are experimenting with new products produced by foreign technologies, with a view to introduce them into the Indian market within the framework of Indian laws.
- Outsourcing of R&D in biotechnology represents a tremendous opportunity for Indian companies for contract research for overseas corporations. The current global spend on outsourced R&D is approximately US$ 7 billion and is expected to grow at 30% per annum for the next 5 years.

- There are around 50 R&D laboratories in the public sector, providing high quality R&D and over 20 conducting research in specific areas of biotechnology. In addition, there are companies in Bangalore with excellent technical manpower and world-renowned institutions such as the Indian Institute of Science (IISc), the National Centre for Biological Sciences (NCBS), Jawahar Lal Nehru Centre for Advanced Scientific Research (JNCASR), which provide high quality R&D services to several organisations throughout the world.
- Global biotech majors including IBM Life Sciences, Incyte Genomics, Affymetrix and AntexBiologics have shown interest in forging partnerships with Indian companies in the biotechnology sector.

2. Biotechnology Institution in India

Below is the list of top biotech colleges in India, both in public sector and private sector.

***(a)* Top Biotech Schools (Public Sector) :** Some of top ten public biotech institutions of India are given below:

(*i*) School of Life Sciences, University of Hyderabad, Hyderabad.

(*ii*) School of Biotechnology, Banaras Hindu University, Varanasi, U.P.

(*iii*) Rajiv Gandhi Centre for Biotechnology, Trivandrum, Kerala.

(*iv*) Institute of Chemical Technology, Mumbai, Maharashtra.

(*v*) University School of Biotechnology, Guru Gobind Singh Indraprasta University, New Delhi.

(*vi*) Animal Biotechnology Research Centre, National Dairy Research Institute, Karnal, Haryana.

(*vii*) Department of Molecular Biology & Genetic Engineering, G.B. Pant University of Agriculture and Technology, Pantnagar, Uttarakhand.

(*viii*) Department of Biotechnology, University of Kashmir, Srinagar, J&K.

(*ix*) School of Biotechnology, JNU, New Delhi.

(*x*) Department of Microbiology and Biotechnology, M.S. University of Baroda, Vadodara, Gujarat.

***(b)* Top Biotech Schools (Private Sector):** Some of top ten private biotech institutions of India are given below:

(*i*) School of Chemical and Biotechnology, Sastra University Thanjavur, Tamilnadu.

(*ii*) Department of Biotechnology, SRM University, Chennai.

(*iii*) Department of Biotechnology and Bioinformatics, Jaypee University of Information Technology, Solan, H.P.

(*iv*) Jaypee Institute of Information Technology, Noida.

(*v*) The Amity Institute of Biotechnology, Noida.

(*vi*) The Oxford College of Science, Bangalore.

(*vii*) PES Institute of Technology, Bangalore.

(*viii*) Department of Biotechnology, R.V. College of Engineering, Bangalore.

(*ix*) Shoolini Institute of Life Sciences and Business Management, Solan, Himachal Pradesh.

(*x*) Shri M & N Virani Science College, Rajkot, Gujrat.

H. ETHICAL ISSUES

Biotechnology is a very controversial subject; therefore, biotech industries are subject to rules and regulations. Legal, social, and ethical concerns have been raised by the ability to genetically alter organisms. Each type of biotech has to deal with its own ethical issues. Therefore, governments have established guidelines for the performance of biotechnology research, and specific requirements have been set to control the introduction of recombinant DNA products into the marketplace. General governmental guidelines for biotech research are published time to time. Some of the issues are briefly discussed below:

(*a*) Stem Cells : The mechanism of cell differentiation and production of specific cells from stem cell lines and cure of various diseases such as cancer, diabetes, etc. can be understood by stem cells research. Only embryonic stem cells can be turned into any cell types. Only embryonic stem cells can be used for these researches; however, embryonic stem cells are derived from frozen, fertilized eggs produced through *in vitro* fertilization.

(*b*) Cloning : Cloning is the process of *in vitro* replication of genes, cells or organisms from a single original entity. There are two types of cloning : *therapeutic cloning* and *reproductive cloning*. Therapeutic cloning is used to develop new tissues that may replace the damaged tissues and cure diseases such as diabetes, Parkinson's, Alzheimer's and various types of cancer and heart diseases. The use of therapeutic cloning techniques would result in the manipulation of tissues and organs genetically identical to the patient; therefore, there will be no problem of transplant rejection from the body. In contrast, reproductive cloning is used to create copies of an existing person born as a transplantable organ bearer. However, if therapeutic cloning is allowed, there may be risk to make reproductive cloning because it would result in psychological and physical problem for the clone and create a black market for sale of organs.

(*c*) Food and Agriculture : Agricultural biotechnology is used to modify plants and animals for better properties, but their consequences are not certain. Species have been crossed for centuries to produce new varieties having better properties; but is it is similar to the genetic crossing of species Still there is an additional problem of contamination: GMO's are spread easily and, if they prove to be harmful to man it will be very difficult to distinguish the natural plants from the GM plants.

(*d*) Use of Animals in Research : A large number of products are tested on experimental animals before testing on humans. Besides, human lives have been saved due to testing of thousands of products on animals. Now-a-days some ecological organisations are fighting against big firms that rely in their research on the excessive use of animals.

QUESTIONS

1. What is biotechnology ? Describe in detail about history of biotechnology.
2. Distinguish between traditional and modern biotechnology.
3. Give a detailed account of the world of biotechnology.
4. Write in detail about public perception of biotechnology.
5. Describe in detail about biotechnology in India.
6. Write short notes on the following :
   (*i*) Synthetic biology,
   (*ii*) Ethical issues of biotechnology,
   (*iii*) Biotechnology market in India,
   (*iv*) Success story of biotech market,
   (*v*) Green biotechnology,
   (*vi*) Global area of biotech crops,
   (*vii*) Red biotechnology,
   (*viii*) White biotechnology,
   (*ix*) Biotechnology institutes in India.

BIBLIOGRAPHY

Bhojwani S.S. and Razdan, M.K. 1996. *Plant tissue culture: theory and practice*. Elsevier. p. 767.

Dubey R.C. 2006. *A textbook of biotechnology*. S. Chand & Co. P Ltd. Ram Nagar. New Delhi-110005, India.

James C. 2011. Global status of commercialized biotech/GM crops: 2011. ISAAA Brief No. 43. ISAAA: Ithaca, New York.

Smith J.E. 1996. *Biotechnology* (3rd Ed.) Cambridge University Press, Cambridge, p. 236.

CHAPTER

2 Cell Biology

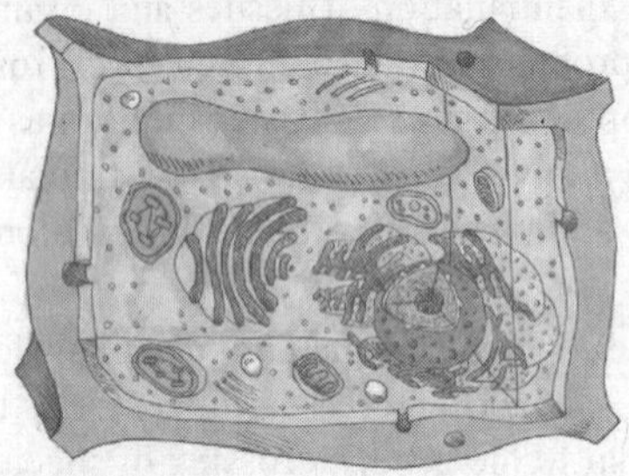

A. Cell structure and function
- A journey from 'molecules to a living cell
- Prokaryotic and eukaryotic cells
- The eukaryotes (cell structure and components)

B. Transport across cell membrane
- Carrier proteins and active membrane transport
- Transport systems in prokaryotes and eukaryotes
- Ion channels

C. Cell signalling and signal transduction
- Types of receptors
- Signal transduction pathway
- Signal amplification

D. Secondary messengers

Cell is a fundamental unit of all living system. It evolved million years ago from the molecules that underwent polymerization. The first cellular organisms evolved were the unicellular ancient prokaryotes called 'archaea'. Such organisms adapted all adverse conditions and thrived well. They bear some unique features that ordinarily lack in true bacteria. The eukaryotes developed after a gap of million years with gradual changes in environmental conditions. In eukaryotes all the cells work in a well coordinated manner.

In its first billion years, the conditions that existed on the earth are still a disputed matter. It is not sure that the surface was initially molten or not, and the atmosphere contained ammonia or methane. However, everyone may agree that the earth consisted of volcanic eruptions, and faced lightning and torrential rains. There would have been a little free oxygen and no layer of ozone to absorb the ultraviolet radiation from the sun. By its photochemical action, the radiation may have helped to keep the atmosphere rich in reactive molecules and far from chemical equilibrium.

Simple organic molecules (*i.e.* molecules containing carbon) are likely to have been produced under such conditions. The best evidence for this comes from laboratory experiments of the Stanley L. Miller and Harold C. Urey in 1953. If water is heated with mixtures of gases such as CO_2, CH_4, NH_3, and H_2 and energized by electrical discharge or by ultraviolet radiation, they react to form small organic molecules such as hydrogen cyanide (HCN) and formaldehyde (HCHO) which accumulate in the U-tube trap (Fig. 2.1). These organic compounds readily undergo further reactions in aqueous solution. Most important, representatives of the major classes of small organic molecules found in cells are generated including amino acids, sugars, and the purines and pyrimidines required to make nucleotides. In 2008, A.P.Johnson and co-workers re-analysed Miller's archived solutions from the original experiments and found that 22 amino acids rather than five were actually created in one of the apparatus used.

Some evidence suggests that Earth's original atmosphere might have contained fewer of the reducing molecules than was thought at the time of the Miller and Urey experiment. There is abundant evidence of major volcanic eruptions 4 billion years ago which would have released CO_2, N_2, H_2S and SO_2 into the atmosphere. More diverse molecules have also been produced through experiments by using these gases. Consequently, a racemic (containing both *L* and *D* enantiomers) mixture is created. In nature, *L* amino acids dominate and later experiments have confirmed that disproportionate amounts of *L* or *D* oriented enantiomers are possible.

Further, the simple organic molecules such as amino acids and nucleotides can associate to form polymers. One amino acid can join with another through a peptide bond. Similarly, two nucleotides can join together by a phosphodiester bond. The repetition of these reactions results in formation of linear polymers which are known as polypeptides and polynucleotides. Large polypeptides, *i.e.* proteins and polynucleotides in the form of both RNA and DNA are found in present day living cells, as the most important constituents of the cell. A set of 20 amino acids constitute the universal building blocks of the proteins, whereas RNA and DNA molecules are constructed from just four types of nucleotides. However, it is uncertain why these particular sets of monomers were selected for biosynthesis in preference to others that are chemically similar. The chemical properties of the corresponding polymers suit them for their specific roles in the cell.

Such experiments cannot exactly reproduce the early conditions on the earth; but the formation of organic molecules is surprisingly easy. The developing earth had immense advantages over any human experimenter; it was very large and could produce a wide spectrum of conditions. In such circumstances at some time and place many of the simple organic molecules (found in present day cells) accumulated in high concentrations.

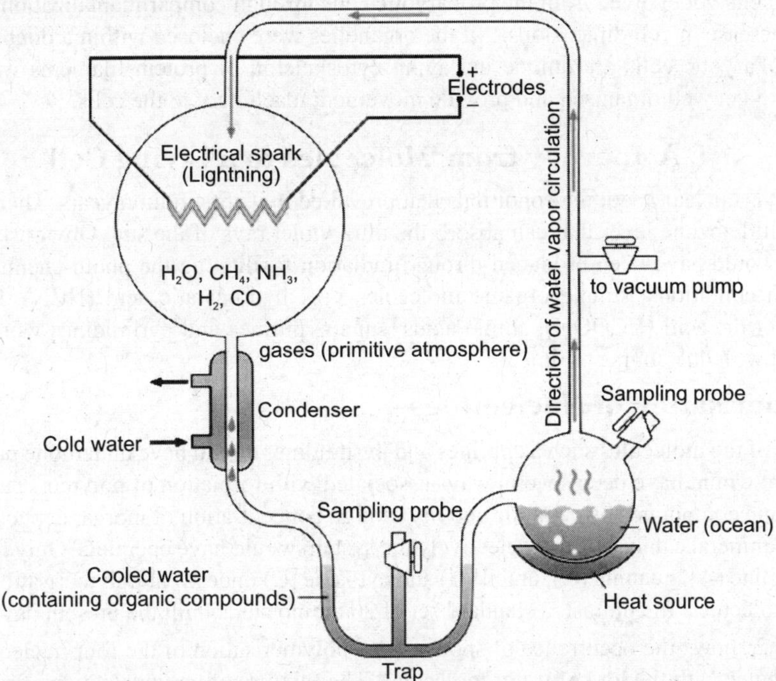

Fig. 2.1. Miller's experiment that demonstrates the conditions which prevailed on primitive earth.

The modern cell biology involvers the distinct areas concerned with the cell : cytology, biochemistry and genetics. Cytology (*cyto-* means 'cell', *logos* means 'study') originated about more than three centuries ago and depended upon light microscope for study. The discovery of electron microscopy helped in understanding the cell components.

Biochemistry helped to unravel the understanding of cell function involving various techniques such as ultracentrifugation, chromatography, electrophoresis, etc. for separation of various cell components and molecules. A significant contribution of biochemistry is the use of radioactively labeled compounds in the study of enzyme-catalysed reactions. Genetics is entirely different which deals with behaviour, storage and transmission of hereditary characters through genetic factors called *genes*. Though the work on genetics was started by Gregor John Mendel about 150 years ago, much understanding came within 75 years of research. Therefore, to understand the present day cell biology one has to study the important contributions and developments made in the above three areas of research. This chapter deals with structural components of the cells and their functioning mechanisms.

A. CELL STRUCTURE AND FUNCTION

About 3.4 billion years ago probably the living cells would have evolved through spontaneous reactions between the molecules present in that environment. Autocatalytic mechanism is the fundamental to all living system. Operation of this system began with the evolution of families of RNA molecules that catalysed their own replication. Thereafter, one of the families of co-operating RNA catalysts developed the ability for direct protein synthesis. Consequently more number of additional protein catalysts would have been accumulated which would have allowed the evolution of more efficient cells.

With the progress of time, RNA was replaced with DNA double helix as a more stable molecule for storing more and more genetic information required by the cells. Gradually biochemically versatile and diverse groups of prokaryotic cells were evolved which had all the metabolic pathways. The eukaryotic cells got evolved from the prokaryotic cells through compartmentalization of organelles and with increase in cell dimensions. All the organelles were enclosed within a double membrane. Besides, eukaryotic cells are unique in having cytoskeleton of protein filaments which help the cytoplasm to get well organised and provide movement machinery to the cells.

I. A Journey from 'Molecules to a Living Cell'

Still it is unclear about the conditions that prevailed in its first billion years. There was no free oxygen or little ozone layer that can absorb the ultra violet rays of the sun. Obviously the reactive molecules would have been produced through radiation facilitating the photo-chemical reactions. Under such conditions, simple organic molecules viz., hydrocyanic acid (HCN), formaldehyde (HCHO), formic acid (HCOOH), amino acids, sugars, purines and pyrimidines would have been synthesised with due course of time.

1. Polymerisation of Nucleotides

Some of the molecules such as purines and pyrimidines would have undergone polymerisation event. There would have been several ways associated with formation of polymers such as heating of dry organic compounds by catalytic activity of high concentration of inorganic phosphates or the other crude mineral catalysts. How selectively the system would have operated? Only four alphabets such as adenine (A), guanine (G), uracil (U) and cytosine (C) underwent polymerisation and formed the RNA molecules. In contrast, a standard set of 20 amino acids built the present day proteins.

Fig. 2.2 shows the occurrence of spontaneous polymerisation of the four nucleotides to form RNA molecules with the loss of water molecules. The mixture of polynucleotide of random length was formed. These polymers would have acted as catalysts and influenced the subsequent reactions.

The origin of life would have required the ability to catalyse the reactions leading to production of more catalyst molecules. Production of catalysts with this self-promoting property would have been favoured. The most efficient molecules that would have added their own production would have diverted raw materials from the production of the other substances. Thus gradually complex chemical system of organic monomers and polymers would have developed which have functioned together to generate more identical molecules.

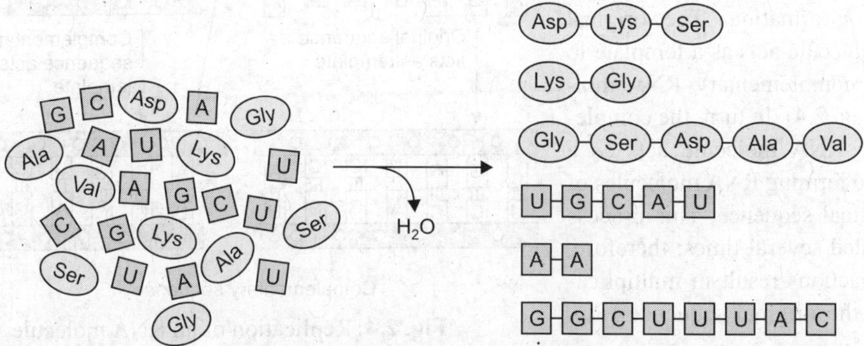

Fig. 2.2. Polymerisation of amino acids and nucleotides to form polypeptides and polynucleotides, respectively (based on Alberts *et al.*, 2002).

Supply of simple raw materials present in the environment has fueled the complex chemical system of organic monomers and polymers. This 'autocatalytic system' would have been evolved as one of the best properties of living matter for: (*i*) random selection of interacting molecules, (*ii*) self-reproduction, (*iii*) competition with other system dependent on the same feedstock, (*iv*) depriving of feed stock that upsets balance of reaction rates, and (*v*) decay and death.

The most versatile catalysts, in the present day living cells, are the polypeptides. These are made up of different amino acids having diverse side chains. Therefore, they adopt various kinds of 3-D forms. In spite of such properties, the polypeptides cannot form another similar sequence. Besides having a limited ability as catalyst, the polynucleotide differs in having ability to guide the formation of exact replica copies of their own sequence. This phenomenon is called replication. This ability of polynucleotide molecules depends on 'complementary pairing' of nucleotide subunits. Because one polynucleotide acts as template to form another copy of polynucleotide having similar sequence if read in 5'→ 3' or 3'→ 5' direction.

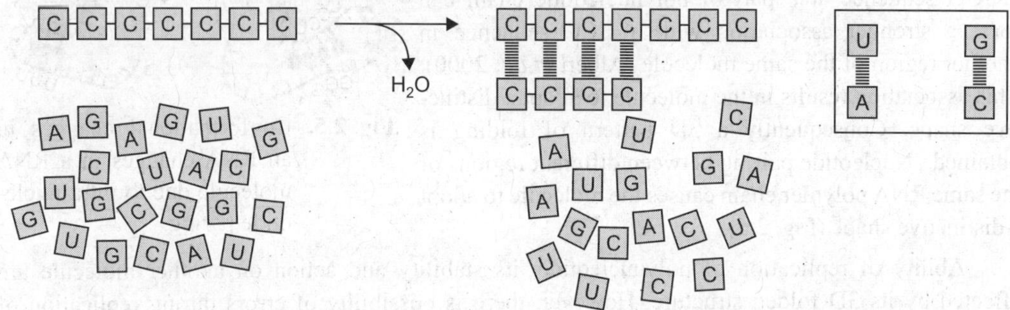

Fig. 2.3. Polynucleotide acting as template (based on Alberts *et al.*, 2002).

For example, polyguanylic acid (or poly G) can line up the subunits that are required to build another polynucleotide (*i.e.* poly cytidilic acid or poly C) along its surface. Hence, poly G promotes the polymerisation of poly C because poly G preferably combines with poly C and vice versa through weak chemical bonds. This pairing enables one polynucleotide to act as template for synthesis of another polynucleotide. In turn poly C synthesises more poly G containing polynucleotide (Fig. 2.3).

Suppose an RNA molecule consists of A, G, U and C in some specific order. When these molecules are added to a mixture of activated nucleotides under suitable conditions, they line up the nucleotides for polymerisation due to complementary pairing.

In biological system the complementary templating mechanisms are involved in transfer process of genetic information. The original RNA molecule acts as a template to form a complementary RNA molecule (Fig. 2.4). In turn, the complementary RNA molecule acts as a template forming RNA molecules of the original sequence. This process is repeated several times; therefore, these reactions result in multiplication of the original sequence (Fig. 2.4).

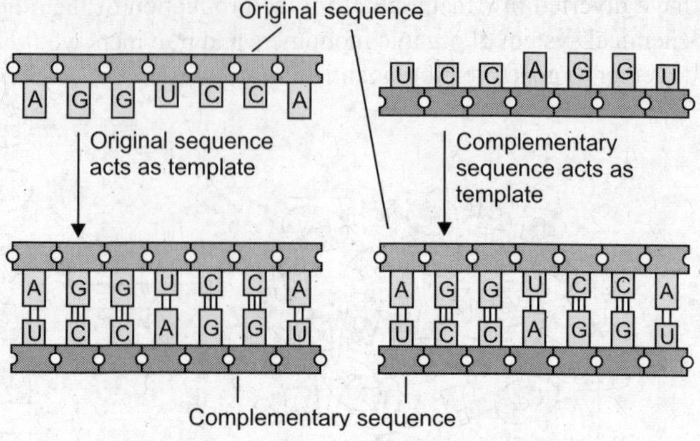

Fig. 2.4. Replication of an RNA molecule

2. Natural Selection of Self-Replicating Molecules

Besides acting as template for their own replication, RNA molecules would have been having different types of catalytic properties. An RNA molecule can act as catalyst for replication of another RNA molecule which further had acted as template. This had resulted in special versatility of RNA molecules which in turn would have played a key role in the origin of life (Alberts *et al.,* 2000).

The chemical nature of nucleotides of RNA molecules affects their behaviour. Its folding in a solution is governed by the specific sequence of nucleotides. As the nucleotides of a template molecule can pair with free complementary nucleotide residues within the polymer itself. For example GGGG sequence at a part of polynucleotide chain can form a stronger association with a CCC sequence in another region of the same molecule (Albert *et al.*, 2000). This association results in the molecule to adopt a distinctive shape. Consequently a 3D pattern of folding is obtained. Nucleotide pairing between different regions of the same RNA polymer chain causes the molecule to adopt a distinctive shape (Fig. 2.5).

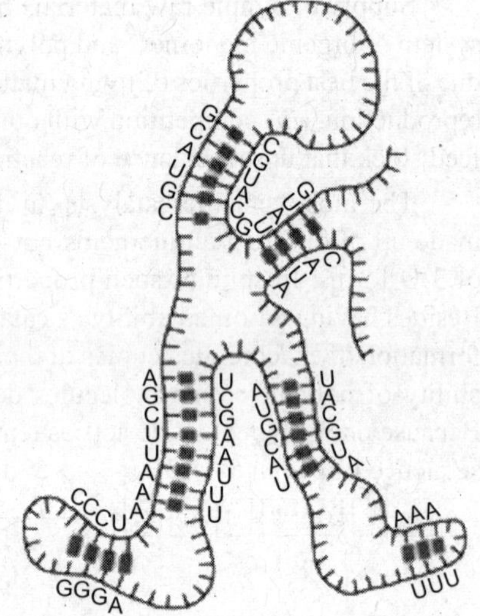

Fig. 2.5. Conformational changes in an RNA changes in a RNA molecule due to intra-nucleotide paring.

Ability of replication of polynucleotide, its stability and action on another molecule are affected by its 3D folded structure. However, there is possibility of errors during replication of RNA. Wrong copies are produced. New variant sequences of nucleotides are produced through repeated replication. In 1990, Elington and Szostak have demonstrated in laboratory conditions a form of natural selection when different sequences predominate according to exact conditions. The RNA molecules can be selected for the ability to bind almost any other molecules specifically.

The two features (informational and functional) are the most important properties required for evolution. The RNA molecules bear these two features *i.e.* it carries information in the form of nucleotide sequence which can pass, and the folded structure which makes it able to interact selectively with the other molecules and its responses to ambient conditions.

3. The Catalytic RNA Molecules – the Ribozymes

The existing environment where the RNA molecules are present governs their natural selection. Moreover, critical component of the environment is the set of other RNA molecules in the mixture that influence the replicating RNA. In 1986, Thomas R. Cech reported that RNA acts as an enzyme. The breakage and formation of covalent bonds between nucleotides can be catalysed by some specialized RNA molecules. They catalyse a change in other RNA molecules and cut a nucleotide at a specific point. Spontaneously, the other types of RNA molecules cut a portion of their own nucleotide sequence and rejoin the cut ends. This process is known as 'self-splicing'. A specific arrangement of atoms forming on the surface of ribozyme governs each RNA-catalysed reaction. Fig. 2.6 shows a 3D structure of the original ribozyme, the self-splicing intron of *Tetrahymena*. Green and blue ribbons indicate the path of the RNA backbone in the two major domains of the RNA, and the red star marks the active site (http://en.wikipedia.org/wiki/Ribozyme).

Some RNAs are Enzymes: For many years it was known that only proteins are the enzymes. The enzymes should be capable of binding the substrate, carrying out chemical reactions, releasing the products and repeating this event many times. Proteins perform all the functions due to attaining 3D (tertiary structure) and opening the binding pockets for substrate (and also co-factor) and active site for catalysis of reaction. But now it is known that some RNAs can attain similar tertiary structure and act as biological catalyst. Such RNA enzymes have been termed as 'ribozymes' (ribonucleic acid enzyme, also called *RNA enzyme* or *catalytic RNA*). The term ribozyme was first introduced by Kelly Kruger and co-workers in 1982. Ribozymes also exhibit similar features of enzymes e.g. presence of an active site, a binding site for substrate and binding site for a co-factor. Ribozymes often have divalent metal ions such as Mg^{2+} which act as cofactors. Such RNAs are one of the first ribozymes which were discovered first. RNase P contains both RNA and protein and also associated with generation of tRNA molecules from the larger precursor RNAs (Watson *et al.*, 2004).

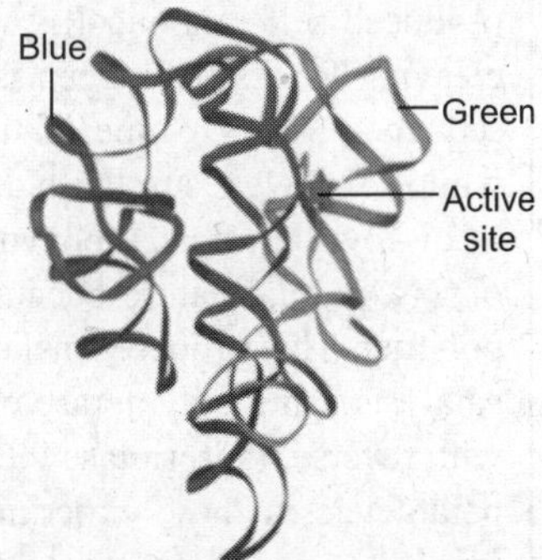

Fig. 2.6. 3D structure of the original ribozyme of *Tetrahymena*.

The first ribozymes were discovered in the 1980s by Thomas R. Cech, who was studying RNA splicing in the ciliated protozoan *Tetrahymena thermophila* and Sidney Altman, who was working on the bacterial RNase P complex. The ribozymes were found in the intron of an RNA transcript, which removed itself from the transcript and in the RNA component of the RNase P complex, which is involved in the maturation of pre-tRNAs.

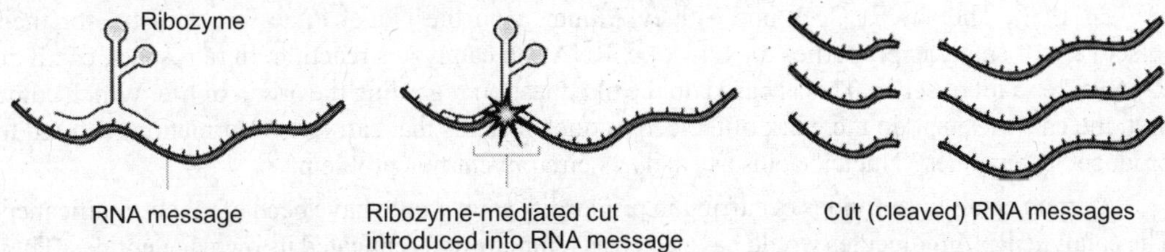

Fig. 2.7. Ribozyme cleavage of RNA. The ribozyme attaches to an mRNA molecule and cleaves into fragments to prevent it from functioning.

Sidney Altman

Thomas R. Cech

Sidney Altman of Yale University (USA) and Thomas R. Cech of University of Colorado (USA) equally shared the Nobel Prize in Chemistry in 1989 for their independent discovery of catalytic properties of RNA. Both of them found that RNA could fold into complex shapes and catalyze biochemical reactions, a function previously thought to be restricted to protein enzymes. Thus, RNA was sometimes an active participant in the chemistry of life, not just a passive messenger. They named these RNA enzymes as 'ribozymes.'

Artificial Ribozymes

There has been interest in the study of new synthetic ribozymes made in the laboratory since the discovery of ribozymes that exist in living organisms. For example, artificially-produced self-cleaving RNAs have been produced which bear good enzymatic activity. Self-cleaving RNAs have been isolatted by *in vitro* selection of RNAs originating from random-sequence RNAs. Some of the synthetic ribozymes that were produced had novel structures, while some were similar to the naturally occurring hammerhead ribozyme. The techniques used to discover artificial ribozymes involve Darwinian evolution. This approach takes advantage of RNA's dual nature because both possess the properties of a catalyst and an informational polymer. This property makes it easy for an investigator to produce vast populations of RNA catalysts using polymerase enzymes. The ribozymes are mutated by reverse transcribing them with reverse transcriptase into various cDNA and amplified with mutagenic PCR. The selection parameters in these experiments often differ. One approach for selecting a ligase ribozyme involves using biotin tags, which are covalently linked to the substrate. If a molecule possesses the desired ligase activity, a streptavidin matrix can be used to recover the active molecules.

In 1989, Thomas R. Cech and Sidney Altman won the Nobel Prize in chemistry for their 'discovery of catalytic properties of RNA', *i.e.* RNA can catalyse a reaction. In retrospect, catalytic RNA makes a lot of sense. This is based on the old question regarding the origin of life. Which come first, the enzymes that do the work of the cell or nucleic acids that carry the information required to produce the enzymes? Nucleic acids as catalysts circumvent this problem.

4. Evolution of Self-Replicating System of RNA Molecules

In 1989, Thomas R. Cech and Sidney Altman won the Nobel Prize in chemistry for their 'discovery of catalytic properties of RNA', *i.e.* RNA can catalyze a reaction. In retrospect, catalytic RNA makes a lot of sense. This is based on the old question regarding the origin of life. Which come first, the enzymes that do the work of the cell or nucleic acids that carry the information required to produce the enzymes? Nucleic acids as catalysts circumvent this problem.

Certain catalytic activities occurring in primordial soup would have been of much significance. The catalytic RNA molecules would have acted as template and replicated its own nucleotides. They would have also influenced its shape (Fig. 2.8A).

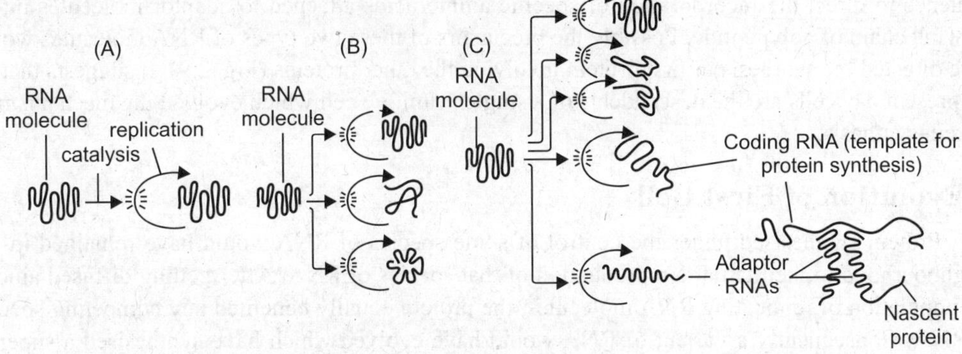

Fig. 2.8. Evolution of a self-replicating system of RNA molecules. A-replication of catalytic RNA; B- families of mutually supportive catalytic RNAs; C- evolution of coding RNA and adaptor RNAs and production of first protein (based on Alberts *et al.,* 2002).

Families of mutually supportive catalytic RNA molecules would have been produced where one molecule would have been catalyzing the reproduction of other molecule (Fig. 2.8B). As a result of which new catalytic RNA molecules got evolved; some of which would have bound the activated amino acids to themselves. By base-pairing to a coding RNA molecule, these RNA molecules allowed an RNA sequence to act as template for the synthesis of amino acid polymers. This would have caused the origin of first genetically-determined protein. Thus they served as the first adaptor between nucleotides and amino acid sequences (Alberts *et al*., 2000) (Fig. 2.8C).

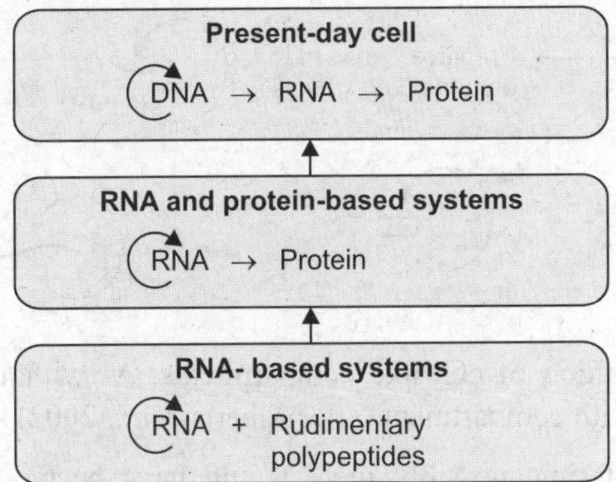

Fig. 2.9. Stages of evolution (based on Alberts *et al*., 2002).

5. Flow of Genetic Information

During the course of evolution the first self-replicating systems of RNA molecules would have mixed with the other organic molecules like simple polypeptides. Probably this would have started the process of evolution. Systems with different sets of polymers would have competed for available precursors to construct their own copies as the present day organisms do. They got success on the basis of accuracy and speed of making copies, and stability of the copies.

The RNA molecules would have collaborated in directing the synthesis of polypeptides (*i.e.* protein synthesis) but this process would have been helped by the previously synthesised proteins. During this process the nucleotides of one RNA molecule would have acted as codes, whereas the other RNA molecule acted as adaptors (for binding a specific amino acid). These two types of RNA molecules formed base pairs complementary to each other. This enabled the nucleotide sequence to direct the incorporation of specific amino acids attached to adaptor molecules into a growing chain of polypeptide. Possibly the precursors of these two types of RNA molecules would have directed the synthesis of first protein involving the other proteins (Fig. 2.9). It suggests that all the present day cells are the descendents of a single primitive cell which evolved the mechanism of protein synthesis.

6. Evolution of First Cell

Protein synthesised under the control of some species of RNA would have remained in the neighbourhood to facilitate the reproduction of that species of RNA. After getting diffused among the population of replicating RNA molecules, the protein equally benefited any competing species of RNA. Consequently a variant of RNA would have evolved which have synthesised a superior enzyme. The RNA molecules could not be selected efficiently according to the quality of the protein before the evolution of some form of compartment that could contain the proteins made by an RNA molecule (Fig. 2.10). Such protein could be available to such RNA for its own use that has formed them. Therefore, the RNA can be selected on the basis of its guiding for production of a better protein.

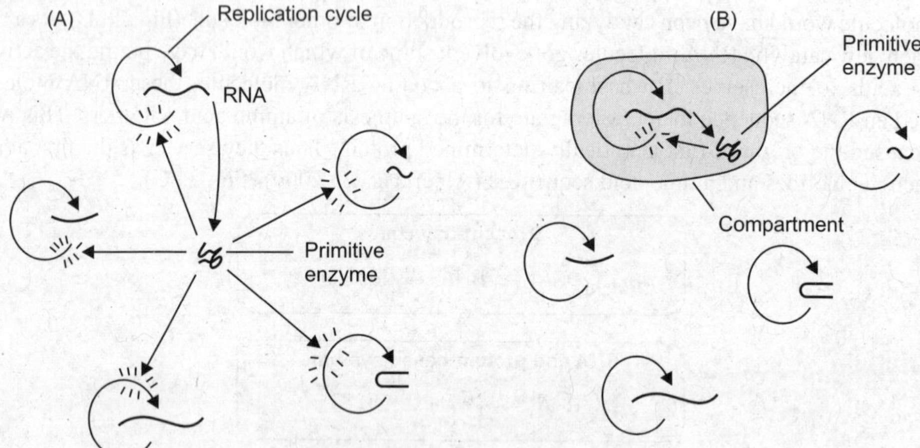

Fig. 2.10. Evolution of cell-like compartments; A- without compartment; B- with compartment (after Alberts *et al.*, 2002)

During the course of time possibly there would have been development of another amphipathic molecules, *i.e.* phospholipids. Amphipathic molecules consist of two parts, one part is hydrophobic (water-insoluble) and the other part is hydrophilic (water-soluble). The amphipathic molecules would have spontaneously aggregated to form bilayer enclosing small vesicles that enclosed the replicating mixture of RNA and other molecules. The aqueous contents of vesicles are separated from the rest of the external medium.

If the RNA molecules got evolved within a membrane, they started to act as carriers of genetic information. It could be presumed that about 3.5 billion years ago the primitive cells would have been produced through prebiotic synthesis of small molecules, the self-replication of catalytic RNAs, translation of RNA sequences into amino acid sequences and assembly of lipids to form double membrane compartments.

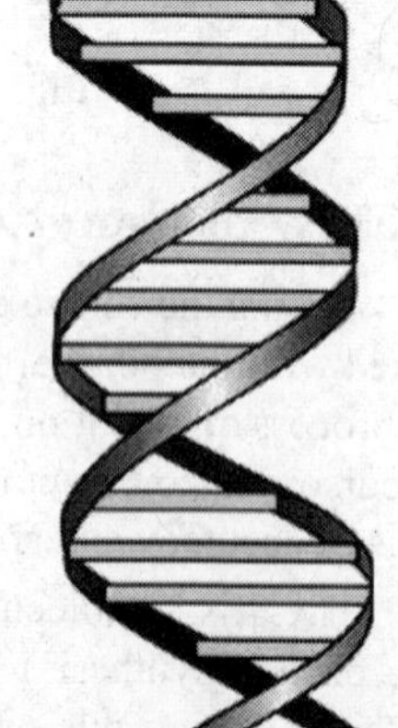

Fig. 2.11. Double stranded DNA helix.

7. The Hereditary Material

Presumably the cells on the earth were not so complicated. In present day cells, hereditary information is stored in DNA instead of RNA. It would have been possible due to evolution of DNA from RNA when stability of cells would have started. Because the polynucleotides must be stable before acting as genetic material and that the RNA molecules were continuously went of changing. However, two different types of work could be assigned for two kinds of polynucleotides where both worked in a collaborative manner.

The double stranded DNA structure (Fig. 2.11) makes it more stable and robust than the RNA and allows it to replicate easily and repair the associated damages. It may be explained that RNA preceded DNA in evolution due to the existence of both catalytic and genetic properties. With the progress of evolution DNA took over the genetic function guiding the RNA to work through expression of protein. The proteins worked as major catalysts, whereas RNA became the intermediate between DNA and protein (Alberts *et al.,* 2002).

II. PROKARYOTIC AND EUKARYOTIC CELLS

On the basis of presence and absence of organised nucleus the cells have been categorised into two broad groups: *prokaryotic cells* and *eukaryotic cells*. Structurally they are very different to each other. One of the major differences is the arrangement of DNA. Prokaryotes lack compartmentalisation of organelles *i.e.* lacking double-membrane bound organelles. For example, DNA of prokaryotes is not surrounded by double membrane. Cells divide by binary fission. However, the prokaryotic nuclear region is called **nucleoid**. Prokaryotic cells (bacteria) contain their chromosome as **circular DNA.** Usually the entire genome is a single circle, but often there are extra circles called **plasmids**. The DNA is packaged by DNA-binding proteins. For detail, *see* Chapter 4 *Molecular biology*.

In terms of numbers of molecules, the most abundant proteins in *E. coli* are the 'histone-like' proteins. There are four main types of proteins that have been well characterised (Table 2.1):

Table 2.1. Four sub-units of histone like proteins and their function.

Protein	*Subunit structure*	*Subunit MW (kDa)*	*Function*	*Gene (s)*
FIS	a2	11.2	DNA bending	*fis*
HNS	a6	15.4	DNA compaction	*hns*
HU	ab	9.5 (a) 9.5 (b)	DNA wrapping	*hupB* (a) *hupA* (b)
IHF	ab	11.2 (a) 10.5 (b)	DNA bending	*himA* (a) *himD* (b)

The eukaryotes are phylogenetically advanced group. Phylogenetic (evolutionary) relationships among the microorganisms have been described on the basis of 16S rRNA sequencing. In 1993, Carl Woese and co-workers presented the three evolutionary lineages among which two are prokaryotes and one is eukaryotes. The three major domains are known as *Bacteria, Archaea* and *Eukarya* (Fig. 2.12).

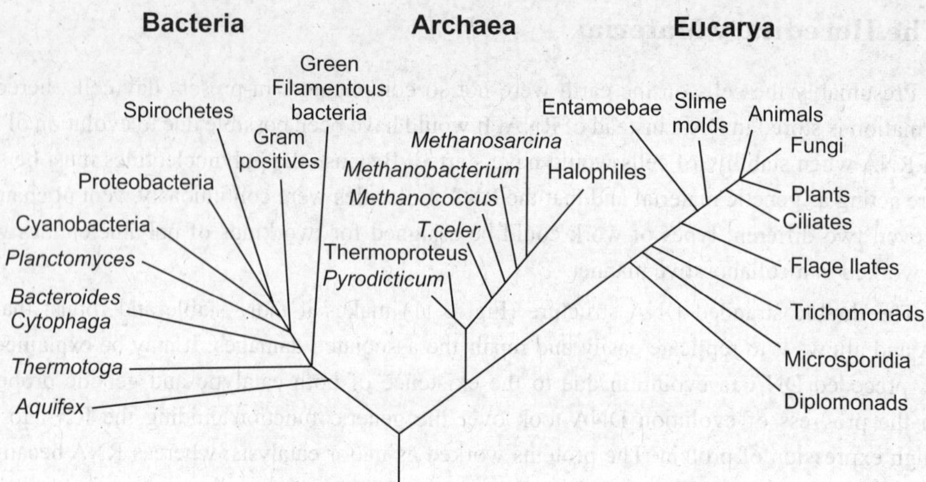

Fig. 2.12. The phylogenetic tree (into three domains) of life based on comparative 16S rRNA sequencing.

The domains Bacteria and Archaea (Gr. *Archae* = ancient, *bact* = rod) include prokaryotes, whereas Eukarya consists of eukaryotes. The phylogenetic tree shows the procaryotes in two domains: Archaea and Bacteria. At a taxonomic level, most organisms at the tips of the Archaeal branches represent a unique Order; most organisms at the tips of the bacterial branches are classified into a unique Phylum. On the Archaeal limb, the three physiological groups are evident in the names: 'thermo' and 'pyro' for the extreme thermophiles; 'methano' for the methanogens; and 'halophiles' for the extreme halophiles. The most important, best known, and diverse groups (phyla) branching of the Bacterial limb are the cyanobacteria, proteobacteria and Gram-positive bacteria.

These organisms lack peptidoglycan in cell walls; however, the cell wall is made up of non-cellulosic polysaccharides or proteins or cell wall is absent. The lipids of the plasma membrane consist of glycerol linked to branched hydrocarbon chains. Archael lipids are glycerol diether isopranoid derivatives, whereas normal lipids are glycerol diester fatty acid derivatives (Fig. 2.13).

Glycerol
Ester bond
CH_2-O-O
Stearic acid
A normal lipid $OH-O-O$
CH_2-OH

Ester bond
CH_2-O
Phytanol
Archaeal lipid OH_2-O
CH_2-OH
Phytenylglycerol diether

Fig. 2.13. Comparison between normal lipid (A) and archaeal lipid (B).

Except mycoplasma, all the members of Bacteria and Archaea have the following major structures: cell wall, cytoplasmic membrane, ribosomes, inclusion bodies and nucleoids.

The Archaea are the most ancient prokaryotes which lived in most extreme environments, for example hydrothermal vents, volcanic walls, hot sulphur springs, the Great Salt Lake, the Dead Sea, etc. (2.14).

The archaea are divided into methanogens (CH_4 evolving bacteria), halophiles (salt-loving bacteria) and thermo-acidophiles (high temperature- and low pH- loving bacteria). Differences between prokaryotes and eukaryotes are given in Table 2. 2.

Fig. 2.14. Unusual environment; A-deep sea hydrothermal vents; B-the Great Salt Lake of Utah.

Table 2.2. Differences between prokaryotes and eukaryotes.

	Prokaryotes	*Eukaryotes*
1.	They are very small from 1 to 10 μm.	They are large sized from 10 to 100 μm.
2.	Cellular materials are not enclosed within double	Double membrane-bound organelles are present.
3.	Cell wall is present in most of prokaryotes.	Cell wall is present in plant cells but not in animal cells.
4.	Cell wall is made up of murein.	Cell wall is made up of cellulose or kitin.
5.	Cell wall component peptidoglycan is present.	No peptidoglycan is found.
6.	Due to small size of cells the rates of metabolic activities are very high.	Rates of metabolic activity are slow due to small size.
7.	They contain primitive nucleus that is called nucleoid.	They contain well developed nucleus.
8.	DNA lacks histone proteins, hence it is not condensed.	Histone proteins are associated with DNA. It gets condensed several hundred times to form chromosomes which are visible at metaphase stage.
9.	DNA is is linear or circular molecule lacking introns; plasmids are present.	DNA is linear molecule which is interrupted by introns. Plasmids are rare.
10.	Ribosomes are 70S.	Ribosomes are 80S.
11.	They multiply by binary fission.	They multiply by mitosis.
12.	They can grow at all extreme environmental conditions.	They fail to grow at extreme environmental conditions.
13.	Respiration occurs in mesosomes.	Respiration takes place in mitochondria.
14.	In some bacteria photosynthetic enzymes are located in chlorosomes/cell membrane.	Photosynthetic enzymes are located in chloroplasts.

1. Structure of Bacterial Cells

Structure of a typical bacterial cell is given in Fig. 2.15. A typical bacterial cell consists of cell envelope (capsule, cell wall and plasma membrane) and cytoplasm that includes proteins, carbohydrates, RNA and DNA molecules, plasmids, ribosomes and several inclusion bodies. There are two types of bacteria based on being stained by Gram's stain: *Gram-negative bacteria* (that cannot be stained by Gram's stain) and *Gram-positive bacteria* (that are stained by Gram's stain). Cell wall composition and structure of these two groups of bacteria differ.

Chemical composition of a bacterial cell may be inferred by chemical analysis of a bacterium such as *E. coli*, which is displayed in Table 2.3. The macromolecules comprise about 96% of the dry weight of the cell. Small molecules and inorganic ions, which are constituents of the cytoplasm, comprise of the remaining 4%.

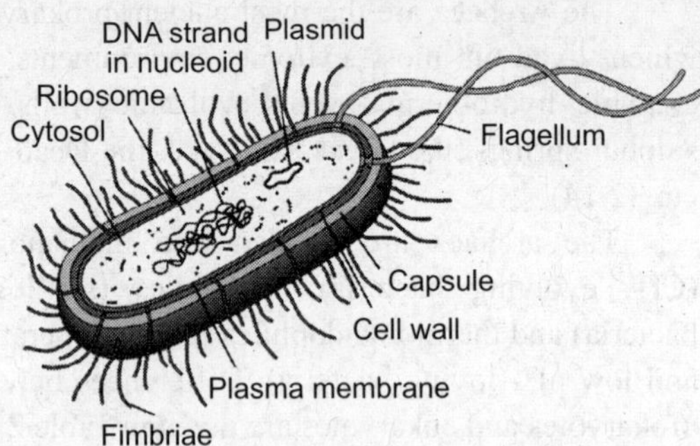

Fig. 2.15. Structure of a bacterial cell.

Table 2.3. Molecular composition of *E. coli* showing balanced growth.

Molecule	***Dry weight (%)***
Protein	55.0
Total RNA	20.5
DNA	3.1
Phospholipid	9.1
Lipopolysaccharide	3.4
Peptidoglycan	2.5
Glycogen	2.5
Small molecules (precursors, metabolites, vitamins, etc.)	2.9
Inorganic ions	1.0
Total dry weight	100.0

(*a*) Capsule: It is also termed as glycocalyx which covers the cell wall. Not all but some bacteria possess capsule. Thickness of glycocalyx varies genus to genus. It is made up of polysaccharides but also contains proteins. It appears like the gum and provides adhesiveness to bacteria. It protects the bacteria from desiccation, phagocytes, toxic chemicals and drugs, viral attack, etc.

(*b*) Flagella (singular *flagellum*)**:** Some bacteria possess flagella for their locomotion. Prokaryotic flagella are smaller than eukaryotic flagella. The flagella are about 20 mm long having a diameter of 20-25 mm. On the basis of the number of flagella per cell, the following four types of arrangements have been described (Fig. 2.16).

Different species of bacteria have different numbers and arrangements of flagella. **Monotrichous** bacteria have a single flagellum (e.g. *Vibrio cholerae*) (Fig. 2.16A). **Lophotrichous** bacteria have two or more flagella at one end of the cell (B). **Amphitrichous** bacteria have a single flagellum on each of two opposite ends (only one flagellum operates at a time, allowing the bacteria to reverse course rapidly by switching which flagellum is active) e.g. *Nitrosomonas*, *Pseudomonas fluorescens* (C). **Peritrichous** bacteria have flagella projecting in all directions (e.g. *Salmonella typhae, Escherichia coli*) (D). Cells devoid of any flagellum (e.g. *Lactobacillus*) are called **Atrichous. Cephalotrichous** bacteria have insertion of flagella in polar tufts.

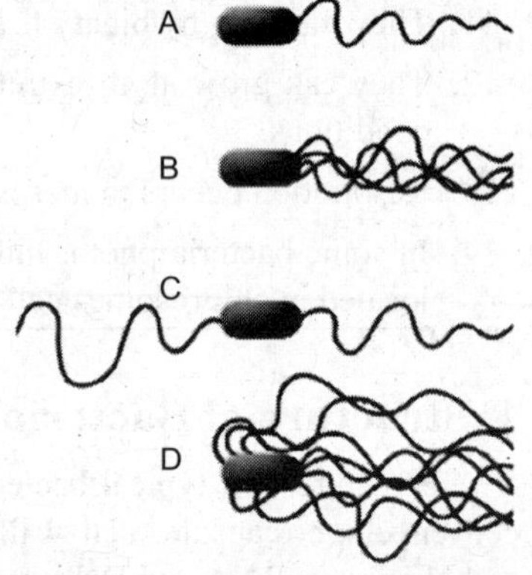

Fig. 2.16. Flagellation in different bacteria.

In some bacteria, such as *Selenomonas*, the flagella are organised outside the cell body, twining about each other into a structure called a 'fascicle'". Other bacteria, such as Spirochates, have a specialised type of flagellum called an 'axial filament' that is located in the periplasmic space, the rotation of which causes the entire bacterium to move forward in a corkscrew-like motion.

***(i)* Structure of flagellum:** Electron microscopic studies show that a flagellum consists of three components: *filament, hook* and *basal bodies* (Fig. 2.17).

- **Filament:** It is the longest part of the flagellum extending out the cell surface. It is 1-70 nm long having 20 nm diametre. Filament is made up of a special protein called *flagellin* where protein molecules are spirally arranged. Filament is attached to the basal body.
- **Hook:** It is a curved and tubular structure which is made up of different protein sub-units. Hook is the thickest part of flagellum.
- **Basal body:** It is embedded inside the cell wall and plasma membrane. Therefore, it is the most complex part of the flagellum. In Gram-negative bacteria it consists of four rings (the L ring associates with the lipopolysaccharide, the P ring associates with peptidoglycan layer, the M ring is embedded in the plasma membrane, and the S ring is directly attached to the plasma membrane) connected to a central rod.

Two rings are embedded in cell wall and peptidoglycan, respectively. The inner two rings are embedded in plasma membrane. The filament ends with a capping protein. Gram- positive bacteria have two of these basal body rings, one in the peptidoglycan layer and one in the plasma membrane. The flagellum rotates at 360° like a whip resulting in forward movement of the bacterium.

The bacterial flagellum is driven by a rotary engine made up of protein which is located at the flagellum's anchor point on the inner cell membrane. The engine is powered by proton motive force *i.e.* by the flow of protons (hydrogen ions) across the bacterial cell membrane due to a concentration gradient set up by the cell's metabolism. In *Vibrio* species there are two kinds of flagella, lateral and polar, and some are driven by a sodium ion pump rather than a proton pump. The rotor transports protons across the membrane, and is turned in the process. The rotor alone can operate at 6,000 to 17,000 rpm, but with the flagellar filament attached usually only reaches 200 to 1,000 rpm.

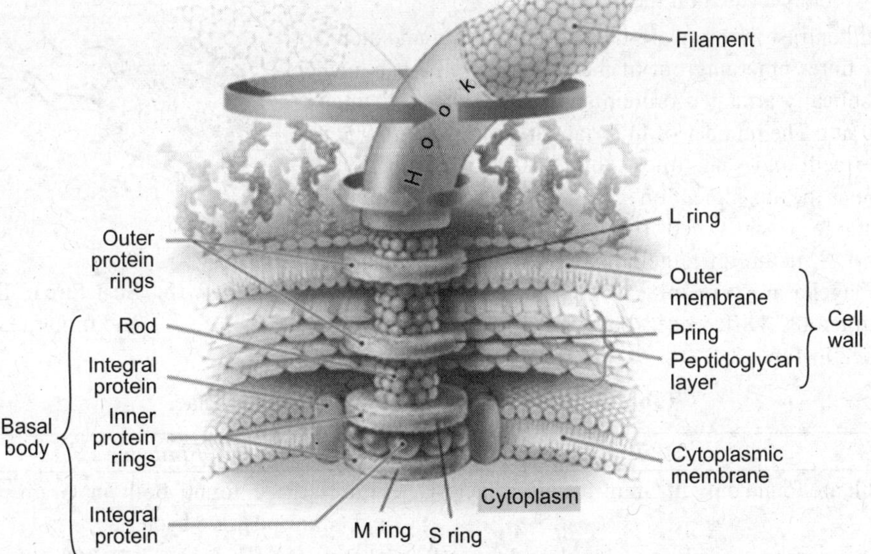

Fig. 2.17. Structure of a bacterial flagellum and its different parts.

Flagella do not rotate at a constant speed but instead can increase or decrease their rotational speed in relation to the strength of the proton motive force. Flagellar rotation can move bacteria through liquid media at speed of up to 60 cell lengths/second. Although this is only about 0.00017 km/h, when comparing this speed with that of higher organisms in terms of number of lengths moved per second, it is extremely fast. When size is accounted for prokaryotic cells, swimming at 50-60 lengths/sec is actually much faster than larger organisms.

(*ii*) Differences between the archaeal and bacterial flagella: Differences between the archaeal and bacterial flagella are as given below:

- Bacterial flagella are motorised by a flow of H^+ ions (or occasionally Na^+ ions); archaeal flagella are almost certainly powered by ATP. The torque-generating motor that powers rotation of the archaeal flagellum has not been identified.
- While bacterial cells often have many flagellar filaments, each of which rotates independently, the archaeal flagellum is composed of a bundle of many filaments that rotate as a single assembly.
- Bacterial flagella grow by the addition of flagellin subunits at the tip; archaeal flagella grow by the addition of subunits to the base.
- Bacterial flagella are thicker than archaeal flagella, and the bacterial filament has a large enough hollow tube inside that the flagellin subunits can flow up the inside of the filament and get added at the tip; the archaeal flagellum is too thin to allow this.

(*c*) Pili and Fimbrae: These are the cell appendages which are not involved in locomotion. Pili (singular pilus) are long, thicker and tubular structure. It is made up of protein molecules called pilin. It bears an adhesive property; hence it gets attached to the surfaces of other bacteria. Thus, pili are involved in conjugation phenomenon in Gram-negative bacteria (but not in Gram-positive bacteria). Therefore, these pili are called sex pili. It also acts as attachment site for bacteriophages like M13 bacteriophage.

The number of pili is genetically governed by the genes of plasmids. If the number of plasmid is 5, the number of pili would be 5. Bacterial plasmid from donor or male cell (designated as F^+ cell) is transferred through the lumen of pilus into the recipient or female cell (designate as F^- cell). Here F refers to fertility factor which is located on the plasmid.

In contrast fimbriae are the brittle like thin and slender tubular fibres emerging out of the bacterial cells. It is made up of helically arranged protein sub-units. It has a diameter of 3-10 nm. The number of fimbriae ranges between 300 and 400 per cell, and sometimes around 1000. The function of fimbriae is the attachment on specific surfaces. Development of fimbriae is governed by genes present on bacterial nucleoid. Structure of flagella and fimbriae is given in Fig. 2.18. Flagella are very long, while the fimbriae are shorter and numerous. Differences between pili and fimbriae are described in Table 2.4.

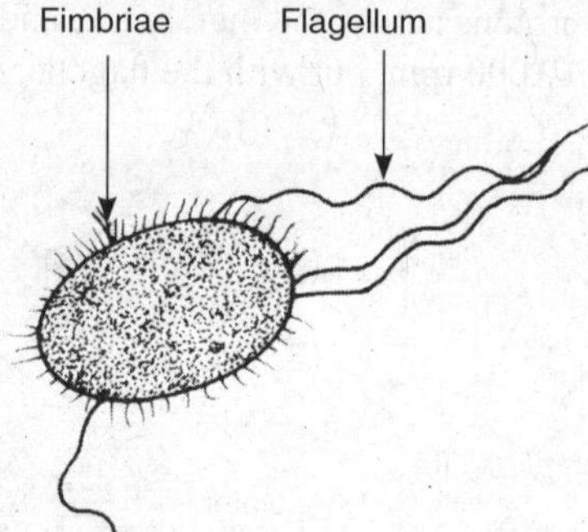

Fig. 2.18. Structure of flagella and fimbriae.

Table 2.4. Differences between pili and fimbriae.

	Pili	*Fimbriae*
1.	Pili are found only in Gram-negative bacteria.	Fimbriae are found both in Gram-negative and Gram-positive bacteria.
2.	The number of pili are very less (generally 3-5) because their formation is genetically governed by genes of plasmid.	The number of fimbriae is 300-400 per cell and their formation is governed by mucleoid genes.
3.	The length and diameter of pili are more than fimbriae.	Fimbriae are shorter and narrower than the pili.

(d) Cell Wall: Except mycoplasma, all prokaryotes contain a rigid cell wall which provides a proper shape to cell. It protects the bacterial cells from environmental shocks. Cell wall is rigid due to the presence of a macromolecule called peptidoglycan (murein or mucopeptide). Peptidoglycan is constituted by the repeating units of N-acetyl glucosamine (NAG) and *N*-acetyl muramic acid (NAM) (Fig. 2.19).

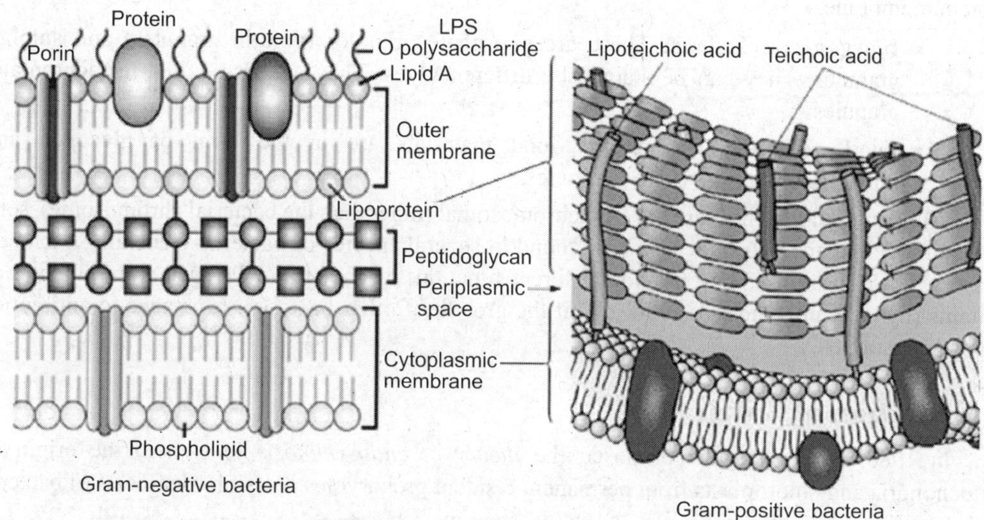

Fig. 2.19. Ultra structure of cell wall of Gram-negative bacteria (A), and Gram-positive (B) bacteria.

NAG and NAM are cross-linked by short peptide chains. It provides flexible framework. This framework can be broken by certain antibiotics (*e.g.* penicillin) and lysozyme (occurring naturally in tears and saliva).

Cell wall of Gram-positive bacteria is thick (20-80 nm), which consists of mainly peptidoglycan and teichoic acid. The cell wall of Gram-negative bacteria is comparatively thin (8-12 nm) and appears two-layered. The outer layer consists of lipopolysaccharides, lipids and proteins. The inner layer consists of peptidoglycan, proteins, lipids, amino acids, etc. The cell wall of *Mycobacteria* and *Nocardia* (Gram-positive bacteria) also contain mycolic acid (a long chain of fatty acid).

(e) Plasma Membrane: An ultra-thin (6-8 nm), semi-permeable, bi-lipid layer encloses the protoplasm. It is made up of phospholipid and proteins that are arranged in a fluid mosaic pattern. A phospholipid molecule comprises of a polar and a flexible tail which is a chain of fatty acids. Proteins are embedded within the lipid layer which serves the important functions of the cells, *i.e.* transportation of nutrients and metabolites across the membrane. Besides, the bacterial membrane also acts as a specific site for attachment of circular DNA which later starts replication.

Plasma membrane invaginates within the cell to form vesicles, tubules and lamellae which are collectively termed as *mesosome*. The mesosomes are commonly found in Gram-negative bacteria. It acts as a site for respiration, increases surface area, help in chromosomal replication. In photosynthetic bacteria also the plasma membrane invaginates to form tubules and vesicles which contain bacterial photosynthetic pigments. These structures are called *chlorosomes*. In some bacteria plasma membrane also consists of photosynthetic pigments.

(f) Cytoplasm: Plasma membrane encloses the cellular materials of the bacteria which carry out all vital functions required for a living cell. It consists of water, carbohydrates, proteins, lipids, RNAs, ribosomes, DNA molecules and several inclusion bodies as reserve food material. They are found freely in cytoplasm e.g. cyanophycean granules in cyanobacteria. In bacteria the inclusion bodies (e.g. poly β–hydroxybutyrate (PHB), sulphur granules and gas vesicles) are covered by a single membrane.

***(i)* Inclusion bodies:** These are aggregates of small hollow vesicles found in cyanobacteria, purple and green photosynthetic bacteria. Each vesicle is surrounded by a single non-unit non-lipid protein membrane.

- **Inorganic inclusions:** These are phosphate granules (volutin granules) and sulphur granules. They can be stained by different dyes. They are also called metachromatic granules.
- **Food reserves:** The reserve food materials are in the form of glycogen and polyhyroxybutyrate (PHB) granules.

***(ii)* Plasmids:** Plasmids are the extrachromosomal (other than the bacterial chromosome), self-replicating, circular DNA molecule. It is found in several species of bacteria. Generally, each cell possesses 3-5 plasmids. They provide different functions to bacterial cells such as production of colicins (by colicinogenic factors to inhibit the growth of other bacteria), resistance to antibiotics (by R plasmid), etc.

2. Evolution of Eukaryotes

In 1980, Lynn Margulis proposed the *theory of endosymbiosis* to explain the origin of mitochondria and chloroplasts from permanent resident prokaryotes. Possible symbiosis of bacteria within early eukaryotic cells was a major step in the evolution of eukaryotic cells. According to this hypothesis some 1.5 billion to 700 million years ago a larger prokaryote (or perhaps early eukaryote) engulfed or surrounded a smaller prokaryote. This resulted in evolution of double membrane system. Steps in endosymbiosis are given in Fig. 2.20.

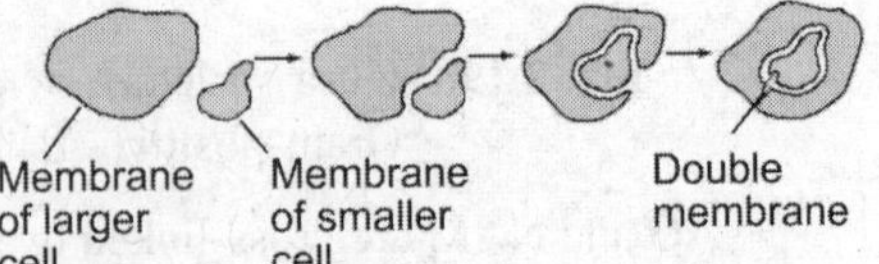

Fig. 2.20. Steps in endosymbiosis.

3. Compartmentalisation of Organelles

Cells are of two types, prokaryotic cells (bacteria, cyanobacteria, mycoplasma) and eukaryotic cells (all fungi, plants and animals). In prokaryotes, all the macromolecules are enclosed within a cell membrane. The cell membrane is covered by a cell wall (except mycoplasma where only three-layered unit membrane is present). In eukaryotes the cell membrane encloses nucleus and cytoplasm. The nucleus is a membrane-bound structure that consists of chromosomes. Chromosomes are the most condensed form of genetic material visible at metaphase. In all prokaryotes, DNA is not bounded by the nuclear membrane, but it remains naked which is called nucleoid. On the other hand, eukaryotic cytoplasm includes various types of double layered membrane-bound organelles such as : (*i*) plastids (only in plants), (*ii*) mitochondria, (*iii*) endoplasmic reticulum, (*iv*) ribosomes, (*v*) Golgi complex, (*vi*) peroxisomes. Plastids are found specially only in plants, whereas all the organelles are found in every types of cells of each organisms. The phenomenon of presence of membrane-bound organelles is called *compartmentalisation of organelles*. Compartmentalisation is important in maintaining the structure of the cell, the ionic balance and electrode potential. Structures are compacted the way because that is how they have adapted to the external environments and the functions they must undertake within the cell.

Instead of digesting the smaller organisms the large one and the smaller one entered into a type of symbiosis known as mutualism, where both organisms benefit and neither is harmed. The larger organism gained excess ATP provided by the 'protomitochondrion' and excess sugar provided by the 'protochloroplast', while providing a stable environment and the raw materials the endosymbionts were required. This is so strong that no eukaryotic cells can survive without mitochondria (likewise photosynthetic eukaryotes cannot survive without chloroplasts), and the endosymbionts cannot survive outside their hosts. Nearly all eukaryotes have mitochondria. Mitochondrial division is remarkably similar to the prokaryotic methods that will be studied later in this course.

The nuclear envelope allows the nucleus to control its contents and separate them from the rest of the cytoplasm where necessary. This is important for controlling processes on either side of the nuclear membrane. In some cases where a cytoplasmic process needs to be restricted, a key participant is removed to the nucleus, where it interacts with transcription factors to downregulate the production of certain enzymes in the pathway.

III. THE EUKARYOTES (CELL STRUCTURE AND COMPONENTS)

The domain Eukarya consists of all eukaryotic (Gr., *eu* = good, *karyotic* = nucleated) organisms. Eukaryotic cells are large sized having a well organised nucleus covered by a double layered membrane called nuclear membrane. All the organelles are covered by a double layered membrane. The eukaryotic cells have different shapes and size. Irrespective of groups of eukaryotes, all the cells are covered by a plasma membrane.

Further, there are two groups in eukaryotes: plants and animals. Plants (from algae to angiosperms) evolved first then animals (from protozoa to mammals) got evolved. Most of the plant cells bear a complex and rigid cell wall, whereas animal cells lack cell wall. It maintains the structure of the cells and protects them from environmental fluctuations. Instead, cell membrane called unit membrane does all the function. Differences between plant and animal cells are given in Table 2.5.

Table 2.5. Differences between plant and animal cells.

Plant cell	*Animal cell*
1. Plant cells are larger than animal cells	Animal cells are generally smaller than the plant cells
2. Each cell consists of a cell wall which is made up of cellulose	Animal cells lack cell wall. Plasma membrane is the outermost covering of the cell
3. Photosynthetic cells consists of chloroplast, hence they appear green	Except *Euglena*, animal cells lack chloroplast
4. Centriole (centrosome) is absent except a few plant cells	Centriole is generally present, except few animal cells
5. Cytokinesis takes place by cell plate formation	Cytokinesis occurs due to formation of furrow in centre of cells
6. Many simple Golgi apparatus is present called dictyosomes	A single and complex Golgi apparatus is present
7. A single large central vacuole is present that pushes cytoplasm towards periphery of the cell	They contain well developed nucleus.

Cell is a small membrane bound unit structure consisting of aqueous solution of chemical compounds. There are many organisms which are unicellular, bicellular, filamentous, coenocytic or septate, branched or unbranched forms with gradually increasing complexity to form tissue, organs or body. Complexity has arisen at anatomical, physiological and reproductive levels. But the basic unit of all such forms is a single cell.

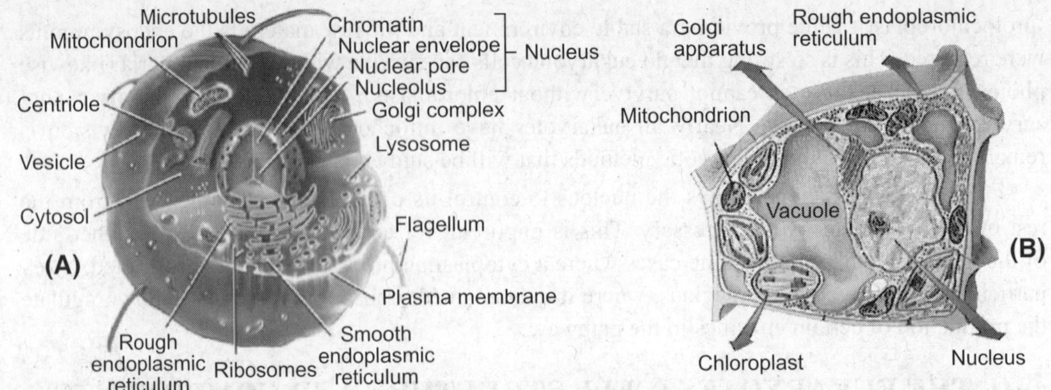

Fig. 2.21. Structure of a typical animal cell (A) and plant cell (B).

In 1665, for the first time Robert Hooke observed the cork cells under a simple microscope. Study of cell and its components is known as *cytology*. Furthermore, several techniques have been developed in recent years which helped to study the structure and function of cells of prokaryotes, plants and animals. In all types of cell there are many common fundamental processes such as macromolecules and their synthesis. Study of these cellular processes is called *cell biology*. Structure of a typical animal and plant cell is given in Fig. 2.21.

1. Cell Wall

The cell wall is a rigid layer that protects the cells of all eukaryotes except the animals and some plants. It surrounds the plasma membrane and governs the proper shape and size of the cell. Therefore, shape of the plant is also governed by the cell wall. On the basis of different types of cells, different tissues are found in plants such as parenchyma (having intercellular spaces), aerenchyma (having air space), sclerenchyma (without intercellular space), etc. The plant cell wall consists of cellulose as one of the major cell wall components. In contrast, the fungal cell wall consists of chitin. Cell wall formation occurs in the form of cell plate at the time of cell division.

***(a)* Structure of Cell Wall:** Cell wall is a non-living complex structure which is differentiated into three layers: primary cell wall, secondary cell wall and tertiary cell wall. Tertiary cell wall may be present occasionally.

***(i)* Primary cell wall:** It is formed in all plants during early growth stages. It is the outermost layer of the cell which is found in parenchymatous and immature meristematic cells. Due to deposition of cutin it becomes impermeable such as epidermal layers of leaf and stem. Primary cell wall is permeable. It is composed of cellulose, hemicellulose and pectin. The cellulose fibrils are linear and interwoven in a gel like matrix. It is 1-3 μm thick.

***(ii)* Secondary cell wall:** The primary cell wall is covered by secondary cell wall after accomplishment of growth of the former. It is present in certain highly specialised and mature cells. Its thickness varies from 5 to 10 μm. In most of the cells protoplast disappears after formation of secondary cell wall. This wall is thick and permeable. It is present near tertiary wall or plasma membrane. It comprises of compactly arranged cellulose macrofibrils. The secondary wall consists of three layers: outer layer, middle layer and inner layer. These three layers get arranged in concentric way and form the secondary cell wall. In this wall hemicellulose, pectin and lignin are sometimes deposited.

***(iii)* Tertiary cell wall:** In some of the plants, a third wall is deposited in the inner layer of secondary wall which is called tertiary cell wall. This layer is very thin and formed in the xylem tracheids of gymnosperms. In addition to cellulose, it possesses another chemical constituent called xylan.

- **Middle lamella:** It is an inter-cellular matrix which cements together the cells of plant tissues. It is made up of proteins, pectin and lignin. But the main constituent of middle lamella is pectin which is a hydrophilic and colloidal substance.
- **Plasmodemata:** Cytoplasmic relationships between the plants cells are maintained by minute apertures which are called plasmodesmata. Hence they act as cytoplasmic bridges or junctions. Intercellular circulation of solutions (containing ions, dissolved gases and other substances) takes place through plasmodesmata.

(*b*) Cell Wall Chemistry: One of the carbohydrates called *cellulose* constitutes the cell wall. In addition, cell wall is also made up of hemicellulose, lignin, pectin, chitin, cutin and proteins. Certain minerals e.g. calcium and magnesium in the form of carbonates and silicates are also found in cell wall.

(*i*) Cellulose: It is a polymer of glucose sub-unit. About 3000 glucose sub-units are linked together by β, 1→4 linkage to form cellulose molecules. Several molecules of cellulose are parallely arranged to form the bundles. It is the most important conmponent of the cell wall of most of the plant groups. The cell wall of fungi and lichen consists of chitin, glucans and mannan.

(*ii*) Hemicellulose: Hemicellulose molecules are made up of monosaccharides e.g. galactose, manose, arabinose, xylose, etc.

(*iii*) Pectin: Pectin is made up of glucuronic acid and galacturonic acid. It is a derivative of carbohydrate which provides rigidity to cell wall. It is found in all the three layers of cell wall. Lignin is added to the cell wall matrix during the process of lignification.

(*iv*) Lignin : Lignin is composed of complex benzene-ring compounds and cutin is comprises of fatty acids. The structural unit of chitin is the glucosamine. The structural proteins are also present that act as inter-locking agent

(*c*) Function of Cell Wall: Cell wall carries out the following important functions:

- It provides a proper shape to the cells.
- It protects plasma membrane and prevents from osmotic bursting of cytoplasm.
- It protects from the infection of pathogens.
- It provides mechanical strength to the cell and tissues and organs.

2. Cell Membrane

In 1855, C. Nugeli and C. Cramer coined the term 'cell membrane' or plasma membrane. It is the outer boundary of the cell. It is 4-5 nm thick constituted by two layers of phospholipids called lipid bilayer. But it is 7.5 nm thick when measured along with protein. Then it appears trilayered or trilaminar. At certain intervals proteins are embedded in lipid bilayer. Cell membrane consists of lipoproteins which are the combination of lipids and proteins. Generally the membrane consists of 40% carbohydrate and 60% proteins on dry weight basis. The carbohydrates are found in glycolipids and glycoproteins.

In 1972, Singer and Nicolson have given the *fluid mosaic model* of cell membrane to explain its major features. This model is widely accepted due to the best explanation of membrane properties. According to fluid mosaic model, the cell membrane is *quasifluid* structure in which lipids and integral proteins are arranged in mosaic manner. It forms a continuous layer of phospholipid molecules in which globular proteins are embedded (Fig. 2.22). The interaction between lipids and proteins results in fluidity to the membrane.

(*a*) Lipids: Lipids found in cell membrane are polar lipids which consist of a polar head (hydrophilic) and two non-polar (hydrophobic) tails. There are three main constituents of lipids that form cell membrane: *phospholipids, glycolipids* and *sterols*. Phospholipids have a phosphate moiety in the hydrophilic head. One of the tails has the saturated fatty acid and the other an unsaturated fatty acid. In a lipid bilayer, head is arranged towards outside and tails are centrally directed. Glycolipid consists of one or more monosaccharide residues linked with lipids. Sterols are steroid alcohols consisting of cyclopentanoperhyrophenanthrene nucleus. The most abundant sterol in animals is the cholesterol and that of plant is phytosterol. Ergosterol is found as the major membrane sterol of eukaryotic microorganisms.

(*b*) Proteins: The membrane proteins are very important as they play a role in governing the function of a cell. There are thousand types of proteins associated with cell membrane. Phospholipids and many intrinsic proteins are amphipatic molecules, *i.e.* they show both hydrophilic and hydrophobic properties (Fig. 2.22). The globular proteins are of two types: *intrinsic* (integral) proteins and *extrinsic* (peripheral) proteins. The intrinsic proteins are insoluble proteins and may be dissociated with difficulty, but the extrinsic proteins are soluble proteins which can easily be dissociated from the membrane.

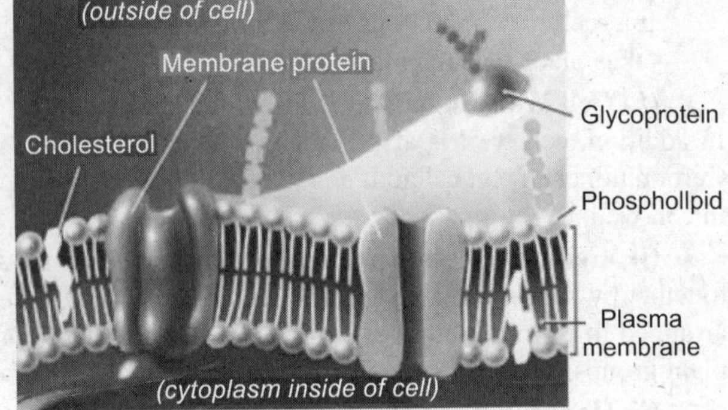

Fig. 2.22. Structure of cell membrane

The intrinsic proteins are partially embedded either out side or inner side of surface of lipid bilayer. The extrinsic proteins are the soluble proteins and penetrate both the layers. The glycoproteins (glycosylated protein, *i.e.* sugar-binding proteins) cement two adjacent cells and thus help in cell-to-cell adhesion. There are basically three classes of proteins present on cell membrane: *structural proteins, carrier proteins* and *enzymes*. The structural proteins form the membrane. The major part of membrane proteins show catalytic properties due to enzymatic nature. Some proteins act as electron carriers, receptor of the other molecules, ligands, recognition sites of viruses, etc. Some proteins act as pump and channel for transporting specific molecules and ions across the cell membrane. They form pores or channels. The carrier proteins transport the substances across the membrane against concentration gradient. Transport across membrane and receptors have been described in detail in the preceding sections.

(*c*) Function of Cell Membrane: The cell membrane carries out several important biological functions : (*i*) it prevents the escape of cellular materials outside the cell, (*ii*) it shows semi-permeability towards molecules/ions, (*iii*) it acts as the site for oxidative phosphorylation or photosynthesis, (*iv*) it bounds the cell organelles in eukaryotes, (*v*) it contains the protein receptors to which various ligands (e.g. antibodies, hormones, etc.) can attach, (*vi*) the exchange of gases and ions takes place through the control by cell membrane, and (*vii*) in bacteria it possesses the attachment site for chromosome and plasmid DNA.

3. Protoplasms

All sub-cellular structures, *i.e.* nucleus and cytoplasm all together are called *protoplasm*. Cytoplasm is the aqueous material between the cell membrane and nuclear membrane. The soluble part of cytoplasm, other than organelles is called *cytosol*. Cytoplasm consists of several organelles, insoluble cellular wastes and storage products.

4. Cell Organelles

As discussed earlier the eukaryotic cells consist of persistent double layered structure called cell organelles. Each organelle is specialised for specific function. Some of the important organelles are described below:

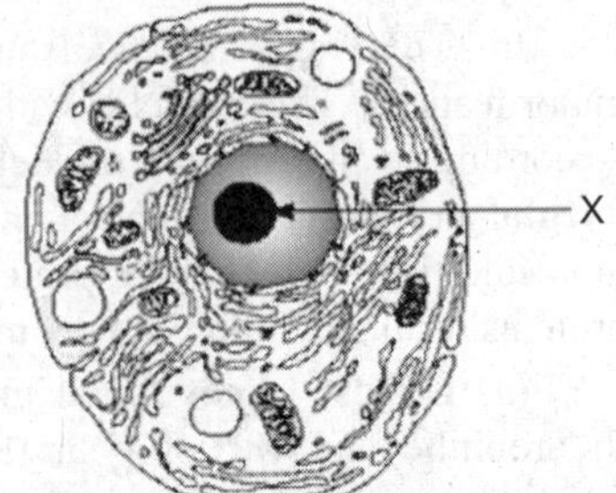

Fig. 2.23. A cell showing nucleus in centre and darkly stained nucleolus (*X*).

(a) Nucleus: The nucleus was the first organelle to be discovered, and was first described by Franz Bauer in 1802. It was later described in more detail by Scottish botanist Robert Brown in 1831 as he was studying orchid microscopically when he observed an opaque area, to which he called the areola or nucleus, in the cells of the flower's outer layer. He did not suggest a potential function. In 1934, J. Hammerling demonstrated that nucleus determines the characters of an organism through grafting techniques of two species of an alga: *Acetabularia crenulata* and *A. mediterranea*. In 1838, M. Schleien proposed that the nucleus plays a role in generating cells; thus he introduced the name "cytoblast" (cell builder). In 1855, Rudolf Virchow who propagated the new paradigm that cells are generated solely by cells (*Omnis cellula e cellula*). The function of the nucleus remained unclear. In 1873, A. Weismann postulated the equivalence of the maternal and paternal germ cells for heredity. The function of the normal nucleus as carrier of genetic information became clear only later after the discovery of nucleus and rediscovery of Mendelian rules at the beginning of the 20th century; thereafter, the chromosome theory of heredity was developed.

Moreover, nucleus is the centrally located, large sized spherical structure of 10 mm diameter. The number of nuclei in a cell is usually one, but the number may be more than one in multinucleate cells or organisms. The polykaryotes of the bone marrow have about 100 nuclei, whereas the RBCs are devoid of any nucleus. Shape of the nucleus varies depending on type of cells and tissues such as disc shaped (in squamous epithelial cells), spherical (in spherical or polyhedral cells), bilobed (in WBCs), C-shaped (*Vorticella*), pyriform *i.e.* kidney shaped (*Paramecium*), etc. A nucleus consists of nuclear membrane, nucleoplasm, chromosomes, and nucleolus (Fig. 2.23).

(i) Nuclear envelope: The nuclear envelope bounds the nucleus. Under electron microscope two membranes can be observed: the outer membrane and inner membrane. Thickness of membranes is 7.5 nm (1 nm = 10Å). Structure of the nuclear envelope and nuclear pores is given in Fig. 2.24. It is made up of two membranes of lipoprotein. At several points the outer membrane remains continuous with the membranes of endoplasmic reticulum and plasma membrane. The ribosomes get attached on the outer side. The two membranes are separated by 15 nm wide perinuclear space which is filled with fluid. Antibodies, lipid droplets and crystalline deposits have also been observed in this site. The nuclear membrane is not continuous but it is interrupted by nuclear pores of varying diameter (30-100Å) through the fusion of outer and inner membrane. The nucleus of a typical mammalian cell will have about 3,000 to 4,000 pores throughout its envelope, each of which contains a donut-shaped, eightfold symmetric ring-shaped structure at a position where the inner and outer membranes fuse. Attached to the ring is a structure called the 'nuclear basket' that extends into the nucleoplasm and a series of filamentous extensions that reach into the cytoplasm. Both structures serve to mediate binding to nuclear transport proteins. Thus, the pores appear circular or octagonal. Nuclear pore complex (NPC) is the sites of exchange of gases and chemical substances between nucleus and cytoplasm. Each NPC consists of about 100 different polypeptides which are called Nup (nuclear pore proteins or nucleoporins).

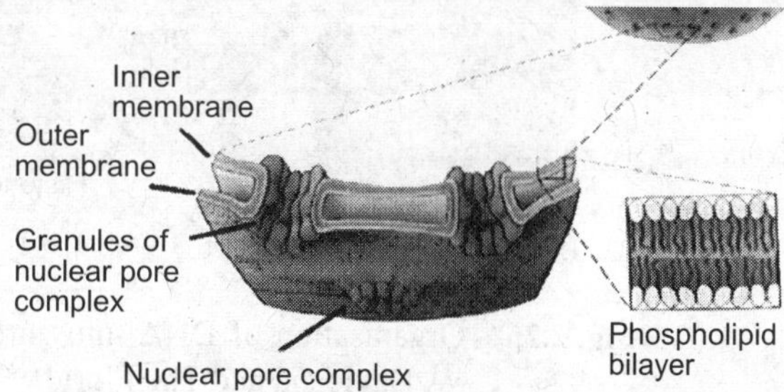

Fig. 2.24. Structure of the nuclear envelope and nuclear pores

There are large numbers of signal sequences which are present on nuclear pore complex (NPC). These sequences facilitate the import and export of proteins and ribonucleoproteins across the NCPs. Biochemically the nuclear membrane is composed of proteins (59-79%) and lipids (17-35%) and low amount of carbohydrates. The nuclear envelope shows ATPase activity also. Besides many transport enzymes, a high concentration of ATPase is found in nuclear pores. One of the well defined examples of RNA export is U snRNAs which consists of hypermethylated cap (m2,3,7 GpppN). It binds to cap-binding protein complex which is involved in export. The complex formed by U snRNAs and cap-binding protein complex is called ribonucleoprotein (RNP). RNP export across the NCP is also mediated by some signal sequences present in the proteins.

***(ii)* Chromosomes:** The space between the nucleolus and nuclear membrane is filled by an aqueous gel-like substance called nucleoplasm. The nucleoplasm consists of dissolved phosphorus, proteins, nucleotides, ribosomes and DNA. DNA takes various shapes such as nucleosomes, fibrils, chromatids and chromosomes (*chroma* = colour; *soma* = thread like). DNA complexed with basic proteins (nucleoproteins) divides first (karyokinesis) followed by division of cytoplasm (cytokinesis). Condensation of chromatin proceeds with progress of cell division from prophase to metaphase. Chromatin is dispersed in nucleoplasm during interphase. There are two types of chromatin, *euchromatin* and *heterochromatin*. Euchromatin is loosely coiled, less compact DNA and less densely stained. It also contains genes that are frequently expressed by the cell. Herterochromatin is the more compact form and contains DNA that is infrequently transcribed.

This structure is further categorised into *facultative* heterochromatin, consisting of genes that are organised as heterochromatin only in certain cell types or at certain stages of development. *Constitutive* heterochromatin consists of chromosome structural components such as telomeres and cetromeres. During interphase the chromatin organises itself into discrete individual patches called chromosome territories. Active genes which are generally found in the euchromatic region of the chromosome tend to be located towards the chromosome's territory boundary.

Length of euchromatin varies with species or organisms. In 1975, E. Strasburger described the chromosomes as a thread like structure that appears during cell division. Helical DNA passes through several organisational stages (DNA → nucleosomes → fibrils → chromatids → chromosomes) to be visible during metaphase which is called chromosomes. The number of chromosomes per cell varies according to vegetative and reproductive tissues.

The chromosomes are the highly condensed form of DNA which becomes visible at metaphase of cell division. A chromosome splits into two chromatids which are joined together at a point called centromere. Organisation of DNA into a typical eukaryotic chromosome is shown in Fig. 2.25.

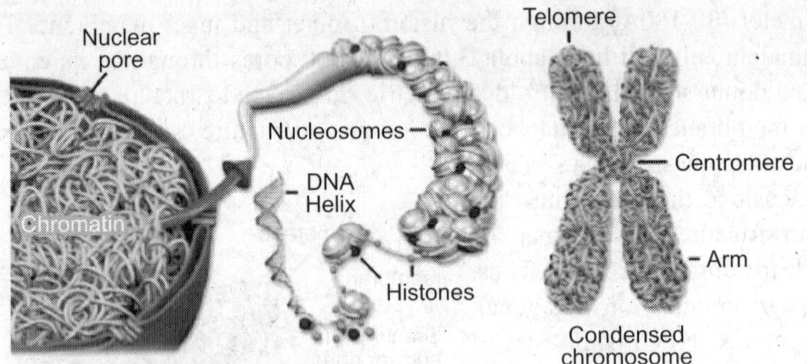

Fig. 2.25. Organisation of DNA into chromosome (*adopted* from Michael W. Davidson and The Florida State University).

***(iii)* Nucleolus:** In 1874, for the first time Fontana recognised the nucleolus. It is a darkly stained spherical and colloidal body present in the nucleoplasm which is associated with nucleolar organiser chromosomes. In this chromosome there lies a special region which is called nucleolar organiser region (NOR); thus the nucleolus gets attached to NOR. It is found in close association with nucleolar organiser region of two or more chromosomes. The cells which are under the process of active synthesis have large nucleoli. Nucleoli are absent in some algae, mammalian cells such as erythrocytes, reticulocytes, and spermatozoa. However, number of nucleoli per cell may be one or many depending on types of cells and organisms.

The nucleolus is composed of large amount of ribosomal proteins and ribosomal RNA. Different types of rRNAs and ribosomal proteins are synthesised by NOR of DNA in the nucleolus. In the less dense region near the centre of nucleus the rRNA becomes folded. The folded rRNA in association with proteins forms ribosome sub-units. The sub-units are transported from nucleus to cytoplasm through nucleolar pore, where it is completely assembled. Ultrastructure of a nucleus showing darkly stained nucleolus in the centre is shown in Fig. 2.26.

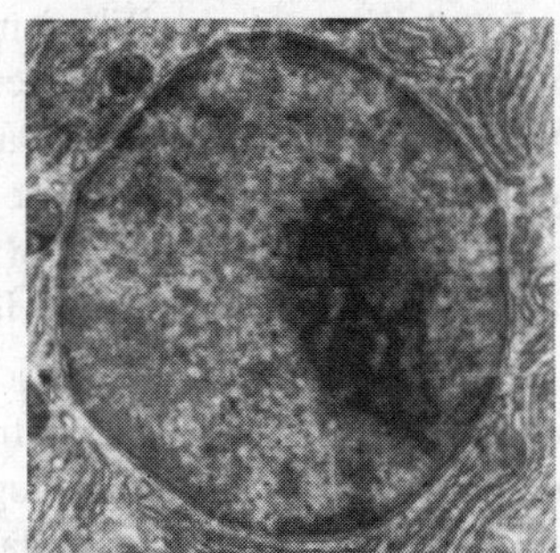

Fig. 2.26. Ultra structure of a typical nucleus and nucleolus in centre.

According to earlier worker, the nucleolus consists of two main components: a continuous coiled filament called *nucleoplasm* and a homogeneous matrix called *pars amorpha*. The filamentous components are embedded in the homogeneous component. But the ultra structure of nucleolus shows four main components: *matrix* or *pars amorpha, chromatin, fibrils* and *granules*. Matrix is homogeneous where granules and fibrils are scattered within it. Chromatin is associated with nucleolus which contains substantial amount of DNA, which acts as template for RNA transcription. *Perinucleolar* chromatin surrounds the nucleolus like a shell and forms septa-like tuberculae inside which is called as intranucleolar chromatin.

As the nucleus consists of small nuclear RNA (U snRNA e.g. Ui, U2, U3, U5, U6, etc.), in the same way nucleolus also consists of more than 50 small nucleolar RNAs (U snoRNA e.g. U3, U7, U8, U14, etc.). The U3 an U14 snoRNAs are ubiquitous and bind to ETS (external transcribed spacer) and/or ITS1 (internal transcribed spacer) to remove them by early processing. Besides, pre-rRNA processing is carried out by seven snoRNA (U3, U7, U8, U14, U22, snR30, RNA component of RNase).

(*b*) Endoplasmic Reticulum (ER) : In 1953, Porter first coined the term ER. It extends through the cytoplasm of eukaryotic cell and occupies a large intracellular space. In light microscope it appears like a net in cytoplasm, hence called as ER. The ER is composed of three types of morphologically distinct structures, *i.e.* cisternae, vesicles and tubules (Fig. 2.27).

(*i*) Cisternae : Cisternae are the long, flattened sac like unbranched tubules of 40-50% µm diameter. Cisternae occur in those cells which have synthetic roles in cells such as pancreas.

(*ii*) Vesicles : The membrane bound vacuolar oval structures of 25-530 µm diameter are called vesicles. They are found in most of the cells.

(*iii*) Tubules : Tubules are the branched structures of 50-190 mm diameter forming a network system along the vesicles and cisternae. The membrane of ER remains continuous with cell membrane, nuclear membrane and Golgi complex. The cavity of cisternae and tubules of ER are developed and act as the passage for transport of the secretory products. The ER is of two types, the smooth ER and rough ER. The smooth ER is more tubular.

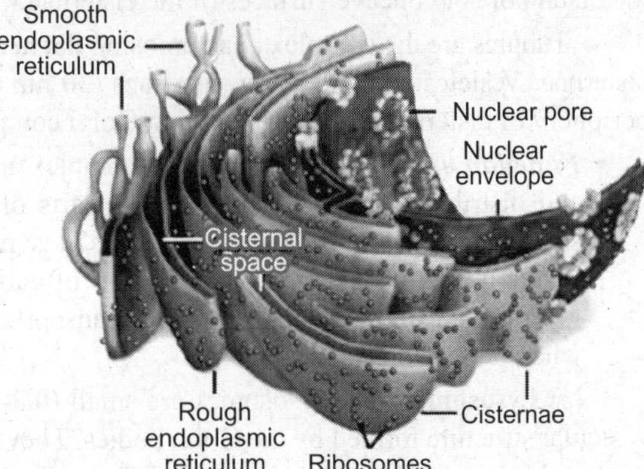

Fig. 2.27. Structure of endoplasmic reticulum.

The membrane of rough ER is covered by ribosomes, therefore, the surface becomes granular. It is the site of protein synthesis. The vesicles covered by ribosomes are called microsomes. The proteins transported from ER to Golgi complex are passed either to the other cell organelles or secreted from the cell.

(c) Ribosomes : Ribosomes are small (about 20 nm diameter) dense, rounded and granular particles of the ribonucleoprotein. Ribosomes were first discovered in plant by Robinson and Brown in 1953, and in animals by Palade in 1955. These are dispersed in cytoplasm and also attached to endoplasmic reticulum. The complete form of ribosomes appears when two different sub-units (small and large sub-units) join together. On the basis of sedimentation rate, ribosomes are classified into two types, 70S and 80S. Here 'S' represents Svedberg unit.

Ribosomes are synthesised in nucleolus. Chemically ribosomes are nucleoproteins, *i.e.* they are made up of almost equal amount of ribosomal rRNA and proteins. Presence of Mg^{++} ions plays a significant role in joining together the two sub-units of ribosomes. When the concentration of Mg^{++} ions decreases in cytoplasm the two sub-units are dissociated. Ribosomes play an important role in protein synthesis. During protein synthesis many ribosomes are aggregated on a common mRNA and form the polyribosomes. Ribosomes move along mRNA in 5' → 3' direction as a zip. The charged tRNA attaches one-by-one as per mRNA codes on A site of ribosome as peptidyl bond between two amino acids are formed one-by-another. Protein synthesis occurs more efficiently in the presence of polyriobosomes.

(d) Golgi Complex: In 1898, Camillo Golgi observed first the parallely arranged flattened membrane-bound vesicles (called cisternae) devoid of ribosomes in nerve cells of owl and cat. After the name of discoverer it was named as Golgi complex, Golgi apparatus, Golgi bodies or dictyosomes. It is found in all eukaryotic cells.

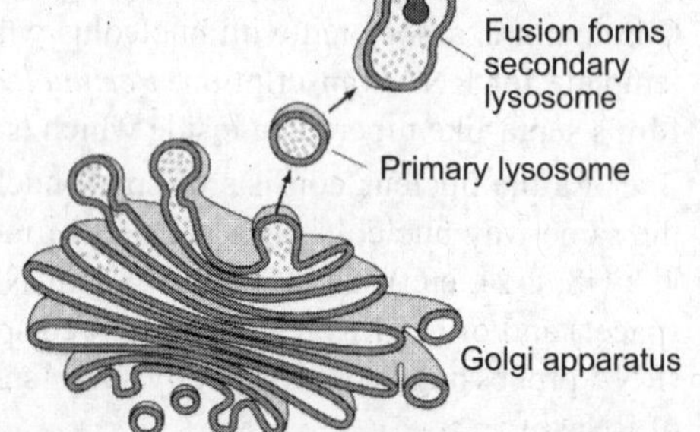

Fig. 2.28. Structure of lysosome formed by Golgi complex.

Golgi complex consists of central flattened plate like *cisternae*, peripheral network interconnecting *tubules*, and peripherally occurring *vesicles* (Fig.2.28). Cisternae are a cavity-filled with fluid within the cell. Its number varies 1-30. Vesicles bud off from the ER and get fused to form cisternae. Cisternae have two poles, one pole has convex surface and the other has concave surface. The process of budding off of vesicles occurs on the inner concave surface of Golgi complex. Formation of cisternae takes place towards the outer concave surface of Golgi apparatus. The secretory vesicles are formed from the distal pole or concave surfaces of the cisternae.

Tubules are the complex anastomosing flat network that arises from the peripheral area of the cisternae. Vesicles are small drop-like sacs (30 nm diameter) which are attached to tubules at the periphery of cisternae. It can be said that Golgi complex originated from the ER.

- ***Function of Golgi complex:*** Golgi complex accepts vesicles from ER, modifies the contents and distributes the products to other parts of the cells. The mnodified vesicles are called secretory vesicles or lysosomes. Thus it synthesises the lysosomes. It also performs the function of glycosylation, *i.e.* addition of carbohydrates moiety to the protein forming the glycoproteins. The glycoproteins are transported in the cisternae. They act as receptor proteins when present on cell surface.

(e) Lysosomes: The lysosomes are small (0.2-0.5 μm diameter), spherical membrane-bound vesicular structure formed by the Golgi bodies. They contain hydrolytic enzymes such as hydrolases that could destroy the cell. About 40 different hydrolases are known such as proteases, lipases, nucleases, phosphatases, α-glucosidase, β-galactosidase, etc. All the enzymes are enclosed within a single lipoprotein membrane which can easily be broken. Enzymes after being synthesised in ER are transported to Golgi complex through its cisternae. They are cleaved from the Golgi complex as vesicles called primary lysosomes. The lysosomes can fuse with the phagosomes (*i.e.* phagocytic vesicles) and form phagolysosomes. Lysosome contents function in the extracellular breakdown of materials.

***(f)* Peroxisomes:** Peroxisomes are small, single membrane bound vesicles found in all eukaryotic cells. They are derived from the endoplasmic reticulum and replicate by fission. Peroxisomes consist of oxidative enzymes catalase that generates and destroy hydrogen peroxide (H_2O_2). They are involved in the catabolism of very long chain fatty acids, branched chain fatty acids, D-amino acids, polyamines, and biosynthesis of plasmalogens, etherphospholipids critical for the normal function of mammalian brains and lungs.They also contain about 10% of the total activity of two enzymes in the pentose phosphate pathway which is important for energy metabolism. It is rigorously debated if peroxisomes are involved in isoprenoid and cholesterol synthesis in animals. Certain enzymes within the peroxisome remove hydrogen atoms from specific organic substrates (labeled as R) by using molecular oxygen producing hydrogen peroxide (H_2O_2) which is toxic.

$$RH_2 + O_2 \rightarrow R + H_2O_2$$

Catalase is another peroxisomal enzyme that uses H_2O_2 to oxidise the other substrates including phenols, formic acid, formaldehyde and alcohol by means of the peroxidation reaction.

$$H_2O_2 + R'H_2 \rightarrow R' + 2H_2O$$

In this way it eliminates the poisonous hydrogen peroxide in the process.

***(g)* Mitochondria:** In 1898, Benda coined the term mitochondria. Mitochondria (Greek *mito* = thread, *chondrion* = granular) are elongated and granular organelles of aerobic cells of plants, animals and microorganisms. They are absent in prokaryotes. Mitochondria are uniformly distributed in cytoplasm. In many cells they are restricted to an area where ATPs are heavily utilised, for example contractile elements of muscles, base of sperm flagellum, etc. The cells of green plants contain less number of mitochondria than the animal cells. The erythrocytes (RBCs) of mammals and other higher animals lack mitochondria. The size of mitochondria varies from cell to cell. They may have 0.2-2.0 μm diameter and 0.3-40 μm length.

Mitochondria contain 65-70% proteins, 25-30% lipids, 0.5% RNA and small amount of DNA. However, the outer membrane consists of 50% proteins and 50% phospholipids. Mitochondria have double membrane of which the inner membrane invaginates to form the folded structure called cisternae (Fig.2.29).

Mitochondria are also called as 'the powerhouse of the cell' because they generate most of the cell's supply of adenosine triphosphate (ATP) which is used as a source of chemical energy. Besides, mitochondria are involved in several other processes such as: cell signalling, cellular differentiation, cell death (apoptosis), and also control the cell cycle and cell growth. Mitochondria are also involved in several human diseases such as mitochondrial disorders cardiac dysfunction, and the aging. Cristae act as the site for oxidative phosphorylation and electron transport. Martix also stores several enzymes required in Krebs cycle.

The inner mitochondrial membrane contains proteins which carry out five types of functions: (*i*) perform the redox reactions of oxidative phosphorylation, (*ii*) ATP synthase, which generates ATP in the matrix, (*iii*) specific transport proteins regulate metabolite passage into and out of the matrix, (*iv*) acts as protein import machinery, and (*v*) mitochondria fusion and fission protein.

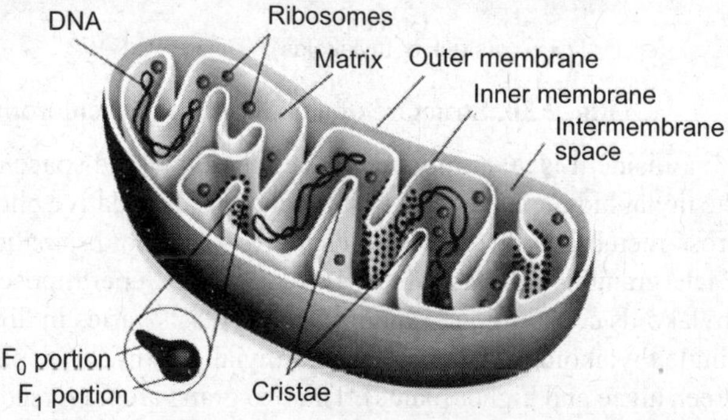

Fig. 2.29. Structure of mitochondria and its components (*adopted* from Encyclopedia Britanica, 1998).

***(h)* Plastids:** In 1885, A.F.W. Schimper first coined the term plastids. These are such organelles that are found only in plants and *Euglena*. Chromatophores (colour bearing bodies) are found inside the plastids in blue-green algae, fungi, plants and certain animals. These are enclosed by a double membrane. Plastids are self-replicating organelles that contain DNA and synthesise proteins.

Plastids perform some of the important biological functions such as synthesis of food and storage of carbohydrates, lipids and proteins. On the basis of structure, pigments and function plastids are of the following types:

***(i)* Chloroplast:** Double membrane-bound and chlorophyll containing organelles found in plant cells and other eukaryotic organisms are called chloroplasts. Chloroplasts are organelles that conduct photosynthesis. Chloroplasts are of various shapes such as filamentous, spherical, ovoid, discoid, club-shaped, collar shaped, branched, stellate, etc. Their size varies from species to species. They are generally 2-3 mm thick having 3-10 mm diameter. The number of chloroplasts per cell varies from 20 to 40 reaching maximum to 1,000 and minimum one per cell. Generally, chloroplasts are considered to be originated from cyanobacteria through endòsymbiosis. In 1883, Schimper observed that chloroplasts closely resemble cyanobacteria.

Fig. 2.30A shows the structure of a typical chloroplast. The chloroplast is surrounded by a double-layered composite membrane with an intermembrane space. It has reticulations or many enfolding filling the inner spaces. The chloroplast has its own DNA which codes for redox proteins involved in electron transport in photosynthesis; this is termed as plastome. Chloroplast genome is considerably reduced compared to that of free-living cyanobacteria, but the parts that are still present show clear similarities with the cyanobacterial genome. Plastids may contain 60-100 genes, whereas cyanobacteria often contain more than 1,500 genes. Many of the missing genes are encoded in the nuclear genome of the host. Fig. 2.30 B shows the structure of a chloroplast genome in liverwort.

The inner matrix of chloroplast is called stroma which also contains grana. Within the stroma are stacks of thylakoids (the sub-organelles) which are the site of photosynthesis. The thylakoids are arranged in stacks called grana (singular granum) which are the flattened sacs filled with fluid and pigment.

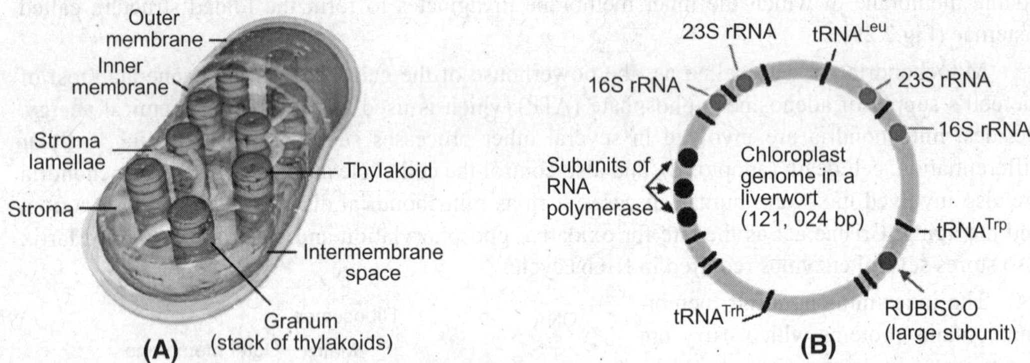

Fig. 2.30. Structure of a chloroplast (A); chloroplast genome in a liverwort (B).

Inside it is an empty area called the thylakoid space or lumen. Photosynthesis takes place on the thylakoid membrane. As in mitochondrial oxidative phosphorylation, it involves the coupling of cross-membrane fluxes with biosynthesis via the dissipation of a proton electrochemical gradient. Each granum consists of 10-100 disc-like superimposed parallel thylakoids. The number of thylakoids constituting granum in chloroplasts varies in different groups of organisms, for example single thylakoid in red algae, paired thylakoids in chrysophyta of algae, in a group of 4 or many (in green algae and higher plants). The two grana are traversed by the lamellae (thylakoid membranes). Thylakoids contain chlorophyll a (75%) and chlorophyll b (25%), carotenoids, electron carriers and enzymes. Dark reaction occurs in stroma and photosystem I (PS I) and PS II takes place inside the grana. Function of PS I and PS II, electron transport chain and photosynthesis are shown in Fig. 2.31.

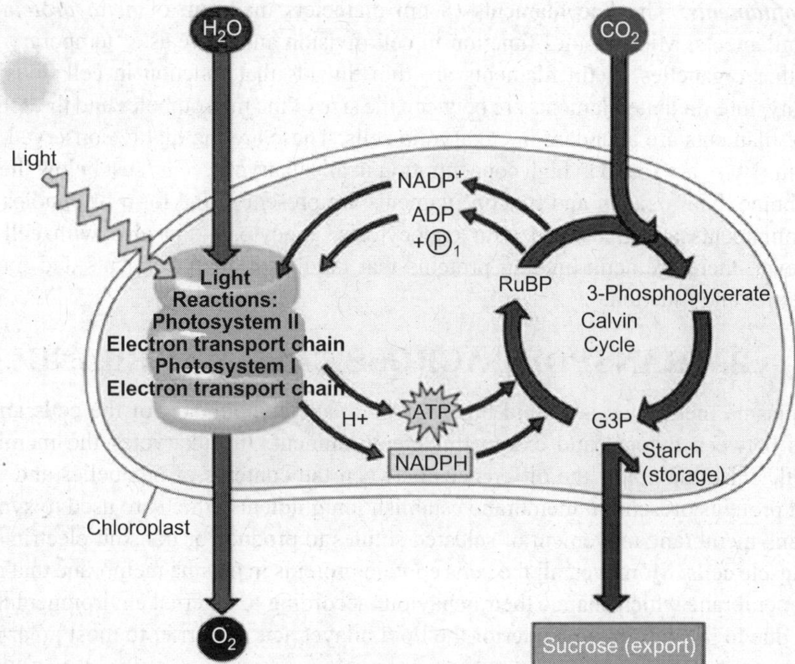

Fig. 2.31. Function of PS I and PS II, electron transport chain in thylakoid interior space, and photosynthesis.

***(ii)* Chromoplasts:** Chromoplasts are the plastids that contain only coloured pigments (red, orange, yellow). These are found in cells of fruits and flowers. Such pigments are non-photosynthetic and they impart only colour to the plants.

***(iii)* Leucoplasts:** The leucoplasts are colourless plastids. These are the sites for storage of food materials. On the basis of types of stored food the leucoplasts of plant cells are named as: *amyloplasts* (which synthesise and store starch), *proteoplasts* (or proteinoplasts which store proteins in seeds), *lipidoplasts* or *elaioplasts* (which store lipids in seeds).

***(iv)* Cytoskeleton:** The shape of living cells is maintained by the arrays of protein filaments The filaments are fibrous proteins that form a network in cytosol. Altogether these fibres are called cytoskeleton. Cytoskeleton maintains the shape of the cell as well as anchoring organelles, moving the cell and controlling internal movement of structures. Cytoskeleton helps in movement of cells or organelles, besides maintaining the shapes. Cytoskeletal filaments are of two types, microtubules and microfilaments.

- ***Microtubules:*** Microtubules are long (25 nm diameter) unbranched cylindrical structures (Fig. 2.32). These are made up of stiff polymers of tubulin (α- and β-) molecules. These extend throughout the cytoplasm and govern the location and movement of cell organelles and other cell components. The two ends of microtubules are different. Microtubules cause rhythmic movement of flagella and cillia. They are also responsible for movement of chromosomes, Golgi complex, etc.

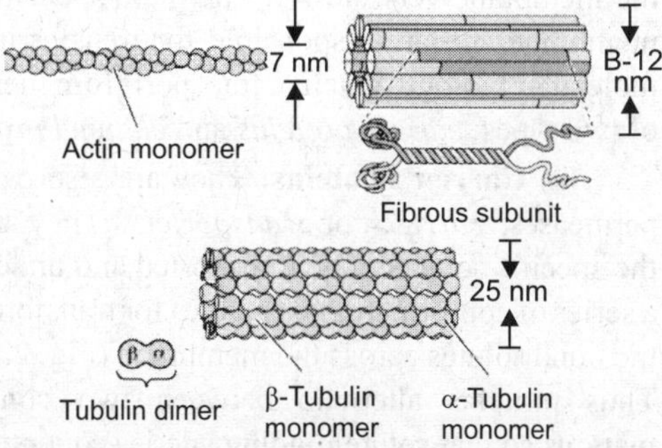

Fig. 2.32. Actin and tubulin components of the cytoskeleton.

- ***Microfilaments:*** The fine filaments (8 nm diameters made up of *actin protein*) are called microfilaments. Microtubules function in cell division and serve as a 'temporary scaffolding' for other organelles. Actin filaments are thin threads that function in cell division and cell motility. Intermediate filaments are between the size of the microtubules and the actin filaments. These filaments are abundant in eukaryotic cells. These are the highly conserved cytoskeletal protein. These are found in high concentration in all eukaryotic cells. Just below the cell surface membrane, both α-actin and β-actin filaments are present in the form of bundles (Fig. 2.32). Microfilaments are associated with endocytosis, exocytosis and also with cell movement. However, there are actin-binding proteins that bind with actin filaments and modulate their properties and function.

B. TRANSPORT ACROSS CELL MEMBRANE

The plasma membrane is a lipid bilayer which acts as boundary of the cells and maintains differences between cytosol and extracellular environment. In eukaryotes the membrane of all the organelles also maintains the differences between the contents of organelles and cytosol. The specialised proteins present on membrane establish ion gradients which are used to synthesise ATP to drive trans-membrane movement of selected solutes to produce or transmit electric signals or in nerve or muscle cells. Moreover, all the cells contain proteins in plasma membrane that act as sensor of plasma membrane which change their behaviour according to external environmental conditions. Moreover, due to its hydrophobic interior the lipid bilayer acts as barrier to most polar molecules.

Due to this barrier the cells maintain concentrations of solutes in cytosol. In 1986, A. Walter and J. Gutknecht reported the permeability of small electrolytes through lipid bilayer membranes. Transport of inorganic ions and small water-soluble organic molecules across the lipid bilayer is achieved by specialised trans-membrane proteins which transfer specific ions or molecules or group of closely related ions or molecules. Cells can also transfer macromolecules selectively.

All molecules diffuse across a protein-free lipid bilayer down to its concentration gradient. The rate of diffusion depends on the size of molecules and its relative solubility in oils. Small monomer molecules like O_2 and CO_2 soon dissolve in lipid bilayer diffusing across the membrane. Uncharged polar molecules also diffuse rapidly if they are small in size e.g. water, ethanol and urea, glycerol diffuses less rapidly, while glucose diffuses hardly. In addition, lipid bilayer is more permeable to charged molecules (ions). Due to the presence of charges and high degree of hydration such molecules fail to enter the hydrocarbon phase of the bilayer.

1. Membrane Transport Proteins

The cell membrane allows the water and vapour molecules to enter through diffusion. However, various polar molecules such as ions, sugars, amino acids, nucleotides and metabolites enter across the membrane very slowly. There are special membrane proteins which occur on all types of cell membrane and are responsible for transferring such solutes. In 1989, H.R. Kaback reported the molecular biology of active transport from membrane to molecule to mechanism. These proteins are of two types: *carrier proteins* and *channel proteins*.

(*a*) Carrier Proteins: They are also called permeases, *carriers* or *transporters*. They bind the specific solutes to be transported and undergo a series of conformational changes for transferring the bond solutes across the membrane (Fig. 2.33). Thus proteins alternate between two conformations so that solute binding site is sequentially accessible on one side of bilayer and then on the other side.

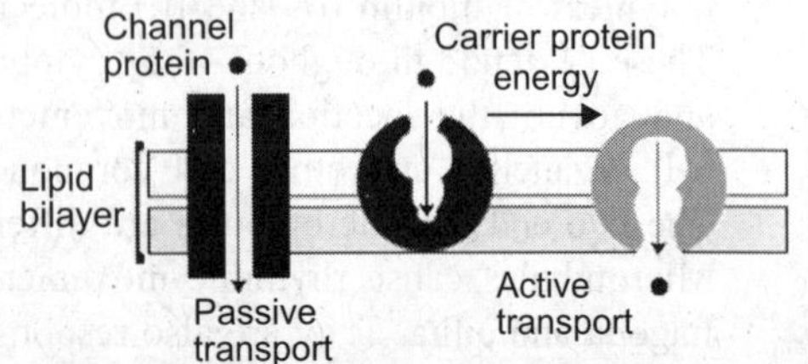

Fig. 2.33. Membrane transport proteins; A carrier proteins; B- channel proteins.

(b) Channel Proteins: These proteins do not bind the solute but they form water-filled hydrophobic pores across the membrane. When these pores are open, specific solutes (e.g. inorganic ions of suitable size and charges) pass through the membrane (Fig. 2.33). Transfer through channel proteins occur very fast than that by carrier proteins.

2. Active and Passive Transport Systems

Many carrier proteins, all channel proteins and solutes passively cross the membrane. This phenomenon is called 'downhill'. It is also a process called *passive transport* or *facilitated diffusion*. If the transport molecule is not changed, it means that the molecule has been transported due to concentration gradient of membrane in concentration of molecules on two sides which drives passive transport and ensures the direction. In contrast, if the solute possesses a net charge, both its concentration gradient and the electrical potential difference across the membrane, the membrane potential affect the transport of solute. The electro-chemical gradient, *i.e.* a net drive force of each charged solute can be calculated by combining the concentration gradient and the electrical gradient. Plasma membrane of all the organisms has an electrical potential difference across them, inside negative and outside positive. This difference favours the entry of positively charged ions into the cell and opposes the entry of negatively charged ions (Fig. 2.34 A).

Active transport is an 'uphill' process which is always mediated by carrier proteins. These proteins actively pump certain solutes across the membrane against their electrical gradient, *i.e.* 'uphill'. The pumping activity of carrier protein is always directional because it is highly coupled to a source of metabolic energy such as ATP hydrolysis or ion gradient (Fig. 2.34 B). Transport by carrier proteins is either active or passive, while that by channel proteins is always passive (Kaback, 1989).

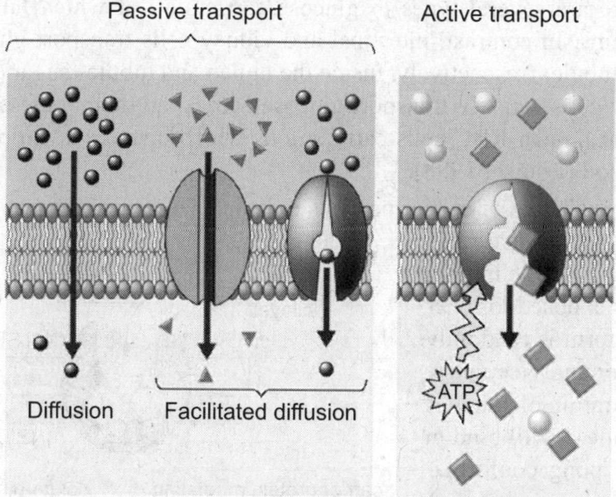

Fig. 2.34. Comparison between passive and active transport across plasma membrane.

Ionophores increase cell permeability: The small hydrophobic molecules that dissolve in lipid bilayer and increase the permeability to certain inorganic ions are called ionophores. Microorganisms synthesise ıonophole as biological weapons. They are of two types: mobile ion carriers (e.g. valinomycin) and channel formers (e.g. gramicidin). They allow net movement of ions only down their electro-chemical gradient because they are not coupled to energy sources.

I. CARRIER PROTEINS AND ACTIVE MEMBRANE TRANSPORT

The carrier proteins present on lipid bilayer behave as specialised membrane-bound enzymes, because each carrier protein consists of one or more specific binding site(s) for substrate. Maximum rate of transport takes place when the carrier is saturated *i.e.* all the binding sites are occupied by the substrate. The rate of carrier-substrate reaction resembles an enzyme-substrate reaction is referred to as V_{max} which is characteristic of the specific carrier.

Each carrier has a characteristic binding constant for its solute designated as K_m which is equal to the concentration of solutes when the transport rate is half its maximum value.

Moreover, the competitive inhibitors (which compete for the same binding site) or non-competitive inhibitor (which bind elsewhere and change the structure of carrier) can block the binding site of the carriers (Kaback, 1989).

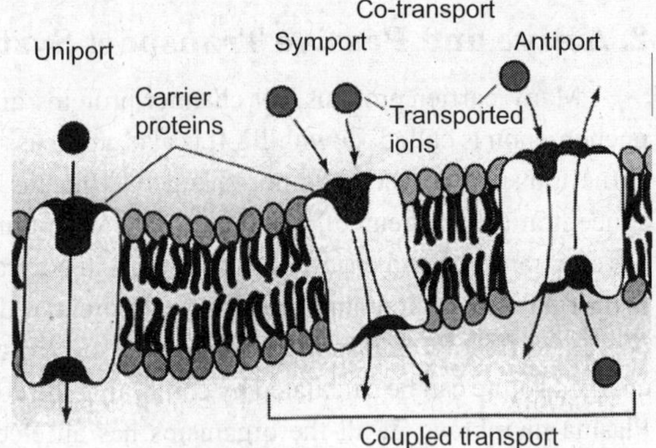

Fig. 2.35. Type of carrier-mediated transporters.

There are three types of carrier-mediated transport: (*i*) some carrier proteins simply transport a single solute from one site of membrane to the other, they are called *uniporter*, (*ii*) some carrier proteins act as coupled (linked) transporters where the transfer of one solute depends on the simultaneous or sequential transfer of a second solute either in the same direction (*symport*) or in the opposite direction (*antiporter*) (Fig. 2.35). .

Most of the animal cells take up glucose from extracellular solution where its concentration is high as compared to cytosol. Glucose carriers operate as uniporter and transfer glucose passively (Lodish, 1988). There are several types of glucose carriers all of which fall in the same family of homologous proteins. In contrast, intestinal and kidney cells transport glucose from lumen of intestine and kidney tubules, respectively. Inside the lumen and tubules sugar concentration is quite low. Therefore, these cells actively transport glucose across the membrane by symport with Na^+. The band 3 protein of human RBC cells carry anions which operate as antiporter for exchanging Cl^- and HCO_3^- in blood (Lodish, 1988).

The conformational state of carrier proteins exist in two forms: *pong* and *ping*. In pong form the binding sites for solutes are exposed outside of bilayer, while in ping form the same site is exposed to inner side of bilayer. One form is randomly changed into the other and vice versa. Hence, if the concentration of solute A is higher on the outside, it will bind in more number to the pong conformation of carrier protein than to ping conformation. Then net transport of solute A will occur down to electrochemical gradient of the bilayer (Fig. 2.36).

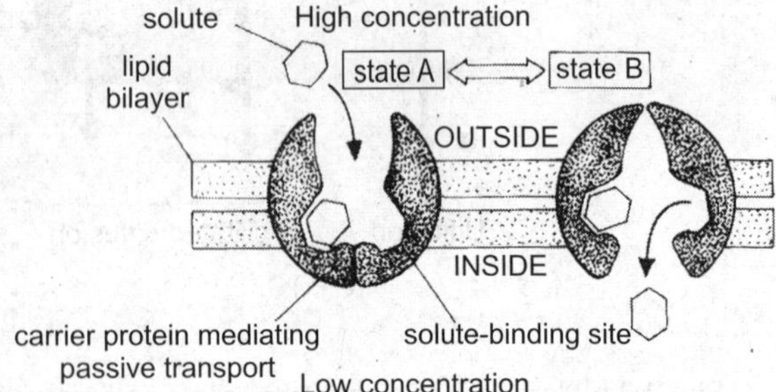

Fig. 2.36. Facilitated diffusion of a solute by a carrier protein mediation (adopted from Alberts, 2002)

1. Na^+/K^+-ATPase

Na^+/K^+-ATPase was discovered by Jens C. Skou in 1957 at the University of Aurhus (Denmark). For this first discovery of ion transporting enzyme (Na^+/K^+-ATPase) he was awarded Nobel Prize in 1997. Na^+/K^+-ATPase (also called Na^+-K^+ ATPase pump) is an enzyme. For maintenance of cell potential a cell keeps low Na^+ concentration and high K^+ ions within the cell, whereas high concentration of Na^+ and low concentration of K^+ is kept out side the cell. This results in diffusion through ion channels in the membrane. Through active transport, it pumps Na^+ out and K^+ in to maintain optimum concentration. Generally the concentration of K^+ remains between 10 and 20 times higher inside the cell than outside, while the concentration of Na^+ is 10-20 times lower inside the cell than outside (Table 2.6).

The Na^+/K^+ pump present in bilayer membrane of all animal cells maintains this concentration difference. This carrier protein actively pumps Na^+ out and K^+ into a cell against their electrochemical gradient. For the hydrolysis of one ATP molecule inside the cell, three Na^+ ions are pumped out and two K^+ are pumped in (Fig. 2.37).

The specific pump inhibitor, *ouabain*, and K^+ compete for the same site on the external side of the ATPase. ATP supplies energy for pumping Na^+ and K^+ ions across the membrane. During pumping cycle, the terminated phosphate group of the ATP is transferred to an aspartic acid residue of the ATPase and is subsequently removed.

The Na^+/K^+ pump has been purified. It consists of a large multi-pass transmembrane catalytic sub-unit of about 1000 amino acid long, and an associated small single pass glycoprotein (Mercer, 1993). The former bears a binding site for Na^+/K^+ and ATP on its cytoplasmic surface and a binding site for K^+ on its external surface. It is reversibly phosphorylated and dephosphorylated during pumping cycle. This function of glycoprotein is not known so far except its involvement in intracellular transport of catalytic sub-unit to the cell membrane (Horisberger, 1991).

Table 2.6. Ion concentration inside and outside a typical mammalian cell (modified after Alberts *et al.*, 2002).

Ions	*Ion concentration within the cell (mM)*	*Ion concentration outside the cell (mM)*
Cations		
Na^+	5–15	145
K^+	140	5
Mg^{++}	0.5	1–2
Ca^{++}	10–4	1–2
H^+	7×10^{-5} ($10^{-7.2}$ or pH 7.2)	4×10^{-5} ($10^{-7.2}$ M or pH 7.4)
Anions		
Cl^-	5–15	110

Na^+/K^+-ATPase is required for maintenance of osmotic balance of the cell. It drives three positively charged ions out of cell for every two ions it pumps in. Therefore, it is electronegative; it means it drives a current across the membrane and creates an electrical potential with inside negative as compared to outside. This contributes 10% membrane potential and the remaining 90% potential depends on indirect pump. In contrast the Na^+/K^+-ATPase controls the solute concentration inside the cell, hence it regulates the osmotic forces which results in shrinking of the cells.

In bacteria and yeasts and many membrane-bound organelles of animal cells, most active transport system driven by ion gradients depend on H^+ but not Na^+ gradients. This shows the predominance of H^+-ATPase and the absence of Na^+/K^+-ATPase in these membranes. The electrochemical H^+-ATPase across the plasma membrane drives the active transport of many sugars and amino acids into the bacterial cells. On the other hand, most cells have one or more types of Na^+-driven antiporters in their plasma membrane. They regulate cytosolic pH and keeps to about 7.2. These proteins use the energy stored in Na^+-gradient and reduce acidity by releasing excess H^+. The H^+ ions either leak in or produce in the cell by acid-forming reaction.

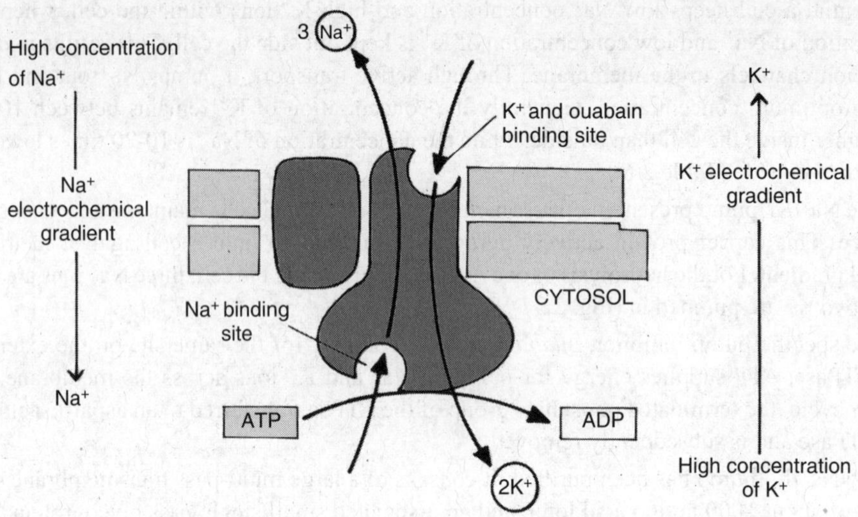

Fig. 2.37. The Na^+/K^+-ATPase carrier (*based* on Alberts *et al.*, 2002).

2. Ca^{2+}-ATPase Pumps

In 1991, E. Carafoli reviewed calcium (Ca^{2+}) pump of plasma membrane. A very low concentration of free Ca^{2+} is maintained by eukaryotic cells in their cytosol (~10^{-7}M) in contrast to higher extracellular Ca^{2+} concentration (~10^{-3}M). The concentration of free Ca^{2+} in cytosol is increased due to small influx of Ca^{2+} ions. The flow of Ca^{2+} down its steep concentration gradient in response to extracellular signals causes the transmission of these signals along the membrane. Therefore, maintenance of a steep Ca^{2+} gradient in the cell is important. The Ca^{2+}-pumps in the plasma membrane maintain Ca^{2+} gradient which actively transport Ca^{2+} out the cell. One of these pumps is an ATPase and the other is antiporter. The later is driven by the Na^+ electrochemical gradient.

The best known Ca^{2+} pump is a membrane-bound ATPase in the sarcoplasmic reticulum of muscle cells which is a special type of endoplasmic reticulum. It acts as an intracellular store of Ca^{2+} ions. Ca^{2+} is released from sarcoplasmic reticulum into cytosol after depolarisation of muscle cell membrane by action potential. This causes contraction of the stimulated muscles. Ca^{2+}-pump accumulate about 90% of membrane protein of organelles. It is also responsible for pumping Ca^{2+} from cytosol to sarcoplasmic reticulum.

It has been shown through gene manipulation studies that carrier proteins belong to a small number of families. Each family consists of proteins of identical amino acids which show that they have evolved from a common ancestral protein and operate by similar mechanism. The family of cation transporting ATPase is a unique example which consists of Na^+/K^+ pump. Both, Na^+/K^{++}-ATPase and Ca^{2+}-ATPase are homologous proteins. In both the cases the large catalytic sub-units exist in several identical forms. It has about 10 membrane spanning α-helices. During pumping cycle the catalytic sub-unit is phosphorylated and dephosphorylated.

II. TRANSPORT SYSTEMS IN PROKARYOTES AND EUKARYOTES

It has earlier been discussed that plasma membrane of all bacteria consists of H^+ gradient across the membrane which pump several kinds of nutrients into the cell. This is done by a family of transport ATPases. Structure of bacterial cell wall has already been discussed in this chapter. Moreover in bacteria e.g. *Escherichia coli*, the transport ATPases are present in the inner membrane. The nutrients are captured and delivered to transporters by an auxillary transport system which is associated with transport ATPases.

Porins (channel-forming proteins) are present in the outer membrane of bacteria. Solutes diffuse through porins into periplasmic space and bind to a substrate-binding protein. As a result of which the substrate binding protein undergoes a conformational change. Therefore, it becomes capable to bind to a transport ATPase in the plasma membrane. Then the transport ATPase picks up the solutes and soon transports it across the lipid bilayer through an ATP hydrolysis reaction. Since peptidoglycan is simple and porous, it allows the solutes and solute binding protein to move across through diffusion.

1. ABC Transport Super Family

In bacterial plasma membrane, the transport ATPases belong to a largest and diverse family of transport proteins called ABC transport superfamily. The reason is that each member of the superfamily consists of a highly conserved ATP-binding cassette. So far about 50 ABC transporters are known and each one is specific for a specific substrate or class of substrate. This superfamily transports a variety of substrates such as amino acids, sugars, inorganic ions, peptides, polysaccharides and proteins. Several family members are described in prokaryotes. They are also described in eukaryotes. One of these proteins is 'multidrug resistance (MDR) protein'. Its expression in human cancer cells makes the cells resistant to many cytotoxic drugs used in chemotherapy of cancer. The cells that over express the MDR transport proteins are selected after treatment with chemotherapeutic drugs. Then the drugs are pumped out of the cell by the transporters and result in reduction in their toxicity and conferring resistance to many drugs. Similarly, malaria caused by *Plasmodium falciparum* is able to amplify a gene then expresses an ABC transporter which pumps the chloroquin out the cell.

III. ION CHANNELS

The channel proteins are present in plasma membrane and form hydrophilic pores unlike the carrier proteins. Most of them present in plasma membrane of plants, animals and microbial cells connect the cytosol to the outer surface of all membrane and form highly selective narrow pores. Since these proteins transport the inorganic ions, they are called *ion channels*. The ion channels are 1000 times more efficient in transporting the ions than the carriers. About 1 million ions per second can pass across the membrane. This transport is always passive (downhill) because the channel is not coupled to an energy source to carry out active transport. The specific ions allow to cross the membrane are Na^+, K^+, Ca^{2+} or Cl^- which diffuse sharply down their electrochemical gradient.

The ion fluxes are controlled by the channels in two ways: (*i*) *through ion selectivity* (ions of suitable size and charge can pass because the pores become narrower enough to transport small sized ions. This limits the rate of passage. The ion flux increases with increase in concentration of ions), and (*ii*) *through gated channels* (the ions channels have gates which open and close in response to certain stimuli). Based on the types of stimuli which cause the ion channels to open, the gated channels are of following types (Fig. 2.38) :

- **Voltage-Gated Channels:** A change in voltage across the membrane causes the gate to open.
- **Ligand-Gated Channels:** Binding of ligand causes the channel to open the gate.
- **Transmitter-Gated Channels:** Ligand can be an extracellular mediator specifically a neurotransmitter.
- **Ion-Gated Channels:** The intracellular ligand such as ions can cause to open the gate of the channel.
- **Nucleotide-Gated Channel:** An intracellular mediator nucleotide can open the gate of the

channel.

There are more than 100 types of ion channels known so far and many more still being discovered. They mediate mostly electrical signals in the nervous system. A single nerve cell can contain 10 types of ion-channels present in different domains in the lipid bilayer. The ion channels which are permeable to K^+ are the most common ones. They are found mostly in all animal cells. Always they remain in a resting or unstimulated cell; hence, K^+ ions are leaked out. Therefore, such channels are also called 'K^+ leak channels'.

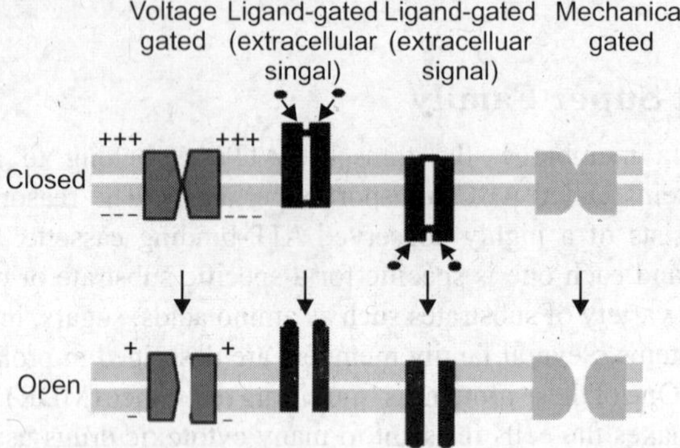

Fig. 2.38. Opening and closing of gated ion channels (diagrammatic).

1. Measurement of Ion Flow

Electrochemical gradient for an anion drives the flow of that ion across the ion channel protein of the plasma membrane. A combination of voltage gradient and the concentration gradient of the ions across the membrane govern the electrochemical gradient. Zero value is achieved when the voltage gradient and concentration gradient balance each other. The voltage gradient is called equilibrium potential at which this equilibrium is achieved. This equilibrium concentration is also called the resting membrane potential. It can be calculated by using the 'Nernst equation' as given below :

$$V = \frac{FT}{zF} \ln \frac{\text{Co}}{\text{Ci}}$$

Where, V = The equilibrium potential in volts (internal potential - external potential)

Co and Ci = Out side and inside concentration of ions, respectively

R = The gas constant (2 cal mol^{-1} °K^{-1})

T = The absolute temperature (°K)

F = Faraday's constant (2.3×10^4 cal V^{-1} mol^{-1})

Z = The valence (charge) of the ion

ln = logarithm to the base e

C. CELL SIGNALLING AND SIGNAL TRANSDUCTION

Several unicellular and multicellular organisms live in a habitat and respond to environmental stimuli via the activation of signal transduction pathways. Each of them communicates to the others by giving signals and receiving signals. For example, slime molds secrete cyclic AMP upon starvation which stimulates individual cells in the immediate environment to aggregate. Yeast (*Sacharomyces cerevisiae*) cells live independently and divide by budding. When a haploid cell is to go for sexual reproduction, it expresses a peptide secreting factors (PSF). The signal in the form of PSF is received by the other mating cells. Then cell proliferation is stopped. The cells prepare themselves to undergo conjugation. After fusion the two haploid cells produce diploid cell which undergoes meiosis to form haploid cells. Later on these cells could proliferate (http://en.wikipedia.

org/wiki/cell_signalling).

Cell signalling is part of a complex system of communication that governs the basic cellular activities and coordinates cell actions. The ability of cells to perceive and correctly respond to their microenvironment is the basis of development, tissue repair, and immunity as well as normal tissue homeostasis. Errors in cellular information processing are responsible for diseases such as cancer, autoimmunity, and diabetes. By understanding cell signalling, diseases can be treated effectively and, theoretically, artificial tissues could be built. Traditional work in biology has focused on studying the individual parts of cell signalling pathways. Systemic biology research helps us to understand the underlying structure of cell signalling networks and how changes in these networks can affect the transmission of information.

Cell signalling has been most extensively studied in the context of human diseases and signalling between cells of a single organism. However, cell signalling can also occur between the cells of two different organisms. In many mammals, early embryo cells exchange signals with cells of the uterus. In the human gastro-intestinal tract, bacteria exchange signals with each other and with cells of human epithelium and immune system.

1. Recognition of Extracellular Signal Molecules

There is a 'biomolecular dialogue' between the signal molecules and signal receptor molecules present on target cells. Cells of higher animals communicate by means of hundred kinds of signalling molecules such as proteins, amino acids, small peptides, nucleotides, steroids, retinoids, hormones and dissolved gases like nitrous oxide and carbon monoxide (Fig. 2.39). Most of these molecules are secreted from cells by exocytosis or released by diffusion through plasma membrane. Irrespective of nature of signal, the target cells that receive signal, respond by means of a protein called receptor which binds the signalling molecules and starts a response in the target cell. The signal receptor molecules are the *trans-membrane proteins* present on the surface of target cells. These molecules transmit signals in the cytoplasm from where it is carried to nucleus or enter itself inside the cell and transmit signals to the nucleus. Accordingly, the target cells transmit the signals.

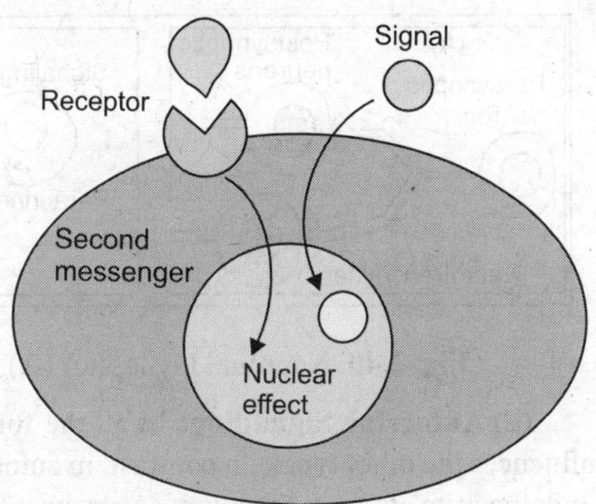

Fig. 2.39. Inter-cellular signaling through secretion of molecules.

2. Forms of Signalling

The signal molecules are either carried to a long distance from its origin or may act as local mediator and affect only cells in immediate environment, or transported to all parts of body through blood stream. Cell signalling requires both extracellular signal molecules and a complementary receptor protein in each cell that binds and responds them. There are following three types of signalling.

(*a*) Synaptic (neuronal) Signalling: In the multicellular complex organisms like mammals short range signalling is not sufficient. Therefore, sets or specialised cells have evolved. These are involved in signalling between the two parts of bodies. The most complex of these are the nerve cells or neurons. Neurons receive, conduct and transmit signals. After being activated by environmental signals, a neuron sends electrical impulses along its axon. Then the terminals after receiving impulse become stimulated and secrete a chemical signal called *neurotransmitter*. In 1982, J. Kistler and R.M. Stroud reported the structure and function of an acetylcholine receptor. Neurotransmitters are ligands that are capable of binding to ion channel proteins resulting in their opening to allow the rapid flow of a particular ion across the plasma membrane. This results in an altering of the cell's membrane potential and is important for processes such as the neural conduction of electrical impulses.The neurotransmitters are delivered to target cells at specialized cell junctions called chemical synapses; therefore it is called synaptic signalling (Fig. 2.40 A).

***(b)* Paracrine Signalling:** The signalling molecules act as local mediator and affect only the cell present in immediate environment (Fig. 2.40 B). The secreted signalling molecules are not allowed to be diffused at a long distance, but are quickly taken up by the neighbouring target cells.

***(c)* Endocrine Signalling:** Body of some organisms possesses endocrine gland the cells of which secrete signalling molecules called hormones. Hormones are secreted into blood stream in animals and cell sap in plants. They carry the signal to target cells which are dispersed throughout the body (Fig. 2.40 C). Moreover, endocrine signalling is slow because it depends on diffusion and blood flow. But the signalling by nerve cells is faster and precise because signal is transmitted over a long distance at the rate up to 100 metres per second. Then the neurotransmitter diffuses less than 100 nm to the target cell. This event is accomplished within one second.

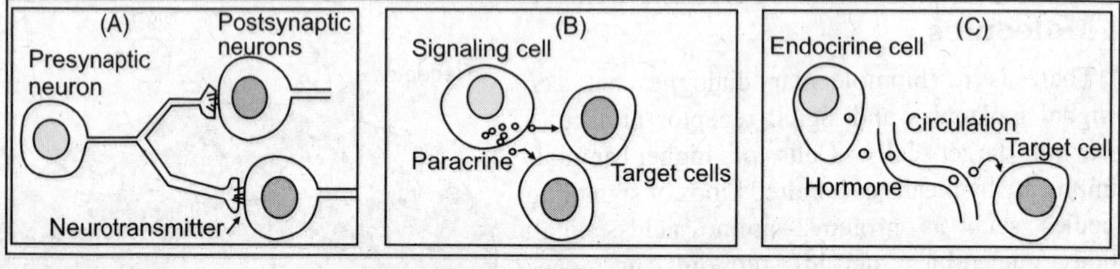

Fig. 2.40. Neuronal (synaptic) (A), paracrine (B) and endocrine (C) signalling.

***(d)* Autocrine Signalling:** In all the forms of signalling as discussed above, one type of cell influences the other types. In contrast, in autocrine signalling a cell secretes signall molecules which binds back to its own receptors. A group of identical cells produces a higher concentration of a secreted signal than does a single cell.

I. TYPES OF RECEPTORS

Receptors can be broadly divided into two major classes: *cell-surface receptors* and *intracellular* receptors. A third *ligand-gated ion channel receptors* are a class of receptor that may occur either at the cell-surface or intracellularly.

1. Cell-Surface Receptors

A cell is surrounded by a membrane which forms a barrier between the inside and outside the cell. This structure consists of double layers of phospholipid molecules interspersed with cholesterol and proteins (many of which are receptors for chemical signals) that float in or are attached to the membrane. Its fat-rich composition provides it with the characteristic of being selectively permeable, allowing the flow of fat-soluble substances across it, while excluding many water-soluble ones. Fat-soluble signals, such as hormones and some vitamins, are thought to simply diffuse across the membrane. Their receptor proteins are usually found within the cell.

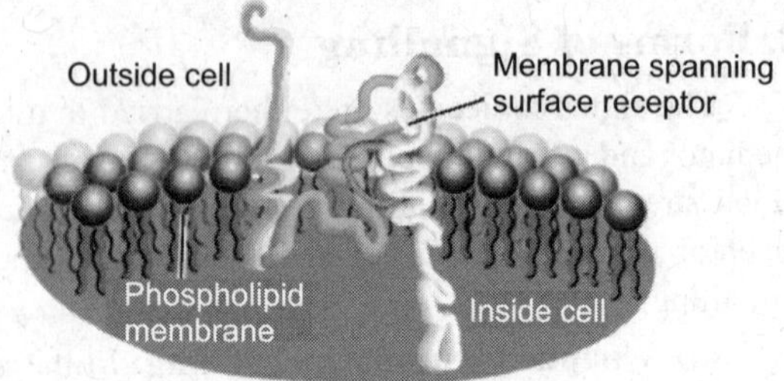

Fig. 2.41. Cell-surface receptors embedded within the membrane.

In contrast, the vast majority of chemical signals are water-soluble and thus unable to cross the fatty lipid layer of the cell membrane. Their receptor proteins must span the membrane, detect their presence on the outside of the cell, and transmit this information into the interior of the cell. These proteins are called *cell-surface receptors* because they span the cell membrane and receive the chemical signals on the outside of the cell and transfer information to the inside of the cell (Fig. 2.41).

Cell-surface receptors are integral trans-membrane proteins and recognise the vast majority of extracellular signalling molecules. Trans-membrane receptors span the plasma membrane of the cell, with one part of the receptor on the outside of the cell (the *extracellular domain*) and the other on the inside of the cell (the *intracellular domain*). Signal transduction occurs as a result of stimulatory molecule or ligand binding to their extracellular domain, the ligand itself does not pass through the plasma membrane prior to receptor binding.

Binding of a ligand to a cell-surface receptor stimulates a series of events inside the cell with different types of receptor stimulating different intracellular responses. Receptors only respond to the binding of a specific ligand. After binding, the ligand initiates the transmission of a signal across the plasma membrane by inducing a change in the shape or conformation of the intracellular part of the receptor. Often, such changes in conformation result in either activation of an enzymatic activity contained within the receptor or expose a binding site for other signalling proteins within the cell. Once these proteins bind to the receptor they themselves may become active and propagate the signal into the cytoplasm.

2. Classes of Cell Surface Receptor Proteins Involved in Signalling

The signalling molecules secreted by a cell are called extra-cellular signalling molecules (ESM). The ESM transduces the signals to the target site. Therefore, the ESM requires a complementary set of receptor proteins in each cell that establishes it to bind and respond them in programmed ways. There are many different classes of *transmembrane receptor* that recognise many different extracellular signalling molecules. Examples of receptors are*: G-protein coupled receptors* (*e.g. chemokine receptors*)*, receptor tyrosine kinases* (*e.g. growth-factor receptors, integrins*)*, toll-like receptors* and *ion-channel-linked receptors.*

***(a)* G-Protein-Linked Receptors:** G-protein-linked receptors (GPLRs) are a family of integral membrane proteins that possess seven membrane-spanning domains, and are linked to a guanine nucleotide binding protein (or G-protein). Many receptors make up this family, including adrenergic receptors, olfactory receptors, chemokine receptors, and rhodopsin.

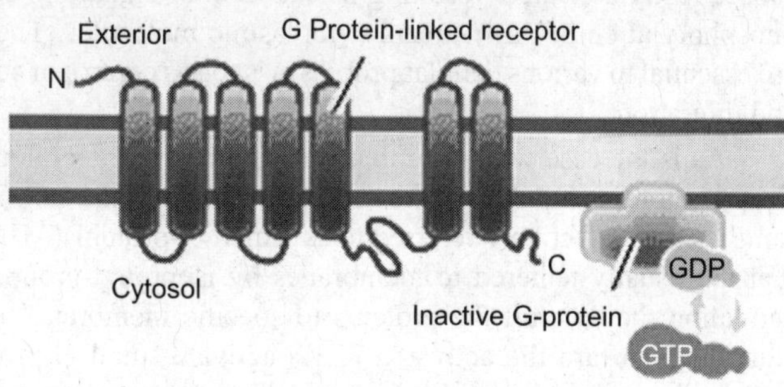

Fig. 2.42. G-protein-linked receptor.

Signal transduction by a GPLR simply begins with an inactive G-protein linked with the receptor (Fig. 2.42). An inactive G-protein exists as a heterotrimer, a molecule composed of three different protein subunits. Once the GPLR recognizes a ligand, the shape (conformation) of the receptor changes to mechanically activate the G-protein, and causes one subunit (Gα) to dissociate from the other two G-protein subunits (Gβ and Gγ). The dissociation exposes sites on the G-protein subunits that interact with other molecules. The activated G-protein subunits detach from the receptor and initiate signalling from many downstream effector proteins, including phosphodiesterase and adenylyl cyclases, phospholipases, and ion channels that permit the release of second messenger molecules such as cyclic AMP (cAMP), cyclic GMP (cGMP), inositol triphosphate (IP3), diacylglycerol (DAG), and calcium (Ca^{2+}) ions. For example, a rhodopsin molecule in the plasma membrane of a retina cell in the eye that was activated by a photon can activate up to 2,000 effector molecules (*i.e.* transducin) per second.

***(b)* Enzyme-Linked Receptors (e.g. Receptor Tyrosine Kinases):** Enzyme-linked receptors are transmembrane proteins with an intracellular kinase domain and an extracellular domain that binds ligand (Fig. 2.43).

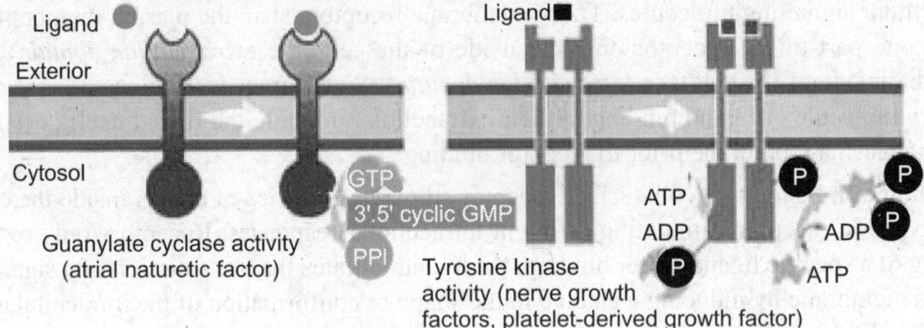

Fig.2.43. Enzyme-linked receptors.

Receptor tyrosine kinases (RTKs) are one of the enzyme-linked receptors. Depending on their structural properties and ligand specificity, RKTs proteins are classified into subfamilies: many growth factor receptors (such as insulin receptor and the insulin-like growth factor receptor), and many other receptors.

Structure of tyrosine is given in Fig. 2.44. RTKs need to form dimers in the plasma membrane for conducting their biochemical signals. Interaction between the two cytoplasmic domains of the dimer is thought to stimulate autophosphorylation of tyrosines within the cytoplasmic tyrosine kinase domains of the RTKs causing their conformational changes. The kinase domain of the receptors is subsequently activated, resulting in initiation of signalling cascades of phosphorylation of downstream cytoplasmic molecules. These signals are essential to various cellular processes, such as control of cell growth, differentiation, metabolism, and migration.

Fig. 2.44. Chemical structure of tyrosine.

As is the case with G-protein coupled receptors, proteins that bind GTP play a major role in transmission of signal from the activated RTK into the cell. The members of the Ras, Rho and Ral families are collectively referred to as 'small G-proteins'. These proteins act as molecular switches that are usually tethered to membranes by isoprenyl groups linked to their carboxyl ends. After activation they recruit the proteins to specific membrane sub-domains where they participate in signalling. In turn the activated RTKs activate small G proteins which in turn activate 'guanine nucleotide exchange factors'. On activation, these exchange factors can activate many more small G-proteins, and amplify the receptors initial signal.

***(c)* Integrins:** Integrins are produced by a wide variety of cell types and play a role in the attachment of a cell to the extracellular matrix (ECM) and to other cells, and in the signal transduction of signals received from extracellular matrix components such as fibronectin, collagen and laminin. Ligand binding to the extracellular domain of integrins induces a conformational change within the protein and a clustering of the protein at the cell surface, in order to initiate signal transduction. Integrins lack kinase activity and integrin-mediated signal transduction is achieved through a variety of intracellular protein kinases and adaptor molecules such as integin-linked kinase (ILK), focal adhesin-kinase (FAK), talin, paxillin, parvins, p130Cas, Src family kinases and GTPases of the Tho family. The main protein coordinating signal transduction is ILK. Cooperative integrin and receptor tyrosine kinase signalling determines the cellular survival, apoptosis, proliferation and differentiation (Fig. 2.45).

Important differences exist between integrin signalling in circulating blood cells and non-circulating blood cells such as epithelial cells. Integrins at the cell surface of circulating cells are inactive under normal physiological conditions. For example cell-surface integrins on circulating leukocytes are maintained in an inactive state in order to avoid epithelial cell attachment. Only in response to appropriate stimuli, leukocyte integrins are converted into an active form, such as those received at the site of an inflammatory response. Similarly, it is important that integrins at the cell surface of circulating platelets are kept in an inactive state under normal conditions, in order to avoid thrombosis. Epithelial cells have active integrins at their cell surface under normal conditions. They help to maintain their stable adhesion to underlying stromal cells, which provide appropriate signals in order to maintain their survival and differentiation (Hehlgens *et al.,* 2007).

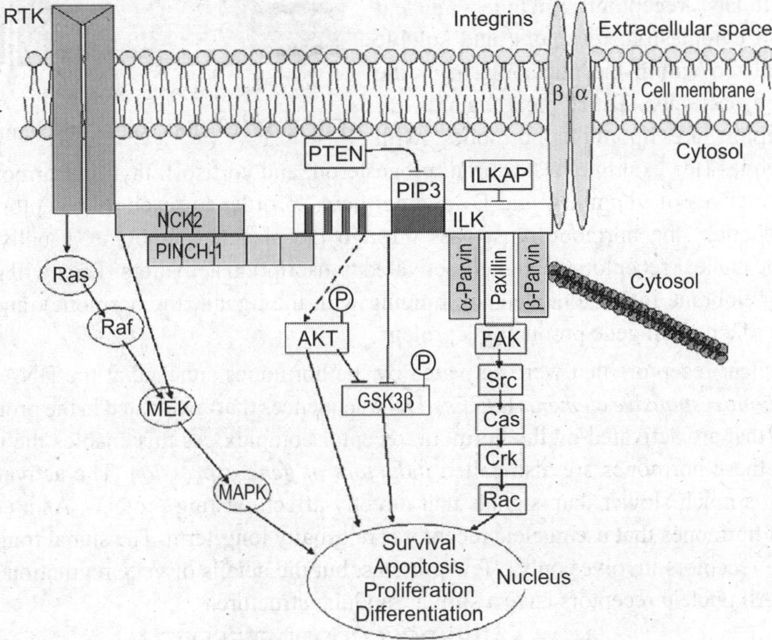

Fig. 2.45. Integrin-mediated signal transduction (after Hehlgens *et al.,* 2007).

(*d*) Toll-Like Receptors: When activated, toll-like receptors (TLRs) recruit adapter molecules within the cytoplasm of cells in order to propagate a signal. Four adapter molecules are known to be involved in signalling. These proteins are known as MyD88, Tirap (also called Mal), Trif, and Tram. The adapters activate other molecules within the cell, including certain protein kinases (IRAK1, IRAK4, TBK1, and IKKi) that amplify the signal, and ultimately lead to the induction or suppression of genes that orchestrate the inflammatory response. In all, thousands of genes are activated by TLR signalling, and collectively, the TLRs constitutes one of the most powerful and important gateways for gene modulation.

(*e*) Ligand-Gated Ion Channel Receptors: A ligand-activated ion channel will recognise its ligand, and then undergo a structural change that opens a gap (channel) in the plasma membrane through which ions can pass. These ions will then relay the signal. An example for this mechanism is found in the receiving cell, or post-synaptic cell of a neural synapse.

In contrast, other ion channels open in response to a change in cell potential, *i.e.* the difference of the electrical charge across the membrane. In neurons, this mechanism underlies the action potential that travel along nerves. The influx of ions that occurs in response to ligand-gated ion channels often induce action potentials by depolarizing the membrane of the post-synaptic cells which results in the wave like opening of voltage-gated ion channels. Chemical signals in the form of neurotransmitters are transduced by ion-channel-linked receptors directly into an electrical signal in the form of a voltage difference across the plasma membrane. This occurs when a neurotransmitter binds to this type of receptor, altering its conformation to open or close a channel (often through or near the receptor) to the flow of Na^{2+}, K^+, Ca^{2+} or Cl^- ions across the membrane. Driven by their electrochemical gradient (*i.e.* one side of the membrane has numerous ions, while the other side has few) the ions rush into or out of the cell, creating a change in the membrane potential due to the positive or negative nature of the ions. This flow of ions through the channel can trigger a nerve impulse, or alternatively stop one from occurring (Fig. 2.46).

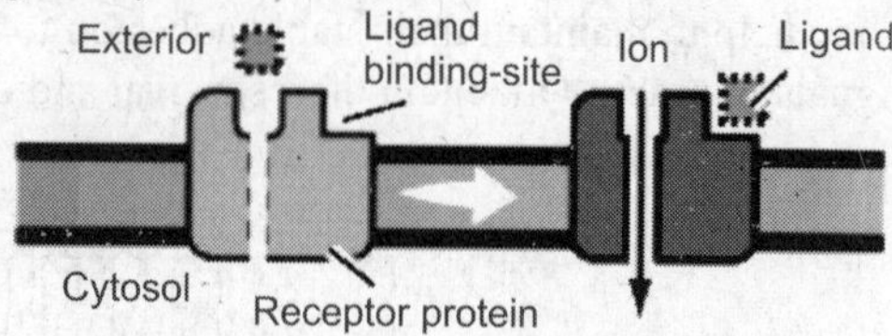

Fig. 2.46. Ligand-gated ion-channel receptor.

3. Intracellular Receptors

Intracellular receptors include nuclear receptors and cytoplasmic receptors, and soluble proteins localized within the nucleoplasm or the cytoplasm, respectively. The typical ligands for nuclear receptors are lipophilic hormones, with steroid hormones (for example, testosteron, progesteron, and cortisol), thyroid hormones, retinoic acid and derivatives of vitamin A and D_3 among them. In order to reach its receptor and initiate signal transduction, the hormone must pass through the plasma membrane, usually by passive diffusion. The nuclear receptors are ligand-activated transcription activators. They will pass through the nuclear membrane into the nucleus on binding with the ligand (the hormone), and enable the transcription of a certain gene producing a protein.

The nuclear receptors that were activated by the hormones attached at the DNA at receptor-specific *hormone responsive elements* (*HREs*), DNA sequences that are located in the promoter region of the genes that are activated by the hormone-receptor complex. As this enables the transcription of the gene, these hormones are also called *inductors of gene expression*. The activation of gene transcription is much slower than signals that directly affect existing proteins. As a consequence, the effects of hormones that use nucleic receptors are usually long-term. The signal transduction via these soluble receptors involves only a few proteins, but the details of gene regulation are not well understood. All nucleic receptors have a similar modular structure.

N-AAAABBBBCCCCDDDDEEEEFFFF-C

Here CCCC is the DNA-binding domain that contains zinc fingers, and EEEE the ligand-binding domain. The latter is also responsible for dimerization of most nuclearic receptors prior to DNA binding. As a third function, it contains structural elements that are responsible for trans-activation, used for communication with the translational apparatus. The zinc fingers in the DNA-binding domain stabilise DNA binding by holding contact to the phosphate backbone of the DNA. The DNA sequences that match the receptor are usually hexameric repeats, either normal, inverted or everted. The sequences are quite similar, but their orientation and distance are the parameters by which the DNA-binding domains of the receptors can tell them apart.

- **Steroid receptors:** Steroid receptors are a subclass of nuclear receptors, located primarily within the cytosol. In the absence of steroid hormone, the receptors cling together in a complex called an aporeceptor complex, which also contains chaperone proteins (also known as heatsock proteins or Hsps). The Hsps are necessary to activate the receptor by assisting the protein to fold in a way such that the signal sequence which enables its passage into the nucleus is accessible. Steroid receptors can also have a *repressive* effect on gene expression, when their transactivation domain is hidden; so it cannot activate transcription. Furthermore, steroid receptor activity can be enhanced by phosphorylation of serine residues at their N-terminal end, as a result of another signal transduction pathway. This behaviour is called cross talk.

II. SIGNAL TRANSDUCTION PATHWAY

Prior to 1977, a term 'signal transmission' or 'sensory transduction' was available in literature. With the widespread use of the term, signal transduction came into existence in a review article published by Rodbell (1980). In biology, signal transduction refers to any process by which a cell converts one kind of signal or stimulus into another, most often involving ordered sequences of biochemical reactions inside the cell, that are carried out by enzymes and linked through second messengers resulting in what is thought of as a 'second messenger pathway'. The second messengers include cyclic nucleotides [such as cyclic AMP (cAMP) and cyclic GMP (cGMP)], phosphatidylinositol derivatives [such as phosphatidylinositol triphosphate (PIP_3), diacylglycerol (DAG) and inositol-triphosphate (IP_3)]. Such processes are usually rapid, lasting on the order of milliseconds in the case of ion flux, to minutes for the activation of protein and lipid mediated kinase cascades. In many signal transduction processes, the number of proteins and other molecules participating in these events increases as the process emanates from the initial stimulus, resulting in a 'signal cascade' and often results in a relatively small stimulus eliciting a large response.

In bacteria and other single-cell organisms, the variety of signal transduction processes of which the cell is capable, influences how many ways it can react and respond to its environment. In multicellular organisms, a multitude of different signal transduction processes is required for coordinating the behavior of individual cells to support the function of the organism as a whole. As may be expected, the more complex the organism, the more complex the repertoire of signal transduction processes the organism possess. Thus, sensing of both the external and internal environment at the cellular level relies on signal transduction. Many disease processes such as diabetes, heart disease, autoimmunity and cancer arise from defects in signal transduction pathways, further highlighting the critical importance of signal transduction to biology as well as medicine. A cell may have many types of receptor molecules, each of which binds with specific signalling molecules. There are two ways through which signals are transmitted, *indirect transduction* and *direct transduction*.

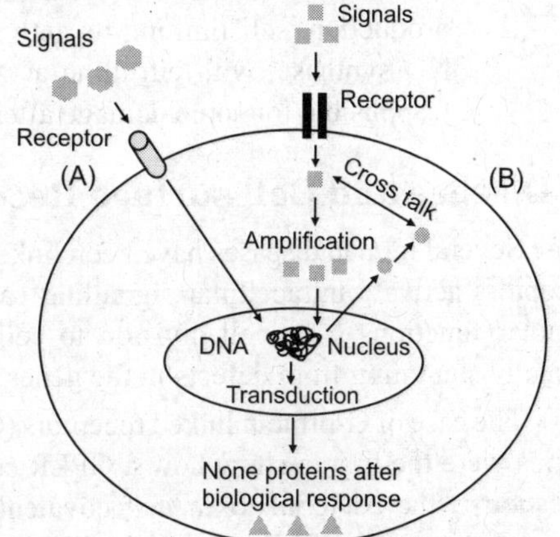

Fig. 2.47. Signal transduction pathway; A-direct, B-indirect transduction.

In the first case, according to stimuli signalling molecules are released which get attached with its specific receptor molecules present on surface of target cell. Then a change in conformation of receptor molecules occurs which is transmitted in cytoplasm to domains of receptor molecules. Further interaction between domain and information relaying molecules (IRM) in cytoplasm takes place resulting in start of a cascade of intracellular signalling which is transmitted to DNA. Genes present on DNA molecules undergo transcription to synthesise mRNA. Specific mRNA synthesises protein and responds according to signal received. The pathway that operates between cell membrane and nucleus involving many proteins is called 'signal transduction pathway' (Fig. 2.47).

In the second type signalling molecules (e.g. interferon, interleukin, etc.) are directly transmitted to the nucleus. Some times the signalling molecules interact with receptor molecules. As a result of interaction the other molecules called 'second messengers' are released. Consequently, cellular activity in terms of protein secretion is stimulated. Not surprising a single cell may have different types of receptor molecules. Each binds with only one specific signalling molecule. Therefore, a cell receives different types of integrated signals from various sources and gives suitable comprehensive response.

1. Role of Second Messenger Molecules in Signal Transduction Pathway

Secondary messengers are a class of small molecules used to transduce extracellular signals to the interior of the cell (http://openwetware.org/wiki/BIO254:2ndMessenger). For detail description, see preceding section.

(*i*) **Ca^{2+} as a second messenger:** *See* preceding section 'Second Messengers'.

(*ii*) **Lipophilic second messenger molecules:** One group of lipohilic second messenger molecules consists of inosite triphosphate and diacylglycerole. Others are ceramide and lysophosphatic acid.

(*iii*) **Nitrogen monoxide (NO) as second messenger:** NO is a univeral messenger molecule, used for regulation in virtually every cell in a multicellular organism. This water-soluble radical can diffuse through the plasma membrane and affect the other nearby cells. NO works mainly by covalent modification of proteins besides the other secondary messenger molecules. These modifications are reversible and work through a redox mechanism. As NO has a halflife of only four minutes, it will cease its function on its own soon after production, self-limiting its activity in time. NO is made from arginine by the enzyme NO synthase, with citrulline as a byproduct. NO is basically toxic, and is said to be responsible for some damage after a stroke. NO serves three main functions.

2. Disease and Cell surface Receptors

Several human diseases have been linked to alterations in cell-surface receptors. These surface receptors activate intracellular signalling cascades. These signalling cascades regulate numerous cellular functions from cell division to cell growth and development. Defects in these proteins (usually stemming from defects in the genes that encode them) can result in disease.

The case of G-protein linked receptors (GPLR) is an illustrative example. Cholera and pertussis toxins were the first to show how a GPLR could be involved in disease. In this case, the GPLR in question is the bacterial toxin that covalently modifies G-protein subunits. After this discovery, many mutations in GPLRs were linked to different disorders. Defective signal of a GPLR can result from many different changes in the normal function of the receptor such as changes in expression levels, post-translational modification or mutations. These mutations can occur as somatic events or they can be inherited. How these mutations affect a person depend on the origin of the mutation. Generalised manifestation occurs in the case of a germline mutation. Somatic mutations result in more focused manifestation. Some germline mutations are incompatible with normal embryonic development.

III. SIGNAL AMPLIFICATION

The principle of signal transduction is the 'signal amplification'. A single or a few hormone molecules can induce an enzymatic reaction that affects many substrates. The amplification can occur at several points of the signal pathway.

1. Signal Amplification at the Transmembrane Hormone Receptor

A receptor that has been activated by a hormone can activate many downstream effector proteins. For example, a rhodopsin molecule in the plasma membrane of a retina cell in the eye that was activated by a photon can activate up to 2000 effector molecules (in this case, transducin) per second. The total strength of signal amplification by a receptor is determined by:

- **The lifetime of the hormone-receptor-complex.** The more stable the hormone-receptor-complex is, the less likely the hormone dissociates from the receptor. Then the receptor will remain active for longer time, hence will activate more effector proteins.
- **The amount and lifetime of the receptor-effector protein-complex.** The more effector protein is available to be activated by the receptor, and the faster the activated effector protein can dissociate from the receptor, the more effector protein will be activated in the same amount of time.
- **Deactivation of the activated receptor.** A receptor that is engaged in a hormone-receptor-complex can be deactivated, either by covalent modification (for example, phosphorylation), or by internalisation.

D. SECONDARY MESSENGERS

Most hormones such as epinephrine and seratonin are too large or too polar to freely pass through the plasma membrane of a cell. For example, second messengers are often free to diffuse to the nucleus, where they can influence gene expression and other processes. The use of a membrane-bound protein with intra- and extracellular domains allows information to pass through the membrane without any exchange of contents. In addition to transducing a signal through the membrane, the use of secondary messengers allows for signal amplification since one bound primary messenger can stimulate the production of many secondary messenger molecules. For instance, enzymes or membrane channels are almost always activated in second-messenger generation; each activated macromolecule can lead to the generation of many second messengers within the cell (http://openwetware.org/wiki/BIO254:2ndMessenger).

Thus, a low concentration of signal in the environment can yield a large intracellular signal. Secondary messengers can be either small organic molecules such as cyclic adensine monophosphate (cAMP), guanosine 3',5'-(cyclic)phosphate (cGMP), diacylglycerol (DAG) and inositol 1,4,5 triphosphate (IP3) or inorganic ions such as Ca^{2+} (Stryer, 2005). Chemical structure of some of the major secondary messengers, such as cAMP, cGMP, calcium ions (Ca^{2+}), inositol 1,4,5-trisphosphate (IP_3), diacylglycerol (DAG) are shown in Fig. 2.48.

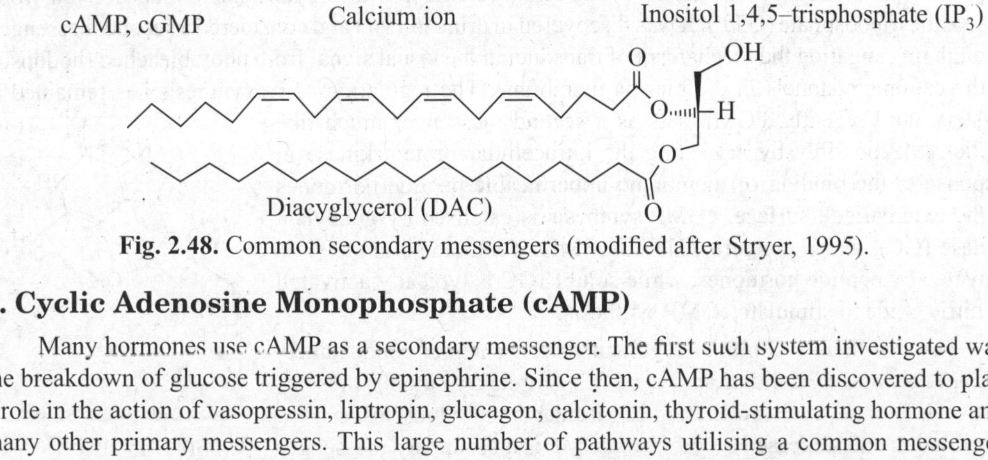

cAMP, cGMP

Calcium ion

Inositol 1,4,5–trisphosphate (IP_3)

Diacyglycerol (DAC)

Fig. 2.48. Common secondary messengers (modified after Stryer, 1995).

1. Cyclic Adenosine Monophosphate (cAMP)

Many hormones use cAMP as a secondary messenger. The first such system investigated was the breakdown of glucose triggered by epinephrine. Since then, cAMP has been discovered to play a role in the action of vasopressin, liptropin, glucagon, calcitonin, thyroid-stimulating hormone and many other primary messengers. This large number of pathways utilising a common messenger increase the potential for 'cross talk' between signalling events. Recent work has indicated the importance of both spatial and temporal control of cAMP concentration. The exact nature of these controls is still an active area of research.

Thereafter, epinephrine is released and travels to the target cell and binds to the extracellular domain of the seven helix transmembrane protein β-andrenergic receptor. The cystolic domain then participates in activation of G_s, a G-protein. Thereafter, the activated G_s releases its α domain along with a bound GTP. Then G_α-GTP stimulates activity of adenylyl cyclase (AC), which catalyses the transformation of ATP into cAMP (Fig. 2.49). AC is primarily bound to the cell membrane, which in this form is responsible for transmitting the epinephrine cascade. The other form of AC is free in solution and localised in the nucleus, mitochondria and microtubules. Soluble AC is activated primarily by bicarbonate and calcium.

Eventually, cAMP triggers the phosphorylation of many other proteins, but this is not accomplished by cAMP itself. Instead, cAMP activates a protein kinase, usually protein kinase A (PKA). Binding of cAMP releases the regulatory domains of PKA and allows the catalytic domains to transfer a phosphoryl group from adenosine triphospate (ATP) onto the target molecule. This reversible phosphorylation directly changes the activity or function of the affected molecule.

Deactivation of the cAMP cascade produced by epinephrine activity is a two step process. First, G-protein (Gs) acts as a GTPase that hydrolyses GTP in the Gα-GTP complex in order to deactivate it and prevents further activity of AC. In the second step, β-andrenergic receptor kinase phosphorylates the cystolic tail of the β-andrenergic receptor to deactivate it. Then β-arrestin caps the phosphorylated tail. In addition to these measures which prevent production of new cAMP, the enzyme cAMP phosphodiesterase also regulates by degrading the existing cAMP.

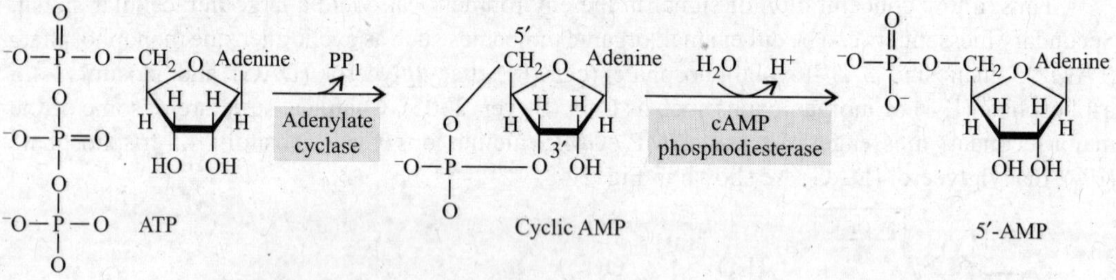

ATP —(PP_1; Adenylate cyclase)→ Cyclic AMP —(H_2O, H^+; cAMP phosphodiesterase)→ 5′-AMP

Fig. 2.49. Adenylyl cyclase (modified after Stryer, 1995).

2. Cyclic Guanosine Monophosphate (cGMP)

The cGMP is very similar to cyclic AMP (cAMP); it is a cyclic nucleotide derived from guanosine triphosphate (GTP). It was discovered in urine in 1963 and considered a second messenger through investigating the mechanism of transducing the visual signal from photobleached rhodopsin to the cationic channels in the plasma membrane. The regulation of its synthesis has remained a mystery until recently. cGMP acts as a second messenger much like cyclic AMP possibly by activating the intracellular protein kinase in response to the binding of membrane-impermeable peptide hormones to the external cell surface. cGMP synthesis is catalised by guanylate cyclase (GC), which converts GTP to cGMP. Membrane-bound GC is activated by peptide hormones, while soluble GC is typically activated by nitric oxide to stimulate cGMP synthesis.

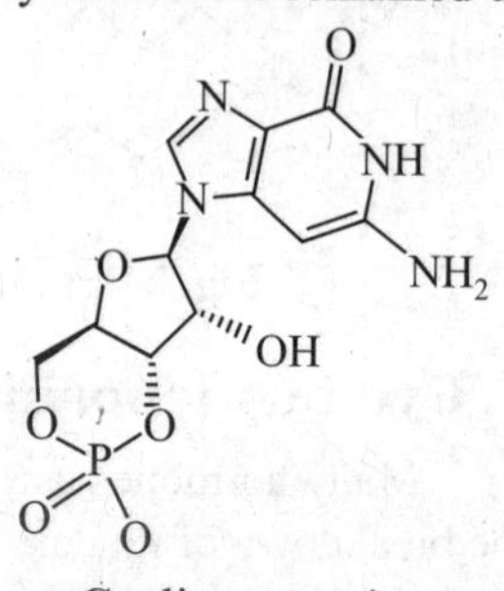

The cGMP is a common regulator of ion channel conductance, glycogenolysis, and cellular apoptosis. It also relaxes smooth muscles tissues. In blood vessels, relaxation of vascular smooth muscles leads to vasodilation and increased blood flow. cGMP is a secondary messenger

Cyclic guanosine monophosphate (cGMP).

in phototransduction in the eye. In the photoreceptors (rods and cones) of the human eye, the absence of light activates photodiesterase, which upregulates cGMP by catalysing its conversion from GMP. The cGMP opens sodium ion channels, which leads to the depolarisation of the photoreceptor's plasma membrane that opens calcium ion channels resulting in the release of neurotransmitters by the photoreceptor cell. In turn it signals the other cells imbedded in the retina and leads to the transfer of information from the eye to the brain and vision.

The cGMP is involved in the regulation of some protein-dependent kinases. For example, protein kinase G is a dimer which consists of one catalytic and one regulatory unit. It blocks the active sites of catalytic domains of regulatory unit. The cGMP binds to sites on the regulatory units of PKG and activates the catalytic units, enabling them to phosphorylate their substrates. Unlike activation of some other protein kinases such as PKA, the PKG is activated but the catalytic and regulatory units do not disassociate.

Transducin is a G-protein that is made up of three different subunits : Tα, (the guanyl nucleotide binding subunit), Tβ, and Tγ (Fig. 2.50). In the dark, with GDP bound, transducin is peripherally associated to the cytoplasmic disc surface. Photoactivated rhodopsin (the photoisomerised protein) forms a transient complex with transducin and catalyses on it a GDP/GTP exchange in the first step of the light triggered cascade.

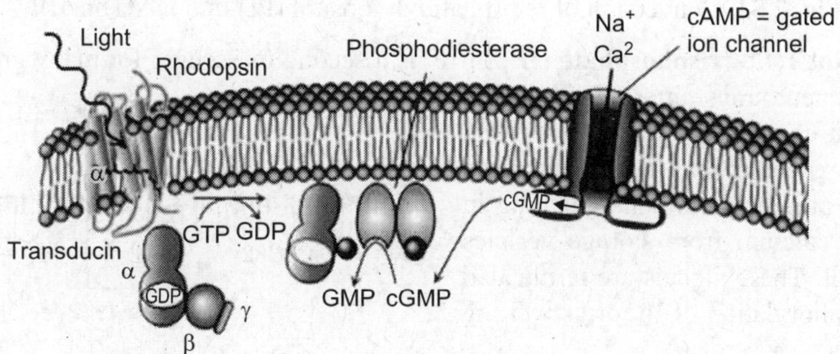

Fig. 2.50. Visual signal transduction (modified after Stryer, 1995).

GTP-bound transducin binds to and activates a cGMP phosphodiesterase (PDE), which converts cGMP to GMP. Cyclic GMP is formed from GTP by guanylate cyclase and is hydrolysed to 5′-GMP by the phosphodiesterase (Fig. 2.51).

GTP —(PP_1; Guanylate cyclase)→ Cyclic GMP —(H_2O, H^+; cGMP phosphodiesterase)→ 5′-GMP

Fig. 2.51. cGMP synthesis (modified after Stryer, 1995).

The concentration of cGMP decreases below; at this stage it is required to open cGMP-gated ion channels of the rod resulting in reduction of flow of ions across the cellular membrane. After hydrolysis transducin can return to the inactive state. It bound GTP molecule into GDP. Because transducin bound to GDP has a low affinity for PDE, it releases the enzyme in an inactive form. Thus it allows cGMP levels to rise again and return the flow of ions across the cell membrane to the 'dark' state. Also, photoreceptor cells possess a protein called regulator of G-protein signalling that accelerates transducin ability to hydrolyse GTP. Photoexcited rhodopsin is deactivated by phosphorylation and then capped by arrestin (He *et al.*, 1998).

3. Inositol trisphosphate (IP_3) and Diacylglycerol (DAG)

A wide variety of G protein-linked receptors link to an intracellular enzyme, phospholipase C (PLC), to hydrolyse phosphatidylinositol-4,5-bisphosphate (PIP_2) and generate inositol 1,4,5-trisphosphate (IP_3) and diacylglycerol (DAG) (Fig. 2.52). These two second messengers in turn regulate intracellular Ca^{2+} concentration and protein kinase C activity.

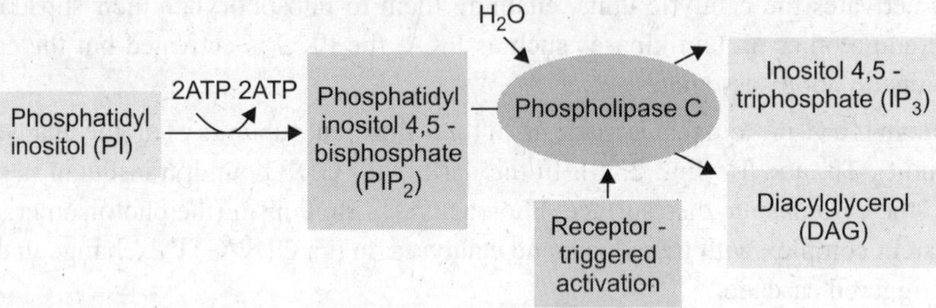

Fig. 2.52. Conversion of phosphatidyl inositol (PI) into DAG and IP_3.

***(a)* Inositol-1,4,5-trisphosphate (IP_3) :** IP_3 is a second messenger for many growth factors, hormones, and neurotransmitters. This soluble molecule diffuses through the cytosol and binds to IP_3 receptors on the endoplasmic reticulum and nuclear membrane resulting in the release of calcium from storage vesicles into the cytosol. These signals are terminated by the dephosphorylation of IP_3 or converting into derivatives.

***(b)* Diacylglycerol (DAG) :** DAG is found in the inner layer of the plasma membrane. It recruits protein kinase C that is a calcium-dependent kinase, which phosphorylates many other proteins that bring about the changes in the cell. Activation of PKC requires calcium ions. These mechanisms are collaborated with IP_3. These signals are terminated by phosphorylating DAG to arachidonic acid or hydrolysing DAG to glycerol or fatty acids. An example of bitter stimuli interaction is shown in Fig. 2.53. T2R/TRB receptors are linked to a heterotrimeric G protein which activates PDE causing decreases in intracellular cAMP and producing DAG and IP_3. Thus it causes the release of Ca^{2+} into the cytosol (Tod, 2001).

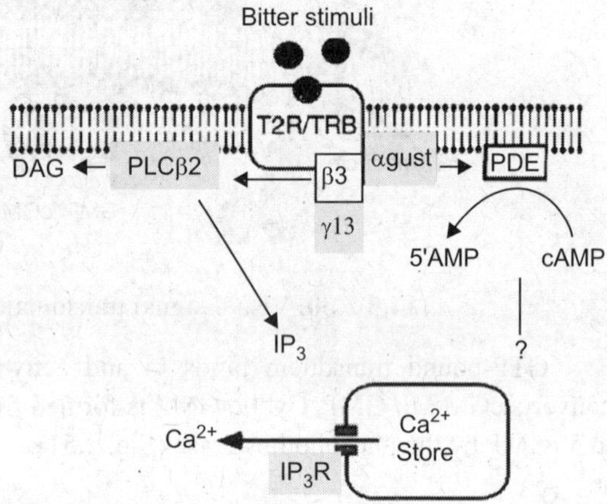

Fig. 2.53. Bitter stimuli interaction (adoped from Tod R.C., 2001)

4. Calcium Ions (Ca^{2+})

Calcium ions are probably the most widely used intracellular messengers. It acts as a signal molecule within the cell. In response to many different signals, a rise of the Ca^{2+} concentration in the cytosol triggers many types of events, such as secretion, contraction, gene expression and apoptosis development. Normally, the level of calcium in the cell is very low (~100 nM). However, its level in the cell can rise dramatically. Ca^{2+} is stored within organelles usually the endoplasmic reticulum where it is bound to molecules like calreticulin. To become active, Ca^{2+} has to be released from the endoplasmic reticulum into the cytosol. There are two combined receptor/ion channel proteins that perform the task of controlled transport of Ca^{2+}:

***(i)* $InsP_3$-receptor :** It will transport Ca^{2+} upon interaction with inosite triphosphate (thus the name) on its cytosolic side. It consists of four identical subunits.

***(ii)* Ryanodin-receptor :** It is named after the plant alkaloid ryanodin. It is similat to the $InsP_3$ receptor and stimulated to transport Ca^{2+} into the cytosol by recognising Ca^{2+} on its cytosolic side, thus establishing a feedback mechanism. A small amount of Ca^{2+} in the cytosol near the receptor will cause it to release even more Ca^{2+}. It is especially important in neurons and muscle cells. In heart and pancreas cells, another second messenger (cyclic ADP ribose) takes part in the receptor activation.

The localised and time-limited activity of Ca^{2+} in the cytosol is also called a 'Ca^{2+} wave'. The building of the wave is done by : (*i*) the feedback mechanism of the ryanodin receptor, and (*ii*) the activation of phospholipase C by Ca^{2+} resulting in production of inosite triphosphate, which in turn activates the $InsP_3$ receptor. Ca^{2+} is used in a multitude of processes. The three main pathways that lead to Ca^{2+} activation are: G-protein regulated pathways, pathways regulated by receptor-tyrosine kinases, ligand- or current-regulated ion channels.

There are two different ways in which Ca^{2+} can regulate proteins: a direct recognition of Ca^{2+} by the protein, binging of Ca^{2+} in the active centre of an enzyme. One of the best studied interactions of Ca^{2+} with a protein is the regulation of calmodulin by Ca^{2+}. Calmodulin itself can regulate other proteins, or become part of a larger protein (for example, phosphorylase kinase). The Ca^{2+}/calmodulin complex plays an important role in proliferation, mitosis and neural signal transduction.

There are two major types of pathways to sense the concentration change of calcium inside the cell.

- After binding of Ca^{2+}, calmodulin is activated when the cytosolic calcium level is raised. Calmodulin is a member of the EF-hand protein family. The EF hand is a Ca^{2+}-binding motif that consists of a helix, a loop, and a second helix, originally discovered in the protein parvalbumin. Calmodulin protein structure changes depending on binding of Ca^{2+} or not. Calmodulin-Ca^{2+} complex stimulate a lot of targets such as CaMKII, which can phosphorylate multiple downstream enzymes and proteins to induce various responses.
- The other example is synaptotagmin which serves as the major Ca^{2+} sensor for regulating exocytosis from neurons. Ca^{2+} triggers two consecutive synaptotagmins interact with each other via coupling its second C2 domain.

QUESTIONS

1. Write a brief note on a journey from molecules to a living cell.
2. Give an illustrated account of structure of a cell wall of Gram-positive and Gram-negative bacteria.
3. Describe the structure of eukaryotic cell wall and cell membrane.
4. Write in detail about transport across cell membrane.
5. Give a brief account of mechanism of active and passive transport systems.
6. Describe the transport system in prokaryotes and eukaryotes.
7. Write cell signalling and signal transduction pathway.
8. What do you know about second messengers and their role in cell signalling.
9. Write short notes on the following :

(*i*) Miller's experiment, (*ii*) polymeris action of nucleotides,
(*iii*) catalytic RNA (ribozyme),
(*iv*) evolution of self catalytic rRNA molecules,
(*v*) differences between prokaryotic and eukaryotic cells,
(*vi*) compartmentalisation of organelles,
(*vii*) differences between plant and animal cells,
(*viii*) structure and function of endoplasmic reticulum,
(*ix*) structure and function of Golgi complex, (*x*) Na^+/K^+–ATPase,
(*xi*) Ca^{++}–ATPase pump, (*xii*) ion channels,
(*xiii*) cell surface receptors and their function, (*xiv*) signal transduction pathways.

BIBLIOGRAPHY

Alberts B. *et al.* 2002. *Molecular Biology of the Cell* (4th Ed.) New York : Garland Science.

He W., *et al.* 1998. RGS9, a GTPase accelerator for phototransduction. *Neuron.* 20: 95–102.

Hehlgans S., Haase M. and Cordes N. 2007. *Signalling via integrins: Implications for cell survival and anticancer strategies. Biochim. Biophys. Acta. 1775 : 163-180.*

Horisberger J.D. *et al.* 1991. Structure-function relationship of Na^+-K^+ ATPase. *Ann. Rev. Physiol.* 53: 565-584.

Kaback H.R. 1989. Molecular biology of active transport from membrane to molecule to mechanism. *Harvey Lect.* **83**: 77-103.

Kistler J., Stroud R.M. et al. (1982). Structure and function of an acetylcholine receptor. *Biophys. J. 37: 371-383.*

Lodish H.F. 1988. Anion exchange and glucose transport proteins-structure, function and distribution. *Harvey Lect.* 82: 19-46.

Mercer R.W. 1993. Structure of the Na, K-ATPase. *Int.Rev.Cytol. 137*C: 139-168.

Rodbell M. 1980. The role of hormone receptors and GTP-regulatory proteins in membrane transduction. *Nature. **284**: 17-22.*

Stryer L. 1995. *Biochemistry.* 4th Ed. WH Freeman & Co., New York.

Tod R. C., Leslie M. S., Robert F. M. and Sue C. K., BMC Neuroscience (2001) **2**, 6

Watson J.D. *et al.* 2004. *Molecular biology of the gene.* 5th Ed. Pearson Education, Singapore.

3 CHAPTER Molecular Biology

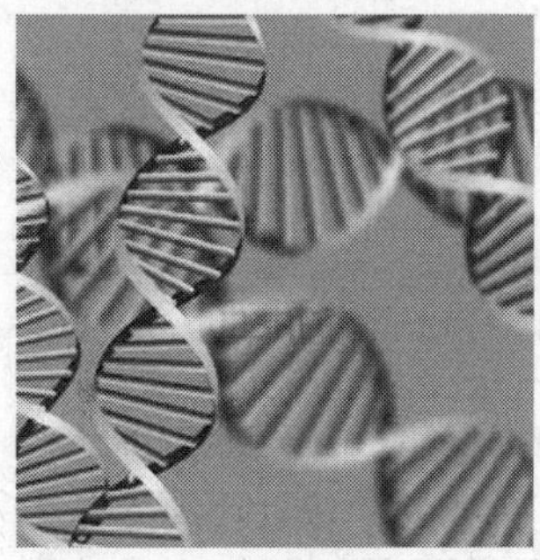

A. Nature of nucleic acids (RNA and DNA)
B. Conformational analysis of nucleic acids
C. Genome organisation
D. Structure of RNA
E. DNA Replication
F. Gene expression (transcription and translation)

The nucleic acids are the first biomolecules which were synthesised on the earth during organic evolution about more than 5 billion years ago. Ribonucleic acid was synthesised first during 'RNA World' than the DNA; the DNA was synthesized during 'DNA World'. Nucleic acids are found in viruses and all living organisms (microorganisms, plants and animals) which carry the genetic information. Nucleic acids are the polynucleotides *i.e.* polymer of nucleotides. Each nucleotide consists of three main constituents: a cyclic five carbon sugar, a purine or pyrimidine base, and a phosphate attached to sugar. The sugar may be ribose or deoxyribose. Based on the types of sugars, the nucleic acids are of two types, ribose nucleic acid (RNA) and deoxyribose nucleic acid (DNA). All the organisms contain DNA but few bacteriophages, and plants and animal viruses contain RNA as genetic material.

A. NATURE OF NUCLEIC ACIDS (DNA AND RNA)

In 1860, for the first time Friedrich Miescher (a Swiss Physician) isolated a precipitate with alkaline solution from white blood cells (WBC). This precipitate contained C, N, H, P and O. He subjected the purified nuclei to an alkaline extraction followed by acidification resulting in a precipitate being formed to which Miescher called *nuclein* (now nucleic acids). Purine and pyrimidines were isolated by Hermann Emil Fischer in 1880. In 1902, he won the Nobel Prize in Chemistry for synthesis of sugar and purine. In 1899, Altaman replaced the term nuclein with nucleic acid. Albrecht Kossel identified the presence of histones and protamines within nucleic acids and was awarded Nobel Prize for demonstrating the presence of two purines (adenine and guanine) and two pyrimidines (thymine and cytosine). During 1910s, P.A. Levene discovered the phosphate and pentose sugars called deoxyribose molecules. In 1953, J.D. Watson (an American biologist) and F.H.C. Crick (a British physicist) presented the double helix model of DNA and they were awarded the Nobel Prize in 1962.

Hermann Emil Fischer (1852-1919)

The DNA is found in all plants, animals, prokaryotes and some viruses. In eukaryotes it is present inside the nucleus, chloroplast and mitochondria, whereas in prokaryotes it is dispersed in cytoplasm. In plants, animals and some viruses the genetic material is double stranded (ds)DNA molecule except some viruses such as ϕX174, M13. In TMV, influenza virus, poliomyelitic virus and bacteriophages the genetic material is single stranded (ss)RNA molecule.

I. DEOXYRIBONUCLEIC ACID

Deoxyribonucleic acid acts as the blue print of life. Genetic information for the charancters of an organism is stored in DNA in the form of nucleotide sequences. Therefore, before discussing anything of the biomolecule one must know the chemical and physical properties of deoxyribonucleic acid.

1. Chemical Composition

DNA isolated and purified from a variety of plants, animals, bacteria and viruses has shown a complex form of polymeric compounds containing four monomers known as deoxyribonucleotide monomers ordeoxyribonucleotides. Each deoxyribonucleotide consists of pentose sugar (deoxyribose), a phosphate group and a nitrogenous base (either purine or pyrimidine). Purines bases (adenine and guanine) are heterocyclic and two ringed bases and the pyrimidines (thymine, cytosine and uracil) are one ringed bases. The following components of deoxyribonucleotide have been described:

(*a*) A Five Carbon Ring: Deoxyribose is a pentose sugar consisting of five carbon atoms. The first and fourth carbon atoms of this sugar combine with one oxygen atom and form a ring. The fifth atom (5′) forms $-CH_2$ group which is present outside this ring. Three -OH groups are attached at position 1′, 3′ and 5′ and the hydrogen atoms combine at position 1′, 2′, 3′ and 4′ of carbon atoms (Fig. 3.1). In ribonucleotides, ribose is the pentose sugar which is similar to deoxyribose except that there is an –H at 2′ carbon atom instead of –OH group. Therefore, the absence of -OH group in DNA at 2′ carbon atom makes it chemically more stable than the RNA.

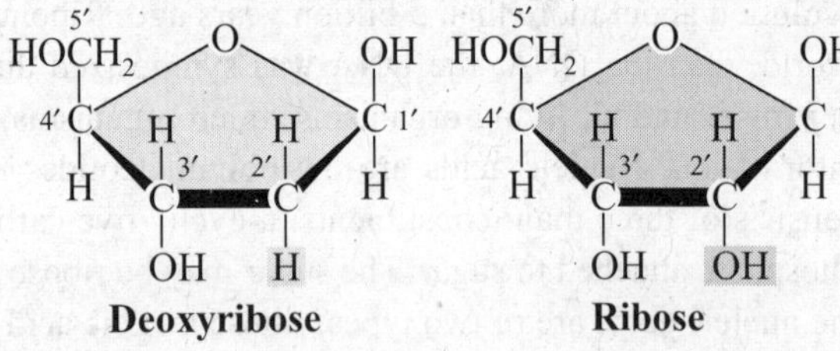

Fig. 3.1. Pentose sugars-deoxyribose and ribose.

(*b*) Nitrogenous Base: There are two nitrogenous bases, purines and pyrimidines. The purines are double ring compounds that consist of 5-membered imidazole ring with nitrogen at 1′, 3′, 7′ and 9′ positions. The pyrimidines are single ring compounds, the nitrogen being at position 1′ and 3′ in 6-membered benzene ring. A single base is attached to 1′-carbon atom of pentose sugar by an *N*-glycosidic bond. Purines are of two types, adenine (A) and guanine (G), and pyrimidines are also of two types, thymine (T) and cytosine (C). Uracil (U) is a third pyrimidine (Fig 3.2). A, G and C are common in both DNA and RNA, but T is found in DNA and U is found only in RNA.

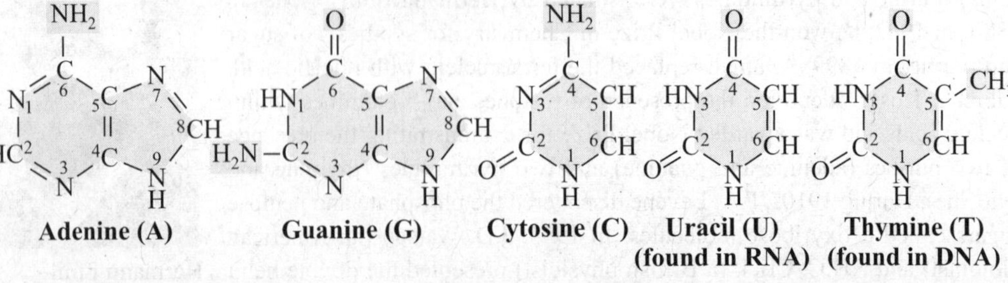

Fig. 3.2. Purines (adenine and guanine), and pyrimidines (thymine, cytosine and uracil).

Purine and pyrimidine are linked together by hydrogen bonds; A and T form double bond, where as G and C form triple hydrogen bond (Fig. 3.3). In RNA A is linked with U.

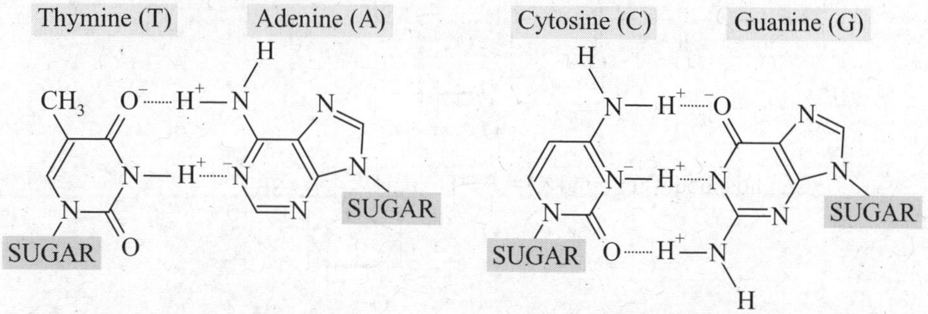

Thymine (T) Adenine (A) Cytosine (C) Guanine (G)

SUGAR SUGAR SUGAR SUGAR

Fig. 3.3. Bond formation between purine and pyrimidine.

(*c*) A Phosphate Group: In addition to the presence of nucleotides in DNA helix, these are also present in nucleoplasm and cytoplasm in the form of deoxyribonucleotide phosphates *e.g.* deoxyadenosine triphosphate (dATP), deoxyguanosine triphosphate (dGTP), deoxycytidine triphosphate (dCTP), deoxythymidine triphosphate (dTTP). The advantage of the four deoxyribonucleotide triphosphate form is that the DNA polymerase uses only triphosphate form of nucleotides during DNA replication. Ribonucleotide triphosphates are the precursors of RNA (Fig. 3.4).

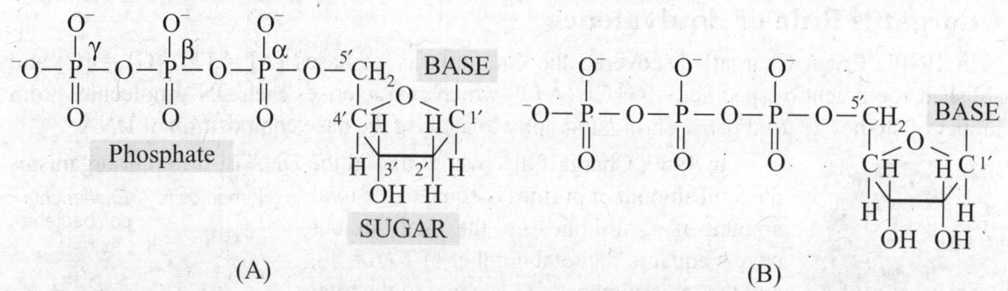

Phosphate BASE SUGAR BASE

(A) (B)

Fig. 3.4. Nucleotides showing base, sugar and phosphate linked with A, B and Y bond; A- a deoxyribonucleotide triphosphate; B- a ribonucleotide triphosphate.

(*d*) Phosphodiester Bond: Two nucleotides of a DNA polymer are linked by a phosphodiester bond (Fig. 3.5). A –OH group is attached to 3′ carbon atom and a phosphate group ($-PO_4^{-3}$) to the 5′-carbon of pentose sugar of a nucleotide. The two nucleotides of a single strand of DNA are linked by a phosphodiester bridge forming a bond between –OH group of 3′ C and $-PO_4$ group of 5′ C of pentose sugar, resulting in a long chain of DNA or RNA molecule. Therefore, each strand contains 3′ end and 5′ end arranged in an alternate manner. Strong negative charges of nucleic acid are due to the presence of phosphate groups.

Similarly, the ribonucleotides contain ribose sugar, nitrogenous bases and phosphate. Except sugar, the other components are similar. However, uracil (U) is found in RNA instead of thymine. Generally, RNA molecule is single stranded besides some exceptions.

The nucleotides undergo the process of polymerisation to form a long chain of polynucleotide. The nucleotides are designated by prefixing 'poly' to each repeating unit also such as poly A (polyadenylic acid), Poly T (polythymidilic acid), poly G (poly guanidylic acid), poly C (polycytidilic acid) and poly U (poly uridylic acid). The polynucleotide that consists of the same repeating units are called homopolynucleotides such as poly A, poly T, poly G, poly C and poly U.

5′-P terminus

$^{-}O-\overset{O^-}{\underset{O}{P}}-O-\overset{O^-}{\underset{O}{P}}-O-\overset{O^-}{\underset{O}{P}}-O-\overset{5'}{C}H_2$ BASE

3′

O

A phosphodiester bond $O=\underset{O^-}{P}-O-\overset{5'}{C}H_2$ BASE

3′

O

$O=\underset{O^-}{P}-O-\overset{5'}{C}H_2$ BASE

3′

OH

3′-OH terminus

Fig. 3.5. Phosphodiester linkage between 3′–OH of a nucleotide and 5′-PO_4 of another nucleotides.

2. Chargaff's Rule of Equivalence

In 1940's Erwin Chargaff discovered the equimolarity of bases [A] = [T], [G] = [C] and established coefficient of specificity [G+C]/[A+T] which characterises each DNA molecule. From a number of studies he used paper chromatography to analyse the base composition of DNA.

Erwin Chargaff

In 1950, Chargaff discovered that in the DNA of different organisms the total amount of purines is equal to the total amount of pyrimidines *i.e.* the total number of A is equal to the total number of T (A = T), and the total number of G is equal to the total number of C (G = C). It means that A/T = G/C, *i.e.* A + T/G + C = 1. In the DNA molecules isolated from several organisms, regularity existed in the base composition (Fig. 3.6).

The DNA molecule of each species comprises of base composition which is not influenced either by environmental conditions or growth stages or age. The molar ratio *i.e.* [A] + [T]/[G]+[C] represents a characteristic composition of DNA of each species. However, in higher plants and animals A=T composition was found generally high and G = C content low, whereas the DNA molecules isolated from lower plants and animals, and bacteria and viruses was generally rich in G = C and poor in A = T contents (Table 3.1).

The two closely related species will have very similar molar % G+C values and vice versa. Thus, the use of base composition has much significance in establishing relationship between two species and in taxonomy and phylogeny of species. For example the two individuals having 95 to 98% similarity in G+C content are supposed to the same organisms. In recent years, it is used as one of the criteria for classifying the microorganisms.

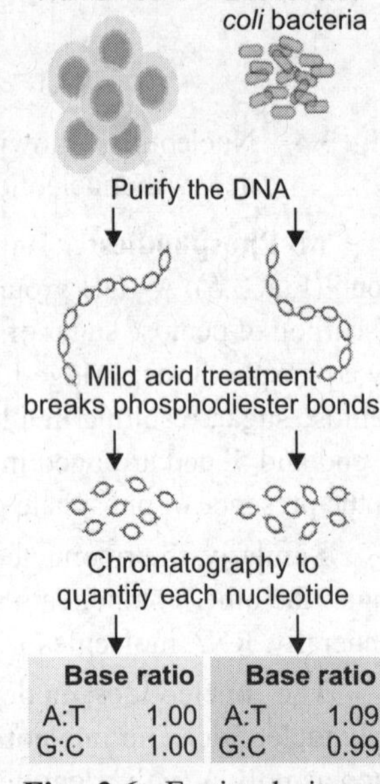

Fig. 3.6. Equimolarity of bases (A: T and G: C) in DNA from human and *Escherichia coli*.

Table 3.1. Relative amount of nitrogenous bases in DNA isolated from different organisms.

Source	*Adenine*	*Guanine*	*Thymine*	*Cytosine*	*Base ratio*
Human sperm	30.9	19.1	31.6	18.4	1.62
Human thymus	30.9	19.9	29.4	19.8	1.52
Sea urchin sperm	32.8	17.7	32.1	18.4	1.85
Wheat germ	26.5	23.5	27.0	23.0	1.19
Yeast	31.3	18.7	32.9	17.1	1.79
Escherichia coli	26.0	24.9	23.9	25.2	1.00
Diplococcus pneumoniae	29.8	20.5	31.6	18.0	1.59
Bacteriophage T2	32.5	18.2	32.6	16.7	1.86

3. X-ray Crystallographic Studies of DNA

Rosalind Elsie Franklin (1920-1958)

In 1920, the basic chemical composition of nucleic acid was elucidated through the efforts of P.A. Levene. In 1938 W.T. Astbury took the first fibre diffraction pictures of DNA, correctly predicted the overall dimensions of the molecule and that the nucleotide bases were stacked at intervals of 3.3Å perpendicular to its long axis. In 1940, for the first time he gave first three dimensional structure of DNA model studied through X-ray crystallography. He concluded that its flat nucleotides get stacked to form polynuceotide. Each nucleotide was oriented perpendicularly to the axis of molecule present at every 3.4 Å along the stack.

Astbury was a founder father of X-ray diffraction studies of biological macromolecules and, in 1961, coined a definition of the new field of 'Molecular Biology'. The chair of biophysics at the University still bears his name, as does the Astbury Centre for Structural Molecular Biology and even the building where the current X-ray diffraction equipment is housed.

Maurice H.F. Wilkins, Rosalind Franklin and coworkers continued to study the work of Astbury and prepared the highly oriented DNA fibres which facilitated for getting photographs of X-ray diffraction. His female associate, R. Franklin prepared a super X-ray diffraction photograph of DNA (Fig. 3.7), that supported Astbury's intermediate distance 3.4 Å and also suggested for helical configuration of DNA.

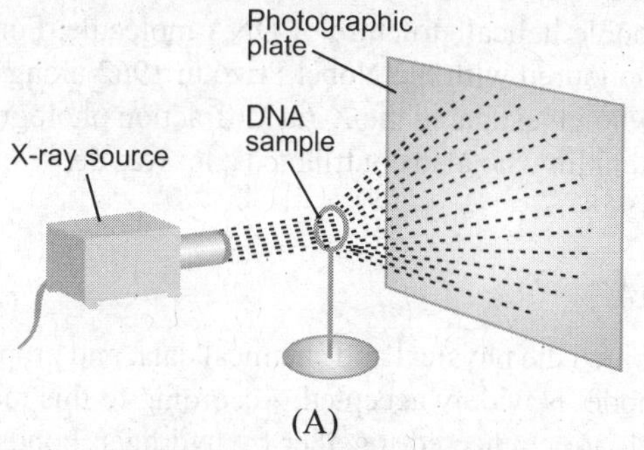

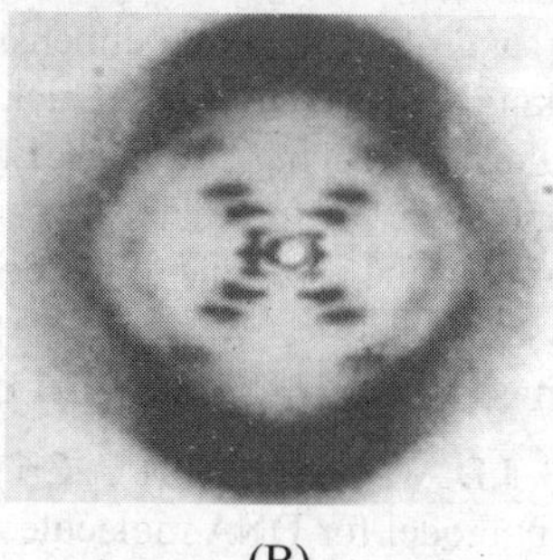

Fig. 3.7. X-ray diffraction studies by Rosalind Franklin and Maurice Wilkins (A); a photograph of X-ray diffraction study of DNA (B).

Some of the points consistently noticed through X-ray diffraction photographic studies made by Wilkins and Franklin were : (*i*) the DNA from different species gave the identical X-ray diffraction besides variation in their base composition, (*ii*) the DNA molecules are about 20 Å thick and more than 30,000 Å long, and (*iii*) the DNA molecules have repeating structure at every 34 Å. However, the pertinent experimental facts were achieved through the study of detailed chemical structure of DNA, Chargaff's base pairing rule and Wilkins-Franklin's X-ray diffraction photographic studies. There was an urgent need to get three dimensional structure of DNA molecule. In February 1953, Pauling and R.B. Corey gave a triple helix model of DNA molecule. However, they could not explain the process of DNA replication.

4. Consideration of Watson and Crick's Model of DNA

In 1953, a young American biologist James D. Watson and an English physicist Francis H. Crick of the Cavendish Laboratory, Cambridge University studied in advance the manuscript of Pauling and Corey, and X-ray diffraction photographs of DNA taken from Franklin. They concluded that : (*i*) polynucleotide chain of DNA has a regular helix, (*ii*) the helix has a diameter of about 20 Å, and (*iii*) the helix makes a full turn at every 34 Å along its length and contains a stack of ten nucleotides per turn because the internucleotide distance is 3.4 Å. They considered the density of DNA molecule and concluded that the density of a cylinder having 20 Å diameter and 34 Å length is too low if it contained a single stack of ten nucleotides and the density will be too high if it contains three or more stacks of ten nucleotides. Therefore, the helix must contain two polynucleotide chains or two stacks of ten nucleotides per turn.

Nobel Prize winners- J.D. Watson, F.H.C. Crick, and M.H.F. Wilkins (left to right)

If the DNA is to contain heredity information which is inscribed as specific sequence of four bases along the polynucleotide chain, the molecular structure of DNA must include any arbitrary sequences along its polynucleotide chain. The dimension of purine ring is more than the dimension of pyrimidine ring; therefore, the two chain helix may contain a constant diameter provided a complementary relationship existed between the two nucleotide stacks, one harbouring purine and the other pyrimidine bases. To provide thermodynamic stability they thought over the formation of hydrogen bonds between amino ($-NH_2$) or hydroxyl (-OH) hydrogens and ketone oxygen or amino-nitrogen of the two bases. On the basis of these considerations they got success in constructing a double helix model for the DNA molecule.

They sent a brief letter to the Journal *Nature* on 2nd April 1953 entitled *a structure for deoxyribose nucleic acid* suggesting the three dimensional double helical structure of DNA molecule. For this ground-breaking work, Watson and Crick were honoured with the Nobel Prize in 1962 along with Maurice Wilkins of the King's College, London who investigated the X-ray diffraction photographs of the DNA molecule. Unfortunately, Rosalind Franklin who also contributed a lot was deprived off possible sharing of the prize due to her death in 1958.

5. Watson and Crick's Model of DNA

In 1953, J.D. Watson and F.H.C. Crick combined the physical and chemical data, and proposed a double helix model for DNA molecule. This model is widely accepted. According to this model, the DNA molecule consists of two strands which are connected together by hydrogen bonds and helically twisted. Each step on one strand consists of a nucleotide of purine base which alternates with that of pyrimidine base. Thus, a strand of DNA molecule is a polymer of four nucleotides, *i.e.* A, G, T, and C. The two strands join together to form a double helix. Bases of two nucleotides form hydrogen bonds, *i.e.* A combines with T by two hydrogen bonds (A = T) and G combines with C by three hydrogen bonds (G≡C). However, the sequence of bonding is such that for every ATGC on one strand there would be TACG on the other strand. Therefore, the two chains are complementary to each other, *i.e.* sequences of nucleotides on one chain are the photocopy of sequences of nucleotides on the other chain. The two strands of double helix run in antiparallel direction, *i.e.* they have opposite polarity (Fig. 3.8 A-B). The left hand strand has $5' \rightarrow 3'$ polarity, whereas the right hand has $3' \rightarrow 5'$ polarity as compared to the first one. The polarity is due to the direction of phosphodiester linkage.

The hydrogen bonds between the two strands are such that they maintain a distance of 20 Å. The double helix coils in right hand direction, *i.e.* clockwise direction and completes a turn at every 34 Å distance (Fig. 3.8 A). The turning of double helix results in the appearance of a deep and wide groove called major groove. The major groove is the site of bonding of specific protein. The distance between two strands forms a minor groove. One turn of double helix at every 34 Å distance includes 10 nucleotides, *i.e.* each nucleotide is situated at a distance of 3.4 Å. Sugar-phosphate (nucleoside) makes the backbone of double helix of DNA molecule (Fig. 3.8 B).

The DNA model also suggested a copying mechanism of the genetic material. DNA replication is the fundamental and unique event underlying growth and reproduction in all living organisms ranging from the smallest viruses to the most complex of all creatures including man. DNA replicates by semiconservative mechanism which was experimentally proved by Mathew Meselson and Frank W. Stahl in 1958. If changes occur in sequence or composition of base pairs of DNA, mutation takes place. Though the presence of adenine, guanine, thymine and cytosine is universal phenomenon, yet unusual bases in DNA molecule are also found. In some bacteriophages 5-hydroxymethylcytosine (HMC) replaces cytosine of the DNA molecule when methylation of adenine, guanine and cytosine occurs. This results in changes in these bases.

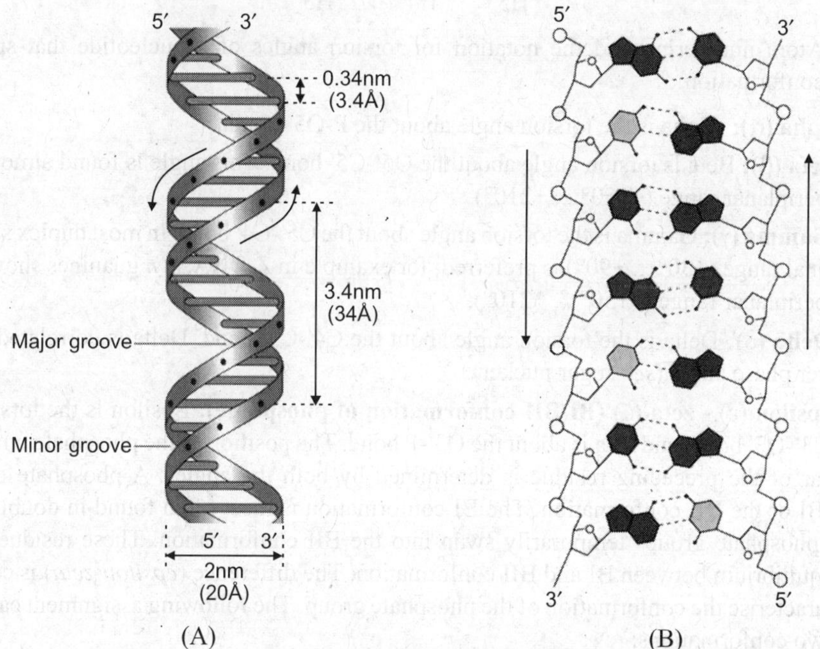

Fig. 3.8. Watson and Crick's model of DNA.

B. CONFORMATIONS OF NUCLEIC ACIDS

Structure of nucleic acids has already been described in previous section. However, this section deals with only conformational analysis of nucleic acids.

I. TORSION ANGLES

The nucleotides are made up of a nitrogenous base, ribose or deoxyribose sugar and three molecules of phosphates attached at 5′ end of the sugar. There developes torsion angles between C-C, C-N and C-PO_4; consequently, they result in changes in corformations.

1. Different Torsion Angles

There are various torsion angles in the backbone of nucleotide units, which are indicated by notations in three sections: *sugar-phosphate backbone* chain, *sugar ring*, and *sugar-base* side chain. The torsion angles are designated with seven Greek letters (*alpha* to *zeta*). Each nucleotide consists of seven torsion angles that specify conformation of every nucleotide in a polynucleotide chain (Fig. 3.9). The sugar-phosphate backbone is characterised by six torsion angles (α, β, γ, δ, ε and ζ), and the orientation of base relative to the sugar is specified by the glycosidic torsion angle chi, χ. The torsion angles govern the conformations of nucleosides and nucleotides for rotation around each bond.

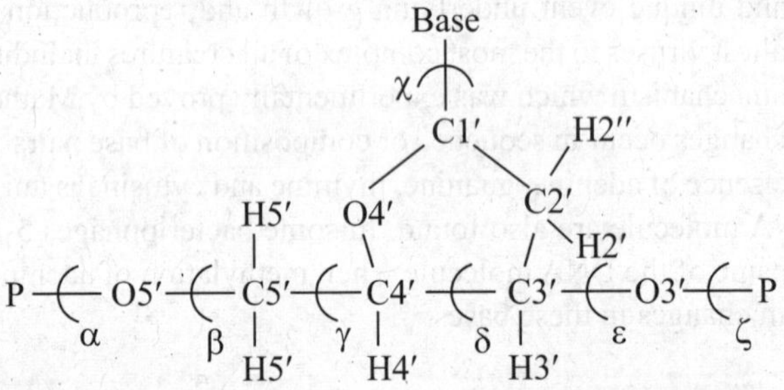

Fig. 3.9. Atom numbering and the notation for torsion angles of a nucleotide that specify its conformation.

(*i*) **Alpha (α):** Alpha is the torsion angle about the P-O5′ bond.

(*ii*) **Beta (β):** Beta is torsion angle about the O5′-C5′ bond. This angle is found almost always in the antiperiplanar range (+150° ... +210°).

(*iii*) **Gamma (γ):** Gamma is the torsion angle about the C5′-C4′ bond. In most duplex structures, the +synclinal range (+30° ... +90°) is preferred; for example in Z-DNA, *syn* guanines show gamma in the antiperiplanar range (+150° ... +210°).

(*iv*) **Delta (δ):** Delta is the torsion angle about the C4′-C3′ bond. Delta is correlated with the sugar pucker phase angle (*see* sugar pucker).

(*v*) **Epsilon (ε) - zeta (ζ) (BI/BII conformation of phosphate):** Epsilon is the torsion angle about the C3′-O3′ bond, and zeta is about the O3′-P bond. The position of the phosphate with respect to the sugar of the preceding residue is determined by both the angles. A phosphate can adopt either the BI or the BII conformation. The BI conformation is most often found in double helices. But some phosphate groups temporarily swap into the BII conformation. These residues show a dynamic equilibrium between BI and BII conformation. The difference (*epsilon-zeta*) is commonly used to characterise the conformation of the phosphate group. The following assignment can be used for these two conformations:

BI: (epsilon-zeta) = – 160º ... + 20º

BI: (epsilon-zeta) = + 20º ... + 200º

A phosphate in the BII conformation forces its neighboring phosphates into the BI conformation. Having two neighboring phosphates in the BII conformation at the same time is an extremely rare event.

Backbone torsion angles in nucleic acid that changes its conformation and structures are shown in Fig. 3.10.

(*vi*) **Chi (χ):** A nucleoside (except pseudouridine) in DNA/RNA contains an *N*-glycosidic bond that connects the base to the sugar. The chi (χ) torsion angle, which characterises the relative base/sugar orientation, is defined by O4′-C1′-N1-C2 for pyrimidines (C, T and U), and O4′-C1′-N9-C4 for purines (A and G).

These angles are very informative for describing the conformation of a double helix. When bond-stretching and bond-angle bending occurs, it change the conformation. But differences in torsion angles also provide the main differences among conformations. The adjacent backbone atoms are in 5′→3′ order; for example beta refers to chain atoms P5′-O5′-C5′-C4′.

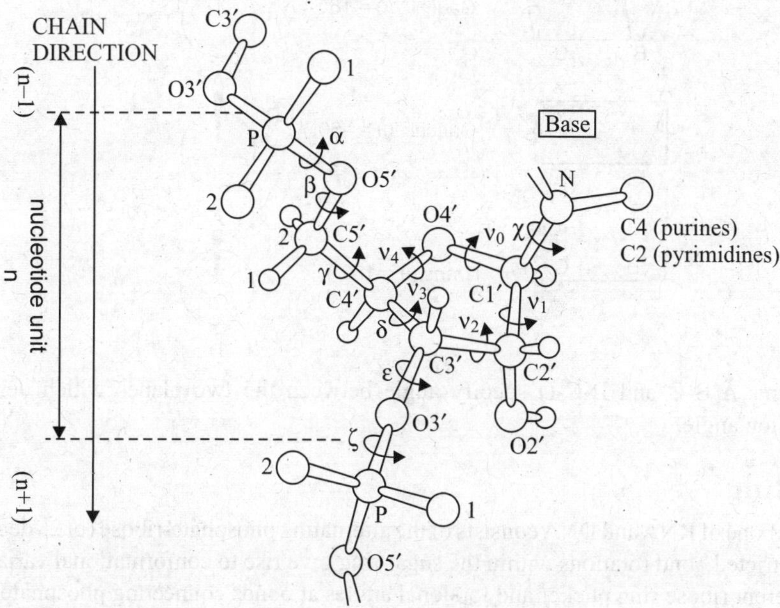

Fig. 3.10. Backbone torsion angles in nucleic acid structures (*source:* Saenger, 1984).

Points to Remember

- Both the backbone torsion angle δ and the endocyclic torsion angle ν3 refer to rotation about the same bond, C4′-C3′. Both angles are needed for complete description of the main-chain and sugar-ring conformations in some studies.
- The notation τ_0–τ_4 previously used to represent torsion angles about the bonds in the sugar ring is superseded by the present notation (ν), which is consistent with polysaccharide nomenclature. The symbol τ is now used to denote a bond angle, which is consistent with polypeptide nomenclature.

The angle between two planes specified by atoms defines the torsion angle. The *cis* conformation (angle 0º) specifies that all atoms are in the same plane with A and B eclipsed. A positive torsion angle is formed upon rotating D clockwise. The *trans* (*t*) planer conformation has a torsion angle of ±180º, whereas Glauchi (*g*) conformations have torsion angles of ±60º or –60º (Fig. 3.11). The conformation of sugar-phosphate backbone is variable, but the conformations with the bases stacked tend to have α (g^-), β (*t*), γ (g^+), ε (*t*) and ζ (g^-). The torsion angle δ is directly related to sugar conformation.

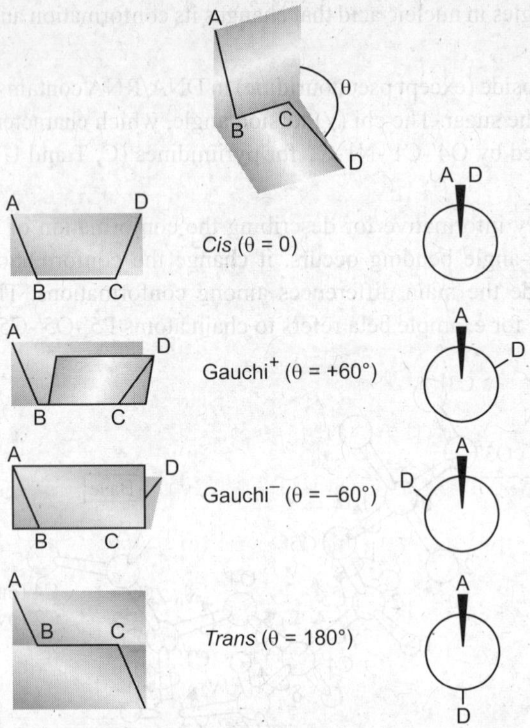

Fig. 3.11. Atoms A-B-C and B-C-D specify angle between the two planes which defines each torsion angle.

2. Sugar Ring

The backbone of RNA and DNA consists of the alternating phosphate-ribose (or 2′-deoxyribose) chain. The restricted bond rotations within the sugar ring give rise to conformational variation. This results in different ribose ring pucker and torsional angles at bonds connecting phosphate to ribose.

***(a)* Endocyclic Torsion Angles:** The pentose sugar ring occupies an important position in a nucleotide forming polypeptide chain because it is a part of both the backbone and the side chain. It is necessary to specify the torsion angles for the ring, the bond lengths and bond angles because it will help to provide a complete description of the ring conformation.

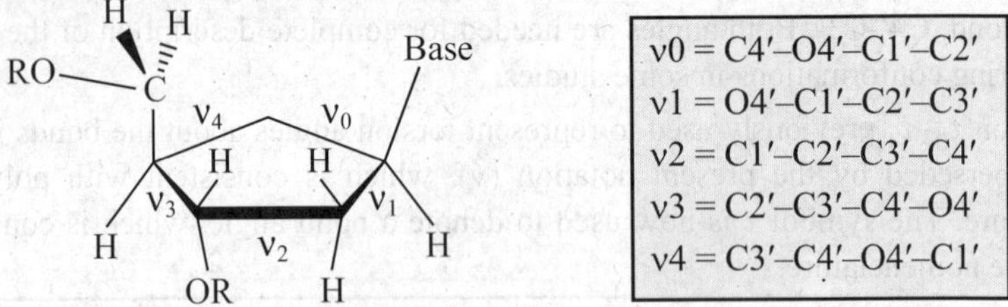

Fig. 3.12. Torsion angles in sugar rings of β-D-nucleosides and nucleotides.

Ribose and deoxyribose are five-membered rings; hence five torsion angles have been specified, which are called 'endocyclic torsion angles'. These five endocyclic torsion angles for the bonds O4′-C1′, C1′-C2′, C2′-C3′, C3′-C4′ and C4′-O4′ are denoted by the symbol *neu*, ν0, ν1, ν2, ν3 and ν4, respectively. The sequence of atoms used defines the each backbone torsion angle, for example ν0 refers to the torsion angle of the sequence of atoms C4′-O4′-C1′-C2′, and so on. The following five endocyclic torsion angles define the ribose sugar geometry (Fig. 3.12).

***(b)* Sugar Puckering:** The conformation of the ribose ring can be described by only two parameters: the sugar pucker 'pseudorotation phase angle' (P) and 'sugar pucker amplitude'; both are calculated from the torsion angles about the five bonds in the ribose ring. The torsion angles in a closed ring are highly correlated. The conformations of the ribose have been named according to the ring atoms which pucker out of the plane given by the other ring atoms. The sugar pucker determines the shape of the α-helix, whether the helix will exist in the A-form or in the B-form. For steric reasons the five-membered ring is not planar because all bonds would be ecliptic. One atom or two are turned out of the plane by 50 pm. The plane C1′-O4′-C4′ is fixed in the ribofuranose.

The ribose ring is not aplanar and usually presents C2′-*endo* (South) or C3′-*endo* (North) conformation (Fig. 3.13). The C2′-*endo* state is the most prominent conformations in B-helix, and C3′-*endo* conformations in A-helix. *Endo*-pucker means that C2′ or C3′ are turned out of this plane into the direction of O5′. *Exo*-pucker describes a shift in the opposite direction. Both conformers (C2′- *endo* and C3′- *endo*) are in a dynamic equilibrium at room temperature. Transitions take place through the O4′-*endo* state. The C3′- *endo* conformation dominates in RNA. DNA is able to attain both the conformations. Several structures obtained from X-ray crystallography or NMR shows the presence of intermediates between C2′-*endo* and C3′-*endo* states. Ring pucker affects the torsional relationships of adjacent C-H bonds in the sugar ring and can be determined from spin coupling in NMR. Ring pucker also affects non-bonded H-H separations which can be measured by NOESY.

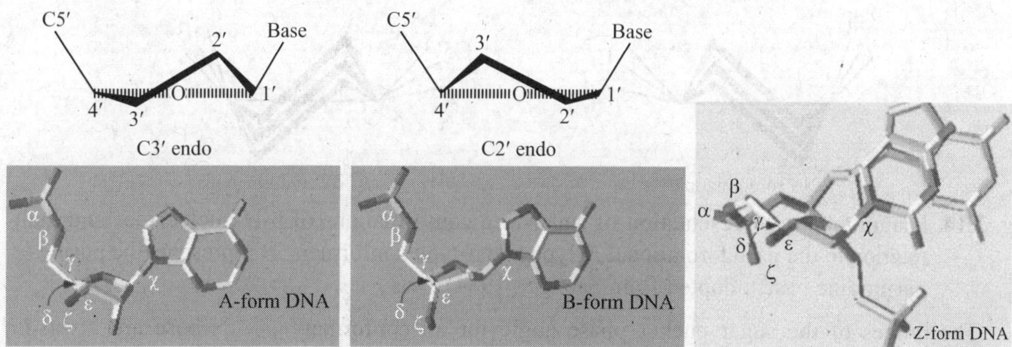

Fig. 3.13. Structures for the C2′-endo (B-form), C3′-endo (A-form) and Z-form conformations of ribose (adopted from Blackburn and Gait, 1996).

If four atoms of the pentose sugar are present in a plane, this plane is called a 'reference plane' and the conformation is called as 'envelope' (*E*). Moreover, if the atoms do not lie in a plane, the reference plane is that of the three atoms which are present closest to the five-atom; this is called 'least-squares plane'. Then the conformation is described as twist (*T*). Atoms lying on the side of the reference plane from which the numbering of the ring appears clockwise are written as superscripts and precede the letter (*E* or *T*).

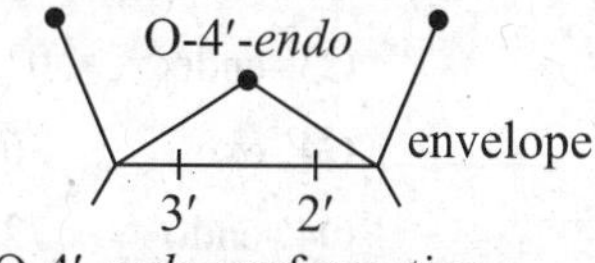

O-4′ *endo* conformation

Points to Remember

- The *E/T* notation has superseded the *endo/exo* description, in which atoms now designated by superscripts were called *endo*, and those now designated by subscripts were called *exo*. C3′-*endo*/C2′-*exo* has become $^3T_2$, and C3′-*endo* has become ^3E.
- Symmetrical twist conformations, in which both atoms exhibit equal displacements with respect to the five-atom plane are denoted by placing the superscript and subscript on the same side of the letter *T*, e.g. $_2^3T$, $_4^3T$, etc.

Atoms lying on the other side are written as subscripts and follow the letter (Fig. 3.14). Atoms on the same side of the plane as C5 in *D*-ribofuranose derivatives are written as preceding superscripts.

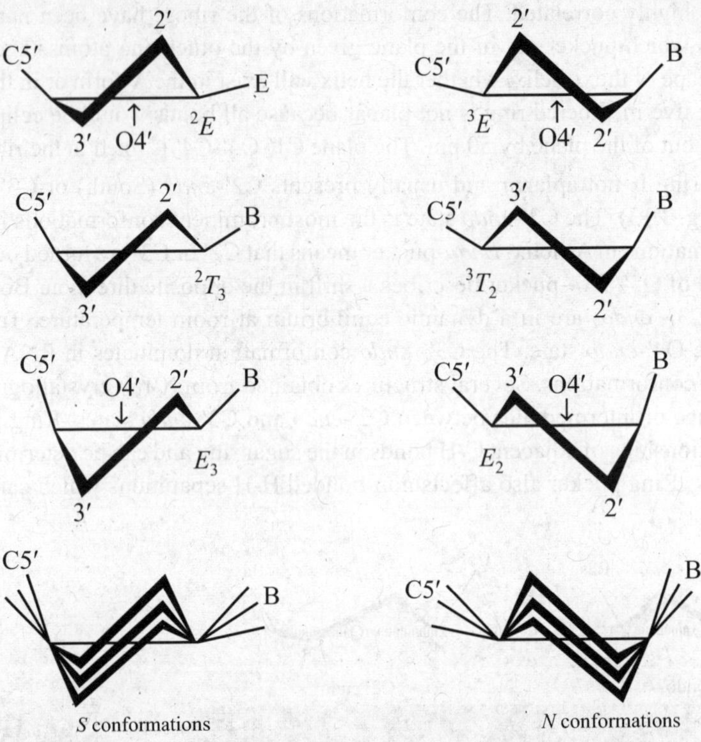

Fig. 3.14. Diagrammatic representation of sugar-ring conformations of β-D-nucleosides and their relation to the pseudorotational *N*-type and *S*-type conformers. B represents the purine or pyrimidine base (adopted from Saenger, 1984).

The ranges of the sugar pucker phase angle for all conformations of ribose are given in Table 3.2.

Table 3.2. Ranges of the sugar pucker phase angle for all conformations of ribose.

Name	Angle range		Occurrence
C2′-exo	–36° ... 0°	324° ... 360°	–36° ... 0°
C3′-endo	0° ... 36°	...	0° ... 36°
C4′-exo	36° ... 72°	...	36° ... 72°
O4′-endo	72° ... 108°	...	72° ... 108°
C1′-exo	108° ... 144°	...	108° ... 144°
C2′-endo	144° ... 180°	...	144° ... 180°
C3′-exo	180° ... 216°	–180° ... –144°	180° ... 216°
C4′-endo	216° ... 252°	–144° ... –108°	216° ... 252°
O4′-exo	252° ... 288°	–108° ... –72°	252° ... 288°
C1′-endo	288° ... 324°	–72° ... –36°	288° ... 324°
C2′-exo	324° ... 360°	–36° ... 0°	324° ... 360°

All the possible conformations of the ribose ring may be obtained with changing the pucker phase angle from 0° to 360°. Fig. 3.15 shows the pseudorotation cycle as a conformational wheel, revealing both systems of designation; for examples C3′-*endo*/C2′-*exo* has become $^3T_2$, and C3′-*endo* has become ^3E. Symmetrical twist conformations, in which both atoms exhibit equal displacements with respect to the five-atom plane' are denoted by placing the superscript and subscript on the same side of the letter *T*, e.g. $_2{}^3T$, $_4{}^3T$, etc. *N*-type conformations correspond to the northern half ($P = 0 \pm 90°$) and *S*-type correspond to the southern half of the pseudorotational cycle. The symbols 'r' and 'd' represent the usual range of *P* values for *N* and *S* conformations of ribo- (r) and 2′-deoxyribo- (d) furanose rings of β-D-nucleosides and nucleotides.

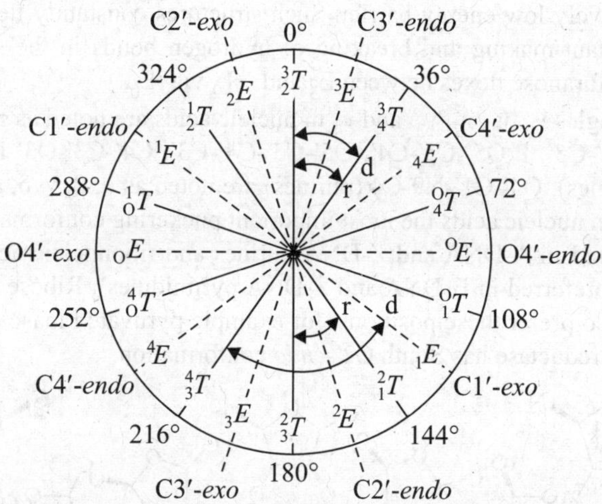

Fig. 3.15. The pseudorotational pathway of the D-aldofuranose ring showing the relation between phase angle of pseudorotation *P* (0-360°), the envelope (*E*) and twist (*T*) notations, and the *endo* and *exo* notations (adopted from C. Altona and M. Sundaralingam, 1972. *J. Am. Chem. Soc.*94: 8205).

(c) Furanose Pseudorotational Angles: The sugar ring conformation has also been described by Altona and Sundaralingam (1973) used the concept of pseudorotation, which has been found advantageous in describing the conformational dynamics of the sugar ring. Each conformation of the furanose ring can be described by two pseudorotational parameters: the *phase angle* of pseudorotation, *P*, and the *degree of pucker*, the t_m. A standard conformation ($P = 0°$) is defined with a maximally positive C1′-C2′-C3′-C4′ torsion angle (*i.e.* the symmetrical $_2{}^3T$ form), and *P* has value 0-360°. Conformations in the upper or northern half of the circle ($P = 0 \pm 90°$) are denoted *N* and those in the southern half of the circle ($P = 180 \pm 90°$) are denoted *S* conformation. The relationship between *P* and the *endo*/*exo* and *T*/*E* notations is given in Fig. 3.15. The symmetrical twist (*T*) conformations arise at even multiples of 18° of *P* and the symmetrical envelope (*E*) conformations arise at odd multiples of 18° of *P*.

Fig.3.16 shows the elaborated form of furanose pseudorotational angles. The example furan is α-*L*-arabinofuranose; *D*-ribose and *D*-deoxyribose are shown top left and right, respectively. The angles v_0, v_1, v_2, v_3, and v_4 are the torsional (dihedral, 3-bond) angles in the clockwise direction looking at the far ring bond along the central ring bond (labeled) relative to the near ring bond. The plus and minus signs show whether the atom lies above (plus) or below the plane formed by the other ring atoms. They denote the torsions C4′-O4′-C1′-C2′, O4′-C1′-C2′-C3′, C1′-C2′-C3′-C4′, C2′-C3′-C4′-C4′, C3′-C4′-O4′-C1′, respectively. The rings are in either the twist (T, relative to a plane formed by three adjacent ring-atoms, so chosen that the *exo*-planar atoms lie on opposite sides of the plane) or envelope (E, containing four adjacent coplanar atoms) conformation. IUPAC recommend that the 0° position is defined with a maximally positive v_2 torsion angle (*i.e.* the symmetrical $^3T_2$ form), shifting angles calculated as above back by 54°.

Alternatively the pucker types may be described as *endo* where the atoms are displaced on the same side as the C5′ or *exo* where the atoms are displaced on the opposite side as the C5′; for example, $^3T_2$ for nucleic acids is C2′-*exo*/C3'-*endo*. The pseudorotational angle, the furan structure and its energy are determined by the displacement of the out-of-plane atoms from the plane given by the remaining ring atoms. Different conformations of the 5-methylhydroxyl group generate different preferred furan structures. There are two minimal conformations in the energy diagram, one in the top half (north, *N*) and one in the bottom half (south, *S*). Both N and S are usually connected by a pathway with a relatively low energy barrier. Such structures constantly flex in aqueous solution, resulting in the constant-making and breaking of hydrogen bonds in the surrounding water; for example α-L-arabinofuranose flexes between E_3 and $^1T_2$ via E_0.

The torsional angles v_0, v_1, v_2, v_3, and v_4 in nucleic acids are noted as τ_0, τ_1, τ_2, τ_3, and τ_4; the sequences, O3′-P-O5′-C5′, P-O5′-C5′-C4′, O5′-C5′-C4′-C3′, C4′-C3′-O3′-P, C3′-O3-′P-O5′, O4′-C1′-N1-C2 (pyrimidines), O4′-C1′-N9-C4 (purines) are noted as α, β, γ, δ, ε, ξ and χ, respectively as described earlier. In nucleic acids the most important puckering conformations are close to north C3'-*endo* (^{3}E; preferred in A-DNA and Z-DNA purines and ribonucleic acid double helices) and south C2′-*endo* (^{2}E; preferred in B-DNA and Z-DNA pyrimidines). Ribose in ATP, ADP and FAD protein complexes also prefer these positions, for example pyruvate kinase possesses a North C3′-*endo* and glutathione reductase has South C2′-*endo* conformation.

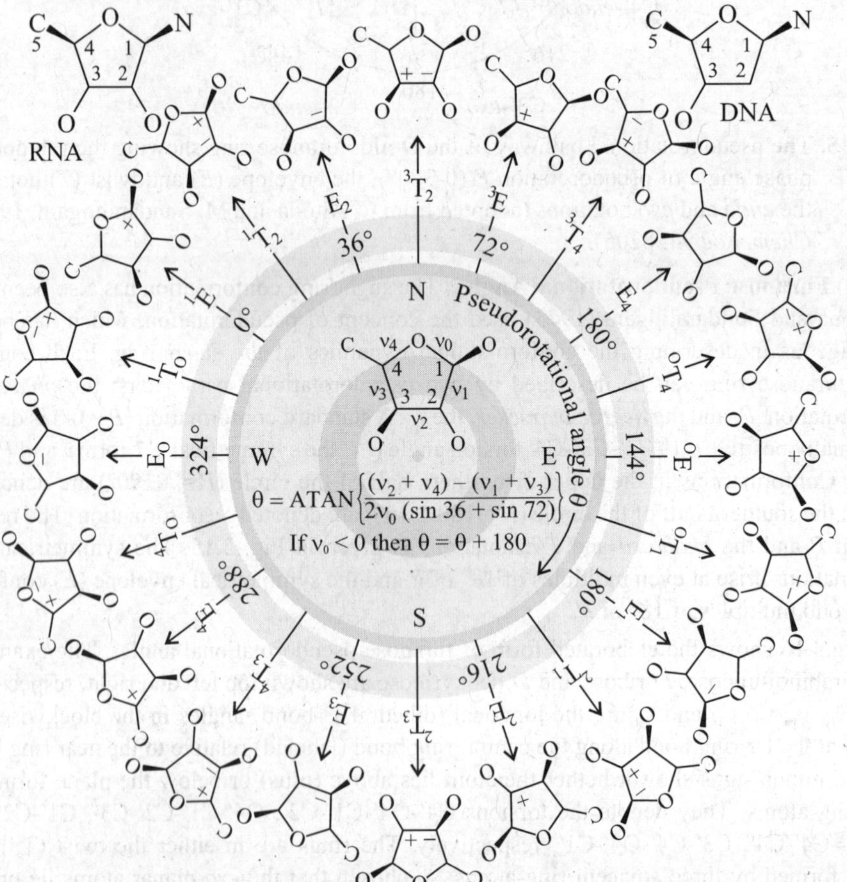

Fig. 3.16. Elaborated form of pseudorotational angles of α-L-arabinofuranose ring.

***(d) Anti-Syn* Orientation about *N*-Glycosidic Bond:** The term *anti* refers to anticlinal (*ac*) and *syn* refers to synclinal (*sc*). The torsion angle about the *N*-glycosidic bond (N-C1′) that links the base to the sugar is denoted by the symbol 'c' which is the same as the notation used to denote side-chain torsion angles in polypeptides. The sequence of atoms chosen to define this angle is O4′-C1′-N9-C4 for purine and O4′-C1′-N1-C2 for pyrimidine derivatives. Thus when chi (χ) = 0° the O4′-C1′ bond is eclipsed with the N9-C4 bond for purine and the N1-C2 bond for pyrimidine derivatives. The definitions of torsion angles of the *N*-glycosidic bond are illustrated, looking along the bond (Fig. 3.17).

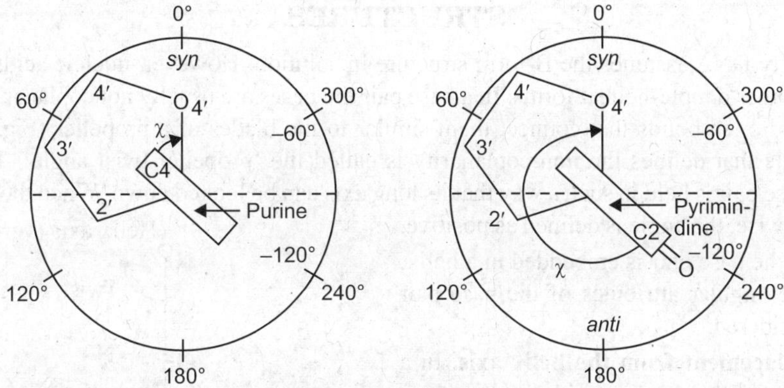

Fig. 3.17. The *N*-glycosidic bond torsion angle chi (χ) and, the *syn* and *anti* regions for purine and pyrimidine derivatives (diagrammatic). The purine derivative (left) is viewed down the N9-C1′ bond and is shown in the +*sc* conformation. The pyrimidine derivative (right) is viewed down the N1- C1' bond and is shown in the -*ac* conformation. The sugar ring is shown as a regular pentagon.

Usually chi (χ) torsion angle falls into the ranges of +90° to +180°, and –90° to –180° (or 90° to 270°) corresponding to the *anti* conformation. But occasionally χ has values between –90° and +90°, which are called *syn* conformation; for example in left-handed Z-DNA with CG repeating sequence, the purine G is in *syn* conformation, whereas the pyrimidine C is *anti*. Rotamers with *chi* values from +90° to +270° is called *anti*.

Fig. 3.17 shows *anti/syn*- orientations of a nucleotide. The χ-related *anti/syn* conformation is a very basic and simple concept. The *N*-glycosidic bond and the corresponding χ torsion angle shows that the base and sugar are two separate entities, *i.e.* there is an internal degree of freedom between them. The sugar edge for classification of base pairs corresponds to the *anti* form only. When a base is flipped over into the *syn* conformation, the sugar edge does not exist, which is in connection with the minor groove side of a nitrogenous base. In Watson-Crick double helix both the bases of a base pair are in *anti* conformation. Bases in *syn* conformation indicate a distortion of the double helix due to base pair opening or mismatched base pairs. The *syn* conformation is also found with guanines in left-handed Z-DNA helix. The glycosidic torsion angle *chi* is correlated with sugar puckering. High *anti* is often found with C2′-endo sugars (B-helix), while C3′-endo sugars (A-helix) are associated with more canonical *anti* values in the range from -150° to -180°.

Points to Remember

- The choice of bond sequence to define c is based on chemical nomenclature. At the same time the use of the terms *syn* and *anti* to describe different conformational regions of c for purine and pyrimidine derivatives is now consistent with accepted chemical nomenclature viz., *syn*, c = 0±90° and *anti*, c = 180±90°.

- In the new convention of IUPAC, most of the conformations formerly described as *syn* and *anti* remain *syn* and *anti,* respectively, except the *high-anti* region which may be described as -synclinal (*-sc*).
- Following the definitions of conformational regions the *anti* conformation of the pyrimidine corresponds to the -anticlinal (*-ac*) conformation and the *syn* conformation of the purine derivative corresponds to the +synclinal (*+sc*) conformation.

II. CONFORMATIONAL VARIATION IN DOUBLE HELICAL STRUCTURES

Normally DNA assumes the B-form structure in solution. However, nucleic acids also occur naturally in other double-helical forms. In a base pair the bases are usually not coplanar, but twisted about the hydrogen bonds that connect them similar to the blades of a propeller (Fig. 3.18). The dihedral angle that defines the non-coplanarity is called the 'propeller twist angle'. To view the angle, the base pair is held in such a way that its long axis can be looked down. When the nearer base rotates clockwise, the angle is defined as positive.

When the base pair is embedded in a helix, several more angular attributes of the base pair must be considered:

- **D-displacement from the helix axis.** In a double stranded nucleic acid, there exists a helix axis which is the average symmetry axes of the base pairs. The base pair 'slips' from this axis and this displacment from the axis is measured as distance from the helix axis.
- **Base pair tilt.** The base pair tilt is the rotation of the plane about the pseudo-dyad axis.

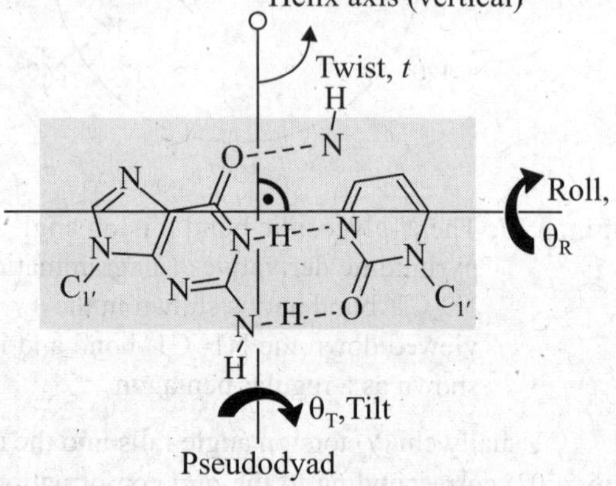

Fig. 3.18. Propeller twist in DNA.

- **Base pair roll.** This is the degree of departure of the mean plane of the base pairs from the perpendicular helix axis on the short axis of the base pairs.
- **Helix twist.** It is the orientation of a base pair with respect to the helix axis.

Propellor twist, base pair roll and displacement are extremely important components in maintaining the stacking interaction of DNA. The bases are coplanar in standard Watson Crick B-DNA, no propeller twist and the bases stack readily upon each other. In such DNA, the helical twist is 36°; it means that there are 10.0 base pair/1 turn of helix. But the stacking of the base pairs is changed upon varying any of these parameters. Thus the base pair tilt, roll, helical twist and propeller all dependent on twist, both from the influence of stacking interaction energies and the van der Waals imposed by different base pairs.

In other helical forms of the nucleic acids, the arrangement of base-pairing remains the same, but the sugar-phosphate groupings that constitute the backbone are inherently flexible and can adopt different conformations. The propeller twist is one of such conformational variations that allows greater overlap between successive bases along a strand of DNA and diminishes the area of contact between bases and solvent water (Fig. 3.19).

A rotation is caused due to successive base pairs with 32° in B-DNA with respect to each other (*i.e.* helical twist) of 36° or so, as observed in the cylindrical axis of the DNA (A). However, the propeller twist (rotation in a different dimension) as in B(2) allows a greater base overlapping within the same strand resulting in reduction of the contact area between the bases and water. Here you can see the edge on to two successive bases in one DNA strand. Dots represent end-on views down the glycosidic bonds. Clockwise rotation has a positive sign (B). The two bases on the left-hand strand of DNA also show positive propeller twist. This motion causes distortion of hydrogen bonds between two bases which remain intact (C).

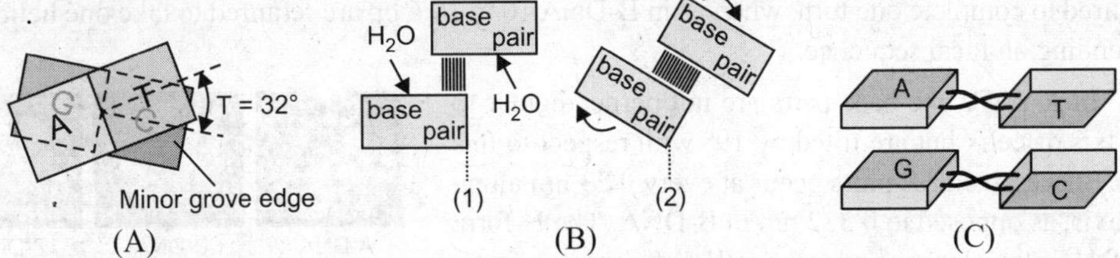

Fig. 3.19. Helical twist and propeller twist in DNA (based on C. R. Callandine and H. R. Drew, 1992. *Understanding DNA: The Molecule and How It Works.* London: Academic Press).

1. Alternative Form of Right-Handed DNA

The most common form of DNA which has right handed helix and proposed by Watson and Crick is called B-form of DNA or B-DNA. In addition, the DNA may be able to exist in other forms of double helical structure. These are A and C forms of double helix which vary from B-form in spacing between nucleotides and number of nucleotides per turn, rotation per base pair, vertical rise per base pair and helical diameter. Structure of B-form of DNA has been proposed by Watson and Crick. It is present in every cell at a very high relative humidity (92%) and low concentration of ions. It has antiparallel double helix, rotating clockwise (right hand) and made up of sugar-phosphate back bone combined with base pairs or purine-pyrimidine. The base pairs are perpendicular to longitudinal axis of the helix. The base pairs tilt to helix by 6.3°. The B-form of DNA is metabolically stable and undergoes changes to A, C, D or Z forms depending on sequence of nucleotides and concentration of excess salts. A comparison of the structural properties of only A-, B-, and Z- forms of DNA double helix is given in Table 3.3, and only A- and Z- forms have been discussed in detail.

Table 3.3. Comparison of the structural properties of A-DNA, B-DNA and Z-DNA of the double helix.

Properties	*Forms of DNA*		
	A-form	*B-form*	*Z-form*
Overall proportions	Short and broad	Longer and thinner	Elongated and slim
Helix sense	Right-handed	Right-handed	Left-handed
Repeating unit	1 bp	1 bp	2 bp
Rotation/bp	33.6°	35.9°	60°/2
Base pairs per turn of helix	10.7	10.0	12
Base-pair tilt from the perpendicular	+19°	-1.2°	-9°
Rise/bp along axis	2.3 Å	3.32 Å	3.8 Å
Helix packing diameter	25.5 Å	23.7 Å	18.4 Å
Pitch/turn of helix	2.46 nm	3.32 nm	4.56 nm
Mean propeller twist	+18°	+16°	0°
Glycosyl bond conformation	*anti*	*anti*	C, *anti*; G, *syn*
Sugar pucker	C3'-endo	C2'-endo	C: C2'-endo, G: C2'-exo
Helix axis location	Major groove	Through base pairs	Minor groove

bp = base pairs; (+), right handed, (–), left handed.

***(a)* A-DNA:** A-DNA molecule is an alternative form of the right-handed double helix, which differs in several ways from B-DNA; for example the pitch (or distance required to complete one helical turn) of B-DNA is 3.4 nm, whereas that of A-DNA is 2.46 nm. In A-DNA 11 bp are required to complete one turn, whereas in B-DNA 10 to 10.6 bp are required to take one helical turn depending on local sequence.

In A-DNA, the base pairs are not perpendicular to the axis of helix but are tilted by 19° with respect to this axis. Successive base pairs occur at every 0.23 nm along the axis, as opposed to 0.332 nm in B-DNA. The B-form of DNA is thus longer and thinner than the short A-form, which has its base pairs displaced around, rather than centered on the helix axis. Fig. 3.20 shows the structural characteristics of A-form, B-form and Z-form of DNA.

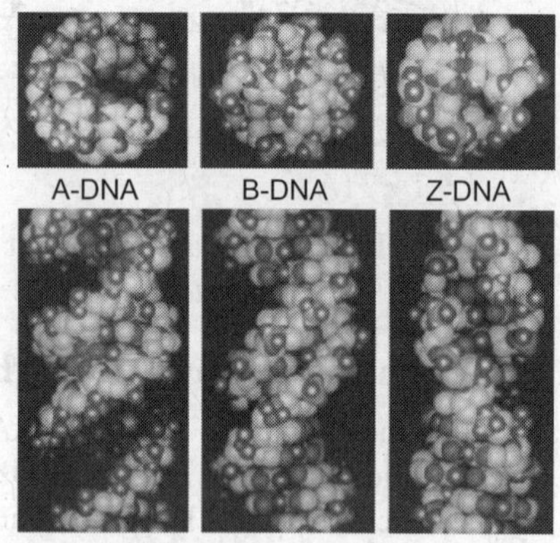

Fig. 3.20. Comparison of the A-DNA, B-DNA, and Z-DNA of the double helix.

In A-DNA the distance required to complete one helical turn is shorter than that in B-DNA. The alternating pyrimidine–purine sequence of Z-DNA is the key to the 'left-handedness' of this helix. The dehydrated DNA fibers adopt the A-conformation under physiological conditions, but it is unclear whether DNA ever assumes this form *in vivo* or not. Probably the double-helical DNA : RNA hybrids have an A-like conformation. In RNA the 2'-OH sterically prevents double-helical regions of RNA chains from adopting the B-form helical arrangement. Moreover, in RNA chains the double-stranded regions assume an A-like conformation with their bases that strongly tilt with respect to the axis of helix.

***(b)* Z-DNA (or Left-Handed Double Helix):** Alexander Rich and his colleagues at MIT first recognised the Z-DNA in X-ray analysis of the synthetic deoxynucleotide dCpGpCpGpCpG, which crystallised into an antiparallel double helix of unexpected conformation. The unusual properties of this oligonucleotide are due to the alternating pyrimidine-purine (Py-Pu) sequence. In this alternating copolymer, *N*-glycosyl bonds of G residues rotate 180° with respect to their conformation in B-DNA. Therefore, the purine ring is in the *syn* conformation rather than *anti* conformation (Fig. 3.20A). In B-DNA, the Cl'–N-9 glycosyl bond is always in the *anti* position due to flipping of the G ring. The C ring should also flip to maintain the normal Watson–Crick base pairing.

In contrast, this bond rotates to adopt the *syn* conformation in the left-handed Z-DNA. Pyrimidine nucleosides do not readily adopt the *syn* conformation because it interferes sterically between the pyrimidine C-2 oxy substituent and the pentose atoms. The cytosine ring does not rotate relative to the pentose; hence the whole C nucleoside (base and sugar) must flip 180°. The change in topological relationships of a six base-pairs from B-DNA to Z-DNA is shown in Fig. 3.21A-B. The purine rings (green) of the deoxyguanosine nucleosides rotate via an *anti* to *syn* change in the conformation of the guanine–deoxyribose glycosidic bond. The pyrimidine rings (blue) are rotated by 180° by flipping the entire deoxycytidine nucleoside (base and deoxyribose) without breaking, and re-forming the G : C hydrogen bonds. Consequently, the B to Z structural transition can take place without disruption of the bonding relationships among the atoms involved.

Dinucleotide is the repeating unit on a given strand in the Z-helix, *i.e.* *n*-1 dinucleotides must be considered for any number of bases (*n*) along one strand. For example, a sequence of GpCpGpC along one strand consists of three successive dinucleotide units: GpC, CpG, and GpC. In the Z-double helix the CpG sequence is conformationally distinct from the GpC sequence along the alternating copolymer chains. The conformational alterations going from B to Z realign the sugar-phosphate backbone along a zig-zag course having a left-handed orientation; therefore, it is called Z-DNA. In GpCpGp subset, the sugar-phosphates of GpC form the horizontal *zig* structure, while the CpG backbone segment forms the vertical *zag* structure. The mean rotation angle around the helix axis is -15° for a CpG step and -45° for a GpC step resulting in -60° for the dinucleotide repeat (here minus sign denotes a left-handed or counter-clockwise rotation about the axis of helix).

Methylation of C at the 5-position favours transition of B-form to Z-form, because in B-DNA these hydrophobic methyl groups would protrude into the aqueous environment of the major groove and destabilise its structure. The sequences that are not strictly alternating Py–Pu can attain Z-form. For example, the hexanucleotide ^{m5}CGAT^{m5}CG [a Py-Pu-Pu-Py-Py-Pu sequence containing two 5-methylcytosines (^{m5}C)], crystallises as Z-DNA. The same methyl groups can form a stabilising hydrophobic patch in Z-DNA. It is obvious that methylation is implicated in gene regulation; the occurrence of Z-DNA may affect the expression of genetic information.

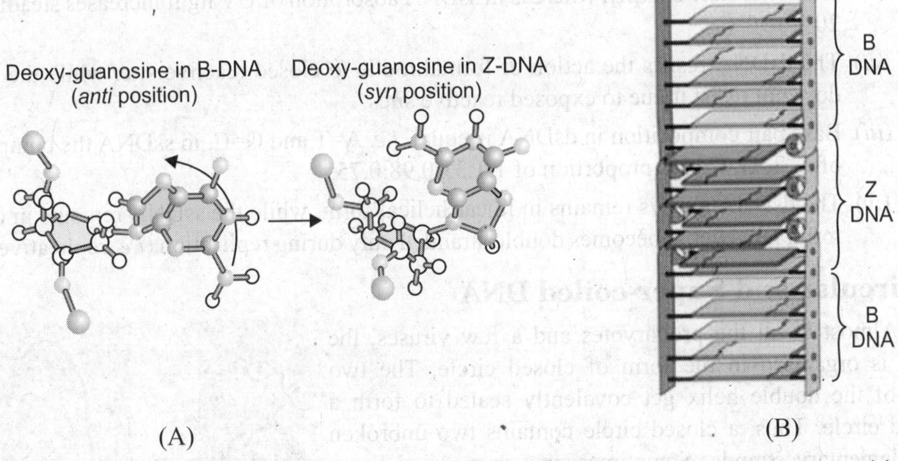

Fig. 3.21. Comparison of the deoxy-guanosine conformation between B-DNA (left, *anti* position) and that of Z-DNA (A); Change in topological relationships of base pairs from B- to Z-DNA.

Differences and Similarities with B-DNA: There are several similarities with B-DNA in having : (*i*) double helix, (*ii*) two antiparallel strands, and (*iii*) three hydrogen bonds between G-C pairing. In addition, the Z-DNA differs from the B-DNA in the following ways:

(*i*) The Z-DNA has left handed helix, while the B-DNA has right handed helix.

(*ii*) The Z-DNA contains zig-zag sugar phosphate back bone as compared to regular back bone of the B-DNA.

(*iii*) The repeating unit in Z-DNA is a dinucleotide due to alternating orientation of sugar residues, whereas in B-DNA the repeating unit is a mononucleotide, and sugar molecules do not have the alternating orientation.

(*iv*) In the Z-DNA one complete turn contains 12 base pairs of six repeating dinucleotide, while in B-DNA one full turn consists of 10 base pairs, *i.e.* the 10 repeating units.

(*v*) Due to the presence of high number (12) of base pairs in one turn of Z-DNA, the angle of twist per repeating unit, *i.e.* dinucleotide is 60° as compared to 36° of B-DNA molecule.

(*vi*) In Z-DNA the distance of twist making one turn of 360° is 45Å as against 34Å in B-DNA.

(*vii*) The Z-DNA has fewer diameters (18Å) as compared to the B-DNA (20Å diameter).

(c) C-Form of DNA (C-DNA): The C-form of DNA is found at 66% relative humidity in the presence of lithium (Lit$^+$) ions. As compared to A-and B-DNA, in C-DNA the number of base pairs per turn is less, *i.e.* 28/3 or 91/3. The base pairs show pronounced negative tilt by 7.8°.

(d) D-Form of DNA (D-DNA): The D-form of DNA is found rarely as extreme variants. Total number of base pairs per turn of helix is eight. Therefore, it shows eight-fold symmetry. This form is also called poly (dA-dT) and poly (dG-dC) form. There is pronounced negative tilt of base pairs by 16.7° as compared to C form, *i.e.* the base pairs are displaced backwardly with respect to the axis of DNA helix.

(f) Single Stranded (ss) DNA: Almost all the organisms contain double stranded DNA except a few viruses such as bacteriophage ϕX174 which consists of single stranded circular DNA. It becomes double stranded only at the time of replication. The differences between ssDNA and dsDNA are given as below:

(*i*) The dsDNA absorbs wavelength 2600 Å of ultra violet light constantly from 0 to 80°C, thereafter rises sharply, whereas in ssDNA absorption of UV light increases steadily from 20° to 90°C.

(*ii*) The dsDNA resists the action of formalin due to closed reactive site, while the ssDNA does not resist it due to exposed reactive sites.

(*iii*) Base pair composition in dsDNA is equal, *i.e.* A=T and G=C, in ssDNA the composition of A, T, G, C is in proportion of 1:1.33:0.98:0.75.

(*iv*) The dsDNA always remains in linear helical form, while the ssDNA remains in circular form; however, it becomes double stranded only during replication (*i.e.* replicative form).

2. Circular and Super-coiled DNA

Almost in all the prokaryotes and a few viruses, the DNA is organised in the form of closed circle. The two ends of the double helix get covalently sealed to form a closed circle. Thus, a closed circle contains two unbroken complementary strands. Sometimes one or more nicks or breaks may be present on one or both strands, for example DNA of phage PM2.

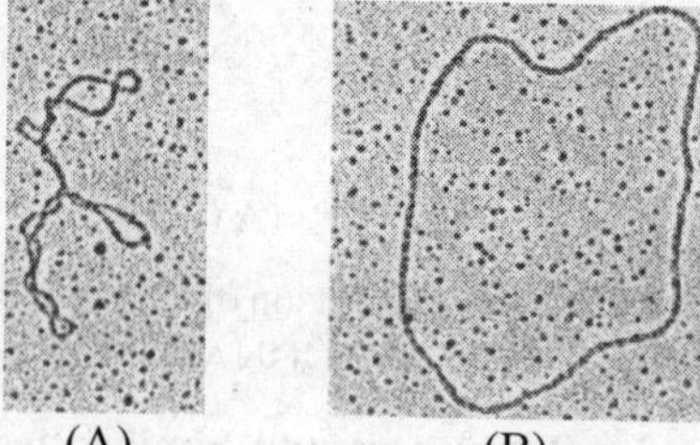

Fig. 3.22. A circular DNA of bacteria; A-super coiled state; B-relaxed state.

Besides some exceptions, the covalently closed circles of DNA (Fig. 3.22A-B) is twisted into super helix or super coils and associated with basic proteins but not with histones found complexed with all eukaryotic DNA.

The two ends of a linear DNA helix can be joined to form each strand continuously. However, if one of ends rotates at 360° with respect to the other to produce some unwinding of the double helix, the ends are joined resulting in formation of a twisted circle in opposite sense, *i.e.* opposite to unwinding direction.

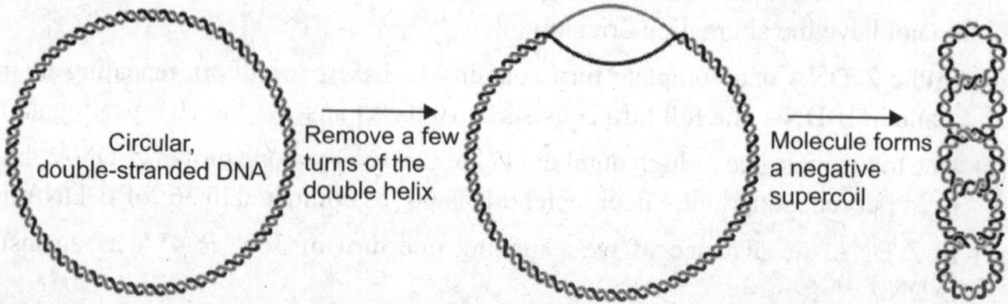

Fig. 3.23. A closed circular bacterial double stranded DNA twisted to form super coil.

Such twisted circle appears as 8, *i.e.* it has one node or crossing over point. If it is twisted at 720° before joining, the resulting super helix will contain two nodes (Fig. 3.23).

3. Palindromic DNA (Palindromes)

A variety of base sequences have been observed in the DNA molecule. Most of them do not have special features. The repeated sequences are of particular interest because they are the site of enzymatic activity and, sometimes give special features to the nucleic acid. For the first time in 1974, D.A. Wilson and C.A. Thomas of the Department of Biological Chemistry (Harvard Medical School Boston, U.S.A.) used the term palindromic DNA. The palindromic DNA or palindromes are the inverted repeats and region of dyad symmetry. The length of palindromes may be short by about 3-10 bases or long by about 50-100 base pairs. As compared to prokaryotic DNA, the eukaryotic DNA contains a large palindrome of about several thousand base pairs. Sometimes a spacer separates the two inverted repeats. The DNA molecules containing palindromes and inverted repeats may exist in alternative forms. After separation of complementary strands, the intramolecular base pairing may result in a double stranded stretch formed between adjacent complementary sequences. This is known as cruciform structure (Fig. 3.24). The cruciform structures have been produced in laboratory but not detected in the DNA isolated from cells.

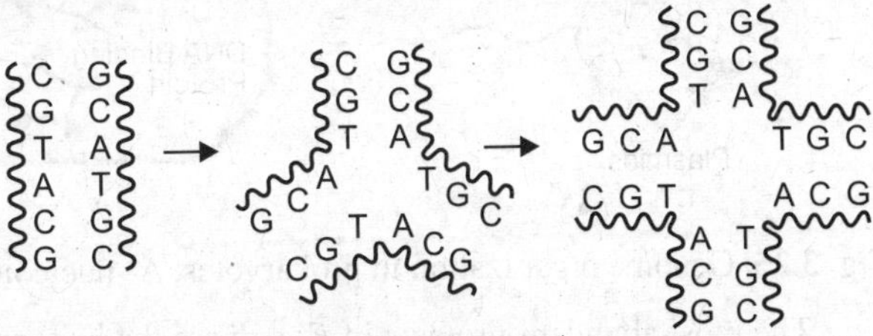

Fig. 3.24. The possible alternative forms of a DNA molecule containing two inverted repeats.

Moreover, if repeats are present in the same orientation, it is called direct repeats. They neither exist in alternative form of dsDNA nor have effect on the single advanced molecule. In addition, if both the palindromes and inverted repeats are present they affect the ssDNA or RNA. They form intra-strand hydrogen bonding between adjacent or nearly complementary sequences. Consquently a palindrome produces an intrastrand double stranded structure which is known as hairpin. The interrupted inverted repeats result in a structure consisting of a double stranded segment with a terminal single-stranded loop. This structure is called stem and loop.

Several possible functions of palindrome have been suggested such as : (*i*) its action as recognition sites of DNA for protein and many bacterial restriction enzymes, (*ii*) its action as bacterial restriction enzyme against destruction by foreign DNA, (*iii*) providing structural strength to the transcribed RNA by hydrogen bonding in the hairpin loops, (*iv*) possible involvement of cruciform structure in genetic recombination, and (*v*) its possession of genes on long palindrome of some lower eukaryotes that code for ribosomal RNA.

C. GENOME ORGANISATION

Genome is haploid set of DNA present in all organisms, including viruses. Genome organisation differs in prokaryotes and eukaryotes.

1. Genome Organisation in Prokaryotes

In prokaryotes The DNA is packaged by DNA-binding proteins. The bacterial DNA is packaged in loops back and forth. The bundled DNA is called the nucleoid (Fig. 3.25). It concentrates the DNA in part of the cell, but it is not separated by a nuclear membrane (as in eukaryotes.) The DNA does form loops back and forth to a protein core attached to the cell wall.

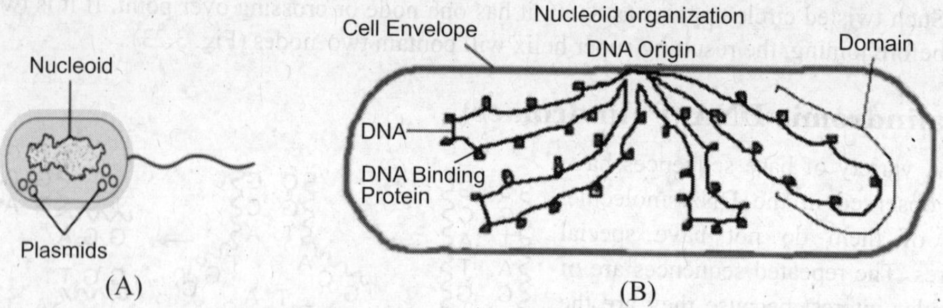

Fig. 3.25. Genome organization in prokaryotes: A- nucleoid and plasmids; B- nucleoid organization.

The most abundant proteins in *E. coli* are the histone-like proteins. There are four main types of histone-like proteins that have been well characterised as shown in Table 3.4. Although the amount of histone-like proteins vary depending on the conditions of the experiment. There may be about 1,00,000 copies of each of these proteins in the bacterial cell. In *E. coli* about 30 to 40% of the total proteins are bound with DNA. In both bacteria and eukaryotes, the DNA is compacted about 1000X, and the isolated chromatin contains roughly equal mass of DNA: protein.

This histone like proteins appear to help the organisation of bacterial DNA into a coiled chromatin structure with the result of nucleosome like structure, folding and super coiling of DNA, and association of DNA polymerase with nucleoids. Several histone-like DNA binding proteins have been described in bacteria (Table 3.4). These nucleoid-associated proteins include HU proteins, IHF, protein H1, Fir A, H-NS and Fis. In archaeobacteria (such as *Archaea*) the chromosomal DNA exists in protein-associated form. Histones like proteins have been isolated from nucleoprotein complexes in *Thermoplasma acidophilum* and *Halobacterium salinarum*. In 1987, Karl Drlica and J. Rouviere-Yuniv have reviewed the the histone-like proteins and the protein associated DNA and nucleosome like structures in a variety of bacteria

If the helix coils clockwise from the axis the coiling is termed as positive or right handed coiling. In contrast, if the path of coiling is anticlockwise, the coil is called left handed or negative coil. The enzyme topoisomerases alter the topological form *i.e.* super coiling of a circular DNA molecule. Type I topoisomerases (*e.g. E. coli* Top A) relax the negatively super coiled DNA by breaking one of the phosphodiester bonds in dsDNA allowing the 3′-OH end to swivel around the 5′-phosphoryl end, and then resealing the nicked phosphodiester backbone. Type II topoisomerases need energy to unwind the DNA molecules resulting in the introduction of super coils. One of type II isomerases, the DNA gyrase, is apparently responsible for the negatively super coiled state of the bacterial chromosome. Super coiling is essential for efficient replication and transcription of prokaryotic DNA. The bacterial chromosome is believed to contain about 50 negatively super coiled loops or domains. Each domain represents a separate topological unit, the boundaries of which may be defined by the sites on DNA that limit its rotation.

Table 3.4. Histones-like proteins of *E. coli* (after Moat and Foster, 1995).

Name	*Subunit Weight*	*Molecular Weight*	*Function*
Hu (hupA)	Hu(a)[HLPIIa]	9,000	Can form nucleosome like structure like H2B
(hupB)	Hu(b)[HLPIIb] linked as heterogenous dimer	9,000	—
HLP 1	—	17,000	Affects RNA polymerase transcription
H1(hns)	—	15,000	May modulate *in vivo* transcription
H	—	28,000	Similar to eukaryotic H2A

2. Genome Organisation in Eukaryotes

Depending on sequence DNA adopts many conformations. The flexibility of the DNA molecules is also affected by sequence; for example A : T sequences are more flexible than G : C sequences. On the basis of structure and flexibility to DNA (not by contacting DNA in sequence specific manner), some proteins bind the DNA.

In eukaryotes the DNA helix is highly organised into the well defined DNA-protein complex. In fact chromosomal DNA is packaged into a compact structure with the help of specialised proteins called histones. The complex DNA plus histones in eucaryotic cells is called chromatin. Among the proteins the most prominent are the histones that can be separated by size in SDS gel. The histones are small and basic proteins rich in amino acids such as lysine and/or arginine. Almost in all eukaryotic cells there are five types of histones, for example H_1, H_2A, H_2B, H_3 and H_4. Histone proteins are abundant in all eukaryotes and are equal in mass to DNA, and evolutionarily conserved. The non-histones proteins are also associated with DNA which differs markedly in type and structure. Eight histone molecules (two each of H_2A, H_2B, H_3 and H_4) form an octamer ellipsoidal structure of about 11 nm long and 6.5-7 nm in diameter which is known as a nucleosome (Fig. 3.26A).

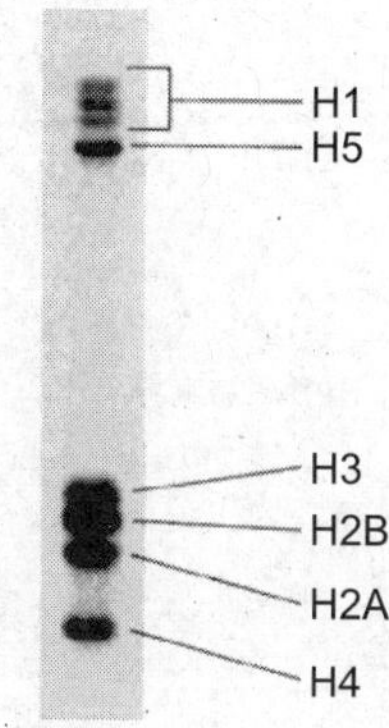

SDS gel of histone proteins

The histones have positively charged (basic) amino acids to bind the negatively charged (acidic) DNA. DNA coils around the surface of ellipsoidal structure of histones, 166 base pairs (about 7/4 turns) before proceeding onto the next and form the nucleosome (Fig. 3.26B). Thus a nucleosome is an octamer of four histone proteins complexed with DNA. About 200 base pairs (bp) of DNA coil around one histone. The coil 'untwists' so as to generate one negative superturn per nucleosome.

The histones play an important role in determining the eukaryotic chromosomes through their conformation. The nucleosomes are the repeating units of DNA which are often termed as beads. The DNA isolated from chromatin looks like string or beads. The stretch of DNA between the nucleosomes is known as 'linker' which varies in length from 14 to over 100 base pairs. The H_1 is associated with the linker region and helps the folding of DNA into complex structure called chromatin fibres which in turn get coiled to form chromatin.

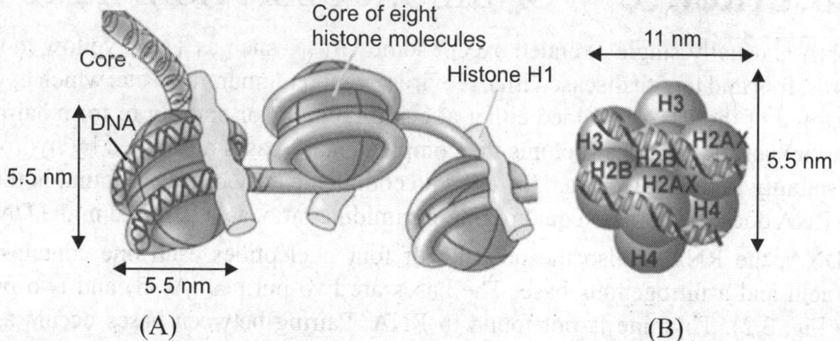

Fig. 3.26. Internal organisation of nucleosomes. A-highly super coiled DNA; B-a single nucleosome.

A second histone (H1) fastens the DNA to the nucleosome core. The total mass of this complex is about 100,000 Daltons. Nucleosomes are packed together with the help of a histone (H1) to form a 30 nm large chromatin fiber. The histone is packed into larger structures called chromatin, condensation and decondensation of which allows or denies proteins involved in transcription and replication of DNA. As a 30 nm chromatin fiber, the typical human chromosome would be about 0.1cm in length and would span the nucleus 100 times (Fig. 3.27).

The extended section of chromosome forms looped domains of 300 nm, and condensed section of chromosome appears 700 nm wide resulting in highly condensed form which becomes visible as chromosome at metaphase of cell division. This suggests higher orders of packaging to give a chromosome the compact structure seen in a typical metaphase cell (Fig. 3.27).

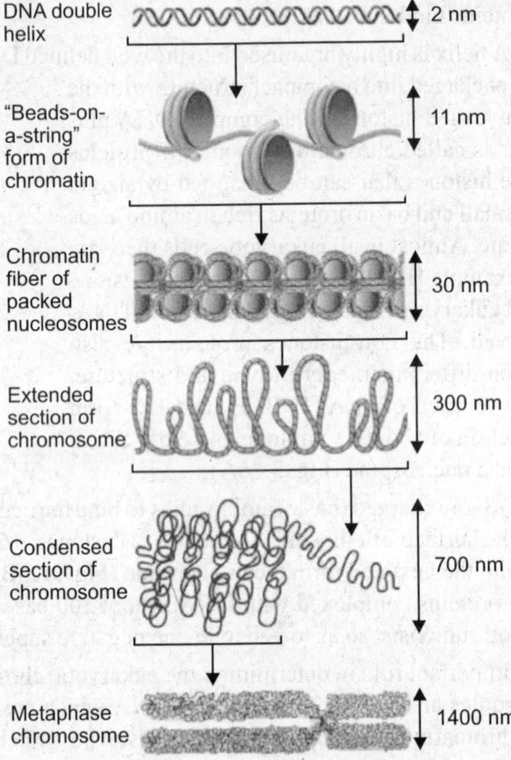

Fig. 3.27. Packaging of DNA into chromosome.

D. STRUCTURE OF RIBONUCLEIC ACID (RNA)

The RNA is usually single stranded except some viruses such as TMV, yellow mosaic virus, influenza virus, foot and mouth disease virus, reovirus, wound tumour virus, etc. which have dsRNA. The single strand of the RNA is folded either at certain regions or entirely to form hairpin-shaped structure. In the hairpin shaped structures the complementary bases are linked by hydrogen bonds which give stability to the molecules. However, no complementary bases are found in the unfolded region. The RNA does not possess equal purine-pyrimidine ratio, as it is found in the DNA.

Like DNA, the RNA is also the polymer of four nucleotides each one contains D-ribose, phosphoric acid and a nitrogenous base. The bases are two purines (A, G) and two pyrimidines (C, U) (*see* Fig. 3.2). Thymine is not found in RNA. Pairing between bases occurs as A-U and G-C. The nucleotides formed by the four bases are adenosine monophosphate (AMP), guanosine monophosphate (GMP), cytosine monophosphate (CMP), and uridine monophosphate (UMP). These are found freely in nucleoplasm but in the form of triphosphates *e.g.* ATP, GTP, UTP and CTP. As a result of polymerization the ribonucleotides form a polynucleotide chain of RNA.

If the RNA is involved in genetic mechanism, it is called genetic RNA as found in plant, animal and bacterial viruses. However, DNA acts as genetic material and RNA follows the order of DNA. In such cells the RNA does not have genetic role, therefore, it is called non-genetic RNA. The non-genetic RNA is of three types: (*i*) ribosomal RNA (rRNA), (*ii*) transfer RNA (tRNA) or soluble

RNA (sRNA), and (*iii*) messenger RNA (mRNA) or template RNA. These three types of RNA differ from each other in structure, site of synthesis in eukaryotic cell and function.

The mRNA and tRNA are synthesised on DNA template, whereas rRNA is delivered from nucleolar DNA. These RNAs are synthesised during different stages. At the time of cleavage most of the mRNAs are synthesised, whereas tRNA is synthesised at the end of cleavage. Synthesis of rRNA occurs during gastrulation. The total population of rRNA is about 90% of all RNAs.

Like DNA, the RNA is not self–replicating but it has to depend on DNA. Therefore, replication of non-genetic RNA is known as DNA-dependent RNA replication. Moreover, the genetic RNA of viruses is self-replicating, *i.e.* it can form its own several replica copies. Differences between the DNA and RNA molecules are given in Table 3.5.

Table 3.5. Differences between RNA and DNA.

RNA	*DNA*
1. RNA is more primitive than DNA	1. DNA originated after RNA
2. RNA is the genetic material of some plant, animal and bacterial viruses.	2. DNA is the genetic material of almost all living organisms
3. Except some viruses (e.g. reovirus) most cellular RNA is single stranded.	3. Except a few (*e.g.* fX174) most of DNA viruses is double stranded.
4. Pentose sugar is ribose.	4. Pentose sugar is deoxyribose.
5. The bases are adenine, guanine, cytosine and uracil.	5. The bases are adenine, guanine, cytosine and thymine.
6. Base pairing occurs between adenine and uracil (A-U)	6. Base pairs are A-T and G-C.
7. Base pairing is seen only in hairpin structure and helical region.	7. Base pairing occurs throughout the length of DNA molecule.
8. RNA contains a few (about 12,000) nucleotides	8. DNA contains millions of nucleotides *e.g.* over 4 millions.
9. The RNA molecules are of three types : rRNA, mRNA, tRNA.	9. DNA is only of one type.
10. The mRNA is found in nucleolus, and tRNA and rRNA are found in cytoplasm. They are formed on the DNA.	10. DNA is found in chromosomes. However, DNA is also found in mitochondria and chloroplasts.
11. RNAs translate the message of DNA into proteins.	11. DNA encodes the genetic messages in the form of mRNA transcripts.
12. Genetic RNA uses the enzyme, reverse transcriptase during replication.	12. This enzyme is not required by DNA. DNA after replication forms DNA and after transcription forms RNA.

There are a number of different types of RNA which play different roles in the cell; but out of many RNAs, only three RNAs (for example rRNA, mRNA and tRNA) are described here with.

Types of RNA found in a Cell

There are a number of different types of RNA which play different roles in the cell:

- **mRNA:** It encodes proteins.
- **rRNA:** It forms the ribosome, including the active site for peptide bond formation.
- **tRNA:** It acts as adaptor which binds amino acids and make link between mRNA and amino acid sequence of protein.

- **snRNA:** It small nuclear RNA forms snRNPs, which process mRNA by removing introns.
- **snoRNA:** The small nucleolar RNA forms snoRNPs which process rRNA mostly by methylation and isomerisation.
- **siRNA:** It is a small interfering RNA involved in gene silencing and regulation.
- **gRNA:** It is guide RNA needed for RNA editing, the removal and insertion of bases into mRNA.
- **tmRNA:** An RNA molecule that disengages ribosome from stalled translation of mRNA in bacteria.
- **Telomerase RNA**: It is an RNA molecule that forms much of the structure and all of the template required by telomerase.
- **hnRNA:** It is rag-bag of unprocessed pre-mRNA transcripts and other heterogeneous nuclear RNAs of less well defined function.
- **Catalytic RNA:** It acts as ribozyme.

1. Ribosomal RNA (rRNA)

The non-genetic RNAs are synthesised on the DNA template and are present in the nucleolus and cytoplasm. The rRNAs are found in ribosomes present in both prokaryotes and eukaryotes. The prokaryotic ribosomes are 70S, whereas eukaryotic ribosomes are 80S. Besides, the rRNA found in mitchondria of vertebrates are 55S. The 70S ribosomes of prokaryotes are made up of two subunits, 50S and 30S. The 50S subunit contains 23S and 5S rRNA, whereas the 30S subunit consists of 16S rRNA. Composition of prokaryotic ribosomes is given in Fig. 3.28.

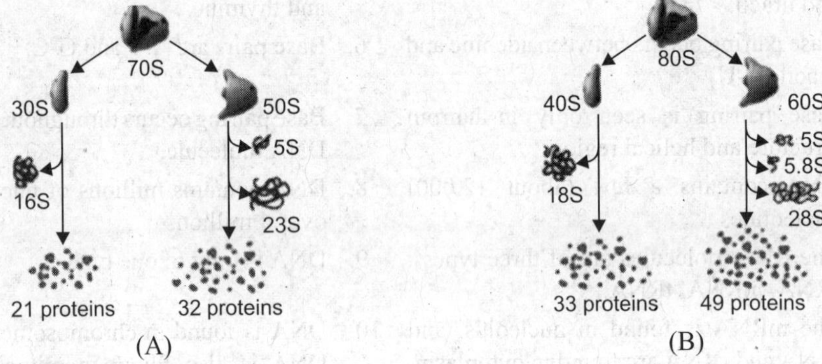

Fig. 3.28. Composition of prokaryotic ribosome (A) and eukaryotic ribosome (B), and proteins of different molecular weight.

The 80S ribosomes consist of 60S and 40S subunits. The rRNA types in both the subunits of plants differ from that of animals (Table 3.6). The base sequences of rRNA and part of DNA where they are synthesised are complementary. In prokaryotes rRNA is formed on a part of DNA called ribosomal DNA, while in eukaryotes these are formed in nucleolus containing the nuclear DNA. The rRNAs are found in ribosomes and accounts for 40-60% of dry weight. In general, it represents about 80% of total RNA of the cell.

Table 3.6. Subunits of rRNA of prokaryotes and eukaryotes.

Source	*Ribosomes*	*Larger subunit*		*Smaller subunit*	
		Size	*RNA*	*Size*	*RNA*
Prokaryotes Eukaryotes	70S	50S	23S and 5S	30S	16S
(*i*) Plants and invertebrates	80S	60S	25S, 5S, 5.8S	40S	16-18S
(*ii*) Vertebrates	80S	60S	28S, 29S, 5S	40S	18S
(*iii*) Vertebrate mitochondria	55S	40S	16-17S, 5S	30S	12-13S

The rRNA is a single stranded molecule which is twisted at certain points to form helical regions. In the helical region most of the base pairs are complementary and linked by hydrogen bonds. But the uncoiled single stranded regions lack the complementary bases (Fig. 3.29). Hence, in rRNA the ratio of purine : pyrimidine is not equal. The rRNA exists in a living cell for about two generations.

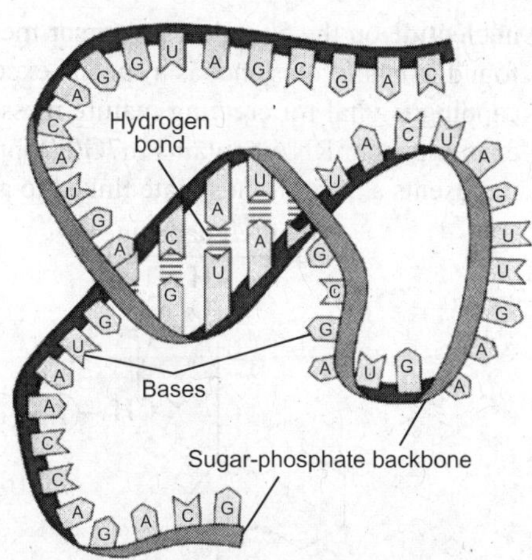

Fig. 3.29. Structure of rRNA.

2. Messenger RNA (mRNA)

The mRNA is transcribed on the DNA template and, therefore, carries the genetic information of DNA. For the first time in 1961, Francis Jacob and Jacques Monod proposed the name mRNA for bearing the transcripts of DNA for protein synthesis on ribosomes. The total population of mRNA in a cell varies from 5 to 10% of the total cellular RNA because the species of mRNA are short lived as these are broken into ribonucleotides by the enzyme ribonuclease. In *E. coli* some of the mRNAs remain alive only for about two minutes. Therefore, the cell does not contain high amount of mRNA at a time. In contrast, the mRNAs of eukaryotes are metabolically stable. The size of mRNA varies. The smallest protein contains about 50 amino acids (50×3=150 nucleotides needed for monocistronic mRNA molecules). Typically protein has 300-600 amino acids (900–1,800 nucleotide long mRNA). In prokaryotes the polycistronic mRNA is more common than monocistronic mRNA and contains 3,000-8,000 nucleotides. Polycistronic mRNA contains usually 10 bases long inter-cistronic sequences called spacers.

The sedimentation coefficient of mRNA is 8S and average molecular weight ranges from 500,000 to 1,00,000. Since they represent a gene, their length and molecular weight change because a gene contains 100 to 1,500 nucleotides. The mRNAs are transcribed by genes; hence individual mRNA represents a single gene. Therefore, in a cell there will be as much mRNAs as genes, and every mRNA differs from each other. Initiation of synthesis of first polypeptide chain of a polycistronic mRNA may begin hundreds of nucleotides from the 5' end. The section of non-translated RNA before coding region is called leader. Untranslated sequences are usually formed at both 5' and 3' ends.

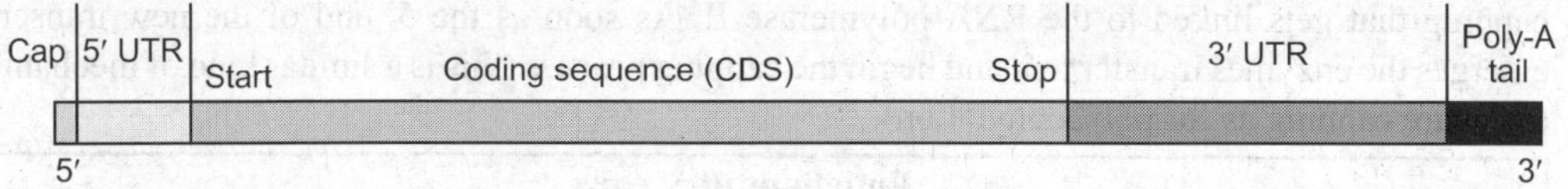

Fig. 3.30. Structure of a mature eukaryotic mRNA consisting of a 5′ cap, 5′ UTR (untranslated region), coding region, 3′ UTR, and poly(A) tail.

As the mRNAs always remain in single stranded form, it may disrupt the biological activity after being coiled. However, the coils lack complementary bases. In 1991, M. Kozak has discussed the structural features in eukaryotic mRNAs that modulate the initiation of translation. The sequence was named after its discoverer, Marilyn Kozak as 'Kozak consensus' or 'Kozak sequence' [(gcc) gccRccAUGG, where R is a purine (A or G) three bases upstream of the start codon (AUG), which is followed by another 'G'].The structure of a mature eukaryotic mRNA is shown in Fig 3.30, and discussed below:

***(a)* The 5' Cap:** In most of the eukaryotes and animal viruses, 5' end of mRNA contains a cap which is formed after methylation of any of four nucleotides. The 5′ cap is a specially altered nucleotide on the 5′ end of precursor messenger RNA and some other primary RNA transcripts as found in eukaryotes and, as a special exception, caliciviruses such as noroviruses. The process of 5′ capping is vital for creating mature messenger RNA which is then able to undergo translation. For example, an mRNA contains m7G(5′)ppp(5′)N where m7G is the methylguanosine and (5′)ppp(5′) represents a 5′-5′ triphosphate linked to a base (N) at 5′ end of an mRNA molecule (Fig. 3.31).

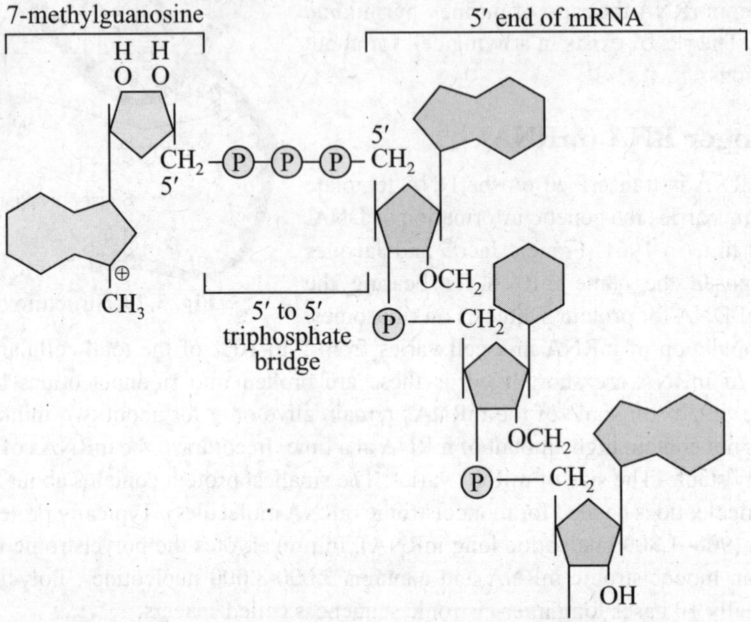

Fig. 3.31. A 7-methylguanosin [m7G (5′) ppp(5′)] cap at 5′ end of mRNA.

The mRNA binds to ribosome with the help of this cap. Therefore, it governs protein synthesis. The bacterial mRNA does not contain 5′ cap; instead they contain a specific (about six) nucleotide long ribosome-binding site, which occurs at several places in the mRNA molecules. These are located at four nucleotides upstream from AUC. In bacterial mRNA there may be multiple ribosome binding sites called Shine- Dalgarno sequences in the interior of an mRNA chain, each resulting in synthesis of a different protein.

Before the start of transcription the nuclear 'capping enzyme complex' (CEC) is required for capping that gets linked to the RNA polymerase II. As soon as the 5' end of the new transcript emerges the enzymes transfer to it and begin the capping process (this is a similar kind of mechanism to ensure capping as for polyadenylation).

Functions of 5′ cap

There are four main functions of the 5' cap: (*i*) regulation of nuclear export, (*ii*) prevention of degradation by exonucleases, (*iii*) promotion of translation, and (*iv*) promotion of 5' proximal intron excision.

(*b*) Non-coding Regions: There are two non-coding regions (NC1) or untranslated region (UTR) first followed by the cap and the second followed by the termination codon. The non-coding region is about 10-100 nucleotides long and rich in A and G residues, whereas the NC2 is 50-150 nucleotides long and contains AAUAAA residues. Both the non-coding regions do not translate protein.

(*c*) Initiation Codon: Both in prokaryotes and eukaryotes the initiation codon (AUG) is present which starts protein synthesis. Bacterial ribosomes, unlike the eukaryotic ribosomes, directly bind to start codons in the interior of mRNA to initiate protein synthesis.

***(d)* Coding Region:** It is the most important region of mRNA which is about 1,500 nucleotides long. This region translates a long chain of protein after attaching with several ribosomes. The combination of mRNA strand with several ribosomes is called polyribosome. Therefore, the bacterial mRNAs are commonly called polycistronic mRNA, *i.e.* they encode multiple proteins that are separately translated from the same mRNA molecule. The eukaryotic mRNAs are typically monocistronic, *i.e.* only one species of polypeptide chain is translated per mRNA molecule.

***(e)* Termination Codons:** The termination codon is required to give the signal to stop protein synthesis. In eukaryotes the termination codons are : UAA, UAG or UGA that terminates the translation process, i.e. the process of protein synthesis.

***(f)* Poly (A) Sequence:** The UTR or NC_2 is followed by poly (A) sequence in the eukaryotic mRNA. The prokaryotic mRNAs lack poly (A). The polyadenylate or poly (A) sequences of 200-250 nucleotides are present at 3′-OH end of mRNA. Poly (A) sequence is added to 3′-OH end of mRNA when it is present inside the nucleus (Fig. 3.32). The function of poly (A) sequence in translation is unknown.

Differences between Prokaryotic and Eukaryotic mRNA

The mRNAs found in prokaryotes differ from that of eukaryotes in the following ways:

- The prokaryotic mRNAs are broken down by the ribonuclease into ribonucleotide; hence these are very short lived, for example 2 minutes in *E. coli*. The short lived nature is due to changes in environmental conditions. In contrast the eukaryotic mRNAs have longer life, *i.e.* they are metabolically more stable than the prokaryotic mRNAs.
- In most of the prokaryotic mRNAs translation starts soon when the mRNAs are being transcribed on DNA template, whereas in eukaryotic mRNAs the translation process begins (in cytoplasm) when transcription process is over (in nucleus).
- Most of the mRNAs of prokaryotes and bacteriophages are polycistronic, *i.e.* contain several sites for initiation and termination of polypeptides due to the presence of several structural genes. Therefore, an mRNA molecule synthesises more than one polypeptide chains. On the otherhand, the mRNAs of eukaryotes are monocistronic, *i.e.* contain only a single initiation and termination codons and synthesise one chain only.
- A very little processing in mRNAs of prokaryotes occurs due to short time between transcription and translation. The mRNAs of eukaryotes undergo a considerable processing after being transcribed. The mRNA processing includes: (*i*) polyadenylation (addition of 200-250 poly (A) at 3'OH end), (*ii*) capping (formation of a 5' cap at 5' end by condensation of guanosine residues, and (*iii*) methylation (addition of methyl, *i.e.* $-CH_3$ group to some of nucleotides).
- Similarly, the mRNAs of prokaryotes do not contain poly (A) at 3' end, whereas the mRNA of eukaryotes contains 200-250 poly (A) residues at 3' end.

3. Transfer RNA (tRNA) or Soluble RNA (sRNA)

Twenty different amino acids required for protein synthesis are present in cytoplasm. Before joining an appropriate amino acid together to form protein, they are activated by attaching to the RNA. Requirement of energy for activation is met from ATP. The RNA which is capable to transfer an amino acid from amino acid pool, possesses capacity to combine with only one amino acid in the presence of an enzyme, aminoacyl tRNA synthetase, and recognises the codon of mRNA, is called tRNA or sRNA. For each amino acid there is different tRNA. It is likely that 20 different tRNAs are present in cytoplasm. However in several cases more than one type of tRNA for each amino acid is present. Therefore, there are more tRNAs than the amino acids. For example, about 100 types of tRNAs are found in bacterial cell.

***(a)* Structure of tRNA:** The tRNA is dissolved in cytoplasm and is too small to be precipitated even at 1,00,000 *g*. Its molecular weight ranges from 25,000 to 30,000 D and sedimentation coefficient is 3.85. It accounts for 10-20% of the total cytoplasmic RNA. They are synthesised in the nucleus on a DNA template by only 0.025% of total DNA at the end of cleavage stage. As the tRNA synthesis is over on DNA template, a part of ribonucleotide (5'CCA3') is added to the 3′ end of each molecule irrespective of amino acid affinity, by an enzyme tRNA phosphorylase. It is the special feature of tRNA as compared to mRNA and rRNA. The tRNA molecules contain about 70-93% nucleotides arranged in a single strand at 5′→3′ end. This sstRNA forms double strands at certain regions with a single stranded loop. The 3′ end terminates with –CCA sequence and the 5' end with G or C.

In addition to bases A, G, C and U present in tRNA, certain unusual bases are also found which are absent in other RNAs. The unusual bases are formed by specific chemical modification such as addition of methyl ($-CH_3$) group to form 3-methyl cytosine or l-methylguanosine, deamination of adenosine to inosine, reduction of uracil to dihydrouracil or rerrangement of uracil into pseudouracil. The other unusual bases are methylguanine (Gme), dimethylguanine (Gme2), methylcytosine (Cme), ribothymine (T), pseudouridine (ψ), dihydrouridine (DHU, H_2U, UH_2), inosine (I) and methylinosine (Ime). Most of the bases pair according to Watson and Crick's model but unusual bases does not because of bringing about changes due to substitution or alterations in those positions that take part in hydrogen bonding. The unusual bases protect the tRNA molecules from break down by RNAase. Consequently several non-base paired loops are formed in tRNA.

***(b)* Clover Leaf Model of tRNA:** For the first time R. Holley (1968) prepared the clover leaf model for yeast tRNA alanine ($tRNA^{ala}$) which includes several known functions of tRNA. This model has been well accepted. A typical clover leaf model is shown in Fig. 3.32.

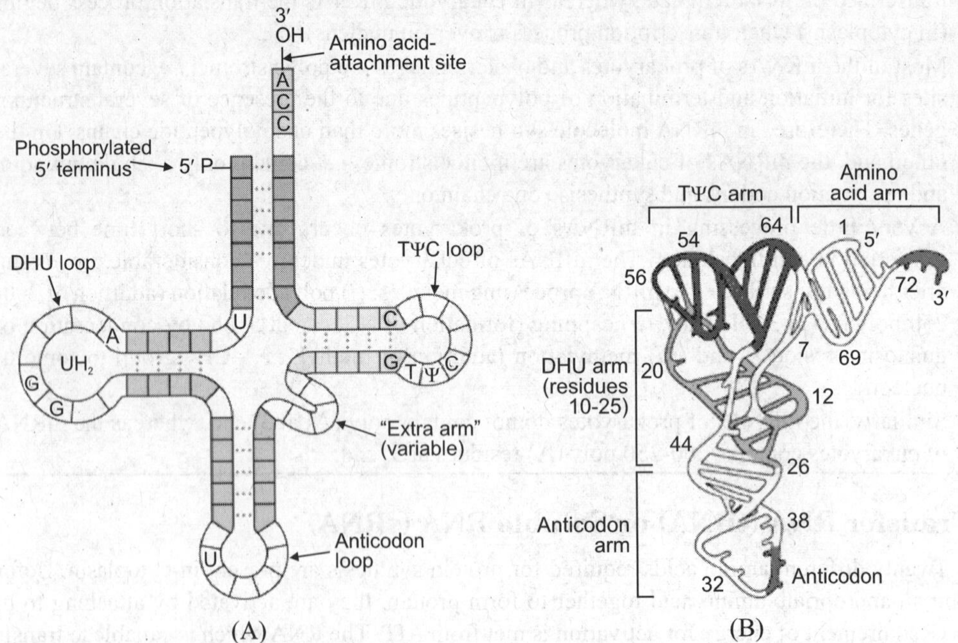

Fig. 3.32. Yeast tRNA; A- clover leaf model, ψ- pseudouridine, DHU- dihydrouridine, T- ribothymidine; B- the three dimensional structure of tRNA.

The clover leaf model reveals the following features:

(i) The single polynucleotide chain of the entire tRNA molecule is folded upon itself to form five arms e.g. acceptor arm, DHU arm, anticodon arm, variable arm, and TΨC arm. An arm consists

of a stem and a loop. Except the acceptor arm, the other arms consist of their respective stem and loop (A).

(ii) The acceptor stem consists of 7 base pairs and 4 unpaired bases, the unpaired bases contain a three -CCA bases and a fourth variable purine (A or G) at 3' end or polynucleotide chain. The last residue adenylic acid (A) acts as amino acid attachment site. The 5' end of tRNA contains either (G) or (C).

(iii) The DHU (dihydrouridine) loop constitutes 7-12 unpaired bases, and acts as the site for recognition of amino acid activating enzyme aminoacyl tRNA synthetase. It consists of a total of 15-18 nucleotides (3-4 base pairs and 7-11 unpaired bases) in the loop. It has two variable regions on both sides of guanine residues. These two regions contain often pyrimidines of 1-3 nucleotides.

(vi) All the tRNA molecules contain different nucleotide triplet codons on anticodon loop. It is also called anticodon or codon recognition site. It is complementary to the corresponding triplet codon of the mRNA molecule. The anticodon stem consists of 5 base pairs, and the anticodon loop contains 7 unpaired nucleotides. The middle three nucleotides act as anticodon which identifies three complementary bases of mRNA molecule. There is a hyper modified purine (HPu) on 3' side chain of anticodon.

(v) The tRNA also possesses a TψC arm that consists of a stem of 5 base pairs and a loop of 7 unpaired bases including pseudouridine. The TψC loop consists of a TψC sequence at 5'→3' direction. The TψC arm has a ribosome recognition site and binds the tRNA molecules to the ribosome.

In some tRNAs with long chain, a variable arm of extra arm is present between the anticodon arm and TψC arm. The variable arm may or may not contain a stem. The three dimensional structure of tRNA reveals a tertiary structure of tRNA where the different limbs are separately formed by the acceptor, TψC and DHU arms and anticodon arm are visible (B). These limbs formed by hydrogen bonds are found between bases and ribose-phosphate backbone, and between the residues of backbone. The tRNA that initates protein synthesis is called initiator tRNA. The initator tRNA of eukaryotes differs from the prokaryotes. The tRNA specifies methionine as the starting amino acid in eukaryotic protein synthesis and *N*-formyl methionine in prokaryotes. Therefore, the two tRNAs specific to these two amino acids are methionyl tRNA (tRNA *f-met*). These two tRNAs differ from each other.

E. DNA REPLICATION

Formation of an additional identical copy is one of the most important properties of the DNA. The process of forming its replica copy is called replication. Replication is the basis of evolution of all morphologically complex forms of life. Howard and Pelc (1953) demonstrated that in eukaryotes replication occurs during interphase between mitotic cycles and also during interphase of meiosis. During interphase of cell division the number of DNA molecules doubles which at anaphase is separated into two daughter cells, and thus equal number of chromosomes is maintained.

Replication does not occur during entire anaphase but is confined only to synthesis (S) phase. There is a post-mitotic gap (G1) between the telophase and S phase. A second premitotic gap (G2) is between the S phase and prophase. Only S phase involves replication process. The G1 phase is most variable and in many eukaryotic cells it is completed within 3 to 4 hours or even months depending on physiological conditions. Mostly DNA synthesis is accomplished in 7 to 8 hours. In bacteria growing at log phase, DNA synthesis occurs from the time a cell originates to give rise to two daughter cells. It is noteworthy that bacteria divide only through binary fission. The DNA carries out two important functions such as heterocatalytic function and autocatalytic function. The heterocatalytic function is protein synthesis directed by DNA, and autocatalytic function is the synthesis of its own DNA into replica copies. However, replication of DNA in prokaryotes differs from that of eukaryotes.

I. WATSON AND CRICK'S MODEL FOR DNA REPLICATION

Each chain of double helix acts as template and is involved in replication of DNA. Watson and Crick proposed that the hydrogen bonds between the base pairs of two strands are broken and separated from each other. Each purine and pyrimidine base of the strands forms hydrogen bonds with complementary free nucleotides to be involved in polymerisation in the cell. The free nucleotides form phosphodiester bonds with deoxyribose residue resulting in formation of a new polynucleotide molecule. This model of Watson and Crick for DNA replication was later on verified experimentally.

1. Meselson and Stahl's experiment for DNA Replication by Semi conservative Method

In 1958, Mathew Meselson and Frank W. Stahl provided the experimental evidence to support the Watson and Crick's model for semi-conservative DNA replication which is also called Meselson-Stahl experiment. They grew *E. coli* cells in medium containing heavy isotopic nitrogen (^{15}N) for several generations (Fig. 3.33). They obtained a population of *E. coli* that contained totally ^{15}N-labelled DNA. Density was measured by density gradient centrifugation in CsCl containing ethidium bromide. Density of ^{15}N- DNA was heavier (1.722 g/cc) than the normal DNA (1.708 g/cc). Then they transferred into a new medium containing less dense isotopic nitrogen (^{14}N). They collected bacterial samples at an interval of 20 minutes starting from 0 minutes up to 40 minutes, and dissolved its DNA in cesium chloride solution and centrifuged rapidly. They found that: (*i*) after zero minute of incubation both strands were labelled, (*ii*) after 20 minutes (F1 generation) one hybrid DNA (one heavy/one light) was produced (in the cells of first generation 50% ^{14}N - DNA and 50% hybrid (^{15}N - ^{14}N) DNA was recorded), and (*iii*) after 40 minutes (F2 generation) one unlabelled molecule and one hybrid (heavy/light) molecule was produced, the ratio of normal and hybrid DNA molecules was 3 : 1. They compared the results of all treatments and found that DNA replicates by semi-conservative method.

This was the semiconservative nature of DNA replication because in the first generation one of the parental strands is converted into progenies and the other complementary polynucleotide strand is replicated. Thus, the two strands were the hybrids. One can postulate for conservative mode of DNA replication, *i.e.* both the original DNA strands act as template for a new duplex but is not separated, and results in an old and new double helix in the first generation. No hybrid $^{15}N-^{14}N$-DNA could be formed. Therefore, such model could not be supported by Meselson-Stahl experiment.

This experiment does not support the dispersive mode of DNA replication in which there is no pattern of replication. The parental strands break randomly at several points during replication. Each segment will replicate and rejoin randomly. This results in varying amount of old and new DNA molecules in daughter cells. After first generation instead of a single ^{15}N- ^{14}N hybrid, a wide spectrum of DNA densities is detected. Therefore, the dispersive mode of replication is also ruled out by Meselson and Stahl experiment.

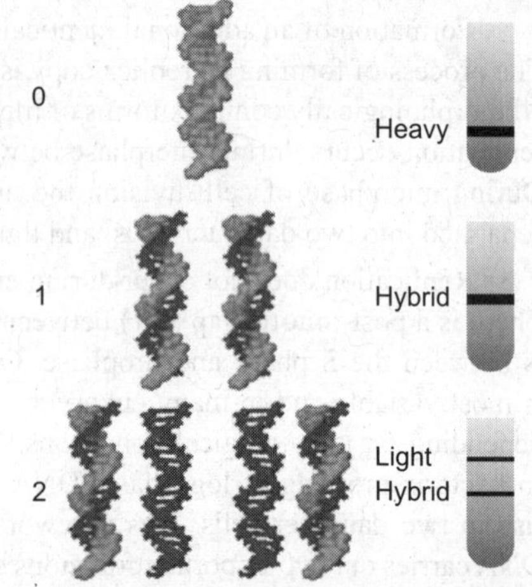

Fig. 3.33. Meselson and Stahl's experiment demonstrating semiconservative replication of DNA.

2. Enzymes Involved in DNA Replication

Both the prokaryotic and eukaryotic cells contain three types of nuclear enzymes that are essential for DNA replication. These enzymes are nucleases, polymerases and ligases.

(*a*) Nucleases: The polynucleotide is held together by 3'→5' phosphodiester bonds. The nucleases hydrolyse the polynucleotide chain into the nucleotides. It attacks either at 3' OH end or 5' phosphate end of the chain. The nucleases are of two types.

(*i*) Exonucleases: The nuclease that attacks on outer free end of the polynucleotide chain is called exonuclease. It breaks phosphodiester bond either in 5'→3' direction or in 3'→5' direction. The enzyme moves in either case stepwise along the chain and removes nucleotides one by one. Thus, the whole chain is digested.

Arther Kornberg (1918-2007)

(*ii*) Endonucleases: The endonucleases attack within the inner portion of one or the double strands. Therefore, a nick is made on double stranded DNA molecule. However, if the polypeptide chain is single stranded (e.g. in DNA viruses), the attack of endonuclease will render the chain into two pieces. On double stranded DNA the nick contains two free ends that in turn act as template for DNA replication. The nicked double helix is distorted due to rotation of free molecules around its intact strand.

(*a*) DNA Polymerases: DNA polymerases carryout the process of polymerisation of nucleotides and formation of polynucleotide chain. This enzyme is called replicase when it replicates the DNA molecules and inherited by daughter cells. In 1959, for the first time A. Kornberg discovered an enzyme in *E. coli* which polymerised the deoxyribonucleotide triphosphate on a DNA template and produced complementary strand of DNA. This enzyme was called DNA polymerase. Later on it was named as Kornberg polymerase or Kornberg enzyme after the name of discoverer, for demonstrating *in vitro* polymerisation of DNA. For catalysis of polymerisation, it requires the four deoxyribonucleotide triphosphates, for example dATP, dGTP, dTTP and dCTP, a DNA template, a primer for initiation of catalytic activity and Mg^{++}.

In prokaryotes, three types of DNA polymerases such as polymerase I (Poly I), polymerase II (Pol II), and DNA polymerase III (Pol III) are found, whereas in eukaryotes three or four polymerases termed as α, β and γ-polymerases and mitochondrial (mt) DNA polymerase are present. Properties of three DNA polymerases of *E. coli* are given in Table 3.7.

Table 3.7. Properties of three DNA polymerases of *E. coli.*

Properties	***DNA polymerase***		
	I	**II**	**III**
Gene for the polymerase subunit	***polA***	***polB***	***polC***
No. of subunits	1	> 4	> 10
Proofreading (3' → 5') exonuclease activity?	yes	yes	yes
5' → 3' exonuclease activity	yes	no	no
Polymerisation rate (nucleotides added per second)	16-20	5 - 10	250-1000
Processivity (nucleotides added before dissociation)	low (3-200)	high (10,000)	very high (500,000)

The molecular weight of α and γ polymerases are over 100,000 and that of β-polymerase is 30,000-50,000. The α and β polymerases are located in the nucleus. The β polymerase copies a poly (A) or poly (C) template. The γ-polymerase copies many polyribonucleotides such as poly (A), poly (C), etc. The mtDNA polymerase is like γ-polymerase.

***(i)* DNA Polymerase I (Pol I):** The Kornberg polymerase is known as Pol I. It is a single peptide chain with a molecular weight of 1,09,000 D. It is the largest single chain of globular protein known so far. One atom of zinc (Zn) per chain is present; therefore, it is metalloenzyme. In *E. coli*, approximately 400 molecules of Pol I are present. Early experiments carried out by Kornberg revealed that when artificially synthesised DNA template strands alternating A and T, *i.e.* poly d(AT) were incubated with polymerase and four radio-labelled nucleoside triphosphate, radioactive DNA containing alternating A and T was synthesised. Though sufficient amount of dGTP and dCTP was present in the solution but these were not synthesised into DNA because the DNA strand contained only poly dAT. This emphasises that DNA Pol I synthesises only complimentary copy of the template.

The shape of Pol I has been studied through electron microscope. It is roughly spherical of about 65 Å diameter (Fig. 3.34) which gets attached regularly to the DNA chain. Pol I possesses several attachment sites such as : (*i*) a template site for attachment to the DNA template, (*ii*) a primer site of about 100 nucleotides contemporary to a segment of RNA on which the growth of newly synthesised DNA occurs, (*iii*) a primer terminus site containing a terminal 3'OH group at the tip, and (*iv*) a triphosphate site for matching the incoming nucleoside triphosphates according to complementary nucleotide of DNA template. The DNA Pol I is a 928 amino acid long polypeptide (MW=103118 D) encoded by the *polA* gene. Pol I plays a significant role in polymerisation as well as degradation (exonucleolytic) process of nucleotides. Polymerisation is a process of synthesis in 5′→3′ direction of short segments of DNA chain from deoxyribonucleoside triphosphate monomers to the 3′–OH end of a DNA strand (Fig. 3.35).

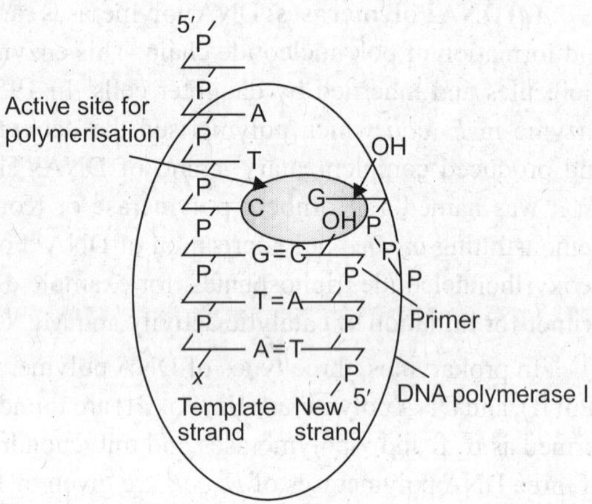

Fig. 3.34. Diagram of DNA polymerase I of *E. coli*.

It is not the main polymerisation enzyme because it cannot synthesise a long chain. It synthesises only a small segment of DNA. It binds only to a DNA and forms nick in dsDNA. Therefore, it takes part in repair synthesis. In *E. coli* DNA Pol I polymerises the nucleotides at the rate of 1,000 nucleotides per minute at 37°C. But the chief enzyme associated with polymerisation is known as polymerase III.

Pol I is broken by trypsin into two fragments, a large fragment (MW 75,000) and a small fragment (MW 36,000). The large fragement shows 3′→5′ exonuclease activity, and the small fragement shows 5′→3′ exonuclease activity. In *E. coli* the following three types of exonuclease activity functions of DNA Pol I have been found.

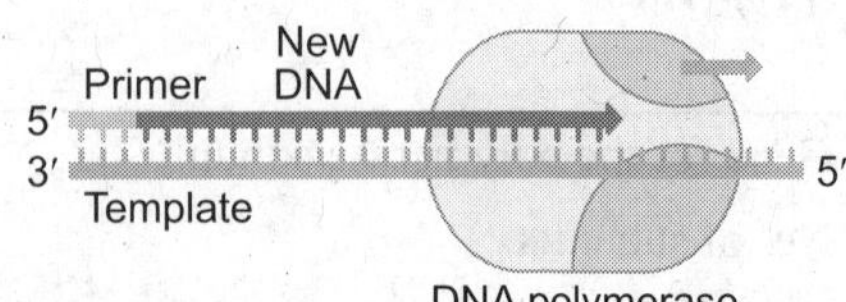

Fig. 3.35. DNA Pol I shows polymerisation in 5′→3′ direction.

3′→5′ exonuclease activity: DNA Pol I contains a 5′ → 3′ nuclease activity located on the small fragment. This activity hydrolyses the phosphodiester bond of bases at or near the 5′ end of double-stranded DNA (or DNA-RNA hybrids). This can then lead to the synthesis of new DNA following the 5′→3′ exonucleolytic activity. It catalyses the breaking of one or two DNA strands in 3′→5′ direction into the nucleotide components, *i.e.* the nucleotides are set free in 3′→5′ direction which is reverse to polymerisation direction; therefore, it is called 3'→5' exonuclease activity (Fig. 3.36.). DNA Pol I corrects the errors made during the polymerisation, and edits the mismatching nucleotides at the primer terminus before the start of strand synthesis. The 3'→ 5′ exonuclease activity of DNA polymerase is not simply due to the catalysis of the reverse polymerase reaction but is a separate and distinct enzymatic activity that has been mapped to its own active site in the enzyme. The function of the 3′ →5′ exonuclease activity is that of proof reading. Any nucleotides such as transiently pairing tautomers, that are incorrectly incorporated, are excised by this activity. Therefore, the function of DNA Pol I is termed as repair synthesis.

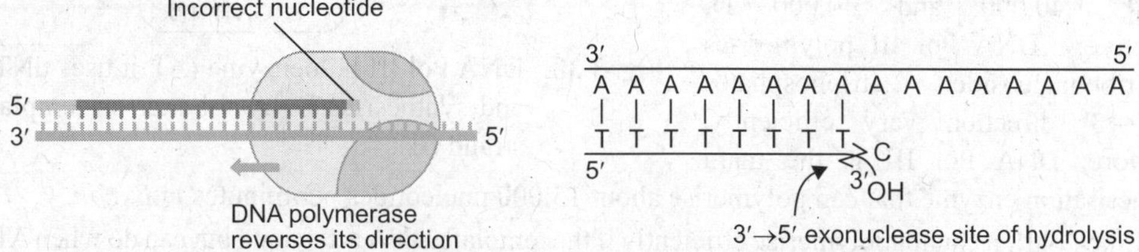

Fig. 3.36. DNA Pol I showing 3'→5' exonuclease activity.

- **5′→3′ exonuclease activity:** DNA Pol I also breaks the polynucleotide chain in 5′→3′ direction with the removal of nucleotide residues. Upon exposure of DNA to the ultraviolet light two adjacent pyrimidines such as thymines are covalently linked forming pyrimidine dimers. These dimers block the replication of DNA. Therefore, removal of pyrimidine dimers *e.g.* thymine dimers (T=T) is necessary. Through 5′→3′ exonuclease activity, Pol I removes pyrimidine dimmers (Fig. 3.37.). Secondly, DNA synthesis occurs on RNA primer in the form of Okazaki fragments. Pol I removes RNA primer and seals the gap with deoxyribonucleotides through 5′→3′ exonuclease activity. Its onward movement results in removal of ribonucleotides from the front portion followed by of deoxyribonucleotides behind it.

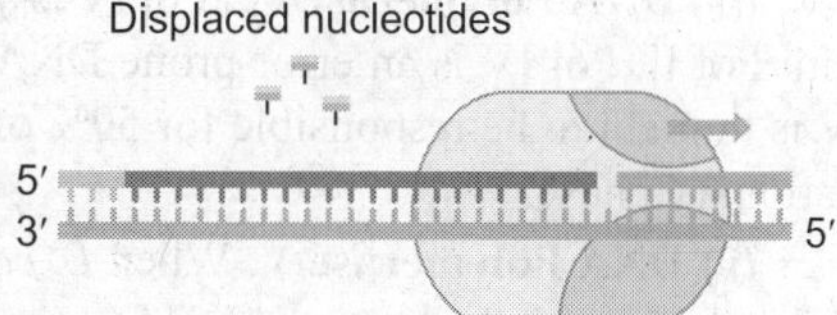

Fig. 3.37. DNA Pol I showing 5′→3′ exonuclease activity.

***(ii)* DNA Polymerase II (Pol II):** For several years Pol I was considered to be responsible for replication in *E. coli,* but the work done during 1970s made it clear that Pol I is associated only with repair synthesis and the other enzymes, Pol II and Pol III are involved in polymerisation process. Pol II is a single polypeptide chain (MW 90,000 D) that shows polymerisation in 5′→3′ direction of a complementary chain. It also shows exonuclease activity in 3′→5′ direction but not in 5′→3′ direction. The polymerisation activity of Pol II is much less than Pol I in *E.coli* cells. About 50 nucleotides per minute are synthesised. *E.coli* cells contain about 40 Pol II molecules. The 3′→5′ exonuclease activity of Pol II shows that it also plays a role in repair synthesis or DNA damaged by U.V. light just like Pol I. In the absence of Pol I, it can elongate the Okazaki fragments. Therefore, Pol II is an alternative to Pol I.

***(iii)* DNA Polymerase III (Pol III):** The core enzyme of DNA Pol III consists of only those subunits that are required for the basic underlying enzymatic activity. In DNA Pol III the core enzyme consists of three subunits: alpha (α), epsilon (ε) and theta (θ) (Fig. 3.38). The α- subunit is coded by the *pol*C gene which is a 1,29,904 kD polypeptide that contains the 5'→3' DNA polymerase activity but no proofreading activity. The epsilon (ε) subunit is coded by the *dna*Q gene which is a 27,099 kD polypeptide that contains the proofreading 3′→5′ exonuclease activity. The 8,846 kD theta (θ) subunit is encoded by the *hol*E gene the function of which is not known. It is not essential for growth and is generally not conserved in bacteria. The holoenzyme is the fully functional form of an enzyme, complete with all of its necessary accessory subunits. The DNA Pol

III holoenzyme consists of the core enzyme, the β sliding clamp and the clamp-loading complex. The model containing two pol III core complexes is the active replicase. DNA polymerase III is several times more active than Pol I and Pol II enzymes. It is a dimer of two polypeptide chains with molecular weight 1,40,000 and 40,000 D, respectively. DNA Pol III polymerises deoxyribonucleoside triphosphates in 5′→3′ direction very efficiently. Therefore, DNA Pol III is the main polymerisation enzyme that can polymerise about 15,000 nucleotides per minutes in *E. coli*.

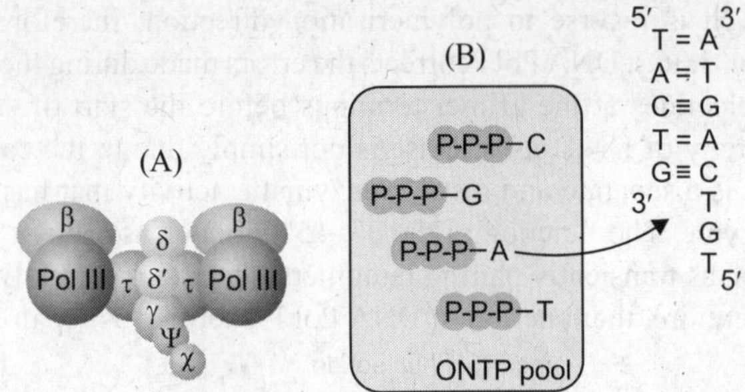

Fig. 3.38. DNA Pol III holoenzyme (A); it uses dNTP and synthesises DNA polymer on template strand (B).

Like Pol II, it cannot polymerise efficiently if the template DNA is too long but can do when ATP and certain protein factors are present. Synthesis of a long template also occurs when an auxillary protein DNA (copolymerase II) is linked with Pol III and produced Pol III-co-Pol II complex. In addition, Pol III also shows 3′→5′ exonuclease activity like Pol II. The 5′→3′ exonuclease activity is absent. All the polymerases *e.g.* Pol I, Pol II and Pol III show 3′→5′ exonuclease activity, whereas the other two polymerases (Pol I and Pol II) lack 5′→3′ exonuclease activity. However, some workers have shown both 3′→5′ and 5′→3' exonuclease activity in Pol III.

***(iv)* DNA Polymerase IV:** Pol IV is also induced in cells of stationary phase where it competes with Pol II. Pol IV is an error-prone DNA polymerase. Because of its error-prone mechanism, Pol IV is thought to be responsible for 50% of the adaptive mutations that are observed to arise during stationary phase.

***(v)* DNA Polymerase V:** When *E. coli* is exposed to high levels of radiation or of a mutagen, major damage to the bacterial DNA can occur. The cell responds by inducing a special ‘last-resort’ repair pathway called the SOS repair pathway. Among the SOS genes that are induced are *umuC* and *umuD*. The products of these two genes form DNA polymerase V (actually, UmuC and two copies of a truncated form of UmuD are required). Pol V replicates past gaps in the DNA. DNA synthesis by Pol V is error-prone; there are none of the proofreading functions of Pol III. It is a relatively poor polymerase that synthesises DNA distributively. Pol V also requires the β subunit and λ complex of Pol III for optimal activity. The β subunit functions as the sliding clamp; it is required for processivity and λ is the clamp loader.

***(c)* DNA Ligases:** The DNA ligases seal single strand nicks in DNA which has 5′→3′ termini. It catalyses the formation of phosphodiester bonds between 3'-OH and 5′-PO_4 group of a nick, and turns into an intact DNA. There are two types of DNA ligases: *E. coli* DNA ligase and T4 DNA ligase. The *E. coli* DNA ligase requires nicotinamide adenine dinucleotide (NAD^+) as cofactor, whereas T4 DNA ligase uses ATP as cofactor for joining reaction of the nick. DNA ligase encircles the DNA double helix. A detailed account of DNA ligases has been given in Chapter 6.

***(d)* DNA Helicase:** DNA helicase play a role in unwinding the DNA helix. It is of two types: DNA helicase A and B. DnaA (the initiator) binds the dsDNA and warps enough to allow helicase (DnaB) to lead on the helix. DNA helicase (DnaB) parts the helices. DNA binding tightens when dTTP (the deoxythymidine analogue of ATP) binds to helicase. Helicase rotates as the dTTP is hydrolysed, screwing itself along the helix, and parting the strands (Fig. 3.39).

Single strand binding proteins (SSBPs) stabilise the parted strands which would anneal themselves (or other strands), forming hairpin loops and other unwanted tangles. Human SSBPs have two fused homologues of *E. coli* SSBPs. SSBPs stabilise the ssDNA and prevent it from re-annealing inappropriately.

(e) Topoisomerase: Topoisomerases stop tangles. Topoisomerase I nicks a single strand ahead of helicase, allowing strain-relief, and preventing snarling up (supercoiling) of the strands as they are forced apart (Fig. 3.40). Topoisomerase II allows DNA molecules to pass through one another.

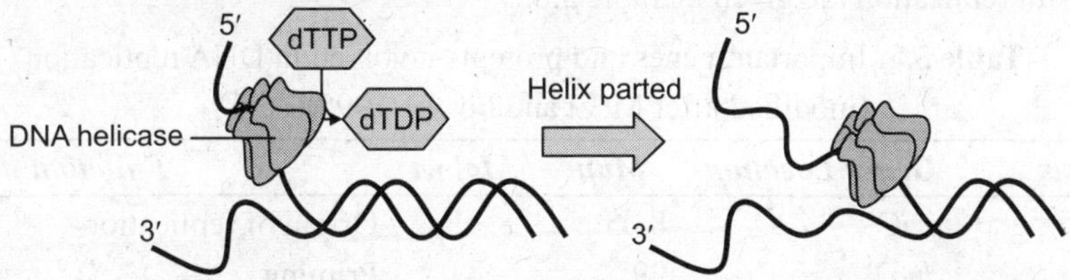

Fig. 3.39. DNA helicase uses the energy of dTTP to screw along a dsDNA molecule, melting the helix and parting the strands.

(f) DNA Gyrase: DNA gyrase, often known as gyrase, is topoisomerase II that introduces negative supercoils (or relaxes positive supercoils) into DNA by looping the template to form a crossing, then cutting one of the double helices and passing the other through it before resealing the break, changing the linking number by two in each enzymatic step. This process occurs in bacteria, whose single circular DNA is cut by DNA gyrase and the two ends are then twisted around each other to form supercoils. The unique ability to introduce negative supercoils into DNA is what gyrase allows bacterial DNA to have free negative supercoils. The ability of gyrase to relax positive supercoils comes into play during DNA replication. The right-handed nature of the DNA double helix causes positive supercoils to accumulate ahead of a translocating enzyme, for example a DNA polymerase in the case of DNA replication. The ability of gyrase (and topoisomerase IV) to relax positive supercoils allows superhelical tension ahead of the polymerase to be released so that replication can continue.

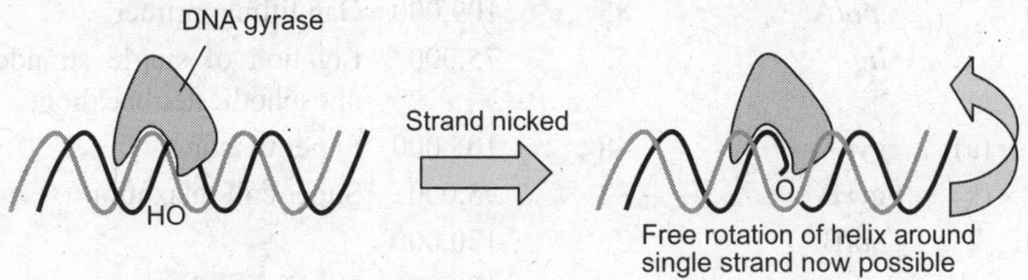

Fig. 3.40. Topoisomerase II allows strain relief ahead of the replication fork.

3. Mechanism of DNA Replication

In 1992, A. Kornberg has nicely discussed the DNA replication. In *E. coli* DNA replication has been investigated most extensively. It was thought that in eukaryotes probably similar mechanism operates. However, it has been found that in *E. coli* replication always starts at a very unique site called the origin. In *E. coli* the replicating apparatus contains an enzyme complex at the point where DNA thread is attached. Through this replicating point DNA thread moves and replication is accomplished. In eukaryotes enzyme moves along the DNA thread. It has earlier been described that *E. coli* possesses three types of DNA polymerases; each of them reads DNA template in 3′→5′ direction and catalyses the synthesis of DNA in 5′→3′ direction. The polymerases read deoxyribonucleotide triphosphates (dATP, dGTP, dCTP, dTTP) as substrate and a DNA template. To thc 3′ end of giowing point, the nucleotides are added after interaction of 3′-OH end of deoxyribose with alpha (first) phosphate group of substrate releasing pyrophosphate as below:

$$\text{dNTPs (dATP, dGTP, dCTP, dTTP)} \xrightarrow[\text{DNA template}]{\text{DNA polymerase III}} \text{DNA + nPPi}$$

Before the replication begins, DNA double helix must be unwounded to give rise to single strand. Unwinding process occurs very rapidly to form a fork that rotates about 75-100 revolutions per second. The unwinding process is facilitated by helicases. Some important genes and proteins associated with replication are given in Table 3.8.

Table 3.8. Important genes and proteins involved in DNA replication (modified after Moat and Foster, 1995).

Proteins	*Genes Location*	*Map*	*Mol wt*	*Function*
–	*ori*C	83.5		Origin of replication
Protein I	*dna*T	99		Priming
Protein n	*pri*B	95		"
Protein n'	*pri*A	89		"
Protein n"	*pri* C	10		"
DnaA	*dna*A	82	54,000	Initiation, binds *ori*C
DnaB	*dna*B	91	55,000	Mobile promoter, helicase prepriming
DnaC	*dna*C	99	25,000	Forms dnaB-dnaC complex
Pol III (a)	*dna*E	4	129,000	DNA Pol III holoenzyme, elongation
Primase	*dna*G	67	60,000	Priming, RNA primer synthesis
g-subunit	*dna*Zx	10	47,500	Synthesis, part of gamma complex
Helix-destabilizing	*ssb*-1	91	20,000	Single strand binding (SSB) protein
Helix-unwinding	*rep*	85	—	Strand separation
E subunit	*dna*Q	5	27,000	Proof reading
Dna Pol I	*pol*A	85	109,000	Gap filling, primer
Ligase	*lig*	52	75,000	Ligation of single stranded nicks in phosphodiester backbone
Dna gyrase (a)	*gyr*A	48	105,000	Super twisting
(b)	*gyr*B	82	95,000	Super coil relaxation
Dna Pol II	*pol*B	2	120,000	
Dna helicase I	—	—	180,000	Unwinding
Dna helicase II	—	—	75,000	Unwinding

Overall DNA replication is accomplished in the following stages (Fig 3.41).

***(a)* Unwinding of Double Helix:** Helicases are responsible for unwinding of double helix. They use energy from ATP to unwind short stretches of helix just ahead the replication fork. After separation of strand it is very necessary to keep them single stranded through single stranded DNA binding proteins (SSB). The SSB is a tetramer with each of four subunits of a molecular weight of 18,500 – 22,000 Dalton. It may bind as a binding site of 8-10 nucleotides. However there is possibility of leading tension and formation of super coils in helix. The relieving of tension and promotion of unwinding process are done by the enzyme topoisomerases which transiently break one of two strands in such a way that it remains unchanged. It ties or unties a knot in DNA strand. DNA gyrase is one of the *E. coli* topoisomerases that removes super coiling of DNA during replication. Thus there is formation of ssDNA template.

***(b)* DNA Replication:** DNA replication is accomplished in several steps. The first step is the RNA primer synthesis on DNA template near origin of replication. Synthesis of RNA primer is very necessary because during DNA replication there is chance of more error in initial lying down of first few nucleotides to pre-existing DNA template. DNA Pol I and Pol II cannot synthesise DNA without an RNA primer; therefore a special RNA polymerase called primase synthesises about 10 nucleotide long short primer. Before priming, preprimer intermediate is formed with the help of six prepriming proteins, for example dnaB, dnaC, n, n', n" and i proteins. For the synthesis of primer, primase needs several accessary proteins which combine with primase. The complex of primase-accessory protein is called primosome. Therefore, DNA Pol III holoenzyme starts synthesis of DNA in 5′→3′ direction at the end of RNA primer (Fig. 3.41).

Several possible functions of palindrome have been suggested such as : (*i*) its action as recognition sites of DNA for protein and many bacterial restriction enzymes, (*ii*) its action as bacterial restriction enzyme against destruction by foreign DNA, (*iii*) providing structural strength to the transcribed RNA by hydrogen bonding in the hairpin loops, (*iv*) possible involvement of cruciform structure in genetic recombination, and (*v*) its possession of genes on long palindrome of some lower eukaryotes that code for ribosomal RNA.

The second step is chain elongation. A new DNA strand starts synthesising by addition of deoxyribonucleoside triphosphates to the 3' end of last nucleotide of RNA primer. DNA synthesis occurs in 5′→3′ direction catalysed by the replisome. The replisome has two DNA Pol III holoenzyme complexes. It is a very large complex containing DNA Pol III and several proteins. The gamma and beta subunits bind the holoenzyme to the DNA template and primer. The a subunit synthesises the DNA. One polymerase continuously copies the leading strand (*i.e.* a strand growing in the direction of replication fork and showing continuous replication). The lagging strand (*i.e.* a strand growing in opposite direction of replication fork and showing discontinuous replication of strand) loops around replisome continuously. There is formation of Y shaped replicating fork at the point where two strands are separated.

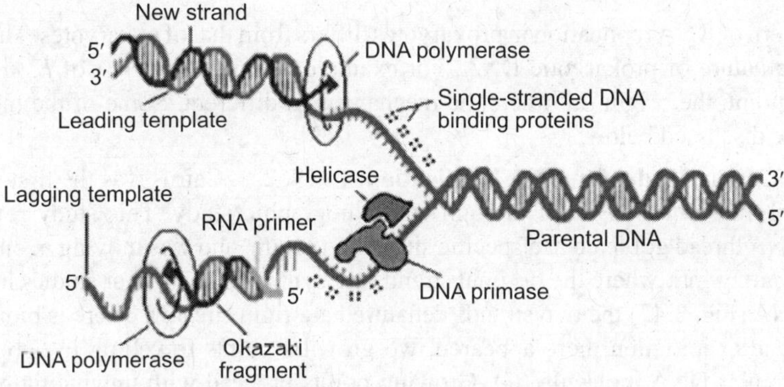

Fig. 3.41. DNA replication A, Overall process of DNA replication in bacteria (diagrammatic).

On leading strand DNA synthesis occurs continuously because there is always a free 3′-OH at the replication fork to which a new nucleotide is added. But on the opposite strand called lagging strand, DNA synthesis occurs discontinuously because there is no 3′-OH at the replication fork to which a new nucleotide can link. On this strand there is free 3′-OH at the opposite end away from growing point. Therefore, on lagging strand a small (11 bases long) RNA primer must be synthesised by primase to provide free 3′ -OH group.

The replication of DNA has two directions, one direction (unidirectional replication) and both the directions (bidirectional replication) from the point of origin. The bidirectional replication is found in most of the bacteria (such as *E. coli*, *Bacillus subtilis*, *Salmonella typhimurium,* etc.), whereas unidirectional replication occurs in *E. coli* bacteriophages (P2 and 186) and mtDNA of mouse LD cells. Finally, about 1000-2000 nucleotides long fragment in bacteria and about 100 nucleotides long fragment in eukaryotic cells are synthesised. These fragments are called the Okazaki fragements after the name of a Japanese discoverer, R. Okazaki (Fig. 3.41).

(*c*) Removal of RNA Primer and Completion of DNA Strand: When the Okazaki fragments are formed, most of lagging strands become duplicated. The RNA primer is removed by DNA Pol I or RNase H. DNA polymerase I synthesises a short segment of complementary DNA to seal the gap. Possibly Pol I removes the primer nucleotide at a time and replaces it with suitable complementary deoxyribonucleotide (Fig. 3.41)

(*d*) Joining of Fragments: At the end, the fragments are joined by DNA ligase that forms a phosphodiester bond between 3'-OH end of growing strand and 5′ end of an Okazaki fragment (Fig. 3.41). In mutants defective ligase is produced; therefore, joining of Okazaki fragements is greatly improved. *E. coli* DNA ligase derives energy from NAD. It is first adenylated by AMP moiety of NAD releasing the nicotinamide mononucleotide (NMN). The adenylated ligase reacts with ssDNA having a nick and forms phosphodiester bond. The complete reaction is as below :

E. coli ligase + NAD ⟶ ligase - AMP + NMN
Ligase - AMP + DNA (with break) ⟶ phosphodiester + ligase + AMP
AMP + NMN ⟶ NAD

DNA replication is a very complex process. If any error occurs during replication, it leads to mutation. *E. coli* makes error about 10^{-6} per gene per generation. The DNA Pol I and Pol III act as proof reader of the newly formed DNA. These move along new DNA synthesised, read mistakes formed due to improper base-pairing and correct those through 3′→5′ exonuclease activity. Despite all these complexity, replication takes place rapidly in bacteria (750-1,000 base pairs per second) and much slower in eukaryotes (50-100 base pairs per second).

4. Models for DNA Replication

The pattern of DNA replication in prokaryotes differs from that of eukaryotes. These differences are due to the nature of prokaryotic DNA. For example, the circular DNA of *E. coli* replicates at a replication point, the origin. In others the mechanism is different. Some of the models for DNA replication are discussed below.

(*a*) The Cairns Model for DNA Replication: In 1962, J. Cairns was the first to visualise the replicating chromosome of *E. coli* through autoradiographic study. This study revealed that the replicating DNA thread got fixed at a specific site called origin, and was moving in one direction and within a replicating fork where the original strands are synthesised. Further studies have shown that in circular DNA (Fig. 3.42) the two strands denatured at origin site (B). There is bidirectional DNA synthesis, *i.e.* after initiation there appeared two growing points travelling in opposite directions around the circular DNA molecule (C). Growing points proceed with unwinding of DNA double helix.

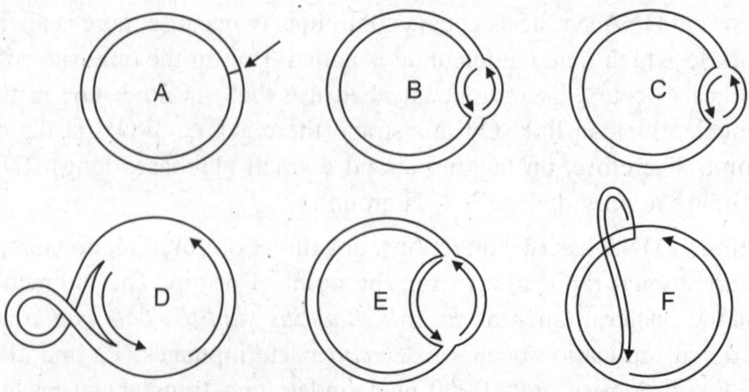

Fig. 3.42. Cairns model for DNA replication (red colour shows newly synthesised DNA in 5′→3′ direction on both the strands).

The process of unwinding creates a torque that is transmitted to the unreplicated part of the DNA molecule resulting in formation of super helix or super twist (D). Super coil prevents its further replication. A temporary nick is made on one of the strands by a swivelling protein (w) which contracts this effect. The nick allows the parental strand for their free rotation on each other and finally freed. At the end swivelling protein seals the nick to continue the replication process (E). Replication process goes on and the two growing point converge on the terminus (F).

(*b*) Rolling Circle Model: In 1969, W. Gilbert and D. H. Dressler described the rolling circle model to explain reactivation in ssDNA viruses (for example ϕX174) (Fig. 3.43) and the transfer of *E. coli* sex factor (for detail discussion, see Chapter 4: *Molecular genetics*).

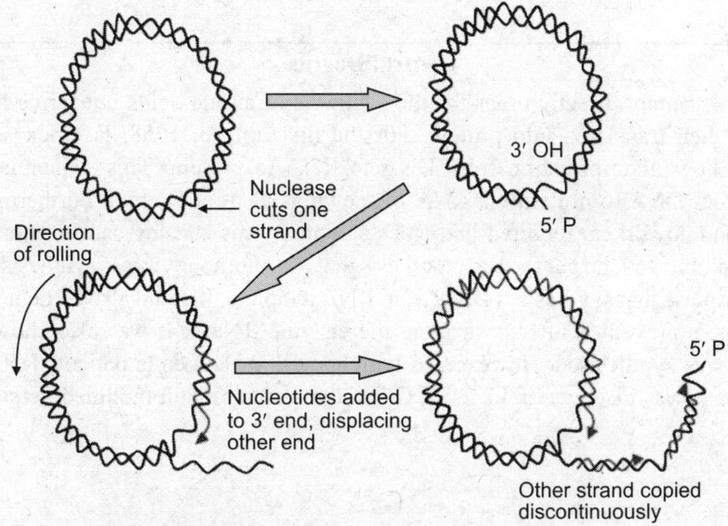

Fig. 3.43. Rolling circle model of DNA replication.

(*c*) Replication in Eukaryotic Chromosome: DNA replication in eukaryotic chromosome is not well understood as compared with prokaryotic chromosome. However, the well accepted model for replication of eukaryotic DNA is the bidirectional model (Fig. 3.44). DNA synthesis starts at a midpoint of replication unit which is called initiation point (*ori*, origin), and progresses in both the directions until reaches the terminal point. The replication fork meets at termination point on the entire chromosome. There may be thousands of initiation points.

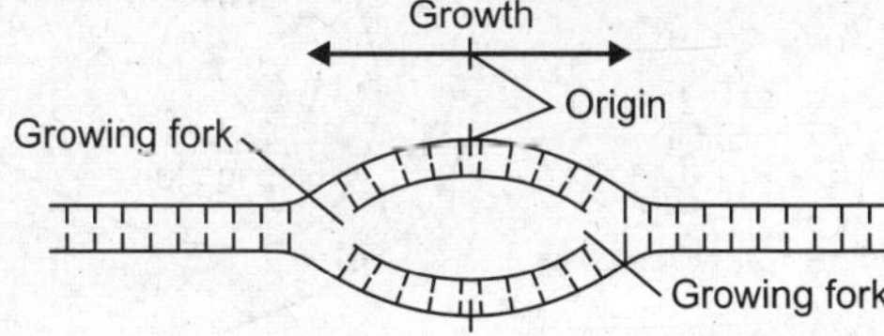

Fig. 3.44. Bidirectional model for replication.

F. GENE EXPRESSION

The process by which information from a gene is used in the synthesis of a functional gene product is called gene expression. These products are often proteins, but in non-protein coding genes such as rRNA genes or tRNA genes, the product is a functional RNA. There are several steps in the gene expression process including the transcription, RNA splicing, translation, and post-translational modification of a protein. Gene regulation gives the cell control over structure and function, and is the basis for cellular differentiation, morphogenesis, versatility and adaptability of any organism. Gene regulation may also serve as a substrate for evolutionary change, since control of the timing, location, and amount of gene expression can have a profound effect on the functions (actions) of the gene in a cell or in a multicellular organism.

During gene expression process DNA is first transcribed into an RNA molecule which is translated into protein, *i.e.* the polypeptides which are polymer of amino acids synthesised as per codes on mRNA. The RNA molecules are synthesised by using a portion of base sequences of single strand of double stranded DNA. This single strand is called template. Hence, formation of an RNA transcript is facilitated by the enzyme RNA polymerase. Therefore, the process of synthesis of an RNA molecule corresponding to a gene is called transcription. By using base sequences and RNA molecule, proteins are synthesised in a definite order. Production of an amino acid sequence from an RNA base sequence is called translation. After completion of translation proteins are synthesised. Therefore, gene expression refers to protein synthesis through two major events, transcription and translation.

Central Dogma

DNA itself cannot directly order for the synthesis of amino acids but forms its transcripts first which is then translated into protein. For the first time in 1958, F. Crick suggested the unidirectional flow of information from DNA to RNA to protein. This sequential transfer of information from DNA to protein via RNA is known as Central Dogma. Furthermore, in 1963 H.M. Temin and coworkers reported that Rous sarcoma virus causing cancer contains RNA as genetic material. In 1964, he put forth a hypothesis that RNA tumour viruses synthesise an enzyme, reverse transcriptase that synthesises DNA from RNA template. It could also be demonstrated that DNA synthesis is prevented after destroying the enzyme RNase. It was also shown that RNA specific DNA was synthesised. However, in lymphocytes of leukemia patients RNA-dependent DNA polymerase was discovered. In 1970, Crick again suggested a modified version of central dogma (Fig. 3.45).

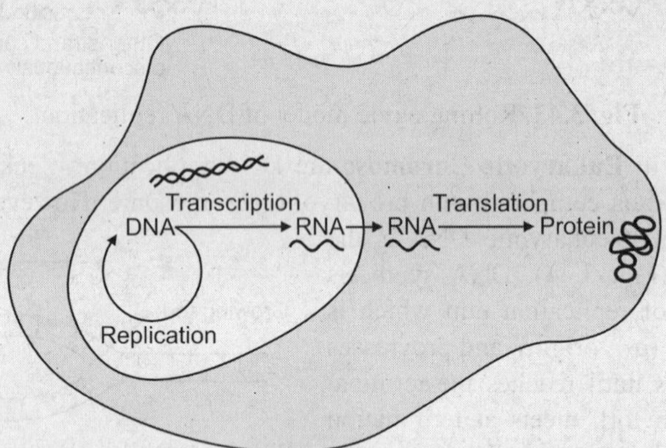

Fig. 3.45. Central dogma (unidirectional flow of genetic information).

I. TRANSCRIPTION (RNA SYNTHESIS)

The tRNA, mRNA and rRNA are involved in the process of transcription. However, an enzyme transcriptase, *i.e.* DNA-dependent RNA polymerase is required for the synthesis of RNA by using ribonucleotide triphosphates, *i.e* ATP, GTP, CTP and UTP. Structure and function of RNA polymeraser are described herewith.

1. RNA Polymerases

On a DNA template elongation of RNA chain at each step is catalysed by RNA polymerase. RNA polymerase is found both in prokaryotes and eukaryotes but the structure and function in these two groups of organisms differ. In prokaryotes only a single enzyme, RNA polymerase governs the synthesis of all cellular RNAs, whereas in eukaryotes, several types of RNA polymerases are involved for the synthesis of a cellular RNA. For example, mRNA, tRNA and rRNA in *E. coli* are synthesised by the same RNA polymerase.

The holoenzyme is a complete RNA polymerase consisting of core, an enzyme and a sigma (σ) factor; hence it is a complex enzyme. The core enzyme of *E. coli* consists of five polypeptide chains, two subunits (αI and aII), one (β), one β′ subunits and one omega (ω) subunit (Fig. 3.46.). There are seven polypeptide chains (such as $\beta\beta'\alpha I\alpha II\delta\omega_1\omega_2$) in core enzyme of *E. coli*. The holoenzyme binds to DNA at specific sites called promoters and transcribes specific length of RNA. Thus σ factor plays a significant role in promoter recognition by RNA polymerase. It can easily be isolated from the holoenzyme. The β-subunit consists of a catalytic site for RNA synthesis and binding sites for substrates and products as well. The β-subunit plays a role in binding RNA polymerase to the DNA template. The two a subunits assemble the two larger subunits into core enzyme (α2ββ'ω). The function of small subunit (ω) is not known in detail; however, it is supposed that it takes part in unwinding. DNA polymerase subunits, structural genes and their functions are summarised in Table 3.9. The *E. coli* RNA polymerase synthesises RNA at the rate of 40 nucleotides per minutes at 37°C.

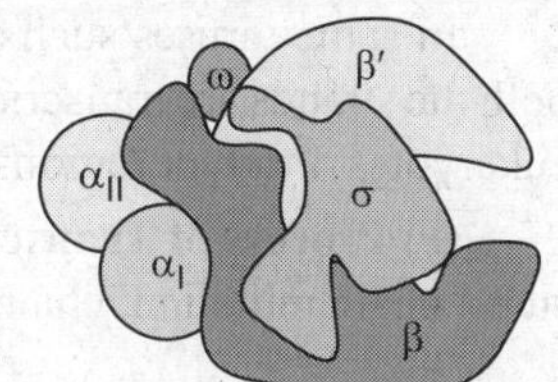

Fig. 3.46. Structure of core enzyme of *E. coli* RNA polymerase.

Table 3.9. RNA polymerase subunits and other transcription factors (modified after Moat and Foster, 1995).

Subunits gene	*Structural position*	*Map weight*	*Molecular*	*Function*
α	rpoA	72	36,000	Initiation
β	rpoB	90	150,000	Initiation, elongation, termination
β'	rpoC	90	160,000	Initiation
σ 70	rpoD	60	83,000	Initiation
σ 32	rpoH	76	32,000	Initiation, heat shock response
Omega (ω)	rpoZ	82	—	Unwinding
CRP	crp	73	22,500	Initiation enhancement
rho (r)	rho	84	50,000	Termination
Nus	nus	69-90	20-69,000	Pausing, transcription, termination
Gre	gre	72	17,000	3'→5' RNA hydrolysis, proof reading.

***(a)* Types of RNA Polymerase:** There are three different eukaryotic RNA polymerases that are transcribed by three different sets of genes (Fig. 3.47). They are distinguished by their sensitivity to a fungal toxin, α-amanitin.

***(i)* RNA polymerase I (RNA Pol I):** It is located in the nucleolus and synthesises precursors of most rRNAs. It is sensitive to α-amanitin.

***(ii)* RNA polymerase II (RNA Pol II):** It is located in the nucleoplasm and synthesises mRNA precursors and some small nuclear RNAs. It is very sensitive to α-amanitin.

***(iii)* RNA polymerase III (RNA Pol III):** It is located in the nucleoplasm. It synthesises the precursors of tRNA, 5S rRNA and other small nuclear and cytoplasmic RNAs. It is moderately sensitive to α-amanitin.

In addition, an antibiotic rifampicin commonly inhibits transcription by interfering with β subunit of prokaryotic RNA polymerase. In 1992, M. Ishihawa presented a current view of RNA polymerase structure and some binding sites for each subunit.

(*b*) Site of Transcription: Transcription begins at the promoter site, *i.e.* cistron localised on DNA molecule. Out of two only one strand of dsDNA is transcribed. It has also been demonstrated that in ϕX174 for any cistron only one strand of DNA is transcribed into RNA. In addition, in phage λ and phage T4 both strands are transcribed, where one strand serves as template for some genes. The rest of genes are transcribed from the other strand. The DNA strand that acts as template is known as anti-sense strand. The nucleotide sequences of anti-sense strand and mRNA transcripts are complementary. The other strand is called sense strand. The bases of sense strand and that of mRNA are exactly the same.

In some viruses such as SP8, ϕX174, etc. out of two only one strand is transcribed. In SV40 both the strands are transcribed for different genes. In contrast in T even phages, phage λ, *E. coli* and eukaryotes different regions of both the DNA strand act as template for RNA synthesis.

(*c*) Process of Transcription: The process of transcription is accomplished in the three main steps: chain initiation, chain elongation and chain termination.

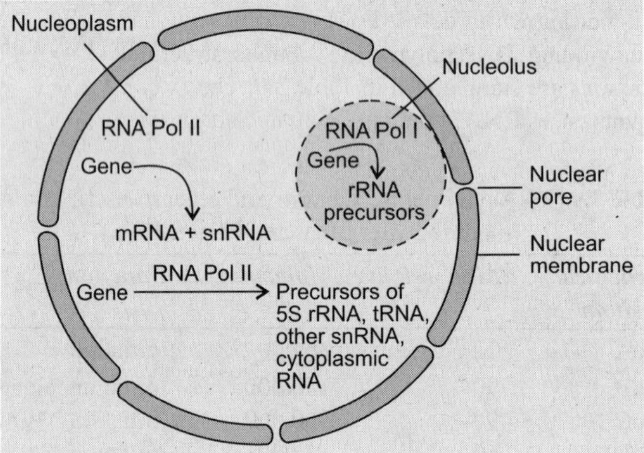

Fig. 3.47. Function of RNA polymerase I, II and III in eukaryotes (diagrammatic).

2. Chain Initiation

During the process of transcription the components required to initiate the RNA chain are the template activated precursors, divalent metal ions (Mg^{++} or Mn^{++}), RNA polymerase and template.

(*a*) Promoter Recognition: The enzyme RNA polymerase plays a key role in recognition and binding of initiation site. Before the formation of complex enzyme, sigma factor interacts with core enzyme at β subunit site. This is required to check transcription of both the strands by core enzyme. The holoenzyme transcribes only one of the two DNA strands. The sigma factor of holoenzyme recognises the promoter region of the DNA (Fig.3.35).

A specific base sequence (20-200 bases long) is called a promoter. Examination of a large number of promoter sequences of different genes from different bacteria has shown that they have many common features. The promoter sequences are centred at –10 and –35 base pairs from the transcription start point and are implicated in normal promoter function. *E. coli* has the following two distinct hexamer components of promoter:

(*i*) Pribnow box: It is an A-T rich sequence of six nucleotides (TATAAT) located 10 base pairs upstream from the start site for transcription. It is called Pribnow box after the name of David Pribnow; the -35 region (5'-TTGACA-3') is called the

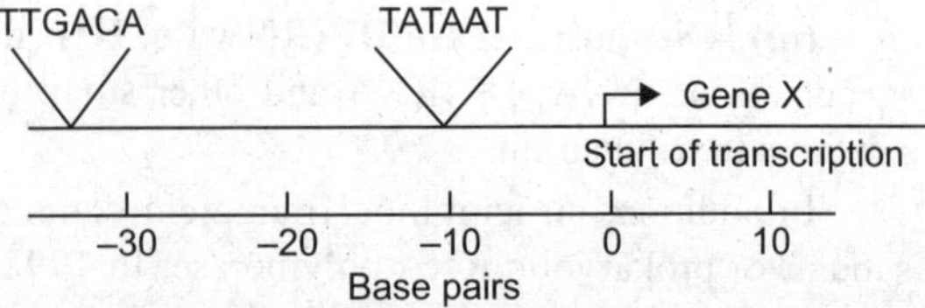

Fig. 3.48. DNA template -35 and -10 promoter sites.

recognition site (Fig. 3.48). It is a consensus sequence, *i.e.* it shows the most frequently occurring base at each position in a large number of promoters analysed. The Pribnow box is found as a part of promoters of all eukaryotes and archaea. The -35 region is located about 35 base pairs upstream (away from the direction of transcription site). The promoter is upstream from the structural gene. Probability of occurrence of each nucleotide is as below:

T	A	T	A	A	T
82%	89%	52%	59%	49%	80%

The RNA polymerase interacts with groups in the major groove and recognises the proper sequence upstream (-35 region) from the Pribnow's box, thereafter forms stable complex by moving laterally to the -10 region. Thus, the sigma factor recognises both the above regions.

***(ii)* TATA box** : The TATA box (also called Goldberg-Hogness box) is a DNA sequence (cis-regulatory element) found in the promoter region of genes in archaea and eukaryotes; about 24% of human genes contain a TATA box within the core promoter. It is the binding site of either general transcription factors or histones, and is involved in transcription by RNA polymerase. The TATA box has the core DNA sequence 5'-TATAAA-3' or a variant, which is usually followed by three or more adenine bases. It is usually located 25 base pairs upstream of the transcription site. The TATA box is usually found as the binding site of RNA polymerase II.

***(b)* Binding of RNA Polymerase to Promoter:** The strength of binding of RNA polymerase to different promoters varies. There are some promoters that have binding sites for proteins rather than RNA polymerase. Therefore, for these promoters the site must be occupied by that protein for RNA polymerase to bind correctly. One of the most common binding sites for this type is one that binds a complex of cyclic AMP receptor protein (CRP). The concentration of AMP governs the activity of these promoters.

***(c)* Unwinding of DNA Double Helix:** The super helical nature of the chromosome may play a role in promoter function. Generally, a negatively super coiled chromosome is a better transcriptional template than relaxed chromosome. The torsional stress imposed by super coiling makes the certain area of DNA easier to separate by RNA polymerase. Hence, super coiling affects the expression of some genes more than the others. Binding of omega (ω) factor of RNA polymerase results in unwinding of a DNA helix. Consequently, a short segment of DNA opens; then the open complex allows tight binding of the RNA polymerase with subsequent initiation of RNA synthesis.

***(d)* Synthesis of First Base of RNA Chain:** The first base of RNA synthesised is always in the form of purine, *i.e.* either triphosphate guanine (pppG) or adenine (pppA). Most of the chains in *E. coli* are started with pppG, whereas in phage T7 and ϕX174 it is initiated with pppA. In contrast to DNA replication, initiation of mRNA synthesis does not require primers. Initiation ends after the formation of first inter nucleotide bond (5'pppPupN) (Fig.3.49).

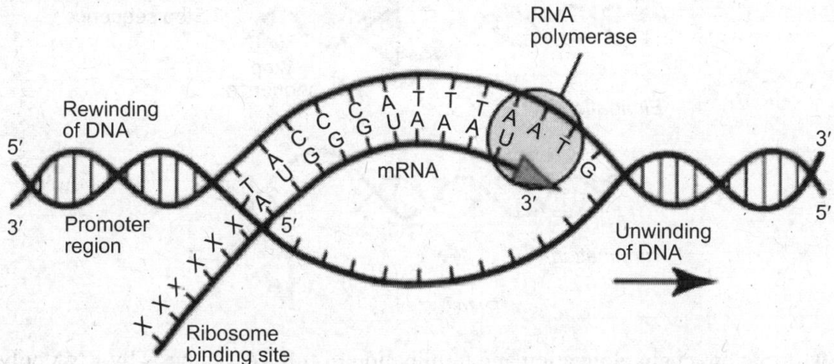

Fig. 3.49. Transcription process started by RNA polymerase.

3. Chain Elongation

Chain elongation occurs by the core enzyme that moves along the DNA template. After beginning the elongation, transcription goes on at a rate between 30 and 60 nucleotides per second at 37°C. In general the elongation reaction includes the steps : (*i*) nucleotide triphosphate binding, (*ii*) bond formation between the nucleotide and the 3′–OH of nascent RNA chain, (*iii*) pyrophosphate release, and (*iv*) translocation of polymerase along the DNA template. The activated ribonucleotide triphosphates (*i.e.* ATP, UTP, GTP and CTP) are added according to nucleotides of one strand of DNA template. The incoming nucleotide forms hydrogen bonds with a DNA base. Reaction occurs between the nucleotides at 3′ end of the RNA molecule and P in the triphosphate group leading to removal of phosphates (PPi). The PPi is soon hydrolysed to inorganic phosphate (Pi).

***(a)* Release of Sigma (σ) Factor:** When 8-9 nucleotide long mRNA transcripts are formed, the σ factor is released abruptly from the complex. Consequently this causes a decrease in the affinity of σ from the RNA polymerase-DNA-nascent RNA complex. The release of σ can combine with any of core enzyme and thus it is reused for the initiation of a new chain.

***(b)* Direction of Transcription:** The ribonucleotide triphosphates attach to the first nucleotide of the template and chain growth takes place in 5′→3′ direction. The region of template that has been transcribed regains its double helical form behind the bubble and the next region of DNA which is to be transcribed unwinds. The RNA transcript does not elongate uniformly along the template. This is due to the presence of pausing sites at certain regions in the template. It has been found that generally the pausing sequence contains GC rich regions about 16-20 bp upstream of 3′-OH end of the paused transcript. Also, the region of dyad symmetry present 16-20 bp upstream of 3′-end causes pausing; however, the mechanism of pausing is not clear.

Moreover, it is believed that after release of sigma factor the NusA and NusG proteins become associated with core enzyme and modulate the rate of elongation. At some sites these proteins enhance pausing. It is interesting to note that the newly formed RNA transcript consists of a triphosphate group at 5′ and free –OH group at 3′ end (Fig.3.50). Moreover, the RNA transcript is larger than the structural gene. Most messages contain a promoter, *i.e.* proximal region of mRNA called the leader sequence, which is the untranslated region before the beginning of translation.

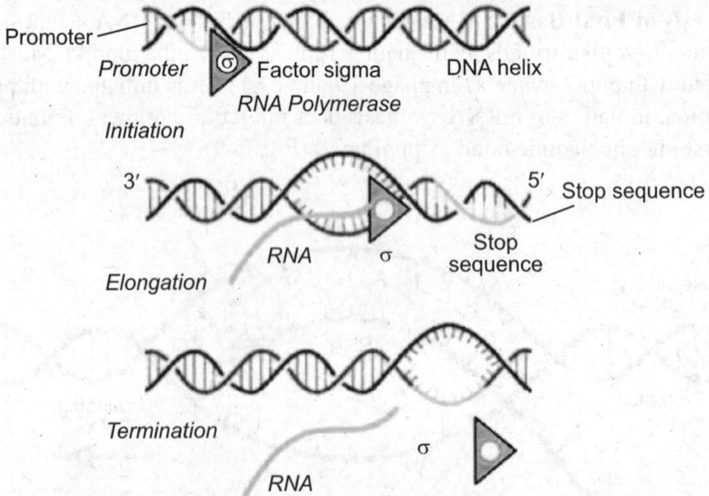

Fig. 3.50. Initiation, chain elongation and termination of mRNA synthesis by RNA polymerase.

***(c)* Proof Reading of mRNA:** The RNA polymerase possesses proof reading activity which is analogous to the 3′→5′ exonuclease activity of DNA polymerase. The two proteins, GreA and GreB enable the RNA polymerase (which is transcriptionally arrested) to back up and cleave 2-3 nucleotides from 3' end of nascent message. This results in release of RNA polymerase which in turn moves through the arrest point at the 3′ end of the gene.

***(d)* Protein-Protein Interaction with RNA Polymerase:** There are several regulatory factors that directly contact the RNA polymerase at the promoter region of template. Such regulatory factors (proteins) can be divided into two classes, I and II. Class I proteins act as activator of transcription that bind upstream from the promoter (for example, CRP, AraC, Fnr and OmpR). Class II transcriptional factors overlap the promoter region; a few of these regulators make contact with C-terminal end of the a subunit.

4. Chain Termination

The process of termination of RNA chain ends with the events: (*i*) cessation of elongation, (*ii*) release of transcript from the tertiary complex, and (*iii*) dissociation of polymerase from the DNA template. There are two types of mechanisms that bring about termination: rho–independent and rho-dependent terminations.

***(a)* Rho-dependent Termination:** In 1983, J.P Richardson gave a model for rho-dependent termination of RNA chain (Fig. 3.51A). Rho-dependent termination requires the rho protein (factor) encoded by *rho* gene. The rho factor is a hexamer. When a strong pause site is apparently located at specific distance from a promoter region the termination associated with rho-factor occurs. However, it must be clear that a strong pause site present close to the promoter will not act as a site for termination. Rho requires ssRNA as a binding region.

Rho factor moves along RNA from its binding site to the enzyme RNA polymerase. The movement is facilitated by the hydrolysis of ATP that provides energy. Hopefully, the ssRNA winds arround the outside of rho factor in the large binding site. This step precedes the arrival of RNA polymerase at the termination site. The RNA polymerase is partially wrapped around the outside of rho factor. Consequently the rho protein comes in the contact of core enzyme of polymerase. This contact is facilitated by NusA or NusG proteins. This step is dependent on the hydrolysis of nucleotide triphosphates. The process of wrapping goes on continuously and disrupts the non-covalent bonds that hold the newly formed RNA to DNA and RNA polymerase. Activation of an RNA/DNA helicase activity causes the mRNA chain and core RNA polymerase to get dissociated from the DNA template. Thus, the rho-protein-RNA complex is set free from the RNA polymerase-DNA complex. Finally, core RNA polymerase is released and interacts with sigma factor that is used for the initiation of second round of transcription. In 1984, D.B. Mose and P.Model has also described a rho-dependent transcription–termination signal in bacteriophage F1.

***(b)* Rho-independent Termination:** DNA itself recognises the rho-independent signal for termination. It consists of GC rich region with dyad symmetry. RNA polymerase reads and extends poly-A sequence on DNA template. It synthesises an RNA transcript that becomes folded to form a stem and loop structure of about 20 bases upstream from the 3′-OH terminus with a terminal stretch of 4-8 poly U residues (Fig.3.51B). This RNA stem-loop structure causes RNA polymerase to pause and disrupt the RNA-DNA hybrid at 5′ end. Poly U residues are present at 3′ end of RNA-DNA hybrid molecule. The A-U hybrid base pairs are relatively unstable; therefore, it causes the 3′ end of the hybrid to break and release the mRNA chain. Thus, the rho-independent termination of transcription is caused by A-U hybrid base pairs.

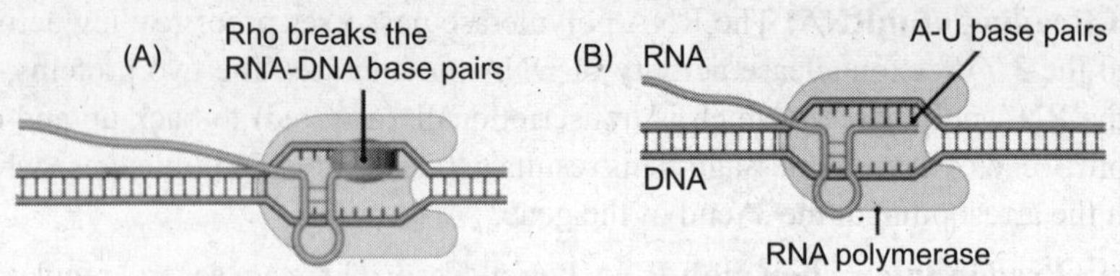

Fig. 3.51. Termination of transcription; A- Rho-dependent termination; B- Rho-independent termination.

II. TRANSLATION (PROTEIN SYNTHESIS)

In protein synthesis the processed RNA molecules take part with the help of ribosomes. All the three types of RNAs are involved in protein synthesis in the following major steps: (*i*) activation of amino acids, (*ii*) transfer of amino acid to tRNA, (*iii*) initiation of protein synthesis, (*iv*) elongation of the polypeptide chain, and (*v*) chain termination. The structure of mRNA and tRNA has earlier been described in detail. During the process of translation the genetic informations are coded in mRNA transcripts in the form of codons which in turn are specifically read by anticodon of tRNA and used to form a polypeptide molecule of defined function.

1. Charging of tRNA

The tRNAs play a great role in translation. Charging of tRNAs takes place in the following steps:

(*a*) Activation of Amino Acids: Only *L*-amino acids take part in protein synthesis. The *D*-amino acids are screened from the all 20 amino acids. In addition, the other amino acids which are not used in protein synthesis are citrulline, alanine, β-alanine, etc. Each amino acid has a specific aminoacyl tRNA synthetase (charging enzyme) and a specific tRNA. At least 32 tRNAs are required to recognise all the amino acid codons, but some cells use more than 32. However, there may be more than one species of tRNA for a specific amino acid but there is only one charging enzyme for each amino acid. Its carboxyl group activates the amino acids through being catalysed by its own specific activating enzyme (aminoacyl tRNA synthetase) in the presence of ATP. Consequently aminoacyl AMP synthetase complex is formed which remains in bound form with the activating enzyme.

(*b*) Transfer of Amino Acid to tRNA: The process of transfer of activated amino acids to tRNA is called charging of tRNA. The tRNAs are specific to their specific amino acid. Therefore, tRNAs are named according to specific amino acid such as $tRNA^{ala}$ (for alanine), $tRNA^{val}$ (for valine), etc. Therefore, the activated amino acid is transferred to its specific tRNA. The aminoacyl-AMP-synthetase complex formed as above is transferred to tRNA as below :

$$AA + ATP \xrightarrow[\text{synthetase}]{\text{Aminoacyl - tRNA}} AA - AMP + PP$$

$$AA - AMP + tRNA \xrightarrow[\text{synthetase}]{\text{Aminoacyl-tRNA}} AA - tRNA + AMP$$

The aminoacyl-AMP synthetase reacts with specific tRNA and forms aminoacyl-tRNA complex by releasing the enzyme aminoacyl-tRNA synthetase. This shows that the enzyme tRNA synthetase has two specific sites. One site recognises the specific amino acids and the other site recognises the specific tRNA molecule. Thus, the tRNA synthetase brings the specific amino acid and tRNA molecule together. However, these recognition properties are essential for making sure that the specific amino acid is charged on the proper tRNA molecule. In the same way the tRNA molecule also consists of two specific sites, one site for recognising its specific aminoacyl-tRNA synthetase and the second (the anticodon) for codon present on mRNA molecule. For the incorporation of an amino acid at proper position in the polypeptide chain, recognition of codon on mRNA by the specific anticodon on tRNA is required.

2. Initiation of Polypeptide Synthesis

There are several specific and complex processes that are involved in the initiation and elongation of polypeptide sequence (Fig. 3.52). The essential components required for initiation are: initiation factors, ribosome, mRNA, guanosine triphosphate (GTP) and aminoacyl-tRNase.

(a) Initiation Factors: There are certain initiation factors (IF) which are required for the initiation of protein synthesis. In prokaryotes three IF (*i.e.* IF-1, 9,000 MW; IF-2, 1,15,000 MW and IF-3,22,000 MW) are involved in the initiation process, whereas in eukaryotes no IF equivalent to IF-1 and IF-2 are found. The, IF-2 is functionally equivalent to eukaryotic eIF-2 and eIF-2′ and IF-3 is equivalent to eukaryotic eIF-3. The IF-1, IF-2 and IF-3 are present in the 30S subunit of the ribosome. The IF-1 and IF-2 help in binding of initiation tRNA ($tRNA^{met}$) to the 30S ribosome subunit.

(b) Formylation of Methionine: Methionine is the starting *N*-terminal amino acid in eukaryotes, whereas in prokaryotes methionine consists of a formyl group (-CHO). Therefore, formyl group containing methionine is called *N*-formyl methionine. In prokaryotes as well as in eukaryotes initiation of protein synthesis occurs through a specific methionyl tRNA which is commonly known as initiation tRNA (*i.e.* $tRNA^{met}$). Binding of initiation tRNA with methionine/formylmethionine occurs as below:

In eukaryotes :

Methionine + tRNA → Methionine tRNA (met-tRNA)

In prokaryotes :

Formyl tetrahydrofolate + N_2- methionyl tRNA $\xrightarrow{\text{Transformylase}}$ *N*-formylmethionyl-tRNA (*N*-fmet-tRN)

(c) Formation of 30S Initiation Complex: The first step in initiation of protein synthesis is the formation of 30S initiation complex. This complex consists of an mRNA, 30S ribosomal subunit, GTP, IF (1, 2 and 3) and the initiator tRNA, *i.e. N*-fmet-RNA. Formation of 30S initiator occurs in the following steps (the actual order of these steps is not known):

(*i*) The initiation factors (IF-1, IF-2 and IF-3) bind to 30S ribosomal subunit in the presence of GTP to form 30S-IF complex (Fig. 3.52). However, when the mRNA is absent IF-1 and IF-3 do not form complex neither with 30S subunit nor 50S subunit.

(*ii*) The second step involves the association of mRNA and initiator tRNA to the 30S subunit. However, the actual order of these steps varies. The IF-3 can bind to both 30S subunit and to mRNA. The 30S-IF complex binds to mRNA at the site containing initiation codon (in the order of preference AUG, GUG, UUG, CUG, AUA or AUU). Each mRNA at its untranslational region consists of a ribosome binding site for every polypeptide in the form of polycistronic message. This ribosome binding site (*i.e.* 5'-AGGAGGU-3') is known as Shine-Dalgarno sequence which is important in the binding of mRNA to the 30S-IF complex (Fig. 3.52). The Shine-Dalgarno sequence pairs to a region present at 3′ end of 16S rRNA. This pairing will results in proper position of initiating AUG codon so that it can combine with an initiator anticodon on tRNA.

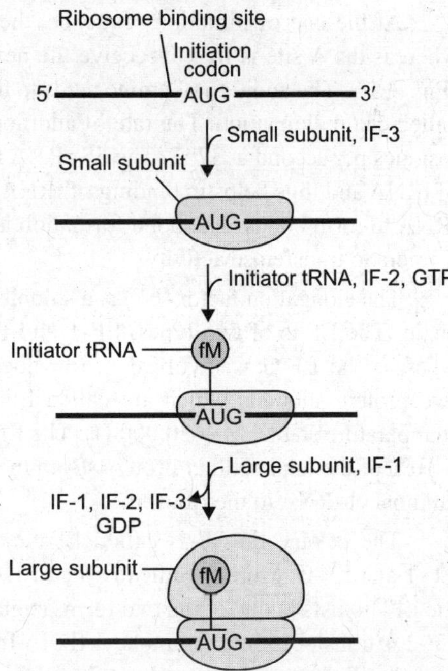

Fig. 3.52. Initiation of translation.

(*iii*) The IF-2 which has combined with GTP, permits the initiator tRNA (*i.e. N*-fmet-tRNA) to bind to the 30S ribosomal subunit (Fig. 3.52). Then it allows to the 30S and 50S subunits to get associated. This binding is followed by removal of IF-3 from the 30S-IF complex. Removal of IF-3 is necessary because its presence inhibits the association of two ribosomal subunits. At this stage the initiation complex consists of mRNA associated with the 30S ribosomal subunits, IF-1, IF-2-GTP and fmet-tRNA.

(*d*) **Formation of the Complete Initiation Com-plex:** The last step in prokaryotes is the union of the 30S initiation complex with 50S ribosomal subunit and forma-tion of a complete 70S initiation complex. This process of union causes the immediate hydrolysis of the bound GTP to GDP + Pi. The process of union is accomplished in the presence of an analogue of GTP (*i.e.* 5'guanyl methylenediphosphate). Therefore, hydrolysis of GTP and subsequent removal of GDP is essential for the IF-1 dependent release of IF-2 from the ribosome (Fig. 3.52).

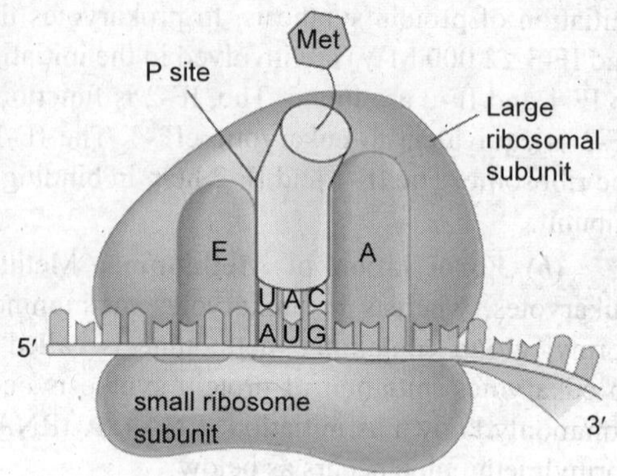

Binding sites (A, P and E) of ribosome

Similarly, in eukaryotes the 40S initiation complex is attached to 60S ribosomal subunit and forms the complete 80S initiation complex. The ribosome has three important binding sites, two are important in protein synthesis. These binding sites are: the aminoacyl-tRNA binding site (A), the peptide (P) binding site and the E site. The A site receives all the incoming charged tRNA, whereas the P site possesses the previous tRNA with the new polypeptide (peptidyl tRNA) attached. The fmet-tRNA (initiation tRNA) directly binds with P site, but not A site. Function of the E site is deacylation.

3. Elongation of Polypeptide Chain

At the end of initiation sequence, the 70S ribosome possesses the fmet-tRNA in the P site, whereas the A site is free to receive the next aminoacyl-tRNA according to the codons on mRNA (Fig. 3.53). The addition of amino acids to the growing polypeptide chain as per codon on mRNA is called chain elongation. The rate of addition of amino acid to the growing polypeptide is about 16 residues per second at 37°C. The 5S rRNA molecule recognises the nucleotide sequence of Ty loop of tRNA and thus helps in binding of tRNA to the A site. The codons direct the specific aminoacyl-tRNA to form bonds. The bond formation is stimulated by an elongation factor T (EF-T) and GTP; T refers to transferase activity.

The elongation factor (EF) is a soluble protein which is required for elongation of polypeptide chain. The EF is of two types, EF-T and EF-G. The EF-T is associated with transferase activity, whereas the EF-G is involved in translocation of mRNA. In prokaryotes the EF-T consists of two protein subunits which are called EF-Tu (temperature unstable, MW 44,000 D) and EF-Ts (temperature stable, MW 30,000 D). The EF-Tu is most abundant protein in *E.coli* that accounts for 5-10% of the total cellular protein. Both the proteins (EF-Tu and EF-Ts) are needed for binding the aminoacyl-tRNA to the ribosome.

The eukaryotic EF is called EF-1 and EF-2 which has resemblance with the prokaryotic EF-T and EF-G. More specifically the EF-1 is like the EF-Tu in its structure and function. At a time the EF-1 exists in one of the two forms (light form, EF-1L and heavy form, EF-1H). The function of EF-2 is translocation of aminoacyl-tRNA from A site to the P site. The GTP is required to drive the process of chain elongation. In 1978, E. Bermerk has discussed the mechanism of chain elongation on ribosome. Elongation of the polypeptide chain is accomplished in the following two steps:

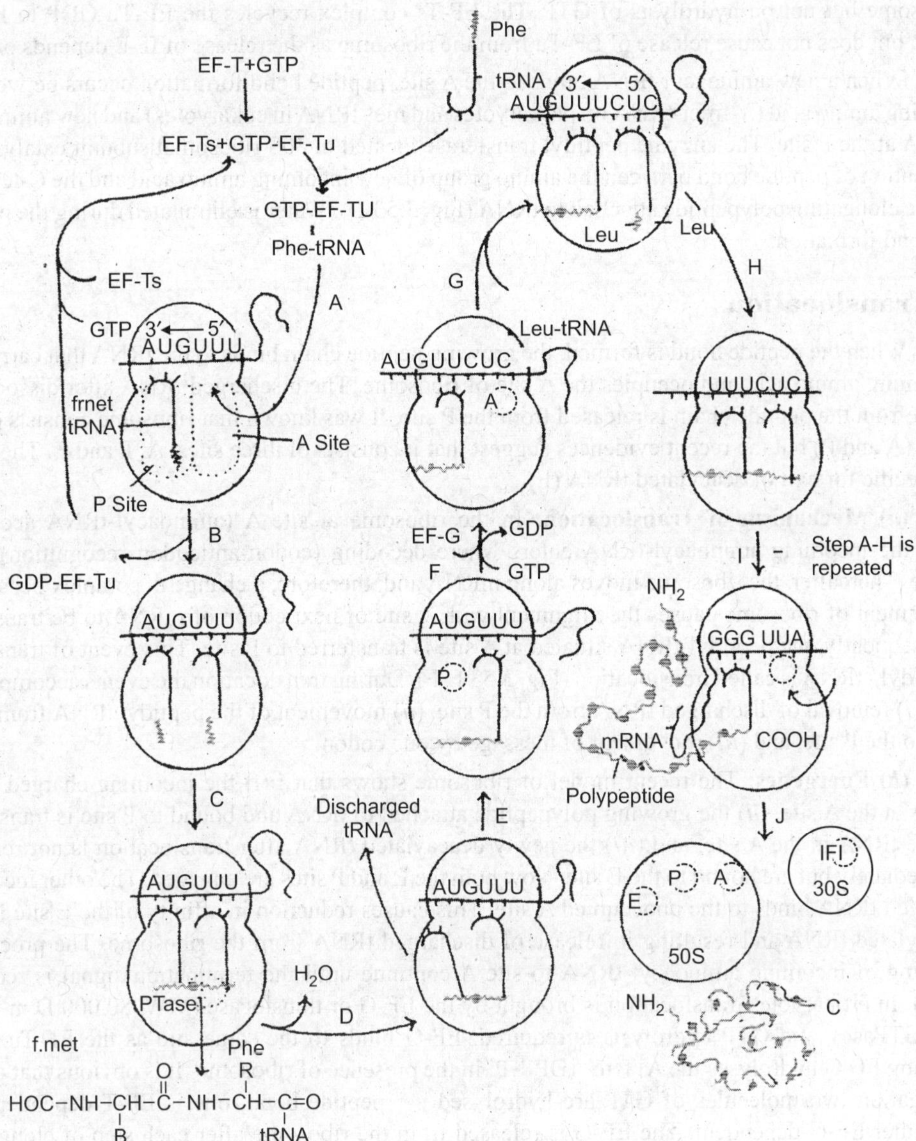

Fig. 3.53. Events of polypeptide chain formation (diagrammatic).

***(a)* Binding of Aminoacyl-tRNA to the A Site:** The GTP binds to EF-T and splits it into EF-Tu-GTP and EF-Ts. The EF-Tu-GTP can bind to all aminoacyl-tRNA (except the initiator tRNA) and results in formation of GTP-EF-Tu-aminoacyl-tRNA complex (Fig. 3.53 A). It is an intermediate complex which is bound to the ribosome. In this step the EF-Ts complex does not play any role. After the aminoacyl-tRNA binds to the A site, GTP is hydrolysed and EF-Tu-GDP complex is released from the ribosome (B). Each aminoacyl -tRNA bound hydrolyses one GDP. The aminoacyl - tRNA may bind to the A site but this binding may not be followed by EF-Tu release from the ribosome. This shows that the purpose of GTP hydrolysis is the release of EF-Tu from the ribosome.

***(b)* Peptide-bond Formation:** Soon the enzyme peptidyl transferase (PTase) catalyses the peptide-bond formation. In fact this is catalysed by the 23S rRNA. This process is called peptidyl transfer (Fig. 3.53 C). However, peptide bond formation depends on release of EF-Tu from the ribosome but not on hydrolysis of GTP. The EF-Ts complex recycles the EF-Tu-GDP to EF-Tu-GTP, but does not cause release of EF-Tu from the ribosome as the release of IF-2 depends on IF-1.

When a new aminoacyl-tRNA binds to the A site, peptide bond formation occurs between the starting amino acid (*N*-fmet-tRNA on prokaryotes and met-tRNA in eukaryotes) and new aminoacyl-tRNA at the P site. The enzyme peptidyl transferase located in 50S ribosomal subunit catalyses the formation of peptide bond between the amino group of new incoming amino acid and the C-terminal of the elongating polypeptide attached to tRNA (Fig. 3.53 D). Water is eliminated during the process of bond formation.

4. Translocation

When the peptide bond is formed, the growing peptide chain binds to the tRNA that carries the incoming amino acid and occupies the A site of ribosome. The discharged tRNA after dissociating itself from the peptide chain is released from the P site. It was known that ribosome consists of two sites (A and P) but the recent evidences suggest that it consists of three sites: A, P and E. The E site is specific for exit of deacylated tRNA (E).

(*a*) Mechanism of Translocation: In the ribosome at site A (aminoacyl-tRNA accepting site) the incoming aminoacyl-tRNA enters where decoding (codon-anticodon recognition) takes place. Thereafter, the ribosome moves along mRNA and, therefore, a change in complex occur. The movement of ribosome causes the alignment with A site of next codon of mRNA to be translated. Consequently, the peptidyl-tRNA situated at A site is transferred to P site. This event of transfer of peptidyl - tRNA is called translocation (Fig. 3.53 E-F). During translocation the events accomplished are: (*i*) removal of discharged tRNA from the P site, (*ii*) movement of the peptidyl tRNA from the A site to the P site, and (*iii*) movement of message by one codon.

(*b*) Energetics: The recent model of ribosome shows that : (*i*) the incoming charged tRNA binds at the A site, (*ii*) the growing polypeptide attached to tRNA and bound to P site is transferred to the tRNA in the A site, and (*iii*) the newly deacaylated tRNA after translocation is not released immediately but are bound to the E site. Now both the E and P sites are engaged. The other incoming charged tRNA binds to the unoccupied A site. This causes reduction in affinity of the E site for the deacylated tRNA and resulting in release of discharged tRNA from the ribosome. The process of binding of incoming aminoacyl-tRNA to site A continue until the termination signal is received (G-I). In prokaryotes translocation is brought by the EF-G or translocase (MW, 80,000 D in which G =GTPase), and GTP hydrolysis is required. EF-G binds to the same site as the EF-Tu. After binding EF-G hydrolyses the ATP to ADP + Pi in the presence of ribosome. It is obvious that during elongation two molecules of GTP are hydrolysed per peptide bond, one is EF-T dependent and the other EF-G-dependent. The EF-G is released from the ribosome after each step of elongation. Since both EF-T and EF-G utilise the same binding site, elongation cannot continue unless EF-G is released.

5. Termination of Polypeptide Chain

The polypeptide chain is terminated in the following two steps :

(*a*) Recognition of Termination Signal: The polypeptide chain continues the elongating until a termination codon on mRNA reaches to ribosome. The termination codons (UAA-ochre, UAG-amber, UGA-opal or umber) are also called as non-sense codon because no tRNA anticodon pairs with them. It is not necessary that the termination codon is the last codon of mRNA. For example, in bacteria and bacteriophages polygenic mRNAs are common and they consist of a number of initiation and termination codons. After translocation of one of the above termination codons into the A site, the ribosome does not bind to an aminoacyl-tRNA-EF-Tu-GTP complex; thereafter, it receives the signal of termination.

***(b)* Release of Polypeptide Chain:** When a termination codon is translocated into the A site, the ribosome instead of binding with a complex containing an amino acid, binds with a peptide release factor (RE) (Fig. 3.53 J). In prokaryotes there are three RF proteins (RF-1, MW 44,000 D; RF-2, MW 47,000 D, RF-3, MW 46,000 D). The RF-1 is active with UAA and UAG codons and the RF-2 is active with UAA and UGA codons. The RF-3 activates the RF-1 and RF-2; therefore, the RF-3 is called stimulatory (S) factor. In eukaryotes there is only one RF protein (MW 56,500 D and 1,15,000 D) which is active with codons UAA, UAG and UGA. The RF protein exists in two units and both of them remain in active form. The ribosome binds either with RF-1 or RF-2; however, the RF protein activates peptidyl transferase which hydrolyses the bond joining the peptide to the tRNA at the site P. This results in release of the peptide chain (Fig. 3.53 J).

6. Post Translational Processing

Some of the processing events occur in the polypeptide chain after release from ribosome. Such modifications occur both in prokaryotes and eukaryotes as given below:

***(a)* Removal of fmet from the Polypeptide Chain:** The formyl group of the *N*-terminal fmet is removed by the enzyme methionine deformylase. The enzyme formyl methionine-specific peptidase (methionyl aminopeptidase or MAP) hydrolyses the entire formyl methionine residues. All the terminal methionines are not removed because there is involvement of discrimination in channelling of different polypeptides through these two alternative steps. The side-chain penultimate amino acid acts as the discriminating factor. Removal of methionine by MAP depends on the length of side chain. The side chain of the longer length has less possibility for MAP to remove the methionine. The other processings are acetylation (of L12 to give rise L7) or adenylation.

***(i)* Loss of signal sequences:** In some polypeptides, about 15 to 30 amino acid residues are present at *N*-terminus. These residues act as signal sequence and direct the protein to its ultimate destination. The signal sequences are cleaved by specific peptidases.

***(ii)* Modification of individual amino acid:** Some amino acid side chains are specifically modified such as : (*i*) enzymatic phosphorylation by ATP of -OH group of certain amino acids (e.g. serine, threonine, tyrosine), (*ii*) binding of Ca^{++} to phosphoresine groups of milk protein, casein, (*iii*) addition of carboxyl (-COOH) group to aspartate glutamate residues of some proteins (*e.g.* blood clotting protein, prothrombin), (*iv*) methylation of proteins (e.g. methylation of lysine residues in cytochrome *c*, calmodulin).

***(iii)* Formation of disulphide cross-links:** Disulphide bridges between cysteine residues of some proteins are formed. Hence, they are covalently cross-linked and attain native from.

***(iv)* Glycosylation:** Attachment of the carbohydrate side chains during or after protein synthesis is called glycosylation, for example glycoproteins.

***(v)* Addition of prosthetic group:** Prosthetic groups get covalently bound to many prokaryotic and eukaryotic proteins. For example, biotin molecule is covalently linked to acetyl-CoA carboxylase.

Protein Synthesis in Prokaryotes and Eukaryotes

The mechanism of protein synthesis in prokaryotes and eukaryotes is the same. However, there are some of the following differences:

- In prokaryotes 70S ribosomes (30S and 50S subunits) and eukaryotes 80S ribosomes (40S and 60S subunits) carryout protein synthesis,
- The bacterial mRNAs are polycistronic (*i.e.* possess many initiation and termination sites), whereas the eukaryotic mRNAs are mainly monocistronic (*i.e.* contain only one functional site for protein synthesis).

- Mostly in bacterial mRNA translation starts while transcription goes on DNA. But in eukaryotes after transcription, mRNAs migrate to cytoplasm through nuclear pore. Translation occurs on ribosomes in cytoplasm. Therefore, during translation mRNA is not associated with DNA.
- In prokaryotes *N*-formyl methionine is the first amino acid which starts translation, whereas in eukaryotes the first amino acid that starts translation is methionine.
- The initiation factors of polypeptide chain in prokaryotes are IF-1, IF-2 and IF-3, whereas in eukaryotes there are several IF factors.
- The elongation factors in prokaryotes are EF-Tu, FF-Ts and EF-G, whereas in eukaryotes the elongation factors are only EF-1 and EF-2. The EF-1 is similar to EF-T (*i.e.* EF-Tu + EF-Ts).
- In eukaryotes the release factor (RF) recognises the three termination codons (UAA, UAG and UGA), whereas in prokaryotes there are three release factors (RF-1, RF-2, RF-3). RF-1 recognises the termination codons UAA and UAG, RF-2 recognises UAA and UGA, and RF-3 is associated with stimulatory activity.

(*b*) Ribosome Editing: During the process of translation certain inappropriate aminoacylated-tRNAs enter the A site of ribosome and remain bound to the ribosome. These outnumber the appropriate amino acylated (aa)-tRNAs. However, the aa-tRNAs remain bound for a long time to the A site for a peptide from P site of ribosome. There are two processes that can reduce the error of surviving polypeptide chain *e.g.* ribosome editing and preferential degradation of polypeptide chain containing erroneous amino acids (Moat and Foster, 1995).

According to the ribosome editing hypothesis the structure of inappropriate peptidyl-tRNA does not correctly complement the structure of mRNA; therefore, it dissociates from the ribosome during protein synthesis. However, the gene *rel*A produces a signal molecule (alarmone), guanosine tetraphosphate (ppGpp) which affects the editing process. The ppGpp interacts with EF-G and results in longer life of peptidyl-tRNA in the A site and enchances the editing process. The gene *pth* synthesises peptidyl-tRNA hydrolase that acts upon the peptidyl-tRNA when it is released. The ribosome free peptidyl-tRNA is hydrolysed by this enzyme. Consequently an intact peptide and an intact tRNA are produced. The defective peptide is degraded by the enzyme.

(*c*) Protein Folding: A protein undergoes spontaneous foldings after the synthesis of polypeptide chain. The secondary folds are formed between the folded regions. Finally, there develops a tertiary structure of polypeptide chain as a result of further folding. Before the terminal amino acids are added in polypeptide chain, the protein reaches to its final shape during the course of chain termination.

According to the recent theories the process of protein folding is complex. Many proteins require assistance to get folded properly. This assistance is provided by the proteins of special kind known as chaperones or chaperonins. These assist polypeptides to self assemble by inhibiting alternative assembling pathway. When chaperones interact with the polypeptide, the chance of incorrect infolding is reduced.

7. Signal Hypothesis (Protein Export)

After synthesis of protein, it is incorporated in membranes or secreted outside by the cell. It is believed that the secretory proteins are synthesised by the ribosomes which are attached to the endoplasmic reticulum and released into it. From endoplasmic reticulum they are transported to various cell organelles (in eukaryotes) where from secreted outside the cell through the process of exocytosis. In 1975, Gunter Blobel (of Rockefeller University) and B. Dobberstein proposed a theory known as 'signal hypotheses' to explain this mechanism. In 1978, Günter Blobel reviewed this hypothesis and postulated that the mRNAs that translate secretory proteins possess a group of signal codons on 3′ side of initiation codon (AUG). For winkling out the details of the translocation mechanism that spirits these proteins into the ER required more than 20 years. Günter Blobel earned the 1999 Nobel Prize in Physiology or Medicine for the discovery that 'proteins have intrinsic signals that govern their transport and localisation in the cell'. The endoplasmic reticulum consists of ribosome receptor proteins. A signal peptide is synthesised by the ribosome as a part of the protein consisting of a special region (signal peptide region). After coming out of the ribosome the signal peptide (with the help of binding proteins) interacts with ribosome receptor protein, results in formation of a tunnel in the membrane and directs the ribosome to a channel in the endoplasmic reticulum. However, the membrane tunnel coincides with the ribosomal tunnel. The growing protein chain penetrates the channel, the signal peptide is cleaved, and the completed protein is released into the lumen of the endoplasmic reticulum. Upon complete synthesis of polypeptide chain, it is released inside the space of endoplasmic reticulum. The protein is subsequently transported out of the cell (Fig. 3.54).

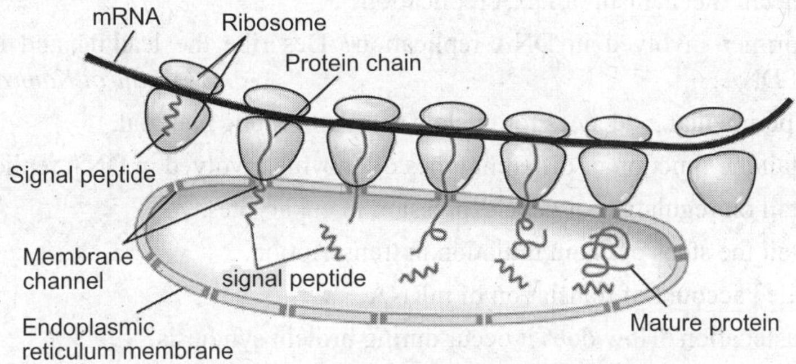

Fig. 3.54. The signal hypothesis.

At the end the ribosomes dissociate from the membrane of endoplasmic reticulum; ribosome receptor proteins get diffused and close the tunnels. This process of entering the proteins into the membranes is also known as 'protein export'.

A significant work has been done in recent years on secretory protein translocation system in *E. coli* and the other bacteria as well, such as *Bacillus subtilis, Salmonella typhimurium, Pseudomonas fluorescens, Vibrio cholerae* etc. Most of the secretory proteins are translated first in a form of precursor containing 15-30 amino acids at *N*-terminus which is called a signal sequence. The signal sequence consists of a hydrophobic region of about 11 amino acid residues and a short stretch of hydrophilic region at *N*-terminus. The signal sequence is involved to bind the nascent polypeptide to the membrane.

In Gram-negative organisms the pathway for export and secretion of signal sequence-containing proteins is called general secretory pathway. The first step is the *sec* gene product dependent translocation of exported protein outside the cytoplasmic membrane. In 1990, L. Brondage and co-workers have presented a model of prompA (an export protein) translocation across the plasma membrane. The protein SecB (a product of *sec*B gene) is a pilot chaperonin. It is associated with the protein that is to be transported, *i.e.* transport protein, for example proOmpA. It is also synthesised on the ribosome. However, in the absence of SecB, the proOmpA aggregates and checks its insertion into the membrane. The protein SecA is basically an ATPase and also forms a part of preprotein translocase in association with integral membrane protein SecY/E. The SecA protein acts as a receptor for proOmpA-SecB complex. Subsequently, the ATP is hydrolysed releasing the proOmpA into the membrane. Once the process of transport has been started at the expense of ATP, further translocation event proceeds through a series of transmembrane intermediates, the energy requirement of which is met by proton motive force rather than ATP hydrolysis.

During the process of translocation, the enzyme signal peptidase (LepB) cleaves the signal sequence of the exported protein. It has to enter the periplasm or translocate across the outer membrane.

QUESTIONS

1. What is the chemical nature of DNA ? Give a detailed account of arrangement of different constituents.
2. Write the X-ray crystallographic study of DNA.
3. With the help of suitable diagrams give Watson and Crick's model of double helix of DNA.
4. Give a detailed account of conformations of nucleic acid.
5. Discuss in detail the different alternative forms of DNA.
6. Write an essay on organization of DNA in eukaryotes.
7. What is the chemical structure of RNA ? Write in brief different types of RNA found in a cell.
8. Discuss in detail mechanism of DNA replication.
9. Why is a primer involved in DNA replication? Describe the leading and lagging strand synthesis of DNA. (*Madurai Kamraj Univ., 2005*)
10. Give the experimental evidences for nucleic acids as genetic material.
11. Write in detail the function of different types of enzyme involved in DNA replication.
12. Write in detail on regulation of gene expression in eukaryotes.
13. Write in detail the steps of chain initiation in transcription.
14. Give a detailed account of translation of mRNA.
15. What is translocation? How does it occur during protein synthesis?
16. Write in detail the post-translational processing of polypeptide chain.
17. Give an illustrated account of signal hypothesis?
18. Write an essay on inhibitors of transcription and translation.
19. Give a detailed account of post-transcriptional processing of RNAs.
20. Discuss in brief about the post translational modification of polypeptide.
21. Write shortnotes on the following :
    (*i*) Chargaff's srule of equivalance, (*ii*) Torsion angle,
    (*iii*) Surgar rings, (*iv*) Sugar puckering,
    (*v*) Furanose pseudorotational angles, (*vi*) *Anti-syn* orientation of N-glycosidic bond,
    (*vii*) Z form of DNA, (*viii*) C form of DNA,
    (*ix*) Genome organization in prokaryotes, (*x*) Circular DNA,
    (*xi*) Palindromic DNA,
    (*xii*) Differences between prokaryotic and eukaryotic mRNA,
    (*xiii*) Structure of rRNA, (*xiv*) Clover leaf model of tRNA,
    (*xv*) Structure of mRNA, (*xvi*) Meselson and Stahl's experiment,
    (*xvii*) DNA polymerases, (*xviii*) Single stranded DNA,
    (*xix*) DNA polymerases, (*xx*) Models of DNA replication,
    (*xxi*) Rolling circle, (*xxii*) Cairns model for DNA replication,
    (*xxiii*) 5′- Capping.
22. Describe the events of post transcriptional modification. (*Madurai Kamraj Univ., 2005*)
23. Define the process of transcription. Where does this process fit into the central dogma of molecular biology? (*Madurai Kamraj Univ., 2005*)
24. Describe organization of operon genes and contrast positive Vs negative control of gene expression. (*Madurai Kamraj Univ., 2005*)

25. Describe prokaryotic protein synthesis with neat sketch. (*Madurai Kamraj Univ., 2005*)

26. What are the roles of mRNA in protein synthesis? (*Guwahati Univ., 2007*)

27. Write short notes of the following :

(*i*) Splicosome, (*ii*) rRNA processing,

(*iii*) mRNA processing, (*iv*) Central dogma,

(*v*) RNA polymerase, (*vi*) Transcription,

(*vii*) Sigma proteins (*viii*) Rho,

(*ix*) Ribosomes, (*x*) Ribosome editing,

(*xi*) Protein synthesis in prokaryotes and eukaryotes.

BIBLIOGRAPHY

Blackburn G.M. and Gait M.J. 1996. DNA and RNA structure. *Nucleic Acids in Chemistry and Biology*. (Eds. Blackburn G.M. and Gait M.J.), Oxford University Press, pp. 28–39.

Bukhari A.I., Shapiro J.A. and Adhya S.L.1977. DNA insertion elements, plasmids and episomes. *Cold spring Harbor Lab.* p. 782.

Burgess R.R. 1971. RNA polymerase. *Ann. Rev. Biochem*. 40:711-740.

Freifelder D.1987. *Microbial genetics.* Jones & Bartlett Publ. Boston, p.601.

Hanna M.M. and Meares C.F. 1983. Topography of transcriptional: path of the leading end of nascent RNA complex *Proc. Natl. Acad. Sci*. 80:4238.

Moat A.G. and Foster J.W. 1995. *Microbiol Physiology.* Wiley-Liss, Inc., New York p.590.

Saenger W. (1984) *Principles of Nucleic Acid Structure*. Springer-Verlag, New York.

Temin, H.M. 1972. RNA directed DNA synthesis. *Sci. Amer*. 226:24.

Travers, A.A. and Burgess,R.P. 1969. Cyclic reuse of RNA polymerase sigma factor. *Nature* 222:537.

4 CHAPTER Molecular Genetics

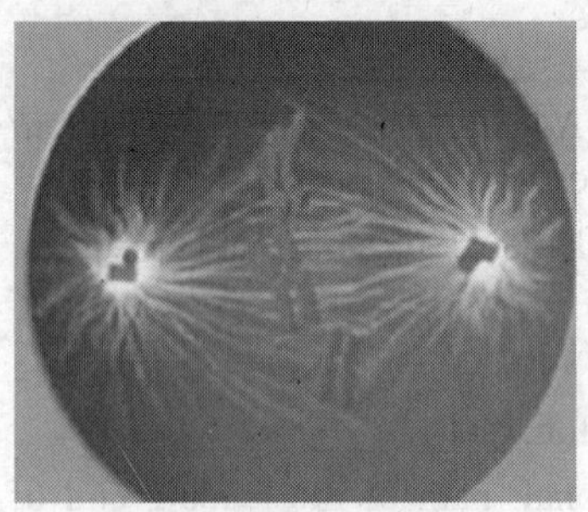

A. Chromosomal basis of Mendelian genetics
B. Cytoplasmic inheritance (extra-nuclear inheritance)
C. Microbial genetics (transformation, conjugation, transduction)
D. Transposable elements
E. Mutations and mutagenesis

Two profound biological breakthroughs were marked in the year 1865 viz., Mendel's publication of his experiments in plant hybridisation, and Charles Darwin's provisional hypothesis of pangenesis. Both of them had no any direct evidence for the theories they proposed. Mendel speculated that cells contained some type of factor that carried traits from one generation to the next, while Darwin proposed that traits could be passed down via units he termed 'gemmules', which he believed travelled from every body part to the sexual organs where they were stored. Because these first attempts to explain the mechanisms of heredity lacked any scientific support. Their profound importance was unrecognised by the scientific community for decades. Nonetheless, Mendel and Darwin's work laid the foundation for formulating a testable and research-based theory of heredity.

The chromosome theory of inheritance was not the work of a single scientist, but rather the collaborative result of multiple researchers working over multiple decades. The seeds of this theory were first planted in the 1860s, when Gregor John Mendel and Charles Darwin each proposed possible systems of heredity. It wasn't until several decades later, following Walther Flemming's discovery of chromosomes and description of their behavior during mitosis, that a probable mechanism for the transmission of traits was identified.

During the middle of the nineteenth century, it could be understood that cells derived from other cells had hereditary information located in the nucleus, but the physical nature of the hereditary material remained unknown. The microscopes provided very poor resolution of cell structures, making it necessary for investigators to treat fixed cells with various stains to enhance the contrast of their contents. Thus, using innovative microscopy techniques and painstaking precision, German anatomist Walther Flemming recognised and explored the fibrous network within the nucleus which he termed 'chromatin' (= stainable material). A few years later the term chromosome was coined by Heinrich Waldeyer. Flemming found that during cell division, the chromatin formed threadlike bodies which he termed 'mitosen' from the Greek word for thread. Based on many observations of cells in various stages of division,

Gregor John Mendel

he correctly deduced the sequence of chromosomal movements during mitosis, which could be confirmed after decades by microscopy of dividing cells. With his characteristic attention to detail, Flemming also made the important observation that chromosomes split along their length during mitosis, and he correctly hypothesised that the split chromosomes were partitioned into different daughter cells at the end of mitosis.

A. CHROMOSOMAL BASIS OF MENDELIAN GENETICS

The idea of a connection between chromosomes and heredity was subsequently strengthened by research conducted by Theodor Boveri and Walter Sutton, but direct evidence in support of chromosome theory didn't come until Thomas Hunt Morgan's experiments with fruit flies at dawn of the twentieth century. Thus, after nearly 50 years of speculation, scientists were finally able to confirm what they had long suspected that chromosomes were indeed the physical carriers of hereditary information. Then, about thirty-five years later, the significance of Mendel's work was emphasised by Walter Sutton, whose observations of chromosome behavior during cell division and gamete formation were consistent with Mendel's findings. Thus, the basis for chromosome theory and the field of cytogenetics was created.

The eukaryotes consist of well organised chromosomes which before mitosis have the different frightening nomenclature such as: homologous chromosomes (identical chromosome present in a cell), non-homologous chromosomes (dissimilar chromosome found in a cell), chromatids (a portion of DNA attached to centromere; one chromosome splits into four chromatids), sister chromatids (two identical DNA molecules of male origin or of female origin), and non-sister chromatids (two chromatids one from father and the other from mother origin).

1. Mendel's Laws of Inheritance

During 1860s, an Australian monk Gregor John Mendel under took the first serious study of genetics of pea. For this pioneering work Mendel is called the father of genetics. He crossed two varieties of pea by selecting the contrast characters and noted the results, *i.e.* changes in height of plant, colour, shape and size of seeds. All the work was published in 1865 under the title '*experiments with plant hybridization*' in Vol. 4 of *Proceedings of the Natural Science Society*. Based on the results of pea his work was formulated as the laws of heredity which is common among all life forms including humans as well.

During the beginning of 20th century three botanists, Hugo de Vries working on *Oenothera*, Carl Correns working on *Xenia*, pea and maize, and Erich von Tshermak working on several flowering plants separately derived the results like Mendel. Later on Mendel's work was reconsidered in 1900 and the original work was republished in *Flora* 89: 364 (1901).

Obviously, genetics is the science of heredity (inheritance) and the variability of the characters of an organism. Heredity is the process by which exact transmission of genetic information takes place from parents to their progenies. The organisms resemble to each other due to the transmission of hereditary information from a common ancestor. Through many generations, species exist simply because the heredity material remains stable. However, genetic diversity exists within most of the species. Hereditary variability is an important factor for evolution which is governed by cell itself or the environment. The genetic makeup adjusts itself according to the environment. Therefore, environment has been constantly influencing the organisms since their origin; that is how evolution took place.

Gregor John Mendel - the Father of Genetics

Gregor John Mendel (1822-1884), an Austrian Monk and a mathematician by training, was born in Sicilian village of Heinzendorf (then a part of Austria). His parents maintained a farm of new varieties of apples. In such farm environment Mendel grew up and completed matriculation at the age of 18 from his village school. He completed two years course in philosophy and tried to become a tutor at Olmutz but did not succeed. This was due to lack of money on part of his father. He returned to his family farm and spent one year. Then he became tutor but could not do further due to combined load of study and teaching; then he left Olmutz (now Brno in Czech). In 1849, he got a temporary post in a school and served for one year. In 1851, he went to the university of Vienna and studied natural history and mathematics. During this period he developed the skill of designing at the monastery at Brno. He studied science and mathematics at University of Vienna. After returning from Vienna to Brno in 1954, he continued to work as priest and as a teacher and stayed for 14 years. During this period the carried out hybridisation works of pea at monastery garden. From 1856 to 1865 he conducted breeding experiment in pea and formulated the principles of heredity; therefore, he is remembered as *the father of genetics*.

Mendel himself did not give any laws; he simply mentioned the conclusions of theoretical and statistical explanations of his experiment. Correns discovered Mendel's work and propounded the whole work into two laws of heredity:

(*a*) The Law of Segregation: In a heterozygote a character remains in contrast paired form which is known as allele. One of alleles remains in dominant form and the other in recessive form or *vice versa*. Dominant and recessive alleles are adhered together throughout the life without getting mixed with each other. During gametogenesis these alleles segregate (separate) from each other with the result that one gamete receives dominant allele and the other recessive allele. After hybridisation (in monohybrid cross) in F1 generation all plants look alike, but their gametes after uniting together in F2 generation give the phenotypic (morphological) ratio of 3 : 1.

(*b*) The Law of Independent Assortment: The law of independent assortment states that the alleles segregate quite independently from each other during the formation of gametes *i.e.* gametogenesis. In F1 generation or dihybrid cross all the plants would be similar but in F2 generation the ratio of characters of plants was 9:3:3:1. Discussion of Mendel's laws in reference to higher plants is not the aim of this chapter, except the brief information.

2. Chromosome Theory of Inheritance (Boveri–Sutton's chromosome theory)

The chromosome theory of inheritance is a fundamental unified theory of genetics which identifies chromosomes as the carrier of genetic material. It explains the mechanism underlying the laws of Mendelian inheritance by identifying chromosomes with the paired factors required by Mendel's laws. It also states that chromosomes are linear structures with genes located at specific sites along them. It states simply that chromosome, which is seen in all dividing cells and pass from one generation to the next, are the basis for all genetic inheritance.

After the discovery of microscope cytological studies were started and meiosis and mitosis were studied. In 1901, Von Winiwarter reported that chromosomes are present in pairs in ovary of half day old rabbit. In 1901, Montgomery studied meiosis in 42 species of insects, and reported that each association of chromosomes in pairs at zygotene stage consists of one maternal chromosome and one paternal chromosome. This was based on the facts that one chromosome was small and the other long.

Theodor Boveri (left) and Walter Sutton (right)

Withelm Roux suggested that the chromosomes look like string of beads and present in nuclei of the cells. The chromosomes duplicate during interphase and separate during meiosis at the time of gamete formation.

In 1902 both Theodor Boveri and Walter S. Sutton independently extended Montgomery's observations and showed a parallelism between the behaviour of chromosomes and Mendelian factors. Boveri was studying sea urchins, in which he found that all the chromosomes had to be present for proper embryonic development to take place. Sutton's work with grasshoppers showed that chromosomes occur in matched pairs of maternal and paternal chromosomes which separate during meiosis.

In 1903, Bovery and Sutton formulated a theory called *chromosomal theory of inheritance*. The important similarities between the two theories are as below:

- The chromosomes occur in pairs as the Mendelian factors (*i.e.* alleles) occur in pairs
- Homologous chromosomes pair during meiosis and segregate into gametes during metaphase I and the Mendelian factors also do
- Each chromosome moves independently from the pairs by spindle fibres as the Mendelian factors assort independently
- The original chromosome number is restored in zygote after fertilisation as Mendelian pairs of alleles are maintained after fertilisation in offspring.

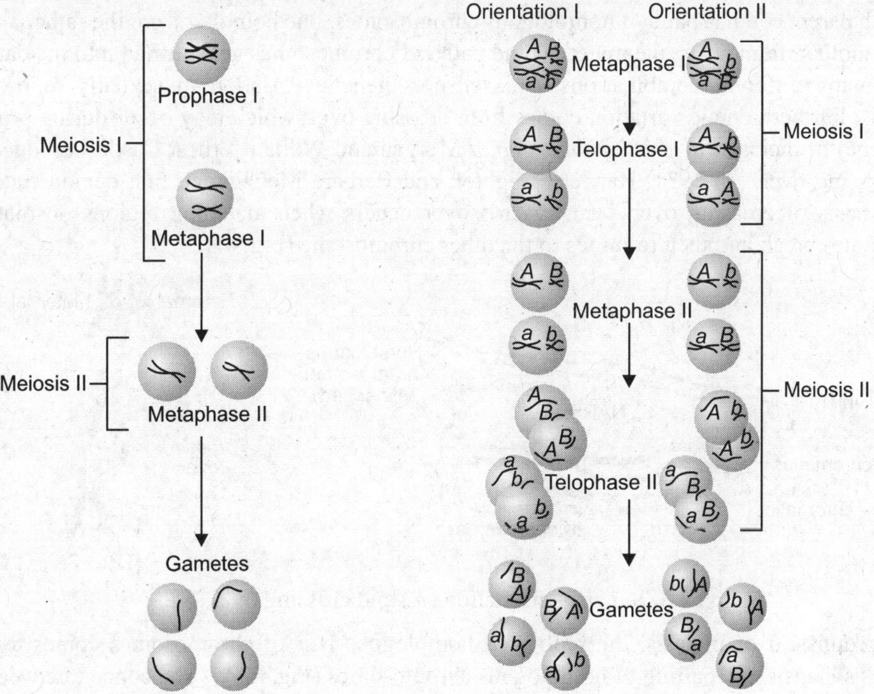

Fig. 4.1. Segregation of homologous chromosomes during meiosis; law of segregation.

Fig. 4.2. Similarity between independent separation of chromosome tetrads of prophase I and Mendel's laws of independent assortment (adopted from McGraw Hill Co).

Fig. 4.1 shows segregation of homologous chromosomes during meiosis which explains Mendel's laws of segregation of alleles. Fig. 4.2 shows similarities between independent separation of tetrads of chromosomes at metaphase I of meiosis in a dihybrid cross and Mendel's laws of independent assortment.

***(a)* Verification of Chromosome Theory of Inheritance:** The proposal that chromosomes carried the factors of Mendelian inheritance was initially controversial, but in 1913 it gained strong support when Eleanor Carothers documented definitive evidence of independent assortment of chromosomes in a species of grasshopper. However, debate continued until 1915 when Thomas

Hunt Morgan's work on inheritance and genetic linkage in the fruit fly *Drosophila melanogaster* provided incontrovertible evidence for the proposal.

(*b*) Sex Chromosomes and Chromosome Theory: The chromosomes that determine the sex are called sex-chromosomes. In 1902, G.E. McClug gave the evidence that chromosomes are associated with sex determination. He found two types of sperms in grasshopper. The male insect consisted of one X chromosome (X0) per cell and the female two X chromosomes (XX) per cell. All the female gametes, *i.e.* eggs were similar in having only X chromosomes. The male gametes, *i.e.* sperms were of two types, having X and O chromosomes in equal number.

Association of X chromosomes with sex determination was suggested. Therefore, Wison and co-workers observed the X chromosomes as dark staining bodies in a specific strain of an insect, *Protenor*. Furthermore, equal number of spermatids with X chromosome and without X chromosome was counted in the testes of *Protenor*. A very strong cytological evidence for chromosome theory of inheritance was given in 1931 by C. Stern in *Drosophila* on crossing over in exchange of chromosome segment.

3. Crossing Over

Each parent cell has pairs of homologous chromosomes, one homolog from the father and one from the mother. In meiosis, the maternal and paternal chromosomes get shuffled into the daughter cells in many different combinations. This ensures genetic variation in sexually repróducing organisms. Further genetic variation comes from crossing over, which may occur during prophase I (diplotene) of meiosis. In 1912, *Thomas Hunt* Morgan and William Ernest Castle introduced the term 'crossing over'. In 1931, Harriet Creighton and Barbara McClintock first demonstrated the physical basis of crossing over. Usually crossover occurs when matching regions on matching chromosomes break and then re-unites to the other chromosome (Fig. 4.3).

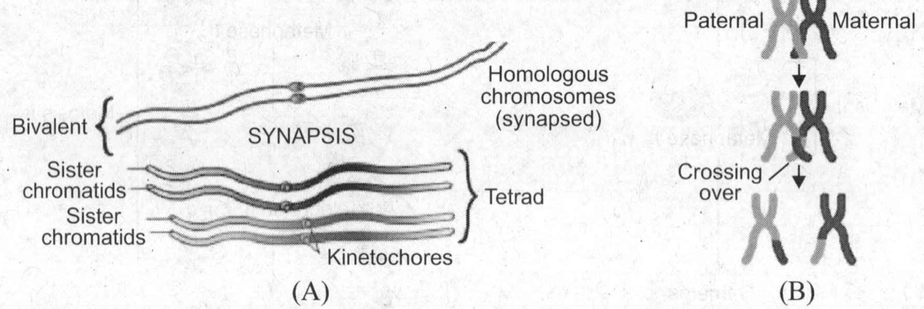

Fig. 4.3. Synapsis formation (A) and crossing over (B).

In prophase I of meiosis, the replicated homologous pair of chromosomes comes together and forms synapsis, *i.e.* pairing of homologous chromosomes (Fig.4.3A). The contact between two chromatids ready to undergo crossing over is called chiasma (Greek letter, X means 'crossing') (Fig.4.3B). Chiasma (plural chiasmata) is the point where two chromatids are intertwined (interwoven) in a cell. Synapsis begins before the synaptonemal complex develops, and is not completed until near the end of prophase I. Consequently sections of the chromosomes are exchanged. After crossing over, the resultant chromosomes are neither entirely maternal nor entirely paternal, but contain genes from both parents. Synapsis and crossing over occur only in meiosis.

Synaptonemal Complex: In 1956, the synaptonemal complex was described by Montrose J. Moses (in primary spermatocytes of crayfish) and by D. Fawcett (in spermatocytes of pigeon, cat and man). During meiosis a protein structure called the synaptonemal complex is formed between two homologous chromosomes (Fig. 4.4); it is thought to mediate chromosome pairing, synapsis, and recombination (crossing over). The synaptonemal complex functions primarily as a scaffold to allow interacting chromatids to complete their crossover activities.

The synaptonemal complex has three specific components: SC protein-1 (SYCP1), SC protein-2 (SYCP2), and SC protein-3 (SYCP3). Moreover, in humans the SYCP1 gene is on chromosome 1p13; the SYCP2 gene is on chromosome 20q13.33; the gene for SYCP3 is on chromosome 12q. The synaptonemal complex is formed by two *lateral elements* (mainly formed by SYCP3 and secondarily by SYCP2), *a central element* (which contains at least two additional proteins and the amino terminal region of SYCP1), and *a central region* (spanned between the two lateral elements containing the transverse filaments composed of the protein SYCP1).

During the pachytene of prophase of meiosis I, this tripartite structure can be observed both in males and in females during gametogenesis. However, during leptotene formation of the lateral elements is started which complete their pairing during the zygotene stage. The synaptonemal complex disappears after completion of pachytene. Formation of the synaptonemal complex reflects the pairing of homologous chromosomes and may be used to probe the presence of pairing abnormalities in individuals carrying chromosomal abnormalities. Among eukaryotic organisms the synaptonemal complex shows very little variation in structure besides some differences in proteins. In many organisms the synaptonemal complex carries one or several recombination nodules associated to its central space (Fig. 4.4). These nodules are thought to correspond to mature genetic recombination events or crossovers.

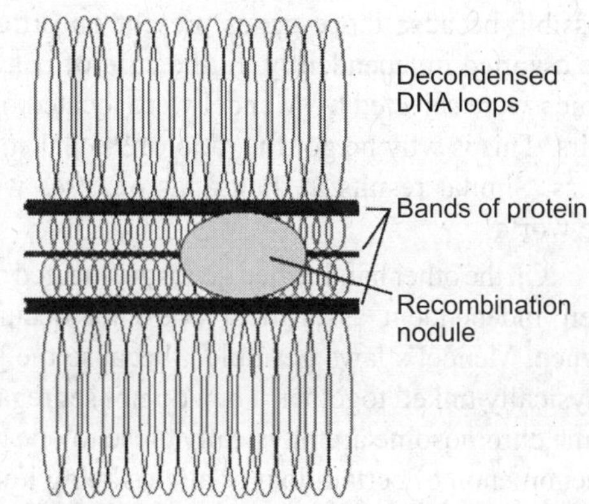

Fig. 4.4. Structure of a synaptonemal complex.

Meiotic recombination initiates with double-stranded breaks that are introduced into the DNA by the Spo11 protein. Then the 5' ends generated by the double-stranded breaks is digested by one or more exonucleases to produce 3′ single-stranded DNA tails. The single-stranded DNA is coated by the meiosis-specific recombinase Dmc1 and the general recombinase Rad51 to form nucleoprotein filaments. The recombinases catalyse invasion of the opposite chromatid by the single-stranded DNA from one end of the break. The 3′ end of the invading DNA primes DNA synthesis, resulting in displacement of the complementary strand, which subsequently anneals to the single-stranded DNA generated from the other end of the initial double-stranded break. The structure formed is called a Holliday junction which is a 'cross-strand exchange' (Fig. 4.5). The Holliday junction is a tetrahedral structure which can be 'pulled' by other recombinases moving it along the four-stranded structure.

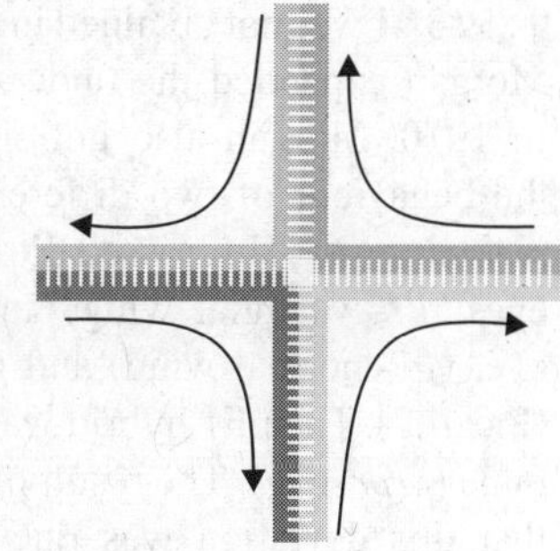

Fig. 4.5. Holliday junction

Crossing over involves breakage and re-union of DNA segments between genes of non-sister chromatids of two homologous chromosomes.This event results in exchange of chromosome segments and formation of new combinations (recombinants) (Fig. 4.6). The recombinants are different from parental chromosomes. The frequency of crossing over

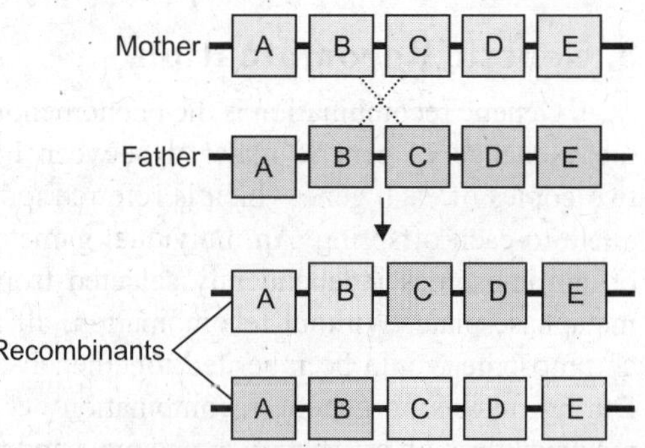

Fig. 4.6. Recombination involves the breakage and re-union of parental chromosomes.

depends on distance between the genes on the same chromosome. Therefore, after analysing the frequency of recombination between the genes, relative distance between the genes on chromosome may be measured. In a population arising from a heterozygous individual, the number of different phenotypes can be analysed. Based on this result the recombination frequency can be calculated.

3. Linkage

Mendel's law of independent assortment of factors (genes) is possible because those genes present on different chromosome pairs are assorted independently. In the case of pea all the seven characters which were selected by Mendel were located on different chromosome pairs. This is why he got the ratio of 9:3:3:1 in F2 progeny of dihybrid cross. Similar results in thousands of cases were reported throughout the world.

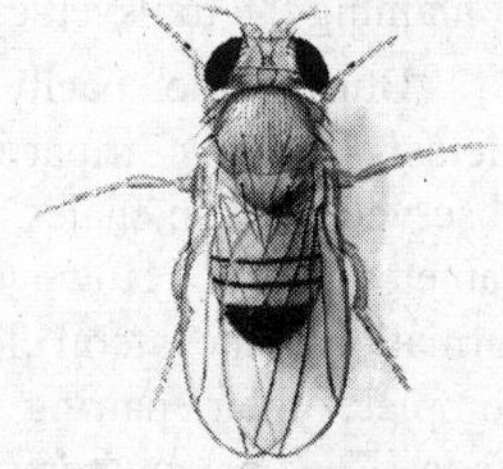

Drosophila melanogaster

On the other hand, when genes are located on the same chromosome their independent assortment does not occur as per Mendel's law. Hence, Mendel's laws get failed, because the genes (alleles) present on the same chromosome are physically linked together. They do not segregate independently during meiosis. Genetic loci on the same chromosome are physically close to one another and tend to stay together during meiosis. The phenomenon of certain loci or alleles being linked together is called linkage.

In 1906, genetic linkage was first discovered by the British geneticists William Bateson and Reginald C. Punnett shortly after rediscovery of Mendel's laws. In 1906, W. Bateson and R.C. Punnett reported a dihybrid experiment in sweet pea where expected Mendelian ratio 9 : 3 : 3 : 1 was not obtained in F2 generation. Thomas Hunt Morgan expanded the understanding of genetic linkage. In 1910, Morgan also got similar result while studying the behaviour of two different gene pairs in *Drosophila melanogaster*. He crossed flies with *PrPrVgVg* (P^r - purple eyes, V^s- vestigial wing) and $Pr^+Pr^+vg^+vg^+$ (Pr^+- red eye, vg^+- normal wing) and got F1 hybrid, $p^{r+}p^{r+}v^{g+}v^{g+}$ (Fig. 4.7). Then F1 hybrid was test-crossed with recessive male (*prprvgvg*). The result of expected test cross ratio 1:1:1:1 was not obtained. Morgan explained that this deviation was due to linkage of genes present on the same chromosome. Morgan's observation that the amount of crossing over between linked genes differs led to the idea that crossover frequency might indicate the distance separating genes on the chromosome.

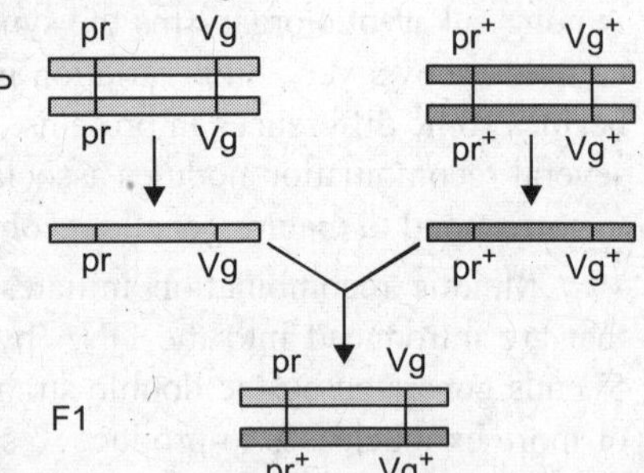

Fig. 4.7. Inheritance of alleles present on the same pair of chromosomes *i.e.* linked genes.

3. Genetic Recombination

Genetic recombination is the phenomenon of formation of new combinations of genes through the exchange of genetic material between homologous chromosomes (Fig. 4.8). A cell carries two copies of each gene which is referred to as an allele in eukaryotes. Each parent transfers one allele to each offspring. An individual gamete inherits a complete haploid complement of alleles on chromosomes independently selected from each pair of chromatids that are lined up on the metaphase plate. Without recombination, all alleles for those genes linked together on the same chromosome would be inherited together. It occurs in nature automatically in the dividing cells. During mitosis no genetic recombination occurs; it takes place during pachytene of prophase of meiosis I through crossing over and male and female gametes are produced. Meiotic recombination allows a more independent selection between the two alleles that occupy the positions of single genes, as recombination shuffles the allele content between homologous chromosomes.

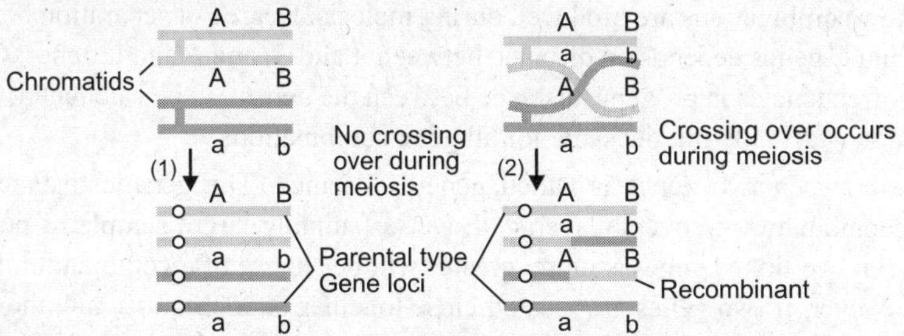

Fig. 4.8. Simples linkage (1); production of recombinants due to crossing over of two non-sister chromatids between the linked genes (2).

After fertilisation new combinations of genes are formed in individuals. Thus varieties are produced that have different genotypes. After every cycle of meiosis and fertilisation between gametes new combinations of genes are produced. Hence, genetic diversity among sexually reproducing organisms has arisen through genetic recombination occurring in nature. Repeated genetic recombination has given rise to thousands of new combinations, *i.e.* allelic forms in organisms. Thus multiple alleles have evolved through this method and also through mutation. This theory of 'independent assortment' of alleles is fundamental to genetic inheritance.

4. Gene Mapping (Genetic Mapping)

Alfred Sturtevant, a student of Morgan, first developed genetic maps which are also known as gene maps or linkage maps. Sturtevant proposed that the greater the distance between linked genes, the greater the chance that non-sister chromatids would cross over in the region between the genes. By working out the number of recombinants it is possible to obtain a measure for the distance between the genes. Genetic mapping is the determination of the relative position of genes on the chromosomes, *i.e.* DNA. In earlier discussion, linkage was explained to occur due to deviation of independent assortment of alleles. The genetic traits are retained on the same chromosome during crossing over. In 1910, T.H. Morgan first demonstrated linkage of genes in *Drosophila melanogaster*. It was explained to occur due to the presence of genes on the same chromosome. Further experiments were done by Morgan and co-workers on mutant *D. melanogaster*. They found that several hereditary characters such as yellow body, white eyes, acute brittles, miniature, etc. are found in one group on sex chromosome. Hence, these are sex-linked characters. There are many groups of traits present on other chromosomes.

In all *D. melanogaster* consists of linked groups of genes each associated with one pair of chromosomes. Similarly, seven linkage groups in sweet pea and ten in maize have been reported. These numbers of linkage groups, *i.e.* four, seven and ten represent the, number of haploid chromosomes in the organisms. Morgan was of the view that there existed a relationship between recombination frequencies of linked genes and their linear distance along the chromosome.

To understand gene mapping, let us consider the presence of three genes *A, B* and *C* in linear order on a chromosome (Fig. 4.9). There are three possibilities for their presence in linear order such as *ABC, ACB* or *BAC*. In the first combination gene B is present in centre, in second gene *C* and in third gene A are located in centrc. The exact position could be found out though three-point test cross. Linear order of gene is searched and linkage map

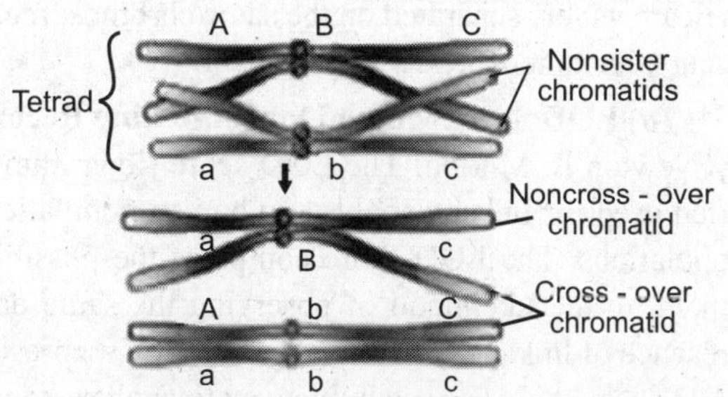

Fig. 4.9. Double crossing over between A and B, and B and C genes and production of recombinants.

is prepared. New combinations are produced during meiosis. Degree of separation between *A* and *B* genes, and B and *C* genes depends on distance between *A* and *B*, and *B* and *C* genes. On the basis of recombination frequencies in progeny, distance between the genes can be calculated. The frequency of recombination is actually not the same for all gene combinations.

Distance between two genes is called genetic distance. The genetic distance, which is a measure of recombination frequency averaged over a (suitably large) sample of pedigrees. If the distance between two linked genes is more, greater will be chance of recombination frequency, and vice versa. However, if two genes are located close together on a chromosome, the possibility for separation of these two genes by a recombination event is less than they were farther apart.

The distance between genes is measured in *map units* (mu). Some scientists refer the map unit as centiMorgan (cM) in the honour of T.H. Morgan. One cM is equal to about 1% recombination frequency between the two genes. For the preparation of genetic map the distance between the genes and the order of their arrangement on chromosomes are required.

Recombination Frequency: Recombination frequency (θ) is the frequency that a single chromosomal crossover will take place between two genes during meiosis. Double crossovers would turn into no recombination. A double crossover is very unlikely if the gene loci are very close (less than 7 cM). But the possibility of a double crossover becomes more when distances between two gene loci become higher. Recombination frequency is a measure of genetic linkage and is used in the creation of a genetic linkage map.

The Mendel's Second Law, *the law of independent assortment,* states that during meiosis the chromosomes assort randomly into gametes such that the segregation of alleles of one gene is independent of alleles of another gene. The law of independent assortment is always true for genes that are located on different chromosomes; but it is not always true for those genes that are present on the same chromosome. When two genes are close together on the same chromosome, they do not assort independently and are said to be linked. Whereas genes located on different chromosomes assort independently and have a recombination frequency of 50%; linked genes have a recombination frequency is less than 50%.

For example consider the crossing of the pure bred homozygote parental strain having genotype *AABB* with a different pure bred strain having genotype *aabb*. *A* and *a,* and *B* and *b* are the alleles of genes *A* and *B*. When these homozygous parental strains are crossed, they result offspring with genotype *AaBb* in F1 generation. The F1 offspring *AaBb* produces gametes (*AB*, *Ab*, *aB*, and *ab*) with equal frequencies (25%); because during meiosis the alleles of gene *A* assort independently of the alleles for gene *B*. Two of the four gametes (50 %) such as *Ab* and *aB* were absent in the parental generation. These gametes are recombinant gametes, which differ from both of the haploid gametes that made up the diploid cell. Since two of the four gametes were recombinant gametes, the recombination frequency is 50%. When two genes are located on different chromosomes or when they are widely separated on the same chromosome, the recombination frequency will be 50%. This occurs due to independent assortment.

(*a*) LOD Score Method for estimating Recombination Frequency: This test was developed by Newton E. Morton. The LOD score [logarithm (base 10) of odds] is a statistical test which is often used for linkage analysis in human populations; but it is also applicable in animal and plant populations. The LOD score compares the possibility of getting the test data if the two loci are linked, to the likelihood of observing the same data by chance. Positive LOD score favours the presence of linkage, whereas negative LOD scores shows less possibility of linkage. Computerised LOD score analysis is a simple way to analyse complex family pedigrees in order to determine the linkage between Mendelian traits (or between a trait and a marker or two markers).

The method is described in greater detail by Tom Strachan and Andrew P. Read in 1999. It works as follows: (*i*) establish a pedigree, (*ii*) make a number of estimates of recombination frequency, and (*iii*) calculate a LOD score for each estimate. The estimate with the highest LOD score will be considered the best estimate. The LOD score is calculated as below:

$$\text{LOD} = Z = \log_{10} \frac{\text{Probability of birth sequence with a given linkage value}}{\text{Probability of birth sequence with no linkage}}$$

$$= \log_{10} \frac{(1-\theta)^{NR} \times \theta^{R}}{0.5^{(NR+R)}}$$

Where, NR is the number of non-recombinant offspring, and R is the number of recombinant offspring. The reason 0.5 is used in the denominator where alleles that are completely unlinked (for example, alleles on separate chromosomes) have a 50% chance of recombination due to independent assortment. Theta is the recombinant fraction which is equal to R / (NR+R). LOD scores are seen in a table which enlists LOD scores of various standard pedigrees and various values of recombination frequency. A LOD score greater than 3.0 is considered evidence for linkage. A LOD score of +3 indicates 1000 to 1 odds that the linkage being observed did not occur by chance. A LOD score less than -2.0 is considered evidence to exclude linkage.

B. CYTOPLASMIC INHERITANCE (= EXTRA-NUCLEAR INHERITANCE)

The term 'extra nuclear' refers to outside the nucleus, *i.e.* in the cytoplasm. The cytoplasmic factors that show replication, independent transmission and expression of characters in organisms are called 'cytogenes' or 'plasmagenes', and all the 'plasmagenes' together constitute 'plasmon' (like genome). They consist of their own genes. The genetic traits are not transmitted in Mendelian fashion. Except a few cases, the traits are transferred by egg cytoplasm of the female plant. However, the traits are detected from the differences obtained in reciprocal crosses. Following are the bases for extra-nuclear inheritance that excludes all possible chromosomal explanations of inheritance.

Genetic traits do not segregate as per Mendelian fashion; therefore Mendalian ratios are not obtained. A trait persists undiminished for many generations. The genetic factors of cytoplasm are also contributed and controlled by chloroplast DNA (in plant) and mitochondrial DNA. Mendelian factors (in the form of mRNA (originated from chromosome) of egg or ovules affect the characters. This affect is called uniparental maternal inheritance. Exceptionally the male parents (through pollens) can also contribute genetic factors to cytoplasm. Such influence is called 'biparental inheritance'. The subject of extra-nuclear or cytoplasmic inheritance is quite vast, yet it has been discussed in the light of some examples.

1. Maternal Effects

In the sexually reproducing organisms female contributes more cytoplasm through eggs or ovules than the male through sperms. So in extranuclear inheritance, contribution of female parent is more. Maternal effects are caused due to effects of genes in cytoplasm. A unique phenotypic effect produced by cytoplasm is called maternal effects. This effect can be exemplified by a fresh water snail (*Limnaea peregra*), in which direction of coiling of shell can be *dextral* (= clockwise, *i.e.* coiling to right direction) or *sinistral* (– anticlockwise, *i.e.* coiling to left). Shell coiling is genetically controlled by a pair of genes. Dextral coiling is done by dominant alleles *DD* and sinistral coiling by recessive alleles *dd*. The phenotype of F1 offspring may be dextral or sinistral in spite of having genotype *Dd* alleles in both cases (Fig. 4.10).

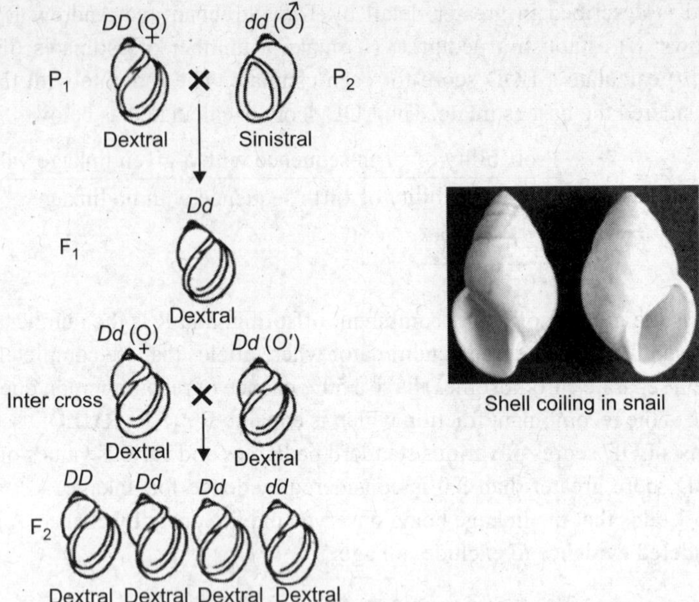

Fig. 4.10. Inheritance of shell coiling in a cross between female dextral (*DD*) and male sinistral (*dd*) snails.

Surprisingly, in the offspring of F1 generation phenotype is determined by genotype of female parent but not of male parent. Snail is hermaphrodite. When dextral female (*DD*) is crossed with sinistral male (*dd*), all the offspring of F_1 generation (*Dd*) have dextral coiling. If sinistral female (*dd*) is crossed with dextral male (*DD*), the offspring have *Dd* genotype but coiling is sinistral. When the latter are self-bred, three types of genotypes are produced- *DD, Dd* and *dd*. However, shell coiling in all is dextral because the mother has *Dd* genotype.

Phenotypes of F1 depend on genotype of female parents. After inter-crossing, all offspring of F2 generation becomes dextral. Similarly F2 offspring in spite of having allele *dd* have dextral coiling if genotype of female parents is heterozygous (*Dd*). The maternal effect lasts only for one generation. In onward generation sinistral coiled offspring are produced from homozygous recessive (*dd*) dextrally coiled parents (Fig. 4.10). However, it is possible that the direction of cell cleavage in fertilised eggs up to a few cell divisions may decide the direction of shell coiling.

2. Extra-Nuclear Inheritance by Cellular Organelles

There are two cell organelles (chloroplasts and mitochondria) which consist of their own DNA and ribosome, *i.e.* protein-synthesising apparatus. It is believed that they were once infectious endosymbiotic prokaryotes that evolved such a dependence on the gene products of the host that they are no longer able to function autonomously. The genetic components of these organelles are often similar to those found in prokaryotes. The chloroplasts of certain algae and *Euglena* contain 70S ribosomes and a naked circular double stranded DNA. Their protein synthesis begins with the amino acid *N*-formyl methionine as found in prokaryotic protein synthesis. Moreover, their DNA-dependent RNA polymerase is sensitive to the inhibitor rifampicin.

Besides, the mitochondrial ribosomes of fungi are larger (77S) than bacterial ribosomes (70S), while mammalian mitochondrial ribosomes are smaller (60S) than bacterial ribosomes. Chloroplast chromosomes consist of more DNA than the average mitochondrial chromosomes and thus can potentially carry more genetic information. The genetic materials of chloroplasts and mitochondria are transmitted to offspring through the egg. Maternal inheritance through chloroplast and mitochondria is discussed giving the following examples :

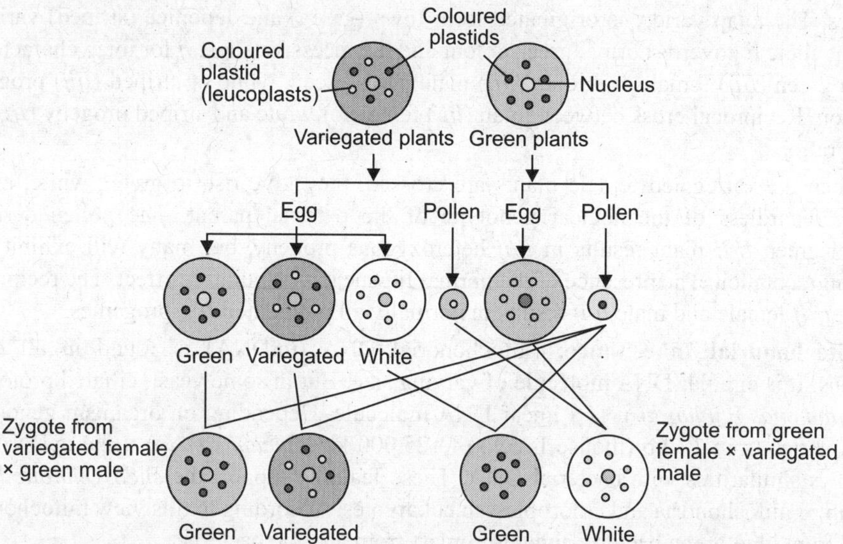

Fig. 4.11. Distribution of plastids and leucoplasts in progeny and results of reciprocal cross between female and variegated male plants.

(a) Plastid (Chloroplast) Inheritance in Plants: Plastids (chloroplasts) are the organelles of the eukaryotic plants cells that contain DNA. The chloroplast DNA (cpDNA) is prokaryotic in nature but also shares eukaryotic characteristics. During mitosis almost equal amount of chloroplasts is distributed into the daughter cells. Each cell has many chloroplasts and each chloroplast has several circular DNA molecules. In lower eukaryotes the total percentage of DNA is lower (about 14%) than the higher plants (about 3%). The cpDNA synthesises rRNA, tRNA, mRNA and proteins.

Four O'clock plant (*Mirabilis jalapa*): In 1909, Carl Correns reported extranuclear inheritance of leaf variegation in four O'clock plant (*Mirabilis jalapa*). The variegated plant contained mixture of leaves, *i.e.* leaf with green and white patches, green leaves and white patches. Accordingly, flowers appeared on the branches. Seeds produced by green branches gave rise to green plants. Seeds from white branches gave rise to white plants (plants died due to lack of chlorophyll). Seeds from variegated branches produced three types of progenies plants with green, white or variegated branches (Fig. 4.11). The cpDNA of female parent is solely responsible for inheritance of variegated phenotype leaves in all progeny. Variegated leaf phenotype does not depend on male parent.

A cross between green female and green/white/variegated male resulted in green plants only in F1 generation (Fig. 4.12). Mating of white female with green/white/variegated male gave rise to white plants in F1. When variegated female was crossed with white/green/variegated male, F1 generation resulted in variegated green or white progeny.

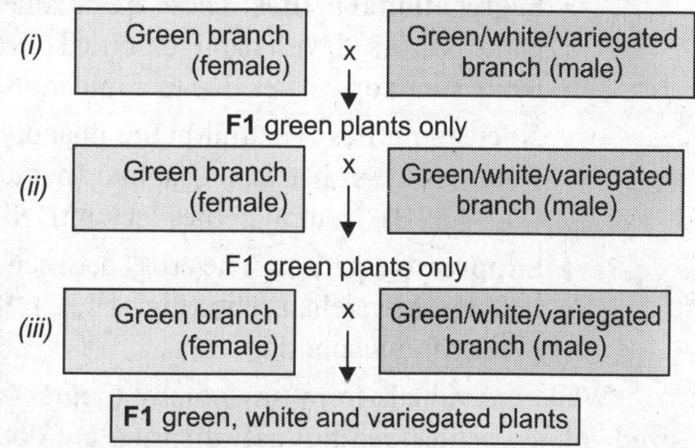

Fig. 4.12. Inheritance of variegated leaves in *Mirabilis jalapa*.

(*ii*) Inheritance by iojap gene of corn: There are other examples also that show uniparental maternal inheritance of characters through cpDNA. The plastid genes control integrity of plastids. A gene in corn plant called lola (ij) has been mapped by in 1946 by M. Rhoades to nuclear chromosomes vii. The *iojap* variety of maize is characterised by contrasting stripes of green and white colours on leaves. The iojap variety is originated from Iowa (green) and Japonica (striped) varieties. The dominant allele Ij governs normal green colour and the recessive allele *ij* for iojap character. A cross between green (*IjIj*) female and iojap (*ijij*) male gives green, white or striped (*Ijij*) progeny in F1 generation. Reciprocal cross between iojap (*ijij*) female of white and striped progeny (*IjIj* and (*IjIj*) in equal ratio.

When the variegated female plants are crossed, they give rise to green, white, and striped progeny, regardless of the nuclear genotype of the paternal parent. The pollen derived from a normal green *Ij/Ij* plant results in *Ij/ij* heterozygous progeny, but many will exhibit abnormal plastid pigmentation. The presence of the normal Ij gene has no curative effect. The reciprocal cross between *Ij/Ij* female and male *ij/ij* results in normally all pigmented *Ij/ij* progenies.

Mitochondrial Inheritance: Mitochondrial DNA (mtDNA) is found in all eukaryotic organisms. It is circular DNA molecule of varying size. But in some yeast, ciliated protozoans and *Chlamydomonas reinhardii* it is a linear DNA molecule. Depending on organism genome size of mtDNA varies from 17 kb (mouse L cells) to 25,000 kb (*Cucumis melo*). Like cpDNA, mtDNA also shows similarities with bacterial DNA. These features support the *endosymbiotic hypothesis* for origin of mitochondria and chloroplast in eukaryotes. According to this view mitochondria have evolved from blue-green bacteria and chroroplast from aerobic bacteria.

(*i*) Inheritance in Yeast: Yeast (= *Saccharomyces cerevisiae*) are single celled ascomycetes fungi. Like chloroplast, mitochondria are also cytoplasmic organelles. They also contain their own DNA and some enzymes for oxidative phosphorylation. However, for proper function, mitochondria need both the nuclear genes and its own genes. The role of mtDNA in cytoplasmic inheritance was first reported in 1955 by Ephrussi and co-workers in petite mutants (petite means small sized) in yeast (*Saccharomyces cerevisiae*). In yeast each mtDNA is 84 kb long.

Certain strains of yeast are comparatively small-sized and slow growing, producing small colonies on agar; therefore, such strains are called 'littles or petites'. The petites grow slowly due to the loss of aerobic respiratory system of the cells and the utilisation by the cell of less efficient fermentation process.

The petite differs from wild type (Grande) to some extent. The petites were insensitive to inhibitors of aerobic pathways, and lacked cytochromes c, a3, b, etc. They contained incompletely developed mitochondria, and did not contain stainable mitochondria. The petite mutants were segregational or non-segregational, *i.e.* extra-chromosomal. There is evidence of spontaneous origin of petite mutants at the rate of one per 500 cell generation. But about 100% mutants may be obtained after treating the wild type with UV light, ethidium bromide or acridine compounds *e.g.* acriflavine. There are three types of petite mutants, the segregational petite, neutral petite and suppressive petite (Fig. 4.13 ABC).

- **Segregational petites:** These are obtained in a ratio 2:2 (petite: wild type) when a haploid petite is crossed with haploid normal (Grande A). This shows Mendelian inheritance. The petite strain originates due to a mutation in nuclear genes (Fig. 4.13A).
- **Nuclear petites:** Neutral petite phenotypes having all progeny wild type are obtained from a cross between haploid petite (mutant) and haploid wild type (Grande) (Fig. 4.13B). Neutral petites lack mtDNA in their mitochondria.
- **Suppressive petites:** The cross between petite and Grande results in 99% petite progeny in a non-Mendelian ratio (Fig. 4.13C). Suppressive petite mutants have mutant mtDNA in their mitochondria.

When individuals from tiny colony (petites) are crossed with the individuals from a normal sized colony, petites and normal offspring are obtained in the ratio of 1 : 1 in F1 generation. In addition to yeast, mitochondrial inheritance has also been found in *Neurospora, Paramecium, Trypanosoma*, maize, etc.

***(ii)* Inheritance in Poky Strain of *Neurospora*:** *Neurospora crassa* is an ascomycetous fungus, in which several mutations of mitochondria are inherited through the female parent. The best studied of these is the poky strain which was first isolated by Mitchell and Mitchell in1952. Pocky strain grows very slowly and results in poorly differentiated mitochondria. The mitochondria deficient in the membrane-bound cytochromes (*b*, *a* and a_3), have excess cytochrome *c*, lack essential proteins of the respiratory electron transport chain, and possess greatly reduced numbers of the small ribosomal units and small (19S) rRNA.

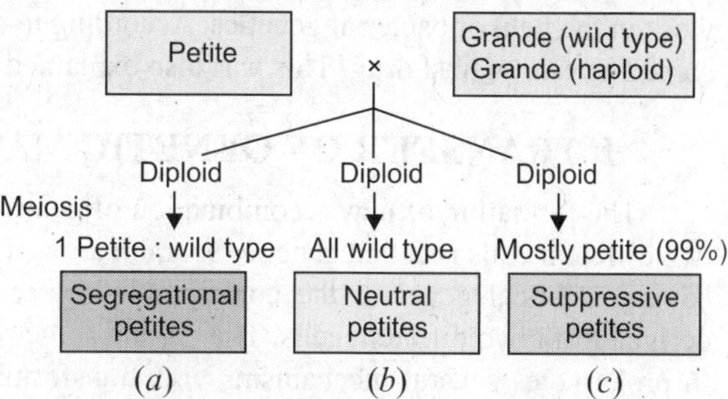

Fig. 4.13. Segregation pattern of petites in yeast.

A four base pair deletion is the primary defect in poky. The membrane-bound, cold-stable, oligomycin-sensitive ATPase activity is present in poky. Poky also contains a second form of mitochondrial ATPase which has an activity of 1.5+/-0.2 mmol/min-mg of protein, is oligomycin sensitive but cold labile, and presumably is attached less firmly to the mitochondrial membrane.

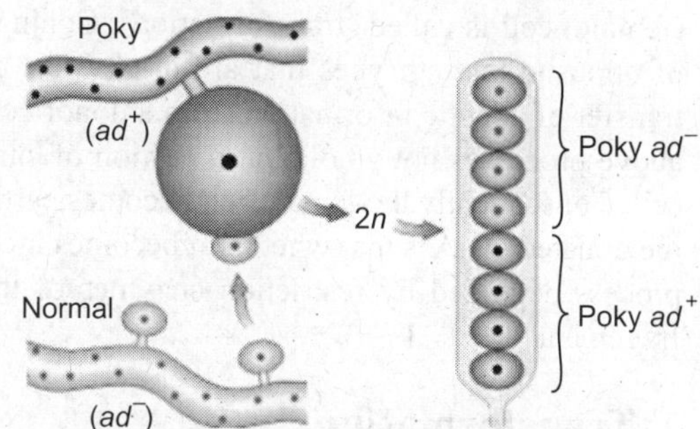

Fig. 4.14. A cross between pocky strain and a normal strain of *Neurospora crossa*.

Thus the poky strain may carry a mutant gene or genes for such mitochondrial deficiencies. But the mitochondrial DNA of poky strain is biochemically similar to mitochondrial DNA of wild type *N. crassa*. Poky mutation is a suppressive mutation where the mutant phenotype suppresses the normal. The poky trait is inherited in a fashion of maternal inheritance. If the proto-perithecium is derived from a poky hypha and the male conidium from a wild type hypha, the resulting eight ascospores are all poky (Fig. 4.14), whereas in reciprocal crosses the ascospores are generally all wild type.

C. MICROBIAL GENETICS

The method of cell division differs in prokaryotes from that of eukaryotes; however, no meiosis or mitosis occurs in prokaryotes because of lack of nucleus. In contrast, most of genetical and molecular works have been carried out in prokaryotes and a few eukaryotic microorganisms because of ease in handling, fast growth, immediate observable characters under any change in conditions and least complex life forms. Bacteria and the other protists transmit characters to their progenies and onward generations. Besides, they also show variations in progenies; these changes are referred to as phenotype (the morphological changes) and genotype (genetical changes). The genotype always remains constant; however, changes occur through mutation which results in morphological changes or phenotypes in progenies.

However, the phenotypic expression depends on the environmental conditions as well. For example, phenotypic changes in *Agrobacterium radiobacter* takes place if it is grown on two different growth media. Mucoid colonies are formed on sucrose salt medium and non-mucoid colonies develop on trypticase soy agar medium. Secondly, yeast produces ethanol when grown in oxygen-deficient conditions, and increases its biomass and does not produce ethanol when grown in the presence of sufficient oxygen.

In 1943, for the first time S. E. Luria and M. Delbruck gave the experimental evidence for acquisition of drug resistance by sensitive bacteria, and this concept formed the basis for modern work in the field of bacterial genetics. According to them resistance occurs as a direct adaptation by some bacteria against drug. This was also explained by spontaneous mutation theory.

I. TRANSFER OF GENETIC MATERIAL IN PROKARYOTES

The formation of new recombination of genes (genotypes) through re-assortment of genes of two different cells is called genetic recombination. This occurs between two different chromosomes having identical genes at the corresponding sites. These are called homologous chromosomes derived from two different cells. The chromosomes of progenies differ from those of parental cells. In prokaryotes several mechanisms viz., transformation, conjugation and transduction have been described that mediate the formation of new recombinants.

Transfer of relatively small segment of naked DNA from a donor cell (male) to the recipient (female) cell is called 'transformation'. Conjugation is the process of gene transfer between cells of opposite mating types that are in physical contact with each other. Transduction involves the transfer of genetic information from a donor cell to a recipient cell through a bacteriophage. In the above processes not whole but a portion of chromosomes is transferred from a donor to recipient cell. Consequently the female cell becomes partially diploid called merozygote. The ultimate fate of the donated DNA is that whether it becomes incorporated in the recipient genome by recombination process degraded by restriction enzymes of the host or maintained as stable extrachromosomal fragement.

1. Transformation

In 1928, an English Health officer, Frederick Griffith first discovered transformation between the two strains of a bacterium *Streptococcus* (*Diplococcus*) *pneumoniae* which causes pneumonia in humans and other animals. The results of Griffith are shown in Fig. 4.15. There were two types of pneumococi, type II and type III. Each type exists in two forms RII, SII, and RIII, SIII forms, where 'R' represents the non-capsulated, and rough colony forming, and non-virulent mutant strain, and 'S' represent the smooth, encapsulated, virulent and wild strain.

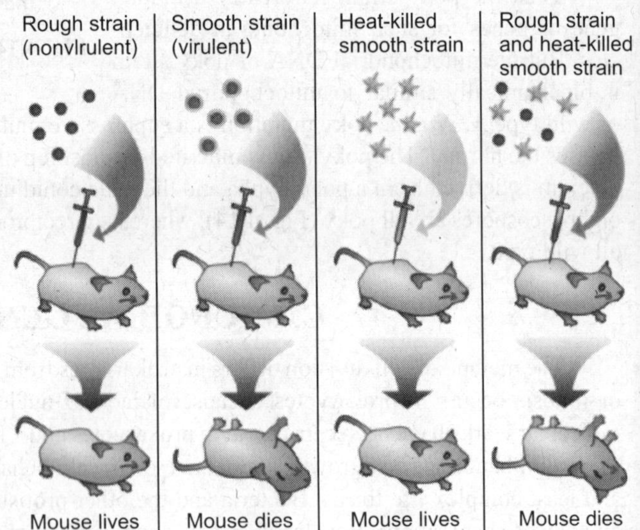

Fig. 4.15. Griffith's transformation experiment.

After injection of smooth (virulent) strain, mice were killed. When boiled smooth strain SIII was injected, the mice were not affected and no pneumococci could be recovered from the mice. Upon injection of a mixture of heat killed smooth strain and live (non-virulent) rough RIII strain, the mice were killed and live virulent bacteria could be recovered from blood of dead mice. Griffith called this change of non-virulent strain into virulent strain as transformation, because the virulent strain transformed the non-virulent strain into the virulent strain. This is called Griffith's transformation experiment (Fig. 4.15). Griffith thought that the transformation would have been caused by a protein.

(*b*) Identification of Transforming Substance: Griffith could not identify the nature of transforming substance. In 1944, Oswald T. Avery, C.M. Mac Leod and M.J. McCarty set out experiments to identify the transforming principle. They demonstrated that heat killed SIII strain transformed RII strain into virulent form which caused death of mice. They destroyed cell constituents in extract of virulent pneumococci SIII using enzyme that hydrolysed DNA, RNA, proteins and polysaccharides. The transforming ability was not altered by treatment with enzyme or by RNase, but was completely destroyed by DNase. They treated the non-virulent strain separately with bacterial extracts as below:

RII Cells + Purified SIII cell polysaccharide	→	R colonies
RII Cells + Purified SIII cell protein	→	R colonies
RII Cells + Purified SIII cell RNA	→	R colonies
RII Cells + Purified SIII cell DNA	→	SIII COLONIES
RII Cells + Purified SIII cell extract + protease	→	SIII colonies
RII Cells + Purified SIII cell extract + RNase	→	SIII colonies

They concluded that a cell-free and highly purified RNA extract of SIII strain could bring about transformation of RII strain into SIII strain. However, this effect was lost when the extract was treated with deoxyribonuclease. Therefore, it proves that the DNA transforms the RIII strain into SIII strain; therefore, DNA is the genetic material and carries sufficient information for transformation. The major steps of transformation are given in Fig. 4.16.

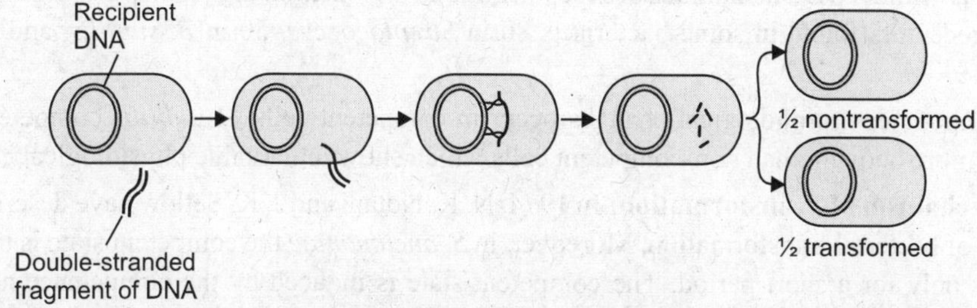

Fig. 4.16. Bacterial transformation.

Thereafter, the transforming material (DNA) was also confirmed in several bacteria such as *Bacillus subtilis, Haemophilus influenzae, Shigella paradysenteriae,* and species of *Neisseria, Xanthomonas, Rhizobium, Bacillus, Staphylococcus,* etc. Generally transformation does not occur in *E. coli* but it does after the treatment with calcium chloride. Possibly it facilitates the entry of DNA into the recipient cells.

(*a*) Factors Affecting Competence: Competence is a physiological state that permits a cell to take up transforming DNA and be genetically changed by it. On the basis of development of competent state the organisms undergoing transformation can be divided into two groups: organisms always present in competent stage (for example, *Neisseria*) and the organisms transiently competent in late exponential phase of growth (for example, *S. pneumoniae*). These differences misrepresent the complex series of regulatory processes that are required to control this mechanism.

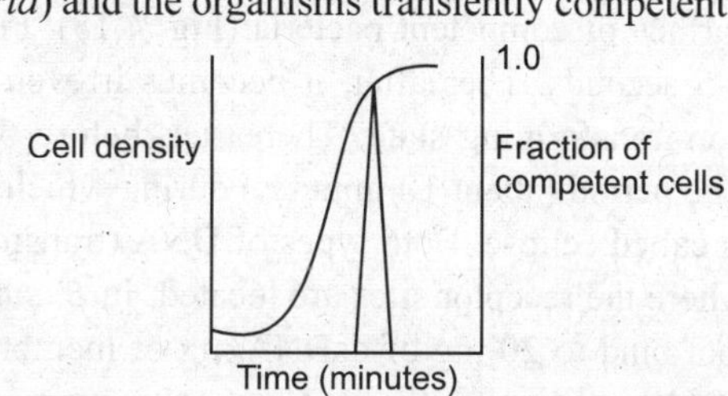

Fig. 4.17. Development of competence in *Streptococcus pneumoniae.*

(*i*) Competence is not a permanent feature of bacteria but occurs only for a short duration during the life cycle. However, there are several factors that affect competence. Commonly competence is observed towards the end of logarithmic phase of growth, *i.e.* just before the establishment of stationary phase (Fig. 4.17). In *S. pneumoniae* duration of competence persists for 7-15 minutes. In *Bacillus subtilis* it persists for several hours.

(ii) Ability of most of the bacteria to receive DNA efficiently is limited. After proper incubation the cell density increases greatly and yields a population of cells in which uptake of DNA is enhanced to 10^4 - 10^6 by a factor. However, the conditions that induce competence and the fraction of competent cells vary from one species to the other. Both the fractions of culture that become competent and the duration of competent state also vary species to species. For example, 100% *S. pneumoniae* cells become competent, while only 20% cells of *B. subtilis* become competent. But the duration of competence in *S. pneumoniae* lasts for a few minutes.

(iii) The growth medium, temperature, degree of aeration, etc. also influence the development of competence in cells. There are three types of competence in *B. subtilis*: nutritional, growth stage-specific and cell type-specific. The best competence develops in glucose minimal medium. Competence develops post exponentially and may involve nitrogen starvation as a key factor. When *Hemophillus influenzae* cells growing exponentially are shifted to the new growth medium, 100% cells become competent.

(iv) Some heat stable/heat labile extracellular competence factors of proteinaceous in nature are involved in development of competence. The competence factor of *S. pneumoniae* has a molecular weight of 10,000 Daltons and of *Streptococcus* strain less than 500 Dalton.

(v) Some kinds of competence receptors are either formed or activated on cell envelope which initially binds with DNA. The number of receptors varies organism to organisms. For example, the number of receptors found in some bacteria is 80 in *Streptococcus,* 50 in *B. subtilis*, and 4 in *H. influenzae.*

(vi) Sometimes morphological changes occur in competent cells. *B. subtilis* competent cells prove to be more buoyant than non-competent cells which show remarkable physiological changes.

(c) Mechanism of Transformation: In 1974, N. K. Notani and J. K. Setlow have described the mechanism of bacterial transformation. Moreover, in *S. pneumoniae* the competent state is transient and persists only for a short period. The competent state is induced by the competence activator protein of molecular weight of 1,000 Dalton. It binds to the plasma membrane of receptor and triggers the synthesis of 10 new proteins within 10 minutes. The competence factor (CF) accelerates the process of transport or leakage of autolysin molecules into the periplasmic space. Moreover, in *H. influenzae* no competence factors have been reported. Only changes in cell envelope accompany the development of competence state. The cell envelope of competent cells contains increased level of polysacccharide as compared to the cells of log phase. Structural changes in competent cells induce numerous vesicles called transformosome buds on the surface that contains protein and mediates the uptake of transforming DNA. Transformation is accomplished in the following steps (Fig. 4.18).

***(i)* DNA binding:** As a result of random collision, DNA comes first in the contact of cell surface of competent bacteria (Fig. 4.18). First the DNA binding is reversible and lasts for about 4-5 seconds. Thereafter, it becomes irreversible permanently. For about 2 minutes it remains in non-transforming state. Thereafter, before 5 minutes it is converted into the transforming state. The period (about 10 minutes) during which no transformation occurs in competent recipient cells is called eclipse. Both types of DNA (transforming and non-transforming) bind to the cell surface where the receptor sites are located. In *B. subtilis* membrane vesicles in competent cells are found that bind to 20 mg of dsDNA/mg of membrane protein. The competent cells show six fold more DNA binding sites than the non-competent cells.

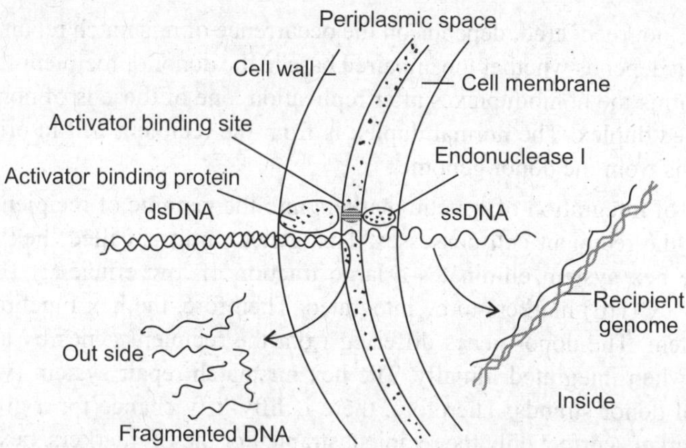

Fig. 4.18. Transformation in streptococci (diagrammatic) (based on Moat and Foster, 1995).

In *H. influenzae* transformosome bud forms the surface and contains proteins that mediate DNA uptake. It binds with conserved sequence (5'AAGTGCGGTCA 3') present at 4 kb interval on DNA. The DNA uptake site contains two proteins of 28 and 52 kilodaltons. After binding, the receptor proteins present the donor DNA to the membrane associated uptake sites. In *S. pneumoniae* the CF induces the ability to bind DNA molecules.

***(ii)* Penetration:** The DNA molecules that bind permanently enter the competent recipient cells. DNA is also resistant to DNase degradation. The nucleolytic enzymes located at the surface of competent recipient cells act upon the donor DNA molecule when it binds the cell membrane. The endonuclease-1 of the recipient cells which is associated with cell membrane acts as DNA translocase by attacking and degrading one strand of the dsDNA. Consequently only complementary single strand of DNA enters into the recipient cells. It has been confirmed by performing the experiments with radiolabelling of donor DNA. The mutant cells of *S. pneumoniae* lack endonuclease-1, therefore, transformation does not occur. Interestingly in *B. subtilis* degradtion of one strand is being delayed. Hence, both the strands enter the recipient cell. The upper limit of peneterating DNA into the recipient cell is about 750 base pairs.

Transformation is affected by the size of donor DNA; however, successful transformation occurs with the donor DNA of molecular weight between 30,00,000 Da and 8 million Da. The number of competent cells increases with increasing the concentration of donor DNA. After penetration the donor DNA migrates from periphery of cell to the bacterial DNA. This movement in different bacteria differs. For example, in *B. subtilis* this movement occurs for about 16-60 minutes. During this movement, DNA is associated with mesosomes which possibly transport it to the bacterial DNA.

***(iii)* Synapsis formation:** Single strand binding (SSB) proteins coat the single stranded DNA and maintains the single stranded region in a replication fork. Then the single strand of the donor DNA or portion of it is linearly inserted into the recipient DNA. The bacterial protein such as *E. coli* RecA protein probably facilitates the DNA pairing during recombination, and causes the local unwinding of dsDNA of the recipient cell from the 5' end. Base pairing, *i.e.* synapsis occurs between the homologous donor ssDNA and the recipient DNA. Unwinding of the recipient DNA continues at the end of assimilated DNA and allows the fraction of invading DNA to increase base pairs. This processs is called branch migration.

***(iv)* Integration:** The unpaired free end of donor DNA or the recipient DNA is cleaved by the endonuclease; this process is called trimming. The nick is sealed by DNA ligase resulting in a heteroduplex region containing mismatched base pairs is formed. In the progenies whether the donor marker is or is not recovered, depends on the occurrence of mismatch repair. If the mismatch repair occurs again, it depends whether the unpaired base in the donor or recipient strand is removed. The heteroduplex forms the homoduplexes after replication, one of these is of normal type and the second is transformed duplex. The normal duplex is from the recipient cell in origin, whereas the transformed duplex is from the donor genome.

The efficiency of integration of genetic markers into the genome of recipient cell varies with different genes that the recipient cell possesses. This genetic trait is called 'hex' (high efficiency of integration). The hex system eliminates a large fraction of low efficiency (LE) markers and permits high efficiency (HE) markers to be integrated. Therefore, the hex function is a mismatch-base correction system. The donor genes differing from the recipient genes by a single base pair create a mismatch when integrated initially. The hex mismatch repair system (with LE markers) can correct either of donor strands. Therefore, there is fifty-fifty chance for a given marker to be retained. The HE markers correct only the recipient strand. For the LE markers, hex mismatch repair system unusually removes the mismatched bases of the donor DNA and the cell retains the recipient genotype, whereas for HE markers the same system removes the mismatched bases of recipient DNA and the cell consists of donor genotype. In the later case, after replication of chromosome and cell division the one progeny cell contains the donor genotype and the other has the recipient genotype. These two types of cells can be differentiated through plating method by using the antibiotic markers.

Hershey-Chase Experiment (evidence from bacteriophages that DNA is genetic material) : In 1952, Alfred D. Hershey and Martha Chase carried out several experiments in bacteriophage T_2 that proved the DNA to be the genetic material of T_2. They prepared a culture medium for the growth of *E. coli* that contained phosphorus and sulphur. Known amount of isotopic phosphorus (^{32}P) and sulphur (^{35}S) was amended to this culture medium. Cell suspension of *E. coli* was poured onto the growth medium. After bacterial growth ^{32}P and ^{35}S were incorporated into the bacterial cell. These cells were allowed to get infected by T_2 bacteriophage. T_2 was multiplied in *E. coli* cells; the ^{32}P was incorporated into DNA molecule and ^{35}S into capsid protein of T_2 phage. Since DNA lacks sulphur, therefore, ^{35}S was not incorporated into DNA. By using this method they labelled differentially the phage DNA and coat protein with the isotopes. These radio-labelled phage particles were recovered from *E. coli* by centrifugation method (Fig. 4.19).

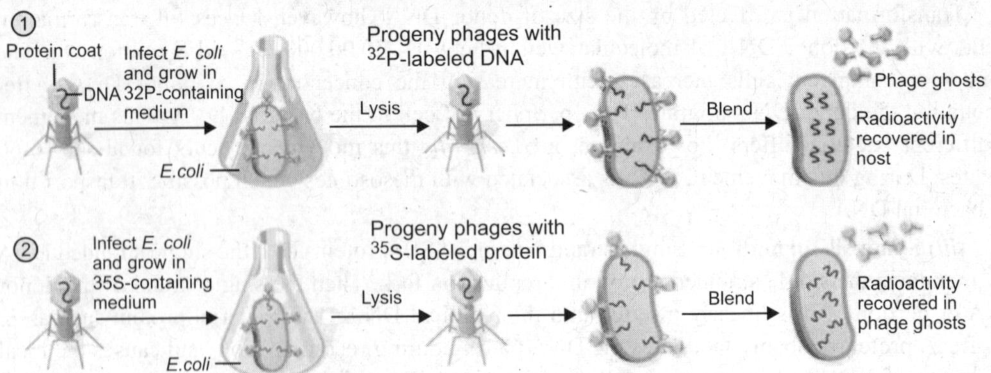

Fig. 4.19. Hershey-Chase bacteriophage experiment done in 1953.

In another experiment, non-radiolabelled *E. coli* cells were allowed to be infected by radiolabelled bacteriophages. After absoption the coat protein remained outside and only radiolabelled DNA was injected. From the infected *E. coli* cells, the viral particles were separated from the host through agitation. The amount of ^{32}P and ^{35}S in bacteriophage, particles, *E. coli* cells and medium was estimated. The ^{32}P was estimated with the bacterial cells and ^{35}S with protein coat left outside the bacterial wall, *i.e.* the growth medium. This experiment shows that the genetic material of T_2 phage resides in DNA but not in their protein. For this work A.D. Hershey shared Nobel Prize in physiology or medicine for 1969 with M. Delbruck and S.E. Luria. Furthermore, this experiment was repeated by several other workers by using different bacteriophages; in all the cases, DNA was found to give evidence of genetic material.

2. Conjugation

In 1946, for the first time, Joshua Lederberg and Edward L. Tatum in their brilliant and remarkable experiment presented the evidence for bacterial conjugation, *i.e.* a process of transfer of genetic material by cell-to-cell contact. They procured the two different auxotrophs (the mutant prototrophs lacking ability to synthesise an essential nutrient and, therefore, obtaining it or precursor from its surroundings) of *E. coli*, mixed them and incubated the two strains for hours in nutrient medium and plated on minimal medium (devoid of biotin, phenylalanine and other amino acids).

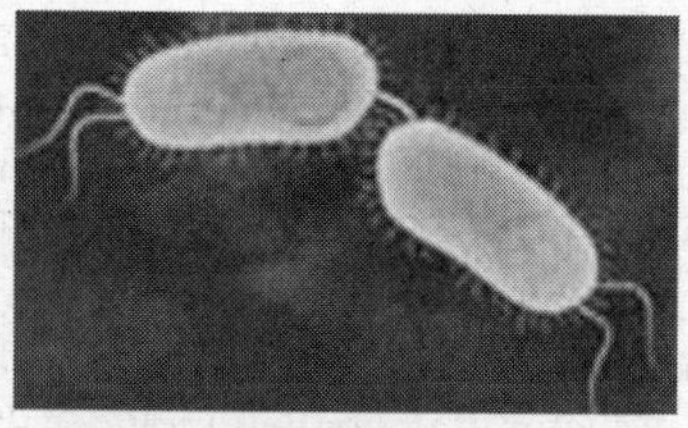

Two bacterial cells showing conjugation.

Fig. 4.20 shows the Lederberg and Tatum's experiment on bacterial conjugation by using triple auxotrophs. They used the triple auxotrophs to rescue from the chance of reversion. One strain (58-161) required biotin (Bio), phenylalanin (Phe) and cystine (Cys) for their growth, hence designated as Bio$^-$ Phe$^-$ Cys$^-$ Thr$^+$ Leu$^+$ Thi$^+$. The second strain (W677) required biotin threonine (Thr), leucine (Leu) and thiamine (Thi) and designated as Bio$^+$ Phe$^+$ Cys$^+$ Thr$^-$ Leu$^-$ Thi. They did not grow on minimal medium. When these two strains were mixed and allowed to grow on minimal medium, after incubation the recombinant prototrophic colonies (*i.e.* microbe requiring the same nutrients as the majority of naturally occurring microbial species) grew on minimal medium. Production of recombinant prototrophs (*i.e.* Bio$^+$ Phe$^+$ Cys$^+$ Thr$^+$ Leu$^+$ Thi$^+$) would have been possible as a result of recombination) between the two auxotrophs. The recombinant prototroph had capacity to synthesise all the six growth factors, *i.e.* amino acids.

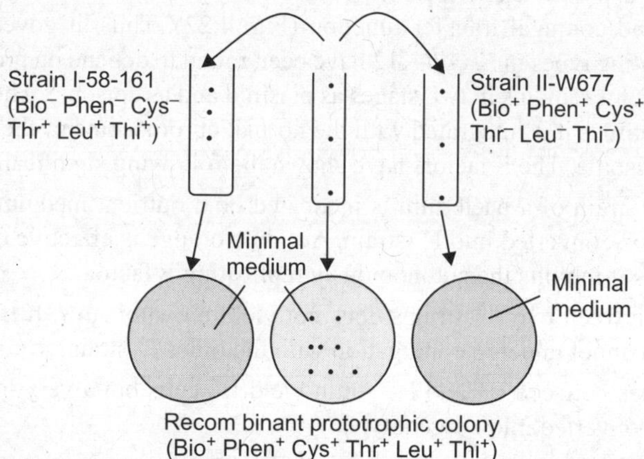

Fig. 4.20. Demonstration of Lederberg and Tatum's experiment on bacterial conjugation by using triple auxotrophs.

In 1946, Lederberg and Tatum could not give the proof for physical contact of cells required for gene transfer. In 1950, the evidence for cell-to-cell contact was provided by Bernard Davis who built a U shaped tube. Two separate pieces of curved glass tubes were prepared and fused at the base to form a U shape with a fritted glass filter between the halves (Fig. 4.21). The filter allows the movement of media but not bacteria from both the ends of U tube. Nutrient medium was inoculated with different auxotrophic strain of *E. coli.* When it was inoculated the medium was pumped back and brought forth from the filter to facilitate the exchange of medium present on either sides of the filter. The bacterial strains from both the halves of U tube were plated after 4 hours of incubation onto minimal medium. The bacterial colonies did not appear on medium because cell-to-cell contact could not be established and, therefore, gene transfer did not occur. This resulted in no production of recombinant prototrophs.

(*a*) Role of Surface Protein in Conjugation: In conjugation physical contact is established between two different strains through a conjugation tube. The genetic material from the donor cell (male) is transferred to the recipient (female) cell. Generally sex pili or F pili (singular, pilus) are present on cell surfaces which form the conjugation tube. The fertility (F) factor enables the cell to act as donor. The F factor of donor cell includes the information of sex pili; the number of F factor varies from 1 to 3 per cell. The cells containing an autonomous F are referred to as F^+ cells; the F factor replicates independently. But under certain specific conditions the number of pili per cell goes to five. The number of pili corresponds to the number of copies of F factor. This explains that each F factor synthesises a single pilus whether it is autonomous replicating conditions (as plasmid) or in integrated conditions (as episome).

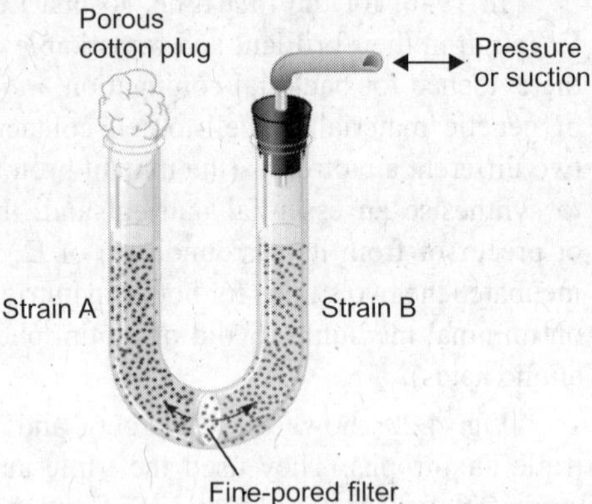

Fig. 4.21. The U-tube experiment of Bernard Davis.

Moreover, the recipient cells possess receptor sites on cell surfaces which are required for conjugation. Certain bacteriophages, such as f_2, MS_2, and Qb act as donor. The donor *E. coli* cells possess sex pili as well as Type I pilus on their cell surfaces. For example phage M12 is adsorbed randomly only on sex pili but not on cell surfaces of recipient bacterial cell.

(*b*) F Factor: The presence of F factor in a bacterial cell determines its autonomous replication, sex pili formation and conjugal transfer function (Fig. 4.22). Thus, it governs the sexuality and conjugation. Two mating types in *E. coli* K12 have been found to depend on presence and absence of the F factor. The F factor remains in two stages as plasmid and as episome. The F plasmid replicates independently; sometimes it is integrated with the normal chromosome of the bacterium; therefore, it is referred to as episome. The F factors have shown the following significant features:

- When F^+ strain of a bacterium is incubated on a nutrient medium mixed with acridine orange, it is converted into F^- strain. Acridine orange is effective only with the growing bacteria as it inhibits the autonomously replicating F factor.
- A cross between two F^- strains does not yield recombinants. It is always sterile as the F^- strain cannot undergo conjugation with the other F^- strain.
- The crosses between F^+ and F^+ strain yield F^+ cells but a very low level. Some of F^+ cells are converted into F^- genotype.
- Transfer of F factor from F^+ to F^- in $F^+ \times F^-$ crosses occurs at a frequency of about 100%, but the production of recombinants occurs at the rate of one per 10^4 to 10^5 cells.
- Among F^+ strains there are certain F^+ substrains that show about 1,000 times more rate of recombination with F^- strains. The sub-strains are called *high frequency recombination* (Hfr) strains. The Hfr strains are produced when F factor integrates with the bacterial chromosomes.

(*c*) Genetic Map of F Plasmid: A genetic map of F plasmid is shown in Fig. 4.22. The F plasmid of *E. coli* is about 100 kb with genes coding for autonomous replication, sex pili formation and conjugal transfer function. The F plasmid contains the transfer (*tra*) region and non-transfer related markers. The non-transfer related markers are the insertion sequences (IS3, d, g and IS2), stable DNA degradation (srn B), inhibition of replication by T7, and II phages (*pif*), and a region for replication (*rep*), incompatibility (*inc*) and origin of vegetative replication (*ori*V).

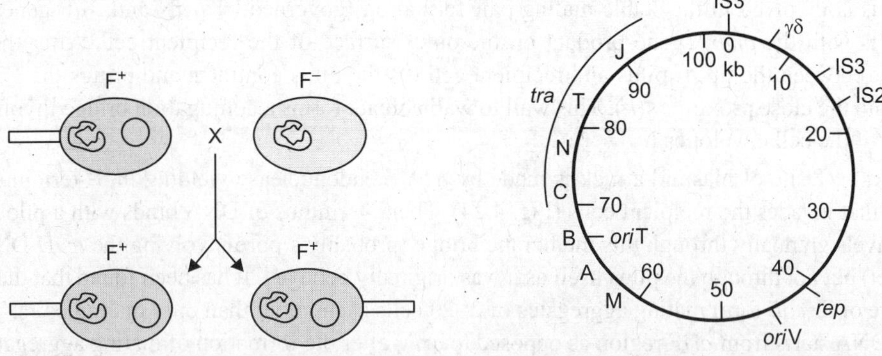

Fig. 4.22. Cross between F$^+$ and F$^-$cells.

Fig. 4.23. Genetic map of F plasmid.

Some genes of R plasmid with their functions are given in Table 4.1. In 1984, N. Willetts and B. Wilkins have given the physical and genetic map of transfer region of F plasmid (Fig. 4.23) which is about 32 kb long consisting of about 25 known transfer genes. Twelve genes are involved in F pilus formation (for example, *tra* A, -L, -E, -K, -B, -V, -W/C, -U, -F, -H, -G). Genes involved in regulation are *fin*P and *tra*J. Stabilisation of mating pairs is done by genes *tra*N and *tra*G, conjugative DNA metabolism by *tra*M, *tra*Y, *tra*D, *tra*I and *tra*Z and surface exclusion by *tra*S and *tra*T. The *tra*M and *tra*J promoter regions have been sequenced and *tra*Y-Z operon possesses its own promoter. Transcription from the promoters for *tra*M and *tra*Y-Z operon is dependent on the product or *tra*J which in turn is negatively regulated by the *Fin*OP repressor. The *tra*I and *tra*Z genes are transcribed continuously from a second promoter at about 18% or the level from the *tra*I induced *tra*Y-Z operon promoter.

Table 4.1. Some genes and sites of F plasmids and their function (after Freifelder, 1987).

Gene	***Function***
*tra*A, *tra*B, *tra*C, *tra*E, *tra*F	Pili formation
*tra*G, *tra*H, *tra*J, *tra*K, *tra*L	
*tra*U, *tra*V, *tra*W	
*tra*J	Structural genes for pilin
*tra*G, *tra*N	Mating aggregation
*tra*I, *tra*M	Initiation of transfer
*tra*O	Operater for *fin*O gene
*tra*Y, *tra*Z	Subunits of endonuclease that nicks
*ori*T, *ori*T	Origin of transfer DNA synthesis.
*ori*V	DNA replication origin.
*tra*S, *tra*T	Surface exclusion (inhibition of mating between F$^-$ containing cells) encodes membrane proteins.
il_z A, il_z B	Lethal zygosis (killing of females by excess Hfr cells)
tp	F replication
inc	Incompatibility of IncF group
*fin*O, *fin*P	Fertility inhibition (found in Col R and F$^-$ like plasmids, but not F)

***(d)* Conjugal Transfer Process:** Specific structural appendages called sex pili are found on cell surfaces of donor cell in the Enterobacteriaceae. The donor cells contain F factor and cell-to-cell contact between F^+ and F^- is established. The steps of F plasmid transfer from F^+ to F^- cell are shown in Fig. 4.24 *a-d*. A pool of preformed subunits is incorporated into mature sex pili. The tip of pilius is involved in the stable mating pair formation (governed by *tra*N and *tra*F genes) when interacts with the *omp*A gene product on the outer surface of the recipient cell. After the initial contact between the tip of pilus and recipient cell (*a*) the pilus contracts and brings the F^+ and F^- cells into the close proximity (*b*). This wall to wall contact forms a conjugation bridge involving the fusion of the cell envelopes (*c*).

At *ori*T site of plasmid a nick is made by *tra*YZ endonuclease yielding in 5'-terminus single strand that invades the recipient cells (Fig. 4.24). The 5′-terminus of DNA binds with a pilot protein and travels gradually through this membrane bridge (probably a pore involving the *tra*D DNA gene product) but not through the pilus itself as it was originally believed. It has been found that the mating mixture of *E. coli* form mating aggregates of 2- 20 cells each rather than only mating pairs. Transfer of F^+ DNA starts from *ori*F region as opposed to *ori*V after the formation of mating aggregates.

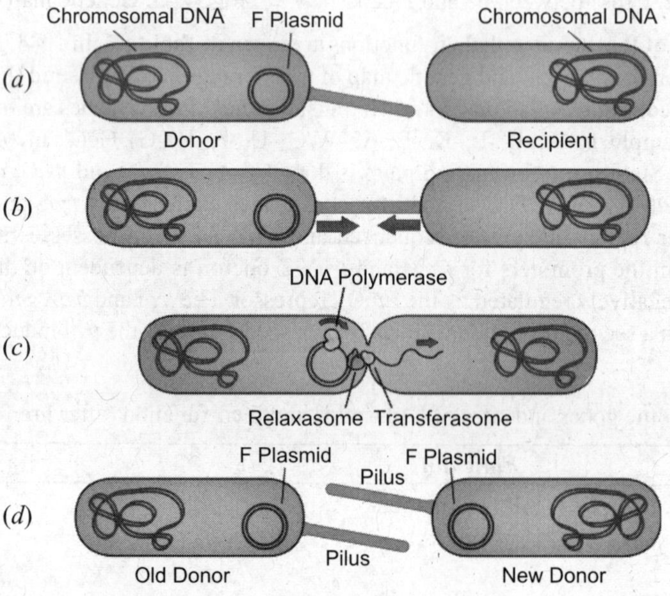

Fig. 4.24. Steps of F plasmid transfer.

When conjugation is initiated by a signal, the relaxase enzyme creates a nick in one of the strands of the conjugative plasmid at the *oriT*. Relaxase may work alone or in a complex of over a dozen proteins known collectively as a relaxosome (*c*). In the F-plasmid system the relaxase enzyme is called TraI and the relaxosome consists of TraI, TraY, TraM and the integrated host factor IHF. Then the nicked strand or T-strand is unwound from the unbroken strand and the 5′ end strand is transferred to the recipient cell. The remaining strand is replicated either independent of conjugative action beginning at the *oriV* or in concert with conjugation replication similar to the rolling circle replication of lambda phage(*c-d*). Conjugative replication may require a second nick before successful transfer can occur.

The *tra*M triggers conjugal DNA synthesis at the nicked *ori*T site by exposing sufficient ssDNA to facilitate the binding of helicase (a *tra*I gene product) or DNA helicase I (Fig. 4.25). The helicase I moves on the other strand which is undergoing transfer for unwinding the plasmid duplex. Helicase I migrates with DNA polymerase III synthesising the replacement strand of the donor DNA. After entering into the recipient cell, the 5′ end strand is attached to the membrane and undergoes replication. Transfer of DNA is associated with synthesis of a replacement strand in the donor cell and of a complementary strand in the recipient cell. Both the processes require *de novo* primer synthesis and the activity of DNA polymerase III holoenzyme. It is assumed that a single strand binding protein coats the DNA and helps the conjugal DNA synthesis. This depends upon the nature of the pore. This protein is also transferred from the donor to the recipient cells. The F plasmid is circularised after the synthesis of complementary strand.

***(e)* Barriers to Conjugation:** Sometimes the cells containing F factor are poor recipient, when conjugative crosses occur. This is due to the presence of surface exclusion. A similar phenomenon (incompatibility) occurs when a F′ element is transferred into a recipient cell that already contained F plasmid. The F′ element renders F plasmid to become unable for fertility. It has been found out that for surface exclusion two genes (*tra*S and *tra*T) are required with *tra*T protein which is an outer membrane-protein. It is hoped that *tra*T protein may block the stabilisation sites of mating pair or inhibits the structural proteins required for stabilisation of mating pair. In most of the conjugative plasmids, such as R 100 transfer of DNA is markedly reduced as compared with F. This is because the fertility inhibition system (FinOP) controls the regulatory system of *tra* genes. The *fin*O and *fin*P gene products interact and form a FinOP inhibitor of *tra* gene expression.

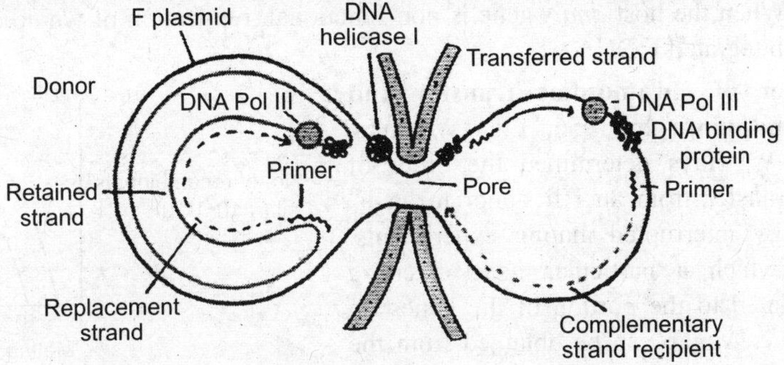

Fig. 4.25. A model for conjugal transfer process of F plasmid (diagrammatic).

***(f)* High Frequency Recombination (Hfr) Strains:** When the chromosome of F^+ cell integrates with F plasmid, it is called high frequency recombination (Hfr) cell. The Hfr cells arise from F^+ cultures (Fig. 4.26). Even after integration of F into chromosome, the chromosome retains a single circular DNA molecule. The F acts as it was a part of the chromosome. The F increases the size of chromosome; however F is capable of transferring the whole chromosome from Hfr cells to the F^+ culture. The frequency of insertion occurs at about 10^5–10^7 per generation, *i.e.* in a bacterial population of 10^7 F^+ cells; there is possibility of 1-100 cells in having an integrated F plasmid with chromosome. Following are the differences between F^+ cells and Hfr cells :

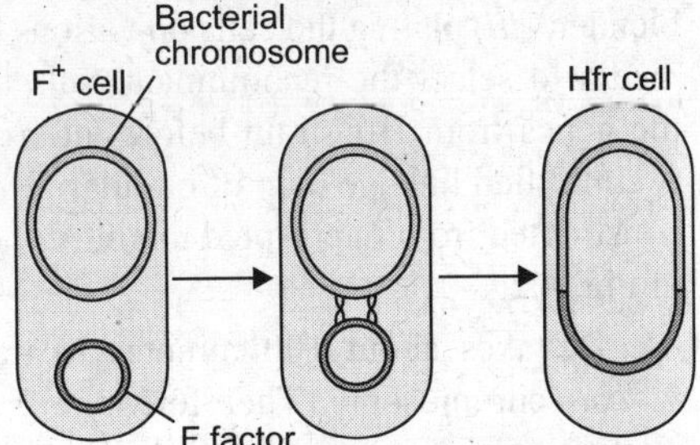

Fig. 4.26. Production of high frequency recombination (Hfr) cell.

- The F factor of Hfr cells is rarely transferred during recombination. In an Hfr × F, the frequency of recombination is high and that of transfer of F factor in low. In contrast in $F^+ \times F^-$ cross, the frequency of recombination is very low and that of transfer of F factor is high.
- It takes about 2 minutes for transfer of F, whereas 100 minutes for entire bacterial chromosome transfer. This difference is mainly due to the relative size of F and the integrated chromosome.
- In a cross between F^- and Hfr cells, F^- cells always remain F^- because of separation of cells before final transfer of ultimate F segment.

- In a mating between an Hfr leu^+ culture and an F^- leu^- culture, F^- leu^+ cells arise. The genotype of the donor is not changed because the concurrent replication in the donor replaces the transferred DNA strand.

***(i)* Mechanism of integration:** Integration is a reciprocal DNA exchange; an integrated F sequence is flanked by two copies of one of the insertion sequences (IS elements) present in F plasmid. Thus, integration involves homologous recombination between two circular DNA molecules resulting in one circular molecule that contains both the DNAs. Both the IS elements (IS2 and IS3) present on F plasmid and IS element on bacterial chromosome (*E. coli* also contain 5 each of IS2 and IS3 sequences) set as the homologous regions for insertion. The integration of F plasmid depends on *rec*A but rarely independent on *rec*A. The F integration also takes place depending on transposition of IS elements. Thus Hfr cells arise due to homologous recombination between the two identical IS elements; one is present in chromosome and the other in F plasmid. Secondly, the Hfr cells also arise by forming a cointegrate mediated by an IS element in F plasmid, and by duplicating a target sequence in the chromosome.

The direction of transfer from *ori*T is such that the *tra* genes are transferred in the last. The orientation of chromosomal IS element is such that the host gene B is as a proximal marker, whereas gene A is transferred in the last. After integration, the F DNA replicates along with the host chromosome. When the host *dna*A gene is non-functional, replication of whole chromosome can begin from an integrated F DNA.

***(ii)* Order of chromosome transfer and conjugation mapping:** In 1956, E.L. Wollman, F. Jacob and W. Hays determined the order of chromosome transfer from an Hfr donor to an F^- recipient cell by interrupted mating experiments. The time at which a particular gene enters a recipient is related to the portion of the genes on the chromosome. A map can be obtained from the time of entry of each gene. The interrupted mating experiment involves: (*i*) mixing of an Hfr strain with F^- strain, (*ii*) interrupting the conjugation at certain intervals by breaking the cells apart in a high speed blender, (*iii*) plating the cells on various types of selective media to select the recombinant cells that had received the genes from Hfr strain before interruption of mating. A simplified linkage map of circular *E. coli* chromosome constructed from interrupted mating experiment is shown in Fig. 4.27.

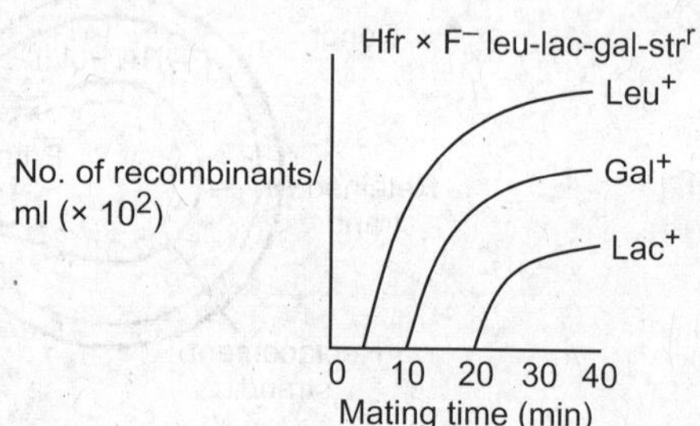

Fig. 4.27. Gene mapping by Hfr interrupted mating experiment.

It takes about 100 minutes to transfer the entire *E. coli* chromosome. Therefore, the genes are mapped relative to the position of the integrated F plasmid by determining the time taken by the gene to be transferred to the recipient cell. The interrupted mating experiment also reveals the frequency of recombination of each marker identified by detectable mutation at a particular locus (Fig. 4.27). The genetic markers are *leu, lac* and *gal*. The donor Hfr cell is wild type, whereas the recipient is *leu*, lac^-, gal^-, *i.e.* the recipient is mutant in *lac* and *gal* but wild type in *leu*. Therefore, the mutant requires lactose or galactose as carbon source.

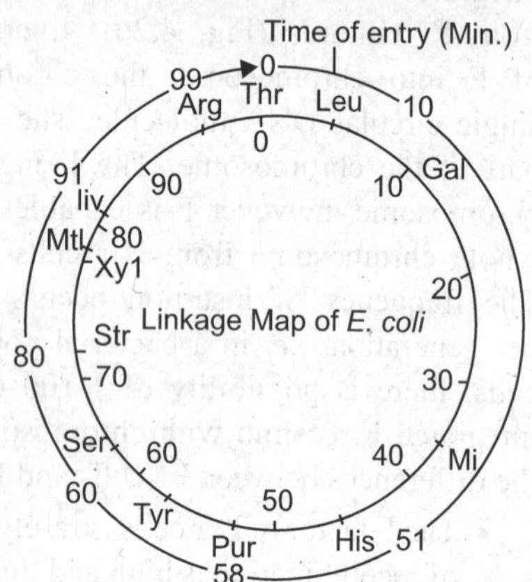

Fig. 4.28. A simplified linkage map of *E. coli* chromosome constructed from interrupted mating experiment.

The Hfr is streptomycin-sensitive (Str^s) and the recipient is streptomycin-resistant (Str^r). After mixing the donor and recipient cells at zero time, the aliquots of mixture are removed at different intervals and mating pairs disrupted by blending. The mixture is plated on minimal media containing: (*i*) glucose to select for Leu^+ recombinant, (*ii*) lactose plus leucine to select for Lac^+ recombinants, and (*iii*) galactose$^+$ leucine to select for Gal^+ recombinants. Colonies growing on these media are the recombinants i.e. the recipient into which the wild type donor gene was transferred and replaced the mutant gene. Thus, the conjugate transfer of Hfr chromosome is time dependent. Each gene enters the F^- cell at a particular time. By measuring different time intervals a graph can be plotted (Fig. 4.27), and linkage map can be constructed (Fig. 4.28).

(*iii*) High resolution mapping: One gets low resolution mapping by interrupted mating experiment. Moreover, this method is not useful for high resolution mapping within a distance of 2 minutes. Hence, study of stable recombinants rather than the gene transfer is required. For example, if abc^+ and thr^+ are the two genes transferred, the frequency of colony can be calculated with thr^+ and thr^- among those with abc^+ genes if abc^+ is more frequent than thr^+. The recombinational distance between abc^+ and thr^+ can be obtained from the proportion of thr^+ (abc^+ thr^-) among total abc^-, *i.e.* [(abc^+ thr^+) + (abc^+ thr^-)] as below:

$$\text{Recombination distance} = \frac{(abc + thr)}{(abc^+thr^+) + (abc^+thr^-)}$$

(*g*) Formation of F⁻ Prime (F′): The Hfr cell can revert to the F^+ state again; hence integration of F factor is a reversible process. When the reversible process occurs the F factor is set free from the chromosome and resumes its autonomously replicating capability. Separation of F factor from the integrated chromosome occurs aberrantly at a low frequency and yields plasmid containing F factor and a small segment of chromosome is called F′ cells. The F′ is of two types, Type I and Type II. Type I F′ has lost some sequence but carries some host DNA located at one or the other side of the integrated F. Type II F′ contains all of F′ plus some host DNA from both sides of the point where F was integrated. In both the condition F′ contains a small segment of chromosome. When such primary F′ cells are crossed with F^- recipients, the F factor is transferred efficiently together in F^- converting them into the secondary F′ cells. The secondary F′ cells are partially diploid; therefore, they called as **merodiploid** or **merozygote** because the recipient cells contained a segment of DNA from the donor F′ cell, besides its own chromosome (Fig. 4.29).

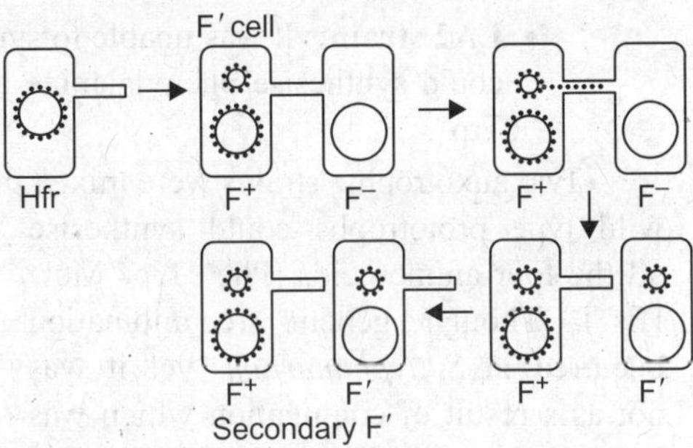

Fig. 4.29. F^- conjugation and formation of F′.

This process of transfer of bacterial DNA from donor cell to recipient cell as a part of sex factor has been called 'sexduction' by F. Jacob and E. L. Wollman in 1961. The F′ conjugation is very important in the study of microbial genetics to find out whether the allele carried by an F′ plasmid in merozygote is dominant or recessive to the chromosomal gene. Since the two neighbour genes are picked up by an F factor, study of F′ plasmid is also useful in mapping the chromosome.

3. Transduction

The phenomenon of transduction was first discovered in 1952 by Norton Zinder and Joshua Lederberg (the Department of Genetics, University of Wisconsin) while searching for sexual conjugation in *Salmonella* species. Transduction is the transfer of genetic material from one cell to another by a bacteriophage. The infection by a bacteriophage is accomplished in several stages such as adsorption, penetration, replication, assembly, lysis and release. In brief the virus particles first attaches to specific receptor site on bacterial cell wall surface. The genetic material penetrates the bacterial cell, and replicates independently by using cell machinery of the host. Consequently, the virus DNA is replicated into multiple copies, and synthesises phage proteins. Complete phage particles are assembled and finally cell is lysed resulting in release of virus particles.

The bacteriophages are of two types depending on mode of reproduction, the virulent phage and the temperate phage. The phages that reproduce by using a lytic cycle are called virulent phages because they destroy the host cell such as T phages, phage lambda (λ), etc. In contrast the temperate phages ordinarily do not lyse the bacterial cell. The viral genome behaves as episome like F factor and becomes integrated into the bacterial chromosome. The latent form of phage genome that remains within the host without harm and integrates with chromosome is called **prophage**. Bacteria containing prophage are known as lysogenic baceria and the relationship between phage and its host is called **lysogeny**. The lysogenic bacteria can produce phage particles under some conditions, and the phage is able to establish the phenomenon of lysogeny and behaves as temperate phage.

Transduction occurs between the closely related species of same genus of a bacterium, *i.e.* intragenic. This preference is due to requirement for specific cell surface receptor for recognition of the phage. In addition, intergeneric transduction has been shown between the closely related enteric bacteria such as *E. coli* and *Salmonella* or *Shigella* species. Several genetic traits for example fermentation potential, antigens and chemical resistance are transducible. Transduction is of two types, generalised transduction and specialised transduction.

(*a*) Generalised Transduction: Generalised transduction was discovered in 1952 by Norton Zinder and Joshua Lederberg. They were repeating the experiments of Lederberg and Tatum on conjugation that occurred in *E.coli* K12 taking another bacterium *Salmonella typhimurium.* They selected the following two strains of *S. typhimurium*.

- **LA22 Strain:** This strain was unable to synthesise the amino acids, phenylalanine and tryptophan (Phen$^-$ Trp$^-$ strain) but could synthesise methionine and histidine (Met$^+$ His$^+$ strain).
- **LA2 strain:** It was unable to synthesise methonine and histidine (Met$^-$ His$^-$ strain) but could synthesise phenylalanine and tryptophan. It is written as LA2 Met$^-$ His$^-$ Phen$^+$ Trp$^+$.

Two auxotrophic strains were mixed that resulted in prototrophs at the rate of $1/10^5$ cells. The wild type prototrophs could synthesise all the four amino acids (Phe$^+$ Trp$^+$ Met$^+$ His$^+$). Though genetic recombination occurred in *S. typhimurium,* yet it was not as a result of conjugation which was confirmed later on. They added each strain in an arm of U tube, of which the two arms were separated by a bacteria proof sintered glass filter which allowed free movement of nutrient media but not bacteria. After applying alternate sunction and pressure the culture medium was allowed to pass from one arm to the other. Thereafter, the two auxotrophs present in two separate arms were grown in the same culture medium. The prototrophs were recovered from LA22 culture but not from LA2 culture. From LA2 strain a genetically active filterable agent was produced which formed prototrophs in LA22 (Fig. 4.30).

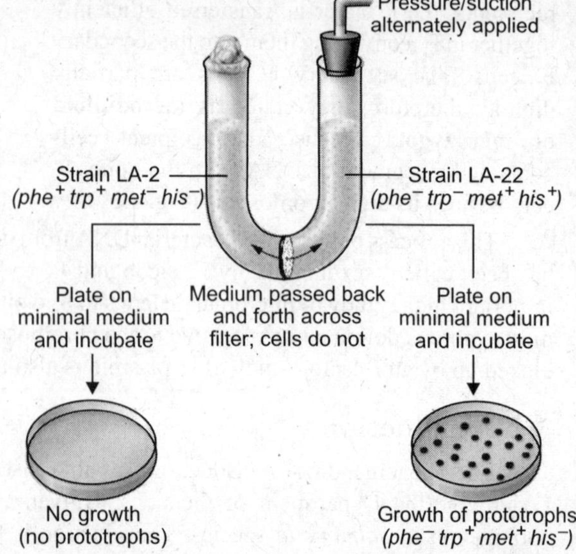

Fig. 4.30. Demonstration of transduction between LA2 and LA22 strains of *Salmonella*.

It was confirmed that the filterable agent was larger than DNA and resistant to DNase. Later on the filterable agent was confirmed as temperate *Salmonella* phage P22. The P22 was carried in one of the parental strain as prophage. Its presence was ascertained by the fact that it could be destroyed by treating with P22 antiserum.

Mostly P22 exists in lysogenic state in LA22 strain of *S. typhimurium*. Generalised transduction occurs during the lytic cycle of virulent or temperate phages. Outline of generalised transduction is given in Fig. 4.31.

After peneteration (A) phage genome multiplies within the cell (B, C), the bacterial chromosome is fragmented into pieces (C). During assembly when virus DNA is packed into protein capsid, by mistake the random fragment of partially degraded bacterial chromosome is also packed (D). Since the upper limit of DNA to be packed in capsid is about 44 kb, some or all viral DNA is left behind. The quantity of bacterial DNA carried by phage DNA depends mainly on the size of capsid. Phage P22 carries about 1% of bacterial chromosome and P1 phage of *E. coli* can carry about 2-25% of bacterial genome. The frequency of such defective phage is about 10^{-5} to 10^{-7} of the total progeny phage produced. This defective phage is also called 'generalised transducing particle'.

The genome is introduced within the host cell when this defective phage infects another bacterium (Fig. 4.31 E). The transferred bacterial DNA (exogenote) is integrated into the recipient bacterial chromosome (endogenote) (F). In addition, 70-90% of transferred DNA is not integrated with the recipient chromosome but survives and expresses itself (G). The mapping of bacterial genes is helped by generalised transduction because the chromosomal segment which has been transferred by bacteriophage contains hundreds of genes.

They are tested by using the genetic markers, for example two temperate (P1 and 363) transfer markers from one strain of *E. coli* to the other. When coliphage P1 is grown in a wild type *E. coli* cells that utilises threonine, leucine and sodium azide (Thr^+ Leu^+ Azi^+) and then used to infect a Thr^- Leu^- Azi^- recipient auxotroph, only 3% of threonine prototrophs (Thr^+ Leu^- Azi^-) are produced. A few recombinants were leucine prototrophs (Thr^+ Leu^+ Azi^-). This shows that only a short segment of bacterial DNA is involved and usually only a single marker locus is produced. The occurrence of multiple transductions (transduction of several gene loci) is rare. This also shows that Thr^+ is closely linked to *leu*$^+$ but not *azi*$^+$. Therefore, the order of linkage is *thr*$^+$ *leu*$^+$ *azi*$^+$.

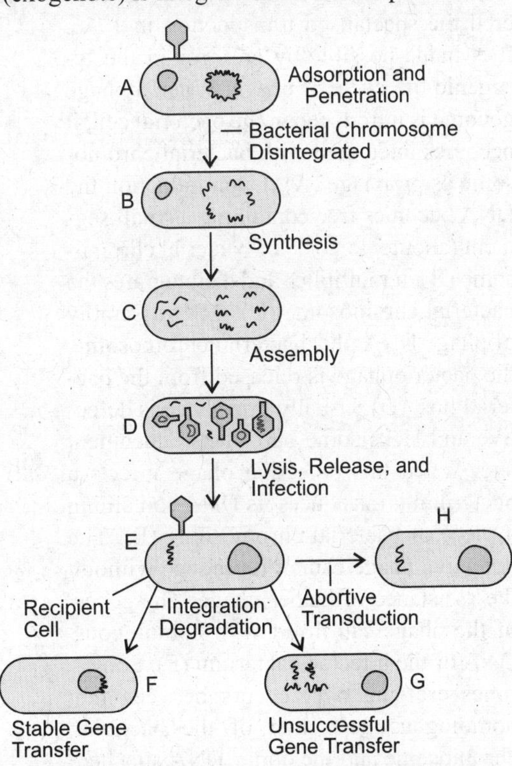

Fig. 4.31. Outline of generalised transduction by bacteriophage.

Abortive transduction: When a segment of bacterial DNA (exogenote) is introduced into another bacterium through a bacteriophage, the exogenote may be integrated into the recipient bacterial chromosome (Fig. 4.31 E-F). This type of transduction is called complete transduction. In contrast, when the exogenote is not integrated into the endogenote and remains free, it is called abortive transduction (H). The recipient bacteria that contain this non-integrated transduced DNA which are partially diploid are called **abortive transductants**. When the abortive transductant divides, only two daughter cells out of two contain exogenote in each generation; the other cells do not contain the exogenote.

The cells containing exogenote synthesise the functional enzymes because the transduced DNA specifies the normal complement by enzymes. However, in non–transduced cells the synthesis of functional enzyme gradually decreases with the progress of cell multiplication. This results in formation of slow growing small colonies or microcolonies.

(*b*) Specialised (Restricted) Transduction: Certain temperate phages can also transfer only a few restricted genes of the bacterial chromosome to the recipient bacterial cell. This transfer of bacterial genes adjacent to prophage only to the recipient chromosome is called restricted or **specialised transduction** (Fig. 4.32A-H). In 1956, for the first time, M.L. Morse and co-workers of the Department of Genetics, University of Wisconsin (U.S.A.) discovered the specialised transduction in 1956. It is made possible by an error in the lysogenic life cycle of phage. When a phage genome is introduced in the bacterial cell, it becomes integrated with bacterial chromosome as prophage (A). Upon induction the DNA becomes free containing a small segment (about 5 to 10%) of bacterial chromosome (B). It multiplies and disintegrates the bacterial chromosome (C). After assembly of phage DNA plus bacterial chromosome, the bacteriophage is released from the bacterial host (D). Usually this phage is defective and lacks some part of its attachment site. When this defective phage infects a bacterium it introduces its DNA containing a piece of bacterial chromosome (E). The defective phage cannot reproduce without the assistance of helper phage. The genes of the phage can insert with homologous DNA of the infected bacterium (F). Sometimes crossing over occurs between the homologous gene loci of the bacterial chromosome and the donor DNA attached with phage genome (G) resulting in integration of the donor DNA with the recipient DNA (H).

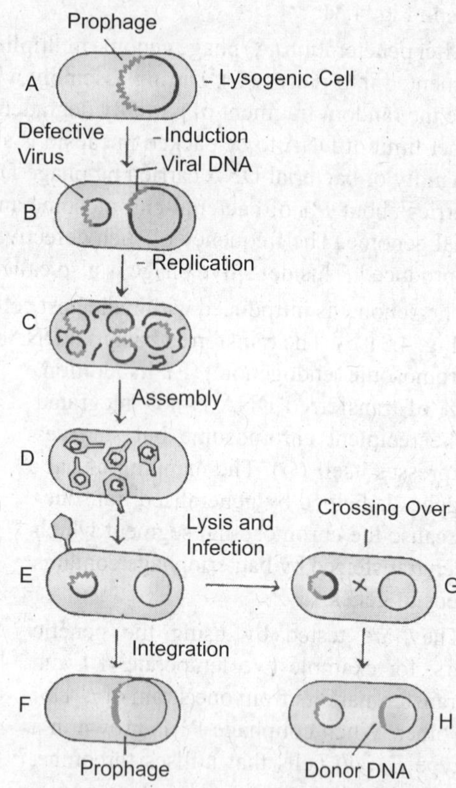

Fig. 4.32. Outline of specialised transduction by a temperate bacteriophage.

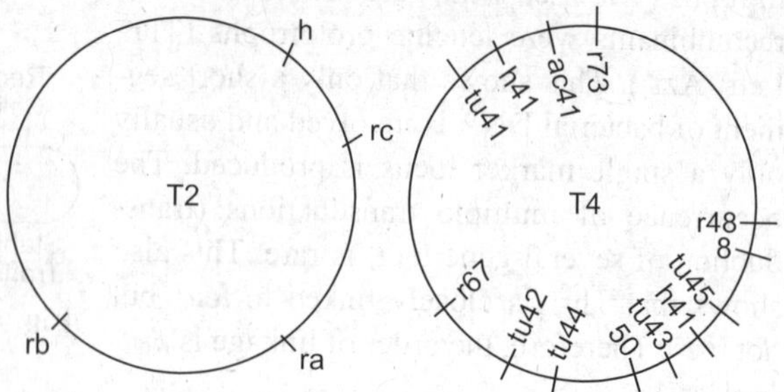

Fig. 4.33. Circular genetic map of bacteriophage T2 (A) showing only four genes and bacteriophage T4 (B); *h* = host range, *ac* = acridine resistance, *tu* = turbid plaque, *os* = resistance to osmotic sock, *e* = lysis defective, *r* = rough colony.

(*c*) Use of Transduction in Genetic Mapping in Viruses: Due to very small in size, viruses are are not considered suitable to study inheritance, because the scorable traits are not available with them. However, some of the bacterial characters have been used for inheritance and recombination studies such as plaque morphology (large or small, fuzzy or sharp), host range and virulence. Plaques are the clear transparent area produced on opaque lawn of bacteria grown on the surface of solid medium in Petri plates. For the first time Alfred D. Hershey attempted to study the inheritance and recombination in viruses by using a cross ($h^-r^+ \times h^+ r^-$, where h = host range, r = rough plaque) in bacteriophage T2. A circular genetic map of phage T2 and T4 is given in Fig. 4.33.

In this cross the genetic traits studied were host range and plaque morphology. The h^+ infects strain 1 and h^- infects both strains 1 and 2. The r^+ lyses slowly and produces small plaques, whereas r^- lyses rapidly and produces large plaques. Both the types of phages were used to infect *E. coli* strain 1 to facilitate **mixed infection** or **double infection.** The lysate was spread over a bacterial lawn having the mixture of strains 1 and 2, and analysed. Four types of plaques were recorded: clear and small plaques (h^-r^+), cloudy and large plaques (h^+r^-), cloudy and small plaques (h^+r^+), and clear and large plaques (h^-r^-). The former two types of plaques have parental phenotypes, while the last two are the recombinants. Out of the four, the recombination frequency (RF) is calculated as below:

$$\text{RF} = \frac{(h^+r^+) + (h^-r^-)}{\text{Total plaques}}$$

D. TRANSPOSABLE ELEMENTS

For a long time it was assumed that the genes have fixed locations on chromosomes. Genetic recombination is a common process of a cell within which exchange between the homologous DNA sequences takes place. Over the past few decades some information of gene rearrangement have come forth that results from illegitimate crossing over between incomplete homologous DNA sequences.

During 1940s, McClintock Barbara (an American scientist and one of the world's most distinguished cytogeneticists) in her genetic experiment with maize found that certain genetic elements regularly jump to new locations, and thereby affect gene expression. Therefore, kernels of maize ear show variation in colours. Her work was followed even after 30 years by the recognition that the bacteria contain mobile DNA sequence. The genetic segments moving from one location to the other were also discovered in bacteria well during 1970s but the frequency was quite low ($10^{-7} - 10^{-2}$ per generation). Several apparent random mutations disrupting gene function in *E. coli* were found to occur due to insertion of a large DNA segments. Similar to maize elements many of these sequences had the jumping properties. Later on the properties of such genes of *E. coli* could be known at molecular levels through the method of gene cloning and DNA sequencing. These moving genetic elements have been called with different names such as *jumping genes, movable genes, transposons* and *transposable elements*. For consistent work on maize and discovery of jumping genes, McClintock Barbara was awarded the1983 Nobel Prize in Physiology or Medicine.

McClintock Barbara (1902 – 1992)

Genome evolves by acquiring new sequences and by rearrangement of existing sequences. The sudden introduction of new sequences results from the ability of a factor to carry information between genome loci. The extra chromosomal element provides information by transferring a small DNA segment. In bacteria plasmids move through conjugation, whereas phages transfer through infection. Both phages and plasmids occasionally transfer host genes along with their own replication. In some bacteria direct transfer of DNA occurs by transformation. In addition, the retroviruses can transfer DNA segment during infection cycle in cukaryotes. However, gene rearrangement in genome may form new sequences by placing them in new regulatory situation. Rearrangement occurs either from recombination under taken by the cellular system for homologous recombination and repair, or from the transportation induced by transposons. Movement of transposable element from one chromosomal site to another is shown in Fig. 4.34.

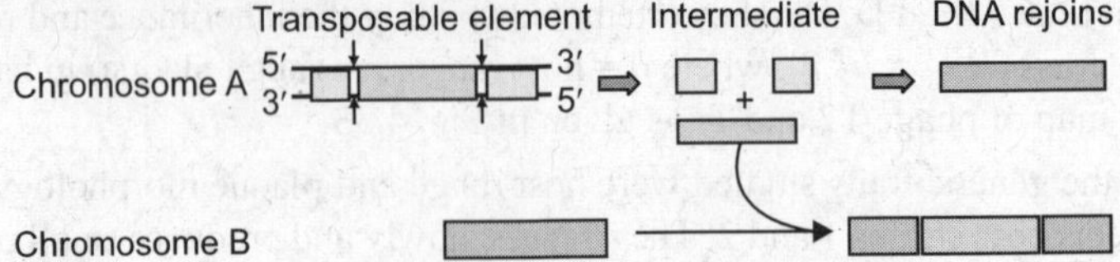

Fig. 4.34. Movement of a transposable element from one chromosomal site to another chromosome. Arrows indicate the site of action of the enzyme transposase and cleaving of DNA strand.

The mobile segments of temperate phages (for example, phage λ and phage Mu) which resemble insertion sequences are called transposable elements, and the mobile sequence of bacterial system called **transposons**. The transposable elements have been defined as *DNA sequences which can insert into several sites in a genome*. The phenomenon of moving genetic segments from one location to the other in a genome is known as transposition. Transposable elements are recognised by the presence of inverted repeat on DNA sequences at their ends. These sequences are necessary for the DNA between them to be transposed by a particular enzyme (transposase) associated with the transposable element.

1. Nomenclature of Transposable Elements

In 1977, A. Campbell and co-workers of the Stanford University (U.S.A.) have described the prokaryotes transposable as simple IS elements, more complex Tn transposons and self-replicating episomes. Definitions and nomenclature rules for these three classes of prokaryotic transposable elements are specified. The transposons discovered first did not contain any known host genes; therefore, they were called insertion sequences or IS elements for historical reasons and designated as IS1, IS2, IS3, etc. Transposons contain easily recognisable bacterial genes especially for antibiotic resistance. The members of this class are designated by the abbreviation, Tn followed by a number such as Tn1, Tn2, Tn3, etc. The number distinguishes different transposons. When it is necessary to refer the genes carried on transposons, these are represented by standard genotypical designations such as Tn1 (*amp*^{r}), where *amp*^{r} refers that the transposon carries the genetic locus for ampicillin resistance. Moreover, transposon present within a particular gene creates mutation in that gene. For example if mutation occurs at position 135 in *gal*T gene by transposon 4, the notation is designated as *gal*T 135::Tn4. In eukaryotes no definite pattern is used for naming of transposable elements. For example the transposable elements in *Drosophila* are named Copia, 497 and P, transposable elements of yeast as Ty1, and transposable element of maize as dissociation element and activator.

2. Classes of Transposable Elements

Transposable elements of prokaryotes are generally called transposons. There are three classes of transposable elements: (*i*) insertion sequences or IS elements also called simple transposons which contain no genetic information except the DNA sequences necessary for transposition, and also lack host gene, (*ii*) transposons which contains antibiotic resistance genes and may or may not be flanked by two identical copies of IS elements, and (*iii*) the transposable phages which are lysogenic phages and employ transposition as a way of life. These have been discussed in detail in the preceeding sections.

(*a*) Insertion Sequences (IS Elements): For the first time transposable elements were identified in the form of spontaneous insertions in bacterial operons. Such insertions inhibited the transcription and/or translation of genes in which it was inserted. These simplest transposable elements were called insertion sequences and designated by the prefix IS followed by numbers of each type e.g. IS1, IS2, IS3, etc. The IS elements are the normal constituents of bacterial chromosome and plasmids. A typical strain of *E. coli* contains more than 10 copies of at least one of the common IS elements. The properties of some IS elements are given in Table 4.2.

Table 4.2. The properties of insertion sequences.

IS elements		*Occurrence (bp)*	*Length terminal repeat (bp)*	*Inverted repeat (bp)*	*Direct No. of copies*
IS1	*E. coli* chromosome	768	23	9	6-10
IS2	" "	1327	41	5	4-13
IS3	" "	1400	38	3.4	5-6
IS4	*E. coli* K12 chromosome	1428	18	11-12	1-2
IS5	" "	1195	16	4	10-11
IS50 R	Tn5 on PJR67	1534	9	9	—
IS101	pSC101	201	37	—	—
IS903	Tn 903 on R6	1057	18	9	—
IS10R	Tn 10 on R100, pSM14	1329	22	9	—

The IS elements are the autonomous units each of which codes only for proteins required for its own transposition. Each IS element differs in sequence size but resemble in organisation in several ways. Mostly all of them are of about 1,000 bp long ending with inverted terminal repeats of about 10-40 bp long. The closely related inverted terminal repeat has two copies of repeats; for example the same sequence of inverted terminal repeat (ITR) of 9 bp is encountered from the flanking DNA on either side of it after proceeding towards the elements. The terminal repeats are believed to serve as recognition sequence for transposition enzymes, the transposases in their role of fusing the ends of IS elements with the target DNA. Besides IS1, all the IS elements possess a single long coding sequence that starts just inside at one end of ITR and terminate just before or within inverted repeat at the other end. It encodes for the transposase. However, IS1 consists of a more complex organisation with two reading frames. The transposase is produced by making a frame shift translation to allow both reading frames to be used (Lewin, 1994).

After transposition by IS element, a sequence of target site of host DNA at the site of insertion is duplicated. The duplicated segment is known as direct repeat, which is repeated in the same orientation. This duplication of sequence of target site is revealed upon comparison before and after insertion of IS element. Therefore, after transposition one copy of sequence of target, *i.e.* direct repeat is present at both the ends of transposon.

A transposon results in different sequences of direct repeats, the length of any IS element remains almost constant that is 9 bp. Insertion of most of IS elements occurs within a host DNA at a variety of sites. The frequency of transposition varies among different elements. The overall rate of transposition is 10^5-10^4 per element per generation (Lewin, 1994). Different IS elements contain different number of bases. The elements contain at least two apparent coding sequences initiated by an AUG and termination with an in-phase stop codon. Study of a large number of bacterial transposable elements indicates that each element encodes at least two proteins (Freifelder, 1987).

The insertion sequences can be detected in two ways : (*i*) interaction and inactivation of genes into which they get inserted, and (*ii*) transcription through their promoters by involving RNA polymerase that turns on adjacent genes. They do not perform any function in bacterial cells except that they may act as natural agent of genetic change by bringing about structure and function. Many structural variations have been observed in IS element regarding the source and host range of IS elements. Two generalisations, can be made : (*i*) the same kind of IS element such as IS1 can be found on plasmid, phage genome and chromosome of different bacteria, and (*ii*) several IS elements can function efficiently in bacteria that do not contain the same IS element in their chromosome. For example, IS10 is active in both *E. coli* K12 and *Salmonella typhimurium*, although their chromosomes do not contain IS10 sequence. Several copies of IS elements may be found on bacterial genome. For example, 8 copies of IS1 and about 5 copies of IS2 are found in *E. coli* chromosome.

(*b*) Transposons : Some additional genetic segment are present on transposons that encoding genetic information such as drug resistance unrelated to transposition process, and may or may not be flanked by IS elements. In 2007, T. Wicker and co-workers from the University Zurich (Switzerland) have described a unified classification system for eukaryotic transposable elements. They propose the first unified hierarchical classification system, designed on the basis of the transposition mechanism, sequence similarities and structural relationships, which can easily be applied by non-experts. The transposable elements are of two types according to their mechanism of transposition : *conservative transposition* (cut-and-paste) and *replicative transposition* (copy-and-paste). The conservative transposition does not involve replication, but simply the elements move from its locus to a new site. When the target site is present within a gene, both types of transposition takes place. In addition, the replicative transposition involves the events of both replication and recombination processes generating the two daughter copies of the original transposable elements, one remaining at the parental site and the other at the target site (Fig. 4.35).

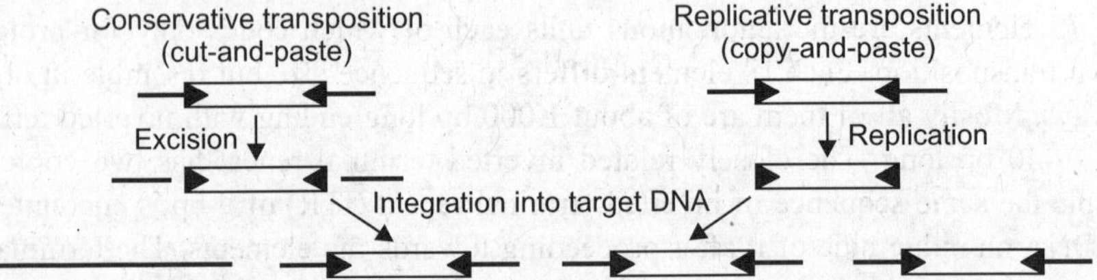

Fig. 4.35. Conservative transposition (cut-and-paste) and replicative transposition (copy-and-paste).

However, there are two classes of transposons, Class I transposons (retrotransposons) and Class II (DNA transposons).

- **Class I (Retrotransposons):** The class I trasposons copy themselves in two stages, first from DNA to RNA by transcription, then from RNA to DNA by reverse transcription. Thereafter, the DNA copy is inserted at a new position into the genome. Reverse transcription is catalysed by an enzyme reverse transcriptase, which is often coded by the *transposable elements* itself (Fig. 4.36 B). The behaviour of retrotransposons is very similar to retroviruses. There are three main orders of retrotransposons (other orders are less abundant): (*i*) retrotransposons with long terminal repeats (LTRs) (encode reverse transcriptase similar to retroviruses), (*ii*) long interspersed elements (LINEs; encode reverse transcriptase, lack LTRs, transcribed by RNA polymerase II), and (*iii*) short interspersed elements (SINEs; do not code for reverse transcriptase, transcribed

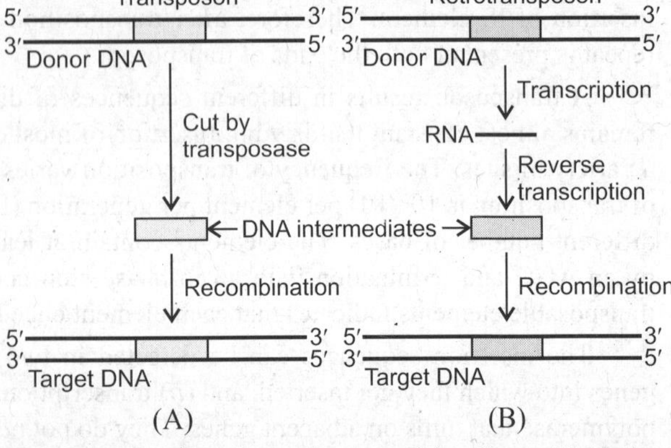

Fig. 4.36. Mechanisms of transpositional recombination; A- for transposons, B- for retrotransposons.

by RNA polymerase III). The LINEs and SINEs are the non-LTR retrotransposons. Retroviruses are also considered as transposable elements. They infect the host cell and convert their RNA into DNA that integrates into the host DNA. The integrated form of DNA (provirus) of the retrovirus acts as a specialized form of eukaryotic retrotransposon. It encodes RNA intermediates that leave the host cells and infect other cells. The transposition cycle of retroviruses is similar to that of prokaryotic transposable elements.

- **Class II (DNA transposons):** An RNA intermediate is not involved in the cut-and-paste transposition mechanism of class II transposable elements. There are different types of transposase enzymes that catalyse the transposition. Some transposases can bind to any non-specific target site, whereas others bind to specific sequence targets. The transposase cuts at the target site generating sticky ends, removes DNA transposon and ligates it into the target site. The resulting gaps from the sticky ends are filled up by a DNA polymerase; there after, DNA ligase seals the sugar-phosphate backbone. This causes duplication of target site; the insertion sites of DNA transposons may be identified by short direct repeats (a staggered cut in the target DNA filled by DNA polymerase) followed by inverted repeats (Fig. 4.36 A). If transposition takes place during S phase of the cell cycle, cut-and-paste transposable elements may be duplicated.

***(i)* Composite transposons:** The transposons that consist of a central region carrying antibiotic resistant genes flanked at both the ends by identical copies of an IS element are called the 'composite transposons'. In addition to transposition mechanisms, the composite transposons carry drug resistance or other markers (Fig. 4.37). These are named with Tn followed by number as described earlier. It is a class of larger transposons. Tn5, Tn9 and Tn10 are the three frequently studied composite transposons.

Tn5 element shows kanamycin resistance (kan^r) and consists of 5,400 base pair segment with 1,450 bp inverted repeats at both ends of the segment. It can be transposed from phage λ to the chromosome of *E .coli* and from a locus of *E. coli* chromosome to another locus. If inserted into genes it causes mutations. Tn9 transposon consists of an R factor derived gene for chloramphenicol resistance (*cam*^{r}). The enzyme that confers drug resistance consists of 2,638 sequences in the middle of Tn flanked by 768 bp long IS1 element on either side. Tn9 IS1 is present in direct order with small inverted repeat at ends. The *cam*^{r} segment is translocated from an R-factor to F-episome as well as phage λ through the phage P1. T9 differs from the other transposons in its instability and a high frequency loss of its antibiotic resistance can be encountered. Tn10 consists of tetracycline resistance genes (*tet*^{r} genes). It is 9,300 bp long consisting of inverted repeats, on either side of 1,400 base pairs. The IS elements is IS10. Tn10 can be translocated from R222 (a drug resistance plasmid) to phage P22.

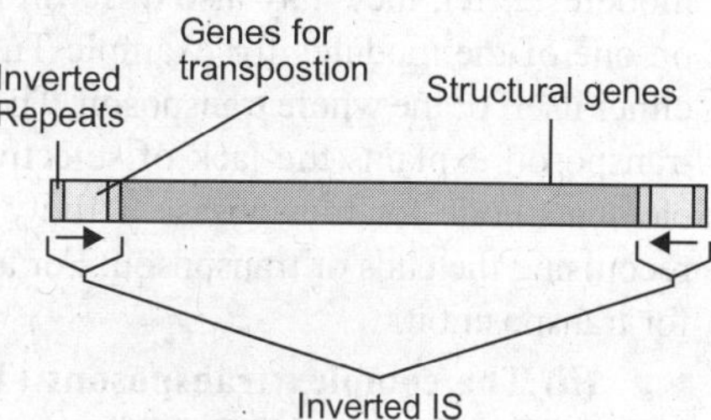

Fig. 4.37. Structure of a composite transposon (diagrammatic).

The IS elements can be in an inverted or direct repeat configuration in composite transposons. The two ends of IS elements are themselves inverted repeats. The relative orientation (direct or inverted) of the flanking IS elements of a composite transposon does not alter its terminal sequences. Thus a composite transposon with arms of direct repeats has the structure: arm L- central region - arm R. If the arms are inverted repeats, thc structure of a composite transposon becomes : left arm, central region, and right arm. The arrows show the orientation of arms according to the orientation of genetic map of the transposon from left (L) to right (R). The features of composite transposons are given in Table 4.3.

Table 4.3. The properties of composite and complex transposons.

Transposons	*Direction of terminal IS elements*	*Length (bp)*	*Target (bp)*	*Terminal IS elements &size (bp)*	*Genetic Marker**
Composite					
Tn5	Inverted	5700	9	IS50 (1500)kan^r	kanr
Tn9	Direct	2638	9	IS1 (768)	can^r
Tn10	Inverted	9300	9	IS10 (1400)	tet^r
Tn204	Direct	2457	—	IS1 (768)	cam^r
Tn903	Inverted	3100	—	IS903 (1050)	kan^r
Complex					
Tn1	Inverted	4957	5	38	amp^r
Tn2	Inverted	4957	5	—	amp^r
Tn3	Inverted	4957	5	38	amp^r
Tn4	Inverted	20,000	—	—	amp^r,str^r, sul^r

* Resistance against kanamycin (Kan), chloramphenicol (Cam), tetracycline (Tet), ampicillin (Amp), streptomycin (Str) and sulfonamide (Sul).

The modules of a composite transposon are identical in some cases, for example Tn9 (direct repeat of IS1 present) or Tn903 (inverted repeats of IS 903 present), whereas the modules are closely related in certain cases. Hence, in Tn10 or Tn5 the modules can be distinguished. When the modules are identical both can sponsor the movement of transposon as found in Tn9 of Tn903. When the modules differ, they may also differ in functional ability. Therefore, transposition depends entirely on one of the modules for example Tn10 or Tn5. Thus an IS module when functional, tranposes either itself or the whole transposon. The ability of a single module to transpose the entire composite transposon explains the lack of selective pressure for both the modules to remain active. The IS elements code for transposase activity that is responsible both for creating a target site and for reconising the ends of transposon. For a transposon only the ends are needed to serve as a substrate for transposition.

***(ii)* The complex transposons (TnA transposon family):** The TnA family of transposons includes Tn1, Tn2, and Tn3. TnA consists of quite large elements. These contain independent units carrying genes for transposition as well as for drug resistance. These are not composite relaying on IS-type transposition modules. The TnA family includes several related transposons of which Tn3 and Tn10 are the best studied ones. TnA family is also known with the name Tn3 family because Tn3 transposon was first discovered in bacterial plasmids in 1974 by R.W. Hedges and A.E. Jacob. Tn3 is an example of a complex transposon. It takes a modular structure as found in Tn10 and is not based on IS elements. Also, it does not have evolutionary links to IS elements. The basic structure of Tn3 is described for TnA family.

One of the unique features of TnA family is that it limits multiple insertions of the same transposon into a plasmid. Most surprising finding is that the transposition frequency to other plasmids in the same cell is unaffected. Transposons of TnA family consist of inverted terminal repeats (of 38 bp long but none are flanked by IS-like elements), an internal *res* site and three known genes, e.g. *tnp*A, *tnp*R and amp^r. The gene *tnp*A encodes for transposase and *tnp*R encodes for resolvase. The amp^r gene (5 bp long) is generated at the target site as direct repeat and codes for β-lactamase that confers resistance to ampicillin. The *res* site consists of three subunits: I, II and III.

***(iii)* Mechanism of transposition:** Analysis of DNA sequences and its junction with target DNA of the bacterial transposon Tn3 provides some clue of the transposition mechanism. Movement of transposons occurs only when the enzyme transposase recognises and cleaves at either 5′ or 3′ of both ends of transposon, and catalyses at either 5′ or 3′ of both ends of transposon and catalyses a staggered cut at the target site. Depending on transposon, a duplication of 3-12 bases of target DNA occurs at the site where insertion is to be done. One copy remains at each end of the transposon sequence. After attachment of both ends of transposon to the target site, two replication forks are immediately formed. From this stage there start two paths for carrying out onward processes.

The first model is the replication path where the transposon replicates and the replicated DNA sealed to flanking sequences generating a co-integrate, that is resolved by the genetic exchange between the two copies of transposon resulting in a simple insertion and regeneration of donor replicon. The second model is the non-replicative path that generates simple insertions without formation of cointegrate. Repair synthesis occurs at the prime termini in the target DNA. The displaced single strand that attaches the transposon to the donor replicon is broken resulting in formation of a simple insertion. It is likely that both the pathways can be used but the frequency of simple insertion and co- integrate formation varies. For transposition the two enzymes, transposase and resolvase coded by *tnp*A and *tnp*R respectively, are required. Transposase recognises the ends of transposon and connects them to the target site. Resolvase provides a site-specific recombination function.

E. MUTATIONS AND MUTAGENESIS

During 1880s, Hugo de Vries for the first time used the term 'mutation' (Latin; *mutare* means to change) to describe the phenotypic changes in evening primrose, *Oenothera lamarckiana.* Mutation refers to any heritable change in nucleotide sequence of a gene of the organisms irrespective of altered phenotypic expression of characters of the organisms. A gene codes for a protein, the physical and chemical properties of which are changed due to alteration in genes. Genes are made up of nucleotide sequences; hence, changes in a gene refer to changes in nucleotides or nucleotide sequence. For example, if the normal sequence was ATT, the change to AAT may lead to entirely different amino acids into a polypeptide chain. Therefore, mutation is the result of stable and heritable changes in nucleotide sequence of DNA. These changes may include alteration of a single base pair of nucleotides or addition/deletion of one or more nucleotides in the coding region of a gene.

An inducible enzyme, β-lactamase is found in *E. coli* which hydrolyses lactose to glucose and galactose, the monosaccharide constituents. When Lac$^+$ *E. coli* cells are grown on nutrient medium devoid of lactose, β-galactosidase cannot be detected within the cell and such cultures are phenotypically lactose sensitive (Lac$^-$). On the other hand, if lactose is substituted for glucose, *E. coli* cells quickly produce large quantity of the enzyme and permease transports lactose into the cell where it is utilised as carbon source. Here lactose does not change the phenotype of *E. coli* cells but induces the necessary enzyme for its utilisation. When lactose is removed from the environment the cells stop producing β-galactosidase and revert to phenotypically Lac$^-$ population. Mutation can occur in genotype of Lac$^+$ cells.

The mutation that arises all of sudden without any effort is called 'spontaneous mutation'; the mutation that arises through induction by addition of chemical mutagens or radiation is called 'induced mutation'.

I. SPONTANEOUS MUTATION

The process of formation of a mutant organism is known as mutagenesis which occurs in an organism by two mechanisms : (*i*) spontaneously or (*ii*) through physical or chemical agents. Any gene of a cell of organisms is vulnerable to mutation. It is not sure which gene will mutate; every gene is subject to mutation. Therefore, mutation occurs in a gene spontaneously and there is possibility of mutating the genes in a cell and more probability of occurrence of mutant allele in a given population of a microorganism. However, it is not sure that mutation will be beneficial; it may be even detrimental too. Thus, the spontaneous mutation arises randomly in a population of organisms ranging from 10^{-6} to 10^{-8} per generation depending on the gene and organism.

1. Random Nature of Mutation

Occurrence of mutation in bacterial population was believed to be due its response to a given selective condition, *i.e.* response to an antibiotic substance present in a medium. In 1943, Salvador E. Luria and Max Delbruck demonstrated the spontaneous and non-adaptive nature of mutation in an experiment, called the 'fluctuation test'. M. Delbrück and Luria won the 1969 Nobel Prize in Physiology or Medicine in part for this work. These experimentations gave birth to microbial genetics. They investigated the origin of mutation in *E. coli* conferring resistance to phage T1 infection.

The number of $T1^{-r}$ (r, resistant) mutant cells arising in different cultures of $T1^{-s}$ (s, sensitive) cells was compared with the number found in repeated samples of the same size taken from a single culture. The result was analysed by a statistical test named as fluctuation test (Table 4.4).

While performing the experiment on fluctuation test, twenty 0.2 ml cultures and one 10 ml culture, each containing 10^3 cells/ml of $T1^{-s}$ bacterium was grown to 21 generation. The number of cells increased to 2.8×10^9 cells/ml. Each of small culture and ten of 0.2 ml of large culture were plated onto individual plates which were already uniformly spread with 10^{10} particles of T1 phage (probably sufficient for destroying all $T1^{-s}$ cells). After incubation the number of $T1^{-r}$ colonies was counted. The number of bacterial cells inoculated in each plate was the same (5.6×10^8) but the number of $T1^{-r}$ colonies depended on whether the cells had grown in small individual culture or in large culture. Out of 20 cultures no $T1^{-r}$ cell was detected in 11 small cultures. In rest of 9 cultures the number of $T1^{-r}$ cells ranged from 1 to 10^7. On the other hand, each of 10 samples of large culture had the same number. This experiment shows that $T1^{-r}$ cells arose by spontaneous mutation at different times in the growth of cultures in the absence of phage T1, therefore the number in different cultures varied greatly. However, if $T1^{-r}$ cells arise in response to phage, these should be about equal numbers in all population of the same size.

Table 4.4. The number of T1 phage-resistant E. coli mutants in small individual cultures and in samples from a large bulk culture (after Luiria and Delbruck, 1943; Genetics. 28: 491 – 511).

Small Individual cultures		***Samples from Large culture***	
Culture	***$T1^{-r}$-colonies/plate***	***Culture***	***$T1^{-r}$-colonies/plate***
1	1	1	14
2	0	2	15
3	3	3	13
4	0	4	21
5	0	5	15
6	5	6	14
7	0	9	20
8	5	10	13
9	0	Variance	15
10	6	Variance/means	0.9
11	107		
12	0		
13	0		
14	0		

15	1
16	0
17	0
18	64
19	0
20	35
Mean	11.4
Variance	694
Variance/mean	60.8

2. Evidence for Spontaneous Mutation

Joshua Lederberg

In 1952, Joshua Lederberg and Esther Lederberg (his wife) gave the direct evidence for the origin of spontaneous mutation in T1^{-r} cells of *E. coli* without exposing to phage. This was presented by a procedure called 'replica plating' (Fig. 4.38). In 1958, J. Lederberg received the Nobel Prize for discovering that bacteria can mate and exchange genes. The steps of replica plating involve: (*i*) plating of bacteria on nutrient medium and proper incubation of plates for growth of bacterial colony, (*ii*) preparation of a solid support mounted with a piece of sterile velvet and gentle pressing it onto surface of plate supporting bacterial colonies, *i.e.* the master plate, (*iii*) again gentle press of velvet (onto which some bacteria from each colony had sticked to fibres) on a Perti plate containing fresh medium that allowed to transfer the bacterial cells on replica plate. Precaution was taken to have the identical position of both the plates and velvet also. The master plate contained 10^7 colonies of bacteria, growing on non-selective medium, whereas the replica plate contained T1 phage particles spread onto medium.

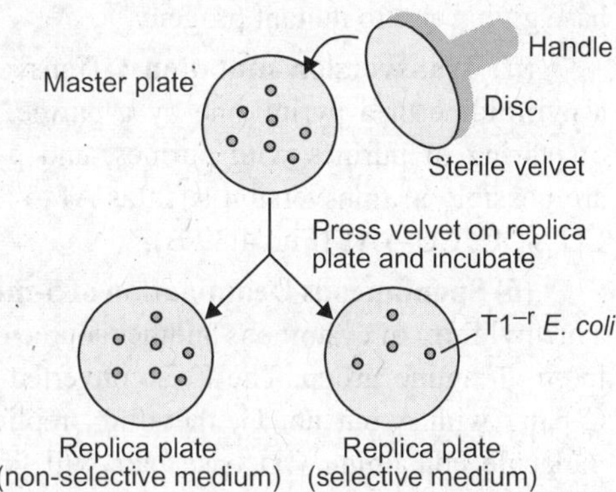

Fig. 4.38. Replica plating technique for selection of T1^{-r} *E. coli* cells

After proper incubation a few T1^{-r} cells formed colonies in the same position on each of replica plate as in master plate. The master plate did not contain T1 phage. Therefore, appearance of T1^{-r} cells on replica plate can be explained to occur due to the mutation for resistance that had taken place by chance in the master plate not exposed to T1 phage particles.

3. Origin of Spontaneous Mutations

There are several mechanisms that bring about alterations in DNA and cause mutation. These modifications may arise from error in DNA replication, damage to DNA from radiation. Errors occur during replication by substitution of frame shift in DNA sequence. Following are the causes of spontaneous mutation.

(*a*) Substitution: Substitution of base pairs is the most common mutation. During replication of DNA repair wrong base pairs are incorporated. A base pair substitution is of two types, transition and transversion.

(*i*) Transition Mutation (due to Tautomerism): The replacement of one purine (A or G) by another purine (G or A) or one pyrimidine (T or C) by another pyrimidine (C or T) is called transition. In transition four types of changes are possible, such as AT→GC, GC→AT, TA→CG,

CG→TA (Fig. 4.39 A). During replication, errors in DNA arise with high frequency. Some of the bases exist in alternative forms having different base pairing properties. Tautomerism is the relationship between two structural isomers that are in chemical equilibrium and readily change into one another. Generally the bases exist in *keto* form, but at a time they can take on either an *imino* or *enol* form. A rare form of adenine can pair with cytosine, and the *enol* form of thymine can pair with guanine. These tautomeric shifts alter the hydrogen bonding characteristics of the bases, and permit for purine substitution or pyrimidine for pyrimidine substitution. These lead to stable modification in nucleotide sequences.

At the time of replication an incorrect base is correctly hydrogen bonded and incorporated to the template strand. Even the editing system does not recognise it as incorrect. When the base assumes its normal function, the mismatch repair system corrects the mismatching bases at this level. However, if the daughter strand containing the incorrect bases is methylated, the mismatch repair system fails to distinguish between the parental and daughter strands. Therefore, the incorrect bases exist in daughter strand and lead to mutation. Such mutation is known as transition mutation. Transition mutations are common, although most of them are repaired by various proof-reading functions. In a strand of DNA molecule a pyrimidine base C is substituted by another pyrimidine base T which is the transition of base pair from one pyrimidine to another. After replication the error is retained in one strand and inherited in the progeny. The other strand remains normal. In the second generation after replication, purine base G is substituted by another purine base giving rise to mutant progeny.

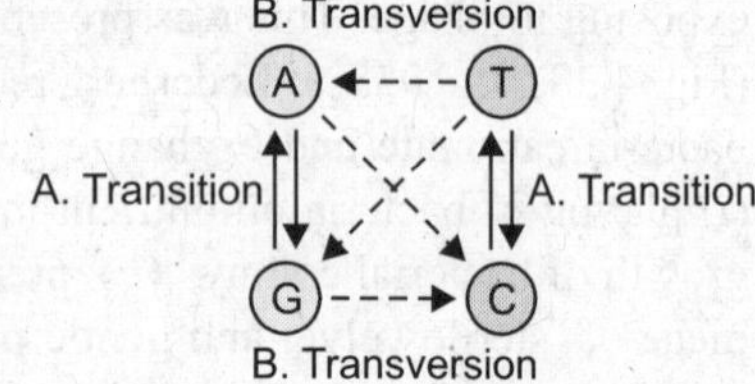

Fig. 4.39. Transition (A) and Transversion (B) mutations.

(*ii*) Transversion mutation : Transversion mutation involves the substitution of purine by a pyrimidine or a pyrimidine by a purine. This type of mutation is rare due to steric problems of pairing of purines with purines, and pyrimidines with pyrimidines. Eight types of changes are possible in transversion such as AT→TA, AT→GC, GC→CG, GC→TA, AT→TA, TA→GC, CG→GC, CG→AT (Fig. 4.39 B).

(*b*) Spontaneous Deamination of 5-methylcytosine : A change in 5-methylcytosine (5-MeC), a methyl form of cystoine is another source of spontaneous mutation. Both MeC and C occasionally loose an amino group. The C is converted to U, and MeC to 5-MeU (thymine) (Fig. 4.40). The C pairs with A but not G, therefore, replication of a molecule containing GU base pair will finally result in substitution of an AT pair for the original GC pair. The process in successive round of replication is GU→AU→AT. The uracil from DNA is removed by uracil glycosylase, hence the conversion of C→U rarely results in mutation. In addition, 5-MeC loses an amino group and converted into 5-methyluracil which is acutally the thymine. There is no removal mechanism; hence GMeC pair becomes a GT pair which may be corrected by mismatch repair system.

5-methyl cytosine —(NH_3, Deamination)→ 5-methyl uracil

Cytosine —(NH_3, Deamination)→ Uracil

Fig. 4.40. Deamination of cytosine and 5-methyl cytosine.

This system does not recognise MeC, therefore, it is present in a methylated strand. It is converted randomly sometimes correct GC pairs and sometimes wrong AT pair. Therefore, MeC acts as mutable site in a gene, and this site is called hot spot. It should be noted that methylated bases are the normal constituents of DNA in some microorganism but not the mutagens. Many organisms contain both C and MeC. Methylation results in protection of DNA against enzymes synthesised by viruses. The bacteriophage T2 contains 5-hydroxylmethylcytosine instead of cytosine.

(c) Frame Shift Mutation : The reading frame of nucleotide sequences is shifted due to deletion or insertion of one or a few nucleotides in the DNA molecule that results in mutation. Therefore, such mutation that results from shifting in reading frame backward or forward by one or more nucleotides is called 'frameshift mutation'. Generally this mutation occurs where there is a short repeated nucleotide sequence.

Fig. 4.41 shows an addition of one G between T.C due to slippage of the normal strand resulting in shifting of nucleotide sequence in genetic code. This causes change in genetic code; the code CGT encodes alanine (Ala), but it gets changed to a new code GCG that encodes arginine (Arg).

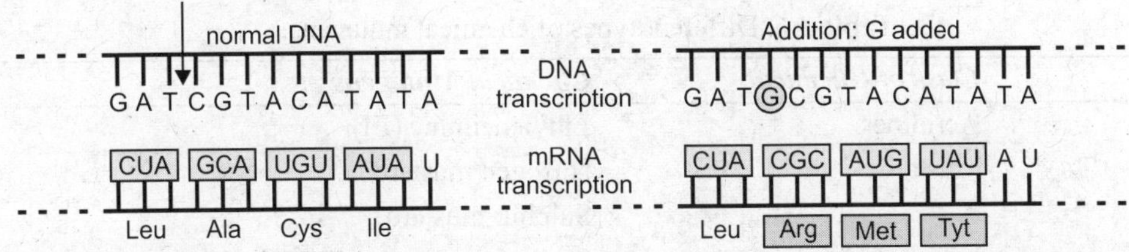

Fig. 4.41. Frame shift mutation resulting from insertion of a GC base pair (see arrow) and sense strand T.

The same can be explained by an analogy of a code of words. If a gene gives the message: THE DOG CAN EAT JAM, each word of three letters represents a codon. After addition of a letter A after G, the new sentence becomes: THE DOG ACA NEA TJA M, which also is meaningless. Similarly, after deletion of C the new sentence becomes THE DOG ANC ATJ AM, which is also meaningless.

II. INDUCED MUTATIONS

Any chemical agent that damages the DNA by altering the base sequence or interfering with repair system will certainly induce mutations in DNA of organisms. The agents causing damage to DNA are called mutagens such as chemicals and radiation. Some common mutagens are listed in Table. 4.5.

Table 4.5. Type of mutagens (after Friefelder, 1987).

Mutagen	*Mode of action*	*Example*	*Consequence*
Base analogue	Substitutes for a standard base during replication, a new base pair appears in daughter cell.	5-Bromouracil	A:T→G:Cand G:C→A:T
		2-aminopurine	A:T → G:C
Chemical mutagen	Alter bases that appear in next generation	Nitrous acid	G:C→A:Tand A:T →G:C
		Hydroxylamine	G:C → AT
		Ethylmethane sulfonate (EMS)	G:C →AT
		G:C → C:G	
		G:C → TA	
		U V light	All single base
		pair changes	
Intercalating agents	Addition or deletion of one or more base pairs	Acridines	frameshifts

Mutator genes	Excessive insertion of incorrect bases or lack of repair of incorrectly inserted bases possible	—	All singlebase pairchange
None	Spontaneous deamination of 5-methylcytosine (MeC)	—	G:MeC→AT

Some of the chemical mutagens and mutagenesis are given in Table 4.6 and described below:

Table 4.6. Different types of chemical mutagens.

Class of Chemical	*Chemical Mutagens*
Acridines	Ethyleneimine (EI)
Mustard	Nitrogen mustard Sulphur mustard
Nitrosamines	Diethylnitrosamine (DMN) Diethylsulphonate (DES) Nitrosomethylurea (NMU)
Epoxide	Ethyleneoxide (EO) Diepoxybutane (DEB)
Alkyl sulphonates	Diethylsulphonate (DES) Methylmethanesulphonate (MMS) Ethylmethanesulphonate (EMS)
Others	Nitrous acid, maleic hydrazide, hydroxylamine

1. Base Analogues

A chemical compound similar to one of the four bases of DNA is called base analogue. It can be incorporated into a growing polynucleotide chain during the normal process of replication. These compounds have base pairing properties different from the bases. They replace the bases and cause stable mutation. A very common and widely used base analogue is 5-bromouracil (5-BrU) which is an analogue of thymine. The 5-BrU functions like thymine and pairs with adenine (Fig. 4.42).

5-BrU (keto) Adenine
(A)

5-BrU (enol) Guanine
(B)

Fig. 4.42. Mutagenesis by base analogue 5-bromouracil. A, the keto form of 5-BrU pairs with adenine; B, 5-BrU is tautomarised to enol form and pairs with guanine rather than adenine.

The 5-BrU undergoes tautomeric shift from *keto* form to *enol* form caused by bromine atom. The *enol* form can exist for a long time for 5-BrU than for thymine. If 5-BrU replaces a thymine, it generates a guanine during replication which in turn specifies cytosine causing G:C pair (Fig. 4.43 A). During the replication, keto form of 5-BrU substitutes for T and the replication of an initial AT pair becomes an A: BrU pair. The rare *enol* form of 5-BrU that pairs with G is the first mutagenic step of replication. Thus, the transition is completed from AT→GC pair. In the next round of replication G pairs with C.

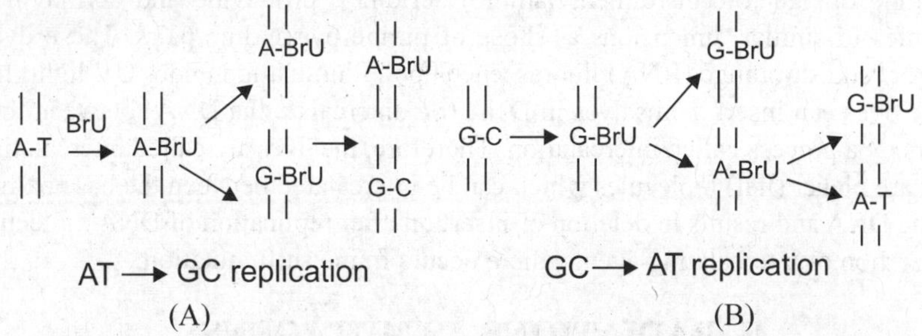

AT → GC replication
(A)

GC → AT replication
(B)

Fig. 4.43. Mechanism of 5-bromouracil (BrU)-induced mutagenesis.

The 5-BrU can also induce the conversion of GC to AT. The *enol* form infrequently acts as an analogue of cytosine rather than thymine. Due to error, GC pair is converted into a G:BrU pair which in turn becomes an AT pair (Fig. 4.43 B). Due to such pairing properties 5-BrU is used in chemotherapy of viruses and cancer. Because of pairing with guanine it disturbs the normal replication process in microorganisms. The 5-bromodeoxyuridine (5-BDU) can replace thymidine in DNA molecule. The 2-aminopurine (2-AP) and 2,6-diaminopurine (2,6-DAP) are the purine analogues. The 2-AP normally pairs with thymine but it is able to form a single hydrogen bond with cytosine resulting in transition of AT to GC. The 2-AP and 2,6-DAP are not as effective as 5-BrU and 5-BrDU.

2. Alkylating Agents

Alkylating agents add an alkyl group to the hydrogen bonding oxygen of guanine (N_7 position) and adenine (at N_3 position) residues of DNA. As a result of alkylation, possibility of ionisation is increased with the introduction of pairing errors. Hydrolysis of linkage of base-sugar occurs resulting in gap in one chain. This phenomenon of loss of alkylated base from the DNA molecule (by breakage of bond joining the nitrogen of purine and deoxyribose) is called **depurination.** Depurination is not always mutagenic. The gap created by loss of a purine can effectively be repaired. Some of the important widely used alkylating agents are : (*i*) dimethyl sulphate (DMS), (*ii*) ethyl methane sulphonate (EMS) $-CH_3CH_2SO_3CH_3$, and (*iii*) ethyl ethane sulphonate (EES) $-CH_3CH_3SO_3CH_3CH_3$.

EMS has the specifity to remove guanine and cytosine from the chain and results in gap formation. Any base (A, T, G, C) may be inserted in the gap. During replication chain without gap will result in normal DNA. In the second round of replication gap is filled by suitable base. If the correct base is inserted, normal DNA sequence will be produced. Insertion of incorrect bases results in transversion or transition mutation. Another example is methyl nitrosoguanidine that adds methyl group to guanine causing it to mispair with thyamine. GC is converted into AT transition after subsequent replication.

Proflavin

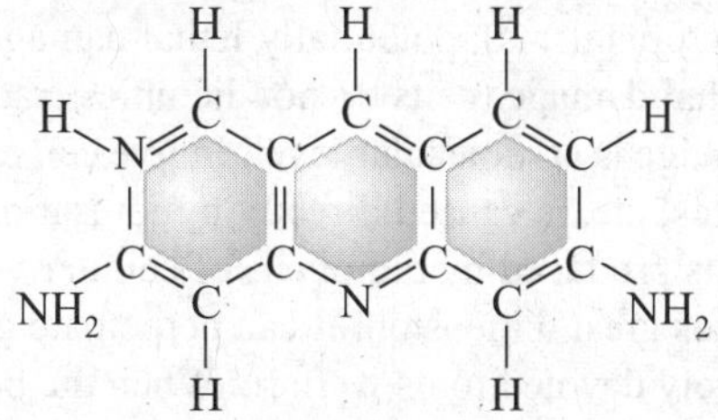

Acridine orange

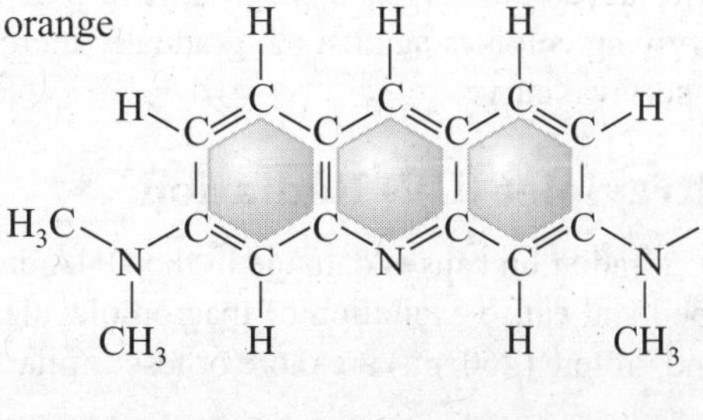

4. Intercalating Agents

Certain dyes used as a probe of nucleic acids in microscopy and related techniques [for example acridine orange [3,6-bis(dimethylamino) acridine], proflavine and acriflavin are three ringed molecules of similar dimensions as those of purine pyrimidine pairs. These dyes yield a green (DNA) or reddish-orange (RNA) fluorescence upon illumination under UV light. In aqueous solution, these dyes can insert themselves in DNA (*i.e.* intercalate the DNA) between the bases in adajacent pairs by a process called intercalation. Therefore, the dyes are called intercalating agents. The acridines are planer (flat) molecules which can be intercalated between the base pairs of DNA, and distorts the DNA and results in deletion or insertion after replication of DNA molecule. Due to deletion or insertion of intercalating agents, there occurs frame shift mutation.

II. RADIATIONS AS MUTAGENS

Radiation or energy is emitted as electromagnetic waves that travel at the speed of light. There are many different types of radiations; each of these types is defined by its wavelength. The wavelength of electromagnetic radiation can vary from being infinitely short to infinitely long. Radiation is the most important among the physical mutagens. The radiations that damages the DNA molecules fall in the wavelength range below 340 nm and photon energy above 1 electrovolt (eV). The destructive radiation consists of ultraviolet (UV) rays, X-rays, γ-rays, alpha (α) rays, beta (β) rays, cosmic rays, neutrons, etc. (Fig. 4.44).

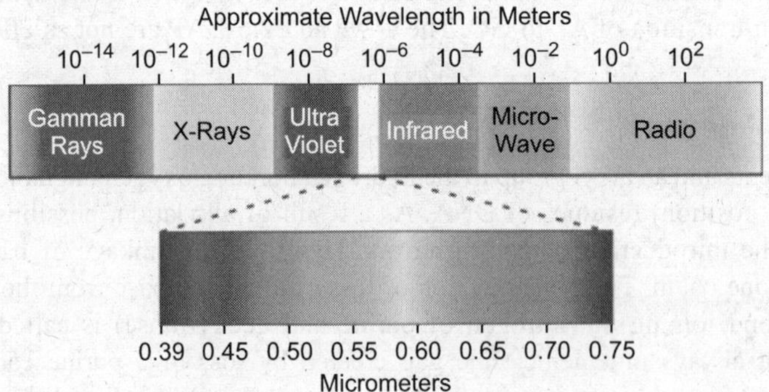

Fig. 4.44. Wavelengths and photon energy of various radiations.

Radiation induced damage can be categorised into the three broad types: lethal damage (killing the organisms), potentially lethal damage (can be lethal under certain ordinary conditions) and sublethal damage (cells do not die unless radiation reaches to a certain threshold value). The effect of damage is at molecular level. In live cell radiation damage to proteins, lipoproteins, DNA, carbohydrates, etc. is caused directly by ionisation/excitation, or indirectly through highly reactive free radicals produced by radiolysis of cellular water. DNA stores genetic informations so damage to it assumes great dimension. It can perpetuate genetic effects and, therefore, the cellular repair system is largely devoted to its welfare. When the bacteria are exposed to radiation they gradually lose the ability to develop colonies. This gradual loss of viability can be expressed graphically by plotting the surviving colonies against the gradually increasing exposure time. This dose-response graph is called survival curve.

1. Ultraviolet (UV) Radiation

UV radiation causes damage in the DNA duplex of the bacteria and phages. The UV rays are absorbed and cause excitation of macromolecules. The absorption maxima of nucleic acid = (280 nm) and protein (260 nm) are more or less similar. The DNA molecule is the target molecule for UV rays but not the proteins. However, absorption spectrum of RNA is quite similar to that of DNA. The excited DNA leads to cross-linking, single strand breaks and base damage as minor lesion and generation of nucleotide dimer as a major one. Purines are generally more radio - resistant than the pyrimidine of the latter, thymine is more reactive than cytosine. The ratio of thymine-thymine (TT), thymine-cytosine (TC), cytosine-cytosine (CC) dimer (Fig. 4.45) is 10:3:3, respectively.

A few dimers of TU and UU also appear. The initial step in pyrimidine dimerisation is known to be hydration of their 4:5 bonds. Formation of thymine-thymine (TT) dimer causes distortion of DNA helix because the thymines are pulled towards one another. The distortion results in weakening of hydrogen-bonding to adenines in the opposing strand. This structural distortion inhibits the advance of replication fork.

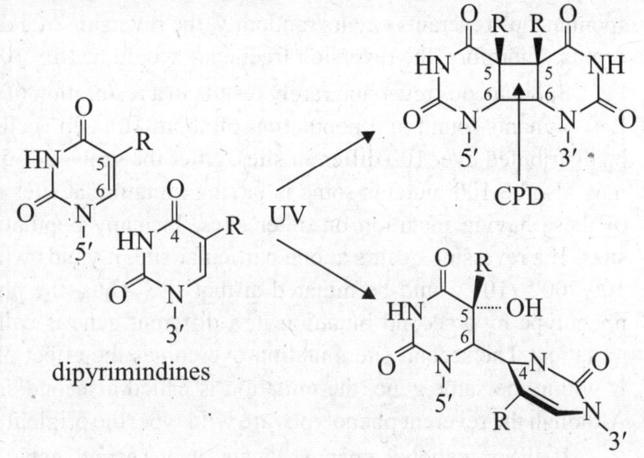

Fig. 4.45. Formation of pyrimidine dimer induced by UV radiation; CPD = cyclobutane pyrimidine dimer.

2. X-Rays

The breaking of phosphate ester linkages in the DNA is caused by X-rays. This breakage occurs at one or more points. Consequently, a large number of bases are deleted or rearranged in the DNA molecule. The X-rays may break the DNA either in one or both strands. If breaks occur in both strands, it becomes lethal. The DNA segment between the two breaks is removed resulting in deletion. Since both the X-rays and UV rays bring about damage in DNA molecule; bacteria and viruses can be sterilised by using X-rays.

III. EXPRESSION OF MUTATIONS

Expression of mutation can be observed only when the detectable and altered phenotypes have been induced. If a mutation occurs from a most prevalent wild type to a mutant form, it is called forward mutation. In forward mutation, a second mutation may take place resulting in appearance of a wild type organism again. This type of mutation is known as back mutation, reverse mutation or reversion. The reverse mutants are called reverants. The phenomenon of reverting of a mutant to the original phenotype characteristics is called reversion, which may occur spontaneously by induction or mutagens as below :

AAA (Lys)	⟶	GAA (Glu)	⟶	AAA (Lys)
Wild type	⟶	Mutant	⟶	Wild type
UCC (Ser)	⟶	UGC (Cys)	⟶	AGC (Ser)
Wild type	⟶	**Mutant**		**Wild type**

The reversion frequency of an organism is a useful criterion for the identification of point mutation. However, if deletion has taken place, reversion will not occur. In bacteria, reversion events can be detected by measuring the ability of formation of bacterial colonies on solid growth medium. For example, if 10^7 Leu^- bacterial cells are plated on solid medium devoid of leucine about 60 colonies are formed. These colonies must be consisting of Leu^+ bacteria which would have reverted spontaneously from Lac^- to Lac^+ cells. Hence, the reversion frequency can be calculated as below:

No. of cells plated = 10^7

No. of colonies formed = 60

Reversion frequency = $60/10^7 = 6 \times 10^{-6}$

This value of frequency is a feature of the reversion of point mutants. Since, the production of spontaneous reverants occurs randomly, the reversion of a double mutant requires two independent events. Therefore, the reversion frequency would be $(6 \times 10^{-6})^2 = 3 \times 10^{-11}$

Spontaneous reversion rarely results in a restoration of the wild type base sequences. Only 100 Leu^- mutants found in a population of about 10^9 Leu^+ cells and mutation causing mutants would be distributed over 100 different sites. After the growth of these cells, in a second population, there may also be 100 mutants some is having mutation at sites present in the first population but most of these having mutation on other sites. In many populations there will be about 500 mutation sites. If a reversion occurs at one particular site, it yields wild type base sequence. On average only $100/500 \times 10^9$ would be mutated at that site. Thus, the phenomenon of reversion into wild type phenotype by a second mutation in a different gene is called second site mutation or suppressor mutation. The second site mutation overcomes the effect of first mutation. If the second mutation is within the same gene, the mutation is called a second site reversion or intragenic suppression. Although the reverent phenotypes are wild type, the original sequences of DNA will not be regained.

If the mutational changes occur in a second gene, it eliminates or suppresses a mutant phenotype which is called suppression or intergenic reversion. This type of suppression has been studied carefully with conditional mutation (described in preceding section) which develops wild type phenotypes on certain conditions and produces mutant phenotype in other conditions. The major class of this mutation is called suppressor sensitive mutation. It acts like wild type when a suppressor molecule is present. For example, a phage mutant can grow in one strain of bacteria but becomes unable to grow in other strain. Suppressor-sensitive mutations are of two types, non-sense (chain termination) mutation and missense (amino acid substitution) mutations.

IV. DETECTION AND ISOLATION OF MUTANTS

Mutation occurring in microorganisms can be detected and efficiently isolated from the parent organisms of other mutants. While studying mutation one must be aware of wild type characters of an organism so that mutants can easily be detected. Since, mutations are rare about one per 10^7 to 10^{11} cells, it is very important to have a very sensitive detection system so that the rare mutant may not be missed from detection. Therefore, to rescue from these difficulties the probability of getting alterations and frequency of mutation (one in 10^3 to 10^6 cells) are increased by inducing mutations. In bacteria and other haploid microorganisms, the detection systems are straight forward because any new allele should be observed immediately. In albino mutation, the deletion is very simple. It requires only change in colour of bacterial colony. The other detection systems are rather complex. Some of the methods of detection have been described as below:

1. Replica Plating Technique

In 1952, Lederberg and Lederberg have given replica plating technique (*see* Fig. 4.38). This technique is used to detect auxotrophic mutants which differentiate between mutants and wild type strains on the basis of ability to grow in the absence of an amino acid.

However, isolation of a leucine auxotroph through replica plating follows the following steps:

- The mutants are generated by treating a culture with a mutagen e.g. nitrosoguanidine.
- A plate containing complete growth medium is inoculated and incubated at proper temperature. Both wild type and mutant survivors will form colonies on the complete medium. This plate containing complete medium is called master plate.

- A piece of sterile velvet is prepared and gently pressed on upper surface of master plate to pick up bacterial cells from each colony. As pressed on master plate, again the velvet gently pressed on replica plates containing complete medium in one set and lacking only leucine in the other set. Thus, the bacterial cells are transferred in replica plates in the same position as in master plate.
- The plates are incubated and the replica plates are compared with master plate for the bacterial colony not growing on replica plate. The leucine auxotrophs (Leu$^-$) will not grow on replica plates that are devoid of leucine. Leu$^-$ cells growing on complete medium are isolated and cultured for further work.

The temperature-sensitive mutants are isolate by using replica plating method. It involves by forming colonies at 30ºC and then transferring these colonies on two plates, one incubated at 30ºC and the other at 42ºC. The colonies growing at 30ºC and absent at 42ºC certainly consist of temperature-sensitive mutation.

2. Resistance Selection Method

Generally the wild type cells are not resistant either to antibiotics or bacteriophages. Therefore, possibly the bacteria can be grown in the presence of the agent (antibiotics or bacteriophage and look for survivors). This method is applied for isolation of mutants resistant to any chemical compounds that can be amended in agar, or phage resistant mutants.

3. Substrate Utilisation Method

This method is employed in the selection of bacteria. Several bacteria utilise only a few primary carbon sources. The cultures are plated onto medium containing an alternate carbon sources. Any colony that grows on medium can use the substrate and are possibly mutants, which can be isolated. Sugar utilisation mutants are also isolated by means of colour indicator plates. A popular medium (EMB agar) is used for this purpose. The EMB agar contains two dyes in the medium, such as eosin and methylene blue. Colour of these dyes is sensitive to pH. This medium also contains lactose sugar as carbon source and complete mixture of amino acids. Therefore, both lactose wild type (Lac$^+$) and lactose mutant (Lac$^-$) cells can grow and form colonies on EMB agar plates. The Lac$^+$ cells catabolise lactose and secrete acids; therefore, local pH of the medium decreases resulting in staining of colony to dark purple. On the other hand, Lac$^-$ cells are unable to utilise lactose and use some of the amino acids as carbon source. After utilisation of amino acid, possibly ammonia is produced that increases the local pH and decolourises the dye resulting in white colony.

4. Carcinogenicity Test

An understanding has developed to identify the environmental carcionogens that cause mutation and induce cancer in organisms can be identified. This method is based on detecting potential of carcinogens and testing for mutagenicity in bacteria. In 1973, B.W. Ames and coworkers developed a method for deletion of mutagenicity of carcinogens which is commonly known as 'Ames test'. It is widely used to detect the carcinogens. The Ames test is a mutational reversion assay in which several special strains of *Salmonella typhimurium* are employed. Each strain contains a different mutation in the operon of histidine biosynthesis.

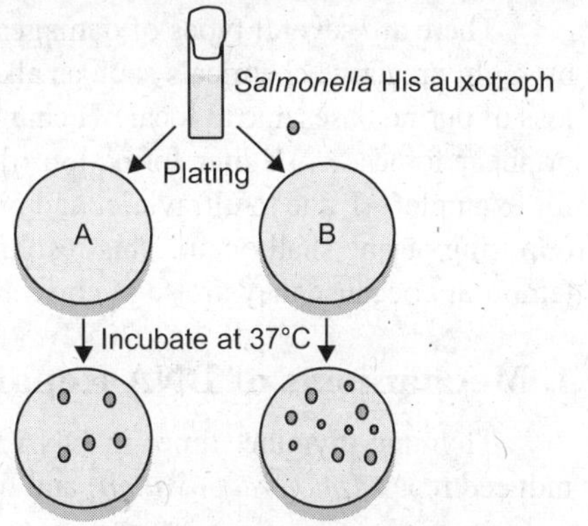

Fig. 4.46. The Ames test for mutagenicity; A, complete medium containing a large amount of histidine; B, medium containing test mutagen and a small amount of histidine.

The steps of Ames test (Fig. 4.46) are : (*i*) preparation of the culture of *Salmonella* histidine auxotrophs (His$^-$), (*ii*) mixing of bacterial cells and test substance (mutagen) in dilute molten top agar with a small amount of histidine in one set along with a check containing the complete medium plus large amount of histidine, and (*iii*) pouring of the molten mix on the top of minimal agar plates and incubation of plates at 37°C for 2-3 days.

All the His$^-$ cells will grow in the presence of test mutagens until depletion of histidine. When histidine is completely exhausted only the reverants (the mutants which have regained the original wild type characters) will grow on agar plate. The number of spontaneous reverants is low, whereas the number of reverants induced by the test mutagen is quite high. In order to estimate the relative mutagenicity of the mutagenic substance the visible colonies are counted and compared with control. The high number of colonies represents the greater mutagenicity.

Before plating a mammalian liver extract is added to the above molten top agar. The extract converts the carcinogens into electrophilic derivatives which will soon react with DNA molecule. In natural way this process occurs in mammalian system when foreign substances are metabolised in the liver. Bacteria do not possess the metabolising capacity as liver does; therefore, the liver extract is added to this test, just to promote the transformation. Therefore, several potential carcinogens that generally are not carcinogenic until modified in liver, for example aflatoxins viz., B_1, B_2, G_1, G_2, etc. The Ames test has now been used with thousands of substances and mixtures such as the industrial chemicals, food additives, pesticides, cosmetics and hair dyes.

V. DNA REPAIR SYSTEMS

The main objective of biological system is to maintain base sequences of DNA from one generation to the other. Changes in DNA sequence arise during replication of DNA damage by chemical mutagens and radiation. During replication if incorrect nucleotides have been added, they are corrected through editing system by DNA Pol I and DNA Pol III. The other systems also exist for correcting the errors missed by editing function. It is called mismatch repair system. Mismatch repair system edits the errors left by DNA Pol I and DNA III and removes the wrong nucleotides. Proof reading by Pol I and Pol III is also known. DNA is always damaged and mutated by several chemicals and radiation as discussed earlier. Only a few errors accumulate in DNA sequence. The stable errors cause mutation and the rest are eliminated. If errors in DNA sequence are corrected before cell division, no mutation occurs. However, there are some DNA damages which cannot be mutated because the damages are not replicated. Therefore, cell death is caused by such damages.

There are several types of damages that occur in DNA: (*i*) modification of one or more bases by highly reactive chemicals such as alkylating agents like nitrosoamine and nitrosoguanidine, (*ii*) loss of purine bases due to local pH change, (*iii*) single strand or double strand break due to bending or shear forces, (*iv*) dimer formation (dimerisation) between two adjacent pyrimidine molecules, for example T-T due to ultraviolet and X-ray radiation. Due to dimerisation no hydrogen bond with opposing purine shall occur. This results in distortion of helix. Most of the spontaneous errors are temporary because they are soon corrected by a process called 'DNA repair'.

1. Mechanisms of DNA Repair

Thymine-thymine dimer in DNA is repaired by involving the four major pathways : light induced repair (*photoreactivation*) and light-independent repair (*dark repair*).

(*a*) Photoreactivation: The damages caused in cells by UV radiation are repaired after exposing the damaged cells in visible light; this is called photoreactivation. In this mechanism an enzyme DNA photolyase cleaves T-T dimer and reverse to monomeric stage (Fig. 4.47). In 1958, Claud S. Rupert and his colleagues were the first to discover what is now known as photolyase. He used the DNA transformation assay to understand the molecular basis of photoreactivation.

Extensive screening revealed that photoreactivation was not universally distributed in the biological world. This enzyme is activated only when exposed to visible light, but the mutant cells lack photolyase. This enzyme absorbs energy, binds to cyclobutane ring to defective sites of DNA and promotes cleavage of covalent bonds formed between T-T. This enzyme is found in several bacteria and placental mammals. Finally, thymine residues are made free and damage is repaired.

Some other photolyases catalyse DNA repair in other ways. The 6-4 photoproduct photolyase repairs the DNA damage, *i.e.* 6-4 photoproduct caused by UV rays. The 6-4 photoproduct is formed due to formation of C4-C6 bond between two adjacent pyrimidines or due to migration of a substituent from C4 position of one pyrimidine to the C6 position of the adjoining pyrimidine. The C4 photoproduct photolyase corrects both the errors.

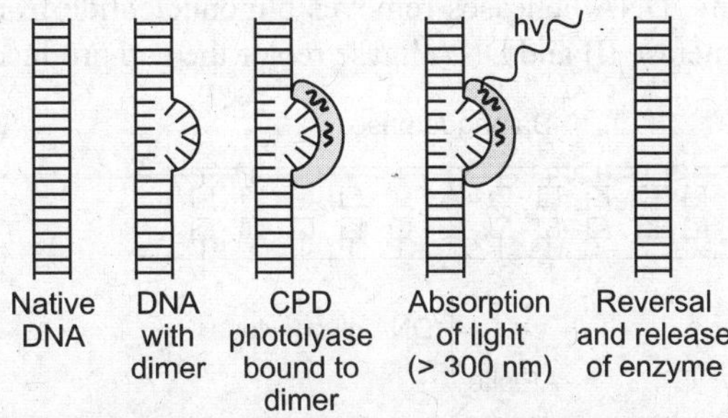

Fig. 4.47. Photoreactivation for repair of thymine dimer.

Photolyase is found in bacteria, archaebacteria, and eukaryotes except in some microbial, plant, and animal species. It is absent in humans, mice, and other placental mammals. In *Bacillus subtilis* a spore photoproduct (5-thyminyl-5, 6-dihydrothymine) is produced after UV radiation, but not cyclobutane dimers. In light-independent reaction, photoproduct lyase is formed which repair C-C bond between the two thymines.

(*b*) Excision Repair: It is an enzymatic process that removes the damaged portion by replacing with new DNA. The second DNA strand acts as template for the synthesis of new DNA fragment. Excision repair involves DNA of different lengths such as: very short patch repair, short patch repair, and long patch repair. The very short patch repair includes the mismatch of a single base, while the latter two deals with mismatches in long patches of the DNA. The short and long patches of damaged DNA molecules are repaired by *uvr* genes for example *uvrA, B C* and *D* which encode repair endonuclease.

(*i*) Base excision repair: The lesions containing non-helix distortion (for example alkylating bases) are repaired by base excision repair. It involves at least six enzymes called DNA glycosylases. Proteins involved in the DNA repairing of *E. coli* are given in Table 4.7. Each enzyme recognises at least bases and removes from DNA strand. The enzymes remove deaminated cytosine, deaminated adenine, alkylated or oxidised base.

Table 4.7. Proteins involved in the DNA repair of *E. coli.*

Repair system	*Enzymes/proteins involved in repair system*
Base excision	DNA glycosylase, AP endonuclease, DNA polymerase I, DNA ligase
Nucleotide excision	Uvr-A, Uvr-B, Uvr-C, DNA polymerase IDNA ligase
Mismatch	Dam methylase, MutS, MutL, MutH, exonuclease, DNA helicase II, SSB protein, DNA polymerase III, DNA ligase

Base excision repair pathway starts with a DNA glycosylation. For example, the enzyme uracyl DNA glycosylase removes the uracyl that has wrongly joined with G which is really deaminated cytosine (Fig. 4.48 A). Then AP-endonuclease (apurinic or apyriminic site) and phosphodiesterase removes sugar-phosphate. AP- sites arise as a result of loss of a purine or a pyrimidine. A gap of single nucleotide develops on DNA which acts as template-primer for DNA polymerase to synthesise DNA and fill the gap by DNA lygase.

***(ii)* Nucleotide excision repair:** Any type of damage having a large change in DNA helix causing helical changes in DNA structure is repaired by this pathway. Such damage may arise due to pyrimidine dimers (T-T, T-C and C-C) caused by sun light and covalently joins large hydrocarbon (e.g. the carcinogen benzopyrene). In *E. coli* a repair endonuclease recognises the distortion produced by T-T dimer and makes two cuts in the sugar phosphate backbone on each side of the damage. The enzyme DNA helicases removes oligonucleotide from the double helix containing damage. DNA polymerase III and DNA ligase repair the gap produced in DNA helix (Fig. 4.48 B).

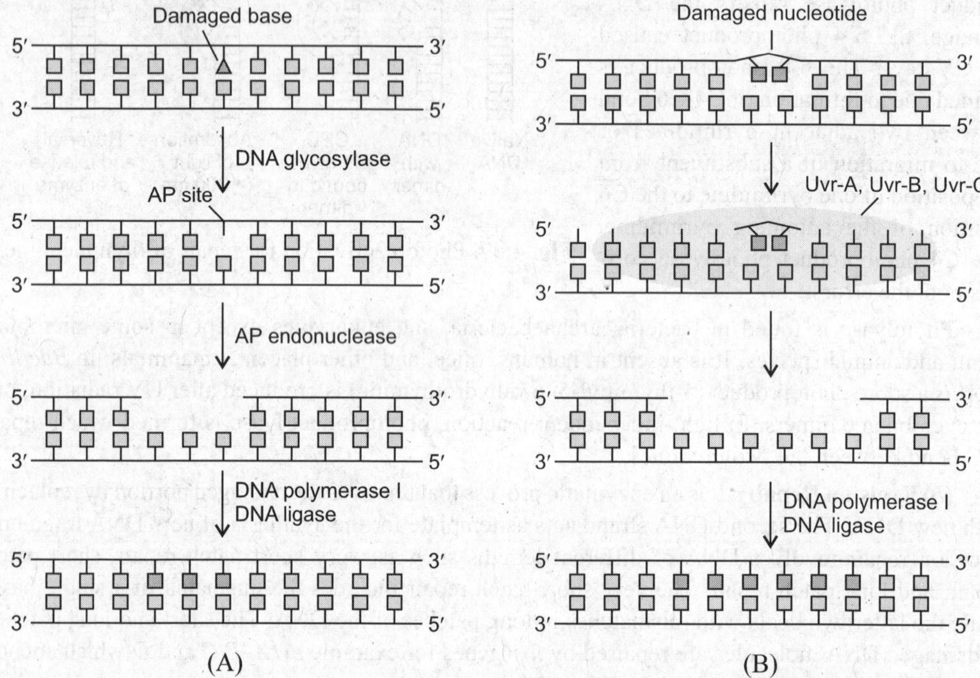

Fig. 4.48. Excision repair pathways; A, base excision repair, B, nucleotide excision repair.

(c) Recombination Repair (Daughter-strand Gap Repair): Recombination repair is required to repair errors when excision repair mechanisms get failed. It operates only after replication; therefore, it is also known as 'post-replication repair'. Homologous recombination and non-homologous end joining mechanisms can repair DNA or result in changes in DNA sequences to varying degrees. The cell has alternative mechanisms for repairing the damaged DNA at the replication fork. The normal action of DNA polymerases cannot copy pyrimidine dimers (T-T or C-C) and other types of lesions; therefore, replication is blocked at the damaged sites. Hence, replication can be initiated again at downstream of the damaged site by the synthesis of an Okazaki fragment, which proceeds along the damaged template strand (Fig. 4.49).

This results in synthesis of a daughter strand having a gap opposite to the damaged site of the parental strand. Such gaps in newly synthesised DNA may be repaired by recombinational repair mechanism or error-prone repair mechanism. Thus, DNA polymerase can bypass the lesion and reinitiate replication at a new site downstream of the dimer resulting in a gap opposite the dimer in the newly synthesised DNA strand. In recombinational repair, this gap is filled by recombination with the undamaged parental strand. One strand of the parental DNA is undamaged; therefore, it can be copied during replication to yield a normal daughter molecule. Although this leaves a gap in the previously intact parental strand, the gap can be filled by DNA polymerase and ligase by using the intact daughter strand as a template. Thus, two intact DNA molecules are formed, and the remaining thymine dimer can be removed by excision repair.

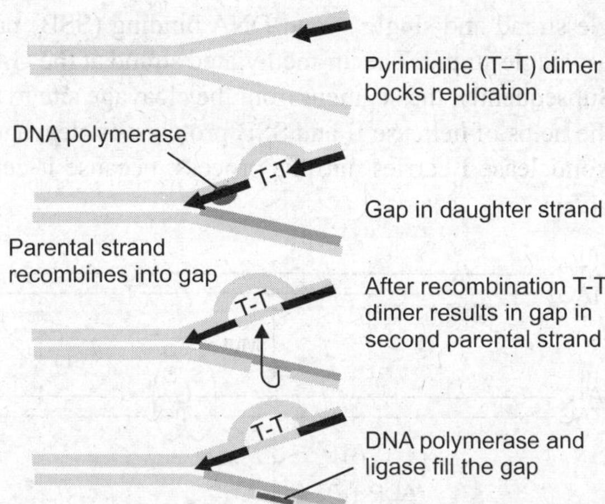

Fig. 4.49. Recombinational repair mechanism.

In *E. coli,* probably RecA protein catalyses DNA strand for sister-strand exchange. Thus a single stranded DNA segment without any defect is excised from a strand on the homologous DNA segment at the replication fork. It is inserted into the gap created by excision of thymine dimer. Then the combined action of DNA Pol I and DNA ligase joins the inserted piece. The gap formed in donor DNA molecule is also filled by DNA Pol I and ligase enzymes.

***(d)* Mismatch Repair System:** During replication, the parent polynucleotide has the correct sequence, but there may be error in the newly synthesised daughter strands. Hence, the repair of error must be made before cell division. It is interesting to know how does the repair process recognise which strand is parent strand and which is the new strand ? In *E. coli* the daughter strand is un-methylated; therefore, it can be distinguished from the parent strand which is fulllly methylated. Methylation of *E. coli* DNA takes place due to DNA adenine methylase (Dam) activities that converts adenines to 6-methyladenines in the sequence 5′-GATC-3′. Similarly, the DNA cytosine methylase (Dcm) converts cytosines to 5-methylcytosines in 5′-CCAGG-3′ and 5′-CCTGG-3′. Methylation of parent strand is not a mutagenic phenomenon, but it is a slight modification of nucleotides where the base-pairs are the same as in the unmodified nucleotides. Thorough methylation property the template and the new strands can be distinguished, and hence the repair system scans the DNA. There is a gradient of methylation along the newly synthesised strand. Least methylation occurs at the replication fork, but the parental strand is uniformly methylated. During this transition period the methylated bases direct the excision mechanisms to the newly synthesised strand containing the incorrect nucleotides (Fig. 4.50).

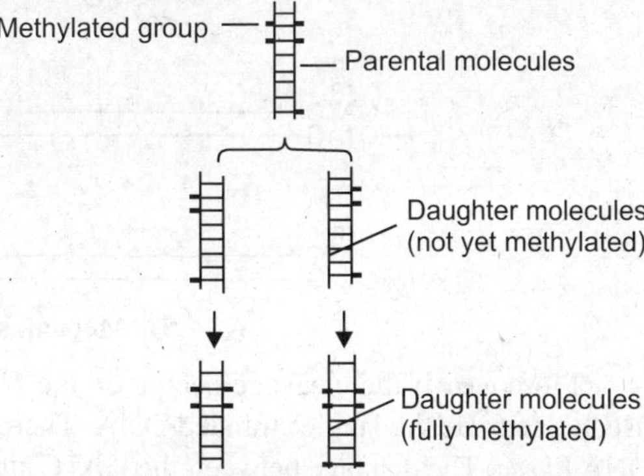

Fig. 4.50. Mismatch repair system in *E. coli.* (methylation-directed)

E. coli has at least three mismatch repair systems which are called *long patch, short patch* and *very short patch*. They are named according to the relative lengths of the excised and re-synthesised segments. The long patch system replaces up to a 1,000 bases or more of DNA and requires the MutH, MutL and MutS proteins, as well as the DNA helicase II. The protein MutS starts the repairing process through binding to mismatched base pairs (Fig. 4.51). Thereafter, MutL is recruited to the complex and activates MutH which binds to 5′-GATC-3′ sequences. MutU supports in unwinding the single strand and single strand DNA binding (SSB) proteins, and maintain the structural topography of single strand. The un-methylated strand at the GATC site is cleaved due to activation of MutH. Subsequently, the segment from the cleavage site to the mismatch is removed by exonuclease with the helps of helicase II and SSB proteins. If cleavage occurs on the 3′ side of the mismatch, then exonuclease I carries out this process, because it degrades a single strand in 3′→5′ direction (Fig. 4.51).

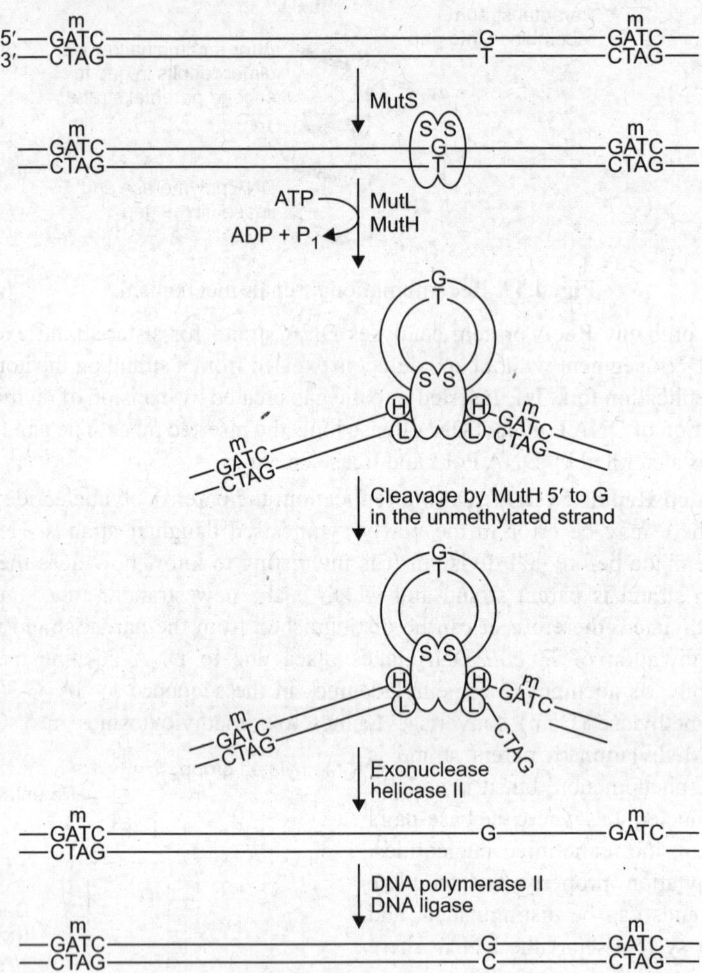

Fig. 4.51. Mechanism of mismatch repair

Moreover, if the cleavage occurs on the 5’ side of the mismatch, exonuclease VII or RecJ is used to degrade the single stranded DNA. Thereafter, the gap is filled by DNA polymerase III and DNA ligase. The distance between the GATC site and the mismatch could be as long as 1,000 base pairs; therefore, mismatch repair is very expensive and inefficient.

The mismatch repair system of eukaryotes is similar to that of *E. coli*. Yeast, mammals, and other eukaryotes also have homologs of MutS and MutL. For example MSH1 to MSH5 of eukaryotes are homologous to MutS; similarly MLH1, PMS1 and PMS2 are homologous to MutL.

***(e)* SOS Repair:** SOS (Save Our Soul) repair is a by-pass repair system, which is also called emergency repair. The damage in DNA itself induces the SOS regulatory system which is a complex cellular mechanism. SOS works where photodimers are formed that lead to cell death. SOS is the last attempt to minimise mutation for survival. It induces a number of DNA repair processes. SOS system works in the absence of a DNA template. Therefore, many errors arise leading to mutation. Generally genes of SOS system remains in repressed condition caused by a protein LexA. Repression of SOS is inactivated by RecA protease. It is formed after the conversion of RecA protein by DNA damage. DNA damage results in conversion of RecA protein into RecA protease (Fig. 4.52).

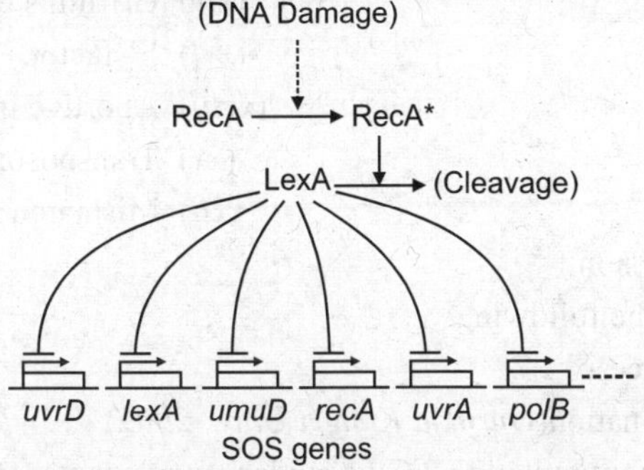

Fig. 4.52. Mechanisms of SOS repair system; LexA- protein expressed by gene *lexA*; RecA- protein expressed by gene *recA*; UvrA- protein expressed by gene *uvrA* (*source:* Kimura *et al.* 2008. *BMC Bioinformatics*.9: 23).

ARecA protease breaks LexA protein. Normally, LexA protein inhibits the activity or *recA* gene and the DNA repair genes (*uvrA* and *umuD*). Finally, DNA repair genes are activated. SOS system does not repair the large amount of damage. When DNA repair is over, RecA protein loses its proteolytic activity. Then LexA protein accumulates and binds to SOS operator and turns off SOS operator. However, repression is not complete. Beside, some RecA protein is also produced that inactivates LexA protein.

QUESTIONS

1. What are the Mendel's laws ? Describe chromosome theory of inheritance.
2. Describe in detail about extra-nuclear inheritance with suitable examples.
3. Why were the experiments of Griffith, Avery and Hershey and Chase important to the development of microbial genetics ? (*Madurai Kamraj Univ., 2005*)
4. What is the mechanism of transformation ? Discuss in brief with suitable diagrams.
5. What is conjugation ? Describe the role of surface proteins and F factors in conjugation.
6. Write a note on mechanism of integration and Hfr formation.
7. How are F plasmids produced ? (*Madurai Kamraj Univ., 2005*)
8. What is transduction ? Describe in detail the types of transduction studied by you.
9. How transduction is used in genetic mapping in viruses ?
10. Describe the use of transformation, transduction and conjugation in genetic engineering. (*Gurukul Kangri Univ., 2009*)
11. What are the transposable elements ? Describe the nomenclature of transposable elements.
12. Give an illustrated account of IS elements.
13. What are transposons ? Write in detail different types of transposons.
14. How does transposition usually occur in bacteria and what happens to the target site? (*Madurai Kamraj Univ., 2005*)
15. Writc short notes on the following :

    (*i*) Crossing over, (*ii*) Linkage,
    (*iii*) Genetic recombination, (*iv*) Gene mapping (genetic mapping),
    (*v*) Recombination Frequency, (*vi*) LOD score method,

(*vii*) Maternal effects, (*viii*) Poky strain of *Neurospora*,
(*ix*) Hershey-Chase experiment, (*x*) Transformation,
(*xi*) Transfection, (*xii*) Conjugation,
(*xiii*) Transduction, (*xiv*) Griffith's experiment,
(*xv*) F^- factor, (*xvi*) F'-factor,
(*xvii*) Hfr strains, (*xviii*) Abortive transduction,
(*xix*) IS elements, (*xx*) Transposons,
(*xxi*) TnA family, (*xxii*) Mismatch repair,
(*xxiii*) Gene conversion,

16. Write short notes on the following:
(*i*) Insertion sequences,
(*ii*) Genetic recombination(*Gurukul Kangri Univ., 2009*)

17. What do you know about mutation ? Give evidences of spontaneous mutation?

18. Write an essay on origin and random nature of spontaneous mutation?

19. What is induced mutation ? Discuss in brief about the role of chemicals in induction of mutation?

20. In what ways radiations act as mutagens ? Discuss in brief the ultra violet light as mutagen?

21. How does mutation express ? Discuss in brief reversion and suppersion?

22. Give a brief account of methods for detection and isolation of mutants.

23. Give a detailed account of repair systems in microorganisms.

24. Write short notes on the following :
(*i*) Mutation rates, (*ii*) Transversion mutation,
(*iii*) Frameshift mutation, (*iv*) Point mutation,
(*v*) Base analogue, (*vi*) Alkylating agents,
(*vii*) Intercalating agents, (*viii*) Reversion,
(*ix*) Suppression, (*x*) Photoreactivation,
(*xi*) Silent mutation, (*xii*) Replica plating,
(*xiii*) Ames test (*xiv*) Spontaneous mutation,
(*xv*) Induced mutation, (*xvi*) Excision repair,
(*xvii*) S0S repair.

25. Describe tautomerism and the way in which this chemical event may lead to mutation. (*Madurai Mamraj Univ., 2005*)

26. Contrast the various types of DNA repair mechanisms known to counter act the effect of UV radiation and other DNA damages. (*Madurai Kamraj Univ., 2005*)

27. Describe how the Ames assay screens for potential environmental mutagens? (*Madurai Kamraj Univ., 2005*)

28. Write an essay on physical and chemical mutagens. (*Gurukul Kangri Univ., 2009*)

BIBLIOGRAPHY

Ames B.W. *et al.* 1973. Carcinogenic arte mutagens: a simple test system combining liver homogenates for activation and bacteri for detection. *Proc. Nathl. Acad. Sci.* 70: 2381.

Avery O.T., Mac Leod C.M. and Mc Carty M. 1944. Studies on the chemical nature of the substance including transformation of pneumococcal types. Induction of transformation by a deoxyribonucleic acid fraction isolated from pneumococcus type III. *J. Exp. Med.* **79**: 137.

Freifelder D. 1987. *Microbial Genetics.* Jones and Bartlett Publ. Inc. Boston.

Griffith F. 1928. The significance of pneumococcal types. *J. Hyg.* **27**: 113.

Hershey A.D. and Chase M. 1952. Independent function of viral protein and nucleic acid in growth of bacteriophage. *J. Gen. Physiol.* **36**: 39.

Jacob F. and Manod J. 1961. Genetic regulatory mechanism in the synthesis of protein *J. Molecular Biol.* **3**: 318-356.

Jacob F. and Wollman E. L. 1961. *Sexuality and the Genetics* of *Bacteria.* Academic Press, New York.

Lederberg J. and Lederberg E. 1952. Replica plating and indirect selection of bacterial mutants. *J. Bact.* **63**: 399.

Lewin B. 1994. *Genes*. Oxford Univ. Press. Oxford. p. 1257.

Luria S.E. and DulbruckM. 1943. Mutations of bacteria from virus sensitivity to virus resistance. *Genetics* **28**: 491.

Moat A.G. and Foster J.W. 1995. *Microbial Physiology.* 3rd ed. Wiley-Liss, New York, p.580.

Perry R.P. 1976. Processing of RNA. *Ann. Rev. Biochem*. **45**: 605-629.

Pettijohn D. E. 1988. Histone -like proteins and bacterial chromosome structure *J. Biol. Chem*. **263**: 12793-12796.

Roth J.R. 1974. Frameshift mutation *Ann. Rev. Genetics* **8**: 319.

Sah N.K. Khan M.A. and Hasnain S.E. 1987. DNA repair mechanism. *Sci. Reptr.* **24**: 279-284.

Singer B. and Kusmierek J.T. 1982. Chemical mutagenesis. *Ann. Rev. Biochem*. **51**: 655-693.

Streisinger G. *et al.* 1966. Frameshift mutation and the genetic code. *Cold Spring Harb. Symp. Quant. Biol.* **31**: 77.

5 CHAPTER Biophysical Methods (= Biophysics)

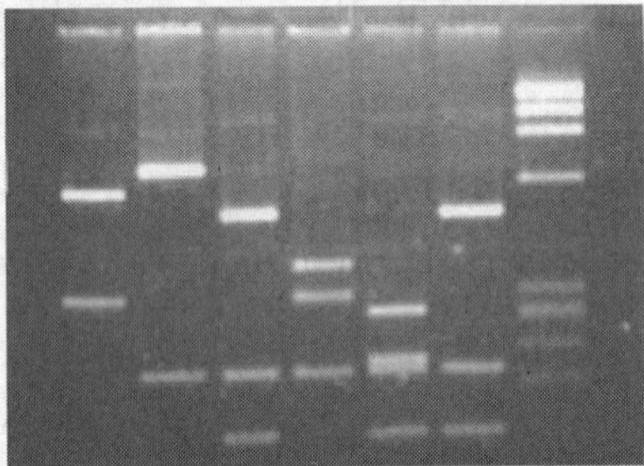

DNA separated as bands

A. Chromatography
B. Electrophoresis
C. Electroblotting
D. Gene sequencing
E. Restriction mapping
F. Site-directed mutagenesis
G. Spectroscopy
H. Polymerase chain reaction
I. Radioactivity
J. Tracer techniques (= radiotracer technique)
K. Laws of thermodynamics

Molecular biology itself grew out of biophysics. The British Biophysical Society defines biophysics as: *an interdisciplinary field which applies techniques from the physical sciences to understanding biological structure and function*. Several principles of physics have helped in understanding the biological processes. In a broader sense, biophysics may be defined as *the study of biology applying physical principles*. The macromolecules such as DNA, RNA, proteins, carbohydrates, etc. are synthesised inside the living cells which vary with each other in respect of molecular weight (size), solubility, presence of charges, absorbance of light wavelength (spectrum), etc. There are many techniques that are used to isolate and characterise the macromolecules on the basis of differentiating features. Size of different types of molecules varies and, therefore, their molecular weight also varies. For example, a macromolecule of small size shows less molecular weight and vice versa. Techniques used on the basis of molecular weight are: gel permeation, osmotic pressure, polarity of charges, X-ray crystallography, nuclear magnetic resonance, laws of thermodynamics, radioactivity, conformations of nucleic acids, and tracer techniques. Some of the biophysical methods are described in this chapter.

A. CHROMATOGRAPHY

Chromatography (Greek *chroma* = color, and *graphein* = write) is the collective term for separation of compounds from a mixture. It includes passing of a mixture dissolved in a 'mobile phase' through a 'stationary phase', which separates the analyte from other molecules present in the mixture on the basis of differential partitioning between the mobile and stationary phases. Slight differences in compounds partition coefficient results in differential retention on the stationary phase and thus changing the separation. Chromatography may be preparative or analytical. The purpose of preparative chromatography is to separate the components from a mixture for further use. Analytical chromatography is done normally with smaller amounts of material for measuring the relative proportions of analytes in a mixture.

TLC shows separated components of chlorophyll

Terminology

Chromatograph: It is an equipment that enables a sophisticated separation, for example gas chromatographic or liquid chromatographic separation.

Mobile phase: The mobile phase consists of the sample being separated and the solvent that moves the sample through the column. The mobile phase moves through the chromatography column (the stationary phase) where the sample interacts with the stationary phase and is separated.

Preparative chromatography: It is used to purify sufficient quantities of a substance for further use rather than analysis.

Retention time: The time taken for a particular analyte to pass through the system (from the column inlet to the detector) under set conditions is called retention time.

Sample: Sample is the matter analyzed in chromatography. It may consist of a single component or may be a mixture of components. During the course of analysis, when the sample is treated, the phase(s) containing the analytes of interest is/are referred to as the sample, whereas everything out of interest separated from the sample is referred to as waste.

Stationary phase: It is the substance which is fixed in place for the chromatography procedure.

an analytical method used for isolation, separation and identification of compounds. Isolation is done on the basis of differences in affinity for a mobile phase and stationary phase. This section describes different methods for isolation of substances.

Thus, chromatography is an analytical method used for isolation, separation and identification of compounds. Isolation is done on the basis of differences in affinity for a mobile phase and stationary phase. This section describes different methods for isolation of substances.

I. GEL PERMEATION

In this technique, small sized ions enter the gel and come out of the column phase, but the bigger molecules cannot enter the gel. The bigger molecules move only with mobile phase through the inner phase of gel material. Different polymeric organic compounds are used to prepare a porous medium. The polymers form a three-dimensional network of pores. The pore size is determined by degree of cross-linking of polymeric chains. Solutes present in the mixture are separated on the basis of their size and shape when they pass through a column consisting of packed gel particles (Fig.5.1). The small sized molecules enter the gel beads, whereas the large-sized molecules remain outside. Solvent usually a buffer is used as an eluent.

In this method a type of sieving of molecules occurs; therefore this method is also called *exclusion chromatography, gel filtration* and *molecular sieve chromatography* or *molecular exclusion chromatography*. In gel permeation chromatography, the gel should be highly rigid and inert. It must contain small amount of ionic groups with similar pore size distributed uniformly. These characteristics are found in agarose, polyacrylamide, silica granules, polystyrene, porous glass, Sephadex (modified starch), etc. Gel chromatography can be carried out by using column chromatography or thin layer chromatography. Size exclusion chromatography (gel filtration) uses crosslinked dextran, agarose or polyacrylamide beads with pores that trap proteins below a particular size. Small molecules are retarded, whilst larger molecules fall through more quickly.

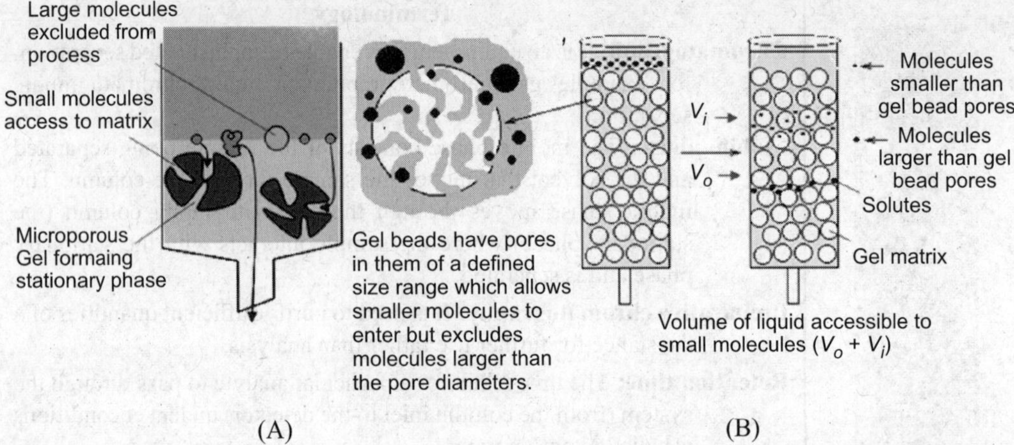

Fig. 5.1. Gel permeation chromatography. A- movement of molecules of different sizes through microporous gel forming stationary phase B- spatial accessibility of molecules during gel filtration.

***(a)* Determination of Molecular Weight of Macromolecules:** The small sized molecules are distributed between the mobile phase (liquid) inside and out side of gel particles. Therefore, different types of molecules in gel are separated from each other. In a given gel, distribution of solutes between inner and outer solvent is called *distribution coefficient* (K_d). The K_d of molecules for a given gel depends upon their size. Very large molecules have no access to mobile phase within the gel; hence their K_d value is 1. The small molecules have accessible inner mobile phase; therefore, their K_d is 1. The K_d values of molecules govern their separation. The relationship of K_d value with elution volume (V_e) is given below:

$$V_e = V_o + K_d \times V_i$$

where, V_e = The elution volume which represents volume of mobile phase (liquid) required to elute the compound from the column

V_o = Void volume or the volume of the mobile phase outside the gel particles

K_d = Distribution coefficient, *i.e.* distribution of solutes between inner and outer solvent

V_i = Volume of mobile phase present inside the gel particles. The volume of inner solvent (V_i) can be calculated by the formula

$$V_i = a \times W_r$$

where, a = dry weight of the gel and W_r = water regain value

Therefore, $V_e = V_o + (K_d \times aW_r)$

Then, $$K_d = \frac{V_e - V_o}{aW_r}$$

Molecular weight of fractioned compound could be known by using the standard graph with distribution coefficient of standards plotted against their molecular weights. Separation of macromolecules by using gel permeation offers many advantages such as: (*i*) separation of labile molecules, (*ii*) recovery of solutes in maximum quantity, (*iii*) short time and expensive equipment, and (*iv*) high reproducibility.

***(a)* Characteristics Desired for Gel Filtration Media:** The characteristics desired of gel filtration media are: (*i*) inertness of gel matrix, (*ii*) presence of low amount of inorganic groups, (*iii*) uniform pore size, (*iv*) wide choice of gel particles and pore sizes, and (*v*) high mechanical rigidity. These criteria can be fulfilled to varying degree by some of the media such as sephadex polydextran gel, polyacrylamide gel, agarose gel, agarose-acrylamide gel and glass beads.

(*b*) Application of Gel Permeation: Gel filtration is used for several purposes as given below:

(*i*) Desalting or group separation: In many experiments (e.g. protein separation) inorganic salts or low molecular weight reagents are used which need to be removed during final preparation by passing through a column of Sephadex G-10 or G-25. The high molecular weight macromolecules (e.g. proteins and nucleic acids) get excluded from the gel particles. Therefore, they are recovered from the column soon after void volume. In contrast, movement of low molecular weight compounds (salts) is impeded due to their entry into beads. This method of desalting is faster than dialysis.

(*ii*) Fractionation or purification: Viral particles and low molecular weight compounds like sugars, proteins and nucleic acids of varying molecular weight can be separated. This method is used for separation or purification of such macromolecules only because of availability of different kinds of gel media (e.g. dextran, agarose, polyacrylamide) the fractionation range of which vary considerably.

(*iii*) Determination of molecular weight: The macromo-lecules are separated on the basis of their relative size.

If molecular weight of a compound is large, it will have lesser elution volume. This method has been used to determine the molecular weight of protein from their elution characteristics (V_e or K_d value). A plot of log molecular weight *vs* elution volume of K_d gives a linear curve (Fig. 5.2). Therefore, a calibration curve can be prepared by using proteins of known molecular weight. The molecular weight of unknown protein can be obtained from the elution volume of K_d.

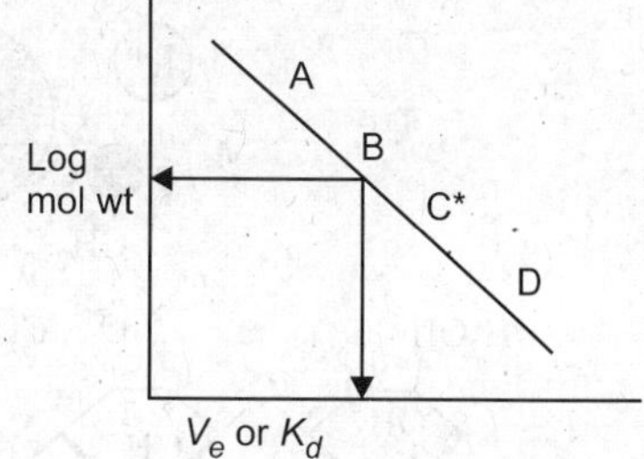

Fig. 5.2. Relation between molecular weight and elution volume during gel filtration. A, B, C are the proteins of known molecular weight; C* is the protein of unknown molecular weight.

II. ION-EXCHANGE CHROMATOGRAPHY (IEC)

Ion methods have been used since 1850, when H. Thompson and J. T. Way (in England) treated different clays with ammonium sulfate or carbonate in solution to extract the ammonia and release calcium. In 1927, the first zeolite mineral column was used to remove interfering calcium and magnesium ions from solution to determine the sulfate content of water. In the early 1970s, ion chromatography was developed by Hamish Small and co-workers at Dow Chemical Company as a novel method of ion-exchange chromatography usable in automated analysis. IEC uses weaker ionic resins for its stationary phase and an additional neutralising stripper or suppressor column to remove background eluent ions. It is a powerful technique for determining low concentrations of ions and is also useful in environmental and water quality studies.

It is a type of adsorption chromatography in which a solute is retained due to irreversible electrostatic interaction with the oppositely charge group on an ion exchanger. Ion exchange chromatography retains analyte molecules on the column based on ionic interactions. The stationary phase surface displays ionic functional groups (R-X) that interact with analyte ions of opposite charge. The ion-exchangers (anion- and cation-exchangers) act as inert and insoluble supporting media. In ion-exchange chromatography the ion-exchangers are packed in columns to separate the substances. When molecules are dissolved in solvent, they dissociate into charged ions and develop polarity.

Generally polystyrene and cellulose are used to prepare ion-exchange resins. But for separation of high molecular weight compounds, Sephadex (modified starch) and sepharose are also useful. Therefore, they can be separated on the basis of polarity, *i.e.* presence of charges. Differences are found in charges, charge densities and distribution of charges on the surface of different molecules. These differences in charge properties of different molecules are taken into account to separate from the other identical molecules. Ion-exchange chromatography is a powerful technique used for separation of two proteins which are very similar to each other, but different from each other in having one charged amino acid (Fig. 5.3). Here the solutes are retained on an ion exchanger due to their reversible interaction with the oppositely charged groups on ion exchanger.

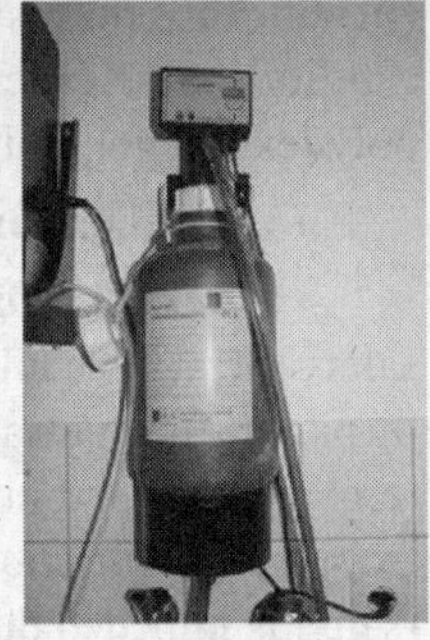

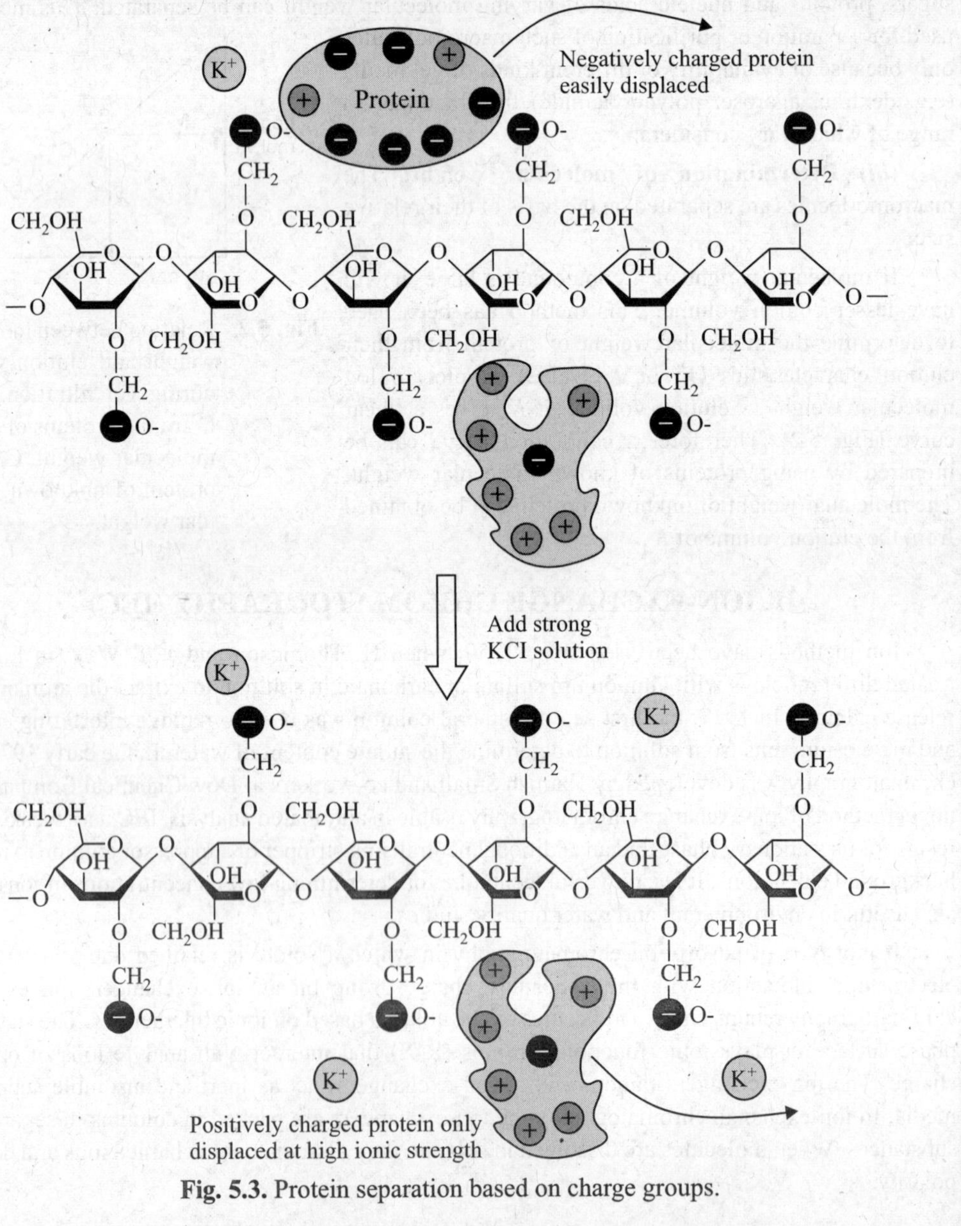

An ion exchanger

Negatively charged protein easily displaced

Protein

Add strong KCl solution

Positively charged protein only displaced at high ionic strength

Fig. 5.3. Protein separation based on charge groups.

Ion exchange separates proteins by charge anion. Exchange groups like DEAE (diethylaminoethyl) are positively charged, so it will reversibly bind the negatively charged anions. Cation exchangers like CMC (carboxymethyl) are negatively charged. These groups are covalently bound onto a cellulose or Sephadex matrix. Proteins can be eluted with increasing salt concentrations, whose ions compete for binding to the stationary phase. The eluate will contain a lot of salt, so it needs to be dialyzed afterwards.

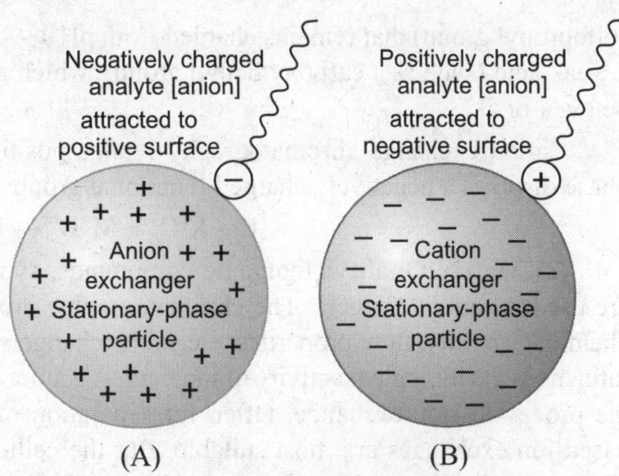

Fig. 5.4. Ion-exchange chromatography. A- anion exchanger; B- cation exchanger.

Ion exchangers are prepared using some synthetic resins which must be insoluble porous organic molecules. The ion exchanger consists of synthetic or normally occurring organic biopolymers as insoluble matrix to which various groups known as fixed ions are covalently attached. These fixed ions are balanced by equal and oppositely charged mobile ions from the solution called *counter ions*. Counter ions are mobile; therefore, they can be exchanged by the other similar charges molecules present in the sample. On the basis of counter ions, the ion exchanger has been grouped into two types: the cation exchangers and anion exchangers.

***(a)* Anionic Exchangers:** These are positively charged exchangers containing negatively charged counter ions (anions) (Fig. 5.4 A). On the basis of strength of charged functional groups, anions exchangers are also divided into three types: *strongly basic, intermediate basic* and *weakly basic* anion exchangers. The charge group on a weak anion exchanger is a weak base, which becomes deprotonated and, therefore, loses its charge at high pH. DEAE-cellulose is an example of a weak anion exchanger, where the amino group can be positively charged below pH ~ 9 and gradually loses its charge at higher pH values.

$$\text{Cellulose}-\underset{\underset{\underset{CH_3}{|}}{CH_2}}{\overset{\overset{\overset{CH_3}{|}}{CH_2}}{\overset{|}{\underset{|}{C}}}}-NH_3 \rightleftharpoons \text{Cellulose}-\underset{\underset{\underset{CH_3}{|}}{CH_2}}{\overset{\overset{\overset{CH_3}{|}}{CH_2}}{\overset{|}{\underset{|}{C}}}}-NH_3 + H^+$$

Diethylaminoethyl (DEAE) cellulose

On the other hand, a strong anion exchanger contains a strong base, which remains positively charged throughout the pH range normally used for ion exchange chromatography (pH 1-14). The anion-exchanger has to be pre-treated first with alkali followed by acid and finally washed with water for its neutralisation. Anion exchange chromatography retains anions using positively charged functional group:

$$R - X^+A^- + M^+B^- \leftrightarrow R\text{-}X^+B^- + M^+ + A^-$$

***(b)* Cation Exchangers:** These are negatively charged exchangers containing positively charged counter ions (cations) (Fig. 5.4 B). On the basis of strength of charged functional groups, cation exchangers are divided into three types: *strongly acidic, intermediate acidic* and *weakly acidic* cation exchangers. The cation exchanger is first treated with acid followed by alkali and finally washed with water till neutral. A strong cation exchanger contains a strong acid (such as a sulfopropyl group) that remains charged from pH 1 - 14; whereas a weak cation exchanger contains a weak acid (such as a carboxymethyl group), which gradually loses its charge as the pH decreases below 4 or 5.

Cation exchange chromatography retains positively charged cations because the stationary phase displays a negatively charged functional group:

$$R - X^{-}C^{+} + M^{+}B^{-} \leftrightarrow R\text{-}X^{-}M^{+} + C^{+} + B^{-}$$

(c) Matrix Quality: Inorganic compounds, synthetic resin or polysaccharide-based matrix, are used in ion exchangers. The chromatographic properties (*e.g.* efficiency, mechanical strength, chemical stability, flow properties, etc.) of exchangers are governed by the matrix. The matrix also influences the biological activity of molecules. Nature of the resin matrix remains unchanged during the process of ion-exchange. Often for separation of low molecular weight biomolecules resin-based ion exchanges are most suitable. But the cellulosic ion-exchangers are used for separation of macromolecules such as proteins and nucleic acids. The basic three groups of materials used in construction of matrix for ion exchangers are cellulose, resins (polystyrene or polyphenolic), and polymers or dextran or acrylamide (Table 5.1).

Table 5.1. Some commonly used ion exchangers.

Type	*Matrix Functional groups*		*Name of functional groups*
A. Cation exchangers			
Strong cation exchangers	Cellulose	$-SO^{3-}$	Sulpho
	Dextram	$-CH_2SO^{3-}$	Sulphomethyl
Weak cation exchangers	Polystyrene	$-CH_2CH_2CH_2SO^{3-}$	Sulphopropyl
	Agarose		
	Cellulose	$-COO^{-}$	Carboxy
	Dextram	$-CH_2COO^{-}$	Carboxymethyl
	Polyacrylamide		
B. Anionic exchanger			
Strong anionic exchanger	Cellulose	$-CH_2N^{+}(CH_3)_3$	Trimethylaminomethyl
	Dextram	$-CH_2CHN^{+}$	Trimethylaminoethyl
Weak anionic exchangers	Polystyrene		
	Agarose		
	Cellulose	$-CH_2CH_2N^{+}H^{3-}$	Aminoethyl
	Dextram	$-CH_2CH_2N^{+}H(CH_2CH_3)_2$	Diethylaminoethyl
	Polystyrene		

(*d*) Separation of Solutes in Ion-exchange Chromatography: The steps for separation in ion-exchange chromatography are given as below :

- The ion-exchanger is first treated with alkali, then acid to neutralize it. Finally it is washed with water. In contrast, the cation exchanger is first treated with acid, neutralized with alkali and finally washed with water.
- The ion-exchanger is packed into a column and equilibrated with counter ions passing buffer of required *p*H.

- When samples containing mixture of compounds is applied onto the top of exchanger bed, the solutes having charges similar to that of counter ions (here negative ions) get exchanged with the counter ions and binds reversely but strongly to the ion exchanger.
- The other solutes (positively charged) and (neutral ones) have no affinity for the stationary phase and are washed down along with the buffer.
- Thereafter, the bound solutes are released step by step by changing *p*H or ionic strength of the elution buffer. This type of elution is called 'gradient elution'.
- Generally a buffer with increasing *p*H and ionic strength is applied with anionic exchanger, whereas with cation exchanger a buffer with decreasing pH and increasing ionic strength gradient is employed for *desorption of bound substances*.
- The ion exchanger is regenerated after passing the buffer continuously.

Gradient Elution

Gradient elution is of two types: *discontinuous gradient* and *discontinuous gradient*. Discontinuous gradient is also called stepwise gradient because in this method a definite volume of mobile phase of a definite pH, ionic strength is passed through the column, followed by elution with fixed volume of mobile phase of different components. In this method, pH or ionic strength of eluent is changed discontinuously in a stepwise manner. This can be changed abruptly at every step. In *continuous gradient* method, a steady and constant rate for change of composition of eluent is followed by using a gradient maker. The gradient maker consists of two vessels of equal diameter to which an outlet tube is connected at their base. The outlet tube is made up of Teflon and equipped with an adjustable screw type pinch cock. The teflon tube connects the recipient chamber and column.

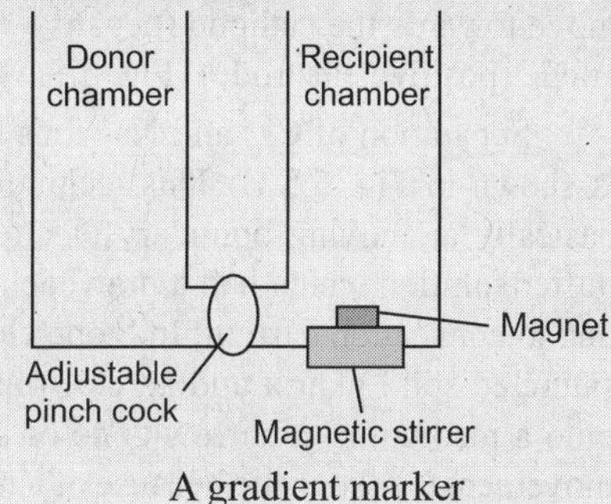

A gradient marker

Initially both the pinch cocks are tightened. Two eluents of different ionic strength and pH are poured into each of two vessels. Generally starting buffer is poured in the recipient chamber with continuously stirring of solution by using a magnetic stirrer. Pick cocks of both the tubes are opened. As the eluent start flowing from recipient vessel to the column, the same volume of eluent enter into the recipient chamber from the donor chamber. It is mixed via stirring by the magnetic stirrer. Thus, the pH and ionic strength of buffer in recipient vessel change depending on the composition of the eluent in the donor chamber.

(*e*) Application of Ion-exchange Chromatography: Ion-exchange chromatography is useful in amino acid analysis so that water could be purified and base composition of amino acids could be detected. Also construction of amino acid autoanalyser is based on the ion-exchange principles. Moreover, biologicals samples are also analysed for detection of concentration of trace metals by using ion-exchange chromatography.

B. ELECTROPHORESIS

Electrophoresis is a method of separation of charged molecules (ions) applying an electric field, under the influence of which the ions or molecules move towards one end of the electrode. For example, sugar molecules are uncharged, therefore, they do not move toward electrode. But proteins move and easily separated because they undergo ionisation due to the presence of -NH_2 and -COOH groups.

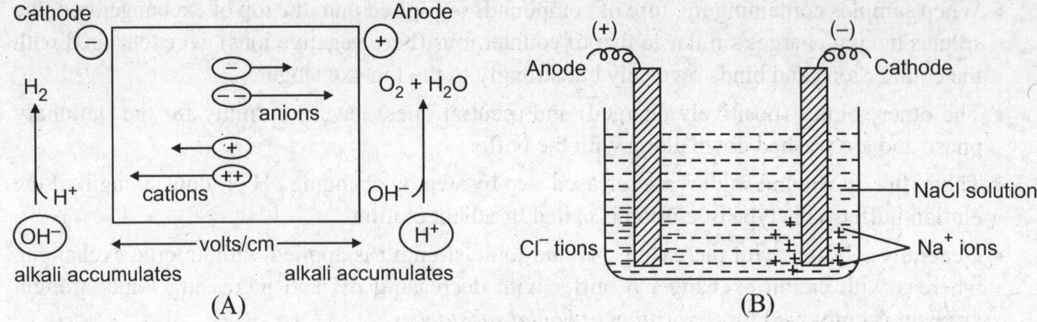

Fig. 5.5. A- the substances dissolved in water dissociate into charged ions (cations and anions) and under the electric field the positively charged ions move towards the cathode (negative electrode) and the negatively charged ions move towards the anode (positive electrode); B- movement of Na^+ and Cl^- ions towards electrodes from NaCl solution.

The basic principle involved in electrophoresis is that any substance dissolved in water dissociate into charged ions (cations and anions). Under the electric field the positively charged ions move towards the cathode (negative electrode) and the negatively charged ions move towards the anode (positive electrode) (Fig. 5.5 A).

Separation of Cl^- and Na^+ ions towards anode and cathode, respectively from NaCl solution is shown in Fig. 5.5 B. This technique was earlier pioneered by A. Tiselius in 1937 which was basically a 'moving boundary electrophoresis'. In this device electrophoresis was performed in buffer solution which had a drawback of diffusion of sample and mixing of components of sample due to convection current. In 'zone electrophoresis' this drawback was overcome by using a buffer saturated solid matrix and an electrophoresis medium (rather than buffer solution). Sample is put onto a medium as a narrow zone or band; the molecules are separated from each other due to its movement to other zone because of having different mobilities. This method is applied for separation of a large number of biomolecules such as proteins, amino acids, peptides, nucleotides, etc.

When the charged molecules are placed in an electric field they migrate depending on their net charges, size, shape and applied current. The velocity of movement of molecules can be represented by the following formula:

$$v = \frac{E \times q}{f}$$

where, v = velocity of migration of molecules
E = electric field in volts/cm
q = net fractional coefficient which is a function of the mass and shape of molecule

Electrophoretic mobility (m) is the movement of charged molecules in an electric field which is defined as 'the velocity per unit of electric field'.

$$(\mu) = \frac{v}{E} = \frac{E \times q}{f}$$

$$= \frac{q}{f}$$

The f value varies with size (but not with shape) for such molecules that have similar conformation. Therefore, electrophoretic mobility (m) of a molecule is proportional to charge density [charge : mass ratio (z/m)]. Under the electric field molecules having different charge : mass ratio (z/m) migrate at different rates resulting in their separation. For separation of RNA, DNA and proteins electrophoresis is commonly used. DNA molecules have negative charges. Therefore, based on their size DNA molecules migrate to anode, *i.e.* small molecules move faster through the pores of matrix than the larger molecules. Protein macromolecules are made up of positively and negatively charged amino acids at specific *p*H. The relative proportion of these two charges governs the net charges of protein macromolecules. If two proteins are of similar size and have identical charges, there shall be a little or no separation of such proteins due to similar charge: mass ratio. The electrophoretic mobility of a molecule depends on charge density (charge/mass ratio).

(*a*) Factors Affecting Migration of Substances: Factors that affect movement of substances from the solution towards the electrodes are as below:

(*i*) Electrodes: Platinum, carbon or Ag/AgCl electrodes are used for electrophoretic studies, but platinum electrodes are preferred over the other types of electrodes.

(*ii*) Electric field: The charged ions of the buffer solution carry out the current applied to the electrodes. Under potential gradient the rate of movement of ions is called mobility. The rate of movement of particles is increased with increase in potential gradient. Besides, the rate of movement increases with increase in electric current and decrease in resistance. The buffer ions carry more charge than the sample and result in slow separation of ions. Therefore, a constant current must be maintained.

(*iii*) Buffer solution: During the process of electrophoresis, buffers play a key role in providing charged ions, optimum pH, electrolytic concentration, ionic strength and viscosity. Most commonly buffers used in electrophoresis are : EDTA, acetate, citrate, phosphate, Tris barbitone, etc. But sometimes mobility of sample components is affected when buffers bind to them. Such buffers are useful in separation of uncharged samples like carbohydrates.

(*iv*) Ionic strength: When ionic strength of a substance is high, their separation would be slow, whereas they are separated fast under low ionic strength.

(*v*) pH: The *p*H governs the degree of ionization of a substance and also affects the rate of their separation. Ionization of acids increases the *p*H and that of bases decreases the *p*H.

(*b*) Supporting Media: There are different types of supporting media used in electrophoresis as given below:

(*i*) Paper: Paper is made up of 95% cellulose which has very less absorption capacity. It can be used as stabilising medium in electrophoresis. However, resolution of separation gets affected by adding the paper as it adsorbs the substances.

(*ii*) Gels: Gels act as a supporting medium for electrophoretic separation of molecules. The nature of gels is porous; hence, mobility of substances is determined by the size of pores in relation to the size of the molecules. Moreover, the charge and the size of molecules govern their separation.

Types of Gels used in Electrophoresis

The following types of gels are used in electrophoresis:

- **Starch:** Pore size of gels made up of starch cannot be controlled due to its high resolving power. It is not suitable to use for protein separation.
- **Agar:** It gets properly dissolved in buffer solution and forms gel with large pore size; therefore, it has no molecular sieving property. However, it is used for separation of macromolecules like nucleic acids and proteins.
- **Polyacrylamide:** Two substances viz., N,N-methylene bisacrylamide, ammonium persulphate and N,N,N,N,-tetramethylene diamine (TEMED) are used to prepare polyacrylamide gel. Pore size of the gel is controlled by using differential concentration of the reagents.
- **Agarose-acrylamide:** This gel consists of mixturc of agarosc and acrylamide where agarose provides physical support to the gel and acrylamide sieving action.
- **Others:** There are other chemical substances also which are rarely used in preparation of gel such as gypsum, pectin, sephadex, polyvinyl chloride, polyvinyl acetate, etc.

I. TYPES OF ELECTROPHORESIS

Basically electrophoresis is of three different types on the basis of mode of operation and separation of substances viz., *boundary electrophoresis, zone electrophoresis* and *immunoelectrophoresis*. Among these three types, the boundary electrophoresis is never used, while zone electrophoresis is commonly used.

***(i)* Boundary electrophoresis**: All types of modern electrophoresis are based on boundary electrophoresis. In this technique all the molecules move towards anode from the sample-buffer mixture forming a boundary. This results in variation in refractive index of the substance. Changes in refractive index can be measured by an optical device.

***(ii)* Zone electrophoresis**: In zone electrophoresis, the charged particles migrate as a zone through a solid medium *e.g.* a paper or a gel. This technique includes paper electrophoresis and gel electrophoresis.

***(iii)* Immunoelectrophoresis**: It works like agar or agarose gel electrophoresis. This technique is a combination of electrophoresis and specific immune reaction. First proteins are separated then these are allowed to react with antigens through diffusion via gel. It is used for separation of proteins based on charge : mass ratio (*z/m*) and their antigenicity.

1. Paper Electrophoresis

Strips of Whatman filter paper or cellulose acetate paper are used as support to separate macromolecules on the basis of their varying sizes. Two beakers containing equal volume and *p*H of same buffer solution are taken and interconnected by a glass plate onto which strips of filter paper of about 3-5 cm are place placed (Fig. 5.6). A spot or line is marked in the centre of filter paper with a pencil. The strips of filter paper are soaked in buffer solution and then blot dried by putting between a blotting paper. Using a micropipette or a capillary tube the sample is put on the spot. The filter strips are placed on the supporting glass plate in such a way that each end should be dipped in buffer solution put in opposite beakers. Two electrodes are taken and one electrode is placed in each beaker. Both the electrodes are connected with a power supply. At a very low voltage (5-10 V/cm) electrophoresis is run for about 45 minutes. After completion of electrophoresis the molecules are separated, dried and stained with indicator stains. For quantitative estimation, the spots developed on filter paper are cut and dissolved in solvent for separation and further work. Photoelectric scanning, colorimetric or densiometric quantifi-cation of the eluted fraction can be carried out.

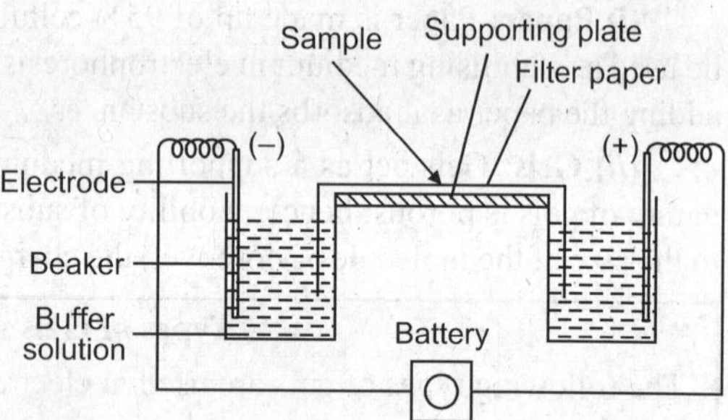

Fig. 5.6. Demonstration of paper chromatography (diagrammatic).

Indicators used for identification of substances

Different types of indicators are used to identify the substances separated through electrophoresis as given below:

Electrophoresed substances	Indicators
Amino acids	Ninhydrin
Carbohydrates	Schiff-periodic acid reagent
Proteins	Azocarmine, naphthalin black, bromophenol blue
Glycoprotein	Diphenylamine

2. Agarose Gel Electrophoresis

Agarose is a linear polymer of D-galactose and 3,6-anhydro-L-galactose which is extracted from seaweeds. By changing the concentration of agarose pore size of the gel can be determined. Gel electrophoresis allows for the separation of DNA molecules based on their size. Agarose gels are more porous than polyacrylamide gels and have comparatively larger pore size. Therefore, it is used for separation of large sized macromolecules such as DNA. Diagram of an electrophoresis is given in Fig. 5.7 A.

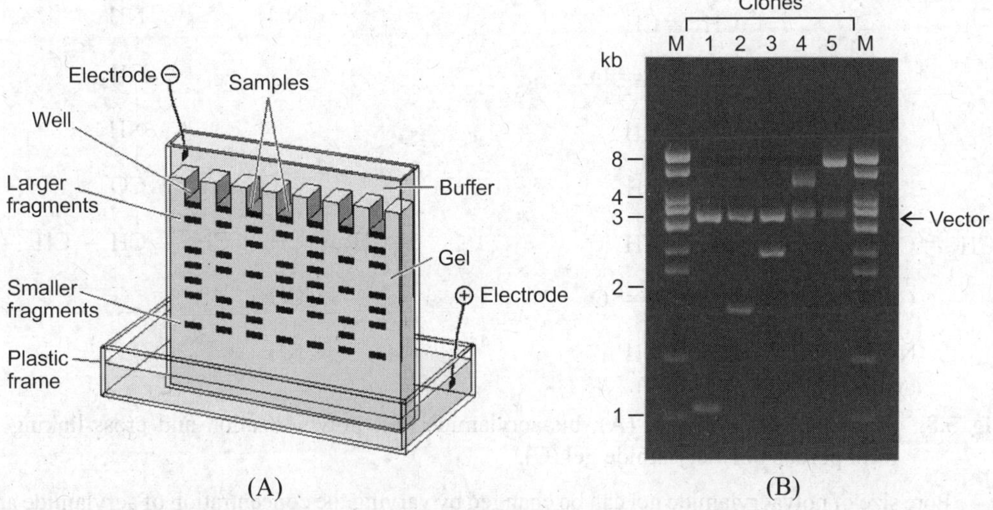

Fig. 5.7. A- an apparatus for slab gel electrophoresis; B- electrophorogram obtained after agarose gel electrophoresis.

Buffers used in electrophoresis

Different types of buffers are used for separation of substances such as phthalate buffer (pH 5.9) for amino acid, borate buffer (pH 10.0) for carbohydrates, and phosphate (pH 7.4) and barbiturate buffers (8.6) for proteins.

The commonly used buffers are Tris-acetate (TAE), Tris-borate (TBE) or Tris-phosphate (TPE) at a concentration of 50mM and a pH range of 7.5 – 8.0.

At neutral pH the negatively charged DNA migrates towards the anode after application of electric field across the gel. It is the standard technique for separation of nucleic acids. Conventional agarose gels cannot resolve fragments larger than 20 kb. Because larger DNA fragments are unable to migrate through pore sieves of gel as compared to small fragments.

Mostly the DNA is stained with ethidium bromide and then visualised with UV light. A typical example of an electrophorogram obtained after agarose gel electrophoresis is given in Fig. 5.7 B. The small-sized samples run faster than the large-sized samples. The bands lower in the gel are less intense than the bands in upper gel. This is because there is less DNA in the smaller bands than the larger bands; in smaller bands the number of molecules is the same but each molecule is smaller; therefore, the total amount of DNA is less.

3. Polyacrylamide Gel Electrophoresis (PAGE)

It is most widely used electrophoresis for separation and characterisation of proteins and nucleic acids. Polyacrylamide gel offers several advantages such as inertness to chemicals, superior resolution, stability over wide range of *p*H, temperature and ionic strength. Polyacrylamide gel is prepared by using the following : (*i*) monomers such as acrylamide, N,N-methylene bisacrylamide, (*ii*) initiator: N,N,N,N,-tetramethylethylene diamine (TEMED), (*iii*) propagators (ammonium persulphate or riboflavin, (*iv*) terminator or inhibitor (oxygen/air). Structure of acrylamide, N,N-methylene bisacrylamide and cross-linked polymers is given in Fig. 5.8. Oxygen is inhibitory for polymerisation.

(A) $CH_2{=}CH-C{=}O-NH_2$

(B) $CH_2{=}CH-C{=}O-NH-CH_2-NH-C{=}O-CH{=}CH_2$

(C) $-CH_2-CH[CH_2-CH]_n CH_2-CH-CH_2-$ (side chains: $CO-NH_2$; $CO-NH-CH_2-NH-CO$ cross-link)
$[CH_2-CH]_n CH_2-CH-CH_2-$ (side chain: $CO-NH_2$)

Fig. 5.8. Structure of acrylamide (A); bis-acrylamide (B); polymerization and cross-linking to form porous polyacrylamide gel (C).

Pore size of polyacrylamide gel can be changed by varying the concentration of acrylamide and bi-acrylamide monomers in a fixed volume of gelatin solution. Concentration of acrylamide should be between 3 to 30% to form gel for separation of molecules of $3{\times}10^6$ daltons. If the concentration of acrylamide is low, gels of larger pore size are formed. Such gels are suitable for separation of high molecular weight compounds; such compounds result in smaller pore size in gels. Therefore, pore size of gels can be maintained by varying the concentration of acrylamide and bis-acrylamide. Free radicals formed by ammonium sulphate activate the acrylamide gel. Long polymer chain is thus produced after reacting with the activated successive acrylamide molecules. Further, bisacrylamide brings about gelation and cross-linking through polymerisation, resulting in formation of complex network of acrylamide.

This apparatus is equipped with a buffer reservoir system and a power pack. The buffer reservoir set up consists of an upper cathode and a lower anode compartments possessing buffer solution. The two compartments are connected with a glass tube (1.4 × 0.5 cm). Each buffer reservoir is connected with a platinum electrode (Fig. 5.9). One end of the glass tube is closed with a rubber cork and the inner volume is filled with 1.1 ml of separation gel (pore solution). A small amount of double distilled water is poured onto separation gel and left for 30 minutes for polymerisation event. Thereafter, water is removed and the space is overlaid with large pore solution called 'spacer gel' followed by distilled water.

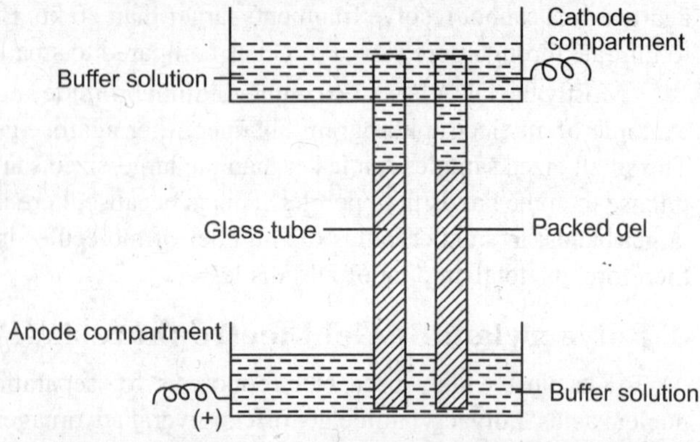

Fig. 5.9. Diagrammatic setup of polyacrylamide gel electrophoresis (PAGE).

Polymerisation occurs in aqueous solution. Small amount of sample (0.05 ml) is added after removing the water followed by addition of marker dye (0.001 ml). Tris buffer is used to fill up the remaining space in the tube.

Now the rubber cork is removed and the tubes are attached to the compartments in such a way that one end of the tube must be connected to the cathode and the other end to the anode compartments. Tris buffer is added to the compartments and current is supplied. The charged molecules migrate on the basis of molecular weight and ionic strength. The electrophoretic run gets stopped after migration of marker dye to the required position in the tube. Then the tubes are taken out from the compartments and gels are soon removed from the tube with needle. The gel is transferred into a tube consisting of acetic acid (70%). The gels are stained by using the suitable dyes. The samples are quantified with the help of densitometer or spectrophotometer.

On the basis of different types of buffer systems used PAGE has been grouped into the four types: (*i*) SDS-PAGE (reducing), (*ii*) SDA-PAGE (non-reducing), (*iii*) native continuous PAGE, and (*iv*) native discontinuous PAGE or native disc gel electrophoresis

4. Sodium Dodecyl Sulphate Polyacrylamide Gel Electrophoresis (SDS-PAGE)

Separation of some of the proteins does not occur due to similar charge : mass ratio(*z/m*). Therefore, such proteins are treated first with an anionic detergent called sodium dodecylsulphate (SDS) before start and during the course of electrophoresis. Therefore, such electrophoresis is called

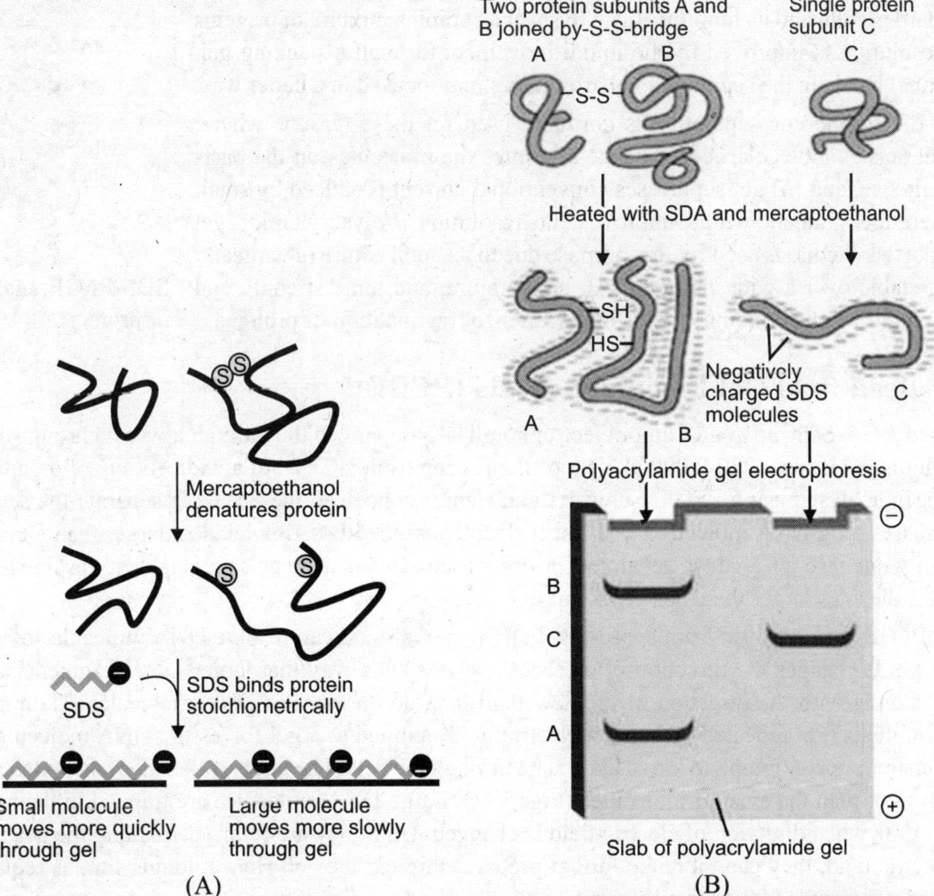

Fig. 5.10. A- effect of mercaptoethanol on separation of proteins; B- SDS-PAGE analysis of proteins.

SDS–PAGE. Identical proteins are denatured by SDS resulting in their sub-units or oligomers. The oligomeric protein is stabilised by hydrogen and disulphide bonds and by hydrophobic interaction also. Protein sub-units need to be opened by solubilising agents. The polypeptide chains get opened and extended after treatment with SDS that breaks hydrophobic interactions and results in negative charge to denatured polypeptides.

Hot mercaptoethanol is used to break disulfide bonds and denature the protein. SDS is then used to swamp the natural charges of protein so that all proteins have same charge to mass ratio (1 SDS binds per 1.4 amino acids) (Fig. 5.10 A). When these proteins are loaded onto a gel and an electric field is applied, the proteins migrate through the pores in the polyacrylamide gel. The gel separates the proteins based on their mobility, which is proportional to the logarithm of their molecular mass alone. Mercaptoethanol reduces inter-polypeptide disulfide bridge and separates the sub-units of a polymeric protein. The molecules are separated on the basis of their mass but not the charges (Fig. 5.10 B).

SDS–PAGE is a high resolution method used universally for analysing the mixture of proteins according to their respective size. SDS solubilised in soluble proteins makes possible the analysis of the other insoluble mixtures also. Separation of the proteins does not occur due to similar charge: mass ratio.

Proteins migrate toward anode at alkaline *p*H through PAGE gel during electrophoresis. The smaller polypeptides moves faster followed by the larger polypeptides. Molecular weight of the separated protein can be analysed by comparing the molecular weight of the standard protein and its mobility. In analysis of a complex mixture of proteins the resolution is improved by the initial movement through a stacking gel. The final bands in the separating gel are sharper and focused in a better way.

Electrophoretic separation is normally used for these reasons where: (*i*) gel acts as molecular sieves hence separates the molecules on the basis of their size, and (*ii*) gel suppresses conventional current produced by small temperature gradient which improves the resolution. Polyacrylamide gel (supporting media) is used for this purpose due to its good nature (chemically inert, stable over a wide range of pH, temperature, and ionic strength and transparent). Polyacrylamide gel is better for size fractionation of proteins.

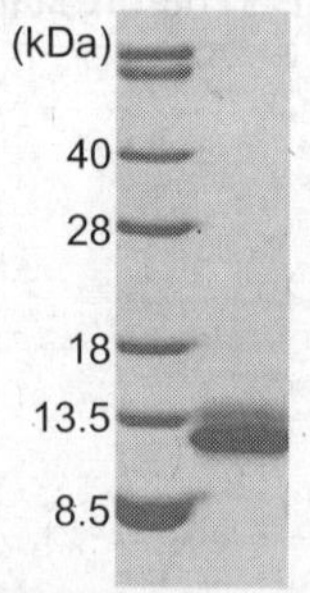

SDS-PAGE analysis of proteins

2. Pulsed-field Gel Electrophoresis (PFGE)

In 1984, Schwartz and Cantor electrophoretically separated the individual yeast chromosomes which ranged in size (230-2,000 kb). Since then chromosomal DNA of a wide variety of organisms has been analysed using PFGE. Schwartz and Cantor conceived the PAGE to separate the several megabases long DNA molecules. All linear double stranded DNA molecules, larger than a certain size, migrate through agarose gel almost at similar rate. But agarose gels are unable to resolve linear DNA molecules larger than 750 kb in length.

PFGE varies a little from agarose gel elctrophoresis and can resolve DNA molecules of 100-1000 kb. It changes the direction of electric field in such a way that larger DNA fragments align more slowly with the direction of the new field than do the smaller DNA molecules. The pulsed electric fields (*i.e.* repeatedly changed electric field) applied to a gel forces the DNA molecules to reorient before continuing to move like snake through the gels. The larger molecules take much time to re-orient than the smaller molecules (Fig. 5.11). Large DNA molecules are trapped in their tube every time when direction of electric field is changed. Until they are re-oriented along the new axis of electric field, they cannot make further progress through the gel. Hence, longer time is required for larger molecules for such alignment.

Therefore, such DNA molecules are functional according to their size whose re-orientation time is less than the period of electric pulses. Unlike a single constant electric field in conventional electrophoresis, two perpendicularly oriented alternating electric fields are applied in PFGE for separation of molecules. The smaller molecules change the direction quickly, move fast and separated from the larger molecules.

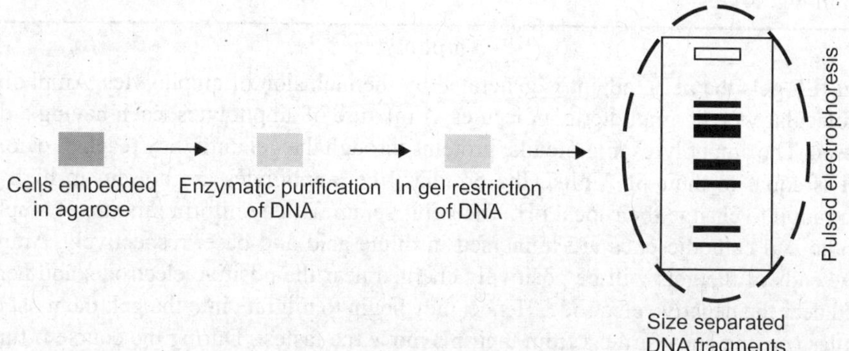

Fig. 5.11. Pulsed field gel electrophoresis.

5. Two-dimensional (2D) Gel Electrophoresis

Electrophoresis techniques can separate up to 100 different proteins on one run. Typically, cell or tissue extracts contain thousands of proteins, most of which will not be resolved into single bands using a separation based on any one parameter such as size or net charge. For any one size range, there is a high probability of more than one protein (out of thousands) falling into this range. On the basis of two parameters, usually size and isoelectric point, separation lowers the probability that two proteins will overlap, and allows the resolution of thousands of protein species on one gel. Two-dimensional gel electrophoresis is very useful and effective technique in separating proteins and resolving thousands of proteins in a mixture. Such separations are carried out by running a one dimensional gel to separate by the first parameter, and then laying this gel or a lane cut from it, across the top of the second dimension gel, the first gel serves as 'sample' for the second. The first dimension usually employs isoelectric focussing (IEP) as the means of separation. In the second dimension, polypeptides are separated according to their molecular weight. The second dimension is sodium dodecylsulphate polyacrylamide gel electrophoresis (SDS-PAGE).

***(a)* Isoelectric Focussing (IEF):** Isoelectric focussing gives a measurement of the isoelectric point (pI) of the protein, *i.e.* the *p*H at which the protein sample will not migrate in an electric field, the protein's net charge being zero. A two dimensional gel can give size and pI information on thousands of proteins in one run. The isoelectric point of a given protein is determined by the relative number of positively charged (*e.g.* lysine) and negatively charged (*e.g.* glutamate) amino acids. The negatively charged amino acids contribute to a high isoelectric point and positively charged amino acids contribute to a low isoelectric point. This method allows for the separation of all cellular proteins on a single large gel. This method often distinguishes between different isoforms of a particular protein, for example a protein that has been phosphorylated.

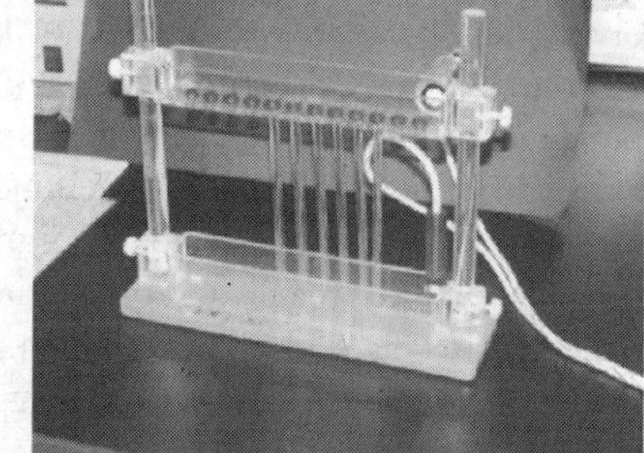

Fig. 5.12. First dimensional isoelectric focusing for 2D-PAGE.

The biomolecules such as proteins, DNA and RNA have electric charges which depend on molecule to molecule and the conditions of the medium (*p*H of buffer in which it is dissolved).

In a capillary tube a protein mixture is subjected to isoelectric focusing on gel. A *p*H gradient is maintained with gradual increase from anode to cathode. Charged molecules can be separated by electrophoresis in gels. Due to the differences in amino acid composition proteins have a unique mass and charge. Hence the proteins have net negative charge and net positive charge or iso-electric point (no charge) at a given *p*H of buffer. The first dimensional isoelectric focusing for 2D-PAGE is shown in Fig. 5.12.

Ampholytes

In IEF gels the pH gradient is generated by the inclusion of ampholytes. Ampholytes are low molecular weight amphoteric molecules. A mixture of ampholytes, each having a different pI, is used. The ampholytes migrate like proteins through the gel until they reach a region where the pH is equal to their pI. Unlike the proteins, the ampholytes are present in high enough concentration to change their local pH. The gel is set up with a uniform mixture of ampholytes. Its anodic and cathodic ends are immersed in dilute acid and base, respectively. Ampholytes near the ends of the gels will be positively charged near the positive electrode, and negatively charged near the negative electrode. Hence, they begin to migrate into the gel; the most charged molecules (*i.e.* the ones furthest from their pI) move the fastest. During the course if time, they separate into zones of defined pH. If the ampholyte system is well designed, a smooth gradient of pH is created with no abrupt charges.

Different mixtures of amphoteric substances have been used as ampholytes, such as amino acids, proteins, and poly acidic poly basic synthetic molecules. Proteins can be good ampholytes, but they interfere with analysis of the sample by introducing new proteins into the mixture. Polycarboxylic acid polyamines are the most commonly used ampholytes. These molecules have excellent buffering capacity and conductivity across a broad pH range, and are usually provided in a molecular weight range of 300 - 500, which is small enough to avoid interference with most subsequent processing. They may bind tightly to the proteins due to ionic interactions, and can be very difficult to remove.

Isoelectric focusing employs a *p*H gradient extending the length of an electrophoresis gel. A protein stops migrating when it enters the zone in which the surrounding *p*H equals its isoelectric point, pI. At any other point in the gradient, the protein acquires a charge which causes it to migrate toward its pI. The amphoteric substances such as proteins, which differ in their isoelectric points, can be separated by IEF. Isoelectric point is a *p*H value at which the net charges on molecules are zero. Ampholytes (*i.e.* complex mixture of synthetic polyaminopolycarboxylic acids) are introduced into the gel to create the *p*H gradient (from 3 to 10, or 7 to 8). Then a potential difference is applied across the gel. The molecules having differences in isoelectric points (by a little as 0.01 *p*H unit) can be separated. Proteins migrate depending on their charge until they reach a region where *p*H corresponds to respective iso-electric point at which *p*H proteins possess no net charge and hence got focused (Fig. 5.13).

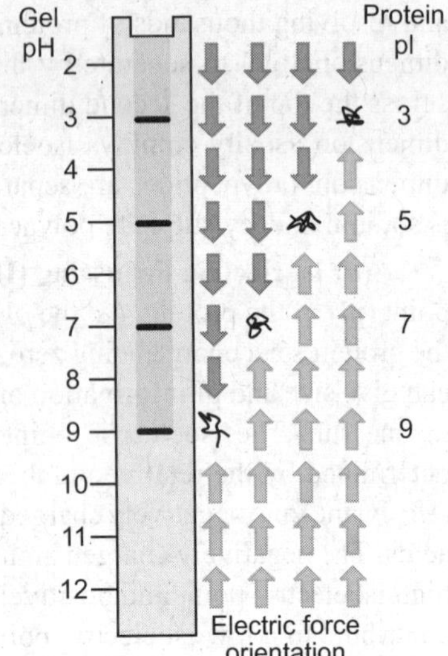

Fig. 5.13. A protein stops migrating at a zone when the surrounding pH equals its isoelectric point (see arrows).

***(b)* SDS-PAGE:** SDS-PAGE is the second dimension of PAGE which is to separate the polypeptides as described earlier. Proteins that have been separated can be cut out of the gel and, thereafter, analysed by mass spectrometry which identifies the protein.

C. ELECTROBLOTTING

In molecular biology and biochemistry, electroblotting is a method to transfer proteins or nucleic acids onto a membrane generally nitrocellulose after gel electrophoresis. Further, protein or nucleic acid can be analysed by using probes e.g. specific antibodies or stains. This technique can be used with all polyacrylamide and agarose gels. Capillary blotting is an alternative technique for transferring proteins from a gel. In 1989, this technique was patented by William J. Littlehales under the title '*electroblotting technique for transferring specimens from a polyacrylamide electrophoresis or like gel onto a membrane*' (http://en.vwikipedia.org/wiki/electroblotting).

Electroblotting technique is based on current and a transfer buffer solution that facilitate the movement of proteins or nucleic acids onto a membrane. After electrophoresis, a standard tank or semi-dry blotting transfer system is set-up. A stack is put together in the order from cathode to anode : sponge - three sheets of filter paper soaked in transfer buffer – gel – nitrocellulose membrane –three sheets of filter paper soaked in transfer buffer – sponge (Fig. 5.14). Membrane is put between the gel and anode because of the movement of current and sample in that direction. Thereafter, the stack is placed in the transfer system, and the current is run for few minutes on the basis of sample being used. After completion of transfer process, electrophoresis gel is stained with coomassie blue, which ensures the transfer of a sufficient quantity of material (http://en.wikipedia.org/wiki/electroblotting).

Fig. 5.14. Schemic presentation of a transfer stack.

D. GENE SEQUENCING

The structure of DNA (such as gene insert, a recombinant plasmid or entire genome) can be analysed by determining the nucleotide sequences. In molecular cloning, the information of nucleotide sequences is essential. In 1965, Robert Holley and his research group at Cornell University completely sequenced the nucleotides of $tRNA^{ala}$ (tRNA for yeast alanine).

I. DNA SEQUENCING

In 1977, two methods were developed for DNA sequencing. Allan Maxam and Walter Gilbert developed a chemical method of DNA sequencing. In the same year, Frederick Sanger and co-workers developed an enzymatic method of DNA sequencing, which is called 'dideoxynucleotide chain termination method' because dideoxynucleotides are used as chain terminator to produce a ladder of molecules. There are four types of deoxynucleotide triphosphates (dNTPs) such as dATP, dCTP, dGTP and dTTP. Similarly, there are four types of dideoxynucleotide triphosphates (ddNTPs) for example, ddATP, ddTTP, ddGTP and ddCTP.

Methods for DNA sequencing are : (*i*) Maxam and Gilbert's method (chemical cleavage), (*ii*) Fred Sanger's method (dideoxy method), (*iii*) automated sequencing (dideoxy, using fluorescent tags), (*iv*) DNA-chip sequencing (currently being developed), and (*v*) sequencing of individual bases (at the atomic level). These are listed roughly in their order of development - the first two became widely available in the late 1970's, both Fred Sanger (at the MRC in England) and Walter Gilbert (at Harvard) got the Nobel Prize for developing this method. However, Sanger's method is faster and the gels look a bit nicer, so not very many people still use the Maxam Gilbert method.

Walter Gilbert

For some DNA sequences (particularly if they are GC rich) it is necessary to resort to Maxam Gilbert Sequencing, but this is quite rare. The automated sequencing machines basically use the Sanger dideoxy method where only fluorescently labelled dideoxy bases are used that allows all four nucleotides to be put in the same lane.

1. Maxam and Gilbert's Method

This is a chemical method of gene sequencing, where the end-labelled DNA is subjected to base specific cleavage reaction before gel separation. This method is not preferred during routine sequencing of DNA.

Walter Gilbert

Walter Gilbert (born on March 21, 1932 in Massachusetts) worked on the theory of elementary particles, *i.e.* the quantum theory of physics. After getting his doctorate degree in 1957, he was trying to identify messenger RNA which serves as a carrier of information from the genome to the ribosomes. After each messenger RNA used a few times to dictate the structure of a protein, it is broken down and recycled to make other messenger RNA molecules. The experiments sought a fleeting new component that they finally managed to pin down.

In 1968, David Dressler and Walter Gilbert invented the rolling circle model which describes one of the two ways how DNA molecules duplicate themselves. In the middle seventies, Allan Maxam and Walter Gilbert developed the rapid chemical DNA sequencing. At this time, he developed some of the recombinant DNA techniques, specifically showing that blunt end ligation was efficient in putting DNA fragments together. He won the Nobel Prize in Chemistry 1980.

Chemicals of different base specificity are used for base alteration, altered base removal and strand cleavage (Table 5.2).

Table 5.2. Chemicals of different base specificity are used for base alteration, altered base removal and strand cleavage.

Base specificity	*Chemical used for base alteration*	*Chemical used for altered base removal*	*Chemical used for strand cleavage*
G	Dimethylsulphate	Piperidine	Piperidine
A+G	Acid	Acid	Piperidine
C+T	Hydrazine	Piperidine	Piperidine
C	Hydrazine + alkali	Piperidine	Piperidine
A >C	Alkali	Piperidine	Piperidine

The two strands of the DNA are separated by using dimethyl sulphoxide and heating to 90°C. Then it is purified by using gel electrophoresis. One of the strands is likely to be heavier than the other due to the presence of more purine nucleotides (A and G) than pyrimidines (C and T) which are lighter). Single-stranded sample is split into separate samples and each is treated with one of the cleavage reagents. This part of process involves alteration of bases (such as dimethylsulphate methylates guanine) followed by removal of altered bases. Lastly, piperidine is used for cleavage of strand at the points where bases are missing.

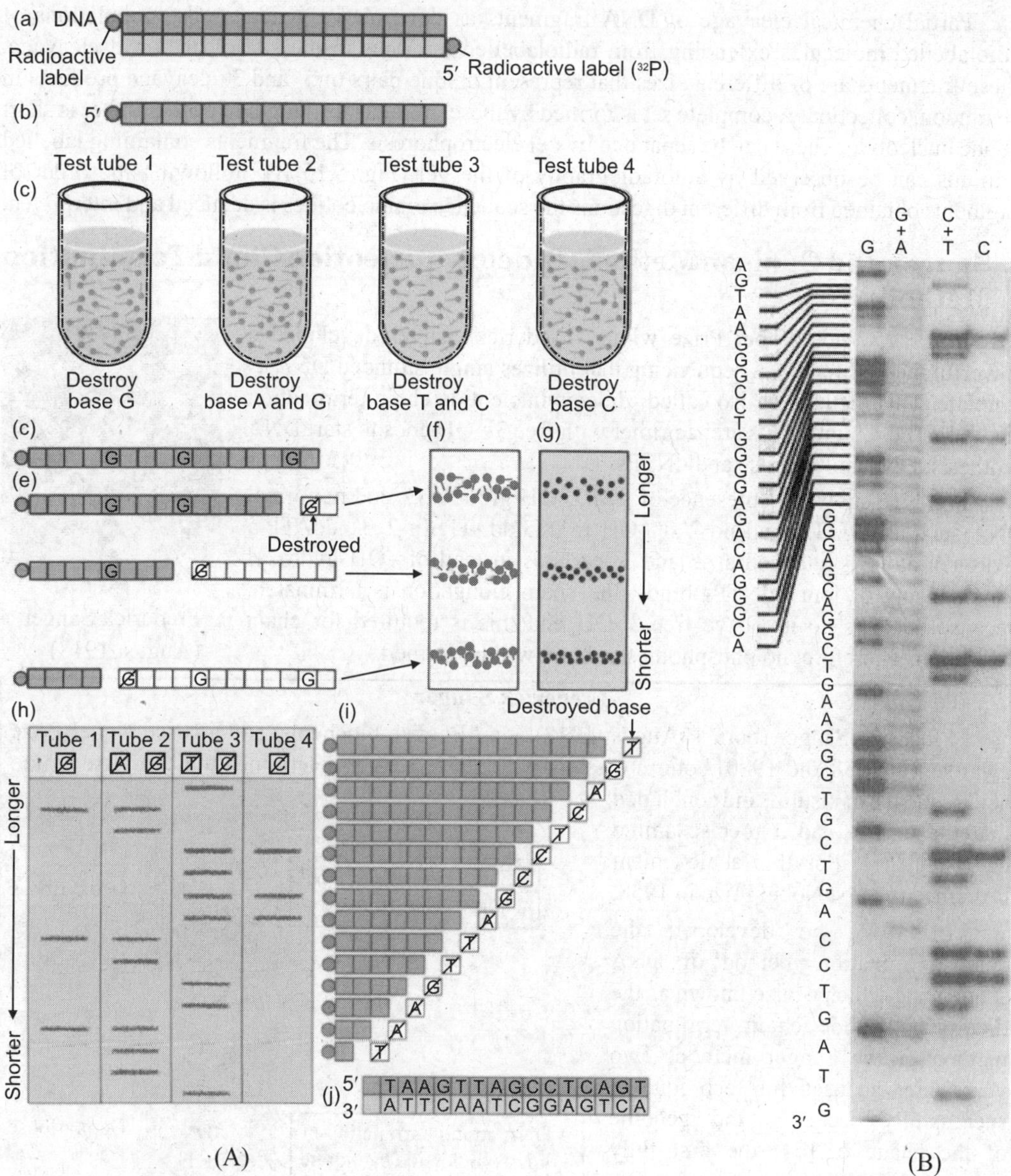

Fig. 5.15. Maxam-Gilbert's chemical degradation method of gene sequencing (A); nucleotide bands on gel (B)

***(a)* Cleavage of Purine:** The mixture is separated in four sets, each set treated with a different reagents which degrade only G or C or A and G or C and T. In one set, DNA is treated with acid followed by dimethyl sulphate. This causes methylation of A (at 3′ position and G (at 7′ position). Subsequently, addition of alkali (-OH) and piperidine results in cleavage of DNA and removal of purines (A or G) (Fig. 5.15 A).

***(b)* Cleavage of Pyrimidine:** Similar to the cleavage of purine, pyrimidine (C or T) is also cleaved in the presence of 1-2 M NaCl solution. It works only with C. Differences between these two indicate the presence of T in the DNA sequence. Cleavage of pyrimidines (C or T) through hydrazine hydrolysis is given in Fig. 5.15 A. If reactions have been arranged to give only one or a few cleavages per DNA molecule, a nested set of end-labelled DNA fragments of different lengths is produced.

Partial chemical cleavage of DNA fragments as done above generates the populations of radiolabelled molecules extending from radiolabelled terminus to the site of chemical cleavage. These fragments are of different sizes that represent unique pairs of 5′ and 3′ cleavage products in the random collection. A complete set is formed by these products the length of each number is short by one nucleotide. These can be separated by gel electrophoresis. The fragments containing labelled terminus can be observed by autoradiography of the gel (Fig. 5.15 B). Following the order of fragments obtained from different digestions the sequence of nucleotides is deduced and interpreted.

2. Sanger and Coulson Method (Dideoxynucleotide Chain Termination Method)

In 1977, twice Nobel Prize winner Frederick Sanger developed a powerful method for DNA sequencing that utilizes single stranded DNA as template. This method is also called 'dideoxynucleotide chain termination method'. The requirements are: a primer with free 3'-OH ends to start DNA synthesis; DNA polymerase and dNTPs.

Frederick Sanger (August 1918)

Fig. 5.16 shows the presence of free 3′-OH group at 3′ end in normal DNA such as 2′dATP, and no 2′,3′ -OH in 2′,3′ddATP. In 2′,3′, ddATP a hydrogen atom is attached at 2′ and 3′ carbons instead of –OH hydroxyl group. If any of four ddNTPs binds, the chain elongation is terminated. Because ddNTPs do not have free 3′-OH end this is required for chain elongation. Therefore, no phosphodiester bond will be formed.

Frederick Sanger

Frederick Sanger (born 13August 1918) is an English biochenist and a twice Nobel Prize awardee (in 1958 and 1980) both in Chemistry. In1955, Sanger determined the complete amino acid sequence of insulin, and concluded that the insulin had a precise amino acid sequence. For this achievement he earned the first Nobel Prize in 1958.

In 1975, he developed the chain termination method of DNA sequencing which is also known as the dideoxynucleotide chain termination method or the Sanger method. Two years later he used his technique to successfully sequence the genome of the phage ϕX174; the first fully sequenced DNA-based genome. For this work he won the second Nobel Prize in Chemistry in 1980, together with Walter Gilbert.

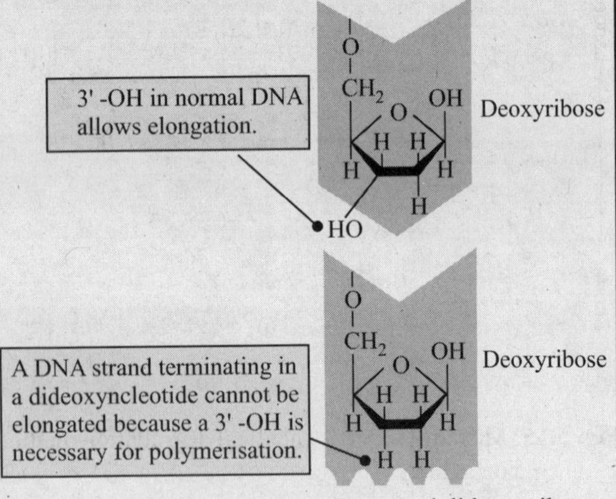

Fig. 5.16. Structure of deoxyribose and dideoxyribose.

Four reaction tubes are labeled with A, T, G and C each containing single stranded DNA template (obtained by NaOH hydrolysis), 5'-radiolabelled DNA primer, and all four radiolabelled dNTPs (dATP, dGTP, dCTP and dTTP) (Fig. 5.17 A). A small amount of ddATP is added to tube 1, ddGTP to tube 2, ddTTP to tube 3 and ddCTP to tube 4. The concentration of ddNTPs should be maintained to about 1% of the concentration of dNTPs. DNA polymerase is added to each tube, DNA synthesis starts and chain elongates. In each tube ddNTP is randomly incorporated and fragments are terminated. The length of each fragment depends on the position of incorporation of ddNTPs.

After completion of reaction, the fragments of each tube are separated by electrophoresis in four different lanes of high resolution polyacrylamide gel. Then the gel is dried and autoradiography is done so that position of different bands (having radiolabelled 5' end) in each lane is observed (Fig. 5.17 B). Ends of DNA fragments present in each lane contain the bases in correspondence to the ddNTPs used. DNA sequence is obtained by reading (from bottom to top of gel) the bands on autoradiogram of four lanes.

3. Automated DNA Sequencers (dideoxy, using Fluorescent Tags)

Automatic DNA sequencing is easier than manual sequencing. In this method, special dyes are used that fluoresce after hitting them by, instead of using radioactive isotopes to label the DNA fragments. Automatic DNA sequencing machines were developed during 1990s. It is an improvement of Sanger's method. Each of the four nucleotides (G, A, T and C) has its own unique color. In this new method a different fluorescent dye is tagged to the ddNTPs. Using this technique a DNA sequence containing thousands of nucleotides can be determined in a few hours. Each dideoxynucleotide is linked with a fluorescent dye that imparts different colours to all the fragments terminating in that nucleotide. All four labelled ddNTPs are added to a single capillary tube. Thus, only one lane per sample is required instead of the four lanes in manual sequencing.

An automatic sequencer.

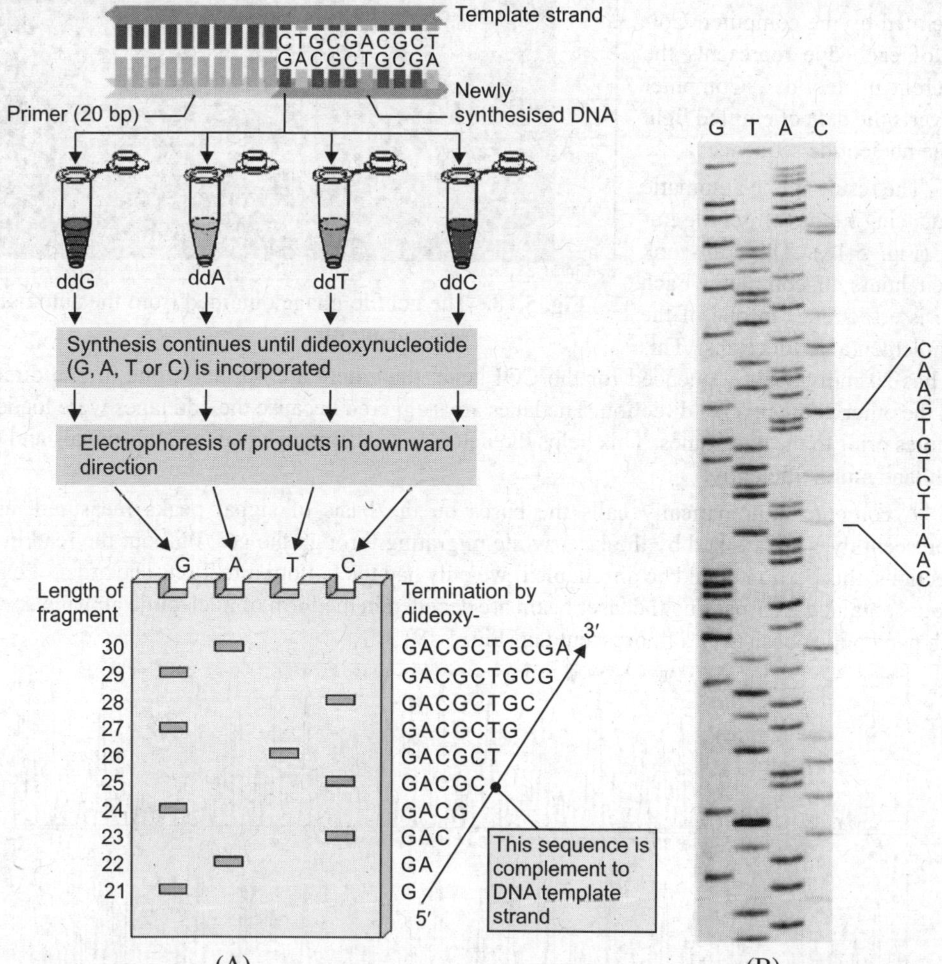

Template strand
CTGCGACGCT
GACGCTGCGA
Newly synthesised DNA
Primer (20 bp)
ddG ddA ddT ddC
Synthesis continues until dideoxynucleotide (G, A, T or C) is incorporated
Electrophoresis of products in downward direction
G A T C
Length of fragment
Termination by dideoxy-
30 GACGCTGCGA 3′
29 GACGCTGCG
28 GACGCTGC
27 GACGCTG
26 GACGCT
25 GACGC
24 GACG
23 GAC
22 GA
21 G 5′
This sequence is complement to DNA template strand
(A)

G T A C
CAAGTGTCTTAAC
(B)

Fig. 5.17. A- Sanger and Coulson method for DNA sequencing, B- DNA bands on autoradiogram of four lanes.

It is a refinement of gel electrophoresis which separates DNA very fast. DNA fragments of different colours are separated by their respective size in a single electrophoretic gel. A current is applied to the gel. The negatively charged DNA strands migrate through the pores of gel towards the positive end. The small sized DNA fragments migrate faster and vice versa. All fragments of a given length migrate in a single peak. The DNA fragments are illuminated with a laser beam. Then the fluorescent dyes are excited and emit light of specific wavelengths which is recorded by a special 'recorder'. The cycle sequenced samples are loaded at the top and the laser scans back and forth while the DNA pieces migrate fast to the bottom.

The DNA sequences are read by determining the sequence of the colours emitted from specific peaks as they pass the detector. This information is directly passed on to the computer which determines the sequence. A tracing electrogram of emitted light of the four dyes is generated by the computer. Colour of each dye represents the different nucleotides. Computer converts the data of emitted light in the nucleotide sequences.

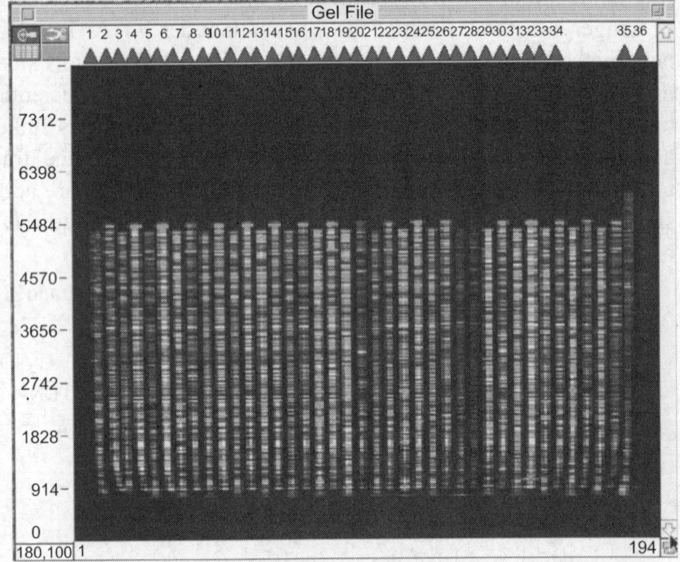

Fig. 5.18. The gel file image obtained from the automatic sequencing run.

The result of one automatic sequencing run is the gel file image (Fig. 5.18). This run took seven hours to complete. Each lane is a sequence in one of the complementary directions. This gel has 20 individuals sequenced for the COI gene; thus there are 40 lanes, one in one direction and the other in the reverse direction. The lanes are staggered because the odd lanes were loaded 10 minutes prior to the even lanes. This helps the detector to distinguish between the signals and track each lane more efficiently.

A computer automatically calls the bases on the basis of signal peaks measured, as the fluorescent dyes are excited by the laser while migrating through the gel. Without the read in both directions, this portion would be unreliable if we only had the bottom direction. The order of excited colours obtained after passing the laser beam are decoded in the form of nucleotide alphabets, where each nucleotide consists of a fluorescent tag (Fig. 5.19).

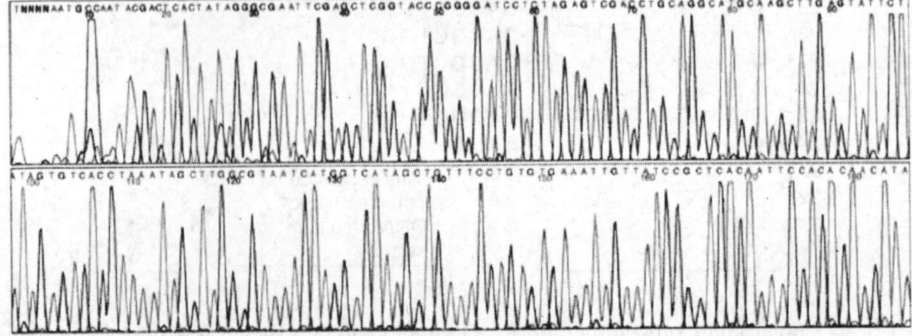

Fig. 5.19. Automated sequencing (dideoxy, using fluorescent tags).

II. RNA SEQUENCING

DNA sequencing is easier than RNA sequencing due to a greater stability. But RNA sequencing some times becomes necessary when one has to determine the position of modified nucleotides present in RNAs. It is achieved by cleaving 5' end labelled RNA by using RNases, that cleave a particular nucleotide at 3' end. RNase T1 cleaves after G, RNase U_2 after A, RNase PhyM after A and U, and *Bacillus cereus* RNase after U and C. The fragments are analyzed by PAGE.

E. RESTRICTION MAPPING

Much has been described about restriction enzymes in Chapter 6, *Recombinant DNA Technology : Enzymes*. Restriction enzymes cut DNA at specific sequences as given below :

*bam*HI	5′GGATCC-3′ 3′-CCTAGG-5′	5′G 3′-CCTAG	GATCC-3′ G-5′
Bg/II	5′-AGATCT-3′ 3′-TCTAGA-5′	5′A 3′-TCTAG	GATCT-3′ A-5′
*Eco*RII	5′-GGATCC-3′ 3′-CCTAGG-5′	5′G 3′-CCTAG	5′-GATCC-3′ G-5′

These enzymes can be used to map a given DNA sequence. A linear DNA is separately cleaved with restriction enzyme 1, enzyme 2, enzyme 1+ enzyme 2. This generates the DNA fragments of varying sizes. Consequently these fragments are passed through gel electrophoresis. Three fragments of varying base pairs are produced with enzyme 1, two fragments with enzyme 2 and four fragments with enzyme 1+ enzyme 2. Thereafter, a restriction map is prepared and combined map is prepared (Fig. 5.20).

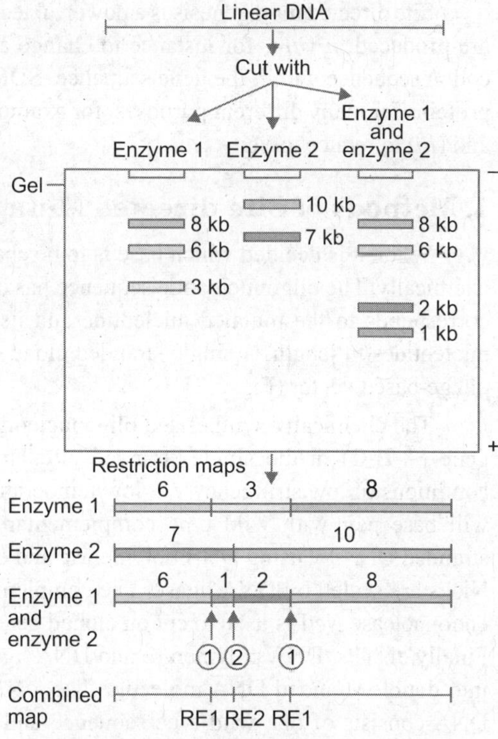

Fig. 5.20. Restriction map of a linear DNA.

F. SITE-DIRECTED MUTAGENESIS (SDM)

In 1978, an important milestone in modern molecular biology came with the first description of site-directed mutagenesis using oligonucleotides for *in vitro* synthesis of mutant DNA. Michael Smith of Canada University of British Columbia (Canada) pioneered this technique which is being used in all biochemistry and molecular biology laboratories. He shared the Nobel Prize in Chemistry in 1993 with Kary B. Mullis who developed polymerase chain reaction (PCR) machine.

In 1987 Kunkel and co-workers improved this technique in which phenotypic selection is not done. The plasmid to be mutated would be transformed into an *E. coli*-deficient in two genes, dUTPase and uracil deglycosidase. The former prevents the breakdown of dUTP, a nucleotide that replaces dTTP in RNA and results in an abundance of the molecule. Deficiency in the latter prevents the removal of dUTP from newly synthesised DNA. As the double-mutant *E. coli* replicates the up-taken plasmid, its enzymatic machinery incorporates the dUTP, resulting in a distinguishable copy. This copy is then extracted and incubated with an oligonucleotide containing desired mutation, which attaches to the complementary wild-type gene sequence, as well as the Klenow enzyme, dNTPs, and DNA ligase.

Michael Smith (1932-2000)

The reaction essentially replicates the dUTP-containing plasmid using the oligonucleotides as primer giving a nearly identical copy. The essential differences being that the copy contains dTTP rather than dUTP, as well as the desired mutation. When the chimeric double-stranded plasmid, containing the dUTP, unmutated strand and the dTTP, mutated strand, is inserted into a wild type *E. coli*, the dUTP-containing strand is broken down, whereas the mutation-containing strand is replicated.

Mutations can be done by directing insertions or deletions to a specific site on the DNA but they are unlikely useful. These minor changes may have some importance if a single amino acid is altered in a protein in order to improve its properties. For such modifications point-mutations, *i.e.* changing of single nucleotide, is done on specific portion in the gene. Thus, the most suitable method that could bring about point-mutation in a gene is known as oligonucleotide-directed mutagenesis. This technique has potential for protein engineering, the engineered enzymes would be better than the wild type ones. Change in a single nucleotide base pair is called 'point mutation'.

Site directed mutagenesis is a powerful technique where site specific changes in DNA sequence are produced *in vitro,* for instance to change an amino acid residue into another by changing the codon sequence within the gene sequence. SDM is also used to 'engineer' commercially important proteins for many different purposes, for example : (*i*) to improve stability, (*ii*) to change specificity, and (*iii*) to reduce toxicity.

1. Methods of Site-directed Mutagenesis

When it is decided which base is to be changed (such as A), an oligonucleotide is synthesised chemically. The oligonucleotide sequence has desired mutation at specific site. The oligonucleotide corresponds to the mutated nucleotide and its neighbouring regions, mainly between 15 and 20 nucleotides in length. A single stranded clone of the wild type gene is produced by using an M13 phage-based vector (Fig. 5.21).

The chemically synthesised oligonucleotide (A′ - C′) is allowed to hybridise with the cloned gene (A-B-C) in the single stranded circular DNA. The hybridisation is performed under the conditions of low stringency, *i.e.* low temperature and high salt concentration. The oligonucleotide will base pair with wild type complementary sequence and acts as primer to produce double stranded DNA by using DNA polymerase and dATP, dGTP, TTP and α-*S*-dCTP (alpha-thio-dCTP). Nicks are sealed by DNA ligase. The complementary strand consists of α-*S*-dCTP. The restriction endonuclease *Nci*I is used to cut on cloned gene and exonuclease removes from the original strand. Finally dNTPs, DNA polymerase and DNA ligase synthesise the single stranded DNA and convert into double stranded DNA molecule. The DNA ligase seals the nick (Fig. 5.21). Double stranded DNA consists of one wild type sequence and the other mutant sequence. The bacterial cells are transformed by introducing the double stranded molecule inside. The duplex DNA replicates in bacterial cell and produces either wild type or mutant plasmids.

Replication of DNA in a bacterial cell will result in a mixture of wild type and mutant double stranded DNA. These can be extracted and used to transform more cells. So far it is not sure whether mutation is present in desired DNA or not. To ascertain this, the DNA is isolated from bacterial cells and sequenced following the sequencing method as described earlier.

Some useful properties in proteins (like stability in subtilisin) can be conferred by point mutation that results in substitution of selected amino acids to improve laundry detergents. Using this technique specific point of a gene can be mutated. Therefore, this method has been used to understand the function of many genes. Moreover, this technique can only be used when nucleotide sequence of gene is known.

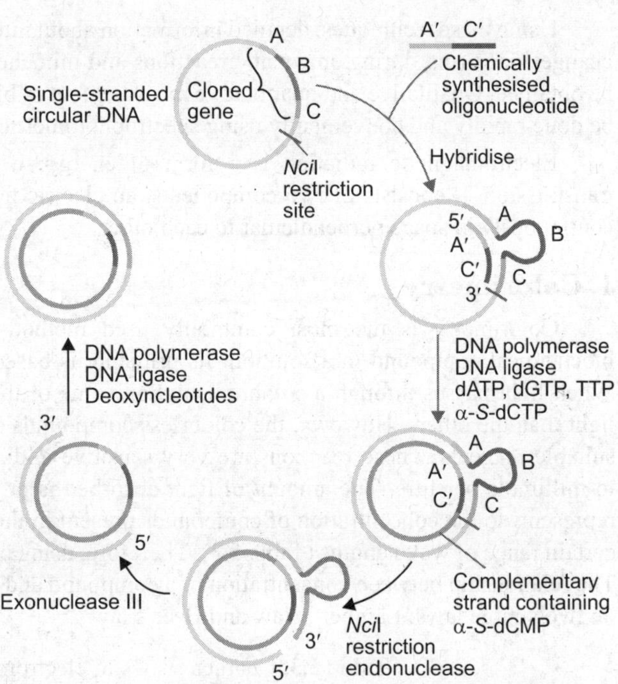

Fig. 5.21. Method of site-directed mutagenesis.

Following site-directed mutagenesis several novel proteins and enzymes have been produced and commercialised such as subtilisin, β-lactamase, tyrosyl-rRNA synthetase, etc. A detailed account of protein engineering in subtilisin protease has been given in Chapter 11: *Proteins: structure, engineering and design*.

G. SPECTROSCOPY

Spectro means production of whole range of continuous wavelength by light source. The study of interaction of electromagnetic radiation with matter is called spectroscopy, which was originally the study of the interaction between radiation and matter as a function of wavelength (λ). Historically, spectroscopy referred to the use of visible light dispersed according to its wavelength by a prism. Thereafter, the concept was expanded to comprise any measurement of a quantity as function of either wavelength or frequency. Thus it also can refer to interactions with particle radiation or to a response to an alternating field or varying frequency (ν). A further extension of the scope of the definition added energy (E) as a variable, once the very close relationship $E=h\nu$ for photons was realised. A plot of the response as a function of wavelength, or more commonly frequency, is referred to as a spectrum. Spectrometry is the spectroscopic technique used to assess the concentration or amount of a given species. The instrument that performs such measurements is called spectrometer or spectrograph. Spectroscopy/spectrometry is often used in physical and analytical chemistry for the identification of substances through the spectrum emitted from or absorbed by them.

Characterisation of chemical compounds on the basis of their spectral properties is done by using spectroscopy. In early days, qualitative estimation and identification of biomolecules were done using visible radiation (light) in spectroscopy. But in recent years spectroscopic techniques have been refined. Therefore, its scope has been broadened. Now-a- days it is used to examine the behaviour of chemical substances after electromagnetic irradiation such as gamma (γ) rays, X-rays, ultraviolet (UV) rays, infrared (IR) rays, radio waves and microwaves.

Using these techniques, detailed information about inter-molecular bonding types or molecular changes occurring during enzymatic reactions and mitochondrial electron transport chain, etc. can be obtained. Qualitative and quantitative measurement of biomolecules even in impure samples can be done rapidly and conveniently using spectrophotometric and flavometric methods.

Electromagnetic radiation is a form of energy; it does not require any medium for its transmission. It consists of two components, an electric field and a magnetic field. They oscillate continuously in space perpendicular to each other.

1. Colorimetry

Colorimetry is the most commonly used method for estimation of concentration of a biochemical compound in a solution. Its principle is based on properties and absorption of light. When light passes through a coloured solution some of the biochemical compounds absorb more light than the others. However, the colourless compounds can be made coloured after mixing with suitable reagents. These reactions are very sensitive and can measure even the concentration up to millimole per litre. The amount of light absorbed is proportional to the colour intensity which represents to the concentration of compounds present in the sample. Coloured filters absorb all but certain range of wavelengths (Table 5.3). Therefore, using coloured filters wavelengths are selected. The relationship between concentration of a compound and the extent of absorption of light is based on two simple laws: Lamberts' law and Beer's law.

Table 5.3. Characteristics of electromagnetic radiation.

Region	*Wavelength* (γ) *(nm)*
Cosmic rays	5×10^5
Gamma rays	10-3 - 0.14
X rays	0.01 - 15
Far ultraviolet	15 - 200
Near ultra violet	200 - 400
Visible	400 - 800
Near infrared	0.8 - 2.5 μm
Far infrared	0.025 - 0.5 mm
Microwave	0.5 - 300 mm

Lambert's law: When a beam of monochromatic light of only one wavelength passes through a homogeneous solution, its intensity decreases exponentially as the length of absorbing medium increases. This law states that the amount of light absorbed (A) is directly proportional to the length or thickness (b) of the solution : $\log_{10} (Io/I) = \varepsilon.b$.

Beer's law: When a ray of monochromatic light passes through an absorbing medium, its intensity decreases exponentially as the concentration of absorbing medium increases. This law states that *the amount of light absorbed is directly proportional to the concentration* (c) *of the solutes in the solution:* $\log_{10} (Io/I) = \varepsilon.b.c$.

These two relationships are combined into one Lambert-Beer law. The combined law states that *at a given wavelength the fraction of incident light absorbed by the solution is related to the length of absorbing medium and the concentration of solutes in the medium:* $(\log_{10}/I) = e.c.l$; where, Io = intensity of incident light, I = intensity of transmitted light after passing through the solution, e = molar absorbance coefficient, c = concentration of solutes, l = length of thickness of liquid in cell or cuvette in cm.

The above equation indicates that the solute molecules and solution are randomly oriented, and the incident light is parallel and monochromatic. The absorbance of light by a solution depends on molar absorbance coefficient and concentration (moles/L) of the absorbing solutes. Molar absorbance or molar extinction coefficient is the absorbance of light by one molar solution of the solute using standard cuvette of 1 cm thickness of absorbing medium (light path). If the light absorbing species is in equilibrium with a state of ionisation (having different ionisation properties), the molar extinction coefficient varies according to the nature of absorbing compounds such as solvent, *p*H and wavelength. The molar extinction coefficient has its maximum value of 10^6; therefore, at least 10^{16} coloured solutes/ml are required to observe the colour of solution.

Colorimeter: Based on above principle, colorimeters have been constructed to measure light absorbance by a chemical substance at a given wavelength. Example of a colorimeter is given in Fig. 5.22. A colorimeter is a device used to measure the concentration of a solution by measuring its absorbance of a specific wavelength of light. Different solutions are prepared along with a control (usually a mixture of distilled water and another solution) is first filled into a cuvette and put inside a colorimeter. It is left for a few minutes to calibrate the machine.

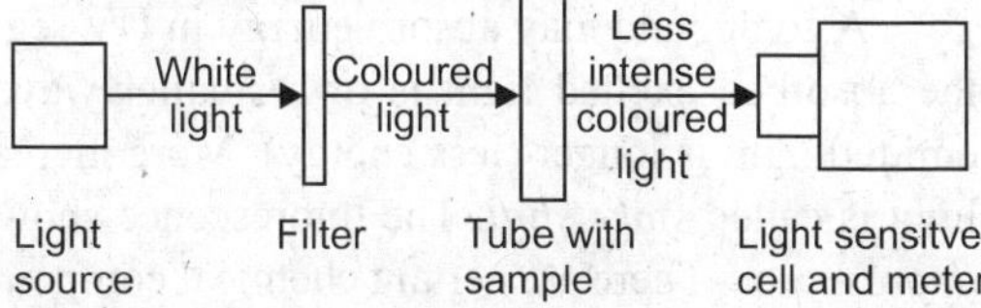

Fig. 5.22. Working principle of a colorimeter.

Only after the device has been calibrated you can use it to find the densities and/or concentrations of the other solutions. You do this by repeating the calibration, except with cuvettes filled with the other solutions. The filter on a colorimeter must be set to red if the liquid is blue. The size of the filter initially chosen for the colorimeter is extremely important, as the wavelength of light that is transmitted by the colorimeter has to be same as that absorbed by the substance.

2. UV-Vis (visible) Spectrophotometry

The principles and techniques of spectrophotometer also are based on Lambert-Beer law. Colorimeters are designed to function only in the visible range (400-700 nm) wavelength). But the spectrophotometers are equipped to operate both in the visible and UV ranges of spectrum, because these are fitted with deuterium or hydrogen (for UV light) and tungsten (visible range) lamps as the light sources. There are several models of spectrophotometer which have varying degree of sophistication. Therefore, a sophistication type of colorimeter is known as spectrophotometer.

This sophistication is in the form of single beam, double beam and recording. In spectrophotometer monochromatic light is produced by a prism. Single beam spectrophotometers utilise a fixed or a continuous wavelength of light source, such as a light-emitting diode (LED), a sample container, and a photodiode detector. Colorimeters consisting of a continuous source are fitted with dispersing element (diffraction grating) and an aperture that selects a single light wavelength which passes through the sample. Then the light scattered is measured by the detector of the colorimeter (Fig. 5.23).

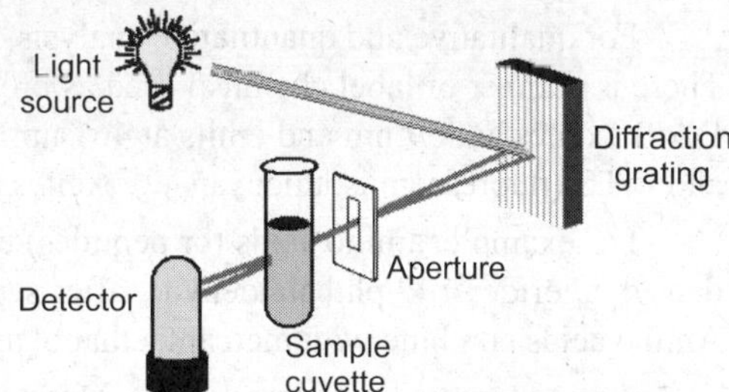

Fig. 5.23. Components of a single beam UV-Visible spectrophotometer.

Spectrophotometer is useful for both quantitative and qualitative estimation of macromolecules such as nucleic acids, proteins, pigments, etc. Even the colourless substances like sugars, amino acids, metal ions can be estimated after reacting them and producing coloured derivatives.

3. Fluorescence Spectroscopy

An electromagnetic spectroscopy which analyzes fluorescence from a sample is called fluorescence spectroscopy or fluorometry or spectrofluorimetry. It involves using a beam of light, usually UV light, that excites the electrons in molecules of certain compounds and causes them to emit light of a lower energy, but not necessarily, visible light. Devices that measure fluorescence are called fluorometers or fluorimeters.

Fluorescence refers to emission of light from a molecule after absorption of radiant energy. When a molecule absorbs radiant energy, its electrons are elevated to excited state. If there is no chemical reaction, dissociation, rearrangement or energy transfer by collision, the molecule return to ground state and excessive energy is lost as heat or light, *i.e.* photons of energy (v). Emission of light is called fluorescence.

A compound may absorb energy in UV region and fluoresce in visible region. In fluorescence the absorbed excited light is of a smaller wavelength (more energy) and the wavelength of the emitted light is longer (less energy). More increase in wavelength of emitted light than the excited light is called *stoke shift*. The fluorescence shown by the natural compounds is known as *intrinsic fluorescence*. There are certain chemical compounds which do not fluoresce. However, fluorescence can be detected after coupling them with a *fluorescence probe* or *fluor*. This phenomenon is called *extrinsic fluorescence*.

The indole-derivative exhibits intense blue fluorescence. Mostly it involves staining with a special type of histological stain called a fluorochrome which is only taken up by specific material or hybridisation of fluorescently labeled complementary DNA sequences. High energy (short wavelength e.g. UV or blue) light is directed onto the preparation and lower energy (longer wavelength e.g. green or red) light is emitted. There are a number of possible routes by which an excited molecule can return to its ground or room temperature state via unstable triplet states. A rapid return results in *fluorescence* and a delayed return results in *phosphorescence* (Fig. 5.24).

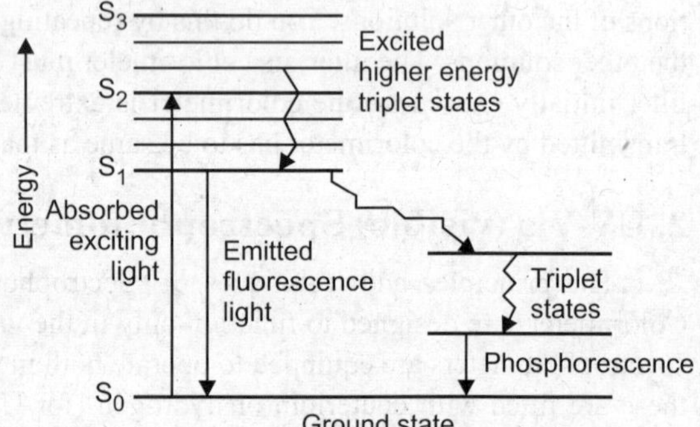

Fig. 5.24. Mechanism of fluorescence and phosphorescence.

For qualitative and quantitative analysis of such chemical compounds, *i.e.* fluor is very useful. There is number of label chemical used as probe which fluoresces at different wavelengths such as DAPI excites at 359 nm and emits at 461 nm, Rhodamine excites at 570 nm and emits at 590 nm, and FITC (fluorescein isothiocyanate) excites at 494 nm and fluorescences at 518 nm.

For example, amino acids (or peptides) can be identified by coupling their primary groups by dansyl chloride or *O*-phthalaldehyde, after separation through electrophoresis or chromatography. Amino acids combine with mercaptoethanol and phthalaldehyde and fluoresce as given below:

Amino acid + Mercaptoethanol + Phthalaldehyde

↓

Indol derivative
(fluorescent)

Basic components of fluorescence spectroscopy are given in Fig. 5.25. It is now usual to use much more efficient epi-illuminators with small 50 or 100 watt super pressure mercury or xenon lamps. Using a matched optical set that consists of an exciter filter, a dichroic beam splitter and a barrier filter, the epi-illuminator allows only light of the correct short, exciting wavelength to strike the preparation and only light of the longer, emitted wavelength to reach the eyepiece, detector or the camera.

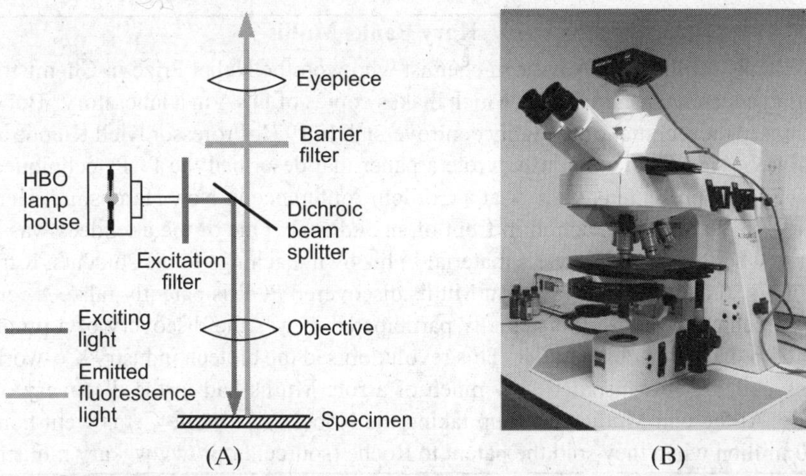

Fig. 5.25. A- basic components of fluorescence spectroscope; B- A fluorescence microscope.

Fluorescence spectroscopy is very useful in quantitative determination of such molecules that are present in very low amount and not detected by the other methods. For example, this technique is very useful in quantitative determination of compounds such as vitamin B1 (in food), hormones, NADH, drugs, carcinogens and some metal ions as well. After adding fluorescent probe, this technique becomes 1000 time more sensitive than the colorimetric method.

H. POLYMERASE CHAIN REACTION

The polymerase chain reaction (PCR) is a technique which is widely used in molecular biology. Its name has been derived from the enzyme, DNA polymerase which is used to amplify a piece of DNA by *in vitro* method. As PCR progresses, the DNA thus generated is itself used as template for replication. It is possible to amplify a single or few copies of a piece of DNA across several orders of magnitude by using PCR, generating millions or more copies of the DNA piece.

Kary Banks Mullis

PCR provides a simple and indigenous method for exponentially amplification of specific DNA sequences by *in vitro* DNA synthesis. In 1983, this technique was developed by Kary Mullis at Cetus Corporation in Emery Ville, California (U.S.A.). Kary Mullis shared the Nobel Prize for chemistry in 1993. This technique has made possible to synthesise large quantities of a DNA fragment without its cloning.

It is ideally suited where the quantity of biological specimen available is very low such as a single fragment of hair or a tiny blood stain left at the site of a crime. Almost all PCR applications employ a heat-stable DNA polymerase, *i.e.* Taq polymerase- an enzyme originally isolated from the bacterium *Thermus aquaticus*. This DNA polymerase enzymatically assembles a new DNA strand from nucleotides by using the single-stranded DNA as template and DNA primers required for initiation of DNA synthesis. The vast majority of PCR methods use thermal cycling, *i.e.* alternately heating and cooling the PCR sample to a defined series of temperature steps. These thermal cycling steps are necessary to physically separate the strands (at high temperatures) in a DNA double helix (DNA melting) used as template during DNA synthesis (at lower temperatures) by the DNA polymerase to selectively amplify the target DNA. The selectivity of PCR results from the use of primers that are complementary to the DNA region targeted for amplification under specific thermal cycling conditions.

Kary Banks Mullis

Kary Banks Mullis is an American chemist who won the Nobel Prize in Chemistry in 1993 for inventing a technique called PCR which makes copies of DNA in a laboratory. But his actual contributions to the technique are highly controversial. In 1971, Professor Kjell Kleppe and Nobel laureate Har Govind Khorana jointly wrote a paper that described the PCR technique. On 18th June 1969 Kleppe presented his work at a Gordon Conference in New Hampshire. Here Kleppe performed a standard PCR reaction in front of an audience. One of the attendees was Professor Stuart Linn, who then used Kleppe's material in his own teachings to his students. Karry Mullis was one of his students. So the idea that Mullis discovered PCR is patently false. A contribution to the PCR technique that Mullis actually participated in was the discovery of a protein which dramatically reduced the cost of PCR. This revolutionised the biotech industry. Co-workers from a company called Cetus disputed how much of a role Mullis had in the discovery of the Taq polymerase. At the end Mullis ended up taking the Nobel Prize while Cetus went home with a cool $300 million when they sold the patent to Roche (source: http://www. kary mulis.htm).

The PCR technique has now been automated and is carried out by a specially designed machine. PCR can be extensively modified to perform a wide array of genetic manipulation. PCR is now a common and often indispensable technique used in medical and biological research laboratories for a variety of applications. An old model of a PCR and a strip of eight PCR tubes are given in Fig. 5.26 A-B.

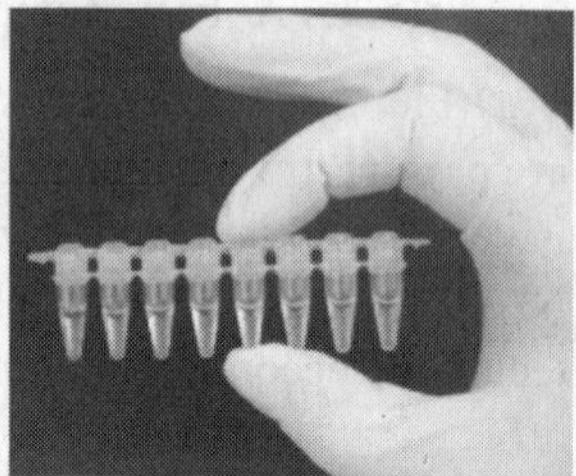

Fig. 5.26. A-an old thermal cycler for PCR; B-a srip of eight PCR tubes each containing 100 µl.

The PCR requires the following :

***(i)* DNA Template:** Any source that contains one or more target DNA molecules to be amplified can be taken as template.

***(ii)* Primers:** A pair of oligonucleotides of about 180-30 nucleotides with similar G+C contents acts as primers. They direct DNA synthesis towards one another. The primers are designed to anneal on opposite strands of target sequence so that they will be extended towards each other by addition of nucleotides to their 3' ends.

***(iii)* Enzyme:** The most common enzyme used in PCR is a thermostable enzyme called *Taq* polymerase. It is isolated from a thermostable bacterium called *Thermus aquaticus*. Taq polymerase has a temperature optimum at around 70°C. But it survives at 95ºC for 1-2 minutes and has a half life for more than 2 hours at this temperature. The other thermostable polymerases can also be used in PCR.

***(iv)* Deoxynucleotide triphosphates (dNTPs):** It is also commonly and erroneously called deoxynucleotide triphosphate), the building blocks from which the DNA polymerase synthesises a new DNA strand.

***(v)* Buffer solution:** It provides a suitable chemical environment for optimum activity and stability of the DNA polymerase.

***(vi)* Divalent cations:** Generally Mg^{2+} is used, but Mn^{2+} can be utilised for PCR-mediated DNA mutagenesis, as higher Mn^{2+} concentration increases the error rate during DNA synthesis.

***(vii)* Monovalent cation**: Potassium ions are also used as monovalent cations.

1. Working of PCR

The PCR consists of a series of 20 to 40 repeated temperature changes called *cycles*; each cycle typically consists of 2-3 discrete temperature steps. Most commonly PCR is carried out with cycles that have three temperature steps. The cycling is often preceded by a single temperature step called hold at a high temperature (>90°C), followed by one hold at the end for final product extension or brief storage. The temperatures used and the length of time applied in each cycle depend on a variety of parameters. These include: the enzyme used for DNA synthesis, the concentration of divalent ions and dNTPs in the reaction, and the melting temperature (*Tm*) of the primers.

The action of PCR involves several cycles. However, there are three steps in one amplification cycles *e.g denaturation* (*melting*), *annealing* and *polymerisation* (*extension*) (Fig. 5.26). Before the start of PCR initialisation step is carried out where the reaction is heated to a temperature 94-96°C (or 98°C if extremely thermostable polymerases are used), and hold for 1-9 minutes. It is required only for DNA polymerases that need heat activation.

(*a*) Denaturation (Melting of Target DNA): The target DNA containing sequence (between 100 and 5,000 bases) to be amplified is heat denatured (94-98°C for 20-30 second) to separate its complementary strands (step 1). It causes melting of DNA template and primers by disrupting the hydrogen bonds between complementary bases of the DNA strands, yielding single strands of DNA. This process is called melting of target DNA. The temperature that results in melting in target DNA is called melting temperature (T_m). After separation each strand acts as template for DNA synthesis.

(*b*) Primer Annealing: The second step is the annealing of two oligonucleotide primers to the denatured DNA strands. Since nucleotide sequence of each of oligonucleotide primer is complementary to 3' end of single stranded template, the primers anneals (hybridises) the each template. The reaction temperature is lowered to 50-65°C for 20-40 seconds allowing annealing of the primers to the single-stranded DNA template. Typically the annealing temperature is about 3-5°C below the *Tm* of the primers used. Consequently the primers form hydrogen bonds, *i.e.* anneal to the DNA on both sides of the DNA sequence (step 2). The polymerase binds to the primer-template hybrid and begins DNA synthesis.The annealing temperature varies, but the too low temperature favours mispairing. The annealing temperature (in °C) can be calculated using the formula: T = 2 (AT) + 4 (G + C).

(*c*) Primer Extension (Polymerisation): Finally, nucleoside triphosphate (dATP, dGTP, dCTP, dTTP) and a thermostable DNA polymerase are added to the reaction mixture. The DNA polymerase acclerates the polymerisation process of primers and, therefore, extends the primers (at 68°C) resulting in synthesis of copies of target DNA sequence (step 3). Only those DNA polymerases which are thermostable, *i.e.* function at the high temperature are employed in PCR technique. For this purpose the two popular enzymes, *Taq* polymerase (of a thermophilic bacterium, *Thermus aquaticus*) and the *vent* polymerase (from *Thermococcus litoralis*) are used in PCR technology. These enzymes exhibit relative stability at DNA-melting replenishment after each cycle of synthesis. Also it reduces the cost of PCR and allows automated thermal cycling.

The concentration of Mg^{++} is maintained between 1 and 4 mM. Thus in the first step the target DNA is copied from the primer sites for various distances until the start of second cycle. However, after completion of step 3 (of one cycle) the targeted sequences on both strands are copied and four strands are produced. The three step cycle (first cycle) is repeated which yields 8 copies from four strands.

The second cycle starts and doubles the DNA molecules synthesised in the first cycle. The second cycle is started with heating of double stranded DNA to result in single stranded DNA. Each single stranded DNA again acts as template, *i.e.* DNA molecules polymerised in first cycle act as template in second cycle. Following the above events all the single stranded DNA molecules of second cycle are converted into the double stranded DNA (Fig. 5.27). Then third and onward cycles are repeated in the same ways to get more DNA products. Always after *n* number of cycles, 2^{11} molecules of DNA are generated using single stranded DNA as template.

Similarly, the third cycle produces 16 strands. This cycle is repeated about 50 times. Theoretically, 20 cycles (each of three steps) will produce about one million copies of the target DNA sequence, and 30 cycles will produce about one billion copies. In each cycle the newly synthesised DNA strands serve as targets for subsequent DNA synthesis as the three steps are repeated up to 50 times. For the working of PCR about 10-100 picomoles of primers are required ml. The PCR machine can carry out 25 cycles and amplify DNA 10^5 times in 75 minutes. Theoretically, the number of target molecules should be doubled after each cycle resulting in exponential increase in the amplified DNA molecules.

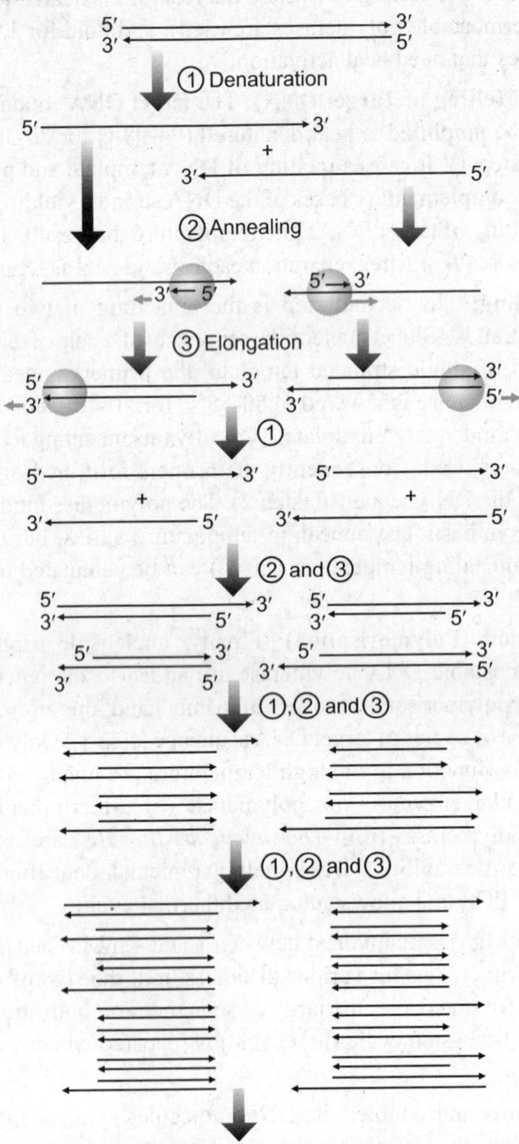

Fig. 5.27. Schematic drawing of the PCR cycle: 1- denaturing at 94-96°C; 2- annealing at ~65°C; 3- elongation at 72°C. Four cycles are shown here. The blue lines represent the DNA template to which primers (red arrows) anneal that are extended by the DNA polymerase (light green circles), to give shorter DNA products (green lines), which themselves are used as templates as PCR progresses.

But the efficiency of exponential amplification is less than 100% due to sub-optimal DNA polymerase activity, poor primer annealing and incomplete denaturation of the template. PCR efficiency is expressed following the formula as given below :

$$\text{PCR product yield} = \text{Input target amount x } (1+\% \text{ efficiency})^n$$

Where, n is the number of cycles.

You can calculate that by using an amplification efficiency value of 70%, about 26 cycles are required to produce 1 mg of PCR product from 1 pg of a target sequence (10^6 amplification [1 mg PCR product = (1 pg target) × $(1 + 0.7)^{26}$]. The PCR technology has been improved in recent years. RNA can also be efficiently used in PCR technology. The *rTth* DNA polymerase is used instead of the *Taq* polymerase. The *rTth* polymerase will transcribe RNA to DNA, thereafter amplify the DNA. Hence, cellular RNA and RNA viruses may be studied when they are present in small quantities. The *rTth* enzyme is the recombinant for of *rTth* polymerase isolated from *Thermus thermophilus*. It can catalyse high temperature reverse transcription of RNA in the presence of $MnCl_2$. The *rTth* is a thermostable DNA polymerase which utilises RNA templates for cDNA synthesis.

2. Application of PCR

After the discovery of PCR, the modern biology has been revolutionised in multifarious ways. Some of the areas of application of PCR have briefly been discussed.

(*a*) Diagnosis of Pathogens: There are several pathogens that grow slowly; therefore, their cells are found less in number in the infected cells/tissues. It is difficult to culture them on artificial medium. Hence, for their diagnosis, PCR-based assays have been developed. These detect the presence of certain specific sequences of the pathogens present in the infected cells/tissues. Besides, it is useful in detection of viral infection before they cause symptoms or serious diseases.

(*b*) Diagnosis of Specific Mutation: In humans there are thousands of genetic diseases. Mutations are also related to genetic diseases. Presence of a faulty DNA sequence can be detected before establishment of disease. By using PCR sickle cell anaemia, phenylketonuria and muscular dystrophy can also be detected. The other diseases can also be diagnosed by using PCR. For example PCR-based diagnostic tests for AIDS, Chlamydia, tuberculosis, hepatitis, human pappiloma virus, and other infectious agents and diseases are being developed. The tests are rapid, sensitive and specific.

(*c*) In Prenatal Diagnosis: It is useful in prenatal diagnosis of several genetic diseases. If the genetic diseases are not curable, it is recommended to go for abortion.

(*d*) DNA Fingerprinting: In recent years DNA fingerprinting is more successfully used in forensic science to search out criminals, rapists, solving disputed parentage and uniting the lost children to their parents or relatives by confirming their identity. This is done through making link between the DNA recovered from samples of blood, semen, hairs, etc. at the spot of crime and the DNA of suspected individuals or between child and his/her parents/relatives.

(*e*) In Research: In addition, DNA fingerprinting of new microorganisms isolated from various extreme environment (soil, water, sediments, air, extreme habitats, etc.) is also carried out to confirm their identity by comparing with the DNA sequences of known microorganisms. Their DNA and RNA can be amplified. Besides, it is also useful to determine the orientation and location or restriction fragments relative to one another.

(*f*) In Molecular Archaeology (Palaentology): PCR can be used to clone the DNA fragments from the mummified remains of humans and extinct animals such as the woolly mammoth and dinosaurs from the remains of ancient animals as recently epitomised in Michel Crighton's Jurassic Park. DNA from buried humans has been amplified and used to trace the human migration that occurred in ancient time.

***(g)* Diagnosis of Plant Pathogens:** It is also applied in diagnosis of plant diseases. A large number of plant pathogens in various hosts or environmental samples are detected by using PCR, for example viroids (associated with hops, apple, pear, grape, citrus, etc), viruses (such as TMV, cauliflower mosaic virus, bean yellow mosaic-virus, plum pox virus, potyviruses), mycoplasmas bacteria (*Agrobacterium tumifaciens, Pseudomonas solanacearum, Rhizobium leguminosarum, Xanthomonas campestris,* etc), fungi (such as *Collectotrichum gloeosporioides, Glomus* spp., *Laccaria* spp., *Phytophthora* spp., *Verticillium* spp), and nematodes (*e.g. Meloidogyne incoginta*).

I. TYPES OF PCR

There are several types of PCR machines which have been designed with certain modifications. Some of the PCRs are described below :

1. Reverse Transcription PCR (RT-PCR)

Amplification of double stranded DNA is not limited only to the genomic DNA. In molecular biology, reverse transcription polymerase chain reaction (RT-PCR) is a laboratory technique for amplifying a defined piece of a ribonucleic acid (RNA) molecule. It can also be applied to double stranded cDNA which is synthesised from mRNA using the enzyme reverse transcriptase. Based on this principle RT-PCR was developed. RT-PCR includes a single application combining the process of cDNA synthesis (by reverse transcription) and PCR amplification. The RNA strand is first reverse transcribed into cDNA, followed by amplification of the resulting DNA using PCR. This can either be a 1 or 2 step process. Using this principle one can isolate very rare mRNA transcripts from cell samples through RT-PCR. The thermostable enzyme *rTth* uses RNA templates from cDNA synthesis and thus allows single enzyme RT-PCR viral reverse transcriptase (RTase) from avian murine virus (AMV RTase) or mammalian leukemia virus (MLV RTase) can also be used for cDNA synthesis. Both the enzymes, *i.e.* RTase and *rTth* DNA polymerase are used alternatively in the RT-PCR.

Application of Reverse-transcription Polymerase Chain Reaction: Amplification of DNA via reverse transcription polymerase chain reaction provides a highly sensitive technique, where a very low copy number of RNA molecules can be detected. Reverse transcription polymerase chain reaction is widely used in the diagnosis of genetic diseases and in the determination of the abundance of specific different RNA molecules within a cell or tissue as a measure of gene expression. Further, Northern blot is used to study the RNA's gene expression.

2. Real-time PCR (RT-PCR)

Real-time polymerase chain reaction, also called 'quantitative real time polymerase chain reaction' (qPCR) or 'kinetic polymerase chain reaction' (kPCR), is a molecular biological laboratory method which is based on the principles of polymerase chain reaction. Although 'real-time quantitative polymerase chain reaction' is often marketed as RT-PCR; it should not be confused with 'reverse transcription polymerase chain reaction' which is also known as 'RT-PCR'.

Real-time polymerase chain reaction is used to amplify and simultaneously quantify a targeted DNA molecule. RT-PCR enables both detection and quantification (as absolute number of copies or relative amount when normalized to DNA input or additional normalising genes) of a specific sequence in a DNA sample. The procedure follows the general principle of polymerase chain reaction.

Activation or deactivation of gene expression regulates the cellular activities of cells in all organisms. This depends on the number of copies of mRNA of a particular gene in a cell. By visualizing the abundance of its mRNA transcript in a sample the expression level of a gene can be estimated with northern blotting technique. In this method, purified RNA is separated by agarose gel electrophoresis, transferred to a solid matrix (such as a nylon membrane), and probed with a specific DNA probe which is complementary to the gene of interest. In this technique relatively large amounts of RNA are required. It provides only qualitative or semiquantitative information of mRNA levels. Amplification of the gene transcript is necessary to detect and quantify gene expression from small amounts of RNA.

DNA is quantified as it accumulates in the reaction in 'real time' after each amplification cycle. Two common methods of quantification are : (*i*) the use of fluorescent dyes that intercalate with double-stranded DNA, (*ii*) and use of modified DNA oligonucleotide probes that fluoresce when hybridised with a complementary DNA. Frequently, real-time polymerase chain reaction is combined with 'reverse transcription polymerase chain reaction' to quantify mRNA molecules present in low amount. Thus it enables to quantify relative gene expression at a particular time or in a particular cell or tissue.

(*a*) Fluorescent Reporter Probe Method: It is the most accurate and reliable method due to use of fluorescent reporter probes, but also it is the most expensive method. In this method a sequence-specific RNA or DNA-based probe is used to quantify only the DNA containing the probe sequence. Hence, the use of the reporter probe significantly increases specificity, and allows quantification even in the presence of some non-specific DNA amplification. By using specific probes with different-coloured labels this allows for multiplexing, assaying for several genes in the same reaction provided the all genes are amplified with similar efficiency.

Commonly it can be carried out with an RNA-based probe with a fluorescent reporter at one end and a quencher of fluorescence at the opposite end of the probe (Fig. 5.28). Detection of its fluorescence is prevented due to close proximity of reporter to the quencher. The Taq polymerase breaks the reporter-quencher proximity by the 5' $\rightarrow$ 3' exonuclease activity, and allows unquenched emission of fluorescence which can be detected (Fig. 5.28). An increase in the product targeted by the reporter probe at each PCR cycle results in a proportional increase in fluorescence due to the breakdown of the probe and release of the reporter.

The steps are : (*i*) preparation of the PCR reaction as used in PCR, and the addition of reporter probe, (*ii*) annealing of both probe and primers to the DNA target during the annealing reaction of the PCR, (*iii*) initiation of polymerisation of a new DNA strand from the primers, and degradation of probe by polymerase reaches the probe, (*iv*) physical separation of fluorescent reporter from the quencher which results in an increase in fluorescence, (*v*) detected of fluorescence and its measurement in the real-time PCR, and (*vi*) determination of threshold cycle (C_T) in each reaction with exponential increase of the product.

(*b*) Applications of Real-time Polymerase Chain Reaction: In laboratory the real-time polymerase chain reaction offers numerous applications. It is commonly used for both diagnostic and research applications. Diagnostically real-time PCR is applied to rapidly detect the presence of genes involved in infectious diseases, cancer and genetic abnormalities. In the field of research, real-time PCR is mainly used to provide highly sensitive quantitative measurements of gene transcription. This method may be used to determine the genetic

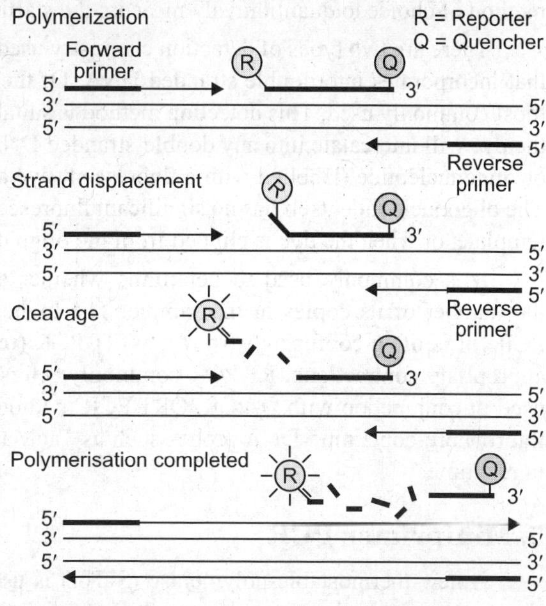

Fig. 5.28. Quenching of reporter fluorescence in intact probes.

expression of a particular gene changes over time, such as in the response of tissue and cell cultures to an administration of a pharmacological agent, progression of cell differentiation, or in response to changes in environmental conditions. This technique is also used in environmental microbiology to quantify resistance genes in water samples.

3. Arbitrary Primed PCR (AP-PCR)

In 1990, J. Welsh and M. McCleland developed the AP-PCR and carried out fingerprinting of genomes with arbitrary primers. It is a type of RAPD where single primers of 10-50 bases are used to amplify genomic DNA in PCR. In the first two cycles, annealing is done under non-stringent conditions. Banding pattern of AP PCR is given in Fig. 5.29.

The AP-PCR differs from RAPD in some respects: (*i*) in AP-PCR amplification occurs in three parts, with its own stringency and concentration of constituents, (*ii*) higher concentrations of primers are used in the first cycles of AP-PCR. Primers of different length are arbitrarily selected *e.g.* M13 universal sequence primer, (*iii*) products of AP-PCR are analysed on acrylamide gels and detected by autoradiography, whereas these of RAPD are analysed on agarose gels, stained with ethidium bromide and visualised under UV light.

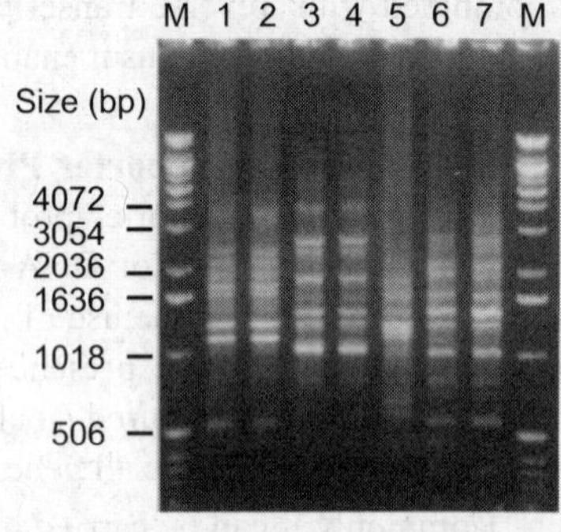

Fig. 5.29. AP PCR banding pattern.

4. Quantitative PCR (Q-PCR)

A PCR product (preferably real-time) is measured by Q-PCR. DNA amplification is monitored at each cycle of PCR. The amount of fluorescence increases above the background when the DNA is in the log linear phase of amplification. The point at which the fluorescence becomes measurable is called the threshold cycle (C_T) or crossing point. A standard curve can be generated of log concentration against C_T by using multiple dilutions of a known amount of standard DNA. The amount of DNA or cDNA in an unknown sample can then be calculated from its C_T value. It is the method of choice to quantitatively measure the starting amounts of DNA, cDNA or RNA.

There are two types of detection chemistry used for Q-PCR. The first uses an intercalating dye that incorporates into double stranded DNA. Of these fluorescent dyes, Sybr® Green I dye, is the most commonly used. This detection method is suitable when a single amplicon is being studied, as the dye will intercalate into any double-stranded DNA generated. The second method uses a primer or oligonucleotide (labelled with a fluorescent dye and quencher) specific to the target of interest. The oligonucleotide itself has no significant fluorescence, but fluoresces either when annealed to the template, or when the dye is clipped from the oligo during extension.

It is commonly used to determine whether a DNA sequence is present in a sample and the number of its copies in the sample. The highest level of accuracy is 'quantitative real-time PCR'. It is often confusingly known as RT-PCR (real time PCR) or RQ-PCR. QRT-PCR is more appropriate contractions. RT-PCR commonly refers to reverse transcription PCR, which is often used in conjunction with Q-PCR. QRT-PCR methods use fluorescent dyes, such as Sybr Green or fluorophore-containing DNA probes such as TaqMan to measure the amount of amplified product in real time.

5. Miniprimer PCR

A new thermostable polymerase (S-Tbr) is used in 'miniprimer PCR' that can extend from short primers (smalligos) as short as 9 or 10 nucleotides, instead of about 20 nucleotides required by Taq. This method allows PCR targeting smaller primer binding regions. It is useful to amplify unknown but conserved DNA sequences such as the 16S (or eukaryotic 18S) rRNA gene. The 16S rRNA miniprimer PCR was used to characterise a microbial mat community growing in an extreme environment. In that study, deeply divergent sequences were discovered with high frequency and included representatives that define two new division-level taxa. Miniprimer PCR may reveal new dimensions of microbial diversity. By enlarging the 'sequence space' that may be queried by PCR primers, this technique may enable novel PCR strategies that are not possible within the limits of primer design imposed by Taq and other commonly used enzymes.

II. PCR-BASED MOLECULAR MARKERS

Following are the PCR-based molecular markers :

1. Random Amplified Polymorphic DNA (RAPD) Marker

RAPD is a PCR-based molecular marker (RAPD analysis) technique. In this method the single short oligonucleotide primers are arbitrarily selected. The scientist performing RAPD creates several arbitrary, short primers (8-12 nucleotides), then proceeds with the PCR using a large template of genomic DNA, hoping that fragments will amplify. These primers amplify a set of discrete regions of DNA which are randomly distributed throughout the genome. On the basis of this principle closely related three techniques were developed at the same time such as RAPD (randomly analysed polymorphic DNA), DAF (DNA amplified fingerprinting) and AP-PCR (arbitrarily primed PCR). Therefore, these techniques are collectively called 'multiple arbitrary amplification profiling' (MAAP). All these methods amplify DNA by using single random primers using the same principle with some differences (http://en.wikipedia.org/wiki/RAPD).

RAPD markers have recently caught the fancy of many individuals in the field of applied plant breeding. This molecular marker is based on the PCR amplification of random locations in the genome of the plant. With this technique, a single oligonucleotide is used to prime the amplification of genomic DNA. Because these primers are 10 nucleotides long, they have the possibility of annealing at a number of locations in the genome. For amplification products, the binding must be inverted repeat sequences generally 150-4000 base pairs apart. The number of amplification products is directly related to the number and orientation of the sequences that are complementary to the primer in the genome. The two genetically distinct individuals differ as they show different polymorphism in pattern of amplified bands. The pattern of amplified polymorphic bands behaves as Mendelian genetic markers and could be used for genome fingerprinting (http://en.wikipedia.org/with/RAPD).

***(a)* Steps of RAPD PCR Amplification of Plant DNA:** The following steps are taken for PCR amplification of plant DNA :

- Isolate total DNA from the individuals of interest.
- Establish the conditions for the amplification of the DNA from specific species. Several considerations are: $MgCl_2$, primer and dNTP concentrations and the quality and concentration of the target DNA. The goal of RAPD experiments is to compare populations. This requires well to well and run to run consistency. Thus, it should be ensured that each well of the cycler will produce the same result at the same given target DNA, primer and dNTP concentration. Without this control you may not be sure that the amplification polymorphisms are the result of population variability or cycler variability.
- The conditions used for the amplification of DNA are: 25 ng DNA, 200 mM dNTP, 200 uM primer, 1.5 mM $MgCl_2$, Taq polymerase buffer, 1 unit Taq polymerase, reaction volume of 10 µl overlaid with 35µl mineral oil.
- The PCR amplification steps are: 94°C for 1 min for denaturation, 35°C for 30 sec for annealing, 72°C for 2 min for extension. This profile is performed 45 times and then the products are completed by a single extension for 7 min at 72°C.

- The amplification products are then separated on a 2% agarose gel, stained and photographed
- Variability is then scored as the presence or absence of a specific amplification product.

For RFLP mapping, one first needs to determine the clone/restriction enzyme combinations that are polymorphic between the parents of your mapping population. The analgous experiment with RAPD markers is to identify the primer and reaction conditions that are polymorphic between the parents. These primers and conditions will be used to amplify and score the products of segregating population.

RAPD amplification can also be carried out by using genomic cDNA (from desired individuals or species) and a single short oligonucleotide primer (about 10 bases). The portion of DNA which is flanked by 10 bp primer in suitable orientation is amplified to get PCR product. The genomic DNA of two different individuals shows different RAPD and represents DNA polymorphism. Hence, these differences can be exploited as genetic marker. An outline of RAPD analysis is given below in Fig. 5.30.

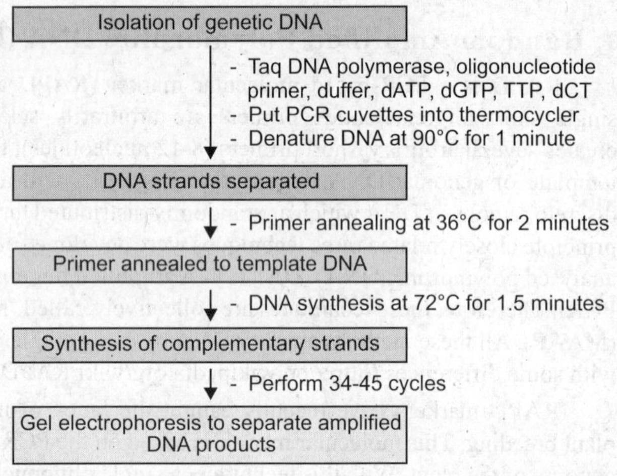

Fig. 5.30. An outline for analysis of RAPD.

RAPD markers have been successfully used for : (*i*) construction of genetic maps, for example *Arabidopsis,* pine, *Helianthus,* etc., (*ii*) mapping of traits, (*iii*) analysis of genetic structure of population, (*iv*) fingerprinting of individuals, (*v*) identification of somatic hybrids, etc.

2. DNA Amplification Fingerprinting (DAF)

In 1994, G. Caetano-Annoles and P.M. Gresshoff described the DAF where a single arbitrary primer (about 5 bases long) is used to amplify DNA by using PCR. In this technique the parameters are carefully optimised. But for early and easy determination of amplified products, automation and fluorescent tagging of primers are extremely amenable. DAF differs from RAPD as: (*i*) it requires higher concentration of shorter primers (5-8 nucleotides), (*ii*) two temperature cycles are used in contrast to three cycles in RAPD, (*iii*) it produces a very complex banding pattern, and (*iv*) it analyses DAF products on acrylamide gels, which are detected by using silver staining.

I. RADIOACTIVITY

Travelling of energy through space is called 'radiation'; sunshine is the most familiar form of radiation. All matter in the universe is made of atoms, and they are made up of even smaller particles called protons, neutrons and electrons. Radiation comes from the activity of these small particles. The arrangement of these particles differentiates one atom from another. Atoms of different types are known as elements. There are over 100 natural and man-made elements. Some of these elements, such as uranium, radium, and thorium share a very important quality, *i.e.* they are unstable. As they change into more stable forms, they release invisible waves of energy or particles. This is called ionising radiation.

Radioactivity

1. Ionising Radiation

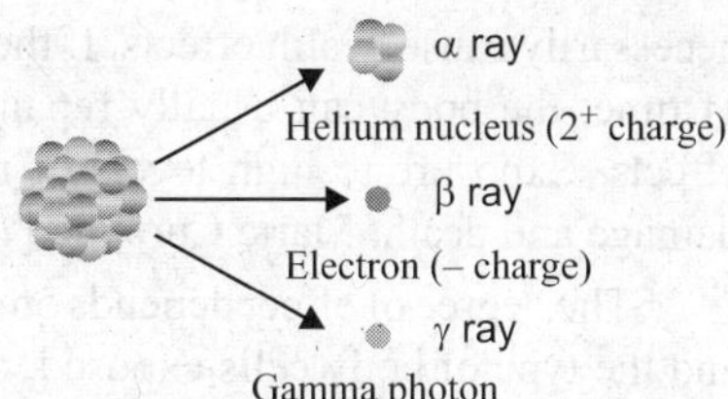

Fig. 5.31. Release of alpha, beta and gamma rays from nuclear fission.

The energy that can ionise or electrically charge an atom by stripping off electrons is called ionising radiation, which can change the chemical composition of many things including living tissue. The amount, type, and intensity of radiation are determined by different units, just as heat is measured in terms of its intensity or its effect using units like degrees and calories. The amounts of radiation can be measured in curies, rems, millirems, and rads.

The unit **curie** (named after Marie and Pierre Curie) is the intensity of the radioactive atoms of the material that decay each second. The unit **rem** is a measurement of the effects of radiation on the body, as degrees Celsius are measurements of the effects of sunlight that heats sand. The millirem (**mrem**) is one-thousandth of a rem, which is used to measure usually very small amount of radiation. The **rad** is the unit of measurement for the physical absorption of radiation. Similar to sunlight, heats pavement by giving up energy to it; radiation gives up rads of energy to objects in its path.

***(a)* Types of Ionising Radiation:** There are three main types of ionising radiation: *alpha, beta* particles and *gamma* rays (Fig. 5.31); but *neutrons* are also included as fourth type :

***(i)* Alpha Particles:** Alpha particles are helium nuclei containing two protons and two neutrons. They are deflected slightly in an electric of magnetic field. They are highly charged (+2), heavy (4 amu), and the most energetic (range 4-6 MeV) of the three types of ionising radiation. But despite their energy, they can travel only a few inches in the air. Alpha particles are not very penetrating and not an external hazard. Alpha particles lose their energy as soon as they collide with anything. A sheet of paper or your skin's surface can easily stop them (Fig. 5.32). Alpha decay is common among nuclides with a high atomic number.

***(i)* Beta Particles**: Beta particles are electrons capable of travelling at speeds approaching the speed of light. They have a charge of -1, and a small mass (1/1836 amu). They are much smaller than alpha particles. They can travel in the air as much as 1,00,000 miles per second for a distance of about 10 feet. Beta particles can pass through a sheet of paper, but may be stopped by a thin sheet of aluminum foil or glass (Fig. 5.32). Low energy betas are not external radiation hazards and may require no shielding (examples are: hydrogen^3, carbon^{14}, sulfur^{35}). Higher energy betas are external hazards and are effectively shielded by 1/4 thick clear plastic laboratory shields; for example phosphorus^{32}, phosphorus^{33}, calcium^{45}.

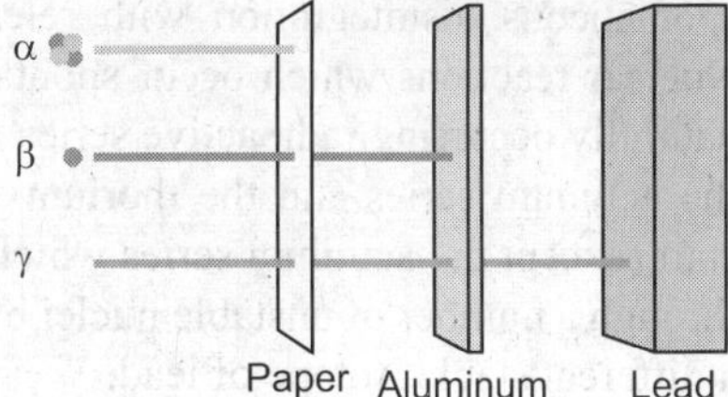

Fig. 5.32. Shielding of alpha, beta and gamma rays.

***(ii)* Gamma ray:** Unlike alpha and beta particles, gammas are high energy electromagnetic radiation with short wavelengths, and they cannot be shielded. Gamma rays travel at the speed of light (1,86,000 miles per second). Gamma radiation is very penetrating and is best shielded by a thick wall of concrete, lead, or steel (Fig. 5.32). Survey meters must be available in laboratories utilising radionuclides having a specific gamma ray emission of more than 0.1 R/h at one meter greater. Examples of gamma emitting nuclides are: iodine^{125}, iodine^{131}, zinc^{64}, cesium^{137}, cobalt^{60}.

***(iii)* Neutrons**: They are classified by energy. Thermal neutrons have same kinetic energy as gas molecules in the same environment. Emission of neutrons accompanies the splitting of nuclei of uranium and plutonium.

***(b)* Effects of Radiation on Health:** Among all natural phenomena radiation is most widely studied. Radiation can chemically change the living cells. However, cell transformation does not necessarily cause health effects. If the radiation dose is low or a person receives it over a long period of time, the body can usually repair or replace the damaged cells without any detectable health effects. Exposure to high levels of radiation can cause serious health effects, such as burns, cell damage and death. Marie Cure died of radiation of isotopes that caused aplastic anaemia.

Marie Curie

The degree of effect depends on the intensity of the dose, the length of exposure and the type of body cells exposed. Sudden large doses exceeding 1,00,000 mrem can cause radiation sickness, with short-term symptoms including nausea, vomiting, extreme tiredness and hair loss. Long-term effects include increased cancer risks. A dose of over 5,00,000 mrem at one time is usually fatal unless medical treatment is available. Besides, there are also man-made sources of radiation, such as medical and dental X-rays, household smoke detectors, and materials released from nuclear and coal-fired power plants.

2. Radioactivity

When the nuclei of unstable atoms are broken up as a result of the imbalance of energy within the nucleus, energetic subatomic particles (such as alpha, beta and gamma) are released. Radioactivity is the release of energy and matter that results from changes in the nucleus of an atom. This phenomenon is observed in the heavy elements such as uranium, thorium and unstable isotopes like carbon^{14}.

The discovery of the phenomenon of radioactivity was purely accidental. In 1896, for the first time radioactivity was discovered by the French scientist Henri Becquerel; after his name the SI unit for radiation, '*Becquerel'* has been given. He found that in presence of uranium salt, photographic plates got heavily fogged even though they were wrapped in opaque paper. He concluded that the uranium salt must be giving off penetrating radiations similar to X-rays that was discovered one year earlier. After a gap of two years, Marie Curie named this phenomenon as 'radioactivity'.

Elements with atomic number above 83 have very heavy and unstable nuclei. These nuclei spontaneously break up to form stable elements with lower atomic numbers. In the process of this breakup, the three types of radiations are released as described earlier. Phenomenon of natural spontaneous disintegration with releasing such radiations is known as 'natural radioactivity'. Nuclear reactions which occur spontaneously are said have natural radioactivity. There are three naturally occurring radioactive series among the elements. These are known as the uranium series, the actinium series and the thorium series; each is named after the element at which the series start (except the actinium series which starts with a different uranium isotope). Each series decays through a number of unstable nuclei by means of alpha and beta emission, until each series ends on a different stable isotope of lead.

Marie Skłodowska Curie (November 1867–July 1934)

Marie Skłodowska Curie was a Polish-born French physicist and chemist famous for her work on radioactivity. She was a pioneer in the field of radioactivity and the first person honored with two Nobel Prizes, one in physics and another in chemistry. She was also the first female professor at the University of Paris. She founded the Curie Institutes in Paris and Warsaw.

Her husband Pierre Curie shared with her the Nobel Prize in Physics. Her daughter Irène Joliot-Curie and son-in-law, Frédéric Joliot-Curie, also shared a Nobel Prize. She was the sole winner of the 1911 Nobel Prize for Chemistry. Curie was the first woman to win a Nobel Prize, and she is the only woman to win the award in two different fields.

She is credited for the creation of a theory of radioactivity (a term she coined), techniques for isolating radioactive isotopes and the discovery of two new elements, polonium and radium. Under her direction, the world's first studies were conducted in the treatment of cancers using radioactive isotopes.

> On 4 July 1934, Marie-Curie died (at the Sancellemoz Sanatorium in Passy, France) of aplastic anaemia, which was certainly contracted from exposure to radiation. The damaging effects of ionizing radiation were not known at that time, and much of her work had been carried out in a shed, without proper safety measures. She had carried test tubes containing radioactive isotopes in her pocket and stored them in her desk drawer.

Transformation theory advanced by Rutherford and Soddy in 1903 states that atoms of radio-elements are unstable and undergo spontaneous disintegration. Succession of transformations (decay) of radium is given in Fig. 5.33. Emission process is unique in that the particles are released without any energy applied to the atom; hence the atom itself is a source of energy. After emitting the particles the atom is changed into an atom of another substance which may be unstable (will undergo further disintegration) or stable (will not emit any more particles). Control of this change is not possible. The rate at which a chemical reaction takes place changes with changing the temperature. However, radioactive decay is independent of temperature, pressure or chemical combination.

Not all nuclear reactions are spontaneous; some reactions occur when stable isotopes are bombarded with particles such as neutrons. This method of inducing a nuclear reaction to proceed is termed as 'artificial radioactivity'. This means new nuclear reactions which would not have been viewed spontaneously, could now be observed. Since 1940, a set of new elements with atomic numbers over 92 (the atomic number of the heaviest naturally occurring element, uranium) have been artificially made. They are called the 'transuranium' elements.

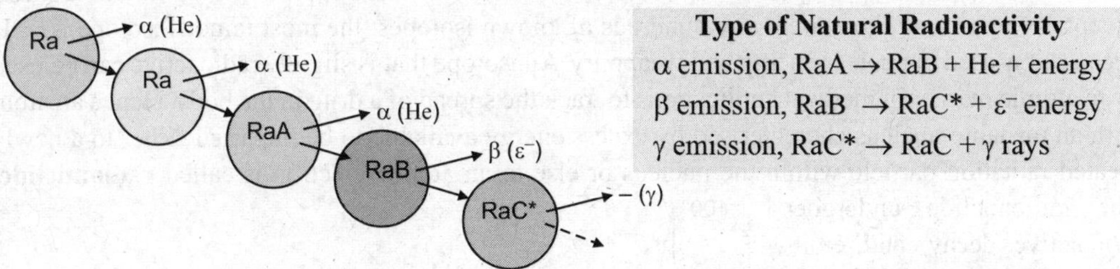

Type of Natural Radioactivity

α emission, RaA $\rightarrow$ RaB + He + energy

β emission, RaB $\rightarrow$ RaC* + ε^- energy

γ emission, RaC* $\rightarrow$ RaC + γ rays

Fig. 5.33. Succession of decay of radium (Ra) and emission energy.

(*a*) Isotopes: Isotopes are such atoms that contain the same number of protons but a different number of neutrons. The number of protons which is called atomic number is the same for each isotope. For example ^{12}C, ^{13}C and ^{14}C each have 6 protons but the number of neutrons in each isotope differs. Except hydrogen, every atomic nucleus is made of both protons and neutrons. Normally, the number of protons and neutrons is the same. In an isotope, this balance is frequently broken, for example ^{238}U has three more neutrons than ^{235}U.

A nucleus becomes unstable due to the lack of necessary neutrons. Protons in the nucleus are positively charged; therefore, they repel each other. The presence of neutrons is necessary to separate these protons slightly that makes the configuration stable. In an unstable configuration, nuclear decay takes place resulting in release of showers of radioactive particles by atoms (Fig. 5.34).

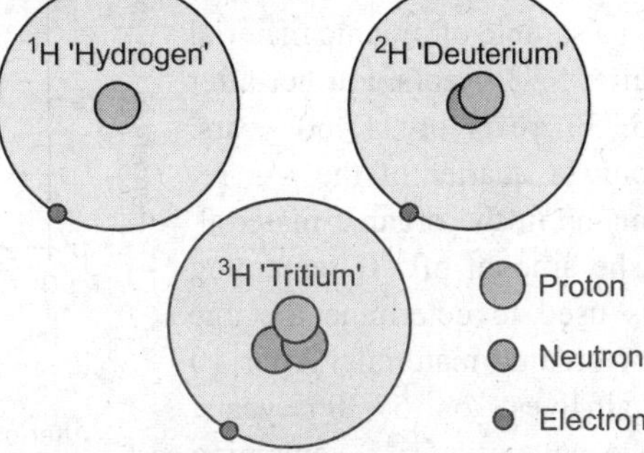

Fig. 5.34. Three isotopes of hydrogen.

(*b*) Beta Decay: The ^{14}C is radioactive; therefore, it decays over time by a process called beta decay. The older artifacts have less ^{14}C than younger ones. During the process, a ^{14}C atom decays into a ^{14}N atom, during which one of the neutrons in the carbon atom becomes a proton. This results in an increase of the number of protons in the atom by one. Consequently a nitrogen atom is created instead of a carbon atom. During this process an electron and an elementary particle are generated which are also called as antineutrino (Fig. 5.35).

***(c)* Half-life:** Each isotope has its own *half-life, i.e.* the time taken for the activity of a given amount of a radioactive substance to decay to half of its initial value. Half-life varies from a fraction of a second to many times longer than the age of the universe. The half life is related to the decay constant as follows:

$$t_{1/2} = \frac{\ln 2}{\lambda} = \tau \ln 2.$$

Where t = time, λ = decay constant, τ = half-life of the isotope.

The relationship between the half-life and the decay constant shows that highly radioactive substances get quickly lost, whereas weak radioactive substances endure longer life. Half-lives of known radionuclides vary from 10^{-23} seconds (for highly unstable isotopes) to more than 10^{19} years (for very nearly stable nuclides e.g. ^{209}Bi).

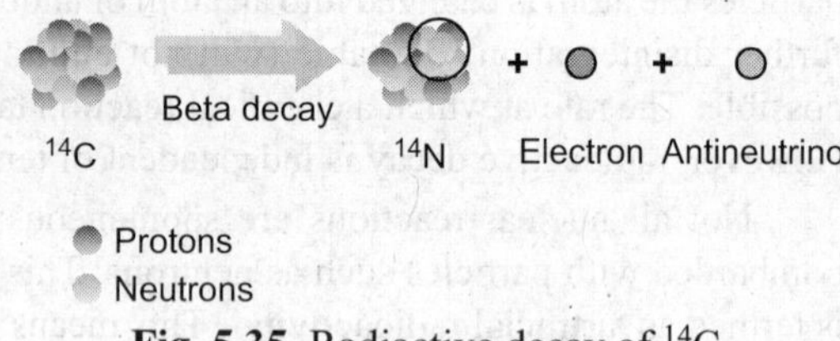

Fig. 5.35. Radiactive decay of ^{14}C.

Some isotopes such as helium^{-3} are not radioactive. An isotope behaves similarly to the natural element. Isotopes can participate in all the same reactions and have approximately the same mass, except in rare cases. There are many hundreds of known isotopes; the most famous isotope is ^{235}U because it is used in nuclear energy and weaponry. An isotope that is slightly radioactive can be used as an atomic marker in medical applications to track the spread of a drug in the body. Hence an atom with an unstable nucleus characterised by excess energy available to be imparted either to a newly created radiation particle within the nucleus or else to an atomic electron is called **radionuclide.** The radionuclide undergoes radioactive decay, and emits a gamma ray (s) and/or subatomic particles. These particles constitute ionising radiation.

The ^{14}C has a half-life of 5,730 years, *i.e.* only half of the original amount of ^{14}C remains in a sample of organic material after 5,730 years. Further after 5,730 years or 11,460 years, only a quarter of the ^{14}C remains in the organic material. The amount of ^{14}C remaining is used to determine the age of organic materials. After 10 half-lives, or 57,300 years, the amount of ^{14}C remaining (<0.1%) becomes very difficult to detect. Thus fossil fuels which are older than 50,000 years have no ^{14}C remaining (Fig. 5.36).

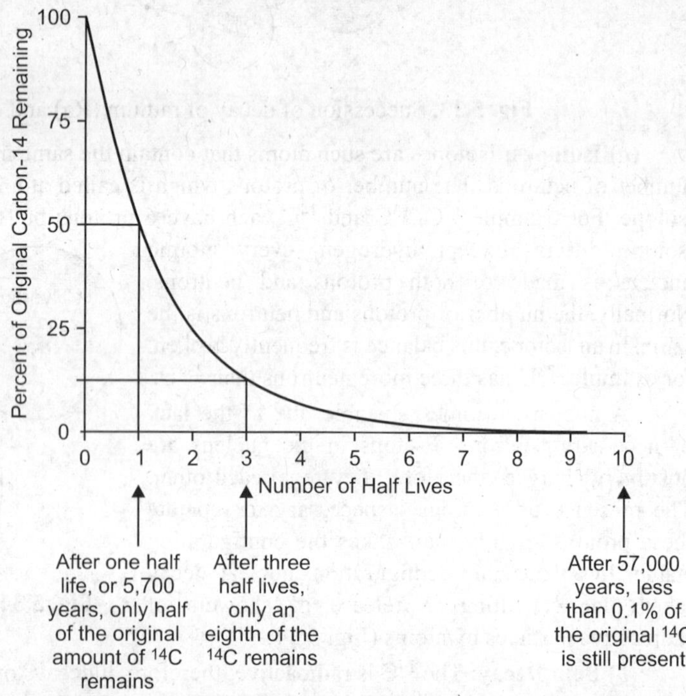

Fig. 5.36. Relationship between ^{14}C remaining after number of half lives.

For about one hundred years, humans have been using radioactivity; it has been added to the natural inventories through its use. The amounts are small compared to the natural amounts. But due to the shorter half-lives of many of the nuclides, they have a marked decrease since the halting of above ground testing of nuclear weapons. A few human-produced radionuclides, which are commercially available are given in Table 5.4.

Table 5.4. List of commercially available human produced radionuclides.

Isotope	*Half Life*	*Energies (KeV)*
Gamma only		
Barium 133	10.7 years	81.0, 356.0
Cadmium 109	453 days	88.0
Cobalt 57	270 days	122.1
Cobalt 60	5.27 years	1173.2, 1332.5
Europium 152	13.5 years	121.8, 344.3, 1408.0
Manganese 54	312 days	834.8
Sodium-22	2.6 years	511.0, 1274.5
Zinc 65	244 days	511.0, 1115.5
Technetium 99 m	6.01 hours	140
Beta only		
Strontium 90	28.5 years	546.0
Thallium 204	3.78 years	763.4
Carbon 14	5730 years	49.5 (average)
Strontium 90	28.5 years	546.0
Alpha only		
Polonium 210	138 days	5304.5

J. TRACER TECHNIQUES (= RADIOTRACER TECHNIQUES)

Radioisotopes are used to trace molecules in cells and organisms. Radioisotope labelling is done to distinguish between two chemically identical molecules but they have different histories; for example they differ in their time of synthesis. Autoradiography radiotracer was used earlier to label various molecules such as amino acids and they were assembled into proteins over the time throughout the cell. This technique was essential to understand: oxidative respiration, photosynthesis, the control of protein synthesis by nucleic acids, the timing of events throughout the cell cycle, the fate of populations of cells, *i.e.* number of cell divisions, migration, relationship of stem cells to the final differentiated phenotypes (by ^{3}H-thmyidine labelling of replicating cells.

The other methods are the analytical techniques, where radioactive labelling of various molecules enables the binding of these molecules (as markers of other molecules) to be accurately monitored by radioisotope cytochemistry, e.g. enzyme inhibitors, antibodies, and nucleic acid probes. In molecular biology experiments S ^{35}, P ^{32} and I ^{125} are widely used to label nucleic acid robes to detect mRNA by *in situ* hybridisation on tissue sections and also for quantitation by Northern analysis on gels. Radioisotope-labelling has great sensitivity but the disadvantage is that hybridisation is performed each time; hence, the probe has to be labelled with fresh radioisotope because it decays rapidly. This is tedious and expensive also. Moreover, radioisotopes are dangerous (especially ^{125}I). Therefore, digoxygenin is now favoured for labelling probes for *in situ* hybridisation studies particularly because digoxygenin-labelled probe is stable for many months. It is detected by an antibody and a colour reaction.

I. RADIOISOTOPIC STUDIES

A radioactive tracer (also called a radioactive label), *i.e.* a substance consisting of a radioisotope is used to measure the speed of chemical processes and to track the movement of a substance through a natural system such as a cell or tissue. A radioactive tracer is identical in chemical composition to the compound of interest and is administered in minute amounts that do not perturb the experimental system. There are number of different radioactive forms of hydrogen, carbon, phosphorus, sulfur, and iodine that are used in biochemical assays, metabolism studies, and medical diagnostics.

Since radioactive elements are used in this method, it is also called 'radioactive tracer technique'. This technique is used to investigate recovery and loss of a microcomponent, in which a radioactively labelled element or compound chemically identical with the microcomponent is added, and its behaviour is followed by sensitive, rapid and selective radioactivity measurements. There are several isotopic elements which are used in different organisms as per requirement. For example, iodine is associated with abnormality of thyroid gland. Hence, radioactive iodine-131 (^{131}I) can be used to study the function of the thyroid gland assisting in detecting disease. Tracers are frequently used to track the localisation of a specific compound or to trace the path of a compound through a series of chemical reactions.

Tracer techniques comprise a set of methods based on the application of stable or unstable (radioactive) isotopes of an element as markers for the naturally occurring form of the element. The tracer isotope may occur as the free element or incorporated into chemical compounds.

Radioactive tracing was developed by a Hungary-born physical chemist George de Hevesy, who won the 1943 Nobel Prize in Chemistry for his pioneering work using radioactive tracers to study metabolic processes in plants and animals. He performed his basic research in this area during the 1930s in Copenhagen. There are many practical applications to the use of radioactivity, such as: (*i*) to study living organisms (microorganisms, plants and animals), (*ii*) to diagnose and treat diseases, (*iii*) to sterilize medical instruments and food, (*iv*) to produce energy for heat and electric power, and (*v*) to monitor various steps in all types of industrial processes.

1. Tracer Isotopes

Depending on the application, a number of different radioisotopes are used as radioactive tracers, as given below :

***(i)* Hydrogen:** Tritium (^{3}H) is a radioactive form of hydrogen which consists of one proton and two neutrons in its nucleus. Tritium emits low energy beta particles and decays into helium-3 (^{3}He). Tritium detected least efficiently by scintillation counting due to low energy of the emitted particles. But tritium is frequently used as a tracer in biochemical studies because of abundance of hydrogen in organic compounds.

***(ii)* Phosphorus:** ^{32}P is a radioactive form of phosphorus that contains 17 neutrons and 15 protons in its nucleus. After emitting a beta particle, ^{32}P decays into ^{32}S, which is frequently used to label amino acids and phosphoproteins, and is commonly used to study protein phosphorylation by kinases in biochemistry. ^{32}P emits a relatively high energy beta particle. On the other hand, ^{33}P is a radioactive form of phosphorus that contains 18 neutrons and 15 protons in its nucleus. After emitting a beta particle, ^{33}P decays into ^{35}S. It emits less energetic beta particles and allows higher resolution assays. ^{33}P is less common and more expensive to produce than ^{33}P.

***(iii)* Sulfur:** ^{35}S is a radioactive form of sulfur that contains 18 neutrons and 16 protons in its nucleus. Upon emitting a beta particle ^{35}S decays into chlorine-35. ^{35}S is frequently used as a tracer in biochemical experiments, where it is used to label amino acids and nucleic acids containing sulfur.

(*iv*) **Carbon:** A radioactive form of carbon is carbon-11 that contains five neutrons and six protons in its nucleus. Due to emission of positron, ^{11}C decays into boron-1. ^{11}C is commonly used as a tracer in positron emission tomography that allows for three-dimensional imaging of functional processes in the human body. Besides, ^{14}C is another radioactive form of carbon that contains eight neutrons and six protons in its nucleus. After emitting a beta particle, ^{14}C decays into ^{14}N. ^{14}C has been used as a tracer in both medical and scientific tests. ^{14}C is frequently used to trace carbons through metabolic pathways. Fig. 5.37 shows the evolution of $^{14}CO_2$ by an experimental plant through respiration after utilising ^{14}C-containing glucose in solution.

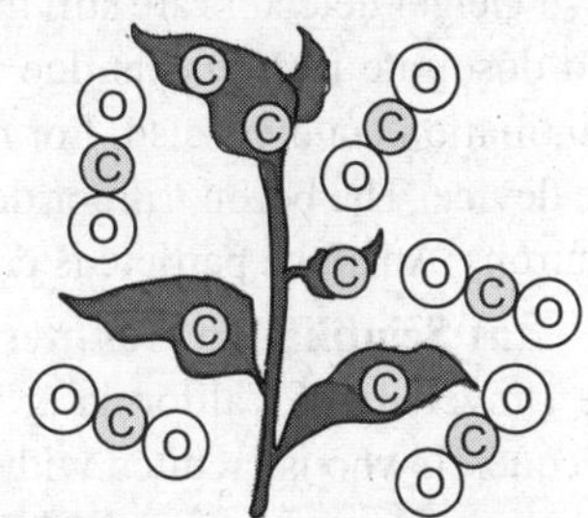

Fig. 5.37. Evolution of $^{14}CO_2$ through respiration.

(*v*) **Iodine:** ^{123}I is a radioactive form of iodine that contains 70 neutrons and 53 protons in its nucleus. By electron capture, ^{123}I decays into tellurium-123 and produces gamma rays. ^{123}I is used in nuclear medicine imaging, specifically to study thyroid function. It has a short half life; hence it is the mostly used isotope in the study of thyroid function. In contrast, ^{125}I is a radioactive form of iodine that contains 72 neutrons and 53 protons in its nucleus. After capturing the electrons, ^{125}I decays into tellurium-125 and produces gamma rays. Due to its long half life and high sensitivity by gamma counters, ^{125}I is frequently used in radioimmunoassay.

2. Detection of Radioactivity

The radiotracers behave exactly in the same way as an unlabeled molecule, but the tracer molecule continually gives off radiation, which cannot be detected by human senses. A variety of handheld and laboratory instruments is available for detecting and measuring radiation. The most common instruments that can detect radioation are: *Geiger counter, scintillation counter* or other type of radiation detection instrument.

(*a*) **Geiger-Müller Counter (or Geiger Counter):** In 1908, Hans Geiger together with Ernest Rutherford developed a device that was later called the 'Geiger counter'. This counter was capable of detecting only alpha particles. In 1928, Hans Geiger and Walther Müller (a Ph.D. student of Geiger) improved the counter to detect more types of ionising radiation. The current version of the Geiger counter is called the 'halogen counter'. It was invented in 1947 by S.H. Liebson. It has superseded the earlier Geiger counter because it has much longer life.

The Geiger-Müller counter is a type of particle detector that measures ionising radiation (usually beta particles and gamma rays, but certain models can detect alpha particles). It is used to detect the objects that emit nuclear radiation. A tube is filled with an inert gas (usually helium, neon or argon with halogens) that conducts electricity when a particle or photon of radiation makes the gas conductive. The tube amplifies this conduction by a current pulse. The pulse is displayed by a needle or lamp and/or audible clicks.

Geiger-Muller counter

Some Geiger counters are used to detect gamma radiation, though sensitivity can be lower for high energy gamma radiation than with certain other types of detector, for examples the X-ray Pancake Geiger tube. Sodium iodide scintillation counter is a better device for detecting gamma rays. Common readout units are roentgens per hour (R/hr), milliroentgens per hour (mR/hr), rem per hour (rem/hr), millirem per hour (mrem/hr), and counts per minute (cpm).

Geiger detectors are still favored as general purpose alpha/beta/gamma portable contamination and dose rate instruments due to their low cost and robustness. There are good alpha- and beta-scintillation counters also. For measuring the neutrons there is a variety of the Geiger tube used in the device. The boron trifluoride gas is used in the tube, and a plastic moderator is used to slow the neutrons. An alpha particle is created inside the detector and neutrons can be counted.

***(a)* Scintillation Counter:** In 1944 Sir Samuel Curran invented the scintillation counter at the University of California at Berkeley. This work is based on the earlier work of Antoine Henri Becquerel, who is credited with discovering radioactivity. A scintillation counter measures ionising radiation. The sensor is called scintillator which consists of a transparent crystal, usually phosphor, plastic containing anthracene, or organic liquid (e.g. liquid scintillation counting). A sensor called a scintillator is fitted in the apparatus that consists of a transparent crystal or organic liquid. When a charged particle strikes the scintillator a flash of light is produced which may or may not be in the visible region of the spectrum. If a flash is produced in a visible region upon being struck by ionising radiation, it can be observed and counted (Fig. 5.38).

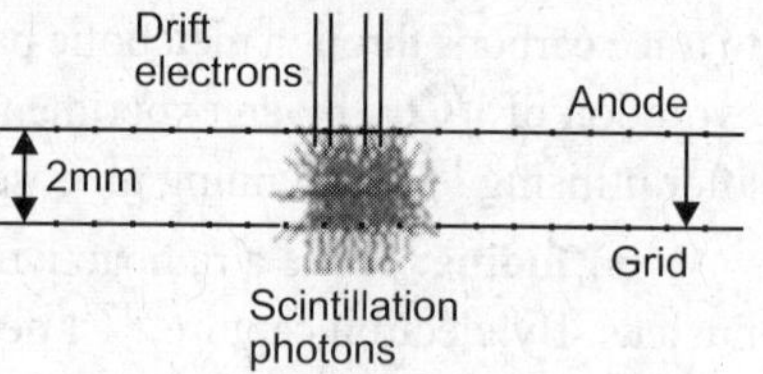

Fig. 5.38. Scintillation detector.

There are three classes of 'solid state scintillation phosphors': *organic crystals, inorganic crystals*, and *plastic phosphors*.

***(i)* Inorganic crystals phosphors:** Inorganic crystals include lithium iodide (LiI), sodium iodide (NaI), cesium iodide (CsI), and zinc sulfide (ZnS). The characteristic features of inorganic crystals are: high density, high atomic number, and pulse decay times of about 1 microsecond. They are highly efficient to detect gamma rays and to handle high count rates.

***(ii)* Organic scintillation phosphors:** Organic scintillation phosphors include naphthalene, stilbene, and anthracene. The decay time of this type of phosphor is approximately 10 nanoseconds. They are efficient to detect the beta particles.

***(iii)* Plastic scintillation phosphors:** Plastic phosphors are made by adding scintillation chemicals to a plastic matrix. The decay constant is the shortest of the three phosphor types, approaching 1 or 2 nanoseconds. The plastic has high hydrogen content; therefore, it is useful for fast neutron detectors.

The density of electrons in the detector of scintillation counters governs the quantum efficiency of a gamma-ray detector (per unit volume). Certain scintillating materials such as sodium iodide and bismuth germanate achieve high electron densities due to high atomic numbers of some of the elements. Moreover, high efficiency of neutron detectors is gained by using of scintillating materials rich in hydrogen that efficiently scatter neutrons.

- **Scintillation counter apparatus:** A suitable scintillation phosphor is coupled to a light-sensitive photomultiplier tube (PMT). Fig. 5.38 shows scintillation counter where a thallium-activated sodium iodide crystal is used. The PMT measures the light from the crystal. The PMT is attached to an electronic amplifier and other electronic equipment to count and possibly quantify the amplitude of the signals produced by the photomultiplier.

  The basis of the scintillation counter apparatus is the association of a scintillator and photomultiplier with the counter circuits. When a charged particle passes through phosphor, some of its atoms get excited and emit photons. The intensity of the light flash depends on the energy of the charged particles. Sodium iodide (NaI) containing a small amount of thallium is used as a scintillator for the detection of gamma waves. Cesium iodide (CsI) in crystalline form is used as the scintillator for the detection of protons and alpha particles.

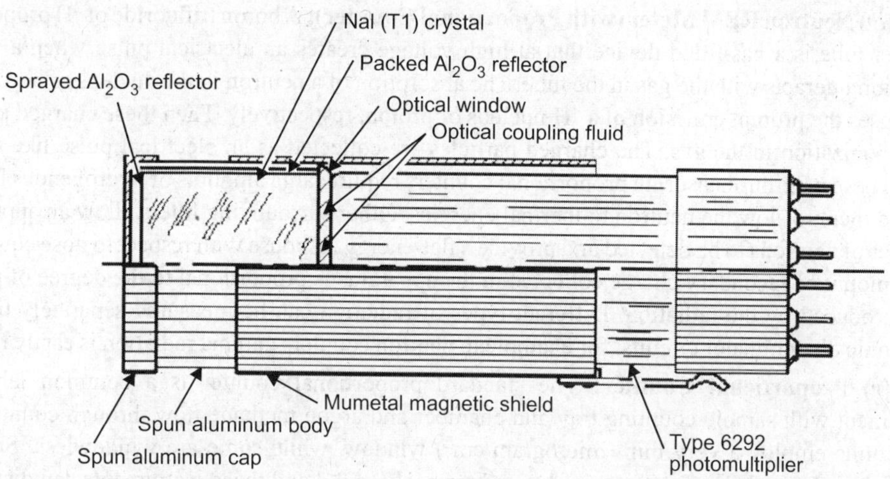

Fig. 5.39. Scintillation counter consisting of sodium iodide (NaI) crystal as scintillator.

In the scintillation counter a layer of phosphor is cemented within one end of the photomultiplier. The inner surface of the apparatus is coated with a photo-emitter, which is called as photocathode. It is connected to the negative terminal of a battery. There are many electrodes called dynodes arranged in the tube at increasing positive potential. Photons are emitted upon striking of a charged particle to the phosphor. In turn, the photon strikes the photocathode in the photomultiplier tube (PMT) resulting in release of an electron. This electron is accelerated towards the first dynode and hits it. Chain of reaction of striking and emission of more electrons continue resulting in intensifying the effect of the first charged particle that generates voltage pulse. The voltage pulse is amplified and recorded by the electronic counter.

- **Liquid Scintillation Counter:** Liquid scintillation counters are an efficient and practical means of quantifying beta radiation. Tritium (^3H) is detectable with liquid scintillation detectors. When a β particle passes through scintillation cocktail, it leaves a trail of energised solvent molecules. These excited molecules transfer its energy to scintillator molecules resulting in light production. Upon activation each scintillator molecule gives off only one photon, and the wavelength of that photon is characteristic of the scintillator, not the β-particle. But multiple scintillators are activated by the energised molecules generated by one β-particle. The path of a β-particle in a cocktail is generally less than 0.1 cm, and the half-life is short which means that the burst of photons from an emission event derives from a small space and reaches the PMT to be read as one pulse of light.

Liquid scintillation counter

The number of photons generated is directly proportional to the path length of the β particle which is determined by its emission energy (the β particle rebounds from solvent molecule to solvent molecule until its energy is exhausted). The intensity of each light pulsc corresponds to the emission energy and the number of pulses per second corresponds to the number of radioactive emissions. A single beta particle gives rise to multiple simultaneous emissions of light when passes through scintillation fluid.

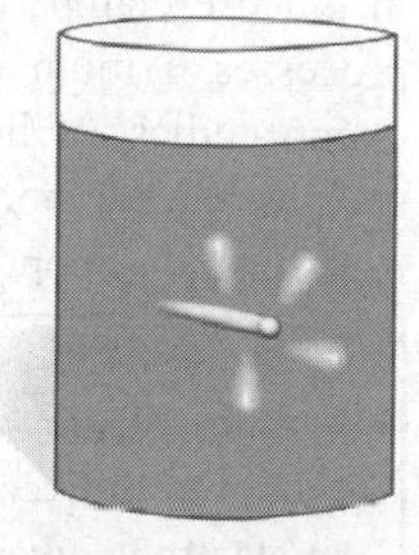

Emissions of multiple simultaneous light by single beta particle in scintillation fluid.

***(iv)* Neutron REM Meter (with Proportional Counter):** A boron trifluoride or ^3H proportional counter tube is a gas-filled device that at high voltage creates an electrical pulse when a neutron radiation interacts with the gas in the tube. The absorption of a neutron in the nucleus of boron-10 or ^3H causes the prompt emission of a ^4H nucleus or proton, respectively. Then these charged particles cause ionization in the gas. The charged particles are collected as an electrical pulse like the GM tube. These neutron-measuring proportional counters require large amounts of hydrogenous material around them to slow the neutron to thermal energies. Other surrounding filters allow an appropriate number of neutrons to be detected and provide a flat-energy response with respect to dose equivalent. The amount of secondary charge collected in the apparatus is proportional to the degree of primary ions produced by the radiation. Different types of radiation can be measured separately by using electronic discriminator circuits; for example in neutron counters gamma radiation is easily rejected.

***(v)* Proportional Counter:** The standard proportional counter is a common laboratory instrument with sample counting tray and chamber and argon/methane flow through counting gas. Most units employ a very thin (microgram/cm^2) window, while some are windowless. Shielding and identical guard chambers are used to reduce background and these instruments can distinguish between alpha and beta radiation. The proportional counters have multiple sample capability and automatic data acquisition, reduction, and storage. Commonly such counters are used to count smear/wipe or air filter samples. Large area gas flow proportional counters are used for counting the whole body and extremities of workers for external contamination while exiting a radiological control area.

3. Applications

The past one hundred years has brought the incredible field of nuclear chemistry into all of our lives. We derive numerous benefits from its applications, yet in most cases, we are not aware of them. Radiotracer techniques are used in many areas of research as briefly described below:

***(a)* Archaeological Dating (= ^{14}C Dating):** Since the discovery of radioactivity and its properties, a significant progress has been made in this field; one such application is ^{14}C dating. One should remember that all biologic organisms contain a given concentration of carbon-14; we can use this information to help solve questions about when the organism died. When an organism dies it has a specific ratio by mass of ^{14}C to ^{12}C incorporated in the cells of its body. The same ratio is in the atmosphere also. At the moment of death, no new ^{14}C-containing molecules are metabolised; therefore, the ratio is at a maximum. After death, the ^{14}C to ^{12}C ratio begins to decrease because ^{14}C is decaying away at a constant and predictable rate. The half-life of ^{14}C is 5700 years, thereafter 5700 years half of ^{14}C remains inside the organism.

***(b)* Metabolic Research:** Radioactive tracers are used frequently in metabolism research in glucose clamps to measure rates of glucose uptake, fatty acid synthesis, and other metabolic processes. Tritium and ^{14}C-labeled water and glucose are commonly used as tracers for metabolic clamp studies. While radioactive tracers are sometimes used in human studies, stable isotope tracers such as ^{13}C are more commonly used in current human clamp studies. Radioactive tracers are also used to study lipoprotein metabolism in humans and experimental animals.

How is ^{14}C used to determine the Age of an Artifact ?

The scientists burn a small piece of an artifact to convert it into carbon dioxide gas so that the amount of radiocarbon left in an artifact can be measured. Radiation counters are used to detect the electrons given off by decaying ^{14}C as it turns into nitrogen. In order to date the artifact, the amount of ^{14}C is compared to the amount of ^{12}C (the stable form of carbon) to determine how much radiocarbon has decayed. The ratio of ^{12}C to ^{14}C is the same in all living things. But at the time of death, the amount of ^{14}C begins to decrease because it is unstable, while the amount of

^{12}C remains constant in the sample. Half of the ^{14}C degrades every 5,730 years as indicated by its half-life. It is possible to determine the age of the artifact by measuring the ratio of ^{12}C to ^{14}C in the sample and comparing it to the ratio in a living organism.

(*c*) Medical Applications: Tracers are applied in a number of diagnostic tests, for example technetium-99 in autoradiography and nuclear medicine, including single photon emission computed tomography (SPECT), positron emission tomography (PET) and scintigraphy. The urea breath test is used to detect the presence of the bacteria *Helicobacter pylori* in the stomach. *H. pylori* causes inflammation, ulcers and atrophy of the stomach. If the labelled urea was metabolised by *H. pylori* in the stomach, the patient's breath would contain labelled carbon dioxide. For this test the patient is given a certain amount of urea with some of the carbon as ^{14}C. If *H. pylori* is present in the stomach, this bacteria will break down the urea and produce CO_2, which will be exhaled by the patient and collected in a balloon. If some of the carbon in the exhaled CO_2 contains is ^{14}C, there must be some bacteria present in the stomach. If more ^{14}C is exhaled, the more *H. pylori* is present. The breath test can be repeated to determine the success of the treatment. The urea breath test is a nice example for determining the total dosage a patient receives. Suppose a 10μCi sample of ^{14}C-urea is given to a 55 Kg patient. The total dosage to his/her body can be determined. The total number of decays occurring in his body can be calculated. Besides, the use of the stable isotope carbon-13 has become the preferred method to minimise patient exposure to radiation in recent years.

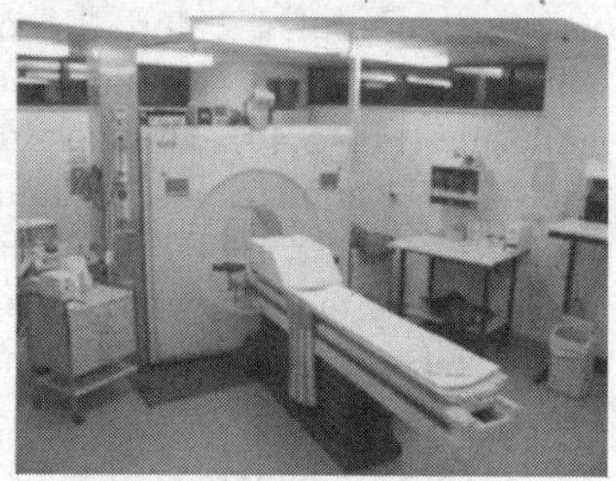

Positron emission tomography.

(*i*) Positron Emission Tomography (PET): Electrons are considered as the matter. Positrons (or anti-electrons) are antimatter-having many of the same properties as matter but with opposite electric charges. An electron is negatively charged while the positron is positively charged (electrons are sometimes called **negatrons** to differentiate them from positrons). When the positrons encounter electrons, they annihilate in bursts of light called gamma rays. These gamma rays are the positron emissions detected by the PET scanners.

There are many types of medical imaging techniques that use slightly different technologies. The X-ray is the simplest and most common technique used to look into the human body. These are high-powered beams of light that pass through some parts of the human body, such as skin, but are stopped by other parts, like bone. This property allows the doctors to easily examine the skeletal structure. Computerised axial tomography (CAT) scans also use X-rays but in a more complicated way. CAT scans image of the body in slices perpendicular to the axis stretching from the feet to the head. These slices fall in the axial plane of the body. This technique of looking at the body in thin slices is called **tomography**.

Positron emission tomography (PET) scans are an important aspect of medical imaging and diagnostics that is used to look into the human body as never before. PET scans differ from other medical imaging techniques because they do not actually look at the body itself. Radioactive tracers are designed to mimic naturally occurring substances

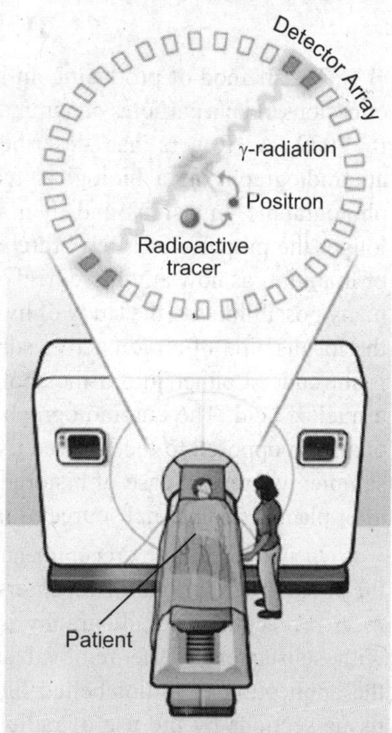

Working of the PET scan

and deliver less radiation than does X-rays. Upon decay of the radioactive tracers inside the body, a positron is released that is equivalent of electrons. Positron encounters one of the billions of electrons inside your body, it annihilates in a flash of light called gamma rays. These gamma rays are the positron emissions which are detected by PET scanners lined with gamma ray detectors.

Before starting of the PET scan, the radioactive tracers are injected into the body of patient. When the radioactive tracers decay, they emit a positron which annihilates with billions of electrons present in your body producing two gamma rays always back-to-back or in opposite direction from each other. The PET scanner detects all rays resulting from body through the sorts of interactions and a computer produces a final image.

About 95 % of PET scans throughout the world use a tracer called 2-fluro-2-deoxy-D-glucose (FDG) which is a fluorine-18 atom attached to a glucose molecule (Fig. 5.40). This positron emitter has a half-life of 109.8 minutes and is the most commonly used radioactive tracer in PET scans. FDG scans can also be used to study the fine detail of sugar metabolism in the brain, and to detect and examine the cancerous tumours throughout the body.

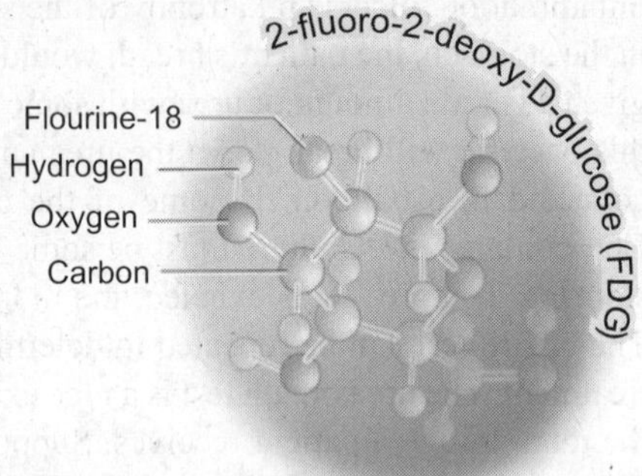

Fig. 5.40. 2-fluro-2-deoxy-D-glucose.

There are hundreds of radioactive tracers available for PET scans but only several dozen are commonly used. The radioactive tracers allow PET scanners to examine different aspects of brain function. For example the oxygen-15 radioisotope is used to study oxygen metabolism in the brain. Oxygen-15 can also be attached to carbon monoxide for the study of blood volume or water for the study of blood flow in the brain.

II. AUTORADIOGRAPHY

The method of producing an image on an X-ray film or nuclear emulsion produced through emissions of beta particles or gamma rays from a radiolabeled compound is called 'autoradiography'. In 1979, A. Rogers has described the techniques of autoradiography. The development of autoradiography as a biological technique started after World War II with the development of photographic emulsions and then stripping film made of silver halide. Radioactivity is now no longer the property of a few rare elements of minor biological interest (such as radium, thorium or uranium) as now any biological compound can be labelled with radioactive isotopes opening up many possibilities in the study of living systems. This technique may be used in biology to determine the localisation of a radioactive substance in the tissue. In this technique, a radioactive substance is introduced either into a metabolic pathway (bound to a receptor or enzyme) or hybridised to a nucleic acid. The autoradiograph (*i.e.* autoradiogram) is obtained by placing the X-ray film or emulsion opposed to the labelled tissue. In autoraduigraphy the radioactive substance is within the sample, whereas in case of historadiography or microradiography X-ray of the sample is obtained after placing an external source of radioactive substance.

In vivo or *in vitro* receptor autoradiography is the use of radiolabelled ligands to determine the tissue distributions of receptors. Generally the ligands are labelled with ^{3}H (tritium) or ^{125}I. *In vivo* receptor autoradiography is the administering of radiolabelled ligand into the circulation with subsequent tissue removal and sectioning, whereas *in vitro* receptor autoradiography is the application of radiolabelled ligand to the tissue sections. Distribution of RNA transcripts in tissue sections by the use of radiolabelled [complementary oligonucleotides or ribonucleic acids (riboprobes)] is called *in situ* hybridisation histochemistry.

The ^3H (thymidine) is radioactive precursors of DNA and ^3U (uridine) that of RNA, which may be introduced to living cells to determine the timing of several phases of the cell cycle. In the same way, RNA or DNA viral sequences can also be located. Usually these probes are labelled with ^{32}P, ^{33}P or ^{35}S.

Liquid photographic emulsion is used in autoradiography where each crystal of silver halide is an independent detector, if insulated from the rest of the emulsion by a capsule of gelatin. Each crystal responds to the charged particle by the formation of a hidden image that is made permanent by the process of development. The record provided by the photographic emulsion is cumulative and spatially accurate. It provides information on the localisation and distribution of radioactivity within a sample. Nuclear emulsion has a very high efficiency for β particles. The effective volume of the detector emulsion in the immediate vicinity of the source may be as little as 100 cubic microns.

1. Method of Autoradiography

The living cells are briefly exposed to a 'pulse' of a specific radioactive compound. The tissue is left for a variable time, and samples are taken, fixed, and processed for light or electron microscopy. Sections are cut and overlaid with a thin film of photographic emulsion, which consists of silver halide crystals in suspension in a clear gelatinous phase, left in the dark for days or weeks (while the radioisotope decays). This exposure time depends on the activity of the isotope, the temperature and the background radiation (this will produce with time a contaminating increase in 'background' silver grains in the film). Following passage through the emulsion of a β-particle or a γ-ray emitted by a radionuclide, the Ag^+ ions are converted to Ag atoms. The photographic emulsion is developed (as for conventional photography). Counterstaining (for example with toluidine blue) shows the histological details of the tissue. The staining must be able to penetrate, but not have an adverse affect on the emulsion. Alternatively, pre-staining of the entire block of tissue can be done with osmium on plastic sections coated with stripping film or dipping emulsion before exposure to the photographic emulsion. This avoids the need for individually (post-) staining each slide. It is not necessary to cover slip these slides. The position of the silver grains in the sample is observed by light or electron microscopy. The grains are in a different plane of focus in the emulsion overlying the tissue section. Often oil with a X100 objective is used for detailed observation with the light microscope. This autoradiograph provides a permanent record. Full details on the batch of emulsion used, dates, exposure time and conditions should be kept for each experiment.

2. Replication of Circular Bacterial Chromosomes

In 1963, J. Cairns first used autoradiography to verify photographically the semi-conservative method of DNA replication in *Escherichia coli.* In this technique the radioactive atoms are exposed to photographic film. The visible silver grains on the film can then be counted to provide an estimate of the quantity of radioactive material present. He radioactively labelled the chromosome by growing the cultures in a medium containing radioactive ^3H-thymidine, a component of DNA nucleotides.

The nucleoside base was incorporated uniformly into the bacterial chromosome. He then isolated the chromosomes by lysing the cells gently, placed them on an electron micrograph (EM) grid and exposed to X-ray film for two months. The emulsion was then developed to produce autoradiograph which was examined under the electron microscope. Each grain of silver represents a radioactive decay. He found when the replication fork moves around the circle, a Greek letter theta (θ)-shaped structure is formed. This experiment clearly demonstrates the theta replication model of circular bacterial chromosomes.

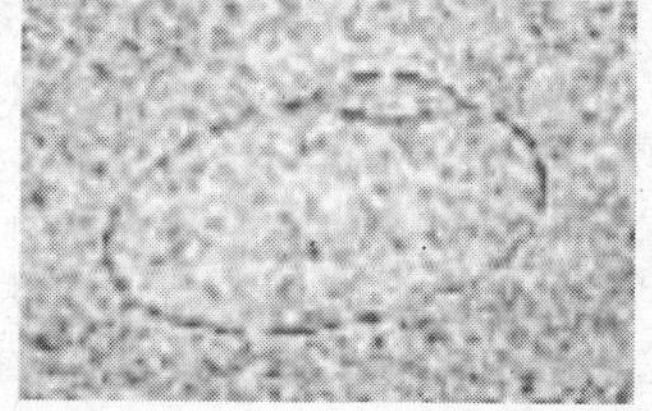

Two replicating strands observed by J. Cairns in *E. coli.*

It was known that the *E. coli* consists of a circle DNA, which replicated maintaining the integrity of the circle *i.e.* the circle, does not break during DNA replication. Formation of intermediate theta structure is due to the replication eye. DNA replicates at one or two moving Y-junctions in the circle replication forks, which supports the semiconservative replication.

It has been confirmed that replication of bacterial chromosome occurs in a bidirectional manner. This was first demonstrated by specifically labelling replicating bacterial chromosomes with radioactive isotopes. The regions of DNA undergoing replication during the experiment were then visualised by using autoradiography and examining the developed film microscopically. This allowed the researchers to observe the point where replication took place. The first conclusive observation of bidirectional replication was provided from studies of *B. subtilis*. Thereafter, the *E. coli* chromosome was also shown to replicate bidirectionally.

III. RADIOIMMUNO ASSAY (RIA)

It take its name from '*radio*' means 'radioisotopes', '*immuno*' refers to antibodies, and '*assay*' mean 'measurement'; hence, the term radioimmunoassay (RIA) has been coined. During the late 1950s, the development of radioimmunoassay by Solomon Sussman Yalow and Aaron Berson Rosalyn is a milestone in the history of the application of radionuclide methodology to biology and to medical investigation. The method offers a technique to assay materials which are unmeasureable or detectable only with difficulty. RIA technique uses radioactive materials to investigate the human body for very small amounts of substances. In 1977, Rosalyn Sussman Yalow received the Nobel Prize in Medicine for the development of the RIA for insulin. The precise measurement of minute amounts of such a hormone was considered a breakthrough in endocrinology. RIA is especially important for the early detection of diseases.

Rosalyn Sussman Yallow

Rosalyn Sussman Yalow (1921- 2011) and Solomon Berson (1918–1972)

Rosalyn Sussman Yalow was born in New York City (U.S.A.). By high school she had changed her career plans to chemistry due to a talented chemistry teacher. Then she switched from chemistry to physics due to receiving more encouragement from her physics Professors than from the chemistry faculty. Life of Marie Curie excited her about nuclear physics. At that time World War II was looming on and many male graduate students were being interviewed for the military. Women students were recruited to take the leaving places of the men. In 1917, Yalow was accepted by the University of Illinois and she arrived as the first woman graduate student in the Department of Physics. There she met and married Aaron Yalow who was also a graduate student in physics.

Solomon Aaron Berson, an American Medical doctor and a scientist, graduated from the City College of New York in 1938, and completed his degree in 1945. After two years in the army he returned to New York to do an internal medicine residency in a Hospital. He started his scientific work in 1950 when he became a member of the Radioisotope Service of the hospital supporting Rosalyn Yalow in her research. In 1950, Berson came to work with Yalow.

Their early laboratory work concerned with iodine and human metabolism of serum albumin, but later they shifted their focus to insulin, a hormone which was difficult to measure in the blood. For the next 22 years Yalow and Berson worked together using radioactive isotopes to study what happens inside the human body. With the success of the insulin RIA, Yalow and Berson extended their success to other hormones, such as corticotropin, gastrin, parathyroid hormone and growth hormone, making significant discoveries in their physiology along the way.

In 1977 Yalow received the Nobel Prize in physiology or medicine for her and Berson's development of RIA applied to tracing hormones in the body. Since Nobel Prizes are only given to the living, Yalow received the award without Berson, who died in 1972. She shared it with two other scientists who were honored for their work on hormone production in the brain.

Radioimmunoassay (RIA) is a very sensitive *in vitro* technique used to measure the concentrations of antigens (e.g. hormone levels in the blood) without bioassay. RIA involves mixing of known quantities of radioactive antigen (labeled with gamma-radioactive isotopes of iodine attached to tyrosine) with antibody to that antigen. Thereafter, unlabelled or 'cold' antigen is added and the amount of displaced labelled-antigen is measured. Initially, the radioactive antigen binds to the antibodies. When unlabelled (*i.e.* quest) antigen is added, the two compete for antibody binding sites. At higher concentrations, more unlabelled antigen binds to the antibody displacing the radioactive variant. The bound antigens are separated from the unbound antigens. Initially, the method of separation was the use of a second 'anti-antibody' directed against the first for precipitation and centrifugation. The use of charcoal suspension for precipitation was extended but replaced later by Werner and Acebedo at Columbia University for RIA of T3 and T4.

1. Method of Radioimmunoassay

A known quantity of an antigen is made radioactive, by labelling it with gamma-radioactive isotopes of iodine attached to tyrosine. Then this radiolabelled antigen is mixed with a known amount of antibody for that antigen. Consequently, antigen and antibody chemically bind to each another; thereafter, a sample from a patient serum containing an unknown quantity of the same antigen is added. The unlabelled (or cold) serum antigen compete with the radiolabelled antigen ('hot') for antibody binding sites (Fig. 5.41).

The unlabelled antigen binds more to the antibody with increasing its concentration, and displaces the radiolabelled variant. This causes a reduction in the ratio of antibody-bound radiolabelled antigen to free radiolabelled antigen. Then the bound antigens are separated from the unbound ones and the radioactivity of the free antigen remaining in the supernatant is measured by using a gamma counter. Known standards are used to prepare a binding curve which allows the amount of antigen in the patient's serum to be derived. The RIA technique is very sensitive and extremely specific. It is the least expensive method but needs specialised equipment to perform the tests.

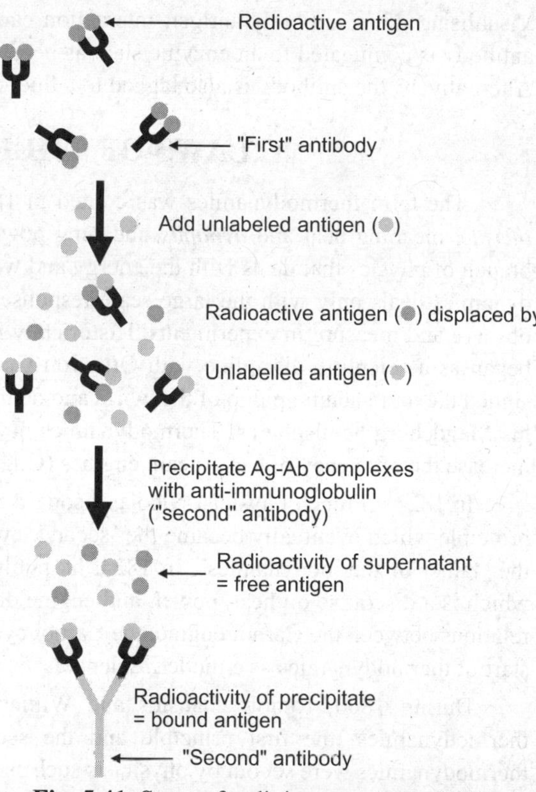

Fig. 5.41. Steps of radioimmunoassay (RIA).

Radioimmunoassay is based on the competition between labelled and unlabelled antigen for the sites of specific antibody, which forms antigen-antibody complexes. At equilibrium, the bound (B) radioactive complex is separated from the free (F) radioactive antigen. The B/F ratio depends on the amount of nonradioactive antigen. Antigen concentration in unknown samples is determined by comparing the B/F ratio to the B/F ratios obtained by incubating varying amounts of known nonradioactive antigen with the same amount of antibody as in the unknown sample under similar assay conditions.

Sensitivity of the order of 10-12 moles/liter may be achieved through the preparation and use of a labelled antigen of high specific activity and the production and selection of antisera with appropriately high affinity constants. Specificity is dependent upon the ability of the antiserum to recognise subtle structural features of the antigen molecule. The ability to assay large numbers of samples with good precision has led to the application of this technique to quantitate substances (like steroids) already measurable. There have been numerous contributions to its further development, refinement and application, since the initial description of competitive binding radioassay techniques.

Since radioactive substances are used in the RIA, it requires special precautions. Now-a-days, it has been supplanted by the ELISA method, where the antigen-antibody reaction is measured using colorimetric signals instead of a radioactive signal. RIA methods are becoming popular because of its robustness, consistent results and low price of testing. An example of radioimmunoassay is the radioallergosorbent (RAST) test which is used to detect the allergen that causes allergy.

IV. IMMUNOHISTOCHEMISTRY (IHC)

The term '*immuno*' refers to antibodies used in the procedure, and '*histo*' means tissue. The immunohistochemistry may be defined as the process of localising proteins in cells of a tissue section exploiting the principle of antibodies binding specifically to antigens in biological tissues. The immunohistochemical staining is widely used in the diagnosis and treatment of cancer. Specific molecular markers are characteristic of particular cancer types. IHC is also widely used in basic research to understand the distribution and localisation of biomarkers in different parts of a tissue. Visualising of an antibody-antigen interaction can be accomplished in several ways. Mostly an antibody is conjugated to an enzyme such as peroxidase that catalyses a color-producing reaction. Alternatively, the antibody is also tagged to a fluorophore such as FITC, rhodamine or Texas Red.

K. LAWS OF THERMODYNAMICS

The term thermodynamics was coined in 1840 from the Greek root *therme,* meaning heat and *dynamis,* meaning power. Thermodynamics is a branch of physics that deals with the energy and work of a system. Thermodynamics deals only with the large scale response of a system that we can observe and measure in experiments. Historically in 1650, thermodynamics began as a scientific discipline with Otto von Guericke who built and designed the first vacuum pump of the world and demonstrated a vacuum using his Magdeburg hemispheres. Thermodynamics developed out of a desire to increase the efficiency of early steam engines (Guggenheim, 1985).

Josiah Villard Gibbs (1839 –1903)

In 1824 a French physicist Nicolas Léonard Sadi Carnot first formulated the thermodynamic principle which eventually became the 'second law of thermodynamics'. Sadi Carnot is known as the 'father of thermodynamics'. In 1824, he published *Reflections on the Motive Power of Fire*, which is a discourse on heat, power, and engine efficiency. The paper outlined the basic energetic relations between the Carnot engine, the Carnot cycle, and motive power (Fig. 5.41). It marked the start of thermodynamics as a modern science.

During 1860, Rudolf Clausius and William Thomson formalised the two principles of thermodynamics, the first principle and the second principle. The foundations of statistical thermodynamics were set out by physicists such as James Clerk Maxwell, Ludwig Boltzmann, Max Planck, Rudolf Clausius and J. Willard Gibbs.

Josiah Willard Gibbs, an American theoretical physicist, chemist, and mathematician, devised the theoretical foundation for chemical thermodynamics as well as physical chemistry. In 1873, Gibbs clearly stated the two absolute laws of thermodynamics, the first law and the second law. In 1901, Gibbs was awarded the highest honor for his greatest contribution of being the first to apply the second law of thermodynamics to the exhaustive discussion of the relation between chemical, electrical, and thermal energy and capacity for external work.

An axiomatic basis of thermodynamics is formed by the thermodynamics laws. It defines the fundamental physical quantities like temperature, energy, and entropy that describe the thermodynamic systems, the transport and conversion of heat and in thermodynamic process.

Thermodynamics defines four laws which do not depend on the details of the systems under study or how they interact. Hence these laws are generally valid and can be applied to systems about which one knows nothing other than the balance of energy and matter transfer. These four laws of thermodynamics are as below:

- The *zeroth law of thermodynamics*: It defines the empirical temperature on the basis of the principle of thermal equilibrium.
- The *first law of thermodynamics*: It defines the conservation of energy and mentions that the flow of heat is a form of energy transfer.
- The *second law of thermodynamics*: It states that the entropy of an isolated macroscopic system never decreases or equivalently perpetual motion machines are impossible.
- The *third law of thermodynamics*: It deals with the entropy of a perfect crystal at absolute zero temperature, and implies that it is impossible to cool a system to exactly absolute zero.

The laws of thermodynamics have become some of the most important fundamental laws in physics and other branches of science. The laws of thermodynamics demonstrate very ordered rules of energy interactions, which the entire universe must obey. Why are the laws of thermodynamics true ? Why are there no exceptions in the entire universe ? This law cannot be an accident, since none of the thermodynamic laws are random. Organization implies an organizer, and design implies a designer. If the laws are not random, then who created them ?

1. Zeroth Law of Thermodynamics

Initially the zeroth law was not recognized as a law, because the importance of the zeroth law was not realized until 1930s. In 1931, Ralph H. Fowler first named the zeroth law of thermodynamics. Moreover, the first, second and third laws found common acceptance by the physicists. After realizing the importance of the zeroth law, the numbers of the other laws were not changed; hence zero was assigned and it was called as the *zeroth law.*

It states that *if two systems are in thermal equilibrium with a third, they are also in thermal equilibrium with each other, i.e. two bodies in thermal equilibrium have the same temperature.* The zeroth law of thermodynamics introduces the concept of 'thermodynamic equilibrium' in which two objects have the same temperature. If you bring two objects into physical contacts that are initially at different temperatures, they exchange heat to establish a thermal equilibrium between each other. Heat is transferred between the objects during the process of reaching thermal equilibrium. The amount of heat transferred DQ is proportional to the temperature difference DT between the objects and the heat

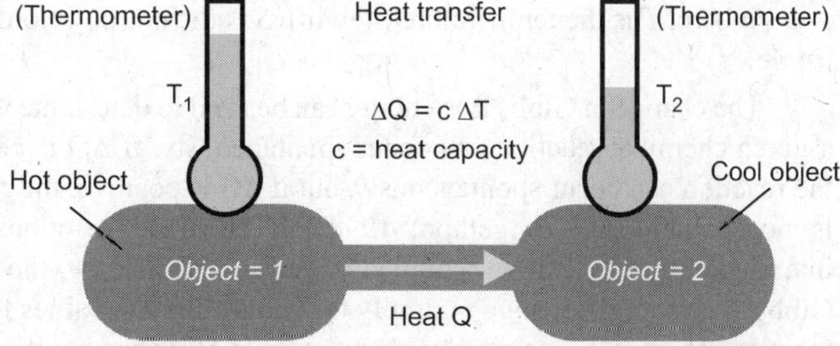

Fig. 5.42. Establishment of thermodynamic equilibrium.

capacity 'c' of the object (Fig. 5.42). The heat capacity is a constant that tells how much heat is added per unit temperature rise.

The value of the constant differs with different materials. Heat is always transferred from the object of higher temperature to the object having lower temperature. If two bodies are at the same temperature, it is not necessary to bring them into contact and measure any changes of their observable properties in time. The law provides a fundamental definition of temperature and justification for the construction of practical thermometers.

Furthermore thermal equilibrium is an equal relation between several numbers of systems. This law also makes a statement about measurability, and establishes an empirical parameter (the temperature) as a property of a system so that systems in equilibrium with each other have the same temperature, for example a gas thermometer can be used as a device to measure the temperature of another system. This law defines temperature in a non-circular logistics without reference to entropy and variables of its conjugates.

(*a*) Thermodynamic Equilibrium: Classical thermodynamics deals with dynamic equilibrium states. The values of its intensive parameters (such as pressure or temperature) determine the local state of a system at thermodynamic equilibrium. Specifically, the minimum of a thermodynamic potential (such as the Helmholtz free energy, *i.e.* systems at constant temperature and volume) characterise thermodynamic equilibrium.

(*i*) The Helmholtz free energy: The Helmholtz free energy was developed by Hermann von Helmholtz, which is a thermodynamic potential that measures the useful work performed from a closed thermodynamic system at a constant temperature and volume. The negative difference in the Helmholtz energy is equal to the maximum amount of work done in a thermodynamic process in which temperature and volume are kept constant. It is minimised at equilibrium under these conditions. Usually the Helmholtz free energy is denoted by a letter A (from the German 'Arbeit' or work), the letter F or as *Helmholtz energy*. The letter F is often referred to as the *Helmholtz function* or simply *free energy*. The Helmholtz free energy is denoted as below:

$$A \cong U - TS$$

Where, A is the Helmholtz free energy (joules), U is the internal energy of the system (joules), T is the absolute temperature (kelvins), and S is the entropy (joules per kelvin).

(*ii*) The Gibbs free energy: The Gibbs free energy (also called Gibbs energy or Gibbs function or free enthalpy) is most commonly used as a measure of thermodynamic potential especially in the field of chemistry. Basically it is such a thermodynamic potential that measures the useful or process-initiating work obtained from an isothermal isobaric thermodynamic system. The Gibbs free energy is the maximum amount of non-expansion work that can be obtained from a closed system, only in a completely reversible process. When a system changes from an initial state to a final state, the Gibbs free energy ΔG equals the work exchanged by the system with its surroundings, minus the work of the pressure forces.

$$\Delta G = \Delta H - T\Delta S \text{ (at constant temperature and pressure)}$$

Where, T is the temperature (kelvin), S is the entropy (joule per kelvin), and H is the enthalpy (joule).

The change in Gibbs free energy can be used to determine whether a given chemical reaction can occur spontaneously. If ΔG is negative, the reaction can occur spontaneously, but if ΔG is positive, the reaction is nonspontaneous (Bergethon, 1998). If chemical reactions share intermediates, they can be 'coupled' together. In this case, the overall Gibbs 'free energy' change is simply the sum of the ΔG values for each reaction. Therefore, an unfavorable reaction (+ΔG1) can be driven by a second, highly favorable reaction (-ΔG2 where the magnitude of ΔG2 >

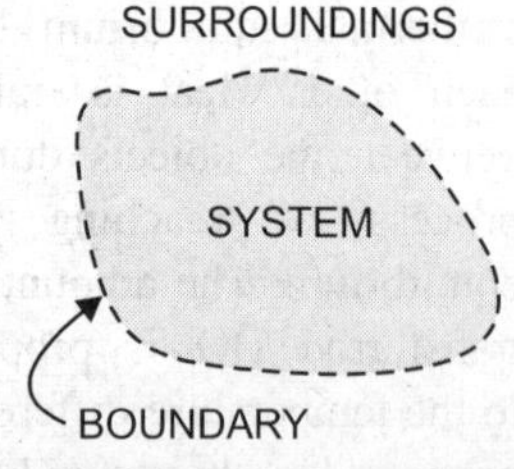

A thermodynamic system

magnitude of ΔG1). For example, the reaction of glucose with fructose to form sucrose has a ΔG value of +5.5 kcal/mole. Therefore, this reaction will not occur spontaneously. The breakdown of ATP to form ADP and inorganic phosphate has a ΔG value of -7.3 kcal/mole. These two reactions can be coupled together, so that glucose binds with ATP to form glucose-1-phosphate and ADP. Thereafter, glucose-1-phosphate is able to bond with fructose producing sucrose and Pi. The ΔG value of the coupled reaction is -1.8 kcal/mole, which indicates the occurrence of spontaneous reaction. This principle of coupling reactions to alter the change in Gibbs 'free energy' is the basic principle behind all enzymatic action in organisms (Alberts, 2009).

The thermalisation process leads to a thermodynamic equilibrium. An example of this is a system of interacting particles that is left undisturbed by outside influences. By interacting, they will share energy/momentum among themselves and reach a state where the global statistics are unchanging in time.

***(b)* Conditions for Equilibrium:** By considering the differential form of thermodynamic potentials, the following relationships can be derived:

- For a completely isolated system, $\Delta S = 0$ at equilibrium.
- For a system at constant temperature and volume, $\Delta A = 0$ at equilibrium.
- For a system at constant temperature and pressure, $\Delta G = 0$ at equilibrium.

The various types of equilibriums are achieved as follows:

- Two systems are in *thermal equilibrium* when their temperatures are the same.
- Two systems are in *mechanical equilibrium* when their pressures are the same.
- Two systems are in *diffusive equilibrium* when their chemical potentials are the same.

2. First Law of Thermodynamics

The principle of conservation of energy is expressed by the first law of thermodynamics. The first law states that '*energy can neither be created nor destroyed, but it can be changed from one form to another.* The law expresses about the transformation of energy from one form to another form. The internal energy of an isolated system is constant which can be transformed into other forms. In any process in an isolated system the total energy remains the same. The change in the internal energy of a system is equal to the amount of heat supplied to the system, minus the amount of work done by the system on its surroundings. For a thermodynamic cycle the net heat supplied to the system equals the net work done by the system.

Hess's law arises from the first law, which states that the heat absorbed or evolved in a given reaction must always be constant and independent of the manner in which the reaction takes place. Some intermediate reactions may be endothermic and others may be exothermic; but the total heat exchange is equal to the heat exchange that directly had occurred in the process. This principle is the basis for the calorimeter, which is used to determine the amount of heat in a chemical reaction. All incoming energy enters the body as food and is ultimately oxidised; hence, the total heat production may be estimated by measuring the heat produced by the oxidation of food in a calorimeter. This heat is expressed in kilocalorie, which is a common unit of food energy (Stacy, 1955).

The first law states that the amount of energy lost in a steady state cannot be greater than the amount of energy gained. This is the statement of conservation of energy for a thermodynamic system. A closed system transfers energy to and from its surroundings by the processes of heat and mechanical work. The rates of these two processes determine the rate of gain or loss in the stored energy of a system. The first law clarifies the nature of energy. It is a stored quantity independent of any particular process path, or any system. If a system undergoes a thermodynamic cycle (whether it becomes warmer, cooler, larger or smaller), then it will have the same amount of energy each time it returns to a particular state (Kondepudi, 2008).

The first law of thermodynamics states that energy is conserved in any process involving a thermodynamic system and its surroundings. A change in the assumed internal energy (U) is regarded as a combination of heat (Q) added to the system and work done by the system (W). Taking ΔU as a differential change in internal energy, one writes.

Increase in internal energy of a system = heat supplied to the system + work done on the system

$$\Delta U = Q + W_{system}$$

The definition of the work is also given in terms of the work performed by a system on its surroundings.

$$\Delta U = Q - W_{surrounding}$$

This is a statement of conservation of energy. The net change in internal energy is such that flows in as heat minus the energy that flows out as the work that the system performs on it environment. This is also stated as a definition of the amount of heat of a process.

For a reversible process, the total amount of heat added to a system can be expressed as $\delta Q = TdS$, where T is temperature and S is entropy. Therefore, for a reversible process:

$$\Delta U = T\Delta S - P\Delta V$$

Since U, S and V are thermodynamic functions of state, the above relation holds for non-reversible changes. The above equation is known as the fundamental thermodynamic relation.

Laws of Thermodynamics in Upanisad

The first law of thermodynamics is just like philosophy. A similar concept has been given in Upanisad as given below:

Na jayate mriyate va vipaschi-,
nnayam kutaschinna babhuv kaschit.
Ajo nityah sasvato'yam purano-,
na hanyate hanyamane sarire

-Kathopanisad. 18/2

(*The everlasting soul is never born, nor it ever dies, nor does it become after being born. For, it is unborn, eternal, everlasting and primeval; though the body is slain even then it is undestroyed*).

3. Second Law of Thermodynamics

In general the total entropy of any system will not decrease by increasing the entropy of some other system. Therefore, the entropy of a system isolated from its environment will not tend to decrease. The heat will not flow from a colder body to a hotter body without applying the work to the colder body. For any device operating on a cycle, it is impossible to produce net work from a single temperature reservoir. The flow of heat from a hotter reservoir to a colder reservoir is required by the production of net work. In a specified process such as a chemical reaction, the increase of entropy reduces, *i.e.* it is energetically more efficient.

The entropy of an unisolated system may not decrease. For example, an air conditioner may cool the air in a room resulting in reduction of the entropy of the air of that system. The heat expelled from the system (the room) involved in the air conditioner always contributes much to the entropy of the environment as compared to the decrease of the entropy of the air of that system. The total entropy of the room (system) plus the entropy of the surroundings increases; thus it agrees with the second law of thermodynamics.

Rudolf Clausius

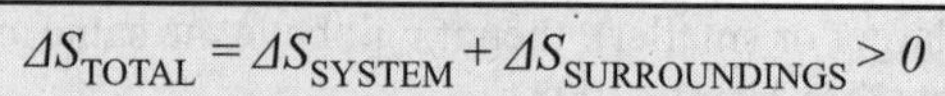

$$\Delta S_{TOTAL} = \Delta S_{SYSTEM} + \Delta S_{SURROUNDINGS} > 0$$

Solid crystals are the most regularly structured form of matter. Solid crystals have low entropy values with considerable predictability of microscopic configuration as well as predictability of bulk behavior. The gases have high entropy values. They behave predictably in bulk even in spite of having their unknown microscopic motions. For a given degree of bulk predictability, the positions of the crystal atoms are more predictable than that of the gas atoms.

When two isolated systems, each in different thermodynamic equilibrium in separate but nearby regions of space are allowed to interact breaking the isolation, they will eventually reach a mutual thermodynamic equilibrium by exchanging the energy. The sum of entropies of the initial isolated systems is less than (or equal to) the entropy of the final exchanging systems. Entropy has increased (or at least has not decreased) in the process of reaching a new thermodynamic equilibrium. The second law states that 'spontaneous natural processes increase overall entropy'. In brief, heat can spontaneously flow from a region of higher temperature to that of a lower temperature'. Nevertheless, energy can be transferred from cold to hot by the expenditure of external work. For example, a refrigerator cools its contents but warms the surrounding air (Kondepudi, 2008).

Since a finite universe is an isolated system, the total entropy of which is constantly increasing. It has been speculated that the universe is fated to a heat death in which all the energy ends up as a homogeneous distribution of thermal energy, so that no more work can be extracted from any source. Entropy of an isolated macroscopic system never decreases. But a microscopic system may exhibit fluctuations of entropy in contrast to that stated by the second law. The second law says that everything moves to a state of order. Everything wants to settle down. According to the second law no natural process can occur unless it is accompanied by an increase in the entropy of the universe.

Entropy and Enthalpy

In 1865, Rudolf Clausius coined the term entropy and originated the concept based on the Greek word *entropía* meaning a turning toward. It is defined as a measure of disorder in the universe or the availability of the energy in a system to do work. It is considered as a measure of energy dispersal at a specific temperature. The measure of the total amount of 'disorder' in the system is given by the following expression:

$$\text{Disorder} = \frac{C_D}{C_I}$$

Where, C_D is the 'disorder' capacity of the system (which is the entropy of the parts contained in the permitted ensemble), C_I is the 'information' capacity of the system.

The living organisms preserve their internal order by taking from their surroundings free energy, in the form of nutrients or sunlight, and returning to their surroundings an equal amount of energy as heat and entropy.

Enthalpy is the total energy of a thermodynamic system. It includes the internal energy required: (*i*) to create a system, and (*ii*) to make room for it by displacing its environment and establishing its volume and pressure. Enthalpy is a thermodynamic potential; it is a state function and an extensive quantity, and the unit of measurement in the International System of Units (SI) is the joule, but it is also measured as calorie. The enthalpy of a system is defined as:

$$H = U + pV$$

Where, H is the enthalpy of the system (in joules), U is the internal energy of the system (in joules), p is the pressure at the boundary of the system and its environment (in pascals), and V is the volume of the system (in m^3). The U is equivalent to the energy required to create the system, and the pV is equivalent to the energy which would be required to 'make room' for the system if the pressure of the environment remained constant.

Living organisms are often mistakenly believed to defy the 'second law' because they increase their level of organisation. This misinterpretation can be corrected, only by referring the definition of 'systems' and 'boundaries'. A living organism is an open system and able to exchange both matter and energy with its environment, for example the virus assembly from its subunits which clearly increases the order of increase. If the virus is considered an isolated system, this process will defy the second law. After virus interaction directly with its environment, the virus assembly results in an increase of entropy in the system as a whole due to the liberation of water of solvation from the components and the resulting increase in rotational and translational entropy of the solvent (Edsall and Gutfreund, 1983).

4. Third Law of Thermodynamics

A chemist Walther Nernst during in the years 1906-1912, developed the third law. Therefore, it is also called Nernst's theorem or Nernst's postulate. The third law of thermodynamics is known less as compared to the other laws of thermodynamics. In its simplest form, the third law of thermodynamics relates the entropy (randomness) of matter to its absolute temperature. According to the third law of thermodynamics, if one could reach absolute zero, all bodies would have the same entropy. In other words, a body at absolute zero could exist in only one possible state, which would possess a definite energy, called the zero-point energy (Kondepudi, 2008)..

As temperature approaches absolute zero, all processes cease and the entropy of the system approaches a minimum. Here absolute zero is −273.15 °C (degrees Celsius), or −459.67 °F (degrees Fahrenheit) or 0 K (Kelvin) at which all activity would stop if it were possible to happen. This state is defined as having zero entropy, because a system at zero temperature exists in its ground state, so that its entropy is determined only by the degeneracy of the ground state. It means that "*it is impossible by any procedure, no matter how idealized, to reduce any system to the absolute zero of temperature in a finite number of operations*".

$S_{T-0} = 0$ (perfect)

The entropy (S) of a pure and perfectly crystalline compound at T=0K is zero (no disorder)

An alternative version of the third law of thermodynamics as stated by Gilbert N. Lewis and Merle Randall in 1923 is as below :

If the entropy of each element in some (perfect) crystalline state be taken as zero at the absolute zero of temperature, every substance has finite positive entropy; but at the absolute zero of temperature the entropy may become zero, and does so become in the case of perfect crystalline substances.

The third law of thermodynamics states that as the temperature of a system approaches absolute zero, its entropy approaches a constant (for pure perfect crystals), that is zero. A pure and perfect crystal is one in which every molecule is identical and the molecular alignment is perfectly even throughout the substance. For non-pure crystals or those with less-than perfect alignment, there will be some energy associated with the imperfections, so the entropy cannot become zero (Kondepudi, 2008).

Mostly the third law of thermodynamics is used in ultra-low temperature chemistry and physics. This law is applied to predict the response of various materials to temperature changes. The third law of thermodynamics is not used directly as much as the others two. Study of the third law of thermodynamics supports the implications of the first two laws. This law demonstrates another detectable, absolute, all-encompassing, and important natural law.

QUESTIONS

1. Write theory and application of gel permeation.
2. Describe principles and application of different types of ion exchange chromatography.
3. With the help of suitable diagram describe electrophoresis.
4. Discuss in detail about principles and method of SDS-PAGE .
5. Describe different types of electrophoresis studied by you.
6. What do you mean about 2D electrophoresis ? Discuss about iso-electric focusing.
7. What is pulsed electric gel elelctrophoresis ? Write in detail its working.
8. What is DNA sequencing ? Write the methods of DNA sequencing studied by you.
9. Describe the principles and methods of site-directed mutagenesis.
10. What is spectroscopy ? Describe the principles of colorimetry.
11. Write the principle and working of UV-Vis spectrophotometer.
12. Give a detailed account of mass spectrometry.
13. What is mass spectrometry ? Give a detailed account of it.
14. What is polymerase chain reaction (PCR) ? Give a detailed account of PCR and its applications.
15. What are the laws of thermodynamics ? Describe second law of thermodynamics.
16. What is ionising radiation ? Describe in detail about radioactivity.
17. Give a detailed account of isotopes and their application in traces techniques.
18. Describe the tracer isotopes and methods of detection of radioactivity.
19. Write various applications of radiotracer techniques studied by you.
20. Write brief notes on the following :

(*i*) Chromatography, (*ii*) Gel permeation,
(*iii*) Electrophoresis, (*iv*) PAGE,
(*v*) SDS-PAGE, (*vi*) Iso-electric focusing,
(*vii*) Colorimetry, (*viii*) Gel permeation,
(*ix*) Agarose gel electrophoresis, (*x*) Ion exchange chromatography,
(*xi*) Electroblotting,
(*xii*) ESI, Maxam-Gilbert method of DNA sequencing,
(*xiii*) Sanger-Coulson method of DNA sequencing,
(*xiv*) Automated DNA sequencers, (*xv*) Restriction mapping,
(*xvi*) Site directed mutagenesis, (*xvii*) Fluorescence spectroscopy,
(*xviii*) PCR, (*xix*) RT-PCR,
(*xx*) AP-PCR, (*xxi*) RAPD,
(*xxii*) DAF, (*xxiii*) Q-PCR,
(*xxiv*) Gibbs free energy, (*xxv*) First law of thermodynamics,
(*xxvi*) Second law of thermodynamics, (*xxvii*) Radioactivity,
(*xxviii*) Tracer techniques, (*xxix*) Autoradiography,
(*xxx*) Radioimmunoassay.

BIBLIOGRAPHY

Alberts B. 2009. *Essential Cell Biology*. New York: Garland Science.

Bergethon P. R. 1998. *The physical basis of biochemistry: the foundations of molecular biophysics.* New York: Springer.

Edsall J.T and Gutfreund H. 1983. *Biothermodynamics : The study of biochemical processes at equilibrium*. Chicheter, West Sussex : Wiley.

Gilby E.D. and Edwards J. 1973. 125-Iodine tracers for steroid radioimmunoassay. *J. Endocrinol.* **58**: 10.

Guggenheim E.A. 1985. *Thermodynamics. An advanced treatment for chemists and physicists*, 7th Ed., North Holland, Amsterdam.

Kondepudi D. 2008. *Introduction to modern thermodynamics*, Wiley, Chichester.

Rogers A. 1979. *Techniques of autoradiography*. 3rd Ed. Elsevier, North Holland, p. 429.

Sambrook J. and Russel D.W. 2001. Molecular Cloning: A laboratory manual, 3rd Ed., Chapter 8: *In vitro* Amplification of DNA by the Polymerase Chain Reaction.

Stacy R.W., David T. *et al.* 1955. *Essentials of biological and medical physics*. New York: McGraw-Hill Book.

Wuthrich K. 1986. *NMR of proteins and nucleic acids*. New York (NY), USA: Wiley- Inter Science.

6 CHAPTER

DNA Modifying Enzymes and Restriction Endonucleases

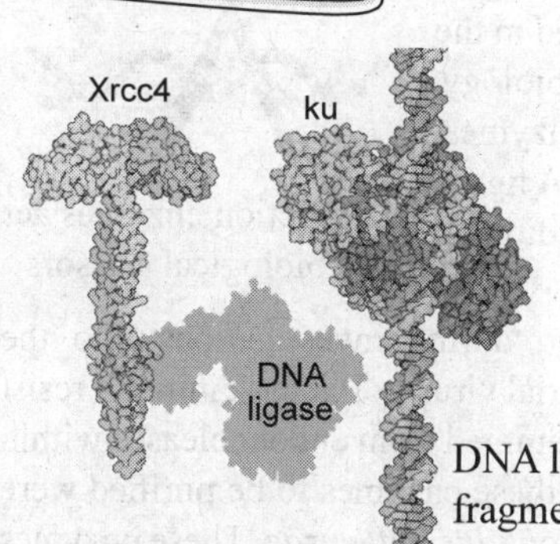

DNA ligase seals two fragments of DNA

A. Exonucleases
B. Endonucleases
C. Restriction endonucleases
D. DNA polymerases
E. Terminal transferase
F. Reverse transcriptase
G. Bacteriophage RNA polymerases
H. Other DNA modifying enzymes

Recombinant DNA technology (= genetic engineering) refers to artificial manipulation, modification and recombination of DNA or other nucleic acid molecules in order to modify an organisms or population of organisms. The term initially meant any of a wide range of techniques for modifying or manipulating organisms through heredity and reproduction. Now the term denotes the narrow field of recombinant DNA technology or gene cloning in which DNA molecules from two or more sources are combined, either within cells or in test tubes, and then inserted into host organisms which are able to reproduce. This technique is used to produce new genetic combinations that are valuable to science, medicine, agriculture or industry. Through recombinant DNA techniques, bacteria have been created that are capable of synthesising human insulin, human interferon, human growth hormone, hepatitis B vaccine, and the other medically useful substances.

Recombinant DNA techniques, combined with the development of a technique for producing antibodies in high quantity have made an impact on medical diagnosis and cancer research. Plants have been genetically adjusted to perform nitrogen fixation and to produce their own pesticides. Bacteria capable of degrading oil have been produced for use in oil-spill cleanups. Genetic engineering also introduces the fear of adverse genetic manipulations and their consequences such as antibiotic-resistant bacteria or new strains of disease.

The first genetically engineered drug was 'human insulin' approved by the Food and Drug Authority of the U.S.A. in 1982. Another early application of genetically engineering was to crease human growth hormones as a replacement for a drug that was previously extracted from human cadavers. In 1986, the FDA approved the first genetically engineered vaccine for humans to control hepatitis B. Since the early uses of technology in medicine, the use of the genetically engineering has expanded to supply many drugs and vaccines.

Proponents of genetically engineering argue that the technology is safe, and it is necessary in order to maintain food production that will continue to match population growth and to feed millions in third world countries more effectively. The others argue that there is more than enough food that may carry some degree of risk.

The late 1950s and the decade of 60s witnessed the enormous breakthrough in DNA enzymology. In 1955, Arthur Kornberg and colleagues isolated DNA polymerase. In 1966, B. Weiss and C.C. Richardson isolated DNA ligase. In 1968, H.O. Smith, K.W. Wilcox and T.J. Kelley isolated and characterised the first sequence-specific restriction endonuclease. These enzymes play roles in synthesis of DNA molecules, the attachment of two or more DNA molecules to one another, and breaking of DNA molecules into fragments. These enzymes make it possible to create entirely the new kinds of DNA molecules and to manipulate the functioning of the genes isolated on these new molecules.

There are various biological tools which are used to carry out the manipulation of genetic material and the cells as well, for example enzymes, foreign DNA, vector or vehicle DNA, gene therapy, etc. Cutting of DNA and pasting its fragments together typically are among the first techniques learned in the molecular biology and are fundamental to all work of molecular biology. Such manipulations of DNA are conducted by a toolkit of enzymes. Restriction endonucleases are used as the molecular scissors, DNA ligase functions to bond piece of DNA together, and a variety of additional enzymes that modify DNA are used to facilitate the process.

Restriction enzymes act as biological scissors

Restriction enzymes were discovered over 30 years ago during investigations into the phenomenon of host-specific restriction and modification of bacterial viruses. Bacteria initially resist transfection by new viruses, and this restriction of viral growth stemmed from endonucleases within the cell that destroy foreign DNA molecules. Among the first of these enzymes to be purified were *Eco*RI from *Escherichia coli*, and *Hin*dII and *Hin*dIII from *Haemophilus influenzae*. These enzymes were found to cleave DNA at specific sites generating discrete, gene-size fragments that could be re-joined in laboratory. Researchers were quick to recognise that restriction enzyme provided them with a remarkable new tool for investigating gene organisation, function and expression.

As the use of restriction enzyme spread among molecular biologists during late 1970s, companies began to search for more enzymes. Except for certain viruses, the restriction enzymes were found only within the prokaryotes. Many thousands of bacteria and archaea have now been screened for their presence. Analysis of sequenced prokaryotic genomes indicates that they are common, and all free-living bacteria and archaea appear to code for them. There are many enzymes which are used in genetic engineering as an important biological tool. Recombinant DNA technology is based totally of different types of enzymes used for cutting and joining of DNA or RNA. Some of the enzymes used in gene cloning are described in detail in this chapter and given in Table 6.1.

Table 6.1. Enzymes used in gene/DNA cloning.

Enzyme	*Use*
Alkaline phosphatase	It removes PO_4 from 5′ end of double stranded or single stranded DNA or RNA.
DNA ligase	It joins sugar-phosphate backbone of double stranded DNA with 5′-PO_4 and a 3′-OH in an ATP-dependent reaction. The two ends of DNA must be blunt or complementary cohesive ones.
DNA Pol I	It synthesises DNA complementary to a DNA template in 5′→3′ direction beginning with a primer with a free 3′-OH. The Klenow fragment is a truncated version of DNA Pol I lacking 5′→3′ exonuclease activity.
Exonuclease III	It cleaves from the ends of linear DNA and digests double stranded DNA from 3′ end only.
Mung bean nuclease	It digests single stranded DNA or RNA but leaves intact double helical region.

Nuclease S1	It behaves as mung bean nuclease and cleaves a strand opposite to a nick on the complementary strand.
Polynucleotide kinase	It adds to 5′-OH end of double stranded DNA or single stranded DNA or RNA in an ATP-dependent reaction. DNA becomes radiolabelled if [g^{32}P]-ATP is used.
Restriction enzymes	These cut both strands of double stranded DNA within a symmetrical recognition site resulting in blunt or sticky ends (5′- or 3′ - overhangs).
Reverse transcriptase	It is a RNA-dependent DNA polymerase which synthesises DNA complementary to an RNA template in 5′→3′ direction by using deoxynucleotide triphosphates (dNTPs) beginning with a primer with a free 3′-OH.
RNase A	It is nuclease which digests RNA, but not DNA.
RNase H	It is also a nuclease which digests the RNA or RNA-DNA heteroduplex.
T3, T7, SP6 RNA polymerases	These are specific RNA polymerases encoded by the bacteriophage T3, T7 and SP6, respectively. These recognise only promoters from its own phage DNA and can be used to transcribe DNA downstream of each promoter.
Taq polymerase	It is a DNA polymerase isolated from *Thermus aquaticus* which operates at 72°C and is stable above 90°C. It is used in PCR.
Terminal transferase	It adds several nucleotides to 3′-ends of linear single or double stranded DNA or RNA molecule.

A. EXONUCLEASES

Exonucleases are enzymes (found as individual enzyme or as part of larger enzyme complexes) that cleave nucleotides one at a time from an end of a polynucleotide chain. These enzymes act upon genome and digest the phosphodiester bonds of base pairs on 5′ and 3′ ends of a single stranded DNA or at single strand nicks or gaps in double stranded DNA. Some examples of exonucleases are deoxyrobonuclease I, deoxyrobonuclease II, deoxyrobonuclease IV, etc. Mechanism of action of exonuclease is given in Fig. 6.1. On left side exonuclease acts on exterior of single strand of double stranded DNA, and cuts A–G nucleotides from 5′-PO_4 end, and C–T–A from 3′-OH end. On the basis of mode of action the exonucleases are grouped into different types.

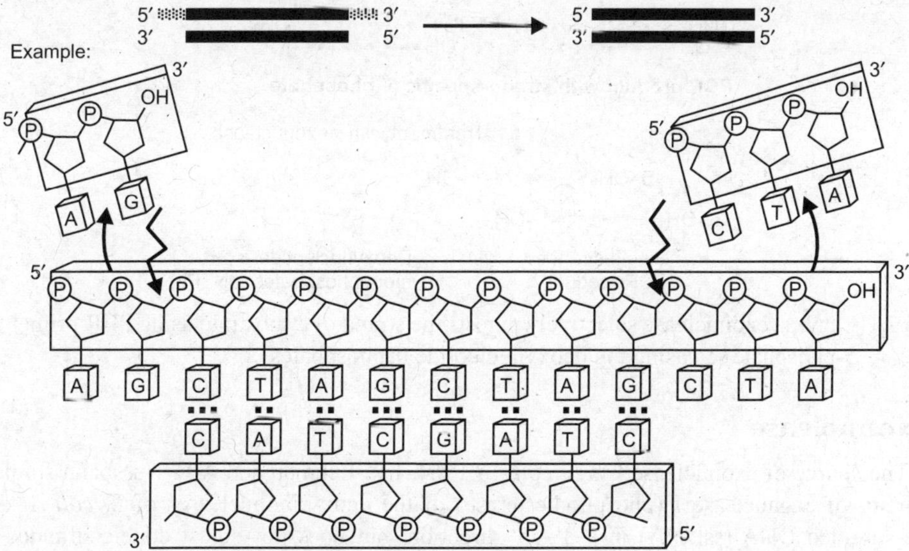

Fig. 6.1. Action of exonuclease on a single strand of double stranded DNA.

1. Lambda (λ) Exonuclease

In 1999, P.G. Mitsis and J.G. Kwagh studied the characteristics of λ exonuclease with DNA ends. It is a simple, elegant and donut-shaped homotrimer processive molecular motor that degrades double-stranded DNA (dsDNA) in the 5′→3′ direction to create long single stranded overhangs used in recombination by the bacteriophage λ. In addition, λ exonuclease does not use an external fuel molecule as an energy source, but uses the energy released in the degradation of its DNA substrate to produce motion. It has been suggested that λ exonuclease may be an example of a Brownian ratchet due to lack of fuel source and structural simplicity. By studying this enzyme, conversion of energy from the hydrolysis of DNA into motion can be understood better. Our approach is to examine λ exonuclease pausing using both single molecule and complimentary bulk biochemistry methods. In 2003, T.T. Perkins and co-workers have described λ exonuclease as an ATP-independent molecular motor.

The preferred substrate is blunt ended 5′-phosphorylated double stranded DNA. It also degrades single-stranded and non-phosphorylated substrates at a highly reduced rate. But λ exonuclease is unable to initiate digestion of DNA at nicks or gaps. This enzyme removes nucleotides from 5′ ends of double stranded DNA. Consequently an improved substrate for terminal transferase is formed. Therefore, this enzyme is used for 5′ end modification. The mode of action of this enzyme is given below :

$$\begin{matrix} 5'\text{-GGAATT}_{OH}\text{-}3' \\ 3'\text{-CCTTAAp-}5' \end{matrix} \longrightarrow \begin{matrix} 5'\text{-GAATT}_{OH}\text{-}3' \\ 3'\text{-Cp-}5' \end{matrix} + \text{Np-}5'$$

(Np = nucleotide phosphate)

Fig. 6.2. shows the digestion of the strand of a substrate by λ exonuclease using a PCR primer with a 5-phosphate, releasing the deoxynucleotide triphosphates (dNTPs).

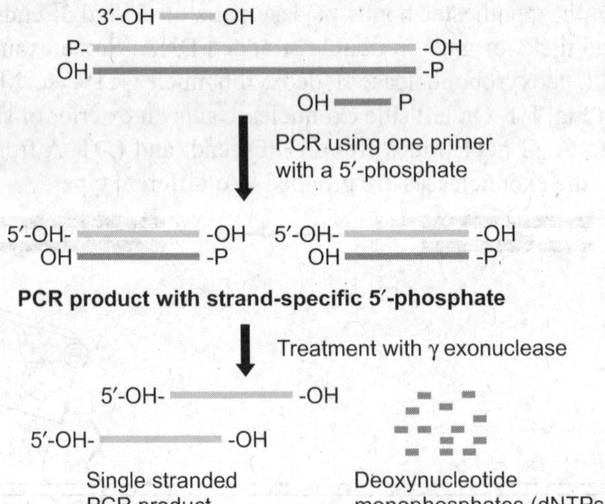

Fig. 6.2. Lambda exonuclease selectively digests the strand of a substrate using PCR primer with a 5-phosphate releasing the deoxynucleotide triphosphates.

2. Exonuclease I

The source of exonuclease I is *E. coli.* In 1964, I.R. Lehman and A.L. Nussbaun studied the specificity of exonuclease I (phosphodiesterase) of the deoxyribonucleases of *E. coli.* It digests single-stranded DNA (ssDNA) in a 3′→5′ direction, but does not digest double-stranded DNA (dsDNA). Although it requires the presence of magnesium and a free 3′-hydroxyl terminus for activity, it is active under a wide variety of buffer conditions and can be added directly into most reaction mixes. Exonuclease I can be heat-inactivated by incubation at 80°C for 15 minutes.

It is used for removal of residual ssDNA including oligos from reaction mixture as well as for removal of ssDNA from nucleic acid mixtures.

One unit of exonuclease I catalyses the release of 10 nano moles of acid-soluble nucleotides from heat-denatured calf thymus DNA in 30 minutes at 37°C under standard assay conditions.

Fig. 6.3 shows digestion of single stranded DNA by exonuclease I. At 1% concentration of agarose gel, exonuclease I completely digested the linear ssDNA oligo (lane 3) leaving the linearised plasmid dsDNA intact (lane 2). Lane 1 has size markers.

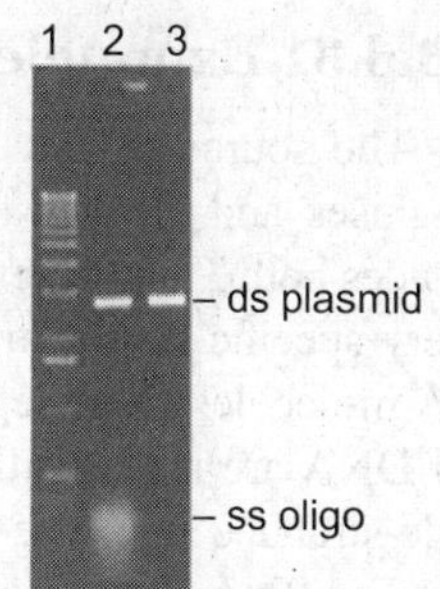

Fig. 6.3. Digestion of single stranded DNA by exonuclease I.

3. Exonuclease III

In 1980, S.G. Rogers and B. Weiss discovered exonuclease III from *E. coli* K-12; hence the source of exonuclease III is *E. coli.* Exonuclease III catalyses the stepwise removal of mononucleotides from 3-OH end of the double stranded DNA. Exonuclease III digests duplex DNA in a 3′→5′ direction from a nick, a blunt or 3′-recessed end, producing single stranded DNA fragments on the opposite strand. DNA degradation by exonuclease III proceeds at a uniform rate under defined reaction conditions, yielding predictable and reproducible digestion results. Consequently an improved substrate is formed which is used in gene manipulation. The preferred substrates are DNAs with blunt ends or protruding ends. It also extends nicks in duplex DNA to create single stranded gaps. It works inefficiently on DNA with 3′ protruding ends, and is active on single stranded DNA. It is commonly used to prepare a set of nested deletions on the termini of linear DNA fragments, *i.e.* used for 3′ end modification of DNA fragments as below :

$$\begin{matrix} 5'\text{-GGAATT}_{OH}\text{-}3' \\ 3'\text{-CCTTAA}_{P}\text{-}5' \end{matrix} \longrightarrow \begin{matrix} 5'\text{-G}_{OH}\text{-}3' \\ 3'\text{-CCTTAA}_{P}\text{-}5' \end{matrix} \quad + \quad N_P\text{-}5'$$

The 3′-protruding termini are resistant to cleavage. The degree of resistance depends on the length of the extension, with extensions four bases or longer being essentially resistant to cleavage. This property is used to produce unidirectional deletions from a linear molecule with one resistant (3′-overhang) and one susceptible (blunt or 5′-overhang) terminus. Exonuclease III is not active on 3′-protruding ends of four bases or more in length, single-stranded DNA or on thioester-linked nucleotides. Therefore, each template must be optimised by using sample digestions to achieve the desired excision rate.

Application of exonuclease III

Exonuclease III is used as below :

- Unidirectional nested deletions.
- Site-directed mutagenesis.
- Production of strand-specific radio-labelled probes.
- Preparation of single-stranded substrates for dideoxysequencing.

Exonuclease III activity depends partially on DNA helical structure and displays sequence-dependence (C >A = T >G). Besides, exonuclease III has also been reported to have intrinsic RNase H, 3′-DNA phosphatase and apurinic DNA endonuclease activities. Temperature, ionic strength, template sequence, and enzyme-to-DNA ratios greatly affect enzyme activity, requiring reaction conditions to be tailored to specific applications. In 1989, J.F. Tomb and G.J. Barcak reported the varying concentrations of NaCl regulate the 3′→5′ activity of exonuclease III.

4. Bal 31 Exonuclease

The source of Bal 31 exonuclease is *Alteromonas espejiana* BAL-31. Bal 31 exonuclease recognises and cleaves the salt-induced conformational junction between B- and Z- DNA and degrades both 3′ and 5′ termini of DNA duplex without generation of internal scissions. This enzyme is very specific single stranded nuclease. It degrades simultaneously both the 3′ and 5′ strands of DNA molecule. Short segments of (dC–dG) left-handed Z-helix comprising approximately 1% of total DNA are specifically detected within two different recombinant plasmids. It also acts as a single-stranded nuclease that cleaves DNA at nicks, gaps and single stranded regions of duplex DNA and RNA. But it does not cleave the internal duplex DNA. This enzyme primarily acts fast as 3′-exonuclease is coupled with a slow acting endonuclease. When Bal 31 exonuclease is present at high concentration, it efficiently shortens DNA molecules by degrading simultaneously both the 3′ and 5′ termini of DNA molecules. Consequently shortened DNA fragments are formed. Therefore, it is used for shortening the DNA fragments. Hence the DNA fragments possess blunt ends at both the termini. Its mechanism of action is as below :

$$\begin{matrix} \text{5′-GGAATT-3′} \\ \text{3′-CCTTAA-5′} \end{matrix} \xrightarrow{\text{Bal 31 exonuclease}} \begin{matrix} \text{5′-GGAA-3′} \\ \text{3′-CCTT-5′} \end{matrix}$$

The Bal 31 enzyme is highly resistant to inactivation by the high concentration of a variety of electrolytes that stabilises left-handed helix, is active at physiological pH, and can be used to probe both linear and circular DNAs. The Bal 31 exonuclease action, at least against duplex substrates, is quasi-processive and removes about 18 and 28 nucleotides per productive enzyme-substrate encounter for the slow and fast-nuclease species, respectively. The Bal 31 exonuclease activities depend on Ca^{2+} and Mg^{2+} concentrations. Additionally, the nuclease cleaves left-handed (dC–dG) only very poorly. Thus Bal 31 exonuclease can be utilised as a probe for helical junctions and consequently for segments of left-handed DNA that might exist within predominantly right-handed naturally occurring genomes.

B. ENDONUCLEASES

The enzymes that cleave the phosphodiester bond within a polynucleotide chain are called 'endonucleases'. There are few classes of endonucleases that cleave only at the specific nucleotide sequences. Moreover, numerous types of DNase and RNase have been isolated and characterised. They differ among other things in substrate specificity, co-factor requirements, and whether the cleave nucleic acids internally (endonucleases), chew in form of the ends (exonucleases) or attack in both of these modes. In many cases, the substrate specificity of a nuclease depends upon the concentration of enzymes used in the reaction, with high concentrations promoting less specific cleavages.

The most widely used nucleases are DNase I and RNase A, both of which are purified from bovine pancreas. They act upon genetic material and cleave the double stranded DNA at any point except the ends, but their action involves only one strand of the duplex.

Cleavage site

5′ C–C–G | C–G–G 3′
3′ G–G–C | G–C–C 5′

Cleavage site   Symmetry of axis

Action of endonuclease

1. S1 Nuclease

The source of S1 nuclease is *Aspergillus oryzae*. In 1973, V.M. Vogt purified and studied the properties of single-strand-specific nuclease from *A. oryzae*. Low concentration of S1 nuclease digests single-stranded DNAs or RNAs, while double stranded nucleic acids (DNA : DNA, DNA: RNA, and RNA : RNA) are degraded by high concentration of enzymes. Moreover, moderate concentrations can be used to digest the double stranded DNA at nicks or small gaps.

The single-stranded DNA or single strand protrusion of double-stranded DNA with cohesive ends is degraded by S1 nuclease. As a result of action of S1 nuclease cohesive ends are converted into blunt ends. Therefore, S1 nuclease is used to remove the incompatible ends so that overlapping ends may be developed for annealing of two fragments of DNA molecules. Commonly it is used to analyse the source of DNA : RNA hybrids (S1 nuclease mapping), and to remove single stranded extensions (overhangs) of DNA fragments to produce blunt ends so that overlapping ends may be developed for annealing of two fragments of DNA molecules. The other uses are : S1 transcript mapping and creation of unidirectional deletions in DNA fragments in conjunction with exonuclease III. In molecular biology, it is used in removing single stranded tails from DNA molecules to create blunt ended molecules and opening hairpin loops generated during synthesis of double stranded cDNA. The mode of action of S1 nuclease is given below :

5′-AG-3′   S1 nuclease   5′-AG-3′
3′-TCTTAA-5′   ⟶   3′-TC-5′ + 3′-TTAA-5′

S1 nuclease is five times more active on DNA than on RNA. The S1 nuclease exhibits 3′-phosphomonoesterase activity.

2. Mungbean Nuclease

Mung bean nuclease is a single strand-specific nuclease derived from sprouted mung beans (*Vigna radiata*). Mungbean nuclease degrades single-stranded DNA or RNA and produces mono- and oligo- nucleotides carrying a 5′- terminus, but it does not digest double-stranded DNA, double-stranded RNA, or DNA / RNA hybrids. It has higher specificity for single-stranded DNA and RNA than S1 nuclease. It is the enzyme of choice for most applications requiring a single strand-specific nuclease. Unlike S1 nuclease, mungbean nuclease will not cleave the intact strand of nicked duplex DNA.

Mungbean nuclease is used for : (*i*) removal of hairpin loops during cDNA synthesis, (*ii*) high-resolution mapping of the termini and exon structures of RNA transcripts (commonly termed Berk-Sharp or S1 Mapping) using either internal-labelled or end-labelled probes, (*iii*) restriction site-modification or removal by digestion of single-stranded protruding ends, (*iv*) cleavage of single-base pair mismatches, (*v*) unidirectional deletion of large DNA (in combination with exonuclease III) to generate ordered deletions for sequencing, (*vi*) removal of 3′ and 5′ extensions from DNA or RNA termini, (*vii*) transcriptional mapping, (*viii*) cleavage of hairpin loops, (*ix*) and excision of gene coding sequences from genomic DNA.

Mungbean nuclease is tested in several functional assays including precise deletion of single-stranded DNA from cleaved restriction sites and non-random cleavage of partially denatured duplex DNA. The enzyme is also tested to ensure that the level of double-stranded nuclease activity is less that 0.05% of the single stranded nuclease activity and is free from phosphatase activity.

3. Deoxyribonuclease I (DNase I)

It is one of the endonucleases that cleaves the double stranded or single stranded DNA. Preferentially its cleavage occurs adjacent to pyrimidine (C or T) residues; therefore, this enzyme is an endonuclease. The major products are 5′-phosphorylated di, tri and tetranucleotides as shown in Fig. 6.4.

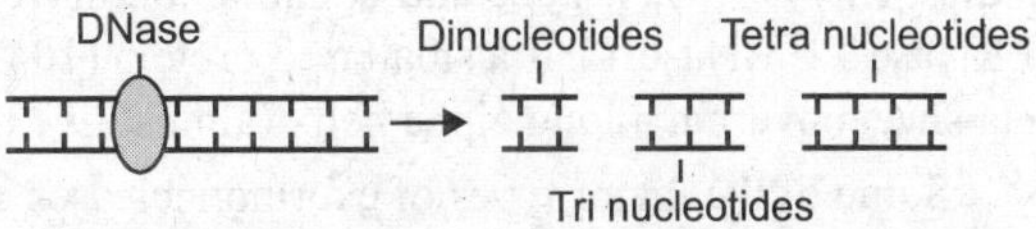

Fig. 6.4. Mode of action of DNase I and generation of 5-phosphorylated di-, tri- and tetra- nucleotides.

DNase I hydrolyzes each strand of duplex DNA independently in the presence of magnesium ions generating the random cleavages. DNase I cleaves both strands of DNA at about the same site in the presence of Mg^{++} ions, resulting in production of blunt ends or fragments with 1-2 base overhangs. DNase I does not cleave RNA, but crude preparations of the enzyme are contaminated with RNase A; RNase-free DNase I is readily available.

Some of the common applications of DNase I are : (*i*) eliminating DNA (e.g. plasmid) from preparations of RNA, (*ii*) analysing DNA-protein interactions via DNase footprinting, and (*iii*) nicking DNA prior to radiolabelling by nick translation.

4. Ribonuclease (RNase)

The nuclease enzyme that catalyses the degradation of RNA into smaller components is called ribonuclease. Basically ribonucleases can be divided into two major groups, endoribonucleases and exoribonucleases. Some major endoribonucleases are described below :

(*a*) Ribonuclease A (RNase A): It cleaves single-stranded RNA at the 3′ end of pyrimidine residues. It degrades the RNA into 3′-phosphorylated mononucleotides and oligonucleotides. Some of the major use of RNase A are : (*i*) eliminating or reducing RNA contamination in preparations of plasmid DNA, and (*ii*) mapping mutations in DNA or RNA by mismatch cleavage. RNAse will cleave the RNA in RNA : DNA hybrids at sites of single base mismatches, and the cleavage products can be analysed.

(*b*) Ribonuclease H (RNase H) : The ribonuclease activity of RNase H cleaves the 3′-O-P bond of RNA in a DNA/RNA duplex resulting in production of 3′-OH and 5′-PO_4-terminated products. In DNA replication, RNase H removes the RNA primer allowing the completion of the newly synthesised DNA. RNase H is a non-specific enzyme that catalyses the cleavage of RNA via a hydrolytic mechanism. Members of the RNase H family can be found in nearly all organisms from archaea to bacteria and eukaryota. RNase H specifically degrades the RNA in RNA-DNA hybrids and will not degrade DNA or unhybridised RNA. Hence, it is commonly used to destroy the RNA template after synthesis of first-strand of complementary DNA (cDNA) by reverse transcription, as well as nuclease protection assays procedures.

(*c*) Ribonuclease P (RNase P) : RNase P is unique from other RNases in that it is a ribozyme (a ribonucleic acid that acts as a catalyst in the same way as an enzyme). It cleaves off an extra or precursor sequence of RNA on tRNA molecules. A form of RNase P has recently been discovered, that is a protein and does not contain RNA. RNase P is one of two known multiple turnover ribozymes in nature; Sidney Altman and Thomas Cech won the Nobel Prize in Chemistry in 1989 for the discovery of ribozymes. In 2006, R. Reiner and co-workers recently demonstrated that RNase P has a new function. It has been shown that human nuclear RNase P is required for the normal and efficient transcription of various small non-coding RNA genes such as tRNA, 5S rRNA, SRP RNA and U6 snRNA genes, which are transcribed by RNA polymerase III.

(*d*) Ribonuclease T1 (RNase T1) : It is a fungal endonuclease that is found in a mold *Aspergillus oryzae*. It cleaves single stranded RNA after guanine residues on their 3′ end. RNase T1 is often used to digest denatured RNA prior to sequencing due to its specificity for guanine. Similar to other ribonucleases such as barnase and RNase A, ribonuclease T1 has been popular for folding studies. In 1991, C.N. Pace and collaborators have described the structure, function, and stability of RNase T1. RNase T1 is a small $\alpha+\beta$ protein (104 amino acids) with a four-stranded, anti-parallel beta sheet covering a long alpha helix (almost five turns).

Some of the major types of exoribonucleases are described below :

(*a*) Ribonuclease D (RNase D) : RNase D is an exoribonuclease which is involved in structured RNA processing. It is one of the seven exoribonucleases identified in *E. coli*. It is a 3′→5′ exoribonuclease which is involved in the 3′ processing of various stable RNA molecules. It adds the 3′ CCA sequence to tRNA during processing of prokaryotic tRNA. In 2005, Yuhong Zuo and co-workers (from the University of Miami, Florida, U.S.A.) have described the crystal structure of *E. coli* RNase D.

***(b)* Ribonuclease R (RNase R) :** In 2005, Zhuan-Fen Cheng and Murray P. Deutscher from the University of Miami (Florida) have described an important role for RNase R in mRNA decay. It is a 3′→5′ exoribonuclease closely related to RNase II, which degrades mRNA in bacteria such as *E. coli*. RNase R has homologues in many other organisms. When a part of another larger protein has a domain similar to RNase R, it is called an RNase R domain.

C. RESTRICTION ENDONUCLEASES

The restriction enzymes are called as *'molecular scissors'*. These act as foundation of recombinant DNA technology. Basically, these are DNA-cutting enzymes found in bacteria. They are often called restriction endonuclease because they cut within the specified sites of DNA molecules. They cleave sugar-phosphate backbone of DNA. In most practical settings, a given enzyme cuts both strands of duplex DNA within a stretch of just a few bases. These enzymes are present in bacteria and provide a type of defense mechanism called the '*restriction-modification system*'. Since the 1950s, it has been known that bacteriophage λ that grows well and efficiently infects one strain of bacteria, *i.e. Escherichia coli* is often unable to grow well and infects the other strain of *E. co*li. In addition phage particles that succeed in infecting the second strain often show the opposite pattern; they are able to efficiently infect the second strain while growing poorly in the original strain. A series of studies showed that phage particles that efficiently grows and infect host cells have DNA molecules that have been chemically modified by the addition of methyl groups to some of their adenine and/or cytosine bases, while the DNA of poorly infecting phage particles does not show this pattern of chemical modification or methylation. Phage particles with unmethylated DNA do not grow and infect efficiently because their DNA molecules are cleaved and degraded by enzymes of host cell, while methylated DNA is protected from this degradation. This phenomenon of degrading unmethylated DNA destroys the growth ability of phage and is responsible for the pattern of growth restriction as described above.

In 1965, Stewart Linn and Werner Arber isolated the two types of enzymes responsible for phage growth restriction in *E. coli*. One of these enzymes methylated DNA, while the other cleaved unmethylated DNA at a wide variety of locations along the length of the molecules. The first type of enzymes was called 'methylase', while the second was called 'restriction nuclease'. These enzymatic tools are important to scientists who were gathering the tools needed to 'cut and paste' the DNA molecules. What was needed now was a tool that would cut DNA at specific sites, rather than at random sites along the length of the molecule, so that scientists could cut DNA molecules in a predictable and reproducible way. However, there was no understanding about cutting and joining of DNA molecules using DNA enzymes before 1970.

Host-control restriction modification system provides defense to *E.coli*. Bacterium protects itself through : (*i*) *restriction mechanism* (*i.e.* identifies the introduced foreign DNA and cuts into pieces called restriction endonucleases, and (*ii*) *modification system* (*i.e.* methylation of certain bases of its own DNA by methylase so that phage λ-expressed restriction endonuclease could not cleave its DNA). Several thousand different restriction endonucleases have been isolated, which collectively exhibit a few hundred different sequence (substrate) specificities.

Werner Arber, a Swiss microbiologist and geneticist, shared the 1978 Nobel Prize in Physiology or Medicine along with American researchers Hamilton Smith and Daniel Nathans for the discovery of restriction endonucleases that recognise the specific sequences of nucleotides in a DNA molecule and cleave the molecule at that particular point. Their work led to the development of recombinant DNA technology.

Werner Arber, D. Nathans and Hamilton Smith (left to right)

I. RESTRICTION-MODIFICATION SYSTEMS

A large majority of restriction enzymes have been isolated from bacteria, where they appear to serve a host-defense role. The idea is that foreign DNA (for example from an infecting virus) will be chopped up and inactivated (restricted) within the bacterium by the restriction enzymes. The presence of restriction enzymes immediately begs the question why they do not chew up the genomic DNA of their host. Almost in all cases, a bacterium that makes a particular restriction endonuclease also synthesises a companion DNA methyltransferase which methylates the DNA target sequence for that restriction enzyme, thereby protecting it from cleavage. This combination of restriction endonuclease and methylase is referred to as *restriction- modification system.*

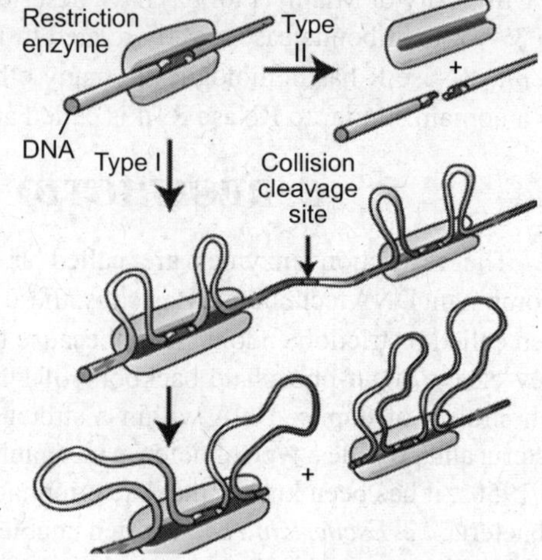

Fig. 6.5. Action of the type I and type II restriction enzymes on DNA (*Source*: David T. F. Dryden. *Nature Struct. Mol. Biol.* 11: 804-806, 2004).

In 2004, David T. F. Dryden has described that the restriction occurs when a target specificity sequence (*see* dark segment on the DNA as a tube) lacking appropriate methylation (*see* yellow spots) on both DNA strands is recognised by the enzyme (Fig. 6.5).

A type II restriction enzyme which is a dimeric protein cleaves the DNA at a specific site within or near the target-specific sequence and then dissociates from the DNA generating two free ends. Unmethylated targets cause the type I restriction enzyme to hold onto the target and trigger DNA translocation (red arrows show direction) by the motor subunits. Multiple contacts with the DNA are made that cause extrusion of expanding loops are extruded from the enzyme as it reels in the DNA. Cleavage at the collision site is triggered by stalling of translocation by collision with another type I restriction enzyme.

Restriction–Modification Systems

Restriction-modification systems enable bacteria to destroy foreign DNA without destroying their own DNA. These systems consist of two enzymic activities : (*i*) a modification methyltransferase that recognises a specific DNA sequence and its recognition site, and a methylates particular bases in that sequence, one in each strand, and (*ii*) a restriction endonuclease that recognises the same sequence and the methyltransferase and proceeds to cleave the DNA if the site is not methylated. The foreign DNA entering the cell will seldom be methylated at the relevant sequences and is usually cleaved. Only if the modification enzyme methylates every site before the restriction enzyme acts at any one site will that DNA survive in the cell. Own DNA of bacteria will be methylated in both strands at every recognition site. After semi-conservative replication, these sites will be methylated in the template strand only, but methylation of one strand is sufficient to block restriction. The modification of enzyme then methylates the newly synthesised strand before the next round of replication.

The enzymes that mediates restriction-modification fall into three types, type I, II and III. In type I and type III systems, both restriction and modification are carried out by a multifunctional enzyme. The enzymes are oligomeric proteins composed of two or more different subunits that are responsible for the restriction and modification activities and/or the recognition of the target sequence. While the type I and III enzymes recognise specific DNA sequences, their restriction subunits do not cleave DNA at the recognition site, but at various distances along the DNA from the site.

After cleavage of DNA, type I restriction enzymes remain attached to the DNA and do not turn over in the restriction reaction. The type I restriction enzyme comprises of a DNA sequence specificity (S) subunit, two methyltransferase (M) subunits and two restriction (R) subunits. The M_2S core is responsible for : (*i*) maintaining methylation of the host chromosome by modifying the hemimethylated specificity sequences produced by DNA replication, and (*ii*) triggering the restriction reaction if recognises unmethylated DNA specificity sequence on both strands and hence from a foreign source. This restriction reaction comprises of ATPase-dependent DNA translocation by the motors, which are the members of the SF2 family of helicases, and DNA cleavage. The restriction reaction is carried out by the R subunits, each of which contains a motor and a DNA cleavage domain.

Type II systems employ separate enzymes for restriction and modification. The modification enzymes are generally monomeric proteins. In each turnover, they methylate the recognition sequence in one strand, and two turnovers are needed to methylate both strands of an unmethylated site. Their primary reaction is the restriction of full methylation following DNA replication. The restriction enzymes are often dimmers of identical subunits that cut both DNA strands in one turnover, although there are several exceptions. In contrast with the type I and type III enzymes, the type II restriction enzymes cut the DNA at specified phosphodiester bonds in or adjacent to their recognition sites. With one exception, Mg^{++} is required for DNA cleavage. Methylated adenine (A^*) protects the host DNA from cleavage by restriction enzymes (Fig. 6.6).

Cleaved	Not cleaved
5′ — GAATTC — 3′	5′ — GA*ATTC — 3′
3′ — CTTAAG — 5′	3′ — CTTAA*G — 5′

A^* = [structure of methylated adenine: H–N–CH_3 (Added methyl group) on the purine ring; N, N, N, N; H; H; deoxyribose]

Fig. 6.6. Protection of host DNA by methylation; A^*- methylated adenine.

The restriction endonuclease *Eco*RV (from *E. coli*) cleaves double stranded viral DNA molecules that contain the sequence 5′-GATATC-3′ but leaves intact host DNA containing hundreds of such sequences. The host DNA is protected by other enzymes called methylase, which methylates adenine bases within the host recognition sequences. For each restriction endonuclease, the host cell produces a corresponding methylase that marks the host DNA and prevents its degradation. These pairs of enzymes are referred to as restriction-modification systems.

Part of the restriction-modification system is found in many bacteria. These enzymes restrict the ability of foreign DNA (such as bacteriophage DNA) to infect/invade the host bacterial cell by cutting it. The host DNA is modified by methylation of the sequences. Methyl groups (CH4) are added to C or A nucleotides in order to protect the bacterial host DNA from degradation by its own enzymes (Fig. 6.7).

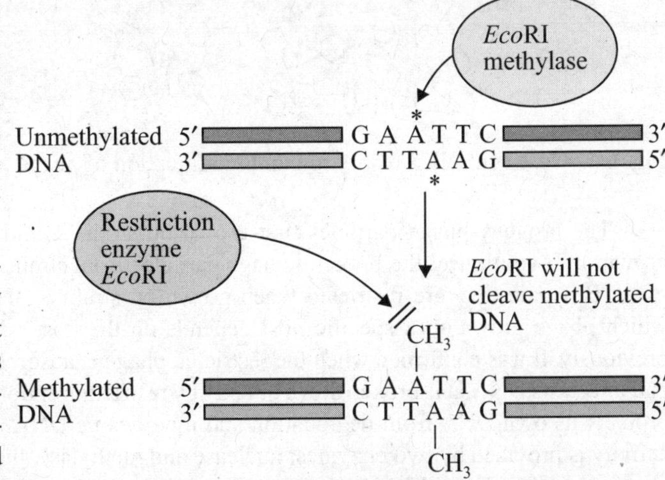

Fig. 6.7. Addition of methyl group (CH_4) to adenine (A) to protect the bacterial host DNA from degradation.

Fig. 6.8 shows the hosts restriction-modification system. If bacteriophage λ propagated on *E. coli* strain K (A) is allowed to infect *E. coli* strain B, the efficiency of infection is very low (B). The progeny phage obtained from plaques of B strain can reinfect *E. coli* strain B with high efficiency (C). But infection efficiency of these progeny phage on strain K is also very low (D). There are three gene loci (*hsd*S, *hsd*M and *hsd*R) that control this system; the *hsd*S encodes a protein which governs the specificity of these systems. The *hsd*M gene product acts as modification enzyme (methylase) which interacts with *hsd*R gene product (restriction enzyme) in the cleavage process.

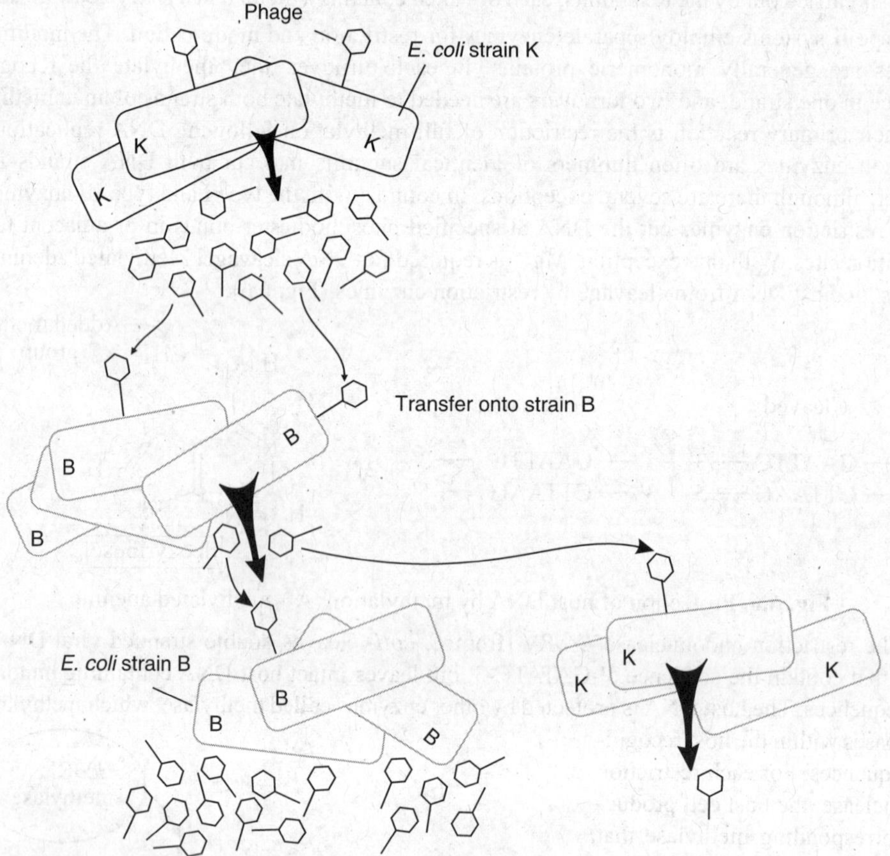

Fig. 6.8. Host-controlled restriction and modification of phage λ in *E. coli* strain K and strain B.

The progeny phage particles first grown on strain K and then strain B are restricted when grown on B strain. But the progeny phage particles from strain B are not restricted when re-infect strain B. Again they are restricted when grown on strain K. This shows that the efficiency with which phage grows on a specific host depends on the host's strain on which it was propagated previously. It was confirmed when the restricted phage is adsorbed on the surface of restrictive host and injects its DNA, it is broken down quickly by restriction endonuclease. Hence, the restrictive host protects its own DNA from degradation and modifies its DNA enzymatically. Thus the restriction activity is provided by two enzymes: nuclease and methylase, the genes of which reside on the host chromosome or plasmid DNA.

II. TYPES OF RESTRICTION ENDONUCLEASES

The important development came in 1968 when H.O. Smith, K.W. Wilcox and T.J. Kelley working at Johns Hopkins University isolated and characterized the first restriction endonuclease whose functioning depended on a specific DNA nucleotide sequence. Working with bacterium *Hemophilus influenza*, this group isolated an enzyme called *Hind*II that always cuts DNA molecules at a particular point within the specific sequence of six base pairs. This sequence is :

5′GT (pyrimidine T or C) (purine A or G) AC 3′

3′CA (purine A or G) (pyrimidine T or C) TG 5′

The *Hind*II enzyme always cuts directly in the centre of this sequence. Whenever this particular sequence of six base pairs occurs unmodified in a DNA molecule, *Hind*II will cleave both DNA backbones between the 3rd and 4th base pairs of the sequence. Moreover, *Hind*II will only cleave a DNA molecule at this particular site. For this reason, this specific base sequence is known as the 'recognition sequence' for *Hind*II.

There are three different types of restriction endonucleases : Type I, Type II and Type III. Each enzyme differs slightly by different mode of action. Type I and Type II are large multi subunit complex which contains both endonuclease and methylase activity. The different bacteria consist of different restriction endonucleases, and in correspondence to these the methylases too are produced. These enzymes occur naturally in bacteria as a chemical weapon against the invading viruses and cut both strands of DNA when certain foreign nucleotides are introduced in the cell. These enzymes cleave a DNA to generate a nick with a 5′ phosphoryl and 3′ hydroxyl termini.

1. Type I restriction Enzyme

The B-K restriction and modification enzyme systems of *E. coli* have been termed as type I endonucleases such as *Eco*B and *Eco*K. They are complex nucleases which function simultaneously as an endonuclease and a methylase. Type I recognizes specific sequences then moves along DNA (~1000 – 5000 bases) before cutting one of the double strands and releasing a number of nucleotides (~75) where the cut is made. A second molecule is required to cut the second strand of the DNA such as *Eco*K. They require Mg^{++}, S-adenosyl methionine and ATP as co-factor. They are single multifunctional enzymes consisting of three different sub-units : (*i*) restrictive sub-unit, (*ii*) modification sub-unit, and (*iii*) specificity sub-unit. The third sub-unit specifies the recognition sitcs. Thc cnzymes show specificity for recognition site but not for cleavage site, *i.e.* they recognise specific sites within the DNA but do not cleave at these sites. Hence, heterogeneous population of DNA fragments is produced.

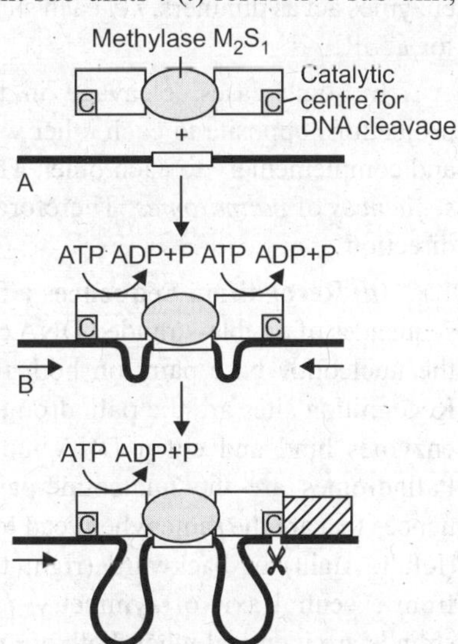

Fig. 6.9. DNA cleavage by type I restriction enzyme.

Recognition site of type I enzyme is 15 bp long and cleavage site is about 1000 bp away from the 5′-end of the recognition sequence TCA which is present within 15 bp long recognition site. The adenine position in the restriction site may be methylated. Such DNA molecules are resistant to type I endonuclease whose recognition sequence is methylated at A residues on both the strands. However, when the recognition sequence is unmethylated, only then DNA is cleaved as below :

*Eco*K 5′- A A A* C N N N N G T C G C -3′

3′- T T T G N′ N′ N′ N′ C A* G C G -5′

Here A* shows methylation at adenine position, N is unspecified bases and N is complementary bases to N. The unmodified DNA molecules are restricted, *i.e.* cleaved by restriction endonucleases.

DNA is shown as an open box representing a recognition site for type I restriction enzyme (Fig. 6.9). This multimeric enzyme complex is composed of the methylase M_2S_1 (grey oval) which mediates binding to the recognition site, and two HsdR subunits (open box). The catalytic centre for DNA cleavage in HsdR is shown as a grey box marked by C. The enzyme binds to its unmethylated recognition site and subsequently translocates non-specific DNA from both directions (see arrows) past the HsdR subunits, irrespective of DNA topology (A).

ATP is hydrolysed to generate energy for this process. When one of the subunits encounters a barrier that can halt the translocation process (hatched box), it introduces a double-stranded break in the DNA substrates at the site of collision indicated by scissors, while the other HsdR subunit can continue in translocation (B). Collision of HsdR subunits of two convergent tracking enzymes would result in introduction of two double-strand breaks in close proximity to each other. The model can also be modified for unidirectional translocation.

2. Type II restriction Enzymes

These restriction enzymes were first isolated by Hamilton Smith. These enzymes are simple and consist of single polypeptide chain and require no ATP for degradation of DNA. They have separate activities for cleaving (endonuclease) and methylation (methylase) of DNA molecules. They are most stable enzymes that require Mg^{++} as co-factor. Daniel Nathans first used these enzymes for mapping and analysing genes and genomes. The first (I) type II restriction enzyme isolated and characterised was from *Escherichia coli* strain RY. Hence, it was named as *Eco*RI. For the discovery of restriction enzymes, W. Arber, H. Smith and D. Nathans were awarded the Nobel Prize in 1978.

dyad axis

G A A | T T C
C T T | C T T

of symmetry

Only type II enzyme is used in gene manipulation. Type I enzyme recognises a specific sequence of DNA molecule but cuts elsewhere, whereas type II make cuts only within the restriction sites and produce two single strand breaks, one break in each strand. Type II enzymes are the most important ones; as a result of their action the broken nucleotides form DNA duplex that exhibit two-fold symmetry around a given point. The dyad symmetry of restriction sites suggests that restriction enzymes act as dimmers, *i.e.* each subunit interacts with one half of the site. This has been confirmed for *Eco*RI.

In some cases, cleavage in two strands are staggerd to produce single stranded short projections opposite to each other with blunt or mutually cohesive sticky ends which are identical and complementary to each other. These complementary sequences are also known as *palindromic sequences* of *palindromes*. Therefore, both strands have the same sequence when read from 5′→3′ direction.

(*a*) Recognition Sequences : The substrates for restriction enzymes are more-or-less specific sequences of double-stranded DNA called 'recognition sequences'. They are specific in recognising the nucleotide base pairs on both the strands and cut exactly within the same recognition site. Recognition sites are the palindromes or palimdromic sequences where most of type II restriction enzymes bind and cut a DNA molecule. Palindromes are the nucleotide-pair sequences that are the same when read forward (left to right) or backward (right to left) from a central axis of symmetry, *i.e.* two strands are identical when both are read in the same polarity, *i.e.* in 5′→3′ direction.

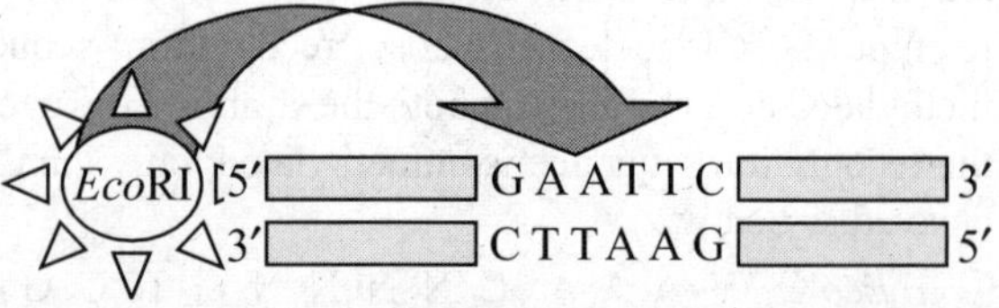

*Eco*RI recognises between G and A of a DNA sequence on both strands.

Recognition sequences of many type enzymes are pallindromic sequences, *i.e.* they exhibit rotational symmetry. The nucleotide sequences are the same when read from 5′→3′ in both the strands. For example, nucleotides of recognition sequences of *Eco*RI are between G and A on both the strands as given in Fig. 6.10.

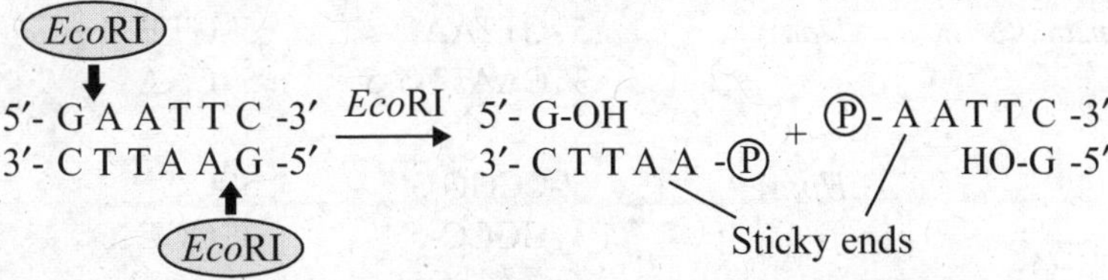

Fig. 6.10. Recognition of restriction site (specific short nucleotide sequence) by *Eco*RI endonuclease and cleaving of both the strands between G and A generating the sticky ends.

In addition, a very useful feature of many restriction nucleases is that they make staggered cuts. For example, *Eco*RI enzyme binds to a region of having specific palindromic sequence.

***(i)* Variation in length of recognition sites of restriction enzymes:** The length of recognition sites of different enzymes varies which can be four, five, six, eight or more nucleotide pairs (Table 6.2). There are many enzymes that recognise a hexanucleotide (a sequence of 6 nucleotides), whereas the others recognise tetra-, penta or octanucleotide sequences present on DNA molecule. The enzymes *Eco*RI, *Sac*I and *Sst*I each recognize a 6 base-pair (bp) sequence of DNA, whereas *Not*I recognises a sequence 8 bp in length, and the recognition site for *Sau*3AI is only 4 bp in length. *Eco*RI restriction enzyme from *E. coli* cleaves DNA at the hexanucleotide 5′GAATTC3′, whereas *Eco*RII isolated from the same bacterium recognises γ-PO_4 end of pentanucleotide 5′CCTGG3′.

***(ii)* Length of the recognition sequence dictates how frequently the enzyme will cut in a random sequence of DNA :** Enzymes with a 6 bp recognition site will cut generally at every 4^6 or 4096 bp; a 4 bp recognition site will occur roughly every 256 bp.

Table 6.2. Sources of type II restriction enzymes, their cleavage sites and cleavage products.

Microorganisms	*Restriction enzymes*	*Restriction sites*	*Cleavage products*	
Bacillus amy-loliquefaciens H	*Bam*HI	↓ 5′-GGATCC-3′ 3′-CCTAGG-5′ ↑	5′-G 3′-CCTAG	GATCC-3′ G-5′
B. globigii	*Bgl*II 3-TCTAGA-5	↓ 5′-AGATCT-3′ 3′-TCTAGA-5′ ↑	5′-A 3′-TCTAG	GATCT-3′ A-5′
Escherchia coli RY13	*Eco*RI 3-CTTAAG-5	↓ 5′-GAATTC-3′ 3′-CTTAAG-5′ ↑	5′-G 3′-CTTAA	AATTC-3′ G-5′
E. coli $R_2$45	*Eco*RII	↓ 5′-GGATCC-3′ 3′-CCTAGG-5′ ↑	5′-G 3′-CCTAG	GATCC-3′ G-5′
	*Hind*II	↓ 5′-GTPyPuAC-3′ 3′-CAPuPyTG-5′ ↑	5′-GTPy 3′-CAPu	PuAC-3′ PyTG-5′

Haemophilus influenzae Rd	*Hin*dIII	↓ 5′-AAGCTT-3′ 3′-TTCGAA-5′	5′-A AGCTT-3′ 3′-TTCGA A-5′
H. parainfluenzae	*Hpa*I	↓ ↑ 5′-GTTAAC-3′ 3′-CAATTG-5′	5′-GTT AAC-3′ 5′-CAA TTG-5′
	*Hpa*II	↓ ↑ 5′-CCGG-3′ 3′-GGCC-5′	5′-C 5′-CGG-3′ 3′-GGC C-5′
Klebsiella pneumoniae OK 8	*Kpn*I	↑ ↓ 5′-GGTACC-5′ 3′-CCATGG-3′	5′-GGTAC C-3′ 3′-C CATGG-5′
Nocardia otitidis-caviarus	*Not*I	↑↓ 5′-GCGGCCGC-3′ 3′-CGCCGGCG-5′	5′-GC GGCCGC-3′ 3′-CGCCGG CG-5′
Providencia stuartii	*Pst*I	↓ ↑ 5′-CTGCAG-3′ 3′-GACGTC-5′	5′-CTGCA G-3′ 3′-G ACGTC-5′
Streptomyces albus G	*Sal*I	↑ ↓ 5′-GTCGAC-3′ 3′-CAGCTG-5′	5′-G TCGAC-3′ 3′-CAGCT G-5′
Staphylococcus aureus 3AI	*Sau*3AI	↓ ↑ 5′-GATC-3′ 3′-CTAG-5′ ↑	5′- GATC-3′ 3′-CTAG -5′

Arrows indicate the recognition sites.

(*iii*) Presence of the same recognition site in different restriction enzymes : Pairs of restriction enzymes specific to the same recognition sequence are called 'isoschizomers', which are isolated from different strains of bacteria at varying conditions. The recognition sites for *Sac*I and *Sst*I are identical. They are isolated from two different sources but possess similar recognition and cleavage sites. Similarly, both *Hin*dIII and *Hsu*I possess 5′AAGCTT3′ recognition sequence. In contrast, some restriction enzymes in spite of having identical recognition sites, differ in cleavage sites. In some cases isoschizomers cut identically within their recognition sites, but sometimes they do not. Isoschizomers often have different optimum reaction conditions and stabilities; for example *Xma*I and *Sma*I have identical recognition site but cut different cleavage sites.

(*iv*) Restriction recognitions sites can be unambiguous or ambiguous : The enzyme *Bam*HI recognises the sequence GGATCC and no others; this is unambiguous property. In contrast, *Hin*fI recognises a 5 bp sequence starting with GA, ending in TC, and having any base between; thus *Hin*fI has an ambiguous recognition site. *Xho*II also has an ambiguous recognition site; thus *Xho*II will recognise and cut sequences of AGATCT, AGATCC, GGATCT and GGATCC.

(*v*) Recognition site for one enzyme may contain the restriction site for another : The *Bam*HI recognition site consists of the recognition site for *Sau*3AI. Therefore, all *Bam*HI sites will cut with *Sau*3AI. Similarly, one of the four possible *Xho*II sites will also be a recognition site for *Bam*HI and all four will cut with *Sau*3AI.

Most of the recognition sequences are palindromes, *i.e.* they read the same forward (5′→3′ on the top strand) and backward (5′→3′ on the bottom strand) sequences. Most recognition sites are palindromes for commonly used restriction enzymes. Often the restriction enzymes bind to their recognition site as dimers (pairs), for example the enzyme *Pvu*II.

(*vi*) Patterns of DNA cutting by restriction enzymes : Restriction enzymes cleave the DNA backbone between deoxyribose and phosphate groups resulting in production of a $-PO_4$ group on the 5′ ends and a –OH on the 3′ ends of both strands. There are some restriction enzymes that cut single stranded DNA usually at low efficiency. The restriction enzymes mostly used in laboratories cut within their recognition sites and generate one of three different types of ends.

- **Overhangs and sticky ends :** An overhang is a stretch of unpaired nucleotides in the end of a DNA molecule. Various overhangs create non-blunt ends. The unpaired nucleotides can be in one strand and create either 3′ or 5′ overhangs. In most cases these overhangs are palindromic in nature.

  Longer overhangs are called 'cohesive ends' or 'sticky ends'. Mostly they are produced upon cutting DNA by restriction endonucleases. Often they cut the two DNA strands four base pairs from each other, creating a four-base 5′ overhang in one molecule and a complementary 5′ overhang in the other. Since they are easily joined back together by a ligase, these ends are called 'cohesive'. Usually different restriction enzymes create different overhangs. Since the overhangs have to be complementary single stranded DNA molecules, they can join in one orientation. The following two 'sticky' ends are compatible to each other :

  ```
  5′-GCT          +      GATGTATG-3′
  3′-CGACTAC                 ATAC-5′
  ```

  Production of a single nucleotide is the simplest example of an overhang. Mostly this is adenosine and is created as a 3′ overhang by some DNA polymerases. Commonly this is used in cloning of PCR products created by such an enzyme. The product is joined with a linear DNA molecule with 3′ thymine overhangs. Since adenine and thymine form a base pair, this helps a ligase to join the two molecules and yield a circular molecule. An example of an A-overhang is given below :

  ```
  5′-ATCTGCTA-3′
  3′-TAGACGA-5′
  ```

  The 5′ or 3′ overhangs generated by enzymes that cut asymmetrically are called 'sticky ends' or 'cohesive ends', because they will readily stick or anneal with their partner by base pairing.

  ◆ **5′ overhangs :** The enzyme cuts asymmetrically within the recognition site in such a way that a short single-stranded segment may be produced extending from the 5′ ends. This is the way of cutting by *Bam*HI. The recognition site is shown below in dark colour and the restriction sites are marked by triangles.

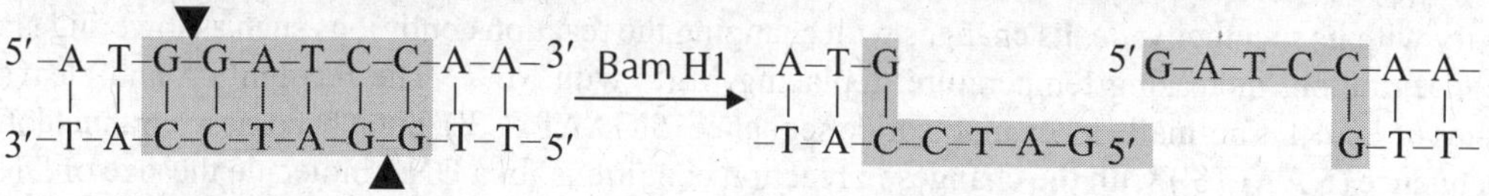

```
      ▼
5′–A–T–G–G–A–T–C–C–A–A–3′  Bam H1   –A–T–G              5′ G–A–T–C–C–A–A–
   | | | | | | | | | |     ------>   | | |                          | | |
3′–T–A–C–C–T–A–G–G–T–T–5′           –T–A–C–C–T–A–G 5′             G–T–T–
                  ▲
```

  ◆ **3' overhangs :** *Kpn*I cleaves within the recognition site asymmetrical and results in production of a single-stranded overhang from the two 3' ends.

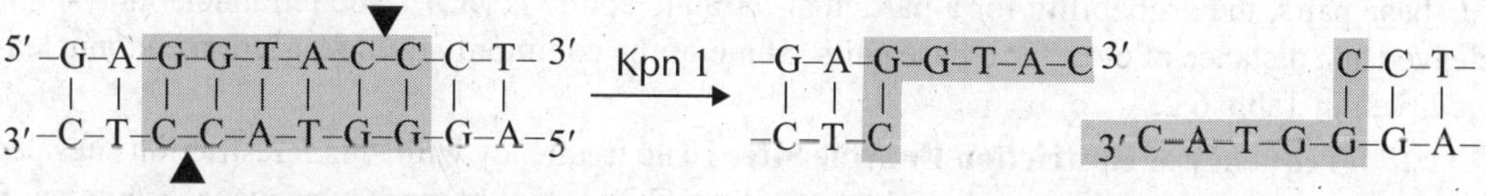

```
                  ▼
5′–G–A–G–G–T–A–C–C–C–T– 3′  Kpn 1   –G–A–G–G–T–A–C 3′               C–C–T–
   | | | | | | | | | |     ------>   | | |                          | | |
3′–C–T–C–C–A–T–G–G–G–A–5′           –C–T–C              3′ C–A–T–G–G–G–A–
      ▲
```

- **Blunts :** There are some other type II restriction enzymes that cleave both strands of DNA at the same base pairs but in the centre of recognition sequence, and results in DNA fragments with blunt ends or flush ends. Blunt end is the simplest DNA end of a double stranded molecule. For example *Hae*III (isolated from *Haemophilus aegypticus*), four nucleotide long palindromic sequence and cuts symmetrically the both DNA strands and form blunt ends. Enzymes that cut at precisely opposite sites in the two strands of DNA generate blunt ends without overhangs. *Sma*I is an example of an enzyme that generates blunt ends as given below :

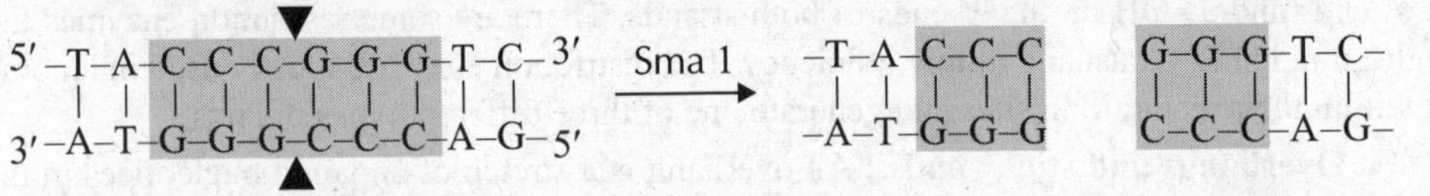

```
          ▼
5′-T-A-C-C-C-G-G-G-T-C-3′   Sma 1    -T-A-C-C-C      G-G-G-T-C-
   | |  | | | | | |  | |    ────→     | | | | |      | | | | |
3′-A-T-G-G-G-C-C-C-A-G-5′            -A-T-G-G-G      C-C-C-A-G-
                ▲
```

Both strands of DNA terminate in a base pair in a blunt-ended molecule. When a DNA ligase is used to join two molecules into one, blunt ends are not always desired due to low yield of recombinants with blunt ends. It also has the disadvantage of potentially inserting the insert DNA in the opposite orientation desired while performing sub-cloning. On the other hand, blunt ends are always compatible with each other.

- **Frayed ends :** Generally it has been found that adenine pair with thymine, and cytosine pair with guanine to form a parallel complementary strand as described below. Two nucleotide sequences which correspond to each other in this manner are referred to as 'complementary'.

  5′-ATCTGACT-3′
  3′-TAGACTGA-5′

  A frayed end is a region of a double stranded (or other multi-stranded) DNA molecule near the end with a significant proportion of non-complementary sequences, *i.e.* a sequence where nucleotides on the adjacent strands do not match correctly :

  5′-ATCTGACTAGGCA-3′
  3′-TAGACTGACTACG-5′

  The term 'frayed' is used because the incorrectly matched nucleotides tend to avoid bonding, thus appears similar to the strands in a fraying piece of rope. Although non-complementary sequences are also possible in the middle of double stranded DNA, mismatched regions away from the ends are not referred to as 'frayed'.

(*vii*) Differences in recognition sites of some restriction enzymes : In spite of differences in recognition sequences some enzymes produce identical sticky ends. For example, recognition sequence of *Bgl*II is 5′AGATCT3′, that of *Bam*HI is 5′GGATCC3′ and *Sau* is 5′GATC3′. All the three enzymes cut recognition sequences and produce 5′GATC3′ sticky ends. Since, fragments generated by each enzyme have identical sticky ends, they can join complementary sticky ends of one another (Table 6.2).

(*b*) Properties of Type II Restriction Enzymes : The properties of type II restriction enzymes vary with its specificity, *i.e.* its changes with changing the reaction conditions such as lowering NaCl concentration, increasing temperature, replacing Mg^{++} with Mn^{++}. The recognition and cleavage sites of *Eco*RI is normally a hexanucleotide sequence (5′GAATTC3′), but it become a tetranucleotide sequence (5′AATT3′) with the changes in reaction conditions. In a DNA molecule the size of DNA, its base composition and the GC content of recognition sequences govern the number of cleavage sites of type II enzymes. For example, if all the four bases are present with 50% AT and 50% GC base pairs, the probability for a particular tetranucleotide is $(X_1)^4$. The tetranucleotide shall be cleaved at a distance of every 256 base pairs. Some of the commonly used restriction endonucleases are given in Table 6.2.

(*c*) Frequency of Restriction Enzyme Sites : The frequency with which restriction sites occur in a random sequence can simply be calculated if the GC content of random sequence is known. For each nucleotide position in the restriction site, determine the frequency with which that position is occupied by the appropriate base. Then multiply the frequencies together to obtain the frequency with which the complete site is observed.

N N N N N N N N N N	**In a random sequence (50% GC)**
G A T C	4-mer sequence will occur
	$1/4 \times 1/4 \times 1/4 \times 1/4 = 1/256$ bp
G G A T C C	6-mer sequence will occur
	$1/4 \times 1/4 \times 1/4 \times 1/4 \times 1/4 \times 1/4 = 1/4096$ bp
G C G G C C G C	6-mer sequence will occur
	$1/4 \times 1/4 \times 1/4 \times 1/4 \times 1/4 \times 1/4 \times 1/4 \times 1/4 = 1/65536$ BP

If the random sequences were 40% GC instead of 50% GC, the 4-mer would cut

$$2/10 \times 2/10 \times 2/10 = 1/278 \text{ bp}$$

Similarly, the 6-mer would cut

$$2/10 \times 2/10 \times 2/10 \times 2/10 \times 2/10 \times 2/10 = 1/6944 \text{ bp}$$

The 8-mer would cut

$$2/10 \times 2/10 \times 2/10 \times 2/10 \times 2/10 \times 2/10 \times 2/10 \times 2/10 = 1/390625 \text{ bp}$$

Average fragment size and frequency of restriction enzyme cutting depend on both the length and the GC bias of the recognition site as well as the GC bias of the genome being cut.

***(d)* Action of *Eco*RI :** *Eco*RI recognises a hexanucleotide sequence 5′GAATTC3′ and cleaves into two fragments leaving four nucleotide-long sticky ends. Two complementary sticky ends again can join after treating with DNA ligase as given in Fig. 6.11.

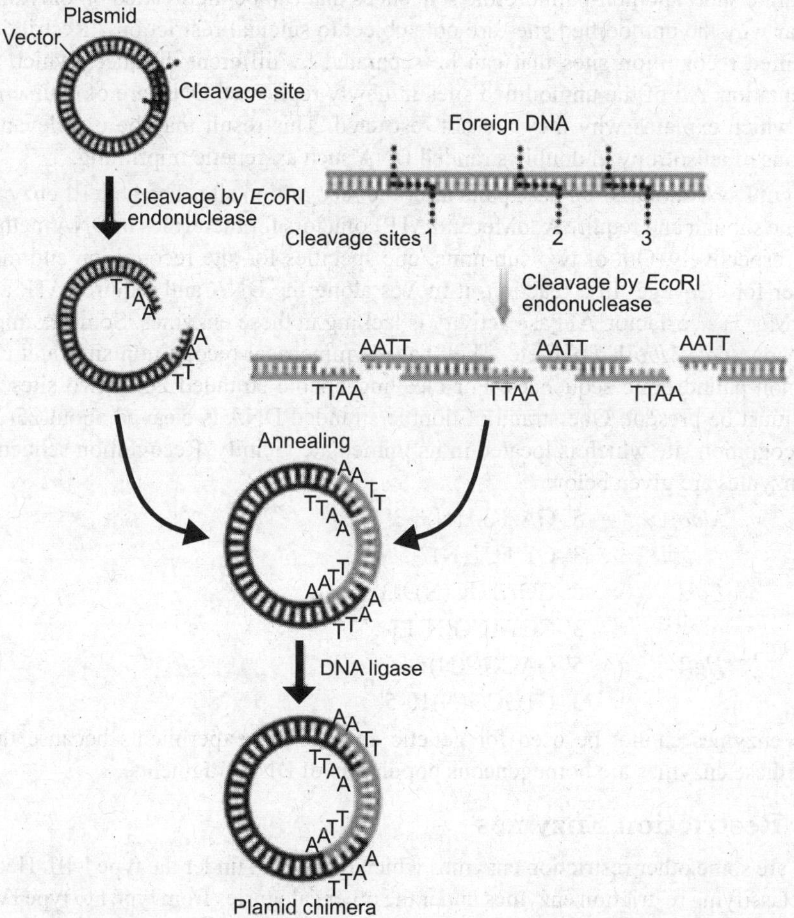

Fig. 6.11. Action of *Eco*RI on both strands of double-stranded DNA and production of recombinant plasmid chimera.

(*e*) Nomenclature of Restriction Enzymes : The restriction enzymes are named based on some of the principles. Name of the organism is identified by the first letter of genus name and the first two letters of the species name to form a three letter abbreviation in the italic, for example, *E. coli* = *Eco* and *H. influenzae* = *Hin*, etc. A strain or type identified is written as subscript *e.g.* *Eco*K for *E. coli* strain K, *Hin*d for *H. influenzae* strain Rd. The extra chromosomal element is identified by a subscript *e.g.* *Eco*RI, *Eco*PI, etc. in such cases where the restriction and modification systems are genetically specified by a virus or plasmid. When a strain has several restriction and modification systems, these are identified by Roman numerals, for example *Hin*dI, *Hin*dII, *Hin*dIII for *H. influenzae* strain Rd. etc. These Roman numerals should not be confused with those in the classification of restriction enzymes into type I.

E	*Escherichia*	Genus
co	*coli*	Species
R	RY13	Strain
I	First identified	Order of identification in a bacterium

4. Type III Restriction Enzyme

In 2003, Meisel and coworkers found that type III restriction-modification enzymes need two inversely oriented recognition sites for DNA cleavage. Type enzymes such as *Eco*P15 recognise the two separate short and non-palindromic sequences that can be methylated on only one strand. It was not clear why the unmodified sites are not subject to suicidal restriction ? Restriction requires two unmodified recognition sites that can be separated by different distances which must be in inverse orientation. All of the unmodified sites in newly replicated DNA are of course in the same orientation which explains why they are not restricted. This result may be of relevance to other manifestations of anisotropy in double stranded DNA such as genetic imprinting.

They cut DNA about 20-30 base pairs after the recognition site. The type III enzymes contain more than one subunit and require AdoMet and ATP cofactors for their roles in DNA methylation and restriction, respectively. Out of two sub-units, one specifies for site recognition and modification, and the other for cleavage. In a reaction, it moves along the DNA and requires ATP as source of energy and Mg^{++} as co-factor. ATPase activity is lacking in these enzymes. Some examples of type III enzyme are *Hpa*I, *Mbo*II, *Fok*I, etc. They have symmetrical recognition sites and cleave DNA at specific non-palindromic sequences. For cleaving double stranded DNA two sites in opposite orientation must be present. One strand of double stranded DNA is cleaved about 25-27 bp away from the recognition site which is located in its immediate vicinity. Recognition sequence of some of type II enzymes are given below :

*Mbo*II	5′-GAAGA(N)8-3′
	3′-CT TCT(N)7-5′
*Fok*I	5′-GGATGC(N)9-3′
	3′-CCTACG(N)13-5′
*Hgl*I	5′-GACGC(N)5-3′
	3′-CTGCG(N)10-5′

These enzymes cannot be used for genetic engineering experiments because the cleavage products of these enzymes are homogeneous population of DNA fragments.

5. Other Restriction Enzymes

There are some other restriction enzymes which do not fall under the type I-III. Hence, there is need for reclassifying restriction enzymes under the other categories from type I to type IV. Examples of the other restriction enzymes are *Eco*571, *Mcr* (modified cytosine reaction), etc. *Eco*571 is made up of single polypeptide. It shows both nuclease modification and nuclease activity. The S-adenosyl methionine stimulates nuclease activity of this enzyme. It bears some other properties that resemble with type II enzymes. Therefore, *Eco*571 should be put under type II to type IV.

In some of the commonly used strains of *E. coli* the DNA of some bacteria and eukaryotes was restricted due to the presence of methyl cytosine in the DNA. This was called modified cytosine reaction. In *E. coli* strain K there are two types of *Mcr* system, McrA and McrBC. The McrA is encoded by a prophage-like element, whereas McrBC is encoded by two genes, *mcr*B and *mcr*C. Both the genes are present adjacent to the genes that encode type I endonucleases in *E. coli* strain K.

Both the strands of DNA are cleaved at multiple sites between the methyl cytosine which is accomplished by GTP. The MerBC system is analogous to the *Rg/B* (restricts glucoseless phage) system which is found in T-even coliphages. T-even bacteriophages consist of glucosylated 5-hydroxymethyl cytosine in their DNA in place of cytosine. The phage DNA is restricted if glucosylation of 5-hydroxymethyl cytosine is inhibited due to mutation of phage or host genes. This results in restriction of phage by the host.

6. Factors Affecting Activity of Restriction Enzymes

Commonly it is difficult to digest DNA with restriction enzymes. The DNA does not appear at time to cut completely, and sometimes it cuts only partially. The restriction sites can be predicted accurately if the sequence is known; but in the laboratory, an enzyme may cut mostly at the wrong sites. Moreover, in some cases, these unexpected results cause a problem unrelated to technique; for example, the sequence may be incorrect, or a restriction map provided by a colleague may be erroneous. However, there are many conditions that influence cleaving of restriction enzymes.

(*a*) Buffer Composition : Different restriction enzymes have differing preferences for ionic strength and major cation such as sodium or potassium. Three to four different buffers handle a large number of available enzymes, though there are some enzymes that require a unique buffer environment. In all the cases, a major function of the buffer is to maintain pH of the reaction (usually at 8.0). Besides, some enzymes are fussy for having their optimal buffer than other enzymes. But the use of wrong buffer can lead to poor cleavage rates.

(*b*) Incubation Temperature : Besides many exceptions, most of restriction enzymes cleave best at 37°C. Enzymes isolated from thermophilic bacteria cleave best at temperatures ranging from 50 to 65°C. Some other enzymes have a very short half life at 37°C; therefore, they should be incubated at 25°C.

(*c*) Effect of DNA Methylation : Almost all strains of *E. coli* used for propagating cloned DNA contain the following two site-specific DNA methylases :

- **Dam methylase :** It adds a methyl group to the adenine in the sequence GATC and yields a sequence which may be written as G^{m}ATC.
- **Dcm methylase :** It methylates the internal cytosine in CC(A/T)GG and generates the sequence C^{m}C(A/T)GG.

Practically there are a number of restriction endonucleases that will not cleave methylated DNA. This concept can be explained by a few examples relative to Dam methylation. The *Mbo*I and *Sau*3AI are isoschizomers that recognise and cut the sequence GATC, which is the sequence recognised by Dam methylase. Digestion of G^{m}ATC by *Mbo*I is completely inhibited, whereas digestion by *Sau*3AI is not affected by methylation.

NNGATCNN / NNCTAGNN —Dam methylase→ NNGATCNN / NNCTAGNN (methylated adenines, H-C-H)

—*Mbo*I→ NNGATCNN / NNCTAGNN

—*Sau*3AI→ NN / NNCTAG + GATCNN / NN

ATCGAT is the recognition site for *Cla*I; this site is not a substrate for Dam methylase. A Dam recognition site is generated and cleavage by *Cla*I is inhibited, if that sequence is followed by a C or preceeded by a G. In this way, half of the *Cla*I recognition sites will not be cut in a random sequence of DNA propagated in most strains of *E. coli*.

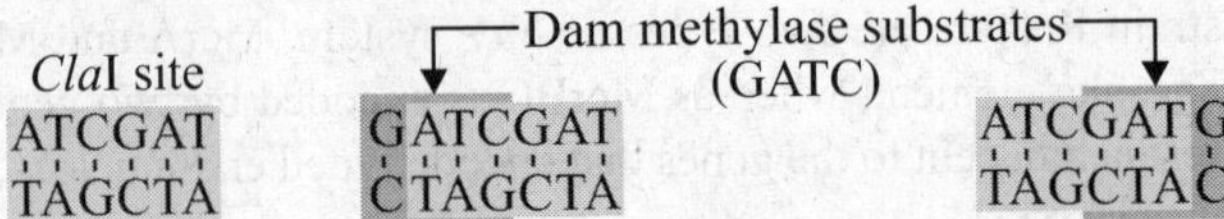

(*d*) Star Activity : Under non-standard conditions, upon digesting the DNA with certain restriction enzymes cleavage can occur at sites different from the normal recognition sequence; such aberrant cutting is called 'star activity'. For example, *Eco*RI enzyme can exhibit star activity that cleaves within a number of sequences differing from the canonical GAATTC by a single base substitution. Factors inducing star activity are : (*i*) high pH (>8.0) or low ionic strength, (*ii*) > 5% glycerol concentrations, (*iii*) very high concentration of enzyme (>100 U/mg of DNA), and (*iv*) the presence of organic solvents such as ethanol and DMSO in the reaction.

(*e*) Digestion with Multiple Enzymes : Digestion of DNA with two enzymes is a commonplace task, and the two enzymes have different buffer requirements. This situation is handled by at least following three ways :

- DNA is digested with both enzymes in the same buffer. Sometimes, a given buffer is not optimal for an enzyme still quite good cleavage rates can be obtained. Usually manufacturer catalogs of enzyme that contain a reference table recommend the use of the best single buffer for conducting specific double digests.
- First the DNA is cut with one enzyme; then the buffer composition is changed and the second enzyme is used to cut the DNA. This provides such situations where one enzyme likes a low salt buffer and the other a high salt buffer. In this case DNA is digested with the first enzyme for a time, and a calculated amount of concentrated NaCl is added followed by cutting with the second enzyme.
- The buffers are changed between digestions with two enzymes. Sometimes, two enzymes have totally incompatible buffers. In such case one digestion is performed, DNA is recovered by precipitation, and re-suspended in buffer suitable for the second enzyme.

D. DNA POLYMERASES

In 1955, Arther Kornberg and co-workers isolated DNA polymerase. Thereafter, different types of DNA polymerases were isolated from various sources that have different function. A detailed account of DNA polymerases has been given in Chapter 3 : *Molecular Biology*.

1. *E. coli* DNA Polymerase

In 1974, Arthur Kornberg reported that this enzyme polymerises the DNA synthesis on DNA or cDNA template), and also catalyses a 5′→3′ and 3′→5′ exonucleolytic degradation of DNA. The DNA polymerase, investigated by A. Kornberg and coworkers in *E. coli* in 1956 is now known as DNA polymerase I (DNA pol I). The other two enzymes are DNA polymerase II (DNA pol II) and DNA polymerase III (DNA pol III). These have almost similar catalytic activity. DNA pol I (mol wt 109,000) has a single polypeptide chain of about 1,000 amino acid residues. The addition of mononucleotide to the free -OH end of a DNA chain is catalysed by this enzyme. Also it catalyses the other two reaction, *i.e.* 3′→5′ exonuclease activity (hydrolysing single nucleotide residues from 5′ - terminus) and 5′→3′ exonuclease activity (hydrolysing single nucleotide residues from 5′- terminus). Function of DNA pol II (mol wt 120,000) is little understood. It catalyses via 3′→5′

exonuclease activity. DNA pol III (mol wt about 140,000) is about several times more active than the other two polymerases. It is a dimer, requires an auxiliary protein DNA co-polymerase III, and after combination yields a DNA pol III–co-pol III complex. Where there is preformed DNA template it produces a parallel strand in the presence of ATP. DNA pol I is susceptible to protease action and can be degraded into fragments.

Nucleic Acid Polymerising Enzymes

Arthur Kornberg (1918–2007) an American biochemist discovered DNA polymerase in the common intestinal bacterium *Escherichia coli.* It is a naturally occurring chemical tool which scientists could use to make copies of DNA. In 1957, Kornberg's group used this enzyme to synthesise DNA molecules. Although the molecules were biologically inactive yet this was an important achievement; it proved that this enzymes catalyse the production of new strands of DNA and it explained how a single strand of DNA acts as a pattern for formation of a new strand of nucleotides acting as the building blocks of DNA. Servo Ochoa (1905-1993) of New York University also discovered another enzyme that catalysed the synthesis of RNA.

Arthur Kornberg

Severo Ochoa

In 1959, Kornberg and Ochoa shared the Nobel Prize in Physiology or Medicine for their discovery of 'the mechanisms in the biological synthesis of deoxyribonucleic acid (DNA)'. Since this discovery, laboratories around the world have used the enzyme to build and study the DNA. This has led to a clear understanding of the biochemical basis of genetics.

2. Klenow (large) Fragment of *E. coli* DNA Polymerase I

The *E. coli* DNA polymerase I is a DNA-dependent DNA polymerase that possesses both 3′→5′ and 5′→3′ exonuclease activities. The *E. coli* DNA Pol I is a single-chain protein (mol wt about 109,000 Da) that requires Mg^{++} as a cofactor.

(*a*) Polymerisation Activity : Each of its three enzymatic activities are encapsulated into distinct domains of the holoenzyme, so that proteolytic deletions can be generated that lack one or more of the activities. The *E. coli* DNA polymerase I makes it unsuitable for many applications due to presence of 5′→3′ exonuclease activities. But this enzymatic activity can readily be removed from the holoenzyme (Fig. 6.12).

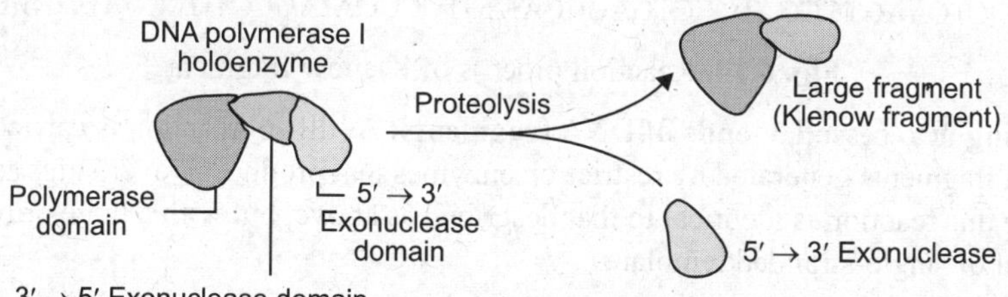

Fig. 6.12. Removal of large (Klenow) fragment of DNA polymerase I holoenzyme.

When DNA polymerase I is exposed to the protease subtilisin it cleaves the molecule into a small fragment, which retains the 5′→3′ exonuclease activity, and a large piece called Klenow fragment (Fig. 6.13). The Klenow fragment is a large fragment that is widely used in recombinant DNA. The Klenow fragment possesses polymerase and 3′→5′exonuclease activity but lack 5′→3′ exonuclease activity. Instead of intact Pol I, this enzyme is used for polymerase activity as below :

5′-G-3′ Klenow polymerase 5′-GGAATT-3′
3′-CCTTAA-5′ ⟶ 3′-CCTTAA-5′

The Klenow fragment of DNA polymerase I has both the DNA polymerase and 3′→5′ exonuclease activities. Hence it is widely used in molecular biology. Besides the generation of Klenow fragment by proteolysis, it can be expressed in bacteria from a truncated form of the DNA polymerase I gene. In the early days of recombinant DNA technology, DNA polymerase I was used frequently for radiolabelling DNA and synthesising cDNA. However, other enzymes have proven to be more effective for these purposes, including a proteolytic fragment of DNA polymerase I called Klenow fragment and T4 DNA polymerase. The holoenzyme DNA polymerase I is no longer frequently used.

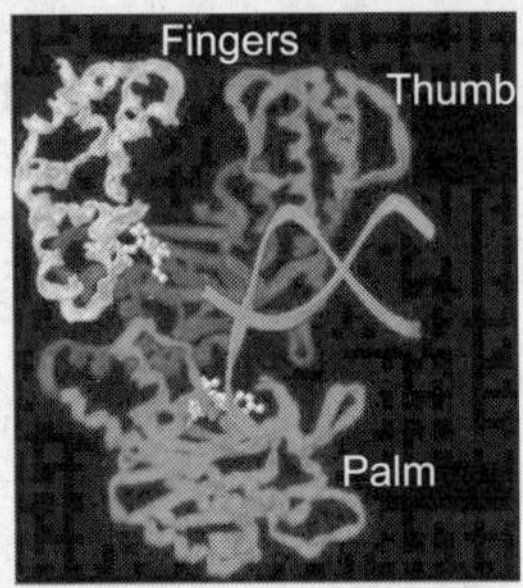

Fig. 6.13. Klenow (large) fragment of *E. coli* DNA polymerase I.

(*b*) Application of *E. coli* DNA Polymerase I : Uses of Klenow fragment are given below :

(*i*) Synthesis of double-stranded DNA from single-stranded DNA templates: Synthesis of complementary strands during DNA replication is the function of DNA polymerases. DNA polymerases require a primer to provide a free 3′-OH group for initiation of synthesis. For most of the *in vitro* polymerisation reactions the primers used are single-stranded DNA, typically 6 to 20 bases in length which are called oligonucleotides. The oligonucleotides must be complementary to some sequences of DNA template. For synthesis of a complementary strand of DNA, single-stranded template, primers and the Klenow fragment are mixed in the presence of a suitable buffer. As Klenow proceeds, it can displace primers downstream and continue synthesising new DNA. The proceeding reaction process of Klenow fragment is given in Fig. 6.14.

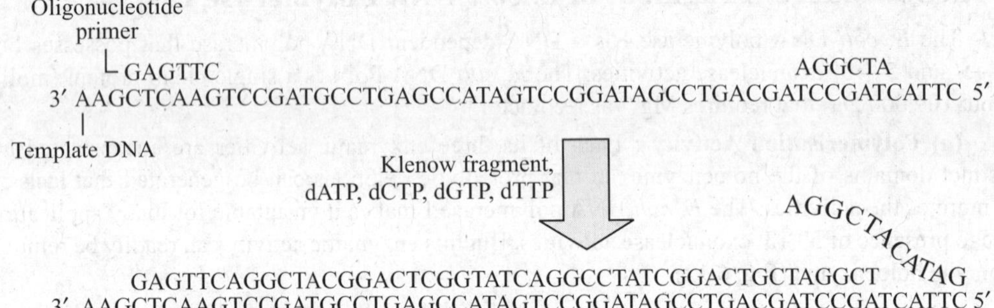

Fig. 6.14. Reaction process of Klenow fragment.

(*ii*) Filling in recessed 3′ ends of DNA fragments: A 'fill-in' reaction is carried out to create blunt ends on fragments generated by restriction enzymes that produce 5′ overhangs as given below. Conceptually this reaction is identical to that described as above, but with a huge primer and a very short segment of single-stranded template.

5′ A–G–G–C–A–G ⟶ (Klenow and nculeotides) 5′ A–G–G–C–A–G–G–A–T–C
3′ T–C–C–G–T–C–C–T–A–C 3′ T–C–C–G–T–C–C–T–A–C

(*iii*) Digestion of 3′ overhangs: In this method the 3′ overhangs generated from restriction enzymes can be digested. The protruding overhang would be digested by the 3′→5′ exonuclease activity of Klenow fragments as given below. Removal of nucleotides from the 3' ends will continue; but the polymerase activity will balance the exonuclease activity in the presence of nucleotides resulting in the blunt ends. This reaction is more efficiently conducted with T4 DNA polymerase because it has much more potent exonuclease activity than the DNA polymerase I activity.

5′ G–A–C–G–A–G–C–T → Klenow → 5′ G–A–C–G
3′ C–T–G–C → 3′ C–T–G–C

***(iv)* Preparation of DNA probes:** The radio-labelled nucleotides can be incorporated into the DNA fragment. Klenow fragment is used frequently to prepare DNA labelled with radionuclides or other markers.

Sometimes, the 3′→5′ exonuclease activity of Klenow fragment is not desirable. Different forms of the enzyme can be expressed by introducing mutations in the gene that encodes Klenow and retains polymerase activity, but lack exonuclease activity. These forms of the enzyme are called 'exo-Klenow fragment'.

3. T4 DNA Polymerase

The T4 DNA polymerase activities are very similar to Klenow fragment of DNA polymerase I. It functions as a 5′→3′ DNA polymerase and a 3′→5′ exonuclease, but does not have 5′→3′ exonuclease activity. Generally, T4 DNA polymerase is used for the same types of reactions as Klenow fragment, especially in producing blunt ends in DNA with 5′ or 3′ overhangs. These two enzymes differ in having different enzyme activites : (*i*) the 3′→5′ exonuclease activity of T4 DNA polymerase is roughly 200 times more than that of Klenow fragment, making it preferred by many investigators for blunting DNAs with 3′ overhangs, and (*ii*) Klenow fragment will displace downstream oligonucleotides as it polymerises but T4 DNA polymerase will not do so. This attribute makes T4 DNA polymerase the more efficient choice for certain types of oligonucleotide mutagenesis reactions.

4. T7 DNA Polymerase

The DNA polymerase of T7 bacteriophage possesses polymerase and 3′→5′ exonuclease activities, but lacks a 5′→3′ exonuclease domain. Therefore, it is very similar to Klenow fragment and T4 DNA polymerase in having these activities. Also T7 DNA polymerase has processivity, *i.e.* the average length of DNA synthesised before the enzyme dissociates from the template is greater than for other enzymes. Therefore, T7 DNA polymerase can be used in DNA sequencing by the chain termination technique.

The 3′→5′ exonuclease activity of T7 DNA polymerase can be abolished by treating it chemically or through genetic engineering technique. These forms of the enzyme are marketed under the name 'Sequenase' and 'Sequenase 2.0', and are widely used for DNA sequencing reactions.

5. Thermostable DNA Polymerases

The original report of this enzyme was published in 1976, by a microbiologist Thomas D. Brock after isolating and purifying from a bacterium *Thermus aquaticus* from a hot spring lake at Yellowstone National Park (U.S.A.). After 10 years of discovery, the polymerase chain reaction was developed and soon *T. aquaticus* polymerase called 'Taq' polymerase became a household word in circles of molecular biology. More recently, a more heat-resistant polymerase has been developed from a hyperthermophilic bacterium *Thermococcus litoralis* growing at 110°C in hydrothermal vent

A hot spring lake at Yellowstone National Park.

ecosystems in the deep ocean; it is called 'Vent polymerase'. The fidelity of Vent DNA polymerase is 5-15-fold higher than that observed for Taq DNA polymerase. This high fidelity derives from an integral 3′→5′ proofreading exonuclease activity in Vent DNA polymerase. More than 90% of the polymerase activity remains even after 1 hour incubation at 95°C.

Like other DNA polymerases, the thermophilic DNA polymerases catalyse template-directed synthesis of DNA from nucleotide triphosphates. To initiate synthesis a primer having a free 3′-OH end and Mg^{++} ions are required. Generally, they have maximal catalytic activity at 75 to 80°C, and reduced activities at lower temperatures. The Taq polymerase has only about 10% of its maximal activity at 37°C.

Besides Taq DNA polymerase, several other thermostable DNA polymerases have also been been isolated and expressed from cloned genes. Three of the most-used polymerases are given in Table 6.3.

Table 6.3. Thermostable DNA polymerases and their properties.

Polymerase	*3′→5′ Exonuclease*	*Source and properties*
Taq	No	Isolated from *Thermus aquaticus*. Its half-life is 1.6 hours at 95°C.
Pfu	Yes	Isolated from *Pyrococcus furiosus*. It appears to have the lowest error rate of known thermophilic DNA polymerases.
Vent	Yes	Isolated from *Thermococcus litoralis*; it is also known as Tli polymerase. Its half-life is about 7 hours at 95°C.

Beside its native polymerases activity, a number of mutants have been generated; for example, a form of Vent polymerase lacks the 3′→5′ exonuclease and is more similar to Taq.

Moreover, presence of error rate is one of the most unique characteristics of thermostable polymerases. Error rates are measured using several different assays. Consequently, the estimates of error rate vary when the assays are performed by different laboratories. A polymerase lacking 3′→5′ exonuclease activity consists of the higher error rates than that of having exonuclease activity. The total error rate of Taq polymerase has been variously reported between 1×10^{-4} and 2×10^{-5} errors per base pair. The Pfu polymerase has the lowest error rate at roughly 1.5×10^{-6} error per base pair, and Vent is intermediate between Taq and Pfu. Error rate is not the only consideration in selecting a polymerase for PCR; otherwise Taq polymerase would not continue to be the most widely used enzyme by far for the myriad of PCR applications.

E. TERMINAL TRANSFERASE

Terminal transferase is a mammalian enzyme that is expressed in lymphocytes. It has been isolated and purified from calf thymus. This enzyme is also known as deoxynucleotidyl transferase. It has a property to add 'oligodeoxynucleotide tail' to 3′-OH end of double-stranded DNA fragments (Fig. 6.17). Thus it extends homopolymer tails this phenomenon is called 'homoplymer tailing'.

Terminal transferase catalyses the addition of nucleotides to the 3′-OH terminus of DNA. It also works on single-stranded DNA, including 3′ overhangs of double-stranded DNA. Thus it is such a DNA polymerase that does not require a primer. It can also add homopolymers of ribonucleotides to the 3′ end of DNA. Protruding 3′ ends are the much preferred substrate for this enzyme, but may also add nucleotides to blunt and 3′-recessed ends of DNA fragments. Cobalt is a necessary cofactor for activity of this enzyme.

1. Application of Terminal Transferase

Following are the application of terminal transferase :

(*a*) Labelling the 3′ ends of DNA : Mostly the substrate for labeling the 3′ ends of DNA is a fragment of DNA generated by digestion with a restriction enzyme that generates a 3′ overhang;

oligodeoxynucleotides can also be used for this purpose. When such DNA is incubated with tagged nucleotides and terminal transferase, a string of the tagged nucleotides will be added to the 3′ overhang or to the 3′ end of the oligonucleotide as below :

```
             5′G–A–T–C–A–C–T–G–C–A
        A–C–G–T–C–T–A–G–T–G 5′
                  ⇩ ── Terminal transferase
                           +
                          dTTP
             G–A–T–C–A–C–T–G–C–A–T–T–T–T–T–T–T
T–T–T–T–T–T–T–A–C–G–T–C–T–A–G–T–G
```

(*b*) Homopolymer Tailing : Addition of complementary homopolymeric tails to DNA is called homopolymer tailing. In the past this procedure was used to clone cDNAs into plasmid vectors; but it has been replaced by more efficient techniques in recent years. Basically, terminal transferase is used to tail a linearised plasmid vector with Gs and the cDNA with Cs. The complementary Gs and Cs anneal to insert the cDNA into the vector when incubated together; then the recombinant plasmid is transformed into *E. coli*. The steps of this technique are given in Fig. 6.15.

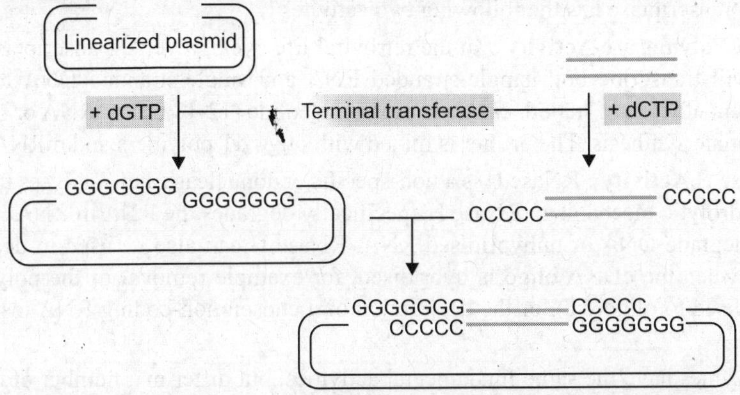

Fig. 6.15. Insertion of cDNA into a linearised plasmid by using terminal transferase.

F. REVERSE TRANSCRIPTASE

Reverse transcriptase is a common name of an enzyme that functions as a RNA-dependent DNA polymerase which transcribes the single-stranded RNA into single-stranded DNA. They are enclosed by retroviruses, where they copy the viral RNA genome into DNA before its integration into host cells. Normal transcription involves the synthesis of RNA from DNA, but reverse transcription is the 'reverse' of this information. In recombinant DNA technology, reverse transcriptase is used to synthesise the copy DNA or complementary DNA (cDNA) by using mRNA as a template. Reverse transcriptase is very useful in synthesis of cDNA and construction of cDNA clone bank.

Passing of the genetic information of DNA to protein through mRNA was known until recently. During 1960s, Howard M. Temin and co-workers at the University of Wisconsin (U.S.A.) postulated that in certain cancer causing animal viruses (Rous sarcoma virus) which contain RNA as genetic material, transcription of cancerous genes (on RNA into DNA) takes places most probably by DNA polymerase directed by virus RNA. Then DNA is used as template for synthesis of many copies of viral RNA in a cell. In 1970, S. Mizutani, H.M. Temin and D. Baltimore discovered that

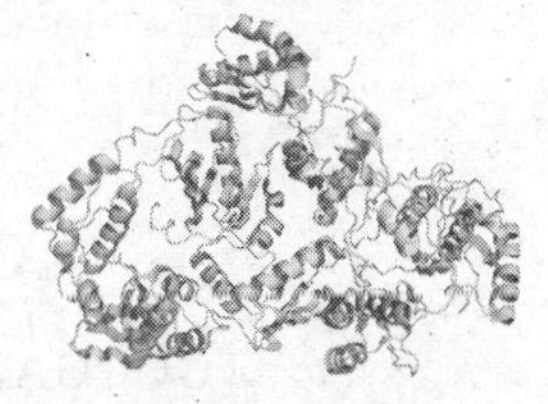

Crystallographic structure of HIV reversed transcriptase

information can also pass back from RNA to DNA. They found that retroviruses (possessing RNA) contain RNA dependent DNA polymerase which is also called 'reverse transcriptase'. This produces single stranded DNA, which in turn functions as template for complementary long chain of DNA. For this major contribution, Temin shared the 1975 Nobel Prize in Medicine and Physiology with David Baltimore and his own former mentor, Renato Dulbecco for the discovery of RNA-dependent DNA polymerase in virions of Rous sarcoma virus RSV).

Retroviral reverse transcriptase has three sequential biochemical activities such as : (*i*) RNA-dependent DNA polymerase, (*ii*) ribonuclease H, and (*iii*) DNA-dependent DNA polymerase. The retroviruses use these activities to convert single-stranded genomic RNA into double-stranded cDNA which can integrate into the host genome, generating a long-term infection. Reverse transcriptase has a domain RNase H family which is vital to replication of retroviruses. The reverse transcriptase has a high error rate when transcribing RNA into DNA; unlike DNA polymerase, it has no proofreading ability. This high error rate allows mutations to accumulate at an accelerated rate relative to proofread forms of replication.

1. Activities of Reverse Transcriptase

Reverse transcriptase has the following two activites :

(*a*) DNA Polymerase Activity : In the retroviral life cycle, reverse transcriptase copies only RNA but it will transcribe both single-stranded RNA and single-stranded DNA templates with easily equivalent efficiency. In both cases, an ologonucleotide (12-18 bases) RNA or DNA primer is required to initiate synthesis. The primer is mixed with oligo dT polymers and mRNA.

(*b*) RNase H Activity : RNase H is a non-specific endonuclease and catalyses the cleavage of RNA via a hydrolytic mechanism. RNase H specifically degrades the RNA in RNA : DNA hybrids and will not degrade DNA or unhybridised RNA. RNase H can also be used to degrade specific RNA strands when the cDNA oligo is hybridised, for example removal of the poly(A) tail from mRNA hybridised to oligo(dT), or the destruction of a chosen non-coding RNA inside or outside the living cell.

Both enzymes have the same fundamental activities, but differ in a number of characteristics including temperature and pH optima. Analysis of the structure-function relationship between the polymerase and RNase H activities indicates that the two activities reside in separate segments of the molecule. While the polymerase domain is contained within the NH_2-terminal 51 kDa, the RNase H activity is located in $COOH^-$ terminal 15 kDa of the p66 subunit. The murine leukemia virus reverse transcriptase has very weak RNase H activity as compared to the avian myeloblastosis enzyme, which makes it the clear choice when trying to synthesise complementary DNAs for long mRNAs. Then reverse trancriptase is added that uses the oligo-dT as a primer to synthesise so called first strand cDNA (Fig. 6.16).

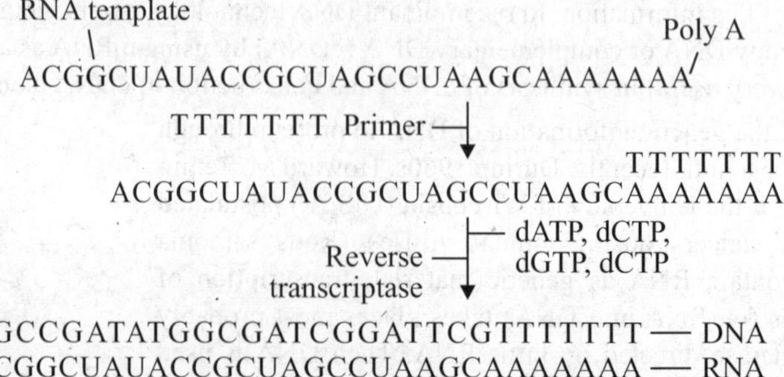

Fig. 6.16. Synthesis of a DNA by reverse transcriptase using mRNA as template and precursor deoxynucleotide triphosphates (dNTPs).

2. Uses of reverse Transcriptase

Following are the uses of reverse transcriptase :

(*a*) Copying RNA into DNA : This task is discussed elsewhere in detail, but the technique is usually initiated by mixing short (12-18 bases) polymers of thimidine (oligo-dT) with mRNA such that they anneal the RNA's polyadenylate tail.

(*b*) Generation of DNA Copies of RNAs before Amplification by PCR : Reverse transcription PCR commonly called RT-PCR is a stupifyingly useful tool for cloning cDNAs, diagnosing microbial diseases rapidly and myriad of other application. In most cases standard preparations of reverse transcriptase are used for RT-PCR, but mutated forms with relatively high thermal stability have been developed to facilitate the process. With the help of reverse transcriptase, RNA can be transcribed into DNA, thus making the PCR analysis of RNA molecules possible.

(*c*) Creation of cDNA Library : Reverse transcriptase is also used to create cDNA libraries by employing mRNA. The commercial availability of reverse transcriptase greatly improved knowledge in the area of molecular biology as it allowed the scientists to clone, sequence, and characterise the DNAs.

(*d*) Preparation of Antiviral Drugs : Since HIV uses reverse transcriptase to copy its genetic material and generate new viruses, specific drugs have been designed to disrupt the process and suppress the growth of viruses during their multiplication cycle. Collectively, these drugs are known as 'reverse transcriptase inhibitors' and include the nucleoside- and nucleotide analogues- such as zidovudine (trade name Retrovir®). Molecular structure of zidovudine that inhibits HIV reverse transcriptase is given in Fig. 6.17.

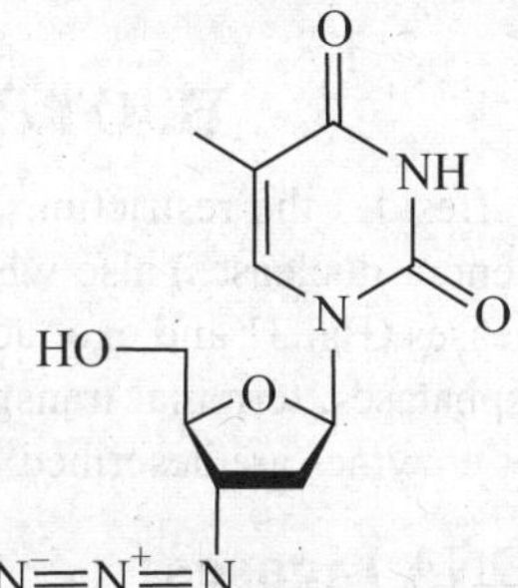

Fig. 6.17. Molecular structure of zidovudine (Retrovir®).

The other drugs are lamivudine (Epivir) and tenofovir (Viread), as well as non-nucleoside inhibitors such as nevirapine (Viramune).

G. BACTERIOPHAGE RNA POLYMERASE

The defined RNAs are generated by using thc DNA-dependent RNA polymerases encoded by bacteriophages for *in vitro* transcription. Mostly the reaction utilises ribonucleotides labelled with radionuclides or some other tag; the resulting labelled RNA is used as a probe for hybridisation. The other applications of *in vitro* transcription, including making RNAs for *in vitro* translation, are to study the structure and function of RNA. Now-a-days, there are several bacteriophage RNA polymerases that are commercially available. They are named after the name of phage that encodes the polymerase. The polymerases are either purified from phage-infected bacteria or produced as recombinant proteins. Some of the RNA polymerases obtained from phage-infected bacteria with their promoter sequences are given in Table 6.4.

Table 6.4. RNA polymerases obtained from phage-infected bacteria with promoter sequences.

Name of polymerases	*Host of encoding phages*	*Promoter sequences*
T7 RNA polymerase	*Escherichia coli*	TAATACGACTCACTATAGGG
T3 RNA polymerase	*E. coli*	AATTAACCCTCACTAAAGGG
SP6 RNA polymerase	*Salmonella typhimurium*	AATTTAGGTGACACTATAGAA

Moreover, there are many plasmids used for carrying cloned DNA that incorporate promoters for bacteriophage RNA polymerases near the cloning site. Consequently, either sense or antisense transcripts of mRNA from the inserted DNA can readily be obtained (Fig. 6.18). This method is often called 'run-off transcription', because the plasmid is cut with a restriction enzyme downstream of the inserted DNA, which results in the polymerase to fall off the template after reaching that site.

It may be assumed that the RNA transc-ribed has the polarity of an mRNA as in Fig. 6.18. Then it is easy to modify the construct to express an antisense RNA by reversing the orientation of the transcribed region. Obviously, most plasmids used have two different phage polymerase promoters flanking the insertion site that are used for *in vitro* transcription. As a result of which transcription of sense RNA with one polymerase and antisense with the other is allowed.

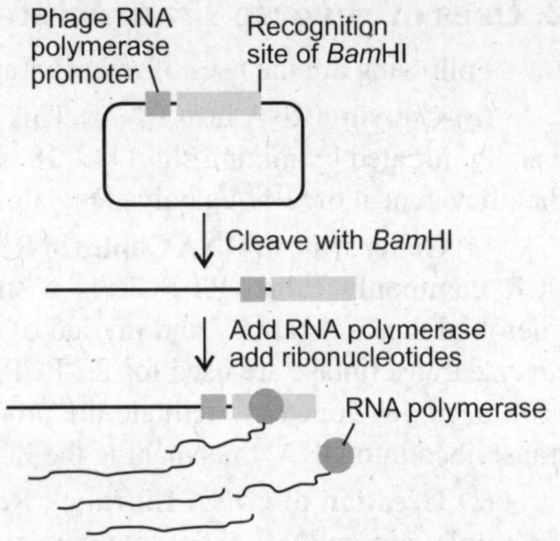

Fig. 6.18. *In vitro* transcription for making RNA by bacteriophage RNA polymerase.

H. OTHER DNA MODIFYING ENZYMES

Besides the restriction enzymes, there are some other DNA modifying enzymes (both exo- and endo- nucleases) also which are used in degradation, alteration, and synthesis of DNA such as nucleases (Bal 31 and exonuclease III, deoxyribonuclease I and SI nuclease), polymerases, ligases, phosphatases, terminal transferases, etc. Some of these enzymes have earlier been described. The other enzymes are described in this section.

1. DNA Ligases

DNA ligases catalyse the formation of a phosphodiester bond between the 5′-PO_4 of one strand of DNA and the 3′-OH of the other. This enzyme is used covalently to ligate together the DNA fragments. The enzyme has also been shown to catalyse the joining of RNA to either a DNA or RNA strand in a duplex molecule but will not join single stranded nucleic acids. Mostly, the reaction involves ligating a DNA fragment into a plasmid vector, which is a fundamental technique of recombinant DNA.

In 1966, B. Weiss and C.C. Richardson isolated DNA ligase, an enzyme that could be used to ligate two DNA strands together. In 1972, J.E. Mertz and R.W. Davis for the first time demonstrated that cleavage of DNA by R 1 restriction endonuclease generates cohesive ends. These cohesive termini of cleaved DNA molecules could be covalently sealed with *E. coli* DNA ligase and were able to produce recombinant DNA molecules. In 1972, Paul Berg used restriction enzyme to cut DNA, and then a ligase to paste two DNA strands together to form a hybrid circular molecule; this was the first recombinant DNA. The DNA ligases used in molecular cloning are either of bacterial origin or bacteriophage-encoded enzyme. All eubacteria (mesophilic or thermophilic) cosist of a single ligase gene which encodes a necotinamide dinucleotide (NAD^+)-depended enzyme.

(*a*) Types of DNA Ligases : Besides the occurrence of DNA ligases in all organisms, they show a wide diversity of molecular sizes, amino acid sequences and properties. In 1974, I. R. Lehnman described the structure, ligation mechanism and function of DNA ligases. DNA ligases can be divided into two broad classes : those requiring NAD^+ as a cofactor (and found only in bacteria), and bacteriophage DNA ligases (which uses ATP as a cofactor and found in eukaryotes, viruses and bacteriophages).

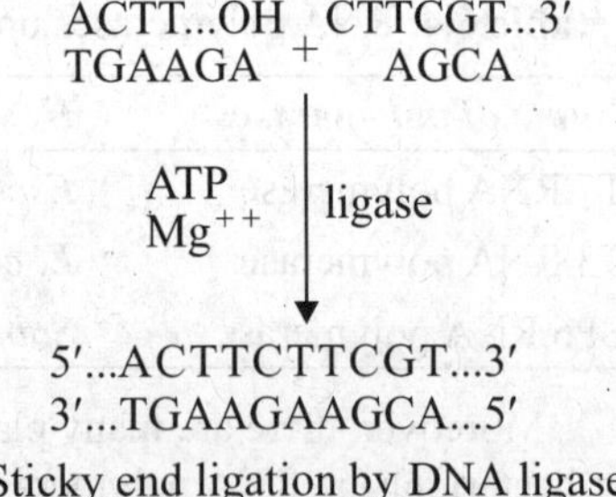

Sticky end ligation by DNA ligase.

The eucaryotic and virally encoded enzymes all require ATP. The ligases in this class range in size from 103 kDa for the human type I enzyme to 41 kDa for bacteriophage T7 DNA. The NAD^+ requiring DNA ligases have only been found in prokaryotic organisms. The amino acid sequences for a number of bacterial DNA ligases are now available. These NAD^+-dependent enzymes are highly homologous and are monomeric proteins of 70-80 kDa but show little homology with ATP-dependent ligases.

***(i) E. coli* DNA ligase :** *E. coli* DNA ligase is an NAD^+–dependent enzyme that catalyses the formation of phosphodiester bonds between complementary 3′-OH (hydroxyl) and 5′-PO_4 (phosphoryl) termini of dsDNA. The enzyme works best with cohesive dsDNA ends and is also active on nicked DNA. *E. coli* DNA ligase is a polypeptide of molecular weight 75,000. It is encoded by *lig* gene of *E. coli*. The nucleotide sequenxce of lig gene shows that *E. coli* DNA ligase consists of 671 amino acids. After cloning and expression of *lig* gene, highly active enzyme preparation could be obtained which could catalyse blunt-end ligation with moderate efficiency. This enzyme is not widely used in molecular cloning.

- **Thermostable DNA Ligase :** In 1991, F. Barany for the first time purified the thermostable DNA ligase from an *E. coli* strain containing the cloned ligase gene from *Thermus aquaticus* HB8. It is the third type of DNA ligase produced by several thermophilic bacteria, and many of the enzymes are commercially available.

  Taq DNA ligase uses NAD^+ as a cofactor and catalyses the formation of a phosphodiester bond between juxtaposed 5′-PO_4 and 3′-OH termini of two adjacent oligonucleotides which are hybridised to a complementary target DNA. The ligation occurs only if the oligonucleotides are perfectly paired to the complementary target DNA and have no gaps between them. Therefore, a single-base substitution can be detected. *Taq* DNA ligase is active at elevated temperatures (45–65°C). They can also catalyse blund-end ligation in the presence of crowding agents even at higher temperature. They are extensively used in ligase amplification reaction for the detection of mutation in mammalian DNA.

***(ii)* T3 DNA Ligase :** T3 DNA ligase is an ATP-dependent dsDNA ligase isolated from bacteriophage T3 that infects *E. coli* strain containing a recombinant gene. It catalyses the formation of a phosphodiester bond between adjacent 5′ phosphate and 3′ hydroxyl groups of dsDNA. Cohesive ends, blunt ends, and nicks are efficiently sealed by T3 DNA ligase. As with T4 DNA ligase, blunt end ligation is enhanced by the addition of PEG 6000 to the reaction. T3 DNA ligase shows 2-fold tolerance of NaCl concentration as compared to T4 DNA ligase making the enzyme a versatile choice for *in vitro* molecular biology protocols requiring DNA ligase activity.

***(iii)* T4 DNA ligase**. The main source of T4 DNA ligase is the bacteriophage T4 that infects overproducing strains of *E. coli,* hence the enzyme is known as T4 DNA ligase. It catalyses the formation of a phosphodiester bond between juxtaposed 5′-PO_4 and 3′ -OH termini in duplex DNA or RNA. This enzyme joins blunt end and cohesive end termini as well as repair single-stranded nicks in duplex DNA, RNA or DNA/RNA hybrids. A typical reaction for inserting a fragment into a plasmid vector (sub-cloning) utilises ATP as a cofactor, sticky ends, and T4 ligase. T4 DNA ligase is a monomeric protein having 487 amino acids. It is most widely used in genetic engineering experiments.

***(iv)* T7 DNA Ligase :** T7 DNA Ligase is an ATP-dependent dsDNA ligase isolated from bacteriophage T7 infecting *E. coli* strains containing a recombinant gene. It catalyses the formation of a phosphodiester bond between adjacent 5′-PO_4 and 3′-OH groups of dsDNA. T7 DNA ligase catalyses the cohesive end-ligation and nick sealing. Unlike T4 and T3 DNA Ligases, T7 DNA ligase does not catalyse blunt end-ligation. However, under typical reaction conditions blunt-end DNA ligation does not occur in the presence of T7 DNA Ligase. This makes it a good choice for applications where blunt and cohesive ends of DNA are present but only the cohesive ends are joined. However, T7 DNA Ligase does not ligate blunt-ends of DNA fragments.

Mammalian DNA Ligases

In mammals, there are four specific types of ligase.

- **DNA ligase I :** It ligates the nascent DNA of the Okazaki fragment (lagging strand) after the ribonuclease H has removed the RNA primer from the Okazaki fragments.
- **DNA ligase II :** It is alternatively spliced form of DNA ligase III found in non-dividing cell.
- **DNA ligase III :** It makes complexes with DNA repair protein XRCC1 to aid in sealing and during the process of nucleotide excision repair and recombinant fragments.
- **DNA ligase IV :** It forms complexes with XRCC4, and catalyses the final step in the non-homologous end-joining DNA double-strand break repair pathway. It is also required for V(D)J recombination in B cells of blood, the process which generates diversity in immunoglobulin and T-cell receptor loci during development of immune system.

(*b*) Ligation Mechanism : Ligation refers to joining linear DNA fragments together with covalent bonds. DNA ligation involves the creation of a phosphodiester bond between the 3′-OH of one nucleotide and the 5′-PO_4 group of another. T4 DNA ligase ligates the DNA fragments having overhanging, cohesive ends that are annealed together, as in the *Eco*RI. This is equivalent to repairing 'nicks' in duplex DNA. T4 DNA ligase also ligates fragments with blunt ends, although higher concentrations of the enzyme are usually recommended for this purpose.

The catalytic activity of the enzyme requires the presence of ATP and Mg^{++} ions. DNA that lacks the required phosphate residues can be rendered capable of ligation by phosphorylation with T4 polynucleotide kinase. An additional reaction of phosphates between pyrophosphate and ATP is also catalysed by this enzyme. *E. coli* DNA ligase is also available, but is not used commonly.

In all DNA ligases, the reaction occurs in three stages : (*i*) formation of a covalent enzyme-AMP intermediate linked to a lysine side-chain in the enzyme, (*ii*) transfer of the AMP nucleotide to the 5′ phosphate of the nicked DNA strand, and (*iii*) attack on the AMP-DNA bond by the 3′-OH of the nicked DNA sealing the phosphate backbone and resealing AMP (Fig. 6.19).

The DNA ends collide by chance and stay together for longer time for joining of ligase. This is the most efficient part of reaction but is easier at low temperature. All molecules move faster at high temperature, hence it is easier for two DNA ends to collide and stay together if they float gently through the solution at low temperature rather than whizzing as they do at higher temperature. Low temperature stabilises the hydrogen bonding between the complementary nucleotides.

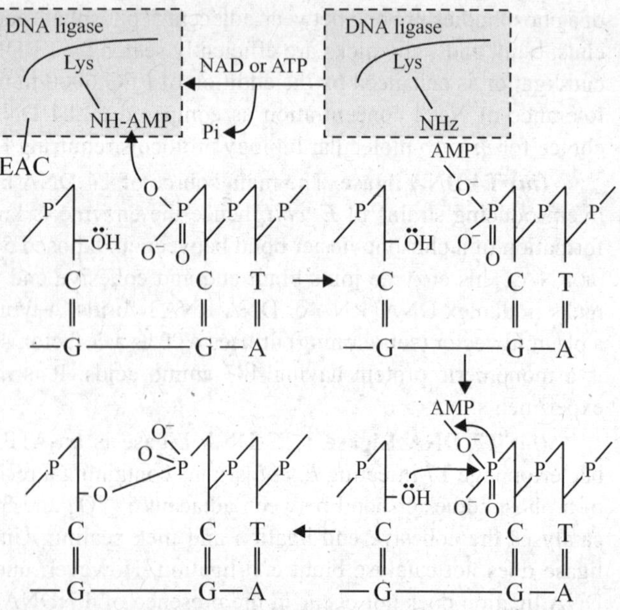

Fig. 6.19. Ligation reaction and bond formation by the enzyme DNA ligase.

The enzymatic reaction allows the DNA ligase to catalyses the joining of 3′-OH and 5′-PO_4 groups. Irrespective of cofactor (NAD^+ or ATP), both types of enzyme contain a $-NH_2$ group on lysine residue (Lys) (Fig. 6.19). The Lys residue acts as the active site for both types of enzymes (NAD^+ and ATP requiring ligases). Then the AMP nucleotide attached to Lys residues is transferred to 5′-PO_4 group. Then the AMP-PO_4 bond is attached by the 3-OH group forming the covalent bond and release of AMP. To allow the enzyme to carry out further reactions the AMP present at the active site of the enzyme must be replenished by ATP.

In both type of ligases, cofactor breaks into AMP (adenosine monophosphate) which in turn adenylates the ligase to form enzyme -AMP complex (EAC). EAC binds to nick containing 3′-OH and 5′-PO_4 ends on a double stranded DNA molecule. The 5′-PO_4 terminus of the nick is adenylated by the EAC with 3′-OH terminus resulting in formation of phosphodiester and liberation of AMP. Nick is sealed after formation of phosphodiester bridge (Fig. 6.20).

T4 DNA ligase has the ability to join the blunt ends of DNA fragments, whereas *E. coli* DNA ligase joins the cohesive ends produced by restriction enzymes. Additional advantage with T4 enzymes is that it can quickly join and produce the full base pairs but it would be difficult to retrieve the inserted DNA from vector. However, cohesive end ligation proceeds about 100 times faster than the blunt end ligation.

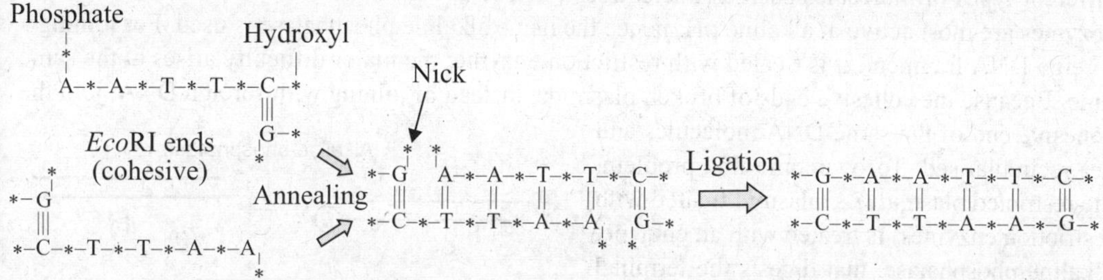

Fig. 6.20. Ligation mechanism in nick on double stranded DNA molecules by DNA ligase (see nick between G and A on both strands).

The optimal incubation temperature for T4 DNA ligase is 16°C and this temperature is recommended when very high efficiency ligation is desired, for example for making DNA libraries. However, ligase is active at a broad range of temperatures; for routine purposes such as sub-cloning, convenience often dictates incubation time and temperature. Ligations usually work well when performed at 4°C overnight or at room temperature for 30 minutes to a couple of hours.

Fig. 6.21 shows the effects of T4 DNA ligase on DNA fragments generated by digestion of a plasmid with two restriction enzymes incubated with different amounts of ligase for varying periods and electrophoresed in 1% agarose. All the fragments are ligated to one another and shifted to higher molecular weights even after 5 minutes treatment with T4 DNA ligase. The plasmid digested with two restriction enzymes, *Pst*I and *Sac*I generate 4 bases and, 3′ overhangs; but the two types of ends are incompatible. Each lane contains an equivalent mass of DNA.

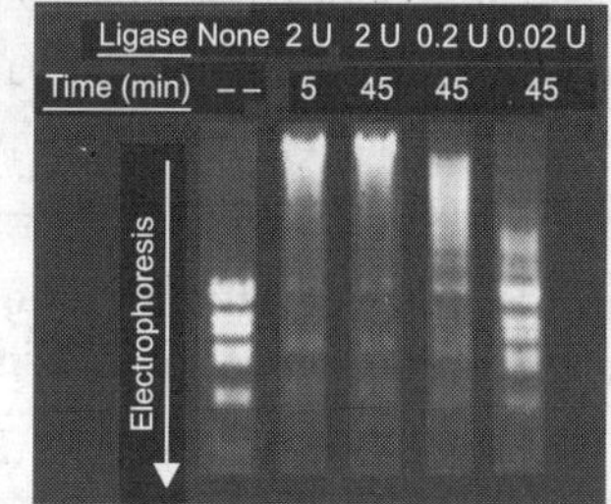

Fig. 6.21. Effect of T4 DNA ligase.

The activity of T4 DNA ligase is very similar to Klenow fragment of DNA polymerase I. It functions as a 5′→3′ DNA polymerase and a 3′→5′ exonuclease, but does not have 5′→3′ exonuclease activity.

Applications of DNA Ligase in Molecular Biology

In modern molecular biology research DNA ligases have become an indispensable tool for generating recombinant DNA molecules; for example, DNA fragments or often genes are inserted into plasmids by using DNA restriction enzymes and ligases.

Controlling the optimal temperature is often atricky aspect to perform successful recombination experiments involving ligase. Mostly T4 DNA ligase (isolated from bacteriophage T4) is used; it is most active at 25°C. However, the optimal enzyme temperature needs to be balanced with the melting temperature T_m (and annealing temperature) of the DNA fragments being ligated to perform successful ligations.

Homologous pairing of the sticky ends will not occur if the ambient temperature exceeds T_m, because hydrogen bonding will be disrupted at high temperature. At the lower T_m, shorter DNA fragments will be generated. Thus ligation experiments are performed at very low temperatures (~4-8°c) for a long time (often overnight) for generation of less than 10 bp-long sticky ends. Some commercially available DNA ligases were originally discovered in bacteriophage T4, *E. coli* or other bacteria.

2. Alkanile Phosphatase

Alkaline phosphatase is a hydrolase enzyme that removes 5′ phosphate groups from many different types of molecules such as nucleotides (DNA and RNA), proteins, and alkaloids. These enzymes are most active at alkaline pH; hence the name alkaline phosphatase is used. For joining a foreign DNA fragment, it is treated with restriction enzyme; the major difficulty arises at the same time. Because the cohesive ends of broken plasmids, instead of joining with foreign DNA, join the cohesive end of the same DNA molecules and get recircularised. To overcome this problem, the restricted plasmid (*i.e.* plasmid treated with restriction enzymes) is treated with an enzyme, alkaline phosphatase, that digests the terminal 5′-PO_4 group as below :

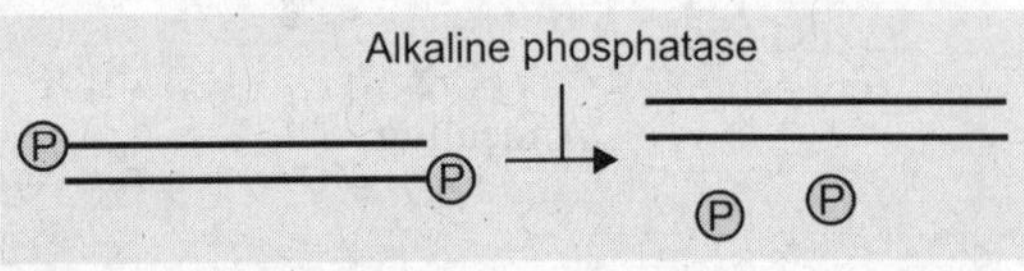

The restriction fragments of the foreign DNA to be cloned are not treated with alkaline phosphatase. Therefore, the 5′ end of foreign DNA fragment can covalently join to 3′ end of the plasmid. The hybrid or recombinant DNA obtained has a nick with 3′ and 5′ hydroxy ends. Ligase will only join 3′ and 5′ ends of recombinant DNA together if the 5′ end is phosphorylated.

Thus, alkaline phosphatase and ligase prevent recircularisation of the vector and increase the frequency of production of recombinant DNA molecules. The nicks between two 3′ ends of DNA fragment and vector DNA are repaired inside the bacterial host cells during the transformation.

The mechanism of action of alkaline phosphatase is given below :

$$\begin{matrix} \text{5'-GGOH-3'} \\ \text{3'-CCAATTAP-5'} \end{matrix} \xrightarrow{\text{Alkaline phosphatase}} \begin{matrix} \text{5'-GG-3'} \\ \text{3'-CCAATTAOH-5'} \end{matrix}$$

***(a)* Sources of Alkaline Phosphatase :** It is a protein that exists in all tissues of the body. There are several sources of alkaline phosphatase that differ in how easily they can be inactivated :

***(i)* Calf intestinal alkaline phosphatase (CIP)** : It is purified from bovine intestine. CIP catalyses the removal of phosphate groups from the 5' end of DNA strands. This is most widely used phosphatase in molecular biology laboratories because it can be effectively destroyed by protease digestion or heat (75°C for 10 minutes in the presence of 5 mM EDTA). The CIP is less active than the BAP. This prevents recircularisation of linearised vector DNA and improves the yield of vector containing the appropriate insert.

***(ii)* Bacterial alkaline phosphatase (BAP)** : It is the most active enzymes from *E. coli*, which may be destroyed with most difficulty at the end of the dephosphorylation reaction. BAP catalyses the hydrolysis of phosphate esters, including those present in nucleic acids and nucleotides. It is more thermal stable than calf intestine alkaline phosphatase (CIAP). Its optimal incubation temperature is about 60°C; however the enzyme remains active from 20°C to 80°C. It is resistant to chemical changes and active over a broad range of buffer conditions. It can be used to remove 5′-phosphates from DNA or RNA prior to 5′-end labeling. It works to remove 5′-phosphates from linearised vector molecules to prevent self-ligation of the vector during cloning procedures.

(*iii*) Shrimp alkaline phosphatase (SAP) : It is derived from a cold-water shrimp. SAP catalyses hydrolysis of 5′-phosphate from DNA (or RNA) regardless of whether the DNA has the 5′ end protruding or recessed, or is blunt-ended. SAP non-specifically catalyses the dephosphorylation of most phosphate monoesters, but it does not degrade diphosphate or triphosphate linkages. It is irreversibly inactivated by heat at 65°C for 15 minutes and denatured.

(*b*) Uses of Alkaline Phosphatas : There are two primary uses for alkaline phosphatase in DNA manipulations:

- **Removal of 5′ phosphates from plasmid and bacteriophage vectors :** The 5′ phosphates are removed from the plasmid and bacteriophage vectors that have been cleaved with a restriction enzyme. In subsequent ligation reactions, this treatment prevents self-ligation of vector and thereby greatly facilitates ligation of other DNA fragments into the vector.
- **Removal of 5′ phosphates from fragments of DNA before $\gamma^{32}PO_4$ radio-labelling :** The 5′ phosphates are removed from DNA fragments before labelling with radioactive phosphate. Polynucleotide kinase is much more effective in phosphorylating DNA if the 5′ phosphate has previously been removed.

Usually it is recommended that dephosphorylation of DNAs with blunt or 5′-recessed ends be conducted using a higher concentration alkaline phosphatase or at higher temperatures than for DNAs containing 5′ overhangs.

3. Polynucleotide Kinase

Polynucleotide kinase (PNK) enzyme catalyses the transfer of a phosphate from ATP to the 5′ ends of either DNA or RNA. It is a product of the T4 bacteriophage; through the expression of cloned phage gene into *E. coli,* commercial preparations can usually be produced. The enzymatic activity of PNK is utilised in two types of reactions as given in Fig. 6.22.

- **Forward reaction :** In the forward reaction, PNK transfers the gamma phosphate (γ^{32}-PO_4) from ATP to the 5′ end of a polynucleotide (DNA or RNA). The target nucleotide is lacking a 5′ phosphate either because it has been dephosphorylated or has been chemically synthesised.
- **Exchange reaction** : In this reaction the target DNA or RNA consisting of a 5′–PO_4 is incubated with an excess of ADP. PNK first transfers the phosphate from the nucleic acid onto an ADP and forms ATP and leaves a dephosphorylated target. Then PNK carries out a forward reaction and transfers a 5′-PO_4 group from ATP onto the target nucleic acid.

The efficiency of phosphorylation via the exchange reaction is less than the forward reaction. PNK also has 3′ phosphatase activity in addition to its phosphorylation activity; though this activity has a little significance.

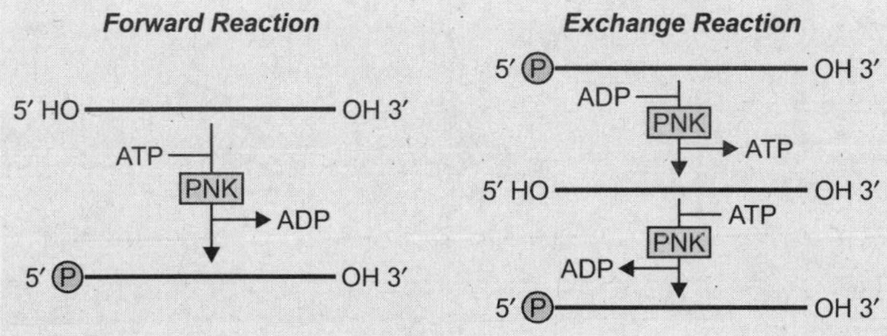

Fig. 6.22. Forward and exchange reactions of polynucleotide kinase enzyme.

PNK shows both phosphorylation as well as phosphatase activities; therefore, two types of reaction can be used accordingly. In phosphorylation activity the γ-PO_4 is transferred to the 5′end of dephosphorylated DNA. When ADP is in excess, it causes the T4 polynucleotide kinase to transfer the terminal phosphate from phosphorylated DNA to ADP. By using radiolabelled gamma-phosphate ($\gamma^{32}PO_4$), it was found that DNA is rephosphorylated by transfer of $\gamma^{32}P$ from $\gamma^{32}P$-ATP. This shows that the enzyme plays a dual role of phosphorylation activity and phosphatase activity. Due to the presence of both properties, this enzyme is used in radiolabelling of the 5′ end of duplex during DNA sequencing, constructing the terminally labelled DNA, phosphorylating the synthetic linkers/oligonucleotide linkers/other DNA fragments lacking terminal 5′ phosphates, etc.

QUESTIONS

1. What is restriction and modification system ?

2. What are the DNA enzymes ? Discuss in detail their mode of action.

3. What are exonuclease ? White in detail about λ exonuclease.

4. Give a detailed account of endonucleases, with special reference to S1 nuclease and deoxyribonuclease I (DNase I).

5. Give a detailed account of restriction endonucleases.

6. Write different types of DNA polymerases.

7. Write short notes on the following :

(*i*) Alkaline phosphatase, (*ii*) exonuclease III,

(*iii*) Terminal transferase, (*iv*) Bal 31 exonuclease,

(*v*) S1 nuclease, (*vi*) Sticky ends,

(*vii*) Palindomes, (*viii*) Linkers and adaptors,

(*ix*) DNA ligases, (*x*) Type II restriction enzyme,

(*xi*) Type III restriction enzyme, (*xii*) Polynucleotide kinase.

BIBLIOGRAPHY

Arber, W. and Linn S. 1969. DNA modification and restriction. *Annu. Rev. Biochem*. **38**: 467-500.

Lodish H.F. 2003. *Molecular cell biology.* 5TH ed., W. H. Freeman and Co., New York, p. 973.

Sambrook J., Fritsch E.F. and Maniatis T. 1989. *Molecular Cloning : A Laboratory Manual*, Cold Spring Harbor, New York.

Cloning Vectors

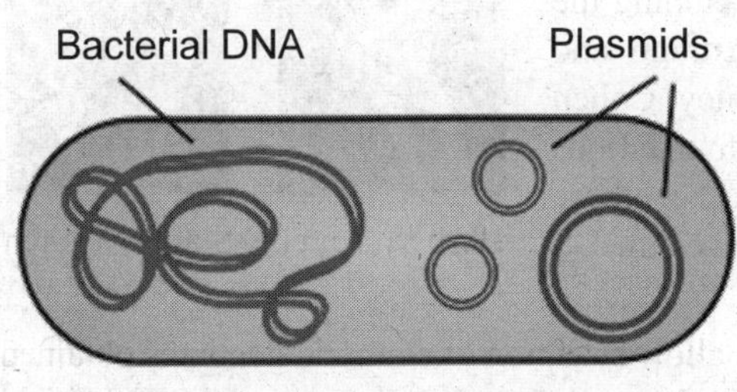

A. Bacterial plasmid vector
B. Bacteriophage vectors
C. Phagemid (phasmid) and fosmid vectors
D. Bacterial artificial chromosome (BAC)
E. Yeast plasmid vectors
F. Yeast artificial chromosomes (YAC)
G. Plant and animal viruses as cloning vectors
H. Transposons as vector
I. Shuttle vectors and expression vectors

In biology cloning is the process of producing populations of genetically-identical individuals that occurs in nature when organisms such as bacteria, insects or plants reproduce asexually. The term 'clone' is derived from the Greek word for 'trunk or branch' which refers to the process through which a new plant can be created from a twig. But in biotechnology cloning refers to processes used to create copies of DNA fragments (the process is called molecular cloning), copies of cells (*i.e.* cell cloning) or copies of organisms.

A remarkable development has been made in recent years in manipulation of prokaryotic and eukaryotic DNA. Similar to DNA enzymes, the other most important requirements for recombinant DNA technology is the availability of cloning vectors, and the molecular analysis of DNA has been made possible by the cloning of DNA. The two molecules required for cloning are the DNA to be cloned and a cloning vector. The recombinant DNA is produced by cloning a foreign DNA isolated either from the genome of an organism or synthesised chemically or prepared as cDNA using mRNA molecule or gene library. However, cloning of DNA can be done only when 'another DNA molecule' is available that may replicate in the transformed host cell. This 'other DNA molecule' used for transferring the foreign DNA is called cloning vectors. The vectors are the DNA molecules that can accept and carry a foreign DNA segment, replicate inside the host cell, and produces many copies of itself and the foreign DNA. Vectors may be plasmids (cloning limit 0.1 to 10 kilobases), a bacteriophage (cloning limit 8-20 kb), cosmids (cloning limit 35-50 kb), phagemids, fosmids, transposons, yeast artificial chromosomes (YAC) (cloning limit 100-1000 kb) or bacterial artificial chromosome (BAC) (cloning limit 75-300 kb), P1 artificial chromosome (PACs), *i.e.* DNA constructs derived from the DNA of bacteriophage (they can carry large amounts of about 100-300 kb of other sequence), or even a virus.

Paul Berg shared Nobel Prize for Chemistry in 1980

Historically, in the early 1970s, Paul Berg, Herbert Boyer, Stanley Cohen and their colleagues at Stanford University used gene cloning to change molecular biology. For cutting DNA and joining to other DNA, *i.e.* production of recombinant DNA molecule, Paul Berg shared half of Nobel Prize in 1980 in chemistry with the team of Walter Gilbert and Frederick Sanger.

Stanley Cohen studied plasmid replication, transfer of plasmids between bacterial strains and mechanism of antibiotic resistance. Stanley Cohen and Herbert Boyer produced the first plasmid vector for cloning purpose which was called pSC101, where 'SC' refers to Stanley Cohen. By genetically engineering cells to produce human substances, Boyer and Cohen invented a quick and easy way to make chemicals like HGH (human growth hormone), synthetic insulin, factor VIII for hemophilia, somatostatin for acromegaly and clot-dissolving agent tissue plasmogen activator (tPA). In 1980, they got patent for pSC101 and developing the gene splicing and cloning technique. Over hot pastrami and corned beef sandwiches, Herbert Boyer and Stanley Cohen opened the door to genetic engineering and laid the foundations for gene therapy and the biotechnology industry. For these outstanding achievements, the two collaborators received the US $500,000 Lemelson-MIT Prize in 1996.

Herbert Boyer

Stanley Cohen

Techniques should also be available which will allow selection of chimeric genomes obtained after insertion of foreign DNA from a mixture of chimeric and the original vector. Another critical desired feature of any cloning vector is that it should possess a site at which foreign DNA can be inserted without disrupting any essential function. Therefore, in each case an enzyme will also have to be selected which will cause a single break.

Sometimes vectors are modified by inserting a DNA segment to create unique site(s) for one or more enzymes to facilitate its use in gene cloning. This inserted DNA with restriction sites for several enzymes is sometimes called a polylinker. Cloning vectors are used to introduce foreign DNA into host cells, where that DNA can be reproduced (cloned) in large quantities.

Herbert Boyer and Stanley Cohen

Herbert Boyer received his M.S. and Ph.D. in bacteriology from the University of Pittsburgh (1963), followed by post-graduate work at Yale. He is currently a director at the leading biotech company Genentech, which he co-founded in 1976.

Stanley Cohen, a native of Perth Amboy, received his M.D. from the University of Pennsylvania. Cohen currently continues research in genetics at Stanford. Both Boyer and Cohen have received the National Medal of Science and the National Medal of Technology, and are members of the National Academy of Sciences.

Boyer and Cohen met in 1972, while presenting papers in Hawaii at a conference on bacterial plasmids. A Stanford University professor, Cohen had been working on ways to isolate specific genes in antibiotic carrying plasmids and clone them individually through introducing them to *E. coli* bacteria. Boyer, hailing from the University of California, had discovered a restriction enzyme that cut DNA strands at specific DNA sequences, producing cohesive ends that could stick to other pieces of DNA.

Boyer and Cohen agreed to collaborate, and they succeeded in splicing a piece of foreign DNA into a plasmid carrier, which then inserted genetic information into a bacterium. When the bacterium reproduced, it copied the foreign DNA into its offspring, acting as a natural factory producing biological substances. Boyer and Cohen have received three patents, from which over 350 licenses have been granted, generating approximately $27 million in royalties.

A vector must possess the following characteristics :

- It must replicate (through *ori* gene) in host cells after its introduction.
- It must contain marker genes such as tet^R (for tetracycline resistance), kan^R (for kanamycin resistance), amp^R (for ampicillin resistance). These help in selection of transformed cells from untransformed cells.
- A unique cleavage site must be present in one of the marker genes. The recombinant molecules can be detected by inactivating it through restriction enzyme. However, many vectors used commonly contain several recognition and cleavage sites for several restriction enzymes in a small region. These are called multiple cloning sites (MCS) or polylinkers. Presence of MCS facilitates the use of restriction enzymes of choice.
- It should contain specific control systems like promoters, terminators, ribosome binding sites, etc. so that the cloned DNA should express properly.

Plasmid vector are the best known of these vectors of which the largest number exists for *E. coli.* During the early 1970s, Stanley Cohen and Herb Boyer selected *Escherichia coli* strain K12 for conducting research work on molecular biology. Since then *E. coli* K12 has been used for various experiments in recombinant DNA technology and production of products.

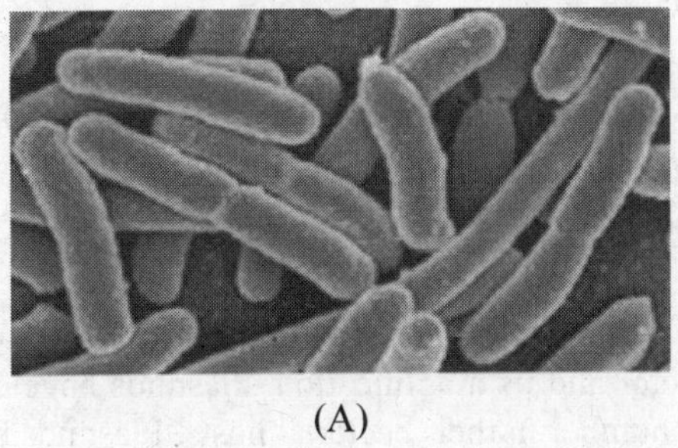

(A)

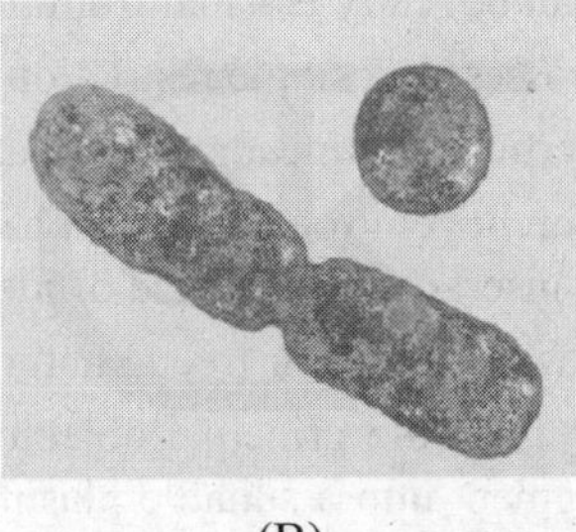

(B)

Fig. 7.1: *E. coli* cells; (A)- electron photomicrograph; (B)- an *E. coli* cell dividing by binary fission.

E. coli is a Gram-negative rod-shaped enteric bacterium of 2.5 m length having peritrichous flagella. An electron photomicrograph of *E. coli* cells is given in Fig. 7.1 . One of the unique features of *E. coli* is that it is found in human intestine and cannot thrive outside the laboratory environment. It grows well on a simple medium containing glucose, ammonia, phosphate and salts such as Ca^{++}, Mg^{++}, Na^{+} and K^{+}. Its generation time is 20 minutes under optimum aerobic condition. It is facultative anaerobe also. Though working with K 12 strain is safe yet one must follow the bio-safety guidelines before working on recombinant DNA.

E. coli strain K12 mutant in restriction-modification system is used in gene cloning. *E coli* strain K12 mutant in *end*h gene (encodes DNA specific endonuclease) increases the yield of plasmid DNA and improves DNA quality. The *rec*A gene is associated with expression of recombination proteins. Therefore, *E. coli* strain K12 mutant in *rec*A gene ensures about the bio-safety of *E. coli* strain K12 because ultra violet light can easily kill them.

A. BACTERIAL PLASMID VECTORS

Plasmids are the extra-chromosomal, self-replicating, and double stranded closed and circular DNA molecules of 100 - 400 kb size present in the bacterial cell (Fig. 7.2). Basically there are three classes of plasmids: virulence plasmids which encodes toxin genes, drug-resistance plasmids which shows resistance against several antibiotics and drugs, and plasmids encoding genes required for bacterial conjugation.

Moreover, there are two types of plasmids, conjugative and non-conjugative plasmids. Conjugative plasmids mediate their own transfer between the bacteria through conjugation specified by *tra* (transfer) and *mob* (mobilizing) regions present on plasmids. The non-conjugative plasmids are not self-transmissible, but they are mobilised by conjugation-proficient plasmid if their *mob* region is functional. Plasmids contain sufficient genetic information for their own replication. A number of host properties are specified by plasmids, such as antibiotic and heavy metal resistance, nitrogen fixation, pollutant degradation, bacteri-ocin and toxin production, colicin factors and phages. Naturally occurring plasmids can be modified by *in vitro* techniques.

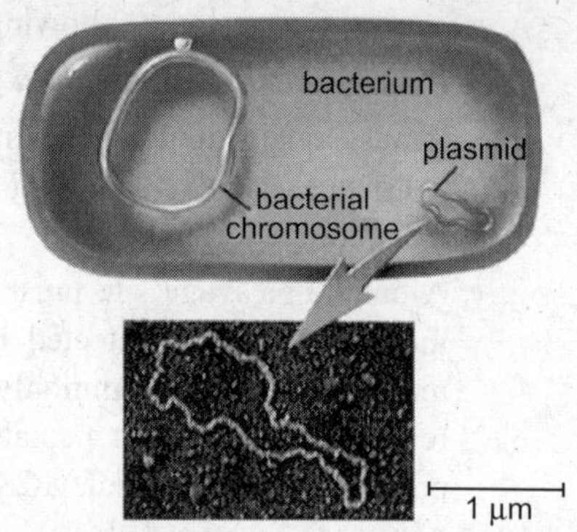

Fig. 7.2. Bacterial cell showing a plasmid with its enlarged view.

1. Some Considerations of Plasmid Vectors

A plasmid can be considered a suitable cloning vector if it possesses the following features :

- It can be really isolated from the cells,
- It possesses a single restriction site for one or more restriction enzyme(s),
- Insertion of a linear molecule at one of these sites does not alter its replication properties,
- It can be reintroduced into a bacterial cell and cells carrying the plasmid with or without the insert can be selected or identified,
- They do not occur free in nature but are found in bacterial cells.

In many cases the principal objective of cloning experiment is the insertion of a particular restriction fragment into a suitable plasmid vector and its amplification. Plasmids have evolved a variety of mechanisms to maintain stable copy number in their bacterial hosts. Plasmids behave as accessory genetic units that replicate and inherited independently of bacterial chromosome.

A plasmid vector is constructed by using the natural plasmids by removing unnecessary segments and adding essential sequences. To clone a DNA sample, the same restriction enzyme must be used to cut both the vector and the foreign DNA to be inserted. Therefore, a vector usually contains a sequence (polylinker) which can recognize several restriction enzymes so that the vector can be used for cloning a variety of DNA samples. For the first time in 1973, Cohen and co-workers reported the cloning of DNA by using plasmid as a vector.

(*a*) Plasmid Numbers: The number of plasmids per cell varies from 1 to 200 depending on maintenance-host proteins and their functions associated with replications. The other way of classification of plasmids is based on the number of copies of plasmids in host cells. This feature is called 'copy number'. The size of plasmids depends on host proteins for replication, maintenance and other functions. They are present in a characteristic number of copies in a bacterial cell, yeast cell, in eukaryotic organelles such as mitochondria. The plasmids are maintained as single copy per cell and they are called single copy plasmid. There are multicopy plasmids also which are maintained as 10–20 copies per cell (Fig. 7.3).

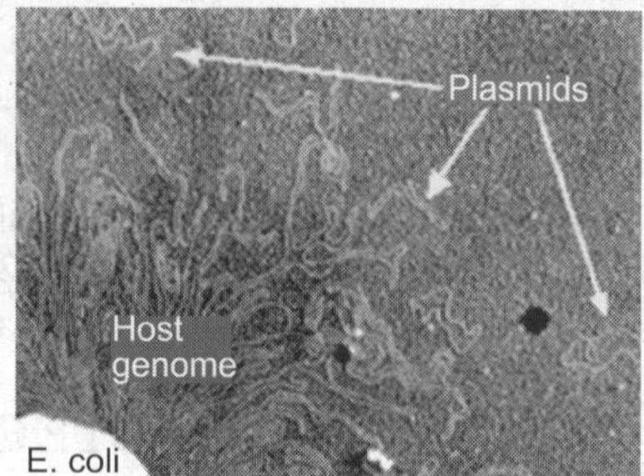

Fig. 7.3. Plasmids and chromosome of *E. coli*.

Apart from this when the plasmids are under the relaxed replication control, they get accumulated in the cell in very large copies, *i.e.* over 1000 copies per cell. The 'low copy number plasmids' tend to exhibit stringent control of DNA replication, with replication of the plasmid DNA closely tied to host's chromosomal DNA replication. The 'high copy number plasmids' are termed as relaxed plasmids whose DNA replication is not dependent on host's chromosomal DNA replication. Due to increased yield potential, such plasmids are used as cloning vectors. Properties of some naturally occurring plasmids are given in Table 7.1.

Table 7.1. Features of some naturally occurring plasmids.

Plasmid	***Conjugative***	***Copy number***	***Selectable markers***
ColE1	No	10-15	Immunity to homologous colicin ($E1^{imm}$)
CloDF13	No	10	Immunity to heterologous colicin ($DF13^{imm}$)
F	Yes	1-2	–
RK6	Yes	10-40	Amp^R, $Strep^R$
RK2	Yes	3-5	Amp^R, Tet^R, Kan^R
pSC101	No	1-2	Tet^R

The ColE1 plasmid replication system of *E. coli* is most extensively studied. It expresses colicins which is an antibacterial protein. The ColE1 transcribes a DNA sequence near the *ori* gene. Due to increased yield potential these plasmids are used as cloning vectors.

A replicon is a genetic unit that consists of an origin of replication and its control systems. A plasmid replicon is the smallest piece of plasmid DNA that is able to replicate autonomously and maintain normal copy number. More than 30 replicons have been identified in plasmids. Mostly all the plasmids used in molecular cloning consists of a replicon from pMB1; it is a plasmid originally isolated from clinical isolates. Its close relative is ColE1 replicon. In each bacterial cell about 15-20 plasmids are maintained under normal growth conditions. The ColE1 or pMB1 replicon relies mainly on long-lived enzyme from host. Replication is primed by an RNA primer and takes place unidirectionally from a specific origin.

E. coli also consists of another plasmid called 'fertility factor' or 'F plasmid' which is associated with conjugation. F plasmid as single copy and encodes the genes which are required for conjugation of F^+ cell of bacteria. The F plasmid is about 95 kb having about 100 genes of which some are required for F plasmid transfer (*tra* gene) into F^- (F-deficient) cells during conjugation. Some of them are needed during the transfer of F factor. F plasmid forms sex pili, the number of which is governed by the number of plasmids. It also encodes cell surface binding proteins for the filamentous bacteriophage M13.

***(b)* Amplification of Plasmids:** The process of increasing the number of a plasmid in a bacterial cell is called amplification. In this process, a cell containing a relaxed plasmid is treated with a drug to inhibit protein synthesis. Consequently, cells stop replicating and the relaxed plasmid pBR322 continues to replicate despite drug treatment. Replication of relaxed plasmid neither depends on cell replication nor requires protein synthesis. For example, addition of chloramphenicol causes pBR322 to get increased about 3000 per cell. Finally the ratio of plasmid DNA to chromosomal DNA is increased which makes easy to isolate the plasmid DNA. In order to replicate and clone a fragment of foreign DNA, it is necessary to possess a sequence of nucleotides which are recognised by the host bacterium as an origin of replication. The origin of replication occurs naturally in plasmid and transferred to onward progenies. During cell division these genes are transferred resulting in gradual decrease in number of such genes. However, due to continuous exchange of genetic materials between plasmid and chromosomal DNA, new genes originate with respect to environmental conditions. Therefore, plasmids have been artificially developed from naturally occurring form and only special features that help in cloning have been preserved.

***(c)* Examples of Plasmids:** There are several plasmid cloning vectors such as pBR320, pSC102, Co1E1, pUC, pRP4, pRK2, pRSFIOlO, pEY, pWWO, Ti- and Ri-DNA plasmids. The plasmid cloning vectors are designated by 'p', some abbreviations and a few numbers. For example, in pBR320, 'p' means plasmid, BR refers to the researchers F. Bolivar and R. Rodriguez who discovered the plasmid and 322 is the numerical designation. Some of the plasmids are given in Table 7.2.

Table 7.2: Example of some cloning vectors.

Cloning vectors	*Natural occurrence*	*Size (Kb)**	*Selective marker***
Plasmids			
pACYC 177	*Escherichia coli*	3.7	Amp^R, Kan^R
pBR322	*E. coli*	4.0	Amp^R, tet^R
pBR324	*E. coli*	8.3	Amp^R, tet^R, $E1^{imm}$
pMB9	*R. coli*	5.8	Tet^R
pRK646	*E. coli*	3.4	Amp^R
pC194	*Staphylococcus aureus*	3.6	Ery^R
pSA0501	*S. aureus*	4.2	Str^R
pBS161-1	*Bacillus subtilis*	3.65	Tet^R
pWWO	*Pseudomonas putida*	117	Kan^R
Cosmids			
pJC74	Derived plasmid from Col EL	16	Amp^R, $E1^{imm}$
pJC720	do	24	$E1^{imm}$., Rif^R
pHC79	Derivative of pBR322	6	Amp^R, Tet^R
Viruses			
SV40	Mammalian cells	5.2	—
Phage $M13^+$	*E. coli*	6.4	—
Phage l	*E. coli*	4.9	—

*, 1 Kb (Kilobase pairs) = 1,000 base pairs = 0.66 mega dalton; **, Resistance to ampicillin (Amp^R), tetracycline (Tet^R), erythromycine (Ery^R), streptomycine (Str^R), Kanamycin (Kan^R), rifampicin (Rif^R), and colicin production (EL imm.)

2. pSC101

In the early 1970, Herbert Boyer and Stanley Norman Cohen produced pSC101 which is the first plasmid vector produced for cloning purposes. In 1973, they successfully cloned two pSC101 plasmids together and created one large plasmid. Thereafter, cloning of genes into plasmids was carried out. In 1980, pSC101 became the first patented commercial DNA cloning vector when patents were awarded to Boyer and Cohen. The 'SC' stands for Stanley Cohen. Although the original pSC101 contained only tetracycline resistance and a restriction site for *Eco*RI, the commercially available pSC101 gained restriction sites for several enzymes including *Hin*dIII, in addition to the *Eco*RI site. The pSC101 is a DNA plasmid that is used as a cloning vector in genetic cloning experiments. In 1973, pSC101 (9.263 kb) was the first cloning vector used by Herbert Boyer and Stanley Norman Cohen (Fig. 7.4). They demonstrated that a gene from a frog could be transferred into bacterial cells and then expressed by the bacterial cells.

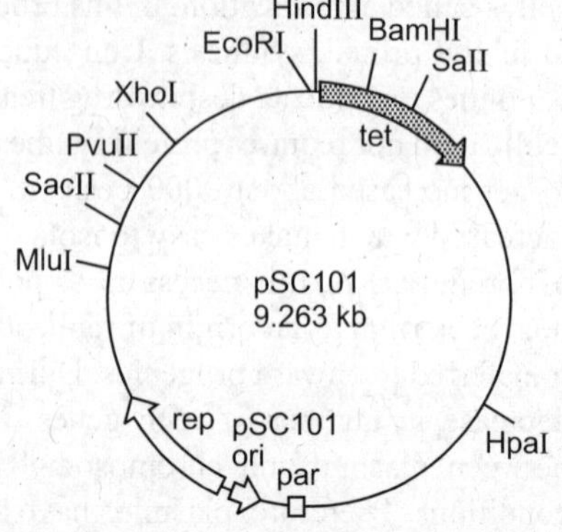

Fig. 7.4. Restriction map of pSC101.

3. pBR322

The naturally occurring plasmids do not bear all the characters that are required for a cloning vector. Therefore, the required genes are constructed in the plasmid for its efficient use. In 1977, F. Boliver and coworkers artificially constructed a new cloning vehicle, pBR322. This plasmid was derived from pBR313 which is a relaxed replicating plasmid sensitive to colicin E1 and carries resistance genes to the antibiotics, ampicillin (*amp*) and tetracycline (*tet*). The antibiotic-resistant genes on pBR322 are not transposable. The vector pBR322 was constructed to have a plasmid with a single *Pst*I site, located in the ampicillin-resistant gene, in addition to four unique restriction sites, *Eco*RI, *Hin*dIII, *Bam*HI and *Sal*I.

F. Boliver and R.L. Rodriguez

The pBR322 is 4.361 kb long and most widely used cloning vector. In the transformed *E. coli* cells its copy number remains about 15-20 molecules per cell. However, its number may be amplified by incubating the transformed *E. coli* cells on a medium containing chloramphenicol.

In order to use antibiotic resistance as a negative as well as a positive selection system, the plasmid vector must carry two different antibiotic resistance genes. An example of such a vector is pBR322 that carries both an ampicllin and a tetracycline resistance gene. The phenotype of bacteria containing the intact plasmid is Amp^R and Tet^R. Insertion of foreign DNA into the *Pst*I site located in the Amp^R gene results in an Amp^S Tet^R phenotype. Conversely, insertion of foreign DNA into the *Eco*RI, *Hin*dIII or *Sal*I sites located in the Tet^R gene results in an Amp^R and Tet^S phenotype.

The complete gene sequencing of pBR322 has been done. It consists of 20 recognition sites; a restriction map of plasmid pBR322 is shown in (Fig. 7.5). The pBR322 is constructed from the

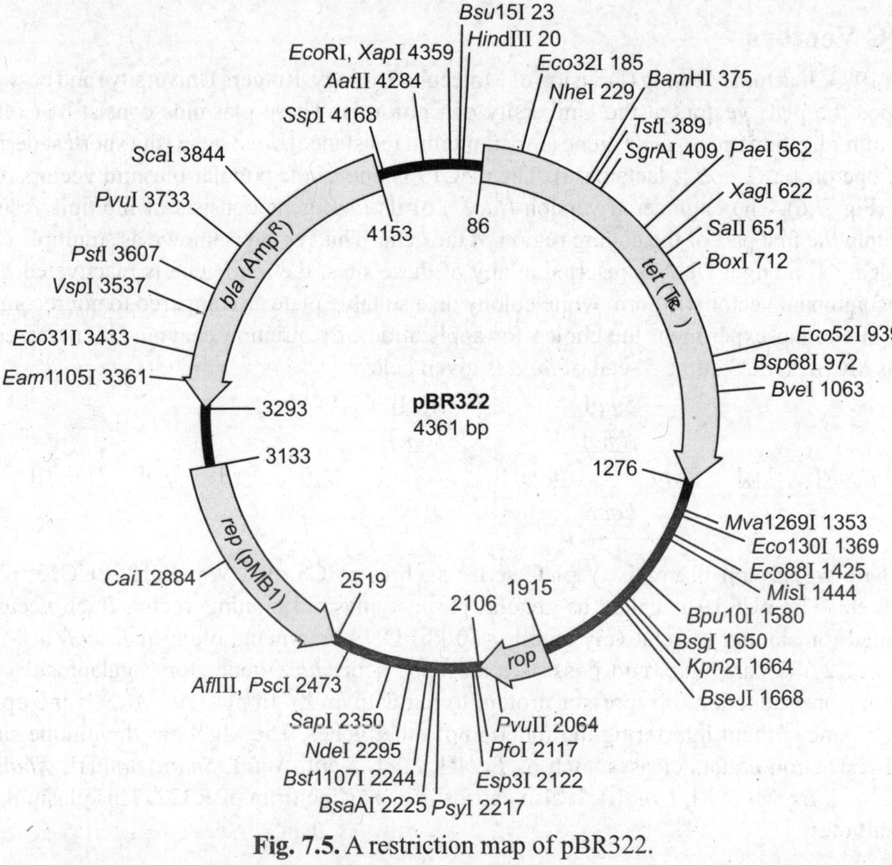

Bsu15I 23
HindIII 20
EcoRI, XapI 4359
AatII 4284
Eco32I 185
BamHI 375
NheI 229
SspI 4168
TstI 389
SgrAI 409
PaeI 562
ScaI 3844
XagI 622
PvuI 3733
4153
86
SalI 651
PstI 3607
BoxI 712
VspI 3537
bla (AmpR)
tet (Tc)
Eco31I 3433
Eco52I 939
Bsp68I 972
Eam1105I 3361
BveI 1063
3293
pBR322
4361 bp
3133
1276
Mva1269I 1353
Eco130I 1369
rep (pMB1)
Eco88I 1425
CaiI 2884
MisI 1444
2519
1915
2106
Bpu10I 1580
BsgI 1650
rop
Kpn2I 1664
AflIII, PscI 2473
BseJI 1668
SapI 2350
PvuII 2064
NdeI 2295
PfoI 2117
Bst1107I 2244
Esp3I 2122
BsaAI 2225
PsyI 2217

Fig. 7.5. A restriction map of pBR322.

plasmids of *E. coli,* pBR318 and pBR320. It contains origin of replication (*ori*) that was derived from a plasmid related to naturally occurring plasmid ColE1. Therefore, its replication may be faster than the bacterial DNA. It also possesses genes conferring resistance to ampicillin (*amp*^{R}) and tetracycline (*tet*^{R}), and unique recognition sites for 20 restriction endonucleases. Six of these 20 sites such as *Eco*RIV, *Bam*HI, *Sph*I, *Sal*I, *Xma*III and *Nru*I are present within the gene coding for tetracycline resistance, two sites (*Hind*II and *Cla*I) are located within the promoter of the tetracycline resistance gene, and the three sites (*Pst*I, *Pvu*I and *Sca*I) within the beta-lactamase gene that provide resistance to ampicillin. If any foreign DNA is cloned into any of these 11 sites, insertional inactivation of any of the antibiotic resistance markers takes place. For example, if a foreign DNA molecule is inserted at *tet*^{R} gene cluster, the property of tetracycline resistance will be lost. The recombinant plasmid will allow the cells to grow only in the presence of ampicillin but will not protect them against tetracycline. The presence of an antibiotic resistant gene in a plasmid of bacterium will confer resistance to that antibiotic.

Therefore, the antibiotic resistant cells can be selected by culturing the cells on nutrient medium supplemented with ampicillin or tetracycline. The circular sequence is numbered such that 1 is the first T of the unique *Eco*RI site GAATTC and the count increases first through the *tet* gene, the pMB1 material, and finally through the Tn3 region. The map shows enzymes that cut pBR322 DNA once. The coordinates refer to the position of first nucleotide in each recognition sequence. The exact position of genetic elements is shown on the map (termination codons included). The *bla* gene nucleotides 4153-4085 (complementary strand) code for a signal peptide. The indicated *rep* region is sufficient to promote replication. DNA replication initiates at position 2533 (+/-1) and proceeds in the direction indicated. Plasmids carrying the pMB1 and ColE1 replicons are incompatible, but they are fully compatible with those carrying the p15A replicon (pACYC177, pACYC184).

4. pUC Vectors

In 1983, Joachim Messing (Division of Molecular Biology, Rutgers University) and co-workers developed the pUC vectors at the University of California. These plasmids consist of ColE1 *ori* (origin of replication) gene, *amp*^{R} gene (for ampicillin resistance), *lacZ* gene (the shortest derivative of *lacZ* operon for *E. coli* β-lactamase). The pUC19 is one of the popular plasmid vectors of pUC family (Fig. 7.6). The engineered version (*lacZ'*) of its *lacZ* gene consists of multiple restriction sites within the first part of the coding region of the gene. This region is known as 'multiple cloning site' (MCS). If a target DNA is inserted at any of these sites, the *lacZ* gene is inactivated. Hence, such recombinant vector will form white colony on a suitable plate as compared to nonrecombinant vector. In cloning experiment the choice for application of restriction enzymes is increased after using as MCS. A MCS at the 5′-end of *lac*Z is given below :

*Sma*I *Acc*II
*Xma*I *Hinc*II
*Eco*RI *Sac*I *Kpn*I *Bam*HI *Xba*I *Sal*I *Pst*I *Sph*I *Hind*III
LacZ
⟶

There are several plasmids of pUC series such as pUC8, pUC9, pUC12, pUC13, pUC18, pUC19, etc. The pUC19 is called as general purpose plasmid cloning vector. Such vectors are designated for cloning of relatively small (<10 kb) DNA fragments often in *E. coli* at MCS. It consists of 2,686 base pairs and possesses an *amp*^{R} gene, *lacZ* gene (for β-galactosidase) and *lac*I gene (for production of repressor protein to regulate *lacZ*). In pUC19, MCS is incorporated into *lac*Z gene without interfering the function of other genes. The MCS has the unique sites for several restriction endonucleases such as *Eco*RI, *Sac*I, *Kpn*I, *Xma*I, *Sma*I, *Bam*HI, *Xba*I, *Sal*I, *Hind*II, *Acc*I, *Bsp*MI, *Pst*I, *Hind*III. It also carries one *ori* gene from pBR322. This plasmid works in prokaryotes.

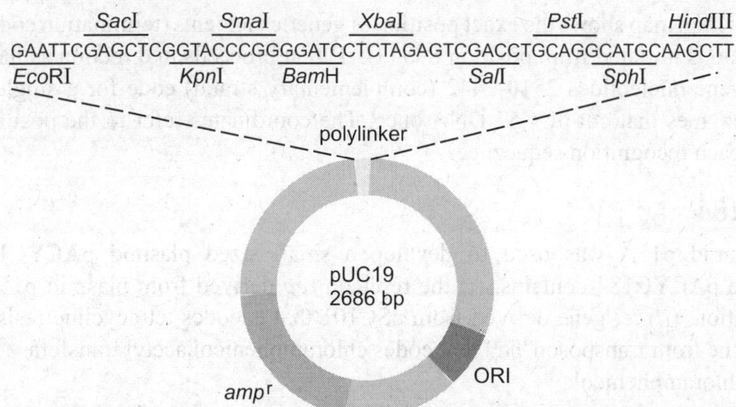

Fig. 7.6. Genetic map of the plasmid cloning vector pUC19. It contains a polylinker which can recognize several different restriction enzymes, an ampicillin-resistance gene (*amp*^R^) for selective amplification, and a replication origin (*ori*) for proliferation in the host cell.

When the cells carrying unchanged pUC19 are grown in the presence of isopropylthiogalactoside (ITG) (an inducer of *lac* operon), the production of *lac*I gene cannot bind to the promoter-operator region of the *lac*Z gene. Therefore, the *lac*Z gene encoded in the plasmid is transcribed. The *lac*Z protein combines with chromosomal DNA-transcribed protein and forms a hybrid β-galactosidase. The MCS is tagged in the *lac*Z gene of pUC19 plasmid which does not interfere the production of hybrid β-galactosidase. On the other hand if a substrate 5-bromo-4-chlorionolyl-β-galactosidase (X-gal) is present in the medium, the hybrid β-galactosidase hydrolyses it to produce a blue product. Therefore, colonies containing unmodified pUC19 appear blue in colour. When the DNA insert is cloned in *lac*Z gene, it will be inactivated and will not express β-galactosidase. Consequently, the colonies formed by *E. coli* cells containing recombinant plasmid will be white in colour. This plasmid vector has been designed for blue-white screening.

5. pUC57

The pUC57 (2710 bp long) is a commonly used plasmid cloning vector in *E. coli* which has been isolated from *E. coli* strain DH5α. Plasmid pUC57 is a derivative of pUC19 of which MCS contains 6 restriction sites with protruding 3′-ends, that are resistant to *E. coli* exonuclease III. This vector is designed for cloning and generation of *Exo*III deletions.

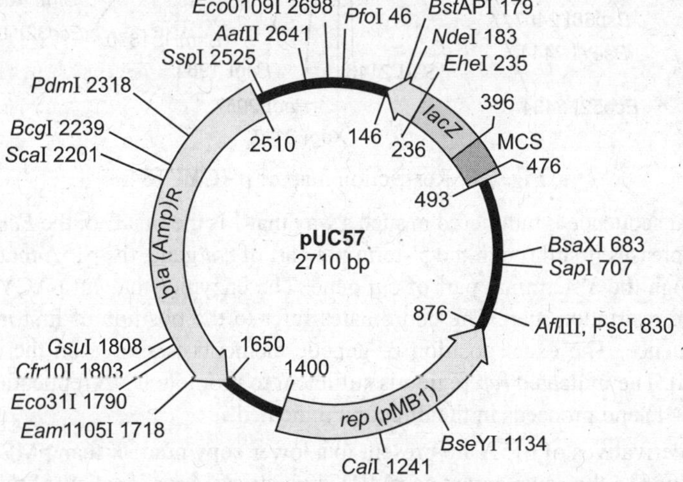

Fig. 7.7. Restriction map of pUC57.

The restriction map shows the exact position of genetic elements (termination codons included). DNA replication is initiated from position 890 (+/- 1) that proceeds in direction as indicated in Fig. 7.7. The *bla* gene nucleotides 2510-2442 (complementary strand) code for a single peptide. The map shows enzymes that cut pUC57 DNA once. The coordinates refer to the position of the first nucleotide in each recognition sequeznce.

6. pACYC184

The plasmid p15A was used to develop a small sized plasmid pACYC184 (4,245kb) (Fig. 7.8). The pACYC184 contains: (*i*) the replicon *rep* derived from plasmid p15A responsible for the replication, (*ii*) *tet* gene derived from pSC101 that encodes tetracycline resistance protein, and (*iii*) *cat* gene from transposon Tn9 that codes chloramphenicol acetyl transferase which confers resistance to chloramphenicol.

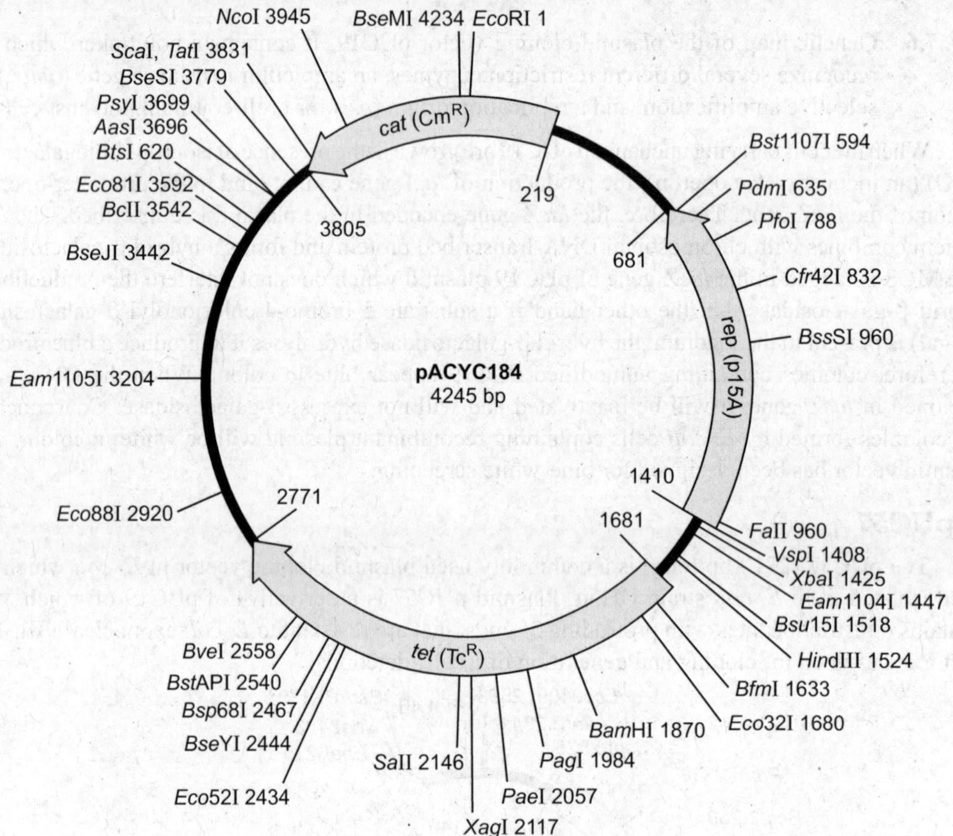

Fig. 7.8. Restriction map of pACYC184.

The circular sequence is numbered in such a way that 1 is the first G of the *Eco*RI site GAATTC and the count increases first through the 5′-terminal part of *cat* gene, the p15A material, the *tet* gene and finally through the 3′-terminal part of *cat* gene. The enzymes that cut pACYC184 DNA once are shown in the restriction map. The coordinates refer to the position of first nucleotide in each recognition sequence. The exact position of genetic elements is shown on the map (termination codons included). The indicated *rep* region is sufficient to promote DNA replication that initiates at position 846 (+/- 1) and proceeds in the direction indicated.

Normally derivatives of p15A are present in a lower copy number than pMB1 derivatives and cannot be amplified to the same extent as pMB1 derivatives. Amplification of this plasmid should be performed using spectinomycin due to the presence of the *cat* gene on pACYC184. The plasmid containing cloned DNA fragments can be detected by insertional inactivation of one of the antibiotic resistance genes. Its number can be amplified unlike pBR322 as the latter is not self-transmissible.

7. pUN121

Nilsson and co-workers developed a positive selection vector in 1983. The pUN121 (4.4 kb) deriving from pBR322 for rapid selection of recombinant plasmid-containing bacteria. This plasmid is used at the time of construction of genomic library. It consists of *amp*^{R} gene P_L promoter of phage λ (exchanging with original promoter region of *tet*^{R}), *cI* gene of phage λ cloning at pUN121. Protein expressed by *cI* gene suppresses the expression of the P_L promoter. Therefore, expression of *tet*^{R} gene is repressed, and the bacterial cells consisting of pUN121 will show resistance against ampicillin but not tetracycline.

The restriction sites for *Eco*RI, *Hind*III, *Bcl*I and *Sma*I are located within the *cI* gene. If a foreign DNA is cloned at any sites of *cI* gene, insertional inactivation of *cI* gene takes place. Hence, suppression of P_L promoter is abolished and the cells become resistant to tetracycline. If such cells are grown on mycelium containing tetracycline, these will form colonies. In these cells *cI* gene produces truncated repressor proteins which cannot repress the P_L promoter (Fig. 7.9).

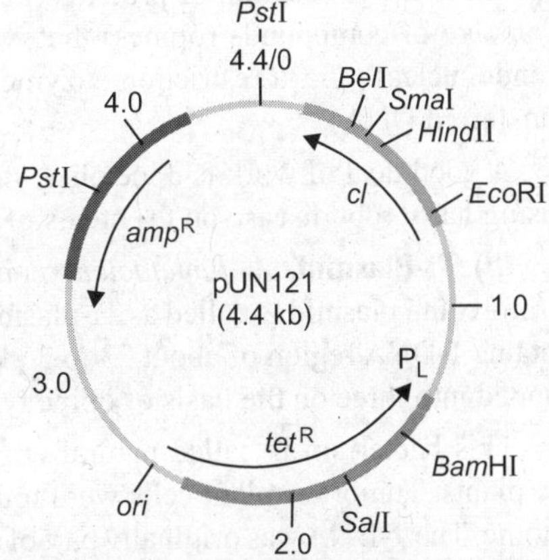

Fig. 7.9. The pUN121 vector without insertion of foreign DNA (A). The pUN121 in which a foreign DNA is cloned in *cI* gene (B).

8. *Agrobacterium*–based Plasmid Vectors

Agrobacterium tumefaciens is found commonly on and around root surfaces- the region termed as rhizosphere where it seems to survive by using nutrients that leak from the root tissues. But it infects only through wound sites, either naturally occurring or caused by transplanting of seedlings and nursery stock, and induces tumour called crown gall (Fig. 7.10). *A. tumefaciens* is a Gram-negative, non-sporing, motile, rod-shaped bacterium which is closely related to *Rhizobium* that forms nitrogen-fixing nodules on leguminous plants. In 1907, Smith and Townsend identified the bacterium as the causative agent of the disease, but it was not until the end of the sixties that a correlation between the tumour and the presence of genetic material of the bacterium was established. Between the 1970s and 1980s, some striking aspects were discovered about the biology, biochemistry, and molecular biology of *Agrobacterium*.

Fig. 7.10. *Agrobacterium tumefaciens* induces crown gall.

Strains of *Agrobacterium* are classified in three biovars based on their utilisation of different carbohydrates and other biochemical tests. The differences between biovars are determined by genes on the single circle of chromosomal DNA. Most of the genes involved in 'crown gall disease' are not borne on the chromosome of *A. tumefaciens* but on a large plasmid called Ti (tumour-inducing) plasmid.

In natural conditions, the motile cells of *A. tumefaciens* are attracted to wound sites by chemotaxis. This is partly a response to the release of sugars and other common root components and it is found even in plasmid-cured strains. However, strains that contain the Ti plasmid respond even more strongly because they recognise wound phenolic compounds such as acetosyringone which are strongly attractive at even very low concentrations (10^{-7} molar). Thus, one of the functions of the Ti plasmid is to code for additional, specific chemotactic receptors that are inserted in the bacterial membrane and enable the bacterium to recognise wound sites. Acetosyringone plays a further role in the infection process, because at higher concentrations (about 10^{-5} to 10^{-4} molar) it activates the virulence genes (*vir* genes) on the Ti plasmid. These genes coordinate the infection process and lead to the production of proteins (permeases) that are inserted in the bacterial cell membrane for uptake of compounds (opines) that will be produced by the tumours, cause the production of an endonuclease, *i.e.* a restriction enzyme that excises part of the Ti plasmid termed as the T-DNA (transferred DNA).

A good deal of work is done on *A. tumefaciens* causing crown gall disease, and *A. rhizogenes* causing hairy root disease on the stems of numerous plants.

(*a*) Ti-Plasmid. *A. tumefaciens* contains a large plasmid which induces tumour in plants; therefore, the plasmid is called as Ti-plasmid. The size of Ti-plasmid ranges between 180–250 kb. It contains T-DNA region of about 23 to 25 kb which is transferred into plant cells. Ti-plasmid can be grouped into three on the basis of opine types, for example, octopine, nopaline and agropine.

This bacterium is called natural engineer and exploited by scientists for gene delivery into new plants. Tumorous plant cells were found to contain DNA of bacterial origin integrated in their genome. The T-DNA was originally part of a small molecule of DNA located outside the chromosome of the bacterium. This DNA molecule was called Ti (tumor-inducing) plasmid (Fig. 7.11 A). The Ti plasmid contains most of the genes required for tumour formation. Wounded plants exude phenolic compounds that stimulate the expression of the virulence genes (*vir* genes) which are also located on the Ti plasmid. The *vir* genes encode a set of proteins responsible for the excision, transfer and integration of the T-DNA into the plant genome. The genes in the T-DNA region are responsible for the tumourigenic process. Some of them direct the production of plant growth hormones that cause proliferation of the transformed plant cells.

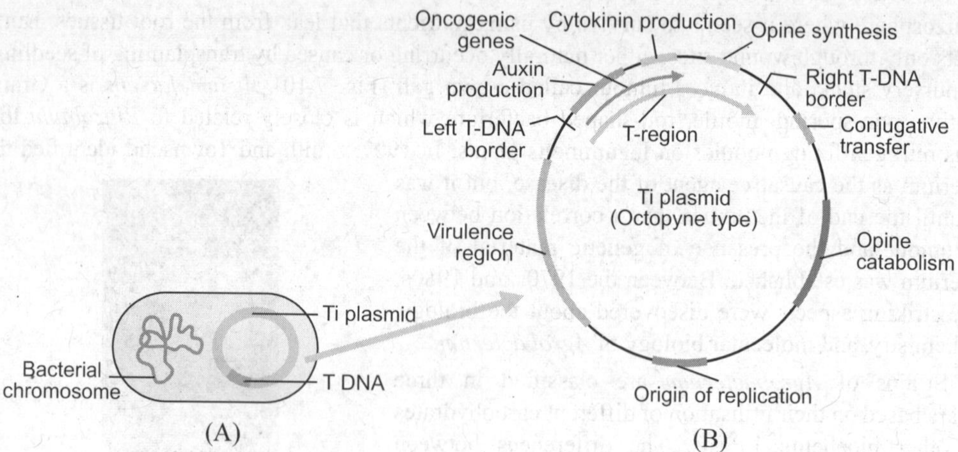

Fig. 7.11. *Agrobacterium tumefaciensis*; (A) - a cell showing bacterial chromosome and a Ti plasmid; (B) - genetic map of Ti plasmid.

The T-DNA region is flanked at both ends by 25 base pairs (bp) of nucleotides called T-DNA borders. Left side of T-DNA is known as TL-T-DNA and right side of the same as TR-T-DNA (Fig. 7.11 B). T-DNA left border is not essential, but the right border is indispensable for T-DNA transfer. The T-DNA border sequences, at least the right border initiates the integration of the T-DNA region into the plant genome. These are sequences of 25 bp imperfect repeats that flank the T-DNA and are required for its transfer. Border sequences encompass the recognition sites for a site-specific endonuclease, which is encoded by the *vir* D operon that is a part of the *vir* genes. The endonuclease cleaves the lower DNA strand of the T-DNA marking the starting point of transfer.

The *vir* genes are required for transfer of the T-DNA region to the plant. The *vir* genes have about 35 *vir* genes map outside the T-DNA (transferred DNA) region and encode products required for excision, transfer, and integration of T-DNA into a plant genome. The *vir* genes act in *trans*, *i.e.* they do not need to be physically attached to the T-DNA to cause integration into the plant genome. Both the essential (T-DNA) and non-essential regions of Ti-plasmids (both octopine and nopaline types) contain many genes. One or more unusual amino acid derivatives encoded by DNA is known as opines. Opine is neither found normally in plants nor required by plants.

A. tumefaciens uses it as a source of carbon and nitrogen for their growth and multiplication. The genes responsible for the biosynthesis (*ocs* gene for octopine and *nos* for nopaline) and degradation (*occ* for octopine and *noc* for nopaline catabolism) of opines are located at the well defined regions. the *arc* genes are required for arginin catabolism. The plasmid also carries transfer (*tra*) gene to help the transfer T-DNA from one bacterium to other bacterial or plant cells, *onc* gene for oncogenecity, *ori* gene for origin of replication and *inc* gene for incompatibility.

T-DNA of plasmid is stably integrated with plant DNA. The sequencing of nucleic acid at the junction between plant DNA and T-DNA has been done. T-region in both octopine and nopaline plasmids is flanked by a direct repeat of 25 base pairs. The *ops* gene encodes enzymes for the synthesis of opines in transformed cells which are required for poliferation of infecting bacteria. TL-T-DNA encodes two enzymes that are involved in biosynthesis of two phytohormones, auxin and cytokinin. This results in disorganised proliferation of cells commonly known as callus, gall or tumour. Structural formulae and diagrams of opine are given in Fig. 7.12.

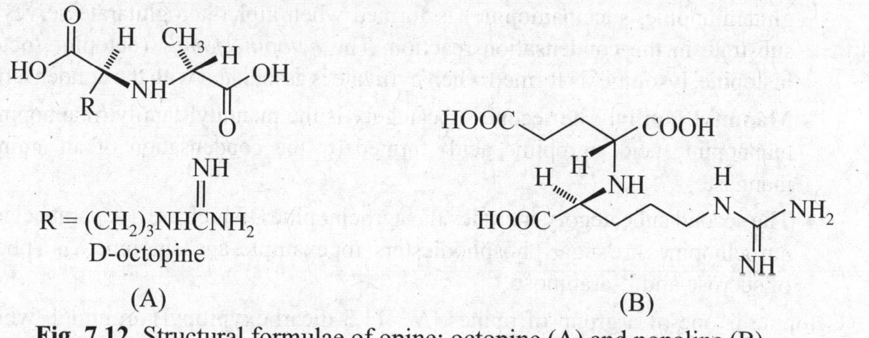

$R = (CH_2)_3NHCNH_2$ (=NH)

D-octopine

(A) (B)

Fig. 7.12. Structural formulae of opine; octopine (A) and nopaline (B).

***(i)* Opines:** Opines are the low molecular weight compounds found in plant crown gall or hairy root tumours formed by species of *Agrobacterium*. The bacterium inserts its T-DNA from Ti plasmid into the plant genome, the genes of which encodes specific enzymes that catalyses opine biosynthesis. In turn the bacterium uses opines as an important source of nitrogen and energy. A specific set of opines is induced and catabolised by each strain of *Agrobacterium*. So far about least 30 different opines are known. Some of examples of opines are given in Table 7.3.

Table 7.3. Examples of some opines and their source.

Opines	*Source*
Agropine	Agropine (1′-Deoxy-*D*-mannitol-1′-yl)-*L*-glutamine, 1′, 2′-lactone) is a member of the mannityl family, which was obtained from crown gall tumours.
Agropinic acid	Agropinic acid (*N*-1-(*D*-mannityl)-*L*-glutamic acid lactam) also belongs to the mannityl family, which is produced by crown gall tumours.

Glutaminopine	Glutaminopine (N^2-(*D*-1,3-dicarboxypropyl) derivative of glutamine) is isolated from crown gall tumour, which is also s a member of the nopaline family.
Heliopine	Heliopine (= vitopine) (N^2-(1-Carboxyethyl)glutamine) was obtained from crown gall tumours.
Histopine	Histopine (*N*-(*D*-1-carboxyethyl) histidine) is a member of the octopine family found in crown gall tumors.
Methopine	Methiopine (*N*-[1-*D*-(carboxyl) ethyl]-*L*-methionine) was found in crown gall tumors.
Nopaline	Nopaline (N^2-(~-1, 3-dicarboxypropyl)-*L*-arginine) was first isolated from crown gall tumors. It is the head member of the nopaline family of opines.
Nopalinic acid	Nopalinic acid (= ornaline) (N^2-(~-1, 3-dicarboxypropy1)-*L*-ornithine) is a member of the nopaline family found in crown gall tumors.
Octopine	Octopine (N^2-(*D*-1-carboxyethyl)-*L*-arginine) is the first opine discovered in 1927 in *octopus* muscle and later in crown gall tumours. It is the principal member of the octopine family of opines.
Octopinic acid	Octopinic acid (N^2-(*D*-1-carboxyethyl)-*L*-ornithine) is a member of the octopine family isolated from crown gall tumours.

On the basis of chemical structure, opines are put into two major subcategories:

- The first subcategory has nopaline and octopine families. The vast majority are secondary amine derivatives that are formed by condensation of an amino acid.
  - **Nopaline family:** The nopaline family (nopaline, nopalinic acid, leucinopine, glutaminopine, succinamopine) is formed when alpha-ketoglutarate serves as the keto substrate in the condensation reaction. The *octopine family* (octopine, octopinic acid, histopine, lysopine) is formed when pyruvate is associated with the condensation reaction.
  - **Mannityl family:** The second subcategory is the mannityl family (mannopine, agropine, mannopinic acid, agropinic acid) formed by the condensation of an amino-acid with mannose.
- The second subcategory includes the agrocinopines, which are a separate class of opines. agrocinopines are sugar-phosphodiesters; for example, agrocinopine A is a phosphodiester of sucrose and L-arabinose.

Octopine is one of a group of opines [N^2- (1,3-dicarboxypropyl) arginine], which is a rare amino acid derivative produced in tissue of certain type of crown galls. The genes responsible for the synthesis of nopaline are part of the T-DNA of the Ti-plasmid. Nopaline is one of theopines. Ti plasmids are classified on the basis of the different types of opines they produce. These may be nopaline plasmids, octopine plasmids and agropine plasmids. These opines are condensation products of amino acids and keto acids or may be from sugar. Opines are used as carbon and nitrogen sources by *Agrobacterium tumefaciens*.

***(b)* Ri-plasmid.** *Agrobacterium rhizogenes* causes formation of adventitious roots. This rhizogenicity has been correlated with the presence of a large plasmid, the Ri-plasmid for root induction parallel to Ti-plasmid. *A. rhizogenes* does not induce enzymes for the biosynthesis of auxin or cytokinin. Plasmid for this species also contains a T-DNA which causes the development of hairy roots. It has also been reported to confer drought tolerance on apple seedlings.

The possible advantage of using Ri-plasmid of *A. rhizogenes* is the production of secondary root systems in plants for : (*i*) better anchorage of plants, (*ii*) ability to resist anoxia from flooding of the soil, (*iii*) increased drought resistance due to high root density, and (*iv*) greater chance of interaction with soil-borne mycorrhizal fungi. It may be applied to establish the secondary root systems in certain plants (fruit trees) by inoculation of soil as well as beneficial for those plants which are propagated through cutting and rooting as stimulants for root formation on cuttings.

(*c*) Properties of Ti-plasmids. *Agrobacterium* plasmids have been used for introduction of genes of desirable traits into plants. The properties of Ti- plasmids are as below:

(*i*) Modified T-DNA region: They do not cause disease when disarmed by removing disease causing gene. A modified T-DNA region of the Ti plasmid, in which the genes responsible for tumour formation are removed by genetic engineering and replaced by foreign genes of diverse origin, e.g. from plants, bacteria, virus. When these genes are removed, transformed plant tissues or cells regenerate into normal-appearing plants, and in most cases the fertile plants.

(*ii*) The *cis*-regulatory regions: They possess sites for insertion of foreign gene which needs to be introduced. Besides, *cis*-regulatory regions that include the right and left T-DNA borders are physically attached to the genes to be transferred into the plant genome. Other *cis*-regulatory regions include promoters and terminators that flank the transgenes and regulate their expression. Commonly used promoters and terminators are the nopaline synthesis gene (NOS) and cauliflower mosaic virus (CaMV) 35S promoter.

(*iii*) Selectable marker genes: They possess selectable markers, *i.e.* genes which help in selecting the transformed cells. In plants selectable marker genes allow the identification and selection of cells with the gene of interest incorporated in their genome. Bacterial selectable markers permit the identification of bacteria transformed with a vector carrying the marker. Examples of plant and bacterial selectable markers are hygromycin phosphotransferase (*hpt*) and kanamycin phosphotransferase (*kpt*), respectively (for detail see Chapter 19: *Transgenic Plants*).

The selectable marker is a powerful promoter, for exampl 35S (for high level of expression of marker) derived from cauliflower mosaic virus (CaMV) and polyadenylation signal (for adding poly A to mRNA). The most commonly used selectable marker is neomycin phosphotrasferase (*npt*II) that confers resistance to neomycin and allows the cells to grow in medium containing neomycin. Thus, a vector is designed after inserting a foreign gene. Then vector is transferred to *A. tumifaciens* or *A. rhizogenes* depending on cases. The transformed bacterium is used to infect the designed host cell where the transformation of gene is to be brought about. For gene transfer, generally explants are used. The explants are protoplasts, suspension of cultured cells, clumps of callus, thin cell layers (epidermis), leaf discs, sliced tissue, floral tissues, stem and root selection.

The explants are incubated or co-cultured with *Agrobacterium* containing vector with modified foreign gene. The transformed cell or colonies are selected due to the selectable marker and used for regeneration of whole plant which are expressed to be transgenic in nature. These regenerated plants can again be tested for the transfer of gene with the help of several scoreable or screenable markers called 'reporter genes' which is also transferred along with desired genes. The most commonly used reporter genes are *npt*II, *cat, gus* and *lux*. In the process of *Agrobacterium* Ti plasmid system, two elements are used in modification. The first element consists of a plasmid vector containing a border sequence such that the vector functions as a T-DNA region. The vector also provides a shuttle that allows the recombinant DNA manipulation to be performed in *E. coli* hosts before subsequent transfer to *Agrobacterium*. Second, the transfer of vector DNA to the plant is mediated by a strain of *Agrobacterium* that contains a full complement of *vir* genes. Besides the two elements, there are considerable number of plant transformation vectors.

9. pTi-based Vectors

Since Ti plasmid cannot be used for cloning of DNA insert in *E. coli* due to large size and absence of unique restriction sites within the T-DNA, modified *E. coli* plasmid can be used for initial gene cloning. Therefore, gene transfer into plants can be done by employing either of the following two approaches, *i.e.* by using co-integrate vector and binary vector.

Disarming of Plasmids

The wild type pTi vector consists of oncogenes (*iaaM, iaaH and ipt*) in T-DNA (that causes callus growth of disorganised mass of cells), is large sized and lack unique cloning sites; therefore, it creates problem in using as cloning vector. These problems have been solved by 'disarming' the PTi vectors. The term disarming refers to deletion of oncogenes (that encode auxin and cytokine) from T-DNA of Ti plasmid. For the first time during 1980s, oncogenes of pTiC58 were replaced by pBR322 sequences but *nos* gene and both right border (RB) and left border (LB) were retained. The modified disarmed T-DNA of pTiC58 was transferred into cells of *Agrobacterium*; this disarmed bacterium did not cause tumour but regenerated planlets and produced nopaline. Thereafter it has become clear that for transferring any DNA insert only LB and RB are necessary. Thus disarmed pTi and pRi are used in gene transfer technology in a routine way.

(*a*) Co-integrate Vectors: The vectors which are derived by integrating the modified *E. coli* plasmid (containing the gene construct) with a resident disarmed Ti plasmid of *Agrobacterium* by homologous recombination via crossing over are called co-integrate vectors (or hybrid Ti plasmids). The co-integrate plasmid normally contains: (*i*) the *vir* genes, (*ii*) the left border and right border of T-DNA, (*iii*) an exogenous DNA sequence between the two T-DNA borders, and (*iv*) plant and bacterial selectable markers (Fig. 7.13). For co-integrate system presence of the following three vectors are necessary:

(*i*) Disarmed pTi of *Agrobacterium*: The oncogenes located in the T-DNA region in pTi plasmids are replaced by foreign DNA (DNA insert); following are the examples of these vectors:

- **SEV series:** In 1985, Robert T. Fraley and co-workers from Monsanto Company (U.S.A.) constructed a SEV system which is a derivative of the octopine type Ti plasmid (pTiB6S3) in which the T_L-DNA oncogenic functions. The T_L-DNA right border sequence and all of T_R-DNA are deleted and replaced with the kanamycin antibiotic resistance marker from Tn903 (601).The right border (RB) of the T-DNA (containing the phytohormone genes coding for cytokinin and auxin) are removed and replaced by a bacterial kanamycin resistance gene (*kan*^{R}). But the left border (LB) and a small part of the left segment (T_L) of the original T-DNA called left inside homology (LIH)) are left intact.

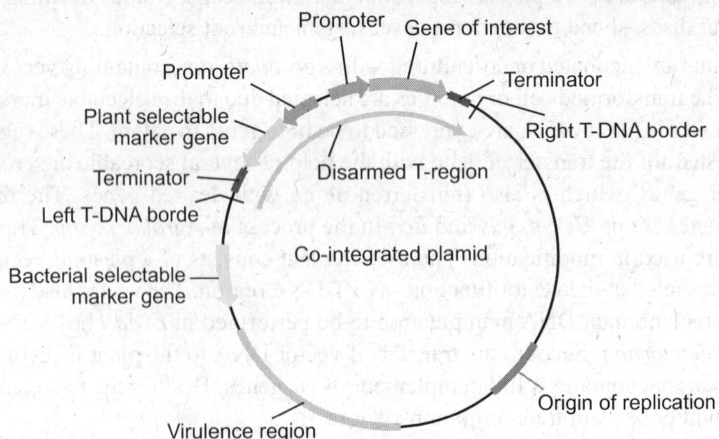

Fig. 7.13. A co-integrate vector.

- **pGV series:** The phytohormone genes located within the T-DNA are excised and substituted by a portion of nucleotide sequence of pBR322. The sequences of LB and RB, and the nopaline synthase gene (*nos*) of the Ti plasmid are conserved. An example of pGV series is pGV3850.

***(ii)* Intermediate vectors:** The small pBR322-based plasmids *(E. coli* vectors) containing a T-DNA region are called 'intermediate vectors'. Since the disarmed Ti plasmids are large sized and lack the unique restriction sites, they create problems in being used as cloning vectors. Therefore, intermediate vectors are used to overcome these problems. An intermediate vector (IV) is produced by modifying pBR322, which consists of *ori* in *E. coli*, pBR322 sequences (located in the T-DNA region of disarmed pTi), T-DNA (devoid of borders) from pTi and suitable selectable markers, such as *neo* gene (for selection of DNA insert) and *kan*^{R} (kanamycin resistance gene for selection co-integrate vector.

Recombinant plasmid IV is produced by propagating plasmid IV in *E. coli*; the DNA insert is readily inserted within the T-DNA present in it. This intermediate vector IV is allowed to replicate in *E. coli* and then is transferred into *Agrobacterium* by conjugation. Vector IV is transferred through a conjugation-proficient plasmid called 'helper plasmid' because IV is non-conjugant. They carry DNA segments homologous to the disarmed T-DNA to allow recombination to form a co-integrated T-DNA structure, because they cannot replicate in *A. tumefaciens*. Examples of an intermediate vectors are pMON120, pMON2, etc.

***(iii)* Helper vectors:** Small plasmids maintained in *E. coli* [that contain transfer (*tra*) and mobilisation (*mob*) genes] are called helper vectors. The transfer of the conjugation-deficient intermediate vectors into *Agrobacterium* is allowed by the helper vectors.

A pTi vector is disarmed by replacing the oncogenes located in T-DNA with suitable sequences of *E. coli* plasmid to be used for cloning of a desired foreign gene. For example, pGV3850 has been constructed by disarming the pTiC58 plasmid via replacing the oncogenes with sequences of pBR322 plasmid of *E. coli*. Recombination takes place within pBR322 region of both intermediate vector and pGV3850 plasmid (Fig. 7.14). The co-integrate pTi vectors are maintained by culture on kanamycin-supplemented medium.

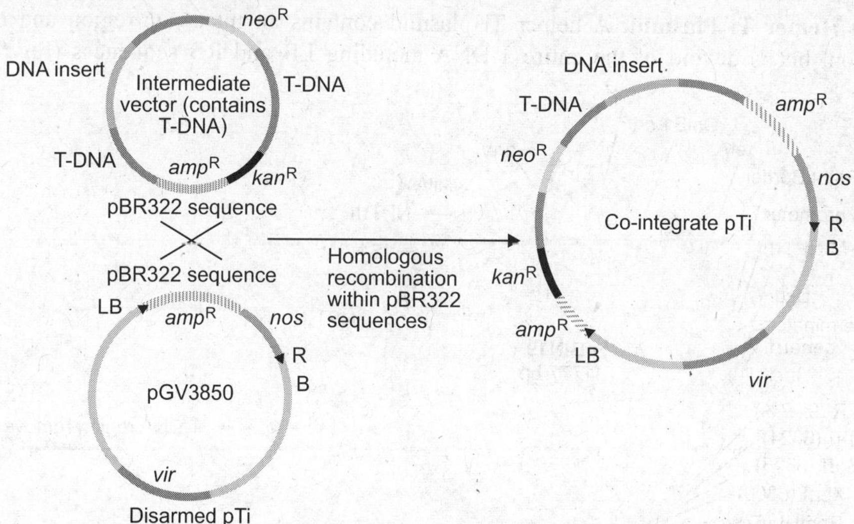

Fig. 7.14. Homologous recombination between an intermediate vector (a recombinant IV) and disarmed pTi plasmid (pGV3850); *neo*, neomycin phophotransferase; *nos*, nopaline synthase; *kan*^{R}, kanamycin resistance; *amp*^{R}, ampicillin resistance; *vir*, virulence regulation; RB, right border; LB, left border.

***(b)* Binary Vectors:** For efficient transfer and insertion of T-DNA into the plant genome, the *vir* genes do not need to be present in the same plasmid with a T-DNA region. This concept gave birth to the construction of a system for plant transformation where the T-DNA region and the *vir* region are present on separate plasmids. Basically, a binary vector is a pair of plasmids consisting of a binary plasmid (*i.e,* a wide-host-range small replicon) and a helper plasmid. Out of the two plasmids, one plasmid consists of disarmed T-DNA (but contains right and left borders of T-DNA), and the helper plasmid consists of *vir* region and does not contain T-DNA and both the borders. The two plasmids are used together to produce genetically modified plants. These vectors are created artificial from the naturally occurring Ti-plasmid of *A. tumefaciens*. The binary vector is able to replicate in both *E. coli* and *A. tumefaciens*. Commonly used binary vectors are: pBin19, pPVP, and pGreen. The two different plasmids employed in the binary vector system are given as below:

***(i)* Disarmed pTi Plasmid:** When the tumour-inducing genes located in the T-DNA are removed, the plasmid is said to be 'disarmed'. Examples of disarmed pTi are pBIN19 (Mini-Ti) and pAL4404 (helper pTi). The 'wide-host-range small replicon' of pBIN19 consists of an origin for replication (*ori*) for maintenance of the plasmid in a wide range of bacteria including *E. coli* and *Agrobacterium*. Typically binary plasmid consists of : (*i*) foreign DNA in place of T-DNA, (*ii*) the left and right T-DNA borders (or at least the right T-border), (*iii*) markers for selection and maintenance in both *E. coli* and *A. tumefaciens*, and (*iv*) a selectable marker for plants (Fig. 7.15 A).

The mini- vector pBIN19 is derived from the broad host range plasmid pRK252. It contains a truncated T-DNA from nopaline plasmid pTiT37 that consists of LB and RB and *nos* and *npt*II genes as a selectable marker. In the mini T-DNA a *lac*Z poly-linker is inserted to provide the cloning sites and colour test. The poly-linker contains galactosidase that results in blue colonies of transformed cells on plates supplemented with IPTG/X-GAL indicator and white colonies if *lac*Z gene is broken due to insertion of a foreign gene. Also the transformants can be selected on kanamycin-containing growth medium due to development of resistance provided by *neo* gene.

In 1995, David A. Frisch and co-workers from Texas A & M University (U.S.A.) have described the complete sequence of the binary vector pBIN19. They found that the non-T-DNA region also contains several superfluous insertions and some functional elements that may not contain optimal sequences.

***(ii)* Helper Ti Plasmid:** A helper Ti plasmid contains an intact *vir* region and origin of replication, but is devoid of the entire T-DNA including LB and RB sequences (Fig. 7.15 B).

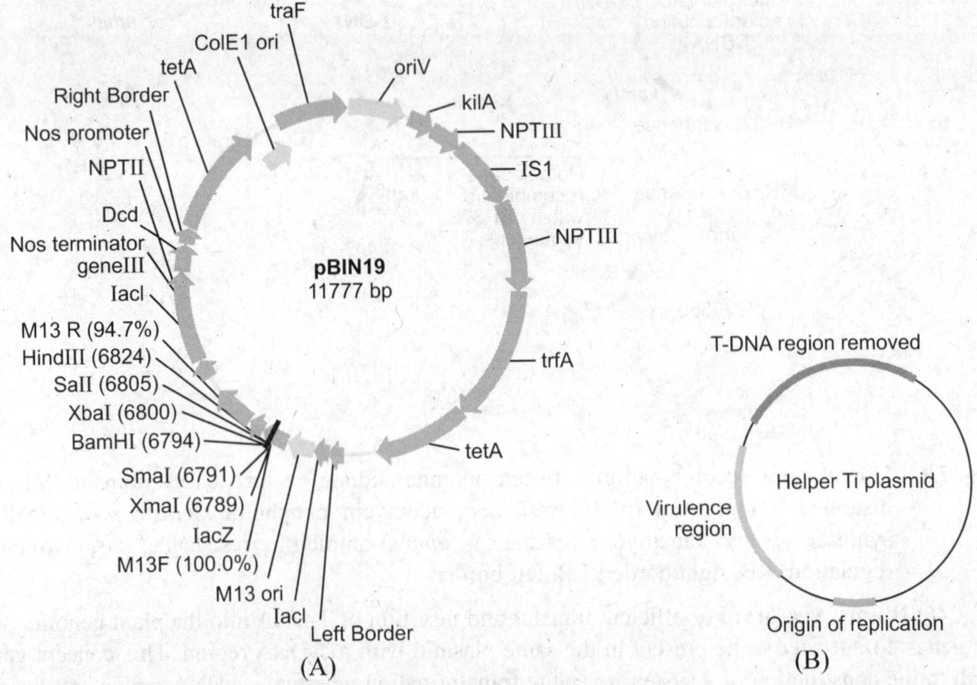

Fig. 7.15. A binary wide host range replicon (A); a helper Ti plasmid (B).

Example of a helper Ti plasmid is pAL4404 which is a derivative of octopine plasmid pTiAch5 where the T-DNA region is completely deleted, but contains the *vir* genes. The recombinant small replicon is transferred directly or through bacterial conjugation. *A. tumefaciens* harboring a helper Ti plasmid, the plant cells are co-cultivated with the *Agrobacterium* to allow transfer of recombinant T-DNA into the plant genome, and the transformed plant cells are selected under suitable conditions.

For most gene transfer studies involving regenerated plants, disarmed strains (e.g. LBA4404) are preferred. This strain can mediate transfer if a T-DNA region is supplied in trains. There are some advantages in using the binary vectors as compared with co-integrated vectors: (*i*) no recombination takes place between the molecules involved, and (*ii*) small vectors are used (instead of a very large, recombinant and disarmed Ti plasmid) which increases transfer efficiency from *E. coli* to *Agrobacterium*. Different types of binary vectors are given in Table 7.4.

The pGreen series are small plasmids of about 3.2 kb size that enable to replicate in *E. coli* but unable to replicate in *Agrobacterium* without the presence of another binary plasmid (pSoup) in the same strain. The pGreen vectors contain : (*i*) a broad host range replication origin (ori pSa) and a ColE1 origin derived from pUC, (*ii*) a pSa replicase gene (*rep* A) that provides replication functions in *trans* and is located in a compatible plasmid (pSoup) in *Agrobacterium*, and (*iii*) multiple cloning sites (*i.e.*, polylinkers) based on the pBlueScript vector, which allow any arrangement of selectable marker and reporter genes (Fig. 7.16).

Table 7.4. Types of binary vectors.

Binary vector series	***Properties***
pGA series	An *ori* derived from RK2 for replication in *E. coli* and *Agrobacterium*, a tetracycline resistance gene (*tet*R), the *cis*-acting factor required for conjugal transfer, the right and left T-DNA borders, a neomycin phosphotransferase (*nptII*) gene and G418 in transformed plants, and a polylinker site (multicloning site).
pCG series	The *ori* of the *A. rhizogenes* plasmid pRiHRI.
pCIT series	The hygromycin (*hph*) resistance gene for plants, lambda *cos* site for cloning long fragments.
pGPTV series	Different plant selectable marker genes close to the left T-DNA border.
pBECK2000 series	Synthetic T-DNA borders and a *bar* gene. The vectors use the phage P1 Cre/*lox*P site-specific recombinase system which permits the transfer and integration of a target and marker genes as a single T-DNA unit into the plant genome.
Binary-BAC (BiBAC) vector	It is based on a bacterial artificial chromosome (BAC) vector and is suitable for transformation of high-molecular-weight DNA. It comprises of low-copy number *ori* for both *E. coli* and *Agrobacterium* to ensure replication of the plasmid as a single-copy in both bacteria, and a helper plasmid carrying *vir*-genes for cloning very large T-DNAs (up to 150 kb) into the plant genome.
pGreen series	It contains a broad host range *ori* gene (*ori* pSa) and a ColE1 *ori* derived from pUC, a pSa replicase gene (*rep* A) for *trans* functions. It is located in a compatible plasmid (pSoup) in *Agrobacterium* and multiple cloning sites based on the pBlueScript vector.

The pGreen/pSoup dual-binary vector system: In 2005, A.S. Afolabi and co-workers from John Innes Centre (U.K.) reported a novel pGreen/pSoup dual-binary vector system in multiple T-DNA co-cultivation as a method for producing marker-free transgenic plants. In 2000, Roger P.

Hellens and coworkers from Norwich research park (U.K.) pGreen as a versatile and flexible binary Ti vector for *Agrobacterium*-mediated plant transformation. The pGreen series of binary Ti vectors are configured for ease of use and to meet the demands of a wide range of transformation methods for several plant species. This plasmid system allows any arrangement of selectable marker and reporter gene at the right and left T-DNA borders without compromising the choice of restriction sites for cloning, since the pGreen cloning sites are based on the well known pBluescript plasmids.

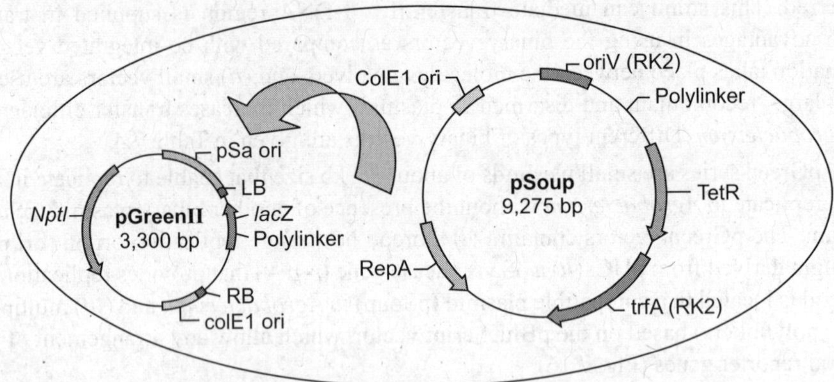

Fig. 7.16. *Agrobacterium tumefaciens* containing pGreenII and pSoup as binary vectors.

pGreen can replicate in *Agrobacterium* only if another plasmid, pSoup, is present in the same strain. pSoup provides replication functions *in trans* for pGreen. Both pGreen and pSoup can be maintained individually in *Escherichia coli*. Versions of pGreen have been used to transform several plant species with the same efficiencies as other binary Ti vectors. Similarly, pClean vectors are compatible with the dual binary vector system pGreen and pSoup, where two binary vectors (*i.e.* pGreen and pSoup) are both present in the same *Agrobacterium*.

B. BACTERIOPHAGE VECTORS

Bacteriophages are the viruses that infect bacterial cells through injection of their genetic material (DNA or RNA) resulting in cell lysis. The viral DNA replicates and expresses inside the bacterial cells, and produces a number of phage particles released after bursting the bacterial cells. This is called lytic cycle of bacteriophage; the released phages re-infect the live cells. The ability of transferring the viral DNA from phage capsid specific bacterial cell gave insight to the scientists to exploit bacteriophages and design them as cloning vectors. The two bacteriophages, for example phage λ and M13 have been modified, extensively studied and commonly used as cloning vectors.

1. Phage Lambda (λ) Vectors

Lambda phage was discovered by Esther Lederberg in 1951. It is a temperate bacteriophage of *Escherichia coli* K12 that consists of icosahedral head and a long tail. DNA is linear and double-stranded (48,502 nucleotides) with 12 bp single-stranded complementary 5'-ends. Phage λ contains a proteinaceous icosahedral (55 nm diam) head and a long tail (180 mm long, non-contractile and without sheath) attached to the head. The head has 20 faces. A 20-faces in 3-D image is called an icosahedron. The head is made of protein of several types and contains a double-stranded DNA. The tail consists of 35 stacked discs and contains a tail fiber at the end (Fig. 7.17 A-B). Upon adsorption on the *lamb* receptor of a host cell, lambda DNA is injected trough the tail, which forms a hollow tube through which the DNA passes. It possesses 50 genes in its genome of which about half of genes are essential. On attachment with tail to cell wall of *E. coli,* it injects its linear DNA into the cell. The linear double stranded DNA molecule cyclises through the single strand of 12 nucleotides commonly known as *cos* sited at its ends. Presence of *cos* (cohesive) site is the key feature of the λ DNA.

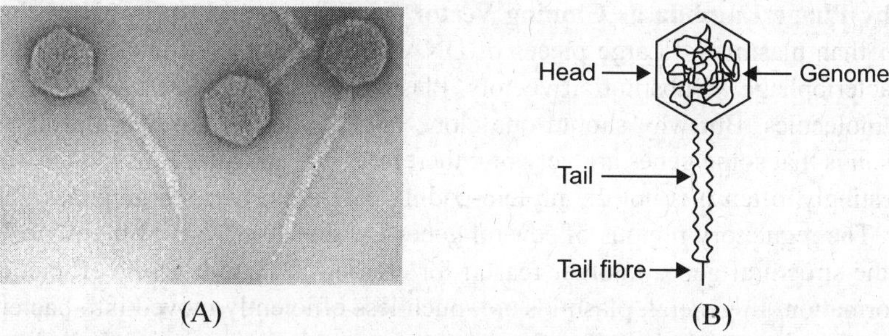

Fig. 7.17. Bacteriophage λ; (A)- an electron micrograph; (B)- structure of head and tail.

The 12 nucleotide long projections (5′GGGCGGCGACCT-3′) show cohesiveness and form the *cos* site. Phage genome has a large non-essential region which is not involved in cell lysis. Taking advantage of it two types of cloning vectors can be produced, either by inserting foreign DNA (insertion vectors) or replacing phage DNA with a foreign DNA (replacement vectors). The upper limit of foreign DNA to be packed is about 23 kb. After the phage particle injects its chromosome into the cell, the chromosome circularises by *cos* end joining as below:

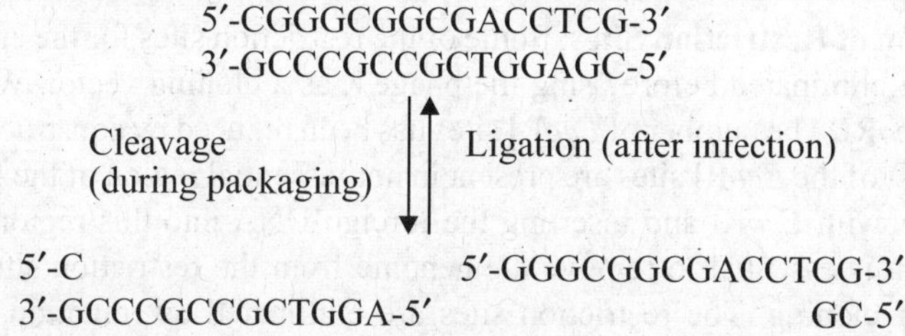

Replication cycle of phage λ is accomplished into two pathways: the lytic and lysogenic pathways. In the lytic pathway, early in the infection sites the circular DNA replicates as theta (θ) forms. By a rolling circle mechanism it produces the long concatemeric molecules joined end to end, and composed of several linearly arranged genomes. At the same time phage DNA directs the synthesis of many proteins required to produce empty heads where DNA is packed after the cleavage of concatemeric DNA at its *cos* site to yield fragments of such sizes as to fit in their heads. Eventually, a tail is attached to the head and finally the mature phage particles are released out of the bacterial cell, where the infected areas appears as clear dots (or plaques) surrounded by live cells. Bacteriophage particles may be recovered from the clear areas. Each plaque originates from a single infected cell.

The lysogenic pathway for replication is another alternative mode of propagation where it becomes stably integrated into the host chromosome and replicated along with the bacterial chromosome. Phage genome integrates by an attachment site (*att*) with a partially homologous site on the *E. coli* chromosome, where it replicates as a chromosomal DNA segment. In this case a protein is produced by *cl* gene which represses all the genes responsible for lytic pathways. In this pathway no phage structural proteins are synthesised. The interactions of two proteins, the *cl* genes expressed protein (by phage genome) and *cro* gene expressed proteins (by *E.coli* chromosomes) decide between the events of the lytic and lysogenic pathways. Phage λ has about 50 genes where only about 50% of these are necessary for growth and plaque formation. Non-essential DNA sequence, therefore, is replaced by donor DNA and the inserted DNA propagated as phage λ recombinant with no detrimental to its replication or packaging. For a detailed description see Λ *Textbook of Microbiology* by R.C. Dubey and D.K. Maheshwari (2010).

Unlike plasmid vectors, the phage vectors are required for cloning of large DNA fragment and, therefore, the gene bank or genomic libraries can be constructed. They are an alternative to bacterial plasmids, and related cosmids.

(*a*) **Why Phage Lambda as Cloning Vector ? :** Why should we use bacteriophage λ as a vector rather than plasmids ? Large pieces of DNA (up to about 20 kilobase pairs) can be easily cloned in bacteriophage λ substitution vectors. Plasmid vectors are less useful for cloning large-sized DNA molecules. But why should one clone large pieces of DNA in the first place ? One obvious reason is that some genes are very big; therefore, it is advantageous to have them all in one piece. Interestingly, often it is not the protein-coding portion of these large genes that contributes to their size. The regulatory regions of several genes are dispersed tens of thousands of base pairs away from the structural gene. Another reason for cloning in lambda is the efficiency it offers in DNA transformation. In general, plasmids are much less efficiently moved into bacterial cells than bacteriophage, which were designed by the nature for their ability to project their chromosome into their hosts.

Advantages of Phage Cloning System

To sum up the advantages of phage cloning system over the plasmids are that : (*i*) DNA can be packed with high efficiency *in vitro* into phage particles and transduced into *E. coli*, (*ii*) foreign DNA up to 20 Kb in length can be inserted into phage vector, and (*iii*) screening and storage of recombinant DNA is easier.

(*b*) **Elimination of Restriction Sites:** Some of the restriction sites for the enzymes, commonly used in cloning, are eliminated before using the phage λ as a cloning vector. Wild type λ has five cleavage sites for *Eco*RI. The number of *Eco*RI sites has been reduced by constructing the derivatives of λ vectors. The rest of the *Eco*RI sites are present in non-essential region of the genome. This helps in cleaving phage λ with *Eco*RI and inserting the foreign DNA into this region. Before using the phage λ as vector, it is essential to remove the genome from the restriction sites for the enzymes commonly used for cloning. The restriction sites are eliminated by mutation or deletion before obtaining a useful cloning vector. Two types of phage cloning vectors have been constructed: insertion vectors (having single restriction site into which foreign DNA can be inserted), and replacement vectors (having two restriction sites where a non-essential segment of vector DNA is replaced with foreign DNA).

Phage λ and cosmid vectors fulfill the requirements of cloning the whole genome in eukaryotes. Phage λ contains about 20–25 kb long segment and cosmid contains 45 kb longsegment. Several phage λ-based cloning vectors have been constructed, for example λgt10 and λgt11, EMBL3 and EMBL 4, charon, etc.

(*i*) **Insertion vectors:** Insertion vectors have unique single cleavage site for one or more restriction enzymes into which relatively small piece of foreign DNA is inserted. This enables DNA fragments to be inserted into lambda genome. Foreign DNA fragment does not affect the function of phage. The upper and lower limit of size of DNA that may be packed into phage particles is between 35 and 53 kb. Therefore, the minimum size of vector must be above 35 kb. It means that the maximum size of a foreign DNA to be inserted is about 18 Kb. Examples of insertion vectors are λgt10 and charon 16A.

- **λgt10:** Phage λ is modified to construct λ gt10 that may clone cDNA fragment. The λgt10 is a dsDNA of 43.3 kb which may clone 7 kb long foreign DNA fragment (Fig. 5.17). It consists of an *Eco*RI site which cleaves left arm of 32.7 kb and right arm of 10.6 kb that accepts about 7.6 kb long insert DNA. *Eco*RI site lies within the *cI* gene (repressor) which forms the basis of selection method based on plaque formation. After inserting DNA, cI^{+} gene is inactivated; therefore, cI^{-} recombinant bacteriophage is formed. The recombinant cI^{-} λgt10 after infecting *E. coli* forms clear plaques which can easily be screened from cloudy plaques formed by non-recombinant λ gt10 cI^{+} vector.

Insertional cloning into the *cI* gene of the λgt10 cDNA cloning vector (DNA inserts of ~1-5 kb) can be selected in hfl (high frequency of lysogeny) mutant strains of *E. coli*. In hflA strains of *E. coli*, expression of the lambda *cII* gene is elevated, resulting in transcriptional induction of the lambda cI repressor gene which promotes lysogeny. Disruption of the lambda *cI* coding sequence by DNA insertion into the unique *Eco*RI site of the λgt10 cDNA cloning vector, blocks the lysogenic pathway leading to cell lysis and plaque formation.

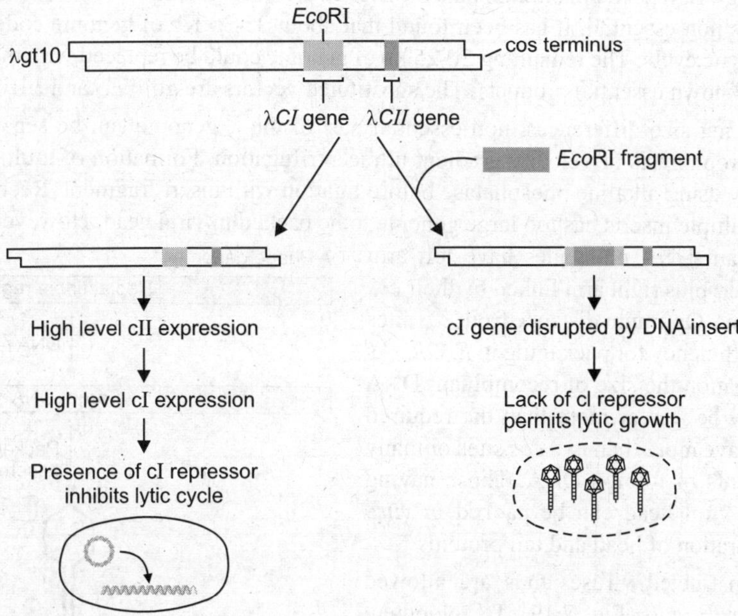

Fig. 7.18. Control of lytic cycle in phage λgt10 cloning vector. See *c*I and *c*II gene and restriction site of *Eco*RI.

- **λgt11:** It is a 43.7 kb long phage dsDNA constructed for cloning mostly cDNA (less than 6 kb). It is an expression vector where the inserted foreign DNA encodes as β-galactosidase fusion protein. After insertion of a foreign DNA, the recombinant λgt11 is converted to λgt11 *gal*$^-$ and non-recombinant λgt11 remains *gal*$^+$. Therefore, by using IPTG (inducer) and X-gal (substrate) with suitable *E. coli* host cells, recombinant λgt11 can be screened. Recombinant λgt11 *gal*$^-$ forms white (clear) colonies while non-recombinant λgt11 *gal*$^+$ forms blue colonies.

cI λgt10
λgt10
Left arm 32.7
Right arm 10.6
EcoRI site

- **Charon 16A:** In 1977, F.R. Blattner and co-workers developed the charon series insertion vectors. Charon 7 consists of *imm*434 and serves as an insertion vector for both *Hind*III and *Eco*RI. Whereas charon 12 contains *lac*Z gene. In charon 16 the arm generated by *Eco*RI digestion are 19.9 kb (left arm) and 21.9 kb (right arm). Fragments of about 9 kb may be cloned. The site for *Eco*RI in charon 16 lies within the β-galactosidase gene (*lac*Z) which helps in detection of recombinants by using X-gal as substrate. Insertion of DNA into the single *Eco*R1 will inactivate the *lac*Z gene of charon 12 and charon 16. The active enzyme results in formation of blue colonies or plaque, whereas after inactivation of *lac*Z colourless colonies are formed. Bacteriophage λ-based insertion vector charon 16A shows *lac*Z gene and recognition site of *Eco*RI.

lac Z
Left arm 19.9
Right arm 21.9
*Eco*RI site

(*ii*) **Replacement or substitution vectors:** The replacement vectors have cleavage sites present on either side of a length of non-essential DNA of phage. As a result of cleavage left and right arms are formed, each arm has a terminal *cos* site and longer a stuffer region, the non-essential region, which can be substituted by foreign DNA fragment. The DNA to be cloned is first inserted into the λ DNA, replacing a nonessential region (Fig. 7.19). Then, by an *in vitro* assembly system, the λ virion carrying the recombinant DNA can be formed. The λ genome is 49 kb in length which can carry up to 25 kb foreign DNA. The maximum size of inserted DNA fragment depends on how much of the phage DNA is non-essential. It has been found that about 25-30 Kb of genome codes for essential products for lytic cycle. The remaining 20-25 kb of genome could be replaced with the foreign DNA fragments of known essential products. The substituted vectors are gt, *WES* and λ1059.

For making size differences, non-essential part of the λ genome can be separated from the arms by electrophoresis or velocity gradient ultracentrifugation. Formation of multiple inserts can be checked by using alkaline phosphatase before ligation with insert fragment. Recombinant DNA formed by multiple inserts has too large genome to be packed in viral head. However, after ligation the recombinant DNA molecules have left arm plus large insert plus right arm linked by their *cos* sites at the arms. Optimum distance from *cos* sites governs the efficiency for packaging in *E. coli.* As a result of ligation the size of recombinant DNA fragment may be less or more than the required size or may have more than two *cos* sites or many small fragments of foreign DNA. Those having the range of viral head can be packed *in vitro* using a preparation of head and tail proteins.

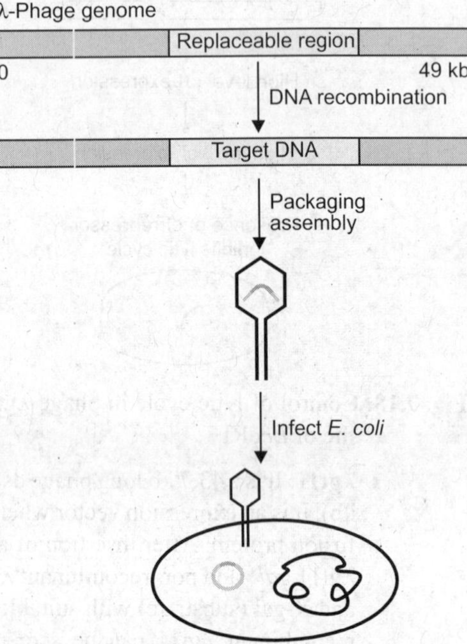

Fig. 7.19. Schematic drawing of the DNA cloning using λ phages as replacement vector.

The constructed viruses thus are allowed to multiply in *E. coli* (Fig. 7.19). Development of plaque turbidity is a useful criterion for the selection of recombinant phages. Plaque turbidity is determined by the presence of non-lysed bacteria. The recombinant phages give clear plaques due to inactivation of *cI* gene. By using these methods the clones containing recombinant DNA can be isolated from the wild type clones *i.e.* turbid plaques. The constructed genome has all the information required for DNA replication and synthesis of the viral protein. The inserts can be identified by colony hybridisation technique.

The phage DNA has been most widely used as phage cloning vector so far, because of ease in handling and screening of large number of recombinant DNA containing phages. That is why it is used as a tool in the building of gene bank, for example, rat, yeast, mouse and human gene banks. Examples of replacement vectors are: EMBL3, EMBL4, charon 40, charon 34, charon 35, and λDASH. In these vectors the central non-essential region of about 44 kb long phage DNA is replaced by a foreign DNA of known function, because 40% of the central region of phage is non-essential for propagation.

- **λ EMBL3 and λ EMBL4 Replacement Vectors:** The EMBL (European Microbiology Laboratory) vectors are the replacement vectors which are used to clone a large fragment (9–25 kb) of genomic DNA. In 1983, Anna-Maria Frischauf and co-workers from European Molecular Biology Laboratory (Federal Republic of Germany) have made a new family of lambda replacement vectors (EMBL1, EMBL2, EMBL3, EMBL4) and derivatives containing amber mutations (EMBL3 *Sam*, EMBL3 *AamBam*, EMBL3 *AamSam*), to simplify the construction and screening of genomic libraries. These vectors have a large capacity and polylinker sequences flanking the middle fragment. The polylinkers allow a choice of cloning enzymes and, especially useful in the case of cloning of Sau3A partial digests, the excision of the entire insert by flanking *Sal*I (EMBL3) or *Eco*RI (EMBL4) sites. EMBL3 and EMBL4 are two such replacement vectors used for preparing genomic libraries in eukaryotes with cloned fragments, 15-25 kb in size. The two vectors have polylinkers with reverse orders of restriction sites with respect to each other.

EMBL3 and EMBL4 vectors are designed in such a way that a central non-essential part of 44 kb long phage can be replaced by a foreign DNA. Cleavage of the phage with an appropriate enzyme generates three fragments (left arm, right arm and a central fragment called stuffer). The central fragment representing 40% of the phage genome is nonessential for propagation of the phage and can be replaced by foreign DNA that may be as long as 20-23 kb. Non recombinant DNA resulting from fusion of left arm and right arm will give a DNA that is too small for packaging, so that there is automatic selection against non-recombinant phages.

Several fragments of the genomic DNA are produce by partial digestion with *Sau*3A1 which has a tetranucleotide recognition site. The fragments are cloned at *Bam*HI site. It should be noted that the DNA fragments produced by *Sau*3A1 and *Bam*HI possess identical cohesive ends. In λ EMBL 3 the *red/gam* genes of λ DNA is replaced with *Bam*HI-generated DNA fragments at the site *trp*E gene. In cohesive sites of these vectors three recognition sites for *Eco*RI, *Bam*HI and *Sal*I are present (Fig. 7.20).

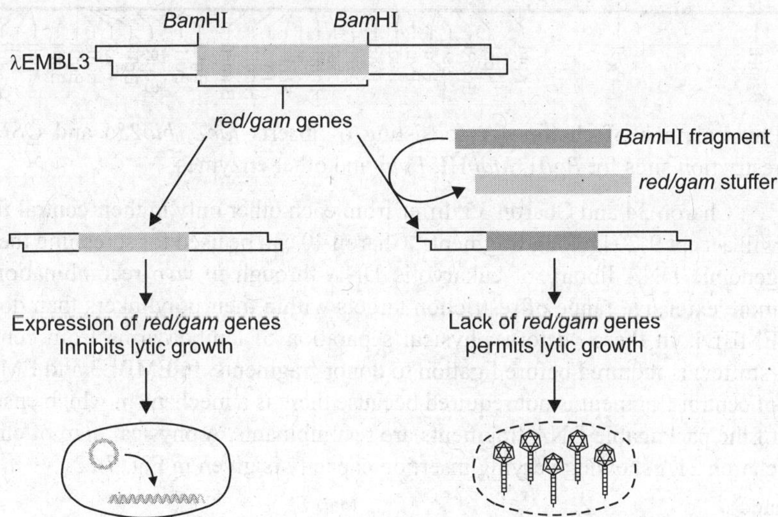

Fig. 7.20. Expression of *red/gam* genes of λ EMBL3 vector.

The inserted foreign DNA can be retrieved by using *Eco*RI or *Sau*I. Suitable *E. coli* cells are infected by bacteriophages containing recombinant DNA and cloned DNA is multiplied. The clones containing recombinant DNA during *in vitro* packaging can be directly selected and easily stored. The phage λ EMBL 3 and λ EMBL 4 contain polylinkers in opposite orientations. They permit the cloning of large fragments generated by any of these enzymes.

Replacement of lambda DNA containing the *red* and *gam* genes in the λ EMBL3 genomic DNA cloning vector with *Bam*HI compatible DNA inserts of ~10-20 kb, permits lytic growth of recombinant phage in *E. coli* strains containing the P2 bacteriophage lysogen.

- **Charons:** Charon is a cloning vector made from the bacteriophage lambda that is used to clone DNA. In 1977, F.R. Blattner and coworkers developed the charon lambda bacteriophages as vectors for cloning. Their construction incorporates mutations that make them simple to use and also greatly increases their safety for the biological containment of cloned recombinant DNA. Three of the Charon vector phages, 3A, 4A, and 16A, have been certified for use as EK2 vector-host systems, when propagated in bulk in a special bacterial host, DP50SupF. Thereafter, a large number of charon vectors have been developed. The replacement vectors have been described to clone large fragment of DNA molecules. Its stuffer consists of DNA molecule polymerised head to tail. It can be cleaved by a single restriction enxyme *Nae*. The other charon series vectors are charon 4A, charon 34 and charon 35. Physical map of charon 4A consisting of inserts *lacZ*, *bio256* and *QSR80* is given in Fig. 7.21.

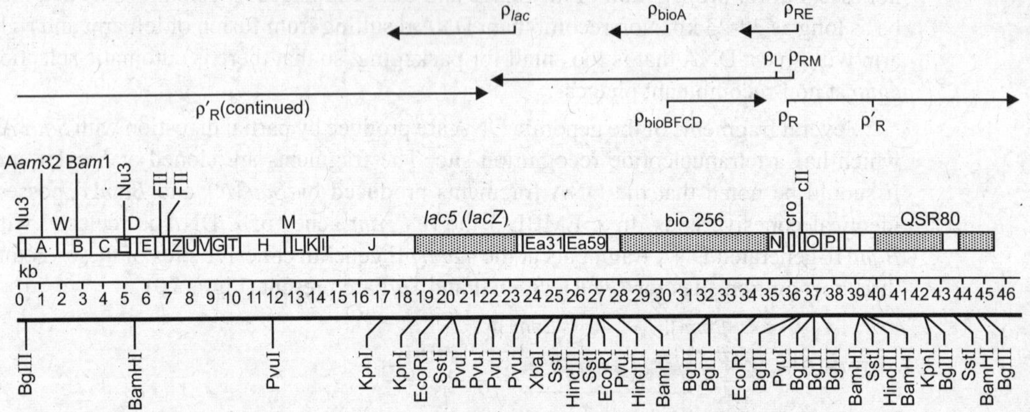

Fig. 7.21. Physical map of charon 4A consisting of inserts *lacZ*, *bio256* and *QSR80*, showing restriction sites for *Bgl*II, *Mab*HI, *Pvu*I and other enzymes.

Charon 34 and Charon 35 differ from each other only in their central fragments and will accept 9-20 kb long fragments. Charon 40 can be used for screening the recombinant genomic DNA library of eukaryotic DNA through *in vivo* recombination. They have more extensive range of restriction targets within their polylinkers than do EMBL3 and EMBL4. In these vectors, physical separation of lambda arms from central fragment (stuffer) is required before ligation to donor fragments. In EMBL3 and EMBL4 removal of central fragment is not required because there is a mechanism which ensures that 98% of the packagable DNA fragments are recombinants. A physical map of charons 3A and charon 21A showing varying insertion capacity is given in Fig. 7.22.

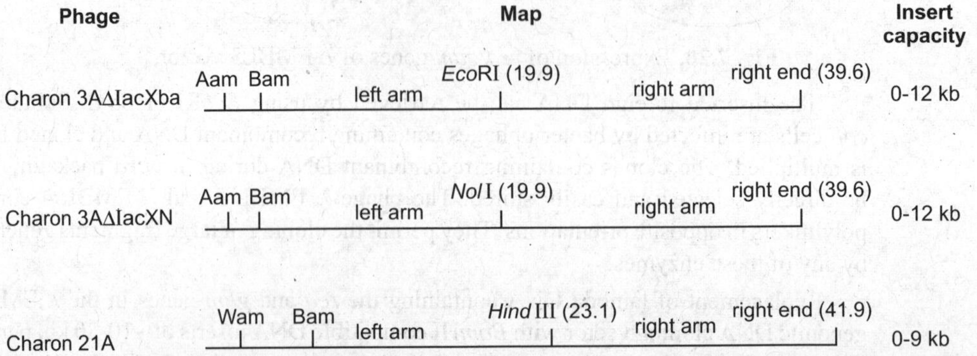

Fig. 7.22. A physical map of charons 3A and charon 21A showing varying insertion capacity.

2. Cosmid Vectors

In 1976, Yasuyuki Fukumaki and co-workers from the Department of Biochemistry, Kyushu University School of Medicine (Japan) showed that the presence of a small segment of phage λ DNA containing cohesive end on the plasmid molecule is a sufficient prerequisite for *in vitro* packaging of this DNA into infectious particles. The cosmids can be defined as the hybrid vectors derived from plasmids which contain *cos* site of phage λ (cosmid = *cos* site + plasmid). In 1978, a small cosmid for efficient cloning of large DNA fragments was developed for the first time by Barbara Collins (Switzerland) and John Hohn (Federal Republic of Germany). They described the production and use of the 6 kb cosmid pHC79, a derivative of pBR322. It can be used for cloning of fragments cleaved by *Eco*RI, *Cla*I, *Bam*HI (also *Bgl*II, *Bcl*I, *Sau*3A and *Mbo*I), *Sal*I (also *Xho*I and *Ava*I), *Eca*I and *Pst*I. Hybrid cosmids containing inserts in the size range of 40 kb are packaged *in vitro* and transduced with an efficiency of $5 \times 10^4 - 5 \times 10^5$ clones/µg of insert DNA.

Cosmids lack genes encoding viral proteins; therefore, neither viral particles are formed within the host cells nor cell lysis occurs. Special features of cosmids similar to plasmids are the presence of : (*i*) origin of replication, (*ii*) a marker gene coding for antibiotic resistance, (*iii*) a special cleavage site for the insertion of foreign DNA, and (*iv*) the small size.

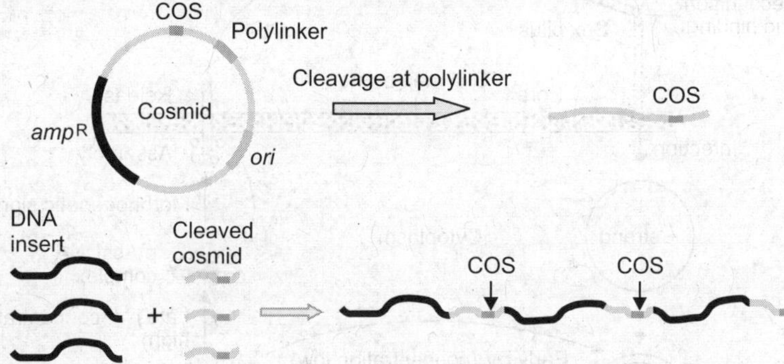

Fig. 7.23. Cloning by using cosmid vectors; A- in addition to *amp*^{r}, *ori*, and polylinker as in the plasmid vector, the cosmid vector also contains a COS site; B- cosmid vectors cleaved with restriction enzyme are ligated with DNA insert. The subsequent assembly and transformation steps are the same as cloning with λ phages.

Character dissimilar to plasmid is the presence of extra phage DNA, the *cos* site, which has about 12 bases. It helps the whole genome in circularisation and ligation. The cosmids have a length of about 5 kb, the upper size limit of the foreign DNA fragment that may be inserted in cosmids and packed into phage particles is, therefore, approximately 45 kb, which is much larger than it would be possible to clone in phage λ or plasmid vector. According to the size of *cos* sites and upper size limit in the head of phage, the recombinant DNA molecules can be packed into bacteriophage particles in *in vitro* packaging system consisting of packaging enzymes, head and tail proteins.

Procedure of DNA cloning by using cosmid vector is shown in Fig. 7.23. Upon transfection of *E. coli* by bacteriophage, the recombinant DNA cyclizes through *cos* sites and then replicates as a plasmid and expresses the drug resistance marker. Recently, based on cosmid vectors, a number of cosmid vectors have been determined from *E. coli*, yeast, and mammalian cells, and gene bank has been constructed.

2. Phage M13 Vectors

M13 is a filamentous phage of *E. coli* which infects only such cells which contain sex pili. It is closely related to the other coliphages such as fd and f1. M13 consists of single stranded circular DNA molecule containing 10 genes arranged in 6,407 base pairs.

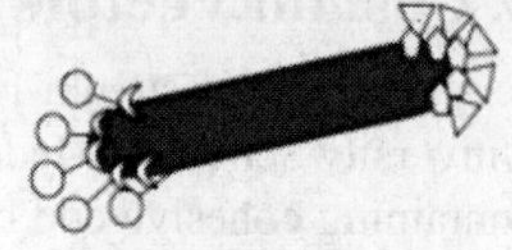

Bacteriophage M13

It lands on sex pilus and binds to receptor proteins present on the surface of sex pili which are specified by F plasmid. Soon it transfects the sex pilus and delivers its DNA into the cell through the lumen of pilus (Fig. 7.24). After delivery, the DNA is converted from single stranded molecule to double stranded molecule. It is called *replicative form* (RF) of DNA. The RF of M13 is double stranded, RF contains both (+) and (–) strands. The RF of M13 replicates and forms 50-100 RF molecules per cell. Finally, single stranded DNA molecules both (+) and (–) strands are produced from RF. Phage M13 encodes single strand binding proteins that bind to the viral single stranded DNA which are synthesised by rolling circle mechanism. Several phage encoded proteins are involved in M13 replication. Finally, the progeny molecules extrude from the cell as M13 particles. The progeny molecules are packed upon extrusion through the cell wall. The cells are not killed but grow slowly.

Foreign DNA of about 500 bp can be cloned into a multiple cloning sequence that forms part of a cloned modified *lac*Z gene on the double stranded RF of M13. Different strategies are followed when large sized foreign DNA (< 2,000 bp) is to be cloned.

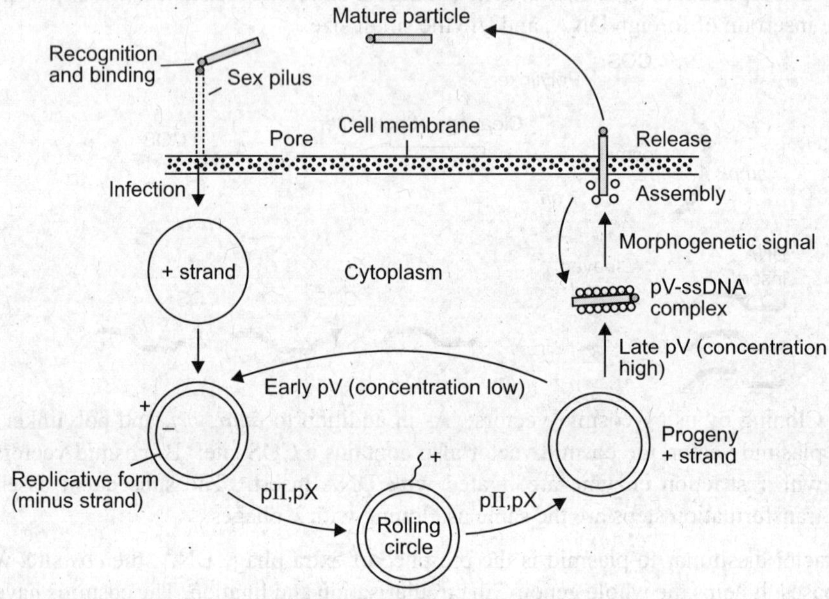

Fig. 7.24. Multiplication of M13 phage in *E.coli* cells.

DNA from the ligation reaction is mixed with the competent *E. coli* cells. Then the cells are plated on nutrient medium containing X-gal as substrate and the plates are incubated for 24 hours. Thereafter, both white (colourless) and blue colonies (M13 plaques) develop on the medium containing X-gal. X-gal turns blue which is hydrolysed by hybrid β-galactosidase where *lac*Z gene is not disrupted, *i.e. lac*Z gene remains functional and lacks any inserted DNA. On the other hand the colourless colonies consist of inserted DNA that has disrupted *lac*Z gene and did not produce β-galactosidase. Therefore, X-gal could not be disintegrated.

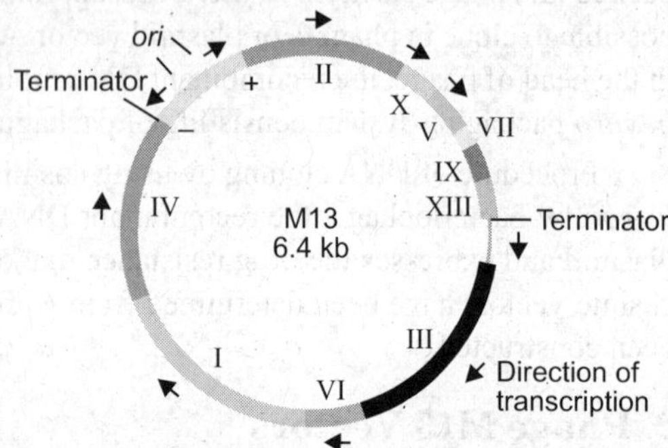

Fig. 7.25. M13 cloning vector.

The M13 particles are isolated from the white colonies. It is used as a source of M13 single stranded DNA which is further employed for DNA sequencing. A primer oligonucleotide is added to the M13 single stranded cloned DNA construct that hybridises the vector DNA near the insertion site of the cloned DNA. The dideoxynucleotide reaction is carried out; the sequence of the insert is read from the autoradiographs.

***(a)* Advantages of using M13 Vectors :** M13 phage has the following advantages :

(*i*) Very large insert can be cloned, as the size of the phage particle is decided by the size of the viral DNA. Hence there are no insert size restrictions like plasmid. M13 host vectors can carry an insert of up to 95 kb size.

(*ii*) The phage DNA is replicated via double stranded circular DNA intermediates. These circular DNA can be isolated and transfected as plasmids (Fig. 7.25).

(*iii*) M13 phages can provide a single stranded DNA molecule which is used in DNA sequencing.

***(b)* M13mp18/M13mp19 Vectors :** In 1983, J. Norrander and co-workers constructed improved M13-based vectors (*i.e.* M13mp18 and M13mp19) by using oligodeoxynucleotide-directed mutagenesis. The M13mp18 and M13mp19 vectors are derivatives of the single-stranded, male-specific filamentous DNA bacteriophage M13 (Fig. 7.26). Both vectors are 7,250 bp in length and have the same MCS inserted in opposing orientations. The vector DNAs contain a region of *E. coli* operon *lacZ* containing CAP protein binding site, promoter P_{lac}, *lac* repressor binding site and 5′-terminal part of *lacZ* gene encoding the *N*-terminal fragment of beta-galactosidase (codons 6-7 of *lacZ* are replaced by MCS).

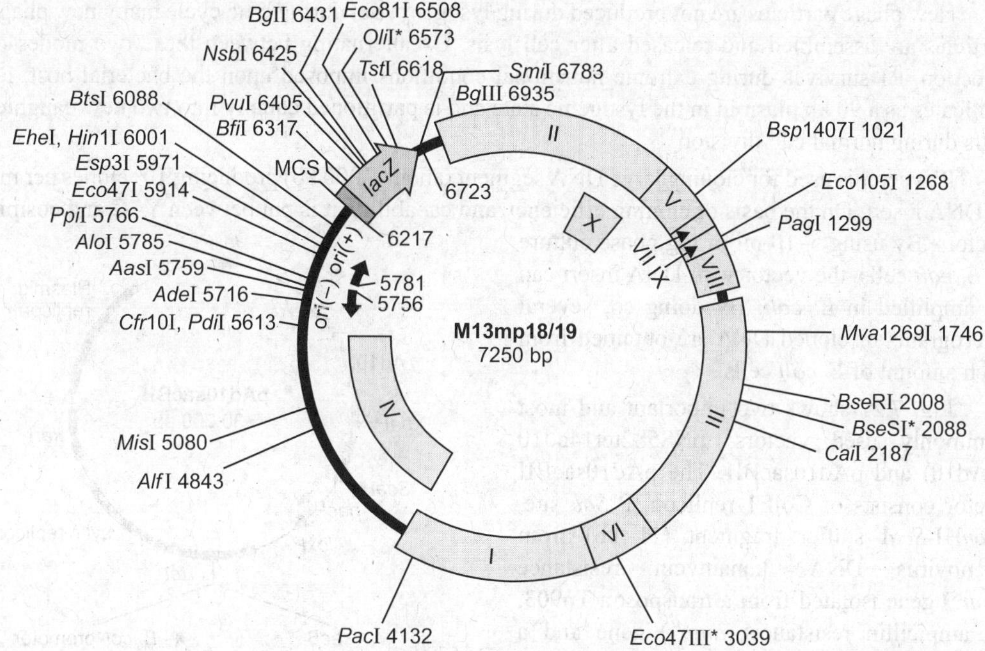

Fig. 7.26. Restriction map of M13mp18/19.

The M13mp18/19 genes are shown on the map (M13 genes are transcribed clockwise). The map shows enzymes that cut M13mp18/19 DNA once. The coordinates refer to the position of first nucleotide in each recognition sequence. The fragment whose synthesis can be induced by IPTG is capable of intra-allelic complementation with a defective form of beta-galactosidase encoded by host (mutation *lacZ*DM15). This results in appearance of blue plaques on media containing

IPTG and X-gal. Recombinant phages containing inserts that destroy the reading frame of *lacZ* are revealed as colorless plaques. Synthesis of viral (plus) single-stranded DNA requires the phage-encoded gene II, X and V proteins. It is initiated at *ori* (+) and proceeds in the indicated direction. The conversion of plus DNA strands to double strands does not require any of the phage genes. DNA synthesis is initiated by a 30 nucleotide-long RNA primer synthesised by host's RNA polymerase and starting at *ori* (–). Double-stranded circular DNA (replicative form, or RF) can be isolated from cells by standard plasmid preparation techniques and used for cloning experiments, while the single-stranded viral DNA (+ strand) can be isolated from phage particles collected from culture medium.

3. Phage P1 Vector

Phage P1 infects a range of enteric bacteria including *E. coli* and *Shigella.* It is a temperate phage, *i.e.* it exists as a prophage within a host cell. The genome of phage P1 is 93.6 kb. However, phage P1 packages about twice as much DNA as lambda whose genome is 48.5 KB. Phage P1 can infect and transduce *Salmonella* if the *Salmonella* O-antigen is removed because the O-antigen blocks the phage absorption sites. Rather, P1 prophage exists as an autonomously replicating plasmid that is maintained at about 1 copy per cell. During lysogeny its genome is not incorporated into the bacterial chromosome as is commonly observed during lysogeny of other bacteriophage. Instead, P1 exists independently within the bacterial cell like a plasmid. In lysogeny, P1 can exist within a bacterial cell as a prophage during lysogeny; in that it exists by replicating with the host chromosome and does not cause cell death.

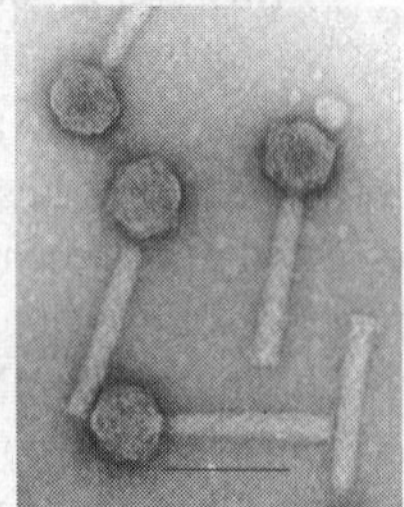

Bacteriophage P1

New phage particles are not produced during lysogeny; but during lytic cycle many new phage particles are assembled and released after cell lysis. By alternating between these two modes of infection, P1 survives during extreme nutritional conditions imposed upon the bacterial host. P1 replicates as a 90 kb plasmid in the lysogenic state and is partitioned equally into two new daughter cells during normal cell division.

P1 vector is used for cloning large DNA segment (about 1,000 kb) producing 10^5 clones per mg of DNA insert. On the basis of cloning efficiency and capability, it is put between YAC and cosmid vectors. By using 5–10 ml of log phase culture of *E. coli* cells, the vector with DNA insert can be amplified in *E. coli.* By doing so, several micrograms of cloned DNA are obtained from such amount of *E. coli* cells.

Fig. 7.27 shows two important and most commonly used vectors: pNS582tet14ad10 (pAd10) and pAd10sacBII. The pAd10sacBII vector consists of ColE1 replicon, P *Sac* site, *Bam*HI-*Sca*I stuffer fragment (11 kb) from adenovirus DNA, kanamycin resistance (*kan*^R^) gene isolated from a transposon Tn903, an ampicillin resistance (*amp*^R^) gene and a recombination site, *lox*P from P1. The IPTG induces the cells containing this vector. If the growth medium contains sucrose, the cells containing this vector die. Therefore, the transformed cells can be selected using sucrose in media. T7 and Sp6 promoters, rare cleavage sites for *Sal*I, *Sfi*I and *Not*I are also present; this

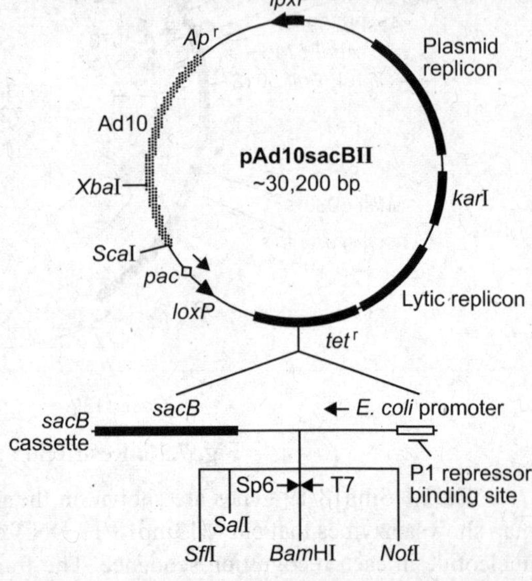

Fig. 7.27. P1 phage cloning vector pAd10sacBII vector.

helps to recover the DNA insert. There are several advantages in using these vectors such as : (*i*) production of multiple copies of genome from representative libraries, (*ii*) production of large amount of DNA clone very easily, (*iii*) separation of cloning process easily, (*iv*) dissection of cloned DNA easily, etc.

In 1990, N. Sternberg introduced *Sac*BII cassette as a mean of selecting for clones with inserts. In this vector system the cloned inserts are introduced between SP6 and T7 RNA polymerase promoters, positioned of the *Bacillus subtilis* genes in the *Sac*BII cassette. During library construction the sequences between the *loxP* sites (containing the Ad10 sequences) are deleted. In 1994, P.A. Loannou and co-workers developed the cloning system through P1-derived artificial chromosome (PAC) for the propagation of large human DNA fragments. This vector consists of characteristics of both P1 and F factor system. The PAC vector contains DNA insert of 100–300 kb and can be transferred into *E. coli* cells through electroporation. The recombinant PAC is stable with *E. coli* cells. This vector consists of genes : for resistance to the antibiotic kanamycin, a plasmid origin of replication that provides a single copy, a lytic origin of replication to allow 20-fold amplification of the plasmid, a *loxP* site (the *cis*-acting site specific recognition signal for the P1 recombinase), and a gene for resistance to tetracycline. Inserts cloned in the vector pNS583tet14Ad10 have been introduced into the *Bam*HI site in the gene that allows tetracycline resistance.

C. PHAGEMID (PHASMID) AND FOSMID VECTORS

A phagemid or phasmid is a type of cloning vector developed as a co-infection of the M13 helper phage and plasmid to produce a smaller version of the virus; an example is PRS314 which is of 4.78 kb size. Phagemids contain an *ori* for double stranded replication as well as an *ori* for single stranded replication, mostly not comprising the entire phagemid (*i.e.* only a small part of the phagemid is copied as a single strand).

Using a single stranded DNA containing bacteriophage M13, a series of cloning vectors were developed. Therefore, a single stranded DNA containing vectors becomes important for DNA sequencing and site-directed mutagenesis. The bacteriophage fl is closely related to M13 which also infects *E. coli*. *E. coli* maintains a plasmid as double stranded DNA due to the presence of *ori* gene. If *E. coli* cells are infected by the helper fl phage, then the fl *ori* is activated. It switches to a mode of replication and produces single stranded DNA which is packed into phage particles as they are extruded from the host cell.

1. Phagemid pBluescript II KS (+/–)

All pBluescript II phagemids are 2961 bp in length, which are the derivative of pUC19 vector. The structure of these phagemids is very similar to that of pTZ19R/U except for the MCSs and the promoters flanking the MCS. pBluescript II phagemids are designed for DNA cloning, dideoxy DNA sequencing, *in vitro* mutagenesis and *in vitro* transcription in a single system. The pBluescript II SK and KS vector series represent two orientations of the MCS within the *lacZ* gene encoding the *N*-terminal fragment of beta-galactosidase (KS represents the orientation of the MCS in which *lacZ* transcription proceeds from *Kpn*I to *Sac*I, while SK - from *Sac*I to *Kpn*I) (Fig. 7.28).

The multiple cloning site (MCS) is flanked by T3 and T7 promoters in opposite directions on the two strands. The MCS is set up for *lac*Z blue-white screening. A *lac*I promoter complements with *lac*Z of *E. coli* upstream of *lac*Z region. This feature helps in selection of recombinant vector based on the criteria of development of blue-white colonies (white colonies are formed if foreign DNA is inserted in the vector). It consists of fl (+) and fl (–) origin of replication obtained from a filamentous phage fl. It is used to get sense (+) and antisense (–) strands of *lac*Z gene after co-infecting the host with a helper phage. But in the absence of the helper phage an origin of replication (ColE1 *ori* from a plasmid) is used. Moreover, for antibiotic selection of recombinant phagemid vector, *amp*^{R} (ampicillin resistance gene) is also used. In these genes a number of restriction sites are present in this vector.

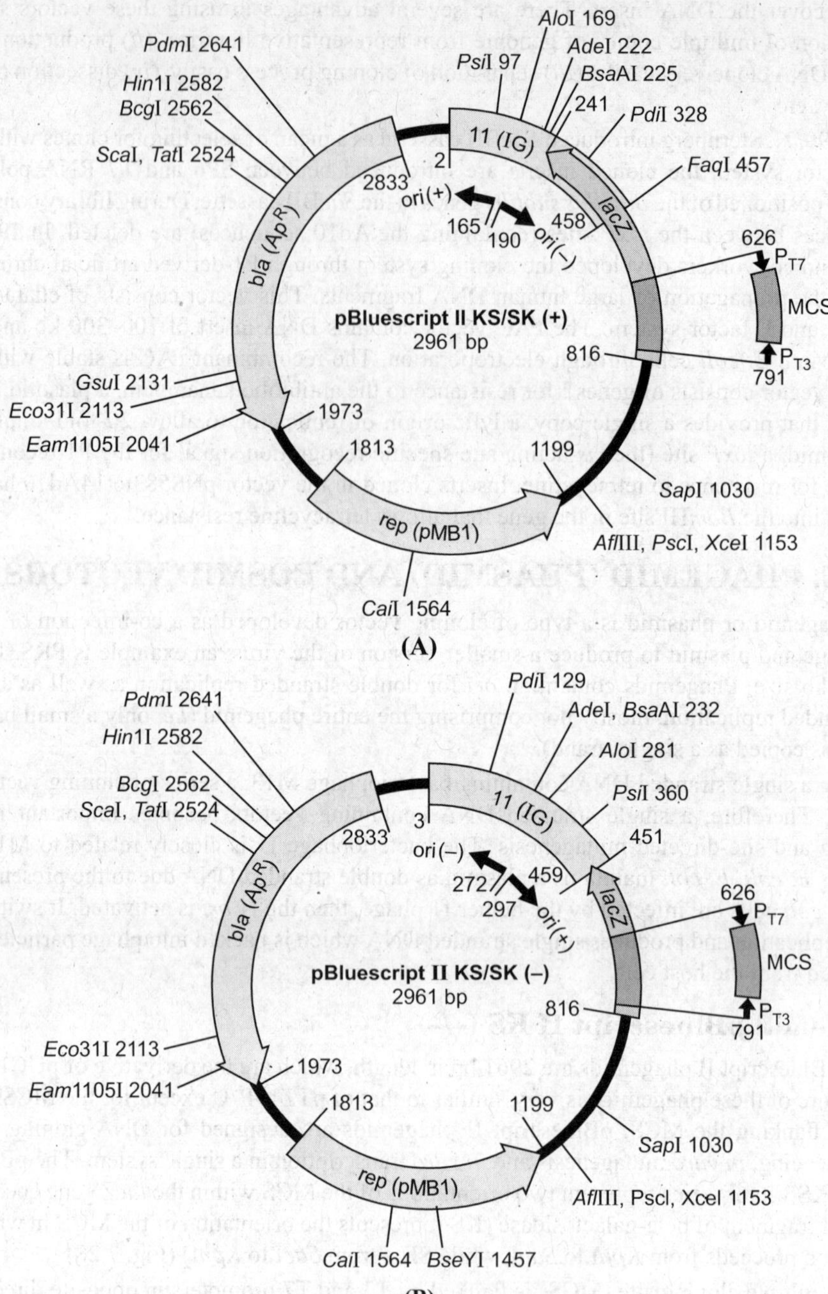

Fig. 7.28. The phagemid vectors; (A)- pBluescript II KS (+) vector; (B)- pBluscript II SK/KS (–) vector.

The (+) and (–) symbols on pBluescript II phagemids indicate the orientation of the cloned phage f1 intergenic (IG) region carrying the sequences required in *cis* for initiation and termination of phage f1 DNA synthesis and for packaging of DNA into bacteriophage particles. Synthesis of single-stranded DNA requires the phage-encoded gene II, X and V proteins and is initiated at *ori* (+). It proceeds in the direction indicated. The conversion of single-stranded DNA strands to double strands does not require any of the phage genes. DNA synthesis is initiated by a 30-nucleotide RNA primer synthesised by host's RNA polymerase and starting at *ori* (–).

The pBluescript II phagemids contain the following : (*i*) f1 (IG) - the intergenic region of phage f1, (*ii*) *rep* (pMB1) - the pMB1 replicon responsible for the replication of phagemid. The indicated region is sufficient to promote replication. DNA replication initiates at position 1213 (+/– 1) and proceeds in the direction indicated, (*iii*) *bla* (ApR) (gene coding for beta-lactamase that confers resistance to ampicillin). Nucleotides 2,833-2,765 (complementary strand) code for a signal peptide, (*iv*) *lacZ* - 5′-terminal part of *lacZ* gene encoding the N-terminal fragment of beta-galactosidase. This fragment allows blue/white screening of recombinant phagemids. The LacZ polypeptide corresponding to *wt* β-galactosidase and essential for blue/white screening ends at nucleotides position 460 (complementary strand). Other codons in the same reading frame come from f1 DNA (Alting-Mees *et al.*, 1992).

The maps show enzymes that cut pBluescript II SK/KS(+) and pBluescript II SK/KS(–) DNA once. The coordinates refer to the position of first nucleotide in each recognition sequence. The exact position of genetic elements is shown on the map (termination codons included) (Fig. 7.28).

2. Phagemids pTZ19R and pTZ19U

The pTZ19R and pTZ19U are the small phagemids of 2862 bp in length (Fig. 7.29). They were constructed by inserting the DNA of phage f1 intergenic region (IG) in pUC19 as well as introduction of T7 promoter sequence near to the MCS of Puc19 (Mead *et al.*, 1986). The phagemids differ in the orientation of cloned f1 IG region. They are designed for DNA cloning, dideoxy DNA sequencing, *in vitro* mutagenesis and *in vitro* transcription in one system. A host strain harboring these phagemids will produce single-stranded DNA if superinfected with the helper phage M13K07.

The pTZ19R/U phagemids contain: (*i*) the pMB1 replicon *rep* responsible for the replication of phagemid (from plasmid pUC19), (*ii*) *bla* gene, coding for beta-lactamase that confers resistance to ampicillin (from plasmid pUC19), (*iii*) region of *E.coli* operon *lac* containing CAP protein binding site, promoter P_{lac}, *lac* repressor binding site and 5'-terminal part of *lacZ* gene encoding the *N*-terminal fragment of beta-galactosidase (this fragment allows blue/white screening of recombinant phagemids in the same way as described in pUC18/19 section), (*iv*) T7 promoter inserted near to MCS of pUC19 allowing *in vitro* synthesis of large amounts of specific RNA, and (*v*) phage f1 intergenic region carrying the sequences required in *cis* for initiation and termination of phage f1 DNA synthesis (+ and – strands) and for packaging of DNA into bacteriophage particles. Synthesis of single-stranded (plus) DNA requires the phage-encoded gene II, X and V proteins. It is initiated at *ori* (+) and proceeds in indicated direction. The conversion of plus DNA strands to double strands does not require any of the phage genes. DNA synthesis is initiated by a 30-nucleotide RNA primer synthesized by host's RNA polymerase and starting at ori (–) (Mead *et al.*, 1986).

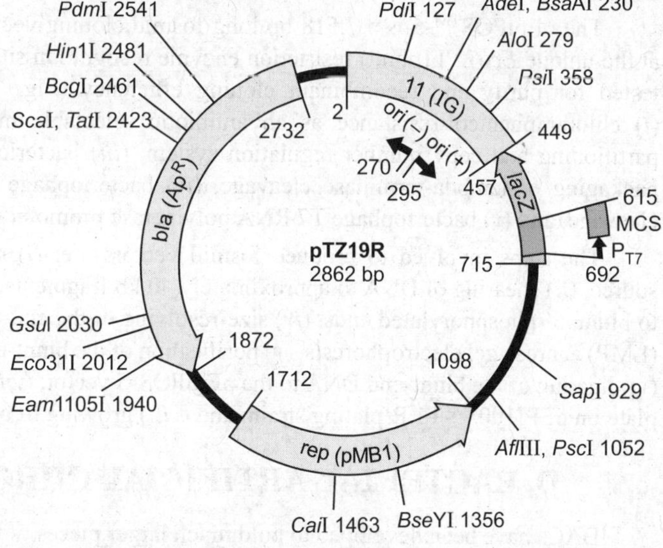

Fig. 7.29. Restriction map of phagemid pTZ19R.

The map shows enzymes that cut pTZ19R DNA once. The coordinates refer to the position of first nucleotide in each recognition sequence. The exact position of genetic elements is shown on the map (termination codons included). In the case of the *bla* gene, nucleotides 2732-2664 (complementary strand) code for a signal peptide. LacZ polypeptide corresponding to *wt* beta-galactosidase and essential for blue/white screening ends at nucleotide position 458 (complementary strand). Remaining amino acids in LacZ reading frame are encoded by f1 DNA. The indicated *rep* region is sufficient to promote DNA replication that initiates in the complementary DNA strand at position 1112 (+/– 1) and proceeds in the indicated direction. Phagemids carrying the pMB1 and ColE1 replicons are incompatible with one another, but are fully compatible with those carrying p15A replicon (pACYC1777, pACYC184). The pMB1-derived plasmids can be amplified by using chloramphenicol (Mead *et al.*, 1986).

3. Fosmid Vectors

Fosmids are plasmids that use the F-plasmid origin of replication and partitioning mechanisms to allow cloning of large DNA fragments. Fosmids are constructed on the basis of bacterial F-plasmid. Since a host (usually *E. coli*) can only contain one fosmid molecule, the cloning vector is limited. Low copy number offers higher stability than comparable high copy number cosmids. Fosmid clones were used to help assess the accuracy of the Public Human Genome Sequence. Fosmid vectors 1-3 provide an improved method for cloning and the stable maintenance of cosmid-sized (35-45 kb) libraries in *E. coli*. The stability of such large constructs *in vivo* is facilitated by the pEpiFOS™-5 vector that maintains the clones at single copy in the cell. Shearing the DNA results in the generation of highly random DNA fragments in contrast to more biased libraries that result from fragmenting the DNA by partial restriction digests.

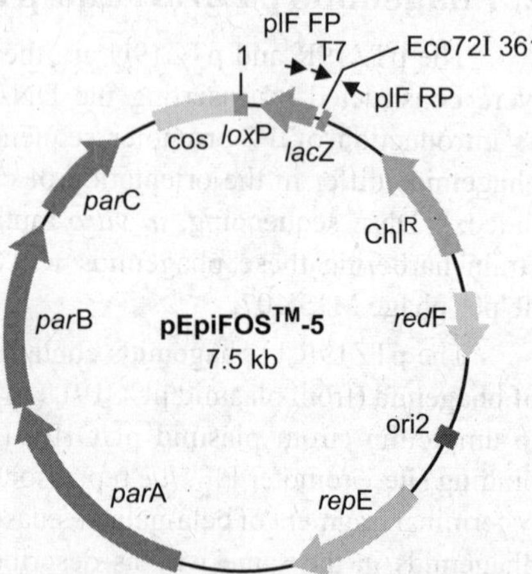

Fig.7.30. EpiFOS™-5 fosmid vector.

The pEpiFOS™-5 is a 7,518 bp long fosmid cloning vector which is provided by linearised at the unique *Eco*72 I (blunt) restriction enzyme recognition site, dephosphorylated and rigorously tested for purity and recombinant cloning efficiency (Fig. 7.30). Features of the vector are: (*i*) chloramphenicol-resistance as an antibiotic selectable marker, (*ii*) *E. coli* F factor-based partitioning and copy number regulation system, (*iii*) bacteriophage lambda *cos* site for lambda packaging or lambda-terminase cleavage, (*iv*) bacteriophage P1 *lox*P site for Cre-recombinase cleavage, and (*v*) bacteriophage T7 RNA polymerase promoter flanking the cloning site.

The steps involved to produce fosmid vectors are: (*i*) purification DNA from the desired source, (*ii*) shearing of DNA to approximately 40 kb fragments, (*iii*) end-repairing of sheared DNA to blunt, 5′-phosphorylated ends, (*iv*) size-resolving of the end-repaired DNA by low melting point (LMP) agarose gel electrophoresis, (*v*) purification of the blunt-end DNA from the LMP agarose gel, (*vi*) ligating of the blunt-end DNA to the pEpiFOS-5 vector, (*vii*) packaging of the ligated DNA and plate on pEPI100™-T1 R plating strain, and (*viii*) growing of clones overnight.

D. BACTERIAL ARTIFICIAL CHROMOSOME (BAC)

BACs have been developed to hold much larger pieces of DNA than a plasmid can. Therefore, it is also called 'super-sized inserts'. BAC vectors were originally created from part of an unusual plasmid present in some bacteria called the F′ plasmid. The F′ plasmid allows bacteria to have 'sex'

(F′ helps bacteria to give its genome to another bacteria but this only happens rarely when bacteria are under a lot of stress). F′ had been studied extensively and it was found that it could hold up to a million base pairs of DNA from another bacteria. Also, F′ has origin of replication and bacteria have a way to control how F′ is copied.

In 1992, Hiroaki Shizuya and co-workers from California institute of Technology discovered the cloning and stable maintenance of 300kb fragments of human DNA in *E. coli* using an F factor-based vector, which is an alternative to YAC vectors. BACs are maintained in *E. coli* as a large single copy. These are used as vectors for cloning very large (>50 kb) sequences of DNA over 300 generations. Sometimes these fragments turn out to comprise non-contiguous (non-adjacent) segments of the genome and frequently lose parts of the DNA during propagation. The BAC system facilitates construction of DNA libraries of complex genomes because of high cloning efficiency, easy manipulation of the cloned DNA, and stable maintenance of inserted DNA.

BAC vectors contain *ori* gene, *rep*E gene for maintenance of F factor, genes (*par*A, B, C) for plasmid copy number, an antibiotic resistance gene, for example chloramphenicol acetyltransferase (chl^R) gene for selection of plasmid and many restriction sites for insertion of foreign DNA. The F factor encodes its own DNA polymerase and is maintained in the cell as one or two copies. BACs are maintained as single copy plasmids in *E. coli* excluding the replication of more than one BAC in the same host cell. The upper limit of foreign DNA to be inserted in BAC is about 300–3500 kb. BACs are alternative to YACs and, therefore, being used in geneome sequencing projects.

The pBeloBAC11 (7,507 bp) is an *E. coli* plasmid cloning vector that has been designed for the construction of BACs (Fig. 7.31). It is maintained in single copy, which allows the cloning and stable maintenance of very large DNA fragments (up to 300 kb). The vector encodes the SopAB functions for active partitioning on the basis of the Ori2 (OriS) replicon of the F factor of *E. coli*. These functions act at SopC to ensure that each daughter cell gets a copy of the plasmid. Initiation factor RepE mediates the assembly of a replication complex at Ori2.

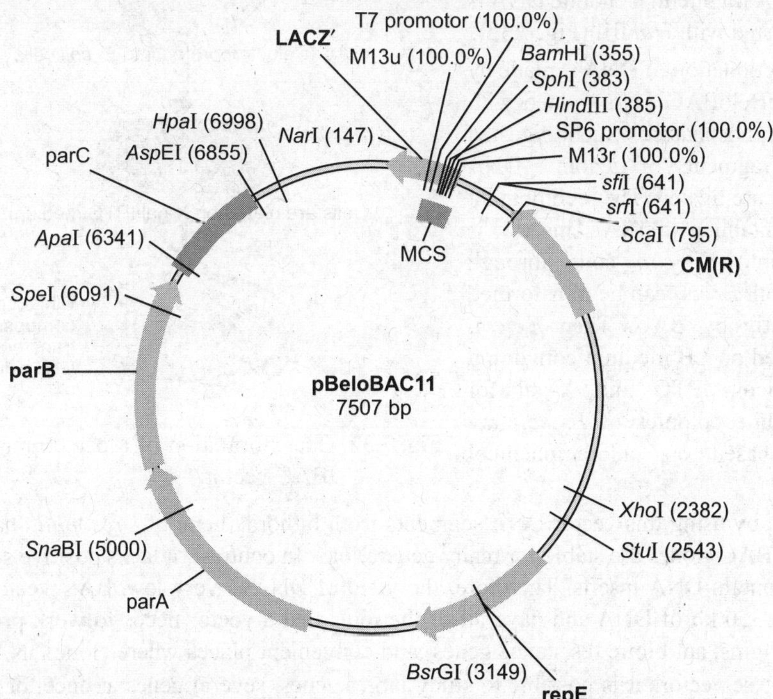

Fig. 7.31. Restriction map of pBeloBAC11 vector.

The cloning region includes: (*i*) unique cloning sites *Bam*HI, *Sph*I and *Hin*dIII in a *lacZα* gene allowing for insert screening by α-complementation, (*ii*) T7 and SP6 phage promoters reading into the cloning sites for generation of RNA probes for blot procedures, (*iii*) several GC-rich restriction sites flanking the cloning segment for removal of the cloned insert, (*iv*) a chloramphenicol selectable marker, (*v*) a lambda *cos* site for packaging into phage lambda particles if desired and also assuring a unique cleavage site for mapping, and (*vi*) a *loxP* site for specific cleavage by Cre recombinase in the presence of *loxP* oligonucleotide. The bold letters shows enzymes with unique restriction sites (Fig. 7.31). Restriction site coordinates refer to the position of the 5′-most base on the top strand in each recognition sequence. Open reading frame (ORF) coordinates are in the form 'translational start – translational stop'; numbers refer to positions on the top (clockwise) strand, regardless of the direction of transcription and include the start and stop codons. Lambda *cos* site coordinates are the boundaries of the *Hin*cII fragment surrounding the annealed 12 base overhangs.

1. Cloning of BAC Vector

The method of preparation of BAC library is the same as for plasmid library; but here the DNA insert is prepared by pulse field gel electrophoresis (PFGE). PFGE has made it possible to separate, map and analyse very large DNA fragments. Large sized globular DNA migrates easily after applying discontinuous electric field. The BAC vector pBeloBAC 11 is partially digested with restriction enzymes during cloning a DNA fragment. Genomic DNA is partially digested with *Hin*dIII (Fig. 7.32).

Size fractionation of DNA is made by PFGE. The pBeloBAC 11 is treated with *Hind*III and phosphatase. Such digested vector and fragments of genomic DNA (100–300 kb) are ligated. The recombinant plasmid containing DNA insert is transferred into *E. coli* cells through electroporation. *E. coli* can be transformed very efficiently by BAC. Then *E. coli* cells are plated on LB medium containing chloramphenicol, IPTG and X-gal for selection. White colonies of *E. coli* are picked up based on chloramphenicol resistance.

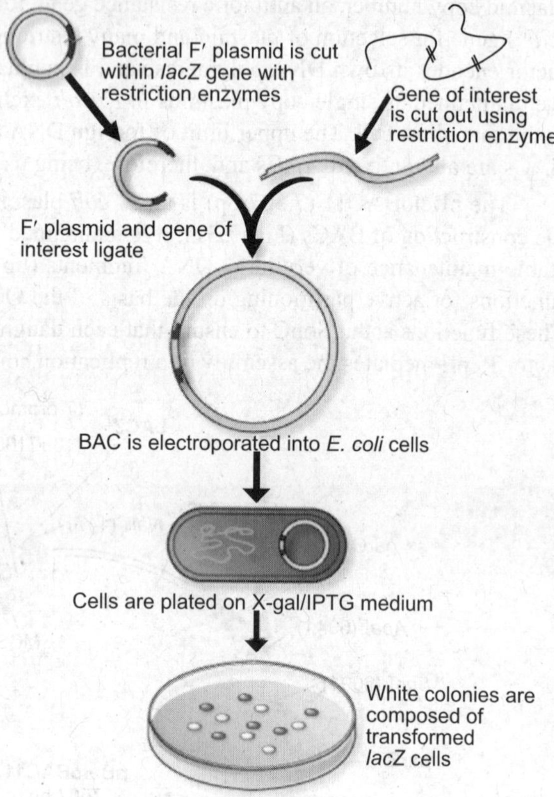

Fig. 7.32. Transformation of a bacterium by using a BAC vector.

In 1989, by using this vector DNA segments from bithorax gene of *Drosophila* have already been cloned. BAC clones are stable for many generations. In contrast, it lacks positive selection for clones that contain DNA inserts. Therefore, the yield of DNA is very low. BAC vectors are able to hold up to 350 kb of DNA and have all of the tools that a vector needs to work properly, like replication origins, antibiotic resistance genes, and convenient places where clone DNA can insert itself. With these vectors it is possible to study larger genes, several genes at once, or entire viral genomes. By using a vector that can hold larger pieces of DNA, the number of clones required to cover the human genome six times theoretically could drop from 1.8 billion to about 50 million.

2. Selection of Transformed Bacteria

The BAC vectors have been modified to make more convenient to use and more useful in specialized situations. A gene was added that enabled the bacteria to turn the colourless substance X-gal/IPTG blue, in addition to the antibiotic resistance gene that was added to identify transfected bacteria. This substance is found in the chicken-soup like media on which the bacteria colonize. This colour-changing gene (*lacZ*) is split apart when the clone DNA is incorporated into the vector, so it is possible to tell not only if a bacteria had been incorporated into the cell, but also if the bacteria was transfected with the vector containing insert DNA or just the vector alone (remembering that if the vector has properly incorporated the clone DNA, it will have lost its ability to change X-gal/IPTG blue) (Fig. 7.33).

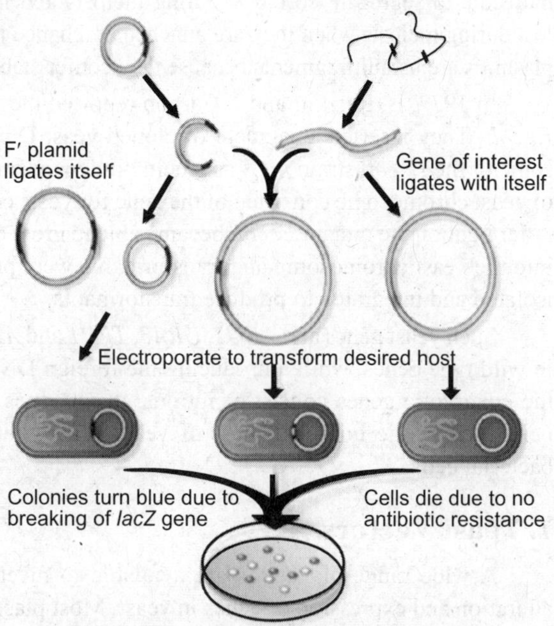

Fig. 7.33. Selection of transformed bacteria.

E. YEAST PLASMID VECTORS

Sacharomyces cerevisiae commonly called yeast is a unicellular eukaryotic microorganism, and has its long historical importance. It has been used for production of alcohol and alcoholic beverages since time immemorial. It can easily be cultured on artificial media. However, its genetic material is well organised into chromosomes present inside the nucleus. Yeast cells contain their own plasmid known as '2 μm plasmid' which is a naturally occuring 6 kb episomal plasmid found in yeasts and used as vector for foreign genes (Fig. 7.34). It is present in many stains in 50–100 copies per cell. The foreign DNA taken up by yeast is integrated by a specific crossing over to yeast DNA containing a homologous sequence on chromosome recognised by the yeast as the origin of replication (*ori*).

Fig. 7.34. A yeast (*Saccharomyces cerevisiae*) cell (A) and its naturally occurring 2μm plasmid (B).

In some cases segments of yeast DNA linked with foreign DNA have been demonstrated to transform very efficiently (about 10^4 cells). The transforming molec-ules replicate autonomously within the cell. DNA sequences of such property are known as *Ars* (autonomously replicating sequence) fragment. *Ars* fragments have also been isolated from a number of eukaryotic organisms such as *Coenhaorabditis elegans, Dictyostelium discoideum, Drosophila melanogaster, Zea mays*, etc. Plasmids have been constructed by using *ars* fragments when joined at centromeric sequences (*cen* region of about 6–10 kb DNA) to ensure the replication and stability in yeast. *Cen* is needed for DNA molecule to segregate correctly during meiosis and mitosis. Sometimes, the minichromosomes are lost during meiosis when they are attached by chance to spindle apparatus. Thus, *cen* regions of yeast plasmids are useful fragments because they confer stability.

In 1977, B. Ratzkin and J. Carbon reported the functional expression of cloned yeast DNA in *E. coli*. They inserted a segment of cloned yeast DNA into ColE1 plasmid of *E. coli*. They found some of the *E. coli* strain *leu*B as leucine-independent transformants. This showed that the segment of yeast chromosome consisted of the gene for yeast enzyme which was absent in *E. coli* strain *leu*B. After ligation the mutant *E. coli* became able to grow on normal medium. The *LEU2* gene was added into the yeast chromosome and transformants were produced. In this way several yeast genes were isolated and integrated to produce transformants.

Such yeast genes are: *LEU2*, *URA3*, *TRP1* and *ARG4*. Thereafter, unique markers were inserted in wild type genes, while introducing the foreign DNA into yeast cells for further selection. Since, the eukaryotic genes consist of introns, these genes rarely function in bacterial cells. In contrast being eukaryotic microbe, most of yeast genes that contain promoters successfully function in bacterial cells.

1. Yeast Vectors

A wide range of vectors are available to meet various requirements for insertion, deletion alteration and expression of genes in yeast. Most plasmids used for yeast studies are shuttle vectors, which contain sequences permitting them to be selected and propagated in *E. coli*, thus allowing for convenient amplification and subsequent alteration *in vitro*. The most common yeast vectors originated from pBR322, and contain an origin of replication (*ori*) promoting high copy-number maintenance in *E. coli*, the selectable antibiotic markers, the β-lactamase gene, *bla* or *Amp*^{R}, and sometime to tetracycline-resistance gene (*Tet*^{R}), conferring resistance to ampicillin and tetracycline, respectively. There are numerous kinds of yeast shuttle vectors, used currently can be broadly classified in either of following three types : integrative vectors (YIp), autonomously replicating high copy-number vectors (YEp), or autonomously replicating low copy-number vectors (YCp) (Table 7.5).

Table 7.5. Components of common yeast plasmid vectors.

Yeast plasmid vectors	***Components***			
Plasmid	YIp	YEp	YRp	YCp
E. coli genes or segments				
ori, bla, tet	+	+	+	+
Yeast genes or segments				
URA3, HIS3, LEU2, TRP1, LYS2	+	+	+	+
leu2-d	0	+	+	0
2μ,; 2μ*m-ori REP3*	0	+	0	0
ARS1, ARS2, ARS3.	0	0	+	+
CEN3, CEN4, CEN11.	0	0	0	+
Host (yeast) markers				
ura3-52, his3-Δ1, leu2- Δ1, trp1- Δ1, lys2-201	+	+	+	+
Stability	++	+	0	+

The difficulty associated with yeast was the transformation at low frequency due to crossing over (integrative transformation). Therefore, several plasmids were developed to solve the difficulty such as YIp, YEp, YRp, YCp, YLp, pYAC, shuttle vectors, retriever vectors, etc. (Fig. 7.35).On the basis of their mode of replication in yeast, yeast vectors can be grouped into five general classes: YIp, YRp, YCp, YEp, and YLp plasmids. With the exception of the Ylp plasmids (yeast linear plasmids), all of these plasmids can be maintained in *E. coli* as well as in *S. cerevisiae*; therefore, these are referred to as 'shuttle vectors'.

1. Yeast Integrative Plasmid (YIp)

The YIp consists of only a yeast marker added to bacterial plasmid. It lacks origin of replication, hence it cannot replicate inside the yeast cell without integrating into a chromosome. Therefore, its integration is must for successful transformation. It results in low frequency transformation of yeast cells yielding a transformant per μg of DNA molecule. The YpI integrative vectors do not replicate autonomously, but integrate into the genome at low frequencies by homologous recombination. Integration of circular plasmid DNA by homologous recombination leads to a copy of the vector sequence flanked by two direct copies of the yeast sequence. The site of integration can be targeted by cutting the yeast segment in the YIp plasmid with a restriction endonuclease and transforming the yeast strain with the linearised plasmid. The linear ends are recombinogenic and direct integration to the site in the homologous genome. In addition, linearisation increases the efficiency of integrative transformation from 10- to 50-fold.

Typically the YIp vectors integrate as a single copy. Multiple integration do occurs at low frequencies, a property that can be used to construct stable strains over-expressing specific genes. YIp plasmids with two yeast segments, such as *YFG1* and *URA3* marker, have the potential to integrate at either of the genomic loci, whereas vectors containing repetitive DNA sequences (such as Ty elements or rDNA) can integrate at any of the multiple sites within genome. Strains constructed with YIp plasmids should be examined by PCR analysis, or other methods to confirm the site of integration.

Strains transformed with YIp plasmids are extremely stable, even in the absence of selective pressure. However, plasmid loss can occur at about 10^{-3} to 10^{-4} frequencies by homologous recombination between tandemly repeated DNA, leading to looping out of the vector sequence and one copy of the duplicated sequence. Integration takes place due to homologous recombination between the inactive gene present on one chromosome and yeast gene carried by YIp as the selectable markers. Crossing over takes place between the chromosomal *LEU2*$^-$ region and *LEU2*$^-$-carrying plasmid. Consequently the flanking region linked to plasmid is integrated to *LEU2*$^-$ region. Autonomously replicating system (*ARS*) is absent in that plasmid; therefore, YIp cannot replicate as a plasmid (Fig. 7.35 A).

2. Yeast Episomal Plasmid (YEp)

The YEp is the 2 μm circle-based vector which replicates autonomously without integrating into a yeast chromosme. The 2 μm plasmid is 6 kb in size with so high frequency for transformation. The 2 μm DNA consists of origin of replication and *rep* genes. This plasmid results in high frequency transformation, *i.e.* about 2 × 10^4 transformants per μg DNA. The number of restriction sites is limited. It is moderately stable in the cell and found in high copy numbers. It will lose its stability if 2 μm sequence is lost. It is moderately stable in the cell (Fig. 7.35 B).

The YEp vectors replicate autonomously due to the presence of a segment of the yeast 2 μm plasmid that serves as an origin of replication (2 μm *ori*) which is responsible for high copy-number and high frequency of transformation of YEp vectors. YEp vectors consist of either a full copy of the 2 μm plasmid or mostly as a region which encompasses the *ori* and the *REP3* gene. The *REP3* gene is required in *cis* to the *ori* form for mediating the action of *trans*-acting *REP1* and *REP2* genes which encode products that promote partitioning of the plasmid between cells at division. Hence, the YEp plasmids containing the region encompassing only *ori* and *REP3* must be propagated in vir$^+$ hosts containing the native 2 μm plasmid.

Most of the YEp plasmids are relatively unstable, being lost in about 10^{-2} or more cells after each generation. Even under conditions of selective growth, only 60% to 95% of the cells retain the YEp plasmid. The copy number of most YEp plasmids ranges from 10 to 40 per cell of vir$^+$ hosts. The plasmids are not equally distributed among the cells, and there is a high variance in the copy number per cell in populations.

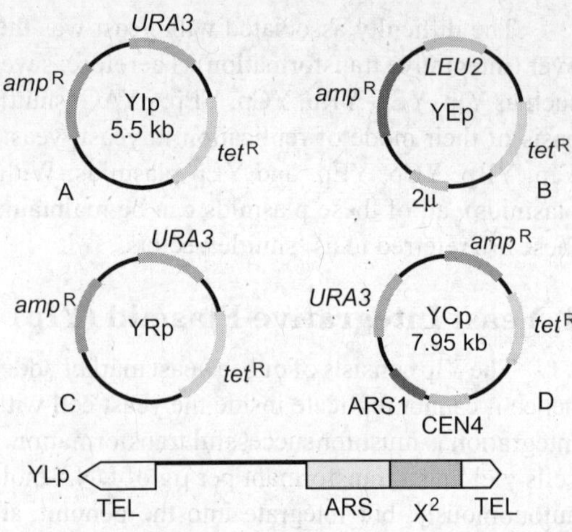

Fig. 7.35. Different types of yeast plasmid cloning vectors.

Several systems have been developed for producing very high copy-numbers of YEp plasmids per cell, including the use of the partially defective mutation *leu2-d*, whose expression is several times less than the wild-type *LEU2*$^+$ allele. The copy number per cell of such YEp *leu2-d* vectors range from 200 to 300, and the high copy-number persists for many generations after growth in leucine-containing media without selective pressure. The YEp *leu2-d* vectors are useful in large-scale cultures with complete media where plasmid selection is not possible. The most common use for YEp plasmid vectors is to overproduce gene products in yeast.

3. Yeast Replicating Plasmid (YRp)

This vector has been developed on the basis of chromosomal elements to support autonomous replication of plasmid in yeast cells. The artificially replicating sequences *(ARS)* was isolated from many regions of genomes of yeast and other eukaryotes and *ARS*-based vectors were constructed. These vectors are also called ARS vectors because they carry *ARS* region of the yeast chromosome. In yeast chromosome, the origin of replication is specified by *ARS* which is about 100 bp long consensus sequence. If any DNA molecule consists of ARS it acts as yeast vector. If it is essential for survival of yeast cell, such a vector will be maintained by the cells. In the absence of such a selection pressure, ARS vectors are soon lost. The YRp vectors are very unstable but can form high copy number inside the cells (Fig. 7.35 C).

One of the examples of YRp vectors is the YRp7 vector which is 5.7 kb in size which is made up of pBR322 plus yeast gene *TRP1*. The chromosomal sequence of YRp7 consists of *TRPI* along with *ARS*. As compared to YEp, it is least efficient in transformation and about 10 times less stable. Transformation rate of YRp vector is 1 to 10,000 transformants per mg plasmid. Copy number of YRp is about 3-100 per cell. Due to these limitations YRp vectors are rarely used in transformation experiments.

4. Yeast Centromeric Plasmid (YCp) or Yeast Centromere (CEN)

It is also yeast-based plasmid consisting of a sequence of chromosomal centromeric DNA of yeast. It acts as true mini-chromosome which becomes circularised. It increases the stability of plasmid and segregates during mitosis and meiosis. The YCp vectors are autonomously replicating vectors containing centromere sequences *(CEN)* and autonomously replicating sequences *(ARS)* (Fig. 7.35 D). The YCp vectors are typically present at very low copy numbers from 1 to 3 per cell, and are lost in about 10^{-2} cells per generation without selective pressure. Copy number of this plasmid is useful when low copy numbers of cloned genes are required.

Mostly the YCp vectors segregate to two of the four ascospores from an ascus, which indicates that they mimic the behavior of chromosomes during meiosis and mitosis. The *ARS* sequences are believed to correspond to the natural replication origins of yeast chromosomes, and all of them contain a specific consensus sequence. The *CEN* function is dependent on three conserved domains, designated as I, II, and III; all three of these elements are required for mitotic stabilisation of YCp vectors. YRp vectors, containing *ARS* but lacking functional *CEN* elements, transform yeast at high frequencies, but are lost at too high a frequency, *i.e.* over 10% per generation, that make them undesirable for general vectors. The stability and low copy-number of YCp vectors make them the ideal choice for cloning vectors, for construction of yeast genomic DNA libraries, and for investigating the function of genes altered *in vivo*. The *ARS1* which is in close proximity to *TRP1* is the most commonly used *ARS* element for YCp vectors, although others have been used. *CEN3*, *CEN4* and *CEN11* are commonly used centromeres that can be conveniently manipulated. For example, the vector YCp50 contains *CEN4* and *ARS1*.

5. Yeast Linear Plasmid (YLp)

In 1992, J. W. Szostak and E. H. Blackburn have constructed a linear yeast plasmid by joining fragments from the termini of a ciliated protozoan, *Tetrahymena* ribosomal DNA to a yeast vector. Structural features of the terminus region of the *Tetrahymena* rDNA plasmid maintained in the yeast linear plasmid include a set of specifically placed single-strand interruptions within the cluster of hexanucleotide (C4A2) repeat units. They used the linear plasmid as a vector to clone chromosomal telomeres from yeast. YLp is similar to *ARS* centromeric plasmid. It is not circular but linear plasmid constructed by adding telomeric (chromosome end) sequences to both ends of a cleaved *ARS*-based plasmid. Telomere is the end of chromosome. This linear plasmid is present in a single copy as a chromosome, and these are very stable and cannot replicate in *E. coli*. Therefore, it has drawn less attention but has raised much possibility to put together the complete set of genes in artificial chromosome of eukaryotes (Fig. 7.35).

F. YEAST ARTIFICIAL CHROMOSOMES (YAC)

The plasmid can accommodate about 10 kb foreign DNA, phage λ can accommodate about 20 kb DNA insert, and cosmid can accommodate about 40 kb long four DNA segments. It was realised that for stable replication and segregation in eukaryotes, particularly in yeast, a small and well defined sequence should be constructed as recombinant chromosome. With this objective David Burke and Maynard Olson at Washington University in St. Louis first developed the YAC cloning system in 1987 by using new approaches where DNA segment of several thousand base pairs (about 1 Mb) can be cloned. Initially it was used for investigation of the maintenance of chromosomes in the cell. Later on it was used as vectors for carrying very large cloned fragments of DNA. It has also been used for physical mapping of human chromosome in 'Human Genome Project'.

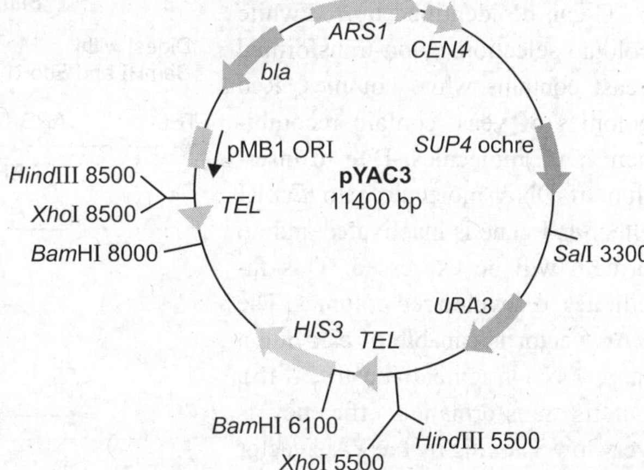

Fig. 7.36. Yeast artificial chromosome (YAC3) vector.

In 1983, A.V. Murray and J.W. Szostak first constructed the 55 kb long yeast artificial chromosomes in yeast that contained cloned genes, replicators, centromeres and telomeres. These molecules have many of the properties of natural yeast chromosomes. Centromere function is impaired on short (less than 20 kb) artificial chromosomes. A typical YAC consists of centromere element (*CEN*) for chromosomal segregation during cell division, telomere, and origin of replication (*ori*) were isolated and joined on plasmids constructed in *E. coli*. Structure of typical YAC3 vector is shown in Fig. 7.36.

The YAC3 vector contains the *E. coli* origin of replication (*ori*^{E}) and a selectable marker (*amp*^{R}), a yeast DNA sequence, genes each for uracil biosynthesis pathway (*URA3*) providing centromeric function (*CEN*), autonomously replication sequence (*ARS*) (which acts as yeast *ori*), tryptophan synthesis pathway (*TRP*) and telomeric (T) sequence (which is extended by the enzyme telomerase inside the yeast cell) (Fig. 7.36). The *TRP1* and *URA3* are yeast selectable markers, one for each end, which ensure that properly reconstituted YACs survive in yeast cells, *Sup*4 is inactivated in recombinants and acts as a basis for red-white selection unlike blue-white selection in *E. coli*. There are recognition sites for restriction enzymes such as *Sma*I and *Bam*HI.

Essential Components of YAC Vectors

The essential components of YAC vectors are given as below:

- Centromers (*CEN*), telomeres (*TEL*) and autonomous replicating sequence (*ARS*) for proliferation in the host cell.
- *amp*^{R} for selective amplification and markers such as *TRP1* and *URA3* for identifying cells containing the YAC vector.
- Recognition sites of restriction enzymes (e.g. *Eco*RI and *Bam*HI)

For cloning purpose, YAC and target DNA are digested with restriction enzyme (*Bam*HIII and *Eco*RI) and ligated by using DNA ligase. Consequently, the recombinants are produced by inserting a large fragment of genomic DNA. This molecule can be maintained in yeast as YAC. The transformants that contain YAC can be identified by red/white colour selection. Non-transformed yeast contains white colonies. Red colonies of yeast contain recombinant YAC molecules. Due to insertion of DNA molecule into *Eco*RI site, *Sup*4 gene is inactivated and no protein will be expressed. This facilitates to develop red colonies. The YAC vector is capable of carrying a large DNA fragment (up to 2 Mb), but its transformation efficiency is very low. Cloning by the YAC vector is shown in Fig. 7.37.

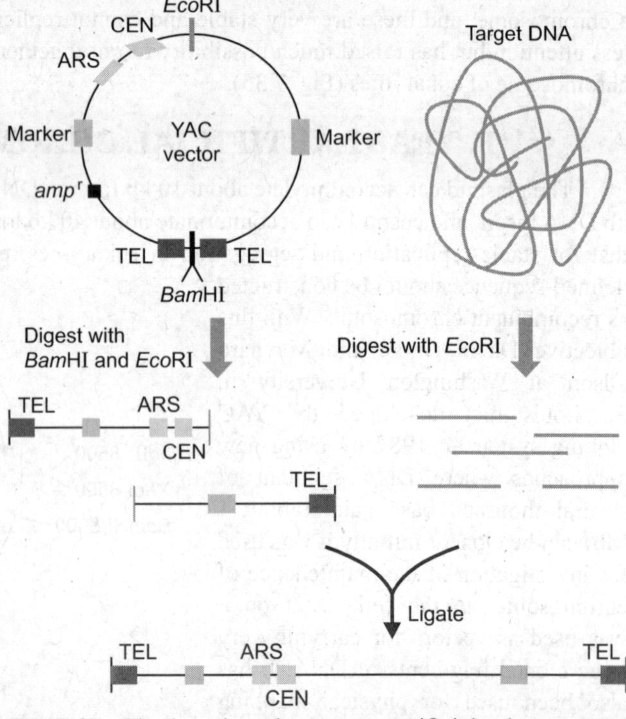

Fig. 7.37. Cloning by the yeast artificial chromosome (YAC) vector.

G. PLANT AND ANIMAL VIRUSES AS CLONING VECTORS

Using gene cloning techniques based on *E. coli* as host, many eukaryotic genes and their control sequences have been isolated and analysed. Many applications of genetic engineering require vectors for the expression of genes on diverse eukaryotic organisms for engineering of new plants, gene therapy and large scale production of eukaryotic proteins. Animal and plant-based vectors designed for this purpose have been discussed in this section.

1. Plant Viral Vectors

Plant virus-based vectors can be used for transfer of genes with high efficiency because virus genome containing DNA insert cause infection in host plant and amplified naturally. However, there are many viruses that cause systemic infection. Consequently their genes are introduced into all the cells. Since virus infection results in delivery of a new genetic material into host cells, they provide a natural example of genetic engineer. The DNA insert linked with viral DNA that is present in host cell replicates and expresses in host cell along with viral genes. For insertion of a desired gene into viral genome, it is necessary to study and characterise the virus into detail so that the DNA linked with virus genome may express in the plant. Construction of viral vector that can express in host cell is based on DNA or RNA molecules.

Plant virus vectors are non-integrative. They are transmitted systematically in plant and assembled in host cell in high copy number. But it never integrates with the host genome. Therefore, to act as viral vector, they should: (*i*) carry extra nucleic acid (DNA or RNA), (*ii*) have broad host range, (*iii*) be easily transmitted, and (*iv*) be preferred to manipulate their genome. There are three groups of viruses (caulimoviruses, geminiviruses and tobamoviruses) which have been worked out in detail for construction of vector.

(*a*) Caulimoviruses: This group includes cauliflower mosaic virus (CaMV) which has been extensively studied. Genome of CaMV is made up of double stranded DNA which was first manipulated by recombinant DNA technology. Viral particles transmitted are isometric, 50 nm in diameter and isolated from inclusion bodies. Viral particles spread rapidly in 3–4 weeks throughout the plant and found at high copy number (about 10^6 virions per cell).

The CaMV promoter (35S) is used in most transgenic crops to activate foreign genes which have been artificially inserted into the host plant. It is inserted into transgenic plants in a form which is different to that found when it is present in its natural brassica plant hosts. This enables it to operate in a wide range of host-organism which would otherwise not be possible. The CaMV DNA remains as linear open circular molecule of 8 kb twisted or knotted forms. A genomic map of cauliflower mosaic virus is given in Fig. 7.38. Its genome consists of tightly packed six major and two minor (total I-VIII) open reading frame (ORF) on one coding strand. It replicates through reverse transcription. After infecting the host cell, DNA enters inside the nucleus where a supercoiled minichromosome is formed after digestion of a single-stranded overhangs and gaps. Inside the nucleus of plant cell, the minichromosome acts as a template for nuclear RNA polymerase II forming RNA transcripts. Then the transcripts are transported to cytoplasm where it undergoes wither translation or replication into (–) DNA strand using reverse transcriptase. The (+) DNA strand is synthesised from two primer-binding sites that are present near

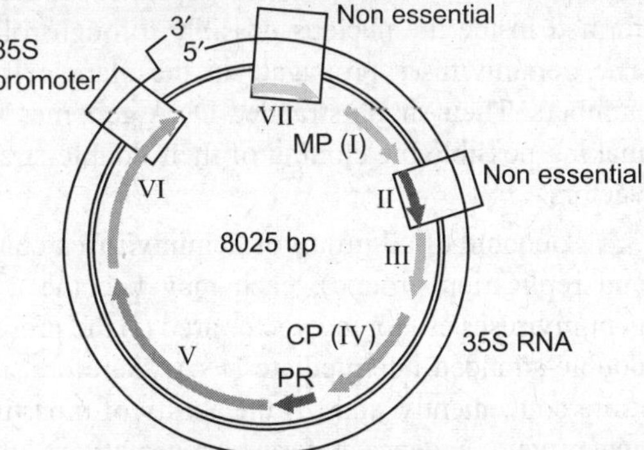

Fig. 7.38. A genomic map of cauliflower mosaic virus.

gaps 2 and 3. The RNA synthesis starts from gap 3′ to 5′ end of (–) DNA strand, whereas synthesis started from gap 3 continues to gap 2. Finally, DNA molecules are packed to form viral particles. These particles re-start the similar events after re-entering the cell of a healthy tissue of other plants.

The properties of the CaMV shows that it could provide a useful vector for gene transfer in higher plants: (*i*) it has a relatively small double-stranded genome that can be easily manipulated *in vitro,* (*ii*) cloned viral DNA is infectious when rubbed onto healthy leaves, and (*iii*) virus spreads throughout the plant and can be found in most cells at high copy number. However, the two genomic regions of the virus (ORF I and ORF VII) are non-essential. ORF II synthesises the factors for insect transmission and ORF VII has un-known function. Therefore, these two regions can be replaced by a foreign DNA and vector constructed. By using this approach, in 1984, N. Brisson and co-workers from the University of Montreal (Canada) transferred CaMV vector containing the bacterial dihydrofolate reductase, into turnip plant cells.

In 1997, Rob J. Noad and co-workers from the Department of Virus Research, John Innes Centre (UK) have described the expression of functional elements inserted into the 35S promoter region of infectious CaMV replicons. They introduced modification and inserts into replaceable domains associated with the 35S promoter. The 35S enhancer (–208 to –56) potentiated the promoter activity in reverse orientation sufficient to establish systemic infection.

(*b*) Geminiviruses: Geminiviruses are plant viruses that consist of single-stranded circular DNA and cause diseases in maize cereals, cassava, tomato, etc. They are characterised by twinned (geminate) icosahedral capsids and depend on DNA-dependent DNA polymerases for their replication. These features differentiate it from the other groups of viruses.

The capsids are icosahedral with 10–20 × 20 nm dimension and enclose covalently closed one or two circular single-stranded DNA molecule(s) (2.6–3.0 kb). Genes diverge in both directions from a virion strand origin of replication (geminivirus genomes are ambisense). After infection DNA molecule enters in the nucleus and the single-stranded DNA is converted into a double-stranded DNA replicative form. Many copies of replicative form of single-stranded DNA are found within in plant nucleus. Thereafter, many copies of replicative forms of DNA are formed inside the nucleus possibly through rolling circle mechanism. The geminiviruses propagate in the plant cell nucleus in high copy numbers. Their single-stranded DNA genomes are small (2.5 to 3 kb), making possible the cloning of their double-stranded replicative intermediates in *Escherichia coli* vectors.

Geminivirus (an electron photomicrograph).

Genome of all group of geminiviruses consists of two elements (*i.e.* an origin of replication and replication protein), each may become a part of vector. The attractive features of using geminiviruses as cloning vector are : (*i*) the presence of single-stranded DNA that replicate through double-stranded intermediate DNA, therefore, it can be manipulated *in vivo* in bacterial plasmid more conveniently, and (*ii*) the ability of bipartite geminiviruses to contain a replacement of virus coat protein sequence by foreign gene without interfering virus genome replication. The geminivirus vectors which have been constructed so far consist of reporter or other desired genes, replacing the coat protein expressing sequence of viral genome. These vectors are allowed to insert foreign gene in protoplasts, cultured cells or leaf discs.

In 1990, Masashi Ugaki and co-workers from the Waksman Institute (U.S.A.) have reported the replication of a geminivirus derived shuttle vector in maize endosperm cells.

(c) Tobamoviruses: Tobacco mosaic virus (TMV) is one of the important and thoroughly worked out viruses of this group which consists of RNA as a linear, large, undivided genome (Fig. 7.39). Virus particles are rod shaped measuring about 280 × 15 mm with mol wt 39 $\times 10^6$ Da.

Virioin contain 2130 protein subunits of identical size which are arranged around a central hole of 4 nm. Each protein subunit is made up of a single polypeptide chain having 158 amino acids. A single-stranded RNA is spirally coiled to form helix inside protein coat.

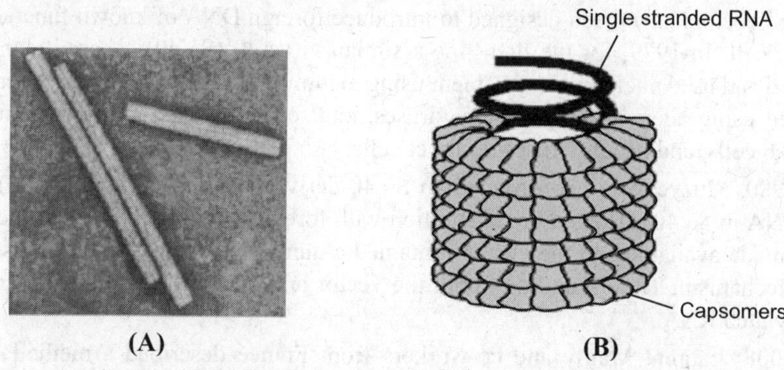

Fig. 7.39. Tobacco mosaic virus (TMV); (A)- electron micrograph of virus particles; (B)- structure of TMV showing RNA as genetic material and protein capsids.

In 1985, M. W. Bevan and co-workers of the Cambridge (U.K.) reported the expression of tobacco mosaic virus coat protein by a cauliflower mosaic virus promoter in plants transformed by *Agrobacterium.* RNA-based vector is constructed by inserting a foreign gene into an intact viral genome. For example, TB2 was the first RNA-based RNA vector which was successfully spread in whole plant (Fig. 7.40).

TB2 vector was constructed by inserting a foreign gene at the 3′ end of the movement protein (MP) ORF. The native subgenome promoter (sgp) drives the expression of the foreign gene and a sgp from the related virus, *i.e.* ORSV (*Odontoglossum* ring spot virus) governs the expression of ORF of coat protein (CP). RNA viral vectors containing gene insert can replicate only in protoplast inoculated leaves or differentiated tissue.

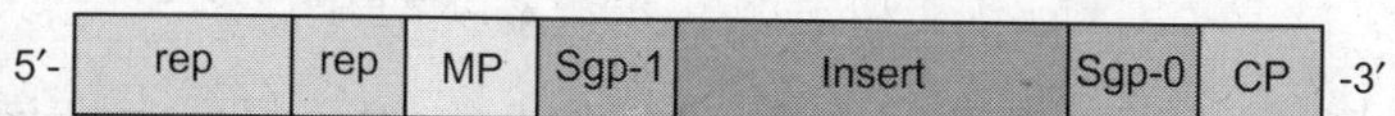

Fig. 7.40. Genomic map of tobacco mosaic virus; TB2 plant RNA viral vector; insert = *npt*II, α-trichosanthin; CP = viral coat protein; MP = viral movement protein; rep= replication.

In 2007, John A. Lindbo from the Department of Plant Pathology, Ohio State University (Ohio) reported TRBO, which is a high-efficiency tobacco mosaic virus RNA-based over-expression vector. He constructed an agroinfection-compatible full-length TMV expression vector called pJL TRBO that expressed all of the TMV genes in addition to a foreign gene. The CP gene sequence was deleted to increase the level of foreign gene expression from a TMV replicon. A map of the T-DNA region of binary plasmid is shown in Fig. 7.41.

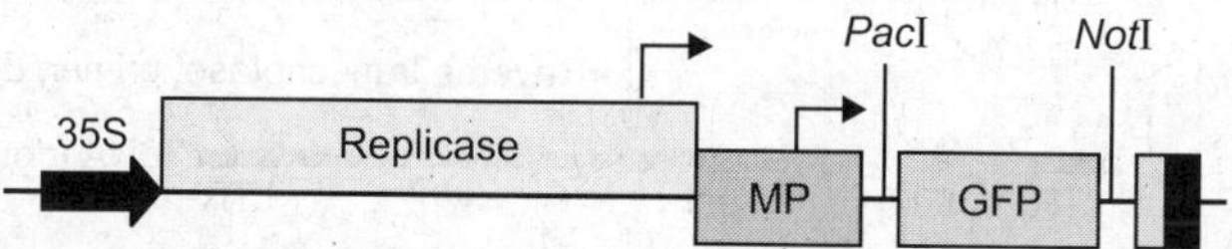

Fig. 7.41. Maps of T-DNA region of binary promoter of plasmid pJL TRBO. Block arrow- CaMV duplicated 35S promoter. Black box, CaMV polyA signal sequence terminator; dark gray box- tobacco etch virus 5′-nontranslated leader sequence. Light gray box, ribozyme; bent arrows- sub-genomic promoters; coloured boxes- ORFs. Identities of ORFs are labeled in white boxes; replicase- TMV 126K/183K ORF; MP- movement protein (based on Lindbo, J.A. 2007. *Plant Physiology,* 145:1232).

2. Animal Viral Vectors

There are several viruses in nature which cause diseases. A virus gets adsorbed at the body surface of a suitable host and infects the cell. This ability of animal viruses has been exploited and virus-based vectors have been designed to introduce foreign DNA of known function into cultured eukaryotic cell. In 1979, for the first time a simian virus 40 (SV40) – based cloning vector was constructed and used in cloning experiment using mammalian cells. Since then several vectors were constructed using adenovirus, papillomaviruses, and retroviruses to clone foreign DNA into the mammalian cells, and baculoviruses in insect cells.

In 1980, Muzyczka N. constructed an SV40-derived cloning vector (dl1142) for cloning of foreign DNA in SV40 virions. For substitution with foreign DNA about 80% of the SV40 genome has been made available and the vector contained a number of unique restriction sites helping the cloning mechanism. They found that both the vector and insert DNA were transcribed mainly as non-adenylated RNA.

In 2008, Laurent Mailly and co-workers from France described a method allowing direct cloning of a transgene in the E3 region of the Ad5 genome that already contained the immediate early CMV promoter upstream of three unique restriction sites. This allowed the construction of recombinant adenoviral genomes in just one step by reducing the time of selection and, production of the corresponding vectors. They directly inserted a transgene in the E3 region of the Ad5 genome that already contained the immediate early CMV promoter upstream of three unique restriction sites. Fig. 3.42 shows restriction map of pAd5CMV/TCS and of the four different inserts such as *Pac*I, *Swa*I, *Bst*BI and *Cla*I. The restriction site for *Cla*I is just near to SV40 pA. Using this vector, they produced recombinant adenoviruses, each giving high-level expression of the transgene in the transduced cells.

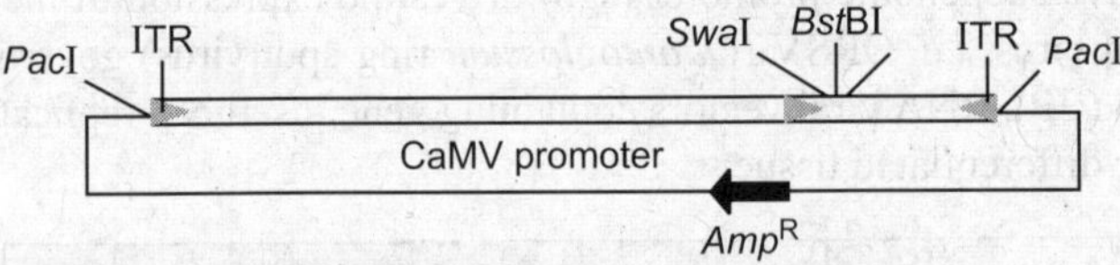

Fig. 7.42. Restriction map of pAd5CMV/TCS and of the four different inserts (based on Mailly, L. *et al.*, 2008. *Virology J.* **5**:73).

In recent years, retrovirus-based vectors are commonly used for gene cloning, because they can infect several types of cells. Since they are single stranded, their high concentrations are required for fully differentiated cells like neurons, hepatocytes, etc. The *gag*, *pol* and *env* genes of retroviruses (which are required for replication and assembly of viral particles) can be replaced with foreign DNA. The recombinant DNA is introduced into mammalian cells in tissue culture that eventually delivers the foreign gene into the desired host cells/tissue (Fig. 7.43).

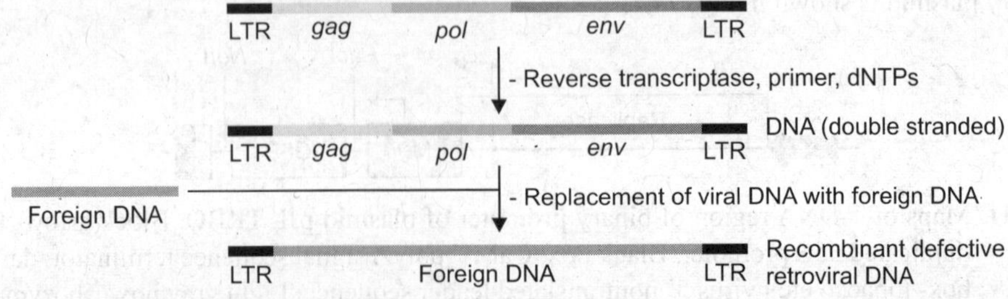

Fig. 7.43. Construction of retroviral genome to be used as cloning vector.

A detailed account of uses of animal viruses as cloning vector in gene therapy has been described in Chapter 9: *Gene Therapy*.

H. TRANSPOSONS AS VECTOR

Different types of transponsons are also used as cloning vectors

1. *Ac-Ds* Elements

The simple transposons of maize are the activator (*Ac*) and dissociation (*Ds*) elements. It has 4.565 kb long autonomous sequence which transcribes 3.5 kb long GC-rich un-translated leader sequence. Four introns are spliced out from the primary transcript and 3.312 nucleotides long mature mRNA is obtained. Consequently 807 amino acids long protein is translated. Within the *Ac* element the terminal inverted repeats (TIR) of 11 bp are located at both the ends which play a key role in transposition. The protein synthesised by *Ac* is called transposase which causes transcription of *Ac* elements (Fig. 7.44). The molecular features of the maize *Ac/Ds* system are that: *Ac* is 4563 bp in length, it contains 11-bp inverted repeats at the ends, it possesses 8-bp direct repeats of target DNA are generated, and *Ds* are the truncated versions of *Ac*

The defective element of *Ac* family is called *Ds* (dissociation). Due to deletion, non-autosomal elements are obtained from the autosomal elements that activate the *trans*-acting transposase. The 11 bp long TIR is common to all *Ds* elements. The enzyme transposase cannot be expressed by the *Ds* element due to deletions. But the transposase produced by *Ac* element is diffused and *trans*acting. It causes transposition of *Ds* elements. By detecting part of the region transposons can be used as vector for cloning a foreign DNA.

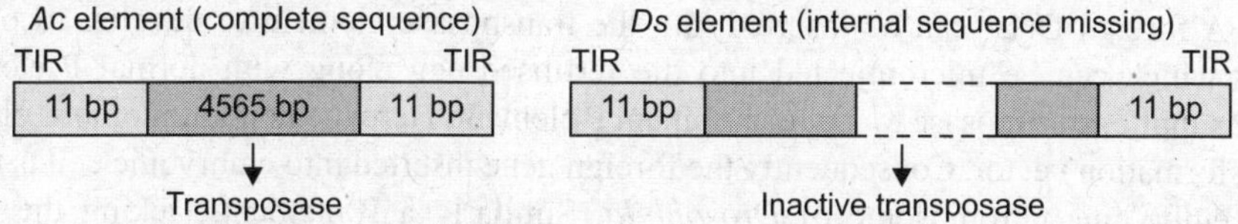

Fig. 7.44. Diagram of *Ac-Ds* elements of transposons in maize.

Excision by transposons is associated with chromosome breaks; generally, host-cell proteins repair this damage, often introducing mutations. Many transposons also use host proteins in the transposition mechanism or in regulation. Transposition in systems lacking host factors that influence the behaviour of these transpositions is useful in determining what those factors are and how they work. In addition, features of transposition and regulation intrinsic to the element itself can be determined. Maize *Ac/Ds* elements transpose in a wide variety of heterologous plants, but their characteristics in these other systems differ from those in maize, including their response to increasing genetic dosage and the types of repair products recovered following excision. The host proteins are involved in *Ac/Ds* activity and potential differences between these proteins among plant species. However, it has also been reported that plant-specific proteins are not essential for *Ac/Ds* transposition. The yeast system is valuable for dissecting the *Ac/Ds* transposition mechanism and identifying host factors that can influence transposition and the repair of DNA damage induced by *Ac/Ds*. Mutations caused by *Ds* excision in yeast suggest the formation of a DNA-hairpin intermediate, and reinsertions occur throughout the genome with a frequency similar to that in plants. The high proportion of *Ac/Ds* reinsertions also makes this system an *in vivo* mutagenesis and reverse genetics tool in yeast and other eukaryotic systems.

2. P Element

A P element is a transposon present in the fruit fly, *Drosophila melanogaster*. It is widely used for mutagenesis and creation of genetically modified flies used for genetic research. There are autonomous and non-autonomous P elements. The autonomous P element encodes one transcript with four exons. It creates an 8-bp target site duplication on transposition. The P element causes a phenotype known as 'hybrid dysgenesis'. Hybrid dysgenesis refers to the high rate of mutation in germline cells of *Drosophila* strains resulting from a cross of males with autonomous P elements (P Strain/P cytotype) and females that lack P elements (M Strain/M cytotype). The hybrid dysgenesis syndrome is marked by temperature-dependent sterility, elevated rates of mutation, chromosome rearrangement and recombination. This is caused by the transposition of P elements within the germ-line cells of offspring of P strain males with M strain females. This transposition occurs only in germ-line cells because a splicing event needed to make transposase mRNA does not occur in somatic cells. P elements are maintained in a dormant state by a factor present in P cytoplasm. Transposition causes a variety of genetic aberrations.

The P element is a class II transposon, *i.e.* its movement within the genome is made possible by a transposon . The complete element is 2,907 bp and is autonomous because it encodes a functional transposase; non-autonomous P elements which lack a functional transposase gene due to mutation also exist. Non-autonomous P elements can still move within the genome if there are autonomous elements to produce transposase. The P element can be identified by its terminal 31-bp inverted repeats, and the 8 bp direct repeats in movement into and out of DNA sequence produces. P Plasmids always contain a *Drosophila* reporter gene, often a red-eye marker (the product of the *white* gene), and the transposase recognition sequences. P elements may also contain: a gene of interest, an *E. coli* reporter gene, often some kind of antibiotic resistance, origin of replication (*ori*) and other associated plasmid 'housekeeping' sequences.

The size of different P elements varies; the complete elements move autonomously due to expression of transposase enzyme. The terminal and sub-terminal sequences cause transposition in them. A foreign DNA can be inserted into the transposase element to develop recombinant P elements which can be microinjected into the fertilised egg along with normal P element. The normal P supplies transposase to the recombinant P element. Here the non-autonomous element acts as a transformation vector. Consequently the foreign gene inserted into embryonic cell is transposed onto an embryonic chromosome of *Drosophila*. Similarly a P element, lacking the ability for integration, was inserted into a plasmid to encode transposase. Both the plasmids were co-injected into the embryo of *D. melanogaster* mutant in rosy gene. It was found that 50% progeny of *D. melanogaster* were rosy progeny. This shows the successful transfer of a foreign gene into a cell through P element.

Structure of a P element, its transposition and cloning vector are given in Fig. 7.45.

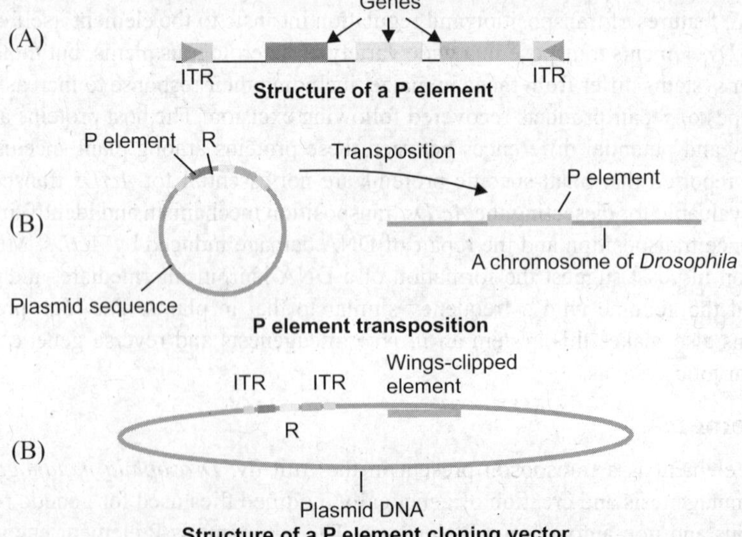

Fig. 7.45. Structure of a P element (A), P element transposition (B) and P element cloning vector (C).

Characteristics of P Elements

Naturally-occurring P elements contain: (*i*) coding sequence for the enzyme transposase, and (*ii*) recognition sequences for transposase action. Transposase is an enzyme that regulates and catalyses the excision of a P element from the host DNA, cutting at two recognition sites, and then reinserts randomly. It is the random insertion that may interfere with existing genes, or carry an additional gene that can be used for genetic research.

The two parts of the P element must be separated to prevent uncontrolled transposition to use this as a useful and controllable genetic tool. The normal genetic tools are, therefore, the DNA coding for transposase with no transposase recognition sequences so it cannot insert a 'P plasmid'.

I. SHUTTLE VECTORS AND EXPRESSION VECTORS

1. Shuttle Vectors

The prokaryotic vectors cannot exist and work in eukaryotic cells because the system of two groups of organisms varies. You know that prokaryotes lack introns, while the eukaryotes consist of introns. Therefore, several vectors have been constructed which may exist both in prokaryotic (*E. coli*) and eukaryotic cells. Such vectors possess two origin for replication *i.e.* ori^E and ori^{Euk} and, therefore, are called 'shuttle vector'. The ori^E functions in *E. coli* and ori^{Euk} functions in eukaryotic cells like yeast. Besides, these vectors also contain the antibiotic resistance genes (such as amp^R) that act as selectable marker gene. One of the examples of shuttle vectors is the yeast episomal plasmid (YEP) (Fig. 7.46). The YEP is a 2 μm circle plasmid-based vector which contain 2 μm origin of replication, *E. coli* shuttle sequence and a selectable marker, *i.e. leu*2 yeast gene. The gene for β-lactamase is an example of a prokaryotic gene that is expressed in yeast. In 1983, C.I. Kado and R.C. Tait have described the properties of shuttle vector as below :

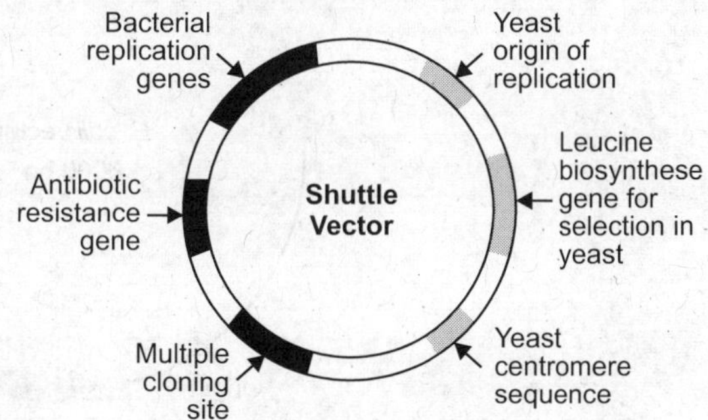

Fig. 7.46. A shuttle vector for yeast.

- The vector must replicate in many organism (*e.g.* bacteria, yeast and plants) to facilitate the isolation and characterisation of genes.
- The vector must be easily recognised by selectable markers.
- The vector should be small in size to accomodate DNA inserts.
- Cloned genes should be easily detected.
- Useful quantities of vector must be easily obtained.
- The vector must be stable, non-pathogenic and non-stress-inducing.

2. Expression Vectors

In addition to incorporation desired gene into the host cell, the objectives of recombinant DNA technology is to produce high amount of proteins encoded by the DNA insert. To achieve this goal, the introduced novel gene must express. Hence, expression of cloned genes is carried out by inserting a 'promoter sequence' (signal for initiation of transcription), and a 'terminator sequence' (that provides signal for termination of transcription). A translation initiation sequence (a ribosome binding site and short codon) is also incorporated near cloning site into the vector. The cloning vectors which contain these signals for protein synthesis are called expression vectors. The basic architecture of an *E. coli* expression vector is shown in the Fig. 7.47.

(*a*) Features of *E. coli* Expression Vector: The special feature of *E. coli* shuttle vector are given below:

(*i*) Selectable marker: In the absence of selective pressure plasmids are lost from the host. Especially in the case of very high copy number plasmids and when plasmid-borne genes are toxic to the host or otherwise significantly reduce its growth rate. The simplest way to overcome this problem is to express from the same plasmid an antibiotic-resistance marker, and supplement the medium with the appropriate antibiotic to kill plasmid-free cells. The use of ampicillin requires special care. The selectable marker, β-lactamase, is secreted into the medium where it hydrolysis all of the ampicillin. This point is already reached when the culture is barely turbid. The cells lacking plasmid will not be killed and could overgrow the culture. Some possible solution are: (*i*) grow cultures overnight at 30°C or less, (*ii*) spin overnight cultures and resuspend the pellet in fresh medium to remove the produced β-lactamase, and (*iii*) use the more stable carbenicillin instead of ampicillin.

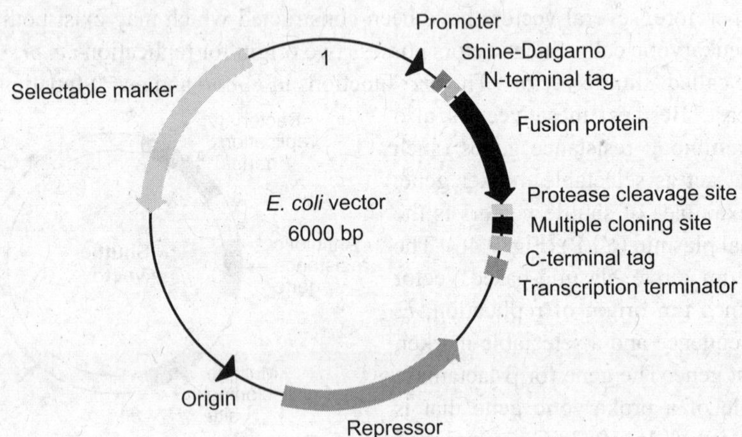

Fig. 7.47. Restriction map of *E. coli* expression vector.

(*ii*) Regulatory gene (repressor): Many promoters show leakiness in their expression, *i.e.* gene products are expressed at low level before the addition of the inducer. This becomes a problem when the gene product is toxic for the host. This can be prevented by the constitutive expression of a repressor protein.

The *lac*-derived promoters are especially leaky. These promoters can be controlled by the insertion of a *lac*-operator sequence downstream the promoter and the expression of the *lac*-repressor by host strains carrying the *lacI^q* allele (for medium copy number plasmids) or from the same or a helper plasmid (for higher copy number plasmids). Alternatively, repression can be achieved by the addition of 1% glucose to the culture medium.

(*iii*) Origin for replication: The origin of replication controls the plasmid copy number.

(*iv*) Promoter. The promoter initiates transcription and is positioned 10-100 nucleotides upstream of the ribosome binding site. The ideal promoter exhibits several desirable features such as: allowing product accumulation up to 50% of the total cellular protein, having a low basal expression level (*i.e.* it is tightly regulated to prevent product toxicity) and ease in induction.

***(v)* Transcription terminator:** The transcription terminator reduces unwanted transcription and increases plasmid and mRNA stability.

***(vi)* Shine-Delgarno sequence:** The Shine-Dalgarno (SD) sequence is required for translation initiation and is complementary to the 3′-end of the 16S ribosomal RNA. The efficiency of translation initiation at the start codon depends on the actual sequence. The consensus sequence is: 5′-TAAGGAGG-3′. It is positioned 4-14 nucleotides upstream the start codon with the optimal spacing being 8 nucleotides. To avoid formation of secondary structures (which reduces expression levels) this region should be rich in A residues.

***(vii)* Start codon:** It is the initiation point of translation. In *E. coli* the mostly used start codon is ATG. GTG is used in 8% of the cases. TTG and TAA are hardly used.

***(viii)* Tags and fusion proteins:** The *N*- or *C*-terminal fusions of heterologous proteins to short peptides or to other proteins offer several potential advantages as given below:

- **Improved expression.** Fusion of the *N*-terminus of a heterologous protein to the *C*-terminus of a highly-expressed fusion partner often results in high level expression of the fusion protein.
- **Improved solubility.** Fusion of the *N*-terminus of a heterologous protein to the *C*-terminus of a soluble fusion partner often improves the solubility of the fusion protein.
- **Improved detection.** Fusion of a protein to either terminus of a short peptide (epitope tag) or protein which is recognised by an antibody or a binding protein (Western blot analysis) or by biophysical methods (*e.g.* GFP by fluorescence) allows for detection of a protein during expression and purification.
- **Improved purification.** Simple purification schemes have been developed for proteins fused at either end to tags or proteins which bind specifically to affinity resins.

***(b)* Protease Cleavage Site:** Protease cleavage sites are often added to be able to remove a tag or fusion partner from the fusion protein after expression. However, cleavage is rarely complete and often additional purification steps are required.

***(c)* Multiple cloning site (MCS):** MCS is a series of unique restriction sites that enables to clone the gene of interest into the vector.

***(d)* Stop codon:** This results in termination of translation. There are three possible stop codons but TAA is preferred because it is less prone to read-through than TAG and TGA. The efficiency of termination is increased by using 2 or 3 stop codons.

QUESTIONS

1. Write in brief on significance of cloning vector in recombinant DNA technology.
2. What is a cloning vector ? Why are they necessary ? Describe the cloning vectors studied by you.
3. What are plasmids ? Write characteristics of plasmids. How do they work as cloning vector ?
4. Describe the characteristics features of some of the plasmids vectors.
5. Describe in detail about *Agrobacterium*-based cloning vectors.
6. Describe in detail about bacteriophage-based cloning vectors.
7. What is phagemid ? Give a brief account of phagemids used as vectors.
8. Describe cloning of BAC vector and selection of transformed bacteria.
9. Discuss in detail the yeast plasmid vectors.
10. What is an expression vector ? Describe the features of expression vectors.

11. Write short notes on the following:

(*i*) Cloning vectors, (*ii*) pBR322,
(*iii*) pSC101, (*iv*) pUC vectors,
(*v*) pUC57, (*vi*) pACYC184,
(*vii*) Ti-plasmid, (*viii*) Ri-plasmid,
(*ix*) Phage λ, (*x*) Binary vectors,
(*xi*) Cosmid and phasmid vectors, (*xii*) Insertion vectors,
(*xiii*) Replacement vectors, (*xiv*) λ EMBL3 replacement vector,
(*xv*) Phage M13 vector, (*xvi*) M13mp18/M13mp19 vectors,
(*xvii*) pBluscrpt II KS (+/–), (*xviii*) pTZ19R and pTZ19U,
(*xix*) BAC vector (*xx*) YAC Vectors,
(*xxi*) Yep, (*xxii*) P1 vector,
(*xxiii*) Plant virus vector, (*xxiv*) Animal virus vectors,
(*xxv*) *AC-DC* elements.

BIBLIOGRAPHY

Alting-Mees, M.A., Sorge, J.A. and Short, J.M. 1992. pBluescript-II : multifunctional cloning and mapping vectors. *Meth. Enzymol,* **216**: 483-495.

Frischauf A.M., Lehrach H., Poustka A. and Murray N. 1983. Lambda replacement vectors carrying polylinker sequences. *J. Mol. Biol.* **15**: 170:827-842.

Lodish H.F. 2003. *Molecular cell biology*. 5 Ed., W. H. Freeman and Co. New York.

Norrander J., Kempe T. and Messing J. 1983. Construction of improved M13 vectors using oligodeoxynucleotide-directed mutagenesis. *Gene.* **26:** 101-106.

Shizuya H. *et al.* 1992 Cloning and stable maintenance of 300-kilobase-pair fragments of human DNA in *Escherichia coli* using an F-factor-based vector. *Proc. Natl. Acad. Sci.* **89**; 8794-8797.

Ung-Jin Kim *et al.* 1992. Stable propagation of cosmid-sized human DNA inserts in an F-factor based vector. *Nucleid Acids Res.* **20**:1083-1085.

Yanisch-Perron C., Vieira J. and Messing J. 1985. Improved M13 phage cloning vectors and host strains: nucleotide sequences of the M13mp18 and pUC19 vectors. *Gene*. **33:** 103-119.

Methods of Gene Manipulation

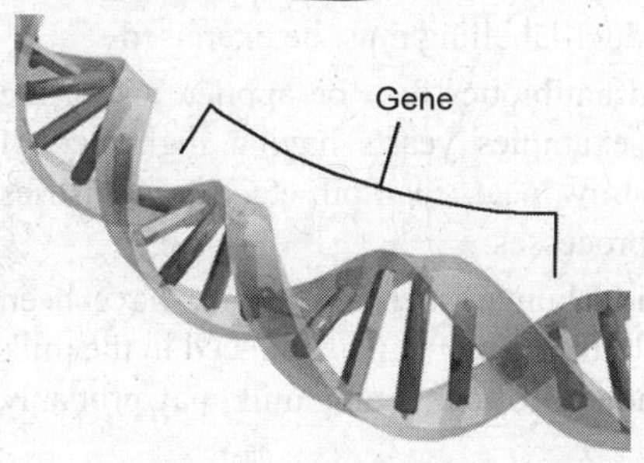

A. The host cells
B. Gene cloning in prokaryotes
C. Blotting techniques
D. Recovery of cells containing recombinant plasmid
E. Detection of nucleic acids
F. Gene cloning in eukaryotes

The DNA fragment to be introduced into the host cell is called 'insert DNA', 'desired DNA', 'target DNA' or 'foreign DNA'. Such gene/DNA fragment is taken out from an organism or gene library, *i.e.* a random collection of cloned DNA fragments in vectors that consist of all genetic information of that organism concerned. However, getting a desired DNA is not so easy because the cellular DNA is very large and the position of desired DNA is not accurately known.

The process of insertion of foreign genes and vectors is used to create recombinant DNA which can be inserted into a host for cloning. The gene to be used must be isolated and well characterised. This would create multiple copies of the gene of interest. This method is relatively a new procedure which allows the analysis of proteins and other parts of most organisms which had previously been very difficult to study due to the scarcity of the gene of interest.

The foreign DNA should be introduced into host cells in such a way that inserted DNA can replicate independently and transferred to progenies during cell division. The transformed cells containing DNA after their characterisation and confirmation can be used commercially for the production of useful compounds such as insulin, interferon, growth hormones, vaccines, proteins, enzymes, etc. Even genetically modified organisms (microorganisms, plants an animals including humans) can performs several beneficial functions and novel chemicals can be produced and new hopes arise for gene therapy in humans in future.

Gene manipulation differs from selective breeding; the later involves only members of the same species. In selective breeding usually only single gene moves into the progeny, because it involves the normal methods of sexual reproduction resulting in transfer of large combinations of genes. Moreover in gene manipulation specific enzymes are used to cut and splice DNA by restriction enzymes that break DNA at specific nucleotide base sequences generating the sticky ends in laboratory. This is done in both the DNAs, from which genes are being taken, and in which genes are being inserted. Then the DNA ligase enzymes are used to rejoin such sections into the other DNA molecule, for example plasmid, cosmid or other cloning vectors of choice. The DNA containing the selected gene for the desired characteristic is then inserted into cells of the target organism by desired vectors. Thus the constructed recombinant plasmids may be transferred into the host cells for purposeful production of products or the genetically modified organism (GMO) may be applied for desired work.

Besides, gene manipulation in plants is rather more drastic because the cell wall forms a barrier; therefore, some other methods can also be used with plant material, such as particle bombardment or ballistic technique, electroporation, pollen-mediated gene transfer, or *Agrobacterium*-mediated gene transfer.

Similarly, various different methods are involved in transfer of novel genes in animal cell for production of desired products, such as microinjection, *in vitro* fertilisation, virus-mediated gene transfer, etc.

There are several biotechnology companies that have attracted the attention of investors excited by the prospect of profits. Similarly, much public distrust has been found, and campaigners on each side have raised the profile of these activities in relation to regulatory authorities. Organisations representing consumer interests want to ban the GMOs from entering the food supply chain or to keep separate from other food supplies and specifically product-labelling must be exercised.

The techniques used in the production of insulin and antibiotics may be applied to use the genetically engineered bacteria in food production. For examples yeasts having high alcohol tolerance, microbes with enhanced ability to digest waste straw, peat, coal, oil, etc. The industries have also favoured the application of genetic modification processes.

Moreover, potential applications of gene manipulation human medical science have been found to provide great opportunities. Production of blood clotting factor can be induced in the milk of sheep, which may be called 'designer milk'. Such low cholesterol-containing milk may probably find a profitable market.

In the earlier chapters two important aspects viz., enzymes and cloning vectors have been described in detail. Different methods of gene cloning have been described in this chapter.

A. HOST CELLS

The success of gene cloning depends on selection and use of a vector and a loving system or host in which the vector can be propagated. This chapter describes various hosts used in cloning experiments and different methods of cloning. Selection of host cells should be done in such a ways where the introduced recombinant vector carrying a DNA insert can multiply. The proper host cells are selected on the basis of purpose of cloning experiments. If the objective of the experiment is the isolation of a desired gene, a simple host cell system can be used that can contribute the required gene of interest.

Table 8.1. Different types of host cells used in gene cloning.

Groups of organisms		*Types*	*Examples*
Prokaryotes			
Bacteria		Gram-negative	*Escherichia coli, Agrobacterium tumifaciens*
		Gram-positive	*Bacillus subtilis, Pseudomonas* spp., *Streptomyces* spp.
Eukaryotes			
	Fungi	Unicellular	*Saccharomyces cerevisiae*
		Filamentous	*Aspergillus nidulans, Neurospora crassa*
	Plants	Protoplasts	Various types like potato, tomato, etc.
		Intact cells	Various types
		Whole organs	Embryo, anther, etc
		Insect cells	*Drosophila melanogaster*
Animals		Mammalian cells	Different types
		Oocytes	Different types
		Whole organisms	Different types

However, a more specific system is required if the aim of the experiment is to express the gene in higher groups of eukaryotes. An ideal host cell is that which should be easy to handle and propagate, and should be available as a wide variety of genetically defined strains. Different types of host cells used in gene cloning are given in Table 8.1.

1. Prokaryotic cells

Prokaryotic cells of various organisms have successfully been used in a variety of experiments. Among the prokaryotes *Escherichia coli* has been used in many cloning experiments due to easy in handling and propagation. The microbial geneticists have isolated many strains of *E. coli* and their genetic mechanisms have been studied. *E. coli* is a Gram-negative rod-shaped enteric bacterium of 2.5 m length having peritrichous flagella (Fig. 8.1).

In 1885 a German pediatrician and bacteriologist Theodor Escherich discovered *E. coli* which is now classified as part of the Enterobacteriaceae family of gamm-proteobacteria. The harmless strains of *E. coli* are part of the normal flora of the gut and can benefit their hosts by producing vitamin K_2, or by preventing the establishment of pathogenic bacteria within the intestine. *E. coli* normally colonizes an infant's gastrointestinal tract within 40 hours of birth, arriving with food or water or with the individuals handling the child. It adheres to the mucus of the large intestine in the bowel. It is the primary facultative bacterium of the human gastrointestinal tract.

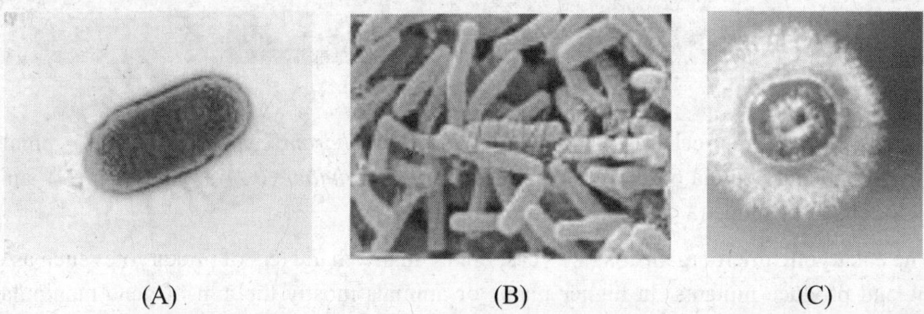

(A) (B) (C)

Fig. 8.1. Prokaryotic host cells; (A)- a cell of *E. coli* showing peritrichous flagella; (B)- rod-shaped cells of *Agrobacterium tumifaciens;* (C)- colony of *Streptomyces* sp.

Genome size of *E. coli* is about 4.6×10^6 bp and the complete sequence of its is now known. It is considered one of the simplest hosts due to coupled process of gene expression (transcription and translation) *i.e.* the newly synthesised mRNA is immediately translated into proteins without post transcriptional processing. Hence it is considered one of the simplest host cells.

The bacteria can also be grown easily and its genetics is comparatively simple and easily-manipulated, making it one of the best-studied prokaryotic model organisms, and an important species in biotechnology. *E. coli* also plays an important role in modern biological engineering and industrial microbiology because of its long history of laboratory culture and ease of manipulation. The work of Stanley Norman Cohen and Herbert Boyer in *E. coli* became a foundation of biotechnology by using plasmid and restriction enzymes to create recombinant DNA.

In addition to *E. coli*, the other bacteria used as hosts in gene cloning are the species of *Agrobacterium, Bacillus, Streptomyces,* etc. Advantages of using *Bacillus subtilis* as a host are: (*i*) safety associated with this bacterium because it has no known pathogenic interactions with man or animals, (*ii*) its wide application in industry for a long time, (*iii*) secretion of many proteins compared to *E. coli* (this property offers an enormous advantage over *E. coli* for the expression of heterologous proteins which are often degraded inside bacteria), and (*iv*) its sporulation and interest as a model of prokaryotic differentiation for bacterial geneticists and physiologists.

2. Eukaryotic Host Cells

One of the disadvantages of using prokaryotes is the lack of membrane bound organelles such as nucleus, mitochondria, etc. Due to this fact certain eukaryotic genes cannot function inside the prokaryotic cells like *E. coli* and this warrants their isolation by selection mechanisms which exclusively depend on gene expression. The reason is that the eukaryotic genes having intervening sequences (intrans) fail to express in intrans-free genomes of prokaryotes. However, if one desires to produce eukaryotic proteins through cloning, it is not easy to ensure that a prokaryotic host will be able to produce the functional proteins as per requirement.

There is large number of eukaryotic organisms, right from microorganisms to large-sized organisms such as yeasts (e.g. *Saccharomyces cerevisiae*), fungi (e.g. *Aspergillus nidulans, Neurospora crassa,* etc.), algae, plants, animal cells and their organs like oocytes, embryos, etc. Fig. 8.2 shows some of the eukaryotic cells that can be used as host cells.

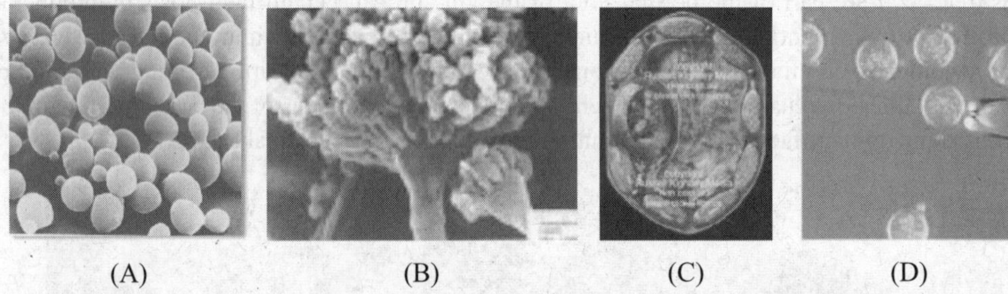

(A) (B) (C) (D)

Fig.8.2. Eukaryotic host cells; (A)- budding cells of *Saccharomyces cerevisiae*; (B)- phialides producing conidia on the vesicle of *Aspergillus nidulans*; (C)- a plant cell; (D)- animal cells to be cloned.

The eukaryotic microorganisms like yeasts share many characters of prokaryotes such as easy to grow and produce mutants. In higher plants or animals mostly the aim of gene manipulation lies on production of transgenic plants or animals rather than isolation and analysis of a gene to produce the desired proteins in large quantity. Transgenic plants and animals have been described in preceding chapters. Plant and animal cells are usually grown as cell cultures which are much easier to manipulate than the cells in whole organism.

Yeasts are unicellular, although some species with yeast forms may become multicellular through the formation of a string of connected budding cells known as pseudohyphae, *i.e.* false hyphae as seen in most of molds. Size of yeast varies greatly depending on the species, typically measuring 3–4 mm in diameter, although some yeasts can reach over 40 μm. The yeast species, *S. cerevisiae,* has been used in baking and fermenting the alcoholic beverages for thousands of years. It is also extremely important as a model organism in modern cell biology research, and is the most thoroughly researched eukaryotic microorganism. Researchers have used it to gather information into the biology of the eukaryotic cell and ultimately human biology.

S. cerevisiae is very common eukaryotic microorganisms used in gene cloning. It has been used for centuries in production of various kinds of beverages, bread, etc. It consists of 3.5 times more DNA than *E. coli*. On 24 April 1996, *S. cerevisiae* was announced to be the first eukaryote to have its genome consisting of 12 million base pairs, fully sequenced as part of the Genome Project. At the time it was the most complex organism to have its full genome sequenced and took 7 years and the involvement of more than 100 laboratories to accomplish.

B. GENE CLONING IN PROKARYOTES

Prokaryotes are efficiently used in cloning experiments as compared to eukaryotic cells for the production of recombinant DNA molecules and functional proteins of choice. In 1980, 'Humulin' was the first recombinant protein pro-duced by applying recombinant DNA technology by using *E. coli* as host cell. Steps of gene cloning are given in Fig. 8.3. Now, genetically engineered hepatitis B vaccine produced by genetically modified yeast cells is available in market.

Bacillus subtilis has been manipu-lated to produce valuable industrially important enzymes such as amylases and alkaline proteases. For example, subtilisin is a serine protease which is used in biological washing powders for increasing the efficiency of stain removal. It accounts for 25% of total world enzyme market. Similarly, species of *Streptomyces* have been manipulated for enhanced production of antibiotics, etc. This section describes the detailed methods of gene cloning in prokaryotes.

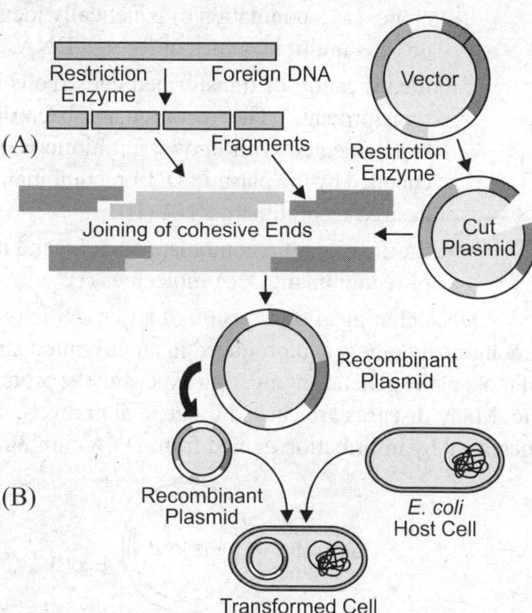

Fig. 8.3. Production of recombinant plasmid (A); transformation of host cell by the recombinant plasmid (B).

I. STRATEGIES OF RECOMBINANT DNA TECHNOLOGY

Recombinant DNA technology involves not a single method but several steps. The procedure for insertion of foreign genes and vectors can be grouped into four categories based on the construction and insertion of recombinant DNA, DNA source, selected vector, restriction enzymes, and host cell transformation.

The gene or sequence of interest is usually important due to its unique function, whether it may be the production of a protein or capsule or regulation of gene expression. Once the DNA source is chosen, it must bc isolated. This can be accomplished with the use of restriction endonucleases. The isolated DNA is then inserted into a vector. Vectors are used as a vehicle by which foreign DNA can be inserted into cells; therefore, the recombinant DNA may be replicated. The most common type of vector is a plasmid. Plasmids are a small circular piece of extrachromosomal DNA found in bacterial cells which are capable of independent replication in the host. Plasmids usually carry a few genes which give them a selective advantage such as genes for antibiotic resistance.

1. Basic Technique of Gene Cloning

In 1976, T. Maniatis and coworkers described the basic techniques of gene cloning the principal steps of which are shown in Fig. 8.4 and given below :

- Isolation of DNA of known function from an organism (A).
- Enzymatic cleavage (B) to generate complementary 'sticky ends', and then joining of insert DNA to cloning vector (C).
- Mixing of opened plasmid and the freed gene with DNA ligase to join two cleaved pieces of DNA and forms a recombinant DNA (*i.e.* vector + insert DNA) molecule (D).
- Transformation of a host cell, *i.e.* transfer and maintenance of this rDNA molecule into a host cell (E).

- Amplification of rDNA to get its multiple copies in a cell and cell multiplication to get a clone, *i.e.* a population of genetically identical cells (F). This facilitates each of clones to possess multiple copies of foreign DNA.
- Identification of transformed cells (cells carrying rDNA) and their selection from non-transformants. The recombinant bacterial cultures are allowed to grow in medium supplemented with known antibiotic (G). All the cells, except those which have been encoded by the plasmid DNA recombinant, are killed leaving a cell culture containing the desired recombinant DNA (H).
- Purification of recombinant bacteria and its multiplication for isolation of a large number of recombinant DNA molecules (I).

DNA cloning allows a copy of any specific part of a DNA (or RNA) sequence to be selected among many others and produced in an unlimited amount. However, this technique is the first stage of most of the genetic engineering experiments: production of DNA libraries, PCR, DNA sequencing, etc. Many diseases are caused by gene alterations. The understanding of genetic diseases could be increased by information gained from DNA cloning.

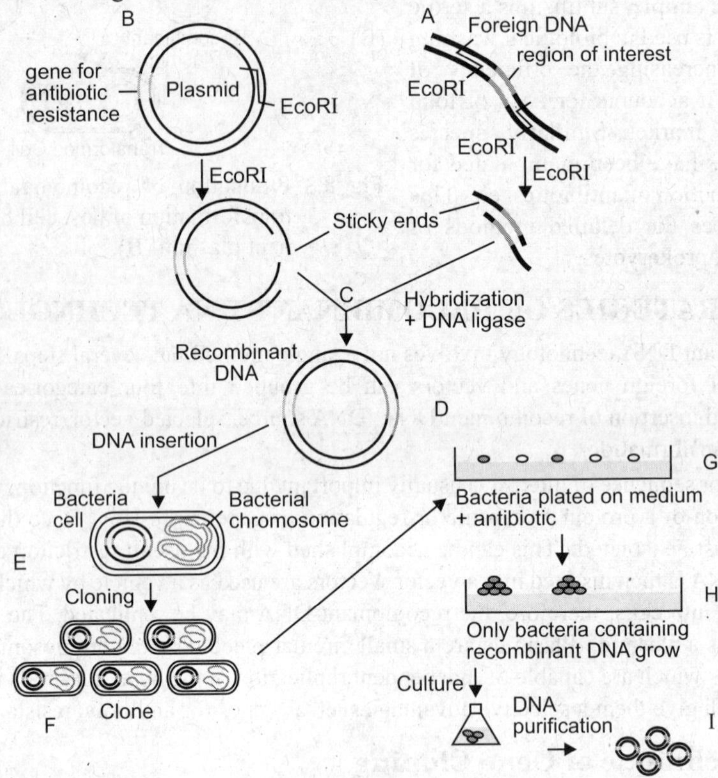

Fig. 8.4. Process by which a plasmid is used to incorporate recombinant DNA into a host cell for cloning.

2. Pros and Cons

Problems that may arise in the process of creating recombinant DNA include: (*i*) difficulty in location and isolation of the sequence of interest, and (*ii*) finding the appropriate vector and successful transformation of the recombinant DNA into the host cell. The recombinant DNA must be made competent by treatment with a $CaCl_2$ solution, but this does not ensure that the procedure will work.

One important concern with using this technique to produce recombinant DNA and inevitably cloned cells is the use of the genetically engineered cells for biological or environmental purposes. Transgenic plants and animals could be created but the uncertainty of their impact on society may not permit their use until further research is conducted.

This method however, has enabled the analysis of many proteins, enzymes and genes which previously had not been possible. This method has a wide variety of applications. It can be used in the medical field to study specific genes that might be linked to genetic disorders. It can be used in the creation of recombinant vaccines. This method can be applied in the agricultural field by creating transgenic plants which exhibit antibiotic resistance.

II. DNA LIBRARY (= GENE LIBRARY OR GENE BANK)

A collection of different DNA sequences from an organism where each sequence has been cloned into a vector for ease of purification, storage and analysis is called 'gene library'. A gene library is also called *DNA library* or *gene bank*. This term also represents the collection of all of the vector molecules, each carrying piece of the chromosomal DNA of the organism, prior to the insertion of these molecules into the population of the host cell.

1. Types of DNA Libraries

There are two types of libraries on the basis of source of DNA used: (*i*) *genomic library* (where genomic DNA is used), and (*ii*) *cDNA library* (where cDNA or complementary DNA is produced using mRNA). A gene library should contain a certain number of recombinants with a high probability of containing any particular sequence. This value can be calculated if you know the genome size and average size of insert in the vector. The most popular prokaryotic vector used is phage λ which can take inserts up to about 20 kb. Lambda vectors can take larger inserts than plasmid vectors and are more stable than cosmid vectors. Now-a-days eukaryotic vectors such as YAC are used which can take enormous inserts which reduces the number of clones that require screening. This is particularly important with large genomes *e.g.* human genome having size up to 2.8×10^6 kilobases (= 2.8×10^9 bases).

(*a*) How many clones are required for a library? : In theory:

$$\text{Minimum number of clones required } (N) = \frac{\text{size of genome (b)}}{\text{mean size of fragments (a)}} \qquad (1)$$

For example, if using human genome and phage λ vector:

$$\text{Minimum number of clones required } (N) = \frac{2.8 \times 10^6 \text{ kb}}{20 \text{ kb}} = 1.4 \times 10^5 \text{ clones}$$

(*b*) Size of Gene library: In practice the number of clones must be greater than this because the process is random. Therefore, some sequences will be included in the library more than once due to overlaps and some sequences will be missed if only the minimum number of clones is used. In 1976, Clarke and Carbon devised a formula which calculates the probability (*P*) of any DNA sequence being included in a library of random fragments.

The size of library can be calculated. The probability of getting a given sequence in a gene library, the number (*N*) of recombinants or clones (in bacterial colonies or viral plaques) can be calculated by using the following formula:

$$N = \frac{\ln (1 - P)}{\ln (1 - f)} \qquad (2)$$

where, ln = natural logarithm, *P* = the desired probability and *f* = the fraction of genome in one insert (*i.e.* = minimum number of clones required (b/a). For a probability of 0.99 with insert sizes 20×10^3 bp, the value of genome for *E. coli* (4.6×10^6 bp) will be:

$$N_{E.\ coli} = \frac{\ln(1-0.99)}{\ln[1-(2\times10^4/4.6\times10^6)]} = 1.1\times10^3$$

Similarly, For a probability of 0.99 with insert sizes 20 × 10³ bp, the value of human genome (3 × 10⁹ bp) will be:

$$N_{\text{human}} = \frac{\ln(1-0.99)}{\ln[1-(2\times10^4/3\times10^9)]} = 6.9\times10^5$$

For example, to achieve 95% (0.95) probability of including a particular sequence in a random human gene library where the number of clones required will be:

$$N = \frac{\ln(1-0.95)}{\ln(1-1/1.4\times10^5)} = 4.2\times10^5 \text{ clones}$$

In other words, 3 times ($4.2 \times 10^5/1.4 \times 10^5$) the theoretical minimum number of clones has to be used to achieve 95% probability of including a given sequence in the library. It should also be noted that to achieve 100% (1.00) probability an infinite number of clones would have to be used!

Another version of the formula:

$$N = \frac{\ln(1-P)}{\ln(1-a/b)} \quad (3)$$

Where, P = probability, a = mean size of fragments, b = size of genome, ln = natural logarithm.

The number of clones required also indicates the number of clones that have to be screened to find a particular sequence (maybe a gene) in a gene library.

cDNA Library vs Genomic DNA Library

The entire sequence of genes is often very large in eukaryotic genomes because coding sequence (exons) is interrupted by introns that do not contain coding DNA sequence. The advantage of a cDNA library is that only the exons, or DNA representing expressed sequence, are templates for the creation of DNA (via reverse transcriptase) to collect the preferred genes. On the other hand, genes from prokaryotic cells does not usually contain introns, and the DNA is, therefore, much shorter. Also, prokaryotic mRNA can be difficult to isolate because of this a genomic library is more preferable for prokaryotic genomic studies.

On the basis of these values on can explain about the possibility of making good libraries from prokaryotes in plasmids, where the insert size is 5 – 10 kb because only a few thousand recombinants are required for this purpose. For larger genomes, the few recombinants are required. For this purpose, bacteriophage λ, cosmid, BAC or YAC vectors have been developed. For construction of good library methods of cloning in phage λ and the efficiency of λ packaging is a good choice. The use of different types of cloning vectors for preparation of human genomic library of different sizes is given in Table 8.2.

Table 8.2. Use of different types of cloning vectors for preparation of human genomic library of different sizes.

Types of vectors	*Insert size (kb)*	*Number of clones**	
		***P* = 95%**	***P* = 99%**
Phage λ replacement	18	5,32,500	8,20,000
Cosmid, fosmid	40	2,40,000	3,70,000
P1-based vector	125	77,000	1,18,000
BAC, PAC	300	32,000	50,000
YAC	6000	16,000	24,500

*Calculation is based on the formula (3)

2. Genomic Library

A collection of clones that represents the complete genome of an organism is called genomic library. All fragments of DNA inserted into vectors for further propagation into suitable host represent the entire genome of an organism. For construction of a genomic library the entire genomic DNA is isolated from host cells/tissues, purified and broken randomly into fragments of correct size for cloning into a suitable vector. There are two basic ways of randomly fragmenting genomic DNA: *physical shearing* (such as pipetting, mixing or sonication) and partial *restriction enzyme digestion* (by using limiting amount of restriction enzyme the DNA is not digested at every recognition sequence). Using these methods genomic DNA is broken randomly into smaller fragments.

A total of 6 random DNA fragments obtained after physical shearing are given in Fig. 8.5. The vector isolated from bacterium is also digested with the restriction enzyme which digests genomic DNA. The fragments of genomic DNA are inserted into a vector. Each vector consists of different fragments of DNA. The recombinant DNA molecules are transferred into bacterial cells, or phage particles are assembled. The genomic library for organisms with smaller genome size (*e.g. E. coli*) can be constructed in plasmid vector. Only 5,000 clones (of the average size of 5,000 bp) results in 99% chance of cloning the entire genome of 4.6 $\times 10^6$ bp. Phage λ-based vectors, cosmid, BAC or YAC vectors have been developed for construction of libraries of organisms containing large-sized genome.

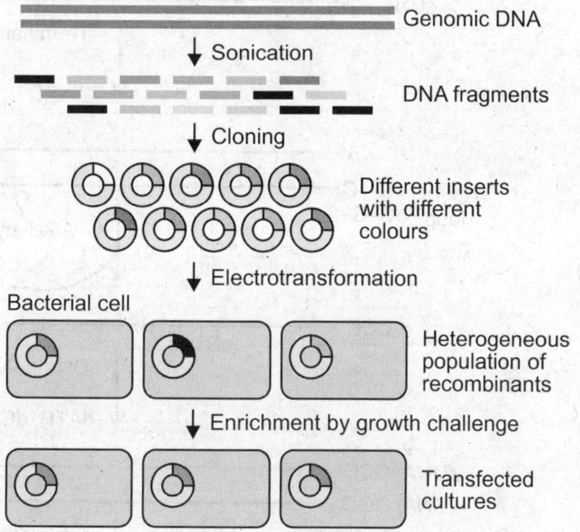

Fig. 8.5. Steps for construction of a genomic library.

One disadvantage of creating this type of library for eukaryotic gene is the presence of introns (non-protein coding pieces of DNA) along with exons (protein coding sequences). Most of DNA in any eukaryotes consists of introns; therefore, many of clones of genomic library consist of introns. Another limitation of genomic library is the difficulties in searching of gene of interest within the large-sized genome in most of the eukaryotes including humans.

3. cDNA Library

There is compartmentalisation of organelles in all eukaryotes. A gene expresses through transcription in nucleus and translation in cytoplasm. After transcription the mRNAs undergo post-transcriptional processing. The mRNAs are highly processed representatives of genes which express under specific conditions. The mRNAs cannot be directly cloned because they are unstable; hence mRNAs are converted into cDNAs. Working with mRNA is difficult because mRNA is unstable and is easily degraded by RNases which can be found even on the skin; therefore, researchers use an enzyme called reverse transcriptase which will produce a DNA copy of each mRNA strand referred to as cDNA. The library made from complementary or copy DNA (cDNA), which is basically reverse transcribed mRNAs and, therefore, it is called 'cDNA library'. The cDNA library represents the DNA of only eukaryotic organisms, not the prokaryotic ones. The genomic DNA of eukaryotes contains introns, regulatory regions, and repetitive sequences. Therefore, establishment of genomic library of eukaryotes is not meaningful. The cDNA library can be constructed by using mRNA because mRNAs are the highly processed, intron-free representatives of DNA having only coding sequence.

A cDNA fragment is prepared directly by using mRNA as template. It follows many biochemical methods; the major steps for production of cDNA are described as below and shown in Fig. 8.6.

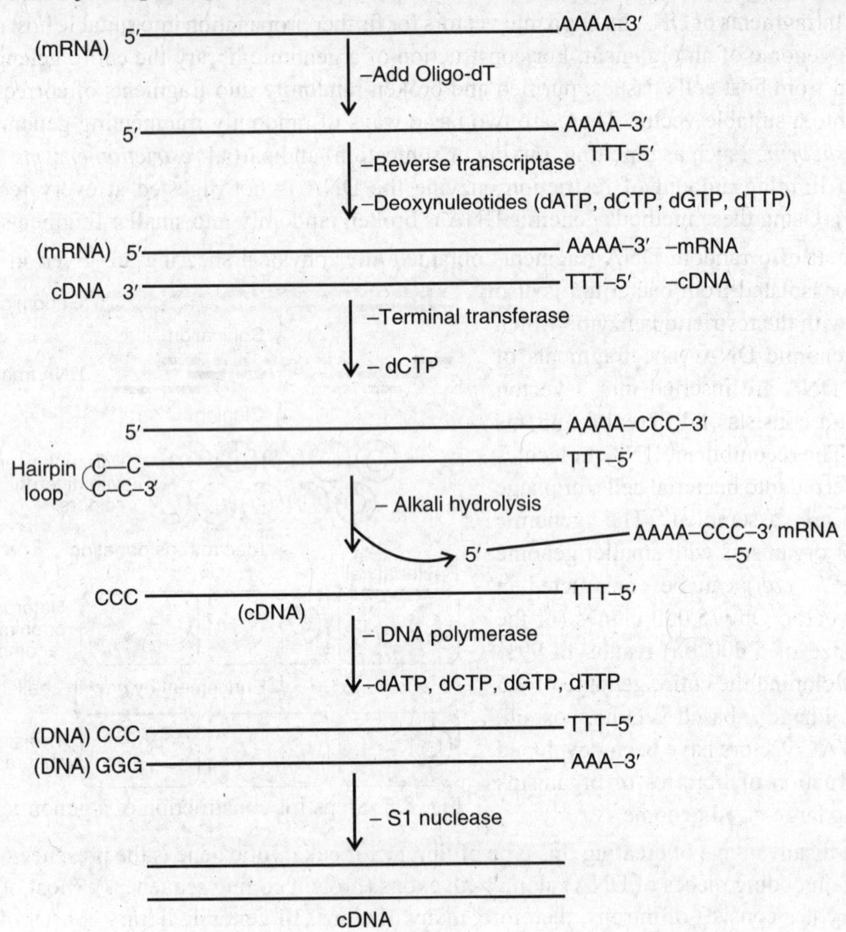

Fig. 8.6. Preparation of cDNA used for cloning.

A cDNA library has several uses : (*i*) a cDNA of an eukaryotic organism (for example, a human) can be cloned into a prokaryotic organism (for example, *E. coli*) and expressed there with limitations, for example post-transcriptional modification; (*ii*) a cDNA library is also important for analysis through bioinformatics, (*iii*) the complete cDNA library of an organism gives the total of the proteins it can possibly express, and (*iv*) comparison of cDNA sequence between libraries constructed from cells derived from different organisms can provide insight into the genetic and evolutionary relationship between organisms through the similarity of their cDNA.

***(a)* Isolation of mRNA:** The mRNA makes up between 1-5% of total cellular RNA. Isolation of specific mRNAs is much difficult if it is in low amount in the cell. It is most frequently used for: (*i*) detection and quantification of extremely rare mRNAs, (*ii*) synthesis of probes for array analysis, and (*iii*) the construction of random-primed cDNA libraries where the use of total RNA would generate rRNA templates that would significantly dilute out cDNAs of interest. The majority of mRNA sequence in eukaryotic cells contains a long polyadenylated tract [*i.e.* about 100 poly adenylated (poly A) residues] at their 3′OH termini due to which they can be separated from the other types of RNAs. Therefore, mRNA binds to an oligo-dT cellulose affinity column or poly-U sepharose from which it can be eluted. By using gel electrophoresis its integrity can be studied, which allows mRNAs to be size-fractionated by recovering chosen regions of the gel lane. Commercial preparations of both the substrates are now available. The mRNA preparations can be enriched for the desired molecules by fractionation according to their size by using sucrose density gradient centrifugation.

***(b)* Reverse Transcription:** Reverse transcriptase was first discovered in RNA viruses called retroviruses. It is required for the synthesis of DNA copy of an mRNA molecule. An oligo–dT primer is required to ensure the synthesis of cDNA at 3′ terminus of mRNA in the polyA tract by the enzyme reverse transcriptase. In case of non-polyadenylated RNA such as ribosomal RNA or influenza RNA, the enzyme poly(A) polymerase from *E. coli* is used to synthesise a poly(A) tract at 3′ terminus. The mRNAs are treated with oligo-dT primer, reverse transcriptase enzyme and four deoxynucleotide triphosphates (dNTPs), *i.e.* dATP, dGTP, dCTP and dTTP. The oligo-dT primer binds to polyadenylated tail that contains –AAAA(n) residues and provides free 3′-OH site for reverse transcription. Reverse transcriptase adds complementary dNTPs one-by-one to free 3′-OH site of primer, forms single strand of DNA resulting in RNA-DNA hybrid. Cellular DNA and total RNA inhibit reverse transcriptase activity; therefore, it is necessary that mRNA must be in pure form before cloning.

The oligo-dT segments contain between 10 and 20 nucleotides that hybridise the poly-A tract on mRNA. The oligo-dT primer ensures the initiation of cDNA synthesis at 3′ terminus of RNA in poly-A tract. The conditions that optimize the synthesis of cDNA are: (*i*) the use of high level of deoxynucleoside triphosphates (dNTPs), (*ii*) addition of sodium pyrophosphate or ribonucleoside triphosphates, and (*iii*) omission of monovalent cations coupled with evaluation of the reaction temperature to 46°C.

***(c)* Oligo-dC Tailing:** The RNA-DNA hybrid is treated with the enzyme terminal transferase and a nucleotide precursor, dCTP. This enzyme adds dCTP one-by-one to 3′-OH group of RNA and DNA strands. Consequently a short oligo-dC tail is produced at 3′ end of both the strands. The residues of cDNA become curved. The curved tail forming a loop like structure is known as 'snapback' or 'hairpin loop'. The hairpin acts as a primer for elongation in the presence of DNA polymerase I or AMV (avian myeloblastosis virus) reverse transcriptase.

***(d)* Alkali Hydrolysis:** The mRNA-cDNA strands are separated into single strand upon hydrolysis by using alkaline (NaOH) sucrose solution.

***(e)* Addition of Oligo-dG Primer:** Oligo-dG primer is added in the reaction mixture and temperature is maintained to about 55°C. This facilitates the binding of dG to oligo-dC tails formed on cDNA. The single stranded cDNA acts as a template for the synthesis of double stranded cDNA in the presence of DNA polymerase I or AMV reverse transcriptase.

***(f)* Synthesis of Second Strand of cDNA:** The dNTP precursors and Klenow DNA polymerase I are added to the reaction mixture. This enzyme adds complementary nucleotides to the cDNA one by one to 3′-OH site of primer. Consequently, double stranded cDNA clone is synthesised from single stranded cDNA, the length of which is similar to that of the mRNA molecule initially used. The aim of using various enzymatic steps is to get the longest possible double stranded cDNA.

***(g)* Cleavage of Hairpin Loop:** Cleavage of phosphodiester bonds formed within the closed loop present at one end of double stranded cDNA and generation of suitable cDNA molecules are necessary before insertion into a cloning vector. Hence, any residual single stranded cDNA which has not been converted into double stranded cDNA is removed from the cDNA molecule by using S1 nuclease. This enzyme is specific to recognize and cleave within very short region of incorrectly matched duplex. This results in loss of hairpin loop from 5′ end of mRNA. A single-stranded cDNA converts very efficiently into double-stranded form when AMV reverse transcriptase is used in the reaction.

***(h)* Generation of Cohesive Ends on Double-stranded cDNA**: It is necessary to generate cohesive ends at both termini of cDNA molecule for efficient formation of recombinant DNA molecules. There are two approaches to generate cohesive ends: (*i*) edition of restriction enzyme linkers, and (*ii*) addition of homopolymer tails. These two approaches have been discussed in the preceding section.

The cDNA is used for cloning and construction of cDNA library and in gene cloning experiments. The double stranded cDNA is identical to the gene which codes for mRNA (only in prokaryotic genes). However, majority of the eukaryotic genes contain introns which interrupt their coding sequences. By using this method intron-free cDNA molecules can be prepared. Moreover, the cDNA can be radioactively labelled by nick translation and used as a hybridisation probe for the identification of DNA fragments containing the gene.

3. Screening of cDNA Clone

There is no need of screening the cDNA clones because cDNA clone is prepared by using a specific mRNA. However, often cDNA clones are the mixture of cDNA molecules. The cDNA clone of desired function may be low or large in number. Therefore, all these clones must be ligated to a suitable vector such as M13 vector, phage λ vectors and transferred into host bacteria. Each bacterial cell possesses a single cDNA clone; hence collection of all recombinant bacteria is called cDNA library. When gene cloning is to be done, the insert DNA is isolated from recombinant bacteria and cut with restriction enzyme. Then the cDNA is separated through electrophoresis, amplified and used for gene manipulation. Screening of recombinant clones has been described in preceding section.

III. INSERTION OF A FOREIGN GENE (DNA INSERT) INTO A CLONING VECTOR

Cloning vector (*e.g.* pBR322) is treated with specific restriction enzymes so that the cohesive ends generated may have complementary residues similar to foreign DNA (Fig. 8.7 A). The insert DNA prepared as above or procured from gene bank is also cleaved by using the same restriction enzyme that forms specific cohesive ends (B). The plasmid contains amp^R and tet^R marker genes to confer resistance against anti-biotics, ampicillin and tetracycline res-pectively. Some time after cleavage by restriction enzymes and ligation of foreign DNA fragment with the plasmid a few important genes are destroyed in the resultant product (C). For example, *Bam*HI destroys tet^R genes. The sticky ends can permanently join if T4 DNA ligase (which forms a phosphodiester linkage between free 5′-PO_4 and 3′-OH group) is added along with ATP at 4-10°C for a long incubation time. A number of products are formed from this reaction which is highly heterogeneous mixture of recombinant DNA molecules together with parental plasmids.

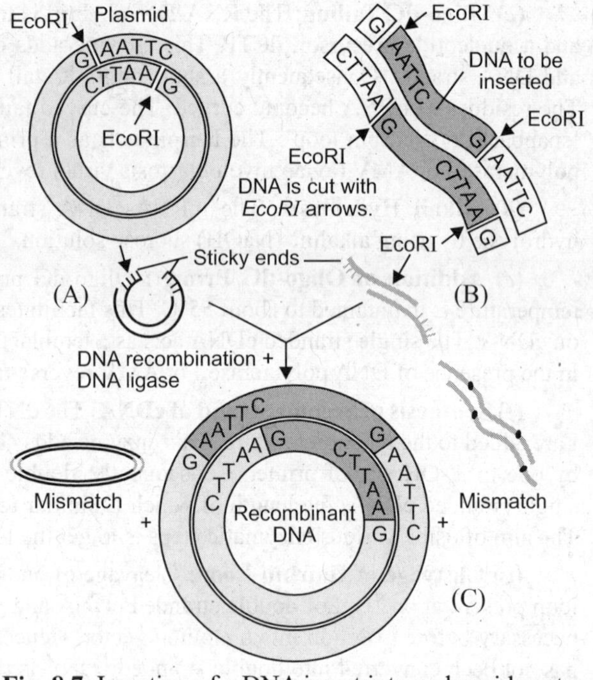

Fig. 8.7. Insertion of a DNA insert into a plasmid vector.

For insertion of double stranded cDNA into a cloning vector it is necessary to add to both termini single stranded DNA sequence which should be complementary to a tract of DNA at the termini of linearlised vector. In order to get efficient formation of recombinant DNA molecules, addition of sticky ends on both termini is necessary. There are two methods for generation of cohesive ends on blunt-ended double stranded cDNA: use of linkers and homopolymer tails.

1. Blunt-End Ligation by Linkers and Adaptors

Joining of DNA molecules with blunt ends by using DNA ligase is called blunt-end ligation. This is done by using *linker molecules* and *homopolymer tailing*. The later approach is described ahead. Chemically synthesised double stranded oligonucleotides are called linkers. Linkers can be defined as *the short tretches of double stranded DNA of 8-14 bp length having recognition site for 3-8 restriction enzymes*. These linkers are ligated to blunt end DNA by ligase. Because of the high concentration of these small molecules present in the reaction, the ligation is efficient when compared with blunt-end ligation of large molecules. Linkers added to double stranded DNA molecules act as recognition site for restriction enzyme. The cohesive ends are generated by digesting the DNA with appropriate restriction enzyme that generates cohesive ends by cleaving in the linkers. The problem with linker is that the sites for the enzyme used to generate cohesive ends may be present in the target DNA fragment. This drawback limits the use of linkers for cloning. For example, the following oligonucleotide possesses recognition sequence for *Eco*RI (asterisk nucleotides).

5′-C C G A*A*T*T*C*G G-3′
3′-G G C T T A A G C C-3′

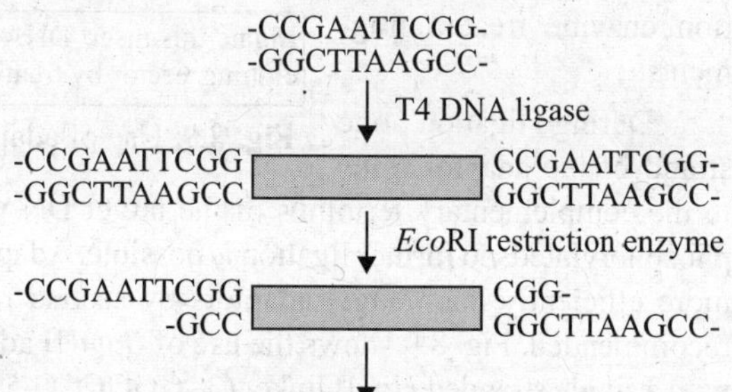

Fig. 8.8. Use of linkers for generation of protruding 5′ termini.

It can be ligated to blunt end of any double stranded DNA and cut by *Eco*RI to generate sticky ends to DNA fragments (Fig. 8.8). The 10-mer 5′-CCGAATTCGG-3′ linker contains the recognition site for *Eco*RI.

On the other hand chemically synthesised DNA molecules containing pre-formed cohesive ends are called 'adaptors'. They are short single stranded non-complementary synthetic oligonucleotides that carry an internal restriction endonuclease recognition site that may be used in conjunction with linkers. Upon annealing, a linker/adaptor with one blunt end and one sticky end is produced, which can be added to the cDNA to provide sticky end cloning without digestion of the linkers. Thus, adaptors are linkers with cohesive ends or a linker digested with restriction enzyme. Example of some of the adaptor sequences are given in Table 8.3.

Table 8.3. Example of some of the adaptor sequences.

Restriction enzyme	***Adaptor sequence***
*Bam*HI-*Pst*I	5′GATCCCTGCAG3′ 3′GGACGTC5′
*Bam*HI-*Sma*I	5′GATCCCCCGGG3V 3′GGGGGCCC5′
*Hin*dIII-*Sma*I	5′AGCTTCCCGGG3′ 3′AGGGCCC5′
*Hin*dIII-*Not*I	5′AGCTTGCGGCCGC3′ 3′GCCGGCG5′
*Eco*RI-*Not*I	5′AATTCGCGGCCGC3′ 3′GCGCCGGCG5′
*Eco*RI-*Sma*I	5′TCGACCCCGGG3′ 3′GGGGCCC5′
*Eco*RI-*Eco*RI	5′AATTCGAATTC3′ 3′GCTTAAG5′

Adaptors are used to replace one type of protruding terminus with another. They are not perfectly double stranded nor single stranded. By adding adaptors to the ends of a DNA, blunt end sequences can be converted into cohesive ends. When recognition site is present within the DNA molecules to be cloned, adaptors are used to develop sticky ends. Adaptor is used in the form of 5′-OH so that self polymerisation could not be done (Fig. 8.9). Thereafter, phosphorylation of foreign DNA ligated adaptor is done at 5′ end. They joined to restriction enzyme treated fragments.

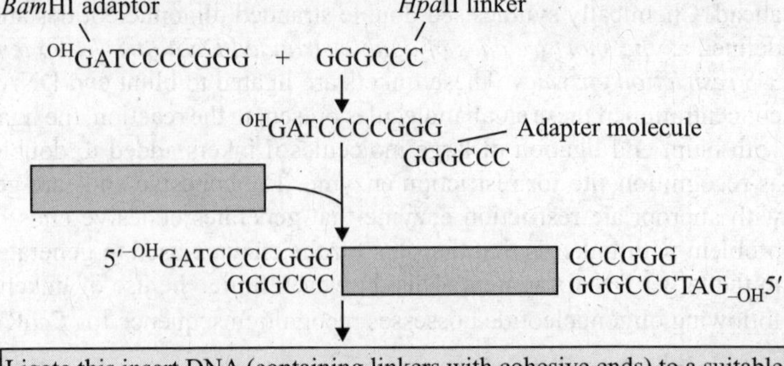

Fig. 8.9. Use of adaptor molecules in formation of cohesive termini.

During ligation one strand of the adaptor joins to the complementary terminus of the target DNA, restoring an *Eco*RI site. Unless the adaptor is phosphorylated, no further ligation is possible. Adaptors can form tandem array, which can be cleaved more efficiently than single adaptors by second restriction enzyme; therefore phosphorylation is recommended. Fig. 8.9 shows the use of *Bam*HI adaptor (5′-GATCCCCGGG-3′) which is annealed with a single stranded *Hpa*II linker (3′-GGGCCC-5′) to get a double stranded sticky-ended molecule. This is added to blunt ended DNA by using DNA ligase. Now the resulting DNA molecule possesses protruding 5′-termini without the need for digestion with a restriction enzyme. Self-ligation can be prevented through dephosphorylation of 5′-termini..

2. Homopolymer Tailing

For the first time D.A. Jackson and coworkers in 1972 applied this method to construct a recombinant DNA by inserting phage λ DNA into SV40 DNA by using dA-dT homopolymers. This is an effective and popular method for cloning cDNA. In this technique an enzyme, terminal transferase, is used to add homopolymer tails of poly dA, dT, dG or dC to a DNA molecule. Earlier this technique was used to produce recombinant DNA molecule by poly-dA tails on one molecule and poly-dT tail on the other molecule. Now this technique is most commonly used for cloning of cDNA into *Pst*I site of a plasmid vector by dG-dC tailing. Homopolymer tailing offers two advantages: (*i*) it provide longer region for annealing two cohesive ends of DNA molecules, and (*ii*) it has specificity as the vector and insert cDNA have different but complementary tails; there is a little chance of self-annealing.

Using terminal transferase the synthesis of homopolymer tails of the defined length at both 3′OH termini of double stranded DNA and vector can be done. In the presence of precursor dGTP, terminal transferase helps to add poly-dG at 3′ termini of vector. Likewise the same enzyme adds poly-C at 3′OH termini of DNA molecule when precursor dCTP is present. The linearised vector having tails is incapable of recircularisation, unless ligated to a double stranded DNA fragment. The vector DNA and insert DNA tail are annealed. The poly dG-dC tails are then ligated by T4 DNA ligase (Fig. 8.10). If poly dG-dC tails are used, instead of poly dA-dT tails, a high initiation temperature (up to 37°C) is required for the annealing reaction. The major advantage in using poly dG-dC tails is that the hybrids are more stable than poly dA-dT hybrids.

In the former method where poly dT-dC tailing is done, the inserted DNA molecule can be retrieved by using restriction enzymes (*e.g. Eco*RI) on the cleavage sites of linker, whereas in the second method, it is very difficult to retrieve the inserted DNA fragment due to the loss of recognition sites when poly dA-dT tails are added. Moreover, recovery of the inserted DNA fragment is possible only when poly dG-dC tails are used, because it provides recognition site for restriction enzymes, for example *Pst*I.

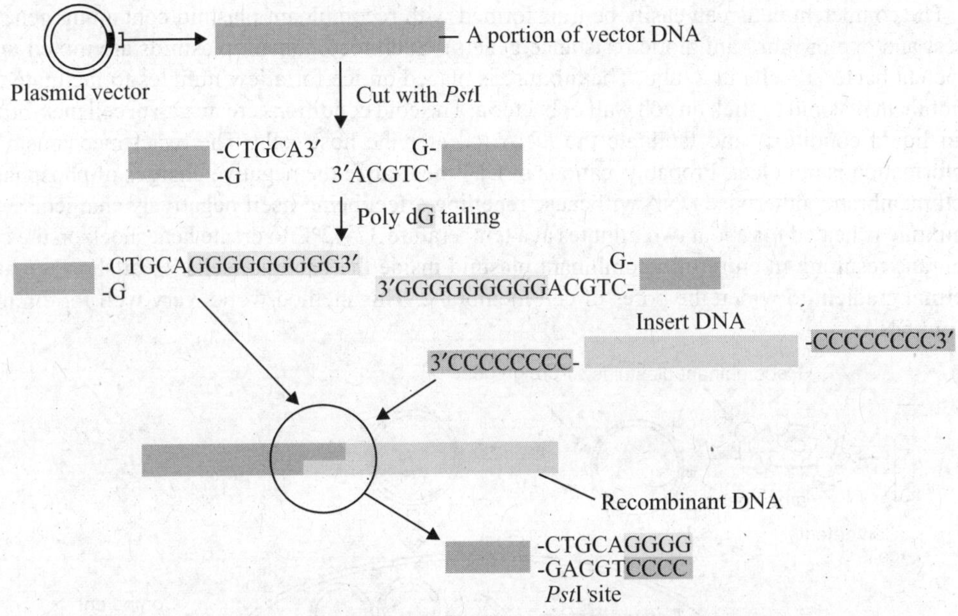

Fig. 8.10. Steps of homopolymer tailing (diagrammatic).

Use of homopolymer tailing is shown in Fig. 8.10. The vector is cut with *Pst*I to generate 3′-OH termini, and then tailed with dG residues by using terminal transferase in the presence of dGTP. This results in dG tail. Similarly, the insert DNA is tailed with dC residues. The dG and dC tails are complementary to each other; therefore, the insert DNA can be annealed with the vector DNA molecule to get a recombinant. At the end of the insert DNA, the *Pst*I sites are regenerated; hence, by using *Pst*I the insert DNA can be cleaved from the recombinant.

IV. TRANSFORMATION OF BACTERIAL CELLS BY RECOMBINANT DNA

There are several methods developed for introduction of recombinant DNA molecule into host cells. The methods are adopted according to types of vectors and host cells. Some of the methods of gene transfer into host cells are briefly discussed herewith.

1. Transformation

The concept of transformation came first in 1928 after the work of an English bacteriologist, Frederick Griffith on *Diplicoccus pneumoniae*. The event of entering the plasmid containing foreign DNA fragment into a bacterial cell is known as transformation. Originally the transformation procedure was developed by M. Mandell and A. Higa in 1979. When a mixture of recombinant DNA plus non-recombinant DNA is obtained, it is allowed to transform the suitable bacterial cells. Most of the bacterial cells do not take up the foreign DNA easily unless the cells are treated with certain chemicals to make them competent (Fig. 8.11).

Competence refers to the state of being able to take up exogenous DNA from the environment. Competence factors develop inside such bacterial cells for a few minutes. The strains of *E. coli* usually do not have restriction systems, hence it degrades foreign DNA. Cohen discovered a technique for treating the competent bacterial cells with ice-cold solution of $CaCl_2$. Ca^{++} ions (cations) disrupt the bacterial cell wall and membrane resulting in formation of small holes though which DNA molecules can enter the cells. Such cells are frozen at ultra low temperature (–80 to –60°C) and used when required in future.

The competent cells can easily be transformed with recombinant plasmid containing gene of interest and one or more antibiotic resistance gene(s). Such recombinant plasmids are mixed with competent bacterial cells in a tube. The mixture is placed on ice for a few minutes to facilitate the recombinant plasmid to stick on cell wall of bacteria. The cold conditions create gap in cell membrane put in liquid condition, and facilitate the DNA to enter the host cells. The exact mechanism of transformation is not clear. Probably, cations of Ca^{++} neutralize the negative charges of phosphates of cell membrane; otherwise DNA will cause repelling effect being itself negatively charged. Soon the mixture is heated for about two minutes at a temperature 37-42°C to create heat shock on the cell membrane resulting in entry of recombinant plasmid inside the bacterial cells. Heat shock creates a thermal gradient to widen the pores of cell membrane. This method works very well for circular plasmid DNAs.

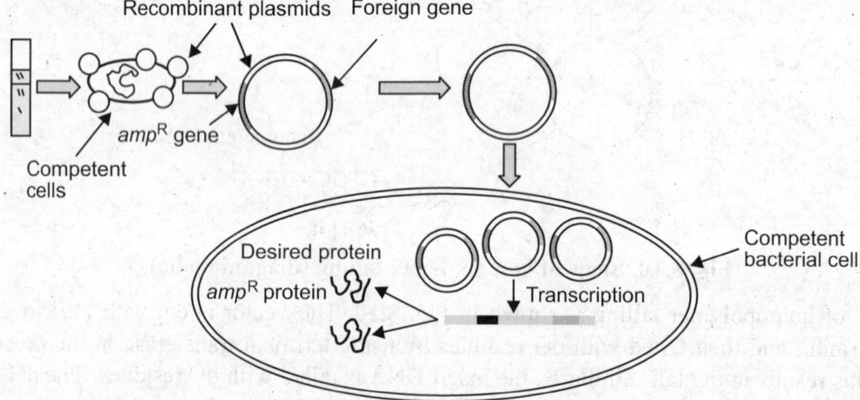

Fig. 8.11. Process of transformation.

Then the transformants are incubated at 37°C for about 60-90 minutes to allow the plasmid properly to get established and express the phenotypic characters. Thereafter, the transformed cells are plated on selective nutrient medium so that the cells could multiply, and also the plasmid could increase its number [*see* the preceding section 'Selection (screening) of recombinants].

An excellent preparation of competent cells will give ~10^8 colonies per microgram of plasmid. A poor preparation will be about 10^4/μg or less. Good non-commercial preparations should give 10^5 to 10^6 transformants per microgram of plasmid. Then the cells are grown on the nutrient agar medium containing antibiotic(s) for selection of recombinants (*see* 'Direct Selection of Recombinants' in proceeding section).

2. Transfection

Phage λ was used to transfer the foreign DNA into *E. coli* cell for the first time; therefore, this mechanism is often termed as *transfection* (a hybrid of transformation and infection). Also, transfer of naked foreign DNA molecules into cultured host cells mediated through chemicals is called transfection. Basically it refers to the process of introducing nucleic acids into eukaryotic cells by nonviral methods. Using various chemical, lipid or physical methods, this gene transfer technology is a powerful tool for studying gene function in the context of a cell. As such DNA does not enter the bacterial cells. Essentially, transfection is a method that neutralises or obviates the issue of introducing negatively charged molecules (e.g. phosphate backbones of DNA and RNA) into cells with a negatively charged membrane.

A mixture of a polycationic lipid and a neutral lipid results in the formation of unilamellar liposome vesicles that have a net positive charge due to the presence of highly positive amine head groups on these molecules. Liposomes have become the method of choice for carrier molecules in routine cell line DNA transfections because liposome-mediated gene delivery is technically easy, highly reproducible and very efficient. A charged chemical substance such as cationic liposomes and calcium phosphate or DEAE-dextran is mixed with DNA molecules. This will neutralise or even create an overall positive charge to the molecule (Fig. 8.12). This makes it easier for the DNA-transfection reagent complex to cross the membrane, especially for lipids that have a 'fusogenic' component, which enhances fusion with the lipid bilayer. The recipient host cells are overlayed by this mixture. Consequently the foreign DNA is taken up by the host cells. For detail see *liposome-mediated gene transfer* in preceding section.

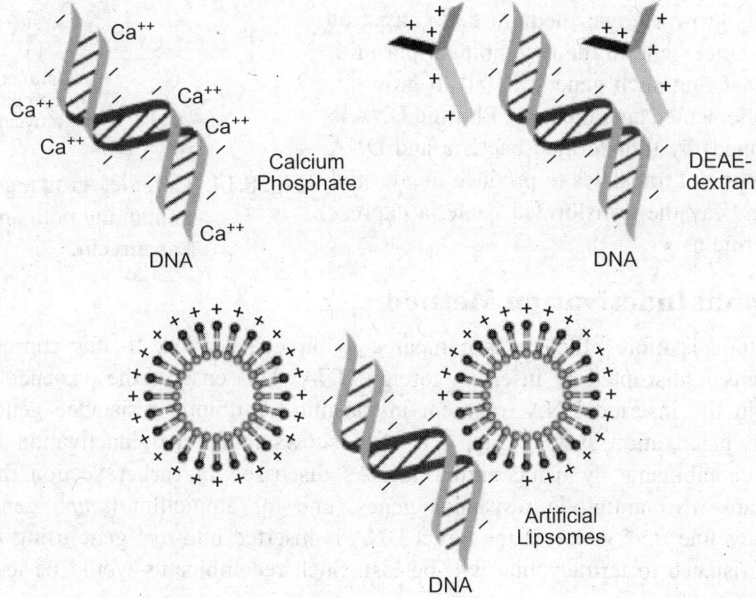

Fig. 8.12. Chemicals like calcium phosphate and DEAE-dextran or cationic lipid-based reagents coat the DNA, and create an overall positive charge to the molecule.

The efficiency of transformation is not high as it is influenced by bacterial strain and size of foreign DNA. It has been found that the efficiency of this process could generate about 10^5 transformants per microgram (μg) of cloned circular plasmid (*e.g.* pBR322). It has not been possible to achieve efficiencies of over 10^8 transformants per g plasmid. If linear DNA is transformed it is almost completely insufficient in transformation.

Fig.8.13 shows transfection by a recombinant plasmid of an animal cell (A); its entry into nucleus (B), and integration with host chromosome (C).

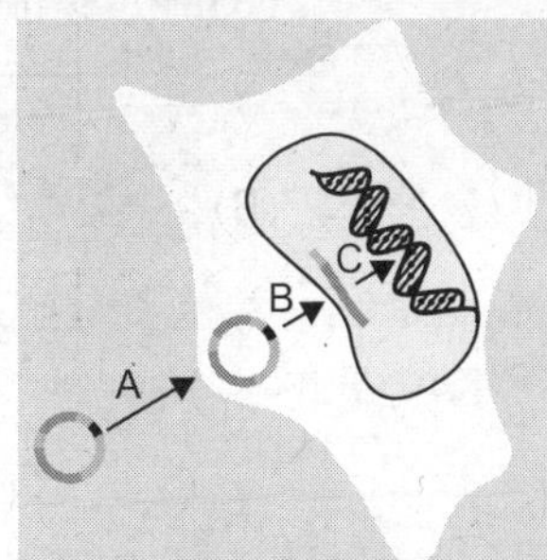

Fig. 8.13. Transfection by a recombinant plasmid of an animal cell.

V. SELECTION (SCREENING) OF RECOMBINANTS

It is essential to identify those cells which have received the recombinant DNA molecules after its introduction into suitable host cells. This process is called *screening* or *selection.* Selection of recombinant DNA cells is based on expression or non-expression of certain characters or traits. The vector or foreign DNA present in recombinant cells expresses the characters, while the non-recombinants do not express the traits. Some of the methods which are used mainly for selection of recombinants in *E. coli* are given as below:

1. Direct Selection of Recombinants

If the cloned DNA itself codes for resistance to the antibiotic(s) ampicillin (*amp*^{R}) and/or and kanamycin (*kan*^{R}), the recombinants can be allowed to grow on minimal medium containing ampicillin and/or kanamycin. Only such recombinants will grow and form colony on medium that contain *amp*^{R} and/or *kan*^{R} gene(s) on its plasmid vector (Fig. 8.14).

Following this procedure we cannot say whether the recom-binants growing on such medium contain religated plasmid vector or contain recombinant plasmid plus foreign DNA fragment because *amp*^{R} and/or *kan*^{R} gene(s) are present in both the recombinants. Only such cells grow on agar medium that consist of antibiotic resistance gene in the recombinant plasmid. The plasmids lacking such gene will fail to grow on antibiotic-supplemented agar medium. Plasmid DNA is replicated (cloned) by transformed bacteria and DNA insert transcribes and translates to produce the desired protein. In this way the transformed bacteria express recombinant proteins.

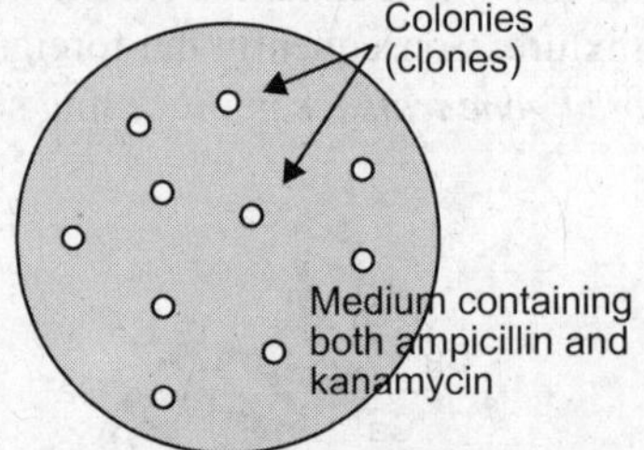

Fig. 8.14. Colonies growing on medium containing both ampicillin and kanamycin.

2. Insertional Inactivation Method

This method is more efficient as compared to direct selection. In this approach one of the genetic traits is disrupted by inserting foreign DNA. This ensures the presence or absence of *tet*^{R} gene in the inserted DNA fragment in plasmid. Antibiotic resistance genes act as a good insertion inactivation system. Fig. 8.15 A-E shows insertional inactivation method for selection of recombinants by using antibiotics. As discussed in earlier section that plasmid pBR322 contains two antibiotic resistance genes, one for ampicillin (*amp*^{R} gene) and the other for tetracycline (*tet*^{R} gene). If the target DNA is inserted into *tet*^{R} gene using *Bam*HI, the property of resistance to tertracycline will be lost. Such recombinants would be test-sensitive.

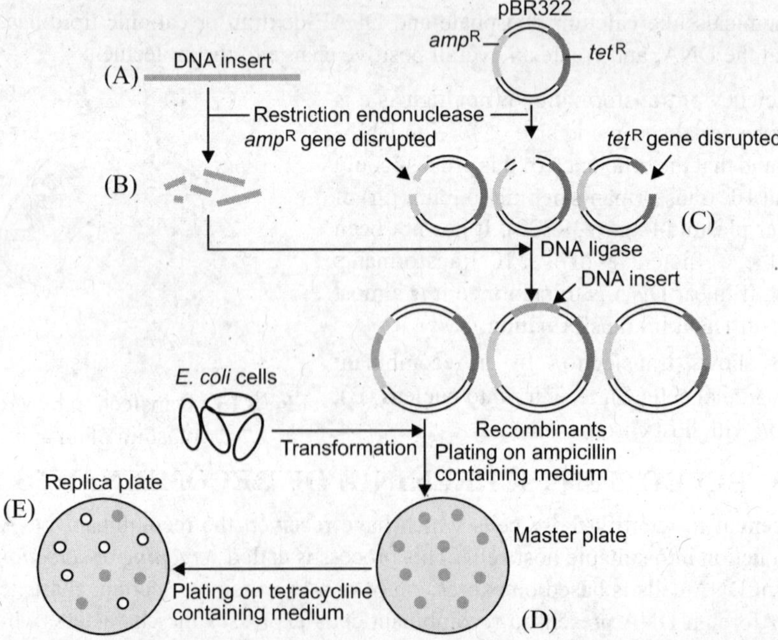

Fig. 8.15. Insertional inactivation method for selection of recombinants by using antibiotics.

When such recombinants (containing target DNA in *tet*^{R} gene) are grown onto medium containing tetracycline, they will not grow because their *tet*^{R} gene has been inactivated. But they are resistant to ampicillin because *amp*^{R} gene is functional. On the other hand the self-ligated recombinants will show resistance to tetracycline and ampicillin. Therefore, they will grow on medium containing both the antibiotics.

Replica plating is done from the master plate for ascertaining the presence or absence of *tet*^{R} gene in the inserted DNA fragment in plasmid. Bacterial colonies on master plate are gently pressed with sterile velvet so that a few cells of each colony may adhere on it which is then pressed on other plate containing the nutrient medium amended with tetracycline. Plates are incubated for the growth of bacterial colonies. The appearance of colonies is compared with the master plate and those colonies that fail to grow on replica plate can be said to have a plasmid which had insert DNA in the *tet*^{R} gene of plasmid and had destroyed *tet*^{R} gene.

3. Blue-White Screening Method

The use of colour selection is a more common method to determine the transformants what so ever contain plasmids with inserts. For *E. coli* this involves the *lac* complex and blue/white screening. This is also a powerful method used for screening of recombinant plasmid by inserting a reporter gene *lacZ* in the vector, for example pUC8 (Fig. 8.16). It must be noted that the Lac complex consists of the *lacI* gene and the *lacZ* gene. The *lacZ* gene produces beta-galactosidase enzyme [that splits a derivative of lactose called 'X-gal' (5-bromo-4-chloro-3-indolyl-beta-D-galactopyranoside)]. This produces a blue dye resulting in the development of blue-coloured colony. Expression of the *lacZ* gene is regulated by the LacI repressor protein (product of the *lacI* gene) in *E. coli* which binds to the promoter and prevents the transcription process. Further transcription starts when another lactose analogue is added. IPTG (isopropyl-beta-D-thiogalactopyranoside) is a non-degradable lactose analogue which binds to the LacI repressor protein and displaces it from DNA it binds in the lac promoter system. Then transcription of *lacZ* can takes place and break down of X-gal occurs.

Moreover, the plasmid pUC8 consists of *amp*^{R} gene for ampicillin resistance, *lacZ* gene and *ori* gene for origin for replication. The *lacZ* encodes the enzyme β-galactosidase and contains several recognition sites for restriction enzymes. The bacteria transformed by recombinant plasmids containing foreign gene of interest are grown on nutrient medium containing ampicillin plus a synthetic substrate X-gal.

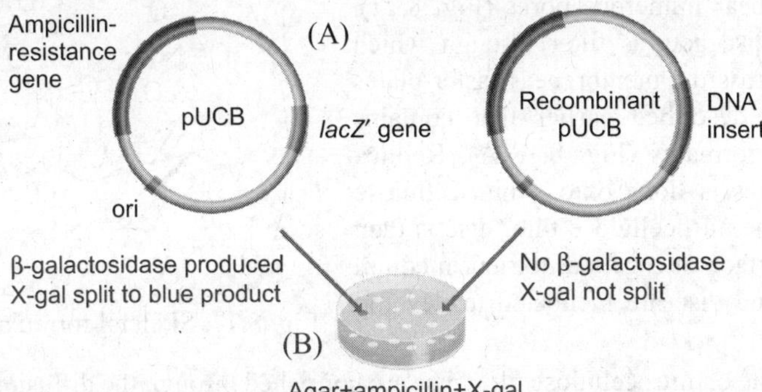

Fig. 8.16. A plasmid pUC8 designed for blue-white screening (A); colonies produced by blue-white screening (B).

X-gal is similar to lactose in structure and turns blue when cleaved by β–galactosidase, due to production of an insoluble blue coloured product. If a foreign gene is inserted into *lacZ*, this gene will be inactivated; therefore, no blue colour will develop because β-galactosidase is not synthesised due to inactivation of *lacZ*. Therefore, the host cells containing recombinant DNA will form white coloured colonies on the medium containing X-gal, whereas the other cells containing non-recombinant DNA will develop the blue coloured colonies because the non-recombinant cells cannot produce β-galactosidase that metabolises X-gal. On the basis of colony colour the recombinants can be selected (Fig. 8.16).

4. Screening by Colony Hybridization (Nucleic Acid Hybridisation) Technique

Colony hybridization technique has been developed by M. Grustein and D.S. Hogness in 1975. After the growth of transformed cells on nutrient medium it is likely that a few cells may have specific DNA as desired among several thousand cells. How is it possible to pick up those cells which have the desired DNA? This method is suitable for use with plasmids. An analogous method, plaque hybridisation technique, has been given for use with phages.

(*a*) DNA Probes: A DNA probe is the radio-labelled ($^{32}PO_4$) small fragment of DNA molecule (20-40 bp) which is complementary at least to one part of desired DNA. Generally probe DNA is labelled with $^{32}PO_4$ or sometimes with ^{125}I. The ^{32}P liberates γ-particles, while ^{125}I emits γ-rays. In addition to using radioactive isotopic elements, isotopic sulphur or fluorescent molecules may also be used. Therefore, a DNA probe recognises A, T, G and C nucleotides and combines with the complementary sequences of the target DNA. A and T of the probe combine with T and A of target DNA and *vice versa.* Similarly, G and C of probe combine with C and G of target DNA and *vice versa*. The colony hybridisation technique is based on the availability of a complementary sequence of a radiolabelled DNA probe. The use of DNA probe in searching the functional target DNA is given in nucleic acid hybridization technique.

The target DNA can easily be caught for hybridisation by the small sized DNA probe than the long sized probe. Besides, a high rate of hybridisation is achieved on maintaining higher concentration of ^{32}P-DNA than the target DNA molecules. Moreover, the DNA probe is very specific to its target DNA; this specificity is called stringency. The probe may be partially pure mRNA, a chemically synthesised oligonucleotide or a related gene which identifies the corresponding recombinant DNA. DNA probes have commercial significance. They diagnose specific DNA sequences (genes) and are used in the diagnosis of diseases, microbiological tests and in research as well.

(*b*) Colony (or Plaque) Hybridization Technique: In 1975, M. Grunstein and D.S. Hogness developed colony hybridisation technique. In this technique the nitrocellulose membrane discs are used which bear numerous pores (Fig. 8.17). Therefore, this disc acts as filter through which nutrients pass across the membrane. Master plates are prepared as described earlier that contains colonies of transformants (Fig. 8.18 A). Replica plating of colonies is done onto a nitrocellulose filter disc (B). The nitrocellulose filter disc is then placed on the surface of a gelled nutrient medium and both plate and disc are incubated to develop colonies (D).

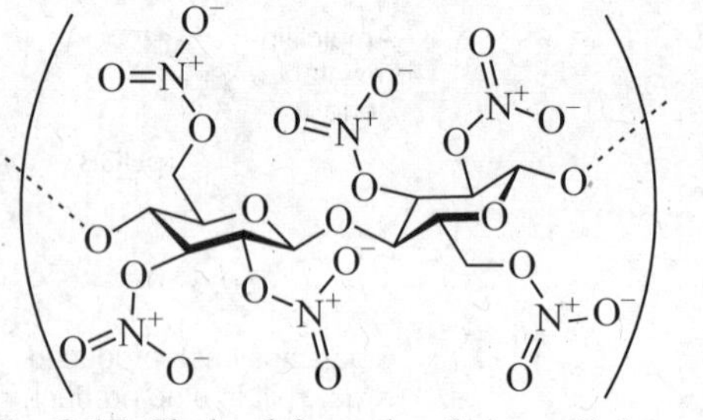

Fig. 8.17. Skeletal formula of nitrocellulose.

Cells growing on nitrocellulose filter disc are nourished through the diffusion of nutrients from gelled nutrient medium (Fig. 8.18). The filter disc is removed and put on blotting paper soaked with 0.5 N NaOH solution. The alkali diffuses into nitrocellulose, lyses bacterial cells and denatures their DNA (E).

Thereafter, the filter disc is neutralised by tris (hydroxymethyl) aminomethane-HCl buffer by keeping high salt concentrations. Then the filter is dipped in proteinase K solution that removes proteins from the filter. This results in binding the DNA with nitrocellulose disc in the same pattern as the bacterial colonies. Then the filter disc is baked at 80°C to fix the cDNA properly (F). The baked filter disc is then incubated putting in a solution containing radioactive chemical labelled probe (^{32}P-DNA) at suitable conditions. The probe will hybridise any bound DNA which contains sequences complementary to the probe. By thorough washing unhybridised (unbound) probes are removed from the hybridised probe (colonies containing sequences complementary to probe) and is identified by autoradiography by exposing the X-rays through nitrocellulose membrane on a photographic film (G).

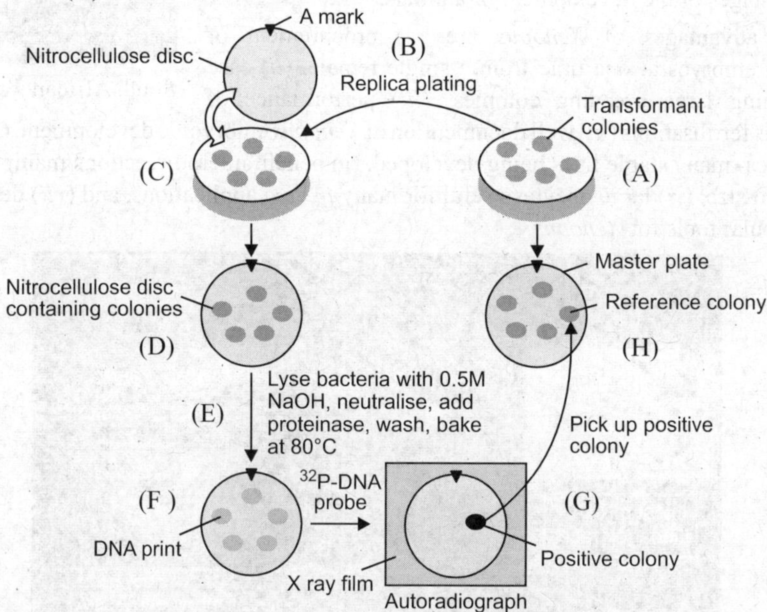

Fig. 8.18. Colony hybridisation technique.

The probe has bound to the filter, radioactivity from radioactive probes or released light (fluorescence or chemilumniscence) from non-radioactive probes will expose the silver grains present in the film. Depending on the number of genes of interest, there may be a few colonies (or plaques) on the filter paper that hybridises to the probe. The autoradiogram developed acts as a permanent record. Autoradiogram (positive X-ray image formed by recombinant colonies) is compared with the master plate to identify which colonies contained recombinant plasmid with gene of interest. These colonies are picked up and multiplied on the nutrient medium to isolate the cloned DNA (H).

For making the replica of plaques developed in the bacterial lawn by bacteriophage recombinants, a similar procedure is followed. Other methods for selection of clones or screening are: *in vitro* translation, and direct immunological techniques.

***(c)* Plaque Lifting Method:** In 1977, W. D. Benton and R. W. Davis devised this technique for screening of bacteriophages from plaque forming units. The recombinant phage particles are isolated and used to construct DNA library. In this method lawn of *E. coli* is prepared on agar medium which is allowed to get infected by recombinant phage particles. They infect *E. coli*, multiply inside cells and lyse the cells resulting in plaque formation. As described in Southern blotting technique, replica plates are prepared from the master plate containing plaques by using nitrocellulose filter paper. This filter is treated with alkali solution so that the phage DNA may be denatured. Then filter is baked at 80°C to fix denatured phage DNA and put in a solution containing ^{32}P-DNA probe. After some times the probe hybridises the desired DNA if complementary bases are present. The filter paper is washed with standard saline citrate solution to remove unhybridised probe and autoradiography of hybridised filter paper is carried out. Dark spots are formed where probes have hybridised the desired DNA. Recombinant phage particles isolated from the plaques are used for further study.

5. Screening by *In Vitro* Translation Method

South African *Xenopus laevis*

Primarily *in vitro* translation method is used as a primary screen for identification of cloned DNA, but now this method is used to confirm the identification of recombinant clones. For a longer time, the South African clawed frog, *Xenopus laevis*, has been the subject of research for developmental biologists to learn the earliest stages of the development of animals.

Some advantages of *Xenopus* are: (*i*) procurement of hundreds of embryos at one time from a single female, (*ii*) ease of maintaining large breeding colonies, (*iii*) performance of simultaneous fertilisations (*iii*) well documention of *Xenopus* embryonic development, (*iv*) possibility of transgenics- many stable lines being developed, (*v*) benefit microdissections/manipulations from large embryo size, (*vi*) large oocytes useful for many *in vivo* applications, and (*vii*) development of many molecular tools for *Xenopus*.

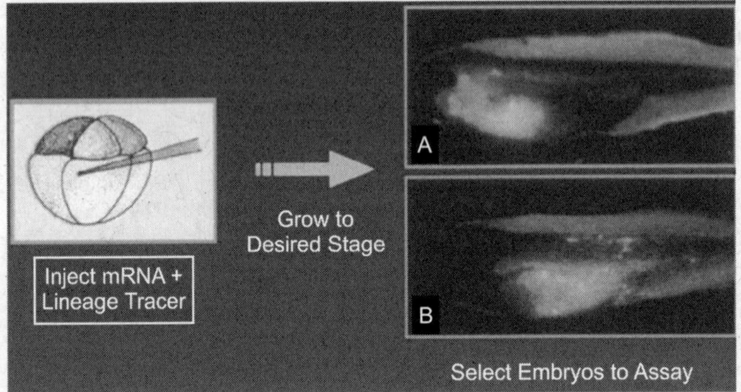

Fig. 8.19. Gene expression can be readily visualised through mRNA microinjection.

In 1980, S. Nagata and collaborators have used primary translation screened by isolating the interferon gene from total leukocyte poly-A^+ RNA. Translation of poly-A^+ RNA is done when it is microinjected into oocytes from *X. laevis* with the result of secretion of interferon into culture medium (Fig. 8.19). Interferon is detected by its antiviral activity. It is supposed that the number of interferon mRNA may be between 10^{-3} and 10^{-4}; therefore, about 10^4 transformants must be screened to get at least single clone. Otherwise, individual colony is screened to get the DNA from different pools of the transformed colonies. A single clone is identified from the pool of DNA which is then immobilised on nitrocellulose or diazobenzyloxymethyl cellulose paper to isolate complementary mRNA by nucleic acid hybridisation as described earlier. The RNA can be eluted from the cDNA-RNA hybrid and microinjected into oocytes and the translation products are analysed by polyacrylamide gel electrophoresis (PAGE) and detected by immunoprecipitation techniques.

6. Direct Immunological Screening Technique

The final tests, analogous to colony hybridisation technique as described earlier, are the immunological techniques. It is an alternative screening procedure which relies on expression, and generally applicable approach to identify a clone synthesizing a particular polypeptide. This is potentially a very powerful method since the only absolute requirement is that the required mRNA encodes a protein for which a suitable antibody is available. Two *in situ* techniques given by S, Broome and W. Gilbert (1978) and H.A. Erlich and coworker in 1978 are in current use.

In the immunological test, instead of radiolabelling of DNA molecules, antibodies (immunoglobulins) are used to identify the colonies or plaques developed on master plates that synthesize antigens encoded by the foreign DNA present in plasmids of the bacterial clones. For this purpose a special vector, known as expression vector, is designed where the foreign DNA is transcribed and translated within the bacterial cell. The growth medium containing specific anti-serum may help in detection of viable immunoprecipitate (precipitin) around the colonies or plaques. The method follows: (*i*) the replica plating of bacterial colonies of master plate on nutrient agar, (*ii*) lysis of cells after their growth by exposure to chloroform vapour, or treatment with high temperature, (*iii*) making gentle contact of a solid support (for example a cellulose filter containing immobilized antibody to solid support with the lysed cells within the colonies to allow absorption of antigen to antibody), and (*iv*) detection of antigen-antibody complex by incubating the cellulose filter with a radio-labelled second antibody. The antibodies that do not react are washed off and position of the antigen-antibody complex is determined by passing the filter through X-ray. It gives the signal of those bacterial cells which synthesize antigen on the master plate (Fig. 8.20).

Master plate
↓ Replica plating
Bacterial colonies on replica plate
↓ Put antibody-immobilized filter paper
Filter paper containing bacterial DNA
↓ Incubate with second antibody
Filter paper
↓ X-rays
Filter paper shows signals for synthesis of antigen

Fig. 8.20. Steps of immunological test.

In 1978, S. Broome and W. Gilbert purified immunoglobulin G (IgG) and bound it to an agar plate so that the antigens released through *in situ* lysis of bacterial colonies can bound to the fixed antibody (Fig. 8.21). The antigen is then detected by using the same IgG preparation which is radio-labelled. This recognises determinants on the bound antigen. In 1978, H.A. Erlich and collaborators used F(ab)½ fragments derived by digestion of pepsin of the immunoglobulin, bound to either polyvinyl cellulose or diazobenzyl oxymethyl cellulose (DBMC) paper. This is incubated first with the antigen and then with undigested antiserum, FC portion of which will bind radiolabelled *Staphylococcus aureus* A protein.

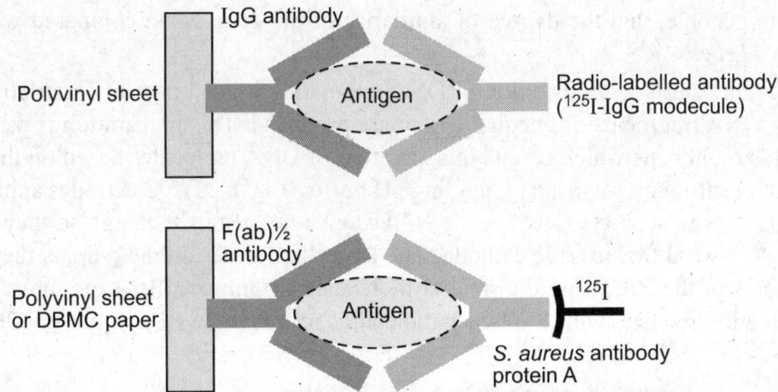

Fig. 8.21. Detection of specific antigen synthesised by a bacterium.

C. BLOTTING TECHNIQUES

After the DNA, RNA, or protein has been separated by molecular weight, it must be transferred to a solid support before hybridisation. Hybridisation does not work well in a gel. This transfer process is called blotting and is why these hybridization techniques are called blots**.** Usually, the solid support is a sheet of nitrocellulose paper. DNA, RNA and protein stick well to nitrocellulose in a sequence-independent manner; hence it is called blots.

DNA, RNA and proteins are separated by blotting techniques. The important properties of the three blotting techniques are given in Table 8.4. A comparative account of these blots is given below:

- **Southern Blots.** DNA is first cut with restriction enzymes and the resulting double-stranded DNA fragments have an extended rod conformation without pre-treatment.
- **Northern Blots.** Although RNA is single-stranded, RNA molecules often have small regions that can form base-paired secondary structures. To prevent this, the RNA is pre-treated with formaldehyde.
- **Western Blots**. Proteins have extensive 2′ and 3′ structures and are not always negatively charged. Proteins are treated with the detergent SDS (sodium dodecyl sulphate) which removes 2′ and 3′ structure and coats the protein with negative charges.

Table 8.4. The important properties of the three blotting techniques.

	Southern blotting	***Northern blotting***	***Western blotting***
What is separated by molecular weight ?	DNA cut with restriction enzymes	RNA is denatured with formaldehyde	Proteins are denatured with SDS
Probe used	Radioactive gene X DNA	Radioactive gene X DNA	Antibody against protein X, labelled with radioactivity or enzyme
What is learnt from it ?	Restriction map of gene X in chromosome	The amount and length of gene X mRNA	Amount of protein X, and length of protein X

1. Southern Blotting Techniques

Southern blotting combines agarose gel electrophoresis for separation of DNA on the basis of its size to transfer the size-separated DNA to a filter membrane for probe hybridisation. The method is named after its inventor, the British biologist Edwin M. Southern (1975), a Professor of Biochemistry at the University of Oxford. Southern blots can easily provide a physical map of restriction sites within a gene located normally on a chromosome, reveal the number of copies of the gene in the genome, and the degree of similarity of the gene when compared with the other complementary genes.

The procedure starts with digestion of DNA population by one or many restriction enzymes. Consequently, DNA fragments of unequal length are produced. This preparation is passed through agarose gel electrophoresis which results in separation of DNA molecules based on their size. The gel is set up for electrophoresis in a tank holding pH buffer (Fig. 8.22). Electrodes apply an electric field. The molecules to separate (DNA and RNA) carry a net negative charge so they move along the electric field toward the positive cathode. The larger molecules are held up as they try to pass through the pores of the gel, while the smaller molecules are impeded less and move faster. This results in separation by size, with the larger molecules nearer the well and the smaller molecules farther away.

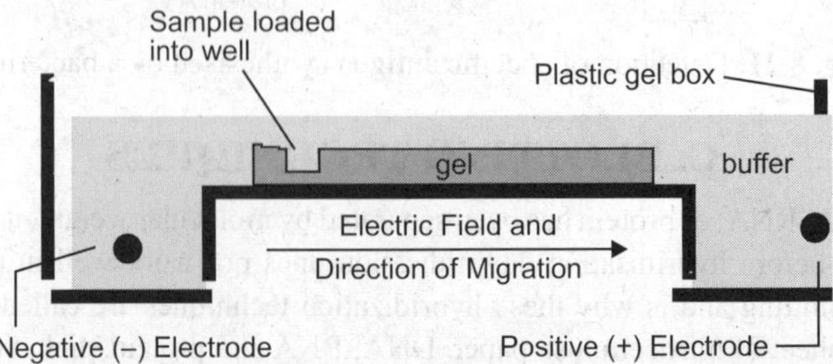

Fig. 8.22. A set up for electrophoresis in a tank holding pH buffer (side view).

DNA restriction fragments present in gel are denatured by alkali treatment. Gel is then put on the top of the buffer saturated filter paper. Upper surface of the gel is covered with nitrocellulose filter and overlaid with dry filter paper. A sheet of nitocellulose (or nylon) membrane is placed on top (or below, depending on the direction of the transfer) of the gel. An even pressure is applied to the gel (either using suction or by placing a stack of paper towels and a weight on top of the membrane and gel), to ensure good and even contact between gel and membrane. Buffer transfer by capillary action from a region of high water potential to a region of low water potential (usually filter paper and paper tissues) is then used to move the DNA from the gel on to the membrane. Ion exchange interactions bind the DNA to the membrane due to the negative charge on the DNA and positive charge on the membrane.

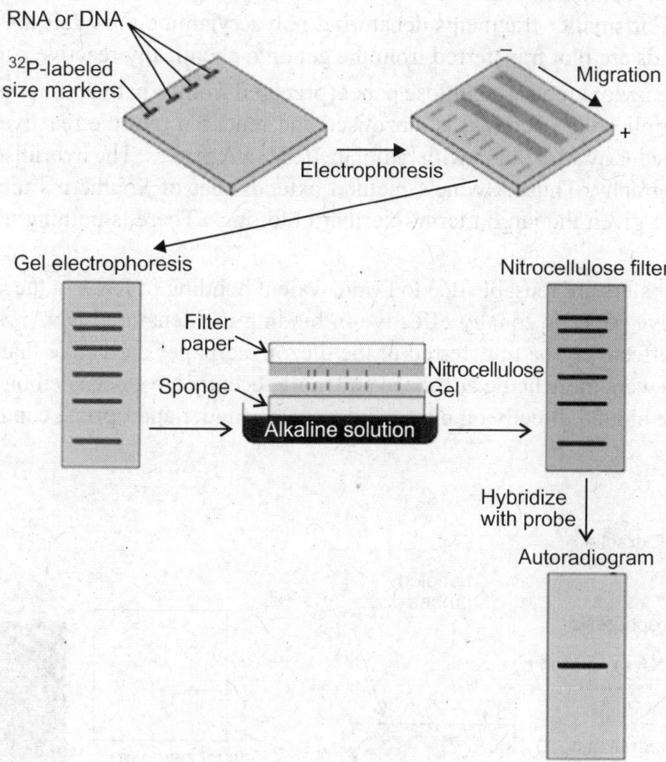

Fig. 8.23. Steps of the Southern blotting.

The dry filter paper draws the buffer through the gel. Buffer contains single stranded DNA. Nitrocellulose filter binds DNA fragments strongly when come in contact of it. The membrane is then baked, *i.e.* exposed to high temperature (60 to 100°C) (in the case of nitrocellulose) or exposed to ultraviolet radiation (nylon) to permanently and covalently crosslink the DNA to the membrane. After baking at 80°C, DNA fragments are permanently fixed to the nitrocellulose filter. Then the filter is placed in a solution containing radio-labelled RNA or denatured DNA probe of known sequences. These are complementary in sequence to the blot-transferred DNA. Thereafter, the membrane is exposed to a hybridisation probe, *i.e.* a single DNA fragment with a specific sequence whose presence in the target DNA is to be determined. In some cases, the hybridisation probe may be made from RNA, rather than DNA.

The radio-labelled DNA probe hybridises the complementary DNA on nitrocellulose filter. After hybridisation, excess probe is washed from the membrane, and the pattern of hybridisation is visualized on X-ray film by autoradiography. The filter is thoroughly washed to remove the probe. The hybridised regions are detected autoradiographically by placing the nitrocellulose filter in contact with a photographic film. The images show the hybridised DNA molecules. Thus, the sequences of DNA are recognised following the sequences of nucleic acid probe.

2. Northern Blotting Technique

The blot transfer of mRNA separated by gel electrophoresis could not be applied directly by Southern blotting technique, because RNA was found not to bind with nitrocellulose filter. Hence, this technique was developed in 1977 by James Alwine, David Kemp and George Stark at Stanford University (U.S.A.). The gels may be run on either agarose or denaturing polyacrylamide gels depending on the size of the RNA to be detected. A notable difference in the procedure in case of agarose gels (as compared with the Southern blot) is the addition of formaldehyde which acts as a denaturant. For smaller fragments denaturing polyacrylamide urea gels are employed. In this method RNA bands are blot transferred from the gel onto chemically reactive paper.

An aminobenzyloxymethyl cellulose paper (prepared from Whatman filter paper No. 540 after a series of uncomplicated reactions) is diazotised and rendered into the reactive paper. Therefore, becomes available for hybridization with radiolabelled DNA probes. The hybridised bands are found out by autoradiography. Thus, Alwine's method extends that of Southern's method, and for this reason it has been given the jargon term 'Northern blotting'. There is nothing northern or western like Southern.

The blot transfers are reusable due to firm covalent bonding of RNA to the reactive paper. The chemically reactive paper is equally effective in binding the denatured DNA. Small fragments of DNA can more effectively be transferred to the diazotized paper derivative than to nitrocellulose. These techniques were more being advanced and more recently have been demonstrated that mRNA bands can also be blotted directly on nitrocellulose paper under appropriate condition (Fig. 8.24).

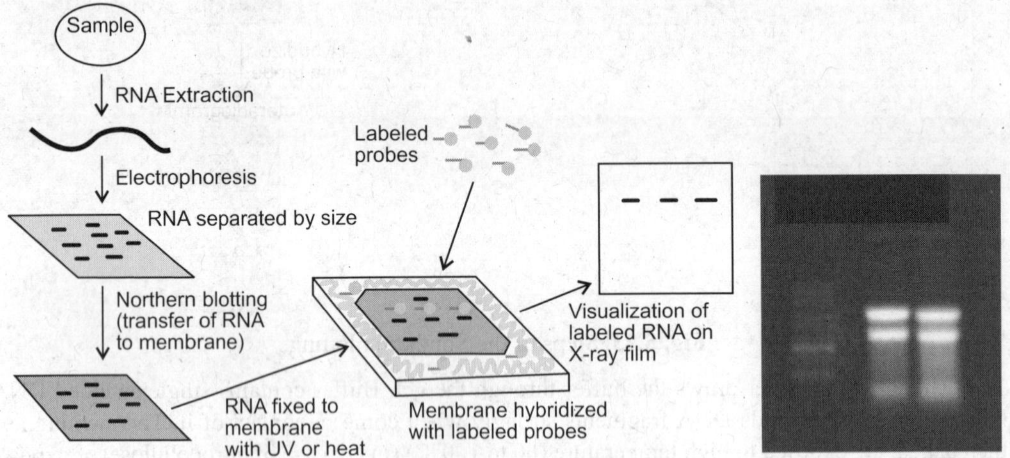

Fig. 8.24. Steps of northern blotting technique (A); RNA run on a formaldehyde agarose gel to highlight the 28S (top band) and 18S (lower band) ribosomal subunits (B).

In this method preparation of reactive paper is not required. The mRNA is isolated from the transformed cells and electrophoresed under such conditions that do not permit the development of secondary structures. The mRNAs separated on the gel are transferred onto nitrocellulose filter which are then hybridised by single stranded probe (RNA or DNA). Thereafter, hybrids are treated with SI nuclease and RNase which digests the single stranded RNA/DNA probe. It does not affect the double stranded nucleic acid formed due to hybridisation of RNA by the complementary sequences of nucleic acid probe. Structure of mRNA is revealed to the extent to which mRNA protects the nucleic acid probe.

***(a)* Applications of the Northern Blot:** Northern blots have been superceded in most areas by Real Time PCR (RT-PCR) and microarray approaches. It is not often used for clinical or diagnostic purposes. The northern blot protocol and its variations are used in molecular biology research to: (*i*) a gold-standard for the direct study of gene expression at the level of mRNA (*ii*) detection of mRNA transcript size, (*iii*) study RNA degradation, (*iv*) study RNA splicing (can detect alternatively spliced transcripts), (*v*) study RNA half-life, (*vi*) study IRES (internal ribosomal entry site) (to remove possibility of RNA digestion vs second cistron translation), and (*vii*) confirm and check transgenic / knockout mice (animals).

***(b)* Advantages of Northern Blotting:** The advantages of northern blots are that: (*i*) it is a widely accepted and well regarded method, (*ii*) northern blotting is a straight-forward method, (*iii*) often it is used as a confirmation or check, (*iv*) often a gold-standard, (*v*) it is a versatile protocol as it can allow the usage of many types of probes (vs real time PCR) including radiolabelled and non-radiolabelled, *in vitro* transcribed RNA and even oligonucleotides such as primers, and (*vi*) sequences with even partial homology, unlike real time PCR or other methods can be used as hybridisation probes (*i.e.* sequence from different species for homology analysis or even genomic fragments can be used).

***(c)* Disadvantages of Northern Blotting:** The disadvantages of using northern blotting are: (*i*) often use of radioactivity (this prevents ease of its performance, use and disposal. New methods of non-radioactive detection have been generated allowing non-radioactive detection), (*ii*) taking the whole process of northern blotting usually a long time from sample preparation through to detection, (*iii*) if RNA samples are even slightly degraded by RNases, the quality of the data and quantitation of expression is quite negatively affected, (*iv*) being the standard northern blot a relatively less sensitive than nuclease protection assays and RT-PCR (the sensitivity of northern blots may be increased with the use of nylon positively-charged membranes, use of a highly specific antisense probe), and (*v*) being problematic while detecting with multiple probes. Often, the membranes must be stripped before hybridisation and detection with a second probe. This is a problem as harsh conditions are required to strip off probes from the blot and is also time consuming. Also, there is a limit to the amount of times a blot may be stripped.

***(d)* Reverse Northern Blot:** Occasionally some differences in the procedure of northern blotting are carried out. Therefore, the northern blot with some differences is called reverse northern blot. In this method, the nucleic acid (*i.e.* substrate to be affixed to the membrane) is a collection of isolated DNA fragments, and the probe is radioactively labelled-RNA extracted from a tissue. The use of DNA microarrays that have come into widespread use in the late 1990s and early 2,000s is more akin to the reverse procedure. They involve the use of isolated DNA fragments affixed to a substrate and hybridisation with a probe made from cellular RNA. Thus the reverse procedure enabled northern analysis to evolve into gene expression profile in which possibly all of the genes in an organism may have their expression monitored.

3. Western Blotting Techniques (Protein Blotting or Electroblotting Technique)

In 1979, H. Towbin and coworkers developed the western blotting technique to find out the newly encoded protein by a transformed cell. Its working principle lies on antigen-antibody reaction; hence, it is an immunodetection technique. In this method radiolabelled nucleic acid probes are not used. This technique follows the following steps :

(*i*) Extraction of protein from transformed cells and its separation by SDS-PAGE (sodium dodecyl sulphate polyacrylamide gel electrophoresis) based on size where SDS acts as detergent for electrophoresis.

(*ii*) Placing the Whatman filter paper on a cathode plate followed by stack of coarse filter, Whatman filter, electrophoresed gel, nitrocellulose filter, Whatman filter paper, coarse filter stack, Whatman filter and anode plate (Fig. 8.25). Putting the complete set up in transfer tank containing sufficient transfer buffer. The negative charge should be on the side of the gel and the positive charge on the side of the nitrocellulose membrane so that the negatively charged proteins can be driven over to the positively charged nitrocellulose membrane. This results in imprinting of the nitrocellulose membrane with the same protein bands as the gel. It should be kept in mind that proteins bind better to nitrocellulose at a low pH. There should be no air bubbles between the nitrocellulose and the gel otherwise proteins will not be transferred. Application of an electric field (30 V overnight for 5 hours) to cause the migration of proteins from the gel to nitrocellulose filter and binding on its surface (the nitrocellulose filter has exact image of pattern of proteins as present in the gel. This type of blotting is called western blotting).

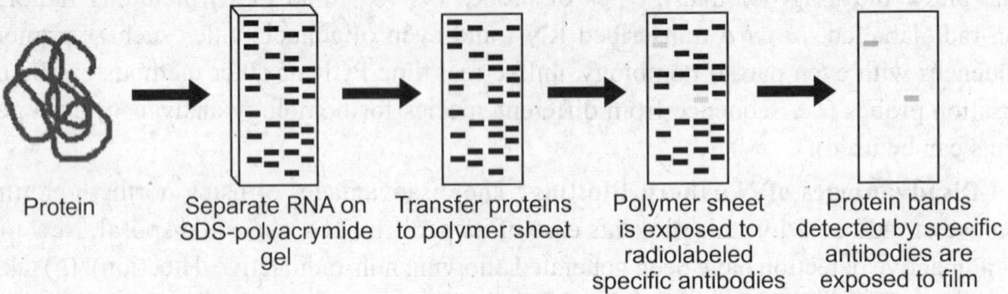

Fig. 8.25. Western blotting by using radiolabelled antibody and detection of bands through autoradiography.

(*iii*) Incubation of nitrocellulose membrane with a primary antibody, which is the specific antibody that sticks to protein and forms an antibody-protein complex with the protein of interest (hybridization of proteins is also done by using radio-labelled antibodies (I^{125}-antibodies) of known structure isolated from the rabbit). Washing of nitrocellulose filter is done with a wash solution (tris-buffered saline + Tween 20) to facilitate the removal of unhybridised antibodies.

(*iv*) Incubation of nitrocellulose membrane with a second antibody, for example a goat-anti-rabbit antibody conjugated to horseradish peroxidase (HRP). The goat-anti-rabbit is the secondary antibody which 'sticks' to the primary antibody) and the HRP is the conjugated enzyme that will allow to visualise the protein (Fig. 8.26). This antibody should be an antibody-enzyme conjugate. The secondary antibody should be an antibody against the primary antibody, *i.e.* the secondary antibody will 'stick' to the primary antibody as the primary antibody 'stuck' to the protein. The conjugated enzyme allows to visualise all of this).

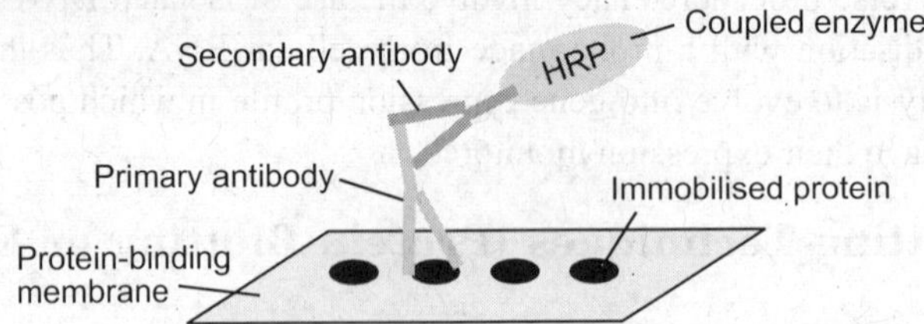

Fig. 8.26. A goat-anti-rabbit antibody conjugated to horseradish peroxidase (HRP) used to detect protein.

Incubation in a reaction mixture specific for the enzyme to see the enzyme action. The bands are seen wherever there is a protein-primary antibody-secondary antibody-enzyme complex. X-ray film is placed on gel to detect a flash of light given off by the enzyme. Usually the reaction runs out in about an hour. The hybridised sequences are detected by autoradiography. The dots show the presence of desired protein (Fig. 8.25).

Secondary Antibody

After rinsing the membrane to remove unbound primary antibody, the membrane is exposed to another antibody directed at a species-specific portion of the primary antibody. This is known as a secondary antibody; due to its targeting properties it is referred to as 'anti-mouse,' 'anti-goat,' etc. The sources of antibodies are animals (or hybridoma cultures); an anti-mouse secondary will bind to just about any mouse-sourced primary antibody. The secondary antibody is usually linked to *biotin* or to a *reporter enzyme* such as alkaline phosphatase or horseradish peroxidase. Several secondary antibodies will bind to one primary antibody and enhances the signal.

Most commonly, a horseradish peroxidase-linked secondary antibody is used in conjunction with a chemiluminescent agent, and the reaction product produces lumniscence in proportion to the amount of protein. A sensitive sheet of photographic film is placed against the membrane, and exposure to the light from the reaction creates an image of the antibodies bound to the blot.

The enzyme can be provided with a substrate molecule that will be converted by the enzyme to a coloured reaction product visible on the membrane.

A third alternative is to use a radioactive label rather than an enzyme coupled to the secondary antibody, such as labelling an antibody-binding protein like *Staphylococcus* protein A with a radioactive isotope of iodine (^{125}I). Since other methods are safer, quicker and cheaper this method is now rarely used.

Uses of Western blotting

This technique determines in which tissues or under which physiological conditions a gene is expressed to produce a protein? This also determines whether a transgenic organism expresses the inserted gene to produce a protein?

4. Dot Blot Technique

Southern blotting, northern blotting, and even western blotting techniques are most often utilised in molecular biology. The whole process is time consuming and requires careful purification of samples containing nucleic acids.

It is expensive also but may be simple if only detection and quantification of any sequence is to be done. The procedure is made simple by excluding purification steps, electrophoresis and blotting of gel. In dot blot the biomolecules to be detected are not separated by chromatography, but a mixture possibly containing the molecule to be detected is applied directly with a pipette onto the surface of nitrocellulose filter paper as a dot (Fig. 8.27). Dotting (putting the spots) may also be done by using an apparatus making as circular or oblong slot. Since dot or slot on the filter paper is done, therefore, this method is also called 'dot and slot blot'. This filter is put in 0.4 M NaOH solution to ensure the binding of denatured DNA on nitrocellulose filter. This technique rapidly detects the nucleic acid sequences and determines the relative amount of DNA or RNA in the sample. This is then immediately followed by detection by either nucleotide probes (northern blot and Southern blot) or antibody (western blot).

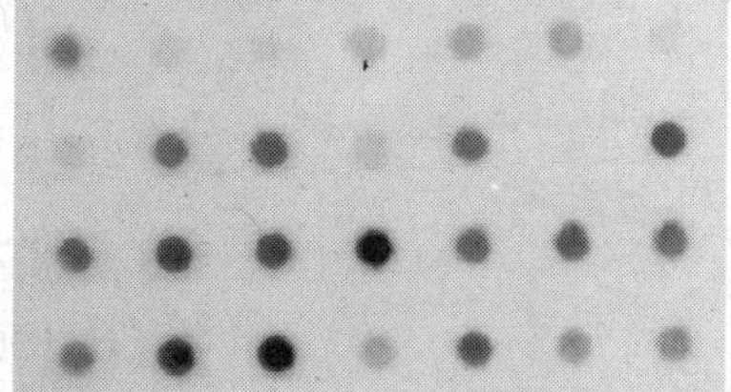

Fig. 8.27. Original dot blot image.

A densitometer is used to scan the autoradiographic signals and quantifies the intensity of a hybridised nucleic acid in the sample. By using this method one can compare the amount of DNA or RNA sequences present in a large number of samples.The technique offers significant savings in time as chromatography, and complex blotting procedure for the chromatography gel is not required. However, it offers no information on the size of the biomolecules. Furthermore, if two molecules of different sizes are detected, they will appear as a single blot. Dot blot can only confirm the presence or absence of a biomolecules or biomolecules which can be detected by the probes or the antibody.

Dot Blot Analysis: The highly specific matching properties of antigen-antibody for detection of antigenic or antibody proteins in complex samples is used in laboratory. The immunodetection method is widely known as dot blot which involves applying the small volumes of antigens or test samples on protein binding membrane; the membrane captures and immobilises proteins. The proteins immobilised on the membrane are probed with a specific antibody and a matching specific detection reagent. The dot blot technique is also applied in clinical testing, forensic, and research laboratories.

D. RECOVERY OF CELLS CONTAINING RECOMBINANT PLASMID

Colonies complementary to nitrocellulose signal are recovered from the master plate and cultured in nutrient broth in order to get the sufficient bacterial clones containing recombinant plasmid. However, in some instances the foreign DNA is cleaved enzymatically from the vector molecules. Sometimes aim lies to achieve expression of gene within the bacterial cell if it encodes for valuable polypeptides.

I. EXPRESSION OF CLONED DNA

The genes of eukaryotes differ in different ways from those of *E. coli.* Therefore, the eukaryotic genes must be modified in such a way that they could resemble the prokaryotic, genes and express in *E. coli.* To begin with transcription in prokaryotes the binding of bacterial RNA polymerase to a promotor region of the DNA immediately before (or upstream of) the gene is essential. Factors taken into account for expression are : (*i*) supply of prokaryotic promoter for expression of eukaryotic genes, (*ii*) supply of ribosomal binding sites for the cloning vector, (*iii*) removal of introns from mRNA obtained from eukaryotic genes, and (*iv*) inhibition of gene responsible for degradation of foreign protein within the bacterium.

1. Shine-Dalgarno Sequence

The mRNA must contain some feature that allows its 5′ end to be recognised by and positioned correctly on the ribosome during the assembly of the protein synthesising apparatus. In bacteria, a special ribosome binding site was identified by the Australian scientists John Shine and Lynn Dalgarno in 1975. They observed that the 3′ end of the 16S rRNA is complementary to a short region just upstream of the start codon in bacterial mRNA (Fig. 8.28).

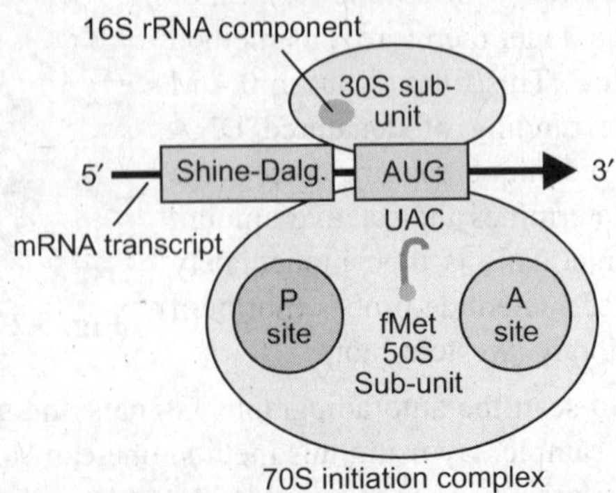

Fig. 8.28. The Shine-Dalgarno sequence recruits the ribosome to the mRNA to initiate protein synthesis by aligning it with the start codon.

What is Shine-Dalgarno Sequence ?

The ribosome binding site in prokaryotes is frequently called the Shine-Dalgarno sequence, which can be located anywhere within an mRNA. In eukaryotes, it is the 5′ cap structure that is added to the mRNA which is required for correct positioning of the ribosome on the mRNA during the initiation phase of protein synthesis.

The Shine-Dalgarno sequence (or Shine-Dalgarno box) was proposed by John Shine and Lynn Dalgarno, hence the name of this sequence was coined after the name of the discoverer. This sequence is a ribosomal binding site generally located 6-7 nucleotides upstream of the start codon AUG. It is also complementary to a sequence near the 3′ terminus of 16S rRNA. This sequence is about 3-9 bases long. The Shine-Dalgarno sequence exists only in prokaryotes. In *E. coli* the Shine-Dalgarno sequence is the six-base consensus sequence AGGAGG (Fig. 8.29). This sequence helps recruit the ribosome to the mRNA to initiate protein synthesis by aligning it with the start codon. The complementary sequence (CCUCCU) is called the 'anti-Shine-Dalgarno sequence' that is located at the 3′ end of the 16S rRNA in the ribosome. The eukaryotic equivalent of the Shine-Dalgarno sequence is called the Kozak sequence.

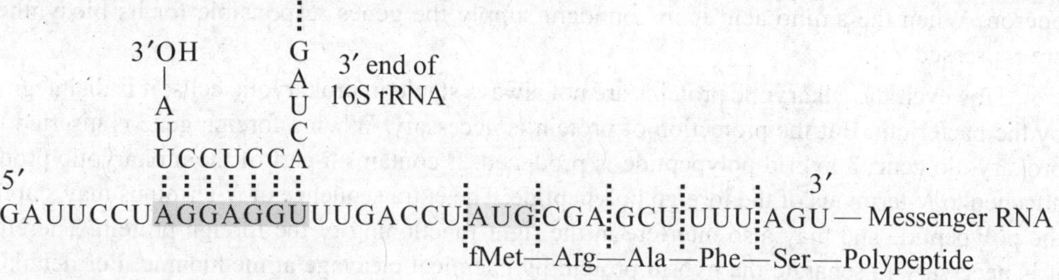

Fig. 8.29. The Shine-Dalgarno sequence (AGGAGGU) and anti-Shine-Dalgarno sequence (UCCUCCA) in bacteria.

Moreover, numerous prokaryotic mRNAs do not possess Shine-Dalgarno sequences at all. The ribosomal protein S1 attracts ribosome to mRNA initiation region and binds to AU-rich sequences found in many prokaryotic mRNAs 15-30 nucleotides upstream of start-codon. The S1 protein is present only in Gram-negative bacteria but is absent in Gram-positive bacteria. The presence of a Shine-Dalgarno sequence is necessary for the expression of eukaryotic gene within *E. coli*.

In the Shinc-Dalgarno sequence mutation can reduce translation due to a reduced efficiency of mRNA-ribosome pairing. It is evidenced by the fact that complementary mutations in the anti-Shine-Dalgarno sequence can restore translation. When the Shine-Dalgarno sequence and the anti-Shine-Dalgarno sequence pair, the translation initiation factors (*IF2-GTP*, *IF1*, *IF3*) as well as the initiator tRNA, fMet-tRNA(fMet) are recruited to the ribosome.

2. Expression Vectors

An expression vector, also called as an expression construct, is generally a plasmid used to introduce and express a specific gene into a target cell. Expression vector allows the production of large amounts of stable mRNA. Once the expression vector is inside the cell, the protein encoded by the gene is produced by the cellular transcription and translation machinery. The plasmid is engineered such that it contains a highly active promoter which causes the production of large amounts of mRNA.

Purification of the protein is required after expression of the gene product. Since the vector is introduced to a host cell, the protein of interest should be purified from the proteins of the host cell. Therefore, the cloned gene should have a tag of histidine (His) or any other marker protein to make purification process easy.

In cDNA, translation is correctly started at the universal initiation codon AUG. The expression vector has been built up to overcome the problem of time consumption in adding a Shine-Dalgarno promoter sequence and initiation codon to every foreign gene. It contains a strong promoter, the activities of which are controlled by temperature, concentration of specific inducer or repressor and a Shine-Dalgarno sequence. In addition to promoter, the expression vectors also have restriction sites, origin of replication and marker genes.

Therefore, it is necessary to replace the eukaryotic promoter with a prokaryotic promoter in order to carry out transcription of a eukaryotic gene. Hence, for the construction of expression vectors a number of bacterial promoters have been used, for example, promoter of *lac* or *trp* operon from *E. coli* and ß-lactamase promoter from pBR322 plasmid. The *tac* promoter is a hybrid promoter constructed between elements of the *trp* and *lac* promoters. These promoters are regulated by their corresponding repressors. The *lac* operon is negatively regulated by the *lac* repressor and can be depressed by addition of an inducer, *i.e.* isopropyl ß-D-thiogalactosidase (IPTG). It is positively regulated by the catabolic activator protein (CAP) complexed with cyclic AMP pathway. Similarly, the *trp* operon encodes five enzymes for the biosynthesis of tryptophan. The complex formed between the *trp* repressor and tryptophan negatively controls the *trp* operon. When the amino acid is in abundant supply the genes responsible for its biosynthesis are repressed.

However, the eukaryotic proteins are not always stable in prokaryotic cells; it is disintegrated by the bacterium. But the protection of protein is necessary. When a foreign gene is inserted in a prokaryotic gene, a hybrid polypeptide is produced. It contains a part of the prokaryotic product attached to *N*-terminus of the foreign polypeptide. The extra sequence of *N*-terminus may stabilise the polypeptide and may also interfere in the right functioning of the foreign protein. Therefore, it is necessary to separate the hybrid protein by chemical cleavage at methionine. For detail, see Chapter 7: *Cloning vectors.*

E. DETECTION OF NUCLEIC ACIDS

There are two approaches for labelling the nucleic acids to be used as DNA probe: radioactive labelling and non-radioactive labelling.

1. Radioactive Labelling

During 1970s, a method for *in vitro* introduction of radioactivity in nucleic acids was developed through metabolic labelling. In this method large amount of radioactivity was required. However, this method was time consuming and laborious. The methods developed later on utilised phage T4 polynucleotide kinase (which transfers γ-PO^4 of ATP to 5′-OH terminus in DNA or RNA) or *E. coli* DNA polymerase I (that replaces a normal nucleotide of double stranded DNA with α-$^{32}PO_4$ labelled nucleotides). The nucleic acids are labelled *in vitro* by using the following methods:

(*a*) Nick Translation: In molecular biology, nick translation technique is used to replace some of the nucleotides of a DNA sequence with their labelled analogues by employing DNA polymerase I (Pol I) resulting in a tagged DNA sequence. Generally, *E. coli* DNA polymerase I is used in this method. These sequences can be used as a probe in 'fluorescence in situ hybridization' or 'blotting techniques'. Nucleotides are eliminated from 5′ end and simultaneously added at 3′ side resulting in movement of the nick along the DNA. Therefore, this method is called nick translation (Fig. 8.30). This is followed by replacement in nicked sites by DNA Pol I which elongates the 3′ OH terminus removing nucleotides by 5′→3′ exonuclease activity replacing them with radiolabelled dNTPs. When DNA polymerase I eventually detaches from the DNA, it leaves another nick in the phosphate backbone. The nick has 'translated' some distance depending on the processivity of the polymerase. This nick is sealed by DNA ligase, or its 3′ OH group could serve as the template for further DNA Pol I activity. This process could cause a break in double-stranded DNA, if DNA polymerase I encounters another nick on the opposite strand. In this case, both phosphate backbones will be incomplete, and the DNA will essentially be cleaved into two shorter dsDNA fragments with blunt ends. This can be reversed by ligation.

One of the incorporated nucleotides provided in the reaction is radiolabelled in the *alpha* phosphate (α-$^{32}PO_4$) position to radioactively label a DNA fragment for use as a probe in blotting procedures. This method is used to label the double stranded DNA probes. If there is no nick, a nick is made on double stranded DNA by using DNase (in a buffered solution containing dNTPs) which exposes 3′-OH group on one strand of DNA.

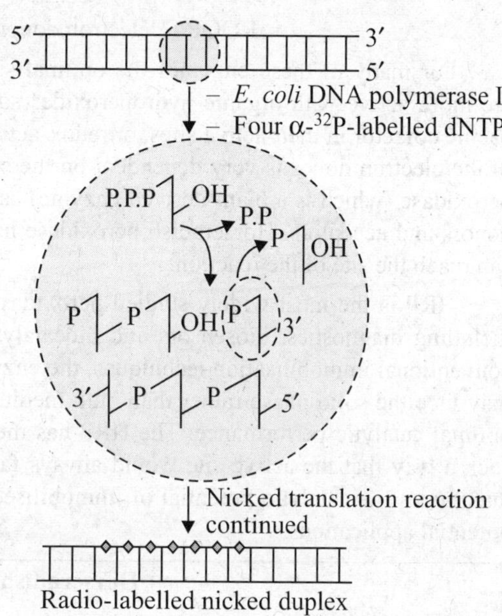

Fig. 8.30. Nick translation by radio-labelling γ-^{32}P-DNA.

Generally only one nucleotide *i.e.* α-^{32}P-dCTP is radiolabelled, whereas the other nucleotides are unlabelled. This enzyme shows 5′→3′ exocatalytic activity and either removes or hydrolyses the nucleotides from 5′ side of the nick. The reaction mixture includes probe DNA to be labelled, α-^{32}P-dCTP, nick translation buffer, the other three dNTPs, DNase 1 and *E. coli* DNA polymerase I. It is incubated at 14°C for 2-4 hours. The hybridised DNA is separated from the rest of dNTPs and used to hybridise the DNA.

***(b)* Random Primed Radiolabelling of Probes:** This technique was developed by A.P. Feinberg and B. Vogelstein in 1983. The restriction fragments purified by gel electrophoresis from denatured linear or circular DNA are used to prepare probes. In this method primers are used for DNA synthesis but DNA nicking is not done. The purified DNA to be labelled is mixed in a buffer and heated at 100°C so that it may be denatured. Then it is put on ice to lower down the temperature. The reaction mixture (containing Klenow fragment of *E. coli*, DNA Pol I, one radiolabelled nucleotide *e.g.* α-^{32}P-dCTP and three unlabelled nucleotide triphosphates) is incubated at 37°C for 30 minutes to 16 hours. Since, Klenow DNA Pol I lacks 5′→3′ exonuclease activity, the radiolabelled probe is synthesised by primer extension but not nick translation as described earlier. The hybridised DNA is separated from unincorporated nucleotides and used as probe.

***(c)* Probes Developed by PCR:** This method is advantageous because very small amount of template DNA can be used. In this method the reaction mixture includes linear template DNA (to be labelled), amplification buffer, *Taq* polymerase, primers, one radiolabelled and three unlabelled dNTPs. The hybridised DNA is isolated and used as probe.

2. Non-radioactive Labelling

Commonly, the non-radiolabelled DNA probes are also used in detection of desired DNA. There are three systems of non-radioactive routes such as using horseradish peroxidase, biotin probe and streptavidin probes as described below. Besides, fluorescent labelling with quantum dots (ODs) has also been described.

***(a)* Horseradish Peroxidase (HRP) Method:** Peroxidases are a large family of enzymes which typically catalyze a reaction :

$$ROOR' + \text{electron donor } (2\ e^-) + 2H^+ \rightarrow ROH + R'OH$$

For many of these enzymes the optimal substrate is hydrogen peroxide (H_2O_2), but others are more active with organic hydroperoxides such as lipid peroxides. Peroxidases can contain a haeme cofactor in their active sites, or redox-active cycteine or selenocysteine residues. The nature of the electron donor is very dependent on the structure of the enzyme. For example, horseradish peroxidase (which is a plant-derived enzyme) can use a variety of organic compounds as electron donors and acceptors. Horseradish peroxidase has an accessible active site and many compounds can reach the site of the reaction.

HRP is the most widely studied peroxidase due to its many diverse uses in biotechnology, including diagnostics, biosensors and biocatalysis. Often, it is used in immobilised form. With conventional immobilisation techniques, the enzyme adheres in random orientation: the active site may face the solid phase rather than bulk medium, impeding substrate access and leading to sub-optimal catalytic performance. The HRP has the ability to immobilise in a directional manner in such a way that the active site would always face outwards from the insoluble matrix. Hence, it maximizes the catalytic potential of immobilised enzyme and increases HRP range of actual and potential applications.

Horseradish Peroxidase (HRP)

Horseradish (*Armoracia rusticana*, syn. *Cochlearia armoracia*) is a perennial plant of the family Brassicaceae which includes mustard, wasabi, and cabbages. Horseradish contains two glucosinolates (sinigrin and gluconasturtiin) which are responsible for its pungent taste.

The enzyme horseradish peroxidase found in the plant is used extensively in molecular biology for antibody detection. It is increasingly important in biochemical research fields. HRP is commonly used for specifically colouring of thin (~5 micrometer) slices of tissue biopsies from patients suspected to have cancer. This is an area of human pathology called immunohistochemistry (IHC). Many molecules of HRP are bound to a polymer together with immunoglobulins that will bind to a primary immunoglobulin that recognises a specific biomarker in cells in the tissue slices. HRP will convert 3, 3-diaminobenzidin (DAB) to a yellowish brown insoluble compound. This compound is visible in a microscope and helps the pathologist to diagnose the cancer. HRP has been employed in materials used to test for the presence of glucose in blood or urine.

HRP is covalently linked to DNA. Chromogenic substrates (such as chloronaphthol) or chemiluminescent substrates (that emit light after reacting with enzyme *e.g.* luminol) are used for detection of HRP-linked DNA. Chloro-naphthol yields an insoluble purple product in the presence of peroxide and HRP. Consequently, the presence of DNA is detected. Similarly, HRP oxidises luminol and emits luminescence by using X ray film; hence luminescence can be detected.

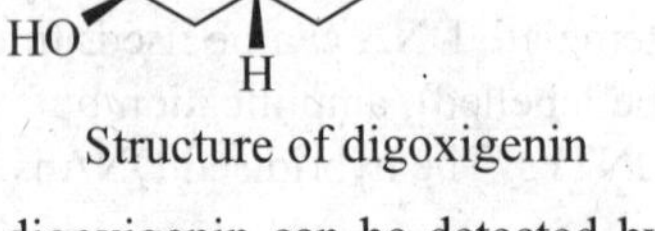

Structure of digoxigenin

***(b)* Digoxigenin (DIG) Labelling System:** Digoxigenin is a cardenolide steroid derived as a hapten from flower and leaves of the plants *Digitalis purpurea* and *D. lanata*. The presence of digoxigenin can be detected by an antibody associated with an enzyme (anti-digoxigenin-alkaline phosphatase conjugate). The probe is labelled with digoxigenin (11)-dUTP which is a nucleotide triphosphate analogue. This analog containing digoxigenin moiety is incorporated into DNA by nick translation or random prime labelling. The DNA acts as probe. Anti-DIG (antibody to digoxigenin) conjugated with alkaline phosphatase detects the DIG-labelled probe through enzyme-linked immuno assay. A chromogenic substrate [such as 5-bromo-4-chloro-3-indolyl phosphate (BCIP) and nitroblue tetrazolium chloride (NBT)] is added which reacts with alkaline phosphatase and produce purple/blue colour (Fig. 8.31).

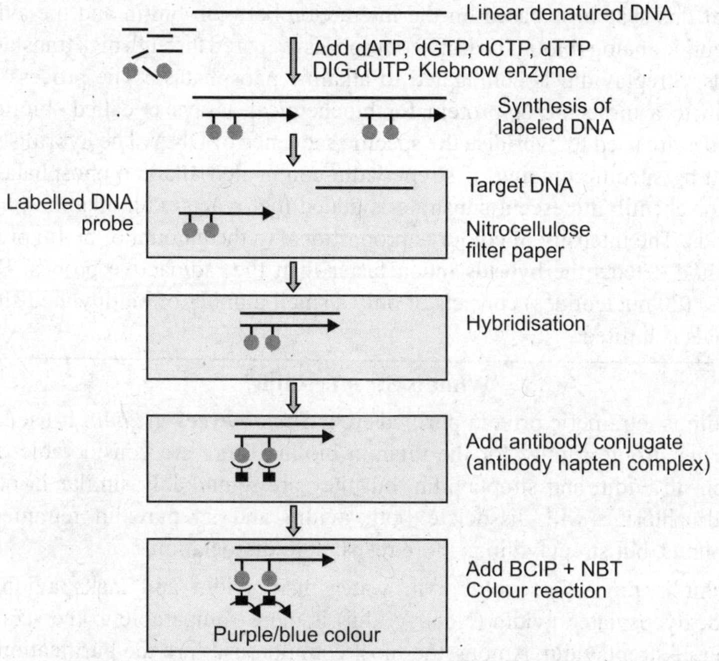

Fig. 8.31. Non-radioactive DIG-DNA labelling.

A chemiluminescent substrate such as dioxetane is also used with digoxigenin or biotin-streptavidin systems. There are several derivatives of digoxigenin which emit light after enzymatic activation by alkaline phosphatase. The denatured DNA present on Southern blots is hybridised by labelled probe. The membrane is placed in detection buffer containing antidigoxigenin (20 μg/μl) and bovine serum albumin (BSA) (5% w/v) and incubated at 37°C for 1 hour. After incubation membrane is washed thrice. Enzyme activity of alkaline phosphatase is detected by using BCIP (0.17 μg/ml) and nitrobutane tetrazolium chloride (NBT) (0.33 μg/ml) as dye. Purple/blue colour develops after enzymatic activation.

What is Digoxigenin ?

Digoxigenin can be attached to biological molecules, along with the availability of antibodies. It makes it a useful tool in biochemistry. Along with fluorescein it has become the standard immunohistochemical marker for *in situ hybridization*. In this case it is conjugated to a single species of RNA nucleotide triphosphate (typically uridine) which is then incorporated into the riboprobe as it is synthesised. It may also be conjugated with sugars, allowing the incorporation of specific sugars into glycoproteins to be imaged in a similar way.

(*c*) Biotin-Streptavidin Labelling System: Avidin (streptavidin) is a common glycoprotein molecule found in egg whites which is used in immunofluorescence labelling, while biotin is vitamin H. The high affinity of avidin for biotin was first exploited in histochemical applications in the mid-1970s. This egg-white protein and its bacterial counterpart (streptavidin) have become the standard reagents for diverse detection methods. Such methods entail applying a biotinylated probe to a sample and then detecting the bound probe with a labelled avidin or streptavidin. These techniques are commonly used to localise antigens in cells and tissues and to detect biomolecules in immunoassays and DNA hybridisation procedures.

Structure of biotin

Principle of this system is based on the interaction between biotin and the avidin. A biotin-containing nucleotide analog (biotin-14dATP) or probe is prepared through nick translation or random priming methods. Streptavidin is conjugated to alkaline phosphatase. The process of chemically tagging of biotin to a molecule or protein for biochemical assays is called 'biotinylation'. The biotinylated probes are used to hybridise the specific sequence of DNA. The hybridised biotinylated DNA is detected by specific binding of streptavidin-conjugated alkaline phosphatase. Thereafter, a chromogenic or chemiluminescent substrate is added that reacts with alkaline phosphatase and detects the hybrids. The intensity of colour is proportional to the amount of biotin present in hybrid DNA. This method detects the hybrids much faster than the radioactive probes. But very small number of probes (20 nucleotides) consists of only a small number of biotinylated sites. Therefore, intensity of signals is limited.

What is Streptavidin ?

Streptavidin is tetrameric protein purified from *Streptomyces avidinii*. It used in molecular biology due to its strong affinity for the vitamin biotin. There are considerable differences in the composition of avidin and streptavidin, but they are remarkably similar in other respects. Guanidine hydrochloride will dissociate both avidin and streptavidin tetramers into their component subunits, but streptavidin is more resistant to dissociation.

Streptavidin is much less soluble in water than avidin and lacks avidin's extensive glycosylation deglycosylated avidin (NeutrAvidin) is more comparable to the size, pI and non-specific binding of streptavidin. Among the most common uses are the purification or detection of various biomolecules. The strong streptavidin-biotin bond can be used to attach various biomolecules to one another or onto a solid support.

DNA is fixed first by digesting it with a restriction exonuclease which produces either a blunt end, a 3′ overhang or a 5′ overhang. Then the DNA is incubated with biotin-11-dUTP (a deoxyribonucleotide analog that is covalently attached to biotin), and the Klenow fragment of the holoenzyme DNA Pol I of *E. coli*. The biotin-11-dUTP is incorporated into the 3′ end of the strand complementary to the 5′ single stranded DNA of the overhang. Because DNA polymerases can only add nucleotides to the 3′-OH of DNA and not the 5′-phosphate.

5′-ACTGGCT*U*-3′

3′-TGACCGAACCGTT-5′

(where *U* is biotin-11-UTP incorporated into the DNA strand)

Care is taken to ensure that only one 5′ overhang with only one possible site is available for dUTP incorporation. This results in production of a DNA strand with a biotinylated end which is used for binding (via non-covalent interactions) to streptavidin coated surfaces. Various DNA hybridisation and immunological assays can be performed with the DNA firmly attached to this substrate. They may also be attached to streptavidin coated agarose microspheres, polystyrene or even paramagnetic beads. These complexes are most commonly used for the purification or isolation of DNA binding proteins. CaptAvidin is an avidin-derivative which is also a biotin-binding protein.

What is Biotin ?

Biotin also known as vitamin H or B_7, has the chemical formula $C_{10}H_{16}N_2O_3S$. It is a water-soluble B-complex vitamin which is composed of an ureido (tetrahydroimidizalone) ring fused with a tetrahydrothiophene ring. A valeric acid substituent is attached to one of the carbon atoms of the tetrahydrothiophene ring. Biotin is often chemically linked or tagged to a molecule or protein for biochemical assays. This process is called biotinylation. Since avidins bind preferentially to biotin, biotin-tagged molecules can be extracted from a sample by mixing them with beads with covalently-attached avidin, and washing away anything unbound to the beads. For example, biotin can be attached to a molecule of interest (e.g. a protein), and this modified molecule will be mixed with a complex mixture of proteins. Avidin or streptavidin beads are added to the mixture, and the biotinylated molecule will bind to the beads. Any other proteins binding to the biotinylated molecule will also stay with the beads. All other unbound proteins can be washed away and a variety of methods can be used to determine which proteins have bound to the biotinylated molecule. Biotinylated antibodies are used to capture avidin or streptavidin in both the ELISPOT and ELISA techniques.

(*d*) **Fluorescent Labelling with Quantum Dots (QDs):** Quantum dots are nanoparticles rather than small molecules which can effectively label the target moleculs at a subcellular level. QDs have the potential to become a new class of fluorescent probes for many biological and biomedical applications. Recently methods for the surface functionalisation of QDs have been developed. In 2002, Xingyong Wu and coworkers at Quantum Dot Corporation (U.S.A.) synthesised specific immunofluorescent probes by linking these QDs to streptavidin and IgGs and conducted comprehensive investigations into the labelling efficiency of these probes at the subcellular level. They detected two different targets in the same cell simultaneously with different colours of QDs or combinations of QDs and organic dyes. They found the QDs as an attractive class of fluorescence labels for biological and biomedical cellular imaging. They used QDs linked to immunoglobulin G (IgG) and streptavidin to label the breast cancer marker Her2 on the surface of fixed and live cancer cells, to stain actin and microtubule fibers in the cytoplasm, and to detect nuclear antigens inside the nucleus.

Fixed breast cancer SK-BR-3 cells were incubated with monoclonal anti-Her2 antibody and goat anti-mouse IgG conjugated to QDs. Her2 was clearly labelled with (A) QD 535–IgG and (C) QD 630–IgG (B, D) (Fig. 8.32). When cells were incubated with normal mouse IgG and QD-IgG, there were no detectable or very weak nonspecific signals on the cell surface. The nuclei were counterstained with Hoechst 33342 (blue). All labelling signals were specific for the intended targets and are brighter and considerably more photostable than comparable organic dyes. Using QDs with different emission spectra conjugated to IgG and streptavidin, it was found that QD-based probes can be very effective in cellular imaging and offer substantial advantages over organic dyes in multiplex target detection.

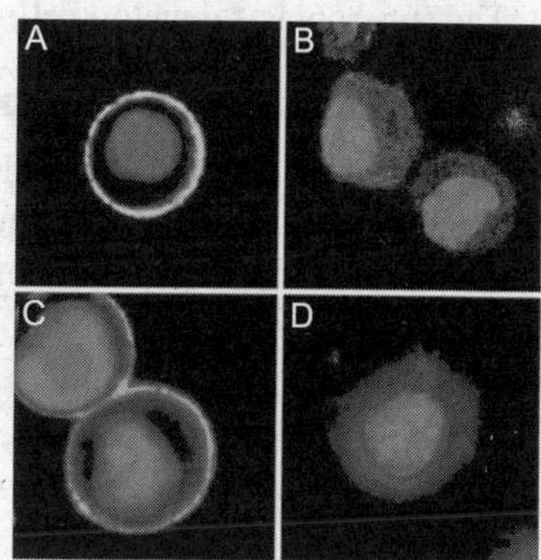

Fig. 8.32. Detection of cancer marker Her2 with QD-IgG conjugates (after X. Wu et al., 2002. *Nat Biotechnol*. 21: 41).

F. GENE CLONING IN EUKARYOTES

In biology the word 'clone' has several different meanings. An identical genetic copy of a piece of deoxyribonucleic acid (DNA), a cell or a whole organism is called as a 'clone'. Identical twins are clones, as are two daughter cells produced by mitosis. But 'to clone' means to produce identical genetic copies of either pieces of DNA, cells, or whole organisms.

Most of studies on gene cloning have been carried out in bacteria. In addition, many difficulties are associated with them for example; (*i*) correction of introns of eukaryotic mRNAs, (*ii*) failure of transfer of equal number of plasmids into daughter cells during cell division and yield of two types of cells, one with plasmid and the second without plasmid, post-translational modification (*e.g.* inhibition of proteolysis, addition of oligosaccharides to specific sites on the polypeptide chain, the glycosylation), (*iii*) threat for hazardous effects, and (*iv*) expression to ensure for the presence of a stable plasmid in the bacterium when applied on large scale for commercial applications. With these prospects the attraction of the use of eukaryotic cells is obvious. In recent years, considerable efforts have been made for the improvement of crop plants through genetic engineering. Many works on various aspects of building up of vectors and expression of insert genes in eukaryotic cells are in progress. Some of them are briefly described in this connection.

I. GENE CLONING IN YEASTS AND MOLDS

1. Gene Cloning in Yeasts

Generally bacterial cultures are suitable for the production of polypeptides on a large scale but not plant or animal cells because of their slow growth rate. But certain limitations with bacteria have forced to develop its alternative and, to some extent, it has been overcome with the use of yeast (*Saccharomyces cerevisiae*). Yeast is a unicellular eukaryotic fungus containing a small well characterised genome. Unlike plant or animal cells, it has rather fast growth rate and itself is a non-pathogenic fungus. Most of its genes contain introns which are spliced during purification of mature mRNA. It appears that introns found in yeast contain sequences for correct splicing as they are totally absent in higher eukaryotes. Moreover, yeasts can carry out post-translational modifications such as removal of a signal sequence from a precursor polypeptide after the secretion of cells. This reveals a major advantage of yeasts over the bacteria.

Success in DNA cloning in yeast depends on uptake of foreign DNA by its spheroplast (naked protoplast when cell wall is enzymatically removed) in the presence of Ca^{++} and polyethyleneglycol (PEG). The spheroplasts develop cell wall after the DNA is taken up.

2. Transformation in Filamentous Fungi

Transformation through plasmid of *E. coli* was first described in protoplasts of *S. cerevisiae* and then in *N. crassa*. However, the situation in filamentous fungi is different from the others as there is no convincing evidence for autonomously replicating plasmids. But sequence having functions similar to ARS have been isolated.

Method of transformation is similar to plant cells. Following are the examples where transformations in protoplasts have been carried out: *A. nidulans, A. niger, A. oryzae, Cephalosporium acremonium, Coprinus cinereus, Glomerella cingulata, Gaeumannomyces graminis, Mucor* sp., *Penicillium chrysogenum, Septoriu nodorum,* etc.

Fig. 8.33 shows *Agrobacterium* tumifaciens containing a Ti plasmid. In 1998, for the first time de Groot and co-workers introduced *Agrobacterium tumefaciens*-mediated transformation of filamentous fungi. Before them the most common method for the genetic transformation of filamentous fungi was treatment of a mixture of fungal protoplasts or spheroplasts and DNA with calcium salt and polyethylene glycol. This was because of the difficulties of protoplast preparation and regeneration, high batch-to-batch variation in the quality of enzyme cocktails used to digest fungal cell wall, and the low yields of transformants for some species. Electroporation .low yields and unstable transformants.

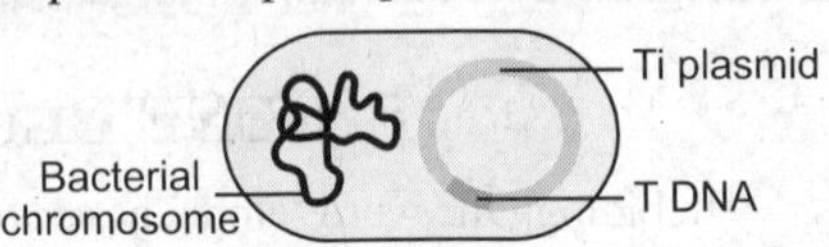

Fig. 8.33. *Agrobacterium tumefaciens* containing Ti plasmid.

Fungal transformation mediated by *A. tumefaciens* consisted of three steps: (*i*) induction of a bacterial culture harbouring appropriate plasmids, (*ii*) co-incubation of the culture with fresh fungal spores on a solid support, and (*iii*) selection of transformants on a medium with a selection agent. Jan Utermark and Petr Karlovsky of the Goettingen University (Germany) have optimised the method extensively and have used it routinely for the transformation of several fungi including *Gliocladium roseum* as given in Fig. 8.34. Conidia of *G. roseum* are harvested from the pure culture of fungus and gene of interest is integreted to the Ti-plasmid. The recombinant plasmid is transferred into the cells of *A. tumefaciens*. The fungal conidia are co-inoculated with the cells of *A. tumefaciens*; eventually the fungal conidia are transformed by the recombinant bacteria. Transformed fungal colonies are picked up and purified for further use.

Routinely this procedure generates 8 - 15 hygomycin-resistant colonies per 10^6 fungal spores per plate of *G. roseum.* Adaptation of the protocol to different fungal species is easy. When a limited number of transformants have to be generated (e.g. transformation of overexpression or silencing constructs), no further optimisation is required. The efficiency of the transformation should be maximized for genome-wide mutagenesis. The most important parameters to optimise are the ratio between *A. tumefaciens* cells and fungal conidia as well as the duration of the co-cultivation.

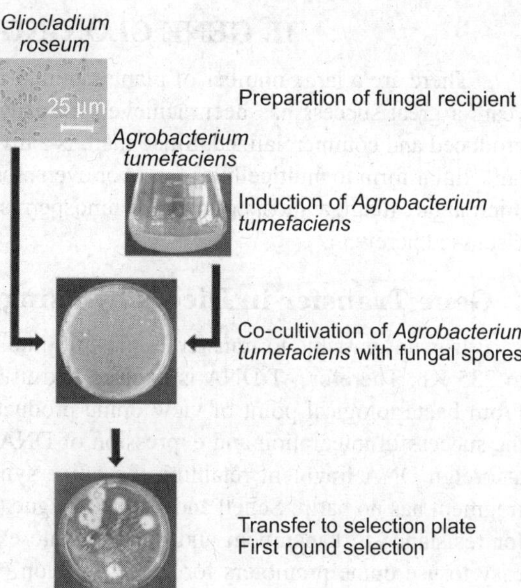

Fig. 8.34. *Agrobacterium tumefaciens* mediated transformation of *Gliocladium roseum* (after Utermark and Karlovsky, 1998).

Application of Transformation of Fungal Protoplasts: Mainly the transformation appr-oaches are used : (*i*) for the analysis of promoters and construction of expression vectors, and (*ii*) biotechnological application in the expression of foreign genes in fungi. Analysis of promoter involves the construction of vectors which have different pieces of promoters. These can be cut by restriction enzymes and joined with DNA ligase. By this method changes in promoter region are also possible and this can be achieved through mutagenesis. Thus, a strain can be improved resulting in overproduction of the products. Factors influencing the fusion frequency of protoplasts have been examined with auxotrophic mutants of *Aspergillus nidulans*. The optimum conditions for protoplast fusion are: a total of 5 to 15 million protoplasts per ml, and 25% polyethylene glycol (PEG) 4000 or 6000 as fusogenic agent in 10 to 100 mM $CaCl_2$ solution.

Adopting these approachcs many mammalian genes have now been expressed in *S. cerevisiae* and aspergilli through expression vector (*i.e.* vector which include sequences promoting high level of transcription of the gene and secretion of the gene products).

Several companies in the U.S.A. and Europe have demonstrated the expression of foreign DNA introduced in *Aspergillus* sp. through transformation. In 1987 A. Upshall and coworkers have constructed a plasmid pM159 for the secretion of human tissue plasminogen activater (tPA) by using *A. nidulans*. Yield of this protein was about 100 mg/litre culture filtrate. Similarly in 1987, D.M. Gwynne and coworkers found the secretion of human interferon by transformed *A. nidulans* mycelia to about 1 mg/litre.

What is tPA ?

tPA is a naturally occurring protein that breaks the inactive zymogen plasminogen into the active enzyme plasmin. Then plasmin cleaves fibrin clots. This process is important in removing clots after completion of the repair of a wound. Often clots form in response to disease conditions such as atherosclerosis, stroke, cancer and some complications during pregnancy. It is used as a powerful drug to dissolve the clots and prevent much of the damage associated with loss of blood flow to vital organs.

II. GENE CLONING IN PLANT CELLS

There are a large number of plants ranging from unicellular to multicellular forms. In recent years a great success has been achieved in many plants, and hundreds of transgenics have been produced and commercialized. Since there is a diverse groups of plants in eukaryotes ranging from unicellular form to multicellular forms or even most complicated and advanced ones such as yeasts, microalgae, fungi, angiosperms and gymnosperms. Some of the most common examples have been discussed herewith.

1. Gene Transfer in Dicots by using *Agrobacterium* Ti-DNA as Vector

It is not possible to transfer Ti-plasmid into plant cells because of its large size, probably up to 235 Kb. Therefore, T-DNA is excised and used as foreign DNA for expression in plant cells. From bacteriological point of view opine production and development of gall is not required for the successful integration and expression of DNA. Hence, replacement of *nos* and *ops* genes with a foreign DNA fragment retaining the opine synthase promoter for expression of inserted DNA fragment has no harm. Schell and Van Montague (1983) undertook some experiments where genes for resistance to kanamycin and methotrexate expressed in cultured callus cells. Generally it is easy to use opine promoters for the expression of foreign DNA. It has become possible that the inserted genes, like other plant genes, will express only at certain stages or specific tissues of plants when they are with specific regulated promoters. There are reports of replacement of opine synthase promoters by regulatory promoters obtained from plants. Murai *et al.* (1983) have introduced gene for bean plant storage protein, the phaseolin (which is supplemented with methionine codon), into the cells of sunflower plant after transformation with Ti-plasmid where phaseolin gene was under the control of its own promoters, the octopine synthase.

2. Gene Transfer in Monocots

In 1985, for the first time a definitive evidence was presented by Ingo Potrykus1nd co-workers at Switzerland for stable gene transfer to cultured cells in a plant of the family Gramineae, *Lolium multiflorum* by using DNA transformation of protoplasts from a non-morphogenic cell culture. A construction consisting of expression signals from gene VI of cauliflower mosaic virus joined to the aminoglycoside (neomycin) phosphotransferase gene (APH(3′)II) from transposon Tn*5* conferred resistance to the antibiotic G-418 to cell colonies arising from transformed protoplasts. By demonstrating a tight correlation between the resistant phenotype, the physical presence of the foreign gene and the presence of the active gene product it has been shown that these colonies are true transformants and that a gene which is expressed well in dicotylodenous plants is also expressed in cells of graminaceous monocots.

The limitation of Ti-plasmid is that it is specific only to the dicot plants. *A. tumefaciens* does not induce tumour in monocot plants, however, most of the major crops including cereals are monocots. In 1984, Hooykaas Van and coworkers discovered that *A. tumefaciens* could transfer T-DNA into certain monocots with the consequences of expression of opine gene within the plant cell without inducing the tumour. This discovery made it possible that T-DNA can be expressed into cells of monocot plants.

3. Plant Cell Transformation

Cell/suspension culture technique has eased the manipulation of plants cells. Hence, plant cell suspension culture coupled with DNA transfer techniques has become a new field of gene manipulation in plants. Cell suspension cultures have been the traditional source of plant material for biochemical selection of plant cell mutants, such as numerous cell lines resistant to various amino analogs. Suspension of protoplasts constitutes an identical plant material from one laboratory to other. The approaches made for plant protoplast transformation are: (*i*) introduction of DNA sequence via integration into T-DNA of Ti-plasmid of *A. tumefaciens*, (*ii*) transfer of naked DNA into protoplast including bacterial spheroplasts and microinjection, and (*iii*) transformation with plant viruses carrying inserts e.g. gemini and caulimoviruses.

It has been found that plant protoplast treated with polyethylene glycol (PEG) used for protoplast fusion, resulted in efficient delivery of DNA to protoplasts. The PEG method has been employed for the successful transformation of protoplasts with Ti-plasmid. Once DNA fragment is introduced into plant cells, it integrates with plant chromosomes and expresses from its signal. Transformation of plant cells with a gene for drug and antibiotic resistance linked to plant promoter has been possible. The gene could show expression in plant cell. Therefore, the transformed cells could be selected from the culture medium amended with drugs and antibiotics. The plants regenerated from cells show drug/antibiotics resistance or any traits according to the introduced gene(s).

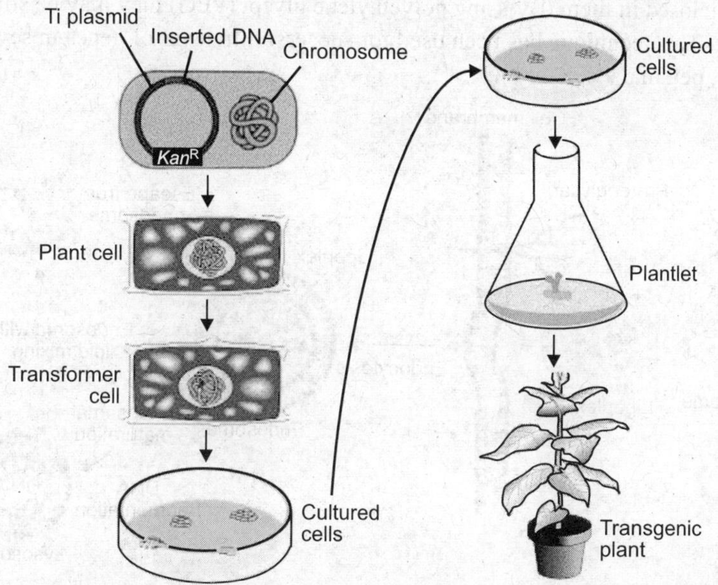

Fig. 8.35. Plant transformation by co-cultivation method.

(*a*) Plant Cell Transformation by Co-cultivation Method: The plant protoplasts/tissues are co-cultivated with *Agrobacterium* carrying T-DNA. Co-transformed plants are selected and the integration of the different T-DNAs was analysed at the genetic and genomic level. Genetic analysis shows that the T-DNAs derived from different bacteria are frequently integrated at the same locus independent of the plant species or transformation method used. In 1997, De Neve and coworkers found 12 out of 27 *Arabidopsis* transformants in Southern analysis that contained the co-transferred T-DNAs linked to each other in all possible configurations but with a preference for those with at least one right border involved in linkage.

(*b*) Plant Cell Transformation by Co-integration Method: The desired gene is inserted into a plasmid containing a segment of the T-DNA. The resulting plasmid called a shuttle vector is then transerred into *E. coli*. Transformants are selected on the basis of antibiotic resistance encoded by genes contained in the plasmid. The plasmid is then transferred from *E. coli* to *Agrobacterium* by a process called conjugation. Once inside *Agrobacterium* the shuttle vector integrates into the Ti plasmid by means of homologous recombination of the T-DNA on both plasmids. The *Agrobacterium* containing the recombinant Ti plasmid is selected and used to infect plant cells. Marker genes contained in the Ti plasmid coding for antibiotic resistance and other physiological traits are used to select transformed plants.

(c) Plant Cell Transformation by Ultrasonication : Ultrasonication is done for various biological experiments. But in some plants this technique has been done for successful gene delivery. The Biotechnology Research Centre, Beijing (China) has used this technique for gene transfer in plants like wheat, tobacco and sugarbeet. When the cultured explants were sonicated with plasmid DNA carrying marker genes such as *cat, npt*II, and *gus,* and sonicated calli transferred to selective medium, shoots were grown successfully. The calli in control set of experiment which were not sonicated with plasmid did not grow on selective medium rather they died. Transgenic tobacco plants were obtained at a frequency of 22%.

(d) Liposome-mediated Gene Transfer: Liposomes are microscopically small sized particles, which contain the phospholipid bilayers that are concentric in nature. Liposomes enclose aqeuous chamber and can entrap water soluble molecules; therefore, they are called lipid bags. Many plasmids are enclosed in them. By using polyethylene glycol (PEG) they may be stimulated to fuse with protoplast. This technique has been used for successful transfer of genes in several plants like carrot, tobacco, petunia, etc.

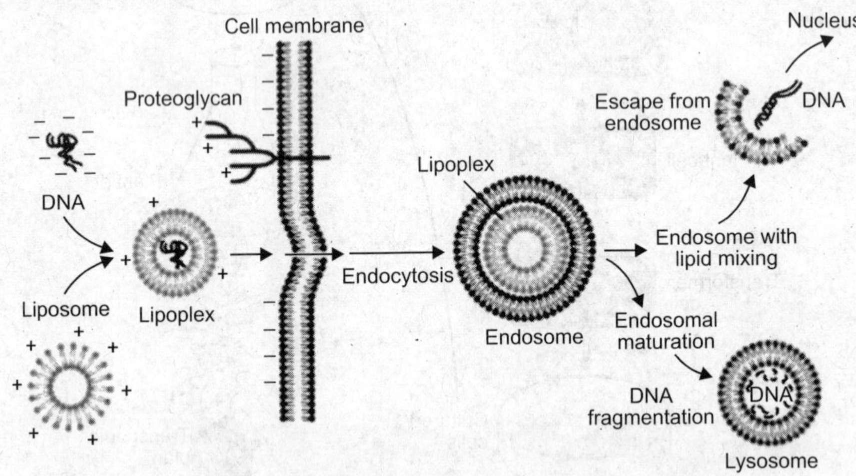

Fig. 8.36. Lipoplex-mediated transfection and endocytosis.

Cationic lipids forming micellar structures called liposomes are complexed with DNA to create lipoplexes (Fig. 8.36). The structures fuse with the cell membrane, after interactions with surface proteoglycans. DNA enters the protoplast due to endocyclosis of liposomes. It gets adhered first to protoplast surface and gets fused with protoplast at the site of surface. This results in the formation of a double-layer inverted micellar vesicle. During the maturation of the endosome into a lysosome, the endosomal wall might rupture releasing the contained DNA into the cytoplasm and potentially towards the nucleus. DNA imported into the nucleus might result in gene expression. Alternatively, DNA might be degraded within the lysosome.

There are several advantages in using this technique such as: (*i*) low toxicity, (*ii*) protection of DNA/RNA from nucleases that lyse them, (*iii*) long stable storage of DNA fragments in liposome, (*iv*) applicability in various cell types, and (*v*) high level of reproductivity.

(e) Electroporation : Electroporation or electropermeabilisation (using electric field) is now used to transfer the foreign DNA into the fragile cells. Therefore, it is a powerful method for introducing cell-impermeant species into living cells. The alternatively cell-impermeant species are modified chemically so that they can passively diffuse through the membrane. Certain polar Ca^{2+}-sensitive dyes are introduced as nonpolar esters for many biologically relevant species such as DNA, RNA, ions, particles, proteins, libraries of compounds, which are inconvenient or impossible.

In addition, electric fields can also induce fusion of cells and liposomes. Electroporation on cell suspensions (referred to as bulk electroporation) gives rise to a statistical distribution of cellular states. The phenomena and parameters described should be considered averages.

Electroporation is done employing the electroporators that create the electric current, and send it through the cell solution (Fig. 8.37). The solution is pipetted into a glass or plastic cuvette (the plastic with aluminium electrodes and a blue lid; they hold a maximum of 400 ml) which has two aluminum electrodes on its sides. A suspension of around 50 microliters is usually used. Before electroporation it is mixed with the plasmid to be transformed. The mixture is pipetted into the cuvette, the voltage is set on the electroporator (240 volts is often used) and the cuvette is inserted into the electroporator.

In medical and biological research, single or small population of cells is used in many different settings and experiments, for example functional cell-based drug screening, stem cell research, flow cytometry, etc. Microfabrication technologies for developing chip-based devices for cell sorting, manipulation and characterisation are likely to be extremely successful. In addition, modern analytical chemistry can provide quantitative data on biological and biochemical phenomena, at the single-cell level. A study of many individual single cells will lead to a more profound understanding than the study of the same cells as an ensemble.

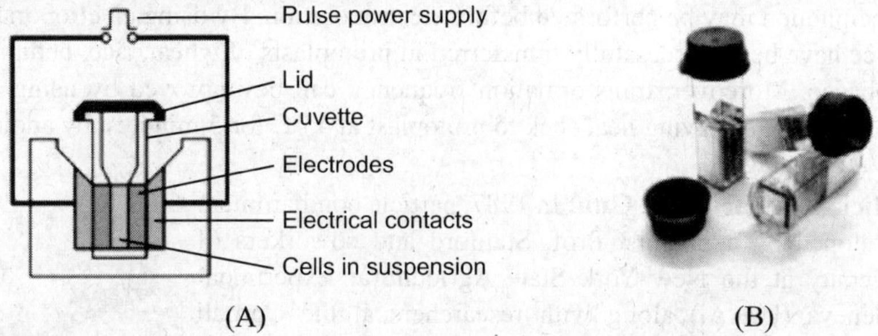

Fig. 8.37. The major components of an electroporator with cuvette loaded (A); cuvettes for electroporation (B).

Brief pulses of high voltage electricity are applied to protoplasts suspension containing naked or recombinant plasmids. The electric pulses induce the formation of large pores in the cell membrane. These pores give a passage through which the foreign DNA can enter into the protoplasts and increase the transformation frequency (Fig. 8.38).

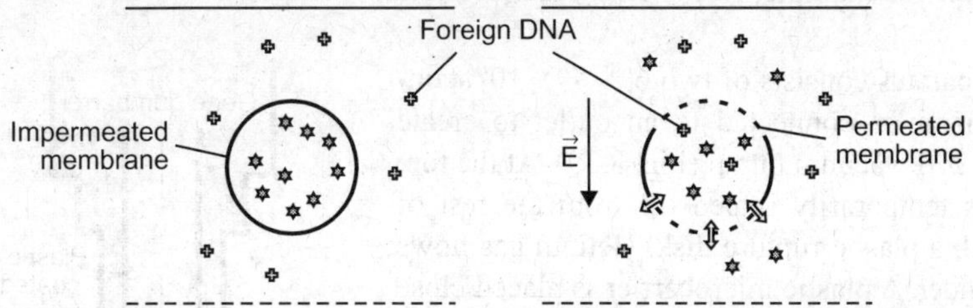

Fig. 8.38. Cell before (left) and after (right) electroporation. Permeated membrane is shown with stylized line and impermeable membrane with solid line.

The transformed protoplasts are cultured for about one month. These develop microcalli which are plated on solid medium containing selective marker (*e.g.* kanamycin). After 37-45 days, the calli are analysed for the presence of differences in transformed cells. The normal assumption in bulk electroporation is that the field in solution is the applied voltage (hundreds to a thousand volts) divided by the inter-electrode spacing (0.1-1 cm). This can be true, but that the magnitude of the effective field may depend on the buffer type.

Electroporation in Other Organisms

The process of electroporation is often used for the transformation of bacteria, yeast, and plant protoplasts, in molecular biology. In addition to the lipid membranes, bacteria also have cell walls which are different from the lipid membranes. These are made of peptidoglycan and its derivatives. However, the walls are naturally porous and only act as stiff shells that protect bacteria from severe environmental impacts. If bacteria and plasmids are mixed together, the plasmids can be transferred into the cell after electroporation. Several hundred volts across a distance of several millimeters are typically used in this process. Afterwards, the cells have to be handled carefully until they have had a chance to divide producing new cells that contain reproduced plasmids. This process is approximately ten times as effective as chemical transformation.

This procedure is also highly efficient for the introduction of foreign gene in tissue culture cells, especially mammalian cells. For example, it is used in the process of producing knockout mice, as well as in tumour treatment, gene therapy, and cell-based therapy.

The success of the elecroporation depends greatly on the purity of the plasmid solution, and especially on its salt content. Impure solutions might cause a small explosion (called arcing), in which case the bacteria are dead and the process needs to be redone. If this happens often, an additional precipitation may be performed before electroporation. By using electroporation method genes of choice have been successfully transferred in protoplasts of wheat, rice, petunia, sorghum, maize and tobacco. Moreover, transformation frequency can be improved by using linear DNA instead of circular one, by giving heat sock to protoplast at 45ºC for 5 minutes, by adding PEG and using 1.25 kV/cm voltage.

(*f*) Particle Bombardment Gun: In 1987, particle bombardment gun was developed by a botanist Prof. Stanford and coworkers of Cornell University at the New York State Agricultural Experiment Station in Geneva (U.S.A.) along with researchers at the Cornell Nanofabrication Facility. As the term denotes, it shoots foreign DNA into plant cells or tissue at a very high speed. This technique is also known as particle bombardment, particle gun method, biolistic process, microprojectile bombardment or particle acceleration.

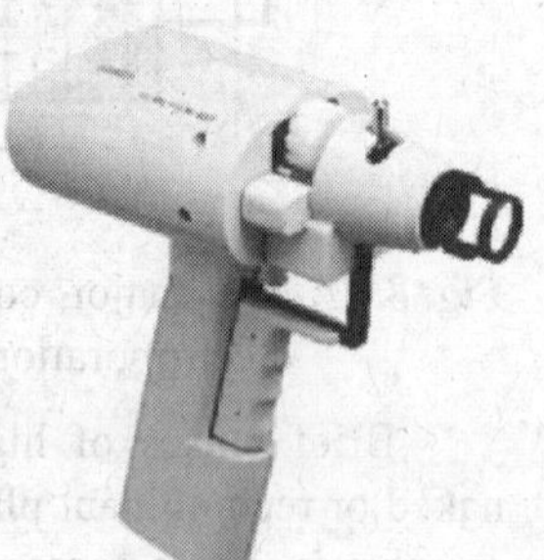

A gene gun

This technique is most suitable for those plants which hardly regenerate and do not show sufficient response to gene transfer through *Agrobacterium* for example, rice, wheat corn, sorghum, chickpea and pigeon-pea.

The apparatus consists of two 6′ × 7′ × 10′ stainless steel chambers connected to an outlet to create vacuum in a 2HP vacuum pump (Fig. 8.39). At the top, a cylinder is temporarily sealed off from the rest of chamber with a plastic rupture disk. Helium gas flows into the cylinder. A plastic microcarrier is placed close to rupture disk. It contains DNA coated tungsten balls (1 micrometer in diameter), the microscopic pellets (*i.e.* coated microprojectiles). When to work the apparatus is placed in laminar flow just to maintain sterile conditions before the start of work. Then the target cells/tissues are placed in the apparatus. A stopping screen is put between the target cells and microcarrier assembly. A blast of helium at 1,000 psi sends the disk shooting forward at about 1,300 feet per second, roughly the

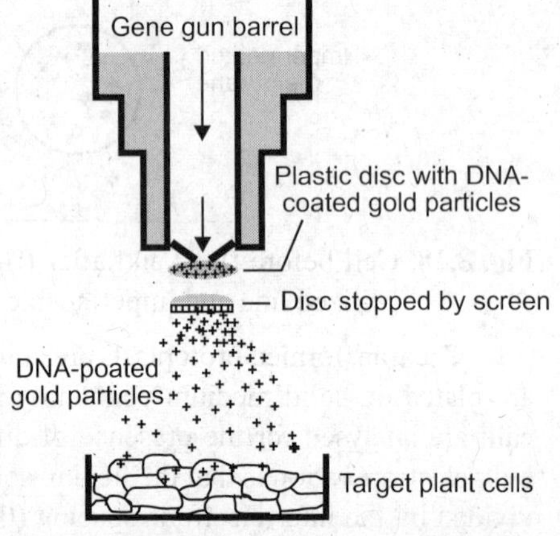

Fig. 8.39. Functioning of gene gun.

same speed as a bullet leaves a rifle. When pressure of cylinder exceeds the bursting point of plastic disk, it gets ruptured. Helium shock waves propel the plastic mircrocarrier containing DNA coated micropellets. The stopping screen allows the micropellets to pass through and deliver DNA into target cells. The balls breech the cell membrane and release the DNA particles. The gene gun utilises recombinant DNA technology to incorporate the expression of the delivered genes. The transformed cells are regenerated onto nutrient medium. The genetically altered cells can be used to make plants that include the desired genetic modification in all of their cells. The regenerated plant tissues are selected over culture media containing either antibiotics or herbicide. The selected plants are then analysed for expression of foreign DNA in the form of polypeptide.

This is a clean and safe technique which enables the scientists to transform the organised tissue of plant species and has a universal delivery system common to many tissue types from many different species. However, it can give rise to un-wanted side effects, such as the gene of interest being rearranged upon entry or the target cell sustaining damage upon bombardment. But it has been quite useful for getting transgenes into organisms when no other options are available.

Using this technique scientist have got success in delivering foreign DNA into epidermal tissues of *Allium cepa*, scutellar tissues of maize and leaf and cell culture of many crops. In addition to bacterial cells, algae, fungi plant organelles (*e.g.* chloroplast and mitochondria), an animal and human cells and fruit fly embryos have been successfully transformed by using this method.

(*i*) Conditions for Bombardment. There are several conditions required for introduction of foreign gene into the desired cell/tissue/pollen, etc. The conditions required for introduction of beta-glucuronidase (GUS) reporter gene for transformation of Basmati Indica rice are given in Table 8.5. In 1996, D. Minhas and coworkers reported for the first time the transient expression of GUS reporter gene through biolistic delivery in mature embryo-derived embryonic callus cultures of *Oryza sative* cv Basmati 370. In 1998, scientists of Plant Transformation Group at ICGEB (New Delhi) have got success in transferring human interferon gamma gene into chloroplasts of tobacco, maize, etc. using a particle gun-mediated gene transformation.

Table 8.5. Bombardment conditions employed on Basmati Indica rice callus (*Source*: Minhas D. 1996. *Curr. Sci.* 71: 1005).

1. Distance between rupture disc and microcarrier	0.5 inch + 15 mm
2. Size of gold particles	1 μm
3. Distance between microcarrier and target	6 and 9 cm
4. Density of particles/shot	9 mg/shot
5. Bombardment medium	MS medium supplemented with 2, 4-D at 2 mg/l (rice callus medium)
6. DNA concentration	1 mg/ml
7. Physical parameters	
(*i*) helium pressure	450-1300 pounds/inch2
(*ii*) vacuum	26 mg Hg
8. Post bombardment culture	On callus medium till GUS assay (48 h)

(*ii*) Pollen transformation through particle bombardment. Genetic improvement in alfalfa has been made by plant tissue culture methods. This method is time-consuming, requires special techniques and efficiency of getting stable regenerants is low. In 1997, S.M. Ramaiah and D.Z. Skinner have produced transgenic alfalfa through direct delivery of DNA into pollen grains by particle bombardment method. Plasmid pBI121 bearing GUS reporter gene was introduced into pollen grains. The microprojectile bombarded liquid pollen suspension was allowed to fertilise the tagged and tripped recipient female flower. Pollinated flowers set seeds in about a month.

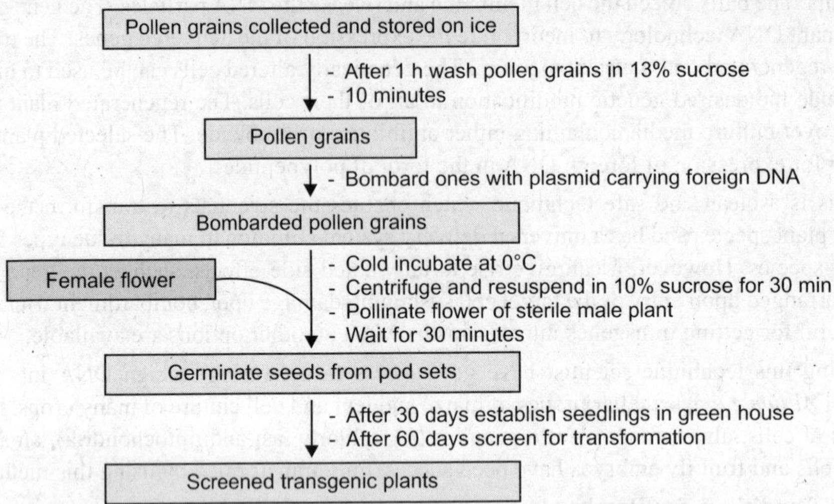

Fig. 8.40. Outline of pollen transformation to obtain transgenic alfalfa plants.

Thirty per cent of plants derived from fertile seeds showed integration of GUS plasmid. It was confirmed by Southern analysis. After 10 vegetative generations some transgenic plants lost the integrated GUS plasmid, whereas in few others number, copies of GUS gene decreased due to unknown reasons. An outline of production of transgenic alfalfa plant through pollen transformation is given in Fig. 8.40.

Fig. 8.41 shows the transient expression of green fluorescent protein (GFP) in a tobacco pollen tube growing in culture medium. Transformation was achieved through the particle bombardment procedure Some of the untransformed pollen grains exhibit auto-fluorescence.

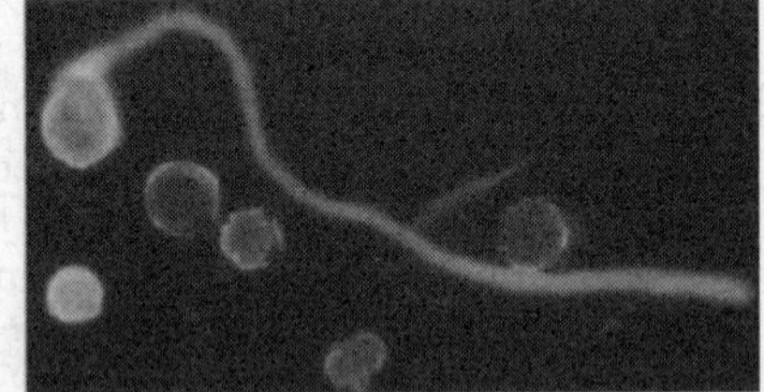

Fig. 8.41. Expression of green fluorescent protein (GFP) in a tobacco pollen tube.

***(f)* Fibre-Mediated DNA Delivery** : Fibre-mediated DNA delivery is a new type of approach of DNA delivery in plant cells. In this approach suspension culture cells were mixed with plasmid DNA having GUS gene and silicon carbide fibres of 0.6 μm diameter and 10-80 μm length; all are suspended in the culture medium. The mixture is vertexed, DNA is delivered into the cell cytoplasm and nucleus through silicon carbide fibres. The cells are assayed for transient GUS expression. The frequency of GUS-transformed cells is one out of 10^{-4} cells.

II. GENE CLONING IN ANIMAL CELLS

Culture of animal cells on a large scale is more difficult and expensive than plant cells, the bacteria, yeast, etc. This is why the commercialisation of natural products of animal origin is limited. However, most of the animal products have been produced by using bacteria as biological tool.

In India, K. Chandrashekharan at the Centre for Cellular and Molecular Biology (CCMB), Hyderabad emphasised the use of recombinant DNA technology for the introduction of foreign DNA into mammalian cells. This would help in curing certain genetic defects.

Since the nucleus of every animal cell contains the entire genome of the animal, it might seem easy enough to clone an animal by placing the nucleus in an egg cell from which the nucleus has been removed. While this was tried many times, it was never successfully accomplished until 1996, in the creation of the sheep Dolly by Ian Wilmut and colleagues in Scotland. Dolly was the first mammal created using the nucleus from a cell of a mature adult mammal. Prior to this feat, it had been thought that normal mammalian development caused irreversible changes in some portion of the DNA that prevented it from acting as embryonic DNA. Amphibians have long been cloned from adult cells, but they invariably die in the tadpole stage. Adult amphibians have been successfully cloned for many years from embryo nuclei. In this technique, nuclei from cells of an early embryo are extracted using a very fine glass pipette and placed in egg cells that have been shed by a female amphibian such as a frog (after removing the unfertilised egg cell nucleus).

In 1998, mice were cloned from adult somatic cell nuclei using the same technique as was used for Dolly. This technique may become especially important for producing large numbers of transgenic animals for use in research or production of specialised proteins. However, cloned mammals are generally not very healthy. Apparently development is not quite normal when it begins with a nucleus that has already existed in another animal, compared to a genome derived from a sperm and an egg.

DNA cloning is usually performed for one of two reasons: either to produce a lot of identical DNA for further study, or to use the DNA in an intact organism to produce useful proteins. In the first case a researcher might want to determine the DNA sequence of the gene or study the factors that control its expression (transcription). In the second case, one might want to produce large amounts of a medically useful protein such as insulin or growth hormone.

Animals that have been cloned

Carp, cats, cattle, dear, dog, ferret, fruit files, goat, gaur, horse, mice, mouflon, mule, pig, rabbit, rat, rhesus monkey, sheep, water buffalo, wolf.

1. Animal Viruses

Different animal viruses are known, such as simian vacuolating virus 40 (SV40), adinovirus, retrovirus (ssRNA), vaccinia virus, etc. They can increase the efficiency of animal cell transformation if they are used as vector. A special feature of viruses is that they contain the strong promoters which can possibly bring about expression of inserted DNA fragment of the viruses. The most commonly used virus is SV40 which contains a circular DNA of about 5.2 kb. In addition to *ori* region, the DNA contains early genes and late genes. Early genes are required for DNA replication, whereas late genes encode coat protein of the virus.

Like phage λ, the amount of DNA to be packed into virus capsid is limited. The genome has 3 identified non-essential regions; hence the use of non-defective viral vectors is limited. The defective virus vectors have missing genes whose infection process has to be helped by viral DNA constituting functional copies of missing genes. If the *cos* cells (the host monkey cells) are used as host, they can be transformed by a plasmid carrying the early region of SV40 with a defective replication origin. *Cos* cells express wild type large T antigen and can support the growth of SV40. Recombinants containing foreign DNA inserted into early region can be propagated without the need of helper virus. In this way, the recombinant and helper can be separately recovered from the host. Adenoviruses carry double-stranded DNA as their genetic material.

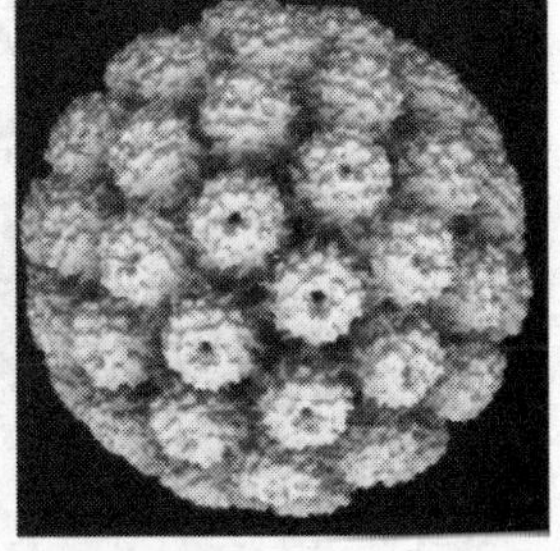

Electron photomicrograph of SV40.

Upon infection their DNA molecule is introduced into the host cells. But their genetic material is not incorporated into the host cell's genetic material. The DNA molecule is left free in the nucleus of the host cell, and the instructions in this extra DNA molecule are transcribed just like any other gene. The only difference is that these extra genes are not replicated when the cell is about to undergo cell division so the descendants of that cell will not have the extra gene. This vector system has shown real promise in treating cancer, and indeed the first gene therapy product to be licenced is an adenovirus to treat cancer (for detail *see* Chapter 9, *Gene Therapy*).

2. Microinjection

Microinjection is a technique of delivering foreign DNA into a living cell (a cell, egg, oocyte, embryos of animals) through a glass micropipette. Cells can be microinjected directly in their culture dish, but often microinjection is combined with microscopy. Typically for microinjection of single cells in culture, primary and tissue culture eukaryotic cells are microinjected with cytoskeletal proteins and antibodies, but the general principles also apply to other molecules such as nucleic acids also. Cells growing on 2- and in 3-dimensional substrates are amenable to microinjection. The sensitivity of a cell to microinjection depends on the cell type. Training in this type of technique relies heavily on hands-on instruction and practice, and empirically determined methods.

It is important to have consideration about choice of cell type. Easy cells to be microinjected are flat tissue culture cells such as Ptk2 cells, NRK cells, most tissue culture fibroblasts, CHO cells, etc. Hard cells to microinjection are round cells, for example mitotic cells and neurons and sensitive cells (e.g. neurons and primary cells including primary fibroblasts). Cells on a two-dimensional substrate are more amenable to microinjection than in a three-dimensional matrix. The cells are plated on glass cover slips (12 mm circles or similar for cells to be fixed or larger circles of about 25 mm for cells to be viewed live on a microscope in an observation chamber). Cells tend to stick less well to glass so usually cover slips are pretreated with acid and poly-L-lysine (PLL) or polyornithine (better for some neurons). One end of a glass micropipette is heated until the glass becomes some what liquefied. It is quickly stretched which forms a very fine tip at the heated end. The tip of the pipette attains to about 0.5 mm diameter which resembles an injection needle. The process of delivering foreign DNA is done under a powerful microscope. Cells to be microinjected are placed in on a cover slip or a container. A holding pipette is placed in the field of view of the microscope. The holding pipette holds a target cell at the tip when gently sucked. The tip of the micropipette is injected through the membrane of the cell (Fig. 8.42). Contents of the needle are delivered into the cytoplasm and the empty needle is taken out.

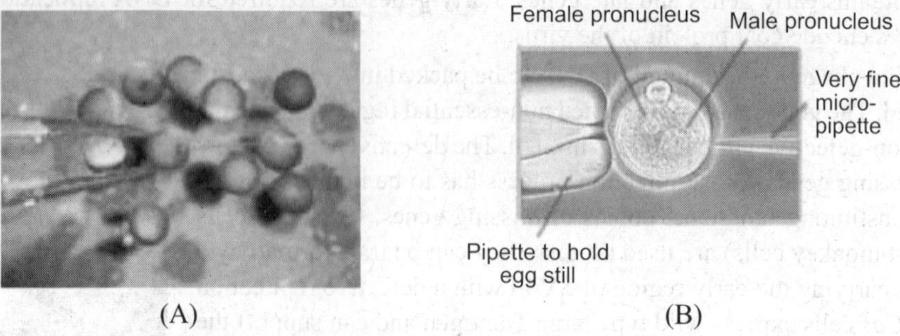

Fig. 8.42. Microinjection; (A)- *Xenopus* oocytes to be microinjected; (B)- delivering foreign gene through microinjection.

Xenopus oocytes have been widely used for the study of transcription by microinjection because oocytes contain between 6,000 and 1,00,000 or more RNA polymerase molecules than somatic cells. Microinjection is technically easy because of large size of oocytes. In 1981, M. Wickens and

R.A. Laskey characterised some of the endogenous pattern of gene regulation during development. The injected DNA integrates randomly with nuclear DNA and its expression could be possible only when the foreign DNA is attached to a suitable promoter sequence. There are many examples where different types of animal cells have been microinjected and successfully transferred.

In 1982, G. Rubin and A. Spradling for the first time introduced *Drosophila* gene for xanthine dehydrogenase into a P-element (parental element) which was microinjected with an intact helper P-element into embryo deficient for this gene. Such embryos later on developed into flies with rosy coloured eyes than mosaic eyes, as in the first parental generation.

QUESTIONS

1. Give an illustrated account of gene cloning in prokaryotes.
2. Describe the methods of construction of cDNA library with suitable diagrams.
3. Give a detailed account of construction of gene bank (DNA library) with suitable diagrams.
4. Describe in brief about linkers and homopolymer tails ? What is their importance in gene cloning ?
5. Discuss in detail about the methods of blunt end ligation.
6. Describe in detail about colony hybridisation technique.
7. Write in detail about the methods of transformation of bacterial cells.
8. Discuss in detail about different methods used for selection of recombinants.
9. Describe different methods of gene cloning in eukaryotes.
10. With the help of illustrated account of radioactive labelling.
11. Give an illustrated account of *Agrobacterium*-mediated gene transfer.
12. Write different methods of plant cell transformation studied by you.
13. Describe the steps of particle bombardment gun method of gene transfer.
14. Describe different methods of gene cloning in animal cells.
15. Write short notes on the following :

(*i*) DNA library (gene bank), (*ii*) Type and size of gene library,
(*iii*) cDNA library (genomic library), (*iv*) Blunt end ligation,
(*v*) Homopolymer tailing, (*vi*) Linkers and adaptors,
(*vii*) Transformation of bactcrial cells, (*viii*) Transfection of a cell,
(*ix*) Insertion selection activation, (*x*) Blue-white screening,
(*xi*) DNA probe, (*xii*) Plaque lifting,
(*xiii*) *In vitro* selection method, (*xiv*) Southern blotting technique,
(*xv*) Western blotting (electroblotting), (*xvi*) Northern blotting method,
(*xvii*) Dot blot technique, (*xviii*) Colony hybridization,
(*xix*) Direct immunological screening tests (*xx*) Shine-Dalgarno sequence,
(*xxi*) expression vector, (*xxii*) Radioactive labelling,
(*xxiii*) Non-radioactive labeling, (*xxiv*) Nick translation,
(*xxv*) Digoxigenin-labelling system, (*xxvi*) Biotin-Streptadin labelling system,
(*xxvii*) Fluorescent labeling with quantum dots, (*xxviii*) Transformation of filamentous fungi,
(*xxix*) Liposome-mediated gene transfer, (*xxx*) Electroporation,
(*xxxi*) Particle bombardment gun,
(*xxxii*) Pollen transformation through particle bombardment,
(*xxxiii*) Microinjection.

BIBLIOGRAPHY

Alwine J.C., Kemp D.J. and Stark G.R. 1977. Method for detection of specific RNAs in agarose gels by transfer to diazobenzyloxymethyl-paper and hybridization with DNA probes. *Proc. Natl. Acad. Sci. U.S.A. 74: 5350–4.*

Drlica, K. 1984. *Understanding DNA and gene cloning: A guide for the curious*. John Wiley & Sons Inc., New York, p. 205.

Dubey, R.C. 2014. *A textbook of biotechnology*. S. Chand & Co. Pvt. Ltd. Ram Nagar. New Delhi, India.

Grange J. M., Fox, K. R. and Morgan N. L. 2009. *Genetic manipulation: techniques and applications.* John Wiley & Sons, p. 416.

http://en.wikipedia.org/wiki/Digoxigenin.

Maniatis T., Fritsch E.F. and Sambrook J. 1982. *Molecular cloning: A laboratory manual.* Cold Spring Harbour Laboratory Press. USA.

Old R.W. and Primrose S.B. 1980. *Principles of gene manipulation: An introduction to genetic engineering*. University of California Press, p. 138.

Primrose S.B, Twyman R.M. and Old B. 2005. *Principles of gene manipulation and genomics.* 7th Ed. Lavoisier S.A.S.

Shine J. and Dalgarno L. 1975. Determinant of cistron specificity in bacterial ribosomes. *Nature.* **254** *(5495): 34–8.*

Stansfield W.D. 1996. *Schaum's outline of theory and problems of molecular and cell biology.* McGraw-Hill, NY, p. 183-200.

CHAPTER

Gene Therapy

A. History of gene therapy
B. Genetic diseases
C. Background of gene therapy
D. Types of gene therapy
E. Method of gene therapy
F. Recent developments in gene therapy
G. Problems and considerations
H. Governmental oversight and public participation

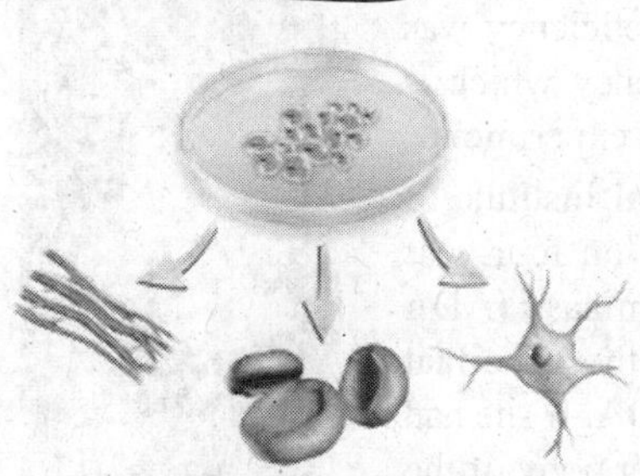

Embryonic stem cells (from extra embryo created during *in vitro* fertilization) differentiating

Gene therapy (= gene transfer and genetic modification) refers to the insertion of genes into the cells and tissues of individuals to treat a disease and hereditary diseases. Gene therapy supplements a defective mutant allele with a functional one. Although the technology is still in its infancy, it has been used with some success.

Basically, gene therapy is a technique for correcting the defective genes which are responsible for disease development. It can be is defined as 'the introduction of a therapeutic gene into a cell whose expression can lead to a cure of a disease or offer a transient advantage for tissue growth and regeneration'. According to the U.S. Food and Drug Administration, human gene therapy is the replacement of faulty genetic material of a person with normal genetic material to cure a disease. The delivery of genes can be undertaken for a number of purposes. Usually it is attempted to enhance or add a function to a cell or a tissue or to delete or reduce another function.

Gene therapy has the potential to mediate the highest possible level of therapeutic specificity. Every normal or diseased cell can switch on or off a gene expression cassette in a tissue-, disease-, and time-dependent fashion, by use of specific transcription factors that are active only in a given unique situation. The delivery of nucleic acids into the target cells is very ineffective and presents a formidable challenge. There are viral and non-viral vector methods of gene therapy. However, future developments include improved targeting, enhanced intracellular uptake, and reduced toxicity of gene vectors. The currently used classes of vectors have complementary characteristics such as high intracellular efficiency of viral vectors and low immunogenicity and greater flexibility of non-viral vectors. In recent years, the merge of viral and non-viral vector technologies is an encouraging strategy for the future, for example chemically modified viral vectors (chemo-viruses) and synthesis of virus-like systems (synthetic viruses) (Boeckle and Wagner, 2006).

A. HISTORY OF GENE THERAPY

There have been tremendous strides in the area of gene therapy in the last couple decades. In 1980, the first unauthorized attempt was made for gene therapy. Martin Cline at the University of

California Los Angeles (UCLA) conducted a recombinant DNA transfer into the bone marrow cells of two patients who had hereditary blood disorders. He made such attempts in direct opposition of 'gene therapy guidelines' of National Institute of Health, and without the approval of the Institutional Review Board, where the work was conducted. The ethical concerns prompted a call for review by a number of organisations including the National Council of Churches, Synagogue Council of America, and the United States Catholic Conference. Consequently, Cline was forced to resign his department chairmanship at UCLA and lost several research grants.

In 1990, adenosine deaminase (ADA) deficiency was the first disease approved for gene therapy treatment by the federal government. Due to lack of ADA enzyme, patients suffering from this disorder have a severely weakened immune system. It can be treated with an expensive drug called PEG-ADA (costing more than $60,000 a year) that requires regular injections throughout the life of patients. Therefore, ADA deficiency was selected for gene therapy because it is a single gene deficiency which is regulated in a simple manner, and involves a relatively simple experiment.

Dr. W. French Anderson and his gene therapy patient Ashanti DeSilva (4 years old).

On September 14, 1990 researchers at the U.S. National Institute of Health performed the first approved gene therapy procedure on four-year old Ashanti DeSilva who was suffering from adenoside deaminase (ADA deficiency. Ashanti DeSilva inherited two broken copies of the gene that contains the instructions for manufacturing a protein called ADA. If she had had just one broken copy, she would have been fine. The other copy of the gene would have made the difference. With two broken copies, her body didn't have the right instructions to manufacture ADA at all. ADA plays a crucial role in our resistance to disease. Without it, special white blood cells called T cells die. Without T cells, ADA-deficient children are wide open to the attacks of viruses and bacteria. These children have severe combined immune deficiency (SCID) disorder which is more commonly known as 'bubble boy disease'.

Ashanti DeSilva, a healthy 14 years girl.

Children with this illness usually develop overwhelming infections and rarely survive to adulthood; a common childhood illness like chickenpox is life-threatening. Ashanti led a cloistered existence: avoiding contact with people outside her family, remaining in the sterile environment of her home, and battling frequent illnesses with massive amounts of antibiotics. Dr. W. French Anderson performed the therapy by infusing white blood cells, with the correct genetic composition, into Ashanti to correct her immunodeficiency. The outside world is threatening to a person with a weak immune system. Everyone you touch, share a glass with, or share the same air with is a potential source of dangerous pathogens. Lacking the ability to defend herself, Ashanti was largely confined to her home.

In Ashanti's gene therapy procedure, doctors removed white blood cells from the child's body, allowed the cells to grow in the laboratory, inserted the missing gene into the cells, and then infused the genetically modified blood cells back into the patient's bloodstream. Laboratory tests have shown that the therapy strengthened Ashanti's immune system. She no longer has recurrent colds; she has been allowed to attend school, and she was immunised against whooping cough. This procedure was not a cure; the white blood cells treated genetically only work for a few months, and the process must be repeated every few months. This therapy did not result in a complete cure but it helped to correct the deficiency so that Ashanti only takes small and weekly doses of the traditional PEG-ADA drug.

Unfortunately, an 18 year old Jesse Gelsinger was the first case who died during gene therapy in 1999. Jesse suffered from OTC disease (ornithine transcarboxylase deficiency) which is a liver disorder that results in poisonous levels of ammonia formed in the body. His disorder was being controlled through diet and medication. Although he was healthy and full of life, he offered to participate in gene therapy study because of his desire to help others with same disease. He died of multiple organ failure four days after starting gene therapy treatment. His death was thought to be due to a severe immune response to adenovirus carrier. The vector was used to carry the altered genes to his liver.

The first attempt for germline gene transfer was started in 2001. As a result of ooplasmic transfer 30 children were borne. Ooplasmic transfer which is also known as ooplasmic or cytoplasmic transplantation can be performed on women who cannot conceive due to defects in their ooplasm, *i.e.* their egg cytoplasm. Healthy ooplasm is inserted from donor eggs into the eggs with defective ooplasm. Moreover, a small amount of mitochondrial DNA is transferred into the mother's egg while inserting the healthy ooplasm from the donor eggs into the mother's defective eggs. Unlike DNA material contained in nucleus, mitochondrial DNA is contained in other cell organelles such as the mitochondria. Since babies borne as a result of this ooplasmic transfer have DNA from their mother, father, and the ooplasm donor (in the form of mitochondrial DNA), they are supposed to have three genetic parents. The DNA of these children, and of their offspring, will contain genetic material from three parents, instead of two parents; therefore, it is considered as germline gene transfer.

In 2000, the first complete gene therapy cures were done. French doctors successfully used gene therapy to treat 10 children with X-linked 'severe combined immunodeficiency disease' (X-SCID. This unprecedented result differed from previous gene therapy treatments (like Ashanti DeSilva) in that it completely cured the children of the disease without requiring additional and/or supplemental medication. Unfortunately, after the treatment two of the 10 cured children developed leukemia.

B. GENETIC DISEASES

More than 4,000 genetic diseases are known so far among which some are very fatal. If the DNA information of a particular gene contains mistakes, the gene may not function properly. Sometimes the malfunction will not be serious but it will cause a severe genetic disease. Examples of some genetic diseases are cystic fibrosis, sickle cell anemia, and hemophilia. Haemophilia is caused by the malfunctioning of the gene that makes the factor that causes blood to clot. After study of human genetics, it has become clear that diseases such as diabetes, cancer, heart diseases, etc. also result due to faulty DNA. There are satisfactory therapies for some genetic diseases. Drugs, blood transfusions, changes in diet, or transplantation of body organs can often help to compensate for the incorrect information from the malfunctioning gene. For example, clotting factor can be administered to patients with haemophilia. Many human diseases are caused by defective genes. All of these diseases are caused by a defect at a single gene locus. The inheritance is recessive, so both the maternal and paternal copies of the gene must be defective. Some common human diseases are given in Table 9.1.

Table 9.1. Some common human diseases.

Disease	*Genetic defect*
Haemophilia A	Absence of clotting factor VIII
Cystic fibrosis	Defective chloride channel protein
Muscular dystrophy	Defective muscle protein (dystrophin)
Sickle-cell disease	Defective beta globin
Haemophilia B	Absence of clotting factor IX
Severe combined immunodeficiency (SCID)	Any one of several genes fail to make a protein essential for T and B cell function

Some of important genetic diseases are given in Table 9.2 and a few are discusses in this section.

Table 9.2. Some common genetic diseases and their origin due to mutation.

Disorder	*Mutation*	*Chromosome*
Alpha-1-antitrypsin deficiency		14q32.1
Alpha-galactosidae A deficiency	P	Xq22.1
Beta thalassemia		
Canavan disease		17p
Ceramide trihexosidase deficiency (Fabry disease)		X
Cerebelloretinal angiomatosis, familial	P	3 (p26-p25)
Color blindness	P	X
Conotruncal anomaly face syndrome	D	22q
Cystic fibrosis	P	7q
Down syndromé (47,XY,+21)	C	21
Down syndrome (47,XX,+21)	C	21
Duchenne muscular dystrophy	D	Xp
Galactosemia	P	9 (p13)
Genetic hypercalciuria		Xp11.22
Haemophilia	P	X
Klinefelter syndrome	C	X
Klinefelter syndrome (47,XXY)	C	X
Myotonic dystrophy	T	19
Phenylketonuria	P	12q
Sickle-cell disease	P	11p
Triple X syndrome (47,XXX)	C	X
Turner syndrome (45,X)	C	X

The term cystic fibrosis (CF) (= mucoviscoidosis or mucoviscidosis) refers to the characteristic 'fibrosis' (tissue scarring) of the biliary tract (*cystic* = biliary vesicle and/or the bladder). It was first recognised in the 1930s. CF is a hereditary disease which affects the exocrine (mucus) glands of the lungs, liver, pancreas and intestines of the patients causing progressive disability due to multi-system failure. It is one of the most common life-shortening, childhood-onset inherited diseases. In the United States, 1 in 3,900 children is born with CF. It is most common among Europeans and Ashkenazi Jews; 1 in 22 people of European descent are carriers of one gene for CF, while Ireland has the highest rate of CF carriers in the world (1 in 19).

Lungs of the patients are frequently infected resulting in production of thick mucus, as well as a less competent immune system. Secretion of pancreatic enzymes is also stopped that results in poor growth, fatty diarrhea and deficiency in fat-soluble vitamins. Due to the conditions of congenital bilateral absence of vas deferens, males can be infertile. Individuals with cystic fibrosis can be diagnosed prior to birth by genetic testing. Newborn screening tests are common and effective (although false positives may occur and children need to be brought in for a sweat test to distinguish disease vs carrier status). Diagnosis of CF may be confirmed with the presence of high levels of salt during a sweat test.

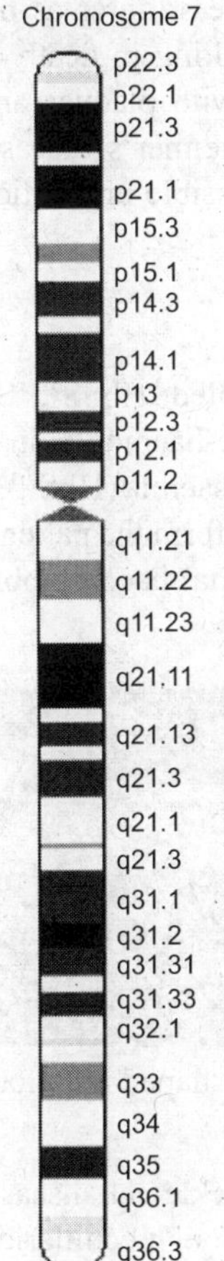

Fig. 9.1. Location of the CFTR gene on chromosome 7

There is no cure for CF, and most individuals with cystic fibrosis die young due to lung failure in their 20s and 30s. However, life expectancy of a person with CF is increasing to ages as high as 40 or 50 with the continuous introduction of many new treatments. CF is caused by a mutation in a gene called the 'cystic fibrosis transmembrane conductance regulator' (CFTR) present on chromosome 7 (Fig. 9.1). The product of this gene is a chloride ion channel important in creating sweat, digestive juices, and mucus. Although most people without CF have alleles of the CFTR gene, only one is needed to prevent cystic fibrosis. CF develops when neither allele can produce a functional CFTR protein. Therefore, CF is considered an autosomal recessive disease.

Women who are pregnant or planning for pregnancy can themselves be tested for CFTR gene mutations to determine the likelihood that their child will be born with cystic fibrosis. Testing is performed first on one or both parents. If the risk of CF is found to be high, testing on the foetus can then be performed. If the functional gene is missing, the doctors advise for abortion of CF-carrier baby. In addition, female infertility may be overcome by assisted reproduction technology, particularly embryo transfer techniques. Male infertility may be overcome with intracytoplasmic sperm injection (Fig. 9.2). Third party reproduction is also a possibility for women with CF.

Gene therapy holds promise as a potential avenue to cure cystic fibrosis. Gene therapy attempts to place a normal copy of the CFTR gene into affected cells. Studies have shown that to prevent the lung manifestations of cystic fibrosis, only 5–10% the normal amount of CFTR gene expression is needed. Many approaches have been theorised and several clinical trials have been initiated.3

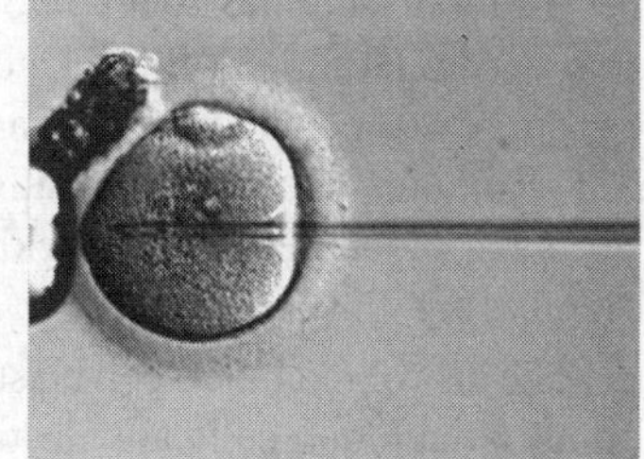

Fig. 9.2. Injection of intra-cytoplasmic sperm to provide fertility for men with cystic fibrosis.

1. Alzheimer Disease

In 2007, R. Brookmeyer and coworkers from Johns Hopkins University (U.S.A.) have forecasted the global burden of Alzheimer's disease. According to them the Alzheimer disease is a degenerative and terminal disease of which there is no cure. Commonly, it afflicts over 65 year old individuals, but there are less prevalent early-onset forms also. Approximately 26.6 million worldwide people were afflicted by 2006.

In 1901, a German psychiatrist, Alois Alzheimer identified the first case of disease in a fifty-year-old woman which is known as Alzheimer's disease. In the following five years, eleven similar cases were reported, some of them already using the term Alzheimer's disease. Diagnosis of Alzheimer's disease was reserved for individuals between the ages of 45 and 65 who developed symptoms of dementia. The term *senile dementia of the Alzheimer type* (SDAT) was used for a time to describe the condition in those over 65, with classical Alzheimer's disease being used for those younger. The term Alzheimer's disease was formally adopted in medical science to describe individuals of all ages with a characteristic common symptom pattern.

The symptom of the disease becomes apparent with loss of memory, *i.e.* difficulty to remember recently learned facts. Alzheimer's develops for an indeterminate period before becoming fully apparent. As the disease advances, progressive symptoms include: confusion, anger, mood swings, language breakdown, long-term memory loss, and the general withdrawal of the sufferer as his or her senses decline. Gradually, minor and major bodily functions are lost resulting in death of the sufferers. Research for the cause and progression of disease is associated with plaques and tangles in the brain. Many preventive measures have been suggested for Alzheimer's disease. Mental stimulation, exercise and a balanced diet are often recommended as a possible prevention and management of the disease.

2. Sickle-cell Anaemia

Sickle-cell anaemia is a specific form of sickle-cell disease which is also called 'HbSS', 'SS disease', or 'haemoglobin S'. Other, rarer forms of sickle-cell diseases are : sickle-haemoglobin C disease (HbSC), sickle beta-plus-thalassaemia (HbS/β^+) and sickle beta-zero-thalassaemia (HbS/β^o). These other forms of sickle-cell disease are compound heterozygous states. In this form the patients has only one copy of the mutation causing HbS and one copy of another abnormal haemoglobin allele. In 1922, Vernon Mason named the disease as 'sickle-cell anaemia'.

Sickle-cell anaemia is caused by a point mutation in the β-globin chain of haemoglobin, resulting in replacement of an amino acid 'glutamic acid' with the less polar amino acid 'valine' at the sixth position. The β-globin gene is found on the short arm of chromosome 11. The association of two wild-type α-globin subunits with two mutant β-globin subunits forms haemoglobin S (HbS). Under low oxygen conditions, the absence of a polar amino acid at position six of the β-globin chain promotes the polymerisation of haemoglobin which distorts red blood cells into a sickle shape and decreases their elasticity (Fig. 9.3).

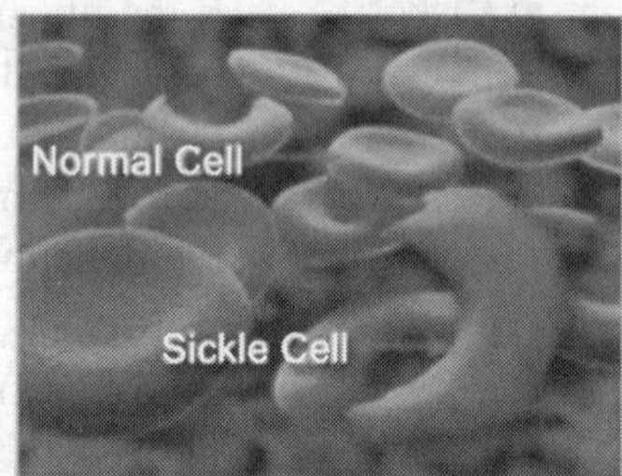

Fig. 9.3. Sickle-shaped red blood cells.

The loss of red blood cell elasticity is the main reason of pathophysiology of sickle-cell disease. Normally red blood cells are elastic which permits the cells to deform to pass through capillaries. In sickle-cell disease, low oxygen tension promotes sickling of RBCs and their repeated sickling damages the cell membrane and decreases elasticity of RBCs. After restoration of normal oxygen tension these cells fail to return to normal shape. Consequently, these rigid blood cells are unable to deform as they pass through narrow capillaries leading to vessel occlusion and ischaemia.

Different approaches are being made to prevent sickling episodes as well as the complications of sickle-cell disease. The other ways to modify haemoglobin switching are being investigated including the use of phytochemicals such as 'nicosan'. Gene therapy is under investigation.

3. Thalassaemia

Thalassaemia (Greek, *thalassa* means 'sea', *haema* means 'blood') is an inherited autosomal recessive disease of the blood. Due to genetic defect the rate of synthesis of one of the globin chains making haemoglobin is reduced which causes formation of abnormal haemoglobin molecules resulting in anaemia which is the characteristic symptom of the thalassaemias. Thalassaemia produces a deficiency of α or β globin, unlike sickle cell disease which produces a specific mutant form of β globin. It is a quantitative problem of too few globins synthesised, whereas sickle-cell anaemia (a haemoglobinopathy) is a problem of synthesis of an incorrectly functioning globin. Thalassaemias usually result in under production of normal globin proteins through mutations in regulatory genes. Haemoglobinopathies imply structural abnormalities in the globin proteins themselves.

Thalassaemias are prevalent in populations that evolved in humid climates where malaria is endemic but it affects all races. This disease is associated with Arabs, people of Mediterranean origin, and Asians. In the world, Maldives has the highest incidence of thalassaemia with a carrier rate of 18% of the population. The estimated prevalence is 16% in people from Cyprus, 1% in Thailand, and 3-8% in populations from Bangladesh, China, India, Malaysia and Pakistan.

Thalassaemias are classified according to affected chain of the haemoglobin molecules. Production of α globin chain is affected in α-thalassaemias, whereas production of β globin chain is affected in β-thalassaemia. A single gene on chromosome 11 encodes β globin chains, while the two closely linked genes on chromosome 16 encode α globin chains. Hence, a normal person having two copies of each chromosome consists of two loci encoding β chain, and four loci encoding α chain. Deletion of one of the α loci has a high prevalence in people of African-American or Asian descent resulting in the development of α thalassaemia. β thalassaemias are common in African-Americans.

Types of Thalassemias

There are three types of thalassemias, α, β and δ thalassemias as given below:

(*i*) Alpha (α) thalassaemias: The α-thalassemias involve the genes HBA1 and HBA2, inherited in a Mendelian recessive fashion. It is also connected to the deletion of the 16p chromosome. The α-thalassaemias result in decreased alpha-globin production, therefore, fewer alpha-globin chains are produced, resulting in an excess of β chains in adults and excess γ chains in newborns. The excess β-chains form unstable tetramers (called Hemoglobin H or HbH of 4 beta chains) which have abnormal oxygen dissociation curves

(*ii*) Beta (β) thalassaemias: Beta thalassaemias are due to mutations in the HBB gene on chromosome 11 also inherited in an autosomal-recessive fashion. The severity of the disease depends on the nature of the mutation. Mutations are characterised as (β^o) if they prevent any formation of β chains; they are characterised as (β^+) if they allow some β chain formation to occur. In either case there is a relative excess of α chains, but these do not form tetramers, rather they bind to the membrane of red blood cell producing membrane damage, and they form toxic aggregates at high concentrations.

(*iii*) Delta (δ) thalassaemia: As well as alpha and beta chains being present in haemoglobin about 3% of adult haemoglobin is made of alpha and delta chains. Just as with beta thalassaemia, mutations can occur that affect the ability of this gene to produce the delta chains. A mutation that prevents formation of any delta chains is termed as delta^0 mutation, whereas one that decreases but does not eliminate production of delta chain is termed as delta$^+$ mutation. When one inherits two delta^0 mutations, no haemoglobin A2 (alpha2, delta 2) can be formed.

About 60-80 million people in the world carry the beta thalassaemia trait. India, Pakistan and Iran are seeing a large increase of thalassaemia patients due to lack of genetic counselling and screening. Thalassaemia may become a very serious problem in the next 50 years that will burden the world's blood bank supplies and the health system. Therefore, genetic counselling is recommended as with other genetically acquired disorders.

4. Duchenne Muscular Dystrophy (DMD)

Duchenne muscular dystrophy (DMD) is a severe recessive X-linked muscular dystrophy. DMD is named after a French neurologist, Guillaume Benjamin Amand Duchenne who first described the disease during 1860s. It is characterised by rapid progession of muscle degeneration leading to loss in ambulation, paralysis, and death. Duchenne muscular dystrophy (DMD) is a X-linked recessive genetic disorder carried on the X-chromosome that results from a mutation in the dystrophin gene. Men are much more likely to get this disease. This affliction affects one in 3,500 males, making it the most prevalent of muscular dystrophies. Generally, females can be carriers but males are only afflicted. The disorder is caused by a mutation in the gene DMD present on human X-chromosome. The DMD gene codes the 'dystrophin' protein which is an important structural component within muscle tissue. Dystrophin provides structural stability to the 'dydtroglycan complex' (DGC) located on the cell membrane.

Structure of a muscle is given in Fig. 9.4 A. In the Duchenne muscular dystrophy, the attachment of muscle fibers to their surrounding endomysium (extracellular matrix) becomes weakened due to mutations in the dystrophin. Dystrophin is part of a large group of proteins that are involved in making contacts between muscle cells and their surrounding extracellular matrix. These protein complexes help muscle cells to withstand the shearing forces of contraction. Thus, the absence of dystrophin leads to increased death and destruction. In turn, scar tissue (a mix of collagen and blood vessels) replaces the muscle fibers, and it can gradually contract leading to increased muscle rigidity. The breakdown of muscle fibers and the infiltration of fibrous scar tissue lead to most of the symptoms observed in DMD.

Usually symptoms appear in males of 6 years old but may be visible in early infancy. First of all weakness of proximal muscle of the legs and pelvis associated with a loss of muscle mass is observed (Fig. 9.4 B). This weakness spreads to the arms, neck, and other areas resulting in pseudohypertrophy (enlargement of calf muscles), low endurance, and difficulties in standing unaided or inability to ascend staircases. By the age 10, a walking aide required in but most patients become wheelchair-dependent by age 12. The later symptom is abnormal bone development that leads to skeletal deformities including curvature of the spine. The average life expectancy of DMD patients varies from early teenage to age of mid 30s.

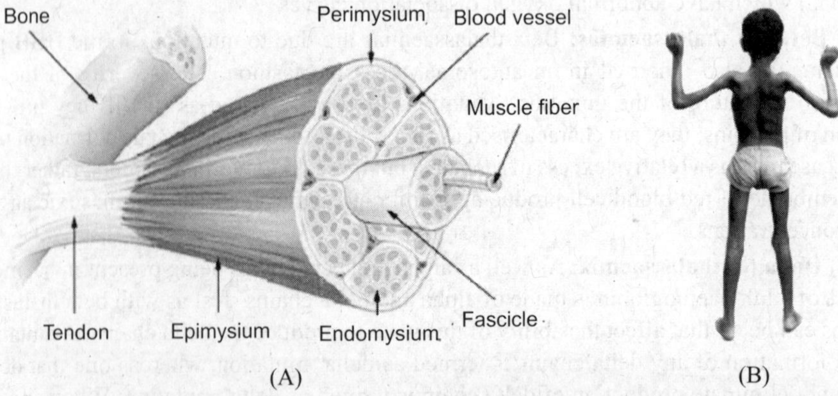

Fig. 9.4. A- structure of a muscle; B- a male child suffering from muscular dystrophy.

The CPK (creatine phosphokinase) test is done to measures the amount of CPK in the blood. In DMD patients CPK leaks out of the muscle cell into the bloodstream; therefore, a high level (50 to 100 times normal) of CPK confirms the damage of muscles. In most cases, diagnosis of DMD is done by DNA testing. The muscle-specific dystrophin gene is composed of 79 exons; DNA testing and analysis identify the specific type of mutation in the affected exon(s).

Prenatal tests are carried out during pregnancy to find out whether the foetus (unborn child) is affected. Different types of prenatal tests are carried out after 10-12 weeks of pregnancy by 'chorion villus sampling' (CVS), and amniocentesis is done at about 14-16 weeks. Placental biopsy and foetal blood sampling are done at about 18 weeks of pregnancy. Women and/or couples decide(s) which test is to be adopted and discuss this with their genetic counsellor.

5. Severe Combined Immunodeficiency (SCID)

Severe combined immunodeficiency (SCID) is such a disease in which the patient has neither the cell-mediated immune responses nor is able to make antibodies. This disease is found in young children because the absence of an immune system results in microbial infections leading to death of patients. About 25% of the cases of SCID are due to homozygous conditions in child that results in defective genes encoding the enzyme 'adenosine deaminase' (ADA). The normal catabolism of purines is deficient in patients and this is particularly toxic for T cells and B cells. The suffering children must live strictly in germ-free environment. All food, water, and air must be sterilised. If a child is getting continuous source of ADA, and T and B cells through a transplant of bone marrow from a normal histocompatible donor, he can survive up to 12 years or more. Since the child has no immune system, he cannot reject the transplant unless the donor was an identical twin. Currently the enzyme ADA is extracted from cows. The injections of ADA are regularly given to the patients. The enzyme ADA is conjugated with polyethylene glycol (ADA-PEG) to delay its breakdown in the blood. The ADA-PEG injections keep the SCID patients healthy; but it must be repeated at frequent intervals as the insulin injections are given to the diabetics. At last report, some of the circulating T cells in these children carried the ADA gene. Since the children continue to receive ADA-PEG, it is unclear how much gene therapy has contributed to their good health.

In 2002, a team of Italian and Israeli doctors treated the two young SCID patients with their own blood stem cells which were transformed *in vitro* with the ADA gene. After a year, both children had fully-functioning immune systems (T, B, and NK cells) and were able to live normal lives without any need for treatment with ADA-PEG or immunoglobulin (Ig). For proper functioning of ADA genes the permanent cure is gene therapy.

C. BACKGROUND OF GENE THERAPY

In the 1980s, advances in molecular biology had resulted in human genes to be sequenced and cloned. Scientists looking for a method of easily producing proteins such as insulin (the protein deficient in diabetes mellitus type I), have investigated the method of introducing human genes to bacterial DNA. The modified bacteria then produced the corresponding protein which can be harvested, purified and injected in people who cannot produce the insulin naturally. Although this simplified explanation of a gene therapy procedure is a boon to sufferers, it is like the first milestone of a long journey.

The biology of human gene therapy is very complex, and there are many techniques that still need to be developed and diseases that need to be understood more fully before using gene therapy. The public debate surrounding the possible use of genetically engineered material in human subjects has been equally complex. Major participants in the debate have come from the fields of biology, government, law, medicine, philosophy, politics, and religion, each bringing different views to the discussion.

Scientists try to introduce genes directly into human cells, focusing on diseases caused by single-gene defects such as cystic fibrosis, haemophilia, muscular dystrophy and sickle cell anaemia. However, it is more difficult than modifying simple bacteria because of the problems involved in carrying large sections of DNA and delivering it into the right site on the genome.

D. TYPES OF GENE THERAPY

There are two major types of cells in the human body, *somatic* (non-reproductive) cells and *germ-line* (reproductive) cells. Most cells in the body are somatic cells that provide all the body structures and perform all the functions except for passing genetic information to the next generation. Germ cells include eggs in women and sperm in men. The genes in sperm and egg cells store information that will go to the next generation to one's children. The genetic information in somatic cells is not passed on to the next generation.

However, it is possible to transform either somatic cells or cells of the germ-line (such as sperm cells, ova, and their stem cell precursors). All gene therapy so far has been carried out in somatic cells, whereas germ-line engineering in humans remains only a highly controversial prospect. For the introduced gene to be transmitted normally to offspring, it needs not only to be inserted into the cell but also to be incorporated into the chromosomes by genetic recombination.

1. Germ-line Gene Therapy

Technically the germ-line gene therapy is more difficult than the somatic gene therapy. Really, some people are of the opinion that it is impossible. Direct manipulation of the reproductive cells or the zygote or embryo is carried out in germ-line gene therapy. The affected cells are taken out and manipulated *in vitro* (outside of the body) so that the unwanted/non-functional gene will be removed and the desired gene will be introduced into the DNA. The healthy and manipulated zygote or embryo is transferred to the uterus of a woman. The developing embryo would contain the altered genetic make-up in all of its cells including its own reproductive cells. In this way the changes will be passed on to subsequent generations.

Alteration in germ-line is more complicated than somatic cell gene therapy. Somatic cell gene therapy can be accomplished theoretically through gene insertion or gene replacement. Germ-line genetic alteration could be performed only by using gene replacement, *i.e.* taking out non-functional gene and introducing a functional gene, which is not possible in human beings at present. If the affected gene is not removed, the disease will pass on to subsequent generations.

Genetic alteration in Germ-line has been performed in animals. In contrast, the objective of this technological advancement has not been the production of similar results in humans. Changes in animal germ-line have been done: (*i*) to produce 'transgenic' breeds that provide an animal model of a human disease, or (*ii*) to produce animals that make commercially valuable proteins. These objectives cannot be applied to humans. However, in humans the failure rate of insertion and transmission is high.

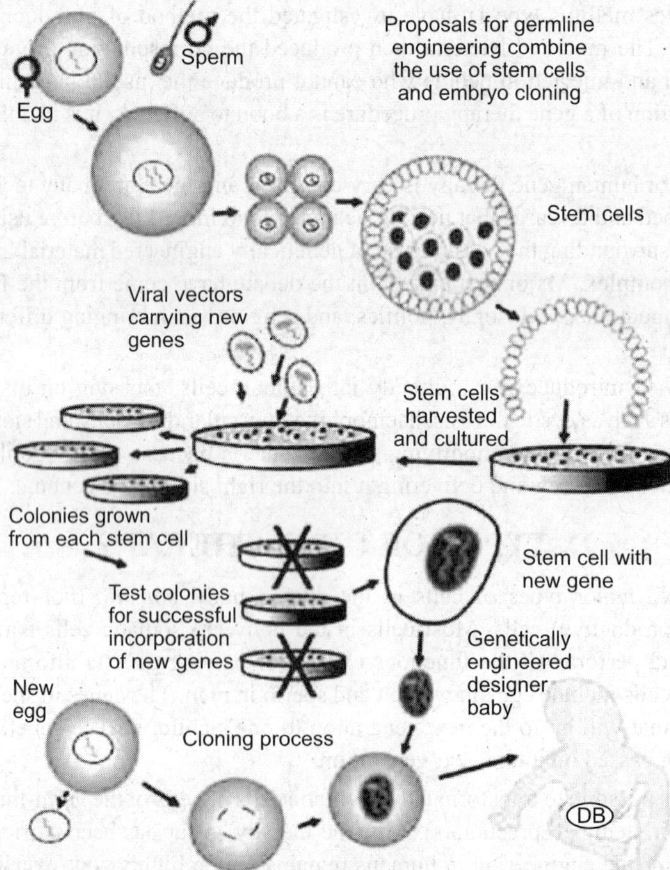

Fig. 9.5. Outline of a proposal for germ-line engineering that combines the use of stem cells and embryo cloning.

At present, the procedure for genetic change in human beings affecting the germ line is totally untested. When a genetic abnormality is identified, the genetic alteration of germ cells, zygotes and embryos can be done. Elimination of affected cells, zygotes or embryos is less risky. All genetic alterations of germ-line must be prohibited to prevent the genetic enhancement, i.*e.* enhancing or improving an already genetically healthy organism. This would be a non-therapeutic use of genetic alteration.

Outline of a proposal for germ-line engineering that combines the use of stem cells and embryo cloning is given in Fig. 9.5. The steps for germ-line engineering involves : (*i*) *in vitro* fertilization of egg with males sperm and incubation of zygote for cell division to form inner mass of cells, (*ii*) culture of inner mass of cells, (*iii*) harvesting and culture of stem cells in synthetic medium, (*iv*) inoculation of cultured cells with recombinant viral vectors containing gene of interest (each stem cell forms colonies), (*v*) screening of stem cells for the presence of new gene of interest, (*vi*) incubation of genetically engineered embryos and its introduction into mother's uterus, and (*vii*) birth of a designer baby.

2. Somatic Gene Therapy

Somatic gene therapy refers to the manipulation of any cells in the body except the reproductive cells. Foreign gene of known function is not passed on to the subsequent generation through somatic gene therapy; therefore, it will not eradicate the disease. In this therapy, a healthy gene is inserted somewhere in the chromosomes of an affected individual. This is not yet possible to insert it within a vector for gene expression. By this way the recessive diseases may be treated. A genetically engineered viral vector containing a gene of choice is introduced into those cells of the affected individual where the disease itself lies. A typical example of somatic gene therapy is cystic fibrosis. The disease symptoms would be relieved through expression of the healthy gene in the respiratory tract.

For diseases it is necessary to search the faulty gene lying in the chromosome and replace it with a healthy one. Only gene insertion through vectors is possible in humans. Somatic cell gene therapy is not expected to be an effective means of therapy because it is still at an early experimental stage. There are several technical problems such as method of insertion, accessibility of the tissue, and regulation of the gene product, etc. The first clinical trials involving human gene therapy began in 1990. Although it has moved rapidly since that time, it is still considered experimental. Although somatic cell gene therapy is still at an early experimental stage, it is not expected to be an effective means of therapy.

As described earlier, on September 14, 1990, the first human somatic gene therapy experiment was conducted on four year-old Ashanti Desilva who suffered from ADA deficiency. White blood cells with the correct genetic composition were infused into Ashanti to correct her immunodeficiency. Though the therapy did not give a complete cure, it helped to correct the deficiency, so that Ashanti may takes small doses of PEG-ADA drug at weekly intervals.

Somatic gene therapy can be grouped into two categories: *ex vivo* and *in vivo* ones. In *ex vivo* somatic gene therapy cells are modified outside the body and then transplanted back again, whereas *in vivo* somatic gene therapy involves changes in genes in cells still present in the body.

Designer Baby

For the first time in Australia approval has been granted to use '*in vitro* fertilisation' (IVF) technology in combination with genetic screening and tissue matching to create a 'designer baby' for a sick sibling. A Melbourne couple has been granted permission to use IVF technology to conceive a genetically screened designer baby in an attempt to cure their terminally ill child.

The two applicants were first given interim approval for the procedure by the Infertility Treatment Authority but it was dependent upon approval from a hospital ethics committee. Approval for the procedure was granted by the Epworth Hospital's ethics committee.

The couple's child was having a terminal illness. Its only chance of survival was to receive a transfusion of umbilical cord blood from a perfectly matched sibling. The technique involves: (*i*) use of fertility drugs for super ovulation to the sufferer woman (*a*), (*ii*) production of multiple eggs, *i.e.* super ovulation (*b*), (*iii*) *in vitro* fertilization of eggs to produce embryos (*c*), (*iv*) screening of embryos to ensure that they are disease-free, and conducting further testing to ensure that the embryo and child's tissue are compatible (*d*), (*iv*) implantation of embryo in the mother's uterus if the embryo meets both tests (*e*), (*vi*) birth of genetically health baby (*f*), and (*vii*) transfusion of umbilical cord blood to sick child (*g*).

One case concerned a four-year-old child with a rare blood disorder called Fanconi's anaemia, while the other child was believed to suffer from a fatal blood condition known as sickle cell anaemia. It is believed the Epworth sought expert opinion from theologians, ethicists and reproductive scientists.

Concerns have been raised that pre-implantation genetic diagnosis might be used in the future to select traits that parents believes are socially desirable. Genetic testing is available in-utero and people have to go at conceiving a baby naturally so that the life of another child could be saved. The novelty in this technique is that it has a combination of IVF technology with the tissue and genetic testing technology.

Cloning of Humans and its Future

Cloning is the production of an organism that has the exactly same genetic make-up of another living or dead organism. In nature this happens all the time. The most common example is the identical twins that arise as a result of an embryo splitting into two haves soon after fertilisation. The other natural example of cloning is asexual reproduction. Many single-celled organisms reproduce by splitting in two. But the current cloning process involves removal of the nucleus from an egg cell and its replacement with the nucleus from a cell of an animal to be cloned. The egg is then forced to get fertilised and proceed to develop an embryo. This process is called 'somatic cell nuclear transfer'. But it does not produce an identical clone, because some genetic information is contained outside of the nucleus, *i.e.* in mitochondria. In somatic cell nuclear transfer, the enucleated egg cell still contains its own mitochondria.

There is a very low success rate in cloning techniques; only 1-2% of embryos survive to develop into the live offspring. But a large proportion of these fail to survive to adulthood. In spite of low success rate, the scientists are hopeful about the potential of cloning in the area of medical sciences and pharma industry. A similar use for these genetic techniques can be done by the pharmaceutical industry. Animals like goats and cattle can be genetically altered and cloned. These would contain genes coding for a desired protein that can be made to be expressed in the animals' milk and then simply extracted from it. But such arguments cannot be made for human cloning. Only a supply of spare organs can be done by human cloning.

E. METHOD OF GENE THERAPY

Normally in gene therapy, an abnormal disease-causing gene is inserted into the genome to replace it with a normal gene. A vector is used as a carrier molecule to deliver the therapeutic gene to the target cells of patients. Mostly, a genetically altered virus is used as a vector that carries the normal human DNA. Viruses encapsulate and deliver their genes to human cells by causing in a pathogenic manner. The disease-causing genes of virus genome are removed and therapeutic genes are inserted into it. The target cells like the liver or lung cells of the patients are allowed to get infected by the vector. Then the vector delivers its genetic material containing the therapeutic human gene into the target cell. The therapeutic gene expresses a functional protein product that restores the target cell to a normal state.

I. VIRAL VECTOR-MEDIATED GENE THERAPY

The viruses infect their suitable hosts and introduce their genetic material (DNA or RNA) into the host cell. This genetic material contains basic 'instructions' for producing more copies of the viral particles by utilising host's normal production machinery. During the course of time, more and more host cells get infected. There are some viruses (*e.g.* retroviruses including HIV)) that physically insert their genes into the host's genome. Such viruses can be used as vehicles to carry functional genes into a human cell. First of all, the genes are removed from the pathogenic viruses. Then the host gene that causes the desired effect (for instance, insulin production in the case of diabetics) is replaced with the functional gene inserted into the virus vector. This procedure must be done in such a way that the genes which allow the virus to insert its genome into its host's genome are left intact. This requires significant research and understanding of the virus genes in order to know the function of each gene.

All this is clearly an oversimplification, and numerous problems exist that prevent gene therapy using viral vectors, such as trouble preventing undesired effects, ensuring that the virus will infect the correct target cell in the body, and the inserted gene doesn't disrupt any vital genes already in the genome. There are several viral vector systems which are used for the delivery of nucleic acid. They have many advantages and limitations as given in Table 9.3. However, this basic mode of gene introduction currently shows much promise, and doctors and scientists are working hard to fix any potential problems.

Table 9.3. Advantages and disadvantages of viral vectors currently used in gene therapy (Boeckle and Wagner, 2006. *AAPS J.* 8: 83).

Vector	*Advantages*	*Disadvantages*
Retrovirus	*Ex vivo* stable transduction of blood cells	Low efficiency *in vivo*; risk of insertional mutagenesis
Adenovirus	High short-term expression *in vivo* (in liver)	Immunogenicity and inflammatory responses
Adeno-associated virus	Long-term expression *in vivo*	Small genome
Viral Vectors in general	High efficiency on particle basis	No repeated administration because of immunogenicity
Naked DNA	Simple (e.g. for vaccination)	Low efficiency
Physically enhanced delivery of DNA (electroporation; gene gun, hydrodynamic delivery)	High expression *in vivo*	Limited localized area, device required
Complex-based gene transfer (lipoplexes, polyplexes)	High expression *ex vivo*; expression *in vivo* (localised and systemic)	Short-term expression; toxicity of cationic carriers
Non-viral vectors in general	High flexibility; lower immunogenicity	Low efficiency on particle basis

Viruses are the good candidates for the development of effective gene delivery because the viruses naturally deliver their genes into host cells. But it is not necessary that of the transduced host cell will survival and expressed genes shall be maintained. Replication-defective viruses are constructed by partly replacing the viral genes with therapeutic genes. Such viruses represent the first generation of viral vectors applied in gene therapy. Vectors and delivery system in gene therapy has been described in detail by Gardlik *et al.* (2005).

Most commonly retroviral vectors and adenoviral vectors have been used in clinical trials (www.wiley.co.uk/genetherapy/clinical). The other viruses developed as viral vectors are adeno-associated virus (AAV), herpes virus, pox virus, and lentivirus. In general viruses are strongly taken up by host cells and efficiency work intracellularly. Therefore, small numbers of viral particles are sufficient to transduce the host cells. Mostly replication-defective viral vectors replicate only in special producer cells but not in the host cells. However, viral vectors can be used to multiply the therapeutic effect. In tumour therapy attenuated replicative viruses were used that can replicate only within tumour cells. The virus replication process directly kills the infected tumour cells. Such viruses are called oncolytic viruses. Consequently the copy number of therapeutic gene carried by the viral vector is amplified.

Viral Vector used in Gene Therapy

There are some viruses that are used as vectors in gene therapy as given below:

- ***Retroviruses :*** A class of viruses that can create double-stranded DNA copies of their RNA genomes. These copies of its genome can be integrated into the chromosomes of host cells. Human immunodeficiency virus (HIV) is a retrovirus.
- ***Adenoviruses :*** A class of viruses with double-stranded DNA genomes that cause respiratory, intestinal, and eye infections in humans. The virus that causes the common cold is an adenovirus.
- ***Adeno-associated viruses :*** A class of small, single-stranded DNA viruses that can insert their genetic material at a specific site on chromosome 19.
- ***Herpes simplex viruses :*** A class of double-stranded DNA viruses that infect a particular cell type such as neurons. Herpes simplex virus type 1 is a common human pathogen that causes cold sores.

1. Retroviruses

The retroviruses consist of a capsid that bears an enzyme reverse transcriptase and RNA as genetic material, in addition to spikes that help in identification and attachment to host cell surface (Fig. 9.6A). But the genetic material of their hosts is the DNA. Generally, after infecting the host cell, a retrovirus introduces its RNA into the cell along with some enzymes. Using host machinery, the retrovirus produces a copy of DNA from its RNA molecule with the help of an enzyme reverse transcriptase and deoxynucleotide triphosphate precursors. The process of producing a DNA copy from an RNA molecule is called 'reverse transcription'. Thereafter, DNA copy is left free in the nucleus of the host cell; then it is inserted into the large DNA molecules in the host's chromosomes. The other enzyme integrase which is carried in the virus carries out this process.

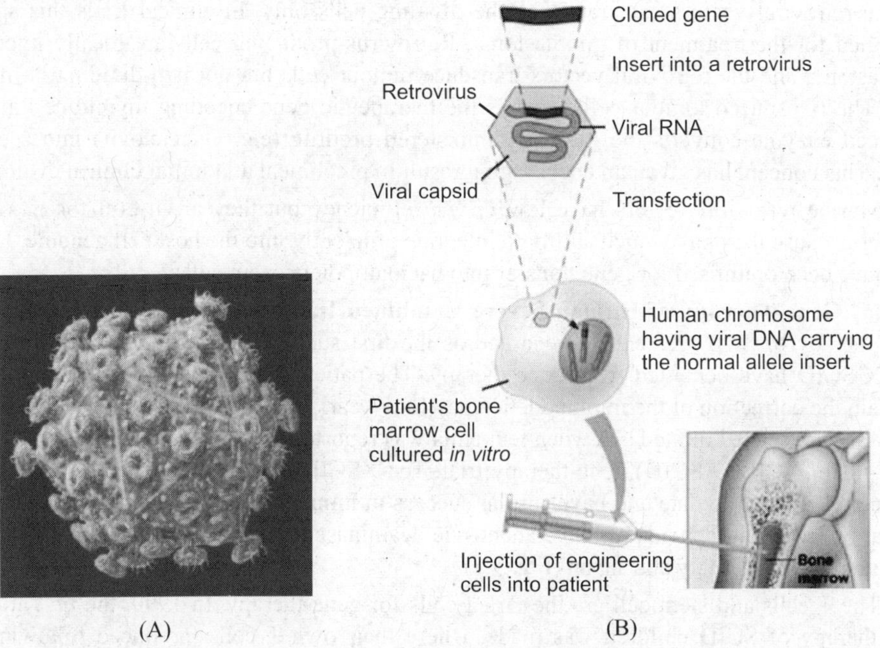

Fig. 9.6. (A)- an electron photomicrograph a retrovirus showing spikes; (B)- steps for introduction of a normal RNA version of functional gene into the bone marrow cells of a patient through a retrovirus.

The normal allele, absent from the cell of a patient is identified and RNA version of allele is produced (Fig. 9.6 A). Then RNA version of normal allele is inserted into the retrovirus to be used as vector for delivery into host cells. Bone marrow cells of the patients are removed and cultured artificially. The genetically modified retroviral particles are allowed to infect the cultured bone marrow cells of the patient. The retrovirus attaches to the surface of host cell and delivers its genetic material that consists of cloned gene of the patient. Thereafter by using reverse transcriptase the RNA version of virus genetic material is converted into cDNA which is integrated into host's chromosome. After incorporation of genetic material of the virus, it becomes a part of genetic material of the host cell. The host cell is now modified to contain a new gene. If the host cell divides later, its descendants will contain the new genes. The genetically engineered bone marrow cells are injected into bone marrow of the patient where the new gene construct of normal allele is delivered that become a part of the patient's body.

One of the problems of gene therapy using retroviruses is that the integrase enzyme can insert the genetic material of the virus in any arbitrary position in the genome of the host. If genetic material happens to be inserted in the middle of one of the original genes of the host cell, this gene will be disrupted (insertional mutagenesis). If the gene regulates cell division, uncontrolled cell division (*i.e.* cancer) can occur. This problem has recently begun to be addressed by utilising zinc finger nucleases or by including certain sequences such as the β-globin locus control region to direct the site of integration to specific chromosomal sites.

Advantages of Using Retroviruses in Gene Therapy

The retroviruses have several advantages for introducing genes into human cells such as : (*i*) their envelope protein enables the virus to infect human cells, (*ii*) RNA copies of the human ADA gene can be incorporated into the retroviral genome using a packaging cell, etc

Gene transfer activity of viral vectors depends on the type of virus used for transduction. Murine retroviral vectors can transduce the dividing cells only. In clinical trials this specificity is applied for the treatment of glioblastoma. Retrovirus-producing cells are locally injected into glioblastoma and the retroviral vectors transduce tumour cells but not non-dividing normal brain cells. The transduced tumour cells express the therapeutic gene encoding thymidine kinase. The produced enzyme converts the locally administered prodrug (e.g. Ganciclovir) into a cytotoxic agent. This concept has given an encouraging result in preclinical and initial clinical evaluation.

Murine retroviral vectors have low *in vivo* efficiency but they are useful for *ex vivo* gene transfer because they have much ability to integrate efficiently into the host cell genome. However, they have been optimised for gene transfer into haematopoietic stem cells.

(*a*) Gene Therapy of Human Severe Combined Immunodeficiency (SCID) Syndrome: Gene therapy of SCID-X1 can be regarded as the first success story. Fifteen out of 16 patients with X-SCID have benefited from gene therapy. The patients treated first were paid attention to maintain the correction of the immunodeficiency for 7 years. In addition, gene therapy trials to treat SCID were restricted in the USA when leukemia was reported in three of eleven patients treated in the X-linked SCID (XSCID) gene therapy trial. Ten XSCID patients treated in England have not presented leukemia to date and have similar success in immune reconstitution. Gene therapy trials to treat SCID due to deficiency of the adenosine deaminase (ADA) enzyme continue with relative success in the USA, Italy and Japan (Fig. 9.7).

The T cells and stem cell are the target cells for gene therapy. In 1990, the first attempt for gene therapy of SCID children was made, where their own T cells, produced following ADA-PEG therapy, were used as the target cells. The T cells were placed in tissue culture, stimulated to proliferate (by treating them with the lymphokine, interleukin 2 (IL-2), infected with the retroviral vector, and returned to the child in a series of treatments. The children developed improved immune function but the injections had to be repeated because T cells live for only 6–12 months in the blood (but there may be a way to make genetically engineered T cells that may live indefinitely by getting them to express telomerase). The children also continued to receive ADA-PEG so the actual benefit of the gene therapy was not known.

Stem cells have also been used as target cells in gene therapy. In 1993, ADA gene therapy was attempted with blood (haematopoietic) stem cells drained from the umbilical cords of three newborn babies who were known to be homozygous for ADA deficiency. The blood stem cells produce (by mitosis) all the types of blood cells, including T and B lymphocytes and also produce more stem cells; this ensures an inexhaustible supply of the cells.

All the requirements for SCID therapy must be met by using a retrovirus as the gene vector. Requirement for SCID gene therapy is as given below :

- A desired gene must be identified and cloned (this has been done for the ADA gene).
- It must be inserted in cells that can take up long-term residence in the patient (this refers to removal of patient's own cells, treating them in tissue culture, and then returning them to the patient).

- It must be inserted in the DNA so that it will be expressed adequately (*i.e.* it must be transcribed and translated with sufficient efficiency).
- Packaging cells are treated so they express: (*i*) an RNA copy of the human ADA gene along with a packaging signal (P) needed for the assembly of fresh virus particles, and inverted repeats (R) at each end to aid insertion of the DNA copies into the DNA of the target cell, and (*ii*) an RNA copy of the retroviral *gag*, *pol*, and *env* genes but with no packaging signal (so these genes cannot be incorporated in fresh viral particles). After being treated with these two genomes, the packaging cell produces a crop of retroviruses with : (*i*) the envelope protein needed to infect the human target cells, (*ii*) an RNA copy of the human ADA gene complete with R sequences at each end, (*iii*) reverse transcriptase needed to make a DNA copy of the ADA gene that can be inserted into the DNA of the target cell, and (*iv*) none of the genes (*gag*, *pol*, *env*) that would enable the virus to replicate in its new host.

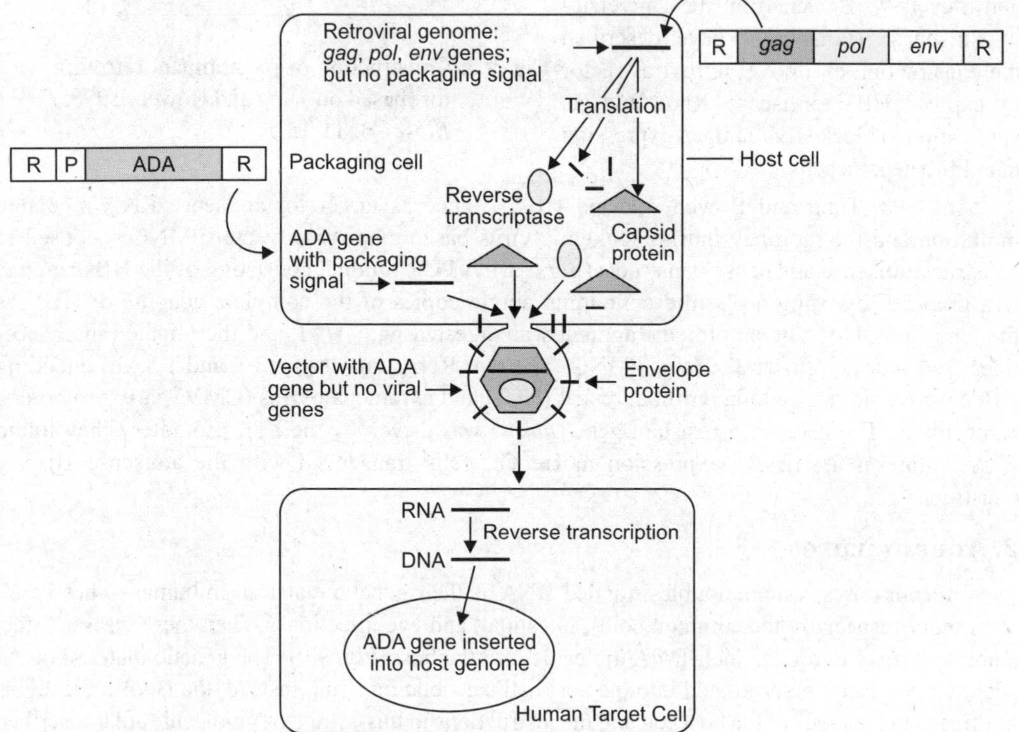

Fig. 9.7. Retroviral gene therapy: delivering adenosine deaminase (ADA) gene into patient through a retrovirus. R, inverted repeats; P, packaging signal.

(*b*) Inhibition of Hepatitis B Virus: HBV is an enveloped hepatotropic DNA virus that causes acute and chronic liver cell injury, inflammation and hepatocellular carcinoma (HCC). Throughout the world, more than 250 million people are chronically infected by this virus, and have a 200-fold greater risk of developing HCC than non-infected ones. Primary HBV infection is usually self-limited, with clearance of viral antigens and infection from liver and blood and the development of lasting immunity to re-infection. But, most of the mortality from HBV infection results in chronic rather than acute disease. Severe chronic hepatitis B leads to premature death from liver failure. There are four major open reading frames (ORFs) which are encoded by minus-strand DNA, ORF S (for HBsAg), ORF C (for HBcAg), ORF X (that encodes a transactivator protein which can transactivate transcription from the HBV core promoter), and ORF P (for viral polymerase).

Virus replication can be completely stopped without killing the host cells by using antisense nucleic acids for inhibition of HBV. Such approach is attractive for cells that have a regenerative ability, such as liver cells. The progeny derived from antiviral gene protected liver cells may replace the damaged cells caused by viral gene expression or host immune responses. Retroviral vectors offer a number of advantages as compared to other gene transfer vectors, such as stable integration of the transferred genes and a lack of retrovirus protein coding sequences in the transfer vector. In 1998, Frank Y.T. Tung and Steven W. Bowen from the University of Pittsburgh (Pittsburgh) have described the construction of a novel retroviral vector to express HBV antisense RNA and its application to block HBV antigen expression in culture hepatic cells.

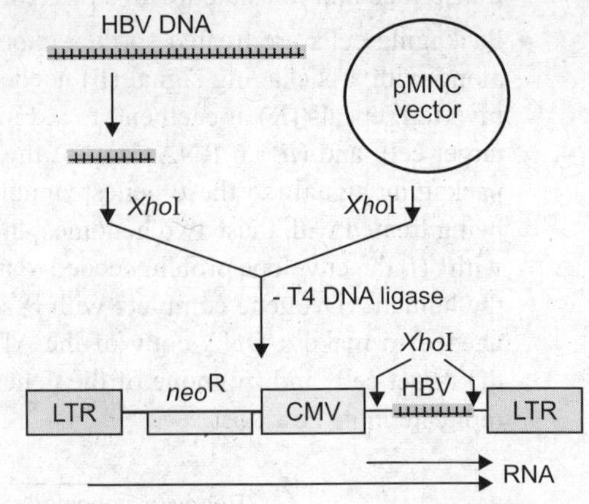

Fig. 9.8. Construction of recombinant retroviral vector (based on Tung and Bowen. 1998. *Front. Biosci.* 3: 11-15).

In 1998, Tung and Bowen selected HBsAg gene as target for antisense RNA mediated inhibition and the moloney murine leukemia virus-based expression vector pMNC was used to construct antisense and sense constructs (Fig. 9.8). A PCR product consisting of the HBsAg gene was prepared by using a plasmid containing single copies of the complete genome of HBV as the template DNA. The amplified fragment was digested with *Xho*I, and then ligated into *Xho*I-linearised murine retroviral vector pMNC. The two RNA transcripts (3.8 and 1.8 kb) encoding HBsAg were driven by long terminal repeat (LTR) and cytomegalovirus (CMV) early promoters, respectively. The neomycin-resistant gene (neo^R) was driven by the LTR promoter. They found 75% inhibition in HBsAg expression in HepG2 cells transduced with the antisense HBsAg construct.

2. Adenoviruses

Adenoviruses contain double-stranded DNA as their genetic material. In humans, they cause respiratory (especially the common cold), intestinal, and eye infections. When these viruses infect a host cell, they introduce their DNA molecule into the host (Fig. 9.9). The genetic material of the adenoviruses is not incorporated into the host cell's genetic material. Instead, the DNA molecule is left free in the nucleus of the host cell and the instructions in this extra DNA molecule are transcribed just like any other gene. Therefore, gene expression is only short term. When the cell is to undergo cell division, these extra genes are not replicated.

Adenoviral vectors efficiently transfer genes into both dividing and non-dividing cells. Both natural immunity against adenovirus (the virus that causes human respiratory diseases) and acute inflammatory and immunological responses limit clinical application to few areas such as localised cancer gene therapy. Systemic administration of a large amount of adenovirus (e.g. into the liver) may cause serious side effects and sometimes results in death. Therefore, to improve viral vector safety the number of remaining adenoviral genes from 'first generation' adenoviruses is reduced (only one early gene segment was removed). In high-capacity 'gutless' adenoviruses, all viral genes are replaced with functional gene of interest. But high-capacity 'gutless' adenoviruses also triggered host responses. Such host responses do not depend on the type of virus and viral gene expression; possibly these responses are related to the viral infection process and to proteins of the vector particle.

Therefore, the descendants of that cell will not have the extra gene. As a result, treatment with the adenovirus will require re-administration in a growing cell population. A new gene is inserted into an adenovirus vector which is used to introduce the modified DNA into a human cell. If the treatment is successful, the new gene will make a functional protein.

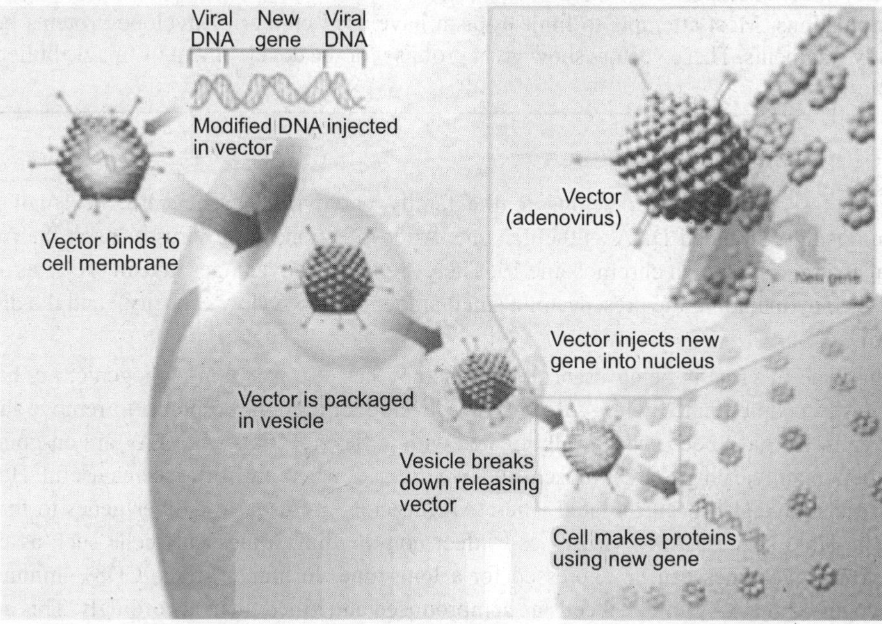

Fig. 9.9. Gene therapy using an adenovirus vector.

Moreover, it has been seen in the SCID trials that the absence of integration into the host cell's genome prevents the type of cancer. This vector system has shown real promise in treating cancer. Indeed an adenovirus is the first gene therapy product to be licensed to treat the cancer.

Envelope Protein Pseudotyping of Viral Vectors

The viral vectors as described above have natural host cell populations that they infect most efficiently. Retroviruses have limited natural host cell ranges, and adenovirus and adeno-associated virus have relatively a broader range of cells. Attachment and entry into a susceptible cell is mediated by the protein envelope present on the surface of a virus. Retroviruses and adeno-associated viruses have a single protein coat on their membrane, whereas adenoviruses are coated with both an envelope protein and fibers that extend away from the surface of the virus. The envelope proteins of each viruses bind to host's cell-surface molecules such as (*i*) *heparin sulfate* (which localizes them upon the surface of the potential host), (*ii*) the *specific protein receptor* that either induces entry-promoting structural changes in the viral protein, or localises the virus in endosomes (where acidification of the lumen induces the refolding of the viral coat). In both the cases, entry into potential host cells requires a favorable interaction between a protein on the surface of the virus and a protein on the surface of the cell.

One should limit or expand the range of cells susceptible to transduction by a gene therapy vector. Therefore, many vectors have been developed in which the endogenous viral envelope proteins have been replaced by either viral envelope proteins, or by chimeric proteins. Such chimera would consist of those parts of the viral protein necessary for incorporation into the virion as well as sequences to interact with specific host cell proteins. Such viruses are referred to as *pseudotyped viruses* in which the envelope proteins have been replaced with viral envelope proteins or chimeric proteins. For example, lentivirus is the most popular retroviral vector used in gene therapy trials.

Simian immunodeficiency virus is coated with envelope proteins, *i.e.* G-protein from vesicular stomatitus virus (VSV). This vector is called VSV G-pseudotyped lentivirus. It infects a universal set of cells. This tropism is characteristic of the VSV G-protein with which this vector is coated. Many attempts have been made to limit the tropism of viral vectors to one or a few host cell populations. Most attempts to limit tropism have used chimeric envelope proteins bearing antibody fragments. These vectors show great promise for the development of 'magic bullet' gene therapies.

3. Adeno-associated Viruses

Adeno-associated viruses belong to the family parvoviridae. They are the small viruses containing single-stranded DNA as the genome. After infection, these viruses insert their genetic material at a specific site on chromosome 19. There are few disadvantages in using AAV as a vector in gene therapy including the presence of a small amount of DNA (low capacity), and the difficulty in its production.

This type of virus is being used in gene therapy because it is non-pathogenic, *i.e.* harmless virus. Mostly people treated with AAV will not develop an immune response to remove the virus and the cells that have been successfully treated with it. Several trials with AAV are on-going or in preparation, mainly trying to treat muscle and eye diseases where the virus seems useful. However, clinical trials have also been initiated where AAV vectors are used to deliver genes to the brain. This is possible because AAV viruses can infect non-dividing (quiescent) cells such as neurons in which their genomes can be expressed for a long time. In human trials, CD8+ immune cells have recognised the AAV infected cells as compromised and killed them accordingly. This action is triggered by a part of the capsid or outer coat of the type 2 virus.

II. NON-VIRAL METHODS

There are several non-viral options for gene delivery also. The simplest method is the direct introduction of therapeutic DNA into target cells. This approach has limited application because it can be used only with certain tissues and requires large amounts of DNA. The other non-viral approach involves the preparation of an artificial lipid sphere with an aqueous core. This liposome which carries the therapeutic DNA is capable of passing the DNA through the membrane of the target cells.

Therapeutic DNA also can enter the target cells by chemically linking the DNA to a molecule that will bind to special cell receptors. After binding to these receptors, the therapeutic DNA constructs are engulfed by the cell membrane and passed into the interior of the target cell. This delivery system tends to be less effective than the other options. Experiments have also been carried out for introducing a 47th (artificial human) chromosome into the target cells. This chromosome would exist autonomously alongside the standard 46 without affecting their workings or causing any mutations. It would be a large vector capable of carrying substantial amounts of genetic code. The body's immune systems would not attack it because of its construction and autonomy. A problem with this potential method is the difficulty in delivering such a large molecule to the nucleus of a target cell.

Non-viral methods present certain advantages over the viral methods such as simple large scale production and low host immunogenicity, etc. Earlier, low levels of transfection and expression of the gene held non-viral methods at a disadvantage. However, recent advances in vector technology have yielded molecules and techniques with transfection efficiencies similar to that of viruses.

1. Direct Injection of Naked DNA

Use of the naked DNA is the simplest method of non-viral transfection. Clinical trials have been carried out of intramuscular injection of a naked DNA plasmid. It resulted in some success. However, the expression has been very low in comparison to other methods of transfection. Besides, there have been trials with naked PCR product which had similar or greater success. However, this success does not compare to that of the other methods leading to research into more efficient methods for delivery of the naked DNA such as electroporation and the use of a gene gun which shoots DNA coated gold particles into the cell using high pressure gas. Gene gun (developed originally for transformation of plant cells) is also being used in animal system for delivery of genes into tissues such as liver, skin, mammary glands and muscle. Direct gene delivery is now being used to map promoter elements that regulate gene expression *in vivo,* in transfection of herpes simplex thymidine kinase genes into melanomas to render them sensitive to Gancyclovir and as a vaccination strategy. Thus, direct gene transfer technique failed to make an impression in treatment of genetic disorders but made enormous impact in the field of vaccine development. The other physical methods *e.g.* microinjection, electroporation, etc. have the limited scope in humans.

Oligodeoxynucleotides: In gene therapy, the use of synthetic oligodeoxynucleotides is to inactivate the genes involved in the disease process. There are several methods to achieve it. One strategy is the antisense therapy specific to the target gene to disrupt the transcription of the faulty gene. In other method small catalytic molecules of RNA called ribozymes is used to cleave specific unique sequences in the mRNA transcript of the faulty gene that disrupts translation of the faulty mRNA and its gene expression. Another strategy uses double stranded oligodeoxynucleotides as a decoy for the transcription factors that activate the transcription of the target gene. The transcription factors bind to the decoys instead of the promoter of the faulty gene which reduces the transcription of the target gene and lowers the expression.

2. Lipoplexes and Polyplexes

For the improvement of delivery of the new DNA into the cell, the DNA must be protected from damage and its entry into the cell must be facilitated. Hence, lipoplexes and polyplexes have been created which have the ability to protect the DNA from undesirable degradation during the transfection process.

Plasmid DNA can be covered with lipids in an organised structure like a micelle or a liposome. When the organised structure is complexed with DNA it is called a lipoplex. There are three types of lipoplexes: anionic (negatively charged), neutral or cationic (positively charged). Initially, anionic and neutral lipids were used for the construction of lipoplexes for synthetic vectors. Although there is little toxicity associated with them; they are compatible with body fluids and there was a possibility of adapting them to be tissue specific. They are complicated and time-consuming to produce; therefore, attention was turned to the cationic versions. The cationic lipids naturally form complex with the negatively charged DNA due to the presence of positive charge. Endocytosis of lipoplex occurs due to presence of their charge. They interact with the cell membrane and the DNA is released into the cytoplasm. The cationic lipids also protect against degradation of the DNA by the cell.

The most common use of lipoplexes has been in gene transfer into cancer cells where the supplied genes have activated tumour suppressor control genes in the cell and decrease the activity of oncogenes. Recent studies have shown lipoplexes to be useful in transfecting respiratory epithelial cells, so they may be used for treatment of genetic respiratory diseases such as cystic fibrosis.

On the other hand, complexes of polymers with DNA are called polyplexes. Most polyplexes consist of cationic polymers and their production is regulated by ionic interactions. One large difference between the methods of action of polyplexes and lipoplxes is that polyplexes cannot release their DNA load into the cytoplasm; therefore, co-transfection must occur with endosome-lytic agents such as inactivated adenovirus. This results in lysis of the endosome that is made during endocytosis, *i.e.* the process by which the polyplex enters the cell. In addition, polymers such as polyethylenimine have their own method of endosome disruption. Liposome-mediated gene transfer has been described in Chapter 8: *Methods of Gene Manipulation*.

3. Receptor-meditated Gene Delivery

Receptor-meditated gene delivery is an attractive strategy that takes advantage of normal physiological pathway. The receptors are exclusively present on hepatocytes. They bind to certain glycoproteins lacking the terminal asialic acid. The concept of delivering genes into specific tissue was examined extensively with the liver asialoglycoprotein receptors. An asialoglycoprotein (orosomucoid) is coupled with poly-L-lysine. This conjugate is then condensed with plasmid DNA via electrostatic interactions (Fig. 9.10). This soluble complex, *i.e.* asialoglycoprotein-poly-L-lysin-DNA complex (ASGP-PL-DNA) is injected intravenously/intraperitoneally which gets entered into hepatocytes via the asialoglycoprotein receptors (ASGPR). DNA is delivered into the nucleus. The transgene expression occurs, therefore, recombination protein is detected for several weeks (Wu and Wu, 1988).

Later on in 1991, Wu and co-workers extended the other cell surface receptors such as transferrin receptor. The major problem related to receptor-mediated gene delivery system has been the degradation of DNA in the endosomes by lysosomal enzymes. Therefore, co-administration of lysosomotropic agents *e.g.* chloroquine is done to enhance gene expression. Moreover, it is an established fact that adinoviruses escape lysosomal degradation by acidification of endosomes.

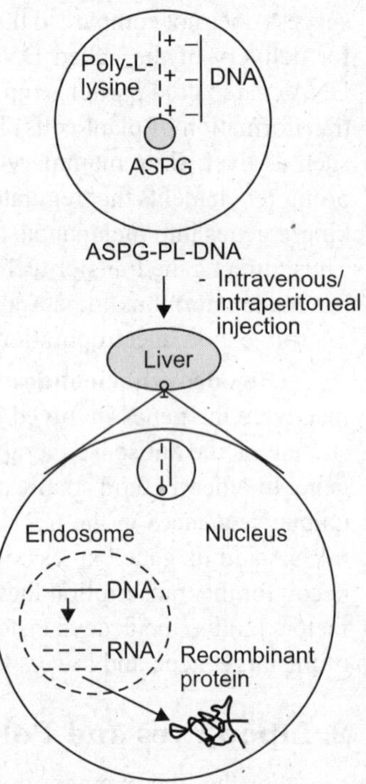

Fig. 9.10. Receptor-mediated gene delivery in liver.

Therefore, methods have been developed to construct adenovirus-linked molecular conjugate vectors such as monoclonal antibodies (against a heterologous epitope on the hexon protein of the adenovirus envelope) that are covalently linked to polylysine. This complex is coupled with plasmid DNA. Delivery of such molecular conjugate has been shown to facilitate more efficient gene transfer than the conventional receptor-mediated gene delivery strategy. Such a conjugate interacts with adenoviruses that express the epitope on the envelope.

It has also been found that positively charged liposomes of cationic lipids can entrap the negatively charged DNA. This has led to the possibility of their use as gene transfer vehicle *in vivo.* Therefore, several catonic lipid formation has been developed and now being used for transfection of cells in culture and a variety of organs *in vivo*. The two potential success of catonic lipids are: the aerosol delivery of genes to lungs, and the development of cancer vaccines.

4. Embryo Therapy through IVF Technology

In 1993, by adopting IVF-technology, Dr. A. Handyside (a team leader of British Researcher at Hammersmith Hospital, London) got success in producing a genetically engineered female baby, whose parents transmitted the genetic disease (cystic fibrosis) in earlier four babies, who died later on. Even after confirmation through prenatal diagnosis that the babies would be sufferer or carrier of this disease they decided to give birth to babies, which failed. This disease clogs the lungs of sufferer and makes them unable to digest the food properly.

Handyside admitted seven couples. Their eggs were collected and fertilised with their respective husbands' sperm in test tubes. After 3 days incubation *in vitro,* when embryo reached the 8 cell stage, a single cell was removed for analysis. The cell was magnified one million times by using a magnifying thermal cycler. The individual defective gene which causes cystic fibrosis was identified and screened out. Thereafter, the embryo was implanted in the womb of mother. Consequently, a female baby (C. O'Brien) was born who is neither sufferer nor carrier of this disease, but a normal one. The birth of this genetically engineered baby has increased the hope of millions of couples who transmit the genetic disease to their offspring, and also of researchers to look upon many more genetic diseases to cure.

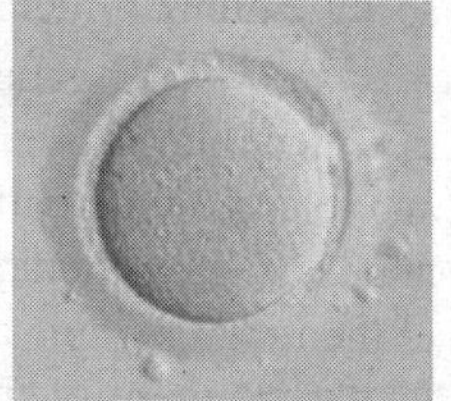

A naked egg

Preimplantation genetic diagnosis (PGD): During 1990s pre-implantation genetic diagnosis and screening was started in conjunction with IVF treatments. Since then hundreds of normal and healthy babies have been born using this advanced reproductive technology. PGD is performed for couples who pass on genetic disease to their offspring. Screening of embryos for chromosomal abnormalities [preimplantation genetic screening (PGS)] improves the possibility of a successful pregnancy and birth for couples with infertility related to recurrent miscarriage or unsuccessful IVF cycles. Patients who also can benefit from PGS/PGD are : (*i*) couples who have a family history of inherited disease, (*ii*) couples who want to use gender selection to prevent a gender-linked disease, (*iii*) women who have had repeated failures with IVF, (*iv*) women with a history of unexplained miscarriage, and (*v*) women who are more than 39 years old.

Numeral chromosomal abnormalities are screened by preimplantation genetic screening (PGS), whereas PGD diagnoses the specific molecular defect of the inherited disease. During the IVF process individual cells from a pre-embryo are analysed in both the cases (PGS and PGD). One or two cells are removed from the pre-embryos (8-cell stage) before the transfer of a pre-embryo back to the uterus of a woman. Thereafter, these cells are evaluated for the presence of normal genes. When the evaluation is completed within 1-2 days, only the normal pre-embryos are transferred back to the uterus of the woman. For implantation fewer embryos are needed; therefore, PGS can reduce the risk of multiple pregnancies.

5. Magic bullets — Cells as Drug Delivery System

Magic bullet is a drug delivery system through a cell loaded with an appropriate treatment molecule which upon injection into a patient migrates through the circulatory system directly to the diseased site. Still, it is only a dream to the patients suffering from genetic diseases. In future it may be possible to cure such diseases such as cystic fibrosis, when a cell can reach to the lung tissues and deliver an enzyme to breakdown the thick gelatinous mucus, and help in breath and food digestion.

(*a*) Genetically Modified Mesenchymal Stromal Cells (MSC) for Therapy: The cells that can be harvested without difficulty, efficiently processed in *ex vivo* conditions, and transplanted into patients are the ideal candidate of cellular therapy for clinical use. MSCs are non-haematopoietic multipotent cells that can self-renew and differentiate into multiple lineages. MSCs are characterised by their plastic adherence in culture, differentiation potential and expression of cell surface markers. MSCs possess all these properties and could be a compatible source for use in human clinical trials. MSCs must express CD105, CD73, and CD90, and lack the expression of CD45, CD34, CD14, or CDllb, Cd79a or CD19, and HLA class II surface markers (Dominici *et al*., 2006).

MSCs can be obtained from bone marrow and other tissues such as peripheral blood, umbilical cord blood, adipose tissue and placenta. The other sources of MSCs are: liver periodontal ligament, hair follicles and amniotic fluid. MSCs reside in the bone marrow in small numbers (0.001–0.01% of nucleated cells but can be easily expanded *in vitro* (*ex vivo*) to yield a sufficient number of cells for clinical issue. Differential gene expression of MSCs as well as different molecular mechanisms involving chemokine receptors and adhesion molecules are associated with MSC migration towards specific tumour microenvironment.

Cell-based gene therapy has gained momentum in the past several years. It is a promising treatment alternative for patients. The genetically modified MSC cells have the efficacy to express and release therapeutic factors. Thus they confirm their ability to serve as an excellent base for cell-mediated gene therapy. Attempts to use MSCs as cellular delivery vehicles have taken advantage of the tumour tropic ability of MSCs, expression of genetically engineered proteins, and utilisation for the delivery of therapeutic gene products. Genetically engineered MSCs have been used for improvement in haematopoietic engraftment following myeloablative transplantation regimens, and the targeted delivery of antitumour factors by secretion of growth factors and cytokines. These studies have led to several clinical trials using MSCs for the treatment of inherited disorders giving promising results.

(*b*) Bone marrow-derived Mesenchymal Stromal Cells: Mesenchymal stromal cells (MSC) are multipotent cells that can self-renew and differentiate into multiple lineages with specific surface marker expression. In 2009, Latha G. Menon and collaborators at the Department of Neurosurgery (Brigham and Women's Hospital, U.S.A.) have reviewed the mesenchymal stromal cells (MSC) as a drug delivery system,

(*i*) Interferon (IFN)– α and β: IFN-β has a wide range of biological activities including potent antiproliferative and proapoptotic effects. But due to toxicity associated with systemic administration its *in vivo* therapeutic efficacy is limited. Engineered human MSCs express interferon β (IFN-β) which have been used for targeted delivery of this potent antiproliferative and proapoptotic agent to metastatic breast and melanoma models and lung metastasis. In 2005, M.Marini Studeny and collaborators from the University of Texas (U.S.A.) found that subcutaneous tumour growth in a SCID mouse xenograft model could be inhibited and prolonged survival of animals was observed only after the co-injection of A375SM melanoma cells with IFN-β-MSCs. They suggested that MSC can be used to target delivery of agents to tumor cells.

(*ii*) Interleukin-2 (IL-2): The use of MSCs genetically modified with interleukin (IL)-2 to exert an effective immune response against the poorly immunogenic B16 melanoma model has been investigated. In 2004, J. Stagg and collaborators from the Lady Davis Institute for Medical Research (Canada) found that IL-2-producing MSCs after being mixed with B16 cells significantly delayed tumour growth in a dose-dependent manner. The results demonstrate that tumour-bearing mice treated with IL-2-producing MSCs developed CD8-mediated tumour-specific immunity and significantly delayed tumour growth, by generating effective immune responses against melanoma in mice with normal immune systems.

(*iii*) Prodrug therapy– cytosine deaminase: In 2009, Lata G. Menon and collaborators have discussed that the prodrug gene therapy refers to the delivery of genes encoding enzymes that convert nontoxic prodrugs into toxic antimetabolites. Three suicide genes are the cytosine deaminase (CD), HSV-1 thymidine kinase and carboxyesterase genes, which confer sensitivity to 5-fluorocytosine 5-FC, ganciclovir (GCV) and camptothecin-11 (CPT-11), respectively. These are being evaluated in clinical trials. The CD gene-directed enzyme prodrug therapy approach relies on the ability of bacterial or yeast CD enzyme to convert the non-toxic substrate antifungal agent 5-FC to antitumor agent 5-fluorouracil (5-FU). 5-FU is capable of non-facilitated diffusion into and out of cells resulting in a significant effect of CD/5-FC.

6. Virosomes in Drug Targeting

Each and every method of gene transfer contains many shortcomings; therefore, some of the hybrid methods have been developed that combine two or more techniques, for example reconstitution of 'virosomes'. Virosomes are small spherical vesicles which are made up of a mixture of synthetic and natural phospholipid bilayer that hold haemagglutinin (HA) and neuraminidase (NA). The HA and NA are the surface proteins of influenza virus or HIV. Therefore, on the basis of source of purified inserted envelop glycoproteins (HA and NA), virosomes are of two types: influenza virosomes and non-influenza virosomes. The non-influenza virosomes are being considered for HIV-1 vaccine research. Electron microscopy and dynamic light scattering are used for estimating the diameter of the vesicles. The mean diameter of the vesicles ranges from 120 to 180 nm. Different components of the virosome are given in Fig. 9.11.

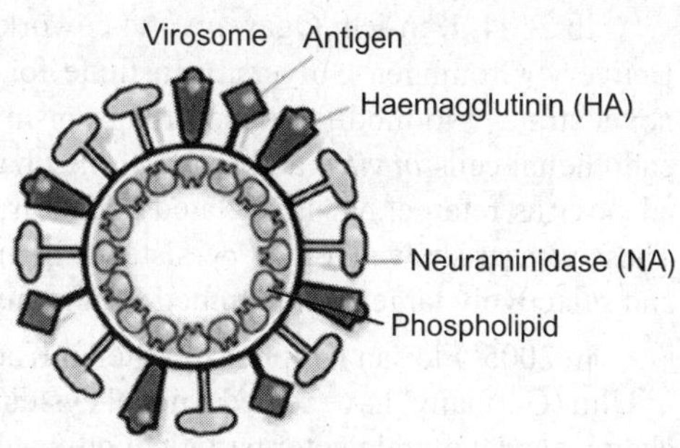

Fig. 9.11. Different components of virosome.

The virosomes are reconstituted fusion-active viral envelopes representing the empty virus envelopes, devoid of the nucleocapsid including the genetic material of the source virus. Antigens can be incorporated into virosomes, adsorbed to the virosome surface, or integrated into the lipid membrane either via hydrophobic domains or lipid moieties cross-linked to the antigen. The HA and NA proteins provide the virosomes the properties to fuse with cells of the immune system. Proteins of other human pathogens can be incorporated into the vesicles. Therefore, the virosomes can deliver these vaccine-specific antigens directly to their targets. Thus virosomes effectively and efficiently stimulate the immune system (http:/www.ncbi.nlm.nih.gov.pubmed/16026906. The virosome concept for influenza vaccines, 2005).

A virosome is a drug or vaccine delivery mechanism which allows it to fuse with target cells. Since, virosomes are pure fusion-active vesicles, they do not replicate. The presence of biologically active influenza HA in their membrane is the unique properties of virosomes. This viral protein contributes to the immunological properties of virosomes and confers structural stability and homogeneity to virosome-based formulations. This is different from the other liposomal and proteoliposomal carrier systems. For the full adjuvant effect of virosomes a physical association between the virosome and the antigen of interest is necessary. Based on the properties of the antigen there are variety of methods that can achieve such physical association.

7. Chemically Modified Viral Vectors (Chemoviruses) and Synthetic Viruses

The currently used classes of viral and non-viral vectors have high intracellular efficiency of viral vectors, and high systemic potential with low immunogenicity of shielded non-viral vectors. However, the merge of viral and non-viral vector research efforts could be an encouraging strategy for the future optimisation of both classes of vector. Viral vectors may greatly benefit from chemical modifications to develop chemoviruses. Non-viral vectors may benefit from modifications mimicking viral intracellular delivery functions to generate synthetic viruses. In addition, hybrid strategies including both non-viral and viral gene vector parts may also provide interesting solutions (Boeckle and Wagner, 2006) (Fig. 9.12).

A good approach may be the combination of chemical and genetic modification of recombinant viruses. All undesired viral functions may be deleted from the viral vector (as described for the development of high-capacity gutless adenoviral vectors), and the original surface of the virus is masked with retargeting ligands and shielding shell/domains by using chemical or recombinant procedures (Fig. 9.12 A).

After coating of adenoviral vectors with hydrophilic polymers (such as polyethylene glycol or poly-N-2-hydroxypropyl-methacrylamide (pHPMA) the virus is shielded from interaction with its native receptor or neutralising antibodies and the incorporated ligands enabled retargeting.

In 2004, Ken-ichi Ogawara and coworkers (at the Department of Pharmacokinetics and Drug Delivery, Groningen University Institute for Drug Exploration, The Netherlands) have reported a novel strategy to modify adenovirus tropism and enhance transgene delivery to activated vascular endothelial cells *in vitro* and *in vivo.* They have described a very encouraging example of systemic adenovirus retargeting. PEGylated adenoviruses were retargeted using anti-E-selectin antibodies. These viruses showed longer persistence in the blood circulation compared with unmodified viruses and selectively targeted inflammed skin in mice resulting in local gene expression.

In 2005, Florian Kreppel and Stefan Kochanek from the Division of Gene Therapy, University of Ulm (Germany) have revied a novel cystein-based vector platform for combination of genetic and chemical adenoviral vector particle modifications. They genetically introduced cysteines at solvent-exposed positions of the adenovirus fibers. The corresponding thiol groups (36 per adenovirus) were highly reactive in controlled chemical coupling to protein ligands such as transferrin. Such particles were efficiently targeted to the transferrin receptor pathway. This targeting concept via cysteine modifications could also be successfully combined with PEGylation of adenovirus amino groups, providing shielding (detargeting) against non-target interactions.

Chimeric viruses (hybrid vectors) are generated by combining viral and non-viral elements. For example, adenovirus can covalently be linked to a separate non-viral formulation (e.g. expression plasmid) (Fig. 9.12 B). The genome of viral vector and the separate non-viral gene expression cassette (outside the virus) may synergise in some applications. For example, a helper plasmid attached to the surface of the virus activates the replication of virus or regulate the viral gene expression cassette. Besides, the non-viral part may be modified with targeting ligands to allow retargeting of the hybrid vector, whereas the virus part may confer greatly improved intracellular efficiency (Boeckle and Wagner, 2006).

In another type of hybrid strategy a whole viral genome is incorporated into a non-viral formulation. A plasmid, encoding for the conditionally replicating oncolytic adenovirus dl1520, was integrated into liposomes. Like the parent virus, after transfection of p53-defective tumour cells this plasmid generates infectious particles. In contrast to the virus, the particles can also transfect coxsackievirus and adenovirus receptor (CAR)-negative tumour cells. The antitumour efficacy of this infectious plasmid was demonstrated with human tumour xenografts in mice after both local and intravenous administration. In contrast to oncolytic viruses, systemic activity was retained, even in the presence of neutralising antibodies (Yotnda *et al.* 2004).

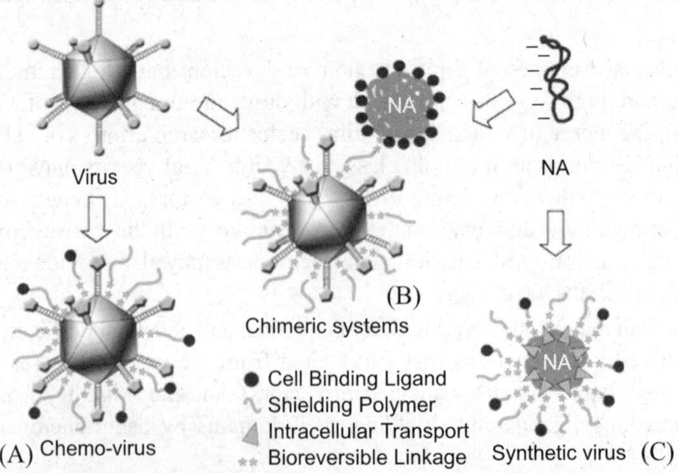

Fig. 9.12. Chemically modified viral vectors (chemo-virus), chimeric systems [hybrid systems containing both viral and non-viral nucleic acid (NA)], and synthetic virus (*source*: Boeckle and Wagner, 2006).

In 1992, C. Plank and coworkers generated one of the first prototypes of synthetic virus-like particles (Fig. 9.12 C). The particles contained a targeting ligand and a membrane-active peptide to induce endosomal release. The system contained a synthetic tetra-antennary carbohydrate ligand for hepatic asialoglycoprotein receptor binding and a synthetic acidic peptide analog derived from the *N*-terminal HA-2 subunit of influenza virus haemagglutinin. Both components were covalently attached to polylysine, a polycationic carrier for DNA binding. The polyplex components were found to compact DNA to small particles with virus-like dimension of ~100 nm. Application of these particles to hepatocytes resulted in efficient, ligand-specific gene expression which was highest in the presence of the endosome-destabilising peptide. This early approach displayed clear limitations. Compared with viral vector particles the system was still very inefficient. The complex formulation was a crude heterogeneous composition that did not contain any measures to shield the particle surface, and the particles aggregated with time.

(*a*) Optimisation of Synthetic Viruses: A synthetic virus-like vector should be available in a defined, purified and stable form. The particles should be purified from all components used for vector preparation. In comparison to a standard non-viral vector, the artificial virus should respond in a more dynamic manner to alterations in their biological micro-environment such as changes in pH or disulfide-reducing conditions, and the vector should undergo programmed structural changes compatible with the different gene delivery steps. This would result in improved shielding and targeting characteristics during systemic circulation and outside the target cell, and in enhanced intracellular trafficking triggered by the intracellular environment. Thus, virus-like functions have to be activated and presented in a bioresponsive fashion.

(*b*) Bioresponsive Elements: Boeckle and Wagner (2006) have described about the optimisation of targeted gene delivery with reference to chemical modification of viral vectors and synthesis of artificial virus vector systems. The merge of viral and nonviral vector technologies is an encouraging strategy for the future. The concepts include chemically modified viral vectors (chemo-viruses) and synthesis of virus-like systems (synthetic viruses). Polyethylene glycol (PEG) acts as a shielding agent because it has a higher degree of shield showing a longer circulation time and better accumulation at the target site but lower gene expression activity in the target cells. However, the intracellular uptake processes is hampered by the shield. Hence, an optimal non-viral approach for systemic application has a more dynamic character. During circulation its surface charge must be neutralised. The membrane-interacting cationic surface should be re-exposed for efficient gene transfer after entering the target cell.

Walker *et al.* (2005) introduced bioresponsive PEG-polycation conjugates with a pH-labile linkage. Pyridyl hydrazones showed acid-dependent hydrolysis. At pH 5 and 37°C about 90% PEG conjugates hydrolyzed within 10 minutes, whereas at pH 7.4 the half-life was 1.5 hours. At neutral pH and 37°C, DNA particles shielded with these bioreversible PEG conjugates remain stable for at least 4 hours but are expected to lose their PEG shield upon reaching endosomal pH (Fig. 9.13). Particles with PEG shield displayed up to 100-fold higher gene transfer activity in cell culture models as compared with polyplexes with the analogous stable PEG shield, and 10-fold higher systemic *in vivo* gene transfer efficiency in a mouse tumour model (Walker *et al.*, 2005). Therefore, in vector design engineering of nanoparticle structures with such dynamic virus-like domains is an encouraging direction.

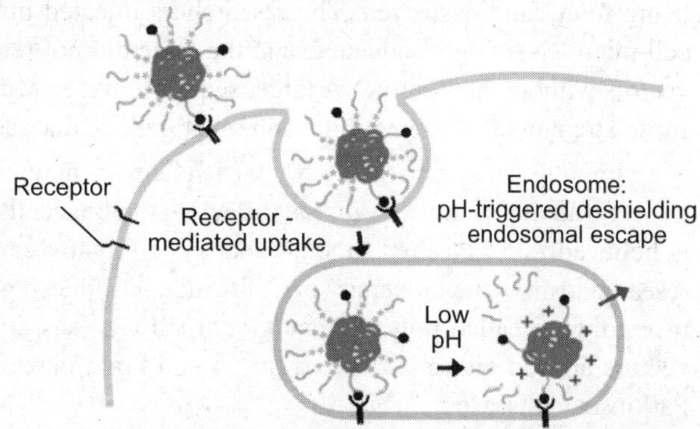

Fig. 9.13. Receptor-mediated delivery of ligandtargeted polyplexes into endosomal vesicles (Walker *et al.*, 2005).

F. RECENT DEVELOPMENTS IN GENE THERAPY

A combination of two tumour suppressing genes delivered in lipid-based nanoparticles drastically reduced the number and size of human lung cancer tumours in mice during trials conducted by researchers from M. D. Anderson Cancer Center (University of Texas) and the University of Texas Southwestern Medical Center. Researchers at the National Cancer Institute (NCI), part of the National Institutes of Health, successfully re-engineered immune cells called lymphocytes to target and attack cancer cells in patients with advanced metastatic melanoma. This is the first time that gene therapy is used to successfully treat cancer in humans. Gene therapy is effectively used to treat two adult patients for a disease affecting non-lymphocytic white blood cells called myeloid cells. Myeloid disorders are common and include a variety of bone marrow failure syndromes, such as acute myeloid leukemia. The study is the first to show that gene therapy can cure diseases of the myeloid system (www.cincinnatichildrens.org).

Gene therapy cures deafness in guinea pigs. Each animal had been deafened by destruction of the hair cells in the cochlea that translate sound vibrations into nerve signals. A gene, called *Atoh1* was delivered to the cochlea by an adenovirus which stimulates the growth of hair cells. The genes triggered re-growth of the hair cells and many of the animals regained up to 80% of their original hearing thresholds. This study is the first to show that gene therapy can repair deafness in animals. It may pave the way to human trials of the gene. New gene therapy approach repairs errors in messenger RNA derived from defective genes. Technique has potential to treat the blood disorder thalassaemia, cystic fibrosis, and some cancers.

Scientists at the National Institutes of Health have successfully treated metastatic melanoma in two patients using killer T cells that retargeted to attack the cancer cells. This study constitutes the first demonstration that gene therapy can be effective in treating cancer. In 2006 a team of scientists led by Luigi Naldini and Brian Brown from the San Raffaele Telethon Institute for Gene Therapy (HSR-TIGET) (Italy) reported a breakthrough for gene therapy in which they developed a way to prevent the immune system from rejecting a newly delivered gene. Similar to organ transplanation, gene therapy has been plagued by the problem of immune rejection.

So far, delivery of the 'good' gene has been difficult because the immune sytem does not recognise the new gene and rejects the cells carrying it. To overcome this problem, the HSR-TIGET group utilised a newly uncovered network of genes regulated by molecules known as microRNAs. Naldini's group reasoned that they could use this natural function of microRNA to selectively turn off the identity of their therapeutic gene in cells of the immune system and prevent the gene from being found and destroyed. The researchers injected mice with the gene containing an immune-cell microRNA target sequence, and the mice did not reject the gene as previously occurred when vectors without the microRNA target sequence were used. This work shows important implications for the treatment of haemophilia and other genetic diseases by gene therapy.

In 2006, an international group of scientists announced the successful use of gene therapy to treat two adult patients for a disease affecting myeloid cells. The study, published in Nature Medicine, is believed to be the first to show that gene therapy can cure diseases of the myeloid system. A research team of the university of California, introduced genes into the brain using liposomes coated in a polymer called polyethylene glycol (PEG). It is a significant achievement because the viral vectors are too big to get across the 'blood-brain barrier'. This method has potential for treating Parkinson's disease.

Researchers at Case Western Reserve University and Copernicus Therapeutics created the tiny liposomes, 25 nanometers across that can carry therapeutic DNA through pores in the nuclear membrane. The success of a multi-center trial for treating children with SCID (severe combined immune deficiency) held from 2000 and 2002 was questioned when two of the ten children treated at the trial's Paris center developed a leukemia-like condition. In 2002, clinical trials were halted temporarily, but resumed after regulatory review of the protocol in the United States, the United Kingdom, France, Italy, and Germany.

1. Gene Silencing (RNA Interference)

DNA carries the information to synthesise all the proteins required for making the life possible. The information of a gene is first transcribed as mRNA, which is then used as a template for synthesising a protein. But unlike DNA, mRNA is single-stranded.

The term 'gene silencing' is generally used to describe the 'switching off' of a gene by a mechanism other than genetic modification. Obviously, genes are regulated at either the transcriptional or post-transcriptional level; hence, there are two types of gene silencing : *transcriptional gene silencing*, and *post-transcriptional gene silencing*. Transcriptional gene silencing occurs as the result of histone modifications that creates an environment of heterochromatin around a gene that makes it inaccessible to transcriptional machinery (RNA polymerase, transcription factors, etc.). Post-transcriptional gene silencing is the result of mRNA of a particular gene which gets destroyed or blocked. The destruction of the mRNA inhibits translation to synthesise a protein. A common mechanism of post-transcriptional gene silencing is use of interference RNA, *i.e.* RNAi. Both transcriptional and post-transcriptional gene silencing are used to regulate endogenous genes. Mechanisms of gene silencing also protect the genome of an organism from transposons and viruses. Gene silencing thus may be part of an ancient immune system protecting from such infectious DNA elements.

2. Sense RNA and Antisense RNA

A DNA molecule of a gene consists of two strands: one strand is the sense strand (or coding strand, plus strand, or non-template strand), and the other is the antisense strand (also called non-coding strand, minus strand, anticoding strand, template strand, or transcribed strand). Antisense is a term that refers to a sequence of DNA or RNA that is complementary to mRNA. Both DNA coding strand and RNA strand are complementary to the template strand; they have the same sequences except that T in the DNA coding strand is replaced by U in the RNA strand. The coding strand contains codons, while the non-coding strand contains anti-codons.

A single-stranded mRNA is produced after transcription; its sequence of nucleotides is called 'sense mRNA' because it results in a gene product called protein (Fig. 9.14). The sense strand of DNA has the same base sequence as the transcribed mRNA except that the base T is replaced by the base U. The tRNA read its 'unpaired nucleotides' as the ribosome proceeds to translate the message. In contrast, the antisense DNA strand consists of nucleotide sequence complementary to the sense strand of DNA. The antisense strand acts as the template for the transcription of mRNA. RNA polymerase polymerises the RNA nucleotides along the sugar phosphate backbone.

5′ ...A T G G C C T G G A C T T C A... 3′ Sense strand of DNA
3′ ...T A C C G G A C C T G A A G T... 5′ Antisense strand of DNA

↓ Transcription of antisense strand

5′ ...A U G G C C U G G A C U U C A... 5′ mRNA

↓ Translation of mRNA

Met — Alu — Thr — Ser — Peptide

Fig. 9.14. Sense and antisense strands of double stranded DNA and its transcription.

The second strand of DNA is called the 'antisense strand' because its sequence of nucleotides is complementary to the sense strand. The second strand of RNA, whose nucleotide sequence is complementary to the first strand of RNA, can form double strand similar to DNA as below :

5′-CUGGA-3′ mRNA

3′-GACCU-5′ Antisense RNA

When mRNA forms a duplex with a complementary antisense RNA sequence, translation is blocked because: (*i*) the ribosome cannot gain access to the nucleotides in the mRNA, or (*ii*) duplex RNA is quickly degraded by ribonucleases in the cell (Fig. 9.15).

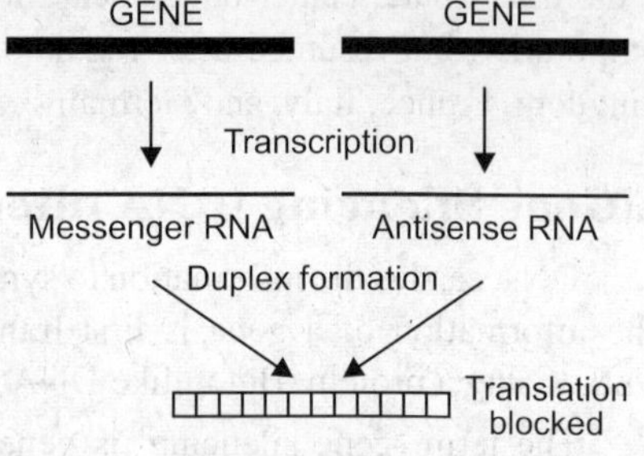

Fig. 9.15. Blocking of translation by double stranded RNA molecule.

With recombinant DNA methods, synthetic genes (DNA) encoding antisense RNA molecules can be introduced into the organism. Sense RNA of the same coding region should be used as a control while testing the effects of antisense RNA. Preparations of sense RNA often becomes as an effective inhibitor as does antisense RNA. Double-stranded RNA (dsRNA) corresponding to a particular gene is a powerful inhibitor of that gene. The suppressive effect of antisense RNA also depends on its ability to form dsRNA (using the corresponding mRNA as a template). The ability of dsRNA to suppress the expression of a gene corresponding to its own sequence is called RNA interference (RNAi); it is also called post-transcriptional gene silencing.

Andrew Z. Fire (left) and Craig C. Mello (right)

Single-stranded RNA is the only RNA molecules that are normally found in the cytoplasm of a cell. If the cell finds the dsRNA molecules, it uses an enzyme called Dicer to cut them into fragments containing ~21 base pairs. Consequently, the two strands of each fragment separate out releasing the antisense strand. With the aid of a protein, it binds to a complementary sense sequence on an mRNA molecule. The mRNA is destroyed if the base-pairing is exactly the same. These RNA fragments are called 'small interfering RNA' (or siRNA) due to their interfering action. The complex of siRNA and protein is called the 'RNA-induced Silencing Complex' (RISC).

In 1998, Andrew Z. Fire and Craig C. Mello injected a cell with the matching mRNA strands and found that the mRNA molecule silenced the expression of the gene, *i.e.* the protein was not synthesised. The matching strand binds to the target RNA and forms a dsRNA molecule as is found in DNA. Later on it was found that this dsRNA is destroyed by a set of proteins as a natural defense mechanism against viruses, as well as a tool to regulate the expression of certain genes.

3. Dicer Enzyme

In 2001, Emily Bernstein (a graduate student at the Cold Spring Harbor Laboratory) first demonstrated the dsRNA 'dicing' activity of the enzyme and named it as Dicer. Dicer is an endoribonuclease in the RNase III family. Dicer contains two RNase III domains and one PAZ domain. The distance between these two regions of the molecule is determined by the length and angle of the connector helix and determines the length of the siRNAs it produces. It cleaves dsRNA and pre-microRNA (miRNA) into short dsRNA fragments called small interfering RNA (siRNA) about 20-25 nucleotides long, usually with a two-base overhang on the 3' end. Dicer catalyses the first step in the RNA interference pathway and initiates formation of the RNA-induced silencing complex (RISC), whose catalytic component argonaute is an endonuclease capable of degrading messenger RNA (mRNA) whose sequence is complementary to that of the siRNA guide strand. The human version of this gene is DICER1.

Fig. 9.16 shows one molecule of the Dicer protein isolated from *Giardia intestinalis* which catalyses the cleavage of dsRNA to siRNAs. The RNase III domains are coloured green, the PAZ domain yellow, the platform domain red, and the connector helix blue. The distance between the RNase III and PAZ domains, determined by the length and angle of the connector helix, has been suggested as the determinant for the length of siRNA molecules produced by a given Dicer variant.

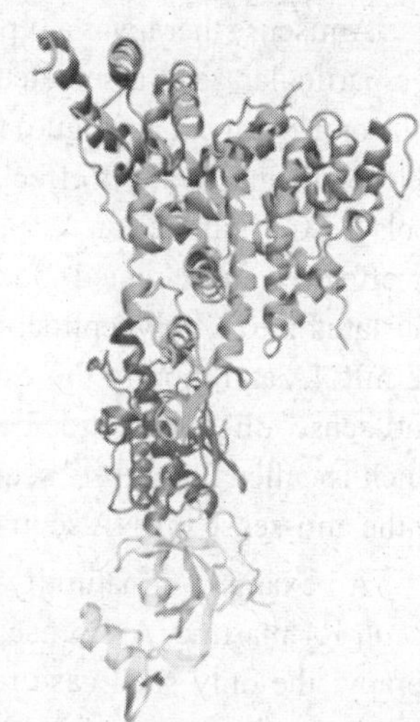

Fig. 9.16. One molecule of the Dicer protein from *Giardia intestinalis.*

RNAi is initiated by the Dicer enzyme which processes dsRNA into ~22-nucleotide small interfering RNAs. Two Dicer molecules with five domains each are shown in Fig. 9.17. On the basis of known mechanisms for the RNase III family of enzymes, Dicer works as a dimeric enzyme. Cleavage into precisely sized fragments is determined by the fact that one of the active sites in each Dicer protein is defective (indicated by an asterisk), shifting the periodicity of cleavage from ~9–11 nucleotides for bacterial RNase III to ~22 nucleotides for Dicer family members. The siRNAs are incorporated into a multicomponent nuclease, RNA-induced Silencing Complex (RISC) enzyme (Fig. 9.16). Recent reports suggest that RISC must be activated from a latent form, containing a double-stranded siRNA to an active form, RISC, by unwinding of siRNAs. RISC then uses the unwound siRNA as a guide to substrate selection.

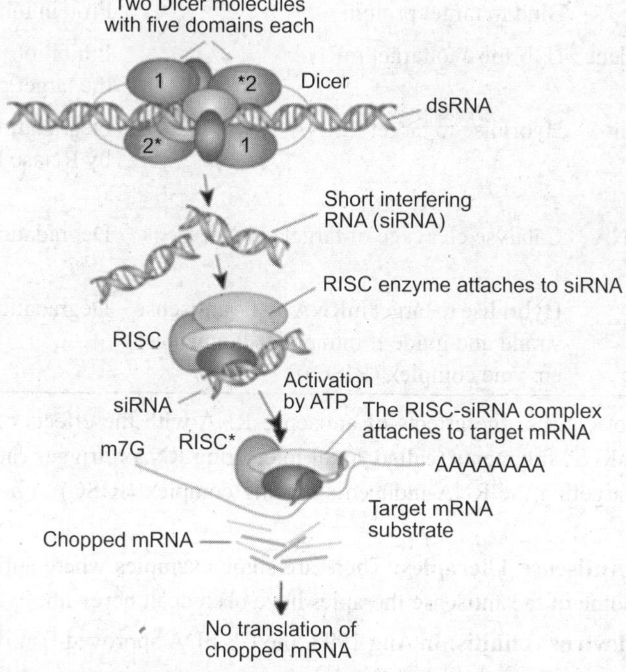

Fig. 9.17. Dicer enzyme initiates the RNAi which processes dsRNA into ~22-nucleotide small interfering RNAs.

4. Antisense Therapy

Antisense RNA is an ssRNA which is complementary to an mRNA strand transcribed within a cell. Antisense RNA may be introduced into a cell to inhibit translation of a complementary mRNA by base pairing and physically obstructing the translation machinery (Weiss *et al.,* 1999).

Antisense therapy is a type of treatment for genetic disorders. If you know the genetic sequence of a particular gene associated with a particular disease, possibly a strand of nucleic acid (DNA, RNA or a chemical analogue) that will bind to the mRNA produced by that gene can be synthesised which will inactivate it effectively turning that gene 'off'. In 2007, Paul A. Morcos from Gene Tools LLC (Summerton Way, U.S.A.) reported the use of steric-block antisense oligos as tools for effective and targeted modification of RNA splicing. BRNA is single stranded which is to be translated into a polypeptide. Alternatively, the strand might be targeted to bind a splicing site on pre-mRNA and modify the exon content of an mRNA. The synthesised nucleic acid is termed as 'anti-sense' oligonucleotide because its base sequence is complementary to the mRNA of the gene, which is called the 'sense' sequence. A sense segment of mRNA (5′-AAGGUC-3′) will be blocked by the anti-sense mRNA segment (3′-UUCCAG-5′).

An example of naturally occurring mRNA antisense mechanism is the hok/sok system of the *E. coli* R1 plasmid. Antisense RNA has long been thought of as a promising technique for disease therapy; the only such case reached the market is the drug 'Fomivirsen'. One commentator has characterised antisense RNA as one of dozens of technologies that are gorgeous in concept but exasperating in commercialisation. Antisense RNA still lack effective design, biological activity, and the efficient route of administration.

Table 9.4. Different gene silencing strategies.

Agent	*Mechanism*	*Result*
Most drugs	Bind to target protein	Protein inhibition
RNase H-independent ODNs	Hybridise to target mRNA	Inhibition of translation of the target protein
RNase H-dependent oligonucleotides (ODNs)	Hybridise to target mRNA	Degradation of the mRNA by RNase H
Ribozymes and DNA enzymes	Catalyse cleavage of target mRNA	Degradation of the mRNA
siRNA	Hybridise to target mRNA by its antisense strand and guide it into endoribonuclease enzyme complex (RISC)	Degradation of the mRNA

One should not confuse the effects of antisense RNA with the effects of RNA interference (RNAi) in which dsRNA fragments (called small interfering RNAs) trigger catalytically mediated gene silencing by targeting the RNA-induced silencing complex (RISC) to bind and degrade the mRNA (Fig. 9.18).

***(a)* Example Antisense Therapies:** There are some examples where antisense therapies are being researched. Some of the antisense therapies have been dealt herewith:

***(i)* Cytomegalovirus retinitis:** In Aug 1998, the US FDA approved Fomivirsen for treatment for cytomegalovirus retinitis, and marketed it as Vitravene.

***(ii)* Cancer:** In 2006, German physicians reported for the compound AP 12009 (a phosphorothioate antisense oligodeoxynucleotide specific for the mRNA of human transforming growth factor TGF-beta2) in patients with high grade gliomas.

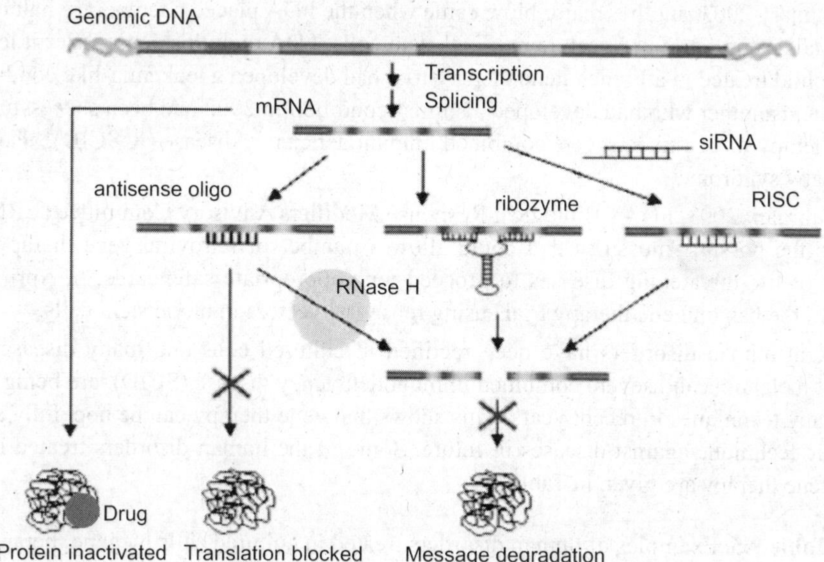

Fig. 9.18. Mechanism of gene silencing.

Now-a-days antisense drugs are being researched to treat cancers (including lung cancer, colorectal carcinoma, pancreatic carcinoma, malignant glioma and malignant melanoma), diabetes, ALS, Duchenne muscular dystrophy, and other diseases such as asthma and arthritis with an inflammatory component. Most potential therapies have not yet produced significant clinical results, though one antisense drug, Fomivirsen®, has been approved by the US Food and Drug Administration (FDA) as a treatment for cytomegalovirus retinitis.

***(i)* HIV/AIDS:** Since 2004, scientists of the United States have been conducting research on using antisense technology to combat HIV. In 2010, they reported the success in reducing HIV viral load using patient T-cells which had been harvested, modified with an RNA antisense strand to the HIV viral envelope protein, and re-infused into the patient during a planned lapse in retroviral drug therapy.

***(ii)* Haemorrhagic fever viruses:** During 2006s, scientists studying the Ebola haemorrhagic fever virus announced a 75% recovery rate after infecting four rhesus monkeys, and treating them with an antisense morpholino drug developed by AVI BioPharma, a U.S. biotechnology firm. The usual mortality rate for monkeys infected with Ebola virus is 100%. In late 2008, AVI BioPharma successfully filed Investigational New Drug (IND) applications with the FDA for its two products for Marburg and Ebola viruses. These drugs, AVI-6002 and AVI-6003 are novel analogues based on AVI's PMO antisense chemistry in which anti-viral potency is enhanced by the addition of positively-charged components to the morpholino oligomer chain.

I. CURRENT STATUS OF GENE THERAPY RESEARCH

The Food and Drug Administration (FDA) has not yet approved any human gene therapy product for sale. Current gene therapy is experimental and has not proven very successful in clinical trials. Little progress has been made since the first gene therapy clinical trial began in 1990. In 1999, gene therapy suffered a major setback with the death of 18-year-old Jesse Gelsinger who was participating in a gene therapy trial for ornithine transcarboxylase deficiency (OTCD). He died from multiple organ failures four days after starting the treatment. His death is believed to have been triggered by a severe immune response to the adenovirus carrier.

In January 2003, another major blow came when the FDA placed a temporary halt on all gene therapy trials using retroviral vectors in blood stem cells. FDA took this action after it learned that a second child treated in a French gene therapy trial had developed a leukemia-like condition. Both this child and another who had developed a similar condition in 2002 had been successfully treated by gene therapy for X-linked severe combined immunodeficiency disease (X-SCID), also known as 'bubble baby syndrome'.

In February 2003, FDA's Biological Response Modifiers Advisory Committee (BRMAC) met to discuss the possible measures that could allow a number of retroviral gene therapy trials for treatment of life-threatening diseases to proceed with appropriate safeguards. In April 2003, the FDA eased the ban on gene therapy trials using retroviral vectors in blood stem cells.

Several human disorders have been rectified in cultured cells and many diseases, such as malignant melanoma and severe combined immunodeficiency disease (SCID), are being treated by gene therapy techniques in recent years. This shows that gene therapy can be hopefully a powerful therapeutic technique against diseases in future. Some of the human disorders treated in cultured cells by gene therapy are given in Table 9.5.

Table 9.5. Examples of human disorders treated in cultured cells by gene therapy.

Disorder	*Affected Gene*
Familial hypercholesterolemia (FH)	LDL receptor
Gaucher disease	Acid β-glucosidase (glucocerebrosidase)
Hemophilia B	Factor IX
Lesch-Nyhan syndrome	Hypoxanthine-guanine Phosphoribosyltransferase (HGPRT)
Phenylketonuria (PKU)	Phenylalanine hydroxylase
SCID	Adenosine deaminase (ADA), purine nucleoside phosphorylase (PNP)
β-Thalassemia	β-Globin

G. PROBLEMS AND CONSIDERATIONS

The Weismann barrier is fundamental in the current thinking for the safety of gene therapy. Therefore, soma-to-germline feedback should be impossible. However, there are indications that the Weissman barrier can be breached.

1. Problems Related to Gene Therapy

The line between germ-line correction of potential health problems and germ-line enhancement is a very blurred and indistinct one. It has been found that people with two copies (alleles) of the long version of a gene for angiotensin-converting enzyme (ACE) have greater muscle efficiency and more stamina than people with a long ACE gene and a short ACE gene, who in turn have more physical endurance than people with two short ACE genes. If there are no negative pleiotropic effects associated with two of the long genes and the germ-line procedure is risk-free, many prospective parents will opt after giving the opportunity for their child to have greater stamina. Similarly, as more generally beneficial single gene effects are discovered the age of the 'designer baby' will inevitable descend upon us. Some of the problems of gene therapy are given below:

***(i)* Short-lived nature of gene therapy:** The therapeutic DNA introduced into target cells must remain functional and the cells containing the therapeutic DNA must be long-lived and stable before gene therapy can become a permanent cure for any condition. Problems with integrating therapeutic DNA into the genome and the rapidly dividing nature of many cells prevent gene therapy from achieving any long-term benefits. Patients will have to undergo multiple rounds of gene therapy.

***(ii)* Immune response:** Any time a foreign object is introduced into human tissues; the immune system is designed to attack the invader. The risk of stimulating the immune system in a way that reduces the effectiveness of gene therapy is always a potential risk. Furthermore, the enhanced response of immune system to invaders makes it difficult for gene therapy to be repeated in patients.

***(iii)* Problems with viral vectors:** Viruses present a variety of potential problems to the patient while the carrier of choice in most gene therapy studies such as toxicity, immune and inflammatory responses, and gene control and targeting issues. In addition, there is always the fear that the viral vector, once inside the patient, may recover its ability to cause disease.

***(iv)* Multigene disorders:** The best candidates for gene therapy are the conditions or disorders that arise from mutations in a single gene. Unfortunately, some of the most commonly occurring disorders, such as heart disease, high blood pressure, Alzheimer's disease, arthritis, and diabetes, are caused by the combined effects of variations in many genes. Multigene or multifactorial disorders would be difficult to treat effectively using gene therapy.

***(v)* Chance of inducing a tumour:** If the DNA is integrated in the wrong place in the genome, it could induce a tumour.

2. Ethical Considerations

These 'designer humans' will also have access to greatly improved somatic techniques and pharmacological technologies allowing for very precise adjustments in hormones and neurotransmitters. Self-made man and woman will become a reality in a very concrete sense. Whether these beings will be more like demi-gods, monsters or something as yet unimagined, no one can predict the future. Such experiments are part of human destiny and seem very probable, assuming that our current civilisation persists sufficiently long into the future.

None of us can know the future with any certainty but if human history is any indicator of what to expect we should be quite concerned. The current state of the world is a direct result of countless actions driven by ancient animal motives. This is not to say that animal motives are necessarily bad or destructive. Our capacity for compassion and empathy is a product of our biology.

We choose changes to make in our body, but we might also be making the choice for our children through the germ-line. Do we have that right and how far should we take our ability? There are several ethical consideration : (*i*) can the argument be used to defend a prohibition of germ-line gene therapy ?, (*ii*) conversely, is it responsible and ethically acceptable to leave the potential of our children to the chance effects of the 'genetic lottery', if we obtain the technological capacity to make positive changes, (*iii*) if genetic engineering became the way of the future, would people whose parents could not afford to genetically 'modify' them while still in an embryo, have a chance of achieving with high standards compared to the people who were 'modified' to be perfect, (*iv*) is it ethical to experiment on embryos that have yet to be born ?, (*v*) how would genetic engineering be used to revolutionise warfare ?, (*vi*) does the superior ability breed a superior ambition ?

3. Social Considerations

Various science fiction writers and futurists have suggested that if an 'artificial intelligence' reached a sufficient level of complexity to achieve self-awareness that such a being would also automatically be incapable of harming other conscious beings. Their reasoning in this is an implicit assumption that a self-aware intelligence would have empathy for other conscious entities. This seems highly improbable. In all likelihood, a self-aware artificial intelligence could know only cold detachment. If the artificial intelligence valued its own existence, it would probably protect itself with absolute ruthlessness regardless of the effect on any innocents that happened to get in the way.

Through genetic engineering, individuals may benefit from non-therapeutic genetic engineering, but some people claim that there may be adverse social implications also. Few resources related to medicine and health care are available to everyone. This allows the most privileged to engineer themselves or their children to have special capabilities to what some call a 'genetic aristocracy'. In addition to some people's desires to alter their physical appearance, numerous enhancements via genetic engineering have been proposed including increased memory, intelligence, and less need for sleep. The advantages created by genetic engineering could lead to new forms of inequality between those with genetic enhancements and those without genetic enhancements, while also exacerbating current inequalities between the rich and poor.

The poor would never have access to such technologies. There is no theoretical limit on how much retrovirus can be cultured in a bioreactor. But the high costs would probably require harsh financing plans. Genetic mortgages of the future could go hand in hand with student loans, spelling disaster for dropouts and increasing the risks of a first class education for a middle or lower class citizen. Many people view genetic engineering as just a way to cure diseases.

On the other hand society could get the point where the government pays for everyone to undergo genetic engineering. This may decrease inequalities, because natural differences in ability would be nearly eliminated leaving socio-economic success or failure entirely to the environment and personal choice.

4. Metaphysical Considerations

The metaphysical or spiritual implications of genetically engineered humans are vast in scope. The individual personality shown to be exclusively the result of genetic information acted upon by the environment; the concepts of the human soul and free will be proven specious. However, could this be shown by the success of actual attempts at genetic engineering any more or less than it is shown by what is already known about the roles of genes and the environment in shaping the phenotypic characteristics including behavior of living things ?

H. GOVERNMENTAL OVERSIGHT AND PUBLIC PARTICIPATION

Oversight of government-funded experiments involving gene therapy for human patients occurs at both the local and national levels. At the local level, facilities at which experiments would take place are required to have two types of committees. First, hospitals and universities involved in experiments with human subjects are required to have Institutional Review Boards (IRB) to ensure that the research complies with Department of Health and Human Services (DHHS) regulations for protection of human subjects. Second, experiments that involve gene insertion must be approved in advance by an Institutional Biosafety Committee (IBC).

These local review boards provide an opportunity for the general public to become involved in the decisions made about research involving gene therapy for human patients. The DHHS regulations require that at least one non-scientist serve as a member of each IRB. The *NIH Guidelines for Research Involving Recombinant DNA Molecules* encourage research facilities to open their IBC meetings to the public. The Director of the NIH must approve each human gene therapy proposal. In making this decision, the Director seeks advice from the Recombinant DNA Advisory Committee (RAC). The initial review of the proposal is performed by the RAC's Human Gene Therapy Subcommittee which is guided by the *Points to Consider in the Design and Submission of Protocols for the Transfer of Recombinant DNA into the Genome of Human Subjects*. The general public is represented on the RAC and the Human Gene Therapy Subcommittee as well as on the local review boards. The membership of the RAC and of the Human Gene Therapy Subcommittee includes scientists, physicians, lawyers, ethicists, and several laypeople.

Since the 1970s, general interest in human gene therapy has increased both here and abroad, along with awareness of the need for oversight and regulation. In 1974, the Department of Health, Education and Welfare (now the DHHS), chartered the RAC to develop recommendations for the regulation of recombinant DNA research. The *Guidelines for Research Involving Recombinant DNA Molecules* were published in 1976. In 1978, the Guidelines were revised, relaxing many of the requirements for recombinant DNA experiments. In 1980, at the urging of the three major religious groups, the President requested that the President's Commission for the Study of Ethical Problems in Medicine and Biomedical and Behavioral Research examine the topic of human genetic engineering. In November 1982, one of the Commission's eleven reports was submitted to the President and Congress. A subcommittee of the United States House of Representatives held three days of hearings on the report. Congressional interest also resulted in a 1984 background paper entitled 'Human Gene Therapy' produced by the Office of Technology Assessment. In response to one recommendation of the President's Commission, the NIH RAC formed a working group in 1984 to specialise in human gene therapy. This is now the Human Gene Therapy Subcommittee.

In 1985, the White House Office of Science and Technology Policy created the Biotechnology Science Coordinating Committee (BSCC). The Committee includes federal officials representing the National Institutes of Health, the Environmental Protection Agency, the U.S. Department of Agriculture, the Food and Drug Administration, and the National Science Foundation. The BSCC provides a forum for discussion of biotechnology issues and an opportunity to make recommendations on the federal regulation of biotechnology.

Legislative interest continues to be expressed through activities of the Subcommittee on Science, Research and Technology of the House Committee on Science and Technology and the Office of Technology Assessment. In 1986, a Congressional Biomedical Ethics Board was formed to oversee research and developments in genetic engineering.

International interest in human gene therapy has resulted in a number of reports and recommendations submitted by foreign government committees. In 1982, the Parliamentary Assembly of the Council of Europe issued a statement including proposals for oversight and recommendations for certain restrictions on human genetic engineering. Thus, the research on human gene therapy is being monitored at both the local and national levels as well as at national and international levels. Members of the general public are represented and are encouraged to participate in the public discussion of this new area of biomedical research.

QUESTIONS

1. Write in brief about the history of gene therapy.
2. Write a brief account of genetic diseases studied by you.
3. What is gene therapy ? Write different types of gene therapy studied by you.
4. Write in detail about different methods used in gene therapy.
5. Give an illustrated account of viral gene therapy with special reference to severe combined immunodeficiency (SCID).
6. Describe in detail about non-viral methods of gene therapy.
7. What is antisense therapy ? Write the method of antisense therapy.
8. What is gene silencing ? Write in detail about methods of gene silencing.
9. Discuss in brief about the problems associated with gene therapy.

10. Write short notes on the following :

(*i*) Sickle cell anaemia, (*ii*) Thalssaemia,
(*iii*) Duchenne muscular dystrophy,
(*iv*) Severe combined immunodeficiency (SCID), (*v*) Germline gene therapy,
(*vi*) Somatic gene therapy, (*vii*) Designer baby,
(*viii*) Vector-mediated gene therapy, (*ix*) Lipoplex and polyples,
(*x*) Receptor-mediated gene therapy, (*xi*) Embryo therapy,
(*xii*) Magic bullet, (*xiii*) Prodrug therapy,
(*xiv*) Genetically modified mesenchymal stromal cell for gene therapy,
(*xv*) Virosomes in drug targeting, (*xvi*) Chemoviruses,
(*xvii*) Synthetic viruses, (*xviii*) Dicer enzyme,
(*xix*) Public perception about gene therapy.

BIBLIOGRAPHY

Boeckle S. and Wagner E. 2006. Optimizing targeted gene delivery: chemical modification of viral vectors and synthesis of artificial virus vector systems. *AAPS Journal*. **8**: 83.

Dominici, M. *et al.* 2006. Minimal criteria for defining multipotent mesenchymal stromal cells. The international society for cellular therapy position statement. *Cytotherapy*. **8**, 315–317.

Gardlik R. *et al.* 2005. Vectors and delivery systems in gene therapy. *Medical Science Monitor.* **11**: *110–121.*

http://en.wikipedia.org/wiki/Gene_therapy; Cystic_fibrosis; Duchenne_muscular_dystrophy; Thalassemia

Kreppel F., Gackowski J., Schmidt E. and Kochanek S. 2005. Combined genetic and chemical capsid modifications enable flexible and efficient de- and retargeting of adenovirus vectors. *Mol Ther*. **12**:107-117.

Walker G.F., Fella C. and Pelisek J. *et al.* 2005. Toward synthetic viruses: endosomal pH-triggered deshielding of targeted polyplexes greatly enhances gene transfer in vitro and in vivo. *Mol. Ther*. **11**:418-425.

Weiss B., Davidkova, G., and Zhou, L-W. 1999. Antisense RNA gene therapy for studying and modulating biological processes. *Cell. Mol. Life Sci.* 55:334-358.

Wu G.Y. and Wu C.H. 1987. Receptor-mediated *in vitro* gene transformation by a soluble DNA carrier system. *J Biol. Chem*. **262**:4429-4432.

Yotnda P, Davis A.R., Hicks M.J., *et al.* 2004. Liposomal enhancement of the antitumor activity of conditionally replication-competent adenoviral plasmids. *Mol Ther*. **9**: 489-495.

10

CHAPTER

DNA Profiling

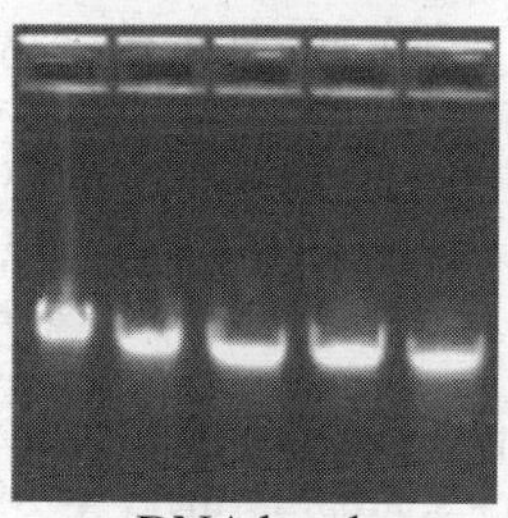

DNA bands

A. DNA fingerprinting (= DNA profiling)
B. Methods of DNA fingerprinting profiling
C. Application of DNA profiling technology
D. Hurdles to DNA profiling technology
E. Disease diagnosis

DNA is the master molecule of all life forms which constitutes the blue prints of every living organism through which characters are passed generation to generations. However, every living organism differs only due to nucleotide sequences of chromosome. The nucleotides form codes; therefore, this coded genetic information can be profiled to produce most authentic 'identity card' of any organism. The complicated technology that facilitates the identification of individuals at genetic level is known as DNA fingerprinting (= DNA testing, DNA profiling, DNA typing, and genetic fingerprinting). DNA fingerprinting was renamed as 'DNA profiling' to stop confusion with fingerprints. This genetic analysis is based on identifying tiny segments of the hereditary material which testify the unique molecular signature which cannot be altered. Identity of burnt or unidentifiable dead body can be done with the help of DNA profiling technology. The basic requirements for DNA profiling is the availability of biological samples which can be anything including blood stains, a piece of hair with its root, few drops of semen, skin cells, a mouth swab, cells of bone marrow of any body tissue.

Large portions of any single person's DNA are the same as every other person's, because we are all human beings and a large chunk of our DNA is dedicated to our species-specific traits. We have feet instead of hooves, skin instead of scales, etc., but other sections or fragments of human DNA are unique to the individual. These fragments are called polymorphic because they vary in shape from person to person. Approximately, 99.9% of human DNA sequences are the same in every person. But much of the DNA is different to distinguish one individual from the other. The number of repeats can vary from about one to thirty and are not the same from person to person. These highly variable repetitive sequences are called 'Variable Number of Tandem Repeats' (VNTRs). DNA profiling uses the VNTRs loci which are very similar between closely related humans but it varies between different individuals who have the same VNTRs. An enzyme can cut the DNA at certain sites within the VNTRs; the location of these sites also varies from person to person. As a result of cutting with the enzyme DNA fragments of different lengths are produced, which are called Restriction Fragment Length Polymorphisms (RFLPs). On the basis of their size these DNA fragments can be separated on an agarose gel. The RFLPs can be visualised by using the complementary radioactive DNA probe that compares different samples of DNA. In the human genome some RFLPs occur only once, while the others occur many times on several chromosomes.

Sir Alec John Jeffreys

In 1984, Sir Alec Jeffreys of the University of Leicester (England) reported the technique of DNA profiling for the first time, which later was commercialised in 1987 (http://en.wikipedia.org/wiki/Genetic_fingerprinting). Essentially, DNA profiling is the process of separating an individual's unique and polymorphic fragments from the common ones. DNA profiling now helps to prove innocence or guilt, resolve immigration arguments and clarify paternity cases. DNA fingerprints have been used in many areas of human health care research, as well as in the judiciary system. DNA fingerprinting is used to diagnose inherited disorders in both prenatal and newborn babies in hospitals all over the world. These disorders include cystic fibrosis, haemophilia, Huntington's disease, familial Alzheimer's, sickle cell anaemia, etc. Early detection of such disorders enables the medical staff to prepare themselves and the parents for proper treatment of the child. In some programmes, genetic counsellors use DNA fingerprint information to help prospective parents to understand the risk of having a disabled child. DNA fingerprints are also vital to developing cures for inherited disorders. Research programmes to locate the genes causing inherited disorders depend on the information contained in the profiles.

The identity of a person at a crime scene can be determined by the police by using the same analysis. From the scene and any suspects the police collect a DNA samples, amplify the VNTRs and digest the DNA with a restriction enzyme in a molecular laboratory. After running the samples on an agarose gel, the bands found at the crime scene are aligned with that of the suspects. DNA fingerprints can either prove someone innocent or guilty.

A. DNA FINGERPRINTING (= DNA PROFILING)

Genes are the functional unit of inheritance that reside at a specific location (locus) within the human genome and include regions involved in regulation of expression and regions that code for specific functional products. DNA fingerprinting involves the generation of a set of distinct DNA fragments from a single DNA sample; these fragments are then used as a source of genotypic information. A wide range of techniques can be used to generate DNA fingerprinting data. Researchers choose among these techniques based on the organism being studied and on the question being addressed.

Nature does not tolerate a lot of variation within genes since changes or mutations can result in genetic disease, if the mutation disrupts the function of a protein. Cancer develops if the changes are in regulatory elements. As important as genes are for medical genetics and cancer, they only account for approximately 2% of the human genome.

Variation is the Cause of Evolution and Basis of DNA Fingerprinting

Humans are 96% genetically identical to the chimpanzee, which are our closest living relative. Humans are even more similar to each other; two people from anywhere on earth are expected to be 99.9% genetically identical. It means that the whole world exists only in 0.1% dissimilarity in DNA. How can DNA of one be unique enough to identify him or her as the perpetrator of a particular crime? The answer is the volume. The human genome is composed of three billion base pairs. Even at 99.9% similarity, any two people will still differ at about three million base pairs. Obviously, no two people on earth have exactly the same genetic sequence, except the identical twins.

Some of these genetic differences influence the unique set of characteristics that makes your eye color, hair color, height, tendency towards heart disease, and numerous other traits. But most of these genetic differences have no discernable effect on your phenotype, or set of physical features. Based on these genetic differences the biologists focus on when they are trying to identify a suspect by using DNA fingerprinting.

Different parts of the genome evolve at different rates. DNA that encodes important traits tends to evolve slowly. This is because most mutations in critical regions of the genome are likely to cause detrimental effects and be selected out of the population rapidly. If a stretch of DNA evolves slowly, few changes in its sequence will occur and many people in the population will likely carry identical sequences. Though important, these regions of the genome will not be very helpful for identification.

Variation in some regions of the genome has little effect on the characteristics of the organisms, and variants are largely invisible to natural selection. Mutations accumulate without much consequence and gene frequencies change through genetic drift. These regions evolve quickly; consequently different individuals in the population carry different sequences in these regions. Within the fast-evolving regions of the genome (hotspots), there may be particular mutation resulting in sequence variation. Many of these regions contain DNA that repeats the same sequence of bases over and over again (e.g. ATGGATGGATGGATGG...). It is thought that cells frequently make mistakes while copying these regions resulting in the production of more or less repeats than in the original DNA sequence, causing a new mutation. Since, they evolve very fast and vary much in term of repeats; therefore, these hotspots are ideal targets for DNA fingerprinting.

DNA profiling is a technique which is employed by forensic scientists to help in the identification of individuals by their respective 'DNA profiles'. DNA profiles are encrypted sets of numbers that reflect the DNA makeup of a person which can also be used to identify a person. For the first time, DNA fingerprinting also allowed the systematic isolation of large numbers of different human minisatellites, an essential prerequisite to any studies of the biology of these loci. Some minisatellites so isolated proved to be extraordinarily variable, with scores or even hundreds of alleles differing in repeat copy number. These loci not only proved to be efficient markers for gene mapping and loss of heterozygosity studies, but also provided the second approach to human identification, namely single-locus DNA profiling, which yields simple two-band hybridisation profiles rather than the complex multi-band phenotypes generated by DNA fingerprinting. DNA profiling saw its forensic debut in 1986 with the Enderby murder case involving two young girls who had been raped and murdered. In 1987, A.J. Jeffrey and co-workers characterized a panel of highly variable minisatellites cloned from human DNA. Semen DNA profiling resulted in the release of a young man falsely accused of the crimes, the first individual to be exonerated by DNA typing, and the successful entrapment of the true murderer following the world's first DNA-based man hunt. DNA profiling is now standard practice in forensic laboratories worldwide, and despite extensive scientific and legal controversy, has maintained its position as a robust and highly reliable DNA typing technology. In 1992, Jeffrey and co-workers identified the skeletal remains of Josef Mengele by DNA analysis and proved the identity.

It is important to note that 99.9% of base sequences are the same in the DNA of all human beings. Only a short stretches of sequences which is about 3 billion base pairs, differ person to person. Out of the total DNA, about one in 1,000 base pairs is a site of variation in the population. This is an unusual sequence of bases of 10-15 base pairs long which is repeated several times. At least 50% of the human genome is made of simple repeat sequences, because these regions of our DNA do not encode proteins or regulatory element. It appears that changes can accumulate with impunity. Genetic elements that are highly variable between individuals are called polymorphic (many forms).

Although they are not genes, yet they are called 'genetic elements'; they reside at specific locations (loci) within the genome. Most forensically important loci are 'tandem repeats'; which are DNA elements that are repeated many times, head to tail. The number of repeats is what varies between individuals. Another important class of repeating elements is the short tandem repeats (STR) which is non-coding DNA sequences (part of the so-called junk DNA). STRs have repeat elements that are 2-5 bp in length. Most STR alleles contain between 5 and 45 repeating elements and range in size from 20-180 bp (the actual size will be larger because of the inclusion of flanking sequences). Because of their relatively small size, STR loci are well suited to PCR amplification. These STRs appear at set points (known as loci) on the human genome, but the precise number of repeated sequences at each locus varies between different people and ranges from four to around forty.

1. Biological Evidence

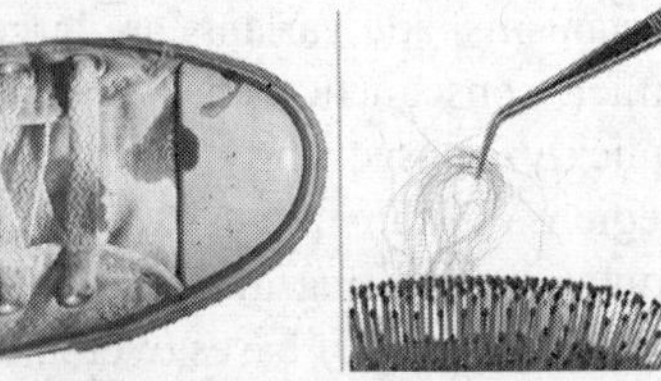

Blood stain and hair pieces act as biological evidence.

A critical challenge in utilising biological evidence is first finding, and then figuring out it exactly. At a crime scene blood may not always be visible (if someone has cleaned up), so investigators apply chemicals such as luminol that changes colour as red blood cells release oxygen. Immunological tests can be performed in the laborator to identify proteins such as haemoglobin that indicates the human blood. Hair samples are examined microscopically for characteristics such as colour, chemical treatment, shaft form (wavy, straight, curly) and pigment distribution. All biological samples have the potential to yield useful DNA evidence. When laboratories perform nuclear DNA testing, examiners look for matches up to 16 short tandem repeat (STR) loci, which vary considerably from person to person. Where nuclear DNA is either degraded or not available, bones, teeth and hair samples that have no roots, the less-exacting mitochondrial DNA (mtDNA) testing process is typically employed.

2. Variable Number of Tandem Repeats

These bases were first isolated and identified by A. Wyman and R. White in 1980 at the University of Utah as hypervariable segments of DNA and termed as 'variable number of tandem repeats' (VNTRs). Variations in VNTR allele lengths in four alleles of individuals are given in Fig. 10.1 A. The repeats are tandem, *i.e.* they are clustered together and oriented in the same direction (Fig. 10.1 B). Hence, the VNTR is a location in a genome where a short nucleotide sequence is organised as a tandem repeat. These can be found on many chromosomes and often show variations in length between individuals. Each variant acts as an inherited allele allowing them to be used for personal or parental identification. The VNTR has repeat units of 9 to 80 bp. Depending on the number of repeat units, the genetic elements can be 1 to 20 kb in size. The length of segment is measured on the basis of number of repeats present in VNTRs. In different person the length of VNTR differs and, therefore, VNTR is the key of DNA fingerprinting (http://en.wikipedia.org/wiki/Variable_number_tandem_repeat).

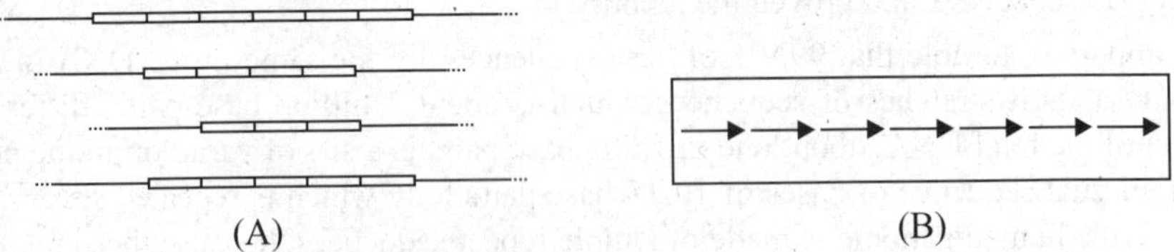

Fig. 10.1. Tandem repeats; (A) a variable number of tandem repeats in four alleles; (B) the arrows represent seven identical elements that are repeated head to tail.

Individual repeats can be removed from (or added to) the VNTR via recombination or replication errors leading to alleles with different numbers of repeats. Flanking the repeats are segments of non-repetitive sequence as shown here as thin lines (Fig. 10.1). This allows the VNTR loci to be extracted with restriction enzymes and analysed by RFLP or amplified by the polymerase chain reaction (PCR) technique and their size is determined by gel electrophoresis.

There are two main families of VNTRs: *microsatellites* and *minisatellites*. The microsatellites are repeats of sequences less than about 5 base pairs in length (an arbitrary cutoff), while the minisatellites involve longer blocks. VNTRs with very short repeat blocks may be unstable, *i.e.* dinucleotide repeats may vary from one tissue to another within an individual, while trinucleotide repeats have been found to vary from one generation to another.

What is Genetic Engineering ?

The structure of every individual is the same but every person has a different sequence of base pairs. Every individual has billions of base pairs, making it too time-consuming to identify individual. Indeed, scientists look at common repeating patterns known to vary among individuals. They can then determine whether two samples are from the related people.

In 1984, for the first time, a British geneticist Alec Jeffreys working at the University of Leicester developed the technique of DNA fingerprinting and DNA profiling. This method has been patented, and being used in Europe and America. Its success is based on identification of small segment of DNA which testifies the unique molecular signature that every person on earth possesses and which cannot be changed in one's life time.

It was Jeffreys who first used VNTRs in forensic science. The VNTRs act as genetic markers. The scheme developed by Jeffreys for measuring the length of fragments of DNA containing the repeated sequences is also known as 'restriction fragment length polymorphism' (RFLP) testing. Lengths of VNTR alleles of different six individuals vary (Fig. 10.2)

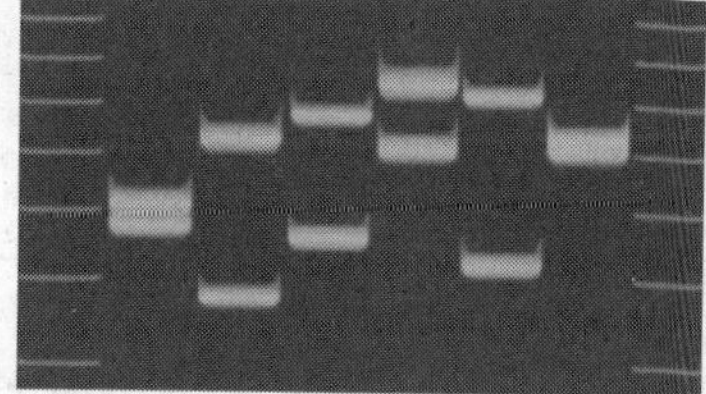

Fig. 10.2. Variations of VNTR allele lengths in six individuals.

3. Microsatellites

Microsatellites (sometimes referred to as VNTRs) are simple sequence tandem repeats (SSTRs) of DNA, such as CACACACA; they occur in non-coding DNA. In some microsatellites, the repeated units (that is CA) are generally di-, tri- tetra- or pentanucleotides; in others it may be seven, two, or thirty. Presence of microsatellite sequence (CA)s in a chromosome is shown in Fig. 10.3. In diploid organisms like elephants, each individual animal has two copies of any particular microsatellite segment. For example, a father might have a genotype of 12 repeats and 19 repeats, a mother might have 18 repeats and 15 repeats, while their first born might have repeats of 12 and 15. On rare occasions, microsatellites can cause the DNA polymerase to make an extra copy of CA. If an individual DNA polymerase adds to the repeated sequence, this slightly larger version can be passed on to offspring who will usually replicate it accurately. As the animals breed in a population, they will recombine their microsatellites during sexual reproduction and the population will maintain a variety of microsatellites that is characteristic for that population and distinct from other populations which do not interbreed.

In birds a common repeat motif is AC_n, where the two nucleotides A and C are repeated in bead-like fashion for a variable number of times (n ranges from 8 to 50). They occur in non-coding regions of the DNA; although a few human genetic disorders are caused by tri-nucleotide microsatellite regions present in coding regions. Flanking regions consisting of unordered DNA are present on each side of the repeat unit. The flanking regions allow to develop locus-specific primers to amplify the microsatellites with PCR. The widely occurring repeat units and locus-specific flanking regions are used as part of strategy for finding and developing microsatellite primers. The primers for PCR will be sequences from these unique flanking regions. A short (100 to 500 bp) locus-specific microsatellite region can be amplified by using a forward and a reverse primer on each side of the microsatellite.

Advantages of using microsatellites as genetic markers are that : (*i*) they are locus-specific (in contrast to multi-locus markers such as minisatellites or RAPDs), (*ii*) they become codominant (*i.e.*, heterozygotes can be distinguished from homozygotes, in contrast to RAPDs and AFLPs), (*iii*) they are PCR-based markers (*i.e.* only a small amount of tissue can be required; they work on highly degraded or ancient DNA), (*iv*) they are highly polymorphic (*i.e.* hypervariable) and provides considerable pattern, and (*vi*) they are useful at a range of scales to fine-scale phylogenies.

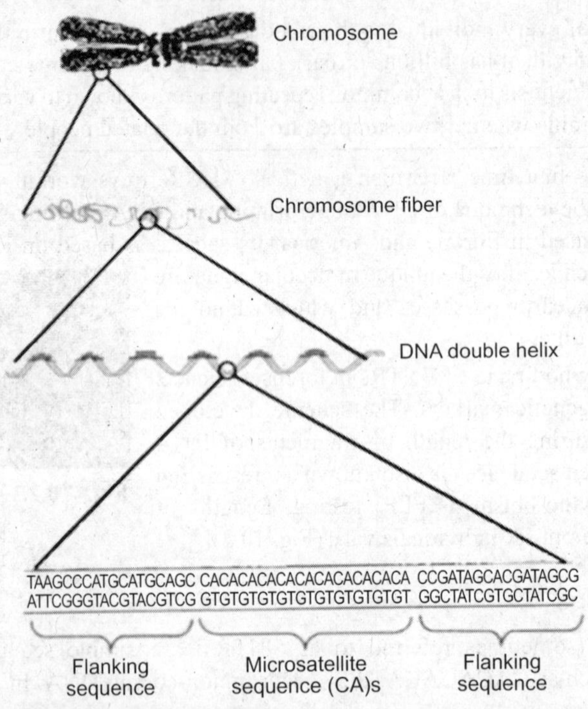

Fig. 10.3. Presence of microsatellite sequence (CA)s in a chromosome.

4. Minisatellites

A section of DNA consisting of a 10–60 bp long series of bases is called minisatellite. In the human genome these occur at more than 1,000 locations. There are some minisatellites that consist of a central (or core) sequence of letters –GGGCAGGANG- (where N can be any base) or more generally a strand bias with purines [adenosine (A) and guanine (G)] on one strand and pyrimidines [cytosine (C) and thymine (T)] on the other. This sequence encourages chromosomes to swap DNA. Moreover, it is the presence of a neighbouring *cis*-acting meiotic double-strand break hotspot which is the primary cause of minisatellite repeat copy number variations. Somatic changes result from replication difficulties. Errors are made when such events occur that cause minisatellites at more than 1,000 locations in the genome of a person. Therefore, such person will have slightly different numbers of repeats. This makes each individual unique. Thus, minisatellites consist of repetitive and generally GC-rich variant repeats that range from 10 to 100 bp. These variant repeats are tandemly intermingled, which make minisatellites ideal for studying the mechanisms of DNA turnover. Minisatellites have also been implicated as regulators of gene expression (e.g. at levels of transcription, alternative splicing or imprint control) or as part of bona fide open reading frames.

In the early 1980s, human gene characterisation revealed the first examples of highly variable minisatellites composed of repeat arrays, typically several kilobases in length and with repeats 10–60 bp long, depending on the locus. The discovery of sequence similarities between repeat units at different loci, the so-called minisatellite 'core' sequence, made it possible to develop hybridisation probes capable of detecting multiple highly variable loci to produce a Southern blot profile that was so variable between individuals that it could be termed a DNA fingerprint. DNA profiling based on highly variable minisatellites in the human genome was developed by a team of scientists led by Peter Gill at the Forensic Science Service in the 1990s. DNA fingerprinting was renamed as DNA profiling to stop confusion with fingerprints.

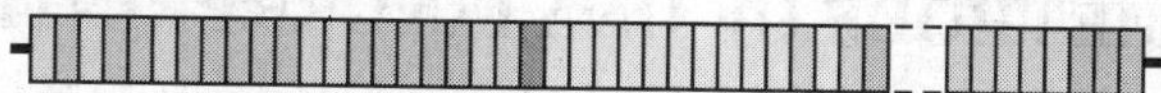

Fig. 10.4. Examples of minisatellites detected in human sperm.

DNA profiling focused on just a few of these highly variable minisatellites, making the system more sensitive, more reproducible and amenable to computer databasing. With highly automated and sophisticated equipment, modern-day DNA profiling can process hundreds of samples each day. This DNA profiling technique was the basis for the U.K. National DNA Database (NDNAD) launched in Britain in 1995. Ten minisatellites plus a marker for sex determination are used with the current system developed for the NDNAD, giving a discrimination power of one in over a billion. Under British law, anyone arrested has their DNA profile stored on a database (whether or not they are convicted), which now contains the DNA information of nearly five million people. Jeffreys has opposed the current use of DNA profiling where the government has access to that database, and has instead proposed a database of all people's DNA whose access would be controlled by an independent third party (http://en.wikipedia.org/wiki/Alec_Jeffreys).

Sir Alec John Jeffreys, FRS

Sir Alec John Jeffreys (born on 9 January 1950 in Oxfordshire) is a British geneticist, who developed techniques for DNA fingerprinting and DNA profiling which are now used all over the world in forensic science to assist police detective work and also to resolve paternity and immigration disputes. He is a Professor of Genetics at the University of Leicester, who became an honorary freeman of the City of Leicester in 1992. In 1994, he was knighted by her Majesty Queen Elizabeth II of England for Services to Science and Technology.

Jeffreys had a 'eureka moment' in his laboratory after looking at the X-ray film image of a DNA experiment on 10 September 1984, which unexpectedly showed both similarities and differences between the DNA of different members of his technician's family. Within about half an hour, he realised the possible scope of DNA fingerprinting, which uses variations in the genetic code to identify individuals.

Jeffreys' DNA fingerprinting was first used when he helped in a disputed immigration case by confirming the identity of a British boy whose family was originally from Ghana. The case was resolved when the DNA results proved that the boy was closely related to the other members of the family. Jeffreys saw the relief in the mother's face when she heard the results. DNA fingerprinting was first used as a police forensic test to identify the rapist and killer of two teenagers, Lynda Mann and Dawn Ashworth, who were both murdered in Leicestershire in 1983 and 1986, respectively. Colin Pitchfork was identified and convicted of murder after samples taken from him matched semen samples obtained from the two dead girls. This turned out to be an important identification method. British Authorities believed that it was inevitable for Richard Buckland, the main suspect at the time, who would have been found guilty. In this case Jeffrey's work proved the real killer, and also exonerated someone who likely would have spent his life in prison.

Differences between individuals in the number of repeats of a minisatellite sequence arise through unequal crossing over between chromosomes during meiosis. Some minisatellite sequences are highly variable in repeat number between individuals. This is the basis of the DNA fingerprinting.

After the discovery of the first human minisatellite in 1980, the extreme polymorphism of minisatellites made them superb for DNA fingerprinting by Alec Jeffreys, this class of repeats has been an intense focus of studies on the turnover mechanisms that provoke their instability. Due to their high level of polymorphism, minisatellites have been extensively used for DNA fingerprinting as well as for genetic markers in linkage analysis and population studies.

B. METHODS OF DNA FINGERPRINTING

In this technique DNA is isolated from blood stains, semen stains or hair roots of disputed children or any suspected person and the same from parents, close associates or relatives of suspected criminals (based on the cases). Since hairs contain less amount of DNA, it can be produced in a large amount by using polymerase chain reaction (PCR). RBCs do not contain DNA; therefore, WBCs act as the source of DNA.

Prof. Lalji Singh

Thus the DNA isolated is cut with restriction enzyme and subjected to Southern blotting. The DNA bands appearing on membrane are hybridised with ^{32}P-DNA probe, washed in water to remove un-hybridised DNA, and passed through X-ray. The hybridised complementary DNA sequences develop images (prints). Identical prints that contain specific DNA sequences appearing on two X-ray films are identified and thus identity is confirmed. Probability of two persons having similar sets of base pairs in the same sequence of VNTR of DNA is one in 30-300 million individuals.

In India, facilities of DNA fingerprinting at international level are available at the Centre for DNA Fingerprinting and Diagnostics (CDFD) (Hyderabad), the then Centre for Cellular and Molecular Biology (CCMB) (Hyderabad). At this centre the Jeffreys technique is not followed. Professor Lalji Singh has developed an indigenous technique where 'BKM'-DNA probe is used for hybridisation of DNA sequences. While he was working on sex determination in snakes in his Ph.D. programme (at the Banaras Hindu University), he found a contrast result. Unlike human and other organisms, the female snake contains XY and the male Y.Y sex chromosomes. A segment of DNA was isolated from sex determining Y chromosome of female banded krait (*Bungarus fasciatus*), an Indian poisonous snake. This unique segment was named as 'banded krait minor satellite' (BKM). It is similar to sex determining chromosome in humans. The probe which is used for this purpose is, therefore, called 'BKM-derived probe'. Different methods of DNA fingerprinting have been described in this section. In 2009, he proved through his research work that the Aryan and Dravidian races in India had the same genetic basis (*Nature*, 2009); this work has been widely acclaimed in the scientific community.

Padmshri Professor Lalji Singh- the Great Son of the Soil

Dr. Lalji Singh has been the Director of the Centre for Cellular and Molecular Biology (CCMB), Hyderabad (India). Besides, he is a respected and widely acclaimed Molecular Biologist in India. Dr. Singh won the Banaras Hindu University Gold Medal for standing first in order of merit in the M.Sc. Zoology in 1966. He earned Ph.D. degree in 1971 from the Banaras Hindu University for his work on 'evolution of karyotypes in snakes'. He was Pool Officer at the Calcutta University in 1974. He was awarded the Dr. S. P. Basu Memorial Medal of Zoological Society (Calcutta) in 1973 for his outstanding contribution in the field of cytogenetics. In 1974 the Indian National Science Academy decorated him with the Science Academy Medal for Young Scientists. He won the two year Commonwealth Fellowship in 1974 to carry out research at the Edinburgh University, U.K.

In June 1987, Dr Singh came back to India and joined the CCMB, Hyderabad as senior scientist. Dr. Singh and his colleagues developed a probe called 'BKM-derived probe' for DNA fingerprinting which brought CCMB to limelight. Since then this probe is being extensively used for forensic investigation, paternity determination and seed stock verification. For the first time DNA fingerprinting evidence was presented in the court of law in India. The Kerala High Court upheld the verdict. Since then this indigenous technique has been used as evidence in more than 500 cases including the assassination of the late Prime Minister of India Shri Rajiv Gandhi, Naina Sahni (Tandoor) murder case, the Beant Singh (the late Punjab Chief Minister) murder case, and the case of the Premananda Ashram in Tamil Nadu, etc. This fetched him and his group the CSIR Technology Award for the year 1992 for biological sciences. This resulted in the formation of an autonomous institution, Centre for DNA Fingerprinting and Diagnostics (CDFD) in 1995. In 1998 he was appointed the Director of the CCMB, Hyderabad. The All India Conference of Intellectuals (AICOI), New Delhi has conferred the prestigious Intellectual's Award '*The great son of the soil*' upon Lalji Singh for his outstanding contributions in developing a universal probe for DNA Fingerprinting technology. In 2004, he was honoured with *Padmashri* by the President of India (http://en.wikipedia.org/wiki/Lalji_Singh). After his retirement, he was appointed as the 25th Vice Chancellor of the Banaras Hindu University, in 2011for five years.

1. RFLP Analysis

Restriction enzyme digestion is the first method used for DNA fingerprinting followed by Southern blotting. More commonly the enzymes and DNA probes were used to analyse VNTR loci. But polymorphism can exist in the restriction enzyme cleavage sites. However, the Southern blot technique is laborious and requires large amounts of undegraded sample DNA. Jeffreys' original technique looked at many minisatellite loci at the same time but making it hard to discern individual alleles. PCR-based assays have supplanted the early techniques.

Until a few years ago, most crime laboratories used a DNA profiling method known as Restriction Fragment Length Polymorphism (RFLP). Although RFLP yielded excellent results in many cases, the method was slow, labour-intensive and required relatively large samples of high quality DNA. In the early days of RFLP, DNA profiles were detected with ^{32}P labelled probes. It would often take 5 or 6 weeks to complete a single case. With the advent of enzyme-linked probes and chemiluminescent detection in the mid 1990s, it was often possible to complete a simple case in 1 or 2 weeks. DNA fingerprinting is given in Fig. 10.5 that requires the following six steps :

(*i*) **Isolation of DNA:** Isolate the DNA from the cells from crime scene evidence and from reference samples. Only a small amount of cells/tissue is needed. For example, the amount of DNA found in one drop of blood is usually sufficient.

(*ii*) **Cutting, sizing, and sorting the DNA:** Special enzymes called restriction enzymes are used to cut the DNA at specific places. For example, an enzyme called *Eco*R1 found in bacteria will cut DNA only when the sequence GAATTC occurs. Usually *Hae*III is used that does not cut within the VNTR elements.

(*iii*) **DNA separation:** The DNA pieces are sorted according to size by a sieving technique called 'electrophoresis' through a gel made from seaweed agarose. Electrophoresis is the DNA separation equivalent to sieving sand through progressively finer mesh screens to determine particle size distribution.

(*iii*) **Transfer of DNA to nylon:** The DNA pieces are transferred to a nylon sheet by placing the agarose gel and nylon next to each other overnight.

(*iv-v*) **Probing:** The DNA fingerprint is generated by adding tagged probes to the nylon sheet. Each probe typically sticks in only one or two specific places, wherever the sequences match (A with T and G with C). The tag allows detection of the probe.

(*vi*) **DNA fingerprint:** The final DNA fingerprint is built from several different probes and resembles the bar codes used at the grocery counter.

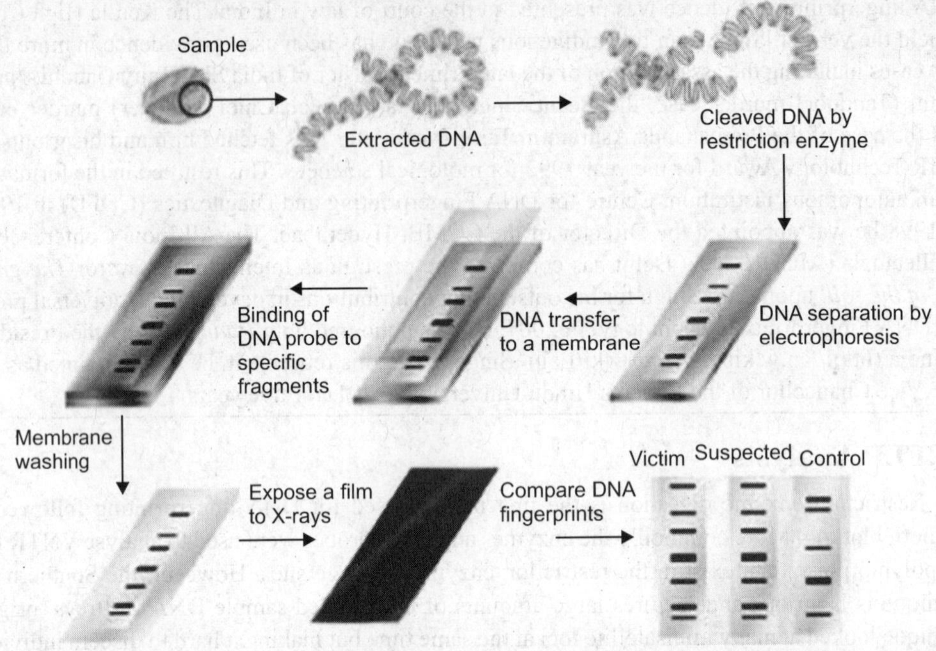

Fig. 10.5. RFLP analysis of DNA from blood or stains from suspects using DNA probe.

RFLP studies can be extremely time-consuming and challenging in the absence of such data. When sequence data is not available, the experimenter must physically clone a region of the genome under study that is large enough to be cut up by enzymes, and this process requires a great deal of time and resources.

2. PCR Analysis

Early assays such as the HLA-DQ alpha reverse dot blot strips became very popular due to their ease of use and the speed to get a result (Fig. 10.6). However, they were not as discriminating as RFLP. It was also difficult to determine a DNA profile for mixed samples such as a vaginal swab from a sexual assault victim. DNA fingerprinting took huge strides forward in both discriminating power and the ability to recover information from very small (or degraded) starting samples with the invention of the PCR technique.

Allele-specific testing can be done on a much smaller and less pure sample of DNA and takes less time than RFLP. In a criminal case this is a distinct advantage in which one is trying to extract DNA from a single strand of hair, a drop of blood or only skeletal remains of the victim. PCR amplifies a specific region of DNA from a small sample by using oligonucleotide primers and DNA polymerase.

Fortunately, the PCR method is readily adaptable for analysing VNTR loci. The FBI of the United States has standardised a set of 13 VNTR assays for DNA typing and has organised the CODIS database for forensic identification of cases like criminals. In the other countries, similar assays and databases have also been set up. Besides, commercial kits are available that analyse single nucleotide polymorphism (SNPs) that use PCR to amplify specific regions with known variations and hybridise them to probes anchored on cards. This results in a coloured spot corresponding to the particular sequence variation.

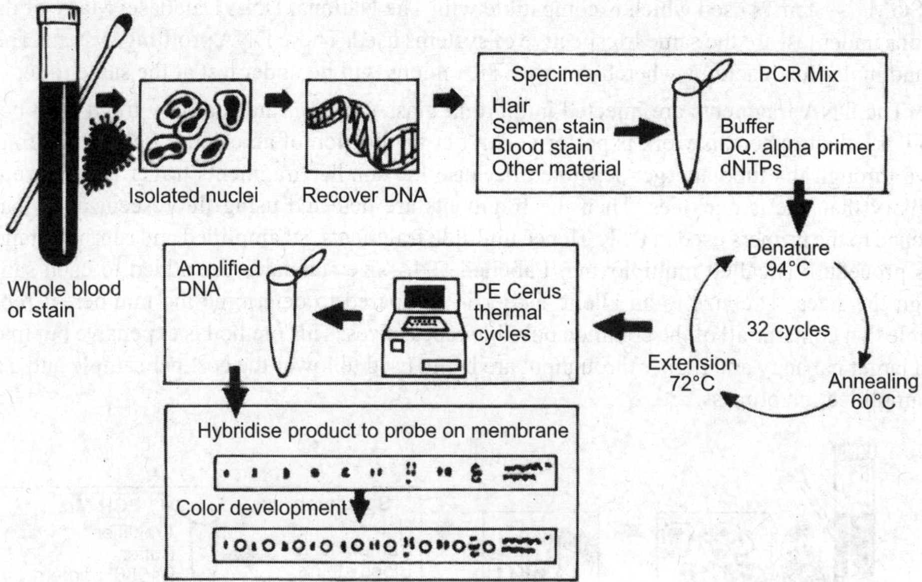

Fig. 10.6. DQ-alpha typing by using primers and hybridisation of product to probes for colour development.

3. Short Tandem Repeat (STRs) Analysis

PCR is one of the most prevalent methods of DNA fingerprinting that uses short tandem repeats (STR) (Fig. 10.7). STR analysis method uses the highly polymorphic regions that have short repeated sequences of DNA. The most common is 4 bases repeated but there are other lengths in use including 3 and 5 bases repeats. Because different people have different numbers of repeat units; hence, these regions of DNA can be used to distinguish between different individuals. These STR loci are targeted with sequence-specific primers and are amplified by using PCR. Thereafter, the DNA fragments are separated and detected using electrophoresis (Fig. 10.8).

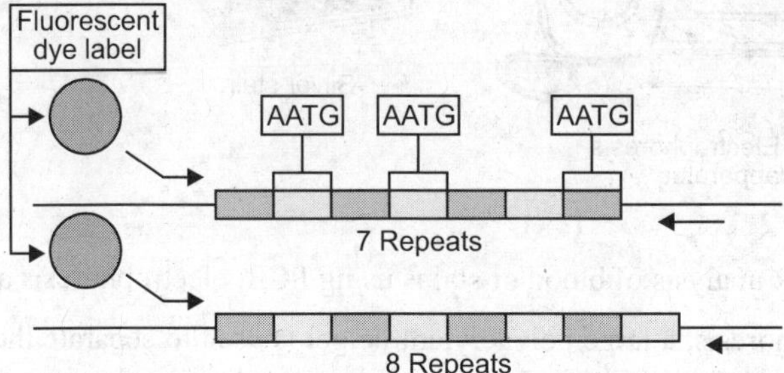

Fig. 10.7. Short tandem repeats (STRs).

There are two common methods of separation and detection: *capillary electrophoresis* and *gel electrophoresis*. Typically each polymorphism is shared by around 5-20% of individuals; hence, the polymorphisms displayed at each STR region are by themselves very common. It is the unique combination of these polymorphisms to an individual that makes this method discriminating as an identification tool. The test becomes more discriminating with the presence of more STR regions in an individual.

Different STR-based DNA profiling systems are used which differ with countries. In North America the CODIS 13 core loci-amplifying systems are most commonly used, while in the UK the SGM+ system is used which is compatible with The National DNA Database. Many of the STR regions under test are the same irrespective of systems used. These DNA profiling systems are based around multiplex reactions whereby many STR regions will be under test at the same time.

The DNA fragments are injected into a thin glass tube, *i.e.* the capillary filled with polymer. Capillary electrophoresis work is performed. After application of an electric field, DNA fragments move through the tube and get separated because the smaller fragments travel faster through the capillary than the larger ones. Then the fragments are detected using fluorescent dyes that were attached to the primers used in PCR. Hence multiple fragments get amplified and run simultaneously. This procedure is called multiplexing. Labelled DNA size standards are added to each sample to assign the sizes. The size to an allelic ladder is compared to determine the number of repeats, a sample that contains all of the common possible repeat sizes. This method is expensive but machines with larger capacity and higher throughput are being used to lower the cost per sample and expedite the amplification process.

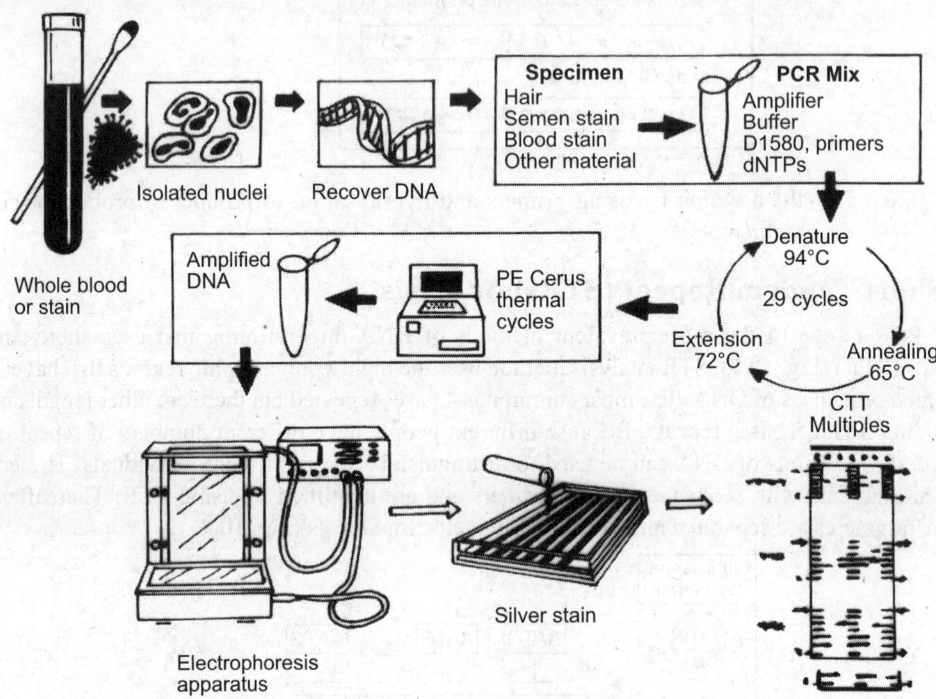

Fig. 10.8. STR analysis of blood or stains using PCR, electrophoresis and silver stain.

In gel electrophoresis, a large polyacrylamide gel is used to separate the DNA fragments. An electric field is applied but instead of running all of the samples by a detector, the smallest fragments are run close to the bottom of the gel. The entire gel is scanned into a computer. This results in production of an image which shows all of the bands corresponding to different repeat sizes and the allelic ladder. Since the allelic ladder is run alongside of samples and serves this purpose, this approach does not require the use of size standards. The DNA bands can be visualised by using fluorescently tagged dyes in the primers or by silver staining the gel before scanning,. It is cost-effective and can be high throughput; hence, silver staining kits for STRs are being discontinued. The statistical power of discrimination is the true power of STR analysis. There are 13 core loci (DNA locations) in the US and these are currently used for discrimination in CODIS because these loci are present independently.

4. Amplified Fragment Length Polymorphism (AFLP)

In 1995, AFLP-PCR was first described by Pieter Vos and his colleagues. This technique involves five major steps as described in the following sections: (*i*) preparing the AFLP template, (*ii*) ligation reaction with restriction fragments and adaptors, (*iii*) selective PCR amplification, (*iv*) electrophoretic separation of amplified DNA fragments, and (*v*) analysis of DNA banding patterns to determine the DNA fingerprint associated with a given sample.

A sample of genomic DNA is isolated and purified from a single individual. Then genomic DNA is digested with a pair of restriction enzymes, often *Mse*I and *Eco*RI. *Mse*I recognises 5′-TTAA-3′ and cleaves after the first 5′-T, whereas *Eco*RI recognises 5′-GAATTC-3′ and cleaves after the 5′-G. The genomic DNA is typically incubated with these enzymes for a period long enough to permit complete digestion of the DNA into restriction fragments. *Mse*I and *Eco*RI generate DNA fragments with 5′ overhangs (5′-TA-3′ and 5′-AATT-3′, respectively) that are distinct from each other and are noncomplementary.

An *Mse*I adaptor and an *Eco*RI adaptor are developed each of which is a short and double-stranded DNA molecule with a 5′ overhang that is complementary to the 5′ overhang generated by *Mse*I (5′-TA-3′) and *Eco*RI (5′-AATT-3′), respectively. The *Mse*I adaptor contains a 3′-C which destroys the *Mse*I recognition site. The *Eco*RI adaptor contains a 3′-G, which destroys the *Eco*RI recognition site once ligated. The *Mse*I and *Eco*RI adaptors each contain their own specific DNA sequence, typically between 19 and 22 base pairs in length located upstream of respective 5′ overhangs. The *Mse*I/*Eco*RI digested genomic DNA is incubated with the two adaptors and with DNA ligase, which covalently links the adaptors with their respective complementary 5′ ends. They prevent re-ligation between *Eco*RI-*Eco*RI and *Mse*I-*Mse*I genomic fragments and will not recognise the ligated adaptor-genomic DNA fragments.

PCR reactions are carried out using primers corresponding to the *Mse*I and *Eco*RI adaptor sequences. Every single genomic DNA fragment is amplified. Primer sets complementary to the *Mse*I or *Eco*RI adaptor sequences starting at their 5′ ends are used to amplify a smaller number of genomic DNA fragments. Amplification of a subset of genomic DNA fragments takes place upon addition of only a small number of base pairs, called selective nucleotides at the end of each primer.

AFLP-PCR reactions are carried out under stringent conditions permitting only the selective amplification of those genomic DNA fragments that are perfectly complementary to the 3′ ends of the primer sequences. The stringent PCR conditions employed in AFLP-PCR reactions lead to highly reproducible results that are readily comparable among different samples. Consequently a wide range of DNA fragment sizes is amplified based on the length of the original *Mse*I-*Eco*RI genomic DNA fragment ligated to the *Mse*I and *Eco*RI adaptors. The DNA fingerprints produced in a given PCR reaction are identified and characterised. In most of the cases one of the two PCR primers is radioactively or fluorescently labelled (typically the *Eco*RI primer), which leads to the production of a labeled PCR product that can be easily detected. A high-resolution electrophoresis is also used to separate the DNA fragments based on their size and overall negative charge; large fragments will migrate more slowly and small fragments will migrate more quickly when exposed to an electrical field.

The resulting DNA banding patterns are analysed to determine the DNA fingerprint associated with a given sample. AFLP-PCR can detect only dominant genetic markers; it cannot report homozygous individuals for a given marker. Amplification of multiple genomic DNA fragments (often between 50 and 100 fragments within a single PCR reaction) with high specificity and reproducibility cannot be done using this technique in the absence of any genome sequence data.

This technique was also faster than RFLP analysis. PCR is used for amplification of DNA samples. Its functioning depends on VNTR polymorphisms for discriminating different alleles that are separated on a polyacrylamide gel using an allelic ladder as opposed to a molecular weight ladder. The DNA bands are formed which are visualised by silver staining of the gel. The D1S80 locus was the popular locus for fingerprinting. Highly degraded DNA or very small amounts of DNA may cause allelic dropout or other stochastic effects as with all PCR based methods. Very high number repeats may bunch together at the top of the gel because the analysis is done on a gel which makes its resolution difficult. AFLP analysis can be highly automated. On the basis of comparing individual samples of DNA a phylogenetic trees can be prepared. Moreover, AFLP methods are popular in third world countries due to its relatively low cost and ease of set-up, and operation.

5. Y-chromosome Analysis

Normally there are 46 chromosomes in the human, a pair of sex chromosomes and 22 pairs of autosomes of which one copy is inherited from each parent. The sex chromosomes are called the X and the Y chromosomes. Presence of at least one X chromosome is must for everyone to survive. Normally females have two X chromosomes, whereas males have one X and one Y chromosome. When the Y chromosome is present, ladies will give birth to male babies. The Y chromosome is different from all of the other chromosomes: (*i*) it contains the fewest number of genes of any chromosome, far fewer than chromosome 21, the next smallest chromosome, (*ii*) the vast majority of the Y chromosome is composed of heterochromatin (a form of DNA that does not contain functional genes), and (*iii*) the genes that are present on the Y chromosome are critically important in sexual development.

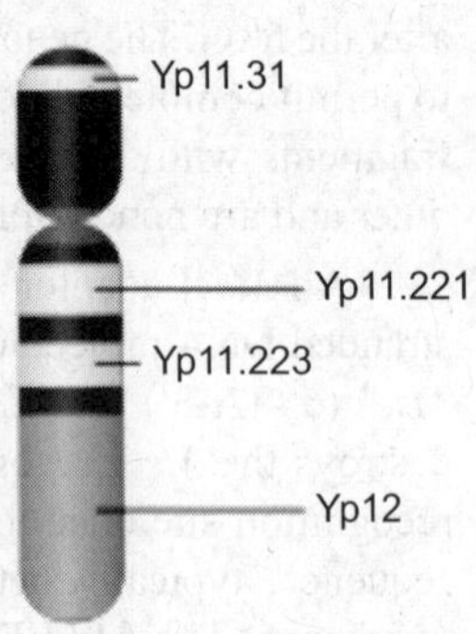

Human's Y-chromatid

Since fathers uniformly transmit the Y chromosome to their sons at conception, and never to their daughters, this allows tracing the inheritance patterns for genes and other markers on the Y chromosome from father to son down through many generations. Because the Y chromosome has so much non-coding DNA, there are many different DNA sequence variants that may be identified on the Y chromosome. These non-coding DNA sequences have a very high rate of mutation, and many potentially informative short tandem repeat (STR) sequences that permit a detailed study of paternity and other forensic testing based on DNA sequences.

Chromosomal analysis for the sex chromosomes can be used to predict the sex of a baby prenatally. After 2005, it is not considered ethical to use chromosome analysis prenatally to facilitate sex selection for parents who desire either a boy or a girl unless there is a risk of sex-linked genetic disease.

The Y chromosome of humans is composed of about 60 million base pairs. Since DNA of the Y chromosome is passed from father to son, Y-DNA analysis may thus be used in genealogy research. Y-chromosomes (Y-STR) are paternally inherited; therefore, Y-STR analysis can help in identification of paternally-related males. Recent innovations have included the creation of primers targeting polymorphic regions on the Y-chromosome (Y-STR). This allows the resolution of a mixed DNA sample from a male and female and/or cases in which a differential extraction is not possible.

The Y chromosome in humans spans about 58 million base pairs and represents about 2% of the total DNA in a male cell. The human Y chromosome contains 86 genes which code for only 23 distinct proteins. The human Y chromosome is unable to recombine with the X chromosome, except for small pieces of pseudoautosomal regions at the telomere (which comprise about 5% of the chromosome's length). These regions are relics of ancient homology between the X and Y chromosomes. The bulk of the Y chromosome which does not recombine is called the 'NRY' or non-recombining region of the Y chromosome. The SNPs present in this region are used for tracing the direct paternal ancestral lines.

Analysis of Y chromosome has many useful applications in several different fields. It can be used to provide evidence of any familial relationship between males, which is particularly valuable in such cases where there is a lack of paper evidence to establish a link. It helps families who have immigrated frequently to many countries and the paperwork is challenging. Y chromosome analysis is also useful for application to families in one area that are not even aware they are related because of surname differences. Therefore, this type of analysis can establish familial links. Y chromosome analysis does not suit much for criminal investigations but provides many benefits for investigating familial relationships, which is particularly valuable for families wishing to identify the other members.

6. Mitochondrial DNA (mtDNA) Analysis

In humans, individuals inherit mitochondrial DNA strictly from their mothers. Therefore, the mtDNA sequences obtained from maternally related individuals, such as a brother and a sister or a mother and a daughter will exactly match each other where there is no mutation in either of cases. This characteristic of mtDNA is advantageous in the cases of missing persons as any maternal relative of the missing individual can supply reference samples. However, mtDNA analysis is limited when compared with nuclear DNA analysis in that it cannot distinguish between individuals of the same maternal lineage or individuals who have the same mtDNA sequence by chance. Structure of the human mitochondrial genome is given in Fig. 10.9.

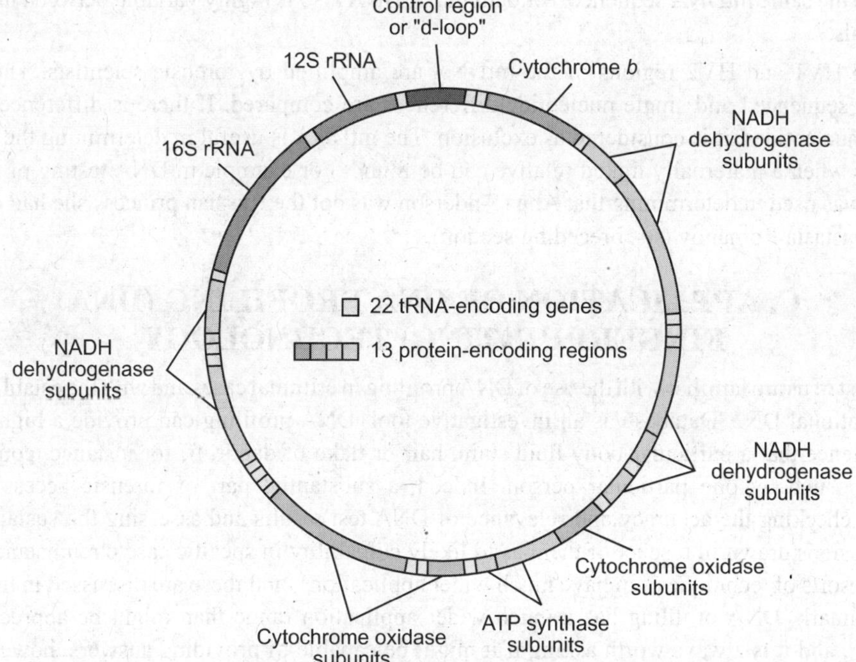

Fig. 10.9. Structure of the human mitochondrial genome.

The mtDNA analysis can offer the law enforcement community some equally noteworthy assistance in solving crimes. The mtDNA is located in mitochondria and found in the outer layer of the cell. A cell may contain hundreds of mitochondria, each of which may have several copies of mtDNA. Thus, mtDNA has a greater copy number than nuclear DNA. This characteristic of mtDNA proves useful in situations where the amount of sample is very limited. Typical sources of evidence suitable for mtDNA analysis include hairs without tissue, bones, and teeth.

Sometimes, it is impossible to get a complete profile of the 13 CODIS STRs for highly degraded samples. Mitochondrial DNA analysis is recommended in certain situations: (*i*) cases with degraded or small samples that have not yielded results using conventional STR analysis (e.g. aged bone or teeth, aged or degraded tissue, and minute quantities of blood or other body fluids), (*ii*) samples such as shed hair which do not contain a root and thus will not work using conventional STR analysis, and (*iii*) samples that have been subjected to the environmental extremes of heat, humidity or burning

The mtDNA is maternally inherited and is passed on to both daughters and sons. Therefore every child has an exact copy of his or her mother's mtDNA. Since mutations occur only once every 50 generations, a maternal line can be traced back many generations using mitochondrial DNA analysis. Every individual within the same maternal lineage will possess the same mtDNA. Mitochondrial DNA sequence analysis is a proven technology useful when standard STR-based approaches have been attempted and found to be uninformative.

Mitochondrial DNA is present deep within the cell and has a circular structure which protects it from deterioration. DNA is bound and protected by a substance called hydroxyapatite found in teeth and bones. The high copy number of mtDNA gives the forensic scientist a better chance of locating and amplifying a piece of un-degraded DNA in a sample. The maternal inheritance of mtDNA can prove advantageous in cases involving missing persons, even though this fact also makes it less discriminating than nuclear DNA because any person who is a member of the same maternal lineage will have the same mtDNA sequence. Mitochondrial DNA also is highly variable between unrelated individuals.

The HV1 and HV2 regions of the mtDNA are amplified by forensic scientists. Then each region is sequenced and single nucleotide differences are compared. If there is difference of two or more nucleotides, it is considered as exclusion. The mtDNA is useful in determining the unclear identities when a maternally linked relative can be found. For example mtDNA testing of missing persons was used in determining that Anna Anderson was not the Russian princess she had claimed to be, Anastasia Romanov (*see* preceding section).

C. APPLICATION OF DNA PROFILING (DNA FINGERPRINTING) TECHNOLOGY

Most of us are familiar with the use of DNA profiling in criminal cases and with the establishment of the National DNA Database as an investigative tool. DNA profiling can provide a high degree of confidence that a particular body fluid stain, hair or flake of dandruff, for instance, could have been deposited by one particular person. Indeed, a substantial part of forensic access's work involves checking the accuracy and relevance of DNA test results and assessing the sustainability of conclusions drawn in respect of them as to likely culpability in specific case circumstances. But the same sorts of techniques can have much wider applications, and these are discussed in this issue of Benchmark. DNA profiling has a much wider application range than might be appreciated at first sight, and it is always worth asking if it might be capable of providing answers, however, old or unusual the circumstances surrounding the question. DNA profiling technology has applications in many more areas as discussed in this section. DBT has established the Center for DNA Finger Printing and Diagnostics in 1996. Since its inception the CDFD has not only developed several new DNA signatures but also provided invaluable services to the criminal justice system of the country in more than 3,500 cases.

In 2002, Nagaraju and co-workers have established the genetic identity and uniqueness of the Indian Basmati rice compared to the Pakistani Basmati was, again by the CDFD, this case had enormous trade and export impact.

Why is DNA used for identification ?

DNA has many remarkable properties; that is why it is suitably used for identification of an individual because:

- DNA remains the same throughout the life of a person,
- DNA does not change with age,
- Irrespective of source of isolation (*i.e.* brain, hair, semen, blood, bone, sputum, urine, skin, kidney or any other tissue) DNA results the same fingerprinting pattern in an individual,
- DNA from one source such as semen can be compared with DNA in the blood in a suspect; hence sources of DNA to be compared for DNA analysis are immaterial,
- DNA is a stable molecule; hence it can be heated, boiled and denatured. But the strand can again form a double helix under optimum conditions, buffer, temperature, etc. Due to its long stability in nature also, it has been isolated from thousands years old remains of bones and even from fossils (several million years old).

1. Setting up of a National DNA Databases

Realising the potential of DNA profiling genetic databank had been/is being set up though out the world. Databases of the genetic fingerprints of criminals have been set up in many parts of the U.S.A. and Britain. Forensic Science Services Birmingham (Britain) has set up the world's first National DNA Database, where about 5 million records would be available. It has collected samples of tissues and DNA of suspected individuals that has been profiled. The Armed Forces Institute of Pathology (U.S.A.) has collected DNA samples from each and every member of the armed services. The samples are stored in freezers in DNA repositories. Each sample consists of blood blots on cards and sealed in individual envelopes. A computer database of all samples has been prepared.

The largest DNA database in the world is located in the United States as the 'Combined DNA Index System'. In 2007, it had more than 5 million records. The United Kingdom maintains the National DNA Database (NDNAD), the size of which is the same. The U.K. National DNA Database was established in April 1995 and currently holds the DNA profiles of over 1,50,000 individuals to provide a powerful tool for identifying re-offenders and for establishing links between apparently unrelated crimes, for example serial rape. The size of this database and its rate of growth are giving concern to civil liberties in the U.K. where police have powers of wide range to take samples and to retain them even in the event of acquittal. A means for the U.S. government is provided by the U.S. Patriot Act of the United States to get DNA samples from other countries. Under the act, the American offices of the company can't divulge to their subsidiaries/offices in other countries.

National DNA Database

A national DNA database refers to a database of DNA samples set up at national levels against which law enforcement agencies can match the suspect DNA. The first national database was set up by the United Kingdom in April 1995. The second national database was set up in New Zealand. France set up the FNAEG in 1998. The FBI has organised the CODIS database in the U.S.A.

It was originally intended for sex offenders, but it has been extended to include almost any criminal offender (http://en.wikipedia.org/wiki/National_DNA_database).

When a match is made from a National DNA Databank to link a crime scene to an offender who has provided a DNA sample to a databank that link is often referred to as a **cold hit**. A cold hit is of value in referring the police agency to a specific suspect but is of less evident value than a DNA match made from outside the DNA Databank.

(A)

(B)

Fig. 10.10. (A) Lalji Singh taking a swab sample from an Onge Tribal; (B) showing DNA fingerprints done in 18 hours.

The 'DNA Profiling Bill 2007' of India drafted by the Department of Biotechnology and sent to all ministries and departments for their feedback, has been modified. The final version has been sent to the Law Ministry, which has sent it to the legal department for final drafting. If the parliament passes the DNA Profiling Bill 2007, India will soon join countries such as the U.S. and U.K. in creating a national DNA database. The present bill is pending in parliament of India. The Bill provides for : (*i*) collection of DNA profiles of individuals for lawful purposes, (*ii*) an authority to create and maintain databases of offenders, suspects, crime scene samples, etc. for detection of, and to serve as deterrents against, repeat offenders, and (*iii*) creation of DNA Profiling Advisory Board of experts to define, establish and modify standards and implement quality control measures including compulsory accreditation of laboratories/agencies, undertaking DNA profiling services also to prevent abuse / misuse of DNA profiling to discriminate against individuals for employment, health/life insurance, etc.

In this bill, six regional laboratories has been proposed to be established in Kolkata, Chandigarh, Hyderabad, Bhopal, Pune and Guwahati that will facilitate the exchange and comparison of forensic DNA profiles created from evidence gathered from crime investigations and convicts. The role of the database would be to compile profiles of convicts, under-trials and suspects as well as from material obtained from crime scenes. At present India has no law that empowers the Government to collect and store DNA profiles of convicts.

Some Most Important Cases where DNA Fingerprinting has been Applied

DNA fingerprinting technology has been used in some of the most known cases :

- Rajiv Gandhi assassination case of 1992
- Identification of Abu Jundal (a famous international terrorist)
- Identification of Saddam Husain (President of Iraq)
- Identification of Osama Bin Laden (Chief of Al Quaida)
- Identification of criminal in Madhumita Shukla murder case (U.P.)
- Identification of murderers in Beant Singh (ex-Chief Minister of Punjab) murder case
- N.D. Tewari paternity case of 2012

2. Reuniting the Lost Children

DNA profiling technology helped in reuniting the lost children with their respective parents or *vice versa* who were separated during war, violence or natural disasters. For example, in Argentina many people were abducted during military rule (1976-1983). They were presumed to be died. But an Arzentinian Human Right Group (the Grandmother's of Plaza de Mayo) was of the opinion that the children are still alive and adopted by military personnel after killing their natural parents. In 1987, a genetic databank storing blood samples of grand parents and relatives of lost children was set up. By matching the DNA profile of one or more grand parents with that of lost children about 40 missing children have been reunited with their families (Chawla, 1998).

3. Identifying Bodies or Body Parts

Over the years identification of human remains has successfully been done by applying sensitive DNA extraction and profiling techniques to bones, teeth and hair in particular. Similar techniques can be used to resolve medical disputes; for instance, in a case where it was suspected that a mastectomy had been performed on the wrong woman. This was confirmed by comparing the DNA profile of a sample of the biopsy tissue taken from the woman in question with the profile of the excised breast tissue. Unfortunately, there had been a critical mix-up of samples in the hospital concerned.

In the 1920s, Anastasia Manahan (usually known as Anna Anderson) claimed that she was Princess Anastasia Romanov of Russia. She died in 1984 at the age of 87 due to pneumonia. After her death, samples of her tissue, that had been stored at a Charlottesville (Virginia hospital) following a medical procedure, were tested using DNA fingerprinting. Anderson's DNA was compared with those of the Romanovs, at the suggestion of Marina Botkin Schweitzer, the daughter of Gleb Botkin. Anderson's DNA sample did not match with that of the Duke of Edinburgh or that of the bones. The DNA analysis showed that Anderson had no relation to the Romanovs. The tissue sample tested belonged to Anderson but could not belong to Anastasia (Godl, 1998).

Anna Anderson

4. Establishing Paternity

DNA analysis has been used in numerous paternity cases and is usually pretty straightforward. But one of the most challenging cases dates back to the 1950s, and has extended the frontiers of forensic DNA analysis. The case concerned a man who discovered that the man that he had always thought was his father might not have been. Both his mother and his putative father were dead by the time his suspicions were aroused and he had no real clues to go on. One possibility for his paternity appeared to be a man living overseas with whom his mother had corresponded. Analysis of DNA in 'saliva' on the 'backs of postage stamps' on mail found amongst his mother's possessions had the potential to hold the key, *i.e.* saliva that had been deposited more than 50 years ago.

Standard techniques produced weak and incomplete profiles of the DNA. It was not possible to exclude either of the men on the basis of these profiles; therefore, a new form of profiling was attempted that analysed DNA on the male-specific Y chromosome only. Y-chromosome profiles were passed on pretty much intact from father to son for generation after generation and so can provide a particularly powerful means of establishing paternity. In this case, the Y-chromosome profiling showed that the recent information was incorrect and indicated strongly that his father had been the man that he always thought.

There are many reasons why paternity testing may be done. These include the following: (*i*) to determine the biological father of the child, (*ii*) to prevent disputes in adoptions, (*iii*) for immigration purposes, (*iv*) to determine financial issues of supporting the child, (*v*) to establish inheritance benefits, (*vi*) to strengthen the bond between father and child, and (*vii*) to obtain medical history.

In India, the courts are slowly considering the importance of DNA test. In June 1988, a four year female baby named Laxmi was stolen from Chennai by a missionary and renamed as 'Merry' by the so called parents. After FIR and taking police help the matter was not solved. Therefore, the real parents went to High Court. The hon'ble High Court sought help from CCMB. Lalji Singh and his team performed DNA fingerprinting of baby and parents (real and false), confirmed the identity of the baby, and submitted the report to the

DNA profiling as evidence in criminal trials.

court. On the basis of report of DNA fingerprinting decision was given by the court in the favour of real parents of Laxmi. Laxmi was handed over to her real parents. This is the first case in India where a court has given decision based on the report of DNA fingerprinting.

In case of Kanchan Bedi vs Gurpreet Singh Bedi, a DNA test of the child was directed where the defendant was denying any marriage had taken place between him and the plaintiff and, therefore, he was not the father of the child.

A DNA test was also conducted in the sensational Premanada Swami's case who was charged with the rape of several teenage girls in his ashram. A DNA test established that 45 year old Swami Premananda was the biological father of the foetus as a result of rape of 19-years old Aruljyothi (*see* preceding section).

The recent Delhi High Court judgment has also given a positive direction to the applicability and acceptance of DNA tests in disputes involving paternity. In order to understand the situation with respect to the second primary issue we have to see another important Supreme Court decision. In Dwarika Prasad Satpatty vs Bidyut Parva Dixit it was held that the refusal to paternity (DNA) test would bar a party from challenging the paternity of the child. This decision was followed in K. Salvaraj vs P. Jayakumari and it was also stated that an adverse inference can be drawn if a party refuses to undergo a DNA test. This seems to be a preferable interpretation and strikes a balance where the court does not have the power to direct for giving of sample, it may draw an adverse inference if it is not given.

In 2010, parentage dispute was on the spark when Rohit Shekhar filed a paternity suit in Delhi High Court in 2008 charging Mr. N.D. Tewari (the Ex-Chief Minister of Uttar Pradesh and Uttarakhand, and Ex-Governor of Andhra Pradesh) as his biological father and demanded for DNA test, to which Mr. N.D. Tewari denied. Mr. Tiwari who has denied having any physical relation with the petitioner's mother (Mrs. Ujjwala Sharma), has also been ordered by the High Court to undergo a DNA test for determination of Shekhar's paternity. DNA test conducted between Rohit and Ujjwala, and Rohit and B.P. Sharma (pre-divorced husband) proved that B.P. Sharma is not Rohit's biological father. Moreover, 'every child has right to know his/her biological father and mother'.

Thus, the court sought the help of a Delhi Government dispensary for determination of modalities to draw blood samples of Mr. Tiwari, Shekhar and his mother for the DNA test. But Mr. Tewari repeatedly denied giving any blood sample to anybody to whom the Delhi High Court took it seriously. Thereafter, the Delhi high court reserved its order on Tiwari's plea contending that he cannot be compelled to give his blood sample for DNA test (Aug. 2, 2011, The Time of India). Then the court issued the notice to N.D. Tewari 'why contempt proceedings should not be initiated against him for his deliberate, willful disobedience of the court's order'. After rejecting his plea and an ultimatum by the Supreme Court, N. D. Tiwari gave his blood sample in May, 2012 at his residence at Forest Research Institute (Dehra Dun) in the presence of district judge, a medical team, a pathologist and a lab technician in front of Joint Registrar of Delhi High Court and lawyers of Tiwari and Shekhar. The blood sample was sealed and handed over to Joint Registrar of Delhi High Court, who submitted it to the court for its DNA profiling. Blood sample was sent to the Centre for DNA Fingerprinting and Diagnostics (CDFD), Hyderabad for DNA analysis. On July 27, 2012 result of the DNA test from the CDFD was de-sealed in the chamber of the judge who declared the result based on positive DNA match that 'N.D. Tiwari is reported to be the biological father of Rohit Shekhar and Ujjwala Sharma the biological mother'. Mr. Tewari has challenged the decision in the Supreme Court.

N.D. Tewari's DNA test confirmed as the biological father of Rohit Shekhar.

The justice administration system needs to assimilate the scientific advancements of genetic profiling and develop procedural techniques for harnessing the emerging juridical challenges. The significant paradigms of DNA fingerprints cannot be left alone to the courts to adjudicate with temporary tailor made solutions. Therefore, in matters of disputed paternity, the legitimacy or illegitimacy of the child cannot be determined solely by Section 112 of the Indian Evidence Act, 1872. DNA technology can conclusively establish the truth in such disputes and, therefore, should be resorted to without any hesitation. It is to be borne in mind when Section 112 was being drafted even the discovery of DNA was not contemplated and, therefore, this section should be amended. An ideal solution could be to provide another outlet apart from the proof of non-access to be provided in the form of evidence of DNA test to rebut the conclusive proof provision in Section 112. Moreover, under CRPC Section 53A, there is provision of considering DNA test as modern scientific method, besides the other biological samples.

5. Solving of Baby Swapping Case

Baby swapping seems to be a relatively new phenomenon but some believe that it will have very dangerous consequences. The baby swapping case saw a mother reunited with her baby son after testing it could be confirmed that she was not the mother of the baby girl. In 2003, Latha Reddy gave birth to a son in the Nayapul Government Maternity Hospital (Hyderabad) but five days later, when she asked to see the child, she was presented with a baby girl by hospital authorities. Then she sat on protest with her husband against the hospital personnel. DNA fingerprinting of baby girl and Latha Reddy was carried out and the DNA test proved that the girl was not her. The employees were believed to be responsible for handing over the newborn son of Latha Reddy to Mehmooda Bai, and baby girl of Mehbooba Bai to Latha Reddy.

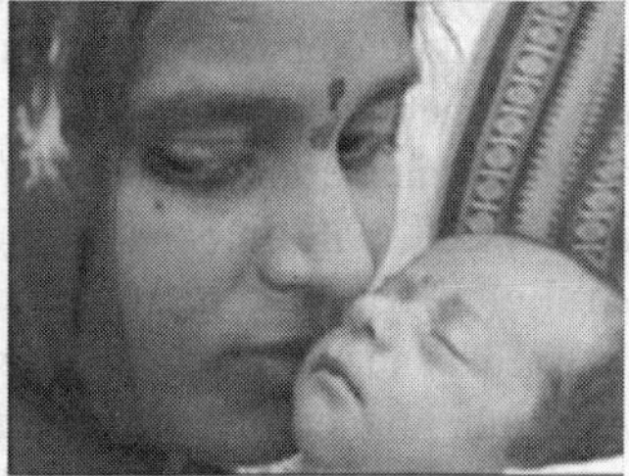

Latha Reddy who got her son back after three weeks.

After DNA test, it was confirmed that the baby son who was with Mehbooba Bai was not actually the son of Mehbooba, but it was the baby son of Latha Reddy. However, Latha Reddy's son was returned to her three weeks later. Nazeer Ahmed, the father of the baby girl, was also arrested. Mehmooda Bai pretended that she and her husband had been forced to take away the male child by the hospital despite their objections. Consequently, four people were taken into custody for their role in the swap, three of them hospital workers and the father of the baby girl.

6. Application in Forensic Diagnostics (Establishing Identity of Criminals and Rapists)

When forensic scientists examine DNA in the laboratory, each sample appears as a unique sequence of dark bars. Patterns of bars are compared to find a match. Fig. 10.11 shows that DNA bands of suspect #2 matches with that of DNA bands of the victim left some at the crime scene, *i.e.* victim.

In 1987, British baker Colin Pitchfork was the first criminal caught using DNA fingerprinting in Leicester, the city where it was first discovered. In 1987, Florida rapist Tommie Lee Andrews was the first person in the United States to be convicted as a result of DNA evidence for raping a woman during a burglary. He was convicted on 6 November 1987 and sentenced to 22 years in prison. In 1988, Timothy Spencer was the first man in the United States to be sentenced to death through DNA testing for several rape and murder charges, he was dubbed 'the South Side Strangler', because he killed all his victims on the south side of Richmond, Virginia. He was later charged with rape and murder, and was sentenced to death. He was executed on April 27, 1994.

In 1994, RCMP detectives successfully tested hairs from a cat known as Snowball, and used the test to link a man to the murder of his wife, thus marking for the first time in forensic history the use of non-human DNA to identify a criminal. In 1998, Dr. Richard J. Schmidt was convicted of attempted second-degree murder when it was shown that there was a link between the viral DNA of the human immunodeficiency virus (HIV) he had been accused of injecting in his girlfriend and viral DNA from one of his patients with full-blown AIDS. This was the first time for using viral DNA fingerprinting as evidence in a criminal trial.

DNA samples
crime scene | suspect #1 | suspect #2 | suspect #3

Fig. 10.11. Comparison of DNA fingerprints of a victim, suspects 1, suspect 2 and suspect 3 where DNA bands of victim and suspect #1 shows matching.

(*a*) Sexual Escapades: In India, for the first time, DNA profiling of aborted foetus has been used as evidence for sexual escapades. In Chennai, many women *ashramites* had undergone repeated abortion in the private clinic run by a local medical practitioner Chandradevi (who was convicted in the case), but the responsibility could not be fixed on Premananda, the accused in the absence of immediate medico-legal evidence. The Ashram had nearly 200 male inmates, including devotees and workers, at any given time. The 19-years old Aruljothi, who escaped to Chennai towards the end of 1994, disclosed to the media in January 1995 that she was pregnant. While charging Premananda with causing her the unwanted pregnancy, Aruljothi alleged that she was impregnated twice before by Premananda and the pregnancies were medically terminated. The DNA profile of the foetus gave unassailable evidence against Premananda that he was using Ashram women for illicit relationship.

The blood sample of Premananda and the tissue from the aborted foetus of Aruljothi were examined at the Centre for Cellular and Molecular Biology, Hyderabad. Lalji Singh of the CCMB in his report confirmed that 'Premananda was responsible for the product of conception whose biological mother was Aruljothi'. When the DNA profiles of Premananda were compared with the foetus, every band present in the foetus tissue was fully accounted for either being inherited from the mother or Premananda, indicating that they were the biological parents. The defence had brought a genetic engineering consultant from the United Kingdom to dispute the DNA profiling, but could not demolish the prosecution evidence and the opinion of the Indian expert. Then Premananda was taken in police custody.

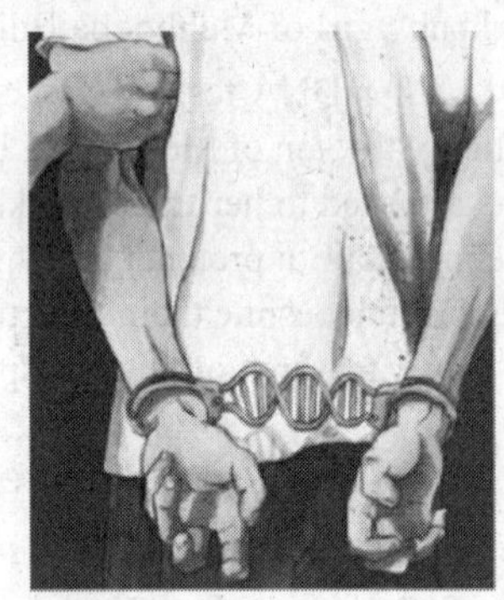

A person under police custody.

(*b*) Tandoor Case: In India more than 500 disputes have been solved through DNA finger-printing. DNA fingerprinting–based decision of identity confirmation of a brutally killed lady was done in Delhi. On the night of July 2, 1995, Delhi police constable and a Home Guard saw the smoke emanating from the open-air Bagiya Restaurant in Ashok Yatri Nivas (now called Indraprastha Hotel). They got suspicious and scaled the hotel's boundary wall to enter the premises. 'Something big' was burning in a 'tandoor' in the restaurant. That 'something big' turned out to be the body of Naina Sahni, who was killed by her husband, the then Delhi Youth Congress president Sushil Sharma.

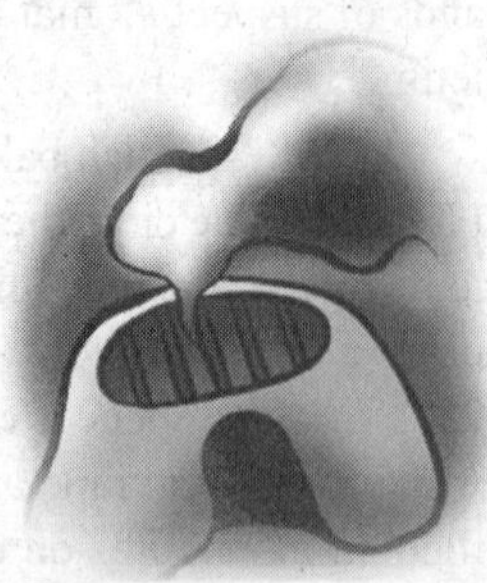

Tandoor case: Murder most foul

According to the 19-page charge sheet submitted by the police, Sharma doubted Sahni's fidelity. Sharma allegedly fired three shots from his licensed revolver at Sahni. One bullet pierced her head, another hit her in the neck while the third missed her and hit an air-conditioner. Sahni died on the spot. Sharma wrapped her body, took it to the restaurant, chopped into small pieces and burnt in his tandoor. Her limbs had been brutally severed to unravel the identity (Chawla, 1998).

After this incidence nobody came forward to receive the dead body neither parents nor relatives. Thanks to DNA fingerprinting technology team let by Lalji Singh that confirmed the identity of brutally killed and roasted lady as Naina Sahni. Naina Sahni case was in lime light during those days and after thorough investigation her husband, who killed her wife Naina Sahni, was given death sentence to which he was trying to convert into life imprisonment.

7. Solving Immigrant Dispute

There is the problem of immigration more-or-less in every country. Ultimately, the identity is not confirmed whether they have crossed the borders or are the resident of that country. Therefore, the problem of immigrants can be solved through DNA databank. India is suffering from problem of immigration from Bangla Desh and Pakistan. Several million illegal Bangla Deshis and Pakistanis are living in India. There is an urgent need to issue identity cards of Indian citizens containing blood stains that can be used further for identification in future.

8. DNA Ancestry Testing

The Focused GeoGenetic Database (FoGG DB™) has been prepared after compiling data from various sources by a firm Indian Biosciences. Then various calculations are performed and certain algorithms are followed to arrive at a result which reveals a person's strongest DNA ancestry. The data from this DNA ancestry test is then compiled into the Ancestral Origins™ DNA Ancestry map which is personalised for each customer. Each Ancestral Origins™ DNA Ancestry map is broken into three major sections: population matches, native region matches, and strength indicators. Individual DNA is measured against every population in the database, and strongly matched profile against each is marked. On the population match map, each population has a shape which tells about the population belonging to a particular anthropological region. A personalised Ancestral Origins DNA Analysis Report together with an account of historical migration is prepared which will explain how different populations have moved around the world. This will give you a more comprehensive insight into your ancestral roots.

9. DNA Testing in IVF Technology

Until now there was no reliable method that could test the viability of human IVF embryos. Recently, fertility researchers have used DNA fingerprinting for the first time to identify which embryos have implanted after *in vitro* fertilisation (IVF) and developed successfully to result in the births of healthy babies. The researchers from Monash University and Monash IVF in Clayton, Victoria, Australia, Genesis Athens Hospital and Athens University in Greece, tested the DNA in blastocysts from *in vitro* fertilization (IVF) embryos before implanting into the womb. When the researchers compared the results with the DNA of the newborn healthy babies, they discovered a cluster of genes that could be used to determine which embryos will be able to make it to full term.

10. Application of DNA Profiling in the Livestock Industry

DNA profiling plays a crucial role in the livestock industry worldwide (Fig. 10.12)). Since the discovery of the technique in the early 1980s, it has proved invaluable in parentage determination. Development of the technology and the adoption of modern approaches have expanded the fields of application. The modern methodology using microsatellites is highly accurate reproducible and automated. A large number of applications for DNA profiling exist in the livestock industry :

(*i*) Artificial insemination (AI) and multi-sire mating systems are widespread practices where DNA profiling can establish the sire.

(*ii*) Embryo transfer is becoming more and more popular and here, too, the technology plays an important role in parentage determination.

(*iii*) The incidence of stock theft is increasing in South Africa and these cases can rapidly and accurately be solved with the aid of microsatellite technology.

(*iv*) DNA profiling can also be used as a diagnostic tool and breeding lines can be selected for certain traits.

(*v*) A further example of the application of the technique is the determination of the zygosity of twins, valuable in a variety of research programmes.

DNA profiling is a valuable tool for the livestock industry in South Africa. In economic terms, both commercial and small scale farmers can benefit by applying this service to select superior breeding animals.

Fig. 10.12. Livestock of cow breeds.

D. HURDLES TO DNA PROFILING TECHNOLOGY

When DNA profiling technology was applied in forensic science, it was challenged in 1987 for its reliability by the defense expert of a murderer in pretrial hearing. Thus, murder trial could not be concluded on DNA testing. The controversial arguments were on methods of collection, labelling and testing of samples.

Certainly, the interpretation of the bar code pattern of a DNA profile is not easy; it needs skill and expertise. For example, a band corresponding to DNA fragment containing 10 repeat sequences may be very close to a fragment with 11 such sequences. Therefore, these two could not be distinguished. Hence, computer scanning device is needed that measures accurate size of each DNA fragment which makes bar code pattern and compares DNA profiles.

Secondly, when blood samples of criminals are collected, it may be contaminated by micro-organisms, the other environmental factors like heat may degrade DNA. If so, there may be misleading matches or mismatches. Therefore, at every step care should be taken in carrying out the analytical work. Since the test depends on comparison of the length of different cut fragments, the DNA samples to be compared should be treated exactly in the same way as same restriction enzyme, same gel, same detective probes, etc. In spite of precautions, some times there occurs 'band shift' which shows still carelessness, because small differences will give wrong result and misinterpretation. Therefore, analysis should be repeated to ensure its validity and reliability. Therefore, for forensic science accurate guidelines should be formulated and strictly followed. In this regard, the National Research Council (USA) recommended a series of safeguards for conducting DNA typing tests. In addition, the Federal Bureau of Investigation (USA) along with the National Institute of Standard & Technology has developed standards for DNA profiling (Chawla, 1998).

E. DISEASE DIAGNOSIS

There are certain microbial and non-microbial diseases which can not be diagnosed due to lack of techniques. Therefore, such diseases can be diagnosed by some other methods. A few methods are given as given below.

I. DIAGNOSIS OF PARASITIC DISEASES

Probes used for diagnosis of pathogens contain the most specific DNA sequences of genetic material of parasite. The specificity lies in such a way that the other related species or strain does not contain those sequences. These unrelated unspecific sequences of parasite are first recognised by using DNA hybridisation technique. Then a DNA sequence, not present in any species, is identified, cleaved by using restriction enzymes, and inserted into a cloning vector (plasmid). The bacterial cells are transformed by the recombinant vector. The transformants are multiplied. Finally, the foreign DNA fragments are retrieved from the host cells. The DNA sequences of the parasite, thus obtained, are labelled with radioisotope and used as a probe. The probe can also be chemically synthesised.

1. Diagnosis of Parasitic Diseases through DNA Probe

Following are the steps for diagnosis of a particular disease by using DNA probe:

(*i*) Isolate the parasite from the infected tissue of the patient. Extract the DNA from parasite and purify it.

(*ii*) Break the DNA by using restriction enzymes to get the DNA fragments of varying size.

(*iii*) Electrophorise the DNA solution of different length by using agarose gel just to get a smear of DNA.

(*iv*) Attach the DNA smear to more firm support by Southern blotting techniques. Thus the filter paper carries the exact replica of the DNA adhered to it.

(*v*) Hybridise the immobilised DNA on filter paper by incubating it with radiolabelled probe. Probe DNA complementary to certain DNA sequences of parasite DNA sticks to it and forms the hybrid.

(*vi*) Wash the filter paper to remove unhybridised probe. Keep the filter in contact of X-ray film. Dark bands appear where DNA probe had hybridised the specific target sequence. Thus, a parasitic disease is diagonised positive. If dark bands on X-ray film do not appear, the absence of parasite is noted.

Specific DNA probes have been designed which have shown good results in diagnosing the infection caused by viruses, bacteria or protozoa. Tuberculosis (T.B.) caused by *Mycobacterium tuberculosis* is one of such diseases which can be diagonised by this method. Moreover, a complete testing system for T.B. is marketed by Genprobe Inc. (California). Similar efforts have also been made for diagnosis of leprosy, Kala azar (caused by *Leishmania donovani*), malaria, African trypanosomes, etc.

(*a*) Identification of African Trypanosomes: DNA probes have been developed for identification of African trypanosomes. Cloned DNA hybridisation probes are used to detect trypanosomes and distinguish within and between species by two approaches: demonstration of the presence or absence of a particular repetitive sequence or detection of restriction enzyme site polymorphism around a sequence. The first approach is more easily applied because blood or tissue samples can be used and propagation of trypanosomes is not necessary for DNA preparation. Repetitive DNA sequences can be easily identified and cloned and can often be used to detect a very small number of parasites, either in the vertebrate host or in the insect vector. For example, repetitive minichromosome sequences cloned from *Trypanosoma congolense* can detect as few as 30 parasites. These probes appear to be specific for subgroups of *T. congolense*. DNA probes that would define important characteristics of the trypanosomes, such as infectivity to man, clinical characteristics or serodeme-specificity, would be more valuable, but are more difficult to identify.

Non-radioactive DNA probes have also been developed of for identification of African trypanosomes. Disadvantages of using radiolabelled probes are : short half life, expense and disposal problems. These problems can be obviated by biotinylation of the probe. The DNA probes can be biotinylated by nick translation using either biotin-11 dUTP or biotin-11 dCTP, both of which are currently available commercially. The hybridisation reaction can then be visualised enzymatically by using avidin biotinylated peroxidase/alkaline phosphatase or streptavidin biotinylated peroxidase/ alkaline phosphatase. The method can be used in all instances where radiolabeling is currently being employed, such as Southern blots, dot blots and colony hybridisation. Since the stability of DNA probes would allow standardisation of the assay, adaptation for field use is possible. This technique can be refined by eliminating the high background and improving sensitivity.

The sensitivity of two techniques, *i.e.* radioactive and non-radioactive probes was similar. However, there are certain problems associated with the use of non-radioactive detection systems as below :

(*i*) The lack of linear relationship between quantity applied and signal obtained when whole trypanosomes are used; this did not apply when purified DNA was used .

(*ii*) Photosensitivity of the reagents need to standardise the reagents in each experiment.

II. GENETIC COUNSELLING

Identification of changes in genes is not always easy. Many of the genes in which changes lead to a condition code for very large messages. Changes can occur anywhere along the length of the DNA segment making up the gene. In addition, a single gene may have many possible changes - some changes make the gene faulty (mutations) and others have no effect on how the gene works. For example, there are over 1000 mutations that have been detected to date at different places along the length of the gene involved in cystic fibrosis. It is also likely that there are other mutations that have not yet been identified. Laboratories often test for only some of the more commonly known mutations in a gene and not for the presence of all of the mutations that occur much more rarely. Laboratories have to rely on a more indirect method to determine whether a mutation is likely to be present or absent in the gene.

Therefore, genetic counselling will provide families considering using indirect gene tracking with an estimate of the chance that the result could be inaccurate. Genetic counselling is a technique which is carried out in detail based on antenatal diagnosis and suggested for the possibilities of future progeny. A genetic disease is first identified by a genetic counsellor who understands the family history of genetic disease, and then suggests for the possibilities of giving birth to the child or aborting the child. The child may be normal, genetically defected or carrier of genetic disease. Take the case of thalassaemia. Alpha thalassaemia is caused by the presence of two chromosomes each of them carries one dysfunctional globin gene or one normal chromosome and other dysfunctional genes. Undoubtedly, a person carrying two chromosomes each with single functional α-genes donates one normal gene to their progenies. It is, therefore, impossible to suffer their progenies from 'hydrops fetalis' as it occurs due to complete dysfunctioning of all α-genes. DNA analysis of thalassaemia has revealed that non-function of both agenes on a single chromosome is only due to deletion of both genes from the chromosome.

Recent techniques have made it possible to detect α-thalassaemic couples for the presence of dysfunctioning of all α-genes gene on chromosome. However, there is risk for giving birth to hydropic foetuses if both of the couples have α-thalassaemia due to no gene chromosome. If there lies any possibility for gene therapy, the patients are also suggested for the same. It has also been suggested that one should go to genetic counsellor before marriage for safe and normal progeny, instead of going to astrologers to see the horoscope.

1. Indirect Gene Tracking

Family test relies on comparing DNA markers from family members who are affected with genetic disease as well as unaffected relatives. If there is larger number of affected family members, it is easier to track the linked markers to the faulty gene in the family. Small family size and the availability of DNA from affected family members can limit the applicability of this test for some families. In addition, it will not be possible to find markers which can differentiate between the family members who have two copies of the faulty gene, no copy or who are carriers of the faulty gene.

While scientists can most often track the markers travelling with the faulty gene copy, sometimes the markers will cross over and travel with the correct gene copy. The markers that have been travelling with the correct gene copy will be associated with the faulty gene in the next generation. This potential for crossing over occurs whenever a sperm or an egg is made. There is a small risk using this method that the diagnosis of a person being affected by a particular condition will be incorrect. However, the more closely marker is linked to the faulty gene, the more likely the presence of the marker will indicate the presence of the faulty gene. In addition, it is some times not possible to find DNA markers that enable the scientists to differentiate between the faulty gene copy and the correct gene copy.

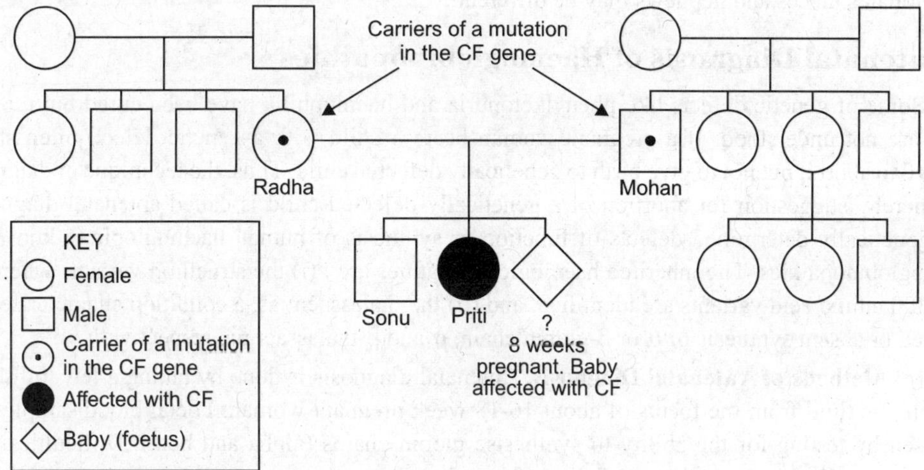

Fig. 10.13. Genetic testing of a family member affected by cystic fibrosis.

Fig. 10.13 shows genetic testing of a family member affected by cystic fibrosis. The DNA from Priti is examined using direct gene testing to see if the mutation in each copy of the gene involved can be identified. If the mutations are detected, testing can be offered to Radha and Mohan in this or future pregnancies. If the mutations causing the gene to be faulty are relatively common, it will be possible to examine the baby's DNA for these common mutations.

The genetic test of Radha and Mohan show that both Radha and Mohan have two bands, as both of them have a faulty gene copy and a correct gene copy (Fig. 10.14). Priti has two copies of the faulty gene and so she is affected by cystic fibrosis. Sonu has inherited both correct copies from his parents. The baby's DNA pattern will indicate whether she or he: (*i*) has cystic fibrosis, (*ii*) is an unaffected faulty gene carrier for CF, or (*iii*) is unaffected with both gene copies having the correct information.

If the mutation in Priti's gene for cystic fibrosis can be identified, Raha's and Mohan's brothers and sisters, who are planning to have children, can be offered direct gene testing to see whether they carry a faulty copy of the gene for cystic fibrosis. If any one of them is a genetic carrier, it may be possible to check their partners to determine whether they are also carriers of the faulty gene.

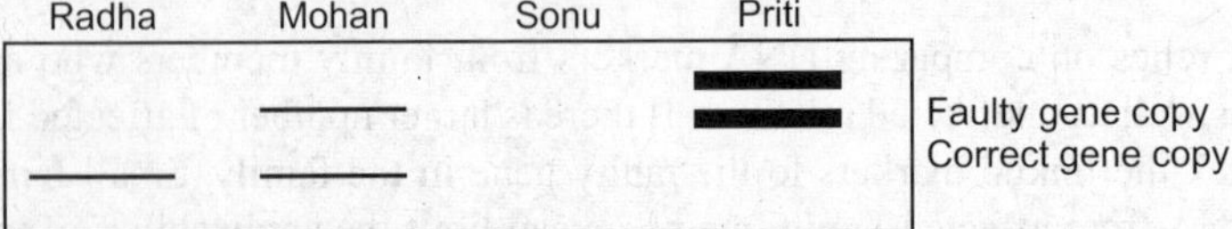

Fig. 10.14. The genetic test of Radha, Mohan, Sonu an Preti for the presence and/or absence of faulty gene for cystic fibrosis.

When Sonu is planning to have children, he will also be able to choose to have carrier testing. If it is not possible to identify the particular mutation in the two gene copies of Priti inherited from Radha and Mohan, they can be offered indirect gene tracking for prenatal diagnosis during this and further pregnancies. DNA markers from Priti, Radha and Mohan will be compared to those from the developing baby. The test will show if the baby's gene sequence is like Priti's, she will have CF. If the pattern is the same as Radha's or Mohan's, the baby will be a genetic carrier of CF just like the parents. However, Radha and Mohan's brothers and sisters may not be able to use this information for prenatal diagnosis because the markers examined are special to Radha and Mohan and their children. The markers in the partners of Radha's and Mohan's brothers and sisters and in Radha's and Mohan's nieces and nephews may be different.

2. Antenatal Diagnosis of Haemoglobinopathies

Some of genetic defects like phenylketonuria and haemophilia have been cured but most of them are not understood. If a pregnant woman bears a child with a genetic defect, often she is advised to abort it but not to give birth to genetically defected child. Thus, the technique of diagnosis and thereby suggestion for abortion of a genetically defected child is called antenatal diagnosis. The genetically determined defects of function or synthesis of human haemoglobin is known as haemoglobinopathies. The inherited haemoglobinopathies are : (*i*) the structural variants where till now 300 amino acid variants are identified, and (*ii*) the thalassaemias, a condition characterised by reduced or absent synthesis of α or ß-globin chain, α and β-thalassaemia, respectively.

***(a)* Methods of Antenatal Diagnosis.** Antenatal diagnosis is done by taking a few millilitres of amniotic fluid from the foetus of about 16-18 week pregnant woman. Foetal blood samples are analysed by testing for the ability to synthesise globin chains (alpha and beta). Difficulties with the antenatal diagnosis are that the foetus of less than 3 months of age synthesises very little α or ß-globin chains. At this stage the foetal blood contains normal level of α and low level of β-chain. Synthesis of ß-chain takes place in low amount in foetus of 8 weeks age and, therefore, it is possible to analyse the α-globin in a foetus but difficult to analyse ß-chains.

The foetus fluid also contains foetally derived fibroblasts which serve, with or without additional culture, as a source of tissue for preparation of foetal DNA. Structural analysis of DNA is done by : (*i*) isolation of all the globin genes as recombinant DNA molecules, and cloning in bacteria, or (*ii*) without cloning the recombinant DNA in bacteria directly by Southern blotting technique, and (*iii*) linked polymorphism. There are about 35 diseases which have been identified by antenatal diagnosis. This method helps in suggesting the abortion of defected baby, and if possible, to cure. It can be recommended for gene therapy as well. Thus, the antenatal diagnosis is applied for both genetic counselling and gene therapy.

3. Foetal Cell/DNA Prenatal Paternity Test

This prenatal paternity test uses only the mother's blood (and samples from the alleged father) to determine the paternity of the child as early as 13 weeks into the pregnancy. This test is noninvasive unlike an amniocentesis or chorionic villus sampling (CVS) test which could cause problems in the pregnancy and/or foetal injuries, induce labour, or terminate the pregnancy. In this test, you do not need to have an amniocentesis or CVS.

The amniocentesis is a test that involves placing a needle in to the womb through the woman's abdomen (tummy) and obtaining a small sample of the fluid called amniotic fluid in which the baby floats. Most of the women say that the test is not painful. This test is usually carried out between 16 and 18 weeks of the pregnancy and can be done in an out-patients clinic without the need for an overnight stay in the hospital. An ultrasound scan is carried out to check whether twins are present and to see where the placenta and the baby are? The fluid contains cells which have come from the baby and these cells are then grown in the laboratory so that genetic tests can be carried out.

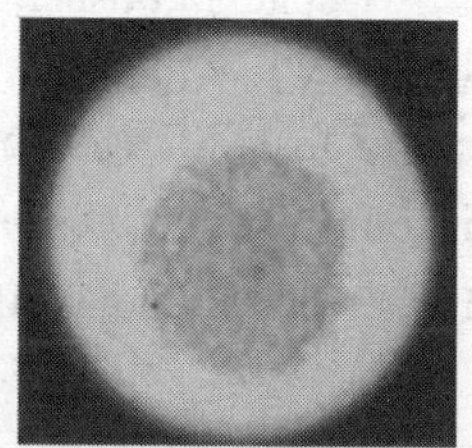

Foetal cells and DNA are in the mother's blood

The collection of all specimens for this test is safe and noninvasive for both the mother and unborn foetus, and no physician is required to collect the sample specimen. From maternal blood both foetal cells and foetal DNA can be isolated. This foetal material is a source of information about the gender and genetic makeup of the developing foetus. Foetal genetic material can be detected in maternal blood early in the pregnancy. During pregnancy, a variety of cell types of foetal origin as well as foetal DNA cross the placenta circulate within maternal peripheral blood. Blood sample is taken from at least 13 weeks pregnant woman. Samples from alleged father(s) may be buccal swab samples, hair samples or forensic samples for discreet testing *e.g.* toothbrush, water bottle, chewing gum, cigarette butt, etc.

QUESTIONS

1. What do you understand by DNA fingerprinting ? Write the role of Sir Alec John Jeffreys in DNA fingerprinting.
2. Explain in brief that variation is the cause of evolution and the basis of DNA fingerprinting.
3. Write in detail about the methods used in DNA fingerprinting.
4. Writc RFLP analysis of DNA.
5. Discuss in detail about application of DNA DNA fingerprinting.
6. Describe DNA fingerprinting and its role in disease diagnosis.
7. Discuss in detail about genetic counselling.
8. Write short notes on the following :

(*i*) Variable number of tandem repeats, (*ii*) Sir Alec John Jeffreys,
(*iii*) Microsatellites, (*iv*) Minisatellites,
(*v*) DNA as blue print of life, (*vi*) Lalji Singh,
(*vii*) PCR analysis of DNA, (*vii*) STR analysis,
(*viii*) Amplified fragment length polymorphism (AFLP),
(*ix*) Y-chromosome analysis, (*x*) Mitochondrial DNA analysis,
(*xi*) Application of DNA fingerprinting in forensic science,
(*xii*) Naina Sahni tandoor case, (*xiii*) Hurdles to DNA profiling,
(*xiv*) Indirect gene tracking, (*xv*) Antenatal diagnosis

BIBLIOGRAPHY

Chawla P. 1998. *The detective DNA*. NISCAIR, CSIR, New Delhi.

Easteal S., McLeod N. and Reed K. 1991. *DNA profiling : principles, pitfalls, and potential*. Harwood Academic Publishers, p. 227.

Godl J. 1998. Remembering Anna Anderson. *European Royal History J.*, Issue VI (August), Arturo Beeche, Publisher, Oakland.

Jeffreys A.J., Allen, M., Hagelberg E. and Sonnberg A. 1992. Identification of the skeletal remains of Josef Mengele by DNA analysis. *Forensic Sci. Int.* **56**: 65–76.

Lincoln P.J. 1998. *Forensic DNA Profiling Protocols*. Humana Press, p. 320.

Mueller U. G. and Wolfenbarger L. L. 1999. AFLP genotyping and fingerprinting. *Trends Ecol. Evol.* **14:** 389–394.

Proteins: Structure, Engineering and Design

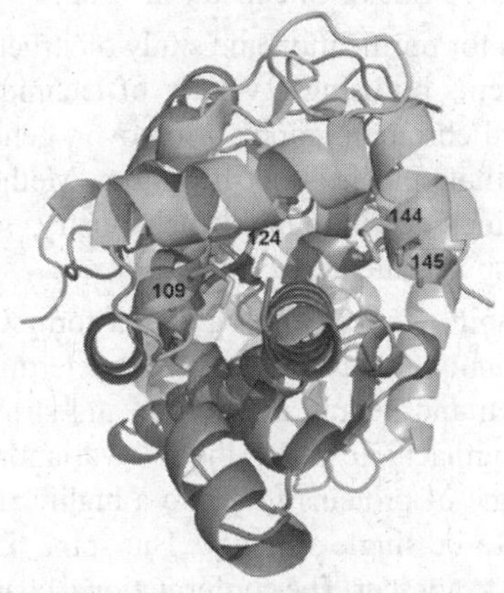

A 3D computer model (molecule of enzyme, adenylate kinase)

A. Chemical bonds
B. Protein structure
C. Conformation analysis of proteins
D. X-ray crystallography
E. Nuclear magnetic resonance (NMR)
F. Mass spectrometry (MS)
G. Protein - DNA interaction
H. Protein engineering and protein design
I. Methods of protein engineering and design
J. Structure-function relationships
K. Some achievements in protein engineering and design

Proteins are the macromolecules which are found in all groups of organisms in varying amount. The life of every organism including viruses is possible only because of proteins. All organisms living on earth function only because of proteins. Structural topography in cell membrane as well as in cell wall is provided by proteins. Muscular contraction takes place only due to proteins. Our heart beats up due to protein, and we are able to see the beautiful world only because of proteins. What we eat, smell, see, feel, think, express, speak or realise, every activity is possible because of proteins. If there had been no proteins on earth, no life would have been possible.

Proteins make possible our body structure, hairs, eyes, height, muscular shape and strength and over all beauty of body. Besides, most of the diseases we suffer from are because of mal functioning of proteins, protein deficiencies, or absence of proteins- for example blood, sickle shaped anaemia, muscular dystrophy, colour blindness, texture and colour of skin, immunity in body to fight against diseases, etc.

Proteins also called polypeptides are the organic compounds which are made up of amino acids arranged in a linear chain and folded into a globular form. A protein is formed by the amino acids joining together by the peptide bonds between the carboxyl and amino groups of adjacent amino acid residues. In a protein the sequence of amino acids is governed by the sequence of a gene, which is encoded in the genetic code represented by mRNA. Moreover, the genetic code specifies 20 standard amino acids; but in certain organisms the genetic code can include selenocysteine, and in certain archaea it contains pyrrolysine. Soon after synthesis, the residues in a protein are chemically modified by post-translational modification resulting in changes in physical and chemical properties, folding, stability, activity, and ultimately, the function of the proteins. Proteins can also work together to achieve a particular function, and they form stable complexes.

In 1838, for the first time the term 'protein' was coined by a Swedish chemist Jöns Jakob Berzelius. In 1926, James B. Sumner showed that the enzyme urease was in fact a protein. Before this report the main role of proteins as enzymes in living organisms was not fully known. Frederick Sanger was the first to sequence the protein insulin; hence he won the Nobel Prize in 1958 for this achievement. In 1958, haemoglobin and myoglobin were the first proteins of which structures were solved by Max Perutz and Sir John Cowdery Kendrew, respectively by determining the three dimensional structures of both proteins through X-ray diffraction analysis. For these discoveries, Perutz and Kendrew were awarded the Nobel Prize in Chemistry in 1962.

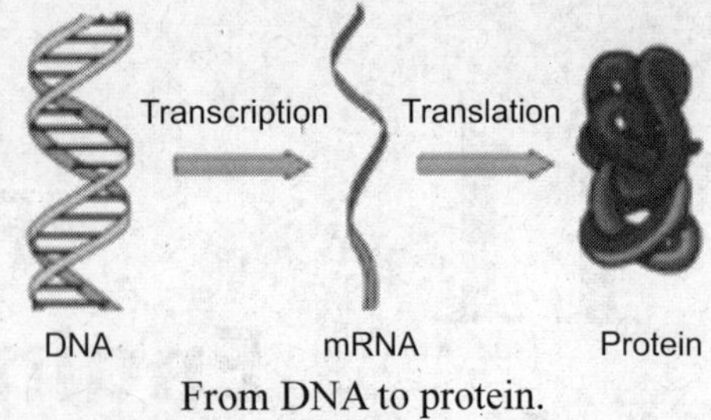

From DNA to protein.

The 20th century witnessed the development of methods for purification and study of structure of proteins. Proteins may be purified from cellular components by using a variety of techniques such as ultracentrifugation, precipitation, electrophoresis, and chromatography. Moreover, genetic engineering has made possible a number of methods to facilitate purification of protein. Methods commonly used to study protein structure and functions include: immunohistochemistry, site-directed mutagenesis, nuclear magnetic resonance and mass spectrometry, etc.

The most distinguishing feature of polypeptides is their ability to fold into a globular form. One or more polypeptide chains twist into a 3D shape forms. This unique shape of the protein determines its function. For instance, structural proteins such as collagen and keratin are fibrous and stringy. But the globular proteins like haemoglobin are folded and compact. However, there is variation in the extent of folding of proteins into a defined structure. Some of proteins fold into a highly rigid structure with small fluctuations; hence they are considered to be single structure. But some other proteins undergo large rearrangements from one conformation to another. The conformational change is often associated with a signalling event. Thus, the structure of a protein serves a medium through which it regulates either the function of a protein or activity of an enzyme. Not all proteins require a folding process in order to function because some function is performed in an unfolded state.

A. CHEMICAL BONDS

The atoms and molecules are interlinked by different types of bonding formed through loss or gain of electrons, proper electron sharing for stable electron configuration, and/or development of slight positive and negative charges. Different types of bonding are described as below:

1. Ionic Bond

An ionic bond is a type of chemical bond that is formed by electrostatic attraction between two oppositely charged ions of a metal and a nonmetal. The metal donates one or more electron, forming a positively charged ion or cation with a stable electron configuration. These electrons then enter the non-metal, causing it to form a negatively charged ion or anion which also has a stable electronic configuration. The electrostatic attraction between the oppositely charged ions causes them to come together and form a bond.

Fig. 11.1 shows the formation of sodium chloride through ionically bonding of sodium and chlorine. The sodium atom loses its outer electron by using ionisation energy (the amount of energy required to remove the outermost electron) resulting in a stable electronic configuration and formation of a cation (Na^+).

$$Na \rightarrow Na^+ + e^-$$

This electron enters the chlorine atom exothermically. The chlorine atom gains an electron forming an anion (Cl^-).

3. Hydrogen Bond

In 1912, Linus Pauling credited T.S. Moore and T.F. Winmill who first mentioned the hydrogen bond to account for the fact that trimethyl ammonium hydroxide is a weaker base than tetramethyl ammonium hydroxide. The attractive interaction of a hydrogen atom with an electronegative atom (coming from another molecule or chemical group) is called 'hydrogen bond'. A hydrogen atom attached to a relatively electronegative atom is a hydrogen bond donor. An electronegative atom such as fluorine, oxygen or nitrogen is a hydrogen bond acceptor. These bonds are formed intermolecularly, *i.e.* between molecules or intramolecularly (within different parts of a single molecule). The hydrogen bond is 5 to 30 kJ/mole stronger than a van der Waals interaction, but weaker than covalent or ionic bonds. This type of bond occurs in both inorganic molecules such as water and organic molecules such as DNA.

The high boiling point of water (100°C) is due to intermolecular hydrogen bonding. Intramolecular hydrogen bonding is partly responsible for the secondary, tertiary, and quaternary structures of proteins and nucleic acids. Ethanol is an example of a hydrogen bond donor where hydrogen is bonded to oxygen. In contrast, diethyl ether is a hydrogen bond acceptor which lacks a hydrogen atom bonded to it. Fig. 11.3 shows hydrogen bond donors and hydrogen bond acceptors.

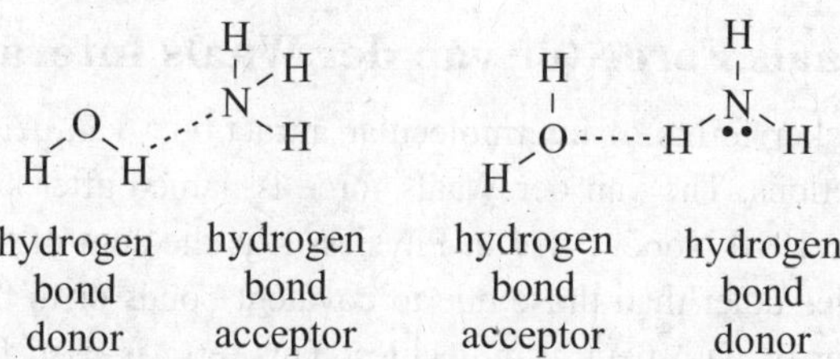

Fig. 11.3. Hydrogen bond donors and hydrogen bond acceptors.

Often the hydrogen bond is described as an electrostatic dipole-dipole interaction. But it has some features of covalent bonding also. It is directional and strong, produces interatomic distances shorter than sum of van der Waals radii, and usually involves a limited number of interaction partners. These covalent features are more substantial when acceptors bind hydrogen from more electronegative donors.

The simplest example of a hydrogen bond is found between water molecules (Fig. 11.4). There are two hydrogen atoms and one oxygen atom in a discrete water molecule. Two water molecules can form a hydrogen bond between them. More bonds are possible in liquid water, because the oxygen of one water molecule has two pairs of electrons, each pair of electrons form a hydrogen bond with a hydrogen atom on another water molecule. This is repeated in such a way that every water molecule is H-bonded with up to four other molecules; two molecules are bonded through its two lone pairs and two molecules through its two hydrogen atoms. The crystal structure of ice is strongly affected by hydrogen bonding. Hydrogen bonding also results in an open hexagonal lattice. The exact number of hydrogen bonds formed by a molecule of liquid water fluctuates with time and depends on the temperature.

Hydrogen bonds are formed in a large number of chemical compounds such as DNA, proteins, polymers, etc. Hydrogen bonds in protein have also been described in the preceding section. For detail *see* Section C, *Protein Structure*.

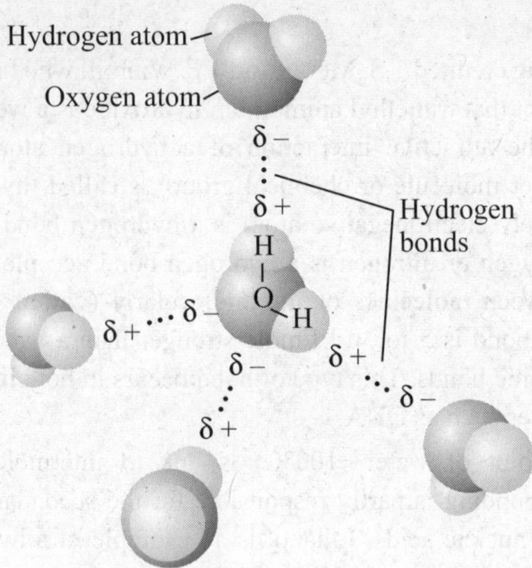

Fig. 11.4. Hydrogen bonds between water molecules.

4. The van der Waals Force (or van der Waals Interaction)

There are two weaker forms of intermolecular attractions: van der Waals dispersion forces and dipole-dipole attractions. The van der Waals force is named after a Dutch scientist J.D. van de Waals, winner of the 1910 Nobel Prize in Physics. It is the sum of the attractive or repulsive forces between molecules other than those due to covalent bonds or to the electrostatic attraction of ions with one another or with neutral molecules. This term is used for the: (*i*) force between two permanent dipoles, (*ii*) force between a permanent dipole and a corresponding induced dipole, and (*iii*) force between two instantaneously induced dipole. The van der Waals forces have been described by International Union of Pure and Applied Chemistry in 1994.

The 'van der Waals radius' of an atom is the radius of an imaginary hard sphere which can be used to model the atom for many purposes (Fig. 11.5). The van der Waals radii are determined by measuring the atomic spacing between pairs of unbonded atoms in crystals. The van der Waals radius is a fixed property of the atom in all circumstances.

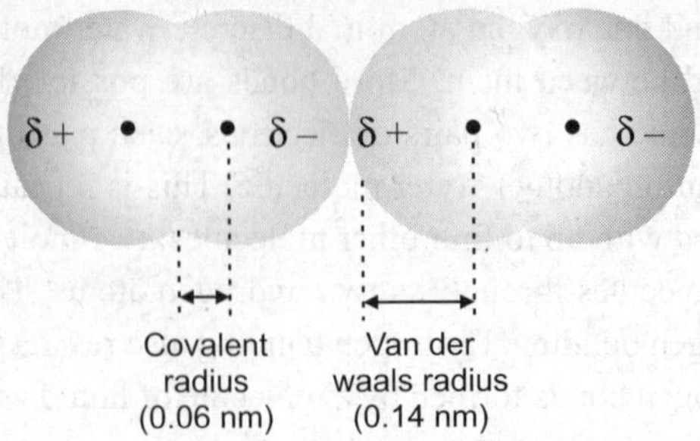

Fig. 11.5. The van der Waals radius of an atom.

As compared to normal chemical bonds, the van der Waals forces are relatively weak; but they play a key role in diverse fields such as supramolecular chemistry, structural biology, nanotechnology, etc. The chemical characters of many organic compounds are defined by the van der Waals forces, such as solubility of organic substances in polar and non-polar media.

5. Peptide Bond

The peptide bonds are actually the amide bonds that covalently link the amino acids together to form peptides (short chains of amino acids) and proteins (longer polypeptide chains that fold into specific three-dimensional conformations with biological function). Peptide bond is formed between two molecules of amino acids when the α-carboxyl group (>COO^-) of one molecule reacts with the α-amine group (-NH_3^+) of the other molecule resulting in release of a molecule of water (H_2O) (Fig. 11.6). This is called condensation reaction or dehydration synthesis reaction which usually occurs between amino acids. The C(O)NH bond is called a peptide bond, and the resulting molecule is an amide molecule. Peptide bonds held together the amino acids forming polypeptides or proteins chains.

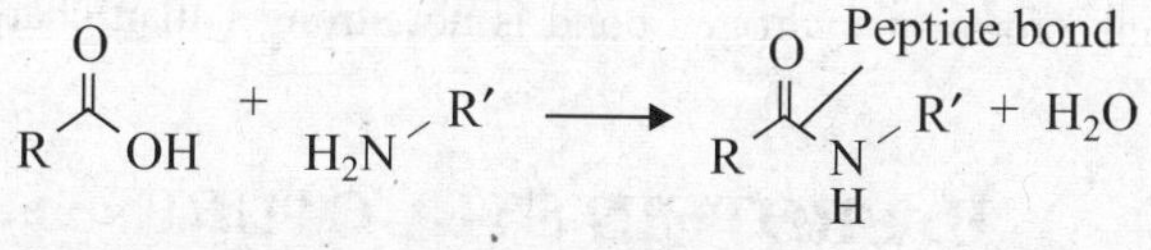

Peptide bond

R–C(=O)–OH + H_2N–R′ ⟶ R–C(=O)–N(H)–R′ + H_2O

Fig. 11.6. Formation of an amide molecule through condensation reaction containing peptide bond.

In a polypeptide chain, R and R′ show Cα atoms and their attached groups. The peptide bond presents a significant barrier to rotation, and hence is central to a roughly planar, rigid group of six atoms. The uncharged, single-bonded form (typically ~60%) is shown on the left, whereas the charged, double-bonded form (typically ~40%) is on the right.

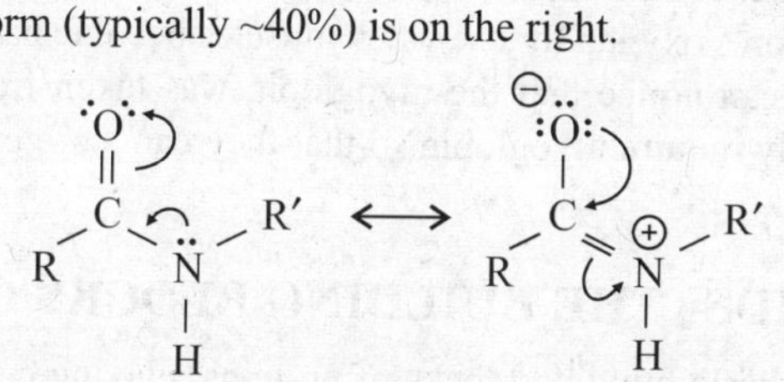

Fig. 11.7. Peptide bond of an amide molecule showing resonance structure.

A resonance structure is invoked to explain the rigidity of the peptide group (Fig. 11.7). Optimal electron delocalisation (π-bonding) within the peptide group depends upon the planar arrangement of these atoms. The partial double-bond character of the peptide bond is reflected in the intermediate length of the peptide bond relative to single and double carbon-nitrogen bonds. Typically the peptide bond has a length of 1.32 to 1.33 Å, while the C–N bond in amines is typically ~1.49 Å, and the C=N bond of an imine is ~1.27 Å.

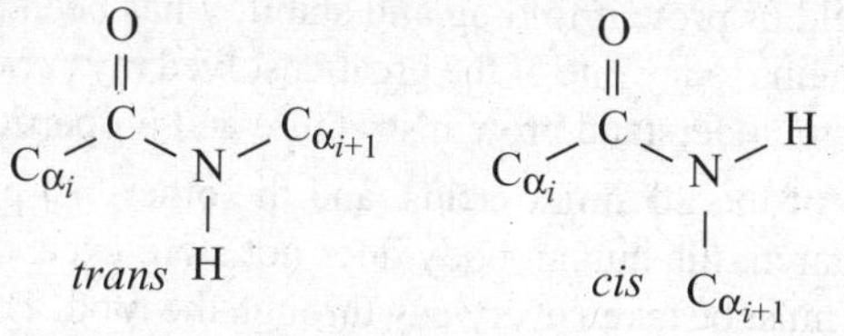

trans *cis*

Fig. 11.8. *Trans* and *cis* configurartion of amide molecule.

Two stable configurations are possible in the planar nature of the peptide group due to the high energetic barrier to rotation about the peptide bond. The *trans* peptide bond is the predominant form in polypeptides, in which the Cα atoms are *trans* to one another (the peptide bond dihedral angle ω is −180°), and the trace of the main chain (*see* bond lines between C and N) is fully extended (Fig. 11.8). The alternative configuration puts the Cα atoms *cis* to one another (the peptide bond dihedral angle ω is 0°). In the case of the *cis* configuration, the trace of the main chain is compacted, *i.e.* the distance between Cα(*i*) and Cα(*i*+1) is less than for the *trans* peptide bond. This is the basis for the energetic preference for the *trans* configuration, the steric strain due to the closer approach of the Cα positions in the *cis* configuration makes it less favorable.

The two resonance forms of amide group confer several important properties. It stabilises the group by 20 kcal/mol and makes it less reactive than similar groups, for example esters. The resonance suggests that the amide group has a partial double bond character, estimated at 40% under typical conditions. At all normal pH values, the peptide group is uncharged; but its double-bonded resonance form gives it an unusually large dipole moment. In certain secondary structures (such as the α-helix), these dipole moments can line up producing a large net dipole.

The characters of the partial double bond can be strengthened or weakened by modifications favouring one resonance form over another. But in hydrophobic environments, the double-bonded form is disfavoured due to its charge. The double-bonded form should be favoured by donating a hydrogen bond to the amide oxygen, or accepting a hydrogen bond from the amide nitrogen. The uncharged single-bonded form of hydrogen bond is not stronger than charged form of hydrogen bond.

B. PROTEIN STRUCTURE

Protein structure really begins with myoglobin because it is the first protein where the science of exploring protein structure really began. In 1960s, John Kendrew and his coworkers determined the atomic structure of myoglobin, laying the foundation for an era of biological understanding. Myoglobin is a small, bright red protein which is very common in muscle cells, and gives meat much of its red colour. It stores oxygen for use when muscles become hard at work. Looking at John Kendrew's PDB file, you can notice that the myoglobin was taken from whale muscles. Marine whales and dolphins greatly require myoglobin so that they can store extra oxygen for use in their deep dives undersea.

I. AMINO ACIDS– THE BUILDING BLOCKS OF PROTEINS

Amino acids are the basic building blocks of proteins, playing central roles both as building blocks of proteins and as intermediates in metabolism. The 20 amino acids forming proteins convey a vast array of chemical versatility. The precise amino acid content and the sequence of those amino acids of a specific protein are determined by the sequence of the bases in the gene that encodes that protein. The biological activity of the protein is determined by the chemical properties of the amino acids. Proteins not only catalyse all of the reactions in living cells but also they control all cellular process. Besides, proteins contain within their amino acid sequences the necessary information to determine how that protein will fold into a three dimensional structure, and the stability of the resulting structure. The field of protein folding and stability has been a critically important area of research for years, and remains today one of the great unsolved mysteries. To understand amino acid structure and properties is to understand protein structure and properties.

Humans produce 10 of the 20 amino acids, and the others 10 proteins are supplied through the food. Unlike fat and starch, the human body does not store excess amino acids for further use; therefore, the amino acids must be taken every day through the food. The 10 amino acids that we can produce are alanine, asparagine, aspartic acid, cysteine, glutamic acid, glutamine, glycine, proline, serine and tyrosine. Tyrosine is produced from phenylalanine; therefore, tyrosine will be required if the diet is deficient in phenylalanine. The essential amino acids are arginine (required for the young, but not for adults), histidine, isoleucine, leucine, lysine, methionine, phenylalanine, threonine, tryptophan, and valine. These amino acids are required in the diet. Plants can make all the amino acids. Humans do not have all the enzymes required for the biosynthesis of all of the amino acids.

Fundamentally, amino acids are joined together by peptide bonds to form the basic structure of proteins. But due to many side groups that are the part of the amino acids the other sorts of bonds may form between the amino acid units. These additional bonds twist and turn the protein into convoluted shapes that are unique to the protein and essential to its ability to perform certain functions within the human body. Upon hydrolysis, proteins yield different types of amino acids. However, the 20 naturally occurring amino acids are polymerised to form polypeptide chain. Thus the proteins are polypeptides containing more than 100 amino acids. Humans can synthesise 10 of them (by inter-conversions) from each other or from other molecules of intermediary metabolism. But the other 10 essential amino acids such as arginine, histidine, isoleucine, leucine, lysine, methionine, phenylalanine, threonine, tryptophan, and valine must be consumed in the diet. This is just one example of the importance of amino acids in the functioning of life. Another cautionary illustration of amino acid power is the gamut of diseases (most notably sickle cell anaemia) that claim the lives of those whose amino acids are out of sequence or malfunctioning.

Amino acids have existed on earth for at least three billion years, long before the appearance of the first true organisms. Amino acids, as ancient and ubiquitous molecules, have been co-opted by evolution for a variety of purposes in living systems. Proteins function as chemical messengers and as intermediates in metabolism.

1. Structure of Amino acids

The term amino acid represents a molecule having two functional groups: *a basic amino* ($-NH_2$) *group* and an *acidic carboxylic* ($-COOH$) *group*. The carbon atom next to the carbonyl group is called the α–carbon, and amino acids with a side chain linked to this carbon are referred to as *alpha amino acids*. The common amino acids are alpha (α)-amino acids because they have a primary amino group and a carboxylic acid group as substitutes of the carbon atoms. In the α–amino acids, the α–carbon is a chiral carbon atom with the exception of glycine. Proline is an exception because it has a secondary amino group (-NH); for uniformity it is also treated as α-amino acid. R is the side chain which differs from each amino acid (Fig. 11.9). The 20 standard amino acids have different structures in their side chains (R groups). All the naturally occurring amino acids have α-amino group (except one which possess α-imino group), and L-configuration at α-C position. The R can be hydrogen atom (H) or methyl group ($-CH_3$) or more complex structure. The general formula of an α-amino acid is $H_2NCHRCOOH$. Different α-amino acids differ where the side chain (R group) is attached to their α-carbon. The side chain varies in size from a hydrogen atom in glycine to a methyl group in alanine, through to a large heterocyclic group in tryptophan.

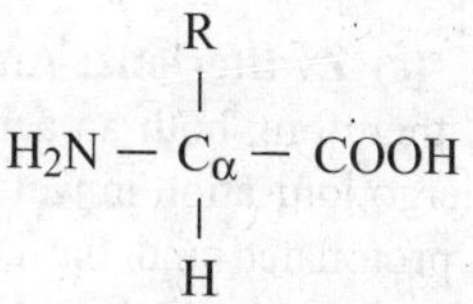

```
       R
       |
H2N — Cα — COOH
       |
       H
```

Fig. 11.9. General structure of L-α-amino acid.

In amino acids that have a carbon chain attached to the α–carbon (such as lysine), the carbons are labelled in order as α, β, γ, δ, and so on (Fig. 11.10). In some amino acids, the aminc group is attached to the β or γ-carbon, and these are referred to as β- amino acid or γ-amino acid.

***(a)* Isomerism:** Among the standard α-amino acids (except glycine) are optically active, *i.e.* they exist in either of two optical isomers called L- (levorotatory) amino acid or D- (dextrorotatory) amino acid. These are the mirror images of each other, *i.e.* they show chirality (Fig. 11.11). L-amino acids show all the amino acids found in proteins during translation in the ribosome. In 2005, K. Pisarewicz and co-workers have found that D-amino acids are present in some proteins that are produced during posttranslational modification after translation and translocation to the endoplasmic reticulum; for example exotic sea-dwelling organisms such as cone snails.

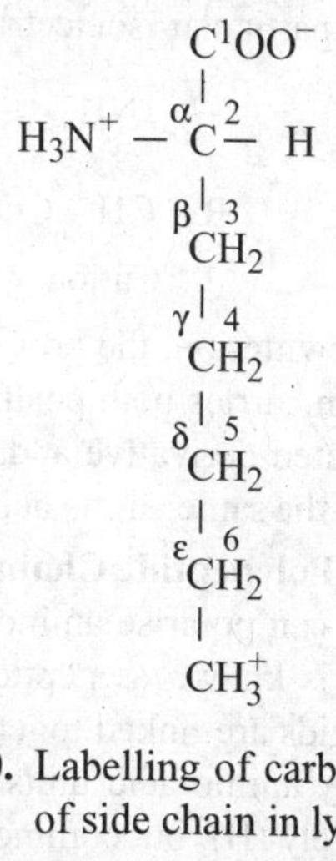

```
          C¹OO⁻
          |
H3N⁺ — α C² — H
          |
        β C³H2
          |
        γ C⁴H2
          |
        δ C⁵H2
          |
        ε C⁶H2
          |
          CH3⁺
```

Fig. 11.10. Labelling of carbon atoms of side chain in lysine.

For amino acid configuration the L and D refers not to the optical activity of the amino acid itself, but to the optical activity of the isomer of glyceraldehyde from which that amino acid can theoretically be synthesised (D-glyceraldehyde and L-glyceraldehyde). In proteins mostly all of the amino acids are (S) at the α-carbon with cysteine being (R) and glycine non-chiral. Since cysteine has a sulfur atom at the first position in its side-chain, it is unusual.

***(b)* Ionic Forms of Amino Acids:** Amino acids bear atleast two ionisable weak acid groups, a – COOH and a $-NH_2$. In solution, two forms of these groups, one charged and another uncharged, exist in protonic equilibrium, where R-COOH and $R\text{-}NH_3^+$ represent the protonated or acidic partners in these equilibria. $R\text{-}COO^-$ and $R\text{-}NH_2$ are the conjugate bases of the corresponding acids.

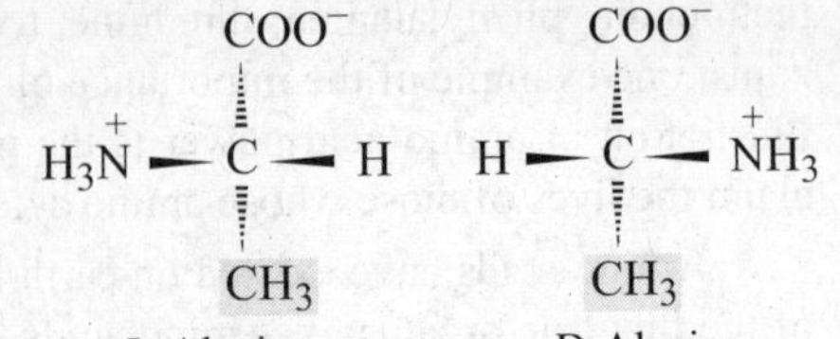

COO^- / $H_3\overset{+}{N}$ — C — H / CH_3 (L-Alanine); COO^- / H — C — $\overset{+}{N}H_3$ / CH_3 (D-Alanine)

Fig. 11.11. L-alanine and D-alanine.

$$R-COOH \rightleftharpoons R-COO^- + H^+$$
$$R-NH_3^+ \rightleftharpoons R-NH_2 + H^+$$

***(c)* Zwitterions:** Amino acids have both amine and carboxylic acid functional groups and are, therefore, both an acid and a base at the same time. Both the $-NH_2$ and the –COOH, groups undergo ionisation in water. At a pH ~7.4 the amino groups are protonated and the carboxyl acid groups are in their conjugate base (carboxylate) form. At a certain pH known as the isoelectric point an amino acid has no overall charge, since the number of protonated ammonium groups (positive charges) and deprotonated carboxylate groups (negative charges) are equal. The amino acids all have different isoelectric points. The ions produced at the isoelectric point have both positive and negative charges of opposite polarity, hence they are known as zwitterions or dipolar ions (German, *Zwitter* means hermaphrodite or hybrid).

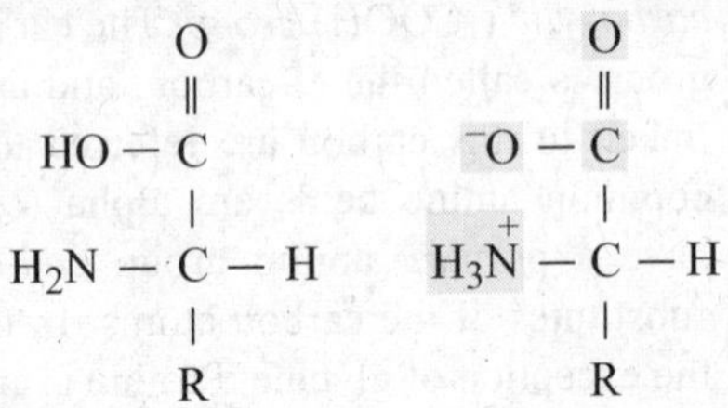

Nonionic form: HO — C(=O) — C(H)(H_2N) — R; Zwitterionic form: ^-O — C(=O) — C(H)($H_3\overset{+}{N}$) — R

Fig. 11.12. Non-ionic and zwitterion forms of an amino acid.

Thus they are amphoteric and overall charges on them are neutral. In solids and in polar solutions such as water amino acids can exist as zwitterions, but not in the gas phase. Zwitterion forms of amino acids are least soluble at their isolectric point. In 2006, M. Remko and B.M. Rode studied the effect of metal ions such as Li^+, Na^+, K^+, Mg^{2+}, Ca^{2+}, Ni^{2+}, Cu^{2+}, and Zn^{2+}, and water coordination on the structure of glycine and zwitterionic glycine. They found that by adjusting the pH to its particular isoelectric point, an amino acid can be isolated by precipitating it from water.

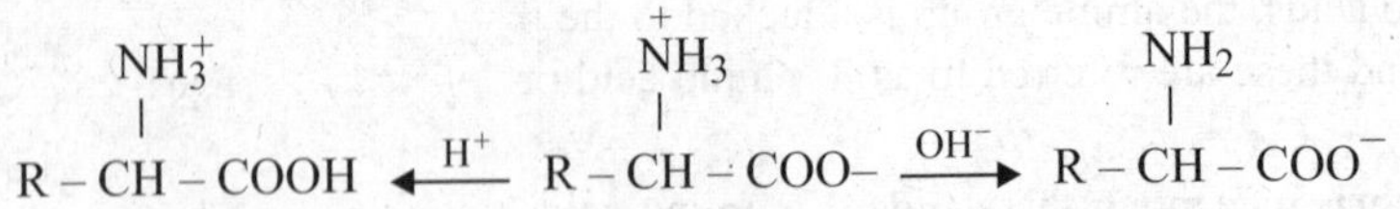

$$\underset{\text{Cation}}{R-CH(NH_3^+)-COOH} \xleftarrow{H^+} \underset{\text{Zwitter ion}}{R-CH(\overset{+}{N}H_3)-COO^-} \xrightarrow{OH^-} \underset{\text{Anion}}{R-CH(NH_2)-COO^-}$$

In zwitterion, the proton from the carboxylic group is transferred to the amino group. Thus a zwitterion carries both positive and negative charges. In acidic solution an amino acid behave like a protonated derivative and, therefore, migrates to the cathode under electric field. In an alkaline medium, the same amino acid behaves like an anion derivative and, therefore, migrates to the anode.

***(d)* Polypeptide Chains:** Amino acids are joined together through a peptide bond. Elimination of water can polarise amino acids to form long chains; the resulting -CO-NH- linkage (an amide linkage) is known as peptide bond (Fig. 11.13 A). A polypeptide chain is formed when several amino acids are linked together by peptide bonds. Polymers composed of two, three, a few (3-10), and many amino acid units are known as dipeptides, tripeptides, oligopeptides, and polypeptides respectively (B); but commonly they are called 'peptides'.

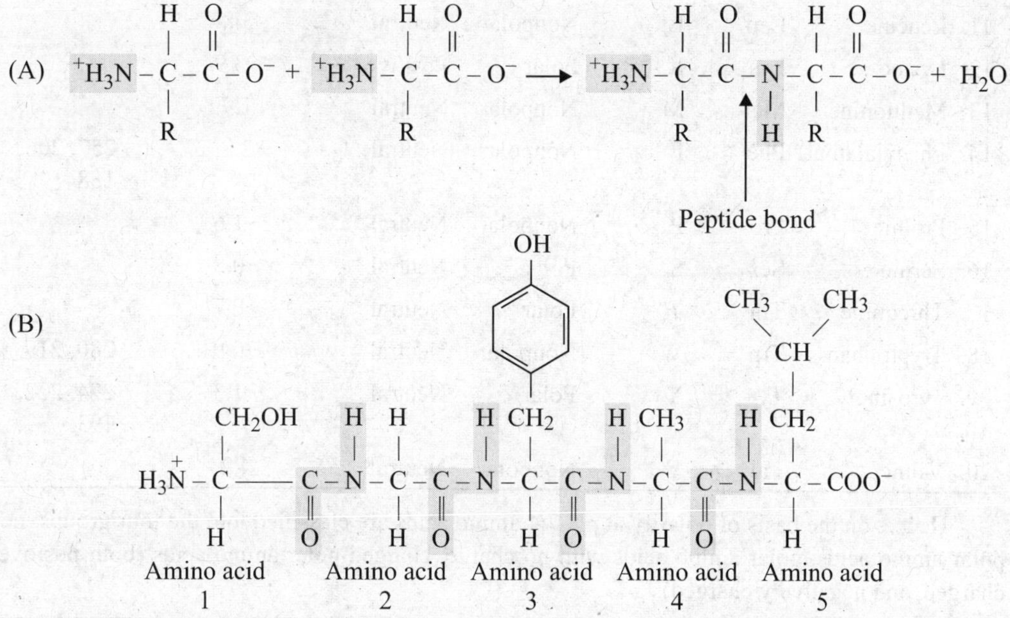

(A)

$$^{+}H_3N-\underset{R}{\overset{H}{C}}-\overset{O}{\overset{\|}{C}}-O^- + {}^{+}H_3N-\underset{R}{\overset{H}{C}}-\overset{O}{\overset{\|}{C}}-O^- \longrightarrow {}^{+}H_3N-\underset{R}{\overset{H}{C}}-\overset{O}{\overset{\|}{C}}-\underset{H}{N}-\underset{R}{\overset{H}{C}}-\overset{O}{\overset{\|}{C}}-O^- + H_2O$$

Peptide bond

(B)

$$H_3\overset{+}{N}-\underset{H}{\overset{CH_2OH}{C}}-\overset{O}{\overset{\|}{C}}-\underset{H}{\overset{H}{N}}-\underset{H}{\overset{H}{C}}-\overset{O}{\overset{\|}{C}}-\overset{H}{N}-\underset{H}{\overset{CH_2-C_6H_4-OH}{C}}-\overset{O}{\overset{\|}{C}}-\overset{H}{N}-\underset{H}{\overset{CH_3}{C}}-\overset{O}{\overset{\|}{C}}-\overset{H}{N}-\underset{H}{\overset{CH_2-CH(CH_3)_2}{C}}-COO^-$$

Amino acid 1 | Amino acid 2 | Amino acid 3 | Amino acid 4 | Amino acid 5

Fig. 11.13. The two amino acids are linked by peptide bond (A); a polypeptide chain formed by five amino acids (B).

2. Classification of Amino Acids

Amino acids are classified into different ways on the basis of polarity, structure, nutritional requirement, metabolic fate, etc. However, a general method of classification is based on polarity. Basically there are three major classifications for amino acids: (*i*) amino acids with non-polar R group, (*ii*) amino acids with uncharged polar R groups, and (*iii*) amino acids with charged polar R group. Table 11.2 shows all 20 amino acids with their codes. The amino and carboxylic acid groups of amino acids readily ionise.

The ionic property of the side chains influences the physical and chemical property of free amino acids, and amino acids in proteins.

Table 11.2. Standard amino acids, abbreviations and their properties.

S.I. No.	*Amino acid*	*3-Letter*	*1-Letter*	*Side chain polarity*	*Side chain charge (pH 7.4)*	*Hydropathy index*	*Absorbance λ_{max}(nm)*
1.	Alanine	Ala	A	Nonpolar	Neutral	1.8	
2.	Arginine	Arg	R	Polar	Positive	−4.5	
3.	Asparagine	Asn	N	Polar	Neutral	−3.5	
4.	Aspartic acid	Asp	D	Polar	Negative	−3.5	
5.	Cysteine	Cys	C	Nonpolar	Neutral	2.5	250
6.	Glutamic acid	Glu	E	Polar	Negative	−3.5	
7.	Glutamine	Gln	Q	Polar	Neutral	−3.5	
8.	Glycine	Gly	G	Nonpolar	Neutral	−0.4	
9.	Histidine	His	H	Polar	Positive Neutral	−3.2	211
10.	Isoleucine	Ile	I	Nonpolar	Neutral	4.5	

11.	Leucine	Leu	L	Nonpolar	Neutral	3.8	
12.	Lysine	Lys	K	Polar	Positive	–3.9	
13.	Methionine	Met	M	Nonpolar	Neutral	1.9	
14.	Phenylalanine	Phe	F	Nonpolar	Neutral	2.8	257, 206, 188
15.	Proline	Pro	P	Nonpolar	Neutral	–1.6	
16.	Serine	Ser	S	Polar	Neutral	–0.8	
17.	Threonine	Thr	T	Polar	Neutral	–0.7	
18.	Tryptophan	Trp	W	Nonpolar	Neutral	–0.9	280, 219
19.	Tyrosine	Tyr	Y	Polar	Neutral	–1.3	274, 222, 193
20.	Valine	Val	V	Nonpolar	Neutral	4.2	

Hence, on the basis of polarity at pH 7.4 amino acids are classified into the four groups: non-polar amino acids, polar amino acids with no charge, charged polar amino acids (both positively charged, and negatively charged).

> Generally, the genetic code specifies 20 standard amino acids; however, in certain organisms the genetic code can include *selenocysteine*, and in certain archaea– *pyrrolysine*.

***(a)* Non-polar Amino Acids:** These amino acids have equal number of amino and carboxyl groups and are neutral. These amino acids are hydrophobic and have no charge on the 'R' group. The amino acids in this group are alanine, valine, leucine, isoleucine, phenyl alanine, glycine, tryptophan, methionine and proline (Fig. 11.14).

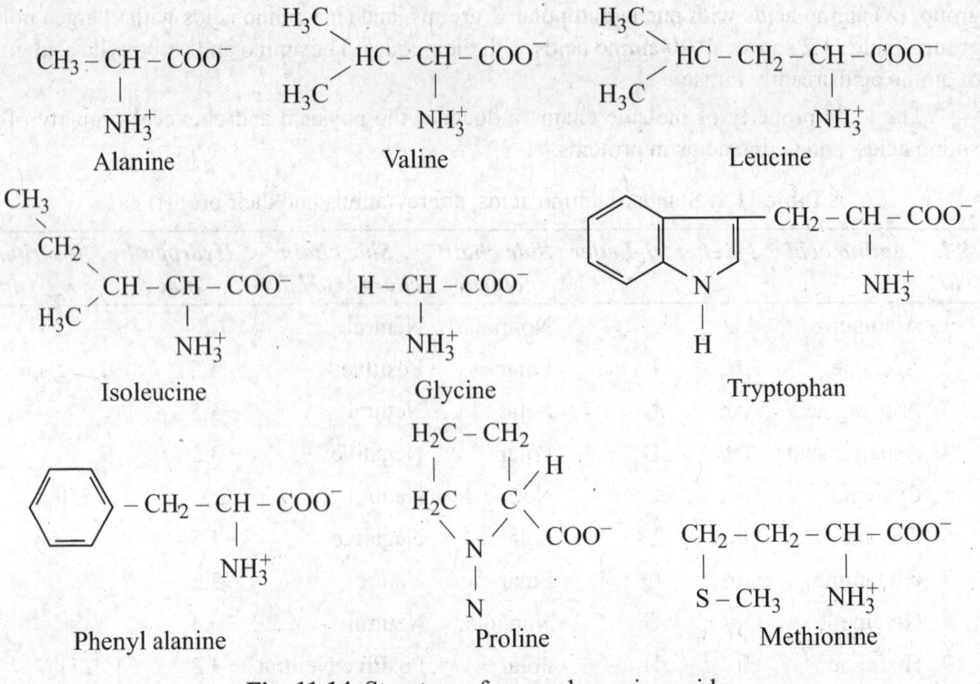

$CH_3-CH(NH_3^+)-COO^-$

Alanine

$(H_3C)_2HC-CH(NH_3^+)-COO^-$

Valine

$(H_3C)_2HC-CH_2-CH(NH_3^+)-COO^-$

Leucine

$CH_3-CH_2-CH(H_3C)-CH(NH_3^+)-COO^-$

Isoleucine

$H-CH(NH_3^+)-COO^-$

Glycine

(Indole ring, N–H) $-CH_2-CH(NH_3^+)-COO^-$

Tryptophan

(Benzene ring) $-CH_2-CH(NH_3^+)-COO^-$

Phenyl alanine

H_2C-CH_2, H_2C, C(H)(COO^-), N, N

Proline

$CH_2-CH_2-CH(NH_3^+)-COO^-$, $S-CH_3$

Methionine

Fig. 11.14. Structure of non-polar amino acids.

Non-polar amino acid side chains have a variety of shapes and sizes. There are basically nine amino acids under this classification. Glycine has the smallest possible side chain, an H atom. Alanine, valine, leucine, and isoleucine have aliphatic hydrocarbon side chains ranging in size from a methyl group for alanine to isomeric butyl groups for leucine and isoleucine. Methionine has a thiol ether side chain that resembles an n-butyl group in many of its physical properties (C and A have nearly equal electronegativities, and S is about the size of a methylene group). Proline has a cyclic pyrrolidine side group. Phenylalanine (with its phenyl moiety) and typtophan (with its indole group) contains aromatic side groups, which are characterised by bulk as well as nonpolarity.

***(b)* Polar Amino acids with no Charge:** These amino acids do not have any charge on the 'R' group. They participate in hydrogen bonding of protein structure. The amino acids in this group are: serine, threonine, tyrosine, cysteine, glutamine and aspargine Fig 11.15.

Uncharged polar side chains have hydroxyl-, amide-, or thiol- groups. There are six amino acids under this category. Serine and threonine have hydroxylic R side chains of different sizes. Asparagine and glutamine have amide-bearing side chains of different sizes. Tyrosine has a phenolic group and is aromatic. Cystein is very unique among all 20 amino acid because it has a thiol group that forms a disulfide bond with other cystein through oxidation.

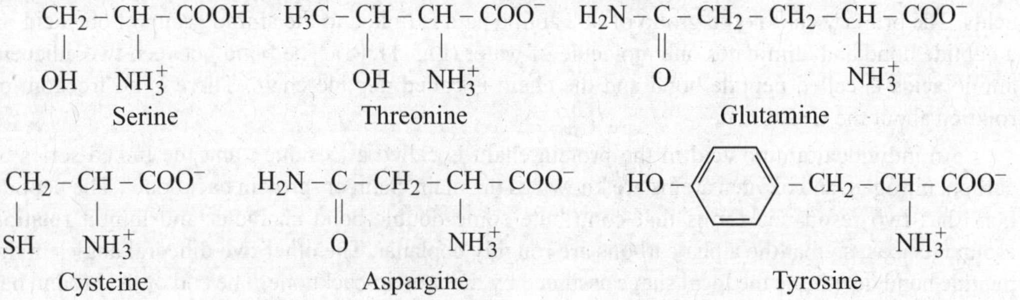

$CH_2 - CH - COOH$ (with OH on CH_2 and NH_3^+ on CH)
Serine

$H_3C - CH - CH - COO^-$ (with OH and NH_3^+)
Threonine

$H_2N - C(=O) - CH_2 - CH_2 - CH - COO^-$ (with NH_3^+)
Glutamine

$CH_2 - CH - COO^-$ (with SH and NH_3^+)
Cysteine

$H_2N - C(=O) - CH_2 - CH - COO^-$ (with NH_3^+)
Aspargine

$HO - C_6H_4 - CH_2 - CH - COO^-$ (with NH_3^+)
Tyrosine

Fig. 11.15. Structure of polar amino acids with no charge.

***(c)* Charged Polar Amino Acids:** Under this groups the amino acids have charged polar side chains which may be positively or negatively charged. There are five amino acids under this type such as lysine, arginine, histidine, aspartic acid and glutamic acid.

***(i)* Positively charged polar amino acids:** Polar amino acids with positive charge have more amino groups as compared to carboxyl groups making it basic. The amino acids, which have positive charge on the 'R' group, are placed in this category. The side chains are positively charged in lysine, arginine and histidine (Fig. 11.16); lysine has a butylammonium side chain, arginine has a guanidine group, and histidine bears an imidazolium moiety.

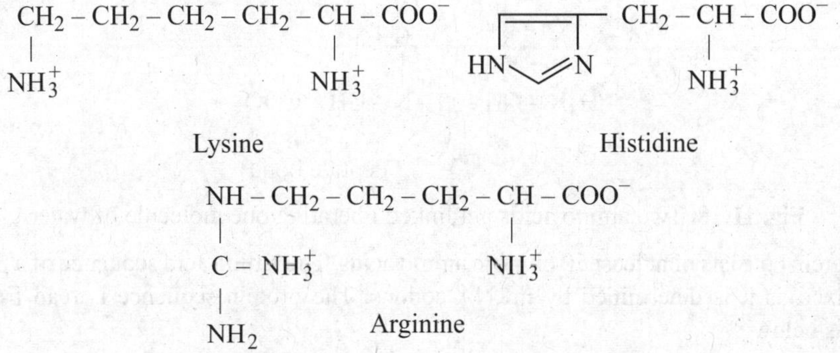

$CH_2 - CH_2 - CH_2 - CH_2 - CH - COO^-$ (with NH_3^+ on the first CH_2 and NH_3^+ on CH)
Lysine

(imidazole ring HN, N) $CH_2 - CH - COO^-$ (with NH_3^+)
Histidine

$NH - CH_2 - CH_2 - CH_2 - CH - COO^-$ (with $C - NH_3^+$, NH_2 on NH and NH_3^+ on CH)
Arginine

Fig. 11.16. Structure of polar amino acids with positive charge..

***(ii)* Negatively charged polar amino acids:** Negatively charged polar amino acids have more carboxyl groups than amino groups. Presence of carboxylic group makes them acidic. The amino acids which have negative charge on the 'R' group are placed in this category, for example aspartic acid and glutamic acid (Fig. 11.17). They are also called as dicarboxylic mono-amino acids.

$$^{-}OOC - CH_2 - \underset{\displaystyle NH_3^+}{\underset{|}{CH}} - COO^- \qquad\qquad ^{-}OOC - CH_2 - CH_2 - \underset{\displaystyle NH_3^+}{\underset{|}{CH}} - COO^-$$

Aspartic acid   Glutamic acid

Fig. 11.17. Structure of polar amino acids with negative charge.

Except bacterial cell wall, D-amino acids rarely occur. On the basis of types of R-groups (whether charged, polar of hydrophobic), α-amino acids are categorized into three groups viz., charged amino acids, polar amino acids and hydrophobic amino acids.

3. Polymerisation of Amino Acids

In a polymer the amino acids are joined together by the peptide bonds between the carboxyl and amino groups of adjacent amino acid residues. Proteins are unbranched linear polymer of amino acids. The primary carboxylic group of one amino acid is linked to the amino group of one acid by a peptide bond and eliminates one molecule of water (Fig. 11.18). The bond between two adjacent amino acids is called peptide bond and the chain is called peptide chain. There is no freedom of rotation about the C−N bond.

An individual amino acid in the protein chain is called a 'residue', and the linked series of carbon, nitrogen, and oxygen atoms are known as the main chain or 'protein backbone'. The peptide bond has two resonance forms that contribute some double-bond character and inhibit rotation around its axis, so that the alpha carbons are roughly coplanar. The other two dihedral angles in the peptide bond determine the local shape assumed by the protein backbone. The end of the protein has a free carboxyl group (-COOH) which is known as the C-terminus or carboxy terminus, whereas the other end with a free amino group ($-NH_2$) is known as the N-terminus or amino terminus.

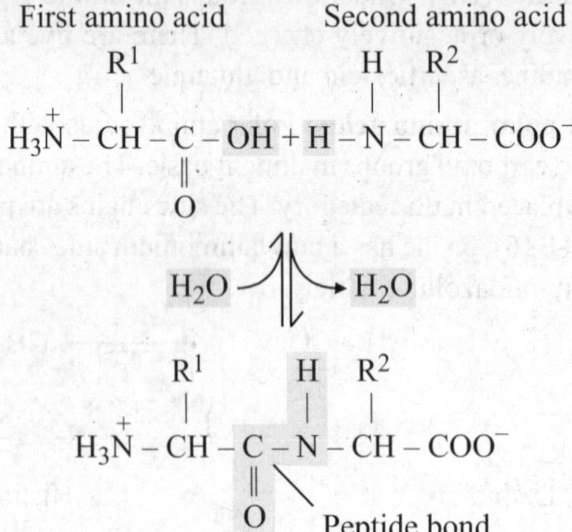

Fig. 11.18. Two amino acids get linked liberating one molecule of water.

A protein consists of at least 50 or more amino acids. The amino acid sequence of a polypeptide chain is fixed as it is determined by mRNA codons. The protein sequence is read from N to C terminus as below:

$\overrightarrow{NH_2} -$ lys – gly – thr – ala ------------- COOH

4. Nomenclature

Table 2 shows abbreviations of a single letter and three letters, most of them are taken from the first three letters, for example Glx refers to Glu or Gln. Care should be taken while writing these notations. One-letter notations are used sequences of several similar proteins are compared. During translation the genetic code in the form of RNA is translated into a protein sequence.

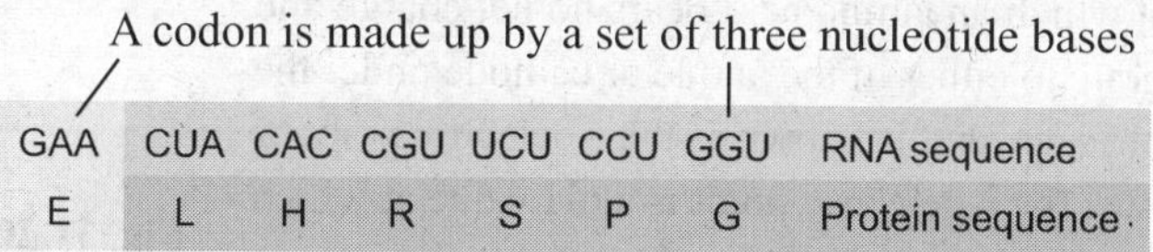

***(a)* IUPAC System of Nomenclature:** The IUPAC system for naming the amino acids is fairly simple. But it is not extensively used. IUPAC method for naming other types of organic compounds is popular. It is very easy to apply to amino acids also. In an amino acid there are: a carboxylic acid with five-carbon chain and with an amino group on the second carbon atom and a methyl group on the third carbon atom (Fig. 11.19). Thus, the IUPAC name for this compound is 2-amino-3-methylpentanoic acid.

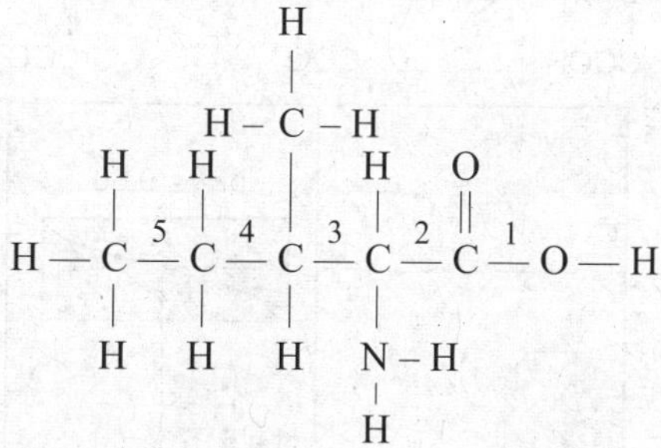

Fig. 11.19. Structural formula of 2-amino-3-methylpentanoic acid.

Non-standard Amino acids

Besides twenty standard amino acids, there is large number of non-standard amino acids also, that are found in proteins formed by post-translational modification. For proper function or regulation of a protein, these modifications are often essential. For example, carboxylation of glutamate allows better binding of calcium cations, and the hydroxylation of proline is critical for maintaining connective tissues. Localisation of the protein can also be determined by such modifications; for example a protein becomes able to bind to a phospholipid membrane after addition of long hydrophobic group.

The non-standard amino acids not found in proteins are: lanthionine, 2-aminoisobutyric acid, dehydroalanine and the neurotransmitter gamma-aminobutyric acid. Mostly these amino acids occur as intermediates in the metabolic pathways for standard amino acids, for example ornithine and citrulline occur in the urea cycle, part of amino acid catabolism. Rarely, the β-amino acid β-alanine (3-aminopropanoic acid), is used in plants and microorganisms in the synthesis of pantothenic acid (vitamin B_5), a component of coenzyme A.

1. Chemical Properties of Amino Acids

Different amino acids have different functional groups such as α-amino groups, carboxylic group, phenolic group (tyrosine), alcoholic group (serine) and sulfhydryl group (cystein). These groups are chemically reactive.

Isoelectric Point (pI): The isoelectric point is the pH at which a particular molecule or surface carries no net electrical charge is present. The net charge (the algebraic sum of all the positively and negatively charged groups present) of an amino acid depends upon the pH, or proton concentration of the surrounding medium. The pH at which an amino acid bears no net charge and hence does not migrate to either of the anode or cathode under the influence of an electric current, is known as the isoelectric point or isoelectric pH. Hence, the isoelectric point is a pH value at which the net charges on molecules are zero (Fig. 11.20).

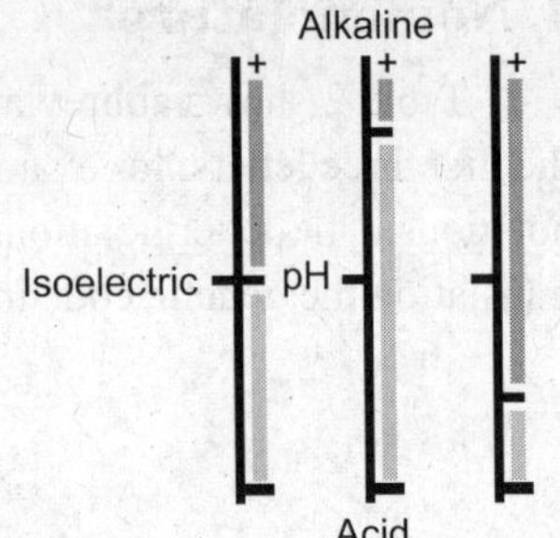

Fig. 11.20. Isoelectric point.

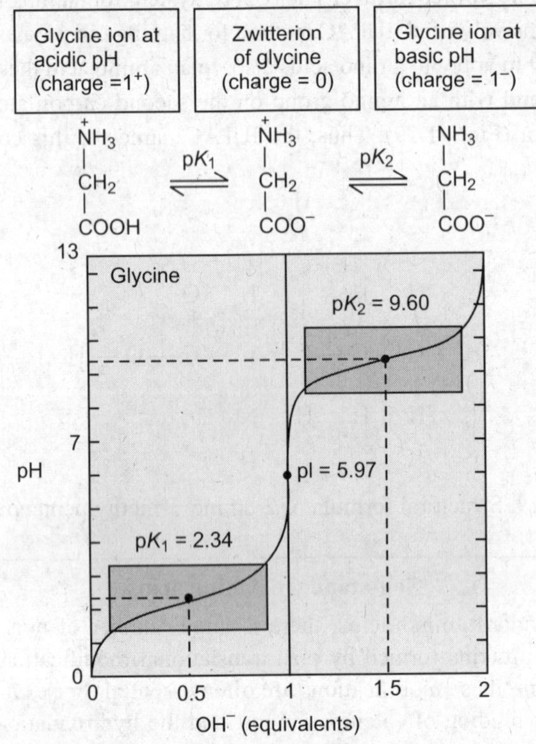

Fig. 11.21. Titration of glycine with NaOH solution. The pK_1 [½ cations *i.e.* $NH_3^+(CH_2)COOH$, and ½ zwitterions *i.e.* $NH_3^+(CH_2)CHCOO^-$]; the pl (all zwitterions) = 5.97; the pK_2 (½ zwitterions, ½ anions) = 9.60; the end of titration [all glycine molecules are in the anionic form: $NH_2(CH_2)CHOO^-$].

It has been discussed earlier that the amphoteric molecules called zwitterions contain both positive and negative charges depending on the functional groups present in the molecule. The net charge on the molecule is affected by pH of their surrounding environment and can become more positively or negatively charged due to the loss or gain of protons (H^+). The pI is the pH value at which the molecule carries no electrical charge or the negative and positive charges are equal.

Proteins are made up of both positively and negatively charged groups; but usually one predominates and they will have an overall negative or an overall positive charge. These charges can balance each other to some degree. At a particular pH, each protein will have an overall charge of zero. This is different for each protein, and is its isoelectric point. For example, during the titration of glycine [$NH_2(CH_2)COOH$] an isoelectric point is established with the pH changes. When titration is started (with NaOH solution as the titrant) from acid solution, relatively small pH changes are observed between pH 2 and 3 (acidic), and again between pH 9 and 10 (alkaline) (Fig. 11.21). These are caused by the buffering capacity of the carboxyl (-COOH) and amine ($-NH_2$) groups as weak electrolytes. The pH for ½ (half) equivalence corresponds to the pK of the acid function (a value related to the equilibrium constant), where one-half of the alanine molecules still carry only a positive charge ($-NH_3^+$), whereas the other half are also negatively charged ($-COO^-$). Hence glycine exists in the zwitterion form.

(*i*) Calculation of pI Values: When the pH is raised above the isoelectric point, the number of charged amino groups is reduced. Consequently the number of charged carboxyl and hydroxyl groups increases which causes it to behave as an acid protein. The number of charged amino groups increases with lowering the pH from the isoelectric point. This results in a decrease in the number of charged carboxyl and hydroxyl groups causing the protein to behave as a basic protein. The pI of an amino acid with only one amine and one carboxyl group can be calculated from the mean of the pKa's of this molecule.

$$\text{pI} = \frac{pK_1 + pK_2}{2}$$

Using this formula, $\text{pI}_{glycine}$ can be calculated by putting the value of pK_1 and pK_2 as given in Fig. 11.21.

$$\text{pI}_{glycine} = \tfrac{1}{2}(2.34 + 9.60) = 5.97$$

The same formula is used for amino acids consisting of more than two ionizable groups, for example lysine. But the two pK values used are those of the two groups that lose and gain a charge from the neutral form of the amino acid. Lysine has a single carboxylic pK_1 and two amine pK_2 values (one of which is on the R-group). Therefore, fully protonated lysine has a 2^+ net charge. For getting a neutral charge, the lysine must be deprotonated twice, and the R-group and pK_1 values of amine should be used. If the amino acid has more than two pKa values like lysine, average of two relevant pK_a values are taken to calculate the pI. Suppose for lysine $pK_1 = 2.18$, $pK_2 = 8.95$ and $pK_R = 10.53$, then:

$$\text{pI} = \tfrac{1}{2}(pK_2 + pK_R)$$
$$= \tfrac{1}{2}(8.95 + 10.53) = 9.74$$

(pK_R is not always higher than pK_2)

COO^-
$H_3\overset{+}{N}-C-H$
CH_2
CH_2
CH_2
CH_2
$^+NH_3$

Lysine

Moreover, the buffer used for gel is applied to determine the pH of an electrophoretic gel. If the pH of the buffer is above the pI of the protein being run, the protein will migrate to the positive pole (negative charged protein migrates to a positive pole). But if the pH of the buffer is below the pI of the protein being run, the protein will migrate to the negative pole of the gel (positive charge is attracted to the negative pole). Besides, the protein will not migrate if it is run with a buffer pH equal to the pI. Table 11.3 shows pKa value of some common amino acids.

Table 11.3. pKa value of some common amino acids.

Amino acids	*a-COOH*	*a-NH_3*	*RH or RH^+*
Ala	2.34	9.69	
Arg	2.17	9.04	12.48
Asn	2.02	8.80	
Asp	2.09	9.82	3.86*
Cys	1.71	10.78	8.33*
Gln	2.17	9.13	

Glu	2.19	9.67	4.25*
Gly	2.34	9.60	
His	1.82	9.17	6.0*
Ile	2.36	9.68	
Leu	2.36	9.68	
Lys	2.18	8.96	10.55
Met	2.28	9.2`	
Phe	1.83	9.13	
Pro	1.99	10.6	
Ser	2.21	9.15	
Thr	2.63	10.43	
Trp	2.38	9.39	
Tyr	2.20	9.11	10.07
Val	2.32	9.62	

* Ionization of R group takes place before the ionization of α-NH_3^+.

(*a*) Reaction with Ninhydrin: Except proline, all amino acids can be distinguished by ninhydrin test. Ninhydrin (triketohydrindene) is a powerful oxidising agent. It causes oxidative dehydration of α-amino acid producing carbon dioxide, ammonia and aldehyde, and reduced form of ninhydrin (that is hydrindantin). The hydrindantin again reacts with ammonia and a mole of ninhydrin to produce a blue or purple coloured Ruhemann's complex (diketohydrin) (Fig. 11.22), which absorbs light at a wavelength of 570 nm. Thus, ninhydrin involves reaction with both carboxylic and amino groups of amino acids. Since proline lacks α-amino group, it gives yellow colour with ninhydrin test. Similarly asparagine having a free amide group reacts to give a brown coloured product.

Besides, the proteins which catalyse the rate of biochemical reactions are called enzymes or biocatalysts. In addition to amino acids, vitamins and co-enzymes are also the building blocks of enzymes and take part in various biological reactions.

Ninhydrin + α-amino acid → Hydrindantin + R-CHO + NH_3 + CO_2
Aldehyde Ammonia Carbon dioxide

Hydrindantin + NH_3 → Ruhemann's purple + H_2O

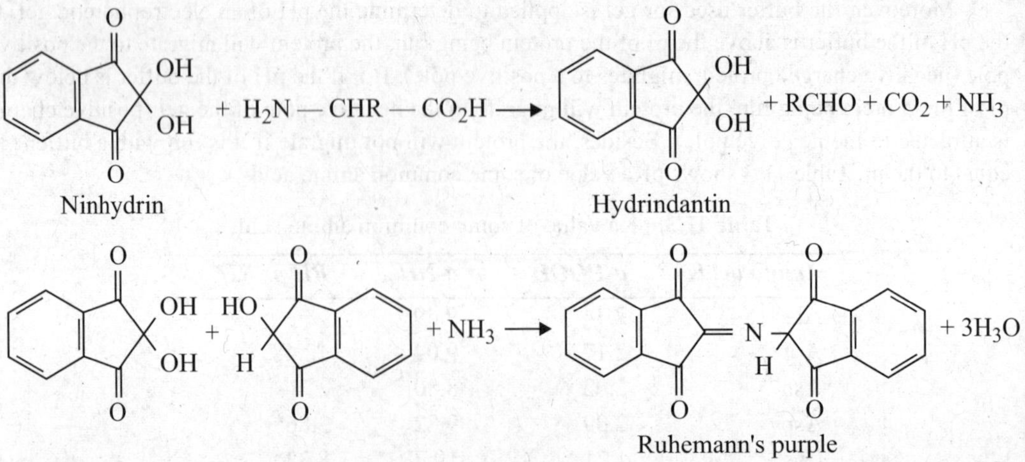

Ninhydrin + $H_2N—CHR—CO_2H$ ⟶ Hydrindantin + $RCHO + CO_2 + NH_3$

+ NH_3 ⟶ Ruhemann's purple + $3H_3O$

Fig. 11.22. Reaction of ninhydrin with an alpha amino acid and production of Ruhemann's purple.

C. CONFORMATION ANALYSIS OF PROTEINS

Proteins are the most abundant intracellular macromolecules. These constitute more than half of dry weight of most organisms. Some proteins provide structural elements of the body such as hairs, collagens, enzymes and oxygen carriers. In the 18th century, Antoine Fourcroy and co-workers recognised the proteins as a distinct class of biological molecules, based on its ability to coagulate or flocculate under heat or acid treatments, for example blood serum albumin, fibrin, albumin from egg whites, and wheat gluten. Gerhardus J. Mulder, a Dutch chemist carried out elemental analysis of common proteins, and found that most of the proteins had the same empirical formula, $C_{400}H_{620}N_{100}O_{120}P_1S_1$. He came to the erroneous conclusion that they might be composed of a single type of (very large) molecule. In 1838, Jöns Jakob Berzelius proposed the term 'protein' to describe these molecules. Protein (Greek, *proteios*) means 'primary', or 'standing in front'. Mulder identified the products of protein degradation such as the amino acid leucine for which he found a molecular weight of 131 Da which was almost correct. The early protein biochemists found it very difficult to purify in large quantities. Therefore, early studies were made to purify large quantities of proteins, such as protein from blood, egg white, and digestive and metabolic enzymes obtained from slaughterhouses, etc. In the 1950s, 1 kg of pure bovine pancreatic ribonuclease A was purified by the Armour Hot Dog Co., which was made available freely to scientists.

Linus Pauling successfully predicted the secondary structures of regular protein on the basis of hydrogen bonding; however, this idea was first given by William Astbury in 1933. The work done by Walter Kauzmann on denaturation contributed an understanding of protein folding and structure mediated by hydrophobic interactions. In 1949 Fred Sanger correctly determined the amino acid sequence of insulin. He demonstrated that proteins consisted of linear polymers of amino acids rather than branched chains, colloids or cyclols. In 1960s the first atomic resolution structures of proteins were solved by X-ray crystallography and by NMR in the 1980s. Up to 2009, the Protein Data Bank (PDB) has over 55,000 atomic-resolution structures of proteins. Recently, the two methods being approached for atomic resolution: cryo-electron microscopy of large macromolecular assemblies, and prediction of computational protein structure of small protein domains.

1. Special Features of Proteins

Some proteins possess the following important features :

- Proteins may be monomeric (made up of single polypeptide chain e.g. myoglobin found in muscles), or multimeric (made up of more than one polypeptide chains e.g. haemoglobin which consists of five polypeptide chains such as 2a and 2β chains, and some other have α, β, γ, δ and ε).
- The metaloproteins contain metal ions (e.g. Zn, Cu, Ag, Hg). For example, enzymes carbonic carbonic anhydrase has Zn^{++}.
- There are some heteroproteins, which are linked with a non-protein portion called prosthetic group which may be either a metal (haem in haemoglobin) or a compound (such as vitamin B-complex called co-enzyme).
- The chromoproteins are linked with coloured pigment, for example myoglobin, cytochromes, flavoproteins, catalase, etc.
- The glycoproteins contain carbohydrates as prosthatic group; examples are serum globulins and serum albumins.
- Phosphoproteins are linked with phosphoric acid, for example casein.
- The nucleoproteins contain nucleic acids and proteins (such as protamines and histone proteins).

2. Primary Structure of Proteins

In 1951, Linderstrøm-Lang first coined the term 'primary structure, which refers to the linear sequence of structural units of proteins, *i.e.* amino acids. The primary structure of a protein starts from the amino-terminal (N) end to the carboxyl-terminal (C). In 1902, two scientists in the same conference at Karlsbad, made a proposal for linear chains of α-amino acids in proteins. In the morning Franz Hofmeister gave the proposal on the basis of his observations of the biuret reaction in proteins, followed by Emil Fischer a few hours later.

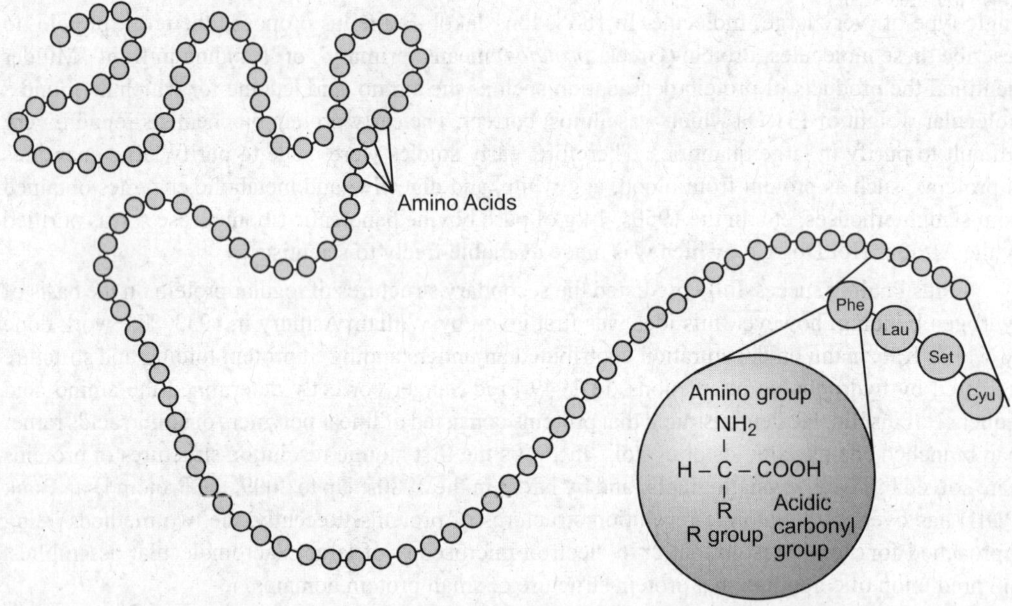

Fig. 11.23. Primary structure of insulin showing a chain of amino acids.

The 'colloidal protein hypothesis' holds that proteins were colloidal assemblies of smaller molecules. In the 1920s Theodor Svedberg disproved this hypothesis based on ultracentrifugation measurements which showed that proteins had a well-defined, reproducible molecular weight and by electrophoretic measurements. Dorothy Wrinch proposed a second hypothesis called the 'cyclol hypothesis' advancing that the linear polypeptide underwent a chemical cyclol rearrangement (C=O + HN→C(OH)-N) that cross linked its backbone amide groups, forming a two-dimensional fabric. Frederick Sanger successfully sequenced insulin and presented the primary structure of protein, and the crystallographic determination of myoglobin and haemoglobin was given by Max Perutz and John Kendrew, respectively. Fig. 11.23 shows the primary structure of insulin consisting of a chain of amino acids.

In general, proteins contain > 40 residues each are linked by peptide bond to results in primary structure of proteins, as described in previous section. A minimum of 100-1000 amino acid residues are needed to fold into tertiary structure, the percent of each amino acid varies. To determine the primary structure, the protein should be separated based on differences in size and composition. Thereafter, proteins are analyzed to determine their structure and function. Fig. 11.24 shows primary, secondary, tertiary and quaternary structure of protein.

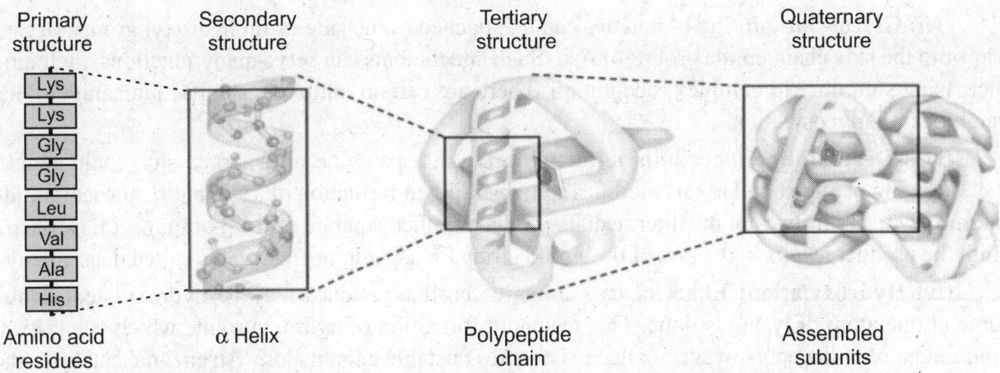

Fig. 11.24. Primary, secondary, tertiary and quaternary structures of protein.

***(a)* Post-translational Modifications:** The N-terminal amino group of a polypeptide can be modified covalently. Modifications in most of the polypeptide occur after synthesis of the protein on the ribosome that occurs in the endoplasmic reticulum. There are many posttranslational modifications that occur in proteins as described below:

***(i)* Acetylation** [$-C(=O)-CH_3$]**:** The positive charge on the N-terminal amino group may be eliminated by changing it to an acetyl group (N-terminal blocking). N-terminal acetylation is shown in Fig. 11.25 A. Acetylation of the lysine amino groups is chemically analogous to the acetylation of the N-terminus. Acetylation of lysine residues is used to regulate the binding of proteins to nucleic acids. The cancellation of the positive charge on the lysine weakens the electrostatic attraction for the (negatively charged) nucleic acids.

***(ii)* Amidation:** Blocking of the C-terminus carboxylate group of a polypeptide takes place that modify it through amidation via neutralising the negative charges (Fig. 11.25 B).

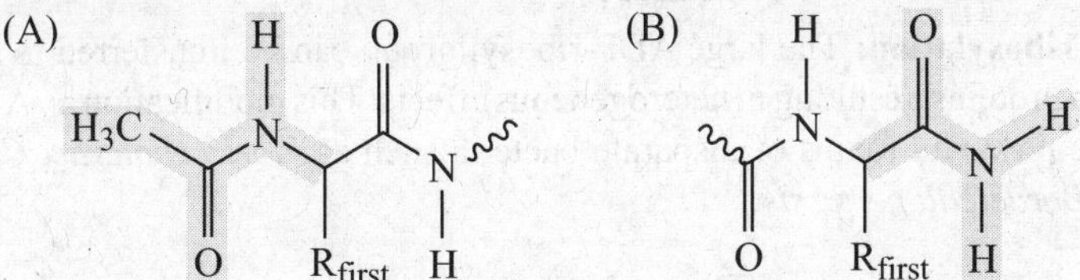

Fig. 11.25. Acetylation (A), and amidation (B) of polypeptide.

***(iii)* Formylation** [$-C(=O)H$]: The N-terminal methionine is usually present after translation. It has an N-terminus blocked with a formyl group (and sometimes the methionine residue, if followed by Gly or Ser), which is removed by the enzyme deformylase.

***(iv)* Pyroglutamate:** An N-terminal glutamine can attack itself forming a cyclic pyroglutamate group resulting in the formation of pyroglutamate from an N-terminal glutamine.

***(v)* Myristoylation [$-C(=O)-(CH_3)_{13}CH_3$]:** It is similar to acetylation. The myristoyl group has a tail of 14 hydrophobic carbons, instead of a simple methyl group which make it ideal for anchoring proteins to cellular membranes.

***(vi)* Phosphorylation:** Phosphorylation is the most important chemical modification of proteins. Phosphorylation refers to attachment of a phosphate group to the side chain hydroxyl group of serine, threonine and tyrosine residues, resulting in addition of a negative charge at that site and production of an unnatural amino acid. The enzyme kinases catalyse such reactions and phosphatases catalyse the reverse reaction. Often the phosphorylated tyrosines are used as 'handles' through which proteins can bind to one another. Conformational changes are induced by phosphorylation of Ser/Thr due to negative charge. Sometimes the effects of phosphorylating Ser/Thr can be simulated by mutating the Ser/Thr residue to glutamate.

***(vii)* Glycosylation:** Sugar moieties can be attached to the side chain hydroxyl groups of Ser/Thr or to the side chain amide groups of Asn. Such attachments can serve many functions, including increasing solubility to complex recognition. There are certain inhibitors such as tunicamycin that can block all glycosylation.

***(viii)* Deamidation (succinimide formation)**: An asparagine or aspartate side chain attacks the following peptide bond in this modification, resulting in formation of a symmetrical succinimide intermediate. Hydrolysis of the intermediate produces either asparate or the β-amino acid, iso(Asp). Both the product results in the loss of the amide group for asparagine, hence it is called deamidation.

***(ix)* Hydroxylation:** Either of two atoms of proline residues may be hydroxylated, as the same at one atom of lysine is done. This results in formation of hydroxyproline which is a critical component of collagen. However, collagen becomes unstable upon its loss. An enzyme catalyses the hydroxylation reaction that requires ascorbic acid (vitamin C). Therefore, deficiencies in ascorbic acid lead to many diseases associated with connective tissue such as scurvy.

***(x)* Methylation**: There are several proteins of which the amino acid residues can be methylated, for example the positive groups of lysine and arginine. At these sites methylation is used to regulate the binding of proteins to nucleic acids. Lysine residues can be methylated once, twice and even thrice; but the positive charge on the side chain cannot be changed by methylation.

***(xi)* Carboxylation:** Carboxylation is used to strengthen the binding to hard metal ions such as calcium. It is a rare modification that adds an extra carboxylate group to a glutamate side chain resulting in production of a Gla residue.

***(xii)* Sulfation:** Sulfation refers to addition of sulphate group on amino acid residues. Sulfation adds a negative charge to a previously neutral site like phosphorylation. Tyrosines may become sulfated on their O^{η} atom. This modification occurs in the Golgi apparatus but not in the endoplasmic reticulum. Like phosphorylated tyrosines, the sulfated tyrosines are used for specific recognition, e.g. in chemokine receptors on the cell surface.

***(xiii)* ADP-ribosylation:** The large ADP-ribosyl group can be transferred to several types of side chains within proteins, resulting in heterogeneous effects. This modification by ADP-ribosylation is a target for the powerful toxins of disparate bacteria such as *Vibrio cholerae*, *Corynebacterium diphtheriae* and *Bordetella pertussis*.

3. Secondary Structure of Proteins

The secondary structure is the general three-dimensional form of local segments of biopolymers such as proteins and nucleic acids. However, it does not describe the specific atomic positions in three-dimensional space, which are considered to be tertiary structure. Hence, the secondary structure of proteins refers to second level of organisation through folding or coiling between neighboring amino acids. In proteins, the secondary structure is defined by the patterns of hydrogen bonds between backbone amide and carboxyl groups. The hydrogen bonding patterns may be significantly distorted, which makes difficult to an automatic determination of secondary structure.

In 1951, L. Pauling and R.B Corey described two types of secondary structures of proteins (α-helix and β-plates) on the basis of nature of hydrogen bonding, *i.e.* intermolecular and intramolecular. The secondary structure may be also governed on the basis of regular pattern of backbone dihedral angles in a particular region of the Ramachandran plot. A segment of residues with such dihedral angles is called a helix, irrespective of the correct hydrogen bonding. Crystallographers also provide the secondary structure in the corresponding Protein Data bank (PDB) file.

***(a)* The alpha Helix (α-helix):** The most common secondary structures are α-helices and β-sheets. The other helices such as the 3_{10} helix and π helix are calculated to have energetically favourable hydrogen-bonding patterns but are rarely if observed in natural proteins except at the ends of α-helix due to unfavourable backbone packing in the center of the helix. The α-helix is a common motif in the secondary structure of proteins. It is a right-handed coiled conformation in which every backbone N-H group donates a hydrogen bond to the backbone C=O group of the amino acid four residues earlier (*i* + 4→*i* hydrogen bonding). This secondary structure is also a classic 'Pauling-Corey-Branson alpha helix'. The α-helix is the most regular, prevalent and predictable. In the 1930s, William Astbury showed drastic changes in the X-ray fiber diffraction of moist wool or hair fibers upon significant stretching. But the unstretched fibers had a coiled molecular structure with a characteristic repeat of ~5.1 Å. Astbury proposed a kinked-chain structure for the fibers that: (*i*) the unstretched protein molecules formed a helix (which he called the α-form), and (*ii*) the stretching caused the helix to uncoil, forming an extended state (which he called the β-form).

Fig. 11.26 shows side view of a α-helix of alanine residues in atomic detail. Two hydrogen bonds to the same peptide group are highlighted in magenta. The protein chain runs upwards, *i.e.* its N-terminus is at the bottom and its C-terminus at the top. The side chains (grey stubs) angle slightly downward, toward the N-terminus, while the peptide oxygen (red) point up and the peptide NHs point down.

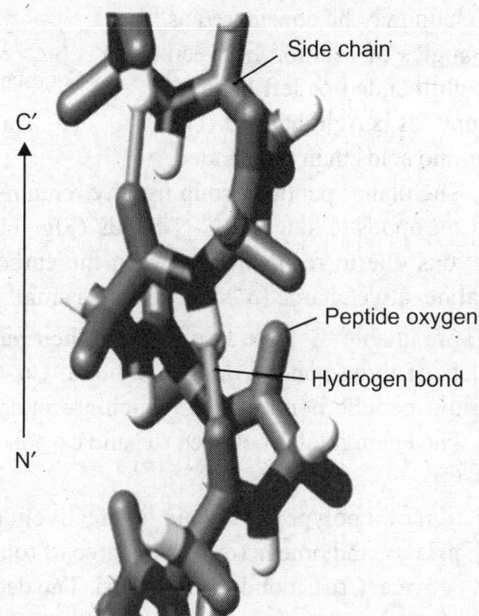

Fig. 11.26. Side view of an α-helix of alanine residues in atomic detail.

Astbury's models of these forms were correct in essence and correspond to modern elements of secondary structure, the α-helix and the β-strand. In 1951 Linus Pauling, Robert Corey and Herman Branson developed the structure in detail showing both right- and left-handed helices. In 1940, Hans Neurath first showed that Astbury's models could not be correct in detail, because they involved clashes of atoms. On the basis of findings of Neurath's paper and Astbury's data, H. S. Taylor, Maurice Huggins and W.L. Bragg proposed the models of keratin that resembles with modern α-helix to some extent.

In an α helix the amino acids are arranged in a right-handed helical structure where each amino acid residue corresponds to a 100° turn in the helix (*i.e.* the helix has 3.6 residues per turn), and a translation of 1.5 Å (0.15 nm) along the helical axis. The pitch of the alpha-helix (the vertical distance between two consecutive turns of the helix) is 5.4 Å (0.54 nm) which is the product of 1.5 and 3.6. The N-H group of an amino acid forms a hydrogen bond with the C=O group of the amino acid, four residues earlier. The most prominent characteristic of an α-helix is this repeated *i* + 4→*i* hydrogen bonding.

The ability of amino acids varies to form the secondary structure. Proline and glycine are called the 'helix breakers' because they disrupt the regularity of the conformation of α-helical backbone. However, both have unusual conformational abilities and are commonly found in turns. Amino acids that prefer to adopt helical conformations in proteins are: *methionine, alanine, leucine, glutamate* and *lysine* (in one letter code of these five amino acids, MALEK). But the large aromatic residues (tryptophan, tyrosine and phenylalanine) and C^{β}-branched amino acids (isoleucine, valine, and threonine) adopt β-strand conformations. On the basis of these preferences the secondary structure can be predicted.

Planarity of the Peptide Bond: The peptide bond is slightly shorter than a standard single bond, reflecting the partial delocalization of pi electrons from the carbonyl group into orbitals shared with the lone pair electrons of the amide nitrogen. This partial double-bond character inhibits rotation around the peptide bond; thus, the four atoms bound to the carbonyl carbon and amide nitrogen form a plane. A polypeptide chain may be considered as a series of planes with two angles of rotation between each plane. A helix can be right handed or left handed. The α-helix of known polypeptides is right handed. The α-helix is most stable for L-amino acids than left handed. It is very common in nature. The planar peptide group usually remains in *trans* configuration, *i.e.* oxygen and hydrogen atoms on opposite sides of C-N bonds (Fig. 11.27). This shows a peptide sharing of two pairs of electrons due to resonance between the carboxyl oxygen and the amide nitrogen. Oxygen has a partial negative charge (δ^+) that sets up a small electric dipole (A).

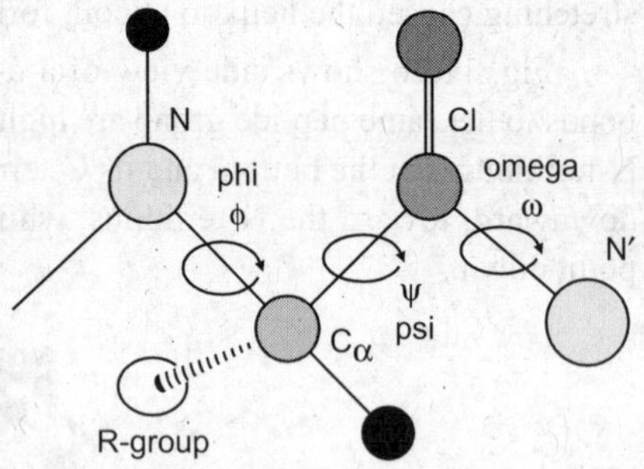

Three dihedral angles

The amide bonds (C-N) are unable to rotate freely due to their partial double bond character. The polypeptide chain consists of three bonds viz., N-Cα bond, Cα-C bond and C-N bond (B). The chain consists of semi-rigid peptide bond planes, which are interconnected to each other by α-carbon of each amino acid. The chain rotates between αC and carbon atoms (N-Cα), and between αC and carbonyl carbon (Cα-C).

The backbone conformation of a polypeptide chain is effectively determined by the values of three dihedral angles: phi (ϕ), psi (ψ), and omega (ω). The degree of rotation at N-αC bond is called as phi, and degree of rotation between Cα-C bonds is called psi. The degree of rotation at C-N bond is called as omega. The ϕ and ψ can have any value between -180° and +180°. Moreover, limitation arises in rotation due to R-group attached to Cα, which may come close to peptide plane (due to steric interference) at ϕ and ψ angles.

(A) $\mathrm{O{=}C{-}N} \longleftrightarrow \mathrm{{}^{-}O{-}C{=}\overset{+}{N}}$ (B) N ------- Cα ------- C ------- N -------
ϕ   ψ   ω   ϕ

Fig. 11.27. Diagrammatic representation of planar peptide bond. A- planar peptide bond in *trans* configuration; B- three bonds in a polypeptide chain, *i.e.* N—αC bond and αC—C bond with y, and C—N bond with ω.

The resonance structures lead to a planar arrangement of the peptide carbonyl C and amide N and the two other atoms connected to each. The two planes can twist around the alpha carbon (αC) (Fig. 11.28). The rotation angles for the two planes (ϕ and ψ) are analogous to the torsion angles in the acyl chains of fatty acid. They can vary from -180 to +180°. The R group substituent attached to the αC can also rotate around the αC and the beta C (βC) of the side chain. This angle is defined as chi. Other rotations also occur within the side chain.

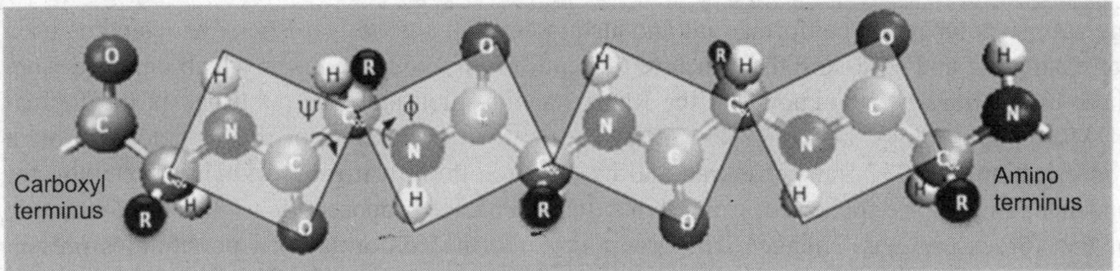

Fig. 11.28. Planes and phi/psi angles of an extended polypeptide chain showing limited rotation around two (N¾Cα and αC¾C) bonds.

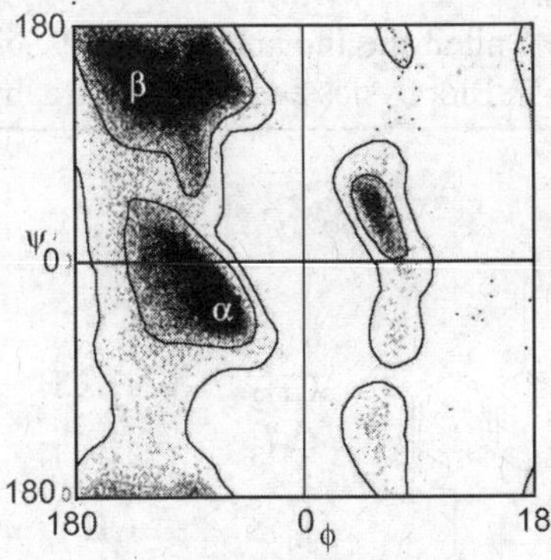

Fig. 11.29. The Ramachandran plot. φ-ψ plot of the torsion angles that specify protein backbone conformation, with the α-helix and β-strand regions labeled.

If the values of φ and ψ for each amino acid are known, the conformation of main polypeptide chain can be determined. In 1963, G.N. Ramachandran and his coworkers collected φ and ψ angles for different α-carbon amino acids of a polypeptide chain. Then he plotted the ψ values against φ which is called 'Ramachandran Plot' (Fig. 11.29). Following this plot he prepared various possible secondary structure of the protein. The left-handed configuration is rarely found in 'collagen', which is a protein of connective tissue. It is constituted by unusual amino acids.

The α-helix is a rod like structure. The α-configuration of planar peptide bonds depends on the intramolecular hydrogen bonding between -NII and -CO groups of peptide bonds. The α-helix constitutes 3.6 amino acids residues per complete turn (360°). It has a pitch of 4.5Å, *i.e.* it rises along central axis with 4.5/3.6=1.5 Å per residue (Fig. 11.30).

Prof. G.N. Ramachandran (October 1922 - April 2001)

Prof. Gopalasamudram Narayana Iyer Ramachandran was born in 1922 in the town of Ernakulam, Kerala, India. But the native place of his forefathers is a village in the old Tirunelveli District of Madras Presidency. In 1942 he joined the Indian Institute of Science, Bangalore in the Electrical Engineering Department. He had interest in physics; hence he switched to the Department of Physics to complete his master's and doctoral thesis under the supervision of Nobel Laureate Sir C. V. Raman. In 1942, he received a master's degree in physics from Madras University. He received his D.Sc. degree in 1947. During his studies he created an X-ray focusing mirror for the X-ray microscope.

IIe was appointed as the Head of the Department of Crystallography and Biophysics, University of Madras, Chennai. In 1954 his interest shifted to the structure of biological macromolecules. Using X-ray diffraction. Ramachandran along with Gopinath Kartha proposed and published the triple helical structure of collagen. Ramachandran decided to use this information to examine the various polypeptide conformations and also to develop a good 'yardstick' that could be used for examining and assessing the structure of peptides. In 1962, the result which emerged from these calculations now is known as the Ramachandran plot, and was published in the Journal of Molecular Biology in 1963. Now it has become an essential tool in the field of protein conformation. In 1957 Ramachandran and his research group earned worldwide reputation for understanding the steriochemistry of polypeptide chains. He proposed the triple helical structure of the fibrous proteins, collagen. His group explored the conformational possibilities present in a polypeptide chain giving the models: (*i*) the polypeptide bond is rigid and will not allow rotation (*ii*) the molecules may be rotated about the other (single) bonds in the backbones, *i.e.* the N-Cα bond and Cα-C bond, (*iii*) all atoms are represented as hard spheres with specific radii, and (*iv*) there are no other attractive of repulsive forces.

In 1970, he founded the Molecular Biophysics Unit at the Indian Institute of Science which was later known as the 'Centre of Advanced Study in Biophysics and Crystallography'.

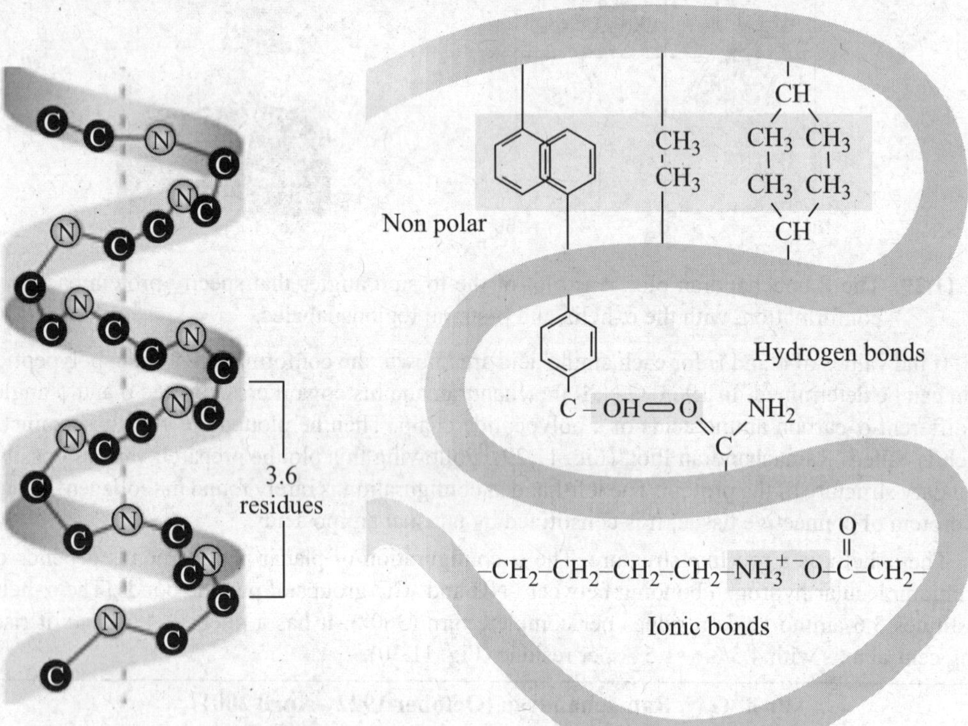

Fig. 11.30. An α-helical configuration of a polypeptide chain.

Many fibrous proteins such as α-keratin have only helical structure. The α-keratin is found in skin, hairs, wool, feathers, and horns. The α-keratin consists of three right-handed coils that are stabilized through cross-linking disulfide bonds (Fig. 11.31). The α-helix is stabilized by hydrogen bonds between carboxy oxygen of one amino acid residue (n, >C=O) and hydrogen atom of amino nitrogen (-NH) of forth amino acid residue (*i.e.* n+4).

HN ---- (1) -------- (2) -------- (3) -------- (4) -------- (5) -------- (6) ---------COO–

O .... H

Fig. 11.31. The 3α-helical coils in keratin.

Do you know ?

Moist hairs are stretched due to reduction in number of intra-strand hydrogen bonds of amino acids; because water itself forms hydrogen bonds with amino groups of hair keratin. Therefore, helix becomes more flexible as compared to dry hairs. This is why barber makes the beard moist before shaving.

***(b)* The β-pleated Sheet:** In the 1930s, the first β-sheet structure was proposed by William Astbury, giving the idea of hydrogen bonding between the peptide bonds of parallel or antiparallel extended β-strands. In 1953, L. Pauling and R.B Corey identified a second type of repetitive structure which they called β-pleated sheet (also the β-sheet) because it was the second structure after the first one, *i.e.* α-helix. The β-sheet is the second form of regular secondary structure in proteins, somewhat less common than α-helix. Beta sheets consist of beta strands connected laterally by at least two or three backbone hydrogen bonds, forming a generally twisted pleated sheet. A beta strand is a stretch of polypeptide chain typically 3 to 10 amino acids long with backbone in an almost fully extended conformation (Fig. 11.32). The higher-level association of β sheets has been implicated in formation of the protein aggregates and fibrils observed in many human diseases such as Alzheimer's disease.

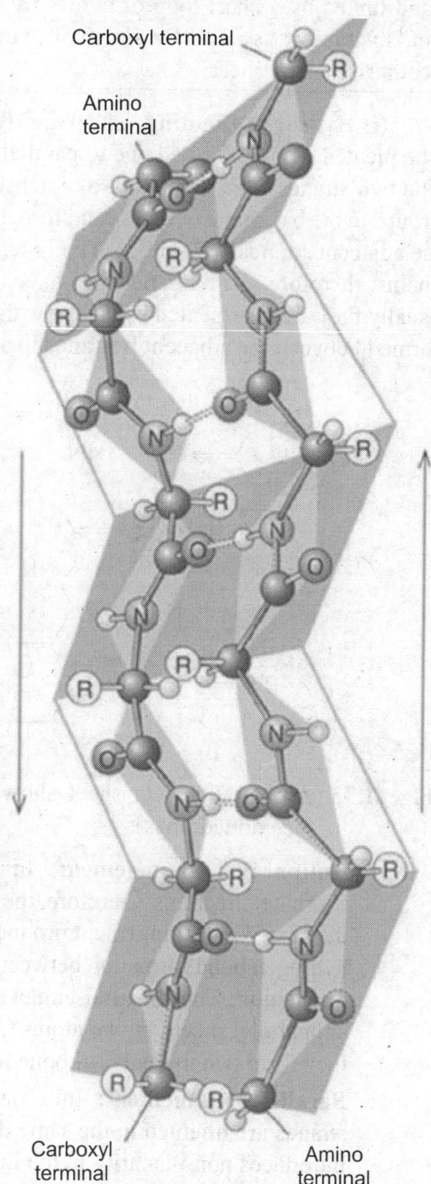

Fig. 11.32. Antiparallel β-pleated sheet.

If you fully extend the β strand, you will find that the successive side chains point straight up, then straight down, then straight up, etc. In a β sheet adjacent β strands are aligned so that their Cα atoms are adjacent and their side chains point in the same direction. The pleated appearance of β strands arises due to tetrahedral chemical bonding at the Cα atom. If a side chain points straight up, then the bond to the Cα must point slightly downwards, because its bond angle is about 109.5°. The pleating causes the distance between C^{α}_{i} and C^{α}_{i+2} to about 6 Å, rather than 7.6 Å (2 × 3.8 Å) expected from two fully extended *trans* peptide virtual bonds. The sideways distance between adjacent Cα atoms is about 5 Å in the hydrogen-bonded β strands. The β strands exhibit a twist due to the chirality of their amino acids. Near (φ, ψ) = (–135°, 135°) (the upper left region of the Ramachandran plot) the dihedral angles diverge from the fully extended conformation (φ, ψ) = (–180°, 180°) (Voet and Voet, 2004).

Often the twist is associated with alternating fluctuations in the dihedral angles to prevent the individual β strands in a larger sheet from splaying apart. The side chains point outwards from the folds of the pleats, more or less perpendicularly to the plane of the sheet; successive residues point outwards on alternating faces of the sheet.

Beta Hairpin.

The simplest protein motif involving two beta strands that look like a hairpin is called beta hairpin (or beta-beta unit) structural motif. The motif consists of two strands adjacent in primary structure oriented in an antiparallel arrangement (where the N-terminus of one sheet is adjacent to the C-terminus of the next) and linked by a short loop of two to five amino acids. Beta hairpins can occur in isolation or as part of a series of hydrogen bonded strands that collectively comprise a beta sheet.

Beta hairpin

(*i*) **Hydrogen Bonding:** The majority of β strands are arranged adjacent to the other strands. The pleated sheet is formed due to parallel or antiparallel running of polypeptide chains in a plane. The two strands form a network of extensive hydrogen bonds with their neighbors where the N-H groups in the backbone of one strand form hydrogen bonds with the C = O groups in the backbone of the adjacent strands(Fig. 11.33). The N-terminus and C-terminus provides directionality to peptide chains; therefore, the β strands also have direction in the same way. In protein topology diagrams usually they are represented by an arrow that points toward the C-terminus. Hydrogen bonds can be formed between the adjacent β strands in antiparallel, parallel or mixed arrangements.

Fig. 11.33. Antiparallel beta sheet showing hydrogen bonding patterns in two different views (*see* dotted lines).

- **Antiparallel arrangement:** In an antiparallel arrangement the successive β strands alternate directions; therefore, the N-terminus of one strand is adjacent to the C-terminus of the next. This arrangement produces the strongest inter-strand stability through inter-strand hydrogen bond formation between carbonyls and amines to be planar. This is their preferred orientation. The dihedral angles (φ, ψ) of peptide backbone are about (–140°, 135°) in antiparallel sheets. If two atoms C_i^{α} and C_j^{α} are adjacent in two hydrogen-bonded β strands, they form two mutual backbone hydrogen bonds to each other's peptide groups.
- **Parallel arrangement:** In a parallel arrangement, all of the N-termini of successive strands are oriented in the same direction. This orientation is slightly less stable because it introduces non-planarity in the inter-strand hydrogen bonding pattern. The dihedral angles (φ, ψ) are about (–120°, 115°) in parallel sheets. In a motif, there are rarely less than five interacting parallel strands which suggest that a smaller number of strands may be unstable. Here if two atoms (C_i^{α} and C_j^{α}) are adjacent in two hydrogen-bonded β strands, they do not form hydrogen bond to each other. Instead, one residue forms hydrogen bonds to the other residues. For example, residue *i* may form hydrogen bonds to residues *j* – 1 and *j* + 1; this is known as a 'wide pair' of hydrogen bonds. By contrast, residue *j* may hydrogen-bond to different residues altogether or to none at all.

- **Mixed bonding arrangements:** An individual strand may exhibit a mixed bonding pattern with a parallel strand on one side and an antiparallel strand on the other. Such arrangements are less common than a random distribution of orientations which indicates that this pattern is less stable than the antiparallel arrangement. The hydrogen bonding of β strands can exhibit localised disruptions known as beta bulges. The hydrogen bonds lay roughly in the plane of the sheet with the peptide carbonyl groups that point in alternating directions with successive residues.

This results in formation of strong and inflexible protein fibres. It is quite common in nature, for example silk and certain synthetic fibres, e.g. nylon and orlon. A portion of two chains of silk fibroin runs in opposite direction. The silk fibre has a repeating sequence of six residues (Gly-Ser-Gly-Ala-Gly-Ala)n.

4. Tertiary Structure of Proteins

The tertiary structure of a protein or any other macromolecule is its 'three-dimensional structure' as defined by the atomic coordinates in biochemistry and molecular biology. The tertiary structure is considered to be largely determined by primary structure of protein or amino acid sequence.

Tertiary Structure of Proteins

The tertiary structure is produced due to coiling of polypeptide chain. Folding of the active amino acids scattered along the chain are brought closer from a distinct cavity or cleft. The structure is bound in three clefts. The tertiary (*i.e.* 3D) conformation determines the shape of such proteins which consists of only one polypeptide chain (monomeric), for example myoglobin (Fig. 11.34). There are a number of bonds which maintain the tertiary structure such as hydrogen bonds, disulfide bonds, ionic bonds and hydrophobic bonds.

The sequestration of hydrophobic amino acid residues in the protein core frequently stabilises the tertiary interactions in globular proteins, from which water is excluded. Stabilisation also takes place by the consequent enrichment of charged or hydrophilic residues present on the water-exposed surface of the proteins. Disulfide bonds between cysteine residues help to maintain the tertiary structure of the proteins that do not spend time in the cytoplasm. In a large number of proteins several common and stable tertiary structures appear that are unrelated in both function and evolution. Another common structure is a highly stable dimeric coiled coil which is composed of 2-7 α-helices.

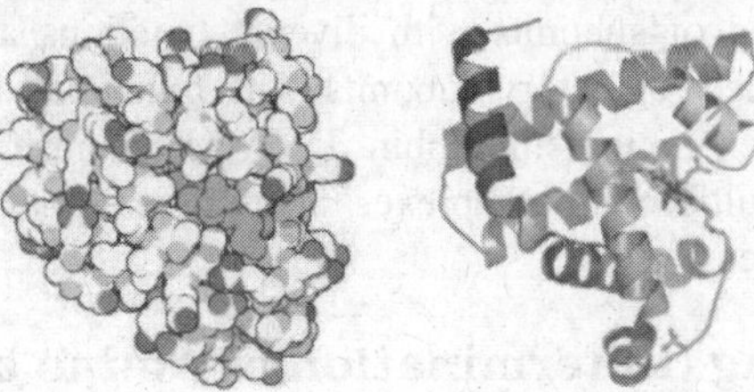

Fig. 11.34. The 3D structure of myoglobin showing coloured alpha helices. Structure of this protein could be solved by X-ray crystallography.

Ribonuclease A (RNase, a globular protein) is an enzyme which is secreted into small intestine by pancreas. RNAse hydrolyses certain bonds of ribonucleic acid present in ingested food. It attains a reniform 3D structure. All the proteins are grouped into two: (*i*) the globular or capsular proteins (these are soluble proteins and their 3D structure appear spherical like globe), and (*ii*) the fibrous or fibrillin proteins (these are insoluble proteins and their 3D structure looks like a long thin fibre).

Christin Anbinsen found that the primary structure (*i.e.* amino acid residues) of the protein governs the folding and 3D structure. Globular proteins are more compact in conformation than the fibrous ones. Anbinsen dissolved RNase into strong solution of chemicals like urea. Consequently

3D structure was destroyed which failed to display hydrolytic activity. However, enzymatic activity for hydrolysis was regained after removal of denaturant chemical that destroyed 3D structure.

X-ray crystallography has solved the majority of protein structures known to date by providing data of high resolution. But it does not provide time-dependent information on the conformational flexibility of proteins. Nuclear Magnetic Resonance (NMR) commonly solves structures of relatively small proteins by providing data of lower resolution; but it provides time-dependent information about the motion of a protein in solution. Dual polarisation interferometry is a time resolved analytical method for determining the overall conformation and conformational changes in surface captured proteins that provides complementary information to these high resolution methods. More is known about the tertiary structural features of soluble globular proteins than about membrane proteins.

5. Quaternary Structure of Proteins

The multimeric proteins (*i.e.* proteins having more than one polypeptide chain) may associate to form quaternary structure. Thus, the quaternary structure is the arrangement of multiple folded proteins or coiling protein molecules in a multi-subunit complex. There are many proteins that are actually assemblies of more than one polypeptide chain; these proteins are called 'multimeric proteins' (*i.e.* proteins having more than one polypeptide chain) or 'multiple-subunit proteins'. Besides the tertiary structure of the subunits, multimeric proteins possess a quaternary structure also, that is the arrangement of the subunits. Conformational changes within individual subunits or reorientation of the subunits relative to each other bring about changes in quaternary structure of proteins.

Fig. 11.35. Quaternary structure of haemoglobin showing α-helix and β-pleated sheets.

The combination of secondary structure elements is called domains or moieties. Subunits of the two polypeptide interact and form covalent (disulfide bond, S–S), and non-covalent bonds, for example ionic and hydrophobic bonds. If the protein has identical subunits, it is called homogeneous quaternary structure (e.g. lactic dehydrogenase). The proteins having dissimilar subunits are called heterogeneous quaternary structure (such as haemoglobin) (Fig. 11.35).

The enzymes composed of subunits with diverse functions are also called holoenzymes, some parts of which are called regulatory subunits. The functional core of the protein is known as the catalytic subunit, for example hemoglobin, DNA polymerase, and ion channels. The other assemblies referred to as 'multiprotein complexes' also possess quaternary structure, for example nucleosomes and microtubules.

6. Protein Sequencing (Determination of Amino acid Sequence)

Molecular weight, amino acid composition and purity of proteins must be known before start of sequencing. The polypeptide chains of multimeric proteins should be separated and molecular weight of each chain should be measured. Molecular weight gives information about the number of amino acids present in polypeptide chain. Generally molecular weight of each amino acid is about 110 Da. If molecular weight of a protein is 55,000 Da, it is supposed that protein is constituted by about 50 amino acids. The following methods are used in protein sequencing.

Do you know ?

The hormone insulin was the first protein to be sequenced by Sanger. More than one gram of insulin purified from large number of pancreatic gland was used. Insulin deficiency in human causes diabetes.

(*a*) **Sanger's Method**: It was widely assumed that proteins were somewhat amorphous. In 1951, Fred Sanger first succeeded in determining the complete amino acid sequence of the two polypeptide chains of bovine insulin. In 1958, Fred Sanger was awarded a Nobel Prize in chemistry for his work on the structure of proteins, especially that of insulin. He showed for the first time that amino acids are covalently linked together by α-amino and α-carboxyl groups.

Polypeptide

NO_2 FDNB NO_2 F

2, 4-Dinitro-phenyl derivative

NO_2 NH $R^1 - CH$ C = O HN $R^2 - CH$ C = O

6M HCl

NO_2 NO_2 NH $R^1 - CH$ COO^-

2, 4-Dinitrophenyl derivative of amino-terminal residue

+ Free amino acids

Fig. 11.36. Sanger's method of protein sequencing.

He suggested stepwise release and identification of amino acid starting from N-terminal. He used a reagent 'fluoro-dinitrobenzene' (FDNB), which is commonly called Sanger's reagent. The FDNB was used to react with the exposed amino groups in the protein and in particular with the N-terminal amino group at one end of the polypeptide chain.

The FDNB reacts with free-NH_2 group of N-terminus. Upon hydrolysis a yellow colored dinitriphenol (DNP)-derivative of N-terminal amino acids is produced (Fig. 11.36). The DNP-amino acid is identified comparing it with a known standard, DNP-amino acid by using gel chromatography. This step is repeated. After each step polypeptide chain is shortened by one amino acid. At present Sanger's method has historical importance.

(*a*) **Edman's Degradation Method:** Pehr Victor Edman, a Swedish biochemist recognised that proteins are distinct entities with a defined molecular mass, electric charge and structure. This inspired Edman to develop a method to determine the sequence of amino acids in the protein. In 1947 he was awarded a travel stipend to go to Rockefeller Institute of Medical Research at Princeton University. In 1950 he published his first paper using the method later known as Edman degradation method to determine the sequence of a protein. To his death he continued to work to improve the method to be able to determine longer stretches with smaller amounts of sample.

Pehr Victor Edman (1916 -1977)

In this method, the amino-terminal residue is labelled and cleaved from the peptide without disrupting the peptide bonds between other amino acid residues. It involves sequential identification of amino acids from N to C termini. Phenyl-isothiocynate (PTC, also called Edman's regent) is used to react with -NH_2 group of first amino acid of polypeptide chain (Fig. 11.37).

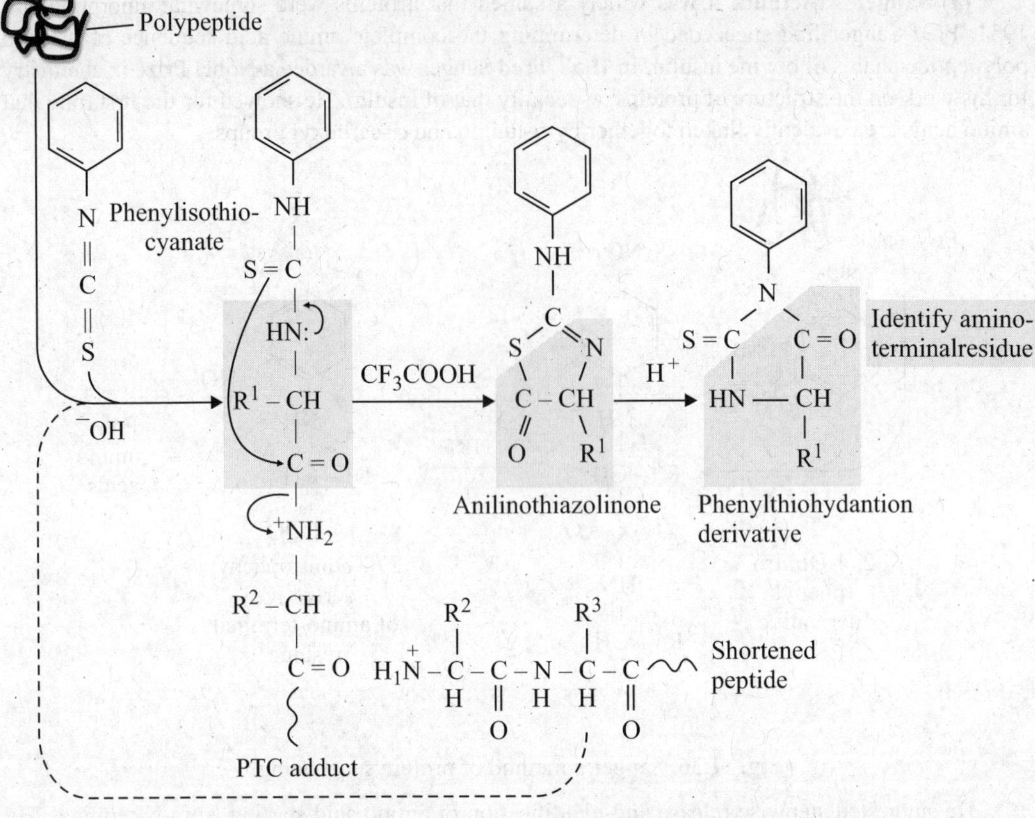

Fig. 11.37. Edmonn's method of protein sequencing.

Under mildly alkaline conditions phenylisothiocyanate reacts with an uncharged terminal amino group to form a cyclical phenylthiocarbamoyl derivative. This derivative of the terminal amino acid is cleaved as a thiazolinone derivative under acidic conditions. Then thiazolinone amino acid is selectively extracted into an organic solvent and treated with acid to form the more stable phenylthiohydantoin (PTH)- amino acid derivative that can be identified by using chromatography or electrophoresis. This procedure can then be repeated again to identify the next amino acid. Phenylthiocynate derivative is released which is then analysed by HPLC and identified using standard. After each step the new amino terminal residue is identified.

In 1973, H.D Niall built up an automated instrument called protein 'sequenator' that was capable of identifying only about 50 amino acid residues by using the same procedure. An advantage of the Edman degradation is that it only uses 10 - 100 picomoles of peptide for the sequencing process. Therefore, large sized proteins (150-15,000 amino acids) must be fragmented before sequencing in the instrument. Using Edman's method proteins up to microgram quantities can be sequenced. Moreover, Edman degradation reaction is automated to speed up the process.

***(b)* Enzymatic Method for Fragmentation of Polypeptide Chains:** The enzymes, trypsin, chymotripsin and pepsin hydrolyse the protein. They are very selective in hydrolysing peptide bonds. For example, trypsin is an endopeptidase and splits peptide bonds in which the carboxylic group is present on lysine or arginine only. Similarly, chymotrypsin breaks peptide bonds in which carbon groups form aromatic residues (e.g. phenol, alanine, tyrosine in tryptophan). Take an example of a dodecapeptide whose composition was found to be Ala_2, Arg, Glu, Gly, Leu, Lys_2, Phe, Try_2, Val. It was determined that N-terminal amino acid of dodecapeptide was valine and the C-terminal amino acid was leucine. Hydrolysis by trypsin resulted in four peptides whose sequence was as below (A to D).

Try - Glu – Lys	Phe - Gly – Arg	Val – Lys	Ala-Try-Ala - Leu	...(1)
A	B	C	D	

Based on C and N termini fragments must represent N-terminal end and D- fragment the C- terminal end of dodecapeptide. To find out the correct order, dodecapeptide was hydrolysed by chymotrypsin. The four peptides obtained were sequenced as given below:

Ala – Leu	Gly – Lys – Ala – Try	Val – Lys – PHe	Gly – Arg – Tyr	...(2)
E	F	G	H	

Comparing sequences of peptides H with E it appears that peptide B must precede peptide A. Therefore, the structure of dodecapeptide determined would be as below:

Val – Lys – Phe- Gly – Arg – Tyr - Glu – Lys – Ala – Try - Ala – Leu ...(3)

***(c)* Chemical Method:** In this method cyanogen bromide (CNBr) is used that breaks selectively the peptide bond following a residue of methionine.

D. X-RAY CRYSTALLOGRAPHY

The method for determining the arrangement of atoms within a crystal by passing the X-rays is called X-ray crystallography. In this method a beam of X-rays strikes a crystal and diffracts into many specific directions. A crystallographer can produce a 3D picture of the electron density within the crystal from the angles and intensities of these diffracted beams. The mean positions of the atoms, their chemical bonds, and their disorder in the crystal can be determined from this electron density.

In 1895, Wilhelm Conrad Röntgen (a Nobel Laureate) discovered X-rays to study the symmetry of crystals. Initially, the nature of X-rays was unknown but soon it was suspected that they were waves of electromagnetic radiation. Maxwell theory of electromagnetic radiation as the wave model of light was well accepted at that time among scientists. It could be shown that X-rays exhibited phenomena of electromagnetic waves, such as transverse polarisation and spectral lines similar to that observed in the visible wavelengths. X-rays are composed of photons; they are not only waves of electromagnetic radiation but also exhibit particle-like properties. In 1905, Albert Einstein introduced the photon concept which was not broadly accepted until 1922 before its confirmation of scattering of X-rays from electrons by Arthur Compton. In 1912, the observation on X-ray diffraction confirmed that X-rays were a form of electromagnetic radiation.

Wilhelm Conrad Röntgen

The salts, metals, minerals, semiconductors, and various inorganic, organic and biological molecules can form crystals. Therefore, X-ray crystallography has developed many scientific fields. X-ray crystallography has helped to determine in the first decade, the atomic size, lengths and types of chemical bonds, and differences among various materials at atomic-scale, such as minerals and alloys.

The structure and function of many biological molecules for example vitamins, drugs, proteins and nucleic acids (e.g. DNA) have been revealed by this method. Still X-ray crystallography is the main method to characterise the atomic structure of new materials and to discern materials that appear similar by other experiments. Unusual electronic or elastic properties of a material can also be accounted by X-ray crystal structures, or serving as the basis for designing pharmaceutical chemicals.

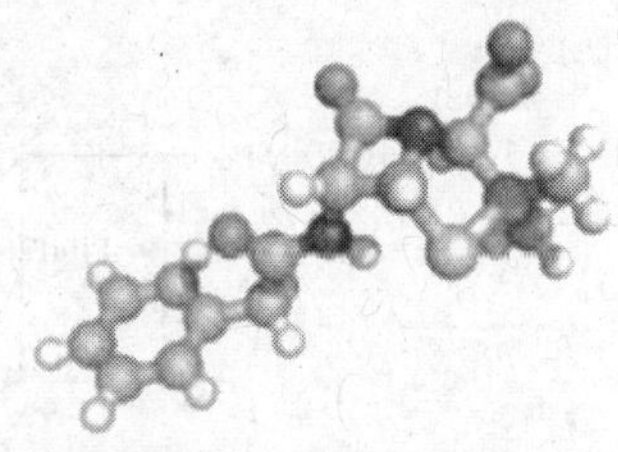

The 3D structure of penicillin

Since the 1920s, X-ray diffraction has been used to determine the arrangement of atoms in minerals and metals, such as silicates, alloy, etc. X-ray crystallography of several biological molecules was carried out by Dorothy Crowfoot Hodgkin, who solved the structures of cholesterol, vitamin B_{12} and penicillin for which she was awarded the Nobel Prize in Chemistry in 1964. In 1969, she solved the structure of insulin and worked for over thirty years.

The other forms of elastic X-ray scattering include powder diffraction, SAXS and several types of X-ray fibre diffraction which was used by Rosalind Franklin in determining the double-helix structure of DNA. The ingle-crystal X-ray diffraction offers more structural information than the other techniques; but it requires a sufficiently large and regular crystal.

1. X-ray Diffraction Analysis of Crystals

In 1895, a German physicist Wilhelm Röntgen discovered the X-rays; therefore, X-ray radiation is also called Röntgen radiation after the name of the discoverer. The traditional unit of X-rays is rad (equal to 10 millijoules of energy deposited per kilogram; 100 rad = 1.00 gray). X-radiation is a form of electromagnetic radiation from outer space (sun, black holes, and supernovae) which is the main source of X-rays. X-rays have a wavelength ranging from 0.01 to 10 nanometers and energies from 120 eV to 120 keV. The wavelengths of X-rays are shorter than that of UV rays and longer than gamma rays. On the basis of their penetrating abilities X-rays from about 10 to 0.10 nm (0.12 to 12 keV) are classified as 'soft X-rays', and from about 0.10 to 0.01 nm (12 to 120 keV) as the 'hard X-rays'. In nature, X-rays are artificially generated by an X-ray tube (vacuum tube) using a high voltage to accelerate the electrons released by a hot cathode to a high velocity. The high velocity electrons collide with the anode and create the X-rays. Electromagnetic spectrum of different rays is shown in Fig. 11.38.

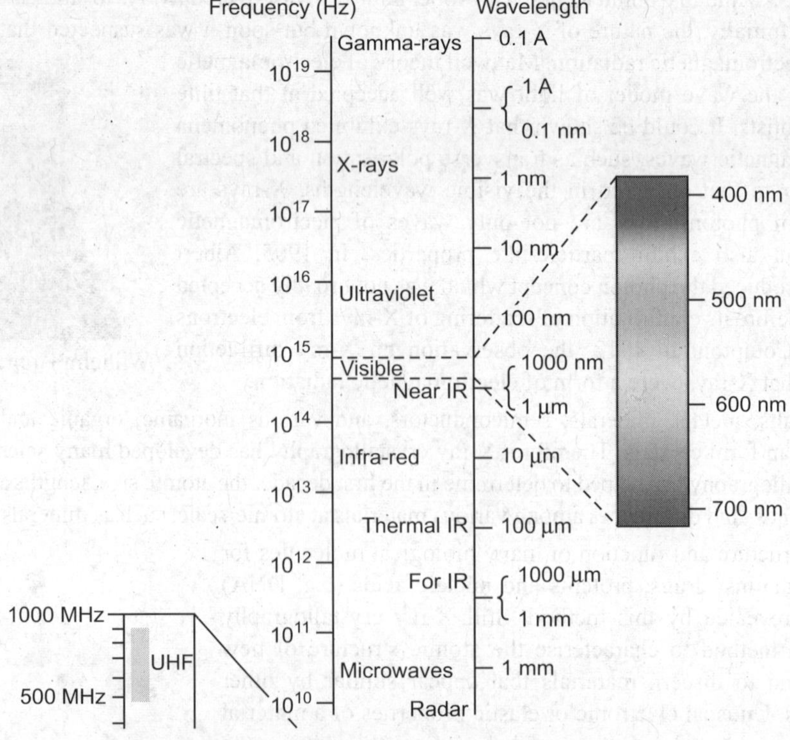

Fig. 11.38. Electromagnetic spectrum.

Besides its potential application in medical science, the X-rays are also used in crystallography in which the pattern produced by the diffraction of X-rays through the closely spaced lattice of atoms in a crystal is recorded and then analysed to reveal the nature of lattice. A related technique, fibre diffraction, was used by Rosalind Franklin to discover the double helical structure of DNA.

The diffraction pattern of crystals is produced by X-rays because their wavelength (λ) is the same order of magnitude (1-100 Å) as the spacing (*d*) between planes in the crystal. In 1665, Francesco Maria Grimaldi predicted first that any wave impinging on a regular array of scatterers produces diffraction. The spacing between the scatterers and the wavelength of the impinging wave should be similar in size for the production of a significant diffraction. In the late 17th century, James Gregory first reported the diffraction of sunlight through a bird's feather. The spacing between lattice planes in a crystal was unknown before the first X-ray diffraction experiments. However, the idea that crystals could be used as a diffraction grating for X-rays arose in 1912 in a conversation between Paul Peter Ewald and Max von Laue. Electromagnetic radiation of a shorter wavelength is required to observe such small spacing. Von Laue suggested that X-rays might have a wavelength comparable to the unit-cell spacing in crystals. He shined an X-ray beam through a copper sulfate crystal and recorded its diffraction on a photographic plate. The plate showed a large number of well-defined spots arranged in a pattern of intersecting circles around the spot produced by the central beam (Fig. 11.39). Von Laue developed a law that connects the scattering angles and the size and orientation of the unit-cell spacing in the crystal, for which he was awarded the Nobel Prize in Physics in 1914.

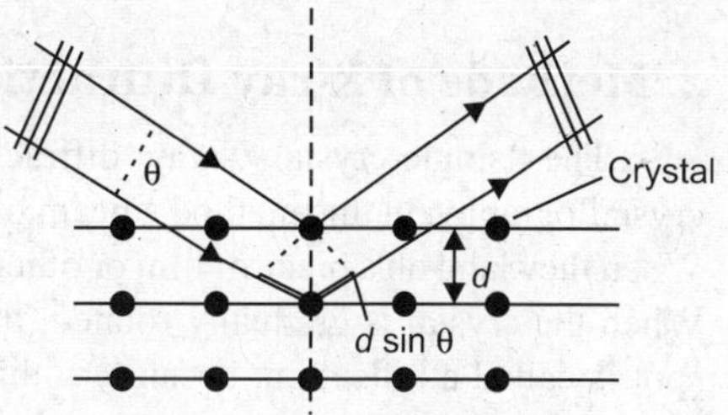

Fig. 11.39. The X-ray diffraction pattern.

For measurement by an X-ray diffraction, a crystal is mounted on a goniometer, bombarded with X-rays and gradually rotated while bombardment is going on (Fig. 11.40). A diffraction pattern of regularly spaced spots is produced which is called 'reflections'. The two-dimensional images are taken at different rotations. Then these images are converted into a 3D model of the density of electrons within the crystal by using the mathematical method of 'Fourier transforms', combined with chemical data of the sample. If the crystals are too small or not uniform in their internal makeup, poor resolution or errors may be obtained.

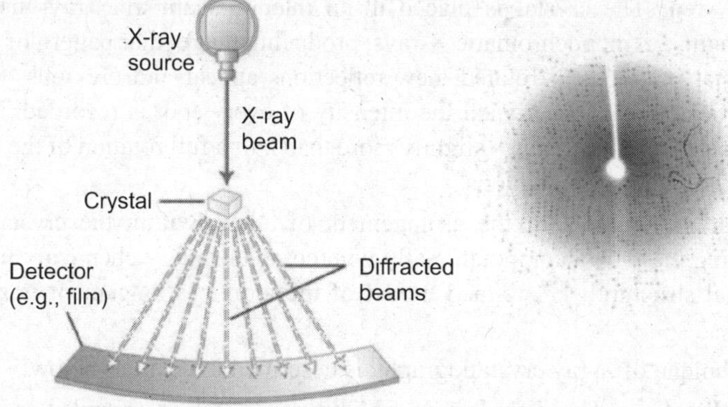

Fig. 11.40. X-ray crystallography of a crystal.

X-rays can be considered waves as the of electromagnetic radiation, and crystals are regular arrays of atoms. Atoms scatter X-ray waves through their electrons, as an ocean wave striking a lighthouse produces secondary circular waves. Similarly an X-ray striking an electron produces secondary spherical waves emanating from the electron. This phenomenon is known as elastic scattering, and the electron (or lighthouse) is known as the 'scatterer' (the repeating arrangement of atoms within the crystal). The X-ray beam coming from upper left causes each scatterer to re-radiate a small portion of its intensity as a spherical wave. A regular array of scatterers produces a regular array of spherical waves. Although these waves cancel one another out in most directions through destructive interference, they add in a few specific directions, which are determined by Bragg's law:

$$2d \sin \theta = n\lambda$$

Where d is the spacing between diffracting planes, θ is the incident scattering angle, n is any integer, and λ is the wavelength of the X-ray beam.

These specific directions appear as spots on the diffraction pattern called *i.e.* reflections. Thus, X-ray diffraction results from an electromagnetic wave (the X-ray) impinging on a regular array of scatterers. If scatterers are arranged symmetrically with a separation d, these spherical waves will add constructively only in directions where their path-length difference $2d \sin \theta$ equals an integer multiple of the wavelength λ. In that case, part of the incoming beam is deflected by an angle 2θ, producing a reflection spot in the diffraction pattern.

2. Methods of X-ray Diffraction

The 'single-crystal X-ray diffraction' is the oldest and most precise method of X-ray crystallography. In this method a beam of X-rays strikes a single crystal, producing scattered beams. When they land on a piece of film or other detector, these beams create a 'diffraction pattern' of spots. When the crystal is gradually rotated, the strengths and angles of these beams are recorded. Each spot is called a reflection, because within the crystal it corresponds to the reflection of the X-rays from one set of evenly spaced planes. X-ray diffraction data can determine the mean chemical bond lengths (of a single sufficient pure and regular crystal) within a few thousandths of an Ångström, and angles within a few tenths of a degree. In a crystal the atoms are not static, but oscillate from their mean positions, by less than a few tenths of an Ångström. X-ray crystallography measures the size of these oscillations.

The technique of single crystal X-ray crystallography has three basic steps.

- **First step:** It is the most difficult step which involves getting an adequate crystal of the material to study. The crystal should be pure, regular in structure, and large-sized (>0.1 mm in all dimensions) with no significant internal defects like twinning or cracks.
- **Second step:** The crystal is placed in an intense beam of X-rays usually of a single wavelength, *i.e.* monochromatic X-rays, producing the regular pattern of reflections. When the crystal is gradually rotated, new reflections appear and previous ones disappear. At every orientation of the crystal the intensity of every spot is recorded. Multiple data sets are collected; each set covers slightly more than half a full rotation of the crystal containing tens of thousands of reflections.
- **Third step:** A model of the arrangement of atoms within the crystal is produced by combining these data computationally with complementary chemical information. Finally a 'crystal structure', *i.e.* refined model of the atomic arrangement is stored in a public database.

The technique of X-ray crystallography is described in detail as below:

***(a)* Crystallisation:** The disorders in an impure crystal or irregular are characterised by crystallography. But a pure crystal of high regularity is required in this method so that the structure of complicated arrangement of atoms could be solved. Sometimes, natural or synthetic materials such as samples of metals, minerals or other macroscopic materials are used to get the pure and regular crystals. However, the regularity of such crystals can be improved by crystal annealing of macromolecules. In many cases, the chief barrier for solving the atomic-resolution structure is to get a diffraction-quality crystal. Generally small molecules have few degrees of conformational freedom; hence they are crystallised by several methods. But there are many degrees of freedom in macromolecules; therefore, their crystallisation can be done for maintaining a stable structure. For example, if the tertiary structure of proteins and larger RNA molecules are unfolded, they cannot be crystallised. Therefore, the conditions of solution in which such molecules remain folded restrict the range of crystallisation conditions.

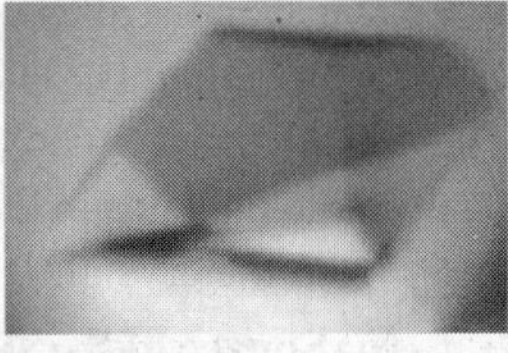
A protein crystal

Always the protein crystals are grown in solution so that the solubility of its component molecules should be lowered very gradually. If this is done very quickly, the molecules will precipitate from solution resulting in formation of a useless dust or amorphous gel on the bottom of the container. Crystal growth in solution is characterised by two steps: 'nucleation' of a microscopic crystallite (containing 100 molecules) followed by 'growth' of that crystallite to a diffraction-quality crystal. The conditions of solution that favour nucleation (first step) are not always the same that favour the subsequent growth (second step). Since larger crystals offer improved resolution of the molecule, solution conditions that favour the development of a single large crystal should be identified. If nucleation is favoured too much, a shower of small crystallites (but not large crystal) will form in the droplet.

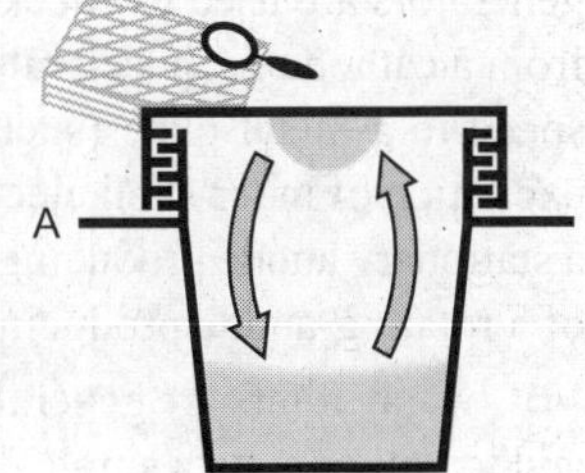

Hanging drop method for preparing crystals.

The good conditions for nucleation (or growth of well-ordered crystals) cannot be predicted. A very large batch of the molecules is prepared, and a wide variety of crystallisation solutions are tested. Before finding the successful results hundreds or thousands of solution conditions are tried such as: change of pH, chemicals lowering dielectric constant of the solution, and large polymers such as polyethylene glycol that affects various temperature ranges supersaturating the solution. It is difficult to get large quantities of crystallisation grade protein. Therefore, for accurately dispensing crystallisation trial drops (in the order of 100 nanoliters in volume) robots have been developed that use 10-fold less protein as compared to crystallisation trials setup by hands.

(*b*) Data Collection: There are different steps of data collection as described below:

(*i*) Mounting the crystal: For measurements the crystal is mounted so that it may be held in the X-ray beam and rotated. There are several methods of mounting the crystal. Earlier crystals were loaded into glass capillaries with the crystallisation solution. In modern approach the crystal is scooped up in a tiny loop made of nylon or plastic, and attached to a solid rod that is flash-frozen with liquid nitrogen. The untreated crystals are pre-soaked in a cryoprotectant solution before freezing to protect is from cracking by flash-freezing. But pre-soaking also results in cracking of crystals. Therefore, successful cryo-conditions must be identified.

The capillary is mounted on a goniometer which positions it accurately within the X-ray beam and rotated. Both the crystal and the beam are often very small, hence the crystal must be centered within the beam to within ~25 mm aided by a camera focused on the crystal. The most common type of goniometer is the kappa (κ) goniometer (Fig. 11.41) which offers three angles of rotation: (*i*) the ω angle which rotates about an axis perpendicular to the beam, (*ii*) the κ angle about an axis at ~50° to the ω axis, and (*iii*) the φ angle about the loop/capillary axis.

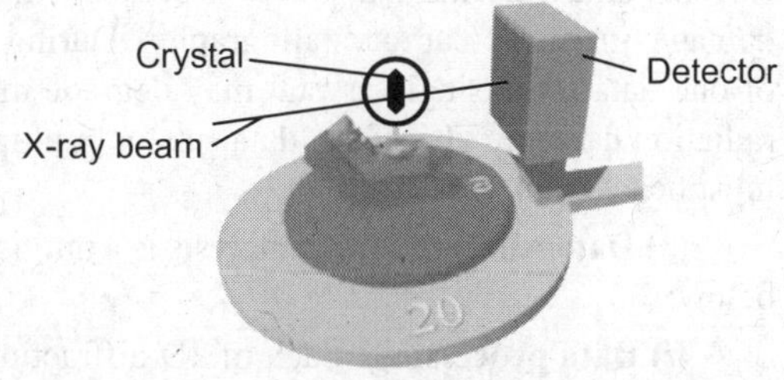

Fig. 11.41. A four-circle kappa goniometer.

The ω and φ axes are aligned when the κ angle is zero. The κ rotation allows mounting of the crystal conveniently. The oscillations carried out during data collection involve the ω axis only. The rotations about each of the four angles φ, κ, ω and 2θ leave the crystal within the X-ray beam resulting in changes in the orientation of crystals. The detector (red box) can be slid closer or further away from the crystal. If closer, it allows higher resolution data.

***(ii)* Source of X-rays:** The mounted crystal is irradiated with a beam of monochromatic X-rays. Synchrotrons are the brightest and most useful X-ray sources; however, their higher luminosity results in better resolution. They also tune the wavelength of the radiation which is useful for 'multi-wavelength anomalous dispersion' (MAD) phasing. Generally synchrotrons are national facilities where data is collected around the clock throughout the week. In laboratories often smaller X-ray generators are used to check the quality of crystals before putting them to a synchrotron. Electrons from a cathode get accelerated through a strong electric potential of ~50 kV, and collide with a high speed to a metal plate (such as copper) emitting some strong spectral lines corresponding to the excitation of inner-shell electrons of the metal. The cheapest variety of sealed X-ray tube consists of a stationary anode producing ~2 kW of X-ray radiation, whereas the more expensive variety consists of a rotating-anode producing ~14 kW of X-ray radiation.

X-ray filters are generally equipped to filtre the X-rays to a single wavelength (monochromatic) and to collimate to a single direction before striking the crystals. The filtering mechanism simplifies the data analysis, and removes radiation that degrades the crystal without giving useful information. Collimation is done either with a collimator (a long tube) or with an arrangement of curved mirrors (preferably for small crystals (0.3 mm).

***(iii)* Reflection recording:** When a crystal is mounted and exposed to a beam of X-rays, it scatters the X-rays into a pattern of spots or reflections (Fig. 11.42) which is observed on a screen behind the crystal. Shining a laser pointer at a compact disc also provides a similar pattern of spots. The information for the arrangement of molecules within the crystal is provided by the relative intensities of these spots in detail, which are recorded with photographic film, an area detector or a charge-coupled device (CCD) image sensor. The peaks at small angles represent to low-resolution data, while peaks at high angles represent high-resolution data. At this point some measures of diffraction quality can be determined such as the mosaicity of the crystal, pathologies of the crystal, etc.

Several images of spots are required for reconstruction of the whole crystal because one image shows only a small piece of information. Therefore, the crystal must be rotated step-by-step through 180° for collection of all the information. But a smaller angular range such as 90° or 45° may be used to record an image at every step for a crystal having higher symmetry. Multiple data sets may be necessary for certain phasing methods. In 2006, R.B. Ravelli and E.F.Garman have described the radiation damage in molecular crystallography. During collection of one data set, a single crystal may degrade much due to radiation damage. Therefore, data sets on multiple crystals must be taken in such cases.

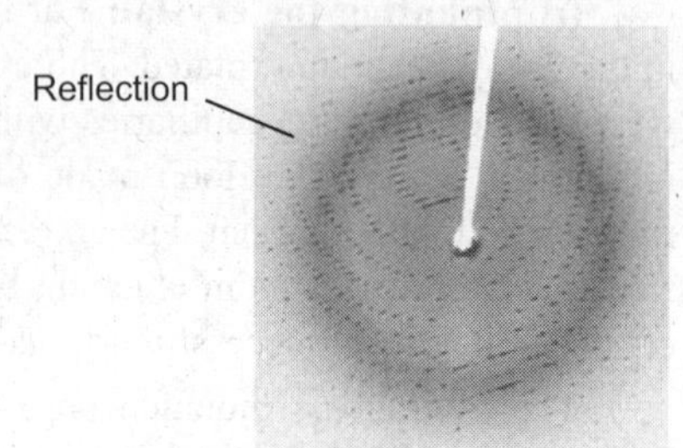

Fig. 11.42. X-ray diffraction pattern of a crystal.

***(c)* Data Analysis:** Data analysis is a multistep approach which involves several steps as given below:

***(i)* Data processing:** Each of 2D diffraction patterns corresponds to the different orientations of a crystal. It is converted into a 3D model of the electron density. Each spot corresponds to a different kind of variation in the electron density. The crystallographer determines *indexing* (the variation corresponding to the spot), *merging and scaling* (*i.e.* the relative strengths of the spots in different images), and *phasing* (*i.e.* yielding total electron density by combining the variations).

- **Indexing:** Indexing of reflections refers to identification of dimensions of the unit cell and image peak corresponding to position in reciprocal space. Some space groups (symmetry of the crystal) can be eliminated from the beginning. The chiral molecules do not show reflection symmetries; therefore, only 65 space groups of 243 possible are allowed for protein molecules which are almost chiral. Indexing is carried out by using an 'autoindexing'. Then data is integrated after assigning the symmetry. A full data set consists of hundreds of separate images taken at different orientations of the crystal. Various images are merged and scaled, to identify peaks appearing in two or more images, and to scale the relative images so that they have a consistent intensity scale. The diffractometer records several times many symmetry-equivalent reflections, calculating the symmetry-related R-factor which is a reliability index.
- **Phasing:** The data collected from a diffraction experiment is a reciprocal space representation of the crystal lattice. The size, shape and the inherent symmetry within the crystal of the unit cell governs the position of each diffraction 'spot'. The intensity of each diffraction 'spot' is recorded. The structure factor is a complex number that consists of information of both the amplitude and phase of a wave. Both amplitude and phase must be known for getting an interpretable electron density map (Fig. 11.43). On the basis of an electron density map, a starting model of the molecule can be built by a crystallographer.

***(ii)* Model building and phase refinement:** An initial model can be built after getting the initial phases, which can be used to refine the phases resulting in an improved model. Then a new model can be fit to the new electron density map and a further refinement can be carried out. Debye-Waller factors or B-factors accounts for the thermal motion of the atom which can be refined to fit the observed diffraction data. This results in a better set of phases. Until the correlation between the diffraction data and the model is maximized, this practice is continued. This agreement is measured by an *R*-factor which is given as below:

$$R = \frac{\Sigma_{\text{all reflections}}\,(F_o - F_c)}{\Sigma_{\text{all reflections}}\,(F_o)}$$

The *R* factor depends on the resolution of the data. The complementary measures of the model quality are the chemical bonding features such as stereochemistry, hydrogen bonding and distribution of bond lengths and angles. Phase bias is a serious problem in such iterative model building which can be checked by a method called 'Omit maps'.

The resulting electron density is an average of all the molecules within the crystal (Fig. 11.43); therefore, it is not possible to observe every atom of the crystallised molecule. Moreover, there is too much residual disorder in some atoms. The resulting electron density of atoms existing in many conformations is smeared to such a way that it is not detectable in the electron density map. The atoms being scattered weakly such as hydrogen are invisible. In an electron density map, a single atom does not appear several times.

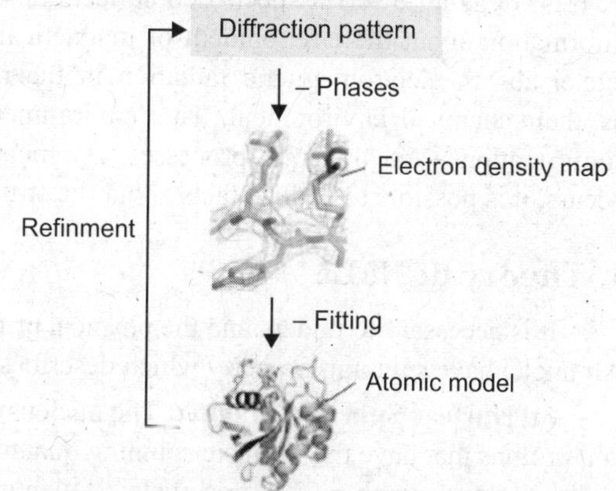

Fig. 11.43. X-ray crystallography of a molecule.

***(iii)* Deposition of the structure of molecules:** After finalising the model of structure of a molecule, it is deposited in a crystallographic database such as the Cambridge Structural Database (for small molecules), the Inorganic Crystal Structure Database (ICSD) (for inorganic compounds), or the Protein Data Bank (for protein structures).

E. NUCLEAR MAGNETIC RESONANCE (NMR)

In 1938, for the first time Isidor Rabi measured the nuclear magnetic resonance in molecular beams. In 1944, Rabi was awarded the Nobel Prize in physics for this work. In 1946, F. Bloch and E. M. Purcell expanded the technique for use on liquids and solids, for which they were awarded the Nobel Prize in physics in 1952. He worked on the production and detection of RF (radio frequency) energy, and absorption of such RF energy by matter.

The magnetic nuclei (such as ^{1}H and ^{31}P) absorb RF energy when placed in a magnetic field of a nucleus-specific strength. The nucleus shows 'resonance' after absorbing the RF energy. The nuclei of different atoms within a molecule resonate at different radio frequencies for the same strength of magnetic field. Therefore, a trained user can discover essential, chemical and structural information about the molecule at such magnetic resonance frequencies of the nuclei of a molecule.

First 1 GHz NMR spectrometer set up in France in 2009

The magnetic nuclei when put in a magnetic field and applied electromagnetic (EM) pulse show the resonance property, which is called 'nuclear magnetic resonance (NMR). Consequently the nuclei absorb energy from the EM pulse and radiate it back. The energy radiated depends on the strength of the magnetic field. Hence the magnetic properties of specific quantum mechanics of an atomic nucleus can be observed. The stable isotopes containing an odd number of protons and/or of neutrons have an intrinsic magnetic momentum and angular momentum, *i.e.* a nonzero spin, whereas all nuclides with even numbers of both have zero spin. The ^{1}H and ^{3}H are the most commonly studied nuclei, whereas the nuclei of ^{2}H, ^{10}B, ^{14}N, ^{15}N, ^{17}O, ^{35}Cl, etc. are also studied by high-field NMR spectroscopy. The NMR phenomenon is exploited through NMR spectroscopy by many scientific techniques to study the molecular physics, crystals and non-crystalline materials. Routinely the NMR is also used in medical imaging techniques, for example in magnetic resonance imaging (MRI).

NMR is a powerful spectroscopic technique used to study atoms and molecules. It reveals information about the environment of magnetically active nuclei. Under proper conditions, such nuclei absorb electromagnetic radiation in the radio-frequency region at frequencies governed by their chemical environment. This environment is influenced by chemical bonds, molecular conformations, and dynamic processes. By measuring the frequencies at which the absorption occurs, it is possible to deduce facts about the structure of the molecule being examined.

1. Theory of NMR

It is necessary to understand the physical principles on which the methods of NMR are based. All nuclei have spinning property which describes the nature of magnetic field of that nucleus.

***(a)* Nuclear Spin and Magnet:** The nucleus of an atom is composed of nucleons, *i.e.* neutrons and protons that have the intrinsic spinning quantum property (Fig. 11.44). Spin is characterised by a spin quantum number (*I*). Some nuclei of elemental isotopes have integral spins (for example I = 1, 2, 3 ....), some have fractional spins (e.g. $I = 1/2, 3/2, 5/2$), and a few have no spin, $I = 0$ (e.g. ^{12}C, ^{16}O, ^{32}S). Isotopes of particular interest and uses are ^{1}H, ^{13}C, ^{19}F and ^{31}P; all of them have $I = 1/2$.

Analysis of this spin state is straightforward. Nuclei whose spin number is equal to zero they have no magnetic field, while nuclei with non-zero spin numbers have magnetic fields. Higher values of spin number imply more complex magnetic fields.

The spin quantum number I determines the overall spin of the nucleus. The even number of both the protons and neutrons in a given nuclide shows the absence of overall spin *i.e.* $I = 0$. When the spins of a proton and neutron are parallel, they will have lower energy as this parallel spin alignment has to do with the quark structure of these two nucleons. Hence the ground state of spin for the deuteron (^{2}H isotope of hydrogen) having only a proton and a neutron corresponds to a spin value of 1, not zero. Besides exceptions, most of nuclides having both even numbers of protons and neutrons also have zero nuclear magnetic moments and zero magnetic dipole and quadrupole moments. Hence, such nuclides do not exhibit any NMR absorption spectra. For example, ^{18}O nuclide has no NMR absorption, whereas ^{13}C, ^{31}P, ^{35}Cl and ^{37}Cl nuclides do exhibit NMR absorption spectra.

The shapes of the magnetic fields become progressively more and more complex with the increase values of I. The shape of magnetic field of a nucleus is described by I. A nucleus with a large magnetogyric ratio has a stronger magnetic field than a nucleus with a small magnetogyric ratio. The strongest magnetic field is conveniently possessed by ^{1}H. This is why hydrogen NMR is the most popular followed by ^{13}C and ^{31}P NMR.

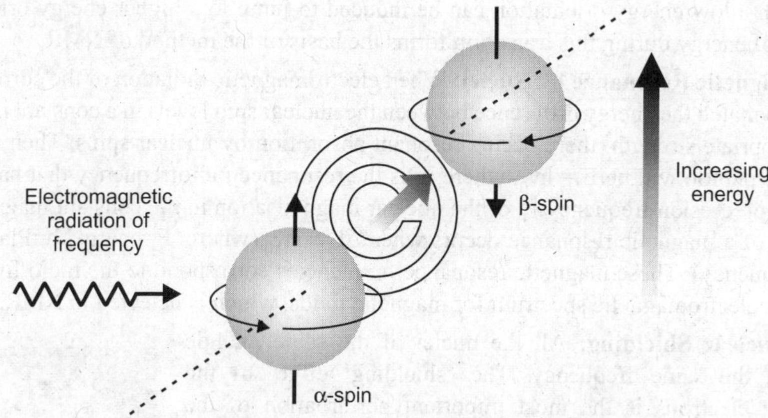

Fig. 11.44. The basis of NMR spectroscopy.

***(b)* Spinning in a Magnetic Field:** When a magnetic nucleus ($I > 0$) is placed between the poles of an external magnet, it is aligned to the externally applied strength of magnetic field (H_o or B_o) (Fig. 11.45). Two magnets can be aligned in an infinite number of orientations. A nucleus can take against an external magnetic field to only a finite number of alignments which depends on the shape of magnetic field of the nucleus. Therefore, it depends on the value of its spin number I. Each possible alignment is assigned a value called I_z which ranges from $-I$ to $+I$ in steps of 1. These orientations are referred to as spin states.

***(i)* The nucleus has two possible spin states:** $I_z = +\frac{1}{2}$ (or alpha, α) or $I_z = -\frac{1}{2}$ (or beta, β) which is also referred to as spin-up and spin-down, respectively (Fig. 11.45). The ^{1}H, ^{13}C or ^{19}F nuclei have a spin of one-half. These spin states are degenerate and, therefore, they have the same energy. Hence at thermal equilibrium the number of atoms in these two states will be more-or-less equal. Upon placing a nucleus in a magnetic field the interaction between the nuclear magnetic moment and the external magnetic field do not has the same energy. Spin states oriented more nearly parallel to the external field are lower in energy than in the absence of an external field. In contrast, spin states whose orientations oppose the external field are higher in energy than in the absence of an external field.

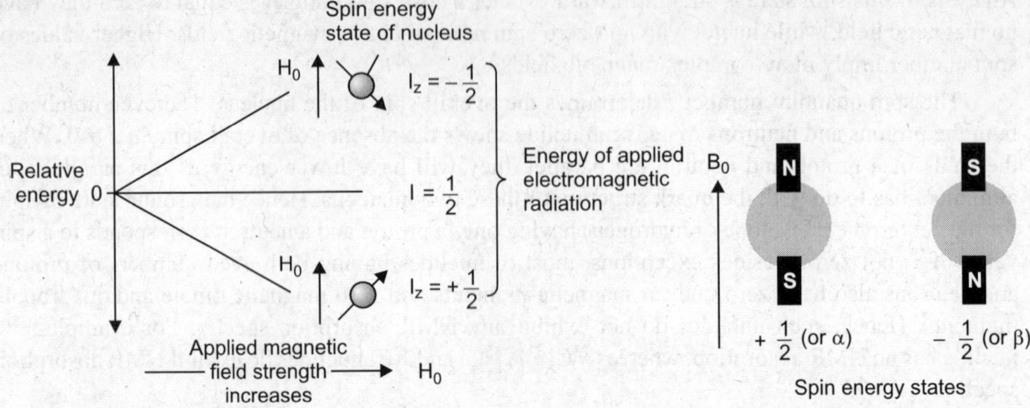

Fig. 11.45. The diagram illustrates the possible spin energy states for a spin 1/2 and a spin 1 nucleus with applied magnetic field strength (H_0).

There is a possibility to induce a transition between various spin states where an energy separation exists. By irradiating the nucleus with electromagnetic radiation of the correct energy, a nucleus with a low energy orientation can be induced to jump to a higher energy orientation. The absorption of energy during this transition forms the basis of the method of NMR.

***(c)* Magnetic Resonance by Nuclei:** When electromagnetic radiation of the correct frequency is applied to match the energy difference between the nuclear spin levels in a constant magnetic field of the appropriate strength, there occurs resonant absorption by nuclear spins. Then the energy of an absorbed photon will be $E = h\nu_0$, where ν_0 is the resonance radiofrequency that has to match to the Larmor precession frequency ν_L of the nuclear magnetisation in the constant magnetic field B_0. Absorption of a magnetic resonance occurs when $\Delta E = h\nu_0$ (where, E=energy, h=Planks constant, and ν = frequency). These magnetic resonance frequencies correspond to the radio frequency (RF) range of the electromagnetic spectrum for magnetic fields, which is detected in NMR.

***(d)* Nuclear Shielding:** All the nuclei of the same nuclide resonate at the same frequency. The 'shielding' effect of the surrounding electrons is the most important perturbation of the NMR frequency for applications of NMR. The magnetic field at the nucleus is reduced by this electronic shielding, which determines the NMR frequency. Consequently the energy gap is reduced and the frequency required to achieve resonance is also reduced. This shift in the NMR frequency is called 'chemical shift'. It is caused due to the orbital coupling of electrons of the molecules to the external magnetic field. The regions in which the induced field supports or adds to the external field are said to be 'deshielded regions', because a slightly weaker external field will bring about resonance for nuclei in such areas (Fig. 11.46).

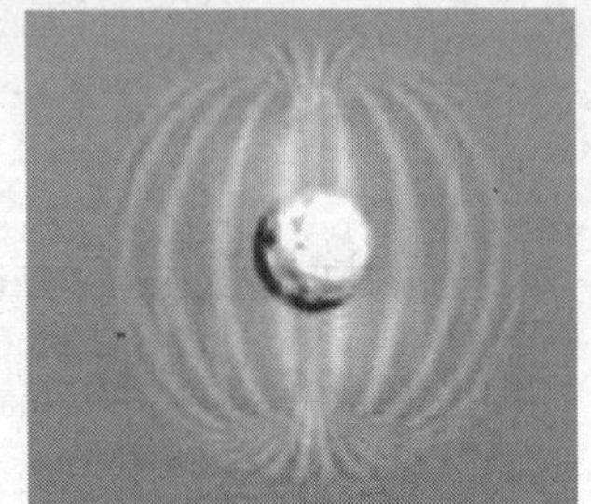

Nuclear shielding

In contrast, the regions in which the induced field opposes the external field are termed as 'shielded regions' because an increase in the applied field is needed for resonance. Shielded regions are designated by a plus sign, and deshielded regions by a negative sign.

The chemical structure of molecules depends on the distribution of electron density in the corresponding orbitals of the molecule. That is why NMR probes the chemical structure of molecules. The NMR frequency of a nucleus is shifted to a lower chemical shift called upfield, when a nucleus is shielded to a higher degree by a higher electron density of its surrounding molecular orbital. But the NMR frequency is shifted to a higher chemical shift called downfield, if it is less shielded by such surrounding electron density.

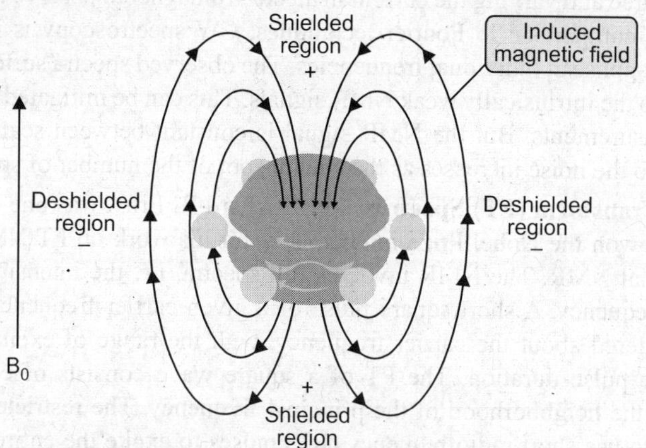

Fig. 11.46. A model showing nuclear shielding.

2. Working Principle of NMR

The principle of NMR usually involves two sequential steps: (*i*) polarisation of the magnetic nuclear spins in an applied constant magnetic field (H_o), and (*ii*) the perturbation of this alignment of the nuclear spins by employing an electro-magnetic pulse usually radio frequency (RF) pulse. The required perturbing frequency depends on both the static magnetic field (H_o) and the nuclei of observation.

A computer controls the apparatus that feeds a signal to a pulse generator. It selects pulses of high frequency radiation generated by a RF generator that are passed to a RF amplifier and then to the coils of the sample probe. The absorbed energy provides a signal which is amplified and passed to the detector, where the RF signal is heterodyned with the original RF source frequency and the frequency difference is amplified by an AF amplifier. An analogue digitises the output from the AF amplifier to digital converter, and the digitised signal is processed by a computer.

3. Types of NMR Spectroscopy

On the basis of application many kinds of spectroscopy has been built up; some of them are given below:

***(a)* NMR Spectroscopy:** The NMR spectroscopy is a technique that exploits the magnetic properties of certain nuclei. NMR is applicable to any nucleus that possesses spin. NMR spectroscopy is used to obtain physical, chemical, electronic and structural information about molecules due to either the chemical shift, Zeeman effect or the Knight Shift effect or a combination of both, on the resonant frequencies of the nuclei present in the sample.

It is a powerful technique that provides detailed information on the topology, dynamics and three-dimensional structure of molecules present in solution and the solid state. The most important applications in organic chemistry are proton NMR spectroscopy and carbon-13 NMR spectroscopy. Analysis of an NMR spectrum provides information on the number and type of chemical entities in a molecule.

The NMR spectroscopy is an invaluable tool in understanding protein and structure and function of nucleic acids. It can also be used to study mixtures of analytes, to understand dynamic effects such as change in temperature and reaction mechanisms, and applied to a wide variety of samples both in the solution and the solid state.

***(b)* Continuous Wave (CW) Spectroscopy:** NMR spectrometers employ a technique called continuous-wave (CW) spectroscopy. Although NMR spectra could be obtained by using a fixed magnetic field and sweeping the frequency of the electromagnetic radiation, it is involved by using a fixed frequency source and varying the current in an electromagnet to observe the absorption signals of the resonant. In comparison to Fourier techniques, CW spectroscopy is inefficient because it probes the NMR response at individual frequencies. The observed spectra suffer from a poor signal-to-noise ratio due to the intrinsically weak NMR signals. This can be mitigated by adding the spectra from repeated measurements. But the NMR signal is constant between scans. Therefore, overall ratio of the signal to the noise increases as the square-root of the number of spectra measured.

(*c*) Fourier Transform (FT) Spectroscopy: Richard R. Ernst was one of the pioneers of FT NMR. In 1991, he won the Nobel Prize in chemistry for his work on FT NMR and development of multi-dimensional NMR. The NMR involves full spectra, *i.e.* the intensity of the NMR signal as a function of frequency. A short square pulse of a given carrier frequency consists of a range of frequencies centered about the carrier frequency, with the range of excitation being inversely proportional to the pulse duration. The FT of a square wave consists of contributions from all the frequencies in the neighborhood of the principal frequency. The restricted range of the NMR frequencies eases to use short radiofrequency (RF) pulses to excite the entire NMR spectrum. All the single-quantum NMR transitions are excited by applying a pulse to a set of nuclear spins. This corresponds to tilting the magnetisation vector away from its equilibrium position. The oscillating magnetisation vector induces a current in a pickup coil nearby and creates an electrical signal oscillating at the NMR frequency.

(*d*) Multi-dimensional NMR Spectroscopy: The spectroscopist extracts different types of information about the molecule by using the pulses of different shapes, frequencies and durations in specifically-designed pulse sequences. Multi-dimensional NMR spectroscopy is types of FT spectroscopy in which there are at least two pulses, and the pulse sequence is varied. However, sequence of pulses and one variable time period are formed in multidimensional NMR. In three dimensions, two time sequences are varied, and so on.

4. Applications of Nuclear Magnetic Resonance

The NMR is applied in many areas of research and medical science as given below:

(*a*) NMR in Structural Biology: The protocol for determination of NMR structure includes the preparation of a homogeneous protein solution, the recording and handling of the NMR data sets, and the structural interpretation of the NMR data. In 1984, the techniques used for the structure determination of bull seminal proteinase inhibitor IIA are listed in Table 11.4. The four steps of the structural interpretation were performed separately, although the result of the first round of constraint collection and structure calculation was subsequently used for additional checks on the sequence-specific resonance assignments as well as on the collection of conformational constraints. Since 1984, the protocol outlined in Table 11.4 has been used in over 3'000 NMR structure determinations of proteins and nucleic acids, and greatly improved experimental techniques have been incorporated into this general scheme (Wuthrich, 2002).

Table 11.4. Standard protocol for NMR structure determination of proteins (after Wuthrich, 2002).

Step	*NMR structure determination*
Sample preparation	Protein isolated from natural source; natural isotope distribution; 16 mM solutions in H_2O and in $2H_2O$, respectively
NMR spectroscopy	2D ^1H NMR
– Resonance assignments	Sequential NOEs
– Conformational constraints	[^1H,^1H]-NOEs, 3*J*HN_, 3*J*__
– Structure calculation	Metric matrix distance geometry
– Structure refinement	Restrained energy minimisation

Major advances in the experimental techniques for NMR structure determination were spurned by the introduction of methods for the production of recombinant proteins labelled with stable isotopes, in particular ^{13}C, ^{15}N and deuterium. This has opened the way for efficient use of heteronuclear NMR techniques with proteins, such as 3D [^{1}H,^{13}C,^{15}N]-triple resonance experiments, 3D ^{13}C- or ^{15}N-resolved [^{1}H,^{1}H]-NOESY, and the use of heteronuclear filters. Important advances have also been made with the methods of structure calculation.

(*b*) **Medicine:** The magnetic resonance imaging (MRI) for medical diagnosis and MR microscopy in research, are the application of nuclear magnetic resonance to the general public. It is widely used in chemical studies in NMR spectroscopy such as proton NMR, carbon-13 NMR, deuterium NMR and phosphorus-31 NMR. Besides, the biochemical information can also be obtained from living tissue such as human brain tumours. In these studies orbiting electrons surround the nuclei. The electrons are the charged particles generating small, local magnetic fields and adding/ subtracting from the external magnetic field. Hence they will partially shield the nuclei.

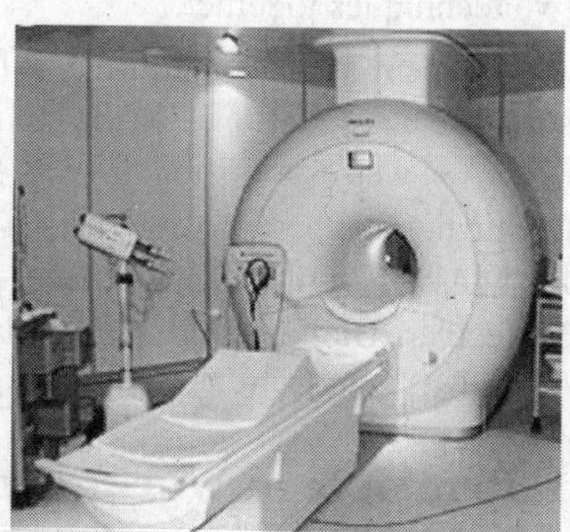

Medical MRI

The exact local environment governs the amount of shielding. For example, shielding of hydrogen bound to oxygen differs from the shielding of hydrogen bound to a carbon atom. If the two hydrogen nuclei are on the same molecule they interact through 'spin-spin coupling' resulting in splitting of the lines of the spectra. NMR is a major spectroscopic technique used in metabolomics. It is used to generate metabolic fingerprints from biological fluids to obtain information about state of diseases.

(*c*) **Chemistry:** The structure of many compounds can be determined by studying the peaks of nuclear magnetic resonance spectra. Many atoms within a molecule or collection of molecules of the same type can be distinguished by NMR. The identity of a compound can be determined by comparing the observed nuclear precession frequencies to known frequencies. The 'spin-spin coupling' is observed to elucidate the structural data. Spin-spin coupling is a process by which the precession frequency of a nucleus can be influenced by the magnetisation transfer from nearby nuclei. It is commonly observed in NMR that involves the common isotopes like hydrogen-1 (^{1}H NMR).

(*d*) **Non-destructive Testing:** The NMR is very useful for analysing samples non-destructively. Radio waves and static magnetic fields easily penetrate different types of matter, for example, expensive biological samples like RNA and DNA or proteins can be studied using NMR before using destructive biochemical experiments. Therefore, NMR is a good choice to analyse several harmful samples.

F. MASS SPECTROMETRY (MS)

In 1900, for the first time J.J. Thomson introduced mass spectrometer which employed fixed magnetic and electric fields to separate ions of different mass and energy. He recognised that charged particles differing in momentum behaved differently in an electric field and used this property for separation of ions with different mass. Its extensive use for research in various field of biological science was started in 1980s. During 1990-2000, mass spectrometry became an important technique for genomic and proteomic research leading to the 2002 Nobel Prize in Chemistry to J.B. Fenn and K. Tanaka.

Nobel Prize Winners in Chemistry in 2002

Koichi Tanaka shared Nobel Prize in Chemistry in 2002 [with J.B. Fenn (Mass Spectrometry) and K. Wüthrich (Nuclear Magnetic Resonance)] for the discovery of Soft Laser Desorption of macromolecules. In his honour Shimadzu Corporation (Japan), the founder of Shimadzu Biotech, has established the Koichi Tanaka Mass Spectrometry Research Laboratory in Kyoto, Japan.

Two-dimensional electrophoresis is more powerful when coupled with MS. The unknown protein spot is cut from the gel and cleaved by trypsin digestion into fragments which are then analysed by mass spectrometer followed by plotting of and the mass of fragments. This mass finger print can be used to estimate the probable amino acid composition of each fragment and tentatively identify the protein. The proteome and its changes can be studied very effectively by employing the two techniques together.

Recent improvement in MS has significantly improved its application in the study of protein structure and function. For the study of macromolecules, the Indian Society for Mass Spectrometry (ISMAS) was established in 1978 with its headquarter at BARC (Bhabha Atomic Research Centre), Mumbai. This society conducts a workshop each year on mass spectrometry. In India, facilities for MALDI-TOF MS have been created at several centres including the Centre for Cellular and Molecular Biology (CCMB) (Hyderabad), Indian Institute of Sciences (Bangalore), BARC (Mumbai), and Rajiv Gandhi Centre for Biotechnology (RGCB) (Thiruvanantpuram), etc.

1. Principle of Working of Mass Spectrometry

The principle of working of mass spectrometry is to create gas phase ions from polar charged biomolecules like peptides, proteins or nucleic acids. In the ionisation source the sample (M) undergoes ionisation upon introduction in the mass spectrometry. Molecular weight more than 1299 Da of the sample results in multiple charged ions *e.g.* M+2H. There are many suitable sites of proteins which undergo protonation. All the backbone amide nitrogen atoms and side chain amino group may be protonated. The charged molecules are electrostatically propelled into analyser. On the basis of m/z ratio analyser separates the ions. The detector detects the ions and transfers the received signals to a computer. The information is stored and processed by the computer.

2. Components of Mass Spectrometry

In recent years, mass spectrometry has become an essential tool for analysis of genome and proteome in its many forms. It is capable of identifying and characterising proteins present even in picomoles (10^{-12}). The MS can also provide valuable information about covalent modification of proteins which can affect their activity. This technique involves: (*i*) the production of ions of the materials in sample, (*ii*) their separation on the basis of their mass : charge (*m*:*z*) ratio, and (*iii*) determination of relative abundance of each ion. Basically it is an analytical device consisting of three components: the *source of ion*, an *analyzer*, and a *detector* (Fig. 11.47). It does not directly measure the molecular mass but detects *m* : *z* ratio. Mass is measured in terms of Da [one Dalton = $1/12^{th}$ of mass of a single atom of isotonic carbon ($^{13}C^{+}$)].

(*i*) Source of Ignition (ions): It involves the gaseous ionisation of analyte (the molecule to be analysed) to be examined. When the molecules gain or lose a charge (through electron ejection, protonation or deprotonation), molecular ions are generated.

(*ii*) Analyser: The ions generated as above are separated according to their mass-to-charge (m/z) ratio.

(*iii*) Detector: The molecular masses of separated ions are determined from the mass spectra obtained after mass spectrometry of analyte.

For identification, protein is separated from the crude extract on a two-dimensional gel. The protein spots are excised and then used as search for mass spectrometry or fragmented into small peptides. Protein fragmentation into peptides is done by proteases (*e.g.* trypsin). The peptides are separated by using liquid chromatography (such anion exchange affinity or reverse phase column chromatography). Moreover, proteins are easily identified if fragmented into peptides.

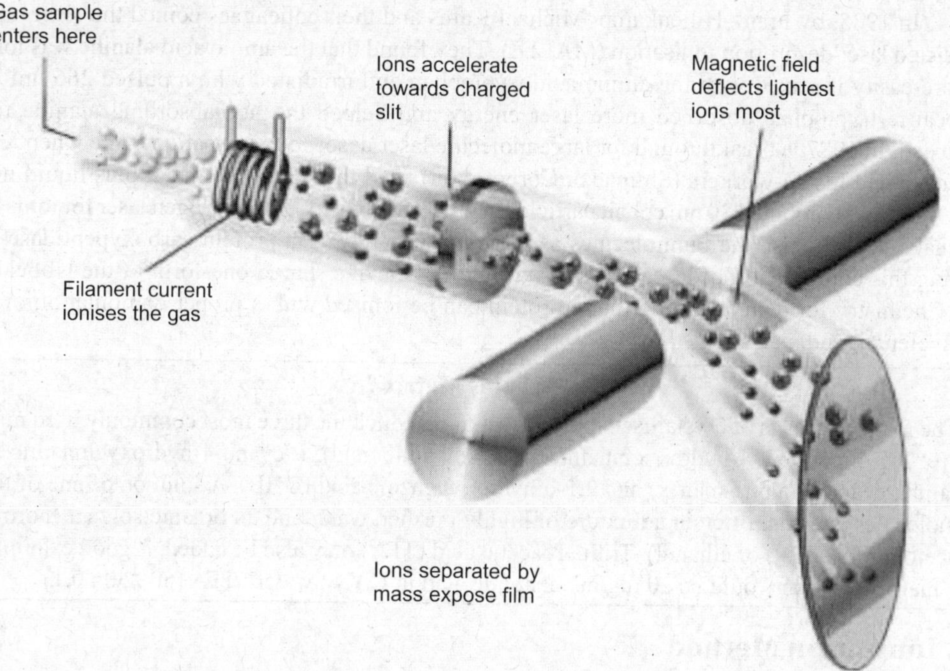

Fig. 11.47. A simplified schematic diagram of mass spectrometry.

3. Uses of Mass Spectrometry

Mass spectrometry is very useful technique which is appropriately used for the analysis of molar masses of molecules for the past 50 years. It is used in identification of unknown compounds, quantification of known compounds and determination of structural and chemical properties of compounds when present in small amount (10^{-6} – 10^{-8} g). However, the application of MS to large biomolecules and synthetic polymers has been limited due to low volatility and thermal instability of these materials. These problems have been solved to a great extent through the development of soft ionisation techniques such as chemical ionisation, secondary-ion mass spectrometry (SIMS), field desorption, fast atom bombardment, and matrix-assisted laser desorption/ionisation mass spectrometry (MALDI-MS). The MALDI-MS technique, in particular, allows for the mass determination of large biomolecules and synthetic polymers of molar mass greater than 200,000 Daltons by ionisation and vaporisation without degradation.

The molecules to be examined in MS should be in the form of gaseous ions. There are several methods available for ionisation of biomolecules but MALDI and electrospray (ES) each coupled with TOF-MS are discussed in this section.

I. MATRIX-ASSISTED LAYER DESORPTION-IONISATION (MALDI)

MALDI is a soft ionisation technique used in mass spectrometry which analyses the biomolecules or biopolymers (such as proteins, peptides and sugars) and large organic molecules (such as polymers, dendrimers and other macromolecules), which tend to be fragile and fragment when ionized by more conventional ionisation methods. MALDI is most similar in character to electrospray ionisation both in relative softness and the ions produced. The ionisation is triggered by a laser beam, usually a nitrogen laser. A matrix is used to protect the biomolecule from being destroyed by direct laser beam and to facilitate vaporisation and ionisation.

In 1985, by Franz Hillenkamp, Michael Karas and their colleagues coined the term matrix-assisted laser desorption ionisation (MALDI). They found that the amino acid alanine was ionised more easily if mixed with the amino acid tryptophan and irradiated with a pulsed 266 nm laser. Because tryptophan absorbed more laser energy and helped the non-absorbing alanine to get ionised. In 1987, a breakthrough for large molecule laser desorption ionisation came when Koichi Tanaka and his co-workers (Shimadzu Corporation) used the 'ultra fine metal plus liquid matrix method' that combined 30 nm cobalt particles in glycerol with a 337 nm nitrogen laser for ionisation. Tanaka could ionise the biomolecules as large as the 34,472 Da protein carboxypeptidase-A by using this combination of laser and matrix. In 2002, Tanaka shared one-forth of the Nobel Prize in Chemistry for demonstrating that a protein can be ionised with a proper combination of laser wavelength and matrix.

What is Matrix ?

The matrix consists of crystallised molecules among which the three most commonly used matrix are: 3,5-dimethoxy-4-hydroxycinnamic acid (sinapinic acid), α-cyano-4-hydroxycinnamic acid (alpha-cyano or alpha-matrix) and 2,5-dihydroxybenzoic acid (DHB). A solution of one of these molecules is made, often in a mixture of highly purified water and an organic solvent (normally acetonitrile (ACN) or ethanol). Trifluoroacetic acid (TFA) may also be added. A good example of a matrix-solution would be 20 mg/ml sinapinic acid in CAN : water : TFA (50 : 50 : 0.1).

1. Ionisation Method

In this method the matrix solution is mixed with the analyte *e.g.* protein sample. The organic solvent allows hydrophobic molecules to dissolve into the solution, while the water allows for water-soluble molecules to get dissolved. The mixed solution (matrix + analyte) is pipetted and spotted onto a MALDI plate which is usually a metal plate designed for this purpose (Fig. 11.48A). The solvents vaporise, leaving only the recrystallised matrix, but now with analyte molecules spread throughout the crystals. The matrix and the analyte are said to be co-crystallised in a MALDI spot. The dried mixture in the MALDI spot is irradiated by laser pulse (337 nm). The irradiation causes energy transfer and desorption resulting in gas phase matrix ions. The matrix absorbs the laser energy and primarily the matrix is ionized. Then the matrix transfers part of its charge to the analyte molecules resulting in ionisation and protection from the disruptive energy of the laser. Ions observed consist of a neutral molecule [M] and an added or removed ion. They form together a quasimolecular ion *e.g.* $[M+H]^+$ in the case of an added proton, $[M+Na]^+$ in the case of an added sodium ion, or $[M-H]^-$ in the case of a removed proton. MALDI is capable of creating singly-charged ions, but multiply charged ions ($[M+nH]^{n+}$) can also be created as a function of the matrix, the laser intensity and/or the voltage used. With the help of matrix ions the non-absorbing analyte molecules also get desorbed into the gas phase and ionised. The charged molecular ions of analyte are detected by MALDI-TOF mass spectrometer (MS) (Fig. 11.48B).

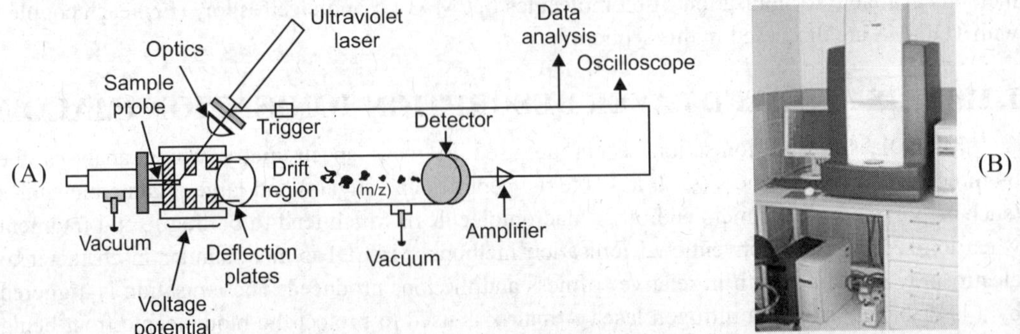

Fig. 11.48. MALDI-TOF mass spectrometer; A- schematic presentation; B- instrument.

Generally, MALDI is used for the study of molecules having mass above 500 Daltons *e.g.* oligonucleotides and peptides. Proteome analysis will be crucial for understanding different biological processes in the post-genomic era. To achieve this objective mass spectrometry is extensively used for peptide sequencing, identification of protein, protein expression in different tissues and conditions, identification and post-transcriptional modification (*e.g.* phosphorylation, glycosylation, etc.) of proteins in response to different stimuli, and characterisation of protein interactions that include protein-ligand, protein-protein and protein-DNA interaction.

MALDI-TOF MS is an emerging technique offering promise for the fast and accurate determination characteristics of a number of polymers. The matrix and polymer are mixed at a molecular level in an appropriate solvent with a ~10^4 molar excess of the matrix. The solvent prevents aggregation of the polymer. The sample-matrix mixture is placed onto a sample probe tip. Under vacuum conditions the solvent is removed, leaving co-crystallised polymer molecules homogeneously dispersed within matrix molecules. When the pulsed laser beam is tuned to the appropriate frequency, the energy is transferred to the matrix which is partially vaporised carrying intact polymer into the vapor phase and charging the polymer chains.

Multiple laser shots are used to improve the signal-to-noise ratio and the peak shapes, which increases the accuracy of the molar mass determination. In the linear TOF analyser (drift region), the distribution of molecules emanating from a sample are imparted identical translational kinetic energies after being subjected to the same electrical potential energy difference (Fig. 11.49).

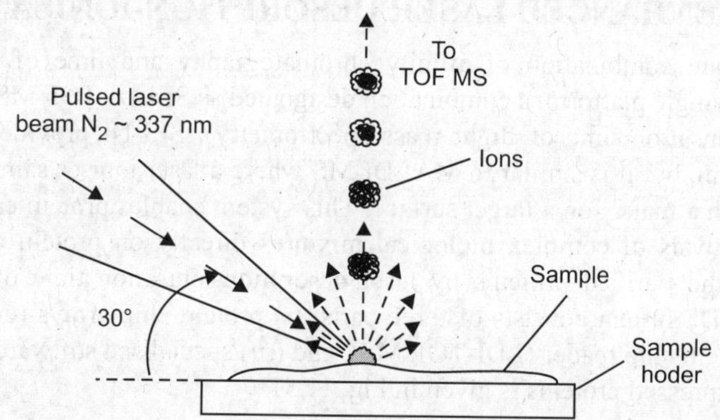

Fig. 11.49. Method of sample ionisation by MALDI-TOF MS.

Thereafter, these ions traverse the same distance down an evacuated field-free drift tube. The smaller ions come to the detector rapidly than the more massive ions. Separated ion fractions arriving at the end of the drift tube are detected by an appropriate recorder that produces a signal upon impact of each ion group (Fig. 11.50). The digitised data generated from successive laser shots are summed yielding a TOF mass spectrum. The TOF mass spectrum is a recording of the detector signal as a function of time. The time of flight for a molecule of mass *m* and charge *z* to travel this distance is proportional to (*m*/*z*)1/2. This relationship (*t* ~ (*m*/*z*)1/2) can be used to calculate the ions mass. Through calculation of the ion's mass, conversion of the TOF mass spectrum to a conventional mass spectrum of mass-to-charge axis can be achieved (Fig. 11.50).

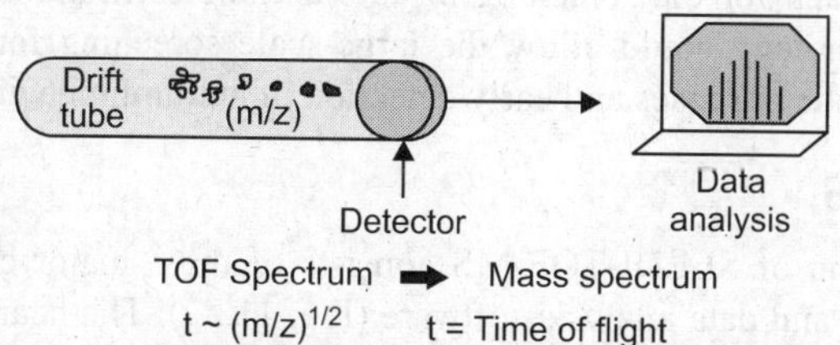

Fig. 11.50. Conversion from TOF spectra to conventional spectra.

2. Application

In recent years MALDI has various applications in different areas as described below:

(*a*) In Biochemistry: MALDI is used in proteomics for identification of proteins isolated through gel electrophoresis, SDS-PAGE, size exclusion chromatography, and two-dimensional gel electrophoresis. One method used is peptide mass fingerprinting by MALDI-MS or with post ionisation decay or collision-induced dissociation. In papers loss of sialic acid has been identified when DHB has been used as a matrix for MALDI MS analysis of glycosylated peptides.

It has been reported that a reduction in loss of some post-translational modifications can be accomplished if IR MALDI is used instead of UV MALDI.

(*b*) In Organic Chemistry: Some synthetic macromolecules *e.g.* catenanes and rotaxanes, dendrimers and hyperbranched polymers have molecular weights extending into the thousands or tens of thousands, where most ionization techniques have difficulty in producing molecular ions. MALDI is a simple and rapid analytical method that can allow chemists to analyze the results of such syntheses and verify their results.

(*c*) In Polymer Chemistry: In polymer chemistry MALDI can be used to determine the molar mass distribution. Polymers with polydispersity greater than 1.2 are difficult to characterise with MALDI due to the signal intensity discrimination against higher mass oligomers.

II. SURFACE-ENHANCED LASER DESORPTION-IONISATION (SELDI)

It is a unique combination of affinity chromatography and time of flight (TOF) mass spectrometry on a single platform, a combination designated as SELDI-TOF MS (surface enhanced laser desoption-ionisation time of flight mass spectrometry). SELDI provides a unique sample preparation platform, but it is similar to MALDI-MS where a laser ionises samples that have been co-crystallised with a matrix on a target surface. This system enables protein capture, purification, ionisation and analysis of complex biological mixtures directly on protein chips array surface and detection of the purified proteins by laser desorption-ionisation time of flight (LDI-TOF) MS analysis. SELDI system consists of three parts: (*i*) protein chip arrays (eight spots of 2 mm diameter), (*ii*) protein chip reader (LDI-TOF MS), and (*iii*) specialised software. The identification of differentially expressed proteins is given in Fig. 11.51.

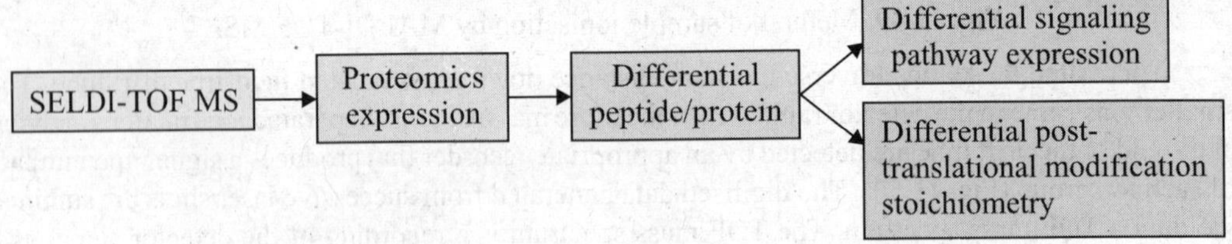

Fig. 11.51. Identification of differentially expressed proteins.

The onchip procedure is quite easy to use, much less time consuming than Western blotting and much more sensitive at least in the low molecular weight range. SELDI-TOF technology is a very promising approach for the screening of auto-antibodies in autoimmune diseases. Due to its versatility, this technology could allow the large-scale screening for complex auto-antibody distributions for diagnostic purposes and early detection of autoimmune diseases could be possible.

1. SELDI-TOF-MS

The instrumentation of SELDI-TOF-MS consists of three major components: protein chip arrays, a mass analyser, and data analysis software (Fig. 11.52). The heart of the SELDI-TOF MS technology is the protein chip arrays that distinguish it from other MS-based systems. The arrays are composed of different chromatographic surfaces that are designed to retain proteins of interest. The protein chip arrays are made of a 10 mm wide × 80 mm long aluminum strip with 8 or 16 2 mm spots composed of a chemically (anionic, cationic, hydrophobic, hydrophilic, or metal ion) or biochemically (immobilized antibody, receptor, DNA, enzyme, etc.) active surface (Fig. 11.52 A).

The mass analyser is a simple TOF mass spectrometer that is equipped with a pulsed, UV, nitrogen laser (Fig. 11.52 B). The chip is irradiated with a laser resulting in ionization of the adherent molecules. . In an electric field the ionized molecules are accelerated into a field-free region under vacuum and their mass-to-charge ratios are calculated from their time of flight through the vacuum chamber (the so called TOF tube) toward an ion detector. The *m/z* of each species is recorded on the basis of the time required for the ions to pass through the TOF tube. The mass analyzer results in low resolution and, therefore, low mass accuracy is obtained.

The data analysis software can convert the MS peak trace into a simulated one-dimensional gel electrophoresis display which is called 'gel view'. The TOF MS analysis generates output is a traces that shows the molecular weights of the detected proteins (Fig. 11.52 C). The software compares these displays to identify unique peaks or those showing a significant abundance difference in one of the samples. The software also carries out cluster analyses to identify significant differences in protein abundances of several samples.

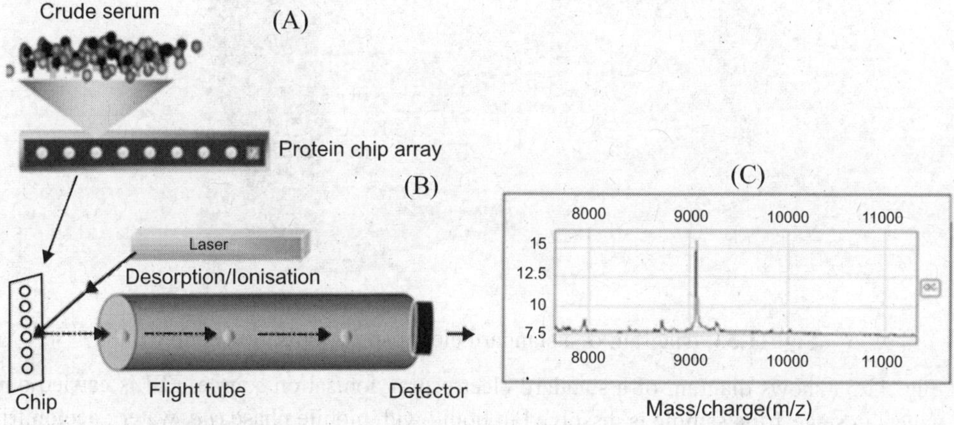

Fig. 11.52. On chip method of SELDI-TOF MS. A, preparation of sample and protein chip arrays; B, mass analyzer; C, mass spectra acquisition.

2. Samples Preparation and Data Processing

Two-dimensional polyacrylamide gel electrophoresis (2D-PAGE) is the proteomic method for discovering disease biomarkers where proteins from two distinct samples are analysed and their expression patterns are compared. Protein spots of interest are excised from the gel, digested proteolytically or chemically, and the resultant peptides are analysed by MS to identify the protein. Steps of sample preparation and data processing are: (*i*) selection of relevant clinical problematic and samples groups composition, (*ii*) pre-treatment samples for removing highly abundant proteins that saturate chips and MS signal, (*iii*) purification or fractionation of sample proteins on SELDI chips, (*iv*) spectra acquisition on SELDI-TOF-MS, (*v*) management of spectra, peaks detection and peak clusters formation, (*vi*) uni- and/or multi- variate statistical differential analysis for groups comparison, (*vii*) purification and identification of the selected potential biomarkers by high resolution mass spectrometry, (*viii*) validation of the selected potential biomarkers by MS or non-MS methods, on independent series of subject samples.

The researchers have discovered potential diagnostic markers by using SELDI-TOF MS for prostate, bladder, breast, and ovarian cancers, Alzheimer's disease and interstitial cystitis. By using this technique several proteins have been characterised, for example phosphorylated and glycosylated proteins, transcription factors, and peptides and proteins secreted by various cancer cell lines.

III. ELECTROSPRAY IONIZATION (ESI)

In the late 1980s, J.B. Fenn developed ESI mass spectrometry and was awarded Nobel Prize in 2002 in chemistry. ESI is a popular technique in mass spectroscopy for ionising samples before they are measured. ESI works well with heavier compounds and is often used in proteomics (analysis of proteins) and petroleomics (analysis of petroleum).

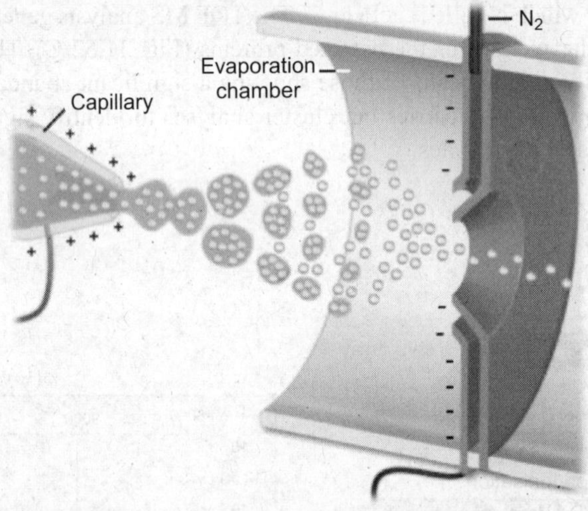

Fig. 11.53. Diagram of a standard electrospray ionisation source.

Fig. 11.53 shows diagram of a standard electrospray ionisation source. ESI is carried out in two stages: in stage 1 the sample is dissolved in liquid with mobile phase (*i.e.* water : acetonitrile : methanol) and pumped through a hypodermic needle at a high voltage into a vessel called a 'capillary' that ends in a very fine tip. A very high voltage is applied to this tip which charges the molecules in the solvent because they are highly charged. About 4,000 V causes strong electric field at the top of nozzle of the metal needle. These molecules with the same charge once are pushed through the nozzle into the 'evaporation chamber', they repel one another violently. When the charged liquid first exits the tip, it briefly forms a cone shape, a 'taylor cone' before the droplets burst away from each other into a fine spray. This disperses electrostatically or electrosprays the highly charged small droplets of about one micrometer in size.

In stage 2, the droplets in the spray go through a series of divisions after being released from the nozzle. This occurs due to gradual evaporation of solvent within these droplets (with the assistance of nitrogen gas pumped into the chamber), that forces the charges in the molecules within these droplets (which again are identical) closer together. After coming closer, the ions repel each other due to presence of equal charges causing the droplets to divide into two smaller droplets. This behavior is known as the coloumb force. This process repeats itself until the solvent is completely evaporated and the droplets have split up to the point that each is a single and charged molecule. One of the advantages of this ionisation method is that the molecules remain intact and will not be broken apart.

The *m/z* of the nucleotide can be analysed in the vacuum chamber. ES produces many more ions than MALDI. These ions are transferred into mass spectrometer for analysis with high efficiency. ES MS can analyse a wide range of molecules including proteins, oligonucleotides, sugar and lipids. Since a MS measures *m/z*, it allows high molecular weight proteins to be observed on low mass range equipment. Thus ESI generates multiple charged molecules. Peptides or proteins are identified on the basis of *m/z* ratio. The m/z ratio of ions is analysed after their generation. The new proteins with human genome sequencing project could be identified through the combination of MS and genomics, and also using the tools of bioinformatics (*see* Chapter 12 : *Bioinformatics*).

IV. DESORPTION ELECTROSPRAY IONIZATION (DESI)

In 2004, this technology was first published by R. Graham Cooks' group. Fig. 11.54 shows the components of DESI and the way of interfacing an ion source with a standard mass spectrometer. DESI is carried out by directing pneumatically assisted electrosprayed droplets onto a surface to be analysed at atmospheric conditions. Ions are produced from the sample originally present on the surface. The resulting mass spectra are similar to normal ESI mass spectra. The contents of the solvent spray, the gas flow rate, the amount of applied voltage, the spray angle and the ion uptake angle as well as the various distances in aligning the spray, sample and mass spectrometer are all variables which can be studied to achieve an optimal mass spectrum for a particular type of sample.

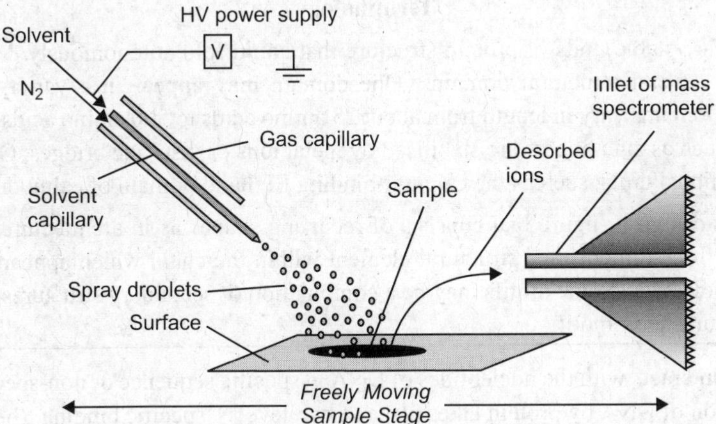

Fig. 11.54. The components of DESI where ions interface with a standard mass spectrometer.

By using DESI, high quality mass spectra have already been obtained for a wide range of molecules by directly interrogating a diverse range of surfaces. For example, explosive agents, chemical warfare simulants, amino acids, peptides, proteins, commercially available drug molecules, alkaloids, terpenoids and steroids have all been successfully ionised using DESI. These molecules are ionised from a diverse range of surfaces (conducting as well as insulating) such as plant tissues, human skin, liquid blood, dried blood spots, urine, as well as from a wide range of inert materials such as polytetrafluoroethylene (PTFE), nitrile gloves, leather belts, pharmaceutical tablets, and many more surfaces that we interact with in our everyday lives.

Benefits of DESI: The benefits of DESI include: (*i*) it is a direct sampling method that requires absolutely no sample preparation, (*ii*) its operation in the atmosphere under ambient conditions, (*iii*) it is a sensitive and gentle ionisation method for the analysis or interrogation of all types of surfaces, (*iv*) it produces electrospray-like spectra, (*v*) it is effective for both organic and biological compounds, (*vi*) it is effective for polar and non-polar molecules, and (*vii*) it allows for *in situ* analysis of biological tissues and other fragile surfaces.

G. PROTEIN-DNA INTERACTION

In the cells of prokaryotes and eukaryotes interaction occurs between DNA and protein, and a DNA-binding domain (DBD) governs the structure and function of DNA. Proteins that bind to DNA have the common folding patterns called 'DNA binding motifs'. A DNA-binding domain (DBD) is an independently folded protein domain that contains at least one motif that recognises double-stranded DNA or single-stranded DNA. A DBD can recognise a specific DNA sequence, called recognition sequence. Hence, each DNA binding motif is composed of a recognition region and a stabilisation region. Some DNA-binding domains may also include nucleic acids in their folded structure.

One or more DNA-binding domains are a larger protein which contains additional domains the function of which differs. The additional domains regulate the activity of the DNA-binding domain. The function of DNA-binding is either structural or functional involving transcription regulation; sometimes the two roles overlap. Such DNA-binding domains have biological roles in the replication, repair, storage and modification of DNA, such as methylation. There are several proteins containing DNA-binding domains which are involved in the regulation of gene expression, for example transcription factors, *i.e.* proteins that regulate transcription by binding DNA. Gene regulation is the final output of most cascades cell signalling.

Terminology

- **Domains** are stable units of protein structure that could fold autonomously. Many proteins consist of several structural domains. One domain may appear in a variety of different proteins. Domains vary in length from about 25 amino acids to 500 amino acids. The shortest domains such as zinc fingers are stabilised by metal ions or disulfide bridges. Often domains form functional units, such as the calcium-binding EF hand domain of calmodulin.
- **Motif** is a design or figure that consists of recurring shapes as in architecture. A structural motif is a three-dimensional structural element within the chain which appears in a variety of other molecules. Some motifs may be a combination of secondary structures, for example the helix-turn-helix motif.

The DBD interacts with the nucleotides of DNA in specific sequence or non-specific sequence; hence, recognition of DNA by protein takes place at two levels: 'specific binding' (between protein side chains and nucleotide bases), and 'nonspecific binding' (between DNA phosphodiester backbone and protein side chains, *i.e.* arginines and lysines or other than these).

The protein α-helices make most of the base-specific interactions in these proteins that recognise and bind the DNA. The DBD recognises at the major or minor groove of DNA, or at the sugar-phosphate backbone of DNA. Mostly base-specific interactions occur in the major groove of the DNA. The hydrogen bonding pattern in the DNA major groove is less degenerate than that of the DNA minor groove providing a more attractive site for sequence-specific DNA recognition. But some sort of complementary molecules must be involved between protein and DNA in non-sequence-specific recognition. Each specific type of DNA recognition governs the function of proteins, for example the DNA-cutting enzyme DNase I cleaves DNA randomly and binds to DNA in a non-sequence-specific manner. But, DNAse I recognises a certain 3D DNA structure yielding a specific DNA cleavage pattern that can be useful for studying the DNA recognition.

Specific DNA sequences are recognized by many DNA-binding domains; for example DBDs of transcription factors that activate specific genes, some enzymes modify DNA at specific sites such as restriction enzymes and telomerase. There are many biochemical and biophysical techniques (such as gel electrophoresis, analytical ultracentrifugation X-ray crystallography, surface plasmon resonance, electron paramagnetic resonance, cross-linking), which are used to study the specificity of DNA-binding proteins.

1. Types of DNA-binding Domains

There are several different types of DNA-binding domains as described below:

***(a)* Helix-turn-helix (HTH) Domain:** The helix-turn-helix domain is commonly found in prokaryotic proteins that regulate gene expression (at the level of transcription of DNA to RNA). The helix-turn-helix motif is found in repressor proteins and is about 20 amino acids long. Examples include the *E. coli* trp repressor, the 434 *cro* repressor (from a bacteriophage), and the *E. coli lac* repressor. The λ repressor of bacteriophage lambda is shown in Fig. 11.55A. The helix-turn-helix is a major structural motif capable of binding DNA. It is composed of two α helices joined by a short strand of amino acids and is found in many proteins that regulate gene expression. It should not to be confused with the 'helix-loop-helix' (HLH) domain.

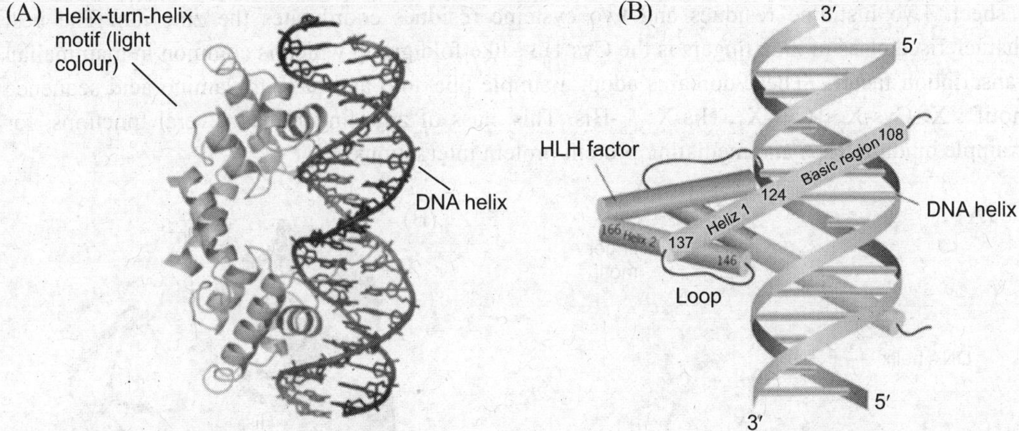

Fig. 11.55. The λ repressor of bacteriophage lambda (left) binds to DNA (right) (A); Helix-loop-helix transcription factor (B).

There are similarities between the genes for Cro, CAP, and λ repressor which share a common sequence of 20-25 amino acid involving DNA recognition. The two α- helices (one occupying the N-terminal end of the motif and the other at the C-terminus) recognise and bind to DNA. Mostly the second helix contributes to DNA recognition; hence it is called the 'recognition helix' as found in the Cro repressor. It binds to the major groove of DNA through a series of hydrogen bonds and various van der Waals interactions with exposed bases. The other α-helix does not play a strong role in its recognition but stabilises the interaction between protein and DNA.

***(b)* Helix-loop-helix (HLH) Domain:** The Helix-loop-helix domain is found in some transcription factors and is characterised by two α-helices connected by a loop (Fig. 11.55 B); typically one helix is smaller in size than the other. Due to the flexibility of the loop it allows dimerisation against another helix by folding and packing mechanisms. In addition, the larger helix typically contains the DNA-binding regions.

***(c)* Zinc Finger Domain:** Originally the name 'zinc finger' was coined to describe the hypothesized structure of the repeated unit of transcription factor IIIA in *Xenopus laevis*. Zinc fingers are small protein structural motifs that can coordinate one or more zinc ions to help stabilise their folds (Fig. 11.56 A).

Generally the zinc finger domain is 23-28 amino acids long and stabilised by coordinating zinc ions with regularly spaced zinc-coordinating residues (either histidine or cysteine). The zinc finger motif consists of a α-helix followed by a turn and two short β-strands. The β-stands form a small loop resembling a finger, and overall the fold sandwiches a zinc ion. The α-helix and the first β-strand contain conserved cysteine and histidine residues that coordinate the zinc metal ion, which is important for stabilsing the protein fold.

The most common class of zinc finger (Cys2His2) coordinates a single zinc ion and consists of a recognition helix and a two-strand beta-sheet. In transcription factors these domains are often found in arrays (usually separated by short linker sequences) and adjacent fingers are spaced at an interval of three base pairs when bound to DNA. They can be classified into different structural families and function as interaction modules that bind to DNA, RNA, proteins, or small molecules. Zinc fingers coordinate zinc ions with a combination of cysteine and histidine residues. They can be classified by the type and order of these zinc coordinating residues, for example Cys_2His_2, Cys_4, and Cys_6.

Fig. 11.56 B shows the Cys2His2 zinc finger motif consisting of an α-helix and an antiparallel β sheet. Two histidine residues and two cysteine residues coordinates the zinc ion. The best characterised class of zinc fingers is the Cys_2His_2-like fold group which is common in mammalian transcription factors. These domains adopt a simple ββα fold and have the amino acid sequence motif : X_2-Cys-$X_{2,4}$-Cys-X_{12}-His-$X_{3,4,5}$-His. This class of zinc fingers has several functions, for example binding RNA and mediating protein-protein interactions.

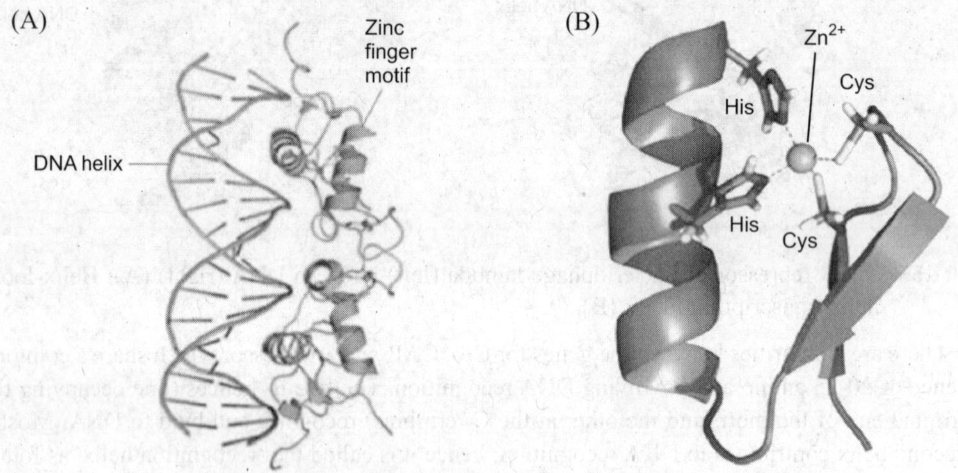

Fig. 11.56. The zinc finger motif (A); Cys2His2 zinc finger motif (B).

***(d)* Leucine Zipper Domain:** A leucine zipper is a common three-dimensional structural motif in proteins which are usually found as part of a DNA-binding domain in various transcription factors. Therefore, leucine zippers are involved in regulation of gene expression. Leucine zippers are found in both eukaryotic and prokaryotic regulatory proteins, but mainly are present in eukaryotes. The leucine zipper domain consists of two α-helices that interact with each other at one end and promote dimerization. The other end of each helix contacts the DNA in the major groove. When binding to the DNA, basic amino acid residues bind to the sugar-phosphate backbone while the helices sit in the major grooves. It regulates gene expression. Fig. 11.57A shows the transcriptional factor AP-1 (protooncogene c-jun) that binds to DNA.

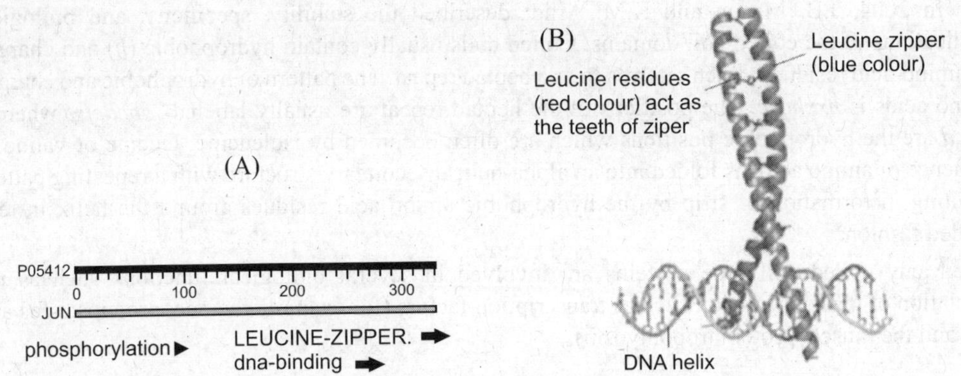

Fig. 11.57. A- transcriptional factor AP-1 (protooncogene c-jun); B- leucine zipper bound to DNA. The leucine residues that represent the 'teeth' of the zipper (see red colour).

The dimerisation region is characterised by the presence of leucine molecules at every seventh position in the helix, forming a tight dimer structure. Like many other transcription factors, the leucine-zipper-containing transcription factors bind DNA as dimers (Fig. 11.57 B). The helices are held together by hydrophobic interactions between leucine residues, which are located on one side of each helix, for examples AP-1, CREB, and Gcn4.

Leucine zipper regulatory proteins are c-fos and c-jun (the AP1 transcription factor) (important regulators of normal development), as well as myc family members including myc, max, and mdx1. When they are overproduced upon mutation, they may generate cancer. These proteins interact with the DNA as dimers (homo- or hetero-) and are also called 'basic zipper proteins' (bZips).

***(e)* HMG-box Domain:** The HMG-box domain consists of three alpha helices separated by loops. The HMG-box domains are found in high mobility group proteins which are involved in a variety of DNA-dependent processes such as replication and transcription.

***(f)* Coiled Coil Motif:** In 1951 by Linus Pauling and coworkers suggested the structure of the alpha helix. In 1952 Francis Crick proposed the possibility of coiled coils for α-keratin and the mathematical methods for determining their structure. A coiled coil is a structural motif in proteins where 2-7 alpha-helices are coiled together that looks like the strands of a rope. However, dimers and trimers are the most common types of coils. The α-helices may be parallel or anti-parallel, and usually adopt a left-handed super-coil. The GCN4 leucine zipper is the classic example of a coiled coil which is a parallel and left-handed homodimer (Fig. 11.58).

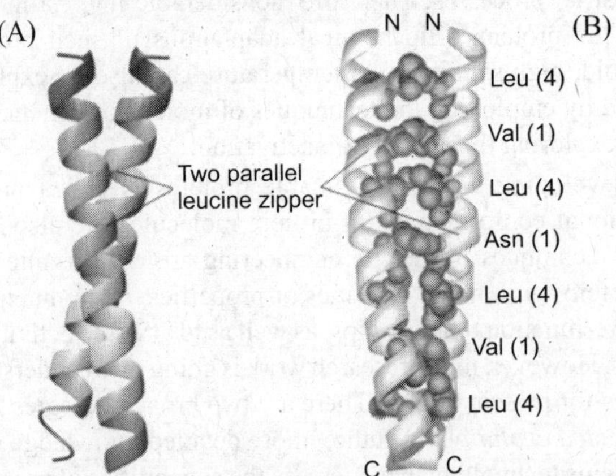

Fig. 11.58. The coiled coil motif of GCN4 leucine zipper is a parallel and left-handed homodimer; arrangement of valine (1) and leucine (4) at varying positions.

In 2004, J.B. Mason and K.M. Arndt described the stability, specificity, and biological implications of the coiled coil domains. Coiled coils usually contain hydrophobic (*h*) and charged (*c*) amino-acid residues which referred to as a heptad repeat. The pattern of hydrophobic and charged amino acids is *hxxhcxc*. The positions in the heptad repeat are usually labeled- *abcdefg*, where *a* and *d* are the hydrophobic positions which are often occupied by isoleucine, leucine or valine. A sequence of amino acids is folded into an alpha-helical secondary structure with a repeating pattern resulting in formation of strip by the hydrophobic amino acid residues around the helix in left-handed fashion.

Many coiled coil type proteins are involved in several biological functions such as the regulation of gene expression such as transcription factors (for example the oncoproteins c-fos and jun, and the muscle protein tropomyosin).

2. Unusual DNA-binding Domains

There are some unusual DNA-binding domains also; some of them are described below:

(*a*) Immunoglobulin Fold: The immunoglobulin domains (*IPR013783*) are usually found in immunoglobulin proteins. It consists of a beta-sheet structure with large connecting loops which recognises either DNA major grooves or antigens. They also present in Stat proteins of the cytokine pathway. Because the cytokine pathway evolved relatively recently and has made use of systems that were already functional.

(*b*) B3 Domain: The B3 DBD (*IPR003340, SCOP 117343*) consists of 100-120 residues. It consists of seven beta sheets and two alpha helices which form a DNA-binding pseudobarrel protein fold. It is found exclusively in transcription factors from higher plants.

(*c*) TAL Effector DNA-binding Domain: TAL effector consists of a central region of tandem repeats of 33-35 residue and each repeat region encodes a single DNA base in the TALE-binding site. TAL effectors are found in bacterial plant pathogens and are involved in regulating the genes of the host plant in order to facilitate bacterial virulence, proliferation, and dissemination.

H. PROTEIN ENGINEERING AND PROTEIN DESIGN

Proteins carryout complex functions in the cells of all living organisms. Many thousands of enzymes have been isolated and characterised biochemically. The reason for large number of enzymes not being used commercially is their activity which occurs naturally but gets failed in highly specialised *in vitro* conditions. Most enzymes are easily denatured due to exposure to adverse conditions such as high temperature, pH changes and presence of organic solvents and reactive chemicals used in industrial processes. Therefore, considerable interest in biotechnology industry exists for engineering the proteins. Biochemical adaptations of such proteins to environmental stresses *e.g.* draught, cold, high salinity, high temperature can also be explored. The stability of a protein can be increased by employing the techniques of molecular genetics (such as gene cloning and mutagenesis) after exploring the causes of inactivation.

The process of developing useful or valuable proteins is called protein engineering. The sizes and three-dimensional conformations of protein molecules are also manipulated by protein engineering. The basic techniques of genetic engineering are used to alter the genes that encode proteins, generating proteins with novel activities or properties. Such manipulations are frequently used to discover structure-function relationships, as well as to alter the activity, stability, localisation, and structure of proteins. However, much research work is going on to understand protein folding and recognition of principles for protein design. There are two broad strategies for protein engineering: *rational design* and *directed evolution*. In future, more detailed knowledge of protein structure and function, and advancements technology may increase the capabilities of protein engineering. Protein engineering is the foremost among the much recent advancement in biotechnology. It reflects some of the exciting possibilities that come out of deeper understanding of living systems.

Protein-engineering and design refer to the possibility of actually changing the structure of a protein at one's desire so that it may function in a better way. The design of new protein molecules either from scratch or by making calculated variations on a known structure is called 'protein design. The use of rational protein design techniques is a major aspect of protein engineering. During 1990s, the design of minimalist computer models of proteins (lattice proteins), and the secondary structural modification of real proteins was started. Thereafter, the *de novo* design of real proteins became possible, and the 21st century has witnessed the creation of small proteins with real biological functions including chiroselective catalysis ion detection and antiviral behavior (Saghatelian *et al.*, 2001). It is hoped that the design of these and larger proteins will have great applications in medicine and bioengineering. Recent computational redesign was capable of experimentally switching the cofactor specificity of *Candida boidinii* xylose reductase from NADPH to NADH.

Based on protein engineering expertise, we can modulate and optimise challenging protein properties such as activity, affinity, selectivity, stability, resistance to proteolytic degradation, and even improve protein expression and solubility. In folded form protein stability is a balance between the stabilising interactions (*e.g.* mainly hydrophobic) and the tendency towards destabilising (which is caused by the loss of conformational entropy as the protein adopts the unfolded form).

A reduction in entropy-difference between folded and unfolded states would require a drastic reduction of the number of possible unfolded conformations. By substituting amino acids the stability of a protein can be changed. The substituted amino acids favour the stabilising interactions in a folded protein or destabilising interactions to some extent in a native protein. Many enzymes and proteins have been attempted to bring about changes so that improvement in their catalytic potency, thermal and pH stability, solvent tolerance, etc. could be made satisfactorily.

1. Differences between Protein Engineering and Protein Design

Protein engineering differs from protein design. In protein engineering an existing protein is changed chemically or genetically so that its function can be changed as per need. In protein design, the aim of the process is to fulfill a desired function through designing a protein. Protein engineering is a multidisciplinary science. The engineering and design flow chart requires the interaction of physicists, biophysicists, molecular biologists, organic chemists and computer scientists (Fig. 11.59).

2. Strategies for Protein Engineering

There are two broad strategies for protein engineering as given below:

***(a)* Rational Design of Proteins:** A detailed knowledge of the structure and function of protein is exploited by the scientist to make desired changes in to it. Since site-directed mutagenesis techniques are well-developed, it is technically easy and inexpensive. But the major drawback is the unavailability of detailed structural knowledge of a protein. However, it is extremely difficult to predict the effects of various mutations even on its availability.

The algorithms of computational protein design identify sequences of novel amino acids of low in energy when folded to the pre-specified target structure. The sequence-conformation space required to be searched is large. The most challenging requirement for computational protein design is a fast energy function that distinguishes optimal sequences from similar suboptimal ones.

***(b)* Directed Evolution:** Random mutagenesis is applied to a protein in directed evolution; a selection regime is used to pick out variants that have the desired qualities. Then further rounds of mutation and selection are applied. This method mimics natural evolution and often produces superior results to rational design. In addition, DNA shuffling technique mixes and matches pieces of successful variants in order to produce better results. This process mimics the naturally occurring recombination during sexual reproduction. Directed evolution there is no need of prior structural knowledge of a protein. Even it is not necessary to predict the effect of a given mutation. Often the results of directed evolution experiments are surprising. Often desired changes are caused by mutations that were not expected to have that effect. Directed evolution requires high-throughput which is not possible for all proteins. Large amounts of recombinant DNA must be mutated and the products must be screened for desired qualities. However, no all desired activities can be easily screened. Often expensive robotic equipment is required to automate the process by the sheer number of variants.

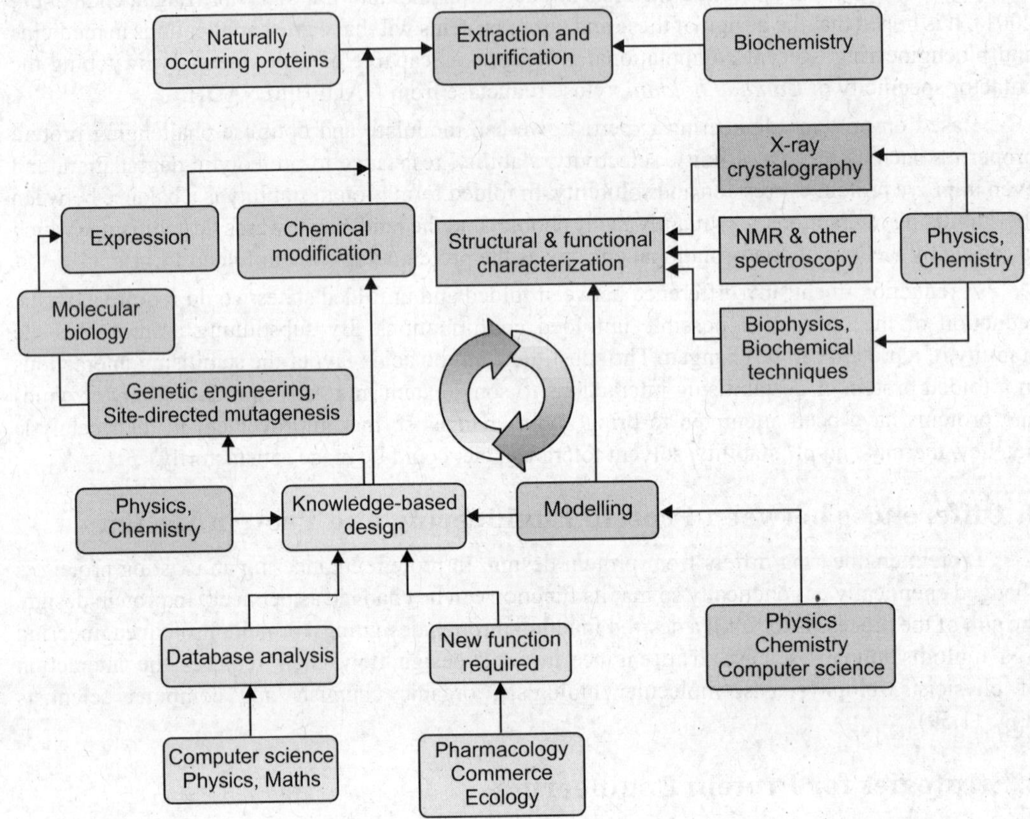

Fig. 11.59. Multidisciplinary approach in protein engineering and design cycle.

***(c)* Examples of Engineered Proteins:** In 2003, B. Kuhlman and co-workers designed a novel protein called Top7 having a novel globular fold with accuracy at atomic level. Similarly, L. Looger and co-workers in 2003 designed the receptor and sensor proteins for unnatural molecules with novel functions using the computational methods. Rilonacept (a pharmaceutical product) produced through engineering of fusion proteins has been approved by FDA (USA) for the treatment of cryopyrin-associated periodic syndrome.

Iterative Protein Redesign and Optimisation (IPRO) is another computational method which successfully engineered the switching of cofactor specificity of *Candida boidinii* xylose reductase. IPRO redesigns proteins to increase or give specificity to native or novel substrates and cofactors. This is done by repeatedly disturbing the backbones of the proteins around specified position of design, identifying the lowest energy combination of rotamers, and determining a lower binding energy by new design. By iterative nature of this process IPRO makes additive mutations to the protein sequences that collectively improve the specificity towards the desired substrates and/or cofactors.

3. Steps of Protein Engineering

The steps of protein-engineering are given in Fig. 11.60 and described below :

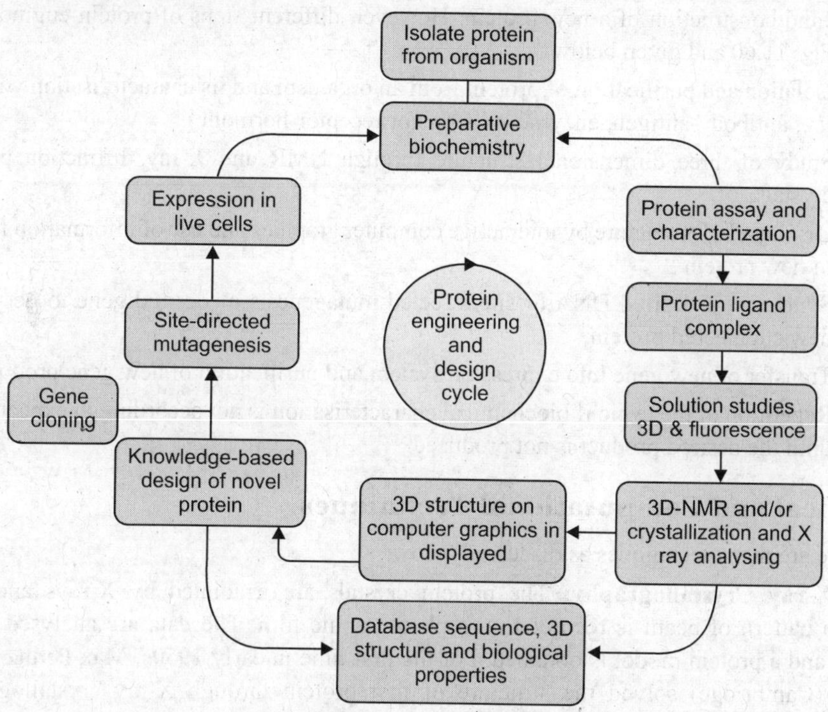

Fig. 11.60. Different steps of protein engineering.

***(i)* Identification:** Protein engineering starts with identification of a protein with a specific function. Depending upon its need and scope for improvement it can be changed accordingly following modern methods of molecular biology and biotechnology.

***(ii)* Isolation and characterisation:** The identified proteins are isolated and characterised biochemically. Then 3D structure and function of protein and relationship between structure and function are found out.

***(iii)* Protein modification:** Based on the above facts and known principles of protein conformation, modification in protein can be suggested which may bring about desired changes.

***(iv)* Incorporation of changes:** Changes are incorporated into the protein by side-directed mutagenesis or molecular techniques. The activity of a new protein must be tested. Protein design starts from knowledge of principles of protein structure. A sequence of amino acid is designed applying these principles with the objectives that it will take up a specific 3D structure and cause the protein for necessary function. Appropriate DNA is synthesised as per amino acid of choice. Then it is cloned into expression vector system by recombinant DNA technology. The expressed protein will undergo tests and next stage of the design process.

The properties possibly required for protein engineering are: (*i*) optimum pH, (*ii*) thermostability, (*iii*) substrate specificity, (*iv*) co-factor requirements, (*v*) protease resistance, (*vi*) kinetic properties and Michaelis constant (Km), (*vii*) allosteric regulation, and (*viii*) molecular weight and structure of subunits.

I. METHODS IN PROTEIN ENGINEERING AND DESIGN

Combined effort of scientists and technologists from many different disciplines are required for design and construction of a new protein. However, different steps of protein-engineering are shown in Fig. 11.60 and given below :

(*i*) Isolation and purification of protein from an organism and its characterisation with ligand (*i.e.* antibody-antigen, enzyme-substrate or receptor-hormone),

(*ii*) Study of three dimensional structure through NMR and X-ray diffraction pattern of crystals,

(*iii*) Display of 3D structure by interactive computer graphics and use of information for design of new protein

(*iv*) Synthesis of desired DNA or site-directed mutagenesis of desired gene to get expected new engineered protein,

(*v*) Transfer of new gene into expression system and purification of new gene product, and

(*vi*) Repetition of the cycle if biochemical characterisation is not according to expected result, until the desired product is not produced.

1. Physical and Computational Techniques

There are many techniques as discussed below:

(*a*) X-ray Crystallography: The protein crystals are irradiated by X-rays and the 3D diffraction pattern of beam is recorded on a photographic film. The data are analysed by using computer and a protein model is obtained. For the first time in early 1950s, Max Perutz and John Kendrew (Cambridge) solved the structure of first protein through X-ray crystallography as described earlier.

(*b*) Nuclear Magnetic Resonance (NMR): The NMR is a technique of studying molecular structure placing the sample in a strong magnetic field. Then it is irradiated by the electromagnetic waves. Each nucleus of the atoms of protein molecule absorbs energy from the electromagnetic waves. Absorption of this energy can be predicted for every nucleus types. For example hydrogen or proton nucleus of H_2 atom would absorb energy different from that of -NH_2 atoms. NMR is very useful in searching out the dynamics of protein structure and understanding the changes in their shape with changing conditions of pH, salt, temperature, etc. Thus after bombarding the protein, the 3D diffraction of waves is measured. Based on this information protein structure is determined as described earlier in this chapter.

(*c*) The Spectroscopic Technique: The UV and spectroscopy can be used when required.

(*d*) Prediction of Protein Structure: Proteins with homologous amino acid sequence are predicted to have similar three dimensional structures. The method of P.Y. Chou and G.D. Fasman (U.S.A.) is most frequently used. This method is based on a statistical analysis of the protein structure database.

The number of occurrences of an amino acid in the α-helix, β-pleated sheet and loop conformations was used to calculate the statistical parameters P(α), P(β) and P(turn) for each residue as a measure of its conformational preference. Specific position of amino acids in the turn is also taken into account. For a given sequence a computer programme is available which prints a list of values of average probability to find a particular residue in a particular secondary structural motifs (*i.e.* super secondary structure of folds).

(*e*) Computer Modelling: There are several computers which use many software packages to implement the prediction of above structure. The Evans and Sutherland range of graphics work stations and the Silicon Graphics range of machines are commonly used by the computers. Generally PCs include the popular software like DTMM (Desk Top Molecular Modeller) and larger machines the FRODO and BioSym. Besides, computer simulation is applied to study the dynamic motion of proteins. However, for the prediction of conformational re-arrangement occurring during protein activity, molecular dynamics algorithms are used. Most common programme packages such as AMBER, CHARMm and X-PLOR are used for computer modelling of proteins (Fig. 11.61).

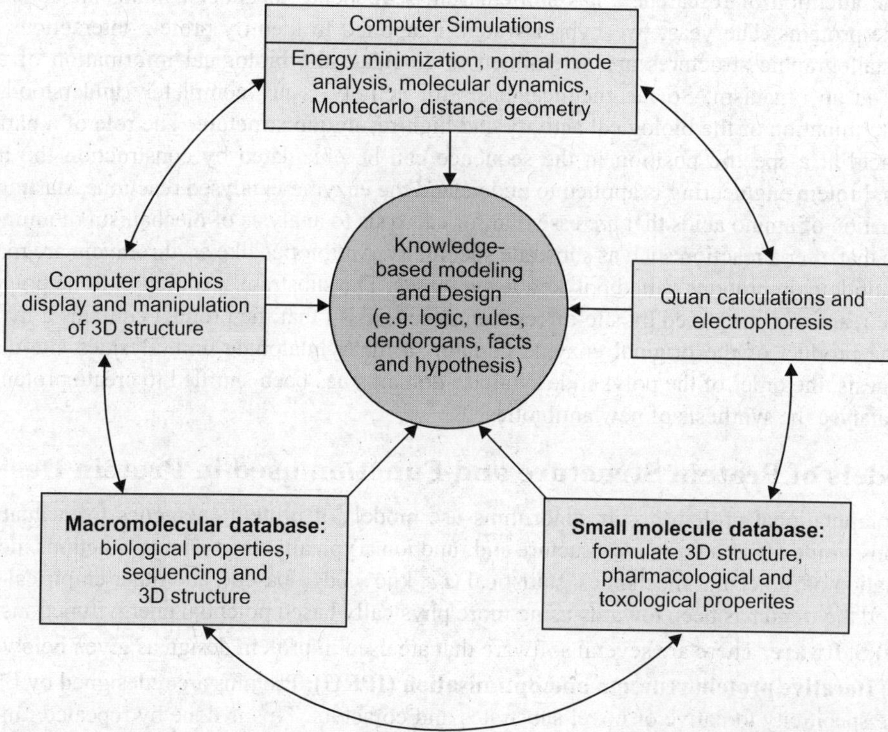

Fig. 11.61. Knowledge-based molecular modelling that involves different components of computer software.

3. Biochemical Techniques and Molecular Biology

(*a*) Site-directed Mutagenesis: By using this technique specific nucleotide sequence of a gene is changed so that the desired protein of altered amino acid should be obtained. When the information of a new DNA sequence coding for a protein is available, the new sequence can also be synthesised chemically and amplified by using PCR (polymerase chain reaction) technique. The new gene is incorporated into an expression system and altered protein is harvested in desired quantity. Method of site-directed mutagenesis is described in detail in Chapter 5 : *Biophysical methods*.

(*b*) Recombinant DNA Techniques: It involves: (*i*) introduction of a fragment of new DNA into vector, (*ii*) transfer of recombinant vector into suitable cells, (*iii*) amplification, and (*iv*) expression of recombinant gene, and getting desired protein of altered amino acid. The methods of recombinant DNA technology are described in Chapter 8 : *Methods of Gene Manipulation*.

J. STRUCTURE-FUNCTION RELATIONSHIPS

There are 20 amino acids which vary in size, shape, charge, hydrogen bonding and chemical reactivity. One amino acid joins to another through peptide bonds. All the amino acids are α-amino acids. The number of amino acids differs in different proteins. The polypeptide chain attains various shapes *e.g.* primary, secondary, tertiary and quaternary structures. Changes in 3D shapes arise due to changes in protein conformation.

The amino acid sequence of a particular protein from many different organisms is now available after sequencing 180 genomes including human. This advancement has created a situation where protein sequences are available without any information of biological activity of the protein. Now, the attention of researchers has shifted from sequencing to understanding the function of all of the proteins. The yeast two-hybrid system is applied to identify protein interactions. Then the crystallographic structures are determined to elucidate the biological information of all the proteins in an organism. So far chemical basis for activity is not completely understood, even after determination of the biological activity and high-resolution structure. The role of a particular amino acid at a specific position in the sequence can be elucidated by construction the mutant proteins. Protein engineering is applied to understand the enzyme-catalysed reactions, starting from identification of amino acids that are essential for catalysis to analysis of mechanism of amino acid changes that alter a reaction such as substrate specificity. Antibiotics like erythromycin are made by large multidomain proteins called polyketide synthases. The substrate specificity of one polyketide synthase reaction is modified by site-directed mutagenesis so that the product contains a malonate unit. The product of the original enzyme contains a methylmalonate unit. Besides site-directed mutagenesis, the order of the polyketide synthase domains has been shuffled to create proteins that could catalyse the synthesis of new antibiotics.

1. Models of Protein Structure and Function used in Protein Design

Computational protein design algorithms use models of protein energetics to evaluate how mutations would affect a protein's structure and function. Typically these energy functions include a combination of molecular mechanics, statistical (*i.e.* knowledge-based), and other empirical terms. However, the trend has been towards using more physically based potential energy functions.

***(a)* Software:** There are several software that ate used in protein design as given below:

***(i)* Iterative protein redesign and optimisation (IPRO):** Proteins are redesigned by IPRO to increase specificity to native or novel substrates and cofactors. This is done by repeated randomly disturbing the backbone of proteins around the positions of specified design, identifying the lowest energy combination of rotamers, and determining a lower binding energy as compared to previous ones. The iterative nature of this process allows IPRO to make additive mutations to the protein sequence that collectively improves the specificity towards the desired substrates and/or cofactors. Experimental testing of predictions by IPRO successfully switched the cofactor preference of *Candida boidinii* xylose reductase from NADPH to NADH.

***(ii)* EGAD- a genetic algorithm for protein design**: A free and open-source software package is available for protein design and prediction of effects of mutation on protein folding stabilities and binding affinities. EGAD can also consider multiple structures simultaneously for designing specific binding proteins or locking proteins into specific conformational states. EGAD can also consider free-moving ligands with or without rotatable bonds. One can use EGAD with single or multiple processors.

***(iii)* RosettaDesign**. A software package RosettaDesign is available under active development and free for academic use which is extensively successful. RosettaDesign is accessible via a web server (http://rosettadesign.med.unc.edu).

***(iv)* SHARPEN**. SHARPEN is a permissive open-source library for protein design and structure prediction. It offers many combinatorial optimisation methods (*e.g.* Monte Carlo, Simulated Annealing, FASTER) and can score proteins using the successful Rosetta all-atom force field or molecular mechanics force fields (OPLSaa). Besides protein modelling library, SHARPEN includes tools for scalable distributed computing.

K. SOME ACHIEVEMENTS IN PROTEIN-ENGINEERING AND DESIGN

There are several proteins where engineering has been tried. For example serine proteases which are present in virtually all organisms functioning both inside and outside the cell. Serine proteases (protein-digesting protein) exist as two families: *trypsin-like* and the *subtilisin-like*. They are called the serine proteases for two reasons: they hydrolyse proteins, and have an essential Ser residue at the active site which is critical for catalysis. In fact this Ser is much more reactive than the other serines in the protein.

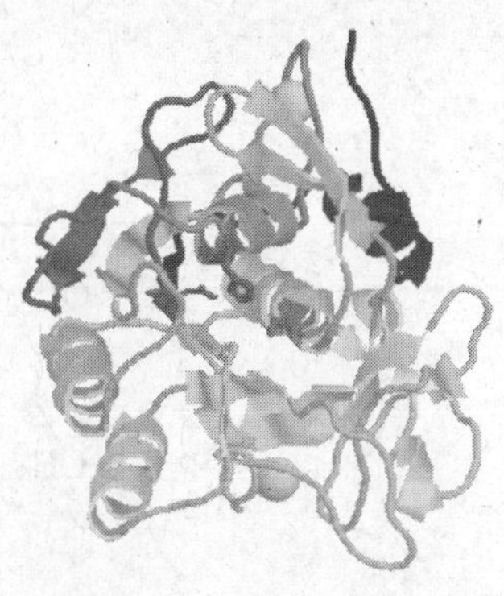

The first crystal structure of serine protease subtilisin from *Bacillus amyloliquefaciens*.

All the mammalian serine proteases have a common 3D structure, besides significant differences in some regions of the surface. Hence they show different physiological functions and, therefore, need for different interactions with different molecules. All of them have the same catalytic mechanism. They have evolutionary relationships as they have the sequence and structural homology. Chymotrypsin, trypsin and elastase have great similarities which show that these proteins evolved through gene duplications of an ancestral serine protease followed by the divergent evolution of the resulting enzymes.

Subtilisin was originally isolated from *Bacillus subtilis*; it is a serine protease of which primary and tertiary structures bear no discernible relationships to those of chymotrypsin. The active site groups of these two enzymes are identical and their relative 3D positions cannot be distinguished. Chymotrypsin and subtilisin are a remarkable example of convergent evolution. In the amino acid sequences of the two enzymes the orders of the corresponding active site residues are quite different (Ser-195, His-57, Asp-102 in chymotrypsin, trypsin and elastase, Ser-221, His-64, Asp-32 in the bacterial subtilisin).

1. How does Serine Protease (Subtilisin) Work ?

Subtilisin uses the same three catalytic residues such as Asp32, His64, and Ser221. The mechanism of catalysis is the same including the positioning of substrate by hydrogen bonding. This shows convergent evolution of serine protease. However, the catalytic process of the proteases is carried out by the catalytic triad: Asp102, His57, and Ser195 (Fig. 11.62).

His57
H–N N····H–O–CH_2 – Ser195
O–C(=O)–CH_2 Asp102

Fig. 11.62. Catalytic triad of Asp102, His57, and Ser195.

All the serine proteases employ this mechanism with some small variations. It can be started with the binding in cartoon form.

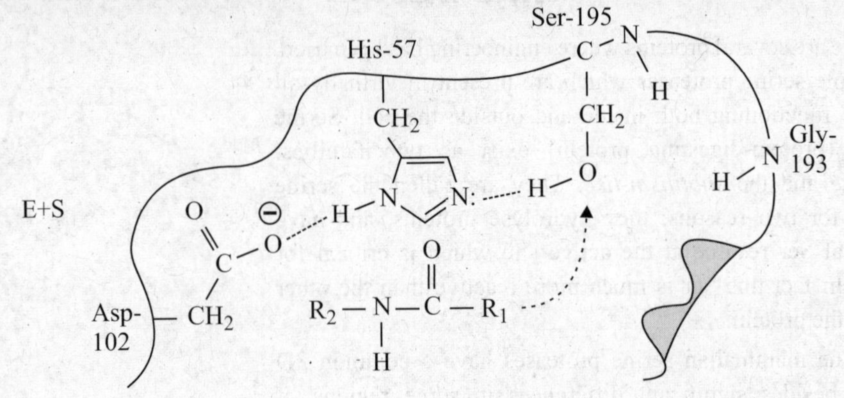

In the next step His57 removes a proton from the Ser105 OH, while the O does a nucleophilic attack on the peptide carbonyl:

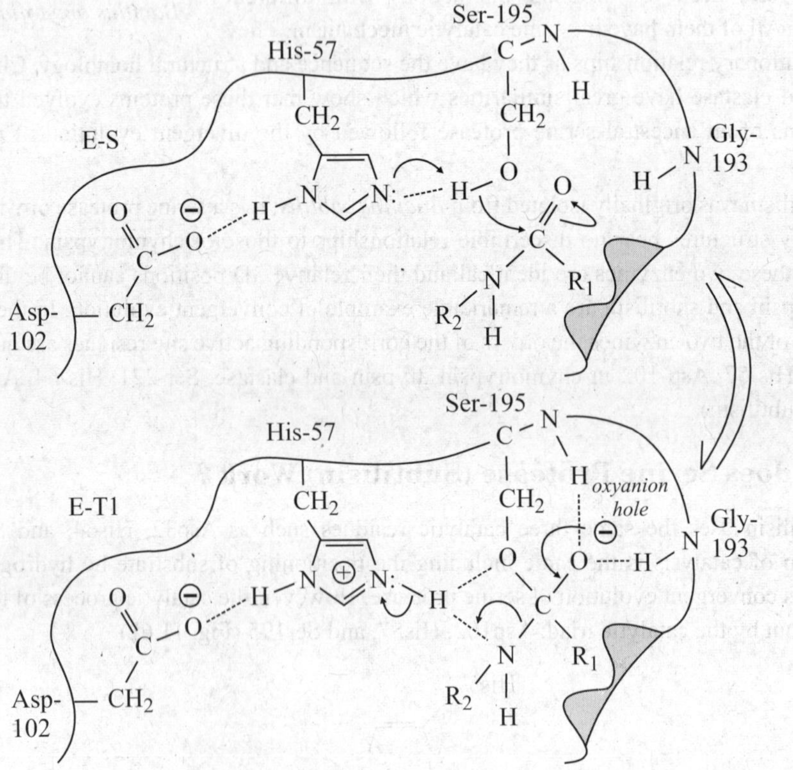

The shift of negative charge onto the carbonyl oxygen of the peptide is facilitated by hydrogen bonding of the oxygen to the backbone -NH groups of Ser195 and Gly193; these form the 'oxyanion hole'. This results in formation of a structure equivalent to the tetrahedral intermediate in the non-enzymatic mechanism. This differs in the sense that the two protons that are part of the -OH groups are merely hydrogen-bonded in the enzyme. A tetrahedral intermediate form was a point of considerable contention in the early investigations of protease mechanisms; biochemists tended to draw nucleophilic acyl substitutions as if they were S_N2 reactions. This results in tetrahedral intermediate that binds to Ser195 with the former carbonyl oxygen in the oxyanion hole.

Ser-195
His-57
Acyl E
+
P1
CH_2
CH_2
Gly-193
Asp-102
CH_2
$R_2 - N$
R_1
Amine product (P_1)

Ser-195
His-57
CH_2
CH_2
Gly-193
Asp-102
CH_2
R_1

Thereafter, the carbonyl double bond is reconstructed in the next step with expulsion of the leaving group. At this stage the protein chain is actually cleaved, producing an 'acyl enzyme': the acyl part of the peptide that was cleaved and bound as an ester to Ser195.

Now the acyl enzyme is cleaved but it is not permanently changed in the reaction. A molecule of water is required for cleaving mechanism. The other end of the original protein is released after restoring the carbonyl double bond; consequently, the catalytic triad is restored to its beginning state.

Ser-195
His-57
CH_2
CH_2
oxyanion hole
Gly-193
Asp-102
CH_2
R_1

CH_2
CH_2
Gly-193
R_1

Upon dissociation of the second protein fragment now the enzyme subtilisin is ready to re-enter in the event of action.

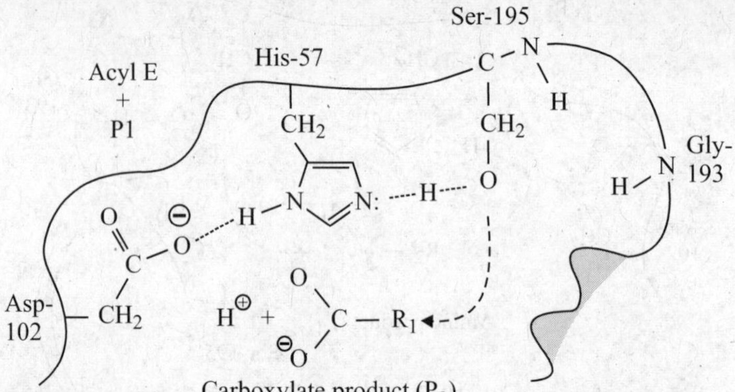

Carboxylate product (P_2)

2. Protein Engineering in Subtilisin- Improvement in a Biological Laundry Detergent

Subtilisin is an alkaline protease of 27.5 kilo Dalton which was originally isolated from a bacterium *Bacillus subtilis*. Bacterial protease subtilisin has a broad specificity for proteins commonly soil clothing. Detergent manufacturers added subtilisin (as advertised with slogans) in 95% of washing to improve the efficiency of laundry detergents as a stain-remover. The protein stains are removed more effectively at lower temperature than are usually required for laundry processing. The ideal requirements for such enzymes are: (*i*) stability up to 70°C and within the pH range of 8-11, (*ii*) resistance to non-ionic detergents and oxidising reagents such as H_2O_2, and (*iii*) the absence of requirement of metal ions. Substilisin acts upon stains by catalytic triad (like chymotrypsin), *i.e.* Asp32, His64 and Ser221. Significantly, enzyme activity is lost if all the three amino acids are replaced either singly or in combinations. By adding bleach about 90% activity of substilisin molecule is lost due to oxidation of an amino acid methionine present at position 222 (Met^{222}) of the polypeptide chain.

Therefore, by using site-directed mutagenesis of the subtilisin gene in *E. coli*, this methionine was changed to many other amino acids, and the subtilisin activity and bleach resistance of the mutated protein was tested. The Ala222 substituted enzyme activity of engineered subtilisin was measured in the presence of bleach (Table 11.5). The alanine222 substituted enzyme showed 53% activity as compared with wild type one. It was found that substitution of methionine222 by alanine222 was the best in terms of activity and stability. At present cloned genetically engineered subtilisin is used in laundry detergent by many detergent industries.

Table 11.5. Site-directed mutagenesis at codon position 222 of subtilisin.

Codon 22	*Abbreviated name*	*% activity wrt wild type*	*Codon 22*	*Abbreviated name*	*% activity wrt wild type*
Cysteine	Cys	138	Glutamine	Gln	7.2
Methionine	Met	100	Phenylalanine	Phe	4.9
Alanine	Ala	53	Tryptophan	Trp	4.8
Serine	Ser	35	Aspartate	Asp	4.1
Glycine	Gly	30	Tyrosine	Tyr	4.0
Threonine	Thr	28	Histidine	His	4.0
Asparagine	Asn	15	Glutamate	Glu	3.6
Proline	Pro	13	Isoleucine	Ile	2.2
Leucine	Leu	12	Arginine	Arg	0.5
Valine	Val	9.3	Lysine	Lys	0.3

The methionine was substituted by either alanine or cysteine through site-directed mutagenesis in subtilisin gene (Fig. 11.63).

The National Library of Medicine maintains a database of enzymes found in household products. However, this database does not indicate the types of enzymes, whether 'wild type' or 'genetically improved version', is used in these products.

Commercial Subtilisin

Subtilisin is an alkaline protease originally produced by *Bacillus subtilis*. It has the largest industrial market as compared to other enzymes. It is supplemented with detergent by manufacturers with catchy slogans on detergent cakes and powders.

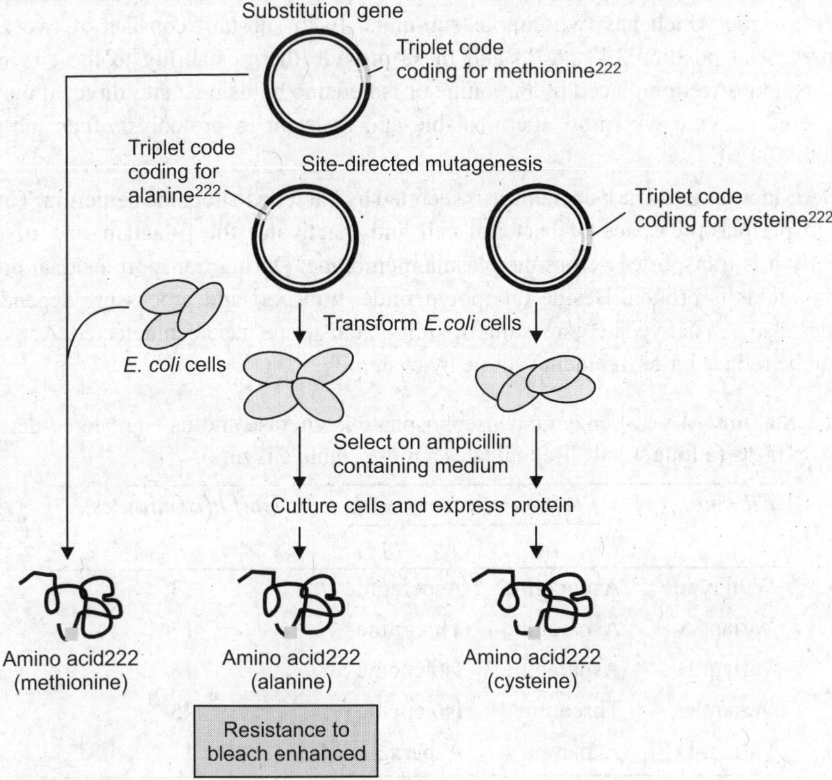

Fig.11.63. Site-directed mutagenesis used to alter the enzyme subtilisin.

3. Recombinant Vaccines

Earlier vaccines were prepared by inactivating the bacteria/viruses or their surface proteins. Such bacterial or viral preparations have been used to generate immunity against specific bacterial or viral diseases. It has been noted that patients often develop fever who were administered the vaccines. It is true that proteins stimulate immune system and cause to secrete specific antibodies. Such specific amino acid sequences in the protein that stimulate immune response are called *epitopes*. Based on selected epitopes recombinant vaccines may be produced on commercial level which can prove more effective and safer than the conventional vaccines.

Working on these lines, a recombinant Hepatitis B vaccine was produced by cloning the synthetic gene (for the surface antigen of the virus) in yeast cells. This gene expressed well in yeast cells and produced 22 nm particles of hepatitis B virus (HBV) surface antigen (as produced in patients) infected with hepatitis B virus. The recombinant vaccine has high immunogenecity. This product has been marketed as a vaccine for protection against HBV infection. Similarly, recombinant vaccine for foot and mouth disease (FMD) virus has been prepared.

4. Protein Engineering in Enzymes

There are many enzymes where protein-engineering has been done. Two enzymes viz., triosephosphate isomerase and β-lactamase have been discussed.

(*a*) Triosephosphate Isomerase: Proteins exposed at high temperature releases ammonia due to deamination of asparagine and glutamine residues. Hence, enzyme activity is lost due to localised changes in folding of protein. For example, triosephosphate isomerase is a yeast (*Saccharomyces cerevisiae*) enzyme which has two similar sub-units. Each sub-unit consists of two asparagine residues present at position 14 and 78 and these provide thermostability to the enzyme. These asparagine residues were replaced by threonine or isoleucine by using a site-directed mutagenesis. The engineered enzyme was more thermostable and resistant to proteolysis than the wild type enzyme (Table 11.6).

(*b*) Beta-lactamase: The β-lactamase is secreted by bacteria to degrade penicillin. This enzyme functions in periplasmic space of bacterial cell and inactivates the β-lactam ring of penicillin. Consequently it is transported across the plasma membrane. During transport a signal polypeptide of 23 amino acids is broken. Beside this polypeptide, transport and processing depend on other polypeptides also. An active site consisting of amino acid serine has been detected. Activity of this enzyme can be reduced after replacing serine by cysteine.

Table 11.6. Stability of yeast enzyme triosephosphate isomerase and its engineered derivatives at 100°C (a longer half life indicates a more stable enzyme).

Enzyme	***Position of amino acids***		***Half life (minutes)***
	14	**78**	
Wild type	Asparagine	Asperagine	13
Variant A	Asparagine	Threonine	17
Variant B	Asparagine	Isoleucine	16
Variant C	Threonine	Isoleucine	**25**
Variant D	Aspartate	Asparagine	11

4. Protein Design

Only a subset of enormous number of possible amino acid sequences will fold reliably and quickly to a single native state. Protein design involves the identification within this subset of novel sequences which is physiologically active in native state. The native state of a protein is the conformational free energy minimum required for the chain. Hence, protein design refers to the search for sequences which have the chosen structure as a free energy minimum. It is the reverse of structure prediction. A tertiary structure is specified and a sequence is identified which will fold to it. Hence it is also referred to as 'inverse folding'.

Similar to mad cow disease prion diseases are helpful examples of designer proteins that possess only one possible stable conformation. However, there is a healthy protein with a fatal weakness in 'mad cow disease'; this protein can comfortably take another conformation. The abnormally folded shape is very stable because it has very little free energy. But, mis-folded prion protein catalyses the other proteins to adopt the mis-folded prion shape resulting in a disease-generating cascade of previously functional proteins that quickly becomes mis-folded. In the new conformation they lose the ability to perform their function; even they form aggregates which are called plaques. Production of these aggregates in the brain results in progressive neuronal death with the consequences of death of the entire organism. Therefore, it is easy to see that a designer protein has only one possible stable tertiary structure, and this remains the case in all environments. Protein design requires an understanding of the molecular interactions that stabilise proteins in specific folded configurations; it does not require an understanding of the dynamical process by which proteins fold.

By using computer models protein design can be accomplished. When folded to the target structure, the computational protein design algorithms search the low energy sequence-conformation space for sequences. This search space is large. Presently the most challenging requirement for computational protein design is an accurate and fast energy function that can distinguish optimal sequences from similar suboptimal ones. A protein with a novel fold as described by Kuhlman and co-worker in 2003, as well as sensors for unnatural molecules as described by Looger and co-workers in 2003 has been designed by using the computational methods. Besides, it is well known that not all possible protein structures are 'designable' *i.e.* there are compact configurations of the chain where no sequences can fold. Particularly the conformations having poor secondary structures are unlikely to be designable.

In order to increase the solubility, stability or ability to crystallise of a given protein, we can use a protein compaction technology based on the deletion of the protein's non-essential floppy loops. The resulting functional mini-protein will display a better therapeutic value (therapeutic proteins, antigens) or a superior ability to crystallise (target proteins).This approach has been successfully applied to both viral antigens for vaccines development and anti-bacterial targets for drug discovery applications. The main objective of protein-engineering is to design a chain of amino acids which may fold to attain the desired structure so that desired function can be done. On the basis of known protein sequences and structure it is easy to design a chain of amino acids for specific structure and function than to predict the specific sequence. Considering these facts a globular water soluble protein without any specific function was designed. It attained definite secondary and tertiary structure. This protein consists of four α-helices which fold to form a compact bundle. This motif is common in proteins. Hydrophobic amino acids are to the interior and hydrophilic amino acids to the exterior. Four 16-residue long sequence of the four helix region was as below:

Gly-Glu-Leu-Glu-Glu-Leu-Leu-Lys-Lys-Leu-Lys-Glu-Leu-Leu-Lys-Gly

This four 16-residue helix sequence was connected by three short loop regions having amino acids: Pro-Arg-Arg-. The corresponding DNA sequence was synthesised and incorporated into *E. coli* by following recombinant DNA technology. It expressed in expression vector system. The second example is the design of an eye lens protein *Crystanova*. Gene encoding this protein was synthesised and incorporated into *E. coli* which expressed crystanova.

5. Improvement in Nutritional Quality of Seeds

The major source of protein of our diet is the seeds and cereal grains. During the development of seeds, the 'seed storage protein's are synthesised and accumulated. When such seeds germinate or being conserved by humans its reserved proteins act as source of amino acids. However, if some amino acids lack in these seeds or grains, these cannot be used as balanced diet. Therefore, those essential amino acids are supplemented to the diets procuring from the other sources.

Out of the proteins found in nature, all are not consumable because some are toxic or allergenic in nature. Therefore, such proteins should be consumed which are nutritious and safe. The highly nutritious food materials are easily contaminated by the microorganisms. During growth and multiplication the microorganisms secrete toxins like mycotoxins by fungi (such as aflatoxins, ochratoxin, rubratoxin ergot, etc.) These toxins cause several health hazards in animals and human too.

However, there are several measures to estimate the nutritional value of food protein which meets the nutritional requirement of an individual. These parameters of protein value are essential amino acid profile (EAP), biological value (BV) or net protein efficiency ratio (PER).

(*a*) **Essential Amino Acid Profile:** The amino acids which are taken up from external food are called essential amino acids. Our body cannot synthesise them. Amount of branched chain amino acids (BCAAs), *i.e.* Ile, Leu, Tys, Thr and Val is highest in whey as compared to the others. BCAAs need to be present in muscle cells so that protein synthesis should be promoted. In the presence of BCAA, bioavailability of complex carbohydrates is increased. When a person performs exercises, BCAAs are released from skeletal muscle. Its carbon part is utilised for energy production and nitrogen part is converted into alanine. Alanine is transported into liver and takes part in energy production.

Therefore, BCAA are the good source of energy for athletes during, before and after exercises and protecting their biomass from breakdown of muscles.

QUESTIONS

1. Write different types of chemical bonds formed in molecules and macromolecules.

2. Describe chemical structure of a protein.

3. Classify different types of amino acids studied by you.

4. Describe different properties of amino acids.

5. Describe in detail about conformational analysis of protein.

6. Give a detailed account of post-transcriptional modification of proteins.

7. With the help of suitable diagram describes secondary structure of proteins.

8. What is protein engineering ? Write different methods of protein engineering.

9. What is X ray crystallography ? Describe the steps of X ray crystallography of a protein crystal.

10. What are the methods of X ray crystallography ? Describe with the help of suitable diagrams.

11. What do you know about NMR ? Write its theory and working principles.

12. What is mass spectrometry ? Write its working principle and application.

13. Discuss in detail about working principle and application of MALDI.

14. Give a detailed account of protein-DNA interaction.

15. Describe structure-function relationship of proteins.

16. Write short notes on the following :

(*i*) Different types of bonds,
(*ii*) Amino acids,
(*iii*) Zwitterions,
(*iv*) Amino acid polymerization,
(*v*) Isoelectric point,
(*vi*) Primary structure of proteins,
(*vii*) Alpha helix of protein,
(*viii*) Planarity of polypeptide bond,
(*ix*) Ramchandran plot of protein,
(*x*) Beta-pleated sheets,
(*xi*) X ray diffraction analysis,
(*xii*) NMR and its application,
(*xiii*) MALDI-TOF,
(*xiv*) SELDI,
(*xv*) Electrospray ionization (ESI),
(*xvi*) G.N. Ramchandran,
(*xvii*) desorption electrospray ionization (DESI),
(*xviii*) Helix-turn-helix domain,
(*xix*) Zinc finger domain,
(*xx*) Leucine zipper domain.

BIBLIOGRAPHY

Pabo C.O. Peisach E. and Grant R.A. 2001. Design and Selection of Novel Cys2His2 Zinc Finger Protein. *Annu. Rev. Biochem.* **70**: 313–40.

Compton A. 1923. A quantum theory of the scattering of X-rays by light elements. *Phys. Rev.* **21**: 483.

Creighton T.H. 1993. In: *Proteins- structures and molecular properties*. San Francisco: W. H. Freeman.

Glick D.M. 1997. Leucine scissors, Ed. *Glossary of Biochemistry and Molecular Biology*. London: Portland Press.

Wuthrich K. 1986. *NMR of Proteins and Nucleic Acids*. Wiley- Interscience USA: New York (NY).

Lilley D.M. J. 1995. *DNA-protein: structural interactions*. Oxford: IRL Press at Oxford University Press.

Maton A, Hopkins J, McLaughlin C.W., Johnson S., Warner M.Q., LaHart D. and Wright J.D. 1993. *Human Biology and Health*. Englewood Cliffs, Michigan, USA: Prentice Hall.

Haner R.L. and Keifer P.A. 2009. *Encyclopedia of Magnetic Resonance*. John Wiley.

Saghatelian A., Yokobayashi Y., Soltani K. and Reza G.M. 2001. A chiroselective peptide replicator. *Nature* 409: 797.

Voet D. and Voet J.. 2004. *Biochemistry* (3rd ed.). Hoboken, NJ: Wiley. p. 227–231.

Wuthrich K. 2002. NMR studies of structure and function of biological macromolecules. *Nobel Lecture,* December 8, 2002.

Chapter 12 Bioinformatics

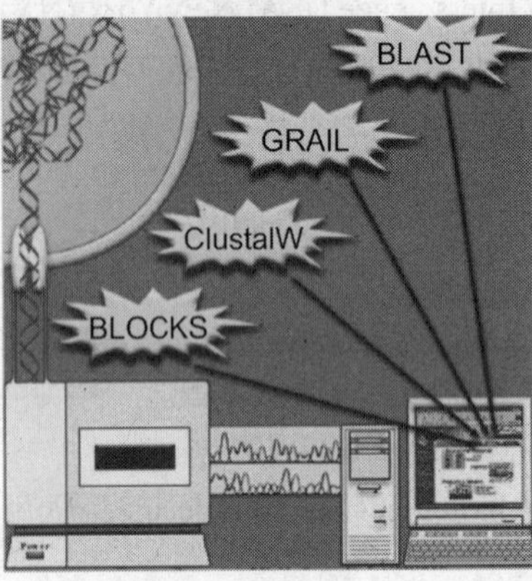

A. Historical background
B. What is bioinformatics ?
C. Databases
D. Sequences and nomenclature
E. Information sources
F. Data retrieval tools
G. Data base similarity searching
H. Resources for gene level sequences
I. Sequence alignment
J. Uses of bioinformatics tools in analysis

In 1943, Paulien Hogeweg (a Dutch theoretical biologist and complex systems researcher) studied biological systems as dynamic information processing systems at many interconnected levels. In 1978, Paulien Hogeweg and Ben Hesper coined the term *bioinformatics* for the study of informatic processes in biological systems. Now the term bioinformatics has become very popular. In her research she stands out as a pioneer to develop new concepts and computational approaches in various areas of biology. Thereafter, the use of bioinformatics has been in genomics involving large-scale DNA sequencing and genetics. A tremendous amount of information related to molecular biology has been produced over the past few decades due to rapid developments in genomic technologies and developments in information technologies.

The end of 20th century witnessed that the laboratory-based biology can be used as a tool for earning money as well as storing, analysing and making the use of all genetic information for further uses in multifarious areas. In future biotechnology will not have to spend much time for cloning genes because they would be able to generate new questions and hypotheses from computer analysis of genome data. Their hypotheses can be tested in laboratory.

Paulien Hogeweg

In 1987, when the Human Genome Project was conceived of, the field of bioinformatics was in its infancy. Today, bioinformatics has become a recognised discipline on its own, born out of the necessity to bring together the information sciences and the biological sciences in understanding the wealth of data that has been created through various projects around the world. The 3 billion characters in the DNA sequence that make up the human genome were translated into biologically meaningful information by using computer. This act gave birth to a new field called 'bioinformatics'. The field of genomics relies upon bioinformatics (*see* Chapter 13: *Genomics and Proteomics*). It holds the key to unlock these data for the next generation of innovations.

In the 21st century biology is being transformed from a purely laboratory-based science to an information science too. The information refers to comprehensive views of DNA sequences, RNA expression and protein interactions. Due to explosion of sequence and structure information available to researchers, people have become optimistic to get answer of fundamental biomedical problems. Similar progress has also been made in computer-based technology also.

A. HISTORICAL BACKGROUND

Historically, the protein databases were prepared first, then nucleotide databases. In 1959, V.M. Ingram first made attempt to compare sickle cell haemoglobin and normal haemoglobin, and demonstrated their homology. In due course of time the other proteins associated with similar biological function were also compared. This resulted in more protein sequencing and accumulation of vast information. Hence, it is realised to have databases so that using computation software the proteins can be quickly compared.

In 1962, using sequence variability, Zuckerkandl and Pauling proposed a new strategy to study evolutionary relations between the organisms which are called 'molecular evolution'. This theory was based on the facts that similarity exists among the functionally related (homologous) protein sequences. Margaret O. Dayhoff found that during evolution protein sequences undergo changes according to certain patterns such as: (*i*) preferential alteration (replacement) in amino acids with amino acids of similar physico-chemical characteristics (but not randomly), (*ii*) no replacement of some amino acids (*e.g.* tryptophan) by any other amino acids, and (*iii*) development of a point accepted mutation (PAM) on the basis of several homologous sequences.

Further work on sequence comparison on the basis of quantitative strategy was carried out. In 1965, Dayhoff and co-workers collected all the protein sequences known at that time and catalogued them as the *Atlas of Protein Sequence and Structure* which was first published by the National Biomedical Research Foundation (Silver Sring MD). Later on collection of such macromolecular sequences was published under the above title from 1965 to 1978. The above printed book laid the foundation for the resources that the entire biotechnology community now depends for day-to-day work in computational biology. The development of computer methods pioneered by Dayhoff and her research group is applicable: (*i*) in comparing protein sequences, (*ii*) detecting distantly related sequences and duplication within sequences, and (*iii*) deducing the evolutionary histories from alignment of protein sequences.

Dayhoff went on to pioneer the development of programmable computer methods for use in comparing protein sequences and deriving their evolutionary histories (in other words, discerning homologies) from their sequence alignments. Though this was before the days of massive outputs of sequence information by automated and other methods, Margaret Dayhoff anticipated the potential of computers to the current theories of Zuckerkandl and Pauling and the method which Sanger had engineered. David J. Lipman has called Dayhoff 'the mother and father' of bioinformatics.

Margaret O. Dayhoff

D.J. Lipman, who is director of the National Center for Biotechnology Information is also the scientist who spearheaded the collaborative project that produced BLAST. His ongoing work in developing better computational methods for molecular biology attests to his inheritance of Dayhoff's legacy.

In 1980, the advent of the DNA sequence database led to the next phase in database sequence information through establishment of a data library by the European Molecular Biology Laboratory (EMBL). The purpose of establishing data library was to collect, organise and distribute data on nucleotide sequence and other information related to them. The European Bioinformatics Institute (EBI) is its successor that is situated at Hinxton, Cambridge, United Kingdom.

In 1984, the National Biomedical Research Foundation (NBRF) established the protein information resource (PIR). The NBRF helps the scientists in identifying and interpreting the information of protein sequences.

In 1988, the National Institute of Health (NIH), U.S.A. developed the National Centre for Biotechnology Information (NCBI) as a division of the National Library of Medicine (NLM) to develop information system in molecular biology. The DNA Databank of Japan (DDBJ) at Mishima joined the data collecting collaboration a few years later.

The NCBI built the GenBank, the National Institute of Health (NIH) genetic sequence database. GenBank is an annotated collection of all publically available nucleotide and protein sequences. The record within GenBank represents single *contig* (contiguous) selection of DNA or RNA with annotations. In 1988, the three partners (DDBJ, EMBL and GenBank) of the International Nucleotide Sequence Database Collaboration had a meeting and agreed to use a common format. All the three centres provide separate points of data submission, yet exchange this information daily making the same database available at large. All the three centres are collecting, direct submitting and distributing them so that each centre has copies of all the sequences. Hence, they can act as a primary distribution centre for these sequences. Moreover, all the databases have collaboration with each other. They regularly exchange their data.

Now sequence data are accumulating day-by-day. Therefore, there is a need of powerful software so that sequences can be analysed. For the development of algorithms [any sequence of actions (*e.g.* computational steps) that perform a particular task] firm basis of mathematics is needed. Now, mathematicians, biologists and computer scientists are taking much interest in bioinformatics. Moreover, biologists are curious to ask reservoir of all such information because they are widely interconnected through network.

Thus bioinformatics is aimed at: (*i*) the development of powerful software for data analysis, and (*ii*) benefit the researchers through disseminating the scientifically investigated knowledge, etc. The nucleotide and amino acid monomers are represented by limited alphabets. The properties of biopolymers, *i.e.* macromolecules (*e.g.* DNA, RNA proteins) are such that they can be transformed into sequences having digital symbols. Genetic data and other biological data are differentiated by these digital data. This resulted in the progress of bioinformatics.

B. WHAT IS BIOINFORMATICS ?

Translation of billions of characters in DNA sequences that make the genome into biologically meaningful information has given birth to a new field of science called 'bioinformatics'. The term bioinformatics has been derived by combining ***biology*** and ***informatics***. The key to biotechnology discoveries is locked in the genomes of organisms. The bioinformatics holds the key to unlock these data for the next generation of innovations.

A more precise definition of bioinformatics is *the application of information sciences (mathematics, statistics and computer sciences) to increase our understanding of biology*. In a broadest sense this term can be considered as information technology applied to management and analysis of biological data. But in view of the recent rapid accumulation of available protein structure, this term is also used to embrace the manipulation and analysis of three-dimensional structure of data. Thus bioinformatics is a multidisciplinary science which aims to use the benefits of computer technologies in understanding the biology of life. Now, as a subject bioinformatics consists of three core areas: (*i*) molecular biology database, (*ii*) sequence comparison and sequence analysis, and (*iii*) the emerging technology of microarrays. In brief, bioinformatics is *the management and analysis of biological information stored in databases*.

Now the formal and practical problems arising from the management and analysis of biological data can be solved through bioinformatics by creating and advancing databases, algorithms, computational and statistical techniques and theory. Over the past few decades a tremendous amount of information on molecular biology have gathered due to rapid developments in genomic and other molecular research technologies and developments in information technologies. However, mathematical and computing approaches made it possible to glean understanding of biological processes.

Mostly the activities in bioinformatics are focused to: map and analyse the DNA and protein sequences, align different DNA and protein sequences so that they can be compared, and 3D models of protein structures can be created and visualised. Main objective of bioinformatics is to increase the understanding of biological processes by focusing the attention to develop and apply computationally techniques such as pattern recognition, data mining, machine learning algorithms, and visualisation. Furthermore, sequence alignment, gene finding, genome assembly, drug design, drug discovery, protein structure alignment, protein structure prediction, prediction of gene expression and protein-protein interactions, genome-wide association studies and the modelling of evolution are the major research efforts in the field.

The most pressing task of bioinformatics is the analysis and interpretation of various types of data, including nucleotide and amino acid sequences, protein domains, and protein structures. The actual process of analysing and interpreting data is referred to as *computational biology*. There are two fundamental methods of modelling a living cell that come under bioinformatic approaches: *static* and *dynamic*. The static approaches are: procuring sequences (e.g. proteins, nucleic acids and peptides), understanding their structures (of proteins, nucleic acids, ligands (including metabolites and drugs) and peptides, and interaction data among the above entities including microarray data and networks of proteins and metabolites. The dynamic approaches are: (*i*) systems biology (including reaction fluxes and variable concentrations of metabolites), and (*ii*) multi-agent-based modelling approaches capturing cellular activities (such as signalling, transcription and reaction dynamics).

C. DATABASES

A database is a repository of sequences (DNA or amino acids) stored in a computer which provide a centralised and homogeneous view of its contents. It can retrieve large amounts of data easily. Every data item in the database is structured according to a *scheme*, defined as a set of pre-specified rules through the data definition language. The contents of database can be accessed through a 'graphical user interface' (GUI) that allows browsing through the contents of the repository very much similar as one may browse through the books in library.

The repository is created and modified through a database management system (DBMS). DBMS can be categorised according to the 'database model' that they support (e.g. relational or XML), the type(s) of computer they support (e.g. a server cluster or a mobile phone), the query language that access the database (SQL or XQuery), performance trade-offs (e.g. maximum scale or maximum speed or others). In these categories more than one entry are covered by some DBMS.

1. Types of Databases

The databases are broadly classified into two categories: *sequence databases* (that involves the sequences of both proteins and nucleic acids), and *structural databases* (that involves only protein databases). In addition, it is also classified into three categories : (*i*) *primary database,* (*ii*) *secondary database,* and (*iii*) *composite database.*

(*a*) Primary Databases: Primary databases contain information of the sequence or structure alone of either protein or nucleic acid e.g. PIR or protein sequences, GenBank and DDBJ for genome sequences. Primary database tools are effective for identifying the sequence similarities, but analysis of output is sometimes difficult and cannot always answer some of the more sophisticated questions of sequence analysis. In 1998, GenBank obtained more than a million of sequences from more than 18,000 organisms.

***(b)* Secondary Databases:** Secondary databases contain derived-information from the primary databases, for example, information on conserved sequence, signature sequence and active site residues of protein families by using SCOP, eMOTIF, etc. It is more useful than the primary databases. Orthology provides an important layer of information when considering phylogenetic relationships between the genes. Depending on the type of analysis method used, relationship may be elucidated in considerable detail including superfamily, family, sub-family and species-specific sequence levels.

***(c)* Composite Database:** The composite database is obviating the need to search multiple resources. The SCOP is structural classification of proteins in which the proteins are classified into hierarchical levels such as classes, folds, super-families. In 1991, Bairoch established a moderate database pertaining to protein sequence and structural correlations on the 'Net'. This database was called PROSIT which later on was strengthened with a database on sequence analysis and comparison of protein sequences known as SEQUANALREE.

***(d)* Other Databases:** The other most important databases are: *flat-file database, relational database,* and *hierarchical database* as given below:

***(i)* Flat-file database:** A flat file database consists of data in a single file in the form of rows and columns, with no relationships or links between records and fields except the table structure. This database describes any of various means to encode a database model like table as a single file. A flat file is a plain or mixed text and binary file containing one record per line or physical record on disc or tape. The single fields can be separated by delimiters e.g. commas. To avoid delimiter collision extra formatting is required; however, no structural relationships exist between the records. Example of a flat file is *name*, *address*, and *phone number* of employees with the fields. A list of names, addresses, and phone numbers written on a sheet of paper is a flat file database.

By using a typewriter or word processor a flat file can be created. For the first time Herman Hollerith gave the idea to represent census data by holes punched in paper cards and tabulated by machine. Therefore, the US Census of 1990 was the first computerised database that consisted of thousands of boxes full of punched cards. Configurable flat-file database computer applications were popular on DOS and the Macintosh during 1980s. Still flat file databases are used internally by many computer applications to store configuration data. Many applications allow users to store and retrieve their own information from flat files, for examples programmes to manage collections of books (Table 12.1). Now XML is a popular format for storing data in plain text files but it allows very complex nested data structures to be represented and contains the definition of the data. It is very different from the flat-file model.

Table 12.1. Record of some book in a library.

Subject	*Accession No.*	*Copies of books*
English	234765-890	125
Hindi	234891-090	200
Microbiology	235090-150	60
Physics	235151-195	45
Chemistry	235196-285	90

The flat model is in the form of a table that consists of a single, two-dimensional array of data elements. Here all members of a given column are assumed to have similar values, and all members of a row are assumed to be related to one another. Each row of the table has specific password associated with an individual user, and columns have a type associated with them that defining as character data, date or time information, integers or floating point numbers.

***(ii)* Relational database:** In 1970, Edgar Codd coined the term relational database at IBM Almaden Research Center. The relational model is a mathematical model defined in terms of predicate logic and set theory. Codd used the relational model to make database management systems more independent of any particular application. The relational databases implement a model that is only an approximation to the mathematical model. A relational database matches data by using common characteristics within the data set. Then resulting groups of data are organised to make easier to understand by people. For example, a data set containing all the real-estate transactions in a town can be grouped by the year the transaction, or it can be grouped by last name of the buyers. Such a grouping uses the relational model. Therefore, such database is called a relational database, and the software used for this grouping is called a relational database management system (RDBMS). At present people prefer to use the relational databases to store personnel data, financial records, medical records, manufacturing and logistical information, etc.

Table is the basic data structure of the relational model where information about a specific entity (*i.e.* an employee) is represented in rows (also called tuples) and columns. The *relation* in 'relational database' refers to the several tables in the database; hence, a relation is a set of tuples. The columns enumerate the various attributes of the entity (the employee's name, address or phone number, etc.), and a row is an actual instance of the entity (a specific employee) that is represented by the relation. Cosequently each tuple of the employee table represents various attributes of a single employee.

Usually a 'tuple' is an object and information about that object; the objects are the concepts or physical objects. Usually a relation is described as a table organised into rows and columns. All the data referenced by an attribute conform to the same constraints and are in the same domain (Fig. 12.1). The set of possible values for a given attribute is described by a domain, which can be considered a constraint on the value of the attribute. Mathematically, attaching a domain to an attribute means that any value for the attribute must be an element of the specified set.

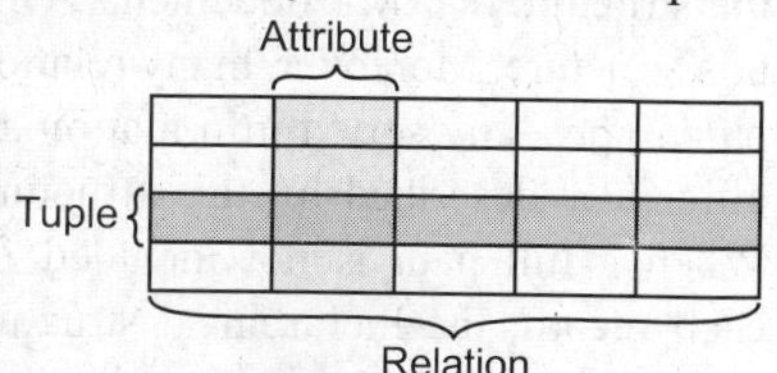

Fig. 12.1. Terminology of relational database.

The relational model specifies that the tuples of a relation have no specific order and impose no order on the attributes. Application accesses data by specifying queries which use operations. For example *select* identifies tuples, *project* identifies attributes, and *join* combines relations. However, by using *insert*, *delete* and *update* operators relations can be modified. Explicit values can be supplied new tuples or be derived from a query. Similarly, tuples can be identified by queries for updating or deleting. For each tuple of a relation it is necessary to be identifiable by some combination of its attribute values. This combination is known as the 'primary key'.

Relational Model

Activity code	Activity name
14	Crack sealing
15	Patching
16	Overlay

Key = 15

Activity code	Date	Route No.
15	01/11/11	I-85
15	02/07/11	I-56

Date	Activity code	Route No.
01/11/11	15	I-85
01/04/11	14	I-485
02/07/11	15	I-56

Fig. 12.2. Relational data model.

In relational database models, three key terms are used extensively: *relations*, *attributes*, and *domains*. A *relation* is a table with columns and rows. The *attributes* are the named columns of the relation, and the *domain* is the set of values to which attributes take (Fig. 12.2).

A primary key is a key that uniquely identify a row in a table, which are commonly used to combine data from two or more tables. For instance, a table of an *employee* may contain a column called *location* which consists of a value that matches the key of a *location* table. Keys create indexes that quickly retrieve data from large tables. Any column may be a key and multiple columns may be grouped together into a compound key.

***(iii)* Hierarchical database:** Hierarchical databases were one of the first data models to be widely accepted and used; it is found on many legacy servers. Hierarchical databases are fast and simple but inflexible as the relationship is restricted to one-to-many; hence it allows one parent segment per child. In some hierarchical database systems multiple hierarchies can be defined over the same data. In this database, records contain groups of parent/child relationships. In modern databases today, the hierarchical model is rarely found. It is primarily used to store information that ranges from geographic, file systems to the Windows registry to XML documents.

In hierarchical model data are organised into a tree-like structure implying a single upward link in each record so that it may describe a sort field to keep the records in a particular order in each list of same level (Fig. 12.3). Hierarchical structures were used in the early database management systems such as the Information Management System (IMS) by IBM. Now it describes the structure of XML documents. One 1 : N relationship between two types of data is allowed by this structure. Moreover, many relationships in the real world, recipes, table of contents, ordering of paragraphs, any sorted information are efficiently described by this structure. When a full path is not included for each record, the hierarchical structure is inefficient for certain database operations. For example, a child may only have one parent but a parent can have many children. Therefore, parents and children are tied together by links called 'pointers'.

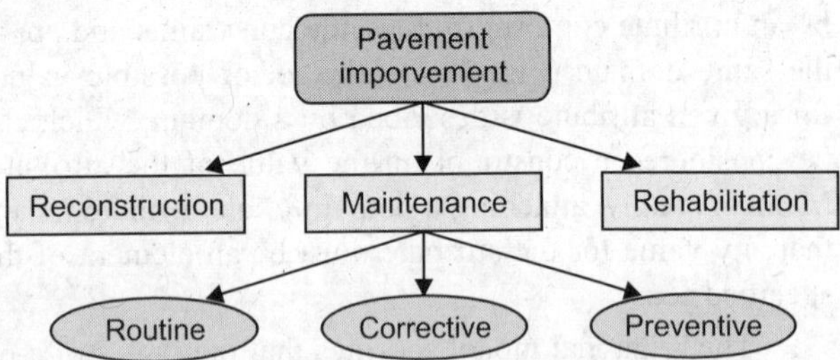

Fig. 12.3. Hierarchical model.

2. Data Structure

Data structures (fields, records, files and objects) are optimised to deal with very large amounts of data stored on a permanent data storage device. A data structure is a way of storing and organising the data in a computer to use it efficiently in future. For different kinds of applications, different types of data structures are suited; some data are highly specialised to specific tasks; for example, B-trees are well-suited for implementation of databases, whereas compiler implementations use hash tables to look up identifiers. Data structures are used almost in every programme or software system. Specific data structures are essential components of many efficient algorithms that possibly manage the huge amounts of data e.g. internet indexing services and large databases.

Data structures are emphasised by some formal design methods and programming languages as the key organising factor in software design. Generally data structures are based on the ability of a computer to catch and store data at any place in its memory specified by an address. Computing the addresses of data items with arithmetic operations governs the record and array data structures; whereas the linked data-structures depend on storing the addresses of data items within the structure itself. Many data structures use both principles, sometimes combined in non-trivial ways. A data structure can be implemented via writing a set of methods that create and manipulate instances of that structure. None can analyse the efficiency of a data structure separately. The theoretical concept of an abstract data type can be motivated by a data structure that is defined indirectly by the operations performed on it, and the mathematical properties of those operations.

3. Database Management Systems (DBMS)

A set of software programmes that controls the organisation, storage, management and retrieval of data in a database is called DBMS. The organisations are allowed to place control of database development in the hands of database administrators (DBAs) and other specialists. DBMSs are categorised according to structures or types of their data. A DBMS is a system software package that helps the use of integrated collection of data records and files known as databases. The DBMS accepts requests for data from an application programme and instructs the operating system to transfer the appropriate data. The queries and responses must be submitted and received according to a format that conforms to one or more applicable protocols. Different programmes of user's application are allowed to access easily the same database. DBMSs use a variety of database models; a DBMS allows users and other software to store and retrieve data in a structured way in large systems. While using a DBMS, information systems can be changed more easily according to change in informational requirements by the organisations. New data can be added to the database without disruption to the existing system.

Database servers are usually multiprocessor computers, with generous memory and RAID disk arrays used for stable storage. Database servers hold the actual databases and run only the DBMS and related software. Hardware database accelerators are connected to one or more servers via a high-speed channel. They are also used in large volume transaction processing environments. Modern DBMSs rely on a standard operating system for good functioning.

During 1960s, several systems were used commercially. Charles Bachman [author of Integrated Data Store (IDS)] founded the Database Task Group within CODASYL (a group that standardised COBOL). In 1971 they delivered their standard that was known as the 'Codasyl approach'. Based on this approach several commercial products were made available. The Codasyl approach was based on the 'manual' navigation of a linked data set formed into a large network. Such operations were too expensive to contemplate anyway because most data was stored on magnetic tape.

In 1968, IBM also had its own DBMS system known as IMS which was a development of software written for the Apollo programme on the System/360. IMS was similar in concept to Codasyl, but used a strict hierarchy for its model of data navigation instead of Codasyl's network model. Thereafter, both concepts could be known as 'navigational databases' due to the way of data accession. IMS is classified as a hierarchical database. IMS and IDMS, CODASYL databases and CINCOMs TOTAL database are classified as network databases.

Some commonly used DBMS are: MySQL, PostgreSQL, Microsoft Access, SQL Server, FileMaker, Oracle, Sybase, dBASE, Clipper, FoxPro, etc. An Open Database Connectivity (ODBC) driver gives rise to every database software that allows the database to integrate with other databases. Most databases also allow querying of its contents through a specialised query language. The data definition language and the query language form the *data model*. A data model not only structures a data, but also defines a set of operations to be performed on the data. Operations such as select, project, and joins are defined by the relational model.

(*a*) Components of DBMS: Most DBMS is a relational model. But the other DBMS systems such as Object DBMS offer specific features for more specialised requirements.

(*i*) Relational DBMS (RDBMS) Components: The relational RDBMS has the following components-

- **Sublanguages:** RDBMS include Data Definition Language (DDL) for defining the structure of the database, Data Control Language (DCL) for defining security/access controls, and Data Manipulation Language (DML) for querying and updating data.

- **Interface drivers:** Interface drivers are code libraries that provide methods to prepare statements, execute statements, fetch results, etc; for example ODBC, JDBC, MySQL/PHP, FireBird/Python.
- **SQL engine:** SQL engine interprets and executes the DDL, DCL, and DML statements. It consists of three major components, for example compiler, optimiser, and executor.
- **Transaction engine:** Transaction engine ensures about success or failure of multiple SQL statements as a group.
- **Relational engine:** This component implements the relational objects such as table, index, and referential integrity constraints, etc.

***(ii)* Storage engine:** This component stores and retrieves data from secondary storage. It also manages transaction commit and rollback, backup and recovery.

***(iii)* Object DBMS (ODBMS) Components:** Object DBMS (ODBMS) consists of transaction and storage components that are analogous to RDBMS components. DDL, DML and update tasks are handled differently by some DBMS. Typically they include a sublanguage and accompanying engine for processing queries with interpretive statements analogous to SQL. For example, the object query languages are OQL, LINQ, JDOQL, JPAQL and others. Instead of relational rows, the query engine returns collections of objects.

SQL is an integral part of a database management system, a language and a tool for communicating with the DBMS. Fig. 12.4 shows some of the components of a typical DBMS and SQL acts as the glue linking them together. The heart of the DBMS is the database engine, which actually structures, stores, and retrieves the data in the database. It accepts SQL requests from other DBMS components, such as forms facility, reports writers, or interactive query facility (from user-written application programs), and even from other computer systems. SQL has emerged as a powerful tool for linking people, computer programmes, and computer systems to the data stored in a relational database.

SQL plays many different roles: (*i*) it is an interactive query language (used to retrieve data and display it on the screen, providing a convenient, easy-to-use tool for ad hoc database queries), (*ii*) it is a database programming language (to access the data in a database, both user written programmes and database utility programmes), (*iii*) it is a database administration language (responsible for managing a minicomputer or mainframe database uses SQL to define the database structure and control access to the stored data), (*iv*) it is a client/server language (to communicate over a network with database servers that stores shared data), (*v*) it is an internet data access language; internet web servers use SQL as a standard language for accessing corporate databases, (*vi*) it is a distributed database language that distributes data across several connected computer systems by using SQL, and (*vii*) it is a database gateway language that allows one brand of DBMS to communicate with another brand.

***(a)* Current Trends:** Database management has affected the field of technology by several ways due to growing up of demand of directory services of organisations. Businesses use directory services that provide prompted searches for company information. Mobile devices store more than contact information of users, and can cache and display a large amount of information. Search engine queries locate data within the World Wide Web. From the developments with data warehousing, the retailers have also derived benefits because they can record transactions of the customers. Now-a-days, for e-business online transactions have become very popular. Consumers and businesses can make payments through some websites of company. All these have been possible through the evolution of database management. Even there is a growing demand for new development with the progress of database management.

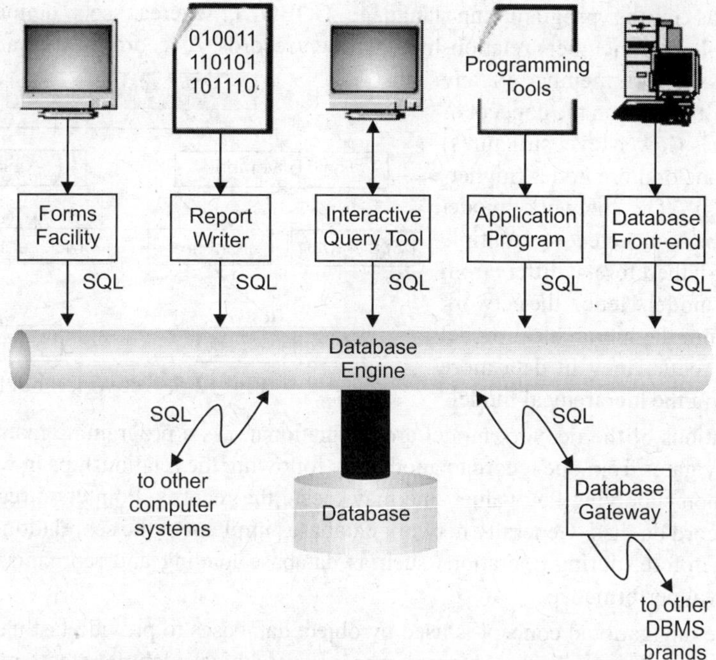

Fig. 12.4. Components of a typical database management system.

Data availability and computing have become more ubiquitous with the increase in speeds of internet connectivity of consumers. Databases are processed by web-based languages such as XML and PHP that allow databases to live in 'the cloud'. Besides Google's Gmail, Microsoft's Office 2010, and Carbonite's online backup services have started to move to web based services due to increasing internet reliability, data storage efficiency and the lack of dedicated IT staff who can manage the hardware. At Rochester Institute of Technology the faculty has planned to add 'cloud-based database' that can compute their curriculum to keep information technology curriculum on the top of technology.

4. Database Models

A database model is a specification that describes to structure and use a database. Hence, it defines the infrastructure offered by a particular database system, for example the relational model. A data modelling language defines the schema of each database hosted in the DBMS. The DBMS are designed to use one of five database structures to provide simple access of information stored in databases; these five database structures are: the *hierarchical model, relational model, network model, multidimensional model,* and *object model*. Moreover, flat model has also been used which does not strictly qualify as a data model as per definition. Both the hierarchical and relational models have been described earlier.

***(a)* Network Model:** The Network Database model was designed to solve some of the more serious problems with the Hierarchical Database Model. The Network model solves the problem of data redundancy by representing relationships in terms of sets rather than hierarchy. The model had its origins in the Conference on Data Systems Languages (CODASYL) which had created the Data Base Task Group to explore and design a method to replace the hierarchical model. The network model is very similar to the hierarchical model. The hierarchical model is a subset of the network model.

The CODASYL specification defines the network model that organises data using two fundamental constructs which are called *records* and *sets*. Records contain fields (organised hierarchically as in the programming language COBOL), whereas sets define one-to-many (*one* = owner, *many* = member) relationships between records. A record may be an owner in any number of sets, and a member in any number of sets. It is built on the concept of multiple branches (lower-level structures) evolving from one or more nodes (higher-level structures). The network model branches can be connected to multiple nodes (Fig. 12.5); therefore, it differs from the hierarchical model. Hence, the network model varies from the hierarchical model and represents redundancy in data more efficiently than in the hierarchical model.

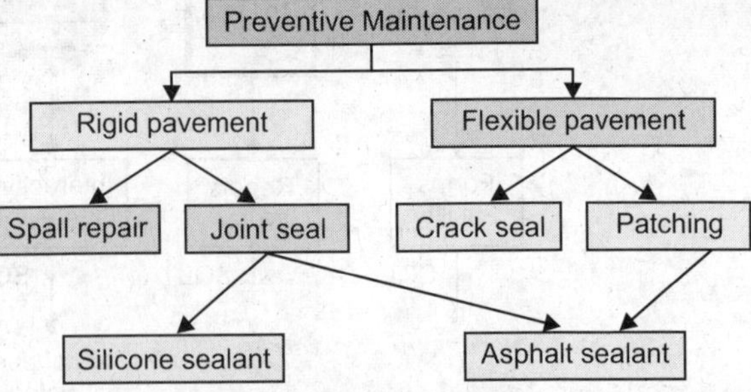

Fig. 12.5. Network model.

The operations of the network model are navigational, *i.e.* a programme maintains a current position and navigates from one record to another by following the relationships in which the record participates. Upon supplying key values one may locate the records. Pointers directly address the location of a record on disk. Generally network databases implement the set relationships by means of pointers. Therefore, during operations such as database loading and reorganisation this gives excellent retrieval performance.

Mostly the navigational concept is used by object databases to provide fast navigation across networks of objects. Generally the object databases uses object identifiers as 'smart' pointers to related objects. Objectivity/DB implements named 1:1, 1: many, many: 1 and many: many named relationships that can cross databases. Many object databases also support SQL combining the strengths of both models. However, instead of using a single-parent tree hierarchy, the network model uses set theory to provide a tree-like hierarchy. This allowed the network model to support many-to-many relationships.

A network database looks like a hierarchical database and appearing like a tree, but the look is more like several trees which shares branches. Therefore, children may have multiple parents and parents may have multiple children. The network model was difficult to implement and maintain. Most implementations of the network model were used by computer programmers rather than real users. Therefore, a simple model was needed which could be used by end users to solve real problems.

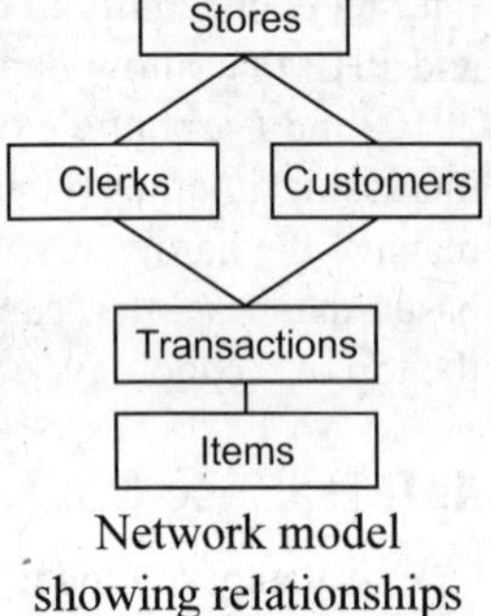

Network model showing relationships

(*b*) Dimensional Model: It is a specialised adaptation of the relational model used to represent data in data warehouses where data are easily summarised by OLAP queries. A database consists of a single large table of facts in the dimensional model that is described by using dimensions and measures. A dimension provides the context of a fact, for example name of participants, date and place of events, and its type. However, it is used in queries to group related facts together. Dimensions are often hierarchical and tend to be discrete; for example, the location includes the building, state, and country. A measure is a quantity that describes the fact e.g. revenue which can be added together from different locations. In an OLAP query, dimensions are chosen and the facts are grouped and added together to make a summary.

Often the dimensional model is implemented on top of the relational model by using a star scheme that consists of one table containing the facts, and surrounding tables containing the dimensions. Complicated dimensions are represented by using multiple tables that result in a snowflake schema. A data warehouse consists of multiple star schemas sharing the dimension tables which allow them to be used together. A standard set of dimensions acts an important part of dimensional modelling.

(c) Object-Oriented Database Models: Object-oriented database technology is a marriage of object-oriented programming and database technologies. Thus object-oriented paradigm has been applied to database technology creating a new programming model known as object databases. Fig. 12.6 shows how these programming and database concepts have come together to provide object-oriented databases.

The most significant characteristic of object-oriented database technology is that it combines object-oriented programming with database technology to provide an integrated application development system. There are many advantages in including the definition of operations with the definition of data: (*i*) the defined operations apply ubiquitously and are not dependent on the particular database application running at the moment, and (*ii*) the data types can be extended to support complex data such as multi-media by defining new object classes that have operations to support the new kinds of information.

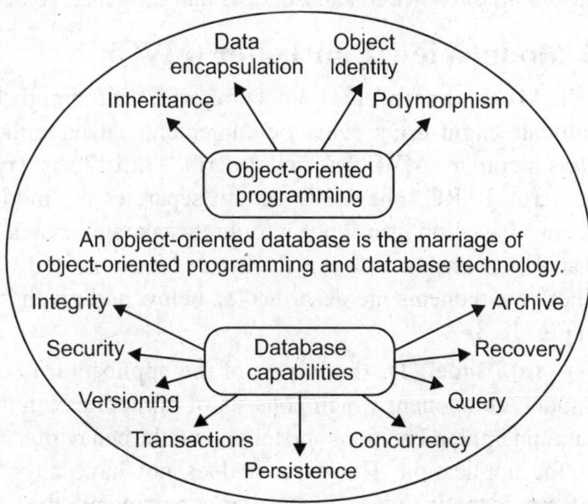

Fig. 12.6. Make up of object-oriented database.

These databases bring the database world and the application programming world closer to each other by using the same type system as the application programme. This avoids the overhead of converting information between its representation in the database (as rows in tables) and that of in the application programme (as objects) (Fig. 12.7). At the same time object databases introduce the ideas of object programming (e.g. encapsulation and polymorphism) into the world of databases.

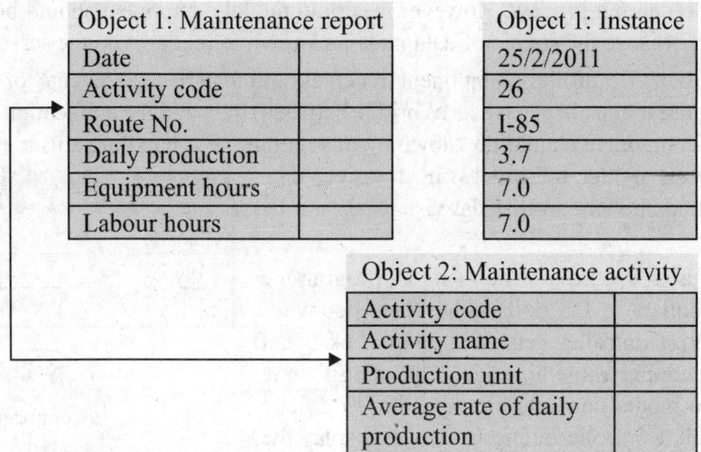

Fig. 12.7. Object-oriented model.

Many ways of storing the objects in a database have been attempted. Addition of some types of query language is required because the conventional programming languages are unable to find objects based on contents of their information. By defining an object-oriented data model and a database programming language for the database, some have attacked the problem from the database end. A database programming language allows full programming capabilities as well as traditional query facilities.

The object databases suffer due to lack of standardisation. Standards were never implemented well enough to ensure interoperability between products in spite of being defined by ODMG. In many applications the object databases (usually specialised applications e.g. engineering databases or molecular biology databases) have been used successfully. However, the relational vendors have picked up the object database ideas and influenced extensions made to these products.

5. Model-View-Controller (MVC)

MVC is a software architecture, which is currently considered as an architectural pattern used in software engineering. It can be implemented even without an object-oriented language or a specific class hierarchy. MVC was first described in 1979 by Trygve Reenskaug, then working on Smalltalk at Xerox PARC. The MVC pattern separates the modelling of the domain, presentation, and the actions based on user input into three separate classes: *model*, *view* and *controller* components so that dependencies between them can be minimised. These components are described as below and given in Fig. 12.8.

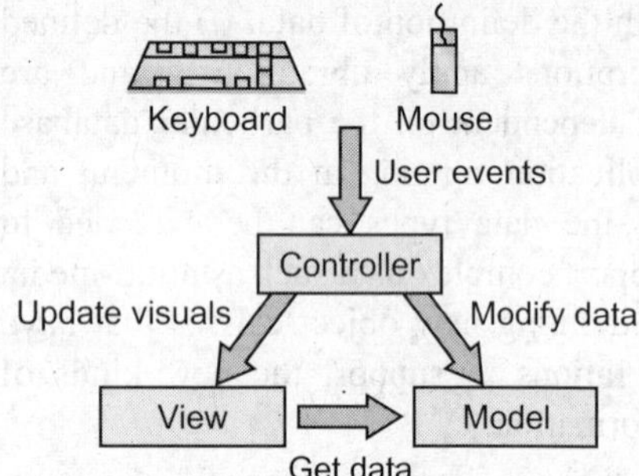

Fig. 12.8. Structural relationship between the three objects: controller, view and model.

***(a)* Model**: The brain part of the application is 'model' component which consists of all application data and implementations and tells about the behaviour of the application. However, it does not have any reference to 'view' or 'controller' component; this shows that model component is purely independent. Controller or view component associated with model is unknown; rather the model component simply processes the requests from any controller or view component.

***(b)* View:** The visual contents are rendered by view component onto the screen. View module is independent on controller; instead upon requests by any controller component to update the visual it performs only rendering process. However, a certain model component should be referenced by view component, because the source of data must be known to render it on the screen.

***(c)* Controller:** Controller component receives and handles all events of users such as keyboard and mouse inputs; hence it acts as bridge between users and the application. Assessment of model and view component should be known by this module. For handling a user event, controller component requests model for processing the new data and telling view component so that the visuals should be updated.

If a user uses a currency conversion programme from Canadian dollars to US dollars, he clicks 'convert' button (Fig. 12.9). Controller gets the button click event first and sends the input value to model requesting for the conversion. Then model converts the input to US dollars and save the result. Controller requests view to display the result. View displays the result on the screen after getting from model.

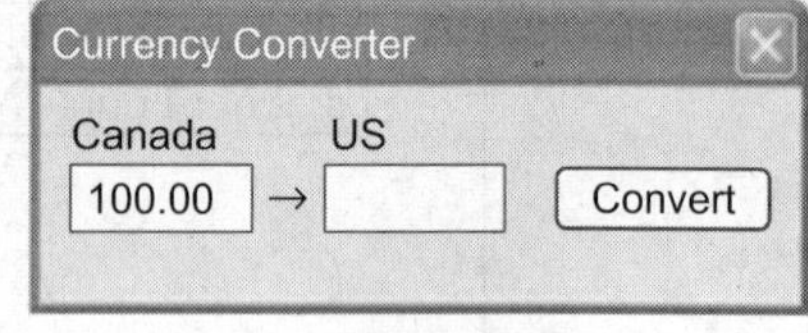

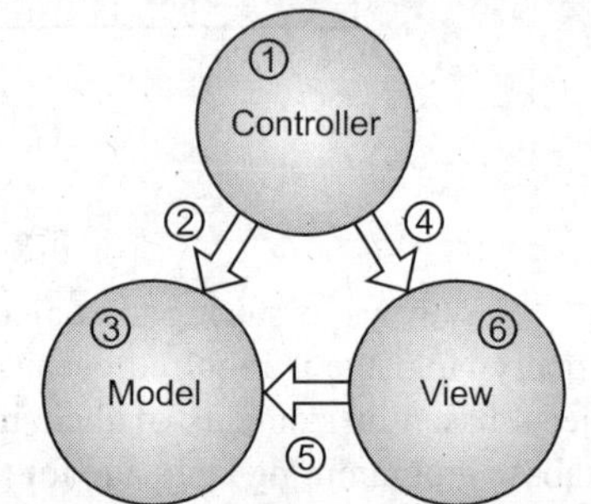

Fig. 12.9. Application of a GUI: currency converter.

Understanding the role of each component is cleaner and easier. It is also easier to maintain each module separately by multiple developers. The components can be interchangeable and scalable due to its efficient modularity. The look-and-feel of view component can be customised without modifying model and controller modules.

Variations of *MVC*

In application programming in Smalltalk-80, how to use MVC (Burbeck92), Steve Burbeck (1992) describes two variations of MVC: *passive model* and *active model*.

***(i)* Passive Model:** The passive model is employed when one controller exclusively manipulates the model. The controller modifies the model and informs the view about the changed model, which should be refreshed (Fig. 12.10). Moreover, the model is completely independent of the view and the controller, *i.e.* there is no means for the model to report changes in its state, for example the HTTP protocol. In the browser there is no simple way to get asynchronous updates from the server. The browser displays the view and responds to user input, but it does not detect changes in the data on the server. Only when the user requests a refresh, the server is interrogated for changes.

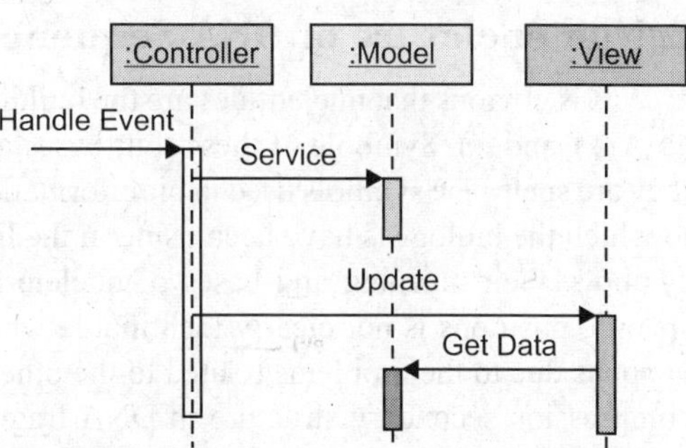

Fig. 12.10. Behavior of the passive model.

***(ii)* Active Model:** When the model changes state without the involvement of controller, the active model is used. This happens when other sources are changing the data and the changes must be reflected in the views. In a stock-ticker display stock data is received from an external source and the views are updated on changing of the stock data. Because changes are detected only by the model to its internal state; the model must notify the views to refresh the display. The way of notification done by the observer about the changed views in model is shown in Fig. 12.11. But in a Unified Modeling Language (UML) sequence diagramme there is no good way to demonstrate the separation of model and view due to representation of instances of objects by the diagram rather than classes and interfaces.

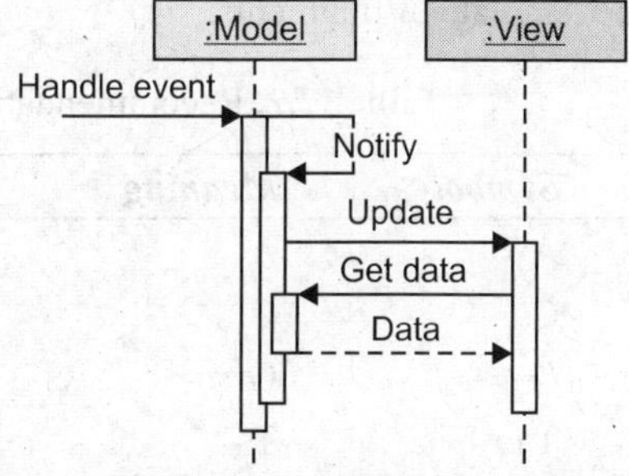

Fig. 12.11. Behavior of the active model.

D. SEQUENCES AND NOMENCLATURE

As mentioned earlier the sequences of digital symbols are the transformed biopolymers. Indirectly the sequence data means the structure of biopolymer, and structure expresses the function. It shows a reductionist approach. Therefore, the sequence data can be used as context free.

1. IUPAC Symbols

The International Union of Pure and Applied Chemistry (IUPAC) has made certain recommendations. The nomenclature system in bioinformatics is based on these recommendations. Different laboratories of the world follow nomenclature system of the IUPAC so that their data set can uniformly and easily be compared. For rapid reproducibility and uniformity, the database institution and editors (who publish journals and research findings) also follow the recommendations of IUPAC.

For routine work, the basic IUPAC nomenclature system of nucleic acids and proteins has been discussed in this section. For detail you should go through the IUPAC web site. Language used in bioinformatics is given below:

Alphabet → Nucleotides

Words → Gene (prokaryotes)
Introns (eukaryotes)

Sentence → Operon (prokaryotes)
Gene (eukaryotes)
Punctuation → Regulatory gene
Chapter → Chromosome

2. Nomenclature of DNA Sequences

It is obvious that nucleotides are the building blocks of DNA, and are constituted by four bases (A, G, T and C). Symbols of these four bases and basis of their nomenclature are used as much as they are spelt. The symbols used in bioinformatics to represent these sequences are the same symbols to which the biologists have been using in the literature for the same purpose. Table 12.2 shows the symbols, their meaning and bases of nucleic acid sequences. Often the identity of sequences at specific positions is not clearly identifiable when sequence data are experimentally determined. It happens due to the problems related to the other secondary structures or 'compression' artifacts. In compression secondary structure in DNA fragments causes them to move in the gel so that more than one size of fragments may migrate to the same position.

Generally by repeating the experiment and sequencing the complementary strand, this problem can be solved. However, if ambiguities persist in some cases, the probable possibility can be deduced from the gel reads, *i.e.* forward and reverse readings give data from opposite strands of DNA. They provide information about the relative orientations of the read pairs (*i.e.* pair of reading) from the same template of fragments.

Table 12.2. Recommendations of the single-letter codes of nucleic acid.

Symbol	*Meaning*	*Base*
A	A	Adenine
C	C	Cytosine
G	G	Guanine
T or U	T	Thymine or Uracil
R	A or G	puRine
Y	C or T	pYrimidine
M	A or C	aMino group
K	G or T	Ketone
S	G or C	Strong interaction (3 H bonds)
W	A or T	Week interaction (2 H bonds)
H	A or C or T	not-G, H follows G in the alphabet
B	G or T or C	not-A, B follows A in the alphabet
V	G or C or A	not-T (not-U), V follows U in the alphabet
D	G or A or T	not-C, D follows C in the alphabet
N	G or A or T or C	aNy nucleotide
X		Masked
–		Gap of indeterminate length

A new symbol 'S' is used when there is doubt for the presence of G or C but there is surety for absence of A or T. N is the common representation for 'unknown', as when a sequencer can't determine a base with certainty. The terminology of the computer science is a little different from that of the biology for the codes as given in Table 12.2. In computer science, these tables define 'two alphabets', *i.e.* finite sets of symbols that can make strings. A sequence of symbols is called a 'string'. A language is a (finite or infinite) set of strings, *i.e.* mainly DNA and protein sequence data. Bioinformaticians often refer to an actual sequence of DNA or protein as a 'string', as opposed to its representation as sequence data. It is a typical example of the terminologies of the two disciplines that cross over into one another.

Table 12.3 shows that data are represented by simple letters, but computers actually use additional codes to represent simple letters. ASCII is a way for computers to store textual (and control) data in their memory. When a programme (e.g. text editor) reads the data and reading ASCII, it actually draws the letters on the screen in a recognisable fashion because it is programmed to know that particular code.

Except a few viruses all the cellular organism consists of double stranded DNA. The two strands are complementary and antiparallel (running from 5′→3′ direction) to each other. This is called Watson and Crick base pairing. When one encounters the symbol, the problem arises due to more than one base at a position. These problems are resolved following IUPAC system of nomenclature. Table 12.3 shows the symbols used in a strand and that of its complementary strand. At certain positions the identical symbols in the strand and its complement are used. This shows that they are the same set of bases (*see* star, *).

Table 12.3. Symbols used in a strand and its complementary strand.

Symbol	A	B	C	D	G	H	K	M	S	A	U	W	N	R	Y
Complement	T	V	G	H	C	D	M	K	S*	A	B	W*	N	Y	R

3. Nomenclature of Protein Sequences

You know that there are 20 amino acids which built the protein. But there are a few symbols which represent more than one amino acid. These symbols representing amino acids are given in Table 12.4.

Table 12.4. Symbols representing the amino acids.

Symbols	***Amino acids***	***Three- letter code***	***Symbols***	***Amino acids***	***Three-letter code***
A	Alanine	Ala	N	Asparagine	Asn
B	Aspartic acid or Asparagine	Asx	O	Glutamine	Gln
C	Cysteine	Cys	P	Proline	Pro
D	Aspartic acid	Asp	Q	Glutamine	Gln
E	Glutamic acid	Glu	R	Arginine	Arg
F	Phenylalanine	Phe	S	Serine	Ser
G	Glycine	Gly	T	Threonine	Thr
H	Histidine	His	V	Valine	Val
I	Isoleucine	Ile	W	Tryptophan	Trp
K	Lysine	Lys	X	Unknown	Xxx
L	Leucine	Leu	Y	Tyrosine	Tyr
M	Methionine	Met	Z	Glutamic acid or Glutamine	Glx

4. Directionality of Sequences

In nucleic acids (DNA and RNA) the nucleotide sequences are synthesised in 5′→3′ direction. The 5′ primer represents the presence of phosphate group at 5th carbon of sugar, and 3′ primer represents the presence of hydroxyl group at 3rd carbon of sugar. Hence, this information is used to collect the data and store it in sequence database. Because data of nucleotide sequences are deposited in the database in the same form as these have been submitted or published.

The nucleotide bases are numbered sequentially starting from 5′ end *i.e.* from 5′ to 3′direction. A word 'C' is indicated for complementary strand which also shows the orientation of chain in 5′→3′ direction. Both the chains run antiparallely, *i.e.* one in 5′→3′ direction and the other in 3′→5′ direction. While depositing sequence data, information on nucleotide sequence of only one strand is submitted in database. The nucleotide sequence of complementary strand is deduced from different web sites or programmes in different packages.

The three letter alphabets of the nucleotide act as codes, and each code represents an amino acid. In nature each cell synthesises proteins from N-terminus to C-terminus (N′→C′) where N′ represents -NH_2 group and C′ represents -COOH group of the amino acids. These fundamental phenomena are universal in all organisms. Hence, this conventional sequence of protein is entered in database. The concept of directionality is a universal fundamental process which is used by different database institutions. If you have a sequence without any label, how will you search out whether this sequence is DNA or RNA or protein ? You will search sequence search programme which scans 20 symbols. If the symbols switch between any of 4 bases, the sequence is DNA. If there is U instead of T, it is RNA sequence. If the symbols switch more than 4 bases, it is proteins sequence.

5. Types of Sequences used in Bioinformatics

There are different types of sequences which are known to have genetic information. Therefore, such sequences are used in bioinformatics; these sequences have been described in this context.

***(i)* Genomic DNA:** The genomic DNA acts as the reservoir of genetic information of all organisms. In recent years it is routinely sequenced in many laboratories of Molecular Biology. Genomic DNA of prokaryotes differs from that of eukaryotes, as the later differs with respect to location and consists of introns. Construction of genomic library is discussed in Chapter 8 : *Methods of Gene Manipulation*.

***(ii)* cDNA:** The double stranded molecules prepared by using mRNA as template and reverse transcriptase are called cDNA. These are expressed genes of the genomic DNA. By using cDNA molecules, substantial number of sequences have been determined and deposited in database. You have to tick at the right position when sequence entry form is to be filled up. This shows that the sequence, which is to be deposited, is the cDNA. Moreover, if you desire to retrieve the sequence this data need to be provided. Details of cDNA production are given in Chapter 8.

***(iii)* Organellar DNA:** Eukaryotic cells consist of different types of organelles *e.g.* chloroplast, mitochondria, Golgi complex, nucleus, etc. In eukaryotes genomic DNA is found in nucleus and organellar DNA molecules are located in mitochondria and chloroplasts.

The DNA molecules of these organelles are usually circular and double stranded of varying size besides a few exceptions. The mitochondrial genomes are relatively small and simple. Human mitochondria genome is made up of 16,569 base pairs. The genome consists of 2 rRNA genes, 22 tRNA genes and 13 protein coding sequences. The size of mitochondrial DNA varies greatly. But the size of chloroplast genome of several organisms has less variation, which are present in many copies. The size of mitochondrial and chloroplast genome of some organisms is given in Table 12.5.

Table 12.5. Mitochondrial and chloroplast genome of some organisms.

Types of DNA	*Genome size (kb)*	*Circular or linear*
1. Mitochondrial genome		
Insects and mammals	16-19	Circular
Homo sapiens (humans)	17	Circular
Saccharomyces cerevisiae (yeast)	33-78	Circular
Aspergillus nidulens (fungus)	32	Circular
Neurospora crassa (fungus)	60	Circular
Chlamydomonas (green alga)	16	Linear
Paramecium (protozoa)	40	Linear
2. Chloroplast genome		
Pisum satium (pea)	120	Circular
Marchantia polymorpha (bryophyte)	121	Circular
Oryza stiva (rice)	134	Circular
Nicotina tobaccum (tobacco)	156	Circular
Chlamydomonas reinhardii (alga)	195	Circular
Higher plants	120-200	Circular

The organeller genomes contain completes genetic system. The chloroplast genome of higher plants contains about 120 genes. The organeller genomes and prokaryotic genomes have several similarities with respect to organisation and functions. The organeller genome codes for their own rRNA, tRNA and some important proteins. The protein synthesising machinery of the organelles especially chloroplast resembles that of bacteria but not the eukaryotes. The ribosomes of chloroplasts closely resemble bacterial ribosomes. The mitochondrial ribosomes show both similarity and differences.

(*iv*) Expressed sequence tag (EST): It was Crag Venter (American biologist) who initiated first the sequencing of cDNA molecules using mRNA. The cDNA is cloned into a vector and cDNA library is constructed. An EST is a short sub-sequence of a transcribed cDNA sequence. It is used to identify gene transcripts and is instrumental in gene discovery and gene sequence determination. The identification of ESTs has proceeded rapidly. At present about 65.9 million ESTs are available in public databases (e.g. GenBank 18 June 2010, all species). Normally each clone has 5′ and 3′ ESTs associated with it. The average length of sequence is of about 400 bases. While the ESTs are short representing only fragments of genes, but do not complete coding sequence. Many sequencing centres have automated the EST production where ESTs are produced rapidly.

John Craig Venter

John Craig Venter (October 14, 1946)

John Craig Venter is most famous for being one of the first to sequence the human genome and for his role in creating the first cell with a synthetic genome in 2010. Venter founded Celera Genomics, The Institute for Genomic Research and the J. Craig Venter Institute, now working to create synthetic biological organisms and to document genetic diversity in the oceans of the world.

In 1975, he completed his Ph.D. degree in physiology and pharmacology from the University of California. He was married with a former Ph.D. candidate Barbara Rae. After working as professor at the State University of New York at Buffalo, he joined the National Institutes of Health in 1984. In Buffalo, he divorced Rae Venter and married his student, Claire M. Fraser and remained married till 2005.

Venter learned a technique at the NIH for rapidly identifying all the mRNAs present in a cell. The short cDNA sequence fragments discovered by this method are called expressed sequence tags (ESTs), a name coined by Anthony Kerlavage at The Institute for Genomic Research. The NIH initially patented these gene fragments in which Venter controversially became involved. Later on the NIH withdrew the patent applications following public outcry. Subsequent court cases declared that ESTs were not directly patentable. But Venter was passionate about the power of genomics to radically transform into healthcare. Then he sought funding from a private sector to fund Celera Genomics where human gene sequencing was started. The goal of the company was to sequence the entire human genome and release it into the public domain for non-commercial use in much less time and less cost than the public human genome project.

Celera used DNA from five demographically different individuals to generate the sequence of the human genome; one of the individuals was Venter himself. In 2001, Celera published the first Human Genome in the journal *Science*, followed Publication of Human Genome Project in the journal *Nature*.

The first complete genome (six billion letters) of 'an individual human' was published on September 4, 2007 by Sam Levy and co-workers where Venter himself donated his own DNA for sequencing. Some of the sequences in Venter's genome are associated with wet earwax, increased risk of antisocial behavior, Alzheimer's and cardiovascular diseases. This publication was especially interesting since it contained a diploid instead of a haploid genome and shows promise for personalised medicine via genotyping.

An EST is produced by one-shot sequencing of a cloned mRNA, *i.e.* sequencing several hundred base pairs from an end of a cDNA clone obtained from a cDNA library. The resulting sequence is a relatively low quality fragment whose length is limited to about 500 to 800 nucleotides. Because these clones consist of DNA complementary to mRNA, and the ESTs represent portions of expressed genes. ESTs may be present in the database as cDNA/mRNA sequence or as the reverse complement of the mRNA, *i.e.* the template strand. Genome Sequencing Centre at Washington University (U.S.A) has been producing over 20,000 ESTs per week. The EST data are used to find out expression pattern using the following formula:

$$\text{Expression pattern} = \frac{\text{No. of ESTs corresponding to each gene}}{\text{Total No. of ESTs}}$$

***(iv)* Gene Sequencing Tags (GSTs):** It has been found that the enzyme mungbean nuclease (Mnase) cleaves between the genes of *Plasmodium falciparum*. Therefore, by digesting *P. falciparum* genome a genomic library was established. It helps in identifying the genes of *P. falciparum*. The approach for construction of GSTs is similar to ESTs. It is constructed by isolating one read of sequence from any of the ends 5′ or 3′. The sequences obtained through this approach are called as GSTs.

***(vi)* Other Biomolecules:** The databases also consist of sequences of tRNA and small sized rRNAs. For example, 16S rRNA sequencing is done in tracing phylogenetic relationship among the species. A similar approach can also be made by using the other molecule. Like mRNAs, rRNA can be copied into DNA, but this practice is rarely done.

E. INFORMATION SOURCES

There are several well developed data repositories that have facilitated the dissemination of genome and protein resources of humans and other organisms. Some of the major biological databases are given in Table 12.6. The most comprehensive resources are the NCBI, genome database (GDB) and Mouse Genome Database (MGD). Each database is informative and very useful. After the components become linked to, for a single integrated resource that generate, the information stored in each database that can be analysed. For example, by integrating various forms of information relative to a particular protein, a researcher may be able to elucidate the previously unknown function. Certain steps within a complex biological pathway become clear which were never understood earlier.

Table 12.6. Major biological databases.

Database name	*Link*	*Information available*
	A. Major Nucleotide Sequence Database	
NCBI's GenBank	http://www.ncbi.nlm.nih.gov/Genbank	All known nucleotides and protein sequences International Nucleotide Sequence Data Collection
EMBL Nucleotide Sequence Database	http://www.ebi.ac.uk/embl/	All known nucleotide and protein sequences International Nucleotide Sequence Data Collection
DNA Data Bank of Japan (DDBJ)	http://www.ddbj.nig.ac.jp	All known nucleotide and protein sequences International Nucleotide Sequence Data Collection
SWISS-PROT	http://www.expasy.ch/sprot	Annotated protein sequence
	B. Major Mutation Databases	
NCBI's dbSNP	http://www.ncbi.nlm.nih.gov/SNP/	Database of single nucleotide polymorphism
OMIM	http://www.ncbi.nlm.nih.gov/OMIM/	Catalogue of human genetic and genomic disorder
	C. Major Gene Expression Databases	
GeneExpression	http://www.ncbi.nlm.nih.gov/GEO	NCBI's repository for gene espression
	D. Major Microbial Genomic Databases	
NCBI's Microbial	http://www.ncbi.nlm.nih.gov/PMGifs/ Geno	?
Genome Gateway	mic/micr./html	
Escherichia coli	http://1 ecoli.aist-nara.ac.jp	?
	E. Major Organism specific Genome Database	
Mouse Genome Database (MGD)	http://www.informatics.jax.org/	Mouse genetics and genomics
Saccharomyces genome database (SGD)	http://genome.www.stanford.edu/ Saccharomyces/	*S. cerevisiae* genome information
Rice Genome Project (RGP)	http://1 rgp.dna.affrc.go.jp/	Reporting current data in rice genome project
	F. Major Protein Database	
Protein Database (PDB)	http://www.rcsb.org/pdb	Structure data determined by X-ray crystallography and NMR

1. National Centre of Biotechnology Information (NCBI)

On November 4, 1988, the NCBI was established as a division of the National Library of Medicine (NLM) for the development of information systems in molecular biology. The NCBI is located in the campus of the National Institute of Health (NIH) in Bethesda, Maryland (U.S.A.). The NLM was selected to host the NCBI of its experience in biomedical database maintenance. It could establish a research programme in computational biology. Since 1992, the NCBI has been maintaining GenBank, the NIH DNA sequence database. Groups of annotators create sequence database records from the literature obtained from the authors, data exchanged with EMBL and DDBJ. The NCBI is the foremost repository of publicly available genomic and proteomic data. After its establishment the services at NCBI have fully expanded. The NCBI makes available the different kinds of biological data, computational resources for analysis of GenBank data and data retrieval system.

More specifically the NCBI has been charged with creating automated systems for: (*i*) storing and analysing knowledge about molecular biology, biochemistry and genetics, (*ii*) facilitating the use of such databases and software by the research and medical communities, (*iii*) coordinating efforts to gather biotechnology information both nationally and internationally, and (*iv*) performing research into advanced methods of computer-based information processing for analysing the structure and function of biologically important molecules.

(*a*) Functions of NCBI: The NCBI does the following to carry out its multifarious responsibilities:

(*i*) It maintains collaboration with several NIH institutes, academia, industry and other governmental agencies.

(*ii*) It fosters scientific communication by sponsoring meetings, workshops and lecture series.

(*iii*) It supports training on basic and applied research in computational biology for post-doctoral fellows through the NIH intra-mutual research programme.

(*iv*) It engages the members of the international scientific community in informatics research and training through the scientific visitors programmes.

(*v*) It develops, distributes, supports and coordinates access to a variety of databases and software for the scientific and medical communities.

(*vi*) It develops and promotes standards for databases, data deposition and exchange, and biological nomenclature.

The NCBI has developed many useful resources and tools. These may be grouped into different types such as database retrieval tools, BLAST family for search of DNA sequences, ePCR, gene level sequences, chromosomal sequences, genome analysis, analysis of gene expression patterns, molecular structure, LocusLink used in genome catalogue information about gene and gene-based markers, OMIM, UniGene etc.

These tools have their own websites which many be used free of cost. You will learn the practical aspects of some of these tools in your practical. Out of the above resources and tools, three sets of resources are discussed in this regard. Using these resources most of the cases of bioinformatics activity can be carried out. While doing advanced studies, the other resources may be used.

(*b*) GenBank: Since 1992, the NCBI has the responsibility for making available the DNA sequence database to the GenBank, which coordinates with individual laboratories and other sequence databases such as those of the European Molecular Biology Laboratory (EMBL) and the DNA Database of Japan (DDBJ). Moreover, the NCBI has grown to provide other databases in addition to GenBank. The NCBI provides Online Mendelian Inheritance in Man, the Molecular Modeling Database (3D protein structures), dbSNP a database of single-nucleotide polymorphisms, the Unique Human Gene Sequence Collection, a Gene Map of the human genome and a Taxonomy Browser. Besides, the NCBI coordinates with the National Cancer Institute to provide the Cancer Genome Anatomy Project. The NCBI assigns a unique identifier (Taxonomy ID number) to each species of organism.

The NCBI has software tools that are available by WWW browsing or by FTP. For example, BLAST is a sequence similarity searching programme. One can use BLAST for sequence comparisons against the GenBank DNA database in less than 15 seconds.

Fig. 12.12 shows the growth in nucleotide sequences submitted to GenBank between 1982 and 2006.The note from each release of GenBank reveal the total number of nucleotides submitted to the database. One data point from each year is used to show the exponential growth rate of nucleotide sequences in the international database. Data from expressed sequence tag and that from sequencing centers are included in Fig. 12.10.

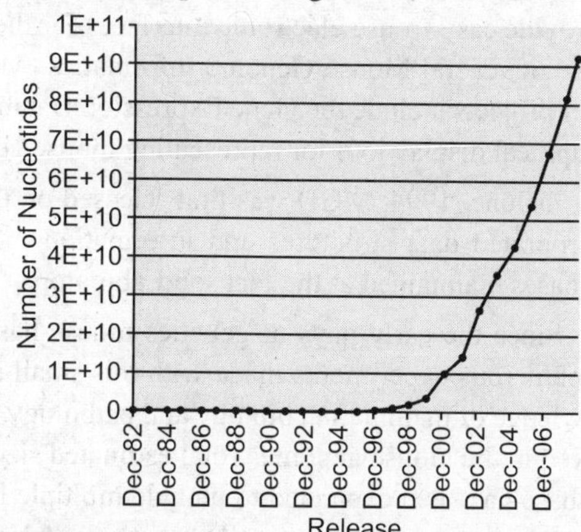

Fig. 12.12. Growth in nucleotide sequences submitted to GenBank between 1982 and 2005 (*Source*: Nature Education, 2008).

GenBank is a comprehensive sequence database that contains publicly available DNA sequences for more than 1,19, 000 different organisms obtained through the submission of sequence data from individual laboratories and batch submissions from large-scale sequencing projects. Mostly data are submitted by using the BankIt (web) or Sequin programmes; thereafter accession numbers are assigned by the staff GenBank. Daily data exchange with the EMBL Data Library in the UK and the DNA Data Bank of Japan helps worldwide coverage.

GenBank is accessible through Entrez which is retrieval system of the NCBI. Entrez integrates data from the major DNA and protein sequence databases along with taxonomy, genome, mapping, protein structure and domain information, and the biomedical journal literature via PubMed.

BLAST provides sequence similarity searches of GenBank and other sequence databases. FTP makes available complete bimonthly releases and daily updates of the GenBank database. One should go to the NCBI home page (http://www.ncbi.nlm.nih.gov.) to access GenBank and its related retrieval and analysis services.

2. Mouse Genome Database (MGD)

The Mouse Genome Database (MGD) is a comprehensive integrated database for genetic and biological data on the mouse, populated both with data from published literature and with data electronically submitted from the research community. The MGD is one component of the Mouse Genome Informatics (MGI) system (http://www.informatics.jax.org), which is a community database resource for the laboratory mouse. The MGD has evolved from a mapping and genetic resource to include sequence and genome information and details on the functions and roles of genes and alleles. MGD provides a

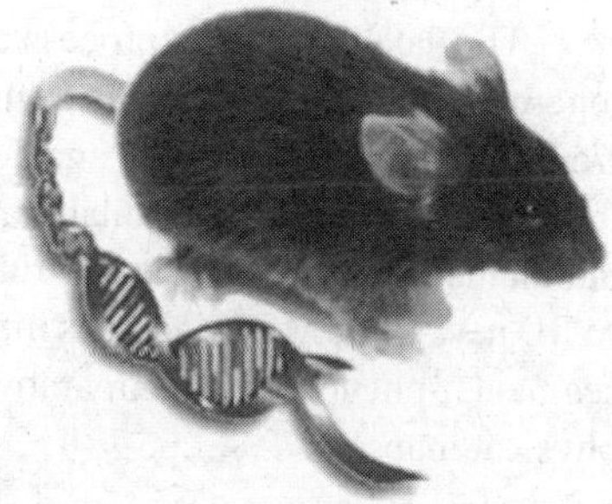

Mouse– a privileged model

comprehensive knowledgebase about the mouse with experiments and data annotated from both literature and online sources. Thus it is the primary public mouse genomic catalogue resource that curates and presents consensus and experimental data representations of genetic, genotype and phenotype information such as detailed reports about genes and gene-products. A relationship between genes, sequences and phenotypes is established to represent integration. MGD collaborates with other bioinformatics groups to curate a definitive set of information about the laboratory mouse and to build and implement the data and semantic standards essential for comparative genome analysis. Recently MGD has facilitated integration of the mouse sequence data and substantial revisions in the presentation, query and visualisation of sequence data.

At the Jackson Laboratory the Mouse Genome Informatics (MGI) project is developing and maintaining the Mouse Genome Database. The focus of MGD is to represent reference-supported elemental genomic data which is suitable for analysis and integration with other biological data, and to provide easy-to-use electronic interfaces for the scientific community to access these data. MGD is one of several Mouse Genome Informatics (MGI) projects going on at the Jackson Laboratory. Other projects include the Gene Expression Database, the Encyclopedia of the Mouse Genome, and a graphical display tool for representing mouse genetic maps.

In June, 1994, MGD was first released on the World Wide Web (WWW). The initial release incorporated data structures and information previously contained in independent, task-specific databases maintained at the Jackson Laboratory.

Since the early days of genetics mouse has been used as a privileged model organism, and the numerous experiments made with this small mammal have regularly contributed to enrich our knowledge of mammalian biology and pathology. MGD helped the annotation of the ~25,000 genes packed in the mouse genome. Total estimated size (NCBI build 37) is 2473 M bp. The scope of the database has been expanded to include multiple links to other biological databases and to enhance the representation of cytogenetic, genetic and physical maps. The Mouse Genome Database (MGD) showing search and other resources is given in Fig. 12.13.

Thus the MGD is an integrated data resource for mouse genetic, genomic, and biological information, which includes many kinds of data ranging from gene characterisation and genomic structures to orthologous relationships between mouse genes (and those of other mammalian species) to maps (genetic, cytogenetic, physical), to descriptions of mutant phenotypes, to characteristics of inbred strains, to information about biological reagents such as clones and primers. Data are accessed via search/retrieval Web forms and displayed as tables, text, and graphical maps, with supporting primary data. A rich set of hypertext links is provided such as those from gene and clone information to DNA and protein sequence databases (GenBank, EMBL, DDBJ, SWISS-PROT), from bibliographic data to PubMed, from phenotypes to OMIM (Online Mendelian Inheritance in Man), and from gene homology records to the genomic databases of other species.

The data integration process of MGD places disparate data in contextual relationship, bringing together information from electronic downloads, the mouse Chromosome Committees, submissions from individual researchers, and actively curated data from the published literature. For facilitating the communication and collaboration among interested scientists electronic bulletin boards are maintained, which provide a forum for community discussions and a communication vehicle for MGD news. The MGD includes information on mouse genetic markers and nomenclature, molecular segments, phenotypes, comparative mapping data, graphical display of linkage, cytogenetic and physical maps.

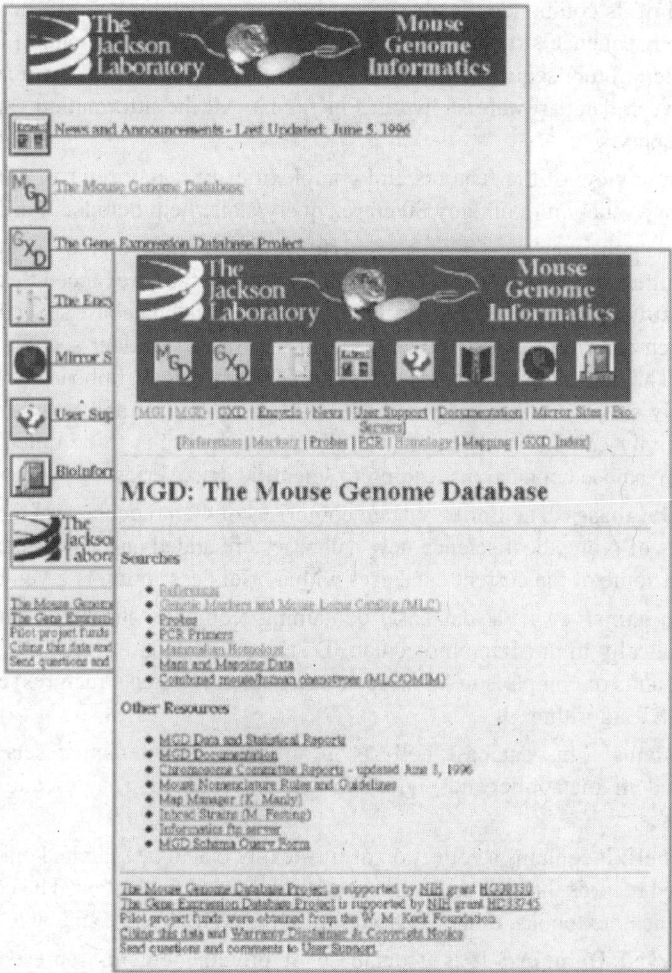

Fig. 12.13. The Mouse Genome Database (MGD) showing search and other resources.

F. DATA RETRIEVAL TOOLS

GenBank contains more than 8 millions sequence record covering 9 million nucleotide bases. Unless the databases are easily searched and entries retrieved in a usable and meaningful format, the biological databases serve a little purpose. Moreover, efforts made on sequencing will not be meaningful if biological community as a whole cannot make use of the information hidden within millions of bases and amino acids. There are several database retrieval tools such as ENTREZ, LOCUSLINK, TAXONOMY BROWSER, etc.

1. ENTREZ

Entrez is primary text search and retrieval system of NCBI that integrates the PubMed database of biomedical literature with 39 other literature and molecular databases including DNA and protein sequence, structure, gene, genome, genetic variation and gene expression. This document is an overview of the Entrez databases with general information on searching and displaying data. The NCBI Help Manual sections on the NCBI bookshelf can help more for the individual Entrez databases. Using Entrez system you can access literature, prepare genome map, sequences (both protein and nucleotides) and get structural data (3D). Entrez is not a database; but it is the interface through which all of its component databases can be accessed and traversed. Entrez has ability to retrieve the related sequence structures and references. The Entrez information includes PubMed records, nucleotide, protein sequence data, 3D structure, information and mapping. The hard link relationships between databases are shown in Fig. 12.12. All the information can be accessed by issuing only one query.

For complete review of the features and complexities one may refer to a tutorial on Entrez system at : http://www.ncbi.nlm.nih.gov:80/entrez/query/static/help/helpdoc.html.

The Entrez search interface features powerful options which include popular configurable limits and preset filters to help focus on specific kinds of results, an Advanced Search interface that facilitates the construction of more sophisticated queries. For each database specialised search fields are available which can be browsed and selected in the Search Builder section of the Advanced Search interface. The other Entrez features are Search History and a Clipboard where search results can be temporarily saved. Entrez integrates data with links within and between databases. This does interconnectivity enhance navigation and allow search results to be quickly focused; these relationships often expose connections leading to scientific discoveries.

(*a*) Entrez Databases: The Entrez system comprises of 40 molecular and literature databases. With the advances of biomedical science new databases are added and new kinds of data become available. A list of some of the current databases with a brief description is given below:

(*i*) **3D Domains:** It is a database containing compact structural domains identified automatically in the macromolecular 3D structure database (MMDB). The 3D domains are the units of comparison for structure neighbour (related structures) calculations using the VAST algorithm.

(*ii*) **BioSystems:** This database collects information on interacting sets of biomolecules involved in metabolic and signalling pathways, disease states and other biological processes.

(*iii*) **Bookshelf:** It contains a collection of full-text books to be searched online. It is linked to PubMed records through research paper citations within the text. The collection includes biomedical textbooks, other scientific titles, the NCBI News, and NCBI help manuals.

(*iv*) **Conserved Domains:** It is a database of protein domains represented by sequence alignments and profiles for protein domains conserved in molecular evolution. The source databases for conserved domains are Pfam, Smart, and COG.

(*v*) **EST:** These are typically short single-pass reads from cDNA libraries often generated as large survey project. The EST database contains sequence records from the bulk EST division of GenBank. Data from EST can be used to catalog expressed genes for a particular organ, tissue or cell type of a species, and compared expression levels of genes in various library sources.

(*iv*) **Gene:** A searchable database of genes is Gene that focuses on genomes which have completely been sequenced and that have an active research community to contribute gene specific data. Information in Gene records includes nomenclature, chromosomal localisation, gene products and their attributes (e.g., protein interactions), associated markers, phenotypes, interactions, and links to citations, sequences, variation details, maps, expression reports, homologs, protein domain content and external databases.

(*vii*) **Genome:** The Genome database consists of sequence and map data from the whole genomes of over 1,000 species or strains. It represents both completely sequenced genomes and sequencing in-progress sequences. Besides many viruses, phages, viroids, plasmids and organelles, all three main domains of life such as bacteria, archaea, and eukaryota are represented in Genome.

(*viii*) **Genome Project:** Genome Projects collects information on complete and in-progress large scale sequencing, assembly, annotation and mapping projects for cellular organisms.

(*ix*) **GEO Datasets:** The Gene Expression Omnibus (GEO) datasets store curated gene expression and molecular abundance data sets assembled by NCBI from the GEO repository of microarray data.

(*x*) **GEO Profiles:** GEO Profiles is a database that stores individual gene expression and molecular abundance profiles assembled from the GEO repository of microarray data.

(*xi*) **GSS:** The GSS database (Genome Survey Sequence) contains sequence records from the bulk GSS division of GenBank. These are the genomic equivalent of EST records; short single pass reads from gDNA libraries. Insert end and other reads from BAC and other large insert genomic libraries identify and assemble candidates for genome sequencing.

(*xii*) **HomoloGene:** The HomoloGene database contains automatically generated sets of homologous genes and their corresponding mRNA, genomic, and protein sequence data from selected eukaryotic organisms. Potential homologs from other organisms are included through sequence similarity to UniGene clusters.

(*xiii*) **Journals:** Journals is a database that contains information for journals cited in any Entrez database record. Journal records provide the journal title, MEDLINE abbreviation, NLM ID, ISO abbreviation or ISSN.

(*xiv*) **Nucleotide:** The Nucleotide database contains all the sequence data from GenBank, EMBL, and DDBJ, the members of the International Nucleotide Sequence Databases Collaboration (INSDC). Nucleotide also includes NCBI-curated Reference Sequences (RefSeqs), submitted assemblies and annotations from the Third Party Annotation (TPA) database, and nucleotide sequences extracted from structure records from the Protein Databank (PDB).

(*xv*) **NLM Catalog:** The NLM Catalog contains records for books, journals, audiovisuals, computer software, electronic resources and the other materials in the National Library of Medicine (NLM).

(*xvi*) **OMIM:** The OMIM (Online Mendelian Inheritance in Man) database contains review articles, human genes, genetic disorders, and other inherited traits. OMIM articles provide links to associated literature references, sequence records, maps, and related databases.

(*xvii*) **Peptidome:** Peptidome is an archive of tandem mass spectrometry peptide and protein identification data generated by the scientific community. Peptidome includes identified proteins, identified peptides for protein identification, mass spectra that support identifications, and supporting documentation.

(*xviii*) **PopSet:** The PopSet database contains related nucleotide sequences that originate from comparative studies: phylogenetic, population, environmental (ecosystem), and mutational. Each record in the database is a set of nucleotide sequences representing the same molecule from the same species (population, mutation), different identifiable species (phylogenetic), or anonymous species from the same biological community (ecosystem).

(*xix*) **Protein Clusters:** Protein Clusters is a collection of related protein sequences (clusters) consisting of Reference Sequence proteins that are encoded by complete prokaryotic genomes as well those encoded eukaryotic organelle plasmids and genomes. The database provides easy access to annotation information, publications, domains, structures, external links, and analysis tools.

(*xx*) **Protein:** The Protein database contains amino acid sequences created from the translations of coding regions provided on nucleotide records in GenBank, EMBL, and DDBJ, the members of the International Nucleotide Sequence Databases Collaboration (INSDC) as well as those from coding regions on NCBI Reference Sequences and the Third Party Annotation (TPA) database records. Protein records are also imported from the outside protein-only data sources Protein Information Resource (PIR), SWISS-PROT, Protein Research Foundation (PRF). Protein sequences are also extracted from structure records from the Protein Data Bank (PDB).

(*xxi*) **PubChem BioAssay:** PubChem BioAssay is a database that contains bioactivity screens of chemical substances described in PubChem Substance. It provides searchable descriptions of each bioassay, including descriptions of the conditions and readouts specific to that screening procedure.

(*xxii*) **PubChem Substance:** The PubChem Substance database contains information on chemical substances including mixtures electronically submitted to PubChem by depositors. This includes any chemical structure information submitted as well as chemical names, comments, and links to the depositor's web site.

(*xxiii*) **PubMed:** PubMed is database of citations and abstracts for biomedical literature from MEDLINE and additional life science journals. Links are provided when full text versions of the articles are available through PubMed Central or other websites.

(*xxiv*) **SNP:** The SNP (Single Nucleotide Polymorphism) database is a central repository for single nucleotide polymorphisms, microsatellites, and small-scale insertions and deletions. Both submitted SNPs and NCBI-produced non-redundant reference records (RefSNPs) that cluster reports of the same polymorphism from different sources are available. SNP also contains population-specific frequency and genotype data, experimental conditions, molecular context, and mapping information for both neutral polymorphisms and clinical mutations.

(*xxv*) **SRA:** The SRA (Sequence Read Archive) contains sequencing data from the next generation sequencing platforms. SRA accepts and presents data from all current next generation sequencing platforms including 454 (Roche), Illumina, SOLiD (Applied Biosystems), HeliScope, and Complete Genomics. Data can include sequence, quality scores, colour values, and intensity graphs depending on the platform involved.

(*xxvi*) **Structure:** The Structure or Molecular Modeling Database (MMDB) contains experimental data from crystallographic and NMR structure determinations. The data for MMDB are obtained from the Protein Data Bank (PDB).

(*xxvii*) **Taxonomy:** The Taxonomy database contains the names and phylogenetic lineages of the more than 160,000 organisms that have molecular data in the NCBI databases. New taxa are added to the Taxonomy database as data are deposited for them. The taxonomy records include links to all molecular data for the organism or group as well as links to outside classification resources. The taxonomy provides the major controlled vocabulary for classifying molecular data across the Entrez system.

(*xxviii*) **UniGene:** UniGene is a database that provides automatically generated non-redundant sets (clusters) of transcript sequences, each cluster representing a distinct transcription locus (gene or expressed pseudogene). UniGene clusters also provide information on protein similarities, gene expression, cDNA clone reagents and genomic location.

(*xxix*) **UniSTS:** UniSTS is a comprehensive database of sequence tagged sites (STSs) derived from STS-based maps and other experiments. STSs are defined by PCR primer pairs and are associated with additional information such as genomic position, genes, and sequences.

2. OMIM (Online Mendelian Inheritance in Man)

OMIM is a non-sequence-based information resource that is very much useful in genomics. It is a web-based electronic version of catalogue that contains thousands of entries for human genes and genetic disorders. It serves as a phenotypic companion to Human Genome Project. In 1998, the information in this database was collected and processed under the leadership of Victor A. McKusick at Johns Hopkins University, Baltimore, U.S.A. assisted by a team of science writers and editors. Relevant articles are identified, discussed and written up in the relevant entries in the MIM database. Concise textual information is provided by OMIM from the published literature on the conditions of human having genetic disorders and full citation information. The online version of OMIM is housed at the NCBI. Links are provided to Entrez from all references cited within each OMOM entry. Internet resource for OMIM is: http://www.ncbi.nlm.nih.gov/omim.

Victor Almon McKusick

Victor Almon McKusick (1921–2008), internist and medical geneticist, was University Professor of Medical Genetics and Professor of Medicine at the Johns Hopkins Hospital, Baltimore, USA. He was the original author and, until his death, remained chief editor of *Mendelian Inheritance in Man* (MIM) and its online counterpart *Online Mendelian Inheritance in Man* (OMIM), a database of heritable diseases and genes. He is widely regarded as the father of Clinical Medical Genetics. McKusick also served as editor-in-chief of the journal *Medicine*.

Victor A. McKusick

On July 22, 2008, McKusick died of cancer at the age of 86 at his home in Towson.

***(a)* MIM Code:** The method of assigning MIM codes is shown in Table 12.7. A six digit number is assigned to every disease and gene, the first number of which classifies the method of inheritance. The initial digit 1 shows the trait autosomal dominant, the 2 shows autosomal recessive, and the digit 3 represents for X-linked traits. In this dictionary wherever a trait defined has a MIM number, the number from the 12th edition of MIM, is given in square brackets with or without an asterisk [asterisks indicate that the mode of inheritance is known; a number symbol (#) before an entry number means that the phenotype can be caused by mutation in any of two or more genes] as appropriate *e.g.* Pelizaeus-Merzbacher disease [MIM #312080] is an X-linked recessive disorder.

Table 12.7. The method of assigning MIM codes.

First Digit	*Range of MIM codes*	*Method of inheritance*
1	100000–199999	Autosomal dominant loci or phenotypes (created before May 15, 1994)
2	200000–299999	Autosomal recessive loci or phenotypes (created before May 15, 1994)
3	300000–399999	X-linked loci or phenotypes
4	400000–499999	X-linked loci or phenotypes
5	500000–599999	Mitochondrial loci or phenotypes
6	600000–	Autosomal loci or phenotypes (created after May 15, 1994)

The OMIM cytogenetic and morbid maps present cytogenetic locations for those genes with published locations and provide an alphabetical list of all the diseases described in OMIM. Therefore, it is necessary to consider the results of web-based biology reported in the scientific literature in order to validate the findings generated through computer-based comparative analysis.

Hence, integration of scientific data with the literature is an important step for creating a unified information resource in the life science. For this purpose, individuals are provided with a direct link from OMIM to PubMed, the NCBI literature system.

It is very easy to perform OMIM searches. A simple query is performed by search engine on the basis of one or more words typed into a search window. Consequently, a list of documents is returned containing the query words. The users can select one or more disorders from the list so as to see the full text of OMIM entry. OMIM statistics showing number of entries for March 10, 2011 (Johns Hopkins University is given in Table 12.8.

Table 12.8. OMIM statistics showing number of entries for March 10, 2011 (Johns Hopkins University).

	Autosomal	*X-Linked*	*Y-Linked*	*Mitochondrial*	*Total*
Gene with known sequence	12605	620	48	35	13308
Gene with known sequence and phenotype	314	18	0	2	334
Phenotype description (molecular basis known)	2725	236	4	28	2993
Mendelian phenotype or locus (molecular basis unknown)	1632	134	5	0	1771
Other, mainly phenotypes with suspected Mendelian basis	1831	130	2	0	1963
Total	19107	1138	59	65	**20369**

3. PubMed (Publishers on Medicine)

PubMed is a Web search interface that provides over 11 million journal citation in MEDLINE and contains links to full text articles at participating publisher's web site. PubMed provides web-based access to over 11 million citations, abstracts and indexing terms for journal articles in the biomedical sciences. It also includes links to full-text journals. At present about 20 million searches are conducted per month and over 1,40,000 users seek information daily through PubMed.

4. Taxonomy Browsers

Diversity of organisms is such that millions of species are known. It is hoped that millions of organisms are also unknown. After a species is known, its various features are studied and information is restored in database. So far information of over 79,000 organisms is restored in database.

The species are classified according to their characters and similarity with others through the information provided by taxonomy browser (http://www.ncbi.nlm.nih.gov/taxonomy/tax.html). Browsers are used to work at web page. Browsers client-server applications that connect to remote site, download the requested information at the site, and display the information on a user's monitor disconnecting from the remote host.

5. LocusLink

LocusLink is an NCBI project to link information applicable to specific genetic loci from several disparate databases. It is principally intended for use by graduate students and professional researchers in the biomedical sciences. LocusLink provides a single query interface to various types of information regarding to a given genetic locus such as phenotypes, map locations, homologies to other genes and information on official nomenclature. Currently LocusLink search space includes information from human, mice, rats, fruitfly and zebrafish. It carries information on mouse homologue of a given human gene, you cannot get. Beginning with LocusLink query simply by typing the name of the gene into the query box which appears on the top of LocusLink home page, you can select the gene of interest from an alphabetical list.

It presents information on official nomenclature, aliases, sequence accessions, phenotypes, EC numbers, MIM numbers, UniGene clusters, homology, map locations, and related web sites. LocusLink links to related web sites including the OMIM. The internet resource for LocusLink is: http://www.ncbi.nlm.nih.gov/LocusLink, as given in Fig. 12.14.

6. Sequence Retrieval System (SRS)

The SRS has been created by Swiss Institute of Bioinformatics and the European Bioinformatics Institute that have also created the Swiss-PROT database. SRS allows retrieval from an extensive catalogue of more than 75 public biological databases. The link button in SRS will allow to get all the entries in one databank which is linked to an entry (or entries) in another databank. Hyperlinks made links between the entries. SRS (http://srs/ebi/ac/uk/) is a homogeneous interface to over 80 biological databases that had been developed at the European Bioinformatics Institute (EBI) at Hinxton, UK [*see* also http://srs.ebi.ac.uk/srs5/man/mi_srswww.html]. The SRS includes databases of sequences, metabolic pathways, transcription factors, application results (such as BLAST, SSEARCH, and FASTA), protein 3-D structures, genomes, mappings, mutations and locus specific mutations. The web page listing all the databases contains a link to a description page about the database including the date on which it was last updated. You select one or more of the databases to search before entering your query. After getting results you choose an alignment algorithm (like CLUSTALW, PHYLIP) enters parameters, and run it. The SRS is highly recommended for use.

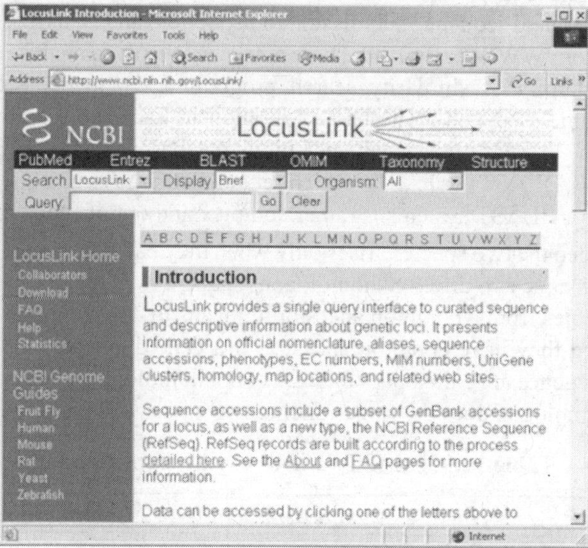

Fig. 12.14. The web site LocusLink.

SRS is one of the most available powerful data browsing/retrieval tools. SRS provides rapid, easy and user friendly access to the large volumes of diverse and heterogeneous 'life science' data stored in more than 400 internal and public domain databases. It can be used to browse the various biological sequence and literature databases. The SRS is premier data integration, analysis and display tool for bioinformatics, genomics and related data of the world. The Computer Centre has recently set up the SRS on the BIOINFO server (http://bioinfo.hku.hk/srs) for supporting large-scale speedy access of the genetics sequence from HKU.

The latest SRS version 7 system has been installed BIOINFO server, which can provide speedy and efficient performance for the searching, query and linkage of sequences from the huge databases. SRS7 provides several enhancements in sequence parsing, sorting and logging, as well as supporting the XML database and EMBOSS applications.

G. DATABASE SIMILARITY SEARCHING

Two software, the Basic Local Alignment Search Tool (BLAST) and FASTA, are described for similarity searching of biomolecules.

1. Basic Local Alignement Search Tool (BLAST)

Due to genome searching projects on a large scale, the flood of DNA sequence data coming into public databases is staggering. The search for sequence similarity is very important; for example, a researcher who has discovered a potentially important DNA or protein sequence wants to know if it has already been identified and characterized by another researcher. Scientists are relying on deducing the function of putative genes through similarity to well characterised proteins.

The Basic Local Alignment Search Tool (BLAST) is one of the most popular software tools in biological research for sequence similarity searching developed at the NCBI. BLAST is an algorithm for comparing primary biological sequence information, such as the amino-acid sequences of different proteins or the nucleotides of DNA sequences. It tests a query sequence against a library of known sequences to find out similarity. Thus, BLAST is a collection of programmes with versions for query-to-database pairs such as nucleotide-nucleotide, protein-nucleotide, protein-protein, nucleotide-protein, etc. It executes sequences searches against the entire DNA database in less than 15 seconds.

Additional software tools provided by the NCBI are: Open Reading Frame Finder (ORF Finder), Electronic PCR (ePCR), and sequence submission tools, Sequin and BankIt. All the databases of the NCBI are available from WWW or by FTP. There are several tools to analyse sequence information among the BLAST family of similarity search programme. Sequence similarity searches use alignments to determine a 'match'. The basic operation in database searching is to sequentially align a query sequence to each subject sequence in the database. Most users prefer BLAST (http://www.ncbi.nlm.nih.gov/BLAST) or FASTA which rely on heuristic strategies to speed up alignment searches. The theory of BLAST systems is rather complex and out of scope of this book.

Similarity between the two species arises only when they had common ancestors. For example, if two species A and B have two different homologous sequences, it means that they have descended into two different species from a common ancestor. It means that two similar sequences may not be homologous, provided they have common ancestry. If a gene duplicates in a genome, it will also show similarity. Such genes are called paralogues, but not homologues; because paralogue differs in function, whereas homologue does the same function.

Table 12.9. BLAST variants for different searches.

Programme	*Query*	*Comparison database*	*Common use*
BLASTp	Protein	Protein	Seeks to align an amino acid query
BLASTn	Nucleotide	Nucleotide	Align to new DNA sequence to a nucleotide sequence database
BLASTx	Nucleotide (translated)	Protein	Analyse new DNA sequence (translated) to find out potential coding regions
TBLASTn	Protein	Nucleotide (translated)	Useful for EST analysis
TBLASTx	Nucleotide (translated)	Nucleotide (translated)	Useful for EST analysis

There are several variants of BLAST, each is distinguished by the types of sequences (DNA or protein) of the query and database sequences (Table 12.9). The principles of working system of BLAST are: (*i*) using matrices of BLAST a given sequence is compared with database sequence. It specifies scores to either *reward* a match or *penalise* a mismatch, (*ii*) based on the criteria the top scoring matches are ranked. It differentiates the similarity either due to ancestry or random chance. Generally these criteria are not changed. But you can change them as you wish, and (*iii*) further examination of true matches are done using the other details which are accessible through Entrez or other tools at NCBI.

The original version of BLAST stretches a longer alignment between the query and the database sequence in the left and right directions, from the position where the exact match occurred. The extension does not stop until the accumulated total score of the HSP begins to decrease. The high-scoring segment pair (HSP) can be listed whose scores are greater than the empirically determined cutoff score. A cutoff score can be determined by examining the distribution of the alignment scores modeled by comparing random sequences (Fig. 12.15).

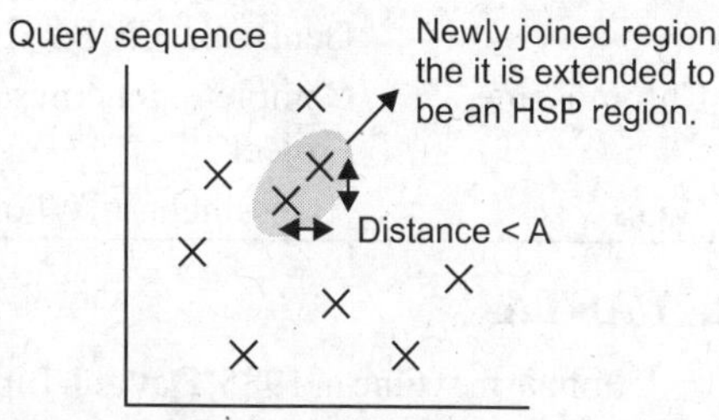

Fig. 12.15. The positions of the exact matches.

(*a*) **Uses of BLAST:** BLAST can be used for several purposes as given below:

(*i*) **Species Identification:** With the use of BLAST, you can possibly correctly identify a species and/or find homologous species. This can be useful, for example, when you are working with a DNA sequence from an unknown species.

(*ii*) **Locating Domains:** When working with a protein sequence you can input it into BLAST, to locate known domains within the sequence of interest.

(*iii*) **Establishment of Phylogeny:** Using the results received through BLAST you can create a phylogenetic tree using the BLAST web-page. It should be noted that phylogenies based on BLAST alone are less reliable than other purpose-built computational phylogenetic methods, so should only be relied upon for 'first pass' phylogenetic analyses.

(*iv*) **DNA Mapping:** When working with a known species, and looking to sequence a gene at an unknown location, BLAST can compare the chromosomal position of the sequence of interest, to relevant sequences in the database(s).

(*v*) **Comparison:** When working with genes, BLAST can locate common genes in two related species, and can be used to map annotations from one organism to another.

One of the important steps to perform BLAST is to select a suitable database for searching the query sequence to carryout matches. By selecting the database from a drop down box, species-specific BLAST search can be made. One can find different types of database in the GenBank, some of them are given in Table 12.10.

Table 12.10. Some of the protein and nucleotide databases available to perform BLAST search.

Database	*Description*
Protein databases	
Refseq	Protein sequence from NCBI reference project
Swissprot	Last major release of SWISSPROT protein sequence database
Pat	Protein from the Patent Division of GenBank
PDB	Sequence obtained from 3'D record structure record from Protein Data Bank
Env_nr	Non-reductant CDS translation from env_nt entries.

Nucleotide databases	
ReSeq _ mrna	mRNA sequences from the NCBI Reference Sequence Project
EST	Databank of GenBank+EMBL+DDBJ sequence from EST division
EST_human	Human subset of EST
GSS	Single-pass genomic data of Genome Survey Sequence
PAT	Nucleotides from Patent Division of GenBank
dBST	Database of Sequence Tag Site Entries from EST division of GenBank+EMBL+DDBJ
Chromosome	Complete genomes and chromosomes from the NCBI Reference Sequence Project
WGS	Assessment of Whole Genome Shotgun sequences

2. FASTA

For the first time in 1985, David J. Lipman and William R. Pearson described FASTA (as FASTP) that is a software package for DNA and protein sequence alignment. FASTA added the ability to do DNA : DNA searches, translated protein:DNA searches, and provided a more sophisticated shuffling programme to evaluate the statistical significance. In this package there are several programmes that allow the alignment of protein sequences and DNA sequences. FASTA was the first widely-used programme for database similarity search that performs optimised search for local alignment using a substitution matrix. It takes sufficient time to apply this strategy exhaustively. This programme uses the observed pattern of word hits to identify potential matches before attempting the more time-consuming optimised search. The *k-tuple* or KTup parameter controls the trade-off between speed and sensitivity.

FASTA is pronounced 'fast A', and stands for 'FAST-All', because it works with any alphabet, an extension of 'FAST-P' (protein) and 'FAST-N' (nucleotide) alignment. The current FASTA package contains programmes for protein:protein, DNA:DNA, protein : translated DNA, and ordered or unordered peptide searches. Recent versions of the FASTA package include special translated search algorithms that correctly handle frameshift errors (which six-frame-translated searches do not handle very well) when comparing nucleotide to protein sequence data.

***(a)* FASTA Format:** FASTA format contains a defimilion line and sequence characters; it may be used as input to many analysis programmes. FASTA format is used in several software suites of molecular biology. In general BLAST tends to be faster and is more sensitive in detecting more alignments, but FASTA returns fewer false hits. The FASTA file format used as input for this software is now largely used by other sequence database search tools (such as BLAST) and sequence alignment programmes (Clustal, T-Coffee, etc).

A sequence in FASTA format begins with a single-line description, followed by lines of sequence data. The description line is distinguished from the sequence data by a greater-than (>) symbol in the first column. It is recommended that all lines of text be shorter than 80 characters in length. An example of sequence in FASTA format is as below:

```
>gi|532319|pir|TVFV2E|TVFV2E envelope protein
ELRLRYCAPAGFALLKCNDADYDGFKTNCSNVSVVHCTNLMNTTVTTGLLLNGSYSE
NRTQIWQKHRTSNDSALILLNKHYNLTVTCKRPGNKTVLPVTIMAGLVFHSQKYNLRL
RQAWCHFPSNWKGAWKEVKEEIVNLPKERYRGTNDPKRIFFQRQWGDPETANLWFN
CHGEFFYCKMDWFLNYLNNLTVDADHNECKNTSGTKSGNKRAPGPCVQRTYVACHI
RSVIIWLETISKKTYAPPREGHLECTSTVTGMTVELNYIPKNRTNVTLSPQIESIWAAELD
RYKLVEITPIGFAPTEVRRYTGGHERQKRVPFVXXXXXXXXXXXXXXXXXXXXXXVQ
SQHLLAGILQQQKNLAAVEAQQQMLKLTIWGVK
```

A multiple sequence FASTA format would be obtained by concatenating several single sequence FASTA files. This does not imply a contradiction with the format as only the first line in a FASTA file may start with a ';' or '>'; hence forcing all subsequent sequences to start with a '>' in order to be taken as different ones (and further forcing the exclusive reservation of '>' for the sequence definition line).

Sequences are expected to be represented in the standard IUB/IUPAC amino acid and nucleic acid codes, with these exceptions: lower-case letters are accepted and are mapped into upper-case; a single hyphen or dash can be used to represent a gap of indeterminate length; and in amino acid sequences, U and * are acceptable letters. Any numerical digits in the query sequence should either be removed or replaced by appropriate letter codes (e.g., N for unknown nucleic acid residue or X for unknown amino acid residue), before submitting a request.

The nucleic acid codes supported are given in Table 12.2. The accepted amino acid codes are given in Table 12.4 for those programmes that use amino acid query sequences (BLASTP and TBLASTN).

***(b)* Method of Searching:** A given nucleotide or amino-acid sequence is taken up by FASTA that searches a corresponding sequence database to find matches of similar database sequences. Searching work is done by using local sequence alignment. A largely heuristic method is followed by FASTA programme which executes at a high speed. Before performing a more time-consuming optimised search, it initially observes the pattern of word hits, word-to-word matches of a given length, and marks potential matches.

The sensitivity and speed of the programme is controlled by the size taken for a word that is given by ktup parameter. The number of background hits decreases with increasing the ktup value. The segments containing a cluster of nearby hits are looked up from the word hits that are returned to the programme. Thereafter, these segments for a possible match are investigated. There are some differences between FASTN and FASTP as far as the use of sequences are concerned, but both use four steps and calculate three scores to describe and format the sequence similarity results as given below:

- It identifies the regions of highest density in each sequence comparison, taking a ktup to equal 1 or 2. In this step all or a group of the identities between two sequences are found using a look up table. The ktup valuc determines the number of consecutive identities required for a match to be declared. Thus the lesser the ktup value: the more sensitive the search. Ktup = 2 is frequently taken by users for protein sequences and ktup = 4 or 6 for nucleotide sequences. Short oligonucleotides are usually run with ktup = 1. Groups of identities with high similarity scores contribute more to the local diagonal score than to identities with low similarity scores. Nucleotide sequences use the identity matrix for the same purpose. Then the best 10 local regions selected from all the diagonals put together are saved.
- The regions taken are rescanned by using the scoring matrices. Trimming the ends of the region includes only those that contribute to the highest score.
- If several initial regions with scores greater than a 'cutoff' value are found in an alignment, the trimmed initial regions can be joined to form an approximate alignment with gaps. A similarity score, *i.e.* the sum of the joined regions penalising for each gap 20 points, is calculated. This initial similarity score (*initn*) is used for ranking the library sequences. Then the score of the single best initial region found in step 2 is reported (*init1*).

- A banded Smith-Waterman algorithm is used to calculate an optimal score (*opt*) for alignment of query sequence to a database (library) sequence. A band of 32 residues centered on the *init1* region of step2 is taken up for calculating the optimal alignment. When all sequences are searched the programme, the initial scores of each database sequence is plotted in a histogram. Then the statistical significance of the 'opt' score is calculated. The final alignment is produced using a full Smith-Waterman alignment for protein sequences, whereas a banded alignment is provided for DNA sequences. FASTA algorithm uses the dynamic programming to optimise the alignment in a narrow band that encompasses the top scoring segments (Fig. 12.16).

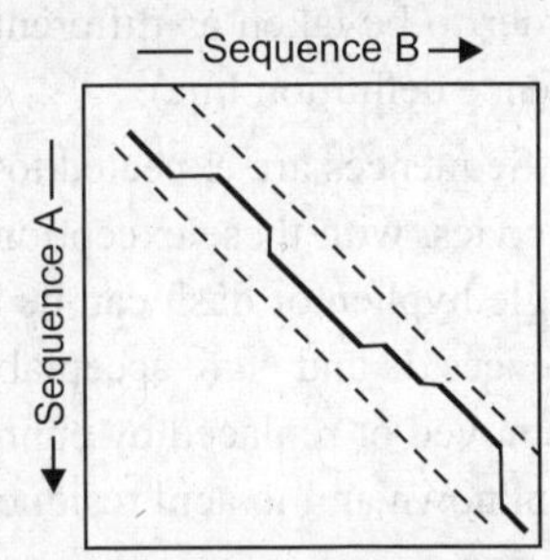

Fig.12.16. A narrow band optimized by FASTA algorithm.

FASTA files can be batch converted to or from MultiFASTA format using tools, some of which are available as freeware. Tools are also available for batch conversion from chromatogram formats (ABI/SCF) to FASTA. A comparison between BLAST and FASTA is given in Table 12.11.

Table 12.11. Comparison between BLAST and FASTA.

Comparison between BLAST and FASTA	
BLAST	***FASTA***
1. It is very fast, and takes 2-3 minutes to perform database search of given sequence.	Relatively it is slow and takes several hours to search database of given sequence.
2. Perfect sequence match is not required by it.	At the initial stage of searching it requires perfect sequence match.
3. It creates unaligned gaps; therefore, it fails to detect longest exons.	It is used for specialised task such as detection of genomic region by using cDNA sequence.
4. It is very sensitive for detecting protein similarity at default setting.	It is less sensitive for detecting protein similarity at default setting.
5. In microsatellite region of DNA, usually it gives erroneous results.	Due to Ktup, it gives better results in microsatellite region of DNA.

H. RESOURCES FOR GENE LEVEL SEQUENCES

There are several tools among the resources for gene level sequences *e.g.* UniGene, HomoloGene, RefSeq, etc.

1. UniGene Database

The ESTs have been described in previous section. Many redundant ESTs are generated during the course of its production. Because several cDNA clones represent the same gene (Fig. 12.17). Therefore, UniGene (one gene) database was developed at NCBI to control redundancy in EST data. UniGene is an NCBI database of the transcriptome and thus, despite the name, not primarily a database for genes. Each UniGene entry is a set of transcript sequences that appear to come from the same transcription locus (gene or expressed pseudogene), together with information on protein similarities, gene expression, cDNA clone reagents, and genomic location.

UniGene clustures ESTs and other mRNA sequences along with coding sequences (CDSs) annotated on genomic DNA into *subsets* of related sequences. The clusters are specific to organisms. At present clustures are available for human, mouse, rat, zebrafish and cattle.

The scheme for clustering ESTs is shown in Fig. 12.17 and steps are given below:

(*i*) First search the sequences for contaminants *e.g.* ribosomal, mitochondrial, and repetitive and vector sequences.

(*ii*) Enter the sequences (those contain about 100 bases) into UniGene. The mRNA and genomic DNA are clustered into gene links.

(*iii*) A second sequence comparison links ESTs to each other and to the gene links. All clusters are anchored and contain either a sequence with a polyadenylation site (poly A) or two ESTs labelled as coming from 3′ end of a clone.

(*iv*) Clone based edges are added by linking the 5′ and 3′ ESTs that derive from the same clone.

(*v*) Finally unanchored ESTs and gene clusters of size 1 are compared with UniGene clusters at lower stringency. The UniGene built is updated weekly. Then the sequence cluster may change.

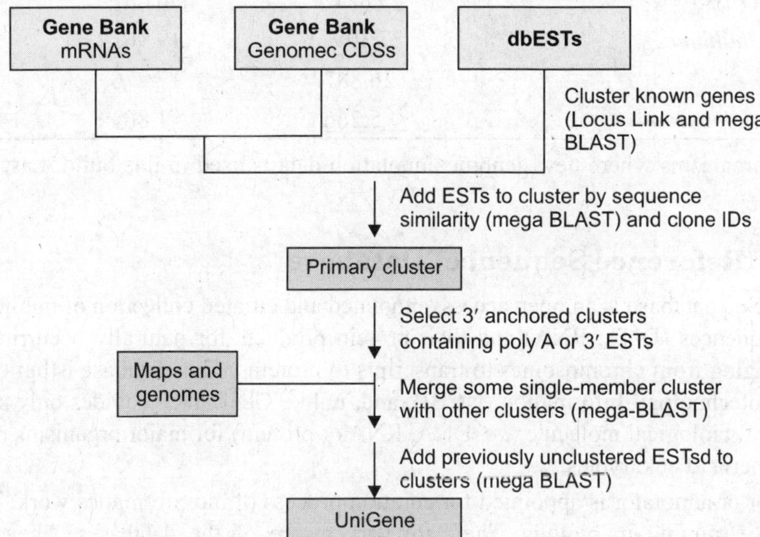

Fig. 12.17. Scheme for clustering ESTs by UniGene.

2. HomoloGene Database

A new UniGene resource has been created which is called 'HomoloGene' (homologous gene). HomoloGene database includes curated and calculated orthologs and homologs for genes from human, mouse, rat, zebrafish and cow. This database is also available in LocusLink. Using HomoloGene, homologous relations can easily be inferred. Homologs are identified as the best match between a HomoloGene cluster in one organism and a cluster in a second organism. When two sequences in different organisms are best matches to another the HomoloGene clusters corresponding to the pair of sequences are considered putative orthologs. HomoloGene is an automated system for detecting homologs among the annotated genes of several completely sequenced eukaryotic genomes (Table 12.12) (http://www.ncbi.nlm.nih.gov/sites/entrez?db=homologene).

Table 12.12. Initial numbers of genes from complete genomes, numbers of genes placed in a homology group, and the numbers of groups for each species.

Species	*Number of Genes*		*HomoloGene*
	Input	***Grouped***	***groups***
Homo sapiens	19,943*	18,981	18,431
Pan troglodytes	25,096	16,850	15,980
Canis familiaris	19,766	16,708	15,951
Mus musculus	25,388	21,766	19,005
Rattus norvegicus	21,991	19,229	17,473
Drosophila melanogaster	13,827*	9,282	7,749
Anopheles gambiae	12,460	8,867	7,541
Caenorhabditis elegans	20,132*	8,678	4,810
Schizosaccharomyces pombe	5,043	3,225	2,935
Saccharomyces cerevisiae	5,880	4,851	4,370
Neurospora crassa	9,821*	6,287	6,144
Arabidopsis thaliana	27,309*	19,961	11,243
Oryza sativa	26,887	17,276	10,627
Plasmodium falciparum	5,266	1,862	799

'*' indicates organisms where new genome annotation data is used in this build. Last updated on: Feb 14, 2011.

3. RefSeq (Reference Sequence) Database

The RefSeq database is an open access, annotated and curated collection of publicly available nucleotide sequences (DNA, RNA) and their protein products for naturally occurring biological molecules ranging from chromosomes to transcripts to proteins. This database is built by National Center for Biotechnology Information (NCBI), and, unlike GenBank, provides only single record for each natural biological molecule (*i.e.* DNA, RNA or protein) for major organisms ranging from viruses to bacteria to eukaryotes.

A curator or annotator is appointed for curation process of bioinformatics work. The curators have extensive training in biology. They are very aware of the databases. They ensure that no sequence data is lost during the process of submission. The curator reviews and checks the newly submitted data and ensures that: (*i*) the biological features are described adequately, (*ii*) the conceptual translation of coding regions follow the universal translation rules, and (*iii*) all mandatory information has been given.

The RefSeq project provides a stable reference sequence for all the molecules corresponding to central dogma of biology, *i.e.* flow of information from DNA → RNA → Protein. RefSeq mRNAs are reference sequence standards for the human genome. To generate a sequence standard, the NCBI investigators first collaborate with external organisation to gather different kinds of information. They assimilate these data using both computational tools and scientific judgment to determine what sequence is all appropriate representation of a gene.

The RefSeq provides separate and linked records, for each model organism, for the genomic DNA, gene transcripts, and proteins arising from those transcripts. RefSeq is limited to major organisms for which sufficient data is available, while GenBank includes sequences for any organism submitted (about 2,50,000 different named organisms).

I. SEQUENCE ALIGNMENT

In an alignment if two sequences share a common ancestor, mismatches can be interpreted as point mutations, and gaps as indels. It means that insertion or deletion mutations introduced in one or both lineages during the time because they diverged from one another. The degree of similarity between amino acids occupying a particular position in the sequence, during sequence alignments of proteins, can be interpreted as a rough measure of a particular conserved region or sequence motif among the lineages. The absence of substitutions or the presence of only conservative substitutions in a particular region of the sequence suggests the structural or functional importance of the region. Conservation of base pairs indicates similarity in structural or functional role, in spite of similarity between DNA and RNA nucleotides more than the amino acids.

The method of arranging the DNA, RNA or protein sequences to identify regions of similarity arising as a result of structural, functional or evolutionary relationships between the sequences is called sequence alignment (Mount, 2004). Within a matrix the nucleotide or amino acid sequences aligned are represented as rows. Between the residues, the gaps are inserted to align the identical or similar characters in successive columns. A sequence alignment of two human zinc finger proteins, produced by ClustalW and identified on the left by GenBank accession number (ClustalW2 FAQs http://www.ebi.ac.uk/Tools/clustalw2/help.html#color) is shown in Fig. 12.18.

Accession No. ***Sequence alignment of two zinc finger proteins of human***

```
AAB24882   TYHMCQFHCRYVNNHSGEKLYECNERSKAFSCPSHLQCHKRRQIGEKTHEHNQCGKAFPT  60
AAB24881   --------------------YECNQCGKAFAQHSSLKCHYRTHIGEKPYECNQCGKAFSK  40
                               **** :   .*** :   * * :** *   :**** .******** ..

AAB24882   PSHLQYHERTHTGEKPYECHQCGQAFKKCSLLQRHKRTHTGEKPYE-CNQCGKAFAQ-  116
AAB24881   HSHLQCHKRTHTGEKPYECNQCGKAFSQHGLLQRHKRTHTGEKPYMNVINMVKPLHNS  98
            ****  * : ****************** :*** :** ::  .******************     : * : :
```

Fig. 12.18. Alignment of zinc finger sequence; key: single letters- amino acids; red- small, hydrophobic, aromatic, not Y; blue- acidic; magenta- basic; green- hydroxyl, amine, amide, basic; gray- others. '*': identical. ':': conserved substitutions (same colour group). '.': semi-conserved substitution.

1. Global and Local Alignments

When the sequences in the query set are similar and of equal size, global alignments are most useful which attempt to align every residue in every sequence. Basically the Needleman-Wunsch algorithm is a global alignment technique that is based on dynamic programming. Local alignments are more useful for dissimilar sequences suspected to contain similar regions or similar sequence motifs within their larger sequence. The Smith-Waterman algorithm is a general local alignment method that is also based on dynamic programming. There is no difference between local and global alignments with sufficiently similar sequences.

Semi-global methods are the hybrid ones which find the best possible alignment that includes the start and end of one or the other sequence. Overlapping of downstream part of one sequence with the upstream part of the other sequence can be especially useful. A global alignment would force the alignment to extend beyond the overlapping region, whereas a local alignment may not fully cover the overlapping region (Brudno, 2003).

Fig. 12.19 shows thc global and local alignments having the 'gappy' quality of global alignments occurring when sequences are similar.

Global FTFTALILLAVAV

F- - TAL-LLA–AV

```
Local   FTFTALILL–AVAV
        - - FTAL-LLAAV- -
```

Fig. 12.19. Global and local alignments showing the 'gappy' quality of global alignments that occur when sequences are similar.

In a global alignment the entire genomic sequence is given in the output even in regions where it does not line up with the cDNA. Scroll down the output until you reach an area of alignment.

```
XLRHODOP    1          ggtagaacagcttcagttgggatcacaggcttcta      35
                       |||||||||||||||||||||||||||||||||||
XL23808     1171  tgggtcatactgtagaacagcttcagttgggatcacaggcttcta  1215
XLRHODOP    36    gggatcctttgggcaaaaagaaacacagaaggcattctttctat   80
                  ||||||||||||||||||||||||||||||||||||||||||||
XL23808     1216  gggatcctttgggcaaaaagaaacacagaaggcattctttctat   1260
XLRHODOP    81    acaagaaaggactttatagagctgctaccatgaacggaacagaag  125
                  |||||||||||||||||||||||||||||||||||||||||||||
XL23808     1261  acaagaaaggactttatagagctgctaccatgaacggaacagaag  1305
XLRHODOP    126   gtccaaatttttatgtccccatgtccaacaaaactggggtggtac  170
                  |||||||||||||||||||||||||||||||||||||||||||||
XL23808     1306  gtccaaatttttatgtccccatgtccaacaaaactggggtggtac  1350
```

The part of the output is very long. There are five aligned regions that represent the five exons as predicted from the dot plot. The intron/exon boundaries have a conserved gt..ag pair of dinucleotides delimiting the splice sites. These boundaries have correctly been aligned as it has no understanding of models of gene structure. The scoring method does not specifically treat splice sites. The programme EST2 genome has an extra scoring factor that allows to do a better job of aligning intron/exon boundaries.

2. Pair-wise Alignment

The best matching local or global alignments of two query sequences are found out by using pair-wise sequence alignment methods. However, pair-wise alignments can be used between two sequences at a time. Pair-wise alignments are often used for methods that do not require extreme precision e.g. searching a database for sequences with high similarity to a query. There are three primary methods used to produce pair-wise alignments: *dot-matrix*, *dynamic programming*, and *word methods* (Mount, 2004). Where the number of repetitions differs in two sequences to be aligned, these methods have difficulty with highly repetitive sequences of low information content.

(*a*) Dot-matrix Methods: The dot-matrix method is a qualitative and simple but time-consuming approach to analyse on a large scale. It produces a family of alignments for regions of an individual sequence. In the absence of noise it can be easy to visually identify certain sequence features, such as insertions, deletions, repeats, or inverted repeats from a dot-matrix plot. Two sequences are written along the top row and leftmost column of a two-dimensional matrix for constructing a dot-matrix plot. A dot is put at any point where the characters in the appropriate columns match. Depending on the degree of similarity of two characters, some implementations vary the size or intensity of the dot to accommodate conservative substitutions. The dot plots of very closely related sequences appear as a single line along the main diagonal of matrix.

However, there are many problems associated with dot plots as an information display technique such as noise, lack of clarity, non-intuitiveness, difficulty in extracting match summary statistics and match positions on the two sequences. Where the match data is duplicated across the diagonal there lays much wasted space. Most of the actual area of the plot is taken up by empty space or noise; dot-plots are limited to two sequences. These limitations do not apply to Miropeats alignment diagrams. Repetitiveness in a single sequence can be assessed by using dot plots. A sequence can be plotted against itself. The regions showing similarities appear as lines off the main diagonal. When a protein consists of multiple domains of similar structure this effect occurs frequently.

(*b*) Dynamic Programming: In 1970, Saul B. Needleman and Christian D. Wunsch first published the algorithm that performs a global alignment on two sequences called *A* and *B*. It is commonly used in bioinformatics to align protein or nucleotide sequences. The Needleman–Wunsch algorithm is an example of dynamic programming, and was the first application of dynamic programming to biological sequence comparison. Needleman and Wunsch described their algorithm explicitly for the case when the alignment is penalised solely by the matches and mismatches, and gaps have no penalty ($d = 0$).

The local alignment is performed via the Smith-Waterman algorithm which is a well-known algorithm for performing local sequence alignment, *i.e.* for determining similar regions between two nucleotide/protein sequences. The Smith-Waterman algorithm compares segments of all possible lengths and optimises the similarity measure. A substitution matrix is used by protein alignments to assign scores to amino-acid matches or mismatches, and a gap penalty for matching an amino acid in one sequence to a gap in the other. A scoring matrix may be used by DNA and RNA alignments; but it often assigns a positive match score, a negative mismatch score, and a negative gap penalty. The 'gap opening' is much larger than the 'gap extension' e.g. -10 for gap open and -2 for gap extension. In an alignment the number of gaps is usually reduced, and residues and gaps are kept together. By using three matrices the Gotoh algorithm implements affine gap costs.

For aligning nucleotide to protein sequences dynamic programming can be useful. A series of global or local pair-wise alignments between a query nucleotide sequence and a search set of protein sequences, are produced by the frame-search method. This method is useful for sequences containing large numbers of indels due to its ability to evaluate frame-shifts offset by an arbitrary number of nucleotides. However, it can be very difficult to align with more efficient heuristic methods. A system having specialised architecture for dynamic programming is needed by this method.

Basic tools for creating translated alignments are provided by the BLAST and EMBOSS suites. Both commercial sources such as FrameSearch (as part of the Accelrys GCG package) and Open Source software (*e.g.* Genewise) make available more general methods.

(*c*) Word Methods: Word methods arc also known as *k*-tuple methods that are heuristic methods not guaranteed to find a solution for optimal alignment but are more efficient than dynamic programming. These methods are useful in large-scale database searches. A large proportion of the candidate sequences have no significant match with the query sequence. Word methods are the best to be used in the database search tools FASTA and the BLAST family (Mount, 2004). In the query sequences a series of short, non-overlapping subsequences called 'words' are identified by word methods. Then these are matched to candidate database sequences. In the two sequences the relative positions of the word compared are subtracted to obtain an offset; if multiple distinct words produce the same offset, this will indicate a region of alignment. Several unnecessary comparisons are eliminated that have no similar sequences.

A value *k* is defined by the user which is used as the word length for searching the database in the FASTA method. At lower values of *k*, this method is slower but more sensitive. However, it is preferred for searching using a very short query sequence. Several algorithms that have been optimised for particular types of queries (e.g. searching for distantly related sequence matches) are provided by the BLAST family of search methods.

BLAST was developed as an alternative to FASTA that is faster and work without sacrificing much accuracy. BLAST uses a word search of length *k* similar to FASTA but only the most significant word matches are evaluated. Instead, FASTA does every word match. A fixed default word length optimised for the query and database type is used by most of the BLAST implementations. However, it can be changed only under special circumstances, e.g. while searching with repetitive or very short query sequences. Several web portals e.g. EMBL FASTA and NCBI BLAST can provide implementations.

3. Multiple Sequence Alignment (MSA)

Multiple sequence alignment is one of the most basic tasks in bioinformatics. It has been used for building phylogenetic trees, locating conserved motifs and domains and predicting secondary and tertiary structures. Three different algorithms have been developed. The progressive algorithm [used in, for example, Pileup, ClustalW, T-Coffee, MUSCLE and MAFFT] starts with the alignment of two sequences, and then adds other sequences one by one according to a predetermined order. A multiple sequence alignment is an extension of pair-wise alignment to incorporate more than two sequences at a time. This method aligns all of the sequences in a given query set. Mostly multiple alignments are used to identify the conserved sequence regions across a group of related sequences. The conserved sequence motifs are used to locate the catalytic active sites of enzymes in conjunction with structural and mechanistic information, which further may help to establish the evolutionary relationships via construction of phylogenetic trees. Three or more sequences of biologically relevant length are difficult and almost always time-consuming to align by hand; therefore, computational algorithms are used to produce and analyse the alignments. In bioinformatics the utility of these alignments resulted in the development of various methods suitable for aligning three or more sequences (Elias, 2006).

***(a)* Dynamic Programming:** Dynamic programming is a method for solving complex problems by breaking them down into simpler sub-problems. The technique of dynamic programming is theoretically applicable to any number of sequences; however, because it is computationally expensive in both time and memory, it is rarely used for more than three or four sequences in its most basic form. Dynamic programming technique is a direct method for producing an MSA to identify the globally optimal alignment solution. For proteins, this method involves two sets of parameters: *a gap penalty* and a *substitution matrix*. On the similarity of chemical properties of the amino acids and the evolutionary probability of the mutation, these parameters assign scores to the alignment of each possible pair of amino acids. A similar gap penalty is used for nucleotide sequences. In the substitution matrix the scores may be either all positive or a mixture of positive and negative (in a global alignment), but must be both positive and negative (in a local alignment).

The naive method requires, for *n* individual sequences, the construction of *n*-dimensional equivalent of the matrix formed in standard pair-wise sequence alignment (where *n* is the number of sequences in the query). In 1989, based on Carrillo-Lipman Algorithm, Altschul introduced a practical method that uses pair-wise alignments to constrain the *n* dimensional search space. First of all standard dynamic programming is used on all pairs of query sequences; thereafter, the alignment space is filled by possible matches or gaps at intermediate positions constructing an alignment. The sum of all of the pairs of characters at each position in the alignment (so called *sum of pair* score) is optimised by the MSA programme. Then these are implemented in a software programme to construct the multiple sequence alignments. Computationally this method is expensive, but globally useful because of being able to accurately align a few sequences. One method has been implemented in the MSA software package to reduce the computational demands of dynamic programming. This relies on the *sum of pair* objective function (Lipman *et al.,* 1989).

***(b)* Progressive Methods:** A heuristic search known as progressive technique (also called hierarchical or tree method) is used by the most widely used approach to multiple sequence alignments. Progressive, hierarchical, or tree methods results in a multiple sequence alignment by first aligning the most similar sequences and then adding less related sequences or groups to the alignment until the whole query set has been incorporated into the solution. The initial tree that describes the sequence relatedness is based on pair-wise comparisons. An efficient clustering method such as neighbor-joining or UPGMA determines the initial 'guide tree'. UPGMA uses distances on the basis of number of identical two letter sub-sequences (as in FASTA rather than a dynamic programming alignment). Like FASTA pair-wise comparisons include heuristic pair-wise alignment methods.

The most related sequences govern the results of progressive alignment. Hence, in the initial pair-wise alignments it may be sensitive to inaccuracies. In the query set most progressive methods of multiple sequence alignment weight the sequences according to their relatedness; hence it improves alignment accuracy.

Progressive alignment methods are efficient for implementing many (100s to 1000s) sequences on a large scale, which are commonly available on publicly accessible web servers. The Clustal family has been the most popular progressive alignment method, for example the weighted variant ClustalW to which access is provided by a large number of web portals such as GenomeNet, EBI, and EMBNet (Higgins and Sharp, 1988).

T-Coffee (Tree-based Consistency Objective Function For alignment Evaluation) is multiple sequence alignment software using a progressive approach. It generates a library of pair-wise alignments to guide the multiple sequence alignment. It can also combine multiple sequences alignments obtained previously and in the latest versions can use structural information from PDB files (3D-Coffee). T-Coffee is the other common method for progressive alignment (Notredame *et al.,* 2000) that is slower than Clustal and its derivatives, which generally produces more accurate alignments for distantly related sequence sets. Pair-wise alignments are calculated by T-Coffee by combining the direct alignment of the pair with indirect alignments, that aligns each sequence of the pair to a third sequence. It uses the output from Clustal and another local alignment programme LALIGN, which finds multiple regions of local alignment between two sequences. Phylogenetic tree and the resulting alignment are used as a guide for producing more accurate weighting factors. Variants of the progressive implementation of Clustal are used for multiple sequence alignment, phylogenetic tree construction, and for prediction of protein structure (Chenna *et al.,* 2003).

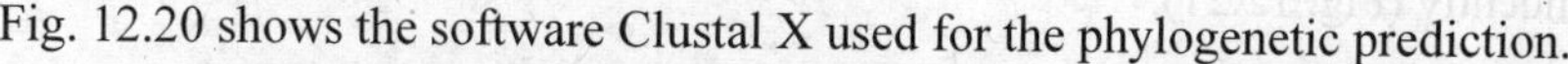

Fig. 12.20 shows the software Clustal X used for the phylogenetic prediction.

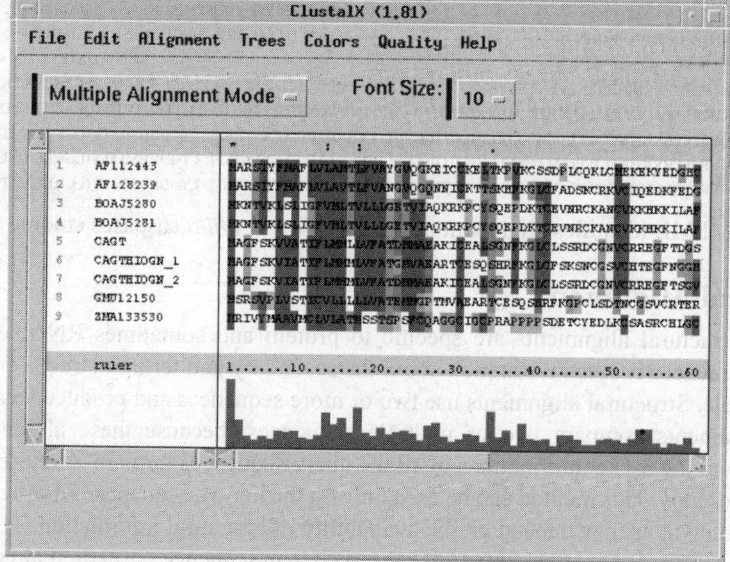

Fig. 12.20. The software Clustal X used for the phylogenetic prediction.

(c) Iterative Methods: These are a set of methods used to produce MSA while reducing the errors inherent in progressive methods. These methods are classified as 'iterative' because they work similar to progressive methods. They realign the initial sequences and add new sequences to the growing MSA. Iterative methods improve on the weak point of the progressive methods, and optimise an objective function on the basis of a selected alignment scoring method. Thereafter, the realigned subsets are themselves aligned for producing the next multiple sequence alignment of iteration. DIALIGN is the other iterative programme that approaches unusually for focusing on local alignments between sub-segments or sequence motifs without introducing a gap penalty. Then the alignment of individual motifs is obtained with a matrix representation as a dot-matrix plot is achieved in a pair-wise alignment. An alternative method is implemented in the CHAOS/DIALIGN suite that uses fast local alignments as anchor points or seeds for a slower procedure of global alignment (Brudno *et al.*, 2003).

MUSCLE (multiple sequence alignment by log expectation) is a third popular iteration-based method that improves on progressive methods by measuring a more accurate distance to assess the relatedness of two sequences. However, the distance measure is updated between iteration stages.

***(d)* Motif Finding (Profile Analysis):** Motif finding (also called profile analysis) is a method to locate sequence motifs in global MSA. It is both a means to produce a better MSA and a scoring matrix that may be used to search the other sequences of similar motifs among the sequences in the query set. There are many methods developed for isolation of the motifs, but all are based on: (*i*) the identification of short highly conserved patterns within the larger alignment, and (*ii*) the construction of a matrix like a substitution matrix that shows the composition of amino acid or nucleotide in the motif. The profile matrix for each conserved region is arranged like a scoring matrix. But the frequency counts of each amino acid/ nucleotide are derived from the character distribution of conserved region rather than from a general empirical distribution. Thereafter, the profile matrices are used to search the other sequences for occurrence of the characterized motif. Pseudo-counts are added to normalise the character distributions represented in the motif where the original data set contained a small number of sequences, or only highly related sequences. Motif positions and sequence alignments are often correlated well but not perfectly when they are generated independently (Fig. 12.21).

```
DAMMfly2R_  : MYLFERTEHCKIERLY ----------------------------- D8NRVN ---------- AEPGOGL ------
DCP1fly2R_  : ------------MTD- ---------------------------ECVTRSIYGVGIHSPNGSINRGS-PIMACNTCAK---GCTPISLVVGG
DRICEf y3R  : MDATNNGESADQVGIRVGN ----------------------PEQPNIHTTALGSV-GSGGAGSSGLVAGSHPY-----GSGAIGOLANG
DECAYfly3R  : MDDTDFSLFGCMNKHR ----------------------------------- MDMADATKIA ------HTITSEL -- --
DRONCfly3L  : MOBRKLPIGMPKRHREHIRNLVENTNYERLAMECVCCGILTVCMLRNTCDILNGE - PENMOKKDVRVEOMRRLLLKITQRGPTAYNLLINA
STRICAfly2  : MGNNSKKSETDRSQPSQELVAQDPRTRVAQDPRTRVQTTBAATETTNTAVQNSTITDNNKQTVIFI-TTRQTVTHTQRALITETTTRRTPSQAELEALFAKI
DREDDPAfly2 : MSASAIYRPFPRVKHKFCIFPIAMAGSNLLIHLDTIDCNDLIYVKRDMNFACXVGIKFL-LYGDDHSDATYILGKLLAMTRSDFPOSDLLIRFAK
DREDDPAfly2 : MSASAIYRPFPRVKHKFCIFPIAMAGSNLLIHLDTIDCNDLIYVKRDMNFACXVGIKFL-LYGDDHSDATYILGKLLAMTRSDFPOSDLLIRFAK
DREDDPAfly2 : MSASAIYRPFPRVKHKFCIFPIAMAGSNLLIHLDTIDCNDLIYVKRDMNFACXVGIKFL-LYGDDHSDATYILGKLLAMTRSDFPOSDLLIRFAK
```

Fig. 12.21. MEME identifies alignment of the seven *Drosophila* caspases colored by motifs.

4. Structural Alignment

Usually structural alignments are specific to protein and sometimes RNA sequences. For aligning the sequences it uses information about the secondary and tertiary structure of the protein or RNA molecule. Structural alignments use two or more sequences and produce local alignments. Structural alignments compare two or multiple sequences; because these alignments rely on information about the 3-D conformations of all the query sequences through X-ray crystallography or NMR spectroscopy. This method can be used only on the known sequences whose corresponding structures are known, as they depend on the availability of structural information. The structure of both protein and RNA is more evolutionarily conserved than sequence. Structural alignments can be more reliable between very distantly related sequences that and diverged sequences.

For those proteins that have low sequence similarity, structural alignment is a valuable tool for their comparison. However, evolutionary relationships between such proteins cannot be easily detected by standard sequence alignment techniques. In the evaluation of alignments for prediction of homology-based protein structure, structural alignments are used as the 'gold standard', because they align structurally similar regions of protein sequences. Highly accurate models of the target protein sequence can be produced for the given structural alignment between a target and a template sequence. Production of structurally accurate alignments is a major stumbling block in homology-based structure prediction (Zhang *et al.,* 2005).

Fig. 12.22 shows the structural alignment of thioredoxins from humans and *Drosophila melanogaster*.

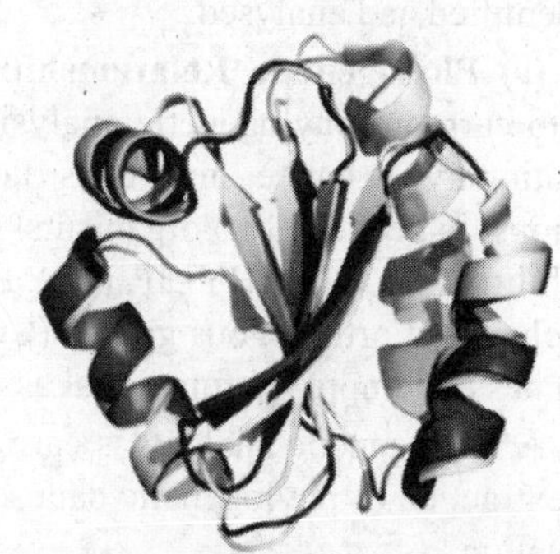

Fig. 12.22. Structural alignment of thioredoxins from humans and *Drosophila melanogaster;* red colour- human protein, yellow colour- fly protein.

(*a*) DALI: The DALI (distance matrix alignment) is a fragment-based method used to construct the structural alignments on the basis of patterns of contact similarity between successive hexapeptides. It produces pair-wise or multiple alignments and identify the structural neighbors of a query sequence in the Protein Data Bank (PDB). It has been used to construct the FSSP structural alignment database (Fold classification based on Structure-Structure alignment of Proteins, or Families of Structurally Similar Proteins). EBI DALI can access a DALI web server, and the FSSP is located at the Dali Database (Holm and Sander, 1996).

(*b*) SSAP (Sequential Structure Alignment Programme): SSAP is a dynamic programming-based method of structural alignment. As comparison points it uses atom-to-atom vectors in structure space. It has been extended to include multiple as well as pair-wise alignments since its original description. It has been used in the construction of the CATH (Class, Architecture, Topology, Homology) hierarchical database classification of protein folds. The CATH database can be accessed at CATH Protein Structure Classification.

(*c*) Combinatorial Extension: A pair-wise structural alignment is generated by the combinatorial extension method of structural alignment. Local geometry is used to align short fragments of the two proteins and then these fragments are assembled into a larger alignment. On the basis of feature such as rigid-body root mean square distance, residue distances, local secondary structure, and surrounding environmental features, local alignments (called aligned fragment pairs) are generated and used to build a similarity matrix. The similarity matrix represents all possible structural alignments within predefined criteria. Then a path is traced from one protein structure to the other through the matrix by extending the growing alignment one fragment at a time. The combinatorial-extension alignment is defined by the optimal path. A web-based server is located at the Combinatorial Extension website that implements the method and provides a database of pair-wise alignments of structures in the Protein Data Bank (Shindyalov and Bourne, 1998).

J. USE OF BIOINFORMATICS TOOLS IN ANALYSIS

By using bioinformatics tools different types of analysis of biological data can be done. Some of the analyses have been briefly discussed in this section.

(*i*) Processing Raw Information: The biological information hidden in DNA/RNA and protein sequences is generated experimentally. These are called raw information. Using bioinformatics tools these informations are processed into genes and proteins and a relationship is established between a gene and a protein. Phylogenetic relationships can also be established among the species of organisms.

***(ii)* Genes:** Using bioinformatics tools such as GenMark (for bacteria), and GenScan (for eukaryotes) gene prediction is carried out in organisms. GenScan can identify introns, exons, promotor sites and poly A signals and other gene identification algorithms.

***(iii)* Proteins:** By using computer programmes protein sequences can be deduced from the predicted genes.

***(iv)* Regulatory Sequences:** Using computer programmes the regulatory sequences can also be identified and analysed.

***(v)* Phylogenetic Relationships:** There are a few web-based applications that will allow you to carryout phylogenetic analysis over the web. By aligning multiple sequences, calculating evolutionary distance and constructing phylogenetic trees, you can establish phylogenetic relationship between two organisms. Internet-based applications that provide phylogenetic analysis capabilities are WEBPHYLIP, PhyloBLAST, GenTree and BLAST2 and Orthologue Search Server. The cluster of orthologous groups (COGs) database simplifies the evolutionary studies of complete genomes and improves functional assignment of individual protein.

A phylogenetic analysis is carried out in four steps: alignment (both building the data model and extraction of phylogenetic data set), determining the substitution model, tree building, and tree evaluation.

***(vi)* Reconstruction of Metabolic Pathways:** Reconstruction of metabolic pathways is one of the indispensable final steps of all genome analyses. It is also a convergence point for the data produced by different methods. The ENZYME database lists name and catalysed reaction for all the enzymes that have been assigned official Enzyme Commission (EC) numbers.

***(vii)* Prediction of Function of Unknown Genes:** By using the bioinformatics tools and databases, you can predict the function of unknown genes. Before going for gene prediction you must be aware of the structure of a gene, transcription process, post-transcriptional processing of RNAs and protein folding, etc. There are two approaches used for gene prediction: by using DNA sequences, and using protein sequences.ONS

QESTIONS

1. Give a historical account of bioinformatics.
2. What is database ? Write different type of databases studied by you ?
3. Write a detailed account of database management system (DMS).
4. Give an account of database models.
5. Write about model vie controller.
6. Write in detail about database similarity searching.
7. Write in detail about sequence alignment methods.
8. List any six resources available from NCBI and their uses.
9. Write the method of nomenclature of protein and DNA sequence.
10. What are the information sources used in bioinformatics ?
11. What are data retrieval tools? Write Entrez map showing the hardlink relationships between databases.
12. Write short notes on the following:

(i) NCBI, *(ii)* Database,
(iii) BLAST, *(iv)* Directionality of sequence,
(v) UniGene, *(vi)* ESTs,
(vii) Entrez, *(viii)* RefSeq database,
(ix) Mouse genome database, *(x)* OMIM,
(xi) FASTA.

BIBLIOGRAPHY

Brudno M. *et al.* 2003. Glocal alignment: finding rearrangements during alignment. *Bioinformatics.* **19** (Suppl) **1**: 154–62.

Chenna R. *et al.* 2003. Multiple sequence alignment with the Clustal series of programs. *Nucleic Acids Res* **31**: 3497–500.

Elias, I. 2006. Settling the intractability of multiple alignments. *J Comput Biol* **13**: 1323–1339.

Higgins D.G. and Sharp P.M. 1988. CLUSTAL: a package for performing multiple sequence alignment on a microcomputer. *Gene.* **73**: 237–244.

Holm L. and Sander C. 1996. Mapping the protein universe. *Sci.* **273**: 595–603.

Lipman D.J., Altschul S.F. and Kececioglu J.D. 1989. A tool for multiple sequence alignment. *Proc. Natl. Acad. Sci. U.S.A.* **86**: 4412–5.

Mount D.M. 2004. *Bioinformatics: Sequence and Genome Analysis* (2nd Ed.). Cold Spring Harbor Laboratory Press: New York.

Notredame C., Higgins D.G. and Heringa J. 2000. T-Coffee: A novel method for fast and accurate multiple sequence alignment. *J .Mol. Biol.* **302**: 205–17.

Shindyalov I.N. and Bourne P.E. 1998. Protein structure alignment by incremental combinatorial extension (CE) of the optimal path. *Protein Eng.* **11**: 739–47.

Zhang Y. and Skolnick J. 2005. The protein structure prediction problem could be solved using the current PDB library. *Proc. Natl. Acad. Sci. U.S.A.* **102**: 1029–34.

CHAPTER

Genomics and Proteomics

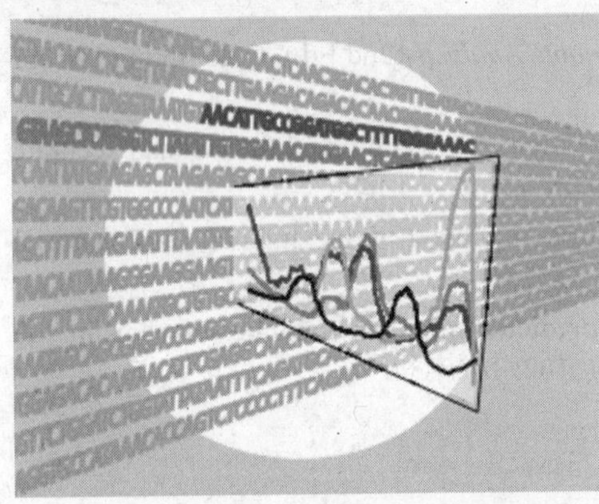

A. The era of genomics and proteomics
B. Genome sequencing projects
C. Human genome project (HGP)
D. Methods of gene sequencing
E. Genome assembly and finishing
F. Gene prediction and gene counting
G. Genome similarity and single nucleotide polymorphisms (SNP)
H. Types of genomics
I. Proteomics

Genomics is an outcome of recent convergence of many sciences including genetics, molecular biology, biochemistry, statistics and computer sciences. These fields were highly developed before the origin of the term 'genomics' by Thomas Roderick in 1986. The history of genetics and molecular biology used in genomics are particularly relevant to the techniques, experimental designs, and intellectual approaches. Ready access to the large information generated throughout the world has been provided to researchers, through the development of computers and the internet. The processes providing the tools for molecular biology, recombinant DNA technology, and the biotechnology industry could be defined up to mid to late of 20th century. The process of DNA replication described the necessary components needed for chain termination method of DNA sequencing. Understanding of replication helped to determine the tools that are necessary for the radiolabelling of DNA. This development was important to support Southern hybridisations and the experiments of early molecular mapping. Understanding of replication also clarified the role of DNA ligase enzyme that is important for DNA cloning.

Restriction enzymes were discovered and used to construct recombinant DNA molecules that contained foreign DNA that could be grown in abundance in bacterial cells. The discovery of reverse transcriptase also enabled cDNA cloning which is essential for the modern EST projects. Cloning is essential for the discovery of gene structure and function. It is also an essential step for all of the genome sequencing projects. The importance of the PCR procedure cannot be emphasised enough. The protein and DNA sequencing launched a new era of phylogenetics; species could now be compared at the molecular level. New methods are developed for the development of phylogenies. The proposed neutral theory of molecular evolution suggests that most mutations are neutral and are fixed by genetic drift and not selection. It is debated whether the evolution of species is driven more by neutral effects or selection.

The age of information is important to genomics. The hallmark of the science is the electronic analysis, distribution and storage of genomic data. The development of both large and small computers which put computing power in the hands of all scientists was critical. The scientists were provided with the tools to study the details of their experiments through free distribution of analytical software. The internet spawned the distribution of information from central databases. E-mail connected scientists and fostered the rapid exchange of ideas. The advent of the WWW provided a new medium for the presentation of information.

A. THE ERA OF GENOMICS AND PROTEOMICS

For the first time whole genomes were sequenced. The gene content is described using ESTs for other species. The first glimpse of global expression patterns was provided by microarray analyses. Proteomics started to describe the protein component of the genome; thereafter, metabolomics was established.

In 1920, H. Winkler coined the term *genome* to describe 'the complete set of chromosomal and extra-chromosomal genes of an organism or virus'. In eukaryotes DNA is also present inside the mitochondria and chloroplasts (in plants only). After a gap of 66 years, Thomas Roderick in 1986 first used the term *genomics* and described it as a scientific discipline of mapping, sequencing and analysing the genome. This definition is unclear but emphasises the systematic exploitation of genome information to answer several queries arising from biology or its related areas. Now the number, location, size and organisation of all genes require making up an organism can be known. Thus *genomics is the study of molecular organisation of genome, their information contents and the gene products they encode*. Significance of genomics was substantially increased when the Human Genome Project was conceived of in 1987. An extensive history of the major developments in these fields is given in Table 13.1.

Table 13.1. An annotated history of genomics and proteomics.

Year	*Who*	*Discovery*
1952	F. Sanger Colleagues	The complete amino acid sequence of insulin is determined, and it is shown that the molecule is a dimer held together by disulphide bonds.
1953	James Watson Francis Crick	A structural model of DNA is presented that states it consists of two anti-parallel chains held together by hydrogen bonds. The model suggests a model of DNA replication.
1955	Newton Morton	LOD score method of determining linkage distance in humans developed.
1956	Arthur Kornberg	DNA polymerase I was purified from *E. coli,* which is an enzyme of DNA replication.
1957	Francis Crick	The central dogma of molecular biology was proposed. This is a first elucidation of the link between the sequences in the DNA molecule and the production of proteins.
1980	Sanger Group	The first complete genome sequence was published. The genome was of the bacteriophage ΦX174 of *E. coli.*
1980	K. Wuthrich	Protein structure was determined using multi-dimensional NMR.
1981	Felsenstein	The maximum likelihood procedure to develop phylogenetic trees was published.

1982	GenBank	GenBank established to be a database of all DNA sequences. Initially housed at the Los Alamos National Laboratory, it was transferred to National Center for Biotechnology Information (NCBI) in 1988.
1983	Kary Mullis	He described the procedure to amplify large amounts of DNA by using Taq DNA polymerase from the bacterium *Thermus aquaticus*.
1986	Leroy Hood Lloyd Smth Michael Hunkapiller Tim Hunkapiller	The first automated DNA sequencer was released. This development, based on fluorescent labeling of nucleotides and the Sanger sequencing technique, made it possible to sequence genomes in a reasonable amount of time. Applied Biosystems released the first sequencer.
1986	SWISS-PROT	The initial version of the SWISS-PROT protein database was released.
1987	Eric Lander et al.	MAPMAKER, a computer program for developing genetic linkage maps from molecular marker data is released.
1988	H.D. Higgins P.M. Sharp	The Clustal multiple sequence alignment approach is published.
1989	Jean Thierr-Mieg Richard Durbin	The ACeDB (a *C. elegans* DataBase) is developed. This is the first database developed for genomic information.
1990	US Government	The 15 year Human Genome Program was launched to 'find all the genes on every chromosome in the body and to determine their biochemical nature'.
1990	Altshul *et al.*	BLAST, a computational approach to aligning two DNA sequences was described. An updated version was released in 1997.
1990	Tim Berners-Lee	Version 1.0 of HTML was published. This language is used to delivery WWW pages.
1991	Craig Venter	The first EST (expressed sequence tagged) sequences are published. These short sequences prove to be important for gene discovery.
1993-4	Genthon	The first high density linkage map of the human genome was published.
1995	Celera Corp Craig Venter	The entire *Haemophilus influenzae* genome sequence (1.8 Mbp) was published. This is the first free-living bacterium that was sequenced by the shotgun genome sequencing method.
1995	Pat Brown Ron Davis	The first microarray system was described. This system uses cDNA sequences printed on glass slides.
1996	Yeast Genome Consortium	The entire *Saccharomyces cerevisiae* (yeast) genome sequence (12.1 Mbp) was published. This is the first eukaryotic genome to be sequenced that contains 6,250 genes, later (2003) revised to 5,700.
1997	deRisi *et al.*	The first study of regulatory pathway using microarray technology was published.
1997	*Escherichia coli* Genome Project	*E. coli* (4.7 Mbp) was sequenced and shown to contain 4,500 genes.
1998	*C. elegans* genome project	The *C. elegans* genome sequence (97 Mbp) was reported. This is the first genome of a multicellular organism to be sequenced. The genome contains 19,100 genes.

1998	Eisen *et al.*	Cluster analysis approaches are described for the analysis of microarray data.
1999	Affymetrix	The oligonucleotide microarray system was developed.
2000	*Arabidopsis* Genome Initiative	The *Arabidopsis thaliana* genome (125 Mbp) was sequenced. This is the first publishing of a flowering plant genome that contains 25,500 genes.
2000	*Drosophila* Genome Initiative	The *Drosophila* genome sequence (123 Mbp) contains 13,400 genes. This is the first shotgun approach to the sequencing of a eukaryotic organism.
2000	Steven Tanksley	The first gene controlling a quantitative trait was cloned.
2001	International Human Genome Sequencing Consortium Celera Corporation	The human genome sequence (2900 Mbp) was published. The genome contains between 35,000 and 40,000 genes. Later (2002) estimates place the number at 30,000 genes.
2001	Sanger Institute Wellcome Trust	The Ensembl Genome Browser provides easy access to human genome data.
2002	Mouse Genome Sequencing Consortium	The mouse genome sequence (2500 Mbp) was published. The number of genes was estimated to be 30,000.
2002	Syngenta, Torrey Pines Research Institute and Beijing Genomics Institute	The draft of the rice (*Oryza*) genome sequence (470 Mbp) was published. The estimates of the number of genes range from 32,000 to 56,000. The sequences of the *indica* and *japonica* strains were published jointly.
2002	Puffer-fish Genome Sequencing Consortium	The puffer-fish (*Fugu rubripes*) genome (365 Mbp) was sequenced. A total of 31,059 genes were observed. This is the smallest animal gene by size but contains a similar number of genes as human and mouse.
2002	Malaria Sequencing Consortium	The sequence of *Plasmodium falciparum* geneome (23 Mb) was published. The genome consists of 5300 genes.
2003	British Columbia Cancer Agency	The SARS-associated coronavirus genome sequence (30 kb) was released. The genome contains 16 open-reading frames. The sequence was released less than five months after the disease began spreading world-wide.
2003	Human Genome Project	Very few people have had their genes analyzed, although one of the most ambitious projects in biological science's history was the Human Genome Project, which mapped the entire DNA that makes up all humans in 2003. Very few people have had their genes analyzed, although one of the most ambitious projects in biological science's *history* was the Human *Genome* Project, which mapped the entire DNA that makes up all humans in *2003*. "These technologies produce thousands of times more data per run and at a fraction of the cost than those technologies/techniques used in the human *genome* project," says Cofactor *Genomics*' Chief Technical Officer, Dr. Jarret Glasscock. "This data will lead to ...
2004	EMBO Conference	Microbial history revealed by comparative genomics

2004	454 Life Sciences	Parallelized version of pyrosequencing was marketed by 454 Life Sciences
2007	Craig Venter	A team led by Craig Venter published his complete DNA sequence unveiling the six-billion-nucleotide genome of a single individual for the first time.
2008	OPHG	In *January 2008*, OPHG hosted a meeting to highlight progress and achievements in genomics and to discuss future directions.
2009	Procter & Gamble	I was reminded after the story published that Synthetic Genomics said it already had sequenced the *Jatropha* Genome in May 2009. Procter & Gamble boosted San Diego-based Somaxon's efforts to launch its new insomnia drug.

Genome, Genomics and Genome Sequencing

A genome is the sum total genes present in haploid set of chromosomes of an individual organism. Thus, genomics is the study of all the genes of a cell or tissue, at the DNA (genotype), mRNA (transcriptome), protein (proteome), or metabolite (metabolome) levels. Obtaining the blocks of DNA sequences and assembling serially into contiguous stretches of sequence and ulatimately into a whole genome sequence using various bioinformatic strategies is called genome sequencing.

Therefore, the current efforts are focused on understanding the protein products encoded by the genes. The corresponding protein becomes defective upon mutating a gene. The emerging field of proteomics aims to understand the biological processes. We need to understand how proteins function in and around cells since they are the functioning units. Moreover, investigators are also turning their attention to the expansive regions of our genome devoid of traditional protein-encoding genes. Gene expresses through production of mRNA; the complete set of mRNAs produced in an organism is called 'transcriptome'. The mRNA molecules ate translated into proteins. Complete set of proteins is called 'proteome' and the study of protein analysis is called 'proteomics'. The mRNA molecules also produce the other metabolites in the cell, in addition to protein. All such metabolites are called 'metabolome' and the phenomenon is 'metabolomics'. Approaches to genomics, proteomics and metabolomics are given in Fig. 13.1.

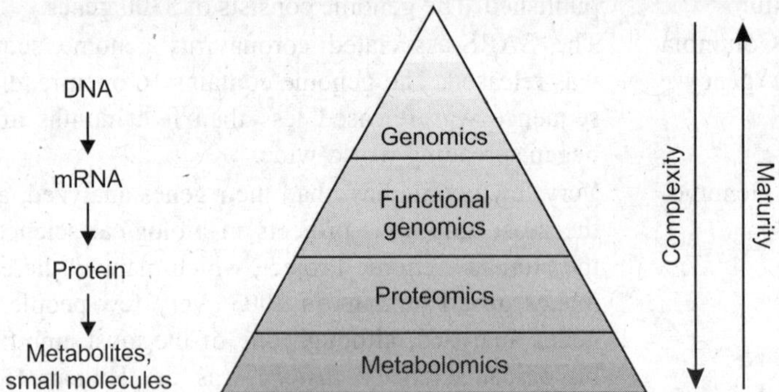

Fig. 13.1. Relation between genomics, proteomics and metabolomics.

B. GENOME SEQUENCING PROJECTS

Throughout the world, scientists were trying to sequence completely the genome of several organisms of important groups. There are many reasons of sequencing the genome such as: (*i*) it provides knowledge of total number of all genes, (*ii*) it shows relationships between genes, (*iii*) it provides opportunities to exploit the genes for desired experimentation, (*iv*) it provides all genetic information about the organisms, (*v*) it expose all type of structural and functional annotated proteins, (*vi*) it show the finest similarity with other organisms, and (*vii*) the genome sequences act as an archive of all genetic information. Now, the number, location, size, organisation of all genes required to make up an organism can be known by genome sequencing.

I. MICROBIAL GENOME SEQUENCING PROJECTS

The complete genome sequencing has been determined for a variety of microbes, including Bacteria, Archaea, and Eukarya. Many additional microbial genome sequencing projects are currently in progress. Lists of current genome sequencing projects are available via: (*i*) DOE Microbial Genome Programme, (*ii*) NCBI Microbial Genomes database, (*iii*) Sanger Genome Centre, and (*iv*) TIGR Microbial database.

Several computer programmes were adopted for data analysis. To carry out sequence assembly, several new computer programmes could also be written. Genome sequencing of many bacteria including those of agriculture, medical and industrial importance, and from extremes of environment (*e.g.* volcano and deep sea vents) were determined at the Institute of Genomics Research (TIGR), Maryland (U.S.A.) which was established by Craig Venter in 1992.

1. Viral Genomics

Virus identification, characterisation and taxonomy are necessarily on the basis of modern technology. However, virus attributes used for identification include measurements of the physical viral structure, pathogenic properties, antibody recognition of capsid proteins (immunochemistry), biochemical and clinical observations, and the genome including its form, hybridization and patterns of restriction enzyme digestion. The Sanger dideoxynucleotide chain terminator method has provided the primary genome sequence. Originally this method was validated by sequencing a DNA virus (ϕX174, 5.4 kb) that led to a continuing cycle of data acquisition and analysis in conjunction with technology development (Seto, 2010).

On a small scale viral genomes are being sequenced and studied; but recent resources allow for a larger genomic view of viruses. The National Institute of Allergy and Infectious Disease (NIAID) and the National Institutes of Health (NIH), through the funding of the Genome Sequencing Center (GSC) for Infectious Diseases at the J. Craig Venter Institute (JCVI), have recently approved 'white papers' and proposed for the large-scale genomic sampling of several groups of viruses such as: arbovirus, adenovirus, coronavirus, influenza, norovirus, paramyxovirus and rotavirus. Recently rhinoviruses were analysed globally and more genomes are being sequenced in this family. Examples of some viral genomes that have been sequenced are given in Table 13.2.

Table 13.2. Examples of some viral genomes that have been sequenced.

Organism	*Year sequenced*	*Genome size (bp)*	*Note*
MS2	1976	3,569	First sequenced RNA genome
SV40	Virus	5,224	-
Φ-X174	Virus	5,386	-
HIV	Virus	9749	-
Phage λ	Virus	48,502	-
Mimivirus	2004	1,181,404	Largest known viral genome

Bacteriophage MS2 (3.6 kb) is the first virus of which genome was completely sequenced in 1976. Genome sequencing of the largest virus (mimivirus, 1.2 Mb) was completed in 2004. A genetic map of bacteriophage MS2 is shown in Fig. 13.2. Human adenoviruses contain DNA genomes of 35 kb, the Poxviridae genomes range from 130 to 375 kb and the coronaviruses consist of the largest RNA genomes (26-38 kb).

Details of high resolution viral genome on molecular basis have revealed biological system and some surprising results. Studies on virus genome sequence data have resulted in development of genome-specific bioinformatic tools, establishment of data sets and databases, implementation of biotechnological tools and applications, new policies and standard, etc. These have continued the growing importance of viral genomics and bioinformatics. Genomes sets from members of certain virus families are available; many of these data sets such as rhinovirus have recently been assembled and reported. Some of them have recently been reviewed e.g. human immunodeficiency virus (HIV), papillomavirus and rhabdoviruses.

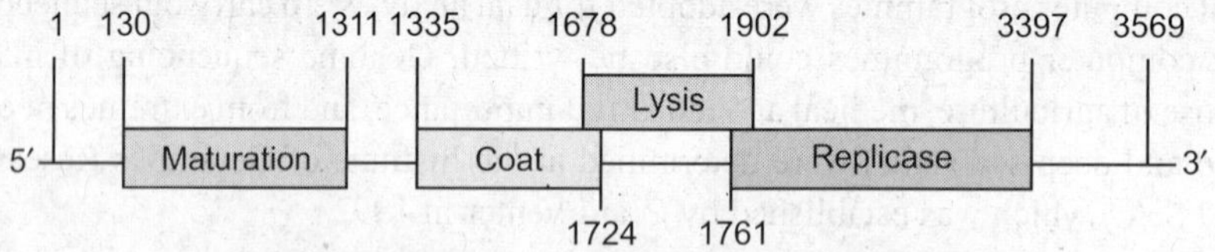

Fig. 13.2. Genetic map of bacteriophage MS2.

***(a)* Standards for Viral Genome and Annotation Data:** Gradually the sets of viral genome data are increasing that prompt the interest to establish the standards for annotation of viral genome sequences. The National Center for Biotechnology Information (NCBI) sponsored to hold the 3rd Prokaryotic Genome Annotation workshop on 26-27 April 2010 at the J. Craig Venter Institute. For viral genomes a virus genome 'working group' was formed and tasked with developing sequence, function, and standards for metadata annotation. About 27,091 full-length virus genomes have been deposited in GenBank up to 2010. Moreover, in the near future a significant of viral genomes may increase. There is need of having 'standards and consistencies' to archive distribute and use these data. The viral genome sequences may be deposited into one or more public databases such as: International Nucleotide Sequence Database Collaboration (INSDC), the DNA Database of Japan (DDBJ), the European Nucleotide Archive (ENA), and GenBank. The universal standards should be developed and imposed on data depositors so that 'cacophony' may be avoided and stress-free effective data mining of these databases may be allowed (Seto, 2010).

***(b)* Biotechnological Applications:** Viral genomics research is important to understand their genome because many deadly viruses mutate quickly after administering vaccines and other antiviral drugs. The application of genomics and bioinformatics to virus studies provides a large amount of data and stimulates applications in many ways. The genomes and the tools used to mine data from them provide unique sequence signatures that may be used to populate microarray chips for surveillance and diagnostics or to develop assays for specific polymerase chain reaction (PCR). Application of viruses used in biomedical and biotechnological processes can be understood in greater detail using these approaches; for example adenoviruses are used in human gene therapy as gene transfer vectors. Genomics and bioinformatics can be used for designing of appropriate and safe vectors (Seto, 2010).

Human health is much affected by viral diseases such as HIV/AIDS and dengue. But the application of genomic technologies to study viral diseases is lagging. The virologists and researchers of infectious disease use the genomic technologies to: (*i*) study the viruses that cause disease, (*ii*) understand their population dynamics and mechanisms of virulence, pathogenesis, host-pathogen interaction, viral emergence, transmission and evolution.

II. BACTERIAL GENOME SEQUENCING PROJECTS

A total of 323 bacterial genome sequences have been retrieved from different databases. The bacterial isolates for which the complete genome sequence was determined belong to 235 different species; 65 sequenced strains are type strains. Thirty-two species are represented by more than one sequenced strain. Complete genome sequencing of *Mycobacterium tuberculosis* was done by several European groups, and that of *Bacillus subtilis* at Pasteur Institute and Sanger Centre. Total whole genome sequences of over 1,000 viruses, over 151 microbes and 9 eukaryotes are known. In 1995, the completely sequenced genomes of the first two smallest bacteria, *Haemophilus influenzae* Rd (1,830 kb) and *Mycoplasma genitalicum* (580 kb) were reported. The sequence data of *Mycoplasma genitalicum* are important because these help to establish the minimal set of gene required f or free-living existence. There are 517 genes, out of which 140 genes code for membrane proteins and 5 genes for regulatory function.

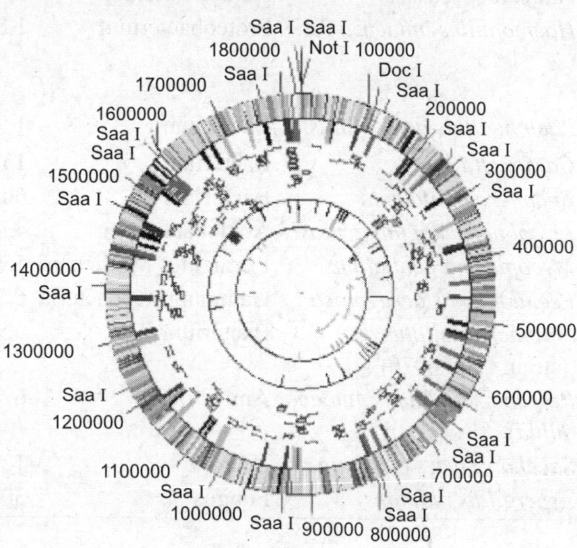

Haemophilus influenzae genome

Recently, hundreds of genome sequencing projects are going on in different parts of the world, and many are being started. Some of the completely sequenced genomes include *E. coli* (4.639×10^6 bp), *Helicobacter pylori* causing stomach cancer (1.67×10^6 bp), and yeast *Saccharomyces cerevisiae* (12.1×10^6 bp). By the end 1999, 97% chromosome 22 (34.5×10^6 bp) was sequenced.

Saccharomyces cerevisiae is the most important fungal species used in biotechnological processes. Its name derives from the fact that it can ferment saccharose (sugar). *S. cerevisiae* genome sequencing project began in 1989, and the sequencing was completed in 1999. Yeast genome is 12.8×10^6 bp and is estimated to contain 6,548 genes.

The NCBI databases also contain complete genomic sequences of eight eukaryotes including: the parasites *Lcishmania major* and *Plasmodium falciparum,* the yeast *S. cerevisiae*, etc. Many more genome sequencing projects are under way in private and public research laboratories that are not yet available on public databases. It is anticipated that the acquisition of new genome sequence data will continue to accelerate. A comparison between genome sizes of different microorganisms is given in Table 13.3.

Table 13.3. Comparison between genome sizes of microorganisms.

Organism	***Organism type***	***Genome size (base pairs)***	***Remark***
Agrobacterium tumefaciens	Bacterium	5,650,368	-
Anabaena nostoc	Cyanobacterium	6,413,771	-
Aquifex aeolicus	Bacterium	1,551,335	-
Bacillus anthracis	Bacterium	5,227,293	-
Bacillus thuringiensis	Bacterium	5,237,682	-
Clostridium tetani	Bacterium	2,799,251	-
Escherichia coli K-12	Proteobacterium	4,639,675	-

Frankia species	Acidobacterium	5,433,628	-
Haemophilus influenza Rd	Proteobacterium	1,830,138	First genome of a living organism sequenced in July 1995
Lactobacillus acidophilus	Bacterium	1,993,564	
Carsonella ruddii	Bacterium	159,662	Smallest non-viral genome
Buchnera aphidicola	Bacterium	600,000	-
Mycobacterium tuberculosis	Acidobacterium	4,403,837	-
Mycoplasma genitalium	Firmicutes Mollicutes	580,076	-
Pseudomonas aeruginosa	Gammaproteobacteria	6,264,403	-
Solibacter usitatus (strain Ellin 6076)	Bacterium	9,970,000	Largest known bacterial genome
Polychaos dubium (*Amoeba dubia*)	Amoeboid	670,000,000,000	Largest known genome
Saccharomyces cerevisiae	Yeast	12,100,000	-
Aspergillus nidulans	Fungus	30,000,000	-

1. *Escherichia coli* Genome Project

Escherichia coli is a natural inhabitant of the lower intestinal tract of animals. In September 1997, a team of scientists led by Frederick Blattner at the University of Wisconsin (U.S.A.) reported the complete 4.6 Mb sequence of *E. coli* K-12 genome.

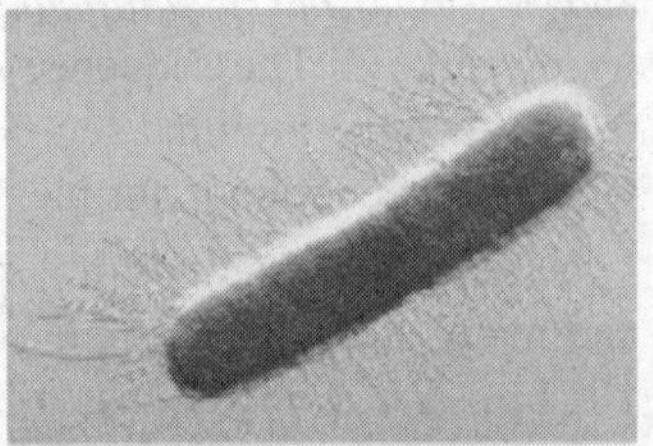

Escherichia coli K-12

Under a variety of physiological conditions knockouts of many genes, archive clones of many ORFs, and an extensive gene expression data set were created. The 4,639,221 base pairs of the genome contain 4,403 genes. The function of third of the genes has yet to be determined. *E. coli* is the best studied bacterium; many of the technological tools available today were developed for the *E. coli* genome sequencing project. The *E. coli* genome sequencing was completed in 1997. The genome size is just over 4.64 $\times 10^6$ bp and contains 4,408 genes. A compact arrangement of genes of *E. coli* K-12 genome separated by only ~120 bp is given in Fig. 13.3.

For more than 70 years, *E. coli* has been one of the most studied organisms for scientists exploring fundamental processes in biochemistry, genetics, and physiology. In recent years it has become the workhorse of biotechnology and serves as a living factory for producing human insulin and other medicines. Due to the presence of similarities in genes across species, this work provides a valuable starting point for identifying and understanding genes in other organisms including humans.

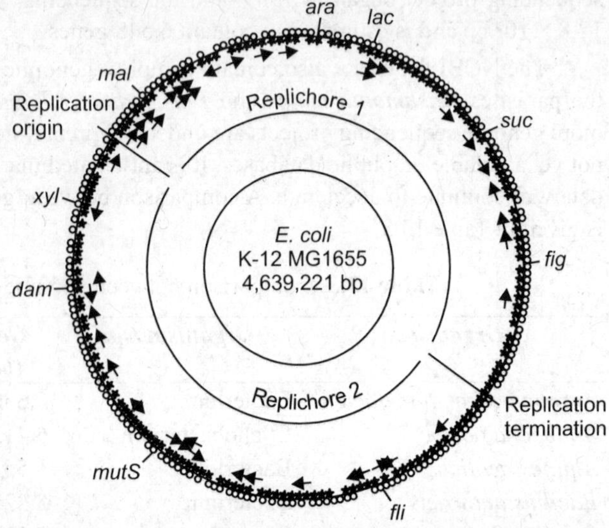

Fig. 13.3. A compact arrangement of genes of *Escherichia coli* K-12 genome separated by only ~120 bp.

2. *Bacillus* Genome Project

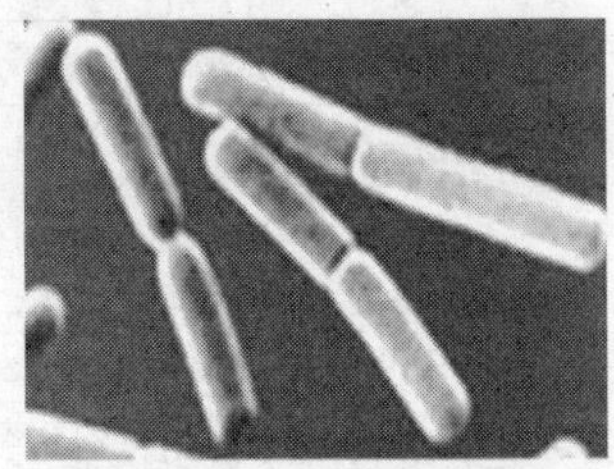

Bacillus subtilis cells

Bacillus is a rod-shaped, Gram-positive bacterium; the members of this bacterium are important organisms for both research and industrial purposes. Since 1995 a major international effort was made to sequence the complete genome of *Bacillus subtilis*. In 1997, Kunst and collaborators published the complete genome sequence of *B. subtilis*.

B. subtilis cells are naturally found in soil and vegetation. *B. subtilis* is much used in industrial processes for both enzyme production and food supply fermentation. It is 'generally regarded as safe' (GRAS). Only one DNA molecule is present in the cells. *B. subtilis* has one circular chromosome. 53% of the protein-coding genes are only seen once, while 25% of the genome relates to families of genes that have undergone gene duplication. It has a genome of 4.21 x 10^6 bp that contains some 4,212 genes. 192 of the 4,100 genes are considered indispensable, and an additional 79 are essential. Most of the essential genes are involved in metabolism. Genome sequences of *B. subtilis* sequenced by different countries, as shown by their respective flags, are given in Fig. 13.4.

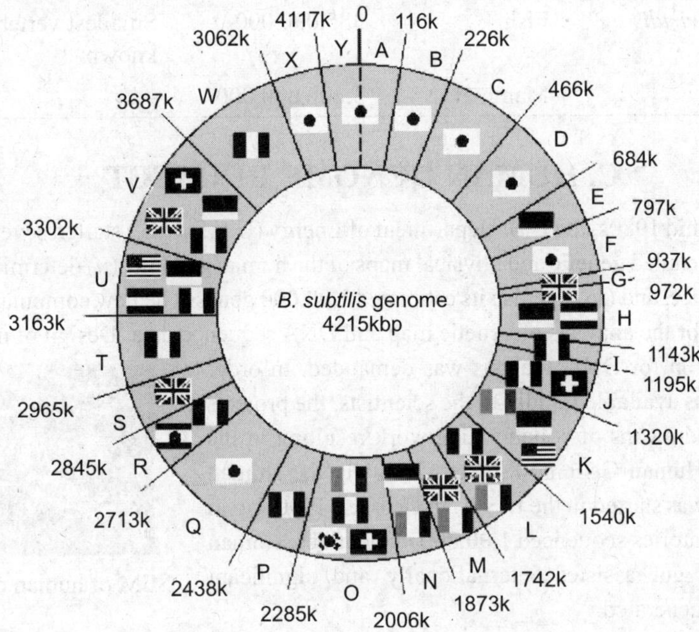

Fig. 13.4. *Bacillus subtilis* genomes sequenced by different countries (*see* flags).

B. sphaericus is natural mosquito pathogen. The strain C3-41 is a highly efficient mosquitocidal bacterium. From 1987 to 2006, more than 1500 tons of this pesticide had been produced and applied in 1.5 million acres in more than 100 cities of China.

III. EUKARYOTIC GENOME SEQUENCING PROJECTS

Some of the completely sequenced genomes of eukaryotes include nematode worm *Caenorhabditis elegans* (100 × 10^6 bp) and the fruitfly *Drosophila melanogaster* (160 × 10^6 bp). Many other genomic sequencing projects are underway including plants, fish, etc. A comparison between genome sizes of different eukaryotes is given in Table 13.4.

Table 13.4. Comparison between genome sizes of eukaryotes.

Organism	*Organism type*	*Genome size (bp)*	*Note*
Arabidopsis thaliana	Plant	157,000,000	First plant genome sequenced, December 2000
Populus trichocarpa		480,000,000	First tree genome sequenced, September 2006
Physcomitrella patens	Moss	480,000,000	First genome of a bryophyte sequenced, January 2008
Caenorhabditis elegans	Nematode	100,300,000	First multicellular animal genome sequenced, December 1998
Drosophila melanogaster (fruit fly)	Insect	130,000,000	-
Bombyx mori (silk moth)	Insect	530,000,000	-
Apis mellifera (honey bee)	Insect	236,000,000	-
Tetraodon nigroviridis	Fish	385,000,000	Smallest vertebrate genome known
Homo sapiens	Mammal	3,200,000,000	-

C. HUMAN GENOME PROJECT

During the mid 1980s, the U.S. Department of Energy (U.S.D.O.E.) started several projects to : (*i*) construct the detailed genetic and physical maps of the human genome, (*ii*) determine its complete nucleotide sequence, and (*iii*) localise its estimated 1,00,000 genes. The new computational methods were developed for the analysis of genetic map and DNA sequence data. Design of new techniques and instrumentation for DNA analysis was demanded. In order to make the results available rapidly to the scientists, the projects used the advanced means of collaborative work resulting in the development of Human Genome Project. Officially the Human Genome Project was started in the U.S.A. in October, 1990. Up to 1990, some laboratories sequenced 1,00,000 nucleotides. Human Genome Project got assisted internationally and significant information was generated.

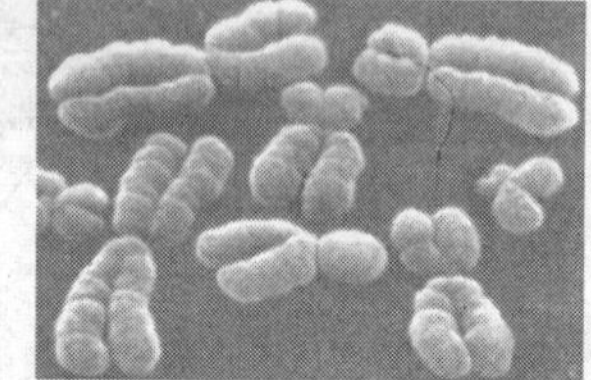

SEM of human chromosomes

In 1991, at the National Institutes of Health, Craig Venter developed the method of finding human genes without sequencing of the entire human genome. Only about 3% of genome is composed of genes that express messenger RNA and others are repetitive sequences or junk DNA. Venter suggested that the most efficient way to find genes would be to use the processing machinery of the cell. Only part of DNA of a cell is transcriptionally active at a given time. These 'expressed' segments of DNA are converted and edited by enzymes into mRNA molecules. Cellular mRNA fragments can be transcribed into complementary DNA (cDNA) by using the enzyme reverse transcriptase. Such stable cDNA fragment is called expressed sequence tag (EST). Computer programmes were used to assemble these cDNA sequences into longer sequences representing large parts or all of many human genes that could match overlapping ends of ESTs.

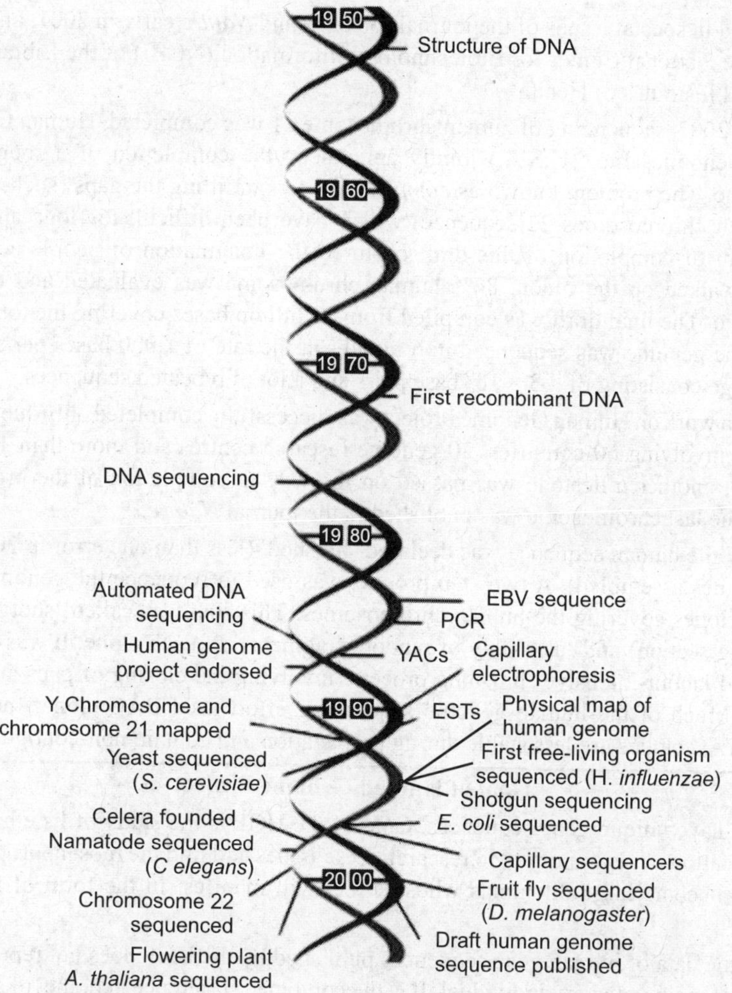

Fig. 13.5. Timeline of Human Genome Project.

In 1992, Venter left NIH to establish The Institute for Genomic Research (TIGR). By 1995 researchers in public and private institutions had isolated over 1,70,000 ESTs which were used to identify more than half of the then estimated 60,000 to 80,000 genes in the human genome 2. In 1998, Venter joined with Perkin-Elmer Instruments (Boston) and formed Celera Genomics (Rockville, MD). Thus, the commercial era of genomics was started with the establishment of Celera Genomics Inc. In 1998, the Human Genome Programme announced to complete the human genome sequence by 2003, on the eve of the 50th anniversary of Watson and Crick's description of the structure of DNA. The objectives of this plan were to: (*i*) achieve coverage of at least 90% of the genome in a working draft based on mapped clones by the end of 2001, (*ii*) finish one-third of the human DNA sequence by the end of 2001, (*iii*) finish the complete human genome sequence by the end of 2003, and (*iv*) make the sequence totally and freely accessible. It was planned to sequence human genome within 15 years, and 5 year's goal was published. J.D. Watson pledged to encode thc genome by 2005 by expending $ 3 billion.

On June 26, 2000, President Clinton met with Francis Collins, the Director of the Human Genome Programme, and Craig Venter of Celera Genomics Inc. (U.S.A.) to announce that the 'working draft' of human genome was completed nearly two years before the schedule. These drafts were published in special issues of the journals *Science* and *Nature* early in 2001, and the sequence is online at the National Center for Biotechnology Information (NCBI) of the Library of Medicine at the National Institutes of Health.

By May 2002, sequencing of human chromosome 21 was completed. Human Genome Project and Celera Genomics Inc. (U.S.A.) jointly announced the completion of a rough draft of the human genome. The problem known as *completion* refers to filling the gaps (such as the 11 small gaps in human chromosomes 22 sequence) which have been difficult to clone and/or sequence. Announcement of completion of this draft captured the imagination of people across the world as man had walked on the moon. 86% human chromosome was evaluated and classified to be highly accurate. The final draft was compiled from 22 billion bases covering the total human DNA sequences. The genome was sequenced in 6 months at the rate of 1,000 bases per second. Human genome is large consisting of $\sim 3 \times 10^9$ base pairs and a lot of repeated sequences.

Research work on Human Genome Project was successfully completed only due to international collaboration involving 60 countries, 20 genome research centres and more than 1,000 scientists. In May 2006, another milestone was passed on the way to completion of the project, when the sequence of the last chromosome was published in the journal *Nature*.

By 2007 the human sequence was declared 'finished' (less than one error in 20,000 bases and all chromosomes assembled). A two step process was used to sequence the genome by analysing the mapped clones covering the human chromosomes. This step was called 'shotgun sequencing' (*see* preceding section) and 'assembly of random fragments' of each clone. It was followed by an expensive and labour-intensive 'finishing process' involving the closing of gaps and resolution of ambiguities. Much of the human genome sequencing effort used BAC clones, not YAC clones, because the YAC clones are less stable during propagation and contain non-contiguous inserts.

Goal of Human Genome Project

All humans have unique gene sequences. Goal of the HGP is the study of its ethical, legal, and social implications. It is important to research these issues and find the most appropriate solutions before they become large dilemmas whose effect will manifest in the form of major political concerns.

Therefore, the data of human gene sequences published by the HGP does not represent the exact sequence of genome of every individual. It is the combined 'reference genome' of a small number of anonymous donors. The HGP genome is a scaffold for future work to identify differences among individuals. Most of the current effort to identify differences among individuals are single-nucleotide polymorphisms (SNP) and the HapMap.

1. Donors of Genome

Researchers collected blood (female) or sperm (male) samples from a large number of donors in the IHGSC international public-sector Human Genome Project (HGP). Out of these samples collected, only a few were processed as DNA resources. Thus the donor identities were protected; therefore, neither donors nor scientists could know whose DNA was sequenced. In the overall project DNA clones from many different libraries were used. In the genomics community it is well known that much of the DNA for the public HGP came from a single anonymous male donor.

The HGP scientists randomly selected 20 male and female donors each and used white blood cells from the blood of two male and two female donors, each donor yielded a separate DNA library. The male samples contained more than half of DNA from the sex chromosomes (one X and one Y chromosomes) as compared to female samples (two X chromosomes). The other 22 chromosomes (the autosomes) were the same for both genders.

The main phase of sequencing of the HGP has been completed but the International HapMap Project has continued the study of DNA variation. It has objectives to identify the patterns of single-nucleotide polymorphism (SNP) groups called haplotypes or 'haps'. For the HapMap the DNA samples were collected from a total of 270 individuals, for example Yoruba people (Nigeria), Japanese people (Tokyo), Han Chinese (Beijing), and the French resource, who were the residents of the United States having ancestry from Western and Northern Europe.

In the Celera Genomics (a private-sector project), DNA from five different individuals were used for sequencing. Craig Venter who was the lead scientist of Celera Genomics acknowledged that his DNA was one of 21 samples in the pool.

On 4th September, 2007 Craig Venter who was the team leader published his complete DNA sequence and unveiled the six billion nucleotide genome of a single individual for the first time.

2. Benefits of Human Genome Project

The human genome project: (*i*) provides clues to the understanding of human biology, (*ii*) provides knowledge of the effects of variation of DNA among individuals that can revolutionise the ways to diagnose, treat and even prevent several diseases that affect human beings.

3. Cost of Human Genome Project

The Human Genome Project was sometimes reported to cost about $3 billion. However, this figure refers to the total projected funding over a 13-year period (1990–2003) for a wide range of scientific activities related to genomics. These activities include studies of human diseases, experimental organisms (such as bacteria, yeast, worms, flies, and mice), development of new technologies for biological and medical research, computational methods to analyse genomes, and ethical, legal, and social issues related to genetics. Human genome sequencing represents only a small fraction of the overall 13-year budget. The DOE and NIH genome programmes set aside 3% to 5% of their respective total annual budgets for the study of the project's ELSI issues.

4. Implications of Genomics on Medical Science

There are hundreds of genetically governed diseases in human, such as sickle cell anaemia, which have very simple, predictable inheritance patterns because each is caused by a change in a single gene. With the vast trove of data about human DNA generated by the Human Genome Project and the HapMap Project, scientists and clinicians have much more powerful tools to study the role of genetic factors that play roles in more complex diseases like cancer, diabetes, and cardiovascular disease constituting majority of health problems. Genome-based research is already enabling medical researchers to develop more effective diagnostic tools to better understand the health needs of people based on their individual genetic make-up and to design new treatments for disease. The role of genetics in health care is the first examples of the era of personalised medicine.

However, it often takes considerable time, efforts and funding to bring the discoveries from the laboratory to medical clinic. Based on genome-based research most of new drugs are estimated to be at least 10 to 15 years away. For a company it usually takes more than a decade to conduct clinical studies needed to receive approval from the Food and Drug Administration.

Pharmacogenomics is a emerging field which involves using information about genetic make-up of a patient to tailor drug therapy better to their individual needs. A deeper understanding of genetics will shed light on more than just hereditary risks by revealing the basic components of cells and explaining how all the various elements work together to affect the human body in both health and disease.

5. Ethical, legal and Social Issues (ELSI)

In addition to identification of about 24,000 genes in the human genome the objective of the project was to address the *ethical, legal and social issues* (ELSI) that may arise from the genetic information. For addressing the ELSI arising from the project, 5% of the annual budget was allocated for this purpose. The burden of genetics education has fallen on the tribes to understand the goal of Human Genome Project and its impacts on their lives. The government has been funding projects to study the indigenous groups without any meaningful consultation.

D. METHODS OF GENE SEQUENCING

DNA sequencing is a powerful technique in molecular biologythat attends the genes at the nucleotide level. This information provides a starting point from which the real research into the thousands of diseases, their regulatory proteins, beneficial genes that have a genetic basis can begin. Therefore, this tool has been applied to several areas of research. PCR is a method which fastly produces many copies of a request piece of DNA, requires first knowing the flanking sequences of this piece. Another important use of DNA sequencing is identifying restriction sites in plasmids. Knowing these restriction sites is useful in cloning a foreign gene into the plasmid. Before the advent of DNA sequencing, molecular biologists had to sequence proteins directly; now amino acid sequences can be determined more easily by sequencing a piece of cDNA and finding an open reading frame.

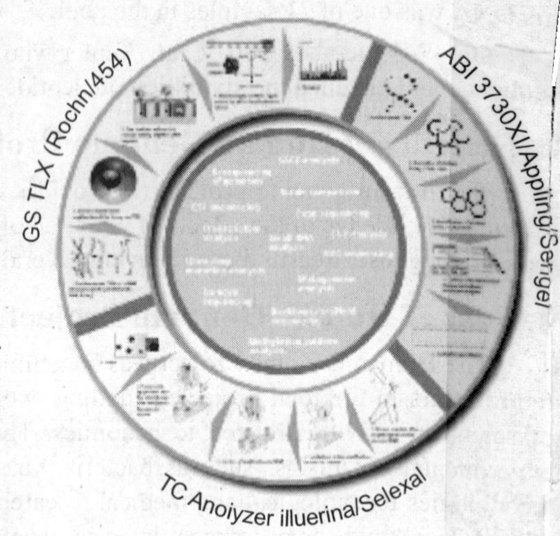

Next generation pyrosequencing

In eukaryotic gene expression, sequencing has allowed researchers to identify conserved sequence motifs and determine their importance in the promoter region. Furthermore, a molecular biologist can utilise sequencing to identify the site of a point mutation. There are only a few examples that illustrate the way in which DNA sequencing has revolutionised molecular biology. Dideoxynucleotide sequencing and pyrosequencing represents the method of sequencing DNA. Dideoxynucleotide sequencing is commonly called as Sanger sequencing or chain termination method.

1. Sanger Sequencing Method

This method generates all possible single-stranded DNA molecules complementary to a given template, and beginning at a common 5' base.This technique utilises 2',3'-dideoxynucleotide triphospates (ddNTPs), molecules that differ from deoxynucleotides by the having a hydrogen atom attached to the 3' carbon rather than an OH group (Fig. 13.6). These molecules terminate DNA chain elongation because they cannot form a phosphodiester bond with the next deoxynucleotide. In this method, the DNA strand to be analysed is used as a template and DNA polymerase is used, in a PCR reaction, to generate complimentary strands using primers. Four different PCR reaction mixtures are prepared, each containing a certain percentage of dideoxynucleoside triphosphate (ddNTP) analogs to one of the four nucleotides (ATP, CTP, GTP or TTP). Synthesis of the new DNA strand continues until one of these analogs is incorporated, at which time the strand is prematurely truncated. Each PCR reaction will end up containing a mixture of different lengths of DNA strands, all ending with the nucleotide dideoxy labelled for that reaction. Gel electrophoresis is then used to separate the strands of the four reactions, in four separate lanes, and determine the sequence of the original template based on lengths of strands ending with types of nucleotide.

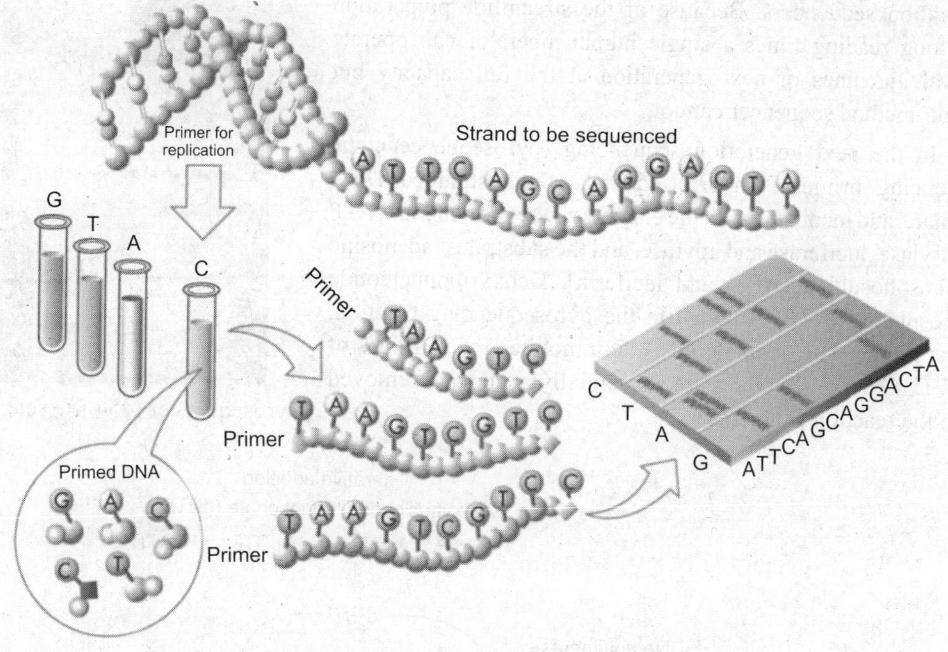

Fig. 13.6 . Steps of sequencing.

In the automated Sanger reaction, primers are used that are labelled with four different coloured fluorescent tags. PCR reactions, in the presence of the different dideoxy nucleotides, are performed as described above. The four reaction mixtures are combined and applied to a single lane of a gel. The colour of each fragment is detected using a laser beam and the information is collected by a computer which generates chromatograms showing peaks for each colour, from which the template DNA sequence can be determined. Typically, the automated sequencing method is only accurate for sequences up to a maximum of about 700-800 base-pairs in length. However, it is possible to obtain full sequences of larger genes and whole genomes by using step-wise methods such as Primer Walking and Shotgun sequencing.

Sequencer (Model: ABI 3130 XL)

2. Pyrosequencing (Next- Generation Sequencing)

In 1996, Pål Nyrén and Mostaf Ronaghi at the Royal Institute of Technology (Stockholm) developed a method of DNA sequencing called pyrosequencing, *i.e.* determining the order of nucleotides in DNA.

The principle of sequencing was based on the 'sequencing by synthesis'. Pyrosequencing differs from Sanger sequencing method. In pyrosequencing method, after incorporating the nucleotides, it detects the release of pyrophosphate, whereas in Sanger's method chain termination takes place with dideoxynucleotides. In this method a single strand of the DNA is sequenced and then its complementary strand along it is synthesised enzymatically. Pyrosequencing method is based on detecting the activity of DNA polymerase with another chemiluminescent enzyme luciferase. The main difference between pyrosequencing and Sanger sequencing is in the throughput of a single run: 96 capillaries (Sanger method) of about 800 bp compared to several thousand to tens of millions (pyrosequencing) shorter reads. Time runs are longer in Sanger method in comparison to next generation sequencers. Because of the streamline preparation and long reading times a single human operator can operate several machines of next generation at the full capacity but Sanger method sequencer cannot.

In the next generation sequencing (pyrosequencer), the sequencing primer is hybridised to a single-stranded DNA template, and incubated with the enzymes, DNA polymerase, ATP sulfurylase, luciferase and apyrase, and the substrates [adenosine 5′ phosphosulfate (APS) and luciferin]. Deoxyribonucleotide triphosphate (dNTP) is added to the pyrosequencing reaction, one at a time. The template DNA is immobile, and solutions of A, T, C, G and nucleotides are sequentially added and removed from the reaction (Fig. 13.7).

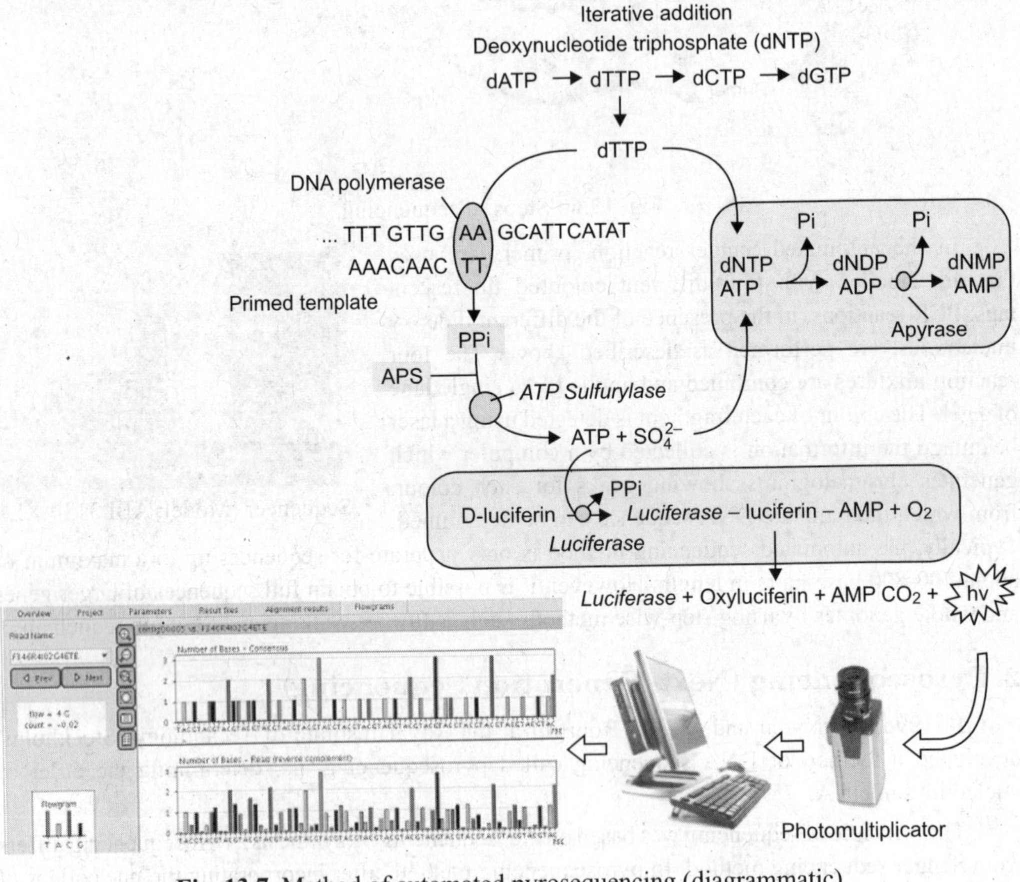

Pyrosequencer (454 life science)

Fig. 13.7. Method of automated pyrosequencing (diagrammatic).

In the case of dATP, deoxyadenosine alfa-thio triphosphate (dATPαS) is added which is not a substrate for a luciferase. Incorporation of a nucleotide is accompanied by release of pyrophosphate (PPi) stoichiometrically. The ATP sulfurylase quantitatively converts PPi to ATP. Light is produced only when the nucleotide solution complements the first unpaired base of the template. The signal light produced by the luciferase-catalysed reaction in presence of ATP is detected by a charge coupled device (CCD) camera and integrated as a peak in a pyrogram. The nucleotide degrading apyrase enzyme continuously degrades ATP excess and unincorporated dNTPs. This process continues with addition of the next dNTP and the nucleotide sequence of the complementary DNA strand is inferred from the signal peaks of the pyrogramme (Ronaghi *et al*., 1996.).

(*a*) Limitation of the Method: The most prior limitation of the pyrosequencing is the cost. Sanger sequencing has low cost in comparison to pyrosequencing. Secondly, the lengths of individual reads of DNA sequence are in the neighborhood of 300-500 nucleotides. It is shorter than the 800-1,000 nucleotides obtainable with chain termination methods. This results in more difficult genome assembly process, especially for those sequences that contain a large amount of repetitive DNA. Therefore, pyrosequencing is most commonly used for resequencing or sequencing of genomes for which the sequence of a close relative is already available and for metagenome (direct from environment genome). The templates for pyrosequencing are made both by solid phase template preparation (streptavidin-coated magnetic beads) and enzymatic template preparation (apyrase+exonuclease).

(*b*) Commercialisation: Pyrosequencing technology has now been commercialised. A company pyrosequencing AB in Uppsala (Sweden) commercialised the machinery and reagents of pyrosequencing for sequencing the short stretches of DNA. In 2003, this Company was renamed as Biotage to which another company named Qiagen acquired in 2008. Further pyrosequencing method was licensed to 454 Life Sciences (Roche), which developed an array-based pyrosequencing technology that emerged as a platform for large-scale DNA sequencing. Most notably it is used for genome sequencing and metagenomics. A single machine of the pyrosequencing platform can generate 400 million nucleotide data within 10 hours.

(*c*) Applications of Pyrosequencing: Pyrosequencing has opened up new possibilities for performing sequence–based DNA analysis. Some of the potential applications of Pyrosequencing are: genotyping of single nucleotide polymorphisms, microbial typing, resequencing, tag sequencing, and analysis of difficult secondary structures.

3. DNA Nanoball Sequencing

The DNA nanoball sequencing is also a high throughput next generation sequencing technology. In this method the entire genomic sequence of an organism can be determined by using the DNA nanoball sequencing technology. This technology is used by the company Complete Genomics to sequence samples that submitted by the researchers throughout the world. Since this technology uses the 'rolling circle replication mechanism' for amplification of small fragments of genomic DNA into DNA nanoballs, it is called DNA nanoball sequencing technology. As compared to the other next generation sequencing platforms, the DNA sequencing technology allows large numbers of DNA nanoballs to be sequenced per run and at low costs (Porreca, 2010). However, by using this method only short sequences of DNA are determined from each DNA nanoball. This makes difficult the mapping the short reads to a reference genome. Many genome sequencing projects are using this technology in recent years.

4. Hierarchical Sequencing (Clone-by-clone Sequencing) Method

Hierarchical sequencing method is also called 'clone-by-clone sequencing method or 'systematic approach' for sequencing the large-sized genome (Fig. 13.8). Genetic linkage map of an organism is prepared in this method. In this method genome is divided into sections of 150 kb, each section is fragmented, sequenced, and assembled. The entire genomic sequence is reconstructed employing the sequenced sections. The public Human Genome sequencing effort exploits this technique.

(*a*) Construction of Physical Map: Entire genome is closed in large insert vector for construction of the physical map. Then all the cloned fragments are arranged in their chromosomal position using genetic markers as probe. For construction of physical maps the basic requirements are: (*i*) large insert cloning vectors, (*ii*) highly saturated genetic and molecular maps, (*iii*) high throughput techniques for clone picking, culturing and selection, and (*iv*) powerful software for constructing minimum tiling paths. Examples of some of the large insert vectors used to construct physical map are: phage λ (insert size 20-30 kb), cosmids (35-45 kb), BAC (100-300 kb), PAC (P1 artificial chromosome, having insert size 100-300 kb), and YAC (insert size 200-1,000 kb). Most often chimeric DNA is present in YAC vectors; therefore, these are not preferred in construction of physical map. Hence, BACs are used on the basis of size of insert DNA.

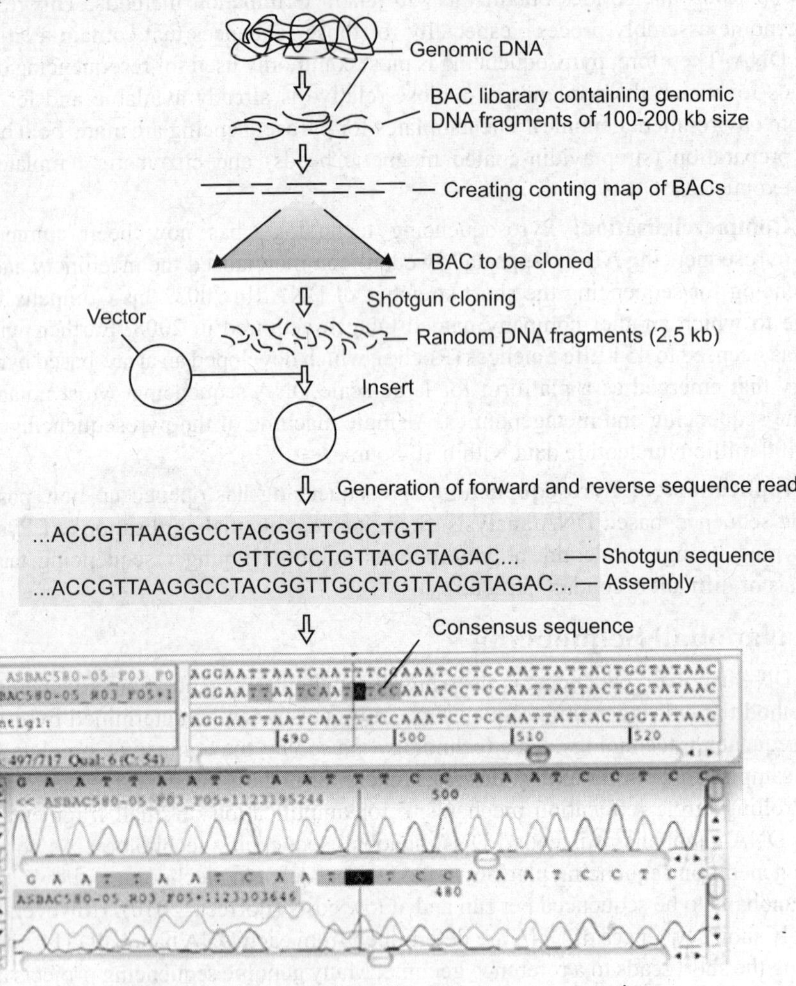

Fig. 13.8. Steps of hierarchical shotgun sequencing.

Consensus Sequence

After alignment of multiple sequences, the most common nucleotides or amino acids present at a particular position are called consensus sequence, i.e. the conserved sequence motifs are called consensus sequences. A consensus sequence can vary in length and can have multiple instances in a single genome (the full set of chromosomes of an individual, or the total number of genes in such a set). The two primary features of any consensus sequence are its conserved residues and common functional roles, despite its genetic location. Because of this, any consensus sequence also serves as a measurement of functional significance over evolutionary time. The results of a multiple sequence alignment are represented by a consensus sequence, where related sequences are compared to each other. The consensus sequence shows the most abundant residues in the alignment at each position.

Consider a DNA sequence: A[CT]N{A}YR, where A means that an A is always found in that position; [CT] means either C or T; N shows any base, and {A} refers to any base except A; Y represents any pyrimidine and R shows any purine. The relative frequency of C or T at that position is not indicated by the [CT]. In addition, a consensus sequence is also represented by a sequence logo which is a graphical representation of the consensus sequence. In a sequence logo the size of a symbol is related to the frequency of a given nucleotide or amino acid occurring at a certain position.

(*b*) Shotgun Cloning: Shotgun libraries are prepared after isolating the high quality BAC DNA. The bacterial cultures containing BAC insert DNA of desired size (2-5 kb) are put in buffer solution and centrifuged to isolate plasmid DNA following different steps such as sonication, nebulisation, hydro-shearing, size selection of fragmented DNA by using electrophoresis, DNA ligation, transformation and template preparation for sequencing.

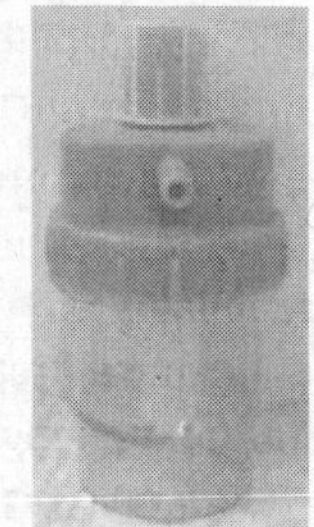

A nebulizer

(*i*) Sonication: A sonicator is used to randomly fragment the DNA of varying size. During sonication process, temperature increases; this affects the size of fragmented DNA. Therefore, for getting good result the temperature of the sonicator should be checked and maintained properly.

(*ii*) Nebulization: DNA fragmentation is important before library construction or subcloning for DNA sequencing. Random fragmentation of DNA using a nebulizer is called nebulisation which is an easy and quick method, and requires only small amounts of DNA (0.5-5 μg). Nebulizer shears the double stranded DNA into 300-800 bp-long fragments. The temperature of the nebulisation buffer does not affect the shearing capabilities, but pressure affects on fragment size. Increasing pressure results in decrease of the average size and vice versa. Time of nebulisation does not affect the size distribution as much as the pressure but longer nebulisation times will result in loss of material.

Buffered DNA (20-50 μg with glycerol (20-50%) is place win a neulizer kept in an ice-cold water bath, followed by passing of inert nitrogen gas through DNA solution at proper pressure. About 2-5 kb sized DNA fragments are obtained at 6-10 pound per inch square pressure for 2-3 minutes. The size of the fragments obtained by nebulization is determined mainly by the speed at which the DNA solution passes through the hole, that alters the pressure of the gas blowing through the nebulizer, the viscosity of the solution, and the temperature. The resulting DNA fragments are distributed over a narrow range of sizes (700-1,330 bp).

(*iii*) Hydroshearing: Shearing of the large DNA fragments by hydrodynamic (water jet) shearing forces using a machine into several little pieces is called hydroshearing. The strands are sheared into many little pieces by a Hydroshear machine because DNA sequencing machines can sequence only small strands of DNA. The sample containing long strands of source DNA is put into the HydroShear machine. The sample passes through a narrow tube where it is forced through a small aperture, and fragments are randomly sheared due to stretching of DNA. Short sheared fragments (2-4 kb) are collected and separated by agarose gel electrophoresis. In the gel lanes the sheared DNA appears as smear. At specific points the lanes bulged out due to accumulation of maximum DNA fragments of desired size. Under UV light gel is sliced that contain DNA fragments of desired size, which are then separated by putting the gel in warm water and ethanol. The purified DNA fragments are used for end repairing which are then cloned in a vector.

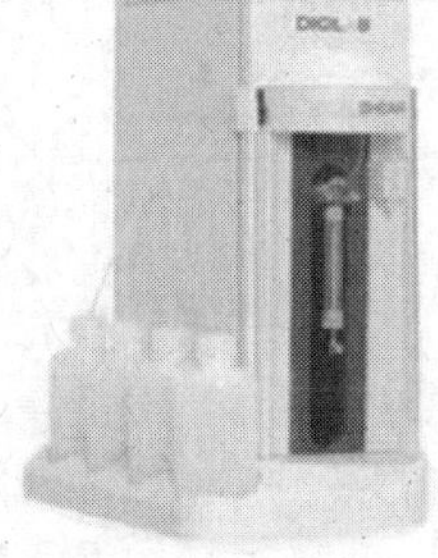

Hydroshear machine

- **DNA Ligation, Transformation and Selection:** The fragmented DNA is ligated into a plasmid, for example pUC18 or pBluscript, by incubating in the presence of T4DNA ligase and rATP. The pUC18 plasmid is transferred into *Escherichia coli* cells for propagation and replication. Each plasmid has a gene for antibiotic resistance as well as a *lac*Z gene that produces blue pigment. The fragments are inserted in the middle of the *lac*Z gene. The transformed *E. coli* cells are selected by blue/white selection plating on LB medium. The white colonies are picked up from each BAC clones and DNA is isolated from it and the templates are used for sequencing.

- **Template Preparation:** By applying the standard protocols the alkaline lyses of bacterial cells are obtained, where from DNA is isolated by using magnetic bead DNA isolation method. Then for setting up a sequencing reaction, quantification of template DNA is carried out on 0.8% agarose gel.

Automated DNA Sequencing : When templates are ready DNA sequencing is carried out via cycle sequencing by using dye terminator protocols. Following the manufacturer's instruction, DNA templates are reacted with Terminator Ready Reaction Mix and plates are kept in PCR machine at optimum reaction conditions. The extension products are purified after 35 cycles by ethanol purification methods.

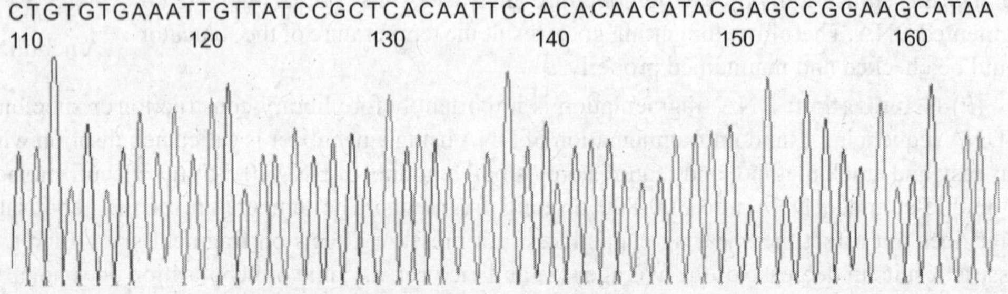

Fig. 13.9. A trace-file view of DNA sequences.

The air-dried DNA pellets after dissolving in double distilled water are loaded in an automated DNA sequencer, which generates a four-color chromatogram showing the results of the sequencing run. About 500 bp read is obtained in the form of a standard chromatogram file in a single sequencing reaction. One can see evenly-spaced peaks each with one color (Fig. 13.9). Peak heights may vary normally three-fold. Baseline peaks with good template and primer may be minimally present.

Difference sequence data of each BAC clone are assembled by using different computational tools. Moreover, before submission of sequence data to the GenBank, it is well polished by using various genome finishing methods. Now many molecular libraries of the world may use this data in functional and comparative genome analyses. An outline of preparation of shotgun libraries and preparation of DNA template for sequencing is given in Fig. 13.10.

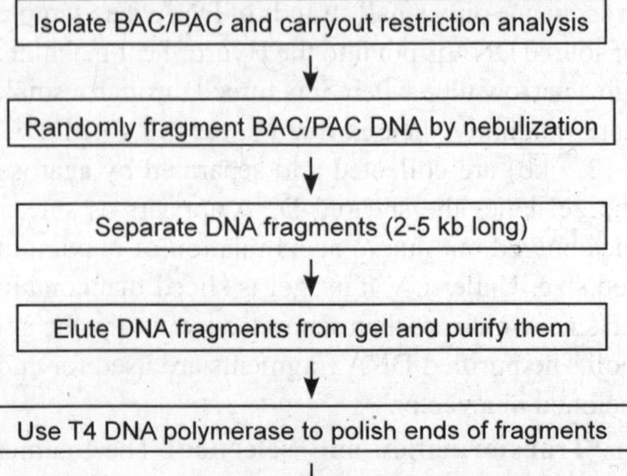

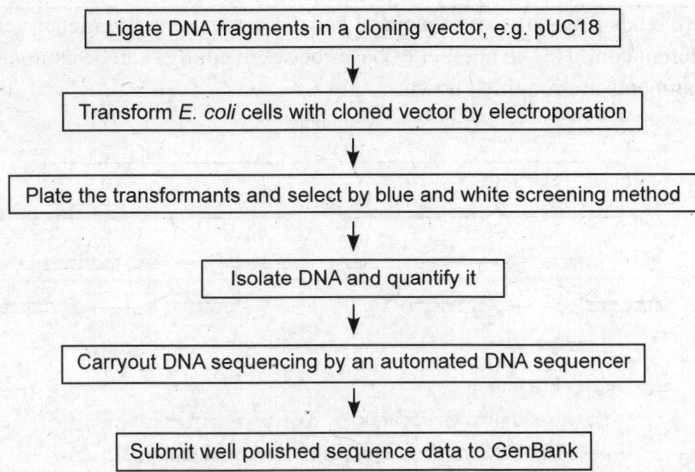

Fig. 13.10. An outline of preparation of shotgun libraries and preparation of DNA template for sequencing.

4. Whole-Genome Shotgun Sequencing

Whole Genome Shotgun (WGS) is a successful method for sequencing the genomes. It is mainly used for sequencing the prokaryotic genomes because prokaryotic genomes are less repetitive in nature; therefore, sequences are assembled by using computational tools. Before 1995, whole genome sequencing was not possible because available computational power was not sufficient to assemble a genome from thousands of DNA fragments. J. Craig Venter and H Smith developed whole-genome shotgun sequencing, and sequenced the genome of two bacteria, *H. influenzae* and *M. genitalium*. This approach may be categorised into the following four stages:

***(a)* Library Construction:** The chromosome is isolated from the desired cells following the methods of molecular biology, and randomly fragmented into small pieces (2-10 kb) using ultrasonic waves. Then the fragments are purified and attached to plasmid vectors to construct 2-10 kb shotgun libraries (Fig. 13.11). Plasmids with single insert are isolated. A library of plasmid clones is prepared by transforming desired host cells with suitable vector that lacked restriction enzymes. Depending on the size of the resulting fragment, various hosts are used to clone these regions that they can accommodate, for example plasmids (2-10 kb), cosmids (40 kb) or bacterial artificial chromosomes (BAC) (70-200 kb).

***(b)* Random Sequencing:** The DNA is purified from plasmid. Thousands of DNA fragments are sequenced using automated sequencer by employing primers labelled with special dyes. Normally with universal primers, thousands of templates were used. These recognise the plasmid DNA sequences next to bacterial DNA insert. The whole-genome is sequenced several times ending with the final accurate results.

***(c)* Fragment-alignment and Gap Closure:** By using special computer programme, the sequenced DNA fragments are clustered and assembled into longer stretches of sequence by comparing nucleotide sequence overlaps between fragments. Two fragments are joined to form a large stretch of DNA if the sequences at their ends overlapped and matched. This 'overlap comparison method' resulted in a set of larger contiguous nucleotide sequence called contigs.

What is Scaffold ?

A portion of the genome sequence reconstructed from end-sequenced whole-genome shotgun clones is called a scaffold. Scaffolds are composed of contigs and gaps. A contig is a contiguous length of genomic sequence in which the order of bases is known to a high confidence level.

Gaps occur where reads from the two sequenced ends of at least one fragment overlap with other reads in two different contigs. The number of bases between contigs can be estimated because the lengths of the fragments are roughly known.

Scaffold

Contig 1 Contig 1

Fragment

Read (known sequence)

Roughly known length but not known sequence

The contigs are aligned in a proper order to form the completed genome sequence. Many bioinformatic tools are available to make the contigs, *i.e.* CAP3 (a bioinformatics software). If there exist gaps between the two contigs, these could be analysed and gap filled in with their sequences. Several other methods were also used to align contigs and filling the gaps. The large fragments overlap the previously sequenced contings. Using oligonucleotide probes, the fragments having overlaps with two contigs allow them to place side-by-side and fill in the gap between them.

(*d*) Proof Reading: Then the proof reading of sequences is done carefully so that any ambiguities in the sequence could be resolved. The sequence is also checked for the presence of any frameshift mutation; if so, the mutation is corrected. This approach required less than four months to sequence the genome (5,00,000 bp) of *M. genitalium*.

The major advantages of this approach are that: (*i*) it is less time-consuming method because it does not require the construction of genetic and physical maps, and (*ii*) in more complex genome, the assembly of sequence reads in the forms of chromosome-specific pseudo-molecules and resolving repeats are difficult.

5. Expressed Sequence Tag (EST) Method

The EST approach was pioneered by J. Craig Venter and co-workers at the National Institute of Health (NIH) (U.S.A.) in the early 1990s. He developed a new method of investing the genes by focusing the attention of the *active* portion of the genome as mRNA. Venter and co-workers isolated mRNA molecules (instead of fragments of genomic DNA) and constructed cDNA molecules. They treated DNA as a part of chromosomal DNA and sequenced to create 'expressed sequence tags' (ESTs). The ESTs were used as handles for isolating the complete genes. Following EST strategy, plenty of databases of nucleotide sequences were generated. Consequently it helped to prepare the transcript map of human genome at preliminary level. The EST technique demonstrated the possibility of sequencing all genes to the highest levels. This attempt boosted up the growth of genomic industry.

Thereafter, Venter switched towards sequencing the entire human genome through whole genome shotgun strategy. In this approach he matched the overlapping ends of DNA fragments, fitted contiguous pieces and attempted for genome sequencing by a 'genome assembly programme'. By sequencing the entire genome of some microorganisms, the validity of this method was also proved.

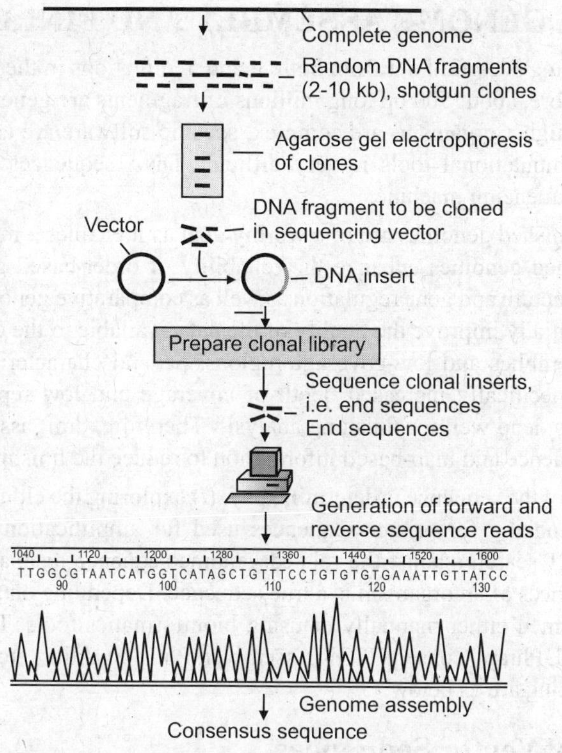

Fig. 13.11. Steps of whole genome shotgun sequencing.

For preparation of ESTs individual's clones are picked up from cDNA library and one sequence is generated from each end of cDNA insert (Fig. 13.12).

The contaminating vectors, mitochondria and bacterial sequences are removed before depositing the ESTs into the public database (dbEST). In database, the ESTs are identified by their clone number and presence of 5′ or 3′ orientation. So far the ESTs that have been submitted to the public sequence databases were created from thousands of different cDNA libraries representing over 250 organisms.

The I.M.A.G.E. Consortium has picked up the individual clone from many libraries and used for EST sequencing and arranged them for easy distribution. Upto 2002, more than 3.8 million cDNA clones have been arrayed from 360 human and 108 mouse cDNA libraries. The I.M.A.G.E. Consortium at present consists of more than half of ESTs in GenBank.

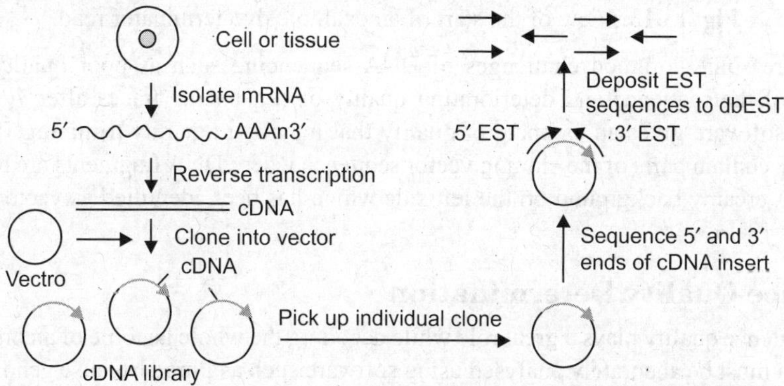

Fig. 13.12. Construction of expressed sequence tag (EST) database.

E. GENOME ASSEMBLY AND FINISHING

DNA sequencing is carried out from both 3′ and 5′ directions in the large genome sequencing programme. Therefore, about 500 bp long millions of fragments are generated from the organisms. Since genomes of higher organisms are complex, specific software are used for genome assembly because use of computational tools is very difficult. DNA sequences are analysed after being produced from a sequencing machine.

Completely finished genomes are of a worthy goal as it enables a much richer set of genomic analysis. The finished genomes enhance the reliability of order-based genomic analysis such as studying operon structure and gene regulation as well as comparative genomic studies. The finishing process can substantially improve the quality of the data available to the community by identifying and fixing mis-assemblies and low-coverage regions. Several characteristics of the new types of sequencing data, specifically increased depth of coverage and low representation biases in the sequencing libraries, lend well for finishing analysis. Therefore, draft assemblies can be combined with additional sequence and map-based information to reduce the finishing effort.

The accuracy of the sequence is determined by: (*i*) exploring the cloning sites of inserts present in the vector, (*ii*) looking for primer sequences used for amplification of the product, and (*iii*) carrying out BLAST search against the related genome sequence databases. Any sequence hit by the unknown sequences of an organism is a true sequence. Depending on the genome length further steps can be performed either manually or using bioinformatics tools. The most commonly used software are : Phred, Phrap, Consed, TIGR Assembler, Phusion, Atlas, etc. Further steps of genome assembly and finishing are as below:

1. Trimming of Vector Sequences

The sequence data produced by sequencing machine consist of vector- borne sequences in some of the trace files, hence such data cannot be used for genome assembly. A software 'cross match' is used to trim the vector sequences, which compares sequence read against the database of vector sequences resulting in vector-masked reads.

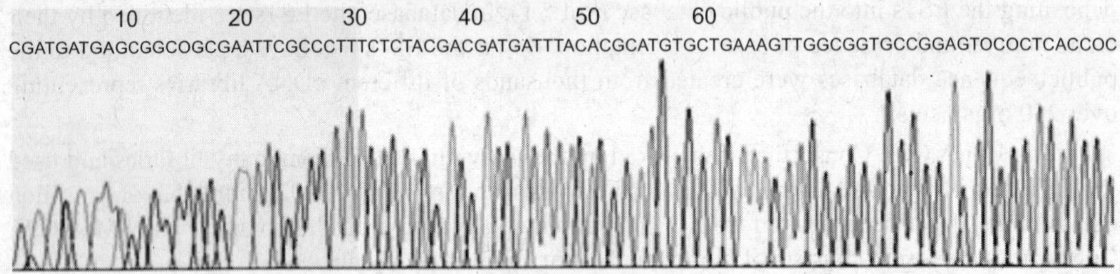

Fig. 13.13. View of the start of an example dye-terminator read.

There are some common challenges of DNA sequencing such as poor quality in the first 15-40 bases of the sequence and deteriorating quality of sequencing traces after 700-900 bases. Base calling software gives an estimate of quality that may be in quality trimming. The resulting sequence may contain parts of the cloning vector sequence where DNA fragments are cloned before sequencing. A creamy background on the left side which has been identified as vector sequence is shown in Fig. 13.14.

2. Sequence Quality Determination

The sequence quality plays a great role while decoding the whole genome of an organism. Raw sequence data must be accurately analysed using software such as Phred because genome sequence data are used in functional and comparative genomics. Phred means Phil's Read Editor developed by

Phil Green and Brent Ewing of the U.S.A. in 1998. On the basis of trace quality in the chromatogram produced by Phred, each nucleotide in the trace file of sequence is assigned a particular statistical score of numerical value called 'base calling'. Therefore, Phred is a 'base calling software' that assigns a probability score to each nucleotide with a high level of accuracy (Fig. 13.14).

A single read from the NCBI sequence read archive (SRA, http://www.ncbi.nlm.nih.gov/Traces/sra/sra.cgih) presented as a FASTA entry are given below:

```
> SRR014849. 1 EIXKN4201CFU84 length = 93
GGGGGGGGGGGGGGGGGTTTTTTGTTTGGAACCGGAAAGG
GTTTTGAATTTCAAACCCTTTTCGGTTTCCAACCTTCCAA
AGCAATGCCAATA
```

The Phred's scores of above sequence are as below:

```
>SRR014849. 1 EIXKN4201CF084 length = 93
18 10 5 3 2 1 1 1 1 1 1 1 1 1 1 1 22 37
31 22 16 11 6 1 26 34 30 11 33 26 30 21
33 26 25 36 32 16 36 32 16 36 32 20 6
24 33 25 30 25 2 24 36 32 15 35 31 17
36 32 20 6 25 29 20 30 25 4 32 26 32 23
32 26 30 24 33 26 35 31 14 28 28 30 22
28 24 27 17 32 23 28 28
```

Fig. 13.14. Scores of each nucleotide in a trace file of sequence assigned by Phred.

The basic steps used by the software to analyse sequences are: (*i*) the use of best reason of the peaks to predict expected peaks, (*i*) identification of observed peaks in the chromatogram, (*iii*) comparison of both expected and observed peaks and separating into two groups, all matched peaks and un-matched peaks, and (*iv*) again analysing the un-matched peaks which might have been missed. The Phred scores can be calculated using the following formula :

$$Q_{(Phred)} = -10 \times \log_{10}(p)$$

Where

$Q_{(Phred)}$ is the quality score

p is the probability of a wrong base call

(*a*) Phred Quality Scores: One of the most innovative features in Phred is the base-specific quality scores. In a chromatogram file the basic criteria for assigning a quality score by Phred are the height, spacing and compression of the peaks. The most important use of Phred quality scores is the automatic determination of accurate, quality-based consensus sequences. Phred examines the peaks around each base call to assign a quality score to each base call. Quality scores range from 4 to about 60; higher Phred score correspond to higher quality. If Phred assigns a quality score of 30 to a base, the chance of an incorrect base is 1 in 1000. The most commonly used method is to count the bases with a quality score of 20 and above. The high accuracy of Phred quality scores makes them an ideal tool to assess the quality of sequences. The quality scores are logarithmically linked to error probabilities as shown in Table 13.5.

The probabilities of Phred's error are very accurate. If Phred assigns a quality score of 40 to a base, the chance of a wrong base is indeed 1 in 10,000. This high accuracy has been observed for sequences generated at different laboratories using a different combination of sequencing enzymes, fluorescent dyes, and gel run conditions.

Phred quality scores are used for: (*i*) assessment of sequence quality, (*ii*) recognition and removal of low-quality sequence (end clipping), and (*iii*) determination of accurate consensus sequences.

Table 13.5. The quality scores logarithmically linked to error probabilities and percent accuracy.

Phred quality score	*Probability that the base is called wrong*	*Accuracy of the base call (%)*
10	1 in 10	90%
20	1 in 100	99%
30	1 in 1,000	99.9%
40	1 in 10,000	99.99%
50	1 in 100,000	99.999%

(*b*) Sequence Assembly: Aligning and merging of a much longer DNA sequence to reconstruct the original sequence is called sequence assembly. DNA sequencing technology cannot read whole genomes at a time, but it can read small pieces between 20 and 1000 bases. The short fragments called reads are produced from shotgun sequencing genomic DNA or gene transcript called expression sequence tags (ESTs). DNA sequence assembly software is required for sequence assembly. Scientists have developed assemblers like Celera assembler, Arachne, Phrap and Phusion assemblers which are able to handle genomes of 100-300 million base pairs.

Phrap is one of the popular and leading programmes for assembling shotgun DNA sequence data because: (*i*) it allows the use of the entire read but not just the trimmed high quality part, (*ii*) it uses a combination of information from user-supplied data and internally computed data quality to improve assembly accuracy in the presence of repeats, (*iii*) it constructs the contig sequence as a mosaic of the highest quality read segments rather than a consensus, (*iv*) it provides extensive assembly information to assist in trouble-shooting assembly problems, and (*v*) it handles large datasets. Phrap is routinely used in some of the largest sequencing projects in the Human Genome Sequencing Project and in the biotech industry.

Characteristic Features of Phrap

- ***Fast assemblies:*** Assemblies of cosmid to BAC sized projects with several hundred to two thousand reads take only minutes to complete on high-powered computers.
- ***Accurate consensus sequences:*** Phrap uses Phred's quality scores and determines the highly accurate consensus sequences. Phrap examines all individual sequences at a given position, and uses the highest quality sequence to build the consensus. Phrap's approach gives more accurate consensus sequences especially in regions of low coverage or regions of systematic errors like compressions.
- ***Consensus quality estimates***: Phrap uses the quality information of individual sequences to estimate the quality of the consensus sequence. Phrap uses available information about sequencing chemistry (dye terminator or dye primer) and confirmation by 'other strand' reads while estimating the consensus quality. Consensus quality estimates can also be very helpful in mutation detection by DNA sequencing.
- ***Ability to assemble very large projects:*** Routinely Phrap is used to assemble bacterial genomes sequenced by the shotgun approach, where each project contained tens of thousands of reads. Smaller bacterial genomes can be assembled in less than three hours.
- ***Improved identification and handling of repeats:*** Phrap uses quality scores to estimate whether discrepancies between two overlapping sequences are more likely to arise from random errors, or from different copies of a repeated sequence. For repeats with 95 to 98% identity (like human Alu sequences) and high quality sequence data, this typically yields correct assemblies.

Phrap uses Phred's quality scores and identifies the most accurate consensus sequences and examines all individual sequences at a given position. On the basis of high quality sequence it builds the consensus sequence. The input file for Phrap is called 'phd' file developed by Phred, whereas the output files of Phrap is called 'ace' file. Phrap also assigns scores to nucleotide bases of the consensus sequence which is called 'Phrap score'. The sum of Phred quality of the overlapping sequences is called Phrap score. Phred and Phrap, and similar programmes enable the assembly of large parts of the human genome (and many other genomes) at an accuracy that was substantially higher (less than 1 error in 10,000 bases) than the typical accuracy of carefully hand-edited sequences that had earlier been submitted to the GenBank database.

***(c)* Assembly View:** There is another graphical tool called 'Consed' that is used for viewing, editing, and finishing the DNA sequence assemblies. In 1998, David Gordon and co-workers at the Phil Green Laboratory developed Consed software and distributed to academic users free of cost. Consed was originally developed as a contig editing and finishing tool for large-scale cosmid shotgun sequencing in the Human Genome Project. Consed was used to check assemblies generated by Phrap, solve assembly problems like those caused by highly identical repeats, and finishing tasks like primer picking and gap closure.

There are three input directories required for it functioning: *chromat_dir* (directory for chromatogram files generated by sequencing machine), *phd_dir* (the output file generated by Phred that contains quality score of each base), and *edit_dir* (directories containing ace files generated by Phrap). Then double click on contig 'myreference' to bring up the Aligned Reads Window. Scroll around a little and right click on a read or two to see the trace. The Aligned Reads window shows the sequence assemblies in the form of graphical interface as in Fig. 13.15.

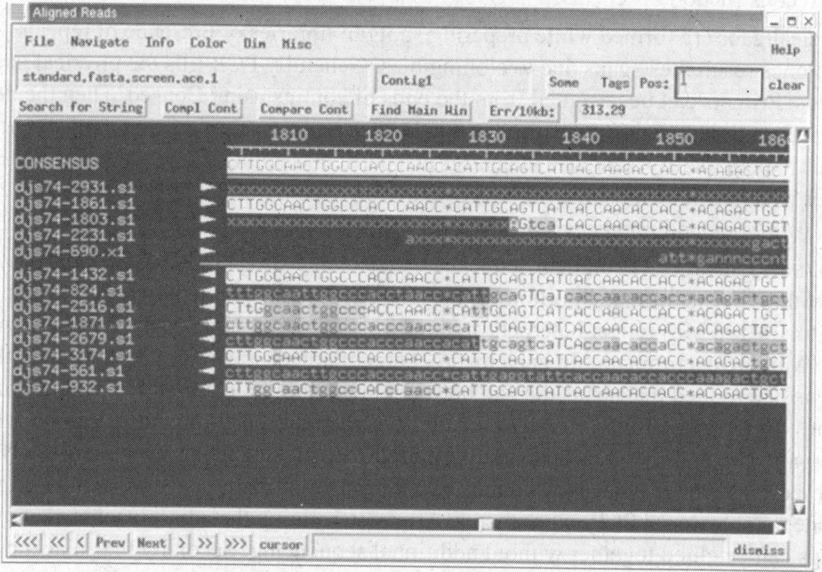

Fig. 13.15. Assembled sequence file shown in Consed window.

Certain sequence quality tags are shown in output of Consed Window. The high quality and mismatched tags are shown in red colour in the white upper case. Navigation of high and low quality discrepancy, comparison and merging of sequence data and picking of primers for finishing can be done by using Consed. The sequences are submitted to GenBank in three different phases as given below after completing the assembly of BAC clones: *phase I* (unoriented contings of about 2 kb), *phase II* (as draft sequence, *i.e.* oriented contigs having delineated insert-vector junctions), and *phase III* (high quality sequence free from any errors and breaks) (Fig. 13.16).

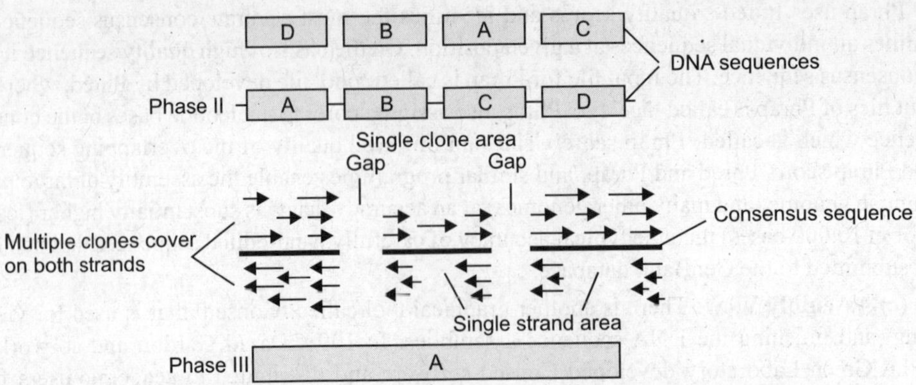

Fig. 13.16. Sequences of different phases submitted to GenBank.

3. Problems Related to Draft Sequence and Genome Finishing

The process of converting the initial draft sequence into a near-complete sequence is referred to as 'finishing'. It is a complex iterative process that proceeds simultaneously at multiple scales, ranging from single nucleotides to the integrity of whole chromosomes. The objective of finishing is to overcome two major limitations of the shotgun sequencing process: (*i*) the output of a genome assembler is fragmented due to difficulties in assembling repeat regions and to cloning/sequencing biases, and (*ii*) the assembled fragments frequently contain errors due to sequencing artifacts or the incorrect reconstruction of repeats.

Physical gaps are formed while preparing shotgun libraries or presence of repeats. Absence of any sub-clones result these gaps due to cloning bias. Generally PCR fills the physical gaps between two contigs. Primers can be designed from the ends of contigs using Consed. Then the desired PCR products are amplified by using the primers and BAC DNA as a template. In contrast, sequence gaps appear due to poor quality sequence in contigs at a particular point. The sequence gaps can be closed by 'primer walking'. The finishing process can thus be decomposed into two steps: *gap closure*, and *assembly validation* and *refinement*.

(*a*) Gap Closure: The human sequence consists of 2,851,330,913 nucleotides, lying almost entirely within the euchromatic portion of the genome. It is interrupted by only 341 gaps, of which 33 gaps (total ~198 Mb) reflect heterochromatin, which was not targeted by the HGP, and 308 gaps (total ~28 Mb) are euchromatic. In gap closure, pairs of adjacent contigs are identified, the genomic sequence is determined that spans the gap between them through directed-PCR and primer-walking approaches. In the availability of mate-pair libraries the adjacency of contigs can be inferred from the mate-pair data and the gaps spanned by paired reads (sequencing gaps) can easily be closed. However, contigs whose adjacency cannot be inferred from mate-pair data require expensive and error-prone combinatorial PCR experiments. After determining the adjacency between two contigs, the contigs can be glued together without additional sequencing. Secondly the adjacency of contigs can easily be determined either through recently developed nextgen mate-pair protocols or through the use of new mapping technologies such as the optical mapping approach from Opgen Inc (see *http://www.opgen.com*).

(*b*) Validation and refinement: The goal of validation and refinement is to correct errors in the assembled sequence, *i.e.* both single-base errors and large-scale errors. The single-base errors are the mis-match bases occurred due to sequencing errors, and large-scale errors are the mis-assemblies formed due to repeats. Mostly the single base errors can be automatically corrected due to a high level of coverage. This works in 454 pyrosequencing also where mostly errors arise within homopolymer tracts. Moreover, assembly software (such as Newbler assembler from 454) has been designed mainly for high-coverage nextgen data that uses conservative algorithms to avoid miss-assemblies. Repeat-induced ambiguities can be resolved through targeted PCR experiments aimed at uncovering the correct adjacency of the assembled contigs. In the assembly the two adjacently determined contigs can be simply glued together.

4. Assembly Verification

The whole sequence is verified for its accuracy after finishing the clones. It is done by Masport analysis, *i.e.* digesting the sequence *in silico* with at least three restriction endonucleases and comparing with actual fingerprints of BAC clones. Both the fingerprints of BAC clones and *in silico*-derived restriction digests are matched. BLAST search of finished sequence of the overlapping clones is performed to determine the north and south coordinates of the overlapping clones. Then the sequence is submitted to GenBank.

F. GENE PREDICTION AND GENE COUNTING

Gene prediction (gene finding) is an area of computational biology concerned with algorithmically identification of stretches of sequence, usually genomic DNA that are biologically functional including protein-coding genes and other functional elements such as RNA genes and regulatory regions. In other words, gene is a region of the DNA which codes for a protein or mRNA. Regulatory regions, *i.e.* regions of DNA that regulate gene expression, are also considered as the functional regions. Gene prediction does not tell us which genes code for which proteins.

Typically a gene consists of: (*i*) an open reading frame (ORF) possessing a well defined start codon ATG, and stop codons (TAG, TGA and TAA), (*ii*) promoter sequences (TATA box, CAAT box) in the 5′ upstream region, (*iii*) poly A tail in the 3′ downstream region, and (*iv*) splicing sites consisting of AG and GT splice signals. Basic characteristics of typical prokaryotic and eukaryotic genes are given in Fig. 13.17.

There are different computational tools that are used for gene predictions; all the tools look for the presence or absence of the above gene sequences and predict the gene accordingly. However, gene prediction is mainly based on codon usage, base composition, splicing site, poly-A signal, di-, tri-, hexa-nucleotide, transcription signals, translation signals and whole genome with accuracy. The general purpose of gene prediction is to predict genes encoding proteins, search for open reading frames (ORFs) and identify promoters and binding sites which regulate gene activity.

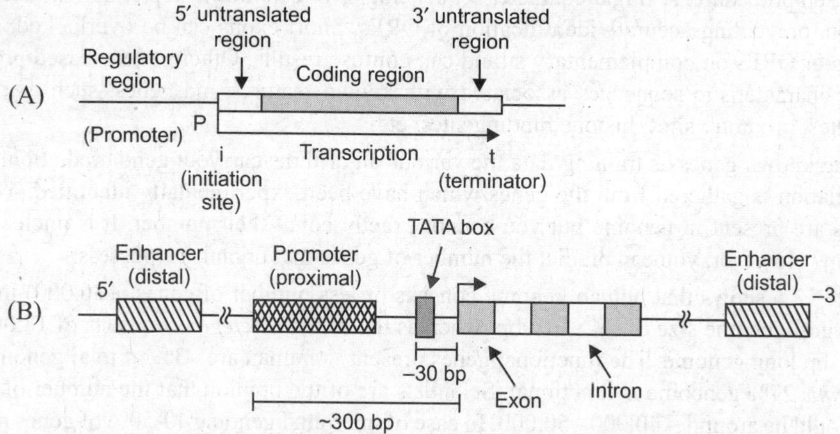

Fig. 13.17. Structure of prokaryotic gene (A) and eukaryotic gene (B).

Terminology

- **Conserved:** It is derived from a common ancestor and retained in contemporary related species.
- **Homologs:** Features in species being compared that are similar because they are ancestrally related.
- **Orthologs:** Homologous features that separated because of a speciation event, they are derived from the same gene in the last common ancestor.
- **Paralogs:** Homologous features that separated because of duplication events.
- **Synteny** : The property of being on the same chromosome.
- **Homology Blocks:** Also defined as conserved synteny; it occurs when the orthologs of genes that are on the same chromosome in one species are also on the same chromosome in the comparison species.
- **Conserved Segments:** Also known as conserved linkages, is a special case of the conserved synteny in which the order of multiple orthologous genes is the same in the compared species.

1. Methods of Gene Prediction

One of the primary tasks in deciphering the functional contents of a newly sequenced genome is the identification of its protein coding genes. There are two primary approaches for predicting genes: intrinsic approach or *ab initio* method, and extrinsic approaches or homology-based methods. The extrinsic methods try to find similarities between genomic sequences and known proteins. A certain limitation of both approaches is that they critically rely on information derived from already known genes, so they tend to be biased towards finding genes that are similar to known genes. The *ab initio* methods use the DNA sequence itself as the only source of information, comparative methods using multiple genomic sequences, and similarity based methods which employ the cDNA or protein sequences of related genes to aid the gene prediction.

(*a*) Content-based Method: Content-based gene finding methods rely on statistical information derived from known sequences to predict unknown genes. Some evaluative measures include: coding potential (based on codon bias), periodicity in the sequence, sequence homogeneity, etc.

(*b*) Pattern-based Method: ORF finding based on start and stop codon frequency is a pattern-based procedure. A single-character sequencing error can hide a stop codon or insert a false stop codon, preventing accurate identification of ORFs. Short exons can be overlooked. Multiple transcripts or ORFs on complementary strand can confuse results. Other pattern-based procedures recognise characteristic sequences associated with known features and genes, such as ribosome binding sites, promoter sites, histone binding sites, etc.

Using known genes as training data the various algorithms carryout gene prediction. Most of the information is gathered from the genes which have been experimentally identified. You know that genes are present in genome but you cannot exactly count their number. It is unclear how to count them? However, you can predict the number of genes that organisms possess.

Table 13.6 shows that human genome consists of less number of genes (~30,000) in spite of having largest genome size of ~3×10^9 bp, whereas the worm *C. elegans* consists of 18,000 genes in 1×10^7 bp long genome. The functional genes present in human are <3% of total genome, while in *C. elegans* 27% geneome is functional. Scientists are of the opinion that the number of genes in human should be around ~ 40,000 - 50,000. In case of microbial genome 40-50% of genes may code for proteins of unknown function. 20-30% genes may encode unknown proteins that are unique to the species.

***(c)* Comparative Methods:** It is extrinsic approach where sequence homology acts as the basis of the comparative methods. The forces of natural selection cause genes and other functional elements to undergo mutation at a slower rate than the rest of the genome, causing negative impact in functional elements of the organisms. Genes can thus be detected by comparing the genomes of related species to detect this evolutionary pressure for conservation. Therefore, comparative study of variation at gene level acts as the principle of comparative approaches. For the first time this approach was applied to the mouse and human genomes by using programmes like SLAM, Twinscan/N-SCAN and SGP.

Table 13.6. Some organisms showing genome size and the number of predicted genes.

Organisms	***Genome size (bp)***	***Number of predicted genes***	***Part of genome encoding proteins***
Escherichia coli	5×10^5	5,000	90%
Saccharomyces cerevisiae	14×10^6	6,000	70%
Drosophila melanogaster	3×10^8	14,000	20%
Caenorhabditis elegans	1×10^8	18,000	27%
Arabidopsis thaliana	1.25×10^8	25,000	20%
Homo sapiens	$\sim 3 \times 10^9$	~ 30,000	<5%

1. Gene Prediction Algorithms

The transformation of raw genomic data into the organised knowledge (which provide new and improved understanding of genome organisation and regulation) is called *genome annotation.* Annotation can be used to identify tentatively many genes. This allows analysing many kinds of genes and functions present in the organism. For computation biologist, genome annotation refers to the process of assigning 'features' or 'label' to raw DNA sequences. It is done by integrating information from the sequence with computation tools, auxiliary data and biological knowledge. Gene prediction requires a combination of algorithms with different types of biological databases. A simplified schematic of algorithmic gene prediction is given in Fig. 13.18.

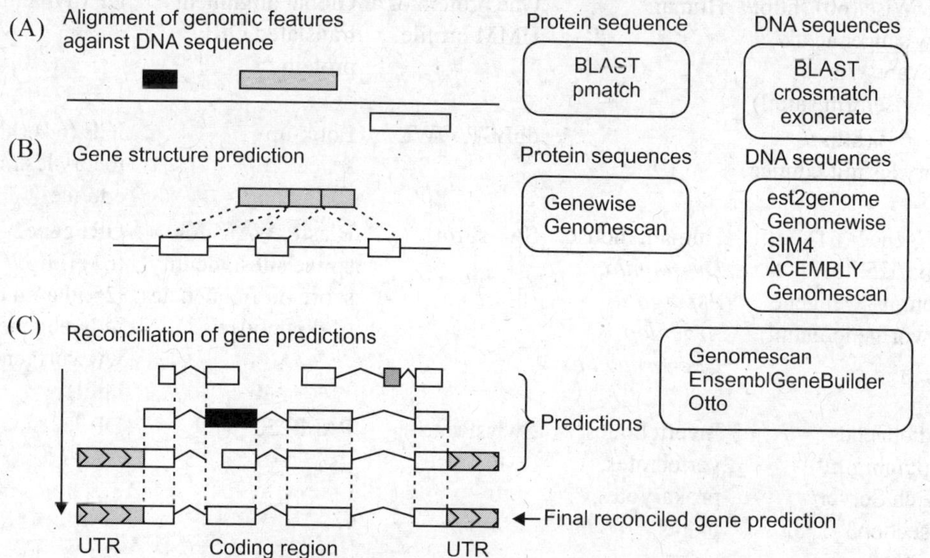

Fig. 13.18. Schematic algorithmic gene prediction.

There are several algorithms for gene prediction as given below:

(*a*) Homology-based Gene Prediction: It is traditionally the first and most commonly used tool to discover new genes. Homology-based gene prediction falls into two categories as below:

(*i*) Gene prediction through detection of homology to know the proteins: This method uses sequence alignment of the translated DNA sequence (using 6 possible reading frames) with databases of known proteins.

(*ii*) Gene prediction through comparison with expressed sequence tags (EST) database: The EST has been described earlier. With the appropriate use of sequence alignment parameters about 90% of the genes annotated on human genomic DNA are detected by ESTs. Homology-based gene prediction programmes are given in Table 13.9.

Table 13.7. Homology-based gene prediction programmes.

Programme	***Organism***	***Databank or required input***	***Alignment***	***Gene reconstruction***
AAT (66) (http://genome.cs.mtu.edu/aat.html)	Primates, rodents, other	cDNA, protein	DDS (improved BLASTX), DPS (improved BLASTN)	NAP, GAP2
EbEST (73) (http://ares.ifrc.mcw.edu/EBEST/ebest.html)	Human, other	dbEST	BLASTN, EST clustering, Smith–Waterman-based gapped alignment	3′-UTR detection, assembly of EST-tagged exons
GeneSeqer (67,68) (http://bioinformatics.iastate.edu/cgi-bin/gs.cgi)	*Arabidopsis*, maize, generic plant	dbEST or EST database or proteins	Spliced alignment, splice recognition with SplicePredictor if missing EST match	Yes
GeneWise (60) (http://www.sanger.ac.uk/Software/Wise2/genewiseform.shtml)	Human	One protein or a HMM profile	Global alignment translated ORF/protein	DP (dynamite)
ICE (64) (http://theory.lcs.mit.edu/ice)		dbEST, OWL	Look-up	ICE (64) (http://theory.lcs.mit.edu/ice)
ORFgene2 (61) (http://l25.itba.mi.cnr.it/~webgene/wwworfgene2.html)	Human, mouse, *Drosophila*, *Aspergillus*, *Arabidopsis*, *Caenorhabditis*	SwissProt	BlastP, WAM for splice sites, identity score on frequencies of dipeptides	ORFgene2 (61) (http://l25.itba.mi.cnr.it/~webgene/wwworfgene2.html)
PredictGenes (http://cbrg.inf.ethz.ch/Server/subsection3_1_8.html)	Invertebrates, vertebrates, prokaryotes, plants	SwissProt	PAM 250	DP

PROCRUSTES (59) (http://www-hto.usc.edu/software/procrustes/wwwserv.html)	Vertebrates	One homologous protein	Protein/protein alignments scored with PAM 120	DP
Pro-Gen (83) (http://www.anchorgen.com/pro_gen/pro_gen.html)		Two genomic sequences	Alignment of translated sequences scored with PAM 120	DP
ROSETTA (80) (http://crossspecies.lcs.mit.edu/)	Human, mouse	Two genomic sequences	GLASS (global alignment system), PAM 20, Genscan method for splice sites	DP
SLAM (85) (http://baboon.math.berkeley.edu/~syntenic/slam.html)	Human, mouse	Two genomic sequences	Generalised pair HMM	DP
SYNCOD (72) (http://125.itba.mi.cnr.it/~webgene/wwwsyncod.html)	Human, mouse, *Drosophila*, *Arabidopsis*, *Aspergillus*, *Caenorhabditis*	BLASTN output	Silent/replacement ratio, Monte Carlo simulations	No
TAP (75) (http://sapiens.wustl.edu/~zkan/TAP/)	Human, mouse, *Drosophila*	dbEST	WU-BLASTN, SIM4	Yes
Utopia (84)	All eukaryotes	Two genomic sequences	Local alignment	Yes

DP, dynamic programming; WAM, weight array matrix; WMM, weight matrix method.

***(b) Ab Initio* Gene Prediction:** In early 1980s, *ab initio* gene prediction has evolved from simple methods based on coding region statistics to sophisticated methodologies that can incorporate biological constrains into computational algorithms. *Ab initio* gene prediction developed due to Human Genome Project. *Ab initio* gene prediction refers to the computation tools and algorithms which are useful in this step of genome annotation. Moreover, gene prediction is still important and widely used of all genome annotations.

It includes the class of 'statistical learning' algorithms which are used for *ab initio* gene recognition. There are several strategies of *ab initio* gene prediction based on: oligonucleotide usage, marker models, statistical pattern recognition and classification, neural networks.

(c) Systenic Gene Prediction: Systenic gene prediction is gene recognition by using cross-species sequence comparisons to identify and align relevant regions. The presence of exonic features at corresponding positions is searched out in both species simultaneously. The reason behind systenic gene prediction is simple. During evolution the exons (*i.e.* functional regions of DNA sequence) tend to more highly conserved than non-functional regions. Hence local conservation identified through comparisons of genomes of related species indicates biological function. Genes in human chromosome systenic to mouse chromosome are shown in Fig. 13.19.

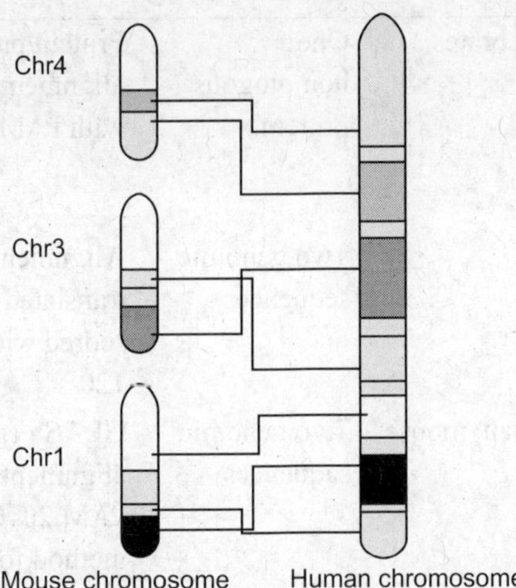

Fig. 13.19. Systeny between human chromosome and mouse genome.

2. Accuracy and Validity of Gene Prediction Algorithms

Although the statistical properties of coding regions allow for a good discrimination between large coding and non-coding regions, the exact identification of the limits of the coding parts of the exons or of gene boundaries remains difficult. In the first case, predicted coding region limits are often incorrect. In the second case, the predicted structure frequently splits a single true gene into several or merges several genes into one. Several teams are currently working on the terminal parts of genes to address these problems. A programme dedicated to predict the 3′-terminal exons and another to identify the 5′-terminal exons have been made available. Such problems are very complex as intergenic and intronic sequences do not differ much.

Moreover, inaccuracies of *ab initio* gene prediction algorithms will travel down the line. This result in errors at the transcriptional level (the proteome level) and could ultimately affect or at least hinder our understanding of biology of species.

3. Gene Prediction Software

There are many different gene finding bioinformatics software packages based on online or offline are available and no one programme is capable of finding everything. Some are briefly describe and listed below:

- **BLAST:** Basic Local Alignment Search Tool (BLAST) finds regions of local similarity between sequences, compares nucleotide or protein sequences to sequence databases and calculates the statistical significance of matches.
- **AGenDA:** AGenDA is a web tool that compares the genomic sequences from evolutionarily related organisms in order to make gene predictions. It takes pairs of genomic sequences as input, aligns the sequences, and makes predictions based on splice signals, start and stop codons, and areas of conserved sequence.
- **AUGUSTUS:** AUGUSTUS is a eukaryotic gene prediction tool for modelling intron length distribution, and searching for motifs and multiple splice variants. It is particularly effective with larger sequences. It can be run through a web interface, or downloaded and run locally.
- **GeneAlign:** GeneAlign is a gene prediction tool that uses conservation of gene structures and sequence homologies between protein coding regions to increase prediction

accuracy. GeneAlign is currently configured to align human and mouse sequences for gene prediction.

- **DNAtools:** DNAtools include predicting DNA curvature; plotting physicochemical, statistical or locally computed paramaters along DNA sequences; producing a 3-D model of a DNA sequence; searching an intron database.
- **Glimmer:** Gene Locator and Interpolated Markov Modeler; this prokaryote-gene finding tool is the primary microbial gene finder used at TIGR; free (including source code) with registration for non-commercial use.
- **SLAM:** SLAM is a comparative-based annotation and alignment tool for syntenic genomic sequences that performs gene finding and alignment simultaneously. SLAM also predicts CNSs (conserved non-coding sequences).
- **TACT:** Transcriptome Auto-annotation Conducting Tool (TACT) is an automated tool for conducting functional annotation of transcripts that integrates sequence similarity searches and functional motif predictions.

G. GENOME SIMILARITY AND SINGLE NUCLEOTIDE POLYMORPHISMS (SNP)

Establishment of genome similarity and single nucleotide polymorphism are the important aspects of bioinformatics. There two are briefly described as below.

1. Genome Similarity

Genomes of all the organisms from prokaryotes to humans are made up of only four alphabets of nucleotides, A, G, T and C. A comparison of genetic blueprints of common chimpanzee with that of the human genome shows the sharing of 96% of our DNA to that of our closest living relatives. The number of genetic differences between humans and chimpanzee is ten times smaller than that between mice and rats. Despite the similarities in human and chimpanzee genomes, about 40 million differences have been identified among the three billion nucleotides in each genome.

When the work on Human Genome Project was in progress, several unknown samples were collected from different regions. During this moment two questions were raised: (*i*) genome of individuals to be sequenced, and (*ii*) the percentage of genome similarity between two individuals. Though all the humans are similar, yet what is the percentage of genome similarity between them? Human genome is about 99.9% similar to each other. The percentage of similarity refers to 'consensus human genome'. It means that 0.1% difference in genome makes them different from each other. There lies a difference of one in thousand nucleotides in the genomes of two different individuals. Only this percentage of DNA brings uniqueness in them.

2. Single Nucleotide Polymorphisms (SNPs)

SNPs are the variations in a nucleotide sequence which occur due to change even in a single base (*e.g.* A, G, T or C). Therefore, certain sites of the sequence nucleotide bases of different individuals differ as below:

```
GTCATAGCATTATTATTATTATTCAGGACTA
CAGTATCGTAATAATAATAATAAGTCCTGAT
1bp            15            ↕ 30
GTCATAGCATTATTATTATTATTCAGGCCTA
CAGTATCGTAATAATAATAATAAGTCCGGAT
```

Single nucleotide polymorphisms

Small variations in DNA are called polymorphisms. Skin type is a common human polymorphism. We have different skin depending on the order in which the nucleotides in our DNA line up. Some polymorphisms are so small they affect the order of just one pair of nucleotides.

These are called 'single nucleotide polymorphisms' or SNPs, which is a genetic polymorphism that involves variation in a single nucleotide. Thus a SNP is defined as a single base change in a DNA sequence that occurs in a significant proportion (more than 1%) of a large population. Over 5 million SNP locations have been identified in the human genome that grows daily. Genetic variation in a DNA sequence that occurs when a single nucleotide in a genome is altered; SNPs are usually considered to be point mutations that have been evolutionarily successful to recur in a significant proportion of the population of a species. In humans scientists have identified about 1.4 million locations where differences occur in single-base DNA. This information revolutionises the process of finding chromosomal locations for disease-associated sequences.

Single-nucleotide polymorphisms (SNPs) are the markers of choice for studies of linkage and historical demography due to: (*i*) the relative abundance of SNPs in the human genome compared with other types of polymorphisms, (*ii*) the efficiency with which they can be assayed, and (*iii*) the ease with which they can be analysed by the tools of population genetics. It is typically assumed that each SNP is the result of a single mutation event and different SNPs segregate independently to one another.

In 2003, the human genome project finished decoding almost the whole genome showing that 99.9% of the base sequences of human genes are common to all people, whereas the remaining 0.1% differ from person to person. This personal variation in the base sequence is called polymorphism. Although there is a wide variety of polymorphisms, the most prevalent are single nucleotide polymorphisms (SNPs) involving only one base (Fig. 13.20).

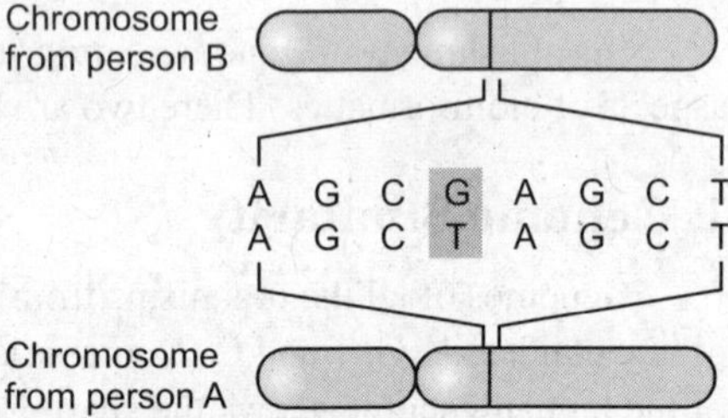

Fig. 13.20. Single nucleotide polymorphism.

SNPs occur normally throughout a person's DNA. They occur once in every 300 nucleotides; which means there are 10 million SNPs in the human genome. Most commonly, these variations are found in the DNA between genes. They can act as biological markers helping scientists to locate genes that are associated with disease. When SNPs occur within a gene or in a regulatory region near a gene, they may play a more direct role in disease by affecting the gene's function. Over half of all known disease mutations come from replacement polymorphisms.

Most SNPs have no effect on health or development. Some of these genetic differences, however, have proven to be very important in the study of human health. Researchers have found SNPs that help to predict an individual's response to certain drugs, susceptibility to environmental factors such as toxins, and risk of developing particular diseases. SNPs can also be used to track the inheritance of disease genes within families. Future studies will work to identify SNPs associated with complex diseases such as heart disease, diabetes, and cancer. Some SNPs are responsible for disease such as sickle cell disease.

Genetic Variations and DNA fingerprinting

DNA fingerprinting of individuals is possible due to these genetic variations in non-coding parts of genome. This technique is used in search of criminals, rapists, solving parentage problem, confirming identity of individuals, etc.

The genetic variations are some time harmful because our body becomes susceptible or resistant to certain diseases, *i.e.* protects from all kinds of pathogens. Besides, genetic variations also govern the severity of illness and the body responses to treatment of medicines. It must be noted that : (*i*) single gene mutation in cystic fibrosis results in autosomal recessive inheritance and occurs due to deletion of 3 bp at codon 508 on chromosome 7 (7q31) that accounts for 70% of all mutations, (*ii*) Huntington disease is also caused by autosomal dominance which occurs due to single mutation, *i.e.* adenine/guanine/cytosine mutation at chromosome 4 (4q16.3), (*iii*) migraine occurs due to polymorphisms which has complex inheritance, (*iv*) Alzheimer's disease occurs due to difference of a single base in *apoE* gene on chromosome 19 (19q13), and (*v*) resistance to HIV and AIDS occurs due to a simple deletion within the chemokine receptor gene *CCR5*.

Databases: Bioinformatics databases exist for SNPs. dbSNP is a SNP database from National Centre for Biotechnology Information (NCBI). SNPedia is a wiki-style database from a hybrid organisation. The OMIM database describes the association between polymorphisms and diseases. The Human Gene Mutation Database provides gene mutations causing or associated with human inherited diseases and functional SNPs, and GWAS Central allows users to visually interrogate the actual summary-level association data in one or more genetic association studies.

H. TYPES OF GENOMICS

In the last few years some interesting findings have been recorded and several new branches have emerged. Consequently, the area of genomics has quietly widened. However, the genomics is broadly categorised into three, *structural genomics, functional genomics* and *comparative genomics.*

1. Structural Genomics

The structural genomics deals with DNA sequencing, sequence assembly, sequence organisation and management. Basically it is the starting stage of genome analysis *i.e.* construction of genetic, physical or sequence maps of high resolution of the organism. The complete DNA sequence of an organism is its ultimate physical map. Due to rapid advancement in DNA technology and completion of several genome sequencing projects for the last few years, the concept of structural genomics has come to a stage of transition. Now it also includes systematics and determination of 3D structure of proteins found in living cells, because proteins in every group of individuals vary and so there would also be variations in genome sequences. With full-genome sequences available, structure prediction can be done more quickly through a combination of experimental and modelling approaches. Thus structural genomics describes the 3D structure of every protein encoded by a given genome.

Since protein structure is closely linked with protein function, the structural genomics provides knowledge of protein function. Besides elucidation of protein functions, structural genomics can be used to identify novel protein folds and potential targets for drug discovery. Structural genomics involves a large number of approaches to structure determination, including experimental methods using genomic sequences or modelling-based approaches on the basis of sequence or structural homology to a protein of known structure, these approaches also depend on chemical and physical principles for a protein with no homology to any known structure.

The next step of genomic studies after yielding of the complete genome sequence of a species is to identify both cellular and molecular function of each gene in the genome. Several pilot projects in structural genomics are in progress. Both approaches (structural and functional genomics analyses) are important for genome research beyond DNA sequencing. Genome map of *Klebsiella pneumoniae* derived through structural genomics is given in Fig. 13.21. Such information helps in establishing gene – protein relationships.

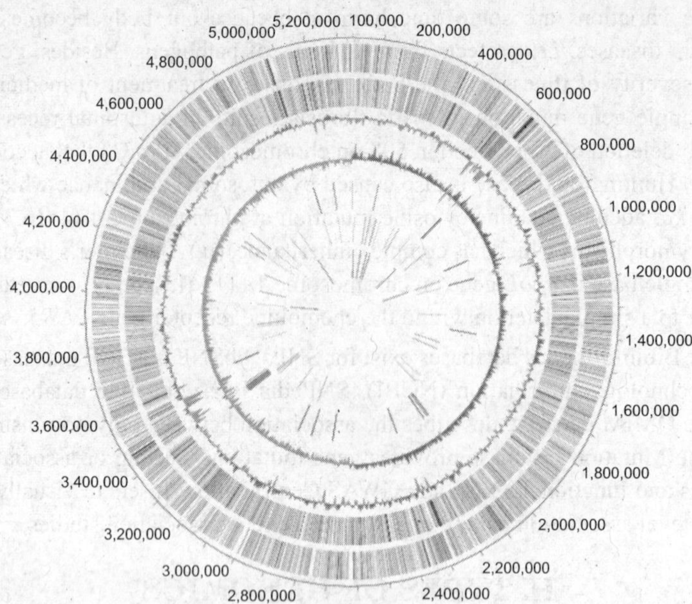

Fig. 13.21. *Klebsiella pneumoniae* structural genomics (*source*: http://kp.life.nthu.edu.tw/).

2. Functional Genomics

Functional genomics deals with the study of function of all gene sequences and their expression in an organism. If you locate a gene and sequence it, the next step would be to know what function it does? Functional genomics includes the approaches under development to ascertain the biochemical, cellular, and/or physiological properties of each and every gene product. J. Pevsner (2009) has defined the functional genomics as 'the genome-wide study of the function of DNA (including genes and non-genic elements), as well as the nucleic acid and protein products encoded by DNA'. Based on the information of structural genomics the next step is to reconstruct genome sequences and to find out the function of the genes. This information also lends support to design experiment to find out the functions that specific genome does. The strategy of functional genomics has widened the scope of biological investigations. Strategy of functional genomics is based on systematic study of single gene/protein to all genes/proteins. Therefore, the large scale experimental methodologies (along with statistically analysed/computed results) characterise the functional genomics. Functional genomics focuses on the dynamic aspects such as gene transcription, translation, and protein-protein interactions, as opposed to the static aspects of the genomic information such as DNA sequence or structures.

Functional genomics includes function-related aspects of the genome itself such as mutation and polymorphism (for example SNP) analysis, and measurement of molecular activities. The molecular activities consist of several '-omics' such as *transcriptomics* (gene expression), *proteomics* (protein expression), *phosphoproteomics* (subsets of proteomics) and *metabolomics*. Mostly multiplex techniques are used in functional genomics to measure the abundance of all gene products, for example mRNAs or proteins within a biological sample. These measurement modalities quantitate the various biological processes and improve our understanding of gene and protein functions and interactions.

Techniques and applications of functional genomics are done : (*i*) at the DNA level (by genetic interaction mapping, the ENCODE project), (*ii*) at the RNA level, *i.e.* transcriptome profiling (by microarrays and SAGE), (*iii*) at protein level via protein-protein interactions (by yeast two-hybrid system and AP/MS), (*vi*) estimating loss of function technique (by mutagenesis and RNAi), (*v*) functional annotations for gene (genome annotations, Rosetta stone approach), etc.

(*a*) Functional Genomics Toolbox: The functional genomics emerged in response to the challenges posed by complete genome sequences. To understand this process the biochemical and physiological function of every gene product and their complexes are necessary to understand. The gene activity manifests at several different levels, including RNA, protein and metabolite levels. Analyses at these levels can provide insight into the possible function of individual gene and the cooperation that occurs between genes and gene products to produce a defined biological outcome. The technology involved in defining functional genomics are DNA or oligonucleotide microarray technology for determining mRNA; 2D gels and mass spectroscopy and other methods for analysing different proteins and GC-MS or LC-MS for identifying and quantifying different metabolites in a cell. High throughput methods for forward and reverse genetics are also integral to functional genomics (Fig. 13.22).

Pre-genomics	**Post-genomics**	**Tools**
Gene	Genome	Bioinformatics, sequencers
↓	↓	
Transcript	Transcriptome	Microarray, SAGE, ALFP, MPSS
↓	↓	
Protein	Proteome	2D gels, MS, 2D-LC, BIA-MS
↓	↓	
Phenotype	Phenotype	Various tools
Forward genetics	Reverse genetics	

Fig. 13.22. Tools of functional genomics.

(*b*) Determination of Function of Unknown Genes: The following approaches are made for determining the function of unknown genes:

(*i*) Computer analysis: Computer plays a significant role to locate the genes in the DNA sequence. The most powerful tools used for this purpose is the *homology searching*. All the related genes have similar sequences. Hence, a new gene is discovered when it is similar to an equivalent already sequenced genes procured from different organisms. Those genes are called homologous genes and they share a common evolutionary sequence similarity. Usually identical sequences lack in a pair of homologous genes due to random changes by mutation. But they possess similar sequences because of operation of random changes on the same starting sequences of the common ancestral gene. Homology searching uses this sequence similarity. The analysis is based on the newly sequenced gene must be similar to a second previously sequenced gene. This shows that both the genes have evolutionary relationship and function of the new gene will be similar to the second gene. If the nucleotide sequences are 80% identical, they will show similarity. Such genes will show 80% similarity rather than 80% homology. Thus homology analysis gives the information on function of all the genes or its segments. But best results are obtained after carrying out amino acid homology. There are several software programmes such as BLAST which are used for such analysis. A homology search is done with a nucleotide or an amino acid sequence.

The homology domains have evolved due to a single nucleotide change or complex rearrangements resulting in new genes inside which the domains are present. For example, *tudor* domain was first identified in *D. melanogaster*. It is 120 amino acid long motifs. A homology search using *tudor* domain has revealed the presence of this domain in several proteins. This is the transport protein expressed during oogenesis in *Drosophila*, RNA metabolism in human, and some other functions involving RNA.

Homology analysis of *S. cerevisiae* genome was done to assign functions to new genes. Out of 6,000 genes known, 30% were identified by conventional method and 70% by homology analysis. Out of these 30% genes of the genome were assigned function and about half of genes were the homologous of earlier known genes and about half shows similarity only to domains, *i.e.* shows less similarity. About 10% genes had homologues in databases and 30% of the total did not have homologues. Of the 30%, about 7% were unreal genes because of having unusual 'codon bias', *i.e.* requirement of a preferred codon for each amino acid by an organism besides coding a single amino acid by several codons.

(*ii*) Experimental analysis: Genes express the visible characters called phenotype. However, the functions of such genes have been analysed by searching the mutant organisms where gene for a specific phenotype has been changed. After locating the position of genes through genetic crosses, they can be identified and sequenced following the methods of molecular biology. Thus the method of assigning functions to unknown genes is based on mutation and identification of the phenotypic changes occurring through mutation.

(*c*) Patterns of Gene Expression: Always all the genes do not express. Only those genes express whose products are required by the cell. The following approaches are made to understand the active genes in a specific tissue at a specific time which may also help to assess the relative degree of activity of the genes.

(*i*) Gene expression array by measuring levels of RNA transcripts: When a gene is switched on, it expresses through formation of an RNA transcript. The gene transcript may be identified to determine the gene. To approach this goal, specific RNA is extracted from the cells. The cDNA is synthesised by using mRNA and used as hybridisation probe. The cDNA copy of the gene being studied is immobilised on a solid support. Then this DNA is hybridised with the hybridisation probe synthesised from RNA transcript. A hybridisation signal is observed when the gene transcripts are present in the extract of RNA. For every coding gene in the genome, this method is repeated and the whole mRNA molecules of a cell is analysed. This is called 'transcriptome'. The DNA chip strategy is employed in such conditions.

(*ii*) Serial analysis of gene expression (SAGE): It is a method for gene analysis that depends on gene sequencing. SAGE is based on PCR which is used to identify differentially expressed mRNA levels in two different samples. In a cDNA pool the relative proportion of gene, specific ESTs shows the relative abundance of the corresponding mRNA transcripts. Therefore, differentially expressed gene transcripts can be identified by determining the sequence of a significant number of ESTs in one cDNA pool. By using many mRNA samples, very short 3′ cDNA sequence tags can be generated. Generation of cDNA tags from one mRNA population is shown in Fig. 13.23.

An mRNA is isolated and used to synthesise cDNA. A biotin-labelled oligo-dT primer starts the synthesis of cDNA. It physically allows to separate 3' ESTs by using the magnetic beads-coated with streptavidin. *Nla*III is used to delineate the tags, which recognises CTAG. Similarly, *Bsm*F1 also delineates the tag and cleaves 10 bp downstream of GGGAC sequence. The restriction fragments are ligated to adaptor and PCR amplified so that a large number of short ESTs (10-15 bp long) may be linked into a single array. This array is subjected to automated DNA sequencing. SAGE technique has been studied in yeast and humans, but not in plants.

Moreover, the wealth of knowledge about these untold stories is being unravelled by the scientists after the development of microarray technology and proteomics (*see* in preceding sections). These two technologies helped to explore the instantaneous events of all the genes expressed in a cell/tissue present at varying environmental conditions like temperature, pH, etc.

3. Comparative Genomics

When the complete genome sequences of cellular life forms become available, the notable findings were recorded. It was found that one third of the genes encoded on each genome had no predictable or known function. In *E. coli* K12 (which is all time favourite model organisms of molecular biologists) about 40% genes have unknown function. The level of evolutionary conservation of microbial proteins is rather uniform with about 70% of gene products from each of sequenced genomes having homologues in distant genomes. The function of these genes can be predicted by comparing different genomes and by transferring functional annotations of protein from the other organisms to their **orthologs** (direct descendants of a sequence in the common ancestor, *i.e.* the 'same' genes in different species that connect vertical evolutionary descendent) as opposed to **paralogs** (*i.e.* genes related by duplication within the genome from less-studied organisms).

Thus comparative genomics can be defined as 'the study of the relationship of genome structure and function across different biological species or strains'. It is an attempt to take the advantage of the information provided by the signatures of selection to understand the function and evolutionary processes that act on genomes. An important application of comparative genomics is the gene finding, *i.e.* discovery of new and non-coding functional elements of the genome. Comparative genomics includes both similarities and differences in the proteins, RNA, and regulatory regions of different organisms to infer the selection that has acted upon these elements. Those elements responsible for similarities between different species should be conserved through time (stabilising selection), while those elements responsible for differences among species should be divergent (positive selection). Finally, the unimportant elements to the evolutionary success of the organism will be unconserved (selection is neutral).

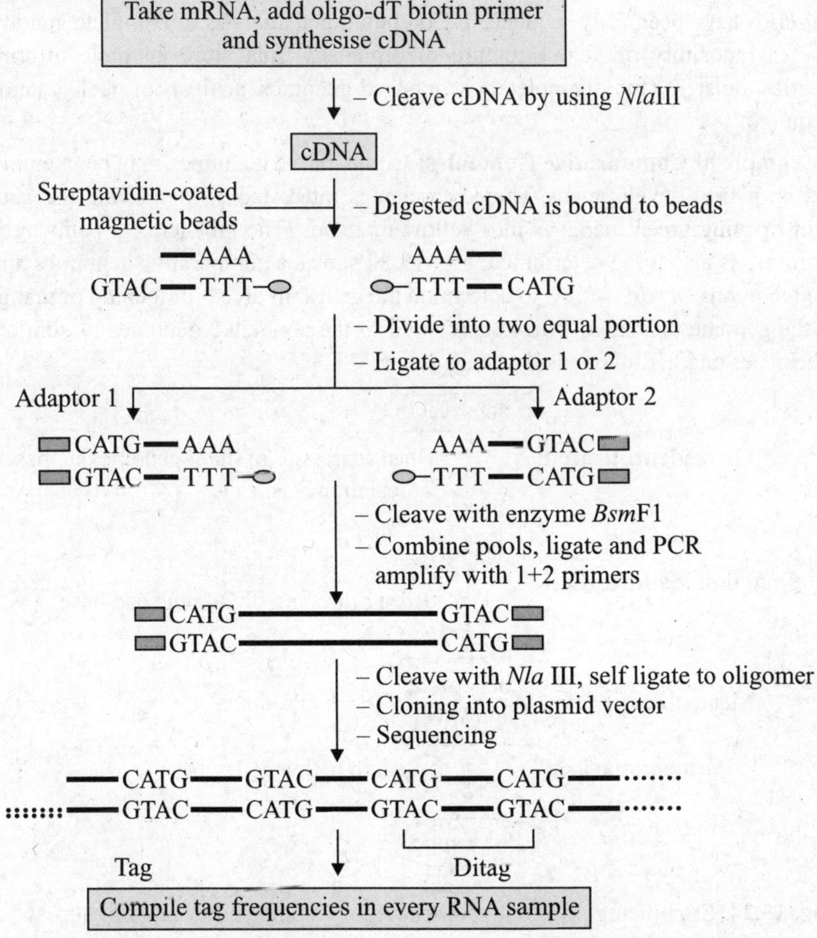

Fig. 13.23. Generation of cDNA tags in a SAGE (diagrammatic).

Biology and organisms make the comparative genomics a powerful approach for better understanding of genomes. Comparative genomics includes several distinct aspects; analysis of protein sets from completely sequenced genomes is one of them. There are several databases (*e.g.* general purpose databases and organisms specific databases) used for comparative genomics.

Genome sequencing projects have made very clear that genomes of different organisms (*e.g.* mouse and man) may be very similar. Systeny between human chromosome 1 and mouse chromosome 1, 3 and 4 has earlier been shown in Fig. 13.19. It has been found that the genomes of human and chimpanzee differ by 1-3%. The working DNA of mouse and human is shared by about 97.5%. About 12% of ~18,000 genes of the worm *C. elegans* show sequence similarity to yeast gene. This datum is based on proteins encoded by this much number of genes.

About 2,000 genes of yeast (one third of ~6,000 genes) are functionally similar to the genes of *C. elegans*. Such similarity between the organisms suggests that in spite of evolution of organisms about 100 million years ago, their genomes have not changed much. That is why some percentage of similarity exists between genes of organisms.

Based on the above facts, study of comparative genomics proved a powerful approach for achieving a better understanding of the genomes and, subsequently of the biology of the respective organisms. Recently, some of the genome of the microorganisms viz., *Haemophilus influnenzae, Mycoplasma genitalum, Methanococcus jannaschii, Saccaromyces cerevisae, Escherichia coli, Bacillus subtilis* have been fully sequenced. Computational analysis of complete genomes requires a database (a repository of gene structure of organisns) that store genomic informations and bioinformatics tools. To study completely sequenced genomes, analysis of nucleic acids, proteins, etc. are required.

***(a)* Example of Comparative Genomics:** Comparative genomics may be exemplified by the genetic basis of fruit development. The researcher is interested in identifying the genes that are involved in ripening green mangoes into yellow mangoes. The biochemical pathway involved in ripening process is also to be determined. Fig. 13.24 shows a comparative genomics approach that can be used in purely *in silico* effort to determine the genes involved in ripening of mangoes. In this approach the genome of mango fruit is compared to the annotated genomes of similar species to identify the genes and the functions that they do.

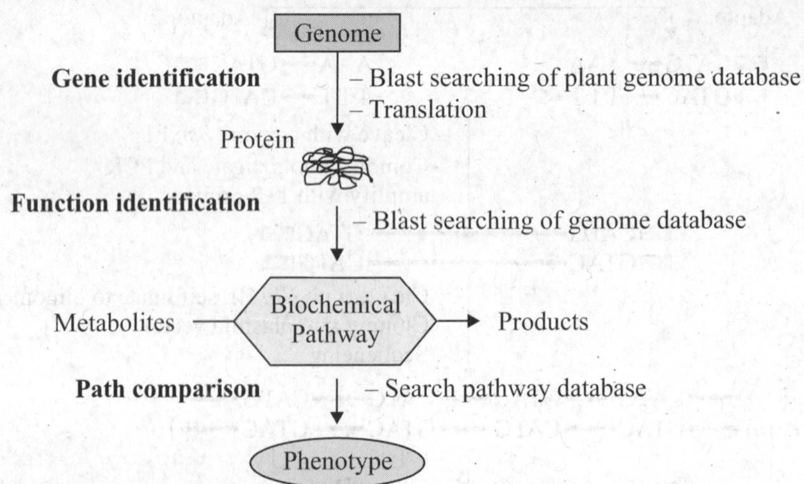

Fig. 13.24. From mango genome to phenotype (colour- yellow and flavour- sweet).

(*b*) Databases for Comparative Genomics: World Wide Web (www) is accessible to anyone by using Internet.

(*i*) PEDANT: This database gives information about the proteins, their three-dimensional structures, enzyme patterns, PROSITE patterns, Pfam domains, BLOCKS and SCOP domains as well as PIR keywords and PIR superfamilies.

(*ii*) COGs: Clusters of Orthologs Groups (COGs) is applicable to simplify evolutionary studies of complete genomes and improve functional assignments of individual proteins. It comprises of ~2,800 conserved families of proteins from each of the sequenced genomes. It contains orthologus sets of proteins from at least three phylogenetic lineages, which are assumed to have evolved from an individual ancestral protein. The functions of orthologs are same in all organisms. The protein families in the COGs database are separated into 17 functional groups that include a group of uncharacterised conserved proteins, as well as a group of proteins for which only a general functional assignment appeared appropriate. In COGs database due to storage of diverse nature of data on proteins, the similarity searches also give some information for those proteins which has no clear information in databases. The databases also act as a tool for a comparative analysis of complete genomes.

(*iii*) KEGG: Kyoto Encyclopaedia of Genes and Genomes (KEGG) Centers on cellular metabolism as proposed by Kaneshisa and Goto (2000). A comprehensive set of metabolic pathway charts both general and specific has been given for the sequenced (genome) organism. In this, enzymes identified in a particular organism are colour-coded, so that one can easily trace the pathways. It also provided the enzymes coded for the orthologous genes. These genes if located adjacent to each other, form like operons; for example, comparison can be made between two complete genomes in which genes are located relatively close or adjacent within five genes. This site is useful to get informations for the analysis of metabolism in various organisms.

(*iv*) MBGD: Microbial Genome Database (MBGD) is situated in University of Tokyo, Japan. This database helps to search the microbial genomes. MBGD accepts several sequences at once (~2000 residues) for searching against all of the complete genomics available displays colour–coded functions of the detected homologs, and their location on a circular genome map. This database also gives information regarding the functions, e.g. degradation of hydrocarbon or biosynthesis of nucleotides, etc.

(*v*) WIT: Similar to KEGG, WIT ('what is there' database) gives informations regarding metabolic reconstruction for completely sequenced genomes. The WIT features arc to provide sequence of reactions bctween two bifurcations besides to include proteins from many partially-sequenced genomes. These features of WIT provide many more information on the sequences of the same proteins/enzymes obtained from different organisms.

4. DNA Chip (DNA Microarray) Technology

A major technological advancement was made in the field of molecular biology during the mid 1990s when DNA chips were produced. It attracted the interest among the biologists throughout the world. There are several synonyms of microarray such as DNA chips, gene chips, DNA arrays, gene arrays and biochips. One can never differentiate each name. In future, some new techniques (*e.g.* nucleotide arrays, carbohydrate arrays, protein arrays, etc.) are expected to be available. DNA chips are high density miniaturised microarrays of large number of cDNA sequences (probes) which are attached in a fixed (spotted) location in a systematic order on a solid support such as glass plates, slides or nylon. DNA fragments from unknown genes are placed using robotic device that accurately deposit nanolitre quantity of DNA solution on a surface measuring only a few square centimetres. The UV light is used to cross link the DNA to the glass slides. The DNA fragments get attached to the surface.

The principle of DNA microarrays lies on the base pairing or hybridisation between the nucleotides, which is the central event in microarray technology. Sequences from a labelled target are recognised by the hybridisation affinities of the oligonucleotide probes in a highly specific and sensitive manner. Using this technology the presence of one genomic or cDNA sequence in 1,00,000 or more sequences can be screened in a single hybridisation. PCR and hybridisation have come together in the development of DNA microarrays which rapidly screens many thousands of genes. The DNA chips contain known oligonucleotides (20-mers) sequences or cDNA of known function. Thus a single DNA chip can give the 'complete picture' of whole genome of an organism. For application in DNA sequencing, DNA chips will have to possess every possible oligonucleotide sequence. Because the maximum sequence 'read' possible is the square root of the number of oligonucleotide sequences on the chip.

(*a*) Types of DNA Chips: There are two types of DNA chips: cDNA-based chips and oligonucleotide-based chips.

(*i*) cDNA-based chips: Since these types of chips are prepared by using cDNA, it is called cDNA chips or cDNA microarray. It was commercialised by Synteni (Freemont, CA, U.S.A.) which are called probe DNA. The cDNAs are amplified by using PCR. Then these are immobilised on a solid support made up of nylon filter of glass slide (1 ×3 inch). The probe DNA samples are loaded into a spotting pin by capillary action. Small volume of this DNA preparation is spotted on solid surface making physical contact between these two. DNA is delivered mechanically or in a robotic manner. When one DNA spotting cycle is completed, the pin is washed and loaded with fresh DNA to start the second cycle. The cDNA microarrays prepared by Synteni consists of about 10,000 groups of cDNA in an area of about 3.6 cm^2.

(*ii*) Oligonucleotide-based chips: The second approach involves the synthesis of oligonucleotides directly on the chip using light-directed solid phase conbinatorial chemistry. It was developed first by Affymetrix (Santa Clara, U.S.A.). A high density of short oligonucleotides (10-20 mers) is found on the chip. Microarrays are prepared in the following ways (Fig. 13.25).

- The glass support is coated with light-sensitive protecting groups, *i.e.* photoprotecting DNA base that prevent random nucleotide attachment.
- The surface is covered with a photomask that holds corresponding to the sites for attachment of the desired nucleotides and deprotects the defined regions of the slide.
- To remove the unexposed protecting group, shine the laser light through the photomask holes.
- The chip is bathed in a solution containing the first nucleoside to be attached. The nucleoside has a light-removable protecting group to prevent addition of another nucleotide until the appropriate time.
- The step 2 is repeated through step 4 with a new mask each time to add nucleoside until all sequences on the chip have been completed.

This method is used for construction of any sequence. The DNA chip available commercially consists of 25 bases long oligonucleotide probes. It is about 1.3 cm on a side and have more than 2,00,000 positions. The probes are often the expressed sequence tags (ESTs) derived from cDNA molecules. Now the chips have probes for every expressed genes or open reading frame in the genome (for example, *E. coli* and the yeast).

The nucleic acids to be hybridised, often called 'targets' (which may be mRNA or cDNA), are isolated and labelled with fluorescent reporter dyes. The chip is incubated with target-mixture to ensure the proper binding to probe with complementary sequences. The unbound target is washed off and chip is scanned with laser beam. Fluorescence indicates that the probe is bound to that particular sequence. The hybridisation pattern shows the genes being expressed. Target samples are labelled with different fluorescent dyes and compared using the same DNA chip. On the basis of gene expression, gene function can be tentatively assigned. Only mRNA can be detected that are currently expressed. If a gene is transiently expressed, its activity may be missed by a DNA chip analysis.

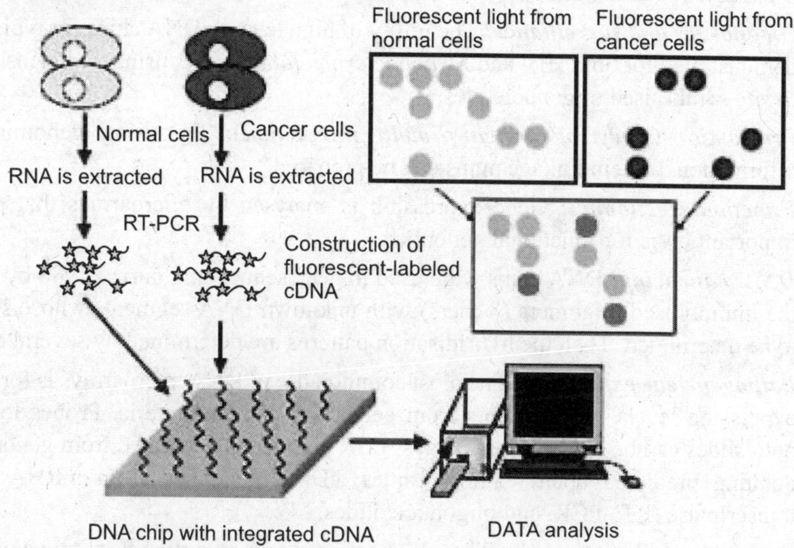

Fig. 13.25. Basic technology for DNA chip.

***(b)* Comparative Microarray Hybridisation:** The major steps of comparative microarray hybridisation are shown in Fig. 13.25. A gene expressed at a given developmental stage can be detected using microarray technology. Different mRNAs produced by two different cell populations can be compared. If you are interested to compare a normal cell with an abnormal cell (*e.g.* mutant or cancerous cells), then separately isolate sufficient amount (1-2 ml) of mRNA from total cell population and purify them. The base pair sequence of cDNA is complementary to that of mRNA. Fluorescent dyes or fluors (*e.g.* Texas red, Rhodamine, etc.) of different colours such as red and green are attached to cDNA of each cell population. The labelled cDNA is called as probe. Dye of different colours enables to distinguish two samples on DNA chips.

The microarrays contain thousands of spots each containing different DNA sequences. The fluorescent cDNAs are mixed and allowed to hybridise to a DNA microarray. The labelled cDNA hybridises the DNA of such spots on microarray which contains complementary sequences.

The spots of nucleotides present on microarray hybridise such labelled cDNAs whose sequences are complementary to that of spotted DNA arrays. Hence, each spot acts as independent assay which determines different cDNA molecules. After removing the unhybridised probes the hybridised arrays are scanned using laser and scanned image is detected. The spots that fluoresce provide a snapshot of all the genes being expressed in the cells at the moment they were harvested.

Those spots show predominantly red or green colour whose mRNA is present in high amount in cell populations. If about equal amount of cDNA from each cell population bounds to spots on arrays, it impart yellow colour. In this way the entire genome on a single DNA chip can be monitored using this technology which can give a better picture of interaction among thousands of genes.

(c) Application of DNA Chips: The DNA chips are used in many areas as given below:

- ***Identification of tissue specific genes:*** Tissue specific genes such as insulin gene in pancreas are identified.

- ***Discovery of drugs:*** The new drugs are discovered for treatment of certain diseases like cancer that interact and combine with suitable altered genes, *i.e.* oncogenes.
- ***Diagnostics and genetic mapping:*** DNA chips have been prepared to identify mutation/multiple mutations e.g. CFTR (cyctic fibrosis), *BRCA-1* (cancer susceptible gene), β-thalassemia mutations in patients, genotyping of hepatitis C virus in blood samples.
- ***Diagnostics and surveillance:*** The utility of high density DNA chips for typing has been demonstrated for both HIV and *Mycobacterium tuberculosis* using high density arrays of *in situ* synthesised oligonucletides.
- ***Proteomics (study of protein-protein interaction):*** By using genomic sequence information, protein linkage maps can be prepared.
- ***Functional genomics:*** Gene expression is analysed by microarrays that provides an important basis for functional genomics.
- ***DNA sequencing:*** DNA chips were used in gene sequencing during 1988 by hybridising the immobilised oligomers (8-mers) with unknown DNA segments whose sequences is to be determined. Then the hybridisation patterns are determined by several devices.
- ***Pathogen gene expression:*** The most common use of DNA microarrays is for monitoring expression levels of transcripts from cells, viruses and bacteria. Probes to be spotted onto slides can be made in three ways: PCR amplicons generated from genomic regions defining predicted open reading frames, cDNA generated from mRNA by reverse transcriptase (RT)-PCR, and oligonucleotides.
- ***Agricultural biotechnology:*** The expression patterns of agricultural crops or transgenic plants under the different environmental conditions can be analysed by using DNAs chips. It is also used to study the DNA polymorphisms when molecular marker-assisted selection is to be studied.

I. PROTEOMICS

The term 'transcriptome' has been coined to describe the complement of mRNA transcribed from the genome of a cell. The mRNA undergoes translation to produce protein within a cell. Proteins are vital parts of living organisms as they are the main components of the physiological metabolic pathways of cells. Human cells have 1,00,000 or more individual proteins and, therefore, scientists were surprised to find that the human genome appears to house only about 40,000 genes. How can this may be when every protein sequence is encoded by our genes? A new science of proteins called 'proteomics' has recently blossomed to answer this questions about how all the proteins in our cells function to sustain life. *Proteomics* is the large-scale study of proteins, particularly their structures and functions. In 1997, the term 'proteomics' was first coined to make an analogy to 'genomics'. The word proteome is a blend of protein and genome, and was coined by Marc Wilkins in 1994 while working on the concept as a Ph.D. student. Wilkins described it as the entire complement of proteins expressed by a genome, cell, tissue or organism. The proteome is the entire set of proteins including the modifications made to a particular set of proteins expressed from the transcriptomes of a cell, systems, or organisms. A cellular proteome is the collection of proteins found in a particular cell type under a particular set of environmental conditions such as exposure to hormone stimulation. It can also be useful to consider the complete proteome of an organism; it can be conceptualised as the complete set of proteins from all of the various cellular proteomes. Proteomics is aimed at examining the whole set of proteins contained by a cell at any one time. There are many areas of modern proteomics such as protein expression, protein structure, protein localisation, protein-protein interaction, etc. (Fig. 13.26).

Proteomics is the direct outcome of advancement made for nucleotide sequencing of different genomes in large scale that helps to identify various proteins. Generation of information about protein is necessary, because, protein governs the phenotypic characters of the cells. Proteomics employs an incredibly diverse range of technologies including molecular biology, chromatography, electrophoresis, mass spectrometry, X-ray crystallography, NMR spectroscopy, robotics and computational biology. Merely genome study cannot provide the understanding of mechanism of disease development and various developmental changes occurring in organisms including humans. Moreover, target drugs for many kinds of diseases can be prepared only after understanding the protein modification and protein functions. In eukaryotes the proteome is larger than the genome because there are more proteins than genes. This is due to alternative splicing of genes and post-translational modifications like glycosylation or phosphorylation.

In different types of biological systems this term has been applied, for example cellular proteome, complete proteome, viral proteome, etc. A 'cellular proteome' is the collection of proteins found in a particular cell type under certain environmental conditions such as exposure to hormones. It may be helpful to consider an organism's complete proteome which can be conceptualised as the complete set of proteins from all of the various cellular proteomes. This is very roughly the protein equivalent of the genome. The term proteome has also been used to refer to the collection of proteins in certain sub-cellular biological systems; for example all of the proteins in a virus can be called a 'viral proteome'.

The genome is constant within the cell, whereas transcriptome and proteome vary: (*i*) within a single cell during cell cycle, and (*ii*) between cell types (*e.g.* brain and liver) or when exposed to various external stimuli that exert changes in gene expression (*e.g.* growth factors, hormones, etc.). However, biochemical machinery of the cell can be modulated in response to internal and external factors. This affects synthesis or degradation of proteins, post-transcriptional modification of protein, alteration in cellular localisation, etc.

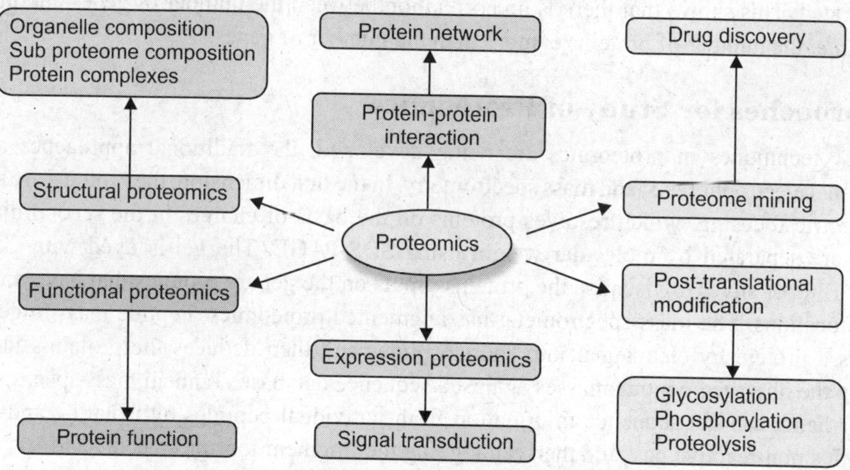

Fig. 13.26. Different area of proteomics.

Proteomics provides information about genome function that mRNA studies cannot do. Always mRNA and protein cannot be correlated because of the post-translational modification of proteins and protein turnover. Measurement of RNA levels can show the changes of gene expression that occurs inside the cell, whereas proteomics discovers what is actually happening. A dream of the proteomicist may be to obtain a dynamic map of an organism's entire proteome that reveals the changes during development and in response to biotic and abiotic challenges. Such a map could have more intrinsic value than the corresponding transcriptome map, which is a more remote indicator of the biochemical activity of the organism. However, proteome map will be more complex than the transcriptome maps, and probably more difficult to achieve because post-transcriptional modification of protein can produce many variants of a single gene product and because proteome map will ultimately need to incorporate information about the location of each protein in a cell.

About one third of gene sequences of organisms (of which genome sequence is known) do not carry out any function. The structural genomic projects can be assisted significantly only identifying the proteins completely. One of the objectives of structural genomics is to prepare 3D structure of all proteins expressed by the genome of the cells. Based on the protein structures possible functions may be assigned to respective proteins.

Proteomics is the identification, analysis and large scale characterisation of proteome expressed by any given cells, tissues and organs under the defined conditions. The major objectives of proteomics are: (*i*) to characterise the post-transcriptional modifications in protein, and (*ii*) to prepare 3D map of a cell indicating the exact location of protein.

1. Relation between Gene and Protein

In eukaryotes transcription occurs inside the nucleus and translation in cytoplasm. DNA is transcribed under transcriptional regulation into pre-mRNAs by RNA polymerase II (Fig. 13.27).

Inside the nucleus, pre-mRNAs undergo various post-transcriptional modifications to increase their stability such as capping at 5′-end, polyadenylation (*i.e.* addition of poly-A at 3′-OH end) and mRNA editing. The introns are spliced out by splisosomes through a process called splicing. Then the mature mRNAs are transported from nucleus to cytoplasm for translation into protein on the ribosomes. Translational regulation of protein occurs in cytoplasm. Then protein undergoes post-translational modifications to about 200 types. From a single gene, various forms of proteins are generated. This shows that there is no correlation between the number of genes and number of proteins, *i.e.* the number of proteins is more than the number of genes.

2. Approaches for Study of Proteomics

New techniques in proteomics are being developed, the traditional approaches are: two-dimensional electrophoresis, and mass spectrometry. In the first dimension, the proteins are separated by isoelectric focusing, which resolves proteins on the basis of charge. In the second dimension, proteins are separated by molecular weight using SDS-PAGE. The gel is dyed with Coomassie Brilliant Blue or silver to visualise the proteins. Spots on the gel are proteins that have migrated to specific locations. The mass spectrometer has augmented proteomics. Peptide mass fingerprinting identifies a protein by cleaving it into short peptides and then deduces the protein's identity by matching the observed peptide masses against a sequence database. Tandem mass spectrometry, on the other hand, can get sequence information from individual peptides by isolating and colliding them with a non-reactive gas, and then cataloguing the fragment ions produced.

3. Types of Proteomics

There are many types of proteomics as shown in Fig. 13.27, but expression proteomics, structural proteomics and functional proteomics have been dealt herewith.

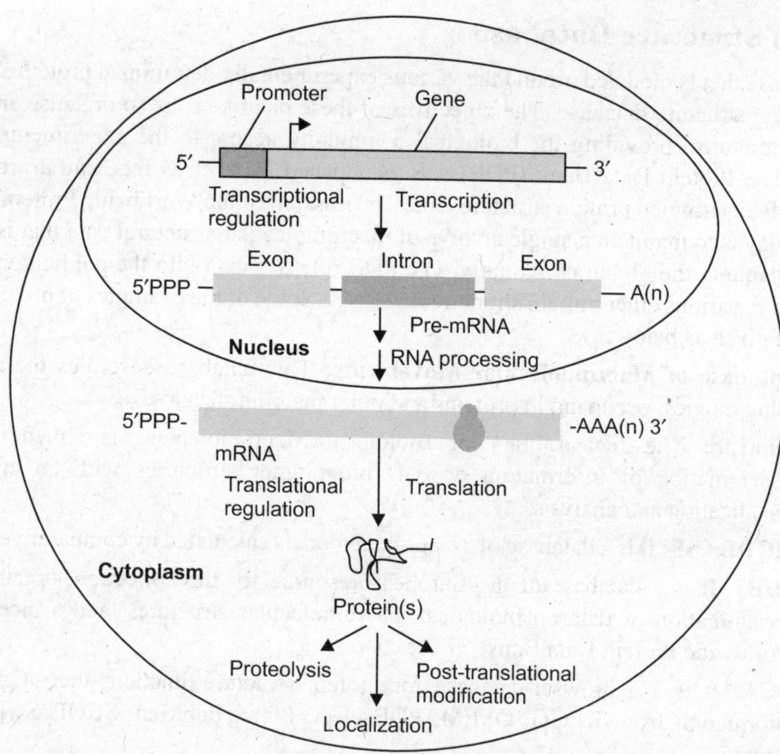

Fig. 13.27. Mechanism of production of multiple gene-products from a single gene.

(*a*) Expression Proteomics: Expression proteomics is the quantitative study of protein expression between the samples which differ by some variables. A comparative study of the whole proteome between the samples can be done using this approach. For example, a patient suffering from mouth cancer develops a small tumour. Thus tumour of cancer patient and similar tissues from a normal person may be taken out and analysed for protein expression following different routes. Using techniques of high resolution protein separation and identification (*e.g.* two-dimensional gel electrophoresis, isoelectric focusing, MALDI mass spectrometry, microarray technique, etc.) under-expressed or over-expressed proteins in cancer patient and normal ones can be characterised and identified. An understanding about the formation of such tumour could be developed on the basis of proteins identified and compared between the two individuals.

(*b*) Structural Proteomics: The structural proteomics deals with the study of structure and nature of protein complexes present in a particular cell organelle. To fulfill this objective specific sub-cellular organelles or all protein complexes are isolated. All proteins present in these complexes are identified and protein-protein interactions occurring between them are characterised. These studies lend support to assemble information about the structural topography of the cells and clues how certain proteins got expressed and gave unique characteristics to the cells.

(c) Functional Proteomics: Functional proteomics is a broad term which embraces all proteomics approaches related to devising its functions. It is defined as the use of proteomics methods for analysis of properties of molecular networks formed in a living cell. In this study molecules are identified which take part in such networks. Recently, some novel proteins have been discovered which transport the important molecules from nucleus to cytoplasm and cytoplasm to nucleus. This functional proteomics is rather a complex process where function of a molecule is found out in the molecular networks.

4. Protein Structure Database

A database that is modeled around the various experimentally determined protein structures is called a protein structure database. The objectives of these databases are to organise and annotate the protein structures, providing the biological community access to the experimental data in a useful way. The Protein Data Bank (PDB) was established in 1971 as the central archive of all experimentally determined protein structure data. The mission of the Worldwide Protein Data Bank (wwPDB) today is to maintain a single archive of macromolecular structural data that is freely and publicly available to the global community. The PDB releases data into the public domain which has been used in various other protein structure databases. Some of the examples of protein structure databases are given as below:

- **Database of Macromolecular Movements:** This database describes the motions by using movies, occurring in proteins and other macromolecules.
- **JenaLib:** The Jena Library of Biological Macromolecules is aimed at a better dissemination of information on 3-D biopolymer structures with an emphasis on visualization and analysis.
- **MODBASE:** It is a database of 3-D protein models calculated by comparative modelling.
- **PDBe:** It is a database of the European resource for the collection, organisation and dissemination of data on biological macromolecular structures, and a member of the Worldwide Protein Data Bank.
- **OCA:** It is a a browser-database for protein structure/function; the OCA integrates information from KEGG, OMIM, PDBselect, Pfam, PubMed, SCOP, SwissProt, and others.
- **OPM:** This database provides spatial positions of 3D structures of protein with respect to the lipid bilayer.
- **PDB Lite:** This database is derived from OCA, PDB Lite was provided to make it as easy as possible to find and view a macromolecule within the PDB
- **PDBsum:** It provides an overview macromolecular structure in the PDB giving schematic diagrams of the molecules in each structure and of the interactions between them.
- **PDBTM :** The Protein Data Bank of Transmembrane Proteins is a selection of the PDB.
- **PDBWiki:** It is a community annotated knowledge base of biological molecular structures.
- **Protein:** It is a NIH protein database : a collection of sequences from several sources including translations from annotated coding regions in GenBank, RefSeq and TPA, as well as records from SwissProt, PIR, PRF, and PDB.
- **Proteopedia:** It is the collaborative 3D encyclopedia of proteins and other molecules. A wiki that contains a page for every entry in the PDB (>50,000 pages) with a Jmol view that highlights functional sites and ligands. It offers an easy-to-use scene-authoring tool so you don't have to learn Jmol script language to create customised molecular scenes. Custom scenes are easily attached to 'green links' in descriptive text that display those scenes in Jmol.
- **SCOP:** It is database of the structural classification of proteins that has a detailed and comprehensive description of the structural and evolutionary relationships between all proteins having known structure.
- **SWISS-MODEL Repository:** It is a database of annotated protein models calculated by homology modelling
- **TOPSAN:** It is the Open Protein Structure Annotation Network. A wiki designed to collect, share and distribute information about protein 3D structures.

- **PMDB:** It is a Protein Model DataBase, which collects three dimensional protein models obtained by structure prediction methods. The database currently stores all models submitted to the last four editions of the CASP experiment. PMDB is a joint project between CASPUR and the biocomputing group of the Department of Biochemical Sciences of the University of Rome 'La Sapienza'.

QUESTIONS

1. What is genomics ? Give a brief account of genome sequencing project.
2. What is the significance of complete genome sequencing ? Name any two organisms whose genome has been sequenced completely.
3. Write in detail bout microbial genome sequencing projects.
4. Give a detailed account of bacterial genome sequencing projects
5. Describe in detail about Human Genome Project.
6. Discuss in detail about different methods of gene sequencing.
7. What is pyrosequencing ? Write the method of pysrosequencing.
8. Give an illustrated account of hierarchical sequencing.
9. Give an illustrated account of DNA microarray technology.
10. Write in detail about whole genome shotgun sequencing method.
11. What is expressed sequence tag (EST) method ?
12. Give a detailed account of genome assembly and finishing.
13. What is gene prediction and gene counting ? Write gene prediction algorithms.
14. Write a note on genome similarity and single nucleotide polymorphism.
15. What is genomics ? Discuss in brief about different types of genomics.
16. What is the microarray technique used to differentiate a normal cell from a diseased one ?
17. What is proteomics ? Write different types of proteomics studied by you.
18. Write short notes on the following:
    (*i*) Viral genome sequencing projects, (*ii*) *E. coli* genome project,
    (*iii*) *Bacillus* genome project, (*iv*) Sanger sequencing method,
    (*v*) Pyrosequencing, (*vi*) Scaffold,
    (*vii*) Sequence quality determination, (*viii*) DNA nanoball sequencing,
    (*ix*) Assembly view, (*x*) Gene prediction algorithms,
    (*xi*) Gene prediction software, (*xii*) Functional genomics,
    (xiii) Comparative genomics, (*xiv*) DNA chip,
    (*xv*) SNPs, (*xvi*) Functional genomics,
    (*xvii*) DNA microarray technology.

BIBLIOGRAPHY

http://en.wikipedia.org/wiki/Structural_genomics

Judith A. Blake *et al.* 2002. MGD: the mouse genome database. *Nucleic Acids Research*. **31**: 193-195.

Pevsner J. 2009. *Bioinformatics and functional genomics* (2nd Ed.). Hoboken, Wiley-Blackwell, NJ.

Porreca J.G. 2010. Genome Sequencing on nanoballs . *Nature Biotechnol.* **28**:43-44.

Ronaghi M. *et al.* 1996. Real-time DNA sequencing using detection of pyrophosphate release. *Analytical Biochem.* **242**: 84.

Seto D. 2010. Viral genomics and bioinformatics. *Viruses*. **2**: 2587-2593.

Venter D. 2993. A part of the human genome sequence. *Science*. **299.** 1183-84.

14

CHAPTER

Metagenomics

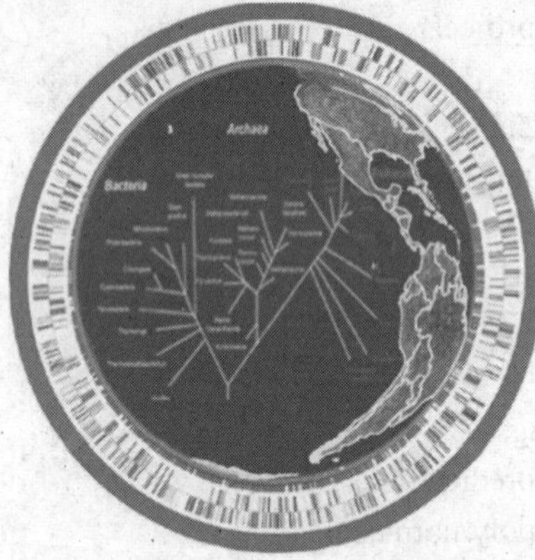

Global metagenomics work

A. Historical background
B. About Metagenomics
C. From genomics to metagenomics
D. Different metagenomics projects
E. Metagenomics at work
F. Technologies for metagenome study
G. Metagenome gene discovery tools
H. Metagenomics via next generation sequencing
I. Metagenomics via bioinformatics
J. Application of metagenomics

During the last 25 years, microbiology has experienced a transformation that has changed the views of microbiologists towards the study of microorganisms. Microbiologists realised that most of the microorganisms cannot grow in pure culture; this belief forced them to answer how can the microbial world be conquered ? This belief could be replaced with our ignorance about the range of metabolic and organismal diversity. An intense effort was made by the microbiologists in the ensuing years to describe the phylogenetic diversity of exotic and normal environments, for example ocean surfaces, deep sea vents, hot springs, soil, animal rumen and gut, human oral cavity and intestine. Many new lineages were classified on basis of their molecular signatures alone. The next challenge was the elucidation of the functions of these new phenotypes and determination whether they represented new species, genera or phyla of prokaryotic life. This challenge spawned the various techniques including metagenomics, *i.e.* the genomic analysis of assemblages of organisms. In a few years, the study of uncultured microorganisms has expanded beyond asking 'who is there' rather to include the difficult question 'what are they doing' ?

The results of the recognition of uncultured microorganisms need to be examined. Metagenomics is one of these outcomes, which is further shaping the microbiology. Metagenomics has already opened the new areas of research by enabling unprecedented analyses of genome heterogeneity and evolution in environmental contexts and provided access to more microbial diversity than has been viewed in the Petri dishes.

The approaches that microbiologists make to solve many problems have been changed by metagenomics. The metagenomics has redefined the concept of a genome and accelerated the rate of gene discovery. There is no end of the potential for application of metagenomics to biotechnology. The potential of metagenomics is realised which depends on the advancement of methods central to library construction and its analysis. For sequence-based approaches, the speed and cost of nucleotide sequencing will be a barrier of rapidly diminishing significance with the continuation of improvement of sequencing technology. Sequence-based assignment of function will also derive benefit from advances in detection of homology, which rely on the tertiary structures of predicted proteins rather than simply on primary sequence. Bioinformatics tools are included for advances that facilitate the management and analysis of large libraries. These tools analyse vast sequence databases and reassemble multiple genomes rapidly and affordable gene chips for library profiling that readily distinguishes clones expressing genes from silent clones.

More innovations in method development are required for functional analysis; most important among these are the strategies to improve heterologous gene expression and approaches for efficient screening of large libraries. Microbiology has relied on diverse methods for analysis; but metagenomics can provide the tools to balance the abundance of knowledge attained from culturing with an understanding of the uncultured microbial life. Myriad environments on earth have not been studied with culture-independent methods other than PCR-based 16S rRNA gene analysis; therefore, their further analysis is needed. Metagenomics may further expand our understanding of a variety of the exotic and familiar habitats attracting the attention of microbial ecologists, for example deep sea thermal vents, acidic hot springs, permafrost, temperate, desert, and cold soils, Antarctic frozen lakes, and eukaryotic host organs such as the human mouth and gut, termite and caterpillar guts, plant rhizosphere and phyllosphere, and fungi in lichen symbioses. Metagenomics will expand and continue to enrich our understanding of microorganisms with improved methods for analysis in the new fields, and attraction of diverse scientists to identify new problems and solve old ones.

Metagenomics Terms

- **Environmental Genomics:** A synonym for metagenomics.
- **Metagenomics:** The sequencing and analysis of DNA from environmental samples without the need for culturing individual, clonal organisms.
- **Microbiome:** The total collection of microbes and their genetic material within an environment.
- **Microbial Diversity:** A type of metagenomic study using a sequencing approach that focuses on a specific hypervariable region of the genome (that is, the V6 region of 16S rRNA) to assess the number of different species within an environment.
- **Random Community Genomics:** Frequently used as a synonym for metagenomics. It refers to sequencing of whole community from the environment with or without cloning, but without screening for any functional component.
- **Virome:** The totality of viruses and their genetic molecules within an environment.

A. HISTORICAL BACKGROUND

During the 1990s, the genomic revolution yielded almost a thousand sequenced microbial genomes. More recently, the explosion of random community genomics or metagenomics, where DNA is sequenced directly from environmental samples, has provided insights into the microbial communities. Currently, two approaches are commonly used to sequence metagenome samples. In the traditional approach, DNA is cloned into BACs or small plasmids, and dideoxy chain termination sequencing (Sanger sequencing) is used to determine the sequences. The alternative approach is DNA sequenced without cloning by using one of the so-called next-generation sequencing techniques which is usually called 'pyrosequencing'.

Jo Handelsman

During the late 1990s, the term metagenomics was first used by Jo Handelsman, Jon Clardy, Robert M. Goodman and others, that first appeared in publication in 1998. Jo Handelsman is a professor of molecular, cellular and developmental biology in Howard Hughes Medical Institute at Yale University. She is also an editor-in-chief of the academic journal *DNA and Cell Biology* and author of many books on scientific education.

Robert M. Goodman has served as the executive dean of agriculture and natural resources at Rutgers working mainly on plant biology and virology in the State University of New Jersey since June 2005. Metagenomics is defined as *the genomic analysis of microorganisms by direct extraction and cloning of DNA from an assemblage of microorganisms*. The availability of 'next generation sequencing technologies' such as 454 pyrosequencing have made it such that a cloning step is no longer essential for metagenomic projects. The National Institutes of Health launched a Human Microbiome initiative with primary goals to determine if there is a core human microbiome to understand the changes in the human microbiome that can be correlated with human health and to develop new technological and bioinformatics tools to support these goals. Advances in genomics have a great impact in the field of microbial ecology.

Metagenomics holds a great promise for accessing and characterising the microbial communities. However, the high diversity and level of complexity present in microbial communities represent an obstacle to understand these assemblages given the current approaches. The integration of microbial community structure with function remains challenging when uncultured microbes in diverse environments are taken into account. Integration of complementing technologies will be required to understand the role of microbes in different ecological habitats (like microarrays, high throughput sequencing, bioinformatics and novel tools, and metagenomics approaches such as metaproteomics, metatranscriptomics and meta-metabolomics). In spite of the many challenges, the field of metagenomics offers the promising perspectives for achieving a more comprehensive view of microbial communities, microbial adaptation and function within their ecosystems.

Robert M. Goodman

B. ABOUT METAGENOMICS

Metagenomics is the study of metagenomes, *i.e.* the genetic material recovered directly from environmental samples and their detailed studies including cloning and sequencing. The broad field may also be referred to as environmental genomics, ecogenomics or community genomics. Random community genomes (metagenomes) are now commonly used to study microbes in different environments. Over the past few years, the major challenge associated with metagenomics shifted from sequence generation to sequence analysis. High-throughput, low-cost next-generation sequencing has provided access to metagenomics to a wide range of researchers. A schematic representation of classical metagenomics is given in Fig. 14.1.

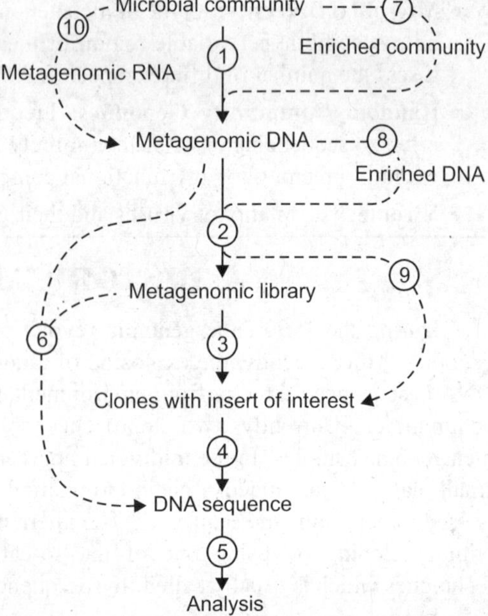

Fig. 14.1. Classical metagenomic approach.

On the earth system, microbes are important for playing a key role in maintaining the well being of our global environment. In addition to importance of microbes, very little is known of their diversity; how many species are present in the environment and what does each individual species do ? There were no appropriate techniques available to answer these questions until recently. Microorganisms growing in very diverse habitat cannot be cultured in the laboratory and, therefore, it was impossible to study by the known methods that have been very successful for known microorganisms throughout 20th century. Thanks to high-throughput technology developed to analyse and sequence DNA procured from various samples of the natural environment. Those information began to accumulate that demonstrated the exceptional diversity of microbes in nature; in fact most of the microorganisms are entirely novel and have not earlier been described.

Survival and sustenance of life on earth is possible only due to microorganisms, but how they do so is still poorly understood. Only 0.1- 1% of the total microbial communities present in most habitats is represented by cultured microorganisms. Until the discovery of metagenomics tools which allow the investigation of non-culturable microorganisms in the laboratory, study of the entire genetic makeup from particular environments was not possible. This resulted in loss of a fruitful source of microbial biodiversity and functional novelties present in them. Metagenomic analyses of biological communities in nature are revolutionising our understanding of the diversity, function, and inter-relationships among organisms in diverse ecological niches. The rapid advancements in metagenomics have been possible due to the technical and analytical methods developed from high throughput platforms for cloning, DNA sequencing, robotics, high-density microarrays, 2D-gel electrophoresis, and mass spectrometry as well as associated bioinformatics software. This approach includes the application of metagenomic techniques to the study the microbial communities directly in their natural environments, bypassing the isolation and cultivation of individual species in laboratory. The metagenomic approaches have its roots in the culture-independent retrieval of 16S rRNA genes, pioneered by Pace and colleagues two decades ago. Since then, metagenomics has revolutionised microbiology by shifting focus away from clonal isolates towards the estimated 99% of non-culturable microbial species.

The influence of ecological factors on microbial life in a wide range of habitats can be assessed through taxonomic and functional differences between metagenomic samples. Traditional microbiology and microbial genome sequencing rely on cultivated clonal cultures. This new field of genetic research enables the study of organisms that are not easily cultured in a laboratory as well as studies of organisms in their natural environment. Early gene sequencing approaches from environmental samples have cloned specific genes (often the 16S rRNA gene) to provide a profile of microbial diversity in a natural sample. It has been found that the vast majority of microbial biodiversity had been missed by culture-based methods. Unbiased samples of all genes from all members of sampled communities have been obtained by using 'shotgun' Sanger sequencing or massively parallel pyrosequencing. A typical metagenomics project begins with the construction of a clone library from DNA sequence retrieved from an environmental sample. Then the clones are selected for sequencing by using either functional or sequence-based screens. In the functional approach, genes retrieved from the environment are heterologously expressed in a host such as *E. coli*, and sophisticated functional screens employed to detect clones expressing functions of interest.

Never the less, this approach has produced many exciting discoveries and spawned several companies aiming to retrieve marketable natural products from the environment and Cubist Pharmaceuticals. In the sequence-based approach, clones are selected for sequencing based on the presence of either phylogenetically informative genes, such as 16S, or other genes of biological interest. The most prominent discovery from this approach is the discovery of the 'proteorhodopsin' gene from a marine community, Recently, facilitated by the increasing capacity of sequencing centers, whole-genome shotgun (WGS) sequencing of the entire clone library has emerged as a third approach to metagenomics (for detail *see* Chapter 13: *Genomics and Proteomics*). Unlike previous approaches, which typically study a single gene or individual genomes, this approach offers a more global view of the community allowing to: better assessment levels of phylogenetic diversity and intraspecies polymorphism, study of full gene complement and metabolic pathways in the community, and reconstruction of near-complete genome sequences. WGS sequencing also has the potential to discover new genes that are too diverged from currently known genes to be amplified with PCR or heterologously expressed in common hosts, and is especially important in the case of viral communities because of the lack of a universal gene analogous to 16S.Nine shotgun sequencing projects of various communities have been completed to date.

1. Biological Insights From Metagenomics

The biological insights from these studies have been well-reviewed elsewhere. Two studies are highlighted here just to exemplify the exciting possibilities of the approach. The acid mine biofilm community is an extremely simple model system consisting of only four dominant species; therefore, a relatively miniscule amount of shotgun sequencing (75 Mbp) was enough to produce two near-complete genome sequences and detailed information about metabolic pathways and strain-level polymorphism. Moreover, the Sargasso Sea community is extremely complex containing more than 1,800 species. Nonetheless, with an enormous amount of sequencing (1.6 Gbp), vast amount of previously unknown diversity was discovered including over 1.2 million new genes, 148 new species, and numerous new rhodopsin genes. These results are especially surprising (because how well the community had been studied previously), and suggest that equally large amounts of biological diversity awaits for future discovery.

The genomic revolution made during 1990s has yielded about a thousand sequenced microbial genomes. More recently, the explosion of random community genomics or metagenomics, where DNA is sequenced directly from environmental samples, has provided insights into the microbial communities. Currently, two approaches are commonly used for sequencing metagenome samples. In the traditional approach, DNA is cloned into BACs or small plasmids, and dideoxy chain termination sequencing (Sanger sequencing) is used to determine the sequences. But in the alternative approach, DNA is sequenced without cloning by using one of the so-called next-generation sequencing techniques, *i.e.* pyrosequencing. Both approaches have advantages and disadvantages. For example, Sanger sequencing generates longer sequence reads but has inherent biases due to the cloning. Pyrosequencing has much higher throughput and a lower error rate per base sequenced compared to Sanger sequencing, but those errors are biased toward certain mistakes.

Regardless of the sequencing approach used to generate the data, the first steps in analysis of any metagenome involve comparing those sequences to known sequence databases. This computationally intensive task provides the basic data types for many subsequent analyses including phylogenetic comparisons, functional annotations, binning of sequences, phylogenomic profiling, and metabolic reconstructions. Recent research is connected to the development of a freely available, fully automated open source system for processing metagenome sequence data to generate these basic elements. A public implementation of this system has been provided for all researchers to analyze their metagenomes. The metagenomics RAST server (*i.e.* mg-RAST) is available over the web to all researchers, and access is not limited to specific groups or data types. About 500 metagenomes have been processed through the beta version of the pipeline so far.

C. FROM GENOMICS TO METAGENOMICS

Genomics came to an age when we began to witness a greater level of microbial diversity within species than previously anticipated. This laid the foundation for generating genomic sequence data from whole environments without culturing them, a field of research now known as metagenomics. In the late 1990s, the term metagenomics was first used and defined as the genomic analysis of microorganisms by direct extraction and cloning of DNA from an assemblage of microorganisms. The availability of next generation sequencing technologies such as Sanger sequencing, Pyrosequencing, etc. have made it such that a cloning step is no longer essential for metagenomic projects. The National Institutes of Health launched a Human Microbiome initiative with primary goals: to determine a core human microbiome, to understand the changes in the human microbiome that can be correlated with human health, and to develop new technological and bioinformatics tools to support these goals.

There has been a great impact of advances in genomics on the field of microbial ecology. Metagenomics holds a great promise for accessing and characterising microbial communities. However, the high diversity and level of complexity present in microbial communities represent an obstacle to understand these assemblages given the current approaches. Integration of microbial community structure with function remains challenging, when taking into account of uncultured microbes in diverse environments. Integration of complementing technologies such as microarrays, high throughput sequencing and bioinformatics, novel tools, and meta-approaches such as metaproteomics, metatranscriptomics and meta-metabolomics, will be required to understand the role of microbes in different ecological habitats.

Progress made in mining the rich genetic resource of non-culturable microbes has led to the discovery of new genes, enzymes, and natural products. The impact of metagenomics is witnessed in the development of commodity and fine chemicals, agrochemicals and pharmaceuticals, where the benefit of enzyme-catalysed chiral synthesis is recognised. Recovery of metabolic pathways and gene clusters involved in biosynthesis of antibiotics and bioactive molecules has increased the prospect of identifying useful natural and synthetic products for drug development. The discovery of biocatalysts operating optimally with high efficiency in conditions amenable to industrial processes requirements are key to successful development of food products, detergent additives, bioactive compounds, fuel alcohol and biodiesel, as well as optically active intermediates for chemical and drug synthesis. An analysis of the basic aspect of culture-independent tools to study microbial communities, the methods available to isolate environmental DNA and to establish metagenomic libraries which can further be used for extensive activity screens are important to learn along with a number of the recent applications of this technology.

D. DIFFERENT METAGENOMICS PROJECT

With the growth of the number and diversity of metagenomics, an appreciation of the challenges has also increased. Many of the challenges are likely to diminish with the development of new technologies and mathematical tools. In addition, many criteria for success of metagenomics studies will remain unchanged by new knowledge or methods. Sometimes we ignore the microbes, but it has colonised almost every millimetre of our body- living in mouths, skin, lungs, gut, intestine, etc. It has been assumed that human intestine bears about 100 trillion bacteria. Indeed, the human body has 10 times as many microbial cells as human cells. Microorganisms maintain our health breaking down undigestible foods, making essential vitamins, and even shaping our immune system. Recent researches suggest that microbes play a role in curing the diseases such as ulcers, heart disease, and obesity.

While microbes make up such an intimate part of us, most of our microbial inhabitants remain a mystery. The bacteria in the human body are very difficult to study, since only about 1 percent of them can be grown in the laboratory. Now a proposed new project to sequence all our microbial residents could change the concept. According to George Weinstock, codirector of the Human Sequencing Center at the Baylor College of Medicine, Houston (USA), 'this is completely unexplored territory that is likely to have a large impact on our understanding of human health and disease. We had not been able to approach it because of the scale of the problem, but now we are finally able to open that door'.

Thanks to ever-improving methods to sequence DNA. Now the scientists can analyse the genomes of entire microbial communities. By comparing microbial communities in people of different ages, origins, and health status, researchers hope to find out precisely how microorganisms prevent or increase risk for certain diseases and whether they can be manipulated to improve the health.

Several metagenomics projects are under way or have been completed including analysis of the microbes living in the human gut and on the skin. But a true snapshot of our microbial menagerie will require a massive effort, along the lines of the Human Genome Project. A total 440 projects are known to be underway. Some of the projects are given below with its brief description.

1. Environmental Microbial Metagenomics Project

This is a megaproject supported by a group of megadiverse countries of Australia. This project includes hypersaline habitats, subsurface thermal (non-volcanic) habitats and unique animals and insects whose guts contain unique microbes. Most of the Australian microbial diversity remains unknown. The project will seek to redress this. Of all the known habitats the most interesting is the deep subsurface thermal aquifer know as the Great Artesian Basin (GAB) of Australia. Fig. 14.2 shows the different environmental regions of Australia.

GAB is a unique deep subsurface thermal non-volcanic aquifer and world's largest underground water reserve. The basin recharge is along the Eastern margins of Australia along the Great Dividing mountain ranges, and the natural discharge occurs as hypersaline mound springs in central Australia. GAB is a 'hot bed' for microbial metagenomics adaptation of new types of microbes like heated by the earth's hot core due to the deep nature of the aquifer- rate of 1°C/30 metre depth, < 95°C at bore source forms a temperature gradient to ambient along drain runoff channels. Aquifer's age is 2 million years (water) but the sedimentary bases are up to 400 million old and pH neutral to 10.

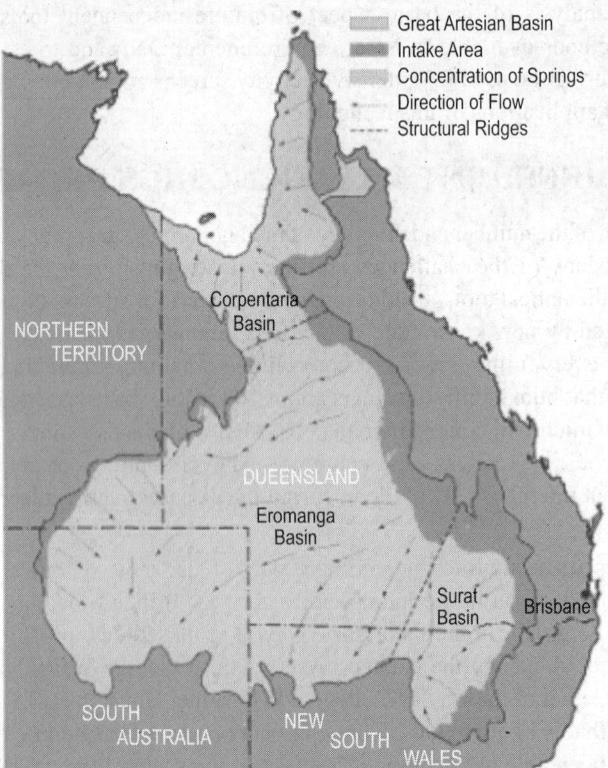

Fig. 14.2. Different environmental regions of Australia.

2. Microbial Metagenomics Project

This project is an integrated consortium of microbiologists who will utilise post-genomics to investigate aquatic microbial assemblages that control biogeochemical cycles. This is a NERC (National Environment Research Council) funded research project investigating the metagenome of marine microbial communities led by project leader, Ian R Joint from the Plymouth Marine Laboratory (UK). The project is built around three specific hypotheses and the study will provide answers to important and current environmental questions. The hypotheses to be tested are:

(*i*) That the microbes in the ocean exist in definable communities and biogeochemical fluxes depend on microbial community structure.

(*ii*) That the oligotrophic ocean is ultimately limited by the availability of nitrogen and the scale of nitrogen limitation is a function of interactions between microbes in the assemblage.

(*iii*) That the marine microbial activity is a major source of atmospheric biogas, much of which involve anaerobic metabolism in an aerobic environment.

NERC is basically the higher metagenomics analyses of virus-driven nutrient regeneration on microbial growth in marine ecosystems and marine microbial communities. The main non-predatory loss route for primary production by phytoplankton in virus-induced cell lysis and as much as one quarter of the photosynthetically fixed organic carbon is recycled back to the dissolved fraction by viral lysis. Virus infection of phytoplankton will lead to cell lysis and release of metagen cell contents (including metagen progeny viruses) that will become available to a microbial community. This system is currently under the stage of development but is building in technology developed for the microbase project. The complete strategy of microbial metagenomics project is given in Fig. 14.3.

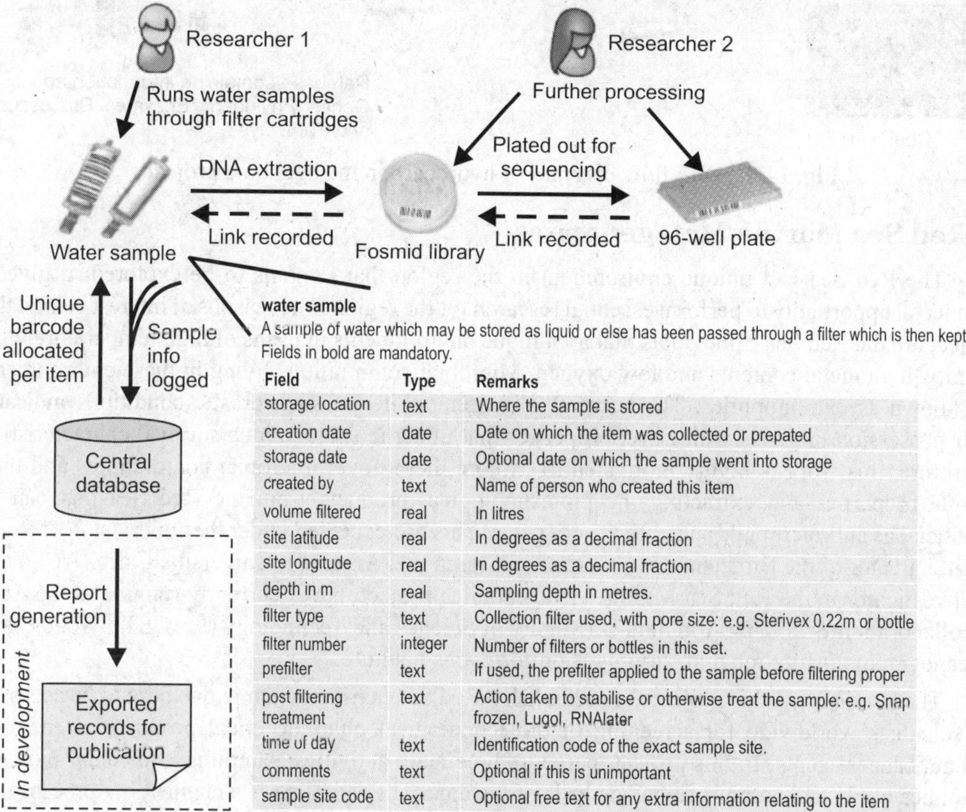

Field	Type	Remarks
storage location	text	Where the sample is stored
creation date	date	Date on which the item was collected or prepated
storage date	date	Optional date on which the sample went into storage
created by	text	Name of person who created this item
volume filtered	real	In litres
site latitude	real	In degrees as a decimal fraction
site longitude	real	In degrees as a decimal fraction
depth in m	real	Sampling depth in metres.
filter type	text	Collection filter used, with pore size: e.g. Sterivex 0.22m or bottle
filter number	integer	Number of filters or bottles in this set.
prefilter	text	If used, the prefilter applied to the sample before filtering proper
post filtering treatment	text	Action taken to stabilise or otherwise treat the sample: e.g. Snap frozen, Lugol, RNAlater
time of day	text	Identification code of the exact sample site.
comments	text	Optional if this is unimportant
sample site code	text	Optional free text for any extra information relating to the item

Fig. 14.3. Strategy for complete microbial metagenomics project.

3. Hydrocarbon Metagenome Project

Microbial activity in oil reservoirs is common; therefore, the microbial fermentation products can lower oil viscosity or interfacial tension, and produced biomass can block undesired flow paths to produce more oil. Due to the high biotechnology benefit from increased understanding of reservoir, microbial ecology through new sequence technologies can help to decrease the environmental impact of oil production. Canada recognises the necessity of transition to global energy production towards renewable resources. It has developed a database to describe and harness the genetic potential of the microorganisms, genes and biological processes that exist naturally in microbial communities in our oil sands and coal beds. By identifying the genes, bioprocesses and bacteria in the old sands and coal beds, they can now improve their understanding of how methane in hydrocarbon resources is generated, and identify the enzymes involved in the natural cracking of hydrocarbons that produce methane and carbon dioxide. In this project the organic bioprocesses can be utilised to decrease the water used and land lost in mining operations and manage the methane emissions from tailings ponds. The publicly accessible knowledge generated by this project will also help other researchers to harness the power of the processes present in deep biosphere of Canada to resolve the other important research questions. Work flow diagram of hydrocarbon metagenome project is given in Fig. 14.4.

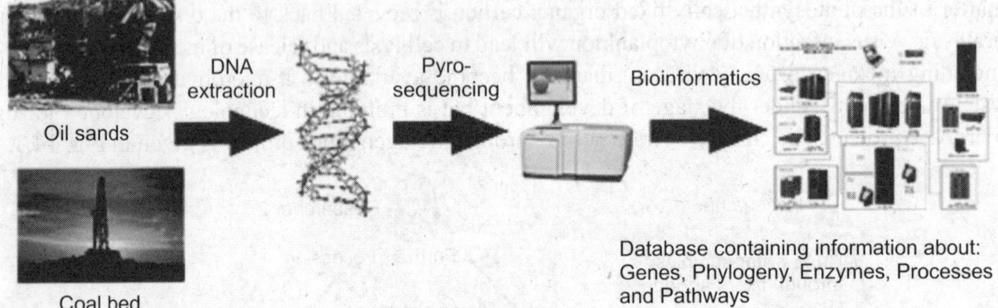

Fig. 14.4. Work flow diagram of hydrocarbon metagenome project.

4. Red Sea Marine Metagenomics

The Red Sea is a unique environment in the region that remains to be explored; it gives a wonderful opportunity to perform essential research for the region. The reasons of interest behind this project are the Red Sea brine pools that are unique environments in terms of high temperature, high salinity, high metal contents and low oxygen. Microbial communities living in these environments are known as extremophiles. The survival of extremophiles in such drastic conditions indicates their possession of genes with novel properties that underlie these unique survival characteristics. Therefore, this work is highly motivated to explore these novel microbial communities and their unique properties and extracting the biotechnological products from the Red Sea that can be beneficial as antimicrobials and anticancer agents. This project is led under the supervision of Rania Siam, Director of the Biotechnology Graduate Programme, American University, Cairo (AUC) and an Investigator in the Red Sea Marine Metagenomics Project that is currently running at the AUC in collaboration with King Abdullah University of Science and Technology (KAUST), Woods Hole Oceanographic Institution and Virginia Bioinformatics Institute.

The main aim of this project is to establish a Red Sea marine genomic database to be accessed by scientists worldwide for screening of biotechnological pharmaceutical products as enzymes and anticancer agents. In this project there are three main activities: sampling, molecular biology/ genomics work, and computational analysis of the data. The sampling is a challenging process that requires rigorous planning where samples should be subjected to proper processing and storage till arrival to the laboratory. The samples are subjected to different procedures starting from DNA extraction, followed by Whole Genome Sequencing (WGS), to identification of unknown or unique genes, and to understand the novel microbial communities. This requires rigorous computational analysis to make sense of the data. Furthermore, fosmid libraries are constructed for isolation and purification of genes of biotechnological interest as lipases and cellulases. In addition, 16S rRNA phylogenetic analysis of the microbial communities in the samples is carried out.

E. METAGENOMICS AT WORK

The endless ability of microorganisms has impact on every part of the biosphere. The microbes convert the key elements of life carbon, nitrogen, oxygen, and sulfur into forms accessible to other living organisms. They also synthesise the necessary nutrients, minerals, and vitamins available to plants and animals. The billions of microbes living in the human gut help them to digest food, break down toxins, and fight the pathogens causing diseases. Microbes also clean up pollutants in the environment, such as oil and chemical spills. All of these activities are carried out not by individual microbes but also by complex microbial communities intricate, balanced, and integrated entities that have a remarkable ability to adapt swiftly to environmental change.

Historically, microbiology has focused on a single species in pure laboratory culture, and thus understanding of microbial communities has lagged behind the understanding of their individual member. Metagenomics is a new tool to study microbes in the complex communities where they live and to understand the work of these communities. Traditional microbiological approaches have already shown how useful microbes can be; the new approach of metagenomics will greatly extend scientist's ability to discover and benefit from microbial capabilities. Although it is still considered 'a youthful emerging form of analysis', metagenomics has already made major contributions to microbiology.

In 1998, the first complete genome sequence of an uncultured bacterium, *Treponema pallidum* causing syphilis, was released that acted a landmark in genome sequencing. Although this infectious agent was discovered over a century ago, yet it has never been isolated in culture. The project led by I. Liu of the Centers of Disease Control utilised the DNA from bacteria that were obtained from the testes of over 400 infected rabbits. Analysis of *T. pallidum* immediately identified as a potential contributor to virulence, that potentially led to the development of DNA-based treatments. Besides, the genomes of *Tropheryma whipple* have also been sequenced using metagenomics; this bacterium is responsible for Whipple disease, and lives inside tsetse flies, fruit-flies, carpenter ants, a nematode and two aphid species.

One such study led to the discovery of proteorhodopsin, *i.e.* a type of light gathering protein in oceanic bacteria. This discovery was surprising not only because the host organisms were believed to depend on organic matter and no light for energy, but also because this family of proteins has never before been found in bacteria. In addition to its contributions to our understanding of the microbial world, metagenomics analysis has resulted in an understanding of why certain bacteria cannot grow in culture, contributing to an improvement in culture techniques, diagnostics, and therapies for uncooperative microbes.

F. TECHNOLOGIES FOR METAGENOME STUDY

In metagenomics, the analysis of the entire DNA present in an organism is applied to entire communities of microbes, bypassing the need to isolate and culture individual microbial species. This new science opens the door to a tremendous amount of scientific exploration. Metagenomics will vastly enhance our knowledge of microbial communities and could lead to practical applications in industry and other human welfare. In Metagenomics, the several methods are available to explore the entire knowledge and productivity of metagenome; some methods are described in this section.

1. Metagenomic Nucleic acid Extraction and Enrichment Strategies

Cell recovery and direct lysis are the two principal strategies for the recovery of metagenomic DNA. Extraction of total metagenomic DNA is necessarily a compromise between the vigorous extraction required for the representation of all microbial genomes, and the metagenomics of DNA shearing and the coextraction of inhibiting contaminants. Mechanical bead beating is used to recover more diversity as compared with chemical treatment. However, chemical lysis is a more gentle method recovering DNA of higher molecular weight. Chemical lysis can also select for certain taxa by exploiting their unique biochemical characteristics.

The methods for recovering RNA from environmental samples are largely similar to those used for DNA isolation and modified to optimise the yield of intact mRNA by minimising single-stranded polynucleotide degradation. Protocols are designed to limit physical degradation and RNase activity, which are the major causes of yield loss. Samples should be processed or frozen at -80°C immediately after harvesting and additional methods are used to minimise the RNA degradation such as the co-precipitation of cellular RNA with proteins (e.g. sulphate salt solution RNAlater, Ambion) and the synthetic capping of the isolated RNA. These techniques provide a feasible route for the construction of metagenomic cDNA libraries for further identification of functional eukaryotic genes. Several strategies have been designed to enrich for the sequences of interest before cloning for increasing the proportion of active clones in a library. The first and essential step is to enrich for the genomes of interest.

(*a*) GC Enrichment by Ultracentrifugation: There are large numbers of microorganisms (e.g. actinomycetes and acidobacteria) that consist of a high GC content in their DNA, they are of particular interest. Their DNA can be extracted from the soil and then subjected to ultracentrifugation to enrich for high GC content DNA. Although this is a fairly crude approach and will not provide a complete separation, yet it will certainly increase the representation of certain genomes in a library.

(*b*) Bromodeoxyuridine (BrdU) Enrichment: The principle of bromodeoxyuridine (BrdU) enrichment (Fig. 14.5) is that metabolically active organisms will incorporate a labelled nucleotide into their DNA, which can then be isolated on the basis of the incorporated label. BrdU can be fed to bacteria in soil samples and the labelled DNA recovered by immunocapture. The 16S rRNA analysis performed on the DNA identifies phylogenetic groups that were metabolically active in the original sample. The addition of selective substrates with the BrdU further discriminates among the members of the microbial community, enriching metagenomic libraries for those that grow on the added nutrient. This strategy could be used to enrich for organisms that grow on xenobiotics or on substrates such as starch, cellulose and proteins to find amylases, cellulases and proteases, respectively, or other enzymes of interest in metagenomic libraries. Limitations of these methods include cross-feeding and recycling of the label within the community resulting in loss of specific enrichment.

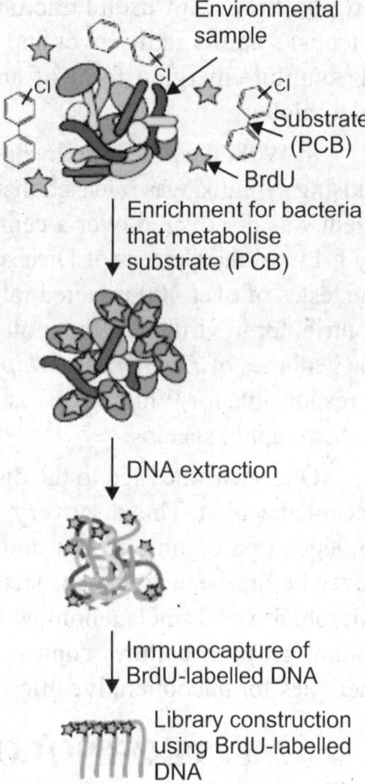

Fig. 14.5. Construction of BrdU-enriched metagenomics library.

(*c*) Stable-isotope Probing (SIP): Stable-isotope probing approach (Fig. 14.6) involves bacteria that can use the substrate incorporating ^{13}C into their DNA, making it denser than normal DNA containing ^{12}C. SIP has been successfully used for labelling and separating microbial DNA and RNA. Density gradient centrifugation separates the labelled nucleic acids from unlabelled nucleic acids, which can then be used either for PCR-based analysis or direct cloning to construct metagenomic libraries. The method has enormous potential for subdividing microbial communities into functional units to simplify analysis and will offer broad opportunity to study community functions if it can be expanded to stable isotopes of other elements such as phosphorus or nitrogen.

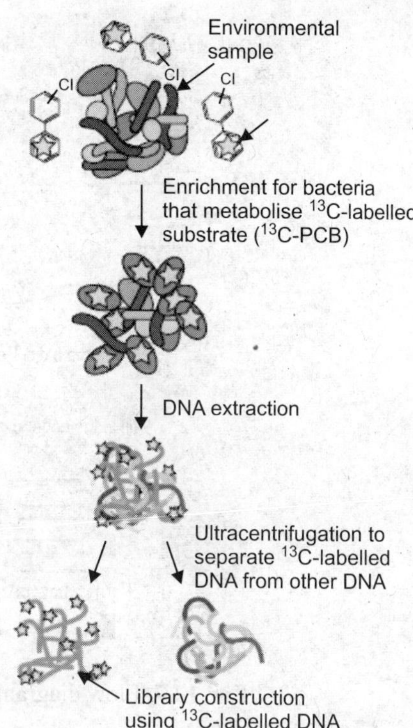

Fig. 14.6. Construction of stable isotope probe enriched metagenomic library.

Although these approaches provide a significant step in refining metagenomic library construction by enriching for DNA from a subset of the community, yet both methods have limitations that need to be addressed by future research. The first potential problem is cross-feeding. Any metabolically active bacterium will take up BrdU and will be represented in a library that is supposed to be enriched for organisms that utilise an added substrate. An additional issue is timing. The more prolonged the substrate feeding, the higher the probability that the substrate will be recycled in the community and the basis of the enrichment will break down. Finally, the immunocapture method and density gradient centrifugation may shear the DNA making it difficult to retrieve pathways encoded on large fragments of DNA.

RNA-based SIP reduces the cross-feeding problems associated with DNA-based SIP, but it has been used only to construct 16S rRNA gene libraries. If the isotope-labelled RNA presents a viable substrate for making cDNA, libraries could be constructed from cDNA representing genes that are actively transcribed in the presence of the labelled compound. Metagenomic clone libraries could be screened in hybridisation arrays using the cDNA clones as probes. This approach has the attractive feature that one library constructed with non-selected DNA from the entire community could be screened repeatedly using different probes, each developed from a distinct subset of the community.

(*d*) Suppressive Subtractive Hybridisation (SSH): Suppressive subtraction hybridisation can identify genetic differences between the microorganisms and, therefore, it is a powerful technique for specific gene enrichment. Adaptors are ligated to the DNA populations and subtractive hybridisation is carried out to select DNA fragments unique to each DNA sample (Fig. 14.7). This method has typically been applied to analyse genetic differences between the two closely related bacteria (e.g. in the identification of genetic elements contributing to pathogenesis). These techniques might be adopted to target specific genes in related metagenomes. For example, genes involved in the bioremediation of an environmental pollutant could be identified by the comparison of a reference metagenome with a 'perturbed' metagenome (*i.e.* affected by a specific pollutant). The relatively crude nature of this approach identifies the total genetic difference between the two bacterial populations and is not specific to genes of interest or genes whose expression was up-regulated on addition of the environmental pollutant or xenobiotic. SSH has successfully been used on complex metagenomes and the sensitivity of the process can be increased by using multiple rounds of subtractive hybridisation.

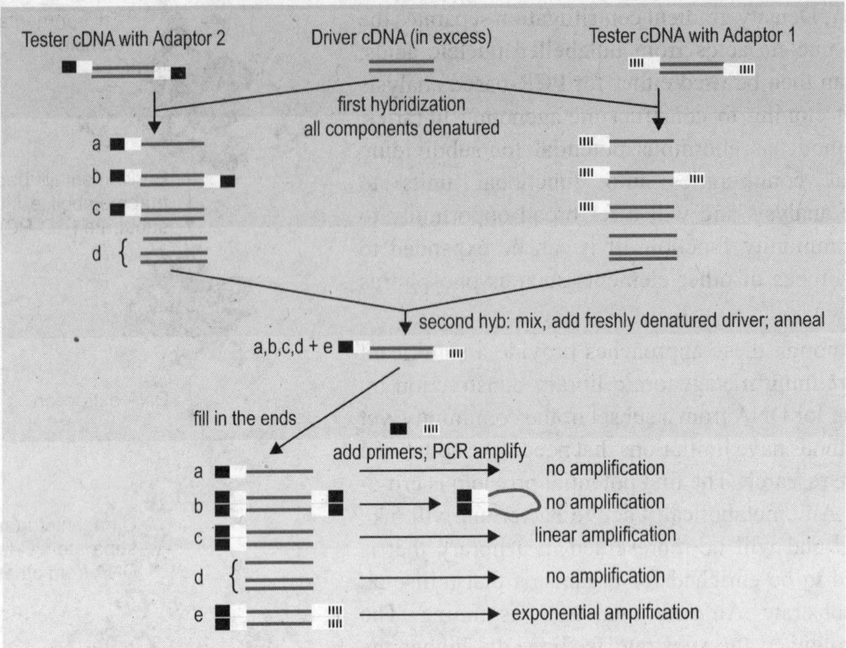

Fig. 14.7. Flow diagram of suppressive subtractive hybridisation.

(*e*) Differential Expression Analysis (DEA): Differential expression analysis (DEA) targets transcriptional differences in gene expression. Several variations in the basic concept exist. These include selective amplification via biotin and restriction-mediated enrichment (SABRE), integrated procedure for gene identification (IPGI), serial analysis of gene expression (SAGE), tandem arrayed ligation of expressed sequence tags (TALEST) and total gene expression analysis (TOGA). These techniques have been effectively applied for eukaryotic gene discovery. Their high sensitivity and selectivity should enable small differences in expression of single copy genes to be detected.

(*f*) Phage Display: Phage-display expression libraries provide a means of isolating a given DNA sequence by affinity selection of the surface-displayed protein to a metagenomic ligand. Biopanning involves repeated cycles of binding that will successively enrich the pool. After several rounds of enrichment, individual clones are characterised by DNA sequencing. This method is efficient and amenable to high-throughput screening, offering the potential to enrich even rare DNA sequences in the metagenome; but current phage technology limits expression of proteins ~50kDa.

(*g*) Affinity Capture: Oligonucleotides covalently binding to a solid support can be used to affinity purify the target genes. The slow kinetics of hybridisation limit this process, but might be improved by using metagenomics mRNA or single-stranded DNA.

(*h*) Microarray: Microarrays represent a powerful high-throughput system for analysis of genes. These are typically used to monitor differential gene expression to quantify the environmental bacterial diversity and catalogue genes involved in key processes. Microarrays of immobilised oligonucleotide gene targets have also been used to select appropriate biotope samples for metagenomic library screening. Such arrays could also be used for the affinity capture of targets as a means of enrichment before construction of metagenomic libraries. Microarray technology could also be used for the pre-selection of genes in metagenomic libraries before shotgun sequencing, thereby reducing the sequencing burden and reducing the proportion of sequences unassigned by database sequence similarity searches. Microarray technology can be applied both as a tool to monitor critical gene activities across a diverse spectrum of genomes or to access the transcriptome of single microbial strains in a complex community. Regardless of the nature of the study, substantial challenges (e.g. efficient RNA extraction, detection of study of signal above background noise for complex samples, and cross hybridisation) need to be overcome before the expression of environmental gene approaches the reproducibility of similar laboratory-based analyses. Finally, gene expression studies are more amenable to time-course studies as compared with large-scale sequencing. The significance of this is that the dynamics of a complex microbial community in a changing environment can be tracked with a single and portable experimental tool. Thus, microarrays are a promising technique for characterisation of genes in environments such as soil and water. Construction of screening of metagenomic clones by using microarray approach is given in Fig. 14.8.

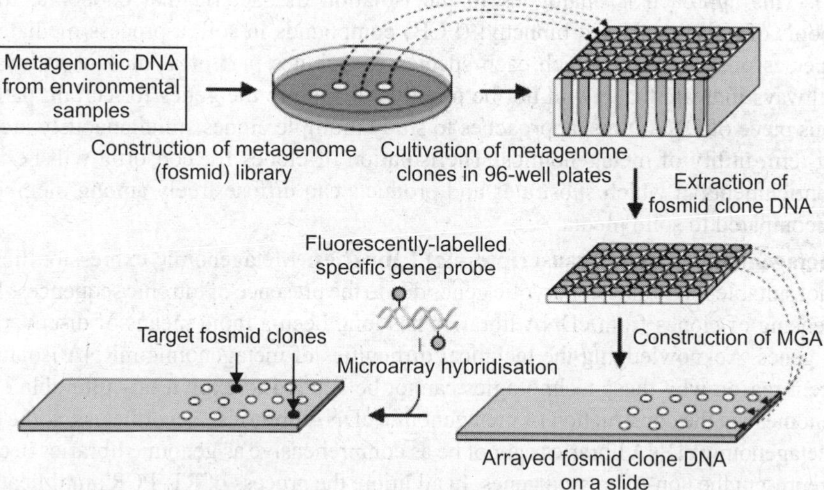

Fig. 14.8. Construction of screening of metagenomics clones by using microarray approach.

DNA microarray platform to screen a metagenomics library with whole microbial genomes and community genomes emerged as an elegant tool to target the unculturable species in the environment (Fig. 14.8). The microarray (COSMO) contains ~1-kb PCR products amplified from the inserts of 672 cosmids along with a set of controls (16S ribosomal DNA probes). Numerous bacterial isolates are obtained from the same sample from which the library was derived. Genomic DNA is purified from each environmental isolate. In addition, metagenomic DNA is purified directly from the mixed population, which is done without cultivation of bacteria. Each test genome is labelled with Cy5-dCTP and probed with COSMO. In order to subtract any signal coming from nonspecific hybridization, two-color hybridization is used in each experiment, in which each test genome is compared to a reference sample of common bacterial DNA. The reference DNA consists of a pooled sample of genomic DNA from several species of microbes. The reference DNA is given a different label (Cy3), and equal amounts of test and reference DNA are combined and hybridised to COSMO. This approach referred to as comparative genomic hybridisation (CGH). CGH is

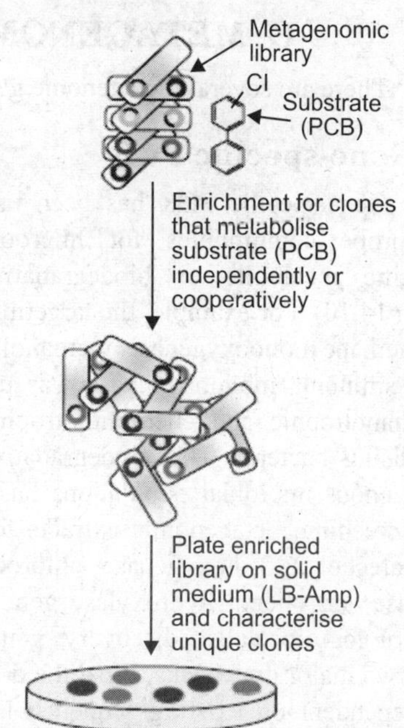

Fig. 14.9. Enrichments technique for metagenomics clones forming metabolic consortia.

repeated for all environmental isolates as well as for the metagenomic DNA sample(s). Positive results are determined based on a Cy5/Cy3 ratio greater than 1. Consequently, a profile for each clone is obtained in the metagenomic library, *i.e.* a graphical representation of its hybridisation to one or more species of bacteria. Clones that are specific to a test strain or community are hybridised only to those DNA samples. Clones that contain a conserved sequence within their corresponding PCR amplicons hybridise to the genomes of multiple species.

***(i)* Enrichments for Metagenomics Clones forming Metabolic Consortia:** Research on metagenomic libraries has focused on activities of individual clones. A key direction for development of this technology is to enrich the consortia of clones that together accomplish a desired function (Fig. 14.9). This approach is analogous to the isolation of bacteria that cooperate to degrade complex pools of polychlorinated biphenyl (PCB) compounds in soil, a process mediated by two or more species of bacteria in which each species contributes part of a 'mosaic pathway'. Such mosaic pathways indicate that it will not be possible to capture the genes for certain pathways on a contiguous piece of DNA; thus, approaches to study multiple clones simultaneously are essential to the long-term utility of metagenomics. The isolation of clones for consortia will be facilitated by liquid enrichments in which substrates and products can diffuse freely among members of the mixture as compared to solid media.

***(j)* Metagenomic cDNA (Transcriptomic) Libraries:** Metagenomic expression libraries are generally not suitable for mining eukaryotic genes due to the presence of intronic sequences. The large-scale sequencing of clones from cDNA libraries has long been a rapid means of discovering novel eukaryotic genes. Acknowledging the technical difficulties of metagenomic mRNA isolation, there is no inherent reason why these technologies cannot be applied to exploit un-culturable eukaryotic enzyme genomes via the construction of metagenomic cDNA libraries. Nevertheless, some caution is advised. Metagenomic cDNA libraries cannot be as comprehensive as genomic libraries because they can never represent the non-expressed genes. In addition, the process of RT–PCR amplification limits the size of inserts and could impose a large sequence-dependent bias on the library.

G. METAGENOME GENE DISCOVERY TOOLS

There are several metagenome gene discovery tools, some of them are described below:

1. Gene-specific PCR

Gene-specific PCR has been used extensively to probe communities for microorganisms with specific metabolic or biodegradative capabilities (Fig. 14.10). For example, the targeting of genes such as methane monooxygenase, methanol dehydrogenase and ammonia monooxygenase was used to identify methanotrophic and chemolithotrophic ammonium-oxidising bacteria. The biodegradative potential of indigenous microbial populations has been assessed by screening metagenomic extracts for the presence of catechol 2, 3-dioxygenase, chlorocatechol dioxygenase and phenol hydroxylase genes. However, as a tool for biocatalyst discovery, gene-specific PCR has two major drawbacks. First, the design of primers is dependent on existing sequence information and skews the search in favour of known sequence types. Functionally similar genes resulting from convergent evolution are not likely to be detected by a single

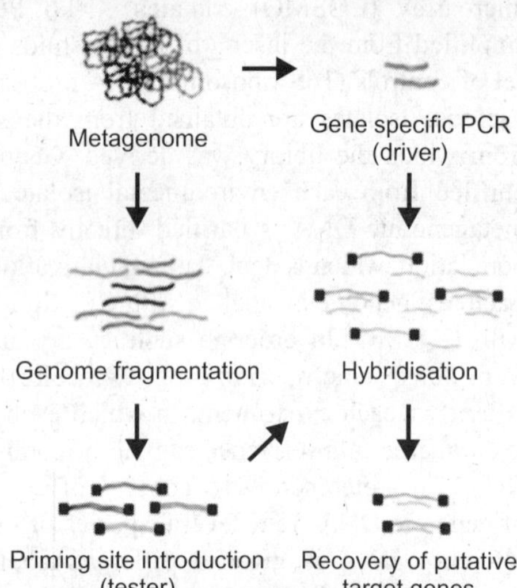

Fig. 14.10. Metagenomic gene discovery by using gene-specific PCR.

gene-family-specific set of PCR primers. Second, only a fragment of a structural gene will typically be amplified by gene-specific PCR, requiring additional steps to access the full-length genes. Amplicons can be labelled as probes to identify the putative full-length gene(s) in conventional metagenomic libraries. Alternatively, PCR-based strategies for the recovery of either the upstream or downstream flanking regions including universal fast walking, panhandle PCR, random primed PCR, inverse PCR and adaptor ligation PCR can be used to access the full-length gene.

These methods are technically more difficult to apply at a metagenomic level due to the increased complexity of a metagenomic DNA sample, but have been used successfully for the recovery of novel gene variants (e.g. 2,5 diketo-D-gluconic acid reductase) from environmental DNA. Because these approaches can be laborious and time-consuming, innovative alternatives have been developed. Cassette PCR has been used to isolate the central fragment of catechol 2, 3-dioxygenase genes from genomic DNA obtained from a phenol- and crude oil-degrading bacterial consortium. The internal fragment of a previously cloned full-length copy of a catechol 2, 3-dioxygenase gene was then replaced with the PCR-derived internal fragment, thus constructing a novel hybrid catechol 2, 3- dioxygenase gene. This approach can be combined with PCR mutagenesis and/or chimeragenesis to generate highly diverse protein variants incorporating random and directed molecular evolution.

2. RNA as Biomarker

The use of RNA might be more effective than DNA for profiling the functional microbial communities because RNA is a more sensitive biomarker due to its high turnover. Reverse transcriptase PCR (RT–PCR) has been used to recover genes from environmental samples, for example in the isolation of naphthalene-degrading enzymes from microorganisms present in a coal tar waste. Although this approach suffers from the technical difficulties associated with mRNA recovery from environmental samples, it benefits from wider genomic access (includes structural genes from lower eukaryotes as well as from prokaryotes) and the facility to select the functional genes in response to alterations in environmental conditions.

3. Integrons

Integrons are naturally occurring gene capture, dissemination and expression systems that have until recently been associated with antibiotic resistant and pathogenic bacteria. They are widely dispersed in nature and could play a significant role in bacterial genome evolution. The key structural features of an integron include a gene cassette integration site (*att*1), an *intI* gene that encodes an integrase and two promoters that drive the exprcssion of the integrase gene and the incorporated genc cassettes. The mobile element in the system is the gene cassette which consists of one or more open reading frame(s) (ORFs) and associated chromosomal attachment sites (*att*C, a 59 base elements). The integrase catalyses the insertion of the gene cassette into the integration site controlled by the strong promoter via site-specific recombination using *att*1 and *att*C as its substrates. Therefore, integrons act as a repository of ORFs coding for many gene products and potentially provide a source of novel genes. Primers designed to target the conserved regions within the 59 base elements have successfully been used to recover novel genes homologous to DNA glycosylase, phosphotransferase, methyl transferase and thiotransferase. The specificity of this system for gene targets could be improved by using a primer specific for the gene of interest and one targeting a flanking 59 base element.

4. Homologous Recombination Cloning

Homologous recombination cloning can be used for single-step gene targeting and screening with only those recombinants containing the gene of interest after transformation. This method requires the design of an *E. coli* host containing the DNA sequence of a vector homologous to the 50- and 30- sequences flanking the gene of interest. The efficiency of bacterial homologous recombination has been improved and commercial systems are now available (Red/ET system, Gene Bridges) (www.genebridges.com).

5. Immobilised Oligonucleotides

Methods requiring only one gene-specific primer impose less sequence-dependent bias compared with standard procedures for twin-primer PCR amplification. An elegant application of this method would be the use of metagenomic oligonucleotides designed to target a specific gene fragment or consensus sequences by affinity binding. Of course, this approach is used routinely for recovery of polyA RNA cDNA library construction. Affinity capture should be equally applicable to either denatured cDNA or genomic DNA fragments and yields could be further enhanced with prior linking of adaptors so that affinity selected DNA fragments could be PCR amplified using linker-specific primers.

6. Microarrays

Microarrays represent a powerful high-throughput system for analysis of genes. They are typically used to monitor differential gene expression, to quantify the environmental bacterial diversity and catalogue genes involved in key processes. Microarrays of immobilised oligonucleotide gene targets have also been used to select appropriate biotope samples for metagenomics library screening. Such arrays could also be used for the affinity capture of targets as a means of enrichment before construction of metagenomic libraries. Microarray technology could also be used for the pre-selection of genes in metagenomic libraries before shotgun sequencing, thereby reducing the sequencing burden and reducing the proportion of sequences unassigned by database sequence similarity searches.

7. Fluorescence *In situ* Hybridization (FISH)

The development of fluorescence *in situ* hybridization (FISH) (Fig.14.11) over the last decade has had a great impact on study of microbial diversity, allowing the fast screening of bacteria present in the sample. In addition, the PCR based approach which samples the total DNA present in a sample cannot distinguish living from dormant or dead bacteria. FISH targets 16S rRNA, thus assessing only metabolically active bacteria which have high cellular rRNA content. Moreover, specific probes can be used to monitor microorganisms at different levels of taxonomic specificity. Since the PCR and FISH methods are not mutually exclusive, both can be used in parallel to catalogue and compare microbial diversity in different soil samples, and to follow dynamic changes in bacterial community structure due to altered environmental factors. FISH has been described in detail in Chapter 13: *Genomics and Proteomics*.

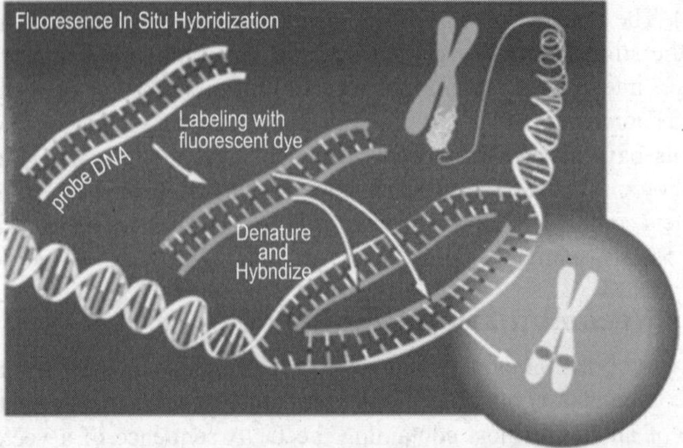

Fig. 14.11. Fluorescence *in situ* hybridization.

8. Substrate-induced Gene Expression Screening (SIGEX)

SIGEX is used to detect metagenomic clones that express desired catabolic genes induced by the presence of suitable substrates using green fluorescent protein (GFP) as a reporter gene. Constitutively expressing clones are first removed by metagenomic-activated cell sorting (FACS), then the expression of desired genes is induced by the addition of substrates and positive clones are again sorted by FACS, separated on agar plates and characterised. A large number of clones can be screened in relatively short time scales by using this approach.

9. Pre-amplification Inverse-PCR (PAI-PCR)

Pre-amplification of the template DNA using phi (ϕ) 29 DNA polymerase coupled with inverse PCR (PAI-PCR), is used to facilitate the isolation of novel genes from small amounts of environmental DNA. This approach could also be useful for marine or water DNA-samples where the amount of extracted DNA is low.

10. Metagenomic DNA Shuffling

Metagenomic DNA shuffling is used to generate novel pesticide-degrading genes in terrestrial systems. The advantages of this method are that the diversity of the metagenome can actually be increased by applying stringent selection steps enzymes suited to industrial applications can be generated. Thus, such an approach could also be useful to find novel enzyme encoding genes in the marine environment.

11. Function-driven Analysis

The function-driven analysis is initiated by identification of clones that express a desired trait followed by characterisation of the active clones by sequence and biochemical analysis. This approach quickly identifies clones that have potential applications in agriculture by focusing on natural products or proteins that have useful activities. The limitations of the approach are that it requires expression of the function of interest in the host cell and clustering of all of the genes required for the function. It also depends on the availability of an assay for the function of interest that can be performed efficiently on vast libraries, because the frequency of active clones is quite low. Many approaches are being developed to mitigate these limitations.

Construction and analysis of metagenomic libraries using function-driven and sequence-driven approaches are given in Fig. 14.12. Improved systems for heterologous gene expression are being developed with shuttle vectors that facilitate screening of the metagenomic DNA in diverse host species and with modifications of *Escherichia coli* to expand the range of gene expression. Although the genes encoding the enzymes required for synthesis of secondary metabolites are usually clustered on a contiguous piece of DNA obtaining fragments of DNA large enough to containing the information required for synthesis of complex antibiotics which can require over 100 genes, presents a challenge. This is a particular problem when the DNA is isolated from soil or other environments contain high concentrations of contaminants that inhibit cloning. Methods to improve the preparation of large fragments of DNA clean enough to clone are being pursued vigorously in many laboratories. To address the challenge of detecting rare but active clones in large libraries, efforts are being directed toward the design of highly sensitive assays and robotic screens that efficiently detect low levels of activity in many samples.

The most convenient traits to study present a selectable phenotype, such as resistance to an antibiotic or growth on an unusual substrate, because selections are orders of magnitude more efficient than screens. Functional screens of metagenomic libraries have identified both novel and previously described antibiotics, an antibiotic resistance gene, lipases, chitinases, membrane proteins, genes encoding enzymes for the metabolism of 4-hydroxybutyrate, and genes encoding the biotin synthetic pathways. Sequencing of the flanking DNA of some clones reveals a gene or a group of genes that can be used to infer the phylogenetic affiliation of the organism from which the DNA in the clone was isolated. In the absence of a highly conserved gene, such as one encoding 16S rRNA, DNA polymerase or RecA-codon usage and sequence alignment of gene clusters with genes in the databases can be used to make phylogenetic inferences, although these are less robust than those based on conserved genes.

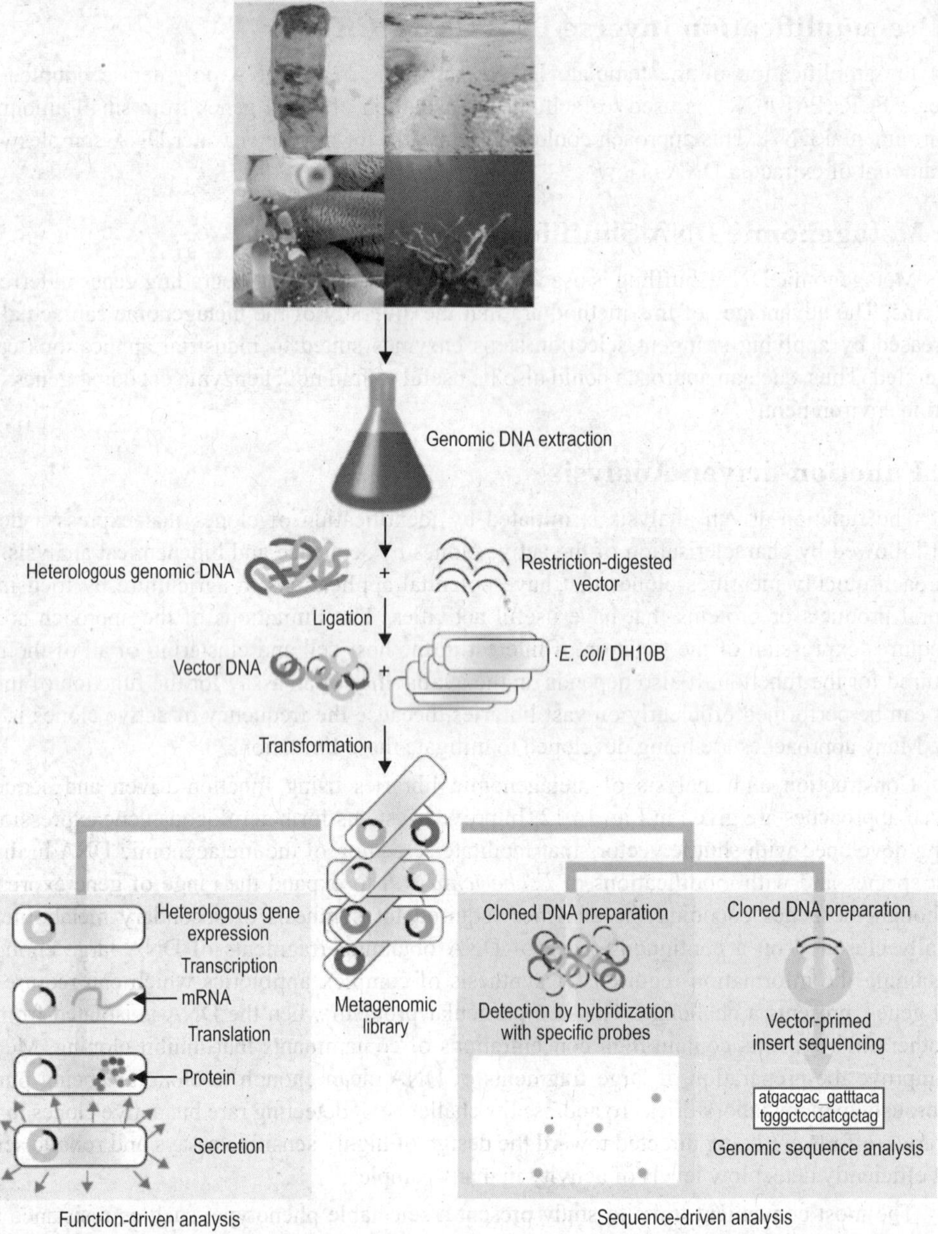

Fig. 14.12. Construction and analysis of metagenomic libraries using function-driven and sequence-driven approaches.

12. Sequence-driven Analysis

Sequence-driven analysis relies on the use of conserved DNA sequences to design hybridisation probes or PCR primers to screen metagenomics libraries for clones that contain sequences of interest. Sequencing of clones carrying phylogenetic anchors, such as the 16S rRNA gene and the archaeal DNA repair gene *radA*, has led to functional information about the organisms from which these clones were derived using 16S rRNA genes to identify clones obtained from the archaea and employing sequencing to compile genomic information about them (Fig. 14.12). The sequence conservation of regions of phylogenetic anchors facilitates their isolation without prior knowledge of the full gene sequence. By contrast, the sequences of most genes of practical importance are far too divergent to make it possible to identify new homologues by PCR or hybridisation. However, a few classes of genes contain conserved regions to facilitate their identification by sequence instead of function.

Two examples are the genes encoding polyketide synthases (PKSs) and peptide synthetases, which contribute to synthesis of complex antibiotics. The PKSs are modular enzymes with repeating domains containing divergent regions that provide the variation in chemical structures of the products. These regions are flanked by highly conserved regions which have provided the basis for designing probes to identify PKS genes among metagenomic clones. Mixing and matching PKS domains from different sources has yielded new antibiotics, stimulating interest in the discovery of new PKS genes. Similar approaches have identified genes encoding light-harvesting complexes from uncultured bacteria and *radA* and DNA polymerase genes from uncultured archaea. Thus, application of this approach is likely to be broadened as patterns emerge that define more gene families containing highly conserved features.

DNA fragmentation is a significant problem when constructing metagenomic libraries. The vigorous extraction methods required for high yields of DNA from environmental samples often result in excessive DNA shearing. This precludes the construction of libraries using cohesive ends because highly sheared DNA (e.g. 0.5–5 kb fragments) cannot be restricted to generate ligatable sticky ends without significant loss of the total gene complement. An alternative approach uses blunt-end or T-A ligation to clone randomly sheared metagenomic fragments.

Cosmid library and bacterial artificial chromosome (BAC) library (Fig. 14.13) have widely been used for the construction of metagenomic libraries. The ability to clone large fragments of metagenomic DNA allows the entire functional operons to be targeted with the possibility of recovering entire metabolic pathways.

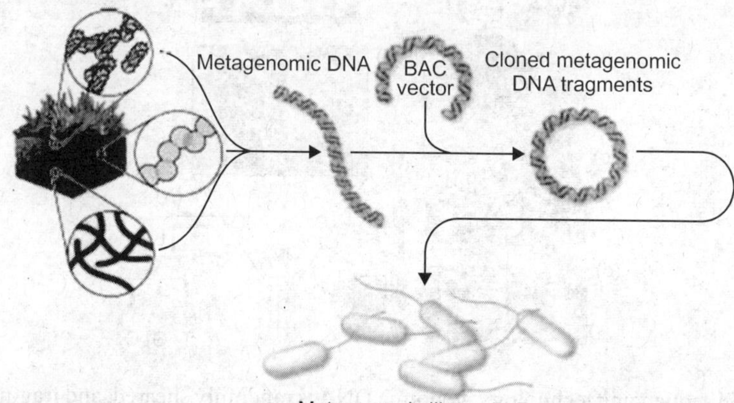

Fig. 14.13. Construction of metagenomic libraries using bacterial artificial chromosome (BAC) vectors.

BAC approach has successfully been applied to the isolation of several multigenic pathways such as that responsible for the synthesis of the antibiotic violacein. Fosmid vectors provide an improved method for cloning and stably maintaining cosmid size inserts (35–45 kb) in *E. coli*.

Phage-display expression libraries provide a means for isolating DNA sequences by affinity selection of the surface-displayed expression products. This method is efficient and amenable to high-throughput screening, offering the potential to enrich even rare DNA sequences in the metagenome. However, phage display is limited by the expression capacity of the bacteriophage, a protein size upper limit of around 50 kDa.

The limitation of *E. coli* as a host for comprehensive mining of metagenomic samples is highlighted by the low number of positive clones obtained during a single round of screening (typically < 0.01%). Moreover, it is virtually impossible to recover translational fusion products owing to the high number of clones (O107) that would need to be screened. Intuitively, expression from native promoters and read-through transcription from the vector-based promoter offer the best chance for recovery of heterologously expressing genes. Statistically, for a small insert (~10 kb) library, between 10^5 and 10^6 clones need to be screened for a single hit. This suggests that without sample enrichment the discovery of specific genes in a complex metagenome is technically challenging.

H. METAGENOMICS VIA NEXT GENERATION SEQUENCING

There are several metagenomic approaches employed for next generation sequencing, as described in this section.

1. 454 Sequencing Technology

The main principles of the 454 technology are an *in vitro* sample preparation and a miniaturisation of sequencing chemistry that enables to generate sequencing data in a parallel manner. Long DNA samples are randomly sheared into small fragments (e.g. by nebulisation) of sizes between 500 and 1000 bp. For library preparation, an adaptor is added to each end of these fragments, respectively (Fig. 14.14). The 'single stranded DNA fragments' (sstDNA) are then used for the subsequent emulsion PCR (emPCR). This compartmentalised enzymatic amplification technique is essential because it allows for a clonal production of templates without the need for host cells like *E. coli.* Hence, the potential loss of sequence coverage due to the cloning bias is avoided.

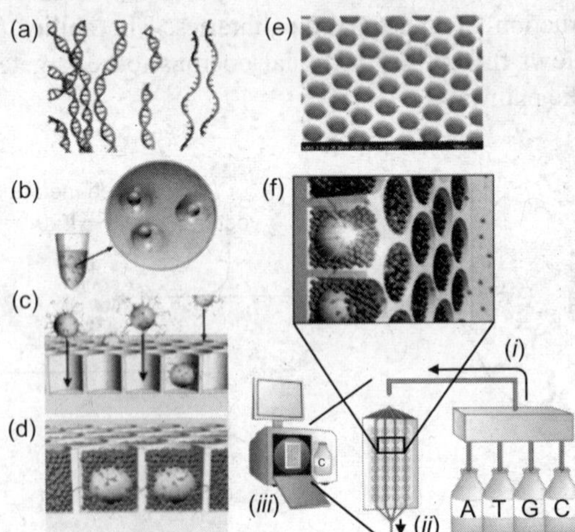

Fig. 14.14. 454 sequencing technology; genomic DNA is randomly sheared, and fragments (sstDNA) are ligated to adaptors (*a*). The single stranded DNA is attached to beads (*b*); an emPCR reaction amplifies the fragments within water-in-oil reactors (*b*); beads are deposited into wells (*c*), and layered with smaller beads carrying enzymes (*d*); scanning electron micrograph showing part of a PicoTiterPlate (*e*); a light signal is emitted by incorporating the sequentially owed nucleotides (*f*); the light signal is captured by a CCD camera.

At the same time, the DNA amplification ensures later to obtain high levels of signal differing from the background noise. For the sequencing phase, the beads carrying sstDNA are deposited into 3.5 million wells of a fibre-optic PicoTiterPlate™. The high amount of wells is a key factor to accomplish the high density and parallelism of the sequencing reactions. The wells are small enough to allow only one bead placement per well (55 μm in depth and a diameter of 44 μm), and such they act as individual reaction vessels where each reaction is isolated from other wells. In addition to the sstDNA beads, smaller beads carrying immobilised enzymes (ATP sulfurylase and luciferase) required for the pyrophosphate reaction are loaded into the wells. The sequencing itself is carried out by owing sequencing reagents (nucleotides and buffers) over the plate. Individual nucleotides are sequently owed (e.g. A-C-G-T-A-C-G-T...). If the current nucleotide is complementary to the template strand, the DNA polymerase incorporates one or more consecutive nucleotides. Incorporation of a complementary base generates pyrophosphate (PPi) which is converted to ATP by the sulfurylase. Luciferase uses the ATP to convert luciferin to oxyluciferin, resulting in light emission (Fig. 14.15).

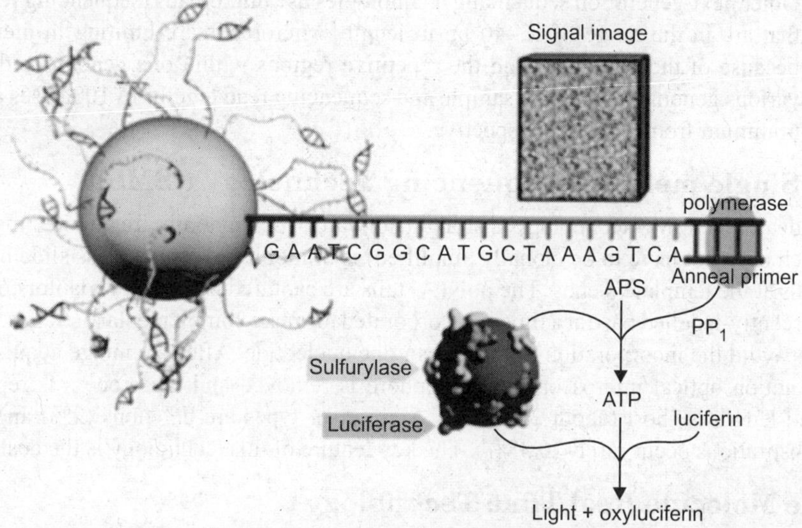

Fig. 14.15: 454 pyrosequencing reaction.

The number of incorporated nucleotides is proportional to the intensity of the emitted light signal generated by the sequencing reaction. The light signal for each flow is captured by a CCD camera. The resulting flow grams for each sstDNA bead in a single well are then translated into sequence space, *i.e.* the base sequence is derived from the intensity of the light signals for each nucleotide flow.

2. Genome Sequencer™ FLX System

The Genome Sequencer™ FLX System from 454 Life Sciences™ and Roche Applied Science is a versatile sequencing platform suitable for a wide range of applications, including *de novo* sequencing and assembly of genomic DNA, transcriptome sequencing, small RNA analysis, and amplicon sequencing. The Genome Sequencer FLX is built upon 454 Sequencing™ technology that allows 200–300 base-pair read lengths and very high single-read accuracy. The Genome Sequencer FLX System provides a comprehensive view of metagenomics samples with high throughput, no cloning bias, and read lengths long enough to allow diversity and functional analysis of microbial communities. This system offers several advantages for metagenomic studies as given below:

(*i*) This tool eliminates the requirement to clone DNA fragments into bacteria which consequently avoids the cloning bias that is introduced in sample preparation observed in Sanger sequencing–based methods.

(*ii*) With over 4,00,000 sequencing reads per instrument run, this system can facilitate extensive surveys to identify large numbers of different genes, metabolic pathways and microbial species that may be present while providing a dramatic reduction in the cost per project. This allows researchers to approach samples and address questions that only major genome centers could manage. As a result, the Genome Sequencer FLX System offers a powerful technology which supports research studies to answer the environmental and ecological questions.

(*iii*) For metagenomic studies sequencing read length is also an important factor. As most genomes in metagenomic samples are unknown and relatively little reference sequence information is available, it is important to have read lengths that allow researchers to both assemble genomes in a *de novo* fashion and uniquely assign reads to a specific gene and/or genome. The Genome Sequencer FLX generates sequencing read lengths of 200-300 bases with read length being dependent on the specific sequence characteristics. Other next-generation sequencing technologies use microreads (sequencing read lengths) that are in the range of 15–40 bp in length. Microreads are limiting in metagenomics because of the homology and the repetitive regions within one genome and across the various genomes within the sample and sequencing read lengths of 100 bases are near the minimum from a utility perspective.

3. True Single-molecule Sequencing Technology (tSMS)

The idea of the tSMS technology is straight-forward. The randomly fragmented metagenomic DNA (which does not have to be clonally amplified) is immobilised onto a glass slide by attaching poly(A) tails at the template's ends. The poly(A) tails are captured by poly(T) adaptors on the slide. One fluorescently-labelled base at a time is incorporated to create complementary strands. Reversible terminators avoid the incorporation of more than one nucleotide. After the image acquisition using a high-resolution optical microscope, the fluophore is removed and the process is repeated. The typical read length is short (about 24-32 bp). Main error types are deletions (2%) and insertions (1.2%); substitutions occur rarely (0.38%). The key feature of this technology is the cost-efficiency.

4. Single Molecule Real Time Technology

The sequencing of metagenome occurs in zeta liter (10^{-21}) wells which contain the immobilised single DNA molecules (Fig. 14.16). Sequencing begins after adding a high concentration of fluorescently labelled nucleotides and the polymerase. As the complementary strand grows, the fluophores are cleaved and the light signal is captured by a highly focused detection system. Due to the small well and the high nucleotide concentration, the sequencing rate is expected to be fast, e.g. 10 bases/second. This technology is able to generate reads that are thousands of nucleotides long, at the expense of overall output. Currently, average read lengths of 568 bp (maximum 2,805 bp) are reported which would really take (meta) genomic research to the next level.

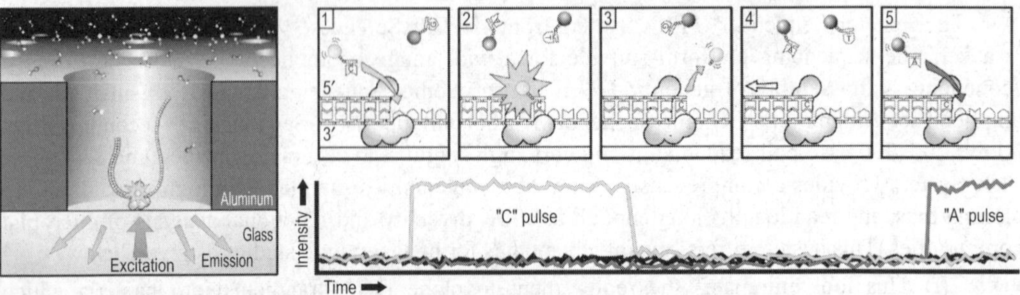

Fig. 14.16. Single molecule real time technology; the sequencing reaction takes place in a zepto liter well enabling the real time monitoring of base incorporations.

5. Oxford Nanopore Technologies

Nanopore sequencing is a fairly different approach to read the DNA molecule. Instead of capturing light signals with CCD cameras to detect base incorporations, the company Oxford Nanopore Technologies modified an α-haemolysin nanopore (1 nm of inner distance) for the detection of individual nucleotide monophosphates. DNA is digested by an exonuclease attached to the pore. Each nucleotide falling through the pore blocks the electrical current that runs through this pore, because each nucleotide has characteristic current amplitude, one is able to read the sequence base-by-base. Similar to Single Molecule Real Time Technology, the nanopore technology has the potential to generate long reads, *i.e.* up to several thousand base pairs. Moreover, read accuracy is quite high (99.8%), and the error correction is expected to be straight-forward. The massive parallelisation of this technology (hundreds of thousands of pores on an array) will be able to produce large-volume sequencing data in short time (http://www.nanoporetech.com).

I. METAGENOMICS VIA BIOINFORMATICS

Development of metagenomics as a branch of science was facilitated by the fact that cultured microorganisms represents no more than 1% of the microorganisms present in the vast majority of environmental habitats. This evidence was derived from analyses of 16S rRNA gene sequences amplified directly from the environment, an approach that avoided the bias imposed by culturing and led to the discovery of vast new lineages of microbial life. Metagenomics facilitates the study of the physiology and ecology of environmental microorganisms. But the recovery of DNA sequences longer than a few thousand base pairs from environmental samples was very difficult until recent advances in molecular biological techniques, particularly related to constructing libraries in BACs, provided better vectors for molecular cloning. Advances in bioinformatics, refinements of DNA amplification, and proliferation of computational power have greatly aided the analysis of DNA sequences recovered from environmental samples. Some brief explanation of available tools used in metagenome analysis is as follows:

1. CAMERA

The Community Cyberinfrastructure for Advanced Microbial Ecology Research and Analysis (CAMERA) is a user-driven site dedicated to provide the scientific community with metagenomic data and analysis tools. The CAMERA enables the microbial ecology community to overcome challenges in sequencing data analysis due to vast amounts of data and the intrinsic complexity of environmental populations. Besides providing computational resources for researchers, the CAMERA Cyberinfrastructure resource includes primary sequence data and associated environmental parameters or metadata. Its computational resources include rich, distinctive bioinformatics tools and a flexible collaborative workflow environment that allows researchers to identify gene type, clean up contaminations, annotated functional genes, assign taxonomy to genes and facilitate researcher's understanding of the biological implications of the experimental data.

CAMERA has integrated 23 billion base pairs of sequencing data along with the available metadata, yielding the largest metagenomic data and metadata resource. It includes all environmental metagenomic projects deposited in NCBI, as well as additional ongoing projects. Researchers may query on diverse metadata categories, such as habitat, sample collection location, time, salinity, pH, and other environmental physicochemical measurements, to match their own needs. In addition, CAMERA incorporates pre-computed annotation results of metagenomic sequences with reference data from PFam, COG, Gene Ontology, etc. CAMERA provides a rich tool environment where users can utilise and create specific analysis pipelines through a scientific workflow system. Users are able to view and download results and browse the step-by-step details of how results are generated (*i.e.* the complete provenance of each analysis is recorded).

CAMERA has developed and integrated tools for sequence quality control, assembly, gene prediction, annotation, and applications of functional and comparative genomics. It also provides high-performance networking access and grid-based computing. The CAMERA data repository contains a sizeable amount of rotating storage and large computational clusters. CAMERA provides a platform for users to upload their own workflows, supporting the tool integration from the broader research communities. CAMERA also provides a data submission tool that allows researchers to share and forward data to other metagenomic sites, and to other community archives such as GenBank to ensure the wider access to this data and any annotations performed on these data (http://camera.calit2.net).

2. CARMA

CARMA (Characterising Short Read Metagenomes) is a software pipeline for characterising the taxonomic composition and genetic diversity of short-read metagenomes. It is a software pipeline for the characterisation of species composition and the genetic potential of microbial samples using short and unassembled reads. The software was originally designed for the analysis of environmental metagenomes obtained by the ultra-fast 454 pyrosequencing system. The recently developed parallel 454 pyrosequencing technique gives the opportunity to rapidly obtain metagenomic sequences at a low cost and without cloning bias. However, the phylogenetic analysis of the short reads produced represents a significant computational challenge. The phylogenetic algorithm CARMA is to predict the source organisms of environmental 454 reads. The algorithm searches for conserved Pfam domain and protein families in the unassembled reads of a sample.

These gene fragments (environmental gene tags, EGTs) are classified into a higher-order taxonomy based on the reconstruction of a phylogenetic tree of each matching Pfam family. The method exhibits high accuracy for a wide range of taxonomic groups, and EGTs as short as 27 amino acids can be phylogenetically classified up to the rank of genus (http://www.cebitec.uni-bielefeld.de/brf/carma/carma.html).

3. CLC bio Workbench

The advanced Workbenches from CLC bio involving integrated framework, is used to support work processes in a number of scientific purposes such as design of cloning strategies, primers, assembly of contigs obtained from clones containing environmental DNA, creation of local databases, BLAST searches in both local, and NCBI databases.

Below is a case study on how software from CLC bio is used by scientists at BioGenia to help in accessing information of microbial biofilm metagenomes. The analyses are performed as shown in the illustration of the work flow that includes the use of CLC software (Fig. 14.17). The work flow alternates between tasks performed in the laboratory and bioinformatics tasks where the CLC Workbench is used.

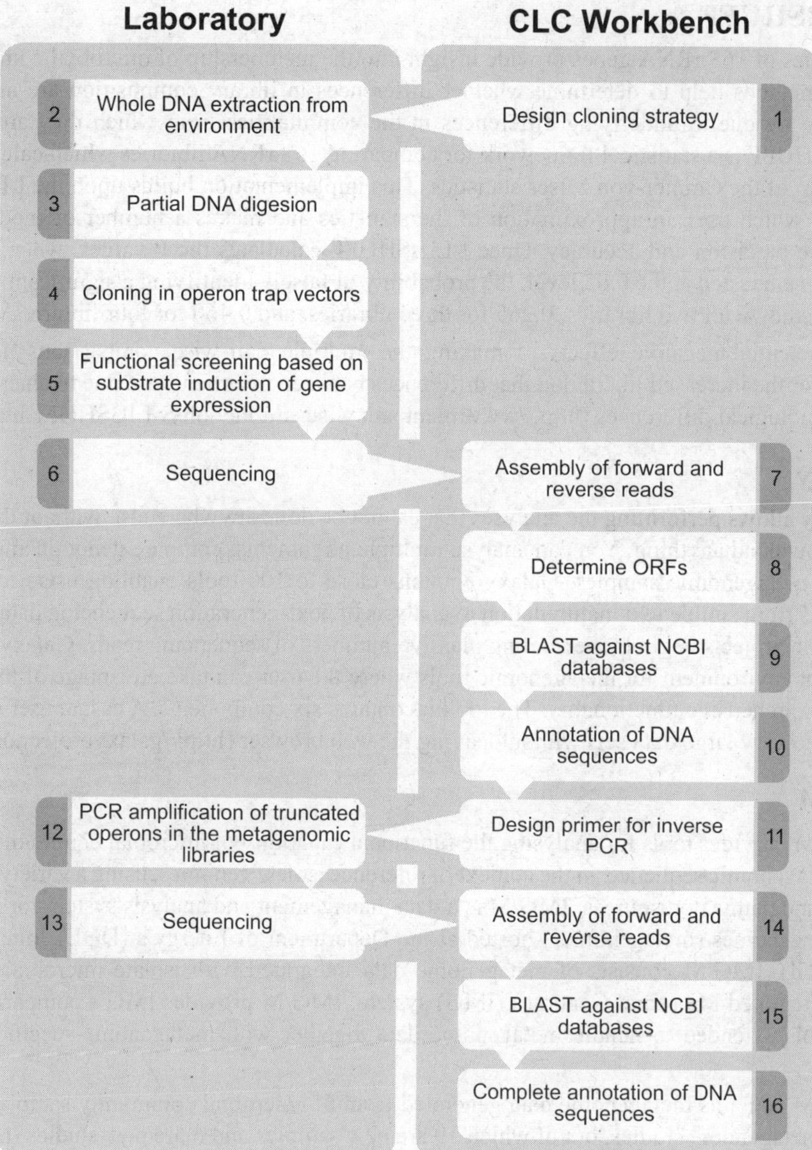

Fig. 14.17. Illustration of typical work flow.

4. DOTUR

DOTUR is a computer programme to define the operational taxonomic units and to estimate species richness. This method overcomes the challenge of assigning sequences to operational taxonomic units (OTUs) based on the genetic distances between sequences. DOTUR assigns sequences to OTUs by using either the furthest, average or nearest neighbor algorithm for each distance level.

DOTUR uses the frequency at which each OTU is observed to construct rarefaction and collector's curves for various measures of richness and diversity (http://www.plantpath.wisc.edu/fac/joh/dotur.html).

5. ∫ -LIBSHUFF

Libraries of 16S rRNA genes provide insight into the membership of microbial communities. Statistical methods help to determine whether differences in library composition are artifacts of sampling or are due to underlying differences in the communities from which they are derived. The ∫-LIBSHUFF is a statistical framework for comparing 16S rRNA libraries which calculates the integral form of the Cramér-von Mises statistics. This implementation builds upon the LIBSHUFF programme which uses an approximation of the statistics and makes a number of modifications that improve precision and accuracy. Once ∫-LIBSHUFF calculates the P values, when pair-wise comparisons are tested at the 0.05 level, the probability of falsely identifying a significant P value is 0.098 for a study with two libraries, 0.265 for three libraries, and 0.460 for four libraries.

The potential negative effects of making the multiple pair-wise comparisons necessitate correction for the increased likelihood that differences between treatments are due to chance and do not reflect biological differences (http://www.plantpath.wisc.edu/fac/joh/S-LIBSHUFF.html).

6. Galaxy

Galaxy allows performing the analyses that cannot be done anywhere else without the need to install or download anything. You can analyse multiple alignments, compare genomic annotations, and profile metagenomic samples. Galaxy contains close to 100 tools enabling users to perform diverse tasks from simple text manipulation to analysis of next-generation sequencing data. As most metagenomic projects begin by generating massive numbers of sequencing reads, Galaxy provides a convenient environment for metagenomic tools where the user can take advantage of the already existing comprehensive functionality. The toolkit contains six components. A casual user can apply this toolkit to very large data sets without leaving the web browser (http://galaxyproject.org).

7. IMG/M

IMG/M provides tools for analysing the functional capability of microbial communities based on their metagenomic sequence, in the context of reference isolate genomes, using a variety of public functional and pathway resources. IMG/M is a data management and analysis system for microbial community genomes (metagenomes) hosted at the Department of Energy's (DOE) Joint Genome Institute (JGI). IMG/M consists of metagenome data integrated with isolate microbial genomes from the Integrated Microbial Genomes (IMG) system. IMG/M provides IMG's comparative data analysis tools extended to handle metagenome data together with metagenome-specific analysis tools.

IMG/M contains metagenome data generated from 65 microbial community samples that are part of 20 metagenome studies, out of which 19 are new samples and 6 are new studies (http://img.jgi.doe.gov/m).

8. Metagenomics RAST Server (MG-RAST)

Metagenomics RAST server (*Metagenome* Rapid Annotation using Subsystem Technology) is a public resource for the automatic phylogenetic and functional analysis of metagenomes. MG-RAST is a fully-automated service for annotating metagenome samples. It provides: (*i*) annotation of sequence fragments, (*ii*) their phylogenetic classification, *(iii)* metabolic reconstructions, and (*iv*) comparison tools. The service is built as a modified version of the RAST server which was originally designed to support high-quality annotation of complete or draft microbial genomes.

The open-source metagenomics RAST service provides a new paradigm for the annotation and analysis of metagenomes. With built-in support for multiple data sources and a back end that houses abstract data types, the metagenomics RAST is stable, extensible, and freely available to all researchers. This service has removed one of the primary bottlenecks in metagenome sequence analysis, *i.e.* the availability of high-performance computing for annotating the data. The MG-RAST server is an open source system based on the SEED framework for comparative genomics. Users can upload raw sequence data in FASTA format; the sequences will be normalised and processed and summaries automatically generated. Genome annotation systems are ever evolving; therefore, in order to accommodate new methods that may be developed, the pipeline was designed with a modular framework that allows the rapid addition of new analysis steps or comparative data at any stage of the analysis. The server provides several methods to access the different data types including phylogenetic and metabolic reconstructions, and the ability to compare the metabolism and annotations of one or more metagenomes and genomes. In addition, the server offers a comprehensive search capability. Access to the data is password protected, and all data generated by the automated pipeline is available for download and analysis in variety of common formats (http://metagenomics.nmpdr.org/).

9. MetaSim

MetaSim (sequencing simulator for genomes and metagenomes) can be used to generate collections of synthetic reads that reflect the diverse taxonomical composition of typical metagenome data sets. Based on a database of given genomes, the programme allows the user to design a metagenome by specifying the number of genomes present at different levels of the NCBI taxonomy, and then to collect reads from the metagenome using a simulation of several different sequencing technologies. A population sampler optionally produces evolved sequences based on source genomes and a given evolutionary tree. The resulting data sets can be used as standardised test scenarios for planning and sequencing projects, or for benchmarking assembler and metagenomic software (http://www-ab.informatik.uni-tuebingen.de/software/metasim).

MetaSim : (*i*) integrates a database for source genome sequences, (*ii*) generates sets of synthetic reads or mate-pairs based on adaptable sequencing error models (e.g. for Sanger chemistry, Roche's 454 and Illumina (former Solexa), (*iii*) enables the user to configure abundance values for each organism to model specific taxon compositions, (*iv*) provides a population sampler to generate evolved sequences based on source genomes and a given evolutionary tree, and (*v*) can be controlled via graphical user interface or in command line mode.

10. MEGAN

MEGAN is a new computer programme that allows laptop analysis of large metagenomic data sets. In a preprocessing step the set of DNA sequences is compared against databases of known sequences using BLAST or another comparison tool (Huson *et al.,* 2007). MEGAN is then used to compute and explore the taxonomical content of the data set, employing the NCBI taxonomy to summarise and order the results. A simple lowest common ancestor algorithm assigns reads to taxa such that the taxonomical level of the assigned taxon reflects the level of conservation of the sequence. The software allows large data sets to be dissected without the need for assembly or the targeting of specific phylogenetic markers. It provides graphical and statistical output for comparing different data sets (http://www-ab.informatik.uni-tuebingen.de/software/megan).

11. MetaStats

MetaStats is a statistical method for handling frequency data to detect differentially abundant features between two populations. This method can be applied to the analysis of any count data generated through molecular methods, including random shotgun sequencing of environmental samples, targeted sequencing of specific genes in a metagenomic sample, digital gene expression surveys (e.g. SAGE), or even whole-genome shotgun data (e.g. comparing the depth of sequencing coverage across assembled genes). It can also be generalised to experiments with more than two populations by substituting the *t*-test with a one-way ANOVA test.

Furthermore, if only a single sample from each treatment is available, a Chi-square test can be used instead of the *t*-test. In the coming years metagenomic studies will increasingly be applied in a clinical setting, requiring new algorithms and software tools to be developed that can exploit data from hundreds to thousands of patients. The methods described above represent an initial step in this direction by providing a robust and rigorous statistical method for identifying organisms and other features whose differential abundance correlates with disease (http://metastats.cbcb.umd.edu).

12. STAMP

STAMP (Statistical Analysis of Metagenomic Profiles) is a software package for analysing metagenomic profiles (e.g. a phylogenetic profile indicating the number of marker genes assigned to different taxonomic units or a functional profile indicating the number of sequences assigned to different biological subsystems or pathways) that promotes 'best practices' in choosing appropriate statistical techniques and reporting results. It encourages the use of effect sizes and confidence intervals in assessing biological importance. A user friendly, graphical interface permits easy exploration of statistical results and generation of publication quality plots for inferring the biological relevance of features in a metagenomic profile. STAMP is open source, extensible via a Plugin framework, and available for all major platforms (http://kiwi.cs.dal.ca/Software/STAMP).

13. SONS

SONS is a tool for operational taxonomic unit-based comparisons of microbial community memberships and structures. SONS implements nonparametric estimators for the fraction and richness of OTUs shared between two communities. SONS is a freely available computer code written in C^{++} that uses the 'list' output file from DOTUR as an input as well as a file containing a paired list of each sequence identifier and a designation from the library from which the sequence was generated. The DOTUR-derived list file can be replaced with an analogous file if sequences were assigned to OTUs by an alternative method, as long as the file formatting is retained. The order of sequence identifiers is used to construct collector's curves of the estimates from equations 1 to 9 for each pair wise comparison and distance level used to define OTUs.

The ordering of sequences can be randomised in the software. Where possible, SONS calculates the 95% confidence interval for the final estimate (http://www.plantpath.wisc.edu/fac/joh/sons.html).

14. SmashCommunity

SmashCommunity is a stand-alone metagenomic annotation and analysis pipeline suitable for data from Sanger and 454 sequencing technologies. It supports state-of-the-art software for essential metagenomic tasks such as assembly and gene prediction. It provides tools to estimate the quantitative phylogenetic and functional compositions of metagenomes, to compare compositions of multiple metagenomes and to produce intuitive visual representations of such analyses (http://www.bork.embl.de/software/smash).

15. TreeClimber

TreeClimber is a test to compare microbial community structures. TreeClimber is adaptation to a method used in population genetics, the parsimony test, to determine the relatedness of communities. TreeClimber is available free as C^{++} source code for compilation in Linux or Mac OSX and as a binary executable for the Windows operating system. Two input files are required to run the programme: a PHYLIP or NEXUS-formatted Newick file containing one or more phylogenetic trees and a file containing the sequence name and its corresponding treatment. The user can select the number of randomly joining trees desired to construct the test distribution. A probability distribution for any number of treatments and replicates can be determined without input of an observed set of phylogenies or set of sequence identifiers.

TreeClimber generates a file containing the probability of observing each parsimony score for the user and randomly joining trees (http://www.plantpath.wisc.edu/fac/joh/treeclimber.html).

16. Treephyler

Treephyler is a tool for fast taxonomic profiling of metagenomes. Assessment of phylogenetic diversity is a key element to the analysis of microbial communities. Tools are needed to handle next-generation sequencing data and to cope with the computational complexity of large-scale studies. Treephyler was evaluated on real metagenome to assess its performance in comparison to previous approaches for taxonomic profiling. Results indicate that Treephyler is in terms of speed and accuracy prepared for next-generation sequencing techniques and large-scale analysis (http://www.gobics.de/fabian/treephyler.php).

J. APPLICATION OF METAGENOMICS

Metagenomics has changed the way that microbiologists approach many problems, redefined the concept of a genome and accelerated the rate of gene discovery. The potential for application of metagenomics to biotechnology seems endless. Functional screens have identified new enzymes and antibiotics and other reagents in libraries from diverse environments. A number of barriers have limited the discovery of new genes that provide insight in to microbial community structure and function or that can be used to solve medical, agricultural or industrial problems. Schematic representation of application of metagenomics is shown in Fig. 14.18.

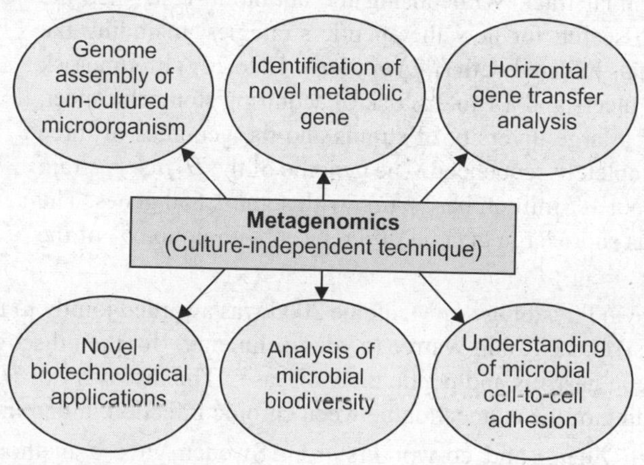

Fig. 14.18. Schematic representation of application of metagenomics.

1. Agricultural Application

There are lots of microorganisms present on the plants, which play an important role in plant growth by providing essential nutrients. Bacteria present in the soil not only increase the soil fertility but also convert the atmospheric nitrogen into ammonia. Some microbes decompose the dead animals and plants, and the others convert the elements like iron and manganese to the nutrients which can easily be used by plants as nutrients.

Suppressive soil is a phenomenon in which the harmful and beneficial microbes live at one place and they do not affect the health of the plants. If the soil is sterilised by killing the harmful bacteria, the health of the plants will be affected and they will die eventually because harmful microbes may penetrate in the soil and cause death of the plants. Scientists have observed that suppressive soil microbes provide protection to the plants and if they are removed, the plants become susceptible to infections. The microbial community of suppressing soil is beneficial in agriculture. Scientists believe that not only species of microbes make the soil suppressed but various species are involved in this process. The science of metagenomics enables the scientists to explore how the microbial species are beneficial for plants. It will help them to produce more healthy crops.

2. Human Health

Human beings are always surrounded with microbes. Microbes not only reside around the humans but they also live within the human body. According to an estimate, there are ten-fold of microbes living in human body as compared to the total number of human cells. Microbes living in the body contain genes more than the human genes. Metagenomics enables the scientists to develop such tools and guidelines in human and animal nutrition and also the drug discovery. It also make possible to study dangerous diseases like cancer and certain immune disorders. Microbes are the essential part of the human body as they help in the digestion of food by living in intestines, and produce certain vitamins and amino acids. Some microbes stay in the stomach and control the pH, and acid level of the stomach. Metagenomics provides opportunities to study both beneficial and harmful microbes. It can also be helpful in identifying the other compounds released or produced by the microbes which are beneficial for the animal and human health.

Helicobacter pylori is a Gram-negative and microaerophilic bacterium found in human stomach. More than 50% of the world's population harbour *H. pylori* in their upper gastrointestinal tract. Infection is more prevalent in developing countries; *H. pylori* is a major cause of certain diseases of the upper gastrointestinal tract. With raising the antibiotic resistance the need is increasing to search for new therapeutic strategies, including the vaccination. Individuals infected with *H. pylori* have 10 to 20% lifetime risk of developing peptic ulcers and a 1 to 2% risk of acquiring stomach cancer. *H. pylori* consists of a large diversity of strains and the genomes of three strains have been completely sequenced. The genome of the *H. pylori* strain 26695 consists of about 1.7 million base pairs, with some 1,550 genes. The two sequenced strains show large genetic differences, with up to 6% of the nucleotides differing.

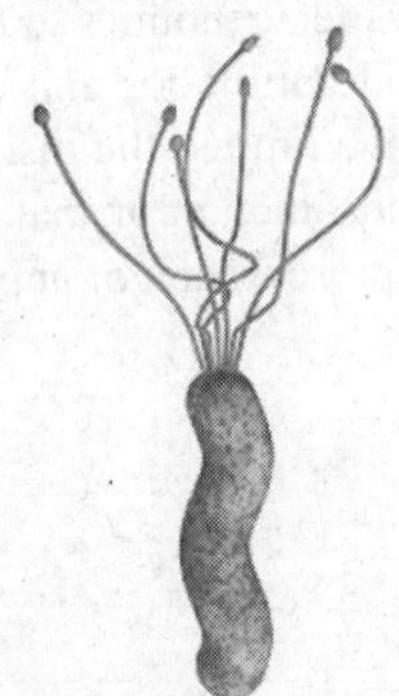

Helicobacter pylori

The Nobel Prize in Physiology or Medicine 2005 was awarded jointly to Barry J. Marshall (an Australian pathologist) and J. Robin Warren (a young clinician) 'for their discovery of the bacterium *H. pylori* and its role in gastritis and peptic ulcer disease'. The discovery of *H. pylori* has led to an increased understanding of the connection between chronic infection, inflammation and cancer.

In 2010, Zongli Zheng and co-workers from Sweden gave a method for metagenomics of *H. pylori* from archived formalin-fixed gastric biopsies permitting Longitudinal studies of carcinogenic risk. Metagenomic profiles of *H. pylori* in two FFPE biopsies (both aged 15 years) from two patients obtained by laser microdissection (LCM) were modified Roche 454 pyrosequencing. After re-culture they compared the frozen homogenised biopsies from these two patients. They found that DNAs from *H. pylori* dissected from FFPE sections had the highest identity with the culture DNAs from the same host (~ 98%) than with other reference genomes (93 ~ 95%).

Barry J. Marshall

J. Robin Warren

3. Environment

Microbes living in the environment are helpful in many ways. They decompose dead materials and are the source of cleaning the environment from pollutants. There are some microbes which use methane gas which seeps through the landfills and some are related with the wastewater treatments.

Microbes are also the source of cleaning the ground water by eliminating gasoline leaked from the petroleum pipes. The analysis of metagenomics enables to identify the particular species of microbes concerned with the water treatment purposes.

4. Industrial Application

Metagenomics has the potential to high impact on antibiotic production. Many of the important products like amidases, oxidoreductases/dehydrogenases production, vitamin biosynthesis, polysaccharide degrading or modifying enzymes, etc. Some recent applications of metagenomics are given in Table 14.1.

Table 14.1. Application of metagenomics in industry.

Target gene/natural product	*Source*	*Vector used for library construction*	*Host strain*
Alcohol oxidoreductase	Meadow, sugar beet field, crop land soil	pBluscript SK	*E. coli* DH5 alpha
Aminoglycoside and tetracycline antibiotic resistance	Soil	pjN105, pCF430	DH10B
Amylase	Soil	pUC19	*E. coli* DH5 alpha
Amidase	Soil and enrichment culture	pZero-2	*E. coli* TOP10
Carbonyl formation	Soil	Not recorded	*E. coli*
Co-enzyme B12-dependent glycerol	Soil	pBluscript SK	*E. coli* DH5 alpha
Esterase	Soil	pCCIFOS	*E. coli* EP1300 TI
Esterase, lipase	Meadow soil	pBluscript SK	*E. coli* DH5
Lipase	Soil	pEpiFOS5	*E. coli* EP1-100
Novel biocatalysts	Unplanted field soil	Cosmid	*E. coli*
Polyketide synthase	Soil	Cosmid	*E. coli*

Source: Brady *et al.* (2004), Ricsenfeld *et al.* (2004).

Two previous studies reported the successful screening of soil metagenomic libraries for indirubin, an active ingredient in a Chinese herbal medicine used to treat chronic myelocytic leukemia. While a range of novel antibiotics have been detected in metagenomic libraries. A clone found in a soil metagenomic library produces deoxyviolacein and the broad spectrum antibiotic violacein. During the screening of seven different soil metagenomic libraries, it was reported that these libraries exposéd 11 clones producing long chain N-acyltyrosine antibiotics, and analysis of their synthases indicated that ten of them were novel (Brady *et al.,* 2004). Metagenomic libraries have also been used for isolating natural antibiotic resistance genes. Riesenfeld *et al.* (2004) identified nine aminoglycoside and one tetracycline antibiotics resistance genes from soil.

QUESTIONS

1. What is Metagenomics ? Write a brief note on history of metagenomics ?
2. Write a rationale on 'genomics to metagenomics' ?
3. Explain the three different metagenomics project with their impact ?
4. Describe different techniques used for metagenomics study.

5. Give a detailed account of metagenomic gene discovery tools.

6. Discuss in detail about metagenomics via bioinformatics by using various tools.

7. With the help of diagram describe any three tools for metagenomics analysis.

8. What is next generation metagenomic analysis ?

9. What is the application of metagenomics ?

10. Write short notes on the following:

(*i*) Microbial metagenomic project, (*ii*) Environmental metagenomic project,
(*iii*) Hydrocarbon metagenomic project, (*iv*) Screening of metagenomics clones,
(*v*) Integrons, (*vi*) Fluorescent *in situ* hybridization (FISH),
(*vii*) Function-driven analysis for metagenomic library,
(*viii*) Sequence-driven analysis for metagenomic library,
(*ix*) 454 sequencing technology, (*x*) MEGAN,
(*xi*) CAMERA, (*xii*) Microarray,
(*xiii*) MG-RAST, (*xiv*) Stable-isotope probing,
(*xv*) *Heliobacter pylori*.

BIBLIOGRAPHY

Brady; S. F.; Chao, C. J. and Clardy, J. 2004. Long chain N-acyltyrosine synthases from environmental DNA. *Appl. Environ. Microbiol.* **70:** 6865 – 6870.

Huson, D.H. *et al.* 2007 MEGAN analysis of metagenomic data. *Genome Res.* **17:** 377-386.

Riesenfeld, C.S.; Goodman, R.M. and Handelsman, J. 2004. Uncultured soil bacteria are a reservoir of new antibiotic resistance genes. *Environ. Microbiol.* **6:** 981 – 989.

15 CHAPTER Enzyme Technology

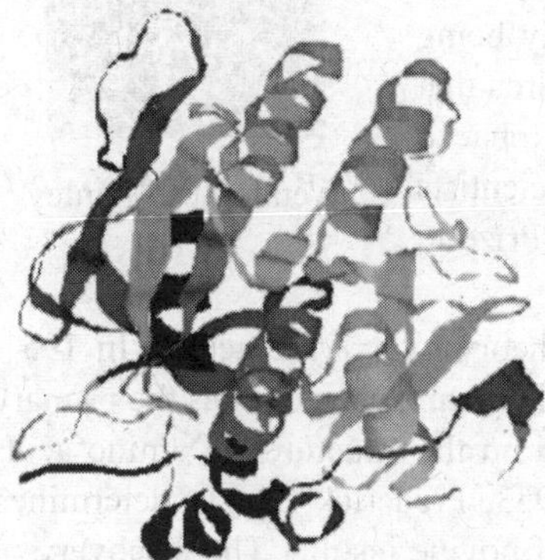

Beta-Subtilisin

A. Enzyme structure
B. Properties of enzymes
C. Novel enzymes : abzymes, synzymes and extremozymes
D. Classification and nomenclature of enzymes
E. Mechanism of enzyme catalysis
F. Enzyme kinetics
G. Immobilization of enzymes
H. Application of enzymes, immobilized enzymes and whole cells
I. Biosensor
J. Enzyme engineering

By the late 17th and early 18th centuries digestion of meat by stomach secretions and the conversion of starch into sugars by plant extracts and saliva were known, but the mechanism of its occurrence was unknown. In the 19th century, when studying the fermentation, Louis Pasteur found that fermentation of sugar to alcohol by yeast was catalysed by a vital force present within the yeast cells to which he called 'ferments'. The ferments were thought to function only within living organisms.

In 1877, Wilhelm Kühne (a German physiologist) first used the term 'enzyme' (Greek, means 'in leaven') to describe leavening process. Later the word *enzyme* was used for nonliving substances for example pepsin, and the word *ferment* was used for chemical activity produced by living organisms. The term enzyme (*en = in; zyme = yeast*; means 'in yeast') was referred to denote the reaction wherein the production of ethyl alcohol and carbon dioxide by the enzyme zymase, present in yeast takes place. The reaction is popularly known as 'alcoholic fermentation'.

Eduard Buchner

In 1897, Eduard Buchner at the University of Berlin found that the yeast extracts alone (without any living yeast cells) fermented the sugar. He conducted several experiments and found the occurrence of sugar fermentation even without the living yeast cells in the mixture. He named the enzyme that carries out sucrose fermentation as 'zymase'. In 1907, he was awarded the Nobel Prize in Chemistry for the discovery of cell-free fermentation. Thereafter, enzymes are named according to the reaction they carry out following Buchner's example. The suffix *-ase* is added at the end of the name of its substrate (for example lactase, *i.e.* the enzyme that break down lactose) or the type of reaction (for example DNA polymerase, *i.e.* the enzyme that polymerises the formation of DNA molecules).

The biochemical nature of enzymes was determined only after studying their function outside a living cell. Several early workers found the association of enzymatic activity with proteins, but many scientists including a Nobel laureate Richard Willstätter reported that proteins were only the carriers of the enzymes. Hence proteins could not carryout catalysis. In 1926, James B. Sumner showed that the enzyme urease was a pure protein which was further crystallised. In 1937, he isolated and crystallised a second enzyme, catalase. In 1929, John Howard Northrop of the Rockefeller Institute obtained the other crystalline enzymes (starting with pepsin) following the similar methods.

Wendell M. Stanley

During 1930s, John H. Northrop and Wendell M. Stanley, working on the digestive enzymes pepsin, trypsin and chymotrypsin, proved that pure proteins can definitively be enzymes. Besides, Stanley being a virologist also isolated the nucleoprotein of tobacco mosaic virus that caused mosaic disease in tobacco, crystallised it and displayed tobacco mosaic virus activity in healthy plants. Hence, these three scientists (Sumner, Northrop and Stanley) were awarded the 1946 Nobel Prize in Chemistry.

In 1948, Linus C. Pauling put forth the transition state theory of enzyme action. In 1951, Pauling, Robert Corey and Herman Branson proposed the alpha helix and beta sheet as the primary structural motifs in protein (enzyme) secondary structure based on the structures of amino acids and peptides and the planar nature of the peptide bond. In 1953, Frederick Sanger determined the complete amino acid sequence of two polypeptide chains of bovine insulin. The discovery of crystallisation of enzymes allowed solving their structures by X-ray crystallography, which was started first with lysozyme by David Chilton Phillips and co-workers in 1965. Lysozyme is an enzyme found in tears, saliva and egg whites that digests the cell wall of some bacteria. The X-ray crystallography helped to get the high resolution structure of lysozyme. This approach marked the field of structural biology and understanding of enzyme working at atomic levels. In 1986, Thomas R. Cech discovered catalytic RNA (ribozyme), while Richard Lerner and Peter Schutlz developed catalytic antibodies (abzymes).

Enzymes are proteins that catalyse the chemical reactions and occur in all living cells. Every cell of the body consists of more than five hundred different enzymes, and each of them helps the cell and the body to work properly. Enzymes increase the rates of such reactions by a factor of 10^3 - 10^{16} that occur naturally at a very slow rate, but they cannot induce a reaction that would not happen normally. Enzyme technology constitutes the use of purified enzymes for generating a useful product or service.

Several enzymes are used in the preparation of many different foods; for example, amylases present in yeast cells are used in brewing and baking to convert sugars into alcohol and carbon dioxide. Proteases are used in tenderising the meat. Rennin is essential to create cheese. Invertase is used to make some soft-centred sweets such as thin chocolate covered mints. A thick and solid paste of sucrose is converted into a mixture of glucose and fructose which creates a soft centre to the chocolate.

Enzymes can be isolated from organisms, purified and then used for various purposes. Application of enzymes in biological washing powders is one of the very common uses at the home. Washing powders contain amylase, lipases and proteases, and break down any stains that contain carbohydrate, fat and protein. Therefore, it digests biological stains from the clothes at a low temperature, saves electricity and makes good for delicate fabrics.

A. ENZYME STRUCTURE

Generally enzymes are globular proteins; they range in size from 62 amino acid residues (monomer of 4-oxalocrotonate tautomerase) to about 2,500 residues (in animal fatty acid synthase). Although the 3-D structure of enzymes determines their function, prediction of a novel enzyme activity from its structure is a very difficult problem that has not been solved so far. A comparison between enzymes and non-biological catalysts is given in Table 15.1.

Table 15.1. Comparison between enzymes and non-biological catalysts.

Feature	*Enzymes*	*Nonbiological catalysts*
Specificity	Highly specific; binding to substrate may be less specific but the catalysed reaction is highly specific	Nonspecific
Rate of reaction	Enhanced by a factor of 10^3-10^6	Enhancement only a fraction of that by enzymes
Regulation	Enzymes are subject to a variety of regulations, which increase/ decrease reaction rate	They are not subject to such regulations
Saturation	Enzymes have a maximum reaction rate with respect to substrate concentration	Most catalysts do not show substrate staturation
Chemical nature	Enzymes are proteins (RNA with enzymatic action is called ribozyme)	Metal and nonmetal inorganic molecules
Temperature, pressure and pH	Mild, biologically compatible	Often high temperature and high pressure
Side reactions	Do not occur	Occur

1. Cofactors and Coenzymes

Enzymes may or may not have a non-protein molecule attached to them. There are some enzymes that consist of carbohydrate groups covalently attached to it which do not affect the catalytic activity, but they may influence enzyme stability or solubility. Many enzymes possess metal ions, whereas some others contain low weight organic molecules called 'cofactors', which are essential for enzyme activity.

(*a*) Cofactors: Cofactors are the non-protein chemical compounds (organic or inorganic) attached to a protein which is required for biological activity. Cofactors can be divided into two broad groups: *organic cofactors* (e.g. flavin or haeme), and *inorganic cofactors* (such as the metal ions Mg^{2+}, Cu^{+}, Mn^{2+} or iron-sulfur clusters). Further the organic cofactors are classified depending on how tightly they bind to an enzyme. The loosely-bound cofactors are termed as *coenzyme* and tightly-bound cofactors as *prosthetic groups*. The term coenzyme refers mainly to the functional properties of a protein. The term 'prosthetic group' denotes the nature of the binding of a cofactor to a protein, *i.e.* it is tightly bound or covalently attached to it. Hence, the prosthetic group refers to the structural property of a protein. Without cofactor an inactive enzyme is called an apoenzyme, whereas the complete enzyme with cofactor is called holoenzyme.

In addition, different sources provide slightly different definitions of coenzymes, cofactors, and prosthetic groups. Some workers consider the tightly-attached organic molecules as prosthetic groups (not as coenzymes), while the others define all non-protein organic molecules needed for enzyme activity as coenzymes, and classify those tightly bound as coenzyme prosthetic groups.

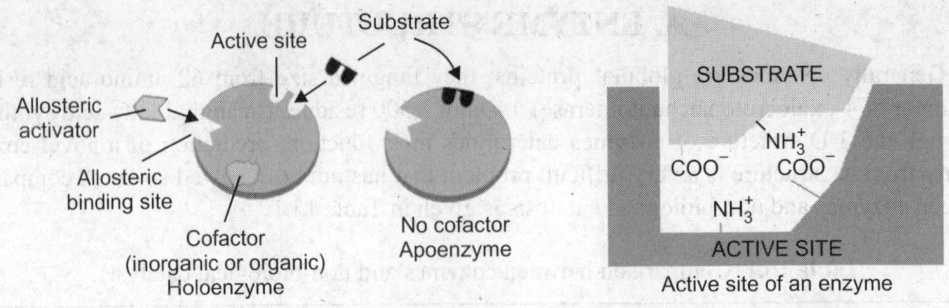

Fig. 15.1. Different sites of enzymes.

An organic cofactor is commonly known as 'coenzyme'. Cofactors and coenzymes may be covalently or non-covalently attached to the protein molecule called apoenzyme, which lacks its catalytic activity in the absence of specific cofactor. When a cofactor is so tightly bound to the apoenzyme that it becomes difficult to remove it without damaging the enzyme, the cofactor is often called a *prosthetic group* (*i.e.* a tightly bound non-polypeptide structure required for the activity of an enzyme or other protein, for example the haem of haemoglobin). Generally both coenzymes and cofactors contribute to enzyme activity as well as stability. The complex of an apoenzyme and the cofactor is known as holoenzyme.

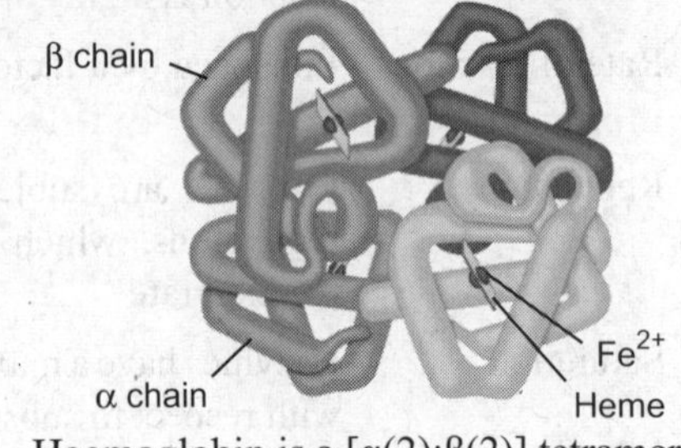

Haemoglobin is a [α(2):β(2)] tetramer

Active Sites and Substrate Interaction: It is very important to understand the nature of the active site in an enzyme. Since enzymes are proteins, the nature of amino acid side chains in the vicinity of the active site is important. The specific amino acid side chains of many enzymes have been determined. The principles involved in the active site for carboxypeptidase A is shown in Fig. 15.2.

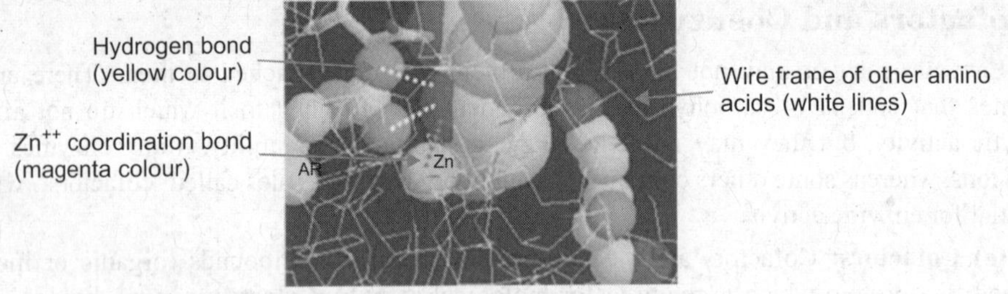

Fig. 15.2. Interaction of substrate with active site of carboxypeptidase A.

The substrate (space filling gray, blue red) can interact with the active site through opposite charges, hydrogen bonding (*see* yellow colour), hydrophobic non-polar interaction, and coordinate covalent bonding to the metal ion activator (*see* magenta colour). The sequence position of the amino acid in the protein is indicated by the numbers behind the amino acids. In the enzyme, white lines represent the wire frames of the other amino acids (Fig. 15.2). The Zn ions interact with the carbonyl bond and activate it resulting in addition of -OH from water to the carbonyl, production of an acid, and rupturing of the C-N bond.

***(i)* Inorganic cofactors (Metalloproteins):** There are many metal ions that are common cofactors. You have studied about essential trace elements; they play a role as cofactors. Example of some metalloproteins is given in Table 15.2. In humans the common metal ions as cofactors are : Fe^{2++}, Mg^{2++}, Mn^{2++}, Co^{2++}, Cu^{2++}, Zn^{2++} and Mo^{2++}. Besides, iodine is also an essential trace element, but it is a part of thyroid hormones rather than as an enzyme cofactor. Ca^{2++} is also a part of human diet, but its role in full activity of many enzymes such as nitric oxide synthase, protein phosphatases, and adenylate kinase, is important. Often Ca^{2++} binds to these enzymes in a complex with calmodulin, and activates them in allosteric regulation. Therefore, calcium is a cell signalling molecule, and not usually considered a cofactor of the enzymes.

Table 15.2. Inorganic metal cofactors.

Ion	*Examples of enzymes containing ion*
Cupric	Cytochrome oxidase
Ferrous or Ferric	Catalase Cytochrome (via haeme), nitrogenase, hydrogenase
Magnesium	Glucose 6-phosphatase, hexokinase, DNA polymerase
Manganese	Arginase
Molybdenum	Nitrate reductase
Nickel	Urease
Selenium	Glutathione peroxidase
Zinc	Alcohol dehydrogenase, carbonic anhydrase, DNA polymerase

There are several other organisms that require additional metals as enzyme cofactors, for example vanadium in the nitrogenase of the nitrogen-fixing bacteria (such as *Azotobacter*), tungsten in the aldehyde ferredoxin oxidoreductase of the thermophilic archaean *Pyrococcus furiosus*, and cadmium in the carbonic anhydrase from the marine diatom *Thalassiosira weissflogii*.

***(ii)* Organic cofactors:** Organic cofactors are small molecules (less than 1,000 Da) that can be loosely or tightly bound to the enzyme, and participate in the reaction. When cofactors are tightly bound, it is difficult to remove them unless the enzyme is denatured; then they are called a prosthetic group. The loosely and tightly bound cofactors cannot be sharply divided. For example, NAD^+ can bind tightly with some enzymes, whereas it is loosely attached to others. Similarly, thiamine pyrophosphate (TPP) tightly binds in pyruvate decarboxylase, whereas it is less tightly bound in pyruvate dehydrogenase. In contrast, the other coenzymes such as flavin adenine dinucleotide (FAD), biotin, and lipoamide are covalently bound. Generally, the tightly-bound cofactors are regenerated during the same reaction cycle, while loosely-bound cofactors can be regenerated in a subsequent reaction catalysed by a different enzyme.

Besides having other functions in the body, some vitamins (such as B_1, B_2, B_6, B_{12}, niacin, folic acid) act as precursor to many organic cofactors, or as coenzymes themselves (e.g. vitamin C). There are many organic cofactors that consist of a nucleotide such as NAD and FAD, and coenzyme A; they carry acyl groups. Some of the cofactors are universal to all forms of life and most of these are found in a huge variety of species. Some of the organic molecules that act as cofactors are given in Table 15.3.

Table 15.3. Vitamins and non-vitamins as cofactors (source: http://en.wikipedia.org/wiki/Enzyme).

Cofactor	*Vitamin*	*Chemical group(s) transferred*	*Distribution*
		A. Vitamins cofactors	
NAD^+	Niacin (B_3)	Electrons	Bacteria, archaea and eukaryotes
Lipoamide	Lipoic acid	Electrons, acyl groups	Bacteria, archaea and eukaryotes
Methylcobalamin	Vitamin B_{12}	Acyl groups	Bacteria, archaea and eukaryotes
Cobalamine	Cobalamine (B_{12})	Hydrogen, alkyl groups	Bacteria, archaea and eukaryotes

Biotin	Biotin (H)	CO_2	Bacteria, archaea and eukaryotes
Coenzyme A	Pantothenic acid (B_5)	Acetyl group and other acyl groups	Bacteria, archaea and eukaryotes
Menaquinone	Vitamin K	Carbonyl group and electrons	Bacteria, archaea and eukaryotes
Ascorbic acid	Vitamin C	Electrons	Bacteria, archaea and eukaryotes
Coenzyme F420	Riboflavin (B_2)	Electrons	Methanogens and some bacteria
		B. Non-vitamins	
Adenosine triphosphate		Phosphate group	Bacteria, archaea and eukaryotes
Coenzyme B		Electrons	Methanogens
Coenzyme M		Methyl group	Methanogens
Coenzyme Q		Electrons	Bacteria, archaea and eukaryotes
Glutathione		Electrons	Some bacteria and most eukaryotes
Haeme		Electrons	Bacteria, archaea and eukaryotes

***(b)* Coenzymes:** Coenzymes are small organic molecules that can be loosely or tightly bound to an enzyme. The tightly bound coenzymes can be called 'allosteric groups'. Coenzymes transport chemical groups from one enzyme to another. Human body does not synthesise vitamins; hence, these are acquired from the diet. The chemical groups include: the hydride ion (H^-) of NAD or $NADP^+$, the phosphate group of ATP, the acetyl group of coenzyme A, formyl or methyl groups of folic acid and the methyl group carried by *S*-adenosylmethionine. Coenzymes are chemically changed due to enzyme action. Therefore, coenzymes can be considered as a special class of substrates which are common to many different enzymes. There are about 700 enzymes that use the coenzyme NADH.

***(i)* Coenzyme-regenerating systems:** Usually coenzymes are continuously regenerated and their concentrations are maintained at a steady level inside the cell; for example, NADPH is regenerated through the pentose phosphate pathway and *S*-adenosylmethionine by methionine adenosyltransferase. This continuous regeneration means very intensive use of even small amounts of coenzymes. Coenzymes (e.g. NAD^+, $NADP^+$, NADH, NADPH, and ATP) are utilised by many oxidoreductases and all ligases.

***(ii)* Immobilised regenerating coenzymes:** Coenzymes represent some of the most useful biological catalyst; their application is presently limited by the high cost and difficulties with their regeneration. If the coenzyme is immobilised together with the enzyme and regenerated *in situ,* these two problems may be overcome. Use of whole cell systems is a simple way of immobilising the regenerating coenzymes that can widely be applied. They have lower efficiency and flexibility than immobilised-enzyme systems. Membrane reactors are used for immobilisation of coenzymes but their pore size should be smaller than the diameter of coenzyme. For adequate immobilisation and regeneration coenzymes must be derivatised. This process activates the coenzymes for attachment to the immobilisation support but does not interfere with its biological function. The most widely applied synthetic routes involve the alkylation of the exocyclic N^6-amino nitrogen of the adenine moiety present in the coenzymes NAD^+, $NADP^+$, NADH, NADPH, ATP and coenzyme A.

B. PROPERTIES OF ENZYMES

Like proteins, enzymes are long, linear chains of amino acids that fold to form a three-dimensional structure. Each amino acid sequence governs a specific structure resulting in unique properties. However, sometimes individual protein chains may group together and form a protein complex. Most of the enzymes can be denatured by heating or chemical denaturants which disrupt the inactivation and folding of 3-D structure. Denaturation may be reversible or irreversible depending on the enzyme structure.

Enzymes are similar to catalysts in the following respects : (*i*) they lower the activation energy of reaction, (*ii*) they do not participate in the reaction, and return to their original form at the end of reaction, and (*iii*) they increase the rate of reaction. But enzymes increase the rate of reaction at a phenomenal scale. They are highly specific to distinguish between the similar parts of the substrate molecule (regiospecificity), and between optical isomers of the substrate (streospecificity). In addition, enzymes are subject to a variety of regulations, and their reaction rates show substrate saturation, which is not the case with catalysts (non-biological). Enzymes are attractive because they operate under mild conditions of temperature, pressure and pH, which save energy

1. Presence of Species Specificity

Usually enzymes are very specific to : (*i*) type of reactions they have to catalyse, and (*ii*) the substrates involved in these reactions. Specificity is determined by complementary shape, charge and hydrophilic/hydrophobic characteristics of enzymes and substrates. Enzymes are species specific because macromolecules including proteins differ in different species. It may be explained that the phylogenetic development, which has given rise to microbiological variation, is caused by variation in these molecules. Different types of enzyme (*e.g.* protease, α-amylase, lactase) found in many species have varying properties; for example protease of two closely related species differs in several ways in spite of certain similarities.

Some of the enzymes show the highest specificity and accuracy and are associated with copying and expression of the genome. In 2002, I.V. Shevelev and U. Hubscher found that these enzymes have 'proof-reading' mechanisms; for example DNA polymerase catalyses a reaction in a first step and then checks and corrects the errors (if any) in the product in a second step. This two-step process results in average error rates of less than 1 error in 100 million reactions in high-fidelity mammalian polymerases. In 2000, M. Ibba and D. Soll have also found the similar proofreading mechanisms in RNA polymerase, aminoacyl tRNA synthetases and ribosomes.

A few enzymes exhibit absolute specificity, *i.e.* they catalyse only one particular reaction. However, the other enzymes are specific for a particular type of chemical bond or functional group. Generally, there are four distinct types of specificity:

- ***Absolute specificity:*** In absolute specificity the enzyme catalyses only one reaction.
- ***Group specificity:*** In this case the enzyme acts only on such molecules that have specific functional groups such as amino, phosphate and methyl groups.
- ***Linkage specificity:*** In linkage specificity the enzyme acts on a particular type of chemical bond irrespective of the rest of the molecular structure.
- ***Stereochemical specificity:*** In this case the enzyme acts on a particular steric or optical isomer.

Since enzymes exhibit the great degrees of specificity, cofactors may serve many apoenzymes. For example, nicotinamide adenine dinucleotide (NAD) is a coenzyme for many dehydrogenase reactions in which it acts as a hydrogen acceptor, such as the alcohol dehydrogenase, malate dehydrogenase and lactate dehydrogenase reactions.

2. Variation in Activity and Ability

Most of microbial enzymes are extracellular; therefore, they are applied in various ways. However, extracellular enzymes are influenced externally by temperature, pH, etc. But their optimum stability and activity are very much close to optimum conditions for microbial growth. For example, optimum temperature and pH for amylase activity of a thermophilic microbial species (e.g. *Bacillus coagulans*) differ from that of mesophilic species of some microbe (*B. licheniformis*). In contrast to extracellular enzymes, the intracellular enzymes are little influenced by external environmental factors.

(*a*) Effects of pH: Enzymes are affected by changes in pH. The point where the enzyme is most active is known as the optimum pH, *i.e.* the most favorable pH value. The optimum pH value varies greatly from one enzyme to another as given in Table 15.4.

Table 15.4. Different pH values for optimum activity of some enzymes.

Enzyme	*Optimum pH*
Lipase (pancreas)	8.0
Lipase (stomach)	4.0 - 5.0
Lipase (castor oil)	4.7
Pepsin	1.5 - 1.6
Trypsin	7.8 - 8.7
Urease	7.0
Invertase	4.5
Maltase	6.1 - 6.8
Amylase (pancreas)	6.7 - 7.0
Amylase (malt)	4.6 - 5.2
Catalase	7.0

The optimum pH for pepsin is 2.0 and for trypsin 8. Generally at a very high or low pH values activity of most of enzymes are completely lost. The pH also acts as a factor in providing stability to the enzymes. There is also a region of pH optimal stability for each enzyme.

The activity and stability of some enzymes also differ. Xylose isomerase is stable at pH range from 4.0 to 8.5 but shows optimum activity at pH between 5.5 and 7.0. Similarly, temperature also influences enzyme activity. On increasing the temperature enzyme activity gradually increases, but at certain stages temperature inactivates the rate of reaction and finally enzyme is denatured (at high temperature) as it is proteinaceous in nature. Thermal stability in the target enzyme may be a useful attribute during enzyme production as heat may be used to destroy the activity of contaminant enzyme. However, for most enzymes the effective pH range is 4-9. Beyond these limits denaturation of enzymes takes place (Fig. 15.3).

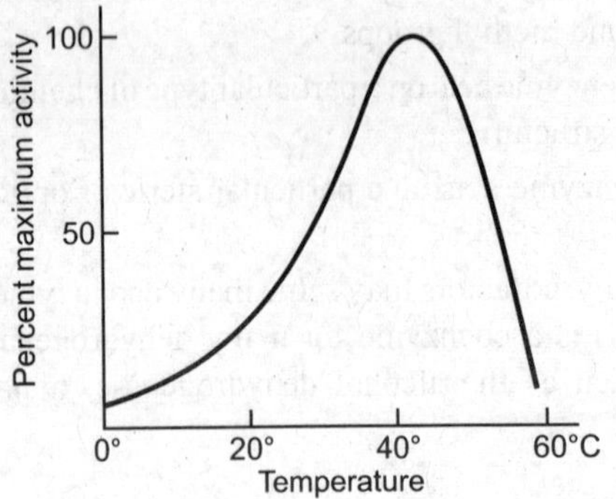

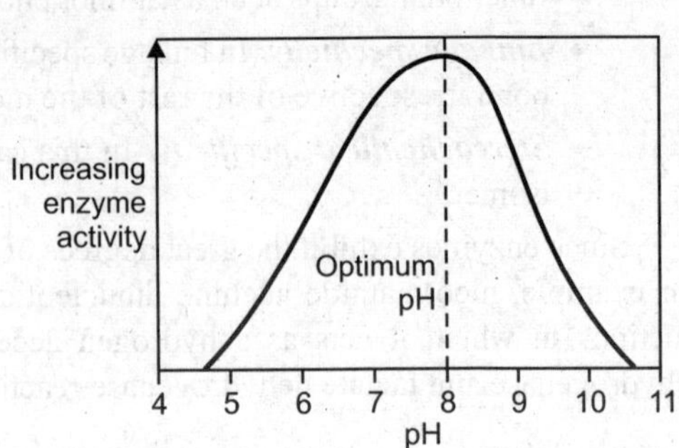

Fig. 15.3. Effect of pH on enzyme activity **Fig. 15.4.** Effect of temperature on enzyme activity.

(*b*) Effect of Temperature: Temperature greatly affects the activity of enzyme. For example, if the temperature is increased by 10°C, the rate of most chemical reactions is doubled. However, there is loss of enzyme activity at 40°C - 60°C, because denaturation of proteins occurs at this temperature (Fig. 15.4).

(*c*) **Effect of Ions**: Enzyme activity is affected by H^+ ion concentration and their ionic concentrations. Some enzymes require a loosely bound cation such as Mg^{++}. In contrast α-amylase is denatured within 4 hours in the absence of Ca^{++}.

(*d*) **Effect of other Factors:** Besides, the stability of enzyme is also increased by many factors such as: (*i*) high concentration of respective enzymes (as protein aggregates and protects them), (*ii*) presence of their substrate and/or product (*e.g.* amylase shows more stability in the presence of starch than in its absence), and (*iii*) reduced amount of water content in reaction mixture (for example, at natural conditions β-glactosidase results in production of glucose and galactose from hydrolysis of lactose in whey. But the same enzyme produces some glucose and galactose and mixture of trisaccharides from the same concentrated whey).

If you plot a graph between initial reaction rate and concentration of a reactant, there are various possibilities depending on the relationship between the concentration and the rate. If the reaction rate is independent of the concentration, this is called a 'zero order reaction'. But if the reaction rate is proportional to the concentration, it is called a 'first order reaction'.

3. Substrate Specificity

Every enzyme is specific for a certain substance called the substrate, and the reaction; here the enzyme can only identify its substrate. However, the characteristics of its active site govern the specificity of enzymes. The active site is the region of the enzyme where it binds to the substrate before carrying out transformations of just one substrate or a family of structurally related substrates. When the enzyme acts only on one substrate, it shows absolute specificity for the substrate. It is the case of succinate dehydrogenase that is specific for succinate or the L-glutamate dehydrogenase specific for glutamate.

The enzyme shows relative specificity over substrate if it binds to some structurally related substrates. For example, the oxidation of different amino acid of the L series can be catalysed by L- amino acid oxidase, but not the oxidation of D- amino acids. Different groups (usually lateral chains of amino acids) related to the active site determines the binding of the enzyme to specific substrate(s) as given below :

- ***Backbone forming groups:*** These groups form the backbone and give an appropriate conformation for binding to substrate.
- ***Orientation groups:*** The orientation groups oblige the substrate(s) to adopt the appropriate orientation for the reaction.
- ***Environmental groups:*** The environmental groups give the hydrophobic, polar negative, or positive environment necessary for warranting an appropriate affinity between the enzyme and the specific substrate.
- ***Catalytic groups:*** The catalytic groups create the necessary tensions for breaking the old linkages and forming new ones among the metabolites implied in the reaction, such as: substrate(s), coenzymes or other prosthetic.

4. Activation and Inhibition

Some enzymes obtained from different sources show different responses to a given activator or inhibitor. For example, β-galactosidase isolated from fungi does not require cobalt, whereas the same of bacterial origin requires cobalt as a cofactor. Cobalt activates β-galactosidase isolated from bacteria and inhibits it when obtained from fungi. Examples of some activators of enzymes used commercially are : proteins (for proteases), starch (for α-amylase), cellulose (for cellulase) and pectin (for pectinase). There are ways through which enzyme activity is inhibited viz., denaturing or destroying the enzyme by heating, changing the pH, or by reaction with several chemicals.

Enzyme inhibitors are substances which alter the catalytic action of the enzyme and consequently slow down, or in some cases, stop catalysis. There are three common types of enzyme inhibition: *competitive*, *non-competitive* and *feedback inhibition*.

(*a*) Competitive Inhibition: Each enzyme binds only with its specific substrate. However, different molecules more-or-less similar to the substrate, are also able to bind to the active site of the enzyme. If such molecule is present in a concentration comparable to that of the substrate, it will compete with the substrate for bonding sites on the enzyme. Consequently it will interfere with the catalytic action of the enzyme and the substrate. This phenomenon is known as 'competitive inhibition' because the enzyme is inhibited by the inactive substrate or competitor molecule.

(*b*) Non-competitive Inhibition: Reaction with the inhibitors causes changes in shape of the active site. The non-competitive inhibitors do not attach directly to the active site, but elsewhere on the enzyme. The inhibitor attaches to a side group in the protein chain, and affects the protein folding into its tertiary structure. In turn it changes the shape of the active site. If the shape of the active site changes, the substrate cannot attach to it. The non-competitive inhibitors attach irreversibly or reversibly to the enzyme; if they attach irreversibly, they stop its working permanently.

Some compounds containing heavy metals such as Pb, Hg, Cu or Ag are poisonous because ions of these metals are non-competitive inhibitors for several enzymes. For example, Ag^{++} reacts with -SH groups in the side groups of cysteine residues of a protein chain. The electronegatively of Ag and S does not differ too much for a full ionic bond; hence they form covalent bond. If the cysteine residue is present in a protein chain in such a way that affect its folding into tertiary structure, the shape of the active site gets changed by changing this group resulting in stop of enzyme working. Similarly, the 2^+ ions from Hg, Cu or Pb also attach themselves to the sulphur in place of the hydrogen, and behave in the same way.

(*i*) Difference between allosteric regulation and non-competitive inhibition: The term allosteric is derived from the word allostery (Greek, *allos* means *other*, and *stereos* means *space*) which refers to the regulatory site of an allosteric protein located separately from its active site. Regulation of an enzyme or protein by binding an effector molecule at the allosteric site of a protein (*i.e.* a site other than active site of the protein) is called 'allosteric regulation'. The effectors that enhance the activity of protein are called 'allosteric activators', while the effectors decreasing the activity of proteins are referred to as 'allosteric inhibitors'. One of the natural examples of feedback control is the allosteric regulation.

Non-competitive inhibition involves an inhibitor molecule binding to the enzyme at a different area, changing the conformation of the enzyme and altering the shape of the active site. As a result the catalytic efficiency of the enzyme is reduced and the shape of the active site may be altered in such a way that stops the reaction. This type of inhibition may or may not be reversible. When the inhibitors are like poison, the inhibition is irreversible; for example penicillin causes irreversible inhibition of an enzyme needed to form the bacterial cell wall resulting in the death of bacteria.

(*c*) Feedback Inhibition (Allosteric Inhibition): There is a second binding site on the enzyme where the inhibitors bind. The structure of inhibitor is not necessarily similar to the substrate. The absence or presence of the inhibitor at this second binding site activates/deactivates the enzyme presumably by changing the conformation of the enzyme so that the active site is made available/unavailable to the substrate. Usually the inhibitor is the product of a reaction in the metabolic pathway. However, the activity of some enzymes involved in metabolic pathways is controlled by a self-regulating mechanism. Some specific substance acts as an inhibitor. Such inhibitor binds to an enzyme at a specific site and modifies the active sites. This inhibits the binding of substrate molecule. Such sites on the enzymes are called allosteric sites and the enzymes as allosteric enzymes.

Comparison between Competitive Inhibition and Feedback inhibition

Competitive inhibition and feedback inhibition differ in the following ways:

- The inhibitor is similar in structure to the substrate in competitive inhibition. The inhibitor binds to the enzyme at the active site and prevents the substrate from binding. But in feedback inhibition, the inhibitor binds to the enzyme at a site away from the active site. Then it acts upon enzyme by causing changes in the shape of the enzyme in such a way that it becomes unable to catalyse the reaction.
- Usually competitive inhibition involves inhibitors, commonly called poisons that do not belong in the cell, while feedback inhibition is a natural process by which an organism regulates the chemical reactions that take place in its cells.

For example, in the citric acid (TCA) cycle acetyl CoA combines with a molecule of oxaloacetic acid and form a molecule of citric acid or citrate. The enzyme citrate synthase (which catalyses this step) is inhibited by ATP. When the cell has a low concentration of ATP, it has low energy. In such condition the TCA cycle operates, feeds the electron transport chain and ATP is produced. When the concentration of ATP increases in the cell (more energy is stored in the cell), the TCA cycle shuts down because ATP inhibits the citrate synthase and stops the formation of citric acid. It means that high amount of ATP accumulation inhibits the TCA cycle.

5. Regulation (Control) of Enzyme Activity

Enzymes activities are regulated by the following five basic ways :

(*a*) Allosteric Control: There are different regulatory and catalytic binding sites in allosteric proteins. Binding of a substrate in one active site of allosteric proteins affects the activity of the rest of the binding sites because allosteric proteins are cooperative proteins. Depending on the biological needs, some substrate binding will favour the protein to be in the active R (relaxed) state, whereas the other substrate binding will favour the protein to be in the inactive state T (tense). The enzymes that are allosterically regulated, do not obey Michaelis-Menten kinetics but they follow the sigmoidal kinetics.

(*b*) Proteolytic Activation. In human body several enzymes are present in their inactive forms; such enzymes are called 'zymogen' or 'proenzyme'. The zymogens cannot be activated without cleaving by a digestive enzyme. Upon cleaving the three dimensional shape of the enzyme is changed resulting in the formation of the active site in the right orientation. Consequently the zymogens become activated in an irreversible reaction.

(*c*) Isoenzymes. There are different amino acid sequences in the isoenzymes but they catalyse the same reaction as enzymes. Usually their K_m and V_{max} values differ and also they have different regulatory mechanisms. The isoenzymes can catalyse the same reaction under the different environments within the different organelles that is an additional advantage. In metabolism isozymes are an important entity; for example lactate dehydrogenase (LDH) has two isozymes having 75% similar amino acid sequence. The H isozyme is expressed in the heart muscle, while the M isozyme is found in the skeletal muscle.

(*d*) Reversible Covalent Modification. When a different group is covalently attached to the active site of an enzyme, its activity can be changed. Due to such modification, binding of the natural substrate to the active site is blocked, for example phosphorylation/dephosphorylation, aceylation/deacylation, etc. However, all forms of covalent modification are not reversible. For example, in some proteins an attachment of a lipid group will inhibit the signal-transduction pathway.

(*e*) Control by Limiting Amount of Enzyme. The amount of enzymes gets produced can be controlled at the transcription level. In double displacement (*i.e.* ping pong reaction), two compounds switch places to form new compounds. Two reactants yield two products (for detail, *see* preceding section).

C. NOVEL ENZYMES

There are number of possibilities for the construction of artificial enzymes. Some of them have been briefly described herewith.

1. Abzymes

An abzyme (derived from antibody and enzyme) are antibodies that function as enzymes by catalysing specific chemical reactions. Basically, they are catalytic monoclonal antibody hence also called *catmab.* Usually abzymes are constructed artificially, but they also occur in normal humans (anti-vasoactive intestinal peptide autoantibodies) and in patients with autoimmune diseases such as systemic lupus erythematosus where they bind and hydrolyse DNA. Antibodies recognise and bind to the ground states of the specific molecules. In contrast, enzymes have binding sites that preferentially bind to the transition state of their substrate molecules. For this reason abzymes are not produced naturally. However, a catalytic antibody is produced in response to molecules having similar structure as proposed for transition state of the substrate of the reaction to catalyse for which the antibody is sought. Production of such catalytic antibodies is the strongest evidence in support of transition theory of enzyme action (http://en.wikipedia.org/wiki/Abzyme).

***(a)* Naturally Occurring Abzymes:** Catalytic activities-containing antibodies were isolated from sera of patients having different types of diseases. The first natural abzymes were obtained from human serum through purification of antibodies. Antibodies showing protease activity against vasoactive intestinal peptide (VIP) were first isolated from the serum of patient suffering from asthma. Besides, antibodies showing similar catalytic activity were obtained by immunising mice with VIP. These monoclonal antibodies demonstrated that the catalytic activity is borne by light chain of the antibody molecule. The other protease activities were characterised for hydrolysing VIII factor in haemophilia patients infused with homologous factor VIII. DNA hydrolysing autoantibodies were also isolated from the sera of patients suffering from systemic lupus erythematosus or rheumatoid arthritis. The DNA-hydrolysing activity could be correlated to the presence of high levels of anti-topoisomerase I antibodies in the sera of patients (http://en.wikipedia.org/wiki/Abzyme).

***(b)* Application of Abyzme Technology:** The specific targeting of cancer cells for its destruction is the most exciting application of abyzme technology. Cancer cells contain tumour cell antigens (on their surface lacking in normal cells) which are the unique determinants. Cancer drugs can be delivered directly to the tumour by utilising antibodies that specifically bind to these tumour cell antigens. In the case of abzymes antibodies with two distinct antigen binding sites have been envisioned (Fig. 15.5); one site binds firmly to a tumour cell antigen and the second site catalyzes the cleavage of a prodrug.

The prodrug is a non-toxic precursor of a cytotoxic drug given to the cancer patients. First, the antibody is administered to patients that bind the tumour cells with high affinity. Secondly, the prodrug is introduced into the bloodstream, but only becomes activated in the vicinity of the targeted antibody. By this technique, tumours are selectively destroyed, while healthy cells are spared from the toxic affect of cancer drugs. In June 2008, S. Planque, Sudhir Paul, and Yasuhiro Nishiyama, of the University of Texas Medical School at Houston, in the journal *Autoimmunity Reviews*, announced that they have engineered an abzyme that degrades the superantigenic region of the gp120 CD4

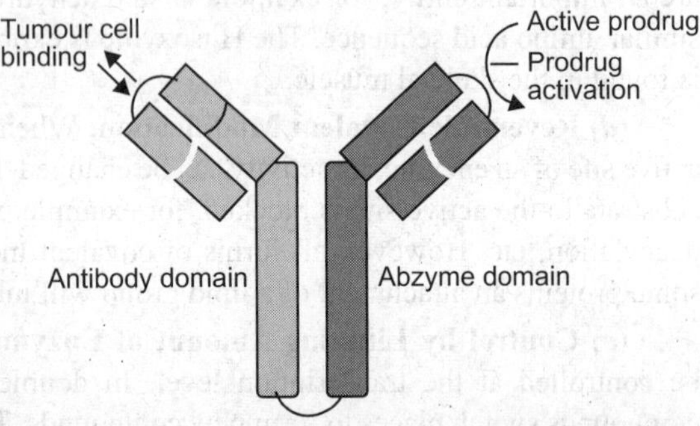

Fig. 15.5. Tumour-targetting anti-cancer abzyme.

binding site, which is one part of outer coating of the HIV that does not change because it is the attachment point to T lymphocytes.

2. Synzymes (= Artificial Enzymes)

Synzymes are synthetic polymers having enzymatic activities and follow the Michaelis-Menten kinetics similar to natural enzymes. Synzymes possess two structural entities: a substrate-binding site, and a catalytically effective site. Production of substrate binding facility is relatively straightforward but catalytic sites are more difficult. Both sites may be designed separately; if the synzyme has a binding site for the reaction transition state, this achieves both the functions. There are some synzymes that are derivatised proteins, for example the derivatised myoglobin by attaching $(Ru(NH_3)_5)^{3+}$ to three surface histidine residues. This converts from an oxygen carrier to an oxidase via oxidising ascorbic acid and reducing molecular oxygen. Synzyme is effective similar to natural ascorbate oxidases.

At present conformations of synzymes are not predictable from their primary structure. Therefore, it is impossible to design protein synzymes from scratch with any probability of success. Such proteins are also sensitive to denaturation, oxidation and hydrolysis, showing the drawbacks of natural enzymes. For example, polylysine binds anionic dyes only 10% as strongly as the natural binding serum protein (albumin), in spite of the many charges and apolar side-chains. Polyglutamic acid shows synzymic properties. It acts as an esterase similar to acid proteases showing a bell-shaped pH-activity and Michaelis-Menten kinetics.

Cyclodextrins (Schardinger dextrins) are naturally occurring toroidal molecules consisting of six, seven, eight, nine or ten α-1, 4-linked D-glucose units joined head-to-tail in a ring (α-, β-, γ-, δ- and ε-cyclodextrins, respectively. Usually synzymic cyclodextrins are derivatised to introduce catalytically relevant groups and many such derivatives have been examined. A C-6 hydroxyl group of β-cyclodextrin was covalently derivatised by an activated pyridoxal coenzyme. The resulting synzyme acts as transaminase and shows stereoselectivity for the L-amino acids.

3. Extremozymes

Extremozymes (the biocatalysts produced by extremophiles) are proteins that function under extreme conditions. Extremophiles are bizarre microorganisms that grow and thrive in extreme environments, which were formerly considered too hostile to support life. The extreme conditions may be high (121°C) or low (-5°C) temperature, high (14) or low (2) pH, high salinity (1-4 M NaCl), high metal concentrations, very low nutrient content, very low water activity, high radiation (5 Mrad), high pressure (38 MPa) and low oxygen tension. Some extremophiles are subject to multiple stress conditions.

Archaea have developed a variety of molecular strategies to survive the often harsh environments in which they exist, hence they are called extremophiles. Extremophiles are structurally adapted at the molecular level to withstand these harsh conditions. Due to their extreme stability extremozymes offer new opportunities for biocatalysis and biotransformation. Examples of extremozymes are : cellulases, amylases, xylanases, proteases, pectinases, keratinases, lipases, esterases, catalases, peroxidases and phytases which have great potential for application in various biotechnological processes. At present only 1–2 % of the microorganisms on the earth have been commercially exploited; there are only few examples of extremophiles. However, the renewed interest that is currently emerging as a result of new developments in the cultivation and production of extremophiles and success in the cloning and expression of their genes in mesophilic hosts will increase the biocatalytic applications of extremozymes (Gomes and Steiner, 2004).

Thermophilic enzymes have attracted most attention during the past four decades. Such enzymes are of great industrial and biotechnological interest due to better suited for harsh industrial processes. There are many advantages of conducting industrial processes at high temperature, such as the increased solubility of many polymeric substrates, resulting in decreased viscosity, increased bioavailability, faster reaction rate and the decreased risk of microbial contamination. These enzymes have also been used as models for the understanding of thermostability and thermoactivity, which is useful for protein engineering.

Thermostable DNA polymerases, isolated from hyperthermophiles, have led to a tremendous advancement in molecular biology due to their capacity to amplify DNA in polymerase chain reaction. A series of thermophilic glycosidases, sold commercially as CLONEZYMEs™, can be used to catalyse the formation of a range of linkages, including Ga-(1-4), Gal-(1-6), Gal-(1-6), D-Fuc-(1-3) and D-Fuc-(1-2) (Gomes and Steiner, 2004).

D. CLASSIFICATION AND NOMENCLATURE OF ENZYMES

There are different kinds of enzymes that are named in different ways. Most often enzymes are named by adding a suffix '*ase*' to the root word of the substrate; for example, lipase (fat hydrolysing enzyme), and sucrase (breaks down sucrose). Sometimes the enzymes are named on the basis of the reaction that they catalyse; for example, polymerase (carries out polymerisation), and dehydrogenase (removes H atoms). Some enzymes have been named on the basis of source from which they were first identified; for example, papayin from papaya. The name of some enzymes ends with '*in*' indicating that they are basically proteins; for example, pepsin, trypsin, etc.

1. Nomenclature Committee of the International Union of Biochemistry and Molecular Biology (NC-IUBMB)

The scheme for enzyme nomenclature was developed in 1955, when the International Congress of Biochemistry set up an Enzyme Commission (EC) in Brussels. The first version was published in 1961. The first Enzyme Commission in 1961 devised a system for classification of enzymes that also serves as a basis for assigning code numbers to them. These code numbers prefixed by EC are now widely used; it contains four elements separated by points with the following meaning : (*i*) the first number shows the six main divisions (classes) to which the enzyme belongs, (*ii*) the second figure indicates the subclass, (*iii*) the third figure gives the sub-subclass, and (*iv*) the fourth figure is the serial number of the enzyme in its sub-subclass.

The current sixth edition published by the International Union of Biochemistry and Molecular Biology in 1992, contains 3196 different enzymes. The NC-IUBMB has recommended the nomenclature and classification of enzymes on the basis of reactions they catalyse. The complete contents of *Enzyme Nomenclature* 1992 with subsequent supplements and other changes are listed in enzyme number order giving just the recommended name. Each entry provides a link to details of that enzyme. If looking for a specific reaction used in the classification of enzymes the broad outline defined by the first two numbers are given. Each of subclass entries is linked to a location where the category is subdivided to sub-subclasses. These are linked to a list of recommended names for each enzyme in the sub-subclass (http://www.chem.qmul.ac.uk/iubmb/enzyme/).

(*a*) Criteria for Inclusion: Direct experimental evidence is required before inclusion of an enzyme in the list about catalysing the reaction claimed by the proposed enzyme. Without evidence for the reaction catalysed close sequence similarity is not sufficient due to requirement of a small change in sequence for altering the activity of an enzyme. Moreover, classification is based mainly on the reaction that the enzymes catalyse; however, in some cases proteins of different sequences catalyse the same reaction.

2. List of Recommended Names for Enzymes

Each enzyme code consists of the letters EC followed by four numbers separated by periods. These numbers represent a finer classification of the enzyme; for example, the tripeptide aminopeptidases have the code EC 3.4.11.4, whose components indicate the following groups of enzymes:

- EC 3 enzymes are hydrolases (enzymes that use water to break up some other molecule)
- EC 3.4 are hydrolases that act on peptide bonds
- EC 3.4.11 are those hydrolases that cleave off the amino-terminal amino acid from a polypeptide
- EC 3.4.11.4 are those that cleave off the amino-terminal end from a tripeptide

The common names of all listed enzymes are given below, along with their EC numbers. Information of an enzyme that has been deleted or transferred to another EC number is also indicated. Each list is linked to either separate entry for each entry or to files having up to 50 enzymes in each file. Common names of all listed enzymes are given in Table 15.5.

Table 15.5. Common names of all listed enzymes.

Common names for	***List linked to***	
EC 1.1 to EC 1.3	Separate	Up to 50
EC 1.4 to EC 1.97	Separate	Up to 50
EC 2.1 to EC 2.4.1	Separate	Up to 50
EC 2.4.2 to EC 2.9	Separate	Up to 50
EC 3.1 to EC 3.3	Separate	Up to 50
EC 3.4 to EC 3.12	Separate	Up to 50
EC 4	Separate	Up to 50
EC 5	Separate	Up to 50
EC 6	Separate	Up to 50

Strictly speaking, EC numbers do not specify enzymes, but enzyme-catalysed reactions. If different enzymes (for instance from different organisms) catalyse the same reaction, then they receive the same EC number. Top-level EC numbers of the enzymes that they catalyse the reactions are given in Table 15.6.

Table 15.6. Top-level EC numbers of enzymes that they catalyse the reactions (Moss, 2006).

Top-level EC numbers			
Group	***Example of enzymes with trivial name***	***Reaction catalysed***	***Typical reaction***
EC 1 *Oxidoreductases*	Dehydrogenase, oxidase	Catalyse oxidation/reduction reactions; transfer H and O atoms or electrons from one substance to another	AH + B → A + BH (reduced) A + O → AO (oxidized)
EC 2 *Transferases*	Transaminase, kinase	Transfer of a functional group from one substance to another. The group may be methyl-, acyl-, amino- or phosphate group	AB + C → A + BC
EC 3 *Hydrolases*	Lipase, amylase, peptidase	Formation of two products from a substrate by hydrolysis	AB + H_2O → AOH + BH

EC 4 *Lyases*	Decarboxylase	Non-hydrolytic addition or removal of groups from substrates. C-C, C-N, C-O or C-S bonds may be cleaved	RCOCOOH → RCOH + CO_2 or [X-A-B-Y] → [A=B + X-Y]
EC 5 *Isomerases*	Isomerase, mutase	Intramolecule rearrangement, i.e. isomerization changes within a single molecule	AB → BA
EC 6 *Ligases*	Synthetase	Join together two molecules by synthesis of new C-O, C-S, C-N or C-C bonds with simultaneous breakdown of ATP	X + Y+ ATP → XY + ADP + Pi

3. Glossary, Reaction Pathways and Links to Other Databases

A start has been done that shows the pathways where enzymes participate. For example, a link under EC 5.3.3.2 (isopentenyl-diphosphate isomerase) leads to the pathway from mevalonate to terpenes. Similarly the links under EC 1.14.99.7 (squalene monooxygenase) and EC 5.4.99.7 (lanosterol synthase) lead to pathways of steroid formation. A glossary entry has been added for the other enzymes which may be a systematic name or a link to a graphical representation. The glossary from *Enzyme Nomenclature* 1992 should be consulted, which is updated with subsequent glossary entries. Each enzyme entry is linked to the other databases.

***(a)* Enzyme Subclasses:** The links list the enzyme sub-subclasses which in turn list the enzymes linked to separate files for each enzyme, or to a list as part of a file with up to 50 enzymes per file. The latest systematic nomenclature system known as the International Enzyme Commission (IEC) system is based upon the type of reaction catalysed. There are six broad groups of enzymes in this system (Table 15.7).

Table 15.7. List of sub-subclasses of some enzymes as proposed by International Enzyme Commission (IEC).

Subclass	*Name*	*Enzyme file type*	
EC 1	**Oxidoreductases**		
EC 1.1	Acting on the CH-OH group of donors	Sub-subclasses	Up to 50
EC 1.2	Acting on the aldehyde or oxo group of donors	Sub-subclasses	Up to 50
EC 1.3	Acting on the CH-CH group of donors	Sub-subclasses	Up to 50
EC 1.4-EC 1.22	Acting on the CH-NH_2 group of donors, etc.	Sub-subclasses	Up to 50
EC 2	**Transferases**		
EC 2.1	Transferring one-carbon groups	Sub-subclasses	Up to 50
EC 2.2	Transferring aldehyde or ketonic groups	Sub-subclasses	Up to 50
EC 2.3	Acyltransferases	Sub-subclasses	Up to 50
EC 2.4-EC 2.9	Glycosyltransferases, etc.	Sub-subclasses	Up to 50
EC 3	**Hydrolases**		
EC 3.1	Acting on ester bonds	Sub-subclasses	Up to 50
EC 3.2	Glycosylases	Sub-subclasses	Up to 50
EC 3.3	Acting on ether bonds	Sub-subclasses	Up to 50
EC 3.4-EC 3.13	Acting on peptide bonds (peptidases), etc.	Sub-subclasses	Up to 50

EC 4	Lyases		
EC 4.1	Carbon-carbon lyases	Sub-subclasses	Up to 50
EC 4.2	Carbon-oxygen lyases	Sub-subclasses	Up to 50
EC 4.34–EC 4.99	Carbon-nitrogen lyases, etc.	Sub-subclasses	Up to 50
EC 5	**Isomerases**		
EC 5.1	Racemases and epimerases	Sub-subclasses	Up to 50
EC 5.2-EC 5.99	*cis-trans*-Isomerases, etc	Sub-subclasses	Up to 50
EC 6	**Ligases**		
EC 6.1	Forming carbon—oxygen bonds	Sub-subclasses	Up to 50
EC 6.2-EC 6.66	Forming carbon—sulfur bonds, etc.	Sub-subclasses	Up to 50

E. MECHANISM OF ENZYME CATALYSIS

The principle of mechanism of enzyme catalysis is similar to other types of chemical catalysis. The enzyme reduces the energy required to reach the highest energy transition state of the reaction by providing an alternative reaction route and stabilising the intermediates. The number of reactant molecules gets increases as the 'activation energy' (Ea) reduces to reach the optimum activation energy and form the product.

1. Induced Fit Theory

The induced-fit theory assumes that the substrate plays a role in determining the final shape of the enzyme which is partially flexible. That is why certain compounds can bind to the enzyme but do not react because the enzyme becomes greatly distorted. The other molecules are very small to induce the proper alignment; hence they cannot react. Only a suitable substrate induces the proper alignment of the active site. The induced fit model is a favoured model for the enzyme-substrate interaction, but the initial interaction between enzyme and substrate is weak. These weak interactions rapidly induce conformational changes in the enzyme that strengthen binding. Due to the stabilising effect of strong enzyme binding the induced fit mechanism has certain advantages. Substrate binding takes place through two different mechanisms : *uniform binding* (having strong substrate binding), and *differential binding* (having strong transition state binding). Both substrate and transition state binding affinity are increased by the stabilising effect of uniform binding, while only transition state binding affinity is increased by differential binding. Enzymes use both the mechanisms and have been chosen to minimise the Ea of the reaction through evolution. The saturated enzymes have a high affinity substrate binding, and require differential binding to reduce the Ea. Differential or uniform binding is used by small substrate unbound enzymes.

Most enzymes are much larger than their substrates, and only a small portion of the enzyme (around 3–4 amino acids) is directly involved in catalysis. Enzymes have an area with a very particular shape. When a molecule of the right chemical for that enzyme comes along it will fit exactly into the shape. The region that contains these catalytic residues binds the substrate and carries out the reaction, is known as the active site. The molecule on which the enzyme will work is called the 'substrate' (Fig. 15.6).

The differential binding mechanism is used to reduce the Ea by most proteins due to these effects. Most proteins have high affinity of the enzyme to the transition state. The induced fit mechanism carries out differential binding, *i.e.* first the substrate binds weakly, and then the enzyme changes the conformation and increases the affinity to the transition state and stabilises it; hence it reduces the activation energy to reach this state. The induced fit concept cannot be used to rationalise catalysis. The chemical catalysis is defined as '*the reduction of Ea (when the system is already in the ES) relative to Ea in the uncatalysed reaction in water (without the enzyme)*'.

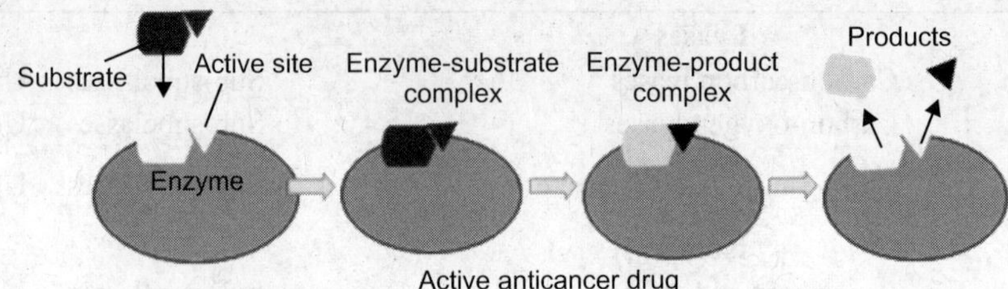

Fig. 15.6. The induced fit theory of enzyme action.

Enzymes can also contain sites that bind cofactors which are needed for catalysis. Some enzymes also have binding sites for small molecules which are the products or substrates of the reaction catalysed. The substrate enters the active site of the enzyme and forms enzyme-substrate complex. Substrate binding causes slight changes in the shape of the enzyme, resulting in the increase or decrease of enzyme activity. It is also a mechanism of feedback regulation. Enzyme-product complex is formed after enzyme action; the product leaves the active site of the enzyme and gets releases from the complex. Then the enzyme becomes ready for another molecule of the chemical. Eenzyme structures may be obtained in enzyme-substrate complex or enzyme-substrate analogs by using 'time resolved crystallography' methods.

2. Mechanisms of Transition State Stabilisation

The catalytic residues present in the active site are brought closer to the chemical bonds in the substrate by the conformational changes. When enzyme-substrate binding is over, the catalysis mechanisms decrease the energy of transition state of the reactions by an alternative chemical pathway. There are six possible mechanisms of transition state stabilisation as given below:

(*a*) Catalysis by Bond Strain: The main effect of induced fit binding is catalysis by bond strain, where the affinity of the enzyme to the transition state is more than the substrate. Consequently, structural rearrangements are induced straining the substrate bonds into a position closer to the conformation of the transition state. Therefore, the difference of energy between the substrate and transition state is lowered resulting in catalysis of reaction. However, the strain effect is a ground state destabilisation effect, rather than transition state stabilisation effect. Enzymes cannot apply large strain effect because they are very flexible. Bond strain may also be induced within the enzyme itself to activate residues in the active site, in addition to bond strain in the substrate.

(*b*) Catalysis by Proximity and Orientation: When enzyme-substrate interactions align the reactive chemical groups and hold them close together, the rate of the reaction is increased due to catalysis by proximity and orientation. The entropy of the reactants is reduced making the addition reactions more favourable. When two reactants become a single product, the overall loss of entropy is reduced. The intra-molecular binding character of the reagents to the enzyme increases the rate of reaction massively.

(*c*) Acid/Base Catalysis (Catalysis involving Proton Donors or Acceptors): In order to stabilise developing charges in the transition state, acids (proton donors) and bases (proton acceptors) donate and accept the protons. This activates the nucleophile and electrophile groups or stabilises the leaving groups. Often histidine is such a residue that is involved in acid/base reactions. Histidine can accept and donate protons because it has a pKa close to neutral pH. There are several mechanisms of reaction involving acid/base catalysis that possibly may change pKa through the local environment of the residue. The environment can modify the pKa significantly in such a way that basic residues in solution may act as proton donors, and acidic residues in solution as proton acceptor. Modification in pKa at different conditions is given in Table 15.8.

Table 15.8. Modification in pKa at different conditions.

Conditions	***Acids***	***Bases***
Hydrophobic environment	Increase pKa	Decrease pKa
Adjacent residues of like charge	Increase pKa	Decrease pKa
Salt bridge (and hydrogen bond) formation	Decrease pKa	Increase pKa

***(d)* Covalent Catalysis:** A transient covalent bond formation by the substrate with residues in the active site or with a cofactor is called covalent catalysis. Thus, an additional covalent intermediate is added to the reaction that reduces the energy of later transition states of the reaction. At a later stage the covalent bond is broken and enzyme is regenerated. Such type of mechanism is found in some enzymes like proteases, for example chymotrypsin and trypsin. An acyl-enzyme intermediate is formed in these cases. The other mechanism is Schiff base formation by using the free amine from a lysine residue as found during glycolysis in the enzyme aldolase. Some enzymes [such as pyridoxal phosphate (PLP) or thiamine pyrophosphate (TPP)] form covalent intermediates with reactant molecules by utilising non-amino acid cofactors. The energy of later transition states is reduced by these covalent intermediates. Cofactors allow enzymes to carryout reactions which cannot be carried out by amino acid side residues. The enzymes that utilise such cofactors are : the PLP-dependent enzyme aspartate transaminase and the TPP-dependent enzyme pyruvate dehydrogenase (Voet and Voet, 2004).

***(e)* Electrostatic Catalysis:** In the active site the residues forming ionic bonds with the intermediate stabilise the charged transition states. The acidic or basic side chains provide these bonds that are found on amino acids (for example, lysine, arginine, aspartic acid or glutamic acid), or the metal cofactors (e.g. Zn). Metal ions effectively reduce the pKa of water and results in an effective nucleophile. In 2006, A. Warshel and co-workers found that electrostatic effects largely contribute to catalysis as has been established by systematic computer simulation studies. Enzyme provides more polar environment than water and the fixed dipoles stabilise the ionic transition states. In order to stabilise ionic and charged states, the water molecules pay reorganisation energy. Thus, pre-organisation of enzyme polar groups is the fact of catalysis.

***(f)* Quantum Tunneling:** In recent years, the physical laws governing life can be refined by using the quantum mechanics. When there is no enough energy in the mix to let the reaction occur by classical routes, quantum tunnelling allows the hydrogen to pass from reactant to product. Quantum tunnelling allows some reactions to occur quickly to sustain life. The majority of studies have been concentrated on understanding how enzymes facilitate passage of the reaction over a static potential-energy barrier. Quantum mechanical phenomena driven by protein dynamics can play a pivotal role in enzyme action.

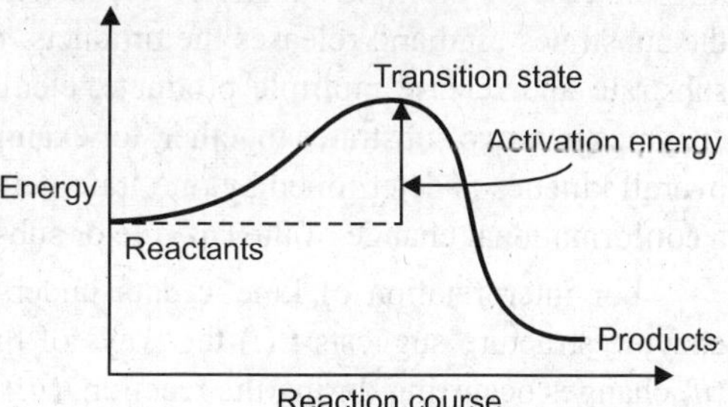

Fig. 15.7. Separation of reactants from products by the activation energy barrier.

During 1972, Volkenshtein and co-workers formulated the first quantum-mechanical model of enzyme catalysis. In some cases, models and observations of 'through the barrier' mechanisms (quantum tunneling) have challenged these traditional 'over the barrier' mechanisms. In 2004, Garcia-Viloca and collaborators reported that a proton or an electron can tunnel through activation barriers in 'through the barrier' models.

The transition state is the top of the barrier (Fig 15.7). However, the reactants surmount the barrier only after putting the right amount of energy into the reaction. The particle can appear in the probability area on the other side of the activation barrier. Hydrogen tunnels through the activation barrier; particles of different energies tunnel at various heights. The thinner barrier at the top promotes tunneling but few particles have sufficient energy at these heights.

Oxidation/Reduction Reactions

In living organisms biochemical reactions are essentially energy transfers. Occurrence of reactions linked together is referred to as oxidation/reduction reactions. Reduction is the gain of an electron; sometimes we also have H ions, so reduction also becomes the gain of H. Oxidation is the loss of an electron (or H). In oxidation/reduction reactions, one chemical is oxidised and its electrons are passed to another (reduced) chemical. Such coupled reactions are referred to as redox reactions. The metabolic processes glycolysis, Kreb's cycle, and electron transport phosphorylation involve the transfer of electrons by redox reactions at varying energy states.

Tryptamine oxidation by aromatic amine dehydrogenase shows quantum tunneling for protons. Since the tunneling contributions are similar in the catalysed and the uncatalysed reactions in solution, quantum tunneling does not appear to provide a major catalytic advantage. The tunneling contribution is likely crucial to the viability of biological organisms. Therefore, the general importance of tunneling reactions has been emphasised.

F. ENZYME KINETICS

The study of the chemical reactions catalysed by enzymes is called 'enzyme kinetics', where the reaction rate is measured with or without varying conditions of the reaction. Enzymes kinetics study reveals the catalytic mechanism of the enzyme, its role in metabolism, control of activity, and enzyme inhibition by a drug or a poison. It has earlier been explained that target molecules of substrate bind to the active site of an enzyme, and through a series of steps are transformed into products. These mechanisms of action can be divided into two: *single-substrate mechanism* (enzymes that bind only one substrate, such as triosephosphate isomerase) and *multiple-substrate mechanism* (enzymes that bind more than one substrate, such as dihydrofolate reductase).

In case of multiple-substrate mechanism, enzyme kinetics displays the sequence in which the substrates bind and releases the products. Proteases are the enzymes that bind a single protein substrate and release multiple products, cleaving one protein into two polypeptides. The others enzymes join two substrates together, for example DNA polymerase links a nucleotide to DNA. The overall kinetics is determined by one 'rate-determining step', which may be a chemical reaction or a conformational change of the enzyme or substrates.

For interpretation of kinetic data understanding of the enzyme structure is necessary. The enzyme structure suggests : (*i*) the ways of binding the substrates and products during catalysis, (*ii*) changes occurring during the reaction, (*iii*) and the role of specific amino acid residues. During the kinetics mechanism some enzymes change their shape which helps to determine the enzyme structure.

Ribozymes and ribosomes carryout several cellular functions e.g. RNA splicing and translation. A more limited set of reactions are performed by ribozymes and their reaction mechanisms and kinetics can be analysed and classified following the same methods.

1. Principles of Enzyme Kinetics

An enzyme-catalysed reaction uses the same reactants and produces exactly the same products as the enzyme-uncatalysed reaction does. The position of equilibrium between substrates and products is not changed by the enzymes similar to the other catalysts. But enzyme-catalysed reactions show saturation kinetics, unlike uncatalysed chemical reactions. The rate of reaction increases linearly with substrate concentration for a given enzyme concentration and relatively low concentrations of substrate. The enzyme molecules freely catalyse the reaction. The rate of reaction increases with increasing the concentration of substrate, *i.e.* the rate at which the enzyme and substrate molecules encounter one another. The theoretical maximum is asymptotically approached by the rate of reaction at high concentrations of substrate. All the active sites of enzyme are occupied and intrinsic turnover rate of an enzyme determines the rate of reaction. K_M value denotes the substrate concentration midway between these two limiting cases. There are two most important kinetic properties of an enzyme : (*i*) quick saturation with a particular substrate, and (*ii*) achievement of its maximum rate. Understanding of these properties suggests the function of an enzyme in the cell and changes its responses in these conditions.

2. Enzyme Assays

The rates of enzyme reactions are measured by the laboratory procedures, *i.e.* enzymes assays because the reactions do not consume the enzymes that they catalyse. For measuring the rate of reaction, changes in the concentration of substrates or products are followed by enzyme assays using several methods. For example, change in absorbance of light between products and reactants are observed by spectrophotometric assays, because rate of the reaction can be measured continuously and conveniently. Moreover, incorporation or release of radioactivity to calculate the amount of product made over time is measured by radiometric assays, which are very sensitive and can measure even very low levels of enzyme activity (Danson and Eisenthal, 2002).

Use of mass spectrometry is an analogous approach for monitoring the incorporation or release of stable isotopes while substrate is converted into product. Use of lasers focused through a microscope to observe changes in single enzyme molecules while catalysing their reactions, is the most sensitive enzyme assays. These measurements use changes in the fluorescence of cofactors during reaction mechanism of an enzyme or fluorescent dyes added onto specific sites of the protein to report movements occurring during catalysis. These studies provide a new view of the kinetics and dynamics of single enzymes in contrast to traditional enzyme kinetics which observes the average behaviour of populations of millions of enzyme molecules.

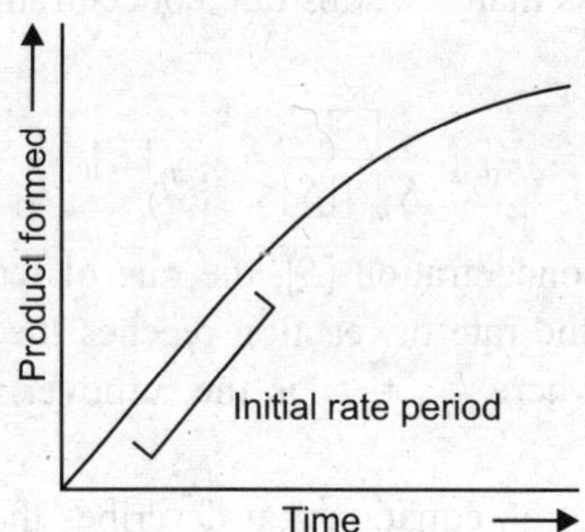

Fig. 15.8. Progress curve for an enzyme reaction.

An enzyme produces product at an initial rate linear for a short period after the start of the reaction (Fig. 15.8). The rate continuously slows with the progress of reaction and consumption of substrate, till the substrate reaches at saturating level. Typically enzyme assays are performed to measure the initial rate, when the reaction proceeds only a few percent. The assay conditions govern the length of the initial rate period that may be from milliseconds to hours. Most studies on enzyme kinetics concentrate on initial linear part of enzyme reactions; but the complete reaction curve is also measured and this data is fit to a non-linear rate equation. Such method of measuring enzyme reactions is called progress-curve analysis. When the initial rate is too fast to measure accurately, this approach is useful as an alternative to rapid kinetics.

3. Single-substrate Reactions

Enzymes that react with a single substrate show single-substrate mechanisms (Fig. 15.9); these include isomerases (e.g. triosephosphateisomerase or bisphosphoglycerate mutase), intramolecular lyases (e.g. adenylate cyclase) and the hammerhead ribozyme (*i.e.* RNAs that self-cleave via a small conserved secondary structural motif), and RNA lyase. In contrast, there are some enzymes that do not fall into this category of mechanisms in spite of having only a single substrate. For example, catalase that reacts with a first molecule of the substrate (*i.e.* hydrogen peroxide) becomes oxidised and finally reduced by a second molecule of substrate. In spite of involvement of a single substrate, a modified enzyme intermediate exists. This shows that the mechanism of catalase is a ping–pong mechanism (*see* preceding section).

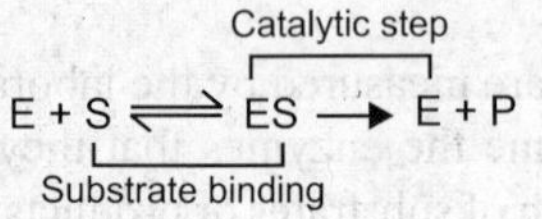

Catalytic step

$$E + S \rightleftharpoons ES \longrightarrow E + P$$

Substrate binding

Fig. 15.9. Single substrate mechanism of an enzyme.

4. Michaelis–Menten Kinetics

In 1913, an American biochemist Leonor Michaelis and a Canadian physician Maud Menten first proposed a model for the most basic enzymatic reactions. It is one of the simplest and best-known models of enzyme kinetics. Suppose there is an enzyme E that reversibly binds to a substrate S and forms a complex ES. In turn this complex is converted into a product P and the enzyme E. This may be written as below:

$$E + S \underset{k_{-1}}{\overset{k_1}{\rightleftharpoons}} ES \xrightarrow{k_2} E + P$$

Where k_1, k_{-1} and k_2 represent the rate constants of the three reactions. However, when the enzyme concentration is much less than the substrate concentration, the rate of product formation is represented as below:

$$V = V_{\max} \frac{[S]}{K_m | [S]} = k_{cat}\,[E]_0 \frac{[S]}{K_m | [S]}.$$

Upon increasing substrate concentration [S], the rate of reaction increases gradually; when all enzymes bound to substrate, the rate of reaction reaches the maximum rate $V_{\max}$. It can also be written as : $V_{\max} = k_{cat}[E]_0$, where $k_{cat} = k_2$ is the turnover number, and $[E]_0$ is the enzyme concentration.

This model takes the form of an equation that describes the rate of enzymatic reactions, by relating the rate of reaction V to [S], the concentration of a substrate S. It may also be represented by the following formula :

$$V = \frac{V_{\max}[S]}{K_m + [S]}.$$

Here, $V_{\max}$ is the maximum rate obtained by the system at maximum saturating substrate concentrations. The Michaelis constant K_m is the substrate concentration at which the reaction rate is half of $V_{\max}$. Michaelis–Menten saturation curve for an enzyme reaction showing the relation between the concentration of substrate and rate is given in Fig. 15.10.

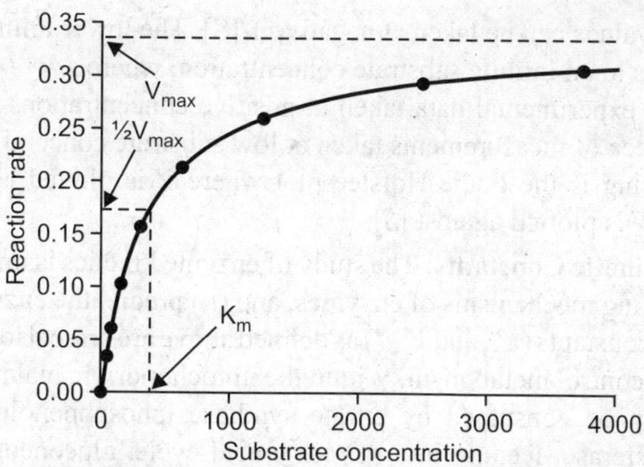

Fig. 15.10. Michaelis–Menten saturation curve for an irreversible enzyme reaction.

Mostly single substrate reactions follow Michaelis–Menten kinetics, where K_m is a measure of affinity of the substrate for the enzyme. When K_m is small, it shows high affinity, *i.e.* the rate will approach V_{max} more quickly (Lehninger *et al.*, 2005).

(*a*) Linear Plots of the Michaelis–Menten Equation: The effects on the behaviour of an enzyme of varying kinetic constants can be explored by using an interactive Michaelis–Menten kinetics. The plot of V versus [S] above is not linear; initially it is linear at low [S] but bends over to saturate at high [S]. Therefore, K_M and V_{max} cannot be accurately estimated due to nonlinearity. Hence, several researchers have developed linearisation of Michaelis–Menten equation; for example the Lineweaver–Burk plot, the Eadie–Hofstee plot, and the Hanes–Woolf plot. For visualising the data these linear representations can be useful but none should be used to determine kinetic parameters. Now-a-days, computer software is available that allows for more accurate determination by nonlinear regression methods.

The Lineweaver–Burk plot or double reciprocal plot is a common method to illustrate the kinetic data. This plot is produced by taking the reciprocal of both sides of the Michaelis–Menten equation. Lineweaver–Burk plot of kinetic data, showing the significance of the axis intercepts and gradient is given in Fig. 15.11.

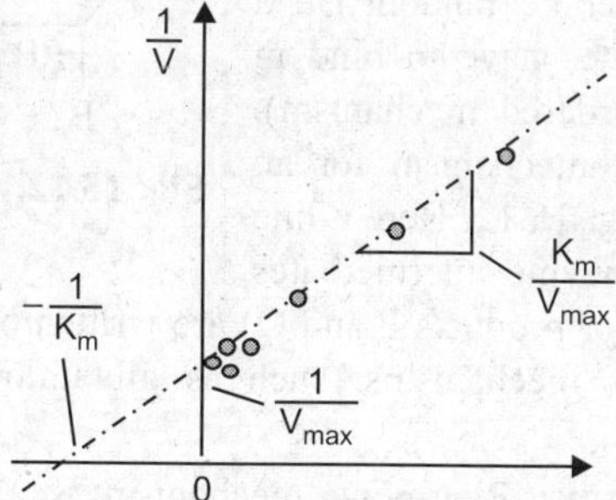

Fig. 15.11. Lineweaver–Burk plot of kinetic data, showing the significance of the axis intercepts and gradient.

This is a linear form of the Michaelis–Menten equation. A straight line with the equation $y = mx + c$ with a y-intercept equivalent to $1/V_{max}$ and an x-intercept of the graph representing -1/K_M, is produced in this plot.

$$\frac{1}{v} = \frac{K_M}{V_{max}\,[S]} + \frac{1}{V_{max}}$$

No experimental values can be taken at negative 1/[S]. The lower limiting value 1/[*S*] = 0 (the *y*-intercept) corresponds to an infinite substrate concentration, where $1/v=1/V_{max}$. The *x*-intercept is an extrapolation of the experimental data taken at positive concentrations. The Lineweaver–Burk plot skews the importance of measurements taken at low substrate concentrations. A more accurate method for linear plotting is the Eadie-Hofstee plot where *V* is plotted against *V*/[*S*], but in the Hanes-Woolf plot, [*S*]/*V* is plotted against [*S*].

Significance of Kinetic Constants: The study of enzyme kinetics is important because it helps to : (*i*) explain the working mechanisms of enzymes, and (*ii*) predict the enzyme behaviour in living organisms. The kinetic constants (K_M and V_{max}) as defined above are critical to understand the working of enzymes together to control metabolism. Within the mitochondrion, malate dehydrogenase forms oxaloacetate which can be consumed by citrate synthase, phosphoenolpyruvate carboxykinase or aspartate aminotransferase, feeding into the citric acid cycle, gluconeogenesis or aspartic acid biosynthesis, respectively. On the basis of knowledge of the concentration of oxaloacetate as well as the concentration and kinetics of each of these enzymes, one can predict how much oxaloacetate goes into which pathway. The mathematical models of entire organisms can be predicted on the basis of huge amounts of kinetic data (generated through metabolic pathways) and gene expression data.

5. Multi-substrate Reactions

The complex rate equations are followed by multi-substrate reactions which describe the methods of binding the substrates in specific sequence. If the concentration of substrate A is kept constant and substrate B varied, the analysis of these reactions is much simpler. At this condition, the enzyme acts like a single-substrate enzyme; an apparent K_M and V_{max} constants for substrate B are obtained after plotting *V* by [*S*]. The mechanism of the reaction can be obtained exploiting these data, by performing a set of these measurements at different fixed concentrations of A. Moreover, there are two types of mechanisms (ternary complex and ping–pong) for an enzyme that takes two substrates A and B and turns into two products P and Q.

(*a*) Ternary-complex Mechanisms: In these enzymes, two substrates A and B bind to the enzyme at the same time to produce an EAB ternary complex. The order of binding can either be random (in a random mechanism) or substrates have to bind in a particular sequence (in an ordered mechanism). Random-order ternary-complex mechanism for an enzyme reaction is shown in Fig. 15.12. Here a line shows the reaction path and enzyme intermediates (containing substrates A and B or products P and Q) are written below the line. There are certain enzymes with ternary-complex mechanisms, such as glutathione *S*-transferase, dihydrofolate reductase, DNA polymerase, etc.

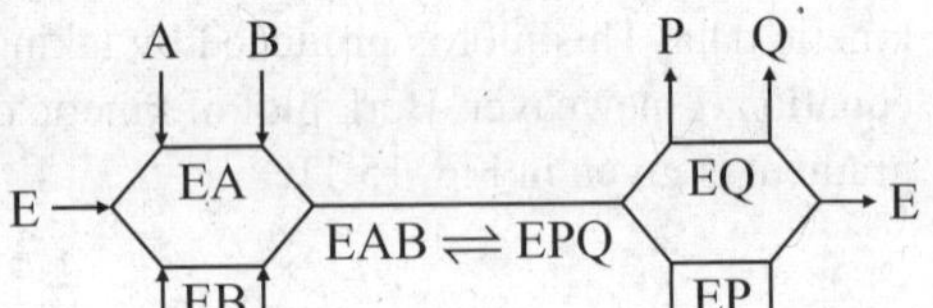

Fig. 15.12. Ternary-complex mechanism for an enzyme reaction.

(*b*) Ping–pong Mechanisms: Ping-pong mechanism is also called double-displacement reaction. Enzymes with a ping-pong mechanism exist in two states, E (an enzyme) and an intermediate E* (a chemically modified form of the enzyme). In such mechanisms, substrate A binds the enzyme E and forms enzyme-substrate complex (Fig. 15.13). E is changed to E* by transferring a chemical group to the active site; thereafter, it is released. When the first product P is released from E*, B binds to modified enzyme E* and regenerates the unmodified E form. B is converted to Q and released as the second product. The process can be repeated. E* contains a fragment of the original substrate A, which can alter the function of the enzyme and gets attached to substrate B, or both. The other characteristic of the ping-pong mechanism is that one product is formed and released before binding the second substrate (Garrett and Charles, 2010).

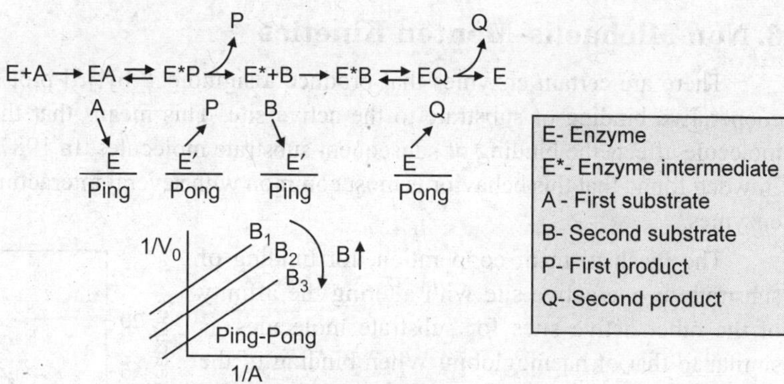

Fig. 15.13. Ping-pong mechanism.

There are certain enzymes that possess ping–pong mechanisms such as some oxidoreductases (e.g. thioredoxin peroxidase), transferases (e.g. acylneuraminate cytydilyltransferase) and serine proteases (e.g. trypsin and chymotrypsin). Serine proteases are a very common and diverse family of enzymes, including several enzymes of the blood clotting cascade, digestive enzymes (trypsin, chymotrypsin, and elastase), etc. In serine proteases, the E* intermediate is an acyl-enzyme species which is formed by the attack of an active site serine residue on a peptide bond.

Fig. 15.14 shows glutamate-aspartate aminotransferase reaction which is a ping-pong mechanism. There is an E′ form of the enzyme which is indicative of a ping-pong mechanism. Pyridoxal is a coenzyme bound to glutamate-aspartate aminotransferase that accepts an amino group from glutamate and becomes pyridoxamine releasing alpha-ketoglutarate. Pyridoxamine bound to the enzyme will then donate its amino group to oxaloacetate to regenrate pyridoxal as well as aspartate. Therefore, it is a ping-pong mechanism.

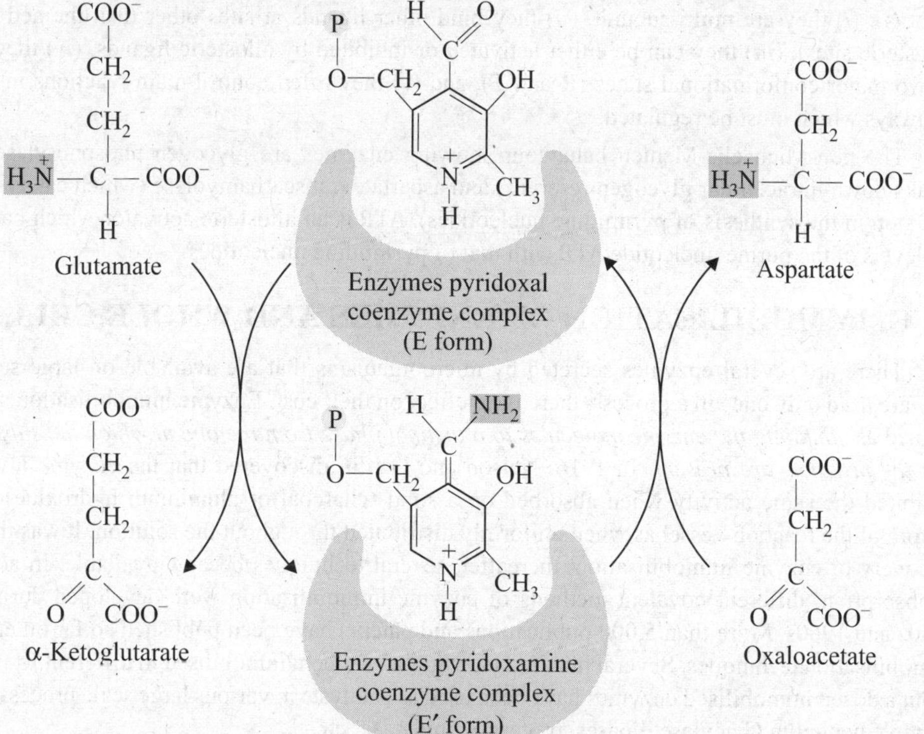

Fig. 15.14. Glutamate-aspartate aminotransferase mechanism.

6. Non-Michaelis–Menten Kinetics

There are certain enzymes that produce a sigmoid *V* by [*S*] plot (Fig. 15.15). This shows a cooperative binding of substrate to the active site. This means that the binding of one substrate molecule affects the binding of subsequent substrate molecules. In 1987, J. Ricard and A. Cornish-Bowden found that this behavior is most common with several interacting active sites in multimeric enzymes.

The mechanism of cooperation, for binding of substrate to one active site with altering the affinity of the other active sites for substrate molecules, is similar to that of haemoglobin. When binding of the first substrate molecule increases the affinity of the other active sites for substrate, it results in positive cooperativity. In contrast, when binding of the first substrate decreases the affinity of the enzyme for other substrate molecules, it results in negative cooperativity. Surprisingly cooperativity is common and regulates the responses of enzymes to changes in the concentrations of their substrates. Enzymes are made more sensitive to [*S*] by positive cooperativity; however, large changes over a narrow range of substrate concentration can be shown by their activities. In contrast, enzymes are made insensitive to small changes in [*S*] by the negative cooperativity. A mammalian tyrosyl tRNA-synthetase (an allosteric enzyme) shows negative cooperativity, and bacterial aspartate transcarbamoylase and phosphofructokinase show positive cooperativity.

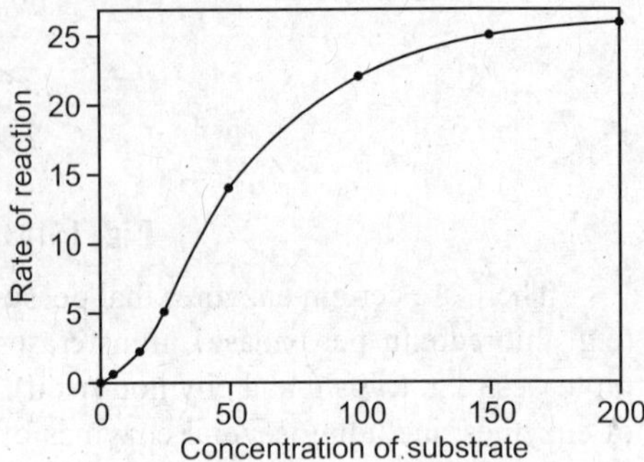

Fig. 15.15. Saturation curve for an enzyme reaction.

Enzymes that display this non-Michaelis-Menten behavior have some common characteristics such as : (*i*) they are multi-subunit, (*ii*) they bind other ligands at sites other than the active site (allosteric sites), (*iii*) they can be either activated or inhibited by allosteric ligands, (*iv*) they exist in two major conformational states (R and T), and (*v*) they often control main reactions in major pathways which must be regulated.

The non-Michaelis-Menten behaviour showing enzymes are glycogen phosphorylase (that breaks down intracellular glycogen reserves) and aspartate transcarbamyolase (which catalyses the first step in the synthesis of pyrimidine nucleotides). ATP is an allosteric activator which balances the levels of the purine nucleotide ATP with that of pyrimidine nucleotides.

G. IMMOBILISATION OF ENZYMES AND WHOLE CELLS

There are several enzymes secreted by microorganisms that are available on large scale; if they are used only once in a process, there is no effect on their cost. Enzyme immobilisation may be defined as *confining the enzyme molecules to a distinct phase from the one in which the substrates and the products are present*. In 1916, Nelson and Griffin discovered that the enzyme invertase exhibited the same activity when absorbed on a solid (charcoal or aluminium hydroxide) at the bottom of the reaction vessel as when uniformly distributed throughout the solution. It was the first discovery of enzyme immobilisation; thereafter, several techniques became available. In addition to absorption, different covalent methods of enzyme immobilisation were developed during the 1950s and 1960s. More than 5,000 publications and patents have been published so far on enzyme immobilisation techniques. Several hundred enzymes have been immobilised in different forms and about a dozen immobilised enzymes have been used as catalysts in various large scale processes, for example penicillin G acylase, lipases, proteases, invertase, etc.

Hence, immobilisation refers to 'the imprisonment of an enzyme in a distinct phase that allows exchange with, but is separated from the bulk phase in which the substrate, effector or inhibitor molecules are dispersed and monitored'. Imprisonment refers to arresting the enzyme by certain means where polymer matrix is formed. The first commercial application of immobilised enzyme technology was realised in 1969 in Japan with the use *of Aspergillus oryzae* amino acylase for the industrial production of L-amino acids. Consequently, pilot plant processes were introduced for 6-amino penicillanic acid (6 APA) production from penicillin G and for glucose conversion to fructose by immobilised glucose isomerase.

Immobead™ caught on a filter (50 μm) (*source*: ChiralVisions, The Netherlands).

From the commercial point of view the enzyme immobilisation is a very important approach to facilitate the enzyme reuse. However, the immobilisation process may or may not change the properties of the free enzyme. For example, the immobilised *Bacillus circulans* levansucrase on wool showed resistance against thermal and chemical denaturation, increased tolerance to pH in a wide range and could be used in the production of levan. This provides a good efficiency for frucosylation to synthesise fructan polymers from sucrose. In 2009, H.A. El-Refai and coworkers found that the produced levan was one form of fructans that offered a variety of industrial applications in the field of cosmetics, food and pharmaceutical products.

The materials used for immobilisation of enzymes are called carrier matrices, which are usually inert polymers or inorganic materials. A wide variety of carriers and methods have been used for the immobilisation of various enzymes. However, limited progress has been made in commercialising immobilised enzyme processes, *i.e.* the high cost of carriers used for immobilisation of enzymes being one of the reasons.

1. Carrier Matrices and its Properties

The most commonly employed matrices are agarose, dextran, celluloses and polyacrylamides. Sepharose (an agarose) is available commercially as beads; it is highly hydrophilic and is generally inert to microbial attack. Sepharose is activated by treating it with chloroformates, carbodiimides, glutaraldehyde or other compounds. Agarose is a polysaccharide (sugar) obtained by purifying agar while agar is directly obtained from red algae and seaweed such as *Gracilaria* and *Gelidium*. Agar consists of a mixture of agarose and agaropectin. Agarose being the predominant component is a linear polymer that is made up of the repeating monomeric unit of agarobiose. Agarobiose is a disaccharide made up of D-galactose and 3,6-anhydro-L-galactopyranose. Agaropectin is a heterogeneous mixture of smaller acidic molecules that gel poorly. Structure of agarose is (1→4)-3,6-anhydro-α-L-galactopyranosyl-(1→3)-β-D-galactopyranan (Fig. 15.16).

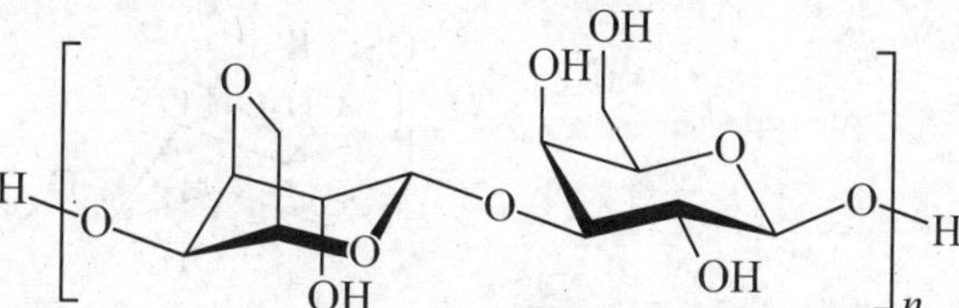

Fig. 15.16. Chemical structure of agarose.

Dextran is a complex and branched glucan (polysaccharide made of many glucose molecules) composed of chains of varying lengths (from 3 to 2000 kD). Structure of dextran is given in Fig.15.17. According to the binding mode of the enzyme, the carrier-binding method can be further sub-classified into: physical adsorption, ionic binding, covalent binding, etc.

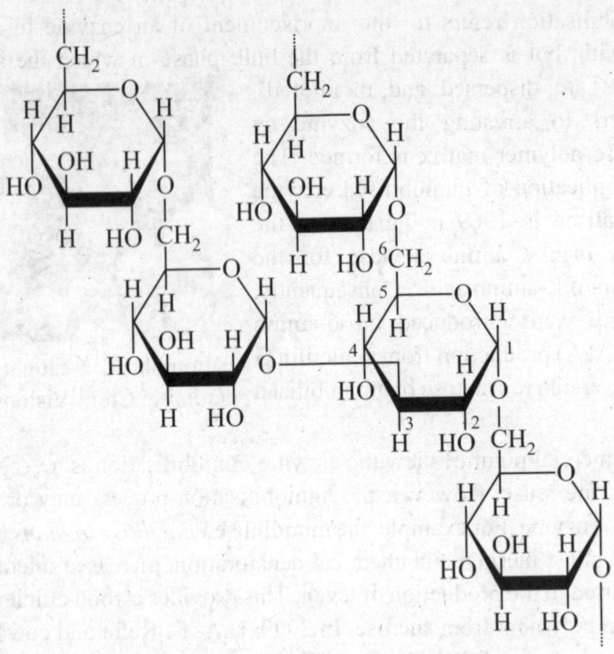

Fig. 15.17. Chemical structure of dextran.

Carrageenans are a family of linear sulfated polysaccharides that are extracted from red seaweeds. Popular sources for carrageenan are the *Chondrus crispus, Eucheuma cottonii* and *Eucheuma spinosum*. *Gigartina skottsbergii* is one of the major species used in the extraction of carrageenan. The basic structure of carrageenan is a linear polysaccharide made up of a repeating dissacharide sequence of α-D-galactopyranose linked 1,3 called the A residue and β-D-galactopyranose residues linked through positions 1,4 (B residues). Carrageenans differ from agar in that the B units in carrageenan are in the D form, whereas they are in the L form in agar. Basic structure of kappa carrageenan is shown in Fig. 15.18.

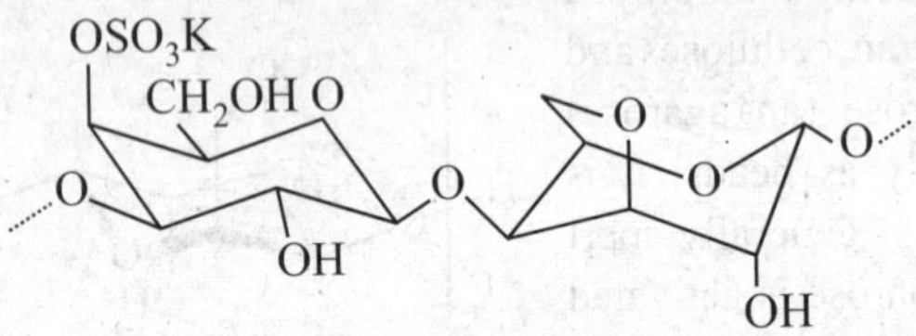

Fig. 15.18. Basic structure of kappa carrageenan.

Carrageenan

There are several varieties of carrageen : (*i*) kappa-carrageenan is used mostly in breading and batter due to its gelling nature, (*ii*) lambda-carrageenan is a non gelling variety that assists in binding, retaining moisture and in contributing to viscosity in sweet doughs, and (*iii*) iota-carrageenan is used primarily in fruit applications and requires calcium ions to develop a heat-reversible and flexible gel.

Carrageenan holds any cell structure and keeps it alive and viable. It is useful for trapping seeds, cells and microorganisms with or without nutrients. One of the chief advantages of carrageenan over the gelling temperature is within the tolerance range of the organism being immobilised. Enzyme activity of immobilised cells is generally high that are entrapped in carrageenan.

Gluteraldehyde is a bifunctional reagent which exists as an equilibrium mixture of monomer and oligomers. It can be used to covalently bind the enzyme molecules or cross link the enzyme molecules. In cross linking, each bifunctional reagent molecule binds to two enzyme molecules producing a network of enzyme molecules linked together. Glutaraldehyde is useful for producing immobilised enzyme membranes for use in biosensors; this is achieved by cross linking the enzyme plus a non-catalytic protein used for dilution within a porous sheet, e.g. lens tissue paper or nylon net fabric.

Several companies including enzyme producers are now offering an increasing range of immobilised enzymes with a wide choice of enzymes and carriers with particle sizes from micrometers to millimeters suiting. Most of the carriers used for enzyme immobilisation are available in the market at low cost and in bulk. The specific knowledge to the market is provided by custom immobilisation services that also make the technique accessible for custom enzymes.

In 1983, H.P.S. Makkar and O.P. Sharma for the first time used eggshell as a carrier for immobilisation of enzymes. They chose eggshell (a by-product of the poultry industry) as a carrier for immobilisation of enzymes due to high cost of carrier matrices. The eggshell is widely available at almost no cost; it has good mechanical strength and is resistant to microbial attack. They immobilised 3-galactosidase on ground eggshell by cross-linking with glutaraldehyde, isolated from *Lactobacillus bulgaricus* and partially purified.

(*a*) Properties of Carrier Matrix : The ideal carrier matrix has several properties such as: (*i*) low cost, (*ii*) inertness, (*iii*) physical strength, (*iv*) stability, (*v*) regenerability after the useful lifetime of the immobilised enzyme, (*vi*) enhancement of enzyme specificity, (*vii*) reduction in product inhibition, (*viii*) a shift in the pH optimum for enzyme action to the desired value for the process, and (*ix*) reduction in microbial contamination and nonspecific adsorption.

There are several enzymes that affect the cost and could not be economical if not reused. Therefore, reuse of enzymes led to the development of immobilisation techniques. It involves the conversion of water soluble enzyme protein into a solid form of catalyst by several methods. Now it is possible to immobilise microbial cells by similar techniques.

(*b*) Carrier Binding : The carrier-binding is the oldest immobilisation technique for enzymes, where the amount of enzyme bound to the carrier and the activity after immobilisation depend on the nature of the carrier. The enzyme binds to the carrier as shown in Fig. 15.19. The carriers are selected according to the nature of the enzyme, as well as the particle size, surface area, molar ratio of hydrophilic to hydrophobic groups, and chemical composition. A higher activity of the immobilized enzymes is achieved with an increase in the ratio of hydrophilic groups and in the concentration of bound enzymes. Some of the most commonly used carriers for enzyme immobilization are derivatives of polysaccharide such as cellulose, dextran, agarose, and polyacrylamide gel.

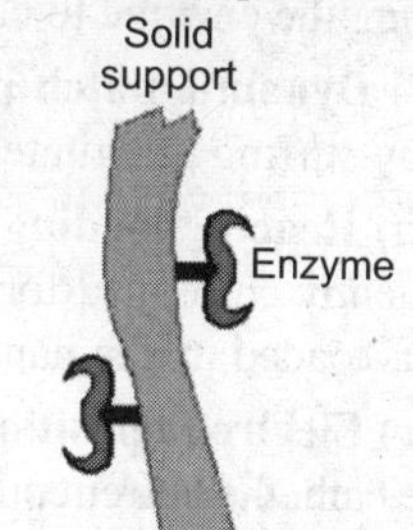

Fig. 15.19. Immobilised enzyme bound to a carrier.

2. Methods of Enzyme Immobilisation

There are five different techniques of immobilising enzymes : (*i*) adsorption, (*ii*) covalent bonding, (*iii*) entrapment, (*iv*) copolymerisation or cross-linking, and (*v*) encapsulation (Fig. 15.20). A membrane or a solid surface is used for the purpose of immobilisation of enzymes carriers, *i.e.* the support materials such as matrix system.

(*a*) Adsorption: Enzymes can be immobilised by bonding to external or internal surface of a carrier or support such as mineral support (aluminium oxide, clay), organic support (starch), modified sapharose and ion exchange resins. In the adsorption, bonds of low energy are involved (*e.g.* ionic interactions, hydrogen bonds, van der Waals forces, etc) (Fig. 15.20 A). The particle size of carriers must be very small to achieve an appreciable surface of bonding while immobilising the enzyme externally. The carrier particles have diameters ranging from 500 Å to about 1 mm. However, no pore diffusion limitations are encountered due to immobilisation of enzymes on external surface.

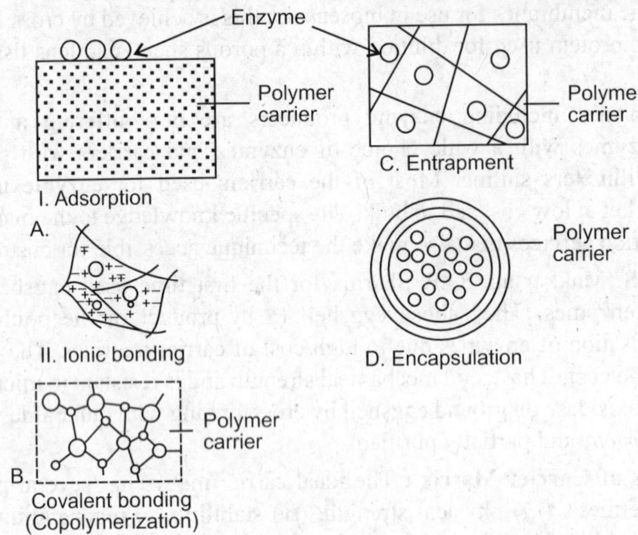

Fig. 15.20. Different types of enzyme immobilisation.

Besides, the enzyme immobilised on an internal surface is protected from abrasion, inhibitory bulk solutions and microbial attack. A more stable and active enzyme system may be achieved due to internal immobilisation. Moreover, the pore diameters of carriers may be optimised for internal surface immobilisation in internal pore immobilisation. Immobilisation by adsorption has the following four procedures :

(*i*) **Static process:** Here enzyme is immobilised on the carrier by allowing the solution containing the enzyme to contact the carrier without stirring.

(*ii*) **Dynamic batch process:** In this process carrier is placed into the enzyme solution and mixed by stirring or agitated continuously in a shaker.

(*iii*) **Reactor loading process:** In this process carrier is placed into the reactor that will be subsequently employed for processing. Then the enzyme solution is transferred to the reactor and carrier is loaded in a dynamic environment by agitating the carrier and enzyme solution.

(*iv*) **Electrode position process:** Here carrier is placed proximal to one of the electrodes in an enzyme bath, the current put on, and the enzyme migrates to the carrier and deposited on the surface.

(*b*) **Covalent Bonding:** A covalent bond is formed between the chemical groups of enzyme and chemical groups present on surface of a carrier. Thus covalent bonding is utilised under a broad range of pH, ionic strength and other variable conditions. Immobilisation steps are attachment of coupling agent followed by an activation process, or attachment of a functional group and finally attachment of the enzyme (Fig. 15.20 B).

Covalent bond formation occurs with the side chains of amino acids of the enzyme. Lysine residues are the most useful in covalent binding since they are usually exposed on the surface. They are highly reactive very rarely occur at active sites of enzymes. Enzyme loading is quite low (Ca. 0.02g/g matrix) only in exceptional cases it may be 0.3 g/g matrix.

Support binding can simply exploit weak hydrophobic and van der Waals interactions, or stronger ones like ionic bonding. Covalent binding of the enzyme is more appropriate for industrial applications because it cannot be leached from the solid support. Fig. 15.21 shows an example of covalent immobilisation of enzymes on a functionalised solid support, where amino groups of lysine side chains are used for the nucleophilic opening of the epoxy rings present on the surface of the support.

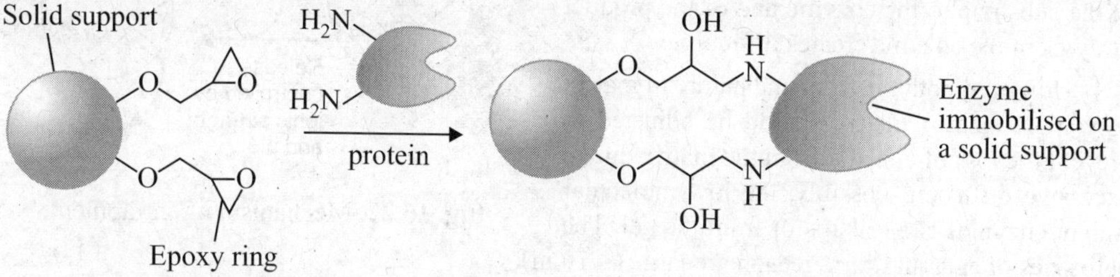

Solid support
Epoxy ring
H_2N
H_2N
protein
OH
H
N
N
H
OH
Enzyme immobilised on a solid support

Fig. 15.21. Covalent immobilisation of enzymes on a functionalised solid support.

Different types of carriers are used in immobilisation such as carbohydrates proteins and amine-bearing carriers, inorganic carriers, etc. Covalent attachment may be directed to a specific group (*e.g.* amine, hydroxyl, tyrosyl, etc.) on the surface of the enzyme. Hydroxyl- and amino groups are the main groups of the enzymes with which it forms bonds, whereas sulphydryl group least involved. There are different methods of covalent bonding as given below :

(*i*) Diazoation: It is bonding between the amino groups of the support *e.g.* aminobenzyle cellulose, aminosilanised porous glass, aminoderivatives and a tyrosyl or histidyl group of the enzyme.

(*ii*) Formation of peptide bond: It is bond formation between the amino or carboxyl group of the support and amino or carboxy group of the enzyme.

(*iii*) Group activation: It refers to use of cyanogen bromide to a support containing glycol group, *i.e.* cellulose, syphadex, sepharose, etc.

(*iv*) Poly functional reagents: Use of a bifunctional or multifunctional reagent *e.g.* glutaraldehyde which forms bonding between the amino group of the support and amino group of the enzyme.

Problem with covalent bonding is that the enzyme may be inactivated by bringing about changes in conformation when undergoes reactions at active sites. But, this problem can be overcome through immobilisation in the presence of enzyme's substrate or competitive inhibitors. The most common activated polymers are celluloses or polyacrylamides onto which diazo, carbodimide or azide groups have been incorporated.

(c) Cross-linking or Co-polymerization: This method is characterised by covalent bonding between the various molecules of an enzyme via a polyfunctional reagent such as glutaraldehyde, diazonium salt, hexamethylene disocyanate, and N-N' ethylene bismaleimide. The demerit of using polyfunctional reagents is that they can denature the enzyme. This technique is cheap and simple but not often used with pure proteins because it produces very little immobilised enzyme that has very high intrinsic activity. It is widely used in commercial preparation.

(*d*) Entrapment: Entrapment is more suited for the immobilisation of whole cells. A matrix (support) of a water soluble polymer such as polyacrylamide type gels and naturally derived gels (*e.g.* cellulose triacetate, agar, gelatin, carrageenan, alginate, etc.) can physically entrap the enzyme inside it (Fig. 15.20 C). Thus, the enzyme molecules are entrapped within suitable gels or fibres and there may or may not be covalent bond formation between the enzyme molcculess and the matrix. Generally entrapment does not involve a chemical or physical/chemical reaction directly with the enzyme molecules. Physical adsorption on active carbon particles and ionic adsorption on ion-exchange resins are important for enzyme immobilisation. A non-covalent entrapment may be viewed as putting the enzyme molecule in a molecular cage. When covalent binding is also to be generated, the enzyme molecules are usually treated with a suitable reagent. Enzyme loading is very high (1 g/g gel or fiber). However, diffusion of the substrate to the enzyme and of the product away from the enzyme creates difficulties.

However, both the form and nature of matrix vary; pore size of matrix should be adjusted to prevent the loss of enzyme from the matrix due to excessive diffusion. Possibly, the low molecular weight enzymes can leakage of from the gel. The pore sizes of agar and carrageenan are large (< 10 m).

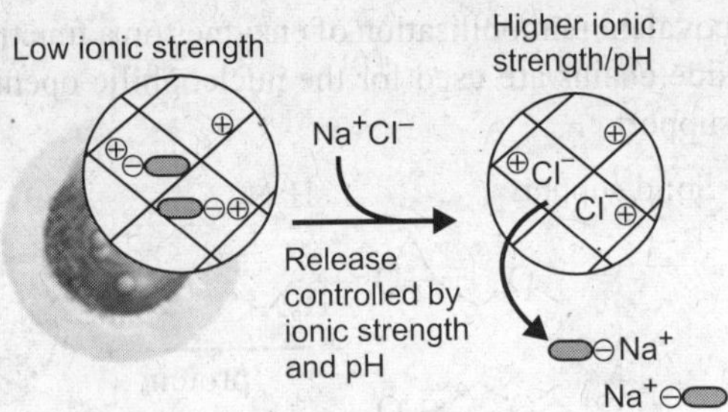

Fig. 15.22. Mechanism of entrapment.

There are two different types of chemical interactions that play a key role to retain molecules, that are ionic bonds and hydrophobic interactions. Entrapment of ionic molecules is particularly stable due to creation of ionic bonds between the central core of glycospheres and the active ingredient molecule. All the entrapped active molecules remain in the particle at low ionic strength; when ionic strength is increased by addition of salt, the ionic bonds split and the active molecules get released (Fig. 15.22).

There are three different commonly used entrapment media : polyacrylamide, calcium alginate, and gelatin. In all the methods, enzymes are mixed well in a solution with monomers/polymers and cross-linking agents. The solution is then exposed to polymerisation promoters to start the process of gel formation. Then the solution is poured into a mold to get the desired shapes. A gel block may be cut into smaller cubes to increase the surface area.

Polyacrylamide is the most widely used matrix for entrapping enzymes among the three gels. It is non-ionic, hence it is advantageous. The properties of the enzymes are modified minimum in the presence of the gel matrix. At the same time, the diffusion of the charged substrate and products is not affected. Dimethylaminopropionitrile is the polymerisation initiator that is highly toxic; hence it must be handled with great care.

Calcium alginate is widely used similar to polyacrylamide; but unlike polyacrylamide gels, gelation of calcium alginate does not depend on the formation of more permanent covalent bonds between polymer chains. Besides, calcium ions cross-link the polymer molecules. Due to this property, calcium alginate beads can be formed in extremely mild conditions; therefore, yield of enzyme activity over 80% can be achieved. As calcium ions can be exchanged for sodium ions, similarly they can also be displaced by other ions. In addition, care must be taken to ensure that the substrate solution does not contain high concentrations of such ions that disintegrate the gel.

Often the entrapment method suffers from three types of difficulties : (*i*) continuous leakage due to wide pore size of the gel [for example, agar, carragcenan, etc. have large pore size (<10 microns)], (*ii*) reduced substrate accessibility to the enzyme, and (*iii*) some loss of enzyme activity due to the free radicals produced during polymerization of the gel.

The entrapment of enzymes has been widely used for sensing application, but not much success has been achieved with industrial process. The major limitations for entrapment of enzymes are their leakage from the matrix. Most workers prefer to use the technique of entrapment for immobilisation of whole cells. Cell entrapment is used for industrial production of amino acids such as L-isoleucine, L-aspartic acid, L-malic acid etc.

(*e*) Encapsulation: Encapsulation is the enclosing of a droplet of solution of enzyme in a semipermeable membrane capsule. The membrane may be polymeric, lipoidal, lipoprotein based or non-ionic in nature. Besides, the capsule is made up of cellulose nitrate and nylon. The method of encapsulation is cheap and simple but its effectiveness largely depends on the stability of enzyme although the catalyst is very effectively retained within the capsule. This technique is restricted to medical sciences only (Fig. 15.20 D). In this method a large quantity of enzyme is immobilised but the biggest disadvantage is that only small substrate molecule is utilised with the intact membrane. Recently encapsulation is being used for immobilisation of enzymes and mammalian cells; for example, pancreatic cells grown in cultures can be immobilised by encapsulation. Hybridoma cells have also been immobilised successfully by these techniques. There are three distinct ways of encapsulation on the basis of : (*i*) special membrane reactors, (*ii*) formation of emulation, and (*iii*) stabilisation of emulation to form microcapsule.

Y. Wang and F. Caruso in 2004 have reported that encapsulation of enzymes in mesoporous silica spheres *via* immobilisation, followed by assembling an organic/inorganic nanocomposite shell on the particle surface leads to high loadings, high enzymatic activity and stability, and protection from proteolysis (Fig. 15.23).

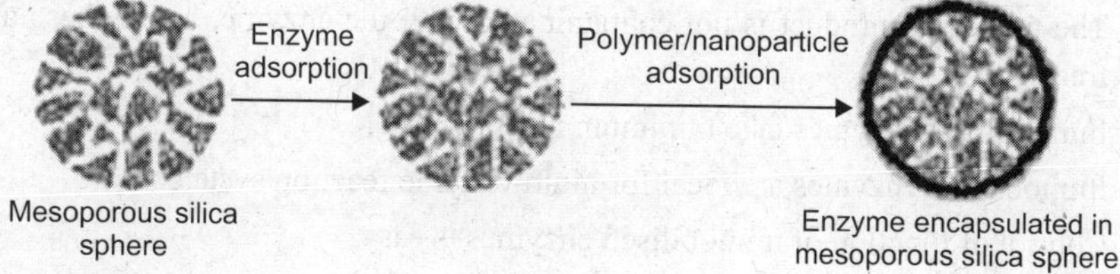

Fig. 15.23. Encapsulation of enzymes in mesoporous silica spheres (*adopted* from Wang and Caruso, 2004).

A comparison of several features of the different immobilisation systems is given in Table 15.9.

Table 15.9. Comparison of different features of immobilisation systems.

Feature	*Immobillisation system*			
	Adsorption	*Covalent binding*	*Entrapment*	*Membrane confinement*
Matrices (examples)	Ion-exchange matrices, clays, glasses, etc.	Sepharose, cellulose, acrylamide, etc.	Acrylamide, cellulose acetate, etc.	Semipermeable membranes, e.g. hollow-fibers, lipsomes, etc.
Preparation	Simple	Difficult	Difficult	Simple
Immobilization mechanism	Hydrophobic effects, salt links	Covalent bonds	Trapping in Gel/fibre, even covalent bonds	Confinement in a semipermeable membrane
Binding force	Variable	Strong	Weak	Strong
Enzyme loading	High (Ca. 19/9 matrix)	Small (Ca. 0.02 gig matrix)	–	–
Enzyme leakage during use	Yes	No	Yes	No
Applicability	Wide	Selective	Wide	Very wide
Problems during operation	High	Low	High	High
Matrix effects on enzyme	Yes	Yes	Yes	No
Diffusion barriers	Absent	Absent	Largc	Large
Protection from microbial attack	No	No	Yes	Yes
Cost	Low	High	Moderate	High

3. Advantages and Disadvantages of Immobilised Enzymes

Immobilised enzymes retain their structural confirmation necessary for catalysis.

(*a*) Advantages: The first immobilised enzymes to be scaled up to pilot plant level and industrial manufacture were immobilised amino acid acylase, panicillin G-acylase and glucose isomerase. Some other industrially important enzymes are aspartase, esterase and nitrilase. There are several advantages of immobilised enzymes as below.

- Immobilised enzymes are more stable and efficient in function than non-immobilised enzymes due to its binding with a matrix.
- The immobilised enzyme can be recovered at the end of the reaction and can be reused several times; it saves the capital cost.
- The processed product is not contaminated with the enzyme, hence these are enzyme-free.
- Immobilised enzymes take minimum reaction time.
- Immobilised enzymes are ideal for multi-enzyme reaction systems.
- Control of function of immobilised enzymes is easy.
- Immobilised enzymes are suitable for industrial and medical use.
- They minimise affluent disposal problems.
- Some manipulations of enzyme catalysed reactions are better in immobilised form.

The immobilised enzymes are preferred over immobilised cells or tissues, because they have a high specificity and yield pure products.

(*b*) Disadvantages: There are certain disadvantages of immobilised enzymes such as possible loss of biological activity of an enzyme during immobilisation or during its use. Moreover, immobilisation is an expensive affair that often requires sophisticated requirements.

- The cost of immobilised enzyme gets increased after immobilisation.
- Immobilisation often adversely affects the stability and/or activity of the enzymes.
- This approach cannot be used when one of the substrates is insoluble.
- Some immobilisation strategies present large problems in diffusion of the substrate to reach the enzyme.

The covalent bonding of the enzyme to a selected carrier has several possible variations. Different types of materials, namely cellulose, glass particles, silica, etc. have been used as carriers for immobilisation. Enzymes immobilised by entrapment and adsorption may lose it activity due to leakage or desorption. Moreover, the chemical treatment may permanently damage some enzyme molecules that form the covalent bond between an enzyme and its carrier. Immobilisation yield can be used to describe the percent of immobilised enzyme activity, *i.e.* % yield = 100′ (activity immobilised/starting activity). However, immobilisation efficiency describes the percent of enzyme activity that is observed : % efficiency = 100′ (observed activity/activity immobilised).

Comparison between Immobilised Enzymes and Immobilised Cells

Generally immobilised enzymes are preferred over immobilised cells due to specificity to yield the product in pure form. However there are several advantages of using immobilised multi enzyme system such as organelles and whole cells over immobilised enzymes. The immobilised cells possesses the natural environment with cofactor availability, also its regeneration capability is particularly suitable for multiple enzymatic reaction.

4. Immobilisation of Whole Cells

In recent years, the technique of whole cell immobilisation has revolutionised the prospects of enzyme application in industry. An alternative to enzyme immobilisation is the whole cell immobilised which is now a well developed method. Several industrial plants have successfully demonstrated the performance of immobilisation. In cell immobilisation technology the main important feature is that enzymes are active and stable for a long period of time. It keeps within the cellular domain together with all cell constituents, whether the cells are dead or viable but in resting state.

The methods of whole cell immobilisation are same as described for enzyme immobilisation, *i.e.* adsorption, covalent bonding, cell to cell cross-linking, encapsulation, and entrapment. Since a long time adsorption of cells to preformed carrier has been done; for example use of woodchips as carrier for *Acetobacter* has been in practice for vinegar production since 1823. Preformed carrier of choice is used (Table 15.10). Cell attachment to the surface of preformed carrier is done by covalent bonding.

(*a*) Immobilisation of Animal Cell Cultures: The microcarrier beads used for cell immobilisation develop charged surfaces, either positive or negative charges; for example, DEAE sephadex (diethyl -aminoethyl type of cross-linked dextran) developed positive charges, polysysteine negative charges, gelatin beads slightly positive or negative charges. The microcarrier beads require surface coating for adhesion of sufficient number of cells because energy is required for cell adhesion. On the other hand, glass and ceramic materials are used as good carriers as they contain high surface energy. Different materials such as glass spheres are packed for cell attachment in different types of bioreactors. These materials increase surface area for cultivation of cells in mass. This type of bioreactor is called packed bed bioreactor. About 50 kg of 3 mm diameter spheres are required for a bioreactor of 100 liter capacity that results in a total surface area of about 20 m^2.

Table 15.10. Immobilisation of whole cells by different methods (Source: Moo-Young, 1985).

Support material	***Cells***	***Reaction***
	A. Adsorption	
Gelatin	*Lactobacilli*	Lactose/lactic acid
Porous glass	*Saccharomyces carlsbergensis*	Glucose/ethanol
Cotton fibers	*Zymomonas mobilis*	Glucose/ethanol
Vermiculite	*Z. mobilis*	Glucose/ethanol
DEAE-cellulose	*Nocardia erythropolis*	Steroid conversion
	B. Covalent Bonding	
Cellulose + cyanuric chloride	*S. cerevisiae*	Glucose/ethanol
Ti (IV) oxide, etc.	*Acetobacter* sp.	Wort/vinegar
Carboxymethylcellulose + carbodiimide	*Bacillus subtilis*	L-histidine/uronic acid
	C. Crosslinking of cell-to-cell	
Diazotized diamines	*Streptomyces*	Glucose/fructose
Glutaraldehyde	*E. coli*	Fumaric acid/L-aspartic acid
Flocculation by chitosan	*Lactobacillus brevis*	Glucose/fructose
	D. Entrapment	
Al alginate	*Candida tropicalis*	Phenol degradation
Ca alginate	*S. cerevisiae*	Glucose/ethanol
Mg pectinate	Fungi	Glucose/fructose
κ-carrageenan	*E. coli*	Fumaric acid/L-spartic acid

Chitosan alginate	*S. cerevisiae*	Glucose/ethanol
	E. Encapsulation	
Cellulose acetate	*Comamonas* sp.	7-ACA production
Ethylcelhilose	*Streptomyces* sp.	Glucose/fructose
Polyester	*Streptomyces* sp.	Glucose/fructose
Alginate-polylysine	Pancreas cells	-
Alginate-polylysine	Hybridoma cells	Monoclonal antibodies

H. APPLICATION OF ENZYMES, IMMOBILISED ENZYMES AND WHOLE CELLS

For the production of antibiotics covalent immobilised penicillin G acylase has been used for over decades. Most of them are mainly used in food processing like immobilised glucose isomerase for the continuous production of fructose syrup. This has forced some chemical producing companies to immobilise enzymes themselves: although both the enzyme and the carrier are available on the market, no immobilised enzyme can be obtained. Only a dozen of immobilised enzymes are used as catalysts in different industrial processes, but a few enzymes are commercially available in immobilised form. The most common is lipase B from *Candida antarctica* (*i.e.* novozymes) absorbed on macroporous beads, which is used in diverse chemical applications.

I. APPLICATION OF ENZYMES AND IMMOBILISED ENZYMES

Enzymes are used in our everyday life, such as cloth washing, lowering the hardness of the water, lowering the alkalinity of the water, etc. Nothing can ultimately compete with enzymes from efficiency, cost and safety perspective. Moreover, the enzymes are broadly used as tools in agriculture, industries and medicine. Table 15.11 shows a list of some enzymes used for commercial activities.

When an enzyme is attached to an inert and insoluble material such as calcium alginate (produced by reacting a mixture of sodium alginate solution and enzyme solution with calcium chloride), it is called 'immobilised enzyme'. An immobilised enzyme provides increased resistance to changes in conditions such as pH or temperature. It holds the enzymes in place throughout the reaction; hence they are easily separated from the products and may be re-used. Therefore, enzyme immobilisation is a more efficient process and so is widely used in industry for enzyme catalysed reactions. In contrast, whole cell immobilisation is an alternative to enzyme immobilisation.

Table 15.11. List of some enzymes used for commercial activities.

Enzyme	*Source*	*Application*	*Use in the Form*
α - Amylase	*Bacillius licheniformis*	Hydrolysis of starch to dextrans	Extracellular, soluble
Glucoamylase	*Aspergillus niger*	Dextran hydrolysis to glucose	Extracellular, soluble
Xylose (glucose) isomerase	*Bacillus coagulans*	Pure glucose to equilibrium mixture of glucose + fructose	Immobilized whole cells
Alkaline protease	*B. licheniformis, B. subtilis*	Detergents (protein digestion)	Extracellular, soluble
Neutral protease	*B. amyloliquefaciens*	Protein digestion in brewing substances	Extracellular, soluble
Acid protease (rennet)	*A. niger*	Milk coagulation and cheese flavour enhancement	Extracellular, soluble

Pectinase	*A. niger, B. subtilis*	Pectin hydrolysis in fruit juices	Extracellular, soluble
Glucanase	*A. niger, B. subtilis*	In fruit juices	Extracellular, soluble
Cellulase	*A. niger, Trichoderma reesei*	Cellulose hydrolysis in fruit juices	Extracellular, soluble
Lipase	*Rhizopus species*	Detergents, lipid hydrolysis	Extracellular, soluble
Lactase	*A. niger*	Milk lactose hydrolysis to glucose	Immobilized on glass beads

Some of the enzyme applications are broadly grouped in the following broad categories.

1. Applications in Pharmaceutical Industry

There are several enzymes which are immobilised for production of a variety of products in pharmaceutical industry.

***(a)* Uses in Semi-synthetic Penicillin Production:** All penicillins consist of an active beta lactam ring, *i.e.* 6-amine penicillanic acid (6-APA) group combined with different side chains (R group). Penicillin G/V acylase removes G/V group from penicillin G/V resulting in separation of 6-APA and R groups. Production of 6-APA by deacylation of the side chain, in either penicillin G or penicillin-V, by using penicillin-G acylase or penicillin-V acylase (penicillin amidase), respectively is one of the major applications of immobilised enzymes in pharmaceutical industry. Finally, new synthetic side chains are coupled with 6 APA to synthesise new semi-synthetic penicillins, for example ampicillin; enzyme reactions are given below:

$$\text{Penicillin-G} \xrightarrow{\text{Penicillin-G acylase}} \text{6-APA + G side chain}$$

$$\text{Penicillin-V} \xrightarrow{\text{Penicillin-V acylase}} \text{6-APA + V side chain}$$

$$\text{6-APA} \xrightarrow{\text{Side chain addition}} \text{Semi-synthetic penicillin}$$

E. coli produces penicillin-G acylase, and fungi and actinomycetes produce penicillin-V acylase, and removers 6-APA. The biosynthesis of penicillin acylase in *E. coli* is controlled by altering the nutrients and culture conditions. The growth substrate, inducers and regulators affect on enzyme formation. Phenyl acetic acid (PAA) (20 mM) has been found to intracellularly stimulate the synthesis of penicillin-G acylase in *E. coli* by 8-10 folds. However, glucose, lactose, sorbitol, acetate and lactate catabolically repress enzyme expression in *E. coli*. But phosphate and yeast extract are essential for growth and enzyme biosynthesis.

Today more than 50% of 6-APA is produced enzymatically by using the immobilised route. A pure product is obtained by employing the immobilised enzymes that minimises the purification costs. During 1970s, for the first time Squibb (USA), Astra (Sweden) and Riga Biochemical Plant (USSR) set up the industrial process for production of 6-APA. Now-a-days, most of the pharmaceutical companies make use of this technology. This has resulted in commercially production of many immobilised systems for penicillin acylase making use of a variety of techniques emploing the isolated enzyme or the whole cells. In India this is also one of the major applications of the immobilised enzyme technology. Similarly 7-aminodeacetoxy-cephalosporanic acid, an intermediate in the production of semisynthetic cephalosporins, has also been produced.

***(b)* Immobilisation of Amino Acid Acylase:** The enzyme amino acid acylase is the first immobilised enzyme used industrially by Tanabe Seiyaku Company (Japan) for the resolution of recemic mixtures of chemically synthesised amino acids. The L form of the *N*-acetyl amino acids is catalysed by amino acid acylase through deacetylation leaving unaltered the *N*-acetyl-d amino acid. It can easily be separated, racemized and recycled. Some of the immobilized preparations are : enzyme immobilised by ionic binding to DEAE-sephadex and the enzyme entrapped as micro-droplets of its aqueous solution into cellulose triacetate fibres by means of fibre wet spinning.

(*c*) Production of High Fructose Syrup: Fructose is the sweetest among the monosaccharides having twice sweetening strength as compared to that of sucrose. Glucose in spite of being the most abundant monosaccharide cannot be a good substitute for sucrose for sweetening because glucose is about 75% sweet as compared to sucrose. Therefore, fructose has a great demand as sweetener due to being very sweet. Similar to glucose or sucrose, fructose also has the same calorific value. Moreover, about equivalent amount of glucose and fructose are present in high fructose syrup; therefore, in the preparation of soft drink, processed food and baking, high fructose syrup is a good substitute for sugar. High fructose syrup can be produced from glucose by employing an immobilised enzyme 'glucose isomerase'. The raw materials such as wheat, potato, corn, etc consist of starch; hence these materials are hydrolysed to produce glucose from starch. Then glucose isomerase isomerises the glucose to fructose (Fig. 15.24). Therefore, the product consists of high fructose syrup, *i.e.* about 50% fructose.

CHO		CH_2OH
CHO		CO
HOCH	Glucose isomerase ⇌	HOCH
CHOH		CHOH
CHOH		CHOH
CH_2OH		CH_2OH
glucose		fructose

Fig. 15.24. Conversion of glucose to fructose by glucose isomerase.

Enzyme		**Starch**		**Products**
Glucose isomerase (*B. coagulans*, *Streptomyces* sp.)	⟶	**Isomerization** pH: 6.5-8.5 Temperature: 60-65°C	⟶	Fructose syrup
α-amylase (*B. licheniformis*)	⟶	**Liquefaction** pH: 5.5-7.0 Temperature: 90°C	⟶	Maltodextrins
α-amylase (*Aspergillus oryzae*)	⟶	**Saccharification** pH: 5.0-7.0 Temperature: 50-55°C	⟶	Maltose syrup in high amount

Glucose isomerase is an intracellular enzyme produced by several microorganisms, such as the species of *Arthrobacter*, *Bacillus* and *Streptomyces*. Isolation of glucose isomerase requires special and costly techniques without loss of biological activity, because it is an intracellular enzyme. Therefore, whole cells or partially broken cells are immobilised and used or production of glucose isomerase.

Table 15.12 shows some of the commercial preparations made through the adsorption or the cross-linking technique. The non-tropical countries have abundant starch raw material; hence application of glucose isomerase technology has gained considerable importance in these countries. In contrast, in tropical countries like India, the high fructose syrups can be obtained by a simpler process of hydrolysis of sucrose using invertase, because sugarcane cultivation is abundant. Invert sugar has a higher solubility and osmotic pressure as compared to sucrose.

Historically, invertase is the first enzyme reported in an immobilised form; several immobilised invertase systems have also been patented. During the end of 1970s, S. F. D'Souza and G.B. Nadkarni demonstrated the possible use of whole cells of yeast as a source of invertase.

Table 15.12. Commercially immobilised glucose isomerase preparations.

Enzyme source	*Methods of immobilisation*	*Name of Companies*
Actinoplanes missouriensis	Cells entrapped in gelatin, and crossed-linked in glutaraldehyde	Gist Brocades
Arthrobacter sp.	Flocculated cells extruded and dried as cylindrical pellets	ICI Americas Inc.
Bacillus coagulans	Enzyme mixed with inorganic diluents and formed into solid spheres or cell lysate cross-linked with glutaraldehyde	Novo Industry
Streptomyces sp.	Heat fixed cells cross-linked with glutaraldehyde	Miles Kali Chemic
Streptomyces olivaceous	Enzyme adsorbed on ion-exchanger	Clinton Corn Processing
S. olivaceous	Glutaraldehyde cross-linked with whole cell granules	Min Car Int
S. olivaceous	Glutaraldehyde cross-linked with whole cell granules	Miles Laboratory
S. oliochromogenes	Enzyme adsorbed on controlled pore alumina	Corning Glass Works
S. oliochromogenes	Adsorption on alumina/other ceramics/ion-exchange resin	CPC Int. Inc

***(d)* Enzymes Used in Production of Glucose Syrup**- Glucose syrup is produced from liquefied starch, which is obtained (by heating to 105°C for 5 minutes) a 30-40% granular starch slurry (containing 20-80 ppm Ca^{2+}) mixed with α-amylase. Due to heating the starch becomes gelatinised; then the temperature is lowered to 90-100°C and incubated for 2 hours. The α-amylase digests the starch to produce soluble dextrin molecules, which is called liquefaction. Small quantities of liquefied starch are spray-dried for use in baby foods and as bulking agents. Liquefied starch is rapidly cooled to 60°C, pH adjusted to 4.0-4.5, and 200 U/kg of glucoamylase is added to it. A debranching enzyme pullulanase from *Bacillus acidopullulyticus* is also added to it at 100 U/kg to digest branched oligosaccharides.

2. Therapeutic Uses (= Medical and Clinical Uses)

Since 19th century pancreatic enzymes have been used in digestive disorders. The digestive enzymes (for example amylases, proteases and lipases) are often used as components of drugs to cure the digestive disorders. Most enzymes are used extracellularly for : (*i*) topical applications (e.g. collagenase), and (*ii*) removal of toxic substances (e.g. rhodonase), or in disorders within blood circulation system (e.g. streptokinase, urokinase, etc.). The enzymes must be in highly pure form and free from any contamination; this is the reason why they are very costly and generally obtained from animal sources. For example, urokinase is isolated from human urine and costs nearly $ 200/mg; the annual market for this enzyme is nearly $150 million. It is used to cure the disorders of the blood circulation system.

All of us are aware of extensive bleeding after surgery; therefore, to stop postoperative bleeding, thrombin from beef plasma is used externally that converts fibrinogen to fibrin which makes blood to clot. Enzymes such as trypsin, ficin, and streptodornase are used as surface disinfectants and for removal of damaged tissue. Trypsin cleans the thick and purulent masses in surface infections and in enlargement of vesicles of lungs associated with pleural pneumonia. Cleaning by digestion heals the wound rapidly than untreated area.

Day-by-day the number of diabetics is increasing throughout the world with increase in population. Moreover, the diabetics regularly determine glucose level in blood calorimetrically by using glucose oxidase in combination with peroxidase.

Humans suffer from stomach ache due to worm infection. Plant proteases like papain from papaya and ficin from fig are used as de-worming medicines for humans as well as domestic animals. These enzymes can digest the worm but the secretions from the worms do not inhibit these enzymes.

Enzyme applications in medicine are as extensive as in industry. The variety of enzymes and their potential therapeutic applications are considerable. A selection of those enzymes which have realised this potential to become important therapeutic agents is shown in Table 15.13. Some of the applications of the enzyme in medicine are discussed briefly.

Table 15.13. Therapeutic applications of enzymes.

Enzyme	*Uses*
Asperaginase	Leukaemia
Collaginase	Skin ulcers
Glutaminase	Leukaemia
Hyaluronidase	Heart attack
Lysozyme	Antibiotic
Rhodanase	Cyanide poisoning.
Ribonuclease	Antiviral
Beta-lactamase	Penicillin allergy
Streptokinase	Blood clots
Trypsin	Inflammation
Uricase	Gout
Urokinase	Blood clots

***(a)* Enzymes Used in Cancer Therapy:** Enzymes have a major potential application in treatment of cancer. Proteolytic enzymes have a long history of use in cancer treatment. In 1906, John Beard, a Scottish embryologist, reported in his book *The Enzyme Treatment of Cancer and its Scientific Basis* on the successful treatment of cancer using a pancreatic extract. Proteolytic enzymes have been promoted by numerous alternative cancer practitioners for many years. After absorption, the body prevents digestion of proteins in the blood and other body tissues producing antiproteases that block the invasiveness of tumour cells and prevent the formation of new blood vessels (angiogenesis). Besides, the proteolytic enzymes exert many other interesting anticancer mechanisms such as inhibition of metastasis (the spread of cancer) and the enhancement of the immune response.

Asparaginase is used in the treatment of lymphocytic leukemia. The tumour cells are unable to synthesise L-asparagine due to an enzyme deficiency; hence they derive this amino acid from the body fluids. Therefore, the intravenous injection of asparaginase drastically reduces the level of L-asparagine in the bloodstream, and creates starvation for this amino acid in tumour cells.

Interestingly the normal cells are not affected at all because they synthesise asparagines on their own. Asparaginase is injected intravenously, shows half-life of about 1 day (in dog), and results in the complete recovery of the cases by 60%.

Proteolytic enzymes can be used effectively with conventional therapy (surgery, chemotherapy and/or radiation). Proteolytic enzymes are not recommended for at least two days before or after a surgery as they may increase the risk of bleeding. Proteolytic enzymes in speed up post-surgical recovery and relieve a complication of surgery and radiation known as lymphedema (Table 15.14).

Table 15.14. Patients given proteolytic enzymes after receiving chemotherapy.

Cancer	*Medication*	*Duration of therapy*	*Effect of enzyme therapy*
Inoperable lung cancer	Fluorouracil, vinblastine, cyclophosphamide plus enzyme	12 months	General improvement in health and life, side effect is fever
Colon cancer	Fluorouracil, levamisol vs. plus enzyme	2-45 months	Adverse effect of chemotherapy reduced, more patients survived
Gastric cancer	Mitomycin, fluorouracil, cytarabine vs picibanil plus enzyme	6-12 months	Increase the ration of T lymphocytes to total lymphocytes
Multiple myeloma	Multicombination chemotherapy vs. plus enzyme	6 months	Survival of patients with stage II multiple myeloma increased by 36 months
Ovarian cancer	Carboplatin, epirubicin, prednimustin plus placebo vs. plus enzyme	6 months	More rapid fall in tumour enzymes

There are some limitations of the therapeutic enzymes that limit its application in medicine because : (*i*) larger molecular size inhibits its distribution among body cells, (*ii*) enzymes can elicit immune response in the patient, especially on prolonged use, because they are, and (*iii*) in the circulatory system most enzymes have short effective life of few minutes.

3. Uses of Enzymes in Foods and Beverages

Most people are already familiar with the domestic application for enzyme technology. Historically, in early biotechnological practices humans have been using enzymes for centuries to produce foods. It was possible to make wine, beer, vinegar and cheeses due to utilisation of enzymes present in the yeasts and bacteria. Now specialised strains have been developed for specific uses that improve the flavour and quality of the products.

By using enzymes the process can be made cheaper and more predictable. Beside, the length of time required for aging can be reduced, which stabilises the products or controls alcohol and sugar contents. Enzymes have also been used for years to turn starch into sugar. Throughout the food industry corn and wheat syrups are used as sweeteners. Production of these sweeteners can be less expensive than using sugarcane sugar. Moreover, there is growing sale of technical enzymes, food and beverage enzymes, and other enzymes (Fig. 15.25).

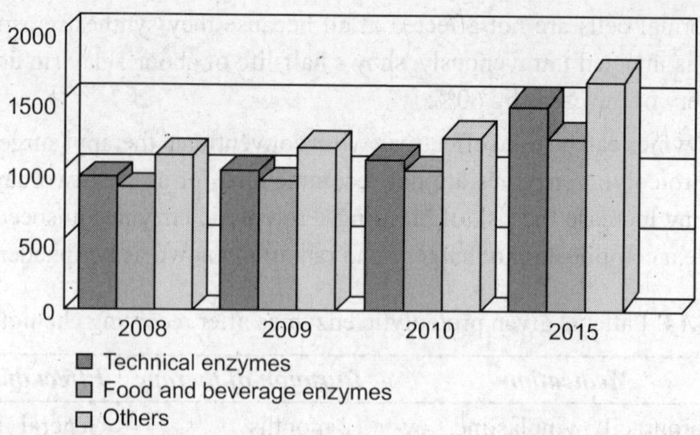

Fig. 15.25. Increasing sale of technical enzymes, food and beverage enzymes, and other enzymes.

4. Uses of Enzymes in Textile Industry

About all textile-manufacturing processes use harsh chemicals that pollute the environment and cause health hazard. Therefore, the demands are increasing to reduce pollution replacing harsh chemicals with enzymes. Now-a-days enzymes are used to prepare the fabrics for making clothing, furniture and other household items. Also enzymes are used to enhance the preparation of cotton for weaving, reducing impurities, minimising pulls in fabric, or as pre-treatment before dying to reduce rinsing time and improve colour quality. These steps results in the process less toxic and eco-friendly. Besides, the costs associated with the production process and consumption of natural resources (water, electricity, fuels) get reduced; the quality of the final textile product is also improved.

5. Use of Enzymes in Removal of Stickies

Stickies are the deposits that consist of synthetic contaminants from the breakdown of adhesives in wastepaper. Stickies are tacky, hydrophobic, pliable organic materials that reduce the quality of the final paper product, can clog the paper mill machinery and cost hours of downtime. Direct breakdown of adhesives in wastepaper results in primary stickies, and/or the interaction of chemicals results in the secondary stickies. Chemical interactions within the system influence the tackiness of stickies particles, which is increased by both wood resins and hydrocarbons from additives. Currently 75% of paper mills utilise the recovered fiber.

The pulp and paper industry use the enzymes for the removal of stickies, the glues, adhesives and coatings introduced to pulp during recycling of paper. The ester bonds held together the stickies; therefore, in pulp the esterase enzymes cut the stickies into smaller water soluble compounds and help to remove them from the pulp. Esterases are effective only at moderate temperature and pH. Their effective temperature, pH ranges, and substrate capabilities can be broadened through the genetic modifications of existing enzymes.

6. Use of Enzymes in Detergents

Traditional use of enzymes in laundry detergents involved those that degrade proteins causing stains. The largest amount of enzymes is used in detergent industries, amounting to about 25-30% of the total sales of enzymes. The enzymes used in detergents must be cost effective, safe to use and capable of performing the task even in the presence of anionic and non-ionic detergents, soaps, oxidants, etc. at pH between 8 and 10.5.

Detergents consist of about 0.4-0.8% crude enzyme by weight. The main enzymes used in detergents are : proteases, α-amylase, lipases, and cellulase. Proteases, in association with dirt, digest proteins present in blood stains, milk, grass, etc. therefore, they remove the dirt also; but only serine proteases are suitable for use in detergents. Serine proteases are produced by *Bacillus licheniformis* and *B. amyloliquefaciens*. Since *B. licheniformis* can grow in alkaline condition, it produces a protease that survives at high pH levels having an optimum pH at around 9 and 10. At this pH it can remove protein-comprised dirt in clothes. Alkaline protease prevents shrinkage and fading colours because it allows low temperatures, which in turn lowers energy usage.

Proteases are packed inside granules coated with paraffin oil or polyethylene glycol plus hydrophilic binders. After washing, the granules disperse and release the enzyme. Coating of proteases protects the users from hypersensitivity to the enzymes. The α-amylase produced by *B. licheniformis* digests the starch present with dirt and stains. Lipases are used in detergent for digestion of lipids present in fat stains and/or dirt, and clean grease traps or other fat-based cleaning applications. In addition, there is a large number of fungi that produce cellulases used for washing cotton fabrics that removes soil particles by digesting the associated cellulose without damaging the major fibers.

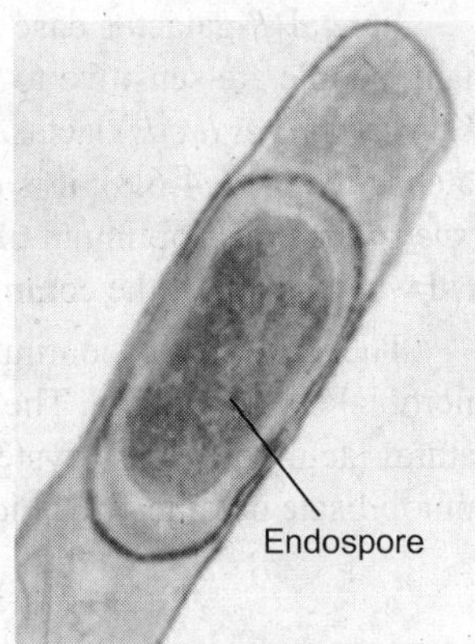

Bacillus licheniformis forms an endospore.

In recent years, recombinant enzymes (engineered proteins) are being sought using different DNA technologies such as site-directed mutagenesis and DNA shuffling for improvement in detergent enzymes.

(*a*) Uses of Lipases in Detergent Industry: Lipase catalyses a series of reactions and cleave the ester bonds of triacylglycerols (hydrolysis); lipases also catalyse the reverse reaction under microaqueous conditions, such as formation of ester bonds between alcohol and carboxylic acid moieties. In addition, lipases consist of much industrial potential such as : use in detergents, leather treatment, controlled hydrolysis of milk fat for acceleration of cheese ripening, hydrolysis and alcoholysis of bulk fats and oils, production of pure compounds, flavours, etc. Lipases are soluble in aqueous phase but the lipids are insoluble. The practical use of lipases in pseudo-homogeneous reactions poses technological difficulties. Many attempts have been made to solve these problems by using immobilised lipases.

7. Application in Dairy Industry

In dairy industry immobilised enzymes (e.g. β-galactosidase) are used to prepare lactose-hydrolysed milk and whey. Hence, the lactose intolerant can consume lactose-hydrolysed milk. In a country like India where lactose intolerance is quite prevalent, this technique has a great significance. The sweetness and solubility of the sugars are enhanced due to lactose hydrolysis; consequently a variety of dairy products can be prepared in future. Lactose-hydrolysed whey may be used as a component of whey-based beverages, leavening agents and feed stuffs. It may also be fermented to produce ethanol and yeast. Thus, it converts an inexpensive byproduct into a highly nutritious, good quality food ingredient. The Centrale del Latte of Milan (Italy) was the first company to commercially hydrolyse lactose in milk through immobilisation of neutral lactase from yeast entrapped in synthetic fibres. Specialist Dairy Ingredients had set up an immobilised β-galctosidase plant in North Wales for the production of lactose-hydrolysed whey. Specialist Dairy Ingredients is a joint venture between the Milk Marketing Board of England, and Wales and Corning.

Fungal β-galactosidase has also been used unlike the milk acidic β-galactosidase. Since many people are sensitive to lactose, lactase is used to remove lactose from milk and whey. Yeast (*Kluyveromyces lactis*) lactase is immobilised in cellulose triacetate fibers which are cut and used in batch (5°C, pH 6.4-6.8); it is used for processing of milk and sweet whey. But fungal lactase (from *Aspergillus niger*, optimum pH 3.0-4.0) is immobilised on porous silica and is used for treatment of acid whey. Some of the commercial lactase systems are given in Table 15.15.

The large-scale continuous processing of milk using immobilised enzyme suffers from microbial contamination. Therefore, during continuous hydrolysis of lactose by the initiation of the natural lacto-peroxidase system in milk, bacterial contamination can be minimised by using a co-immobilisate obtained by binding of glucose oxidase on the microbial cell wall.

Table 15.15. Commercially immobilised lactase preparations.

Source of enzymes	*Methods of immobilization*	*Company*
Aspergillus niger	Enzyme adsorbed/cross-linked to phenol form-aldehyde resin	Valio laboratory
A. niger	Enzyme covalently bound to silanised Mn-Zn ferrite	Sturge
A. oryzae	Enzyme covalently bound to microporous PVC-silica sheet	Amerace Corporation
A. oryzae	Enzyme covalently bound to macroporous amphoteric ion exchange resin	Sumitomo
Kluyveromyces lactis	Enzyme entrapped in cellulose triacetate fibres	Snampogetti
Saccharomyces cerevisiae	Enzyme entrapped in cellulose triacetate fibres	Gist Brocades

8. Uses of Enzymes in Food Industries

Several bioactive peptides are being used in food industries as sweeteners, flavourings, antioxidants and nutritional supplements. Over the last two decades immobilised proteases have emerged as powerful catalysts for the synthesis and modification of peptides. The synthesis of peptide sweetener using immobilised enzymes like the thermolysin is one of the important large scale applications. Proteolytic enzymes are used (Bayer AG) to resolve A *N*-acyl-DL-phenylglycine ester racemate, yielding *N*-acyl-D-esters or *N*-acyl-D-amides and *N*-acyl-L-acids; these proteolytic enzymes include subtilisin, *a*-chymotrypsin, papain, ficin or bromelain which have been immobilised by covalent binding, adsorption or cross-linking to polymeric supports. DL-phenylgycinamide racemates can be separated by the immobilised aminopeptidases. The immobilised hydropyrimidine hydrolase have been used to prepare D-carmamyl amino acids and the corresponding D-amino acids from various substituted hydantoins.

9. Uses of Enzymes in Biosensors

There are many types of biosensors, for example phosphate detection biosensors consist of alkaline phospatase immobilised on glass fiber for detection of phosphate in water samples, milk and shrimp. This attachment occurs by covalent bond or physical adsorption. Phosphate biosensor has widely been used as the traditional enzyme immobilisation technology. Glass provides desirable mechanical strength and thermo-chemical stability, but it is difficult to achieve high enzyme loading as required for efficient bioprocess application. Some of the enzymes/whole cells are used in biosensosors; it has been dealt with in detail in preceding section.

II. APPLICATION OF IMMOBILISED WHOLE CELLS

Enzyme immobilisation is a more efficient process; therefore, it is widely used industrially. In contrast, whole cell immobilisation is an alternative to enzyme immobilisation. Immobilised cells are used for the production of a large number of products where normally they are not produced, for example industrial enzymes, alcohols, acids, biosensors, etc. Some of the applications are described herewith.

1. Enzyme Produced by Immobilised Cells

A substantial amount of useful enzymes are produced by microorganisms. However, cell immobilisation technology is suited to produce extracellular enzymes. Interest in applying cell immobilisation techniques is gradually increasing for the continuous production of enzymes. The starch degrading enzymes (*i.e.* α-amylase and glucoamylase) have been studied extensively among the microbial enzymes. Several researchers have attempted the production of α-amylase by immobilised cells. Entrapment of cells in polyacrylamide, calcium alginate, agar, and several other polymer supports have been done. Immobilisation of *Bacillus cereus* in calcium alginate is known, and packed-bed and fluidised-bed reactors are employed to continuously synthesise thermostable α-amylase. Alginate fibres are spun by pultrusion technique to reduce the diffusion-resistances in the gel matrix, resulting in a 24-fold increase in the productivity compared to batch fermentation with free-cells.

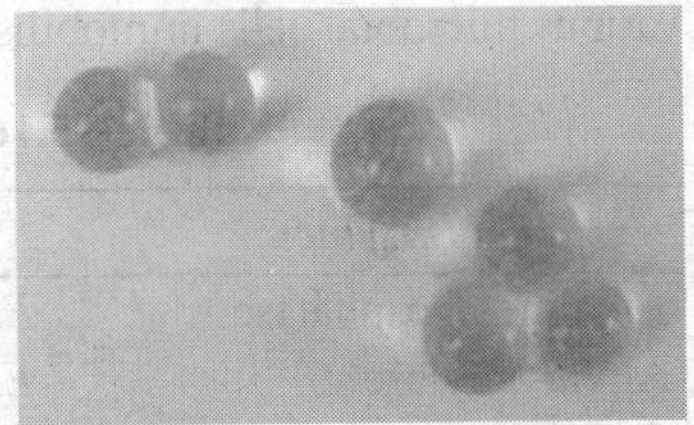

Sodium alginate beads.

Thermophilic α-amylase is also being produced from transformed cells of *E. coli* immobilised in kappa-carrageenan with supplementation of glycine. However, silicone foam has been found to be superior for entrapment of the recombinant *E. coli* to produce thermostable α-amylase; this results in a five-fold increase in production of the enzyme.

Alginate

Alginate (= alginic acid or algin) is an anionic polysaccharide widely distributed in the cell wall of brown algae, including the giant kelp *Macrocystis pyrifera*, *Ascophyllum nodosum*, and species of *Laminaria*. It is also produced by two bacterial genera, *Pseudomonas* and *Azotobacter*. In cell wall it binds to water and forms a viscous gum. In extracted form it absorbs water quickly, and is able to absorb 200-300 times its own weight in water. Its colour appears from white to yellowish brown. It is sold in form of filament, granules or powdered. The chemical compound sodium alginate is the sodium salt of alginic acid. Its empirical formula is $NaC_6H_7O_6$. Structure of sodium alginate is given below:

COONa H H COONa
H O H H O
H O OH HO H O–
–O OH HO H O OH OH
H O H
H H COONa H H

Enzymatic hydrolysis of cellulosic materials has been the focus of attention in recent years, because these materials can be converted into fuels. There are several investigations on immobilised *Tricoderma reesei* for the continuous production of cellulase. Immobilisation of the fungi on polyester cloth, nönwoven material, and cellulosic fabric has been attempted. In 1992, L.Z. Xin studied the effect of surface property of the matrix for the immobilisation of *Treesei*. The carrier materials were pre-coated with hydrophilic and hydrophobic monomers by using co-polymerisation technique, and the treated materials were used for passive immobilisation of fungi. Pretreatment with hydrophilic matrices allows the increased growth of immobilised organisms, thereby resulting in higher activity.

Lipase-catalysed inter esterifications have been superior to chemical processes in recent years. Immobilisation of various microbial cells such as *Rhizopus chinensis*, *A. niger*, *Candida rugosa*, and *Sporotrichum thermophile apinis* has been done for the production of lipases. Various biomass support particles made from urethane, nylon, polyester, and stainless steel were tried using *Rhizopus* sp. to enhance the activity of lipase several folds as compared to free-cells. Enzyme production by various microbial cells immobilised on different carrier support systems is given in Table 15.16.

Table 15.16. Enzyme production by immobilized cells.

Enzymes	***Microorganisms***	***Carrier support***
Alkaline phosphatase	*Escherichia coli*	Kappa-carrageenam
Alpha-amylase	*E. coli*	Kappa-carrageenam
Cellulase	*Trichoderma reesei*	Cellulosic fibres
Chitinase	*Micromonospora chalcae*	Calcium alginate
Glucose oxidase	*Aspergillus niger*	Wheat, rye
Invertase	Yeast	Polymer
Pectinase	*Aspergillus niger*	Wheat
Penicillin acylase	*E. coli*	Polyacrylamide
Xylanase	*Aspergillus sydowii*	Calcium alginate

2. Alcohol Production by Immobilised Cells

Ethanol fermentation is one of the widely studied systems using immobilised cells of yeast, *Saccharomyces cerevisiae*. All the methods of immobilisation were tried for alcohol fermentation, such as entrapment, adsorption, and cross linking. In 1992, H. Nellaiah and P. Gunasekaran used immobilised cells of *Zymomonas mobilis* have also been used for alcohol production. *In situ* generation of CO_2 takes place in large quantities during continuous fermentation that usually encounters problems in alcohol fermentation, resulting in severe operational difficulties. These problems are associated with gas generation, for example bed compaction, breakage of the gel particles, and improper CO_2 ventilation. Attempts have been made to overcome these problems. In 1984, M. Nagashima and co-workers have operated a pilot plant of 4.0 kilo litre capacity by using alginate-entrapped cells of yeast for alcohol fermentation for a period of 4,000 hours with a constant alcohol production rate of 8.5 to 9.0%.

Several studies have been carried out on physiology of the yeast under immobilised state, which have provided information on process intensification during immobilisation. It has been reported that alcohol can be produced from starch directly in a single-step process by co-immobilising an aerobic amylolytic fungus *A. awamori* with an obligate anaerobic bacterium *Z. mobilis*. The O_2 gradients in the gel divide the growth of mixed cultures into two zones: the peripheral region of the gel bead occupied by the fungi and the core-central region by the bacteria. This is a typical case of exploiting the diffusional behaviour of gel matrix (Jamuna and Ramakrishna, 1989).

3. Use of Cell Immobilisation in Meat Industry

The meat industry is potentially benefited by cell immobilisation through : decreasing the costs of starter culture, ensuring uniform distribution of bacteria, and simplifying recovery and recycling procedures. Meat fermentation is contributed by immobilizing cells. Cell mobility is restricted by the solid/semisolid nature of the meat substrates. Production of a continuous cell line may be possible by using immobilised cells in meat starter cultures, which ensures retention of stable cell lines. One disadvantage of incorporating immobilised cells into fermented meats might be a decrease in substrate availability as compared with fluid products. Immobilized cell technology can improve ecological competence of a starter culture.

I. BIOSENSORS AND ITS APPLICATIONS

A biosensor is an analytical sensing device containing an immobilised biological material (*e.g.* enzyme, antibody, nucleic acid, hormone or an organelle/whole cell) which can specifically interact with an analyte and produce physical, chemical or electrical signals that can be measured. An analyte is the compound (*e.g.* glucose, urea, drug, pesticide) whose concentration has to be measured. Biosensor provides selective identification of different types of substances at ultra-trace levels in industrial products, chemical compounds, environmental samples (air, soil and water) or biological systems (bacteria, virus or tissue components) or for biomedical diagnosis. Biosensor is used for the detection of a specific analyte that combines an immobilised biological sensing material (bioselector, biocatalyst) with a physico-chemical detector component, *i.e.* transducer. The biological material may be tissue, microorganisms, organelles, cell receptors, enzymes, antibodies, nucleic acids, etc., physico-chemical transducer may be optical, electrochemical, calorimetrical, conductimetrical, thermometric, piezoelectric or magnetic substances.

II. CHARACTERISTICS AND TYPES OF BIOSENSORS

Biosensors possess several characteristic features, based on which various types of biosensors have been prepared. Some of the biosensors are described in this section.

1. Components of a Biosensor

The biological component (biocatalyst) consists of immobilised biological sensing materials that detects the test analyte or converts the substrate into product. An interface which is a polymeric film or membrane links the biological component to the transducer. The transducer determines the biological reaction and converts it into an electrical signal. It is the key part of a biosensor and makes use of a physical change accompanying the reaction. This may be : *(i)* the heat produced (or absorbed) by the reaction (calorimetric biosensor), *(ii)* change in the ionic concentration resulting in the production of an electrical potential (potentiometric biosensor), *(iii)* electron flux produced in an enzyme catalysed redox reaction (amperometric biosensor), *(iv)* conductance varying due to the change in ionic environment (conductimetric biosensor), *(v)* the catalytic and/or affinity reactions causing a change in fluorescence or absorbance (optical biosensor), and *(vi)* variations in resonance frequency being observed in piezoelectric crystals due to change in the mass of reactants or products (piezo-electric biosensor).

The output electrical signal from the transducer is often low and superimposed upon a relatively high and noisy baseline. This output signal and also a reference baseline signal produced from a similar transducer which is not in contact to a biocatalytic membrane are amplified by an amplifier. Both the amplified signals are transmitted to the microprocessor, which then processes both the signals that involves subtracting the reference baseline signal from the sample signal. The resultant signal difference is electronically filtered to make it free from unwanted noise. This signal is converted to a digital signal in terms of concentration units and is proportional to the concentration of material being analysed. The digital signal thus generated is passed to an output device. It is a digital panel for displaying and/or storage of data. A biosensor comprises of following main components (Fig. 15.26) :

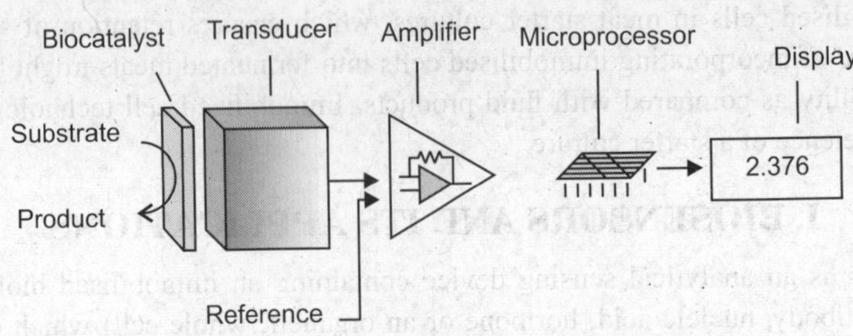

Fig. 15.26. Components of a biosensor.

2. Characteristics of Biosensors

Generally the biological component is immobilised near or onto the surface of the transducer to facilitate reuse, to minimise interference and to maximise response. In each case it is the ability of the biological component to react or respond specifically to an analyte(s) that makes the biomolecule suitable for use as the sensing element of a biosensor. For example, an antibody will only bind to the specific antigen under suitable conditions. The biological signal produced is converted by a suitable transducer into a quantifiable electrical signal.

A wide range of analytes can be detected by biosensors such as pollutants, explosives, viruses, biochemical and pharmaceutical products, vitamins, amino acids, heavy metals, ions, gases, etc. Some remarkable characteristic features of biosensor are given below :

(*i*) **Specificity:** Biosensor has the remarkable ability to distinguish between the analyte of interest and other similar substances. An analyte can be detected using biosensor with great accuracy. The biocatalyst must be highly specific for the purpose of the analyses, be stable under normal storage conditions, and show good stability over a large number of assays.

(*ii*) **Response time:** The response should be accurate, precise, reproducible and linear over the useful analytical range without dilution or concentration. Analytic tracers or catalytic product can be detected directly and instantaneously with the help of biosensors. Thus they are very much suitable for on-line monitoring.

(*iii*) **Simplicity**: The sensing molecule and the transducer are integrated on to a single probe. Thus handling of biosensors is very easy.

(*iv*) **Continuous monitoring ability:** Biosensors can be regenerated and reused. Thus they are suitable for continuous monitoring purposes which can never be done with conventional analysis methods.

(*v*) **Reproducibility:** Biosensors are very reliable; reproducibility of the values is also a great advantage.

(*vi*) **Portability:** Portability of such sensors has many advantages which can never be with a spectrophotometer.

(*vii*) **Cost:** The cost of the biosensor limits its application to some extent. The complete biosensor should be cheap, small, portable and capable of being used by semi-skilled operators.

(*viii*) **Marketability:** There should be a market for the biosensor. There is clearly a purpose for developing a biosensor if other factors (such as government subsidies, the continued employment of skilled analysts, or poor customer perception) encourage the use of traditional methods and discourage the decentralisation of laboratory testing.

The biocatalytic membrane determines the biological response of the biosensor by accomplishing the conversion of reactant to product. There are number of advantages in immobilised enzymes; that is why they are used in such systems. A similar catalytic activity is present for a series of analyses; therefore they may be re-used.

3. Type of Biosensors

Biosensors are of different types based on the use of different biological material and sensing devices; some of the biosensors constructed for various purposes are described below:

(*a*) Glucose-electrode Biosensor: A glucose-electrode is built by immobilising the glucose oxidase in polyacrylamide gel around a platinum oxygen-electrode separated by a teflon membrane. Potassium chloride (KCl) solution is placed around the platinum-oxygen electrode. Glucose oxidase is intimately covered from the upper surface by a cellulose acetate membrane (Fig. 15.27).

Glucose solution
Cellulose acatate membrane
Immobilized glucose oxydase
Permeable teflon membrane
Ring seal
Saturated KCl solution
Platinum O_2 electrode

Fig. 15.27. A glucose electrode.

The enzymatic and detection parts are closely combined in one solid state electrode. The electrode surface is renewable; hence it overcomes the limitations of membrane-immobilised enzyme systems. An electronic box controls the sensor that automatically exudes a determined quantity of carbon paste from the cartridge. Then the electrode surface is polished and a new clean reaction surface is provided for each measurement. When the electrode is placed in a solution containing glucose and polarised, the reaction takes place as shown in Fig. 15.28.

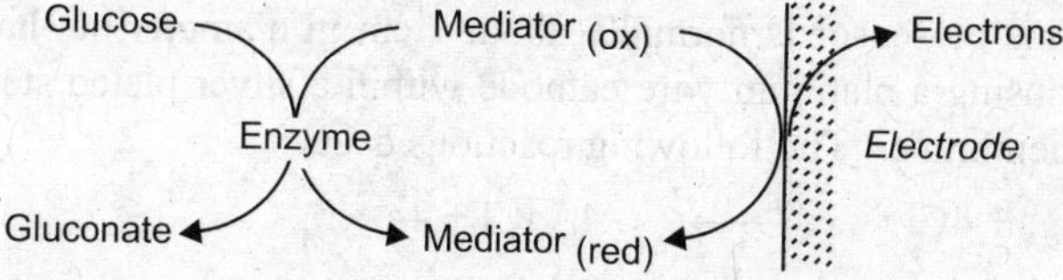

Fig. 15.28. Biochemical mechanism of glucose electrode biosensor.

Upon putting the glucose solution into the contact of membrane, glucose and oxygen pass through membrane into the enzyme layer. As a result of oxidation-reduction reactions, glucose is converted into gluconate (*i.e.* gluconic acid) and hydrogen peroxide in the presence of water, oxygen and glucose oxidase. Consequently, oxygen concentration in the gel around the electrode is lowered down. Hydrogen peroxide (H_2O_2) brings a change in current, *i.e.* measurable signal. Electrode records the rate of reactions; the rate of diminition of oxygen concentration is proportional to glucose concentration of the sample. It responds linearly to glucose concentrations over a range of 10^{-1}-10^{-5}mol dm^{-3} with a response time of 1 minute.

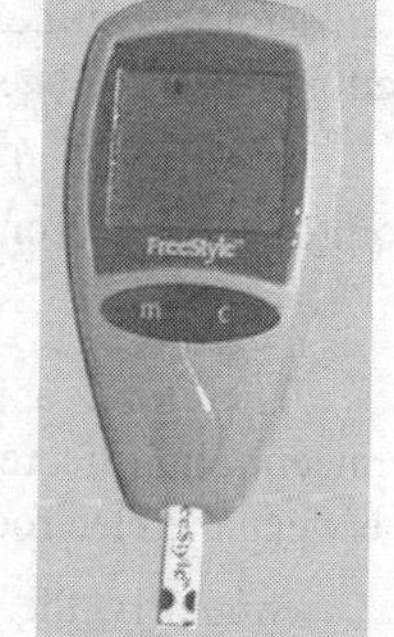

A glucose meter

In 1987, for the first time Yellow Springs Instruments Co. (USA) developed a biosensor for diagnostic purposes for measuring glucose in blood plasma. It is a hand - held machine which measures six components of blood plasma for example, glucose, urea, nitrogen, sodium, potassium and chloride. In India, an indigenous glucose sensor has been developed by the scientists at Central Electrochemical Research Institute (CECRI), Karaikudi. It gives electrical signal for a glucose concentration as low as 0-15 millimoles.

(b) Electro-chemical Biosensor. This type of biosensor has been developed using electronic devices such as field effect transmitors or light emitting diode. The effect transmitors measure the charge accumulation on their surface and the light emitting diode measures photoresponse generated in silica-based chip as an alternating current. Thus, a biochemical reaction is measured by the field effect transmitor at the surface, which is then induced into current. Moreover, the field effect transmitors can be modified to ion-sensitive, enzyme-sensitive or antibody-sensitive transmitors by using selective ions, enzymes or antibodies, respectively.

(c) Amperometric Biosensor. The biosensor that measures the reaction of anylate with enzyme and generates electrons directly or through a mediator is called 'amperometric biosensor'. When a potential is applied between two electrodes, amperometric biosensors function by the production of a current. Similar to the potentiometric biosensors, generally they have response times, dynamic ranges and sensitivities. The simplest amperometric biosensors involve the Clark oxygen electrode (Fig. 15.29).

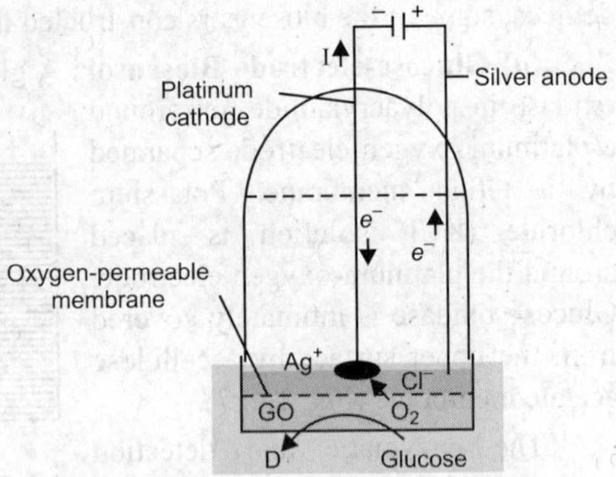

Fig. 15.29. A simple amperometric biosensor.

An amperometric biosensor consists of a platinum cathode (at which oxygen is reduced) and a silver/silver chloride electrode. After applying a potential of -0.6 V (relative to the Ag/AgCl electrode) between the central platinum cathode and the annular silver anode, a current (I) proportional to the oxygen concentration is generated which is carried between the electrodes by means of a saturated KCl solution. Normally both electrodes are bathed in a solution of saturated KCl and separated from the biocatalyst (for example glucose oxidase, GOD) by a plastic membrane (form example teflon, polytetrafluoroethylene) permeable only to an oxygen. This biosensor is normally about 1 cm in diameter but has been scaled down to 0.25 mm diameter using a platinum wire cathode within a silver plated steel needle anode and utilising dip-coated membranes. The following reactions occur:

Ag anode $4Ag^0 + 4Cl^- \longrightarrow 4AgCl + 4e^-$

Pt cathode $O_2 + 4H^+ + 4e^- \longrightarrow 2H_2O$

The amperometric biosensors comprises of either enzyme-electrode or without a 'mediator' or chemically modified electrodes. The enzyme-electrode biosensors are the oxygen- and peroxide-based biosensors as discussed earlier. By using a mediator some advancement has been made into this type of biosensors. In addition, the direct electron transfer systems are more advanced types. Principle of a mediated biosensor is shown in Fig. 15.30. A redox reaction catalysed by an enzyme is directly coupled to an electrode in a mediated biosensor, where enzyme is presented with the oxidisable substrate. The electrons are transferred from the substrate to the electrode via enzyme and redox mediator. The oxygen requirement of the enzymes is replaced by the oxidase.

In presence of the enzyme glucose oxydase, the glucose that enters the membrane is oxidised to produce glucuronic acid and hydrogen peroxide.

$$\text{Glucose} + O_2 \longrightarrow \text{Glucuronic acid} + H_2O_2$$

Therefore, as the oxygen moves up through the membranes to reach the cathode, its concentration decreases. This decrease in oxygen concentration results in a decrease in current between the electrodes.

Similarly, the decrease in H_2O_2 concentration can also be used to find the concentration of glucose, by changing the voltage applied between the electrodes to +0.68 V relative to the Ag/AgCl electrode causing the reactions:

Pt anode H_2O_2 $O_2 + 2H^+ + 2e^-$

Ag cathode $2AgCl + 2e^- \longrightarrow 2Ag^0 + 2Cl^-$

(*i*) Enzyme electrode biosensor. It is a new type of biosensors which has been designed for the amperometric assay or potentiometric assay of substrates such as glucose, urea, amino acid, alcohol, and lactic acid. In this biosensor, the electrode consists of an electrochemical sensor in close contact with a thin permeable enzyme membrane that reacts with a given substrate. The membrane encloses the enzyme that produce O_2, H^+, NH_4^+, CO_2 or other small molecules depending on enzymatic reactions. The specific sensor attached with this device detects it. The substrate concentration is determined by the magnitude of the response. A simple biosensor combining on electrochemical electrode and an enzyme immobilised on to a semi-permeable membrane is given in Fig. 15.31.

Properties of Mediators

The biosensors depend on the concentration of dissolved oxygen. Hence, this problem may be overcome by the use of 'mediators' which transfer the electrons directly to the electrode bypassing the reduction of the oxygen co-substrate. These mediators possess several useful properties as given below:

- The mediators must react rapidly with the reduced form of the enzyme.
- The reduced form of the mediator should not readily react with oxygen.
- Both in the oxidised and reduced forms they must be soluble to get diffused rapidly between the active site of the enzyme and the electrode surface. Generally the mediator should be non-toxic.
- The over potential for the regeneration of the oxidised mediator, at the electrode, should be low and independent of pH.

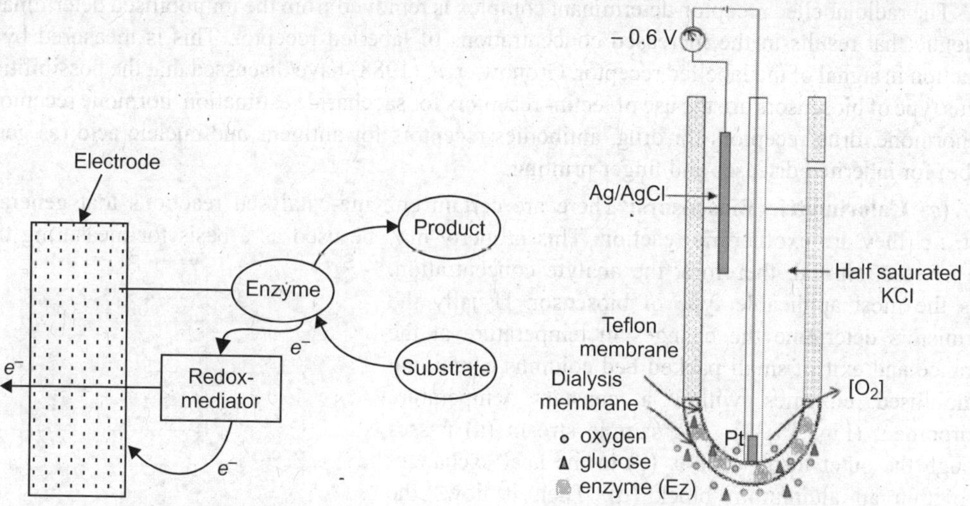

Fig. 15.30 Mediated biosensor.

Fig. 15.31. An enzyme electrode biosensor.

(*d*) Thermistor-containing Biosensor: There occurs fluctuation in temperature during biochemical reactions. A thermistor is attached with this biosensor to record even a small temperature change (0.1 – 0.001°C). Different types of thermistors have been developed by immobilising the enzymes such as cholesterol oxidase, glucose oxidase, invertase, tyrosinase, etc. (Gronow *et al.,* 1988). Moreover, in case of thermometric enzyme linked immunoabsorbant assay (ELISA), thermistors are also employed for the study of antigen-antibody with very high sensitivity (10^{-13} mol dm^{-3}).

(e) Split-flow Enzyme Thermistor (Cholesterol Biosensor): In 1991, Satoh and Teramura described a split-flow enzyme thermistor for biosensing of cholesterol in Japan. When a cholesterol sample is passed through a column containing three kinds of enzymes (cholesterol esterase, cholesterol oxidase and catalase) covalently co-immobilised onto porous glass beads, the heat of oxidation and decomposition evolves which is measured by this device. The calibration graph for standard cholesterol was found linear up to 400 mg/dl for 50 μl samples. The assay took about three minutes. This method can be applied to determine cholesterol in human blood sera and can be compared with a conventional spectrophotometric method.

(f) Bioaffinity Biosensors: Recently, bioaffinity bisensors have been developed, that measures the concentration of the determinants, *i.e.* substrates based on equilibrium binding. This shows a high degree of selectivity. Bioaffinity bisensors are of diverse nature because of the use of radiolabelled substance, enzyme-labeled substance, or fluorescence-labeled substance. Fig. 15.32 shows the principle of bioaffinity biosensor. A receptor is radiolabelled in this biosensor binds with determinant analogue immobilised onto the surface of a transducer. The labelled-receptor forms an intimately bound complex with determinant upon increasing the concentrations of a determinant.

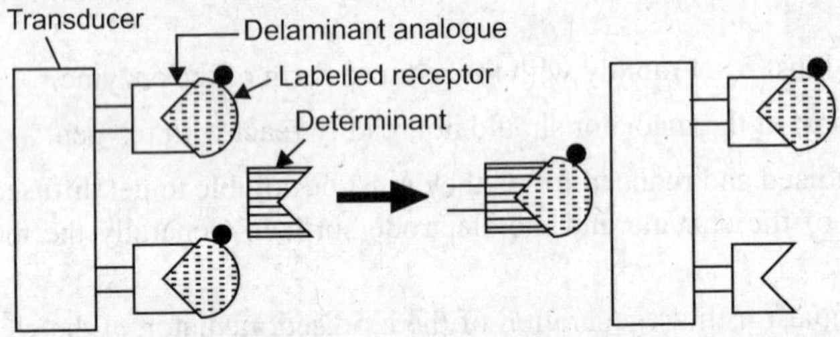

Fig. 15.32. Principle of bioaffinity biosensor.

The radiolabelled receptor-determinant complex is removed from the immobilised determinant analogue that results in the increased concentrations of labelled receptor. This is measured by a reduction in signal of the labelled receptor. Gronow *et al.* (1988) have discussed that the possibilities of this type of biosensors are the use of lectin- receptors for saccharide estimation, hormone receptors for hormone, drug receptors for drug, antibodies receptors for antigens and nucleic acid (as gene probe) for inherited diseases and finger-printing.

(g) Calorimetric Biosensors: There are certain enzyme-catalysed reactions that generate heat, *i.e.* they are exothermic reaction. This property may be used as a basis for measuring the rate of reaction and, therefore, the analyte concentration. It is the most applicable type of biosensor. Usually the thermistors determine the changes in temperature at the entrance and exit of small packed bed columns containing immobilised enzymes within a constant temperature environment (Fig. 15.33). The sample stream (a) passes through the outer insulated box (b) to the heat exchanger (c) within an aluminium block (d). Then it flows the reference thermistor (e) and into the packed bed bioreactor (f) containing the biocatalyst, where the reaction takes. A change in temperature is determined by the thermistor (g) and the solution passed to waste (h). The difference in the resistance is determined by the external electronics (l), and hence temperature between the thermistors.

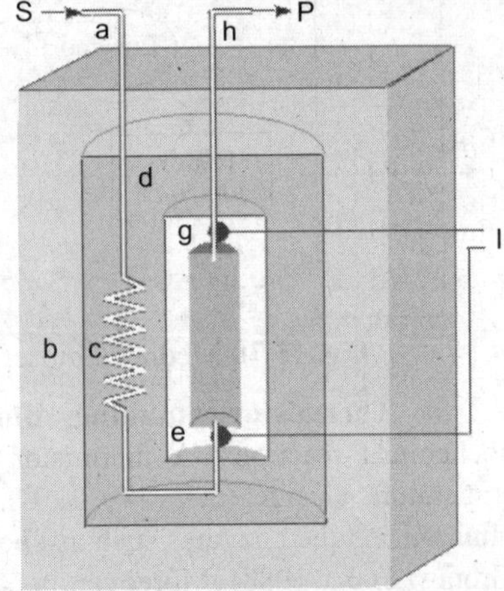

Fig. 15.33. A calorimetric biosensor (diagrammatic).

Up to 80% of the heat generated in the reaction under closely controlled conditions may be registered as a temperature change in the sample stream. This may be calculated from the enthalpy change and the amount reacted. If a 1 mM reactant is completely converted to product in a reaction generating 100 kJ mole^{-1}, each ml of solution generates 0.1 J of heat. This will cause a change in temperature of the solution to about 0.02°C at 80% efficiency. For a useful biosensor, the temperature change commonly necessitates a temperature resolution of 0.0001°C.

***(h)* Opto-electronic Biosensor.** In these biosensors either enzymes or antibodies are immobilised on the surface of a membrane. For measuring colour, biosensor with enzyme and dye is immobilised to a membrane. When a substrate is catalysed to yield product, changes in pH of the medium occur. This results in changes in dye - membrane complex. These changes in colour are measured by using a light emitting diode and a photodiode (Gronow *et al,* 1988).

The most common use of this technology is the whole-blood monitoring in diabetes control. Here, the strips include glucose oxidase, horseradish peroxidase and a chromogen (such as *o*-toluidine or 3,3',5,5'-tetramethylbenzidine). The hydrogen peroxide, produced by the aerobic oxidation of glucose, oxidises the weakly coloured chromogen to a highly coloured dye as given below :

$$\text{Chromogen (2H)} + H_2O_2 \xrightarrow{\text{Peroxidase}} \text{dye} + 2H_2O$$

Evaluation of the dyed strips is best achieved by the use of portable reflectance meters, although direct visual comparison with a coloured chart is often used. At present a wide variety of test strips involving the other enzymes are commercially available.

***(i)* Piezo-electric Biosensors:** Under the influence of an electric field Piezo-electric crystals (e.g. quartz) vibrate acoording to the strength of electric field. The frequency of this oscillation (f) depends on their thickness; each crystal has a characteristic resonant frequency (Hz) that changes as molecules adsorb or desorb from the surface of the crystal.

$$\Delta f = \frac{Kf^2 \, \Delta m}{A}$$

Where, Δf = the change in resonant frequency, Δm = the change in mass of adsorbed material (g), K = a constant for the particular crystal dependent on such factors as its density and cut, and A = the adsorbing surface area (cm^2)

For any piezo-electric crystal the change in frequency is proportional to the mass of absorbed material to about a 2% change. Relatively unsophisticated electronic circuits can easily detect this frequency change. For example, a formaldehyde biosensor utilises a formaldehyde dehydrogenase coating immobilised on a quartz crystal and sensitive to gaseous formaldehyde. The demerit of these devices is that they can be interfered by atmospheric humidity and they are difficult to be used for the determination of material in solution. However, they are inexpensive, small and robust, and can give a rapid response.

***(j)* Whole Cell Biosensors (Microbial Biosensors).** In this device, either immobilised whole cell of microorganisms or their organelles are used. These react with a large number of substrates and show generally slow response. Immobilised *Azotobacter vinelandii* coupled with ammonia electrode shows sensitivity range between 10^{-5} and 8×10 mol dm^{-3}. It measures the concentration of nitrate within 5-10 min^{-2}. Examples of microbial biosensors are given in Table 15.17.

Table 15.17. Microbial whole cell biosensors containing oxygen electrode (Source : Kumar and Kumar, 1992).

Microorganism	***Sensoring for***	***Responsc time(min)***	***Range***
Brevibacterium lactofermentum	Assimilase sugars	1-10	Linear above 1 m mole dm^{-3}
Bacillus subtilis	Mutagen screening	90-100	1-6m g cm^{-3}

Methylomonas flagellata	Methane	1	upto 6.6m mol dm^{-3}
Trichosporon brassicae	Acetate	6-10	Linear upto 22.5 mg dm^{-3}
T. brassicae	Ethanol	10	below 22.5 mg dm^{-3}

III. APPLICATIONS OF BIOSENSOR

In the beginning biosensor was applied in the field of medicine and industry. But in recent years, biosensors are becoming popular in many areas due to the small size, rapid and easy handling, low cost, and greater sensitivity and selectivity. Application of biosensor in some of the areas is described below:

1. Uses in Medicine and Health

In the field of medical science biosensors have tremendous potential for its application. In 1979, the first glucose analyser was commercialised by Yellow Springs Instruments Co. (USA) by using biomolecule for the detection of blood glucose. A minipump filled with insulin has been constructed to deliver insulin to diabetics based on glucose levels of blood. The device delivers accurate amount of insulin required by the diabetics after getting information by biosensor.

2. Uses in Industry

Estimation of substrates utilised and the products formed in the fermented broth is done by using spectrophotometer and autoanalyser; but certain problems are associated with this approach. Therefore, the biosensors are constructed for measuring the fermentation products : (*i*) to improve the feedback control, (*ii*) to carry out rapid sampling and rejection of below standard raw materials, and (*iii*) to improve the efficiency of workers. Isaokarube and coworkers of Tokyo University Research Centre of Advance Science & Technology have recently developed an ion sensitive field effect transistor. This device is highly sensitive to change the ion concentration. Using this biosensor, it is possible to measure the odor, freshness and taste of foods. ATPase, aminoxidase or putrescine oxidase is used to measure fish freshness. ATPase detects the presence of ATP in fish muscle. Signals do not occur when ATP is absent in staled food. Cransfield Institute of Technology (U.K.) has developed a biosensor recently which measures cholesterol levels in butter. When the enzyme cholesterol oxidase is immobilised on the electrodes, it reacts with cholesterol of food.

3. Biosensor in Military

The darker side of biosensor application is to provide support to military with such a biosensor that can detect toxic gases including chemical warfare agents. Such biosensors have advantages over the traditional methods of sensing of chemicals.

A recent U.S. Arms Control and Disarmament Agency report on worldwide arms control compliance states that both Syria and Egypt have offensive biological warfare programmes. Protection of a military unit or community from dangerous bacteria means to detect them before they reach to intended victims. In such situations people can be warned to leave an area or at least wear protective gear. Hazardous bacteria can be detected using biosensors that detect and measure concentrations of specific bacteria or hazardous chemicals. It can also measure the acidity levels (pH). Genetically engineered bacteria have been designed to provide a detectable signal (such as light) in the presence of a specific pollutant.

4. Uses of Biosensors in Environmental Monitoring

Biosensors are very helpful in environmental monitoring and pollution control, since they can be miniaturised and automated. As far as quality control of drinking water is concerned, the biosensors are successful in monitoring of pesticides in water. In Japan, a biosensor coupled with oxygen electrode and immobilised *Trichosporon cutaneum* is used for measuring biological oxygen demand (BOD) (Gronow *et al,* 1988).The whole cell biosensor developed by immobilising *Salmonella typhimurium* and *Bacillus subtilis* in conjugation with oxygen electrode has been used to measure mutagenicity and carcinogenecity of several chemical compounds. Some of the important biosensors used in environmental pollution monitoring are described below :

(*a*) Gas Biosensors: Microbial biosensors have been developed in order to detect gases such as sulphur dioxide, (SO_2), methane, carbon dioxide, etc. *Thiobacillus*-based biosensor detects SO_2, whereas methane (CH_4) can be detected by immobilised *Methalomonas*. A particular strain of *Pseudomonas* is used to monitor carbon dioxide levels.

(*b*) Immunosensors: Immunoelectrodes as biosensors are used to detect the low concentrations of pollutants. Pesticide specific antibodies can detect the presence of low concentrations of triazines, malathion and carbamates, by using immunoassay methods. Recently, ELISA techniques have been combined with biosensors to form immunosensors to increase their range, speed and sensitivity. Fig. 15.34 shows a simple immunosensor configuration where the biosensor merely replaces the traditional colorimetric detection system. More advanced immunosensors are being developed which rely on the direct detection of antigen bound to the antibody-coated surface of the biosensor. Piezoelectric and FET-based biosensors are particularly suited to such applications.

A tube is coated with immobilised antigen (A). An excess of specific antibody-enzyme conjugate is placed in the tube and allowed to bind it properly. After some time any unbound material is washed off (B). The analyte antigen solution is passed into the tube, binding and releasing some of the antibody-enzyme conjugate dependent upon the antigen's concentration. The amount of antibody-enzyme conjugate released is determined by the response from the biosensor (C).

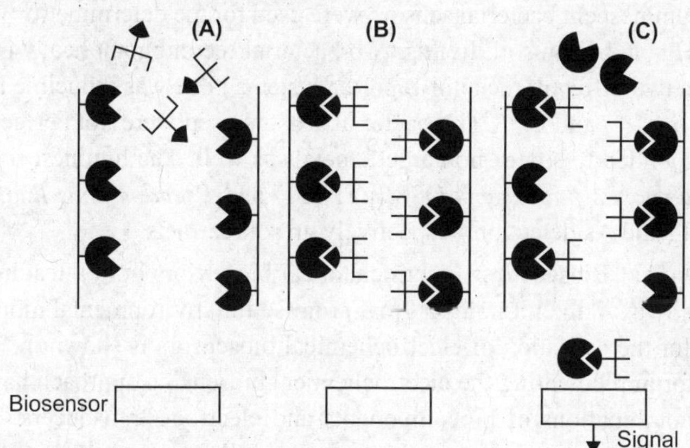

Fig. 15.34. Principles of immunosensors.

(*c*) BOD Biosensor: Biological oxygen demand (BOD) is an indicator of the amount of biodegradable organic compounds found in waste water. The BOD value determination conventionally requires five days incubation of the water sample with an inoculum of microorganisms typically present in the waste treatment system to yield an end-point measurement for oxygen consumption. The use of indicator microorganisms interfaced to signal transducers allows the measurement of the rate of metabolism of organic compound in a significantly shorter time (*e.g.* 15 minutes to 1 hour), rendering this technology highly advantageous for process control applications. Although these biosensors appear to work well for *in situ* monitoring of industrial waste waters that result in high BOD values, they currently require improvements for better response. Biological oxygen demand (BOD) is widely used as a test to detect the levels of organic pollution. This requires five days of incubation but a BOD biosensor using the yeast *Trichosporon cutaneum* with oxygen probe takes only 15 minutes to detect the organic pollution.

(*d*) Luminescent Bacterial Biosensors: Luminescent bacterial biosensors have been developed for monitoring several heavy metals. Bioluminescence is produced by the immobilised bacterial cells in response to arsenite exposure. The intensity of the bioluminescence is proportional to the arsenite concentration which can be recorded after predefined incubation periods in a luminometer. Bioreporter cells carry a plasmid with the genes for bacterial luciferase (*luxAB*) under expression control of the ArsR transcriptional repressor protein. Cellular entrance of arsenite (or antimonite) causes the release of transcriptional repression and subsequent synthesis of luciferase by the cells.

Moreover, the measurement of the bioavailability of mercury is quite difficult with traditional analytical methods. Hence, a luciferase-based biosensor using *Escherichia coli* MC1061 immobilised in a latex copolymer film was developed for the determination of bioavailable mercury. It utilises firefly luciferase gene as a reporter under the control of the mercury-inducible *mer* promoter from transposon Tn21 and *Escherichia coli* MC1061 as a host organism. The luminescence-based sensor was evaluated for the selectivity and sensitivity of the detection of mercury. The *mer* operon encodes for resistance to mercury, which is a nonessential and toxic metal for bacteria. The luciferase catalyses the following reaction:

$$\text{ATP} + O_2 + \text{D-luciferin} \longrightarrow \text{AMP} + \text{PPi} + CO_2 + \text{oxyluciferin} + \text{light (560 nm)}$$

The firefly luciferase gene *(lucFF)* was cloned and has become a widely used reporter gene in prokaryotic as well as in eukaryotic systems because it provides sensitive and simple detection of the gene regulation. The quantification of the luminescence is possible with a liquid scintillation counter, luminometer or even X-ray film.

Recombinant luminescent bacterial sensors were used for the determination of the bioavailable fraction of Cd, Zn, Hg, and Cr in soil. In this, two bacterial recombinant heavy metal sensors were constructed based on two different receptor-reporter systems : one was inducible by Zn^{2+}, Cd^{2+}, and Hg^{2+}, and the other by Cr^{6+} and Cr^{3+}. The bacterial sensors used were not perfectly specific to one heavy metal, but responded to some "nontarget" metals as well. The luminescence-based bacterial sensor strains *Pseudomonas fluorescens* OS8 (pTPT11) and *Pseudomonas fluorescens* OS8 have also been used for Hg and As detection, respectively, in soil extracts.

(*e*) Electrochemical Biosensors: Electrochemical biosensors hold a leading position among the bioprobes currently available; it holds a great promise for environmental monitoring. A general schematic diagram for the operation of electrochemical biosensors is shown in Fig. 15.35. A great variety of schemes for implementing the electrochemical biosensing approach have been suggested based on different combinations of biocomponents and electrode transducers. These rely on the immobilisation of enzymes, antibodies, receptors or whole cells onto amperometric or potentiometric electrodes. Enzyme electrodes have the longest tradition in the field of biosensors. Such devices are usually prepared by attaching an enzyme layer to the electrode surface which monitors changes occurring as a result of the biocatalytic reaction amperometrically or potentiometrically.

Amperometric enzyme electrodes rely on the biocatalytic generation or consumption of electroactive species. A large number of hydrogen-peroxide generating oxidases and NAD^+-dependent dehydrogenases have been particularly useful for the measurement of a wide range of substrates The liberated peroxide or NADH species can be readily detected at relatively modest potentials (0.5-0.8V vs Ag/AgCl), depending upon the working electrode material. Lowering of these detection potentials is desired for minimising interferences from coexisting electroactive species.

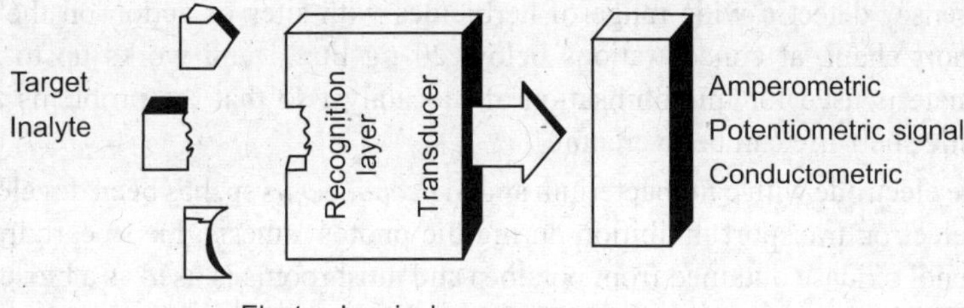

Fig. 15.35. An electrochemical biosensors showing biorecognition and signal transduction (diagrammatic).

Potentiometric enzyme electrodes rely on the use of ion- or gas-selective electrode transducers, and thus allow the determination of substrates whose biocatalytic reaction results in local pH changes or the formation or consumption of ions or gas (e.g. NH_4 + or CO_2). The resulting potential signal thus depends on the logarithm of the substrate concentration. Proper functioning of enzyme electrodes is greatly dependent on the immobilisation procedure.

***(f)* Enzyme Electrodes:** There are many enzyme electrodes that have proved usefulness for environmental monitoring, for example highly sensitive amperometric biosensors for phenolic compounds. Such devices rely on the immobilisation of tyrosinase (polyphenol oxidase) onto carbon- or platinum transducers, and the low potential detection of the liberated quinone product (Fig. 15.36). Assays of industrial wastes or natural water have been documented including possible remote phenol ensuing and single-use on-site sensing. Similarly, low potential biosensing of organic peroxides or hydrogen peroxides can be accomplished at peroxidase-modified electrodes. Class-selective enzyme electrodes, based on tyrosinase or peroxidase, can be used for semiquantitative field screening. They can also be used as detectors for liquid chromatography, hence provide quantitation of the individual substrates. The organic- phase activity of these enzymes should be useful for chromatographic separations and rapid solvent extraction.

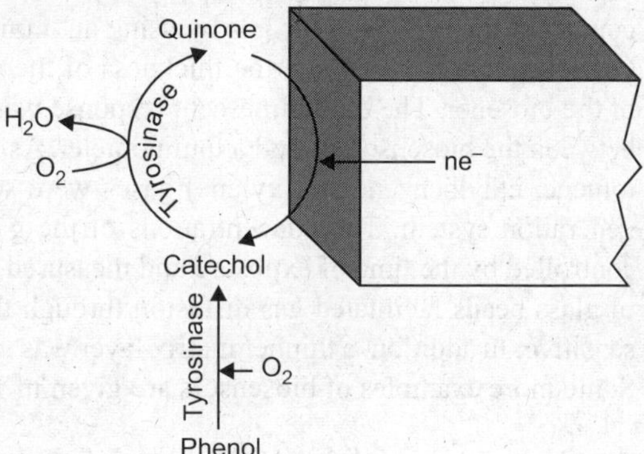

Fig. 15.36. Enzyme (tyrosinase) electrode for monitoring the phenolic compounds.

Another environmentally important microbial sensor offers rapid estimate of BOD (biochemical oxygen demand), hence replacing the long (5 days) conventional BOD test.

(g) Whole Cell Biosensors: Risk of pollution to drinking water from industry and agriculture is growing; hence there is need of continuous on-line monitoring of water. There is increasing use of living organisms as the sensitive agent to detect the presence of pollutants. The whole-cell biosensors have particular advantages in such environmental monitoring. One of the attempts the development of a mediated amperometric biosensor by immobilising the cells of a cyanobacterium called *Synechococcus* sp. as the biocatalyst herbicide monitoring (Fig. 15.37).

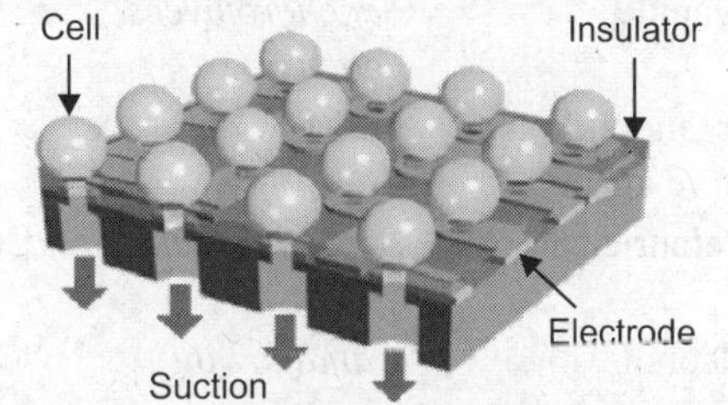

Fig. 15.37. A whole-cell biosensor made of a cyanobacterium *Synechococcus* used for environmental monitoring.

This biosensor detect a wide range of herbicides with sites of action on the photosynthetic electron transport chain, at concentrations below 20 μg $litre^{-1}$ and works up to 7 days. In this biosensor alginate is used for immobilisation of biocatalyst so that the problems associated with getting a realistic shelf-life can be overcome.

A graphite electrode with cynobacterium and *Synechococcus* sp. has been developed to measure the degree of electron transport inhibition during the photosynthesis due to certain pollutants like herbicides. Phenol oxidase obtained from potatoes and mushrooms is used as a biosensor for phenol detection. Such biosensors have also been developed for the detection of polychlorinated biphenyls (PCBs), chlorinated hydrocarbons and certain other organic compounds. Biosensors employing acetylcholine esterase obtained from bovine RBC can be used for the detection of organophosphorus compounds in water.

In 2001, Geun Cheol Gil and coworkers immobilised a genetically engineered bioluminescent bacterium (*lac::luxCDABE*) to develop a whole cell biosensor for the detection of toxic gaseous chemicals. The toxicity of chemicals can be evaluated through bioluminescent reaction as it reduces in intensity when the cells experience toxic or lethal conditions.

This whole cell biosensor was fabricated for the measurement of toxicity through direct contact of the cells with the gas by using an immobilisation technique utilising solid agar medium. Glass beads were used and the thickness of the agar layer was reduced to enhance the sensitivity of the biosenor. The bioluminescent response was measured by using a fiber optic probe connected between the biosensor kit and a luminometer. As sample gaseous toxic chemicals, BTEX (benzene, toluene, ethylbenzene and xylene) gases were selected and their vapors were produced by a gas generation system. The concentrations of the gaseous chemicals injected into the chamber were controlled by the time of exposure and measured by using a portable gas chromatograph. Additions of glass beads facilitated gas diffusion through the solid medium which made the biosensor more sensitive. In addition, a thinner matrix layer was more advantageous for the detection of gas toxicity. Some more examples of biosensors are given in Table 15.18.

Table 15.18. Some of examples of whole cell biosensors.

Analyte	*Microorganism*	*Transducer/immobilization*	*Detection limit*
BOD	*Trichosporum cutaneum*	Miniature O_2 electrode	0.2-28 mg/L
BOD	activated sludge (mixed microbial consortium)	O_2 electrode/flow injection system	>3.5 mg/L
Phenolic compounds	*Pseudomonas putida*	O_2 electrode (reactor with cells adsorbed on PEI glass)	100 μM
	Nitrobactor vulgaris	O_2 electrode (adsorption on Whatman paper)	>10 μM
Cyanide	*Saccharomyces cerevisiae*	O_2 electrode (PVA)	0.15-15 nM
Organophosphate nerve agent	GEM *E. coli*	Potentiometric	0.055-1.8 mM
Mercuric chloride	*Synechococcus sp.* PCC 7942	Photoelectrochemical	0.2 and 0.06 μM
Alcohol	*Candida vini*	O_2 electrode (porous acetyl cellulose filter)	0.2-0.02 mM
CO_2	CO_2 utilizing autotrophic bacteria	O_2 electrode (bound on cellulose nitrate membrane)	0.2-5 mM

Toxicity of Zn, Cu and Cd, alone or in combination	*E. coli* HB101 and *P. fluorescens* 10586 genetically modified with *lux*CDABE
Polycyclic aromatic hydrocarbons	*P. fluorescens* HK44 genetically modified with *lux*CDABE

J. ENZYME ENGINEERING

Enzyme engineering has already been described in Chapter 11: *Protein: Structure, Engineering and Design.*

QUESTIONS

1. Write the structure of an enzyme.
2. Write different factors that affect enzyme activity.
3. Give a detailed account of activation, inhibition and regulation of enzyme activity.
4. Write a note of abzymes, synzymes and extremozymes.
5. Write classification and nomenclature of enzymes.
6. Describe in detail about mechanism of enzyme action.
7. What do you know about enzyme kinetics ? Write principles of enzyme kinetics.
8. What is Maechilis-Menten kinetics ? Derive a linear plot of Maechilis-Menten equation.
9. What do you know about enzyme immobilisation ? Write the use of carrier matrices and its properties.
10. Give a detailed account of methods of enzyme immobilisation.
11. Write in detail about application of enzyme and immobilised enzymes.
12. Discuss in detail about application of immobilised enzymes in pharma industry.
13. Describe therapeutic uses of immobilised enzymes.
14. Write uses of enzymes in food, beverages and detergent industry.
15. Give a detailed account of immobilised whole cells producing enzymes and alcohols.
16. Describe the characteristic features and types of biosensors.
17. Write application of different biosensors.
18. Write short notes on the following :
    (*i*) Cofactors and coenzymes, (*ii*) Active sites of an enzyme,
    (*iii*) Substrate specificity, (*iv*) Induced fit theory,
    (*v*) Mechanism of transition state stabilization,
    (*vi*) Quantum tunneling, (*vii*) Michaelis-Menten constant,
    (*viii*) Ternary complex mechanism, (*ix*) Ping-pong mechanism,
    (*x*) Non- Michaelis-Menten constant, (*xi*) Carrier matrices,
    (*xii*) Enzyme entrapment, (*xiii*) Covalent bonding,
    (*xiv*) Enzyme encapsulation, (*xv*) Advantages and disadvantages of enzyme immobilisation,
    (*xvi*) Enzymes used in cancer therapy, (*xvii*) Amperiometric biosensor,
    (*xviii*) Glucose-electrode biosensor, (*xix*) Enzyme-electrode biosensor,
    (*xx*) Bioaffinity biosensor, (*xxi*) Opto-electronic biosensor.

BIBLIOGRAPHY

Chakraborty S. 2009. *Microfluidics and Microfabrication*. Springer. p. 357.

Danson M. and Eisenthal R. 2002. *Enzyme assays: a practical approach*. Oxford: Oxford University Press.

Garrett R. and Charles M. G. 2010. Enzymes-Kinetics and Specificity. *Biochemistry (4rth Ed.)*, Belmont, CA: p. 406-407.

Gomes J. and Steiner W. 2004. Extremophiles and extremozymes. *Food Technol. Biotechnol.* **42:** 223–235.

Gronow M. *et.al.* 1988. Biosensors. In: *Molecular biology and biotechnology.* (Eds. J.M. Walkerand E.B. Gingold), p. 323–47.

Jamuna R. and Ramakrishna S. V. 1989. In *Physiology of Immobilized Cells* (Eds. de Bont J.A.M., Visser J., Mattiasson B. and Tramper J.), Elsevier Sci. Publ., Amsterdam, p. 522–538.

Lehninger A.L., Nelson D.L. and Cox M.M. 2005. *Lehninger principles of biochemistry*. New York: W.H. Freeman.

Moo-Young M. 1985 (Ed.) *The Principles of Biotechnology: scientific fundamentals.* Pergamon Press, Oxford.

Metzler D.E. and Metzler C.M. 2001. *Biochemistry: the chemical reactions of living cells* (2nd Ed.). Academic Press, San Diego/ Harcourt.

Voet D. and Voet J. 2004. *Biochemistry*. John Wiley & Sons, Inc.

16 CHAPTER Immunology

RBCs- the oxygen carrier

A. History of immunology
B. Branches of Immunology
C. Antigen
D. Innate Immunity
E. Adaptive or acquired immunity
F. Humoral immunity
G Cell-mediated immunity
H. Antibody diversity
I. Complements
J. Histocompatibility
K. Hypersensitivity
L. Autoimmunity

The term immunity is coined from a Latin word *immunis* which means 'free of'. Before the development of the concept immunity, several human organs were characterised which later were regarded as the parts of the immune system. For example, thymus and bone marrow are the main primary lymphoid organs of the immune system. But the secondary lymphatic tissues are spleen, tonsils, lymph vessels, lymph nodes, adenoids, and skin. There are several cells associated with immune system which are not associated with any specific organ; but these circulate in various tissues located throughout the body. Thus, the study of all aspects of the immune system in all organisms is called immunology. It deals with: (*i*) the physiological functioning of the immune system, (*ii*) malfunctions of the immune system in immunological disorders (*e.g.* autoimmune diseases, hypersensitivities, immune deficiency, transplant rejection), and (*iii*) the physical, chemical and physiological characteristics of the components of the immune system.

Immunology is the study of the molecular and cellular components comprising of the immune system, including their function and interaction. The immune system has been divided into a more primitive *innate immune system*, and *acquired* or *adaptive immune system*, the latter is further divided into *humoral* and *cellular* components.

The humoral (antibody) response is defined as the interaction between *antibodies* and *antigens*. Antibodies are specific proteins released from a certain class of immune cells (B lymphocytes). Antigens are defined as anything that elicits generation of antibodies; hence they are 'antibody generators'. Immunology itself rests on an understanding of the properties of these two biological entities. However, equally important is the cellular response, which can not only kill infected cells in its own right, but is also crucial in controlling the antibody response. But simply, both systems are highly interdependent.

Edward Jenner who discovered vaccination.

A major triggered enzyme plasma system is called 'complement system'. It causes coating of microbes with molecules which provides susceptibility to engulfment by phagocytes. The permeability of the capillaries is increased by vascular permeability mediators which allow more flowing of plasma and complement fluid to the site of infection. Phagocytes die after carrying out their work and pus is formed their corpses, pockets of damaged tissue and fluid.

The normal flora of our body is constituted mostly by bacteria that live in and on our body without any harmful effects. In our body about 10^{14} bacteria are found, among which most of them reside in large intestine. There are 10^3–10^4 microbes per cm^2 on the skin *e.g. Staphylococcus aureus*, *Staph. epidermidis*, diphtheroids, streptococci, *Candida*, etc. Many bacteria reside in nose and mouth. Lactobacilli live in the stomach and small intestine. The upper intestine has about 10^4 bacteria per gram; the large bowel has 10^{11} per gram, of which 95–99% are anaerobes. Various bacteria and diphtheroids lightly colonise the urogenital tract. After puberty vagina is colonised by *Lactobacillus aerophilus*. This bacterium ferment glycogen and maintains an acidic pH. Almost all of the available ecological niches harbour normal flora in the body which produce bacteriocidins, defensins, cationic proteins, and lactoferrin. These microbial products kill the other bacteria that compete for their niche in the body.

Upon invasion of these resident bacteria a new place in the body they can become problematic also; for examples: (*i*) *Staphylococcus* lives on skin and can enter into body through small cuts, (*ii*) some antibiotics *e.g.* clindamycin kill some of the bacteria in our intestinal tract resulting in an overgrowth of *Clostridium difficile*. This causes pseudomembranous colitis where the inner lining of the intestine cracks and bleeds. This is a painful condition.

The major histocompatibility complex (MHC) is a set of molecules present on cell surfaces that are responsible for lymphocyte recognition and 'antigen presentation'. The MHC molecules control the immune response through recognition of 'self' and 'non-self', and serves as targets in transplantation rejection. The Class I and Class II MHC molecules belong to a group of molecules known as the 'immunoglobulin super-gene family', which includes immunoglobulins, T-cell receptors, CD4, CD8, and others. Several genes located on human chromosome 6 encode the MHC.

A. HISTORY OF IMMUNOLOGY

Smallpox has been reported in ancient Asian cultures as early as 1,122 BC in China and is mentioned in ancient Sanskrit texts of India in Vedic literature (7,000-9,000 BC). It is believed to have appeared around 10,000 BC, at the time of the first agricultural settlements in north-eastern Africa. From early writings, it is clear that primitive man knew about disease and its ravages. One finds in the Babylonian Epic of Gilgamesh (2000 BC, Mesopotamian hero) records of the presence of pestilence and disease. In other, more recent writings from old dynasties of ancient Egypt, one finds even more descriptions of disease. Further, one can even identify the disease of which they spoke. The earliest evidence of skin lesions resembling those of smallpox is found on faces of mummies from the time of the 18th and 20th Egyptian Dynasties (1570–1085 BC).

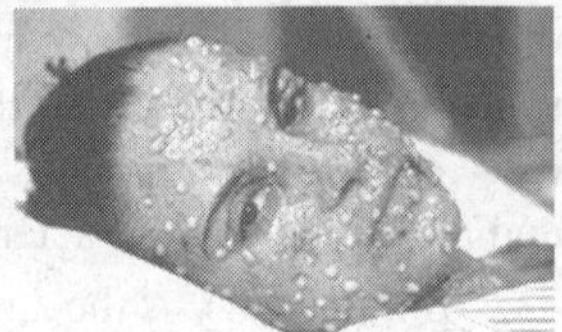

Smallpox caused by variola virus

For many centuries smallpox devastated mankind. Edward Jenner (1749–1823) deserves for thanks for development of vaccination. In recent decades, the historic origins of immunisation are often forgotten with the rapid pace of vaccine development. Unfortunately, since the attack on the World Trade Center on September 11, 2001, the threat of biological warfare and bioterrorism has reemerged. Smallpox has been identified as a possible agent of bioterrorism.

In those days, disease and pestilence was punishment rendering as a result of 'bad deeds' or 'evil thoughts'. The man knew that once he had been afflicted with disease, if he survived, he was normally not able to contract it again.

The science of immunology grew from the common knowledge that those who survived many of the then common infectious diseases rarely contracted that disease again. This was an observation that was made long before the establishment of the germ theory of disease. Louis Pasteur and Robert Koch were instrumental in defining microorganisms as the etiological agents of a large number of diseases.

In 430 BC, Thucydides recorded that while the plaque was raging in Athens, the sick and dying would have received no attention had been paid for those individuals who had already contracted the disease and recovered and recognised their immune status.

Around 1000 AD, the ancient Chinese practiced a form of immunization by inhaling dried powders derived from the crusts of smallpox lesions. Around the 15th century, a practice of applying powdered smallpox crusts and inserting them with a pin or poking device into the skin became common place. The process was referred to as variolation and became quite common in the Middle East. However, the primary intent of variolation was that of preserving the beauty of their daughters and no mention was made of saving lives.

The process of variolation can be traced from Turkey to the inhabitants of a country called Arcassia. The people were poor but were blessed with large number of beautiful women, which unfortunately was the chief trade and very important to the Arcassian economy. Most sales were to the Sultan of Turkey. Eventually, the process was popularized in Great Britain, largely through the efforts of Mary Wortley Montagu. It was vehemently opposed by the Church. There was no standardisation of the inoculum engrafted; the practice occasionally resulted in death/disfigurement from smallpox. This coupled with the widespread acceptance of herbal medicine resulted in it not being widely accepted. In 1721, Mary Wortley Montagu's daughter was the first person to be engrafted in Great Britain. The first step to a safer procedure was to substitute material derived from the lesion of cowpox (vaccinia) for the inoculation. Cowpox is a benign disease due to infection with a virus closely related to the smallpox (variola) virus.

Variolation in China

Inoculation against smallpox in China did not become widely known and practiced until the period 1567-72. Vivid descriptions of the practice are recorded by Yu Chang (1643) in his book *Miscellaneous Ideas in Medicine*. Ming dynasty (1368-1644) started variolation which is insertion of scabs from patients under the skin of healthy individuals or blowing dried scab material up the noses of the individuals with a silver tube (Gan Miao: dry vaccine) or using water to make a paste from scabs to insert into the nostrils (Shi Miao: wet vaccine). A more systemic summary of the Chinese smallpox vaccination was done by Chang Yen in 1741 in his 'Zhong Miao Xin Shu' (a new book about vaccination).

In 1774, Benjamin Jesty, a farmer, inoculated his wife with the vaccinia virus obtained from farmer Elford of Chittenhall, near Yetminster. This is the first record of anyone using vaccinia virus to protect against smallpox.

In 1798, Jenner inoculated a young boy named James Phipps with material obtained from a cowpox lesion. The results were conclusive but were met with great resistance by the Church. For this feat, Jenner received a cash prize of 30,000 pounds. The acceptance of Jenner's thesis was strengthened when 70 of the principal physicians and surgeons of London threw their weight behind him.

To further advance the fledgling science of immunology required the development of the 'germ theory of disease'. Louis Pasteur played a pivotal role in the evolution of the science. Pasteur's

Louis Pasteur

work was concerned with bacterial infectious disease. He was concerned with the prevention of diseases caused by bacteria and how the human body was changed subsequent to infection so as to resist further. Louis Pasteur became the first experimental immunologist.

In 1886, Theobold Smith (an American microbiologist) demonstrated that 'heat killed' cultures of chicken cholera bacillus were also effective in protection from cholera as reported by Pasteur. This demonstrated that the microorganisms did not have to be viable to induce the protection.

Pasteur proceeded to do the same with anthrax. Robert Koch found that the disease was due to a bacterium called *Bacillus anthracis*. He maintained the culture under adverse laboratory conditions (incubation at 42-43°C). He produced the vaccine that was used for the famous demonstration at Pouilly-le-Fort.

Louis Pasteur (1822-1895)

Louis Pasteur contributed in the start of microbiology working in detail on fermentation, pasteurisation, immunisation of experimental animals, etc. Also he contributed on 'germ theory of disease'. Thus he became the first experimental immunologist. He developed attenuated vaccines for cholera, anthrax and rabies. On July 4, 1886, a 9- year old boy, Joseph Meister was bitten repeatedly by a rabid dog. Pasteur treated him with his attenuated rabies vaccine two days later. Joseph Meister survived who later became a gatekeeper of the Pasteur Institute. In 1940, when he was ordered by the German occupiers to open Pasteur's crypt, Joseph Meister refused and committed suicide

On May 5, 1881 Pasteur vaccinated 24 sheep, 1 goat and 6 cows with five drops of the living attenuated anthrax bacillus. On May 17, he inoculated all of the animals with a less attenuated strain. On May 31, all of the animals received 'viable virulent' anthrax bacilli. Additionally, 24 more sheet, 1 goat and 3 cows received the virulent microorganism without the protection of the vaccination. On June 2, 1881 all of the non vaccinated animals had died, while only two sheep of the protected group had succumbed. One of the sheep had died due to pregnancy complications. A third means of virulence attenuation was soon found to involve the passage of the microorganism in an unnatural host.

The first human trial was on July 6, 1885. A nine-year old lad named Joseph Meister had been severely bitten by a rabid dog two days earlier. His parents knew that he 'was a goner' and were desperate for any ray of hope. They heard of Pasteur's work and traveled to Paris in hopes that Pasteur would 'work a miracle'. Pasteur injected the attenuated virus into young boy and waited. You can guess the rest of the story. Joseph Meister, after receiving the immunisation, survived rabies. This was the first known case of an individual being bit and surviving rabies. Joseph Meister was later to become the gate porter of the Pasteur Institute in Paris and served as guard to Pasteur's crypt. Within a year, over 350 people bitten by rabid animals had been treated with no fatalities.

Emil Adolf von Behring (1854 – 1917)

In 1888, Roux and Yersin discovered the diphtheria toxin. Two years later, in 1890, Von Behring (a Student of Koch) and Kitasato demonstrated the presence of anti-toxin (*i.e.* the antibodies that can neutralise the toxin) in the blood of individuals recovering from diphtheria. Von Behring was the first to use this antiserum in treating active disease to what we call serotherapy. Von Behring with Kitasato and Wernike, discovered anti-toxin for Diphtheria and Tetanus and applied as therapy. In 1901, he was awarded first Nobel Prize in Physiology.

The ideas of circulating neutralising anti-toxins predominated the early immunological thoughts and the French and German schools dominated immunological research. At the turn of the century, immunology developed into two schools of thought:

Humoralists were of the opinion that immunity was due to *humoral* substances, *i.e.* antibodies. A pioneer in this area was Paul Ehrlich. He proposed what was the most plausible humoral theory of antibody formation, the 'side chain theory'. Emil Von Behring (worked at the Koch Institute in Germany) used serum to treat diseases. Germans were big proponents of **humoral immunity**.

Cellularists were of the opinion that immunity was due to the existence of 'phagocytic' cells within our bodies. The pioneer here was Eli Metchnikoff and he became the strongest proponent of **cellular immunity** after observing water Daphnia phagocytose smaller materials and examining blood cells devour foreign bacteria in blood samples. Metchnikoff was co-worker with Louis Pasteur (he worked at the Pasteur Institute) and he had many a vitriolic fights with the Germans who were proponents of humoral immunity.

F.M. Burnet (1899-1985)

In 1903, Maurice Arthus, described the localising allergic reaction called the Arthus response. In 1905, a Frenchman by the name of Von Pirquet shocked the world when he provided evidence that immune responses can be deleterious. He was studying serum sickness, a form of hypersensitivity or allergy.

In 1930, American scientists Landsteiner and Kabat described detailed experiments of the specificity of the immune response by chemically altering antigens. In 1944-45, another American named Peter Medawar provided the immunological basis of transplantation immunology.

In 1958, Sir McFarlane Burnet and Neils K. Jerne proposed 'clonal selection theory' to explain tolerance. In 1960, he was awarded Nobel Prize for the discovery of acquired immunological tolerance. Rejection of donor grafts was due to an immunological reaction and that tolerance can be built up by injections into embryos.

Niels K. Jerne

In 1960, the first real demonstration of a cellular basis for humoral and cellular immunity was carried out. The terms T and B cells became vernacular. The late 60's to early 70's have been referred to as the beginning of modern immunology. The molecular/genetic techniques of the 70's have revolutionised our understanding of how the immune system works.

Niels K. Jerne (1912-1994)

Niels K. Jerne worked on antibody avidity maturation. Pre-existing repertoire (in host DNA) theory helped the formation of clonal selection theory. Host MHC is the driving force for the maturation and selection of T cells in the thymus. He was awarded Nobel Prize 1984, for theories concerning 'the specificity in development and control of the immune system' and the discovery of 'the principle for production of monoclonal antibodies'.

Table 16.1. History of immunology.

Year	*Development*
1798	First demonstration of vaccination smallpox vaccination by Edward Jenner
1877	Mast cells discovered by Paul Ehrlich
1880- 1881	Theory that bacterial virulence could be attenuated by culture *in vitro* and used as vaccines. Used to make chicken cholera and anthrax vaccines (Louis Pasteur)
1883 - 1905	Cellular theory of immunity via phagocytosis by macrophages and microphages (polymorhonuclear leukocytes) by Elie Metchnikoff

1885	Introduction of concept of a therapeutic vaccination. First report of a live attenuated vaccine for rabies by Louis Pasteur
1890	Demonstration of antibody activity against diphtheria and tetanus toxins. Beginning of humoral theory of immunity by Emil von Behring and Shibasaburo Kitasato.
1891	Demonstration of cutaneous (delayed type) hypersensitivity by Robert Koch
1896	An antibacterial, heat-labile serum component (complement) was described by Jules Bordet
1901	Blood groups given by Karl Landsteiner
1938	Antigen-antibody binding hypothesis given by John Marrack
1940	Identification of the Rh antigens by Karl Landsteiner and Alexander Weiner
1944	Hypothesis of allograft rejection
1949	Immunological tolerance hypothesis
1951	Vaccine against yellow fever discovered
1957	Clonal selection theory given by Frank Macfarlane Burnet
1957	Discovery of interferon
1958-1962	Discovery of human leukocyte antigens by Jean Dausset and co-workers
1959-1962	Discovery of antibody structure independently by Gerald Edelman and Porter.
1960	Discovery of lymphocyte blastogenic transformation and proliferation in response to mitogenic lectins-phytohemagglutinin (PHA) by Peter Nowell
1961-1962	Discovery of thymus involvement in cellular immunity by Jacques Miller
1963	Development of the plaque assay for the enumeration of antibody-forming cells *in vitro* by Niels Jerne, and Albert Nordin
1968	Passenger leukocytes identified as significant immunogens in allograft rejection by William L. Elkins and Ronald D. Guttmann
1971	Peter Perlmann and Eva Engvall at Stockholm University invented ELISA
1974	T-cell restriction to major histocompatibility complex by Rolf Zinkernagel and Peter C. Doherty
1975	Generation of the first monoclonal antibodies by G. Köhler and C. Milstein
1976	Identification of somatic recombination of immunoglobulin genes by S. Tonegawa
1980-1983	Discovery and characterisation of the first interleukins, 1 and 2 IL-1 IL-2 by K. A. Smith
1986	Hepatitis B vaccine produced by genetic engineering
1988	Discovery of biochemical initiators of T-cell activation: CD4- and CD8-p56lck complexes by C. E. Rudd
1990	Gene therapy for severe combined immunodeficiency (SCID)

1991	Role of peptide for MHC Class II structure by Sadegh-Nasseri and Germain
1994	Danger model of immunological tolerance by P. Matzinger
1995	Regulatory T cells by S. Sakaguchi
2001	Discovery of FOXP3 - the gene directing regulatory T cell development
2005	Development of human papillomavirus vaccine by I. Frazer

B. BRANCHES OF IMMUNOLOGY

In the 21st century, immunology has broadened its horizons with much research being performed in the more specialised areas. This includes the immunological function of cells, organs and systems not normally associated with the immune system, as well as the function of the immune system outside classical models of immunity. There are several branches of immunology as given below:

***(i)* Clinical Immunology:** Clinical immunology deals with the study of diseases which are caused by disorders of the immune system such as failure, aberrant action, and malignant growth of the cellular elements of the system. It also involves diseases of other systems, where immune reactions play a part in the pathology and clinical features.

The diseases caused by disorders of the immune system are grouped into two broad categories: *immunodeficiency* (in which parts of the immune system fail to provide an adequate response (*e.g.* chronic granulomatous disease), and *autoimmunity* (in which the immune system attacks its own host's body *e.g.* systemic lupus erythematosus, rheumatoid arthritis, Hashimoto's disease, etc.). The other disorders of the immune system are hypersensitivities in which the system responds inappropriately to harmless compounds (asthma and other allergies) or responds too intensely.

AIDS is an immunodeficiency caused by HIV which is characterised by the lack of CD4+ (helper) T cells and macrophages, which are destroyed by HIV. The immune system attempts to destroy allografts or xenografts; therefore, the clinical immunologists study the ways to prevent the transplant rejection.

***(ii)* Immunotherapy:** Immunotherapy deals with the use of components of immune system to treat a disease. Immunotherapy is most commonly used in the treatment of cancers along with chemotherapy and radiotherapy. Often the immunotherapy is used in the immunosuppressed patients such as HIV patients, and in the sufferers of the other immune deficiencies or autoimmune diseases.

***(iii)* Diagnostic Immunology:** Specific bond are formed between antibody and antigen. Hence, specificity of the bond has made it an excellent tool in the detection of substances in a variety of diagnostic techniques. Therefore, antibodies specific for a desired antigen can be conjugated with a radiolabel, fluorescent label, or color-forming enzyme and are used as a 'probe' to detect the antigens.

***(iv)* Immuno-biotechnology:** The immune system many of dual contrast properties such as self/non-self recognition, general/specific, natural/adaptive (innate/acquired), cell-mediated/ humoral, active/passive, primary/ secondary. Parts of the immune system are: (*a*) *antigen-specific* (they recognise and act against particular antigens), (*b*) *systemic* (not confined to the initial infection site, but work throughout the body), and (*c*) have *memory* (recognise and mount an even stronger attack to the same antigen the next time). Self/non-self recognition is achieved by every cell due to the presence of a marker based on the major histocompatibility complex (MHC). If a cell does not display this marker, it is treated as non-self and attacked. Even the undigested proteins are treated as antigens.

Due to breaking down of the immune system, the process breaks down and attacks self-cell. Such situation occurs in autoimmune diseases like multiple sclerosis, systemic lupus erythematosus, and some forms of arthritis and diabetes. In some cases the immune response to innocuous substances is inappropriate such as allergies where the simple substance called an allergen elicits the responses.

C. ANTIGENS

In 1899, Ladislas Deutsch (Detre) (1874-1939) named the hypothetical substances halfway between bacterial constituents and antibodies as *immunogenes* or *antigenes*. He was of the opinion that the substances to be precursors of antibodies like zymogen are the precursor of zymase. In 1903, he could know that an antigen induces the production of immune bodies (antibodies) and wrote that the word 'antigen' was a contraction of 'antisomatogen = immunkorperbildner'. An antigen (Ag) (*antibody-generating*) or immunogen is a substance that induces the generation of antibodies and causes an immune response. Moreover, the immune system does not consist of only antibodies, but it includes all substances that are recognised by the adaptive immune system. The substances that elicit a response from the immune system are called immunogens, whereas antigens are the substances that bind to specific antibodies. All immunogens are antigens, but not all antigens produce an immunogenic response.

Do You Know?

- **Tolerogen** is a substance that invokes a specific immune non-responsiveness due to its molecular form. If its molecular form is changed, a tolerogen can become an immunogen.
- **Immunogen** is a substance that induces a specific immune response.
- Cells present their antigens to the immune system via a histocompatibility molecule. Depending on the antigen presented and the type of the histocompatibility molecule, several types of immune cells can become activated.
- **Allergen** is a substance that causes the allergic reaction resulting after exposure via ingestion, inhalation, injection or contact with skin.

Basically antigens are proteins which include parts of bacteria (such as capsules, cell wall, flagella, fimbrae and toxins), viruses and other microorganisms. Lipids and nucleic acids combined with proteins and polysaccharides also act as antigen. Non-microbial exogenous antigens include pollen, egg white and proteins from transplanted tissues and organs or on the surface of transfused blood cells.

A small molecule which elicits an immune response only when attached to a large carrier *e.g.* a protein is called 'hapten'. The carrier molecule by itself may not elicit an immune response. The small-molecule hapten may bind to the antibody when the body has generated antibodies to a hapten-carrier adduct; but it will usually not initiate an immune response. Only the hapten-carrier adduct can initiate an immune response. Sometimes the hapten can prevent adduct from binding to the antibody, and thus block the immune response to the hapten-carrier adduct. There are some haptens which can induce autoimmune disease such as hydralazine. It is a drug which lowers blood pressure and produce drug-induced lupus erythematosus in certain individuals. The other haptens commonly used in molecular biology are fluorescein, biotin, dinitrophenol and digoxigenin.

Thus, haptens have the property of antigenicity but not immunogenicity. A portion of an antigen that combines with the products of a specific immune response is called epitope or antigenic determinant (Fig.16.1).

Antigenic determinants are created by the primary sequence of residues in the polymer *i.e.* sequence determinants, and/or by the secondary, tertiary or quaternary structure of the molecule *i.e.* conformational determinants. Generally, the antigenic determinants are small and limited to about 4-8 residues (amino acids and or sugars). The combining site of an antibody will accommodate an antigenic determinant of approximately 4-8 residues. Although, each 4-8 residues can constitute a separate antigenic determinant and the number of antigenic determinants per antigen is much lower than expected number. Usually the antigenic determinants are limited to those portions of the antigen that are accessible to antibodies

1. Types of Antigens

Basically antigens are of the following two types:

(*a*) T-independent Antigens: The antigens which can directly stimulate B cells to produce antibody without the help of are called T-independent antigens *e.g.* polysaccharides which are T-independent antigens. However, the responses to these antigens differ from the same of other antigens. These are characterised by the same antigenic determinant repeated several times (Fig.11.1). Moreover, many of these antigens can activate B cell clones which are specific for other antigens. Then it is termed as *polyclonal activation*. On the basis of their ability to polyclonally activate B cells, T-independent antigens are put into two subgroups, Type 1 and Type 2. Type 1 T-independent antigens are polyclonal activators, while Type 2 antigens are not. Generally, T-independent antigens are more resistant to degradation; therefore they persist for longer time and stimulate the immune system continuously.

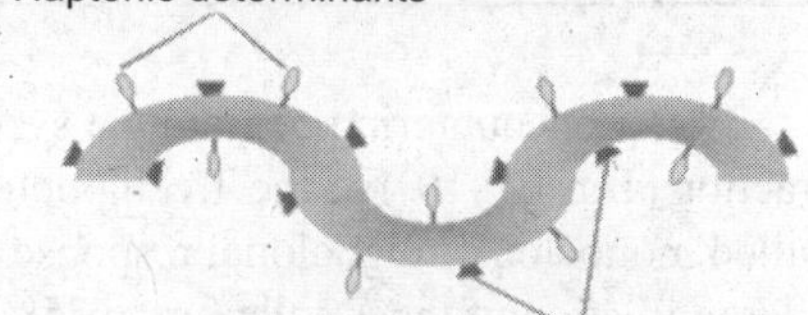

Fig.16.1. Haptenic determinants present on hapten-carrier conjugate of an antigen.

(*b*) T-dependent Antigens: T-dependent antigens are those that do not directly stimulate the production of antibody without the help of T cells. Proteins are T-dependent antigens. These antigens are characterised by a few copies of many different antigenic determinants (Fig. 16.1).

2. Origin of Antigens

Antigens can be classified according to their origins such as *exogenous antigens, endogenous antigens, autoantigens, tumor antigens* and *native antigens*.

(*a*) Exogenous Antigens: The antigens which have entered the body from the outside are called exogenous antigens. They can enter by inhalation, ingestion or injection. Further these antigens are taken into the antigen-presenting cells (APCs) through endocytosis or phagocytosis. Then APCs present the processed fragments to T helper cells ($CD4^+$) by using the class II histocompatibility molecules present on their surface. There are certain T cells which are specific for the peptide-MHC complex. They become activated and start to secrete cytokines which activate cytotoxic T lymphocytes (CTL), antibody-secreting B cells and macrophages.

(*b*) Endogenous Antigens: The antigens that are generated within the cell due to normal cellular metabolic activities or due to viral or intracellular bacterial infection are called endogenous antigens. Then the fragments are presented on the cell surface in the complex with MHC class I molecules. After being recognised by activated cytotoxic $CD8^+$ T cells, the T cells begin to secrete various toxins that cause the lysis or apoptosis of the infected cell.

Self-reactive T cells are deleted from the repertoire as a result of tolerance to keep the cytotoxic cells from killing cells just for presenting self-proteins. They include heterologous, autologous and idiotypic or allogenic (homologous) antigens.

(c) Autoantigens: A normal protein or complex of proteins, DNA or RNA which is recognised by the immune system of patients suffering from a specific autoimmune disease is called autoantigen. Under normal conditions these antigens are not the target of the immune system. The normal immunological tolerance for such an antigen has been lost in the patients due to genetic and environmental factors.

(d) Tumour antigens: The antigens presented by MHC I or MHC II molecules on the surface of tumour cells are called tumour antigens or neoantigens. Sometimes, these antigens are also presented by tumour cells and never by the normal ones. Therefore, they are called tumour-specific antigens (TSAs) because they are produced due to tumour-specific mutation. The antigens that are presented by tumour cells and normal cells are called tumour-associated antigens (TAAs). Tumour antigens may also be present on the surface of the tumour as a mutated receptor where they are recognised by B cells.

(d) Native Antigen (Nativity): An antigen that is not yet been processed by an APC to smaller parts is called a native antigen. The B cells can be activated by native antigens. But the T cells cannot bind to native antigens but they are processed by APCs.

Superantigens

After encountering of immune system to a conventional T-dependent antigen, only a small fraction (1 in 10^4-10^5) of the T cell population recognises the antigen and becomes activated. It is called monoclonal/oligoclonal response. But there are some antigens which polyclonally activate a large fraction of the T cells (up to 25%). These antigens are called superantigens.

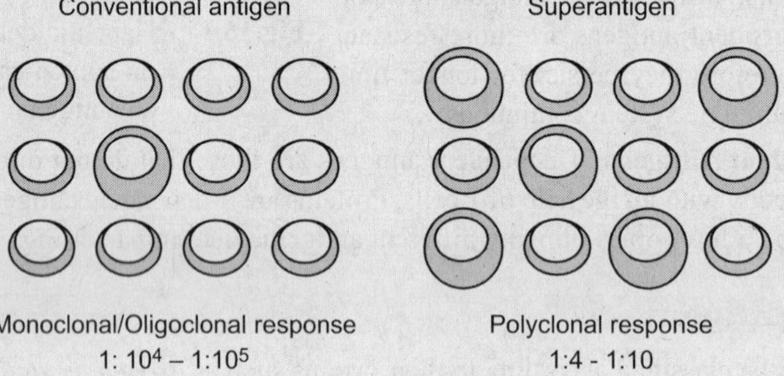

The examples of superantigens are: staphylococcal enterotoxins (food poisoning), staphylococcal toxic shock toxin (toxic shock syndrome), staphylococcal exfoliating toxins (scalded skin syndrome) and streptococcal pyrogenic exotoxins (shock). The bacterial superantigens have been studied in detail. But superantigens associated with viruses and other microorganisms are also known. When superantigens are exposed, there starts hyper-activation of the immune system and subsequent release of biologically active cytokines by activated T cells resulting in development of disease.

D. INNATE IMMUNITY

The cells and mechanisms that defend the host from infection by other organisms, in a non-specific manner are called 'innate immunity'. The cells of the innate system recognise and respond to pathogens in a generic way. Unlike adaptive immune system, it does not confer protective immunity for a long time to the host and immediate defense against infection is provided by the innate immune systems in all types of animals. The innate immunity system is present in us since birth and it is nonspecific. It is inherited generation to generations. Differences between innate and adaptive immunity are given in Table 16.2.

Table 16.2. Difference between innate and adaptive immunity.

	Difference between innate and adaptive immunity	
	Innate immunity	***Adaptive immunity***
Action time	Early (within hours)	Late (after a few days)
Cell involved	Macrophages, dendritic cells, neutrophils	B, T Lymphocytes
Receptors	Fixed in genome *e.g.* toll-like receptor	Gene rearrangement is necessary *e.g.* B cell receptor, T cell receptor
Recognition	Conserved molecular patterns *e.g.* lipopolysaccharide	Wide variety of molecular structure (proteins, peptides)
Evolution	Evolutionarily conserved (in plants, animals)	Only vertebrates (jawed fish-human)

In vertebrates the functions of the innate immune system are: (*i*) recruitment of immune cells to sites of infection, through the production of chemical factors such as specialised chemical mediators called cytokines, (*ii*) activation of complement cascade to identify bacteria, activate cells and to promote clearance of dead cells or antibody complexes, (*iii*) identification and removal of foreign substances present in organs, tissues, blood and lymph, etc. by specialised white blood cells, and (*iii*) activation of the adaptive immune system through a process known as antigen presentation.

I. CELLS OF INNATE IMMUNE RESPONSE

Blood is a vascular connective tissue consisting of 52–62% liquid plasma and 38–48% cells. The plasma is mostly water (91.5%) and acts as a solvent for transporting other materials [7% protein (consisting of 54% albumins, 38% globulins, 7% fibrinogen (and 1% other assorted stuff)] and 1.5% other stuff. Blood is slightly alkaline (pH = 7.4).

Table 16.3. Counts of blood cells in a normal adult.

Counts of blood cells in a Normal adult	
Red Blood Cells	$5.0 \times 10^6/mm^3$
Platelets	$2.5 \times 10^5/mm^3$
Leukocytes	$7.3 \times 10^3/mm^3$
Neutrophil	50-70%
Lymphocyte	20-40%
Monocyte	1-6%
Eosinophil	1-3%
Basophil	<1%

The stem cells manufacture all blood cells via a process called haematopoiesis which live mostly in the bone marrow (Fig. 16.2).

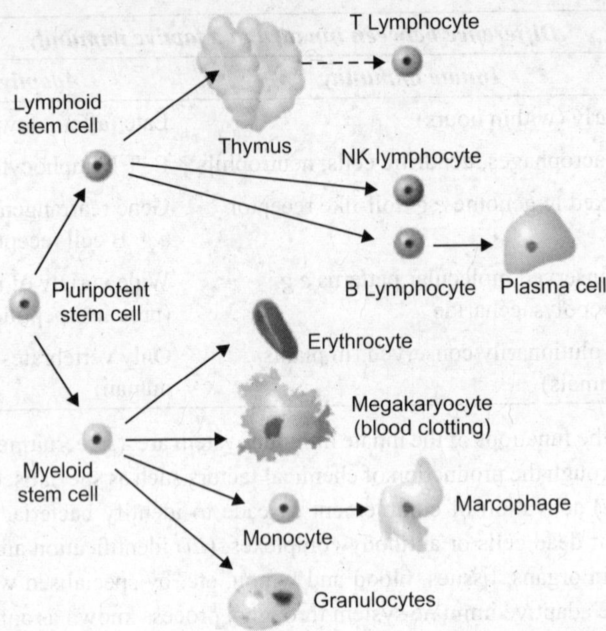

Fig. 16.2. Generation of blood cells from pluripotent stem cells in bone marrow.

Haemocytoblasts produced by the stem cells differentiate into the precursors for all the different types of cells, and mature into three types of blood cells: *erythrocytes* (red blood corpuscles or RBCs), *leukocytes* (white blood cells or WBCs), and *thrombocytes* (platelets). Further the leukocytes are subdivided into *granulocytes* (containing large granules in the cytoplasm) and *agranulocytes* (without granules). The granulocytes consist of neutrophils (55–70%), eosinophils (1–3%), and basophils (0.5–1.0%). The agranulocytes are *lymphocytes* (consisting of B cells and T cells) and *monocytes*. Lymphocytes circulate in the blood and lymph systems. Blood cell counts in a normal adult are given in Table 16.3.

(*a*) Erythrocytes: The red blood corpuscles (RBCs) are known as erythrocytes. They are the carriers of oxygen. The rate of production of RBCs is about 2,400,000 per second; each RBC lives for about 120 days. Before their death they migrate to the spleen to die. A healthy male has about 5 million RBCs per mm^3, while females have less than the males. In addition, about 5000–10,000 WBCs per mm^3 are found in a healthy man that live for 5-9 days.

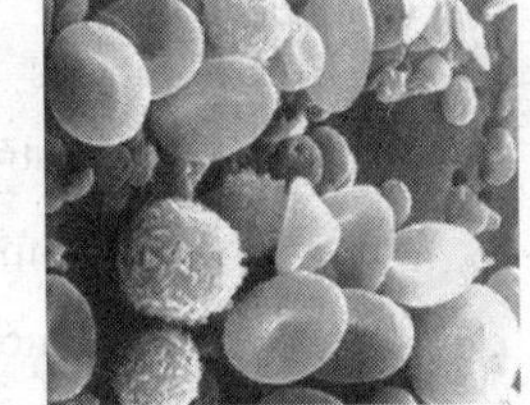

SEM showing normal circulating human blood cells

The antigen present on RBCs is responsible for the ABO blood grouping. Each group is characterised by the presence or absence of A and/or B antigens on the surface of the RBCs. Blood group AB contains both antigens, and group O lacks both antigens. Type A blood has A antigen and type B blood has B antigen.

(*b*) Leukocytes: The white blood cells (WBC) of all types are called leukocytes. Leukocytes are not tightly associated with a particular organ or tissue. Therefore, they work independently as the single-celled organisms. Leukocytes move freely, interact with and capture cellular debris, foreign particles or invading microorganisms. They are produced by pluripotent haematopoietic stem cells present in bone marrow. The innate leukocytes include: natural killer cells, mast cells, eosinophils, basophils and the phagocytes cells including macrophages, neutrophils and dendritic cells. They are able to identify and eliminate the pathogens that cause infection (Charles *et al.,* 2001).

(*c*) **Mast cells:** Mast cells are found in connective tissue and mucous membranes. They are intimately associated with defense against pathogens, wound healing. Often, they are also associated with allergy and anaphylaxis. Upon activation, mast cells soon release characteristic granules which are rich in histamine and heparin. Different types of hormone mediators and chemokines or chemotactic cytokines are also released into the environment. Inflammation is caused due to dilation of blood vessels by histamine called vasodialation.

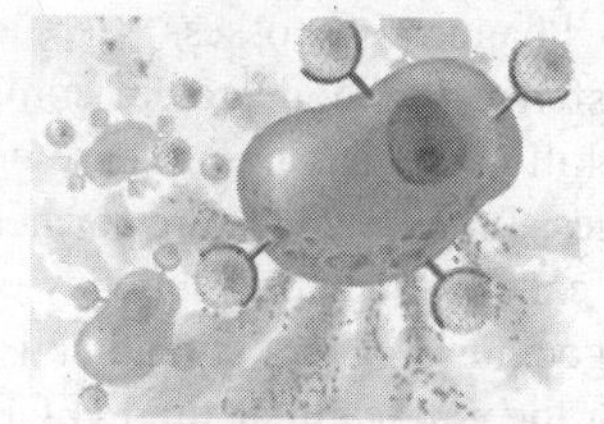

A mast cell releases histamines after encountering with allergens.

(*d*) **Phagocytes**: The cells that eat upon the foreign particles or pathogens are called phagocytes. The phenomenon of eating of pathogens is called phagocytosis. The phagocytic cells of the immune system include macrophages, neutrophils, eosinophils and dendritic cells. These cells provide immunity. These cells phagocytose *i.e.* engulf pathogens or particles. A phagocyte extends portions of its plasma membrane and wraps the membrane all around the particle. After being enveloped the invading pathogen are taken up inside an endosome which merges with a lysosome (Fig. 16.3). The lysosome consists of acids that kill organisms. They also contain enzymes that digest the pathogens or particles. Generally the phagocytes patrol the body searching for pathogens. Besides, the phagocytes react to a group of highly specialised molecular signals produced by other cells which are called cytokines.

Phagocytosis

The cellular process of phagocytes is called phagocytosis. Bacteria, dead tissue cells, and small mineral particles may be phagocytosed. The process is only homologous to eating at the level of single-celled organisms. The phagocytes engulf solid particles by the cellular mechanism to form an internal phagosome. Phagocytosis is a specific form of endocytosis that involves the vesicular internalisation of solid particles such as bacteria. Therefore, it differs from other forms of endocytosis such as **pinocytosis** which is the vesicular internalisation of various liquids. It is involved in the acquisition of nutrients for some cells. In the immune system it is a major mechanism which uses to remove pathogens and cell debris.

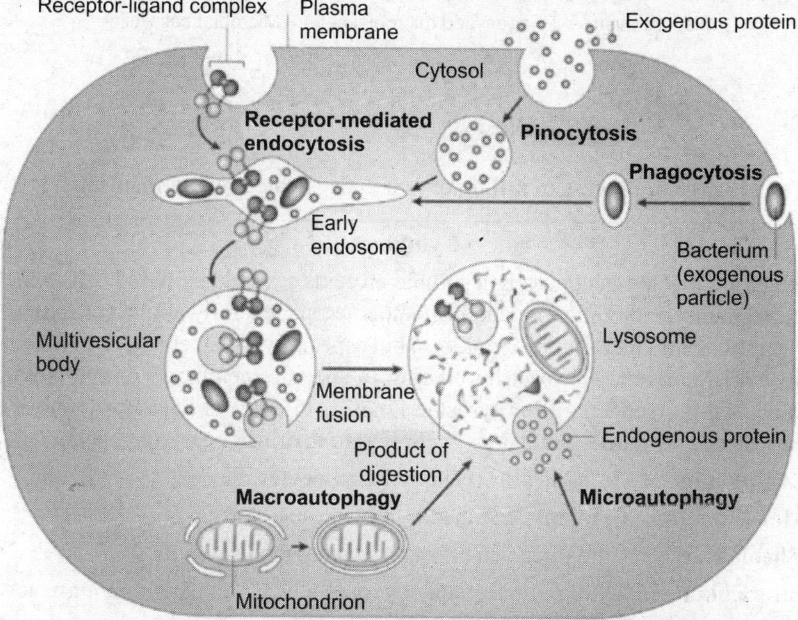

Fig. 16.3. Mechanism of endocytosis- phagocytosis and pinocytosis.

Phagocytes possess receptors (*e.g.* CR3, FcgR, mannose receptors, scavenger receptors, transferrin (a blood protein complexed with iron) on their cell surface to facilitate phagocytosis and kill pathogens by protease enzymes. Phagocytes attract, adhere, engulf and ingest the foreign bodies. Promonocytes are formed in the bone marrow which later on are released into the blood and called circulating monocytes. In turn they get matured into macrophages *e.g.* 'big eaters'. Phagocytosis occurs sporadically and results in the ingestion of particulate matter (*e.g.* bacteria) from the extracellular fluid (ECF). When cell receptors encounter a molecule of transferrin, they bind tightly to it.

The complex of transferrin and its receptor is then engulfed by endocytosis. The endosome of macrophages is so large that it is called a phagosome. During the course of time, phagosomes deliver their contents to lysosomes. The membranes of the two organelles fuse. Once inside the lysosome, the contents of the phagosome, e.g. ingested bacteria, are destroyed by the degradative enzymes of the lysosome.

Phagocytosis is accomplished in three steps: (*i*) unbound phagocyte surface receptors do not trigger phagocytosis, (*ii*) binding of receptors causes them to cluster, and (*iii*) phagocytosis is triggered and the particle is taken-up by the phagocyte.

(*i*) Macrophages: The macrophages (Greek, '*macro*' means 'large' and '*phage*' means 'eating cell') are large phagocytic leukocytes which move outside of the vascular system across the cell membrane of capillary vessels and enter the areas between cells to search the invading pathogens. In tissues, organ-specific macrophages are differentiated from phagocytic cells present in the blood called monocytes. Macrophages are the most efficient phagocytes and can phagocytose substantial numbers of bacteria or other microbes. The binding of bacterial molecules to receptors on the surface of a macrophage triggers it to engulf and destroy the bacteria through the generation of a 'respiratory burst'. Pathogens also stimulate the macrophage to produce chemokines which communicate the other cells to the site of infection. Extracellular non-encapsulated microorganisms are easily phagocytosed and killed within macrophages. But the encapsulated microorganisms require the production of antibody in order to be effectively phagocytosed. However, they are easily killed after being engulfed by the phagocytes (Fig. 16.4).

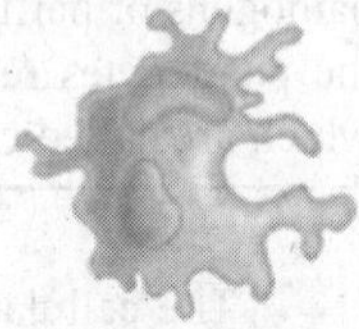

A macrophage

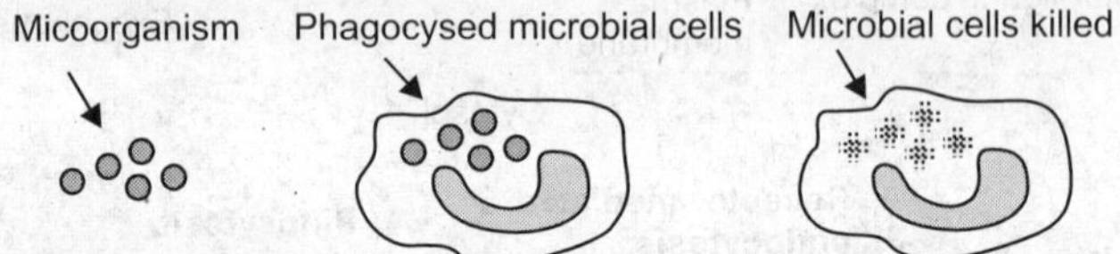

Fig. 16.4. Process of killing of non-encapsulated microorganisms.

Cytokines

Cytokines are low molecular weight soluble proteins secreted by blood cells which mediate and regulate immunity, inflammation, and haematopoiesis. Generally all the cells associated with innate and adaptive immunity secrete cytokines, but especially T-helper (T_H) lymphocytes secrete them in substantial amount. They act by binding to specific membrane receptors which signal the cell via second messengers, often tyrosine kinases, to alter its behavior (gene expression). Cytokines are named with different names according to their origin as given below:

- **Lymphokine** : Cytokines secreted by lymphocytes
- **Monokine** : Cytokines secreted by monocytes
- **Chemokine** : Cytokines which possess chemotactic activities
- **Interleukin** : Cytokines secreted by one leukocyte which in turn act on other leukocytes

The cells that secrete them are also affected by cytokines; such action is called 'autocrine action'. When the other nearby cells are affected, it is called 'paracrine action'. When the action of cytokines takes place on distant cells, it is called 'endocrine action'.

Cytokines act on their target cells by binding the specific 'membrane receptors'. The receptors and their corresponding cytokines have been divided into several families on the basis of their structure and function as given below:

- **Haematopoietin family:** Receptors are dimers or trimers with conserved cysteines in their extracellular domains and a conserved Trp-Ser-X-Trp-Ser sequence. Examples are receptors for IL-2 through IL-7 and GM-CSF.
- **Interferon family:** Receptors have the conserved cysteine residues but not the Trp-Ser-X-Trp-Ser sequence, and include the receptors for IFNa, IFNb and IFNg.
- **Tumour necrosis factor family:** Receptors have four extracellular domains; they include receptors for soluble TNFa and TNFb as well as membrane-bound CD40 (important for B cell and macrophage activation) and Fas (which signals the cell to undergo apoptosis).
- **Chemokine family:** Receptors have seven transmembrane helices and interact with G protein. This family includes receptors for IL-8, MIP-1 and RANTES. Chemokine receptors CCR5 and CXCR4 are used by HIV to preferentially enter either macrophages or T cells.

Haematopoietin cytokine receptors are the best characterised. They generally have two subunits, one cytokine-specific and one signal transducing.

***(ii)* Neutrophils:** Neutrophils, along with two other cell types, eosinophils and basophils, are known as granulocytes due to the presence of granules in their cytoplasm or due to presence of lobed nuclei. They have no mitochondria and get their energy from stored glycogen. They are nondividing, short-lived (half-life of 6–8 hours, 1–4 day lifespan) and have a segmented nucleus. Many types of toxic substances are found in granules of neutrophil that inhibit the growth of bacteria and fungi. Like macrophages, neutrophils also attack pathogens through activation of respiratory burst. The main products of the neutrophil respiratory burst are strong oxidising agents such as hydrogen peroxide, free oxygen radicals and hypochlorite. Neutrophils represent by 50 to 60% of the total circulating leukocytes. Usually they are the first cells to come at the site of an infection. In a normal healthy adult, the bone marrow produces about 100 billion neutrophils per day. But during acute inflammation more than 10 times neutrophils per day are produced. The neutrophils provide the major defense against pyogenic (pus-forming) bacteria and first fight to infection. About three to four hours later, they are followed by the wandering macrophages.

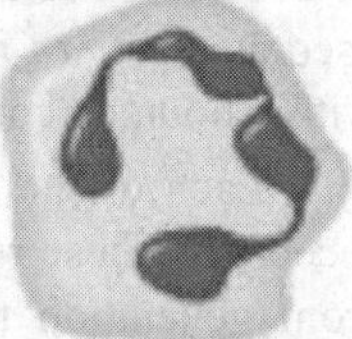
A neutrophil

***(iii)* Dendritic cells:** Dendritic cells are covered with the membranous processes that look like nerve cell dendrites. Most of them are highly efficient antigen presenting cells. There are four basic types: Langerhans cells, interstitial dendritic cells, interdigitating dendritic cells, and circulating dendritic cells. Langerhans cells are found in the epidermis and mucous membranes especially in the anal, vaginal and oral cavities. These cells make a point of attracting antigen and efficiently presenting it to T helper cells for their activation. Dendritic cells are in contact with the external environment especially the skin and the inner mucosal lining of the nose, lungs, stomach and intestines. These cells are not connected to the nervous system. Dendritic cells are very important in the process of antigen presentation. They act as a link between the innate and adaptive immune systems.

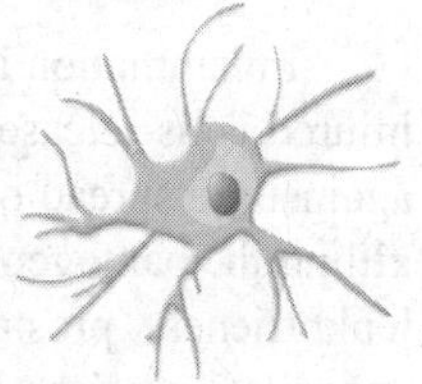
A dendric cell

***(e)* Basophils and Eosinophils:** Basophils and eosinophils are the cells that have relation to the neutrophil. About 13% of the WBCs are eosinophils. Their lifespan is about 8–12 days. Eosinophils are attracted to cells coated with complement C3B, where they release major basic protein (MBP), cationic protein, perforins, and oxygen metabolites. All of which work together to burn holes in cells. Upon interaction with a pathogen basophils release histamine that is important in defense against parasites. It plays a role in allergic reactions *e.g.* asthma (Janeway *et al.,* 2001). When eosinophils are activated, they secrete several highly toxic proteins and free radicals which effectively kill bacteria and parasites. Besides, they damage the tissue during allergic reactions. Therefore, activation and toxin release by eosinophils is tightly regulated so that any inappropriate tissue destruction may be prevented (Stvrtinová, *et al.,* 1995).

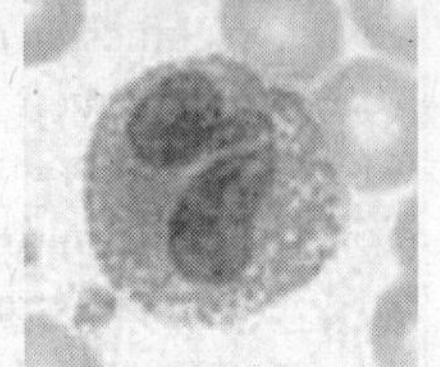
An eosinophil

***(f)* Natural Killer Cells:** The natural killer (NK) cells are a constituent of the innate immune system which does not directly attack the invading microbes. NK cells move in the blood and lymph to lyse cancer cells and virus-infected body cells. They are large granular lymphocytes that attach to the glycoproteins on the surfaces of infected cells and kill them. They recognise the cells by a condition called 'missing self' *i.e.* the cells with low levels of a surface marker called MHC I (major histocompatibility complex). This situation arises in viral infections of host cells. The natural killer cells do not require activation to kill cells having missing self.

(g) Gamma-Delta (γδ) T Cells: The γδ T cells show some characteristics on the basis of which they are placed at the border between innate and adaptive immunity. However, γδ T cells may be considered as a component of adaptive immunity, because they rearrange TCR genes to produce junctional diversity and develop a memory phenotype. Moreover, various subsets may also be considered as a part of innate immune system where a restricted TCR or NK receptors may be used as a pattern recognition receptor. For example, a large numbers of Vγ9/Vδ2 T cells respond within hours to common molecules produced by microbes.

Each of the cells in the innate immune system binds to antigen using pattern-recognition receptors. These receptors are encoded in the germ-line of every person. This immunity is passed from generation to generation. During the course of human development these receptors for pathogen-associated molecular patterns have evolved through natural selection. Therefore, they are specific to certain characteristics of infectious organisms. There are hundreds of receptors which recognise bacterial lipopolysaccharide, peptidoglycan, DNA, dsRNA and other substances.

II. RESPONSES OF INNATE IMMUNE SYSTEM

There are many molecules and mechanisms which are involved in the innate immune system. Some of the mechanisms are discussed below:

1. Inflammation

Inflammation is one of the first responses of the immune system to infection or irritation. The injured cells release chemical factors that stimulate inflammation and establish a physical barrier against the spread of infection. These chemicals also promote healing of the damaged tissue after killing the pathogens (Stvrtinová, *et al.,* 1995). Chemical factors (histamine, bradykinin, serotonin, leukotrienes, prostaglandins) produced during inflammation sensitise pain receptors, cause vasodilation of the blood vessels, and attract phagocytes *e.g.* neutrophils. Thereafter, neutrophils release the factors that activate the other leukocytes and lymphocytes, and trigger the other parts of the immune system. The symptom quintet of inflammatory response is: *rubor* (redness) *calor* (heat), *tumor* (swelling), *dolor* (pain) and *functio laesa* (involvement of possible dysfunction of organs or tissues) (Stvrtinová, *et al.,* 1995).

2. Complement System

The complement system is a biochemical cascade of the immune system that helps or 'complements' the ability of antibodies to clear pathogens or mark them for destruction by other cells. Complement proteins exist in blood serum. They coat pathogens and form pores on membranes causing their lysis with the help of antibodies. Complement fragments called opsonins adhere to pathogen surface (opsonisation) and promote phagocytosis by leukocytes. The cascade is composed of many plasma proteins synthesised in the liver mainly by hepatocytes. Complement system has been described in detail in preceding section.

Thus the proteins work together to: (*i*) trigger the recruitment of inflammatory cells, (*ii*) tag pathogens for destruction by other cells by opsonisation (*i.e.* adhering to pathogen surface) or coating, the surface of the pathogen, and (*iii*) disrupt the plasma membrane of an infected cell resulting in cytolysis of the infected cell causing the death of the pathogen.

Opsonisation

Adherance of opsonins (complement fragments) to the surface of pathogen and promotion of phagocytosis by leukocytes is called opsonisation. An opsonin is any molecule that acts as a binding enhancer for the process of phagocytosis by coating the negatively-charged molecules on the surface of membrane. Both the membrane of a phagocytosing cell as well as its target have negative charges, which make the two cells difficult to come closer together. During the process of opsonisation, antigens are bound by antibody and/or complement molecules. Phagocytes express receptors that bind opsonin molecules which include the Fc receptors. Binding of the antigen to the phagocyte is greatly enhanced with the antigen coated in these molecules. Most of the phagocytic binding cannot occur without opsonisation of the antigen. Opsonisation of the antigen and subsequent binding to an activated phagocyte cause increased expression of complement receptors on neighboring phagocytes.

3. Cytoplasmic Viral Receptors

Proteins present inside the cells detect the viral products directly such as double-stranded RNA. As a result of the detection of viruses by these receptors, production of interferons (anti-viral proteins) or programmed cell death *e.g.* PKR, RIG-I can be initiated.

4. Antimicrobial Peptides

The antimicrobial peptides such as defensins, cathelidins, etc. are the small cationic molecules secreted from animals and plants that kill pathogens such as bacteria, fungi or viruses (some antimicrobial peptides even kill cancer cells). In humans, antimicrobial peptides are secreted to the body surface such as the skin or intestinal epithelium. They kill pathogens by disrupting their membranes.

5. Apoptosis (Programmed Cell Death)

Animal cells infected by virus have a 'suicide' programme called programmed cell death or apoptosis. After death of virus-infected cells, proliferation of viruses in host cells can be inhibited and their propagation to the other cells can be reduced.

***(a)* Innate Immune Response:** Recognition of pathogens (or Pathogen-Associated Molecular Patterns, PAMPs) by receptors such as TLRs or NBD-LRR proteins results in activation of antigen-presenting cells (APCs). This activation results in the production of inflammatory cytokines. At the same time co-stimulatory molecules are expressed on the cell surface (Fig. 16.5).

***(b)* Adaptive Immune Response:** After activation of innate immune response, the MHC molecules on APCs present the antigens to T lymphocytes. This is not enough to activate T lymphocytes. An additional signal from co-stimulatory molecules is required. Macrophages, dendritic cells and epithelial cells have a set of transmembrane receptors that recognize different types of PAMPs. These are called toll-like receptors (TLRs). Hence, TLR (or NBD-LRR) stimulation induces the expression on APCs. Further the activated T lymphocytes are stimulated by cytokines *e.g.* IL-12, and become differentiated to effector T lymphocytes (Fig. 16.5).

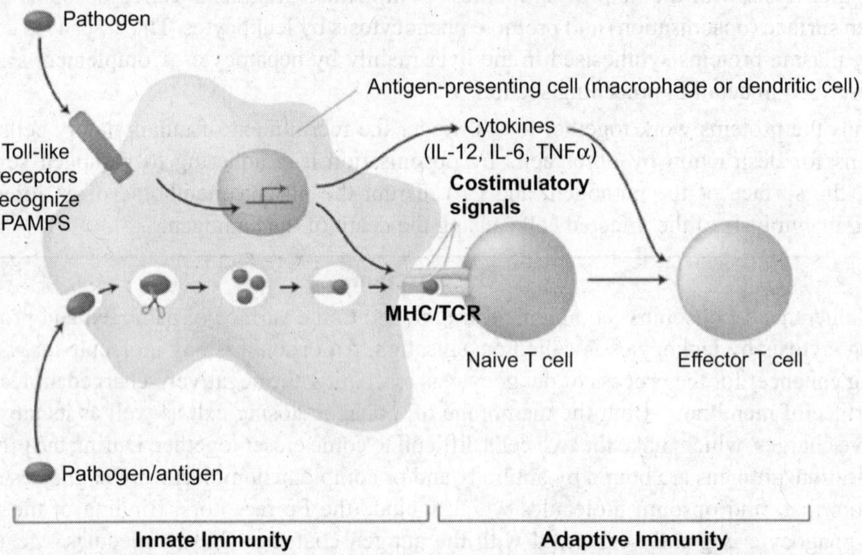

Fig. 16.5. Mechanism of innate immunity and adaptive immunity.

E. ADAPTIVE OR ACQUIRED IMMUNITY

Lymphocytes are of two main types: B cells and T cells. The peripheral blood contains 20–50% of circulating lymphocytes, while the rest move in the lymph system. About 80% of them are T cells, 15% B cells and remaining are undifferentiated cells. Among total WBC of the body, lymphocytes constitute 20–40%. The stem cells in the fetal liver and the bone marrow produce B cells which produce antibody and provides humoral immunity (Fig. 16.6).

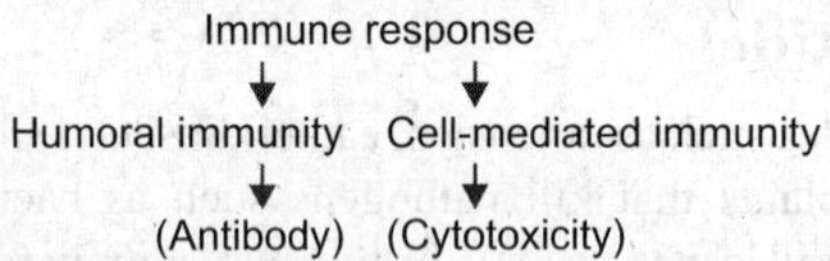

Fig. 16.6. Types of adaptive immune response.

The other type of lymphocytes are also produced in the bone marrow but sensitized in the thymus; therefore these are called T cells. They are nonantibody-producing cells which constitute the basis of cell-mediated immunity.

B lymphocytes are thymus-independent that express intrinsically produced immunoglobulins on their external membranes and upon stimulation by antigen differentiate into plasma cells that produce and secrete large numbers of antibody molecules. Pre-B cells, the immediate precursors of B cells, are restricted to the bone marrow and are characterised by the presence of cytoplasmic μ chains (H chains for IgM) but no L chains. Non-stimulated mature B cells express monomeric IgM antibodies, MHC class II molecules, CD19- CD20, the Epstein-Barr virus/C3d (CR2) (CD21) receptor, and T cell interaction molecules, B7-1, B7-2 and the CD40 ligand, CD39 (Fig. 16.7). Function of markers of B cells is given in Table 16.4. B cells account for 10-15% of blood lymphocytes. They and antibody-producing cells reside mainly in peripheral lymphoid organs.

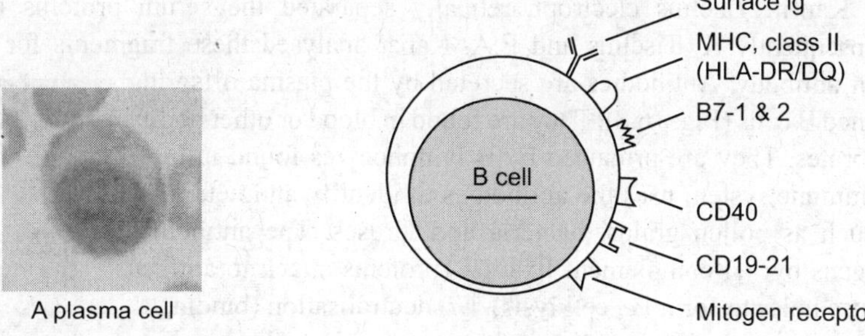

Fig. 16.7 . Surface markers on a B cell.

The immune system is changeable and can adapt to attack the invading antigen. There are two fundamental adaptive mechanisms: *humoral immunity* and *cell-mediated immunity*.

Table 16.4. Function of markers of B cells.

B cell markers	*Function*
Surface Ig	Antigen specific recognition
MHC class II (HLA-DR/DO)	Antigen presentation
B7-1&2	T-cell activation
CD40	Receives activation signal from T cells
CD19-21	Unique B cell markers
Mitogen receptor	Binding and activation by bacterial products

F. HUMORAL IMMUNITY

When an antigen binds to the surface receptors of an immunocompetent immature B-lymphocyte, the later is stimulated to maturity in the presence of a T helper cell nearby (to release a cytokine). This sensitises the B cell and it undergoes clonal selection. Thus B-lymphocyte reproduces asexually by mitosis. Most of the family of clones becomes plasma cells which after an initial lag produce highly specific antibodies at a rate of about 2000 molecules per second for four to five days. The other B cells become long-lived and behave as memory cells.

I. ANTIBODY – STRUCTURE AND DIVERSITY

An antibody is a soluble globular protein (glycoproteins) which is produced in response to a specific antigen and binds to it via non-covalent interactions. The term 'immunoglobulin' is often used interchangeably with 'antibody'. The term immunoglobulin is used to describe *any* antibody, regardless of specificity, and the term 'antibody' to describe an *antigen-specific* 'immunoglobulin'.

Arne Wilhelm Kaurin Tiselius (1902 –1971)

Arne Wilhelm Kaurin Tiselius was a Swedish biochemist who won the Nobel Prize in Chemistry in 1948. He studied at the University of Uppsala specialising in chemistry. In 1925, he became research assistant in the Svedberg's laboratory and obtained his doctor's degree in 1930 on the 'moving-boundary method of studying the electrophoresis of proteins'. He published a number of papers on diffusion and adsorption in naturally occurring base-exchanging zeolites. He took interest in proteins and the application of physical methods to biochemical problems. This led to a much-improved method of electrophoretic analysis which he refined in subsequent years. Therefore, he is known as the 'father of electrophoresis'.

In 1937, Kaurin Tiselius electrophoretically separated the serum proteins into different fragments. Consequently K. Tiselius and E.A. Kabat analysed these fragments for the presence of properties of antibody. Antibodies are secreted by the plasma offspring (clones) of primed B cells (Fig. 16.8). They are found in blood or other bodily fluids of vertebrates. They are produced by B lymphocytes found in bone marrow. The immune system uses the antibodies to identify and neutralise the antigens such as pollen grains, bacteria and viruses. The antibodies inactivate antigens by: (*i*) complement fixation (proteins attach to antigen surface and cause holes to form *i.e.* cell lysis), (*ii*) neutralisation (binding to specific sites to prevent attachment—this is the same as taking their parking space), (*iii*) agglutination (clumping), (*iv*) precipitation (forcing insolubility and settling out of solution), etc.

A.W.K.Tiselius

Elvin A. Kabat (1914–2000)

Elvin Abraham Kabat was an American biomedical scientist who is considered as one of the founding fathers of modern quantitative immunochemistry together with his mentor Michael Heidelberger. Elvin Kabat was engaged in the discovery of the structural and genetic basis for the remarkable specificity of antibodies. Kabat first demonstrated that antibodies are gamma globulins. He characterised the size and stunning heterogeneity of antibody-combining sites. Kabat and his coworkers determined the structures of the major blood group antigens, the embryonic-stage-specific carbohydrate antigens and the carbohydrate markers of leukocyte (white blood cell).

Kabat's work was basic research with many applications to clinical medicine. It greatly advanced our knowledge of inflammation, autoimmunity, and blood transfusion medicine.

Kabat did postdoctoral work with Arne Tiselius and Kai Pederson in Sweden to learn the new methods of ultracentrifugation and electrophoresis. He returned to New York City to take up a position at Cornell University Medical College.

The membrane-bound form of an antibody may be called a 'surface immunoglobulin' (sIg) or a 'membrane immunoglobulin' (mIg). It is part of the B cell receptor (BCR) and secreted by specialised cells called plasma cells that are derived from the B cells. When a specific antigen is present in the body, the mIg allows a B cell to detect it and also trigger B cell activation.

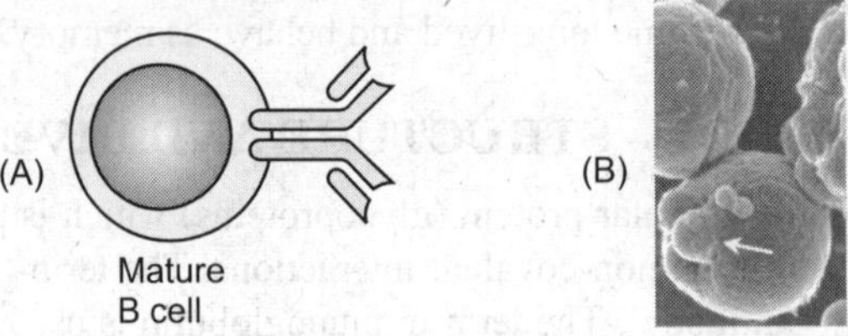

Fig. 16.8. A mature B lymphocyte containing an antibody molecule (A); lymphocyte producing antibodies (B).

What is Antitoxin ?

An antitoxin is an antibody which neutralises a specific toxin. These are produced by certain animals, plants and bacteria. In addition to neutralising the toxins, they can kill bacteria and other microorganisms. Organisms synthesise the antitoxins; but it can be injected into other organisms including humans. When an animal is injected with a safe amount of a particular toxin, the animal's body makes the antitoxin required to neutralise the toxin. Thereafter, blood is withdrawn from the toxin-injected animal. The antitoxin obtained from the blood is purified and injected into a human or other animal for inducing passive immunity.

1. Antigen Binding

Antigen binding by antibodies is the main function of antibodies which results in protection of the host. Antibodies bind mainly to specific antigenic determinant of one or a few closely related antigens. The valency of antibody refers to the number of antigenic determinants that an individual antibody molecule can bind. The valency of all antibodies is at least two and in some examples it is more than one.

There is no direct biological effect on frequently binding of an antibody to an antigen. But, the important biological effects are the result of secondary effector functions of antibodies. The antibodies mediate different effector functions. Antigen-antibody binding determines the ability to carry out a specific effector function. All the effector functions are not mediated by every immunoglobulin. Such effector functions include: *complement fixation* (which results in cell lysis and release of biologically active molecules), and *binding to various cell types* (such as phagocytic cells, lymphocytes, platelets, mast cells and basophils which have receptors that bind antibodies).

2. Structure of Antibodies

Immunoglobulins are composed of four polypeptide chains: two 'light' chains and two 'heavy' chains. The four polypeptide chains are held together by covalent disulfide (-S-S-) bonds (Fig. 16.9). There are two types of disulfide bonds: *inter-chain disulfide bonds* and *intra-chain disulfide bonds*. The inter-chain disulfide bonds held together the heavy and light chains and the two heavy chains by non-covalent interactions. The number of inter-chain disulfide bonds varies among different immunoglobulin molecules. There are also intra-chain disulfide bonds within each of the polypeptide chains.

Fig. 16.9 shows the structure and different binding sites present on an immunoglobulin molecule. Light chains are composed of 220 amino acid residues (V_L 110 amino acids plus C_L 110 amino acids), while heavy chains are composed of 440-550 amino acids (V_H 110 amino acids plus C_H 330-440 amino acids). Each chain has 'constant' (C) and 'variable' (V) regions. Variable regions are contained within the amino (NH_2) terminal end of the polypeptide chain (amino acids 1-110). The arms of the antibody molecule forms a Y shaped structure at a region called the hinge region because at this point there is some flexibility in the molecule.

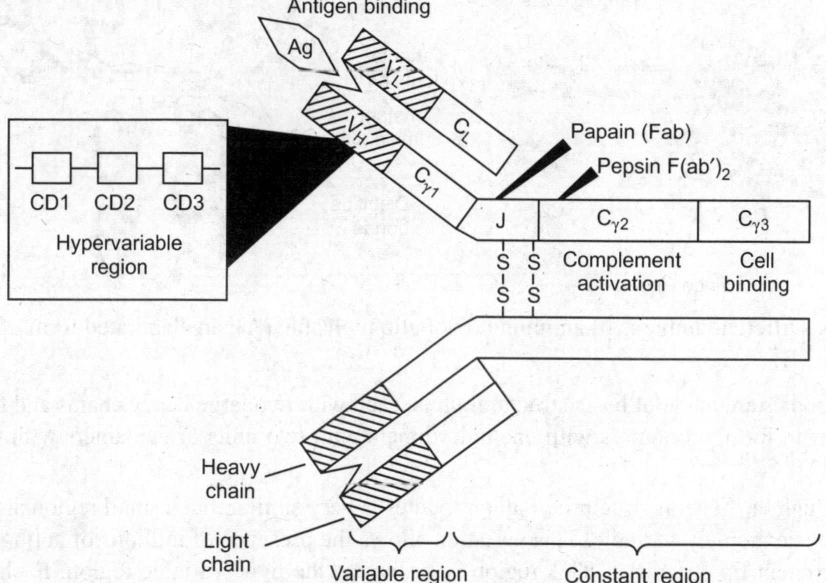

Fig. 16.9. Structure and different binding sites present on an immunoglobulin molecule.

Variable (V) regions of H and L chains show great variability in the sequence of amino acids. Localised areas of these hypervariable regions of H and L chains interact to form antigen binding sites (*i.e.* CD1, CD2 and CD3 (Fig. 16.9). In contrast, C regions of H chains dictate other functions of immunoglobulins such as binding to cell surface receptors. Eight immunoglobulin isotypes *e.g.* IgG1, IgG2, IgG3, IgG4, IgA1, IgA2, IgM, IgD, and IgE are produced by B cells as a result of rearrangements of V genes for H chains (VH), D genes for H chains (DH), J genes for H chains (JH), V genes for L chains (VL), J genes for L chains (JL), and C region genes (vide infra).

When two antibodies are compared, these amino acid sequences are quite distinct. Constant regions comprising amino acids 111-220 are rather uniform from one antibody to another within the same isotype. Hypervariable regions or 'complementarity determining regions' (CDRs) are found within the variable regions of both the heavy and light chains. These regions are called 'antigen-binding site' which recognise and bind to specific antigens.

The immunoglobulin molecule is not straight but it is folded into globular regions each of which contains an intra-chain disulfide bond. These regions are called domains (Fig. 16.10). There are two domains: light chain domains (for example V_L and C_L) and heavy chain domains (V_H, C_{H1} - C_{H3} or C_{H4}). In most immunoglobulins, carbohydrates are attached to the C_{H2} domain. However, carbohydrates may also be attached at other locations in some cases.

Immunoglobulins can also be classified by the type of light chain that they have. Light chain types are based on differences in the amino acid sequence in the constant region of the light chain. These differences are detected by serological means. The types of light chains are: kappa (κ) light chain and lambda (λ) light chain. The light chains can also be divided into subtypes based on differences in the amino acid sequences in the constant region of the light chain; for example lambda subtypes are: λ1, λ2, λ3, and λ4.

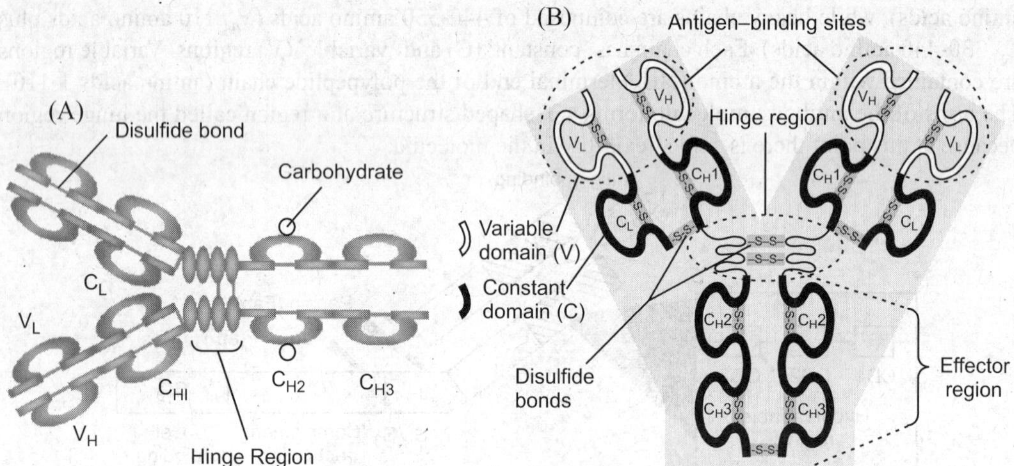

Fig. 16.10. Different domains of an immunoglobulin molecule (A); an elaborated form of domains (B).

Antibodies are made of basic structural units, each with two large heavy chains and two small light chains to form, monomers with one unit, dimers with two units or pentamers with five units (Fig. 16.11).

Although the general structure of all antibodies is very similar, but a small region at the tip of the protein is extremely variable. This variation allows the presence of millions of antibodies with slightly different tip structures. This region is known as the hypervariable region. Each of these variants can bind to a different target known as an antigen. This huge diversity of antibodies allows the immune system to recognise an equally wide diversity of antigens. The specific proteins of the antigen recognised by an antibody are called as epitope.

These epitopes bind with their antibody in a highly specific interaction called 'induced fit' that allows antibodies to identify and bind only their unique antigen in the midst of the millions of different molecules that make up an organism. The upper part of Fab portion of the antibody attaches to the epitopes of the antigen. Thus antibody recognises the epitope and not the entire antigen. The Fc region is fraction crystallisable and is responsible for effectors functions, *i.e.* the end to which immune cells can attach.

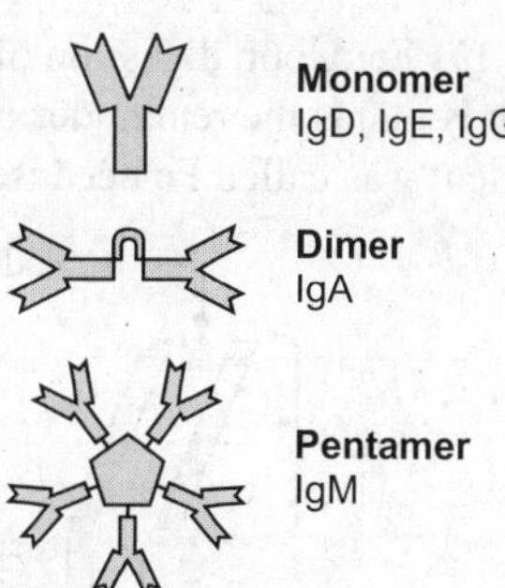

Fig. 16.11. Antibodies - monomer, dimer and pentamer.

Epitopes

An epitope is also called *antigenic determinant* which is the part of a macromolecule. It is recognised by the immune system specifically by antibodies, B cells or T cells. The part of an antibody that recognises the epitope is called a paratope. The epitopes are derived from nonself proteins, but the sequences derived from the host that can be recognised are also classified as epitopes. Most epitopes recognised by antibodies or B cells may have three-dimensional surface features of an antigen molecule. These features fit precisely and thus bind to antibodies.

T cell epitopes are presented on the surface of an antigen-presenting cell where they are bound to MHC molecules. T cell epitopes presented by MHC class I molecules are typically peptides between 8 and 11 amino acids in lengths, whereas MHC class II molecules present longer peptides, and non-classical MHC molecules also present non-peptidic epitopes such as glycolipids.

Epitopes are sometimes cross-reactive. This property is exploited by the immune system in regulation by anti-idiotypic antibodies. If an antibody binds to an antigen's epitope, the paratope can become the epitope for another antibody that will bind to it. If this second antibody is of IgM class, its binding can upregulate the immune response. But if the second antibody is of IgG class, its binding can downregulate the immune response.

2. Immunoglobulin Fragments- Structure-Function Relationships

Upon proteolytic digestion fragments of immunoglobulin are produced which have proven very useful in elucidation of structure-function relationships in immunoglobulins.

***(a)* Fab:** Upon digestion with papain, the immunoglobulin molecule is broken (in the hinge region before the H-H inter-chain disulfide bond) into two identical fragments which contain the light chain and the V_H and C_{H1} domains of the heavy chain. These fragments are called the Fab (fragments antigen binding) because they contained the antigen binding sites of the antibody (Fig. 16.12 A). Each Fab fragment is monovalent, whereas the original molecule was divalent. The divalent fragment was called **F(ab')$_2$** (Fig. 16.12B). The combining site of the antibody is created by both V_H and V_L. An antibody is able to bind a particular antigenic determinant because it has a particular combination of V_H and V_L. Different combinations of a V_H and V_L result in antibodies that can bind different antigenic determinants. The F(ab')$_2$ binds antigen but it does not mediate the effector functions of antibodies.

Immunoglobulins are normally very heterogeneous because they are composed of different classes and subclasses each of which has different types and subtypes of light chains. Besides, different immunoglobulin molecules can have different antigen binding properties because of different V_H and V_L regions.

***(b)* Fc:** Upon digestion of immunoglobulin molecule with papain also a fragment is produced which contains the remainder of the two heavy chains each containing a C_{H2} and C_{H3} domain. This fragment was called Fc because it was easily crystallized *i.e.* fragment crystalizable (Fig. 16.12 A).

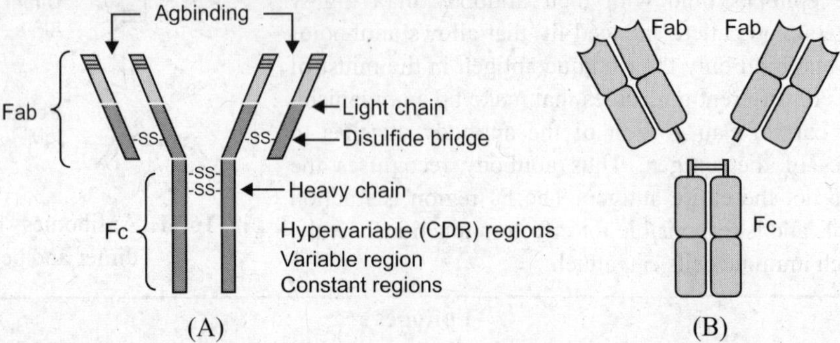

Fig. 16.12. Various regions of immunoglobulin (A); digestion of immunoglobulin into two Fab fragments, and one Fc fragment (B).

The effector functions of immunoglobulins are mediated by this part of the molecule. Different functions are mediated by the different domains in this fragment. Normally the ability of an antibody to carry out an effector function requires the prior binding of an antigen; however, there are exceptions to this rule. When immunoglobulins are digested in different enzymes, different types of fragments are obtained. After digestion with papain, immunoglobulin molecule is broken into three fragments, two identical fragment (Fab) which can bind with the antigen, and a third fragment which can easily be crystallized, *i.e.* Fc fragment (Fig. 16.12 B).

In addition, the proteolytic enzyme papain cleaves the immunoglobulin into a fragment that contains both of antigen binding arms $(Fab')_2$ and several pieces of the Fc fragment. Fab interacts with the antigen, and Fc binds to Fc-receptors on different cells.

Epitope Mapping

Epitope mapping is the process of identification and characterisation of the minimum molecular structures that are able to be recognised by the elements of immune system mainly T and B cells. A collection of *in vivo* and *in vitro* methodologies are used for epitope mapping. The most commonly used methods are binding assay, ELISPOT, HLA transgenic mice and prediction software. Epitope mapping of proteins is used for development of new vaccines and diagnostic approaches.

II. IMMUNOGLOBULIN ISOTYPES

Antibodies can come in different varieties known as isotypes or classes. They are named with an 'Ig' prefix which refers for immunoglobulin. The two light chains (lambda or kappa), and two heavy chains (alpha, delta, gamma, epsilon or mu) that are linked by disulphide bridges. The type of heavy chain determines the five different classes of immunoglobulin isotypes *viz.,* IgA (α heavy chain, 15%), IgD (δ heavy chain, 1%), IgG (γ heavy chain, 76%), IgE (ε heavy chain, 0.002%), IgM (μ heavy chain, 8%), respectively. Antibody differs in its biological properties, functional locations and ability to deal with different antigens as given in Table 16.5.

These antibodies are responsible for autoimmune responses such as such as allergies and diseases like arthritis, multiple sclerosis, and systemic lupus erythematosus. They perform different roles and direct the appropriate immune response for each different type of foreign object to which they encounter. IgG is the only antibody that can cross the placental barrier to the foetus and it is responsible for the 3 to 6 month immune protection of newborns that is conferred by the mother.

The B cell receptor (BCR) is composed of surface-bound IgD or IgM antibodies and associated Ig-α and Ig-β heterodimers which carry out signal transduction. A typical human B cell binds 50,000 to 100,000 antibodies to its surface. They cluster in large patches of 1 μm in diameter after binding of antigen on lipid rafts that isolate the BCRs from most other cell signaling receptors. These patches may improve the efficiency of the cellular immune response.

Generally, immunoglobulins carryout one of the two roles: (*i*) they may act as plasma membrane bound antigen receptors on the surface of a B-cell, or (*ii*) they are present freely in cellular fluids and eliminate antigenic determinants. In both the cases, function of antibody is intimately related to its structure.

Table. 16.5. Properties of serum immunoglobulins.

Immunoglobulins	*Heavy chains*	*Other chains*	*% Total*	*Ig placental transfer*	*Location*	*Function*
IgM	Mu	J, pentamer	5-10	No	Blood serum	Very effective agglutinator, produced early in immune response
IgG	Gamma	Monomer	80-85	Yes	Blood serum, extracellular body fluid, on killer lymphocytes	Most abundant Ig of internal body fluid, combat with microbes and their toxins
IgA	Alpha	J, dimer	15	No	Blood serum and lymphocyte surface	Major Ig in seromucus, secretion, and on defend external body helper
IgE	Epsilon	Monomer	0.002	No	Blood serum fixed to basophils and mast cells	Raised in infection, responsoble for allergy symptoms
IgD	Delta	Monomer	0.2	No	Blood serum and lymphocytes in new born	Transient present on lymphocyte surface

1. Immunoglobulin Gamma (IgG)

IgG antibody is a large monomeric molecule of about 150 kDa composed of four peptide chains (Fig. 16.13). It contains two identical heavy chains gamma (γ) of about 50 kDa and two identical light chains of about 25 kDa; thus it provides tetrameric quaternary structure. The two heavy chains are linked to each other and to a light chain each by disulfide bonds. The resulting tetramer has two identical halves which together form the Y-like shape. Each end of the fork contains an identical antigen binding site.

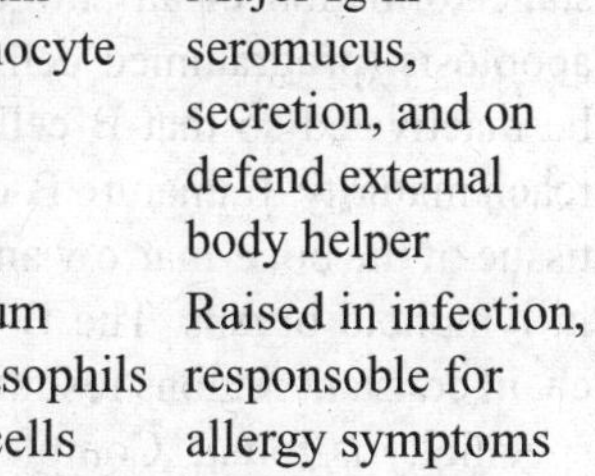

Moreover, there are four subclasses of IgG (IgG1, 2, 3 and 4) in humans which are named on the basis of their abundance in serum; for example, IgG1 predominates among the other IgG subclasses. IgG is the most abundant immunoglobulin which is equally distributed in blood and in tissue liquids constituting 75% of serum immunoglobulins in humans. The plasma B cells synthesise the IgG

Fig. 16.13. Structure of IgG.

molecules and secrete outside the cell. IgG antibodies are predominately involved in the secondary antibody response which occurs about one month after antigen recognition. Thus the presence of specific IgG generally corresponds to maturation of the antibody response. Pro-inflammatory cytokines, especially IL-4 and IL-2, play a critical role in activation of the IgG antibody response.

This is the only isotype that can pass through the human placenta; therefore, it provides protection to the developing foetus in uterus. Along with IgA secreted in the breast milk, residual IgG absorbed through the placenta provides the neonate with humoral immunity before the development of its own immune system. The IgG antibodies can bind to several pathogens such as viruses, bacteria and fungi. It protects the body against them by agglutination and immobilisation, complement activation, opsonisation for phagocytosis and neutralisation of their toxins. Besides, IgG is also associated with Type II and Type III hypersensitivity.

2. Immunoglobulin Delta (IgD)

Structure of immunoglobulin delta (δ) isotype is given in Fig. 16.14. The IgD makes up about 1% of proteins in the plasma membranes of immature B-lymphocytes where it is co-expressed with IgM, the other cell surface antibody. IgD is also produced in a secreted form in very small amounts in blood serum as a monomer with two heavy chains of the delta (δ) class and two light chains. When the young B cells in the spleen are ready to be activated, the IgD functions to signal. Such activated B cells take part in the immune system to provide defense to the body.

The specificity of an antibody is selected by V(D)J recombination in an immature B-cell. Thereafter, this specificity is refined via the process called 'affinity maturation'. IgM antibodies are produced often at the immature stage. At this stage formation of an antigen-antibody complex results in apoptosis (programmed cell death) of B-cells. Also it may be inactivated so that B-cells expressing antibodies do not reach maturity. Immature B-cells move from haematopoietic tissue of the bone-marrow and their maturation is completed in lymphoid organs. The B-cells upon reaching the mature state, co-express both μ and δ heavy chain constant regions resulting in IgM and IgD antibody isotypes. No apparent change in antibody specificity is found. Complementarity determining regions (CDRs) have already been selected in the immature stage of B-cells. An antigen-antibody complex triggers a specific immune response at mature state causing the proliferation of B-cell to progeny cells *i.e.* plasma cells and memory cells.

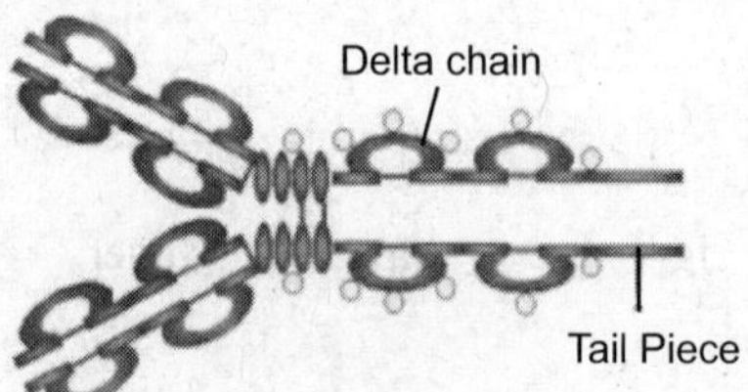

Fig. 16.14. Structure of IgD.

3. Immunoglobulin Alpha (IgA)

Immunoglobulin A consists of alpha (α) heavy chain. It is a dimeric molecule (Fig. 16.15). In blood, its population remains to about 15% of all immunoglobulins. IgA provides mucosal immunity; more IgA is produced in mucosal linings than all other types of antibody combined. In its secretory form, it is the main immunoglobulin present in mucous secretions including tears, saliva, intestinal juice, vaginal fluid and prostate and respiratory epithelial secretions. It is also found in small amounts in blood. It is resistant to degradation by enzymes; hence, secretory IgA can survive in harsh environments (such as the digestive and respiratory tracts) to provide protection against microbes multiplying in body secretions. IgA poorly activates the complement system and opsonises it weakly.

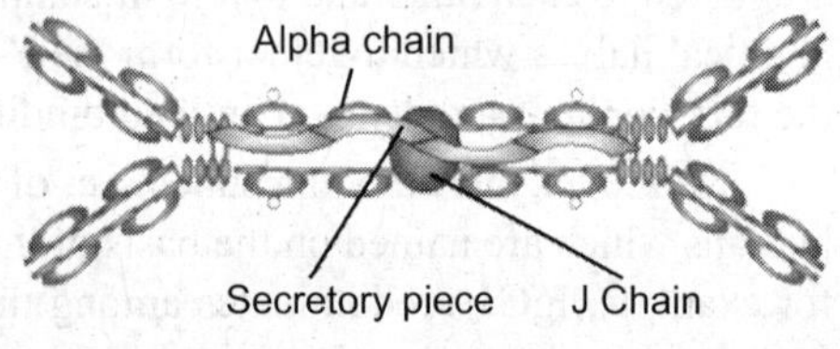

Fig. 16.15. IgA forming dimer.

4. Immunoglobulin Mu (IgM)

Immunoglobulin M is a basic antibody present on B cells which consists of μ heavy chain. It is the primary antibody against A and B antigens expressed on the surface of red blood cells. IgM is the largest antibody found in the human circulatory system. The main antibody involved in primary response is IgM. IgM forms polymers where multiple immunoglobulins are covalently linked together with disulfide bonds mostly as a pentamer (Fig. 16.16) but also inked as a hexamer. IgM has a molecular mass of about 900 kD in its pentamer form, because a pentameric IgM consists of 10 antigen binding sites. The large size of most antigens hinders the binding to nearby sites. Therefore, IgM cannot bind 10 antigens at the same time.

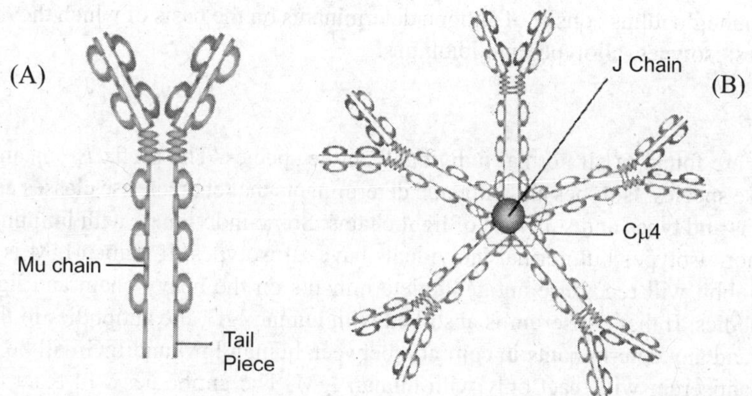

Fig. 16.16. Structure of immunoglobulin M (IgG); A- monomer, B- pentamer.

In contrast, the hexameric form of IgM lacks J chain due to constraints of space in the complex, whereas the pentameric IgM consists of J chain because IgM cannot diffuse well due to its large size. Hence, it is found in very low quantities mostly in serum. IgM possesses high avidity due to its polymeric nature.

The gene segment, encoding the μ constant region of the heavy chain in germ line cells, is positioned first among the other constant region. This is the reason why IgM is the first immunoglobulin expressed by mature B cells. Also, it is the first immunoglobulin expressed in about 20 weeks of the foetus. IgM antibodies do not pass across the human placenta. Phylogenetically this antibody develops early. This antibody, appears early soon after infection and usually reappears after further exposure. Therefore, these two biological properties of IgM make it useful in the diagnosis of infectious diseases.

5. Immunoglobulin Epsilon (IgE)

Structure of immunoglobulin epsilon (ε) (IgE) is given in Fig. 16.17. The IgE molecule is a class of isotype that is found only in mammals. In 1966, the Japanese couple Teruka and Kimishige Ishizaka discovered IgE. It plays a major role in allergy, and is associated with hypersensitivity Type 1. Although IgE is typically the least abundant isotype of blood serum, its concentration in a normal individual is only 0.05% of the IgG concentration. IgE elicits an immune response by binding to Fc receptors which are found on the surface of mast cells and basophils. Also it is found on eosinophils, monocytes, macrophages and platelets in humans.

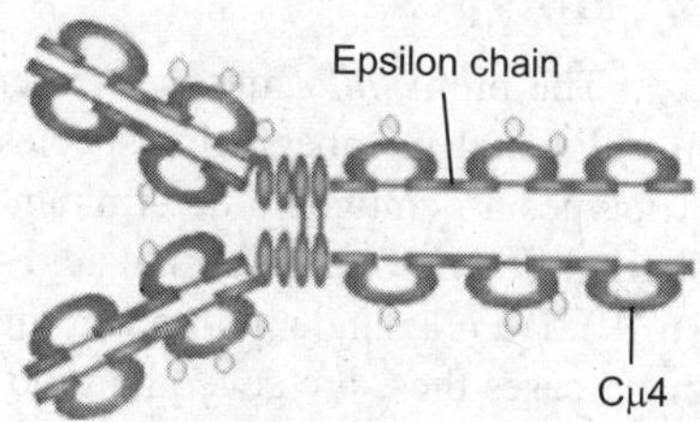

Fig. 16.17. Structure of IgE.

IgE is implicated in immune system responses to most parasitic worms and are important during immune defense against certain protozoan parasites like *Plasmodium falciparum*. Besides, IgE can recognise an allergen. The allergens are typically a protein such as dust mite DerP1, cat FelD1, grass or ragweed pollen, etc. It has a unique long-lived interaction with its high affinity receptor (*i.e.* Fc εRI) so that basophils and mast cells become primed. IgE may be an important target in treatments for allergy and asthma. IgM is the dominant antibody produced in primary immune responses, while IgG dominates in secondary immune responses. IgM is physically much larger than the other immunoglobulins.

III. ANTIGEN DETERMINANTS OF IMMUNOGLOBULINS

The immunoglobulins consist of antigen determinants on the basis of which they are classified into three types: isotypes, allotypes and idiotypes.

1. Isotypes

Isotypes are found in all normal individuals in the species. The prefix *iso* means same in all members of the species. Isotypes are antigenic determinants that characterise classes and subclasses of heavy chains and types and subtypes of light chains. Some individuals with immunodeficiencies lack one or more isotypes but normal individuals have all isotypes. If human IgM is injected into a rabbit, the rabbit will recognise antigenic determinants on the heavy chain and light chain and produce antibodies. If that antiserum is absorbed with human IgG, the antibodies to the light chain determinants and any determinants in common between human IgM and IgG will be removed and the resulting antiserum will react only with human IgM. The antibodies will react only with the constant region of the μ chain. Antibodies to the variable region are rare because only a few copies of each different variable region are represented in the IgM; therefore, effective immunisation does not occur.

The determinants that are recognised by such antibodies are called 'isotypic determinants' and the antibodies to those determinants are called 'anti-isotypic antibodies'. Each class, subclass, type and subtype of immunoglobulin has its unique set of isotypic determinants. Heavy chain isotypes are found on the Fc portion of the constant region of the molecule while light chain isotypes are found in the constant region. The location of isotypic determinants is shown in Fig. 16.18. Antibodies to isotypes are used for the quantisation of Ig classes and subclasses in various diseases, in the characterisation of B cell leukemia and in the diagnosis of various immunodeficiency diseases.

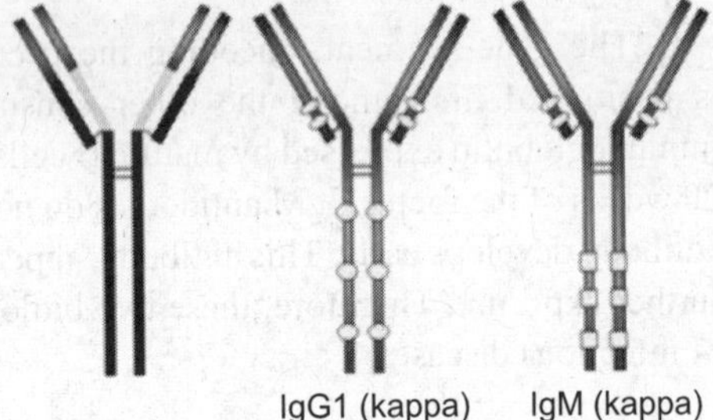

Fig. 16.18. Immunoglobulin isotypes.

2. Allotypes

The prefix *allo* means different in individuals of a species. Individual allotypes are found in individual members of a species. All allotypes are not found in all members of the species. Allotypes are antigenic determinants specified by allelic forms of the Ig genes. Allotypes shows slight differences in the amino acid sequences of heavy or light chains of different individuals (Fig. 16.19). Even a single amino acid difference can give rise to an allotypic determinant, although in many cases there are several amino acid substitutions. Allotypic differences are detected by using antibodies directed against allotypic determinants. These antibodies can be prepared by injecting the Ig from one person into another. In practice anti-allotype antisera from women are obtained who have had multiple pregnancies or who have received blood transfusions from people or from some patients with rheumatoid arthritis. In man the allotypic differences are localised to the constant region of the heavy and light chains as shown in Fig. 16.19.

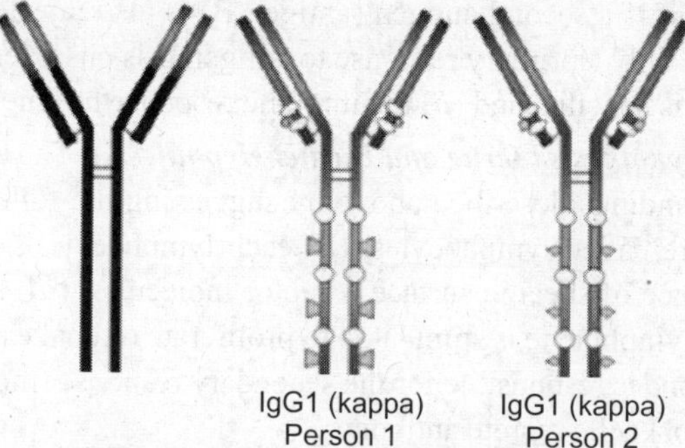

Fig. 16.19. Immunoglobulin allotypes.

3. Idiotypes

There are some unique antigenic determinants which are present on individual antibody molecules or on molecules of identical specificity, *i.e.* all antibodies molecules have the exact same hypervariable regions. The idiotypes are the antigenic determinants created by combining the hypervariable regions of an antibody and the antibodies elicited to the idiotypes are called 'anti-Id antibodies'. The anti-idiotypic antibodies are directed against the hypervariable regions of an antibody. Idiotypes are localised on the Fab fragment of the Ig molecules (Fig. 16.20). Idiotypes are localised at the hypervariable regions of the heavy and light chains. In several cases individual antigenic determinant (*i.e.* idiotype) consists of some framework residues near the hypervariable region.

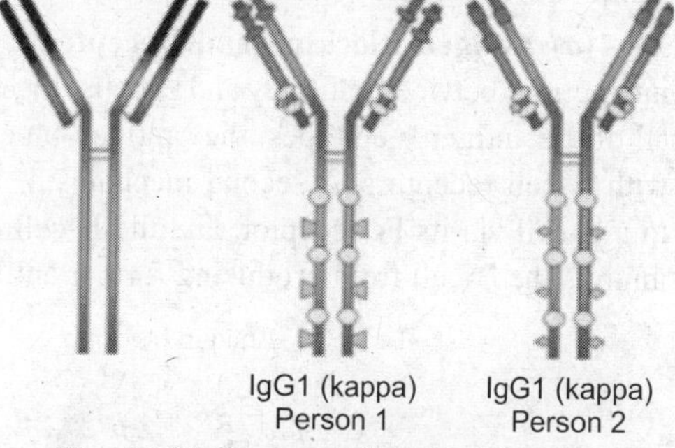

Fig. 16.20. Immunoglobulin idiotype.

Idiotypes are a useful marker for a particular variable region. Immune responses are regulated by anti-Id antibodies directed against our own Id antibodies. In some cases anti-idiotypic antibodies stimulate B cells to produce antibodies and they can be used as a vaccine. This method is being tried to immunise against highly virulent pathogens which cannot be safely used as a vaccine

IV. ANTIBODY KINETICS

Production of antibody in response to antigenic substances is given in Fig. 16.21. An animal was injected with antigen A at day 0 which invokes a primary response after 4 days as indicated by a rise in the specific antibody titer, *i.e.* measure of the amount of antibody in the animal's serum per unit volume. In the beginning, this antibody is mostly IgM (and some IgG).

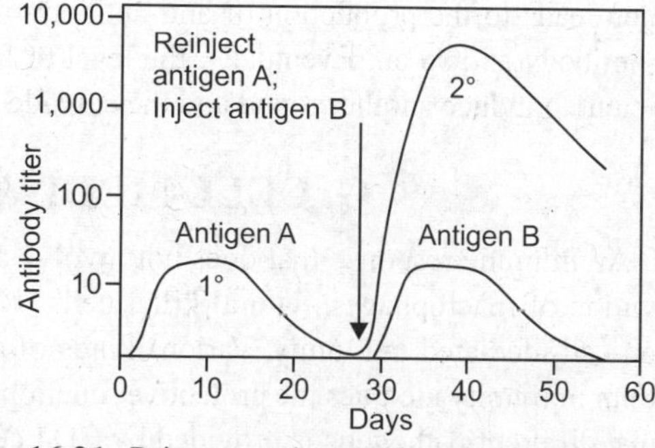

Fig. 16.21. Primary and secondary responses invoked by antigens A and B, respectively.

After 7 to 10 days of a peak titer, the response decreases rapidly. If the animal is then re-injected with antigen A at 28th day, production of antibody begins almost immediately and reaches a level 1000-fold greater than that was measured in the primary response. This is known as the secondary response and the principal antibody produced is IgG. If a second antigen (antigen B) is also injected at the same time as the re-injection of antigen A, only a primary response to antigen B is observed. These results show that: (*i*) the immune response is specific, and (*ii*) the immune response has memory.

Clonal Selection Hypothesis (Jerne and Burnet Hypothesis): The clonal selection hypothesis attempts to explain the findings described above by suggesting the following: (*i*) animals contain numerous cells which are called lymphocytes, (*ii*) each lymphocyte is responsive to a particular antigen due to the presence of specific surface receptor molecules, (*iii*) after making contact to its appropriate antigen, the lymphocyte is stimulated to proliferate (clonal expansion) and differentiate, and (*iv*) the expanded clone is responsible for the secondary response (more cells to respond), while the differentiated (effector) cells secrete antibody.

1. Regulation of Humoral Response

The immune response is regulation possibly in several ways. First, a specific group of T-cells (*i.e.* suppressor T-cells) are involved in turning down the immune response. Like helper T-cells the suppressor T-cells are stimulated by antigen. But suppressor T-cells release factors that suppress the B-cell response, instead of releasing lymphokines that activate B-cells (and other cells). It is more complicated than the activation pathway, and possibly involves the additional cells.

(*a*) Antigen Blocking and Receptor Cross-linking: The other ways of regulation involve interactions between antibody and B-cells (Fig. 16.22A-B). When high doses of antibody interact with all of the antigen's epitopes, the 'antigen-blocking' mechanism occurs, which inhibits interactions with B-cell receptors. A second mechanism, 'receptor cross-linking' results when antibody binds to a B-cell via its Fc receptor, and the B-cell receptor combines with antigen. This 'cross-linking' inhibits the B-cell from producing further antibody.

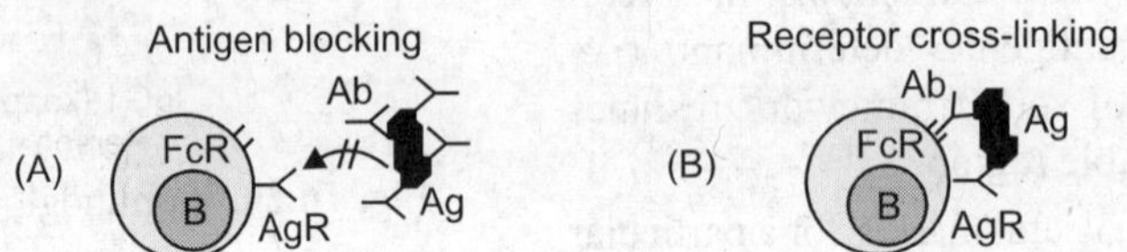

Fig. 16.22. Antigen blocking (A); receptor cross-linking (B).

(*b*) Idiotypic Networking: The idiotypic network hypothesis suggests that the idiotypic determinants of antibody molecules are very unique and act to the immune system like foreign molecules; therefore, they are antigenic in nature. Hence, antibody production in response to antigen leads to the production of anti-antibody in response, and anti-anti-antibody and so on. Eventually, the level of $[\text{anti}]_n$-antibody is not sufficient to induce another round and the cascade ends (Fig. 16.23).

Ag anti-Abs →

Ab1 Ab2 Ab3 Ab4

Fig. 16.23. Idiotypic networking.

G. CELL-MEDIATED IMMUNITY

An immune response that does not involve antibodies or complement but rather involves the activation of macrophages, natural killer cells (NK), antigen-specific cytotoxic T-lymphocytes, is called cell-mediated immunity. Various kinds of cytokines are released in response to an antigen. Cellular immunity includes the protective function of immunisation associated with cells. Protection against different pathogens is provided by CD4 cells or helper T cells. The human immune system has about two trillion lymphocytes, among which about one trillion are T-cells. T lymphocytes develop from precursor stem cells in foetal liver and bone marrow and differentiate into mature cell types during residence in the thymus. Mature T lymphocytes (antigen responding, response control and response mediating cells) are present in thymus, spleen, lymph nodes, throughout skin and other lymphatic organs, and in the bloodstream (Fig. 16.24).

T cells are processed in the thymus where they undergo two selection processes. The first positive selection process weeds out only those T cells with the correct set of receptors that can recognise the MHC molecules responsible for self-recognition. A negative selection process starts where T cells that can recognise MHC molecules complexed with foreign peptides are allowed to pass out of the thymus.

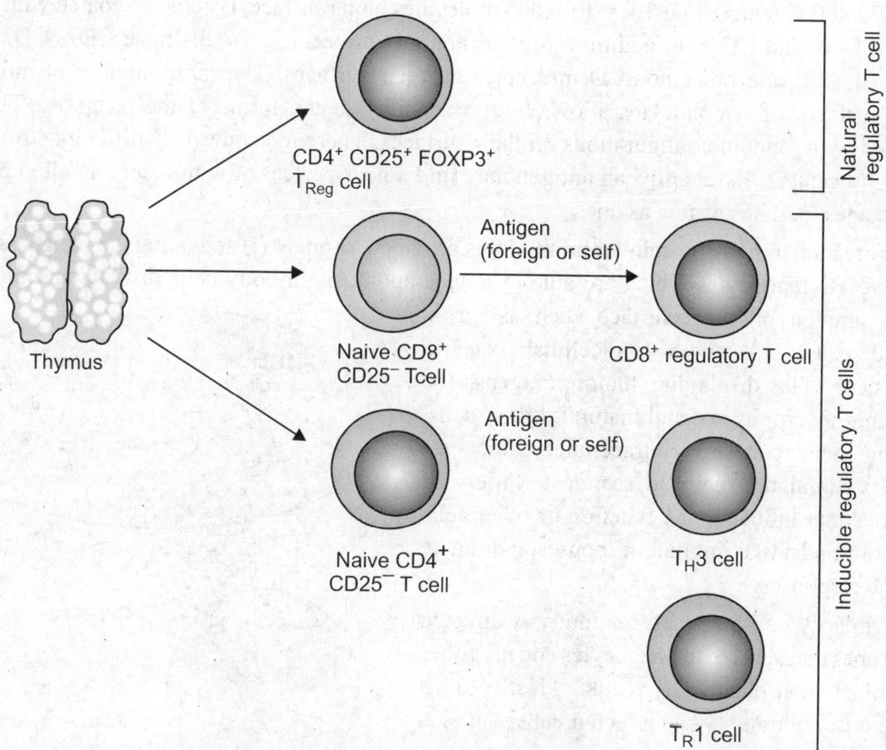

Fig. 16.24. Processing of T cells.

Natural regulatory T cells express the cell-surface marker CD25 [CD = **c**luster of **d**ifferentiation) and the transcriptional repressor FOXP3 (forkhead box P3)]. There are more than one hundred and sixty clusters, each of which is a different chemical molecule that coats the surface. Processing of T cells is shown in Fig. 16.24. T cells mature in thymus, migrate from it and constitute 5–10% of peripheral T cells). Other populations of antigen-specific regulatory T cells can be induced from naive $CD4^+CD25^-$ or $CD8^+CD25^-$ T cells in the periphery under the influence of semi-mature dendritic cells, interleukin-10 (IL-10), transforming growth factor-β (TGF-β) and possibly interferon-α (IFN-α).

The populations of 'inducible regulatory T cells' include distinct subtypes of $CD4^+$ T cell: *T regulatory 1* (T_R1) cells which secrete high levels of IL-10, no IL-4 and no or low levels of IFN-γ; and *T helper 3* (T_H3) cells which secrete high levels of TGF-β. Although $CD8^+$ T cells are normally associated with cytotoxic T-lymphocyte function and IFN-γ production, these cells or their subtype can secrete IL-10 and have been called $CD8^+$ regulatory T cells.

Cytotoxic or killer T cells ($CD8^+$) release lymphotoxins which cause cell lysis. Helper T cells ($CD4^+$) serve as managers and direct the immune response. They secrete chemicals called lymphokines that stimulate cytotoxic T cells and B cells to grow and divide, attract neutrophils and enhance the ability of macrophages to engulf and destroy microbes. Suppressor T cells inhibit the production of cytotoxic T cells. Memory T cells are programmed to recognise and respond to a pathogen once it has invaded and been repelled.

Macrophages engulf antigens/microbial cell/viral particle, process them internally and show some parts on their surface together with some of their own proteins. This sensitises the T cells to recognise these antigens. All cells are coated with various substances. $CD8^+$ is read 'CD8 positive'. Every T and B cell has about 10^5 = 1,00,000 molecules on its surface. B cells are coated with CD21, CD35, CD40, and CD45 in addition to other non-CD molecules. T cells have CD2, CD3, CD4, CD28, CD45R, and other non-CD molecules on their surfaces. The large number of molecules on the surfaces of lymphocytes allows huge variability in the forms of the receptors. They are produced with random configurations on their surfaces. There are some 10^{18} different structurally different receptors. Essentially, an antigen may find a near-perfect fit with a very small number of lymphocytes, perhaps as few as one.

Protection to body by cellular immunity is provided through: (*i*) activating the antigen-specific cytotoxic T-lymphocytes which are able to induce apoptosis in body cells displaying epitopes of foreign antigen on their surface such as virus-infected cells, cells with intracellular bacteria and cancer cells displaying tumor antigens, (*ii*) activating macrophages and natural killer cells, enabling them to destroy intracellular pathogens, and (*iii*) stimulating cells to secrete a variety of cytokines that influence the function of other cells involved in adaptive immune responses and innate immune responses.

Primarily cell-mediated immunity is directed at microbes that survive in phagocytes and microbes that infect non-phagocytic cells. Mostly it is effective in removing virus-infected cells, but also participates in defending against fungi, protozoans, cancers, and intracellular bacteria (Fig. 16.25). It also plays a major role in transplant rejection.

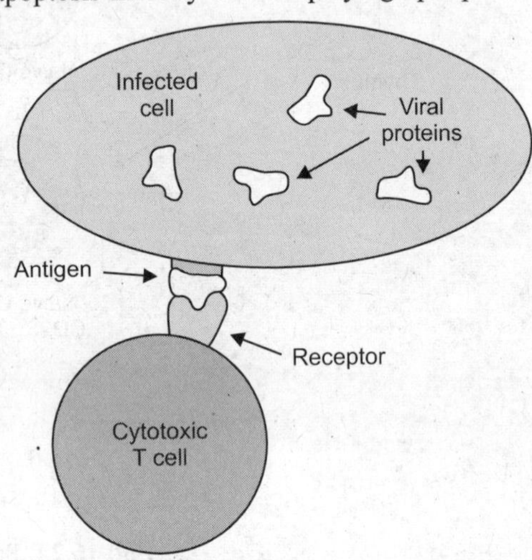

Fig. 16.25. A T-cell having receptors.

H. ANTIBODY DIVERSITY

The genes for antibody (globular proteins) are constructed in B-cells by recombining some regions of DNA from several different locations. Millions of different permutations can be constructed to create a random library of antibody molecules. The chance of recognising a new antigen such as virus by these randomly constructed antibody molecules would be very high. How do antibodies recognise so many different antigens was unknown until the discovery of 'gene rearrangement by Susumu Tonegawa who received the Nobel Prize in Physiology or Medicine in 1987 for working out the mechanism of generating antibody diversity. This is one of the most significant Nobel Prizes of the 20th century.

S. Tonegava

Susumu Tonegawa (1939-)

S. Tonegawa, a Japanese researcher, was born in Nagoya, received the Nobel Prize in Physiology or Medicine in 1987 for working out the mechanism of generating antibody diversity. This was one of the fundamental problems in immunology. How do antibodies recognise so many different antigens was unknown until this discovery.

In Japan he developed an interest in chemistry during high school. Hence, he appeared in the entrance test in the Department of Chemistry of the University of Kyoto. After having failed once, he was admitted to this University in 1959. This happened to be one year before the first ten-year term of the defence treaty between Japan and the United States expired.

Being influenced with the work of Jacob and Monod, he decided to pursue graduate study in molecular biology and was accepted by Professor Itaru Watanabe's laboratory at the Institute for Virus Research at the University of Kyoto. He went to the United States for a few years of postdoctoral study after obtaining the Ph.D. in Japan. He studied in the laboratory of Professor Masaki Hayashi, carrying out a thesis project on the transcriptional control of phage lambda and received Ph.D. in molecular biology in 1968. He changed his interest from molecular biology to immunology after coming in the contact of immunologists. By the end of 1971, he was introduced the great debate on the genetic origins of antibody diversity. He decided to resolve this question by applying the recently invented techniques of molecular biology such as restriction enzymes and recombinant DNA. Later on his research group got expanded and worked involving many capable postdoctoral fellows and devoted technical assistants.

In 1981 Professor Salvador E. Luria, Director of the Cancer Center (USA) gave an offer to him of Professorship to work further. He worked on two major problems, the role of somatic rearrangement in the activation of the rearranged antibody gene and the antigen receptor of T cells. He contributed to the understanding of both problems by discovering a tissue-specific transcriptional enhancer in the immunoglobulin heavy chain gene and by identifying, cloning, and sequencing genes coding for the polypeptide subunits of the T cell receptor. An intriguing development made during the latter study was the identification of a gene that led to the discovery of a new T cell receptor, gamma delta. While the function of the T cells bearing this receptor is currently unknown, data accumulated during the past year in his laboratory as well as many other laboratories suggested that these T cells may be involved in an entirely new aspect of immunity. Further, he got success in solving the question that around one thousand genes are able to create billions of different forms of antibodies.

1. Genetic Basis of Antibody Diversity

If an antigen appears that binds to one of the randomly generated membrane antibodies of a B-cell, that clone can be selected to proliferate and differentiate into memory and antibody producing plasma cells. It has earlier been described that an immunoglobulin (Ig) consists of two identical light chains (L) and two identical heavy chains (H). The 3D structure of Ig chain shows one N-terminal variable domain (V) and one (for an L chain) or several (for an H chain) C-terminal constant domain (C).

The B line cells synthesise immunoglobulins which are produced either on the surface of B-lymphocytes or are secreted by the plasmocytes. The antibody-producing chromosome includes many coding regions called V for variable, D for diversity and J for joining (Fig. 16.26). These are passed from generation to generation by the germ cells.

In B cell, DNA segments for both heavy (VDJ) chain and light (VJ) chain are randomly combined. Specific antibodies are produced as a result of immunoglobulin gene rearrangement, *i.e.* recombination of V, D, J and C genes. Consequently millions of different antibody molecules are generated from the pool of V genes. Separate sets of V genes encode the variable domains of immunoglobulin H and L chains. The two chains are produced separately but the mechanisms to achieve their diversity are similar in principle.

Do You Know ?

The immunoglobulin genes are located on different chromosomes which get rearranged in B germ line cell before maturation.

- Light chain kappa (κ) gene is located on chromosome 2
- Light chain lambda (λ) gene is located chromosome 22
- Heavy chain gene is located on chromosome 14

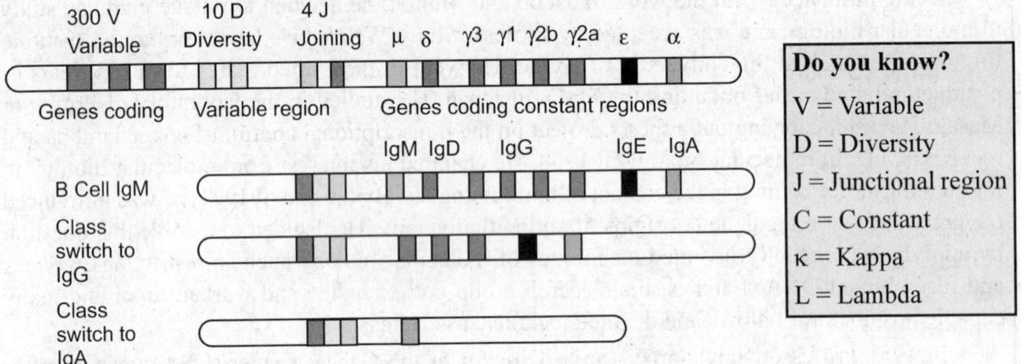

Fig. 16.26. The antibody gene rearrangements occur prior to exposure to antigen.

***(a)* Light Chain (kappa or lamba) Formation:** The light chain may be kappa or lambda chain. The light chains possess numerous V and J genes but lack D genes (Fig. 16.27).

***(i)* Kappa chain– V-J rearrangement:** Mostly the antibody molecules use kappa (κ) light chain. The Igκ (kappa) genes are located at 2p11 on chromosome 2. The κ gene cluster consists of several hundred (~300) VL genes, a few J genes (5) and one C gene. The V, J and C genes are separated in the DNA of germ line of the Ig genes and are transcriptionally inactive.

As the B cell matures, genes are rearranged so that one V gene is joined to one J gene. The rearranged VJ segment together with the C gene is transcribed. The portion of DNA between the joined segments is deleted, and the transcripts are processed by splicing to produce the messenger RNA for the L chain. V-J rearrangement occurs at the recombination signals (RS), which include a heptameric sequence (7 nucleotides) and a nanomeric sequence (9 nucleotides), separated by a spacer. Each IgκV gene is followed downstream (in the 3' position) by an RS consisting of a CACAGTG heptamer, and by a 12 bp spacer, and then an ACAAAAACC nanomer.

The κ chains are encoded by a separate cluster of V, J and C genes, but the rearrangement and transcription are similar to that of the L chain. Any given B cell uses only one type of L chain to produce the immunoglobulin molecule. The L chain combines with the H chain during their transport from polyribosomes to the membrane.

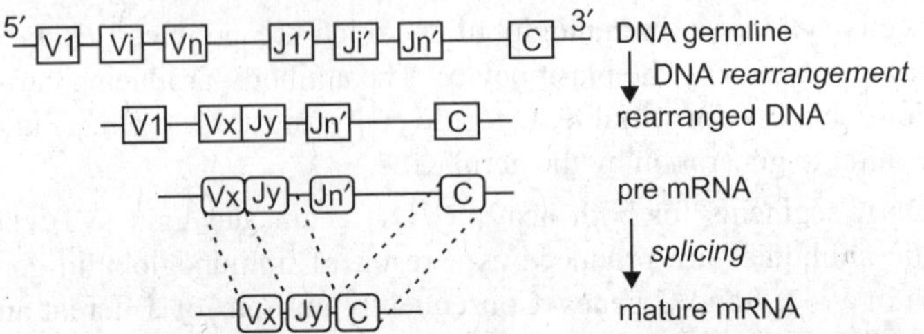

Fig. 16.27. V-J gene rearrangement in light (kappa) chain.

(*ii*) **Lambda chain – V-J rearrangement:** The IgL (lambda) genes are located at the 22q11 position on chromosome 22. The mechanism of V-J rearrangement is the same as described for the Igκ genes. Rearrangement takes place between one of the 29 to 33 functional IgLV genes and a J gene. There are 4 to 5 functional IgLC genes each of which is preceded by an IgLJ gene.

(*b*) **Heavy (H) Chain Formation:** The genes of heavy chain are located on chromosome 14. The H chain gene is similar to that of light chain but it is slightly more complex than the light chain gene. In addition to ~1,000 VH genes, there are >10 D genes and ~4 J genes also. Furthermore, this genetic cluster has nine C genes that encode different immunoglobulin isotypes. The mature B cell rearranges its immunoglobulin genes, joins them together and deletes the DNA between the joined segments. The rearranged VDJ gene segment is transcribed together with a Cμ or Cδ gene, and this long transcript is spliced into VDJCμ or VDJCδ messages resulting in the expression of IgM and IgD, respectively on the B cell surface. Both immunoglobulin molecules use the same VDJ segment and, therefore, possess the same immunological specificity. The B cell is now ready to bind to a specific antigen and become further differentiated.

(*i*) **V(D)J Rearrangement:** In the human genome multiple genes are encoded for the variable regions that contain three distinct types of segments: V, D and J. The immunoglobulin H chain region contains 65 variable (V) genes plus 27 diversity (D) genes and 6 joining (J) genes (Fig. 16.28).

DNA rearrangement takes place between one of the 38 to 46 functional variable IgHV genes, one of the 23 functional diversity IgHD genes, and one of the 6 functional junction IgHJ genes. There are also some RSs, which are located downstream (in position 3′) of the V genes on both sides of D genes and upstream (at 5′) of the J genes. During V-D-J rearrangement, a junction is first formed between 1 D and 1 J, and then one between 1 V and the D-J complex (http://en.wikipedia.org/wiki/V(D)J_recombination).

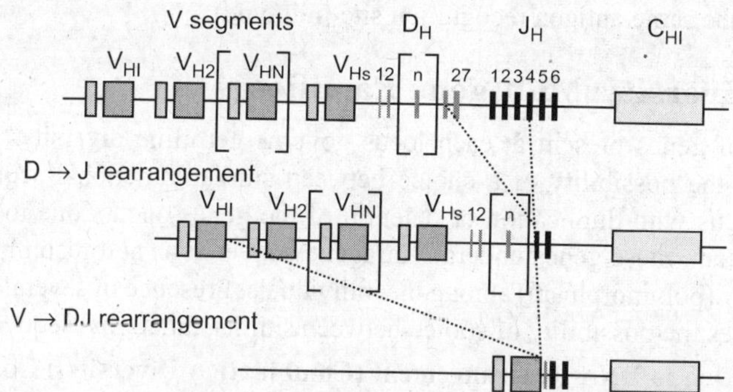

Fig. 16.28. Rearrangement of V-D-J region in immunoglobulin heavy (IgH) chain gene.

There are 2 or 3 open reading frames of D genes, each of which codes for 2 or 3 different peptides. The V-D-J junctions are also characterised by nucleotide deletions by using an exonuclease and by the random addition of nucleotides [by using TdT (terminal deoxynucleotidyl transferase)]. The V regions increase the diversity of the V-D-J junctions of the variable domains of the heavy chains of the immunoglobulins. Through the mechanism of DNA rearrangement of these regional genes it is possible to generate a large antibody repertoire (for example 65 V genes × 27 D genes × 6 J genes = 10,530 combinations, with additional variability through light chains).

2. Isotype Switching

An IgM (mu) chain is first synthesised in pre-B lymphocyte because the constant IgHM gene (CÊ) is located near to the V-D-J rearrangement. The IgM chain is associated with the pseudo-light chain and the combination constitutes the pre-B receptor. The first complete Ig synthesised by the B-lymphocyte is an IgM in which the M chain is combined with a light kappa or lambda chain.

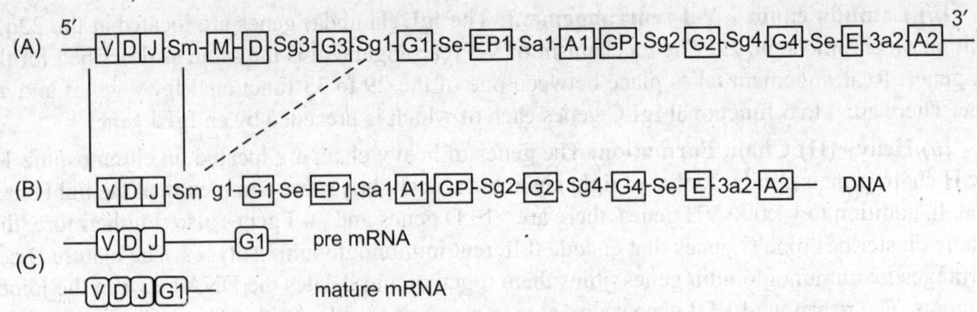

Fig. 16.29. Mechanism of isotype switching.

DNA recombination takes place between the switch sequence of the IgHM gene and another IJHC in 3' (Fig. 16.29A). Genes IgHM, IgHD, IgHG3, IgHG1, IgHA1, IgHG2, IgHG4, IgHE and IgHA2 code for the constant region of M, D, G3, G1, A1, G2, G4, E and A2 chains. There is a pseudogene IgHEP1 and an ORF (open reading frame) IgHGP. However, IgHGP which has an ORF cannot be used because there is no switch.

The B lymphocyte can express some other isotype or sub-isotype of Ig during its differentiation. This involves the replacement of an IgHC gene by another as the result of 'isotype switch' (*i.e.* DNA recombination) with the excision of the whole deletion loop. This excision occurs at the switch sequences (Fig. 16.29B). Then the normal sequence carries out the function as synthesis of pre-mRNA (B), splicing of the introns resulting in mature mRNA (Fig. 16.29C), and protein synthesis.

This is why : (*i*) a B-lymphocyte can at first synthesise an IgM and during its differentiation, an IgG (IgG1, IgG2, IgG3 or IgG4), an IgA (IgA1 or IgA2) or an IgE, and (*ii*) it retains the same V-D-J rearrangement and the same antigen recognition site (idiotype).

3. Germline Diversity: Multigene Families

The number of genes present at each locus governs germline diversity. These are families of genes that offer the possibility of a choice between similar functional sequences. Long-term evolution of the locus with duplication or deletion of the genes occurs due to possible intergene recombination. Further, these genes undergo intragene conversions and recombination that lead to mixing and diversity (polymorphism) among the individuals. Presence of several ORFs *e.g.* in IgHD genes again increases the possibility of choice between similar functional sequences.

***(a)* Diversity Due to DNA Rearrangement (Combination Diversity):** Combination diversity allows the synthesis a million of immunoglobulins. The IgH genes synthesise about 6,000 heavy chains, IgK or IgL genes of about 160 light chains, which is equal to about a million possible combinations 6 $\times$ 10 ^{3} $\times$ 160). Possible combination diversity of immunoglobulin genes is given in Table 16.6.

During the rearrangements of the IgH of the heavy chains, the acquisition of the N regions, and using one or other of the reading frames for the D genes at the V-D-J junctions, and during the IgK or IgL rearrangements of the light chains, flexibility of the V-J junctions. These mechanisms contribute to increasing the diversity by a factor 10^3 to 10^4 (potential synthesis of 10^9 Ig chains).

***(b)* Diversity as a Result of Somatic Hypermutations:** There are several somatic mutations that produce much targeted characterisation of the rearranged V-J and V-D-J genes of the immunoglobulins. But the mechanism of start of mutation is unknown. Activation-induced cytidine deaminase (AID) may be involve both in the occurrence of the mutations and the switch mechanism. In the lymph glands, mutations occur during differentiation of B lymphocyte. This results in the increase of immunoglogulin diversity by a factor of 10^3, which possibly achieves a potential diversity of 10^{12} different immunoglobulins. These different mechanisms of diversity make it possible to obtain 10^{12} different immunoglobulins which become capable to respond to the several million known antigens.

Table 16.6. Combination diversity imunoglobulins genes.

	Number of functional genes					*Combination diversity*
Locus	***V***	***D***	***J***	***C***	***Total***	***V-J or V-D-J***
IgK (2p11)	31-35	0	5	1	37-41	31 × 5 = 155 (minimum) 35 × 5 = 175 (maximum)
IgL (22q11)	29-33	0	4-5	4-5	37-43	29 × 4 = 116 (minimum) 33 × 5 = 165 (maximum)
IgH (14q32)	38-46	23	6	9	76-84	38 × 23x6 = 5244 (minimum) 46 × 23x6 = 6348 (maximum)
					= 150-168	

I. COMPLEMENT

The complement system is a biochemical cascade that helps to eradicate the pathogens from an organism. It is part of the larger immune system as it belongs to the innate immune immunity. However, it can be recruited and brought into action by the adaptive immune system.

In the late 1890s, Paul Ehrlich introduced the term 'complement' as part of his theory of immune system. This theory advocates that the immune system consists of cells which possess specific receptors on their surface. These receptors recognize the antigens. When a person is immunised with an antigen, more receptors are formed. In turn these are shed from the cells and get circulated in the blood. Ehrlich called these receptors as 'amboceptors' to emphasise their bifunctional binding capacity. These were later on called as the 'antibodies'. The antibody recognises and binds to a specific antigen. At the same time, they also recognise and bind to the heat-labile antimicrobial component. Ehrlich named this heat-labile substance (proteins) as 'complement' because it complements the cells of the immune system.

The complement system consists of several small proteins that are found in blood. These proteins are generally synthesised by the liver and normally circulate as inactive zymogens. Upon stimulation by one of several triggers, proteases in the system cleave specific proteins and release cytokines and initiate an amplifying cascade of further cleavages. The result of this activation cascade is massive amplification of the response and activation of the cell-killing membrane attack complex. The complement system consists of more than 20 proteins and protein-fragments including serum proteins, serosal proteins, and cell membrane receptors. These proteins are synthesised mainly in the liver and account about 5% of the globulin fraction of blood serum.

Function of Complement

The complement system is complex and has multicomponent system which consists of about 26 proteins. The complement cascade is constitutive and non-specific but it must be activated in order to function.

The functions of complement are: (*i*) making bacteria more susceptible to phagocytosis, (*ii*) opsonize for promotion of phagocytosis of particulate antigens, (*iii*) lysing some viruses, bacteria and foreign cells, (*vi*) producing chemotactic substances, (*v*) increasing vascular permeability, (*vi*) binding to specific complement receptors on the cells of the immune system that triggers specific cell functions, inflammation and certain immunoregulatory molecules, (*vi*) causing smooth muscle contraction promoting degranulation of mast cells, (*viii*) binding to specific complement receptors on the cells of the immune system, triggering specific cell functions, inflammation and certain immunoregulatory molecules, and (*ix*) immune clearance which removes immune complexes from immune system and deposits them in the spleen and liver.

1. Activation of Complement Cascade

There are three biochemical pathways that activate the complement system: the *classical complement pathway*, the *alternative complement pathway*, and the *mannose-binding lectin pathway* (Janeway *et al*., 2005). Most of the complement components are numbered C1, C2, C3, etc. but some are referred to as 'factors'. Some of the components are cleaved enzymatically to activate their function, and the others combine to form active complexes. The components and their functions are given in Table 16.7.

Table 16.7. The function of its components of complements.

Native component	***Active component(s)***	***Function(s)***
		Components of the classical pathway
C1(q, r, s)	C1q, C1r, C1s	C1q binds to antibody that has bound antigen and activates C1r. C1r cleaves C1s to activate protease function. C1s cleaves C2 and C4.
C2	C2a, C2b	Function of C2a is not known. Active enzyme of classical pathway is C2b which cleaves C3 and C5.
C3	C3a, C3b	C3a mediates inflammation. C3b binds C5 for cleavage by C2b. Binds cell surfaces for opsonisation and activation of alternate pathway.
C4	C4a, C4b	C4a mediates inflammation. C4b binds C2 for cleavage by C1s, and binds cell surfaces for opsonisation.
		Components of the alternate pathway
C3	C3a, C3b	C3a mediates inflammation; anaphylatoxin. C3b binds cell surfaces for opsonisation and activation of alternate pathway.
Factor B	B, Ba, Bb	B binds membrane bound C3b. Cleaved by Factor D. Ba unknown. Bb cleaved form stabilised by P produces C3 convertase.
Factor D	D	Cleaves Factor B when bound to C3b.
Properdin	P	Binds and stabilises membrane bound C3bBb.
		Components of the membrane-attack complex
C5	C5a, C5b	C5a mediates inflammation; anaphylatoxin, chemotaxin. C5b initiates assembly of the membrane-attack complex (MAC).
C6	C6	Binds C5b, forms acceptor for C7.
C7	C7	Binds C5b6, inserts into membrane, forms acceptor for C8.
C8	C8	Binds C5b67, initiates C9 polymerisation.
C9	C9n	Polymerises around C5b678 to form channel that causes cell lysis.

***(a)* Classical Pathway:** C1 is a recognition complex composed of 22 polypeptide chains in three subunits viz., C1q, C1r, and C1s. Actually C1q is the recognition portion, *i.e.* a glycoprotein containing hydroxyproline and hydroxylysine that looks like a tulip flower. The classical pathway is triggered by activation of the C1-complex (C1q, two molecules of C1r, and two molecules of C1s thus forming C1qr^2s^2). Classical pathway starts with C1q binding to immunoglobulin Fc (primarily IgM or IgG complexed with antigens) or directly binding of C1q to the surface of the pathogen (Fig. 16.30). Such binding causes a conformational change in C1q molecule that activates two molecules of C1r *i.e.* a serine protease.

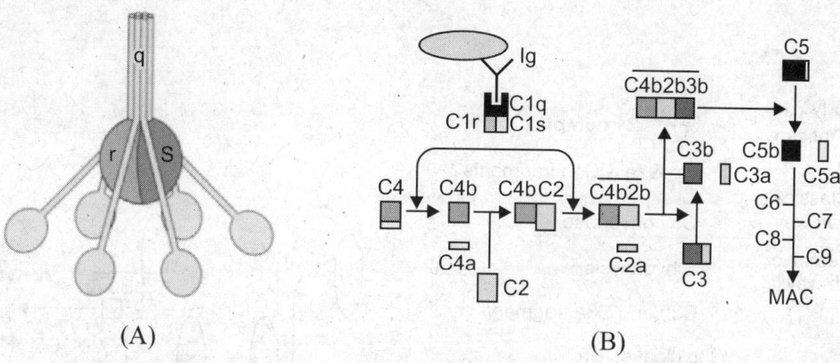

Fig. 16.30. C1 protein showing subunits C1r, C1s and the C1q tails (A); mechanism of classical pathway (B).

Then C1r is activated to become a protease that cleaves C1s to a form that cleaves both C2 and C4 to C2a and C2b, and C4a and C4b. The C2b and C4b combine to form C3 convertase (C4b-C2a complex, *i.e.* C3 activating enzyme).

This enzyme promotes the cleavage of C3 into C3a and C3b. Thereafter, C3b joins with C4bC2a (the C3 convertase) to make C5 convertase (C4bC2aC3b complex). C3b is also able to bind directly to cells and make them susceptible to phagocytosis. C5 is converted by C5 convertase (*i.e.* C4b2b3b) to C5a and C5b. C5a possesses anaphylatoxic and chemotaxic activities, whereas C5b functions as an anchor on the surface of target cell forming the lytic membrane-attack complex (MAC) *e.g.* C5b, C6, C7, C8 and C9 (Fig. 16.31). Lysis takes place when C9 polymerises to form a hole in the cell wall. C1-inhibitor controls the inhibition of C1r and C1. C4a has anaphylactic activity (inflammatory response).

***(b)* Alternative Pathway:** IgA is associated with activation of alternative path. Besides, IgE or non-immunological means for example lipopolysaccharide (LPS) initiates the alternate pathway. C3 is mildly unstable in aqueous environment. Due to breakdown of the thioester bond via condensation reaction, the alternative pathway is directly triggered by spontaneous C3 hydrolysis to form C3a and C3b. A small amount of C3b is always present in circulation (Fig. 16.31 A). Like the other pathways, it does depend on a pathogen-binding antibody. Then C3b binds covalently to the surface of a pathogen and can become protected after binding to sugars on a cell surface. In the absence of a pathogen in blood, the C3a and C3b protein fragments rejoins and get deactivated. After binding to a cellular membrane C3b is bound by factor B to form C3bB. This complex is cleaved into Ba and Bb in presence of factor D.

Bb remains covalently linked to C3b to form C3bBb which is the alternative pathway C3-convertase. Protein C3 is produced in the liver. Bb contains the active site for a C3 convertase. The C3bBb complex, which is hooked onto the surface of the pathogen, hydrolyses C3 into C3a and C3b (B).

This positively affects the number of C3bBb hooked onto a pathogen. Next properdin binds to C3bBb to stabilize the C3bBb convertase on cell surface leading to cleavage of C3. Upon hydrolysis of C3, C3b forms a complex (C3bBbC3b complex which is a C5 convertase) which splits C5 into C5a and C5b. After linking of C5b with C6, C7, C8, and C9 (C5b6789) a membrane attack complex (MAC) is formed which is inserted into the cell membrane, punches a hole and lyses the cells (B).

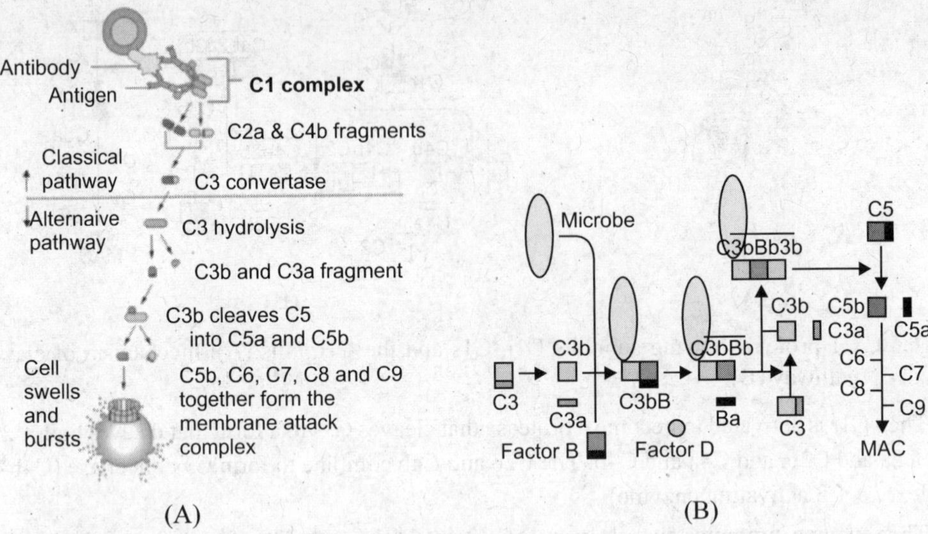

Fig. 16.31. Classical and alternative complement pathways (C2a read C2b about C2 fragments) (A); alternative pathway (B).

Regulation of Cascade of Complement

Because both the classical and alternate pathways depend upon C3b. Reegulation of the complement cascade is mediated via three proteins which affect the levels and activities of this component.

- **C1 Inhibitor** inhibits the production of C3b by combining and inactivating C1r and C1s. This prevents formation of the C3 convertase, C4b2b.
- **Protein H** inhibits the production of C3b by inhibiting the binding of factor B to membrane-bound C3b, thereby preventing cleavage of B to Bb and production of the C3 convertase, C3bBb.
- **Factor I** inhibits the production of C3b by cleaving C3b into C3c and C3d, which are inactive. Factor I works only on cell membrane bound C3b, mostly on red blood cells (*i.e.* non-activator surfaces).

Generally, only Gram-negative cells can be directly lysed by antibody plus complement, while Gram-positive cells are mostly resistant. However, phagocytosis is greatly enhanced by C3b binding because phagocytes have C3b receptors on their surface. Antibody is not always required for such binding. In addition, complement can neutralise virus particles by direct lysis or preventing viral penetration of host cells.

(*c*) **Lectin Pathway [Mannose-binding Lectin (MBL) Pathway]:** Lectins are the sugar-binding proteins found in most of organisms including human. Lectin pathway is homologous to the classical pathway; but instead of C1q, it operates with the opsonin, mannose-binding lectin (MBL) and ficolins. This pathway is activated by binding mannose-binding lectin to mannose residues present on the surface of pathogens. This activates the MBL-associated serine proteases (MASP-1, and MASP-2), which is very similar to C1r and C1s, splitting C4 into C4a and C4b and C2 into C2a and C2b. Similar to classical pathway, C4b and C2a bind together to form C3-convertase. Ficolins are homologous to MBL and function via MASP in the same way.

J. HISTOCOMPATIBILITY

Histocompatible refers to the individuals who have the same tissues *i.e.* identical twins. The term histocompatibility is used to determine how identical two unrelated individuals are? In case of two histocompatible individuals, a tissue or organ from a donor (the person giving the organ or tissue) that will not be rejected by the recipient (the patient in whom the tissue or organ is transplanted). Histocompatibility is the property of having the same or mostly the same alleles of a set of genes called the 'major histocompatibility complex'. These genes are expressed in most tissues as antigens to which the immune system makes antibodies.

I. MAJOR HISTOCOMPATIBILITY COMPLEX (MHC)

The major histocompatibility complex (MHC) is a large cluster of genes found on the short arm of chromosome 6. The complex consists of four million base pairs of DNA and possesses 128 genes and 96 pseudogenes. In this complex all the genes play important roles in the immune system. Traditionally, the MHC is divided into the class I, II and III regions each contains groups of genes with related functions. Due to the codominance of the MHC alleles and the extreme polymorphism of the MHC genes, a great number of variant molecules is seen on the cells of one individual and throughout the whole population.

MHC Proteins

The MHC I and II proteins carry out the important function of antigen presentation. Through this mechanism the immune system searches out what is happening inside our cells even though it can only survey them from the outside. Proteins inside the cell are broken into short fragments and displayed as peptide antigens by MHC proteins on the surface. This helps the immune system to discriminate between normal (self) antigens and potentially dangerous foreign antigens.

Class I molecules are encoded by the BCA region of the chromosome, whereas class II molecules are encoded by the D region. On chromosome 6, a region between these two genes encodes class III molecules (Fig. 16.32).

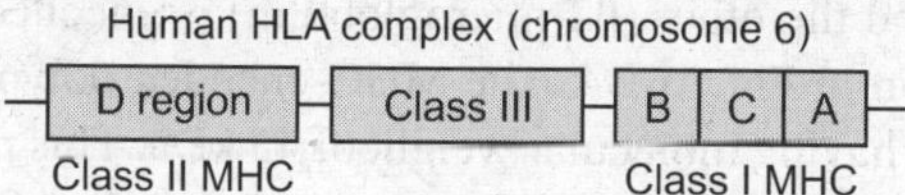

Fig. 16.32. Arrangement of MHC I, II and III genes on human chromosome 6.

Class I molecules are found on all nucleated cells but not red blood cells, whereas class II molecules are found only on antigen presenting cells (APCs) which include dendritic cells, macrophages, B cells and a few other types (Fig. 16.33).

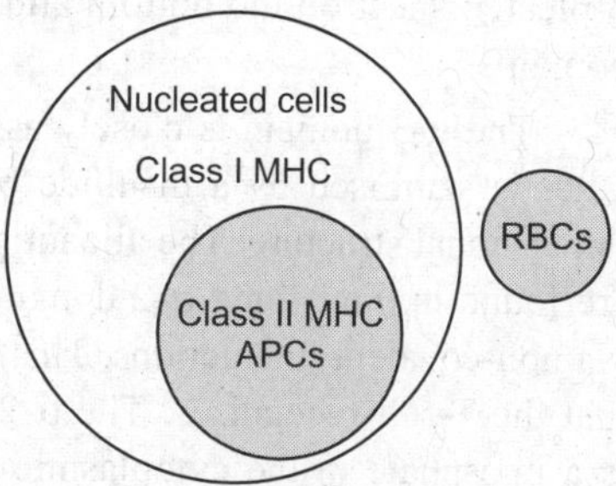

Fig. 16.33. Presence of class I and class II MHC molecules on human cells.

The human leukocyte antigens (HLAs) present on cell surface are encoded by class I and II MHC genes which define types of tissue of an individual. Each HLA exists as a large number of varieties; therefore, there are many possible tissue types among the population. Immune system of every one is tolerant to its own HLAs. If foreign HLAs are detected, the cells showing them are attacked and destroyed. This is the reason why the body rejects grafts and transplants from donors which have not been matched for tissue type.

The T cell receptor (TCR) recognises antigenic peptides in association with MHC molecules. T cells recognize portions of protein antigens bound non-covalently to MHC gene products. Cytotoxic T cells (Tc) recognise peptides bound to class I MHC molecules and helper T cells (Th) recognise peptides bound to class II MHC molecules. Function and expression of subgroups of MHC molecules are given in Table 16.8.

Table 16.8. Function and expression of subgroups of MHC molecules.

Name	*Function*	*Expression*
MHC class I	Encodes heterodimeric peptide-binding protein, as well as antigen-processing molecules such as tapasin.	MHC class I proteins contain an α chain and β2-microglobulin (not a part of the MHC). They present antigen fragments to cytotoxic T-cells and bind to CD8 on cytotoxic T-cells.
MHC class II	Encodes heterodimeric peptide-binding proteins and other proteins that modulate antigen loading onto MHC class II proteins in the lysosomal compartment	On antigen-presenting cells MHC class II proteins contain α- and β- chains and they present antigen to T-helper cells by binding to the CD4 receptor on the T-helper cells.
MHC class III	Encodes for other immune components such as complement components (*e.g.* C2 anzd C4) and some that encode cytokines (*e.g.* TNF-α).	Variable

1. MHC Class I Proteins

MHC I molecules consist of two polypeptide chains: one chain is encoded by the BCA region of chromosome 6 and the other (ß2-microglobulin) is encoded elsewhere. The structure of the class I MHC is shown in Fig. 16.34. The MHC-encoded polypeptide is glycosylated and about 350 amino acids long having molecular weight of 45 kDa. This polypeptide folds into three separate domains called α-1, α-2 and α-3. The β2-microglobulin (β_2m) has molecular weight of 12 kDa which is non-covalently associated with the α-3 domain. A region is bounded by a β-pleated sheet on the bottom and two alpha helices on the sides lies between the α-1 and α-2 domains.

The α-3 domain is closely associated with the non-MHC -encoded β-2 microglobulin. The latter is stabilised by a disulfide bridge and is similar to an immunoglobulin domain in three-dimensional structure. The alloantigenic sites which carry determinants specific to each individual are found in the α- and α-2 domains. This region binds a small peptide of about 10 amino acids via non-covalent interactions. This peptide is presented to a T-cell and defines the antigen epitope that the T-cell recognises. The α-2 domain also consists of a carbohydrate chain (CHO). There is a phosphate in the cytoplasmic domain. Papain cleaves near the outer surface of the plasma membrane.

MHC I proteins are found on all cell types. They present fragments of proteins synthesised inside the cell. The peptide antigens presented in this manner are checked by killer T-cells which have receptors for the class I MHC proteins.

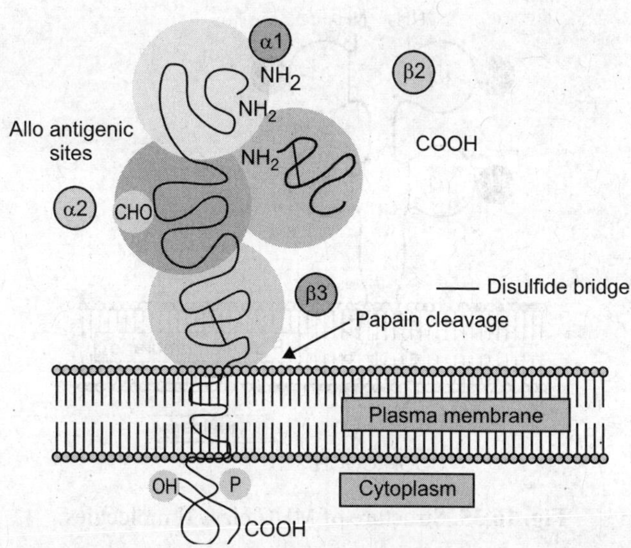

Fig. 16.34. Structure of MHC class I molecule showing three globular domains- α-1, α-2 and α-3.

2. MHC Class II Proteins

MHC II proteins which are found only on immune cells such as phagocytes are composed of two polypeptide chains both encoded by the D region. The α-polypeptide is glycosylated and about 230 amino acids long with molecular weight about 33 kDa. The β polypeptide is also glycosylated and about 240 amino acids long having molecular weights of 28 kDa. These polypeptides fold into two separate domains: α-1 and α-2 for the alpha polypeptide and β-1 and β-2 for β-polypeptide. Between α-1 and β-1 domain there lies a peptide binding region very similar to that present on class I MHC molecule. This region is bounded by a β-pleated sheet on the bottom, and by two α-helices on the sides. It is capable of binding a small peptide of about 10 amino acids via non-covalent interactions. This small peptide is presented to a helper T-cell and defines the antigen epitope. The helper T-cells has receptors for class II MHC proteins which help to recognise the epitope of antigen. The MHC-encoded polypeptides are shown in Fig. 16.35.

MHC class II molecules consist of two non-identical peptides (α and β) which are non-covalently associated and traverse the plasma membrane with the N terminus to the outside of the cell. The domains closest to the membrane in each chain are structurally related to immunoglobulins. With the exception of the α-1 domain, all domains are stabilised by disulfide bridges. Both the α- and β- chains are glycosylated. The β-chain is shorter than the α-chain and contains the alloantigenic sites. There is some polymorphism in the α-chain of some MHC II molecules.

MHC II proteins are found only on immune cells such as phagocytes which engulf foreign particles including bacteria. These cells present peptide antigens from digested particles. The purpose of this surveillance system is to stop the immune system running out of control and attacking the body's own cells.

Within the MHC there are 5 loci that encode class II molecules each of which contains a gene for an α- chain and at least one gene for a β-chain. The loci are designated as HLA-DP, HLA –DQ, HLA-DR, HLA-DM, and HLA-DO. Among these, HLA-DP, HLA–DQ and HLA-DR are the most important and are most polymorphic.

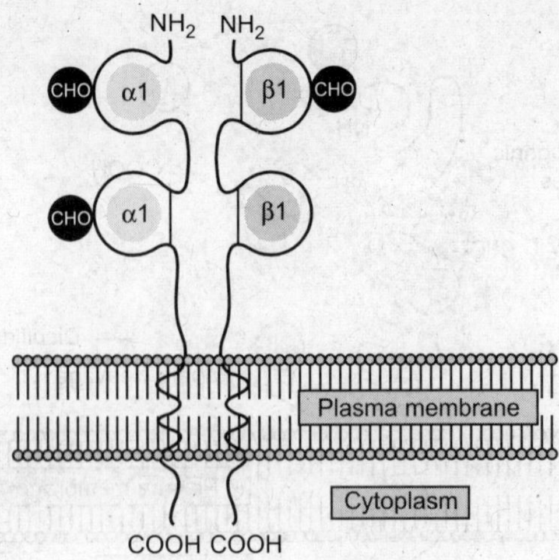

Fig. 16.35. Structure of MHC class II molecules.

3. MHC Class III Proteins

Several components of the complement system are encoded by MHC III genes. The components are a collection of soluble proteins found in the blood that targets foreign cells and breaks open their membranes. A group of genes that controls inflammation lays adjacent to the class III region.

The MHC shows a high degree of polymorphism, *i.e.* 10% difference between any two unrelated individuals. Many of these polymorphisms are associated with either increased or decreased susceptibility to a wide range of infectious diseases like hepatitis, AIDS, malaria, tuberculosis, leprosy and typhoid. Defects in certain MHC genes lead to autoimmune disorders in which the body fails to recognise self-antigens such as multiple sclerosis, some forms of arthritis and diabetes.

K. HYPERSENSITIVITY

The term hypersensitivity is used to describe immune responses which are damaging rather than helpful to the host. Some times, the immune system responds inappropriately in the presence of antigen. These responses are referred to as hypersensitivities. Gell and Coombs (1963) proposed a scheme of classification which defined four types of hypersensitivity reactions (Table 16.9). The first three are mediated by antibody, and the fourth by T cells. These types are classified as: Type I (immediate hypersensitivity), Type II (cytotoxic hypersensitivity), Type III (immune complex hypersensitivity), and Type IV (delayed hypersensitivity),

Table 16.9. Types of hypersensitivity as classified by Gell and Coombs (1963).

Type	*Alternative names*	*Mentioned disorders*
I	Allergy (immediate)	Atopy, anaphylaxis, asthma
II	Cytotoxic, antibody-dependent	Autoimmune hemolytic anemia, Thrombocytopenia, erythroblastosis fetalis Goodpasture's syndrome
III	Immune complex disease	Serum sickness, Arthus reaction,
IV	Delayed-type hypersensitivity (DTH), cell-mediated immune memory response, antibody-independent	Contact dermatitis, Mantoux test, chronic transplant rejection, multiple sclerosis

1. Type I Hypersensitivity

It is the rapid immediate allergic reaction. The symptoms produced by exposure of a sensitised person to antigen depend upon the site of contact. Allergic rhinitis, eczema, asthma and urticaria result from type I hypersensitivity. It is caused after contact with antigen, against which the host has pre-existing IgE antibody. IgE is present in very low levels in serum in most people. In serum its half life is only 2-3 days but much of the IgE in the body is bound to high affinity receptors (Fc εRI). In the bound state the half-life is ~3 weeks. The high affinity Fc epsilonεRI (Fc εRI) receptors are found on mast cells and basophils. Each cell has a high density of these receptors (40-250,000 per cell) so that a wide spectrum of antigen specificities is represented. The cells are activated by the cross-linking of the Fc εRI receptors via antigen binding to the bound IgE molecules (Fig. 16.36).

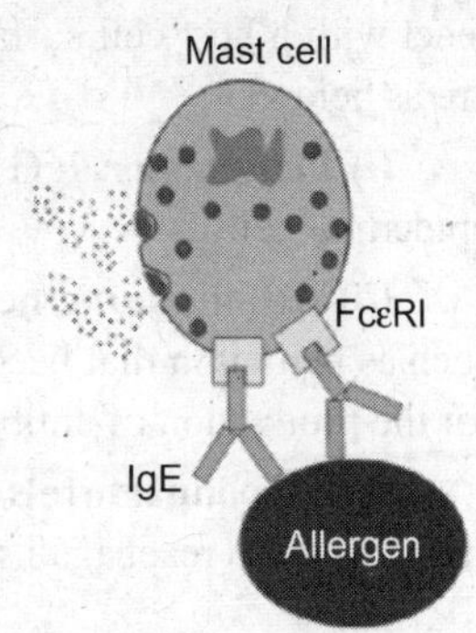

Fig. 16.36. Type I hypersensitivity.

This cross-linking results in rapid degranulation (60-300 seconds) of the mast cells and the release of primary inflammatory mediators which are stored in granules. These mediators cause an acute inflammatory reaction *e.g.* increased vascular permeability, smooth muscle contraction, granulocyte chemotaxis, extravasation, etc. Mast cell activation through Fc εRI also results in production of two other types of mediators.These secondary mediators comprise arachadonic acid metabolites (prostaglandins and leukotrienes) and proteins (cytokines and enzymes). They must be synthesised *de novo* unlike the stored granule contents.

Common Antigens Causing Type I Hypersensitivity

Some common antigens that cause Type I hypersensitivity are given below :

Pollen grains	:	Oil seed rape, birch, ragweed
Food	:	Eggs, seafood, nuts (e.g. peanut)
Drugs	:	Penicillins, salicylates (e.g. aspirin)
Animal hair	:	Cat hair
Insect products	:	House dust mite, bee venom

After an initial introduction of antigen Type I hypersensitive reaction produces an antibody response. Moreover, the type of antigen induces the synthesis of IgE antibody. Immunoglobulin IgE binds very specifically to receptors on the surface of circulating mast cells. Reintroduced antigen interacts with IgE on mast cells causing the cells to degranulate and release large amounts of histamine, lipid mediators and chemotactic factors that cause smooth muscle contraction, vasodilation, increased vascular permeability, broncoconstriction and edema. These reactions take place very quickly causing death.

Type I hypersensitivities are: allergies to penicillin, insect bites, molds, etc. Sensitivity of persons to these allergens can be tested by a cutaneous reaction. If the specific antigen is injected subcutaneously and the patient is sensitive, a specific reaction known as 'wheal and flare' can be observed within 15 minutes. Therefore, the persons who are hypersensitive to such allergens must avoid contact with large inocula to check the anaphylactic shock.

2. Type II Hypersensitivity

Type II hypersensitivity is also called 'cytotoxic hypersensitivity' which involves antibody-mediated reactions (Fig. 16.37). However, the immunoglobulin isotype is generally IgG. This process involves K-cells rather than mast cells. K-cells are involved in antibody-dependent cell-mediated cytotoxicity. Type II hypersensitivity may also involve complement that binds to cell-bound antibody. The antibodies are specific for 'self' antigens. When these circulating antibodies react with a host cell surface, tissue damage may result. The examples of Type II hypersensitivity are as below:

(*i*) Pemphigus: IgG antibodies that react with the intracellular substance are found between epidermal cells.

(*ii*) Autoimmune hemolytic anaemia (AHA): This disease is generally inspired by a drug such as penicillin that becomes attached to the surface of red blood cells (RBC) and acts as hapten for the production of antibody which then binds the RBC surface leading to lysis of RBCs.

(*iii*) Goodpasture's syndrome: Generally it manifests as a glomerulonephritis *i.e.* IgG antibodies that react against glomerular basement membrane surfaces can lead to kidney destruction.

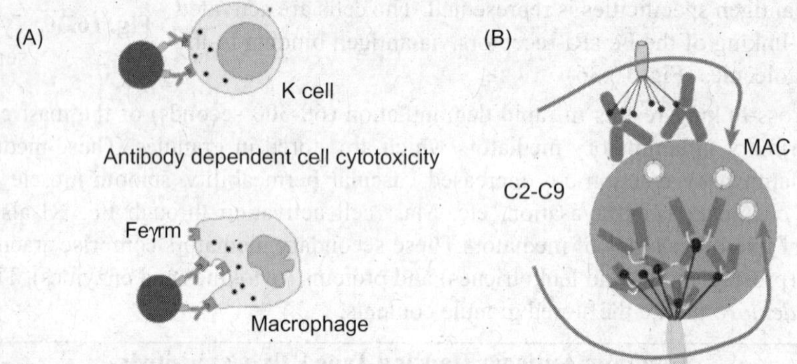

Fig. 16.37. Type II hypersensitivity (A); classical pathway showing complement activation (B).

3. Type III Hypersensitivity

Type III hypersensitivity reactions arise with circulating antibody that reacts with free soluble antigens. The pathology is caused by the deposition of antigen-antibody aggregates or immune complexes at certain sites of tissue (Fig. 16.38). Immune complexes are generated in all responses of antibody but their pathogenic potential is determined by their size and the amount, affinity and isotype of the responding antibody. Type III or 'immune complex hypersensitivity' includes the circulating antibody that reacts with free soluble antigen. Upon deposition of circulating complexes on tissues, onward reactions with complement results in tissue damage.

Thus this type of hypersensitivity develops as a result of systematic exposure to an antigen which dependent on: (*i*) the type of antigen and antibody, and (*ii*) the size of the resulting complex. Too small complexes remain in circulation, too large complexes are removed by the glomerulus, and intermediate complexes may become lodged in the glomerulus which damages the kidney.

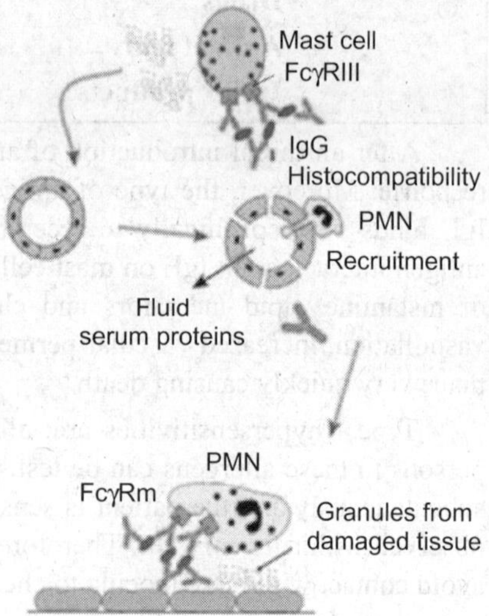

Fig. 16.38. Type III hypersensitivity.

One example of Type III hypersensitivity is serum sickness which is a condition that may develop when a patient is injected with a large amount of antitoxin that was produced in an animal. After about 10 days, anti-antitoxin antibodies react with the antitoxin forming immune complexes that deposit in tissues. Type III hypersensitivities can be ascertained by intradermal injection of the antigen, followed by the observance of an 'arthus' reaction (swelling and redness at site of injection) after a few hours.

4. Type IV Hypersensitivity

Type IV hypersensitivity is also known as delayed type hypersensitivity which are mediated by antigen-specific effector T cells. This is the only class of hypersensitive reactions to be triggered by antigen-specific T cells which is called as T_H1 cells but they were originally termed T_{DTH} after the alternative name, *i.e.* delayed type hypersensitivity (T_{DTH}). A delayed type hypersensitivity is achieved when an antigen presenting cell (mainly a tissue dendritic cell picking up antigen), processes and displays the suitable peptide fragments bound to class II MHC contacted by an antigen specific T_H1 cell patrolling the tissue. The activation of T cell produces cytokines *e.g.* chemokines for macrophages (Fig. 16.39), other T cells and, neutrophils as well as TNFbeta and IFN-γ. Consequently, *e.g.* T cells and macrophages predominate in a cellular infiltrate reaching maximal in 48-72 h.

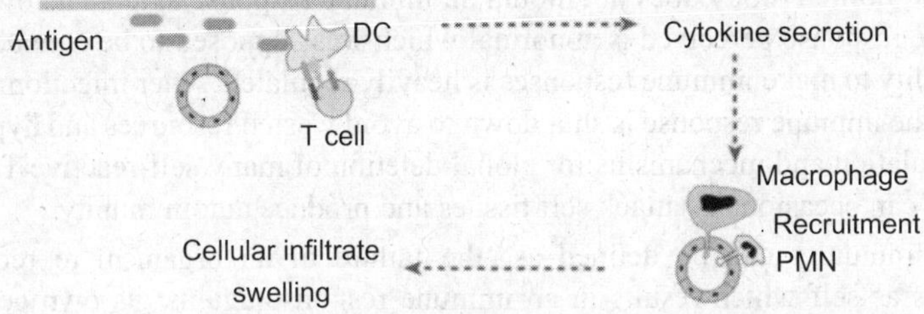

Fig. 16.39. Type IV hypersensitivity.

The classical example of delayed type hypersensitivity is in tuberculosis. The tuberculin test is used to determine whether an individual has previously been infected with *Mycobacterium tuberculosis*. Small amounts of tuberculin (a complex mixture of peptides and carbohydrates derived from *M. tuberculosis*) are injected intradermally. A local T cell-mediated inflammatory reaction evolves over 24–72 hours in individuals who have previously been exposed to the bacterium either by infection with the pathogen or by immunisation with BCG, *i.e.* an attenuated form of *M. tuberculosis*. A summary of hypersensitive diseases has been given in Table 16.10.

The response is mediated by T_H1 cells which enter the site of antigen injection, recognise complexes of peptide-MHC class II molecules on antigen-presenting cells, and release inflammatory cytokines, such as IFN-γ and TNF-β. A more familiar example is contact hypersensitivity which results from exposure of certain individuals to metal salts and small reactive chemicals. It is important to note that in such cases the antigen must be a complex of the hapten and a self peptide and T cells specific for Ni^{++} and reactive chemicals *e.g.* dinitrofluorobenzene.

Table 16.10. Summary of hypersensitivity classes.

Type	Descriptive name	Initiation time	Mechanism	Example
I	IgE-mediated hypersensitivity	2-30 min	Ag induces cross-linking of IgE bound to mast cells with release of vasoactive mediators	Systemic anaphylaxis, Local anaphylaxis, Hay fever, asthma, eczema
II	Antibody-mediated	5-8 h	Ab directed against cell-surface antigens mediates	Blood transfusion reactions,

	cytotoxic hypersensitivity		cell destruction via ADCC or complement	Haemolytic disease of the newborn, autoimmune haemolytic anaemia
III	Immune-complex mediated hypersensitivity	2-8 h	Ag-Ab complexes deposited at various sites induces mast cell degranulation via FcgammaRIII, PMN degranulation damages tissue	Arthus reaction (localised); systemic reactions disseminated rash, arthritis, glomerulonephritis
IV	Cell-mediated hypersensitivity	24-72 h	Memory TH1 cells release cytokines that recruit and activate macrophages	Contact dermatitis, tubercular lesions

L. AUTOIMMUNITY

In the beginning of 20th century, Paul Ehrlich, proposed the concept of **horror autotoxicus**. In such condition a 'normal' body does not mount an immune response against its own tissues. Thus, any autoimmune response perceived is abnormal which was supposed to be connected with human disease. The ability to make immune responses is heavily regulated. After infection when pathogens are eliminated, the immune response is shut down to avoid wasted resources and hypersensitivity. In spite of this regulation and mechanisms for clonal deletion of many self-reactive T and B cells, the immune system can occasionally attack self tissues and produce autoimmunity.

Thus autoimmunity can be defined as 'the failure of an organism to recognise its own constituent parts as self which results in an immune response against its own cells and tissues'. Hence, deficiencies in the ability to make an adequate immune response can result in life-threatening infections (http://en.wikipedia.org/wiki/Autoimmunity).

1. Causes of Autoimmunity

Autoimmunity is caused by an adaptive immune response against 'self' antigen. Clonal deletion and anergy to self-specific lymphocytes greatly reduce the possibility of low affinity self-specific responses. Transient autoimmune responses are common but usually cause no lasting damage. Because self antigens are continually present in the body; when autoimmune responses are prolonged, the resulting tissue damage can be life-threatening. Certain individuals are genetically susceptible to autoimmune diseases. This susceptibility is associated with multiple genes along with other risk factors. Genetically-predisposed individuals do not always develop autoimmune diseases. There are three main sets of genes suspected in many autoimmune diseases. These genes are related to immunoglobulins, T-cell receptors, and major histocompatibility complexes (MHC). Risk factors for autoimmune disease include the presence of certain HLA alleles, sex hormone levels, infection, and other environmental factors. Autoimmune diseases are classified by corresponding type of hypersensitivity *i.e.* type II, type III or type IV (no type of autoimmune disease mimics type I hypersensitivity).

Paul Ehrlich

Risk factors for autoimmune diseases are genetically linked to the presence of specific class I or class II HLA alleles. There are two possible models that explain HLA linkage to autoimmunity. The first model explains that certain HLA alleles are better at presenting pathogen peptides which resemble self peptides to mature T cells. The HLA B27 (class I) allele is associated with an 80-fold increased risk of alkylosing spondylitis characterised by inflammation and damage to the spine. B27 binds the peptides in the absence of tapasin. This is an advantage in certain virus infections where the virus interferes with production of tapasin. However, the peptides bound in the absence of tapasin are different from those bound in the presence of tapasin. They might include some self peptides that could induce autoimmunity. Therefore, people with the B27 Class I MHC allele possibly will respond more to virus infections by producing CTL that can recognize self peptides and kill uninfected cells.

Paul Ehrlich

Paul Ehrlich (14 March 1854 – 20 August 1915) was a German Nobel Prize winner together with I.I. Mechnikov for Medicine in 1908. He worked in the fields of haematology, immunology, and chemotherapy, and. He is noted for his research in autoimmunity, calling it *horror autotoxicus*. He coined the term 'chemotherapy' and popularised the concept of a 'magic bullet'. The magic bullet concept comes from the experience of 19th century German chemists with selectively staining tissues for histological examination, and selectively staining bacteria. Ehrlich reasoned that if a compound could be made that selectively targeted a disease-causing organism, then a toxin for that organism could be delivered along with the agent of selectivity. Hence, a *magic bullet* would be created that killed only the target organism. The concept of a magic bullet was fully realised with the invention of monoclonal antibodies.

He worked with his friend Emil Adolf von Behring on the development of the diphtheria serum. The side-chain theory of Behring explained the effects of serum and enabled measurement of the amount of antigen. In 1906 he discovered the structural formula of atoxyl, a chemical compound used to treat sleeping sickness. He tried to create a less toxic version of the medicament. In 1909 he and his student Sahachiro Hata developed Salvarsan which was an effective treatment against syphilis.

The second model for autoimmunity proposes that certain HLA alleles are less efficient at presenting self peptides to developing T cells in the thymus so that negative selection fails, for example 'insulin-dependent diabetes mellitus' (IDDM) or type I diabetes. IDDM is strongly linked to the presence of HLA DR3 and DR4 alleles.

When autoimmune disease is caused by autoantibodies, the disease mechanism is classified as Type II or Type III hypersensitivity. In autoimmune haemolytic anaemia, antibodies to red blood cell antigens initiate complement lysis and phagocytosis of RBC in the spleen reticuloendothelial system (RES). Antibodies to platelets and neutrophils can also cause depletion of these cells in the RES. Leukocytes are more resistant to complement lysis than erythrocytes. Autoimmune haemolytic anaemia can be treated simply by removing the spleen. Table 16.11 summarises the information about some human autoimmune diseases.

Table 16.11. Autoimmune diseases of humans.

Human Autoimmune Diseases			
Disease	***Autoantigen***	***Symptoms***	***Extent****
Type II: antibodies to cell surface molecules			
Autoimmune hemolytic anaemia	Rh blood group antigens, I antigen	Lysis of RBC by complement and FcR^+ cells, anemia	O
Hypoglycemia	Insulin receptor (agonist)	Low blood glucose	O
Insulin-resistant diabetes	Insulin receptor (antagonist)	High blood glucose, ketoacidosis	O
Pernicious anemia	Intrinsic factor, gastric parietal cells	Anemia	O
Rheumatic fever	Streptococcal cell wall antigens; antibodies cross-react with heart muscle	Arthritis, myocarditis, heart valve scars	O
Spontaneous infertility	Sperm antigens	Infertility	O

Type III: Immune complex disease			
Ankylosing spondylitis	Immune complexes	Damage to vertebrae	S
Mixed essential cryoglobulinemia	Rheumatoid factor IgG complexes	Systemic vasculitis	S
Rheumatoid arthritis	Rheumatoid factor IgG complexes	Arthritis	S
Systemic lupus erythematosus (SLE)	DNA, histones, ribosomes, snRNP, scRNP	Glomerulonephritis, vasculitis, rash	S
Type IV: T cell-mediated disease			
Hashimoto's thyroiditis	Thyroid antigen(s)	Thyroid under activity	O
Insulin-dependent (Type I) diabetes mellitus (IDDM)	Pancreatic B cell antigen(s)	B cell destruction	O
Rheumatoid arthritis	Unknown synovial joint antigen	Joint inflammation and destruction	S

*O = organ-specific, S = systemic

Antibodies to cell surface receptors may stimulate or inhibit receptor function. Antibodies to thyroid stimulating hormone (TSH) receptor in the thyroid stimulate thyroid hormone production and induce hyperthyroidism in Grave's Disease. Antibodies to the receptor for acetylcholine (which is a neurotransmitter) block the signals of nervous system to muscle cells and leads to myasthenia gravis, a progressive muscle weakness. Antibodies to insulin receptor mimic insulin function and cause low blood sugar and in other cases block insulin function and cause high blood sugar.

IgG autoantibodies may be transferred across the placenta and cause transient symptoms in the newborn infant. Damage is usually not permanent; removal of maternal antibodies can be accomplished with plasmapheresis. T cell-mediated damage results in several autoimmune diseases such as multiple sclerosis, rheumatoid arthritis (which also has some Type III characteristics) and insulin-dependent diabetes. It is more difficult to identify autoimmune T cells and the antigen to which they are responding than to identify antibodies and their antigens. Since autoimmune disease is mediated by normal immune mechanisms, controlling the disease without making the patient susceptible to infection is the greatest challenge.

2. Tolerance

Immunological tolerance is the ability of an individual to ignore 'self' while reacting to 'non-self'. Natural tolerance is the inability to make an immune response to an antigen. It is caused by several factors. Clonal deletion of self-reactive T and B cells during development removes lymphocytes with high avidity receptors for ubiquitous self antigen present in the thymus and marrow. The particular HLA alleles available to present self-antigen to developing T cells influence self peptides which are presented with high avidity to induce clonal deletion, and which are presented with low avidity. When immature B cells encounter soluble antigen that cross-links BCR and T cells encounter unprocessed antigen or processed antigen in the absence of co-stimulatory signals, clonal anergy occurs in the periphery. Clonal anergy maintains tolerance to some (but not all) self antigens that are not available for clonal deletion in the thymus and marrow. Specific autoimmune diseases like those described above are probably not due to general failure of clonal deletion or clonal anergy.

Immunological ignorance refers to the common occurrence of low numbers of low avidity self-specific T cells existing in the presence of low levels of presented self peptides without becoming activated to respond. Humans make at least 10^5 proteins (average size 300 amino acids) which are processed to produce 3×10^7 peptides for presentation to T cells. Each APC has a maximum of

10^5 MHC molecules per cell with which to present these peptides. T cells must bind to about 10-100 identical peptides on an APC to become activated. Hence, most peptides presented on APC would be below the threshold for T cell detection. The best chance of stimulating autoimmune T cells would be with tissue-specific antigens which are less likely to induce clonal deletion. There are relatively few autoimmune diseases and all people with a given disease respond to the same antigen(s). It means that that autoimmunity occurs only under rare circumstances. The level of antigen presentation depends on availability of MHC alleles, and certain MHC alleles are linked with certain autoimmune diseases.

Tolerance is also maintained by regulatory (suppressor) T cells (*e.g.* $CD4^+$ $CD25^+$ T cells), which can transfer tolerance to another animals. For example, neonatal rats injected with bone marrow cells from allogeneic rats become tolerant to skin grafts of the same allotype but not other allogeneic skin grafts. T cells transferred from the tolerant rat to another rat make the recipient tolerant to the same MHC allotype as long as allogeneic cells are transferred with the suppressor T cells. Tolerance can be eliminated if cells are transferred from an animal immunised to the allogeneic cells. Probably these cells kill the allogeneic cells.

(*a*) Origin of Immunological Tolerance: The exact genesis of immunological tolerance is still elusive but several theories have been proposed to explain its origin. Three hypotheses have gained widespread attention among the immunologists:

(*i*) Clonal deletion theory: It was proposed by F.M. Burnet which explains that self-reactive lymphoid cells are destroyed during the development of the immune system in an individual. For discovery of acquired immunological tolerance, Frank M. Burnet and Peter B. Medawar were awarded the Nobel Prize in Physiology or Medicine in 1960.

(*ii*) Clonal anergy theory: This theory was proposed by Nosaal, in which self-reactive T- or B-cells become inactivated in the normal individual and cannot amplify the immune response.

(*iii*) Idiotype network theory: This theory was proposed by N. Jerne, wherein a network of antibodies capable of neutralising self-reactive antibodies exists naturally within the body.

(*b*) Genetic Factors: Certain individuals are genetically susceptible to developing autoimmune diseases. This susceptibility is associated with multiple genes plus other risk factors. Genetically-predisposed individuals do not always develop autoimmune diseases. Three main sets of genes are suspected in many autoimmune diseases. These genes are related to immunoglobulins, T-cell receptors, and major histocompatibility complexes (MHC).

3. Treatment of Autoimmune Diseases

Traditionally treatments of autoimmune diseases have been the immunosuppressive, anti-inflammatory or palliative. The non-immunological therapies, for example hormone replacement in Hashimoto's thyroiditis, are the result of autoaggressive response. Dietary manipulation results in decline in the severity of celiac disease. Steroidal treatment limits inflammatory symptoms of many diseases. Specific immunomodulatory therapies like the TNFα antagonists e.g. etanercept, the B cell depleting agent rituximab, anti-IL-6 receptor tocilizumab and the costimulation blocker abatacept have been useful. Some of these immunotherapies may be associated with increased risk of adverse effects such as susceptibility to infection.

QUESTIONS

1. Give a historical account of immunology.
2. What are antigens ? Write origin and types of antigens.
3. What do you know about innate immunity ? Describe different types of cells associated with innate immunity.
4. Discuss in detail about responses of innate immunity.

5. Write structure and diversity of antibodies.
6. Whay is an immunoglobulin ? Describe its isotypes.
7. Give an illustrated account of antigen-determinants of immunoglobulins.
8. Describe in detail about cell-mediated immunity.
9. Write genetic basis of antibody diversity.
10. Explain complement system describing in detail about activation of complement cascade.
11. What is major histocompatibility complex ? Describe MHC proteins.
12. Give a detailed account of different types of hypersensitivity. Describe in brief about causes of autoimmunity and treatment of autoimmune diseases.
13. Write short notes on the following :
    (*i*) Variolation, (*ii*) Antigens,
    (*iii*) Autoantigens, (*iv*) Innate immunity,
    (*v*) Phagocytosis, (*vi*) Antibody,
    (*vii*) Immunoglobulin fragment structure and function relationship,
    (*viii*) Susuman Tonegawa, (*ix*) Immunoglobulin isotypes,
    (*x*) Allotypes, (*xi*) Antibody kinetics,
    (*xii*) Regulation of humoral response, (*xiii*) Antibody diversity,
    (*xiv*) Isotype switching, (*xv*) Autoimmunity.

BIBLIOGRAPHY

Aderem A. and Underhill D.. 1999. Mechanisms of phagocytosis in macrophages. *Ann. Rev. Immun.* **17**: 593-623.

Atkinson J.P. 2009. Complement system. *Kelley's Textbook of Rheumatology* (Eds Firestein G.S. *et al.*) Philadelphia, PA: Saunders/Elsevier, pp. 323-336.

Black C.A. 1999. Delayed type hypersensitivity: Current Theories with a Historic Perspective *Dermatol. Online* J. 5 (1):7 at http://dermatology.cdlib.org/DOJvol5num1/reviews/black.html

Charles J., Travers P., Walport M. and Shlomchik M. 2001. *Immunobiology.* V. Ed. New York and London: Garland Science.

DeFranco A.L., Locksley R.M. and Robertson M. 2007. *Immunity: The Immune Response in Infectious and Inflammatory Disease*, New Science Press, Sunderland, MA: London;

Edwards J.C., Cambridge G, Abrahams VM 1999. Do self perpetuating B lymphocytes drive human autoimmune disease ? *Immology* .**97**: 1868-1876.

Edwards J.C. and Cambridge G. 2006. B-cell targeting in rheumatoid arthritis and other autoimmune diseases. *Nature Rev. Immunol.* **6** (5): 394-403.

Gell P.G.H. and Coombs R.R.A. 1963. *Clinical Aspects of Immunology*. 1st Ed. Oxford, England: Blackwell.

Goldsby R.A., Kindt T.K., Osborne B.A. and Kuby J. 2003. *Immunology*, 5th Ed. W.H. Freeman and Company, New York.

Janeway C.A.Jr. *et al.* 2005. *Immunobiology*. (6th Ed.), Garland Science.

Jerne N. 1974. Towards a network theory of the immune system. *Ann Immunol* (Paris) **125C**: 373–89.

Pike B., Boyd, A. and Nossal, G. 1982. Clonal anergy: the universally anergic B lymphocyte. *Proc Natl Acad Sci U.S.A.* **79** (6): 2013–7.

Stvrtinová V., Jakubovský J. and Hulín I. 1995. *Inflammation and Fever from Pathophysiology: Principles of Disease.* Computing Centre, Slovak Academy of Sciences: Academic Electronic Press.

17 CHAPTER Immunotechnology

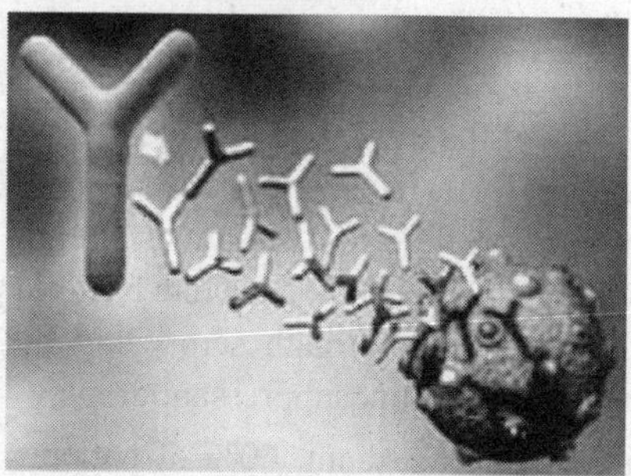

Immunoglobulin acts on a virus

A. Monoclonal antibodies (hybridoma technology)
B. Interferons
C. Vaccines
D. Transplantation (transplant immunology)
E. Tumour immunology
F. Cancer immunotherapy

Immunotechnology is the technology based on applications of cells/tissues and molecules of the immune system. It is an important arm of biotechnology which constitutes the industrial scale application of immunological procedures to produce vaccines for mass immunisation to prevent prevalent diseases and/or to produce immunological therapeutic agents to cure the afflicted. For a long time protein vaccines have been produced on a large-scale, and the current trend is to develop the more specific DNA vaccines.

Now scientists are able to obtain the substances in quantity secreted by cells of the immune system, both antibodies and lymphokines through a strategy known as hybridoma technology. The ready supply of these materials has not only revolutionised immunology but has also created a resounding impact throughout the medicine and industry. Moreover, genetic engineering also called recombinant DNA technology allows the scientists to pluck genes from an organism and combine them with genes of a second organism. In this way, relatively simple organisms such as bacteria or yeast, or even mammalian cells in culture and mammals such as goats and sheep, can be induced to make quantities of human proteins including hormones such as insulin as well as lymphokines and monokines. Microorganisms can also be changed to manufacture proteins from infectious agents such as the hepatitis virus or the AIDS virus for use in vaccines.

During 1770s, Edward Jenner heard that a milkmaid never have the disease smallpox because she had already had cowpox. The latter has a very mild effect in humans, whereas the smallpox is fatal. In 1796, Jenner took pus from the hand of a milkmaid suffering from cowpox and inoculated an 8 years old boy. After a gap of 6 weeks Jenner variolated the arm of the boy with smallpox. He found that the boy did not receive smallpox. Further experimentation showed the efficacy of the procedure on an infant. Vaccination with cowpox was much safer than smallpox inoculation; even then smallpox inoculation was banned in 1840 in spite of being widely practiced in England. Jenner's idea was generalised by Louis Pasteur through developing a rabies vaccine (now termed an antitoxin).

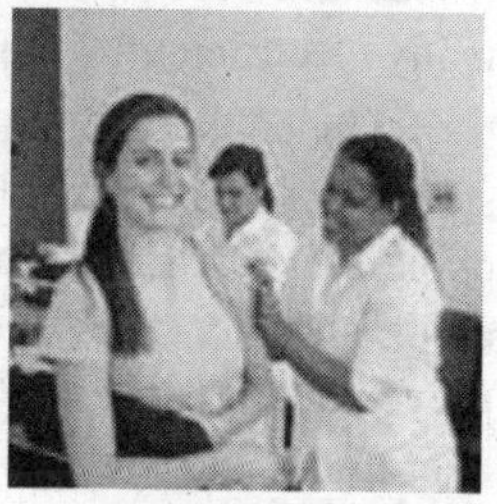

A woman receiving rubella vaccine.

In 19th century vaccines were considered a matter of national prestige. The 20th century witnessed the introduction of several successful vaccines such as the vaccines against diphtheria, measles, mumps, rubella, etc. Major achievements are the development of polio vaccine in the 1950s and the eradication of smallpox during the 1960s and 1970s. Now, the vaccines have become more common than earlier. However, vaccines remained elusive for many important diseases, including HIV and malaria.

In recent years xenotransplantation is becoming popular because xenotransplants are potentially a more effective alternative. Xenotransplants can save thousands of patients who are waiting for donated organs. The animal organ, probably from a pig or baboon could be genetically altered with human genes to trick the immune system of a patient into accepting it as a part of its own body. They have re-emerged due to the lack of organs available and the constant struggle to keep immune systems from rejecting allotransplants.

When a transplanted organ/tissue is not accepted by the body of the transplant recipient, there occurs transplant rejection because the immune system of the recipient attacks the transplanted organ or tissue and rejects it. Obviously, the purpose of immune system is to distinguish foreign material within the body and destroy it similarly as it destroys the infectious pathogens like bacteria and viruses. Moreover, it is possible to reduce the transplant rejection through serotyping by determining the most appropriate donor-recipient match via using the immunosuppressant drugs.

There is a worldwide shortage of organs for clinical implantation. About, 60% of patients die who are waiting for replacement of organs. Moreover, recent advances in understanding the mechanisms of transplant rejection have brought science to a stage where it is reasonable to consider that organs from other species, probably pigs, may soon be engineered to minimise the risk of serious rejection and used as an alternative to human tissues. This may possibly solve the problem of organ shortage. The use of cells or tissues from other species is the other strategy which is being investigated in early clinical trials. It aims to treat the life-threatening and debilitating illnesses such as cancer, diabetes, liver failure and Parkinson's disease. If perfectly proved, vitrification could allow for long-term storage of xenogenic cells, tissues and organs. This will help the society in making the transplant more readily available.

We have learnt in previous chapter that immunology is a very complex and sophisticated area of biology. Now it has become one of the most versatile research tools in biology and medicine, and a powerful weapon in prevention and management of several viral and bacterial diseases. Immunology has been benefited from advancements done in the area of human physiology, biochemistry, medicine, molecular biology (especially protein structure and chemi-stry), computers and a many electronic instruments.

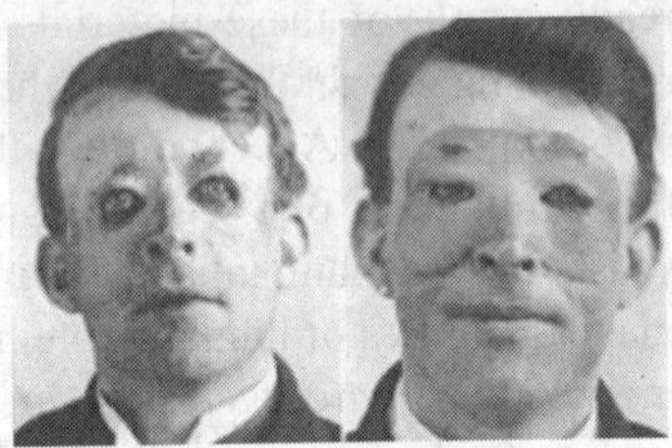

In 1917, a British soldier Walter Yeo was the first person to get benefit from plastic surgery employing tubed pedicle flaps [photos- before operation (left) and after operation (right)].

Undoubtedly, immunology has helped us in eradicating the smallpox worldwide. It has saved million of sufferers from polio, cholera, hepatitis, tetanus, rabies, and several other potentially debilitating or fatal diseases. It is the hope of mankind, to prevent many other diseases such as malaria, tuberculosis, and even certain types of cancer, through appropriate vaccines. The significance of immunology in human is reflected by the award of large number of Nobel Prizes to this area of research (Table 17.1). These award winning discoveries show the mile stones in the development of modern immunology. Moreover, it constitutes more than 10% of all Nobel awards for Physiology or Medicine.

Table 17.1. Nobel prizes awarded for discoveries in immunology/immunotechnology.

Year	*Scientists*	*Discovery*
1908	Ilya I. Mechnikov and Paul Ehrlich	Immunity
1913	Charles R. Richet	Anaphylaxis
1919	Jules Boidet	Immunity related processes
1930	Karl Landsteiner	Human blood groups
1960	Sir F. M. Burnet and Sir Peter B. Medawar	Acquired immunological tolerance
1972	Gerald M Edelman and Rodney R Porter	Chemical structure of antibodies
1980	Baruj Benacerraf, Jean D. and George D. Snell	Genetically determined structures on the cell surface that regulate immunological reactions
1984	Niels K. Jerne, George K.J.F. Kohler and Cesar Milstein	Method for producing monoclonal antibodies
1987	Susuma Tonegawa	Genetic principle for generating antibody diversity
1996	P.C. Doharty and R.M. Zinkernagel	Specificity of cell mediated immune defence

A. HYBRIDOMA TECHNOLOGY (MONOCLONAL ANTIBODIES)

Monoclonal antibodies (MoAb) are produced by one type of immune cell which are all clones of a unique parent cell. Basically these are monospecific antibodies; however, it is possible to create monoclonal antibodies that specifically can bind to a given substance. Therefore, they are useful to detect or purify that substance. Now, this principle has become an important tool in biochemistry, molecular biology and medicine.

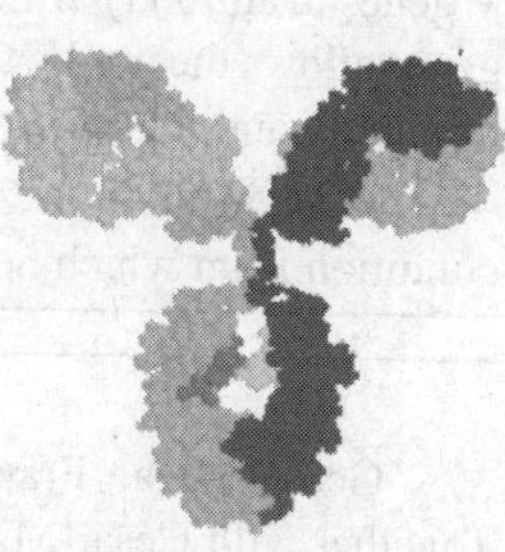

Computer-generated model of a MoAb

In the beginning of the 20th century, Paul Ehrlich first proposed the idea of a 'magic bullet' by postulating the concept that 'if a compound could be made that selectively target a disease-causing organism, a toxin for that organism could be delivered along with the agent of selectivity'.

In the 1970s, it could be understood that the cancerous B-cells (multiple myeloma) produce a single type of antibody, *i.e.* a paraprotein. This was used to study the structure of antibodies, but it was not possible to produce identical antibodies specific to a given antigen. In 1973, Jerrold Schwaber described the production of monoclonal antibodies involving 'human–mouse' hybrid cells (Schwaber and Cohen, 1973). But its claims to priority have been controversial. A science paper on the subject gave some credit to Schwaber for inventing a widely cited technique, but stopped suggesting that he had been cheated. The invention was conceived by Professor Pieczenik along with Professor John Sedat. This technology was discovered in 1975 by two scientists, Georges Kohler of West Germany and Cesal Milstein of Argentina, who jointly with Niels Jerne of Denmark were awarded the 1984 Nobel Prize for Physiology or Medicine. The main idea was: (*i*) to use a myeloma cell line that had lost their ability to secrete antibodies, (*ii*) to fuse these cells with healthy antibody-producing B cells, and finally (*iii*) to select successfully the fused cells.

In 1988, Greg Winter and co-workers pioneered the techniques to humanise monoclonal antibodies by removing the reactions that is caused in some patients by many monoclonal antibodies.

César Milstein

César Milstein (8 October 1927 – 24 March 2002) was an Argentine biochemist. In 1984, he shared the Nobel Prize in Physiology or Medicine with Niels K. Jerne and Georges Köhler. Milstein was borne in Argentina in a Jewish family. He graduated from the University of Buenos Aires and obtained a Ph.D. under Prof. Stoppani (a biochemist) in the Medical School on kinetic studies with the enzyme aldehyde dehydrogenase. In 1958, he joined the Biochemistry Department at the University of Cambridge to work for a Ph.D. under Malcolm Dixon on the mechanism of metal activation of the enzyme phosphoglucomutase. During this work he collaborated with Frederick Sanger whose group he joined with a short-term Medical Research Council appointment.

The major part of Milstein's research was devoted in studying the structure of antibodies and the mechanism by which antibody diversity is generated. In 1975 he together with Georges Köhler, developed the hybridoma technique for the production of monoclonal antibodies, a discovery recognised by the award of the 1984 Nobel Prize for Physiology or Medicine. This discovery led to an enormous expansion in the exploitation of antibodies in science and medicine.

Milstein's early work on antibodies focused on the nature of their diversity at the amino acid level as well as on the disulphide bonds by which they were held together. Part of this work was done in collaboration with his wife, Celia.

The development of the hybridoma technology coupled to advances in nucleic acid sequencing then allowed Milstein to chart the changes that occurred in antibodies following antigen encounter. He demonstrated the importance of somatic hypermutation of immunoglobulin V genes in antibody affinity maturation. In this process, localised mutation of the immunoglobulin genes allows the production of improved antibodies which make a major contribution to protective immunity and immunological memory.

Milstein died on 24 March 2002 in Cambridge, England at age 74 as a result of a heart condition from which he had suffered for many years.

Georges Jean Franz Köhler

Georges Jean Franz Köhler (March 17, 1946 – March 1, 1995) was a German biologist. Together with César Milstein and Niels K. Jerne, Köhler won the Nobel Prize in Physiology or Medicine in 1984, for work on the immune system and the production of monoclonal antibodies. A portion of this research was performed at the Basel Institute for Immunology. In 1984 he became the director of the Max Planck Institute for Immunobiology where he worked until his death.

1. Hybridoma Technology (Production of Monoclonal Antibodies)

Hybridoma technology is the technique for production of monoclonal antibodies in specialised cells. A rabbit is immunised through repeated injection of a specific antigen for the production of specific antibody, facilitated due to proliferation of the desired B cells (Fig. 17.1). Tumours are produced in a mouse or a rabbit. Mouse spleen cells (spleen cells are rich in B cells and T cells) producing specific antibodies, and myeloma cells producing tumours are cultured separately. The myeloma cell line used is unusual in two ways: it has stopped synthesising antibodies and it is a mutant called HGPRT$^-$ that can not synthesise the enzyme hypoxanthine-guanine phosphoribosyl transferase (HGPRT).

By using polyethylene glycol (PEG), fusion of spleen cells to myeloma cells was induced to produce hybridoma. The hybrid cells are grown in selective hypoxanthine aminopterin thymidine (HAT) medium. HAT medium contains a drug aminopterin (a folic acid analogue which inhibits dihydrofolate reductase, DHFR), which blocks one pathway for nucleotide synthesis making the cells dependent on another pathway that needs HGPRT enzyme absent in myeloma cells. Therefore,

HGPRT⁻ myeloma cells die which do not fuse with B cells because they are mutant. B-cells also will die that do not fuse because they lack tumourigenic property of immortal growth. Therefore, hybridoma cells are selected by using HAT medium which have inherited the HGPRT gene from B-cells and tumourigenic property from myeloma cells.

What is Hybridoma ?

The term hybridoma is applied to fused cells produced through the fusion of following two types of cells :

(*i*) an antibody producing lymphocyte cell (e.g. a spleen cell of mouse immunised with red blood cells from sheep), and

(*ii*) a single myeloma cell (bone marrow tumour cell) which is capable of multiplying indefinitely.

These fused hybrid cells or hybridoma have the antibody producing capability inherited from lymphocytes and have the ability to grow continuously (immortal) like malignant cancer cells.

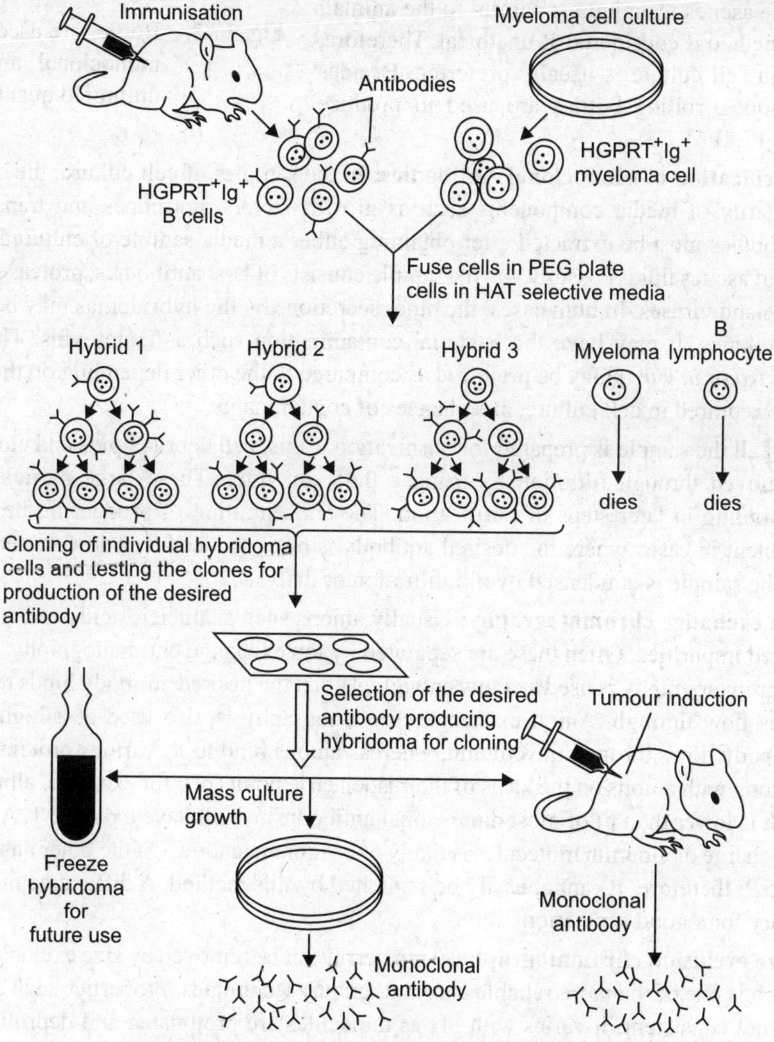

Fig. 17.1. Production of monoclonal antibodies.

The desired hybridoma is selected for cloning and antibody production. This is facilitated by preparing single cell colonies that will grow and can be used for screening of antibody producing hybridomas. Only one cell in several hundred cell hybrids will produce antibodies of the desired specificity. Then the selected hybridoma cells are cultured for the production of monoclonal antibodies in large quantity. These hybridoma cells may be frozen for future use and may also be injected in the body of an animal to produce antibodies which can be recovered from the body fluid later on.

The hybridomas can be grown indefinitely in a suitable cell culture media, or they can be injected into mice (in the peritoneal cavity *i.e.* the gut). Consequently, they produce tumours containing an antibody-rich fluid called 'ascites fluid'. During selection, the medium must be enriched so that they can further favour the growth hybridoma. This can be achieved by using a layer of feeder fibrocyte cells or supplement medium such as briclone. The ascites technique is painful to the animal; hence, this method is considered as unethical. Therefore, production in cell culture is usually preferred. Besides other techniques, rolling bottles are used to produce MoAb (Fig. 17.2).

Fig. 17.2. Bottles are used to produce monoclonal antibodies in unlimited quantities.

(*a*) Purification of Monoclonal Antibodies: In the samples of cell culture, the contaminants consist primarily of media components such as growth factors, hormones and transferrins. The desired antibodies must be extracted after obtaining either a media sample of cultured hybridomas or a sample of ascites fluid. Possibly, *in vivo* sample consists of host antibodies, proteases, nucleases, nucleic acids and viruses. In both cases, the other secretions by the hybridomas may be present, for example cytokines. It may have the bacterial contamination such as endotoxins. Therefore, one method (*in vivo* or *in vitro*) may be preferred as compared to the other depending on the complexity of the media required in cell culture, and chances of contaminants.

First of all the sample is prepared for purification. Cells, cell debris, lipids and clotted material are first removed through filtration by using a 0.45 μm filter. These large particles can cause membrane fouling in later steps of purification. The concentration of product in the sample may not be sufficient in cases where the desired antibody is one produced by a low-secreting cell line. Therefore, the sample is condensed by ultrafiltration or dialysis.

(*i*) Ion exchange chromatography: Usually anions such as nucleic acids and endotoxins act as the charged impurities. Often these are separated by ion exchange chromatography. Either cation exchange chromatography is used at a low enough *p*H that the desired antibody binds to the column, while anions flow through. Anion exchange chromatography is also used at a high *p*H that the desired antibody flows through the column, whereas anions bind to it. Various proteins can also be separated along with anions on the basis of their isoelectric point (pI); for example, albumin has a pI of 4.8 which is lower than pI of most monoclonal antibodies which have a pI of 6.1. At a given pH, the average charge of albumin molecules is likely to be more negative. On the other hand transferrin, has a pI of 5.9; therefore, it cannot easily be separated by this method. A difference in pI of at least 1 is necessary for a good separation.

(*ii*) Size exclusion chromatography: Transferrin can be removed by size exclusion chromatography which is one of the more reliable chromatography techniques. Properties such as charge and affinity are not consistent; it varies with *p*H as molecules are protonated and deprotonated. It has drawbacks such as low resolution, low capacity and low elution times.

***(iii)* Protein A/G affinity chromatography:** Protein A/G affinity chromatography is a much quicker, single-step method of separation. The antibody selectively binds to protein A/G; therefore, a high level of purity (generally >80%) is achieved. This method may be problematic for antibodies that are easily damaged, as harsh conditions are generally used. The bonds can be broken at a low *p*H to remove the antibody from the column. Low *p*H can cause protein A/G itself to leak off the column and appear in the eluted sample. Gentle elution buffer systems which employ high salt concentrations are also available to avoid exposing sensitive antibodies to low *p*H. Since, immobilised protein A/G is a more expensive resin, cost is also considered with this method.

***(iv)* Affinity purification:** In affinity purification an antigen is used to provide exquisite specificity for the antibody so that maximum purity can be achieved in a single step. In this method, the antigen is covalently attached to an agarose support to generate the antibody. If the antigen is a peptide, it is commonly synthesized with a terminal cysteine which allows the selective attachment to a carrier protein during development and to the support for purification. The antibody-containing medium is then incubated with the immobilized antigen, either in batch or as the antibody is passed through a column, where it selectively binds and can be retained while impurities are washed away. An elution with a low *p*H buffer or high salt buffer is done to recover purified antibody from the support.

For further selection, the antibodies can be precipitated using sodium sulfate or ammonium sulfate. Antibodies precipitate at low concentrations of the salt, whereas most of the other proteins precipitate at higher concentrations. Suitable concentration of salt is added to achieve the best separation. Then excess salt is removed by a desalting method *e.g.* dialysis. Finally, the purity of antibody is analysed by using a chromatogram. Any impurities will produce peaks, and the volume under the peak indicates the amount of the impurity. Alternatively, gel electrophoresis and capillary electrophoresis are carried out. Bands of impurities in varying intensity are formed on the basis of amount of impurities present.

***(b)* Production of Recombinant Monoclonal Antibodies:** Recombinant monoclonal antibodies are produced by involving the technologies called 'repertoire cloning' or 'phage display' or 'yeast display'. By using viruses or yeast recombinant antibody engineering is carried out to produce antibodies rather than mice. These techniques depends on rapid cloning of immunoglobulin gene segments to prepare libraries of antibodies with slightly different amino acid sequences from which antibodies with desired specificities can be selected. The phage antibody libraries are a variant of the phage antigen libraries first invented by Professor Pieczenik (Schwaber and Cohen, 1973). These techniques are used to enhance the specificity with which antibodies recognise antigens, their stability in various environmental conditions, their therapeutic efficacy, and their delectability in diagnostic applications. For the production on a large scale, fermentation chambers have been used to produce these antibodies.

Phage Display Library

Phage display library is the construction of V_H and V_L gene libraries and expression of them on a filamentous bacterophage. The phage expresses an antigen-binding domain specific for a particular antigen to screen the MoAbs.

2. Applications of Monoclonal Antibodies

Monoclonal antibodies are used in many areas as disussed below:

***(a)* Diagnostic Tests:** Once monoclonal antibodies for a given substance have been produced, they can be used to detect the presence of this substance. The protein present on a membrane is detected by the Western blot test and immuno dot blot tests. Also these are used in immunohistochemistry to detect antigen in fixed tissue sections. The substance present in a frozen tissue section or in live cells is detected by the immunofluorescence test.

(*i*) **In diagnostic histopathology:** With the help of monoclonal antibodies, tissues and organs can be classified based on their expression of certain defined markers, which reflect tissue or cellular genesis. Prostate specific antigen, placental alkaline phosphatase, human chorionic gonadotrophin, α-fetoprotein and others are organ-associated antigens. Production of monoclonal antibodies against these antigens helps in determining the nature of a primary tumour. Monoclonal antibodies are especially useful in distinguishing morphologically similar lesions like mesothelioma and adenocarcinoma, and in the determination of the organ or tissue origin of undifferentiated metastases. Selected monoclonal antibodies help in the detection of occult metastases by immuno-cytological analysis of bone marrow, other tissue aspirates, as well as lymph nodes and other tissues (Nelson *et al.*, 2000).

A study was done on a sensitive immuno-histochemical assay on bone marrow aspirates of 20 patients with localised prostate cancer. Three monoclonal antibodies (T16, C26, and AE-1) were used in the assays which were capable to recognise the membrane and cytoskeletal antigens expressed by epithelial cells for detecting the tumour cells. Bone marrow aspirates of 22% of patients with localised prostate cancer (stage B, 0/5; Stage C, 2/4), and 36% patients with metastatic prostate cancer (Stage D1, 0/7 patients; Stage D2, 4/4 patients) had antigen-positive cells in their bone marrow. The immuno-histochemical staining of bone marrow aspirates is very useful to detect occult bone marrow metastases in patients with apparently localised prostate cancer (Bretton *et al.*, 1994).

Although immuno-cytochemistry using tumour-associated monoclonal antibodies has led to an improved ability to detect occult breast cancer cells in bone marrow aspirates and peripheral blood, further development of this method is necessary before it can be used routinely. One major drawback of immuno-cytochemistry is that only tumour-associated and non-tumour specific monoclonal antibodies are used; consequently some cross-reaction with normal cells can occur (Kvalheim, 1998).

(*b*) **Monoclonal Antibodies for Cancer Treatment:** Monoclonal antibodies are used as one of the possible treatments for cancer which involves the binding to only cancer cell-specific antigens. The monoclonal antibodies induce an immunological response against the target cancer cell. A toxin, radioisotope, cytokine or other active conjugate can be delivered by such modified MoAb. However, it is also possible to design bispecific antibodies that can bind with their Fab regions both to target antigen and to a conjugate or effector cell. Obviously, every intact antibody can bind to cell receptors or other proteins with its Fc region.

There are four major types of antibody which have been developed as a result of progress of monoclonal drug development *i.e* murine MoAb, chimeric MoAb, humanised- MoAb and human MoAb.

Types of MoAb Designed

There are four types of monoclonal antibodies designed as given below:

- **Murine source MoAb:** Rodent MoAbs with excellent affinities and specificities generated using conventional hydrioma technology. Clinical efficacy compromised by HAMA (human anti-murine antibody) response lead to allergic or immune complex herpersensitivities.
- **Chimeric MoAb:** Chimeras combine the human constant regions with the intact rodent variable regions. Affinity and specificity are unchanged. Also it causes human anti-chimeric antibody response (30% murine resource).
- **Humanised MoAb:** It contains only the CDRs of the rodent variable region grafted onto human variable region framework.
- **Human MoAb:** They are produced by transferring human immunoglobulin genes into murine genome resulting in the production of fully human antibodies

***(i)* Murine monoclonal antibodies (*suffix* omab):** Kohler and Milstein received a Nobel Prize for producing murine antibodies by hybridoma technology. The dissimilarity between murine and human immune systems resulted in the clinical failure of these antibodies with some exceptions. The major problems associated with murine antibodies are: reduced stimulation of cytotoxicity, and formation of complexes after repeated administration which results in mild allergic reactions and anaphylactic shock.

***(ii)* Chimeric monoclonal antibodies (*suffix* ximab) and Humanised monoclonal antibodies (*suffix* zumab):** The standard procedure of production of monoclonal antibodies yields mouse antibodies. In spite of many similarities between murine antibodies and human antibodies there are some differences also. Hence the human immune system recognises mouse antibodies as foreign particle; therefore, they rapidly remove them from circulation and cause systemic inflammatory effects. Such responses are recognised as producing 'human anti-chimeric antibodies' (HACA) or 'human anti-mouse antibodies' (HAMA).

Therefore, this problem can be overcome if the human antibodies can directly be generated from humans. Primarily this is not easy because it is generally not seen as ethical to challenge humans with antigen to produce antibody. It is not easy to generate human antibodies against human tissues. However, the ethics of doing the same to non-humans is a matter of debate.

Since the late 1980s, various attempts have been made to overcome this problem by using recombinant DNA technology. The DNA that encodes the binding portion of monoclonal mouse antibodies is taken and merged with human antibody-producing DNA. Then the mammalian cell cultures are used to express this DNA and to produce these half-mouse and half-human antibodies. Since bacteria produce this kind of glycoprotein, they cannot be used for this purpose. On the basis of how big a part of the mouse antibody is used, it is called chimeric antibodies or humanised antibodies.

The murine molecules are engineered to remove immunogenic content and to increase their immunologic efficiency. Initially this was produced through the production of chimeric and humanised antibodies. Chimeric antibodies are composed of murine variable regions fused with human constant regions. Human gene sequences (obtained from kappa light chain and the IgG1 heavy chain) results in antibodies which are about 65% human. This reduces immunogenicity and increases half-life of serum.

Humanised antibodies are produced upon grafting of murine hypervariable amino acid domains into human antibodies. This results in a molecule of about 95% human origin. It has been found that humanised antibodies bind to antigen more weakly than the parent murine monoclonal antibody. This results in decrease in several hundredfold affinity. Increases in strength of antibody-antigen binding have been obtained by bringing about mutations in the complementarity determining regions (CDR) by using chain-shuffling, randomisation of complementarity determining regions, and generation of antibody libraries with mutations within the variable regions by error-prone PCR, *E. coli* mutator strains and site-specific mutagenesis.

***(iii)* Human Monoclonal Antibodies (*suffix* mumab):** After the discovery that monoclonal antibodies could be generated *in vitro*, scientists have targeted to create the 'fully' human monoclonal antibodies so that some of the side effects of humanised and chimeric antibodies could be avoided. Two successful attempts were made: phage display-generated antibodies and genetically engineered mice to produce more human-like antibodies.

Human monoclonal antibodies are produced by using transgenic mice or phage display libraries. These are produced by transferring human immunoglobulin genes into the murine genome, after which the transgenic mouse is vaccinated against the desired antigen. This results in the production of monoclonal antibodies allowing the transformation of murine antibodies *in vitro* into fully human antibodies (Janeway *et al.,* 2005).

Cambridge Antibody Technology (CAT) was one of the most successful commercial organisations behind therapeutic monoclonal antibodies. At CAT scientists showed that phage display could be used such that variable antibody domains could be expressed on filamentous phage antibodies (McCaffertyet *et al*., 1990). Further CAT developed their display technologies into several patented antibody discovery and/or functional genomics tools which were named Proximol^{TM} and ProAb^{TM}. ProAb was announced in December 1997 and involved high throughput screening of antibody libraries against diseased and non-diseased tissue, whereas Proximol used a free radical enzymatic reaction to label molecules in proximity to a given protein.

Transgenic mice can be modified to produce human antibodies. There are a number of commercial organisations that has exploited this approach, for example: Medarex who market their UltiMab platform, Abgenix who marketed their Xenomouse technology, Abgenix were acquired in April 2006 by Amgen, Regeneron's VelocImmune technology, etc. Monoclonal antibodies have been generated and approved to treat: cancer, cardiovascular disease, inflammatory diseases, macular degeneration, transplant rejection, multiple sclerosis, and viral infection (see monoclonal antibody therapy).

In 2006 the Pharmaceutical Research and Manufacturers of America reported that U.S. companies had 160 different monoclonal antibodies in clinical trials or awaiting approval by the Food and Drug Administration (Table 17.2) (Anonymous, 2006).

Table 17.2. Example of some of monoclonal antibodies in clinical trials (Anonymous, 2006).

Monoclonal antibodies			
Type	***Mode***	***Mechanism***	***Application***
Abciximab	Chimeric	Inhibits the receptor GpIIb/IIIa on platelets	Prevent coagulation in coronary angioplasty
Aemtuzumab	Humanized	Targets an antigen CD52 on T- and B-lymphocytes	B cell leukemia
Adalimumab	Human	Inhibits TNF-α	Rheumatoid arthritis, Crohn's disease, Ulcerative Colitis
Basiliximab	Chimeric	Inhibits IL-2 on activated T cells	Acute rejection of kidney transplants
Bevacizumab	Humanized	Inhibits VEGF	Anti-angiogenic cancer therapy
Daclizumab	Humanized	Inhibits IL-2 on activated T cells	Acute rejection of kidney transplants
Etanercept	Fusion protein	Contains TNF receptor	Rheumatoid arthritis
Gemtuzumab	Humanized	Targets myeloid cell surface antigen CD33 on leukemia cells	Relapsed acute myeloid leukaemia
Infliximab	Chimeric	Inhibits TNF-α	Rheumatoid arthritis, Crohn's disease, Ulcerative Colitis
Rituximab	Chimeric	Targets phosphoprotein CD20 on B lymphocytes	Non-Hodgkin's lymphoma
Palivizumab	Humanized	Inhibits an RSV fusion (F) protein	RSV infections in children
Trastuzumab	Humanized	Targets the HER2/neu (erbB2) receptor	Anti-cancer therapy for a specific kind of breast cancer

3. Monoclonal Antibody Therapy

In 2003, T.A. Waldmann has described the past, present and future of immunotherapy. The use of monoclonal antibodies (MoAb) to specifically bind to target cells is called 'monoclonal antibody therapy'. Then it stimulates the immune system of patients by attack the target cells. Possibly a MoAb specific to almost any extracellular target can be created. A large amount of research and development currently are in progress to create monoclonals for many serious diseases such as multiple sclerosis, rheumatoid arthritis and cancers. MoAbs can be therapeutically used in many ways, *e.g.* MoAb therapy can be used to destroy malignant tumour cells and to inhibit the growth of tumour by blocking the specific cell receptors. Variations also exist within this treatment *e.g.* radioimmunotherapy where a radioactive dose localises on target cell line delivering lethal chemical doses to the target.

(*a*) Origins of Monoclonal Antibody Therapy: Immunotherapy was developed with the discovery of the structure of antibodies and the development of hybridoma technology. The first reliable source of monoclonal antibodies was provided by hybridoma technology. Further advances in these techniques allowed the specific targeting of tumours both *in vitro* and *in vivo*.

Initial therapeutic antibodies were simple murine analogues. These antibodies have a short half-life *in vivo*, limited penetration into tumour sites, and inadequate recruitment of host effector functions. In modern therapeutic antibody applications, chimeric and humanised antibodies have generally replaced murine antibodies. Hybridoma technology has been replaced by recombinant DNA technology, transgenic mice and phage display.

(*b*) Function of Human and Therapeutic Antibodies: The immune system responds according to the environmental factors it encounters on the basis of discrimination between self and non-self. Since tumour cells are the patient's own cells, one's own immune system specifically does not target the tumour cells. Tumour cells are highly abnormal and most of them produce unusual antigens that are either inappropriate for the cell type or are only normally present during the development of organisms *e.g.* fetal antigens (Janeway *et al.,* 2005).

Some of the tumour cells have cell surface receptors that are rare or absent on the surfaces of healthy cells. These cell surface receptors are responsible for activation of cellular signal transduction pathways that cause the unregulated growth and division of the tumour cell. For example ErbB2 is an active cell surface receptor that is abnormally produced at high levels on the surface of about 30% of tumour cells causing breast cancer. Such breast cancer is known a HER2 positive breast cancer. Antibodies play a central role in both in the recognition of foreign antigens and the stimulation of an immune response to them. Therefore, they are a key component of the adaptive immune response. The advent of monoclonal antibody technology has made it possible to raise antibodies against specific antigens presented on the surfaces of tumours (Janeway *et al.,* 2005).

4. Mechanism of Action

The therapeutic effects of monoclonal antibodies are achieved by various mechanisms. They directly can cause apoptosis or programmed cell death. They can block growth factor receptors arresting the proliferation of tumour cells. They can bring about anti-idiotype antibody formation in cells that express monoclonal antibodies.

Indirect effects of monoclonal antibodies are the recruitment of cells that have cytotoxicity, for example monocytes and macrophages. This type of antibody-mediated cell killing is called antibody-dependent cell mediated cytotoxicity (ADCC). Monoclonal antibodies also bind complement, leading to direct cell toxicity which is called complement dependent cytotoxicity (CDC) (Fig. 17.3).

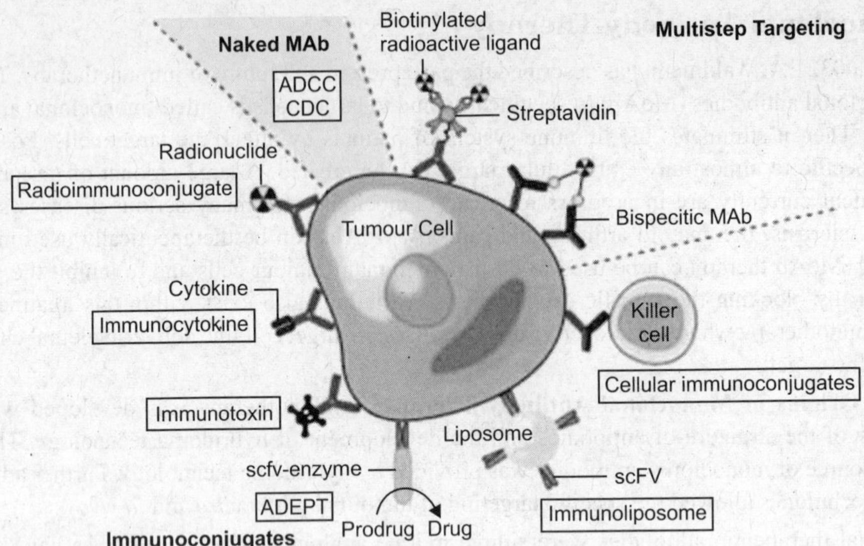

Fig. 17.3. Monoclonal antibodies for cancer treatment. ADEPT- antibody directed enzyme prodrug therapy; ADCC- antibody dependent cell-mediated cytotoxicity; CDC- complement dependent cytotoxicity; MoAb- monoclonal antibody; scFv- single-chain Fv fragment (modified from Carter, 2001).

Human antibodies especially IgG_1 and IgG_3 direct the killing of tumour cells by antibody-dependent cellular cytotoxicity (ADCC) or complement-dependent cytotoxicity (CDC). ADCC is triggered by an interaction between the Fc region of an antibody (that has bound to a tumour cell through its antigen-binding region) and the Fcg receptors (Fcg Rs) particularly Fcg RI and Fcg RIII (on immune effector cells such as neutrophils, macrophages and natural killer cells) (Fig. 17.4).

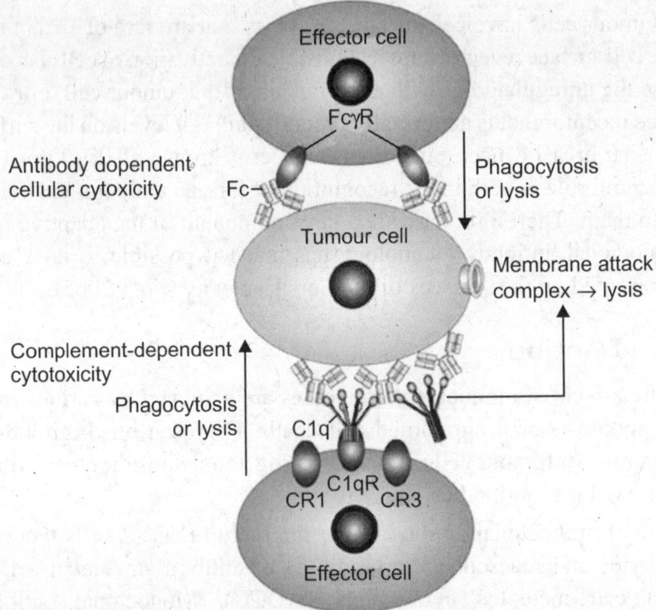

Fig. 17.4. Mechanism of antibody-dependent cellular cytotoxicity (ADCC) and complement-dependent cytotoxicity (CDC).

Depending on the type of mediating effector cell, the tumour cell is eliminated by phagocytosis or lysis. The first step for CDC is the recruitment of complement component C1q by IgG bound to the tumour cell surface. This triggers a proteolytic cascade to activate complement. This can lead to the formation of a membrane attack complex that kills the target cell by disrupting its cell membrane. Tumour-cell-bound C1q can bind to complement receptors such as C1qR, CR1 (CD35) and CR3 (CD11b/CD18) on effector cells such as neutrophils, macrophages and natural killer cells. This can trigger cell-mediated tumour-cell lysis or phagocytosis depending on the type of effector cell.

5. Antibody-directed Enzyme Prodrug Therapy (ADEPT)

Fig. 17.3 shows the ADEPT involves the application of cancer associated monoclonal antibodies. The MoAbs are linked to a drug-activating enzyme. After systemic administration, a non-toxic agent results in its conversion to a toxic drug that poses a cytotoxic effect targeting at malignant cells. So far the clinical success of ADEPT treatments is limited. However, it holds a great promise in playing a key role in oncological treatment in future.

6. Drug and Gene Therapy (Immuno-liposomes)

Antibody-conjugated liposomes are called 'immunoliposomes'. Liposomes can carry drugs or therapeutic nucleotides. When conjugated with monoclonal antibodies, they may be directed against malignant cells. Still this technique is in its infancy. *In vivo* immunoliposomes have been successfully used to achieve the targeted delivery of tumour-suppressing genes into tumours by using an antibody fragment against the human transferrin receptor. Tissue-specific gene delivery by using immunoliposomes has also been achieved in the tissues of brain and breast cancer.

7. Approved Therapeutic Antibodies

In 1986, the first FDA-approved therapeutic monoclonal antibody was a murine IgG2a CD3 specific transplant rejection drug, OKT3 that is muromonab. This drug is used in recipients of solid organ transplant who became steroid resistant. There are hundreds of therapies which are under clinical trials; however, most of therapies are concerned with immunological and oncological targets (Table 17.3).

Table 17.3. FDA approved therapeutic monoclonal antibodies (source: T. A. Waldmann, 2003, *Nature Med.*9: 269).

Antibody	*Brand name*	*Year*	*Type*	*Target*	*Treatment*
Abciximab	ReoPro	1994	Chimeric	Inhibition of glycoprotein IIb/IIIa	Cardiovascular disease
Adalimumab	Humira	2002	Human	Inhibition of TNF-α signaling	Several auto-immune disorders
Alemtuzumab	Campath	2001	Humanised	CD52	Chronic lymphocytic leukemia
Basiliximab	Simulect	1998	Chimeric	IL-2Rα receptor (CD25)	Transplant rejection
Bevacizumab	Avastin	2004	Humanised	Vascular endothelial growth factor (VEGF)	Colorectal cancer
Cetuximab	Erbitux	2004	Chimeric	Epidermal growth factor receptor	Colorectal cancer, Head and neck cancer

Certolizuma pegol	Cimzia	2008	Humanised	Inhibition of TNF-α signaling	Crohn's disease
Daclizumab	Zenapax	1997	Humanised	IL-2Rα receptor (CD25)	Transplant rejection
Eculizumab	Soliris	2007	Humanised	Complement system protein C5	Paroxysmal nocturnal hemoglobinuria
Efalizumab	Raptiva	2002	Humanised	CD11a	Psoriasis
Infliximab	Remicade	1998	Chimeric	Inhibition of TNF-α signaling	Several autoimmune disorders
Muromonab CD3	Orthoclone OKT3	1986	Murine	T cell CD3 Receptor	Transplant rejection
Natalizumab	Tysabri	2006	Humanised	Alpha-4 (α4) integrin	Multiple sclerosis and Crohn's disease
Omalizumab	Xolair	2004	Humanised	Immunoglobulin E (IgE)	Mainly allergy-related asthma
Palivizumab	Synagis	1998	Humanised	An epitope of the RSV F protein	Respiratory syncytial Virus
Panitumumab	Vectibix	2006	Human	Epidermal growth factor receptor	Colorectal cancer
Ranibizumab	Lucentis	2006	Humanised	Vascular endothelial growth factor A (VEGF-A)	Macular degeneration
Rituximab	Rituxan, Mabthera	1997	Chimeric	CD20	Non-Hodgkin lymphoma
Tositumomab	Bexxar	2003	Murine	CD20	Non-Hodgkin lymphoma
Trastuzumab	Herceptin	1998	Humanised	ErbB2	Breast cancer

8. Problems with Successful Therapy

Successful therapy with monoclonal antibodies is associated with a number of problems. The antigen of malignant cells is heterogeneously distributed. Therefore, some cells express tumour antigens, whereas the others do not. Moreover, flow of tumour blood is not always optimal. If monoclonal antibodies are delivered through the blood, it may be difficult to get the therapy at the site.

Within the tumour high interstitial pressure prevents the passive monoclonal antibodies from binding. Sometimes, tumor antigen is even released, hence the antibody binds to a free-floating antigen and not the tumor cell takes place. Since monoclonal antibodies are derived from mouse cell lines, there is possibility for existence of an immune response to the antibodies. This response decreases the efficacy of monoclonal antibody therapy and eliminates the possibility of re-treatment. Very rarely there occurs cross-reactivity with normal tissue antigens. The general target antigens that are not cross reactive with normal tissue antigens are chosen. In spite of these problems, there has been tremendous success in application of monoclonal antibodies in haematologic malignancies and solid tumours.

B. INTERFERONS

In 1957, Briton Alick Isaacs and Swiss Jean Lindemann both from London discovered interferon. When they were analysing the effects of viral infection on cells in a tissue culture, they found that already virus-infected cells became resistant to infection by the other viruses for certain period of time. The first infection was found to *interfere* (inhibit) with the second infection. Therefore, the inhibitory protein isolated from these infected cells (which was absent in uninfected cells) was termed as 'interferon' (IFN).

Interferons are a group of proteins belonging to the large class of glycoproteins known as cytokines. INF is known primarily for their role in inhibiting viral infections and in stimulating the entire immune system to fight disease in response to infection by pathogens. In contrast to antibodies which are virus specific, interferons are not virus-specific but host specific. Thus, viral infections of human cells are inhibited only by human interferon (Janeway *et al.,* 2005).

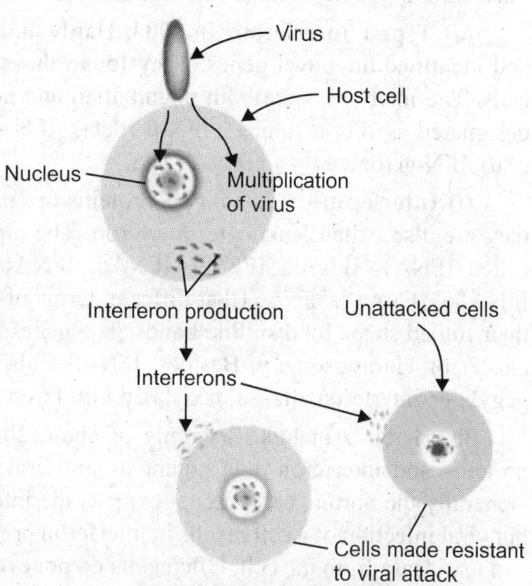

Fig. 17.5. Induction and action of interferon.

In 2007, S. Pestka reviewed 50 years after the discovery of interferons. Until 1978, purification of interferons did not occur. A series of publications from the laboratories of Sidney Pestka and Alan Waldman between 1978 and 1981, describe the purification of the type I interferons IFN-α and IFN-β.). By the early 1980s, the genes for these interferons were cloned. It was proved that interferons really were responsible for interfering with viral replication. In 1980, Nagata and coworkers reported the synthesis in *E. coli* of a polypeptide with human leukocyte interferon activity. They varied out gene cloning experiment and confirmed that IFN-α was encoded by not one gene but a family of related genes.

Now it is known when the body is exposed to attack by viruses, other microorganisms or tumor cells these inhibitory proteinaceous substances are produced by white blood cells (leucocytes and T-cells) as part of the body's natural immune response. After its production the interferon diffuses into uninfected neighbouring cells and induces these cells to produce another antiviral protein which makes these cells resistant to viral infections. As a result, the concerned person exhibit increased resistance to viral infections. Thus interferon induces antiviral state in non-infected cells. Interferon is considered to be the body's first line of defense against infection by viruses.

Interferons are part of the non-specific immune system. They are induced at an early stage in viral infection, *i.e.* before the specific immune system has responded. Thus interferons play an important role in the first line of defense against viral infections. It is known to decrease the virulence in chicken pox, rabies, rubella, eye infections, etc.

1. Types of Interferons

Interferons are classified primarily through their amino acid sequence. Interferon-alpha, -beta, -tau, and -omega, which have relatively similar amino acid sequences, are classified as Type I interferons. Type I interferons are known primarily for their ability to make cells resistant to viral infections. Interferon gamma is the only Type II interferon, classified as such because of its unique amino acid sequence. This interferon is known for its ability to regulate overall immune system functioning.

In addition to their structural makeup, type I and type II interferons have other differences. Type I interferons are produced by almost every cell in the body, while the type II interferon, *i.e.* interferon-gamma is produced only by specialised cells in the immune system. The two classes also bind to different kinds of receptors which lie on the surface of cells and attract and combine with specific molecules of various substances.

On the basis of their structure, these inhibitory proteins are be classified into three groups of interferon: alpha (α), beta (β) and gamma (γ).

***(a)* Type I Interferon:** In 2004, Hardy and coworkers characterised type I interferon locus and identified the novel genes. They found that type I interferons are produced by virus-infected cells. The interferons are both mammalian and non-mammalian types. The mammalian types are designated as IFN-α (alpha), IFN-β (beta), IFN-κ (kappa), IFN-δ (delta), IFN-ε (epsilon), IFN-τ (tau), IFN-ω (omega) and IFN-ζ (zeta).

***(i)* Interferon-α:** The IFN-α proteins are produced by virus-infected leukocytes; therefore, they are also called leukocyte interferon. The alpha group alone consists of at least 15 subtypes called IFNA1, IFNA2, IFNA4, IFNA5, IFNA6, IFNA7, IFNA8, IFNA10, IFNA13, IFNA14, IFNA16, IFNA17, IFNA21that differ in terms of their amino acid sequence and are maintained in their folded shape by disulfide bonds. The genes for these IFN-α molecule are found together in a cluster on chromosome 9. Besides, IFN-α is also made synthetically as medication, for example pegylated interferon alfa-2a, pegylated interferon alfa-2b, etc.

Interferon-α includes a family of about 20 related proteins and interferon-β is potent as antiviral agents. Generally the normal cells do not express the interferon, but viral infection of a cell results in interferon production and its release from the cell. Often cells do not eventually die due to the infection. The interferon induces an antiviral state after binding to target cells. Both DNA viruses and RNA viruses induce interferon but RNA viruses induce higher levels of interferons than the DNA viruses. During viral infection double-stranded (ds)RNA are produced. This dsRNA may be an important inducing agent for interferon production. The other stimuli will also cause the interferon induction and their production *e.g.* exogenous dsRNA, lipopolysaccharide, other components of certain bacteria.

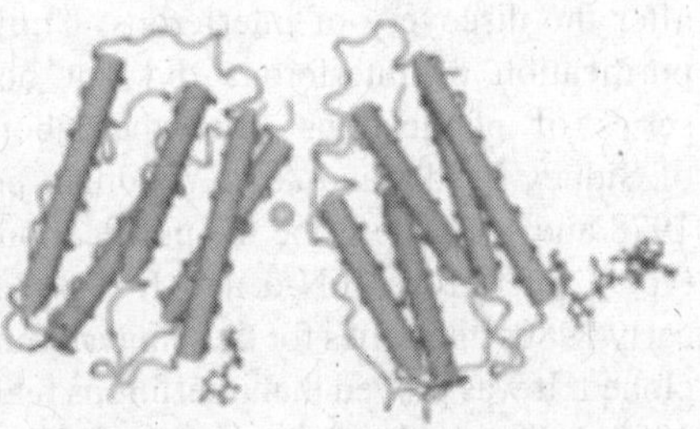

Structure of IFN- β

Interferon α2a is a protein consisting of 165 amino acids without a glucose unit that is maintained in its three-dimensional loop structure by two disulfide bridges.

***(ii)* Interferon-β:** The IFN-β proteins are produced in large quantities by fibroblasts or virus-infected epithelial cells; hence it is also called 'fibroblast interferon'. They have antiviral activity mainly involved in innate immune response. There are two types of IFN-β which have been described: IFN-β1 (IFNB1) and IFN-β3 (IFNB3) (a gene designated IFN-β2 is actually IL-6).

***(iii)* IFN-ε (epsilon), IFN-κ (kappa), IFN-τ (tau), IFN-δ (delta), IFN-ζ (zeta, also known as limitin):** IFN-ε, –κ, -τ, and –ζ appear to come in a single isoform in humans, *IFNK*. Only ruminants encode IFN-τ, a variant of IFN-ω. So far, IFN-ζ is found only in mice, while a structural homolog, IFN-δ is found in a diverse array of non-primate and non-rodent placental mammals. Most of the placental mammals encode functional IFN-ε and IFN-κ genes.

***(iv)* IFN-ω:** IFN-ω has only one functional form described to date (*IFNW1*). Moreover, it has several pseudo-genes (*IFNWP2*, *IFNWP4*, *IFNWP5*, *IFNWP9*, *IFNWP15*, *IFNWP18* and *IFNWP19*) in humans. Many non-primate placental mammals express multiple IFN-ω subtypes.

***(v)* IFN-ν:** This subtype of Type I IFN was recently described as a pseudo-gene in human but potentially functional in the domestic cat genome. In all other genomes of non-feline placental mammals, IFN-ν is a pseudo-gene. In some species the pseudo-gene is well preserved, whereas in others it is badly mutilated or is undetectable. In cat genome, the IFN-ν promoter is deleteriously mutated. It is likely that the IFN-ν gene family was rendered useless prior to mammalian diversification. Its presence on the edge of the Type I IFN locus in mammals may have shielded it from obliteration resulting in its detection.

Viral replication stimulates the infected host cell to produce Type I interferons (Fig. 17.6A). Many viruses produce dsRNA as a part of their life cycle. In response to infection, immune-activated cells or virus-infected cells produce interferons, which upon binding to uninfected neighboring cells exert their antiviral activity and induce them to produce enzymes that degrade mRNA (B). Interferon combines to interferon receptor protein present on cell membrane of host cell, inhibits translation of viral mRNA into viral protein and eventually kills the host cell. Body's defense against viruses is also enhanced by interferons by enhancing the activities of CTLs, macrophages, NK cells, and antibody-producing cells.

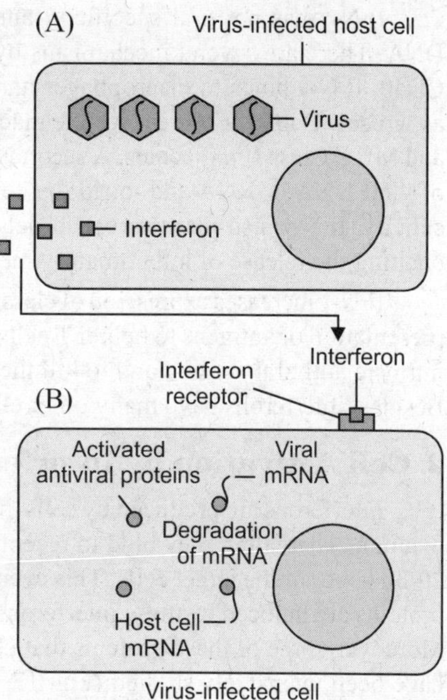

Fig. 17.6. Antiviral action of interferon. A, viral replication stimulating interferon production; B, induction for production of viral mRNA degrading protein.

(*b*) Type II Interferon: Type II interferons are produced by certain activated cells and antigens. It is only one type called 'interferon-γ' or 'immune interferon'. IFN-γ is produced by certain activated cytotoxic T cells, NK cells, dendritic cells, progenitor Th0 cells and Th1 cells. IFN-γ has been located on the human chromosome 12q15. Interferon-γ is produced in response to antigen (including viral antigens) or mitogen stimulation of lymphocytes.

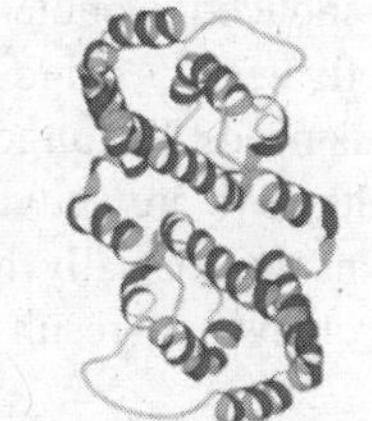

Structure of IFN-γ

As a cytokine, IFN-γ is a highly versatile homodimer glycoprotein formed by antiparallel association of two 21-24kD subunits. Each subunit has 6 α-helices held together by short non-helical sequences. There are no ß-sheets. The subunits have a flattened elliptical shape, whereas the overall structure of the dimmer is globular.

INF-γ plays an essential role in cell-mediated immune responses to viral and mycobacterial infections. The IFN-γ gene codes for a 17 kDa protein, which then undergoes post-translational glycolysis converting the IFN-γ to a 20-25 kDa glycoprotein (http://www.anticancer.net/resan/gamma_interferon.html accessed 2000 Feb 19). The molecule that binds to the IFN-γ receptor is a homodimer and has an intricate degree of linkage between the two subunits. Each subunit contains six alpha helices and low amounts of beta pleated sheets. It is a secondary function of IFN-γ to decrease the concentration of intracellular tryptophan, thereby starving intracellular parasites. Other tryptophan inhibitors have been reported that share the same characteristic subunit interlocking and high degree of alpha helices found in IFN-γ.

Any antigen can cause the secretion of IFN-γ. It is enhanced by IL-2 and IL-12 which induce the NK-cells and Th cells to form IFN-γ. B-lymphocytes need IL-1 to produce IFN-γ which influences on the secreting cells as well as the cells the IFN-γ-receptors. The first step in the functioning of the IFN-γ pathway is its interaction with receptors located on the surface of the cells. The receptor for IFN-γ is composed of two structurally homologous polypeptides called IFN-γ-Ra and IFN-γ-Rb. Signal transduction is mediated by Jak1, Jak2 and STAT1a finally initiating the transcriptional (Janeway *et al.*, 2005).

IFN-γ has no viral specificity and inhibits the spread of viruses containing either RNA or DNA. There are several mechanisms by which IFN-γ reacts to viral infection. In conjunction with CD40, IFN-γ binds to macrophages and activates them by killing the intracellular pathogens such as viruses. Bound IFN-γ causes the macrophage to produce elevated amounts of both MHC Class I and MHC Class II molecules. A second antiviral mechanism of INF-γ is to shutdown the replication of viral DNA or RNA and to rid the cell of pathogen without killing it. In addition to its antiviral activity, IFN-γ also plays a role in delayed type hypersensitivity by binding to macrophages and eliciting the release of inflammatory mediators.

IFN-γ increases expression of class II MHC molecules on antigen-presenting cells and promotes presentation of antigens to helper T cells. IFN-γ can also activate the ability of macrophages to resist intrinsic antiviral activity and to kill the other cells if they are infected (extrinsic antiviral activity). Besides, interferons have many other effects on gene expression.

2. Cell Activation by Interferon- alpha and Interferon-beta

Interferons are produced by cells in response to a stimulus, and are released into the surrounding medium. Thereafter, they bind to receptors present on target cells and induce transcription of about 20-30 genes of the target cells. This event results in an anti-viral state in the target cells. About 20-30 proteins are induced by these interferons the function of most of the proteins is not fully understood. Moreover, three of these proteins that play an important role in the induction of the anti-viral state, have been intensively studied (Fig. 17.7). Expression of a protein 2′5′ oligo A synthase causes an activation of the second protein (a ribonuclease) which break down mRNA. Expression of the third protein a 'protein kinase' inhibits the initiation step of protein synthesis. These activities target viral protein synthesis, but also result in inhibition of host protein synthesis. These proteins are synthesised and activated as per requirement.

Upon treatment with interferon the synthesis of inactive form of these proteins in the target cell is induced. Double-stranded RNA is required for activation of these proteins. The dsRNA directly activates 2′5′ oligo A synthase and protein kinase R, and indirectly activates ribonuclease L (since this needs 2′5′ oligo A, the product of 2′5′ oligo A synthase, for activation). These potentially toxic pathways are activated only in the interferon-treated cell if double-stranded RNA is synthesised. This will usually happen if actually virus infection occurs. Sometimes, the activation of these proteins may result in the death of the cell. But at least the progress of the infection is inhibited.

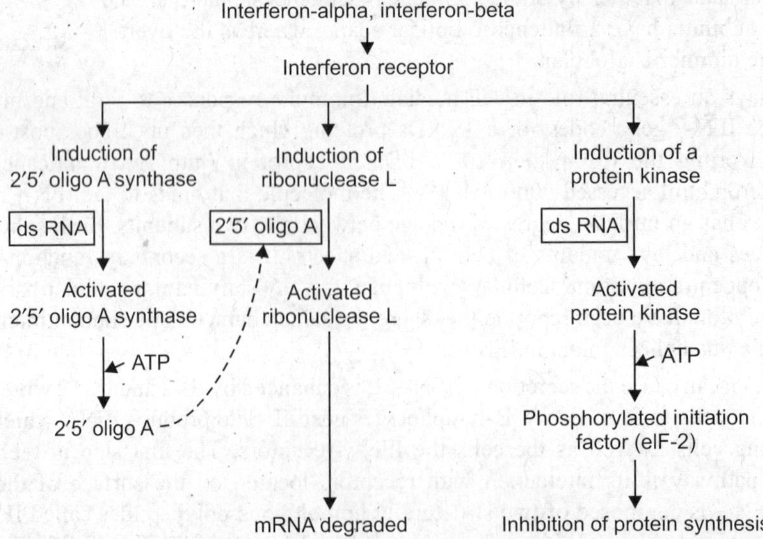

Fig. 17.7. Induction of the anti-viral state by interferon-α and interferon-β.

The pathway described as above is not the only the final way for providing protection by interferons to cells against viruses and other pathogens. All three interferons increase expression of class I MHC molecules and thus promote recognition by cytotoxic T cells. All three interferons can activate NK cells which can then kill virus-infected cells.

3. Role of STAT Protein

The 'signal transducers and activator of transcription' (STAT) protein, also called 'signal transduction and transcription' protein regulates cell growth, its survival and differentiation by regulating the expression of certain immune system genes. Domains and covalent modification sites of STAT proteins are given in Fig. 17.8.

The transcription factors of this family are activated by the Janus kinase (JAK). Dys-regulation of this pathway is frequently observed in primary tumours and leads to increased angiogenesis, enhanced survival of tumours and immunosuppression. Knockout studies have provided evidence that STAT proteins are involved in the development and function of the immune system and play a role in maintaining immune tolerance and tumor surveillance.

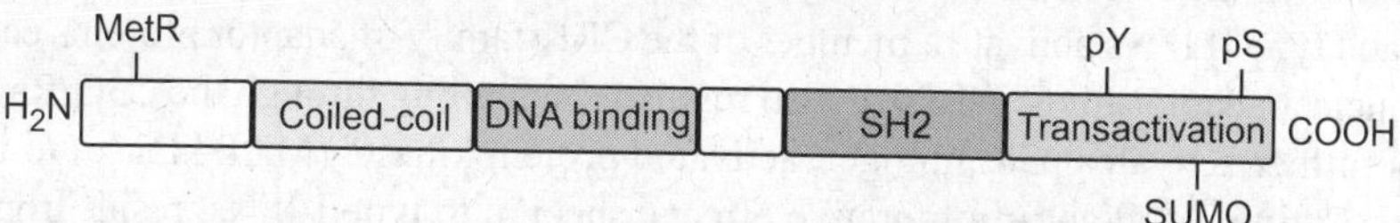

Fig. 17.8. Domains and covalent modification sites of STAT proteins.

***(i)* STAT Family:** The first two STAT proteins were identified in the interferon system. The seven mammalian STAT family members identified are: STAT1, 2, 3, 4, 5A and B, and 6. STAT1 homodimers are involved in type II interferon signalling, and binds to the 'interferon-gamma activated sequence' (GAS) promoter to induce expression of 'interferon stimulated genes' (ISG). In type I interferon signalling, STAT1-STAT2 heterodimer combines with 'interferon response factor' (IRF9) to form 'interferon stimulated gene factor (ISGF3), which binds to the 'interferon stimulated response element' (ISRE) promoter to induce ISG expression.

***(ii)* Activation:** Activation of the intracellular Janus kinase (JAK) is induced by extracellular binding of cytokines (IFNs) *i.e* the classical Janus kinase-STAT (JAK-STAT) signalling pathway. It results in phosphorylation of a specific tyrosine residue in the STAT protein which promotes the dimerization of STAT monomers via their SH2 domain (Fig. 17.9).

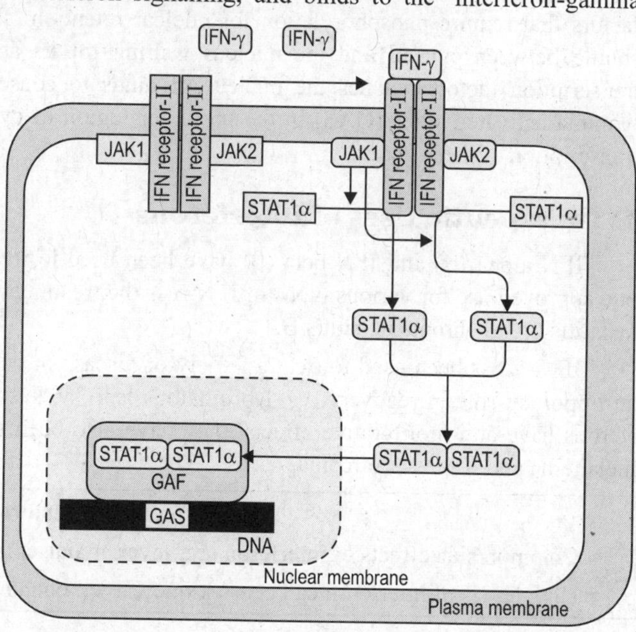

Fig. 17.9. Signal transduction pathway of interferrons.

Then the phosphorylated dimer is actively transported in the nucleus via importin a/b and RanGDP complex. After entering inside the nucleus, the active STAT dimer binds to cytokine inducible promoter regions of genes containing gamma activated site (GAS) motif and activate transcription of these genes. The STAT protein can be dephosphorylated by nuclear phosphatases which lead to inactivation of STAT. Consequently, the transcription factor is transported out of the nucleus by exportin crm1/RanGTP.

IFN-γ Signal Transduction Pathway: Like IFN-γ, IFN-γ receptor is a homodimer with a binding dissociation constant of 10^{-11} to 10^{-10} M^3. IFN-γ receptors are found in high amounts on the surface of many different types of cells *e.g.* T cells, B cells, macrophages, NK cells and fibroblasts. The receptor consists of an alpha subunit to which IFN-γ binds to and a beta subunit necessary for signal transduction. The alpha subunit is associated with Janus kinase 1 (JAK1) and the beta subunit is associated with Janus kinase 2 (JAK2). After binding of IFN-γ to its receptor the alpha subunit, JAK1 and JAK2 are tyrosine phosphorylated (Fig. 17.9). Then JAK1 phosphorylates a tyrosine on signal transducers and activators of transcription alpha (STAT alpha). In 1994, Ke Shuai and coworkers found that after phosphorylation, two STAT alpha molecules dimerise via interaction between their SH2 domains. This dimmer *i.e.* IFN-γ activator factor (GAF) enters into the nucleus and binds to IFN-γ activator sequence. This results in an elevated transcription of IFN-γ.

In addition to the JAK-STAT pathway, IFNs can activate several other signalling cascades. Both type I and type II IFNs activate a member of the CRK family of adaptor proteins called CRKL. CRKL is a nuclear adaptor for STAT5 that also regulates signalling through the C3G/Rap1 pathway. Type I IFNs further activates p38 'mitogen-activated protein kinase' (MAP kinase) to induce gene transcription. Antiviral and anti-proliferative effects, specific to type I IFNs, result from p38 MAP kinase signalling. The phosphatidylinositol 3-kinase (PI3K) signalling pathway is also regulated by both type I and type II IFNs. PI3K activates P70-S6 kinase 1, an enzyme that increases protein synthesis and cell proliferation.

***(iii)* Function:** Mainly the STAT proteins were described as latent cytoplasmic transcription factors that require phosphorylation for nuclear retention. The unphosphorylated STAT proteins shuttle between cytosol and the nucleus waiting for its activation signal. When the activated transcription factors reaches the nucleus, it binds to consensus DNA-recognition motif called 'gamma activated sites' (GAS) in the promoter region of cytokine inducible genes and activates transcription of these genes.

4. Therapeutic Uses of Interferons

IFN alpha (α) and IFN beta (β) have been used to treat various viral infections. Currently one approved use for various types of IFN-α is the treatment of certain cases of acute and chronic hepatitis C and chronic hepatitis B.

IFN--γ has been used to treat a variety of disease in which macrophage activation may play an important role in recovery *e.g.* lepromatous leprosy, leishmaniasis, toxoplasmosis. Since interferons have anti-proliferative effects, they have also been used to treat certain tumours such as melanoma and Kaposi's sarcoma.

Side Effects of Interferon

- Common side effects of interferon are: fever, malaise, fatigue and muscle pains.
- High levels of interferon can cause kidney, liver, bone marrow and heart toxicity.

5. Pegylation of Interferons

Polyethylene glycols (PEGs) are nontoxic and amphophilic, *i.e.* soluble both in water and in most organic solvents. PEGs are being used in recent years to improve the drug efficacy. Chemical structure of PEG reagent is given in Fig. 17.10. Pegylation is a method of altering the intravenously administered therapeutic proteins to improve their stability, biological half-life, water solubility and immunologic characteristics. Rearrangement of molecular structure with the help of PEG allows the changes in molecular properties of a drug. IFN α-2a is pegylated to improve the efficacy of the drug.

$$CH_3O(CH_2CH_2O)n - O - \overset{\overset{\displaystyle O}{\|}}{C} - NH$$

mPEG

$$\varepsilon\,(CH_2)_4$$

Lysine

$$\alpha\,CH$$

$$CH_3O(CH_2CH_2O)n - O - \underset{\underset{\displaystyle O}{\|}}{C} - \underset{H}{N} \qquad \underset{\underset{\displaystyle O}{\|}}{C} - O - N$$

mPEG

NHS

Fig. 17.10. Chemical structure of PEG reagent

Generally protein pegylation is done via stable covalent bonds between an amino or sulfhydryl group of the protein and a chemically reactive group (carbonate, ester, aldehyde, or tresylate) on the PEG. As a result of pegylation, the structures may be linear or branched. The reaction can be controlled by protein type and concentration, reaction time, temperature, and pH value. Environmental factors influence electrostatic binding properties and protein charge, form and size.

A PEG mass of 40-50 kDa is sufficient to increase the size of a small molecule to such an extent that it is less readily excreted through the kidneys and persists in the body for longer. Pegylated proteins are less rapidly broken down by the body's enzymes than are unmodified proteins. These modifications can reduce the frequency of injections by increasing the biological half-life and improving the efficacy of drugs in the body. They also reduce the rapidity and intensity of the body's immune reaction against the interferon molecules.

Fig. 17.11 shows the 3D structure of standard (unpegylated) and pegylated interferon. Almost any molecule can be modified in this way. The therapeutic proteins and peptides such as growth factors and blood products are now being pegylated for medicinal purposes. Due to the safety of the method, substances used in foods and cosmetics such as liposomes are also being pegylated. Human beings could find themselves coming into daily contact with PEGs.

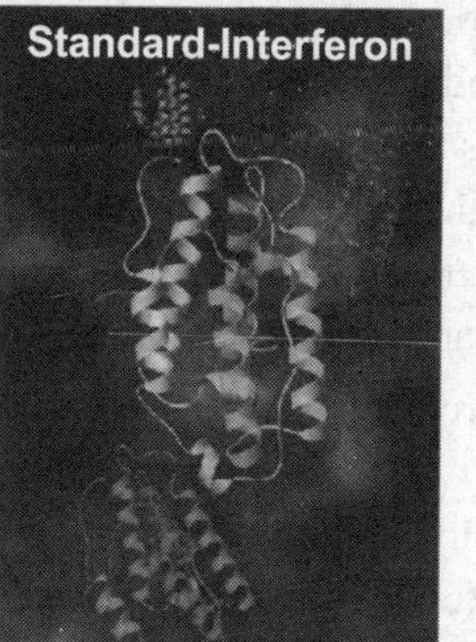

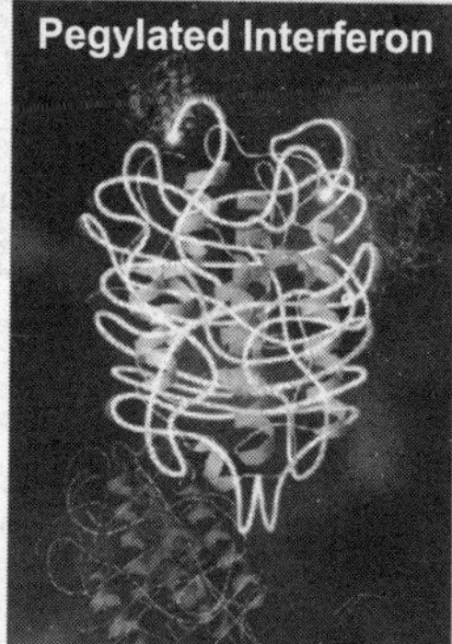

Fig. 17.11. 3-D structure of interferon.

Interferon alfa-2b is an antiviral drug originally discovered in the laboratory of Charles Weissmann at the University of Zurich, developed at Biogen, and marketed by Schering-Plough under the trade name Intron-A®. It has been used for a wide range of indications including viral infections and cancers. This drug is approved around the world for the treatment of chronic hepatitis C, chronic hepatitis B, hairy cell leukemia, chronic myelogenous leukemia, multiple myeloma, follicular lymphoma, carcinoid tumor and malignant melanoma. Some of the interferon alfa-2b products are given in Table 17.4.

Table 17.4. Interferon alfa-2b products (Source: Dmitrij *et al.*, 2007).

Interferon alfa-2b products			
Product	**Manufacturer**	**Features**	**Special uses**
Pegasys	Roche	Pegylated	
Pegintron	Schering-Plough	Pegylated	
Infagel	Vector-Medica	ointment	
Altevir	Bioprocess subsidiary	Liquid, free of HSA	
Viferon	Feron	Suppository or ointment with vitamins C, E	Viral, bacterial, chlamydiosis
Grippferon	Firn-M	Nasal formulation	
Opthalamoferon	Firn-M	with dimedrol	Eye infections
Gerpferon	Firn-M	with acyclovir, lidocaine ointment	Herpes

Pegylated IFN-α/-2a (PEG-INF, Pegasys) shows better results in the treatment of chronic hepatitis C than does the unmodified interferon (IFN). For production of second generation of this drug, a branched polyethylene glycol (PEG) molecule with an average molecular weight of around 43 kDa was attached to it. In animals this pegylation produced a 12 to 135-fold increase in effect against viruses and an 18-fold increase in effect against tumor cells. The modified drug has a biological half-life about ten times more than 8.5-hour half-life of the unmodified drug. Therefore, it needs to be injected only once weekly as compared with three times weekly.

6. Production of Recombinant Interferon (alfa-2a)

For a long time about 60,000 liters of human blood were required in order to produce one gram of interferon. In recent years, it is produced by 'recombinant DNA technology'. In this method such cells are utilised that can be grown in large amounts and can produce high concentrations of interferon. For this purpose, the DNA sequence coding the desired protein is isolated and inserted into the genetic material of an organism.

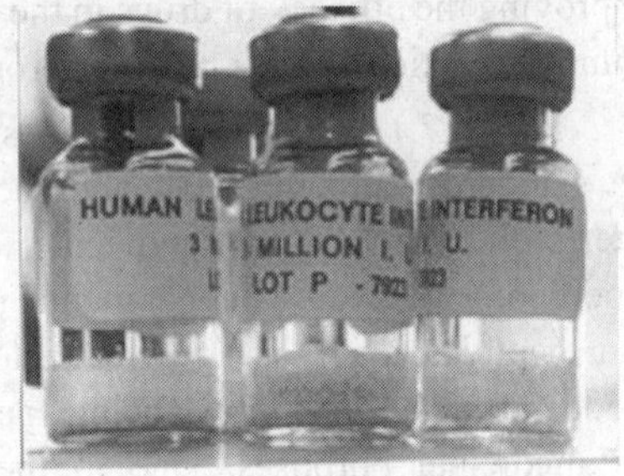

Human leukocyte interferon (IFN-α)

Interferon was scarce and expensive until 1980 when the interferon gene was inserted into bacteria using recombinant DNA technology, allowing mass cultivation and purification from bacterial cultures or derived from yeast (*e.g.* Reiferon Retard is the first yeast derived interferon-alpha 2a).

The genetic information that is required to be mass-produced in the case of interferon is a human leukocyte interferon gene. The rapidly replicating microorganism used for production is a non-pathogenic *Escherichia coli*. The bacteria that are made 'recombinant' by insertion of the human genetic information become able to produce their own proteins as well as human interferon. By employing this method, a purified product containing only the desired interferon subtype alfa-2a can be obtained. The purity of the product is brought to about 99% by using several chromatographic and filtration steps.

Side Effects of Interferons

The most common side effect of interferon alfa-2a or alfa-2b therapy is a flu-like reaction with fever, fatigue, irritability, chills, headaches, and muscle aches. These effects should become less severe and less frequent as the patients continue the therapy. Regular advice is taken from the physician if any of these problems continue.

In 1986, the first recombinant interferon alfa-2a was produced by Roche (Roferon A®). It was licensed in the USA and in Switzerland for the treatment of hairy cell leukemia. Sidney Pestka and his coworkers at the Roche Institute for Molecular Biology developed the actual manufacturing process.

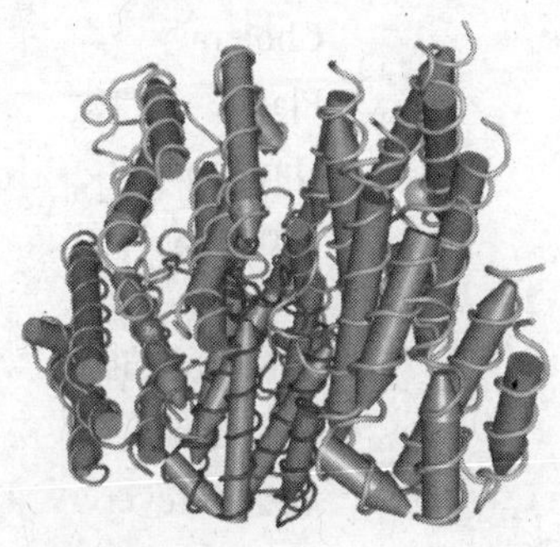

Recombinant human INF-α-2b

Recombinant human interferon α -2b has the role of broad-spectrum anti-viral, anti-tumor, inhibiting cell proliferation and enhancing immune function, etc. Interferon combines with cell surface receptors, induces cells to generate a variety of anti-viral protein, inhibits reproduction of virus in the cell and improves immune function including increasing phagocytosis of macrophage enhancing the function of natural killer cell and toxicity of target cell which is caused by lymphocytes.

Interferon alfa-2a or alfa-2b is injected subcutaneously or intramuscularly. Such treatment is done to prevent tumour cells or viruses from growing inside your body. It does not work for all patients; however, some patients respond to the drug better than others. For the first 10-24 weeks, the patients are injected daily; thereafter, they are injected three times a week. Generally, therapy lasts for at least 6 months. The physician may measure the effectiveness and side effects of the treatment using physical examinations and laboratory tests before and during the treatment. The length of treatment depends on the response to the medication.

C. VACCINES

The history of immunology has been described in the previous chapter Immunology. Moreover, in ancient India and China, there was an effective but highly dangerous practice of introducing the fluid from the pustules of smallpox (*variola, variolae)* patients into healthy individuals through dermal incisions. This practice was called variolation. By using live smallpox virus the patients could be protected from contracting the disease.

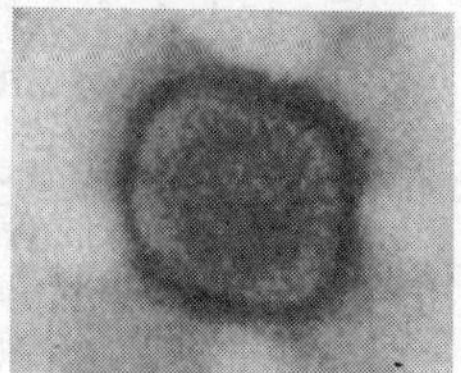

Smallpox virus

The term vaccine is derived from the Latin word *vacca* which means cow; the introduced substance is called the 'vaccine' and the process of introducing the substance is 'vaccination'. Vaccines may be defined as *the harmless molecules (usually but not necessarily proteins) that elicit an immune response providing the protective immunity against a potential pathogen*. The pathogen may be a bacterium, a eukaryotic protozoan, and a virus. Most successful vaccines have been raised against viruses (Table 17.5).

In 1979, the World Health Organization declared smallpox an eradicated disease. This was the result of coordinated public health efforts by many people, but vaccination was an essential component.

Table 17.5. Development of vaccine to prevent diseases.

Disease	***Year***	***Disease***	***Year***
Small pox	1798	Measles	1963
Rabies	1885	Mumps	1967
Typhoid	1896	Rubella	1969
Cholera	1896	Anthrax	1970
Plague	1897	Meningitis	1975
Diphtheria	1923	Pneumonia	1977
Pertussis	1926	Hepatitis B	1981
Tetanus	1927	Haemophius influenza type b	1985
Tuberculosis	1927	Japanese encephalitis	1992
Influenza	1945	Hepatitis A	1995
Yellow fever	1953	Varicella	1995
Poliomyelitis	1955	Rotavirus	1998

Future Application of Interferons

IFN is a powerful weapon against viruses, but acts as a blunderbuss, not a magic bullet. As its cell-regulatory activities are becoming better understood, it is finding increasing use as a cancer treatment *e.g.* IFN-α/HCL. The long term prospects for antiviral use are uncertain.

I. TYPES OF VACCINES

Several approaches are made to designing the vaccines against a microbe. These choices for design of vaccine are based on basic information about the microbe, for example how it infects cells and how the immune system responds to it, and also the countries where the vaccine would be used. Different types of vaccines produced so far are: live attenuated vaccines, inactivated vaccines, subunit vaccines, toxoid vaccines, conjugate vaccines, DNA vaccines, and recombinant vaccines. A brief description of these vaccines is given in this section.

1. Live Attenuated Vaccines

A version of the live and weakened microbe *in vitro* that has lost its disease causing ability is called 'live attenuated vaccines'. A live, attenuated vaccine has similarity to the natural infection. The germs used in these vaccines still can infect people, but they rarely cause serious disease. Such vaccines are good for the immune system. Often they confer lifelong immunity with only one or two doses due to eliciting strong cellular and antibody responses.

Comparatively, live attenuated vaccines are easy to create for certain viruses. Vaccines against measles, mumps and chickenpox are prepared by using this approach. Viruses contain a small number of genes; therefore, their characteristics can be controlled easily. Often viruses are attenuated by growing them for several generations in such cell cultures in which they do not reproduce very well. Such hostile environment forces them to adapt a new environment. As a result of which they become weaker with respect to human beings *i.e.* their natural host. Vaccines are made from viruses whose disease-causing ability has been deteriorated from multiple passages.

Advantages and Disadvantages of Attenuated Vaccines

Advantages

- They elicit humoral IgG and local IgA, and also activate all phases of immune system

- They raise an immune response to all protective antigens (but inactivation, by formaldehyde in the case of the Salk vaccine, may change antigenicity
- They stimulate antibodies against multiple epitopes which are similar to those elicited by the wild type virus
- They are not more costly
- In majority of vaccines, they provide quick immunity
- They can easily be administered as in cases of polio and adenovirus vaccines
- They can easily be transported
- They can eliminate from the community the wild type virus

Disadvantages

- A major disadvantage is that they be mutated and thus lead to reversion to virulence.
- In tropics, sometimes there is poor take
- Spread of the vaccine virus that is not standardized and may be mutated
- In patients of immunodeficiency disease, live viruses are a problem

It is more difficult to create live, attenuated vaccines for bacteria; because, bacteria consist of thousands of genes which become more difficult to control. Therefore, the researchers use recombinant DNA technology to remove several key genes to make a live vaccine for a bacterium. By using this approach a vaccine against *Vibrio cholerae* causing cholera has been created. The other examples of live attenuated vaccines are: measles vaccine, mumps vaccine, rubella (German measles) vaccine (all are found in a common MMR vaccine), oral polio vaccine (OPV), and varicella (chickenpox) vaccine.

In spite of several advantages the live attenuated vaccines have several demerits also. Being a nature entity, it may undergo certain changes or mutation. But there may be a remote possibility to revert an attenuated microbe in the vaccine to a virulent form and being able to cause disease. The live attenuated vaccines cannot safely be given to every patient. Also such patients cannot be given live vaccines that have damaged or weakened immune systems due to HIV infection or have undergone chemotherapy. Another limitation with live attenuated vaccines is that it must always be kept in refrigerated conditions to save its potentiality. The countries where there is no refrigeration a live vaccine may not be used.

Since attenuation was achieved empirically, molecular basis of attenuation is not known in most cases. The empirical method of foreign-cell passage cause several mutations in a virus. It is difficult to determine the important mutations. Many attenuated viruses are temperature-sensitive (*i.e.* they grow better at 32 – 35°C than 37°C) or cold adapted (*i.e.* they grow at 25°C). There are 57 nucleotide changes in the genome in type 1 polio virus attenuated vaccine strain resulting in 21 amino acid changes. One third of the mutations are found in the VP1 gene (*i.e.* in only 12% of genome). Attenuation results from changes in surface proteins of the virus

Synthetic Vaccine

A vaccine that consists of mainly synthetic peptides, carbohydrates or antigens is called a 'synthetic vaccine'. It is usually considered to be safer than vaccines from bacterial cultures. In 1982, Louis Chedid (from the Pasteur Institute, Paria) and Michael Sela (from the Weizmann Institute) created the world's first synthetic vaccine from diphtheria toxin. In 1986, Manuel Elkin Patarroyo created the SPf66, the first version of a synthetic vaccine for Malaria.

2. Inactivated (killed) Vaccines

The inactivated vaccines are produced by killing the disease-causing microbe with chemicals such as formaldehyde, heat or radiation. Such vaccines are more stable and safer than live vaccines; because there is no risk of mutating the dead microbes and converting back to their disease-causing form. Usually there is no need of refrigeration for inactivated vaccines; therefore, they can easily be stored and transported in a freeze-dried form. It makes the vaccines accessible to people in such countries where there is nor refrigeration facility. However, mostly inactivated vaccines stimulate a weaker immune system response than the live vaccines. Therefore, to maintain the immunity of person, several additional doses or booster doses are given. Where people don't have regular access to health care, they can be forfeited to get booster doses in time. Examples of inactivated (killed) vaccines are: inactivated polio vaccine (IPV) which is the shot form of the polio vaccine, inactivated influenza vaccine, etc.

The flu virus strain 1 consists of eight segments of genes and that of strain contains six gene segments. The haemalutinin (HA) gene and nuraminidase (NA) gene of strain 1 is cleaved by restriction enzyme and separately inserted into different plasmids (Fig. 17.12 A). Additional recombinant plasmids are constructed by inserting the remaining six genes from flu virus strain 2 (B). Animal cell culture is grown in laboratory. The cells are transformed *in vitro* by recombinant HA and NA plasmids from flu virus strain 1 and six plasmids carrying genes from flu strain 2 (C). The genes present in plasmids instruct the animal cells to make desired new strain of flu virus (D-E). The combined genes of new strain of flu virus can grow in eggs. It is harmless in humans.

Advantages and Disadvantages of Inactivated Vaccines

Advantages

- Inactivated vaccines provide sufficient humoral immunity after giving booster doses
- There is no mutation or reversion
- They can be used with immuno-deficient patients
- They do better protection in tropical areas sometimes

Disadvantages

- It is costly
- Some vaccines do not induce immunity
- Regular booster doses are must
- There is a shortage of monkeys in the case of polio
- There have been failures in inactivation leading to immunization with virulent virus like smallpox

3. Subunit Vaccines

During 1980s, new strategies were developed for the production of subunit vaccines with the advent of modern molecular biology techniques. These are vaccines comprised of proteins derived from pathogenic viruses, bacteria or parasites. In general the proteins are produced not by the pathogens, but by expression of the gene encoding the protein in a 'surrogate organisms'. Scientists have used the yeast to produce the Hepatitis B surface Antigen (HBsAg) which is now used in almost all commercial Hepatitis B vaccines. The antigens that stimulate the immune system best are called the *subunit vaccines*. But in some cases, the subunit vaccines contain the epitopes, *i.e.* the specific parts of the antigen to which antibodies or T cells recognise and bind. Subunit vaccines contain only the essential antigens, not all the other molecules that make up the microbe. There are least chances of adverse reactions to the vaccine. Anywhere they can contain 1 to 20 or more antigens. It is a tricky and time-consuming process that identifies the best antigens that stimulate the immune system.

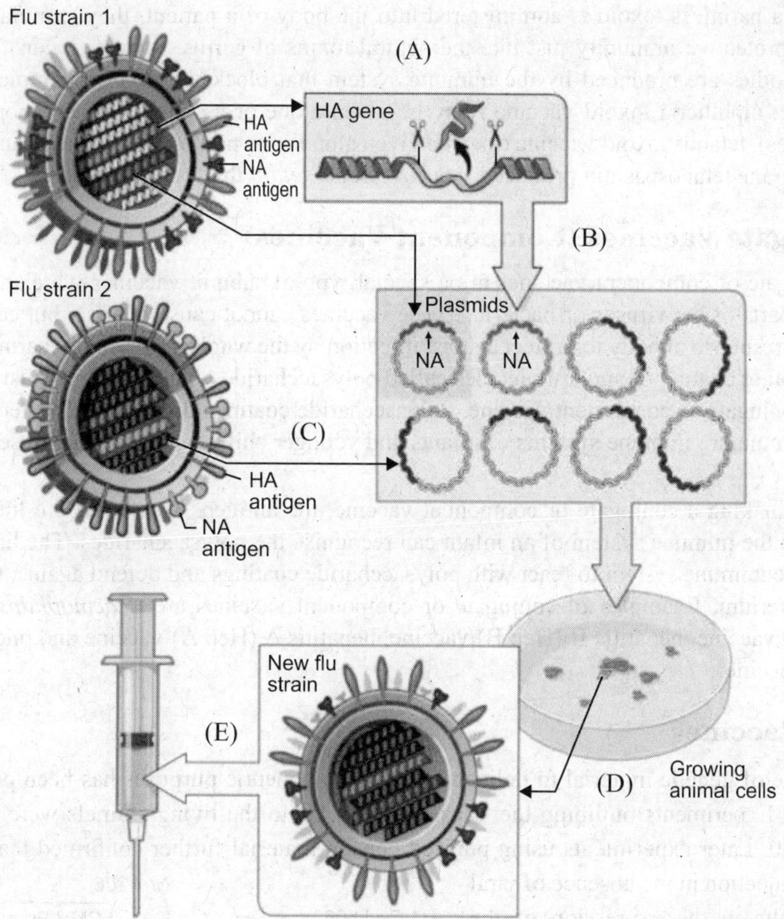

Fig. 17.12. Avian flu vaccine development by using reverse transcriptase (*source:* the National Institute of Allergy and Infectious Diseases).

Subunit vaccines are made in one of two ways: (*i*) the desired microbe is grown in the laboratory and then chemicals are used to break it apart and gather the important antigens, (*ii*) the antigen molecules from the microbe can be manufactured by using recombinant DNA technology. Thus the vaccines produced through this way are called 'recombinant subunit vaccines'.

A recombinant subunit vaccine for the hepatitis B virus has been made. Hepatitis B virus genes that code important antigens are inserted into yeast cell *i.e. Saccharomyces cerevisiae*. Then the genetically modified yeast produced the antigens, which are collected and purified for use in the vaccine. Further work is going on to produce a recombinant subunit vaccine against hepatitis C virus also.

4. Toxoid Vaccines

A toxoid is a bacterial toxin (usually an exotoxin) whose toxicity has been weakened or suppressed, while the other properties *e.g.* typically immunogenicity are maintained. These vaccines are used when a disease is caused mainly by a bacterial toxin. Moreover, the toxins produced by bacteria can be inactivated by treating them with heat or formalin (a solution of formaldehyde and sterilised water) to destroy their disease causing ability. Such detoxified toxins are called 'toxoids' which are safe and used as vaccines as they induce an immune response to the original toxin or increase the response to another antigen.

When a harmless toxoid is administered into the body of a patient, they stimulate the body to produce protective immunity just like the natural toxins of germs. It fights against the natural toxin. Antibodies are produced by the immune system that block the toxin. Examples of toxoid vaccines are: diphtheria toxoid vaccine (may be given alone or as one of the components in the DTP vaccines), tetanus toxoid vaccine (may be given alone or as part of DTP). The tetanus toxoid is derived from the tetanospasmin produced by *Clostridium tetani* that causes tetanus.

5. Conjugate Vaccines (Component Vaccines)

Conjugate or component vaccines are a special type of subunit vaccines which are made by using only parts of the viruses or bacteria. These vaccines cannot cause disease, but can stimulate the immune response of body to protect against infection by the whole germ. Many harmful bacteria possess an outer coating of sugar molecules called polysaccharides. Then attempt must be made to prepare a conjugate or component vaccine. Polysaccharide coatings disguise the bacterial antigens so that the immature immune systems of infants and younger children cannot recognise or respond to them.

When making a conjugate or component vaccine, the antigens or toxoids are linked from a microbe that the immune system of an infant can recognise the polysaccharides. The linkage helps the immature immune system to react with polysaccharide coatings and defend against the disease-causing bacterium. Examples of conjugate or component vaccines are: *Haemophilus influenzae* type b (Hib) vaccine, hepatitis B (Hep B) vaccine, hepatitis A (Hep A) vaccine and pneumoccocal conjugate vaccine.

6. DNA Vaccines

The use of genetic material to deliver genes for therapeutic purposes has been practiced for many years. Experiments outlining the transfer of DNA into the living animals were reported as early as 1950. Later experiments using purified genetic material further confirmed that the direct DNA gene injection in the absence of viral vectors results in the expression of the inoculated genes in the host. There have been additional experiments that purified nucleic acids could be directly delivered into a host, and proteins would be produced. In 1992, Tang and Johnson reported that the delivery of human growth hormone in an expression cassette *in vivo* resulted in production of detectable levels of the growth hormone in host mice. They also found that these inoculated mice developed antibodies against the human growth hormone; they termed this immunisation procedure as genetic immunisation, which describes the ability of inoculated genes to be individual immunogens.

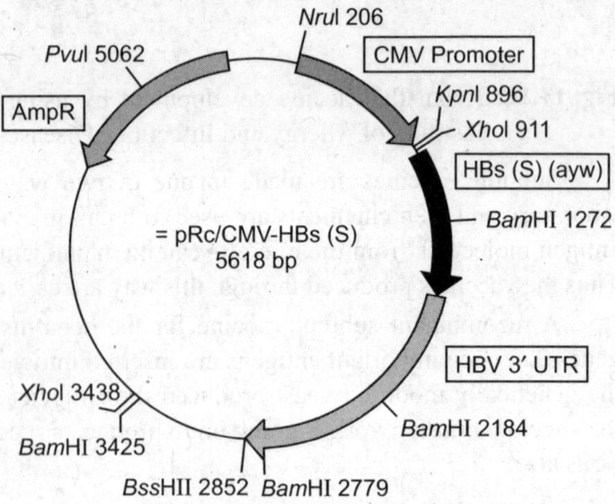

Fig. 17.13. A plasmid pRc/CMV-HBs(S) (5618 bp) used to construct a DNA vaccine.

A DNA vaccine can be created against a microbe only after analysing its genes. These vaccines show great promise; several types of vaccine are being tested in humans. But it is still in the experimental stage.

(*a*) **Construction of DNA Vaccines:** DNA vaccines are composed of bacterial plasmids (Fig. 17.13). In 2001, R. Schirmbeck and R. Jorg realised the potential of DNA-based vaccination based on the success from hepatitis B virus surface antigen. Expression plasmids used in DNA-based vaccination normally contain two units : (*i*) the antigen expression unit composed of promoter/enhancer sequences followed by antigen-encoding and polyadenylation sequences, and (*ii*) the production unit composed of bacterial sequences necessary for plamid amplification and selection.

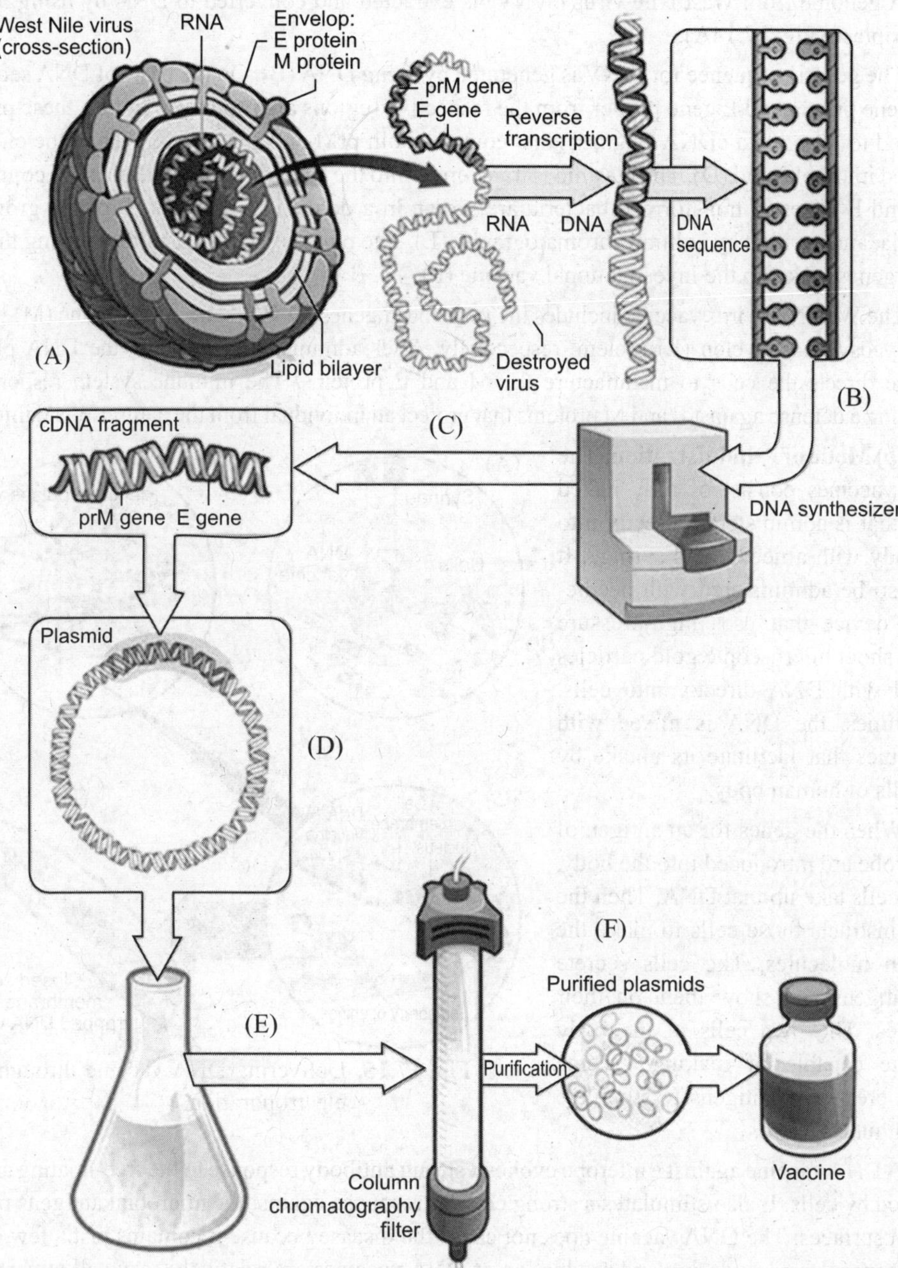

Fig. 17.14. Construction of a DNA vaccine against West Nile virus. (*Source:* National Institute of Allergy and Infectious diseases).

Construction of bacterial plasmids with vaccine inserts is accomplished using recombinant DNA technology. The constructed vaccine plasmid is transformed into bacteria, where bacterial growth produces multiple plasmid copies. Then the plasmid DNA is purified from the bacteria by separating the circular plasmid from the much larger bacterial DNA and other bacterial impurities. This purified DNA acts as the vaccine (http://www.asmusa.org/acasrc/Colloquia/dnareprt).

Construction of a DNA vaccine has been discussed giving the example of West Nile virus. The ssRNA genome from West Nile virus (WNV) is extracted and converted to DNA by using reverse transcriptase (Fig. 17.14A).

The genetic sequence for WNV is generated by using DNA (B). On the basis of DNA sequence prM gene primer and E gene primer from the respective regions are produced. In turn these primers are used to generate a cDNA fragment that contains both prM and E genes (C). Then the cDNA is inserted into a plasmid (D). The plasmid is transferred into the bacterial cells. The plasmid containing prM and E genes in transformed bacteria are grown in a defined medium used for the growth of bacteria, and purified by column chromatography (E). The purified DNA plasmids carrying the prM and E genes make up the investigational vaccine (F).

The West Nile virus vaccine includes the prM and E genes for WNV transmembrane (M) protein and glycosylated envelop (E) protein, respectively. After administering in body, the DNA plasmid vaccine directs the cells to manufacture the M and E proteins. The immune system responds by mounting a defense against E and M proteins that protect an individual from the natural WNV infection.

***(b)* Mode of Administration:** The DNA vaccines consist of only naked DNA that is administered directly into the body with a needle and syringe. It can also be administered with needle-fewer device that uses high-pressure gas to shoot microscopic gold particles coated with DNA directly into cells. Sometimes, the DNA is mixed with molecules that facilitate its uptake by the cells of human body.

When the genes for an antigen of a microbe are introduced into the body, some cells take up that DNA. Then the DNA instructs those cells to make the antigen molecules. The cells secrete the antigens and show them on their surfaces. The own cells of the body become capable of making vaccine which create the antigens to stimulate the immune system.

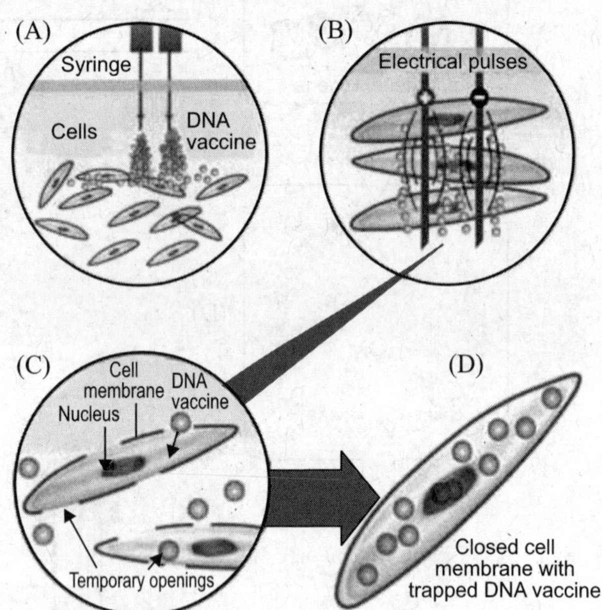

Fig. 17.15. Delivering DNA vaccine through electroporation.

A DNA vaccine against a microbe evokes a strong antibody response to the free-floating antigen secreted by cells. It also stimulates a strong cellular response against the microbial antigens present on cell surfaces. The DNA vaccine does not cause the disease because it contains just a few copies of its genes. Besides, design and production of DNA vaccines are relatively easy and inexpensive. Naked DNA vaccines being tested in humans include those vaccines against the viruses that cause influenza and herpes.

- ***Electroporation-mediated DNA delivery technology:*** DNA vaccination differs from traditional vaccines. The DNA coding for a specific component of a disease-causing organism is injected into the body. The DNA can be administered either in a saline solution injected through a hypodermic needle or on DNA-coated gold beads propelled into the body using gene guns. The actual production of the immunising protein takes place in the vaccinated host. This eliminates any risk of infection associated with some live and attenuated virus vaccines.

  The method of electroporation is the introduction of a needle-less vaccine via electronic pulses through the skin or soft tissue. The purpose of electroporation is to help the uptake of DNA vaccine into a cell. The biological material is first injected into or applied to the surface of the target tissue (Fig. 17.15A), followed by the application of brief and controlled electrical pulses directed to that tissue (B). The millisecond electrical pulses temporarily create the enhanced permeability of pores in the cell membrane (C). After a short period of time the pores reseal leaving the cells undamaged. During this period when these pores remain open, a substantial amount of the injected DNA is taken up and then trapped inside the cell (D). The electroporation technology increases the cellular uptake *i.e.* transfection of an agent by 1,000 times or more. The trapped DNA enables the cells to produce antigen designed to control cancer and chronic infectious diseases like HIV. The antigen also triggers the antibody production and prevents the disease. This method increases levels of gene expression *i.e.* production of the coded protein by 100 times or more compared to plasmid DNA delivered without other delivery enhancements.

(*c*) Mechanisms of Action: A plasmid vector containing a gene of interest (*e.g.* viral protein gene) along with an appropriate promoter is injected into the skin or muscle of the host. The plasmid is taken up; the protein is produced endogenously and processed intracellularly into small antigenic peptides by the host proteases. Then the peptides enter the lumen of the endoplasmic reticulum (ER) by membrane-associated transporters (Fig. 17.16). In the ER peptides bind to MHC class I molecules. These peptides are presented on the cell surface at MHC class I. Consequently, $CD8^+$ cytotoxic T cells (CTL) are stimulated and they evoke cell-mediated immunity. CTLs inhibit virus through both cytolysis of infected cells and noncytolysis mechanisms e.g. cytokine production. The MHC class II pathway presents the foreign protein by APCs which elicit helper T cells ($CD4^+$) responses.

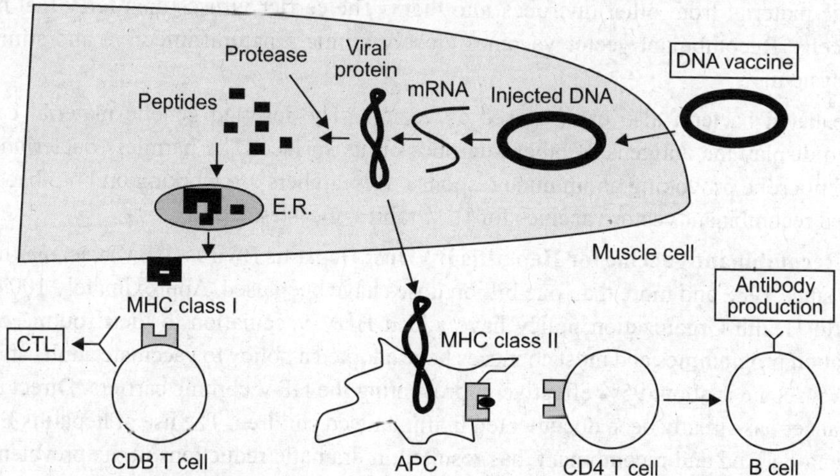

Fig. 17.16. Mechanism of action of DNA vaccine.

These $CD4^+$ cells recognise the peptides formed from exogenous proteins that were endocytosed or phagocytosed by APC, then degraded to peptide fragments and loaded onto MHC class II molecules. B cells are stimulated and antibodies are produced depending on the type of $CD4^+$ cell that binds to the complex. In 2001, R. Schirmbeck and R. Jorg found that the traditional vaccines also work in the same manner.

***(d)* Advantages of DNA Vaccines over Classic Vaccine:** Unlike many component vaccines that require multiple inoculations to maintain immunity, DNA vaccination provides long-lived immune responses. Vaccines for multiple diseases can be given in a single inoculation. Unlike other conventional vaccines, DNA vaccines can be stored under a vast array of conditions either dried or in a solution. This eliminates the need of refrigerators required to maintain a vaccine during distribution. This will greatly improve the ability to deliver vaccines to remote areas in developing countries. Candidate vaccines can be recovered from diseased tissue. Microbial DNA can be isolated from the tissue of an infected animal, purified, amplified and screened for vaccine candidates.

- **Limitations:** Although DNA can be used to raise immune responses against pathogenic proteins, certain microbes have outer capsids that are made up of polysaccharides. This limits the extent of the usage of DNA vaccines because they cannot substitute for polysaccharide-based subunit vaccines (Anonymous, 1996).
- **Future:** Recently it has been discovered that the transfection of myocytes can be amplified by pretreatment with local anesthetics or with cardiotoxin which induce local tissue damage and initiate myoblast regeneration. Full understanding of this mechanism of DNA uptake could prove helpful in improving applications for gene therapy and gene vaccination. Both improved expression and better engineering of the DNA plasmid may enhance antibody response to gene products and expand the applications of the gene vaccines.

7. Recombinant Vector Vaccines

Recombinant vector vaccines are experimental vaccines similar to DNA vaccines, but they use an attenuated virus or bacterium to introduce microbial DNA to cells of the body. Virus or bacterium used as the carrier for delivering the gene construct is used as vector. Scientists have figured out to take out the genomes of certain harmless or attenuated viruses and insert portions of the genetic material from other microbes into them. The carrier viruses then ferry that microbial DNA to cells. Recombinant vector vaccines closely mimic a natural infection and stimulate the immune system.

Attenuated bacteria also can be used as vectors. The inserted genetic material causes the bacteria to display the antigens of other microbes on its surface. The harmless bacterium mimics a harmful microbe provoking an immune response. Researchers are working on both bacterial and viral-based recombinant vector vaccines for HIV, rabies, measles, etc.

***(a)* Recombinant Vaccine for Hepatitis B Virus:** Hepatitis B virus (HBV) vaccines have been available since 1982 and more than one billion doses have been used. Approximately 100 countries with World Health Organization policy have added HBV vaccination to their routine childhood immunisation programmes and most countries have adopted a policy to vaccinate adults at increased risk. HBV vaccine is about 95% effective in preventing the HBV chronic carriers. Direct reduction of liver cancer has already been documented in immunised children. The use of hepatitis B vaccine, both plasma-derived and recombinant, has resulted in dramatic reductions in the prevalence of the carrier state in many areas.

Hepatitis B virus (HBV) is wide spread in man and produces several chronic liver disorders such as fulminant chronic hepatitis, cirrhosis and primary liver cancer. HBV DNA is a double stranded circular molecule of about 3Kb size and has a large single stranded gap which must be required with an endogenous polymerase before digestion with restriction enzyme for DNA cloning. After infection in human being HBV fails to multiply and infect a large number of cells and even does not grow in cultured cells. This property has been explained to be due to hindrance of its molecular characterisation and development of vaccines. Human plasma has been detected to have varying amount of antigens. Its capsid has different surface proteins from the rest of the virus which act as antigens. These antigens are recognised by antibody proteins that bind specifically to one of these surface proteins. Three types of viral proteins are recognised to be antigenic: (*i*) viral surface antigen (HBsAg), (*ii*) viral core antigen (HBcAg), and (*iii*) the endogenous antigen (HBeAg). Genome organisation of hepatitis B virus is given in Fig. 17.17.

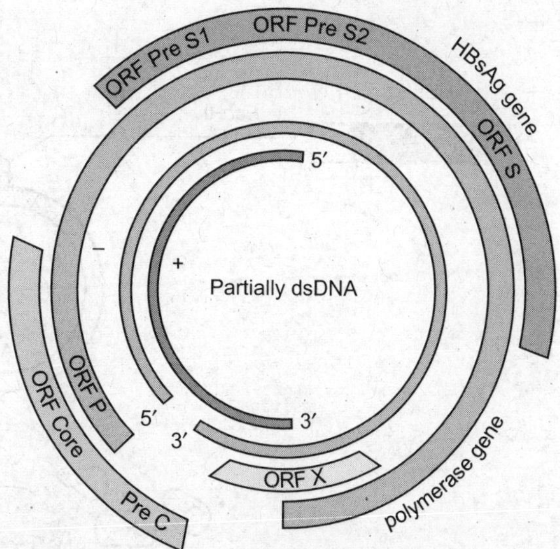

Fig. 17.17. Genome organisation of hepatitis B virus (genes overlap, green ORF S encodes HBsAg).

Four Known Genes of HBV

- **C :** gene for core protein (HBcAg) which is antigen
- **P :** gene for polymerase
- **S :** the 3 polypeptides of the surface antigen (preS1, preS2 and S - produced from alternative translation start sites (HBsAb shows antigenecity)
- **X :** a transactivator of viral transcription. HBx is conserved in all mammalian hepadnaviruses. Though not essential in transfected cells, it is required for infection *in vivo*

Since the whole viral genome has been cloned and sequenced, there is limited information about amino acid sequence of surface and core antigens. Today these antigen-proteins can be a genetically manufactured (*e.g.* transgene *E. coli*) to produce material for a simple antigen test, which finds out the presence of HBV. Recently, HBV DNA has been successfully cloned in *E. coli* and mammalian cells, and synthesis of HBsAg and HBcAg particles has been done in the cells. HBcAg genes are inserted in PBR322 near β-glactosidase gene. Production of these genes is needed in order to get production of vaccines on a large scale. In yeast or mammalian systems, these antigens are synthesised more efficiently than in prokaryotes.

Recombinant vaccine for HBV was produced by cloning HBsAg gene of the virus in yeast cells. The yeast system has its complex membrane and ability of secreting glycosylate protein. This have made it possible to build an autonomously replicating plasmid containing HBsAg gene near the yeast alcohol dehydrogenase (ADH) 1 promoter (Fig. 17.18). The HBsAg gene contains 6 bp long sequences proceeding the AUG that synthesises N-terminal methionine. This is joined to ADH promoter cloned in the yeast vector PMA-56. The recombinant plasmid is inserted into yeast cells.

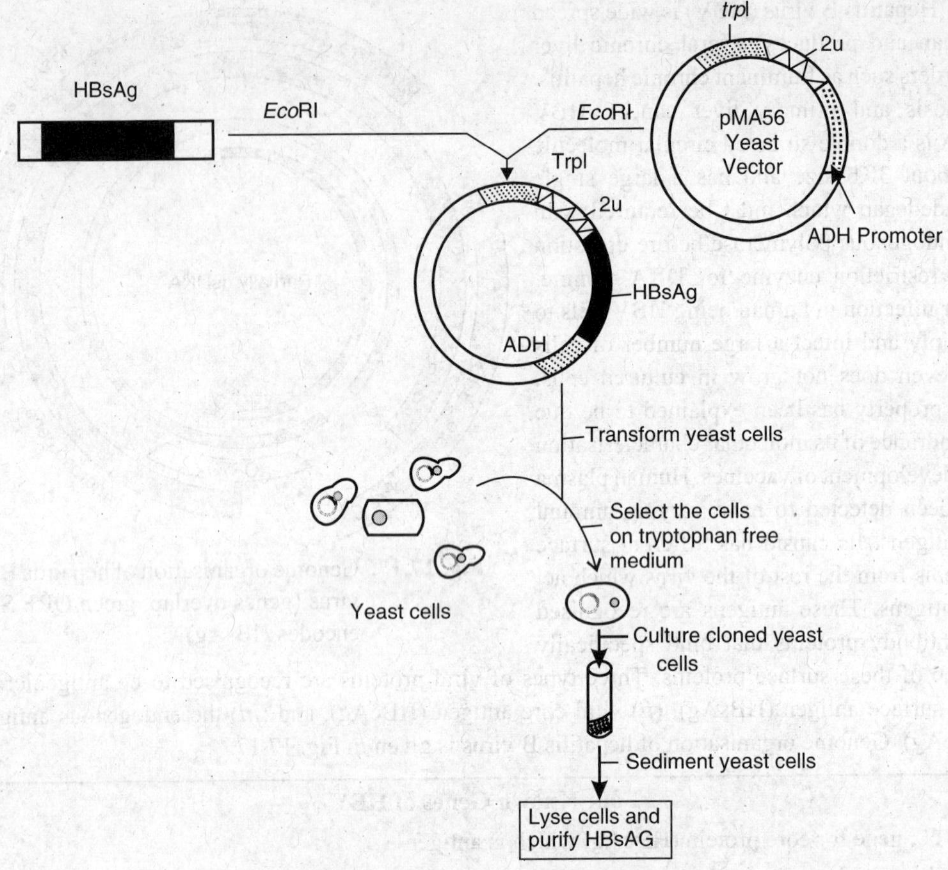

Fig. 17.18. Expression of HBsAg gene in yeast.

The transformed yeast cells are multiplied in trytophan-free medium. The transformed cells are selected. The cloned yeast cells are culture for expression of HBsAg gene. This inserted gene sequence expresses and produces particles similar to the 22 μm particle of HBV as these particles are produced in serum of HBV patients. The expressed HBsAg particles have similarity in structure and immunogenicity with those isolated from HBV-infected cells of patients. Its high immunogenicity has made it possible to market the recombinant product as vaccine against HBV infection.

***(b)* Indigenous Hepatitis-B vaccine.** India's first genetically engineered vaccine (*Guni*) against HBV, developed by a Hyderabad based laboratory (Shantha Biotechnics Pvt. Ltd.) was launched in 1997. India is the fourth country (after the U.S.A., France and Belgium) to develop this highly advanced vaccine. The indigenous yeast-desired HBV vaccine is one third the cost of the imported vaccine. This new vaccine had undergone human clinical trials at Nizam's Institute of Medical Sciences, Hyderabad and K.E.M. Hospital, Mumbai. The clinical trials clearly proved that the seroprotection is about 98%. It was found more effective than the imported vaccine. The Drug Controller General of India has permitted it for commercial manufacture.

II. POLIO VACCINE

Poliovirus is a human enterovirus and member of the family of *Picornaviridae* which causes poliomyelitis or polio. In 1936, Maurice Brodie, a research assistant at New York University,

Polio victimised children

attempted to produce a formaldehyde-killed polio vaccine from ground-up monkey spinal cords. In 1948, a breakthrough came when a research group headed by John Enders at the Children's Hospital Boston successfully cultivated the poliovirus in human tissue in the laboratory. Throughout the world two polio vaccines are used to combat poliomyelitis or polio. The first was developed by Jonas Salk and tested first in 1952. Salk announced to the world on April 12, 1955. It consists of an injected dose of inactivated (dead) poliovirus. Secondly, an oral vaccine was developed by Albert Sabin using attenuated poliovirus. Human trials of Sabin's vaccine began in 1957 which was licensed in 1962. Because there is no long term carrier state for poliovirus in immunocompetent individuals, polioviruses have no non-primate reservoir in nature.

Moreover, survival of the virus in the environment for an extended period of time appears to be remote. The last cases of paralytic poliomyelitis caused by endemic transmission of wild virus in the United States occurred in 1979. A global effort to eradicate polio, led by the World Health Organization, UNICEF, and The Rotary Foundation, began in 1988. Polio has been entirely eradicated in the America by 1994. In 2000, polio was officially eradicated in 36 Western Pacific countries, including China and Australia. Europe was declared polio-free in 2002. As of 2008, polio remains endemic in only four countries namely Nigeria, India, Pakistan and Afghanistan.

Interruption of transmission of the virus person-to person by vaccination is the critical step in global polio eradication. The two vaccines have eradicated polio from most countries in the world, and reduced the worldwide incidence from an estimated 350,000 cases in 1988 to 1,652 cases in 2007. Although transmission of poliovirus has been interrupted much in the world, transmission of wild poliovirus does continue and creates an ongoing risk for the importation of wild poliovirus into previously polio-free regions. Outbreaks of poliomyelitis may develop especially in areas with low vaccination coverage and poor sanitation. As a result, high levels of vaccination coverage must be maintained.

Poliovirus Serotypes

- There are three poliovirus serotypes: type 1 (PV1 or Mahoney), type 2 (PV2, Lansing) and type 3 (PV3, Leon).
- Unfortunately, infection with one type does not protect an individual against infection with the other two types.
- The development of effective vaccines to prevent paralytic polio was one of the major medical breakthroughs of the 20th century.
- Two different kinds of vaccine are available:
  (*i*) A live attenuated (weakened) oral polio vaccine (OPV) developed by Albert Sabin in 1961. OPV is given orally.
  (*ii*) An inactivated (killed) polio vaccine (IPV) developed in 1955 by Jonas Salk. Unlike OPV, IPV is injected by a trained health worker.
- Both vaccines are highly effective against all three types of poliovirus. There are significant differences in the work of each vaccine.

Immunity against polio can be stimulated in two ways: (*i*) through immunisation, or (*ii*) following natural infection with poliovirus. Poliovirus infection provides lifelong immunity against the disease, but this protection is limited to the particular type of poliovirus involved (Type 1, 2, or 3).

Poliovirus enters into mouth and established in pharynx, gastro-intestinal tract and local lymphatic tissue. It spreads through blood to lymphatic and central nervous system. Also viral spread takes place along nerve fibers where it reproduces and multiplies resulting in destruction of motor neurons. Consequently there develops polio.

1. Oral Polio Vaccine (OPV)

Oral polio vaccine (OPV) is a 'live-attenuated vaccine' produced by the passage of the virus through non-human cells at a sub-physiological temperature which produces spontaneous mutations in the viral genome. Oral polio vaccines were developed by several groups, one of which was led by Albert Sabin in 1960. In 1958, the National Institutes of Health (U.S.A.) created a special committee on live polio vaccines. The various vaccines were carefully evaluated for their ability to induce immunity to polio while retaining a low incidence of neuropathogenicity in monkeys. On the basis of good results, the Sabin strains were chosen for worldwide distribution.

In 2005, O. Kew and coworkers have reviewed on vaccine-derived polioviruses and the endgame strategy for global polio eradication. So far there are 57 nucleotide substitutions which distinguish the attenuated Sabin 1 strain from its virulent parental strain *i.e.* the Mahoney serotype. The two nucleotide substitutions attenuate the Sabin 2 strain, and 10 substitutions are involved in attenuating the Sabin 3 strain. The primary attenuating factor common to all three Sabin vaccines is a mutation located in the internal ribosome entry site of the virus which alters stem-loop structures and reduces the ability of poliovirus to translate its RNA template within the host cell.

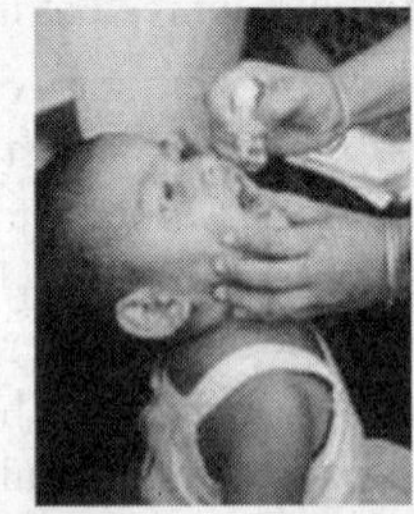

A child receiving OPV

Gut is the primary site of infection and replication where the attenuated poliovirus in Sabin vaccine replicates very efficiently. But it fails to replicate efficiently within nerve tissue. OPV is superior in its entry inside the body where no syringe is required and is more suitable vaccine for mass vaccination campaign. The immunity provided by OPV is also long-lasting as compared to the Salk vaccine. The monovalent oral poliovirus vaccine (MOPV) type 1 and 2 was licensed in 1961, and type 3 MOPV was licensed in 1962. In 1963, trivalent OPV (TOPV) was licensed which became the vaccine of choice in the U.S.A. and the other countries of the world, which replaced the inactivated polio vaccine. A second mass immunisation movement led to a decline in the number of polio cases. Between 1962 and 1965 about 56% of the population received the Sabin vaccine.

Usually OPV is available in vials containing 10-20 doses of vaccine. A single dose (usually two drops) of OPV contains 1,000,000 infectious units of Sabin 1 (effective against PV1), 100,000 infectious units of the Sabin 2 strain, and 600,000 infectious units of Sabin 3. Besides, the vaccine contains small traces of antibiotics (neomycin and streptomycin) but does not contain preservatives. One dose of OPV produces immunity to all three poliovirus serotypes in approximately 50% of recipients. Three doses of live-attenuated OPV provides protective antibody to all three poliovirus types in more than 95% of recipients.

The intestine is the primary entry site of wild poliovirus. OPV produces excellent immunity preventing the infection of wild virus in areas where the virus is endemic. The live virus used in the vaccine is passed in the stool. Moreover, it can spread to the others within a community resulting in protection against poliomyelitis even in individuals who have not been directly vaccinated.

(*a*) Advantages of OPV: OPV is an orally applicable vaccine. It does not have to be administered by a trained health worker. It can be given by volunteers and does not require sterile syringe and needles. The vaccine is relatively inexpensive, which is a major consideration when Governments have to purchase large quantities of vaccine. The recently immunised children shed short-term of vaccine virus in the stools which means the highest incidence in areas where hygiene and sanitation are poor. Immunisation with OPV can result in the passive immunisation of persons within close contact. The ability of OPV to induce intestinal, *i.e.* local immunity is probably responsible for the extraordinary effect of OPV mass campaigns in interrupting wild poliovirus transmission. Hence, OPV is of great choice for the eradication of polio which would not be feasible with inactivated polio vaccine (IPV).

***(b)* Disadvantages of OPV:** OPV is effective and safe but in rare cases the live attenuated vaccine virus causes paralysis either in the vaccinated child or in a close contact. The reason is because of immune deficiency of the recipient. Rarely, a strain of poliovirus in OPV may genetically change and circulate among a population. These are known as 'vaccine-derived polioviruses' (VDPV). This risk of vaccine-associated polio (VAPP) is well known and accepted by most public health programmes in the world because without OPV hundreds or thousands of children would be crippled every year. Now some countries are considering the combined immunisation schedules using both OPV and IPV, where the risk of wild-virus caused polio has come down to zero.

Iatrogenic (Vaccine-induced) Polio

A major draw back about the OPV is its reversion to a form that can infect the nerves and cause paralysis. Clinical paralysis caused by vaccine-derived poliovirus (VDPV) cannot be distinguished from that caused by wild polioviruses. It occurs rarely, but outbreaks of vaccine-associated paralytic poliomyelitis (VAPP) occur in areas of low coverage by OPV, because the OPV is itself protective against the related outbreak strain.

After the widespread use of polio vaccines in the mid-1950s, the incidence of poliomyelitis declined rapidly in many industrialised countries. The use of OPV was discontinued in the U.S.A. in 2000, and in 2004 in the U.K. but its usage is continued throughout the world.

The rate of VAPP varies with region but it is generally about 1 case per 7,50,000 recipients of OPV. VAPP is more likely to occur in adults than in children. The risk of VAPP is about 7,000 times higher in immunodeficient-children, particularly for persons with B-lymphocyte disorders e.g. agammaglobulinemia and hypogammaglobulinemia, which reduce the synthesis of protective antibodies.

The World Health Organisation considers the benefits of vaccination to far outweigh the risk of vaccine derived polio. In order to immunise the entire population, outbreaks of vaccine-derived polio have been stopped by multiple rounds of high-quality vaccination.

2. Inactivated Polio Vaccine (IPV)

In 1952, Jonas Salk at the University of Pittsburgh developed the first effective inactivated polio vaccine (IPV). But its testing was done for years. Salk went on CBS radio to report a successful test on a small group of adults and children on March 26, 1953 just to encourage patients. The IPV is injected and works by producing protective antibodies in the blood *i.e.* it provides serum immunity; hence it prevents the spread of poliovirus to the central nervous system. However, it induces only very low levels of immunity to poliovirus inside the gut. Consequently, it provides protection against polio paralysis to individuals; but unlike OPV, it fails to prevent the spread of wild poliovirus.

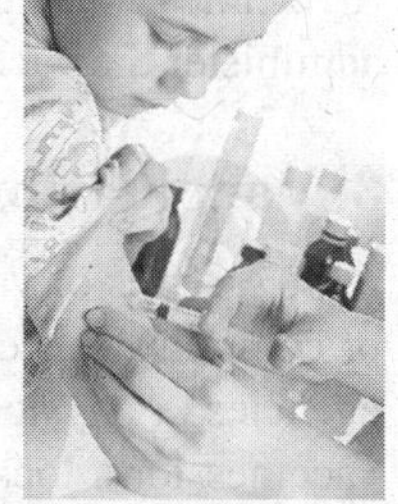

A child is being injected with IVP

The Salk vaccine is based on three wild virulent reference strains *i.e.* Mahoney (type 1 poliovirus), MEF-1 (type 2 poliovirus) and Saukett (type 3 poliovirus). In 2005, Kew and coworkers found that the wild virulent virus types are grown in Vero cell line (a type of monkey kidney tissue culture) then inactivated with formalin. The Salk vaccine is injected into the individuals that provide IgG-mediated immunity in the bloodstream. It prevents the infection of polio from progressing to viremia and protects the motor neurons. Therefore, it eliminates the risk of bulbar polio and post-polio syndrome.

In 1954, Salk vaccine was used in a test called the Francis Field Trial which is the largest medical experiment in the history. This test was done on 4,000 children at Franklin Sherman Elementary School in Virginia, and 1.8 million children in 44 states from Maine to California. About 4,40,000 children received one or more injections of the vaccine, about 2,10,000 children received a placebo (consisting of harmless culture media), and 1.2 million children received no vaccination serving as a control group. The control would then be observed to examine the contracted polio. It was found that the Salk vaccine was 60 - 70% effective against PV1 (poliovirus type 1), over 90% effective against PV2 and PV3, and 94% effective against the development of bulbar polio. The Salk vaccine was licensed in 1955. In the U.S, after a mass immunisation campaign the annual number of polio cases felt to 5,600 by 1957; and only 161 cases could be recorded in the United States by 1961.

In 1987, the IPV of enhanced-potency was licensed in the U.S.A. and is the best choice till date. The first dose of polio vaccine is given usually between 1–2 months of age of a child; a second dose is given at 4 months of age. The third dose should be given between 6–18 months of age. A booster vaccination is given at 4 to 6 years of age. Thus a total of four doses must be given to the child before entering the school.

In 2002, a vaccine of pentavalent (5 component) combination containing IPV was approved in the U.S.A. This vaccine also contains combined vaccines of diphtheria, tetanus and acellular pertussis (DTaP) and a pediatric dose of hepatitis B vaccine. In the U.K., IPV is combined with tetanus, diphtheria, pertussis and *Haemophilus influenzae* type b vaccines. About 90% or more of individuals develop protective antibody to all three serotypes of poliovirus after two doses of the current formulation of the (IPV, and at least 99% are immune to poliovirus after getting three doses).

(*a*) Advantages of IPV: IPV is not a live vaccine but the polio virus is inactivated. There is no risk of vaccine-associated polio paralysis by immunisation with IPV. It triggers an excellent immune response in most of IPV recipients.

(*b*) Disadvantages of IPV: The IPV produces less gastrointestinal immunity than the OPV. Therefore, the persons who receive the IPV may more easily be infected by wild poliovirus. The IPV is supposed to be a good vaccine in regions where no wild poliovirus is found. Live vaccine is still used in regions with higher incidence of polio.

The IPV confers only very little immunity in the intestinal tract unlike the oral vaccine. When an IPV-immunised person is infected with wild poliovirus, virus still multiplies inside the intestine and discharged in stools. Hence, the risk continues. The other disadvantages of IPV are: the price (five times more than OPV), the cost of the syringe, and the need for trained health workers to administer the vaccine using procedures for sterile injection.

3. Vaccination Schedule

A series of vaccinations including the timing of all doses which may be either recommended or compulsory is called 'vaccination schedule'. Vaccine schedules are developed by governmental agencies or physicians groups to achieve maximum effectiveness using required and recommended vaccines for a locality, while minimising the number of health care system interactions. Over the past two decades, the recommended vaccination schedule has grown rapidly and become more complicated as many new vaccines have been developed. Vaccination schedule in developing countries is given in Table 17. 6.

Some vaccines are recommended only in certain areas (countries, sub-national areas or at-risk populations) where a disease is common. For example, yellow fever vaccination is the routine vaccine schedule of French Guiana. It is also recommended in certain regions of Brazil; but in the U.S.A. it is only given to travelers heading to countries. In developing countries, vaccine recommendations also have to consider poor health care access, high vaccine cost and issues with vaccine availability and storage.

Table 17.6. A childhood vaccination schedule for infants in developing countries with Yellow Fever.

Vaccination schedule in developing countries		
Vaccine	***Age of immunization***	***Notes***
BCG (tuberculosis)	Birth	
Polio	6, 10 and 14 weeks	At birth, in endemic countries
Diphtheria, Tetanus, Pertussis	6, 10 and 14 weeks	
Hepatitis B	6, 10 and 14 weeks	Vaccine not widely available.
Haemophilus influenzae	6, 10 and 14 weeks	Vaccine not widely available.
Yellow Fever	9 months	In countries where Yellow fever poses a risk.
Measles	9 months	A second vaccination should be provided.

Source: State of the World's Vaccines and Immunizations from the World Health Organization (2003 edition).

Sample vaccinations schedules discussed by the World Health Organisation show that a developed country uses a schedule for the first five years of a child's life and vaccines costs over $700 including administration costs; while a developing country uses a schedule providing vaccines in the first 9 months of life and costing only $25. This difference shows that health care costs are much lower in developing nations and more expensive vaccines are not used. In 1900, the only smallpox vaccine was administered to children. By the 1960s, children routinely received five vaccines (against diphtheria, pertussis, tetanus, polio, and smallpox), and eight shots by two years of age (Table 17.6).

III. EDIBLE VACCINES (PLANT-DERIVED VACCINES)

Green plants can also be used as the 'surrogate production organism' to produce antigens of human pathogens (including HBsAg) and these proteins can elicit priming and boosting immune response in humans when given orally. Unlike all cell lines used for production of vaccines, components of plant cells have always been an important part of the normal human diet. Therefore, plants offer significant new opportunities for making safe and effective oral vaccines. The following discussion will address issues relating to the commercial development of 'plant-derived vaccines' and their usefulness in preventing infectious diseases in poor populations of developing countries.

GM banana produces vaccines against diarrhea, Hepatitis B and measles

1. Edible Vaccines

The plants are capable of producing vaccines in large quantities at low cost but the purification may require more cost. Therefore, attention has been paid to produce such antigens that stimulate mucosal immune system to produce secretary IgA (S-IgA) at mucosal surface such as gut and respiratory epithelia because of their effectiveness on sites as most of the pathogens invade these regions. For example, bacteria and viruses are transmitted via contaminated food or water and cause diseases such as diarrhea, whooping cough, etc. In 1990, the first report of the production of edible vaccine (a surface protein from *Streptococcus*) in tobacco at 0.02 per cent of total leaf protein level was published in the form of a patent application under the International Patent Cooperation Treaty. Thereafter, expression of a number of antigens in plants was successfully made and reported

(Table 17.7).

Table 17.7. Antigens produced in transgenic plants.

Proteins	***Plants***
Hepatitis B surface antigen	Tobacco, banana
Rabies virus glycoprotein	Tomato
Norwalk virus capsid protein	Tobacco
E. coli heat-labile enterotoxin β-subunit	Potato
Cholera toxin β-subunit	Potato, tobacco
Mouse glutamate dehydrogenase	Potato
VP1 protein of Foot & mouth disease virus	*Arabdiopsis*
Insulin	Potato
Glycoprotein of swine-transmissible gastroenteritis coronavirus	*Arabdiopsis*

In 1997, edible vaccines were first tested on humans, when scientists asked volunteers to eat anti-diarrheal transgenic potatoes produced by the Boyce Thompson Institute at Cornell University. After consuming the potatoes, almost all the volunteers produced antigens in their bodies just as if they had received a traditional anti-diarrheal vaccination. And they experienced no adverse side effects. Volunteers also tested raw potatoes engineered to produce a hepatitis B antigen at the Roswell Park Cancer Institute in Buffalo, New York.

Acute watering diarrhea is caused by enteroxigenic *Escherichia coli* and *Vibrio cholerae* that colonise the small intestine and produce enterotoxin. Cholera toxin (CT) is very similar to *E. coli* toxin. The CT has two subunits, A and B. Attempt was made to produce edible vaccine by expressing heat-labile enterotoxin (CT-B) in tobacco and potato.

In 2001, T. Arakawa and coworkers from Loma Linda University reported the transgenic plant-based cholera vaccines. A tobacco plant was produced that expressed CT-A or CT-B subunits of the toxin CT-A. CT-A produced in plant was not cleaved into CT-A1 and CT-A2 subunits which generally happens in epithelial cells. Similarly, CT-B subunit when expressed in potato was processed in natural way, the pentameric form (the naturally occurring form) being the abundant form. Even after boiling transgenic potato tubers till they became soft, about 50% of the CT-B was present in the pentameric GM1 ganglioside-binding form.

Similarly, a rabies virus coat glycoprotein gene has been expressed in tomato plants. Orally administered protein elicited protective immunity in animals. It was hoped that after further effort an edible oral vaccine against rabies can be developed. The value of vaccine can be improved by providing other adjuvants which either enhance the immunogenic potential or reduce degradation of the active ingredients by the microorganism of gut.

The obtainment of transgenic edible plants carrying recombinant antigens is a desired issue in search for economic alternatives viewing vaccine production. A strategy for genetic transformation of lettuce plants (*Lactuca sativa*) using the surface antigen HBsAg of hepatitis B virus has been reported in 2008 by Marcondes and Hansen. Transgenic lettuce seedlings were obtained through the application of a regulated balance of plant growth regulators. Genetic transformation process was acquired by co-cultivation of cotyledons with *Agrobacterium tumefaciens* harboring the recombinant plasmid.

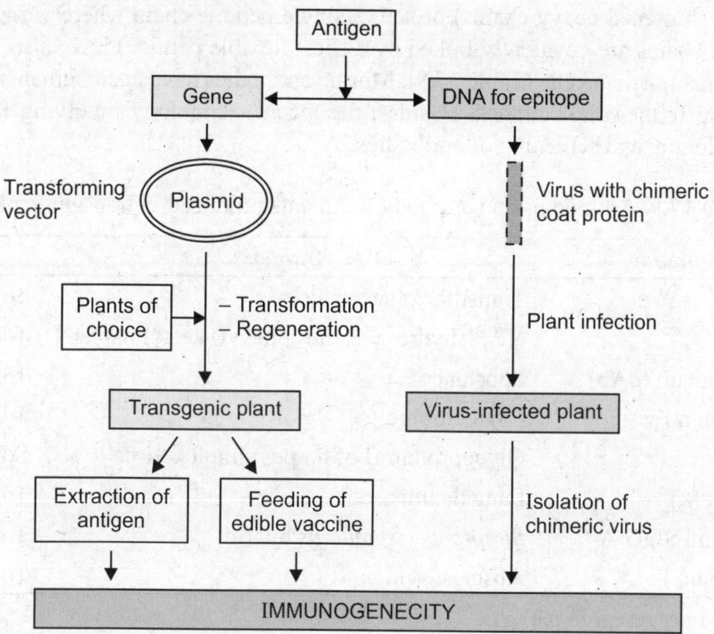

Fig. 17.19. Outline of production of candidate vaccine through transgenic plants.

In transgenic tobacco plants the hepatitis B surface antigen (HBsAg) accumulates to 0.01% of soluble protein level. The HBsAg was recovered in virus like particles of 22 nm diameter (similar to yeast-derived HBsAg-based vaccine) which is known to be a prerequisite for better immunogenicity (Fig. 17.19). A crude extract from plant was used for immunisation in mice. Immune response included all IgG subclasses and IgM against hepatitis B.

One of the alternative strategies of producing a plant-based vaccine is to infect the plants with recombinant virus carrying the desired antigen that is fused to viral coat protein. The infected plants have been reported to produce the desired fusion protein in large amounts in a short duration. The technique involved either placing the gene downstream a subgenomic promoter, or fusing the gene with capsid protein that coats the virus (Fig. 17.19; Table 17.8).

Table 17.8. Transient production of antigens in plants after infection with plant viruses expressing a recombinant gene.

Protein	*Plant*	*Carrier*
Influenza antigen	Tobacco	Tobacco mosaic virus
Murine zona pellucida antigen	Tobacco	Tobacco mosaic virus
Rabies antigen	Spinach	Alfalfa mosaic virus
HIV-1 antigen	Tobacco	Alafalfa mosaic virus
Mink enteritis virus antigen	Black eyed bean	Cowpea mosaic virus
Colon cancer antigen	Tobacco	Tobacco mosaic virus

2. Edible Antibodies (Plantibodies)

Transgenic plants are being looked up as a source of antibodies which can provide passive immunisation by direct application. They provide as a tool for drug targeting. Gene technology has provided impetus to the utility of antibodies. The genes coding for both light and heavy chains have been expressed. Moreover, the modified genes capable of expressing Fab fragments (assembled light chain and shortened heavy chains) or scFV (single peptide chain where variable domains of heavy and light chains are covalently linked by a short flexible peptide) have also been expressed in bacteria and mammalian cells (Table 17.9). Murine antibodies have been humanised by changing the constant and framework domains. Besides, the recent technology involving PCR and phage display allow cloning and screening of antibodies.

Table 17.9. Antibodies and antibody fragments produced in transgenic plants.

Antibody	*Antigen*	*Plant*
IgG (k)	Transition stage analogy	Tobacco
IgM (l)	NP (4-hydroxy-3-nitrophenyl) acetyl hapten	Tobacco
Single domain (dAb)	Substance P	Tobacco
Single chain Fv	Phytochrome	Tobacco
IgG	Glycoprotein B of herpes simplex virus	Soyabean
Fab : IgG (k)	Fungal cutinase	Tobacco
IgG (k) and SIgG/A	*Streptococcus mutans* hybrid	Tobacco
Single chain Fv	Abscisic acid	Tobacco

Source : Many research papers.

In 1995, a molecular biologist Shengwu Ma of the London Health Sciences Center and his team successfully produced multimeric secretary IgA (SIgA) molecules in plants which represent the predominant form of immunoglobulin in mucosal secretions. SIgA contained light and heavy chains, and domainised by a J-chain. The chains were protected by a fourth polypeptide. Thus, four transgenic tobacco plants were produced by genetic engineering which produced a murine monoclonal antibody like κ-chains, hybrid IgA-G antibody heavy chain, murine J-chain and rabbit secretary component. A series of sexual crosses were carried out to allow expression of all the four proteins simultaneously. The progenies produced a functional secretary immunoglobulin very efficiently. This demonstrates the potential of plants in assembly of antibodies, and flexibility of system (Fig. 17.20).

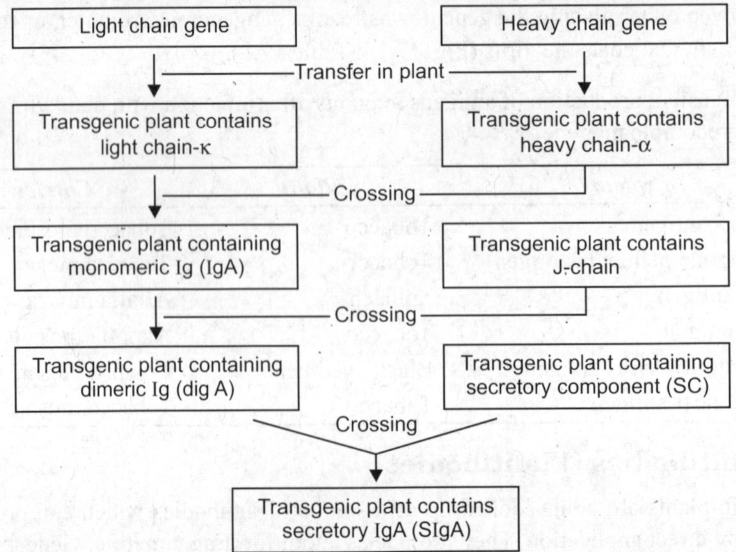

Fig. 17.20. Outline for production of secretory antibody in plant.

It has similar potential to combat other autoimmune diseases such as multiple sclerosis, rheumatoid arthritis, and lupus, and even transplant rejection. Plants are ideal because they can synthesise and assemble proteins to provide huge quantities of soluble proteins at relatively low cost. Many traditional drugs are difficult to make and are costly. However, once this technology is perfected, growing transgenic potatoes will cost very little.

In 1998, Ma and co-workers successfully used a hybrid monoclonal antibody (IgA/G) having constant regions of IgG fused with IgA against human dental carriers caused by the bacterium, *Streptococcus mutans.* The secretary antibody generated SIgA/G in transgenic tobacco and the original mouse IgG was compared. It is interesting to know that both had similar binding affinity to surface adhesion protein of *S. mutans.* SIgA/G survived for 3 days in the oral cavity, whereas IgG survived only for one day. The plant antibody provided protection against the colonisation of *S. mutans* for at least four months.

3. Edible Interferon

Scientists at ICGEB have successfully produced transgenic tobacco and maize plants that secrete human interferon (IFN-γ). It was produced by transformation of nucleus and chloroplast through particle bombardment method. It has been found that interferon was 10-15 folds greater in chloroplast-transformed plants than nucleus-transformed plants.

What are Plantibody ?

A plantibody (plant and antibody) is an antibody produced by genetically modified crops. Antibodies are part of animal immune systems and are produced in plants by transforming them with antibody genes from animals. For the first, A. Hiatt, R. Caffferkey and Ka Bowdish a paper in *Nature* (342) in 1989 on production of antibodies in transgenic plants. They got success in producing a mouse antibody by tobacco plants. Plants expressing single gamma or kappa immunoglobulin chains were crossed to yield progeny in which both chains expressed simultaneously. A functional antibody accumulated to 1.3% of total leaf protein in plants expressing full-length cDNAs containing leader sequences. The results demonstrate that production of immunoglobulins and assembly of functional antibodies occurs very efficiently in tobacco. Assembly of subunits by sexual cross might be a generally applicable method for expression of heterologous multimers in plants.

Plants do not naturally make antibodies; but plantibodies have been shown to function in the same way as normal antibodies. A company Biolex has trademarked the concept and the term plantibody. Theré are two potential major applications of plantibodies. Besides Biolex, a number of other companies such as Planet Biotechnology (of California) and Medicago (of Canada)

D. TRANSPLANTATION (TRANSPLANT IMMUNOLOGY)

The phenomenon of moving of organs/tissues/fluid from one body to another or from a donor to the patients to replace the damaged or absent organ of the later is called 'transplantation', and the transplanted organs/tissues/fluids are called transplants. The emerging field of 'regenerative medicine' may soon allow organs to be re-grown from the own cells (stem cells or cells extracted from the failing organs) of the patients. The organs to be transplanted are: heart, kidneys, liver, lungs, pancreas, intestine, thymus, and skin. Tissues to be transplanted include: bones, tendons, cornea, heart valves, and veins. The kidneys are the most commonly transplanted organs throughout the world. Similarly blood cells from one person to the other person are also transferred.

Allotransplantation (Greek: *allo-* means 'other') is the transplantation of cells, tissues, or organs, sourced from a genetically non-identical member of the same species as the recipient. The transplant is called an allograft e.g. most human tissue and organ transplants. Besides, a transplant from the other species is called a xenograft. A transplanted organ or tissue from a genetically identical donor, *i.e.* an identical twin is termed an isograft. When a tissue is transplanted from one site to another on the same patient it is termed an autograft. In bone marrow transplantation, the term for a genetically identical graft is syngeneic, whereas the equivalent of an autograft is termed autologous transplantation.

I. ORGAN TRANSPLANTS

Successful human transplants, *i.e.* allotransplants have a relatively long history; the operative skills were present long before the discovery of necessities for post-operative survival. It has been reported that Pien Chi'ao (Chinese physician) exchanged hearts between a man of strong spirit but weak will with one of a man of weak spirit but strong will to get a balance in each man.

In the 2nd century BC, the Indian surgeon Sushruta gave the first reasonable account, who used autografted skin transplantation in nose reconstruction rhinoplasty. Success or failure of these procedures is not well documented. In 3rd century AD, Saints Cosmas and Damian miraculously transplanted the black leg of a dead slave onto an elderly servant.

In 1837, the first successful corneal allograft transplant was performed in a gazelle model. In 1905, the first successful human corneal transplant was performed by Eduard Zirm in Czech Republic. In the early 1900s the French surgeon Alexis Carrel and Charles Guthrie carried out the first pioneering work in the surgical technique of transplantation of arteries or veins. Their skillful new suturing techniques, laid the groundwork for later transplant surgery and Carrel won the Nobel Prize in Physiology or Medicine in 1912.

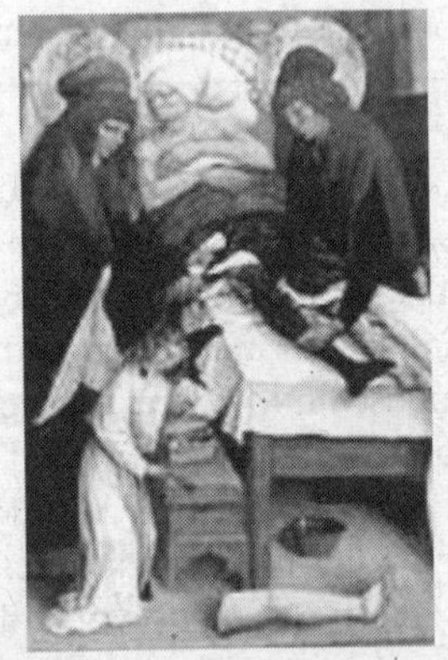
Transplant done in 3rd century

From 1902 Carrel performed transplant experiments on dogs. Surgically he was successful in moving kidneys, hearts and spleens; he was one of the first to identify the problem of rejection. During World War I, the major steps in skin transplantation were taken up; during World War II the work on reconstructive surgery was carried out. In 1962 the first successful replantation surgery, *i.e.* re-attaching a severed limb and restoring (limited) function and feeling was performed.

In 1926, a Russian surgeon Peter Vasil'evič Kolesnikov transplanted a testis from a living donor to a recipient. The donor was a convicted murderer whose death sentence was commuted to 20 years imprisonment, and he was led to believe that it was done because he had donated his testis to an elderly medical doctor. Both the donor and the receiver survived, but charges were brought in a court of law by the public prosecutor against Kolesnikov, not for performing the operation, but for lying to the donor. In 1954, Joseph Murray performed the first successful kidney transplant between identical twins, and it was successful because no immunosuppression was necessary in genetically identical twins. In 1963, James Hardy in Mississippi performed a successful deceased-donor lung transplant into a sufferer of lung cancer. The patient survived for 18 days before death of kidney failure. Cyclosporine altered transplants from research surgery to life-saving treatment. In 1968 surgical pioneer Denton Cooley performed 17 transplants including the first heart-lung transplant. His 14 patients died within six months. By 1984 two-thirds of all heart transplant patients survived for five years or more.

There is substantive research on xenotransplantation (transgenic organs). Although these forms of transplant are not being used in humans, yet clinical trials involving the use of specific cell types have been conducted such as porcine islets of Langerhans to treat diabetes type one. Still there are many problems that need to be solved before patients requiring transplants. Researchers are looking into steroid-free immunosuppression. This type of immunosuppression is being pioneered on a large scale in the U.S.A. Besides, calcineurin inhibitor-free immunosuppression is currently undergoing extensive trialing, the result of which will allow sufficient immunosuppression without nephrotoxicity associated with standard regimens that include calcineurin inhibitors.

In 2008, Professor Paolo Macchiarini and his team of the University of Barcelona performed the first tissue engineered trachea (wind pipe) transplantation. Adult stem cells were extracted from the patient's bone marrow, grown into a large population, and matured into cartilage cells (chondrocytes) by using an adaptive method originally devised for treating osteoarthritis. The newly grown chondrocytes and epithelial cells were seeded into a decellularised (free of donor cells) tracheal segment donated from a 51 year old donor who died of cerebral hemorrhage. After four days of seeding, the graft was used to replace the main bronchus of the patient.

One of the most challenging and complex area of modern medicine is the 'transplantation medicine'. The key areas for medical management are the problems of transplant rejection, during which the body has an immune response to the transplanted organ, leading to transplant failure and the need to immediate removal of the organ from the recipient. In 2001, C. Frohn and co-workers reported the effect of HLA-C matching on acute renal transplant rejection. Transplant rejection can be reduced through serotyping to determine the most appropriate donor-recipient match and through the use of immunosuppressant drugs.

There is a shortage of suitable organs for transplantation in most countries. However, transplantation also raises a number of bioethical issues including the definition of death, when and how consent should be given for an organ to be transplanted and payment for organs for transplantation. A chronological order of timeline of successful transplants is given below :

1905 - First successful cornea transplant by E. Zirm
1954 - First successful kidney transplant by J. Murray (U.S.A.)
1966 - First successful pancreas transplant by R. Lillehei and W. Kelly (Minnesota, U.S.A.)
1967 - First successful liver transplant by T. Starzl (U.S.A.)
1981 - First successful heart/lung transplant by B. Reitz (U.S.A.)
1983 - First successful lung lobe transplant by J. Cooper (Canada)
1986 - First successful double-lung transplant by J. Cooper (Canada)
1995 - First successful laparoscopic live-donor nephrectomy by L. Ratner and L. Kavoussi (U.S.A.)
1998 - First successful live-donor partial pancreas transplant by D. Sutherland (U.S.A.)
1999 - First successful tissue engineered bladder transplanted by A. Atala (U.S.A.)
2005 - First successful partial face transplant (France)
2006 - First jaw transplant to combine donor jaw with bone marrow from the patient by E. M. Genden (New York)
2008 - First baby born from transplanted ovary.
2008 - First transplant of a human windpipe using a patient's own stem cells by P. Macchiarini (Spain).

1. Types of Transplants

There are different types of transplants used by persons as and when required as given below:

(*a*) Autograft: Transplant of tissue to the same person is called 'autograft'. Sometimes, it is done with surplus tissue or tissue that can regenerate or tissues more desperately needed elsewhere (for example skin grafts, vein extraction, etc.) An autograft is also carried out to remove the tissue and then treat it or the person before (*e.g.* stem cell autograft and storing blood in advance of surgery).

***(b)* Allograft:** Transplant of an organ or tissue between two genetically non-identical members of the same species 'allograft' or allogeneic transplant or homograft. Mostly the human tissue and organ transplants are allografts. Due to the genetic difference between the donor and the recipient, the immune system of the recipient identifies the organ as foreign, and attempt to destroy it causing transplant rejection. Possibly, transplant rejection can be reduced through serotyping to determine the suitable donor-recipient match and through the use of immunosuppressant drugs. This affects the whole immune system and making the body susceptible to infection by pathogens.

***(c)* Isograft:** It is a subset of allografts in which organs or tissues are transplanted from a donor to a genetically identical recipient (*e.g.* an identical twin). Isografts are differentiated from other types of transplants because they are anatomically identical to allografts and do not trigger an immune response.

***(d)* Split Transplants:** Sometimes a deceased-donor organ (usually a liver) is divided between two recipients especially an adult and a child. Usually this method is not preferred because the transplantation of a whole organ is more successful.

***(e)* Domino Transplants:** When a patient is suffering from cystic fibrosis, this operation is usually performed because both lungs need to be replaced. Technically it is an easier operation to replace the heart and lungs at the same time. When the recipient's heart is healthy, it can be transplanted into the other whom a heart needs transplant. Domino transplant is also used for a special form of liver transplant in which the recipient suffers from familial amyloidotic polyneuropathy, a disease where the liver slowly produces a protein that damages other organs. Liver of such patient can be transplanted into an older patient who is likely to die from other causes.

***(f)* Xenograft or Xenotransplant:** Xenograft or xenotransplant (Greek, *xeno* means 'foreign') is a transplant of living organs, tissue or cells from one species to another such as from pigs to humans. The other examples are: porcine heart valve transplants, piscine-primate (fish to non-human primate) transplant of islet tissue, *i.e.* pancreatic tissue, etc. However, xenotransplantion is an extremely dangerous type of transplant because of the increased risk of non-compatibility, rejection and disease present in the tissue.

Human xenotransplantation offers a potential treatment for end-stage organ failure, a significant health problem in parts of the industrialised world. It also raises many novel medical, legal and ethical issues. A continuing concern is that pigs have shorter life spans than humans: their tissues age at a different rate. Some successful cases of xenotransplantation have also been reported.

Why Pigs are used in Xenotransplants ?

Pigs are currently thought to be the best candidates for organ donation. The risk of cross-species disease transmission is decreased because of their increased phylogenetic distance from humans. They are readily available, their organs are anatomically comparable in size, and new infectious agents are less likely since they have been in close contact with humans.

***(i)* Future of xenotransplants:** Worldwide there is a shortage of organs for clinical implantation. About 60% of patients waiting for replacement organs die on the waiting list. Recently the mechanisms of transplant rejection has been understood and reasonably considered that organs from other species (probably pigs) may be engineered to minimise the risk of rejection and used as an alternative to human tissues. Possibly it will end the organ shortages. Secondly, the other methods being investigated for clinical trials are the use of cells or tissues from other species to treat life-threatening illnesses such as cancer, diabetes, liver failure and Parkinson's disease. The perfected vitrification can allow for long-term storage of xenogenic cells, tissues and organs to make it more readily available for transplant. The animal organ (*e.g.* pig or baboon) could be genetically altered with human genes so that the immune system of a patient can be tricked and it can accept as a part of its own body. Xenotransplants are thus potentially a more effective alternative. In 2010, P.E.

Bols and coworkers carried out xenotransplantation in immunodeficient mice or SCID mice to study ovarian follicular development in domestic animals. Mature follicles have developed even after use of cryopreserved ovarian tissue. Both host and graft vessels contribute to the revascularisation of xenografted human ovarian tissue in a mouse.

Chimpanzees were originally considered to be the best option since their organs are of similar size and they have good blood compatibility with humans. Since, chimpanzees are listed as an endangered species; they cannot be used as donor. Baboons are more readily available, but they are also not potential donors due to their smaller body size, infrequency of blood group O (the universal donor), their long gestation period, and production of few offspring. Since nonhuman primates are closely related to humans, there is a major problem of the increased risk of disease transmission.

Major Organs and Tissues to be Transplanted

- **Thoracic organs:** Heart, lung, heart/lung
- **Abdominal organs**, Kidney, liver, pancreas, intestine, stomach, testis
- **Tissues, cells, fluids:** Hand, cornea, skin including face replant (autograft) and face transplant (extremely rare), islets of Langerhans, bone marrow/adult stem cell, blood transfusion/blood parts transfusion, blood vessels, heart valve (porcine/bovine), bone.

2. Types of Donor

There are two types of donor such as living and deceased donors as given below:

(*a*) Living or Deceased: In living donor, the donor remains alive and donates a renewable tissue, cell or fluid (*e.g.* blood, skin). An organ/part of an organ is donated by the donors in whom the remaining organ can regenerate or take on the workload of the rest of the organ (mainly a single kidney donation, partial donation of liver, small bowel). One day regenerative medicine may allow for laboratory-grown organs by using patient's own cells (stem cells, or healthy cells extracted from the failing organs).

(*b*) Deceased (formerly Cadaveric): Deceased donors are such donors who have been declared brain-dead and whose organs are kept viable by ventilators/other mechanical mechanisms until these are excised for transplantation. Brain-stem dead donors have formed the majority of deceased donors for the last twenty years. Besides, there is an increasing use of 'donation after cardiac death' (DCD) donors (formerly non-heart beating donors) to increase the pool of donors to meet the demand of transplants. But these organs do not give good results as compared to the organs from a brain-dead donor.

3. Reasons for Donation

Following are the reasons for donating the organs:

(*a*) Living Related Donors: Living related donors donate to family members or friends in whom they have an emotional investment. The risk of surgery is offset by the psychological benefit of not losing someone related to them or not seeing them suffer the ill effects of waiting on a list. A 'paired-exchange' is a technique of matching willing living donors to compatible recipients using serotyping (Fig. 17.21). For example a spouse desires to donate a kidney to his/her partner but cannot donate because it cannot match biologically. The kidney of willing spouse is

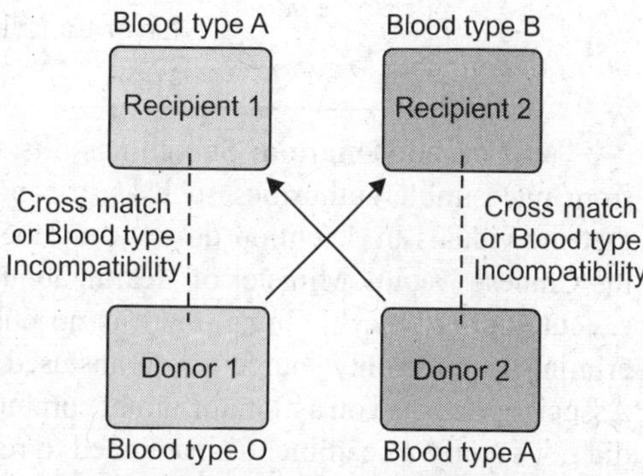

Fig. 17.21. Paired-exchange technique.

donated to a matching recipient who also has an incompatible but willing spouse. The second donor must match the first recipient to complete the pair exchange. In 1997, paired exchange programmes were popularised by L.F. Ross.

A paired exchange is the simplest case of a much larger exchange registry programme where willing donors are matched with any number of compatible recipients.

(*b*) Compensated Donation: In this donation, the donors are provided money or other compensation in exchange for their organs. Whether legal or not, but this practice is common in some parts of the world, and is one of the many factors driving medical tourism. The National Organ Transplant Act of 1984 of the U.S.A. has made organ sales illegal. The Human Organ Transplant Act 1989 of the United Kingdom first made the organ sales illegal, and has been superseded by the Human Tissue Act 2004. In 2007, two European conferences recommended that the 'sale of organs' as illegal. Two books, *Kidney for sale by owner* by Mark Cherry (Georgetown University Press, 2005), and *Stakes and kidneys: why markets in human body parts are morally imperative* by James Stacey Taylor, Ashgate Press, 2005) advocate using markets to increase the supply of organs available for transplantation.

(*i*) Comparative cost of organ: In 2006, Iran became the only country to legally allow individuals to sell their kidneys, and the market price was of the order of US$ 2,000 to US$ 4,000. In Pakistan, 40 to 50 % of the residents of some villages have only one kidney because they have sold the other for a transplant into a wealthy person of the other country. Pakistani donors are offered $ 2,500 for a kidney but receive only about half of that because middlemen take so much (WHO, March 30, 2007). In Chennai, southern India, poor fishermen and their families sold kidneys after their livelihoods were destroyed by the Indian Ocean tsunami in December 26, 2004. About 100 people, mostly women, sold their kidneys for ₹ 40,000-60,000 ($900–$1,350) (http://www.reuters.com/article/healthNews/idUSDEL21432720070116.). A comparative cost of organ transplant in some countries is given in Table 17.10.

Table 17.10. Comparative cost of organ/organ transplantation.

Comparative cost of organs		
Country	***Organs/ organ transplantation***	***Cost***
Latin America	Human kidney	< $ 10,000
South Africa	Human kidney	< $ 20,000
China	Human kidney	$ 70,000
	Liver transplantation	$ 1,60,000
	Heart transplantation	$ 1,20,000
U.S.A.	Kidney transplantation	$ 1,00,000
	Liver transplantation	$ 2,50,000
	Heart transplantation	$ 8,60,000

(*c*) Forced Donation: Sometimes it is accused that certain authorities are harvesting organs from undesirable authorities such as prison populations. The World Medical Association states that individuals in detention do not give their free consent to donate their organs. According to the Chinese Deputy Minister of Health, about 95% of all organs used for transplantation are from executed prisoners. In China, there is no public organ donation programme; hence the executed criminals are the only source for organs used in transplants. In October 2007, the Chinese Medical Association agreed on a moratorium of commercial organ harvesting from condemned prisoners, but did not specify a deadline. China agreed to restrict transplantations from donors to their immediate relatives.

4. Organ Transplant Laws

Various policies have been forged by both the developed and developing countries so that the safety and availability of organ transplants to their citizens could be increased. Brazil, France, Italy, Poland and Spain have framed rules for all adults to opt as potential donors. Moreover, Iran is the only country in the world that has constituted the law for the citizens to sell an organ to another for transplantation.

The Government of India has difficulty in tracking the flourishing of 'organ black market' in the country and officially condemns such incidence occurring in the country. The other countries which have been victimised by illegal organ trade have implemented legislative reactions.

Since July 2006, China has made selling of organs illegal and claims that all prisoner organ donors have filed consent. Starting from May 1 (2007) doctors involved in commercial trade of organs face fines and suspensions in China. In order to curb illegal transplants, only a few certified hospitals will be allowed to perform organ transplants. Harvesting organs without donor's consent is also supposed to be a crime (http://news.bbc.co.uk/2/hi/asia-pacific/6534363.stm.).

However, doctors in the United Kingdom have accused China of abusing its high capital punishment rate. Despite these efforts, illegal organ trafficking continues to thrive and can be attributed to corruption in healthcare systems, which has been traced as high up as the doctors themselves in China, Ukraine and India. The blind eye economically strained governments and health care programmes must turn to organ trafficking. Besides, some organs are also shipped to Uganda and the Netherlands. This was a main product in the triangular trade in 1934.

5. Ethical Concerns

In spite of benefits to the organs receivers, there are many ethical concerns in developing countries regarding the procedures for the existence and distribution of organ transplantation. The major ethical issues and the notion of distributive justice are both the sources and method of obtaining the transplantable organs. The World Health Organization argues that transplantations promote health, but the notion of 'transplantation tourism' has the potential to violate human rights or exploit the poor.

In the developing countries this might be coercive irrespective of 'gift of life'. The coercion practice may be exploitative to the poor, and violating basic human rights according to Articles 3 and 4 of the Universal Declaration of Human Rights. Also there is a powerful opposition that organ trade is a mutually beneficial transaction between two consenting adults if it is properly and effectively regulated to ensure that the seller is fully informed of about the consequences of donation. Prohibition of organ trade would itself be a violation of Articles 3 and 29 of the Universal Declaration of Human Rights.

In 2005, L. Whetstine and coworkers have dealt with the pro/con ethics debate. They found the concern in the developed countries that enthusiasm may trample for the right to life for increasing the supply of organs. The question is made even more complicated by the fact that the 'irreversibility' criterion for legal death cannot be defined and can easily change with changing technology.

II. HAEMATOPOIETIC STEM CELL TRANSPLANTATION

Transplantation of blood stem cells derived from the bone marrow or blood is called haematopoietic stem cell transplantation (HSCT). Mostly it is performed in such patients who are suffering from diseases of the blood, bone marrow or certain types of cancer.

In 2008, R.K Burt and coworkers have reported the clinical applications of blood-derived and marrow-derived stem cells for nonmalignant diseases. After the availability of stem cell growth factors (GM-CSF and G-CSF) most haematopoietic stem cell transplantation are performed by using stem cells collected from the peripheral blood but not the bone marrow. Peripheral blood stem cells provide a bigger graft. It does not require that the donor be subjected to general anesthesia to collect the graft, results in a shorter time to engraftment and may provide a lower long-term relapse rate. It is a risky procedure having many possible complications. Traditionally it is reserved for those patients who suffer from life-threatening diseases.

Between 1950 and 1970, a team led by E. Donnall Thomas (at the Fred Hutchinson Cancer Research Center) pioneered stem cell transplantation by using bone-marrow-derived stem cells. Their work was later recognised with a Nobel Prize in Physiology or Medicine. Thomas showed that bone marrow cells infused intravenously could repopulate the bone marrow and produce new blood cells. His work also reduced the likelihood of developing a life-threatening complication called graft-versus-host disease.

In 1968, Robert A. Good at the University of Minnesota was the first physician to perform a successful human bone marrow transplant. Candidates for HSCTs are: pediatrics who have an inborn error such as severe combined immunodeficiency or congenital neutropenia with defective stem cells, children or adults with aplastic anaemia who have lost their stem cells after birth. Other conditions treated with stem cell transplants include sickle-cell disease, myelodysplastic syndrome, neuroblastoma, lymphoma, Ewing's sarcoma, desmoplastic small round cell tumour and Hodgkin's disease.

1. Types of HSCT Graft

There are two types of HSCT grafts as described below :

***(a)* Autologous Graft:** Autologous HSCT requires the extraction (apheresis) of haematopoietic stem cells (HSC) from the patient and storage of the harvested cells in a freezer. Then a high-dose chemotherapy with or without radiotherapy is given to the patient to eradicate the malignant cell population of the patients at the cost of partial or complete bone marrow ablation. Thereafter, stored stem cells of the patient are returned to their body, where they replace the destroyed tissue and resume the normal blood cell production of the patients.

There is lower risk of infection in autologous transplants during the course of treatment because of rapid recovery of immune function. There is very rare incidence of rejection also because the donor and recipient both are the same individual. Autologous HSCT has become one of the standard second-line treatments for diseases such as lymphoma. In children and adults, non-myeloablative haematopoietic stem cell transplantation is being attempted as a possible treatment for type I diabetes (insulin dependent).

***(b)* Allogeneic Graft:** Two people are required for allogeneic HSCT: the donor (healthy) and the recipient (patient). HLA tissues of allogeneic HSC donors must match with that of the recipient. Matching is done based on the variability at three or more loci of the (HLA) gene and a perfect match is preferred. Allogeneic transplant donors may be: (*i*) *syngeneic* (a monozygotic or identical twin of the patient because they offer a source of perfectly HLA matched stem cells), or (*ii*) *unrelated* (sometimes few unrelated donors have a very close degree of HLA matching).

In 2005, I, Nivison-Smith and coworkers have described haemopoietic stem cell transplantation in Australia and New Zealand in 1992-2001 progress report from the Australasian Bone Marrow Transplant Recipient Registry. They have described the performance of allogeneic transplants by using umbilical cord blood, because it acts as a source of stem cells. By transplanting healthy stem cells to the immune system of recipients, allogeneic HCSTs improves the chances for cure or long-term remission once the immediate transplant-related complications are resolved.

Additional HLA-testing should be done from the blood of potential donors just to find out a compatible donor. There are two categories of HLA genes, Type I and Type II. Generally, the risk of graft rejection increases with mismatches of the Type-I genes *i.e.* HLA-A, HLA-B, or HLA-C.

In contrast the risk of graft-versus-host disease increases with a mismatch of an HLA Type II gene *i.e.* HLA-DR or HLA-DQB1. The perfect matches require knowledge of the exact DNA sequence of these genes for both donor and recipient.

2. Sources of HSCT

In allogeneic HSCT, the donor should have the same human leukocyte antigens (HLA) as the recipient for minimising the risks of transplanted stem cell rejection or the severe graft-versus-host disease. About 25 to 30 % of allogeneic HSCT recipients have an HLA-identical sibling. Even the perfect match has mismatched minor alleles that contribute to graft-versus-host disease.

(*a*) Bone Marrow: In case of a bone marrow transplant, the HSCs are removed from a large bone of the donor (giving general anesthesia) especially the pelvis by using a large needle which can reach the centre of the bone. This method is known as 'bone marrow harvest.

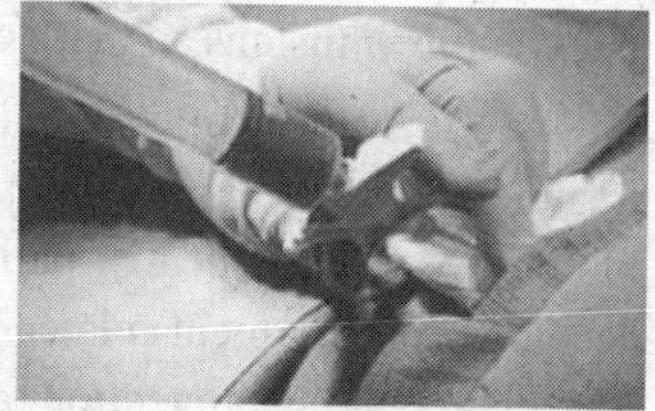

Harvesting the cells from bone marrow

(*b*) Peripheral Blood Stem Cells: Now the most common source of stem cells for allogeneic HSCT is peripheral blood stem cells (PBSCs). The PBSCs are collected from the blood through a process known as apheresis. A sterile needle is inserted in one arm of the donor and blood is withdrawn by passing through a machine that removes white blood cells. Again the red blood cells are returned to the donor. After injecting subcutaneously the granulocyte-colony stimulating factor daily, the peripheral stem cell yield is boosted. It mobilises stem cells from the donor's bone marrow into the peripheral circulation.

(*c*) Amniotic Fluid: Possibly the haematopoietic stem cells can be extract from amniotic fluid which can be used for both autologous or eterologous.

(*d*) Umbilical Cord Blood: When a mother donates the umbilical cord and placenta after the birth of her infant, the umbilical cord blood is obtained. Umbilical cord blood has a high amount of HSC as compared to that of a normal adult blood. However, for transplantation into small children a small quantity of blood (about 50 ml) obtained from an umbilical cord makes it more suitable than the adults.

Storage of HSC

Bone marrow cells can be frozen for a prolonged period through cryopreservation (at -196°C) without damaging too many cells. For autologous HSC it is necessary because the cells must be harvested for months in advance of the transplant treatment. But fresh HSC are preferred for allogeneic transplants to avoid cell loss that may occur during the freezing and thawing process. Allogeneic cord blood is stored frozen at a cord blood bank because it is obtained only at the time of childbirth. For cryopreservation of HSC, DMSO must be added as a preservative and the cells must be cooled very slowly in a control rate freezer so that osmotic cellular injury could not take place during ice crystal formation. HSC may be stored for many years in a cryofreezer which utilizes liquid nitrogen because it is non-toxic and very cold (-196°C).

3. Complications and Side Effects

A high treatment-related mortality in the recipient (10% or higher) has been found in cases of HSCT, therefore, its use becomes limited which themselves are life-threatening. The major complications are: veno-occlusive disease, mucositis, infections (sepsis) and graft-versus-host disease.

***(a)* Veno-occlusive Disease:** The phenomenon of severe liver injury is called hepatic veno-occlusive disease (VOD). The clinical hallmarks of this condition are elevated levels of bilirubin, hepatomegaly and fluid retention. The generalised cellular injury and obstruction in hepatic vein sinuses a greater appreciation; it has been referred to as 'sinusoidal obstruction syndrome' (SOS). A high rate of mortality has been found with severe cases. However, severity of VOD can effectively be reduced by using anticoagulants or defibrotide; but it may increase bleeding complications.

***(b)* Mucositis:** The injury of the mucosal lining of the mouth and throat is a common regimen-related toxicity following ablative HSCT regimens. Usually, it is not life-threatening but is very painful which prevents eating and drinking. Mucositis is treated with pain medications plus intravenous infusions so that dehydration and malnutrition can be prevented.

***(c)* Infection:** In bone marrow transplantation (usually myeloablation) the recipients undergo treatment for several weeks before starting engraft-ment without appreciable numbers of white blood cells. Consequently, patient remains at high risk of infections, sepsis and septic shock, and accounts for treatment-related mortality. The immunosuppressive agents employed in allogeneic transplants for the prevention or treatment of graft-versus-host disease increase the risk of infection. After transplantation immunosuppressive drugs are given for at least 6-months or more if required for the treatment of graft-versus-host disease. Transplant patients lose their acquired immunity *e.g.* immunity to childhood diseases such as measles or polio. For this reason transplant patients must be re-vaccinated with childhood vaccines.

***(d)* Graft-versus-host Disease (GVHD):** It is named graft-versus-host disease (GVHD) because bone marrow transplantation is the only transplant method in which the transplanted cells must accept the body rather than the body accepting the new cells. GVHD is an inflammatory disease which arises in allogeneic transplantation. In case of GVHD the new immune cells of bone marrow attack against the immune tissues of the recipients. Even if the donor and recipient are HLA-identical this can occur because the immune system can recognise the other differences between their tissues. Acute GVHD occurs in the first 3 months after transplantation and involves the skin, intestine or the liver, and is often fatal. A standard treatment is high-dose corticosteroids such as prednisone; but this immuno-suppressive treatment often leads to deadly infections. Chronic GVHD may also develop after allogeneic transplant.

III. TRANSPLANT REJECTION

When a transplanted organ or tissue is not accepted by the body of the recipient, the phenomenon of transplant rejection takes place, because immune system of the recipient attacks the transplanted organ or tissue. Such situation occurs in the transplant recipients, because the immune system distinguishes the foreign material within the body and destroys it, as it destroys the infectious organisms such as viruses and bacteria. Possibly, transplant rejection can be reduced via serotyping to determine the most appropriate 'donor-recipient match' and via using immunosuppressant drugs (http://ndt.oxfordjournals.org/cgi/content/full/16/2/355).

1. Types of Rejection

There are different types of transplant rejection as given below:

***(a)* Hyperacute Rejection:** A complement-mediated response in recipients with pre-existing antibodies to the donor (*e.g.* ABO blood type antibodies) is called 'hyperacute rejection'. It occurs within minutes. The transplant must be immediately removed so that a severe systemic inflammatory response may be prevented. In kidney transplants rapid risk of blood agglutination occurs. Therefore, before kidney transplantation a prospective cytotoxic cross-match is required to ensure that antibodies to the donor are not present. Hyperacute rejection is unlike blood transfusion reaction. Therefore, the problem of hyperacute rejection can be prevented by transplanting only ABO-compatible grafts. In liver allografts and cellular transplants hyperacute rejection is not significant because its tissues have remarkable ability for regeneration. In non-immunosuppressed recipients, hyperacute rejection is the outcome of xenotransplanted organ.

(*b*) Acute Rejection: Usually acute rejection starts after one week of transplantation. However, the risk of acute rejection is highest in the first three months after transplantation. Acute rejection can also occur months to years after transplantation. In all transplants, acute rejection occurs to some extent except identical twins. It is caused by mismatched HLA present on all cells of the body. There are a large number of different alleles of each HLA. Therefore, a perfect match between all HLA in the donor tissue and the body of the recipients is very rare. One week after transplantation, T-cells are involved in the rejection mechanism. The T-cells differentiate before starts of rejection. In the transplanted tissue, T-cells force the cells to lyse or to produce cytokines that results in breaking down of the transplanted tissue.

(*c*) Chronic Rejection: The case where the rejection is due to a poorly understood chronic inflammatory and immune response against the transplanted tissue is called 'chronic rejection.

2. Rejection Mechanisms

Transplant rejection is an adaptive immune response. It is mediated by both T cell-mediated and humoral (antibodies) immune mechanisms. The speed and magnitude of the rejection response is determined by the number of mismatched alleles. There are different mechanisms which act against different grafts (Table 17.11).

Table 17.11. Different mechanism involved in transplant rejection.

Organ/tissue	*Mechanism*
Blood	Antibodies (iso-haemagglutinins)
Kidney	Antibodies, CMI
Heart	Antibodies, CMI
Skin	CMI
Bonemarrow	CMI
Cornea	Usually accepted unless vascularised, CMI

CMI = Cell mediated immunity

3. Treatment of Rejection

Chronic transplant rejection cannot be treated effectively because it is irreversible. Re-transplantation is the only definitive treatment if donors are available. Moreover, acute transplant rejection can be treated using chemotherapeutic drugs designed to suppress the immune system such as high-dose corticosteroids, corticosteroid plus a calcineurin inhibitor and an anti-proliferative agent. In bone marrow transplant, the recipient's immune system can be replaced with the donor's immune system. This helps the recipient's body to accept the new organ without risk of rejection. Moreover, immunosuppressive drugs can be used to treat transplant rejection.

Monoclonal anti-T cell antibody *e.g.* OKT3 was used in treatment of severe acute rejection. But the recipient has fallen out of common use due to the severe 'cytokine release syndrome' and late 'post-transplant lymphoproliferative disorder'. Both of these are commonly associated with use of OKT3.

E. TUMOUR IMMUNOLOGY

Each and every normal cell follows the laws of chemistry. Cell division is a physiological process that occurs in almost all tissues and under many circumstances. Normally the balance between proliferation and programmed cell death is maintained by tightly regulating both processes to ensure the integrity of organs and tissues. Growth and proliferation of normal cells is carefully regulated. Mutations in DNA lead to cancer by disrupting these orderly processes via disrupting the programming that regulates the processes. Only certain mutations can lead to cancer and the majority of potential mutations will have no bearing. Cells may also undergo mutations when exposed to chemical carcinogens, irradiation and certain viruses resulting in their transformation into abnormal cells that show uncontrolled growth producing a tumour or neoplasm.

A tumour may be of two types– benign or malignant. If a cell does not grow and proliferate indefinitely and the host survives, it is called **benign tumour**. Benign tumours do not spread to other parts of the body or invade the other tissues. They are rarely a threat to life unless they compress vital structures or are physiologically active. Some types of these may turn into malignant tumours. On the other hand, if the tumour continues to grow indefinitely and spreads to distant locations (metastasises) eventually killing the host, it is called **malignant tumour** (cancer). The uncontrolled growth may be due to up regulation of oncogenes (*i.e.* cancer-causing genes) and/or down regulation of tumour suppressor genes that normally inhibit tumour growth often by inducing cell death. Thus, carcinogenesis is caused by this mutation of the genetic material of normal cells which upsets the normal balance between proliferation and cell death. This results in uncontrolled cell division and the evolution of those cells by natural selection in the body.

Cells infected with oncoviruses *e.g.* EBV and HPV produce abnormal proteins. Virus infected cells contain latent viral DNA which is transcribed and the resulting protein produces an immune response.

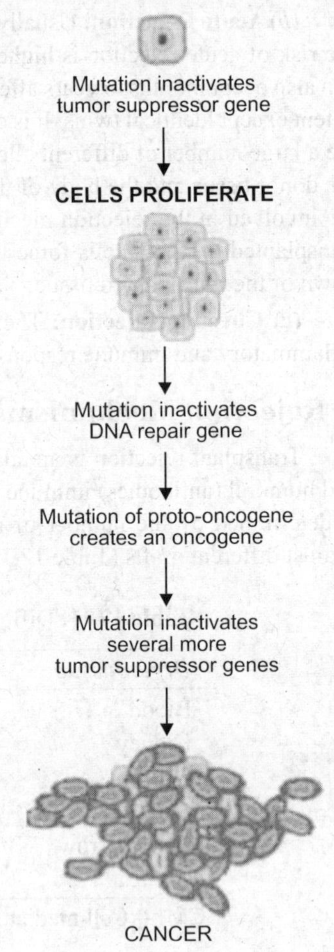

A series of mutations causes cancer.

I. TUMOUR ANTIGENS

A substance (produced in tumour cells) that triggers an immune response in the host is called tumour antigen. Tumour antigens are the proteins produced in a tumour cell; they have an abnormal structure due to mutation. Such abnormal proteins are produced due to mutation of the concerned gene. They are useful to identify tumour cells and are potential candidates for use in cancer therapy. The tumour must have antigens that are recognised as foreign, to react the immune system against a tumour. Many alterations occur in gene expression in cells during tumourigenesis. Tumourigenesis may lead to expression of new antigens (*i.e.* neoantigens) or alteration in existing antigens found on normal cells. The tumour antigens include: membrane receptors, regulators of cell cycle and apoptosis, or molecules involved in signal transduction pathways.

Modern Classification of Tumour Antigens

The older classification of tumour antigens is imperfect because many antigens thought to be tumour-specific are also expressed on some normal cells as well. Hence, the modern classification of tumour antigens is based on their molecular structure and source. Accordingly they can be classified as below:

- Products of mutated oncogenes and tumour suppressor genes
- Products of other mutated genes
- Overexpressed or aberrantly expressed cellular proteins
- Tumour antigens produced by oncogenic viruses
- Oncofoetal antigens
- Altered cell surface glycolipids and glycoproteins
- Cell type-specific differentiation antigens

Normally proteins are produced in very low quantities but its production is increased in tumour cells which trigger an immune response. For example, normally the enzyme tyrosinase is required in minute quantities for melanin production, but its levels are very much increased in melanoma cells. There are two main types of tumour antigens: *tumour-specific antigens*, and *tumour-associated antigens*.

1. Tumour-specific Antigens

The causes of the tumour are mutation of protooncogenes and tumour suppressors resulting in production of abnormal proteins which are called 'tumor-specific antigens'. Examples of tumour-specific antigens are the abnormal products of *ras* and *p53* genes.

Tumour-specific transplantation antigens (TSTA) are unique to tumour cells which are present only on tumour cells and not on any other cell; they are not expressed on normal cells. They are responsible for rejection of the tumour. Although chemical- induced tumours, UV- induced tumours or virus-induced tumours express neo-antigens. Most of these tumours are often weakly immunogenic or non-immunogenic. Mostly TSTAs cannot be identified easily. Some of these antigens may be secreted, while the others may be membrane-associated antigens.

In addition to proteins, other substances like cell surface glycolipids and glycoproteins may also have an abnormal structure in tumour cells and could thus be targets of the immune system.

Chemically-induced tumours are different from virally-induced tumours in that they are extremely heterogeneous in their antigenic characteristics. Thus, any two tumours induced by the same chemical (even in the same animal) rarely share common tumour specific antigens (Fig. 17.22). These unique antigens on chemically-induced tumours are referred to as tumour specific transplantation antigens (TSTA).

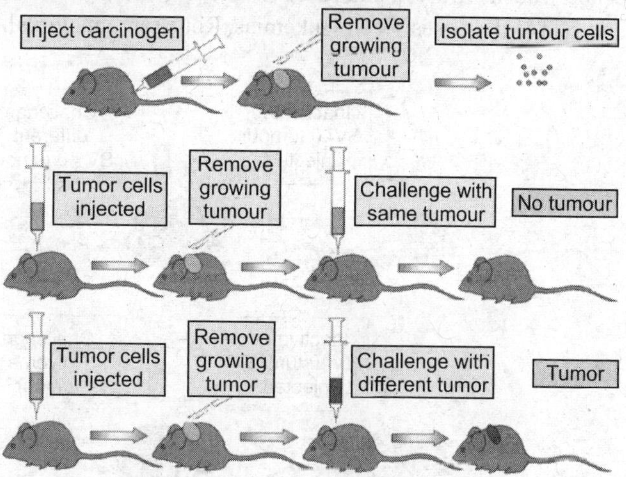

Fig. 17.22. Tumour development in mice challenged with same or different tumour.

2. Tumour-associated Antigens

Mutation of other genes unrelated to the tumour formation may lead to synthesis of abnormal proteins which are called tumour-associated antigens.

(*a*) Tumour-associated Transplantation Antigens (TATA): The majority of tumour antigens present on some tumour cells and also on some normal cells are referred to as 'tumour associated transplantation antigens'. They are expressed at higher levels on tumour cells as compared to normal cells. They may also be expressed only during development of cells and lost during adult life.

(*b*) Tumour-associated Developmental Antigens or Onco-foetal Antigens: Oncofoetal antigens are another important class of tumour antigens, for examples alphafoetoprotein (AFP) and carcinoembryonic antigen (CEA) found secreted in the serum. These proteins are normally produced in the early stages of embryonic development and disappear after full development of the immune system. Thus self-tolerance does not develop against these antigens.

AFP is found in patients with hepatocellular carcinoma as a secreted protein. In humans, the normal concentration of AFP is 0-20 ng/ml. This level rises considerably in patients with hepatomas and non-seminal testicular carcinoma. For monitoring hepatomas and testicular cancers, a 5-fold of this protein or more is used. AFP level may also be raised in some non-malignant conditions *e.g.* cirrhosis, hepatitis and other forms of liver damage.

In contrast, CEA is found in colon cancer present both on cell membranes and in secreted fluids. Since secreted antigens contribute little toward immunity against tumours, the role of these neo-antigens in immuno-surveillance is questionable. Normally CEA level in humans is found 2.5 ng/ml, but it increases in certain malignancies such as colo-rectal cancers. It may also rise in some non-malignant conditions such as chronic cirrhosis, pulmonary emphysema and heavy smoking.

3. Tumour-associated Transplantation Antigens on Viral Tumours

Many viruses cause different types of tumours in animals (*e.g.* SV-40 virus, adenovirus). Viruses that cause human tumours include: DNA viruses [*e.g.* papova viruses - papilloma viruses and polyoma viruses, papilloma virus (causes cervical cancer), hepatitis B virus (causes hepatocellular cancer), adenoviruses], and RNA viruses [*e.g.* retroviruses- human T-lymphotropic viruses (HTLV-I and HTLV-II) causes T cell leukemias, Rous sarcoma virus].

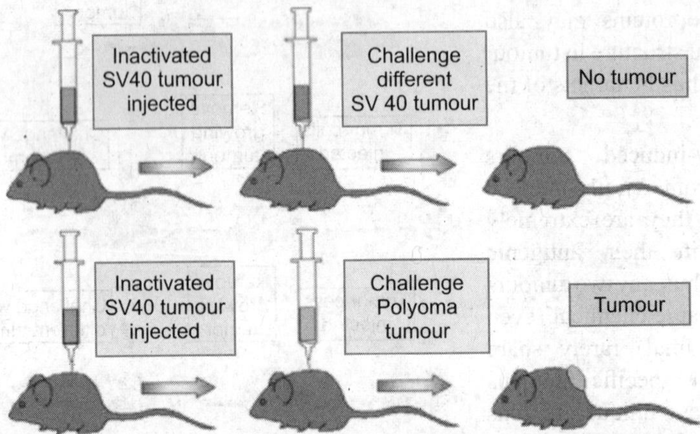

Fig. 17.23 . Antigens are characteristics to tumour-inducing virus.

Viruses are involved in some human malignancies (HTLV-1 in leukemia, hepatitis-B virus in hepatic carcinoma, papilloma virus in cervical cancer). Virus-induced tumours express cell surface antigens (distinct from antigens of the virion itself) which are shared by all tumours induced by the same virus. These antigens are characteristic of the tumour-inducing virus, irrespective of origin of tumour tissue or animal species in which the tumour exists (Fig. 17.23).

4. Importance of Tumour Antigens

Tumour antigens are useful in identifying specific tumour cells due to their relative abundance in tumour cells. Certain tumours have certain tumour antigens in abundance (Table 17. 12). Thus certain tumour antigens are used as tumour markers. In 1990, M. Hareuveni and collaborators have reported the use of tumour antigens in cancer therapy as tumour antigen vaccines.

Table 17.12. Tumour antigens present in different types of tumours.

Tumour antigen	*Tumor in which it is found*	*Remarks*
Alphafetoprotein (AFP)	Germ cell tumours	
Carcinoembryonic antigen (CEA)		
CA-125	Ovarian cancer	
MUC-1	Breast cancer	
epithelial tumor antigen (ETA)	Breast cancer	
Tyrosinase	Malignant melanoma	Normally present in minute quantities; greatly elevated levels in melanoma
Melanoma-associated antigen (MAGE)	Malignant melanoma	Also normally present in the testis
Abnormal products of *ras*, *p53*	Various tumours	

F. CANCER IMMUNOTHERAPY

The presence of neo-antigens on tumour cells has been exploited for both diagnostic and therapeutic purposes. Monoclonal antibodies labelled with radioisotope have been used for *in vivo* detection of relatively small tumour foci. Antibodies have also been used *in vitro* to identify the cell origin of undifferentiated tumours especially of lymphocytic origin. Immuno-histological staining is used to confirm suspected metastatic foci especially in bone marrow.

Immunotherapy has been used as a novel means of treating cancer. Both active and passive means of stimulating the non-specific and specific immune systems have been employed in some cases.

1. Active Immunotherapy

In this approach the host actively participates in mounting an immune response. Specific activation is achieved by using vaccines, such as hepatitis B vaccine, human Papilloma virus (HPV) vaccine (Table 17.13). Similarly non-specific activation is achieved by immunisation with Bacillus Calmette-Guerin (BCG), and *Corynebacterium parvum*. These activate macrophages to be tumouricidal.

2. Passive Immunotherapy

Passive immunity involves the transfer of preformed antibodies, immune cells and other factors into the hosts (Table 17.13).

Table 17.13. Immunotherapy of tumours.

Active	Non-specific	BCG, *Propionibacterium acnes*, levamisole, cytokine genes, etc.
	Specific	Killed tumour cells or their extract, recombinant antigens, idiotype, co-stimulatory molecule genes, etc.
	Non-specific	LAK cells, cytokines
Passive	Specific	Antibodies alone or coupled to drugs, pro-drug toxins or radioisotope; bispecific antibodies; T-cells
	Combined	LAK cells and bispecific antibody

* BCG: Bacillus Calmette Geurin is a bovine strain of *Mycobacterium tuberculosis*

Passive immunity is of the following two types :

***(a)* Specific Passive Immunotherapy:** Antibodies work against tumour antigens, *e.g.* Her2/Neu is used for treatment of breast cancer. Antibodies against IL-2R for human T lymphotropic virus (HTLV-1)-induced adult T cell leukemia.

***(i)* Antibodies against CD20:** Antibodies against CD20 are expressed on non Hodgkin's B cell lymphoma. These antibodies bind to tumour antigens on the cell surface and activate complement (C) to mediate tumour cell lysis. In addition, Fc receptor bearing cells such as NK cells, macrophages and granulocytes may bind to the antigen-antibody complexes on the tumour cell surface and mediate tumour cell killing through antibody-dependent cell-mediated cytotoxicity.

***(ii)* Conjugated antibodies:** Antibodies conjugated to toxins, radioisotopes and anti-cancer drugs have also been used. These enter the cells and inhibit protein synthesis *e.g.* anti-CD20 conjugated to *Pseudomonas* toxin or ricin toxin.

Problems in Using Antibodies

There are several problems with the use of antibodies

- Antibodies are not efficient because the tumour antigens are associated with class I MHC antigens.
- Tumours may shed antigen or antigen-antibody complexes. Therefore, immune cells cannot mediate tumour destruction.
- Some antibodies may not be cytotoxic.
- Antibodies may bind non-specifically to immune cells expressing the Fc receptors which include NK cells, B cells, macrophages and granulocytes without binding to tumour cells.

***(b)* Non-specific passive immunotherapy:** Non-specific passive immunotherapy involves the following:

- Adoptive Transfer of lymphocytes:
  - Lymphokine-activated killer (LAK) cells which are IL-2 activated T and NK cells.
  - Tumor-infiltrating lymphocytes (TIL)
- Dendritic cells pulsed with tumour antigens may induce tumour-specific T cell responses. Tumour lysates are used because tumour antigens are usually unknown.
- Cytokines
  - **IL-2:** It activates T cells/NK cells expressing IL-2 receptors. This is used in the treatment of renal cell carcinoma and melanoma.

- **IFN-alpha:** This induces MHC expression on tumours and used in the treatment of hairy B cell leukemias.
- **IFN-gamma:** This increases class II MHC expression, and used in the treatment of ovarian cancers.
- **TNF-alpha:** This kills tumor cells.

• Cytokine gene transfected tumour cells may also be used which can activate T or LAK cell-mediated anti-tumor immunity.

A number of cytokines have been used to potentiate the immune function of the host since the discovery that these cytokines have potent and selective effects on certain components of the immune system (Table 17.14).

Table 17.14. Cytokine therapy of tumours.

Cytokine	*Tumour type and result*	*Anti-tumour mechanism*
IFN-alpha, beta	Remission of hairy cell leukemia, weak effect on some carcinomas	Increased expression of class I MHC, possible cytostatic anti-tumour effect
IFN-gamma	Remission of peritoneal carcinoma of ovary: ineffective systemically	Increased MHC antigens; macrophage, Tc and NK cell activation
IL-2	Remission in renal carcinoma and melanoma	T-cell proliferation and activation, NK cells activation
TNF-alpha	Can reduce malignant ascites	Macrophage and lymphocyte activation

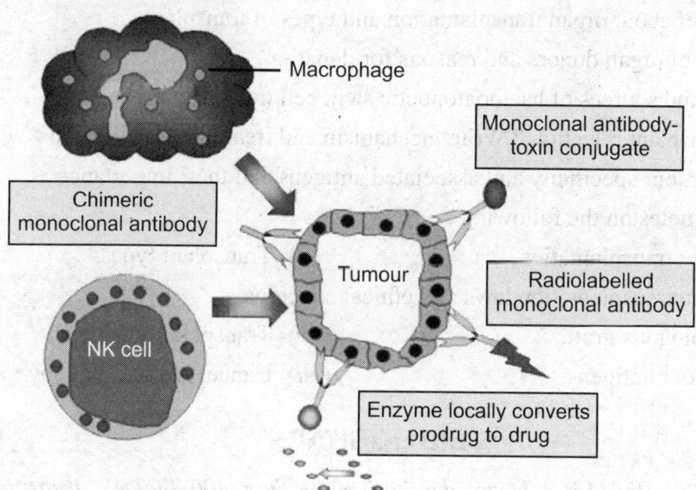

Fig. 17.24. Different forms of antibodies used in cancer therapy.

3. Monoclonal Anti-tumour Antibodies

For the treatment of cancer, monoclonal anti-tumour antibodies have been used in different forms because of their direct effect or as vehicles to target anti-cancer drugs, toxins and the non-specific components of the host's immune system to the site of tumour (Fig. 17.24). Such specific antibodies are also used in the diagnosis of metastatic lesions; It is not detectable by conventional radiologic means.

QUESTIONS

1. What are monoclonal antibodies ? Write the method of their production.
2. Describe production, purification and application of monoclonal antibodies.
3. Write in brief about monoclonal antibodies therapy and action of monoclonal antibodies.
4. What are interferons ? Write different types of interferons, cell activation and potential application.
5. Describe different types of vaccines and their role in immunity.
6. Discuss construction, administration and mechanism of action of DNA vaccines.
7. Discuss in detail about polio vaccines, advantages and disadvantages and vaccination schedule.
8. Describe edible vaccines and edible antibodies.
9. Write short notes on the following :

   (*i*) Monoclonal antibodies, (*ii*) Hybridoma technology,
   (*iii*) Application of MoAb, (*iv*) Chimeric MoAb,
   (*v*) Humanised MoAb, (*vi*) Human MoAb,
   (*vii*) MoAb therapy, (*viii*) Interferons,
   (*ix*) ATAT proteins, (*x*) Recombinant interferons,
   (*xi*) Interferon pegylation, (*xii*) Toxoid vaccines,
   (*xiii*) Conjugate vaccines, (*xiv*) DNA vaccines,
   (*xv*) Recombinant DNA vector vaccine, (*xvi*) Oral polio vaccine,
   (*xvii*) Edible vaccine, (*xviii*) Plantibodies.
10. Write in brief about organ transplantation and types of transplants.
11. Write types of organ donors and reasons for donation.
12. Write type and sources of haemoatopoetic stem cell transplantation.
13. What is transplant rejection ? Write mechanism and treatment of rejection.
14. Describe tumour specificity and associated antigens and their importance.
15. Write short notes on the following :

   (*i*) Oragn transplantation, (*ii*) Transplant types,
   (*iii*) Organ transplantation laws and ethical concerns,
   (*iv*) Autologous graft, (*v*) Transplant rejection,
   (*vi*) Tumour antigens, (*vii*) Cancer immunotherapy.

BIBLIOGRAPHY

Anonymous, 2006. *PhRMA Reports identifies more than 400 biotech drugs in development.* Pharmaceutical Technology, August 24, 2006.

Bretton P.R., Melamed M.R., Fair W.R., Cote R.J. 1994. Detection of occult micrometastases in the bone marrow of patients with prostate carcinoma. *Prostate* **25** (2): 108–14.

Carter P. *et al.* 1992. Humanization of an Anti-p185HER2 antibody for human cancer therapy. *Proc. Natl. Acad. Sci. USA*; 89: 4285-4289.

Carter P. 2001. Improving the efficacy of antibody-based cancer therapies. *Nat Rev Cancer*. 1:118-129.

Dmitrij I.B. and Rabinovich M.L. 2007. Russia through the prism of the world biopharmaceutical market. *Biotechnol. J* 2: 1–17.

Janeway C.A., Travers P., Walport M., Shlomchik M.J. 2005. *ImmunoBiology*, 6th Ed. New York: Garland Science Taylor and Francis Group.

Kvalheim G. 1998. Diagnosis of minimal residual disease in bone marrow and blood in cancer patients--methods and clinical implications. *Acta oncologica (Stockholm, Sweden)* 37: 455–62.

Leboeuf R., Langlois M.F., Martin M., Ahnadi C.E. and Fink G.D. 2006. Hook effectin calcitonin immunoradiometric assay in patients with metastatic medullary thyroid carcinoma: case report and review of the literature. *J. Clin. Endocrinol. Metab.* **91**: 361–4.

McCafferty J., Griffiths A., Winter G. and Chiswell D. 1990. Phage antibodies: filamentous phage displaying antibody variable domains. *Nature.* 348: 552–554.

Nelson P.N., Reynolds G.M., Waldron E.E., Ward E., Giannopoulos K., and Murray P.G. 2000. Monoclonal antibodies. *Molecular Pathology: MP* 53: 111–7.

Rapaport F.T. 1986. The case for a living emotionally related international kidney donor exchange registry. *Transplant. Proc.* **18** (3 Suppl 2): 5–9.

Raz E. 1999. *Gene Vaccination: Theory and Practice*. Springer-Verlag, Heidelburg, p. 169.

Schirmbeck R. and Jorg R. 2001 Revealing the potential of DNA-based vaccination: lessons learned from the hepatitis B virus surface antigen. *Biol. Chem*. 382: 543-552.

Schwaber J. and Cohen E.P. 1973. Human × mouse somatic cell hybrid clone secreting immunoglobulins of both parental types. *Nature.* 244 (5416): 444–7.

CHAPTER

18 Plant Biotechnology

Organogenesis in a plant tissue culture

A. History of plant cell, tissue and organ culture
B. Laboratory techniques
C. Methods of plant cell, tissue and organ culture
D. Meristem culture
E. Callus culture
F. Cell (suspension) culture
G. Protoplast culture
H. Protoplast fusion and somatic hybridization
I. Somatic embryogenesis and encapsulated seeds
J. Micropropagation
K. Somaclonal variation
L. Anther and pollen culture
M. Mentor pollen technology
N. Ovary culture and embryo culture
O. Embryo rescue
P. Cryopreservation

During the last two decades plant cell, tissue and organ culture have developed rapidly and become a major biotechnological tool in agriculture, horticulture, forestry and industry. Those problems which were not feasible through conventional techniques, have now been solved via these techniques, for example, inter- and intra-specific crosses, micropropagation, somaclonal variation, encapsulated seeds, etc. Plant tissue culture is used to propagate plants under sterile conditions. Different techniques in plant tissue culture offer many advantages over the traditional methods of propagation such as : (*i*) production of exact copies of plants that produce particularly good flowers, fruits, etc., (*ii*) quick production of mature plants, (*iii*) production of multiples plants without seeds, (*iv*) regeneration of whole plants from genetically modified cells, (*v*) production of pest and/or pathogen-free plants, (*vi*) production of plants from such seeds that have very low chances of germination efficiency like orchids, and (*vii*) production of virus-free plants.

The success of plant tissue culture depends of on the cells having the regeneration ability into a whole plant that is called 'totipotency'. A new plant can also regenerates from a single cell without a cell wall (protoplasts), leaf pieces, apical meristem or roots on culture media consisting of necessary nutrients and plant hormones.

The concepts of plasticity and totipotency are important to understand plant cell culture and regeneration. Plants have developed a greater ability to tolerate the extreme conditions due to their sessile nature and long life. Several processes associated with plant growth and development also adapt the environmental conditions. Due to this plasticity the plants alter their metabolism, growth and development that can suit them best to their environment. The important aspects of this adaptation are the abilities to initiate cell division from any tissue of the plant and to regenerate the lost organs or undergo different developmental pathways in response to particular stimuli. *In vitro* cultured plant cells and tissues generally show a very high degree of plasticity resulting in initiation of one type of tissue or organ from the other type. Subsequently the whole plants regenerate; however, regeneration of whole organisms depends upon the concept that all plant cells can express the total genetic potential of the parent plant. This maintenance of genetic potential is referred to as 'totipotency'. In fact the most compelling evidence for totipotency is provided by plant cell culture and regeneration.

Totipotency

Totipotency is the basis of plant cell and tissue culture. Each living cell of a multicellular organism is capable of independent development, when provided with suitable conditions. In 1901, T.H Morgan coined the term 'totipotency' to denote the capacity of cell to develop into an organism by regeneration. However, the concept of totipotency is important in tissue culture. Use of multicellular organisms in research as biological units, is rather difficult; therefore, study of an organism using its cultured cells as basic organism is of fundamental importance.

The profound effects of the composition of the medium, especially the plant hormones and the nitrogen source (nitrate versus ammonium salts or amino acids), have been found on the morphology of the tissues that grow from the initial explant. For example, excess of auxin results in root proliferation, whereas that of cytokinin causes shoot proliferation. Therefore, a balance amount of both auxin and cytokinin often produces an unorganised mass of cells that is called callus; however, the morphology of the outgrowth depends on both plant species and the medium composition. After the growth of cultures, its some pieces are cut and transferred to new medium (sub-cultured) to allow for growth or to change the morphology of the culture. The skill and experience of the tissue culturist are important in judging that which pieces should be cultured and which not. As shoots emerge from a culture, they may be sliced off and rooted with auxin to produce plantlets; after maturation, these shoots can be transferred to potting soil for further growth in the greenhouse as normal plants (Bhojwani and Razdan, 1996) (Fig. 18.1).

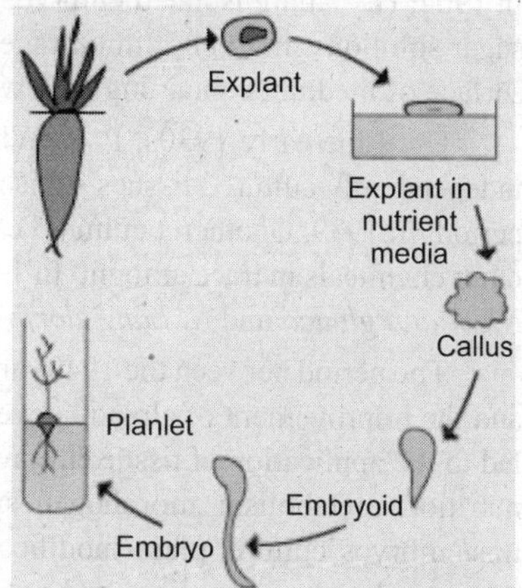

Fig. 18.1. Tissue culture and totipotency.

The International Centre for Genetic Engineering and Biotechnology (ICGEB) provides a scientific and educational environment of the highest standard and conducts innovative research in life sciences for the benefit of developing countries. It strengthens the research capability of its member countries through training and funding programmes and advisory services and represents a comprehensive approach to promoting biotechnology internationally. This centre is dedicated to advanced research and training in molecular biology and plant biotechnology and holds out the application of the latest techniques in the field of biomedicine, crop improvement, environmental protection and remediation, biopharmaceuticals and biopesticide production.

A. HISTORY OF PLANT CELL, TISSUE AND ORGAN CULTURE

The term 'tissue culture' can be applied to any multicellular culture growing on a solid medium (or attached to substratum and nurtured with a liquid medium) that consists of many cells in protoplasmic continuity. Thus plant tissue culture is the aseptic culture of cells, tissues, organs, and their components under defined physical and chemical conditions *in vitro*. But in organ culture (*e.g.* excised roots) the cultured plant material maintains its morphological identity, more or less, with the same anatomy and physiology as *in vitro* of the parent plants. It is an important tool in both basic and applied studies as well as in commercial application.

Historically Henri-Louis Duhamel du Monceau in 1756 pioneered the experiments on wound healing in plants through spontaneous callus (unorganised mass of cells) formation on decorticated region of elm plants. But the science of cell and tissue culture could be advanced after propounding the cell theory by M.J. Schleiden and T.W. Schwann in 1839. In 1853, A, Trecul observed callus formation in a number of plants. In 1878, Vochting suggested the presence of polarity as a key feature that guide the development of plant fragments. He observed that the upper portion of a piece of a stem always produced buds and the basal region produced callus or roots.

During 19th century the idea of development of callus (a disorganised proliferated mass of actively dividing cells) from isolated stem fragments and root apices came into existence. Callus could also be developed from buds, and root and shoot fragments of about 1.5 mm in size without using nutrient medium.

From 1902 to 1930 attempts were made for organ culture. It owes its origin to the ideas of a German Botanist Gottlieb Haberlandt (in Berlin) at the begining of the 20th century. In 1902, he developed the concept of culture of isolated cells of *Tradescantia* in artificial condition. Though his experiment failed to induce the cells to divide; he did not succeed because by that time even auxin was not discovered. But he lent a foundation to plant physiology. He described the cultivation of mesophyll cells of *Lamium purpureum* and *Eichhornia crassipes*, epidermal cells of *Ornithogalum* and hair cells of *Pumonaria*. Cell survived for 3–4 weeks. Due to this endeavour, Hamberlandt is regarded as the 'father of tissue culture'.

Gottlieb Haberlandt

The early studies led to root cultures, embryo cultures, and the first true callus/tissue cultures. In 1904, B. Hannig isolated embryos of some crucifers and successfully grew on mineral salts and sugar solutions. In 1908, Simon regenerated a bulky callus, buds, roots from a poplar trees on the surface of medium containing IAA which proliferated cell division.

Until the early 1930s, P. R. White (USA), R.J. Gautheret (France) and Nobercourt (France) independently cultured tissues excised from several plants on the defined nutrient media for a long period. In 1939, Gautheret cultured cambium tissue of carrot on Knop's solution supplemented with other chemicals in trace amount. In 1939, P.R. White cultured tobacco tumour tissue from the hybrid *Nicotiana glauca* and *N. Langsdorffii.*

The period between the 1940s and the 1960s was marked by the development of new techniques and the improvement of already developed methods. It was the availability of these techniques that led to the application of tissue culture to five broad areas namely, cell behavior (including cytology, nutrition, metabolism, morphogenesis, embryogenesis, and pathology), protoplasts/ anthers/ roots tips/embryos culture, plant modification and improvement, pathogen-free plants and germplasm storage, clonal propagation, and product (mainly secondary metabolite) formation, starting in the mid-1960s.

During 1940 to 1970, suitable nutrient media were developed for culture of plant cells and tissues. *In vitro* morphogenesis (*i.e.* regeneration of complete plant from cultured tissue) of plants was always done successfully. In 1941, van Overbeek and co-workers used coconut milk (embryo sac fluid) for embryo development and callus formation in *Datura.*

In the 1950s several important achievements were made in the field of plant physiology. The understanding of plant growth hormone in rapid multiplication of totipotent cell was developed. In 1958, F.C. Stewart and J. Reinert obtained regeneration in callus tissue culture of *Duccus carrota.* The foundation of commercial plant tissue culture was laid in 1960 with the discovery of G.M. Morel for a million fold increase in clonal multiplication of an orchid, *Cymbidium.* At the University of Wisconsin, Skoog and co-workers found out the role of cytokinins in tissue culture. Consequently, several chemicals were tested which stimulated callus. Adenine in the presence of auxin was found to induce callus growth and bud formation in tobacco cultures. In 1955, C.O. Miller and co-workers eventually isolated and identified a potent cell division factor from degraded DNA preparations, and named as 'kinetin'. The term cytokinin was given to this group after substituting the aminopurine compounds that stimulate cell division in cultured plant tissue and behave physiologically similar to kinetin. Later on other cytokinins (the naturally occurring plant hormones) such as zeatin, and isopentyl adenin were discovered. In 1957, F. Skoog and C.O. Miller advanced the hypothesis of organogenesis in cultured callus by varying the ratio of auxin and cytokinin in the growth medium. The shoot was formed with keeping the ratio of kinetin higher and root developed when ratio was lower.

When fragments of callus are transferred into a liquid medium and aerated on a shaker, it gives a suspension of single cell and aggregate of cells. A cell can be propagated by sub-culturing. In 1953, W.H. Muir developed a successful technique for the culture of single isolated cells which is commonly known as paper-raft nurse technique (placing a single cell on filter paper kept on an actively growing nurse tissue). Later on, attempts were also made for single cell culture by hanging drop and agar plate method. During 1965s, phenomenon of totipotency was fully developed by V. Vasil and A.C. Hilderbrant who demonstrated that a single isolated cell can divide and regenerate a whole plant.

In 1952, the Pfizer Inc., New York (U.S.A.) got the US Patent and started producing industrially the secondary metabolites of plants. The first commercial production of a natural product shikonin by cell suspension culture was obtained.

Cell wall creates a barrier in plant protoplast culture. In 1960s, the role of enzymes such as cellulase and pectinase in dissolution of cell wall in buffer solution at suitable pH, and isolation and culture of protoplast was developed. During this period extensive study of plant tissue culture was done by using the explants (excised plant parts) from different parts of gymnosperms and angiosperms. In 1966, Sipra Guha Mukherjee and S.C. Maheshwari developed techniques for the production of vast numbers of embryos from cultures of pollens and sporogenous tissues of anther. In 1974, J.P. Nitsch gave methods to double the chromosome number in microspors of *Nicotiana* and *Datura,* and collected seeds from the homozygous diploid plants within five months.

However, the benefits from protoplast culture currently availed are : (*i*) intraspecific, interspecific and intergeneric protoplast fusion and sometimes between plant and animal, (*ii*) transfer of mitochondria and plastids into protoplast, (iii) uptake of certain beneficial genes of blue green algae, bacteria and viruses by protoplasts, and (*iv*) transfer of genetic information into isolated protoplasts.

During 1980, recombinant DNA technology made possible to transform artificially cultured plant cells by introducing foreign genes. Thus several transgenic plants of special characteristics have been produced (*see* Chapter 19: *Transgenic Plants*). The gene revolution has made the second green revolution. Now it is possible to develop plants of desired genetic characters.

In 1982, I.K. Vasil has emphasised to develop the following techniques of cell culture and somatic genetics of mainly grasses and cereals: (*i*) rapid clonal propagation, (*ii*) regeneration from single cell and protoplast (*iii*) androgenetic haploids, (*iv*) mutant/variant cell lines and plant regeneration, (*v*) somatic hybridisation, (*vi*) transformation, and molecular biology.

The 1990s witnessed the continued expansion in the application of the *in vitro* technologies to an increasing number of plant species. Cell cultures have remained an important tool in the study of basic areas of plant biology and biochemistry and have assumed major significance in studies in molecular biology and agricultural biotechnology. Thorpe (2007) has described a comprehensive account of the history of plant tissue culture. The historical development of these *in vitro* technologies and their applications are given in Table 18.1.

Table 18.1. History of milestones in the field of plant tissue and cell culture.

Year	*Authors*	*Important Discoveries*
1838 and 1839	M. J. Schleiden and T. W. Schwann	Cell theory suggesting totipotency of cells
1902	Haberlandt Gottlieb	First but unsuccessful attempt of tissue culture was made using monocots.
1904	A. Hannig	First attempt in embryo culture of selected crucifers.
1909	E. Kuster and Dtsch Ber	Fusion of plant protoplasts though the products failed to survive
1924	F. Blumenthal and P. Z. Meyer	Callus formation on carrot root explants by use of lactic acid
1925	F. Laibach	Embryo culture for inter-specific crosses in *Linum* spp.
1934	P. R White	Successful long-term culture of tomato roots
1934	F., Kogl	Identification of the first plant hormone IAA leading to cell enlargement
1939	R. J. Gautheret P. R. White	Successful continuously growing cambial cultures of carrot and tobacco
1941	J. van Overbeek	Coconut milk used for growth and development of very young *Datura* embryos
1942	R. J. Gautheret	Observation of secondary metabolites in plant callus cultures
1943-1950	Braun A. C	Tumour-inducing principle of crown gall tumours identified
1944	F. Skoog	First *in vitro* culture of tobacco used to study adventitious shoot formation
1946	E. Ball	First whole plants of *Lupinus* and *Tropaeolum* from shoot tips were carried out
1950	G. Morel	Monocots were successfully cultured first using coconut milk
1951	J. P. Nitsch	Culture of excised ovaries *in vitro*
	F. Skoog	Chemical control of growth and organ formation in culture demonstrated.
1952	G. Morel and C. Martin	Virus-free *Dahlia* through meristem culture

1953-1954	W. R. Tulecke	Haploid callus from pollen grain of *Ginkgo biloba,* First calli produced from a single cell by use of nurse cultures
1957	F. Skoog and C. O. Miller	Discovery that root or shoot formation in culture depends on auxin : cytokinin ratio
1958	P. Maheshwari and N. S. Rangaswamy	Regeneration of somatic embryos from nucellus of *Citrus* ovules
	F. C. Steward and co-workers	Growth and development in suspension cultures.
1959	W. Tulecke and L. G. Nickell	Production of large amounts (134 L) of plant tissue by submerged culture
1960	K. Kanta	First test tube fertilisation in *Papaver rhoeas*
1964	S. Guha and S. C. Maheshwari	First haploid plants from *Datura* androgenesis
1969	T. Ericksson and K. Jonassen	Protoplast isolation from suspension culture of *Hapopappus gracilis*
1972	P. S. Carlson and co-workers	Interspecific hybridisation of *Nicotiana* spp. using protoplasts
1976	M. Seibert	Shoot induction from cryo-preserved shoot tips of carnation
	G. Bomhoff and co-workers	Octopine and nopaline synthesis and break-down regulation by Ti plasmid.
1977	M. D. Chilton and co-workers	Successful integration of T-DNA in plants
	M. Noguchi and co-workers	Cultivation of tobacco cells in 20,000 L bioreactors
1978	G. Melchers	Somatic hybridisation of tomato and potato
1982	F. A. Krens and co-workers	Naked DNA transformation of protoplasts.
	U. Zimmermann	Electrofusion of protoplasts
1981	P. J. Larkin and W. R. Scowcroft	Introduction of the term somaclonal variaion
1983	Mitsui Petrochemicals	First industrial production of secondary metabolites by suspension cultures of *Lithospermum* spp.
1985	R. B. Horsch	Infection and transformation of leaf discs with *Agrobacterium tumefaciens* and regeneration of transformed plants
1986	A. Crossway and coworkers	Transformation of tobacco protoplasts by direct DNA microinjection
1987	T. M. Klein and co-workers	Use of micro-projectile gun for particle bombardment for genetic transformation and recovery of individuals
1990	G. Neuhaus	Plant transformation by microinjection of intact plant cells.
	R. A. Dekeyser	Electroporation of intact plant tissues for direct DNA delivery
1991	P. T. Lynch and E. E Benson	Cryopreservation of alkaloid-producing cell culture of *Catahranthus*. The cells retain the property of alkaloid synthesis even after thawing

	Y. Huang and co-workers	Production of first transgenic plants of a conifer by *Agrobacterium rhizogenes*
1992	D.J. Yun	Metabolic engineering of *Atropa elladonna* for increased alkaloid production
	S. K. Dutta and co-workers	Herbicide resistant rice plants through PEG mediated transformation of protoplasts
1994	-	Marketing of genetically engineered tomato 'Flavr-Savr'
1996	G. Hansen and M. D. Chilton	Development of 'agrolistic' method of plant transformation
	C. M. Hamilton	Development of a binary bacterial artificial chromosome vector for *Agrobacterium*-mediated transformation
2000	Ingo Potrykus	Golden rice was created by transforming rice with two beta-carotene biosynthesis genes
2005	International Rice Genome Sequencing Project	Rice genome sequenced under International Rice Genome Sequencing Project

1. Research Work on Plant Tissue Culture in India

In India, research work on tissue culture was started during mid 1950s at the Department of Botany (University of Delhi) by Panchanan Maheshwari who is regarded as 'father of embryology' in India. Different tissue culture methodologies were involved for morphogenic studies involving ovary, embryo, endosperm, ovules, etc. At the University of Delhi, Sipra Guha and S.C. Maheshwari (1964-67) for the first time developed the haploids through anther and pollen culture. Discovery of haploid production was a land mark in the development of plant tissue culture.

The advancement made in cell and tissue culture technology is due to the development in composition of culture media. Based on the success of plant cell culture techniques many recent advances have been done in the area of micropropagation, production of secondary metabolites and pathogen-free plants, genetic manipulation (e.g. *in vitro* pollination, somatic hybridisation/cybridisation and induction of haploid, genetic transformation and production of transgenic plants).

Panchanan Maheshwari (1904-1966)

In 1983, the International Centre for Genetic Engineering and Biotechnology (ICGEB) was launched with the signing of its Statutes (the international treaty sanctioning its existence, deposited with the Secretary General of the United Nations) by 26 countries during a Meeting, held in Madrid. In 1984, the representatives of the Signatory Countries met in Vienna and decided that ICGEB would be structured with two Components, one located in Trieste, Italy (which would also be the Headquarters of the Centre), and one in New Delhi, India, as well as a network of Affiliated Centres.

In 1987, the ICGEB started operations as a special programme of the United Nations Industrial Development Organization (UNIDO). The implementation of the programme was made possible through funds provided by the Host Governments of Italy and India. In 1994, the statutes of the centre entered into force on 3 February, and the ICGEB became an autonomous and intergovernmental organisation in accordance with Article 1. In 2007, the third component of the ICGEB was established in Cape Town (South Africa). Up to 2011, the ICGEB has 61 full Member States, a further 22 countries are still pending ratification of, or accession to, the Statutes of ICGEB.

To boost up these areas ICGEB at the Trieste and New Delhi recommended the need of more research in developing countries on plant cell culture, differentiation, regeneration and transformation in tropical grain legumes, woody legumes and cereals. The emphasis was laid to improve growth under stress condition, pest and disease resistance, improved nutritional quality, nitrogen fixation and the control of partitioning within the plants. A detailed account of plant cells, tissue and organ culture is given in this context.

Several certified plant tissue culture laboratories have been established in different states in India that produce commercially the micropropagated plants. For example there is a plant tissue culture laboratory in Hyderabad, approved by the Department of Biotechnology (India) that produces 8 million Tc banana plants. This laboratory is looking for growers in U.P., M.P., Chattisgarh and West Bengal who would be interested in planting banana plants.

Banana rooting

Similarly, KIIT-Technology Business Incubator (KIIT-TBI) at KIIT University, Bhubaneswar (Orissa, India) provides the facilities for incubating ideas and innovation having commercial potential in Biotech. The KIIT-TBI can provide infrastructure requirements including seed funding for initial development of business.

B. LABORATORY TECHNIQUES

Plant cell, tissue and organ culture is carried out *in vitro*; therefore, a tissue culture laboratory requires many facilities as described below:

I. A TISSUE CULTURE LABORATORY

Culture of plant cells and tissues *in vitro* is not an easy task. It requires all the nutrients and physicochemical factors as maintained in a laboratory. For the tissue culture, a good laboratory, similar to a microbiological laboratory, is required which must have facilities for : (*i*) nutrient medium preparation, sterilisation, cleaning and storage of supplies, (*ii*) aseptic condition for working the living materials, (*iii*) a controlled environmental conditions for growth and development of cultures, (*iv*) observation and evaluation of culture as hoped, and (*v*) recording the observations made during the experiment.

1. Basic Facilities

The most basic facility that an individual needs for tissue culture requires the following :

(*a*) Washing and Storage : A separate area is required which should have large sink with provision for hot and cold running water, distillation apparatus, washing machine, pipette washer, drier and cleaning brushes, keeping and weighing the chemicals, and putting glassware.

(*b*) Media Preparation Room : An area is required for preparation of media. There should be provision for bench space for chemicals, lab ware, culture vessels, closures and miscellaneous equipment required for media preparation and dispensing. In this room provision is also made for placing hot plates or stirrers, pH meter, balance, water bath, burners, oven, autoclave, culture, vessel refrigerator, etc. Vitamins and growth hormones are carefully weighed. Stock solutions of chemicals are kept in refrigerator to avoid contamination.

(*c*) Transfer Area: Earlier transfer of plant tissue was done in open in laboratory bench under clean and dry atmosphere conditions. Later on closed plastic box was constructed which consisted of UV tube. UV light from the tube and use of 95 % ethanol helped to maintain sterile conditions inside the space of plastic box.

Different techniques are applied for sterilisation of glassware, working tables, nutrient media and plant materials. The glassware (culture tubes or Erlenmeyer conical flasks) containing sterilised nutrient medium after inoculation with plant material, are plugged with non-absorbent sterile cotton, and finally kept in growth chamber. Environmental conditions such as temperature, moisture and light are controlled as needed. For cell suspension culture, orbital shaker or aeration instrument is placed in growth chamber. Generally temperature is set between 25–27°C. Environmental conditions may vary according to plant species and nature of experiment. Light intensity, quality and photoperiod (light and dark cycles) are also regulated. Other laboratory accessories are dissection microscope, compound microscope, centrifuge, first aid kit, etc. which are required during *in vitro* culture of plant material.

II. NUTRIENT MEDIA

Vital activity of a cell is the absorption of nutrients through cell membranes and rapid proliferation into innumerable cells. In 1934, P.R. White observed the unlimited growth of isolated root tissues when provided with nutrient medium containing inorganic salts, sucrose, vitamins, growth hormone and a few amino acids. Culture media used for *in vitro* cultivation of plant cells are composed of following basic components: (*i*) complex mixture of salts (essential elements or mineral ions), (*ii*) organic supplement (e.g. vitamins and/or amino acids), (*iii*) carbon source (usually sugar sucrose), (*iv*) gelling agents, (*v*) plant growth regulators, and (*vi*) antibiotics. Some of the routinely used culture media are : MS (Murashige and Skoog) medium, ER (Eriksson) medium, B5 (Gamborg and co-workers) medium, etc.

1. Complex Salt Mixture

Complex mixture of salts includes the essential elements or mineral ions that are required for plant nutrition and their physiological function. Furthermore, these essential elements can be divided into: (*i*) macroelements (or macronutrients), (*ii*) microelements (or micronutrients), and (*iii*) iron source.

(*a*) Macroelements: The microelements are required in large amounts for plant growth and development. Nitrogen, phosphorus, potassium, magnesium, calcium and sulphur (and carbon, which is added separately) are regarded as macroelements. These elements comprise at least 0.1% of the dry weight of plants.

(*b*) Microelements: These elements are required in trace amounts for plant growth and development. Microelements are manganese, iodine, copper, cobalt, boron, molybdenum, iron and zinc; but in some formulations the other elements such as aluminum and nickel are frequently present.

(*c*) Iron Source: Usually iron is added in the medium as iron sulphate, or sometimes as iron citrate. Ethylenediaminetetraacetic acid (EDTA) is used in conjunction with the iron sulphate that forms EDTA-iron complexes. This complex slowly and continuously releases the iron into the medium. The uncomplexed iron precipitates out of the medium as ferric oxide.

A concentrated stock solution of macronutrients and microelements is prepared in advance and finally added to medium as per requirement. But there arises the problem of solubility. Therefore, a stock solution of iron is prepared in a chelated form as sodium salt of ferric ethylenediamine tetra acetate (Fe-EDTA) to overcome the problem of solubility.

2. Organic Supplements

The organic compounds serve a source of carbon and energy. These include various kinds of simple sugars, complex organic carbons compounds, vitamins and amino acids.

***(a)* Carbon Source:** Sugar is very important part of nutrient that acts as a source of energy. Most of plant cultures fail to carryout photosynthesis effectively due to inadequately developed cells and tissues, lack of chlorophyll, limited gas exchange and carbon dioxide in tissue culture vessels, etc. Therefore, the auxotrophic ability lacks in them; thus, they need external carbon for energy. The most preferred carbohydrates acting as carbon and energy source are sucrose and D-glucose used at a concentration of 20-60 g/L; but glycerol and myoinosital are also the principal source of carbon. Sucrose is hydrolysed to glucose and fructose after autoclaving the medium; then the latter are used up for growth. Other mono or disaccharide and sugar alcohols like glucose, sorbitol, raffinose, etc. may also be used depending upon plant species. Carbohydrates also provide osmoticum and, therefore, higher concentration of sucrose (6-12%) is used in anther culture.

***(i)* Complex Organics:** The complex organic compounds are the undefined supplements such as casein hydrolysate, coconut milk, yeast extract, orange juice, tomato juice, etc. When no other combination of known defined components produces the desired growth, these compounds can be used for the said purpose. A significant success in tissue culture has been achieved with casein hydrolysate, and potato extract has been found useful for anther culture. The composition of these natural extracts is unknown and varies from one lot to the other lot and also affects reproducibility of results with age, therefore, these natural extracts should be avoided from use.

***(ii)* Activated Charcoal:** The activated charcoal promotes and also inhibits the growth of culture depending upon plant species. For example, it stimulates growth and differentiation in orchids, carrot, ivy and tomato, whereas inhibits the same of tobacco, soybean, etc. Moreover, it absorbs brown-black pigments and oxidises phenolic compounds produced during culture resulting in reduction in toxicity. Besides, it also absorbs the other organic compounds such as vitamins which may inhibit the growth. But one drawback is that the activated charcoal causes darkening of medium and helps root formation. Other complex organic compounds are peptone, yeast extract, malt extract, coconut water, tomato juice, etc.

***(b)* Vitamins:** Vitamins are required in trace amount as they catalyse the enzyme system of the cells. Vitamin B_1 (thiamine) is essential for the culture of plant cells *in vitro*; therefore, it is the most commonly used for all plant tissue cultures. However, other vitamins are often added to for historical reasons because they also stimulate growth and development of cultured tissue, such as niacine (nicotinic acid) vitamin B_2 (riboflavin) vitamin B_6 (pyrodoxin), vitamin C (ascorbic acid), vitamin H (biotin) and vitamin B_{12} (cyanocobalamin).

***(c)* Amino acids:** Although nitrogen sources are present in the inorganic salts, yet various amino acids and amides are used in plant tissue culture media. Glycine is the most commonly used amino acid. The most widely used amino acids are L-aspartic acid, L-asparagin, L-glutamic acid, L-glutamine, alanine, glutamic acid, and proline and L-arginin. Casein hydrolysate can be used as a source of a mixture of amino acids.

3. Growth Hormones

Several growth hormones are known that stimulate the biological activity in cultured plant cell, tissue and organs. But there are five main classes of plant growth regulators used in plant cell culture such as : auxins, cytokinins, gibberellins, abscisic acid, and ethylene as described below:

***(a)* Cytokinins (CK) :** Cytokinins are a class of plant hormones that promote cell division, or cytokinesis in plant roots and shoots. Cell growth and differentiation occur due to cytokinins besides its effect on apical dominance, axillary bud growth, and leaf senescence. In the 1940s, its effect was first discovered through the use of coconut milk by F. Skoog and co-workers at the University of Wisconsin (U.S.A.).

There are two types of cytokinins: adenine-type cytokinins (e.g. kinetin, zeatin, and 6-benzylaminopurine), and phenylurea-type cytokinins (e.g. diphenylurea and thidiazuron). The cytokinin zeatin was discovered first in maize (*Zea maydis*); hence it is named after the name *Zea*. The majority of adenine-type cytokinins are synthesised in the roots, cambium and other actively dividing tissues. The natural occurrence of phenylurea cytokinins in plant tissues is not known. Cytokinins are involved in both local and long distance signalling. Cytokinins act in concert with auxin that stimulates cell expansion, and promotes cell division regulating growth and development like kinetin (6-furfuryl aminopurine). The effect of cytokinin on plant growth is governed by the auxin : cytokinin ratio.

(*b*) Auxins: Auxins are the plant growth hormones which have a cardinal role in coordination of many growth and behavioral processes in the life cycle of plants and are essential for development of plant body. For the first time a Dutch scientist Ken Thimann isolated an auxin and determined its chemical structure as indole-3-acetic acid (IAA); he also described its role in plant growth. IAA is the most abundant auxin that occurs natively and functions in the intact plants.

Auxin resembles indole acetic acid (IAA) and stimulates shoot elongation. Gibberellins are of less importance; however, GA_3 is used in apical meristem. The most widely used cytokinins are adenin, kinetin, zeatin, benzyladenin; the auxins are IAA, NAA (α -naphthalene acetic acid), 2,4-D. For the induction of callus the amount of kinetin should be 0.1 mg/litre. Structural formulae of cytokinins and auxins are given in Fig. 18.2.

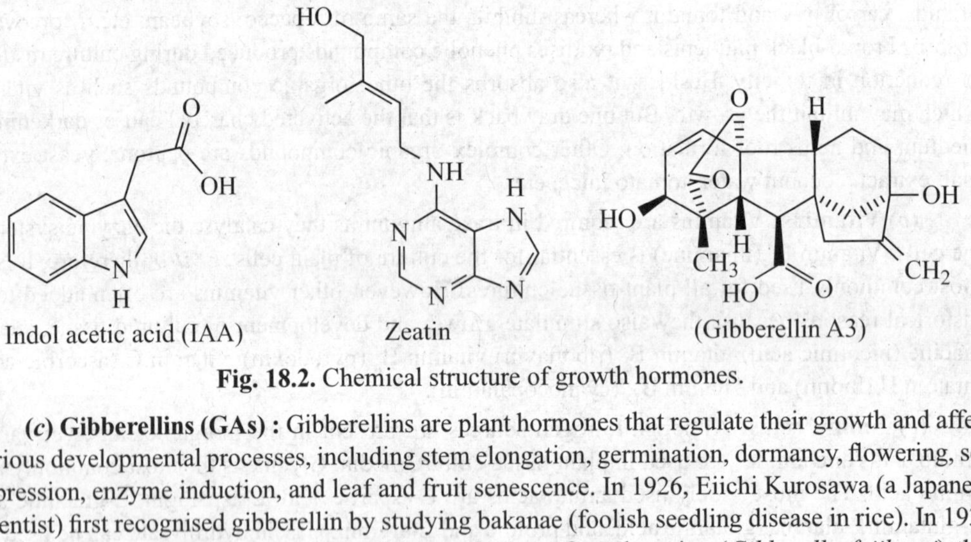

Indol acetic acid (IAA) Zeatin (Gibberellin A3)

Fig. 18.2. Chemical structure of growth hormones.

(*c*) Gibberellins (GAs) : Gibberellins are plant hormones that regulate their growth and affect various developmental processes, including stem elongation, germination, dormancy, flowering, sex expression, enzyme induction, and leaf and fruit senescence. In 1926, Eiichi Kurosawa (a Japanese scientist) first recognised gibberellin by studying bakanae (foolish seedling disease in rice). In 1935 Teijiro Yabuta first isolated this growth hormone from fungal strains (*Gibberella fujikuroi*) that was provided by Kurosawa and named it as the gibberellin. When the plant is exposed to cold temperatures, gibberellins are produced in high amount. The gibberellins are named as GA1....GAn in order of discovery. The first structurally characterised gibberellin was the gibberellic acid, GA3 (Fig. 18.2). As of 2003, about 126 GAs have been identified from plants, fungi and bacteria.

(*d*) Abscisic Acid (ABA) : It is used in plant tissue culture to promote distinct developmental pathways such as somatic embryogenesis. Besides, abscisic acid inhibits cell division and seed germination, and in antagonism with gibberellin. ABA also prevents loss of seed dormancy, but it is most commonly used in plant tissue culture to promote distinct developmental pathways such as somatic embryogenesis. Structure of abscisic acid is given in Fig. 18.3.

(*e*) Ethylene : Ethylene is a gaseous, naturally occurring plant growth regulator which is most commonly associated with control of fruit ripening in climacteric fruits. Its use in plant tissue culture is not widespread. Some plant cell cultures produce ethylene which can inhibit the growth and development of the culture, if accumulated in sufficient amount. Some plant cell cultures produce ethylene, which inhibit the growth and development of the culture. The type of culture vessel used affects the evolution of ethylene, *i.e.* a closure affects the gaseous exchange between the culture vessel and the outside atmosphere; thus the level of ethylene present in the culture is regulated accordingly. Structure of ethylene is given in Fig. 18.3.

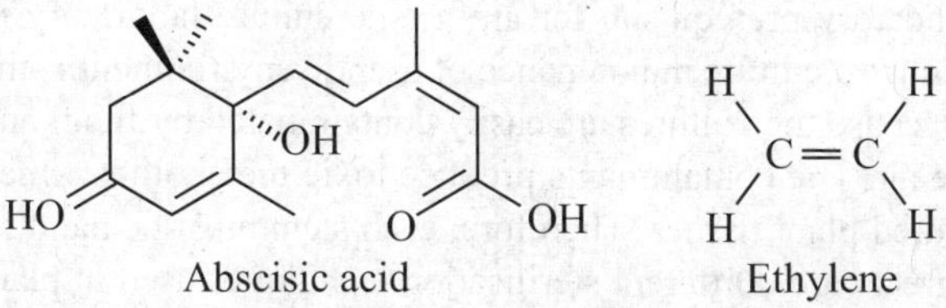

Abscisic acid Ethylene

Fig. 18.3. Structure of abscisic acid and ethylene.

4. Antibiotics

Antibiotics are secondary metabolites produced by certain microorganisms that suppress the growth of other microorganisms and eventually kill them. Antibiotics are used for suppression of bacterial contamination and infections in plant cell and tissue culture. It also suppresses mould and yeast infections in cell cultures, and eliminates *Agrobacterium* species after the transformation of plant tissue.

5. Solidifying Agents

Most commonly agar (a high molecular weight polysaccharide obtained from seaweed, *i.e.* a red alga *Gelidium amansii*) is used as solidifying or gelling agent. Agar gels do not react with constituents of media and not digested by plant enzymes. Generally 0.5 -1% agar is used to form gel. Before use of agar, gelatin (10%) had been used as gelling agent. Demerit of gelatin is that it melts at low temperature (25°C). Media for plant cell culture *in vitro* can be used in either liquid or 'solid' forms, depending on the type of culture being grown. For any culture types that require the plant cells or tissues to be grown on the surface of the medium, it must be solidified (more correctly termed gelled). Agar, produced from seaweed, is the most common type of gelling agent, and is ideal for routine applications. Since agar is a natural product, its quality varies from supplier to supplier and from batch to batch.

Agarose is a purified extract of agar used for protoplast culture. Alternative gelling compounds like gelrite, etc. form clear translucent gels unlike agar; hence it is easier to detect contamination which develops during the growth of culture. However, mechanical support for cell or tissue growth can also be provided without using any gelling agent by filter paper bridge, perforated cellophane, polyurethane foam, etc.

6. pH

Uptake of ions and solidification of gelling agent are affected by pH. Optimum pH between 5.8 and 6.0 is required for growth and development of cultured tissues. But values of pH less than 4.5 or more than 7.0 significantly inhibit growth and development of *in vitro* cultured tissues. After autoclaving the pH of culture media generally decreases by 0.3 to 0.5 units due to oxidation and differential uptake and secretion of substances by growing tissue. Therefore, optimum pH of the medium should be maintained before its sterilisation.

III. MAINTENANCE OF ASEPTIC ENVIRONMENT DURING CULTURE

Plant cells divide slowly as compared to that of bacteria, fungi, etc. and even minor contaminants can easily overgrow the plant tissue culture. Therefore, all the materials like glassware, instruments, medium, explants, etc. to be used in culture work must be freed of microbes. To avoid contamination, laminar flow is a mandatory prerequisite for any tissue culture laboratory in recent years. During *in vitro* culture maintenance of aseptic environment is the most difficult task. Because the cultures are easily contaminated by fungi and bacteria present in the air. The contaminants produce toxic metabolites which inhibit growth of cultured plant tissues. Therefore, each stem must be handled aseptically and with great care. Different sterilisation techniques used in plant tissue culture are enlisted in Table 18.2.

Laminar air flow

Table 18.2. Methods of sterilisation in plant tissue culture.

Methods	*Materials sterilised*
1. *Steam sterilisation/autoclaving* (121°C at 15 psi for 20-40 minutes)	Nutrient media, culture vessels, glassware and plastic wares
2. *Dry heat* (160-180°C for 3 hours)	Instruments (scalpel, forceps, needles etc.), glassware, pipettes, tips and other plastic wares
3. *Flame sterilisation*	Instruments (scalpel, forceps, needles, etc.), mouth of culture vessel
4. *Filter sterilisation* (membrane filter made of cellulose nitrate or cellulose acetate of 0.45-0.22 μm pore size)	Thermolabile substances like growth factors, amino acids, vitamins and enzymes.
5. *Alcohol sterilisation*	Worker's hands, laminar flow cabinet
6. *Surface sterilisation* (sodium hypochlorite, hydrogen peroxide, mercuric chloride, etc.)	Explants

Following are some of the sterilisation methods for aseptic manipulation of plant tissues.

(*a*) Sterilisation of Glassware: Glassware (Petri plates, vials, culture tubes, flasks, pipettes, etc.), metallic instruments are sterilised in a hot air oven at 160-180°C for 2-4 hours.

(*b*) Sterilisation of Instruments: The metallic instruments (*e.g.* forceps, scalpels, needles, spatulas, etc.) are flame sterilised, *i.e.* dipping them in 75% ethanol followed by flaming and cooling. It is called incineration.

(c) Sterilisation of Culture Room and Transfer Area: Floor and walls of culture room are washed first with detergent then 2% sodium hypochlorite or 95% ethanol. Larger surface area is sterilised by exposure to UV light. The cabinet of laminar airflow is also sterilised by exposing UV light for 30 minutes and 95% ethanol 15 minutes before beginning of work inside the cabinet of laminar airflow. The UV radiation is harmful to eyes. Therefore, UV light should be exposed when no experiment is being done, and no person is allowed to see UV light.

(*d*) Sterilisation of Nutrient Media: Culture media are properly dispensed in glass container, plugged with cotton or sealed with plastic closures and sterilised by autoclaving (steam sterilisation) at 15 psi (that gives 121°C) for 30 minutes. Minimum time required for autocalving of nutrient media depends of size of volume of the vessel; therefore, it varies from vessel to vessel as given in Table 18.3.

During autoclaving vitamins, plant extracts, amino acids and hormones are denatured. Therefore, the solution of these compounds are sterilised by using millipore filter paper which has 0.2 mm pore diameter.

Table 18.3. Minimum time required for autoclaving of nutrient media at 15 psi.

Volume (ml)	*Sterilisation time (minutes)*
1-200	15
200-1,000	30
1,000-2,000	40

(*e*) Sterilisation of Plant Materials: Surface of all plant materials has microbial contaminants. Therefore, disinfectants (such as sodium hypochlorite, hydrogen peroxide, mercuric chloride, or ethanol) should be used to make plant materials sterile. Then the chemicals must be washed 6–8 times using sterile distilled water. Then explants are transferred aseptically on nutrient medium inside the cabinet of laminar airflow.

C. METHODS OF PLANT CELL, TISSUE AND ORGAN CULTURE

There is a little variation in the methods of plant cell, tissue and organ culture, but the basic steps are almost the same.

1. Basic Steps

The basic steps for regeneration of whole plants form an explant or cells as shown in Fig. 18.1and described below :

(*a*) Preparation of Suitable Nutrient Medium: Suitable nutrient medium as per objective of culture is prepared and transferred into suitable containers (*e.g.* flasks, Petri plates, culture tubes) and autoclaved at 15 psi (pound per inch square) for 30 minutes. The hormones and vitamins are sterilised by using millipore filter.

(*b*) Selection of Explants: Explants are any excised part of plant to be used in tissue culture, for example axillary bud, leaf and stem segments, root tip, shoot tip, anther, ovary, endosperm, etc. The explants can be obtained either from fields or from controlled cultured plants. Always young and healthy parts of the plants are selected as explant. The regeneration of plants *in vitro* conditions is largely dependent upon the maturity of explants.

The explant cultures are generally used to induce callus or plant regeneration. There are a variety of forms of seed plants viz., trees, herbs, grasses, which exhibit the basic morphological units, *i.e.* root, stem and leaves. These again vary with differences in cells and tissues, and their topography.

Parenchyma is the most versatile of all types of tissues. They are capable of division and growth. However, development of a tissue is characterised by three types of cell growth : cell division, cell elongation and cell differentiation. For this purpose, the explant from healthy and young part of the plant is used. Presence of parenchyma is first consideration in a particular species. Parenchyma from stems, rhizomes, tubers, roots is easily accessible and will generally respond quickly to culture conditions *in vitro*. However, parenchyma cells have an identical morphogenetic potential.

(*c*) Sterlisation of Explants: Explants are sterilised properly to kill the microbial contaminants on its surface; hence it is called 'surface sterilisation'. Collected explants were thoroughly washed with 2-4 drops of surfactant (Tween 20 or any soft liquid swap) for 10 min under running tap water.

Surfactants reduce the surface tension of material being cleaned resulting in the disinfectant solution more effective. After washing the explants are subjected to chemical sterilisation ($HgCl_2$ or NaOCl) in the laminar airflow. It is then washed 3-4 times with sterilised double distilled water. Now the required medium is inoculated by the explants.

(*d*) Inoculation of Explant (Explant Transfer): The inoculation of sterilised explants is carried out in the cabinet of laminar airflow. Laminar airflow is sterilised with UV for 30 minutes followed by 15 minutes flow of motor before inoculation. The forceps are further sterilised by flame with repeated dipping in ethanol. Finally the explants are transferred in the medium. The mouth of the culture vial is immediately re-plugged. The sterile explant is inoculated on the surface of solidified nutrient medium under aseptic conditions. The cabinet of laminar airflow provides the sterile conditions

(*e*) Incubation: Incubation of culture vials is carried out in incubation room under proper photoperiod (*i.e.* light conditions of 16 hours photoperiod and 8 hours dark period created artificially in growth chamber) at a temperature of 25+3°C and relative humidity of 50 to 60%. After defined period callus develops on the medium or shoots/ roots develop from explant. The cultures are observed after one week and after wards with proper growth sub-cultured. Fig. 18.4 shows the growth of explants in culture tubes.

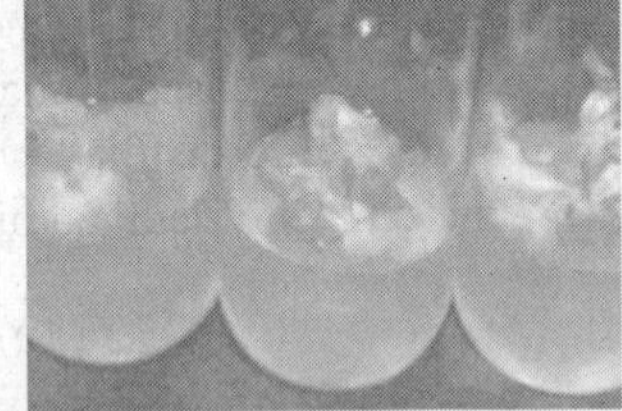

Fig. 18.4. Growth of explants in culture tubes.

(*f*) Regeneration: In culture protoplasts show rapid cytoplasmic streaming and increase in size, and most of the cell organelles (particularly the chloroplasts), aggregate around the nucleus. The rate of cell wall formation depends on the state of differentiation of the donor cells, conditions of isolation of protoplasts and the plant species. The process of cell wall formation begins within few hours after isolation and may be completed in two to several days. Soon after developing a wall around the protoplast the reconstituted cells considerably increase in size, and generally the first division occurs within 2 to 7 days. Subsequent divisions give rise to small cell colonies. Thus, there is a direct relationship between wall formation and cell divisions. Upon completion of wall formation, protoplasts lose their characteristic spherical shape. Plantlets regenerate after transferring a portion of callus onto another medium and induction of roots and shoots or directly from explants. The callus may be induced to undergo organogenesis or whole plant regeneration.

(*g*) Hardening: Hardening is the gradual exposure of plantlets for acclimatisation to environmental conditions. Considerable efforts have been made to optimise the conditions for *in vitro* stages of micropropagation. The physiological and anatomical characteristics of micropropagated plantlets should be gradually acclimatised to the environment of the greenhouse or field. Before transplanting, acclimatisation and hardening should be accomplished as a one-step procedure within a short period of time. A significant impact on the success of hardening for micropropagation has been achieved with optimisation of time for harvesting of microshoots, shoot size, soil pH (4.0–6.4), humidity conditions, plant growth regulator treatment (IBA 500 mg l^{-1} for 30 minutes), CO_2 enrichment and light (15 μmol m^{-2} s^{-1}) in specially designed hardening chambers.

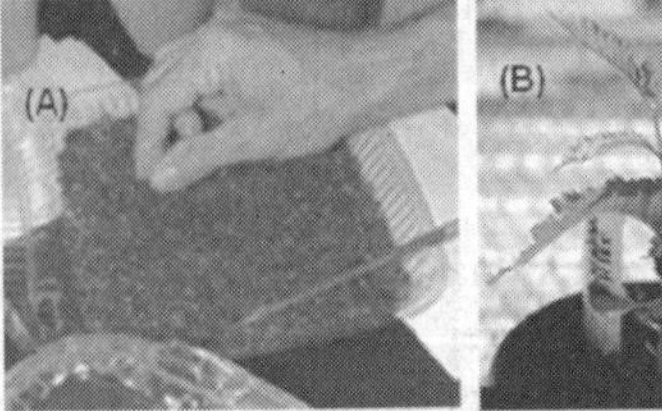

Fig. 18.5. Transplantation of plantlets into potting mix (A), and then in soil (B).

(*h*) Plantlet Transfer to Greenhouse and Field: After complete developments, cultured plants are acclimatised/ hardened in the greenhouse. After hardening process the healthy and tolerating plantlets are transferred to green house and finally to fields. When the plantlets are large, they can be removed from the tissue culture container and transferred into pots containing potting soil (Fig. 18.5 A). The young plants are grown in a greenhouse similar to any young seedling or cutting until they acclimatise to the ambient environment. After acclimation, the young plants can be transplanted in pots in a greenhouse to produce new plants (B). The greenhouse and field have substantially lower relative humidity, higher light level and septic environment that are stressful to micropropagated plants as compared to *in vitro* conditions. The benefit of any micropropagation system can only be fully realised by the successful transfer of plantlets from tissue-culture vessels to the ambient conditions found *ex vitro.* Most species grown *in vitro* require an acclimatisation process in order to ensure that sufficient number of plants survive and grow vigorously when transferred to soil.

D. MERISTEM CULTURE

Apical meristem is the region of shoot apex lying distal to leaf primordium, which can be cultured *in vitro*. It clearly differs from shoot apex by having shoot apex and a few leaf primordia (Fig. 18.6). Because of culture of relatively large tips (5-10 mm long sections), this technique is also known as meristem culture, meristemming and mericlones. In 1950s Morel and co-workers pioneered the work of shoot apex culture of orchid, *Cymbidium,* for obtaining virus -free plants from already virus-infected plants. In 1960, he tried this method in a virus-infected *Cymbidium* and succeeded in obtaining the plantlets free of infection. Thereafter, meristem culture technique was developed, in which a single detached meristem is dissected into several bits and each bit is sub-cultured. The bids are then sterilised with 2 % calcium hypochlorite solution and placed in a liquid nutrient medium in a flask under sterile conditions. The flask is then placed on a rotary shaker and rotated at a given speed. This prevents the formation of any polarity to the developing meristem.

The meristems turn green within a week. It starts developing several lumps of tissue. Now the protocorn is taken out, washed and cut up into as many pieces as there are balls of tissue. Each bit is dipped in a sterilant and put in a tube containing fresh liquid medium and agitated. The procedure is repeated until many plantings as required are produced. Then the tiny balls of tissue are transferred to normal nutrient agar medium and kept static as in seedling culture. After reaching 4 mm in length each ball of tissue differentiates into an individual plant.

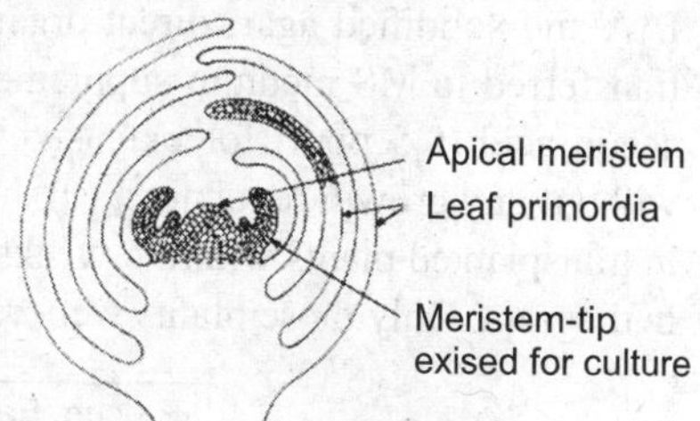

Fig. 18.6. Structure of an apical meristem.

Meristem culture–based technique of propagation provided a great boost to the orchid industry. Rare plants which could not be propagated by seeds due to genetic factors could be multiplied easily at a fantastic rate. This naturally brought down the prices of these orchids which were otherwise too expensive for most people. Precious plants could also be salvaged which fell victim to certain virus diseases.

The use of cytokins for promoting dormant buds to develop into young plants is one of the more recent developments in vegetative multiplication of orchids. Experiments are in progress on several genera but the greatest success has been achieved in *Phalaenopsis.* The mericloning technique is widely used by large establishments like commercial nurseries, for the amateur growing a few plants and desiring to increase them. Thus a large number of plantings could be obtained from a single protocorn within a year by repeating this several times.

I. PRODUCTION OF DISEASE-RESISTANT PLANTS

Many plant species, which propagate vegetatively are systematically infected by virus, bacteria, fungi, and nematodes. Their inoculum is carried over several generations resulting in continued adverse effect of productivity and quality of crops. In order to ensure highest possible yield and quality, it is necessary to provide disease free stock plants to growers. Tissue culture techniques have solved the problem and minimised the time of biological testing. Unless large scale population of pure inoculum of test pathogens is available, it is difficult to peruse the establishment of pathogenecity and crop loss-assessment as it is done in field condition. Now it has become possible to carry out such experiments in laboratory within short span of time by using tissue culture technologies.

1. Production of Virus-free Plants from Meristem Culture

About 10 per cent of viruses are transmited through seeds. In some cases, they are confined to seed coat (*e.g.* TMV) or internally seed-borne (in legumes). Moreover, viruses result in great loss, for example, potato leaf roll virus or potato virus X can cause up to 95% reduction in tuber yield and potato virus X between 5 and 75% depending on virus strain and host cultivar.

Tissue culture technique can be utilised for the production of virus-free plants either through meristem culture or chemotherapy or selective chemotherapy of larger explants from donor plants or dormant propagules or a combination of the two. Above plants were made virus-free by meristem culture by 1983.

Sood and Palni (1992) got early flowering in tissue culture raised and virus tested Easter lily plant (*Lilium longoflorum*). They have described various steps for production of virus-free plants of Easter lily (Fig. 18.7) : (*i*) bulb scale segments from commercially available bulbs were tested positive (virus indexing); segments were subjected to hot water treatment (40°C for 24 h), (*ii*) surface sterilised segments transferred to MS medium supplemented with 2.5 mg/L kinetin and 0.5 mg/1 IAA and solidified agar; direct organogenesis (shoot bud formation) was observed, (*iii*) explants transferred to MS medium supplemented with 6-benzylaminopurine (2.5 mg/l and α-naphthalene acetic acid (0.5 ppm) for extensive rooting, (*iv*) rooted plants transferred after 6 weeks to pots without any growth regulators, (*v*) the plantlets transferred to soil. Normal flowering was observed in transplanted plants within 6 weeks. Virus indexing was carried out at various stages as indicated in diagram. Only those plants were retained which tested negative.

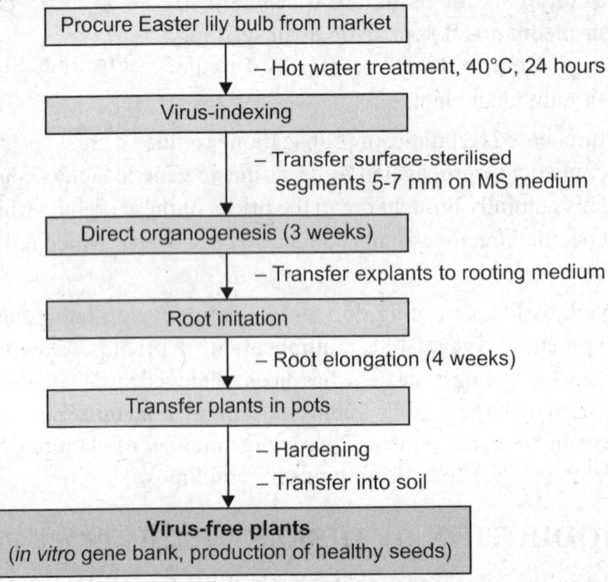

Fig. 18.7. Outline for production of virus-free plants of Easter Lily (based on Sood and Palni, 1992).

In 1952, G. Morel and C. Martin for the first time successfully obtained a virus-free plant through shoot meristem culture. Later on, several virus-free plants were regenerated *in vitro*. In 1983, C.Y. Hu and P.J. Wang produced virus-free plants such as *Pisum sativum, Trifolium repens* and *Citrus* sp., *Dactyl is glomerata* (from mild mosaic mottle viruses) and *Lolium multiflorum* (from ryegrass mosaic virus).

E. CALLUS CULTURE

There are different types of cultures which are produced through cultured plant material such as explant culture, callus culture, cell or suspension culture, protoplast culture, organ culture. Callus refers to an unorganised (undifferentiated) mass of cells (generally parenchymatous in nature) derived from explants for use in biological research and biotechnology. In plant biology, callus cells are those cells that cover a plant wound. In nature, callus develops by infection of microorganisms from wounds due to stimulation by endogenous growth hormones, the auxins and cytokinins. Auxins are added to culture medium for callus induction but the nature and quantity of auxin depend on the nature and source of explant and its genotype. For prolonged periods callus cultures can be maintained by repeated sub-culturing.

Callus can be artificially developed by adopting tissue culture techniques. A 2-5 mm sterile segment, excised from a stem, tuber or root is transferred into nutrient medium and incubated at 25-28°C in an alternate light and dark regime of 12 h. Nutrient medium supplemented with auxins induces cell division. Soon the upper surface of explant is covered by callus. A callus is an amorphous mass of loosely arranged thin walled parenchyma cells developing from proliferating cells of the parent tissue (Dodds and Roberts, 1985). The unique feature of callus is that the abnormal growth has biological potential to develop normal root, shoots and embryoids ultimately forming a plant (Fig. 18.8).

Callus formation is governed by the source of explant, nutritional composition of medium and environmental factors. Explants of meristematic tissues develop cells more rapidly than thin three walled and lignified cells of tissue. Callus is formed through three developmental stages : induction, cell division and differentiation.

During induction metabolic rate of cells is stimulated, duration of which depends on physiological status, and nutritional and environmental factors. Owing to increased metabolic rate, cells accumulate high contents of nutrients and finally divide to form many number of cells. Cellular differentiation and expression of certain metabolic pathways start in the third phase leading to secondary products. Sometimes callus appears of different colours, for example, yellow, white, green or red. Within the cell population of callus, the genetic instability results in variations in phenotypes which may be attributed to developmental (epigenetic) or genetic basis. Epigenetic changes involve selecting gene expression. They are stable and heritable at cellular level.

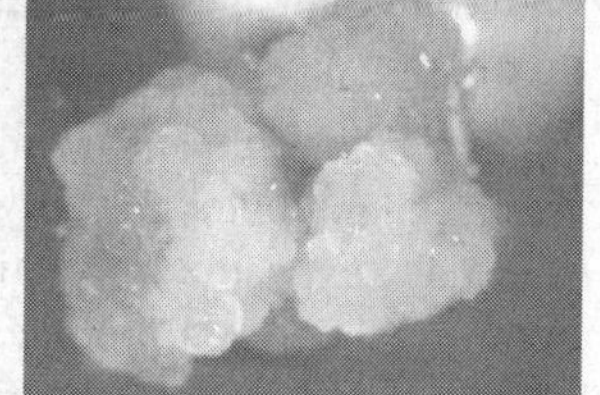

Fig. 18.8. Callus formation in *Pteris* sp.

When callus has grown on nutrient medium after a long time it becomes essential to subculture it within 28 days on a fresh medium. Otherwise there develops nutrient depletion in original medium which results in paucity of water and accumulation of toxic metabolites.

***(a)* Factors Affecting Callus Culture:** The factors that affect callus culture are : (*i*) the source and the genotype of the explants, (*ii*) composition of the medium (most commonly used-MS medium), (*iii*) temperature (22-28°C suitable for callus formation), (*iv*) growth regulators such as auxins and cytokinins alone or in combinations, (*v*) age of the plant, (*vi*) location of the explants, and (*vii*) physiology and growth conditions of the plant.

Callus cultures are used for: (*i*) plant regeneration, (*ii*) preparation of single cell suspensions and protoplasts, and (*iii*) genetic transformation studies.

I. PLANT REGENERATION

The steps of organogenesis and somatic embryogenesis result in regeneration of plants in the cultured tissues.

1. Organogenesis

Many aspects of plant growth and development are regulated by plant hormones. For a long time, both auxin and cytokinin have been known to act either synergistically or antagonistically to control several significant developmental processes in plants. Root, shoot and leaves (but not embryos) are the organs that are induced in plant tissue culture. Since embryo is an independent structure and does not have vascular supply, it is not supposed to be the plant organ. Organogenesis (development of organs) starts with stimulation caused by the chemicals of medium, substances carried over from the original explants and endogenous compounds produced by the culture.

In 1944, F. Skoog for the first time indicated that the organogenesis could be chemically controlled. He observed root initiation (rhizogenesis) and shoot inhibition (caulogenesis) after addition of auxin to the medium. Further, Skoog and co-workers gave the concept of regulation of organogenesis by a balance between cytokinin and auxin. Thus the auxin and cytokinin (CK) are important regulators of the developmental fate of pluripotent plant cells.

In 1957, F. Skoog and C.O. Miller demonstrated that a high ratio of auxin: cytokinin stimulated the formation of root in tobacco callus, but a low ratio of the same induced shoot formation. The hypothesis of organogenesis was advanced and propounded that organogenesis in callus starts with the development of a group of meristematic cells, *i.e.* meristemoids that can respond to the factors within the system to initiate a primordium which, depending on kinds of factors, induces root, shoot or embryoid (Table 18.4).

A fine balance between synthesis and catabolism spatially and temporally regulates the cytokinin homeostasis. Cytokinin modulates auxin-induced organogenesis via regulation of the efflux-dependent intercellular auxin distribution. Auxin (not cytokinin) is capable of triggering organogenesis in certain explants. Cytokinin affects differential distribution of auxin, and the cytokinin-mediated modulation of organogenesis is simulated by inhibition of transport of polar auxin.

A callus culture system demonstrates the quantitative control of root- and shoot- organogenesis by auxin and cytokinin, or growth regulating substances; for example roots can be inducted by a four days exposure to relatively high kinetin and low 2,4-dichlorophenoxyacetic acid (2,4-D). In contrast, shoots are produced after a 4 day induction period in high 2,4-D concentrations and low kinetin concentrations. On induction medium differentiation does not occur; but regeneration takes place upon transfer of tissue to growth regulator-free regeneration medium. Organogenesis of both roots and shoots in mulberry is shown in Fig. 18.9.

Table 18.4. *In vitro* control of organogenesis by auxins and cytokinins (source : Nandi and Palni, 1992).

Auxin (mg/1)	*Cytokinin (mg/1)*	*Organogenesis*
0.0	0.2	No growth
0.03	1.0	Shoots
3.0	0.02	Roots
3.0	0.2	Callus

***(a)* Cytokinin–auxin Cross-talk Controls Shoot Meristem Development:** Exciting progress has been made over the past few years to reveal the molecular mechanisms underlying the auxin–cytokinin action and interaction. Su *et al.* (2011) have reviewed the regulation of meristem development through auxin–cytokinin interaction. During embryogenesis apical meristem of shoot arises and generates all of the aerial parts of a plant. Studies on the classical chemical regulation of growth and organ formation in *in vitro* plant tissues have indicated that excess of cytokinin over auxin promotes shoot formation from callus. This suggests a positive action of cytokinin on activity of apical meristem of shoot. Cytokinin deficiency reduces size and activity of shoot meristem.

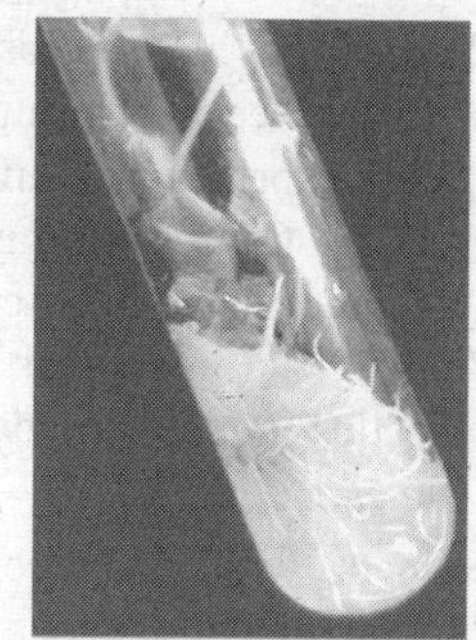

Fig. 18.9. Organogenesis in mulberry: rooting of *in vitro* regenerated shoot (*source*: CSRTI, Mysore, India).

The mutual regulation of cytokinin and stem cell-related genes can explain the regulation of cytokinin production and signaling in the shoot apical meristem. Cytokinin triggers an increase in mRNA levels of the KNOXI genes indicating a positive feedback loop between shoot apical meristem and cytokinin signalling. KNOX transcription factors repress the biosynthesis of the gibberellins (GA) to maintain the normal function of meristem. KNOX proteins are essential for meristem development as the key growth-regulators by simultaneously activating cytokinin and repressing GA biosynthesis. Besides, the type-A ARRs (ARR7/15) is the main response genes of cytokinin signalling required for CLV3 expression. Since CLV3 limits the expression of WUS which in turn down-regulates the expression of ARR5, ARR7 and ARR15, a negative regulation loop exists between type-A ARRs and WUS in the shoot apical meristem. WUS expression is also partially regulated by cytokinin signalling through this CLV-dependent pathway. In the clv mutants the induction of WUS transcripts after cytokinin treatment reveals a CLV-independent mechanism of cytokinin-induced WUS expression. Therefore, the cytokinin-induced increase of WUS transcripts is mediated mainly via an AHK2/AHK4- dependent pathway (Su *et al.*, 2011).

Auxin also plays a critical role in the maintenance of shoot meristem. Auxin produced by *Yucca* genes accumulates in the central zone of meristem at an optimal level to stimulate downstream auxin-induced genes through the Aux/IAA–ARF signalling pathway. Increase in ARR7 and ARR15 expression was observed in the shoot apical meristem of the certain plants indicating that ARR7 and ARR15 activation can be directly induced by the loss of local auxin accumulation (Zhao *et al.*, 2010). Since auxin suppresses the expression of STM [a member of Class IKNOTTED-like homeobox (KNOXI) proteins] that promotes cytokinin biosynthesis in the shoot meristem, it might function on the repression of ARR7 and ARR15 through the STM-mediated pathway. Another pathway has been identified through the analysis of ARR7 and ARR15 expression in the certain mutant plants (Zhao *et al.*, 2010).

The development of root meristem is controlled by the auxin–cytokinin cross-talk. Cytokinin and auxin mutually regulate their signalling pathways or their metabolisms through certain integrators, which are the basis of interaction between these two hormones to determine a specific developmental output in root meristem. Recent studies on a genetic framework have shown that the cell division and cell differentiation in the root meristem is controlled by antagonistic interaction between cytokinin and auxin, which occurs through a regulatory circuit via the SHORT HYPOCOTYL 2 SHY2/IAA3 that is a member of the Aux/IAA gene family (Su *et al.*, 2011).

***(b)* Auxin–cytokinin Interaction Regulates *in vitro* Organogenesis:** Plants have a profound capacity to regenerate organs from their differentiated somatic tissues through the manipulation of hormones. It has already been reported that a high auxin : cytokinin ratio induces root regeneration, while a low ratio promotes shoot induction. This shows that during *in vitro* organogenesis auxin and cytokinin might have a cross-talk. Cytokinin triggers the induction of ectopic WUS expression within the callus, because during *in vitro* establishment of the shoot meristem from the cultured root explants WUS expression is sufficient for inducing shoot regeneration. During shoot induction the increased AHK4 transcription is required for WUS activation. The roles of ARR7 and ARR15 in shoot regeneration have been characterised. Shoot regeneration is suppressed by over expression of ARR7 and ARR15, whereas loss-of-function of these two genes stimulates callus formation on auxin-rich callus-inducing medium and promotes shoot induction on cytokinin-rich shoot-inducing medium (Su *et al.,* 2011).

Treatment with exogenous auxin triggers the organogenic processes accompanied by the production of endogenous cytokinin and tissue-specific activation of cytokinin signalling. Endogenous cytokinin modulates this auxin-induced organogenesis through a negative regulation on the expression of PIN proteins, resulting in an optimal auxin distribution for the induction of root-like organ. It is suggested that auxin is capable of inducing organogenesis in hypocotyl explants. It is critical to identify the crucial signalling components that integrate the auxin and cytokinin signalling for callus induction and shoot regeneration. Auxin-regulated WUS expression is necessary for formation of shoot apical meristem during somatic embryogenesis. The establishment of auxin gradients is correlated with induced WUS expression and subsequent shoot apical meristem formation. These results shed new light on the cross-talk between the two hormones in regulating shoot and root stem-cell formation during somatic embryogenesis (Su *et al.,* 2011).

II. ROOT CULTURE

In 1934, for the first time P.R. White reported the successful organ culture, *e.g.* potentially unlimited growth of excised tomato roots. Subsequently, roots of several species of gymnosperms and angiosperms have been successfully cultured. Generally excised root is cultured in liquid medium. It has several advantages over the solid media. The techniques of root culture give certain important information such as : (*i*) nutritional requirements, (*ii*) infection by *Rhizobium* and nodulation, and (*iii*) physiological activities, for example, production of alkaloids, nicotine, etc. (Dodds and Roberts, 1985).

Root cultures can make a significant contribution in the production of secondary metabolites. Most of the research efforts that use differentiated cultures instead of cell suspension cultures have focused on transformed roots (Srivastava and Srivastava, 2007).

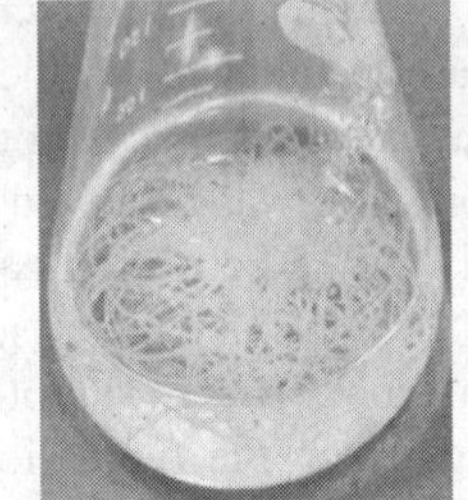

Fig. 18.10. Root culture in a plant.

Ghosh *et al.* (2002) excised root cultures of *Gloriosa superba* that reached 7.5 g dry weight l^{-1} and accumulated 240±40 μg colchicine g^{-1} cell dry weight after 4 weeks growth. While all precursors (except *trans*-cinnamic acid) enhanced colchicine content of root cultures without adversely affecting root growth, treatment with *p*-coumaric acid + tyramine (each at 20 mg l^{-1}) increased colchicine content to 1.9 mg g^{-1} cell dry weight.

Since roots are infected by a naturally occurring soil bacterium (*Agrobacterium rhizogenes*) that develops hairy root disease, such hairy roots can also be cultured *in vitro*.

1. Hairy Root Culture

Hairy root culture is a type of plant tissue culture used to study plant metabolic processes or to produce valuable secondary metabolites with plant genetic engineering. *A. rhizogenes* contains root inducing plasmids, such as Ri plasmids can infect plant roots and cause them to produce opines (a food source for the bacterium) and to grow abnormally. Particularly the abnormal roots are easy to culture in artificial media because hormones are not needed, and they are neoplastic (cancerous) with indefinite growth. The neoplastic roots produced by *Agrobacterium rhizogenes* infection have a high growth rate as compared to untransformed adventitious roots. These roots also have the genetic and biochemical stability. At present the main constraint for commercial utilisation of hairy root culture is the development and scaling up of appropriate bioreactors for the delicate and sensitive hairy roots (Georgiev *et al.*, 2010). So far, the only company based on hairy root cultures as an industrial level is a Swiss company called ROOTec bioactives AG (www.rootec.com).

(*a*) Application of Hairy Root Culture: Hairy root cultures can be used for phytoremediation; they are especially valuable for studies of the metabolic processes involved in phytoremediation. The engineering Ri plasmids containing T-DNA can be used for genetic transformation (biotransformation) of the plant cells.

Srivastava and Srivastava (2007) have reviewed the nature, applications, and perspectives and scale up of hairy root cultures for the production of valuable secondary metabolites. *Agrobacterium rhizogenes* causes hairy root disease in plants. The neoplastic roots produced by *A. rhizogenes* infection are characterised by high growth rate, genetic stability and growth in hormone-free media. Compared to intact plants, the genetically transformed root cultures can produce high levels of secondary metabolites. Hairy root culture offers promises for high production of valuable secondary metabolites (used as pharmaceuticals, pigments and flavors) in many plants. However, the main constraints for commercial exploitation of hairy root cultivations is the development and scaling up of appropriate reactor vessels (bioreactors) that permit the growth of interconnected tissues normally unevenly distributed throughout the vessel. Therefore, emphasis has been paid on designing the bioreactors suitable to culture the delicate and sensitive plant hairy roots. Recent reactors used for mass production of hairy roots can roughly be divided as liquid-phase, gas-phase or hybrid reactors (Srivastava and Srivastava, 2007).

Thus the hairy root cultures can be applied for detailed studies of fundamental molecular, genetic and biochemical aspects of genetic transformation and of hairy root induction. Hairy root culture can also be used for regeneration of whole plants and production of artificial seeds.

F. CELL (SUSPENSION) CULTURE

A fragment of callus (about 500 mg) or any other part of the plant is transferred to the liquid medium (500 ml) and agitated aseptically to make the cells free; thus the suspension of single cells including cell clumps is called 'cell suspension'. Single cells are isolated by using both mechanical and enzymatic methods. In mechanical methods, the tissue is grinded to a fine suspension in a buffered medium followed by filtration and centrifugation so that cell debris can be removed. The enzymatic method involves pectinase or macerozyme to dissolve the middle lamella between the cells. It is difficult to have suspension of single cell. The suspension includes single cell, cell aggregates (varied number of cells), residual inoculum and dead cells. A good suspension consists of a high proportion of single cells than small cluster of cells. It is more difficult to have a good suspension than to find optimum environmental factors for cell separation.

For separation of cells from callus, no standard technique has been recommended; therefore, cells are separated by changing the nutritional composition of medium. The cells are cultured by batch cultures or continuous cultures. To overcome this problem of creation of anaerobic conditions, the suspension cultures are agitated by a rotary shaker which disperses the cells and expose them to air. The cell density should be determined before sub-culturing. Growth of culture depends on a critical cell density below which culture will not grow. For instance, $9–15 \times 10^3$ cells/ml are required for a clone of *Acer pseudoplatanus*. Generally cultures are agitated on orbital shaker or magnetic stirrer for good result. In liquid medium free cells divide after lag phase (no cell division) and linearly increase their population. After some time, based on nutrient level, the rate of cell division decelerates until it comes to stationary phase. At this stage, it is essential to subculture the cells to keep the cells viable. By using plate technique, cell lines can be raised where a mass of cells is spread over medium. Further, growth of plated cells depends on cell density.

The cultured cells of a higher plant possess necessary genetic information for existence on cell level (reproduction, growth, mature state, programmed death). On the other hand, the cultured cells retain supplementary information that determines the production of substances which are important for integrative functions and biocoenotic interactions of the plants. The information determining the progression to a programme for the development of a whole plant is also redundant for existence at cell level. Cell culture systems have been employed in numerous morphological analyses by varying the origin of cells and physicochemical factors.

Common Features of Cultured Plant Cells and Microbial Cells in Suspensions Culture

The properties of cultured plant cell suspensions are common with culture of microorganisms as : (*i*) they grow in sterile environment, (*ii*) they are homogeneous in size, (*iii*) they have a doubling time which is longer than that of microorganisms but considerably shorter than cells *in situ*, and (*iv*) they can be grown on a large scale.

1. Cell (Suspension) Culture in Mass

Plant cells are cultured *en masse* in specially designed 'plant bioreactors' which essentially lack a stirrer because the plant cells are sensitive to shearing. Therefore, gas is gently bubbled in place of stirrer, which provides stirring and also meets the demand of a higher oxygen supply. The suspension cultures are broadly classified into mainly two types, batch culture and continuous culture:

(*a*) Batch Culture : A culture flask is used for batch culture of cells. The culture medium and the cells produced are retained in the culture flask (100-250 ml) containing 20-75 ml of the culture medium. The biomass growth in batch cultures follows a fixed pattern. Cultures are maintained continuously by sub-culturing, *i.e.* transfer of a small aliquot of inoculum from the grown culture to fresh medium at regular intervals. The cell number of a batch culture follows a typical sigmoidal curve as the bacterial cells do pass through 'lag phase' (a phase during which cells do not divide and their number remains constant), followed by brief exponential or 'log phase' (where cell number increases rapidly because of cell division). After 3-4 generations of the cultured cells their growth decreases which is called 'deceleration phase', and the culture enters the 'stationary phase' (where cell number again becomes static). Finally, the cell division stops due to depletion of nutrients and accumulation of cellular wastes. In batch cultures cell density and metabolism undergo change constantly; therefore, batch culture is not used to study cell behaviour. But it is convenient to maintain batch cultures; hence, it is used to initiate cell suspension and scale-up of continuous cultures. For detailed description see Chapter 21 : *Industrial microbiology*.

(*b*) Continuous Culture : In continuous culture the steady state of cell density is maintained by regular replacing of a portion of the used up medium with fresh medium. There are two types of continuous cultures as given below:

(*i*) Closed continuous suspension cultures : In closed type continuous cultures the used medium is replaced with fresh medium; therefore, the cells from used medium are mechanically taken out and added back to the culture. This results in increase in the cell biomass.

(*ii*) Open continuous suspension cultures : In open type continuous cultures both cells and used medium are replaced with fresh medium resulting in maintenance of culture at constant and sub maximal growth rate. For detailed description see Chapter 21 : *Industrial microbiology*.

Furthermore, there are two types of open continuous suspension culture- *turbidostat* and *chemostat.* In turbidostat, cells grow up to certain turbidity (based on optical density) when a predetermined volume of the culture is replaced by fresh culture.

Synchronisation of Suspension Cultures

Generally cell suspension is asynchronous; different cells have different size, shape, DNA, nuclear content and also in different stages of cell cycle (G1, S, G2, M). In cell metabolism studies on these features are undesirable Therefore, it is necessary to obtain synchrony in suspension cultures which can be achieved by employing the following methods :

- **Starvation:** Cells are starved of a nutrient like a growth regulator which is necessary for cell division resulting in arrest of cell growth during G1 or G2. After some time, all arrested cells enter divisions synchronously after supplying the nutrients.
- **Inhibition:** After using an inhibitor of DNA synthesis (such as 5-aminouracil), cells are arrested at G1 phase of cell cycle; hence removal of inhibitor leads to synchronous cell division.
- **Mitotic arrest:** You know that colchicine is always used for mutation by geneticists. It arrests cells at metaphase but only for short duration. Ccolchicine treatment for longer may induce abnormal mitosis and chromosome stickiness resulting in mitotic arrest.

In addition, in chemostat the fresh medium to be added has one nutrient kept at a concentration so that it is depleted rapidly and becomes growth-limiting, while the other nutrients are still in concentration higher than required. Increase or decrease in the concentration of growth limiting factor is correspondingly expressed by increase or decrease in growth rate of cells. Therefore, the level of concentration of growth-limiting factor with respect to that of other constituents is adjusted to maintain the desired rate of cell growth. The effect of individual nutrients on cell growth and metabolism can be determined by using chemostat.

(*c*) Benefits from Cell Culture: Cell suspension cultures have many advantages over the callus cultures as below:

- The suspension can be pipetted easily.
- Suspension culture offers the opportunity to confirm single cell origin of somatic embryos and the asymmetry of the first cell division which starts the process.
- They are less heterogeneous and cell differentiation is less pronounced.
- They can be cultured in volumes up to 1,500 litres.
- They can be subjected to more stringent environmental controls.
- They can be manipulated for production of natural products by feeding precursors.
- The other advantage of cell suspension culture is the use of single embryogenic cells and somatic embryos as a source for direct gene transfer via particle bombardment, transit gene expression and confocal microscopy observation.

2. Culture of Isolated Single Cells

Cell suspensions are grown as single cells under *in vitro* conditions to obtain single cell clones by a process called cell plating which is a most popular technique developed by M.K. Razdan in 1993. In this method the free cells are suspended in liquid medium at a two-fold density as compared to finally desired plating cell density. Here equal volumes of the nutrient media containing single cells and a melted (30-35°C) agar medium (0.6-1%) are mixed. Soon the two mixed medium is spread in the Petri dishes in such a way that the cells should be fixed in 1 mm thick cooled and solidified agar layer.

3. Application of Cell Suspension Culture

The cell suspension cultures can be used for : (*i*) induction of somatic embryos/shoots, (*ii*) *in vitro* mutagenesis and mutant selection, (*iii*) genetic transformation, (*iv*) production of secondary metabolites, etc.

Cell suspension cultures are particularly suitable for physiological, biochemical and molecular studies of the process of somatic embryogenesis and its different stage, such as induction development, maturation and conversion. The development of a genome and proteome database of model *Medicago truncatula* species together with the presence of protocol of cell suspension will provide the opportunity to identify and characterised genes involved in the whole process of somatic embryogenesis.

(*a*) Production of Secondary Metabolites : The most important chemicals produced by using cell cultures are the secondary metabolites, *i.e.* the cell constituents which are not essential for plant survival. The secondary metabolites are : alkaloids, glycosides (steroids and phenolics), terpenoids, latex, tannins, etc. During plant growth the cells undergo morphological differentiation and maturation, where some of the specialised cells produce secondary metabolites. Production of secondary metabolites *in vitro* is higher in differentiated tissues than that in non-differentiated tissues. The cell cultures produce secondary metabolites through several ways such as : (*i*) a new route of synthesis of products, for example codeine, quinine, pyrethroids, (*ii*) a route of synthesis to a novel product from plants difficult to grow e.g. thebain from *Papaver bracteatum*, (*iii*) a source of novel chemicals in their own right such as rutacultin from culture of *Ruta*, and (*iv*) as biotransformation systems either on their own or as part of a larger chemical process e.g. synthesis of digoxin.

(*i*) Advantages of *in vitro* production of secondary metabolites : There are many advantages in production of secondary metabolites *in vitro* as given below :

- The cell cultures and cell growth are easily controlled for the formation of improved product.
- The recovery of the product is easy.
- Increase in the production of the required metabolite is easy because the cell culture systems are independent of environmental factors, seasonal variations, pest and microbial diseases, geographical location constraints.
- Mutant cell lines can be developed for the production of novel and commercially useful compounds.
- As per demand of market the compounds are produced under controlled conditions.
- Minimal labour is involved; therefore, it takes less production time and also is cost effective.

(*ii*) Applications of secondary metabolites : Secondary metabolites are very important because of its uses in many ways such as immense use of alkaloids in medicine. In cell culture yield of these chemicals is generally lower than in whole plants, but it can be substantially increased through manipulation of physiological and biochemical conditions of the medium. Cells from *Lithospermum erythrorhizon* produce a dye shikonine on a commercial scale. In addition, a large number of secondary metabolite products that are being used for various purposes. For example, vincristine is used as anticancer compound, digoxin controls cardiovascular disorders, pyrithrins is an insecticide, etc. In recent years, the production of specialty chemicals by some plants has become a multibillion industry. Some of the secondary metabolites and their uses are given in Table 18.5.

Table 18.5. Secondary metabolites obtained from some plants through tissue culture techniques.

Product	***Plant source***	***Uses***
Artemisin	*Artemisia* spp.	Antimalarial
Azadirachtin	*Azadirachta indica*	Insecticidal
Berberine	*Coptis japonica*	Antibacterial, anti- inflammatory
Capsaicin	*Capsicum annum*	Cures rheumatic pain
Codeine	*Papaver* spp.	Analgesic
Camptothecin	*Campatotheca accuminata*	Anticancer
Cephalotaxine	*Cephalotaxus harringtonia*	Antitumour
Digoxin	*Digitalis lanata*	Cardiac tonic
Pyrethrin	*Chrysanthemum cinerariaefolium*	Insecticide (for grain storage)
Morphine	*Papaver somniferum*	Analgesic, sedative
Quinine	*Cinchona officinalis*	Antimalarial
Scopolamine	*Datura stramonium*	Antihypertensive
Taxol	*Taxus brevifolia*	Anticarcinogenic
Vincristine	*Cathranthus roseus*	Anticarcinogenic

(*b*) Production of Stress-tolerant Plants : Rai *et al.* (2011) have reviewed the developing stress tolerant plants through *in vitro* selection. The efforts to develop stress-tolerant plants are of immense importance to increase crop productivity. Tissue culture-based *in vitro* selection has emerged as a feasible and cost-effective tool for developing stress-tolerant plants. Plants tolerant to both the biotic and the abiotic stresses can be acquired by applying the selecting agents such as NaCl (for salt tolerance), PEG or mannitol (for drought tolerance) and pathogen culture filtrate, phytotoxin or pathogen itself (for disease resistance) in the culture media. Only the explants capable of sustaining such environments survive in the long run and are selected. *In vitro* selection is based on the induction of genetic variation among cells, tissues and/or organs in cultured and regenerated plants. The selection of somaclonal variations appearing in the regenerated plants may be genetically stable and useful in crop improvement. Under stress conditions plants have evolved many biochemical and molecular mechanisms for its survival.

The cultured cells operate biochemical mechanisms determining resistance to biocide chemicals and provide promise for selection *in vitro*. In 1974, M.H. Zenk has reported cell suspension tolerant to 2, 4-D when the cells were subcultured for 6 months in liquid medium supplemented with increasing amounts of herbicides. The cells were able to grow in 1mM (milli mol) 2,4-D, while the control suspension was completely inhibited at 0.3 mM 2, 4-D. The mechanism of tolerance involves an enhancement in the metabolism of 2, 4-D. In 1977, J.M Widholm observed that carrot cells lost the tolerance to the herbicides when transferred to 2, 4-D free medium. The tolerance was the result of induction of enzymatic systems responsible for the degradation of 2, 4-D.

In 1978, R.S. Chaleff and M.F. Persons reported for the first time that regenerated plants from stable cell lines were resistant to herbicides. They also obtained mutant tobacco plants with monogenetic dominant resistance to picloram. The promising possibility of selecting for herbicide resistance *in vitro* was supported by many workers in many plants, such as *Citrus,* barley, tomato, etc.

Uses of protoplasts as a system to select cell lines tolerant to herbicides have not been extensively explored. Protoplast technology can be used to increase the possibility to obtain monoclonal lines and offer the opportunity for intraspecific transfer of cytoplasmic factor of resistance to some type of herbicides. A potential application of transfer of herbicide tolerance factor is the transplant of cytoplasrnic organelles, for example chloroplasts.

G. PROTOPLAST CULTURE

Plant cell wall acts as physical barrier and protects cytoplasm from microbial invasion and environmental stress. Protoplasts are the naked plant cells (*i.e.* cell without cell wall) that can be isolated by using enzymes (like cellulases, pectinases) from leaf, seedling, calli, pollen grains, embryo sacs, etc. Protoplasts are the biologically active and most significant material of cells and the recent researches are based on it. Protoplast consists of a complex mixture of cellulose, hemicellulose, pectin, lignin, lipids, protein, etc. Therefore, it is essential to have the respective enzymes for dissolution of different components of the cell wall. Upon *in vitro* culture the protoplasts regenerate cell wall, undergo cell division, and form callus, which can also be sub-cultured. Some of the examples of plant species that have been regenerated from protoplasts are : *Cucumis sativus, Capsicum annum, Ipomoea batata, Glycine max, Chrysanthemum* sp.

1. Isolation of Protoplasts

Until suitable methods were developed, protoplasts were isolated by cutting the plasmolysed plant tissues and releasing protoplast through deplasmolysis of cells. In 1960, E.C. Cooking for the first time isolated the protoplasts of plant tissues by using cell wall-degrading enzymes *viz.,* cellulase, hemicellulase, pectinase, and protease extracted from a saprophytic fungus *Myrothecium verrucaria.* By 1968, commercial preparations of purified cell wall degrading enzymes such as macerozyme, cellulase and hemicellulose became available that gave further progress to enzymatic isolation of protoplasts. However, the basic technique for isolation of protoplast is described briefly as below and given in Fig. 18.11.

***(a)* Surface Sterilisation of Leaf Samples:** Fully expanded mature leaves are collected from 10 weeks old healthy plants such as tobacco, washed in tap water to remove adhering soil particles and sterilised by first dipping them into 70% ethyl alcohol for about a minute and then treating with sodium hypochlorite solution for 20-30 minutes.

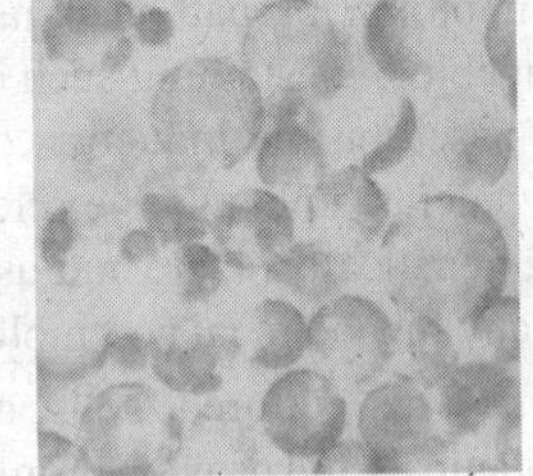

Protoplasts of *Petunia* sp.

***(b)* Rinsing in Suitable Osmoticum :** After 10 min, sample is properly washed with sterile distilled water or MS medium adjusted to a suitable pH and buffer to maintain osmotic pressure. Subsequent operations are carried out under aseptic conditions such as under laminar air flow chamber. Washing should be done for about 6 times to remove the traces of sodium hypochlorite.

(c) Plasmolysis of Cells : The lower epidermis covered by thin wax cuticle is removed with a forceps. Stripping should be done from midrib to margin of lamina. The stripped surface of leaf is kept in mannitol solution (13% w/v) for 3 hours to allow plasmolysis of cells.

***(d)* Peeling of Lower Epidermis :** The lower epidermis of the sterilised leaves is carefully peeled off and the stripped leaves are cut into small pieces. The peeling operation is easy, if the water supply is limited before excision or the leaves are allowed to become flaccid after sterilisation. Mesophyll protoplasts can be obtained from these peeled leaf segments, while those for epidermis are obtained from the peeled epidermis. In cereals, it is difficult to peel off the epidermis; therefore, leaves are cut in long strips and used with enzyme mixture. The sterilisation step can be omitted when the starting material is an *in vitro* shoot culture, and leaves can be cut into pieces under aseptic conditiosn and used directly for enzymatic treatment.

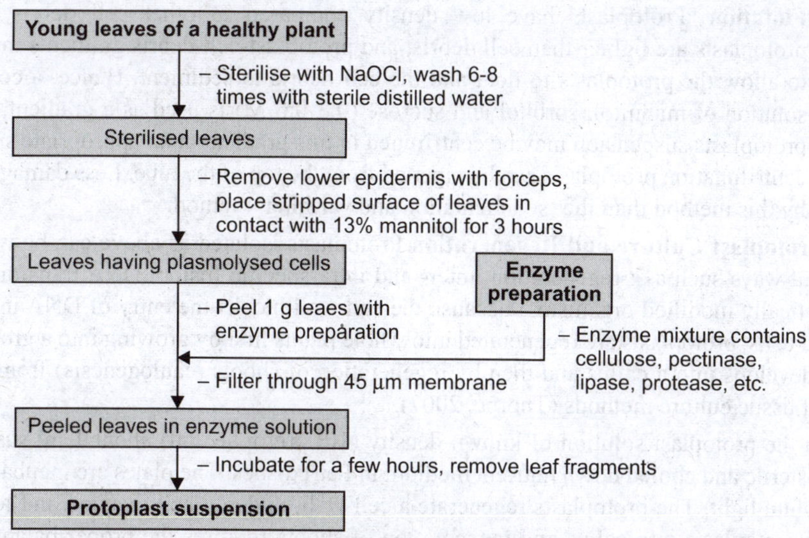

Fig. 18.11. Isolation of protoplasts.

About 1 g peeled off leaves is transferred into enzyme mixture already sterilised through a Seitz filter (0.45 µm). This facilitates the penetration of enzyme into tissue within 12-18 hours at 25°C. From the peeled leaf segments, the protoplasts can be isolated using any one of the two methods :

- **Mixed (one-step) enzymatic method:** This involves simultaneous separation of cells and degradation of their walls to convert protoplast by immersing plant tissues in mixture of macerozyme (or pectinase) and cellulases.
- **Sequential (two-step) enzymatic method:** This involves two steps where first macerated plant tissues are incubated with pectinase to get single cells followed by cellulase treatment to get protoplast.

In both the methods, the peeled leaves are put in such a way that their lower surface should be downwards in a Petri dish containing the enzyme mixture. In one-step method the enzyme mixture consists of 0.5% macerozyme + 2% cellulase in 13% sorbitol or mannitol at pH 5.4, while in sequential method, two enzyme mixtures are used one after the other. But the protoplast yield is higher in one step method because both palisade and spongy mesophyll tissues are used in this method, while only palisade is used in sequential method. However, the protoplasts isolated by sequential method are better.

(*e*) Purification of Protoplasts: The protoplasts isolated as above are present in the media together with a range of cell debris and broken cell organelles. There are many methods for purification of protoplasts from this mixture. Leaf debris is removed with forceps, and enzyme solution containing protoplasts are filtered with a nylon mesh (45 µm). There are two methods commonly used for purification of protoplasts as described below :

- **Sedimentation and washing**: In this method, the crude protoplast suspension is centrifuged first at low speed (50-100 ***g*** for 5 minutes), then at high speed (7,000 rpm). The intact protoplasts form a pellet, and supernatant containing cell debris can be pipetted. Again a fresh MS medium plus 13% mannitol and rewashed and centrifuged. Repeated washing with nutrient medium, centrifugation and decantation are done for about three times. Finally specific concentration of protoplast suspension is prepared, and transferred on specific medium. The steps for purification and regeneration of protoplasts are given in Fig. 18.12.

- **Flotation**: Protoplasts have low density compared to other cell debris; therefore, protoplasts are lighter than cell debris, and protoplast-cell debris gradients may be used to allow the protoplasts to float and the cell debris to sediment. Hence, a concentrated solution of mannitol, sorbitol and sucrose (0.3-0.6 M) is used as a gradient, and crude protoplasts suspension may be centrifuged in this gradient at an appropriate speed. After centrifugation protoplasts can be pipetted from the top of the tube. Less damage is caused by this method than the 'sedimentation and washing' method.

***(f)* Protoplast Culture and Regeneration** Protoplasts isolated as above can be widely used for different ways such as : regeneration, inter- and intra-specific fusion, DNA transformation (to make genetically modified organisms) because the cell wall blocks the entry of DNA into the cell. In the plant cells, protoplasts are regenerated into whole plants first by growing into a group of plant cells that develops into a callus and then by regeneration of shoots (caulogenesis) from the callus using plant tissue culture methods (Thorpe, 2007).

From the protoplast solution of known density (10^5 protoplast/ml) about 1 ml suspension is poured on sterile and cooled down nutrient medium in Petri dishes. The plates are incubated at 25°C in a dim white light. The protoplasts regenerate a cell wall, undergo cell division and form callus. Growth of protoplasts into callus and regeneration of shoots requires the proper balance of plant growth regulators in the tissue culture medium that must be customised for each species of plant. The callus can also be sub-cultured. Embryogenesis begins from callus when it is placed on nutrient medium lacking mannitol and auxin. The embryo develops into seedlings and finally mature plants (Fig. 18.12).

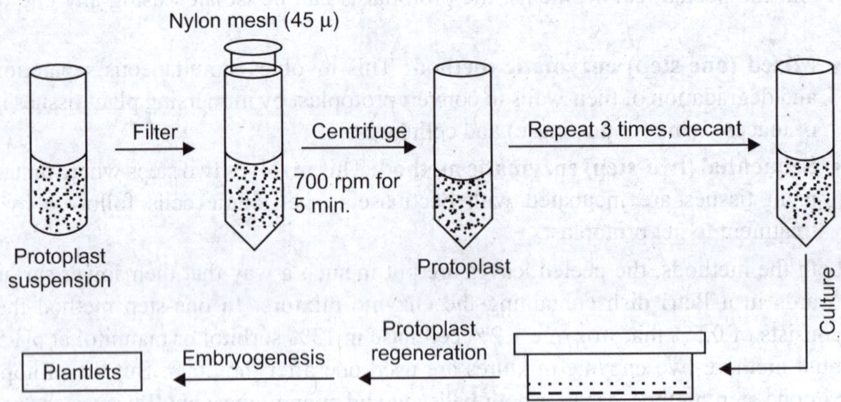

Fig. 18.12. Purification and regeneration of protoplasts.

Examples of plant species that have regenerated from protoplasts are *Cucumis sativus, Capsicum annum, Ipomoea batata, Beta vulgaris, Helianthus annuus, Glycine max, Rosa sp., Chrysanthemum* sp., etc. The isolated protoplasts are used for various purposes such as : (*i*) biochemical and metabolic studies, (*ii*) plant breeding using a technique called protoplast fusion, *i.e.* fusion of two different somatic cells to get somatic hybrids, (*iii*) fusion of nucleated (containing nucleus) and enucleated (without nucleus) cells to produce cybrid (cytoplasmic hybrid), (*iv*) genetic manipulation, (*v*) drug sensitivity, etc.

Protoplast viability and Plating Density

Protoplast viability and plating density must be tested before culturing protoplast. Protoplast viability is estimated by staining with fluorescein diacetate (FDA). Under UV the viable protoplasts exhibit green fluorescence. It should be examined within 5-15 min after the FDA treatment when FDA dissociates. There are some protoplasts which have leaky membrane through which metabolites are leached out. The metabolites can be absorbed back if there is sufficient density of protoplasts. Therefore, a minimum plating density is required to begin the growth and cell density is estimated by haemocytometer.

Single or few protoplast cultures is required for genetic engineering. In conditional media plant cells are already grown and metabolites are leached into it. Therefore, this medium is used after filtering for growing the isolated protoplasts. Feeder layer is prepared by plating solid media with protoplasts followed by irradiation. This results in inactivation of the nucleus with viable protoplasts. The protoplasts at lower density can be plated on this feeder layer.

H. PROTOPLAST FUSION AND SOMATIC HYBRIDISATION

With the development of techniques for enzymatic isolation of protoplasts and subsequent regeneration, a new tool of genetic manipulation of plants has now become available. Moreover, the fusion of protoplasts of genetically different lines or species has also been possible. For example, some plants that show physical or chemical incompatibility in normal sexual crosses may be produced by the fusion of protoplasts obtained from two cultures of different species. Somatic hybridisation of crop plants represents a new challenge to plant breeding and crop improvement. In the field of pest and disease resistance and transfer of C_3 photosystems into C_4 crop plants somatic crosses show most interesting promises (Dodds and Roberts, 1985).

The results obtained so far from somatic hybridisation represent that it is possible to recover fertile and stable amphidiploid somatic hybrids after protoplast fusion. There are four major aspects of protoplast fusion as : (*i*) production of fertile amphidiploid somatic hybrids of sexually incompatible species (induced fusion of protoplasts from two genetically different lines of species must result in a variety of homo-as well as hetero-karyotic fusion products (heterokaryon or heterokaryocytes. Selection of a few true somatic hybrid colonies from the mixed population of regeneration protoplast is a key step in successful somatic hybridization technique), (*ii*) production of heterozygous lines within one plant species which normally will be propagated only vegetatively *e.g.* potato, (*iii*) the transfer of only a part of genetic information from one species to another using the phenomenon of chromosome elimination, and (*iv*) the transfer of cytoplasmic genetic information from one to a second line or species. It has been possible to transfer useful genes (such as *nif* genes, disease resistance genes, rapid growth genes) from one species to another, thereby, to widen the genetic base for plant breeding.

1. Hybridisation and Cybridisation

Fusion of cytoplasm of two protoplasts results in coalescence of cytoplasms. The nuclei of two protoplasts may or may not fuse together even after fusion of cytoplasms. The binucleate cells are known as heterokaryon or heterocyte (Fig. 18.13). When nuclei are fused the cells are known as hybrid or synkaryocyte, and when only cytoplasms fuse and genetic information from one of the two nuclei is lost, it is known as cybrid, (cytoplasmic hybrid or heteroplast) (Doods and Roberts, 1985). Production of cybrid (through fusion of nucleated and enucleated different somatic cells) is called **cybridisation,** while production of hybrid (by fusion of two nucleated cells) is called **hybridisation.** There are some genetic factors which are carried in cytoplasmic inheritance, instead of nuclear genes, for example male sterility in some plants. Susceptibility and resistance to some of the pathotoxins and drug are controlled by cytoplasmic genes. Therefore, production of cybrids which contain the mixture of cytoplasms but only one nuclear genome can help in transfer of cyloplasmic genetic information from one plant to another. Thus, informations of cybrid can be applicable in plant breeding experiments.

In China cybrid technology in rice is a great success. Such plants are very useful in producing hybrid seeds without *emasculation.* However, cybrid technology has successfully been applied to carrot, *Brassica* sp., *Citrus* sp., tobacco and sugar beet.

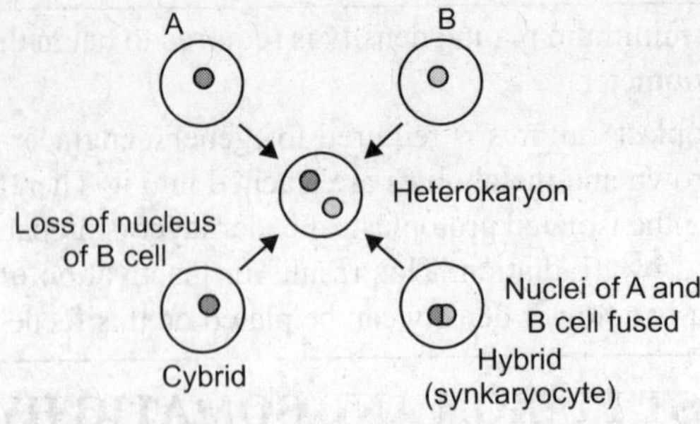

Fig. 18.13. Hybrid/cybrid production through protoplast fusion.

(*a*) Cybrids: Cybrids are referred to the cytoplasmic hybrids where the nucleus is derived from only one parent and the cytoplasm is derived from both the parents. The process of formation of cybrids is called 'cybridisation'. During cybridisation process and heterokaryon formation, the nuclei are stimulated for segregation; therefore, one protoplast contributes to the cytoplasm while the other contributes nucleus alone. The protoplasts can be made inactive and non-dividing by irradiation with gamma rays and X-rays, and use of metabolic inhibitors. You know that there are certain genetic traits in some plants that are cytoplasmically controlled, for example male sterility, resistance to certain antibiotics and herbicides. Hence, the cybrids are important for transferring the genetic traits of cytoplasmic male sterility, antibiotic- and herbicide- resistance in some plants useful in agriculture. Cybrids of *Brassica raphanus* have been developed that contain nucleus of *B. napus*, chloroplasts of atrazin-resistant *B. capestris* and male sterility from *Raphanus sativas*.

2. Methods of Somatic Hybridisation

The steps for successful somatic hybridisation are : (*i*) isolation of protoplasts from suitable plants, (*ii*) mixing of protoplasts in centrifuge tube containing fusogenic chemicals, *i.e.* chemicals promoting protoplast fusion, such as polyethylene glycol (PEG) (20%, w/v), sodium nitrate ($NaNO_3$), maintenance of high pH 10.5 and temperature 37°C (as a result of fusion of protoplasts viable heterokaryons are produced. PEG induces fusion of plant protoplasts and animal cells and produces heterokaryon), (*iii*) wall regeneration by heterokaryotic cells, (*iv*) fusion of nuclei of heterokaryon to produce hybrid cells, (*v*) plating and production of colonies of hybrid cells, (*vi*) selection of hybrid, subculture and induction of organogenesis in the hybrid colonies, and (*vii*) transfer of mature plants from the regenerated callus.

A pomato plant

In 1987, O.L. Gamborg and M.W. Nabors have described a number of variations in cell fusion such as : (*i*) fusion of two protoplasts each from cells with normal parental ploidy, (*ii*) fusion of two protoplasts each obtained from haploid cell sources (the fusion of two diploid cells of *Solanum* sp. would yield a tetraploid which is the normal ploidy for potato, and (*iii*) fusion of a protoplast with an enucleated protoplast. The result of this type of fusion would bring about cybrid with a single nucleus of one parent and the respective ploidy.

In 1972, P.S. Carlson succeeded in carrying out protoplast fusion of *Nicotiana glauca* × *N. langsdorffii*, and produced the first inter-specific somatic hybrid between them. In 1978, G. Melchers, Luther Burbank and co-workers developed first intergenic somatic hybrids between potato (*Solanum thuberosum*) and tomato (*Lycoperscon esculentum*). The hybrids are known as '**pomato** or topato' which has a combination of two members of the Solanaceae family (the potato and the tomato). It is a small tomato-like fruit, with white flesh, edible either raw or cooked. Potato+tomato cross is shown in Fig. 18.14.

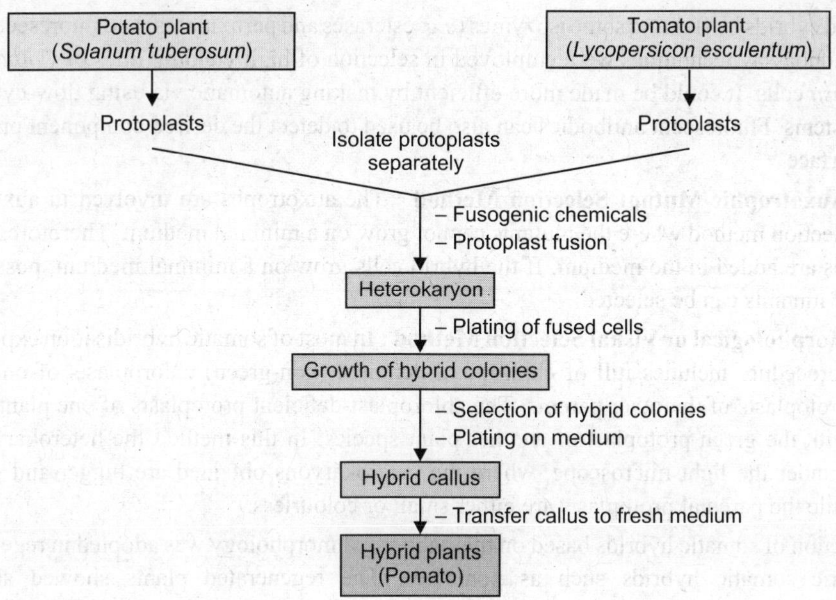

Fig. 18.14. Fusion of potato and tomato protoplasts and production of pomato plant.

In 1985, similarly A.A. Kuchko achieved somatic hybrid through inter-specific cross of wild and cultivated variety of potatoes, *i.e. Solanum tuberosum* × *S. chacoense.* Intergeneric hybrids which may be difficult to achieve by sexual crosses have been obtained through protoplast fusion, for example tomato + potato, *Datura* + *Atropa*, barley + wheat, barley + rice, wheat + oat, and sugarcane + sorghum.

With regard to inter familial cell hybrids, there have been very few successful experiments to obtain such hybrids, for example, hybrids of *Glycine* sp. × *Nicotiana* species. Based on detailed studies on inter familial hybrids, it has been concluded that inter familial hybrids of plant cells are genetically unstable and show species specific elimination/ reconstruction of chromosome belonging to one of the parents. This was usually the parent represented by the mesophyll cells in hybridisation process. These hybrids are incapable of regeneration and morphogenesis into plants. Similarly, intertribal hybrids were studied more widely due to their higher genetic stability; the first intertribal cell hybrid was obtained by Y.Y. Gleba and F. Hoffmann in 1978 from callus of *Arabidopsis thaliana* crossed with mesophyll protoplasts of turnip (*Brassica compestris*).

3. Selection of Somatic Hybrids and Cybrids

After protoplast fusion cell regenerate cell walls and undergo mitosis. This results in mixed population of parent cells, homokaryotic fusion product and hybirds. Hybrid cells should be differntiated from other cells. There are various selection methods used for selection of fusion products. Selection methods are dependent on : (*i*) physical properties of fused cells, (*ii*) biological properties of fused cells, and (*iii*) biological properties of colonies formed from fused cells. The somatic hybrids cannot be identified. Hence, biochemical markers are required for purpose.

The methods used for the selection of hybrid cells are : biochemical, visual and cytometric methods using fluorescent dyes.

***(a)* Biochemical Methods :** In this method, biochemical compounds are used in the medium for selection of hybrid cells. The drug sensitivity method is useful for the selection of hybrids of two plants species, if one of them is sensitive to a drug. Biochemical method is used for selection of hybrids and cybrids by staining some isozymes (*e.g.* esterases and peroxidases) and fluoresecent dyes. Radioimmunoassay techniques were employed in selection of high yielding lines of *Catharanthus* and *Anchusa* cells. It could be made more efficient by making automatic via using flow-cytometric sorting systems. Fluorescent antibodies can also be used to detect the desired component present on the cell surface.

(*b*) Auxotrophic Mutant Selection Method : The auxotrophs are involved in auxotrophic mutant-selection method where the mutants cannot grow on a minimal medium. Therefore, specific compounds are added in the medium. If the hybrid cells grow on a minimal medium, possibly the auxotropic mutants can be selected.

(*c*) Morphological or Visual Selection Method : In most of somatic hybridisation experiment, selection procedure includes full of chlorophyll-deficient (non-green) chloroplasts of one parent with the protoplasts of the other parents. The chloroplast-deficient protoplasts of one plant species is fused with the green protoplast of another plant species. In this method the heterokaryons are identified under the light microscope, where the heterokaryons obtained are bigger and green in colour, while the parental protoplasts are either small or colourless.

Selection of somatic hybrids based on their abnormal morphology was adopted in regenerating inter-generic somatic hybrids such as pomatoes. The regenerated plants showed abnormal morphology and proved to be somatic hybrids by analysis of chromosomal and fraction-1-protein, the ribulose1, 5-bisphosphate carboxylase. Intermediate morphology of the callus also determined the inter-generic somatic hybrids between *Vicia faba* and *Petunia hybrida*.

(*d*) Cytometric Method : The cytometric method uses flow cytometry and flourescent-activated cell sorting (FACS) techniques for the analysis of plant protoplasts. Flow cytometry technique counts and examines the microscopic particles by suspending them in a stream of fluid and passing them by an electronic detection apparatus, for example cells and chromosomes. This technique employs the principles of hydrodynamic focusing for presenting cells to a light excitation source like laser. The sample is injected into the center of a sheath flow. The combined flow is reduced in diameter; therefore, it enforces the cell into the center of the stream. This follows the principle of passing one cell at a time through the laser. A flow cytometer is like a microscope except the flow cytometry offers 'high-throughput' (for a large number of cells) automated quantification of set parameters, instead of producing an image of the cell. First a single-cell suspension should be prepared to analyse solid tissues. A flow cytometer consists of the following five main components :

- **Flow cell :** It is a liquid stream (sheath fluid) which carries and aligns the cells so that they pass single file through the light beam for sensing.
- **Measuring system** : The commonly used measuring system are measurement of impedance (or conductivity) and optical systems e.g. lamps (mercury, xenon), high-power water-cooled lasers (argon, krypton, dye laser), low-power air-cooled lasers [argon (488 nm), red-HeNe (633 nm), green-HeNe, HeCd (UV)], and diode lasers (blue, green, red, violet) resulting in light signals.
- **Detector and analogue to digital conversion (ADC) system :** The detector is connected that generates FSC and SSC as well as fluorescence signals from light into electrical signals to be processed by a computer.
- **Amplification system :** An amplification system is also connected to it.
- **Computer :** A computer is used for analysis of the signals.

There are many laboratories that use techniques of flow cytometry and fluorescent activated cell sorting (FACS) to analyse plant protoplasts when maintaining their viability. For sorting and selection of heterokaryons these techniques have also been applied by many researchers.

(*i*) **Fluorescence-activated cell sorting (FACS):** Fluorescence-activated cell sorting (FACS) is a specialised type of flow cytometry. It provides a method for sorting a heterogeneous mixture of biological cells into two or more containers, one cell at a time, on the basis of the specific light scattering and fluorescent characteristics of each cell. It provides fast, objective and quantitative recording of fluorescent signals from individual cells and also physical separation of the particular cells. For the first in 1965, Mack Fulwyler invented cell sorter by using the Coulter principle. Len Herzenberg expanded this technique and coined the term FACS; therefore, he won the Kyoto Prize in 2006.

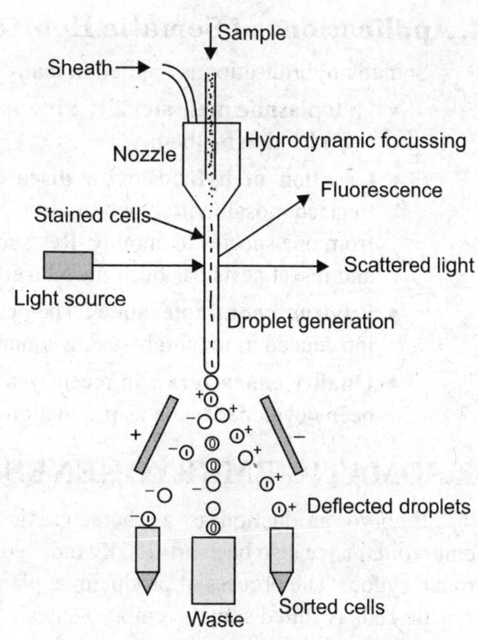

Fig. 18.15. Steps of flow cytometry.

The cell suspension is passed through the center of a narrow rapidly flowing stream of liquid. The stream of cells is broken into individual droplets by a vibrating mechanism. There is a low probability of passing of more than one cell per droplet because of adjustment of the system. The flow passes through a fluorescence measuring station just before breaking of the stream into droplets; therefore, the fluorescent character of interest of each cell is measured at this stage. An electrical charging ring is placed at the point of breaking of stream into droplets. This is done on the basis of prior fluorescence intensity measurement. As the opposite charge gets broken from the stream, it is trapped on the droplet. On the basis of presence of charges, the charged droplets fall through an electrostatic deflection system that diverts droplets into containers (Fig. 18.15).

A universally applicable method for electronic sorting of heterokaryons formed by fusing the protoplasts of two parents labelled with different vital fluorescent dyes, such as rhodomine isothiocyanate and floreceine isothiocyanate has been described by Galbraith co-workers in 1989. A cell sorter machine sorts the fused and unfused products on the basis of the presence or absence of fluorescence of both dyes in the fusion products.

Limitations of Somatic Hybridisation

Somatic hybridisation has many limitations as given below:

- Somatic hybridisation does not always produce plants that give fertile and visible seeds.
- Somatic hybridisation between two diploids results in the formation of an amphidiploid which is not favourable, therefore haploid protoplasts are recommended in somatic hybridisation.
- There is genetic instability associated with protoplast culture.
- Protoplast fusion between different species/genus is easy, but the production of viable somatic hybrids is not always possible.
- There are limitations in the selection methods of hybrids, as many of them are not efficient.
- It is not certain that a specific character will get expressed in somatic hybridisation.
- Regenerated plants obtained from somatic hybridisation are often variable due to somaclonal variations, chromosomal elimination, organelle segregation etc.

4. Applications of Somatic Hybridisation

Somatic hybridisation is applied in many areas as given below :

- **Cytoplasmic male sterility :** It was possible to transfer cytoplasmic male sterility using cybridisation method.
- **Creation of hybrids with disease resistance:** Many disease resistance genes (e.g. tobacco mosaic virus, potato virus X, club rot disease) could be successfully transferred from one species to another. Resistance against diseases such as TMV, spotted wilt virus and insect pests has been introduced in tomato.
- **Environmental tolerance :** The genes conferring tolerance for cold, frost and salt were introduced in tomato by using somatic hybridisation.
- **Quality characters :** In recent years, somatic hybrids with certain characteristics have been developed, such as production of high nicotine content in tobacco.

I. SOMATIC EMBRYOGENESIS AND ENCAPSULATED SEEDS

Embryo production is a characteristic feature of the flowering plants. Such structures (embryoids) have also been artificially induced in cultured plant tissues, besides embryo production from a zygote. The process of producing a plant or embryoid from a single somatic cell or group of somatic cells is called somatic embryogenesis.

1. Somatic Embryogenesis

For the first time somatic embryogenesis was discovered in 1958 by F.C. Steward and co-workers and in 1959 by J. Reinert in carrot cell suspension cultures. The physiological and biochemical events occurring during plant development can be explored through somatic embryogenesis.

(*a*) Types of Encapsulated Seeds: In 1978, H.W. Kohlehbach proposed the following classification of embryos :

(*i*) Zygotic embryos: In zygotic embryogenesis two haploid cells combine to form one diploid cell called zygote.

(*ii*) Non-zygotic Embryos: It is formed from the cells other than zygote. It is of following types :

- **Somatic Embryos:** Somatic embryos are mainly induced from somatic cells *in vitro* by using either solid or liquid nutrient media containing plant growth regulators, *i.e.* auxins and in a smaller amount cytokinin also. The example of somatic embryogenesis occurring in nature is ovules in *Paeonia.* Shoots and roots are monopolar, whereas somatic embryos are bipolar that allows to form a whole plant without culturing on multiple media types.
- **Parthenogenetic Embryos:** Parthenogenesis is a form of apogamy in which the embryo comes from an unfertilised egg. Thus the embryos produced through this process are parthenogenetic embryos.
- **Androgenic Embryos:** Androgenesis refers to male parthenogenesis in which the embryo contains only paternal chromosomes due to the failure of the egg nucleus to participate in fertilisation. Thus the androgenic embryos are formed by pollen grains (see androgenesis in preceding section).

Generally, somatic embryos called embryoids are similar to zygotic embryos (or seed embryos) except they originate from somatic cells and are larger in size. Somatic embryogenesis can be initiated in two ways, directly or indirectly. Indirect embryogenesis occurs through induction of embryogenic cells from unorganised tissue (callus) within the preformed callus. Direct embryogenesis occurs from pre-embryonic determined cell (without callus) of explant tissue which are ready to differentiate into embryoids and then an identical clone.

In the first case, embryoids are initiated in callus from superficial cell aggregates where cells contain a large vacuole, dense cytoplasm, large starch granules and nucleus. Two nutritional media of different composition are required to obtain embryoids. First medium contains auxin to initiate embryogenic cells. Second medium lacks auxin or reduced level of auxin is needed for subsequent development of the embryonic cells into embryoids and plantlets. In both the cases reduced amount of nitrogen is required. The embryogenic cells pass through three different stages to form embryoids, for example globular, heart shaped and torpedo shaped embryoids.

***(b)* Steps of Somatic Embryogenesis :** Regeneration of plant through somatic embryogenesis occurs in the five steps : (*i*) initiation of embryogenic cultures, (*ii*) proliferation of embryogenic cultures, (*iii*) prematuration of somatic embryos, and (*iv*) maturation of somatic embryos, and (*v*) plant development on nonspecific media.

A medium rich in auxin induces differentiation of localised meristematic cells and results in initiation and proliferation of embryogenic cultures. Typically 2,4-D is used as auxin. The embryogenic cells can develop into mature embryos when transferred to a medium with low or no auxin. The somatic embryo germinates only when it is mature and consists of functional root and shoots apices.

These embryoids can be separated and isolated mechanically by using glass beads. When embryoids reach torpedo stage they are transferred to filter paper bridge (a sterile and plugged culture tube containing about 10 ml MS liquid medium supplemented with Kinetin (0.2 mg/L) and sucrose (2% w/v) on which Whatman No.1. filter paper is placed to make a bridge (Dodds and Roberts, 1985). Some plants in which somatic embryogenesis has been induced *in vitro* are *Atropa belladona, Brassica oleracea, Carica papaya, Coffea arabica, Citrus cinensis, Daucus carrota, Nicotiana tabacum, Pinus ponderosa* and *Saccharum officinarum.* The somatic embryos produced in soybean are shown in Fig. 18.16.

***(b)* Factors Influencing Somatic Embryogenesis :** There are certain compounds excreted by plant tissue cultures in media that coordinate cell division and morphological changes in the developing tissue. These compounds include various polysaccharides, amino acids, growth regulators, vitamins, low molecular weight compounds and polypeptides. Several signalling molecules affect or control the formation of somatic embryos. These include extracellular proteins, arabino-galactan proteins and lipochito-oligosaccharides. The maturation of the somatic embryo can also be affected by temperature and light.

Fig. 18.16. Somatic embryos of soybean.

2. Encapsulated Artificial Seeds (= Synthetic Seeds, Somatic Seeds)

In 1977, T. Murashige of the U.S.A. for the first time gave the concept of artificial seeds at a Symposium in Belgium. Artificial (encapsulated) seeds are the somatic embryos covered with a protecting gel. These seeds are compared to the true seeds; artificial seeds are a novel analogue to the plant seeds. In these seeds, the gel acts as seed coat and artificial endosperm providing nutrient as in true seeds (Fig. 18.17). Plant propagules may consist of shoot buds or somatic embryos that have been grown aseptically in tissue culture. These plant propagules can easily grow into individual plants in culture as we have the capacity to control its growth providing chemicals in the culture media.

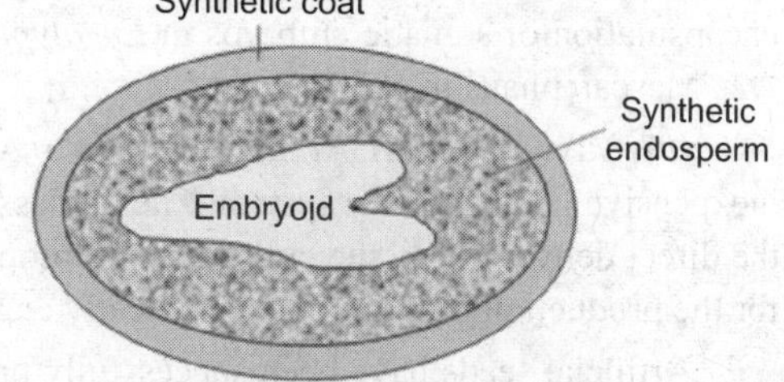

Fig. 18.17. Diagram of an encapsulated seed.

Water soluble gels (hydrogel) must be used as the protective gel. Usually Na/Ca- alginate (a product of brown algae) is selected for encapsulation purpose because it is less toxic to embryos and easy to handle. The production of artificial seeds is especially useful for plants which do not produce viable seeds. It represents a method to propagate these plants. Artificial seeds are small sized and this provides further advantages in storage, handling, shipping and planting.

While producing the artificial seeds, an artificial endosperm can be created within the encapsulation matrix. The encapsulation matrix is a hydrogel made of natural extracts from seaweed (agar, carageenan or alginate), plants (arabic or tragacanth), seed gums (guar, locust bean gum or tamarind) or microbial polymers (dextran, gellan or xanthan gum). These compounds form gel when mixed with or dropped into an appropriate electrolyte (copper sulphate, calcium chloride or ammonium chloride). Ionic bonds are formed to produce stable complexes. Useful adjuvants such as nutrients, plant growth regulators, pesticide and fungicide can be supplied to the plant propagule within the encapsulation matrix. In most cases, a second coat covering the artificial endosperm is required to simulate the seed coat.

(*a*) Methods of production : The steps for the production of encapsulated seeds are : (*i*) induction of pseudoembryos (artificial embryos) from cell suspension culture, (*ii*) mixing of embryos well with 2% Na- alginate, (*iii*) dropping of embryos in a bath of calcium salt, *e.g.* solution of $Ca(NO_3)_2$ for 30 minutes (this results in very quick complex formation at surfaces due to exchange of ions, *i.e.* Na^+ and Ca^+ Consequently, individual embryo is enclosed into a clear and hardened beads of about 4 mm), (*iv*) sieving of beads through a nylon mesh [$Ca(NO_3)_2$ solution can be recycled], and (*v*) testing of growth vigour of beads by plating in sand or soil amended with pesticides.

Artificial seeds

In 1985, S.L. Kitto and J. Janick produced *Citrus* embryos *in vitro* and tested eight compounds for their synthetic coating properties on embryos. Out of the chemicals tested, a polyethyleneoxide (polyox WSR-N75) revealed the good encapsulating properties. It was selected for use in further research with *in vitro* produced carrot embryos. Later on polyox coated embryos were kept in dry condition and then allowed for germination at suitable conditions. The germination percentage of the encapsulated embryos was low. The uncoated embryos did not survive after dry treatment. Then S.L. Kitto and J. Janick applied a number of presumptive hardening treatments *viz.,* (*i*) high inoculum density (0.8 embryo suspension per 25 ml medium), (*ii*) 12% sucrose instead of 2%, (*iii*) chilling at 4°C during the last three days of the embryo-induction phase, and (*iv*) amending 1 mm abscisic acid in the nutrient medium at the time of embryo induction. These treatments should be combined with polyox coating. All the treatments increased the survivability of the coated embryos. Similarly, production of artificial seeds by encapsulation of somatic embryos in *Eucalyptus* sp. has been reported by E.M. Muralidharan and A.F. Mascarenhaus in 1989.

(*b*) Advantages of Artificial Seeds : Artificial seeds have the potential for providing an inexpensive plant delivery system. The process of plánting can be easily mechanised, which allows the direct delivery of tissue cultured plant propagules to the field. It also provides rapid bulking up for the production of individual genetically engineered plants.

Artificial seeds have been successfully produced in the following crops for commercial use : *Alfalfa, Apium graveolens, Cymbidium* (orchid), *Daucus carota, Dendrobium* (orchid), *Dioscorea ftoribunda, Gossypium hirsutum, Medicago sativa, Morus indica, Picorrhiza kursoa, Pogostemon patchouli, Phalaenopsis* (*orchid*), *Rheum emodi, Santalum album, Spathoglottis plicata* (orchid), and *Azadirachta* (neem).

The concept of artificial seed is in its infancy. There is an immense potential for the use of this technique for the propagation and preservation of elite germplasm. Once the technique is perfected it will be a boon to the seed production industry. Use of artificial seeds will become popular for green house production. Additional studies, however, will be needed to understand hardening treatments, coating, duration of storage, embryo conversion frequency, and most important the economics of artificial seeds as compared to natural seeds.

J. MICROPROPAGATION

The practice of rapidly multiplying stock plant material using modern plant tissue culture methods to produce a large number of progeny plants is called micropropagation. There are the following four developmental stages of micropropagation :

Stage I: Selection of a suitable explant and inoculation into nutrient medium.

Stage II: Multiplication and growth of culture take about 2 months followed by repeated subcultures. At this stage rootless (with apical dominance, AD) cultures are obtained. In contrast, in fast multiplying cultures rootless multishooted (MS) cultures are also developed.

Stage III: At this stage cultures are obtained by changing the medium to planting out in the following three sub-sub-stages :

- **Microcutting stage (MC):** In this stage, culture is transferred to multishoot inducing medium so as to get longer shoot and to harvest by MC. Finally MC is sub-cultured on medium for rooting. Stage II AD cultures also respond to MC procedure.
- **Stage-III MS culture stage:** At this stage, stage II MS cultures are divided into individual shoots which are induced to form apical dominant shoots with roots.
- **MS culture stage:** Small clumps of culture from stage II MS culture are transferred to AD inducing medium to get a bushy plant. Generally this type of plant is preferred for many ornamental horticulture plants.

- **Stage IV:** This is the planting out stage where plantlets are aseptically removed from test tube environment to natural and harsh environment. At this stage roots should be fully functional in potting mix (the soil environmental where plantlets are transplanted).

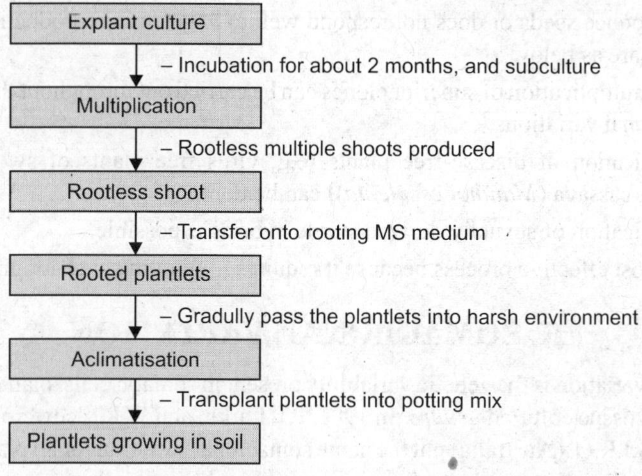

Fig. 18.18. Micropropagation involving different stages.

During this procedure generally plantlets fail to survive because of desiccation (from 100% humidity of test tubes to low humidity under ambient conditions), harsh environment, invasion of soil microorganisms, unadjustibility from dependent (artificial medium) to independent nutrition (by photosynthesis). These procedures generate the concept that a single medium is not sufficient for multiplication and regeneration of plants. Establishment of explant is influenced by the tissue of explants, constituents of nutrients and environmental factors. The region of young plant having cells (shoot tip, auxillary buds) is the suitable explant. Mature tissue impregnated with high amount of phenolic compounds is difficult to culture. Therefore, oxidation of these phenolics is necessary before culturing them.

In 1983, C.Y. Hu and P.J. Wang have suggested for : (*i*) addition of antioxidants to the medium, (*ii*) pre-soaking of explants in antioxidant solutions before culture, (*iii*) sub-culturing to a fresh medium, and (*iv*) providing light or no light during the start of culture.

In addition to nutritional composition, light and temperature are equally important in propagation. For morphogenesis and biosynthesis of chlorophylls, fluorescent light is necessary. For stage I and II light intensity of 1,000 lux has been found the best, and for stage III it was recorded between 3,000-10,000 lux.

A medium that does not initiate callus formation is thought to be suitable because the genetic instability of callus will lead to a high degree of genetically aberrant plants (Dodds and Roberts, 1985). Chemical constituent that favour callus development should be avoided. The source of auxin is usually IAA, α-NAA and IBA (indole-3-butyric acid). The auxin and 2, 4-D should be discarded as they stimulate callus formation and inhibit organogenesis. Exogenous GA3 (0.1 m1/L) is required for the development of shoot tips of isolated potato, carnation, *Chrysanthemum* and *Dahlia* and their subsequent micropropagation.

At planting out stage, it is necessary to develop acclimatisation capability in plantlets before removing them from the test tubes. This can be done by : (*i*) induction to develop some normal and functional leaves, (*ii*) induction of functional roots, and (*iii*) exposing the *in vitro* cultures to harsh environment before first two weeks of planting out. The successful potting mix are peat alone, vermiculite alone, mixture of loam-peat, peat-perlite-vermiculite-ash, perlite-pulverised pine bark-peat-river sand, and perlite-vermiculite-sand. Micropropagation is used to multiply novel plants, for example those plants that have been genetically modified through conventional plant breeding methods. It is also used to provide a sufficient number of plantlets for planting from a stock plant which does not produce seeds or does not respond well to vegetative reproduction. The benefits of micropropagation are as below :

- Rapid multiplication of superior clones can be carried out throughout the year, irrespective of seasonal variations.
- Multiplication of disease-free plants [e.g. virus free plants of sweet potato (*Ipomea batata*), cassava (*Manihot esculenta*)] can be done.
- Multiplication of sexually derived sterile hybrids is possible.
- It is a cost effective process because it requires minimum space for growing the cultures.

K. SOMACLONAL VARIATION

Somaclonal variation is the genetic variability present in somatic cells, plants or plant progenies derived from cells/tissue cultured *in vitro*. In 1981, P.J. Larkin and W.R. Scowcroft at the Division of Plant Industry, C.S.I.R.O. (Australia) put the name somaclonal variation for all plant-variants derived from any form of cell or tissue culture. Some variants are obtained in homozygous condition in the plants regenerated from cells cultured *in vitro* (R0 generation) but mostly variants are recovered in the selfed progeny of tissue culture regenerated plants (R1 generation). This variation consists of aneuploids, sterile plants and morphological variants, some of which may include traits of economic importance for crop plants. But the somaclonal variation may be genetic or epigenetic. Only gametic variations are important for crop improvement because they follow Mendelian inheritance pattern and transmitted to next generation. Hence, R0, R1 and R2 progenies are analysed in several crops for transmission of variant trait to sexual progeny (R1) and 3:1 segregation that results in isolation of true breeding variants (R2).

The significance of somaclonal variation in crop improvement was first demonstrated in sugarcane and potato where few somaclones were recovered with disease resistance against Fiji, downy mildew (sugarcane,) and late and early blight of potato. The genetic heterogeneity of cells in a population represents continuity of genotypes, whereas phenotypically the population is represented as a discrete sum of sub-clones. After cloning a single cell, from the population of a strain of *Dioscorea deltoidea*, there developed a subpopulation that differed in their growth rate and sapogenin (diosgenin and yamogenin) content. The traits varied independently. The genetic nature of variability in cultured cells explains the appearance of a large number of somaclonal variants with heterosis among the regenerated plants (Table 18.6).

Table 18.6. Somaclonal variation in some plants.

Species	***Explants***	***Variant traits***
Brassica spp. (Brassicaceae)	Anther, embryo, meristem	Flowering time, growth habit, wax
Nicotiana tabacum (Solanaceae)	Anther, protoplast, leaf callus	Plant height, leaf size, alkaloids
Oryza sativa (Poaceae)	Embryo	Plant height, tillering number, panicle size, seed fertility, flowering date
Saccharum officinarum (Poaceae)	Various	Auricle length, sugar yield, pathogenic disease
Solanum tuberosum (Solanaceae)	Leaf callus, protoplast	Plant habit, disease resistance and shape, yield and maturity date of tubers
Triticum aestivum (Poaceae)	Immature embryo	Plant height, spike shape, maturity tillering, α-amylase, leaf wax
Zea mays (Poaceae)	Immature embryo	Endosperm and seedling mutant, mtDNA sequence rearrangement

Source : Several research papers.

Somaclonal variations generated during tissue culture may be due to : (*i*) reflection of heterogeneity between cells and explant tissue, (*ii*) a simple representation of spontaneous mutation rate, or (*iii*) activation by culture environment of transposition of genetic materials. The mass occurrence of somaclonal variants increases the resistance, productivity and vital force of the plant (the heterotic effects). This would be due to dominance of nonlethal mutations that lead to heterozygosis with a wild type allele which is thus phenotypically expressed as hybrid heterosis.

One of the most complex and difficult problems involving the population of cultured cells is the retention and expression of the trait of totipotency, *i.e.* the ability to realise the programme of development from cell to plant. The genetic variability of cultured cells is the basis for getting somaclonal variants of the plants that have valuable traits. But this variability makes it difficult to realise the results of cell and genetic engineering as the trait may be lost in the row of cell generations leading to the regeneration of the plant. Stability is an essential requirement for clonal micropropagation and stabilising selection in plant breeding.

Most significant achievement of somaclonal variation was made by James E. Shepard and co-workers in old variety of potatoes in 1980. They screened about 100 somaclones produced from leaf protoplasts of Russet Burbank and found a significant and stable variation in compactness of growth habit, maturity, date, tuber uniformity, tuber skin colour and photoperiodic requirements. The characters of greater tuber uniformity and early onset of tuberisation were agronomic improvements over the parent variety. Moreover, somaclonal variation is applicable for seed propagated plants only such as rice, wheat, maize, tobacco, etc. not for vegetative propagated species. Thus, somaclonal variation has proved an alternative tool to plant breeding for generating new varieties that can exhibit disease resistance and improvement in quality and yield in plants such as cereals, legumes, oil seeds, tuber crops, fruit crops, etc.

In 1984, P.J. Larkin and co-workers have performed a detailed genetic analysis of wheat somaclonal variation, coupled with qualitative and quantitative assessment for yield and other attributes of commercial importance. The main findings are: (*i*) encountering of heritable variation for characters such as height, awns, tiller number, grain colour, gliadin protein and amylase regulation, (*ii*) occurrence of numerical and structural aberration in chromosome, besides morphological variation, (*iii*) correlation of morphological variations with chromosomal variation, (*iv*) the major scope occurrence of translocations for exploitation by introgression of genes from wild relatives by culture of immature embryos of wide crosses, and (*v*) a great scope of somaclonal variation for utilisable germplasm for crop improvement.

In Hawaii, tissue culture is integrated in the breeding programme and somaclonal variation has been exploited for the isolation of clones resistant to certain diseases, for example, viruses, downy mildew and eye spot disease. Some of the resistant clones have advantages in having higher sugar content and yield than parental ones.

1. Causes of Somaclonal Variation

The major causes of somaclonal variation are physiological, genetic, and biochemical as described below :

(*a*) Physiological Causes : There are certain variations induced by physiological factors in culture medium for example prolonged exposure to plant growth regulators such as 2,4-D, 2,4,5-T, etc. These variations results in variability among the regenerants. Mostly such variations are epigenetic; therefore, these do not follow Mendelian inheritance.

(*b*) Genetic Causes : The genetic causes of variation observed in regenerants include all the alterations at chromosomal level. The chief source of somaclonal variation is the chromosomal rearrangements such as deletion, duplication, translocation, inversion, polyploidy and aneuploidy. Variation among somaclones is also caused by meiotic crossing over involving symmetric and asymmetric recombination responsible. Transposable elements in maize like Ac-Ds also get activated in *in vitro* culture. The main cause of somaclonal variation in maize and broad beans (*Vicia faba*) is 'late replicating heterochromatin'. Moreover, the single gene mutations in cultures also result in variations, which are not detected in plants regenerated *in vitro* from any cell or tissue (R0 plants) but express in R1 plants (after selfing R0 plants).

(*c*) Biochemical Causes : Change in carbon metabolism leading to failure of photosynthesis *viz.* albinos in rice is the most common biochemical variation. Somaclonal varaiation is also resulted in by any change in other cell processes such as starch biosynthesis, carotenoid pathway, nitrogen metabolism, antibiotic resistance, etc.

2. Application in Crop Improvement

As compared to the conventional mutation breeding, cell culture systems allow plant breeder to have greater control on selection process as they have the option to select from large amount of genetically uniform material. Hence, Somaclonal varaiation is the only approach for genetic improvement in perennial species limited by narrow germplasm and long regeneration cycle, asexually propagated plants like bananas, for isolation of biochemical mutants like auxotrophs.

Observations over a period of three tuber generations suggest stable changes in tuber shape, yield, and maturity date, in photo-period requirements for flowering, and in plant morphology. Somaclonal variants have been isolated for variety of valuable traits like disease resistance [such as enhanced resistance to early blight (*Alternaria solani*) and late blight (*Phytophthora infestans*)], stress (salt, low temperature) resistance, improved yield and efficient nutrient uptake etc.

Pusa Jai Kisan, with bolder seeds and higher yield developed at Indian Agricultural Research Institute (IARI), New Delhi is another successful application of somaclonal variation of *Brassica juncea* variety 'Varuna'. Somaclonal variants in *Lathyrus* with low toxin level have also been developed at IARI.

3. Examples of Somaclonal Variants

There are several examples of use of somaclonal variation in releasing new varieties; some of the examples are given below :

- In India, Sugarcane Breeding Institute (Coimbatore) has released varieties produced through the process of somaclonal variation.
- **Bio-13.** In India, a somaclonal variant of a medicinal plant *Citronella java* has been released as 'Bio-13' for commercial cultivation by Central Institute for Medicinal and Aromatic Plants (CIMAP) (Lucknow). Bio-13 yields 37% more oil and 39% more citronellon than the control variant.
- **Super-tomatoes.** Heinz Co. and DNA Plant Technology Laboratories (USA) have developed 'super-tomatoes' by screening somaclones. Super-tomatoes have high solid component which reduced shipping and processing costs. In 1986, according to an estimate about 100 million dollars per annum can be saved in the U.S.A. alone by 1% increase in tomato solid processors. New 'super-tomatoes' have been developed by the British researchers which have been fortified with minerals that can improve our immune system and prevent cancer. The new varieties have been enriched with selenium (a powerful anti-oxidant which not only boosts the immune system but also helps to prevent cancer). The mineral is found naturally in foods such as Brazil nuts, shellfish and liver; it is also important for the thyroid gland which quickly determines the energy usages of body and also produces proteins
- **Oxalate removal:** The nutritional stress factors have been estimated in some vegetables through selection of somaclonal variants. First application of this technique was shown in a leafy vegetable called *Amaranthus gangeticus*.

L. ANTHER AND POLLEN CULTURE (PRODUCTION OF HAPLOID PLANTS)

The male reproductive organ is called anther, which is diploid in chromosome numbers. As a result of microsporogenesis, tetrads of microspores are formed from a single spore mother cell. They are known as pollen grains after release from tetrads. The aim of anther and pollen culture is to get haploid plants by induction of embryogenesis. Haploid plants have single complete set of chromosomes that in turn may be useful for the improvement of many crop plants. Moreover, chromosome set of these haploids can be doubled by mutagenic chemicals (*e.g.* colchicine) or regeneration technique to obtain fertile homozygous diploids.

In 1951, W.R. Tulecke cultured pollen grains of *Ginkgo biloba* (gymnosperm) and succeeded to induce the development of haploid callus. In 1964 S. Guha Mukherjee and S.C. Maheswari from University of Delhi (India) made a remarkable discovery by culturing pollen grains of an angiospermic plant, *Datura innoxia* on the nutrient agar medium and also developed torpedo-shaped embryoids that metamorphosed into plantlets.

In 1979, N. Sunderland has described that the anthers to be cultured should be one of the three categories *i.e.* premitotic, mitotic and postmitotic. In premitotic anthers, where the microspores have completed meiosis but not started first pollen division, the best response is achieved, such as *Hordeum vulgare.* In mitotic anthers where first pollen division has started the optimum responses are achieved, such as *N. tabacum* and *D. innoxia.* In post mitotic anthers, the early bicellular stage of pollen development is the best time to culture *e.g. Atropa belladona.*

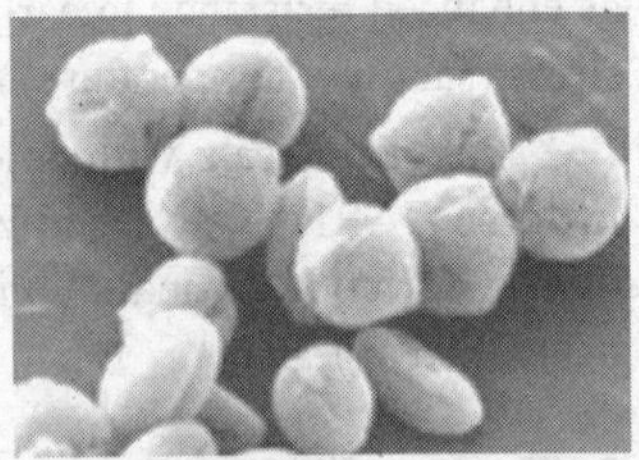

SEM of Jasmine pollen grains.

I. HAPLOID PRODUCTION

Haploid plants are very useful in : (*i*) direct screening of recessive mutation because in diploid or polyploid screening of recessive mutation is not possible, and (*ii*) development of homozygous diploid plants following chromosome doubling of haploid plant cells.

In China, the most widely grown wheat is a doubled haploid produced through homozygous diploid lines. Anther culture of rice is also successfully grown. Haploid plants have been produced in tobacco, wheat and rice through pollen culture. These are used for the development of disease resistant and superior diploid lines. The Institute of Crop Breeding and Cultivation (China) has developed the high yielding and blast resistant varieties of rice Zhonghua No. 8 and Zhonghua No.9 through transfer of desired alien gene.

At present, more than 247 plant species and hybrids belonging to 38 genera and 34 families of dicots and monocots have been regenerated using anther culture technique. They include economically important crops and trees such as rice, wheat maize, coconut, rubber trees, etc.

1. Method of Pollen Culture

Method of anther culture is shown in Fig. 18.19. Anthers are superficially sterilised and washed with double distilled sterile water. They are excised from the flower buds and their proper developmental stages are determined under microscope. On confirmation of a stage, the anthers are directly transferred on nutrient agar or liquid medium where induction of embryogenesis occurs, or the pollen grains are aseptically removed from the anthers and cultured on liquid medium.

(*a*) *In Vitro* Androgenesis : The formation of sporophyte from the male gametophyte on artificial medium is called *in vitro* androgenesis. It is most commonly found in family Solanaceae and Poaceae (Graminae). Success of *in vitro* androgenesis is based on adjustment of developmental stage of pollen, minerals of culture medium and growth regulators as well as thermal shock or other treatments.

Method of *in vitro* androgenesis is given in Fig. 18.19. The pollens are extracted by pressing and squeezing the anthers with a glass rod against the sides of the beaker. The debris of anther tissue is removed by passing the pollen suspension through a filter; the well collected large and healthy pollens are washed with sterile distilled water, centrifuged, decanted and collected. The pollens are inoculated in liquid medium, and then sub-cultured in solid MS medium. In culture pollen grains can be induced to produce callus or embryo. The callus or embryo is transferred to a suitable medium to regenerate a haploid plant in one month. This technique is most successful in *Brassica, Datura, Petunia,* etc. The haploid are identified by looking at their morphological features or by using genetic markers, for example a marker for brown coloured aleurone to detect haploids in maize plant.

Anthers of the Solanaceae respond best to excision and culture by production of embryoids. Several workers have demonstrated the direct transformation of microspores into embryoids by different species of the genera of Solanaceae, for example *Capsicum, Atropa, Datura, Nicotiana, Lycium, Petunia, Solanum,* and *Hyoscyamus.* Before final elongation and formation of shoot and root meristems, embryoids go through the globular, heart-shaped and torpedo stages similar to typical ontogeny of normal diploid zygotic embryos. Cultured anthers of certain cereals like rice also produce embryoids. There are some other plants that do not produce haploids directly from the microspores, but they do through the intervention of a callus. In such plants the microspores first develop into multicellular bodies which later on form dense callus which upon subculturing in an embryogenesis-inducing medium (containing auxins and a cytokinin) initiates haploid seedlings.

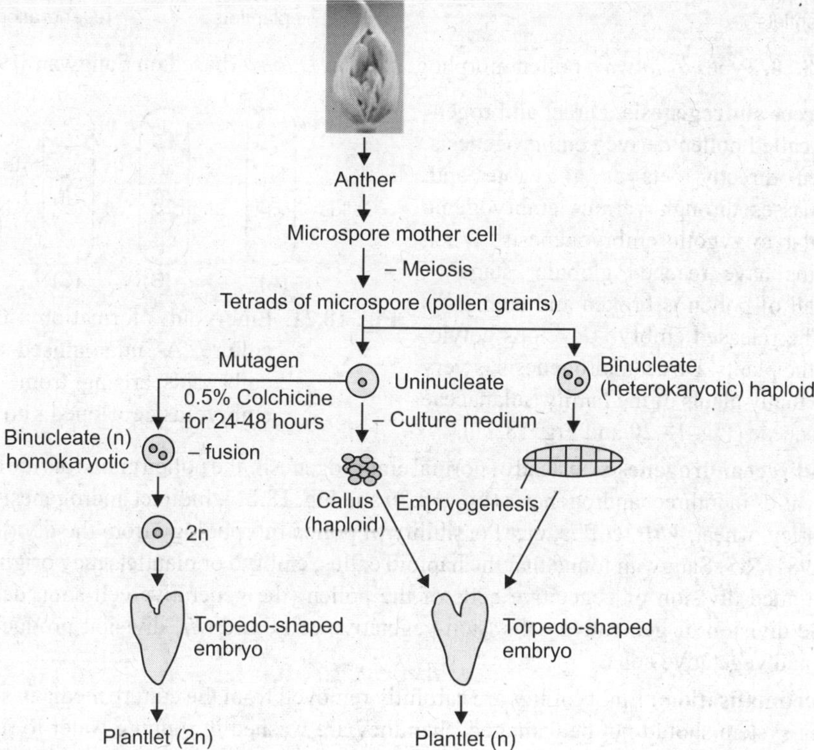

Fig. 18.19. Methods of anther culture, and production of haploid and diploid plants *in vitro.*

Therefore, the following alternative pathways exist for haploidy in anther cultures :

- Microspore ⟶ callus ⟶ embryogenesis ⟶ plantlet e.g. Solanaceae
- Microspore ⟶ callus ⟶ embryogenesis ⟶ plantlet *e.g. Oryza, Brassica, Hordeum, Coffea, Populus.*
- Microspore ⟶ callus ⟶ plantlet *e.g. Datura mtelloides.*

Mainly there are two ways of *in vitro* morphogenesis of pollen grains, direct and indirect (Fig. 18.20).

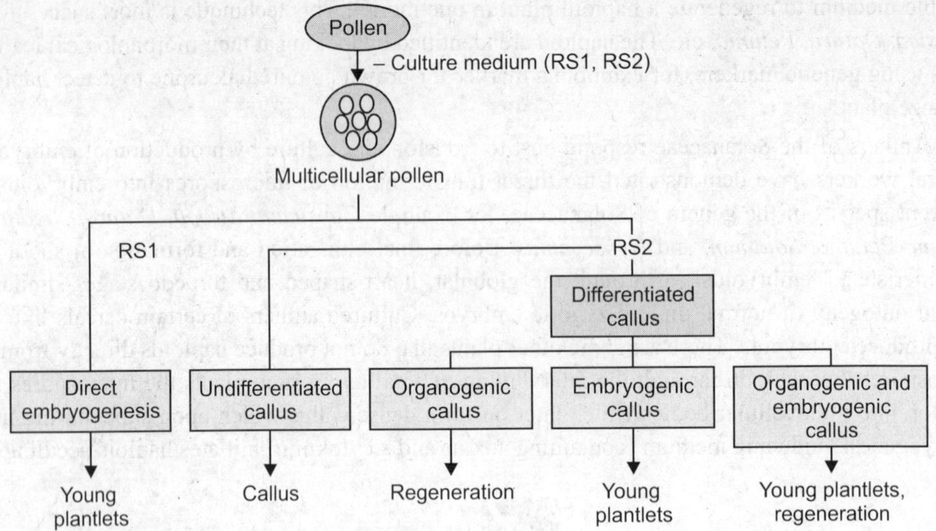

Fig. 18.20. Types of *in vitro* pollen morphogenesis in *Datura* (based on Sangwan, 1981.)

(i) **Direct androgenesis.** Direct and rogenesis is also called pollen derived embryogenesis. Here pollen directly acts as a zygote and, therefore, passes through various embryogenic stages similar to zygotic embryogenesis. When pollen grains have reached globular stage of embryo, wall of pollen is broken and embryo is released. The released embryo develops cotyledons, then the plants. Direct androgenesis is very common in many plants of the family Solanaceae and Brassicaceae (Fig. 18.20 and Fig. 18.21).

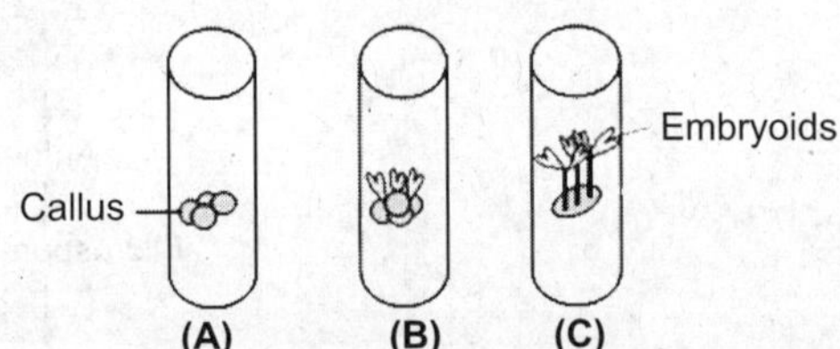

Fig. 18.21. Embryoid formation in tissue culture: A- unorganised callus, B- embryoids arising from callus; C- embryoids developed into plantlets.

(ii) **Indirect androgenesis.** Instead of normal embryogenesis, the pollen grains divide erratically to develop callus in indirect androgenesis (Fig. 18.20 and Fig. 18.21). Indirect androgenesis has been found in barley, wheat, *Vitis,* coffea, etc. Possibility of pollen morphology from the dividing pollen varies. In 1981, R.S. Sangwan found that the haploid callus, embryo or plantlets may originate from: *(i)* the continued division of vegetative cells of the pollen (the generative cell soon degenerate), *(ii)* multiple division of generative cells (non-vegetative cell), and *(iii)* division products of both generative and vegetative cells.

(b) **Acclimatisation :** Embryolings are carefully removed from the culture media in such a way that the root system should not be damaged; then they are washed in running water to remove the traces of media. Clean embryolings are carefully transplanted into small pots filled with soil mixture after dipping in an aqueous solution of 0.5 % Benlate. The soil mixture is prepared by mixing sand, peat moss and humus in the ratio 1: 1: 2 and aired for two days after autoclaving at 121°C and 15 lb pressure for 45 minutes. After planting in the soil mix, embryolings are kept in the growth room at 25± 2°C and with 3000 lux light intensity under 16 h photoperiod for 20 days before transferring to the greenhouse. By reducing the relative humidity in the incubating containers embryolings are gradually acclimatised to the greenhouse environmental conditions.

(c) **Factors Affecting Androgenesis :** The factors that affect androgenesis are the genotype and the physiological state of the donor plants, composition of the culture medium, the method adopted in the pretreatment of anthers to develop in to haploid plants and the stage of the microspore or pollen selected for culture.

(*d*) Limitations in Haploid Production : Due to low frequency of production of the haploids its selection is very difficult. Those haploids develop frequently in cultures that have deleterious traits. Hence, sometimes it is difficult to isolate haploids from the culture since the polyploids outgrow haploids. The doubling of haploids (diploidisation) does not always lead to the formation of homozygous plant. Often the embryos derived from haploids get aborted.

(*f*) Use of Haploids : Haploid plants are very useful in : (*i*) direct screening of recessive mutation because screening of recessive mutation is impossible in diploid or polyploidy plants, (*ii*) development of homozygous lines in a short span of time, (*iii*) the generation of exclusive male plants by the process of androgenesis using techniques to double the chromosome numbers, (*iv*) production of disease-resistant plants by introducing disease resistant genes, for example barley accession Q 21681 is resistant to stem rust, leaf rust, and powdery mildew, and (*v*) cytogenetic research, for example production of aneuploids, determination of origin and basic number of chromosome numbers, induction of genetic variability.

II. PRODUCTION OF MONOPLOID PLANTS

When breeders want to induce and select new favorable gene mutations and to find new combinations of favorable alleles at different loci, diploidy becomes an inherent nuisance. Unless new recessive mutations are homozygous, they cannot be detected. In heterozygotes meiosis breaks up favorable combinations of alleles. In some plant species, monoploids can be artificially derived from the products of meiosis in the anthers. A cold treatment induces a cell destined to become a pollen grain into an embryoid (*i.e.* a small dividing mass of monoploid cells) when placed on agar plates containing certain plant hormones. Under these conditions, haploid embryoids will grow into 'monoploid plantlets'. When these plantlets are placed on medium containing different plant hormones, they will grow into mature monoploid plants with roots, stems, leaves, and flowers. A monoploid plantlet can be potted in soil and allowed to mature (Fig. 18.22).

1. Uses of Monoploids

Monoploid plants can be used in many ways; for example, they are first examined for favorable allelic combinations arisen from heterozygosity either already present in the diploid parent or induced in the parent by mutagens. Therefore, from a parent *A*/*a*, *B*/ *b* might come a favorable monoploid combination *a, b*. Then the monoploid can be subjected to chromosome doubling caused by colchicines through inhibiting the microtubules. Thus the cells produced are homozygous diploid, *a*/ *a*, *b*/ *b* and reproduce normally.

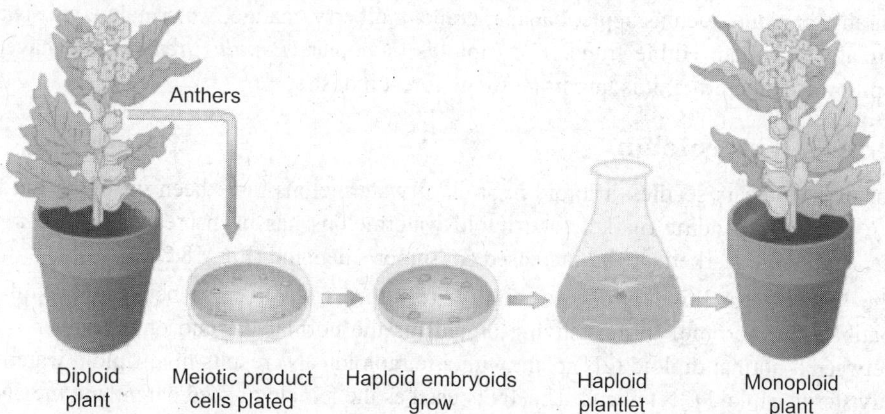

Fig. 18.22. Production of monoploid plants by tissue culture (*adopted from*: Griffiths *et al.*, 1999).

In another approach a population of monoploid cells is isolated, their walls are removed enzymatically, and they are treated with mutagen. Basically the monoploid cells are treated as a population of haploid organisms in a mutagenesis-and-selection procedure. Then the monoploid cells are plated on a selective medium that selects for some desirable phenotype. By using this approach, certain traits are selected, for example resistance to toxic compounds (produced by certain plant parasites), and resistance to herbicides (being used by farmers to kill weeds). Eventually the resistant plantlets grow into haploid plants; thereafter, there chromosome number can be doubled (by colchicine) into a pure-breeding, diploid, resistant type. These powerful techniques have been successfully applied to several important crop plants, such as soybeans, tobacco, etc.

III. TRIPLOID PRODUCTION

An angiosperm ovule is shown in Fig. 18.23. The egg cell and polar nuclei are contained within the embryo sac. The sperm nuclei are derived from the pollen grains. A pollen grain that lands on a compatible stigma can send out a pollen tube to deliver the sperm to the female gametophyte inside the ovule. Double fertilisation occurs in a majority of individuals except the families, Orchidaceae, Podostemaceae and Trapaceae. This results in two fusion products : (*i*) the diploid zygote (through fusion of an egg and one of the male gametes), and (*ii*) the triploid primary endosperm (fusion of the second male gamete with two polar nuclei). The zygote gives rise to embryo and the latter forms endosperm (also called secondary embryo). The endosperm nurses the developing zygotic embryo.

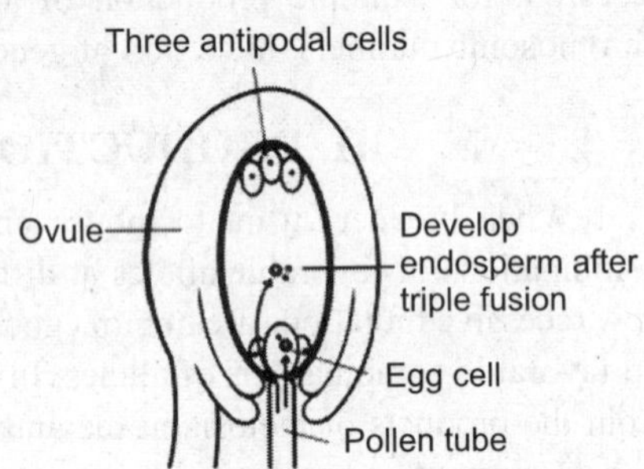

Fig. 18.23. An ovule of angiosperm showing double fertilisation and development of embryo and endosperm.

In 1933, for the first time L. Lamp and C.O. Mills grew maize endosperm on nutrient medium 10-20 days after pollination. The endosperm proliferated slightly. In 1949, LaRue succeeded to produce callus from immature endosperm. In 1965, B.M. Johri and S.S. Bhojwani at the University of Delhi (India) reported the endosperm culture. Some examples of triploid plants raised from endosperm cultures are : *Asparagus officinalis,* barley (*Hordeni vulgare*), rice (*Oryza sativa*), maize (*Zea mays*), *Prunus persica, Pyrus malus, Citrus grandis,* sandle palnt (*Santulum album*). The triploid plants are self-sterile and usually seedless. This characteristic increases edibility of fruits and desirable in plants such as apple, banana, grape, mulberry, mango, watermelon, etc. These are commercially important edible fruits. The triploids of poplar (*Populus tremuloides*) have better quality pulpwood; therefore, it is important to the forest industry.

1. Seedless Watermelon

For over 40 years seedless triploid hybrids of watermelons have been grown in the United States. A rapidly expanding market for triploid watermelons has been created due to improved varieties, aggressive marketing, and increased consumer's demand (Fig. 18.24).

The seedless condition is due to sterility resulting from a cross between two plants of incompatible chromosome. In most living organisms the normal chromosome number is 2N. A cross between a normal diploid (2N) plant with a tetraploid (4N) results in a triploid watermelons on highly sterile triploid (3N) plants. The 4N is used as the female or seed parent and the 2N is the male or pollen parent.

Sterile triploid plants normally do not produce viable seed. However, small, white rudimentary seeds develop within the fruit which are eaten along with the fruit as in cucumber. However, the number and size of these rudimentary seeds (*i.e.* pips) vary with variety. In triploid melons an occasional dark, hard, viable seed may be found. Seedless watermelons have 10 or less mature seeds, excluding pips/caplets on the face of the melon which has been cut into four equal sections.

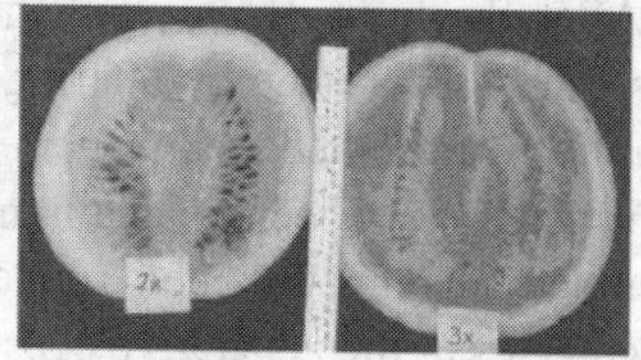

Fig. 18.24. Watermelon produced through embryo rescue approach: normal (2X) and seedless triploid (3X).

2. Triploid Papaya

De-Quan *et al.* (2011) obtained triploid papaya (*Carica papaya*) plants by immature endosperm culture. A maximum frequency of callus induction was obtained when immature endosperm with embryo was cultured on (MS) medium containing 2,4-D, α-naphthaleneacetic acid (NAA) and 6-furfurylamino purine (Kn). Shoot buds were produced on subculturing of callus on a medium supplemented with 6-benzyladenine or thidiazuron along with NAA. Shoots were detached from the callus and transferred to the elongation medium supplemented with indole-3-acetic acid and 6-benzyladenine. The combination of IAA and 6-benzyladenine was the best to produce axillary shoot proliferation by 93.8%.

The addition of indole-3-butyric acid to the 1/2 MS medium was most effective in inducing root formation, where 90% of the shoots developed four to five roots. Healthy rooted plantlets were transferred to pots containing sterilised bed soil and perlite (3:1) mixture in the greenhouse and 78% of the micropropagated plants survived transplantation. The leaves from endosperm-derived plants showed larger stomata and more chloroplasts in guard cells than that from the parent plants. Over 75% of the endosperm-derived plants were triploid.

M. MENTOR POLLEN TECHNOLOGY

Pollination is a natural phenomenon which occurs in many plant species. However, there are certain plants where pollen grains remain viable but unable to set seed on its own pistil, *i.e.* they are self incompatible. Similarly, in several other plants, besides being viable they even fail to germinate on stigmas of another species, *i.e.* they are interspecific incompatible.

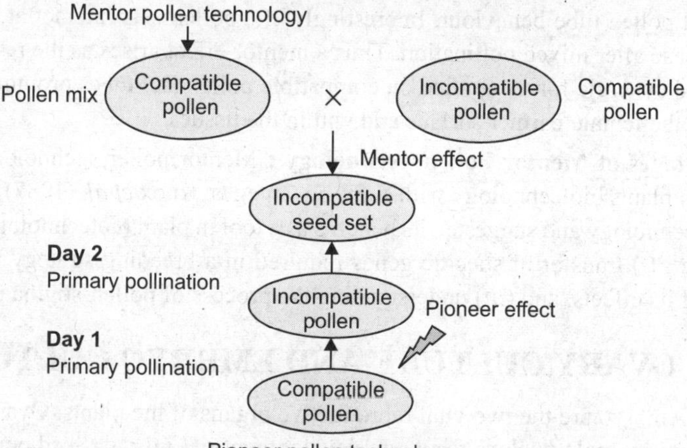

Fig. 18.25. Steps involved in mentor pollen technology and a comparison between mentor pollen technology and pioneer pollen technology.

After the birth of recent mentor pollen technology the incompatibility problem of pollen grains can be overcome by altering the fertilisation ability. The pollens which have been treated purposefully are called *mentor pollen.* Hence, the mentor pollen may be defined as the compatible pollens which have been treated in many ways to inhibit its fertilisation ability and retain its power to stimulate incompatible pollen to accomplish fertilisation. The pollen signals are received by the male gametes or the generative cells and affect their behavior during fertilisation.

Fig. 18.25 shows the steps involved in mentor pollen technology and a comparison between mentor pollen technology and pioneer pollen technology. The pioneer pollen technology is based on the unexpected consequences of double pollination of the same stigmas at successive time intervals. The mentor pollen itself may or may not germinate on stigmas. As a result of this treatment, hybridisation between closely related species or genera, and itself pollination in cross-pollinating plants could be achieved. Some mentor pollen experiments are given in Table 18.7.

Table 18.7 Some experiments of mentor pollens.

Pollen	*Treatment*	*Barrier overcome*
Brassica oleracea	Ether on pollen	Sporophytic self-incompatibility
Malus sp.	Methanol	Gametophytic self-incompatibility
Petunia hybrida	Gamma radiation	Gametophytic self-incompatibility
Populus sp.	Gamma radiation	Interspecific self-incompatibility
Populus sp.	Methanol	Interspecific self-incompatibility
Populus sp.	Ethyl acetate	Interspecific self-incompatibility
Populus sp.	Freezing-thawing	Interspecific self-incompatibility

For example, a major barrier to hybridisation of some *Ipomoea* species is located at the interface of stigma-style. In five inter-specific combinations examined with the aid of fluorescence microscopy, pollens have been found to germinate and penetrate the stigma surface without passing into the style. In the cross, *I. hederacea* × *I. purpurea,* pollen tubes are found in the style and capsules were set without producing full seeds. Thus, the use of mentor pollen + foreign pollen mixes was not successful; even attempts to facilitate hybridisation among *Ipomoea* species got failed.

The mentor effect has been investigated in several plant species including poplars. In poplars, for compatible and incompatible crosses, interspecific incompatibility are analysed *in situ* by pollen germination and pollen tube behaviour. Interestingly, the pollen sets interact at different levels of the progamic phase after mixed pollination. Thus a mentor effect arises as the result of interactions of antagonist and cynergic forces applied on compatible pollen and tubes promotion of pollen tube growth, both on the female partner surface and within the tissues.

***(a)* Advantages of Mentor Pollen Technology :** Mentor pollen technology has become an important tool in plants biotechnology with some advantages. Knox *et al.* (1987) have reviewed the mentor pollen technology and suggested its benefits as a tool in plant biotechnology. The advantages that it offers are : (*i*) transfer of specific genes required in a breeding strategy through seeds, (*ii*) quantification of its effects, and (*iii*) understanding the process of pollen-stigma recognition.

N. OVARY CULTURE AND EMBRYO CULTURE

Ovary and embryo are the two vital reproductive organs of the plants. Ovary produces ovules where fertilisation by male nucleus results in development of embryo. Seed contains embryo that after germination gives rise to plant seedling. These two important organs have been cultured *in vitro*.

I. OVARY CULTURE

Ovary culture refers to the culture of unfertilised ovaries to obtain haploid plants from egg cell or other haploid cells of the embryo sac. The process of producing haploid plants from ovary is termed as 'gynogenesis'. In 1976, San Naom report gynogenesis for the first time in case of barley. Success has been obtained with many species namely maize, wheat, tobacco, rice, sugar beet, petunia, sunflower, rubber, Gerbera, etc. The success rate of gynogenesis is low; about 0.2-6% of the cultured ovaries show gynogenesis and each ovary gives rise to one or two (rarely 8) plantlets.

The success rate of ovary culture varies with species and also affected by genotypes of the explant. Mostly the stage far ovary culture is nearly mature embryo sac, but the free nuclear embryo sac stage is the most responsive in rice ovaries. Moreover, when the whole flowers, ovary and ovules are attached to placenta, their culture responds better; but isolated ovules of Gerbera and sunflower show better response.

In gynogenesis the growth hormones play a key role; they may induce callusing of somatic tissues and even suppress gynogenesis at higher levels. Requirement of growth hormones depends on species; for example, growth hormone-free medium is the best in sunflower, and even a low level of 2-methyl-4-chlorophenaxyacetic acid (MCPA) induces somatic calli and somatic embryos. In rice, the optimum requirement of MCPA is 0.125-0.5 mg/l for gynogenesis.

Sucrose level also plays a key role in ovary culture; for example in sunflower 12% sucrose results in gynogenic embryo production, whereas somatic calli and somatic embryos were also produced at lower levels. Generally ovaries and ovules are cultured in light but at least in same species. Dark incubation favours gynogenesis and minimises somatic callusing in sunflower and rice. Moreover, gynogenesis occurs either through embryogenesis or through regeneration of plantlet from callus. There are two or many stages of gynogenesis, and each stage may have distinct requirements. In rice, there are two stages *viz.*, induction and regeneration. Ovaries are floated on a liquid medium having low auxin during induction and kept in dark; but ovaries are transferred on to agar medium with higher auxin concentration and incubated in light for regeneration. In most of the species haploid plants generally originate from egg cell (*in vitro* parthenogenesis); whereas in some species such as rice, haploid plants develop mainly from synergids. But in *Allium tuberosum* even antipodal produce haploid plants; the phenomenon of producting such haploid is called '*in vitro* apogamic'.

In contrast, sugarbeet develops embryos, while sunflower embryos regenerate from the callus. Regeneration from a callus phase is easier than direct embryogenesis. There are mainly two limitations in ovary culture: (*i*) it has been successful only in less than two dozen species so far, and (*ii*) the frequency of response of ovaries (1-5%) and the number of plantlets/ovary (1-2) is very low. That is why anther culture is preferred over the ovary culture. But ovary culture is recommended only in those cases where anther culture fails, for example sugarbeet.

II. EMBRYO CULTURE

Embryo culture has also been carried out for the production of haploid plants. Embryo culture is used for the recovery of plants from distinct crosses. Embryo culture is useful where embryo fails to develop due to degeneration of embryonic tissues. It is being used extensively in the extraction of haploid barley (*Hordeum vulgare*) from the crosses *H. vulgare* × *H. bulbosum.* Embryo culture is also a routine technique employed in orchid propagation and in breeding of those species that show dormancy. The embryo callus produces somatic embryoids within 8 weeks of culture in the second medium which differentiated into buds after 2 weeks. After 16 weeks of culture several shoots with 4-6 leaves developed.

1. Culturing Method

The general method of embryo culture follows the following steps :

(*i*) Pluck healthy and mature fruits from the field and wash thoroughly in running water for about an hour.

(*ii*) Surface sterilise with 0.01% between-20 for 15 min, rinse seeds several times with distilled water and finally treat with 0.01% $HgCl_2$ solution for 10-15 min. Finally rinse it for six times with sterilised distilled water.

(*iii*) Break seeds aseptically and isolate the embryo.

(*iv*) Culture embryo on callus proliferation medium. Supplement the basal medium with different combinations and concentrations of sugar, vitamins, hormones and other growth adjuvant for callus proliferation and shoot regeneration.

(*v*) Incubate the cultures at 22-25°C under a 16 hours photoperiod of 2000 lux luminous intensity.

(*vi*) After two weeks of inoculation the embryo begins to swell on callus proliferation medium. Distinct callus growth is observed after 4 weeks.

(*vii*) After 8 weeks of inoculation transfer the callus on shoot regeneration medium. Within 4 weeks of transfer into second medium the callus turns green and produces soft spongy tissue. Some of these tissues are differentiated into embryoids.

(*viii*) The embryoids produce cluster of budlets when subcultured onto shoot regeneration medium. The budlets grow into shoots and produce 2-3 leaf appendages within 12 weeks. Thereafter, they are separated into individual shoots and then subcultured into a fresh medium of the same composition until shoots develop.

O. EMBRYO RESCUE

Viable hybrids are produced as a result of sexual crosses between two varieties of the species. However, if sexual crosses are done between the species of the same genus or between two different genera, production of hybrid is rather difficult because of several barriers arising either during pollination, fertilization or embryogenesis. It has been observed that in some cases embryo does not develop in spite of successful pollination and fertilisation. This is due to inherent deficiencies or incompatibility between the developing embryo and endosperms. Therefore, many distant crosses fail due to degeneration of endosperm. In such cases, immature embryos are dissected out from the fruit (seed) and grown artificially on medium which differentiate into shoot, root and plantlets. This technique of growing immature embryo is termed as 'embryo rescue'. Embryo culture has been widely used for this purpose.

Embryo rescue is one of the earliest and successful forms of *in vitro* culture techniques that are used to assist in the development of plant embryos that do not survive to become viable plants. In the 18th century, embryo rescue was first documented when Charles Bonnet excised *Phaseolus* and *Fagopyrum* embryos, planted them in soil and the cross resulted in dwarf plants. During 1890 to 1904, study of embryo rescue became systematic by applying nutrients containing salts and sugars employing aseptic technique. In 1904, E. Hanning performed the first successful *in vitro* embryo culture of Raphanus and Cochlearia on a mineral salt medium containing sugars, amino acids and plant extracts. He also described the problems of precocious embryos that resulted in small immature, and embryos that produced weak and often unviable plantlets.

1. Method of Embryo Rescue

Methods of embryo rescue were developed in the early 1900's that enabling unripe seed or embryos from adult plants to be rescued to form small plants. This technique was used to save embryos from ovules, which had been fertilised but had never developed into viable seed on the mother plant. Complete ovaries were put in tissue culture where seedlings were obtained from embryos which died in a later stage of development. Mostly a combination of these techniques is used where parts of the ovaries are placed in tissue culture followed by application of ovule and/or embryo culture.

Ovary of lily.

Ovaries containing immature embryos are gently taken out from the flower, surface sterilised with sodium hypochlorite and rinsed six to eight times with sterile distilled water. Embryo is gently scooped from below the pericarp with a surgical blade tip or forceps. Then the embryo is immediately placed on embryo rescue medium (modified MS medium). The embryo is kept in the dark at about 28ºC in a growth chamber until it germinates (2-3 days). Then it is transferred to a suitable photoperiod regime at 28ºC. At the 2-3 leaf stage when the root system has developed, transplant seedling into appropriate soil mixture. Immediately pot is placed on a deep tray, and kept covered with plastic wrap during the first days after transplanting so that the seedlings may be prevented from wilting. Holes are made on wrap and finally removed acclimatisation of seedlings. Thereafter, seedlings are transferred to a greenhouse.

Recent examples are the recovery of hybrids from *H. vulgare* × *Secale creale*, *T. aestivum* × *Agropyron repens*, *H. vulgare* × *Triticum* sp. including *T. aestivum,* etc. Hybrid seeds from *T. aestivum* × *H. vulgare* are not obtained, but hybrid seedlings have been obtained by embryo culture. Hybrid embryos from 9-12 day old caryopses were removed and placed on 14-18 days old barley endosperm cultured on a nutrient medium. When the hybrid embryos were cultured, a high proportion of them developed into seedlings.

2. Application of Embryo Rescue

The embryo-rescue techniques are used in the acquisition of interspecific and intergeneric hybrids. Interesting products have been acquired in potatoes, vegetables and various ornamental crops using this method. Many of the new lily hybrids have been obtained from this technique. It is impossible to imagine modern plant breeding without this technique. Embryo rescue is applied for : (*i*) breeding of incompatible interspecific and intergeneric species, (*ii*) overcoming seed dormancy, (*iii*) determination of seed viability, (*iv*) recovery of maternal haploids that develop as a result of chromosome elimination following interspecific hybridisation, and (*v*) studies on the physiology of seed germination and development.

Embryo rescue technique is very useful in wide hybridisation, complete growth of embryo in plant and breaking dormancy of certain seeds where dormancy period is very long. By using embryo rescue technique wild varieties can be crossed with cultivars. As compared with cultivars, the wild species have greater resistance to pests and pathogens, and produce grains of better quality. At the International Crop Research Institute for Semi Arid Tropics (ICRISAT), Hyderabad (India), this technique has been used to improve groundnut, pigeon pea and chick-pea. A new hybrid variety of pigeon-pea (ICPH8) has been developed at ICRISAT that matures in 100 days instead of 200 days. This variety is resistant to pathogens and pests and yields 20% higher than original cultivars. However, it can grow in a wide range of conditions.

P. CRYOPRESERVATION

Cryopreservation (Gr. *Kryos* means frost) refers to preservation in the frozen state of live plant, animals and microbial cells, tissue and organs. Basically it refers to storage at very low temperature such 77 K or –196ºC created by liquid nitrogen that is the boiling point of liquid nitrogen. Cryopreservation of fowl sperm was started by a team of scientists in the U.K. led by Christopher

Polge in 1957. In the 1950s the process was applied in humans with pregnancies obtained after insemination of frozen sperm. The rapid immersion of the samples in liquid nitrogen did not produce the necessary viability to make them usable on thawing for certain samples, such as types of embryos, bone marrow and stem cells. Cryopreservation has also been applied in storing plant tissues and organs as well as microbial cells. Moreover, cell damages also occur during cryopreservation at the freezing stage due to solution effects, extracellular ice formation, dehydration and intracellular ice formation; but many of these effects can be reduced upon addition of cryoprotectants.

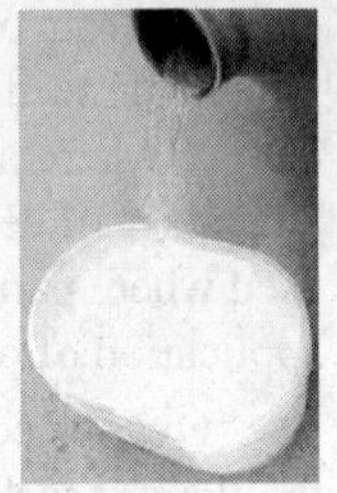
Liquid nitrogen.

Besides, the biologists realised the necessity for storage of genetic resources of plant realm in general and agricultural plants in particular due to gradual disappearance of economic and rare plant species. The conventional methods of storage fail to prevent from losses caused by : (*i*) attack of pathogens and pests, (*ii*) climatic disorders, (*iii*) natural disorders, and (*iv*) political and economic causes. However, the conventional methods could not save the viability of short lived seeds of economic plants, for example oil palm (*Elaeis guineensis*), rubber (*Hevea brasiliensis*), *Citrus* sp. and *Coffe* sp.

Generally the plant material is preserved and maintained in liquid nitrogen. Growth rate of cells retards due to storage of these materials at low temperature; consequently biological activities are conserved for long time. The study of metabolic activities and their responses in plant materials (microbial and animal cells also) stored at low temperature (–196°C) by using liquid nitrogen in the presence of cryoprotectants is called **cryobiology**.

Dodds and Roberts (1985) have discussed three principal methods used for growth suppression in plant tissue culture : (*i*) the alteration of physiological conditions of culture *i.e.* temperature or gas composition within the vessel, (*ii*) changing the composition of basal medium such as use of sub or supra-optimal concentrations of nutrients (some factors essential for normal growth may be either omitted or employed at a reduced level), and (*iii*) supplementing the medium with growth retardants (*e.g.* abscisic acid) or osmoregulatory compounds (mannital, sorbital, etc.).

Liquid Nitrogen

Nitrogen in a liquid state at a very low temperature is called 'liquid nitrogen'. It is produced by fractional distillation of liquid air in an industry. Liquid nitrogen is a colourless clear liquid with density of 0.807 g/mL at its boiling point and a dielectric constant of 1.4. Liquid nitrogen boils at 77 K (-196°C, -321°F) at atmospheric pressure. It is a cryogenic fluid causing rapid freezing on contact with living tissue and leads to frostbite. It can be stored and transported after well insulation, for example in vacuum flasks. By slow boiling of the liquid very low temperature is held constant at 77 K that results in the evolution of nitrogen gas. The holding time of vacuum flasks ranges from a few hours to a few weeks depending on the size and design.

1. Method of Cryopreservation

A standard protocol for cryopreservation of apical meristem is given in Fig. 18.26. The method of cryopreservation is freezing-storage-thawing cycle that is an external procedure being accomplished in the following basic stages :

***(a)* Selection of Materials:** A number of factors are taken into account for selecting the plant materials; the important ones are nature and density of cells in the vials/ampoules to be cryopreserved because the cryoability of the cell cultures depends on these. Young meristematic, highly cytoplasmic and small cells, which are non-vacuolated and thin walled and in small aggregates, are the good materials to be selected for this purpose. Cell density in vials or ampoules should be high, as it shows prolonged survival at high cell density. The ability of explant to survive at –196°C is influenced by the ability of their morphological and physiological conditions. Different types of explants used in cryopreservation are apical meristem and plant organs, ovules, anther/pollen, embryos, protoplasts, etc. Log phase cell suspension of many plant species has successfully been cryopreserved. The cultured cells are not the ideal system that can be cryopreserved. In contrast the organised structures (e.g. shoot pieces, young plantlets, embryos) are preferred for cryopreservation.

(*b*) Addition of Cryoprotectors: The two strong sources cause cell damage during freeze preservation : (*i*) formation of large ice crystals inside the cells that rupture cell organelles and cells itself, and (*ii*) intracellular concentration of solutes that increases to toxic levels due to dehydration before or during storage. Cryoprotectors are the chemicals which decrease cryodestruction. These are sugars, glycols, sugar alcohols, alcohols, polyvinylpyrrollidone, polyethylene glycol (PEG), polyethylene oxide (PEO), dextrans, hydroxystarch, glycerine, sucrose, and some amino acids (*e.g.* proline). Bajaj (1987) has advised to use a mixture of two or three cryoprotectants at low concentrations rather than a single cryoprotectant at a high concentration as it may be toxic. During treatment, the cultures should be maintained in ice to avoid deleterious effects.

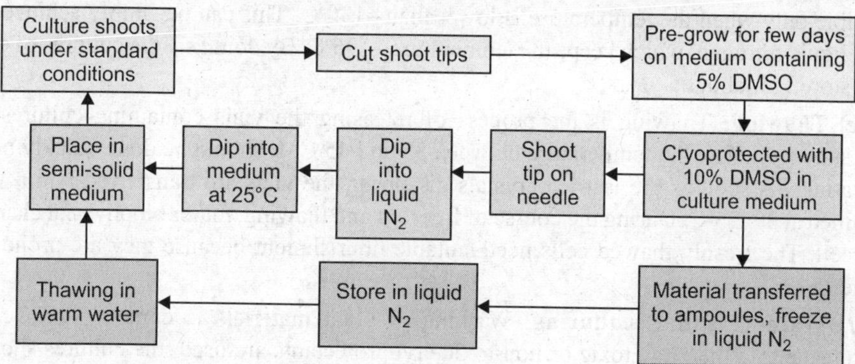

Fig. 18.26. Steps of cryopreservation of shoot tips.

Dimethyl sulfoxide (DMSO), sucrose, glycerol and proline are most frequently used cryoprotectants. DMSO has proved an excellent cryoprotectant because : (*i*) it has low molecular weight, (*ii*) it is easily miscible with solvent, (*iii*) it is toxic at low concentration, (*iv*) it is easily permeable into cells and easily washable. The material to be preserved is put in culture medium and treated with a cryoprotectant. Such material is transferred to sterile cryovials or ampoules which are made up of polypropylene. Gradually 5-10% cryoprotectant is added into the ampoules. It is tightly closed with a screw cap.

(*c*) Freezing: Freezing should be done in such a way that it does not cause intracellular freezing and crystal formation. To avoid this problem, regulated rate of cooling or pre-freezing is done. Moreover, freezers have also been developed which allow the uniform temperature decrease at a desired rate, commonly not less than 1°C per minute. In 1987, the Institute of Cryobiology and Cryomedicine of the Ukrainian Academy of Sciences (earlier in U.S.S.R.) devised the programme freezer which envisaged lower rate (0.5°C per minute) of temperature decrease. The following types of freezing can be done :

(*i*) Rapid freezing: Rapid freezing method is simple and easy to handle. After placing the plant material, the cryovials are put into liquid nitrogen which causes a decrease in temperature. Freezing is done quickly so that there should be least change or development of intracellular crystals. Ultra cooling prevents ice crystals. To achieve this objective, dry ice (CO_2) can be used instead of nitrogen. Rapid freezing of several plant materials has been done, for example, somatic embryos and shoot tips of *Brassica napus,* strawberry, potato, etc.

***(ii)* Slow freezing :** In this method the rate of freezing is slow, *i.e.* 0.1 –10°C per minute. This facilitates the flow in water from inside to outside. Therefore, extracellular ice crystals are formed but not intracellular crystals. Consequently, cell cytoplasm gets concentrated to cause cellular dehydration. Such dehydrated cells survive for a longer duration. Generally slow freezing has been done by using computer-controlled freezers. Meristems of potato, cassava, strawberry, etc. have successfully been cryopreserved.

***(iii)* Stepwise freezing:** It includes the earlier described two methods. In this method temperature gets lowered by –20 to –40°C. It allows protective freezing of the cells. Further, freezing is stopped for 30 minutes; thereafter, it is rapidly freezed in liquid nitrogen to get –196°C. By doing such stepwise decline in temperature, formation of big crystal is increased and good results are obtained. Excellent results have been obtained with suspension cultures and strawberry by adopting this method.

***(d)* Storage in Liquid Nitrogen:** If the cells are not stored at sufficiently low temperature, an additional injury to the cultures may be caused. The storage temperature should be such that it stops all metabolic activity and prevents biochemical injury. Prolonged storage of frozen materials is possible only when the temperature is lower than –130ºC. This can be simply achieved with the help of liquid nitrogen, which keeps the temperature –196ºC. By doing so, the cultures of carrot cells can be stored for about 5 years.

***(e)* Thawing:** Thawing is the process of releasing the vials containing cultures from the frozen state to elevate the temperature between 35 and 45ºC. It should be done quickly but without overheating. As soon as the last ice crystals disappear, the vials are transferred into a water bath maintained at 20–25ºC. During the course of freezing and thawing, major biophysical changes occur in the cell. The freshly thawed cells need suitable nourishment because they are prone to further damage.

***(f)* Washing and Reculturing:** Washing of plant materials is done to remove the toxic cryoprotectants. When low toxic or non-toxic cryoprotectants are used, the cultures should not be washed, but simply recultured. Washing becomes necessary only when cryoprotectants have toxic effects on cells. Washing follows the following procedure : dilution, resuspension, centrifugation and removal of cells. It is, however, possible that some cells die due to storage stress and the most stable ones survive. Therefore, determination of cell viability by culturing them on growth medium is essential. Cell viability test can be done by using FDA staining and growth measurements. The parameters of growth measurements are counting cell number, dry and fresh weight, mitotic index, etc. Mitotic index (MI) is counted by using the following formula:

$$MI = \frac{\text{Number of cells destined to cell division } (i.e.\ P + M + A + T)}{\text{Total number of cells both dividing and undivding}}$$

Where, P, cells in prophase; M, metaphase; A, anaphase and T, telophase.

***(g)* Regeneration of Plantlets:** The viable cells are cultured on non-specific growth media to regenerate into plantlets. Bajaj (1987) has given an extensive list of works on cryopreservation of cells, tissue, and organ culture of various plants *e.g.* potato, cassava, sugarcane, soyabean, groundnut, carrot, cotton, citrus, coconut, etc.

2. Risks-associated with Cryopreservation

During the freezing stage there are certain risks associated with cryopreservation such as: solution effects, extracellular ice formation, dehydration and intracellular ice formation. Many of these effects can be reduced by adding cryoprotectants. In freezing water solutes are excluded with growth of ice crystals that results in the increase of solute concentration in the remaining liquid water. However, high concentrations of some solutes can be very damaging. Water migrates from the cells with slow cooling of tissues, and ice is formed in the extracellular spaces. High amount of extracellular ice causes mechanical damage to the cell membrane. Besides, cellular dehydration is also caused by migration of water forming extracellular ice. There are certain organisms and tissues which can tolerate some extracellular ice. In addition, there are certain stresses associated on the cell that directly cause cell damage.

Methods to prevent damages caused by cryopreservation are a well established combination of 'controlled rate' and 'slow freezing' as well as *vitrification* (a newer flash-freezing process). Usually vitrification requires the addition of cryoprotectants before cooling.

3. Achievement through Cryopreservation

Various forms of plant materials *viz.*, cell suspensions clones, callii, tissues, somatic embryos, root/shoot tips propagules (tubers) pollen grains, etc. have been preserved in liquid nitrogen for prolonged time and tested for their survival and regeneration potential. No doubt, in most of the cases, the cells/tissues, organs regenerated into plants.

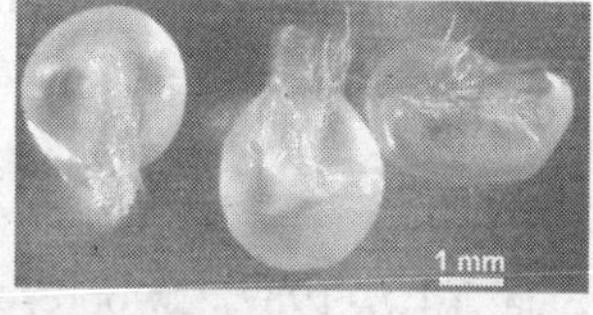

Cryopreservation used to establish Germplasm Bank.

Bajaj (1987) has described a number of plant species that have been successfully cryopreserved. Some of the observations made are as below :

(*i*) Cryopreservation of cell lines : For example, cell suspensions (soybean, tobacco, dhatura, carrot) and somatic hybrid protoplasts (rice × pea, wheat × pea).

(*ii*) Cryopreservation of pollen and pollen embryos : For example, fruit crops, trees, mustard, carrot, peanut, etc.

(*iii*) Cryopreservation of excised meristems : For example, potato, sugarcane, chickpea, peanut, etc.

(*iv*) Cryopreservation of germplasm of vegetatively propagated crops: potato, sugarcane, etc.

(*v*) Cryopreservation of recalcitrant seeds and embryos : large sized seeds that are short-lived and abortive, such as oil palm, coconut, walnut, mango and cocoa.

QUESTIONS

1. Write a brief history of development of *in vitro* techniques of plant cell culture.
2. What are the requirements for establishing a tissue culture laboratory ?
3. Give a brief account of methods of sterilization.
4. How will you prepare the culture medium? Write the formulation and nutritional component of culture medium.
5. What is an explants ? How will you induce callus from it ?
6. Write an essay on organogenesis and its application.
7. Give a brief account of benefits from root culture.
8. How will you prepare cell suspension ? Discuss the benefits of using aqueous medium over the solid one.
9. What do you know about micropropagation ? How does it differ from vegetative propagation?
10. Write an essay on somatic embryogenesis.
11. How will you isolate protoplasts from the plant cells? Give the applications of protoplast culture.
12. What are the benefits from pollen and anther culture ?

13. Why does a cultured anther permit pollens to develop embryos but not the cultured pollen grains?

14. Write short notes on the following ?

(*i*) Explant,
(*ii*) Callus,
(*iii*) Totipotency,
(*iv*) Organogenesis,
(*v*) Cell culture,
(*vi*) Somaclonal variants,
(*vii*) Hybrid and cybrid,
(*viii*) Fugigenic chemicals,
(*ix*) Anther and pollen culture,
(*x*) Micropropagation,
(*xi*) Protoplast fusion in Fungi,
(*xii*) *In vitro* androgenesis,
(*xiii*) Mentor pollen technology,
(*xiv*) Embryo culture,
(*xv*) Embryo rescue.

BIBLIOGRAPHY

Bajaj Y.P.S. 1987. Cryopreservation of pollen and pollen embryos and the establishment of pollen bank. *Intl. Rev. Cytol.* 107: 397-420.

Bhojwani S. S. and Razdan, M. K. 1996. *Plant tissue culture: theory and practice.* Elsevier, pp.767.

De-Quan, S.; Xin-Hua, L.; Guo-Lu, L.; Qi-Gao, G.; Yi-Wei, M.; Jiang-Hui, Xie. 2011. Production of triploid plants of papaya by endosperm culture.*Plant Cell, Tissue and Organ Culture.* 104: 23-29.

Dodds J.H. and Roberts, L.W. 1985. *Experiments in plant tissue culture*, 2nd Ed. Cambridge University Press, Canbridge, U.K.

Ghosh B.; Mukherjee, S.; Jha, T.B. and Jha, S. 2002. Enhanced colchicine production in root cultures of *Gloriosa superba* by direct and indirect precursors of the biosynthetic pathway. *Biotechnology Letters*. 24: 231-234.

Griffiths A.J.F.; Gelbart, W.M.; Miller, J.H. *et al.* 1999. *Modern Genetic Analysis*. New York: W. H. Freeman.

Hazarika B. N. 2003. Acclimatization of tissue-cultured plants. *Current Sci.* **85**: 1704-1712

Knox R.B. *et al*. 1987. Mentor pollen techniques. *Int. Rev. Cytol*. **107**: 315-332.

Kurz WG, Constabel F. 1979. Plant cell cultures, a potential source of pharmaceuticals. *Adv. Appl. Microbiol.* 25: 209-40.

Rai M.K., Kalia, R.K. *et al.* 2011. Developing stress tolerant plants through *in vitro* selection- an overview of the recent progress. *Environ. Experimental Botany*. **71**: 89-98.

Sangwan R.S.1981. *Ph.D. Thesis*, Univ. of Paris, Paris.

Shepard F., Bidney, Dennis and Shahin, Elias. 1980 Potato Protoplasts in Crop Improvement. *Science*.**208**: 17-24.

Sood A and Palni, L.M.S. 1992. In *Microbial Activity in the Himalaya.* (ed. R.D. Khulbe), Shree Almora Book Depot, Almora, pp. 375-383.

Srivastava S. and Srivastava, A.K. 2007. Hairy root culture for mass-production of high-value secondary metabolites. *Crit Rev Biotechnol*. **27**:29-43.

Su Y.H.; Liu, Y.B. and Zhang, X.S. 2011. Auxin–cytokinin interaction regulates meristem development. *Molecular Plant.* **4**: 616–625.

Thorpe T.A. 2007. History of plant tissue culture. *Mol. Biotechnol.* **37**: 169–80.

Zhao Z. *et al.* 2010. Hormonal control of the shoot stem cell niche. *Nature*. **465:** 1089–1092.

CHAPTER 19

Transgenic Plants

A. Gene construct
B. Methods of transformation
C. Herbicide- resistant transgenic plants
D. Virus- resistant transgenic plants
E. Fungus- resistant transgenic plants
F. Insect- resistant transgenic plants
G. Modification in nutrient quality
H. Molecular farming
I. Improvement of seed protein quality

GM tomato

Methods of plant cell, tissue and organ culture have been described in detail in Chapter 18 *Plant biotechnology* discussing different types of culture. However, transgenic technology has been developed by combining the two technologies: recombinant DNA technology and plant tissue culture. You know that the genetic variation is generated through hybridisation, mutagenesis and polyploidy which is necessary for crop improvement. In recent years, somaclonal variation and somatic hybridisation have also been used for crop improvement. But, the transfer of specifically constructed gene assemblies through various transformation techniques is the most potent biotechnological approach for creating the desired plant; really this approach is called genetic engineering (*see* Chapters 6, 7 and 8). Thus, the plants constructed through genetic engineering consist of a gene(s) from the other organism. Such genes are called 'transgenes,' and the plants in which a functional foreign gene has been incorporated by biotechnological methods that is generally not present in plant, are called 'transgenic plants'.

In dicotyledons transgenic plants have been produced to improve the crops in certain traits so that they may become resistant to herbicides, insects and viruses, abiotic stresses such as drought, salinity, etc. and also be improved in nutritional qualities or may contain certain new properties like antibody production or act as vaccines. Normally photosynthesis or biosynthesis of amino acids are affected by herbicides. However, resistant plants can normally be developed either by making target molecules insensitive to herbicide or by overproducing the target proteins. A second method is the introduction of such a pathway that detoxifies the herbicide(s).

About 134 million hectares worldwide are cultivated with transgenic crops after 14 years (1996 to 2009). The most commonly grown cultivars contain enhanced herbicide tolerant and insect resistance traits (James, 2009). Although developed countries were the primary users of transgenic crops, developing countries (such as Argentina, Brazil, India, China and South Africa) now account for about 46% of the global transgenic crop area. Of these, farmers are growing mainly transgenic cotton, maize or soybean (James, 2009), which have been developed and made available by companies such as Monsanto and Syngenta. Seeds of insect-resistant cotton and maize (Bt crops) were distributed by Monsanto to smallholders in South Africa between 1998 and 2002. Similarly, small-scale farmers are growing Monsanto's insect-resistant cotton in China, India and Mexico, and transgenic herbicide-tolerant soybean in Argentina and Brazil.

Professor Ingo Potrykus

Naturally the rice plant can produce β-carotene which is a carotenoid pigment occurring in the leaves and is involved in photosynthesis. Moreover, the rice plant does not normally produce the pigment in the endosperm because the photosynthesis does not occur in the endosperm. But the people eat the rice endosperms, not the leaves. This has resulted in the problems of vitamin A deficiency that is most prevalent throughout the world because the number of rice consumers is gradually increasing with increase in world population. Therefore, this deficiency is very critical because it can cause blindness and affect the severity of many diseases including diarrhea and measles. This is a severe problem affecting more than 100 million children worldwide. A simple solution to this problem is to distribute vitamins to the affected children. Unfortunately, many countries having chronic deficiency lack the necessary infrastructure to deliver the vitamin tablets to the neediest persons.

The solution to this problem was to improve the vitamin content in rice to be widely consumed and readily available to the consumers. In 2,000, Professor Ingo Potrykus and his team at Swiss Federal Institute of Technology created a transgenic plant called 'Golden rice' by inserting genes from daffodil plant rich in vitamin A. It is called 'Golden Rice' because its colour is yellow rather than white due to presence of abundant amount of β-carotene (a yellow precursor to vitamin A) in the seed endosperm. Therefore, it contains elevated levels of the precursor to vitamin A. In 2005 a new variety called Golden Rice 2 was announced which produces up to 23 times more β-carotene than the original variety of Golden rice.

The most important and accepted biotech product is the 'edible vaccines', the mode of working of which is the same as that of normal vaccines. A protein enters the body in some manner and the human immune system produces antibodies against that protein. The human immune system is turned on and destroys the pathogen after exposing the edible vaccine to the viral pathogen(s).

Toxin gene (*bt2* or *cry2*) is isolated from *Bacillus thuringiensis* for inducing resistance in insects. Since long time the Bt (Cry) toxins produced by this bacterium has been used as a biological insecticide. The *bt*2 gene was used in *Agrobacterium* Ti plasmid-mediated transformation of tobacco, cotton and tomato plants. The transgenic plants were resistant to a pest of tobacco called *Manducta sexa*. Cowpea (*Vigna unguiculata*) has been used to demonstrate the induction of resistance to storage pests. In cowpea resistance to attack by the major storage pests and insects is developed by a trypsin inhibitor gene, *cpTi*. The *cpTi* gene was transferred by using binary vectors for inducing resistance in insects. Similarly, this approach was also applied for the production of transgenic tobacco plants expressing high level of CpTi, and making them resistant to a variety of insect pests.

Thus the transgenic plants are applicable in many ways, for example: (*i*) they are extremely valuable tools in studies on plant molecular biology, regulation of gene action, identification of regulatory/promotary sequences, etc., (*ii*) their agronomic and other features have been improved by transferring specific genes into plants, (*iii*) transgenic plants resistant to insects, viruses, etc. have been generated by engineering the genes for resistance to various biotic stresses, (*iv*) they have improved produce quality by transferring several genes such as Golden rice, (*v*) certain novel biochemicals like hirudin can be produced by transgenic plants which are not normally produced by natural plants, (*vi*) transgenic plants can be used as vaccines for immunization of humans against pathogens.

Transgenic crops have tremendous potential to benefit producers, processors and consumers. As advances in biotechnology are made it is important to learn and understand the technology to avoid confusion and misinformation that could prevent the potential benefits of the technology from being realised.

Transplastomic Plants

A genetically modified plant in which the new genes have been inserted in the DNA of the chloroplasts but not in the nuclear DNA is called 'transplastomic plant'. In many plant species plastid DNA is not transmitted through pollen, which prevents gene flow from the genetically modified plant to other plants. This is the major advantage of this technology. The most common and suitable method for plastid transformation is the particle bombardment, but the transformation efficiency is lower than *Agrobacterium*-mediated transformation.

Plastid transformation is suitable only for certain crop species; reliability of this method has only been proven for tobacco. In 2007, a team of researchers led by Ralph Bock (at the Max Planck Institute of Molecular Plant Physiology, Germany), studied genetically modified tobacco in which the transgene was integrated in chloroplasts. The Co-extra researchers analysed more than two million seedlings; they found that less than 20 in 10,00,000 inherited the transgene. This reduction is because of loss of some parts of the seedlings during their development into mature plants. Tobacco has a strong tendency towards self-fertilisation, hence the reliability of transplastomic plants is assumed to be even higher under field conditions.

A. GENE CONSTRUCTS

There is need of a promoter (start signal) and a terminator (stop signal) for gene expression. However, in some cases the additional sequences are required (e.g. marker genes) accompanied by a promoter and a terminator. The name 'construct' is used because the sequences normally do not exist in this combination but they are constructed by putting together. A gene construct consists of all the information needed for expression of the desired gene in an organism. This includes an antibiotic resistance marker, a tissue-specific promoter, the transgene/gene of interest containing a reporter gene (such as GUS or GFP) and a stop sequence.

I. ORGANISATION OF A TYPICAL PLANT GENE

In recent years structural genes of a large number of plant species have been analysed. During 1971, splitting of plant genes into several distinct segments by intervening segments of DNA was discovered by many researchers. The coding segments are called exons and that of intervening segments are called introns. Thus the genes in higher groups of organisms are discontinuous; therefore, these genes are also referred to as mosaic or split genes. The introns are spliced out before start of transcription. A typical plant gene is well organised encoding a polypeptide chain. The adjacent genes are often separated by an 'insulator' which helps them to avoid cross-talk between each other's promoters and enhancers (and/or silencers).

A gene is read from 5′→3′ direction; hence the different regions of a protein-coding gene starting with the 5′-end consists of promoter, enhancer/silencer, cap site or transcriptional start, an untranslated region (5′ UTR) or leader sequence, initiation codon, exons, introns, the stop codon(s), a second untranslated region (3′ UTR), and a poly (A) tail which is typically G/AATAA(A)n.

The initiation codon for translation of mRNA transcript is ATG (AUG in mRNA); therefore, it marks the beginning of coding sequence of the gene. The sequence located between the cap site and ATG consists of untranslated region (5′ UTR), hence it forms the 5′-leader sequence of mRNA. The end of coding sequence of the gene contains the 'chain termination codon', TAG (in mRNA, UAG) or one of the two nonsense codons (TAA and-TGA). TAG is followed by a stretch of untranslated region (3′ UTR) and the end contains polyadenylation site (poly A).

Proteins synthesised in the cell move across the membranes or used for building the cell components and being involved in metabolisms. Such genes have a coding sequence after ATG, *i.e.* the initiation codon at their 5′ ends which encodes a signal peptide. After translocation of proteins to the desired location, the signal peptides get cleaved to generate the mature functional proteins. The different regions of a gene are briefly described as below:

1. Promoters

A region of DNA facilitating the transcription of a particular gene is called promoter that is located typically near upstream (towards the 5′ region of the sense strand) of the genes on the same strand. The promoter contains specific DNA sequences that are recognised by proteins known as 'transcription factors'. The transcription factors bind to the promoter sequences and recruit RNA polymerase, *i.e.* the enzyme that synthesises the RNA from the coding region of the gene. The proximal sequence upstream of the gene that tends to contain primary regulatory elements is called 'proximal promoter' which is approximately -250 bp upstream of the start site. It is also the specific transcription factor binding sites. In contrast, the distal sequence upstream of the gene that may contain additional regulatory elements (often with a weaker influence than the proximal promoter) is called 'distal promoter.

Promoter regions have a consensus sequence that is the most common sequence at this position, but some genes have 'strong' promoters that bind the transcription machinery well, and the others have 'weak' promoters that bind poorly. Promoters are the critical elements that work in concert with the other regulatory regions such as enhancers, silencers, boundary elements/insulators to direct the level of transcription of a given gene. The promoter is associated with transcription initiation providing the RNA polymerase binding site. The CAAT and TATA boxes are the consensus sequences present within promoters of RNA polymerase II. The 'basal or core promoter' is located within about 40 base pairs (bp) of the start site, and an 'upstream promoter', may extend over as many as 200 bp farther upstream.

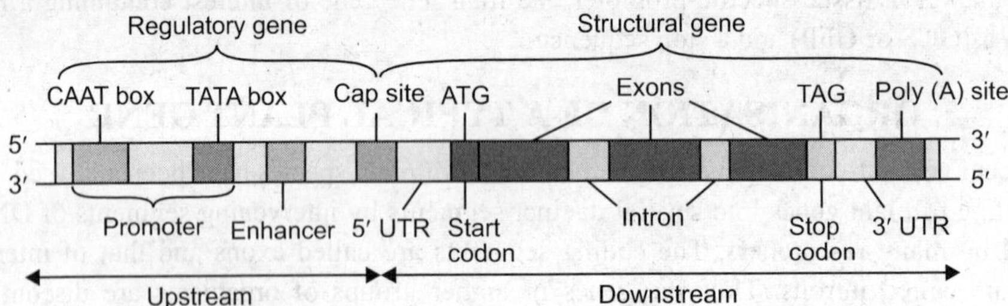

Fig. 19.1. Structural organisation of a typical nuclear plant gene.

***(a)* TATA Box:** One or two short sequence element(s) upstream from cap site is commonly found in eukaryotic genes which are involved in transcription (Fig. 19.1). TATA box (or Goldberg Hogness box) is one of the best characterised sequences. Almost in all plant and animal genes, this sequence is normally located -25 to -30 upstream of the cap site. The TATA box lies very close to the transcriptional start site (within 50 bases). The eukaryotic genes express correctly by involving this sequence.

***(b)* CAAT Box:** Consensus sequence CAAT box or GG (CT) CAATCT is the other sequence involved in transcription of some of the eukaryotic genes, present -40 and -80 nucleotides upstream of the cap site (Fig. 19.1). However, the sequence of this element varies greatly but mostly present as CAAT; therefore, this sequence is called CAAT box. But this sequence is also called AGGA box in plants.

***(i)* Promoter sequences used to drive gene expression in transgenic plants:** Foreign genes (especially prokaryotic genes) must consist of an appropriate promoter (5′-leader and 3-terminator sequences) so that they can expression efficiently in plant cells. In response to a specific stimulus or specific developmental stage a suitable enhancer sequence will be required if the gene is needed to express in a specific tissue. Several promoter sequences are used to drive gene expression in plant cells such as the 35S RNA gene promoter of cauliflower mosaic virus (CaMV), and *nos* (nopaline synthase), *ocs* (octopine synthase) and *mas* (mannopine synthase) promoters from *Agrobacterium*. Both in dicots and monocots 35S promoter is the most commonly used constitutive promoter. As compared to the *nos* promoter, the 35S promoter is 10-40-fold more efficient in expression. Expression in roots (*i.e.* anaerobic induction) is shown by the maize alcohol dehydrogenase 1 (Adh1) promoter, the activity (in monocots) of which is higher than that of the 35S promoter. For the monocots the 35S promoter is not an efficient promoter.

Difference between Eukaryotic and Prokaryotic Promoters

- **Prokaryotic Promoters**

  The prokaryotic promoter consists of two short sequences at -10 and -35 positions upstream from the transcription start site.

  - The sequence present at -10 upstream is called the Pribnow box or the -10 element which usually consists of the six nucleotides TATAAT. The Pribnow box is absolutely essential to start transcription in prokaryotes.
  - The other sequence is found at -35 upstream, and hence called the -35 element which usually consists of the six nucleotides TTGACA. It allows a very high rate of transcription.

- **Eukaryotic Promoters**

  The eukaryotic promoters are very diverse, therefore, they are difficult to characterise. Mostly they are located upstream of the gene and have regulatory elements several kilobases away from the transcriptional start site. Their transcriptional complex can cause the DNA to bend back on itself allowing for placement of regulatory sequences far from the actual site of transcription. Many eukaryotic promoters consist of a TATA box (sequence TATAAA) which in turn binds a TATA binding protein that helps in the formation of RNA polymerase transcriptional complex.

In maize two proteins have been identified viz., alcohol dehydrogenase I (ADH-I) and alcohol dehydrogenase II (ADH-II), of which ADH-I is the most important because its gene promoter has been widely used. Anaerobic regulatory elements and an intron present in the 5′ untranslated regions of the maize ADH genes are important for driving gene expression in monocots. The regulatory elements are present within a 247 bp upstream of the CAP site, *i.e.* between positions -140 and -99 of the maize Adh-1 promoter. There are two essential regions of around 15 bp present within this 40 bp segment, which are required for expression under low oxygen conditions. Moreover, for driving the expression of genes of interest the whole Adh-1 promoter has been used in cereals such as rice, oat and barley and in dicots such as tobacco. Very low level of expression is provided in dicot crops. The Adh-I promoter along with the intron has been found much superior for transformation of some cereals like rice.

For driving gene expression in transformed monocots plants the two regulatory transcription regions of the maize *adh* 1 gene currently in use are: (*i*) the 'anaerobic regulatory elements' (AREs) of the ADH-1 promoter, and (*ii*) the first intron of the maize *Adh*-1 gene. The sequences of region II of ARE of maize *Adh*-1 is 5′-CGTGGTTGCTTGCC-3′, and the sequence of *Adh*-2 is 5′-CGAGCCTTTCTTCCC-3′.

The two elements are the basis of a recombinant promoter construct called pEmu that is formed in 5′ → 3′ direction by: (*i*) a truncated ADH-1 promoter having multiple copies of ARE, (*ii*) enhancer elements from the octopine synthase (OCS) gene of *A. tumefaciens*, (*iii*) a TATA box from a promoter expressed in plants, and (*iv*) the maize Adh-1 intron 1. The promoter construct has been tested in various monocot crops, driving a very high level of gene expression as compared to 35S CaMV promoter. It has also been used for the development of transgenic rice and sugarcane.

Types of Promoters

According to the type of control of gene expression different type of promoters are used in biotechnology. They can be divided into the following four types:

(*i*) Constitutive promoters. Constitutive promoters direct the expression in all tissues independent of environmental and developmental factors. Constitutive promoters are active across species because their expression is not conditioned normally by endogenous factors.

(*ii*) Tissue-specific or development-stage-specific promoters. Expression of a gene in specific tissue or at certain stages of development is directed by the tissue-specific promoters. Those promoter elements can be found in heterologous promoters, *i.e.* distantly related species that are expressed or affect the expression of genes in photosynthetic tissues, tubers, roots, seeds, and other vegetative or reproductive organs. But specificity is achieved with homologous promoters, *i.e.* from the same species, genus or family. However, the coordinated expression of transcription factors is important to regulate the activity of the promoter.

(*iii*) Inducible promoters. Performance of inducible promoter is not conditioned to endogenous factors but to environmental conditions and external stimuli to be controlled artificially. There are such promoters that can be modulated by abiotic factors such as light, oxygen, heat, and cold. However, promoters that respond to chemical compounds but not found naturally in the organism of interest are of particular interest. Such promoters that respond to antibiotics, copper, alcohol, steroids, and herbicides have been adapted and refined to allow the induction of gene activity independently of other biotic or abiotic factors.

(*iv*) Synthetic promoters. Synthetic promoters are made by bringing together the primary elements of a promoter region from diverse origins.

On the basis of trans-activating proteins there are regulatory expression systems that regulate the expression of genes of interest irrespective of their physical position to the target genes. Several chemical-inducible promoters incorporate trans-activating proteins and constitutive promoters as part of the regulatory system.

2. Transcription Start Site (TSS)

The site where transcription of the gene into RNA begins is called the transcription start site. A RNA polymerase II (Pol II) binds to this site. RNA Pol II is a complex of 12 different proteins. Eukaryotic promoters are very diverse and difficult to characterise. They are present upstream of the gene and consist of regulatory elements several kilobases away from the transcriptional start site (enhancers). In eukaryotes, the transcriptional complex causes the DNA to bend back on itself; this allows for placement of regulatory sequences far from the actual site of transcription.

Several eukaryotic promoters (10-20% of all genes) contain a TATA box (sequence TATAAA), which in turn binds a TATA binding protein assisting in the formation of the 'RNA polymerase transcriptional complex'. Typically the TATA box is located very close to the transcriptional start site (often within 50 bases). Eukaryotic promoter sequences bind the proteins called 'transcription factors' which form the transcriptional complex, for example the E-box (sequence CACGTG) which binds transcription factors in the basic-helix-loop-helix (bHLH) family.

3. Enhancers/Silencers

The other classes of elements located within a few hundred nucleotides upstream of start of transcription at 5′ end of the gene, that increases the activity of promoter of a gene and show the transcription at the high level, are called 'enhancers' (Fig. 19.1) that can compensate for a weak promoter. Usually the weak promoters permit a lower rate of transcription than the strong promoters, because the transcription machinery binds to them and initiates transcription less frequently. In addition to flanking of these sequences at 5′ ends, they may also be found at other regions of the genes. Thus the enhancers can be located upstream, downstream, or even within the gene they control. Initially enhancers were discovered in viruses but they are also found in plants. The core sequence of an enhancer identified by sequence comparison is written as GGTGTGGAAAG (or generally GTGGT/AT/AT/AG). Here T/A may also be written as W. Thus the sequence GTGGT/AT/AT/AG can be written as GTGGWWWG.

DNA sequences suppressing promoter activity are called silencers; they are the control regions of DNA. Similar to enhancers, they may be located thousands of base pairs away from the gene they control. However, when transcription factors bind to them, expression of the gene they control is repressed. Thus enhancer/silencer sequences are concerned with gene regulation activity by regulating the promoters but they are not themselves involved in the promoter activity. Tissue-specific gene expression may also be stimulated by enhancer/silencer-like sequences in plants.

Silencer sequence is associated with the *adh*1 gene of maize which is an example of a negative regulating element. Under aerobic conditions *adh*1 expression is suppressed by the ADH1 enhancer sequence. Thus, the first intron (of adh1gene and shrunken 1 gene) of maize acts as enhancers of 35S promoter of CaMV when these two are placed between the promoter and the gene. The shrunken gene intron 1 is about 10-fold more effective in enhancing the efficiency of CaMV 35S promoter than the Adh1 intron.

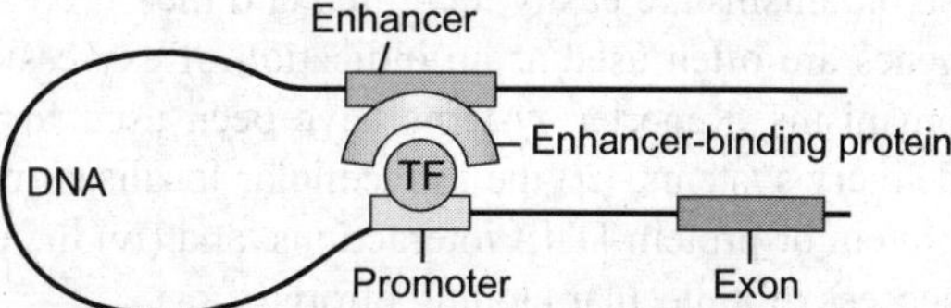

Fig. 19.2. Binding of transcription factors (TF) to sites of enhancer-binding proteins.

There are certain transcription factors (enhancer-binding proteins) that bind to regions of DNA located thousands of base pairs away from the gene that they control. The rate of transcription of the gene gets increased due to their binding to a gene. How does the binding of a protein to an enhancer regulate the transcription of a gene thousands of base pairs away ? In addition to their DNA-binding site, the enhancer-binding proteins have sites that bind to transcription factors (TF) assembled at the promoter of the gene. Consequently, the DNA is drawn into a loop as shown in the Fig. 19.2. The same protein complex called 'cohesion' that holds sister chromatids together during mitosis stabilises these loops.

4. Introns and Exons

The non-coding DNA sequence of a gene that intervenes the coding sequence and interrupts transcription process is called an intron. The introns can be detected and characterised by comparing the complete nucleotide sequences of DNA to the mRNA using genomic and cDNA clones. The cloned genomic DNA is annealed with RNA and the products are examined under an electron microscope. The intron sequences are not present in the RNA; hence they form single stranded loops in the heteroduplex molecule. These loops can be seen and their positions in the gene mapped from an electron micrograph only if they are large-sized, *i.e.* more than 50 nucleotides in length.

It is argued that 'the cell puts a huge amount of its energy into the creation of these introns' then discards them. Why 'most genes of higher organisms are broken up into short exons separated by huge stretches of seemingly useless DNA' is particularly puzzling because 'the introns are all transcribed into RNA only to be snipped out by snRNPs and thrown away immediately'.

A sequence of intron has much evolutionary and functional significance. In 1981, W. Gilbert (Harvard University) put forward originally an appealing hypothesis that the presence of introns is related to an evolutionary process by which new genes arises from recombination events that juxtapose the previously existing genes. The number of introns varies much in both plant and animal genes. However, there are many plant genes that lack introns, for example the zein storage protein genes of maize, two other soybean protein genes, one rare class leaf gene and most cab genes (nuclear genes encoding the chlorophyll a/b-binding protein of the photosystem). There are other plant genes that have 2-3 introns such as French bean (*Phaseolus*) gene, soybean leghaemoglobin gene, a soybean actin gene, soybean glycinin gene, a rare class leaf gene and RuBP carboxylase small subunit genes. In addition, there are few plants that possess 5-9 introns with longer sequences such as maize *adh* genes (9 introns) and phytochrome genes (5 introns). All the plant genes are similar in having first and last bases in the introns, *i.e.* GT and AG, respectively where the intervening sequences are present.

5. Reporter Genes (Marker Genes)

There are some genes that upon transferring into the cells/tissue of another organism produce enzymes the activities of which can easily be detected. Therefore, these genes can be used as a basis for selection of transformed cells such as GUS gene because the characteristics they confer on organisms are easily identified and measured as they are selectable markers. Thus, the reporter genes are often used as an indication of expression of gene that has been taken up in the cell or organisms. Reporter systems have been used for: (*i*) the determination of the efficiency of gene delivery systems, (*ii*) the intracellular localisation of a gene product, (*iii*) the detection of protein-protein or protein-DNA interactions, and (*iv*) the efficiency of translation initiation signals and the success of molecular cloning efforts.

Usually the reporter gene is in the form of a circular DNA molecule called plasmid for bacteria or eukaryotic cells in culture. For introduction of a reporter gene into an organism, the reporter gene is fused with the promoter of the gene of interest and placed in the same DNA construct to be inserted into the living cells/organism (Fig. 19.3). The transformed cells can be selected easily among the thousands of non-transformed cells. Reporter genes present in transformed cells express specific protein, which are easily detectable and quantifiable even without destroying the tissue. Therefore, it is very useful for monitoring and detection of plant transformation, for studying the activities of regulatory elements such as promoter and enhancer. Since the expression of the reporter is being used as a marker for successful uptake of the gene of interest, it is important to use a reporter gene that is not natively expressed in the cell/organism.

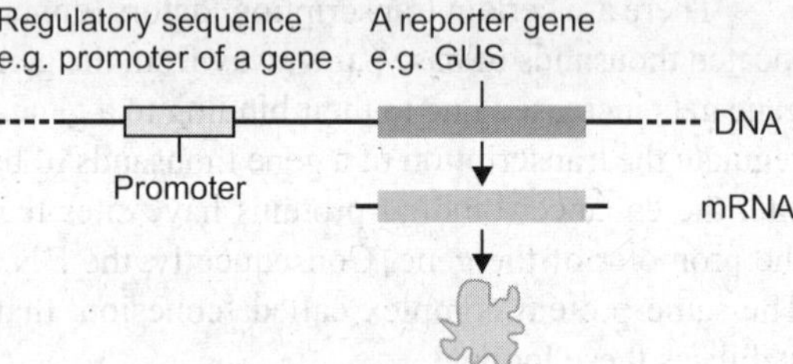

Fig. 19.3. Study of a regulatory sequence by using a reporter gene.

The marker genes can be used to identify a promoter, to study the expression pattern and strength of the promoter. They are used for the study of regulation of different cellular processes, for example to study gene expression, protein localisation and intracellular protein movement without destroying the plants.

Reporter genes are mainly of two types: *scorable markers* and *selectable markers*. Most genes can be tagged with reporter gene to detect its expression easily either through highly sensitive enzyme assays (scorable reporter genes) or through expression of resistance to a toxin (selectable reporter genes). Most common easily detectable enzyme producing genes are: *nos* (nopaline synthase from *Agrobacterium*), *lux* (luciferase from bacteria or firefly that emits light), *cat* (chloramphenicol acetyltransferase from bacteria that confers resistance to the antibiotic chloramphenicol), β-glucuronidase (GUS from bacteria), etc. A common reporter in bacteria is the *LacZ* gene which encodes the protein β-galactosidase. This enzyme causes bacteria expressing the gene to appear blue when grown on a medium containing X-gal as substrate analogue.

An Ideal Reporter Gene

The ideal reporter gene product should be biologically inert and not influence the cell population under investigation. Products of an ideal reporter gene should be: (*i*) easily quantifiable, (*ii*) able to rapidly degrade the enzyme, (*iii*) able to cause low endogenous background (high signal-to-noise ratio), (*iv*) non-toxic to cells, (*v*) resistant to the chemicals used in the processing, and (*vi*) sensitive and reliable to assay. Moreover, there is no such reporter gene that can be ideal for all the plants; some have advantages and some disadvantages.

One of the examples of an ideal reporter gene is the luciferase gene that provides temporal and spatial information on promoter activity in eukaryotes. Luciferase activity has its short half-life. Therefore, dynamic changes in gene expression of individual seedlings can be assayed repeatedly for several days. Such transgenic plants are required that contain a luciferase gene under the control of a promoter of interest. Low light detecting charge-coupled device (CCD) cameras or scintillation counters is used to detect even the low level of luminescence produced by luciferase gene in plants.

Similarly GUS enzyme acts as marker and its assay is very sensitive to gene expression taking place within a cell. Therefore, slight changes in expression can be monitored.

On the basis of extensive information, the selectable marker genes *npt*II, *hpt*, *bar/pat*, and *man*A, and the scorable marker gene *uid*A, all have little risk in terms of biosafety. These appear to represent the safest options for use in plant biotechnology. Some commonly used selectable markers are given in Table 19.1.

(*a*) Scorable (Screenable) Markers: When a scorable marker gene present in gene construct in transformed cells/tissue of a transgenic plant expresses, it results in a quantifiable phenotype, *i.e.* it makes the transformed cells different in look. As compared to selectable markers, the scorable markers results in distinct phenotypes that can easily be identified in background on non-transformed cells (e.g. GUS, GFP, etc.). The objective of using the reporter genes to study molecular processes in living cells is to simplify the detection of the product or to distinguish it from similar genes in the genome by introducing the synthetically modified gene construct in natural genes. Thereafter, these reporter genes are assayed at the level of protein. The scorable reporter gene produces an enzyme the activity of which is determined: *in situ* in the transformed tissues, by *in vitro* assays using the extract from plant tissue, or employing immunological methods.

Table 19.1. Some commonly used marker (reporter) genes.

Genes	*Substrates*	*Enzymes*	*Sources*
		Scorable Markers	
lacZ	X-gal	Beta-galactosidase	*E. coli*
gus (*uidA*)	X-gluc,4-mu gluc	Beta-glucuronidase	*E. coli*
luc	Luciferin/coelenterazine	Luciferase	*Photinus pyralis/ Renilla reniformis*
gfp	No substrate	Green fluorescent protein	*Aequorea victoria*

		Selectable Markers	
cat	Chloramphenicol + acetyl CoA	Chloramphenicol acetyl transferase	Tn9 of *E. coli*
spt	Streptomycin	Streptomycin phosphotransferase	Tn5
nptII	Neomycin (kanamycin or paramomycin G418)	Neomycin phosphotransferase II	*E. coli* K12
ble	Bleomycin	Bleomycin resistance	*E. coli* Tn5
bar	Phosphinothricin (herbicide)	Phosphinothricin acetyl transferase	*Streptomyces hygroscopicus*
EPSP synthase	Glyphosate (herbicide)	5-Enolpyruvylshikimate -3- phosphate synthase	*Petunia hybrida, Zea mays*
bnx	Oxynils (herbicide)	Bromoxynil nitrilase	*Klebsiella pneumoniae* subspecies *ozanaenae*
		Other Markers	
noc, ocs	Arginine + pyruvate + NADH	Nopaline synthase, octopine synthase	*Agrobacterium tumifaciens*
		Biotic- and Abiotic- stress-based Markers	
hsp101	47oc, 60 minutes	Plant heat sock protein	*Oriza sativa, Nicotiana tabacum*
pflp	*Erwinia carotovora*	Ferredoxin-like protein	*Capsicum annum*

An ideal scorable marker gene possesses many such characters that it should: (*i*) be easily quantifiable, (*ii*) results in relatively rapid degradation of enzyme, (*iii*) has low endogenous background (high signal-to-noise ratio), (*iv*) be non-toxic to cells, (*v*) be resistant to the chemicals used in the processing, and (*vi*) be sensitive and reliable to assay. Moreover, there is no such reporter gene that can be ideal for all the plants; some have advantages and some disadvantages.

Some commonly used reporter genes that induce visually identifiable characteristics involve fluorescent and luminescent proteins, for example the gene that encodes green fluorescent protein (GFP) (from the jellyfish which causes the transformed cells to glow green under blue light), the enzyme luciferase (which catalyses a reaction with luciferin to produce light), and the red fluorescent protein from the gene dsRed. The GUS gene has been commonly used in plants but luciferase and GFP are becoming more common.

Scorable markers can also be used as reporter genes that can be placed under the control of a specific promoter. In the presence of the necessary cell machinery the promoter will be activated to transcribe the mRNA of the reporter gene followed by translation of mRNA resulting in polypeptide production. For determination of gene activation in plant system, the cells/tissue is treated with the suitable substrate and expression is monitored. The expression pattern of the promoter is determined based on expression of the reporter gene.

(*i*) GUS or Beta-glucuronidase (*uidA*): During 1980s, Richard Anthony Jefferson originally developed the GUS reporter system during his doctoral research work at the University of Colorado at Boulder. He adopted the technique for the use in plants as he worked in the *Plant Breeding Institute* of Cambridge between 1985 and 1987. The enzyme β-glucuronidase commonly known as GUS breaks down glucuronides that imparts a coloured reaction. Hence, it can be detected inside the plant tissue by cutting thin sections. Fig. 19.4 shows GUS expression in rice embryos. In 1996, for the first time the *E. coli uidA* gene was used as a positive selection approach in tobacco encoding I2-glucuronidase enzyme. The Transformed cells turned blue in the presence of substrate, X-gluc (Fig. 19.4). When it was paired with 7.5–15 mg L^{-1} benzyladenine N-3 glucuronide, the transformation efficiency scored by shoot transformation was increased by 1.7-2.9 times more than what was achieved by *nptII* gene in control.

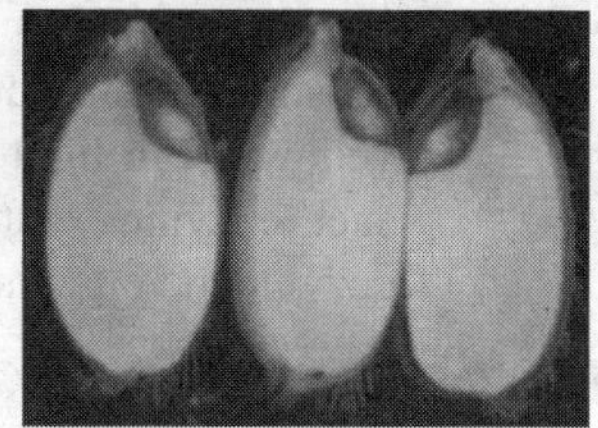

Fig. 19.4. GUS expression in rice embryos.

There are different types of glucuronides that can be used as substrates for the β-glucuronidase depending on the type of detection required (histochemical, spectrophotometrical, fluorimetrical). The most common substrate for GUS histochemical staining is 5-bromo-4-chloro-3-indolyl glucuronide (X-Gluc); in this case the product of the reaction is a clear blue colour. The other commonly used substrates are: *p*-nitrophenyl β-D-glucuronide (PNPG) (for the spectrophotometrical assay) and 4-methylumbelliferyl-β-D-glucuronide (MUG) (for the fluorimetrical assay). The other substrates for glucuronidase are: 5-bromo, 4-chloro, 3-indolyl/glucuronide (BCIG), naphthol AS-B1 glucuronide (NAG) and resorufin glucuronide (REG).

Advantages and Disadvantages of using GUS System

The enzyme GUS is coded by a gene *uidA* which was first isolated from *E. coli*. The major advantage of using this scoreable marker (reporter gene) lies in its assay which requires no DNA extraction, electrophoresis or autoradiography.

Since there is no detectable GUS activity in higher plants, mosses, algae, ferns, fungi and most bacteria, the assay is perfectly suited for these organisms. Thus it is widely used in plant science.

The disadvantage in using the GUS assay is the destruction of plant material. Hence, several kinds of GUS reporter gene assay are actually available depending on the substrate used.

If an organism has no β-glucuronidase or the activity is very low (*i.e.* background activity), it is suitable for a GUS assay. For this reason the assay is not useful in most vertebrates and many molluscs.

(*ii*) **Luciferase (Luc, Lux):** Luciferase offers several advantages as a reporter including the capability of monitoring patterns of gene expression non-destructively in real time with great sensitivity. For example, this allows the continuous monitoring of gene activity during development. There are two types of luciferase markers, firefly luciferase (Luc) and bacterial luciferase (Lux), but the bacterial luciferase differs from firefly luciferase. In bacteria, luciferase Lux consists of two peptide subunits encoded by gene *luxA* and gene *luxE*, but in firefly it is made up of single polypeptide encoded by a gene *luc*. In these two organisms, the enzyme luciferase uses different substrates for luminiscence. Therefore, the assay for detection of presence of luciferase also differs depending upon the source of luciferase gene. Moreover, both *luc* and *lux* are conditional non-selectable marker genes (Li *et al.* 1992).

- **Firefly luciferase (Luc):** The light emitted by fireflies function for communication during courtship, involving the emission of a flash by one sex as a signal, to which the other sex responds in a species-specific pattern. The *luc* gene responsible for expression of the enzyme luciferase is isolated from *Photinus pyralis*. Luciferase is a generic term for the class of oxidative enzymes used in bioluminescence. In *Photinus* the luciferase Luc is encoded by a transcript of about 1,800 nucleotides, which is synthesised from a single copy *luc* gene containing six introns. Luciferase Luc enzyme is a 62 kDa protein.

Luciferase catalyses the production of light at 560 nm. The enzyme luciferase oxidises its substrate (firefly luciferin) in the presence of Mg^{++}, ATP and molecular oxygen emitting yellow-green light (Fig. 19.5). Luc luciferase shows sequence homology with several such enzymes that utilize ATP in adenylation reactions. Due to the presence of Luc enzyme the organisms become capable of glowing in dark.

$$\text{d-Luciferin} + \text{ATP} + O_2 \xrightarrow[Mg^{++}]{\text{Firefly Luciferase}} \text{Oxyluciferin} + PP_i + \text{AMP}\ CO_2 + \text{light (560 nm)}$$

Fig. 19.5. Bioluminescent reactions catalyzed by firefly luciferase.

The firefly Luc catalyses the two-step bioluminescent oxidation of luciferin to oxyluciferin in the presence of magnesium, ATP and oxygen. In the first step of reaction, the substrate is activated in the presence of ATP and Mg^{++} to form an adenylate intermediate. In second step, luciferyl adenylate reacts with oxygen to form oxyluciferin emitting yellow-green light:

Luciferin + ATP+ Mg^{++} ⟶ Luciferyl adenylate +PPi

Luciferyl adenylate + oxygen ⟶ Oxyluciferin + AMP + yellow-green light

When excess substrate is added to firefly Luc, the reaction produces a flash of light that is proportional to the quantity of luciferase in the reaction mixture. Then the light emission decays to 10% of the peak level within 1 minutes.

The nucleotide sequence of the luciferase gene from the firefly *Photinus pyralis* was determined from the analysis of cDNA and genomic clones. There are six introns in luciferase gene; all are less than 60 bases in length. Due to its short half life *in-vivo,* this gene is not destructive to the plants and, therefore, reflects real-time gene expression status of the transgenic tissue under investigation. Luciferase reporter vectors have been constructed where each vector contains a *cis* acting DNA binding element that is recognised by a specific transcription factor (Fig. 19.6).

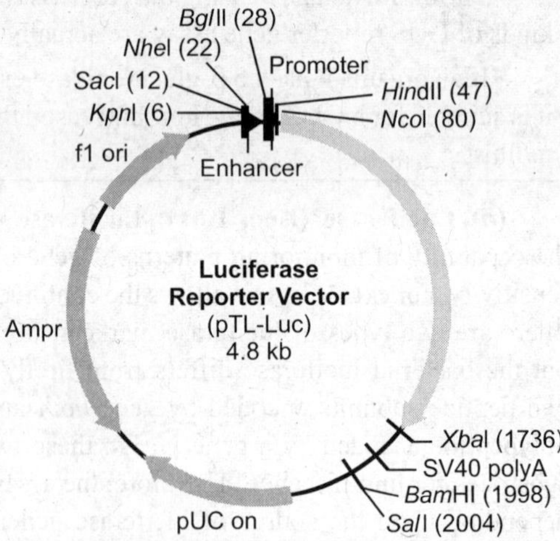

Fig. 19.6. Luciferase reporter gene containing cloning vector.

The binding activity of transcription factors can be monitored and quantified. At this site binding results in the expression of firefly luciferase. Light emitted from the chemical reaction is directly proportional to the amount of expressed enzyme and thus the binding activity of the targeted transcription factor. This reporter vector is an ideal for validating the results obtained by using protein/DNA arrays. The use of the luciferase reporter vectors allows the activity of transcription factor to be monitored in living cells.

For the study of gene expression luciferase reporter gene acts an important tool that allows the *in vitro* and *in vivo* measurement of gene expression from any endogenous genetic control element. Commonly luciferase is used as a reporter to assess the transcriptional activity in cells transfected with a genetic construct containing the luciferase gene under the control of a promoter. Luciferase is also used to detect the level of cellular ATP in cell viability assays or for kinase activity assays. Additionally proluminescent molecules that are converted to luciferin upon activity of a particular enzyme can be used to detect enzyme activity in coupled or two-step luciferase assays.

By using the firefly luciferase gene as a means of monitoring, promoter activity would offer several advantages over the bacterial chloramphenicol acetyltransferase (CAT) gene that is commonly used for this purpose in eukaryotes.

In 1986, the scientists of Iowa State University inserted a genetic structure from fireflies into a tobacco plant which started to glow without exposing to UV light (Fig. 19.7). Glow was caused by the reaction of the enzyme luciferase and pigment luciferin. Unlike the green fluorescent protein, the firefly-derived glow does not require UV light to fluoresce.

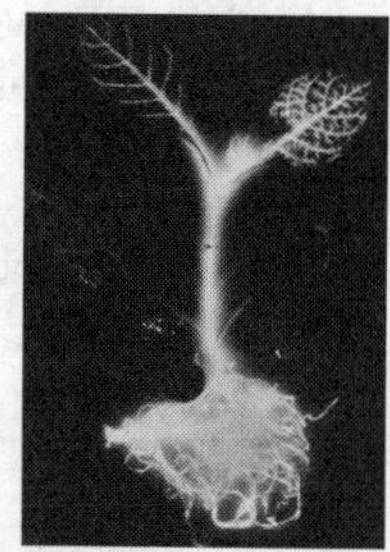

Fig. 19.7. Firefly luciferase gene introduced into tobacco DNA via a virus.

- **Bacterial luciferase (Lux):** Luminous bacteria (e.g. *Vibrio harveyi, V. fisheri*, and *Photobacterium phosphoreum*) continuously emit blue-green light. However, a single bacterium may emit 10^4 or 10^5 photons/second. Luminous bacteria are ubiquitous in the oceans and can be isolated from samples of seawater, from surface to about 1,000 metre depth. But they do not grow in seawater because it is a poor growth medium. The *lux* gene isolated from *Vibrio harveyi* has also been tested successfully in plants. Often *lux* gene is used with other marker genes as a control and also used as a visual marker for selection of transgenic material.

Reduced flavin mononucleotide is the substrate (luciferin) that reacts with oxygen in the presence of Lux to produce an intermediate peroxyflavin. Thereafter, this intermediate reacts with a long-chain aldehyde (tetradecanal), and forms the acid and the luciferase-bound hydroxy flavin. Since the flavin forms the emitter, it acts as substrate luciferin. The enzyme systems maintain the supply of tetradecanal and the oxidised flavin is recycled. Pathway of the bacterial luciferase reaction is shown in Fig. 19.8.

Substrate ⟶ NADH ⟶ Cytochromes ⟶ O_2

Flavin reductase

$FMNH_2$

O_2

AMP + PP

$NADP^+$

RCHO

Luciferase

Myristic acid reductase

ATP

NADPH

RCOOH

R, N, N, O, $E_{\tilde{x}}$, NH, N, OH, O (*)

Light + H_2O + FMN

Fig. 19.8. Bacterial luciferase reaction pathway.

Luciferase reacts with reduced flavin from the electron transport pathway by using molecular oxygen and forms a flavin-peroxy intermediate. Then in the excited state, oxidation with long-chain aldehyde hydroxy-FMN occurs resulting in light emission. Thereafter, acid and FMN products are recycled.

The *lux* genes cloned from several different bacterial species exhibit sequence similarities that indicate evolutionary relatedness. In *Vibrio fischeri*, the *lux* operon has five structural and two regulatory genes with established functions (Fig. 19.9).

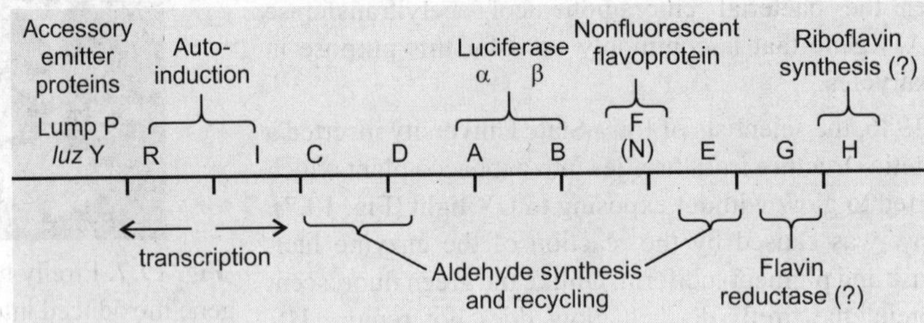

Fig. 19.9. Organisation of the *lux* genes in *Vibrio fischeri*.

These include *luxA* and *luxB* that code for α- and β- subunits of luciferase. The *luxC*, *D* and *E* encode for the reductase, transferase and synthetase components of the fatty acid reductase complex. In the same operon, the regulatory gene *luxI* is upstream and adjacent to it but transcribed in the opposite direction is *luxR*, whose product is responsible for the transcriptional activation of *luxA* through *E*.

***(iii)* Green fluorescent protein (GFP):** GFP is a small protein of 238 amino acids (26.9 KDa), first isolated from the jellyfish *Aequorea victori*, which is commonly used as a marker in biotechnology research. The protein readily fluoresces green under blue light without altering the cell. This led the researchers to insert the gene into DNA and use it for multifarious purposes. Originally GFP was characterised as part of a publicly funded research grant by Osamu Shimomura of Princeton. In 1994 Martin Chalfie from Columbia University explained the method of inserting green GFP gene into several cell types and its expression. Thereafter, the GFP gene was inserted into several different species to observe the patterns of gene expression or cell development. GFP has played a great role in research in all the three systems (plants, animal and microbes). In 2008, O. Shimomura, Martin Chalfie, and Roger Y. Tsien got Noble Prize for the discovery of GFP.

O. Shimomura, Martin Chalfie, Roger Y. Tsien (left to right) who won Nobel prize in 2008.

This reporter gene is used for the study of sub-cellular localisation of proteins inside the cell and to determine the zygosity of transgenic plants. GFP fluorescence is a reliable and quantitative reporter of underlying differences for gene expression in individual eukaryotic cells when fluorescence is measured by flow cytometry. This conclusion is supported by two evidences that: (*i*) GFP fluorescence increases in direct proportion to the GFP gene copy number delivered to cells by a replication-defective adenovirus vector, Ad.CMV-GFP, and (*ii*) the intensity of GFP fluorescence is directly proportional to GFP mRNA in cells. The induction of GFP gene expression from two inducible promoters (the TRE and ICP0 promoters) is readily detected by flow cytometric measurement of

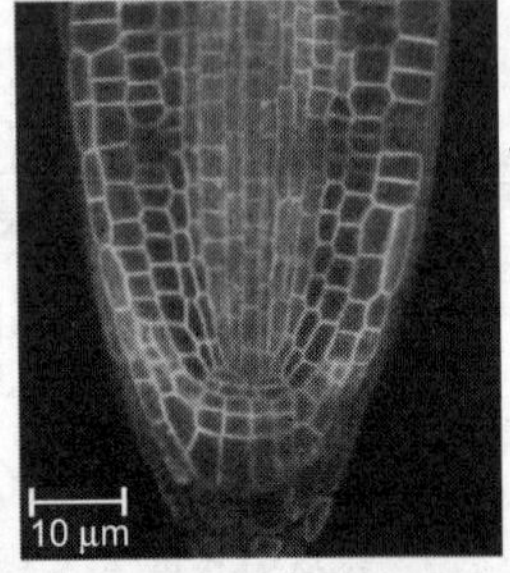

Fig. 19.10. Expression of a plasma membrane protein by *Arabidopsis* root tips.

GFP fluorescent intensity. Expression of a plasma membrane protein fused to GFP by *Arabidopsis* root tips is shown in a confocal image as given in Fig. 19.10.

(*a*) Selectable Markers: A selectable marker is a type of reporter gene introduced into a cell (a bacterium or cells in culture) that confers a trait suitable for artificial selection. They indicate the success of a transfection or introduction of a foreign DNA into a cell. Often the selectable markers are antibiotic resistance genes. Bacteria containing such foreign DNA plus a selectable marker are grown on a medium containing an antibiotic; such bacterial colonies that contain foreign genetic material express and grown in antibiotic-containing medium. The selectable markers allow the transgenic cells/tissue to survive and multiply even at lethal concentrations of selective chemicals such as antibiotics. The marker gene-containing cells are capable to survive under selective conditions where the other cells lacking that specific gene die. Only few cells are transformed by the foreign DNA among the millions of cells. Every individual cell cannot be checked, therefore, a selective agent is required to select the desired cells only, eliminating the non-transformed cells.

The commonly used selectable marker genes include those conferring resistance to the antibiotics such as kanamycin, chloramphenicol and hygromycin, and broad range herbicides such as glyphosate (modified versions of the enzyme EPSPS, 5-enolpyruvate shikimate-3-phosphate synthase isolated from *E. coli* or *Salmonella typhimurium*), phosphinothricin (*bar*, isolated from *Streptomyces hygroscopicus* that codes for phophinothricin acetyltransferase), etc.

Measurement of Expression of Reporter Gene

Enzyme activity assay of the expressed enzyme is encoded by the reporter gene using chromo, fluoro, luminogenic substrate. Immunological assay of the expressed protein is encoded by the reporter gene. Histochemical staining of cells or tissue to localise enzymatic activity is expressed from reporter gene construct cells.

A marker should be selected on the basis of the plant species and other considerations, because each selectable marker presents some favourable and some unfavourable characteristics. The desired marker gene is fused with a suitable plant promoter along with the gene being transferred and placed in a suitable vector. The selectable markers are of the following two types:

(*i*) Antibiotic resistance markers: The *CAT*, *nptII*, and *hptII* antibiotic resistance markers have been discussed as below:

- **Chloramphenicol acetyl transferase (*CAT*) gene:** The transposon Tn9 of *E. coli* is the source of this gene that codes the CAT enzyme; this enzyme catalyses the transfer of the acetyl groups from acetyl coenzyme A to chloramphenicol. It has been used for selection of tobacco transformants with the *cat* gene driven by the *nos* promoter. Selection on kanamycin conferred by the *npt*II gene is more efficient than that on chloramphenicol. Due to its inefficiency the use of the *cat* gene as a selectable marker becomes limited; therefore, this enzyme is no longer widely used as a reporter gene. Moreover, the *CAT* gene containing the *nos* promoter has also been introduced in the tobacco chloroplast genome by *Agrobacterium*-mediated transformation under selection with chloramphenicol.
- ***nptII* gene:** The neomycin phosphotransferaseII (*nptII*) gene (also known as *neo*) isolated from transposon Tn 5 (*E. coli* K12 strain) is most commonly in plant transformation. The gene *nptII* encodes for aminoglycoside 3′ phosphotransferase enzyme which confers resistance by inactivating through phosphorylation of a range of aminoglycoside antibiotics such as kanamycin, neomycin, puromycin, G418, etc. The recipient cell becomes able to grow in the presence of the antibiotic due to the presence of this gene; therefore, the cells become kanamycin-resistant. Here the NPT II controlling elements are bacterial in origin; therefore, for plant transformation *nptII* gene requires

eukaryotic controlling elements. The *npt*II was first used to construct chimeric genes for constitutive expression in plants by fusing it with the 5′- and 3′- regulatory sequences of the *Agrobacterium tumefaciens* T-DNA gene nopaline synthase (*nos*). In addition, the tolerance level differs plant to plant; for example, the selection of transformed tobacco cells is done on a medium containing 50-350 mg/l of kanamycin sulphate. Due to *npt*II, kanamycin resistance has been widely used as a selectable marker in many dicot plants; but in monocots the value of this gene is limited due to their tolerance to the high levels of antibiotics (Miki and McHugh, 2004).

- ***hptII* gene:** Hygromycin B is an aminocyclitol antibiotic inhibitor of protein synthesis; it has a broad spectrum activity against prokaryotes and eukaryotes. This antibiotic is very toxic to plants. The *E. coli* gene *aphIV* (*hph, hpt*), coding for hygromycin B phosphotransferase (HPT), confers resistance to bacteria, fungi, animal cells and plant cells by detoxifying hygromycin B via an ATP-dependent phosphorylation of a 7-hydroxyl group. The hygromycin phosphotransferase gene (*hpt*) was isolated from *E. coli*, which codes enzyme hygromycin phosphotransferase (HPT) (also called aminoglycoside 4-phosphotransferase) that inactivates hygromycin B. HPT catalyses the phosphorylation of hydroxyl group in the antibiotic hygromycin B and inhibits the binding to ribosomes. Hygromycin B is more toxic than kanamycin and kills the sensitive cells more quickly. Therefore, hygromycin B is preferred for selection of a wide range of plants including monocots. For selection of plant cells, its concentration varies from 20 to 200 mg/ml. After kanamycin, hygromycin B is the second most widely used antibiotic for selection, and *hpt* is the second most prevalent antibiotic selectable marker listed in the database of US field trials (Miki and McHugh, 2004).

(*ii*) Herbicide resistance marker: There are many herbicide-resistance markers used for research, a few of them are discussed below:

- ***Bar* gene:** The *bar* gene was originally isolated from *Streptomyces hygroscopicusi* that confers resistance to the herbicide bialaphos (bar). This gene also confers resistance to the herbicide phosphinothricin. This gene encodes the enzyme phosphinothricin acetyl transferase (PAT) which acetylates phosphinothricin (PPT) that is a component of bialaphos. Generally, glutamine synthetase (GS) incorporates ammonia into protein in the normal cells and, therefore, maintains ammonia level in cells. Phosphinothricin is a competitive inhibitor of GS, hence it blocks the GS activity. Ammonia accumulation takes place due to inhibition of GS that reach to the lethal level. Therefore, the plants become resistant to PPT through the transfer of *bar* gene in cells. The transformed plant cells that express *bar* gene grow normally on lethal concentrations of phosphinothricin, whereas the non-transformed cells do not grow, become necrotic and die within 10-21 days. For the selection of transformed cells phosphinothricin is used at a concentration of 1- 10 mg/l.
- **EPSP synthase gene:** The plastid enzyme 5-enolpyruvyl shikimate-3-phosphate (EPSP) synthase is the target of glyphosate (*N*-[phosphonomethyl]glycine) that is a broad-spectrum herbicide and the active ingredient of the commercial Roundup® formulations. It is involved in aromatic amino acid biosynthetic pathway. Glyphosate is an inhibitor of EPSP synthase which is essential in the shikimate pathway for the biosynthesis of the aromatic amino acids. This enzyme is synthesised in petunia flowers at a high level and is rarely found in its leaves. But it is synthesised both in flowers and leaves in tomato. Sequences between -1800 and -800 of EPSPS gene are responsible for tissue-specific expression of this gene in petunia.

Miki and McHugh (2004) have reviewed the selectable marker genes used in transgenic plants describing a number of mechanisms occurring for glyphosate resistance for example: (*i*) over expression of a petunia EPSP synthase gene using the 35S promoter that results in glyphosate tolerance in transformed petunia, (*ii*) expression of mutant forms of the EPSP synthase gene *aroA* in *Salmonella typhimurium* or *E. coli* targeted to chloroplasts that confer glyphosate resistance to tobacco, (*iii*) a naturally-glyphosate-resistant EPSP synthase gene from the *A. tumefaciens* strain CP4 fused to the transit peptide sequence of *Arabidopsis* EPSP synthase for chloroplast targeting that has conferred glyphosate resistance to several crop species, and (*iv*) catabolism of glyphosate to glyoxylate and minomethylphosphonic acid (AMPA) by bacterial glyphosate oxidoreductase (GOX) targeted to the chloroplast confers glyphosate resistance to several different plants.

- **Acetolactate synthase (ALS) gene:** The *als* gene was isolated from *Arabidopsis thaliana* that encodes for the enzyme acetolactate synthase and provides resistance to the herbicide sulfonylurea. When *als* gene is transferred to crop of interest, it will become resistant to sulfonylurea. A mutant acetolactate synthase (ALS) gene (*csr1-1* gene) isolated from sulfonylurea herbicide-resistant *Arabidopsis thaliana*, was placed under control of a cauliflower mosaic virus 35S promoter (35S). Rice protoplasts transformed with the 35S/ALS chimeric gene regenerate into fertile transgenic rice plants. The northern blot hybridisation analysis demonstrates the effective expression of 35S/ALS gene in transformed cells, and provided about 200-fold greater resistance (at extremely low chlorsulfuron concentrations of 10 nm) to transformed calli than the non-transformed calli. It was found that the 35S/ALS gene is an alternative selectable marker for transformation of rice protoplast and production of fertile transgenic rice. But the mutant form of *Arabidopsis* ALS enzyme operates normally in rice cells. The mechanism of protein transport to chloroplast and ALS inhibition by chlorsulfuron is conserved among plant species as diverse as *Arabidopsis* (dicotyledons) and rice (monocotyledon) (Li *et al.* 1992).

B. METHODS FOR DEVELOPING GENETICALLY MODIFIED (GM) PLANTS

Methods of plant transformation have been described in Chapter 8: *Methods of Gene Manipulation*. Some new methods exclusively on plant transformation are described in this section. Techniques used to manipulate the genetic composition of an organism by adding specific useful genes are called 'genetic modification', and the plants developed through this method are referred to as genetically modified (GM) plants, transgenic plants or genetically engineered (GE) plants.

1. Steps for Developing GM Plants

The main steps involved in the development of GM plants are as below:

(*i*) Isolation of gene(s) of interest: The genes responsible for the desired trait in an organism, for example drought tolerance or insect resistance, are identified and isolated by using the existing knowledge about the structure, function or location on chromosomes.

(*ii*) Insertion of gene(s) into a vector: A vector is a gene transfer tool into the cells of desired hosts. The most commonly used vector for plant transformations is Ti plasmid derived from *A. tumefaciens*. The gene(s) of interest is inserted into Ti plasmid by using recombinant DNA technology. For detail, *see* Chapter 8: *Methods of Gene Manipulation*.

(*iii*) Plant transformation: The transformed *A. tumefaciens* cells containing the modified plasmid with transgene are mixed with plant cells, leaf discs or explants. Some of the cells take up a piece of the plasmid called T-DNA (transferred-DNA). The *A. tumefaciens* inserts the desired genes into one of the plant's chromosomes to form transgenic or GM cells. The other most commonly used method to transfer DNA is particle bombardment (gene gun) where small particles coated with DNA molecules are bombarded into the cell. For detail *see* Chapter 8: *Methods of Gene Manipulation.*

***(iv)* Selection of transformants:** The genetically modified plant cells are differentiated following various methods after transformation, from great majority of cells that have not incorporated the desired genes. Mostly selectable marker genes are used to favour growth of the transformed cells as compared to the non-transformed cells. Genes that confer antibiotic or herbicide resistance are inserted into the vector; the recombinant vector is transferred to the plant cells conferring desired traits. Upon exposing the cells to the antibiotic or herbicide, only transformed cells, *i.e.* cells containing and expressing the selectable marker gene, will survive and the non-transformants fail to grow. By using tissue culture methods, the transformed cells are then regenerated to form whole plants.

***(v)* Regeneration into whole plants:** Whole plants via tissue culture involve the explants placed onto nutrient media that induce development of the cells into various plant parts to regenerate into whole plants (for detail description, *see* Chapter 18: *Plant Biotechnology*). The plantlets are transferred to pots after developing the roots and grown under controlled environmental conditions.

***(vi)* Confirmation of transformation:** Inheritance and normal function of the gene inserted into transformed plants is verified by determining the number of copies inserted, and interference (if any) of the insertion with other genes effecting adversely. Gene expression (*i.e.* formation of mRNA and proteins) is done to ensure proper function of the gene.

***(vii)* Plant performance testing:** Generally plant performance is tested first in greenhouse to determine the desired new trait present in the modified plant, and to ensure that the plant does not have any new unwanted traits. Further field testing of the well performing plants is carried out. The plants are first grown in confined field trials to test the expression of desired traits in the open environment. After confirming the good working of this technology the plants are tested in multi-location field trials to ensure that the crop performs well in different environmental conditions also. The GM crop is considered for commercial production on the basis of good performance in all the trials.

***(viii)* Bio-safety assessment.** In conjunction with testing of plant performance the food and environmental safety assessment are also carried out to ensure that the GM crop does not have any bad impact on human health and environment.

2. Agroinfection and Trasformation

Generally plant viruses are transmitted by insect vectors, but an alternative route for viral infection of plants has been searched out in recent years. Agroinfection is a phenomenon of infecting a host mediated by T-DNA of Ti plasmid of *Agrobacterium*. In 1986, N. Grimsley and coworkers used agroinfection as an alternative method for viral infection of plants by using *Agrobacterium tumefaciens* to transfer infectious viral DNA as a part of T-DNA of Ti plasmid to plants. Eventually, viral DNA is integrated into the T-DNA and delivered into plant cells as *Agrobacterium* T-DNA transfers normally. After infection viral DNA is released to form a functional virus. Agroinfection results in the production of transgenic plant containing integrated viral DNA. Agroinfection with maize streak virus has been done in maize successfully. *Agrobacterium*-based vector system can be used for genetic engineering in cereals. Agroinfection results in production of transgenic cereals, but not better than *Agrobacterium*.

For developing this method, cauliflower mosaic virus (CaMV) was selected because it is the best characterised plant DNA virus and can be introduced into plants via aphids, virus particles, viral DNA or cloned DNA. Systemic infection can be caused in turnip after creation of wound and inoculation with *A. tumefaciens* containing Ti plasmid where more than one genome of CaMV has been placed tandemly in the T-DNA. The viral genome is escaped from such constructed T-DNA present inside the plants. The use of combined Ti plasmid and viral DNA has opened a new way to molecular biology to exploit these approaches.

3. Satellite RNA and its use in Transformation

The species of RNA associated with specific strains of some plant RNA viruses are called 'satellite RNA'; their replication is not necessary. Replication and transmission of satellite RNA depend on the virus, though it is unrelated to viral genome. Severity in disease symptoms is reduced due to the presence of satellite RNA; therefore the satellite RNA can be used to develop resistance against specific viruses. By using this principle, transgenic tobacco plants have been produced which carry a DNA fragment that reduces the severity of disease after transcribing a species of satellite RNA.

In one case, transgenic tobacco plants were produced by using DNA for RNA satellite of cucumber mosaic virus (CMV). The presence of satellite RNA (transcribed from DNA) in transgenic plants led to reduction in disease symptoms when satellite-free aggressive CMV strains were used for infection. These transgenic plants are also protected against tomato aspermy virus. The other transgenic tobacco plants could be produced by involving tobacco ringspot virus, which carried DNA version of satellite RNA of tobacco ringspot virus. The deleterious effect of subsequent tobacco ringspot virus infection does not arise in these transgenic plants due to protection provided by satellite RNA. Use of satellite RNA strategy is advantageous over coat protein because satellite RNA is not constitutively expressed like coat protein, but is expressed only after virus infection.

C. HERBICIDE RESISTANT TRANSGENIC PLANTS

The shikimic acid pathway is present in plants, fungi, and bacteria, but not in animals. The products of this pathway are precursors to the synthesis of the three aromatic amino acids (such as phenylalanine, tyrosine, and tryptophan) as well as other essential aromatic compounds in bacteria, fungi, and plants. Animals have no way to synthesise these three aromatic amino acids, which are essential nutrients in animal diets.

1. EPSP Synthase

The enzyme 5-enolpyruvylshikimate 3-phosphate synthase (EPSPS) is involved in shikimate pathway that catalyses phosphoenol pyruvate (PEP) and shikimate-3-phosphate (S3P) to form 5-enolpyruvylshikimate-3-phosphate (EPSP) and inorganic phosphate. The EPSP synthase (EPSPS) enzyme is critical in the metabolic pathway leading to development of the three aromatic amino acids-phenylalanine, tyrosine and tryptophan. All plants synthesise the amino acids needed for their growth and function (Fig. 19.11).

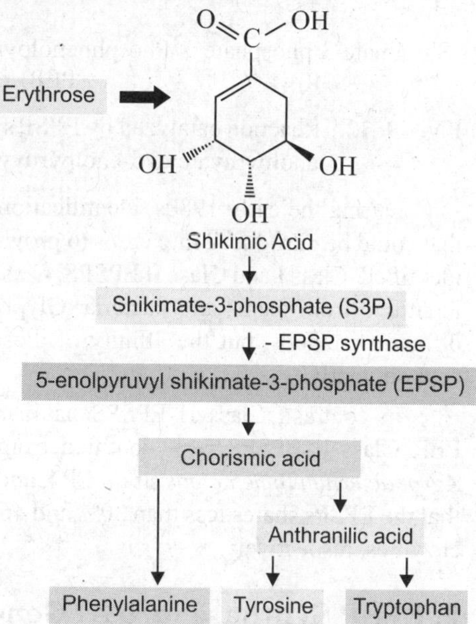

Fig. 19.11. Shikimic acid pathway produces three amino acids- phenylalanine, tyrosine and tryptophan.

EPSP synthase is synthesised at a high level in the flowers of petunia and is hardly found in leaves, but in tomato it is synthesised both in flowers and leaves.

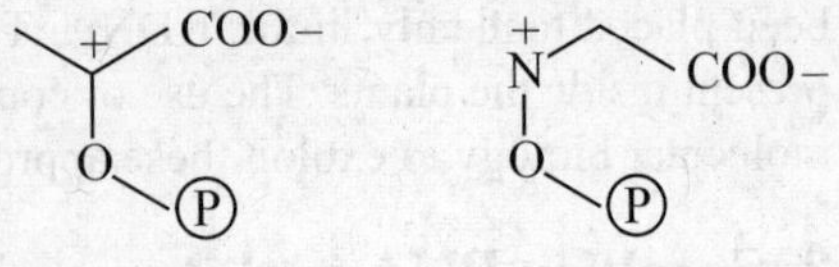

On the other hand, there are a large number of herbicides used to kill the obnoxious weeds that competes with the crop plants. One of the most widely used, broad-spectrum and safest herbicides to humans and the environment is the synthetic compound glyphosate (N-phosphonomethylglycine) (marketed as 'Roundup™ and Touchdown™)

PEP oxonium ion Glyphosate

Inhibition of EPSP synthase by glyphosate.

that has been extensively used for control of weeds throughout the world. The glyphosate is a competitive inhibitor to this enzyme, effectively shutting down aromatic amino acid biosynthesis and also synthesis of other aromatic compounds derived from these amino acids. Consequently, there is an unregulated flow of carbon into the pathway causing high levels of accumulation of S3P. But glyphosate binds to 5-enolpyruvyl shikimate-3-phosphate synthase (EPSPS), inhibiting EPSP synthase by mimicking the PEP oxonium ion and occupying the PEP-binding site (Fig. 19.12). Thus it blocks the pathway and kills all plants including food crops.

The enzyme EPSPS is the target of the non-selective herbicide glyphosate that is an analogue of PEP. It is a competitive inhibitor of EPSPS and binds in a ternary complex of EPSPS–S3P–glyphosate. A blocked metabolic pathway is critical for synthesis of the amino acids and eliminates the plant's ability to make required proteins and enzymes resulting in plant death. Therefore, much attention was paid to search the glyphosate-resistant enzymes produce transgenic crops.

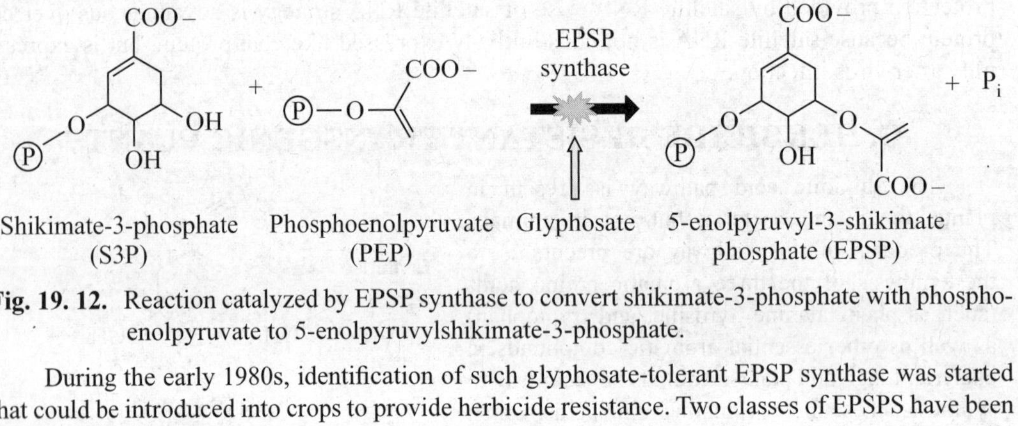

Shikimate-3-phosphate (S3P) + Phosphoenolpyruvate (PEP) —EPSP synthase (inhibited by Glyphosate)→ 5-enolpyruvyl-3-shikimate phosphate (EPSP) + P_i

Fig. 19. 12. Reaction catalyzed by EPSP synthase to convert shikimate-3-phosphate with phosphoenolpyruvate to 5-enolpyruvylshikimate-3-phosphate.

During the early 1980s, identification of such glyphosate-tolerant EPSP synthase was started that could be introduced into crops to provide herbicide resistance. Two classes of EPSPS have been identified: Class I and Class II EPSPS. Class I EPSPS are naturally sensitive to glyphosate, generally identified from plants and bacteria. Glyphosate-tolerant Class I enzymes have been produced by different methods, but the affinity of those glyphosate-tolerant enzymes for PEP often decreases concomitantly.

In contrast, Class II EPSPS has a natural tolerance to glyphosate and a high affinity for PEP. Class II EPSPS was identified from some bacteria (e.g. *Pseudomonas* sp. strain PG2982, *Agrobacterium tumefaciens* strain CP4, and *Staphylococcus aureus*). Phylogenetic analysis revealed that the EPSPs shares less than 30% and 46% amino acid similarity with Class I and Class II EPSPS enzymes, respectively.

2. EPSP Synthase (*aro A*) Gene

Rice EPSP synthase gene resides 45-kb downstream of the waxy (*wx*) gene on the short arm of chromosome 6. The EPSP synthase gene is located at 45-kb downstream of the *wx* gene (Fig. 19.13). Both genes are encoded in the same strand. This fragment also includes the ribosomal protein small subunit 40 (*rps40*) gene and an unknown transcription product. Genomic hybridisation patterns show that the EPSP synthase gene exists as a single copy in the genome of Japonica rice.

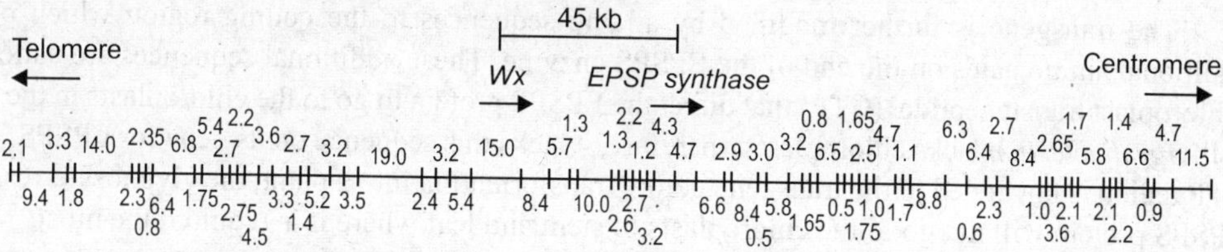

Fig. 19.13. Location of EPSP synthase gene in the rice genome on the short arm of chromosome 6. The short vertical lines are HindIII sites and number correspond to length (kb) between nearest sites (source: T. SATO and co-workers of Hokkaido University, Japan).

3. Microbial EPSP Synthase (*aro A*) Genes

In the early 1980s, several enzymes were identified through selective evolution, site-directed mutagenesis, and microbial screens. Glyphosate-insensitive EPSP synthases were explored and introduced into crops to provide herbicide resistance. However, an increased tolerance for glyphosate in EPSP synthase was often accompanied by a decrease in the affinity of enzyme for PEP, resulting in decreased catalytic efficiency. More favourable kinetic characteristics were observed in some enzymes with substitutions including Pro-101–Ser and Thr-97–Ile (numbering according to *Escherichia coli* EPSP synthase). The naturally occurring glyphosate-tolerant microbes were identified, including *Agrobacterium* sp. strain CP4, *Achromobacter* sp. strain LBAA, and *Pseudomonas* sp. strain PG2982. The enzymes isolated from these bacteria were designated as Class II EPSP synthases on the basis of their catalytic efficiency in the presence of glyphosate and their substantial sequence variation compared with EPSP synthases from plants or *E. coli*. The *Agrobacterium* sp. strain CP4, isolated from a waste-fed column at a glyphosate production facility yielded a glyphosate-resistant EPSP synthase suitable for the production of transgenic crops tolerant to glyphosate. The other Classes II EPSP synthases were from Gram-positive bacteria, such as *Streptococcus pneumonia* and *Staphylococcus aureus*.

Similar to EPSP synthases from *E. coli* and *S. pneumoniae*, the CP4 EPSP synthase exists in an open, unlinked state and a S3P liganded state, suggesting an induced-fit mechanism with binding of S3P as a prerequisite for the enzyme's interaction with PEP. The overall structure of the CP4 EPSP synthase is more similar to the enzyme from *S. pneumoniae* than to the enzyme from *E. coli*. In the binary complex with S3P, the structure of active site is highly conserved, even between the CP4 and *E. coli* enzymes. A single residue in the active site (Ala-100) renders the CP4 EPSP synthase insensitive to glyphosate, whereas a highly conserved Gly residue is found at this position in known natural plant and bacterial enzymes.

4. EPSPS-resistant GM Plant (Roundup Ready)

Agrobacterium sp. strain CP4 was discovered that contained an alternate form of the EPSPS enzyme called CP4-EPSPS that is not inhibited by glyphosate. The two genes that code for the enzyme differ between slightly. A single point mutation in the gene switched the nucleotide guanine for cytosine, which causes the amino acid alanine to be substituted for glycine and prevents glyphosate from binding the enzyme; this allows the normal function of shikimate pathway. The bacterial *CP4-EPSPS* gene that codes for the enzyme CP4-EPSP synthase was modified by adding a promoter that is recognised by plants. A transgene has a viral 35S promoter (from cauliflower mosaic virus), combined with EPSP synthetase gene (Fig. 19.14).

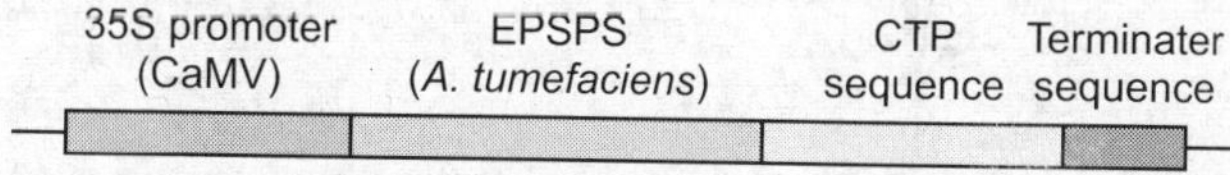

Fig. 19.14. Ti plasmid of *Agrobacterium tumefaciens* 'A transgene containing 35S promoter of CaMV, CTP (chloroplast transit peptide) sequence, and a terminator sequence

The transgene is further modified by adding sequences to the coding region which places additional amino acids on the end of the EPSPS enzyme. These additional sequences are called the 'chloroplast transit peptide' (CTP) that direct the EPSPS protein to go to the chloroplasts in the plant cell. *Agrobacterium* lacks chloroplasts; therefore, it lacks this sequence on its gene for EPSPS. If the CTP region is not added to the transgene before transformation into a plant, the glyphosate-resistant EPSPS protein will not go to the chloroplasts of stem and leaf where it is required the most.

Then the transgene is inserted into a modified Ti plasmid of *A. tumefaciens* at multiple cloning site (MCS) to get a recombinant plasmid (Fig. 19.15). The modified Ti plasmid consists of Amp^r region required for selection on ampicillin-containing medium. Then the recombinant Ti plasmid is transferred back into *A. tumefaciens* cells. Thus the transformed cells of *Agobacterim* are co-cultured with leaf discs of the desired plant (e.g. soybean, tobacco, potato, petunia, maize, cotton, etc.) in tissue culture medium. Eventually, *A. tumefaciens* infects cells of cultured plant discs and transforms the desired plant cells. Thereafter, the transformed discs are grown for about 24 hours on the selective medium impregnated with antibiotic. The selective medium allows only those plant cells to grow that have acquired DNA from the bacteria. Integration of the transgene into the plant chromosome via *Agrobacterium* allows selection of glyphosphate-resistant calli. Plants regenerated from these calli are glyphosphate-resistant, periodic spraying of fields with the herbicide kills weeds but does not affect plants. Protein levels and glyphosate resistance in transformed cells are tested. The rooted seedlings are grown in potting mix that contain transgene which resist the herbicide glyphosate.

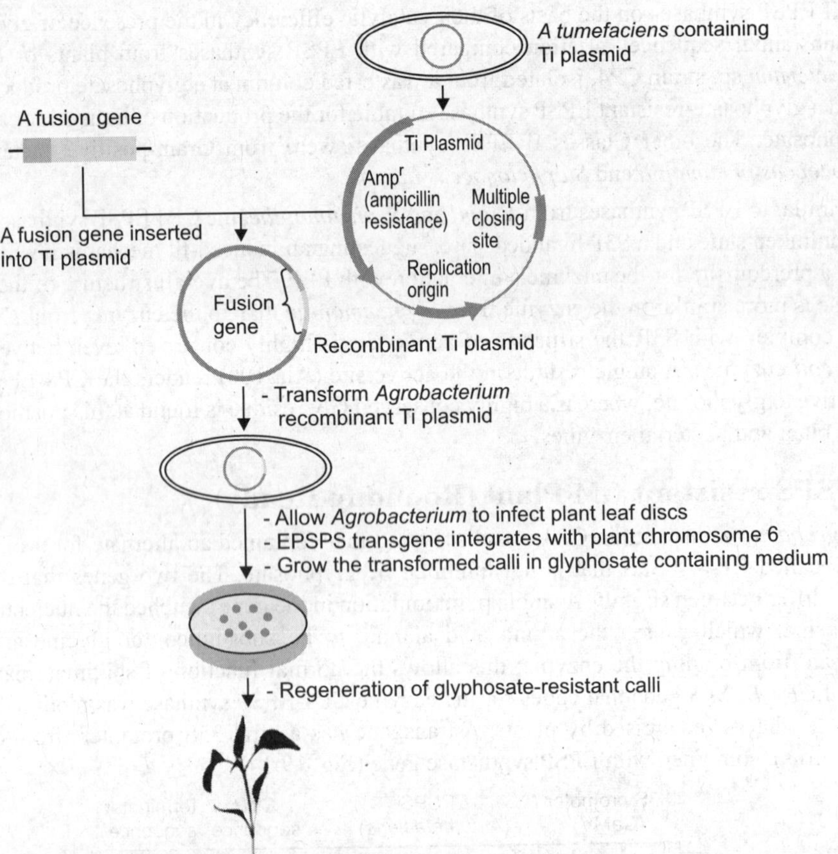

Fig. 19.15. Production of transgenic herbicide glyphosphate-resistant plant.

In 1999, L.H. Chen and co-worker of the China Agricultural University (Beijing) reported the transformation of common wheat (*Triticum aestivum*) with herbicide-resistant *EPSPS* gene. They transferred the herbicide-resistant *EPSPS* gene into about 1,000 young spikes and 800 young embryos of wheat variety (Jinghua 1) by using gene gun method. The regenerated plants were screened with glyphosate, and analysed by PCR and/or Southern blotting. It was found that *EPSPS* gene got integrated stably into the genome of Jinghua 1, and some of the transformants were fertile. Therefore, herbicide-resistant *EPSPS* gene could be used as selective marker in the transformation of monocotyledon cereal crops like wheat.

D. VIRUS-RESISTANT TRANSGENIC PLANTS

Many agricultural crops seriously suffer from viruses resulting in reduced growth, and yield of products and their quality. Infection of a plant with a strain of virus producing only mild effect can be protected against infection that is called cross-protection. Though the mechanism of cross protection is not fully known, it can be assumed that a particular viral encoded protein is responsible for the protective effect. Transgenic virus resistance is the most advanced applications of biotechnology for the management of plant pathogens. Research work on improving virus disease resistance in papaya (*Carica papaya*), tomato (*Solanum lycopersicum*), pepper (*Capsicum annuum*) and citrus (*Citrus sinensis*), is underway at various stages of development. Viruses have some attractive features to act as vectors: (*i*) they infect cells of adult plants, (*ii*) they produce large number of copies per cell leading to gene amplification, and production of the recombinant protein in large quantities, and (*iii*) they spread systematically throughout the plant.

Moreover, the virus genomes do not integrate into the plant genome; therefore, they cannot be used to produce stable and heritable transformations. But they can be used to express transgenes with a view to either improve the phenotypic performance of host plants or to produce large quantities of valuable proteins.

I. VIRUSES AS CLONING VECTOR

Rapid advances in techniques of molecular biology have resulted in the cloning and sequence analysis of the genomic components of many plant viruses. A majority of plant viruses have (+) ssRNA as genomes; there are two such viruses that have great potential as vectors are, brome mosaic virus (BMV) and tobacco mosaic virus (TMV). In addition, the greatest progress has been made with the two groups of viruses having DNA as genetic material, for example caulimoviruses and geminiviruses.

Depending of sources of genes used, there are mainly two approaches for developing genetically engineered resistance. The genes can either be from the pathogenic virus itself, or non-pathogen-derived sources. The first strategy is called *pathogen-derived resistance* and the second as *non-pathogen-derived resistance*. Comparative performance of transgenic virus-resistant plants is given in Table. 19.2.

Table 19.2. Comparative performance of transgenic virus-resistant plants.

Crops	***Viruses***	***Transgene***	***Yield increase (%)***
Papaya	Potato ringspot virus (PRSV)	PRSV CP	90
Potato	Potato virus X (PVX), Potato virus Y (PVY)	PVX + PVY CP	38
Squash	Zucchini yellow mosaic virus (ZYMV), watermelon mosaic virus 2 (WMV2)	ZYMV CP	77

Squash	" "	ZYMV + WMV2 CP	90
Squash	" "	CMV + ZYMV + WMV2 CP	97
Tomato	Tobacco mosaic virus (TMV), cucumber mosaic virus (CMV)	TMV CP	40
Tomato	TMV, CMV	CMV satellite	14

1. Caulimoviruses

Caulimoviruses are double stranded DNA containing viruses that usually have unenveloped particles which can be either bacilliform or isometric. The viral particles are about 35-50 nm in diameter and 900 nm in length (for the bacilliforms) or 45-50 nm in diameter and icosahedral symmetry (for the isometrics) (Fig. 19.16). The virion shell is covered with virion associated protein (VAP). Caulimoviruses are DNA pararetroviruses that replicate in plants via a RNA intermediate evolved from LTR retroelements. In 1985, Howard M. Temin coined the term 'pararetrovirus' to define animal viruses (*Hepadnaviridae*) and plant viruses (*Caulimoviridae*) that differ from retroviruses on the basis of their DNA genome and no regular integration into the host genome for replication.

In 1995, A.R. Mushegian and R.J. Shepherd of the University of Kentucky (U.S.A.) have reviewed the genetc elements of plant viruses as a tool for genetic engineering. Caulimovirus genome is a monopartite, open circular dsDNA of about 8,000 bp in size flanked by direct terminal repeats internally in an inverted form, contain an intergenic poly (A) region and single stranded discontinuities or gaps at specific sites of both strands. One (-) strand of dsDNA contains a single interruption, whereas the other (+) strand can be interrupted at one, two or three sites. These interruptions are site-specific and are thought to arise as a result of the reverse transcription mechanism. The breaks in DNA are repaired upon entry of the virus into the host nucleus. Consequently, a covalently closed virus minichromosome transcribed by a host RNA polymerase II-like enzyme is formed.

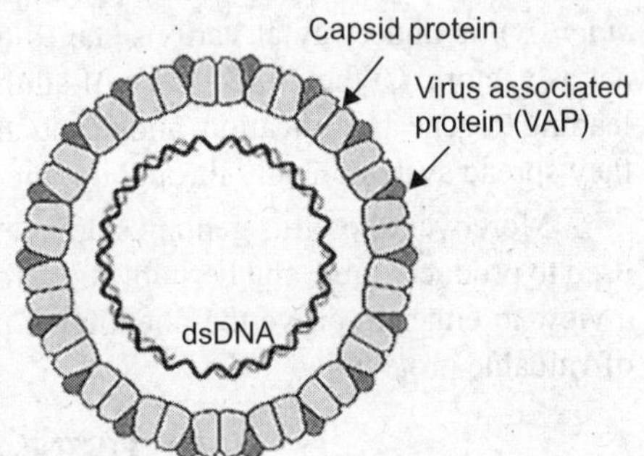

Fig. 19.16. Morphology of caulimovirus showing capsid protein, virus associated protein and dsDNA (*source*: Swiss Institute of Bioinformatics, 2011).

Their genomes are flanked into seven open reading frames (ORFs) and thus encode for 6 to 7 proteins. The other genes code for the movement (MP) protein, the aphid transmission factor (ATF), the virus associated protein (VAP) and the transactivator protein (TAV) as well as diverse additional genes that vary depending on the lineage (Fig. 19.17). This shows that the most likely origin of caulimoviruses was chimeric, *i.e.* a hybrid between LTR retrotransposons and other RNA viruses.

Caulimovirus virion interacts with a cellular receptor, and enters the cell. The dsDNA genome is transported to the nucleus, where it is transcribed by host RNA polymerase II. Replication involves two phases: transcription of an RNA intermediate from the viral DNA in the nucleus and then reverse transcription of this 35S RNA in the cytoplasm into new dsDNA genomes. The genome of pararetroviruses also contains a sequence complementary to a plant $tRNA^{Met}$ that corresponds to the initiation site of DNA replication. Usually this site is located inside or downstream of the large intergenic region (non-coding region) and is generally designated nucleotide 1. Plant pararetroviruses do not require their integration into the host genome for their replication unlike retroviruses; therefore, their genome does not encode the integrase protein.

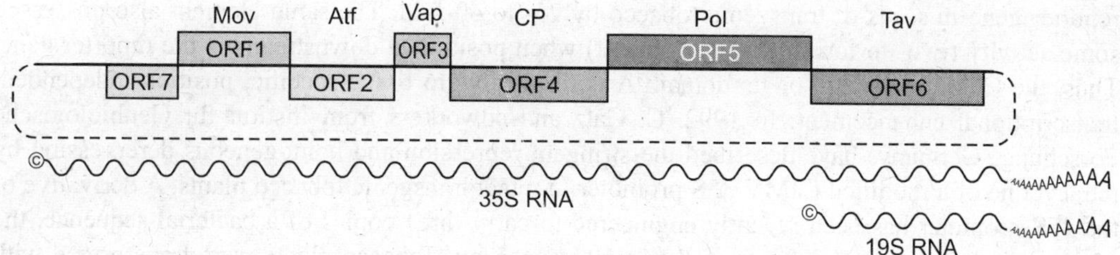

Fig. 19.17. Organisation of CaMV genome (8 kb size). ORF 1 (open reading frame 1), encodes for movement protein (MP); ORF 2, encodes aphid transmission factor (ATF); ORF 3, encodes virus associated protein (VAP); ORF 4, encodes coat protein (CP); ORF 5, encodes polyprotein containing aspartic protease; ORF 6, encodes reverse transcriptase and RNase H domains (Pol) and the transactivator (TAV) (see the positions of two virus promoters (35S RNA and 19S RNA) and polyadenylation signal).

After infection by CaMV, only two major virus transcripts are found *in vivo*. The 35S pregenomic RNA spans the whole genome, while the 19S subgenomic RNA is the messenger for the synthesis of a single polypeptide, the product of CaMV gene VI (Fig. 19.17). The 35S RNA and its spliced derivatives (e.g. 19S RNA) serves as polycistronic mRNA for viral proteins. Interactions between the two promoters (*i.e.* 35S RNA and 19S RNA) depend on their relative position in the DNA. Polycistronic translation depends on Tav which is beleived to be a translation reinitiation factor. Tav is produced from the monocistronic 19S RNA which is the second major viral transcript. Translation of 35S RNA is initiated by ribosome shunt in which scanning ribosomes bypass leader sequence and multiple short ORFs. Genomes are encapsidated with capsid protein, releasing new viral particles.

***(a)* 35S Promoter of CaMV:** In 1990, Philip N. Benfey and Nam-Hai Chua from the Rockefeller University, New York (U.S.A.) have reported the combinatorial regulation of transcription in plants in the presence of 35S CaMV promoter. The 35S promoter of CaMV was initially characterised as a 350 bp DNA fragment spanning the very 39 end of gene VI, *i.e.* the transactivator (TAV) and part of the large intergenic region (Fig. 19.17). This segment confers strong expression in tobacco calli regenerated from the transformed cells when fused to various reporter genes. The sequence from 246 to 18 contains the conventional proximal element of eukaryotic promoters (the TATA box) and is sufficient for the accurate initiation of transcription, at low level as compared with the full-length promoter. The sequence from 290 to 18 (domain A) enables strong expression of a reporter gene in the roots of transgenic tobacco and much weaker expression in the aerial parts of a plant.

In 1995, Mushegian and R.J. Shepherd described that the 35S promoter of CaMV contains modules useful for engineering differentially expressed gene fusions, but the optimal combination of subdomains must be reestablished for each plant species. Similarly, the full-length 35S promoter is constitutive, but this might not be altogether true for every plant or tissue. Further modifications of the 35S promoter of CaMV have created new useful promoters. A DNA sequence element of about 170 bp has been isolated from the promoter of a 9 subunit of the soybean beta-conglycinin gene. This segment was inserted upstream of the 290 segment of the 35S promoter. The resulting chimeric promoter was able to enhance the low expression of a reporter gene in seeds of transgenic tobacco by 25- to 40-fold. The same element also possessed some activity when positioned downstream of the reporter gene. Thus, the CaMV promoter or its A domain is responsive for tissue-specific, position-independent transcriptional enhancement.

***(b)* Production of New Promoters:** New useful promoters can be created by modifying the 35S promoter of CaMV. A DNA element of about 170 bp has been isolated from the promoter of a 9 subunit of the soybean beta-conglycinin gene, which was inserted upstream of the 290 segment of the 35S promoter. The resulting chimeric promoter enhanced the low expression of a reporter gene in seeds of transgenic tobacco by 25- to 40-fold. The same element also possessed some activity (two- to four-fold enhancement) when positioned downstream of the reporter gene. Thus, the CaMV promoter or its domain A is responsive to tissue-specific, position-independent transcriptional enhancement. In 1992, C. Gatz and coworkers from Institut für Genbiologische Forschung (Germany) have described the stringent repression and homogeneous derepression by tetracycline of a modified CaMV 35S promoter in intact transgenic tobacco plants. A derivative of the 35S promoter has been recently engineered to carry three copies of a bacterial sequence, the 19-bp palindromic *tet* operator from the Tn10 transposon. Tobacco plants were transformed with the GUS reporter gene under control of this chimeric promoter and with the bacterial *tet* repressor gene. In such double transgenic plants, the 35S promoter is repressed in the presence of the inducer tetracycline, the promoter is de-repressed and an impressive 500-fold increase in reporter activity is observed.

Similarly, 35S promoter activity in tobacco protoplasts was enhanced by inserting an animal virus-derived DNA element into a plant expression cassette after providing a transcriptional activator in *trans*. The 2343 to 290 activator sequence was tandemly repeated and inserted upstream of the full-length CaMV 35S promoter. The resulting promoter directed 10-fold higher expression of a reporter gene in transgenic tobacco plants than the native 35S promoter. Now, versions of the CaMV 35S promoter with the double enhancer are widely used whenever strong constitutive expression of a transgene is desirable.

2. Geminiviruses

Geminiviruses are being used as autonomously replicating vectors for foreign gene amplification in plants. They are used to analyse the regulation of gene expression in a wide range of hosts, including both mono- and dicotyledonous species by using tissue culture techniques. Geminivirus symptom-induction has been used as a sensitive marker for DNA uptake in monocotyledonous plants that are particularly recalcitrant to transformation.

Geminiviruses are plant viruses that are non-enveloped, about 38 nm in length and 22 nm in diameter, twinned (geminate) incomplete icosahedral capsid made of 110 capsid proteins (CP). Each geminate particle contains only a single circular ssDNA encoding genes that diverge in both directions from a virion strand origin of replication (*i.e.* geminivirus genomes are ambisense) (Fig. 19.18).

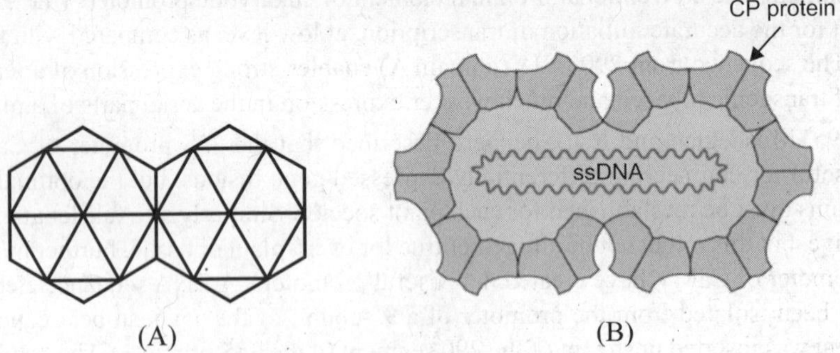

Fig. 19.18. Morphology of geminivirus; (A)- bipartite incomplete icosahedral capsid; (B)- arrangement of capsid protein and ssDNA (*source*: Swiss Institute of Bioinformatics, 2009).

Geminiviruses cause a significant amount of crop damage worldwide, resulting in epidemics due to a number of factors, including the recombination of different geminiviruses co-infecting a plant, which enables novel, possibly virulent viruses to be developed. The other factors that contribute to disease are: the transport of infected plant material to new locations, expansion of agriculture into new growing areas, and the expansion and migration of vectors that can spread the virus from one plant to another.

The genome replicates through rolling-circle mechanism forming double-stranded intermediates. The intergenic region includes a conserved sequence (TAATATTAC) where from ssDNA synthesis is initiated. The genome can either be a single component between 2500-3100 nucleotides or two similar-sized components each between 2,600 and 2,800 nucleotides (e.g. some begomoviruses).

Geminiviruses replicate inside the nucleus; a viral particle penetrates into the host cell, un-coats the virion, and release of the viral genomic ssDNA into the nucleus. Then the ssDNA is converted into dsDNA utilising the cellular factors. Then bidirectional transcription from the IR (inverted repeats) promoter produces viral mRNAs and translation of viral proteins. The (+) strand is cleaved off to initiate the replication by rolling circle mechanism like bacteriophages such as M13 producing ssDNA genomes. New single-stranded DNA form of the virus genome (plus-sense) is formed by interaction of the coat protein with replicating DNA intermediates, as genomes lacking a CP gene do not form ssDNA. The newly synthesised ssDNA can either : (*i*) be converted to dsDNA and serve as a template for transcription/replication (*ii*) be encapsidated by CP and form virions released from the cell, or (*iii*) be transported outside the nucleus to a neighbouring cell through plasmodesmata (cell-cell movement) with the help of viral movement protein.

3. Cucumovirus (CMV)

CMV is a (+) RNA virus consisting of a divided genome. It is non-enveloped, icosahedral, 29 nm in diameter, composed of 180 coat proteins (Fig. 19.19 A)). It has segmented, tripartite linear ssRNA (+) genome which is composed of RNA1, RNA2 and RNA3. RNA4 [1 kb long segmented genome of RNA (sgRNA4)] acts as genome for coat protein (CP). Each genomic segment has a 3' tRNA-like structure and a 5′cap.

RNA 1 is about 3,400 nucleotides long that encodes the 1a protein involved in replication; it has the motifs of a methyl transferase (presumed to provide the 5′ cap structure to the genomic RNAs) and a helicase. RNA 2 is about 3000 nucleotides that encode the 2a protein with a RNA-dependent RNA polymerase motif, and the 2b protein involved in suppression of host gene silencing. RNA 3 (about 2200 nt) encodes the movement protein (MP) required for cell-to-cell movement, and the coat protein that encapsidates the RNA. Sometimes CMV strains harbour a parasitic RNA called the satellite RNA (satRNA), which are small (about 330-400 nucleotides) and do not appear to encode any proteins. They are highly structured having a variety of effects on the virus and the host (Fig. 19.19 B).

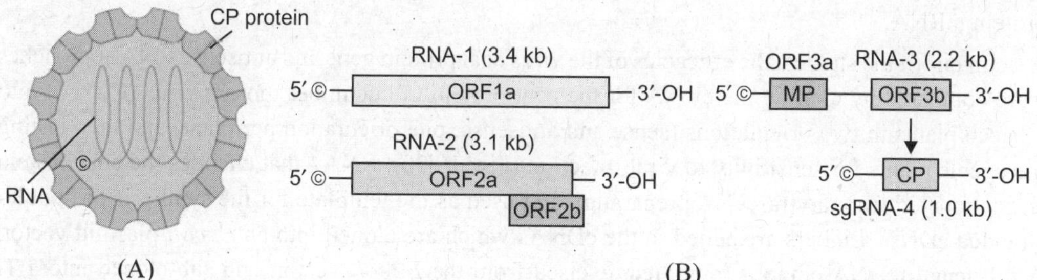

Fig. 19.19. Cucumber mosaic virus; (A)- morphology of viral capsid; (B)- genome organisation.

Cucumovirus shows cytoplasmic replication; virus particle attaches to host receptors, un-coats, and release the viral genomic RNA into the cytoplasm. Expression of protein 1a and 2a takes place that promotes the replication of genomic RNAs. Replication of viral RNA (a negative-sense complementary ssRNA) is synthesised by using the genomic RNA as a template. New genomic RNAs are synthesised by using the negative-sense RNAs as templates. Subgenomic RNA4 is transcribed from RNA3 antigenome, which is translated producing the capsid proteins. Then the new virions are assembled and released after the cell lysis.

II. EXPRESSION OF VIRAL GENES

Several transgenic crops have been developed by introducing a sequence of viral genome in desired plants, such as viral CP gene, MP (movement protein) gene, replicase (*rep*) gene, and satellite RNA, and post-transcriptional gene silencing. Some of these approaches have been discussed in this section.

1. Expression of Coat Protein (CP)

Production of virus-free transgenic plants by using CP gene is one of the greatest successes in crop biotechnology. Several plants have been produced to express CP and show high level of virus-resistance in transgenic crops as compared to non-transgenic crops (Table. 19.3). In 1986, for the first time P. Powell-Abel and co-workers produced a transgenic tobacco by using TMV CP gene that showed resistance against TMV. The transgenic tobacco showed delayed appearance of symptoms and reduced titre of virus in the infected transgenic plants as compared to non-transgenic ones. The resistance against TMV using TMV CP in tobacco was also reported to be effective against the other tobamoviruses whose CP was closely related to that of TMV, but not effective to the other viruses that are distinctly related to TMV. Similarly transgenic potato expressing CP of Potato virus X (PVX) also shows resistance against PVX. Hence, the viral coat protein gene approach has been used to confer tolerance to a number of different plant viruses. With this approach, several virus-resistant transgenic plants have developed for a number of crops. Although complete protection is not usually achieved, high levels of virus resistance have been reported. Moreover, a coat protein gene from one virus sometimes provides tolerance to a broad spectrum of unrelated viruses.

An RNA molecule complementary to a normal gene transcript (mRNA) is called 'antisense RNA'; while the mRNA to be translated is known as sense RNA. The presence of antisense RNA can decrease the synthesis of the gene product by forming a duplex molecule with the normal sense mRNA; hence, its translation is prevented. Thus the antisense RNA—mRNA duplex is also rapidly degraded. This response diminishes the amount of that particular mRNA in the cell. Therefore, it is possible to inhibit plant viruses from replicating and subsequently damaging the plant tissues by producing the transgenic plants which synthesise antisense RNA complementary to virus coat protein mRNA.

In one of the studies, the efficacies of the virus coat protein gene and antisense RNA approaches were compared by cloning the cDNA for the coat protein of cucumber mosaic virus (CuMV) into tobacco plants in two orientations (sense and antisense; one orientation per plant) and then testing transgenic plants for sensitivity to viral infection (Fig. 19.20). RNA4 that encodes the coat protein is fractionated from a viral RNA preparation and used as the template for the synthesis of double-stranded cDNA. Linkers are added to the cDNAs which are cloned into an *E. coli* plasmid vector. A full-length cDNA clone is identified, excised from the *E. coli* vector, and sub-cloned into a Ti plasmid cloning vector between the 35S promoter CaMV and the transcription terminator from the gene for the small subunit of ribulose bis phosphate carboxylase (tRBC). This cloning step creates two orientations for the cDNA of RNA4: in one case, the RNA transcribed is translated into coat protein (sense RNA), and in the other the transcribed RNA is complementary to the mRNA for the coat protein (antisense RNA).

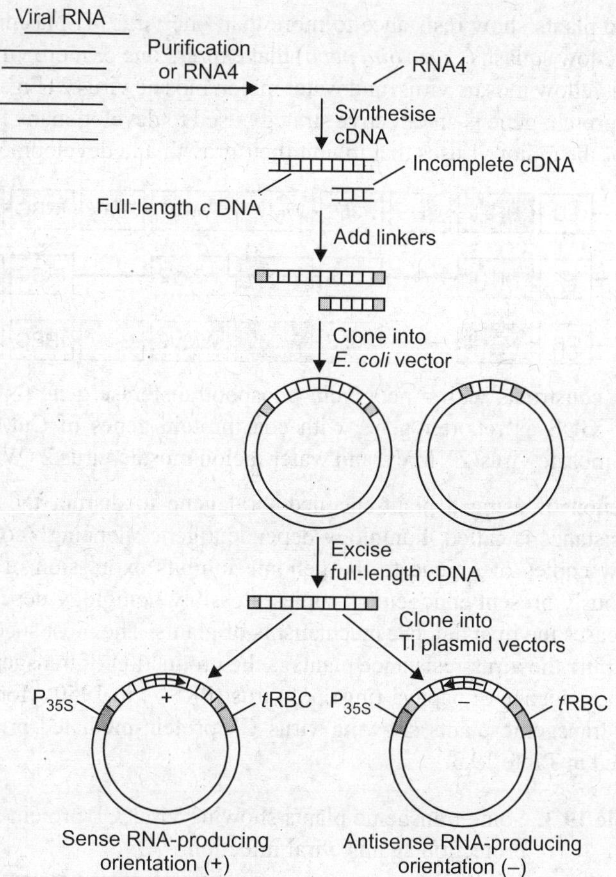

Fig. 19.20. Procedure for putting cucumber mosaic virus (CuMV) coat protein into cucumber.

Both protein–producing sense and antisense RNA producing cDNA sequences are transferred to separate tobacco cells through Ti plasmid binary vector system, from which transgenic plants are regenerated (Fig. 19.21). The transgenic tobacco plants that expressed CuMV coat protein are protected from viral particle accumulated in the cells and do not show any viral infection, irrespective of high or low concentration of the virus inoculum, whereas the cDNA construct having antisense orientation protects the transgenic plants only against low viral doses.

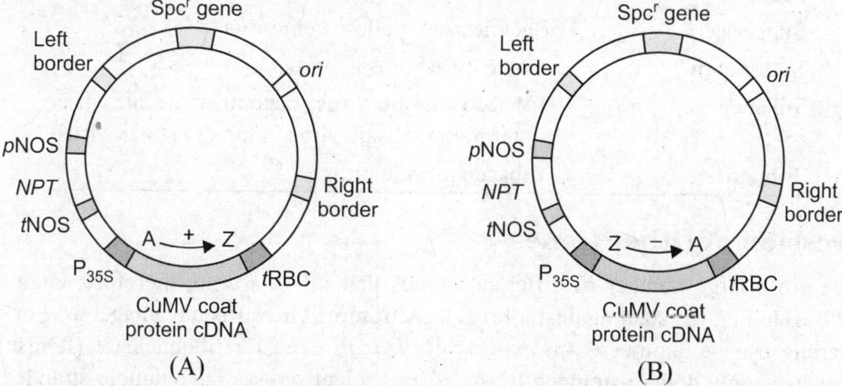

Fig. 19.21. Ti plasmid binary vectors containing CuMV coat protein cDNA; (A)- protein producing sense orientation; (B)- antisense RNA producing orientation.

The transgenic plants show resistance to more than one virus. Ti plasmid binary vectors are used to transform yellow squash (*Cucurbita pepo*) that express one or more viral coat protein genes for CuMV, zucchini yellow mosaic virus, and water melon mosaic virus 2 (Fig. 19.22) . Use of more than one viral coat protein gene is an effective strategy used to develop many transgenic plants that are resistant to all of the major viruses that inhibit their growth and development.

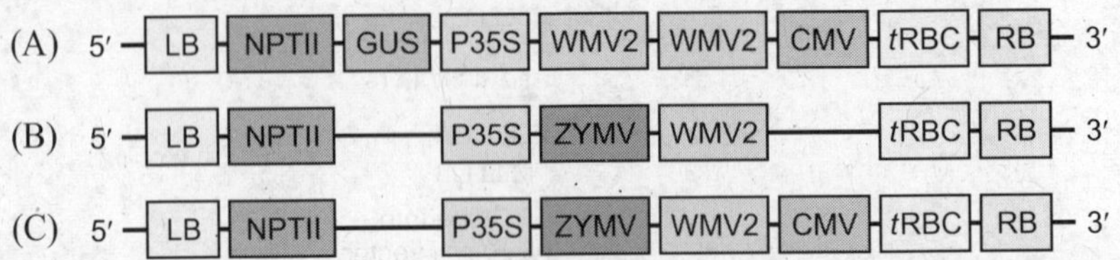

Fig. 19.22. T-DNA constructs with a neomycin phosphotransferase gene (NPTII) as a selectable marker, GUS as reporter gene, with coat protein genes of CuMV (CMV), zucchini yellow mosaic virus(ZYMV), and water melon mosaic virus 2 (WMV2).

The phenomenon of using a plant-encoded viral gene to disrupt the virus life cycle and conferring viral resistance is called 'homology-dependent gene silencing' (co-suppression), where the addition of new copies of a gene to the genome inhibits expression of both the introduced gene and the previously present endogenous copies. Possibly homology-dependent gene silencing is present in some cases the own defense mechanisms of plants. The most successful story of using CP gene for developing the virus-resistance plants is the production of transgenic papaya in Hawaii that suffered a high incidence of papaya ring spot virus (PRSV) in 1950s (for detail see preceding discussion). Some transgenic plants showing virus CP protein-mediated protection against viral infection are enlisted in Table 19.3.

Table 19.3. Some transgenic plants showing virus CP protein-mediated protection against viral infection.

Transgenic plants	***Sources of viral coat protein***
Alfalfa	Alfalfa mosaic virus
Cucumber	Cucumber mosaic virus
Muskmelon	Zucchini yellow mosaic virus
Papaya	Papaya ringspot virus
Potato	Potato virus X, potato virus X
Rice	Rice stripe virus
Sugar beet	Beet necrotic yellow vein virus
Sweet corn	Maize dwarf mosaic virus
Tobacco	Tobacco mosaic virus, cucumber mosaic virus, watermelon mosaic virus 2, tomato streak virus
Tomato	Tobacco mosaic virus

2. Expression of Other Genes

There are a large number of different viruses that infect a crop; therefore, engineered the plants for developing resistant against a broad spectrum of viruses is advantageous. For example, a plant variety can be engineered to express the *E. coli* gene for ribonuclease (RNase) III (*i.e. rnc*) that cleaves only double-stranded RNA, as most plant viruses have double-stranded RNA as their genetic material. After testing, transgenic plants express the *rnc* gene resulting in resistance to several RNA plant viruses. But the *rnc* gene expressing plants become stunted due to the interaction between the plant RNA and the enzyme. This problem can be solved by using a mutant gene (*rnc70*) encoding mutant RNase III that binds to double-stranded RNA without cleaving the substrate. If the mutant *rnc70* gene is introduced into wheat under the control of a corn ubiquitin gene by micro-projectile bombardment, it expresses mutant RNase III. Fig. 19.23 shows a portion of the gene construct used to transform wheat plants with the gene for the mutant form of *E. coli* RNase III. The construct includes the maize ubiquitin promoter, first exon and intron in front of a mutant form of *E. coli* RNase III and a transcription terminator region from a nopaline synthase gene.

5′ — *Ubi* promoter - *Ubi* exon - *Ubi* intron - *Rnc70* gene - TT — 3′

Fig. 19.23. A portion of the gene construct used to transform wheat plants with the mutant RNase III gene of *E. coli*; *Ubi*- maize ubiquitin.

It has been found that the transgenic wheat plants expressing mutant RNase III develop normally and exhibit a high level of resistance to infection by barley stripe mosaic virus. The mutant RNase III was found to bind the replicating barley stripe mosaic virus; consequently viral replication was inhibited.

This is an effective approach to eliminate viroid infection also. Viroids are the circular, single-stranded RNA disease-causing agents. Viroids have a highly base-paired RNA molecules forming double strand-like structure through intra-strand base pairing. Since plant viroids do not encode any proteins, they are difficult to control.

3. Anti-viral Proteins of Other Plants

In addition to expression of viral proteins in the plant cells, protection can also be conferred by antiviral plant proteins. For example, there are three antiviral proteins in the cell wall of pokeweed (*Phytolacca americana*): (*i*) PAPI (pokeweed antiviral protein I) found in spring leaves, (*ii*) PAPII is found in summer leaves, and (*iii*) PAP-S found in seeds. Both PAP and PAPII are ribosome-inactivating proteins and remove a specific adenine residue from the large ribosomal RNA of the 60S subunit of eukaryotic ribosomes. Depending on the season PAP or PAPII is synthesised upon infection of pokeweed plants by viruses, inactivating the ribosomes in the infected cells. Therefore, PAP and PAPI can be said to be a good candidates for developing transgenic plants resistant to various kinds of plant viruses. Thus a cDNA encoding PAP has been isolated and introduced into tobacco and potato plants with binary Ti plasmid vectors under the transcriptional control of the 35S promoter. Less number of lesions could be observed in transformants expressing PAP as compared to non-transformed plants.

III. TRANSGENIC PLANTS

There are several transgenic virus-resistant plants produced through transfer of virus CP protein, movement protein or any other proteins. Some of the transgenic plants have been described as below:

1. GM Papaya

Papaya is consumed in many tropical and subtropical regions, but it has also entered in international markets in recent years. In Jamaica commercial production of papaya has exceeded 20,000 metric tonnes since the mid-1980s. But the papaya industry is threatened by papaya ringspot virus (PRSV) disease transmitted by the aphids (*Aphis gossypii* and *Myzus persicae*). Papaya fruits produce ring-spotting blemishes and the leaves develop mosaic patterns. Consequently, tree vigour and fruit production have got declined.

The CP gene of PRSV has been transferred to papaya to confer virus resistance by using biotechnological methods. In the late 1990s, papaya cultivars were field-tested and the transgenic trees have exhibited varying levels of resistance against virus infections. The selected nutritional components (e.g. vitamins, minerals, acidity and sugars) of the fruits are comparable to those of conventional cultivars of papaya. Safety studies with animal models show no detrimental effects associated with the consumption of the transgenic papaya. The country has prepared to ratify the Cartagena Protocol with the development of the National Biosafety Framework, National Biosafety Policy and National Biosafety Act.

In 1992, papaya ringspot virus (PRSV) was discovered on Hawaii Island, where 95% of Hawaii's papaya was grown. But timely commercialisation of PRSV resistant transgenic papaya trees has revived Hawaii's papaya industry and provides the challenges and opportunities for horticultural biotechnology. In the same year, two cultivars were developed and designated as 'SunUp' and 'Rainbow'. Sun-Up is homozygous for the coat protein gene, while Rainbow is an Fl hybrid of SunUp and the non-transgenic 'Kapoho'. Unfortunately, by October 1994, PRSV had spread throughout the region but a field trial in 1995 showed that 'SunUp' and 'Rainbow' were resistant under heavy pressure of disease.

The ringspot virus resistant genetically engineered papaya introduced in Hawaii in 1998 had been a commercial failure resulting in collapse of papaya industry with a decline of papaya yield by 28%. The transgenic papaya was recently deregulated in Canada, which is a relatively small market for Hawaii. The main challenge is deregulation of transgenic papaya in Japan, where about 30% of Hawaiian papaya was sold. In December 2000, Japan's Ministry of Agriculture, Forestry and Fisheries approved line 55-1. Anticipated approval of transgenic papaya in Japan allowed Hawaiian growers to expand their transgenic papaya markets and to eliminate excessive costs associated with segregating transgenic and non-transgenic fruits.

GM papaya resistant to 'papaya ringspot virus' (PRSV) due to over expression of the virus coat protein (see arrow)

The truth about the failures in Hawaii was also due to the pressure that the US exerted on other countries to permit GE papaya on their land. GE papaya promoters such as the United States Agency for International Development (USAID) and the genetic engineering industry association the International Service for the Acquisition of Agribiotech Applications (ISAAA) claim that it has been a success and they urged countries such as Thailand, the Philippines, and Indonesia to open their farms to the genetically engineered fruit.

In 1997, farmers received an average of US $1.23 per kg, before selling of ringspot virus-resistant GE papayas. In 1998, that figure immediately crashed to $0.89 when traditional buyers of Hawaiian papayas (Japan and Canada) rejected the GE papaya. Hawaiian papaya exports to Japan fell from $10.3 million (in 1998) to $4.6 million (in 2005). Despite Canada's approval of the GE fruit in 2003, the price has never recovered. For the past two years (2004-2005), it has averaged less than $0.80 per kg, at or only marginally above the production cost for many farmers.

Conventional and organic papayas are worth more at market and receive a price premium over the low-valued GE papaya. For example, organic papayas fetched US $3.20 per kg versus $0.99 for the GE papaya in Maui in November 2005. Much of Hawaii's conventional (non-GE) papaya is sold in Japan.

Hawaii's Papaya

2. GM Tomato

Potato virus X (PVX) contains a single infectious genomic RNA of 2×10^{6} daltons; the 3′-end region of the PVX RNA has been cloned and sequenced. This region contains the coat protein gene, which codes for a protein that is 237 amino acid residues in length. A gene construct comprising the coat protein coding sequence of potato virus X (PVX CP) is prepared employing a process similar to that used for engineering TMV constructs.

Potex viral RNAs are polyadenylated, which makes possible an alternative approach to cDNA synthesis by using oligo dT as a primer for first-strand synthesis and DNA polymerase or avian myoblastosis virus reverse transcriptase for the second-strand synthesis. The double-stranded cDNA can be manipulated for isolation and expression in plants of the coat protein sequence as described earlier. The copies of cDNA designated as 'clone p3a', is cloned into the *Pst*I site of pBR322 via the dG-dC tailing method. The final construct contains the cDNA of 18 bp of 5′ non-coding region, 657 bp of coding region of the coat protein sequence (including TAA, the translation termination codon), 72 bp of 3′ non-coding region and 40 bp of dA-dT stretch in pEMBL12(+). The sequence of the PVX coat protein gene is shown in FIG. 15. A horizontal arrow indicates the 5′ boundary of the PVX sequence in p3a. The region derived from synthetic DNA is marked with a wavy line above the sequence. Restriction sites used in construction are underlined.

The full-length cDNA of the PVX coat protein gene is inserted (in both orientations) into expression vectors derived from pMON505, utilising either the CaMV35S promoter (pMON9818) or the ssRUBISCO promoter (pMON9819) and the rbcS-E9 3′ end. The vectors were made to express the PVX coat protein gene, and they were transferred to tobacco plants using the *Agrobacterium*-mediated transformation system.

The transgenic tomato plants expressing the coat protein (CP) genes of cucumovirus have also been tested under field conditions to assess the level of resistance and agronomic performance. Trials were performed in different areas in Italy and the target virus in the majority of tests was spread naturally by the indigenous aphid populations. Twenty-three homozygous lines of variety UC82B, transformed to contain four different CP genes of CMV, were evaluated. The lines were preselected for CP expression, single gene copy, and virus resistance in growth chamber experiments.

In 1999, L. Tomassoli and coworkers demonstrated the resistance of transgenic tomato to cucumber mosaic cucumovirus under field conditions. CMV resistance was confirmed under field conditions though resistance in the field was less effective than what was observed in growth chamber experiments. The resistance observed in multi-year and multi-location experiments is of commercial value for several of the most resistant lines. Engineered resistance upon transfer to Italian varieties by breeding or direct transformation has been used in tomato production in Italy or elsewhere.

E. FUNGUS - RESISTANT TRANSGENIC PLANTS

During the last decade, our knowledge of molecular events occurring during plant–pathogen interactions has expanded. Several strategies have emerged for developing crop varieties resistant to pathogens, for example manipulation of resistance by expression of pathogenesis-related proteins (PRs), antifungal peptides and biosynthesis of phytoalexins. But in these cases the observed resistance was not absolute and was restricted to a limited number of fungi. For long term success of strategies of the antifungal compounds, level of resistance in transgenic plants should be increased and its range should be broadened by isolating new genes and testing new combinations of genes. Interactions between disease resistance (*R*) genes in plants and their corresponding pathogen avirulence (*Avr*) genes are the key determinants of whether a plant is susceptible or resistant to a pathogen attack. Genes involved in *R* – *Avr* interaction have been isolated from several crops and fungus-resistant transgenics are being produced by incorporating the *R* genes in susceptible plants within a genus or a family. The use of transgenic plants, including development of mutants with altered *R* genes, has provided new insights into the mechanisms involved in pathogen perception, signal transduction and subsequent resistance to disease in plants (Gachomo *et al*., 2003).

1. Pathogenesis-related Proteins (PRs)

One of the most efficient defense mechanisms in nature is the hypersensitive reaction to a pathogen that leads to the induction of numerous plant genes encoding defense proteins. These proteins include: (*i*) structural proteins that are incorporated into the extracellular matrix and participate in the confinement of the pathogen, (*ii*) enzymes of secondary metabolism, *i.e.* enzymes of the biosynthesis of plant antibiotics, and (*iii*) pathogenesis-related (PR) proteins which represent major quantitative changes in soluble protein during the defense response.

The PRs have physicochemical properties that enable them to resist to acidic pH and proteolytic cleavage, and to survive in harsh environments such as vacuolar compartment, cell wall or intercellular spaces. Since the discovery of the first PRs in tobacco many other similar proteins have been isolated from tobacco and also other plant species, including dicots and monocots. In 1993, A. Stintzi and coworkers have discussed the plant pathogenesis-related proteins and their role in defense against pathogens. The tobacco PRs have been classified in five major groups (PR-1 to PR-2) on the basis of first serological properties and sequence data. Group PR-1 contains the first discovered PRs of 15-17 kDa molecular mass, whose biological activity is still unknown, but some members have antifungal activity. Group PR-2 contains three structurally distinct classes of 1,3-beta-glucanases, with acidic and basic counterparts, with different specific activity towards linear 1,3-beta-glucans and with different substrate specificity. Group PR-3 consists of various chitinases-lysozymes belonging to three distinct classes; they are vacuolar or extracellular and show differential chitinase and lysozyme activities. Some of them (alone or in combination with 1,3-beta-glucanases) have antifungal property *in vitro* and *in vivo* (transgenic plants) by hydrolysing their substrates as structural components in the fungal cell wall. Group PR-4 contains four members of 13-14.5 kDa of unknown activity and function. Group PR-5 contains acidic-neutral and very basic members with extracellular and vacuolar localisation, respectively; all members show sequence similarity to the sweet-tasting protein thaumatin. Several members of the PR-5 group from tobacco and other plant species inhibit hyphal growth or spore germination of various fungi *in vitro* probably by a membrane permeabilising mechanism. Recently, ten pathogenesis-related protein (PRs) families in plants and their putative functions have been described by E.W. Gachomo *et al.* (2003) as shown in Table 19.4.

A great deal of research has been focused on the isolation, characterisation, and regulation of expression of pathogenesis-related proteins. The constitutive or inducible expression of PR-proteins at effective levels could be used as a tool to enhance yield in areas where pathogens and pests are endemic. PRs in plants analyse the practical aspects of employing PR-proteins for plant protection, in a possible role as the first or last line of defense against pathogens and pests.

Table 19.4. Pathogenesis-related protein (PRs) families and their functions (*Source:* Gachomo *et al.*, 2003).

Protein family	*Reporter protein activity*	*Targeted pathogen sites or components*
PR-1	Pathogenesis-related protein 1 precursor	Membrane
PR-2	1,3-ß-glucanase	Cell wall glucan
PR-3	Endochitinase	Cell wall chitin
PR-4	Endochitinase	Cell wall chitin
PR-5	Osmotin	Membrane
PR-6	Proteinase inhibitor	Proteinase
PR-7	Proteinase	Not defined
PR-8	Endochitinase	Cell wall chitin
PR-9	Peroxidase	Cell wall protein and phenolics
PR-10	RNase	Pathogen-RNA
PR-11	Endochitinase	Cell wall chitin

2. Barnase–barstar Component System

Barnase (**Ba**cterial **R**ibo**N**ucle**ase**) is an extraceullular cytotoxic protein with RNAse activity having a 110-residue that is synthesised and secreted by *Bacillus amyloliquefaciens*. It shows a ribonuclease activity but is lethal to the cell when expressed without its inhibitor barstar, a 90-residue polypeptide. The ribonuclease active site is bound and occluded by the inhibitor, preventing barnase from damaging the RNA of cells before its secretion. The barnase-barstar complex has extraordinarily tight protein-protein binding (Hartley, 1988).

Barstar acts as the inactivator of barnase; it is a small protein synthesised by the bacterium *B. amyloliquefaciens*. It inhibits the ribonuclease activity of its binding partner barnase, and forms a tightly bound complex within the cell until barnase is secreted (Fig. 19.24). Barstar expression is necessary to counter the lethal effect barnase (Hartley, 1988).

(*a*) Mechanism of Catalysis: Barstar binds tightly to barnase through salt bridges and hydrogen bonds, resulting in the inhibition of its RNase activity. After binding the active site of barnase is sterically blocked by barstar with an alpha helix and the loop segment connecting it to the adjacent helix. Formation of the barstar-barnase complex depends on pH as evidenced by the presence of >35 water molecules within 4.5 angstroms of both protein molecules. Small changes occurs in the structure of barnase after binding to barstar. Barstar functions by blocking the active site rather than affecting conformational changes. The changes that occur in barnase are limited to the localized region around the barnase-barstar complex. The two lobes of barstar expand slightly to accomodate barnase after binding.

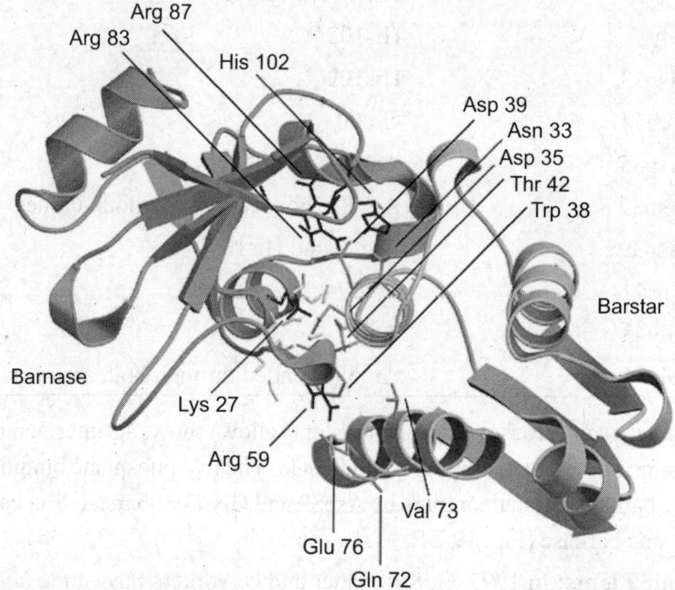

Fig. 19.24. Structure of the barnase–barstar complex (*source:* Lee and Tidor, 2001).

In 1998, D.E. Mossakowska and coworkers characterised the kinetics of recombinant ribonuclease from *Bacillus amyloliquefaciens* (barnase) and investigated the key residues in catalysis by site-directed mutagenesis. They found that Barnase catalyses hydrolysis at diribonucleotide GpN sites. Cleavage occurs in two steps using a general acid-base mechanism: during the first transesterification step a cyclic intermediate is formed, which is then hydrolysed to release the cleaved RNA. The two most important residues involved in catalysis are Glu73 and His102, which are both essential for enzymatic activity. Glu73 is the general base, whereas His102 is the general acid. Although it is not directly involved in acid-base catalysis, Lys27 is also critical for activity; it has been implicated in transition-state substrate binding.

(*b*) Electrostatic Optimisation of Binding Complex: Electrostatic interactions are important for molecular binding and recognition. Structures obtained from analysis of atomic resolution have shown rich protein–protein interfaces in polar and charged groups; these groups generally make interactions between hydrogen bonds. Lee and Tidor (2001) used a novel charge optimisation technique to study the small ribonuclease barnase and to analyse its interaction with a natural tight binding inhibitor, the protein barstar. A continuum model was used in this approach to determine the charge distributions leading to the most favourable electrostatic contribution to binding when competing desolations and interaction effects are included. Table 19.5 shows electrostatic interaction between barnase and barstar through 14 hydrogen bonds between residues.

Barstar is electrostatically optimised for tight binding to barnase when compared with mutants where residues have been substituted with one of the 20 common amino acids. Thus the natural proteins appear to use optimisation of electrostatic interactions as a strategy for achieving tight binding (Lee and Tidor, 2001).

Table 19.5. Electrostatic interaction between barnase and barstar.

Barstar Residue	*Barnase Residue*
Tyr29	Asn84 and Arg83
Gly31	His102
Asn33	His102
Leu34	Glu60
Asp35	Arg59
Asp39	Arg 83 (bonded through both amines)
Asp39	Arg87 and His102
Thr42	Lys27
Gly43	Arg83
Glu76	Arg59 (bonded through both amines)

The complex between barnase (blue) and barstar (yellow) shows 12 interfacial water molecules (grey); the side chains important in binding are indicated. The two phosphate binding sites in barnase are occupied in the barnase-barstar complex by Asp39 and Gly43 of barstar. You can see two barstar proteins for every one barnase (Fig. 19.24).

(c) Transgenic Plants: In 1995, G. Stritmatter and coworkers placed the *barnase* gene under the control of pathogen-inducible potato *prp-1-1* promoter so that *barnase* activity kills the cells at the site of infection. The *barstar* gene was constitutively expressed in all tissues to avoid cell death due to unwanted expression of the *barnase* gene. Cells are killed only if barnase activity is higher than barstar activity. Transgenic potato plants showed severe local necrosis of leaf tissue upon inoculation with *Phytophthora infestans* spores. Symptom development was greatly reduced. However, this strategy was not tested at the field level.

F. INSECT-RESISTANT TRANSGENIC PLANTS (BT TECHNOLOGY)

Since the pests and insects cause substantial damage to our crops, the use of pesticides is a common measure in plant protection programmes. Mostly these pesticides are chemically synthesised. However, during the last 30 years Bt toxins produced by *Bacillus thuringiensis* (Bt) is an exception, where its spore preparation has been used as a biological insecticide (or biopesticide). The genes coding for these proteins were isolated from the bacteria, modified with promoters that would be recognised by plants and inserted into the crop species. Then the plant synthesises the particular *Bt* protein encoded by the gene inserted into that crop. A corn hybrid with a *Bt* gene encodes crystaline proteins from the bacteria that are toxic to insect larvae. When the European corn borer eats the crystaline Cry proteins, it binds to the insect's mid-gut and cause water imbalance in the cells resulting in bursting of cells and killing the corn borer. Moreover, in different crops multiple *Bt* genes can be used to target different insect pests.

1. *Bacillus thuringiensis* (Bt)

Bacillus thuringiensis is a Gram-positive, soil-residing bacterium, which has commonly been used as a bio-pesticide. *B. thuringiensis* also occurs naturally in the gut of caterpillars of various types of moths and butterflies, as well as on the dark surface of plants. During sporulation, many strains of *B. thuringiensis* produce proteinaceous crystal inclusion bodies, which are called δ-endotoxins (Fig. 19.25). The δ-endotoxins, also called Cry (crystal) toxin, and Cyt (cytotoxin) toxin, are pore-forming toxins; hence the Cry toxin crystals may be extracted and used as a pesticide because it has insecticidal action. Therefore, Cry toxin crystals are used as insecticides.

Upon sporulation, *B. thuringiensis* forms crystals of proteinaceous insecticidal δ-endotoxins which are encoded by *cry* genes, located on the plasmid in certain bacterial strains. Cry toxins have specific activities against insect species of the orders Lepidoptera (moths and butterflies), Diptera (flies and mosquitoes), Coleoptera (beetles), Hymenoptera (wasps, bees, ants and sawflies) and nematodes. Thus, *B. thuringiensis* serves as an important reservoir of Cry toxins for production of biological insecticides and insect-resistant GM crops. When insects ingest toxin crystals; the alkaline pH of their digestive tract activates the toxin. The Cry toxin crystals enter into the cell membrane of insect gut and form a pore that results in cell lysis and death of the insect.

(A)

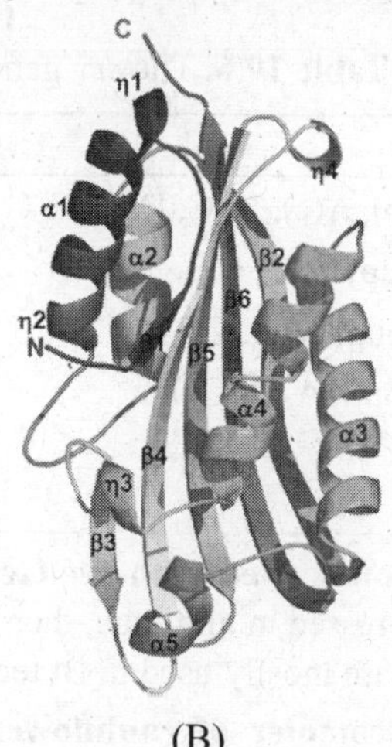

(B)

Fig. 19.25. *Bacillus thuringiensis* subsp. *israelensis* δ-endotoxin; (A)- bipyramidal crystals; (B)- structure of activated Cyt2B(a) monomer.

Different types of *B. thuringiensis* subspecies have been isolated from dead or dying insects mostly from the orders Coleoptera, Diptera and Lepidoptera; but many subspecies have also been isolated from soil, leaf surfaces and other habitats. The carcasses of dead insects often contain large quantities of spores and insecticidal crystal protein that enter the environment. The coleopteran-active and lepidopteran-active Bt subspecies are mainly associated with the soil and leaf surfaces, whereas the dipteran-active Bt subspecies are commonly found in aquatic environments (WHO, 1999).

(*a*) Mechanism of Action of Cry Protein Crystals: When insects ingest toxin crystals, the crystal dissolves in the guts. The alkaline pH and reducing environment in the insect mid gut helps to dissolve the crystal into proto-toxin subunits of ~130,000-Da. The proto-toxin is then digested by pancreatic trypsin-like proteases to form the activated toxin which is half the size of the proto-toxin and is resistant to further trypsin digestion. This trypsin-resistant core binds to specific receptors on the insect mid gut (the central section of the digestive tract in which the processes of digestion and absorption take place) and forms a pore lysing the epithelial cells of the mid gut. These membranes maintain the ion balance of the gut. But the ion balance is destroyed due to disruption of mid gut; therefore, the insect is paralysed and soon dies.

(*b*) Genetics of *Bacillus thuringiensis:* *B. thuringiensis* produces two types of toxins namely: Cry (crystal) toxin encoded by different *cry* genes, and cytolytic (Cyt) toxin. More than 50 genes have been recorded that encode for Cry toxin and they are sequenced for various studies. The various genes and their properties are given in Table 19.6. The crystal (*cry*) gene types encode insecticidal crystal protein of *B. thuringiensis* that is specific to either Lepidoptera (*cryI*), Diptera and Lepidoptera (*cryII*), Coleoptera (*cryIII*), Diptera (*cryIV*), or Coleoptera and Lepidoptera (*cryV*). The commercial Bt cotton available today contain genes from the isolate *B.* thuringiensis, ssp. *Kurstaki* that produces Cry I A(a), Cry I A(b), Cry I A(c), Cry IIA proteins.

In addition, a non-specific cytolytic factor is encoded by the cytolytic (*cyt*) genes present in Bt insecticidal crystal protein and some other Bt subspecies. However, a new nomenclature based on insecticidal crystal protein sequences has been proposed due to the increasing number of characterised insecticidal crystal protein genes and inconsistencies in the existing *cry* gene nomenclature. This new system for identification of Bt employs both the identity of the *cry* genes (defining the host range), and the H-serovars (that define the subspecies). Recently DNA fingerprinting has been used for further characterisation of subspecies (WHO, 1999).

Table 19.6. The *cry* genes and Cry proteins showing insect actively.

Gene	*Crystal shape*	*Protein size (KD)*	*Insect actively*
cry I A(a), A(b), A(c), B, C, D, E, F, G	Bipyramidal	130 – 138	Lepidopteran larva
cry II (subgroup) A, B, C	Cuboidal	69 - 71	Lepidoptera, diptera
cry III (subgroup) A, B, C	Flat irregular	73 – 74	Coleoptera
cry IV (subgroup) A, B, C, D	Biopyrimidal	73 – 134	Diptera
cry V – IX	Various	35 – 129	Lepidopteran and coleopterans larvae

(*c*) Promoters used with *cry* Gene: If a promoter is used with the *Bt* gene inserted into a crop that is expressed in all tissue, then the trait is effective in all plant parts. There are two type of promoters that are mostly used in Bt technology.

(*i*) 35S promoter of cauliflower mosaic virus (or CaMV35S promoter): For most *Bt* transgenic events, the 35S promoter of cauliflower mosaic virus (or CaMV35S promoter) is used, which expresses in all tissue including the pollen (Fig. 19.26 A). When pollen from *Bt* corn drifts to other plants, it causes the death of non-target insects.

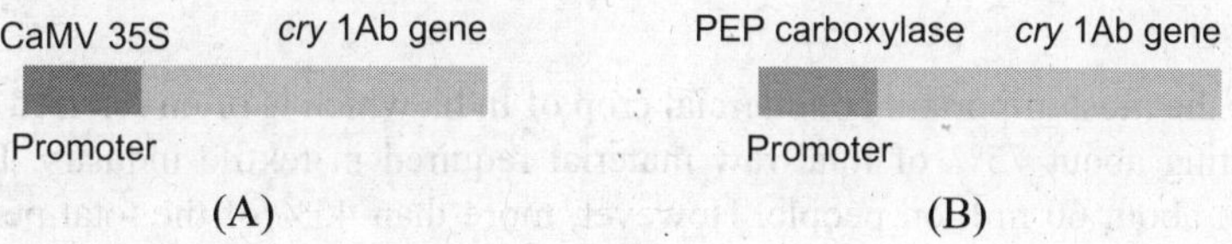

Fig. 19.26. Promoter controls differential expression of *Bt* gene.

(*ii*) Phosphoenolpyruvate (PEP) carboxylase promoter: Phosphoenolpyruvate (PEP) carboxylase promoter from a plant gene is an alternate promoter to CaMV35S that encodes a photosynthetic enzyme. Hence, the *Bt* transgene containing PEP carboylase promoter will express and produce the protein only in green cells where photosynthesis is carried out. Therefore, in Bt corn the Bt trait is not expressed in the root, tassel, or ear tissue where photosynthesis does not occur (Fig. 19.26 B).

(*d*) Transformation of *A. tumefaciens* Cells and Creation of Transgenic Plants: The *cry* gene tagged with desired promoter, for example CaMV 35S, is ligated with Ti-plasmid of *Agrobacterium* to result in recombinant plasmids. Then the plasmids are used for transformation o *A. tumefaciens* cells. The transformants consist of plasmid that contains *cry* gene (Fig. 19.27). The tissue of the desired plant is co-cultured with recombinant *A. tumsfaciens* cell. Eventually, the bacterium infects the cells and delivers *cry* gene which is taken up by the plant cells. Such transgenic plants show resistance towards insect pests, but the normal plants get damaged severely (Fig. 19.28).

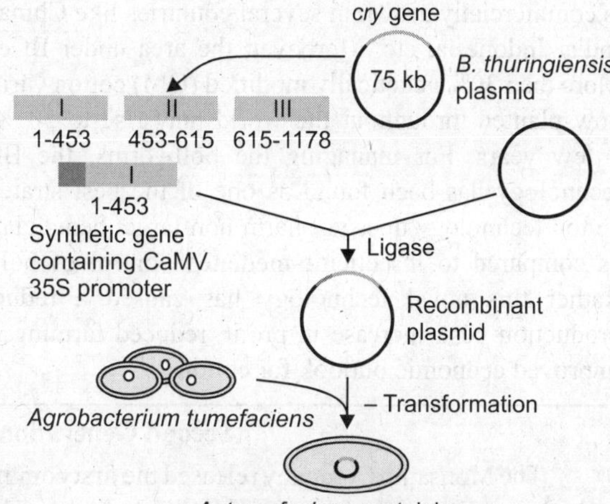

Fig. 19.27. Construction of a recombinant *A. tumefaciens* containing *cry* gene of *B. thuringiensis*.

Therefore, the insect pests attacking these tissues are not controlled because the expression is slow. Moreover, when the plant is completing its life cycle and photosynthesis is reduced, the expression stops toward the end of the season. The above toxin gene (*bt2*) from *B. thuringiensis* has been isolated and used for *Agrobacterium* Ti plasmid-mediated transformation of tobacco, cotton and tomato plants (Fig. 19.28).

The transgenic plants are resistant to the *Manducta sexta*, a pest of tobacco. Experiments of feeding the leaves of these plants to larvae of *M. sexta*, showed 75%-100% mortality of the larvae, while the control plants carrying no transgenes are severely damaged. The presence of the gene, bt2 as well as that of the toxin protein synthesised under its control can also be demonstrated by appropriate experiments.

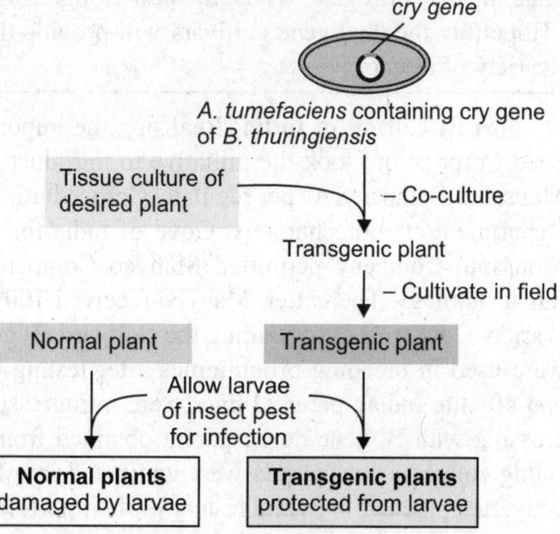

Fig. 19.28. Creation of a transgenic plant: *Agrobacterium*-mediated transfer of *cry* gene into a desired plant and its field trials involving larvae of insect pest.

2. Bt Cotton

Cotton is the most important commercial crop of India which is often referred to as the 'white gold' contributing about 75% of total raw material required in textile industry. It also provides employment to about 60 million people. However, more than 45% of the total pesticides used in our country are consumed by cotton. The important insect pests attacking cotton crop are classified into two broad groups: (*i*) *sap sucking insect pests* (for example, aphids, jassids and white fly), and (*ii*) *leaf chewing insect pests* (such as bollworms, leaf eating caterpillars, etc.). Under severe incidence the boll worms alone reduce the yield up to 40 – 70 %.

The Monsanto Holdings P Ltd (a US based multinational agricultural biotechnology corporation) of the U.S.A. first released the commercial variety of Bt cotton (named Bollgard), which contains *cry1A(c)* gene of *B. thuringiensis*. This was the 'first generation Bt-cotton'. Bt cotton is commercially grown in several countries like China, Australia, Mexico, South Africa, Argentina, India, Indonesia, etc. Moreover, the area under Bt cotton is increasing worldwide year-by-year. More than 30% genetically modified (GM) cotton varieties are now planted throughout the world may rise to 50 % within a few years. For managing the bollworms, the Bt cotton technology has been found as one of the best strategies. Bt cotton technology does not harm non-target beneficial insects as compared to insecticide-mediated control of bollworms. Rather this novel technology has caused a reduction in production cost, increase in profit, reduced farming risk and improved economic outlook for cotton.

Bollworm chewing cotton leaves.

Second Generation Bt Cotton

The Monsanto Company released the first commercial variety of Bt cotton named 'Bollgard', which contains *cry1A(c)* gene of *B. thuringiensis*. This was the 'first generation' Bt-cotton. Then this Company developed the 'second generation' of improved Bt-cotton by introducing the two Bt-genes, which is designated as 'Bollgard II'. Bollgard II cotton was developed by using particle acceleration plant transformation procedures to add the *cryIA(b)* gene to the cotton that already had the *cry1A(c)* gene. Thus, the new Bollgard II cotton contains *cryIIA(b)* and *cry1A(c)* genes. Hopefully the dual gene cultivars will provide the growers with a broader control over a wide variety of insects.

(*a*) Bt Cotton in India: Realising the importance of Bt cotton Mahyco (Maharastra Hybrid Seed Corporation) took the initiative to introduce this technology into India in collaboration with Monsanto Company. As per regulatory procedure, Mahyco sent its application in March 1995 to the Department of Biotechnology, Govt. of India for getting permission to introduce this technology. Monsanto Company permitted Mahyco Company in March 1996 to use this technology. Thereafter, Mahyco received 100 grams of Bt cotton seeds (variety Cocker 312) containing the Bollgard Bt gene (*cryIAc*). These seeds were used in breeding programmes after testing the efficacy of this gene, and 40 elite Indian parental lines were introgressed with *cryI A*(*c*) gene by crossing with Bt gene donor parent obtained from Monsanto. Some of the ruling conventional hybrids were converted into Bollgard cotton using the converted parental lines and tested for their performance.

Bollgard cotton

Nearly 500 field trials were carried out in different agro-climatic regions between 1998 and 2001 to assess the efficacy of MON 531 against bollworms and the concomitant agronomic benefits. Simultaneously, the Indian Council of Agricultural Research (ICAR) conducted 55 multi-location field trials through their network of All India Coordinated Cotton Improvement

Project (AICCIP) for assessment of insecticidal efficacy and economic benefit of Bt cotton (AICCIP, 2002). AICCIP trials clearly showed that the first generation Bt cotton hybrids provided effective control of bollworms, requiring no or fewer applications of insecticides and also provided high economic benefits to farmers (Table 19.7). It was also found that *cry* gene incorporation into Indian cotton did not have any negative effect on fibre quality parameters. On the strength of comprehensive testing, three hybrids namely MECH-12, MECH-162 and MECH-184 were approved by GEAC (Genetic Engineering Approval Committee) in April 2002 for commercial cultivation in central and southern cotton-growing zones of India. This approval was proceeded by a large number of laboratory studies to demonstrate the safety and benefit of Bt cotton as per regulatory requirements (Karihaloo and Kumar, 2009).

The annual value of the biotech cotton crop, based on the sale price of Bt cotton seeds in 2008-09 was estimated to be Rs 1,500 crore (http://www.biospectrumindia.commakesections.asp/0911061.asp). In India, three Bt cotton hybrids have been approved for commercial cultivation during 2002. Moreover, the area for cultivation of such hybrids has sharply increased from 72,000 acre in 2002-03 to 30,00,000 ha in 2006-07. GEAC has approved 59 Bt hybrids in different cotton growing zones so far after extensive field trials by All India Coordinated Cotton Improvement Project Centres and RCGM (Review Committee on Genetic Manipulation) trials in farmer's fields.

In recent years, the effects of Bt crop cultivation on non-target organisms including insect predators, parasitoids and pathogens have been investigated quite extensively. These studies indicate that Bt is rarely directly harmful to beneficial invertebrates and the effects on non-target organisms are negligible in comparison to those of conventional insecticides (Karihaloo and Kumar, 2009).

Table 19.7. Performance of Bt hybrids in ICAR field trials (*Source:* Karihaloo and Kumar, 2009).

Variety/hybrid	*Yield q/ha*	*Gross Income/ ha (Rs)*	*Insecticide Cost/ha (Rs)*	*Additional cost of Bt seeds/ha (Rs)*	*Net Income/ ha (Rs)*
MECH-12 Bt	11.67	21,006	1,727	2,425	16,854
MECH-162 Bt	13.67	24,606	1,413	2,425	20,768
MECH-184 Bt	14.00	25,200	1,413	2,425	21,362
Local control	8.37	15,066	2,845	—	12,221

(*b*) Use of *cry* Gene(s) in Cotton: The currently deployed *cry* genes are from two sources. Monsanto developed and deployed the *cryIA*(*c*) gene in its Bollgard varieties, which are the most widely used in all nine countries that grow cotton. The second source is the Bt-fused gene that was developed by the public sector, Chinese Academy of Agricultural Sciences (CAAS) in Beijing (China). Moreover, genes utilised for the development of transgenic cotton hybrids in India are listed in Table 19.8.

Table 19.8. Genes utilised for the development of transgenic cotton hybrids in India.

Company's Name	*Gene(s) utilised*
Mahyco (India)	*cry1Ac*
Monsanto (U.S.A.)	*cry1Ac+2Ab*
Nath seeds	*cry1Ab+Ac* fusion (China)
JK seeds	*cry1Ac* modified (I.I.T. Khargpur)
Syngenta	*vip3A +cry1Ab*
Dow Agri. Science	*cry1Ac + cry1F,*
Metahelix	*cry1C*
I.C.A.R. (New Delhi)	*cry1Aa3, cry1F, cry1 Ia5, cry1 Ab, cry1 Ac*
N.B.R.I. (Lucknow)	*cry1Ec*

(*c*) **Biosafety Tests and Assessment of Toxicity to Non-target Organisms:** Biosafety tests of Bt cotton were carried out by the National Dairy Research Institute (Karnal), Central Avian Research Institute (Bareily), Industrial Toxicological Research Centre (Lucknow), National Institute of Nutrition (Hyderabad), Central Institute of Fisheries Education (Mumbai), and G.B. Pant University for Agriculture and Technology (Pantnagar, Nainital). These studies have resulted in absolute safety to goats, cows, buffaloes, fish and poultry. Feed-safety studies with Bt cotton seed meal were carried out with goats, buffalos, cows, rabbits, birds and fish. The animals fed with Bt cotton seed meal were comparable to the control animals in various tests and showed no adverse effects.

Bt-cotton expressing Cry1A(c) is non-toxic to all other non-target organisms such as beneficial insects, birds, fish, animals and human beings, because the Cry1A(c) is mainly toxic to the bollworms (cotton bollworm, pink bollworm and spotted bollworm), semiloopers and hairy caterpillars. Laboratory and field studies carried out in India have shown that the Cry1A(c) protein of Bt-cotton did not have any direct effect on any of the non-target beneficial insects. In addition, there has been a reduction in numbers of predators and parasitoids which specialise on the Bt controlled bollworms, but also of increases in numbers and diversity of generalist predators such as spiders. Decrease in population of the parasitoid and predator was due to a decrease in the densities of the pest populations of Bt-cotton. Therefore, farmers had to spray pesticides 3-5 times on Bollgard as compared to 6-8 times on non-Bt cotton. Moreover, there was no increase in numbers of green vegetable bug, aphid or whitefly on Bt cotton. Such adverse effects are very small as compared with the side effects of the spraying of conventional insecticides (Tamil Nadu Agriculture University Publication, Coimbatore, 2008).

(*d*) **Economic Impact of Bt Cotton in Developing Countries:** The most extensive studies of transgenic crop adoption in developing countries have been conducted for insect-resistant (IR) cotton in Argentina, China, India, Mexico and South Africa. Table 19.9 summarises the results from the most comprehensive economic studies of the farm-level impacts of IR cotton in developing countries. Each of the studies was based on data from two or three seasons of commercial farm production. The average percent difference between IR and conventional cotton for all farmers over all seasons covered in the study is shown in Table 19.9.

The economic studies were based on farm-level field trial data, and as such did not reflect the actual farm experience with commercial cultivation. These studies estimated potential yield benefits of 80%. This study found large net gains from IR cotton adoption at the national level in India, although significant variation was observed across states and one state, Andhra Pradesh, experienced negative results. It may be speculated that the lack of locally adapted cultivars was the main reason for poor performance in this state. At the time of the study, the Indian Biosafety authorities had approved only four IR cotton varieties for use throughout the country. By 2005, that number had increased to 20, and the area of the country planted with IR cotton almost tripled from the previous year.

Table 19. 9. Performance advantage of insect-resistant (IR) over conventional cotton expressed as a percentage (*source:* Raney, 2006).

	India	*Argentina*	*China*	*Mexico*	*South Africa*
Yield	34	33	19	11	65
Revenue	33	34	23	9	65
Pesticide costs	- 41	- 47	- 67	- 77	- 58
Seed costs	17	530	95	165	89
Profit	69	31	340	12	299

3. Bt Brinjal

Brinjal (eggplant/aubergine) (*Solanum melongena*) is an important and indigenous vegetable crop of India, where about 2,000 varieties of brinjal are grown throughout the country that have genetic variability (Fig. 19.29). The major brinjal producing states are: Uttar Pradesh, Orissa, Bihar, Assam, West Bengal, Maharashtra, Gujarat, Andhra Pradesh, and Karnataka. The production of brinjal in the year 2005-06 was 9,136,300 metric tons from an area of 553,300 hectares (>1.4 million acres). About 1.4 million farmers cultivate it annually. Brinjal is crucial to India's Rs 72,000 crore vegetable economy.

Currently, India is the second largest grower of brinjal in the world after China with 26% market share. China is ahead by four percent. In India most of the crops are destroyed by the insect pests resulting in huge loss of farmers. More than 40% of the India's brinjal crop is lost due to repeated pest attacks by the insect *Leucinoda orbonalis* or fruit and shoot borer which increases the input costs and keeps the production down. Therefore, the scientists have introduced Bt brinjal to save its yield from bugs which is capable to fight with bugs. The Bt brinjal can minimise this loss to single digits and make the country the largest grower of brinjal in the world. Mahyco is number one seeds company of India that, in collaboration with American multinational company Monsanto, produced genetically modified brinjal by injecting BT gene (http://www.biospectrumindia.com).

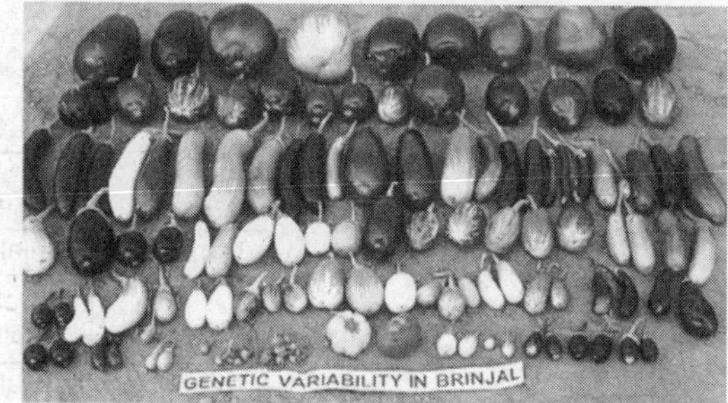

Fig.19.29. Genetic variability in brinjal [*Source*: Report of National Consultations on *Bt* brinjal (http://moef.nic.in/index.php)].

***(a)* Methods of Transformation:** The Bt brinjal technology developed by Monsanto was taken up by Mahyco and the two other public institutions: the Tamil Nadu Agricultural University (Coimbatore) and University of Agricultural Sciences (Dharwad, Karnataka). Bt brinjal was developed by transforming the brinjal proprietary line of Mahyco. The Bt brinjal contains three genes inserted via genetic engineering techniques. First, the *cry1A*(*c*) gene, which encodes for an insecticidal protein Cry1A(c), was derived from the common soil bacterium *B. thuringiensis* subsp. *kurstaki* (Btk). The *cry1A*(*c*) gene is driven by enhanced CaMV 35S promoter. Secondly, the *nptII* gene derived from the prokaryotic transposon Tn5 encoding the selectable marker neomycin phosphotransferase II (NPTII). The *nptII* gene was used to identify the transformed cells that contained the Cry1AC protein. It has no pesticidal properties. The third approach is the use of AAD gene isolated from transposon Tn7 which encodes for the bacterial selectable marker 3-(9)-O-aminnoglycoside adenyl transferase (AAD). It was allowed for the selection of bacteria containing the pMON10518 plasmid on media containing spectinomycin or streptomycin. The *aad* gene is under the control of a bacterial promoter and hence not expressed in Bt brinjal.

Fig. 19.30 shows the steps for the production of Bt brinjal by using *Agrobacterium*-mediated transformation system. The *A. tumefaciens* strain LBA4404 carrying the vector pMON 10518 (which contains *cry1Ac, nptII* and *aad* genes) was used to transform the cultured brinjal cells *in vitro*. Consequently, the T-DNA (that carries *cry1Ac, nptII* and *aad* genes) was transferred into the genome of brinjal cells. After transformation, transgenic brinjal plants were regenerated using kanamycin as the selection agent. At Mahyco all the methods were employed for *Agrobacterium*-mediated brinjal transformation. The regenerated plantlets from tissue culture medium containing kanamycin for selection were selected and analysed for the presence of Cry1AC protein by using ELISA technique. The plants expressing Cry1AC proteins were used further and analysed in subsequent generations to identify lines in which the transgene segregated in the following Mendelian rules. The selected lines were also analysed by Southern blot (http://www.biospectrumindia.com).

***(b)* Mode of Action of Cry1A(c) Protein:** In the transgenic Bt brinjal the *cry1A(c)* gene behaved as a single dominant Mendelian factor indicating stable inheritance of the gene in the plant chromosome. The Cry1A(c) protein must be ingested by lepidopterans insects (brinjal fruit and shoot borer) to be active against them. The Cry1A(c) protein is active in the midgut of insect due to alkaline pH; hence, the protein binds to specific receptors present on the gut cells, inserts into the membrane and forms ion-specific pores. Consequently, the digestive process is disrupted resulting in death of the insect. The Cry1A(c) protein produced in Bt brinjal is non-toxic to non-lepidopteran insects, birds, fish and mammals because these species lack receptors for the proteins on the surface of their gut cells. The gut cells of these organisms remain in acidic medium that makes Cry1AC protein inactive. Besides, NPTII and AAD proteins are used as a selectable marker and have no pesticidal activity to any species.

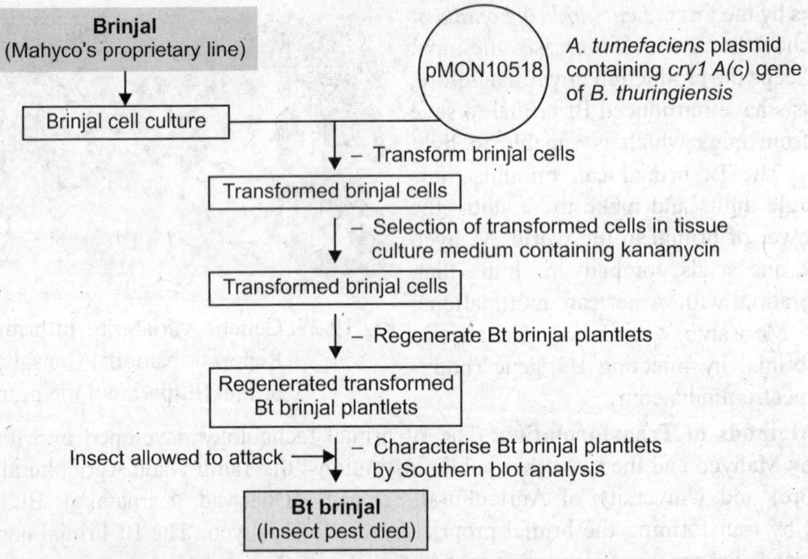

Fig. 19.30. Steps for producing Bt brinjal.

***(c)* Field Trials:** Mahyco regularly conducted multi-location field trials from 2004 to 2006, under the guidance of the Department of Biotechnology and the Ministry of Environment and Forests (MoEF), to evaluate the efficacy of Bt brinjal in controlling brinjal fruit and shoot borer. Field trials of Bt brinjal were conducted at five places in Karnataka, Goa, and Maharashtra. Six popular varieties such as Kudchi, Malapur, Udupi, Gulla, Rabakali, and Goa 112, cultivated locally were genetically modified of which field trials were conducted. The Indian Council of Agriculture Research (ICAR) (New Delhi) independently extensively conducted field trials of Bt brinjal during 2004-05 and 2005-06. A Monitoring and Evaluation Committee (MEC) was framed to evaluate the protocols and supervision of field trial conducted by Mahyco. The experts of MEC regularly visited these trials and submitted its reports to RCGM/GEAC. The insect reactions, yield parameters, insecticide usage, and the economics of Bt brinjal were assessed in these trials (http://www.biospectrumindia.com).

No differences were observed between Bt brinjal and non-Bt brinjal fruit with respect to shape, size and colour. The Bt brinjal was prevented from pest damage as compared with the non-Bt brinjal and yielded higher number of marketable fruits. On October 14, 2009, the GEAC gave the approval to Bt brinjal in a meeting in the MoEF; though three out of 28 members of GEAC were against the approval of commercial release of Bt brinjal variety. But anti-GM activists created the stormy atmosphere throughout the country. Though the GEAC has approved the commercial use of the Bt brinjal, the commercial cultivation is on hold for some more time.

***(d)* Bt Brinjal Controversy:** Monsanto that promoted GM crops in India through Mahyco found itself in the center of this debate. In February 2010, the BT Brinjal was at a stage of controversy. The Union Minister for Environment put the decision of commercial use of BT Brinjal on hold, due to safety concerns attached with the BT brinjal after the failure of genetically modified cotton seeds. Some environmentalists are of the opinion that the use of Bt brinjal on rats could be fatal. Therefore, it would be better to go through research before experimenting it on humans.

Bt brinjal

India is a party to the Convention on Biodiversity and has ratified the Cartagena Protocol (CP); therefore, it is committed to the safe handling of the GMOs. CP provides a broad framework on biosafety especially focusing on trans-boundary movements of GMOs and also covers seeds that are meant for intentional release into the environment, as well as those GMOs that are intended for food, feed or used in food processing.

The first transgenic crop, to be released in India in the year 2002, was the Bt cotton. After its introduction there has been a lot of controversy surrounding Bt cotton. Its performance, impact on the environment, biodiversity and health of cattle has been widely debated. In September 2007, the GEAC announced its approval for large scale field trials for Bt brinjal in India. Some analysts opined that Bt brinjal would have a significant economic impact on farmers, while some raised doubts on its safety and environmental implications.

The Union Minister of Environment and Forest faced a huge outcry from activists, scientists, farmers, and general public. Finally, in March 2010, he put a moratorium on the release of this genetically modified vegetable. It shall be released after more scientific studies and field trials. It shows the crucial failure of the Indian biotechnology sector due to its inability to communicate to the public the pros and cons of the GM technology in an unbiased manner.

Bt brinjal protest in Chandigarh (India)

G. MODIFICATION IN NUTRIENT QUALITY

Transgenic crops with improved nutritional quality have been produced by introducing genes involved in the metabolism of vitamins, minerals and amino acids. A transgenic *Arabidopsis thaliana* has been developed that can produce ten-fold higher vitamin E (alpha-tocopherol) than the native plant. In *A. thaliana* the biochemical machinery that produces a compound close in structure to alpha-tocopherol is present. A gene that can finally produce alpha-tocopherol is also present, but is not expressed. This dormant gene was activated by inserting a regulatory gene from a bacterium which resulted in an efficient production of vitamin E. Similarly, the glycinin gene from soybean has been introduced into rice and successfully expressed because glycinin is a lysine-rich protein of soybean. Therefore, the transgenic rice plants produced glycinin with high contents of lysine in plants.

1. Golden Rice (*Oryza sativa*)

Golden rice is genetic engineered rice produced in 2,000 that biosynthesises beta-carotene (a precursor of pro-vitamin A) in its edible parts, *i.e.* the endospersm. Golden rice was developed for those countries where people suffer from vitamin A-deficiency. But none of these varieties is currently available so far for human consumption. Soon after its development environmental and anti-globalisation activists have started protest. The edible part of rice grains consists of the endosperm, filled with starch granules and protein bodies, but it lacks several essential nutrients for the maintenance of health such as carotenoids that exhibits activity of provitamin A. Using genetic engineering Prof. Potrykus and Dr. Peter Beyer in 2000 have developed rice by introducing three genes involved in the biosynthetic pathway for carotenoid, the precursor for vitamin A. Rice plants synthesise β-carotene in vegetative tissues but not in the grain. All but two steps of the biosynthetic pathways are present in the grain. By addition of only two genes, phytoene synthase (*psy*) and phytoene desaturase (*crt* I), the pathway is reconstituted and β-carotene is consequently accumulated in the endosperm that is the edible part of the grain. The two encoding genes (phytoene synthase and lycopene-beta-cyclase) were from plant (*Narcissus pseudonarcissus,* commonly known as wild daffodil), and phytoene desaturase from a bacterium. Rice plant lack these three enzymes. The first transgene encodes phytoene synthase (*psy*), which utilises the endogenously synthesised geranylgeranyl-diphosphate to form phytoene, a colorless carotene. The second encodes a bacterial carotene desaturase (*crtI*) that introduces conjugation by adding four double bonds. The combined activity of *psy* and *crtI* leads to the formation of lycopene, which is a red compound due to its undecaene chromophore. Lycopene has never been observed in any rice transformant and different genetic backgrounds.

The cDNA coding for phytoene synthase (*psy*) and lycopene ß-cyclase (ß-*lcy*) from wild daffodil was introduced in a single combined transformation effort to obtain a functioning provitamin A (ß-carotene) biosynthetic pathway in rice endosperm. Both the genes were under the control of the endosperm-specific glutelin promoter together with a bacterial phytoene desaturase (*crt*I, from *Erwinia uredovora* under constitutive 35S promoter control). This combination covers the requirements for ß-carotene synthesis and yellow ß-carotene-bearing rice endosperm was obtained in the T_0-generation. The presence of ß-*lcy* is not necessary because *psy* and *crt*I alone are able to drive ß-carotene synthesis and the formation of downstream xanthophylls. These downstream enzymes are constitutively expressed in rice endosperm or are induced by the transformation. Here *trans*-lycopene or a *trans*-lycopene derivative acts as an inductor as it stimulates the endogenous carotenogenic genes. Gene construct used to generate Golden Rice is shown in Fig. 19.31.

Fig. 19.31. Gene construct used to generate Golden Rice. RB, T-DNA right border sequence; *Glu*, rice endosperm-specific glutelin promoter; *tpSSU*, pea ribulose bis-phosphate carboxylase small subunit transit peptide for chloroplast localisation; *nos*, nopaline synthase terminator; *Psy*, phytoene synthase gene from *Narcissus pseudonarcissus* (GR1) or *Zea mays* (GR2); *Ubi1*, maize polyubiquitin promoter; *Pmi*, phosphomannose isomerase gene from *E. coli* for positive selection (GR2); LB, T-DNA left border sequence.

Alpha- and beta-carotene are found together with variable amounts of oxygenated carotenoids such as lutein and zeaxanthin. The carotenoid pattern observed in the endosperm revealed that the pathway proceeded beyond the end point expected from the enzymatic action of the two transgenes alone (Fig. 19.32).

Thus, rice got enriched in pro-vitamin A. The aim was to help millions of people who suffer from night blindness due to Vitamin A deficiency, especially whose staple diet is rice. In 2000, X. Ye and coworkers from Swiss Federal Institute of Technology (Switzerland) including I. Potrykus and P. Beyer (University of Freiburg, Germany) engineering provitamin-A (beta-carotene) biosynthetic pathway into (carotenoid-free) rice endosperm to improve its nutritional value. The presence of β-carotene in the rice gives a characteristic yellow/orange colour, hence this provitamin-A enriched rice was named as 'Golden Rice'. In 2005, Jacqueline A. Paine and coworkers hypothesised that the daffodil gene encoding phytoene synthase (*psy*) was the limiting step in β-carotene accumulation.

Through systematic testing of other plant *psys*, they identified a *psy* from maize that substantially increased carotenoid accumulation in a model plant system. They developed 'Golden Rice 2' by introducing this *psy* in combination with the *Erwinia uredovora* carotene desaturase (*crtI*) that was used to generate the original Golden Rice. Golden Rice 2 produced about 23 times more (maximum 37 μg/g) β-carotene than the original variety of golden rice.

***(a)* Protest against Release of Golden Rice:** The activists oppose the release of any genetically modified organisms into the environment. They feel that golden rice is a Trojan horse that will open the door to more widespread use of GMOs. Vandana Shiva, an Indian anti-GMO activist, argued the problem was not that the crop had any particular deficiencies, but that there were potential problems with poverty and loss of biodiversity in food crops. These problems are aggravated by the corporate control of agriculture based on genetically modified foods. By focusing on a narrow problem (vitamin A deficiency), Shiva argued that the Golden rice proponents were obscuring the larger issue of a lack of broad availability of diverse and nutritionally adequate sources of food. Other groups have argued a varied diet containing foods rich in vitamin A such as sweet potato, leafy green vegetables and fruit would provide children with sufficient vitamin A.

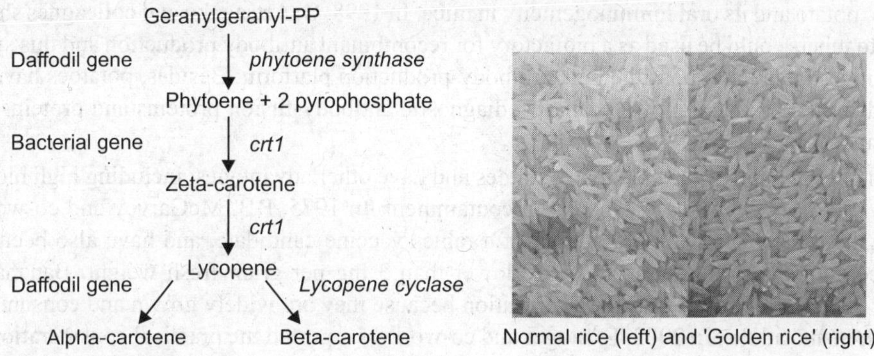

Fig. 19.32. Complete beta-carotene pathway after introducing three genes in rice.

H. MOLECULAR FARMING

Plants have many advantages over other production systems, particularly in terms of practicality, economy and safety. However, several constraints that hinder the widespread use of plants as bioreactors remain to be addressed. Important factors include quality and homogeneity of the final product, the challenge of processing plant-derived pharmaceutical macromolecules under good manufacturing practice conditions and concerns about biosafety. In recent years, several plant-derived biopharmaceutical protein products are reaching for commercial development, which include antibodies, vaccines, human blood products, hormones and growth regulators. For these products, plants offer practical and safety advantages as well as lower production costs compared with traditional systems based on microbial or animal cells, or transgenic animals. Molecular farming in plants is finally coming of age with an increasing number of products.

I. MOLECULAR FARMING OF PLANTBODIES (ANTIBODIES IN PLANTS)

Vaccines and antibodies play a key role in healthcare sector, but its full potential has been hampered by the cost of production and maintaining a chain for vaccine distribution. Expression of antigens as vaccines and of antibodies against antigens of pathogens in transgenic plants is a convenient and inexpensive source for these immunotherapeutic molecules. Various antigens and antibodies have already been expressed successfully in plants and have been shown to retain their native functional forms. Edible plant vaccine against diarrhoea, expressed in potato, and antibody against dental caries, expressed in tobacco, is already in pre-clinical human trials. Attempts are being made to express many proteins of immunotherapeutic use at high levels in plants and to use them as bio-reactors of the modern era. A major advantage of protein expression in fruit and vegetable crops is that the edible organs can be consumed as uncooked, unprocessed or partially-processed material, making them ideal for the production of recombinant subunit vaccines, nutriceuticals and antibodies designed for topical application. Potatoes are the major system for vaccine production and transgenic potato tubers have been administered to humans in at least three clinical trials.

In 1989, A. Hiatt and co-workers from the Research Institute of Scripps Clinic (California) reported the production of antibodies in transgenic plants using cDNAs derived from a mouse hybridoma messenger RNA by transforming the tobacco leaf segments followed by regeneration of mature plants. They found the accumulation of a functional antibody to 1.3% of total leaf protein in plants expressing full-length cDNAs containing leader sequences. Specific binding of the antigen recognised by these antibodies was similar to the hybridoma-derived antibody. They found very efficient occurrence of immunoglobulins and assembly of functional antibodies in tobacco.

In 2003, J. Yu and W. Langridge reported the expression of rotavirus capsid protein VP6 in transgenic potato and its oral immunogenicity in mice. In 1998, O. Artsaenko and colleagues showed how potato tubers could be used as a biofactory for recombinant antibody production and this system has been investigated as a possible bulk antibody-production platform. Besides, potatoes have also been used for the production of glucanases, diagnostic antibody-fusion proteins and proteins from human milk.

Tomatoes are more palatable than potatoes and have other advantages, including high biomass yield and use of greenhouses for increased containment. In 1975, P.B. McGarvey and co-workers first used the tomatoes for the production of a rabies vaccine candidate, and have also been used to produce antibodies, although yields are lower than 3 mg per gram fresh weight. Bananas are attractive vehicles for edible vaccine distribution because they are widely grown and consumed by both children and adults. In 2002, E. Stoger and co-workers reported the practical considerations for pharmaceutical antibody production in different crop systems.

In 2003, Richard M. Twyman from the University of York (U.K.) are of the opinion that plants offer a unique combination of advantages for the production of pharmaceutical antibodies. The main benefit is the low production costs which reflect that traditional agricultural practices and unskilled labour are sufficient for maintaining and harvesting antibody-expressing crops. Scale-up is rapid and efficient, requiring only the cultivation of more land.

The structure of complex glycans differs a little such as the presence of the residues $\alpha(1,3)$-fucose and $\beta(1,2)$- xylose in plants, but not in mammals. These residues are immunogenic in several mammals including humans, but not in mice. The glycan profile of proteins produced in plants can be 'humanised'. In 2005, Richard M. Twyman and co-workers reported the successful production in plants of full-size antibodies and various functional antibody derivatives including Fab fragments, scFvs, bispecific scFvs, single domain antibodies and antibody fusion proteins. Several other therapeutic proteins have also been produced like haemoglobin and erythropoietin in tobacco plants, lactoferrin in potato, trypsin inhibitor in maize etc. The first proteins/enzymes that were produced in transgenic plants (maize) are avidin and beta-glucuronidase; these are used in diagnostic kits.

1. Different Plant-based Expression Systems

The most widely used strategy for antibody production in plants is the nuclear transgenic system, where the antibody transgenes are transferred to the plant nuclear genome. The advantages of this approach include: (*i*) *Agrobacterium*-mediated transformation and microprojectile bombardment for delivering DNA-coated metal particles, (*ii*) use of a stable transgenic line as a permanent genetic resource, (*iii*) the simplest way of to maintaining transgenenics among the various plant systems, and (*iv*) establishment of seed master banks. Disadvantages are : (*i*) relatively long development time (required for transformation, regeneration, analysis of transgenics, selection and bulking up of the producer line), (*ii*) the unpredictable impact of epigenetic events on transgene expression (e.g. post-transcriptional gene silencing, position effects), and (*iii*) the potential for transgene spread from some crops throughout crossing.

Recombinant therapeutic or diagnostic recombinant antibodies produced by molecular farming in plants are listed in Table 19.10. Many antibodies in commercial development remain undisclosed until IP rights have been secured. There are some antibodies that have the properties of phytomodulation or the prevention of plant diseases.

2. Downstream Processing

Downstream processing is an integral part of every bio-manufacturing process which refers to the isolation and purification of the recombinant product. Downstream processing represents up to 80% of overall production costs irrespective of production system this also depends on purity and is highest for clinical grade materials. Though certain classes of product can be isolated using a standardised approach (e.g. affinity chromatography to isolate recombinant antibodies), it is necessary to develop specific processing steps for each product. Many aspects of downstream processing need to be customised for plant systems, such as the removal of fibres, oils and other by-products from certain crops, and process optimisation for the treatment of different plant species and tissues. The steps of downstream processing should meet the standards set for other biopharmaceutical production systems for the production of clinical grade antibodies. The standards must also include a strict regime of quality assurance and quality control so that the regulatory agencies must accord the approval.

Table 19.10. Recombinant therapeutic/diagnostic antibodies/antibody-segments produced through molecular farming in plants.

Antigen	*Antibody format*	*Production system*	*Remark*
B cell lymphoma, murine 38C13	scFv	Virus vectors in tobacco leaves	Maximum yield 30.2μg/g leaves
Carcinoembryonic	scFv, IgG1	Tobacco agroinfiltration	Maximum yields 5μg scFv/g leaves, 1μg IgG/g leaves
Antigen	scFv	Rice, rice cell	Maximum yields 3.8μg/g callus, 29 μg/g leaves, 32 μg/g seed
Hepatitis B virus surface antigen	IgG	Tobacco leaves	Up to 25 mg antibody per kg biomass
Herpes simplex virus	IgG1	Soybean	Secreted to apoplasts
HIV antibodies in blood	scFv-fusion	Tobacco leaves, barley grains, potato tubers	Maximum yield 150 μg/g
Human	IgG1	Alfalfa	Secreted to apoplast. Maximum yield 1% TSP.
Interleukin-4	scFv	Tobacco roots	Maximum yield 0.18% TSP41.
Salmonella enterica lipopolysaccharide	scFv	Tobacco	7 μg purified scFv per g leaf tissue
Tetanus toxin C	IgG2a fused to tetanus toxin C	Tobacco	Animals immunised with recombinant immune complex without adjuvant were fully protected against lethal challenge

3. Regulatory Landscape

The regulatory landscape is one of the greatest uncertainties surrounding the use of plants for the production of pharmaceuticals. When plants are grown in glasshouses and in enclosed bioreactors, the production of pharmaceuticals is regulated similar to the other production systems, and also comes under the Regulatory Authority. Moreover, the production process is made more complex by involving multiple regulatory agencies, because the jurisdiction of each authority is not clear, and only draft guidelines are available (FDA 2002). Consequently, the market is suppressed. When full guidelines become available, hopefully more companies may be interested in plant-derived pharmaceuticals.

II. EDIBLE VACCINES

For production of vaccine antigens the use of plant-based systems offers many practical advantages that make them an attractive option and a practical methodology for the propagation of mucosal vaccines on a global scale. A major advantage of protein expression in fruit and vegetable crops is that the edible organs can be consumed as uncooked, unprocessed or partially-processed material, making them ideal for the production of recombinant subunit vaccines, nutriceuticals and antibodies designed for topical application. It would be less expensive than to produce antigens *via* industrial fermentation; there would be no need to purify the antigen if it is expressed in plant edible tissue. The plant expression system minimises risks arising from contamination.

Crop plants offer cost-effective bioreactors to express antigens which can be used as edible vaccines. Genes encoding antigenic proteins from the pathogens are isolated and then expressed in plants. Such transgenic plants or their tissues producing antigens can be eaten for vaccination/immunisation (edible vaccines). The expression of such antigenic proteins in crops like banana and tomato are useful for immunisation of humans since banana and tomato fruits can be eaten raw. Transgenic plants (tomato, potato) have been developed for expressing antigens derived from animal viruses e.g. rabies virus, herpes virus, etc.

GM banana produces vaccine against hepatitis B virus.

In 1990, the first report of the production of edible vaccine (a surface protein from *Streptococcus*) in tobacco at 0.02% of total leaf protein level was published in the form of a patent application under the International Patent Cooperation Treaty. The concept of edible vaccines got impetus after the work of C.J. Arntzen and co-workers who found expression of hepatitis B surface antigen in tobacco plants in 1992. The first clinical trials in humans, using a plant derived vaccine were conducted in 1997 and limited success was achieved. This involved the ingestion of transgenic potatoes with a toxin of *E. coli* causing diarrhea.

Potatoes are the major system for vaccine production, and transgenic potato tubers have been administered to humans in at least three clinical trials. Recently the production of rotavirus VP6 capsid protein in transgenic potatoes for vaccination against acute viral gastroenteritis has been described.

Tomatoes are more palatable than potatoes and have other advantages, including high biomass yield and use of greenhouses for increased containment. Tomatoes were first used for the production of a rabies vaccine candidate, and have also been used to produce antibodies, although yields thus far are lower than 3 mg per gram fresh weight.

Bananas are attractive vehicles for edible vaccine distribution because they are widely grown and consumed by both children and adults.

The edible vaccines have much importance because they are cost-effective, easy to administer, easy to store, fail-safe and socio-culturally easily acceptable vaccine delivery system, especially for the poor developing countries. The selected desired gene is introduced into plants and then these altered plants are induced to manufacture the encoded proteins. It has been used in prevention of autoimmune diseases, birth control, cancer therapy, etc. Currently edible vaccines are being developed for a number of human and animal diseases. There is growing acceptance of transgenic crops in both industrial and developing countries. Resistance to genetically modified foods may affect the future of edible vaccines. They have passed the major hurdles in the path of an emerging vaccine technology. Various technical obstacles, regulatory and non-scientific challenges need to be overcome.

1. Methods of Production of Edible Vaccines

There are the following two approached used for the production of edible vaccines by the GM plants:

(*a*) *Agrobacterium*–mediated Gene Transfer Method: The process of making of edible vaccines involves the incorporation of a plasmid carrying the antigen gene and an antibiotic resistance gene, into the bacterial cells e.g. *Agrobacterium tumefaciens*. The small pieces of plant leaves are exposed to an antibiotic which can kill the cells that lack the new genes. The surviving cells with altered genes multiply and form a callus. This callus is allowed to grow into plantlets, which are subsequently transferred to soil to form a complete plant. Plant-expressed foreign antigen is extracted and tested on experimental animal. Within a few weeks, the plants produce antigen vaccines. If you have selected potato as experimental plant, its tubers produce antigen vaccines (Fig. 19.33 A).

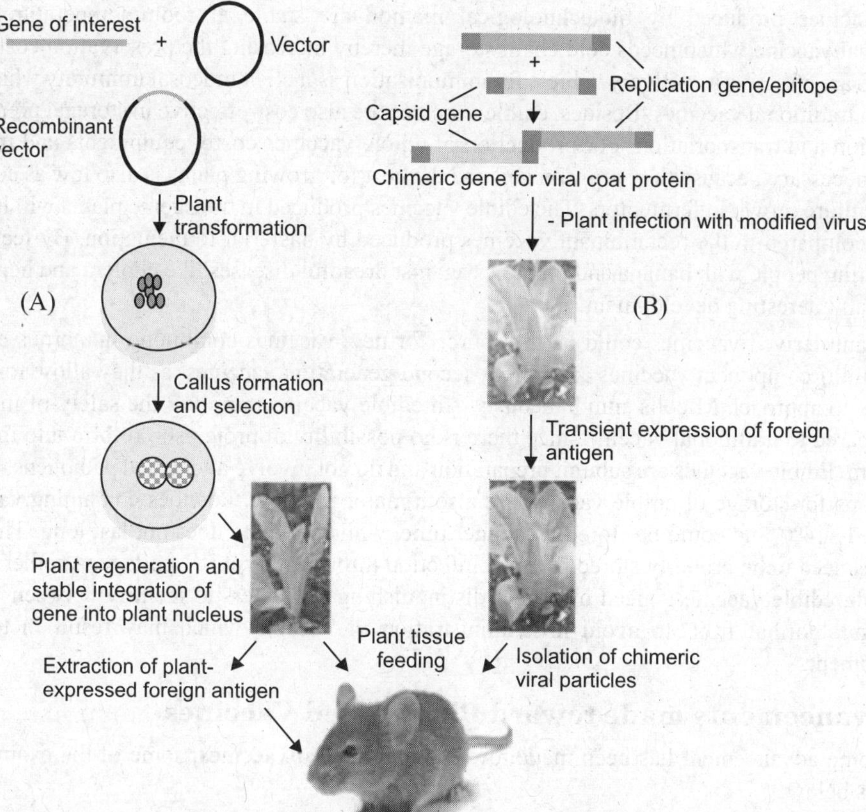

Fig. 19.33. Steps for production of candidate vaccine antigens in plant tissues.

(*b*) **Virus- mediated Gene Transfer Method:** The other strategies to produce a plant-based vaccine are to infect the plants with recombinant virus carrying the desired antigen fused to viral coat protein. In the first approach, the entire structural gene is inserted in to a plant transformation vector downstream to a sub-genomic promoter between 5 'and 3' regulatory element. This will allow the transcription and accumulation of coding sequence in plants. In the second approach, epitope within the antigen is identified, and DNA fragment encoding these is used to construct genes by fusing the coats protein gene from plant virus e.g. TMV or CMV. Thereafter, the genetically engineered virus is allowed to infect the healthy test plant. The infected plants are reported to produce the desired fusion protein in large amounts in a short duration. The antigen protein can be used for further immunological studies employing the experimental animals (Fig. 19.33 B).

2. Advantages of Edible Vaccines

Plants can be used as the cheap chemical factories because they require only water, minerals, sun light and carbon dioxide to produce thousands of sophisticated chemical molecules with different structures. By transferring the right genes, plants can serve as bioreactors to modified or new compounds such as amino acids, proteins, vitamins, plastics, pharmaceuticals (peptides and proteins), drugs, and enzymes for food industry. The transgenic plants as bioreactors have some advantages, such as low cost of production, unlimited supply, safe and environmental friendly, and no scare of spread of animal borne diseases. Tobacco is the most preferred plant as a transgenic bioreactor because it can be easily transformed and engineered. Tobacco is an excellent biomass producer with about 40 tons of fresh leaf production as against rice with 4 tons. The seed production is very high (about one million seeds per plant) and it can be harvested several times in a year.

Vaccines produced by biotechnological method are stable at room temperature, unlike traditional vaccine which needs cold chain storage thereby increasing the preservation cost. Edible vaccines are effective as delivery vehicle for immunisation as it elicit mucosal immunity which is not found in traditional vaccines. Besides, edible vaccines are also cost-effective in storage, preparation, production and transportation. For production of edible vaccines costly equipments and machines are not necessary because they grow on soil and the cost for growing plants is also low as compared to cell culture grown in fermentors. The edible vaccines produced in transgenic plants will have low cost as compared to the recombinant vaccines produced by bacterial fermentation. By feeding the vaccinating people with banana and tomatoes against dreadful diseases like cholera and hepatitis B, will be an interesting development.

Plant derived vaccines could be the source for new vaccines combining numerous antigens. These multi-component vaccines are called 'second generation vaccines' as they allow for several antigens to approach M cells simultaneously. An edible vaccine improves the safety of individual as compared to traditional vaccine since there is no possibility of proteins to reform into infectious organism. Edible vaccines are subunit preparation and do not involve attenuated pathogens. Variable conditions for storage of edible vaccine are also a major problem. Potatoes containing vaccine to be stored at 4°C and could be stored for longer time, while a tomato does not last long. Thus these vaccines need to be properly stored to avoid infection through microbial spoilage. Another concern regarding edible vaccine is need of proper distinguishing characters to identify between 'vaccine fruit' and 'normal fruit' to avoid misadministration of vaccine which may result in tolerance development.

3. Advancements made toward Plant-based Vaccines

Some advancement has been made towards plant-based vaccines; some of the examples are given as below:

(*a*) **Hepatitis B:** HB virus infects liver and causes liver cancer that may also provoke cirrhosis and liver insufficiency. By means of genetic engineering techniques. Three subunit vaccines have been developed based on Hepatitis B virus surface antigen (HBsAg). They are produced in yeast cells (for example, Recombivax HB and Engerix-B) and in mammalian cell cultures (for example, GenHevac B) and proved to be safe, immunogenic and efficacious in preventing the HBV infection.

Many edible plant species such as banana, lettuce, lupin, tomato and potato were employed as hosts for the expression of HBsAg and several approaches in animal model have shown a great success. In 1999, J. Kapusta and coworkers of the Thomas Jefferson University (USA) performed the first study in humans. Volunteers who ate raw transgenic lettuce leaves developed specific-IgG response. Two out of the three human volunteers mounted a significant immune response after a second ingestion of transgenic lettuce. In 2005, Y. Thanavala and coworkers carried out a clinical trial with previously vaccinated individuals. At an interval of two weeks volunteers were fed with 100 g of raw potato containing 8.5 μg of HBsAg/g of transgenic potato. Sixty three percent individuals fed with three doses and 53 % individuals fed with 2 doses exhibited marked increases in serum antibody titers with boosting effects. The results obtained in this study showed that an orally administered plant-based vaccine could be successful in promoting an immune response. Therefore, Thanavala proposed that this type of vaccine should be included as a tool during a global immunisation programme.

(*b*) Influenza: Influenza is a highly contagious viral disease resulting in fever and respiratory symptoms leading to hospitalisation and death, particularly in young children and adults over 65 years age. Haemagglutinin (HA) and neuraminidase (NA) are influenza viral proteins; antibodies to these proteins play a key role in the protection of influenza vaccines. Therefore, HA and NA are the best candidates for influenza subunit vaccine development. In 2008, V. Mett and co-workers transiently expressed the stem and globular domains of HA in *Nicotiana benthamiana* plants, as fusions to a carrier protein. They immunised ferrets with purified antigens. These plant-produced antigens were highly immunogenic and conferred complete protection against infection in the ferret challenge model. When they added the plant-produced NA, an enhancement of the immune response was observed.

In transiently transformed plants, virus like particles (VLPs) have also been obtained and assayed as potential immunogens against influenza. In 2007, R.A. Bright and co-workers in preclinical trials with influenza VLPs have demonstrated their capacity to induce both humoral and cellular immune responses at low antigen doses of antigens. The production of VLPs does not require the protective measures imposed by the manipulation of highly infectious pandemic virus particles as any recombinant protein production methodology. Currently influenza VLPs are produced by using the cell culture methods which may not meet the cost. Therefore, M.A. D´Aoust and co-workers in 2008 developed an alternative method for the production of influenza VLPs at lower costs by utilising transgenic plants. They found that influenza VLPs can form and accumulate in leaves of agroinfiltrated *Nicotiana benthamiana* producing HA from either H5N1 or H1N1. They also demonstrated the induction of a strong and protective immune response in mice after parenteral administration of purified H5-VLPs with alum.

In 2009, Y. Shoji and co-workers described the production of HA from a strain of H5N1 influenza virus by transient expression in *N. benthamiana* plants. Immunisation of mice and ferrets with plant-derived HA elicited serum haemagglutinin-inhibiting antibodies and protected ferrets against challenge infection with a homologous virus.

(*c*) Rotavirus: One of the most common causes of diarrhea and severe infection (such as rotavirus gastroenteritis) is rotavirus that causes acute dehydrating diarrhea in infants and young children. This virus is transmitted through faecal-oral route that lines the small intestine after infection and produces an enterotoxin. Consequently, it induces gastroenteritis leading to severe diarrhea and death through dehydration. Several developments on plant-based vaccines against this agent have been reported. In 2005, J.L. Dong and co-workers synthesised a codon-optimised gene (sVP6) encoding the VP6 protein of human group rotavirus and inserted into the alfalfa genome using *Agrobacterium*-mediated transformation. Female mice were given weekly with 10 mg of transgenic alfalfa extract CpG-rich oligodeoxynucleotides, a mucosal adjuvant. The immunised animals developed high titers of anti-VP6 serum IgG and mucosal IgA. In addition, their offspring developed less severe diarrhea after challenge with simian rotavirus SA-11; this indicates that antibodies generated in the dams provided passive protection to the newborn mice.

Moreover, rotavirus protein VP7 was also tested by using plant expression systems. In 2003, Y.Z. Wu and co-workers found that rotavirus VP7 could maintain its neutralizing immunity when it was expressed in transgenic potato plants. Mice immunised with the transformed tubers successfully elicited serum IgG and mucosal IgA.

In 2006, S. Saldaña and co-workers demonstrated the successful expression of rotavirus capsid proteins VP2 and VP6 in fruits of transgenic tomato plants. They recovered rotavirus VLPs from tomato fruit, even though only a small proportion of VP2/VP6 assembled into VLPs. For investigation of immunogenicity, the adult mice were immunised three times with a protein extract from transgenic tomatoes in adjuvant. Transgenic tomato extract induced detectable levels of anti-rotavirus antibodies in serum. But the contribution of either the free rotavirus proteins or the VLPs to the induction of the antibody response could not be determined.

(*d*) Tuberculosis: *Mycobacterium tuberculosis* is the causal agent of tuberculosis (TB) that affects the lungs of the humans. The disease is endemic and related to poverty. The Bacillus Calmette-Guérin (BCG) is the only vaccine that works against this disease. It is derived from *Mycobacterium bovis* and induces high levels of protection in animal models but its efficacy varies in humans. One of the main antigen of *M. tuberculosis* that is being targeted for the development of new subunit vaccines is ESAT 6, a 6kDa secretory protein inducing a potent cellular response and is absent in *M bovis* BCG. In 2004, M.M. Rigano and co-workers successfully expressed this protein in *A. thaliana* as a fusion protein with the enterotoxigenic *E. coli* heat-labile toxin B subunit (LTB) and proved to be immunogenic when orally administered to mice. ESAT 6 was also expressed in *N. benthamiana* by the *Agrobacterium*-mediated delivery of TMV-based vectors along with Ag85B, other TB antigen. It was also expressed in *N. tabacum* using a potato virus X (PVX)-derived vector as a fusion protein with the coat protein (CP).

4. Commercial use of Transgenic Plants

The main goal of producing transgenic plants is to increase the productivity. In 1995-96, transgenic potato and cotton plants were used commercially for the first time in U.S.A. By the year 1998-99, five other major transgenic crops cotton, maize, canola, soybean, and potato were introduced to the farmers. These accounted for about 75% of the total area planted by crops in USA. There are still a lot of concerns regarding the harmful environmental and hazardous health effects of transgenic plants. The major areas of public concern are : the development of resistance genes in insects and generation of a super weed by mutation, etc. Certain other legal and regulatory hurdles pertaining to commercial use of transgenic plants needs to be addressed.

I. IMPROVEMENT OF SEED PROTEIN QUALITY

The nutritional quality of cereals and legumes has been improved by using biotechnological methods. For improvement in seed protein quality two genetic engineering approaches have been used. In the first approach, a transgene (for example, gene for protein containing sulphur rich amino acids) was introduced into pea plant (which was deficient in methionine and cysteine, but rich in lysine) under the control of seed-specific promoter. In the second case, the endogenous genes are modified so as to increase the essential amino acids like lysine in the seed proteins of cereals.

In the seed storage proteins of several crop plants these transgenic routes have helped to improve the essential amino acids contents, for example overproduction of lysine by de-regulation. The four essential amino acids namely lysine, methionine, threonine, and isoleucine are produced from a non-essential amino acid, *i.e.* aspartic acid. The formation of lysine is regulated by feedback inhibition of the enzymes aspartokinase (AK) and dihydrodipicolinate synthase (DHDPS). The lysine feedback-insensitive genes encoding the enzymes AK and DHDPS have been isolated from *E. coli* and *Cornynebacterium*, respectively. These genes were introduced into soybean and canola plants after appropriate genetic manipulations. The transgenic plants produced in this way had high quantities of lysine.

QUESTIONS

1. What do you mean by gene construct ? Describe the organization of a typical plant gene.
2. Briefly describe the reporter genes.
3. What are scoreable and selectable markers ? Describe them in detail.
4. Write a note on antibiotic- and herbicide- resistance markers.
5. Discuss virus-resistant transgenic plants and their benefits.
6. Explain the expression of virus coat protein gene into a plant.
7. Describe transgenic plants, benefits of their use and risks after release into environment.
8. What is Barnase-barstar component system ? Describe its catalysis mechanism, binding and some transgenic plants.
9. What is Bt technology ? Give a comprehensive account of Bt genes, mechanism of protein action with suitable examples.
10. What is the fate of Bt technology in India ? Discuss biosafety measures and economic impact of Bt plants in India.
11. Describe the fate of Bt brinjal with reference to transformation, mode of action of *cry* genes and field trials.
12. What is Golden rice ? How was it created ? Write the benefits of using Golden rice.
13. What is molecular farming ? Describe molecular farming of antibodies.
14. Write in detail about edible vaccines, its production methods and advantages.
15. Describe advances made towards production of plant-based vaccines.
16. Write short notes on the following :

(*i*) Promoter sequences, (*ii*) Type of promoters,
(*iii*) Enhancers, (*iv*) Reporter (marker) genes,
(*v*) Scoreable markers, (*vi*) Selectable markers,
(*vii*) Luciferase *luç* and *lux* genes, (*viii*) GFP and its expression,
(*ix*) EPSP synthase gene, (*x*) Agroinfection,
(*xi*) Roundup Ready, (*xii*) CaMV 35S promoter,
(*xiii*) Pathogenesis-related Proteins (PRs), (*xiv*) Bt cotton,
(*xv*) Bt brinjal, (*xvi*) Molecular farming,
(*xvii*) Edible vaccines.

BIBLIOGRAPHY

Gachamo E.W. *et. al*. 2003. The molecular initiation and subsequent acquisition of disease resistance in plants. *African J. Biotech*. 2:26-32.

Hartley R.W. 1988. Barnase and barstar. Expression of its cloned inhibitor permits expression of a cloned ribonuclease. *J. Mol. Biol.* **202**: 913–915.

http://www.biospectrumindia.com

James C. 2009. ISAAA 41: Global status of commercialized biotech/GM crops: 2009. The first fourteen years, 1996 to 2009. *International service for the acquisition of agri-biotech applications* (*ISAAA*), Ithaca, NY, USA.

Karihaloo J.L. and Kumar P.A. 2009. Bt cotton in India – A status report (2nd Ed.). Asia-Pacific Consortium on Agricultural Biotechnology (APCoAB), New Delhi, India. p. 56.

Lal P., Ramachandran V.G., Goyal R. and Sharma R. 2007. Edible vaccines: current status and future. *Indian J. Medical Microbial*. **25**: 93-102.

Lee Lee-Peng and Tidor B. 2001. Barstar is electrostatically optimized for tight binding to barnase. *Nature Structural Biology:* 8: 73 – 76.

Li Z., Hayashimoto A. and Murai N. 1992. A sulfonylurea herbicide resistance gene from *Arabidopsis thaliana* as a new selectable marker for production of fertile transgenic rice plants. *Plant Physiol.* **100**: 662–668.

Miki B. and McHugh S. 2004. Selectable marker genes in transgenic plants: applications, alternatives and biosafety. *J. Biotechnol.* **107:** 193–232.

Raney T. 2006. Economic impact of transgenic crops in developing countries. *Curr. Opinion Biotech.* **17**:1-5

World Health Organization. 1999. *Environmental Health Criteria* 217- Microbial Pest Control Agent *Bacillus thurigiensis.* p. 125.

Transgenic Animals (= Animal Biotechnology)

CHAPTER 20

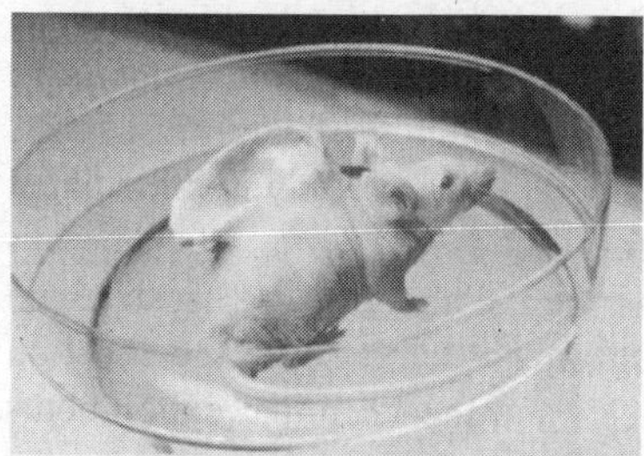
A hairless mouse with a human ear on its back

A. Gene construct
B. Vectors
C. Transfection methods
D. Embryonic stem cell-mediated gene transfer
E. Targeted gene transfer
F. Rescue and recovery of transgenes from animal cells
G. Detection of transgenic animals
H. Breeding with transgenic livestock
I. Transgenic animals
J. Molecular pharming
K. Bioethics

In 1974, Rudolf Jaenisch (from California, U.S.A.) and Beatrice Mintz (from Pennsylvania, U.S.A.) created the first genetically modified animal by inserting a DNA virus into a mouse embryo of an early-stage; further they showed the presence of inserted genes in every cell of mice. But, the mice did not pass the transgene to their offspring. In 1981, Frank Ruddle from Yale, and Frank Constantini and Elizabeth Lacy from Oxford microinjected the purified DNA into the pronucleus of a single-cell embryo of a mouse and demonstrated the transmission of the genetic material to subsequent generations. In 1981, J.W. Gordon and F.H. Ruddle first used the term 'transgenic'; since then a rapid development took place to produce the genetically engineered animals. When an organism is engineered to carry a foreign gene of interest (= transgene, a gene construct) as part of its own genetic material, it is called 'transgenic animal'; the process of gene manipulation in the desired cells of animals is called as 'transgenesis'.

By using recombinant DNA technology, foreign DNA is introduced into the animal, and then transmitted through the germ line so that the same modified genetic material should be introduced in every cell of the desired animal. The genetic material integrates randomly on a chromosome in some of the eggs and becomes part of the genetic material of those cells of the animal. Due to the presence of a foreign gene such animals are called the transgenic animals. Transgenic organism is produced by introducing cloned genes (from microbes, animals or plants) into plant and animal cells. Thus the transgenic technology includes such methods that allow the transfer of genes between different species. The transgenic animals are very useful for delineating the function of newly discovered genes and producing useful proteins in large animals.

Observation of inherited characteristics or spontaneous mutations was the only way of studying the regulation and function of mammalian genes before the development of molecular genetics. Selective breeding was a common practice among farmers for the enhancement of chosen traits such as increased milk production before Mendel.

In 1974, Ralph Lawrence Brinster (an American geneticist) at the University of Pennsylvania, produced the first chimeric mice by combining together the cells of two different embryos of different strains at an early stage of development (eight cells) to form a single embryo. This embryo subsequently developed into a chimeric adult that showed the characteristics of each strain. These techniques form the foundation for all experimentation with the mammalian embryo, including transgenic, embryonic stem cell, human *in vitro* fertilisation, mammalian cloning, and knockout technology. This 'Brinster Method' of embryo manipulation is widely used in modern biology.

Rudolf Jaenisch- creator of first GM animal.

Brinster first demonstrated that adult 'chimeric' mice can be formed by combining the teratocarcinoma cells with blastocyst cells. This method showed the possibility for changing the genetic character of mice. He was the first scientist who developed microinjection technique to fertilised eggs and generate transgenic mice. Thereafter, DNA microinjection technique has been successful used in a large number of animals such as rats, rabbits, sheep, pigs, birds, and fish. Then the two other main techniques were also developed, such as retrovirus-mediated transgenesis and embryonic stem (ES) cell-mediated gene transfer.

Richard D. Palmiter of the University of Pennsylvania was the collaborator of Brinster who pioneered the methods to transfer foreign genes into mammals and to demonstrate the function of genes. In the 1980s, they developed the first 'transgenic mice', and their seminal experiments catalysed a worldwide revolution in genetic engineering. For the first time they showed that new genes introduced into the mammalian germ line have the potential for increasing disease resistance, enhancing growth, and producing vital proteins like blood-clotting factors required by haemophilic patients. Thus, they provided the first proof of : *(i)* expression of transgenes, *(ii)* the first example of cancer arising from a transgene, and *(iii)* the first proof of the targeted integration of DNA by egg injection.

In recent years, treatment of many serious diseases has been possible by using the genetically modified animals. Certain therapeutic proteins can be developed simply by bringing about changes in their DNA or transferring the desired DNA to an animal. Several human proteins have been stably expressed in many animals, such as rats, pigs and sheep. Human-α-1-antitrypsin has also been developed in sheep, and used for treatment of this deficiency in humans.

Ralph L. Brinster

Richard D. Palmiter

The most important use of transgenic animals in food sector is the production of novel products. Much progress has been made in production of pharmaceutical proteins in the milk of transgenic animals. Since milk contains tens of grams of proteins like casein, β-lactoglobulin and whey acidic protein per litre, the transgenic cattle offer the potential to produce a high amount of milk. You know that the production of these proteins is regulated by promoters that regulate gene expression only in the cells of mammary gland of animal. Keeping in view, the regulatory sequences from the genes of milk-specific proteins have been cloned with the gene of interest in transgenic animals such transgenes control the genes expressions.

Ralph Lawrence Brinster and Richard D. Palmiter

Ralph Brinster was an officer in the US Air Force (1953–1956) and served during the Korean War. He returned from military service and earned his V.M.D. (1960) and Ph.D. in Physiology (1964) from the University of Pennsylvania. In the 1960s, Brinster developed the first reliable *in vitro* culture system for early mammalian embryos. These techniques form the foundation for all experimentation with the mammalian embryo including transgenic, embryonic stem cell, human *in vitro* fertilisation, mammalian cloning, and knockout technology. The 'Brinster Method' of embryo manipulation is so ubiquitous in modern biology that other scientists rarely cite the work in current publications.

He is acknowledged as the founder of mammalian transgenesis, and known throughout the scientific community for his research in embryonic-cell differentiation, developmental mechanisms of gene control, and stem cell physiology. Brinster pioneered the methods for manipulating mouse embryos; consequently, the mouse embryo acted as the major genetic model for understanding the basis of human biology and disease. In 2011, he was honoured with International Society of Transgenic Technology Award.

Richard D. Palmiter is a Professor of Biochemistry at the University of Washington (U.S.A.). He received his Ph.D. degree from Stanford University. Prior to moving to the University of Washington, he was a postdoctoral fellow at Stanford University with Robert Schimke, at G.D. Searle Research Laboratories in England.

Richard Palmiter used genetic techniques to disable neurotransmitter functions in mice to understand how and where these signalling molecules act in the brain to regulate various behaviours. Throughout 1980s, Palmiter and Brinster together developed several first animal models of human disease. Their partnership resulted in the first transgenic rabbits, sheep, and pigs. Their continued transcontinental collaboration formed the foundation for generation of scientific progress in genetic modification via transgenesis, homologous recombination (knock-out techniques) and cloning.

Changing of the milk constituents is the prime the goal of transgenesis of dairy cattle. The amount of cheese produced from milk is directly proportional to the K-casein protein. However, production of K-casein can be increased through over-pressed K-casein transgenic. Besides, lactose-free milk can also be produced though heterologous expression of a lactose transgene in the mammary gland of cattle.

Since 1990s, transgenic livestock have been used as bioreactors to secrete continually the high levels of desired proteins such as insulin into their milk that can be harvested simply by milking the animals and purifying it through standard chromatographic procedures. A bioactive recombinant human lysozyme was expressed in the milk of cloned transgenic cattle in March 2011. Some of the human genes expressed in transgenic animals include: anti-thrombin III (to treat intravascular co-agulation), collagen (to treat burns and bone fractures), fibrinogen (used for burns and after surgery), human fertility hormones, human haemoglobin, human serum albumin (for surgery, trauma, and burns), lactoferrin (found in mother milk), tissue plasminogen activator, and specific monoclonal antibodies (including one that is effective against a particular colon cancer). Animals mostly used for expression of these proteins are pigs, cows, sheep and goats. A timeline of animal biotechnology and transgenic animals is given in Table 20.1.

Table 20.1. Timeline of animal biotechnology and transgenic animals.

Year	*Events*
1972	A mouse was born from a stored frozen embryo.
1973	A calf born from a stored frozen embryo.
1978	Birth of Louise Brown (25th July), the first 'test-tube' baby born from *in vitro* fertilisation (IVF).
1979	Twin lambs born through artificial embryo splitting.
1980	First transgenic mouse borne by microinjection of DNA to zygote pronuclei.
1982	Giant mouse produced by transferring growth hormone genes from a rat. A calf was born after IVF technology.
1983	R. Palmiter and R. Brinster placed a gene for human growth hormone in a mouse embryo.

1984	Sheep-goat chimaera was created, and dubbed a 'geep'.
1985	First transgenic domestic animal, a pig was borne.
1986	Cloning of embryo cells from sheep was done.
1987	Transgenic mice was produced whose milk had an artificially altered composition caused by the expression of genes coding for sheep beta-lactoglobulin and human alpha-1-antitrypsin.
1988	The Harvard University patented the first animal called 'Oncomouse', which is a transgenic mouse genetically engineered to develop cancer.
1990	Transgenic sheep 'Tracy' was produced with the human gene for alpha-1-antitrypsin which is expressed in the milk.
1991	A first transgenic cow was born that produced human lactoferrin in its milk, with the intention to harvest lactoferrin for the treatment of septicaemia.
1995	The first baby was produced through fertilisation of an ovum with a spermatid (spermatozoa precursor).
1997	A lamb (Dolly) was cloned by nuclear transfer using a mammary cell from a sheep as nucleus donor and an enucleated ovum as recipient. The first transgenic cloned sheep 'Polly' was born.
1998	Lee Bo-yon of Kyunghee University (South Korea) claimed to have been the first to carry out human cloning by nuclear transfer.
2000	1. Successful treatment of severely combined immune-deficiency (SCID) was done in France. 2. Successful treatment of heart muscle blood supply was done using vascular endothelial growth factor gene.
2001	First human 'embryo' cloned. Ádvanced Cell Technologies U.S.A. used somatic adult cell nuclear transfer method to create a clone.
2002	World's first cloned cat was released.
2004	The first human embryos were created at Seoul National University in South Korea.
2005	The first cloned dog Snuppy (**S**eoul **N**ational **U**niversity pu**ppy**) was born.
2006	South Korean Scientists cloned two females of an endangered species of wolf named Snuwolf and Snuwolffy.
2008	Snuppy become a father after the world's first successful breeding involving only cloned canines
2009	The first buffalo calf was born at NDRI, Karnal (India).
2011	A South Korean team claimed to have cloned coyotes for the first time.
2012	World's first chimeric monkeys were born at Oregon Health and Science University (U.S.A).

The amounts of milk needed to provide a national supply of these pharmaceuticals are really very reasonable. Assuming that the animals produce 1 g of the protein per liter milk and one can purify efficiently about 30% (*i.e.* 30% of the protein is recovered in the pure sample), then a pig can produce 75 g of protein per year, a goat 100 g, a sheep 125 g, and a cow 3 kg. As the national need of blood-clotting factor IX is 2 kg /year, one cow per country can do the job. For other proteins the demand is larger (for example, 75 kg per year tissue plasminogen activator and 1,000 kg /year human serum albumin), but there is a limited number of animals which are expected to meet the national demand for pharmaceutical proteins.

In February 2009, the U.S. Food and Drug Administration (FDA) approved the first biological product called ATryn®, produced by genetically engineered animals. ATryn® is a recombinant anticoagulant antithrombin manufactured by GTC Biotherapeutics U.S.A. ATryn® is made from the milk of transgenic goats that produce human antithrombin; it is a plasma protein having anticoagulant properties. According to GTC Biotherapeutics, one GM goat can produce the same amount of antithrombin in a year as 90,000 blood donations. Goats reproduce more rapidly than cattle and produce more protein than rabbits or mice; therefore, they are preferred over the other animals. ATryn® can be used for treatment of patients with hereditary antithrombin-deficiency who are undergoing surgical or childbirth procedures.

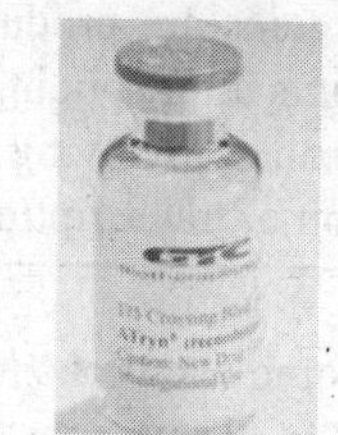

ATryn®- the first drug made by transgenic goat protein.

A. GENE CONSTRUCTS

A DNA segment containing a gene sequence that has been isolated from one organism and is introduced into a different organism by using genetic engineering techniques is called 'transgene'. It can either be a cDNA segment or the gene itself present in its original region of genomic DNA.

In addition to the gene of interest, a promoter and a terminator are also required for efficient expression. Mostly a promoter and terminator also accompany these additional sequences (e.g. marker genes). Since, the sequences do not normally exist in this combination but are artificially constructed; the name 'construct' is used to such sequences. A gene construct contains all the information needed for transfection into a cell and expression of the desired gene in an animal. This includes an antibiotic resistance marker, a tissue-specific promoter, the transgene/gene of interest and a stop sequence. The construct must have a bacterial promoter, ribosomal binding site, and a transcription terminator. The trasgene associated with sequences for transcription and translation must be present in proper orientation in the host cell. The sequences required for efficient gene expression are : promoter/enhancer, ribosome binding site, ATG as initiation codon in DNA (AUG in mRNA), termination codons (for example TAG in DNA), and transcription termination sequences such as polyadenylation sequence (Fig. 20.1). The transgene is usually excised from the plasmid vector before injection so that the interference from the bacterial vector sequences can be minimised. After injecting the transgene into the pronuclei of fertilised mouse eggs, they are randomly incorporated into loci as head-to-tail concatemers consisting of varying numbers (1 >1000) of copies. Mostly transgene is integrated at a single locus and is present in all cells of the mouse, but presence of multiple integration sites in mice is rare.

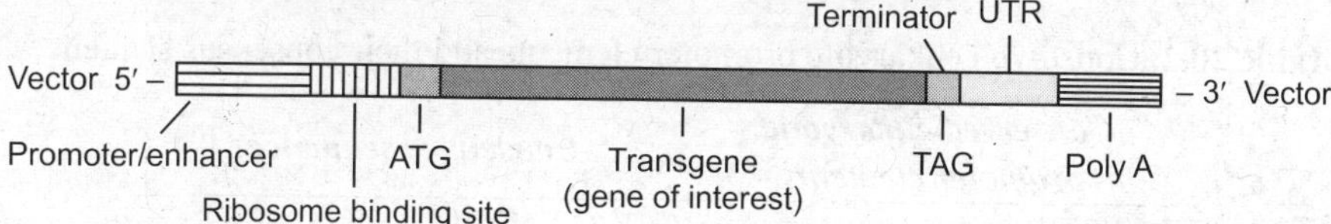

Fig. 20.1. A typical gene construct used for expression in animal/plant cells. ATG, initiation codon; TAG, stop codon; UTR, untranslated region; Poly A, poly adenylate.

Generally, a transgene contains a promoter, a cDNA, an intron, and a polyadenylylation signal. When a cDNA is used, an intron of some sort is added to stimulate the transport of mRNA out of the nucleus, since this process is coupled to the splicing process. The artificial intron should be placed at the 3′ end of the cDNA; some variations in transgenes may also occur.

All the genes are not needed all the times, but the genes required in a particular cell at a particular time are 'switched on' to expression a protein. Hence, performance of a cell depends on expression of particular genes. Thus, an important part of molecular biology is the characterisation of gene expression profiles. Moreover, transgene expression varies depending on the way of foreign gene integrated into the genome.

Proteins are produced by a cell via two general steps: *transcription* (production of mRNA transcripts after splicing), and *translation* (translation of mRNA into a polypeptide). Transcription is controlled involving several DNA regions, where the promoter is the main determinant for the initiation of transcription and timing of gene expression.

Variations in Transgenes

Some variations may also occur in transgenes as below:

- **BAC Transgenes:** BAC transgenes allow insertion of entire genes, including all introns and promoter and enhancer elements with 'insulator' sequences that allow expression of the gene even after integration of transgene into heterochromatin.
- **Promoterless Transgenes:** The promoterless transgenes can act as gene-trap or promoter-trap vectors and mostly consists of cDNA for a reporter gene, for example GFP or *lac*Z.
- **Internal Ribosome Entry Sites (IRES):** The IRES are added along with a second cDNA to achieve the expression of two proteins under the control of the same promoter. The second protein could be a reporter, for example GFP or β-galactosidase.
- **Fusion Proteins:** Fusion proteins get fused with either a reporter gene or an affinity tag, and result in easily detection of the protein products.

1. Promoter

In eukaryotes RNA polymerase II transcribes a gene into RNA. It works in association with the other transcription factors that recognise signals located in promoter region. There are at least seven different factors necessary for the binding of an RNA polymerase II to the promoter. Promoters is located upstream of the gene and have regulatory elements several kilobases away from the transcriptional start site (enhancers). It consists of specific DNA sequence that provides a binding site for RNA polymerase and proteins, which is called transcription factors. Promoters work in association with the other regulatory regions (enhancers, silencers, boundary elements/insulators) and direct the transcription of a given gene. RNA polymerase II first binds at the TATA region and travels along the DNA until it reaches the CAP site where the actual RNA synthesis starts. Transcription process takes place only in the downstream direction from 5' → 3' direction.

Many eukaryotic promoters consist of a TATA box (TATAAA sequence) which lies very close to the transcriptional start site (often within 50 bases). The TATA box binds to a TATA-binding protein that helps in the formation of the RNA polymerase transcriptional complex. These elements are normally regarded as constituents of the promoter region. Example of the conserved eukaryotic promoter elements and their respective consensus sequences are given in Table 20.2.

Table 20.2. Conserved eukaryotc promoter elements and their consensus sequences.

Conserved eukaryotic promoter elements	***Consensus sequence***
CAAT box	GGCCAATCT
TATA box	TATAA
GC box	GGGCGG
CAP site	TAC

Different types of gene expression can be achieved by different promoters. Some promoters are active in all cells at all times, whereas the other promoters are specific to different organisms or type of tissues (e.g., mammary gland tissue-specific promoter, Islets of Langer Hans cell-specific promoter). There are some others promoters that are sensitive to external signals such as temperature, certain chemicals, etc.

All genes must have a promoter for proper expression; therefore, genes transferred by genetic engineering technique must be accompanied by a promoter. The selection of promoters for a transgene depends on the objectives of the research. However, there are four basic types of promoters used in expression vectors for gene expression : (*i*) *minimal promoters* (used to study gene regulatory elements such as enhancer elements), (*ii*) *constitutive promoters* (used to direct expression of gene products), (*iii*) *cell-specific promoters* (used to specify expression to target cells), and (*iv*) *regulated promoters* (used to control the on/off expression of cloned genes). Gene promoters commonly used for expression in cultured cells and transgenic animals are given in Table 20.3.

The *constitutive promoters* [promoters from the human cytomegalovirus (CMV), SV40 or the X-linked phosphoglycerate kinase-1 locus (PGK)], and immediate early promoter/enhancers fused to a minimal chicken beta-actin gene promoter (CAG) are often used to drive expression in all tissues. *Tissue-specific promoters* can be used to limit the spatial expression pattern, while *inducible promoters* can be used to control the timing of expression. Development *regulated promoters* can be used for control of timing.

Table 20.3. Gene promoters commonly used for gene expression in cultured cells and transgenic animals.

Gene name	*Promoter name*	*Application in molecular genetics*
		Minimal promoters
adh	Deleted *Drosophila* alcohol dehydrogenase promoter	A ubiquitous low level promoter that is used to construct reporter genes
HSVtk	Herpes simplex virus thymidine kinase promoter	A low level activity promoter used to direct the synthesis of marker genes in mammalian cells
		Constitutive promoters
CMV	Cytomegalovirus immediate early promoter	High level gene expression in mammalian cells
SV40	Simian virus 40 early enhancer/ promoter	Moderately high level gene expression in mammalian cells
Tuba1	*Drosophila* tubulin α-1promoter	Moderately high level gene expression in *Drosophila* cells
		Cell-specific promoters
WAP	Whey acidic protein promoter	Targeted expression of genes to mammary tissue in animals
lck	Lymphocyte-specific tyrosine kinase promoter	Targeted expression of genes to mouse thymocytes for immunological studies
		Regulated promoters
MMTV	Mouse mammary tumor virus long terminal repeat enhancer/ promoter	Steroid-regulated gene expression in mammalian cells
hsp	*Drosophila* heat shock 70 promoter	Heat-regulated gene expression in *Drosophila* cells

Enhancers: Enhancers are elements located at variable distances from the promoter and contain several closely arranged sequence elements that bind to transcription factors. These elements enhance the activity of a promoter and are orientation-independent with respect to the promoter and can be upstream or downstream of a promoter (e.g., within introns of a gene). Enhancer of simian virus 40 (SV40) is a 72 bp long sequence of tandem repeats located near its origin (Fig. 20.2). The SV40 enhancer contains binding sites for a number of transcription factors. The DNA sequences of

SV40 controlling transcription of the early promoter include a TATA box (an upstream sequence element containing binding sites) and an enhancer. Sequences within the enhancer are involved in both early and late transcription to study the function of their regulatory sequences. The enhancer also contains sequences involved in late transcription. The early gene promoter contains strong enhancer elements which activate it in newly infected cells. The SV40 late promoter is very strong and is activated by binding of large T-antigen to the 72 bp repeats upstream of the transcription start site.

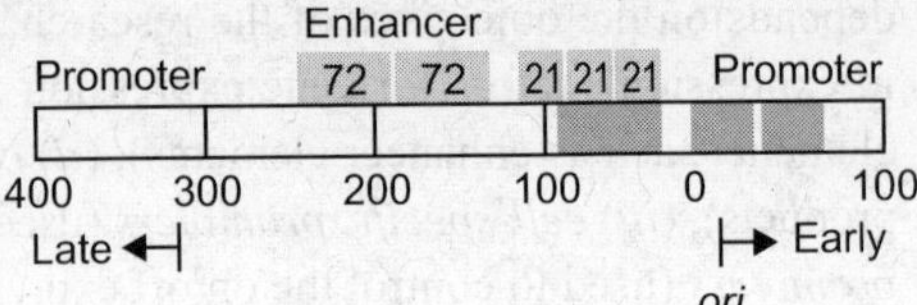

Fig. 20.2. Enhancer elements of early and late gene promoters of SV40; *ori*, origin for replication.

In 1988, Ondek, Brian and co-workers reported that the SV40 transcriptional enhancer is composed of separate 15 to 20 bp long enhancer elements that cooperate with one another to enhance transcription. These elements are bipartite and composed of subunits called 'enhansons' that can be duplicated to create new enhancer elements. Enhansons differ from the enhancer elements because they are very sensitive to change spacing.

Enhancers can be successfully attached to heterologous promoter regions so that the transcriptional activity can be increased and in some cases additional levels of control can be provided to confer tissue-specific or stage-specific expression of a gene.

The human cytomegalovirus (CMV) and Rous sarcoma virus (RSV) promoters have been widely used to provide high levels of expression in many cell types, such as skeletal muscle cells of different maturity, the mammalian embryo, cultured nodose sensory neurons, human lung fibroblasts, etc. Moreover, transgene expression obtained with either the CMV or the RSV promoter reflects the limitation of cell-type-specific infection. Selection of the promoter is critical for the success of the viral vector to express a transgene in specific cell types.

In 1999, A.A. Zarrin and co-workers from the University of Toronto (Canada) compared the activity of the CMV, RSV and SV40 viral promoter/enhancers in different cell lines. They found that the CMV and RSV promoter/enhancers contained stronger regulatory elements than do the SV40 for gene expression in lymphoid cell lines.

Enhancer Trap

A transgenic construction for the identification of enhancers is called an 'enhancer trap'. The enhancer trap consists of a mobile element (that is inserted in the genome randomly) usually P element and a reporter gene. The spatial regulation by enhancers is necessarily identified by the reporter gene. P(lacZ) from *E. coli* and P(GAL4) from yeast are the most common and basic enhancer traps.

2. Reporter Gene (=Marker Gene)

A gene or segment of DNA used to determine the successful insertion of nucleic acid sequence (on the basis of phenotypic characteristics) into the DNA of an organism is called a marker gene. There are two types of marker genes: a *selectable marker* and a *scorable marker* (for screening). Reporter gene is also called scoreable marker, whereas marker gene is called selectable marker. Examples of some of the scorable markers are : *E. coli* chloramphenicol transferase (CAT) gene, *E. coli* transposon Tn9, *E. coli* β-galactosidage gene and the bacterial luciferase gene (*luc*) (for detail see Chapter 19: *Transgenic plants*) . The scorable genes are used to analyse transient gene expression, such as monitoring of promoter activity, identification of transcription-regulating sequences, etc.

Moreover, a selectable marker protects the organism from a selective agent such as antibiotics that would normally kill it or prevent its growth. In mammals, antibiotics are used as selectable markers because it kills the mitochondria and develop resistance to antibiotics. The marker gene is transferred along with the gene of interest into the desired cells which can help to identify even the small number of cells transformed with the new gene as compared to non-transformed cells. Some of the marker genes for mammalian cells are given in Table 20.4.

Table 20.4. Markers system for mammalian cells.

Marker gene	***Selective agent***	***Marker gene protein acting on selective agent***
ada (adenine deaminase)	Xyl-A	Deaminates Xyl-A
ble (bleomycin binding protein)	Bleomycin	Protein binds to bleomycin
neo (neomycin phosphptransferase)	Neomycin, G418	This protein phosphorylates G-418
gs (glutamine synthatase)	MSX	Cells producing excess glutamine synthatase survive
dhfr (dihydrofolate reductase)	Methotrexate (MTX)	Cells producing excess dihydrofolate reductase survive
codA (cytosine deaminase)	*N*-phosphonacetyl-*L*-aspartate (PALA)	Lowers concentration of cytosine in the medium through conversion of cytosine into uracil

Reporter genes have been used to study gene regulation by cloning *cis*-acting test enhancer/ promoter sequences into a multiple cloning site (MCS) upstream of the reporter gene. Besides, reporter cassettes are becoming available in which the reporter genes are placed under the control of a variety of specific transcription factor binding sites, which then report on the induction of *trans*-acting events. Such reporter gene cassettes can give transcriptional readouts in response to many external stimuli, such as receptor binding, protein-protein interactions and virus infections. Consequently, they induce specific signal transduction cascades and transcription factor binding.

The marker genes are described in detail in Chapter 19- *Transgenic plants*. Besides, some other selectable reporter genes used in animals are described below :

(*a*) Chloramphenicol Acetyltransferase (CAT): It is a very popular selectable reporter gene and protein. *cat* is a bacterial gene that neutralises chloramphenicol antibiotic by transferring the acetyle groups to chloramphenicol and changing its shape into a harmless form. Therefore, this enzyme is used as a reporter protein in eukaryotic cells. For determination of promoter activation, the promoter of interest is placed upstream of the CAT coding sequence and either mRNA or expressed protein is measured. The amount of acetylated chloramphenical can be measured either by scanning the density of the dark spots on the X-ray film or by measuring the radioactivity in each spot on the TLC sheet. In both the conditions, CAT assays offer an indirect but quantitative way to measure the amount of transcription driven from any given promoter.

In 2010, Stephen T. Smale described the CAT assay. The activity of reporter protein or fluorescence within a transfected cell population is proportional to the level of steady-state mRNA. In this method, cells transfected with an *E. coli* transposon CAT reporter plasmid are lysed and cellular debris is removed by centrifugation. The lysate is incubated with ^{14}C-chloramphenicol and acetyl-coenzyme A; CAT catalyses the acetylation of chloramphenicol. The acetylated products and the unmodified reactants are separated from the aqueous solution, and acetylation is monitored by autoradiography using thin-layer chromatography (TLC) when the acetylated form is separate from the unacetylated form. The percent conversion of ^{14}C-chloramphenicol to acetyl-(^{14}C)-chloramphenicol can be measured by densitometry analysis of an autoradiograph.

Several *cat* genes from *E. coli, Proteus mirabilis, Staphylococcus aureus, Bacillus pumilus, Clostridium difficile, Clostridium perfringens, Campylobacter coli* and *Streptomyces acrimycini* have been sequenced, and all showed similarity around the active site. In 1992, Roger Parent and Paul H. Roy from Canada sequenced the gene coding for the CAT of Tn2424 of plasmid NR79. The

Tn2424 *cat* gene is part of a 4.5-kb region that codes for a protein of 23,500 Da. It also contains the *aacAla* aminoglycoside-6'-*N*-acetyltransferase gene within 4.5 kb region. Tn2424 is a multi-resistance integron which has evolved by site-specific integration of discrete units, principally containing antibiotic resistance genes between two conserved segments, *i.e.* a 5' conserved segment containing a common antibiotic resistance promoter (and, in the opposite direction, a gene encoding a site specific recombinase of the phage integrase family) and a 3' conserved segment containing ORF4, *sul*l, and ORF5 (Fig. 20.3).

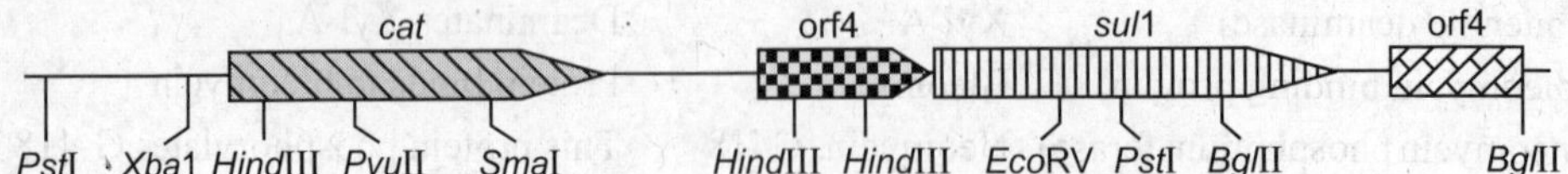

Fig. 20.3. Restriction map of sequenced regions of Tn2424 of plasmid NR79 showing *cat* gene, ORF4, *sul*1 and ORF5.

***(b)* Thymidine Kinase (TK):** Thymidine kinase isa phosphotransferase enzyme encoded by thymidine kinase (*tk*) gene. TK is expressed in most mammalian cells and some viruses (herpes viruses, pox viruses, etc.). In mammalian cells, it is present in two forms of isoenzyme, TK1 and TK2. Deoxythymidine (Thd) reacts with ATP to give deoxythymidine monophosphate (TMP) and adenosine 5'-diphosphate (ADP); thymidine kinase catalyses the reaction as below:

$$\text{Thd} + \text{ATP} \rightarrow \text{TMP} + \text{ADP}$$

Thymidine kinases have a key function in the synthesis of DNA and also in cell division, because they are part of the unique reaction chain to introduce deoxythymidine into the DNA. It is used as a selectable marker that is why tk^- cells and viruses do not survive in medium containing hypoxanthine - aminopterin - thymidine medium (HAT medium); such cells co-transfected with a vector carrying a tk gene (tk+) can be rescued. The complete nucleotide sequence of the human thymidine kinase cDNA is 1,421 bp in length with a large open reading frame of 702 bp capable of specifying a protein with a molecular weight of 25,504. It has two highly conserved promoter elements: the CCAAT element located at -40 and -70, and a TATA element at -21.

In 1960, H.A. Johnson and co-workers for the first time labelled human tumour cells *in vivo* by tritiated thymidine and measured the radioactivity to identify the dividing cells. For this purpose tritiated thymidine is included in the growth medium. In spite of errors in the technique, it is still used to determine the growth rate of malignant cells and to study the activation of lymphocytes in immunology.

In 1996, Lonfye Lye (from University of California, U.S.A.) and Ching C. Wang (from National Taiwan University, Taiwan) tested the herpes simplex virus *tk* gene. They found it as a potential positive and a negative selectable marker in *Trypanosoma brucei* for isolating glycosome-deficient *T. brucei* mutants in the procyclic form and to identify genes involved in glycosomal biogenesis. They have established the methodology for both positive and negative selection of potential glycosome-deficient mutants of *T. brucei*427/*ptk*-SKL.

***(c)* Dihydrofolate Reductase (DHFR) (Resistant to Methotrexate):** The DHFR found in all organisms regulates the level of tetrahydrofolate in the cell. Bacterial species possesses distinct DHFR enzymes, but mammalian DHFRs are very similar to it. In humans the DHFR enzyme is encoded by the *DHFR* gene, located in the q11 → q22 region of chromosome 5. Tetrahydrofolate is essential for synthesis of purines and thymidylate, which in turn are important for cell growth and proliferation. DHFR plays a key role in the synthesis of precursors of nucleic acid; the mutant cells completely lacking DHFR require glycine, purine, and thymidine for their growth. DHFR reduces the NADPH-dependent reduction of 7,8-dihydrofolate (H_2F) to 5,6,7,8-tetrahydrofolate (H_4F) by using NADPH as electron donor (Fig. 20.4); thus it maintains the intracellular levels of H_4F. Tetrahydrofolate acts as an essential cofactor required during synthesis of purines, thymidylate and many amino acids.

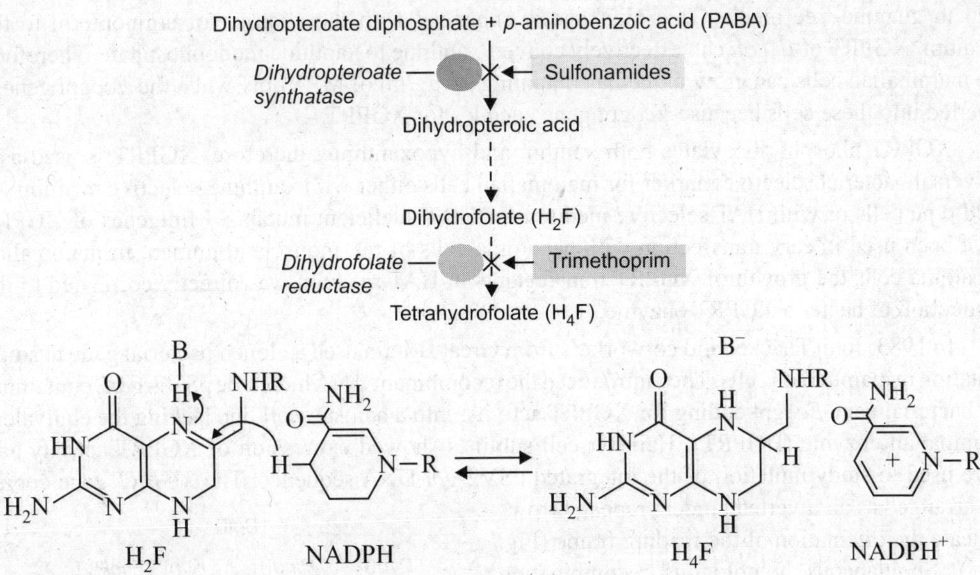

Fig. 20.4. Reduction of dihydrofolic acid to tetrahydrofolic acid by DHFR using NADPH.

The plasmid-encoded DHFR shows the maximum resistance to an antibiotic trimethoprim, which is a homotetramer with an unusual pore that contains the active site. Structure of trimethoprim differs from that of chromosomal DHFRs. Since the rapidly dividing cells require folate to make thymine, the use of antifolate may be advantageous to chemotherapy. For example, methotrexate (4-amno-10-methylfolic acid) inhibits the division of neoplastic cells and, therefore, it is used in chemotherapy of cancer. It must be noted that the DHFR is also required by the bacteria for their growth and multiplication; hence inhibitors selective for bacterial vs. host DHFR have found application as antibacterial agents.

The methotrexate and trimethoprim are the inhibitors of the enzyme dihydrofolate reductase, where trimethoprim is an antibiotic and methotrexate is a chemotherapeutic agent of cancer. This differential behaviour is because of a selective affinity of methotrexate for mammalian dihydrofolate reductase, and that of trimethoprim for the bacterial DHFR.

In 1992, Masahiro Iwakura and Teruo Tanaka reported the dihydrofolate reductase gene (*DHFR*) as a versatile expression marker. They used *E. coli DHFR* gene as a genetic marker specifying trimethoprim resistance (Tmp^R). They constructed three types of plasmids (promoter cloning vector, terminator cloning vector, and the plasmid) containing the *DHFR* gene cassette. In these systems, the selection of recombinant plasmids was carried out just by examining the Tmp^R phenotype of the transformed cells. They evaluated the efficiency of promoters and terminators in the fused DNA fragment on the basis of levels of the enzymatic activity of DHFR. Here the most important aspect of gene expression studies is choosing suitable reporter for experimentation. Each and every reporter differs in providing the information about gene expression. There are two general types of reporter gene constructs, transcriptional reporters and translational reporters.

***(d)* Bacteria Xanthine-Guanine Phosphoribosyl Transferase (XGPRT) (Resistant to Mycophenolic acid):** Inosine monophosphate dehydrogenase is an enzyme which is required in guanosine monophosphate (GMP) synthesis pathway. But the activity of inosine monophosphate dehydrogenase is blocked by an antibiotic mycophenolic acid (MPA). Therefore, MPA can be used as a selective agent in mammalian systems because of its differential effects on the activities of mammalian HPRT (hypoxanthine phosphoribosyl transferase) and bacterial XGPRT. Both enzymes, mammalian HPRT and bacteria XGPRT, mediate an alternative pathway for GMP synthesis. During GMP synthesis mammalian cells use a small amount of xanthine, which is blocked by mycophenolic acid in guanine-free media. This blockage is enhanced with the addition of aminopterin to the medium. XGPRT of *E. coli* can effectively convert xanthine to xanthine monophosphate. Therefore, the mammalian cells can grow in media containing 25 μg/ml of MPA only when the *Eco*gpt gene is inserted into these cells because *Eco*gpt gene encodes for XGPRT.

XGPRT phosphoribosylates both xanthin and hypoxanthine; therefore, XGPRT is used as a universal bacterial selective marker for mammalian cells either with xanthine-selective medium for wild type cells or with HAT-selective medium for HPRT-deficient mutants. Minigenes of XGPRT have been used in eary transfection with hepatoma cells of rat, mouse and human. In mammalian hepatoma cells the growth of XGPRT transfectants in HAT medium was directly correlated to the production of bacteria XGPRT enzyme.

In 1983, John Thacker and co-workers from Great Britain used a cloned bacterial gene to study mutation in mammalian cells. They introduced the recombinant DNA molecule pSV2-*gptI,* containing the bacterial gene *Eco*gpt coding for XGPRT activity, into a hamster cell line lacking the equivalent mammalian enzyme (HGPRT). Hamster cell sublines showed expression of XGPRT activity and were used to study mutation of the integrated pSV2-*gpt* DNA sequence. The *XGPRT* gene copies are arranged as an inverted tandem repeat; arrows indicate the orientation of the reading frame (Fig. 20.5). Spontaneous homologous recombination between these copies can reconstitute a functional *XGPRT* gene. A sequence of 460 bp uninterrupted homology mediates recombination. An early SV40 promoter (*Pvu*II-*Hind*III fragment) controls the intact *XGPRT* gene. The 5' gene copy contains a 5' deletion of 460 bp which eliminates the SV40 promoter and part of the bacterial promoter region; deletion locations are indicated by black bars. The 3' copy is inactivated by a deletion of the 3' gene region (475 bp).

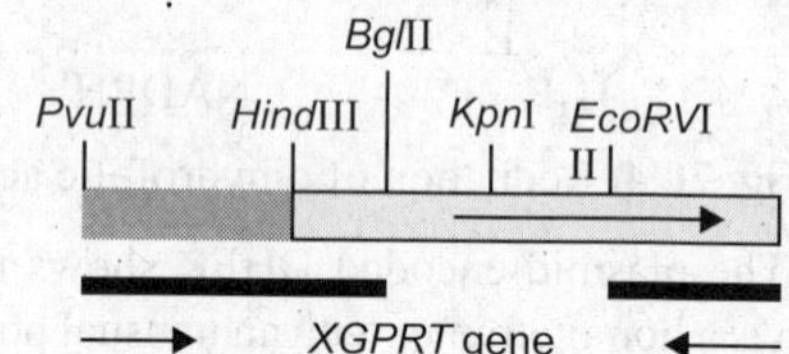

Fig. 20.5. Bacterial *XGPRT* gene constructs under control of an early SV40 promoter (*Pvu*II–*Hind*III fragment) (the arrow indicates the protein coding sequence within the gene).

***(e)* Neomycin Phosphotransferase (NPT) (Resistant to Neomycin, G-418)**: G418 is an aminoglycoside antibiotic produced by *Micromonospora rhodorangea* that inhibits the elongation step in both prokaryotic and eukaryotic cells, and thus blocks protein synthesis. Resistance to G418 is conferred by the neomycin resistance gene (*neo*) from bacterial transposon Tn5 encoding an aminoglycoside 3′-phosphotransferase [APH (3′) II] or neomycin phosphotransferase (NPTII). It is of two types, NPTI and NPTII; probably the NPTII is the most widely used selectable marker for plant transformation. In different organisms it is also used in gene expression and regulation studies because N-terminal fusions can be constructed that retain enzymatic activity (see Chapter 19 : *Transgenic Plants*). In animal cells, G418 and neomycin are used as selectable agents. In mammalian cells markers is usually achieved within three to seven days with concentrations ranging from 400 to 1,000 μg/ml. The dividing cells are affected more quickly than un-dividing cells.

***(f)* Transcriptional and Translational Reporters:** In recent years reporter genes have become standard genetic tools used to evaluate either the transcriptional or the translational activity associated with genes of interest, whose products cannot be easily assayed. The *lacZ* gene from *E. coli* has been used very effectively to quantify such regulated activities in many different organisms. Transcriptional reporters consist of a promoter fragment from a gene of interest driving GFP (Fig. 20.6 A).

Typically, promoter fragments of a few kilobases immediately upstream of the start codon contain a significant portion of the *cis*-regulatory information necessary to provide a tentative expression pattern of the endogenous gene under study. For example, one of the translational reporters is in frame gene fusions between GFP and a gene of interest (B). Ideally, a translational reporter includes the entire genomic locus of a gene (5′ upstream region, exons, introns, 3′ UTR). A marker gene such as GFP can be inserted at any point in the open reading frame, preferably at a site that does not disrupt protein function.

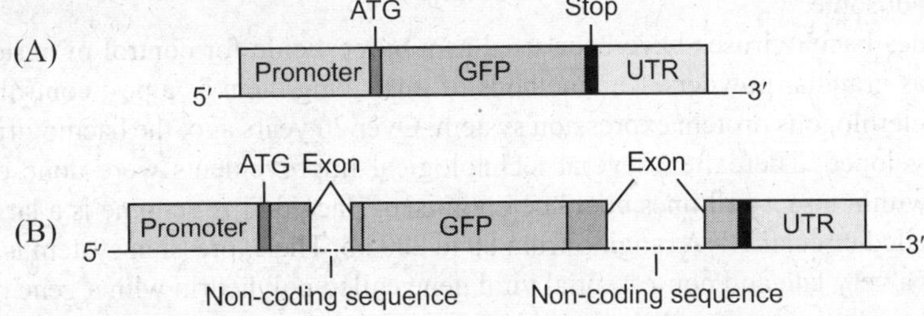

Fig. 20.6. Structure of reporter gene constructs; A- transcriptional reporter; B- translational reporter.

In 2001, Allen L. Honeyman and co-workers from the University of South Florida (U.S.A.) have reported the construction of a pair of plasmids to be used for transcriptional or translational *lacZ* gene fusions in *Streptococcus mutans*. They compiled the complete nucleotide sequences of these gene reporter plasmids. The translational *E. coli* β-galactosidase gene (*lacZ*) fusion plasmid (pALH109) and the transcriptional *lacZ* gene fusion plasmid (pALH122) were used successfully in *S. mutans* to measure the activity of various phosphotransferase system (PTS) genes. Both plasmids employ fusions with the *E. coli lacZ* gene that can be easily quantified using standard *O*-nitrophenyl-β-*D*-galactopyranoside (ONPG) based enzyme assays using 4-methyl-umbelliferyl β-*D*-galactopyranoside (MUG) as the substrate.

Most reporter genes are engineered so that it may contain an artificial eukaryotic translation initiation sequence and a polyadenylation signal to improve their translational efficiency. For this a small intron is added in the transcript, because full RNA processing has been found to contribute to message stability and transport.

B. CLONING VECTORS

Construction of animal vectors is based on the other viruses such as adenoviruses, adeno-associated viruses, retroviruses, bacculoviruses, bovine papilloma viruses, simian vacuolating virus 40 (SV40), chemoviruses (chemically modified viruses), synthetic viruses, and transposable elements e.g. P element of *Drosophila*. Cloning vectors of some of these viruses are described in detail in Chapter 7: *Cloning Vectors* and Chapter 9: *Gene Therapy*. These viruses have been used for the purposes of gene therapy. In biotechnology, animal cloning vectors are also required for the synthesis of recombinant protein from such genes that do not express properly in *E. coli* or yeast. This section deals with some of the cloning vectors not described in previous chapters.

1. Baculovirus Expression Vectors

Though baculoviruses are known for hundreds of years due to infection of Chinese silkworms, significant information could be gathered during the mid of 20th century. Baculoviruses are capable of entering mammalian cells in culture but it is unknown whether they can replicate within cells of mammals or other vertebrates. Baculoviruses are a family of large rod-shaped viruses that can be divided to two genera: nucleopolyhedroviruses (NPV) and granuloviruses (GV). Further, there are two groups within NPV : (*i*) multiple-embedded nucleopolyhedrovirus, e.g. *Autographa californica* MNPV (AcMNPV), and (*ii*) single-embedded nucleopolyhedrovirus, e.g. *Trichoplusia ni* SNPV (TnSNPV). AcMNPV is the most widely studied NPV.

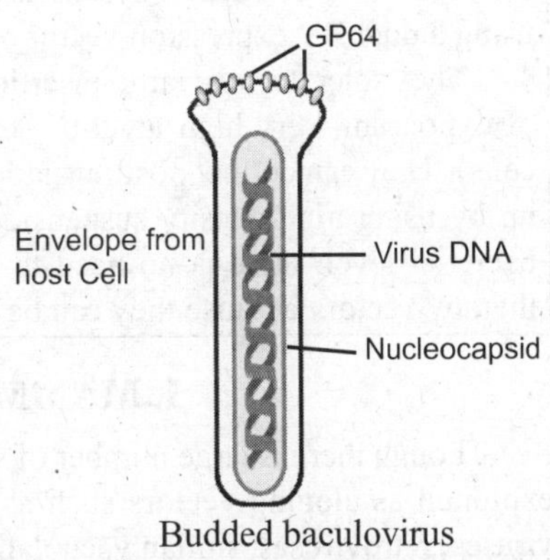

Budded baculovirus

Gp64 is a homotrimeric membrane glycoprotein present polarly on the rod-shaped virion. During infection, it is required for efficient budding of the virion, for cell-to-cell transmission cycle and for binding to cell surface. The major function of the gp64 is to cause pH-mediated envelope fusion to the endosome.

For decades baculoviruses have been used as a biopesticide for control of agricultural pests and marketed as granular powders. The methods of improving them as a pest control have lead to their use as a heterologous protein expression system. Over 20 years ago, the baculovirus expression vector was developed. Thereafter, several technological improvements were done expression of foreign genes within insect cell lines could be optimised. Baculovirus genome is a large covalently closed, supercoiled circular DNA ranging from 80 to 220 kb. The expression system is based on the replacement of a very late and non-essential viral gene called polyhedrin with a gene of interest.

Recently, baculovirus vectors have been developed in such a way that may contain mammalian cell-active promoters and enhancers to permit stable gene expression within higher eukaryotic cell lines, such as human hepatocytes and Chinese hamster ovary cell lines. Baculoviruses can be an attractive and useful tool to study gene expression and gene function within mammalian systems and cell lines. Baculovirus expression vector system is based on the introduction of a foreign gene into nonessential region of virus genome via homologous recombination with a transfer vector containing target gene. The resulting recombinant baculovirus lacks one of nonessential gene replaced with foreign gene encoding heterologous protein which can be expressed in cultured insect cells and insect larvae. There are several baculovirus expression systems including Bac-to-Bac, flashback, BacMam, etc.

***(a)* Bac-to-Bac Method:**The Bac-to-Bac baculovirus system (Invitrogen) rapidly generated recombinant baculoviruses. The basis for Bac-to-Bac system is site-specific transposition of an expression cassette into a baculovirus shuttle vector (bacmid). Bacmid DNA is propagated in *E. coli* DH10Bac as a large plasmid, which confers resistance to kanamycin. The bacmid results in a *lacZ* deletion on the chromosome to form colonies, that are blue (Lac^-) in the presence of a chromogenic substrate and the inducer IPTG. The *lacZ* sequence is disrupted due to transposition of gene of interest from the pFastBac1 donor plasmid into the bacmid. Consequently the recombinant bacmids cause the appearance of white colonies. When the recombinant bacmid DNA is transfected into insect cells it generates recombinant particles of baculovirus. These particles can be amplified during further infection and used to study protein expression.

Advantages of Baculovirus Vectors

The baculovirus expression vector system offers several desirable features such as efficient expression of high levels of soluble recombinant proteins exceeding 600 μg/ml, easy to scale-up production, proper protein folding, oligomerisation and post-translational modifications (glycosylated, fatty acid acylated phosphorylated and methylated) similar to that occurring in mammalian cells. The BV system is particularly well suited to the expression of multiple proteins using a dual BV expression vector or co-infection.

They tolerate very large insertions (15 kb, but theoretically up to 100 kb). They are capable of expressing very high level of heterologous protein (~1 mg of protein per 1×10^6 of infected cells). They can correct posttranslational modification of heterologous proteins. It is easy to scale-up by using high-density suspension culture. Since the bacmid DNA is already selected at the bacterial level, there is no need to make plaque purification. Baculoviruses are potential gene therapy vectors because they can be modified for transduction of mammalian cells.

I. MAMMALIAN VIRUS VECTORS

Though there is large number of viruses causing diseases in mammals, some of them have been exploited as cloning vectors such as adenoviruses, adeno-associated viruses, lentiviruses, sendai viruses, retroviruses, simian vacuolating virus 40 (SV40), etc.

1. SV40-based Vectors

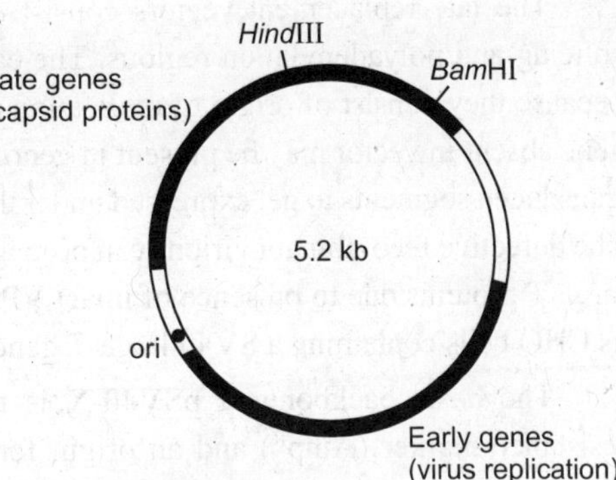

Fig. 20.7. SV40 genome.

In 1979, for the first time D.H. Hamer and P. Leader constructed the mammalian viral vector to express the mouse *maj-globin* gene cloned simian virus 40. SV40 is a double stranded DNA containing tumour virus having 5243 bp long genome. SV40 encodes two early gene products (small and large T antigens) and four late gene products (VP1, VP2, and VP3; VP = viral protein).

SV40 occurs naturally in rhesus monkeys, but kidney cells of African monkeys are preferred for laboratory tests. After infection, the virus replicates within the nucleus of the host cells. SV40 leads two life cycles which depend upon host cell line used, for example a productive lytic cycle that occurs in permissive cells (monkey cells), and a non-permissive life cycle (in rat or mouse cells) where host cells are transformed. Recombinant SV40 vectors are good candidates due to the presence of several key features : (*i*) a well-known virus, (*ii*) easy to make non-replicative vectors, (*iii*) production in titers of 10(12) IU/ml, (*iv*) efficient transduction of both resting and dividing cells, (*v*) delivering persistent transgene expression to a wide range of cell types, and (*vi*) presence of nonimmunogenic property. Three types of vectors (plasmid vectors, transducing vectors and transforming vectors) have been developed by using SV40 genome (Fig. 20.7).

(*a*) SV40 Transducing Vectors: SV40 should be looked as a potential vector for delivering the foreign DNA into mammalian cells, because discrete segment of viral DNA can be replaced with foreign DNA of known function. Then such reconstructed genome can be transduced into suitable cells. The transduced segment cannot exceed half of size of viral DNA so that enough DNA should be retained to maintain viral function. For example 2 kb late region segment encoding VP1, VP2, and VP3 can be replaced with cDNA of almost same size. The transduced segment is supposed to replicate if *cis*- and *trans*- acting transcription elements are retained properly, that consist of early- and late- region promoters, *ori* region, poly-A site, and large T-protein expressing ability.

The vectors where VP1, VP2, and VP3 encoding region (late expression gene) replaced with a DNA insert (e.g. gene X), is called '*late replacement vector*', such as SVGT-5 which is an improved SV40-based transducing vector. Similarly, the vectors where VP1, VP2, and VP3 encoding region is present but the early gene (large-T gene) is replaced with a DNA insert (e.g. gene X), is called '*early replacement vector*', such as COS CV-1 vector. The method of construction of both late and early replacement vectors are shown in Fig. 20.8 A-B.

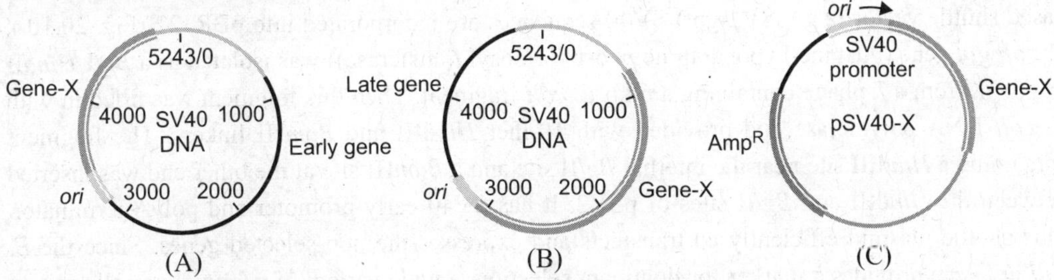

Fig. 20.8. The SV40 replacement vectors; A- late replacement vector; B- early replacement vector; C- plasmid pSV40-X.

The late replacement vectors consist of : (*i*) ori gene, (*ii*) early region of SV40, and (*iii*) splicing and polyadenylation regions. The early replacement vectors do not need any helper virus because they consist of origin of replication and genes encoding VP1, VP2 and VP3. The essential gene absent in vector may be present in genome of host cells. The recombinant segment permits the transduced segments to get expressed under the control of the early region promoter and Poly-A site. The defective recombinant virions can propagate if co-infected with a helper virus (that supplies the large-T proteins due to presence of intact VP1, VP2 and VP3 region) or special primate cells (such as CHO cells containing a SV40 large-T gene in its genomic DNA).

The basic backbone of pSV40-X is the pBR322 plasmid sequence containing antibiotic resistance marker (Amp^R) and an origin for replication (*ori*) that helps the plasmid to replicate with *E. coli* cells. The desired coding sequence (gene-X) is flanked by the promoter of SV40 early region at 5′ (Fig. 20.8 C; see arrow) in such way that transcription may proceed toward gene-X. The small-T intron and the SV40 early region Poly-A site follows the coding sequence so that splicing and polyadenylation of the gene-X mRNA may be permitted.

In 1975, for the first time Tom Maniatis and coworkers from Cold Spring Harbor Laboratory (New York) cloned the cDNA of rabbit β-globin gene at *Eco*RI site of pMB9 using homopolymer tails, and inserted the modified cDNA into SVGT-5. Then β-globin cDNA was inserted into pBR322 to construct SVGT-5-Ra-*β*G (Fig. 20.9.). The monkey kidney cells were infected with a mixture of SVGT-5-Ra-*β*G DNA and DNA from mutant SV40 so that SVGT-5-Ra-*β*G genome could be propagated.

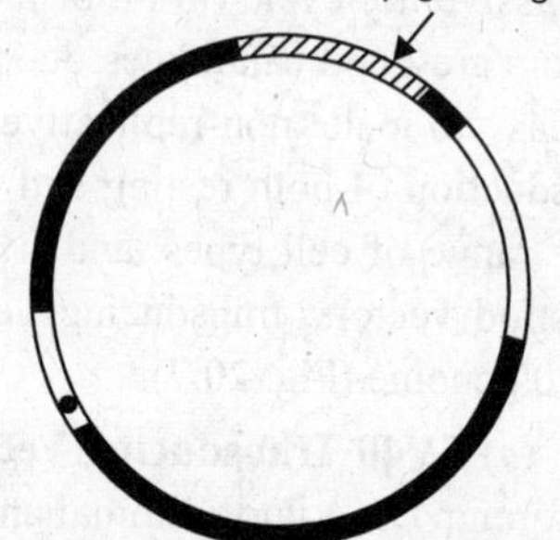

Fig. 20.9. SVGT-5 with inserted rabbit β-globin gene.

In 1985, Masahide Asano and co-workers from the Kanazawa University (Japan) have constructed an early replacement SV40 vector (pSV40-Muβ) by replacing the part of early virus genes with mouse interferon-β (IFN-β) cDNA. After transfection of COS-7 cells with this DNA, the transducing viral particles were produced which infected various cells and produced mouse IFN-β. After virus infection, the IFN production was very high in monkey kidney cells, less in human epithelial cells, but least in mouse, pig, hamster cells and in human lymphocytes. The efficiency of introduction of DNA to monkey kidney cells was compared with that by the calcium phosphate precipitation method. The viral vector was ten to hundred times more efficient.

(*b*) SV40 Plasmid Vectors: These vectors replicate in mammalian tissue culture cells without being released as virions as packaging is not done within host cells. Several phasmid shuttle vectors have been constructded such as pSV2, pSV3, pSV2-gpt, etc. For construction of SV40 plasmid-based shuttle vector (e.g. pSV2-gpt) SV40 sequences are incorporated into pBR322 (Fig. 20.10.). *E. coli gpt* gene (*g*uanine-hypoxanthine *p*hosphoribosyl *t*ransferase)) was isolated as a *Bgl*I-*Hind*II fragment from a λ phage containing a 6 kb *E. coli* fragment. Then this fragment was filled in with *E. coli* DNA polymerase and provided with 10-mer *Hin*dIII and *Bam*HI linkers. The fragment containing a *Hin*dIII site near the internal *Bgl*II site and a *Bam*HI site at the other end was inserted between the *Hin*dIII and *Bgl*II sites of pSV2. It has SV40 early promoter and polyA terminator. This shuttle plasmid efficiently co-transfects and expresses the non-selected genes. Since the *E. coli gpt* gene provides a marker for dominant selection, a wide variety of mammalian cells can be used. Plasmid-containing cells use xanthine as a guanine nucleotide precursor. The hosts are *E. coli*, mammalian cells, COS cell lines and SV40 permissive cells.

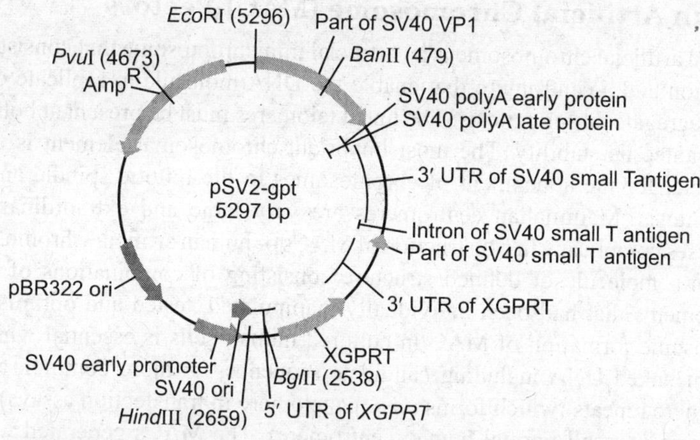

Fig. 20.10. A SV40 plasmid-based shuttle vector- pSV2-gpt.

2. Bovine Papilloma Virus (BPV) Vectors

Expression of these genes requires their replication within eukaryotic cells in culture. The papillomavirus-based shuttle vectors are capable of replication in both bacterial and eukaryotic cells; therefore, they have provided several interesting findings about the relationship between eukaryotic gene structure and function. The episomal nature of the BPV vector system makes it a useful tool for looking the biological phenomena pertaining to gene replication and regulation, without the interference by host chromosomes. In culture the host cells are transformed by the viral genome resulting in piling up of host cell colonies rather than forming the typical monolayer. Here the transformed host cells behave like tumour cells. For the selection of recombinant DNA, the eukaryotic DNA is first cloned in *E. coli* cells, its plasmid (*i.e.* pBR322) is deleted from recombinant DNA, and the remaining DNA is inserted into the animal cells. The viral vector becomes circularised and behaves like a plasmid.

The ability of the BPV system to allow plasmids to be shuttled between bacteria and animal cells could be further utilised in two ways : (*i*) genes that affect cellular phenotypes can be analysed, and (*ii*) gene libraries made in BPV expression vectors could be introduced into animal cells and cloned cell lines could be isolated, that produce the desired phenotype. Then plasmid DNA could be rescued from these cells and the desired gene is isolated.

Fig. 20.11 shows a linear map of BPV1 genome opened at *Hpa*I site; there are 10 open reading frames (ORF) in BPV, out of which eight ORFs (E1 to E8) are required for early viral function, replication and transformation, and two are late ORFs (L1 to L2). L1 expresses the major capsid proteins and L2 the minor capsid protein.

In 1981, N. Sarver and coworkers from the National Cancer Institute (Maryland) demonstrated that a novel eukaryotic vector, derived from the transforming region of bovine papilloma virus, was highly effective for introducing foreign genes into animal cells.

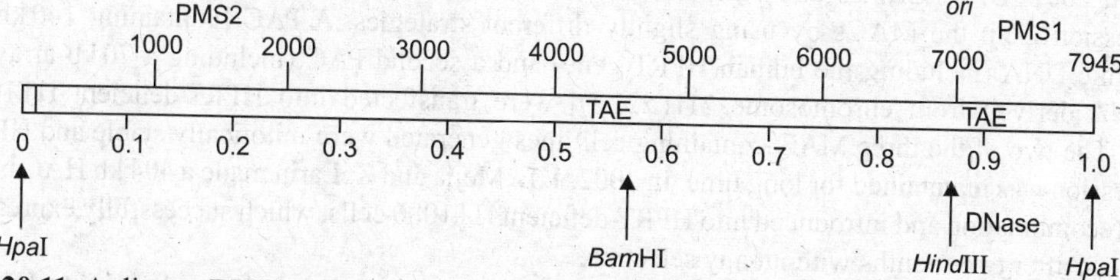

Fig. 20.11. A linear BPV1 genome map opened at *Hpa*I site; T- TATA promoter elements, A- polyadenylation recognition sequences, A-enhancer elements; PMS- plasmid maintenance sequence.

2. Mammalian Artificial Chromosome (MAC) Vectors

A mammalian artificial chromosome is an artificial minichromosome that consists of telomeres, origin of replication and a centromere that enable the DNA molecule to replicate during S phase and to correctly segregate during mitosis. The lineal telomeres must be present at both ends of DNA molecule to guarantee its stability. The most important chromosomal element is the centromere because it is crucial for the attachment of chromosomes to the mitotic spindle and their correct segregation in mitosis. Mammalian centromeres are very large and extraordinarily difficult to engineer; this causes complicacy in construction of MACs or human artificial chromosomes (HACs).

The MAC is a molecule of defined structure consisting of combinations of several cloned chromosomal elements that had been individually manipulated, tested and optimised. For proper function, chromosome formation of MAC in cultured human cells is essential which is followed by transfection of naked DNA including candidate sequences. A more common approach is the inclusion of telomere repeats (which forms *de novo* telomere in transfection assays) and β-satellite DNA (which is present at all normal human centromere). The MACs generated were mitotically stable, associated with functionally important cetromere proteins and were in the 1–10 Mb size range.

A human artificial chromosome (HAC) vector is a microchromosome that can act as a new chromosome in a population of human cells. In 1997, J. Harrington and coworkers for the first time described the human artificial chromosomes (HAC) by combining portions of alpha satellite DNA with telomeric DNA and genomic DNA into linear microchromosomes in human HT1080 cells called 'artificial chromosomes'. HAC is about 6-10 Mb in size and able to carry new genes. They are useful in expression studies as gene transfer vectors and are a tool for elucidating human chromosome function. They have been found mitotically and cytogenetically stable for up to six months when grown in HT1080 cells.

In 1998, H. Masumoto and collaborators have showed how two different alphoid DNA loci from the same chromosome behave in a completely opposite manner. It was demonstrated that no telomere sequences are needed when the input DNA is circular. They found that alphoid-derived YAC unable to form HACs do not have CENP-B boxes in addition to a more diverged alphoid repeats (Fig. 20.12). Thus MACs can be formed in HT1080 cells from either linear or circular input DNA, with circular alphoid constructs being very effective at forming MACs, whether or not human telomere sequences are added. Indeed, the presence of these arrays did not favour MAC formation. By contrast, it was ascertained how linear DNA molecules require telomere sequences to maintain and protect minichromosome ends.

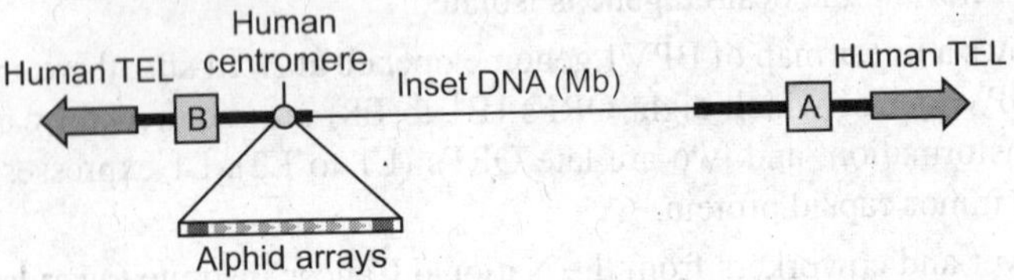

Fig. 20.12. Human artificial chromosome (HAC) vector; TEL- telomere.

In 2001, B. R. Grimes and other collaborators demonstrated the entire human HPRT gene expression from the MACs by using slightly different strategies. A PAC [containing 140 kb of genomic DNA (including the human HPRT gene) and a second PAC (including a 70 kb array of α-DNA derived from chromosome 21 (α21-I)] were transfected into HPRT-deficient HT1080 cells. The two of the three MAC-containing cell lines generated were mitotically stable and HPRT expression was maintained for long time. In 2002, J. E. Mejía and Z. Larin made a 404 kb HAC by *in vivo* recombination and introduced into HPRT-deficient HT1080 cells, which successfully expressed the gene after two months without any selection.

In a second study, a 404 kb HAC made by *in vivo* recombination and introduced into HPRT-deficient HT1080 cells successfully ensured expression of the gene after two months without any selection. Human artificial chromosomes (HACs) have several advantages as gene therapy vectors including stable episomal maintenance, and the ability to carry large gene inserts. HAC vectors have been developed from the normal human chromosomes through chromosome engineering.

In 2011, Y. Kazuki from Tottori University (Japan) and coworkers from the other institutions reported a refined human artificial chromosome called 21HAC, which is based on a stripped copy of human chromosome 21 producing a 5 Mb long chromosome. In 2011, Masaharu Hiratsuka from Tottori University (Japan) and other coworkers used this platform to create the integration-free pluripotent stem cells from mouse embryonic stem cells. Thus, the HAC vector could generate uniform, integration-free induced pluripotent stem cells (iPS cells) with a built-in safeguard system.

C. TRANSFECTION METHODS

Transfection refers to a process of deliberately introducing the nucleic acids into cells. Originally this term was used for the introduction of DNA (or RNA) from a virus into cells causing infection. But in animal cell biology the term is generally used for non-viral methods in eukaryotic cells, referring to a change in cell properties due to deliberate introduction of DNA. Consequently, transient pores in the cell membrane get opened during transfection permitting the uptake of DNA molecules. This results in morphological changes and abnormalities in target cells.

1. Types of Transfection

Transfection is of two types, stable transfection and transient transfection.

(*a*) Stable Transfection. Stable transfection refers to the production of a population of cells in which the gene(s) of interest are stably expressed into it. It means that a gene introduced into the cell gets integrated into the host DNA and carried over several generations during cell division. In contrast, the plasmid initially introduced into a cell is often episomal; that is why it can frequently be degraded or lost. At this time plasmid expression is transient because the DNA is only transiently present within the cells. Therefore, a marker gene is co-transfected to accomplish stable transfection, which gives the cell some selectable advantage such as resistance to certain antibiotics. For isolation of such cells, they are fist transfected with the reporter construct along with the other plasmids carrying a selectable marker like neomycin resistance gene ($G418^R$). Upon co-transfection with two plasmids, mostly cells will integrate either both or none of these plasmids. Such cells that stably express a reporter construct can be isolated by eliminating the other cells that have not stably integrated the DNA construct. The cells stably integrated with the DNA construct can be selected by using G418, and a population of cells expressing the reporter construct of interest can be raised. Then the cell population can be divided to study gene expression under a variety of conditions.

Moreover, the flanking sequences likely have strong influences on the expression of DNA of interest depending on the site of integration. In stable transfection the DNA is integrated into the host chromosome, which may influence either to increase or decrease the expression of the gene construct. Therefore, it is necessary to study not single isolates but rather populations of isolated cells or multiple isolates to eliminate this problem.

(*b*) Transient Transfection. Transient analysis is the second approach for transfection analysis. Here DNA is introduced into a cell population by transfection but stable cell lines are not isolated. Gene expression is studied immediately after transfection within 24-72 hours.

1. Methods of Transfection

The cell membrane is a barrier to any molecule and does not permit to enter into the cell. A large and highly charged molecule such as DNA gets a little success to enter the cell. There are many ways to overcome the permeability barriers. Several methods are used to introduce the DNA into animal cells/embryos as given below :

***(a)* Calcium-phosphate Precipitation.** In 1973 for the first time F.L. Graham and A.J. van der Eb originally discovered a new technique using calcium phosphate for the assay of infectivity of human adenovirus 5 DNA. First of all the DNA preparation to be used for transfection is dissolved in HEPES [4-(2-hydroxyethyl)-1-piperazineethanesulfonic acid] (a zwitterions)-buffered saline solution (HeBS) followed by mixing of calcium chloride to the DNA solution. This results in the formation of insoluble calcium phosphate that co-precipitates with the DNA construct as calcium phosphate-DNA complex combining with calcium and phosphate. Then calcium phosphate-DNA co-precipitate is added to the desired cells to be transfected. When this complex is added to cell culture, the cells take up the co-precipitate particles through phagocytosis and the particles are frequently internalised through endocytosis. The DNA is integrated into cell genome, and then expressed (Fig. 20.13). About 20% of the cells are transfected by this method. The varying size of the particles affects the transfection efficiency.

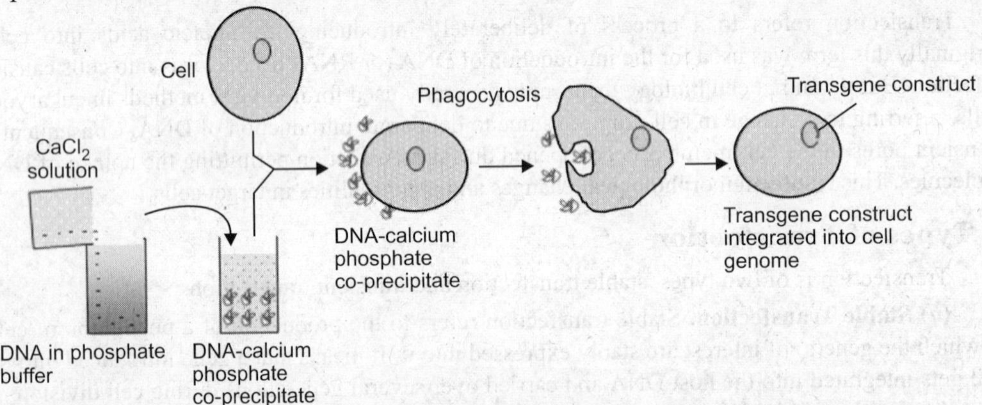

Fig. 20.13. Transfection by calcium-phosphate precipitation method.

***(b)* DEAE-dextran Mediated Transfection.** DEAE-dextran (dimethylaminoethyl-dextran) is solubilised well in water; it is a polycationic organic chemical compound that contains multiple positive charges on its surface. DNA construct is complexed with DEAE ion exchange resin. Positively charged DEAE-dextran interacts with the negatively charged DNA molecules; the cell surface components play an important role in transfection. DEAE-dextran brings about DNA uptake by the cells through endocytosis which is then internalised into the cells. This method is mostly used for transient transfection particularly by using COS cell lines; but this approach is inefficient for stable transfection. For detail, see Chapter 8 : *Methods of gene manipulation*.

***(c)* Viruses-mediated Transfection.** Virus-mediated gene transfer has been described in detail in Chapter 9 : *Gene therapy*.

***(d)* Liposome-mediated Transfection (Lipofection):** Liposome-mediated delivery of DNA into cells is called lipofection. Cationic lipids are used to make liposomes because DNA efficiently complexes with these liposomes to form detergent-DNA complexes. Lipofection is considered to be an effective process of gene transfer because : (*i*) it has improved transfection efficiencies, (*ii*) it is highly cost effective, (*iii*) it poses no significant toxicity, and (*iv*) it has successfully been used on many cell lines. For detail, see Chapter 8 : *Methods of gene manipulation*.

***(e)* Electroporation.** For detail, see Chapter 8 : *Methods of gene manipulation*.

***(f)* Impalefection:** Insertion of DNA into a cell after binding to a surface of a nanofiber is called impalefection. Cells to be transfected are impalled by the arrays of elongated nanostructures such as carbon nanofibers or silicon nanowires which have already been functionalised with plasmid DNA. Arrays of nanofibers can be introduced into large numbers of cells and intact tissue by using this approach.

(*g*) Microinjection. For detail, see Chapter 8 : *Methods of gene manipulation.*

3. Production of Transgenic Cattle

Transgenic cattle are produced by direct microinjection of DNA which is a widely used technique for gene transfer. In 1980, J.W. Gordon and coworkers were the first to demonstrate the microinjection of recombinant plasmids into mouse oocytes and detection of the injected DNA in a fraction of the fully developed mice by Southern blotting technique. In additional, several experiments have demonstrated the integration of DNA into the genome and expression of this newly integrated DNA.

The purpose of gene transfer is to create such organisms in which many cells can carry the new foreign gene. Therefore, transfer of the new gene into the organisms should be done at very early stage of its development. Hence, microinjection of DNA is carried out either with fertilised oocytes or with 2-cell stages. The transgenic cattle can be produced into five main stages : (*i*) preparation of donor animals, isolation of fertilised oocytes, and visualisation of pronuclei, (*ii*) preparation of DNA solution for injection, (*iii*) picroinjection of DNA solution into pronuclei, (*iv*) transfer of injected zygotes into recipients, and (*v*) screening of animals born for integration of the transferred gene.

Isolation of fertilised oocytes from cattle is an extremely laborious process due to unavailable of non-surgical techniques. Surgically the abdominal cavity is opened by making an incision in the median line, and the oviducts are flushed to collect oocytes. An alternative method of getting embryos is the slaughtered superovulated animals, where reproductive organs may be obtained easily during the process of slaughtering. After 12 to 24 hours of fertilisation oviduct is flushed to isolate the oocytes, which are washed and kept in culture medium.

In fertilised oocytes of cattle and some other domesticated mammals, the pronuclei are not visible because they are covered by lipid-containing dark granulae. Therefore, oocytes are centrifuged at 15,000 ***g*** for three minutes so that granulae should migrate to one pole of the oocyte and the pronuclei should remain in the middle of oocyte cell resulting in separation of cellular components and visualisation of the pronuclei. The transgene construct is precipitated and dissolved in injection buffer after proper washing. The solutions to be used for DNA injections must be filtered aseptically so that any contaminating particles should be removed; because the injection pipette may be obstructed by impurity present in DNA preparations. The concentration of the DNA should be adjusted in such a way that about hundreds of the gene copies per picolitre should be present.

Use of an inverted microscope and micromanipulators for microinjection is the most suitable set-up. One drop of culture medium is placed on a microscope slide and covered with paraffin oil; soon it is placed under the microscope. One micromanipulator holds the holding pipette at one side, while the injection pipette is fitted to the other manipulator. Then the DNA solution is poured into an injection pipette. The oocytes are placed in the drop of culture medium held by the holding pipette; it is properly turned to keep the pronuclei in the correct position. Then the injection pipette is used to penetrate the cell membrane of oocyte called *zona pellucida*, the membrane of one of the pronuclei and to inject the DNA. An increase in volume of the pronucleus indicates that transgene has been successfully injected. The injected-oocytes are transferred into a culture dish for inspection and selection of degenerated oocytes. The zygotes that survive are surgically transferred into the oviducts of intermediate recipient animals. After a period of about six days, embryos are recovered by flushing the uterus of the intermediate recipients and evaluated for their development. The bovine embryos (morulae - blastocysts) that have developed normally are transferred to synchronised recipients by using the conventional non-surgical methods.

The animals that have developed from microinjected oocytes are used to collect their tissue samples. However, the best suitable method for calves is the collection of blood samples because leukocytes are a good source of DNA as they contain nuclei; the presence of injected DNA in leukocytes can be tested by Southern blot analysis.

D. EMBRYONIC STEM CELL–MEDIATED GENE TRANSFER

The cells found in all multicellular organisms which can divide and differentiate into numerous specialised types of cells and can self-renew to produce more stem cells are called the 'stem cells'. There are two broad types of stem cells in mammals: *embryonic stem* (ES) cells of blastocysts and *adult stem cells* found in different tissues. In the beginning, scientists worked mainly with two types of stem cells obtained from animals and humans, embryonic stem (ES) cells and non-embryonic somatic stem cells. During 1981, methods for deriving embryonic stem cells from early mouse embryos were discovered. In 1981, M.J. Evans and M.H. Kaufmann from the University of Cambridge (U.K.) were the first to isolate totipotent cell lines directly from mouse embryos cultivated *in vitro* (EK-cells).

In 1998, the biology of mouse stem cells was studied in detail that led to the discovery of a method of deriving stem cells from human embryos *in vitro*. Human embryos reach the blastocyst stage after 4–5 days of fertilisation. Therefore, embryonic stem (ES) cells are derived from the inner cell mass (ICM) of the blastocyst after 4–5 days of fertilisation *in vitro* after donation by a female; but the ES cells are not derived from eggs fertilised in a woman's body. The inner cell mass gives rise to the entire body of the organism, including all specialised cell types and organs such as the heart, lung, skin, sperm, eggs and other tissues. Moreover, Thomson and co-workers derived blastocyst embryonic stem cell lines from human in 1993. Blastocyst consists of 50–150 cells. After isolation of embryoblast or inner cell mass (ICM) the fertilised human embryos get destructed; that is why it raises ethical issues. In 2006, a breakthrough was made when those conditions were identified that allow some specialised adult cells to be genetically 'reprogrammed' to assume a stem cell-like state. These stem cells are called 'induced pluripotent stem cells' (iPSCs). ES cell-mediated gene transfer is the method of choice for gene inactivation, *i.e.* the so-called knock-out method.

Fig. 20.14 shows the stages of zygote development. Soon after fertilisation a zygote passes through several developmental stages to develop an organism. The stages from zygote to the morula are totipotent, because the cells of these stages can generate a complex organism. But at blastocyst stage, only the cells of the inner cell mass (ICM) are capable of forming the three primary germ layers, namely ectoderm, mesoderm and endoderm, as well as the primordial germ cells (PGC). The male and female gametes develop from PGC.

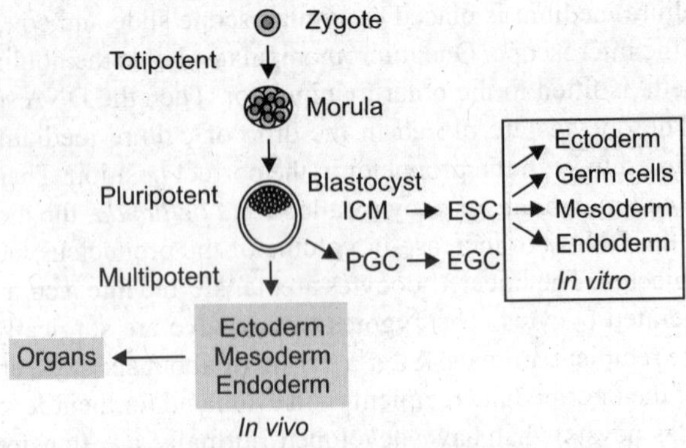

Fig. 20.14. Stages of zygote development.

Due to unique regenerative abilities, stem cells offer new potentials for treatment of some diseases such as diabetes, and heart disease. However, much work is to be done for clinicians to understand the use of these cells for cell-based therapies to treat disease; such treatment is also referred to as regenerative medicine.

Potency of Cells

The potential to differentiate the stem cell into different cell types of cells is called cell potency as given below:

- **Totipotent:** A single cell capable of dividing and producing all the differentiated cells in an organism including extraembryonic tissues is a totipotent cell. Totipotent cells are produced from the fusion of an egg and sperm cell.
- **Pluripotent:** A stem cell capable to differentiate into any of the three germ layers: endoderm (interior stomach lining, gastrointestinal tract, the lungs), mesoderm (muscle, bone, blood, urogenital), or ectoderm (epidermal tissues and nervous system) is a pluripotent cell.
- **Multipotent:** Multipotent stem cells can differentiate into a number of cells, but only those of a closely related family of cells, such as a hematopoietic cell, *i.e.* a blood stem cell.
- **Oligopotent:** Oligopotent stem cells can differentiate into only a few cells, such as lymphoid or myeloid stem cells.
- **Unipotent:** Unipotent cells can produce only one cell type (e.g. hepatocytes which form liver) but have the property of self-renewal, which distinguishes them from non-stem cells (e.g. muscle stem cells).

1. Embryonic Stem Cell Transfer Technology

The ES cell based technology is called as the ES technology. So far the ES cell technology is confined only to mice because the embryos do not survive in other mammal; even ES cell lines are difficult to be produced. But in recent years ES cell lines have been produced in other animals also including humans. Steps of embryonic stem cell transfer technology are briefly described here and given in Fig. 20.15.

(*a*) Generation of Embryonic Stem Cells: The ES cells are generated by transferring the cells from a preimplantation-stage embryo into a plastic culture dish containing a nutrient broth. At proper growth conditions, the cells divide and spread over the surface of the dish in culture. The inner surface of the culture dish is coated with a feeder layer, *i.e.* the coating layer of cells (e.g. mouse

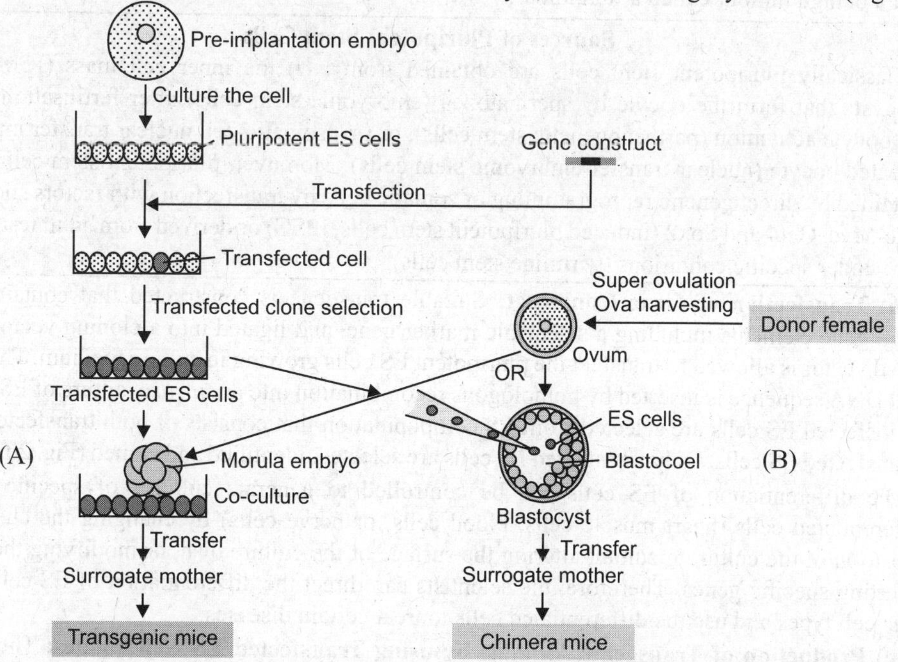

Fig. 20.15. Steps of embryonic stem (ES) cell transfer method (diagrammatic).

embryonic skin cells). The cells of feeder layer are treated to inhibit their division. On the bottom of the culture dish the cells of feeder layer provide a sticky surface attachment, and release nutrients into the culture medium for their growth. Growth of embryonic stem cells without mouse feeder cells has also been done. This checks the spread of viruses from mouse to human cells.

If the plated cells survive, divide and multiply to form cell population in the dish, they are sub-cultured by removing gently and plating into several fresh culture plates. Cell sub-culturing is repeated several times. The original cells produce millions of ES cells when the cell line is established. The ES cells that have proliferated in cell culture for a long time without differentiation are pluripotent. Such ES cells that have not developed genetic abnormalities are referred to as 'embryonic stem cell line'.

(*i*) Characterisation of generated ES cells: During the process of generation of embryonic stem cell line, the cells are characterised by testing at various points for the presence of fundamental properties of embryonic stem cells. It includes several kinds of tests as given below :

- Long-term growth and self-renewal abilities,
- Presence of two important transcription factors, the Nanog and Oct4 (they turn genes on and off at the right time), which is an important part of the processes of cell differentiation and embryonic development. Both Oct 4 and Nanog maintains the stem cells in an undifferentiated state, hence they are capable of self-renewal.
- Determination of the presence of specific cell surface markers produced by undifferentiated cells.
- Examining the chromosomes to assess any damage in the chromosomes.
- Determining of ability of cells for re-growth or sub-culture after freezing, thawing, and re-plating.
- Testing of pluripotent nature of ES cells by : (*i*) allowing them to differentiate in culture, (*ii*) manipulating the cells to differentiate to form cells characteristic of the three germ layers, or (*iii*) injecting the cells into a mouse with a suppressed immune system to test for the formation of a benign tumour called a 'teratoma'.

Sources of Pluripotent Stem Cells

Classically pluripotent stem cells are obtained from : (*i*) the inner cell mass (ICM) of blastocysts that form the oocyte by spermatozoa (embryonic stem cells) after fertilisation, (*ii*) direct oocyte activation (parthenogenetic stem cells), or (*iii*) somatic-cell nuclear transfer into an enucleated oocyte (nuclear-transfer embryonic stem cells). Moreover, pluripotent stem cells can be obtained by direct genetic reprogramming of somatic cells by transfection with factors such as Klf4, c-Myc, Oct4 and Sox2 (induced pluripotent stem cells, iPSC) or derived from adult testis by culture under specific conditions (germline stem cells).

(*b*) Transfection of Gene Construct: Suitable transgene is constructed that contains the essential gene elements including a selectable marker gene, and ligated into a cloning vector. The modified vector is allowed to transfect the pluripotent ES cells growing in culture medium. Thus the desired DNA sequence is inserted by homologous recombination into an *in vitro* culture of ES cells. The transfected ES cells are selected from ES cell population that consists of both transfected and non-transfected ES cells. Only transfected ES cells are selected, identified and cloned (Fig. 20.15A).

The differentiation of ES cells can be controlled to generate cultures of specific types of differentiated cells (heart muscle cells, blood cells, or nerve cells) by changing the chemical composition of the culture medium, altering the surface of the culture dish, or modifying the cells by inserting specific genes. Therefore, the scientists can direct the differentiation of ES cells into specific cell types and use the differentiated cells to treat certain diseases.

(*c*) Production of Transgenic Animals by using Transfected ES Cell Clones: There are several approached for the production of transgnic animals; therefore, any of following methods can be used for this purpose.

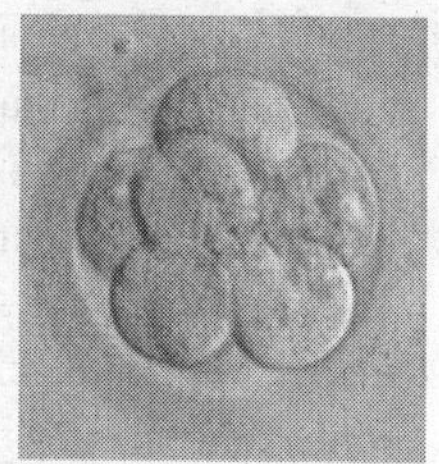

Eight-cell stage of an embryo.

(*i*) Co-culture of ovum with transfected ES cells: From a donor female ova are collected through superovulation method. The embryo starts growing after mating. The zona-pellucida of 8- cells to morula stage embryo is removed; the zona-free morullae are co-cultured with ES cells. Eventually, the ES cells enter into the inner cell mass in blastocoels of the developing embryo.

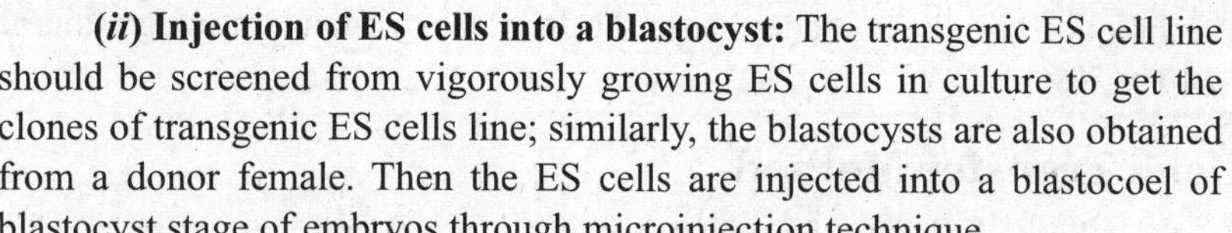

(*ii*) Injection of ES cells into a blastocyst: The transgenic ES cell line should be screened from vigorously growing ES cells in culture to get the clones of transgenic ES cells line; similarly, the blastocysts are also obtained from a donor female. Then the ES cells are injected into a blastocoel of blastocyst stage of embryos through microinjection technique.

(*iii*) Transplantation of ES nucleus into enucleated fertilised ovum: Nuclear transplantation technique is not very popular because it is tedious methods as done for production of Dolly in 1996. For detail see Chapter 8 : *Methods of gene manipulation*.

The embryo transfected through co-culture or microinjection is implanted into the uterus of a surrogate mother (mice) for development of complete offspring. Consequently, non-chimeric transgenic mice of variable frequency from co-culture technique (Fig. 20.15 A) and chimera mice of variable frequency from the ES cell-injection embryo (B) are borne. Tissues of about 30% chimera mice are derived from the ES cells. Moreover, the pluripotent cells are able to give rise to germ cells also. Secondly, the frequency of targeted gene transfer (integration of transgene at specific site) is low because of random integration of transgene in host genome.

E. GENE TARGETING (TARGETTED GENE TRANSFER)

There are several methods for insertion of DNA sequences into the organisms to create recombinant DNA. The genetically modified organisms use the process of homologous recombination. This method is also called 'gene targeting', *i.e.* integration of transgene at a specific site of an allele of the chromosome. Gene targeting is most commonly applied to study yeast and mouse genetics. In knockout mice (*i.e.* such mouse whose gene has been inactivated or eliminated) targeted gene transfer is done by using mouse embryonic stem cells to deliver a transgene (mostly therapeutic gene). Consequently the target gene of the mouse is repressed through homologous recombination, *i.e.* a biological process where two homologous chromosomes pair during prophase I of meiosis, exchange some distal portion of their DNA of non-sister chromatids through crossover (Fig. 20.16). For detailed description see Chapter 4: *Molecular Genetics*.

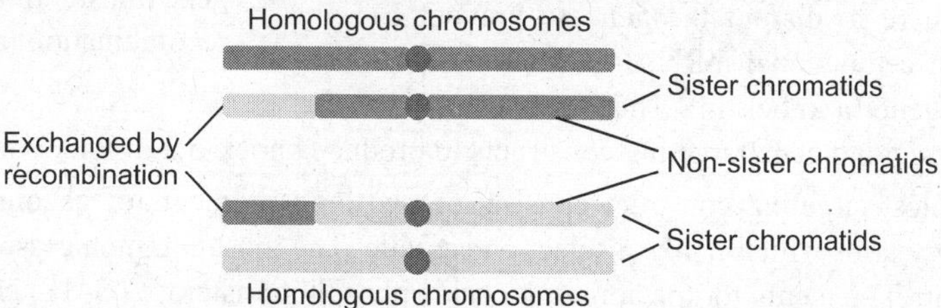

Fig. 20.16. Homologous recombination between non-sister chromatids of two homologous chromosomes.

Mouse acts as a working model to understand the effects of a specific mammalian gene. For the first time principles for introduction of specific gene modifications through homologous recombination was discovered in mice by Mario Capecchi, Martin Evans and Oliver Smithies by using embryonic stem cells; for this significant discovery they were awarded the 2007 Nobel Prize for Physiology or Medicine.

In recent years, Gene targeting in mouse ES cells has become the 'gold standard' for determining gene function in mammals. This technology has revolutionised the study of mammalian biology and human medicine since its inception. Gene integration by homologous recombination is most common in yeasts, but in mammals gene integration occurs more frequently than homologous recombination. There is about 0.1–1% frequency of random integration by homologous recombination; therefore, recovery of transgenics (that representing targeted gene transfer) is very difficult. But in recent years some refinements in techniques have been made; hence the frequency of random integration has been increased to about 10–50%.

1. Targetted Gene Transfer Method

The use of ES cells is limited only to the mouse because it is the only animal from which ES cell lines have been established. However, further work is needed before making available the totipotent ES cell lines of other species. In 1998, J.A. Thomson and coworkers isolated ES cells from human blastocysts for the first time. After 2000, many other researchers have also isolated human ES cells. In 2003, T.P. Zwaka and J.A. Thomson achieved gene targeting in human ES cells through homologous recombination. Some necessary experiments required to pursue this goal would be ethically unacceptable. Thus, the ES cell route presently remains closed against human germline gene therapy.

Advancement in gene targeting technologies has been possible only because of homologous recombination mechanics of cells. Since free ends of transfecting DNA promote homologous recombination between the transgene and endogenous chromosomal genes, a linearised simplest gene construct should be used for transfection of ES cells. A normal transgene flanked by the same DNA sequences as the defective mutant gene on the host chromosome is replaced after transfection. The homologous regions on a specific locus selected (*i.e.* flanking sequences) undergo crossing over and transgene replaces the mutant gene resulting in production of the normal gene in host chromosome and mutant vector (Fig. 20.17). The aim lies here to disrupt the gene to prevent transcription of a functional mRNA (a knock-out), or mutate the gene (a knock-in). There are several methods for designing gene targeting constructs to produce knock-out or knock-in mice.

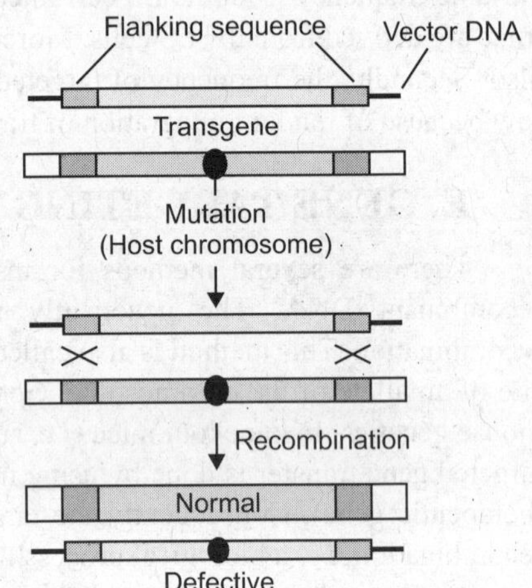

Fig. 20.17. A generalised scheme for targetted gene transfer through homologous recombination (diagrammatic).

The simplest targeting construct consists of two long segments of genomic DNA (gDNA) called homology arms which flanks a selection cassette. The most common cassette consists of the cDNA and control elements for the aminoglycoside phosphotransferase (G418) gene or kanamycin resistance (*neo*) Gene. After transfection of transgene into mouse ES cells (by electroporation) the gDNA homology arms carrying the selection cassette undergo a double-reciprocal recombination event with their matching sequences on one chromosome. Consequently, the gDNA between homologous regions is replaced by the selection cassette. The sequence to be replaced contains the TATA box, the start codon, and one or more initial exons; moreover, such sequence results in a complete knockout. Deletion of downstream exons is often desirable to avoid disruption of promoter.

Use of Positive–Negative Selection Cassettes in Targeting Constructs

The function of any cloned gene can be disrupted by using bacterial *neo* gene; consequently the cells containing disrupted gene construct shows resistance to antibiotic G418. Therefore, transgene containing cells can be selected by using G418. Hence this method is called *positive selection*.

Similarly, use of a herpes simplex virus thymidine kinase (*HSV-tk*) gene is one of the negative selection cassettes. The *HSV-tk* gene-containing cells are sensitive to a toxic nucleotide analogue 'ganciclovir' which is an antiviral drug. The HSV-tk enzyme permits the cells to incorporate ganciclovir into their DNA; hence it helps to select against certain cells. The *HSV-tk* cassette is cloned into the targeting construct outside of the homology arms; therefore, it will not be incorporated into mouse chromosome during homologous recombination. But during non-homologous recombination it is incorporated randomly. Moreover, the cells that randomly incorporate recombinant DNA retain the entire DNA construct including the virus *HSV-tk* gene. An exchange of homologous DNA sequences is involved in cells that display homologous recombination between the recombinant construct and cellular DNA. The non-homologous *HSV-tk* gene at the end of the construct is eliminated. Cells expressing the *HSV-tk* gene are killed by the antiviral drug ganciclovir; this process is called *negative selection*. Therefore, those cells are unique that undergo homologous recombination because they are resistant to both the antibiotic and ganciclovir, and allow the effective selections. Hence, such selection method is called *positive-negative selection*. In addition, the gene for diphtheria toxin A (DTA) is the other negatively selectable marker. The A subunit of toxin inhibits protein synthesis but the other cells cannot take it. The DTA is advantageous over *tk* because it works independently without using a second drug to the culture medium.

There are two major advantages of targeted gene transfer : *gene disruption* and *gene replacement*.

(*a*) Gene Disruption: The knockout mice are produced through homologous recombination method by disrupting or removing the specific gene present in their chromosome. Fig. 20.18 shows the steps for disrupting gene A present on any chromosome by inserting a bacterial *neo* gene. A specific gene that is to be disrupted (e.g. gene A) is isolated, cloned and a selectable marker gene (bacterial *neo* gene and *HSV-tk*) is inserted within the coding sequence of the gene. The mice ES cells are transfected by this gene A construct containing *neo* and *HSV-tk* genes. Two types of recombination will take place, random integration and homologous recombination. In random integration, the intact whole gene A construct integrates randomly with any mouse chromosome at any site in a large number of ES cells. The transfected ES cells will show resistance to G418 due to presence of *neo* gene, and sensitivity to ganciclovir due to *HSV-tk* gene.

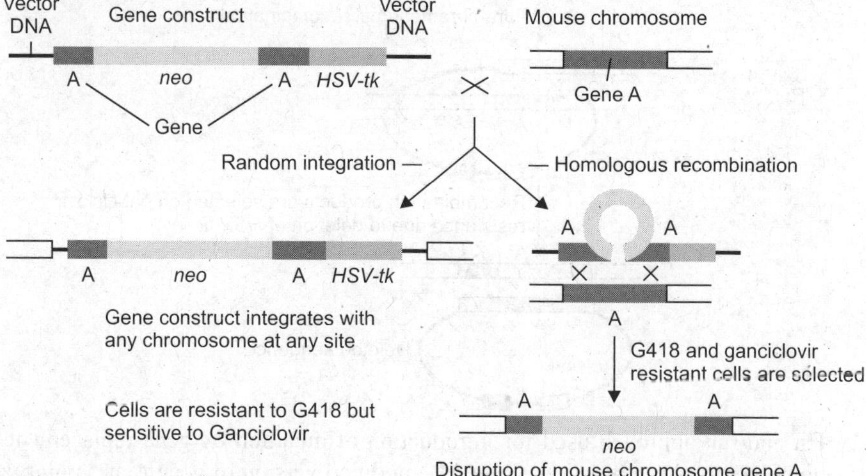

Fig. 20.18. A method of disrupting gene A of any chromosome by inserting a bacterial *neo* gene.

In some ES cells homologous recombination will occurs between the gene A of chromosome and gene A construct. The two loci of gene A located on both sides of *neo* gene will integrate into mouse chromosomal gene A and disrupt it resulting in loss of function of gene A in such ES cell. Therefore, these ES cells will show resistance to G418 through the *neo* gene, and resistance to ganciclovir due to absence of *HSV-tk* gene. Hence, such transfected ES cells can be selected by using G418 and ganciclovir in growth medium.

Only those ES cell will survive and grow on such medium in which homologous recombination has occurred and mouse chromosomal gene A has been disrupted by *neo* gene. Such ES cells are selected by positive-negative selection method, and introduced into young mice embryos, as a result of which knocked out mice are borne. In such mice the function of gene A has been knocked out. Thus a chimeric mouse has agouti coat colour and offspring are homozygous for the agouti gene.

Chimeric mouse borne via gene targeting.

***(b)* Gene Replacement:** The 'hit-and-run' approach is used to for gene replacement by homologous recombination. This method is also called `in and out' because the vector sequence is first integrated into the genome and then removed from genome. In the first step `hit' or 'in' involves homologous recombination by using a targeting vector of insertion type that consists of the *neo* and *HSV-tk* genes, and a gene of interest, for example gene A* (Fig. 20.19). The ES cells are allowed to be transfected by this recombinant vector. Within the ES cells homologous recombination occurs between the gene A* and gene A of host chromosome. After integration of recombinant vector with host chromosome inside the transfected ES cells, gene A duplicates carrying the two selection markers (e.g. *neo* as the positive marker and *HSV-tk* as the negative marker). Therefore, these cells can be selected by using the antibiotic G418.

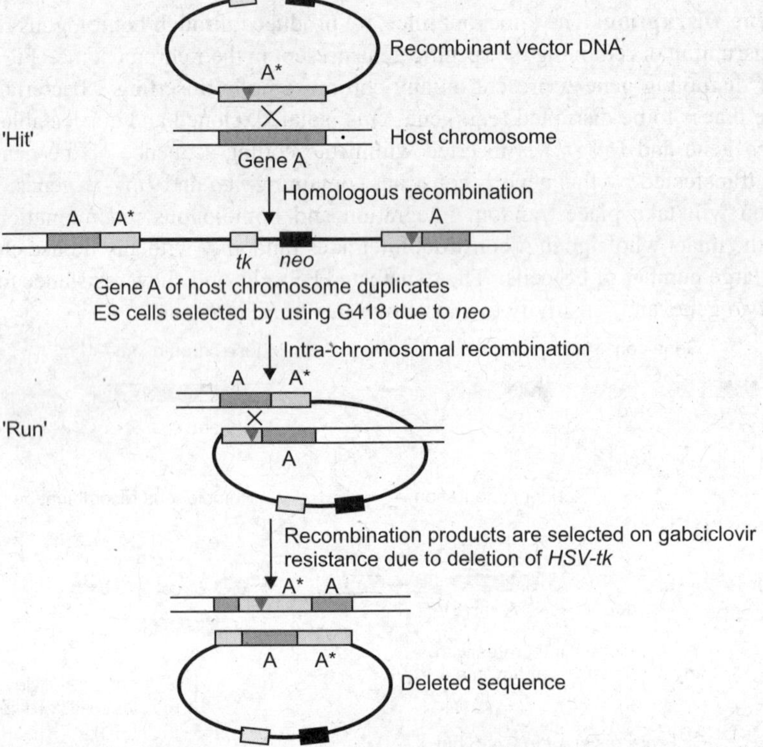

Fig. 20.19. Hit-and-run approach used for introduction of mutation by gene replacement through homologous recombination. Gene A*- modified version of gene A at * point; ×- site of crossing over.

In the second step (`run' or `out') in transgenic ES cells a rare intra-chromosomal recombination occurs between the two copies of gene A (*i.e.* gene of interest) including the negative selection marker *HSV-tk*. Duplication event will spontaneously resolve via intra-chromosomal recombination or exchange of sister chromatids resulting in loss of both *neo* and *HSV-tk* genes (Fig. 20.19). On the basis of actual position of crossover, mutation introduced in the first step will remain incorporated in the genome or the wild-type allele shall be restored. Hence the reverent ES cells in which endogenous genes are replaced by homologous recombination can be selected by using nucleoside analogue ganciclovir. The ES cells that express *HSV-tk* show sensitivity to ganciclovir. It means that the transgenic ES cell line will show sensitivity to ganciclovir, and those produced by intra-chromosomal recombination will show resistance to ganciclovir.

(*c*) Tag and Exchange: The `tag and exchange' strategy involves two sequential gene targeting steps by using two different replacement vectors. In the first targeted replacement step of homologous recombination, the gene is 'tagged' with positive and negative selectable marker *neo/HSV-tk* gene cassette using vector I (Fig. 20.20). The 'tag' event leads to the replacement of exon 2 with a positive (*neo*) and a negative (*tk*) selection marker. Then the G418 resistant recombinants (positive selection) are selected by using neomycin. In the second targeted replacement event, vector II is used that carries a subtle mutation in the flanking region of homology. The *neo*R clones that have undergone homologous recombination are subjected to a second round of gene targeting (`exchange'). Here the inserted markers (*neo* and *HSV-tk*) are replaced by exon 2 having a point mutation (*see* triangle). Embryonic stem cell clones modified by exchange with the mutation vector II are selected for loss of *HSV-tk* gene by using ganciclovir. Then the clones are screened and identified by PCR and Southern blot analysis. Replacement of the sequences introduced in the first round is enriched by negative selection approach. The 'tag and exchange' strategy may save the time if a series of mutants of the same locus is envisaged.

In 1993, G. R. Askew and co-workers from the University of Cincinnati (Ohio) used sequential gene targeting to introduce point mutations into one alpha-2 isoform Na,K-ATPase homolog in mouse ES cells. They altered the endogenous alpha 2 isoform Na,K-ATPase gene to encode the other two amino acids so that the enzyme may become resistant to inhibition by cardiac glycosides while maintaining its transmembrane ion-pumping function. This method is useful for testing specific mutations introduced into the genomes of tissue culture cells and animals and for developing animal models encompassing the mutational variability of known genetic disorders. In 1995, R.C. Moore and co-workers generated a series of mouse strains with different alterations of the prion protein by using this double replacement strategy.

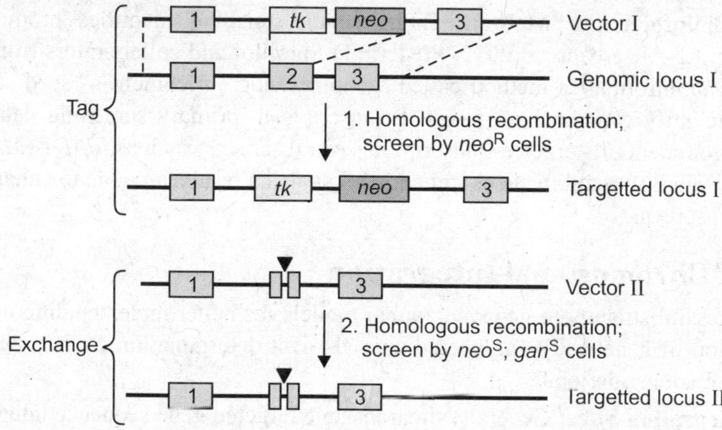

Fig. 20.20. The 'tag and exchange' strategy for targeted gene transfer.

2. Identification of Homologous Recombination

For deciphering gene functions and demonstrating disease causality, chromosome engineering by random integration (to create transgenic animals) or by homologous recombination (to replace/remove specific genes) has been widely used. But, these techniques have certain limitations. The concatenation of transgenes integrating into the genome makes it difficult to determine copy number and distinguish homozygotes from heterozygotes. Besides, gene-targeting events are infrequent and identification of correctly targeted embryonic stem (ES) cells can be laborious. Homologous recombination can be identified by many approaches as given below:

(*a*) Negative Selection Method: In 1990, Takeshi Yagi and coworkers from Japan developed an efficient method for isolation of homologous recombinants of mice at the c-*Jfyn* locus through the use of the diphtheria toxin A-fragment (DT-A) gene for the negative selection.

(*b*) PCR and Southern Blot Methods: PCR and Southern blot are the traditional screening methods that are not compatible with homology arms >10 kb. Alternative screening methods aiming to solve this problem attempt to either quantify a two-fold difference in the copy number of a targeted allele via quantitative PCR and multiplex ligation-dependent probe amplification (MLPA) or to visualise integration sites by fluorescence *in situ* hybridisation (FISH).

(*c*) Recombineering Method: In 2001, Hilary M. Ellis and coworkers from the Bethesda (M.D.) developed the recombineering (recombination-mediated genetic engineering) method. It is a technique of genetic and molecular biology that is based on homologous recombination systems (in contrast to commonly using restriction enzymes and ligases to combine DNA sequences). Recombineering is widely used for bacterial genetics, in the generation of target vectors for making a conditional mouse knockout, and for modifying DNA of any source often contained on a bacterial artificial chromosome (BAC). The BAC recombineering method can uses extremely long homology arms (>10 kb), increases homologous recombination efficiency, and reduces the need of isogenicity between targeting constructs and recipient ES cells

(*d*) Reference-query Pyrosequencing (RQPS) Method: In 2009, Zhenyi Liu and coworkers from the U.S.A. and Austria developed the reference-query pyrosequencing (RQPS) method, which is based on quantitative pyrosequencing and uniquely designed probes containing single nucleotide variations (SNVs). It offers a simple genotyping solution that identifies homologous recombinants independent of the homology arm size, determines the micro-amplification status of endogenous human loci, and quantifies transgene copy number in experimental animal. The simple pyrosequencing-based protocol could be used for the enrichment of homologous recombinant embryonic stem (ES) cells.

(*e*) Bioinformatics Method: Bioinformatics method identifies prominent off-targeted transcripts in RNAi screens. In 2012, Frederic D Sigoillot and collaborators from the U.S.A have developed a bioinformatics method called 'genome-wide enrichment of seed sequence matches' (GESS) to identify candidate off-targeted transcripts in primary screening data. GESS analysis revealed a prominent off-targeted transcript in several screens, such as *MAD2* (*MAD2L1*) for genes required for spindle assembly checkpoint. GESS analysis results are able to enhance the validation rate in RNAi screens.

3. Site of Chromosomal Integration

A successful strategy to generate animal models for better understanding of gene function is the production of genetically engineered animals. But determination of the integration sites of a transgene is challenging job.

(*a*) Integration Sites: Generally the transgene/targeted gene sequence integrates at a specific or random site in any chromosome of host genome but the mechanism of integration of a transgene is unknown. However, several characteristics of this process have been described so far, for example:

(*i*) random integration of transgenes to the chromosome, and (*ii*) more efficient integration of linear molecules into the host genome than the circular molecules. All the molecules integrate on the same chromosome at the same site, and usually the multiple integrated molecules get arranged tandem.

The site of integration is important because : (*i*) the expression of the transgene is affected by the neighboring regions of chromatin, and (*ii*) interference of inserted transgene on the expression of endogenous genes. Depending on the delivery system, there are typically one insertion site for some transgene animals, though multiple transgene copies are often found in a tandem array at that integration site. For other transgene animals, there are often multiple integration events with random transgene insertions on several chromosomes.

(*b*) Expression of Transgene: In the first case the expression of the transgene may be low or undetectable if the transgene is incorporated in heterochromatic regions of the genome. Therefore, such line of transgenic mice should be selected that show the highest expression of the transgene in the tissues. The transgenes must be insulated by linking transgenes to DNA sequences containing matrix-attached regions so that it could be protected from the negative effects of the flanking sequences of the genome. Hopefully these sequences help in the re-organisation of the linked transgene into active chromatin. It also permits the transgene to function independently from neighboring sequences. Such sequences should not be routinely used to produce transgenic mice because the use of sequences containing matrix-attached regions in the production of transgenic mice is not well-characterised. The transgenes that randomly integrate in the proximity of heterochromatin can show variable levels of expression. The effect of repeat sequences on transgene expression in several animals has led to the term homology-dependent gene silencing (HDGS).

(*c*) Effect of Transgene on Function of other Genes: Permissive location in the host genome allowing the expression of endogenous genes without disturbance is also an important problem. The expression of endogenous genes could be altered due to the integration of the transgene into the mouse genome. Therefore, the ability of a researcher to relate the function of the transgene to a phenotype gets limited due to occurrence of such problem.

Hence, it is necessary to study at least 2-3 lines of transgenic mice generated from different founders so that the problem of permissive integration could be eliminated. The phenotype would not be observed in other transgenic lines due to the site of chromosomal integration of the transgene, *i.e.* changes in the expression of endogenous genes and not the transgene.

The effect on the phenotype of interacting genes when their relative positions on the chromosome are altered as by translocation is called 'position effect'. It has been described well in *Drosophila* with respect to eye colour and is known as 'position effect variegation' (PEV). In biotechnology, the position effect refers to variation of expression by identical transgenes inserted into different regions of a genome. The enhancers that regulate the neighbouring genes cause differences in expression pattern. The expression pattern of the transgene can be affected by these local enhancers. Each transgenic organism has the transgene at different location; therefore, expression pattern of each transgenic organism is unique.

You know that telomeres also influence the expression of adjacent genes, a process called telomere-position effect. In 2007, Q. Gao and coworkers from the University of California (U.S.A.) reported that telomeric transgenes are silenced in adult mouse tissues and embryo fibroblasts but are expressed in embryonic stem cells.

(*d*) Determination of Transgene Integration Sites: Though determination of transgene integration sites is challenging, several PCR-based methods (known as chromosome walking techniques) have been developed to isolate DNA fragments adjacent to known sequences; some of these methods include : inverse PCR, ligation-mediated PCR (LMPCR), randomly primed PCR (RP-PCR), and T-linker PCR. In 2006, Elizabeth C. Bryda and collaborators from the University of Missouri (USA) developed a new method that enables quick and inexpensive determination of multiple independent transgene integration sites in founder animals and their offspring that have been created by using lentivirus. The transgene delivery systems involving retroviruses mostly result in the integration of multiple copies of the transgene throughout the host genome. The identification of the number and chromosomal location of the multiple copies of the transgenes is important to correlate inheritance of the transgene at a particular chromosomal site with a specific and robust phenotype. They identified multiple chromosomal integration sites in transgenic founder animals by using a technique that combines restriction endonuclease digest and several rounds of PCR amplification followed by nucleotide sequencing.

F. RESCUE AND RECOVERY OF TRANSGENES FROM ANIMAL CELLS

In some cases, it is desirable to isolated DNA after its transfer into cells and organisms, either as a means to isolate that DNA after functional selection or to access the chromosomal DNA adjacent to the integrated site. The transgene can be isolated from the transfected cells by using transgenic cell line and then selected. Since the transgene expresses in the phynotypic characteristics, it can be identified on the basis of phenotypic effects in cultured cells. One of the retrieval strategies is plasmid rescue, where cyclised fragments of a cellular genome consisting of integrated bacterial plasmid DNA are used to transform *E. coli*. Initially, plasmid rescue was applied in yeast and flies but it was also used in mammalian cells at low frequency. There are variations in different strategies of plasmid rescue, but a simple method is described for rescue of chicken *tk*$^+$ gene from transfected cells in Fig. 20.21.

In 1980, Manuel Perucho and co-workers from the Cold Spring Harbor Laboratory, New York (U.S.A.) isolated the chicken thymidine kinase (*tk*) gene by using plasmid rescue strategy. They used the bacterial plasmid pBR322 as a vehicle to isolate genes coding for selectable markers from higher eukaryotes. They obtained the chicken *tk*$^+$ gene as a 2.2-kilobase *Eco*RI/*Hind*III insert in pBR322. The cloned gene transformed the Tk$^-$ animal cells with the efficiency equal to that of the cloned herpes *HSV-tk* gene.

The chicken cells/tissue is procured and the genomic DNA containing *tk*$^+$ gene is digested with a restriction enzyme *Hind*III to generate fragment having sticky ends. Similarly, pBR322 plasmid containg a selectable marker (*amp*^{R} gene) is cleaved by using *Hind*III which does not cleave the *tk* gene (Fig. 20. 21). Thus both types of fragments containing cohesive ends are ligated by using T4 DNA ligase to concatenated DNA containing chicken *tk*$^+$ gene-tagged to *amp*^{R} gene. Some transformants were expected to contain chicken *tk*$^+$ gene linked to pBR322 sequences. This concatenated DNA was used to transform mouse Tk$^-$ cells to *tk*$^+$ phenotype which later on are selected on HAT medium *in vitro*. Thereafter, the DNA from transgenic cells containing *tk*$^+$ gene is isolated, and used to transfect the chicken Tk$^-$ cells second time. Similarly, the Tk$^+$ transgenic cells are selected on HAT medium *in vitro*. Most of the non-selected recombinant DNA integrated at first round of transfection is eliminated during the second round of transfection. Again the genomic DNA containing *tk*$^+$ gene is isolated from chicken Tk$^+$ transgenic cell. Then it is cleaved with *Eco*RI which does not cleave the chicken *tk*$^+$ gene but cleaves pBR322 only once on either side of the single *Hind*III site of the circular pBR322 molecule. The cleaved DNA ligated in cyclisation conditions are used to transform *E. coli* cells to ampicillin resistance. Such *E. coli* cells are selected on ampicillin-containing medium. The recombinant *E. coli* cells that contain chicken *tk*$^+$ gene-linked to *amp*^{R} gene will be rescued and recovered for examination and analysis.

In 1980, D. R. Lowy and coworkers adopted an alternative strategy to isolate hamster *aprt* gene. In this method the intact plasmid function is not maintained while passing in animal cells. They cleaved hamster genomic DNA using *Hind*III which leaves the *aprt* gene intact; the fragmented genomic DNA was ligated to a *Hind*III-cleaved plasmid pBR322 DNA. This DNA was transfected into hamster Aprt$^+$ cells to generate a primary Aprt$^+$ transformant which has integrated multiple plasmid sequences. Then the DNA from the primary transformant was transferred to recipient Aprt$^-$ cells to construct secondary Aprt$^+$ transformants containing *aprt* gene linked to a single plasmid sequence. The DNA from the secondary transformants was partially cleaved with *Bam*HI. The fragmented DNA (20 kb long) were collected and used to construct a library of 10^6 independent recombinant phage by using charon 4A vector.

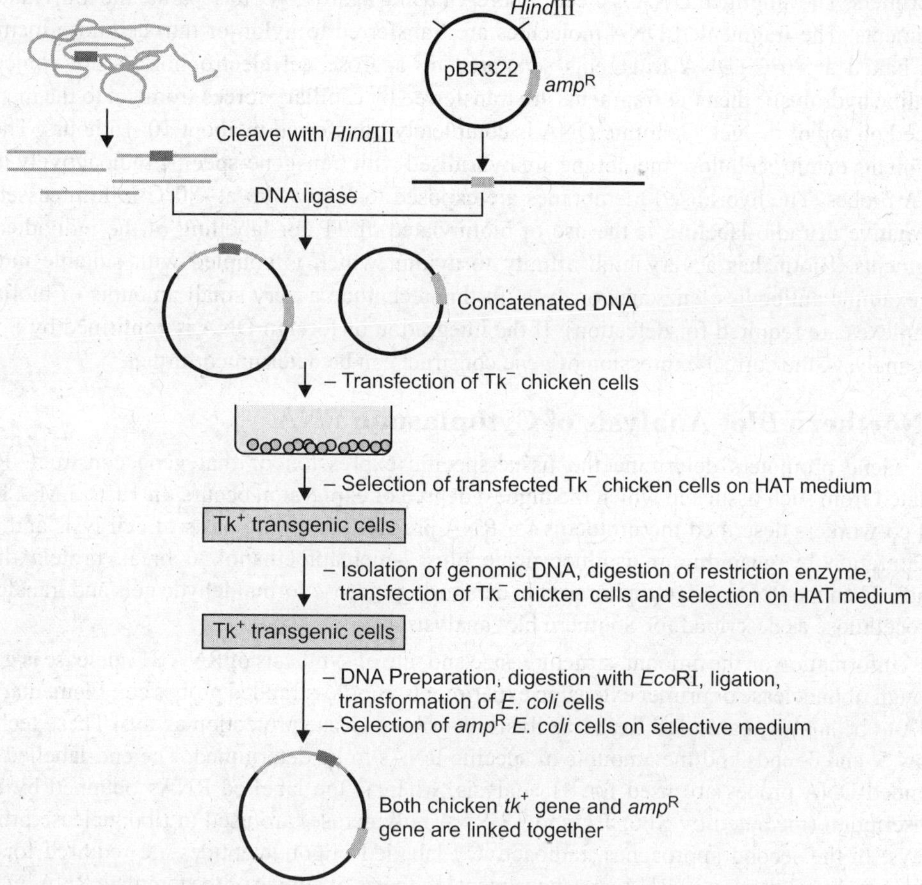

Fig. 20.21. Steps of plasmid rescue for isolation of chicken *tk*$^+$ gene (associated with a phenotypic character) from the cultured cells.

The recombinant phage library was screened using a highly radioactive pBR322 DNA as a hybridisation probe. They screened 6×10^5 plaques to identify a single clone containing plasmid sequences which generated Aprt$^+$ colonies when added to hamster Aprt$^-$ cells with a frequency of one colony per 400 pg/5×10^5 cells. It was five-fold greater than that obtained with total genomic DNA. Preliminary mapping data localised the functional *aprt* gene to an 8-kb *Hin*dIII fragment.

G. DETECTION OF TRANSGENIC ANIMALS

Identification of transgenic animals is important because it proves the stable integration of transgene into the genome and its proper expression to produce the desired protein. Generally dot-blot technique is used for such purposes (for detail see Chapter 8 : *Methods of gene manipulation*). However, the stable integration and expression of a gene construct in a host animal can be detected by using the following methods.

1. Southern Blot Analysis of Transfected Gene Construct

DNA is isolated from tissue or blood specimens after the birth of a transgenic animal. Specific DNA of high-molecular weight is obtained and genomic libraries are constructed. For detection of transgenes a mean molecular weight of 1.65×10^7 Da is quite sufficient. The restriction endonucleases are used to digest the isolated DNA which is then amplified by PCR using unique transgene sequences as primers. The amplified DNA is electrophoresed using agarose gel to separate the individual DNA fragments. The fragmented DNA molecules are transferred to nylon or nitrocellulose membranes and baked at 80°C. DNA fragments separated by agarose gel electrophoresis are subjected to alkaline hydrolysis; then the fragments are transferred by capillary forces from gel to the membrane placed on top of the gel. Genomic DNA is completely transferred in about 10–16 hours. The DNA fragments on nitrocellulose membrane are hybridised with transgene-specific radioactively labelled DNA probes. The hybridised membranes are exposed to X-ray film at -80°C in film cassettes (an alternative of radio-labeling is the use of biotinylated dUTP for labelling of the individual DNA fragments. Biotin has a very high affinity to avidin, which is coupled with suitable indicators for example antibodies, enzymes or dyes. In this technique a very small amounts of biotin/DNA complexes are required for detection). If the integration of foreign DNA is confirmed by Southern blot analysis, the correct expression of gene construct can be determined further.

2. Northern Blot Analysis of Cytoplasmic RNA

Gene promoters determine the tissue-specific expression of that gene construct. RNA is isolated from such tissues in which the highest degree of expression occurs. In 1979, J.M. Chirgwin and co-workers described the protocols for RNA preparation on the basis of cell lysis and protein denaturation by guanidinium isothiocyanate plus 2-mercaptoethanol to break protein disulfide bonds. Then the RNA is electrophoresed in denaturing agarose formaldehyde gels and transferred to nitrocellulose as described for Southern blot analysis.

Information on the amount, structure, size and rate of synthesis of RNA S1 nuclease is gathered through ribonuclease or primer extension experiments. Single-stranded probes complementary to the RNA to be analysed are used for S1 analysis and ribonuclease protection assays. These techniques allow 5′ and 3′ ends and the amounts of specific RNAs to be determined. The end-labelled single-stranded DNA probes are used for S1 analysis, whereas the labelled RNAs obtained by *in vitro* transcription (mediated by SP6, T7, and T3 RNA polymerase) are used in ribonuclease protection assays. In the second approaches, radioactively labelled oligonucleotides are required for primer extension, hybridisation to RNA and subsequent synthesis of copies of the template RNA by reverse transcription. The synthesis products are analysed on polyacrylamide sequencing gels. The RNA molecules are transferred to a nylon membrane and hybridised with a radioactive-labelled DNA probe. After hybridisation the filter is washed and exposed to an X-ray film for 24 hours.

All the three techniques (*i.e.* RNA S1 nuclease analysis, primer extension analysis, reverse transcription analysis) allow analysis of the size and amount of specific RNAs. Primer extension analysis differs in one decisive point from the other two methods. Reverse transcriptase yields a complete copy of RNA irrespective of the presence of any splice sites. Therefore, primer extension analysis complements S1 analysis because it allows exact determination of the 5′ ends of transcripts.

3. Translation Analysis

Translation analysis provides the correct transcription of a microinjected gene construct. This is done by measuring the amounts of protein expressed in specific tissue. There are several techniques used to measure protein production such as : (*i*) colorimetric method for quantitative estimation of proteins (Bradford technique, Lowry technique), (*ii*) gel electrophoretic methods for analysis and subsequent staining with Coomassie blue or silver stain, (*iii*) western blotting method by transferring the proteins to nitrocellulose or nylon filters and western blot analysis, and (*iv*) clinical techniques such as radioimmunoassay, ELISA.

H. BREEDING WITH TRANSGENIC LIVESTOCK

The final stage in the process of developing transgenics is to study animals carrying the transgene. Transgenic cattle which contain the gene construct in their germ cells pass on the transgene to the offspring. Since integration takes place at one locus in the chromosome, inheritance follows that of a Mendelian gene. Integration occurs only at one site within a chromosome under normal circumstances. Therefore, these primary transgenic animals are called hemizygote transgenics. Use of the term 'heterozygous' is not suitable because the transgenic allele misses from the normal homologous chromosome. About 50 % of the offspring of a transgenic animal will inherit the transgene whose gonads are not mosaics. Normally mosaics occur only in the F_o generation. Offspring of transgenic F_1 and the following generations contain the gene construct in all somatic and gonadal cells. It has been observed that a transgene is not transmitted through several generations because it is not stably integrated; hence a transgene can be lost from the genome. If the gonads contain transgenic and non-transgenic cell lines, the frequency of transgenic offspring may be between 0 and 50% depending upon the degree of contribution of transgenic cell lines to the composition of the gonads. If the cells of the gonads do not contain transgenes, they may not be passed on to the offspring. About one third of all transgenic animals generated by microinjection will be mosaics which will not pass on the transgene. This problem exists only in the offspring derived from the primary transgenic animal. In G1 generation, the transgenic offspring will harbour the gene construct in all the somatic and germ cells. It may be assumed that the integrated foreign DNA may be removed from the genome by repair mechanisms.

In rare cases there are more than one integration sites for a transgene. If the distance between these sites is more for random recombination, more transgenic offspring will be produced. A transgenic animal consisting of two independent integration sites will transmit the transgene to its 75 % offspring; 25 % of these will inherit both transgenes, 25 % will contain one or other of the two integration sites, and 25 % offspring will be non-transgenic.

After mating two hemizygous transgenic animals one can get 25% homozygous transgenic, 50% hemizygous transgenic, and 25% normal offspring. There will be no problem till an intact functional allele is retained on the homologous chromosome. If integration occurs in one of these genes, exceptionally the additive genetic effects will be obtained. This will result in a reduction in its allele expression.

After mating the two hemizygous animals carrying an insertion mutation, no transgenic offspring will be produced because the gene affected by the integration event is essential for foetus development. In 1983, E.F. Wagner and coworkers reported the prenatal lethality in mice homozygous for human growth hormone (hGH) gene sequences integrated in the germ line. They observed this lethality in the offspring of transgenic mice carrying the hGH gene. They described two lines of transgenic animals characterised by prenatal mortality of homozygous animals due to an insertion of the gene construct into an essential DNA region.

If the aim is to introduce a specific gene into a cattle population, it will not be sufficient to produce only one transgenic animal. However, there will be too low chances of establishing a functional transgenic line with a single integration of a transgene. It must be expected from a transgenic breeding line that : (*i*) the transgene will be integrated in all somatic cells, (*ii*) stable transmission of the transgene will occur to all offspring, (*iii*) expression of the transgene (DNA » mRNA» protein) will take place, and (*iv*) biological activity of the protein expressed from the transgene will occur properly.

1. Advantages of Transgenic Breeding over Selective Breeding

Transgenic technology is still experimental and very expensive. Its use has many advantages over the traditional methods. Transgenic breeding is more trait-specific, faster, flexible and less costly. So far researches are in progress to introduce new traits involving one or a few genes. However, there is need of more research before manipulating the complex traits. Transgenic breeding offers a number of advantages over the traditional methods. Some of the advantages of transgenic breeding are given below:

- **Trait-specific :** Transgenic breeding is more trait-specific; hence, the traits can be chosen with greater accuracy. The undesired traits can be kept to a minimum.
- **Faster method :** Transgenic breeding is faster because it takes only one generation to establish the trait as compared with the traditional selective breeding that requires several generations.
- **Flexible and cheaper :** Transgenic breeding is more flexible because the traits that will not be available in some animals may be achieved by using transgenic methods. Besides, it is less costly as much cost and labour involved in chemical treatments and feed supplements to animals could be avoided.
- **Eco-friendly :** Transgenic goat producing fibres, *i.e.* BioSteel® is eco-friendly with respect to its composition and production process. Hence, it can be used for as military uniforms, medical sutures, and strings of tennis rackets, biodegradable fishing lines, and soft body armour.

2. Applications of Transgenic Animals

Traditionally, farmers have been using the selective breeding to produce animals that show the desired traits (such as increased milk production, high growth rate), etc. But the conventional breeding is a time-consuming and difficult method. Use of molecular biology techniques resulted in the development of traits in animals easily and shorter time. Transgenic animals can be used in many ways such as for disease resistance, organ production, and production of proteins or drugs. In 1980, Gottfried Brem from the Ludwig Maximilians Universität (Germany) has highlighted the possibilities of application of gene transfer in livestock. Applications of transgenic animals can be categorised in the following three groups:

(*a*) Agricultural Application: Agricultural application includes the following important areas of transgenic animals:

(*i*) Growth: Attempts are being made to increase growth and body composition of animals through the transfer of these genes which are responsible for regulation of growth-hormones. Experiments with the respective proteohormones in growing animals have shown success in achieving the goal especially in pigs.

(*ii*) Disease resistance: Attempts are being made to produce disease-resistant animals such as influenza-resistant pigs; but in farm animals a very limited number of genes are known at present to be responsible for resistance to diseases (www.actionbioscience.org/biotech/margawati.html). A model for such work is the influenza resistance caused by the Mx-gene. Production of Mx protein is strongly induced by α–interferon and β–interferon, but less strongly by γ-interferon. In 1986, Peter Staeheli and coworkers from California (U.S.A.) have reported that α- interferon stringently activates Mx-gene expression at transcription level of the mouse Mx gene. The authors attempted to produce influenza-resistant pigs through the transfer of three Mx-gene constructs.

(*iii*) Product quality: The improvement in the quality and composition of animal products through the transfer of gene constructs provides a new prospect for animal production. In 1987, Laura Mercier proposed a model for reduction of the lactose content of milk. In the milk of transgenic sheep and cattle lactose splits into galactose and glucose. The lactose gene is combined with an udder specific promoter that is involved in lactose splitting. The large population of the world suffering from lactose intolerance can also consume such milk.

In 1986, K.A. Ward, J.D. Murray and C.D. Nancarrow from the CSIRO Division of Animal Production (Australia) have tried to promote synthesis of cysteine in the mammal organism through gene transfer and to influence positively wool growth. Transgenic cows produce more milk or milk with less lactose or cholesterol, transgenic pigs and cattle have more meat on them, and transgenic sheep produces more wool. Moreover, some transgenic cows produce less cholesterol in their milk and sheep produce more wool.

In 1996, Sami Damak and co-workers from New Zealand have reported the improved wool production in transgenic sheep expressing insulin-like growth factor-1. This is the first report on improvement in production trait by genetic engineering of a farm animal without adverse effects on health or reproduction.

***(iv)* Gene Farming:** In the mid-1980s, the basic concept of gene farming approach was developed which gained new impetus with the successful generation of cloned farm animals by nuclear transfer. The recent success achieved by the production of polyclonal human immunoglobulins from transchromosomic calves has shown the potential of this technology. One of its attractive prospects is the production of large amounts of specific human immunoglobulins after vaccination. In 1987, K. Gordon and coworkers have reported the production of human tissue plasminogen activator (tPA) in transgenic mouse milk. In the same year, J.P. Simons, M. McClenaghan and A.J. Clark reported the alteration of milk quality by expression of sheep β-lactoglobulin in milk of transgenic mice.

***(b)* Medical Application:** There are many areas in medical science where transgenic animals are very useful in solving the problems of the patients.

***(i)* Xenotransplantation :** Every year human patients die due to lack of a replacement heart, liver or kidney. For instance about 5,000 organs are needed each year in the United Kingdom alone. Therefore, the transgenic pigs have hope in providing the transplant organs required by the patients. But recently, it has been found that a pig protein can cause donor rejection that hampers xenotransplantation. However, researches are in progress how to remove the pig protein and replace it with a human protein.

***(ii)* Pharmaceuticals and nutritional supplements:** Many animal products, for example insulin, growth hormone, and blood anti-clotting factors have already been obtained from the milk of transgenic cows, sheep, or goats. Research is also in progress to manufacture the milk through transgenesis that can be used for treatment of debilitating diseases like phenylketonuria (PKU), cystic fibrosis, and hereditary emphysema.

Rosie, the first transgenic cow.

In 1997, the first transgenic cow, Rosie (from the Roslin Institute and PPL, Scotland) was created through artificial manipulation. She was designed to be a factory to produce the protein *en masse*, which can be used to make drugs. Her body consisted of a human gene, α-lactalbumin that produced a protein normally made only by the human body. Rosie's milk was human protein-rich that consisted of 2.4 g/litre of human protein. This transgenic milk is more nutritionally balanced product than natural bovine milk and could be given to babies or young with special nutrition. Moreover, human protein products such as insulin, growth hormone and blood anti-clotting factor have already been produced in the milk of cow, sheep, or goat.

***(iii)* Human gene therapy:** Human gene therapy involves adding a normal copy of a gene (transgene) to the genome of a person carrying defective gene. There are 5,000 genetic diseases known so far; the transgenic animals could play a great role in curing the genetic diseases. For example, human RBC-growth promoting substance by a transgenic calf gene has been reported by the A. I. Virtanen Institute (Finland). For detail *see* Chapter 9 : *Gene therapy*.

***(c)* Industrial Applications:** Transgenic breeding can also been exploited industrially as given below :

Transgenic spider-goats; Webster (left) and Peter (right)

***(i)* Material fabrication:** In 2001, two scientists at Nexia Biotechnologies Inc. (Canada) spliced spider genes into the cells of lactating goats. Consequently, the goats started producing silk along with milk and secreted small silk strands from their body. The scientists can create a light, tough, flexible material by extracting polymer strands from the milk and weaving them into thread. Nexia is developing recombinant spider silk, trade named BioSteel®, for applications in the medical, military and fiber industry. The offspring of GM herd can produce spider silk protein in their milk that can be collected, purified and spun into the fibres. These fibres can be used for as military uniforms, medical sutures, and strings of tennis rackets, biodegradable fishing lines, and soft body armour. BioSteel® is eco-friendly with respect to its composition and production process. BioSteel® is expected to be biodegradable in the presence of water. Thus it makes a promising material for applications as medical sutures and fishing lines. It has the potential to be non-polluting and eco- friendly as compared to the other synthetic fibers.

***(ii)* Increased amount of sheep wool:** Breeding transgenic sheep can grow better wool without need of dietary supplements of sulphur-containing amino acids. This aspect of research is in progress.

***(iii)* Increased meat production:** Meat production can be increased by an abnormally high quantity of growth hormone in the transgenic animal for example pigs and cattle have more meat on them. Earlier farmers used growth hormones to spur the development of animals but this technique posed problems due to presence of hormones remained in the animal product.

I. TRANSGENIC ANIMALS PRODUCED

In the creation of transgenic animals, mice have been the pioneers and very popular transgenic animals because they have been studies extensively and are relatively inexpensive and easy to maintain unlike the other mammals. In addition, some other animals have also been genetically altered, such as rats, rabbits, pigs, sheep, goats, fishes, flies, monkeys, cows, birds, chickens, etc. (Table 20.5).

Table 20.5. Medicine production by transgenic animals (*source*: Breekveldt and Jongerden, 1998).

Animal	*Drug*	*Disease/Target*
Cow	Alpha-lactalbumin	Anti-infection
Cow	Collagen I	Tissue repair
Cow	Collagen II	Theumatoid arthritis
Cow	Lactoferrin	GI tract infection, infectious arthritis
Cow, sheep	Fibrinogen	Wound healing
Cow, sheep, pig	Factor IX	Haemophilia
Goat	Antithrombin 3 (ATIII)	Thrombosis
Mouse	Msp-1	Malaria
Mouse, cow	Human serum albumin (HSA)	Maintains blood volume
Mouse, goat	Glutamic acid decarboxylase	Type 1 diabetes
Mouse, goat	Tssue plasminogen activator tPA)	Thrombosis
Pig, sheep	Factor VIII	Haemophilia
Pig, sheep	Human protein C	Thrombosis
Rabbit	Human calcitonin	Osteoporosis
Rabbit	Alpha-glucosidase	Pompe disease
Sheep	Alpha1 anti-trypsin (AAT)	Deficiency leads to emphysema
Sheep, mouse	CFTR	Cystic fibrosis

There are many reasons for creating the transgenic animals for their use in science and society. Thus the transgenic animals can be grouped under the five basic categories, such as scientific models, disease models, transpharmers, xenotransplanters and food sources (Table 20.6).

In the human genome the function of every gene is not known and over-expression of newly discovered proteins on biological development can be studied by creating the animal models. The scientists can learn better about symptoms, pathways, causes, treatments, and possible cures of human diseases by creating transgenic disease models. Through this approach involvement of many genes in disease development can also be understood. Secondly, the transgenic animals (or transpharmers) are used to study the production of a pharmaceutical protein or antibiotics in their milk. It must be kept in mind that all transpharmers are always females because males cannot yield milk. Organ rejection occurs frequently due to its non-compatibility with the organs of patients. Therefore, baboon and pig organs are looked up by the scientists for this type of transplants due to physiological similarity with human organs. Such genetically modified animals that have their blood physiology similar to humans and better prepare their organs for transplantation are called 'xenotransplanters', such as pigs. These animals are modified in such way that may grow human-compatible or more human-like organs for transplantation from animal to human. Certain transgenic animals have been produced that yield more meat and protect the other animals from killing to get similar amount of meat; they are called 'super' animals and grouped under food sources. Some of the transgenic animals have been described in this section.

Table 20.6. Example of five basic groups of transgenic animals with examples.

Groups of transgenic animals	*Examples*
Scientific (biological) models	Smart mouse (= Doogie mouse)
Transgenic models	Alzheimer's mouse, OncoMouse™, ANDi, and Super Fly
Transpharmers	Genzyme goats, alpha-1-antitrypsin sheep, Lactoferrin baby herman
Xenotransplanters	Pig xenotransplanters
Food sources	Superpig, Superfish

1. Mouse

For the first time in 1981, J.W. Gordon and F.H. Ruddle reported the integration and stable germ-line transformation of genes injected into mouse pronucleus. In this method a chosen gene construct (a single gene or a combination of genes) from another member of the same species or from a different species is directly microinjection into the pronucleus of a fertilised ovum. The introduced DNA may be integrated with other host gene randomly resulting in over- or under-expression of other gene(s). However, the introduced gene probability will not insert itself into such a site on the host DNA that will allow its expression. A recipient female or foster mother is mated with a vasectomised male just to induce to act as a recipient. The manipulated fertilised ovum as above is transferred into the oviduct of an induced foster mother (pseudo-pregnant) mouse. Such mouse mother gives birth to transgenic offspring.

In 1982, Richard Palmiter (University of Pennsylvania) and R.L. Brinter isolated the rabbit growth hormone (α–globin) gene, human growth hormone (α-globin) gene as well as thymidine kinase gene and linked separately to the promoter region of mouse associated with the metallothionein I (*MT*) gene (a gene which encodes a metal binding protein). This was joined to pBR322 plasmid to produce the recombinant plasmids. On the other hand, mature eggs from adult mouse were recovered surgically and fertilised with sperms *in vitro*. Immediately the fertilised eggs were microinjected with recombinant plasmids before fertilisation of egg nucleus by a sperm and formation of a diploid zygote. The plasmids generally combine homologously with each other within the egg forming a long repeated concatemer which then integrates randomly to give repeated genes at a single chromosomal site. The engineered embryos were then implanted into the uterus of a host mouse mother for further development.

The resulting mice are called 'transgenic mice' since part of genome comes from another genetically unrelated organism. Due to introduction of foreign gene before nuclear fusion, chromosomal integration takes place early and progeny contains new genes. Twenty one mice were borne; among them seven contained the fusion gene and six were two-fold larger in body weight than the others. The level of growth hormones increased many times (between 200 and 800 times) more than the normal mice. In addition, the level of growth hormone mRNA was also increased in liver cells (Fig. 20.22).

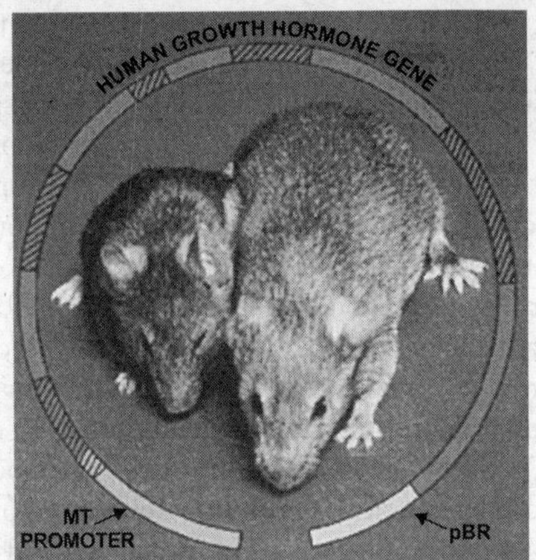

Fig. 20.22. Transgenic mice produced by Richard Palmiter and R.L. Brinter.

(*a*) Alzheimer's Mouse: In 1980 for the first time transgenic mouse was successfully created by transferring gene that helped to study the role of genes in development and physiology of mouse. However, during the past two decades a real model for human pathology has increased. Today, several innovative technologies have been developed based on transgenic mice. Recently, a biotechnological company created the transgenic mouse that is a most powerful tool in transgenesis boosting the research activities.

In 1995, Professor David Adams and collaborators at the Athena Neurosciences Corporation created an Alzheimer's Mouse by artificially creating Alzheimer's disease in them; consequently, mice did not normally acquire Alzheimer's disease. The purpose of this study was to create a vaccine for this disease to cure millions of people.

(*b*) OncoMouse™: Cancer is a deadly disease developed due to accumulation of mass of cells that have lost the control over cell division. The cancer patients have such cells that never respond to signals that control the ability of cell division and multiplication. Philip Leder at Harvard Medical School created OncoMouse™, a transgenic disease model, patented and later licensed to DuPont Pharmaceuticals.

Leder and his collaborators inserted the active cancer-causing genes (oncogenes) into special strains of mice which were passed into their offspring; consequently, the offspring developed cancer. As compared to traditional animal models, transgenic models are more precise for cancers development. By using transgenic oncomice it has been found that some human cancers develop due to working of two or more oncogenes together. Thus OncoMouse is useful model to study human cancers such as breast (leukemias) and blood cancers (lymphomas).

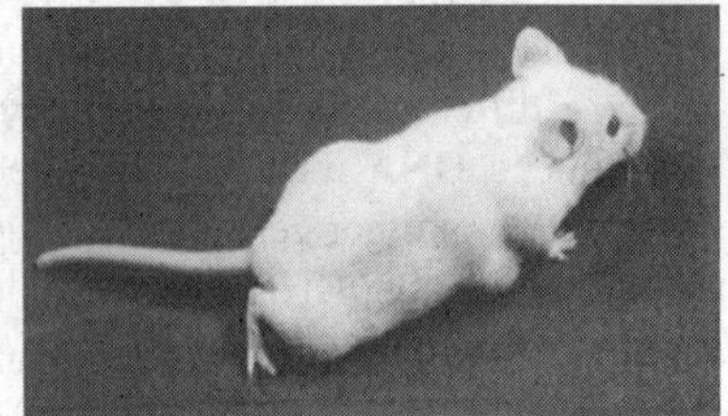
OncoMouse™ with a tumour (Leder, 2003) (see its right front leg).

In India, a team of scientists led by Vijay Kumar at the International Center for Genetic Engineering and Biotechnology (ICGEB) in association with the National Institute of Immunology (NII), New Delhi (India) have engineered and tested a new mouse model that develops liver cancer in 12 to 20 weeks, which is much faster than any of the existing mouse models. This is the first transgenic animal from India for medical research that has got a US Patent. This new transgenic mouse will speed up drug development for liver cancer induced by the hepatitis-B virus (HBV). A protein called 'x-protein' of the hepatitis B virus plays a key role in development of liver cancer.

Since HBV cannot be propagated in cell cultures, analysis of HBV-infected patients and chimpanzees will help to study disease development.

2. Fish

In recent years, transgenic fish are being engineered by the artificial transfer of rearranged genes into newly fertilised eggs by microinjection technique. Many fusion genes have been developed to enhance integration of transgenes. Mass gene transfer methods have also being developed, such as lipofection, particle bombardment, and electroporation of embryos and sperm cells. Applications of fish transgenesis must be focused on the development of fish with accelerated growth, tolerance to low temperature, disease resistance, and eco-friendly that do not disturb genetic diversities of the natural population. However, the possible impact on the environment must be assessed before the release of transgenic fish into the environment.

(*a*) GloFish® : Scientists got success in bringing about changes in genetics of several types of fish; moreover, these fish can be purchased as pets. The GloFish is a Zebrafish that has been crossed with a jellyfish.

GloFish fluorescent fish

In 2000, P.B. Hackett and M.C. Alvarez have reported that growth-enhanced GM fish consist of promoters that drive an over-production of growth hormone of 'all fish'. This resulted in growth enhancement of several species. Fish are the first GM animals that have been made available publicly as a pet. The GloFish is a patented brand of genetically modified fluorescent zebrafish with bright red, green, and orange fluorescent color. However, GloFish fluorescent fish are available for purchase in five beautiful colours: Starfire Red®, Electric Green®, Sunburst Orange®, Cosmic Blue™ and Galactic Purple™. GloFish fluorescent fish can be bred from the earlier developed offspring of fluorescent fish. Offspring of every new GloFish fluorescent fish inherits its colour directly from their parents and pass the genetic traits of colour to offspring.

3. Cows

Transgenic cows are produced by using several steps of techniques such as DNA cloning, restriction enzyme digests, ligation, polymerase chain reaction (PCR), transfection, nuclear transfer, and *in vitro* embryo production. The steps for creation of transgenic cows are : (*i*) isolation of a human gene responsible for production of a desired protein, (*ii*) hormonal treatment to an animal given to produce several embryos, and their collection from the oviduct, (*iii*) insertion of human gene into the pronucleus of fertilised egg by microinjection, (*iv*) implantation of a week-old transgenic embryo into a surrogate cow which gives birth to the transgenic calf, and (*v*) testing of offspring for the presence of new gene (Fig. 20.23).

The cow gives birth to a transgenic calf; the calf is transgenic or not, it can be confirmed by using : (*i*) PCR to determine the presence or absence of the transgene, (*ii*) quantitative PCR (q-PCR) to determine the number of copies of the transgene, and (*iii*) fluorescent *in situ* hybridisation (FISH) to visualise where the transgene is in the chromosome and has integrated into more than one chromosome. If the milk protein gene has been transferred, the expression of the transgenic protein is checked in the milk of lactating calf.

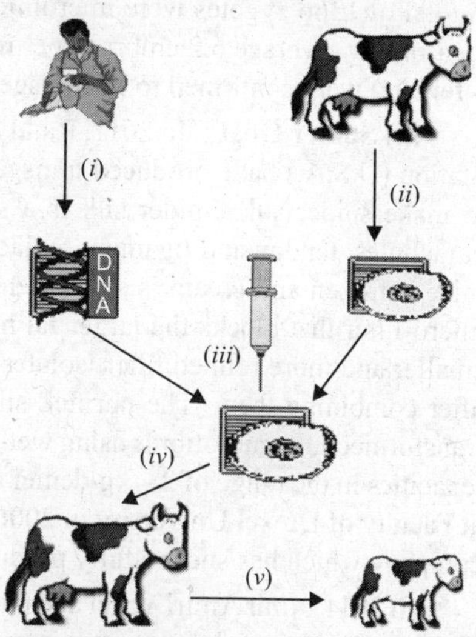

Fig. 20.23. Steps of creation of transgenic cows.

(*a*) Low Cholesterol Cow Milk: Some transgenic cows produce more milk and the others produce milk with lower cholesterol. Casein constitutes about 80% of the total protein in milk. Structure of casein proteins varies due to four different variants, such as alpha-s1, alpha-s2, beta and kappa casein. Amino acid composition, genetic variations and functional properties of each variant differs. All the four casein variants combine and form a structure called a 'micelle', which is dispersed in the water phase imparting white colour to milk. Casein is one of the most valuable components of milk due to its nutritional value. It is also important in making cheese. In 2000, their first transgenic cows were generated which produced modified milk. These cows had extra kappa casein genes artificially inserted in their genome that resulted in increased kappa casein protein in their milk because the transgenic cow had their two naturally occurring kappa casein genes along with the additional inserted kappa casein gene(s).

(*b*) Disease-resistant Cows: About 30% of mastitis in dairy cows is caused by a bacterium *Staphylococcus aureus* that destroy milk-secreting cells of mammary glands. In 2000, Kevin Wells of United States Department of Agriculture produced a mastitis-resistant transgenic cow named 'Annie', which resisted such cellular attacks by bacteria that secrete an added protein called lysostaphin. Only 15% of cows infected with *S. aureus* can be treated with antibiotics. Therefore, lysostaphin offers an alternative defense bioengineered into the cells.

It has also been reported that transgenic technology can be used to modify milk composition in cows, for example transgenic cows can be used to produce therapeutic proteins in bulk.

(*c*) Human Breast Milk in Cows: Human breast milk genes have successfully been introduced into 300 dairy cows to produce milk with the same properties as human breast milk. High quantities of key nutrients are found in human milk that can trigger the baby's immune system. This milk also reduces the risk of infections due to presence of human protein lysozyme. Now it is believed that milk from transgenic cows could provide an alternative to human breast milk and formula milk for babies. Moreover, some animal welfare groups are opposing the use of such milk questioning the safety of milk from GM animals and its effect on the health of cattle.

4. Goat

In 2003, H. Baldassarre and co-workers at Nexia Biotechnologies Inc. (Canada) produced transgenic goats by pronuclear microinjection of *in vitro* produced zygotes derived from oocytes. A total of 1366 zygotes were microinjected and transferred into 219 recipient goats by midventral laparotomy (average 6.2 embryos per recipient). A total of 150 kids were born, of which 9 (6 male : 3 female) were confirmed to be transgenic by PCR and Southern blotting analyses.

(*a*) Spider Goat : In 2011, Randy Lewis at the Utah State University Agricultural Experiment Station (U.S.A.) have produced transgenic goat that produces two proteins in their milk to be used to make spider silk. Spider silk is a strong and versatile fibre that could be the key to stronger parachutes, tendon and ligament replacements and safer airbags. The goats are milked twice a day, milk is frozen and cream is separated removing most of the fats. Then new milk is pushed into a micro-filter that blocks the larger fat molecules allowing the smaller proteins through it. Further a smaller and more refined filter isolates the silk proteins. The two proteins are pushed into a needle after combining them. The purified silk proteins could be dried and dissolved using solvents and transformed into microfibres using wet-spinning fiber production methodologies. The spun fibres had tenacities in the range of 2 - 3 g/denier and elongation range of 25-45% by Milind Ramesh Gandhiin at Faculty of Drexel University in 2006. Lewis has spun 10 metres of fibres so far. Nexia is the only company which has successfully produced fibres from recombinant spider silk commercially.

In 2011, Amir Amiri Yekta and co-workers from Royan Institute for Reproductive Biomedicine (Tehran, Iran) produced a transgenic goat carrying the human coagulation factor IX (hFIX) transgene. They electroporated goat fetal fibroblasts with a linearised marker-free construct in which the transgene was juxtaposed to β-casein promoter designed to secrete the recombinant protein in goat milk. They used two different lines of transfected cells as donors for nucleus transfer into enucleated oocytes. Two transgenic goats were live-born. DNA sequencing of the corresponding transgene locus confirmed authenticity of the cloning procedure. This study provided a promising approach for large-scale production and therapeutic application of hFIX expressed in transgenic goats.

This transgenic goat has a transgene that codes for a human protein under the control of a promoter region that targets the expression specifically to the mammary gland. The human protein is secreted in the goat's milk but not in other animals.

***(b)* Therapeutic Milk Proteins:** In 1999, the world's first cloned transgenic goats were borne through a joint research programme by LSU Agricultural Center and Genzyme Transgenic Corporation. The anti-thrombin III gene was attached to a promoter gene (usually casein gene) and then microinjected into the male pronucleus of the newly fertilised egg. The gene got attached to the genetic material of embryo. The transgene was incorporated into every cell of the developing goat embryo. The embryo was first cultured and then transfered to the surrogate mother of goat, which developed normally. Only the female goat produced milk containing anti-thrombin III protein; this protein could be extracted from milk for pharmaceutical use, for example to make a drug for patients undergoing coronary bypass surgery. The drug works in conjunction with heparin, which prevents blood from clotting.

A transgenic goat

In 2006, the first human therapeutic protein called ATryn® (an antithrombin) made in a transgenic animal was approved for use in Europe and the USA. ATryn® is used to prevent blood clots in patients with a rare blood disease. About 80 goats could produce sufficient protein to meet the demand in whole Europe.

5. Sheep

Domestic sheep are multi-purpose and small ruminants having more than 200 breeds created purposefully. Besides, sheep meat and milk were one of the earliest staple proteins consumed by humans. Domestic sheep provide wool and the meat, the sale of which is the most profitable enterprise in the sheep industry. But less sheep meat is consumed as compared to chicken, pork or beef. The Roslin Institute of Edinburgh (Scotland) used sheep for genetic research producing the groundbreaking results. In 1995, two ewes named Megan and Morag were the first mammals cloned from differentiated cells. In 1996, a Finnish Dorset sheep named Dolly was the first mammal to be cloned from the somatic cell of an adult. In 1997, Polly and Molly were the first transgenic mammals to be simultaneously cloned at the Roslin Institute.

Tracy- a transgenic sheep.

***(a)* Tracy:** A transgenic ewe named Tracy was born in 1990 after human DNA was inserted into fertilised sheep embryos at the Roslin Institute, Edinburgh (Scotland). In an attempt to reproduce animals such as Tracy, the Roslin Institute later cloned Dolly sheep. Tracy was created to supply a chemical in her milk that would hopefully help the patients suffering from cystic fibrosis and emphysema.

***(b)* Dolly and Bonny:** In 1997, Dolly was borne through nuclear transfer technique at the Roslin Institute, Edinburgh (Scotland); she was named after a famous singer called Dolly Parton. In 1998, Dolly gave birth to a lamb called Bonny; this experiment proved that clones are able to produce healthy offspring.

(c) Polly and Molly: Like Dolly, Polly and Molly were cloned by the same nuclear transfer method. Polly is the first transgenic sheep created by A.E. Schnieke and co-workers in 1996 at the

Roslin Institute; Polly and Molly consisted of a human gene. They were created by using fibroblast cells containing a human gene (*i.e.* blood clotting factor IX). This could produce pharmacological and therapeutic proteins in her milk for treatment of human diseases.

Geep- sheep-goat chimera

The biotech companies are patenting their work to get maximum profits. This raises the issue of animal rights; whether these animals will be treated as sentient beings or they will simply be treated as walking factories.

(*d*) Geep (Sheep-goat Chimera) : In 1984, Carole B. Fehilly and coworker from ARC Institute of Animal Physiology, Cambridge (U.K.) reported the production of sheep–goat chimaeras by embryo manipulation and the use of inter-specific chimaerism to allow successful inter-specific embryo transplantation in sheep and goats. The chimeras were a mosaic of goat and sheep tissue, showing the parts that grew from the sheep embryo were woolly and the part that grew from the goat embryo were hairy. A sheep-goat chimera differs from a sheep-goat hybrid; the former is produced by mating of goat with a sheep. The chimera contains cells from two different genetic individuals, each of these arising by normal mating; it has four parents. But, a hybrid has only two parents. Sometimes, the term 'shoat' is used for sheep-goat hybrids and chimeras.

(*e*) 'Peng Peng' Sheep: In 2012, the world's first transgenic sheep was produced with a simplified technique called 'handmade cloning' in Xinjiang Uygur Autonomous Region (China). This transgenic sheep was named 'Peng Peng' (after the identical names of the two cloners). Due to genetic modification Peng Peng has improved meat quality through increasing the contents of unsaturated fatty acids. According to Yutao Du (BGI Ark Biotechnology Co., Ltd.), the gene associated with 'omega-3 poly unsaturated fatty acid' (ω-3PUFA) was successfully transferred into Peng Peng. ω-3PUFAs serve as essential fatty acids for humans reducing the risk of coronary heart disease and supporting the normal development of the brain, eye and neurons. People can absorb ω-3PUFAs by drinking milk or eating meat. Industrial-scale development of ω-3PUFAs needs to be done.

Peng Peng sheep

(*f*) Organ Donor Sheep: Esmail Zanjani at the University of Nevada (Reno) are of the opinion of using sheep organs that can be transplanted into humans or cells that can be injected into human foetuses *in vitro* for treatment of genetic diseases. Human stem cells taken from the bone marrow of an adult or from approved embryo lines are injected into the sheep foetus before the development of immune system of unborn lamb that rejects the human cells. Such hybrid transgenic animals created by cell mixing are known as chimeras. Thousands of people die each year due to non-availability of organ transplants.

6. Pigs

Transgenic pigs have also been produced by fertilising the normal eggs with sperm cells containing a transgene; this method is called 'sperm-mediated gene transfer'. In recent years transgenic pigs are being produced with several objectives, some of them are described below:

(*a*) Geen Pig (= FluoroPig) : In 2006, Taiwan scientists used green fluorescent protein (GFP) gene from a jelly fish and implanted into pig embryos. By injecting fluorescent green protein (GFP) gene (from jellyfish, *Aequorea victoria*) into embryonic pigs, Wu Shinn-Chih and coworkers from the National Taiwan University bred three male transgenic pigs. The GFP gene integrated with chromosomes of embryo entered in almost all cells during cell division and cell development. Offspring pigs glow bright green in the

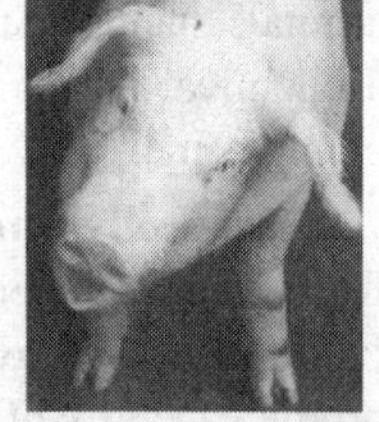
FluoroPig for food production

dark; but during daylight they have a tinge of green on their skin, snout and teeth. The whole body of pigs glows green including their internal organs. The transgenic pigs were produced to study human diseases but it would help to monitor and trace changes of the tissues during the physical development.

***(b)* EnviroPig™ :** Generally the normal pigs cannot break phosphorus; hence it comes out in their faeces. The phosphorus pollutes water, increases algal blooms and destroys fish habitats. This is why the EnviroPig™ was engineered which discharges a very little phosphorus in their faeces (Fig. 20.24). In 1999, Cecil W. Forsberg and coworkers at Molecular and Cellular Biology (University of Guelph) created the first phytase pig and called it 'Wayne'. University of Guelph patented the knowhow and sold the license to a private enterprise, Ontario Pork which is the only private investor that owns the trademark EnviroPig™. The EnviroPig™ is a genetically modified line of Yorkshire pigs consisting of edited DNA from a pig and genetic material from mice.

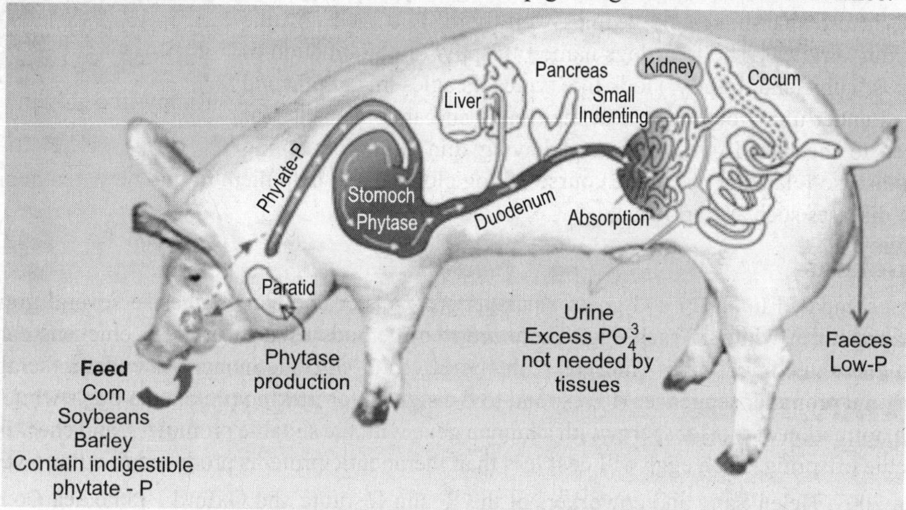

Fig. 20.24. EnviroPig™- phytate digestion.

EnviroPig™ produces the enzyme phytase in the salivary glands which in turn is secreted in saliva. When the pigs consume cereal grains, the phytase mixes with feed in their mouth. After swallowing the feed the phytase is activated in the acidic cnvironment of the stomach and degrades the indigestible phytate in the feed accounting for 50 to 75% of the grain phosphorus. As a result of phytate degradation, phosphate is released that is readily digested by the pig. Therefore, EnviroPig™ is able to break down phosphorus more efficiently than the normal pigs.

Phosphorus is the major nutrient that enables algal growth, blooming of which results in fish mortality due to anoxic conditions and reduced water quality. Therefore, EnviroPigs release less phosphorus into freshwater ponds, streams and rivers resulting in reduced environmental pollution in areas where soil phosphorus exceeds the desirable levels.

7. Dogs

Transgenic dogs can be studied as models of human disease such as narcolepsy, certain cancers and blindness to show the application of cloning and transgenesis to a wide range of mammals.

(a)* Snuppy:** In 2005, Lee and stem cell researcher Woo Suk Hwang created the first cloned dog 'Snuppy' (S**eoul **N**ational **U**niversity pu**ppy***). They used somatic cell nuclear transfer method, the same technique that was used to create Dolly. Hwang isolated nuclei from ear cells of his three years old father (Afghan hound) and implanted into enucleated eggs from female dogs.

Snuppy- a transgenic dog

The research work carried out by Hwang on human cells turned out to be fraudulent, but Snuppy was not. Hwang's team transferred modified eggs into 123 surrogate mothers resulting in only three pregnancies- one miscarried, and two others went to term. One of the clones died from pneumonia at 22 days old and only one survived. Later on, DNA fingerprinting of Snuppy was done, which confirmed that it was genetically identical to his father. Snuppy is the latest mammal to be cloned after sheep, mice, cats, rats, cows, goats, pigs, horses, rabbits and a mule.

***(b)* Ruppy (Glowing Dogs):** In 2007, Lee Byeong-chun from Seoul National University cloned the world's first transgenic dogs carrying fluorescent genes that express red colours. They can develop cures for human diseases using cloning techniques. The researchers took skin cells from a beagle, inserted fluorescent genes into them and put them into eggs before implanted them into the womb of a surrogate mother. Through cloning with a gene producing a red fluorescent protein, six female beagles were born; two died, but the others four survived. The four dogs named 'Ruppy' (a combination of the words '**ru**by' and 'pu**ppy**') look like typical beagles in daylight and glow red under ultraviolet light. But their nails and abdomens appear red even to the naked eye because of having thin skin. Implanting human disease-related genes in the course of dog cloning will help them to find new treatments for genetic diseases such as Parkinson.

Ruppy- the glowing dogs

8. Chickens

As compared to sheep and goats chickens grow faster and can synthesise several grams of human protein in white of each egg. There are two methods used to produce chickens carrying foreign genes : (*i*) infection of embryos with a viral vector carrying human gene (for a therapeutic protein) and promoter sequences (to respond to the signals for making proteins in egg's white), and (*ii*) transformation of rooster sperm with a human gene with the suitable promoters, and checking the transgenic offspring. Such eggs will cost less than therapeutic proteins produced in culture vessels.

In 2007, Helen Sang and coworkers at the Roslin Institute and Oxford Biomedica Company produced transgenic hens by using an infectious anaemia lentivirus as a vector to insert the genes for desired pharmaceutical proteins into the hen's gene for ovalbumin. They inserted the human genes into chicken embryonic stem cells and then blended these cells with that of a normal chicken embryo to create a chimaera, a blend of GM and normal cells. The cells in the oviduct (which lays eggs) consisted of GM cells and passed on the implanted gene so the egg could make the drug protein. This protein is present about in 54% of egg whites (around 2.2 g per egg) which is a sufficient amount as per standards of biotechnology. This method offers the future prospect for mass production of the high priced drugs that cost thousands of pounds per patient per year. This shows advancement in the use of farm animals for the production of pharmaceuticals.

GM hen.

J. MOLECULAR PHARMING IN TRANSGENIC ANIMALS

Molecular pharming (=biopharming, gene pharming) refers to the production of human pharmaceuticals using genetically modified plants or farm animals. At present pharming is in the developmental stage with possible commercialisation in future. Molecular pharmng is gaining application since the development of transgenic 'super mice' in 1982 and the first mice to produce a human drug called tissue plasminogen activator (tPA) for treatment blood clots in 1987. Transgenic organisms have been modified to contain foreign DNA of known function by genetic engineering. The pharmaceutical companies will raise these transgenic animals.

Production of pharmaceuticals in GM animals is a separate area of research. More than 20 pharmaceuticals have been synthesised in the milk of mammals so far. Pharmaceutical agents have also been produced in chicken's eggs, blood, urine, and sperm. Since 2008 the first drug produced by transgenic animals is available in the European Union. Antithrombin can be extracted from the milk of genetically modified goats; it is such a chemical that blocks blood clotting. Some of the pharming products currently in development are given in Table 20.7.

Table 20.7. Pharming products currently under development.

Animal	*Drug/protein*	*Use*
Chicken, cow, goat	Monoclonal antibodies	Other vaccine production
Cow	Alpha-lactalbumin	Anti-infection
Cow	Factor VIII	Treatment of hemophilia
Cow	Fibrinogen	Wound healing
Cow	Collagen I, collagen II	Tissue repair, treatment of rheumatoid arthritis
Cow	Lactoferrin	Treatment of GI tract infection, Treatment of infectious arthritis
Cow	Human serum albumin	Maintains blood volume
Goat	Human protein C	Treatment of thrombosis
Goat	Antithrombin 3	Treatment of thrombosis
Goat	Glutamic acid decarboxylase	Treatment of type 1 diabetes
Goat	Pro542	Treatment of HIV
Pig	Factor VIII, IX	Treatment of hemophilia
Pig,	Tissue plasminogen activator	Treatment of thrombosis
Sheep	Alpha1 anti trypsin	Deficiency leads to emphysema
Sheep	CFTR	Treatment of cystic fibrosis
Sheep	Tissue plasminogen activator	Treatment of thrombosis
Sheep	Factor VIII, IX	Treatment of hemophilia
Sheep	Fibrinogen	Treatment of wound healing

Milk is presently the most mature system to produce recombinant proteins from transgenic organisms. A common application of these proteins is its expression in the milk of cows, sheep or goats, because milk is produced in substantial amount and protein purification from milk is easy. Due to faster breeding, hamsters and rabbits have also been used in preliminary studies. Creation of a transgenic mammal producing the biopharmaceuticals in its milk is an approach to this technology. An animal produced through microinjection method is able for use in cloning technology so that additional offspring can be created that should carry the required transgene. In 2006, the European Medicines Agency granted marketing permission for ATryn (antithrombin protein purified from the milk) that is the first drug produced in genetically modified livestock, *i.e.* goats. In 2009, the U.S. FDA also granted marketing approval for ATryn (Houdebine, 2009).

A transgenic animal to be exploited for pharmaceutical production must : (*i*) produce the desired drug at high levels without endangering its own health, and (*ii*) pass its ability to produce the drug at high levels to its offspring. The current strategy to achieve these objectives is to couple the protein drug gene with a DNA signal that directs milk production in the mammary gland. Then this DNA is injected into fertilised cow, sheep, goat, or mouse embryos following microinjection method by using a micromanipulator and a microscope. Though the new gene is present in every cell of the animal, it functions only in the mammary gland; therefore, the protein drug is produced only in milk. There is no harm to the animal in making the 'foreign' protein drug because the mammary gland and milk are found outside of main life support systems of the animal.

Success rate of creating a transgenic animal that can produce the drug is low. About 10–30% mouse embryos can produce transgenics, but less than 5% goats, sheep or cows can produce transgenics. When an animal attains maturity and bred, production of drug is measured during lactation. Currently development of transgenic animals is very expensive, because of the long time involved in production of transgenic animals and low success rates. Animal pharming is 5-10 times more economical and 2-3 times cheaper in startup costs than cell culture production technology. Some useful drugs currently made/being developed in transgenic animals are given in Table 20.8.

The Edinburgh Company has produced transgenic sheep that produce alpha-1-antitrypsin protein which is used to treat cystic fibrosis. Moreover, transgenic sheep produce proteins used for treatment of clotting disorders, such as haemophilia, including fibrinogen, factor VII and factor IX.

In 1987, Lothar Hennighausen and Heiner Westphal at the National Institutes of Health, in collaboration with Katy Gordon and her colleagues at Integrated Genetics, started a mini-revolution after reporting the production of a pharmaceutical in mammary glands of transgenic animals. In 1992, a new industry was formed. Transgenic animal bioreactor technology moved speedily from laboratories to industry for production of pharmaceuticals. The dairy industry could not derive benefits from transgenic technology with a speed as the pharmaceutical industry did. Consequently, many companies have been set up to exploit transgenic technology. The U.S. market for bioreactor products exceeded $3 billion/annum.

Transgenic casein cows.

The target species will be determined on the basis of demand products. Presumably at production level of 1 mg/ml, you can calculate the number of animals required to produce some of the proposed pharmaceutical bioreactor products. An organisation may consider generating more than 27,000 rabbits necessary to produce 150 kg fibrinogen. About 17 cows may produce all of the fibrinogen required to satisfy the current need of the world. The cloned cows produce milk having high amount of protein. The herd has extra copies of genes for two types of casein, the major protein in milk and the main component of curd.

Recombinant bovine growth hormone (rBGH) is also known as recombinant bovine somatotropin (rBST) has been produced using cow gene. It is a synthetic growth hormone which is injected into cows to increase growth rates and milk production. The rBST protein which induces cattle to produce more milk can be synthesised by an engineered bacterium. The use of rBST drug is used in the USA for twenty years but not permitted in Europe. The arguments for protest of this drug are that : (*i*) it is 'naturalness', and (*ii*) it casts ill effects on health of cattle.

The anti-thrombin III protein is now in the third phase of human clinical trials. The product may be in the market within few years. One of the first applications of cloning to produce transgenic goats for the pharmaceutical industry is to clone the three Genzyme goats. Genzyme takes the gene for anti-thrombin III from one species and implanting it into the genome of another. The transgene containing embryo is implanted into surrogate mother to give birth of new transgenic animal. When you breed the anti-thrombin III female offspring, 50% offspring will possess this gene. Genzyme is trying to develop technologies to reproduce these transgenic animals quickly.

Table 20.8. Current market price of certain drugs from transgenic animals and supply produced by an animal.

Drug	*Animals*	*Application*	*Value/animal/ year ($%)*
Alpha-1 antitrypsin	Sheep	Inherited deficiency leads to emphysema which is a respiratory complication	15,000
CFTR (cystic fibrosis transmembrane conductance)	Sheep, Mouse	Treatment of cystic fibrosis	75,000
Blood clotting factor VIII	Sheep	Treatment for haemophilia	37,000
Blood clotting factor IX	Sheep	Treatment for haemophilia	20,000
Haemoglobin	Pig	Blood substitute for human transfusion	3,000
Human protein C	Pig	Anticoagulant, treatment of blood clots	10,00,000
Lactoferin	Cow	Infant formula additive	20,000
tPA (tissue plasminogen activator)	Goat	Treatment for blood clots	75,000

In 1991, G. Wright and coworkers reported high level expression of active human alpha-1-antitrypsin in the milk of transgenic sheep, and patented this technology for large scale preparation of alpha-1 proteinase inhibitor and use thereof. They reported this method suitable for processing scaled-up amounts of source material in the range of tens of kilograms for the purification of alpha-1 proteinase inhibitor (API) from a mixture of unpurified proteins (US Patent 7879800).

In 2000, success was reported for inserting a transgene into a specific gene locus. The gene was the human gene for alpha1-antitrypsin. The human protein was expressed in large quantities in milk of two animals, goat and sheep. In 2006, GTC Biotherapeutics won preliminary approval to market a human protein and antithrombin in Europe. Antithrombin III is plasma protein that inhibits the formation of thrombin by binding and inactivating prothrombin, factor 9 and factor 10.

K. BIOETHICS

Many people are rightly concerned about animal welfare and relationship between humans and animals. Transgenic animal researches are essentially human endeavours to improve the availability, quality, and safety of drugs, to improve animal health, and to enhance human health. Alteration in the animal DNA has made a revolution in farm animals and cattle.

However, many ethical issues are raised for pharming regarding the treatment of animals. It has been found that many transgenic offspring have birth defects and gross abnormalities, or do not produce the desired protein. For production of transgenic animals, a large number of eggs are used and one egg is found out that can be implanted. Mostly the egg donor is slaughtered in the process of donation.

A biosafety and ethical problems arise while using transgenic animals as a source of pharmaceutical proteins. Unlike microorganisms, the escape of the animals in environment is impossible, and mostly animals do not suffer from expression of a foreign protein in their milk or eggs. But some specific problems arise from many proteins. For example, alteration in rabbit health is reported to be due to expression of human erythropoietin in rabbit milk. Moreover, least effective side effects have been observed in rabbits while they were producing human growth hormone in their milk.

It has been reported that rBST manufactured by Monsanto is sold to dairy farmers with the trade name, Posilac™. Injection of Posilac™ hormone forces cows to increase about 10% milk production. Monsanto and the Food and Drug Administration (FDA) insist that rBGH milk is safe to cows and consumers. But rBGH milk is supercharged with high levels of a natural growth factor (IGF-1). Moreover, rBGH milk is contaminated by pus from mastitis induced by rBGH, and antibiotics used to treat the mastitis. Excess levels of rBGH have been the major causes of cancers in breast, colon and prostate of cows.

Besides several benefits as described earlier, decision-making ethical issues cannot be ignored by the biotechnology industry, scientists, policy-makers, and the public. These ethical issues include several questions. In general the scientific community agrees that regulators should review the existing policies and guidelines made for transgenic animals. To ensure safety for animals and health of humans and animals new policies regarding transgenic and cloned animals must be prepared. The public debate on transgenic technologies will decide further research demanded by animal producers, regulators, environmentalists and the public.

QUESTIONS

1. Give a historical account of transgenic animals.
2. What are the reporter genes ? Discuss with suitable examples.
3. Describe the characteristic features of some animal cloning vectors with suitable diagrams.
4. What is transfection ? Write different types and methods of transfection.
5. What is stem cell ? Discuss the methods of embryonic stem cell-mediated gene transfer.
6. Describe the steps of targeted gene transfer.
7. Write the steps of rescue and recovery of transgenes from animal cells.
8. How are the transgenic animals detected ? Write the method of detection.
9. Write a note on breeding of transgenic livestock and its advantages.
10. Write an essay on transgenic animals produced.
11. What is molecular farming ? Describe it in brief.
12. Write short notes on the following :
    (*i*) Gene construct, (*ii*) Promoters and enhancers,
    (*iii*) *CAT* gene, (*iv*) *DHFR* gene,
    (*v*) Bacterial *XGPRT* gene, (*vi*) *NPT* gene,
    (*vii*) Baculovirus expression vector, (*viii*) SV40 transducing vector,
    (*ix*) SV40 plasmid vector, (*x*) MAC vectors,
    (*xi*) Embryonic stem cells, (*xii*) Gene targeting,
    (*xiii*) Gene disruption, (*xiv*) Gene replacement,
    (*xv*) Sites of chromosomal integration, (*xvi*) Transgenic cow,
    (*xvii*) Transgenic goat, (*xviii*) Spider goat,
    (*xix*) EnviroPig, (*xx*) FluroPig,
    (*xxi*) Bioethics.

BIBLIOGRAPHY

Breekveldt J. and Jongerden J. 1998. Transgenic animals in pharmaceutical production. *Biotechnol. Development Monitor.* No. 36, p. 19-22.

Clark A.J., Simons P. *et al.* 1987. Pharmaceuticals from transgenic livestock. *TibTech.* **5**: 20-24.

Federation of European Laboratory Animal Science Associations (FELASA) September 1992, revised February 1995. Transgenic Animals - Derivation, Welfare, Use and Protection.

Houdebine L.M. 2009. Production of pharmaceutical by transgenic animals. *Comparative Immunol. Microbiol. Infect. Dis.* **32**: 107–121.

Lodish H. *et al.* 2000. *Molecular Cell Biology* (4th Ed.). W. H. Freeman and Company.

Smith R. L. *et al*. 2000. Characterisation of promoter function and cell-type-specific expression from viral vectors in the nervous system. *J. Virol.* **74**: 11254–11261.

Zarrin A.A. *et al.* 1999. Comparison of CMV, RSV, SV40 viral and vlambda1 cellular promoters in B and T lymphoid and non-lymphoid cell lines. *Biochim Biophys Acta*. **1446**: 135-139.

21 CHAPTER Industrial Microbiology

An industrial fermentor

A. Microbial growth
B. Technique of microbial culture
C. Fermentor: designing and construction
D. Fermentation processes
E. Production of microbial products
   (i) Strain improvement
   (ii) Fermentation process
   (iii) Alcohol production
   (iv) Malt beverages
   (v) Organic acid production
   (vi) Antibiotic production
   (vii) Amino acid production

Industrial microbiology deals with the exploitation of microbes for use in industrial processes, for examples fermentation, biotechnology industry, brewing, etc. Since the time immemorial microorganisms have been used by humans in a variety of ways such as preparation of curd and cheese, production of alcohol and alcoholic beverages, vinegar, fermented foods like *idli, dosa, jalebi*, tempeh, etc. but these activities were purely an art, not a science. The science of industrial microbiology is only about 150 years old. After the discovery of microscope the mystery of microbial world could be unraveled and different types of microorganisms were discovered. Thereafter, the beginning of 20th century witnessed the production of antibiotics, solvents, vitamins, enzymes, and ended with the methods of production of therapeutic recombinant proteins by genetically engineered microorganisms. All these achievements have been possible after exploitation of different metabolic routes of microorganisms. Low cost technology for commercial production of products was developed when microbiology was combined with biochemistry, chemistry, physics, biochemical engineering, economics and technology. The scientists made endeavour to exploit the full capabilities of microorganisms. Hence, the statistical calculations helped a lot in designing the experimental models.

Though Antoni von Leeuwenhoek (a Dutch tradesman and scientist commonly known as 'the Father of Microbiology') observed the microorganisms first in 1677, the experiments of Spallanzani in 1799 and that of Schwan in 1837 provided a means of sterilisation of liquids (by heat) and air (by heat), respectively. Schwan's findings also suggested that alcoholic fermentation was due to yeast, and inoculation resulted in quick fermentation. But microbiology began in 1857 when Louis Pasteur reported his studies on lactic acid fermentation including the microscopic features of the microorganisms and a suitable medium for the process. The scientific basis of industrial microbiology began thereafter.

Antonie van Leeuwenhoek (1632-1723)

In 1860, Pasteur reported the first synthetic medium for cultivation of microorganisms and used for alcohol fermentation. In 1861, Pasteur showed that growth and physiology of yeast and the accumulation of alcohol differ depending on the presence or absence of oxygen. This phenomenon is known as Pasteur Effect. In 1878, Lister described dilution technique for isolation of the first pure culture of lactic acid bacterium. In 1881, Robert Robert Koch described a simpler and more effective technique for obtaining pure cultures from separate colonies developed on solidified medium. In 1897, Buchner demonstrated alcohol fermentation by cell-free yeast juice; he suggested that a proteinaceous enzyme was responsible for fermentation. In 1901, Wildiers demonstrated that yeast required growth factors (vitamins) for growth, especially at low inoculum levels; vitamins are used in fermentation even today.

In 1929, Alexander Fleming accidentally discovered penicillin produced by *Penicillium notatum* growing as contaminant in a Petri plate of *Staphylococcus*. Fleming developed the technique for assay of antibacterial activity of penicillin using bacteria and showed its low toxicity to man and animals. This was followed by an intensive search for antibiotics during the Second World War leading to the discovery of streptomycin, chloramphenicol, tetracyclines, etc. Later developments have resulted in the use of metabolically blocked mutants of microorganisms, which accumulate large amounts of metabolic intermediates.

A. MICROBIAL GROWTH

Growth is defined as an orderly increase in cellular components. The 'balanced growth' is doubling of every biochemical unit of the cell within the time duration of a single division without a change in the rate of growth. There are number of methods available for measuring microbial growth the choice of which depends on the objectives and usefulness of the available techniques. In some cases of industrial fermentations that contain complex media, indirect methods for estimation need to be used; however, irrespective of methods used considerable care is required while interpreting the results. Furthermore, bacterial growth can be measured either by : (*i*) colony counting or cell counting, (*ii*) weighing the cells (cell mass measurement), (*iii*) ATPase activity, or (*iii*) cell activity (turbidity method) measurement.

1. Growth Parameters

There are several parameters used to measure microbial growth, as given below:

(*a*) Viable Plate Count (Standard Plate Count): Viable plate count is a common method for enumeration of bacteria. Serially diluted suspension of bacteria is spread over the surface of growth medium, and the plates are incubated at proper temperature to develop the bacterial colonies. Each bacterial colony is supposed to arise from an individual viable cell. The total number of colonies is counted and multiplied by the dilution factor to calculate the total number of cells present in the original sample. By using this method only those cells are counted that grow on the given medium under optimum conditions. However, some cells are viable but non-culturable, hence non-culturable cells cannot grow. The viable plate count method is an indirect measurement of cell density. Normally, the bacterial sample is diluted by factors of 10 and plated on agar. Generally, the number of colony forming units (CFUs) per ml of sample is determined. For example, if the number of colonies is 300 colonies, it is multiplied by the dilution factor of the plate counted (*i.e.* the number of times the original volume of bacteria was diluted).

(*b*) Cell Counting by Direct Microscopic Count Method: Microbial cells can be counted directly; direct microscopic counts are rapid but limited for their inability to distinguish between the living and dead cells unless differentiated by use of a vital staining technique. Bacterial cells can be accurately counted by using Petroff-Hausser counting chamber (Fig. 21.1 A). The chamber includes a glass slide, a cover slip which is framed and kept 1/50 mm above the slide so that bacterial suspension is present in each ruled square of the slide (B). The area of square is 1/400 mm^2; glass cover slip rests 1/50 mm above the slide hence volume over a square is 1/20,000 mm^3 or 1/20,000,000 cm^3, for example, if in one square an average of five bacteria are present, these are $5 \times 20,000,000$ or 10^8 bacteria per ml.

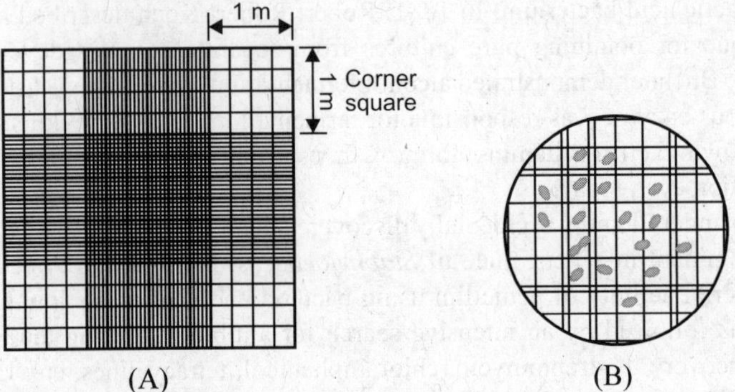

Fig. 21.1. The central square millimeter of Petroff-Hausser counting chamber (A); bacterial cells in a square (B).

(*c*) Viable Counting (Plate Counting) Technique: Viable plate count is a common method for enumeration of bacteria. Serially diluted suspension of bacteria is spread over the surface of growth medium, and the plates are incubated at proper temperature to develop the bacterial colonies. Each bacterial colony is supposed to arise from an individual viable cell. The total number of colonies is counted and multiplied by the dilution factor to calculate total number of cells present in the original sample. By using this method only those cells are counted that grow on the given medium under optimum conditions. However, some cells are viable but non-culturable, hence non-culturable cells cannot grow. Serial dilution of soil sample is shown in Fig. 21.2.

A measured amount of the sample of bacterial suspension is mixed in the agar medium having temperature of 40 – 45±°C. It is plated after mixing thoroughly. Each organism grows, reproduces and forms a visible mass in the form of colony. These are counted either by using colony counter or with the aid of large magnifying lens. If many colonies appear and overlap each other, the sample is diluted so that the colonies could be accurately counted. This method is called 'pour plate method'. The plate count is also performed by 'spread plate method'. In this method 0.1 ml sample containing bacteria is spread over the surface of an agar plate using a sterile glass spreader. In both pour plate and spread plate methods the plates containing bacterial suspension are incubated until the colonies appear and counted. As stated earlier the sample must be diluted to obtain the appropriate colony number. Serial dilutions of the sample are usually carried out. To make a 10-fold (10^{-1}) dilution, 10 ml sample is mixed with 90 ml diluents.

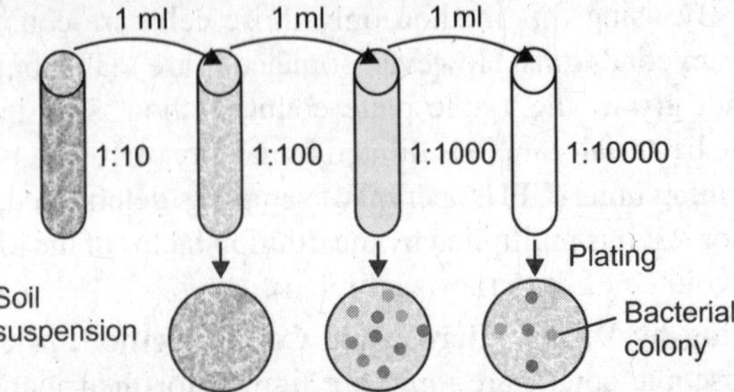

Fig. 21.2. Serial dilution technique for isolation of microorganisms.

In most cases, serial dilutions are performed to obtain the final dilution. Demerits of the above methods are not only the suitability of the culture medium and incubation conditions but sometimes bacterial cells deposited on the plate, does not show their visibility in the form of colony if incubation period is short. Further, viable counts and preparation of dilutions of the sample also give wrong information. To get correct information, viable counts are often expressed as the number of colony forming units (CFU) per millilitre rather than number of viable cells. This method is adopted in counting microorganisms in soil.

***(d)* Measurement of Cell Mass:** Cell mass is directly proportional to cell numbers. This can be obtained after centrifugation of a known volume of culture and weighing the pellet obtained. Measuring cell mass is an easy step of cell growth measurement. A known volume of culture sample from the fermentor is withdrawn and centrifuged. Wet weight of pellets is measured by using pre-weighed filter paper. A pre-weighed filter paper of similar size is used to subtract the weight of wet filter paper. Thus wet weight of cells is calculated.

Dry weight measurement of cell material is similar to that of wet weight. Dry weight of pre weighed filter paper containing pellets of microbial cells is measured. Dry weight of filter paper is nullified by subtracting the dry weight of only filter paper of similar size. Thus dry weight of pelleted microbial cells grown at 90-100°C overnight can be obtained.

***(e)* Absorbance (Optical Density Turbidity) Measurement Method:** The cell mass and numbers are also obtained by using optical density method. Due to the presence of cells, turbidity develops in the liquid medium that makes cloudy appearance to the eyes. When sample is more turbid it means that more cells are present. Hence more light is scattered. Turbidity can be measured with a photometer or a spectrophotometer device that detects the amount of un-scattered light recorded in photometer unit. This is indirect method; hence, some direct measurements of cell number should also be determined. Since optical density (OD) is proportional to cell mass and thus also to cell number, therefore, turbidity reading acts as an estimate of cell number or cell mass. The optical density is directly proportional to the cell concentration. If the absorbance is greater, the number of bacteria will be greater. Light entering a cloudy solution will be absorbed. A clear solution will allow almost all of the light through.

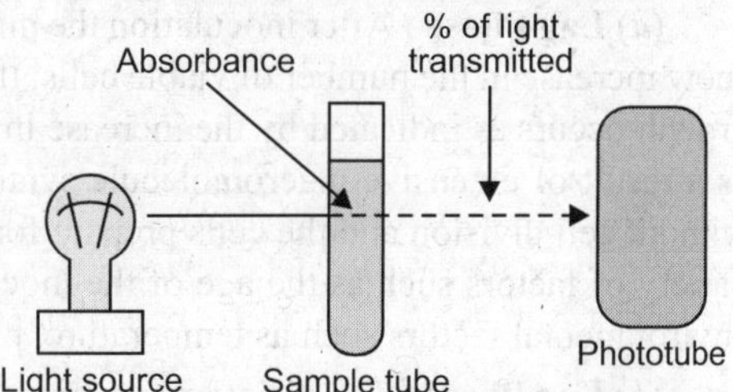

Fig. 21.3. Principle of photometer.

A spectrophotometer consists of : *(i)* a light source, *(ii)* a filter (that allows only a single wavelength of light to pass through), *(iii)* the sample tube containing the bacterial suspension, *(iv)* and a photocell (that compares the amount of light coming through the tube with the total light entering the tube). A light passes through a filter, then through the sample and onto a light-sensitive phototube (Fig. 21.3) resulting in production of an electrical current. The absorbance meter measures the amount of light that has been blocked by the sample and prevented from striking the phototube. A clear tube of water or other clear solution has zero absorbance. The amount of substance in the solution is directly proportional to the reading of absorbance. A graph of absorbance vs. concentration will produce a straight line. As the number of bacteria in a broth culture increases, the absorbance increases.

The plotting between semi-logarithmic versus time, growth rate of microbial cultures is obtained and used to calculate the 'generation time' of the growing culture. Such a curve can contain data for both cell number and cell mass, allowing for an estimate of both parameters from a single turbidity reading.

There are certain demerits of this method also. If the concentration of the cells in sample is high, light scattered away from the detecting unit by one to one cell can be scattered back by another. Hence, one-to-one correspondence between cell number and turbidity does not follow linearity. Secondly, dead cells also interfere during measurement. This method is reasonably accurate only for measurement of microbial growth before early log phase.

2. Growth in Batch Culture

Batch culture is the simplest method for growing a desired microbe in a closed culture system on a limited amount of medium. The laboratory grown microorganisms in ordinary flask is basically a batch culture. When inoculum of a microorganism is transferred from a slant culture to a known volume of liquid medium the population undergoes a characteristic sequence with its rate of increase in cell number. After incubation of medium, the microbe (yeast, bacteria or fungal inoculum) passes through five growth stages such as lag phase, log (or exponential growth) phases, deceleration phase, stationary phase and death phase (Fig. 21.4).

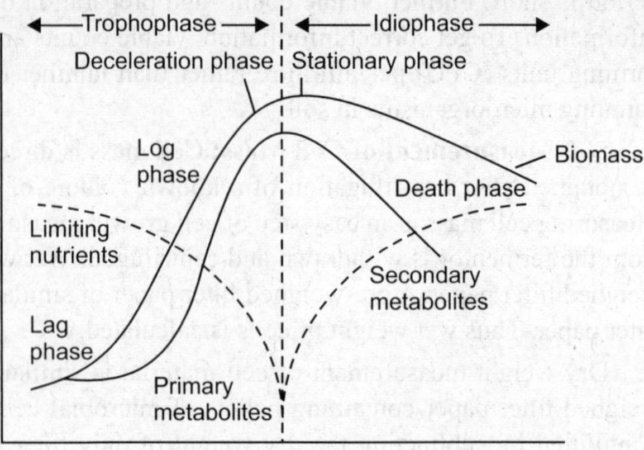

Fig. 21.4. Microbial growth in batch culture (dark line). Dotted lines show primary and secondary metabolic phases.

Due to gradual consumption of medium and production of metabolites the growing environment goes on changing. This influences the growing cells. Therefore, fresh medium is used in each batch.

***(a)* Lag Phase :** After inoculation the microorganism adapts the new environment and does not show increase in the number of viable cells; this period is called lag phase (Fig. 21.4). However, cell growth occurs as indicated by the increase in cell mass. During this period the cells increase in size as a result of extensive macromolecule synthesis. This stage represents a period of active growth without cell division and the cells prepare for division. The length of the lag phase depends upon a variety of factors such as the age of the inoculum, the composition of the growth medium and the environmental factors such as temperature, pH, aeration, etc.

***(b)* Log Phase :** The cells use nutrients of the growth medium and grow exponentially until the nutrients of medium are not depleted. The period of exponentially dividing cells is called log phase or exponential growth phase. During log phase there occurs an increase in cell mass and cell number with gradual decline in nutrients. Cell population increases logarithmically and the cells divide at the maximum rate permitted by the composition of medium and environmental conditions. The cell number during this period increases as a function of the exponent ($2^1, 2^2, 2^3, 2^4, . . 2^n$). Growth rates are measured during this period since growth occurs at a maximal rate. The growth rate can be expressed in time. For example, a culture which has a double time of 60 minutes will give a growth rate of one.

The growth rate is influenced by a variety of factors such as the composition of medium, environmental and the inherent properties of the culture. The length of log period is determined mainly by the composition of the medium and the rate of accumulation of inhibitory products. Although in the log phase, the growth is maximum; a culture in log phase represents a population of cells of different ages, some have just divided and others are in the intermediate stages of the division cycle. Therefore, in physiological terms the population is not homogeneous.

***(c)* Deceleration Phase :** Cell mass and cell number increase utilising the nutrients during log phase resulting in gradual decline of nutrients in growth medium. There is no microbial growth at all because the growth environment gets changed due to gradual consumption of nutrients and accumulation of metabolites in fermentor. Therefore, this stage is called deceleration phase.

***(d)* Stationary Phase :** The period of deceleration phase is followed by stationary phase where the total number of viable cells remains constant. At stationary phase microbial growth gets decreased to zero. The stationary phase is achieved when the viable cell number does not increase. The duration of stationary phase varies with the microorganism and environmental conditions. Some organisms may display long stationary phase lasting for several days, while the others may show a very short stationary phase of only few hours before the next phase begins.

Moreover, the sporulating bacteria may form endospores as they reach the stationary phase of growth and, therefore, they resist to lysis. The number of viable cells will remain constant after attaining the stationary phase and the phase of decline may not be seen.

***(e)* Death Phase :** Accumulation of secondary metabolites exceeds and no sufficient nutrients are left in the fermentor; therefore, the cells gradually start to die. This stage is called death phase, which is characterised by an exponential decrease in number of the viable cells. A cell is considered to have died or to have to become non-viable when it is no longer capable of multiplying.

When a population of cell from a stationary phase or death phase is used to inoculate fresh growth medium, the cells will not continue to die but re-enter the lag phase and initiate new growth. Cells from the death phase may, however, show a longer lag in contrast to cells transferred from either the stationary phase or the logarithmic phase.

Bacteria divide by binary fission. During the growth cycle cells double its mass and also the amount of all the cell constituents. Prior to division cell wall and membrane is synthesised and the parent cell divides into two identical cells, each of which grows exactly as the original cell. Doubling time for most bacteria is reported to be as fast as 15-20 minutes under optimal conditions (Table 21.1).

Table 21.1. Generation time of some Gram-positive and Gram-negative bacteria.

	Bacteria	***Generation Time (minutes)***
Gram-positive	*Bacillus megaterium*	25
	Streptococcus lactis	26
	Bacillus subtilis	28
	Staphylococcus aureus	30
	Lactobacillus acidophilus	66-87
	Escherichia coli	20
	Pseudomonas aeruginosa	35
Gram-negative	*Bradyrrhizobium japonicum*	344-461
	Mycobacterium tuberculosis	792-932
	Treponema pallidum	1980

Yeasts divide by budding (exceptions to include yeast that grow by fission or by forming hyphae). Budding involves the formation of a bud on the mother cell. The bud grows until it becomes same in the size of the mother. At this point the daughter cell separates. Under optimal conditions, yeast may divide in as little as 45 minutes However, 90-120 minutes is optimum.

3. Microbial Growth Kinetics and Specific Growth Rate

During log phase when cells utilise nutrients and grow to increase biomass, the growth behaves similar to autocatalytic reaction. At this phase growth rate is proportional to cell mass of that period. During this time the rate of 'cell mass increase' (*dx/dt*) is equal to the specific growth rate (μ) and cell concentration.

$$\frac{dx}{dt} = \mu . x \qquad ...(1)$$

Where, x = cell concentration (mg/cm^{-3}), t = time of incubation (hours), μ = specific growth rate (h^{-1})

Equation (1) may be written as

$$\mu = \frac{dx}{dt} \cdot \frac{1}{x} \qquad \ldots (2)$$

The specific growth rate (μ) acts as an index of rate of cell growth in that very environment. If you plot a graph between *dx/dt* and *x*, and determine the slope of straight line, you can calculate the specific growth rate. The value of specific growth rate can be converted into doubling time (*i.e.* time required by a cell to divide and double its number). It gives a better appreciation of the meaning of these values.

In the beginning of log phase (t = 0) the cell biomass is x_0. Assume that all the cells of batch culture require the same time to double their number (doubling time). Then on integration equation (2) you can calculate doubling time as below:

$$\mu\, dt = \frac{1}{X} . dx$$

or $\log x = \mu.t + K$ (K = constant of integration) ...(3)

When t = 0 from equation (3), then

$$\log x_0 = K$$

Putting the value of K in equation (3), we get

$$\log X - \log X_0 = \mu.t$$

$$\log \frac{X}{X_0} = \mu.t \qquad \ldots (4)$$

$$\therefore \quad t_d = \frac{1}{\mu} \ln \frac{X}{X_0}$$

When $t = t_d$ (doubling time)

Then, $X = 2X_0$

$$\therefore \quad t_d = -\ln 2 \frac{1}{\mu} \qquad \ldots (5)$$

$$\text{or} \quad t_d = \frac{\ln 2}{\mu}$$

If $t = t_d$ (doubling time), and X = double of X_0, then

$$\ln 2 = \mu.t_d$$

The μ is inversely proportional to t_d. If the microbial culture has high t_d, it will have low μ and *vice versa*. Specific growth rate of 1 h^{-1} is equivalent to t_d of 0.693. Therefore,

$$t_d = \frac{0.693}{\mu} \qquad \ldots (6)$$

You can calculate the doubling time of the culture by calculating and substituting in equation (6)

Specific Growth Rate

The specific growth rate is related to the doubling time (*td*) in terms of hours of the biomass. The *td* is equal to the generation time of a cell. The specific rates are very important parameter because it is related to cell productivity. Therefore, specific productivity of a given metabolite directly indicates the capacity of the cells to synthesise this metabolite. Furthermore, one can calculate the capacity of the biomass production per bioreactor volume (*i.e.* volumetric productivity) simply by multiplying the biomass concentration. In a kinetic model the specific rates are specified as functions of the variables in the systems, *i.e.* substrate concentrations.

There is another method of calculating the specific growth rate of a microbial culture. Number of microbial cells at different time is counted. If the medium is inoculated with X_0 cells at the time t_0, the number of cells after the time t_t becomes X. The equation (1) may be written as

$$\ln \frac{X}{X_0} = \mu.t \quad \text{or} \quad \ln X - \ln X_0 = \mu.t$$

After converting the natural logarithms (ln) to logarithm to the base 10, you get

$$\frac{\mu}{\dfrac{\log_{10} X - \log_{10} X_0}{2.303}} = -(t_t - t_0) \qquad \text{...(7)}$$

or $$2.303\{\log_{10} X - \log_{10} X_0) = \mu\,(t - t_0)$$

or $$\mu = \frac{(\log_{10} X - \log_{10} X_0)\ 2.303}{t - t_0}$$

This equation is true for biomass concentration (X) as well as cell number (N). If at the time t_o the microbial culture consists of 10^4 cells/ml and attains 10^{10} cells/ml after 3 hours, the specific growth rate (μ) of the culture can be calculated using equation (3) as below:

$$(t - t_0)\ 2.303$$

$$\mu = \frac{(t - t_0)\ 2.303}{\log_{10}(10^{10} - 10^{4})}\ (\log_{10} = 0.301)$$

$$= \frac{(10 - 4)\ 2.303}{3} = \frac{6 \times 2.303}{3} = 2 \times 2.303 = 4.606\ \text{h}^{-1}$$

The specific growth rate of microbial culture is 4.606 h^{-1}. Using the value of μ (*i.e.* 4.606 h^{-1}) in equation (6) you can derive t_d of the culture:

$$t_d = \frac{0.693}{\mu} = \frac{0.693}{4.606} = 0.15 \text{ hour} = 4 \text{ minutes}$$

The t_d of the microbial culture is 4 minutes.

It shows a reciprocal relationship between specific growth rate and doubling time t_d. If μ is high, the t_d will be low and vice versa. The μ is the capacity of the microbial culture to grow in a specific environment. The μ of a microbial culture is measured in log phase of growth during which balanced cell growth occurs. In a batch culture the values of μ varies having maximum value at exponential (log) phase of growth. The environmental factors (*e.g.* pH, temperature, medium composition, aeration, etc.) that affect microbial growth also affect the specific growth rate of the culture. Representative values of μ_{max} of some microorganisms are given in Table 21.2.

Table 21.2. Representative values of μ_{max} of some microorganisms.

Microorganisms	μ_{max} (h^{-1})
Benechea natriegens	4.24
Methylomonas methanolytica	0.53
Aspergillus nidulans	0.36
Penicillium chrysogenum	0.12

If a bacterial culture contains 10^5 cells/ml at t_0 and 10^{10} cells/ml after 4 hours, calculate its specific growth rate and doubling time.

Specific growth rate (m)

$$m = 2.303 \log_{10} (t - t_o)/\text{time interval}$$
$$= 2.303 \log_{10} (10^{10} - 10^5)/4$$
$$= 2.303\ 5/4$$
$$= 2.88/\text{h}$$

Doubling time (t_d) of bacterial cells

$$= 0.693/m = 0.693/2.88$$
$$= 0.240\text{h} \times 60 = 14.43 \text{ minutes}$$

(*a*) Synchronous Growth : In log phase of growth cells are present in various stages and division cycle. Analysis of such population, therefore, yields only average value of any parameter. To understand the properties of individual cell, during the course of its division cycle, it is necessary to analyse each cell, which is practically not possible. A system that closely resembles and amplifies the behaviour of single cell is a synchronous culture, which contains cells that are physiologically identical and are in the stage of division cycle. A synchronous population can be generated either by physically separating the cells in the same stage of division or by forcing a cell population to attain an identical physiological condition by a change in the environment. To obtain synchronised population, cells may be inoculated into a medium at a sub-optimal temperature. After sometime, they will metabolise slowly but will not divide. When the temperature is raised to optimum, the cells will undergo a synchronised division.

In the synchronised culture, the cells are physiologically identical, cell division occurs periodically at constant intervals. The dry mass of the cell, optical density, total proteins or RNA contents increase at a constant rate. The amount per cell will increase in population to the cell number. On the other hand, the pattern of DNA synthesis can be either periodic or continuous depending upon how fast the culture is growing. *E. coli* grows in a medium with a generation time greater than 40 minutes, will show a period when no DNA synthesis occurs. Recently, to obtain synchronous culture, the exponentially grown culture is centrifuged either in sucrose, glycerol or sorbitol gradients in order to separate cells based on their densities which is directly related to their age.

(*b*) Diauxic Growth : In a culture medium containing two carbon sources bacteria display a different growth pattern called diauxic growth (bi-phasic growth). In 1941, Jacques Monod discovered diauxic growth during his experiments with *E. coli* and *Bacillus subtilis*. He was testing the effects of combinations of sugars as nutrient sources on bacteria. He found that bacteria grown with two different sugars often displayed two phases of growth in batch culture. For example, if both glucose and lactose are provided, glucose shall be metabolised first (growth phase I) followed by lactose (growth phase II) (Fig. 21.5). Even in the first two panels, the sugars are consumed sequentially with no diauxic lag. Focusing on the well defined diauxic growth of *E. coli* on the mixture of glucose and lactose, his later work led to the *lac* operon model of gene expression and subsequently a Nobel Prize.

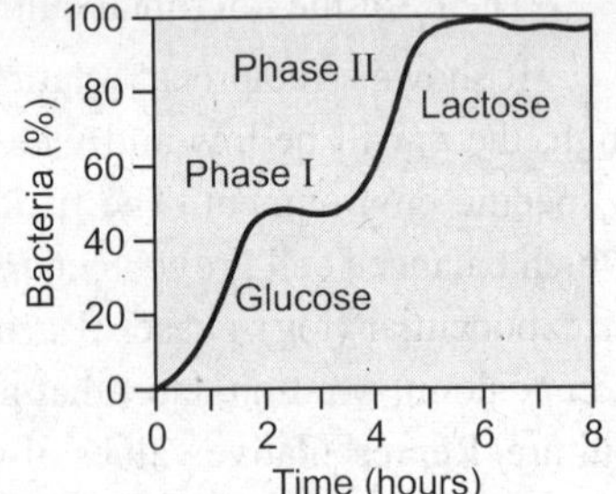

Fig. 21.5. Diauxic growth of *E. coli*; phase I-growth on glucose; phase II-growth on lactose.

Microbial cells commonly metabolise the two available sugars simultaneously in a sequential pattern resulting in two separate growth phases. Under this condition if glucose and lactose are supplemented in medium having *E. coli,* it will utilise glucose first, and thereafter lactose is utilised. In between there exists a short lag period. This led us to conclude that *E. coli* preferentially utilises certain carbon sources.

(*i*) Genetic regulation of diauxic growth: Metabolism of lactose does not occur during the first part of the diauxic growth curve because β-galactosidase is not synthesised when both glucose and lactose are present in the medium. The reason of diauxie growth is the use of operons of microorganisms to differently control the expression of enzymes needed to metabolise different sugars. The organisms regulate their genetic control mechanisms to express those genes resulting in the fastest growth rate. *Lactococcus lactis* when grown in the presence of both glucose and maltose produces enzymes to metabolise glucose first. It changes its gene expression to use maltose only after exhausting the glucose. When yeast grows aerobically on glucose, commonly it shows diauxic growth in batch culture. During the growth phase I the yeast prefers glucose fermentation to aerobic respiration when glucose and oxygen are available. Yeast has evolved the choice due to the intracellular genetic regulatory mechanisms. Fermentation provides a faster growth rate for the yeast cells than the aerobic respiration of glucose. In the presence of oxygen, ethanol is oxidised in a slower second growth phase after depletion of glucose.

(*ii*) Mechanism of diauxic growth: Microbial cells prefer to metabolise the glucose faster during the growth phase I. After depletion of glucose the cells switch to the growth phase II. Often there is a lag period at the time of the 'diauxic shift' during which cells produce the enzymes needed to metabolise the second sugar. Joshua Lederberg isolated β-galactosidase during his work on the *lac* operon of *E. coli.* He isolated it in high amount from colonies grown on lactose rather than the other sugars. Thus, the idea of enzyme adaptation was replaced with the concept of enzyme induction. In the case of enzyme induction a molecule induces expression of an operon. An inducer binds to a repressor protein and prevents it from attaching to the operator.

During bacterial diauxic shift from glucose to lactose metabolism, initially glucose inhibits the enzyme adenylate cyclase to synthesise cyclic AMP (cAMP). The catabolite activator protein (CAP) needs cAMP to bind to DNA and activate the transcription of the *lac* operon. The presence of lactose is sensed through the activity of the *lac* repressor, which inhibits transcription of the *lac* operon unless it is bound by lactose. The cAMP levels remain low in the presence of glucose; hence CAP does not activate the transcription of the *lac* operon, irrespective of the presence or absence of lactose. When glucose is exhausted, cAMP levels rise that allow the CAP to activate the genes required to metabolise the other substrates, including lactose also.

2. Growth in Fed-batch Culture

Fed-batch culture is basically a batch culture which is continuously fed with nutrient medium in the fermentor without removing the growth culture or growth products. Consequently volume of the medium is increased. The nutrients should be added at the same rate as they are consumed by the growing cells. Therefore, excess of nutrient addition should be avoided. In batch culture when high concentration of substrate inhibits microbial growth, the fed-batch culture is preferred over the former. Hence, in a fed-batch culture substrate is fed at such a concentration that remains below the toxic level. This activity accelerates the cell growth. A high cell density is achieved in batch culture as compared to fed-batch culture.

Fed-batch culture is an ideal process for production of intracellular metabolites in maximum amount. For example, alkaline protease used in biological detergents is produced by the species of *Bacillus*. Batch feeding of nitrogen sources (*e.g.* ammonia, ammonium ions and amino acids) keeps these substrates at low concentration and induces protease synthesis.

3. Growth in Continuous Culture

When culture becomes old in batch culture, the composition of the culture medium is drastically changed. To keep the cultures in constant environments for longer durations continuous culture method is adopted. Continuous culture is an open process in which microbial cultures grow continuously in log phase. A continuous culture essentially requires a flow of constant volume to which medium is added continuously and from which continuous removal of medium along with culture can occur. When such a system is in equilibrium, cell number and nutrient status remain constant and system is in steady state. The rate of supply of nutrients in the form of raw material and removal of products/cells should be volumetrically the same, *i.e.* volume added is equal to volume removed. Hence the volume always remains constant.

The fresh culture medium flows into a growth chamber at a carefully controlled rate, the volume of culture is maintained by controlling the rates of inflow and out flow, the rate of growth is then controlled by regulating the inflows rates. The rate of loss of cells through overflow can be expressed as below :

$$DM = \frac{DM}{dt} = \frac{F}{V}$$

Where flow rate F is measured in the culture volume V/h. The expression F/VM is called the dilution rate (D). In a continuous culture, $KM = DM$.

$K = D$, *i.e.* growth rate (K) will equal the dilution rate in a stabilised continuous culture. Under such conditions growth of the culture is linear rather than exponential, since the amount of mass/unit volume (growth rate) remains constant. The continuous culture does not represent synchronous growth since these do not contain cells that are physiologically identical.

(*a*) Chemostat for Growth Measurement: Chemostat (*kee-mo-stat*, *kem.o.stat*) is a device for maintaining organisms in continuous culture; it regulates the growth rate of the organism by regulating the concentration of an essential nutrient by a pump connected to a medium reservoir (Fig. 21.6). The growth rate and growth yield can be controlled independently on each other by adjusting the dilution rate and by varying the concentration of nutrients present in a limiting amount. At steady state the specific growth rate (μ) of the microorganism is equal to the dilution rate (D). The dilution rate is defined as the rate of flow of medium over the volume of culture in the bioreactor.

$$D = \frac{\text{Medium flow rate}}{\text{Culture volume}} = \frac{F}{V}$$

The high dilution rate does not allow the organism to grow fast enough to keep up with its dilution rates, a large fraction of the cells may die from starvation due to non-availability of nutrients required for maintenance of cell metabolism. Hence, a minimum amount of energy is required to maintain cell structure. Such energy is called maintenance energy.

To achieve a steady state, parameters other than the dilution rate and culture volume must be kept constant (for example, temperature and pH). The fermentor is stirred to provide a homogeneous suspension in which all individual cells in the culture come into contact with the growth-limiting nutrient immediately, and to achieve optimal distribution of air (O_2) in the fermentor when aerobic cultures are in use. The number of cells/ml is called cell density which is controlled in the chemostat by the nutrients. If the concentration of the nutrient in the incoming medium is increased with the constant dilution rate, the cell density will increase but growth rate will remain the same. Usually the laboratory chemostats contain 0.5 to

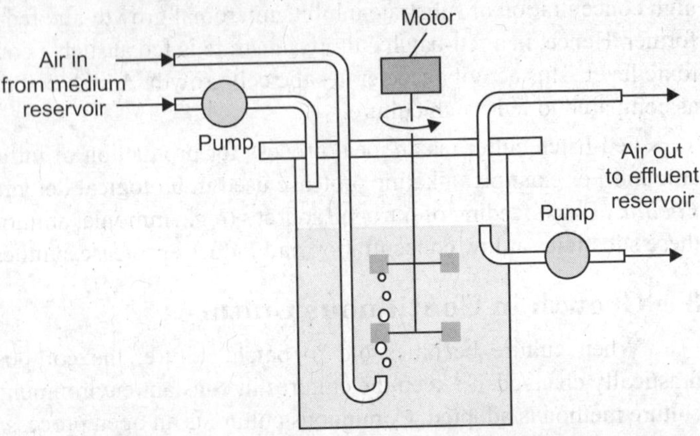

Fig. 21.6. Diagram of a chemostat apparatus.

10 litres of culture, but industrial chemostat can involve volumes up to 1,300 m^3 for continuous production of microbial biomass.

B. TECHNIQUES OF MICROBIAL CULTURE

The technique of microbial culture is a multistep process and requires media formulation, sterilisation, environmental control, operation of bioreactor, etc. These steps are discussed in this section.

1. Sources of Nutrition

Carbon and nitrogen present in medium act as energy sources (besides using carbon for cell mass and production of products). Energy is needed for growth of microorganisms. There are different sources of nutrients required by different types of microorganisms; these are given below:

(*i*) Carbon Sources: Choice of carbon sources has a major effect on design of the fermentor. Because it provides sufficient amount of oxygen from complete utilisation of carbon sources and sufficient heat exchange area for removal of heat. The carbon sources used for large scale microbial culture in fermentor are sugarcane molasses, beet molasses, vegetable oil, starch, cereal grains, whey, glucose, sucrose, lactose, malt, hydrocarbons, etc.

(*ii*) Nitrogen Sources: The nitrogen sources are corn steep liquor, slaughterhouse wastes, urea, ammonium salts, nitrate, peanut granules, soybean meal soya meal, yeast extract, distilled soluble, etc.

(*iii*) Growth Factors: There are certain microorganisms which are not capable of synthesising vitamins or amino acids. Therefore, to achieve optimum growth, media are supplemented with the growth factors.

(*iv*) Trace Elements: Microorganisms also require certain trace elements (such as Zn, Mn, Mo, Fe, Cu, Co, etc.) in trace amount. Because these are associated with stimulation of metabolism or enzymes (metallo-enzymes) and proteins (leg-haemoglobin).

(*v*) Inducers, Repressors and Precursors: The catabolic enzymes are induced only in the presence of inducers; for example, yeast extract induces streptomycin fermentation. Production of catabolic enzymes and secondary metabolites is repressed by the presence of certain compounds in culture medium. For example, rapidly utilised carbon sources repress the formation of amylases, griseofulvin, penicillin, etc. Besides, production of certain products is increased if precursor metabolite is supplemented in medium, for example addition of phenyl acetic acid enhances penicillin fermentation.

(*vi*) Antifoams: Protein sources in culture media (*i.e.* products of medium or produced by microorganisms) cause foaming. Mostly foaming creates problem in microbial process. Therefore, to check the foaming problem, some antifoam, *i.e.* fatty acids (such as sunflower oil or olive oil) are added in the culture medium.

(*vii*) Water: Water is the most important component of the living cells. Because all metabolic activities occur in cytosol. Water-soluble ionic forms of nutrients are absorbed by the cell. In laboratory single or double distilled water is used for preparation of culture media. But for large stage industrial production clean water of consistent composition is required. Dissolved chemicals and pH of water are measured. Water is also needed for ancillary activities for example, cleaning, washing, rinsing, cooling, heating, etc.

1. Culture Media

The culture media consist of chemicals which support the growth of culture or microorganisms. Microbes can use the nutrients of culture media as their food is necessary for cultivating them *in vitro*. The medium should neither be acidic nor alkaline. It should contain all kinds of nutrients in suitable amounts. It must be sterilised before use. Microorganisms require several nutrients (e.g. carbon, nitrogen, phosphorus, minerals) and oxygen for growth and yield. The nutrient formulations which support optimum microbial growth and yield are called growth media. On the basis of purity of chemical compounds used, media are of different types.

(a) Types of Media: The first medium prepared was meat-infusion broth. As most pathogenic microbes require complex food similar in composition to the fluids of the animal body. It was Robert Koch and his colleagues who used meat infusion and meat extracts as basic ingredients in their culture media for the isolation of pathogenic microbes, while one of his assistants named Petri designed and developed glass dishes, known today as Petri dishes, are used in microbiological work.

On the basis of chemical composition, the media are classified into three types:

***(i)* Synthetic or chemically defined media:** These media are prepared by mixing all the pure chemicals of known composition, for example Czapek Dox medium.

***(ii)* Semi-synthetic or undefined media:** Such are those media, where exact chemical composition is unknown, for example potato dextrose agar or MacConkey agar medium.

***(iii)* Natural Media:** The media prepared by using the natural complex compounds are called natural media e.g. soybean seed extract broth, V8 juice broth, soil extract broth, etc.

On the basis of consistency, the media are of three types:

***(i)* Solid medium or synthetic medium:** When 5-7% agar or 10-20% gelatin is added the liquid broth, it is solidified. Such media are used for making agar slants and agar stab.

***(ii)* Liquid medium or broth:** In such cases no agar is added or used while preparing the medium. After inoculation and later incubation, the growth of cells becomes visible in the form of small mass on the top of the broth.

***(iii)* Semi-solid or floppy agar medium:** Such media are prepared by adding half quantity of agar (1/2 than required for solid medium), *i.e.* about 0.5% in the medium. This type of medium may be selective which promotes the growth of one organism and retards the growth of the other organism. On the other hand, there are differential media which differentiate organisms growing together.

***(b)* Preparation of Media:** The liquid medium or broth is prepared by dissolving the known amounts of chemicals in distilled water; the pH is adjusted by adding N/10 HCl or 1N NaOH. The liquid medium is dissolved into either Erlenmeyer flasks or clean test tubes. In 15 ml capacity of test tube, 5 ml medium should be poured while in flask of 250 ml capacity, the amount of the medium should be 100 ml. These are then plugged with non-adsorbent cotton plugs. The plugged tubes or flasks should be wrapped by brown paper and placed for sterilisation by autoclaving at a pressure of 15 lbs/inch^2 (at temperature 121°C) for 15 min. The heat sensitive substances (such as protein, enzymes, etc.) should be sterilised by using membrane filters. The agar is dissolved separately and dispensed after dissolving all ingredients of the medium. All the glassware should be sterilised in oven at 170°C for 3 h before using them. Such sterilised glassware are needed for pouring the medium used for culturing of microorganisms.

Every biological process requires energy for their vital activities. The basic cell building requirements are supplied by the nutrition, which is manipulated according to its requirement. Nutrition provides energy and also acts as precursors for growth of microorganisms. The nutritional requirement of an organism depends upon the biochemical capacity. If an organism is capable of synthesising its own food using various inorganic components, it requires a simple nutritional diet; but organism unable to meet such synthesis requires complex organic substances.

C. FERMENTORS – DESIGNING AND CONSTRUCTION

Production of products has been carried out throughout the world since time immemorial.

Fermentation could have been done in a well designed vessel called fermentors. One should be very clear about the difference between fermentor and fermenter.

Fermentor Vs Fermenter

In some books both fermentor and fermenter are used as synonym. As per Webster's dictionary the word 'fermenter' (*fer·men·ter*) means 'an organism that causes fermentation', and the term 'fermentor' (*fer·men·tor*) means 'an apparatus for carrying out fermentation'. Similarly, according to The Free Dictionary by Farlex, *fer-men-ter* means an organism that causes fermentation; *fer-men-tor* means an apparatus that maintains optimal conditions for the growth of microorganisms used in large-scale fermentation and in the commercial production of antibiotics and hormones.

Different types of microorganisms are exploited in fermentation process for the production of products by building up different types of fementors. In 1944, De Becze and Liebmann used the first large scale (above 20 litre capacity) fermentor for the production of yeast. During the World War I, a British scientist named Chain Weizmann (1914-1918) developed a fermentor for the production of acetone. Since the importance of aseptic conditions was recognised, steps were taken to design and construct piping, joints and valves in which sterile conditions could be achieved and manufactured when required.

During 1930s, large scale aerobic fermentors were used for the first time in central Europe for the production of compressed yeast. The fermentor consisted of a large cylindrical tank with air introduced at the base via network of perforated pipes. In later modifications, mechanical impellers were used to increase the rate of mixing and to break up and disperse the air bubbles. This process led to the compressed air requirements. Baffles on the walls of the vessels prevented a vortex forming in the liquid. In 1934, Strauch and Schmidt patented a system in which the aeration tubes were introduced with water and steam for cleaning and sterilisation.

The decision to use submerged culture technique for penicillin production, where aseptic conditions, good aeration and agitation were essential, was probably a very important factor in forcing the development of carefully designed and purposefully built fermentation vessels. In 1943, when the British Government decided that surface culture was inadequate, none of the fermentation plants were immediately suitable for deep fermentation. In 1950, the first pilot fermentor was established at Hindustan Antibiotic Ltd., Pimpri (Pune) in India.

1. Basic Function of Fermentors (Bioreactors)

The purpose of using the fermentors is to provide a controlled environment for growth of microorganisms (or a defined mixture of microorganism) to obtain a desired product while bioreactors refer to production units of mammalian and plant cell culture. Certain criteria must be used in designing and constructing a fermentor. The vessel should be capable of being operated aseptically for several days and should be reliable for long term operation. The adequate aeration and agitation should be provided to meet the metabolic requirements of the microbes. However, the mixing should not damage the microorganism. The power consumption should be low and temperature and pH control system should be provided. The evaporation losses from the fermentor should not be excessive. The vessel should be designed to require the minimal use of labour in operation, harvesting, cleansing and maintenance. It should have proper sampling facility. The vessel should be constructed to infuse instead of flange joints. The cheapest and best material should be used and there should be adequate service provisions for individual plants.

A bioreactor should provide the following: (*i*) agitation (for mixing of cells and medium), (*ii*) aeration (aerobic fermentors for O_2 supply), (*iii*) regulation of factors like temperature, pH, pressure, aeration, nutrient feeding, liquid level, etc., (*iv*) sterilisation and maintenance of sterility, and (*v*) withdrawal of cells/medium (for continuous fermentors). Modem fermentors are usually integrated with computers for efficient process monitoring, data acquisition, etc.

2. Construction of Fermentors

The criteria considered before selecting materials for constructions of a fermentor are : (*i*) the materials that have no effect of sterilisation, and (*ii*) its smooth internal finish, discouraging lodging of contamination. The internal surface should be corrosion-resistant. There are two types of such materials : (*i*) stainless steel, and (*ii*) glass which are used in fermentor. According to American Iron and Steel Institute (AISI) if steel contains 4% chromium, it is called stainless. The long and continuous use of stainless steel sometimes shows pitting. It is also important to consider the materials used for aseptic seal. Sometimes it is made between glass and glass, glass and metal and metal joints between a vessel and detachable top or base plate. On pilot scale, any material to be used will have to be assessed on their ability to withstand pressure, sterilisation, corrosion and their potential toxicity and cost.

(*a*) Temperature Control. Since heat is produced by microbial activity and mechanical agitation, it is sometimes necessary to remove it. On the other hand, in certain processes extra heat is produced by using thermostatically controlled water bath or by using internal heating coil or jacket meant for water circulation.

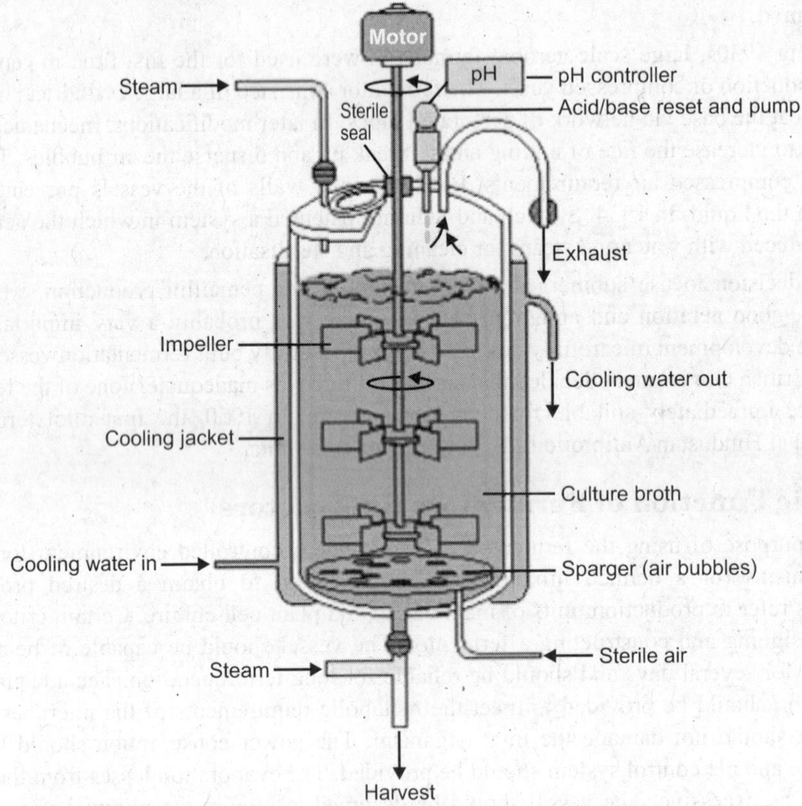

Fig.21.7. Basic design of a fermentor with multi-bladed impeller.

(*b*) Aeration and Agitation. The main purpose of aeration and agitation is to provide oxygen required to the metabolism of microorganisms. The agitation should ensure a uniform suspension of microbial cells suspended in nutrient medium. There are following necessary requirements for this purpose : (*i*) the agitator (impeller) for mixing, (*ii*) stirrer glands and bearings for aseptic sealing, (*iii*) baffles for checking the vortex resulting into foaming, (*iv*) the sparger (aeration) for introducing air into the liquid (Fig. 21.7).

(*i*) Agitator (Impeller): The size and position of the impeller in the vessel depends upon the size of the fermenter. In tall vessels, more than one impeller is needed if adequate aeration is to be obtained. Ideally, the impeller should be 1/3 or 1/2 of the vessel diameter above the base of the vessel. The number of impeller may vary from size to size to the vessel (Fig. 21.7).

***(ii)* Stirrer gland and bearing:** Four basic types of seals assembly have been used, the packed gland seal, the simple bush seal, the mechanical seal and the magnetic drive.

***(iii)* Baffles:** The baffles are normally incorporated into agitated vessel of all sizes to prevent a vortex and to improve aeration efficiency. They are metal strips roughly one-tenth of the vessel diameter and attached radially to the walls.

***(iv)* Sparger:** A sparger may be defined as a device for introducing air into the liquid in a fermentor. It is important to know whether sparger is to be used on its own or with mechanical agitation as it can influence equipment design to determine initial bubble size. Three basic types of sparger have been used and may be described as the porous sparger, the orifice sparger and the nozzle sparger.

***(v)* Microbial biosensors:** A microbial sensor consists of a microorganism immobilised on a membrane and an electrode. The principle of working of biosensor is the change in respiration or the amount of metabolites produced as a result of the assimilation of substrate by the microorganism. A wide range of thermophilic microbes have been used for the manufacturing of microbial sensors as given in Table 21.3.

Table 21.3. Use of microorganisms in biosensor for measuring substances in industry.

Name of Microorganism	*Detected Substance*	*Substance to be measured*
Citirobacter freundii	IT	Cephalosporin
Lactobacillus arabinosus	H^+	Vitamin-nicotinic acid
Escherichia coli	CO_2	L-glutamate
Streptococcus faecalis	NH_3	L-arginine
Trichosporon brassicae	O_2	Alcohol- ethanol
Clostridium butyricum	Fuel Cell	Formate
Yeast cells	O_2	Nystatin

Microbial biosensors have several uses in clinical analysis, such as general health care monitoring, veterinary and agricultural applications, industrial product processing and monitoring besides control of environmental pollution. Biosensors are low cost, small sized, sensitive and selective in nature. A large number of bacteria and eukaryotic cell cultures are used in manufacturing the new products. The monitoring of these items is essential by using microbial biosensors, this minimises the production cost.

4. Design and Operation

Fermentors are designed to support the best possible growth and biosynthesis of industrially important cultures, and to allow ease of manipulation for all operations associated with the use of the fermentors. These vessels must be strong enough to resist the pressure of large volume of agitating medium. The product should not corrode the material nor contribute toxicity to the growth medium. This involves a meticulous design of every aspect of the vessel parts and other openings, accessories in contact, etc.

The size of fermentors ranges from 1-2 litres laboratory fermentors to 500,000 litres or even more; fermentors of up to 1.2 million litres have also been used. Generally, 20-25% of fermentor volume is left unfilled with medium as 'head space' to allow for splashing, foaming and aeration. The fermentor design varies greatly depending on the type of fermentation for which it is used.

In fermentations, provisions should be made for the control of contaminating organisms for rapid incorporation of sterile air into the medium in such a way that the oxygen of air should be dissolved in the medium. Therefore, oxygen should be readily available to the microorganisms and CO_2 produced from microbial metabolism is flushed from the medium. Some stirring devices should be available for mixing the organism through the medium so as to avail the nutrients and oxygen. The fermentor has a possibility for the intermittent addition of antifoam agent. Some form of temperature control for efficient heat transfer system should also be provided for maintaining a constant temperature in the fermentor during the growth of an organism. The pH should be detected. The other accessories in the fermentor consist of additional inoculum tank or seed tank in which inoculum is produced and then added directly into it without employing extensive piping because it may cause contamination problem (Fig. 21.8).

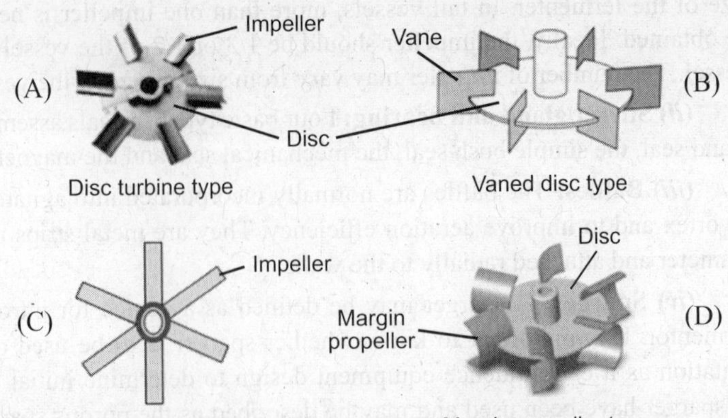

Fig. 21.8. Different types of impellers; A- disc turbine type; B- vaned disc type; C- open turbine types; D- margin propeller type.

4. Types of Fermentors

Bioreactors are extensively used for food processing, fermentation, waste treatment, etc. Fermentors can be classified into two main types based on shape : (*i*) tabular and (*ii*) stirred tank. Cooling coils are provided to maintain constant temperature inside the bioreactor. It can be operated asepticaly for many days and simple in construction. The disadvantages are high power requirement, shearing on the organisms caused by vigorous agitation and inhibition exercised by the product. In terms of process requirements, bioreactors are of the different types such as: (*i*) aerobic, (*ii*) anaerobic, (*iii*) solid state, and (*iv*) immobilised cell bioreactors.

But according to the requirements and various kinds of uses, the fermentors are classified as below:

(*a*) Fluidised Bed Bioreactor: During the 1920s, Fritz Winkler developed the first fluidised bed gas generator in Germany. A fluidised bed bioreactor is a type of device used for variety of multiphase chemical reactions. It is a new tool in the field of chemical engineering. In this device a fluid (gas or liquid) is passed through a granular solid material (usually a catalyst at high velocities so that the solids may be suspend (Fig. 21.9).

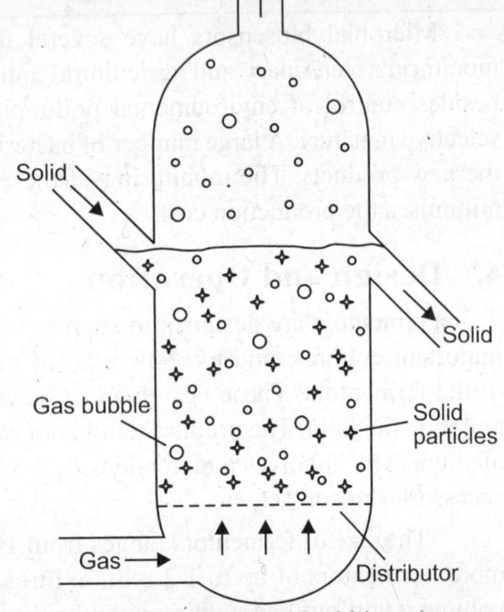

Fig. 21.9. A fluidised bed bioreactor (diagrammatic).

This process is called fluidisation that imparts many important advantages to the FBR. The solid substrate material, *i.e.* the catalytic material upon which chemical species react, in the fluidised bed reactor is supported by a porous plate called a distributor. Then the fluid is forced through the distributor. At low fluid velocities as the fluid passes through the voids in the material, the solids remain at its place (Howard, 1989). These are mostly used in conjunction with immobilised cells or enzyme system and are operated continuously.

The fluidized bed reactor is used in many industries for many ways. However, it is more popular in chemical industry rather new to biochemical industries. In 1942, the Standard Oil Company of New Jersey (U.S.A.) used the Catalytic Cracking Unit as one of the first fluidized bed reactors in the petroleum industry.

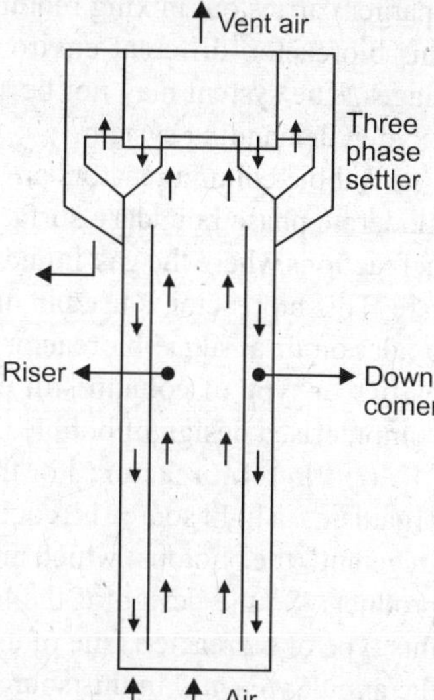

Fig. 21.10. An air lift reactor.

***(b)* Air-lift Bioreactor:** In the conventional bioreactor, oxygen is supplied by vigorous agitation of the bioreactor content. The heat is generated which is a problem in conventional type. In air-lift fermentor, cooling becomes simpler due to the position of inner or outer loop. In air-lift bioreactors gas is injected to perform mixing circulation and aeration if needed by additional external liquid circulation to obtain the required mixing pattern (Fig. 21.10). Air lifts results in more recirculation for the same air flow but lower oxygen transfer rates than bubble columns. Small bubbles can be used in aeration to limit shear, but this may cause excessive foaming and requires more energy for their generation at porous distributors.

***(c)* Membrane Bioreactor:** Membrane bioreactor (MBR) is the combination of a membrane process like ultra-filtration with a suspended growth bioreactor. The membrane is semi-permeable that is made up of cellulose acetate or other polymeric materials. The primary purpose of the membrane is to retain the cells within the bioreactor, thus increasing their density. At the same time it allows metabolic products to pass through the membrane (Fig. 21.11).

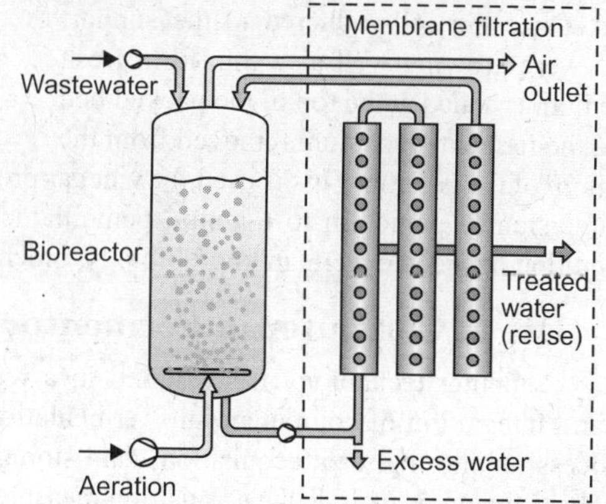

Fig. 21.11. A membrane bioreactor (diagrammatic).

Membrane bioreactor is widely used for treatment of municipal and industrial wastewater having the plant sizes up to 80,000 populations. In the membrane bioreactor, the membranes are immersed in and integral to the biological reactor. Membrane bioreactor processes mixed liquor suspended solids at higher concentrations as compared to conventional settlement separation systems. Therefore, the reactor volume is reduced to achieve the same loading rate.

***(d)* Pulsed Column Bioreactor.** The essential component of a pulsed column bioreactor is a column bioreactor generator connected to the bottom of the column. A pulsed column bioreactor can be utilised as an aerobic bioreactor, enzyme bioreactor or as a separation unit since its original successful application was in the extraction of uranium.

***(e)* Bubble Column Bioreactor.** A bubble column reactor is an apparatus used for gas-liquid reactions. It consists of vertical arranged cylindrical columns. The gas is introduced at the bottom of the column that causes a turbulent stream. Also it enables an optimum gas exchange. The gas sparger carries out mixing requiring less energy than mechanical stirring. In this bioreactor, different environmental conditions are provided in various stages. The system may not be suitable for fungal fermentation due to high oxygen demanding system.

Bubble column reactors are characterised by a high liquid content and a moderate phase boundary surface. The bubble column is particularly useful in reactions where the gas-liquid reaction is slow in relation to the absorption rate. This bioreactor is used in different kinds of chemical reactions like wet oxidation or as algae bioreactor (Fig. 21.12). Experience in the choice of a particular type of column still plays an important role due to restriction in computerised design of bubble columns to a few partial processes.

Fig. 21.12. A algae bioreactor.

(*f*) **Photobioreactor:** For the growth and production of photosynthetic organisms, a light source is required. In photobioreactor, there is an important 'reactant' (the photons) which must be absorbed in order to react and produce products. So the design of the light source is critical in the performance of this type of bioreactor. One of the most interesting photochemical reactors is the annular reactor. In this source of radiation is a cylinder with an annular section, which encloses the lamp completely. The nutrient passing from the product is removed from the top. This is used for *Spirulina* (SCP) and other algal protein production. A diagram of tubular photobioreactor is given in Fig. 21.13.

(*g*) **Packed Tower Bioreactor:** It consists of cylindrical column packed with inert material like wood shavings, twigs, cake, polyethylene or sand. Initially, both medium and cells are fed into the top of the packed bed. Once the cells adhered to the support and were growing well as a thin film, fresh medium is added at the top of the packed bed and the fermented medium removed from the bottom of the column. This is used for vinegar production, sewage effluent treatment and enzymatic conversion of penicillin to 6-amino penicillanic acid. The design of fermentor involves the co-operation between experts in microbiology, biochemistry, mechanical engineering and economics.

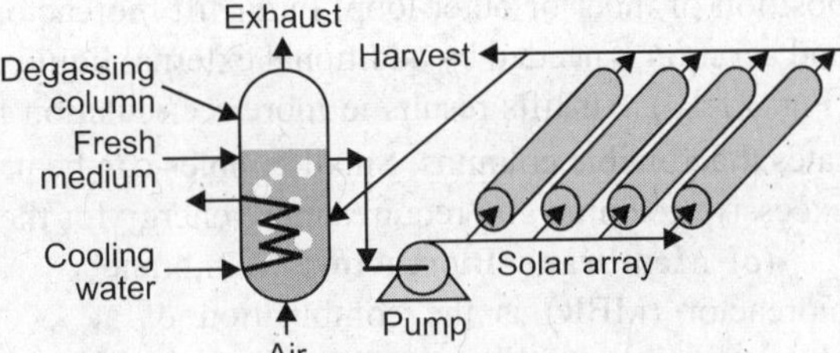

Fig. 21.13. Schematic tubular photobioreactor.

5. Use of Computer in Fermentor

Computer technology has resulted in a remarkable impact in fermentation work in recent years. Integration of computers into fermentation systems is based on the computer capacity for process monitoring, data acquisition, data storage, and error-detection. Some typical on-line data analysis functions include the acquisition measurements, data verification, filtering, unit conversion, calculations of indirect measurements, and differential integration calculation of estimated variables, data reduction, and tabulation of results, graphical presentation of results, process stimulation and storage of data.

In alcoholic fermentation using *S. cerevisiae*, the stoichiometry of glucose conversion to ethanol and CO_2 is given as below :

$$\underset{\text{(Glucose)}}{C_6H_{12}O_6} \longrightarrow \underset{\text{(Ethanol)}}{2\,C_2H_5OH} + \underset{\text{(Carbon dioxide)}}{2CO_2}$$

This equation shows that 0.511 g of ethanol and 0.489 g of CO_2 are produced from each gram of glucose. The real stoichiometric yields are 90-95% of these values because some of the glucose is used for the production and synthesis of secondary products and cell components. It is possible to follow the kinetics of fermentation by measuring the mass of CO_2 released. Moreover, at high glucose concentrations the glucose consumption is inhibited. Therefore, the fed-batch fermentation is preferred to avoid this inhibition. In this process, fermentation is started batch-wise with a low glucose concentration. When all the initial substrate is consumed, a new addition restores glucose concentration to just below the point where it produces inhibitory effects. The fed-batch fermentation is a sequence of batch fermentations of increasing volumes. The apparatus for this experiment has a magnetically stirred 2-liter flask (Fig. 21.14).

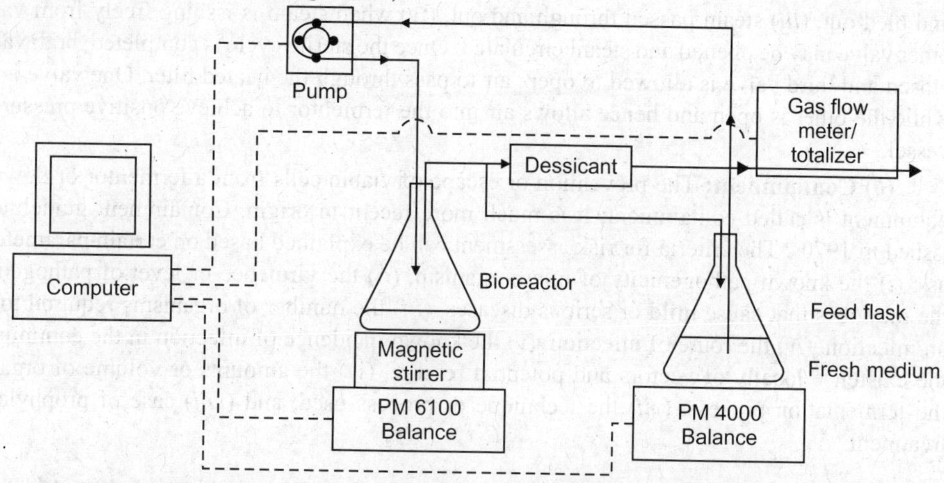

Fig. 21.14. Use of computer in monitoring of fermentation process.

Fresh fermentation medium is supplied from another 2-liter flask through a computer controlled peristaltic pump. The computer controls the pump and collects data. Both the fermentation and medium flasks are on top-loading balance which is equipped with serial computer interfaces. It enables on-line calculation of mass balances and productivity for the fermentation. A meter vents the fermentation flask and independently measures the CO_2 released during sugar fermentation and production of ethanol.

In a fermentation pilot plant, the aim of computers is to prepare and present data for the analysis of fermentation processes. The hardware is normally bought from computer manufacturers with some additional installations. The biotechnologists develop application software; therefore, the software structure is the most important part of the computer application. Data storage is divided into three parts, short-time memory, long-time storage, and a medium memory mainly to be used in process analyses and process control. There are four types of programmes used: main schedule tasks, low-priority on-line tasks, sense-switch routine and different off-line programmes. For demonstrating the software structure a table of all programmes is presented describing in detail the main schedule task.

6. Maintenance of Aseptic Conditions

It is necessary to sterilise the fermentor and its contents throughout the complete growth. The following operations need to be performed to achieve and maintain aseptic conditions during fermentation : (*i*) sterilisation of fermentor, (*ii*) sterilisation of air supply, (*iii*) aeration and agitation, (*iv*) addition of inoculum, nutrients and other supplements, (*v*) sampling, (*vi*) foam control, (*vii*) monitoring and control of various parameters.

(*a*) Sterilisation: The fermentors are sterilised by steam under pressure. The medium may be sterilised in the vessel or separately and subsequently added aseptically. As every point of entry and exit from the fermentor is a potential source of contamination, steam should be introduced through all the entry and exit points except the air outlet from which steam should be allowed to leave. Sterile air will be required in very large volumes in many aerobic fermentation processes. Although, there is a number of ways of sterilising air, only three have found permanent application. These are heat filtration through fibrous material and filtration through granular material. Heat is generally too costly for full scale operation. The chosen filter must remove microbes to a high level of efficiency, to be relatively cheap, robust and have a low pressure drop. Air filter must also be sterilised in association with the fermentor. Use of a steam packed air filter is an alternate approach. At the beginning of sterilisation cycle the valve should be : (*i*) closed, (*ii*) steam passed through valves and bled out, (*iii*) steam passed through and out, (*iv*) when steam is issuing freely from valve, the other valve may be opened and steam circulated. Once the sterile cycle is completed, both valves are closed and third valve is allowed to open, air to pass through the heated filter. One valve is closed, while the other is open and hence allows air into the fermentor to achieve positive pressure in the vessel.

(*b*) Containment: The prevention or escape of viable cells from a fermentor or downstream equipment is called containment; it is much more recent in origin. Containment guidelines were issued in 1970s. The criteria for risk assessment can be explained based on certain parameters such as : (*i*) the known pathogenicity of microorganism, (*ii*) the virulence or level of pathogenicity of the microbes that cause mild or serious diseases, (*iii*) the number of organisms required to initiate an infection, (*iv*) the route of infection, (*v*) the known incidence of infection in the community and the existence locally of vectors and potential reserve, (*vi*) the amounts or volume of organism in the fermentation process, (*vii*) the technique or process used, and (*viii*) ease of prophylaxis and treatment.

D. FERMENTATION PROCESS

Fermentation processes occur according to fermentors used, for example batch fermentation, fed-batch fermentation and continuous fermentation.

1. Batch Fermentation

Batch fermentation is a process in which a large volume of nutrient medium is inoculated to proceed for the harvest and recovery of the product. This ends the batch fermentation as the vessel is cleansed and re-sterilised for the subsequent batches. This is a close system in which all the nutrients are initially added to the vessel and inoculated. Subsequent treatments include maintaining adequate aeration by providing stirrers and pH control by the addition of an acid or alkali. In aerated systems, antifoam agents such as palm oil or soybean oil are added. Large scale growth of microorganisms tends to generate heat in the system. Temperature control is maintained by providing water circulation system around the vessel for heat exchange.

In a large-sized fermentor, the fed-batch method is employed to control the concentration of critical nutrients. For example, penicillin production is controlled by maintaining low glucose level in the medium because its accumulation in the medium results into catabolite repression of secondary metabolite. In batch fermentation, the growth of microorganism follows the characteristics growth curve with a lag phase followed by a log phase, finally reaching the stationary phase due to limitation of nutrients and other factors.

2. Fed-batch Fermentation

In such (fed-batch) cases, substrate is added in increments as the fermentation progresses. The formation of many secondary metabolites is subject to catabolite repression by high concentration of glucose, other carbohydrates, or nitrogen compounds. In such situations, in fed-batch method the critical elements of the nutrient solution are added in small concentration in the beginning of the fermentation and these substances (substrates) continuously are added in small doses during production phase.

It is difficult to measure the substrate concentration directly and continuously during fermentation. Indirect parameters, which are correlated with the metabolism of the critical substrates, have to be measured in order to control the feeding process. For example, in the process of organic acids the pH is to be used to determine the rate of glucose feeding. Sometimes dissolved O_2 or the CO_2 content in the exhaust air is monitored.

3. Continuous Fermentation

During continuous fermentation, some parts of the components (including media and inoculum) of upstream process are withdrawn intermittently and replacement or withdrawn substances are made by adding the fresh medium or nutrients. The withdrawn part from the fermentor is used for recovery of the products. During continuous fermentation, the equipment is always in use, and secondly inoculum is not required in subsequent addition of the nutrients. It is carried out in three different processes as given below :

(*a*) Single Stage Fermentation : In such fermentation a single fermentor is inoculated then kept in continuous fermentation. 'Recycle' fermentation in which, a portion of the withdrawn culture or the residual unused substrate including withdrawn culture, is recycled in fermentor. During this process, the substrate is further utilised for product formation and inoculum can also be recycled.

(*b*) Multiple-stage Fermentation : It involves two or more stages with the use of two or more fermentors in sequence. In such instances, the microbial growth occurs in first stage fermentor followed by a synthetic stage in the next fermentor. This happens in such cases where metabolities are not related with growth. The microbial activity in continuous fermentation can be controlled.

(*c*) Turbidostat : In this case the total cell population is held constant by employing a device that measures the culture turbidity so that it regulates both the nutrient feed rate and the culture withdrawn rate from the fermentor. Sometimes, the inoculum added is multiplied so quickly that its level gets increased too much; hence medium is added further so as to dilute the inoculum. However, the microbial growth should be maintained in log phase by adding the nutrient. During this process unused nutrient is lost from the harvested culture. On the other hand, in the chemostat the nutrient and harvest culture withdrawal rate remains at constant values. This growth rate can be controlled by toxic product formation during fermentation, pH and even the change of temperature. Actually, it is necessary to maintain a constant cell population in fermentor. The flow rate is also related to the growth of organism. If it is low, it can allow the culture to go into the maximum stationary phase. Too high flow rate can adversely affect a number of microorganisms namely, *Penicillium, Saccharomyces, Torula,* etc. which are being used in continuous process of fermentation. Some of the products such as beer have been commercialised.

Fermentation media sometimes require strong mixing if the medium is viscous. To overcome the problem of adequate mixing, mathematical assumptions are necessary while processing for continuous fermentation. It is also necessary to evaluate the microbial cells that do not adhere/attach to the surface of the fermentor. Sometimes, filamentous fungi grow in piping and valve. This is immobilised state and cultures do not expose to nutrient added. This condition makes difficult the proper regulation of inflow and outflow for the fermentation. It is also observed that conditions for the microbial growth and product formation differ. Hence, the optimum conditions for microbial growth do not align with the optimum conditions required for product formation.

In single stage fermentation, a compromise must be laid in the nutrient and physical conditions during the fermentation, growth in the first stage and product formation in the second or succeeding stages are required. The multistage chemostat is useful in the utilisation of multiple C sources in the production of secondary metabolites. Similar to continuous brewing it has limited applications in industries due to complexity in this system.

I. STOICHIOMETRY

It is a part of chemistry which deals with composition of substances, especially with determination of combining proportion of chemical equivalents. To start bioprocess operation, a medium is formulated for the growth of selected cells and production of desired products. Such constituents should be used in medium which must meet the elemental requirements for cell biomass and metabolite production. Therefore, there must be an adequate supply of energy for biosynthesis and cell maintenance. The economically designed medium shows minimal wastage of components. During biochemical reactions new groups of atoms and molecules are formed through their rearrangement.

The optimum chemical reaction occurs at suitable pH, temperature, solvent and atmospheric pressure. Such biochemical reactions also occur inside the cells or microorganisms. Therefore, microorganisms also need optimum conditions (such as pH, temperature, etc.) and growth media containing nutrients in particular substrate. The microbes utilise the nutrients at optimum conditions and convert into desired products (Fig. 21.15). Utilising a major amount of substrate the microbe also increases their biomass and growth and multiplication. However, the conversion efficiency of substrate into product falls in the range of 20-30% as compared to 80-90% conversion efficiency of chemical reaction.

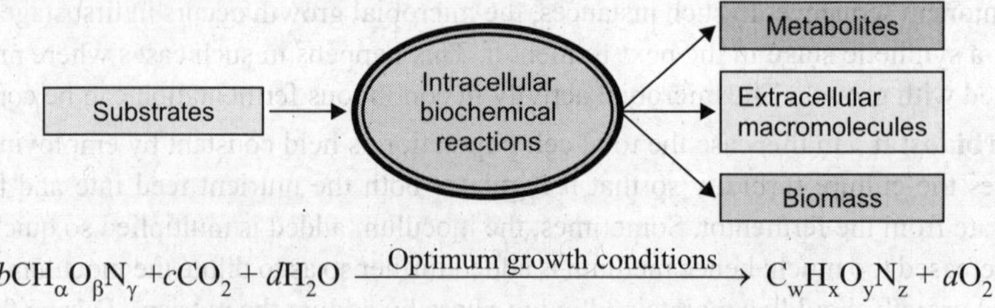

$$bCH_\alpha O_\beta N_\gamma + cCO_2 + dH_2O \xrightarrow{\text{Optimum growth conditions}} C_wH_xO_yN_z + aO_2$$

Fig. 21.15. Cellular growth and production of product.

Bioconversion is carried out in a specially designed fermentor. The product is taken out from the fermentor, then isolated, purified and identified following physical and chemical procedures. By applying stoichiometric calculation the relationships of mass and moles between the reactants consumed and products formed can be measured. However, this calculation is possible only when equation and atomic weights of reactants are correctly written. Hence, the first step is to consider an equation based on stoichiometry for growth and product formation. For example:

Carbon and energy sources + Nitrogen sources + Other requirements → Cell biomass + Product + CO_2 + H_2O+ Heat

Equation for all alcoholic fermentation is written as below:

$$C_6H_{12}O_6 \xrightarrow{\text{Yeast cells}} 2C_2H_5OH + 2CO_2 + \text{Heat}$$

The following two quantities are conserved during biochemical reactions:

- *Total mass*

  Total mass of reactants = total mass of products

- *Number of atoms of each elements*

  Total number of C, H, O, N, P atoms in reactants = number of C, H, O, N in products

Knowledge of elemental composition of medium and compounds of cellular biomass and products is necessary. It should be noted that the nutrients are not supplied exactly similar to mathematical equation. Some reactants are provided more than what are required by the growing cells such as P and K, while the others are near the limiting value such as Zn and Cu. The concentration of P is raised in many media to increase the buffering capacity. Here the C has dual role, biosynthesis and energy generation.

The C requirement under aerobic condition: It may be estimated from the cellular yield coefficient (*Y*) which is written as below:

$$Y = \frac{\text{Quantity of cell dry mass produced}}{\text{Quantity of carbon substrate utilised}} \qquad \text{...(23)}$$

In 1979, C.L. Cooney has calculated the theoretical conversion yield in penicillin G biosynthesis on the basis of material and energy balance by using a biosynthetic pathway based on reaction of stoichiometry. The stoichiomentry for overall synthesis is:

$$aC_6H_{12}O_6 + bNH_3 + cO_2 + dH_2SO_4 + e\text{PAA} \longrightarrow n\text{Penicillin G} + pCO_2 + qH_2O$$

Where, *a,b,c,d,e,n,p,* and *q* are the stoichiometric coefficients and PAA is phenylacetate. It was calculated that the theoretical yield was 1.1 gram penicillin G /gram of glucose.

It has also been found that sometimes the reactants are consumed by the cells and some other side reactions take place. Therefore, some side products are produced which are not desired in main reaction equation. Finally, mixture of products is produced. Therefore, knowledge of additional information on biosynthesis of side products is also necessary in stoichiometric calculations of amount of reactants used and products formed.

II. PROCESS FLOW

Bioprocess is a large-scale operation in a bioreactor where the raw materials (biological or abiological) are transformed into a product by any biological agency (such as microorganisms, animal or plant cell cultures or materials e.g. enzymes, organelles, etc.). The products obtained so have a little commercial value (e.g. wastes) or a very high commercial value, and sold in market such as alcohols, organic acids, enzymes, single cell proteins insulin, antibiotics (penicillin), streptomycin), etc. The cells are provided with carbon sources, nitrogen sources, inducers for the production of desired useful products. In most of the bioprocess reactions the carbohydrates act as source of carbon and energy to the growing cells and production of bioproducts. Some of the carbohydrates are used for commercial production of various useful products (Table 21.4).

Table 21.4. Bioconversion of carbon sources into various products.

Cells of microorganism, plants, animals, enzymes	***Carbon sources***	***Products***
Acetobacter suboxidans	Glucose	Vitamin
Aspergillus niger	Molasses	Citric acid through Koji fermentation
Animal cell	Glucose	Erythropoitin
Escherichia coli	Glucose	Insulin
Glucose isomerase	Sucrose	High fructose syrup
Hybridoma (hydrid animal cells)	Glucose	Monoclonal antibodies
Lactobacillus lactis	Whey	Lactic acid
Plant cells	Sucrose	Shikonin
Streptomyces griseus	Starch hydrolysis	Streptomycin
Xanthomonas compestris	Sucrose/glucose	Xanthum gum

A bioproces is completed in the four stages : (*i*) *pretreatment*, (*ii*) *bioreaction*, (*iii*) *downstream processing*, and (*iv*) *formulations* (Fig. 21.16). Biochemical engineering involves both upstream and downstream operations. **Upstream process** is the pretreatment of raw materials so that it may be suitable for biological reactions. The bioreactions occur in a bioreactor which forms the core of the bioreaction steps. Environmental conditions of the bioreactor are controlled according to the requirement. Biorection results in biotransformation of substrate into biomass and /or some other products. Whole cells (as suspension or immobilised) or enzymes are used as biocatalyst. The **downstream process** is extraction and purification of desired end products obtained from the bioreactor so that it may be converted into usable forms. Different types of carbon and nitrogen sources, and suitable inducers are used in fermentation medium (Table 21.5).

Table 21.5. Some carbon and nitrogen sources and inducers commonly used in fermentation media.

Carbon sources	*Nitrogen sources*	*Inducers*
Starch	Ammonia	Various proteins
Glucose	Ammonium salt	Starch
Malt	Nitrate	cellulose
Beet molasses	Corn-steep liquor	Pectin
Cane molasses	Peanut granules	Phenylacetate
Vegetable oil	Soybean meal	
Hydrocarbons	Dried blood	
	Yeast extract	
	Pharmamedia	

Microorganisms also utilise precursors and trace elements for the production of products, because precursors are the pre-requisite and trace elements act as a component of enzymes for accelerating the rate of reaction. Some of the precursors and trace elements are given in Table 21.6.

For getting the required products the design and efficient operation of downstream processing are the important steps. Over 90% of 200 staff is associated with the recovery process. Therefore, it is the major part of overall costs of most processes. Initially the bioreactor broth is separated into a liquid phase and solid phase, and subsequently concentration and purification are carried out. Processing event involves more than one stage. The final product of the downstream purification stages should have some degree of stability for commercial distribution. It is done through spray-drying, fluidized- bed drying or freeze-drying. The products which are not supplied easily in dried form are sold in liquid preparation.

Table 21.6. Utilisation of precursors and trace elements by some microorganisms for production of products.

Microorganisms	*Precursors*	*Trace elements*	*Products*
Lactobacillus bulgaricus	Propionate	Fe, Co	Riboflavin
Penicillium chrysogenum	Phenylacetate	Fe, Zn, Cu	Penicillin
Propionibacterium	Cyanide	Co	Vitamin B_{12}
Streptomyces griseus	Chloride	Zn	Griseofulvin

Care should be taken to save the products from contamination and denaturation. Thus downstream processing is only physical operation aimed at concentration and purification of products in different forms (e.g. liquid, slurry, powder, crystalline).

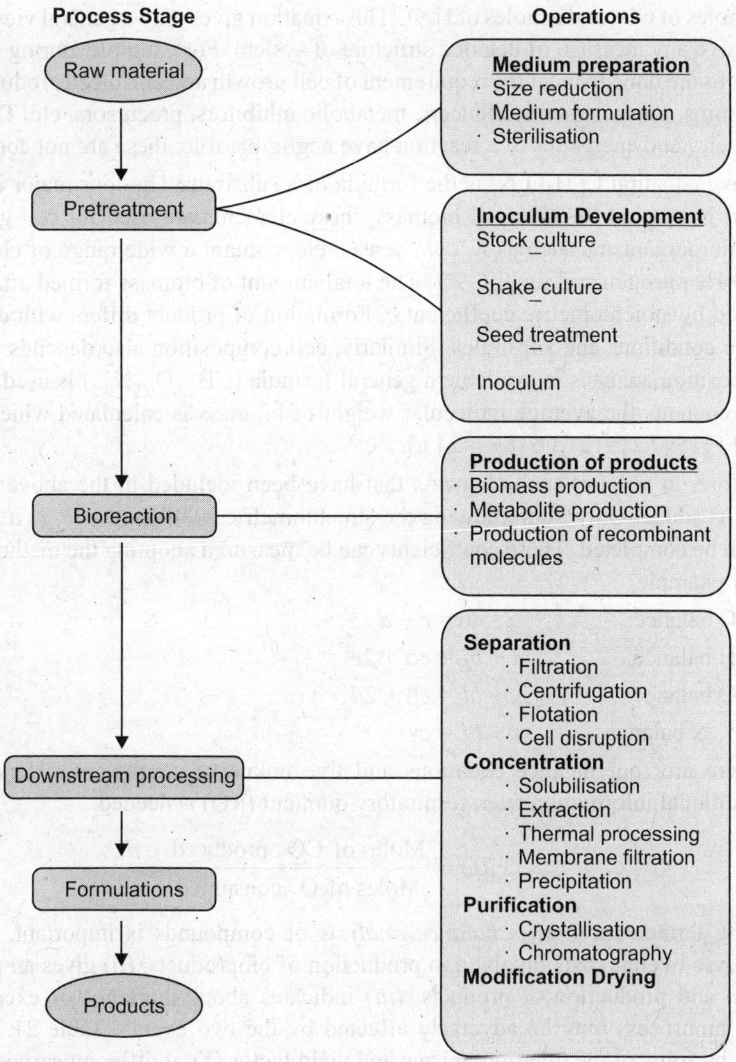

Fig. 21.16. A flow diagram showing bioprocess – stages and operations.

III. MATERIAL BALANCE (MASS BALANCE)

In fermentor the growing cells utilise substrates and convert them into products; however, to discuss mass balance in aerobic growth the following equation may be written:

$$C_wH_xO_yN_z + aO_2 + bN_iH_jC_kO_l \xrightarrow{\text{Cells}} cCH_\alpha O_\beta N_\gamma + dCO_2 + eH_2O \qquad ...(24)$$

where,

$C_wH_xO_yN_z$ = chemical formula of substrate having a carbon source; for example, for glucose, w = 6, x = 12, y = 6, z = 0

$N_iH_jC_kO_l$ = Chemical formula of nitrogen source

$CH_\alpha O_\beta N_\gamma$ = Chemical formula of dry biomass

a, b, c, d, and *e* = Stoichiometric coefficients

The above equation is called *mass balance equation*. It is written on the basis of consumption of 1 mole of substrate, *a* moles of oxygen, *b* moles of nitrogen sources, and production of *c* moles of product, *d* moles of CO_2 and *e* moles of H_2O. This equation gives only a general view of metabolism. It does not give any message of detailed structure of system. For example, during catabolism some other additions are done to meet the requirement of cell growth and enhanced production of products such as vitamins, trace elements, inducers, metabolic inhibitors, precursors, etc. Their contribution in stoichiometry and energetics of a reaction have negligible role; these are not considered.

In above equation $C_wH_xO_yN_z$ is the formula of a substrate. The four major elements such as C, H, O and N account 90-95% cell biomass; these elements are used for cell growth. Different groups of microorganisms such as *E. coli*, yeasts, etc. contain a wide range of elements. Bacteria require 11-14% nitrogen and fungi 6-9%. The total amount of biomass formed after bioconversion is represented by stoichiometric coefficient *c*. Formation of product differs with different types of cells, culture conditions and substrates. Similarly, cell composition also depends on these factors. When composition analysis is available a general formula ($CH_{1.8}O_{0.5}N_{0.2}$) is used. On the basis of C, H, O, N contents the average molecular weight of biomass is calculated which comes to 24.6 [12+(2x0.5}+{15x0.2}=12+3.6+8+3=24.6].

Therefore, to account such elements that have been included in the above formula, 5-10% residual ash is added. Only after knowing the stoichiometric coefficient *a, b, c, d* and *e*, the above equation will be completed. These coefficients can be measured adopting the methods for elemental balance. For example

C balance: $w = c + d$ ...(25)

H balance: $x + bj = c\alpha + 2e$ ...(26)

O balance: $y + 2a + bl = c\beta + 2d + e$ ...(27)

N balance $z + bi = c\gamma$ ...(28)

There are four balance equations and five unknown coefficients. Hence, to solve the equation additional information from respiratory quotient (RQ) is needed.

$$\text{RQ} = \frac{\text{Moles of } CO_2 \text{ produced}}{\text{Moles of } O_2 \text{ consumed}} = \frac{d}{a} \quad \text{...(29)}$$

During fermentation *mass balance analysis* of compounds is important. Because it: (*i*) helps to analyse overall costs involved in production of bioproducts, (*ii*) gives an idea of substrate consumption and production of products, (*iii*) indicates about shortage or excess of nutrients. The overall bioprocess may be adversely affected by the two events. Table 21.7 shows oxygen requirement by some of the microorganisms and yield factor (*Y*) of different carbon sources.

Table 21.7. Oxygen requirement by some microorganisms and yield factor (*Y*) of different carbon sources.

Microorganisms	***Substrate***	***O_2 requirement (g O_2/g dry weight)***	***Y (g biomass/g substrate)***
Escherichia coli	Glucose	0.4	0.53
Pseudomonas C	Methanol	1.2	0.54
Pseudomonas sp.	Octane	1.6	1.06

1. Some Calculations

(*i*) *Bacillus polymyxa* under ideal conditions occupies 5% of the total volume of a fermentor. Under which conditions the culture medium (after bacilli are filtered) has a concentration of 1 mg/ml of the secreted protein trehalose?

(*i*) Calculate the volume of culture filtrate required to purify 1 kg of trehalose.

(*ii*) What is the total number of *B. polymyxa* cells which produce this amount of enzyme?

Assume that each cell is a cylinder which has dimension of 1 um diameter and length 2 μm.

Diameter of a cell = 1 um *i.e.* radium (*r*) = 0.5 μm

Length (1) = 2 μm

Volume of the single cell $= \pi r^2 h = 22/7 \times 0.5 \times 0.5 \times 10–12 \times 10–6 \times 2$

$= 1.57 \times 10^{-18}\ m^3$

Concentration of trehalose = 1 mg/ml

For 1 kg mass there must be 10^6 mg

In comparison to same ratio, volume = 10^6 ml = 103 L = 1 m^3

So, No. of *B. polymyxa* cells to produce trehalose

= Total volume/single cell volume = $1/1.57 \times 10^{-18}$

$= 10 \times 10^{17}/1.57 = 6.37 \times 10^{17}$ cells.

(*ii*) *Bacillus polymyxa* is a rod shaped bacterium having 1 μm diameter and 2 μm length. The average density of a cell is 1.28 g/ml. About 13.5% of the wet weight of *B. polymyxa* is soluble protein. Estimate weight of 100,000 Da enzyme represents 0.1% of the total soluble protein.

Diameter = 1 μm, hence radium (r) = 0.5 μm, length = 2 μm

Density = 1.28 g/ml *i.e.* $1.28 \times 10^6 g/m^3$

Volume of each cell $= \pi r^2 h = 22/7 \times 0.5 \times 0.5 \times 10^{-12} \times 2 \times 10^{-6}$

$= 1.57 \times 10^{-18}\ m^3$

Mass = volume × density

$1.57 \times 10^{-18} \times 1.28 \times 10^6 = 2.0096 \times 10^{-12}$

Soluble protein = 13.5% of mass = $13.5/100 \times 2.0096 \times 10^{-12}$

Weight of enzyme = 0.1% of soluble protein

$= 0.1/100 \times 13.5/100 \times 2.0096 \times 10^{-12} = 2.7129 \times 10^{-16} g$

1,00,000 g of enzyme = 6.023×10^{23} molecules

1g " " $= 6.023 \times 10^{23}/105$

$2.7129 \times 10–^{16} g = 6.023 \times 10^{23}/10^5 \times 2.7129 \times 10^{-16}$

$= 16.34 \times 10^2$ = 1,634 molecules

(*iii*) Bacterial cells harvested from a 500 L capacity fermentor yield 5 g of a protein of molecular weight 15,000 Da. If each bacterial cell contains 2,500 molecules of the given protein, calculate the total number of bacterial cells harvested. Also calculate the percentage volume of the fermentor occupied by the cells assuming that each cell is 1 μm in diameter and 2 μm length.

Size of fermentor = 500 L, protein harvested = 5 g,

Molecular weight of protein = 15,000 Da

Each cell contains 2500 molecules,

Diameter of a cell = 1 μm [radius (r) = 0.5μm], length = 2μm

15,000 Da enzymes = 6.023×10^{23} molecules

5 g " " $= 6.023 \times 1023 \times 5/15,000$

No. of bacterial cells $= 6.023 \times 10^{23} \times 5/15,000 \times 2500 = 80 \times 10^{21}$

Volume of a single cell $= \pi r^2 h = 22/7 \times 0.5 \times 0.5 \times 10^{-12} \times 10^{-6} \times 2 - 1.57 \times 10^{-18}\ m^3$

Total volume occupied by bacterial cells $= 80 \times 1021 \times 1.57 \times 10^{-18}$

Total " " " " $= 125.6 \times 1^{-3}\ m^3 = 125.6$ L

Percent volume occupied by cells in a fermentor = 125.6 × 100/500 = 25.12%

IV. SCALE-UP OF FERMENTATION

Scale-up is the determination of the proper incubation conditions to be employed with large scale production tanks as based on information obtained with various sized smaller tanks. This process allows carrying out laboratory procedure at industrial scale. It is the best way to obtain fermentation information for production tanks directly in large tanks. However, this is not practical for : (*i*) new fermentation, (*ii*) variation in studies on fermentation already in production, and (*iii*) valid experiment cannot be carried out with only a single tank; one or more tanks are required as experimental controls. Besides these considerations, costs and media also affect the scale-up process.

- **Use of Erlenmeyer Flasks:** The conventional methods provide a poor production because of the poor aeration characteristics associated with this vessel.
- **Use of Baffle Flasks:** Flask with glass baffles projecting into the medium from bottoms or sides provide better aeration than the Erlenmeyer flask.
- **Use of Small Laboratory Fermentors:** One to ten or twelve litre size fermentors are most ideal for this type of studies, since their aeration and agitation conditions can be varied and the overall fermentation conditions of these tanks more closely resemble with those of the larger production tanks. These tanks allow fermentation studies on a scale that has meaning in relation to production tank but without too high expenses for media, labour, power input, *etc.*

Experience with particular fermentation equipment, previous fermentations is the only real guide in translating scale up.

(*a*) Sterilisation of Gases and Nutrient Solutions : In all fermentation processes, it is mandatory for a cost-effective operation to have contamination-free seed culture at all stages from the preliminary culture to the production fermentor. A fermentor can be sterilised by destroying the organism with some lethal agents or by removing the viable organisms by a physical process such as filtration.

Initially prepared nutrient media contain a variety of different cells and spores derived from the constituents of the culture medium, the water and the vessel. These must be eliminated by a suitable means before inoculation. A number of means are recommended but in practice heat is the main mechanism used for large scale sterilisation.

(*b*) Stock Cultures: It is important to maintain microorganisms for extended periods in viable conditions, and in situation which do not alter their desired product formation capacity. This condition is also true for strains used in biological assays. Thus, microbial species procured from various culture collection centres are maintained in viable conditions and known as stock-culture collection. The stock culture generally retains all the characteristics initially described. Stock culture collection centres have been established throughout the world to help microbiologists in obtaining cultures for various studies. These centres also help in classifying a newly isolated organism. There are two types of stock cultures: working stocks and primary stocks.

(*i*) Working Stocks: These stocks are used frequently and they must be maintained in vigorous and uncontaminated conditions on agar slants, agar stabs, spore preparations, or broth culture and one held under refrigeration. They must be checked constantly for possible changes in growth characteristics, nutrition, productive capacity and contamination.

(*ii*) Primary Stocks: The cultures that are held in reserve for presently practical or new fermentation for comparative purposes, for biological assays, or for possible later screening programmes. These are not maintained in a state of physiological activity. Transfers from these cultures are made only when a new working stock culture is required, or when the primary stock culture is sub-cultured to avoid death of the cells. Thus, primary stock cultures are stored in such a manner as to require the least possible numbers of transfers over a period of time. Further, these are stored at room temperature, maintained in sterile soil, or in agar or broth overlaid with sterile mineral oil.

Agar and broth culture without mineral oil are refrigerated. The cultures in milk of agar are maintained frozen at low temperature. Finally, primary stock cultures are lyophilised or frozen-dried, and stored at low temperature. The culture of *Blakeslea trispora* used in β-carotene production cannot be stored at refrigeration temperature because they die relatively quickly. However, at room temperature transfers are being made to fresh medium when the cultures become nearly dried out.

E. PRODUCTION OF MICROBIAL PRODUCTS

Before the production of products, microbial strains and fermentor are required. Wild strains of microorganism produce low amount of commercially important products; therefore, investigation needs to be done to increase the productivity of selected organisms. To achieve the increased yield, optimisation of culture medium and growth conditions are the pre-requisites but the potential productivity of the microorganisms is controlled or regulated by its genome; therefore, genome must be modified so as to enhance the productivity. The modified microorganisms have different nutritional requirements with different culture conditions such as temperature, pH, etc. These factors provide benefits to the process of strain improvement involving genome modification of the microorganisms followed by reappraisals of its cultural requirements.

I. STRAIN IMPROVEMENT

Microbial strains are improved for better production of products. There are number of methods to improve the microbial strains, some of which are briefly described as below:

1. Methods of Strain Improvement

A mutant of *Streptomyces fradiae* requiring oleic acid for neomycin formation shows a decrease in the intracellular level of neomycin precursors in the mutant. The other example is the supersensitive mutants of β-lactam antibiotics. Recent approaches towards strain improvement are given below:

***(a)* Plasmid Improvement:** Plasmid genes are involved in antibiotic production in *Streptomyces* spp. Plasmids are involved in genetic characteristics on curing experiments. Involvement of plasmids in biosynthesis of aureothricin and kasugamycin in *S. kasuaensis* was demonstrated more than decades ago by M. Okanishi and co-workers in 1970. The genetic study using *S. venezuelae* ISP 5230 [a chloramphenicol (CM) producer] contains most of the structural genes for the CM biosynthetic steps treated between *met* and *ilu* on the chromosome and the plasmid played role in increasing CM production. A linear plasmid like DNA (pSLA2) of 11.2×10^6 dalton molecular weight from *Streptomyces* sp. produced antibiotics.

***(b)* Protoplast Fusion:** Protoplast fusion is one of the useful techniques for obtaining hybrids or recombinants of different microorganism strains. Various studies have been carried out by using protoplast fusion in *Streptomyces, Saccharomyces,* and fungi. Protoplast fusion in *Sterptomyces* was first reported by M. Okanishi and his team in the year 1966. Further, they have worked on formation, stabilisation and regeneration of protoplast of *Str. griseus* and *Str. venezuelae.* Fusion of yeast protoplasts has been reported with *Sacchromyces cerevisiae.* Technique for protoplast fusion in *Brevibacterium flavum* has been used for strain improvement.

***(c)* Mutation:** Screening after major subjection of a parent strain to physical or chemical mutagen greatly increased the probability of finding improved strain.

***(i)* Major mutations:** It involves the selection of mutants with a pronounced change in a biochemical character of practical interest. Such variants are commonly used in genetic studies and are generally 'low mutants'. They are isolated routinely from population surviving after prolonged exposure to a mutagen, for example, selection of non-pigmented *Penicillium chrysogenum* strains with high penicillin production. The initial strain of *Sterptomyces griseus* (a streptomycin producing organism) synthesised the small amount of streptomycin but its variant was isolated which produced greater amount of streptomycin. For further improvement it is also necessary to study the biosynthetic pathways which contribute to the identification of precursors as in case of a modified tetracycline synthesised by a mutant strain of *S. aureofacies.* The molecule got changed at the C-5 position and was devoid of antibiotic activity. The other mutant strain S-604 synthesised 6-dimethyl tetracycline, a new antibiotic (not elaborated by the parent strains) proved to have several advantages. Today it is one of leading commercial forms of tetracycline.

***(ii)* Minor mutations:** It plays a dominant role in strain improvement. By definition, such mutation affects only the amount of product synthesised. Such variants are usually phenotypically similar to the parent with rapid and abundant mycelial and conidial development. A 10 to 15% increase in conidial population exposed to moderate doses of a mutagen has been obtained after repeated isolation of minor (positive) variants and using each succeeding strain for further mutation and selection. Such increases have also been obtained by repeated selection without introduction of mutagen. In this case, the population to be tested must be large and assay for the desired product also must be accurate and specific. This technique fetched importance in improving *P. chrysogenum.* For example, Wisconsin series were the famous Q-176 culture with significantly improved antibiotic titres, and strains BL3-D10, which does not produce the characteristic and troublesome chrysogenin pigment. All further mutant selections over the next decade were derived from Q-176.

2. Isolation of Mutants and their Use in Microbial Processes

The mutants are isolated from the mixed population of strains and can be used further for various microbiological processes.

***(a)* Localised Mutagenesis and Computation:** Localised mutagenesis affecting the small selected regions of the chromosomes, offers a promising new approach. Mutation programmes can be directed to maximise mutations in any marked area on the chromosome, specially the areas known to affect the formation of end products. Isolation of strains in unknown loci linked to the revertant site can be done by a heterokaryon method or by the use of temperature sensitive mutants.

***(b)* Sexual and Parasexual Processes:** In fungi, the vegetative mycelium is haploid and can be propagated almost indefinitely by serial transfer of hyphal fragments and can also be propagated by asexual spores/conidia. Two strains of opposite mating types (A or B) are required to initiate the sexual cycle and allow to mate by mixing the conidia of mating type A with mycelia on appropriate media. After a period of nuclear division and migration, fusion between A and B nuclei takes place. Each fused nucleus (diploid) undergoes meiosis to form four haploids, which divide mitotically into eight nuclei contained in ascus.

Few of the industrially important fungi form heterokaryon in which rare diploid nuclei result from the fusion of two haploid nuclei. This process is called 'parasexuality'. Although, recombination in few fragments in the parasexual cycle compared to the meiotic process can occur by mitotic crossing over or by other mechanism. The importance of mitotic crossing over is that it makes possible genetic analysis and controlled breeding in organisms with no sexual cycle. Strain improvement through parasexual cycle has been reported in *P. chrysogenum* in a study.

II. FERMENTATION PROCESSES

Industrial fermentation processes require two most important stages to complete product fermentation. These are upstream process and downstream process.

1. Upstream Process

The upstream process includes procurement and maintenance of inoculums, strain improvement to enhance and yield, and continuous development of selected strains to increase the economics efficiency. It also includes the maintenance of strain purity, preparation of inocula and further efforts to improve the economic efficiency of the process.

There are two distinct processes involved in upstream process but both the processes run simultaneously. The medium is prepared for the culture (inoculum) preparation as well as for the production of desired product. The vessel of the fermentor is filled with production media and inoculated with suitable microbial culture for the industrial product formations as shown in Fig. 21.17.

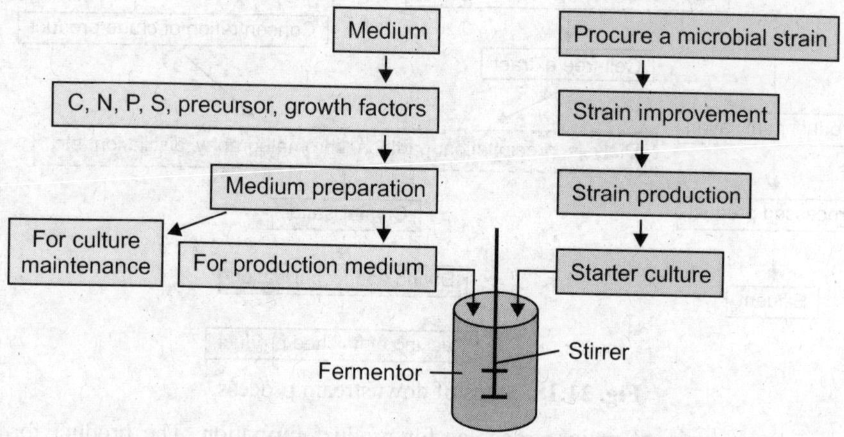

Fig. 21.17. Steps of upstream process.

2. Downstream Process (DSP)

The downstream process includes the selection of cost-effective suitable fermentation media, optimisation of important fermentation condition to give rise maximum yield and profits, and fermentation conditions to optimise the growth of microorganism or the production of a desired product. The main objective of this process is to get reproducibility of result as far as possible besides safe recovery of the target product. Further processing is carried out depending upon whether the product is intracellular or present inside the cell or membrane bound or release outside the cell or extracellular in nature. This leads to recovery of the product in pure form.

The presence of undesirable products (bye products), impurities in media, antifoam agents, etc. affect the steps of downstream process. Therefore, each step requires thorough check up in both upstream as well as in downstream processes. For example, a cheap carbon and energy source may increase the cost of downstream process. At commercial level use of existing equipments, ancillaries, etc. are more advantageous than that of creating new facilities because the later step requires more investment which may lead to escalation of cost of the product and processes. The downstream process consists of series of distinct unity or processes which are connected to each other for getting the final product as shown in Fig 21.18.

For the precautionary measures it is necessary to keep this unit process minimum. This will not only save the cost of the plant but will also ensure not to lose the product during individual steps. Now-a-days, most preferred DSP is to link with the fermentation. This increases productivity and reduce time and cost.

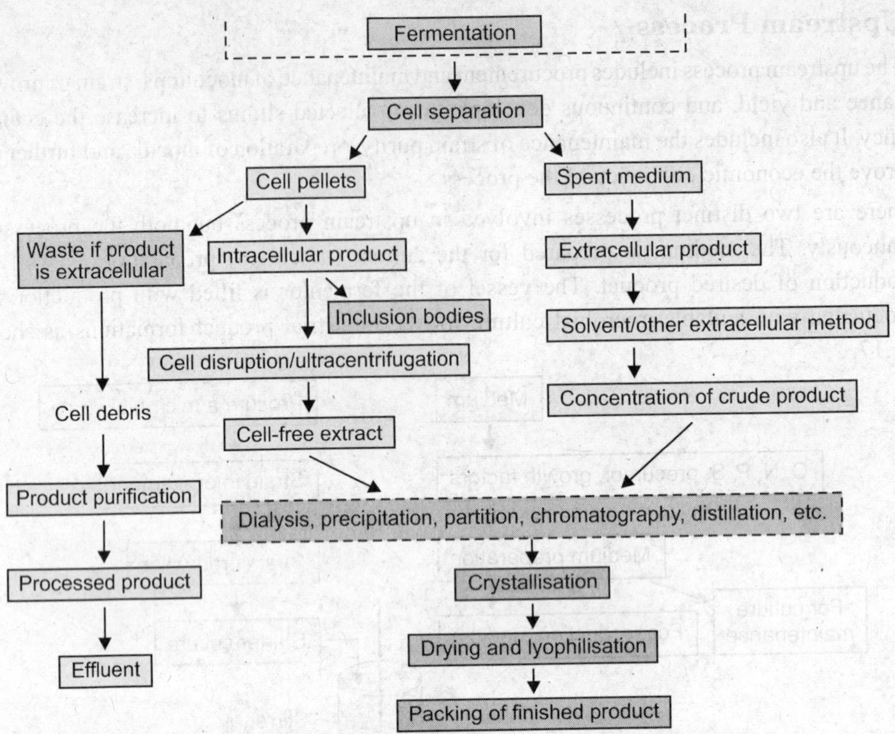

Fig. 21.18. Steps of downstream process.

There are two types of major processes for product formation. The product formation is coupled with growth, *i.e.* primary metabolites and another where product is formed after growth as in case of antibiotics. In case of primary metabolites, higher productivity is achieved by using integrated system of fermentation with DSP. This kind of system maintains a high cell density through cell retention of recycling. If product is inhibitory, various methods have been employed to partition fermentor as in case of formation of glucose by breakdown of cellulose. This allows the rapid *in-situ* removal of product by extraction, adsorption or stripping. The process can be *ex-situ* where the product is removed outside the fermentor and the processed medium is returned to the fermentation. Such processes have been successfully used for removing alcohols, solvents, proteins, etc. as shown in Fig. 21.18. The product recovery is carried out after downstream processing is over.

3. Product Recovery

The following steps are employed for the recovery of the product(s) obtained after downstream processing :

(*a*) Cell Harvesting : Since, microbial cells are in liquid medium or broth; solid-liquid separation method is influenced by the size and morphology of the microorganisms (single cells, aggregates or mycelia). Further, the process also depends upon the location of product whether it is intracellular or extracellular it has been secreted into the medium. Besides, other factors such as specific gravity, viscosity, and rheology of the medium also influence cell harvesting method. Rheology is the study of the flow of matter (primarily in the liquid state but also as soft solids) under conditions in which they respond with plastic flow rather than deforming elastically in response to an applied force.

(*b*) Broth Conditioning Technique : Cells are separated from large volume of liquid medium after formation of floccules or precipitate by using chemical, physical and biological treatments. Coagulating materials such as simple electrolytes, acids, bases, salts, multivalent ions, etc. are added to form small flocs from dispersed colloids or suspended materials. Certain flocs precipitation methods are also used at the end of many traditional beer and wine fermentation processes for the precipitation of yeast cells.

(*c*) Sedimentation: This process is widely used for yeast separation during alocohol production and in wastewater treatment. The rate of particle sedimentation is a function of both size and density. If larger the size, greater its density, and the faster rate of sedimentation. Therefore, for quick separation the difference in density between the particle and the medium must be large with low viscosity.

(*d*) Centrifugation : This process allows small particles or cells to separate from liquid. It is also suitable for some liquid-liquid separations. Similar to sedimentation, its effectiveness also depends upon cell size, density difference between the cells and the medium, and its viscosity. Higher speed of centrifugation requires for the separation of smaller microorganisms such as bacteria. In case of yeast cells present in beer, relatively low centrifugation effectively recovers residual yeast cells.

(*e*) Filtration : There are certain filters such as clothes, glass wool or cellulose that retain the solids and pass on the liquid. The solids accumulate above the filter. Such technique is useful in case of separation of fungal mycelia. There are two types of filters used in industrial process: plate and frame filters or filter press, and rotatory lvacuum filter, which help in harvesting the fungal mycelia during antibiotic manufacture, baker's yeast production and in dewatering of sewage sludge.

(*f*) Disruption of Microbial Cells : When the product is intracellular, it is necessary to disrupt the cells/cell membrane so as to release the product. Cell disruption can be achieved by mechanical and non-mechanical methods. Some of the methods used are ultrasonication and application of lysozyme. The *'French press'* is often used in laboratories, while Manton and Gaulin homogeniser (high pressure homogeniser) is employed for pilot scale cell disruption. These are used for disruption of bacterial, yeast cells and fungal mycelia. The non-mechanical cell disruption is carried out by cell permeation. This can be accompanied by autolysis, osmotic shock ruptures by ice crystal or heat shock treatment. Some organic solvents such as acetone, butanol, chloroform and methanol have been used to release enzymes from microorganisms. Simple treatment with sodium dodecyl sulfate (SDS) or Triton X-100 is also found effective. Lysozyme is useful for Gram-positive organisms but addition of EDTA improves the effectiveness of lysozyme of Gram-negative bacteria. Some antibiotics namely penicillin and cycloserine may also be used to lyse the actively growing bacterial cells. Chitosan is effective for yeast cells.

III. ALCOHOL PRODUCTION

Simple organic compounds act as feed stock for chemical industry. Microbial production of one of the organic feed stocks from sugarcane molasses is presently used for ethanol production. This alcohol was produced by fermentation in the early days but for many years by chemical means through the catalytic hydration of ethylene. In modern era, attention has been paid to produce ethanol for chemical and fuel purposes by microbial fermentation. Ethanol is now-a-days produced by using sugar beet, potatoes, corn, cassava, and sugarcane (Fig. 21.19).

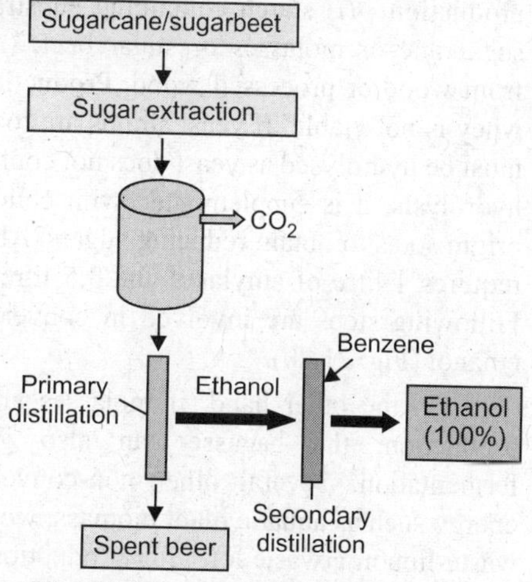

Fig. 21.19. Ethanol production from molasses.

Both yeasts (*Saccharomyces cerevisiae, S. uvarum S. carlsbergensis, Candida brassicae, C. utilis, Kluyveromyces fragilis, K. lactis*) and bacteria (*Zymomonas mobilis*) have been employed for ethanol production in industries. The commercial production is carried out with *S. cerevisiae.* On the otherhand, *S. uvarum* has also largely been used. The *Candida utilis* is used for fermentation of waste sulphite liquor since it also ferments pentoses. Recently, experimentation with *Schizosaccharomyces* has shown promising results. When whey from milk is used, strain of *K. fragilis* is recommended for the production of ethanol. It has also been found that *Bacillus* and *Pachysolen tannophilus* (yeast) can transform pentose sugars to ethanol.

Theoretically, it is interesting to note that fermentation process retains most of the energy of the sugar in the form of ethanol. The heat of combustion of solid sucrose is 5.647 MJ mol^{-1}, the heat of combustion of glucose is 2.816 MJ mol-1 but the heat release is 1.371 MJ mol^{-1}. The equations are given below:

1st equation :	Because	$C_{12}H_{24}O_{11}$ (Sucrose)	$4C_2H_5OH + 4CO_2$ (Ethanol)
	Hence,	5.647	$4 \times 1.371 = 5.184$ (97% conversion)
2nd equation :	In this case,	$C_6H_{12}O_6$ (Glucose)	$2C_2H_5OH + 2CO_2$ (Ethanol)
	Hence,	2.816	$2 \times 1.371 = 2.742$ (97% conversion)

Thus, the above reactions show that 97% sugar transforms into ethanol. But in practice, the fermentation yield of ethanol from sugar is about 46% or one hundred grams of pure glucose will yield 48.4 grams of ethanol, 46.6 g of CO_2, 3.3 grams of glycerol and 1.2 g of yeast. The biosynthesis of ethanol is given in Fig. 21.20.

It is noteworthy that the ethanol at high concentration inhibits the yeast. Hence, the concentration of ethanol reduces the yeast growth rate which affects the biosynthesis of ethanol. It can produce about 10-12 % ethanol but the demerit of yeast is that it has limitation of converting whole biomass derived by their ability to convert xylulose into ethanol. The *Zymomonas* has a merit over yeast that it has osmotic tolerance to higher sugar concentration. It is relatively having high tolerance to ethanol and have more specific growth rate.

1. Preparation of Medium

Three types of substrates are used for ethanol production : (*i*) starch containing substrate, (*ii*) juice from sugarcane or molasses or sugar beet, (*iii*) waste products from wood or processed wood. Production of ethanol from whey is not viable. If yeast strains are to be used, the starch must be hydrolysed as yeast does not contain amylases. After hydrolysis, it is supplemented with cellulases of microbial origin so as to obtain reducing sugars. About 1 ton of starch requires 1 litre of amylases and 3.5 litre of glucoamylases. Following steps are involved in conversion of starch into ethanol (Fig. 21.20).

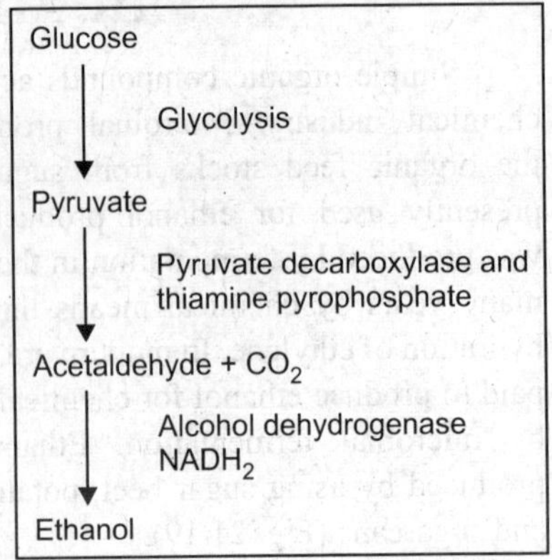

Fig. 21.20. Ethanol biosynthesis.

On the other hand, if molasses are used for ethanol production, the bagasse can also give ethanol after fermentation. Several other non-conventional sources of energy such as aquatic plant biomass, wood after hydrolysis with cellulases gives ethanol. Sulphite waste-liquor, a waste left after production of paper, also contains hexose as well as pentose sugar. The former can easily be converted microbially.

2. Fermentation

Ethanol is produced by continuous fermentation; hence, large fermentors are used for continuous manufacturing of ethanol. The process varies from one country to another. India, Brazil, Germany and Denmark have their own technology for ethanol production.

The fermentation conditions are almost similar (pH 5, temperature 35°C) but the cultures and culture conditions are different. The fermentation is normally carried out for several days but within 12h, it starts production. When fermentation is over, the cells are separated to get biomass of yeast cells which are used as single cell protein (SCP) for animal's feed. The culture medium or supernatant is processed for recovery of ethanol (Fig. 21.20). Ethanol is also produced by batch fermentation as no significant difference is found both in batch and continuous fermentation. Within 12 h *Saccharomyces cerevisiae* starts producing ethanol at the rate of 10% (v/v) with 10-20 g cells dry weight/lit. The reduction in fermentation time is accomplished through cell recycling continuously in fermentation.

3. Recovery

Ethanol can be recovered up to 95% by successive distillations. To obtain 100% it requires forming an azeotropic mixture containing 5% water which is removed from azeotropic mixture of ethanol, water and benzene after distillation. In this procedure, benzene-water ethanol and then ethanol-benzene azeotropic mixture are removed so that absolute alcohol is obtained.

Yeasts utilise pyruvate during fermentation resulting in the formation of an intermediary product, acetaldehyde. This is trapped by hydrogen sulfite to yield the acetaldehyde in precipitated form and fluid product formation is glycerol as shown below :

$$CH_3\,CHO + NaHSO_3 \longrightarrow CH_2\text{-}CHOH\text{-}SO_3Na$$

In place of acetaldehyde, dihydroxyacetone phosphate acts as a hydrogen acceptor which is reduced to glycerol-3-phosphate. After removal of phosphate, *i.e.* dephosphorylation, it gives glycerol as given below :

$$C_6H_{12}O_6 + H_2SO_3 \longrightarrow CH_2\text{-}CHOH\text{-}SO_3Na + \text{Glycerol} + CO_2$$

Neuberg's fermentation process is categorised as reward and third fermentation. The first fermentation equation is given below :

$$2\text{Glucose} + H_2O \longrightarrow C_2H_5OH + \text{acetate} + \text{glycerol} + 2CO_2$$

IV. MALT BEVERAGES

Man has been using microorganisms to produce alcohol (beer) from starch of barley grains, or wine from sugar in grapes for more than 6000 years.

1. Beer Production

Brewing refers to the production of malt beverages; it originated from Mesopotamia where about 40% of total cereal production was used for brewing purpose. It is likely that beer was a later discovery than wine. Although, most of the beer is produced by using barley and other cereals are occasionally used. Wheat beers such as Berliner Weisee and the gueuze lambic beers of Belgium are notable exceptions. One reason for barley's pre-eminence is that the grain retains the husk which affords protection during storage and transport and also acts as an aid to filtration during wort separation.

(*a*) Malting Process: The barley (*Hordeum vulgare*) kernels separated from the stalk and chaff, are allowed to germinate under controlled temperature and humidity to generate enzyme systems which partially degrade endospermic starch and protein. For this, the barley grains are soaked in water for 1-2 days and after sprouting kernels stopped by drying in hot air. During this process, some flavour and colour components are also formed. The dried malt (source of amylase and proteinase) is crushed before use (Fig. 21.21).

(*b*) Mashing Process: The crushed malt and adjuvent such as corn grits or rice are made into a mash with brewing water, and combined for further enzymatic degradation and subsequent extraction of sugars and proteins. The process of making malt as soluble as possible by using enzymes, adjuncts, etc. is called *mashing.* This process allows the malt amylases and proteinase to degrade starch into maltose, and protein into peptone and peptides, respectively. The enzymatic hydrolysis is influenced by temperature and pH. Therefore, both of the factors are adjusted to get partially and totally degraded enzymatic products. The remaining extract (liquid) is called *wort.* The wort is boiled with flower of hops (*Humulus lupulus*) which imparts longer durability, preservation during beer processing. The wort now is used for fermentation. The boiled wort is cooked, inoculated or 'pitched' with a pure yeast culture (*Saccharomyces carlsbergensis* : bottom type) and fermented. Each beer producing industry or brewery maintains its own strain.

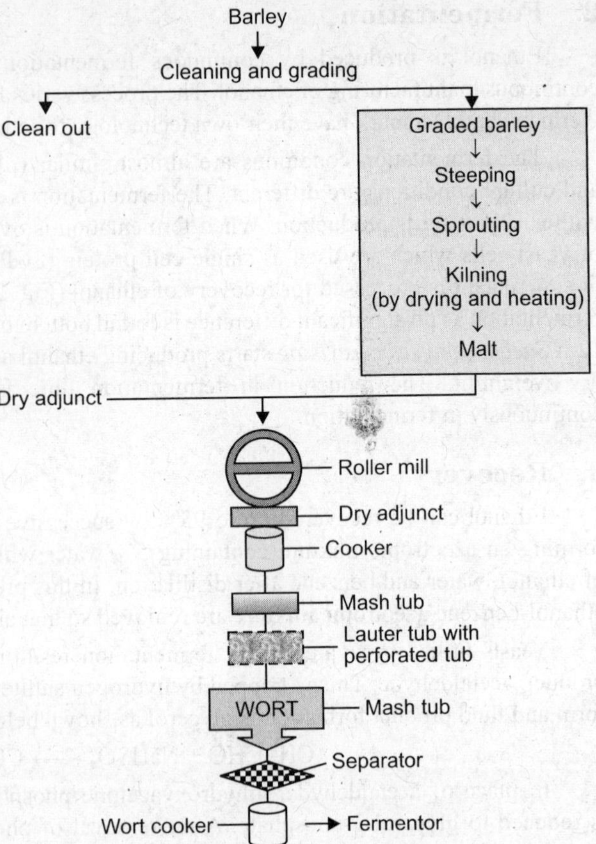

Fig. 21.21. Different steps of brewing process.

The fermentable sugars are converted to ethyl alcohol and trace quantities of flavour ingredients while the non-fermentable carbohydrates are also left in the beer. The fermentation process is completed within 8-14 days. During the later stages, the bottom yeast 'breaks' flocculate and settle. The bacterial growth is not desired during the fermentation. The yeast generated during the fermentation is separated from the beer by centrifugation or natural setting. After storage for several weeks, recarbonation and two or more filtrations are necessary to remove the yeast. The finished product is kept for aging or maturing for several months to years. The 'green' beer is stored or 'laged' in vats at 0°C for several weeks, the proteins along with remaining yeast and other substances precipitate. The flavour is added due to ester formation and aroma develops. The body changes from harsh to smooth (Fig. 21.22).

(*c*) Finishing: The beer is later on carbonated to CO_2 content (45-52%). The beer is now cooled, clarified or filtered and packed in bottles, consor barrels; such process is called *"finishing"*.

Sake is the other malt product which is of the Japanese origin rice beer or wine with an alcohol content varying from 4-17%. A starter or *Koji,* for sake is made by *Aspergillus oryzae* grown on soaked and steamed rice mash until a maximum yield of enzymes is obtained. The koji contains amylases which cause the hydrolysis of starch to sugars available to yeast plus other hydrolytic enzymes such as proteins. The liquor obtained after filtration is called *sake.* The rice beer is also manufactured in India but called 'sonti' which is produced by *Rhizopus sonti* and yeast.

2. Wine Production

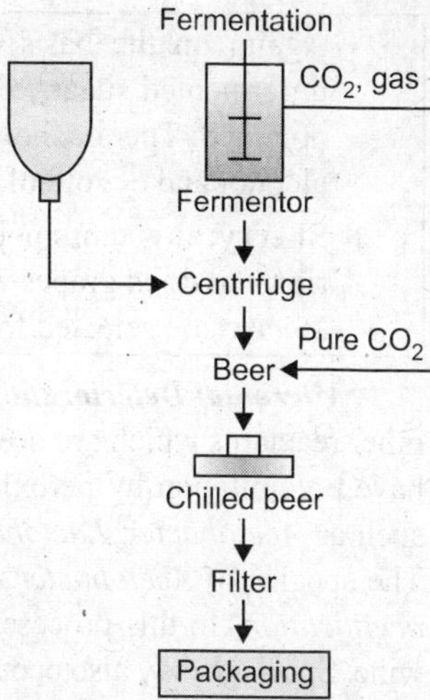

Fig. 21.22. Steps of brewing process.

Wine is a microbial product of grapes obtained after normal alcoholic fermentation by yeast. Wine production and recovery are carried out under the following steps :

(*a*) Microbial Process: Wine is basically the transformation under anaerobic conditions of grapes sugars by yeast into ethanol, carbon dioxide and small amount of byproducts, D-glucose and D-fructose. The two principal sugars of grape juice yield ethanol and carbon-dioxide. The natural yeast flora of crushed grapes and fermentation equipment are less commonly used. In such fermentation, a succession of yeast and bacterial population is found during fermentation of grapes. Moreover, wine is produced by fermentation of the fruit juices, berries, honey, etc.

(*b*) Wine from Grapes: Two kinds of grapes are cultivated and used for wine production. The red grape (where the skin of grape is red) gives rise red wine, while the white grape juice after fermentation gives white wine. Sometimes, the red pigments may be added in juice.

When the grapes have the derived sugar content by 14-20 %, it should be harvested for the extraction of juice. For this, the grapes must be stemmed, cleaned and crushed and then sodium or potassium meta-bisulphite should be added to check the undesirable organisms. The crushed grape is known as *must*. Sometimes, it is difficult to control the natural microbial population present in the *must*. It is, therefore, recommended to use the pure culture of wine yeast of proven quality. Such strains are usually used in conjunction with proper concentration of sulphur dioxide to which the wine yeast is resistant and wild yeast and bacteria are sensitive.

(*c*) Fermentation: This process is carried out by adding 2-5% of wine yeast, namely *Sacharomyces cerevisiae* var. *ellipsoideus* in the *must*. The whole contents are mixed twice a day by punching the 'cap' of floating grapes so as to allow profuse growth of yeast due to increase in aeration in the beginning. This process aids in the extraction of colour (white or red) depending upon the quality of grapes used. Thereafter, mixing should be stopped and anaerobic fermentation allowed to be continued. The temperature is maintained at 24-27°C for red wine for 3-5 days, and 10-21°C for white wine for 7-12 days. The heat generated during fermentation increases the temperature. Sometimes lactobacilli may grow at such a temperature which contaminates the product. Hence, artificial cooling is required.

(*d*) Recovery: The fermented juice is drawn off from the residue (Pomace) and stored under atmosphere of carbon dioxide for carrying out further fermentation for about 7-11 days at 21-29°C. If *dry wine* is to be obtained, whole residual sugar should be fermented to yield ethanol.

The wine may be pasteurised before aging. During pasteurisation, protein is precipitated and removed. It is cooled, filtered and transferred in the wooden tanks made up of white oak or red wood or plastic concrete tank for aging. Aging imparts flavour, aroma, sanctity and colour to the wine. After aging, the wine is clarified, barreled or bottled. The final alcohol content in wine varies from 6 to 9% by weight, or 8 to 13% by volume.

Types of Wine

On the basis of production of carbon dioxide, wines are of two categories.

- **Still wine:** The wine in which no carbon dioxide is produced during the fermentation is called still wine.
- **Carbonated wine:** The wine in which considerable quantity of carbon dioxide is produced is called carbonated wine.

Again, on the basis of presence of sugars, these are called *dry wine,* which have no unfermented sugar, while *sweet wines* contain sugar. Upon distillation of wine '*brandy*' is obtained. The alcohol percentage is increased up to 21% in brandy. *Table wines* have low alcohol and devoid of sugar.

- **Sherry:** It is more popular in France; hence, it is called French dry sherry. It is made from ripe and dried grapes with high sugar contents. Sometimes, the grapes infected with *Botrytis cinerea* are selected for sherry production.

***Microbial Deterioration and Spoilage of Wine* :** Defects in wine are mainly due to metal and other reactants which are used for wine processing. White wines may turn brown and red wine may have precipitation by peroxidases or oxidising enzyme of certain molds. Various microorganisms such as *Acetobacter, Lactobacillus, Leuconostoc*, wild yeasts, molds, etc. cause spoilage of wine. The species of *Acetobacter aceti* or *Gluconobacter oxydans* impart an undesirable process called *acetification*. In this process, glucose may oxidise and may give a `mousy' or sweet sour taste to wine. Spoilage may also occur due to abnormal fermentation by natural microflora that results in low alcohol content, high volatile acidity, undesirable flavour and cloudiness in wine. Some common defects in wine are either due to bacterio-fermentative process by lactobacilli that result in acid production from sugars. Such spoilage is called *tourne*. On the other hand, bitter taste in wine also occurs due to the fermentation of glycerol. Such process is called *amerture*. The liberation of carbon dioxide or gassiness results due to the action of heterofermentative lactis that is called *pousse*.

Distilled Beverages or Liquors

- **Rum:** It is obtained by distillation of alcoholically fermented sugarcane juice, syrup, or molasses.
- **Whiskey:** It is obtained after distillation from saccharified and fermented grain mashes. Whiskey is prepared by using wheat, corn, rye, etc. Whiskeys are made by novel strains of *S. cerevisiae* var. *ellipsoideus*.
- **Brandy:** It is obtained from grape wine after distillation. The brandy can also be obtained by using apple, peach and apricot.

V. ORGANIC ACID PRODUCTION

A large number of organic acids is produced by using various microorganisms; some of organic acids are discussed below :

1. Vinegar Production

The vinegar is derived from the French '*vinaigre*' means sour wine. The vinegar mainly depends upon the type of local alcohol production, for example rice vinegar in Japan, or sugarcane vinegar in India.

$$\mathrm{H{-}\overset{\displaystyle H}{\underset{\displaystyle H}{C}}{-}\underset{\displaystyle \| \atop O}{C}{-}OH}$$

Acetic acid

There are mainly two steps in vinegar production. In the first stage yeast converts sugar into ethanol anaerobically, while in the second step ethanol is oxidised to acetic acid aerobically by *Acetobacter* and *Gluconobacter* through a process is called acetification. Both the bacteria are Gram-negative, catalase positive, oxidase negative and strictly aerobic. *Acetobacter* species are the better acid producers and are commonly used in industries.

Vinegar is an alcoholic fermentation product and if enough acetic acid is present, it is legal vinegar. The *S. cerevisae* var. *ellipsoideus* carried out the conversion of sugar into ethanol in the first step, while the species of *Acetobacter aceti* or *Gluconobacter* convert ethanol into acetic acid.

***(a)* Substrate:** Vinegar is produced by using fruit juices, starchy vegetables, malted cereals such as barely, rice, wheat, corn, sugar cane syrups, molasses, honey, alcohol, etc. From the fruit juice such as apple juice, the vinegar produced is known as cider vinegar. If it is produced by using Ale (light-yellow coloured fermented product of hops), then it is called Alegar. While malted grain made malt vinegar and ethanol produced sprit vinegar.

(*b*) Methods of Production: Mainly there are two methods : (*i*) slow or 'let home' process, and (*ii*) quick or 'French orleans' process. Besides, some other old methods are used in many countries.

(*i*) Natural fermentation: It can be made easily by fermenting fresh sap into plastic or earthen jar until it becomes sour. Then packed into plastic bottles and placed under the heat of sun for few days. The very common package is the use of 1.5 liters coke bottle. Sugar palm and coconut sap are some of the common examples.

(*ii*) Slow process: In the slow process, the alcoholic liquid (substrate) is not allowed to move during acetification and fermented juice or malt liquors are used for acetic acid production.

Old method of vinegar production

Apple juice is specially used for alcoholic production by batch fermentation process. The barrel is partially filled with the fermented juice and is allowed to undergo the process of acidification. The vinegar bacteria are supplemented from the previous batch of vinegar. A film of vinegar bacteria called 'mother of vinegar' should grow on the surface of the liquid which indicates that ethanol is oxidised into acetic acid and formed vinegar. Demerits of this process are the absence of productive strain, and poor yield of ethanol gives inferior quality of vinegar.

(*iii*) Continuous quick process: In this process about 1/4 barrel is filled with raw vinegar from previous run supplemented with active vinegar bacteria. This is acetified to check competing microbes. The barrel is filled by alcoholic hydrolysate up to half filling and examined for growth of bacteria in a film on the top of the liquid. Generally, this process takes weeks to months at 21 to 29°C. The recovery is made in part by withdrawing and replacing by equal quantity of alcoholic liquor. This is a continuous fermentation. In this process the major difficulty is dropping of the gelatinous film of vinegar bacteria which results in retardation of the acetification. To avoid this, a raft or floating frame-work is provided to support the filter Another important process is called quick generator process.

(*iv*) Generator process: This involves the movement of ethanol liquid during acetification. This ethanol liquid is trickled over the film of vinegar bacteria. A simple generator or cylindrical tank made up of wood is used (Fig. 21.23).

In starting a new generator, the slime of vinegar bacteria must be established before vinegar is produced. Hence, in the beginning the middle section of the tank is filled with raw vinegar that contained active vinegar bacteria. To inoculate the shaving with the desired bacteria an alcoholic liquid acetified with vinegar is slowly trickled through the generator to build up bacterial growth on the shaving and then is recirculated.

The fringe generator is also recommended for vinegar production which contains a large cylindrical airtight tank equipped with a sprinkler at the top. The cooling coils are fitted in the lower part of the middle section. Recirculation of the vinegar is done from the bottom collection chamber. The generator gives high yield of acetic acid and leaves little residue (*i.e.* alcohol).

(*v*) Makin's process: A fine mist of a mixture of vinegar bacteria and nutrient alcoholic solution is sprayed through jet nozzles into a chamber. The mist is kept in circulation by filtered air for a while and is allowed to fall to the bottom for circulation, collection, etc.

***(vi)* Submerged process:** A stirred medium containing 8-12% of ethanol is inoculated with *Acetobacter acetigenum* and is held at 24-29°C. The bacteria grow in a suspension of fine air bubbles fermenting liquid.

***(f)* Finishing:** It is the slow process; vinegar produced is less harsh, while in quick process improved vinegar is produced as far as its taste, and body and flavor are concerned. The unit of vinegar is in grams, and it is commercially and internationally defined in terms of grain which is 10 times more than that of gram.

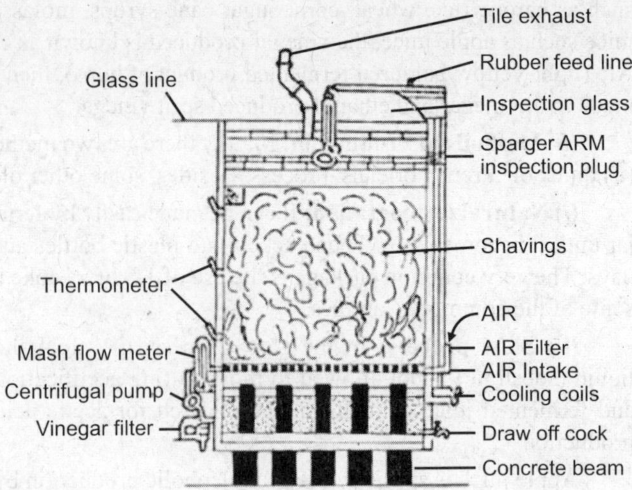

Fig. 21.23. Vinegar production through generator process.

If the storage vessel is made up of steel or iron, the ferric ion is oxidised into ferrous ion. The tannin, phosphate or proteins are formed which make the vinegar hazy in appearence. If the barrel is made up of tin or copper, cloudiness develops resulting in darkening of vinegar. In addition, the animal pests such as mites, fruit flies, vinegar eel (*Anguillula aceti*) attack the film of acetic acid bacteria which cause to sink and deteriorate vinegar. Sometimes, *Leuconostoc* and *Lactobacillus* contaminate the vinegar imparting off flavour.

Types of Vinegar

- **Wine Vinegar**: It is made from wine, contains citric and tartaric acid, and a small portion of acetic ether, which communicates an agreeable flavour. It is initiated by adding acetic ether and colouring matter to vinegar made from whiskey.
- **Cider Vinegar**: It includes all fruit vinegars and contains malic acid.
- **Malt or Corn Vinegar**: It is made from weak worts, contains phosphates of lime and magnesia, gum, and extractive matter.
- **Wood Vinegar or Pyroligneous Acid**: When crude, contains tar and wood spirit.
- **Distilled Vinegar:** It is obtained from vinegar by distillation; distilled vinegar is weaker than the common vinegar but is used sometimes in pickles.

2. Lactic Acid Production

Lactic acid was first discovered by Scheele (1789) from sour milk. Later on, L.Pasteur (1857) identified the microorganism involved in lactic acid production. In the year 1881, first commercial production was started by M/S Clinton Processing Company, Iowa (U.S.A.) through the fermentation process.

Lactic acid production by using chemical process was not economical and recovery and purification were also not up to the mark; hence continuous efforts were made to improve the process. Moreover, the requirement in plastic industry was of very high purity.

$$\begin{array}{c} COOH \\ | \\ HCOH \\ | \\ CH_3 \end{array}$$

Lactic acid

***(a)* Fermentation:** Lactic acid is produced by several microorganisms which differs in their ability to produce either D (–) lactic acid, L (+) lactic acid or the racemic mixture. The particular acid formed seems to be characteristic of the individual microorganism. The racemic mixture is formed due to the production of an enzyme called 'recemase'. The lactic acid recovered is optically active but becomes inactive due to the action of the enzyme.

Various microorganisms are involved in the production of lactic acid. *Rhizopus oryzae* produces only L(+) lactic acid. However, the production is quite slow and yield is also low. There are mainly two important processes based on end product formation.

***(i)* Homofermentative process:** This process involves certain bacteria namely, *Lactobacillus delbruckii, L. bulgaricus, L. pentosus, L. leichmanii, L. casei, Streptococcus lactis,* etc. These bacteria utilise the EMP pathway to produce pyruvic acid which is then reduced by the lactate dehydrogenase to lactic acid. All the microbes are considered to be anaerobic, although they can withstand some oxygen. The end product is lactic acid with traces of others.

***(ii)* Heterofermentative process:** This process involves the action of *Leuconostoc mesenteroides* which produces lactic acid, carbon dioxide, ethanol, acetic acid, water and few other products.

***(b)* Medium and Manufacturing Process:** The culture medium contains semirefined sugar, (molasses or whey contains semi-refined sugar), molasses or whey starch, maltose, lactose, sucrose, calcium carbonate with ammonium hydrogen phosphate. The malt sprouts are mixed and pH is kept between 5.5 and 6.5.

Lactic acid is quite corrosive, hence metals are avoided, and consequently wooden fermentors are used. The thermophilic clostridia results in the production of some butanol and butyric acid which are the major contaminants in lactic acid production.

The colonies of *L. delbruckii* are transferred into large culture vessel kept at 45-55°C. Each stage of culture building requires 16-18h. A slight excess of calcium carbonate is present in each stage. The inoculum volume is usually 5% and the fermentation is carried out for 5-10 days. The sugar is reduced to 0.11% or less during the fermentation because residual sugar makes the recovery of better quality of lactic acid difficult. Aeration and agitation are required. Lactic acid fermentation allows NAD^+ to be recycled and glycolysis to continue. Lactic acid is oxidised back to pyruvate which then goes through the Krebs cycle when vigorous activity ceases (Fig. 21.24).

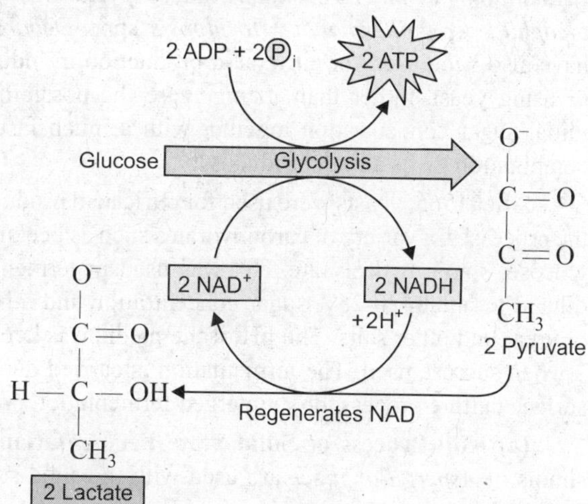

Fig. 21.24. Anaerobic lactic acid production.

***(c)* Recovery:** To the fermentation medium, $CaCO_3$ is added; pH adjusted to 10, broth is heated and filtered. Lactic acid is converted to calcium lactate. It decomposes residual sugar which kills bacteria. The H_2SO_4 is added to remove Ca as $CaSO_4$. Lactic acid is recrystallised as calcium lactate. The activated charcoal is added to remove impurities and lactic acid is recovered.

In the solvent extraction procedure free lactic acid is extracted with isopropyl ether directly. It is washed with cold water in a centrifuge to get lactic acid. In another process, methyl ester of free lactic acid is prepared. The fermentation broth is distilled to get ester. It is boiled with water (hydrolysis) which allows the ester to decompose. The product is distilled to get aqucous solution of lactic acid and distilled product is methanol. The lactic acid can also be recovered as tertiary alkaline salt. Hence, the fermentation broth, organic solvent is added to get salts of lactic acid which is decomposed and free acid is released.

***(d)* Uses:** Since it is a weak acid with good solvent properties, it polymerises readily for the production of polymers. It provides acidity in foods and beverages and served as a preservative in food stuff. The delining of hides in leather industries is also carried out by its utilisation. Certain other industries such as textile and laundry use lactic acid in fabric treatment. Calcium lactate is employed in baking powder and as a source of calcium in pharmaceutical industries. The pure form of lactic acid is used in plastic industry.

2. Citric Acid Production

This acid was first produced commercially by John and Edmund Sturage Company in UK in the year 1826. Scheele (1789) reported the isolation and crystallisation of the four constituents of lemon juice. Grimoux and Adams (1880) synthesised citric acid from glycerol. Wehmner (1893) observed the occurrence of citric acid as a microbial product by using *Penicillium* and *Citromyces*. Millard in 1922 recorded accumulation of citric acid in culture of *Aspergillus niger* under condition of nutrition deficiency. In 1923, Pfizer began operation of fermentation based process in the U.S.A.

Aspergillus niger has been the choice for the production of this primary metabolite citric acid for several decades. A large number of other microorganisms (fungi and yeast) such as *Aspergillus clavatus, A. wentii, Penicillium luteum, P. citrinum, Mucor pyriforms, Candida lipolytica, C. oleophila, C. guillermondis, Torulopsis* spp., *Debaromyces daussenii,* etc. have also been used for citric acid production in industries. The advantages of using yeast, rather than *A. niger* are the possibility of using very high initial sugar concentration together with a much faster fermentations. This combination gives a high productivity

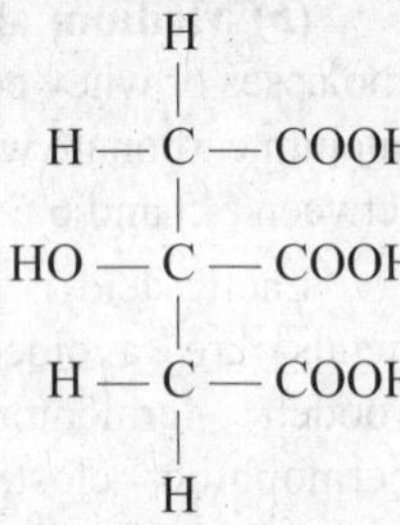

$$\begin{array}{c} \text{H} \\ | \\ \text{H}-\text{C}-\text{COOH} \\ | \\ \text{HO}-\text{C}-\text{COOH} \\ | \\ \text{H}-\text{C}-\text{COOH} \\ | \\ \text{H} \end{array}$$

Citric acid

After 1965, yeasts were used for citric acid production using carbohydrate and *n*-alkanes. In all the processes, a variety of carbohydrates such as beet molasses, cane molasses, sucrose, commercial glucose, starch hydrolysate, etc. was used in fermentation medium. The starchy raw material is diluted to obtain 20-25% sugar concentration and mixed with a nitrogen source (ammonium salts or urea) and other salts. The pH of the medium is kept around 5 when molasses is used, and at pH 3 when sucrose used. The fermentation is carried out by any of the processes, koji process, liquid surface culture process, or submerged fermentation process.

(*a*) Koji Process or Solid State Fermentation : It is a Japanese process in which special strains of *Aspergillus niger* are used with the solid substrate such as sweet potato starch. Mold is used in the preparation called koji to which wheat bran was substituted in the sweet potato material. The pH of the bran is adjusted between 4 and 5, and additional moisture is picked up during steaming so as to get the water content of the mash around 70-80%. After cooling the bran to 30-60°C, the mass is inoculated with a koji which was made by a special strain of *A. niger* which is probably not as possible to the presence of ions of iron as the culture strains used in other process. Since bran contains starch which on saccharification by the amylase enzyme of *A. niger* induces citric acid production. After inoculation the bran is spread in trays to a depth of 3-5 cm and kept for incubation at 25-30°C. After 5-8 days, the koji is harvested and citric acid is extracted with water.

(*b*) Liquid Surface Culture Process : In this case, *A. niger* floats on the surface of a solution. Aluminium or stainless steel shallow pans (5-20 cm deep) or trays are used. The sterilised medium usually contains molasses and salts. Fermentation is carried out by blowing the spores of *A. niger* over the surface of the solution for 5-6 days, after which dry air is used. Spore germination occurs within 24 hours and white mycelia grow over the surface of the solution, eight or ten days after inoculation. The initial sugar concentration (20-25%) is reduced to range of 1-3%. The liquid is drained off and any portion of mycelial mat left becomes submerged and inactivated. The small quantity of citric acid is produced during the growth phase, which is called primary metabolite. The mycelia can also be reused.

During preparation of fermentable sugar from molasses, sucrose is the main carbohydrate along with some glucose as well as protein, peptide, amino acids, and inorganic ions. This is subjected to heat; so it contains saccharic acids and related compounds in traces. The initial sugar concentration is about 20-25%. Removal of metallic ions or reduction in quantity of undesirable ions in sucrose syrup by adsorption with a combination of $CaCO_3$, colloidal silica, tricalcium phosphate and starch are the other important steps. Addition of calcium ferrocyanide results in iron precipitation.

Initially, the pH remains in the range of 5-6, but after spore germination, pH approaches the range of 1.5-2 as ammonium ions are removed from the solution. It is important to mention that at initial pH of 3-5 some oxalic acid is also produced. The presences of iron also favours oxalic acid production, and of yellow or yellow green pigments in the mycelia sometimes secreted into the culture solution and are difficult to remove during product recovery and purification. Since citric acid is a primary metabolite, it is produced during growth (trophophase) of the microorganism as a result of interruption of TCA cycle *i.e.* defective Kreb's cycle. In this process a good amount of sugar transport through EMP pathway takes place which results in acetyl CoA production, which condenses with oxaloacetic acid to yield citric acid. Further, citric acid cycle is interrupted by inhibition of aconitase and isocitric dehydrogenase by copper or H_2O_2. Fe is an essential cofactor and the Cu or Fe ratio is 0.3: 2 (mg/lit) which interrupt the activity of these enzymes to pH 2.0 (Fig. 21.25).

***(c)* Submerged Culture Process:** It is a process in which the fungal mycelia grow throughout the solution in a deep tank. This process is quite economical. In this case, the organism (*Aspergillus japonicus*) is slowly bubbled in a steam of air through a culture solution of 15 cm depth. Since the organism shows subsurface growth and produces citric acid in the culture solution, the yield is inferior as compared to liquid surface culture fermentation.

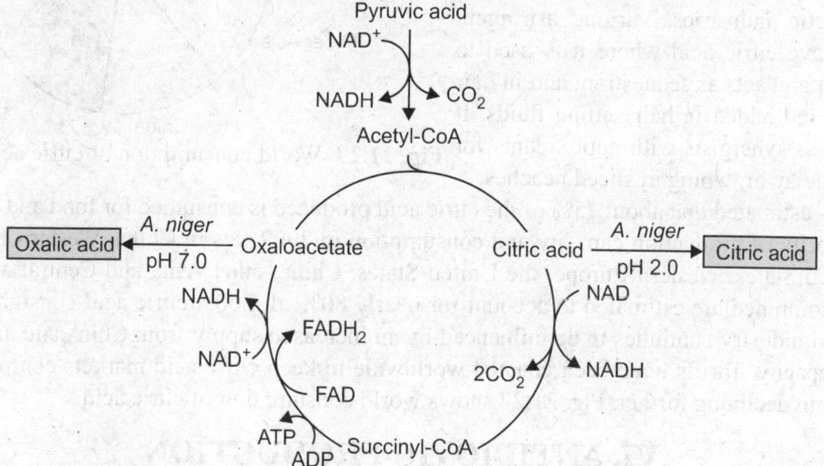

Fig. 21.25. Biochemical pathway of oxalic acid and citric acid.

The earlier workers used shaking culture and extracted Mollard's phosphate deficiency concept to submerged fermentation. But they could not realise the role of metallic ions which commonly occur as impurities in phosphate salts. Aeration is required for the continuous fermentation. The addition of copper ion is must to ensure the right biochemical process. Addition of the antifoam agents is necessary; such agents must be free from iron, cobalt or nickel. Continuous culture technique is not considered suitable for citric acid product.

***(d)* Recovery:** The culture filtrate used to be hazy due to the presence of residual antifoam agents, mycelia and oxalate. The $Ca(OH)_2)$ slurry is added to precipitate calcium citrate. After filtrations the filtrate is transferred and treated with H_2SO_4 to precipitate Ca as $CaSO_4$. This is subjected to the treatment with activated carbon. It is demineralised by successive passages through ion exchange beds and the purified solution is evaporated in a circulating crystallisers. The crystals are removed by centrifugation. The remaining mother liquor is returned to the recovery stream. The solvent extraction can also be performed by adding 100 parts tri-*n*-butyl phosphate and 5-30 parts *n*-butyl acetate or methyl isobutyl ketone which are to be mixed with the filterate. The solvent is then extracted with water at 70-90°C. Citric acid is further concentrated, decolourised and crystallised (Fig. 21.26).

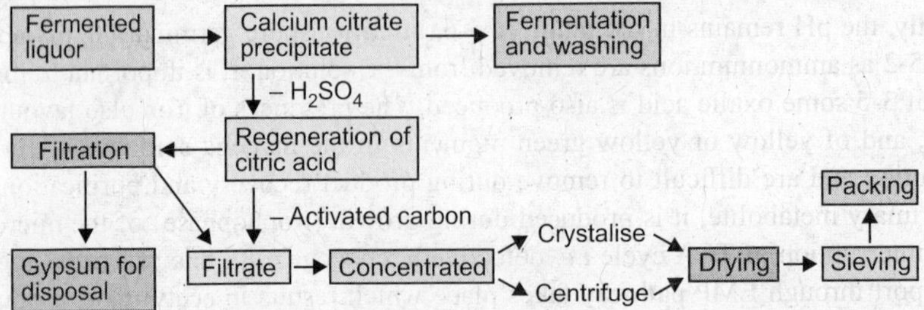

Fig. 21.26. Steps of recovery of citric acid.

***(e)* Uses:** Citric acid is sold in the market as an anhydrous crystalline chemical, as crystalline monohydrate, or as a crystalline sodium salt. Citric acid is used in the soft drinks, jams, jelly, wines, candies and frozen fruits. It also used in artificial flavours. In medical applications, it is used in blood transfusion and as effervescent product. In cosmetic industries, various astringent lotions have citric acid where it is used to adjust pH and acts as sequestrant and in hair dressing and added in hair setting fluids. It also acts as synergists with antioxidants for oils and delay browning in sliced peaches.

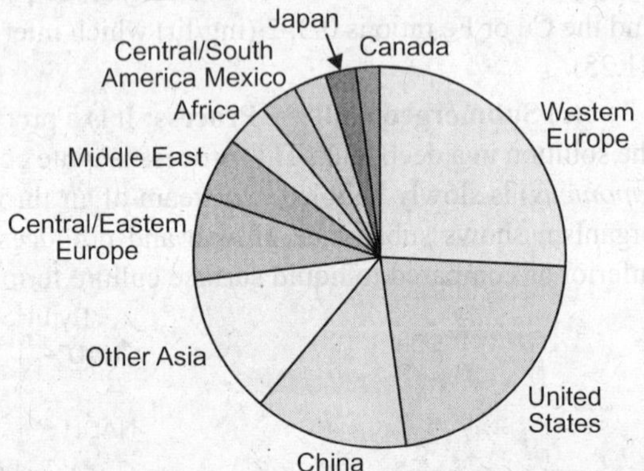

Fig. 21.27. World consumption of citric acid in 2009.

It is estimated that about 75% of the citric acid produced is consumed for food and beverages. The majority of production capacity and consumption in 2009 was in China, Western Europe and the United States. Western Europe, the United States, China, other Asia, and Central and Eastern Europe combined are estimated to account for nearly 80% of global citric acid consumption. The citric acid industry continues to be influenced by an increased supply from China and an abundant global capacity. Tariffs have been enacted worldwide to keep citric acid markets competitive and prices from declining further. Fig. 21.27 shows world consumption of citric acid.

VI. ANTIBIOTIC PRODUCTION

Antibiotics (Fr. *anti* = against; *bios* = life) are chemical substances secreted by some microorganisms which inhibit the growth and development of other microbes. Most of them are produced by actinomycetes, specially the genus *Streptomyces*, and filamentous fungi. The study of antibiotics began by the discovery of penicillin in 1929, when Alexander Fleming proved that the filtrate of a broth culture of *Penicillium notatum* has antibacterial properties in relation to Gram-positive bacteria. In 1940, E. Chain and H. Florey obtained a relatively stable preparation of penicillin. In 1932, Clutterbuck studied the chemistry of penicillin and observed that it can be extracted from organic acid solvents from aqueous solution of low pH, and it was heat labile. But its activity was lost during evaporation of solution to dryness. By keeping low temperature

Alexander Fleming

during extraction, its curative property was demonstrated. In1941, Florey and Heatley shifted from U.K. to the U.S.A. during World War II and received aid from the American Government to carry out further research work. By the time American troops entered in France in 1944, and sufficient amount of penicillin was available for saving the wounded soldiers.

Strain improvement for Secondary Metabolites Production: The most wide spread method or increasing product yield in antibiotic producing microorganisms is still random mutation by treatment with mutagenic agents. Morphological differentiation has been studied in relation to cephalosporin synthesis in *Cephalosporium acremonium* by causing mutation for elucidation of the biosynthesis pathways for cephalosporin synthesis.

In streptomycete, the ability to produce aerial mycelium is sometimes associated with production. Another approach is by raising blocked mutants for product cosynthesis. An end product is synthesised by two such blocked mutants (secretor and converter) when they are propogated together, *i.e.* cosynthesis. These studies successfully elucidated the biosynthetic pathway for butirosins using blocked mutants of *Bacillus circulans*. Similarly, for new antibiotic production, biosynthetic blocked mutants for secondary metabolites have been evolved. This is called mutational synthesis e.g. macrolide antibiotics production.

Sometimes, mutant's resistance to antibiotics or precursors is effective for strain improvement. Corynesin (chloramphenicol analogue) is produced by *Corynebacterium hydrocarboclastus.* Another example is the growth inhibition of *Streptomyces hygroscopicus* by valine, which produced more maridomycin than the parent strain. A mutant requiring oleic acid for neomycin formation by *Streptomyces fradiac* showed a decrease in the intracellular level of neomycin precursors in the mutant.

1. Penicillins

Penicillins are a group of antibiotics such as penicillin G, procaine penicillin, benzathine penicillin, and penicillin V. Penicillins are derived from the fungus, *Penicillium chrysogenum* and *P. notatum*. Penicillins differ from one another in the side chain attached to its amino group. Most of these penicillins are 6-aminopenicillanic acid derivatives and all have β-lactam ring which is responsible for the antibiotic activity (Fig. 21.28). A beta-lactam ring is a lactam (*i.e.* a cyclic amide) with a heteroatomic ring structure consisting of three carbon atoms and one nitrogen atom. Pencillin acts against Gram-positive bacteria and inhibit their cell wall synthesis. These are non-toxic to mammals except for certain allergic reactions. More than 100 penicillins have been synthesised so far. β-Lactam antibiotics inhibit the formation of peptidoglycan cross-links in the bacterial cell wall and results in cell wall lysis.

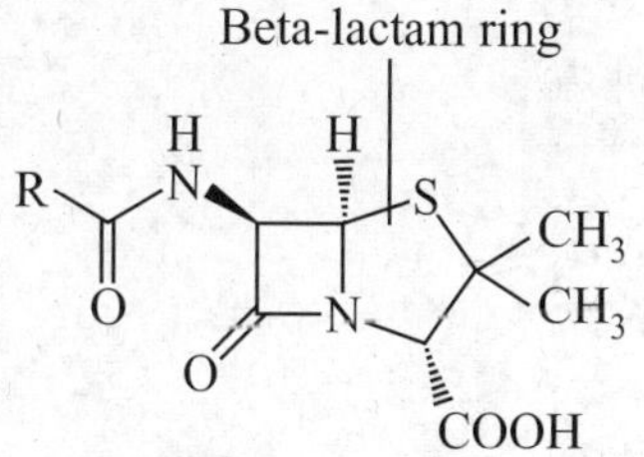

Beta-lactam ring

Fig. 21.28. Penicillin G, a natural product.

Penicillium species required the production medium which contains lactose (1%), calcium carbonate (1%), corn steep liquor 8.5%, glucose (1%), sodium hydrogen phosphate (0.4%) and phenyl acetic acid (0.5 g). The pH is kept between 5 and 6, and temperature for incubation is 23-25°C. Aeration and agitation are necessary.

***(a)* Fermentation:** Penicillin is produced by fed-batch culture using *P. chrysogenum* Q-176, a fungus that can be grown in stirred fermentors. The inoculum under aerobic condition (seed) can be produced when glucose is present in the medium in sufficient amount. If particular penicillin is produced, specific precursor (substance added prior or simultaneously with the fermentation which is incorporated without any major change in the molecules) is added in the medium (for example phenyl acetic acid or its derivatives such as ethanol amide) to get penicillin G (Fig. 21.28). An antifoam agent such as vegetable oil (corn or soybean oil) is added to the medium before sterilisation.

The spore suspension is inoculated in flasks, each containing 15 g barley seeds. These flasks are vacuum-dried, to which sterilised quartz is added. The inoculum is pressed on barley seeds. The flask containing 15 g barley seeds are mixed with the mother culture, and incubated at 25°C for 7 days. The spores developed on barley seeds are suspended in distilled water to make spore suspension. After testing the antibiotic activity, seeds containing flasks are ready for seeding in fermenter. Three growth phases are differentiated during cultivation of *Penicillium chrysogenum.*

- **First phase.** In this phase mycelial growth occurs, hence antibiotic yield is quite low. Lactic acid present in corn steep liquor is utilised at a maximum rate by the microorganism. Lactose is used slowly. Ammonia is liberated into the medium resulting into rise in pH.
- **Second phase.** In this phase intensive synthesis of penicillin occurs due to rapid consumption of lactose and the ammonium nitrogen (NH_3N). The mycelial mass increases, and the pH remains unchanged.
- **Third phase.** The concentration of antibiotic decreases in the medium. The autolysis of mycelium results in liberation of ammonia with slight rise in pH.

(*b*) Biosynthesis of Penicillin G (Benzylpenicillin): In 1998, A.A. Brakhage has described that three amino acids viz., L-α-aminoadipic acid, L-cysteine and L-valine condensed into a tripeptide in the first step (Fig. 21.29). The amino acid L-valine undergoes epimerisation and becomes D-valine. They undergo condensation and become tripeptide, which is named as δ-(L-α-aminoadipyl)-L-cysteine-D-valine (or ACV). During this process a catalytic enzyme δ-(L-α-aminoadipyl)-L-cysteine-D-valine synthetase (ACVS) is added, which is required for the activation of the three amino acids before condensation and the epimerisation of transforming L-valine to D-valine. In the second step a very weak intermediate enzyme isopenicillin N synthase is used to change ACV into isopenicillin N, resulting in oxidation process and formation of bicyclic ring. In the last step the side-chain group, *i.e.* the alpha-aminoadipyl side-chain of isopenicillin N is exchanged for a phenylacetyl side-chain to form penicillin G.

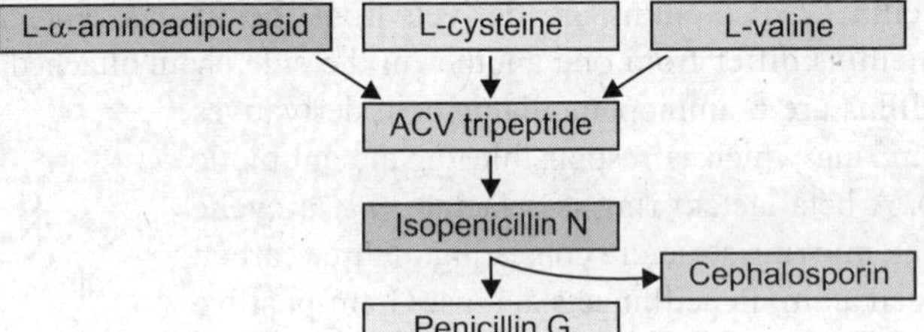

Fig. 21.29. Biosynthesis of penicillin G.

(*c*) Recovery: When fermentation cycle (7 days) is completed, the whole batch is harvested for recovery. Its activity disappears on evaporation to dryness, hydrolysed to penicilloic acid. Penicillin has tendency that it remains in aqueous phase at normal pH and in solvent phase at acidic pH. This property of penicillin is used in recovery of potassium penicillin from natural solutions. Once the fermentation is completed the broth is separated from fungal mycelium and processed by absorption, precipitation and crystallisation to yield the final product (Fig. 21.30). This basic product can then be modified by chemical procedures to yield a variety of semisynthetic penicillins such as ampicillin, amoxycillin, etc.

Fig. 21.30. Structure of sodium penicillin G.

(*d*) Units of Antibiotics: The potency of antibiotics is normally expressed in terms of units/ml of solution or in 1 mg (U/mg). In most of the cases, the antibiotic vial shows the quantity expressed in terms of weight as well as activity for example 1 mg of the penicillin G (benzyl penicillin) is equivalent to 1,665 units.

Semisynthetic Penicillins

Semi-synthetic penicillins are the derivatives of penicillins. The main objective of producing semisynthetic penicillin is to generate compounds with improved properties e.g. acid stability, resistance to enzymic degradation, broader spectrum of activity, etc. In such cases, side chain is removed to form 6-amino penicillanic acid (6-APA) via immobilisation in a column of penicillin acylase. For example, penicillin G is converted to 6-APA and phenylacetic acid. The 6-APA is then chemically acylated to produce semi-synthetic penicillin.

2. Streptomycin

It is effective against tuberculosis causing organism, *Mycobacterium tuberculosis* and Gram-negative bacteria. The prolonged use of streptomycin in mass can produce neurotoxic effects and loss in hearing. Since its discovery in 1944 by Schatz, Burgie and Waksman most of the strains of *S. griseus* are genetically improved.

(*a*) Structure of Streptomycin: The commercially available streptomycin is basically hydrochloride of streptomycin ($C_{21}H_{39}N_7O_{12}.3HCl$) with calcium chloride. During the production of streptomycin, mannosidostreptomycin or hydroxystreptomycin is also produced in the early fermentation. This salt is not economical and is easily converted to streptomycin by the action of *S. griseus* (Fig. 21.31). No precursor is reported to increase the yield.

(*b*) Production: The contents of medium for streptomycin production are : soybean meal (10 g), glucose (10 g), peptone (5 g), meat extract (5 g) and sodium chloride (5 g). The pH is kept at 7.6-8.0 before sterilisation and after inoculation the culture is incubated at 28°C. Streptomycin is produced by *S. griseus* which is grown in stirred fermentation; it strongly requires high aeration and agitation. Therefore, continuous fermentation process is done to produce streptomycin. The spores are produced on media which provide enough sporulated growth to initiate liquid culture of mycelia. The optimum fermentation temperature is approximately 28°C, and the whole process completes within 5-7 days. There are three main steps in the production process of streptomycin.

(*i*) First phase: Growth of mycelium occurs, and the proteolytic activity of *S. griseus* releases ammonia from the soybean meal. Carbon from soybean meal induces growth but glucose is utilised at a minimum rate. The yield of streptomycin is low. The pH rises during the first phase of fermentation process.

(*ii*) Second phase: Streptomycin synthesised at a rapid rate in this phase due to rapid utilisation of ammonia and glucose. The total incubation period lasts from 24 h to 6 - 7 days. No mycelial growth occurs in this phase. The pH remains between 7.6 and 8.

(*iii*) Third phase: The sugar depletes from the medium resulting in stop of streptomycin production. The cells lyse, releasing ammonia resulting in rise in pH. Before lysis, fermentative material is harvested for recovery of streptomycin.

(*c*) Recovery: After filteration, the broth is treated with activated carbon and streptomycin is eluted with dilute acid. The eluted streptomycin is then precipitated by solvents, filtered, and dried before further purification. In another process, culture filtrate is acidified, filtered and neutralised. It is passed through cation exchange column to absorb streptomycin. Later it is dissolved in methanol, after filtration acetone is added to yield

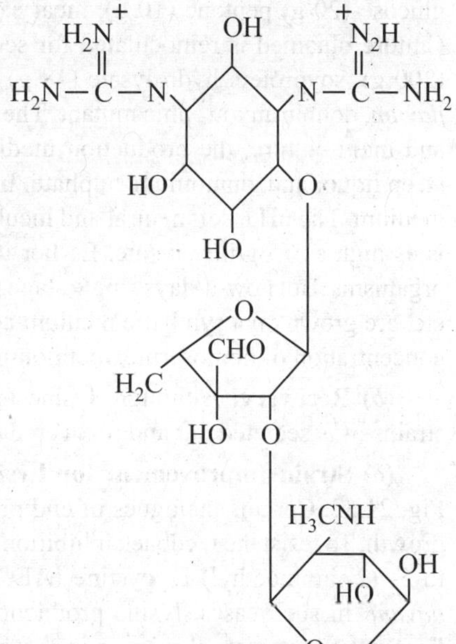

Fig. 21.31. Structure of streptomycin.

streptomycin. About 1,200 µg/ml yield is obtained. The yield is not affected by contamination except transformation by actinophages which are sometimes found during penicillin production.

(*d*) Unit of Streptomycin: One unit of streptomycin activity is equivalent to the microgram of free base.

VII. AMINO ACID PRODUCTION

Some amino acids have been produced in Japan by bacterial fermentation since 1950. Their biosynthesis is now possible by genetically manipulating and physiologically altering the microorganism. *Corynebacterium glutamicum* and *Brevibacterium flavum* are involved in the synthesis of L-lysine and L-threonine from a common intermediate, aspartic acid. The regulatory mutants of bacteria produce lysine in sufficient amount. Such bacteria eliminate the dependence of enzyme formation upon addition of inducer.

The second aspartokinase is repressed by 1-methionine as well as by 1-threonine and isoleucine in addition to repression by L-lysine. The dihydrodipicolinate synthase is also inhibited by end product L-lysine.

1. L-Lysine Production

Generally, *Escherichia coli* and *Enterobacter aerogenes* are used for the formation of diaminopalmelic acid and the decarboxylation of the diaminopalmelic acid by an enzyme DAP decarboxylase (*E. aerogenes* origin), respectively. *E. aerogenes* is an auxotroph that requires L-homoserine or a mixture of L-threonine and L-methionine. The lysine-histidine, double auxotrophic mutant of *E.coli* (ATCC 13002) produces diaminopalmelic acid on a molasses medium with a yield of 19-24 g/litre. The entire fermentation solution including the cell material is subsequently incubated with *E. aerogenes* (ATCC 12409) at 35°C. After 20h, the DAP is quantitatively decarboxylated to L-lysine (C6H14N2O2).

(*a*) Fermentation: For industrial production of lysine, the seed culture is prepared by using glucose (20 g) peptone (10 g), meat extract (5 g), sodium chloride (2.5 g) in 1 litre of tap water. Culture obtained is reinoculated for second seed culture in media containing sugarcane molasses (200 g), soyprotein hydrolysate (18 g) mixed in tap water (1 litre), inoculated by *Brevibacterium flavum*, double auxotrophic mutant. The acetate is used as C source. After preparation of first, second and main culture, the production medium is used. Fermentation media consist of glycerol, corn steep liquor and ammonium sulphate. In addition, calcium carbonate is employed in the production medium. The pH is left neutral and incubated for 72 h at 28°C with high aeration. The yield of lysine is as high as 75 g/litre lysine. Earlier it was produced by two stage processes using two different organisms. But now-a-days single stage process using mutants of *Conynebacterium, Brevibacterium,* etc. are grown on a synthetic medium containing glucose, an inorganic nitrogen source and a small concentration of homoserine, methionine, etc. in addition to a small concentration of biotin.

(*b*) Recovery: Although, lysine is bound in the cell but due to mutations in the producing strains, it is secreted out and recovered from the medium.

(*c*) Strain Improvement for Lysine Production: The biosynthesis of lysine is depicted in Fig. 21.32. Certain analogues of end products act as false feedback effectors and inhibit microbial growth. To resist the feedback inhibition of aspartate kinase by lysine and threonine, mutants resistant to 5- (2- aminoethyl)-L- cystine (AEC) a lysine analogue, has been reported in *Brevibacterium flavum.* In such cases, lysine production may be increased to the extent of 57mg/ml. The strain does not accumulate threonine in the medium because normal feedback inhibition of homoserine dehydrogenase still operates. Some other amino acids such as arginine, histidine, proline, valine, leucine, citrulline, etc. are also improved by isolating analog-resistant mutants (regulatory mutants).

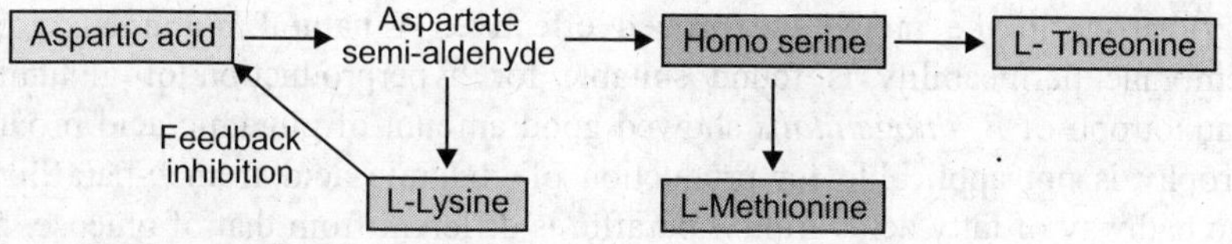

Fig. 21.32. Biosynthesis of L-lysine in *Corynebacterium glutamicum.*

2. L-Glutamic Acid

In 1957, S. Kinoshita, S. Udaka and M. Shimono observed that L-glutamic acid ($C_5H_9O_4N$) is produced by various microorganisms, including *Micrococcus glutamicus* (syn: *Corynebacterium glutamicum*). Although, this organism is not an auxotroph, yet it requires biotin for growth in the medium. This amino acid is present intracellular but leaks out in the medium subjected to optimum biotin level available for fermentative production of L-glutamic acid. If excess of biotin is present in the medium, there is heavy cell growth, but there starts lactic acid production. Some microorganisms, such as *Corynebacterium herculis, C. lilium, Arthrobacter globiformis, Microbacterium salicinovorum, Brevibacterium divaricatum, B. aminogenes, B. flavum, Bacillus megaterium* are other glutamic acid producing species.

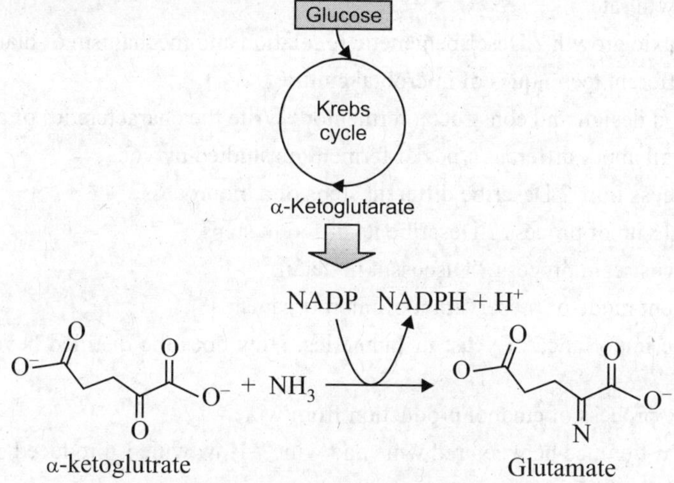

Fig. 21.33. Biosynthesis of L-glutamic acid.

Modification in the permeability of a microorganism is provided by the glutamic acid fermentation. The permeability of *Corynebacterium glutamicum* may be controlled by the composition of the culture medium if supplemented with biotin (Fig. 21.33).

(*a*) Fermentation: The medium contains glucose (121 g), ammonium acetate (5 g), molasses (6 g), potassium hydrogen phosphate (1.2 g), potassium sulphate (6 μg), manganese sulphate (6 μg), antifoam agent (0.1 ml) in one litre distilled water. The size of the inoculum remains 6%. The fermentative microorganism is *Brevibacterium divaricatum* (NRRL B-231). The incubation was carried out for 16 hours at 35°C. In the beginning of the fermentations olive oil is added at a concentration of 0.65 ml per litre. The pH is set at 8.5 with ammonia and is automatically maintained at 7.8 during fermentation. After growth of the culture (about 14 h), the temperature is increased from 32-33°C to 38°C. The glucose feeding is done until the fermentation is completed.

(*b*) Recovery: Glutamic acid content is analysed at one hour interval. The fermentation is stopped after 30-35 hours with a yield of 100 g per litre. If molasses from starch saccharification is substituted for glucose, the glutamic acid yield is 94 g per litre after 36 hours.

(*c*) Strain Improvement for Glutamic Acid Production: Sometimes elimination of the permeability barrier in the membrane proved effective. A natural biotin auxotroph that has altered membrane permeability is found suitable for hyperproduction of glutamic acid. The oleic acid auxotroph of *B. thiogenitalis* showed good amount of glutamic acid production. Oleic acid auxotrophy is not applicable for production of glutamic acid from *n*-paraffin because the biosynthetic pathway of fatty acids from *n*-paraffin is different from that of glucose. Mutants with altered permeability in the membrane to glutamic acid led to the success in obtaining a glycerol auxotroph from *Corynebacterium alkanolyticum.* Protease production by kabacidin resistant strains of *Fusarium* sp. is another example of participation of permeability barrier in production.

(*d*) Uses: It is used as a flavour enhancer in the form of monosodium glutamate. It also helps in meat tenderisation.

QUESTIONS

1. What is difference between an industrial fermenter and Erlenmeyer flask or culture vessel ?

2. What are the parameters of microbial growth ? How is it measured ?

3. Write different steps of microbial growth in batch culture, and microbial growth kinetics and specific growth rate.

4. What is diauxic growth ? Describe genetic regulation and mechanism of diauxic growth.

5. Describe different techniques of microbial culture.

6. How will you design and construct a fermentor? Write the characterstics of a fermentor.

7. Write in detail about different types of fermentors studied by you.

8. What is process flow ? Describe different steps of a bioprocess .

9. What is Scale-up of process ? Describe its different steps.

10. What is downstream process ? Discuss it in detail.

11. Write different mode of preservation of microorganisms.

12. Describe the importance of yeast in Industries. How does the distilled beverage differ from wine?

13. Describe the process of ethanol production from wine.

14. What is the difference between red wine and wine ? How wine is produced commercially ?

15. What is brewing ? How beer is produced?

16. Write with suitable diagrams about organic acid production.

17. What are antibiotics ? How is penicillin produced industrially ?

18. What are amino acids ? How are they produced industrially ?

19. Describe in brief about the lysine or glutamic acid production by using microbes.

20. Define microbial metabolites. How are these regulated? Discuss overproduction of antibiotics by feed-back resistant mechanism. (*Gurukul Kangri Univ., 2007*)

21. What is batch culture ? Discuss microbial growth kinetics in batch. (*Gurukul Kangri Univ., 2007*)

22. Describe production of penicillin in industry. (*Manonmanian Sundernar Univ., 2006*)

23. If a broth of *Brevibacterium flavum* contains 108 cells/ml at zero hour and after 3 h incubation its number raised to 1012 cells/ml. Calculate its specific growth rate and generation time? (**Ans:**1.15/h and 84 minutes)

24. In a 1000 litre fermentor, 1.0 kg of antibiotics of molecular weight of 17,000 D manufactured after harvesting of the cells. If each cell contains 2800 molecules of such antibiotic, calculate the total number of bacterial cells required ? (**Ans :** 99.2 × 1024)

25. Write short notes on the following :

(*i*) Synchronous growth,
(*ii*) Diauxic growth,
(*iii*) Chemostat,
(*iv*) Turbidostat,
(v) Impellers,
(*vi*) Use of computer in fermentor,
(*vii*) Stoichiometry,
(*viii*) Mass balance,
(*ix*) Strain improvement,
(*x*) Downstream processing,
(*xi*) Vinegar production,
(*xii*) Citric acid production,
(*xiii*) Koji fermentation,
(*xiv*) Idiophase and trophophase,
(*xv*) Lyophilisation,
(*xvi*) Baker's yeast.

BIBLIOGRAPHY

Buchta K. 1983. Lactic acid, In *Biotechnology* 3 (Eds. H.J. and G. Reed), Verlag Chemic, Weinheim, pp. 40–417.

Cruegar W. and Cruegar A.1990. *Biotechnology: A Textbook of Industrial Microbiology*. (Ed. T.D. Brock and M. Sunderland), Sinauer Associated.

Dubey R.C. and Maheshwari D.k. 2012. *A Textbook of Microbiology*. S. Chand & Co. New Delhi-5, India.

Florent J. 1986. Vitamins, In *Biotechnology* 4, (Eds. Relm, H.J. and G. Reed), Verlag chemic, pp. 115-158.

Glazer A.N and Nikaido H. 1995. *Microbial Biotechnology. Fundamentals of Applied* Microbiology W.H. Freeman, New York.

Godfrey T. and West S. (Eds). 1996. *Industrial Enzymology-The application of enzymes in Industry.* 2nd eds. Stockhom, New York.

Howard J. R. 1989. Fluidized bed technology: principles and applications. New York, NY: Adam Higler.

Lockwood L.B. 1979. Production of organic acids by fermentation, pp. 355-387. (Eds. Peppler, H.J. and D. Perlman. *Microbial Technology*. Vol. 1 Academic Press, New York.

Perlman D. 1978. Vitamins. In *Economic Microbiology* : *Primary products of metabolism*, Vol-2 (Ed. A.H. Rose), Academic Press, London, pp. 303–326.

Primrose S.B. 1991. *Molecular Biotechnology*. 2nd Ed. Oxford: Blackwell Scientific publishers.

Sikyata B. 1995. *Techniques in Applied Microbiology - Progress in Industrial Microbiology.* Vol. 31. Elsevir Science, Amsterdam.

Vining L.C. and Stuttard C. 1994. *Genetics and biochemistry of Antibiotic production*. Butterworth. Heinmann, Stoneham, MA.

22 CHAPTER

Environmental Biotechnology

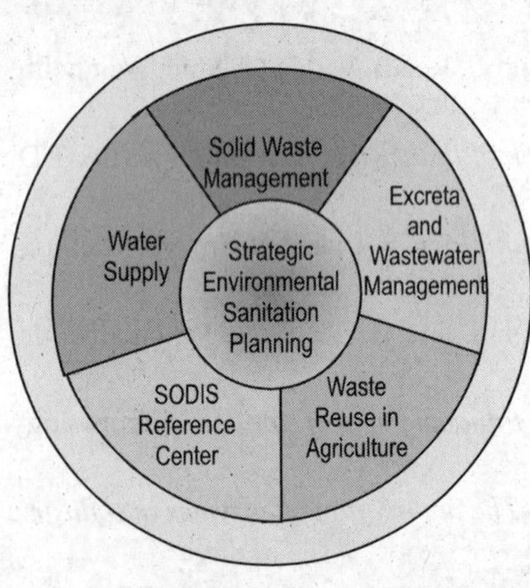

A. Biosensors in environmental monitoring
B. Bioremediation
C. Composting
D. Vermicomposting (vermitechnology)
E. Biogas (*gobar* gas) production technology
F. Biomining (=Bioleaching)

About 100 million people are added to the world population each year. Therefore, more and more industries, food, health care, medicines, vehicles, etc. are required to meet the demand of people. Consequently, a large volume of organic matter as wastes is generated from agricultural activities, dairy farms and animal shelters which usually is dumped in a corner where it putrefies emanating a foul smell. In addition, the other forms of wastes are solids, liquids and gases released from multifarious ways into the environment. The solid waste as organic materials is valuable resource and can be utilised by composting it into a value added end product called manure. The objective to compost organic wastes should not be for the disposal of solid organic wastes but to produce superior quality manure to feed nutrient-poor soil.

According to Third World Network reports more than one billion pounds (450 million kilograms) of toxins are released into atmosphere and hydrosphere. The contaminants posing environmental troubles causes discrepancy in natural world is of worldwide anxiety. The environmentalists around the world are trying to prevail over it by numerous ways.

Oil spill

Moreover, for enhanced crop production farmers have used chemical fertilizers on free hands for the last forty years. Chemical fertilizers have been responsible for deterioration of soil friability, and destruction of beneficial soil life such as earthworms, bacteria and microarthropods. Moreover, plants grown on artificial fertilizers have lower nutrient value than their organically grown counterparts. The most harmful impact of chemical fertilizers arises their leaching into the subsoil that permanently contaminated the soil and the groundwater. Moreover, they also enter the food chain and harm organisms including man. This upsets ecological balance and has a negative impact on the beneficial birds and insects.

Composting imitates nature's way of rebuilding soil by encouraging the decomposition of organic substances, but it does so more rapidly because heat, microbes, and then the worms combine to speed up the process. Compost prepares the soil for plant roots to penetrate into it. Besides, it is the cheapest, most practical and environment-friendly method of disposing organic wastes.

Besides adequate nutrients, plants need helpful microorganisms like bacteria, fungi, actinomycetes and protozoans, and other soil symbionts. These provide a good structure to the soil to allow the root zone to breathe and absorb moisture.

To speed up the rate of decomposition and rendering the biodegraded products involving potential organisms present in the environment, the recently developed new methods can be exploited. This is possible by involving biotechnology. However, it is probable that the modern biotechnology based on the sophisticated art of genetic engineering could solve all these problems. The environmental biotechnology employs the application of genetic engineering to improve the efficiency and cost which are central to the future widespread exploitation of microorganisms to reduce the environmental burden of toxic substances. It is hoped that in future the application of microorganisms coupled with genetic engineering techniques will make a major contribution tó improve the quality of our environment. However, is there risk associated with the release of genetically engineered microorganisms into the environment ?

This chapter deals with the application of biosensors for monitoring the environmental pollution, uses of microorganisms in environmental clean up with special reference to bioremediation types, processes and methods with several examples (*e.g.* hydrocarbons, heavy metals, dyes, xenobiotics, etc), utilisation of sewage and agro-wastes, composting, vermicomposting (vermitechnology), bioleaching and biogas.

A. BIOSENSORS IN ENVIRONMENTAL MONITORING

The substances or energy liable to cause hazards to human health, harm to living resources and ecological damage, are gradually increasing into the environment. The changes occurring in the environment caused by either natural processes or human activities involve many large parameters. These changes range from a slow increase in global temperature over the years to the rapid accumulation of heavy metals and xenobiotics. Some of the changes may require careful monitoring over the long periods. Some of the elements and compounds that accumulate in the environment are present in such low concentrations that they may hardly be detected.

The main groups of environmental pollutants which can contaminate land, water, and air are: inorganic compounds (such as metals), organic compounds (such as sewage, petrochemicals), synthetic xenobiotic compounds (like phenol), and microorganisms (such as pathogens) and genetically engineered organisms (GMOs). The environmental pollutants can originate from contaminated sites of defunct industries such as gasworks, from existing domestic and industrial sources (e.g. sewage) and metals from industries like electroplating, from accidents, and illegal dumping.

Hence, there is a need to monitor the environmental pollution by various methods and abate the pollutant accordingly. One of such methods is the application of biosensors for monitoring the environmental pollution. Biosensors are currently being considered for detection of environmental pollutants such as phenols, genotoxins, and pesticides like organophosphates, 2,4-D, etc. Biosensor technology lends itself to fast, economical and continuous monitoring capabilities; therefore the development of these systems to complement classical analytical measurements is expected to result in a substantial cost benefit, especially when sample turnaround time and cost per analysis are important issues.

A detailed account of different types of biosensors has already been given in Chapter 15: *Enzyme Technology.*

B. BIOREMEDIATION

Industrialisation has provided mankind a comfortable life and may be quoted as a characteristic of the present age. A serious aspect which had earlier been overlooked but recently attracted much attention in the course of modernisation which was increased pollution levels in the environment. The hazardous pollutants from most of the industries as effluents are frequently released in the surroundings.

It is highly desirable to either stop the continuous use of these noxious xenobiotic chemicals or to find some way to detoxify them as far as environment and health hazards are concerned. But the application of chemical fertilizers and pesticides can not be stopped immediately because protection of crops from damage by insect pests and pathogen is must to meet the demand of food for increasing population. The other alternatives left are: (*i*) their partial replacement by biological control agents (biopesticides) in the integrated pest management (IPM) system, and (*ii*) their detoxification through degradation by biological means *i.e.* the process known as bioremediation. Technologies using microorganisms with extensive biodegradative capacities have been developed for use in both pollution prevention and site remediation. Bioremediation as defined by the American Academy of Microbiology is '*the use of living organisms to reduce or eliminate environmental hazards resulting from accumulation of toxic chemicals or other hazardous wastes*'. It is evident in that humankind has practiced composting, sewage treatment and fermentation since the beginning of human civilization.

In general the approaches to bioremediation are: *intrinsic bioremediation* (by applying naturally occurring microbial population), *biostimulation* (stimulating viable native microbial population through application of nutrients and aeration), *bioaugumentation* (artificial introduction of viable population of microorganisms), and *phytoremediation* (use of pollution tolerant plants) (Fig. 22.1). The microorganisms degrade the complex form of pollutants and use as a source of carbon and energy. Also they accumulate some of the metals inside their cells through the process of bioaccumulation (using live cells) or biosorption (use of dead microbial biomass).

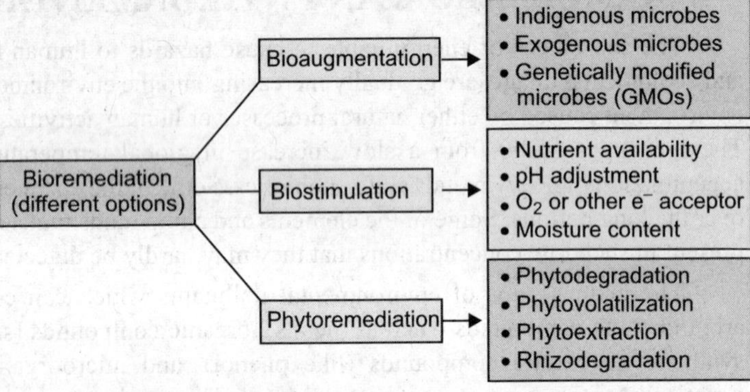

Fig. 22.1. Different approaches for bioremediation.

Bioremediation process has several advantages and disadvantages. The advantages includes: minimal exposure of the contaminant to on site workers, long term protection of public health, the cheapest of all methods of pollutant removal. The process can be done on site with a minimum space and equipment, eliminating the need to transport of hazardous material in the natural process, transforming the pollutants instead of simply moving them from one media to another, performance of the degradation in an acceptable time frame.

The disadvantages are: poor management, inability to estimate the length of time that it takes which may vary from site to sites, taking a few months to few years, unbiodegrability of all organic compounds.

I. TYPES OF BIOREMEDIATION

Bioremediation has been grouped into two types, *in-situ* bioremediation and *ex-situ* bioremediation as described in detail as below :

1. *In- Situ* Bioremediation

In situ techniques do not require excavation of the contaminated soils; therefore, it is less expensive, creates less dust, and causes less release of contaminants than *ex situ* techniques. Also, it is possible to treat a large volume of soil at once. *In situ* techniques may be slower than *ex situ* techniques; it may be difficult to manage, and is most effective at sites with permeable soil. The goal of aerobic *in situ* bioremediation is to supply oxygen and nutrients to the microorganisms present in the soil. Aerobic *in situ* techniques supply oxygen to the organisms that degrade the contaminants. Such methods are *biostimulation* and *bioaugmentation.* Oxygen can be provided by pumping air into the soil above the water table (bioventing) or by delivering oxygen in liquid form as hydrogen peroxide.

In situ bioremediation may not work well in clays or in highly layered subsurface environments because oxygen cannot be evenly distributed throughout the treatment area. *In situ* remediation often requires years to reach cleanup goals, depending mainly on specific biodegradable contaminants. Less time is required for easily degraded contaminants (Environmental Protection Agency, 1996). There are two types of *in situ* bioremediation: *intrinsic bioremediation* and *engineered in situ bioremediation.*

(*a*) Intrinsic Bioremediation: Intrinsic bioremediation refers to conversion of environmental pollutants into the harmless forms through the innate capabilities of naturally occurring microbial population. However, there is increasing interest on intrinsic bioremediation for control of all or some of the contamination at waste sites. The intrinsic *i.e.* inherent capacity of microorganisms to metabolise the contaminants should be tested at laboratory and field levels before use for intrinsic bioremediation. Through site monitoring programmes progress of intrinsic bioremediation should be recorded time to time. The conditions of site that favour intrinsic bioremediation are ground water flow throughout the year, carbonate minerals to buffer acidity produced during biodegradation, supply of electron acceptors and nutrients for microbial growth and absence of toxic compounds. The other environmental factors such as pH, concentration, temperature and nutrient availability determine whether or not biotransformation takes place. Bioremediation of waste mixtures containing metals such as Hg, Pb, As and cyanide at toxic concentration can create problem.

The ability of surface bacteria to degrade a given mixture of pollutants in ground water is dependent on the type and concentration of compounds, electron acceptor and duration of bacteria exposed to contamination. Therefore, ability of indigenous bacteria degrading contaminants can be determined in laboratory by plate count and microcosm studies. The microbes release the enzymes that break down the contaminant into digestible products. The microbial cells consume the contaminants as food (Fig. 22.2).

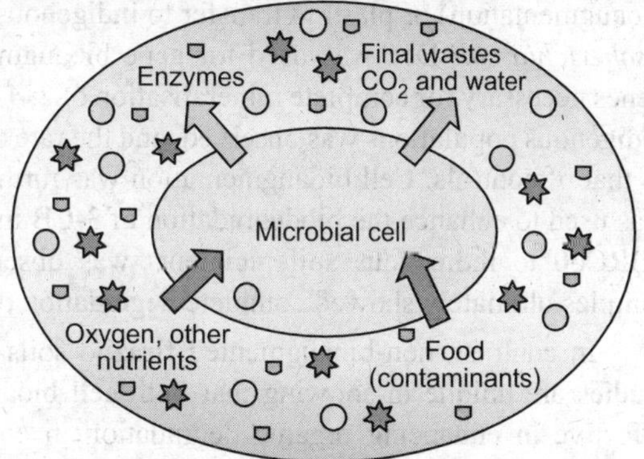

Fig. 22.2. Secretion of enzymes by microbial cell during bioremediation.

(*b*) Engineered *in situ* Bioremediation: At some places the intrinsic bioremediation is satisfactory, but it is slow process due to poorly adapted microorganisms, limited ability of electron acceptor and nutrients, cold temperature and high concentration of contaminants. When site conditions are not suitable, bioremediation requires construction of engineered systems to supply materials that stimulate microorganisms. Engineered *in situ* bioremediation accelerates the desired biodegradation reactions by encouraging growth of more microorganisms via optimising physico-chemical conditions. Oxygen and electron acceptors (such as NO31⁻ and SO42⁻) and nutrients (*e.g.* nitrogen and phosphorus) promote microbial growth in surface.

When contamination is deeper, amended water is injected through wells. But in some *in situ* bioremediation systems both extraction and injection wells are used in combination to control the flow of contaminated ground water combined with above-ground bioreactor treatment and subsequent re-injection of nutrients spiked effluent are done.

Lagoon Treatment: Lagoon treatment is an *in situ* bioremediation process that degrades a mixture of organic and inorganic hazardous substances. It requires a combined treatment both aerobic and anaerobic biodegradation methods depending on characteristics of contaminant. For wastes containing volatile chlorinated hydrocarbons, anaerobic processes are applied followed by an aerobic phase. Solids distribution and slurry mixing techniques including air sparging, pumping, mechanical and dredging must be used in the lagoon treatment in order to maintain high bioremediation rates. Lagoon treatment can remediate galvanising wastes, phenols, pickling acids, polychlorinated biphenyl (PCB), heavy metals, polycyclic aromatic hydrocarbons (PAHs), and benzene, toluene, ethylbenzene, and xylene (BTEX). This method has been effectively used to treat hazardous substances discharged into lagoons by industrial or other sources. Lagoon treatment requires less cost than alternate technologies.

(*i*) Bioaugmentation: Bioaugmentation deals with introduction of specially selected or genetically engineered strains of microbes to a contaminated site. If site assessments reveal that species of indigenous microorganisms are unable to degrade target contaminants, exogenous microorganisms with the required biochemical capabilities can be introduced to successfully degrade specific waste compounds. Bioaugmentation is a rational rearrangement of microbial richness leading to dominance of microbial group(s) with specific catabolic traits which are necessary for cleaning up the pollutants. Bioaugmentation enhances the performance of indigenous bacteria through the addition of bacterial strains with specific activities for several folds over a relatively short time. Bioagumentation studies have focused on exogenous introduction of efficient pollutant-degrading strain(s) or bacterial consortium to contaminated sites.

Few studies have been carried out to evaluate the performance of *in-situ* biodegradation. The enhanced degradation takes place due to survival of the introduced microorganism (cell bioaugmentation) or plasmid transfer to indigenous microbial populations (gene bioaugmentation). *Escherichia coli* D11 was used for gene bioaugmentation in Madera soil. It lacks chromosomal genes necessary for complete mineralisation of 2, 4-D. Significant gene transfer of the plasmid to the indigenous populations was observed, and the rate of 2, 4-D degradation was enhanced as compared to that of controls. Cell bioaugmentation was further demonstrated when *Comamonas testosteroni* was used to enhance the biodegradation of 3-CB in Madera soil. In this case no transfer of plasmid pBRC60 to indigenous soil recipients was observed. For the Madera soil, non-bioaugmented samples ultimately showed complete degradation of 2, 4-D.

In contrast, non-bioaugmented Brazito soils showed incomplete 2, 4-D degradation. These studies are unique in showing that both cell bioaugmentation and gene bioaugmentation can be effective in enhancing organic degradation in co-contaminated soils. Bioaugmentation strategy depends on the degree of contamination and the time frame available for remediation. Besides, a few other variants of bioagumentation have also been developed and implemented successfully for remediation purpose.

Co-bioaugmentation is one such variant wherein the process is rendered effectively by exogenous introduction of multiple microbial strains with different metabolic potentials. Bioaugmentation is a straightforward approach yet the use of this technology has not become common due the limited success with isolation of efficient pollutant degrading microorganisms. Further the bioaugmentation process needs to address several other issues such as optimisation of the process before its application.

(*ii*) Biodegradation: Degradation of organic contaminants in soil and/or groundwater by indigenous or transplanted/acclimated microorganisms such as bacteria and fungi is called biodegradation. Organic contaminants are converted into carbon dioxide, water, and microbial cell mass under aerobic conditions (Fig. 22.3).

Fig. 22.3. Aerobic biodegradation of oil in soil (EPA, 1996).

Biodegradation can be used to degrade environmental pollutants such as pentachlorophenol (PCP), lindane (hexachlorocyclohexane), dichlorodiphenyl trichloroethane (DDT), and 2, 4-dichlorophenol. Fungi, such as *Fusarium oxysporium* and *Phanerochaete chyrysosoporium* as well as bacteria such as *Pseudomonas cepacia* and *Pseudomonas putida* are used to degrade these pollutants. Microorganisms eat oil or other organic contaminants. They digest the oil and convert into CO_2 and water resulting in release of CO_2. Aerobic biodegradation of oil in soil is given in Fig. 22.4.

Oxidative chanelling pathway

Oxygenolytic cleavage

2 hydroxy muconic semialdehyde

Dioxygenase O_2

Meta cleavage pathway

Ortho cleavage pathway

cis, cis muconic acid

hydroxylated ring

Monoxygenase O_2

Rearrangement

arenc oxide

Epoxide hydrolase pathway

trans dihydrodiol

Fig. 22.4. Aerobic ring cleavage.

But under anaerobic conditions, methane, small quantities of carbon dioxide and hydrogen, and occasionally intermediate species are produced which may be less or more hazardous than the metabolised compound (Fig. 22.5).

The biodegradation of high molecular weight polycyclic aromatic hydrocarbon (PAH) aromatic compounds have been well described. The initial step in the aerobic catabolism of a PAH molecule occurs via oxidation of a PAH by a multi-component enzyme system. Thereafter, intermediates may proceed through either an *ortho* cleavage type pathway or a *meta* cleavage pathway, leading to central intermediates which are further converted to intermediates of tricarboxylic acid cycle, and finally ring is opened and end products are produced.

Carboxyl chanelling pathway

Hydrolytic reductive cleavage

CO_2

+ Energy

COOH

SCoA

e^-

H_2O

COOH

Unknown enzyme single or multiple steps?

Hypothetical analogue to monoaromatic pathways

Fig. 22.5. Anaerobic ring cleavage.

***(iii)* Biostimulation:** Biostimulation refers to the addition of oxygen and/or inorganic nutrients to indigenous microbial populations in soils and groundwater. *In situ* or *ex situ* methods can be employed to stimulate biodegradation of contaminants. Biostimulation involves the modification of the environment to stimulate existing bacteria capable of bioremediation. This can be done by addition of various forms of rate-limiting nutrients such as phosphorus, nitrogen, oxygen or carbon (*e.g.* in the form of molasses) and electron acceptors/donors (acetate, nitrate, sulphate, glutamate, etc.) and gaseous formulations to contaminated environment (Fig. 22.6). Alternatively, pollutant- removal rates have also been stimulated by generating an optimal balance of physical factors such as aeration, temperature and buffering of environmental pH by altering the redox state and electrokinetics state of contaminated samples. Additives are usually added to the subsurface through injection wells; although injection well technology for biostimulation purposes is still emerging. Removal of the contaminated material is also an option. Biostimulation can be enhanced by bioaugmentation. This process is referred to as bioremediation and is an EPA-approved method for reversing the presence of oil or gas spills. It is more effective if used in combination with bioaugmentation methods. To evaluate the performance of the biostimulation methods in comparison of bioaugmentation and natural attenuation, the kinetic efficiency of biostimulation was found to be relatively slow as compared to bioaugmentation process.

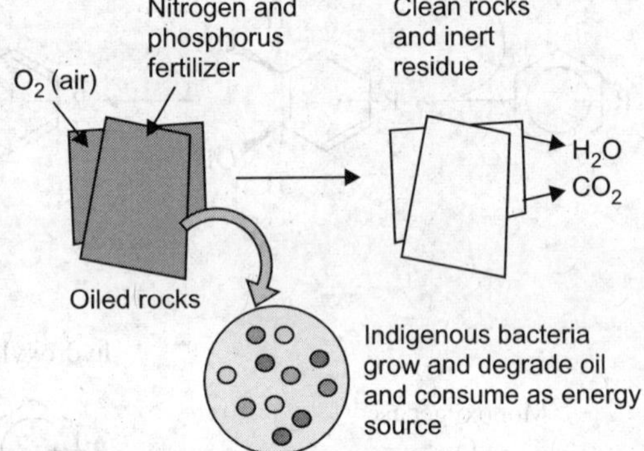

Fig.22.6. Biostimulation of indigenous bacteria by adding nitrogen and phosphorus fertilizers that speed up biodegradation of complex chemicals present in oil.

***(iv)* Bioventing:** Bioventing is a process of motivating the natural *in situ* biodegradation of contaminants in soil by providing air or oxygen to existing soil microorganisms. Oxygen is most commonly supplied through direct air injection into residual contamination in soil. In addition to degradation of adsorbed fuel residuals, the volatile compounds are biodegraded as vapours move slowly through biologically active soil. Bioventing is applicable to any chemical that can be aerobically biodegraded. It is successfully used to remediate soils contaminated by petroleum hydrocarbons, non-chlorinated solvents, some pesticides, wood preservatives, and other organic chemicals.

***(v)* Biosorption:** Microorganisms remove heavy metals and radionuclides present outside their cell through either bioaccumulation or biosorption. Retention and concentration of a substance by an organism is called 'bioaccumulation', where metals move from outside of microbial cell into cytoplasm through their cell membrane. Inside the cell the metal ions are sequestered and immobilosed. But in biosorption phenomenon, the positively charged metal ions are sequestered first through adsorption to the negatively charged groups present on cell surfaces such as capsules. Usually the binding sites present on the cell surfaces of microorganisms are: phosphate residues, carboxyl residues, sulfahydro- groups (SH- group), or hydroxyl group. Depending on the charge density on the cell surface, the amount of metal biosorbed to bacterial cells often exceeds. The charged functional groups serve as nucleation sites for deposition of various metal bearing precipitates.

Biosorption is a passive metabolism-independent physico-chemical interaction between heavy metal ions and microbial surface. It is defined as '*a non-directed physico-chemical interaction that occurs between microbial cells and metal/radionuclides.* Most biosorption phenomena are combination of processes such as electrostatic interactions, ion exchange complexation, ionic band formation, precipitation, nucleation, etc. (Fig. 22.7).

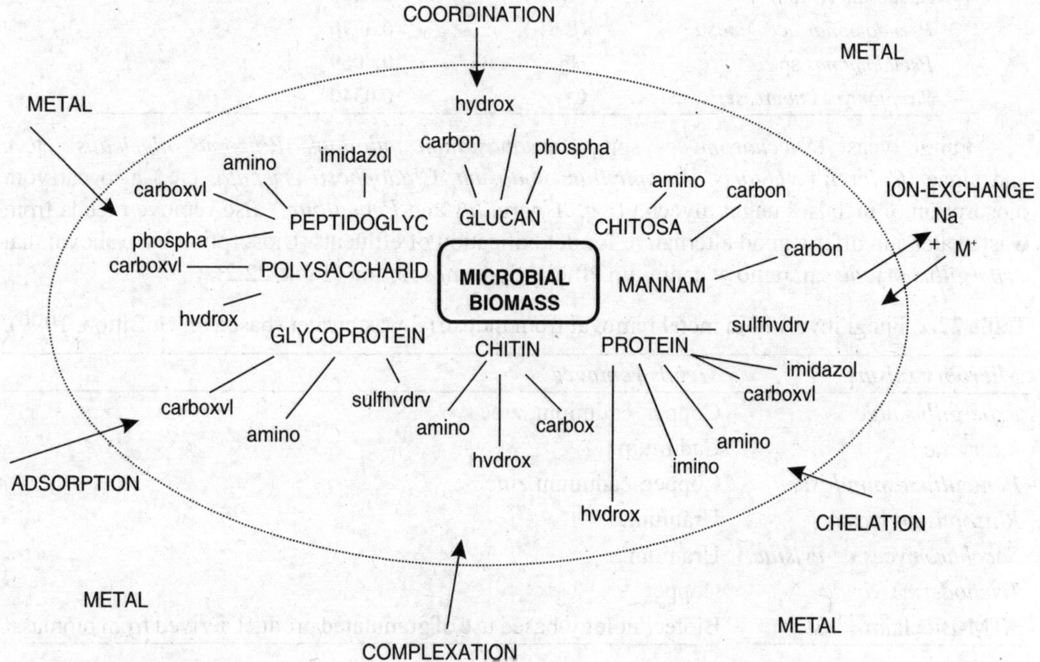

Fig.22.7. Microbial biomass showing various ligands involved in metal biosorption (Panikar *et al.,* 2003).

These interactions occur due to complexity of microbial surfaces and chemical/physical properties of metal ions. The active state of cells is not prerequisite for biosorption. The process can occur even with inactivated/dead cells. Biosorption offers several advantages, for examples,

- Process is not governed by physiological constraints of microbial cells. There is no need of costly nutrients for growth and aseptic operation of cells. It operates at a wider range of conditions such as pH, temperature and metal concentrations.
- Inactivated biomass works as an ion exchanger. Therefore, the process is rapid, for example *Streptomyces nouressi* carries out biosorption of chromium at a high speed (0.0340 mmol/g). Metals can be desorbed readily from the biosorbent and recovered.

***(i)* Organisms involved in biosorption:** There are several microorganisms such as bacteria (*Arthrobacter viscosus, Pseudomonas syryngae, P. aeruginosa, P. putida, Bacillus subtilis, E. coli, Streptomyces nouresei, S. pimprina,* etc.) whose sorption capacity for various heavy metals varies (Table 22.1).

Table 22.1. Sorption capacity of bacteria for various heavy metals (*source:* many research papers).

Bacteria	*Metals*	*Sorption capacity (mmol/g)*
Bacillus amyloliquifaciens	Cu	0.0437
B. amyloliquifaciens	Hg	0.0437
B. amyloliquifaciens	Pb	0.0437
B. subtilis	Cu	2.9900
B. subtilis	Mn	0.8010
B. subtilis (cell wall)	Au	0.4010
B. subtilis (walls)	Cu	0.5300
B. subtilis (walls)	Ni	0.1070
B. subtilis (walls)	Pb	0.5430
Pseudomonas aeruginosa	(LaIII)	0.3970
Pseudomonas sp.	Hg	0.0089
Streptomyces nouressi	Cr	0.0340

Fungi, yeast (*Saccharomyces* spp., *Aureobasidium pullulans, Rhizopus nigricans,* etc.), and algae (*Chlorella valgaris, Ascophyllum nodosum, Cladophora crispata,* etc.) also carryout biosorption of metals. Fungal mycelia (*e.g. Aspergillus* and *Penicillium*) also remove metals from wastewater and offer a good alternative for detoxification of effluents. Biosorption have shown that *Aspergillus oryzae* can remove cadmium efficiently from solution (Table 22.2).

Table 22.2. Fungi involved in metal removal from industrial wastewater (based on G. Bitton, 1999).

Microorganisms	*Metals removed*
Aspergillus niger	Copper, cadmium, zinc
A. oryzae	Cadmium
Penicillium spinulosum	Copper, cadmium zinc
Rhizopus arrhizus	Uranium
Saccharomyces cerevisiae	Uranium
Trichoderma viride	Copper
ATM-Bioclaim^{TM}	Biotechnology-based use of granulated product derived from biomass

Biosorption of some organisms sorbing various metals at different conditions (such as pH, temperature (T), initial metal concentration (Co), biomass concentration (B) and value of metal uptake (Q) are given in Table 22.3.

Table 22.3. Biosorption of metals by some microorganisms under different conditions.

Organism and metals		*Operating conditions*				*Q (mg/g)*
		pH	*T(°C)*	*Co (mg/ml)*	*B (g/1)*	
Arthrobacter sp.	Cd	7	20	1	0.6	1.4
	Cu	3.5-6	30	180	0.4	1.48
	Pb	5-5.5	30	250	1.4	130
Streptomyces nouresei	Cd	6	30	1-110	3.5	3.4
	Cu	5.5	30	0.6-65	3.5	3.4
	Pb	6.1	30	2-207	3.5	36.5
	Ag	6	30	1-100	3.5	38.6
	Zn	5.8	30	0.6-65	3.5	1.6

Rhizopus arrhizus	Cd	3.5	26	10-400	NA	25
	Cr	1-2	25	25-400	NA	4.5
	Cu	5.5	25	0.6-25	NA	9.5
	Pb	5	25	150	1	6.5
	Zn	6-7	N.A.	10-600	3	13.5
Penicillium digitatum						
	Cd	5.5	25	10-50	6.5	3.5
	Cu	5.5	25	10-50	6.5	3
	Pb	5.5	25	10-50	6.5	3

Source : K.M. Paknikar *et al.* (2003); NA- not available.

***(ii)* Factor affecting biosorption of metals:** Biosorption depends on various factors like type of biomass, metal chemistry, temperature, pH, biomass concentration, initial metal concentration, composition of biomass, etc.

***(iii)* Metal bisorption technologies:** Several technologies for metal removal have been commercialised and employed as briefly given below:

- **ATMBIOCLAIMTM process:** The Advance Mineral Technology (ATM) Inc. (U.S.A.) developed a wastewater treatment process with *Bacillus* sp. (immoblised in beads) pretreated with caustic solution. It is specific for metal cations in the order: $Cu^{2+} > Zn^{2+} > Cd^{2+} = Ni^{2+} > Pb^{2+}$.
- **AlgaSORBTM process:** Biorecovery systems, Inc. (U.S.A.) developed this proprietary based-material which consists of several types of living and nonliving algae. Algal cultures are immobilised in silica gel in the form of beads and then desorption of metal takes place.
- **BIO-FIXTM process:** The Bureau of Mines (U.S.A.) developed this process that consists of biomass immobilised in polysulfone. It consists of thermally killed biomass of *Sphagnum* peat moss, algae, yeast, bacteria and /or aquatic flora. The beads are suitable for practical application in stirred tank reactor, fixed and fluidized-bed columns.

2. *Ex- Situ* Bioremediation

Ex situ bioremediation refers to abating the pollution outside the source of wastes. Methods of bioremediation are faster, easier to control and used to treat a wider range of contaminants and soil types than the *in situ* techniques. *Ex-situ* techniques require excavation and treatment of the contaminated soil before the start of actual bioremediation. The techniques include *slurry-phase bioremediation* and *solid phase bioremediation.*

The *ex-situ* bioremediation is carried out aboveground by physically extracting the impacted medium. It is commonly applied to dissolved-phase contamination via pumping and treatment with aboveground bioreactors. Soils are treated aboveground through land-farming, biopiling and composting. The primary advantage to these e*x-situ* approaches is the degree of control that can be exerted over the processes being used to manipulate the system. But the primary disadvantage is the expense and disruption associated with removal, treatment and disposal or replacement of the impacted medium. *Ex- situ* bioremediation at contaminated sites involves various treatments.

(*a*) Slurry-phase Bioremediation: Contaminated soil is combined with water and other additives in a large tank called bioreactor, and mixed to keep the microorganisms already present in the soil in contact with the contaminants in the soil (Fig. 22.8). Nutrients and oxygen are added, and conditions in the bioreactor are controlled to create the optimum environmental conditions for the microorganisms to degrade the contaminants. When the treatment is completed the water is removed from the solids which are disposed off or treated further if they still contain pollutants. Slurry-phase biological treatment can be a relatively rapid process as compared to other biological treatment processes particularly for contaminated clays. The success of the process depends on the specific soil and chemical properties of the contaminated material. This technology is particularly useful where rapid remediation is a high priority (EPA, 2003).

***(i)* Slurry-based Reactors:** Slurry-phase biological treatment is used to remediate a mixture of water and excavated soil. Bioreactors represent highly controlled methods of treating contaminated soils and groundwater. Because temperature, pH, nutrient levels and agitation can be controlled in constructed batch- or continuously-fed reactors, microbial activity, and thus contaminant degradation can be optimised.

The slurry contains between 5 and 40% solids by weight depending on the nature of the biological reactor. The soil is suspended in a reactor vessel and mixed with nutrients and oxygen. Microorganisms may be added depending on treatment requirements. When biodegradation is complete, the soil slurry is dewatered. Biological treatment in slurry-phase reactors has been successfully used to remediate soils, sludge and groundwater contaminated by hydrocarbons, petrochemicals, solvents, pesticides, wood preservatives, and other organic chemicals.

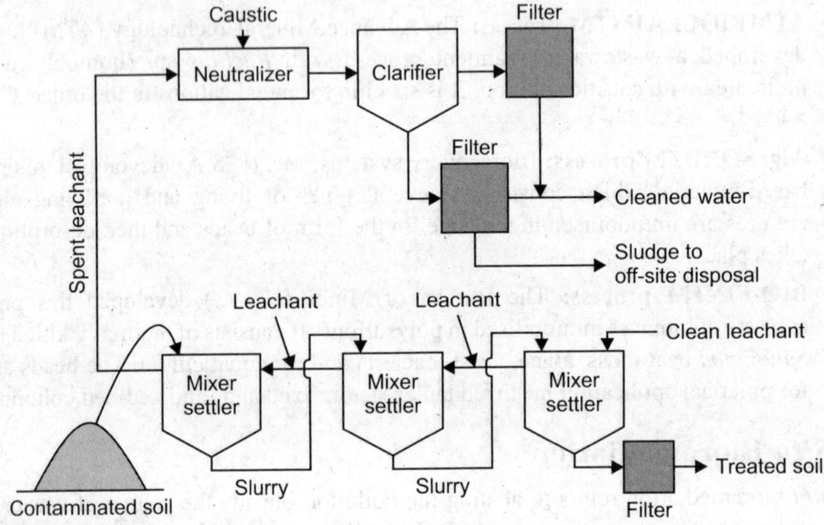

Fig. 22.8. Slurry- phase bioreactor (FRTR, 2000).

***(b)* Solid-phase Bioremediation:** In solid-phase bioremediation soils are treated in above-ground treatment areas equipped with collection systems that prevent any contaminant from escaping the treatment. Moisture, heat, nutrients or oxygen are controlled to enhance biodegradation for the application of this treatment. Solid-phase systems are relatively simple to operate and maintain, require a large amount of space, and cleanups require more time to complete than with slurry-phase processes. Solid phase soil treatment processes include *landfarming, soil biopiles,* and *composting*.

***(i)* Landfarming:** This is relatively a simple treatment method where the contaminated soils are excavated and spread on a pad with a built-in system to collect any leachate or contaminated liquids that seep out of contaminant-soaked soil (Fig. 22.9). The soils are periodically turned over to mix air into the waste. Moisture and nutrients are controlled to enhance bioremediation. If the nutrients, oxygen or temperature are not properly controlled, the length of time for bioremediation will be longer. In some cases, reduction of contaminant concentrations may be attributed more to volatilisation than biodegradation. But volatilisation losses are minimised when the process is conducted in enclosures controlling escaping volatile.

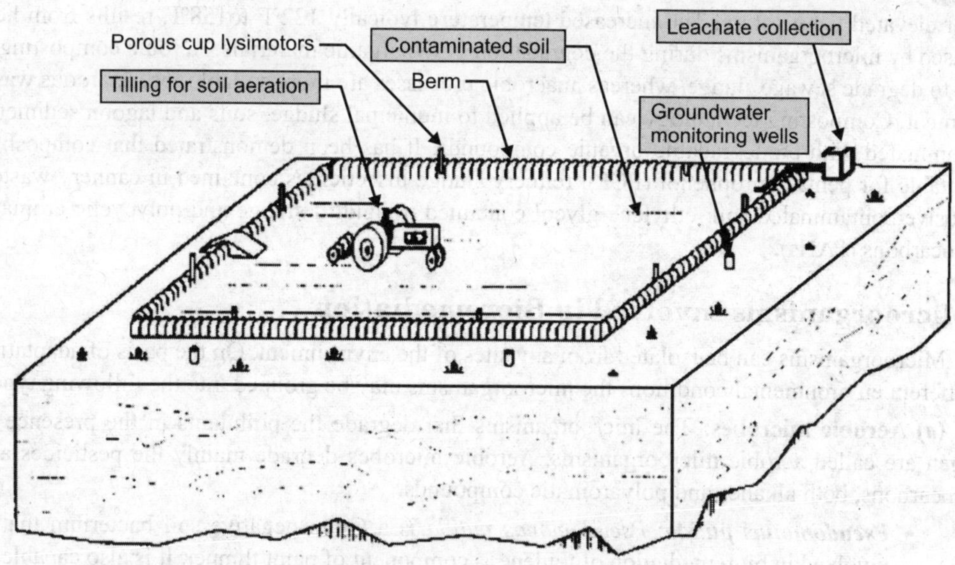

Fig. 22.9. Landfarming (FRTR, 2000).

(ii) **Soil Biopiles:** Contaminated soil is piled in heaps several meters high over an air distribution system. Aeration is provided by pulling air through the heap with the help of a vacuum pump. Moisture and nutrient levels are maintained at such levels that maximise bioremediation process. The soil heaps can be placed in enclosures. Since volatile contaminants are the part of the air stream being pulled through the pile, they can easily be controlled (EPA, 2003).

(iii) **Composting**. Composting is a controlled biological process to convert biodegradable hazardous materials to harmless and stabilised byproducts by using microorganisms under elevated temperature. The increased temperature results from the heat released by microorganisms during the degradation of the organic materials in the waste. Aerobic composting is used to degrade sewage sludge and anaerobic processes are more suitable for hazardous waste treatment. The composting of contaminated soil utilises a bulking agent such as saw dust, animal wastes to increase porosity of the media. The composting efficiency is impacted by moisture content, pH, oxygen, temperature, and C : N ratio. The composting technology can be applied to municipal sludge, soils, and lagoon sediments contaminated with biodegradable organic compounds. It has been demonstrated that composting is suitable for pentachlorophenol (PCP), refinery sludge, insecticides in cannery wastes, explosive contaminated soil, ethylene glycol in landfill sludge and polycyclic aromatic hydrocarbons (PAHs). Biodegradable wastes are mixed with a bulking agent such as straw, hay, or corn cobs to make it easier to deliver the optimum levels of air and water to the microorganisms. In 2000, C.J. Cunningham and J.C. Philip have described the three common designs of composting as given below :

- **Static pile composting:** Compost is formed into piles and aerated with blowers or vacuum pumps.
- **Mechanically agitated in vessel composting:** Compost is placed in a treatment vessel where it is mixed and aerated.
- **Windrow composting:** Compost is placed in long piles known as windrows and periodically mixed by tractors or similar equipment.

Composting has been described in detail in the preceding section.

Compost-based Reactors: A compost-based reactor is a biological approach to convert biodegradable hazardous materials to harmless and stabilised byproducts by using microorganisms under elevated temperature. The increased temperature typically 122ºF to158ºF, results from heat released by microorganisms during the degradation of the organic materials. Aerobic composting is used to degrade sewage sludge, whereas anaerobic processes are more suitable for hazardous waste treatment. Composting technology can be applied to municipal sludge, soils and lagoon sediments contaminated with biodegradable organic compounds. It has been demonstrated that composting is suitable for pentachlorophenol (PCP), refinery sludge insecticides contained in cannery wastes, explosive-contaminated soil, ethylene glycol contained in landfill sludge and polycyclic aromatic hydrocarbons (PAHs).

3. Microorganisms Involved in Bioremediation

Microorganisms can be isolated from any sites of the environment. On the basis of adaptation at different environmental conditions the microorganisms may be grouped into the following types:

(*a*) Aerobic microbes: The microorganisms that degrade the pollutants in the presence of oxygen are called aerobic microorganisms. Aerobic microbes degrade mainly the pesticides and hydrocarbons, both alkanes and polyaromatic compounds.

- ***Pseudomonas putida***: *Pseudomonas putida* is a Gram-negative soil bacterium that is involved in bioremediation of toulene, a component of paint thinner. It is also capable of degrading naphthalene, a product of petroleum refining, in contaminated soils.
- ***Dechloromonas aromatica:*** It is a soil bacterium which is capable of degrading perchlorate and aromatic compounds.
- ***Nitrosomonas europaea, Nitrobacter hamburgensis,*** and ***Paracoccus denitrificans***: Industrial bioremediation is used to clean wastewater. Most treatment systems rely on microbial activity to remove unwanted mineral nitrogen compounds (*i.e.* ammonia, nitrite, and nitrate). The removal of nitrogen is a two stage process that involves nitrification and denitrification. During nitrification, ammonium is oxidised to nitrite by organisms like *Nitrosomonas europaea*. Then, nitrite is further oxidised by microbes like *Nitrobacter hamburgensis*. In anaerobic conditions, nitrate produced during ammonium oxidation is used as a terminal electron acceptor by microbes like *Paracoccus denitrificans* resulting is gas. Through these processes two pollutants, *i.e.* ammonium and nitrate production of responsible for eutrophication in natural waters are remediated.
- ***Methylibium petroleiphilum:*** *Methylibium petroleiphilum* (formally known as PM1 strain) is capable of methyl tert-butyl ether (MTBE) bioremediation. PM1 degrades MTBE by using the contaminant as the sole carbon and energy source.
- ***Deinococcus radiodurans:*** *D. radiodurans* is a radiation-resistant extremophilic bacterium which is genetically engineered for the bioremediation of solvents and heavy metals. An engineered stain of *D. radiodurans* has been shown to degrade ionic mercury and toluene in radioactive mixed waste environments.
- **Methylotrophs:** Petroleum products and oils constitute hydrocarbons, and if present in the surroundings causes pollution. An oil slick caused by oil tankers and petrol leakage into the marine environment is currently a frequently event.

  A number of microorganisms can consume oil as a basis of food and many of them produce effective surface-active compounds that can emulsify oil in water and make possible its subtraction. Unlike chemical surfactants, the microbial emulsifier is non-hazardous and biodegradable. The microorganisms able to degrade petroleum include pseudomonads, various *Corynebacteria, Mycobacteria* and some yeast.

  Methylotrophs are aerobic bacteria that utilise methane for carbon and energy for their growth. They produce an enzyme methane monooxygenase which degrade the chlorinated aliphatic trichloroethylene and 1, 2-dichloroethane. Microbes are the natural scavengers. Their ability to convert natural and artificial chemicals into sources of power and untreated resources for their own development suggests that costly chemical or physical remediation processes might be replaced with natural processes that are lesser in expenditure and more environmentally pleasant. Therefore, microbes symbolise a promising, largely available reserve for latest environmental biotechnologies.

(*b*) Anaerobic microbes: They degrade the pollutants in the absence of oxygen. Anaerobic bacteria are not as frequently used as aerobic bacteria. Mainly used for bioremediation of polychlorinated biphenyls (PCBs) in river sediments, dechlorination of the solvent trichloroethylene (TCE), and chloroform.

(*c*) Lignolytic fungi: In 1995, Andrzej and R.L. Crawford reported the potential for bioremediation of xenobiotic compounds by the white-rot fungus *Phanerochaete chrysosporium*. The lignin-degrading fungi have the ability to degrade lignin containing substrate for example : straw, saw dust, and corn cobs, wood, etc. *P. chrysosporium* exhibits strong potential for bioremediation of pesticides, polyaromatic hydrocarbons, PCBs, dioxins, dyes, TNT and other nitro explosives, cyanides, azide, carbon tetrachloride, and pentachlorophenol. White rot fungi degrade lignin with nonselective extracellular peroxidases which can also facilitate the degradation of other compounds containing similar structure to lignin within the proximity of the enzymes released.

4. Bioremediation Process

Recently it has been discovered that *Geobacter metallireducens* is potential to remove uranium (a radioactive waste) from drainage waters in mining operations and from infected groundwater. Additional investigation of microbial diversity is likely to lead to the detection of a lot of supplementary organisms with distinctive properties valuable in bioremediation. The exploitation of microorganisms is not restricted to individual field of study of bioremediation, but it has a widespread application.

Microbes also have the ability to remove industrial wastes, reduce the toxic cations of *heavy metals* (such as selenium) to a much less toxic soluble form. For example, plants like locoweed remove large amounts of the toxic element selenium. The selenium is stored in plant tissues where it possesses no harm until and unless the plant is eaten. Many algae and bacteria release secretions that absorb metals toxic in high levels. Metals are removed from the food chain. Degradation of dyes is also brought about by some anaerobic bacteria and fungi. Due to increase in population the major problem towards world is to increase the food production. To increase food production rate several pesticides have been used. Utilisation of these pesticides in crops has lead to accumulation of artificial complex compound called 'xenobiotics', and these can be degraded by genetically altered microbes.

(*a*) Bioremediation of Heavy Metals: Heavy metals are usually defined as elements with metallic properties (ductility, conductivity, stability as cations, ligand specificity, etc.) and atomic number >20. The most common heavy metal contaminants are Cd, Cr, Cu, Hg, Pb and Ni. Metals are usual components in soil with a number of heavy metals being requisite by plants as micronutrients. Pollution of biosphere by noxious metals has accelerated since the beginning of the industrial revolution such as mining and smelting of metals, electroplating, gas exhaust, energy and fuel production, fertilizer, sewage and pesticide application, municipal waste generation, etc. Metal pollution has become one of the most rigorous environmental troubles at present. Excessive buildup of heavy metals is toxic to the majority of plants. Heavy metal ions when present at a high level in the environment extremely absorbed by roots and translocated to shoot lead to impaired metabolism and reduced growth. Heavy metal contamination in water and soil poses a key environmental and human health trouble. Excessive metal absorption in polluted soil results in decreased soil microbial action and soil productiveness and yield losses.

Cadmium is a non-essential, toxic heavy metal to plants which inhibit root and shoot growth, affect nutrient uptake and homeostasis and is accumulated by agriculturally significant crops. When Cd-enriched crop products are consumed by animals and humans, it causes diseases. Upon increasing the concentration of Cd the soil may become unusable for crop production. Contamination of soil with Cd can negatively affect biodiversity and the action of soil microbial communities. Canola seeds developed normally in the presence of up to 1 mili mol/litre nickel chloride, but plant root and shoot elongation were inhibited at higher levels.

Heavy metals cannot be destroyed biologically but are only transformed from one oxidation state or organic complex to another. Excavation and land fill, thermal treatment, acid leaching and electroreclamation are not suitable practically due to their high cost, low efficiency, destruction of soil structure and fertility and soil properties etc. Thus the development of phytoremediation strategies for heavy metals contaminated soils is necessary.

Application of Bioremediation

In 1989, bioremediation was employed to treat the Exxon Valdez oil spill in Prince William Sound, Alaska. Hydrocarbon degrading microbes exist in marine systems because natural sources of hydrocarbon exist as a result of geological seeps and other sources. In 2008, J.G. PH and coworkers described that during the Exxon cleanup effort, the activity of these microorganisms was enhanced through the addition of nitrogen and phosphorus to oil laden beaches.

***(i)* Detoxification of organomercurials:** The organomercurials, phenylmercuric, methylmercuric, and ethylmercuric are first cleaved to produce benzene, methane, and ethane, respectively (Fig. 22.10). Methane and ethane are volatile to a lesser extent. Upon subsequent reduction the mercury released from the organomercurials is volatilised. These processes have been followed both by volatilisation of radioactive (^{14}C and ^{203}Hg)- labelled materials and also by gas chromatographic identification of benzene, methane, and ethane. The enzymatic cleavage of *p*-hydroxymercuribenzoate into benzoic acid and Hg^{2+} has been shown for the organomercurial-resistant soil isolates of *Pseudomonas* K62. The broad-spectrum plasmids of *E. coli* determine the ability to cleave *P*-hydroxymercuribenzoate *in vivo*. The release of volatile mercury from thimerosal (ethylmercurithiosalicylate) was found with broad-spectrum organomercurial resistance plasmids in *E. coli*, *P. aeruginosa* and *S. aureus*. The two remaining organomercurials, merbromin and fluoresceinmercuric acetate, show no evidence for enzymatic destruction.

***(ii)* Enzymes associated with mercury and organomercurial detoxification:** Mercuric reductase and organomercurial lyase are the two important enzymes associated with detoxification process. The crystal structure of the enzyme mercuric reductase from *Bacillus* by X-ray diffraction has been studied. Mercuric reductase is roughly a globular dimer; each active site requires cooperation from both subunits. Similarly, elimination of Cys558 of the Tn501 reductase (equivalent to *Bacillus* position Cys628 which is involved in metal binding in the crystal structure) had a much greater effect on *in vitro* enzyme activity than did elimination of Cys559 (equivalent to Cys629 which is not directly involved in metal binding). However, elimination of either cysteine resulted in hypersensitivity to Hg^{2+} by growing cells showing that both cysteines are essential *in vivo*.

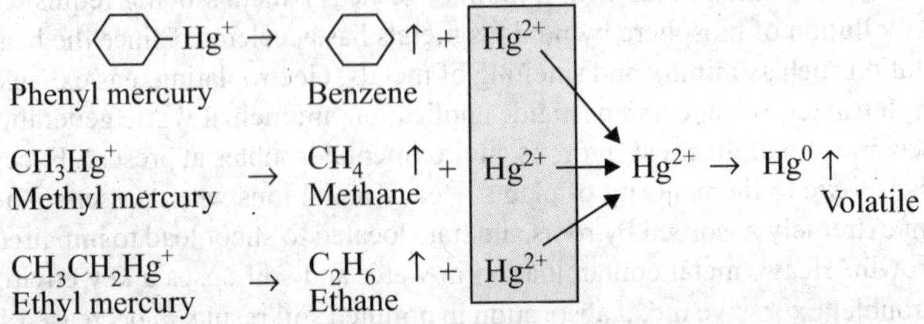

Fig. 22.10. The detoxification of Hg^{2+} and organomercurials.

(*b*) Polycyclic aromatic hydrocarbons (PAHs): Polycyclic aromatic hydrocarbons (PAHs) are the organic compounds that consist of multiple benzene rings. These are the components of asphalts, fuels, oils and greases. They are also called polynuclear aromatic hydrocarbons. The bacterial degradation of PAHs with more than three rings (Fig. 22.11) often referred to a high molecular weight (HMW) PAHs are challenging to be mineralised by microorganisms. This is related to their low solubility and high affinity to solid surfaces and organic liquids, which generally limits their bioavailability to microorganisms present in soils. Therefore, surfactant-enhanced bioremediation of soils containing low soluble hydrocarbons is a promising biodegradation technique.

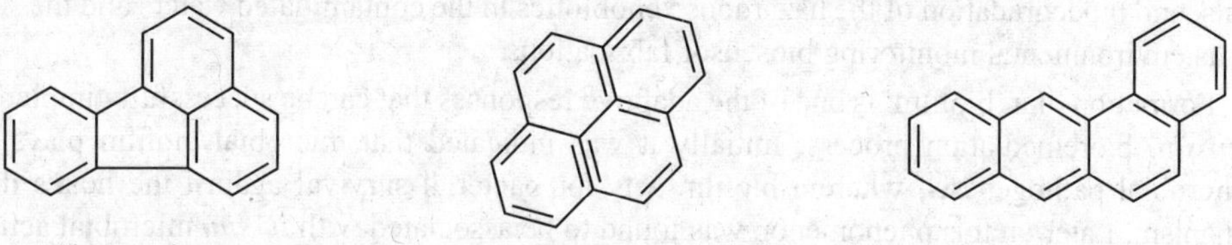

Fluoranthene Pyrene Benz(a)anthracene

Fig. 22.11. Structure of high molecular weight PAH compounds.

5. Factors Affecting Bioremediation

There are several biotic and abiotic factors that affect biodegradation, influence rate of microbial growth and enzymatic activities, affect the rate of petroleum hydrocarbon degradation. The persistence of petroleum pollutants depends on the quantity and quality of the hydrocarbon mixture and on the properties of the affected ecosystem. Some of the factors influencing bioremediation are given below:

(*a*) Bacterial Chemotaxis: Chemotaxis allows the increased bioavailability of the pollutants and helps in maximisation of degradation of pollutants. A variety of bacteria belonging to diverse group have been identified that shows chemotaxis towards environmental pollutants such as petroleum-associated hydrocarbons, explosive, nitroaromatic compounds, polycyclic aromatic hydrocarbons. Some of the common soil bacteria such as *Rhizobium* sp., *Bradyrhizobium* sp., *Pseudomonas* sp., and *Azospirillum* sp. show chemotaxis attracted toward different aromatic hydrocarbons. Many of these compounds are present in soils, sediments and rhizosphere, and serve as sources of carbon and energy for the microorganisms. In 2002, G. Pandey and R.K. Jain reported that aromatic acids such as benzoate, *p*-hydroxybenzoate (PHB), methylbenzoates, *o*-, *m*-, and *p*- toluates, salicylate, DL-mandelate, ß-phenylpyruvate, and benzoylformate act as attractants for *Pseudomonas putida* PRS2000.

Chemotaxis is divided into two subcategories based on metabolism: metabolism–dependent chemotaxis and metabolism-independent chemotaxis. The first one is associated with pollutants that are used as source of metabolic energy. The second one is associated with pollutants that are co-metabolically transformed to generate lesser toxic product. Electron donors/acceptors act as important components of the metabolic machinery. Therefore, bacterial moments towards electron acceptors/donors may be called metabolism-dependent chemotaxis. Generally microbiologists consider positive chemotaxis for detoxification of hydrocarbons. Negative chemotaxis phenomenon could be utilised to explore the alternative methods for control of the microbial fouling by coating surfaces with less universal toxic chemical at non lethal concentrations which prevent fouling by repelling a potential primary microbial film. Positive chemotaxis could be used to enhance the *ex-situ* bioremediation. However, negative chemotaxis may be utilised for developing new approaches for controlling the non-specific microbial fouling.

(*b*) Biofilm: Microbes adhere to the environmental surfaces and their association with these surfaces involves the synthesis of extra cellular homo- or hetero-polymers of sugars called exo-polysaccharides (EPS). These extracellular polymers are fabricated and extended from the cells to the surfaces in a very organised manner forming a tangled and channelised network of the polymeric fibres. The whole assemblage including the resident microbes and the channelised network of the EPS is termed as 'biofilm'. Because of its polyelectrolyte nature the microbial biofilms are highly absorptive and can collect significant quantities of silt, clay hydrocarbons and other detritus from their immediate environment. In 1993, J.D. Brayers is of opinion that the unique physical, biological and chemical properties of biofilms make them a very useful tool for environment cleanliness. Hence the microbial biofilms are implicated in the natural and modulated environmental cleanliness systems for purification of the drinking water, detoxification of the oil spills, and removal of heavy metals, and biodegradation of the hazardous xenobiotics in the contaminated waters and the soils, as well as environmental monitoring biosensor fabrications.

Formation of a 'biofilm' is one of the adaptive responses that can be successfully implemented for *in-situ* bioremediation process. Initially it was indicated that microbial biofilm plays a role in microbial pathogenesis, wherein biofilm acts for bacterial survival against the host's defense mechanism. Later on this phenomenon was found to be associated with *in-situ* microbial activities.

(c)* Biosurfactants:** The word surfactant is a contraction of ***Surface ***Act***ive ***Agent***. Biosurfactants are microbial-derived surface active agents that can lower surface and interfacial tension of liquids. Biosurfactants strongly influence the *in-situ* bioremediation of chemical and metal pollutants.

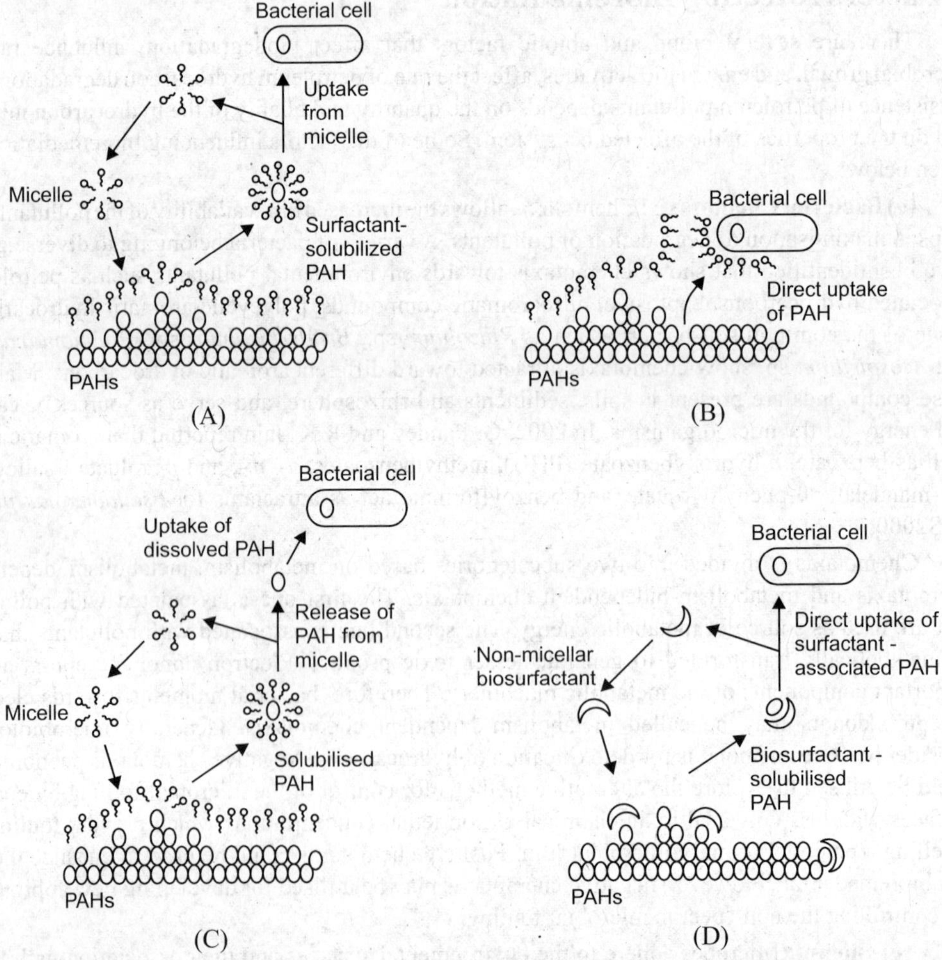

Fig. 22.12. Biosurfactant's mechanism enhancing the bioavailability of PAH (adopted from G.A. Ivey, 2000).

Biosurfactants are produced by a variety of microorganisms belonging to the species of *Pseudomonas*, *Flavobacterium*, *Bacillus*, *Rhodococcus*, *Nocardia*, *Corynebacterium*, *Candida*, *Mycob-acterium* and *Arthrobacte*r. Biosurfactants are amphiphilic molecules consisting of hydrophilic and hydrophobic domains, which facilitate the uptake of hydrocarbons into living cells. Due to the presence of surface active properties, microbial surfactants can accumulate at interfaces form micelles, lower surface tension, and enhance the solubilisation in water of poorly soluble compounds. Mechanisms for enhancement of bioavailability of pollutants by using biosurfactants are as follows (Fig. 22.12) :

- PAH is taken up by bacterial cells directly from the partial micelle (Fig. 22.12A).
- PAH is taken up by bacteria from aqueous phase after release from partial micelle (B).
- The direct uptake of PAH is facilitated via bacterial cell- surfactant- PAH contact (C).
- Non-micellular biosurfactant enhances PAH solubilisation where direct uptake of surfactant-associated PAH takes place (D).

***(d)* Abiotic Factors:** Abiotic factors play important role in the majority of environmental phenomena. The metabolic reaction involved in the bioremediation of hydrocarbons also follows the principal of 'enzyme catalysis' and exhibits optimal performance only over a very narrow range of phsio-chemical parameters. A successful bioremediation requires: (*i*) the availability of specific nutrients (N, P, K, etc.) to support growth of these bacterial species, (*ii*) abundance of electron acceptors (*i.e.* dissolved oxygen and secondary electron acceptors), and (*ii*) contact with the dissolved and adsorbed contaminants (Fig. 22.13).

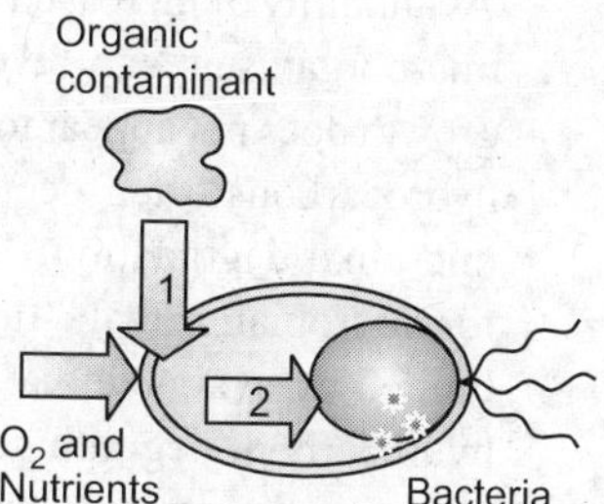

Fig. 22.13. Bioremediation of organic contaminants by bacteria using oxygen and nutrients.

Some of the abiotic factors that influence bioremediation are described below:

- **Temperature:** Biodegradation of hydrocarbons can occur over a wide range of temperatures. Many psychrotrophic, mesophilic, and thermophilic hydrocarbon-utilising microorganisms have been isolated. Hydrocarbon degradation at below 0°C as well as at 70°C has been reported in 1960s. In 1974, G.J. Mulkins-Philips and J.E. Stewart found that 9 months after the spillage of Bunker C fuel oil into Chedabucto Bay, the bacterial population isolated from contaminated areas showed rates of degradation at 5°C that were 21 to 70% less during 14 days of incubation than during 7 days at 10°C. The evidence suggests a seasonal shift to a microbial community capable of low temperature hydrocarbon degradation.

The effects of temperature differ depending on the hydrocarbon composition of a petroleum mixture. Low temperature retards the rates of volatilisation of low molecular weight hydrocarbons, some of which are toxic to microorganisms. The presence of such toxic components delays the onset of oil biodegradation at low temperatures. At 20°C, lighter oils have greater abiotic losses and are more susceptible to biodegradation than heavier oils. The light crude oils had toxic volatile components which evaporated only slowly, inhibiting microbial degradation of these oils at 10°C. A significant lag phase before the onset of biodegradation was found for the lighter oils. No toxic volatile fractions associated with the heavy oils were tested. Paraffinic, aromatic and asphaltic fractions were also subjected to biodegradation.

- **Nutrients:** There is a mass of carbon available for microbial growth within a limited area. Since microorganisms require nitrogen and phosphorus for incorporation into biomass, the availability of these nutrients within the same area as the hydrocarbon is critical. Rates of diffusion may be inadequate to supply sufficient nitrogen and phosphorus to establish optimal C/N and C/P ratios for microbial growth and metabolism. Researchers examining the fate of large oil spills have thus concluded in many cases that concentrations of N and P are limiting with respect to rates of hydrocarbon biodegradation.

- **Physical state of oil pollutants:** The physical state of petroleum hydrocarbons has a marked effect on their biodegradation. The degree of spreading of the oil spill determines the surface area of oil available for microbial colonisation by hydrocarbon degrading microorganisms. The degree of spreading is reduced at low temperatures because of the viscosity of the oil. In soils, petroleum hydrocarbons are absorbed by plant materials and soil particles that limit its spreading. Liquid aromatic hydrocarbons are utilised by bacteria at the water - hydrocarbon surface, whereas solid aromatic hydrocarbons are not metabolised. At 30°C diphenylmethane is a liquid and can be degraded easily but at 20°C its solid form can not be utilised by a *Pseudomonas* sp. Similarly naphthalene can not be utilised in the solid form but can be utilised if dissolved in liquid hydrocarbon.

  Availability of increased surface area accelerates biodegradation. Hydrocarbon degrading microorganisms can be observed growing over the entire surface of an oil droplet; its growth does not appear to occur within oil droplets. After extensive weathering, petroleum hydrocarbons often occur in the environment as tar balls, becoming quite resistant to microbial degradation. The surface area-to-volume ratio of a tar ball is not favourable for microbial growth. Tar balls often accumulate on areas where microbial activities are limited by available water, which is needed to support microbial growth and enzymatic hydrocarbon- degrading activities. Thus, dissolution and emulsification of hydrocarbons appear to have positive effects on degradation rates.
- **Salinity and Pressure:** These two factors are specific features of particular ecosystems such as saline lakes and deep seas (high hydrostatic pressure), which represent specialised environment where petroleum contamination can occur. In 1978, D.M. Ward and T.D. Brock have examined hydrocarbon degradation in hypersaline environments. When hydrocarbons were added to natural samples of various salinities (ranging from 3.3 to 28.4%) from salt evaporation ponds of Great Salt Lake (Utah) rates of metabolism of these compounds decreased with increase in the salinity. Microorganisms in saline waters containing more than 20% of salinity could not use mineral oil as sole source of carbon and energy. Thus proving that biodegradation is not possible in hyper-saline environment.
- **Oxygen:** The reports of existence of microorganisms which are capable of anaerobic hydrocarbon metabolism in nature have been provided by many researchers including that of some microorganisms capable of alkane dehydrogenation. These microorganisms have an enzymatic mechanism which should permit addition of water across the double bond forming a secondary alcohol and permit anaerobic growth. A strain of *Pseudomonas* consumed oxygen when growing on heptane even though it had an *n*- heptane dehydrogenase enzyme. Mineralisation of hydrocarbons in estuarine sediments depends on oxygen availability at pH 5 and 8. Rates of hydrocarbon degradation decreases with decreasing oxygen reduction potential, *i.e.* with increasing the anaerobic condition.

6. Mass Balance

The remediation processes depend on mass balance, *i.e.* the amount of a contaminant existing in the subsurface and removal of this mass by the proposed remediation system. In aerobic bioremediation, the mass balance is a function of the stoichiometry of the oxidation-reduction reactions that govern the biological utilisation of a particular compound. For example, oxidation-reduction reactions for the aerobic utilisation of benzene are as follows:

- **Oxidation reaction:** $C_6H_6 + 12\ H_2O \rightarrow 6\ CO_2 + 30\ H^+ + 30\ e^-$
- **Reduction reaction:** $7.5\ O_2 + 30\ H^+ + 30\ e^- \rightarrow 15\ H_2O$
- **Overall Reaction:** $C_6H_6 + 7.5\ O_2 \rightarrow 6\ CO_2 + 3\ H_2O$

In this reaction 1 mole of C_6H_6 requires 7.5 moles of O_2, and converting to the appropriate mass ratio results in the following:

1 unit C_6H_6 : 7.5 × (32/78) units of O_2, or 1 unit C_6H_6 requires 3 units of O_2

Therefore, aerobic biological degradation requires 3 kg of dissolved oxygen to degrade 1 kg of benzene present in the subsurface.

Oxygenating a 10-gpm flow of water with 45 mg/l of dissolved oxygen, a bioremediation unit can deliver about 6 pounds of dissolved oxygen to the subsurface per day, *i.e.* about 200 ponds of dissolved oxygen per month. This will support the aerobic biological degradation of about 750 pounds of benzene per annum.

Many sites have thousands of pounds of contamination in the subsurface; hence dissolved oxygen alone cannot provide the necessary mass of electron acceptor. To solve this problem, a nutrient amendment can be done that also provides secondary electron acceptors (nitrate, sulfate) to support facultative degradation of some of the constituents. By combining the dissolved oxygen and secondary electron acceptor delivery, degradation of certain constituents can stimulated resulting in cleanup levels within 1-2 years.

7. Superbug- A New Concept of Bioremediation of Petro-pollutants

Each year, over 10 million metric tons of petro-pollutants enter the sea ecosystem and kill birds, shell fish, fishes, invertebrates and other sea animals and planktons. This poses more adverse effect on marine life and food chain. For more than two decades, scientists have been heralding the arrival of genetically modified organisms for bioremediation. But so far, none have actually reached the toxic sites where they are supposed to clean up.

Anand Mohan Chakrabarty

In the 1970s, the most famous incident occurred when Ananda Chakrabarty, the India-born American scientist, doing work for General Electric succeeded in producing a bacterial **superbug** (the oil eating bug) by genetically engineered strain of *Pseudomonas putida*. It is capable of utilising complex chemical compounds like hydrocarbons. The superbug was supposed to eat up oil spills.

Prof. Ananda Mohan Chakrabarty

Prof. Ananda Mohan Chakrabarty (generally called 'Al' by scientific colleagues) is an Indian-American microbiologist, scientist and researcher. He is most notable for his work in directed evolution and his role in developing a genetically engineered organism using plasmid transfer. In 1971, he genetically engineered a new species of *Pseudomonas* bacteria commonly called the *oil-eating bacteria* while working for the Research & Development Center at General Electric Company in New York. Now he is a professor at the College of Medicine, University of Illinois, U.S.A.

When he applied for a first-ever patent for living organism, the bacteria drew international attention. His patent was initially denied by the Patent Office because it was thought that the patent code precluded patents on living organisms. The United States Court of Customs and Patent Appeals overturned the decision in Chakrabarty's favor writing ...*the fact that microorganisms are alive is without legal significance for purposes of patent law.*

Sidney A. Diamond, Commissioner of Patents and Trademarks then appealed to the Supreme Court. The Supreme Court case was argued on 17 March 1980 and decided on 16 June 1980.

This patent was granted by the U.S. Supreme Court (Diamond vs Chakrabarty) in a 5-4 decision when it determined that…*A live, human-made microorganism is patentable subject matter under [Title 35 U.S.C.] 101.* Respondent's microorganism constitutes a 'manufacture' or 'composition of matter' within that statute.

His landmark research has paved the way for many patents on genetically modified microorganisms and other life forms, and catapulted him into the international spotlight. The 'oil-eating bacteria' has been used to clean up many toxic oil spills including the one caused by the Exxon Valdez disaster.

The Company said that the organisms sold in products eat the oil, and then eat each other. When there is no more oil to eat they eat them, and then there is just one microbe left, and he dies of starvation.

Pseudomonas is an aerobic and Gram-negative bacterium belonging to the family *Pseudomonadaceae*. *P. aeruginosa* and *P. fluorescens* are fluorescent pigment-forming bacteria. This superbug utilises a number of toxic chemicals like chlorobenzenes, 2.4-D, 2.4.5-T, and petrochemicals like octane, camphor, etc. This discovery of superbug really opens the door for checking oil-pollution in sea ecosystem.

Petro-products, including oils, generally contain octane, xylene, camphor and naphthalene. No single strain of *P. putida* is able to digest all these four petro-chemicals. There was need of four strains of *P. putida* with four different plasmid DNAs. Chakrabarty introduced four plasmid DNAs from four different strains into one single cell of *Pseudomonas putida* and formed a superbug.

During that time four known species of oil-metabolising bacteria were known to exist. But after introduction of bacteria into an oil spill, they competed with each other and they degraded the crude oil. It is possible because of the presence of the genes on plasmids, such as *OCT* (octane) gene, *XYL* (xylene) gene, *CAM* (camphor) gene and *NAH* (naphthalene) gene that are required to degrade the oil. These plasmids were transferred among the bacterial species (Fig. 22.14).

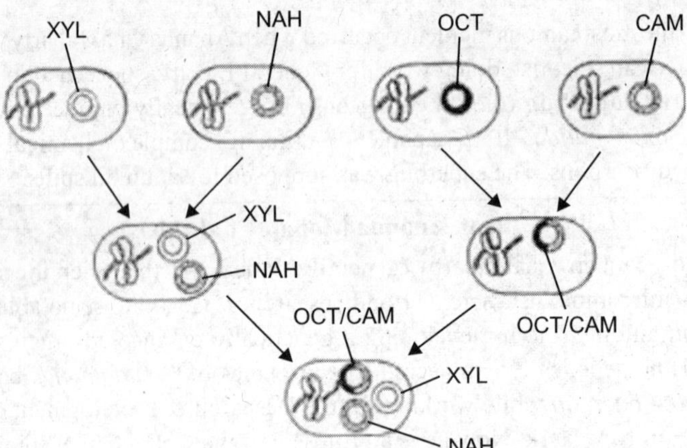

Fig. 22.14. Steps of gene transfer and creation of superbug *Pseudomonas putida*.

After plasmid transfer, Chakrabarty irradiated the transformed bacteria with UV light and discovered a method for genetic cross-linking that fixed all four plasmid genes in place and produced a new and stable bacterial species called *Pseudomonas putida*. This bacterium was capable of consuming oil one- or two- fold faster than the previous four strains of oil-eating bacteria. The new was called 'multi-plasmid hydrocarbon-degrading *Pseudomonas*' which could digest about two-thirds of the hydrocarbons found in a typical oil spill.

Commercial Microbial Products for Bioremediation

As the field of microbial biology continues to advance, scientists have found that some oil-eating microbes naturally occur at oil spills. Current bioremediation efforts focus on encouraging those microbes to grow faster by feeding them fertilizer or oxygen. Natural microbes that like to eat oil are also added to sites to speed up the process.

Today, commercial products used to clean up industrial oil spills, such as BioZorb (developed by Oppenheimer Biotechnology in Austin, Texas) and EcoSolve, contain oil-eating microorganisms, but their manufacturers have not brought genetically modified microorganisms into the marketplace.

8. Monitoring of Bioremedation

To monitor the bioremedation potential of a soil one can probe for the existence of specific degradation pathways in the soil community or monitor for specific enzymes involved in the process. There are two common ways to test for functional genes involved in the degradation of a compound: (*i*) use of specific DNA hybridisation probes, and (*ii*) use of specific RNA hybridisation probes. The specific DNA hybridisation probes can be used to indicate potential for the organisms to degrade the desired compound. The specific RNA hybridisation probes are used to indicate the expression of the functional genes in the environment.

The actual change in pollutant concentration or degradation byproducts can also be monitored to determine the amount of pollutant removal. Controlled laboratory experiments are used to determine whether the degradation of a desired compound is the result of abiotic or biotic activity. The concentration of a pollutant in a non-sterile microcosm containing soil from the environment of interest is compared to a sterile control. The sterile control shows the non-biological contribution to the disappearance of the pollutant due to adsorption to clay particles or volatilisation. The non-sterile microcosm simulates the microbial contribution to the degradation of the pollutant in the natural environment, but also includes other abiotic mechanisms. The microbial contribution to pollutant disappearance is the difference between removal in the biologically active bottle and removal in the sterile control. This helps to quantify whether the disappearance of the pollutant is the result of biological or non-biological mechanisms.

9. Biosafety from GMOs

Public fears that a new bacterium or plant may spread into the environment like some type of superbug that cannot be controlled. After eating the oil, what will it eat next ? But scientists say that getting a genetically modified organism or any organism sufficient nutrients to grow enough to do the job that you want it to do is the more immediate challenge. Unless you engineer an organism that does not need oxygen or water, the self-limiting factors in the environment will keep it from gobbling up a whole oil spill. The scientists hope that public fears about genetically modified organisms will not keep their discoveries in the laboratories and that they can at least go out into the field and test them.

10. Phytoremediation by Superfund Superplant

In Aug 28, 2003 in the 19th issue of the *Proceedings of the National Academy of Sciences* (U.S.A.) a report was published by a group of biologists at the University of California about a kind of Superfund superplant, *i.e.* a leafy organism designed to devour the hazardous waste left in the ground due to human carelessness and industrial machinations. The researchers found that a modified mutant variety of *Arabidopsis* (a relative of the mustard plant) plant can detoxify and suck up several common heavy metals into their stems and leaves. Four of the top six heavy metals such as lead, arsenic, mercury and cadmium are found at Superfund sites to those the plants can ingest. The genetically modified mutant plants suck heavy metals such as lead and arsenic from soil into their roots, stems and leaves resulting in cleansing the contaminated soil.

Moreover, the scientists have found that phytochelatins (chemicals produced by an enzyme expressed by a gene) allow the metals to move through the plant up from the roots to the shoots for easier harvesting.

C. COMPOSTING

Compost is the organic materials derived from plants and animal that has been decomposed through aerobic decomposition. Compost is rich in nutrients, and is used in gardens, landscaping, horticulture and agriculture. It is beneficial for the land in many ways such as a soil conditioner, a fertilizer to add vital humus or humic acids, and as a natural pesticide for soil. In ecosystems, compost is useful for control of soil erosion, reclamation of land and stream, construction of wetland, and as landfill cove. Composting is one of the easiest ways of recycling, and can be practiced by individuals in their homes, farmers on their land, and industrially by industries as well.

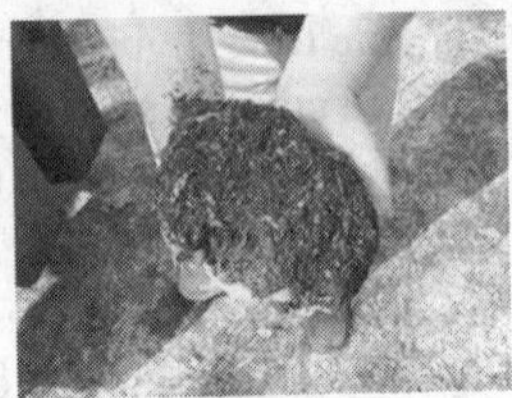

Compost

In *Rig Veda* (7,500-9,000 B.C.) it has been mentioned that animal dung (from goat, sheep, cows, etc.), and water provide healthy growth to the plants. A similar effect of manure has also been described in *Vriksayurveda.* Increase in crop yield upon adding several kinds of manures, oil cakes, animal excreta in soil has also been mentioned during *Mahabharat* period (5,000 B.C.). Even today, Indian farmers practice the composting process for enhanced crop yield.

Traditionally, composting was piling of the organic materials and leaving them for about a year until the next planting season comes, and applying the materials in crop field. In the beginning of 1920s, composting was modernised in Europe as a tool for organic farming. The first industrial station for the transformation of urban organic materials into compost was set up in Austria in the year 1921. The early personages most cited for propounding composting within farming are Rudolf Steiner, the founder of a farming method called biodynamics for the German-speaking world is known to work of composting. In 2006, J. Heckman wrote a history of organic farming transitions. It was Sir Albert Howard who worked extensively in India on sustainable practices and Lady Eve Balfour was a huge proponent of composting. Composting was much boosted up in America also by various followers of early European movements. By chance some of these personages met briefly in India, all were quite influential in the U.S. from the 1960s to the 1980s.

Now-a-days several modifications have been made to speed up the traditional composting. Rapid composting rectifies some of the perceived problems associated with traditional slow composting. By doing so compost can be made in 2 to 3 weeks by incorporating a few changes in traditional methods including smaller and more homogenised pieces in the compost, controlling carbon to nitrogen ratio (C:N) at 30:1 or less, and monitoring the moisture level more carefully.

Generally, compost is recommended as an additive to soil to supply humus and nutrients. It provides a rich growing medium and a porous absorbent material that holds moisture and soluble minerals. It provides the support and nutrients to soil in which plants can grow well. It is rarely used alone; hence it is used to produce loam primarily by mixing with soil, sand, grit, bark chips, vermiculite, perlite or clay granules. In 2004, P. Morel and G. Guillemain have asssessed the possible phytotoxicity of a substrate by using an easy and representative biotest. Most often, direct seeding into a compost should not be done due to rapidly drying of seeds and the possible presence of phytotoxins which may inhibit seed germination, and tie up of nitrogen to incompletely decomposed lignin. Therefore 20–30% compost should be mixed up in soil for transplanting seedlings at cotyledon stage.

Composting can also destroy pathogens or unwanted seeds. Unwanted living plants are destroyed by covering with mulch and/or compost. In 1992, A. Tränkner reported that the use of agricultural and municipal organic wastes to develop disease suppressiveness to plant pathogens. Several thermophilic (high-temperature) and mesophilic microorganism reside in compost. Hence such compost consists of the 'microbial pesticides'. Theromophilic composting is well known to completely destroy all seeds and pathogens. Compost also includes certain composting detritivores (such as black soldier fly larvae) and redworms that also reduce many pathogens. Therefore, where there is a high possibility for the presence of pathogens, the sanitising qualities of thermophilic composting are useful.

Benefits from Compost

- It helps your local environment to keep hygienic and pollution-free.
- It reduces the amount of agricultural, garden and kitchen wastes going to landfill.
- It returns the nutrients locked up in biomass again into soil and increases soil fertility.
- It saves money in terms of reducing the uses of chemical fertilizers, that imbalance the soil nutrients an turn into barren land.
- It helps garden plants and agro-crops to grow naturally.

1. Method of Composting

A composter is located on bare soil nearly by the house or somewhere in garden that is easily accessible all year round. Green and Brown materials are evenly mixed when composting is to be carried out. Greens are soft, wet materials like grass cuttings, vegetables and fruit scraps. Browns are harder, dryer materials like hedge trimmings and strips of cardboard, etc. Some of the wastes used for composting are given in Table 22.4. A lot of materials are put in all at once when composting is carried out. Large items should be chopped into smaller pieces to help the process. Compost must be kept moist but not wet. When materials are too dry some amount of water should be added and covered; however, dry material is added if it is too wet. To speed up the composting process, compost accelerator is added. Adding of materials is continued and the contents are agitated for every couple of weeks throughout spring and summer seasons to keep airs flowing through. The compost becomes ready when it turns into dark colour and smells like earth. Depending on the materials used and the time of the year this can take about 6 to 18 months.

A compost pile

Table 22.4. Wastes used in vermicomposting.

Wastes	*Types*
Agricultural waste	Agricultural fields: stubble waste, husk, straw, and farmyard manure. Stems, leaf matter, fruit rind, pulp and stubble. But be careful while handling an all-citric waste.
Animal waste	Dung, urine and biogas slurry
Urban solid waste	Kitchen waste from household and restaurants, waste from market yards and places of worship, and sludge from sewage treatment plants.
Agro industries	Food processing units: Peel, rind and unused pulp of fruits and vegetables, fine bagasse, pressmud and seed husk, stems, leaves and flowers after extraction of oil

2. Factors Affecting Composting

The time of composting depends on the types of ingredients of composting materials. Basically the organic materials used for composting are made up of carbohydrates, proteins, and cellulose; therefore, the building block of organic materials such as carbohydrates, proteins, and cellulose really govern the rate of transformation of organic materials by microorganisms. To work effectively, composting organisms require the following four important components:

- **Carbon**: It is required for energy; the microbial oxidation of carbon produces the heat. High carbon materials tend to be brown and dry.
- **Nitrogen:** Nitrogen is needed to grow and reproduce more organisms to oxidise the carbon. High nitrogen materials tend to be green (or colorful, like fruits and vegetables) and wet.
- **Oxygen:** It is required for oxidising the carbon, the decomposition process.
- **Water:** Water is an essential part of all life form. The metabolic activities occur in aqueous condition. The right amount of water is required to maintain the microbial activity without causing anaerobic conditions.

(*a*) Carbon : Nitrogen Ratios: All organics have both carbon and nitrogen, but the amount of these vary widely with characteristics noted above (dry/wet, brown/green). Fresh grass clippings have an average C:N ration of about 15:1 and dry autumn leaves about 50:1 depending on species. Mixing equal parts by volume approximates the ideal C:N range. A carbon : nitrogen ratio of about 30:1 results in the most efficient composting.

Certain ratios of these elements provide beneficial bacteria with the nutrients to work at a rate that will heat up the pile. In that process much water is released as vapor and the oxygen is quickly depleted. It explains the need to actively manage the pile. The air/water balance is critical to maintain high temperatures until the materials are broken down. Excessive amount of air or water also slows the process as does too much carbon or too little nitrogen.

(*c*) Urine: The urea is found in urine which can be toxic to plants or can create toxic substances if not captured by sorbents and allowed to compost aerobically.

When urine of a healthy person is added into the mulch/compost, it may increase temperatures resulting in killing of pathogens and unwanted seeds. Use of fresh urine from a healthy person is generally much safer than fresh faeces. Urine neither attracts disease-spreading flies (such as house flies or blow flies), nor it harbours the pathogens such as parasitic worm eggs. Usually urine does not stink for long time particularly when fresh and deposited under sorbents.

(*b*) Microorganisms: When the water, oxygen, carbon, and nitrogen are properly mixed, microorganisms break down the organic matter and form the compost. For the composting process microorganisms are absolutely necessary; the organic matter in compost heap cannot undergo the composting process without them. There are five types of microorganisms found in good quality of compost.

- **Bacteria:** Bacteria found in compost are the most common of all the microorganisms.
- **Actinomycetes:** Actinomycetes are very important for breaking down of paper products such as newspaper, bark, etc.
- **Fungi:** Molds and yeast help the break down of organic materials that bacteria cannot. Some of the organic material consists of lignin which is decomposed mainly by lignocellulosic fungi falling under the class Basidiomycetes.
- **Protozoa:** They ingest bacteria and fungi, therefore, balance the composting cycle.
- **Rotifers:** Rotifers also help break down organics in the compost and also ingest bacteria and fungi.

The difficulty of breaking down of organic materials in landfills is the lack of microorganisms.

What to Compost and What not ?

- **Composting materials**
  Grass cuttings, hedge cuttings, vegetable peelings, tea bags, coffee grounds, ripped cardboard and paper, fruit cuttings, etc.
- **Not to compost the following:**
  Meat, fish or cheese, coal ash, cooked left over, metals, glass or plastic, nappies, etc.

3. Types of Composting

On the basis of presence and absence of oxygen, the composting practices are grouped into two types: *anaerobic composting* and *aerobic composting.* In India, anaerobic composting has been practiced from the past where pits are made into which the wastes are dumped and the pit closed. On completion of six months the manure is excavated and used according. In contrast, the aerobic composting is a scientific process where the duration of composting is drastically reduced. There are several aerobic methods designed time to time. Some of them are: biodung composting vermicomposting, etc.

(*a*) Biodung Composting: It is a very interesting method which includes partially aerobic and anaerobic processes. This method is generally applied as a prerequisite to vermicomposting, because the biodung method can destroy parasites, pests and viable seeds of weeds due to increased temperature in the heap or pile. The temperatures may reach to about 65º to 75ºC. The pile can be turned over with the material one or two times. After 30 days the materials can be used for vermicomposting.

(*b*) Compost Tea: The liquid suspension made by steeping compost in water is called compost tea. It is used as fertilizer and also to prevent plant diseases. The liquid is sprayed to non-edible plant parts or as a soil-drench (root dip) such as seedlings or as a surface spray to reduce incidence of harmful phytopathogenic fungi in the phyllosphere *i.e.* immediate vicinity of the leaf.

(*c*) Vermicompost: Vermicompost (also called Vcompost, compost, vermicast, worm castings, worm humus or worm manure) is the product of composting by using different kinds of species of worms, usually red wigglers, white worms and earthworms to produce a heterogeneous mixture of decomposing vegetable or food waste, bedding materials, and vermicast. Vermicompost is the end-product of the breakdown of organic matter by some species of earthworm. Vermicompost is one of the best nutrient-rich organic fertilizers and soil conditioners that contain water-soluble nutrients, microbes including fungi, protozoa and bacteria. The process of producing vermicompost is called vermicomposting (*see* preceeding section).

(*d*) Bokashi Compost: A method of intensive composting is called Bokashi. The compost is produced through an aerobic or anaerobic inoculation. Like yogurt culture, a starter culture can be used to extend the culture indefinitely. Bokashi compost consists of molasses, water, effective microorganisms (EM), and wheat bran. The EMs are the natural lactic acid bacteria, yeast and phototrophic bacteria that act as a microbe community that ferment and accelerate the breakdown of the organic matter.

At homes the kitchen waste is placed into a container sealed with an air tight lid. Then the scraps are inoculated with a Bokashi EM mix. Usually it takes the form of a carrier such as rice hulls, wheat bran or saw dust that has been inoculated with composting microorganisms. The user would place alternating layers of food scraps and Bokashi mix until the container is full.

(*e*) Industrial Composting Systems: Since composting is a process for converting decomposable organic materials into useful stable products, a concern is growing worldwide interest in recycling by means of composting. Industrial scale composting in the form of in-vessel composting, aerated static pile composting, and anaerobic digestion takes place in most Western countries. In many areas it is mandated by law.

Gradually, the industrial composting systems are increasingly and being installed as a waste management alternative to landfills, along with other advanced waste processing systems. Mechanical sorting of mixed waste streams combined with anaerobic digestion or in-vessel composting is called 'mechanical biological treatment', This kind of treatment is used in developed countries due to regulations controlling the amount of organic matter allowed in landfills. If the biodegradable wastes are treated before entering a landfill, global warming from fugitive methane is reduced. In a landfill the untreated wastes are broken down anaerobically resulting in the production of landfill gas that contains methane which is a potent greenhouse gas.

The world's largest municipal solid waste (MSW) co-composter is the Edmonton Composting Facility in Alberta (Canada) which turns 2,20,000 tonnes of residential solid waste and 22,500 dry tones of biosolids per year into 80,000 tonnes of compost. The facility is 38,690 meters^2 and the operating structure is the largest stainless steel building in North America.

D. VERMICOMPOSTING TECHNOLOGY (VERMITECHNOLOGY)

The term 'verm' refers to a Greek word which means any cylindrical and elongated limbless animal having a tendency to crawl or burrow. In this context *verms* or *vermi* refer to earthworms. Cultivation of a biological material or its propagation in a substrate under semi natural condition is known as *culture*. Vermiculture is propagation of earthworms producing a byproduct called vermicompost and the process of producing vermicompost is called 'vermicomposting'. Vermicomposting is a method of using earthworms and microorganisms to convert organic waste into black, earthy-smelling and nutrient-rich humus.

Earthworms are divided into two groups: humus formers and humus feeders. The humus formers dwell on the surface and feed on nearly 90% organic materials. They are generally darker in colour, and are also called epigeic or detritivorous earthworms. Generally these worms are exploited for vermicomposting. The humus feeders are the burrowing worms. They are useful in both compost preparation as well as making the soil porous. Generally the burrowers help in mixing and distributing humus through the soil. Vermiculture biotechnology helps in bioconversion of substance from one form to another through a biological process.

The earthworms can degrade organic wastes speedily and efficiently. But to increase the efficiency of vermicomposting care must be taken for worms that thrive well on organic matter and breed faster adapting to moisture and climatic fluctuations. Vermicomposting is a fully aerobic system; therefore, its most beneficial feature is to eliminate foul smell of decaying organic wastes.

The concept of vermiculture became well known in the 50s of the 20th century when facilities were set up in industrialised countries of Western Europe for the mass breeding of earthworms. Subsequently, the other countries such as the U.S.A, England and France conducted several experiments related to vermiculture technology for efficient disposal of organic wastes.

1. Working Mechanism of Vermiculture

Earthworms feed on organic waste and consume two to five times more wastes than their body weight. They use a relatively small amount of waste for their growth. They process the organic materials in their alimentary canal and excrete the mucus coated undigested matter as vermicasts. Vermicasts consist of organic matter that has undergone physical and chemical breakdown through muscular gizzard that grinds the material with enzymes releasing amino acids, sugars and organic molecules. The nutrients present in the vermicasts are readily soluble in water which in turn are taken up by plants. Therefore, vermicast is a rich source of macro- and micronutrients, vitamins, enzymes, antibiotics, growth hormones and microflora.

Vermicomposting results in the production of a better end product. Vermicomposting differs from composting in several ways: (*i*) it is a mesophilic process which utilises microorganisms and earthworms that are active at 10–32°C (this temperature remains within the pile of moist organic material), (*ii*) the process of vermicomposting is faster than composting because the material passes through the earthworm gut where it is transformed resulting in earthworm castings (worm manure) which are rich in microbial activity and plant growth regulators, and fortified with pest repellence attributes, and (*iii*) the earthworms transform garbage into eco-frienly vermicompost (Vermi Co, 2001).

Tropical soils are deficient in all necessary plant nutrients but large quantities of such nutrients contained in domestic wastes and agricultural byproducts are wasted. It is estimated that in cities and rural areas of India nearly 700 million tones of organic waste is generated annually which is either burned or land filled. Such large quantities of organic wastes pose a problem for safe disposal. Most of these organic residues are burned currently or used as land fillings. In nature there are a number of micro- and macro- organisms that convert organic waste into valuable resources containing plant nutrients and organic matter. These maintain soil productivity. Microorganisms and earthworms are important biological organisms that help nature to maintain the nutrient flows from one system to another and also minimise environmental degradation. The earthworm population is about 8–10 times higher in uncultivated area. This shows that earthworm population decreases with soil degradation and thus can be used as a sensitive indicator of soil degradation.

2. Importance of Vermicompost

The earthworms consume different kinds of organic wastes and reduce the volume by 40–60%. Each earthworm weighs about 0.5 to 0.6 g, eats waste equivalent to its body weight and produces cast equivalent to about 50% of the waste it consumes in a day. These worm castings have been analysed for chemical and biological properties. The moisture content of castings ranges between 32 and 66% and the pH is around 7.0. The worm castings contain nearly twofold of both macro- and micronutrients than the garden compost (Table 22.5).

Table 22.5. Nutrient composition of vermicompost and garden compost.

Nutrient element	*Vermicompost (%)*	*Garden compost (%)*
Organic carbon	9.8–13.4	12.2
Nitrogen	0.51–1.61	0.8
Phosphorus	0.19–1.02	0.35
Potassium	0.15–0.73	0.48
Calcium	1.18–7.61	2.27
Magnesium	0.093–0.568	0.57
Sodium	0.058–0.158	<0.01
Zinc	0.0042–0.110	0.0012
Copper	0.0026–0.0048	0.0017
Iron	0.2050–1.3313	1.1690
Manganese	0.0105–0.2038	0.0414

3. Physical Quality of Vermicast

Vermicast is one of the best forms of organic manure and is practically 100% like humus. It is dark brown-black humus like coarse material, soft in feel and free from any foul smell, live weed seeds and other contaminations. Mucus type substance is coated on each particle that increases aeration in the soil, provides excellent water retention properties and improves the drainage in heavy soils. Vermicast contains sufficient moisture (25-35%) at the time of packing to ensure high level of biological activities. Beneficial bacteria and fungi present in vermicast introduce very high level of biological activities in the soil (Table 22.6).

In India, Concept Biotech (Gujarat) has taken the initiative by involving the lowly creatures as versatile natural bioreactors for : (*i*) cleaning up the environment with cost effective waste management technology, and (*ii*) development of sustainable agriculture. Concept Biotech undertakes the setting up of plants for solid waste management especially Industrial sludge generated from textile industry, oil refinery, dairy, distilleries, pulp and paper mills, food processing units, mushroom waste and other allied agricultural waste with the aim of treating toxic organic waste into non-toxic manure. It is the first in Gujarat to introduce the industrial waste treatment by using vermitechnology.

Table 22.6: Presence of beneficial bacteria and fungi in vermicast.

Parameters	***Vermicast normal range***
Nitrogen fixing bacteria (cells/g)	2.5 - 10^6
Phosphate solubilizing microbes (bacteria and fungi)	In sufficient quantities
Beneficial fungi (spores and surviving structures)	In sufficient quantities

4. Vermicomposting by Using Earthworms

There are different types and species of earthworms used in vermicomposting as below:

(*a*) Use of Local Varieties: Approximately, 2,500 species of earthworms have been identified throughout the world of which more than five hundred species of earthworms have been identified in India. These earthworms can be cultured in composting material in pits, crates, tanks, concrete rings or any containers. Organic material to be used is recommended only after pre-processing or pre-digestion of respective material through partial anaerobic phase (done under black polythene cover or with a clay seal layer). The biodung composting technology is highly recommended as a pre-digestion mechanism.

Diversity of earthworm species varies with different types of soils. Hence, selection of a local or native species of earthworm for the local soil and for vermicomposting is an important aspect. Local species of earthworms generally used in India are *Perionyx excavatus* and *Lampito mauritii*.

(*b*) Use of Exotic Varieties: Internationally there are three species of earthworms that have received acclaim for vermicomposting, for example *Eisenia foetida* and *Eudrilus eugeniae* are exotic, and *Perionyx excavatus* is endemic. However, the earthworm species most often used are Red wigglers (*Eisenia foetida* or *Eisenia andrei*), but European nightcrawlers (*Eisenia hortensis*) also may be used. Blueworms (*Perionyx excavatus*) may be used in the tropics. These species commonly are found in organic-rich soils throughout Europe and North America and live in rotting vegetation, compost, and manure piles.

In India exotic species of earthworms have also been used for vermicomposting. Since 1982, *E. eugeniae* has been promoted for waste degradation. *Eudrilus eugeniae* is reported to have surpassed both in feeding and reproductive rates as compared to other species of earthworms. However, *Eisenia foetida* is used in certain areas in India for domestic waste composting.

(*a*) Collection of Native Earthworms: Worm-inhabited soil marked by visible earthworm castings on the soil surface is identified. About 500 g jaggery (native sugar) and 500 g fresh cattle dung are dissolved in 2 litres of water and mixture is sprinkled on an area 1×1 m^2 surface soil. Then straw lumps are covered with an old gunny bag. Water is sprinkled for about 20 to 30 days. A combination of epigeic and anecic native worms will aggregate in the spot that could be collected, identified and used.

(*b*) Preparation of Vermipit: Vermipits pit of any convenient dimension can be constructed in the garden or in a field. It may be single pit, two pits or tank of any sizes made up of brick and mortar with proper water outlets. The most convenient pit or chamber of easily manageable size is $2 \times 1 \times 0.75$ m^3 (Fig. 22.15). The size of the pits and chambers should be determined according to the volume of biomass and agricultural waste. A thin layer (5 cm) of broken bricks and coarse sand is prepared above the bottom of the pit. Then the surface is sealed with 25 mm thick layer of soil paste. Holes of 50 mm diameter are made at every 300 mm distance and sticks are introduced into them for two days which are removed later on. This provides the track for air circulation. The pits can also have holes of 15 mm diameter all round the walls at 300 mm apart to provide cross ventilation. The well-aerated material does not emit any foul smell. To combat the ants from attacking the worms, it is good to have a water column in the centre of the parapet wall of the vermipits. After two weeks the earthworms are released on the soil surface. They enter into the organic matter and mix through the crevices left on the surface.

Red wriggler (*Eisenia fetida*)

(A)

(B)

Fig. 22.15. Construction of vermipits; A- four tank vermipit made up of bricks and concrete showing holes on the wall, B- three tank vermipit

(*c*) Preparation of Vermibed: Vermibed (*vermes* = earthworms; *bed*= bedding) is the actual layer of good moist loamy thick soil of about 15 to 20 cm placed at the bottom.

(*i*) The compost along with 150 earthworms is introduced into the loamy soil kept in a compost pit of about 2 m × 1 m × 0.75 m, with a vermibed of about 15 to 20 cm thick. Earthworms inhabit into loamy soil as their home.

(*ii*) Then handful lumps of fresh cattle dung are placed randomly over the vermibed. The compost pit is then layered to about 5 cm with dry leaves or preferably chopped hay, straw or agricultural waste biomass. The set-up is left undisturbed for six weeks. During the composting process water is sprinkled occasionally on the surface.

(*iii*) The bed should neither be dry or soggy. The pit may then be covered with coconut or Palmyra leaves or an old jute (gunny) bag to discourage birds. Plastic sheets on the bed are to be avoided as they trap and retain heat.

(*iv*) After the first 30 days, wet and pre-digested organic waste of animal and/or plant origin from the kitchen, hotels, hostel or farm is spread over it to a thickness of about 5 cm. This can be repeated twice a week.

A crate (600 mm × 45 mm × 600 mm) can have a population of 1,500 adult *Eudrilus eugeniae* and 3,000 - 5,000 of *Eisenia foetida* and *Perionyx excavatus*. Frequent harvesting of *Eudrilus eugeniae* is necessary to reduce population pressure and to enable continuous growth of earthworm population.

(*d*) Harvesting of Compost: The black, granular, lightweight and humus-rich compost is ready for use when the material is moderately loose and crumbly. Between 60 to 90 days (depending on the size of the pits) the compost becomes ready as indicated by the presence of earthworm castings (vermicompost) on the top of the bed. Vermicompost can now be harvested from the pit. During separation of the worms from the compost two to three days before emptying the beds, watering should be stopped. The worms can also be separated by using sieves. The earthworms and the thicker material which remains on top of the sieve are put back in the pit and the process starts again. The smell of the compost is earth-like. Any bad smell indicates the incomplete fermentation and the bacterial processes are still going on. A musty smell indicates the presence of mold or overheating which leads to loss of nitrogen. If this happens, aerate the heap better or start again adding more fibrous material and keeping the heap drier. The compost is then sieved before being packed. The harvested material should be placed in a heap in the sun so that most of the worms move down to the cool base of the heap.

Vermicompost prepared in multitak

In the two or four pit system, watering should be stopped in the first chamber so that worms will automatically move to another chamber where the required environment for the worms are maintained in a cyclic manner and harvesting can be done continuously in cycles.

5. Benefits of Vermicompost

Organic wastes can be broken down and fragmented rapidly by earthworms resulting in a stable nontoxic material with good structure. It has a potentially high economic value and also acts as soil conditioner for plant growth. Vermicompost or worm compost is enriched with the earthworm intestine microbes, ferments, enzymes, vitamins and other biologically active elements that are essential to promote healthy plants. Worm compost is rich in humus, essential nutrients such as nitrogen, phosphorus, potassium, sulphur, and other trace elements; hence it acts an excellent natural fertilizer. Vermicompost is well balanced nutrient content that promotes healthy growth of plants. It has high organic content that helps to improve nutrients and moisture retention, drainage and aeration. Microbial activities on soil help to improve the soil structure and convert nutrients in soil to form readily available to plants. Vermicompost improves water-holding capacity, thus a superb choice for potted plants such as flowers, small fruits trees and other ornamental plants. It is natural organic fertilizer produced from organic materials through vermicomposting. Hence, it is a safe, healthy and green choice. Vermicomposting involves great reduction in populations of pathogenic microorganisms, thus not differing from composting from this point of view.

Vermicomposting also leads to decrease the environmental problems arising from their disposal without needing in many cases to complete the process. It should be realised that vermicomposting can be a useful cottage industry for the underprivileged and the economically weak persons as it can provide them a supplementary income. If every village can formulate a cooperative society of unemployed youth/women group, it could be a wise venture for them to produce vermicompost and sell it back to the village at a recommended price. The youth will not only earn money, but also aid society by providing excellent quality organic manure for sustainable agro-practices.

Vermicompost ready for use

6. Vermicompost Manufacturers and Suppliers

There is large number of companies dealing with vermicompost. Names of some of the companies along with their profiles are given in Table 22.7.

Table 22.7: Manufacturers and suppliers of vermicompost in India.

Name of manufacturers	*Products*	*Location*
Shree Namo Fertilizer	Vermicompost, agricultural vermicompost, natural vermicompost, organic vermicompost, vermi fertilizer, natural plant growth nutrient, plant booster	Patel Nager, Ghaziabad, U. P.
S. K. International	Bulk supplier of vermicompost fertilizers, crops Vermicompost fertilizers, plants vermicompost fertilizers, natural vermicompost fertilizers	Nakur, District Saharanpur, U. P.
Concept Biotech	Black vermicompost, organic vermicompost, vermicompost fertilizers and plant growth vermicompost fertilizers	Karelibaug, Vadodara, Gujarat
Sarva Mangalam Natural Products	Organic vermicompost, herbal vermicompost, multipurpose cleaning liquid and herbal pesticides	Bhatoli Village, Gwarighat, Jabalpur, M. P.
Total Agri Care Concern	Vermicompost, premium vermi compost, organic fertilizer, plant nutrients and organic nutrients	Netaji Subhas Road, Kolkata, West Bengal
Shri Vinayakk, Coimbatore	Vermi compost, earthworm vermi compost, organic vermi compost and bio-organic vermi compost	Uppilipalayam Post, Coimbatore, T. N.
Gujarat Chemicals and Fertilizer Trading Co	Vermicompost, bio-vermicompost and agricultural bio- vermicompost	Karelibaug, Vadodara, Gujarat
Akshat Farms	Vermicompost fertilizers, organic vermicompost, bio-vermicompost, earthworm compost, vermi organic wash, vermi bio wash, vermiwash	Hiran Magri, Udaipur, Rajasthan
Swati Enterprises	Vermicompost, earth worms, undeveloped earth worm, worm castings, bedding materials, organic wastes	Pocket A, Dilshad Garden, New Delhi,
Ruchi Biochemicals	Vermi compost, vermicompost fertilizer, earthworm vermicompost, commercial vermicompost, eartworm vermiwash and vermi wash	Bimbisar Nager, Goregaon East, Mumbai, Maharashtra
Suriya Farms	Organic compost, vermi culture, bio-organic fertilizers, compost fertilizer, natural organic manure, press mud	M. R. Road, Madurai, T. N.
Sushil Corporation	Vermicompost, vermicompost fertilizers, agricultural vermicompost, bio-fertilizers, organic fertilizers, organic manure and organic neem manure	Vikram Tower, Sapana Sangita Road, Indore, M. P.
Baba Bio Tech	Organic fertilizers, organic manure, vermicompost	M. M. D. A., Maduravoyal, Chennai, T. N.
G. D. Vermicompost	Organic manure,vermicompost	Saraiyaganj, Muzaffarpur, Bihar

E. BIOGAS (*GOBAR* GAS) PRODUCTION TECHNOLOGY

Energy is the basis of all live forms on the earth. Everyone needs energy in one form or the other form, whether it is human beings, animals or plants. Energy is needed not only for the survival of the mankind but also for the growth of a country. Higher the consumption of energy, higher would be the growth of the country, because socio-economic development of a nation depends on the availability of energy and energy consumption.

At present the most commonly used energy resources all over the world are: oil, gas, coal, etc. Several countries have been able to utilise nuclear energy to meet their energy requirements. India has to use the available conventional sources judiciously but also harness the non-conventional sources of energy. The most commonly available non-conventional sources of energy are biogas (*gobar* gas), solar, wind, biomass, etc.

For the first time in 1776, an Italian Physicist, Volta demonstrated methane in the marsh gas, generated from organic matter in bottom sediments of ponds and streams. Under anaerobic conditions, the organic materials are converted through microbiological reactions into gases (fuel) and organic fertilizer (sludge). The mixture of gases is composed of 63% methane, 30% CO_2, 4% nitrogen and 1% hydrogen sulphide and traces of hydrogen, oxygen and carbon monoxide. Methane is the main constituent of biogas. It is also referred to as biofuel, sewerage gas, Klar gas, sludge gas, will-o-the wisp of marsh lands, fool's fire, *gobar* gas (cow dung gas), bioenergy and fuel of the future. About 90% of energy of substrate is retained in methane. Biogas is used for cooking and lighting purposes in rural sector. It is devoid of smell and burns with a blue flame without smoke.

In our country biogas has been developed and is being utilised to the maximum extent. The largest cattle population in the world and also the warm climate in major part of the country for most of the year make the simple biogas technology highly successful in our country.

Rural people depend upon biomass fuels such as fire wood, animal waste and crop residues for meeting the requirements of cooking and heating energy. These fuels are burnt in traditional *chulhas*, causing economic loss to the nation and health problems to women. Technologies have been developed indigenously to produce modern biomass fuel, such as biogas from cattle dung and other organic wastes and to burn biomass in an efficient manner in improved *chulhas*.

Biogas technology is an alternate source of energy in India to meet the basic need for cooking fuel in rural areas. By using local resources viz., cattle waste and other organic wastes, energy and manure are derived. In the late 1970s, realisation of this potential and the fact that India supports the largest cattle wealth led to the promotion of National Biogas Programme. Therefore, The Ministry of Non-Conventional Energy Sources continues to implement a centrally sponsored scheme, National Biogas and Manure Management Programme (NBMMP), a modified version of the Ninth Plan Scheme on National Project on Biogas Development (NPBD), with the objectives to promote family type biogas plants and biogas power stations.

Cattle as source of biomass used in biogas (*gobar* gas) production.

1. Biogas Technology in India

Many developing countries are encouraging for installation of biogas plants to meet the demand of fuel. India is one of the pioneer countries in biogas technology where biogas research and plant construction has been carried out over the past 30 years. In 1939, the experiments were initiated at the IARI, New Delhi. However, there are many other institutions where research and development programmes are carried out such as Khadi and Village Industries Commission (KVIC) (Bombay), the Gobar gas Research Station (Ajitmal, Etawah), the Planning Research and Action Institutes (Lucknow), and National Environmental Engineering Research Institute (Nagpur).

In 1951, for the first time, biogas plants were constructed with the target of 8,000 units before 1973. KVIC played a major role in construction of biogas plants during 1960s. Government of India launched an All India Coordinated Project with the target of 1,00,000 units by 1978, but the number could reach to 50,000 units only.

Up to 1993, Non-Conventional Energy Development Agency (NEDA) of Uttar Pradesh has installed about 100 night soil based biogas plant throughout the state. One of the unique examples is the Rajapurwa biogas plant of Kanpur where about 1400 kg human waste from 50-seat toilet complex is pooled through underground pipelines at one place. The waste is processed through a digester into about 55 m^3 methane gas. This quantity is enough to run an 8 HP engine and pump set and to provide fuel to some biogas lamps and cooking burners to about 3500 dwellers. For maintenance of the plant, each family pays some nominal charges.

The sewage plant at Okhla (New Delhi) has 15 digesters of 5,665 m^3 capacity each, and produces 17,000 m^3 gases per day. The gas generated is equivalent to about 10,000 litres of kerosene per day, where 1 m^3 biogas has energy potential equal to about 2/3 litres of kerosene. The Dadar Sewage Treatment Plant (Mumbai) produces about 2,800 m^2 biogas per day from sewage. From this plant pipe lines are connected to about 500 houses, located about 5 km away from the station.

In Sonepat District of Haryana, over 300 biogas units of 2 to 10 m^3 capacity have been set up for cooking and lighting purposes. The Haryana Government provided 25% subsidy per plant for setting up the units. National Sugar Institute (Kanpur) has developed methods for production of biogas from bagasse and other agricultural residues. This plant was built up in 1960s. In this plant, 12 steel digesters, each having capacity of 50 m^3, are set up to which about 14 tones of mixture of agricultural wastes and cattle manure and 28 tones of water are fed to begin the biogas production. Himachal Pradesh Government set up the National Project on biogas, under which about 12, 292 units in 1987-89, and 3,505 units in 1989-90 were setup.

Government policy is to provide direct subsidy of 25% of the cost of plants and encourage banks to pay loans for remaining costs to aspirants. High construction cost, and cost of cement and steel digesters and rates of interest on loans are some of the factors that discourage villagers to setting biogas plants. It has been pointed out that in villages, support of minimum number of cattle (at least 5) is necessary to run a small home digester plant. Five cattle generate dung to produce about 2 m^3 biogas plant to meet the demand of cooking and lighting for a family of 4–5 people. In addition, attempts are also being made to develop a small engine powered with methane gas.

(*a*) Benefits from Biogas Plants: Biogas is used mainly for cooking and lighting purposes. Besides, there are many other advantages in installing the biogas plants. It is used in internal combustion engines to power water pumps and electric generators. It is also used as fuel in fuel type refrigerators. Sludge is used as fertilizer. The most economical benefits are minimising environmental pollution and meeting the demand of energy for various purposes.

(*b*) Feedstock Materials: There are two sources of biomass for biogas production, *i.e.* plant and animal. Biomass from plant origin is aquatic and terrestrial ones derived from various sources. Biomass from animals is cattle dung, manure from poultry, goats and sheep, and slaughter house and fishery wastes. In most of biogas plants cattle dung is used for gas production. There are many cases where *gobar* gas plants are in operation, and sullage gas production is started. Beside cattle dung (*gobar*), agricultural wastes containing cellulosic and lignocellulosic materials are also being used.

(*c*) Construction of a Biogas Plant: India is a leader country in *gobar* gas technology. There are perhaps millions of gobar gas plants in India. A long-term goal is to have one at each of our homes. Different states have started installation of *gobar* gas plant at village level.

First a pit of 10 feet deep is dug. Then a water-tight cement cylinder (with brick or gravel) is constructed (Fig. 22.16 A). A wall is built across the middle, extending up from the bottom, not quite to the top. Intake and outgo pipes are installed. The whole unit is made water-tight.

Fig. 22.16. Types if biogas plant; A- training for construction of fixed dome biogas plant; B- a fixed dome biogas plant; C- a floating drum biogas plant.

A second type of biogas plant is also installed where floating metallic tank is prepared (C). A pit of defined size is dug and cemented side walls are constructed. A metallic tank is fabricated which has inlet and outlet knobs. Cow dung slurry is prepared and transferred into the pit. The manure is mixed with water in the intake basin to make slurry which goes down the pipe to the bottom of the left side. This side of the cylinder gradually fills and overflows to the right side. Meanwhile, the whole mass bubbles methane up to the top. It collects under the large metal bell-like cover. The gas builds pressure, and can be taken off through a rubber tube to a gas stove in a kitchen.

When both sides of the cylinder are full, the effluent flows out from the bottom of the right side. Each time more raw manure is added to the left. What comes out on the right is of more value as fertilizer than the raw manure. In this process, organic fraction of the wastes is segregated and fed to a closed container (biogas digester) where, in the presence of methanogenic bacteria and under anaerobic conditions, it undergoes bio-degradation producing methane-rich biogas and effluent.

2. Biogas Production: Anaerobic Digestion

Anaerobic digestion is carried out in an air tight cylindrical tank which is known as digester. A digester is made up of concrete bricks and cement or steel. It has a side opening (charge pit) into which organic materials for digestion are incorporated. There lies a cylindrical container above the digester to collect the gas. In India, single stage digester is set up in gobar gas plant. However, in other countries single-stage, two stage and multistage digester(s) are set up to accomplish digestion at high rate. A diagram of a single stage digester for biogas/*gobar* gas production is shown in Fig. 22.17.

In biogas plant, fresh cattle dung is deposited into a charge pit, which leads into the digestion tank. Dung remains in tank. After 50 days, sufficient amount of gas is accumulated in gas tank, which is used for house-hold purposes. Digested sludge is removed from the basin and is used as fertilizer. Usually digesters are buried in soil in order to benefit from insulation provided by soil. In cold climate, digester can be heated by the installations provided from composting for the agricultural wastes.

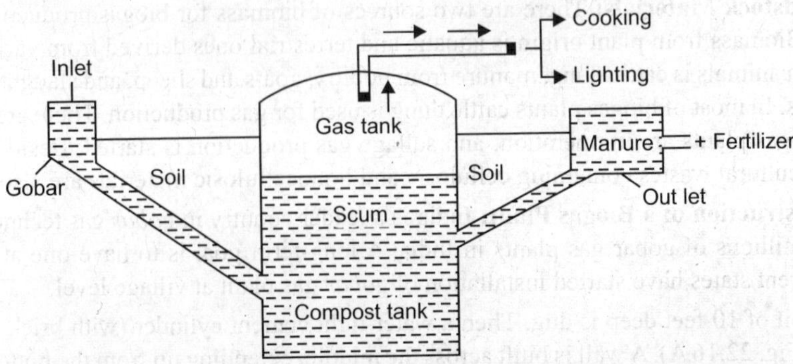

Fig. 22.17. Diagram of a Janata Vayu Gas (*gobar* gas) plant.

There are several groups of bacteria that perform each step; in total dozens of different species are needed to degrade a heterogeneous stream completely. Most of these bacteria adhere to the substrate prior to extensive hydrolysis. Anaerobic digestion is accomplished in three stages, *solubilization, acidogenesis* and *methanogenesis*. These stages are characterised by thee groups of bacteria.

(*a*) Solubilisation/Hydrolysis: It is the initial stage, when feedstock is solubilised by water and enzymes. The feedstock materials (cattle dung and other organic polymers) containing high weight organic molecules such as proteins, carbohydrates, fat and cellulose are dissolved in water to make slurry. The complex polymers are hydrolysed and broken down into smaller molecules like sugars, amino acids, fatty acids, water and alcohols by hydrolytic fermentative bacteria which are mostly anaerobes (Fig. 22.18).

Cattle dung is used to make slurry.

(*b*) Acidogenesis: During this stage, the second group of bacteria, *i.e.* facultative anaerobic and H_2 producing acidogenic bacteria convert the simple organic material via oxidation / reduction reactions into organic acids, carbondioxide, hydrogen sulfide and ammonia. These substances serve as food for the final stage.

(*c*) Acetagenesis: The products from the acidogenesis are used for the production of acetates, carbon dioxide and hydrogen via acetagenic dehydrogenation by obligate H_2 producing acetagenic bacteria. There is other group of acetagenic bacteria which produce acetate and other acids from H_2 and CO_2 via acetagenic hydrogenation (Fig. 22.18).

(*d*) Methanogenesis: Methane-producing bacteria are called methanogens. This is the final stage of anaerobic digestion where the products of acidogenesis and acetagenesis (*i.e.* acetate, H_2 and CO_2) are converted by methane producing bacteria (methanogens) into methane, carbon dioxide, water and other products (Fig. 22.18).

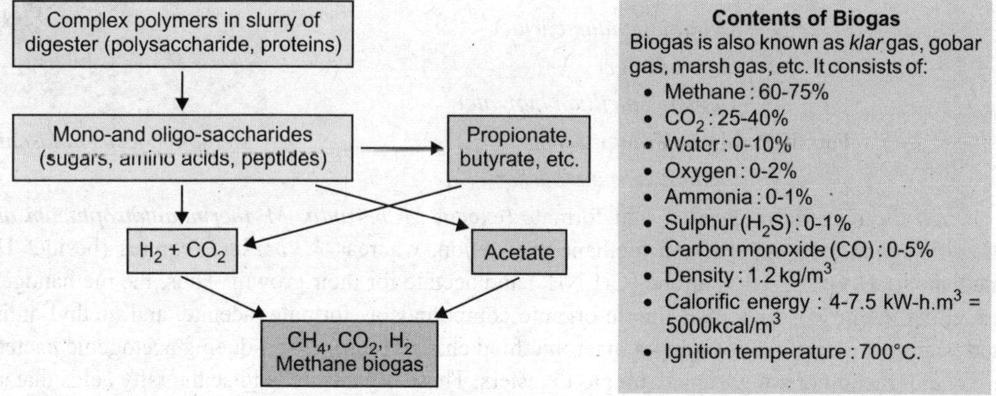

Fig. 22.18. Metabolic pathway of methane production *i.e.* anaerobic digestion of organic matter and methane production.

Different species of methanogens are involved in breakdown of complex organic matter into acetate or other organic acids. Acetate is one of the substrates of methanogenic bacteria. Hydrogen with CO_2 is a general substrate for methanogenesis. Numbers of these bacteria differ with type of substrates. For example, counts of $10 - 10^6$ per ml and $10^5 - 10^8$ per ml of hydrogen utilising bacteria were determined from the pig waste and sewage sludge digesters, respectively.

(*i*) Methanogens. Methanogens are a unique group of bacteria. They are obligate anaerobic microorganisms living in a variety of oxygen-free environments such as sediments of lakes, swamps, rice fields, rumen of cows, the intestine of termites or the anaerobic stage of sewage plants. They gain their energy for ATP synthesis by producing methane from either CO_2/H_2 or from acetate.

They are obligate anaerobes and have slow growth rate. They play a major role in breakdown of substrate into gas form. They are the only organisms which can anaerobically catabolise acetate and H_2 to gaseous products in the absence of exogenous electron acceptors other than CO_2 or light energy. In their absence, effective degradation would cease because of accumulation of non-gaseous, reduced fatty acid and alcohol products of the fermentative and other H_2 using bacteria that have almost the same energy content as the original organic matter. In morphology, they are of different types such as cocci, bacilli, spirilli and sarcina. In 1979, W.E. Balch and coworkers have classified the methanogens as below:

Order 1. Methanobacteriales

Family. Methanobacteriaceae

Methanobacterium formicicum

M. bryantii

M. thermoautotrophicum

Methanobrevibacter ruminantium

M. arboriphilus

M. smithii

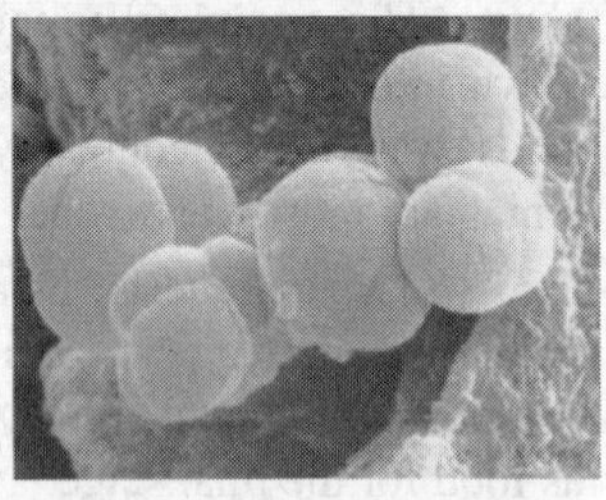

Methanosphaera stadtmanae

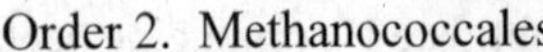

Order 2. Methanococcales

Family. Methanococcaceae

Methanococcus vanielii

M. voltae

Order 3. Methanomicrobiales

Family 1. Mathanomicrobiaceae

Methanomicrobium mobile

Methanogenium cariaci

M. marisnigri

Methanospirillum hungatei

Family 2. Methanosarcinaceae

Methanosarcina barkeri

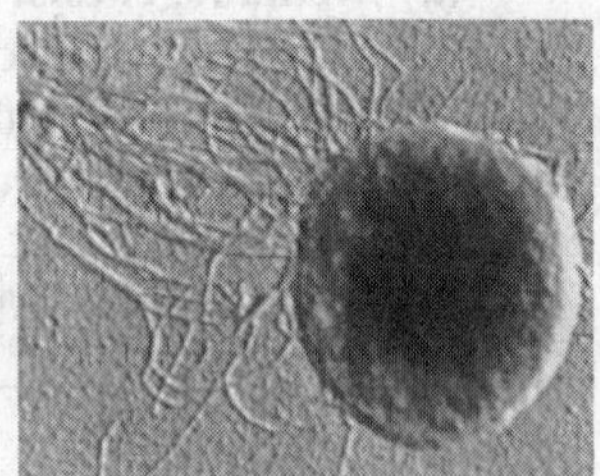

Methanococcus janaschii

All the bacteria require H_2 and formate (except *M. bryantii, M. thermoautotrophicum and M. arboriphilus*) for growth and methane production, whereas *M. barkeri* requires (besides H_2) methanol (CH_3OH), methyl amine (CH_3NH_2) and acetate for their growth. Thus, the methanogens are either autotrophs or utilise simple organic compounds as formate, acetate, and methyl amine and occupy the terminal position in anaerobic food chain. Hydrogen-producing acetogenic bacteria are one of the important groups in biogas digesters. These organisms oxidise the fatty acids that are longer than acetate and thereby release energy from the substrate in the form of methane.

Methanogens possess very limited metabolic repertoire, using only acetate or C_1 compounds (H_2 and CO_2, formate, methanol, methylamines or CO). Methane is the end product of the reaction. Of the methanogenic genera, *Methanosarcina* sp. and *Methanosaeta* sp. form methane by aceticlastic reaction. While faster-growing *Methanosarcina* sp. is predominant in high-rate, shorter-retention digesters wherein acetate concentration is higher, *Methanosaeta* sp. are predominant in low-rate, and slow-turnover digesters. Pathway of methane formation is given in Fig. 22.18.

***(ii)* Mechanism of methane formation.** Metabolically the methanogens are very peculiar. Carbon dioxide fixation, Calvin cycle, serine or hexulose pathway are absent in them. The mechanism of methane formation is not well understood. In 1979, Ralph S. Wolfe has found the involvement of several new coenzymes which are not present in any other group of bacteria. These coenzymes are methyl coenzymes M reductase (MCR), hydroxy methyl coenzyme M, coenzyme F420, coenzyme F430, component B, corrinoids, methanofuran or carbon dioxide reducing factor and methanopterin and formaldehyde activating factor.

The key enzyme in this process is methyl coenzyme M reductase (MCR). Its active site contains the hydroporphinoid nickel complex coenzyme F430. The nickel center of coenzyme F430 is given in Fig. 19A, which catalyses the unusual reaction (B) as given below :

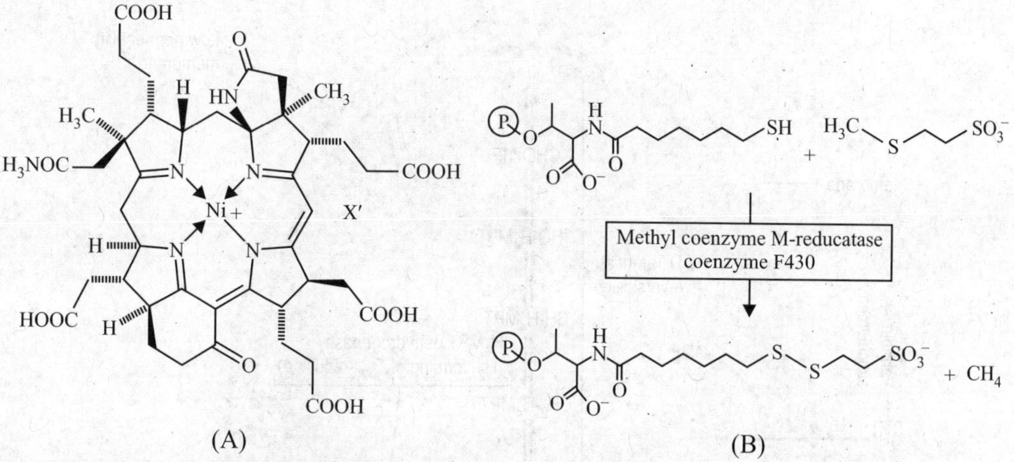

(A) (B)

Fig. 22.19. Coenzyme F430 of MCR containing Ni centre (A); catalysis of reaction by MCR and formation of CH_4 (B).

A detailed description of biochemistry and pathways of methane formation is out of scope of this text. Therefore, only reactions are given in this connection. Primary reaction in which carbon monoxide takes part is as below:

$$CO + H_2O \longrightarrow CO_2 + H_2$$

The secondary reaction takes place in the presence of sufficient hydrogen:

$$CO_2 + 4H_2 \longrightarrow CH_4 + 2H_2O$$

Other reactions showing methane formation from various substrates are given below:

$$4CH_3OH \longrightarrow 3CH_4 + CO_2 + 2H_2O$$

(Methanol)

$$4HCOOH \longrightarrow CH_4 + 3CO_2 + 2H_2O$$

(Formate)

$$CH_3COOH \longrightarrow 12\ CH_4 + 12\ CO_2$$

(Acetate)

***(i)* Adaptations by Methanogens:** The different adaptations that allow methanogens to thrive in conditions of chronic energy stress using the example of the H_2/CO_2 catabolic pathway are shown in Fig. 22.20. There are seven adaptations included in this figure as given below:

- The use of isoprenoidal, ether-linked lipids to reduce futile ion cycling (A).
- The regulation of isoenzyme production in the catabolic pathway as a function of H_2 (Gibbs free energy) availability (B).
- The frequently low quantity of Gibbs free energy available from catabolism (as low as ~10 kJ mol^{-1} of methane produced) (C).

- The inferred capacity to thrive with the net extrusion of one ion per round of catabolism (D).
- The dynamic stoichiometry of proton extrusion during heterodisulphide (CoB-S-S-CoM) reduction as controlled by H_2 concentration (E).
- The reversibility of the first five steps of the catabolic pathway which allow for near-equilibrium conditions and high efficiency (F).
- The variations in stoichiometry of proton-coupled ATP synthesis that probably allow variability in the BEQ (G).

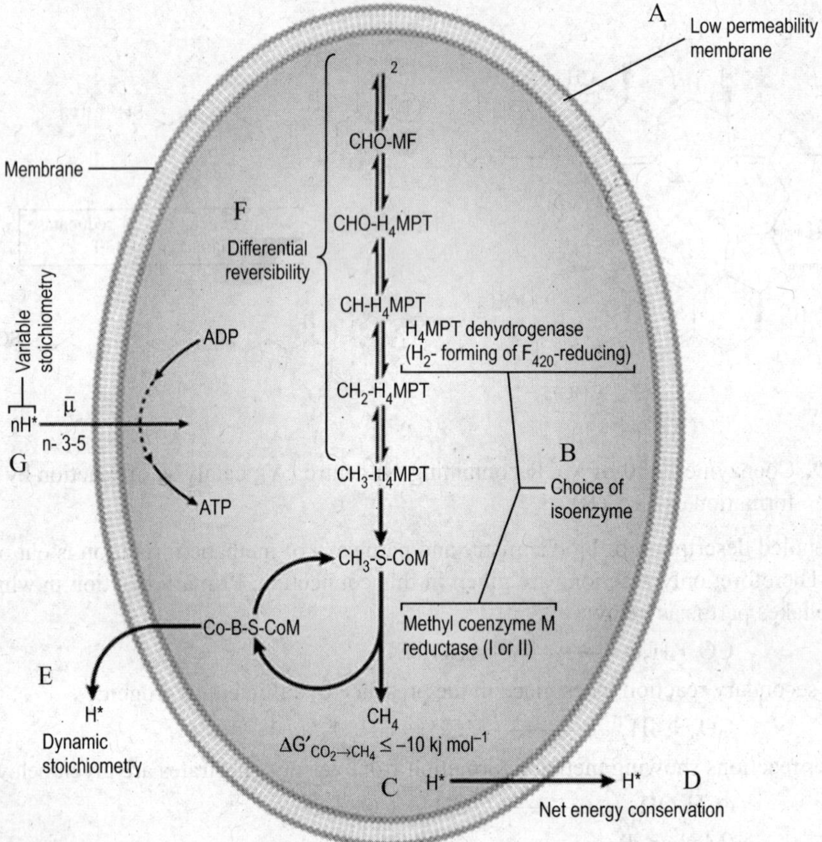

Fig. 22.20. Different adaptations that allow methanogens to thrive in chronic energy-stress conditions. BEQ, biological energy quantum; MF, methanofuran; H_4MPT, tetrahydromethanopterin; CoM, coenzyme M; CoB, coenzyme B.

3. Biogas Production from Different Feedstocks

Although there are various types of biomaterials differing in chemical composition, yet *gobar* (cow dung) is most popular in India. Besides *gobar*, other materials which would be successful in the coming years are agricultural wastes, municipal wastes, industrial wastes and some of the aquatic biomass.

***(a) Salvinia*:** *Salvinia* is a water fern and can be a feedstock material for biogas production. Fermentation of *Salvinia* starts within 7-9 days on putting under water in suitable container. Biogas yield is about 0.1 litres /kg fresh weight for 4 weeks. Air dried weed produces about 1 litre/kg for 90 days and fermentation is continued for 3 months. Thereafter, gas yield gradually declines.

Daily requirement of gas is 0.4 litre per capita. Two tonnes of air dried *Salvinia,* therefore, can meet the fuel requirement of a small family for 3–5 months. Special advantage of using *Salvinia* is that unpleasant odours do not come out. This will certainly provide cheap and clean burning fuel and reduce the increasing pressure on forests.

(*b*) Water Hyacinth: This weed can be used for the production of laboratory and domestic fuel. Its decomposition rate is higher than that of cow dung for gas production. Water hyacinth is totally decomposed within 3 days in summer, while cow dung takes 8 days.

The ratios between total gas evolved by water hyacinth and cow dung under the identical conditions in summer and winter are 5:3 and 5:1, respectively. Methane production under varied conditions is given in Table 22.8.

Table 22.8. Composition of methane in biogas produced by water hyacinth under different conditions.

Experimental design Wet mass: water ratio (g/ml)	*Temperature (°C)*	*Biogas (ml)/g wet weight*	*Methane (%) in total biogas*	*Methane (ml)/g wet weight*
300: 800*	25 ± 5	11.9	61.5	11.0
754: 800*	25 ± 5	11.1	57.2	6.4
800 : 800**	25 ± 5	12.9	61.1	7.9
500 : 350*	36	12.9	69.2	8.9
500 : 350*	36	12.4	91.2	11.3

*, Chopped water hyacinth; **, water hyacinth blended into slurry; (amendment of Ni (3.95 mg/ kg wet mass) and Cd (12.66 mg/kg wet mass).

Works done at the National Space Technology Laboratory, Florida (U.S.A.) revealed that 1 kg dried water hyacinth produced 374 litres of biogas in which the proportion of methane was 60–80%. Moreover, one hectare area can produce 600 kg of dry biomass per day which can generate 2,29,400 litres of biogas/ha/day.

In India, works on biogas production by water hyacinth has gained much momentum. Research works have been carried out at the Hissar Agricultural University (Hissar), IIT (Madras), R.R. Laboratory (Jorhat) and Central Mechanical Engineering Research Institute (CMERI) (Durgapur). CMERI has designed a continuous feed type plant of capacity of 3,000 litres gas/day, in which initially 1,800 kg coarsely chopped semi-dried water hyacinth is used. Thereafter, plant is required to feed 40 kg raw materials. Peak of biogas generation period arrives within 13–18 days. Total available energy from water hyacinth is estimated as 1.334 × 109 Kcal per year plus Rs. 2.25 million worth complex.

Gas produced from water hyacinth by methanogens in an air tight tank contains 70% methane and 30% CO_2. Carbon dioxide can be removed before its use for cooking, heating, lighting and source of power for engines, tractors and cars. Increase in recreational value of water bodies from which hyacinth is to be removed can be an additional advantage of exploiting it for gas production. The hyacinth biomass contains substantial amounts of nutrients, the removal of which may keep a water body clean and less eutrophic.

(*c*) Municipal Sewage: New techniques have been developed to make available sullage gas for cooking purpose or industrial activities. The Okhla Sewage Disposal Works (New Delhi) receives 25 m^3 sewage per second, and generates 18,000 m^3 of gas. In energy terms, it is equivalent to about 4.161 million litres of kerosene/year. From the plant, gas pipe lines are given to about 700 families of the surrounding villages. It is hoped, if whole sewage generated from the city is digested anaerobically to yield biogas, it could meet about 20% fuel demand of the city.

4. Microbial Diversity in Biogas Digesters

Microbial diversity in biogas digesters is as great where several fermentative bacterial species have been reported to play an important role for biogas production. The nature of the substrate determines the type and extent of the fermentative bacteria present in the digester. Presence of high number of proteolytic organisms in cow dung-fed digesters has been found as compared to the other animal's waste-fed digesters. The cow dung-fed digesters support higher amylolytic microorganisms and poultry waste-fed digesters show higher proteolytic population. Among fermentative organisms, *Bacteroides succinogens*, *Butyrivibrio fibrisolvens, Clostridium cellobioparum*, *Ruminococcus albus* and *Clostridium* sp. remain predominant.

There is clear differentiation in the distribution of type of cellulolytic bacterial in rumen and biogas digester. In rumen, *Ruminococcus* sp. alone accounted for 60% of the total population, whereas in biogas digester the predominant species belong to the genera *Bacteroides* and *Clostridium* rather than the genus *Ruminococcus*. *Ruminococcus flavefaciens, Eubacterium cellulosolvens, Clostridium cellulosolvens, Clostridium cellulovorans, Clostridium thermocellum, Bacteroides cellulosolvens* and *Acetivibrio cellulolyticus* are some of the other predominant fermentative bacteria present in cattle dung-fed digesters.

Most of these bacteria adhere to the substrate before substrate hydrolysis. The digester slurry contains higher cellulolytic population, but the outlet of the digester records the least cellulolytic population.

5. Factors Affecting Methane Formation

Following are the factors affecting methane production:

(*i*) **Slurry:** Proper solubilisation of organic materials (the ratio between solid and water) should be 1:1 when it is house hold things.

(*ii*) **Seeding:** In the beginning, seeding of slurry with small amount of sludge of another digester activates methane evolution. Sludge contains acetogenic and methanogenic bacteria.

(*iii*) **pH:** For the production of sufficient amount of methane, optimum pH of digester should be maintained between 6-8 as the acidic medium lowers methane formation.

(*iv*) **Temperature:** Fluctuation in temperature reduces methane formation, because of inhibition in growth of methanogens. In case of mesophilic digestion, temperature should be between 30°C and 40°C but in case of thermophilic ones, it should be between 50°C and 60°C.

(*v*) **Nitrogen concentration:** Excess amount of nitrogen inhibits growth of bacteria, and thereby lowers methane production. Therefore, use of such materials should be discouraged.

(*vi*) **Carbon-nitrogen (C : N) ratio:** Improper C:N ratio lowers methane production. Maximum digestion occurs when C : N ratio is 30:1. Amendment of nitrogen or carbon substrates should be done exogenously according to chemical nature of substrates used in fermentation.

(*vii*) **Creation of anaerobic conditions:** It is obvious that methane production takes place in strictly anaerobic condition; therefore, the digesters should be totally airtight. In Indian conditions, digesters are burried in soil.

(*viii*) **Addition of algae:** In 1989, S. Ramamoorthy and N. Sulochana have found an enhancement in biogas production from cow dung on addition of the green alga, *Zygogonium* sp. The amount of biogas produced from the alga was twice (344 ml/g dry algae) of that obtained from cow dung (179/g dry cow dung) alone. Also, the duration of gas evolution increased with increasing the proportion of slurry. The caloric value of the gas was 4,800 KCal/ m^3 and the percentage of methane was 55.4%. Generally in winter season gas production is lowered to some extent. Therefore, addition of algae holds promise to get biogas in sufficient amount even in winter season also.

6. Uses of Biogas and Sludge

Biogas can be utilised either for cooking / heating purposes or for generating motive power or electricity through dual-fuel, gas engines, low pressure gas turbines or steam turbines (Fig. 22.21).

The sludge from anaerobic digestion can be used as a soil conditioner after stabilisation or as manure depending upon its composition, which is determined mainly by the composition of the input waste. Digester slurry contains higher cellulolytic population of bacteria, while the outlet of the digester recorded the least cellulolytic bacterial population. Increased crop yield has been found after use of sludge as manure. The physical, chemical, and biological characters of the soil get improved with application of slurry into the soil. Moreover, increased grain yield of blackgram by 80.5% has been recorded with addition of 40 t/ha slurry together with recommended dose of NPK. Moreover, the spent slurry can be used as a carrier for *Rhizobium*. Shelf -life of *Rhizobium* in immobilised spent slurry has been found for about three months.

Fig. 22.21. Anaerobic digesters powering generators.

F. BIOMINING (= BIOLEACHING)

Biomining is a technique used to recover the various types of minerals from ore using microorganism. Biomining encompasses two related microbial processes that are useful in the extractive metallurgy of heavy metals: *bacterial leaching* (also known as *bioleaching*), and *biooxidation*. Leaching is the solubilisation of one or more components of a complex solid by contact with a liquid phase. In bioleaching, the solubilisation is mediated by bacteria. So bacterial leaching is a process by which the metal of interest is extracted from the ore by bacterial action. On the other hand, biooxidation implies the bacterial oxidation of reduced sulfur species accompanying the metal of interest, for example biooxidation of refractory gold minerals.

Methods of biomining are cheap, reliable, efficient and also environment friendly. Using biotechnology efficiency of biomining can be increased by using genetically modified microorganisms. High grade ore deposits are easily accessible, hence these become rapidly depleted. Therefore, it is necessary to recover mineral resources from 'low grade ore deposits'. Different compositions of metals in different ores are given in Table 22.9.

Table 22.9. Metal characterization of the mining ore.

Metal	*Concentration (mg/kg residue)*
Copper	7,245
Iron	26,470
Zinc	201
Nickel	27

Still no appropriate technology is available for recovery of metals from low grade deposits. Interestingly, some microorganisms can do it efficiently. Recently this potential of microbes could be realised and efforts are being made to use them for enhanced recovery of mineral resources from natural deposits. In recent years, microorganisms have been used for recovery of two important natural resources such as metals and petroleum.

In 1957, American microbiologists recognised a relationship between the presence of *Thiobacillus ferrooxidans* and the dissolution of metals in copper leaching operation. *Thiobacillus ferrooxidans* and *T. thiooxidans* are the thermoacidophilic archaebacteria. They are autotrophs and grow in acidic and hot environments. Commercial use of *T. ferrooxidans* for metal extraction is widespread. It has been found that the *Thiobacillus* spp. can be used for extraction of copper and uranium from insoluble minerals. The approach is less expensive and has fewer environmental impacts than conventional processes.

This implication of microbial activity in weathering, leaching and deposition of mineral ores could develop into a recent field of biotechnology, *i.e.* biohydrometallurgy. Biohydrometallurgy refers to inter-disciplinary field involving the processes that: (*i*) use microbes usually bacteria and archaea (*bio*), (*ii*) mainly take place in aqueous environment (*hydro*), and (*iii*) deals with metal production and treatment of metal containing materials and solutions (*metallurgy*).

It may be defined as a branch of biotechnology that deals with the study and application of the economic potential of the interactions between the microbial world and the mineral kingdom. Biohydrometallurgy is used to perform processes involving metals, for example, microbial mining, oil recovery, bioleaching, water-treatment and others. It is mainly used to recover certain metals from sulfide ores, and usually utilised when conventional mining procedures are too expensive or ineffective in recovering a metal such as copper, gold, lead, nickel and zinc.

Biomining Efficiency

Bioming efficiency or mineral recovery from its ore depends on several factors such as choice of bacteria, crystal lattice energy, and surface area, composition of the ore, acidity, temperature and also surfactants used in biomining. Copper, gold and oil are extracted from its ore or reservoir by biomining method.

One of the first reports where bioleaching might have been involved in the mobilisation of metals is given by the Roman writer Gaius Plinius Secundus (23 - 79 A.D.). He described the method of recovering copper minerals by using a leaching process. The translated version is as follows : Chrysocolla is a liquid in the before mentioned gold mines running from the gold vein. In cold weather during the winter the sludge freezes to the hardness of pumice. It is know from experience that the most wanted (chrysocolla) is formed in copper mines, the following in silver mines. The liquid is also found in lead mines although it is of minor value. In all these mines chrysocolla is also artificially produced by slowly passing water through the mine during the winter until the month of June; subsequently, the water is evaporated in June and July. Timeline of bioleaching is given in Table 22.10.

Table 22.10. Timeline of bioleaching.

Time	*Developments*
Old Testament	Red and green organisms [fungi? lichens?] on the walls of houses create pits and cavities.
23 - 79 A.D.	Roman writer Gaius Plinius Secundus described that copper minerals are obtained using a leaching process.
166 A.D.	The scientist Galen described *in-situ* leaching in an old, cypriotic mine.
1494 – 1555	Georgius Agricola described techniques for copper winning based on the leaching of copper-containing ores. Agricola also described roasting pyrite to prepare for leaching and produce $FeSO_4$
1572	Industrial heap leaching of copper-sulfides in Rio Tinto, Spain. Roasting of copper sulfide and iron sulfide, and leaching.

1890	Efforts were made to establish bioleaching to the Rio Tinto mines. Heaps of low-grade ore, left for one to three years for natural decomposition. The heap leaching in the Rio Tinto mines continued successfully until the 1970's.
1922	Oxidation of metal sulfides was reported.
1940	Millions of tones of sulfuric acid were discharged into the Ohio river because of the involuntary leaching of sulfides in carbon obtained from the carbon mines in Pennsylvania.
1947	*Thiobacillus ferrooxidans* was identified as part of the microbial community found in acid mine drainage.
1950's	Leach dumps (predecessor of leach heaps) developed.
1958	A first patent was granted on cyclic process where a ferric sulfate and sulfuric acid lixiviant solution is used for metal extraction, regenerated by aeration (ferrous iron oxidation by iron oxidizing microorganisms), and re-used in a next leaching stage.
1961	*Thiobacillus ferrooxidans* (reclassified as *Acidithiobacillus ferrooxidans*) was identified in the leachates.
1965	Discovery of the first iron- and sulfuroxidising archaea, *Acidianus brierleyi* from an acidic, thermal spring in Yellowstone National Park (USA).
1995	Bioleaching of chalcopyrite concentrate was developed and evaluated on a commercial scale.
2000	Commercial scale applications with archaea.
2003	Demonstration of potential commercial applications - the [BioCop^{TM}] process by BHP. Billiton employed acidophilic, iron-oxidizing archaea for bioleaching of chalcopyrite concentrates in aerated stirred tanks.

Examples of mixed populations of microbes useful for bioleaching or biooxidation (other than chalcopyrite ores and concentrates) are given in Table 22.11.

Table 22.11. Mixtures of microbes used as inoculants for bioleaching.

Company	*Microorganisms*	*Growth parameters*
Gencor Limited	*Thiobacillus ferrooxidans, Thiobacillus thiooxidans* and *Leptospirillum ferrooxidans*	Mesophilic bacteria which operate in the temperature range of 35°C to 45°C
Gencor Limited	*Thiobacillus ferrooxidans, Thiobacillus thiooxidans* and *Leptospirillum ferrooxidans*	pH of preferably 3 with an ore preferably crushed to below 6 mm
Bactech Pty Ltd	"M4"	Moderately thermophilic bacterial culture operating in the temperature range of 46°C to 50°C
Mintek	*Thiobacillu ferrooxidans/Leptospirillu ferrooxidans*	Mesophilic mixed bacterial culture

1. Identification of Bacteria

Bacteria used in biomining are identified from the other bacteria using techniques such as immunofluoresence, dot immunoassay and dot-blot hybridisation as given below:

(*a*) Immunoflurosence: Immunofluurosence technique is used to identify the specific antibody or antigen present in the biological fluid. Fluorescent antibodies are used to identify the biomining bacteria.

(*b*) Dot Immunoassay: The bacteria such as *T. ferroxidans* and *T. thioxidans* adhered to ore can be identified using dot immunoassay technique. In this technique bacteria are placed as dots on nitrocellulose paper, antigen-antibody reaction is done on the film, and at last secondary antibodies are added to produce visible color. Comparing the sample with test sample can be used to approximate the sample.

(c) Dot-blot Hybridisation: The biomining bacteria such as *T. ferroxidans* can be identified by using dot-blot hybridisation technique. Dot-blot hybridisation is a DNA-based technique. First the bacteria are isolated from the ore, and then bacterial cells are treated with some chemicals which help in the extraction of DNA. Extracted DNA are then purified and fixed on nitrocellulose membrane via Southern blotting technique. The biomining bacteria are identified by using DNA or RNA probes.

Characteristic Features of Bacteria used in Biomining

Bacteria are the most suitable microorganism for the extraction of minerals from low grade ores.

- Thermophilic bacteria which can survive at high temperature are selected for biomining as the mineral extraction involves the high temperature procedures.
- Chemophilic bacteria are also used in biomining in strong acid and alkali.
- Autotropic bacteria are selected for biomining. They have the ability to produce energy from inorganic compounds through photosynthesis or chemosynthesis.
- Bacteria which have the capacity to form biofilms are also selected for biomining purpose. Techniques such as genetic engineering and conjugation are used to produce the bacteria with all the desired characteristics to increase the biomining yield, *i.e.* to increase the extraction of minerals from low grade ores.

2. Methods of Biomining

Minerals are extracted from the ores using microorganisms by two methods such as: (*i*) direct or indirect oxidation, or (*ii*) reduction of the ore. The principal commercial methods are *in situ*, dump, heap and vat leaching.

(*a*) *In situ* Leaching: *In situ* leaching involves pumping of solution and air under pressure into a mine or into ore bodies made permeable by explosive charging. The resulting metal-enriched solutions are recovered through wells drilled below the ore body. Generally three types of ore bodies are considered for *in situ* leaching, for example surface deposits above the water table, surface deposits below the water table and deep deposits below the water table.

(*b*) Dump Leaching: Dump leaching involves uncrushed waste rock which is piled up. Some of these dumps are huge, containing in excess of 10 million tons of waste rock. Generally dumps contain about 0.1-0.5% copper, too low to recover profitably by conventional procedures. First commercially applied in the 1950s, dump leaching is an important process for the copper mining industry. At many open pit operations a very large amount of material is too low grade to sustain the cost of flotation and smelting. Therefore, this marginal grade ore (typically less than 0.5% copper) is fractured by blasting in the pit and hauled as large rock fragments to dumps. Dumps contain millions of tons of run-of-mine ore and are often several hundred feet deep. Water containing a small amount of sulfuric acid is applied to the top surface of the dump using sprinklers or drippers. As the acidic solution percolates through the dump, favorable conditions develop for the growth of naturally occurring microorganisms, which catalyse the oxidation of the copper sulfide minerals.

Dump bioleaching is a highly economic method of recovering copper from very low-grade ores because of the large tonnages of copper-containing rocks that can be processed and due to its low production costs. In 2006, the dump bioleach operation was initiated by BHP Billiton at the Escondida Mine in Chile. It includes technical enhancements to improve microbial activity and increase the rate and overall recovery of the copper. The Escondida dump bioleach is expected to produce 180,000-200,000 metric tons (198,450 – 220,500 tons) of cathode copper per year over the next 40 years, making Escondida the largest dump bioleach operation in the world.

A stacker forming big heaps

(c) Heap Leaching: Heap bioleaching is widely practiced around the world for the extraction of copper from 'secondary copper ores' (chalcocite and covellite). First the soil is compacted on a slightly sloping ground and then covered with an impermeable pad such as an asphalt layer or a flexible plastic sheet. The ore is crushed to about ¾ inch or less and piled on the pad in the form of big heaps. The pads are lined with high-density polyethylene and perforated plastic drain lines are placed on the pad to improve the drainage of copper containing solution from the bottom of the ore heap. A coarse rock layer is placed above the drain lines and within this rock layer a network of perforated plastic air lines is arranged. Air is forced through the air lines and directed to the microorganisms in the heap by blowers external to the heap. The ore is stacked to a depth of six to 10 metres (20 to 33 feet) most often with automated stackers.

(d) Vat Leaching: Vat leaching as currently applied to oxide ores involves the dissolution of crushed materials in a confined tank. For enhanced recovery of metals conditions can be controlled by using bioreactors; but the use of bioreactor is more costly. However, they are being considered actively for precious metals and ore concentrates.

Heap bioleaching- Some Points for Consideration

- Aeration of the heap is crucial and also affects the temperature of the heap.
- Sometimes, thermofilms are used to cover the heap to reduce the heat loss from exothermic reactions. Usually a high temperature will increase the leaching rate.
- The most common bioleaching process mainly used for bioleaching of copper and refractory gold-bearing ores is the heap bioleaching.
- Although one can rely on the natural movement of microbes to eventually inoculate the heap, initial rates of bioleaching can be improved by effective heap inoculation but this is difficult to achieve.

(i) Use of reactors in biomining: In recent years, microbial processes applied to mining operations are gaining increasing interest. Potential and current applications include the mining of gold, copper and other heavy metals, desulfurisation of coal and oil, tertiary recovery of oil and biosorption of metal ions. Moreover, for a biomining process the selection of a suitable reactor and its design should be based in the physical, chemical and biological characteristics of the system. Much attention should be paid on the complex nature of the reacting sludge which is composed of an aqueous liquid, suspended and attached cells, suspended solids, and air bubbles. Due to very large volumes of material to be processed, bioleaching and biooxidation are best performed in a continuous mode of operation in which volumetric productivity is high and reactor volumes can be kept low. Considering the kinetic characteristics of microbial growth, a continuous stirred tank reactor (CSTR) appears as the first choice for biomining processes (Acevedo, 2000). A stirred tank bioreactor (5 L capacity) used for bioleaching is shown in Fig. 22.22.

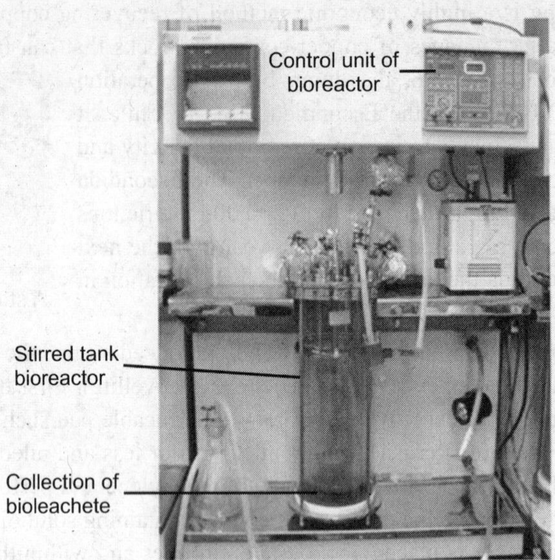

Fig. 22.22. A stirred tank bioreactor (5 litre capacity) for bioleaching.

An important consideration in selecting a suitable reactor refers to the autocatalytic nature of microbial growth. This is common to all fermentation operations, but in bioleaching there is an important difference. In industrial fermentations the nutrients are chosen by their high affinity with the microbial population, where as in biomining the mineral species involved are usually recalcitrant to microbial action, which means that there is quite low affinity.

The substrate-microorganism affinity is related to Monod's saturation constant, K_S. High affinities are reflected in low K_S values, of the order of a few milligrams per liter as in the case of most sugars. Some minerals have saturation constants as high as 3 to 6 g/l, *i.e.* thousands of orders of magnitude higher. This situation affects the selection of the reactor. If there is need of a high degree of conversion, a single agitated tank will require a very large volume; therefore, an arrangement of reactors will be more suitable. A CSTR followed by a tubular plug flow reactor gives the minimum reaction volume to attain a certain conversion. Because the need of aeration and the presence of solid particles, the PFRs become unpractical. But its performance can be approximated by using a series of CSTRs.

Other types of reactors that have been studied for their application in biomining are the percolation column, the Pachuca tank, the air-lift column, and some special designs such as rotary reactors.

***(ii)* Gas mass balance in bioreactor:** A gas mass balance around the bioreactor gives:

$$\frac{dC_{Li}}{dt} = k_L a_i \, (C^*_i - C_{Li}) - N_i \qquad ...(1)$$

where i is the oxygen or carbon dioxide, N is the gas demand (g gas/l · h), $k_L a_i$ is the volumetric gas transfer coefficient (h^{-1}), C^*_i is the gas equilibrium concentration (g/l), C_{Li} is the dissolved gas concentration, and t is time (h).

The gas supply, $k_l ai(C^*_i - C_{Li})$, must equal the gas demand in order to avoid growth limitation, hence

$$N_i = k_L a_i \, (C^*_i - C_{Li}) \qquad ...(2)$$

The gas demands can be calculated as

$$N_i = \frac{\mu X}{T_i} \qquad ...(3)$$

where μ is the specific growth rate of the cells (h^{-1}), X is cell mass concentration (g/l), and Y_i is the gas cell yield (g cells/g gas).

The bioleaching process can be represented by a stoichiometric equation. In the case of a leaching organism, *Thiobacillus ferrooxidans* growing in a simple defined culture media with ferrous iron as the energy source, the following equation can be written:

$$1 \cdot CO_2 + m \cdot NH_3 + n \cdot FeSO_4 + o \cdot H_2SO_4 + p \cdot O_2 \rightarrow C_5H_7O_2N + r \cdot Fe_2(SO)_3 + s \cdot H_2O \quad ...(4)$$

$C_5H_7O_2N$ represents the biomass with an elemental composition of 53.1% C, 6.2% H, 28.3% O, and 12.4% N. Elemental mass balances on C, H, O and N, together with the experimental value of the ferrous ion cell yield of 0.0086 g cells/g Fe^{2+}, allows for the calculation of the stoichiometric coefficients:

$$5.0\ CO_2 + NH_3 + 241.37\ FeSO_4 + 120.67\ H_2SO_4 + 55.34\ O_2 \rightarrow C_5H_7O_2N + 120.67\ Fe_2(SO_4)_3 + 118.67\ H_2O \quad ...(5)$$

The oxygen and carbon dioxide cell yields can be calculated from equation (5):

$$Y_{O_2} = \frac{1 \cdot M_c}{55.34 \cdot M_{O_2}} = 0.064 \text{ g cells/g consumed oxygen} \quad ...(6)$$

$$Y_{CO_2} = \frac{1 \cdot M_c}{5.0 \cdot M_{CO_2}} = 0.51 \text{ g cells/g consumed carbon dioxide} \quad ...(7)$$

In which M_c, M_{O2} and M_{CO2} are the molecular masses of the cells, oxygen and carbon dioxide, respectively.

In an actual bioleaching operation, a similar stoichiometric representation can be made. For instance, for the biooxidation of enargite (Cu_3AsS_4) from a refractory gold concentrate the following equation applies, considering an experimental value of 0.6 g cell/g Cu (Garcia, 1997):

$$5.0\ CO_2 + NH_3 + Cu_3AsS_4 + 4.5\ H_2O + 3.75\ O_2 \rightarrow C_5H_7O_2N + 3.0\ CuSO_4 + H_3AsO_4 + H_2SO_4 \quad ...(8)$$

In this case the oxygen and carbon dioxide cell yields are:

$$Y_{O_2} = \frac{113}{3.75 \cdot 32} = 0.94 \text{ g/g} \quad ...(9)$$

$$Y_{CO_2} = \frac{113}{5 \cdot 44} = 0.51 \text{ g/g} \quad ...(10)$$

As could be expected, the carbon dioxide yield, related only with cell growth, is the same in the defined soluble medium and in the bioleaching of a mineral.

Equations 2, 3, 6, 7, 9 and 10 can be used to estimate the required mass transfer coefficients. However, the required coefficient for oxygen can also be estimated from the stoichiometry equation of the main oxidation reaction. Oxygen transfer coefficients estimated by such method are included for chalcocite (Cu_2S), covellite (CuS) and chalcopyrite (CuFeS).

I. BIOLEACHING OF METALS BY BACTERIA

Bioleaching and biomining of some of the metals such as iron, copper, uranium, and gold are described in this section.

1. Iron Bioleaching

In nature, pyrite (FeS_2) is one of the most common forms of iron and sulphur, which is a highly insoluble crystalline structure. The oxidation of the ferrous ion (Fe^{2+}) to the ferric ion (Fe^{3+}) is an energy producing reaction for some microorganisms. As only a small amount of energy is obtained, large amounts of (Fe^{2+}) have to be oxidised. Fe^{3+} forms insoluble precipitate in H_2O. Many Fe^{2+} oxidising microorganisms also oxidise sulfur, and are obligate acidophiles that further acidify the environment by the producing H_2SO_4. At neutral pH, Fe^{2+} is rapidly oxidised chemically in contact with the air. In these conditions there is no enough Fe^{2+} to allow significant growth. However, at low pH, Fe^{2+} is much more stable. This is the reason why most of the Fe^{2+} oxidising microorganisms are found only in acidic environments and are obligate acidophiles. The chemistry of oxidation of pyrites, *i.e.* the production of ferrous ions and subsequently ferric ions is very complex. The best studied Fe^{2+} oxidising bacterium is *Thiobacillus ferrooxidans*, an acidophililic chemolithotroph.

The sulfur containing ore pyrite (FeS_2) is at the start of this process. Pyrite is an insoluble crystalline structure that is abundant in coal- and mineral ores. Normally pyrite is shielded from contact with oxygen and not accessible for microorganisms. Upon exploitation of the mine, pyrite is brought into contact with air (oxygen) and microorganisms, and oxidation starts thereafter. This oxidation relies on a combination of chemically and microbiologically catalysed processes. A general equation for this process is:

$$2FeS_2(s) + 7O_2(g) + 2H_2O \rightarrow 2Fe^{2+}(aq) + 4SO_4^{2-}(aq) + 4H^+(aq)$$

This reaction acidifies the environment and the Fe^{2+} formed is rather stable in the presence of oxygen. In such an environment *Thiobacillus ferrooxidans* will be able to grow rapidly. Upon further acidification *Ferroplasma* also develops and further acidifies. As a result of the microbial activity, the oxidation of the sulfide to sulfate solubilises the ferrous iron (Fe^{2+}), which in turn is oxidised to ferric iron (Fe^{3+}):

$$4Fe^{2+}(aq) + O_2(g) + 4H^+(aq) \rightarrow 4Fe^{3+}(aq) + 2H_2O$$

Either of these reactions can occur spontaneously or can be catalyzed by microorganisms that derive energy from the oxidation reaction. The ferric irons produced can also oxidize additional pyrite and oxidize into ferrous ions:

$$FeS_2(s) + 14Fe^{3+}(aq) + 8H_2O(l) \rightarrow 15Fe^{2+}(aq) + 2SO_4^{2-}(aq) + 16H^+(aq)$$

The net effect of these reactions is to release H^+ which lowers the pH and maintains the solubility of the ferric ion. The Fe^{2+} ions are again oxidised by bacteria to Fe^{3+} which again reacts with more pyrite and results in rapid oxidation of pyrite at low pH. Naturally some of the some of Fe^{2+} produced by bacteria leaches away and carried by ground water into surrounding streams which are oxidised to Fe^{3+} ions by bacteria aerobically resulting in precipitation of insoluble salt (Fig. 22.23).

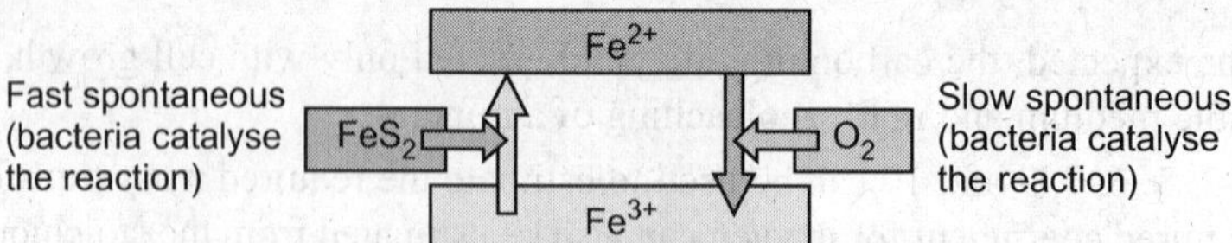

Fig. 22.23. Oxidation of pyrite by iron-oxidising bacteria.

2. Copper Bioleaching

Metals can be extracted economically from low grade sulphide or sulphide-containing ores. Low-grade copper ore contains 0.1 - 0.4% Cu. The pregnant leaching solution may contain 1 to 3 g of Cu^{++}. In laboratory, about 97% of the copper in low-grade ores has been recovered under optimal conditions, by exploiting metabolic activities of *Thiobacillus ferrooxidans*; but such high yields are not achieved in actual mining operations. The process is at present commercially used for recovery of copper and uranium from low-grade ores.

In copper leaching operations, *Thiobacillus* involves both, direct oxidation of *Covellite* (CuS) and indirect oxidation of CuS via generation of ferric ions from ferrous sulphide present in the important copper ores such as chalcopyrite ($CuFeS_2$). In the latter case, copper replaces iron. In 1980s, various firms began to utilise the bioleaching for extracting copper from low-grade ores.

***(a)* Mechanism of Copper Leaching:** The thermophilic sulfide-oxidising archaea is exploited in copper extraction and has attracted interest due to the efficient extraction of recalcitrant metals from sulfide ores. Microbial leaching is useful for copper ores because copper sulfate, formed during the oxidation of copper sulfide ores, is water-soluble. About 25% of all copper mined worldwide is now obtained from leaching processes. The acidophilic archaea *Sulfolobus metallicus, and Metallosphaera sedula* tolerate up to 4% of copper; therefore they are being exploited for mineral biomining. In primary reactors about 40 to 60% copper is extracted and more than 90% extraction is done in secondary reactors in about 6 days.

Chalcopyrite ($CuFeS_2$) ore

Thiobacillus ferrooxidans oxidises the Cu^+ in chalcocite (Cu_2S) to Cu^{2+}; thus it removes some of the cupper in the soluble form as Cu^{2+}, and forming the mineral covellite (CuS). This oxidation of Cu^+ to Cu^{2+} is an energy yielding reaction. Covellite can then be oxidised releasing the sulfate and soluble Cu^{2+} as products.

A second mechanism, and probably the most important in most mining operations, involves chemical oxidation of the copper ore with ferric (Fe^{3+}) ions formed by the microbial oxidation of ferrous ions (derived from the oxidation of pyrite).

There are three possible reactions involved in copper oxidation as given below:

$$Cu_2S + O_2 + H^+ \longrightarrow CuS + Cu^{2+} + H_2O$$

$$CuS + O_2 \longrightarrow Cu^{2+} + SO_4^{2-}$$

$$CuS + 8Fe^{3+} + 4H_2O \longrightarrow Cu^{2+} + 8Fe^{2+} + SO_4^{2-} + 8H^+$$

The copper metal is then recovered by using Fe^0 from steel cans:

$$Fe^0 + Cu^{2+} \longrightarrow Cu^0 + Fe^{2+}$$

Besides, chalcopyrite dissolves in the presence of ferric ion (Fe^{3+}) according to the following reaction:

$$CuFeS_2 + 4Fe^{3+} \longrightarrow Cu^{2+} + 5Fe^{2+} + 2S^0$$

$$CuFeS_2 + 4Fe^{3+} + 3O_2 + 2H_2O \longrightarrow Cu^{2+} + 5Fe^{2+} + 2H_2SO_4$$

Chalcopyrite ($CuFeS_2$) follows the two stages of bioleaching. The chemistry of the bioleaching process is relatively straightforward. The bioleach microorganisms catalyse the oxidation of ferrous iron and sulphur, to produce ferric iron and sulphuric acid as below:

$$Fe^{2+} + O_2 + H^+ \longrightarrow Fe^{3+} + H_2O$$

$$SO_2 + H_2O \longrightarrow H_2SO_4$$

Often the temperature inside the leach dump rises due to microbial activities. Thus, thermophilic iron-oxidising chemolithotrophs such as thermophilic *Thiobacillus* species and *Leptospirillum* and at higher temperatures the thermoacidophilic archaeon *Sulfolobus* (*Metallosphaera sedula*) are the important bacteria required in leaching process above 40°C.

***(b)* Microbiology of Copper Bioleaching:** A consortium of microorganisms is applied in bioleaching comprise that are characterised by their varying range of operating temperature. The mesophilic bacteria operate well between 30 and 42°C. The acidophilic bacteria are: *Acidithiobacillus ferrooxidans*, *Acidithiobacillus caldus* and *Leptospirillum ferrooxidans*. Other *Acidithiobacillus* species also thrive in this temperature range. The moderate thermophiles are mostly bacteria that exist within a relatively narrow temperature range of 45 to 55°C, and include *Sulfobacillus*, *Acidimicrobium* and *Ferroplasma* species, as well as *Leptospirillum ferriphilum* and

Acidithiobacillus caldus. The extreme thermophiles dominated by archaea (as opposed to bacteria) thrive at elevated temperatures between 60ºC and 90°C. It includes various species of *Sulfolobus*, *Acidianus* and *Metallosphaera*.

Most of these organisms are autotrophic (grow by fixing CO_2 from the atmosphere) and acidophilic tolerating a low pH environment). Many of these bacteria oxidise both iron and sulphur, whereas some can tolerate remarkably high concentrations of a wide range of metal ions. *Thiobacillus ferrooxidans* (= *Acidithiobacillus ferrooxidans*) and its proposed role in bioleaching is given in Fig. 22.24.

In 2008, Jorge Valdés of Chile and collaborators reported that *Acidithiobacillus ferrooxidans* as a major participant in consortia of microorganisms used for the industrial recovery of copper. They found that the bacterial cells form biofilm on mineral surface. The chemolithoautotrophic metabolism of *A. ferrooxidans* results in the oxidation/reduction of iron and sulfur compounds. The solubilisation of copper and other commercially valuable metals takes place via bioleaching. It also results in the production of acidified solutions in pristine environments and acid mine drainage in bioleaching operations.

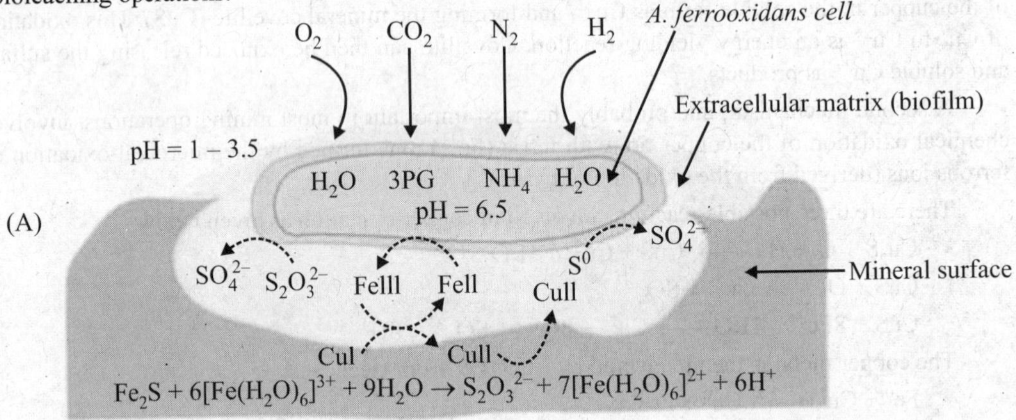

Fig. 22.24. *Acidithiobacillus ferrooxidans* and its proposed role in bioleaching. A- model of copper bioleaching by *A. ferrooxidans*; B- oxidation/reduction reactions carried out by *A. ferrooxidans* (based on Valdés *et al.*, 2008)

***(c)* Biomining of Copper:** Bacterial leaching of copper is usually performed in heaps of ground ore or in dumps of waste or spent material. Big heaps are prepared as described earlier. Fine particles of ore are agglomerated with water containing sulfuric acid to condition the ore for the microorganisms and also to affix fine particles to the larger rock particles so that its permeability may be increased.

Copper leachate from heap to pond.

Permeability is critical for the recovery rates from a heap (as well as from dump and *in-situ* operations). If the heap material is too compact, the leaching operation time increases as the reagent liquid will only slowly percolate through the heap. If permeability in the heap is uneven, the leaching solution might cause channelling leaving parts of the heap untreated. Therefore, fine particles are often agglomerated to ensure a good permeability. The heap is sprayed with leaching reagent, *i.e.* lixiviant fluid over the surface to leach metal from the heap. The dilute sulphuric acid sprinkled on top percolates down through the dump, lowering the pH and promoting the growth of acidophilic microorganisms. The required metals are solubilised as the reagent percolates through the heap. The leachate (metal containing solution) drained from the heap is collected at the bottom of the dump in a pond (Fig. 22.25), and then pumped from the pond to a recovery station.

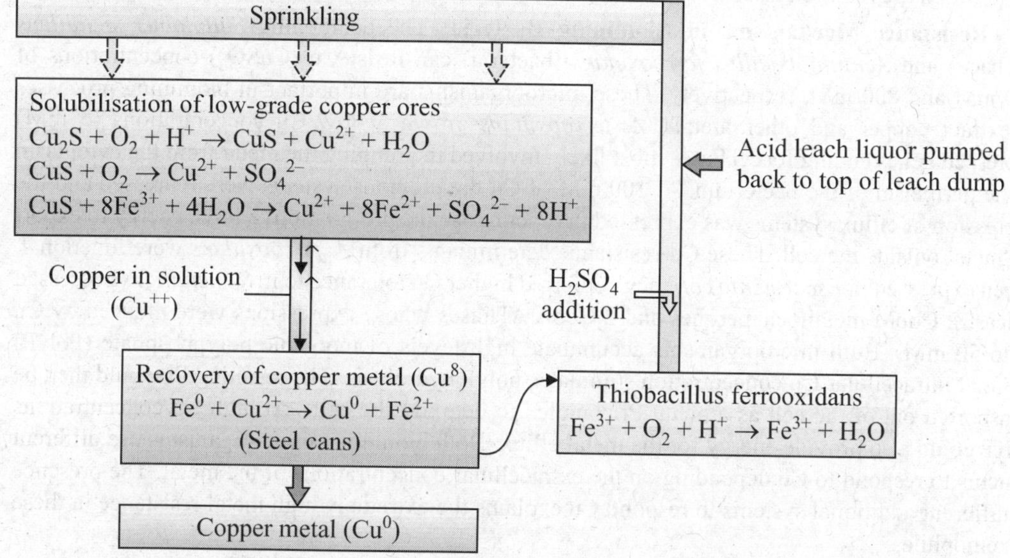

Fig. 22.25. Copper heap bioleaching (flow diagram).

After getting the desired concentration of the metal, the rich liquor is pumped to the solvent extraction (SE) section and then sent to electrowinning (EW), where the fine metal is recovered (Fig. 22.26). The raffinate from the SE section is recycled to the heap or dump and the spent liquor of the EW section is recycled to the SE operation.

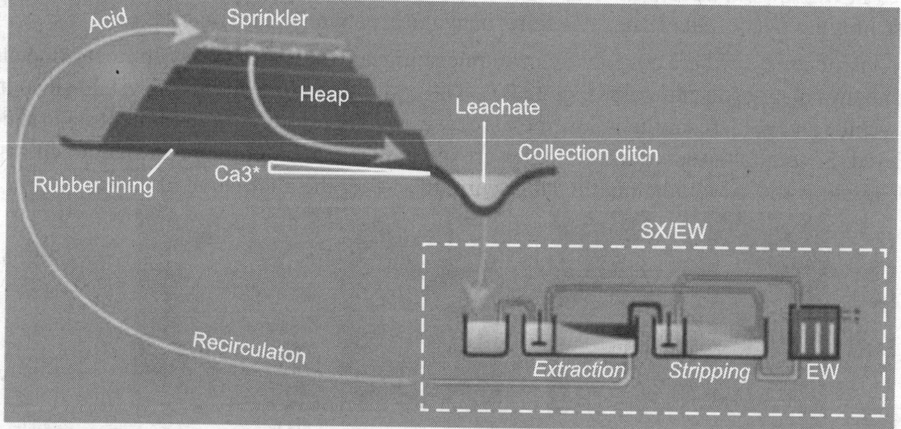

Fig. 22.26. Heap leaching followed by solvent extraction and electrowinning (SX/EW).

Heap leaching takes months rather than years as for dump leaching. The maximum copper leached from heap bioleach operations is 80-85% requiring 250- to 350-days of on-pad leaching to achieve this recovery. The main advantages of heap bioleaching is the rapid start-up and commissioning of operations, low capital and operating costs, the absence of any toxic emissions and the minimisation or complete elimination of any water discharges because all solutions are recycled.

Heaps and dumps present a number of advantages such as simple equipment and operation, low investment and operation costs and acceptable yields. But the operation suffers from some severe limitations: the piled material is very heterogeneous and practically no close process control can be exerted, except for intermittent pH adjustment and the addition of some nutrients. Moreover, the rates of oxygen and carbon dioxide transfer that can be obtained are low, and extended periods of operation are required in order to achieve sufficient conversions (Acevedo *et al.*, 1999).

Resistance Mechanisms in Biomining Bacteria: The acidophilic *Sulfolobus metallicus* (archaea) and *Acidithiobacillus ferrooxidans* (bacteria) can resist Cu ($CuSO_4$) concentrations of 200 mM and 800 mM, respectively. These microorganisms are important in biomining processes to extract copper and other metals. *A. ferrooxidans* grown at low Cu concentrations (5 mM) expressed genes coding for ATPases most likely involved in pumping the metal from the cytoplasm to the periplasm of the bacterium. At 100 mM of Cu the previous systems were repressed and the expression of efflux systems was expressed in order to use the proton motive force energy to export the metal outside the cell. These Cu-resistance determinants from *A. ferrooxidans* were functional. When expressed in *Escherichia coli* they conferred higher Cu tolerance to it. *S. metallicus* possessed at least 2 CopM metallochaperones and 2 CopA ATPases whose expressions were induced by Cu (5 to 50 mM). Both microorganisms accumulate high levels of inorganic polyphosphate (PolyP) and that intracellular Cu concentration stimulates polyP hydrolysis. The resulting Pi would then be transported out of the cell as a metal-Pi complex to detoxify the cells. At high Cu concentrations polyP could also provide energy for the metal efflux. Both biomining microorganisms use different systems to respond to Cu depending on the extracellular concentrations of the metal. The presence of different additional systems to respond Cu explains the extremely high metal resistance in these extremophiles.

3. Gold Biomining

Gold occurs as a native metal usually alloyed to a greater or lesser extent with silver (as electrum) or sometimes with mercury (as an amalgam). Native gold is found as sizeable nuggets, fine grains or flakes in alluvial deposits, or as grains or microscopic particles embedded in other rocks. Ores in which gold occurs in chemical composition with other elements are comparatively rare. These ores are pyrite, calaverite, sylvanite, nagyagite, petzite and krennerite.

Small particles of gold are covered by insoluble sulfides in ores known as refractory impeding the contact between cyanide and gold (Fig. 22.27). These sulfides make difficult to establish contact between cyanide and gold, resulting in low metal recoveries. Therefore, a pre-treatment stage should be considered. Several alternative technologies are available such as pressure oxidation, chemical oxidation, roasting and bio-oxidation, the latter currently being the alternative of choice.

(A)

(B)

Fig. 22.27. A- A very good gold-quartz ore (*source*: A quartz vein near Alma, Colorado); B- a gold grain occluded within the sulfide mineral, pyrite (*Source:* Newmont Mining Corporation).

(*a*) Biooxidation of Gold: In biooxidation process, the gold grain occluded within the ores for example sulfide mineral (pyrite) must first be oxidised using ferric iron produced by microorganisms to dissolve efficiently this gold grain with dilute cyanide or other gold-dissolving reagent. The, gold is recovered from the solution.

Iron-and sulfur-oxidising microorganisms are used to release occluded copper, gold and uranium from mineral sulfides. Most of industrial plants for biooxidation of gold-bearing concentrates have been operated at 40°C with mixed cultures of mesophilic bacteria of the genera *Thiobacillus* or *Leptospirillum ferrooxidans*. The dissimulatory iron-reducing archaea that is *Pyrococcus furiosus* and *Pyrobaculum islandicum* reduce gold chloride to insoluble gold. In the biooxidation process, bacteria partially oxidize the sulfide coating covering the gold microparticles in ores and concentrates. Microorganisms belonging to the *Thiobacillus* and *Leptospirillum* genera are commonly used, although an increasing interest exists in thermophilic archeons (Dew *et al.*, 1999).

In this way, gold recovery from refractory minerals can increase from 15-30% to 85-95%. In 2000, Fernando Acevedo from the Universidad Católica de Valparaíso Avenida (Chile) found that potential applications include the mining of gold, copper and other heavy metals, desulfurisation of coal and oil, tertiary recovery of oil and biosorption of metal ions. Currently, bacterial leaching of copper and biooxidation of refractory gold concentrates are well-established large-scale processes carried on using heaps and tank reactors. During the last 15 years at least ten large-scale commercial gold processing units have been established in South Africa, Brazil, Australia, Ghana, Peru and USA, and eight of them use bioreactors.

In first stage bacteria catalyse the breakdown of the mineral arsenopyrite (FeAsS) by oxidising the sulphur and metal to higher oxidation states whilst reducing dioxygen by H and Fe. This allows the soluble products to dissolve.

$$FeAsS \longrightarrow Fe^{2+} + As^{3+} + S^{6+}$$

This process actually occurs at the cell membrane of the bacteria. The electrons pass into the cells and are used in biochemical processes to produce energy for the bacteria to reduce oxygen molecules to water. In second stage, bacteria oxidise Fe^{2+} to Fe^{3+} and reducing O.

$$Fe^{2+} \longrightarrow Fe^{3+}$$

Then they oxidise the metal to a higher positive oxidation state. With the electrons gained they reduce Fe^{3+} to Fe^{2+} to continue the cycle:

$$M^{3+} \longrightarrow M^{5+}$$

The gold is now separated from the ore and in solution. Gold recovery from refractory minerals can increase from 15-30% to 85-95% after biooxidation. A major drawback in the biooxidation compared with the physicochemical pretreatment processes is that in many cases the consumption of cyanide in the subsequent gold-recovery process is considerably higher following biooxidation.

(*i*) Stirred-Tank Mineral Biooxidation: The other major form of bioleaching is known as stirred tank processes. When the ore body contains high concentrations of arsenopyrite (AsFeS), gold extraction is carried out by stirred tank biooxidation process. Arsenopyrite is extremely recalcitrant to chemical reactions and effectively shields the gold from the cyanide. It is used to dissolve it. The stirred tank processes are used as a pretreatment to remove low value sulphide material from higher value minerals, such as gold. The solid particles leaving the last aeration tank are then moved for chemical processing. This pretreatment can increase the yield of gold-bearing ore by twenty times over the traditional crushing and immediate chemical extraction. These reactors can be enormous with one such facility consisting of a series of 24 tanks, each holding 1,000,000 litres of crushed ore and liquid.

The process requires the construction of the large aerated tanks arranged in a series so that runoff from one tank serves as raw material for the next (Fig. 22.28). Generally, the ore is crushed to get very small-sized particles to ensure that the solids remain suspended in the liquid medium. The slurry of crushed ore and liquid ore is moved from tank to tank to ensure adequate time for oxidation of arsenopyrite (AsFeS). The last tank is a settling tank where the solids containing gold are removed and extracted by using cyanide. In this way, the reactor operates in continuous flow mode with fresh ore being added to the first tank, while the runoff from the final tank is removed and treated. Mineral nutrients in the form of $(NH_4)_2SO_4$ and KH_2PO_4 are also added to the tanks to ensure the maintenance of maximal microbial density.

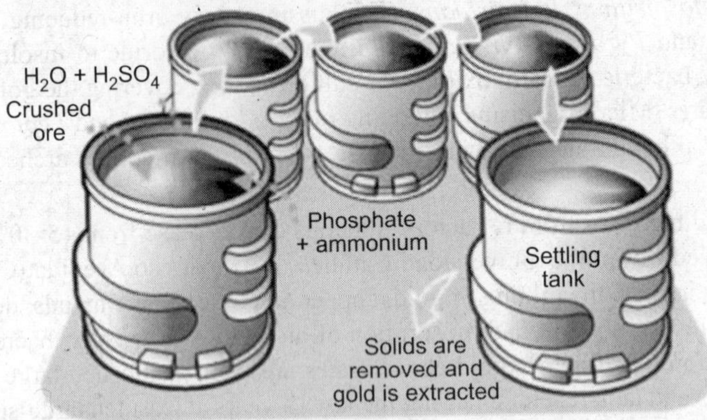

Fig. 22.28. Stirred tank biooxidation (diagrammatic).

Usually biooxidation /bioleaching in an aerated continuous stirred tank reactor (CSTR) is applied to mineral concentrates, due to the capital and operating costs associated with this technology. CSTR technology is carried out in a series of large bioreactors made up of stainless steel, each of capacity of 1,380 m³ (Fig. 22.29). Each bioreactor is equipped with agitators that keep the finely ground sulfidic-refractory gold concentrate in suspension and ensure that oxygen and carbon dioxide are efficiently transferred into the solution for the microorganisms present over a billion per milliliter of solution. Once the CSTRs are inoculated with the microorganisms, no additional inoculation is needed, because the process is continuous. Air provided by blowers is introduced below the impeller of agitator. Internal coils through which cooling water is circulated are mounted along the inside walls of the tanks. It takes about three to five days for the biooxidation of the concentrate of reactors.

Fig. 22.29. An aerated stirred tank bio-leach plant (*Source:* Kasese Cobalt Company, Uganda).

Some large-scale plants used for biooxidation of refractory gold minerals are given in Table 22.12. The tank leaching plants which use several tanks connected in series. The Fairview plant operates with a total residence time of four days and a gold recovery of 94%. The Sao Bento plant was originally designed for pressure oxidation. A biooxidation reactor was added afterwards that allowed for a significant improvement in overall performance. The Ashanti plant has three biooxidation modules, each one consisting in six 900 m³ reactors (Dew *et al.*, 1999). The Youanmi plant consists in two 500 m³ tanks operated in parallel as a first stage followed by a series of two 500 m³ tanks. Although the oldest plants were designed using disk turbine agitators, the newer ones use hydrofoil axial impellers, resulting in a better solids suspension and reduced power requirement (Dew *et al.*, 1999).

***(b)* Future Applications:** The future of bioreactors in mining appears promising. Gold biooxidation operations increased in number and size in several countries. Mostly the use of reactors will be extended to the bioleaching of other metals such as copper. Currently studies are being carried on for the development of processes for the bioleaching of copper concentrates. The experience gained in the heap leaching of copper and in the biooxidation of gold concentrates is being used in these studies. The bioleaching of chalcopyritic copper concentrates in the next few years will constitute a big breakthrough in biomining. The application of these technologies to the processing of Ni, Zn and other heavy metals may also become a reality in the near future.

Table 22.12. Large-scale plants for the biooxidation of refractory gold minerals (*source:* Brierley *et al.*, 1997).

Plant	Start-up	Technology	Capacity
Fairview, South Africa	1986	Tank leaching	35 tons gold concentrate/day
Sao Bento, Brazil	1990	Tank leaching	150 tons gold concentrate/day
Harbour Lights, Australia	1992	Tank leaching	40 tons gold concentrate/day
Wiluna, Australia	1993	Tank leaching	115 tons gold concentrate/day
Ashanti, Ghana	1994	Tank leaching	1,000 tons gold concentrate/day
Youanmi, Australia	1994	Tank leaching	120 tons gold concentrate/day
Sansu, Ghana	1994	Tank leaching	1,000 tons gold concentrate/day
Tamboraque, Peru	1999	Tank leaching	260,000 tons of zinc flotation tailings
Mount Leyshon, Australia	1992	Heap leaching	1,370 tons of copper-gold ore/day
Newmont-Carlin, USA	1995	Heap leaching	10,000 tons gold ore/day

4. Uranium Recovery

Uranium recovery proceeds in very similar way to copper recovery, although less amount of uranium is obtained than copper. For getting one tonne of uranium, a thousand tonne of uranium ore must be handled. *In situ* uranium leaching is gaining vast acceptance. However, uranium leaching from ore on a large scale is widely practised in the USA, South Africa, Canada and India.

Much like copper, uranium is recovered by the conversion of insoluble tetravalent uranium oxides to soluble sulfates though the action of hot ferric iron and sulfuric acid produced by microbes. Consequently, hexavalent uranium sulfate is produced at pH 1.5–3.5 and temperature 35°C

$$UO_2 + Fe_2(SO_4)_3 \longrightarrow 6UO_2\ SO_4 + 2FeSO_4$$

$$UO_3 + H_2SO_4 \longrightarrow 6UO_2SO_4 + H_2O$$

Bacterial activity is limited to oxidation of uranium leaching is indirect process. *T. ferrooxidans* does not directly attack on uranium ore, but on the iron oxidant. The pyrite reaction is used for the initial production of Fe^{3+} leach solution, pyrite and ferrous ion. Biomining has been successfully used to obtain uranium from waste gold ore.

$$2FeS + H_2O + 7\tfrac{1}{2}\,O_2 \xrightarrow{T.\ ferrooxidans} Fe_2(SO_4)_3 + H_2SO_4$$

In addition, contamination of soil and water by uranium is problematic because uranium takes hundreds of millions of years to decay radioactively. For example in the Department of Energy (DOE) (U.S.A.) more than 3,000 contaminated disposal sites have radioactive uranium in groundwater or soil as contamination.

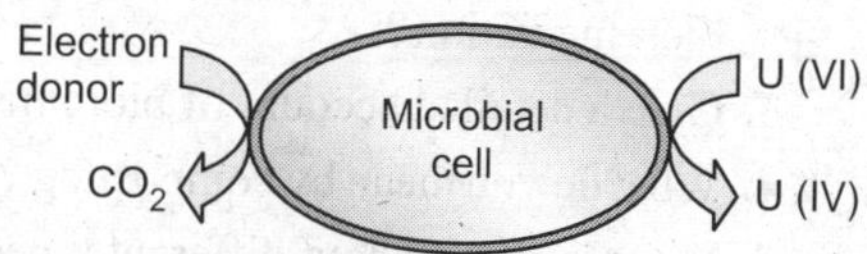

Fig. 22.30. Microbial uranium reduction (diagrammatic).

Oxidation state of uranium governs its transportation which is hexavalent [U(VI)] or tetravalent (U(IV)] (Fig. 22.30). U(VI) state is very soluble which travels with water but U(IV) state is insoluble and immobile. Certain bacteria are able to reduce uranium from its hexavalent state to its tetravalent state. This results in its decrease in solubility and trapping it in the soil underground. As a result, microbial activity can be stimulated *in situ* to remove U from groundwater.

II METAL BIOLEACHING WITH FUNGI

Several species of fungi can be used for bioleaching. Fungi can be grown on many different strata as with electronic scrap, catalytic converters, and fly ash from municipal waste incineration. It has been found that two fungal strains viz., *Aspergillus niger* and *Penicillium simplicissimum* mobilize Cu and Sn by 65%, and Al, Ni, Pb, and Zn by more than 95%. So it can be used for bioleaching sulfides. *A. niger* exhibits good potential in generating a variety of organic acids such as citric acid which are effective for metal solubilisation. Organic acid effectiveness may be enhanced with addition of sulfuric acid to the medium. Different agricultural wastes such as potato peels, and different auxiliary processes can be evaluated in order to elevate the efficiency and reduce costs. Maximum solubilisation of Cu (68%), Zn (46%) and Ni (34%) could be achieved following this process.

Fig. 22.31 shows the formation of copper oxalate, and crystals of calcium oxalate around the hyphae.

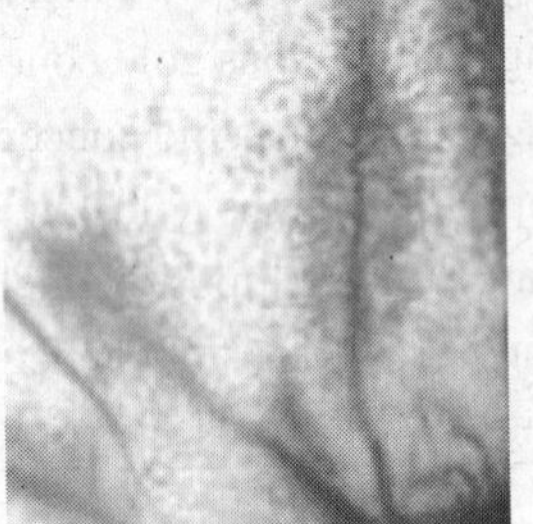

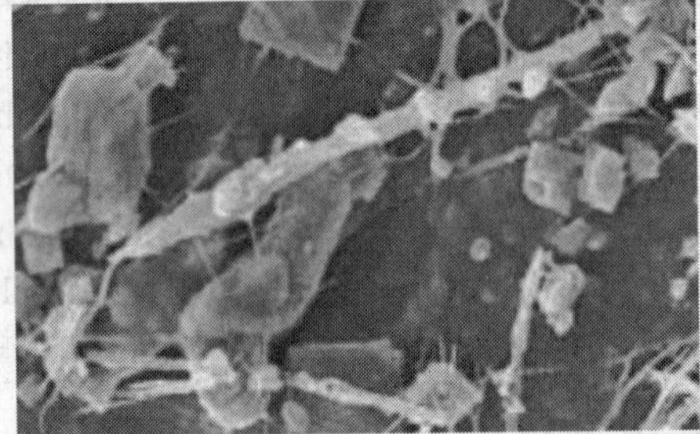

Fig. 22.31. Biomineral formation by fungi; A- light micrograph of moolooite crystals (copper oxalate, $CuC_2O_4.xH_2O$) around the fungal hyphae; B- a crust of calcium oxalate (weddelite and whewellite) crystals and tubular crystalline sheath around fungal hyphae (ESEM dry mode) (Gadd *et al.,* 2006).

In 2002, D.V. Rao and co-workers reported bioleaching of copper from its low-grade chalcopyrite ore by using 26 isolates of acidophilic fungi, for example the species of *Aspergillus*, *Penicillium* and *Rhizopus*. Maximum leaching of copper (78 mg/L) was observed with *Aspergillus flavus* (DSF-8) and *Aspergillus niger* (DOF-1) in Czapek Dox minimal medium upon shaking condition at room temperature for 20 days. Nutritional and environmental conditions for optimum bioleaching were standardised.

QUESTIONS

1. What are biosensors ? How are they used in environmental monitoring ?
2. Discuss in detail about microbial bioremediation. What are the factors that affect bioremediation?
3. Give a detailed account of bioremediation of polycyclic aromatic hydrocarbon.
4. What do you mean by composting ? Write in detail about different factors affecting composting.
5. Write in detail about different types of composting.
6. What is vermicomposting ? Write the working mechanism of vermicomposting.
7. Write in detail about vermicomposting by using earthworms and benefits from vermicompost.
8. Give an illustrated account of biogas technology in India.
9. What is biomining ? Write different methods of biomining.
10. Discuss in detail about copper bioleaching.

11. Write short notes on the following:

(*i*) BOD biosensor,
(*ii*) Electrochemical biosensor,
(*iii*) enzyme biosensor,
(*iv*) Whole cell biosensor,
(*v*) In situ bioremediation,
(*vi*) Ex situ bioremediation,
(*vii*) Super bug,
(*viii*) Bio-safety from GMOs,
(*ix*) Composting
(*x*) Types of composting,
(*xi*) Vermicomposting technology,
(*xii*) Benefits from vermicompost,
(*xiii*) Biogas
(*xiv*) Biomining,
(*xv*) Gold biomining,
(*xvi*) Copper biomining.

BIBLIOGRAPHY

Acevedo F. 2000. The use of reactors in biomining processes. *Electronic S. Biotech.* **3**:184-194.

Acevedo F., Canales C. and Gentina J.C. 1999. Biooxidation of an enargite-pyrite gold concentrate in aerated columns. In: *Biohydromatallurgy and the environment toward the mining of the 21st Century,* Part A. Eds. Amils, R. and Ballester, A., Elsevier, Amsterdam: The Netherlands. pp. 301-308.

Bell N. and Quan, L. 1999. The application of Bactech (Australia) Ltd. technology for processing refractory gold ores at Youanmi gold mine. Paper M2.3. In: *IBS'97-BIOMINE'97: Biotechnology comes of age.* Conference Proceedings. Sydney, Australia, 4th-6th August.

Bitton G. 1999. *Wastewater microbiology.* New York: Wiley-Liss.

Brierley C.L. 1997. Mining biotechnology: research to commercial development and beyond. In *Biomining: Theory, Microbes and Industrial Processes.* (Ed. D.E. Rawlings), Springer Verlag, Germany, pp. 3-17.

Dew D.W., Lawson E.N. and Broadhurst J.L. 1997. The BIOX process for biooxidation of gold-bearing ores or concentrates (Ed. Rawlings, D.E.). *Biomining: Theory, Microbes and Industrial Processes*. Springer Verlag, Berlin, Germany. pp. 45-80.

FRTR (2000). *In-situ Biological Treatment.* Remediation technologies screening matrix and reference guide, Version 4.0. Federal Remediation Technologies Roundtable.

Gadd G.M. 2007. Geomycology: biogeochemical transformations of rocks, minerals, metals and radiounuclides by fungi, bioweathering and bioremediation. *Mycol. Res.* **111:** 3-49.

Ivey G.A. 2000. Surfactant enhanced remediation (SER) using Ivey-sol® surfactant technology.

Sylvia D.M., Fuhrmann J.F., Hartel P.G., and Zuberer D.A 2005. *Principles and applications of soil microbiology*. New Jersey, Pearson Education Inc.

Tränkner A. 1992. Use of agricultural and municipal organic wastes to develop suppressiveness to plant pathogens. In: *Biological Control of Plant Diseases*. (Eds. E. C. Tjamos, G. C. Papavizas, and R. J. Cook). Plenum Press, New York.

Glossary

Abortive transduction : Transduction in which the genetic fragment from the donor bacterium is not integrated in the genome of the recipient bacterium. When the recipient bacterium divides, the genetic fragment from the donor bacterium is transmitted to only one of the daughter cells.

Abzyme [= antibody and enzyme, *catmab* (catalytic monoclonal antibody)] : It is a monoclonal antibody with catalytic activity.

Accessory cell : Cell required for, but not actually mediating, a specific immune response. It is often used to describe antigen-presenting cells (APC).

Active immunity : It is a type of acquired immunity where the body produces antibodies after infection or vaccination.

Active ingredient : It is the active agent in a drug that is responsible for its medicinal properties.

Active site : The specific portion of an enzyme that attaches to the substrate.

Active transport : A kind of transport wherein ions or molecules move against a concentration gradient, which means movement in the direction opposite that of diffusion or movement from an area of lower concentration to an area of higher concentration. Hence, this process will require expenditure of energy, and the assistance of a type of protein called a carrier protein.

Adaptive immunity (= acquired immunity) : A type of immune response. It is a specific reaction to a particular antigen that becomes stronger when the individual is exposed to the antigen again. Compare with innate immunity.

Adenylate cyclase : The enzyme that changes adenosine triphosphate (ATP) into adenosine 3′, 5′ cyclic monophosphate (cAMP). It has an important role in controlling mechanisms within a cell.

Affinity : A measure of the binding constant of a single antigen combining site with a monovalent antigenic determinant.

Amplified Fragment Length Polymorphism (AFLP) : A highly sensitive method for detecting DNA polymorphism. Following restriction enzyme digestion of DNA, a subset of the DNA fragments is selected for PCR amplification and visualisation.

Agarose gel electrophoresis : It is a method used to separate DNA fragments by molecular weight. Fragments move by electric current at different speeds through the agarose.

Agglutination : The aggregation of particulate antigen by antibodies. Agglutination applies to red blood cells as well as to bacteria and inert particles covered with antigen.

***Agrobacterium tumefaciens* :** The soil bacterium containing the Ti plasmid is able to form crown galls on a number of dicotyledonous plants.

Agroinfection : A technique for infecting plant cells with DNA from a plant virus by using the T-DNA portion of a Ti plasmid.

Algorithms : A step-by-step procedure for calculations used for calculation, data processing, and automated reasoning.

Alleles : Alternative forms of a genetic locus; a single allele for each locus is inherited separately from each parent (e.g. at a locus for eye color the allele might result in blue or brown eyes).

Allergen : A substance that causes the allergic reaction resulting after exposure via ingestion, inhalation, injection or contact with skin.

Allogeneic transplant : In an allogeneic transplant, the patient receives bone marrow or peripheral blood stem cells from another person, usually a sibling, but sometimes an unrelated donor.

Allograft : A tissue transplant (graft) between two genetically non-identical members of a species.

Allosteric regulation : Regulation of an enzyme or other protein by binding an effector molecule at the protein's allosteric site (*i.e.*, a site other than the protein's active site).

Allosteric site : A specific receptor site of an enzyme molecule remote from the active site. Molecules bind to the allosteric site and change the shape of the active site, making it either more or less receptive to the substrate.

Allotransplantation : Transplantation of cells, tissues, or organs, sourced from a genetically non-identical member of the same species as the recipient.

Allotypes : Antigenic determinants present in allelic (alternate) forms. When used in association with immunoglobulin, allotypes describe allelic variants of immunoglobulins detected by antibodies raised between members of the same species.

Alpha (α) globin : A protein chain that forms part of normal adult haemoglobin, haemoglobin Gower II, fetal haemoglobin, and haemoglobin A_2. It is produced at low levels during early foetal life and at increased levels during adulthood.

Alpha (α) thalassaemia : A hereditary blood disease in which the production of alpha globin chains in haemoglobin is reduced.

Alpha hairpin : This motif comprises two sequential alpha helices linked by coil, which lie adjacent in and run approximately antiparallel.

Alpha helix : A common secondary structure of proteins which is right-handed coiled or spiral conformation (helix), in which every backbone N-H group donates a hydrogen bond to the backbone C=O group of the amino acid four residues earlier ($i + 4 \rightarrow i$ hydrogen bonding).

Ames test : It is a biological assay to assess the mutagenic potential of chemical compounds. A positive test indicates that the chemical is mutagenic and therefore may act as a carcinogen.

Amniocentesis : A procedure in which a needle is inserted into the amniotic fluid around the embryo in the womb. The needle is used either to remove a sample of amniotic fluid or to inject a substance.

Amniotic fluid : The liquid that surrounds the embryo of reptiles, birds and mammals. The liquid is contained in a cavity or sac, which is lined with membrane.

Amphipathic : Molecules that have both hydrophilic and hydrophobic parts are termed amphipathic.

Ampholyte : It is a molecule or ion that can react both as an acid as well as a base.

Amplification : A process where a gene or fragment of DNA is duplicated many times.

Amplified Fragment Length Polymorphism (AFLP) : A PCR-based tool used in genetics research, DNA fingerprinting, and in the practice of genetic engineering developed by Keygene.

Anaemia : A condition where there are fewer red cells in the blood or a lower haemoglobin concentration than normal.

Androgenesis : The formation of sporophyte from the male gametophyte on artificial medium.

Antenatal : Occurring before birth. See prenatal diagnosis.

Anterior : Situated towards the front or near the head.

Antibiotic resistance marker gene : Genes (usually of bacteria origin) used as selection markers in genetic engineering, because their expression allows cell survival in the presence of normally toxic antibiotic agents. These genes were commonly used in the development and release of first generation genetically engineered crop plants.

Antibiotic : A substance that is produced by one microorganism and which inhibits or kills another. Antibiotics vary widely in structure. They can act either on a particular microorganism (*i.e.* they are specific) or a group of microorganisms (*i.e.* they are broad spectrum).

Antibody (Ab) : A protein that is produced by the lymphoid tissue to fight against antigens. All antibodies are immunoglobulins.

Antigen (= immunogen) : A substance that induces a specific immune response.

Antigen (Ag) (= immunogen) : Any substance that the body identifies as potentially dangerous, e.g. a virus or bacterium. Its presence stimulates the body to produce antibodies.

Antigen presenting cell (APC) : A type of cell in the immune system. It presents antigen to lymphocytes, e.g. T cells. Antigen presenting cells include macrophages and dendritic cells.

Antigen receptor : The specific antigen-binding receptor on T or B lymphocytes; these receptors are transcribed and translated from rearrangements of V genes.

Antigen-binding site : The part of an immunoglobulin molecule that binds antigen specifically.

Antigenic determinant : A single antigenic site or epitope on a complex antigenic molecule or particle.

Antigen-presenting cell (APC) : A specialised type of cell, bearing cell surface class II MHC (major histocompatibility complex) molecules, involved in processing and presentation of antigen to inducer or helper T cells. Examples: macrophage, dendritic cells.

Antihistamine: A drug that acts against histamine. Antihistamines may be given during an allergic reaction, in which histamine is released by certain cells of the immune system.

Anti-oncogene (= tumor suppressor gene) : A gene that protects a cell from one step on the path to cancer. When this gene is mutated to cause a loss or reduction in its function, the cell can progress to cancer, usually in combination with other genetic changes.

Antiparallel beta sheets : This is a beta-pleated sheet in which adjacent beta strands in the structure point in opposite directions, for example amino-terminal to carboxy-terminal, carboxy-terminal to amino-terminal, amino-terminal to carboxy-terminal, etc. Such sheets have narrowly spaced hydrogen bond pairs that alternate with widely spaced pairs. In each residue the main chain -NH and -CO groups point in the same direction, but which alternates with successive residues

Antiporter (= **exchanger** or **counter-transporter**) **:** It is an integral membrane protein involved in secondary active transport of two or more different molecules or ions (*i.e.,* solutes) across a phospholipid membrane such as the plasma membrane in opposite directions.

Antisense RNA : A single-stranded RNA complementary to a mRNA strand transcribed within a cell.

Antisense therapy : A form of treatment for genetic disorders or infections by using a strand of nucleic acid (DNA, RNA or a chemical analogue) that binds to the mRNA produced by that gene and inactivate it turning that gene "off".

Antitoxin : It is an antibody with the ability to neutralise a specific toxin. Antitoxins are produced by certain animals, plants and bacteria.

Anti-viral proteins : Proteins induced by human or animal cells to interfere with viral replication.

Apoptosis : A process where a cell kills itself. It is a kind of programmed cell death.

Arbitrary primed PCR (AP-PCR) : It is a deoxyribonucleic acid (DNA) fingerprinting technique in which one short arbitrary primer is used to amplify multiple DNA fragments of different length, which yield a fingerprint after separation in gel electrophoresis.

Arthritis : A condition where a joint becomes inflamed.

Artificial ribozymes : They are artificially-produced self-cleaving RNAs bearing good enzymatic activity.

Assay : It refers to a test that is used to measure the concentration of a substance.

ATCC : American Type Culture Collection, which is global bioscience organisation that provides biological products, technical services and educational programmes to private industry, government and academic organisations.

Attenuated vaccine : A vaccine created by reducing the virulence of a pathogen but still keeping it 'live'.

Autograft: It is the transplant of tissue to the same person.

Autoimmune disease : Any disease that results from such an aberrant immune response, such as Coeliac disease, diabetes mellitus type 1.

Autoimmunity : The failure of an organism to recognise its own constituent parts as 'self', which allows an immune response against its own cells and tissues.

Autoradiography : A phenomenon of producing a radiograph by means of the radiation (e.g. X rays) emitted from the specimen being photographed.

Avidity: The summation of multiple affinities, for example when a polyvalent antibody binds to a polyvalent antigen.

B cell/ B lymphocyte : A type of white blood cell that is formed by lymphoid tissue. It produces circulating antibodies. It is also known as a B lymphocyte.

B cells : Lymphocytes (one of the five kinds of white blood cells or leukocytes) that play a major role in the humoral immune response (as opposed to the cell-mediated immune response, which is governed by T cells).

B lymphocyte (B cell) : The precursors of antibody-forming plasma cells; these cells carry immunoglobulin and class II MHC (major histocompatibility complex) antigens on their surfaces.

Bacterial artificial chromosome (BAC) : A length of DNA with sequences that allow it to be maintained and reproduced in bacteria. DNA libraries for genome sequencing can be made using BACs.

Barnase–Barstar : Barnase and barstar are the extracellular ribonuclease and its intracellular inhibitor produced by *Bacillus amyloliquefaciens*. Both are small single-chain proteins suitable for application to the study of how a protein's sequence directs its fold.

Base analogue : It is a chemical that can substitute for a normal nucleotide sequence in nucleic acids.

Base pair (bp) : Two nitrogenous bases (adenine and thymine or guanine and cytosine) held together by weak bonds. Two strands of DNA are held together in the shape of a double helix by the bonds between base pairs.

Base sequence : The order of nucleotide bases in a DNA molecule.

Basic local alignement search tool (BLAST) : An algorithm for comparing primary biological sequence information, such as the amino acid sequences of different proteins or the nucleotides of DNA sequences.

Batch culture : A large-scale closed system culture in which cells are grown in a fixed volume of nutrient culture medium under specific environmental conditions up to a certain density in a tank or airlift fermentor, harvested and processed as a batch, especially before all nutrients are used up.

Benign tumour : Harmless tumour that is not capable of indefinite growth and the host survives.

Beta pleated sheet : *See* beta sheet.

Beta (β) globin : A protein chain that forms part of normal adult haemoglobin. It is produced at low levels as early as six weeks of foetal life and at higher levels throughout adulthood.

Beta helix : In this the polypeptide chain winds up into a large right handed coil (or super helix). There are two different types of beta helix, one with two beta sheets and one with three beta sheets.

Beta sheet (= beta pleated sheet) : A secondary protein structure in which two or more extended polypeptide chains are hydrogen-bonded to one another in a planar array; the 'pleats' are formed by angles of bonds in the peptide chains. The beta strands in a sheet can be arranged to form parallel, antiparallel or mixed sheets.

Beta strand: A beta strand describes a single length of polypeptide chain that forms part of a beta sheet.

Beta thalassaemia : A blood abnormality where β globin chains are produced very slowly or not at all. The condition is inherited.

Beta turn : A beta turn occurs where the polypeptide chain makes a sharp reversal by 180 degrees within 4 residues. A turn is defined by 4 consecutive residues (i to i+3) if the C-alpha(i) to C-alpha(i+3) distance is less than 7 angstroms, and if the central two residues are not helical. There is often a hydrogen bond bewteen the CO(i) and NH(i+3).

Binary vector : A pair of plasmids consisting of a binary plasmid and a helper plasmid. The two plasmids are used together to produce genetically modified plants.

Bioaugmentation : The introduction of specially selected or genetically engineered strains of microbes to a contaminated site.

Bio-business : It refers to business of biotech products through national and international markets.

Biodegradation : Degradation of organic contaminants in soil and/or groundwater by indigenous or transplanted/acclimated microorganisms, primarily bacteria and fungi.

Bioethics: The study of ethical issues raised by medical and biological research and the applications of that research.

Biofilm : Microbes adhere to the environmental surfaces and synthesise extra cellular homo- or hetero polymers of sugars called exo-polysaccharides (EPS). The whole assemblage including the resident microbes and the channelised network of the EPS is termed as biofilm.

Biogas : A gas (methane, hydrogen, and carbon monoxide) produced by the anaerobic digestion or fermentation of biodegradable materials such as manure, sewage, municipal waste, green waste, plant material. It is a renewable energy source, like solar and wind energy.

Biogas digesters : A device that maintains anaerobic condition for anaerobic breakdown of organic matter.

Bioinformatics : Bioinformatics is the application of computer technology to the management of biological information.

Bioleaching : Bioleaching is the solubilisation of one or more components of a complex solid mediated by bacteria.

Biolistics (= gene gun) : A biolistic particle delivery system designed for plant transformation, is a device for injecting cells with genetic information. The payload is an elemental particle of a heavy metal coated with plasmid DNA.

Biomarker : A molecule that allows for the detection and isolation of a particular cell type, for example, the protein Oct-4 is used as a biomarker to identify embryonic stem cells.

Biomining : Biomining is a technique used to recover the various types of minerals from ore-using microorganism.

Bioremediation : The use of living organisms to reduce or eliminate environmental hazards resulting from accumulation of toxic chemicals or other hazardous wastes.

Biosafety : The safe handling, transfer and use of living organisms modified through biotechnology.

Biosensor : An analytical device, used for the detection of an analyte, that combines a biological component with a physicochemical detector.

Biosorption : A physiochemical process that occurs naturally in certain biomass which allows it to passively concentrate and bind contaminants onto its cellular structure.

Biostimulation : The addition of oxygen and/or inorganic nutrients to indigenous microbial populations in soils and groundwater.

Biosurfactants : The microbial-derived surface active agents that can lower surface and interfacial tension of liquids.

Biotechnology : The integrated use of biochemistry, microbiology and engineering sciences in order to achieve technological applications of capabilities of microorganisms, cultured tissues, cells and part thereof.

Bioventing : A process of motivating the natural *in situ* biodegradation of contaminants in soil by providing air or oxygen to existing soil microorganisms.

Blastocoel : The fluid-filled cavity inside the blastocyst, an early preimplantation stage of the developing embryo.

Blocks : Multiply aligned ungapped segments corresponding to the most highly conserved regions of proteins.

Blotting technique : The techniques used for separation of DNA, RNA and proteins using different steps.

Blue (Marine) biotechnology : It is the application of molecular biological methods to marine and freshwater life, using marine organisms, and their derivatives, to increase seafood supply and safety, control proliferation of noxious water-borne organisms, and develop new drugs.

Bokashi compost : A method of intensive composting is called Bokashi. The compost is produced through an aerobic or anaerobic inoculation.

Brookhaven Protein DataBank: *See* PDB.

Bt cotton : Genetically cotton modified that contains a natural toxin created by a bacteria *Bacillus thuringiensis* (Bt).

Bursa of fabricius : An out pouching of the cloaca in birds; site of development of B cells in birds.

Ca^{2+}–ATPase pump : The Ca^{2+}–pump in the plasma membrane maintains Ca^{2+} gradient which actively transport Ca^{2+} out the cell. It is very important for cells to maintain low concentrations of Ca^{2+} for proper cell signaling.

CAAT box : A distinct pattern of nucleotides with GGCCAATCT consensus sequence that occur upstream by 75-80 bases to the initial transcription site.

Callus : A mass of unorganised parenchyma cells derived from plant tissue (explants) for use in biological research and biotechnology.

Callus culture : A technique of tissue culture; it is usually on solidified medium and initiated by inoculation of small explants or sections from established organ or other cultures (the inocula). Callus culture is used as the basis for organogenic (shoot, root) cultures, cell cultures or proliferation of embryoids.

CAMP (adenosine 3′, 5′ cyclic monophosphate) : A molecule that is formed by the enzyme adenylate cyclase from adenosine triphosphate. It is important in controlling mechanisms within a cell.

Cancer : Uncontrolled growth of the cells of a tissue or an organ in a multicellular organism.

Cancer immunotherapy : The use of the immune system to reject cancer. The main premise is stimulating the patient's immune system to attack the malignant tumour cells that are responsible for the disease.

Capillary : The smallest blood vessel (5-20 micrometres in diameter). It has a wall only one cell thick, so oxygen, carbon dioxide, salts etc. can be exchanged between the blood and body tissue.

Carcinogen : A substance that causes cancer.

Carcinogenesis (= the creation of cancer) : The process by which normal cells are transformed into cancer cells.

Carrier DNA : DNA of undefined sequence content which is added to the transforming (plasmid) DNA used in physical DNA-transfer procedures. This additional DNA increases the efficiency of transformation in electroporation and chemically mediated DNA-delivery systems. The mechanism responsible for this effect is not known.

Carrier proteins : A protein that transports specific substance through intracellular compartments, into the extracellular fluid, or across the cell membrane.

Carrier : A large immunogenic molecule or particle to which an antigenic determinant is attached, allowing the determinant to become immunogenic.

CAT box : An element of eukaryotic promoters, lying 70-80 nucleotides upstream of the transcription start site.

Catalyst : A substance that promotes a chemical reaction by lowering the activation energyof a chemical reaction, but itself remains unaltered at the end of the reaction.

Catalytic triad : It refers to any set of three residues that function together and are directly involved in catalysis.

cDNA clone : A double-stranded DNA molecule that is carried in a vector and was synthesised *in vitro* from an mRNA sequence by using reverse transcriptase and DNA polymerase.

cDNA library : A collection of cDNA clones that were generated *in vitro* from the mRNA sequences isolated from an organism or a specific tissue or cell type or population of an organism.

cDNA: The double-stranded DNA complement of an mRNA sequence; synthesised *in vitro* from a mature RNA template using reverse transcriptase (to create a single strand of DNA from the RNA template) and DNA polymerase (to create the double-stranded DNA).

Cell culture : The *in vitro* growth of cells derived from multi-cellular organisms. The cells are usually of one type.

Cell differentiation : Continuous loss of physiological and cytological characters of young cells, resulting in getting the characters of adult cells. The unspecialised cells become modified and specialised for the performance of specific functions. Differentiation results from the controlled activation and de-activation of genes.

Cell fusion : Formation of a single hybrid cell from two cells of different species cultured *in vitro*. The cells fuse and coalesce, but their nuclei may remain separated.

Cell signalling : A part of a complex system of communication that governs basic cellular activities and coordinates cell actions.

Cell suspension culture : The culture of cells in medium without adherence to a stationary substrate.

Cell-mediated cytotoxicity (CMC) : Killing (lysis) of a target cell by an effector lymphocyte.

Cell-mediated immunity (CMI) : A form of immunity where antibodies have no role or play only a secondary role. It includes the processes where cytotoxic cells kill pathogens.

Cell surface receptors : Cell surface receptors (membrane receptors, transmembrane receptors) are specialised integral membrane proteins that take part in communication between the cell and the outside world. A transmembrane receptor may be subdivided into three parts or domains, extracellular, transmembrane and intracellular.

Cellular immunity: A condition where lymphocytes and macrophages are able to fight against viruses, parasites, fungi and possibly cancer cells. They acquire the ability after they are exposed to antigens or pathogens.

Cendritic cell : A type of cell that takes up antigen and presents it to other cells of the immune system. It is a phagocyte, which is present in lymphoid and other tissues. Dendritic cells in lymphoid tissues stimulate T cell responses.

Centimorgan (cM) : A unit of measure of recombination frequency. One centimorgan is equal to a 1% chance that a marker at one genetic locus will be separated from a marker at a second locus due to crossing over in a single generation. In human beings, 1 centimorgan is equivalent, on average, to 1 million base pairs.

Central dogma : Flow of genetic information from DNA/RNA to polypeptide via mRNA within a cell. It was first stated by Francis Crick in 1958

Central nervous system (CNS) : The brain and spinal cord.

Centromere : A specialised chromosome region to which spindle fibers attach during cell division.

Channel proteins : In some cases the channel proteins simply act as a passive pore. Molecules will randomly move through the opening in a process called diffusion. This requires no energy, molecules move from an area of high concentration to an area of low concentration.

Chaperone : A protein that helps other proteins to fold up into their correct 3D structure.

Chemokine : A small protein that binds to specific receptors on the surface of immune system cells, particularly phagocytic cells and lymphocytes. Chemokines play a crucial role in inflammatory responses.

Chemostat : A bioreactor to which fresh medium is continuously added, while culture liquid is continuously removed to keep the culture volume constant and **Chem**ical environment **stat**ic.

Chemotherapy : The use of chemicals or drugs to treat disease.

Chiasma : The contact between two chromatids ready to undergo crossing over is called chiasma.

Chimera (= chimaera) : An individual exhibiting two or more genotypes in patches derived from two or more embryos. An individual derived from two embryos by experimental intervention.

Chimeric monoclonal antibody (= ximab) : A combined form for chimeric monoclonal antibody, that usually elicits a neutralising antibody response in a patient who receives them.

Chirality : Chemical version of handedness.

Chromatography : It is the collective term for a set of laboratory techniques for the separation of mixtures dissolved in a fluid called the mobile phase from the stationary phase.

Cladogram : A dichotomous phylogenetic tree that branches repeatedly, suggesting the classification of molecules or organisms based on the time sequence in which evolutionary branches arise.

Clinical diagnosis : A process where a disease is identified from history taking and physical examination.

Clonal deletion : The loss of lymphocytes of a particular specificity due to contact with either 'self' or artificially introduced antigen.

Clonal selection theory : The prevalent concept that specificity and diversity of an immune response are the result of selection by antigen of specifically reactive clones from a large repertoire of preformed lymphocytes, each with individual specificities.

Clone : Generation of identical copies of a region of a DNA molecule or to generate genetically identical copies of a cell, molecule or organisms.

Clone bank : *See* genomic library.

Clones : A group of cells derived from a single ancestor.

Cloning vector : DNA molecule originating from a virus, a plasmid, or the cell of a higher organism into which another DNA fragment of appropriate size can be integrated without loss of the vectors capacity for self-replication; vectors introduce foreign DNA into host cells, where it can be reproduced in large quantities. Examples are plasmids, cosmids, etc.

Cloning : Incorporation of a DNA molecule into a chromosomal site or a cloning vector.

cM : See centimorgan.

Coagulation factor : A protein used by the body to prevent bleeding.

Codon : A group of three DNA or RNA nucleotide bases. Together they specify the choice of an amino acid residue during peptide synthesis. If one or more of the bases changes, a different amino acid may be specified.

Coenzyme : *See* cofactor.

Cofactor : An organic molecule, such as vitamin, that binds to an enzyme and is required for its catalytic activity.

Coil : In a protein secondary structure is defined to include 'helix', 'sheet' and 'turn' residues. 'Coil' residues can be considered as any section of chain that does not fall into these three categories

Colony hybridization : A technique that uses a nucleic acid probe to identify a bacterial colony with a vector carrying a specific cloned gene or genes.

Complement : A group of 20 proteins found in blood. They are involved in the destruction of foreign cells, after the cells have been coated with antibodies.

Complement : A series of serum proteins involved in the mediation of immune reactions. The complement cascade is triggered by the interaction of antibody with specific antigen.

Complementary DNA (cDNA) : DNA that is synthesised from a mRNA template using the enzyme reverse transcriptase; the single-stranded form is often used as a probe in physical mapping.

Complementary sequences : Nucleic acid base sequences that can form a double stranded structure by matching base pairs; the complementary sequence to G-T-A-C is C-A-T-G.

Compost : Compost is the organic materials derived from plants and animal that has been decomposed through aerobic decomposition, which is rich in nutrients.

Compost tea : The liquid suspension made by steeping compost in water is called compost tea.

Composting : A controlled biological process to convert biodegradable hazardous materials to harmless and stabilised byproducts by using microorganisms under elevated temperature.

Conjugate : Certain bacteria have polysaccharide outer coats that are poorly immunogenic. By linking these outer coats to proteins (e.g. toxins), the immune system can be led to recognise the polysaccharide as if it were a protein antigen. This approach is used in the *Haemophilus influenzae* type B vaccine.

Conjugate vaccines (= component vaccines) : Vaccines created by covalently attaching a poor antigen to a carrier protein (preferably from the same microorganism), thereby conferring the immunological attributes of the carrier on the attached antigen.

Conjugation : The formation of a connection between two bacteria for the exchange of genetic material.

Consensus sequence : The calculated order of most frequent residues, either nucleotide or amino acid, found at each position in a sequence alignment.

Conserved sequence : A base sequence in a DNA molecule (or an amino acid sequence in a protein) that has remained essentially unchanged throughout evolution.

Contig map (= contiguous map) : A map depicting the relative order of a linked library of small overlapping clones representing a complete chromosomal segment.

Contig : A group of overlapping clones that provide a physical map of a portion of a chromosome.

Continuous culture : A suspension culture continuously supplied with nutrients by the inflow of fresh medium. The culture volume is normally constant.

Continuous fermentation : A process in which cells or microorganisms are maintained in culture in the exponential growth phase by the continuous addition of fresh medium that is exactly balanced by the removal of cell suspension from the bioreactor.

Cosmid : Artificially constructed cloning vector containing the *cos* gene of phage lambda. Cosmids can be packaged in lambda phage particles for infection into *E. coli;* this permits cloning of larger DNA fragments (up to 45 kb) than can be introduced into bacterial hosts in plasmid vectors.

Crossing over : The breaking during meiosis of one maternal and one paternal chromosome, the exchange of corresponding sections of DNA, and the rejoining of the chromosomes. This process can result in an exchange of alleles between chromosomes.

Cry proteins : A class of crystalline proteins produced by strains of *Bacillus thuringiensis* and engineered into crop plants to give resistance against insect pests. These proteins are toxic to certain insects but are harmless to mammals and most beneficial insects.

Cryogenic : At very low temperature.

Cryopreservation (= freeze preservation) : The preservation of germplasm resources in a dormant state by cryogenic techniques, as currently applied to storage of plant seeds and pollen, micro-organisms, animal sperm, and tissue culture cell lines.

Cryoprotectants : Compounds preventing cell damage during freezing and thawing processes. Cryoprotectants are agents with high water solubility and low toxicity, e.g. glycerol, DMSO, dextran, ethylene glycol, polyvinyl pyrolidone and hydroxyethyl starch.

Culture : A population of plant or animal cells or microorganisms that is grown under controlled conditions.

Cybrid : A cytoplasmic hybrid, originating from the fusion of a cytoplast (the cytoplasm without nucleus) with a whole cell, as in nuclear transfer. The nucleus and cytoplasm of the fused cell products are from different genetic sources.

Cybridisation : A phenomenon of combining the somatic cells by fusing the membranes of cells and mixing of cytoplasm and nuclear content of the parental cells to hybrids with a summation of the chromosomes.

Cybrids (= cytoplasmic hybrid) : A eukaryotic cell line produced by the fusion of a whole cell with a cytoplast.

Cytokine : A soluble molecule that is involved in signalling between cells during an immune response, e.g. tumour necrosis factor.

Cytoskeleton : The network of structural proteins inside a cell. It gives the cell its shape, flexibility and motility.

Cytotoxic (= cytolytic) T cell : Cell that kills target cells bearing appropriate antigen within the groove of an MHC class I molecule that is identical to that of the T cell.

Data source: An electronic collection of biological data, usually in the form of a database (such as relational, object-oriented or flat-file database), but sometimes in the form of a programme.

Database : An organised collection of data that is typically organised to model relevant aspects of reality in a way that supports processes requiring this information.

Database management systems (DBMS) : A set of programmes that enable storing, modifying, and extracting information from a database; it also provides users with tools to add, delete, access, modify, and analyse data stored in one location.

Database models : The theoretical foundation of a database and fundamentally determines in which manner data can be stored, organised, and manipulated in a database system. It thereby defines the infrastructure offered by a particular database system

DEAE-dextran : A positively charged resin used in ion exchange chromatography used in protein and nucleic acid purification/separation.

Denature : To induce structural alterations that disrupt the biological activity of a molecule. Often it refers to breaking hydrogen bonds between base pairs in double-stranded nucleic acid molecules to produce in single-stranded polynucleotides or altering the secondary and tertiary structure of a protein, destroying its activity.

Desorption electrospray ionisation (DESI) : An ambient ionisation technique that can be used in mass spectrometry for chemical analysis. It is an atmospheric pressure ion source that ionises gases, liquids and solids in open air under ambient conditions.

Determinant : Part of the antigen molecule which binds to an antibody-combining site or to a receptor on T cells.

Diagnosis : A process where a disease is identified by signs and symptoms, a history and laboratory tests.

Diauxic growth : Diauxic growth means doubling of growth caused by the presence of two sugars on a culture growth media, one of which is easier for the target bacterium to metabolise.

Dicer enzyme : An endoribonuclease in the RNase III family that cleaves dsRNA and pre-microRNA (miRNA) into short double-stranded RNA fragments called small interfering RNA (siRNA) about 20-25 nucleotides long, usually with a two-base overhang on the 3' end. Dicer contains two RNase III domains and one PAZ domain

Digoxigenin : A steroid found exclusively in the flowers and leaves of the plants *Digitalis purpurea.* It is a hapten, a small molecule with high immunogenicity used in many molecular biology applications similarly to other popular haptens such as DNP (dinitrophenol), biotin, and fluorescein.

Disulphide bridge : This is a covalent link formed between the sulphur atoms of two cysteine residues. These residues can be anywhere in the polypeptide chain, and connections are 'determined' by the tertiary structure of the protein.

DNA Amplification Fingerprinting (DAF) : It is a novel strategy to detect genetic differences among organisms using a thermostable DNA polymerase directed by usually one short (≥5 bp) oligonucleotide primer of arbitrary sequence to amplify short segments of genomic DNA and generate a range of DNA extension products, which can be analysed by polyacrylamide gel electrophoresis and silver staining.

DNA chip (= DNA microarray) : A collection of microscopic DNA spots attached to a solid surface that is used to measure the expression levels of large numbers of genes simultaneously or to genotype multiple regions of a genome.

DNA fingerprinting (= DNA profiling) : Studies of variations in the lengths of pieces of DNA that are used to establish identity. These fragments are obtained from different individuals due to genetic polymorphism of repeated DNA, minisatellites or microsatellites, in the genome.

DNA library : See genomic library.

DNA probes : *See* probe.

DNA repair : A collection of processes by which a cell identifies and corrects damage to the DNA molecules that encode its genome. There are different types of repair stems.

DNA replication : The use of existing DNA as a template for the synthesis of new DNA strands. In humans and other eukaryotes, replication occurs in the cell nucleus.

DNA sequence : The relative order of base pairs, whether in a fragment of DNA, a gene, a chromosome, or an entire genome. See base sequence analysis.

DNA sequencer : A scientific instrument used to automate the DNA sequencing process.

DNA sequencing : The process of determining the precise order of nucleotides within a DNA molecule. It includes any method used to determine the order of the four bases, *i.e.* adenine, guanine, cytosine, and thymine in a DNA strand.

DNA vaccine : A preparation of naked DNA which upon injection protects an organism against disease producing an immunological response.

DNA/RNA polymerase : Enzymes that catalyse the synthesis of nucleic acids on pre-existing nucleic acid templates, assembling RNA from ribonucleotides or DNA from deoxyribonucleotides.

Domain : A discrete portion of a protein with its own function. The combination of domains in a single protein determines its overall function.

Dot blot technique : A technique in used to detect biomolecules, and for detecting, analysing, and identifying proteins .

Double helix : The structural form of the DNA molecule consisting of two coiled polynucleotide chains held together by hydrogen bonds.

Downstream processing : The process of recovery and purification of biosynthetic products, particularly pharmaceuticals, from natural sources such as animal or plant tissue or fermentation broth, including the recycling of salvageable components and the proper treatment and disposal of waste.

Downstream : The region extending in a 3′ direction from a gene.

Duchenne muscular dystrophy (DMD) : An X-linked hereditary disease characterised by the absence of a protein known as dystrophin which causes degeneration and necrosis of skeletal muscle fibres.

EBI : European Bioinformatics Institute in Cambridge, England.

Ectoderm : The outermost germ layer of cells derived from the inner cell mass of the blastocyst; gives rise to the nervous system, sensory organs, skin, and related structures.

Edible vaccines : Antigenic proteins that are genetically engineered into a consumable crop.

Electroblotting : It is a method used in molecular biology/biochemistry/immunogenetics for transferring the proteins or nucleic acids onto a membrane by using PVDF or nitrocellulose, after gel electrophoresis.

Electrophoresis : It is an analytical method frequently used in molecular biology and medicine, which is applied for the separation and characterisation of proteins, nucleic acids as well as viruses and small organelles.

Electrospray ionisation (ESI) : A technique used in mass spectrometry to produce ions especially in producing ions from macromolecules because it overcomes the propensity of these molecules to fragment when ionised.

EMBL : European Molecular Biology Laboratory in Heidelberg, Germany.

Embryo : An early stage in the development of an animal. The embryo is contained in an egg, egg membranes, or the mother. The human embryo develops into a foetus.

Embryo culture : A component of *in vitro* fertilisation wherein resultant embryos are allowed to grow for some time in an artificial medium before being inserted into the uterus.

Embryo rescue : One of the earliest and successful forms of *invitro* culture techniques that is used to assist in the development of plant embryos that might not survive to become viable plants.

Embryonic stem cell : Pluripotent stem cells derived from the inner cell mass of a blastocyst, an early-stage embryo.

Encapsulated artificial seeds (= synthetic seeds) : Artificially encapsulated somatic embryos, shoot buds, cell aggregates, or any other tissue that can be used for sowing as a seed and that possess the ability to convert into a plant under *in vitro* or *ex vitro* conditions and that retain this potential also after storage.

Encapsulation : The condition of being enclosed as in a capsule.

Endocrine : Relating to cells or glands that secrete a substance directly into the bloodstream. The substance is delivered by the blood to anywhere in the body it is needed.

Endocytosis : A process by which a cell takes up material from its environment. The cell membrane folds inwards to take in material. It includes pinocytosis and phagocytosis.

Endogenous : Developing within an organism. Compare exogenous.

Endonuclease : An enzyme that cleaves its nucleic acid substrate at internal sites in the nucleotide sequence.

Endoplasmic reticulum : A structure within a plant or animal cell. It is a highly folded arrangement of membranes within the cytoplasm. There are two types, called rough and smooth endoplasmic reticulum.

Endosymbiosis : This theory holds that mitochondria, plastids (e.g. chloroplasts), and possibly other organelles of eukaryotic cells originate through symbiosis between multiple microorganisms.

Enhancer : A regulatory DNA sequence that greatly enhances the transcription of a gene.

Enthalpy : A measure of the total energy of a thermodynamic system. It includes the internal energy, which is the energy required to create a system, and the amount of energy required to make room for it by displacing its environment and establishing its volume and pressure. Enthalpy is a thermodynamic potential.

Entrapment : Trapping of enzymes in insoluble beads or microspheres, such as calcium alginate beads, that hinders the arrival of the substrate, and the exit of products.

ENTREZ : A Global Query Cross-Database Search System is a powerful federated search engine, or web portal that allows users to search many discrete health sciences databases at the National Center for Biotechnology Information (NCBI) website.

Entropy : It is a mathematically defined function that helps to account for the flow of heat through the Carnot cycle. It has been found to be more generally useful and it has several other definitions

Enucleated : Having had its nucleus removed.

EnviroPig™ : The trademark for a genetically modified line of Yorkshire pigs with the capability to digest plant phosphorus more efficiently than ordinary unmodified pigs that was developed at the University of Guelph.

Enzyme : A protein or an RNA molecule that acts as a catalyst, speeding the rate at which a biochemical reaction proceeds but not altering the direction or nature of the reaction.

Enzyme immobilisation : Attachment of an enzyme to an inert, insoluble material such as calcium alginate (produced by reacting a mixture of sodium alginate solution and enzyme solution with calcium chloride); this can provide increased resistance to changes in conditions such as pH or temperature.

Enzyme-linked immunosorbent assay (ELISA) : An assay in which an enzyme is linked to an antibody and a coloured substrate is used to measure the activity of bound enzyme and, hence, the amount of bound antibody.

Epitope : Also known as allergenic determinant, is the part of a macromolecule that is recognised by the immune system specifically by antibodies, B cells or T cells.

Epitope mapping: The process of identification and characterisation of the minimum molecular structures able to be recognised by the immune system elements, mainly T and B cells.

Epitope mapping : The process of matching surface proteins to antibodies that will bond to them.

EPSP (5-enolpyruvylshikimate-3-phosphate) synthase : An enzyme that catalyses the chemical reaction:

***Escherichia coli* :** Common bacterium that has been studied intensively by geneticists because of its small genome size, normal lack of pathogenicity, and ease of growth in the laboratory.

EST : Expressed sequence tag. See sequence tagged site.

Exogenous DNA: DNA originating outside an organism.

Exon : The coding sequence of a gene.

Exonuclease : An enzyme that cleaves nucleotides sequentially from free ends of a linear nucleic acid substrate.

Explant : Tissue taken from the body and grown in an artificial medium.

Expressed gene : *See* gene expression.

Expressed sequence tag (EST) : A short sub-sequence of a cDNA sequence used to identify gene transcripts and are instrumental in gene discovery and gene sequence determination.

Expression library : A library of genes or cDNAs whose encoded proteins are expressed by specialised vectors.

Expression vectors : A plasmid or virus designed for protein expression in cells which is used to introduce a specific gene into a target cell, and can commandeer the cell's mechanism for protein synthesis to produce the protein encoded by the gene.

Expression : It is shown by production of RNA or protein.

Extremozyme : An enzyme created by extremophiles that function under extreme environmental conditions such as very high pH, very low pH, high temperature, high salinity or other factors.

F⁻ prime (F′) : A bacterium possessing a plasmid containing the F factor and a part of the bacterial genome is called F′.

F(ab′)2 : A fragment of an antibody containing two antigen-binding sites generated by cleavage of the antibody molecule with the enzyme pepsin which cuts at the hinge region C-terminally to the inter-H-chain disulphide bond.

Fab : Fragment of antibody containing the antigen-binding site, generated by cleavage of the antibody with the enzyme papain, which cuts at the hinge region N-terminally to the inter-H-chain disulphide bond and generates two Fab fragments from one antibody molecule.

FASTA : A DNA and protein sequence alignment software package first described (as FASTP) by David J. Lipman and William R. Pearson in 1985. Its legacy is the FASTA format which is now ubiquitous in bioinformatics.

FASTA format : A text-based format for representing either nucleotide sequences or peptide sequences, in which nucleotides or amino acids are represented using single-letter codes.

Fatty acids : A group of molecules that have a chain structure and contain a carboxyl (COOH) group. They make up lipid molecules and are an essential component of cell membranes

Fc : Fragment of antibody without antigen-binding sites, generated by cleavage with papain; the Fc fragment contains the C-terminal domains of the heavy immunoglobulin chains.

Fermenter : An organism that causes fermentation.

Fermentor (= bioreactor) : A container in which fermentation takes place.

Fertilisation : The process during which a sperm enters an egg to form a zygote.

Fibrin : The protein that causes blood to clot.

Fibrinogen : A plasma protein. It is changed to fibrin in the blood by thrombin.

FISH (fluorescence *in situ* hybridisation) : A physical mapping approach that uses fluorescein tags to detect hybridisation of probes with metaphase chromosomes and with the less-condensed somatic interphase chromatin.

Flanking region : The DNA sequences extending on either side of a specific locus or gene.

Flow cytometry : Analysis of biological material by detection of the light-absorbing or fluorescing properties of cells or sub-cellular fractions passing in a narrow stream through a laser beam. An absorbance or fluorescence profile of the sample is produced. Automated sorting devices sort successive droplets of the analysed stream into different fractions depending on the fluorescence emitted by each droplet.

Fluorescence spectroscopy (= fluorometry or spectrofluorometry) : A type of electromagnetic spectroscopy which analyses fluorescence from a sample, by using a beam of ultraviolet light that excites the electrons in molecules of certain compounds and causes them to emit visible light .

Fluorescence-activated cell sorting (FACS) : A specialised type of flow cytometry that provides a method for sorting a heterogeneous mixture of biological cells into two or more containers, one cell at a time, based upon the specific light scattering and fluorescent characteristics of each cell.

Fluorescent antibody : An antibody coupled with a fluorescent dye, used with a fluorescence microscope to detect antigen on cells, tissues or microorganisms.

Fluorescent *in situ* hybridisation : See FISH.

FluoroPig : The GM pigs the entire body of which fluoresces due to presence of GFP gene.

Foetus : The unborn child in the womb during the later part of pregnancy.

Fourier transform (FT) spectroscopy : A measurement technique whereby spectra are collected based on measurements of the coherence of a radiative source by using time-domain or space-domain measurements of the electromagnetic radiation.

Frameshift : A process where nucleotides in DNA are inserted or deleted. As a result there is a change in the sequence in which triplets of nucleotides are read as codons.

G protein : A protein that has a role in controlling mechanisms within a cell. G proteins interact with adenylate cyclase in controlling the level of adenosine 3′, 5′ cyclic monophosphate (cAMP) within cells.

Gamma (γ) globin : A protein chain forming part of foetal haemoglobin. It is mainly produced in the foetus, but also at low levels after birth.

Gamma interferon (IFNγ) : A cytokine that is produced by T cells in response to antigen. It makes macrophages active.

Geep (sheep-goat chimera) : A chimera produced by combining the embryos of a goat and a sheep, and contain the cells of both sheep and goat origin.

Gel electrophoresis : It is a method for separation and analysis of macromolecules (DNA, RNA and proteins) and their fragments, based on their size and charge.

Gel filtration chromatography : It is a technique used for separation of proteins, peptides, and oligonucleotides on the basis of size through a bed of porous beads.

GenBank : A genetic database that collects all known DNA sequences.

Gene : A functional unit of heredity that is a segment of DNA found on chromosomes in the nucleus of a cell. Genes direct the formation of an enzyme or other protein.

Gene amplification : The presence of multiple copies of a gene.

Gene construct (= transgene) : A gene or genetic material transferred naturally or by any of a number of genetic engineering techniques from one organism to another.

Gene deletion : A process where nucleotides change or are rearranged. As a result, the codon changes.

Gene expression : The process by which a gene directs the synthesis of mRNA, which in turn directs the synthesis of protein, thereby affecting the phenotype of an organism.

Gene families : Groups of closely related genes that make similar products.

Gene finding (= **gene prediction) :** The process of identifying the regions of genomic DNA that encode genes; it includes protein-coding genes as well as RNA genes, but may also include prediction of other functional elements such as regulatory regions.

Gene gun : *See* biolistics.

Gene library : *See* genomic library.

Gene mapping : Determination of the relative positions of genes on a DNA molecule (chromosome or plasmid) and of the distance, in linkage units or physical units between them.

Gene marker : A gene that determines an easily recognisable phenotype and is used as a reference point in the study of inherited characteristics.

Gene pharming : A technique of using transgene for making human proteins of medicinal value.

Gene prediction : *See* gene finding.

Gene product : The biochemical material, either RNA or protein, resulting from expression of a gene. The amount of gene product is used to measure how active a gene is; abnormal amounts can be correlated with disease-causing alleles.

Gene silencing (RNA interference) : Loss of gene expression either through an alteration in the DNA sequence of a structural gene, or its regulatory region; or because of interaction between its transcript and other mRNAs present in the cell.

Gene targeting : A genetic technique that uses homologous recombination to change an endogenous gene.

Gene therapy : The transplantation of normal genes into cells in place of missing or defective ones in order to correct genetic disorders.

Genetic counselling : A service that is performed by a trained counsellor. The counsellor combines information from individuals' medical and family history with knowledge of human genetics. This provides information about whether a hereditary condition is possibly present. The counsellor provides information about the treatment available and how to prevent the condition.

Genetic distance : Distance between two genes is called genetic distance.

Genetic engineering : The deliberate modification of the characteristics of an organism by manipulating its genetic material.

Genetic engineering technologies : *See* recombinant DNA technologies.

Genetic map (= **linkage map) :** A map of the relative positions of genetic loci on a chromosome, determined on the basis of how often the loci are inherited together.

Genetic mapping (= genome mapping) : The creation of a genetic map assigning DNA fragments to chromosomes.

Genetic recombination : 1. Any process that results in offspring inheriting a different combination of genes to both parents. 2. A process that occurs during meiosis when homologous chromosomes pair up and exchange segments of DNA.

Genetically modified foods (GM foods) : These are foods produced from genetically modified organisms (GMOs) that that have had their genome altered through genetic engineering or foods that contain ingredients from GMOs.

Genetically modified organism (GMO) : A life form that has been genetically modified using techniques of modern gene technology.

Genetics : The study of the patterns of inheritance of specific traits.

Genome : The total number of genes in a single set of chromosomes, *i.e.* in a haploid nucleus.

Genome library : A collection of the complete DNA of an organism. Compare expression library.

Genome projects : Research and technology development efforts aimed at mapping and sequencing some or all of the genome of human beings and other organisms.

Genome similarity : Similarity between small portions of the genomes of two different organisms.

Genomic library : A collection of clones made from a set of randomly generated overlapping DNA fragments representing the entire genome of an organism. Compare library, arrayed library.

Genomics : The study of the genome of humans and other organisms.

Genotype : All of the genes possessed by an individual; in practice it refers to the particular alleles present at the loci in question.

Germ layers : After the blastocyst stage of embryonic development, the inner cell mass of the blastocyst goes through gastrulation, a period when the inner cell mass becomes organised into three distinct cell layers, called germ layers. The three layers are the ectoderm, the mesoderm, and the endoderm.

Germ line: Refers to genes in germ cells as opposed to somatic cells, that is, genes in their unrearranged state rather than those rearranged for production of a protein.

GloFish® : A patented and trademarked brand of GM fluorescent fish; a variety of different GloFish are currently in the market. Zebrafish were the first GloFish available in pet stores; now they are sold in bright red, green, orange-yellow, blue, and purple fluorescent colors.

Glycoprotein : A protein that has carbohydrate molecules attached to it.

Glycosylation: The addition of sugar side groups to a polypeptide.

Golden rice : A variety of *Oryza sativa* rice produced through genetic engineering to biosynthesise beta-carotene, a precursor of vitamin A, in the edible parts of rice called endosperm.

Green biotechnology (= plant biotechnology) : The use of biotechnology for agricultural processes and products.

Green fluorescent protein (GFP) : Traditionally GFP refers to the protein first isolated from the jellyfish *Aequorea victoria.* It is a protein composed of 238 amino acid residues (26.9 kDa) that exhibits bright green fluorescence when exposed to light in the blue to ultraviolet range.

Guanosine 3′, 5′ cyclic monophosphate (cGMP) : A molecule that is formed by the enzyme guanylate cyclase from guanosine triphosphate. It is important in controlling mechanisms within cells.

Guanosine triphosphate (GTP) : A source of cellular energy. Energy is stored in the chemical bonds of the three phosphate groups. It is released when GTP is split into its diphosphate and monophosphate forms.

Guanylate cyclase : The enzyme that changes GTP into cGMP.

GUS (=beta-glucuronidase) : A reporter gene system, particularly useful in plant molecular biology.

Haem : A pigment which contains iron and porphyrin.

Haematology : The study of the blood and blood forming tissues.

Haematopoietic stem cell : The multipotent stem cells that give rise to all the blood cell types from the myeloid (monocytes and macrophages, neutrophils, basophils, eosinophils, erythrocytes, megakaryocytes/platelets, dendritic cells), and lymphoid lineages (T-cells, B-cells, NK-cells).

Haemoglobin : A protein consisting of a pigment (haem) and polypeptides (globin). Haemoglobin gives blood its red colour. It takes up oxygen as the blood passes through the lungs and carries it to the body's tissues.

Haemoglobinopathy : A inherited disorder occurring in the globin chain part of haemoglobin. It includes structural abnormalities such as sickle haemoglobin and reduced production of globin chains as in the thalassaemias.

Haemophilia : An inherited condition where the blood's ability to clot is reduced. Affected individuals tend to bleed spontaneously and when they suffer minor trauma.

Haemopoietic : Relating to the production of blood cells.

Hairpin : This is a term used to describe a motif of two sequential secondary structures joined by a loop which folds back on itself to form a 'hairpin' structure. e.g. beta hairpin, alpha haipin.

Hairpin loop : A loop formed when single-stranded DNA or RNA base pairs to itself.

Hapten : A low molecular weight that is not itself immunogenic but that, after conjugation to a carrier protein or cells, becomes immunogenic and induces antibody, which can bind the hapten alone in the absence of carrier.

Hardening : Adapting plants to outdoor conditions by gradually withholding water, lowering the temperature, increasing light intensity, or reducing the nutrient supply.

HbS (sickle haemoglobin) : An abnormal type of haemoglobin found in sickle cell disease and sickle cell trait. The abnormality occurs when there are changes in the beta globin gene. These changes then produce an abnormal beta globin chain.

Heavy chain (H chain) : The larger of the two types of chains that comprise a normal immunoglobulin or antibody molecule.

Helix : A chain or coil that carries the genetic material in DNA. It is composed of nucleotides. DNA has two chains/coils that are wound around each other. This is called a 'double helix'.

Helix-turn-helix : A secondary polypeptide structure, found in several DNA-binding proteins, in which two adjacent alpha helices are oriented at approximately right angles to one another.

Helper cell : A functional subclass of T cells that can help to create cytotoxic T cells and cooperate with B cells to produce an antibody response. They usually recognise antigens in association with class 2 major histocompatibility complex (MHC) molecules.

Hematopoietic stem cell : A stem cell that gives rise to all red and white blood cells and platelets.

Hematopoietic stem cell transplantation (HSCT) **:** The transplantation of blood stem cells derived from the bone marrow (in this case known as bone marrow transplantation) or blood.

Heredity : Transmission of genetic material or characteristics from parent to offspring.

Heterogeneous nuclear ribonucleoproteins (hnRNPs) **:** Complexes of RNA and protein present in the cell nucleus during gene transcription and subsequent post-transcriptional modification of the newly synthesised RNA (pre-mRNA).

Heterologous : Derived from the tissues or DNA of a different species.

Heterozygous : Inheriting different alleles at a particular point on the chromosome. Compare homozygous

Hfr (High-frequency recombination strain) : In such *Escherichia coli* strains the F episome is integrated into the bacterial chromosome.

HGP : Human Genome Project.

Hinge region : A flexible, open segment of an antibody molecule that allows bending of the molecule. The hinge region is located between Fab and Fc and is susceptible to enzymatic cleavage.

Histamine : A molecule released from mast cells during inflammation. It makes blood vessels more permeable.

Histocompatibility : The ability of tissues to get along. The antigens are collectively referred to as histocompatibility antigens.

Histocompatibility complex : The collection of genes coding for peptides present on the surface of nucleated cells; these peptides are responsible for the differences between genetically non-identical individuals that cause rejection of tissue grafts between such individuals.

Histocompatibility : The degree to which tissue from one organism will be tolerated by the immune system of another organism.

Homeobox : A short stretch of nucleotides whose base sequence is virtually identical in all the genes that contain it. It has been found in many organisms from fruit flies to human beings. In the fruit fly, a homeobox appears to determine when particular groups of genes are expressed during development.

Homolog : Any member of a set of genes, DNA sequences or protein sequences whose nucleotide sequences show a high degree of one-to-one correspondence.

Homologous recombination : The exchange of DNA fragments between two DNA molecules or chromatids of paired chromosomes (during crossing over) at the site of identical nucleotide sequences.

Homozygote : An individual who has the same alleles at a particular point on the chromosome map. Compare heterozygote.

Homozygous: Inheriting the same alleles at a particular point on the chromosome map.

Hormone : A substance produced by a gland. Hormones are important in controlling mechanisms within the body.

HTS : High-throughput screening.

Human embryonic stem cell (hESC) : A type of pluripotent stem cells derived from early stage human embryos, including the blastocyst stage, that are capable of dividing without differentiating for a prolonged period in culture, and are known to develop into cells and tissues of the three primary germ layers.

Human gene therapy : Insertion of normal DNA directly into cells to correct a genetic defect.

Humanised monoclonal antibodies (zumab) : Antibodies from non-human species whose protein sequences have been modified to increase their similarity to antibody variants produced naturally in humans.

Humoral immunity : Any immune reaction that can be transferred with immune serum is termed humoral immunity (as opposed to cell-mediated immunity).

Hungtington's disease : Huntington's chorea. An autosomal dominant genetic disease, whose symptoms of involuntary muscle movements (chorea) and mental dysfunction usually appear at the age of 35-45, leading to a rapid death.

Hybrid arrested translation : A method used to identify the proteins encoded by a cloned DNA sequence. A crude cellular mRNA preparation, composed of many individual types of mRNA, is hybridised with cloned DNA. Only mRNA molecules homologous to the cloned DNA will anneal to it.

Hybridisation : A laboratory process where a double-stranded molecule is produced from two complementary strands of DNA or a strand of DNA and a complementary strand of RNA.

Hybridoma : A hybrid cell that results from the fusion of an antibody-secreting cell with a malignant cell; the progeny secrete antibody without stimulation and proliferate continuously both *in vivo* and *in vitro*.

Matrix-assisted layer desorption-ionisation (MALDI) : A soft ionisation technique used in mass spectrometry, allowing the analysis of biomolecules (such as DNA, proteins, peptides and sugars) and large organic molecules (such as polymers, dendrimers), which tend to be fragile and fragment when ionised by more conventional ionisation methods.

Megabase (Mb) : Unit of length for DNA fragments equal to 1 million nucleotides and roughly equal to 1 cM.

Meiosis : The process of two consecutive cell divisions in the diploid progenitors of sex cells. Meiosis results in four rather than two daughter cells, each with a haploid set of chromosomes.

Memory cell : A type of T or B cell that has been exposed to antigen. The cell remembers the antigen and will respond more effectively if exposed again.

Mentor pollen : The compatible pollen which has been treated in many ways to inhibit its fertilisation ability and retain its power to stimulate incompatible pollen to accomplish fertilisation.

Meristem : A region of plant tissue consisting of actively dividing cells forming new tissue.

Messenger RNA (mRNA) : RNA that serves as a template for protein synthesis.

Metabolism : The physiological and chemical processes that enable the body to function and to grow.

Metabolite : A substance that is produced or takes part in a chemical process within a living organism.

Metagenomics : The study of metagenomes, genetic material recovered directly from environmental samples. The broad areas are referred to as environmental genomics, ecogenomics or community genomics.

Methanogens : The obligate anaerobic microorganisms living in different oxygen-free environments. They gain their energy for ATP synthesis by producing methane from either CO_2/H_2 or from acetate.

Methylotrophs : Methylotrophs are an aerobic bacterium that utilises methane for carbon and energy for their growth.

Microinjection : The process of using a glass micropipette to insert substances at a microscopic or borderline macroscopic level into a single living cell.

Micropropagation : The practice of rapidly multiplying stock plant material to produce a large number of progeny plants by using modern plant tissue culture methods.

Microsatellites : Very short DNA motifs (1-10 base pairs) which occur as tandem repeats at numerous loci throughout the genome. Also known as simple sequence repeats (SSR), simple tandem repeats or simple repetitive sequences.

Microtubule : A hollow tube found in the cytoplasm of some cells. It is mainly composed of the protein tubulin. Microtubules help cells keep their shape and divide.

Mitochondrial DNA : The genetic material occurring within mitochondria.

Mitogen : A substance that stimulates the proliferation of many different clones of lymphocytes.

Mitosis : The process of nuclear division in cells that produces daughter cells which are genetically identical to each other and to the parental cells.

Mixed beta sheet : A sheet which has both parallel and antiparallel beta strands hydrogen bonded together.

Model-view-controller (MVC) : A software architecture pattern that separates the representation of information from the user's interaction with it.

Modern biotechnology : The application of *in vitro* nucleic acid techniques, including recombinant deoxyribonucleic acid (DNA) and direct injection of nucleic acid into cells or organelles.

Molecular characterisation : DNA sequence data and the mapping of particular functions on the plant chromosome and on the inserted DNA.

Molecular farming : A portmanteau of farming and pharmaceutical and refers to the use of genetic engineering to insert genes that code for useful pharmaceuticals into host animals or plants that would not express those genes, thus creating a genetically modified organism (GMO).

Monoclonal : A preparation of antibody that is monogenous or cells of a single specificity.

Monoclonal antibody : An antibody that is produced artificially from a cell clone. It consists of only one immunoglobulin, which comes from its parent lymphocytes.

Monozygotic : Coming from one fertilised egg, e.g. identical twins. Compare dizygotic.

Multiple cloning site (MCS) (= polylinker) : A short segment of DNA which contains many (up to ~20) restriction sites, a standard feature of engineered plasmids. MCSs are commonly used during procedures involving molecular cloning.

Multiple sequence alignment (MSA) : A sequence alignment of three or more biological sequences, generally protein, DNA or RNA.

Mutation : A permanent change in the DNA sequence of a gene caused spontaneously or induced chemically resulting in alteration in a gene sequence / amino acid sequence of the protein encoded by the gene.

Myeloma : A tumour of plasma cells, generally secreting a single species of immunoglobulin.

Na^+/K^+–ATPase (=Na^+/K^+ pump) : It is an antiporter enzyme, *i.e.* an electrogenic transmembrane ATPase located in the plasma membrane of all animal cells. It maintains the resting potential, avails transport, and regulates cellular volume.

Nanopore sequencing : A small hole of the order of 1 nanometer in internal diameter; this method is used for determining the order in which nucleotides occur on a strand of DNA.

National Centre of Biotechnology Information (NCBI) : Advances science and health by providing access to biomedical and genomic information.

Natural killer (NK) cell : A type of cell that is formed by lymphoid tissue. It kills cells that are infected with pathogens.

Neurotransmitter : A chemical substance that is released from nerve endings. Such substances carry impulses across the spaces between nerves.

Neutrophil : A white blood cell containing a lobed nucleus and fine granules. These cells fight bacterial infection.

Nick translation : A tagging technique in which DNA polymerase I is used to replace some of the nucleotides of a DNA sequence with their labelled analogues, creating a tagged DNA sequence which can be used as a probe in fluorescent *in situ* hybridisation or blotting techniques. It can also be used for radiolabelling.

NIH : National Institutes of Health, Bethesda, MD.

Northern blotting technique : A method used to study gene expression by detection of RNA (or isolated mRNA) in a sample.

Nuclear magnetic resonance (NMR) : A physical phenomenon in which nuclei in a magnetic field absorb and re-emit electromagnetic radiation.

Nucleotide : A subunit of DNA or RNA consisting of a nitrogenous base (adenine, guanine, thymine, or cytosine in DNA; adenine, guanine, uracil, or cytosine in RNA), a phosphate molecule and a sugar molecule (deoxyribose in DNA and ribose in RNA).

Oligonucleotide : A short chain of nucleotides.

Oligosaccharide : A carbohydrate made up of a small number of monosaccharides (usually between two and ten).

OMIM : The On-line Mendelian Inheritance of Man database.

Oncogene: A gene, one or more forms of which is associated with cancer. Many oncogenes are involved directly or indirectly in controlling the rate of cell growth.

Oncogenic : Cancer-causing.

OncoMouse™ : A type of laboratory mouse that has been genetically modified using modifications designed by Philip Leder and Timothy A Stewart.

Open reading frame (ORF) : A reading frame in a sequence of nucleotides in DNA that contains no termination codons and so can potentially translate as a polypeptide chain. It is the primary transcript RNA transcript immediately after transcription in the nucleus, before RNA splicing or polyadenylation to form the mature mRNA.

Operator : A prokaryotic regulatory element that interacts with a repressor to control the transcription of adjancent structural genes.

Operator gene : A gene that activates the production of messenger RNA by adjacent structural genes cistron, gene, a segment of DNA that is involved in producing a polypeptide chain; it can include regions preceding and following the coding DNA as well as introns between the exons.

Operon : A segment of DNA containing adjacent genes including structural genes, an operator gene and a regulatory gene

Opsonin : A substance, usually antibody or complement component, which coats a particle such as a bacterium and enhances phagocytosis by phagocytic cells.

Opsonisation : Literally means 'preparation for eating'. The coating of a bacterium with antibody and/or complement that leads to enhanced phagocytosis of the bacterium by phagocytic cells.

ORF : *See* open reading frame.

Organ transplant is the moving of an organ from one body to another or from a donor site on the patient's own body, for the purpose of replacing the recipient's damaged or absent organ.

Organelle : A structure that is always present within a cell and has a particular function or functions. Organelles include the mitochondrion and the Golgi body.

Organogenesis : Induction of organs such as root, shoot and leaves (but not embryo) in plant tissue culture.

Orthologue : Functionally equivalent genes across species.

Overhangs : Unpaired nucleotide(s) in the end of a DNA molecule which may be produced by a staggered cut on the DNA using a restriction enzyme.

P Element : A transposon present specifically in the fruit fly *Drosophila melanogaster* which is used for mutagenesis and the creation of genetically modified flies for genetic research.

Palindrome : A palindrome is a sequence that may be read the same way in either direction, with general allowances for adjustments to punctuation and word dividers.

Palindromic sequence : A DNA sequence whose $5' \rightarrow 3'$ sequence is identical on each DNA strand. The recognition sequences of many restriction enzymes are palindromic.

Paraprotein : An immunoglobulin or immunoglobulin light-chain that is produced in excess by the clonal proliferation of plasma cells.

Paratope : A part of an antibody that recognizes the epitope.

Passive immunisation : Immunisation by the administration of preformed antibody into a nonimmune individual.

Passive transport : A process where a molecule moves down its electrochemical gradient without requiring energy.

Pathogen : A microorganism or other substance that causes disease, e.g. a bacterium.

Pathogenesis : The way a disease develops and its effects.

Pathogenesis-related proteins : Proteins produced in plants in the event of a pathogen attack; they are induced as part of systemic acquired resistance.

PDB : The Protein Data Bank consisting of 3D structures biological macromolecules.

Peng Peng' sheep : The transgenic sheep is named 'Peng Peng' (after the identical given names of the two cloners).

Peptide : A short segment of a protein formed from two or more amino acids are linked by a peptide bond.

pH : A measure of the acidity or alkalinity of a solution. Pure water has a pH of 7; acidic solutions have a pH less than 7, and alkaline solutions a pH greater than 7. pH is measured on a logarithmic scale from 0 to 14.

Phage : A virus for which the natural host is a bacterial cell.

Phage display : A laboratory technique for the study of protein–protein, protein–peptide, and protein–DNA interactions that uses bacteriophages to connect proteins with the genetic information that encodes them.

Phagocyte : A cell that can surround and destroy foreign particles, e.g. cell debris or bacteria.

Phospholipid : A molecule containing phosphate and lipid. It is an important part of biological membranes.

Phylogeny : The evolutionary history of a molecule such as gene or protein, or a species.

Physical map : A map of the locations of identifiable landmarks on DNA (e.g., restriction enzyme cutting sites, genes), regardless of inheritance.

Phytoremediation : The efficient use of plants to remove, detoxify or immobilise environmental contaminants in a growth matrix (soil, water or sediments) through the natural biological, chemical or physical activities and processes of the plants.

PIR: The Protein Identification Resource, a database of protein sequences.

Plantibody : An antibody produced by genetically modified crops. Antibodies are part of animal immune systems, and are produced in plants by transforming them with antibody genes from animals.

Plantlet : Young or small plants used as propagules; they are usually grown from clippings of mature plants.

Plaque lifting method : A technique used in recombinant DNA technology for screening DNA libraries constructed in bacteriophages such as lambda.

Plasma : The straw-coloured fluid in which the blood cells are found. It is made of a solution of inorganic salts and protein.

Plasma membrane : A very thin membrane that is composed of lipids and protein. It surrounds the cytoplasm of a cell and controls the passage of substances into and out of the cell.

Plasmid : Autonomously replicating, extrachromosomal circular DNA molecules, distinct from the normal bacterial genome and nonessential for cell survival under nonselective conditions.

Plasmolysis : The process in plant cells where the cytoplasm pulls away from the cell wall due to the loss of water through osmosis.

Platelet : A cell fragment that does not have a nucleus. Platelets help blood to form a clot.

Pluripotent cell : The state of a single cell that is capable of differentiating into all tissues of an organism, but not alone capable of sustaining full organismal development.

Polyacrylamide gel : A substance that is used for electrophoresis of proteins. It is used in the western blot diagnostic test.

Polyclonal antibody : Specific antibodies that have been obtained by immunisation of an animal. They are the products of different clones of antibody-producing cells.

Polymerase chain reaction (PCR) : A procedure where a piece of DNA is copied many times to do genetic analysis. It can be used to amplify very small amounts of DNA.

Polymerisation of nucleotides : A long chain of nucleotides where each nucleotide is linked to the other with posphodiester bond.

Polymorphism : A detectable difference at a particular gene or marker occurring among individuals.

Polypeptide: A molecule that is made up of linked amino acids.

Polysaccharide : A type of carbohydrate. It contains many molecules of a simple sugar, e.g. glucose, which are joined together. When a polysaccharide is digested, the smaller sugar units are released. Polysaccharides include starch and glycogen.

Positron emission tomography (PET) : A method that is used to record internal body images. Gamma rays are detected from body tissues that are isotopically labelled.

Post-transcriptional processing : A process in cell biology by which, in eukaryotic cells, primary transcript RNA is converted into mature RNA.

Post-translational modification : The addition of specific chemical residues to a protein after it has been translated e.g. sugars (glycosylation), methyl groups (methylation).

Precursor : Anything that develops into a specified thing. An example is an undifferentiated cell from which a certain type of cell develops.

Prenatal diagnosis : Testing for diseases or conditions in a foetus or embryo before it is born.

Primary structure : The amino acid sequence of a protein.

Primer : A short sequence of nucleotides. Primers are used in PCR to start copying of the DNA that is under analysis.

Primer : Short pre-existing polynucleotide chain to which new deoxyribonucleotides can be added by DNA polymerase.

Probe : Single-stranded DNA or RNA molecules of specific base sequence, labelled either radioactively or immunologically, that are used to detect the complementary base sequence by hybridisation.

Prokaryote : Cell or organism lacking a membrane-bound, structurally discrete *nucleus* and other sub-cellular compartments, for example bacteria.

Promoter : A site on DNA to which RNA polymerase will bind and initiate transcription.

Promoter region : A region of genomic DNA that binds RNA polymerase. It is the site that regulates the synthesis of mRNA.

Promoter : A region of DNA to which RNA polymerase binds before initiating the transcription of DNA into RNA.

Prophage : A phage chromosome inserted as part of the linear structure of the DNA chromosome of a bacterium.

PROSITE : A method for determining what is the function of uncharacterised proteins translated from genomic or cDNA sequences. It consists of a database of biologically significant sites, patterns and profiles that help to identify to which known protein family (if any) a new sequence belongs.

Protein : A large molecule composed of one or more chains of amino acids in a specific order

Protein engineering : The process of developing useful or valuable proteins by bringing slight changes into it.

Proteomics : The large-scale study of proteins, particularly their structures and functions.

Proto-oncogene : A normal gene that can become an oncogene due to mutations or increased expression. Proto-oncogenes code for proteins that help to regulate cell growth and differentiation.

Protoplast culture : The isolation and culture of plant protoplasts by mechanical means or by enzymatic digestion of plant tissues or organs, or cultures derived from these. Protoplasts are utilised for selection or hybridisation at the cellular level and for a variety of other purposes.

Protoplast fusion : The coalescence of the plasmalemma and cytoplasm of two or more protoplasts in contact with one another. Initial adhesion of two protoplasts is a random process but coalescence can be induced.

Prototroph : An organism such as a bacterium that will grow on a minimal medium.

Pseudogene : A pseudogenes are defined by their possession of sequences that are related to those of the functional genes, but that cannot be translated into a functional protein.

Pulsed-field gel electrophoresis (PFGE) : A method used to separate high molecular weight linear DNA fragments that are greater than 50 kb in size, by pulsing an electrical field across an agarose gel rather than applying the constant field used in conventional electrophoresis.

Pyrosequencing : A method of DNA sequencing (determining the order of nucleotides in DNA) based on the sequencing by synthesis principle.

Quantitative : Relating to characteristics that are based on numbers.

Quantitative PCR (Q-PCR) : The abbreviation used for real-time PCR.

Quantitative trait locus (QTL) : Location of a specific gene that affects a measurable or quantifiable trait. These traits are typically affected by more than one gene, and also by the environment; for examples plant height and body weight.

Quantum tunnelling : The quantum mechanical phenomenon where a particle tunnels through a barrier that it classically could not surmount.

Quaternary structure : A protein may consist of several polypeptide chains. This is called the quaternary structure of the protein.

Radioactive labelling : A means or agent by which certain substances or structures can be identified or followed.

Radioactivity : The property of certain nuclides of spontaneously emitting particles or gamma radiation, or of emitting X-radiation following orbital electron capture, or of undergoing spontaneous fission.

Radioimmunoassay (RIA) : A technique for measurement of primary antigen-antibody interactions, and for the determination of the level of important biological substances in mixed samples. It takes advantage of the specificity of the antigen-antibody interaction and the sensitivity that derives from measurement of radioactively labelled materials.

Radioisotopes : A radioactive isotope which has been directly exposed to nuclear energy either naturally or artificially.

Ramachandran : A diagram of the sterically allowed regions for a dipeptide unit in phi-psi space. The conformation of the dipepetide is defined by two dihedral angles phi (the angle of rotation around the N-C-alpha bond in a peptide unit) and psi (the angle of rotation around the C-alpha-C' bond in a peptide unit). The diagram is named after G.N Ramachandran who was the first to make calculations of the allowed regions in protein structures.

Random amplification of polymorphic DNA (RAPD) : A widely-used technique for amplifying anonymous stretches of DNA using PCR with arbitrary primers.

Real-time PCR (RT-PCR) : A quantitative real time polymerase chain reaction (qPCR) or kinetic polymerase chain reaction. It is a laboratory technique based on the PCR used to amplify and simultaneously quantify a targeted DNA molecule.

Receptor : A molecule on the surface of a cell that binds to certain chemicals or proteins.

Recognition sequences : The site of any DNA-binding protein motif that exhibits binding specificity, refers to the DNA sequence (or subset thereof), to which the domain is specific.

Recombinant : A chromosome, cell or organism that result after recombination either naturally or artificially.

Recombinant clones : Clones containing recombinant DNA molecules.

Recombinant DNA molecules : A combination of DNA molecules of different origin that are joined using recombinant DNA technologies.

Recombinant DNA technologies : Procedures used to join together DNA segments in a cell-free system. Under appropriate conditions, a recombinant DNA molecule can enter a cell and replicate there, either autonomously or after it has become integrated into a cellular chromosome.

Recombinant DNA technology : The technology of preparing recombinant DNA in vitro by cutting up DNA molecules and splicing together fragments from more than one organism.

Recombination : A process by which genetic information is rearranged during meiosis. This process also occurs during the somatic rearrangements of DNA that occur in formation of genes that encode antibody molecules.

Recombination frequency : Recombination frequency (θ) is the frequency that a single chromosomal crossover will take place between two genes during meiosis.

Red biotechnology : Biotechnology applied to the pharmaceutical, medical and healthcare field, which deals with pharmacogenomics, designing organisms to produce antibiotics and vaccines, clinical research and trials, gene therapy and diagnostics.

Regeneration : The process of renewal, restoration, and growth that makes genomes, cells, organs, organisms, and ecosystems resilient to natural fluctuations or events that cause disturbance or damage.

Regulator gene : A gene involved in controlling the expression of one or more other genes. A regulator gene may encode a protein, or it may work at the level of RNA, as in the case of genes encoding microRNAs.

Regulatory regions or sequences : A DNA base sequence that controls gene expression.

Relaxosome : During conjugation the relaxase enzyme creates a nick in one of the strands of the conjugative plasmid at the *oriT* and work in a complex of over a dozen proteins known collectively as a relaxosome.

Replica plating : A technique in which one or more secondary Petri plates containing different selective growth media (lacking nutrients or containing chemical growth inhibitors such as antibiotics) are inoculated with the same colonies of microorganisms from a plate, reproducing the original spatial pattern of colonies.

Replication : A process where an exact copy of genetic material (DNA or RNA) is made.

Reporter gene : A gene that researchers attach to a regulatory sequence of another gene of interest in bacteria, cell culture, animals or plants.

Repressor gene : A gene that prevents a non-allele from being transcribed.

Resolution : Degree of molecular detail on a physical map of DNA, ranging from low to high.

Restriction enzyme : A bacterial enzyme that breaks DNA at a short, specific sequence. It is also known as a restriction endonuclease.

Restriction fragment length polymorphism (RFLP) : Variation in the lengths of restriction fragments of DNA. These fragments are obtained from different individuals due to genetic polymorphism of the restriction sites in the chromosomes.

Restriction map : Map of DNA showing the position of sites recognised and cut by various restriction endonucleases.

Restriction site : A specific nucleotide sequence of DNA at which a particular restriction enzyme cuts the DNA, most typically 4-6 bases long. Some sites occur frequently in DNA (e.g., every several hundred base pairs), others much less frequently.

Reverse transcriptase : An enzyme that synthesises a complementary cDNA strand from an RNA template.

Reverse transcription PCR (RT-PCR) : It is one of many variants of PCR which is commonly used to detect RNA.

Reverse turn : A beta turn occurs where the polypeptide chain makes a sharp reversal by 180 degrees within 4 residues. A turn is defined by 4 consecutive residues (*i* to *i*+3) if the C-alpha(i) to C-alpha(*i*+3) distance is less than 7 angstroms, and if the central two residues are not helical. There is often a hydrogen bond bewteen the CO(i) and NH(*i*+3).

Ribonucleic acid (RNA) : A type of nucleic acid containing ribose, phosphate, and bases that contain nitrogen. Cellular forms include ribosomal RNA, messenger RNA and transfer RNA. It is also the genetic material in some viruses, e.g. HIV.

Ribosomal RNA (rRNA) : A class of RNA found in the ribosomes of cells.

Ribosomes : Small cellular components composed of specialised ribosomal RNA and protein; site of protein synthesis. See ribonucleic acid (RNA).

Ribozymes : Some RNAs that show catalytic activities.

RNA polymerase : An enzyme that catalyses the production of RNA from a DNA template.

Satellite RNA : A small, self-splicing RNA molecule that accompanies several plant viruses, including tobacco ringspot virus.

Scale up : Conversion of a process, such as fermentation of a microorganism, from a small scale to a larger scale.

Scanning electron microscope (SEM) **:** An electron-beam-based microscope used to examine, in a three dimension

Scintillation counter : A counter in which light flashes produced in a solid scintillator (usually a sodium iodide [NaI] crystal) by ionising radiation are converted into electrical pulses by a photomultiplier tube. The amount of light emitted by the crystal is directly proportional to the amount of energy deposited in the crystal by the radiation. Usually referred to as an NaI probe or NaI detector.

SCOP : Structure classification of proteins aims to provide a detailed and comprehensive description of the structural and evolutionary relationships between proteins whose 3D structure is known.

Scorable (screenable) marker : They are also called reporter genes that cause a colour change or other visible change in the tissue of the transformed organism. This allows the investigator to quickly screen a large group of cells for the ones that have been transformed.

Second messengers : Second messengers are often free to diffuse to the nucleus, where they can influence gene expression and other processes. Secondary messengers can be either small organic molecules such as cAMP, cGMP, DAG and IP3 or inorganic ions such as Ca^{2+}.

Secondary metabolism : The production by living organisms of substances not essential for primary metabolic functions or physiology. Their role is associated with interaction with the environment, for example for defence, as elicitors, or as attractants.

Secondary metabolites : Organic compounds that are not directly involved in the normal growth, development, or reproduction of an organism.

Secondary structure : The localised and repetitive folding of the polypeptide backbone of a protein due to hydrogen bond formation (alpha helix and beta sheet).

Selectable marker : A gene introduced into a cell, especially a bacterium or to cells in culture, that confers a trait suitable for artificial selection.

Sequence analysis : A technique in which the sequences of molecules, e.g. DNA, from different organisms are compared. More closely related organisms have more similar sequences.

Sequence motif : Biologically related sequences often share some common properties. The sum of these common properties between related sequences is called a sequence motif.

Sequence tagged site (STS) : Short (200 to 500 base pairs) DNA sequence that has a single occurrence in the human genome and whose location and base sequence are known. STSs are useful for localising and orienting the mapping and sequence data reported from many different laboratories and serve as landmarks on the developing physical map of the human genome.

Sequencing : Determination of the order of nucleotides (base sequences) in a DNA or RNA molecule or the order of amino acids in a protein.

Severe combined immunodeficiency (SCID) : A condition where there is a severe lack of T and B lymphocytes. SCID patients usually have a very weak immune system.

Sex chromosomes : The X and Y chromosomes in human beings that determine the sex of an individual. Females have two X chromosomes in diploid cells; males have an X and a Y chromosome. The sex chromosomes comprise the 23rd chromosome pair in a karyotype.

Sheet : These structures are formed from residues in an extended conformation with psi-phi bond angle pairs in the wide allowed region in the upper left hand corner of the Ramachandran plot.

Shine-Dalgarno sequence : A ribosomal binding site in the mRNA, generally located 8 basepairs upstream of the start codon AUG. the Shine-Dalgarno sequence exists both in bacteria and archaea, being also present in some chloroplastic and mitochondial transcripts. The six-base consensus sequence is AGGAGG.

Shotgun method : Cloning of DNA fragments randomly generated from a genome.

Shuttle vectors : A vector (usually a plasmid) constructed so that it can propagate in two different host species.

Sickle cell anaemia : A group of inherited blood diseases. Individuals with one of these diseases produce (abnormal) sickle haemoglobin and experience a wide range of clinical problems. It is also known as sickle cell disease.

Signal amplification : It is the principle of signal transduction. The amplification can occur at several points of the signal pathway.

Signal hypothesis : The major mechanism whereby proteins that insert into or cross a membrane arc synthesised by a membrane-bound ribosome. The first thirteen to thirty-six amino acids synthesised, termed a signal peptide, are recognised by a signal recognition particle that draws the ribosome to the membrane surface by interaction with a docking protein. The signal peptide may later be removed from the protein.

Signal sequence : A 16-30 amino acid sequence, in a secreted polypeptide, that directs the protein to its target compartment or membrane.

Signal transduction: The processes by which an extracellular signal (typically a hormone or neurotransmitter) interacts with a receptor at the cell surface. This causes a change in the level of a second messenger (for example cyclic AMP) and ultimately effects a change in the cell's functioning.

Signal transmission : Signal transmission refers to any process by which a cell converts one kind of signal or stimulus into another by involving ordered sequences of biochemical reactions inside the cell, that are carried out by enzymes and linked through second messengers resulting in what is thought of as a 'second messenger pathway'.

Silencers : The control sections of DNA that, like enhancers, may be located thousands of base pairs away from the gene they regulate.

Single nucleotide polymorphism (SNP) : A common but minute variation that occurs in DNA sequences of a genome. These variations can be used to track inheritance in families or species.

Single-gene disorder : Hereditary disorder caused by a mutant allele of a single gene(e.g., Duchenne muscular dystrophy, retinoblastoma, sickle cell disease). Compare polygenic disorders.

Site-directed mutagenesis (SDM) : A technique used to make specific and intentional changes to the DNA sequence of a gene and any gene products.

Sodium dodecyl sulfate polyacrylamide gel electrophoresis (SDS-PAGE) : A method used to separate proteins according to their electrophoretic mobility (a function of the length of a polypeptide chain and its charge) and no other physical feature.

Somaclonal variation : The variation seen in plants that have been produced by plant tissue culture; an important source of this variation is chromosomal rearrangements.

Somatic (adult) stem cells : A relatively rare undifferentiated cell found in many organs and differentiated tissues with a limited capacity for both self renewal (in the laboratory) and differentiation. Such cells vary in their differentiation capacity, but it is usually limited to cell types in the organ of origin.

Somatic cells : Any cell in the body except gametes and their precursors.

Somatic embryogenesis : A process where a plant or embryo is derived from a single somatic cell or group of somatic cells.

Somatic embryos : The somatic embryos formed from plant cells that are not normally involved in the development of embryos, *i.e.* ordinary plant tissue. No endosperm or seed coat is formed around a somatic embryo.

Somatic hybridisation : Hybridisation by induced fusion of cells (protoplasts) from two contrasting genotypes for production of hybrids or cybrids which contain various mixtures of nuclear and/ or cytoplasmic genomes, respectively.

Somatostatin : Growth-hormone-inhibiting hormone.

Southern blotting : Transfer by absorption of DNA fragments separated in electrophoretic gels to membrane filters for detection of specific base sequences by radiolabeled complementary probes.

Spectrophotometry : It is the quantitative measurement of the reflection or transmission properties of a material as a function of wavelength.

Spectroscopy : It refers to the study of the interaction between matter and radiated energy.

Spliceosome: A structure within a cell in which splicing of messenger RNA takes place.

Splicing site: The join between a noncoding sequence and a coding sequence of RNA. This is where the noncoding sequences come apart from one another during the process of transcription.

Splisosome : It is a complex of snRNA and protein subunits that removes introns from a transcribed pre-mRNA (hnRNA) segment. This process is generally referred to as splicing.

Stem cells : Cells with the ability to divide for indefinite periods in culture and to give rise to specialised cells.

Stoichiometry : The quantitative description of the proportions by moles of the substances in a chemical reaction.

Stop codon : Any of three mRNA sequences (UGA, UAG, UAA) that do not code for an amino acid and thus signal the end of protein synthesis.

Strain : A group of organisms coming from a particular source that shares a set of distinguishing characteristics.

Stringency : Reaction conditions such as notably temperature, salt concentration(s) and pH, that dictate the annealing of single-stranded DNA/DNA, DNA/RNA and RNA/RNA hybrids. At high stringency, duplexes form only between strands with perfect one-to-one complementarity; lower stringency allows annealing between strands with some degree of mismatch between bases.

Stringent plasmid : A plasmid that only replicates along with the main bacterial chromosome and is present as a single copy, or at most several copies, per cell.

Structural gene : A DNA sequence that forms the blueprint for the synthesis of a polypeptide.

Structure motif : A commonly occurring substructure, usually comprising two to three secondary structures, e.g. beta-hairpin.

Sub-cloning : The process of transferring a cloned DNA fragment from one vector to another.

Submerged culture : A method for growing pure cultures of aerobic bacteria in which microorganisms are incubated in a liquid medium subjected to continuous, vigorous agitation.

Subtilisin (= serine protease) : Enzymes that cleave peptide bonds in proteins in which serine serves as the nucleophilic amino acid at the active site of an enzyme.

Sugar puckering : It has to do with changes to the shape of the DNA helix. The sugar pucker determines the shape of the α-helix, whether the helix will exist in the A-form or in the B-form. The five-member ring is for steric reasons not planar because all bonds would be ecliptic. One atom or two are turned out of the plane by 50 pm. *Endo*-pucker means that C2' or C3' are turned out of this plane into the direction of O5'. *Exo*-pucker describes a shift in the opposite direction

Super antigens : Those antigens which polyclonally activate a large fraction of the T cells (up to 25%).

Superbug : The oil eating bug produced first by Ananda Mohan Chakrabarty, the India-born American scientist, by genetically engineered strain of *Pseudomonas putida*. It is capable of utilising complex chemical compounds like hydrocarbons.

Supercoil : A DNA molecule that contains extra twists as a result of overwinding (positive supercoils) or underwinding (negative supercoils).

Surface-enhanced laser desorption-ionisation (SELDI) : An ionisation method in mass spectrometry that is used for the analysis of protein mixtures.

Suspension culture : A type of culture in which (single) cells and/or clumps of cells grow and multiply while suspended in a liquid medium. *See* cell suspension.

Swiss-Prot : A curated protein sequence database which strives to provide a high level of annotations (such as the description of the function of the protein, its domains structure, post-translational modifications, variants, etc.), a minimal level of redundance and high level of integration with other databases.

Symporter : An integral membrane protein involved in movement of two or more different molecules or ions across a phospholipid membrane such as the plasma membrane in the same direction; hence it is a type of co-transporter.

Synapse : The gap between one nerve cell and another. Nerve impulses are carried by neurotransmitters across the gap.

Synaptonemal complex : During meiosis a protein structure called the synaptonemal complex is formed between two homologous chromosomes; it mediates chromosome pairing, synapsis, and recombination.

Synchronisation : The coordination of events to operate a system in unison.

Synchronous culture : A microbiological culture or a cell culture that contains cells that are all in the same growth stage.

Synteny : The occurrence of two or more loci on the same chromosome, without regard to the distance between them.

Synthetic biology : It is the design and construction of new biological entities such as enzymes, genetic circuits, and cells or the redesign of existing biological systems is called synthetic biology.

Synzymes (= artificial enzymes) : Substances with catalytic capabilities. Synzymes consist mainly of organic molecules tailored in such a way that they catalyse certain kinds of reactions.

T cell : A lymphocyte which undergoes a developmental stage in the thymus.

Tandem repeat sequences : Multiple copies of the same base sequence on a chromosome; used as a marker in physical mapping.

Technology transfer : The process of converting scientific findings from research laboratories into useful products by the commercial sector.

Telomere : The ends of chromosomes. These specialised structures are involved in the replication and stability of linear DNA molecules.

Thalassaemia : A group of conditions where there is a reduction in the rate at which globin chains of haemoglobin are formed.

Thawing : It refers to a process when something frozen is heated and then it gets so its no longer frozen but not cooked.

Thermophile: An organism which grows at high temperatures.

Thrombin : An enzyme that converts fibrinogen into fibrin.

Tissue plasmogen activator (tPA) : A protein involved in the breakdown of blood clots. It is a serine protease found on endothelial cells, *i.e.* the cells that line the blood vessels.

Titre : The reciprocal of the last dilution of a titration giving a measurable effect, e.g. if the last dilution giving significant agglutination is 1:124, the titre is 124.

Tolerogen : A substance that invokes a specific immune non-responsiveness due to its molecular form. If its molecular form is changed, a tolerogen can become an immunogen.

Tomography : Imaging by sections or sectioning, through the use of any kind of penetrating wave.

Torsion angles : The 3D structure of DNA calculates a set of nucleic acid backbone parameters, including the six main chain torsion angles (α, β, γ, δ, ε, and ζ) around the covalent bonds, χ about the glycosidic bond, and the sugar pucker.

Totipotency : The ability of a single cell to divide and produce all the differentiated cells in an organism.

Totipotent : Having the ability to give rise to all the cell types of the body plus all of the cell types that make up the extraembryonic tissues such as the placenta.

Toxin : A substance that can cause adverse health effects under typical circumstances of exposure.

Toxoid : An inactivated toxic compound in cases where these (rather than the microorganism itself) cause illness. Examples of toxoid-based vaccines include tetanus and diphtheria.

Tracer isotope : The isotope or non-natural mixture of isotopes of an element which may be incorporated into a sample to make possible observation of the course of that element, alone or in combination, through a chemical, biological, or physical process.

Transcription : The synthesis of an RNA copy from a sequence of DNA (a gene); the first step in gene expression.

Transcriptome : The set of all RNA molecules, including mRNA, rRNA, tRNA, and other non-coding RNA produced in one or a population of cells.

Transduction : It is a process by which DNA is transferred from one bacterium to the another by a virus. It also refers to the process whereby foreign DNA is introduced into another cell via a viral vector.

Transfer RNA (tRNA): A class of RNA having structures with triplet nucleotide sequences that are complementary to the triplet nucleotide coding sequences of mRNA. The role of tRNAs in protein synthesis is to bond with amino acids and transfer them to the ribosomes, where proteins are assembled according to the genetic code carried by mRNA.

Transformation : A process by which the genetic material carried by an individual cell is altered by incorporation of exogenous DNA into its genome.

Transgene (= gene construct) : *See* gene construct.

Transgenesis : The process of introducing an exogenous gene called a transgene into a living organism so that the organism will exhibit a new property and transmit that property to its offspring.

Transgenic (= genetically modified, GM) plants : Plants the DNA of which has been modified using genetic engineering techniques to resist pests and agents causing harm to plants and to improve the growth of these plants to assist in farmers efficiency.

Translation : The process in which the genetic code carried by mRNA directs the synthesis of proteins from amino acids.

Transplant : An implant of an organ or tissue from one part of the body to another or from one person to another.

Transplant rejection : It occurs when a transplanted organ or tissue is not accepted by the body of the transplant recipient. This is explained by the concept that the immune system of the recipient attacks the transplanted organ or tissue.

Transplastomic plants : A genetically modified plant in which the new genes have not been inserted in the nuclear DNA but in the DNA of the chloroplasts.

Transport Proteins : These proteins refer to as a transmembrane pump, transporter protein, escort protein, fatty acid transport protein, cation transport protein, or anion transport protein) is a protein which serves the function of moving other materials within an organism.

Transposons (= jumping genes, movable genes and transposable elements) : Transposons are DNA segments that can move around to different positions in the genome of a single cell resulting in mutations.

Triploid plant : The plants that really grow with only n chromosomes.

Tumour : An abnormal swelling or growth. It can be benign or malignant. It is also known as a neoplasm.

Tumour antigen : A substance produced in tumour cells that triggers an immune response in the host. They are used to identify tumour cells and are potential candidates for use in cancer therapy.

Turbidostat : A continuous culture device, similar to a chemostat or an auxostat which has feedback between the turbidity of the culture vessel and the dilution rate.

Turn : A beta turn occurs where the polypeptide chain makes a sharp reversal by 180 degrees within 4 residues. A turn is defined by 4 consecutive residues : (i to i+3) if the C-alpha (i) to C-alpha(i+3) distance is less than 7 angstroms, and if the central two residues are not helical. There is often a hydrogen bond bewteen the CO(i) and NH(i+3).

Ultraviolet (UV) radiation : Electromagnetic radiation with wavelengths between soft X-rays and visible violet light, often broken down into UV-A (315-400 nm), UV-B (280-315 nm), and UV-C (100-280 nm).

Uniporter : It is an integral membrane protein that is involved in facilitated diffusion; they can be either a channel or a carrier protein.

Upstream : The region extending in a 5′ direction from a gene.

Upstream processing : All those steps related with inoculum development, media development, improvement of inoculum by genetic engineering process, optimisation of growth kinetics so that product development can improve tremendously.

USDA : U.S. Department of Agriculture.

Vaccination : A process where individuals are given a substance, such as a vaccine, to make them resistant or less susceptible to a disease. The vaccine can be given by injection or orally.

Vaccine : A biological preparation usually but not necessarily proteins that elicit an immune response thereby providing protective immunity against a potential pathogen.

Variable number of tandem repeats (VNTR) : A location in a genome where a short nucleotide sequence is organised as a tandem repeat; these can be found on many chromosomes and show variations in length between individuals.

Variolation : A method of inoculating or immunising an individual against the smallpox virus (variola) through deliberate exposure to a mild form of the virus in order to create a localised infection and generate immunity against further infection.

Vector : *See* cloning vector.

Vermibed : Vermibed (*vermes* = earthworms; *bed*= bedding) is the actual layer of good moist loamy thick soil of about 15 to 20 cm placed at the bottom.

Vermicast : The excreta of earthworms produced through processing of organic materials in their alimentary canal which consists of the mucus coated undigested matter. It consist of organic matter that has undergone physical and chemical breakdown through muscular gizzard that grinds the material with enzymes releasing amino acids, sugars, organic molecules.

Vermicompost : The end-product of the breakdown of organic matter by some species of earthworm. It is the best nutrient-rich organic fertilizers and soil conditioners that contain water-soluble nutrients, microbes including fungi, protozoa and bacteria.

Vermicomposting : The process of composting by using various worms, such as red wigglers, white worms, etc. to create a heterogeneous mixture of decomposing vegetable or food waste, bedding materials, and vermicast.

Vermiculture (= worm farming) : The science of raising and managing the worms.

Vermitechnology : *See* vermicomposting.

Virome : The genomes of all the viruses that inhabit a particular organism.

Virosomes : A drug or vaccine delivery mechanism consisting of unilamellar phospholipid membrane vesicle incorporating virus derived proteins to allow the virosomes to fuse with target cells. Virosomes are not able to replicate but are pure fusion-active vesicles.

Virus : A noncellular biological entity that can reproduce only within a host cell. Viruses consist of nucleic acid covered by protein; some animal viruses are also surrounded by membrane. Inside the infected cell, the virus uses the synthetic capability of the host to produce progeny virus.

Warehouse: A single, physical database that is constructed by merging a number of physically separate data sources.

Western blotting techniques (= protein blotting or electroblotting technique) : An analytical method used to detect specific proteins in the given sample of tissue homogenate or extract that uses gel electrophoresis to separate native proteins by 3-D structure or denatured proteins by the length of the polypeptide.

White biotechnology : Application of biotechnology for industrial purposes, including manufacturing, alternative energy or bio-energy, and biomaterials.

White blood cell : A mature blood cell that contains a nucleus. White blood cells, e.g. monocytes, lymphocytes and granulocytes, are found in the blood or tissue fluid. They are part of the body's immune system. It is also known as a leucocyte.

Whole-genome shotgun sequencing : A method used for sequencing long DNA strands. It is named by analogy with the rapidly expanding, quasi-random firing pattern of a shotgun.

Xenograft (= xenotransplant) : A transplant of organs or tissue from one species to another; for example porcine heart valve transplants, which are quite common and successful.

Xenotransplantation : The process of grafting or transplanting organs or tissues between members of different species.

X-rays : Penetrating electromagnetic radiations having wavelengths shorter than those of visible light. They are usually produced by bombarding a metallic target with fast electrons in a high vacuum, as in an X-ray machine.

Yeast artificial chromosomes (YAC) : A vector used to clone DNA fragments larger than 100 kb and up to 3000 kb. YACs are useful for the physical mapping of complex genomes and for the cloning of large genes.

Zinc finger : A structural motif of DNA-binding proteins in which finger-like loops of amino acids are stabilised with zinc atoms.

Zipper : *See* leucine zipper.

Zwitterion : A neutral molecule with a positive and a negative electrical charge at different locations within that molecule.

Zygote : A diploid cell formed in the fusion of haploid gametes; a fertilised ovum.

Zymotechnology : Application of the art of fermentation is called zymotechnology. It refers to the use of enzyme 'zymase' in fermentation by yeast or other fungi.

Index

A

Ab Initio gene prediction, 603
Absorbance (optical density turbidity) measurement method, 995
Abzymes, naturally occurring, 666
  application of, 666
Acetagenesis, 1079
Acidogenesis, 1079
Activation (of enzymes), 663
Activation of complement cascade, 752
Active and passive transport systems, 51
Active membrane transport (and carrier) proteins, 52
Adaptive or acquired immunity, 732
Adeno-associated viruses (in gene therapy), 400
Adenoviruses (in gene therapy), 400
A-DNA, 87
Affinity capture, microarray,559
A-form of DNA, 87
Agarose gel electrophoresis, 193
Agrobacterium Ti-DNA as vector, 291, 292
Agroinfection and trasformation, 902
Alcohol production (by immobilised cells), 700
Alcohol production, 1025
  fermentation of, 1027
  preparation of medium for, 1026
  recovery of, 1027
Alkaline phosphatase, 278
  sources of, 276
  uses of, 279
Alkylating agents, 169
Allogeneic graft, 816
Allograft, 812
Allosteric inhibition (feedback inhibition), 664
Allotypes, 742
Alpha (**α**) helix, 472
Alternative pathway, 753
Alzheimer disease, 387
Alzheimer's mouse, 988
Amino acid production, 1040
  L-glutamic acid, 1041
  L-lysine, 1040
Amino acids (building blocks of proteins), 458
  chemical properties of, 465
  classification of, 461
  nomenclature, 405
  non-polar amino, 462
  polar, 463
  polymerisation of, 464
  structure of, 459
Amperometric biosensor,
Ampholytes (*see* box), 198
Amplification of plasmids, 285
Amplified fragment length polymorphism (AFLP) analysis, 433
Androgenesis (*in vitro*), 870
  factors affecting, 872
Animal biotechnology, Chapter 20, 941–991
Animal cells, 378
  gene cloning in, 378
Anionic exchangers, 187
Antenatal diagnosis (of haemoglobinopathies), 448
  methods of, 448
Anther and pollen culture, 869
  method of pollen culture, 870

Anti–syn orientation about *N*-glycosidic bond, 85
Antibiotic production, 1036
Antibody diversity, 746
  genetic basis of, 747
Antibody kinetics, 743
Antibody, structure and diversity, 733
  antigen binding, 735
  structure of, 735
Antigen binding, 735
Antigen determinants of immunoglobulins, 742
Antigens, 722
  origin of, 723
  types of, 723
Antimicrobial peptides, 731
Antisense RNA, 411
  therapy, 413
Anti-viral proteins, 915
Apoptosis, 731
Application (of DNA profiling) in the livestock industry, 443
Application of
  biosensor, 708
  cell suspension culture, 852
  DNA profiling (DNA fingerprinting) technology, 436
  enzymes and immobilised enzymes, 649
  immobilised whole cells, 699
  metagenomics, 651
  somaclonal variation (in crop improvement), 868
  transgenic animals, 876
Approved therapeutic antibodies, 779
Arbitrary primed PCR (AP-PCR), 218
Artificial enzymes (synzymes), 667
Artificial ribozymes, 27
Assembly verification, 599
Assembly view, 597
Autograft, 811
Autoimmune diseases, 765
  treatment of, 765
Autoimmunity, 762
  causes of, 727
  tolerance, 729
Autologous graft, 816
Automated DNA sequencers, 203
Automated DNA sequencing,
Autoradiography method of, 232

B

Baby swapping case, solving of, 441
BAC vector, cloning of, 315
Bacillus genome project, 579
Bacillus thuringiensis (Bt), genetics of, 922
Bacteria xanthine-guanine phosphoribosyl transferase (XGPRT), 952
Bacterial artificial chromosome (BAC), 314
Bacterial biosensors, luminescent, 710
Bacterial genome sequencing projects, 577
Bacterial plasmid vector (some considerations of), 283
Bacteriophage RNA polymerases, 273
Bacteriophages vectors, 300
Baculovirus expression vectors, 953
Bac-to-Bac method, 954
  bar gene, 900
Barnase–barstar component system, 919
  electrostatic optimisation of binding complex, 920
  mechanism of catalysis, 919
Base analogues, 168
Basic local alignement search tool (BLAST), 554
  uses of, 555
Batch culture (cell in mass), 850
Batch culture, growth in, 996
Batch fermentation, 1012
Beer production, 102
Benefits of vermicompost, 1074
Beta-pleated sheet, 477
Binary vectors, 297
Bioaffinity biosensors, 706
Bioaugmentation, 1048
Bio-business in India, 17
Biodegradation, 1049
Biodung composting, 1069
Bioethics, 989
Biofertilizers, 9
Biofilm, 1059
Biogas (*gobar* gas) production technology in India, 1076

Biogas digesters, microbial diversity in, 1084
Biogas production (anaerobic digestion), 1078
    from different feedstocks, 1082
Biogas, bases of, 1084
Bioinformatics tools, use in analysis, 567
Bioinformatics, 12, 524, 569
    historical background of, 525
    what is, 526
Bioinformatics, types of sequences used in, 537
Bioleaching (biomining), 1085
    copper, 1092
    iron, 1091
    metal (bioleaching) with fungi, 1100
    of metals by bacteria, 1091
Biomining (bioleaching), 1085
    gold, 1069
    methods of, 1088
    of copper, 1092
    uranium recovery, 1099
    use of reactors in, 1089
Biomining bacteria, resistance mechanisms in, 1096
Biooxidation, of gold, 1097
stirred-tank mineral, 1097
Biophysical methods, 182, 244
Biopesticides, 9
Bioreactors (*see* fermentors), 1004
Bioremediation, 1046
    biosafety from GMOs, 1065
    ex- situ, 1053
    factors affecting, 1059
    in- situ, 1047
    mass balance, 1062
    microorganisms involved in, 1056
    monitoring of, 1065
    of heavy metals, 1057
    of petro-pollutants, 1063
    process, 1057
Biosensor, 701
    BOD, 709
    characteristics of, 701
    components of, 701
    gas, 709
    immunosensors, 907
    in environmental monitoring, 708, 1045
    types of, 703
    uses in environmental monitoring, 708
Biosensors (type of), 703
    amperometric, 704
    bioaffinity, 706
    calorimetric, 706
    electro-chemical, 704, 710
    enzyme electrode, 711
    glucose-electrode, 703
    opto-electronic, 707
    piezo-electric, 705
    thermistor-containing, 705
    whole cell biosensors, 707
Biosorption, 1051
    factor affecting, 1053
    organisms involved in, 1051
Biostimulation, 1050
Biosurfactants, 1060
Biotech (global area of crops), 9
    consumption of products, 19
    top schools, 20
    turn-over, 17
Biotech market (success story of), 17
    booming, 18
Biotechnology
    blue (marine), 11
    definition of, 1
    green, 8
    history of, 2
    in India, 16
    industrial, 10
    institution in India, 20
    market in India, 16
    pharmaceutical, 7
    public perception of, 15
    red, 7
    timeline of, 3
    traditional, 5
    white, 10
    world of, 7
Biotin-streptavidin labelling system, 367
Bioventing, 1050
Blotting techniques, 355
    northern blotting, 358

Southern blotting, 356
western blotting, 359
Blue biotechnology, 11
Blunt end, 344
ligation (by linkers and adaptors), 344
BOD biosensor, 709
Bokashi compost, 1069
Boveri–Sutton's chromosome theory (chromosome theory of inheritance, 130
Bovine papilloma virus vectors, 957
Breeding with transgenic livestock, 975
Bromodeoxyuridine (BrdU) enrichment, 632
Bt brinjal, 627
controversy of, 929
field trials of, 928
methods of production of, 927
mode of action of Cry1A(c) protein, 928
Bt cotton, 924
economic impact of, 926
in India, 924
Bt technology (insect-resistant transgenic plants), 924

C

Ca^{2+}-ATPase pumps, 52, 54
CAAT box, 888
Calcium ions, 54
Callus culture, 845
factors affecting, 845
Calorimetric biosensors, 706
CAMERA, 645
cAMP (cyclic adenosine monophosphate), 58
CaMV (35S promoter of), 909, 922
Cancer immunotherapy, 823
active, 823
passive, 823
Carcinogenicity test, 174
CARMA, 646
Carrageenan (see box), 682
Carrier matrices and its properties, 681
carrier binding, 681
properties of, 681
Carrier proteins and active membrane transport, 52
Carrier proteins, 50
CAT (chloramphenicol acetyl transferase) gene, 899
Catalytic RNA molecules, 27
Cation exchangers, 18
Caulimovirus as vector, 323, 908
cDNA, 342
library, 341
screening of (clone), 343
c-DNA, 89
Cell biology, 20, 70
Cell (suspension) culture, 859
application of, 852
benefits from, 852
in mass, 850
synchronisation of (*see* box), 851
Cell counting (by direct microscopic count) method, 993
Cell immobilisation, use in meat industry, 700
Cell mass measurement, 994
Cell membrane, 41
function of, 42
Cell membrane, transport across, 50
Cell organelles, 42
Cell signalling, 56
Cell structure and function, 24
Cell wall, 37, 40
chemistry, 40
structure of, 40
Cell-mediated immunity, 744
Cells as drug delivery system, 465
bone marrow-derived mesenchymal stromal cells, 406
genetically modified mesenchymal stromal cells (MSC), 405
Cells of innate immune response, 730
Cells
eukaryotic, 31
prokaryotic, 31
structure and components, 39
structure of bacterial, 33
Cell-surface receptors, 59-62
classes of, 59
enzyme-linked, 60
G-protein-linked, 59
ligand-gated ion channel, 61
toll-like, 61
Central dogma, 112

C-form of DNA, 89
cGMP (cyclic guanosine monophosphate), 59
Chain termination, 117
  rho-dependent, 117
  rho-independent, 117
Channel proteins, 50
Chargaff's rule of equivalence, 74
Charging of tRNA, 118
Chemical bonds, 452
  covalent bond, 453
  hydrogen bond, 455
  ionic bond, 452
  peptide bond, 457
  van der Waals force, 456
Chemically modified viral vectors (chemoviruses) and synthetic viruses, 407
Chemostat, 1002
Chemotaxis, bacterial, 1059
Chemoviruses and synthetic viruses, 407
Chloramphenicol acetyltransferase (CAT) 949,
Chloroplast, 48
Chromatography, 182
Chromosomal basis (of Mendelian genetics), 129
Chromosome theory of inheritance, 130
Chromosomes, 44
Circular and super-coiled DNA, 90
Citric acid production, 1034
  Koji process or solid state fermentation of, 1034
  liquid surface culture process of, 1034
  recovery of, 1035
  submerged culture process of, 1035
Classical pathway, 752
CLC bio workbench, 646
Clone-by-clone sequencing, 316
Cloning of BAC vector, 316
Cloning vectors, 281, 953
Coat protein (CP) gene, expression of, 912
Coenzymes, 660
Cofactors, 6
  inorganic, 658
  organic, 659
Coiled coil motif, 507
Co-integrate vectors, 296
Colony (or plaque) hybridisation technique, 352
  screening by, 352
Colorimetry, 208
Combinatorial extension, 567
Commercial use of transgenic plants, 938
Comparative genomics, 610
  databases for, 613
  example of, 612
Comparative microarray hybridisation, 615
Compartmentalization of organelles, 38
Complement, 751
  activation of cascade, 752
Complement system, 695
Compost tea, 1069
  harvesting of compost, 1073
Composting, 1055, 1066
  biodung, 1069
  compost tea, 1069
  factors affecting, 1068
  method of, 1067
  types of, 1055, 1069
Conformation analysis (of proteins), 469
Conformational variations in double helical structures, 86
Conformations of nucleic acids, 77
Conjugal transfer process, 150
Conjugate vaccines, 794
Conjugation, 147
  barriers to, 151
  conjugation mapping, 152
  role of surface protein in, 148
Construction of fermentors, 1006
  aeration and agitation in (fermentor), 1007
  temperature control (of fermentor), 1006
Construction of physical map, 587
Containment, 1012
Continuous culture, growth in, 850
Continuous fermentation, 1013
  multiple-stage, 1013
  single stage, 1013
Copper bioleaching, 1092, 1094
  mechanism of, 1093
  microbiology of, 1093
Cosmid vectors, 307
Covalent bond, 453

Cows (GM), 981
Criminals and rapists, establishing identity of, 441
Crossing over, 132
Cry protein crystals, mechanism of action of, 92
Cryopreservation, 879
    achievement through, 883
    method of, 880
    risks-associated with, 882
Cryoprotectors, 881
Culture media 1003
    preparation of, 1004
    types of, 1004
Cybridisation, 858
Cybrids, 858
    selection of, 859
Cyclic adenosine monophosphate (cAMP), 58
Cyclic guanosine monophosphate (cGMP), 59
Cytokinin–auxin cross-talk, 847
Cytoplasmic inheritance (extra-nuclear inheritance), 137
    by cellular organelles, 138
Cytoplasmic viral receptors, 596

D

DALI, 567
Data retrieval tools, 547
Data structure, 530
Database management systems (DBMS), 531
    components of, 531
    current trends, 532
Database models, 533
    dimensional model, 534
    network model, 533
    object-oriented models, 535
Database similarity searching, 554
    basic local alignment search tool (BLAST), 554
    FASTA,556
Database, 527
D-DNA, 89
Deoxyribonucleic acid, 75
Design and operation (of fermentors), 1007
Designing and construction (of fermentors), 1006
Desorption electrospray ionization (DESI), 503
Detection of nucleic acids,71
Detergent industry, uses of lipases in, 697
D-form of DNA, 89
Diacylglycerol (DAG), 60
Diagnosis of parasitic diseases through DNA probe,, 445
Diauxic growth, 1006
Dicer enzyme, 412
Dideoxynucleotide chain termination method, 202
Differential expression analysis, 634
Digoxigenin (DIG) labelling system, 366
Dihydrofolate reductase (DHFR), 950
Direct immunological screening technique, 354
Directionality of sequences (of nucleotides), 540
Disease-resistant plants, production of, 843
DNA amplification fingerprinting (DAF), 220
DNA ancestry testing, 443
DNA chip (DNA microarray) technology, 613
    application of DNA chips, 615
    types of, 614
DNA delivery, fibre mediated, 378
DNA fingerprinting, 422
    amplified fragment length polymorphism (AFLP) analysis, 433
    application of, 436
    method of, 428
    mitochondrial DNA (mtDNA) analysis, 436
    PCR analysis, 430
    RFLP analysis, 429
    short tandem repeat (STRs) analysis, 431
    Y-chromosome analysis, 434
DNA gyrase, 107
DNA helicase, 106
DNA libraries, 339
    size of, 339
    types of, 339
DNA ligases, 106
DNA ligases, 274
    E. coli DNA ligase, 275
    ligation mechanism of, 276
    T3 DNA ligase, 275
    T4 DNA ligase, 275
    T7 DNA ligase, 275
    thermostable DNA ligase, 275

DNA microarray technology, (*see* DNA chip), 613
DNA polymerase, 266
  E. coli, 266
  T4, 269
  T7, 269
  thermostable, 269
DNA modifying enzymes and restriction endonucleases, 245, 280
DNA probe, diagnosis of parasitic diseases through, 445
DNA probes, 353
DNA profiling (DNA fingerprinting) technology, 436
  application of (in the livestock industry), 443
  application of, 436
  hurdles to, 444
DNA profiling, chapter, 10, 421–450
DNA repair systems, 174
  base excision repair, 175
  excision repair, 175
  mechanisms of repair, 174
  mismatch repair, 175
  nucleotide excision repair, 175
  recombination repair, 176
  SOS repair, 178
DNA replication, 101
  enzymes involved in, 103
  in eukaryotic chromosome, 111
  mechanism of, 107
  Meselson and Stahl's experiment for, 102
  models for, 110
  Watson and Crick's model for, 102
DNA sequences, nomenclature of, 538
DNA sequencing, 199
  Maxam and Gilbert's method for, 200
  Sanger and Coulson method for, 202
DNA testing in IVF technology, 443
DNA vaccines, 794
  advantages of, 738
  construction of, 795
  mechanisms of action of, 797
  mode of administration of, 796
  recombinant vector for, 738
DNA, circular and super-coiled, 90
DNA, palindromic (palindromes), 91
DNA, single stranded, 90
DNA, Watson and Crick's model of, 17
DNA, X-ray crystallographic studies of, 75
DNA-binding domains (types of), 505
  coiled coil motif, 507
  helix-loop-helix (HLH), 505
  helix-turn-helix (HTH), 505
  HMG-box, 507
  leucine zipper domain, 506
  unusual, 508
  zinc finger, 505
Dolly and Bonny, 983
Domino transplants, 612
Donors of genome, 582
Dot blot analysis, 361
Dot-blot hybridization, 1088
Double helical structures, conformational variations in, 50
Downstream process, 1016
Downstream processing, 933
Draft sequence and genome finishing, problems related to, 594
Drug delivery system, cells as, 405
Drug targeting, virosomes in, 406
Duchenne muscular dystrophy, 389
Dump leaching, 1088

E

E. coli DNA polymerase, 269
  application of, 270
  Klenow (large) fragment of, 269
Edible antibodies (plantibodies), 807
Edible interferon, 809
Edible vaccines, 805, 934
  advancements made toward plant-based, 936
  advantages of, 936
  methods of production of, 935
Edible vaccines, 934
Electroblotting technique, 675
Electroblotting, 189
Electrochemical biosensor, 704, 710
Electrophoresis, 189
  2D gel, 197
  agarose gel, 193
  paper, 191

PFGE, 190
polyacrylamide gel, 191, 193
SDS-PAGE, 195
types of gels used in, 191
types of, 192
Electroporation, 374
Electroporation, 374
Electrospray ionization (ESI), 502,
Embryo (plant) culture, 878
culturing method, 878
Embryo rescue,878
application of, 878
method of, 879
Embryo therapy through IVF technology, 404
Embryonic stem cell (transfer technology), 963
generation of, 963
production of transgenic animals by using transfected, 964
transfection of gene construct into, 964
Embryonic stem cell–mediated gene transfer, 962
technology, 963
Encapsulated artificial seeds, 863
advantages of, 864
methods of production of, 864
Encapsulated seeds, 863
factors influencing, 864
steps of, 864
types of, 864
Endocyclic torsion angles, 103
Endonucleases, 103
Enhancers, 891
ENTREZ, 547
Entropy and enthalpy, 241
Environmental biotechnology, chapter 22, 1044-1101
Environmental microbial metagenomics project, 528
Environmental monitoring, biosensors in, 708
uses of biosensors in , 708
EnviroPig™, 985
Enzyme assays, 675
Enzyme electrode, 711
biosensor, 701
Enzyme engineering, 713
Enzyme kinetics, 674
principles of, 674
Enzyme prodrug therapy, antibody-directed, 779
Enzyme produced by immobilised cells, 699
Enzyme structure, 657
co-factors, 657
Enzyme technology, Chapter 15, 655-714
Enzyme-linked receptors, 60
Enzyme
advantages and disadvantages of, 688
classification and nomenclature of, 668
immobilization, 660
methods of enzyme, 663
subclasses, 669
Epitope mapping (*see* box), 737
Epitopes (*see* box), 737
EPSP synthase gene, 900
microbial, 903
EPSP synthase, 1, 903
EPSPS-resistant GM plant, 905
Escherichia coli genome project, 578
Ethical issues, 20
Eukaryotic genome sequencing projects, 579
Eukaryotic host cells, 336
Ex- situ bioremediation, 1053
Examples of plasmids, 286
Exons, 891
Exonuclease, 247
Bal 31 exonuclease, 250
dcoxyribonuclcasc I (DNAsc I), 249
exonuclease, 248
exonuclease, 249
lambda (l), 248
mungbean, 251
ribonuclease (RNase), 252
S1 nuclease, 250
Expressed sequence tag (EST), 541
gene sequencing tags, (GSTs) 542
Expression of cloned DNA, 363
Expression of coat protein (CP) gene, 912
Expression of mutation, 171
Expression proteomics, 699
Expression systems (different plant-based), 932
Expression vectors, 329, 363
baculovirus, 953

features of *E. coli,* 330
protease cleavage site, 330
Extracellular signal molecules, recognition of, 57
Extra-nuclear inheritance (cytoplasmic inheritance), 137
Extremozymes, 667

F

F Factor, 148
F Plasmid, genetic map of, 148
F^- Prime, formation of F′, 153
Factors affecting bioremediation, 1059
FASTA, 556
format, 556
method of searching, 557
Fed-batch culture, 1001
fermentation, 1001
growth in, 1001
Feedback inhibition (allosteric inhibition), 664
Fermentation process, 1021
batch, 1012
continuous, 1013
downstream, 1023
fed-batch, 1013
product recovery, 1024
upstream, 1023
Fermentation, scale-up of, 1020
Fermentors, 1004
basic function of, 1005
construction of, 1006
design and operation of, 1007
designing and construction of, 1006
types of, 1008
use of computer in, 1010
Fibre-mediated DNA delivery, 378
Fimbrae, 36
Finishing, 1028
Flagella, 34
differences between the archaeal and bacterial, 36
structure of, 35
Flat-file database, 529
Fluorescence-activated cell sorting (FACS), 861
Fluorescent *in situ* hybridization (FISH), 603
Fluorescent labeling, 369
FluoroPig (= green pig), 984
Forms of signaling, 57
Fosmid vectors, 314
Fragment-alignment and gap closure, 591
Frayed ends, 262
Functional genomics, 608
toolbox, 609
Functional proteomics, 619
Furanose pseudorotational angles, 83

G

Gas biosensors, 709
GC enrichment by ultracentrifugation, 597
Geep (sheep-goat chimera), 984
Geiger-Müller counter, 227
Gel permeation, 163
application of, 185
characteristics desired for, 184
Geminivirus as vector, 910
Gene and protein, relation between, 61
Gene cloning
basic technique of, 337
in animal cells, 378
in filamentous fungi, 370
in yeasts, 370
prokaryotes, 337
Gene constructs, 887, 945
Gene delivery, receptor-meditated, 404
Gene disruption, 967
Gene expression, 111
Gene farming, 977
Gene mapping (genetic mapping), 135
Gene prediction algorithms, 601
Ab Initio gene, 603
accuracy and validity of, 604
homology-based, 602
systenic gene prediction, 603
Gene prediction and gene counting, 599
gene prediction algorithms, 601
methods of, 600
Gene prediction software, 604
Gene replacement, 968
Gene sequencing tags (GSTs), 541

Gene sequencing, methods of, 504
Gene silencing (RNA interference), 411
Gene splicing, 7
Gene targeting (targetted gene transfer), method, 965
Gene therapy, Chapter 9
 current status of research in, 415
 ethical considerations for, 417
 germ-line, 392
 history of, 383
 metaphysical considerations for, 418
 method of, 395
 non-viral methods, 402
 of human severe combined immunodeficiency syndrome (SCID), 396
 problems related to, 416
 recent developments in, 410
 social considerations for, 417
 somatic, 393
 types of, 392
 viral vector-mediated, 395
Gene transfer
 embryonic stem cell–mediated, 963
 in dicots, 372
 in monocots, 372
Genetic counselling, 446
Genetic diseases, 385
Genetic engineering, plant, 7, 8
Genetic information, flow of, 29
Genetic mapping (gene mapping), 135
Genetic recombination, 134
Genetically modified (GM) plants, 901
 methods for developing, 901
 steps for, 901
Genetically modified mesenchymal stromal cells (MSC) for therapy, 405
Genome assembly and finishing, 594
Genome organisation, 91
 in eukaryotes, 92
 in prokaryotes, 91
Genome sequencer™ FLX System, 643
Genome sequencing projects, 574
 eukaryotic, 575
 microbial, 572
Genome similarity, 605
Genomic DNA, 540
Genomic library, 340
Genomics and proteomics, chapter 13, 570-621
Genomics
 comparative, 610
 functional, 608
 structural, 607
 types of, 607
Germline diversity, 750
Germ-line gene therapy, 391
GFP (green fluorescent protein),
Gibbs free energy, 238
Global and local alignments, 561
GloFish®, 981
Glowing dogs (Ruppy), 986
Glucose-electrode biosensor, 703
GM cows, 981
GM hen, 986
GM papaya, 915
GM tomato, 916
Gold biomining, 1096
Golden rice (*Oryza sativa*), 929
 protest against release of, 931
Golgi complex, 46
G-protein-linked receptors, 59
Green biotechnology, 8
Green pig (= FluoroPig), 984
Growth kinetics (microbial) and specific growth rate, 997
Growth measurement, chemostat for, 1002
Growth parameters, 993
GUS (beta-glucuronidase), 894
 advantages and disadvantages of (*see* box), 894

H

Haematopoietic stem cell transplantation, 815
 complications and side effects, 817
 sources of HSCT, 817
 types of HSCT graft, 816
Haemoglobinopathies, antenatal diagnosis of, 440
Hairy root culture, 848
Haploid plants, 870
 limitations in, 873

production of, 870
use of haploids, 873
Hardening, 842
Heap leaching, 1089
Heavy (H) chain formation, 749
Helix-loop-helix (HLH) domain, 509
Helmholtz free energy, 238
Hepatitis B virus, inhibition of, 399
Hepatitis-B vaccine (indigenous), 936
Herbicide resistance marker, 900
Herbicide resistant transgenic plants, 903
Hershey-chase experiment, 146
Heterofermentative process, 1033
Hierarchical sequencing, 551
High frequency recombination (Hfr) strains, 151
Histocompatibility, 755
History of biotechnology, 2
History of plant cell, tissue and organ culture, 830
HMG-box domain, 507
Homofermentative process, 1033
HomoloGene database, 559
Homologous recombination (identification of), 970
Homologous recombination cloning, 602
Homopolymer tailing, 346
Horseradish peroxidase (HRP) method, 360
Human gene therapy, 967
Human genome project, 580
benefits of, 583
cost of, 583
ethical, legal and social issues, 583
implications on medical science, 583
Human monoclonal antibodies (*suffix* mumab), 775
Humoral immunity, 733
Humoral response, regulation of, 744
Hybridoma technology, 769
Hydrocarbon Metagenome Project, 595
Hypersensitivity, 738
type I, 759
type II, 759
type III, 760
type IV, 761

I

Identification of homologous recombination, 970
Idiotypes, 743
Idiotypic networking, 744
Immigrant dispute (solving), 443
Immobilisation, 680
animal cells culture, 680
of amino acid acylase, 691
of enzymes (*see* enzymes immobilization), 683
of whole cells, 689
Immobilised cells
alcohol production by, 700
enzyme produced by, 699
Immuno-biotechnology, 721
immunoelectrophoresis, 192
Immunoglobulin fragments, structure-function relationships, 757
Immunoglobulin isotypes, 738
• Immunoglobulin isotypes, 738
alfa (IgA), 740
delta (IgD), 741
epsilon (IgE), 741
gamma (IgG), 739
mu (IgM), 741
Immunohistochemistry (IHC), 236
Immunology, branches of, 721
clinical, 721
diagnostic, 721
immuno-biotechnology, 721
immunotherapy, 72
Immunology, Chapter 16, 715-766
branches of, 721
history of, 719
Immunosensors, 709
Immunotechnology, Chapter 17, 769-827
Immunotherapy, 721
Impalefection, 960
Improvement in nutritional quality of seeds, 521
In- situ bioremediation, 104
In vitro translation method, screening by, 354
Inactivated (killed) vaccines, 792
Inactivated polio vaccine (IPV), 803
Indirect gene tracking, 447
Induced fit theory, 671

Industrial biotechnology, 10
Industrial microbiology, 992-1043
Inhibition (of enzymes), 663
- competitive inhibition, 664
- feedback inhibition (allosteric inhibition), 664
- non-competitive inhibition, 664

Inhibition of hepatitis B virus, 399
Initiation complex, formation of, 120
Initiation factors, 119
Initiation of polypeptide synthesis, 119
Innate immune response, cells of, 725
Innate immune system, responses of, 725
Innate immunity, 724
Inositol trisphosphate (IP_3), 68
Insect-resistant transgenic plants (Bt technology), 921
- *Bacillus thuringiensis* (Bt), 921

Insertion sequences (IS elements), 158
Integrins, 160
Integrons, 637
Intercalating agents, 170
Interferon (edible), 809
Interferon (IFN), 406
Interferons, 781
- cell activation by, 784
- herapeutic uses of, 786
- production of recombinant, 788
- types of, 781

Interleukin-2 (IL-2), 400
Intracellular receptors, 62
Introns, 891
Ion channels, 56
Ion flow, measurement of, 56
Ion-exchange chromatography, 185
- application of, 189
- separation of solutes in, 188

Ionising radiation, 221
Ionophores, 55
Iron bioleaching, 1091
IS elements, 185
Isoclectric focussing (IEF), 198
Isoelectric point (pI), 466
Isograft, 812
Isomerism, 459
Isotype switching, 749
Isotypes, 744
IVF technology, 443
DNA testing in, 443
embryo therapy through, 404

K

Kappa chain (V-J rearrangement), 748
Koji process or solid state fermentation, 1034

L

Lactic acid production, 1032
- by homofermentative process, 1033
- by heterofermentative process, 1033

Landfarming, 1054
Laundry detergent, improvement in, 518
Leucine zipper domain, 506
L-Glutamic acid production, 104
Library (gene) construction, 591
Ligand-gated ion channel receptors, 61
Light chain (kappa or lamba) formation, 749
Linkage, 134
Linkers and adaptors (blunt-end ligation by), 344
Lipofection, 960
Lipoplexes and polyplexes, 403
Liposome-mediated gene transfer, 374
Liposome-mediated transfection, 960
Liquid surface culture process (for citric acid production), 1034
L-lysine production, 1040
LocusLink, 552
LOD score method for estimating, 136
Luciferase
- bacterial (Lux), 897
- firefly (Luc), 895

M

Magic bullets, cells as drug delivery system, 405
Maintenance of aseptic environment during culture (sterilisation), 840, 1011
Major histocompatibility complex (MHC), 755
- class I proteins, 756
- class II proteins, 757
- class III proteins,758

Malt beverages, 1027
Malting, 1027
Mammalian artificial chromosome (MAC) vectors, 958
Mammalian virus vectors, 954
Marine biotechnology, 11
Marker gene, 892
  herbicide resistance, 900
  scorable (screenable) marker, 893
  selectable marker, 893
Mashing, 1028
Mass spectrometry (MS), 495
  components of, 496
  principle of working of, 496
  uses of, 497
Material balance, 1017
Maternal effects, 137
Matrix quality, 188
Matrix-assisted layer desorption-ionisation (MALDI), 497
  application of, 500
  ionisation method of, 498
Maxam and Gilbert's method (for DNA sequencing), 200
Measurement of
  cell mass, 200
  ion flow, 56
Meat industry, use of cell immobilisation in, 700
Mechanism of
  action of Cry protein crystals, 922
  enzyme catalysis, 671
  methane formation, 1080
Membrane transport proteins, 50
Mendelian genetics, chromosomal basis of, 129
Mendel's laws of inheritance, 129
Mentor pollen technology, 875
Meristem culture, 843
  production of virus-free plants from, 843
Messenger RNA (mRNA), 97
  differences between prokaryotic and eukaryotic, 32
  proof reading of, 117
Metagenome gene discovery tools, 636
Metagenomic
  cDNA (transcriptomic) libraries, 636
  DNA shuffling, 639
  nucleic acid extraction and enrichment strategies, 632
Metagenomics project
  environmental microbial, 628
  hydrocarbon 630
  microbial, 629
Metagenomics
  RAST Server (MG-RAST), 648
Metagenomics, chapter, 14, 622-654
  application of, 681
  at work, 631
  biological insights from, 626
  from genomics to, 626
  historical background of, 623
Methane formation
  factors affecting, 1084
  mechanism of, 1080
Methanogenesis, 1079
Methanogens, 1079
  adaptation by 1081
Method of cryopreservation, 880
Methods of
  bioleaching, 1088
  enzyme immobilization, 688
  gene prediction, 601
  gene manipulation, 333-382
  gene sequencing, 604
  plant cell, tissue and organ culture, 841
  production of edible vaccines, 805, 934
  strain improvement, 1021
  transfection, 959
Michaelis–Menten kinetics, 676
Microarrays, 630, 634
Microbial biosensors (whole cell biosensors), 707
Microbial culture, techniques of, 1003, 1088
Microbial diversity in biogas digesters, 1088
Microbial genetics, 141
Microbial genome sequencing projects, 579
  bacterial genome, 580
  viral genomics, 579
Microbial growth, 993
  in batch culture, 996
  kinetics and specific growth rate, 997
  microbial growth kinetics and specific growth rate, 997

parameters, 993
Microinjection, 38
Microorganisms involved in bioremediation, 105
Micropropagation, 865
Microsatellites, 425
Miniprimer PCR, 218
Minisatellites, 426
Mitochondrial DNA (mtDNA) analysis, 435
Mitochondrial inheritance, 140
Models for DNA replication, 110
Cairns model, 110
rolling circle model, 111
Model-View-Controller (MVC), 536
Molecular biology, 71-127
Molecular genetics, 128-181
Molecular farming, 931
of plantbodies, 931
Molecular markers, 219
DNA amplification fingerprinting (DAF), 220
random amplified polymorphic DNA (RAPD), 219
Molecular pharming (in transgenic animals), 986
Monitoring of bioremediation, 1065
Monoclonal antibodies, 769
applications of, 773
cancer treatment by, 774
diagnosis by, 773
diagnostic tests by, 773
mechanism of action, 777
production of recombinant, 777
purification of, 772
therapy, 779
Monoclonal antibody therapy, 777
Monoclonal anti-tumour antibodies, 825
Monoploid plants, production of, 873
Mouse Genome Database (MGD), 545
Multiple sequence alignment (MSA, 564
Multi-substrate reactions, 678
ping–pong mechanisms for, 678
ternary-complex mechanisms for, 678
Mutants, detection and isolation of, 172
Mutation
and mutagenesis, 163
expression of, 169
frame shift, 166
induced, 167
random nature of, 164
spontaneous, 163
transition, 165
transversion, 166

N

Na^+/K^+-ATPase, 52
National Centre of Biotechnology Information (NCBI), 544
Natural selection (of self-replicating molecules), 26
Neutron REM meter, 230
NMR spectroscopy, 490
applications of, 494
types of, 493
Nomenclature of
DNA sequences, 538
protein sequences, 539
Non-Michaelis–Menten kinetics, 680
Non-radioactive labeling, 365
biotin-streptavidin labeling, 367
digoxigenin (DIG) labelling, 366
fluorescent labeling, 369
horseradish peroxidase (HRP) method, 365
Non-viral methods (used in gene therapy), 402
direct injection of naked DNA, 403
lipoplexes and polyplexes, 403
Northern blot analysis of cytoplasmic RNA, 974
Northern blotting technique, 358
advantages of, 359
applications of, 359
disadvantages of, 359
reverse northern blot, 359
Novel enzymes, 666
abzymes, 666
extremozymes, 667
synzymes, 667
nptII gene, 899
Nuclear magnetic resonance (NMR), 490
theory of, 490
working principle of, 493
Nucleic acid hybridisation technique, screening by, 352

Nucleic acids
conformations of, 77
nature of, 71
Nucleoid, 92
Nucleosome, 93
Nutrient media, 83
Nutritional quality of seeds, improvement in, 521

O

OMIM (online Mendelian inheritance in man), 551
OncoMouse™, 980
Optical density turbidity (absorbance) measurement method, 995
Opto-electronic biosensor, 707
Oral polio vaccine (OPV), 802
advantages of, 802
disadvantages of, 803
Organ transplant laws, 815
Organ transplants, 810
Organellar DNA, 540
Organic acid production, 1030
Organogenesis, 846
auxin–cytokinin interaction regulates *in vitro, 84*
Origin of antigens, 723
Ovary (plant) culture, 877
Overhangs, 261
Oxford nanopore technologies, 648

P

Pair-wise alignment, 562
dot-matrix methods, 562
dynamic programming, 563
word method for, 563
Palindromic DNA (palindromes), 91
Paper electrophoresis, 191
Parasitic diseases, diagnosis of (through DNA probe), 445
Particle bombardment gun, 376
conditions for, 377
pollen transformation through, 377
Passive immunotherapy, 823
non-specific, 824
specific, 824
Paternity establishing, 439
Pathogenesis-related proteins (PRs), 918
Patterns of DNA cutting (by restriction enzymes), 610
PCR, 211
PCR (gene specific), 636
Peng Peng sheep, 984
Penicillins,656
biosynthesis of, 656
production, 656
Peptide bond, 121
planarity of, 140
Peroxisomes, 47
Phage λ vectors, 302
charons, 306
insertion vectors, 302
replacement vectors, 303
λ EMBL3 and λ EMBL4, 304
Phage display, 634
Phage M13 vectors, 307
advantages of using, 309
M13mp18/19 vectors, 309
Phage P1 vector, 310
Phagemid (phasmid), 311
pBluescript II KS (+/–), 311
pTZ19R and pTZ19U, 313
Pharmaceutical biotechnology, 7
Pharmacogenomics, 7
Photoreactivation, 174
Phred quality scores, 595
Phytoremediation (by superfund superplant), 1065
Piezo-electric biosensors, 707
Pili, 36
Ping–pong mechanisms (of enzyme reaction), 644
Plant and animal viruses as cloning vectors, 323
Plant biotechnology, chapter, 18, 828-884
Plant cell transformation, 372
by co-cultivation method, 373
by co-integration method, 373
by ultrasonication, 374
liposome-mediated gene transfer, 374
Plant cell, tissue and organ culture, 830
history of, 830
research work in India, 834
Plant gene, organisation of, 887

Plantibodies (edible antibodies), 807
Plaque lifting method, 353
Plasma membrane, 40
Plasmid numbers, 284
Plasmid(s), 284
 amplification of, 285
 disarming of (*see* box), 296
 examples of, 286
 pACYC184, 286
 pBR322, 287
 pSC101, 286
 pUC, 288
 pUC57, 289
 pUN121, 291
 Ri-plasmid, 294
 Ti-plasmid, 292
Polio vaccine, 800
 inactivated, 803
 oral polio, 802
Pollen culture, method of, 870
Pollen transformation (through particle bombardment), 377
Polly and Molly, 983
Polyacrylamide gel electrophoresis (PAGE), 194
Polymerase chain reaction (PCR), 211
 application of, 215
 types of PCR, 216
 working mechanism of, 212
Polymerization of nucleotides, 24
Polynucleotide kinase, 279
Polypeptide synthesis, 120
 elongation of, 119
 initiation of, 112
 termination of, 112
Polyplexes and lipoplexes, 403
Pomato, 858
Positron emission tomography (PET), 231
Post-translational processing, 23
Potency of cells (*see* box), 963
Preimplantation genetic diagnosis (PGD), 405
Prenatal paternity test, foetal cell/DNA, 439
Pribnow box, 114
Problems related to draft sequence and genome finishing, 598
 gap closure, 598
 validation and refinement, 598
Process flow, 1015
Prodrug therapy, 1015
Product recovery, 1082
Production of
 cattle, 961
 disease-resistant plants, 843
 recombinant interferon alfa-2a, 788
 recombinant monoclonal antibodies, 773
 secondary metabolites, 852
 stress-tolerant plants, 853
 virus-free plants from meristem culture, 844
Prokaryotes and eukaryotes, transport systems in, 50
Prokaryotic host cells, 335
Promoter, phosphoenolpyruvate (PEP) carboxylase, 923
Promoter(s), 888,946
 35S promoter (of CaMV), 909
 difference between eukaryotic and prokaryotic (*see* box), 889
 transcription start site, (TSS), 890
 types of (*see* box), 890
Proof reading (of mRNA), 592
Properties of enzymes, 660
Protein blotting (*see* Western blotting techniques technique), 359
Protein design, 508
 models of protein structure and function used in, 514
 some achievements in, 515
Protein engineering, 508
 in enzymes, 520
 methods in, 512
 steps of, 512
 strategies for, 509
Protein export (signal hypothesis), 724
Protein folding, 125
Protein sequences, nomenclature of, 539
Protein sequencing, 480
 Edman degradation method for, 481
 enzymatic method for, 482
 Sanger's method for, 481
Protein structure database, 607
Protein structure, 458

Protein synthesis [in prokaryotes and eukaryotes (*see* box)], 123
Protein-DNA interaction, 504
Proteins
conformation analysis of, 469
primary structure of, 470
quaternary structure of, 480
secondary structure of, 472
special features of, 496
tertiary structure, of, 479
Proteomics, 616
approaches for study of, 616, 618
expression, 619
functional, 619
structural, 619
types of, 619
Protoplast culture, 854
and regeneration, 858
isolation of, 854
purification of, 855
Protoplast fusion and somatic hybridization, 857
pTi-based vectors, 295
Public perception of biotechnology, 15
PubMed (Publishers on Medicine), 552
Pulsed-field gel electrophoresis (PFGE), 552
Pyrosequencing (next- generation sequencing), 549

Q

Quantitative PCR (Q-PCR), 218
Quantum tunneling, 638

R

Radiations as mutagens, 170
Radioactive labeling, 364
nick translation, 364
of probes (random primed), 365
Radioactive labelling, 364
Radioactivity, 220
detection of, 227
Radioimmuno assay (RIA), 234
method of, 235
Radiotracer techniques, 225
Ramachandran plot, 475
Random amplified polymorphic DNA (RAPD) marker, 219
Random primed radiolabelling of probes, 365
Random sequencing, 591
Real-time PCR (RT-PCR), 216
Receptor-meditated gene delivery, 404
Receptors, intracellular, 62
Recognition of extracellular signal molecules, 57
Recognition sequences, 258
Recombinant DNA technology, strategies of, 337
Recombinant DNA, transformation of bacterial cells by, 347
Recombinant vaccines, 519
Recombinant vector vaccines, 798
Recombinants, 349
blue-white screening of, 351
direct selection of, 350
insertional inactivation of, 350
selection of, 349
Recombination frequency, 136
Recombineering method, 970
Red biotechnology, 7
Red sea marine metagenomics, 630
RefSeq (reference sequence) database, 560
Regeneration (plant), 846
Regulation of enzyme activity, 630
Regulation of humoral response, 709
antigen blocking and receptor cross-linking, 709
idiotypic networking, 710
Relation between gene and protein, 618
Replica plating technique, 172
Reporter gene, 892
Resources for gene level sequences, 558
HomoloGene Database, 559
RefSeq (reference sequence) database, 560
UniGene database, 558
Responses of innate immune system, 725
Restriction endonucleases, 256, 245-280
types of, 257
Restriction enzymes
factors affecting activity of, 265
nomenclature of, 265
patterns of DNA cutting by, 261
type I, 258
type III, 264

Restriction mapping, 205
Restriction- modification system, 254
Retrotransposons, 160
Retroviruses (in gene therapy), 397
Reverse northern blot, 359
Reverse transcriptase, 271
Reverse transcription PCR (RT-PCR), 216
RFLP analysis, 249
Ribonucleic acid (RNA), structure of, 94
Ribosome editing, 124
Ribozymes, (catalytic RNA molecules), 94
    artificial, 27
Right-handed DNA, alternative form of, 87
Ri-plasmid, 294
RNA as biomarker, 631
RNA interference (*see* gene silencing), 411
RNA polymerases, 112
    types of RNA, 113
RNA sequencing, 203
RNA
    sense and antisense, 411
    structure of, 94
    types of, 95
Roundup Ready, 905
Ruppy (glowing dogs), 986

S

Sanger and Coulson method (DNA sequencing), 202
Sanger sequencing method, 584
Satellite RNA (and its use in transformation), 903
Scaffold, 555
Scale-up of fermentation, 1020
Scintillation counter, 228
Scorable (screenable) markers, 893
Screening by *in vitro* translation method, 351
Screening of cDNA clone, 343
Screening technique, direct immunological, 354
SDS-PAGE, 195
Second messenger molecules, 65
    role in signal transduction pathway, 65
Secondary antibody (*see* box), 361
Secondary metabolites, production of, 852
Seed protein quality, improvement of, 929
Seedless watermelon, 874
Selectable markers, 899
Selection of recombinants, 349
    auxotrophic mutant selection method for, 860
    biochemical methods for, 860
    cytometric method for, 860
    fluorescence-activated cell sorting (FACS) for, 861
    morphological selection method for, 860
    somatic hybrids and cybrids, 859
Self-replicating molecules, natural selection of, 26
Semisynthetic penicillins (*see* box), 1039
Sense RNA and antisense RNA, 411
Separation of solutes in ion-exchange chromatography, 158
Sequence alignment, 561
    global and local alignments, 561
    pair-wise alignment, 562
Sequence assembly, 594
Sequence quality determination, 594
Sequence Retrieval System (SRS), 553
Sequence-driven analysis, 641
Sequences and nomenclature, 537
454 Sequencing technology, 642
Serine protease (subtilisin), 518
Setting up of a national DNA databases, 437
Severe combined immunodeficiency (SCID), 390
    gene therapy of, 396
Sheep-goat chimera (Geep), 984
Shine-Dalgarno sequence, 98
Shoot meristem development, cytokinin–auxin cross-talk controls, 847
Short tandem repeat (STRs) analysis, 43
Shotgun cloning, 589
Shuttle vectors, 32
Sickle-cell anaemia, 388
Signal amplification, 56
    at transmembrane hormone receptor, 57
Signal hypothesis (protein export), 124
Signal transduction pathway, IFN-g, 786
Signal transduction, 56
    role of second messenger molecules in, 57
Signalling, forms of, 57
Single molecule real time technology, 644
Single nucleotide polymorphisms (SNPs), 605

Single stranded DNA, 90
Single-substrate reactions, 676
Site of chromosomal integration, 970
Site-directed mutagenesis (SDM), 205
Snuppy (dog), 944
Sodium dodecyl sulphate polyacrylamide gel electrophoresis (SDS-PAGE), 195
Solid state fermentation (Koji process), 1034
Solid-phase bioremediation, 1054
Soluble RNA (sRNA), 95
Somaclonal variation, 866
    application in crop improvement, 868
    causes of, 868
    examples of, 869
Somatic embryogenesis, 862
Somatic gene therapy, 393
Somatic hybridisation, 859
Somatic hybridization, limitations of (*see* box), 861
Somatic hybridization, 857
    applications of, 862
    limitations of (*see* box), 861
Somatic seeds, 863
    advantages of, 864
    methods of production of, 864
Sources of HSCT, 817
Southern blot analysis (of transfected gene constructs), 974
Southern blotting techniques, 356
Specific growth rate, 998
Spectrophotometry, UV-Vis, 999
Spectroscopy, 207
    fluorescence, 209
Spider goat, 982
Split transplants, 812
Spontaneous mutation, 163
    evidence for, 164
    origin of, 164
SSAP (sequential structure alignment programme), 567
Stable transfection, 959
Stable-isotope probing (SIP), 632
Stem cell therapy, 8
Sterilisation (maintenance of aseptic environment during culture), 1011
Sticky ends, 261
Stoichiometry, 1014
Strain improvement, 1021
    isolation of mutants for, 1022
    methods of, 1021
Strategies of recombinant DNA technology, 337
Streptomycin, 1039
    production, 1039
    recovery, 1039
    structure of streptomycin, 1039
Structural alignment, 566
Structural genomics, 607
Structural proteomics, 699
Structure of antibodies, 736
Structure of ribonucleic acid (RNA), 94
Structure-function relationships, 513
Submerged culture process (for citric acid production), 1035
Submerged culture process, 1035
Substrate specificity, 663
Substrate-induced gene expression screening (SIGEX), 639
Subtilisin (serine protease), 518
    protein engineering in, 518
Subunit vaccines, 792
Success story (of biotech market), 16
Sugar puckering, 81
Superbug, 1063
Super-tomatoes, 869
Suppressive subtractive hybridization (SSH), 633
Surface-enhanced laser desorption-ionisation (SELDI), 500
Suspension (cell) culture, 849
    application of, 851
    benefits from, 851
    in mass, 850
    synchronisation of (*see* box), 851
SV40-based vectors, 955
    plasmid vectors, 956
    transducing vectors, 955
Synaptonemal complex, 132
Synchronous growth, 1006
Synthetic biology, 13
Synthetic seeds, 863
    advantages of, 864
    methods of production of, 864

Synthetic vaccine, 791
Synthetic viruses, 409
Synzymes (artificial enzymes), 667

T

T4 DNA polymerase, 269
T7 DNA polymerase, 269
Tag and exchange, 969
Targetted gene transfer (gene targeting), 965
Targetted gene transfer method, 966
TATA box, 1/5, 888
Taxonomy browsers, 552
Techniques of microbial culture, 1003
Terminal transferase, 270
    application of, 270
Termination of polypeptide chain, 122
Thalassaemia, 388
Therapeutic antibodies, approved, 779
Therapeutic uses of interferons, 786
Thermistor-containing biosensor, 705
Thermodynamic equilibrium, 238
    conditions for, 239
Thermodynamics (laws of), 236
    first law of, 239
    second law of, 240
    third law of, 242
    zeroth law of, 237
Thermostable DNA polymerases, 269
Thymidine kinase (TK), 95
Ti-plasmids, 292
Tissue culture laboratory, 836
TnA transposon family, 162
Tobamovirus as vector, 324
Toll-like receptors, 61
Topoisomerase, 107
Torsion angles, 78
    endocyclic, 80
Totipotency (*see* box), 829
Toxoid vaccines, 793
Tracer isotopes, 226
Tracer techniques, 225
Tracy (GM cow), 981
Traditional biotechnology, 5
Transcription (RNA synthesis), 112
Transcription start site (TSS), 890
Transcriptional and translational reporters, 952
Transcriptomic (metagenomic cDNA) libraries, 636
Transduction, 153
    abortive transduction, 155
    generalized, 154
    specialized (restricted), 156
    use in genetic mapping of viruses, 156
Transfected gene constructs
    Southern blot analysis of, 156
Transfection methods, 959
Transfection, 348
Transfer RNA (tRNA), 95
Transformation, 142
    of bacterial cells by recombinant DNA, 347
    plant cell, 373
    use of satellite RNA in, 903
Transgene, variation in (*see* box), 946
Transgenes (rescue and recovery of from animal cells), 972
Transgenesis, 943, 978
Transgenic animals, 941-991
    Alzheimer's mouse, 980
    applications of, 976
    cows, 981
    detection of, 973
    Dolly and Bonny, 983
    EnviroPig™, 985
    FluoroPig, 984
    Geep (Sheep-goat chimera), 984
    GloFish®, 981
    glowing dogs (Ruppy), 986
    GM hen, 986
    OncoMouse™, 980
    Peng Peng Sheep, 984
    Polly and Molly. 983
    Ruppy (glowing dogs), 986
    Sheep-goat chimera (Geep), 984
    Snuppy (dog), 984, 985
    Spider goat, 982
    Tracy (GM cow), 981
Transgenic breeding, advantages of, 976
Transgenic cattle, production of, 961

Transgenic livestock, breeding with, 975
Transgenic plants, chapter 19, 885-940
  Barnase–barstar component system in, 919
  Bt brinjal, 927
  Bt cotton, 924
  commercial use of, 938
  fungus–resistant, 917
  herbicide resistant, 903
  virus-resistant, 907
Transient transfection, 959
Translation (protein synthesis), 122
Translation analysis, 122
Translocation 3/43, mechanism of, 122
Transplant rejection, 818
  types of, 818
  rejection mechanisms, 819
  treatment of, 819
Transplantation antigens, tumour-associated, 821, 822
Transplantation, 809
  ethical concerns for, 818
  organ, 815
  types of, 813
Transplastomic plants (*see* box), 887
Transport
  across cell membrane, 50
  systems in prokaryotes and eukaryotes, 50
Transport systems, active and passive, 51
Transposable elements, 157
  classes of, 158
  nomenclature of, 158
Transposons as vector, 327
Transposons, 160
  complex, 167
  composite,161
  mechanism of, 162
Trasformation, 901
Treatment of autoimmune diseases, 765
Trimming of vector sequences, 594
Triploid production, 874
  papaya, 875
tRNA (transfer RNA), 95
  charging of, 118
  clover leaf model of, 100
  structure of, 100
True single-molecule sequencing technology (tSMS), 609
Tumour antigens, 820
Tumour immunology, 820
Tumour-associated transplantation antigens, 821, 822
Turbidostat, 1013
Two-dimensional (2D) gel electrophoresis, 197
Type II restriction enzymes, properties of, 256
Type III restriction enzyme, 258
Types of (organ) donor, 81
Types of antigens, 726
  T-dependent, 726
  T-independent, 726
Types of bioreactors (fermentors), 1008
  air-lift, 1009
  bubble column, 1009
  fluidised bed, 1008
  membrane, 1009
  packed tower, 1010
  photo bioreactor, 1010
  pulsed column, 1009
Types of composting, 1069
Types of databases, 527
Types of donor, 813
Types of electrophoresis, 192
  boundary electrophoresis, 192
  immunoelectrophoresis, 192
  zone electrophoresis, 192
Types of gene therapy, 391
Types of interferons, 781
Types of NMR spectroscopy, 493
  continuous wave, 493
  Fourier transform, 493
  multi-dimensional, 494
Types of restriction enzymes, 257
Types of sequences used in bioinformatics, 540

U

Ultraviolet (UV) radiation, 170
UniGene database, 558
Upstream process, 1016
Uranium recovery, 1099

Use of
- biogas, 1076
- bioinformatics tools in analysis, 567
- cell Immobilisation in meat industry, 700
- computer in fermentor, 1010
- haploids, 873
- monoploids, 873

UV-Vis (visible) spectrophotometry, 209

V

V(D)J rearrangement, 748
Vaccination schedule, 456
Vaccines, 789
van der Waals Force, 804
Variable number of tandem repeats (VNTR), 424
Variations of MVC, 537
Variolation (*see* box), 717
Vat leaching, 1089
Vector sequences, trimming of, 594
Vectors, 953
Vermibed, preparation of, 1073
Vermicast, physical quality of, 1071
Vermicompost, 1069
- benefits of, 1074
- manufacturers and suppliers of, 1072

Vermicomposting by using earthworms, 1072
Vermicomposting technology (vermitech-nology), 1070
Vermiculture 1074
- importance of, 1071
- working mechanism of, 1070

Vermitechnology, 1070
- physical quality of vermicast, 1071
- vermicomposting by using earthworms, 1072
- working mechanism of vermiculture, 1070

Viable plate count, 992
Vinegar production, 1030
Viral genomics, 575
Virosomes in drug targeting, 406
Virus-resistant transgenic plants, 907

W

Watson and Crick's model of DNA, 17
Western blotting techniques (protein blotting or electroblotting technique), 359
White biotechnology, 10
Whole cell biosensors (microbial biosensors), 707
Whole cell biosensors, 711
Whole-genome shotgun sequencing, 599
Wine
- production, 1029

X

Xenograft, 812
Xenotransplants, future of, 812
X-ray crystallographic studies of DNA, 75
X-ray crystallography, 483
X-ray diffraction analysis of crystals, 484

Y

YAC (yeast artificial chromosomes), 321
Y-chromosome analysis, 434
Yeast
- artificial chromosomes (YAC), 321
- centromeric plasmid (YCp) or yeast centromere (CEN), 320
- episomal plasmid (YEp), 319
- integrative plasmid (YIp), 319
- linear plasmid (YLp), 321
- plasmid vectors, 317
- replicating plasmid (YRp), 320

Z

Z-DNA, 87
Zinc finger domain, 505
Zwitterions, 460
Zymotechnology, 2